P9-BZO-423

EXCITING NEW FEATURES FOR 1995

MERRILL

LIFE SCIENCE

Daniel
Ortleb
Biggs

AN ACTIVE, REAL-LIFE APPROACH THAT CONNECTS SCIENCE TO THE EVERYDAY WORLD

An Educational Program from

GLENCOE

INVOLVE YOUR STUDENTS IN ACTIVE, REAL-LIFE LEARNING

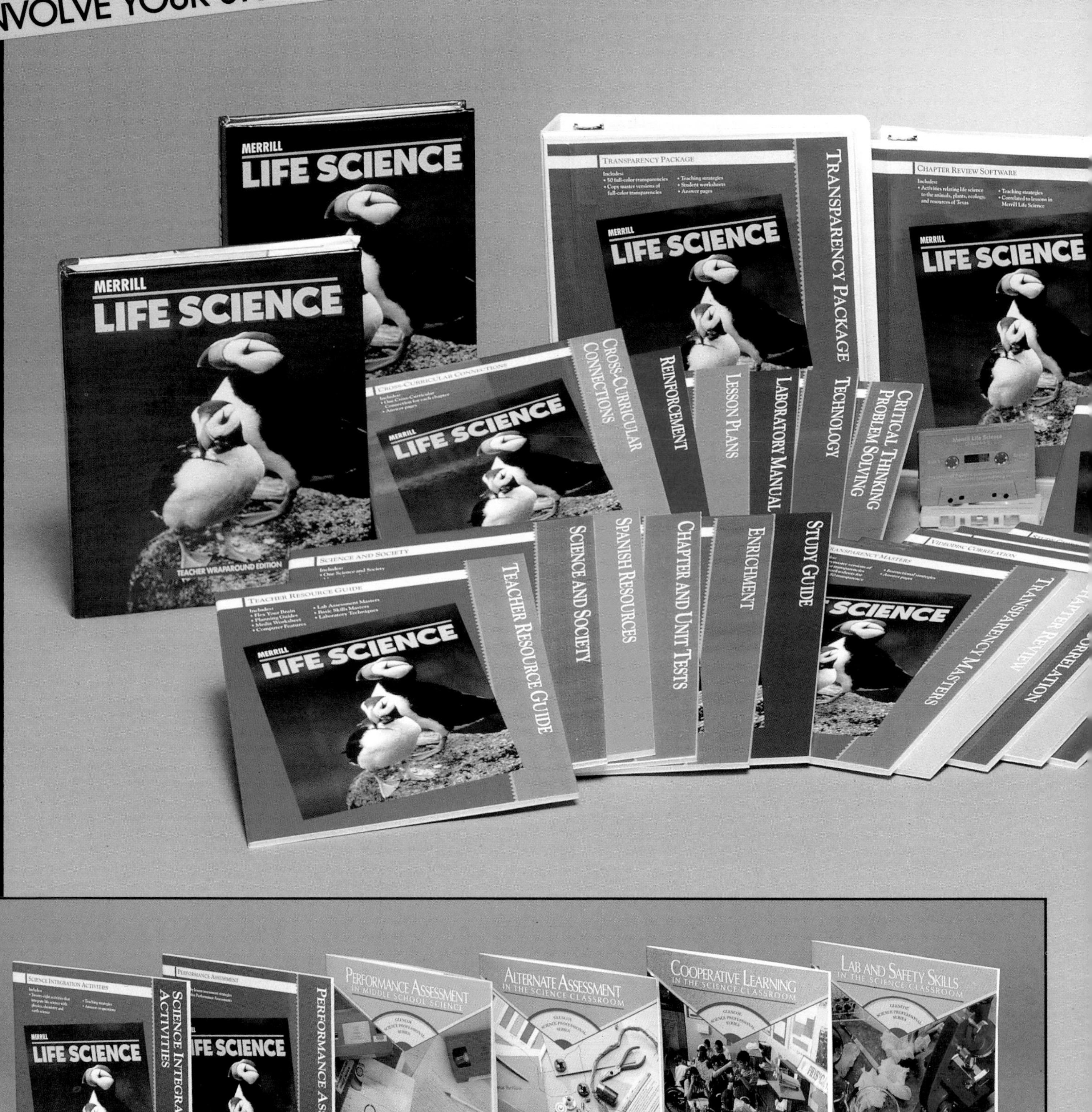

MERRILL
LIFE SCIENCE
TEACHER WRAPAROUND EDITION
TRANSPARENCY PACKAGE
CHAPTER REVIEW SOFTWARE
CROSS-CURRICULAR CONNECTIONS
REINFORCEMENT
LESSON PLANS
LABORATORY MANUAL
TECHNOLOGY
CRITICAL THINKING PROBLEM SOLVING
SCIENCE AND SOCIETY
SPANISH RESOURCES
CHAPTER AND UNIT TESTS
ENRICHMENT
STUDY GUIDE
TEACHER RESOURCE GUIDE
TRANSPARENCY MASTERS
CHAPTER REVIEW
CORRELATION

SCIENCE INTEGRATION ACTIVITIES
LIFE SCIENCE
PERFORMANCE ASSESSMENT
PERFORMANCE ASSESSMENT IN MIDDLE SCHOOL SCIENCE
ALTERNATE ASSESSMENT IN THE SCIENCE CLASSROOM
COOPERATIVE LEARNING IN THE SCIENCE CLASSROOM
LAB AND SAFETY SKILLS IN THE SCIENCE CLASSROOM

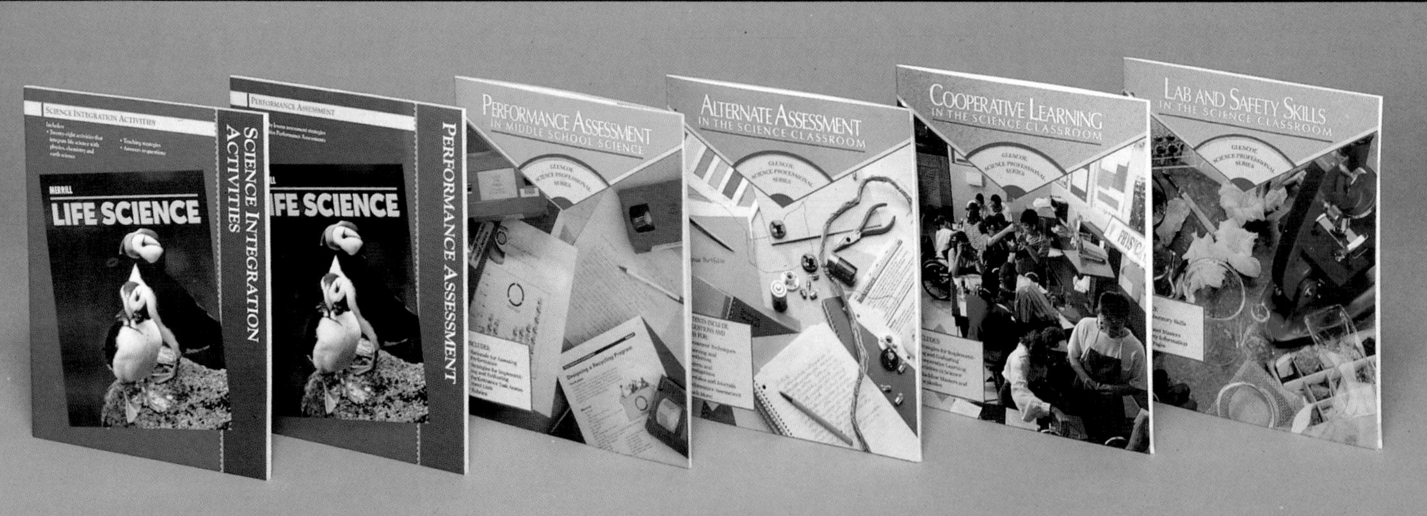

*From the moment your students become involved with the new **Merrill Life Science,** they will experience life in a whole new way. Walking down a city street will no longer be viewed as a simple, everyday act. Rather, your students will see it as a phenomenon that involves their muscles, nervous system, and circulatory system, a phenomenon that is based on the principles of life science.*

*The **Merrill Life Science** comprehensive learning and teaching program provides everything you need to make life science an **active learning experience.** Your students will learn through:*

- ***Solid concept development,** thematically structured to focus on the principle ideas of life science.*
- ***Hands-on activities** that actively involve students in the subject matter.*
- ***Strong skill development** that leads to student success in science and the real world.*
- ***Fascinating, real-world applications** that will make the study of life science come alive in your classroom.*

*As for teachers' classroom support, the new **Merrill Life Science** provides a wealth of resource materials in the **Teacher Wraparound Edition** and **Teacher Classroom Resources.** Teaching suggestions, applications, extension activities, plus all new assessment options, and Mr. Wizard's **Science and Technology Videodisc Series** components provide you with all the valuable tools to help you enrich your program and reduce your preparation time.*

STUDENTS LEARN BY OBSERVING, EXPERIMENTING, AND ASKING QUESTIONS

Understanding life science is more than memorizing facts. It's learning to adapt the basic principles of life science to everyday applications. The new ***Merrill Life Science*** *offers numerous activities that give students the chance to further explore and investigate the scientific topics covered in the textbook. This not only helps students relate life science to their everyday world, it also helps to build skills that will make them responsible decision makers and critical thinkers.*

New open-ended activities provide students with options to design their own experiments in every chapter.

ACTIVITY 4-2
Making a Model

Problem: What does a model of DNA look like?

Materials
- tracing paper
- scissors
- heavy paper
- crayons

Procedure
1. Trace the four DNA parts shown below.
2. Cut out the four tracings. **CAUTION:** *Always be careful when using scissors.*
3. Copy each of the tracings onto heavy paper six times. Label each tracing as shown below.
4. Color the phosphate brown. Color the sugar orange. Color the A bases red, the T bases blue, the C bases green, and the G bases yellow.
5. Cut out all 24 parts.
6. Use the following order of bases to model a DNA molecule: ATCCGT. Use your desk as a work space.
7. Separate the bases in the molecule and replicate the original molecule.
8. Use any combination of DNA parts to create other model molecules. Compare your molecules with those of your classmates.

Analyze
1. What does each DNA part represent?
2. What does the A base pair with?
3. What does the C base pair with?
4. What is the order of bases in the second strand of the DNA molecule in Step 6?
5. Were the molecules you created in Step 8 the same as your classmates'? Explain.

Conclude and Apply
6. Using the model, explain why T only matches with A.
7. Based on your observations of the DNA model, infer why a DNA molecule seldom copies itself incorrectly.
8. Explain why scientists use models.

P S G P S C P S T P S A

94 CELL REPRODUCTION

16 Animals

370

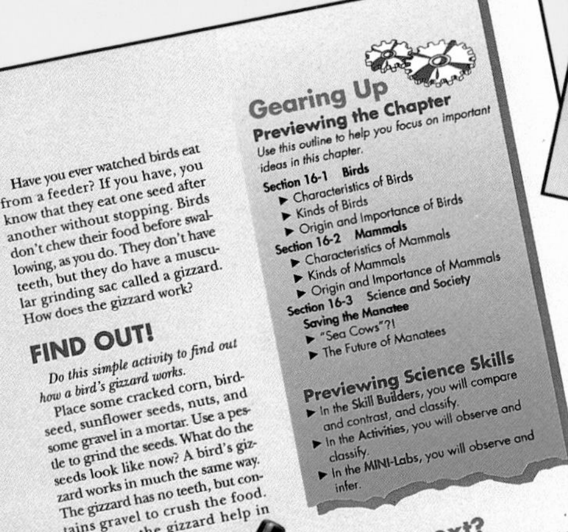

Have you ever watched birds eat from a feeder? If you have, you know that they eat one seed after another without stopping. Birds don't chew their food before swallowing, as you do. They don't have teeth, but they do have a muscular grinding sac called a gizzard. How does the gizzard work?

FIND OUT!

Do this simple activity to find out how a bird's gizzard works.

Place some cracked corn, birdseed, sunflower seeds, nuts, and some gravel in a mortar. Use a pestle to grind the seeds. What do the seeds look like now? A bird's gizzard works in much the same way. The gizzard has no teeth, but contains gravel to crush the food. How does the gizzard help in digestion?

Gearing Up
Previewing the Chapter
Use this outline to help you focus on important ideas in this chapter.

Section 16-1 Birds
- Characteristics of Birds
- Kinds of Birds
- Origin and Importance of Birds

Section 16-2 Mammals
- Characteristics of Mammals
- Kinds of Mammals
- Origin and Importance of Mammals

Section 16-3 Science and Society
Saving the Manatee
- "Sea Cows"?!
- The Future of Manatees

Previewing Science Skills
- In the Skill Builders, you will compare and contrast, and classify.
- In the Activities, you will observe and classify.
- In the MINI-Labs, you will observe and infer.

What's next?
In this chapter, you'll learn more about birds. You'll learn how they've adapted for flight and how they live. You'll also learn about mammals, the class to which you belong. You will learn how mammals are classified and how they live.

371

MINI-Lab
What are the parts of a bird's egg?

Open a chicken egg into a shallow bowl. Observe the shell and the contents of the egg. Identify the parts of the egg from the diagram below. What do you think is the function of each part?

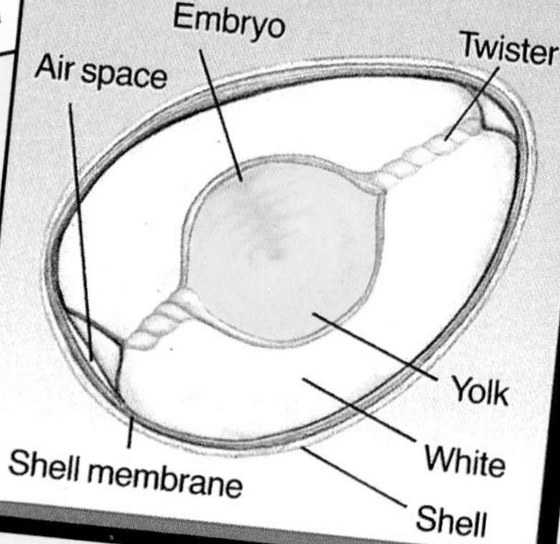

Chapter Openers offer fun, interesting ways to get your students excited about the upcoming lessons.
- A unique **Find Out** activity entices students to make observations and raise questions about upcoming content.
- **Gearing Up** previews key concepts and skills.
- **What's Next** provides an intriguing transition into the chapter's main ideas.

Mini-Labs are quick hands-on activities that give your students additional opportunities to practice important process skills. Great for in-class instruction or take-home exercises.

PROBLEM SOLVING

The Blood Type Mystery

Detective Johnson was on to something. He had been assigned to a confusing robbery case. He had received conflicting testimony from the people involved. Now he had the last piece of evidence he needed to solve the crime.

A person had come home to find someone he knew robbing his apartment. The victim struggled with the robber. He broke a vase over the robber's head. The victim later identified the robber in a police line-up, but the robber said he had an alibi.

While investigating, Detective Johnson found blood on the vase and carpet. He sent the blood to a lab to be analyzed. The blood of the victim and robber also were analyzed, even though the robber offered proof that he had just received a blood transfusion with type O blood.

When Detective Johnson received the blood analysis report, the blood on the vase and carpet was found to be type A. The victim's blood type was B. The robber's blood type was A. The robber was found guilty. The case was solved.

Think Critically: What did the robber know about the blood he received?

The Rh factor is another inherited substance in blood. If present in a person's blood, the person has Rh-positive (Rh+) blood. If it is not present, the person is said to be Rh-negative (Rh−). An Rh− person receiving blood from an Rh+ person will produce antibodies against the Rh+ factor. Antibodies are proteins that destroy or neutralize foreign substances, such as pathogens, in your body.

A problem also occurs when an Rh− mother carries an Rh+ baby. Close to the time when the baby is about to be born, antibodies from the mother can cross the placenta and destroy the baby's red blood cells. If this happens, the baby can receive a blood transfusion before or right after birth. At 28 weeks and immediately after the birth, the mother can receive an injection that destroys any Rh+ antibodies she has made. Thanks to the work of Dr. Landsteiner, blood groups and Rh factor now are checked before transfusions and during pregnancies.

Did You Know?

Rh, or Rhesus, factor was first discovered in the rhesus monkey in 1940. From 85 to 90 percent of the people in the United States have this antigen.

20-2 BLOOD 475

Problem Solving features are real-life stories about kids. A critical thinking question engages the student in solving an everyday problem linked to the chapter content.

Flex Your Brain blends critical thinking and problem solving. Students use a step-by-step method to explore a topic while they learn to develop good problem solving skills.

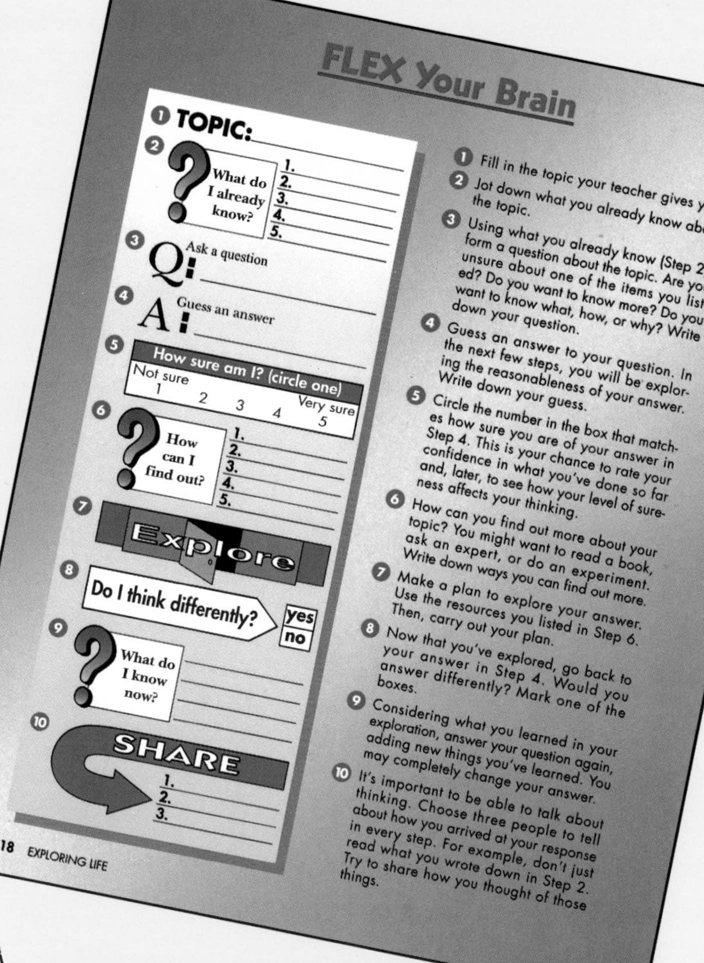

FLEX Your Brain

1 TOPIC:

2 What do I already know? 1. 2. 3. 4. 5.

3 Q: Ask a question

4 A: Guess an answer

5 How sure am I? (circle one) Not sure 1 2 3 4 5 Very sure

6 How can I find out? 1. 2. 3. 4. 5.

7 Explore

8 Do I think differently? yes no

9 What do I know now?

10 SHARE 1. 2. 3.

1. Fill in the topic your teacher gives you.
2. Jot down what you already know about the topic.
3. Using what you already know (Step 2), form a question about the topic. Are you unsure about one of the items you listed? Do you want to know more? Do you want to know what, how, or why? Write down your question.
4. Guess an answer to your question. In the next few steps, you will be exploring the reasonableness of your answer. Write down your guess.
5. Circle the number in the box that matches how sure you are of your answer in Step 4. This is your chance to rate your confidence in what you've done so far and, later, to see how your level of sureness affects your thinking.
6. How can you find out more about your topic? You might want to read a book, ask an expert, or do an experiment. Write down ways you can find out more.
7. Make a plan to explore your answer. Use the resources you listed in Step 6. Then, carry out your plan.
8. Now that you've explored, go back to your answer in Step 4. Would you answer differently? Mark one of the boxes.
9. Considering what you learned in your exploration, answer your question again, adding new things you've learned. You may completely change your answer.
10. It's important to be able to talk about thinking. Choose three people to tell about how you arrived at your response in every step. For example, don't just read what you wrote down in Step 2. Try to share how you thought of those things.

18 EXPLORING LIFE

In Your JOURNAL

Choose one of the rain forest plants, animals, or products listed here to research: toucans, squirrel monkeys, mahogany, kapok, curare, sloths, and bromeliads. Make a drawing of your choice **in your Journal**, and describe its location and its importance.

In Your Journal features help students assess their progress and keep records of their learning.

Skill Builder

Observing and Inferring

A corn plant produces thousands of pollen grains on top of the plant in incomplete flowers that have no odor or color. The pistils grow from the cob lower down on the plant. Explain how a corn plant is probably pollinated. If you need help, refer to Observing and Inferring in the **Skill Handbook** on page 682.

266 THE SEED PLANTS

A **Skill Builder,** at the end of each section, challenges students to use basic process skills. The activity also directs the students to the **Skill Handbook,** located at the back of the text, for a step-by-step overview on how to accomplish that particular skill.

STUDENTS DISCOVER HOW LIFE SCIENCE IMPACTS PEOPLE AND PLACES AROUND THE WORLD

Reading a passage from "The Sea Around Us," which vividly describes life in the sea as seasons change, does wonders to impact your student's learning experiences. All of the features on these pages show real world application your students will get with the new ***Merrill Life Science.*** *And that's what makes this program more enriching than any other life science program currently available.*

Connect to... features in every chapter integrate life science with physics, chemistry, and earth science.

Connect to... Chemistry

Insect-eating plants, such as pitcher plants and sundews, often grow in areas where heavy rains wash nitrogen from the soil. Few other plants can grow there. *Infer* how these plants get the nitrogen they need if nitrogen is not available from soil.

The **Science and Society** lesson in each chapter challenges students to make decisions and formulate their own viewpoints about issues facing them and their community.

SCIENCE & SOCIETY

16-3 Saving the Manatee

New Science Words

manatees
poaching

Objectives

- Identify the characteristics of manatees.
- Explain the major threats to manatees today.

"Sea Cows"?!

Have you ever thought about running into a cow in the water? Unless you live in Florida, you probably didn't realize that "sea cows" exist. Sea cows, or **manatees,** are large mammals belonging to the order Sirenia that live in salt waters. Adults weigh about half a ton. They are known to live in the rivers, estuaries, and coasts in the tropical and subtropical regions. They cannot tolerate cold temperatures.

Unfortunately, manatees are now on the endangered species list. The slow, curious "sea cows" are not afraid to approach and investigate visiting humans, making them very easy to hunt. In the past, they were heavily hunted for their meat, oil, and hides.

Today, manatees are still dying. Manatees are in danger from barges and motorboats. They often collide with these boats and are injured or cut by propellers. In fact, collisions are so common with manatees that researchers are able to identify more than 900 individual manatees just by using their distinctive scars caused by these collisions. Manatees are herbivores and feed on aquatic plants that clog waterways. Many manatees are poisoned by herbicides used to kill these plants.

EcoTip

Ask if your classroom can "adopt" an endangered animal at your local zoo. You will be helping to preserve wildlife.

The Future of Manatees

Researchers are very concerned about the future of these huge, quiet mammals. Although the United States protected them by law in 1983, the state of Florida cannot employ enough law enforcers to prevent illegal hunting called **poaching.** In 1988, 133 manatees were killed in Florida. By 1989 only about 1200 manatees were believed to be living in United States waters.

Manatees have a very slow rate of reproduction. Manatees that are killed by accident are not quickly replaced. Combine this steady decrease in population with the loss of habitat due to development (nearly 1000 people move to Florida each year) and extinction does not look to be very far away.

What can be done to save the manatees? Some suggest developing more refuges where boats would not be allowed to travel. Because Florida real estate is very valuable, this is not likely to be popular with coastal landowners. Another option is to enforce a slower speed limit on the boats that travel in the Intracoastal Waterway where many manatees are injured.

Whatever the solution, manatees need more protection soon if we hope to save them from their current fate of extinction.

Figure 16-16. Warning signs are being placed in areas where manatees are found to help protect them from being injured.

SECTION REVIEW

1. What are manatees and where do they live?
2. What two major factors account for the gradual decrease in the population?

SCIENCE & SOCIETY

You Decide!

In Miami, Florida, injured manatees are taken by a special ambulance to Miami's Seaquarium where they can be treated. Scientists are hoping to release the treated manatees and help the population become more stable. The estimated cost of treatment for one injured manatee is about $18 000. When returned to its environment, there is no guarantee that it will escape another, perhaps fatal, injury. Some people believe that it's not right to spend this much money on saving the manatees. What do you think?

386 WARM-BLOODED ANIMALS

16-3 SAVING THE MANA

Two **Careers** are featured in every unit to expose students to the interesting and wide range of career choices available for those with a knowledge of science.

CAREERS

ZOOLOGIST

A zoologist is a scientist who studies animals. These scientists are interested in the animal's habitat, behavior, diseases, and life processes. Zoologists who study birds are called ornithologists. Those who study reptiles are called herpetologists. Mammalogists study mammals and ichthyologists study fish.

A person who wants to become a zoologist should like working with animals and be willing to spend long and irregular hours with them. Zoologists need a background in science, particularly biology, zoology, and animal behavior. A trained zoologist will need to have a bachelor of science degree for non-research work. If they plan to do research, they will need to have a master's degree.

For Additional Information

Contact the American Society of Zoologists, 104 Sirius Circle, Thousand Oaks, CA 91360.

VETERINARIAN'S TECHNICIAN

A veterinarian's technician likes animals and enjoys working with them. He or she should be calm and reliable, as well as willing to work hard. Some of the work that veterinarian's technicians do is not easy. Cages have to be cleaned, animals groomed, and excited animals must be calmed. Most veterinary technicians are trained on the job. It is possible to start as a part-time animal attendant and advance to technician. Experienced technicians may also find work in kennels, small zoos, pet stores, as assistant laboratory animal technicians, or with the Humane Society.

Some veterinarian technicians want to open their own kennels and raise, groom, or board small animals.

For Additional Information

Contact Animal Caretakers Information, The Humane Society of the U.S., Companion Animals Division, Suite 100, 5430 Grosvenor Lane, Bethesda, MD 20814.

UNIT READINGS

- "Zebras Are Coming." *Popular Mechanics,* February 1991.
- Schaller, George G. "Pandas in the Wild." *National Geographic,* December 1981.
- Clutton-Brock, T.H., "Red Deer and Man." *National Geographic,* October 1986.

414

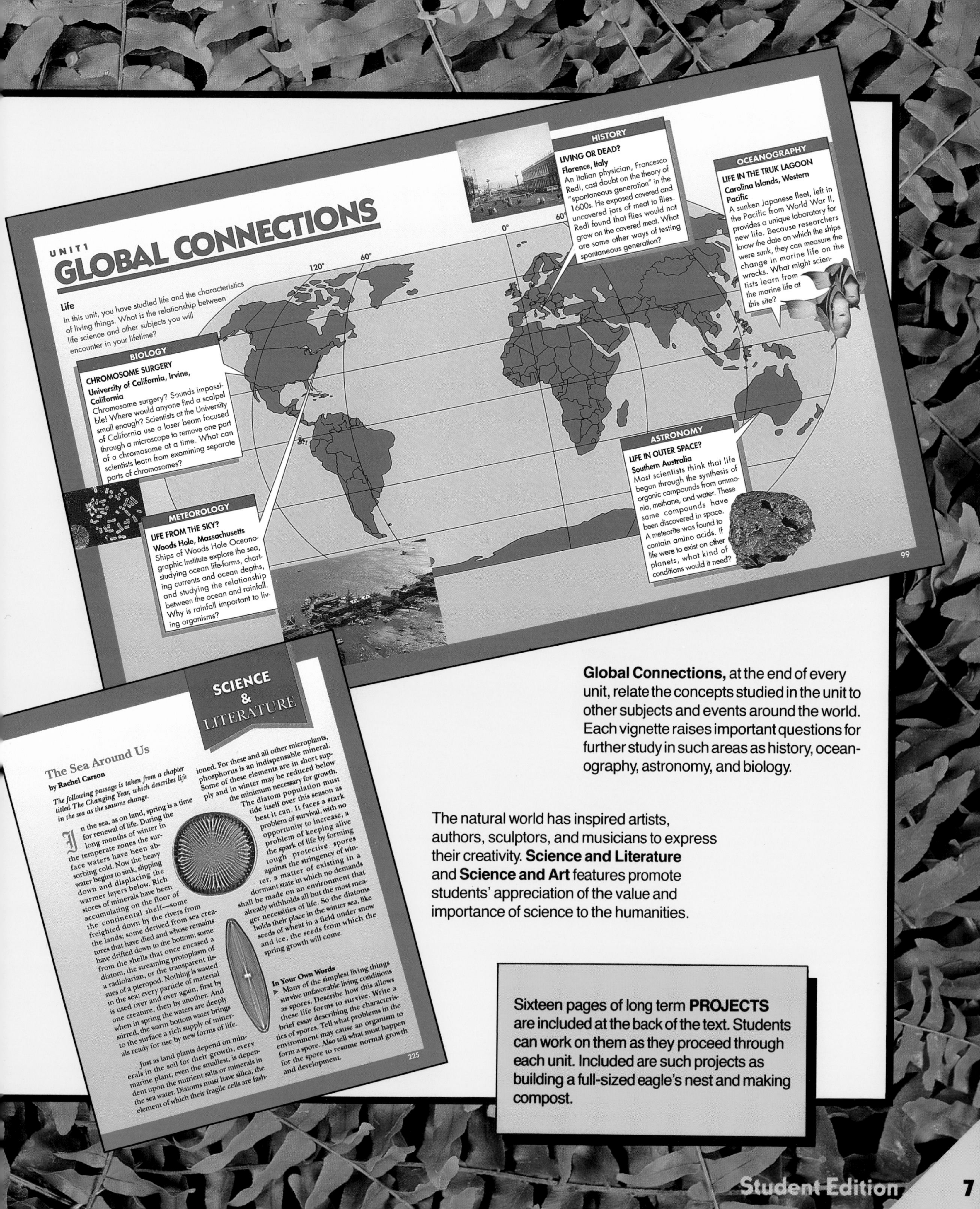

UNIT 1

GLOBAL CONNECTIONS

Life

In this unit, you have studied life and the characteristics of living things. What is the relationship between life science and other subjects you will encounter in your lifetime?

BIOLOGY

CHROMOSOME SURGERY
University of California, Irvine, California
Chromosome surgery? Sounds impossible! Where would anyone find a scalpel small enough? Scientists at the University of California use a laser beam focused through a microscope to remove one part of a chromosome at a time. What can scientists learn from examining separate parts of chromosomes?

HISTORY

LIVING OR DEAD?
Florence, Italy
An Italian physician, Francesco Redi, cast doubt on the theory of "spontaneous generation" in the 1600s. He exposed covered and uncovered jars of meat to flies. Redi found that flies would not grow on the covered meat. What are some other ways of testing spontaneous generation?

OCEANOGRAPHY

LIFE IN THE TRUK LAGOON
Carolina Islands, Western Pacific
A sunken Japanese fleet, left in the Pacific from World War II, provides a unique laboratory for new life. Because researchers know the date on which the ships were sunk, they can measure the change in marine life on the wrecks. What might scientists learn from the marine life at this site?

METEOROLOGY

LIFE FROM THE SKY?
Woods Hole, Massachusetts
Ships of Woods Hole Oceanographic Institute explore the sea, studying ocean life-forms, charting currents and ocean depths, and studying the relationship between the ocean and rainfall. Why is rainfall important to living organisms?

ASTRONOMY

LIFE IN OUTER SPACE?
Southern Australia
Most scientists think that life began through the synthesis of organic compounds from ammonia, methane, and water. These same compounds have been discovered in space. A meteorite was found to contain amino acids. If life were to exist on other planets, what kind of conditions would it need?

99

SCIENCE & LITERATURE

The Sea Around Us

by Rachel Carson

The following passage is taken from a chapter titled The Changing Year, which describes life in the sea as the seasons change.

In the sea, as on land, spring is a time for renewal of life. During the long months of winter in the temperate zones the surface waters have been absorbing cold. Now the heavy water begins to sink, slipping down and displacing the warmer layers below. Rich stores of minerals have been accumulating on the floor of the continental shelf—some freighted down by the rivers from the lands; some derived from sea creatures that have died and whose remains have drifted down to the bottom; some from the shells that once encased a diatom, the streaming protoplasm of a radiolarian, or the transparent tissues of a pteropod. Nothing is wasted in the sea; every particle of material is used over and over again, first by one creature, then by another. And when in spring the waters are deeply stirred, the warm bottom water brings to the surface a rich supply of minerals ready for use by new forms of life.

Just as land plants depend on minerals in the soil for their growth, every marine plant, even the smallest, is dependent upon the nutrient salts or minerals in the sea water. Diatoms must have silica, the element of which their fragile cells are fashioned. For these and all other microplants, phosphorus is an indispensable mineral. Some of these elements are in short supply and in winter may be reduced below the minimum necessary for growth. The diatom population must tide itself over this season as best it can. It faces a stark problem of survival, with no opportunity to increase, a problem of keeping alive the spark of life by forming tough protective spores against the stringency of winter, a matter of existing in a dormant state in which no demands shall be made on an environment that already withholds all but the most meager necessities of life. So the diatoms hold their place in the winter sea, like seeds of wheat in a field under snow and ice, the seeds from which the spring growth will come.

In Your Own Words

▶ Many of the simplest living things survive unfavorable living conditions as spores. Describe how this allows these life forms to survive. Write a brief essay describing the characteristics of spores. Tell what problems in the environment may cause an organism to form a spore. Also tell what must happen for the spore to resume normal growth and development.

225

Global Connections, at the end of every unit, relate the concepts studied in the unit to other subjects and events around the world. Each vignette raises important questions for further study in such areas as history, oceanography, astronomy, and biology.

The natural world has inspired artists, authors, sculptors, and musicians to express their creativity. **Science and Literature** and **Science and Art** features promote students' appreciation of the value and importance of science to the humanities.

Sixteen pages of long term **PROJECTS** are included at the back of the text. Students can work on them as they proceed through each unit. Included are such projects as building a full-sized eagle's nest and making compost.

STRATEGIES AND RESOURCES ARE RIGHT AT YOUR FINGERTIPS

As a teacher, you will thoroughly enjoy all the valuable information ***Merrill Life Science*** *provides in our* ***Teacher Wraparound Edition.*** *Everything is well organized, highly visible, and positioned at the point of instruction, to give you the most teaching value.*

The **Three-Step Teaching Cycle** includes **Motivate, Teach,** and **Close.** It provides various strategies for developing your individual lesson plans, and also highlights optional activities and program resources for enhancing your presentation.

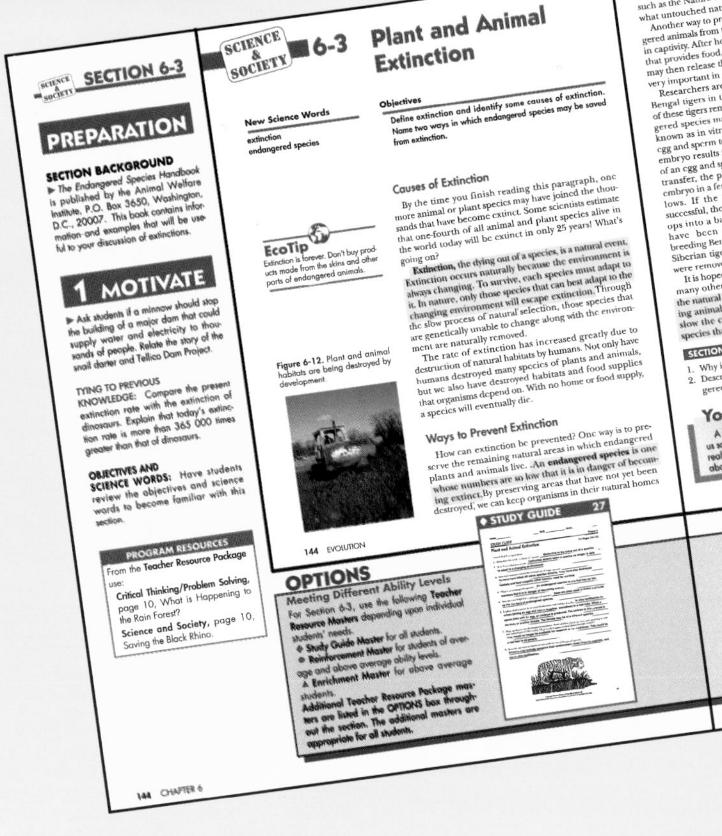

SECTION 6-3

PREPARATION

SECTION BACKGROUND

▶ *The Endangered Species Handbook* is published by the Animal Welfare Institute, P.O. Box 3650, Washington, D.C., 20007. This book contains information and examples that will be useful to your discussion of extinctions.

1 MOTIVATE

▶ Ask students if a minnow should stop the building of a major dam that could supply water and electricity to thousands of people. Relate the story of the snail darter and Tellico Dam Project.

TYING TO PREVIOUS KNOWLEDGE: Compare the present extinction rate with the extinction of dinosaurs. Explain that today's extinction rate is more than 365 000 times greater than that of dinosaurs.

OBJECTIVES AND SCIENCE WORDS: Have students review the objectives and science words to become familiar with this section.

PROGRAM RESOURCES

From the Teacher Resource Package use:

Critical Thinking/Problem Solving, page 10, What is Happening to the Rain Forest?

Science and Society, page 10, Saving the Black Rhino.

6-3 Plant and Animal Extinction

New Science Words: extinction, endangered species

Objectives: Define extinction and identify some causes of extinction. Name two ways in which endangered species may be saved from extinction.

EcoTip: Extinction is forever. Don't buy products made from the skins and other parts of endangered animals.

Causes of Extinction

By the time you finish reading this paragraph, one more animal or plant species may have joined the thousands that have become extinct. Some scientists estimate that one-fourth of all animal and plant species alive in the world today will be extinct in only 25 years! What's going on?

Extinction, the dying out of a species, is a natural event. Extinction occurs naturally because the environment is always changing. To survive, each species must adapt to it. In nature, only those species that can best adapt to the changing environment will escape extinction. Through the slow process of natural selection, those species that are genetically unable to change along with the environment are naturally removed.

The rate of extinction has increased greatly due to destruction of natural habitats by humans. Not only have humans destroyed many species of plants and animals, but we also have destroyed habitats and food supplies that organisms depend on. With no home or food supply, a species will eventually die.

Figure 6-12. Plant and animal habitats are being destroyed by development.

Ways to Prevent Extinction

How can extinction be prevented? One way is to preserve the remaining natural areas in which endangered plants and animals live. An endangered species is one whose numbers are so low that it is in danger of becoming extinct. By preserving areas that have not yet been destroyed, we can keep organisms in their natural homes

144 EVOLUTION

and let them rebuild their populations. Organizations, such as the Nature Conservancy, are working hard to save what untouched natural areas remain.

Another way to prevent extinction is to remove endangered animals from their natural habitats and breed them in captivity. After helping animals reproduce in a setting that provides food, shelter, and protection, researchers may then release them into the wild. Zoos have become very important in these types of programs.

Researchers are battling the extinction of Bengal tigers in this way. Only 3000 to 5000 of these tigers remain in the wild. This endangered species may be saved by a technique known as in vitro fertilization, joining an egg and sperm together in a test tube. An embryo results from the joining of an egg and sperm. Embryo transfer, the placing of the embryo in a female, then follows. If the procedure is successful, the embryo develops into a baby tiger. Zoos have been successful in breeding Bengal tigers by placing the embryo in a female Siberian tiger. Through surgery, the developed kittens were removed from the foster mother.

It is hoped that in vitro fertilization can be used to help many other endangered animal species. By protecting the natural areas that are not yet destroyed and by breeding animals in captivity, preservationists are working to slow the current rate of extinction and preserve the species that remain.

What is an endangered species?

Figure 6-13. Bengal tigers are being bred in captivity to help save them from extinction.

SECTION REVIEW

1. Why is extinction not always a bad event?
2. Describe two ways in which we can work to save endangered species.

SCIENCE & SOCIETY

You Decide!

A lot of energy and money is put into programs that help us save endangered species from extinction. Are these species really worth our time and effort? Should we be concerned about saving an animal or plant that we'll never see?

2 TEACH

Key Concepts are highlighted.

▶ The cheetah is endangered in much of Africa. Cheetahs have very little genetic variability, and this is the apparent cause of their present problem. How does this natural situation differ from the mass extinctions occurring in the tropical rain forests? The cheetah is going extinct naturally, whereas other organisms in the rain forest are going extinct because of human activity.

CHECK FOR UNDERSTANDING

Ask questions 1 and 2 in the Section Review.

RETEACH

Have students list the effects of human activity in rain forests, the Arctic tundra, and other fragile environments.

EXTENSION

For students who have mastered this section, use the **Reinforcement** and **Enrichment** masters or other OPTIONS provided.

3 CLOSE

▶ Have each student present a brief report on a different endangered species. Students should mention numbers, if known, and what is being done to save the species.

SECTION REVIEW ANSWERS

1. Extinction is a natural event that can open habitats to new species.
2. Areas where endangered species live can be preserved. Endangered species can be removed and bred in captivity. In vitro fertilization can be used to help endangered species.

YOU DECIDE!

Answers will vary. Some students may agree that the government should preserve all or some species. Others may think preservation should be left up to nature. Still other students may not think that extinction is important. Answers should be reasonable and logical.

CHAPTER 6 145

OPTIONS

Meeting Different Ability Levels

For Section 6-3, use the following **Teacher Resource Masters** depending upon individual students' needs.

◆ **Study Guide Master** for all students.

● **Reinforcement Master** for students of average and above average ability levels.

▲ **Enrichment Master** for above average students.

Additional Teacher Resource Package masters are listed in the OPTIONS box throughout the section. The additional masters are appropriate for all students.

◆ STUDY GUIDE 27

● REINFORCEMENT 27

▲ ENRICHMENT 27

144 CHAPTER 6

CROSS CURRICULUM

▶ **Reading:** Read the poem "The Chambered Nautilus" by Oliver Wendell Holmes. Provide a nautilus shell for students to examine. Have them draw the shell and write a report on the shell and its function.

Preceding each chapter is a two-page **Planning Guide** that gives you quick access to chapter content, features, activities, skill exercises, and all other program components.

Cross Curriculum strategies provide unique ways to connect life science to other sciences and other disciplines such as math, reading, writing, fine arts, and health.

ALTERNATE TEACHING STRATEGIES AND BUILT-IN ASSESSMENT KEEP YOUR STUDENTS ON THE RIGHT TRACK.

Student masters designed to help you address different levels of learning abilities are shown in reduced form in the bottom margins. They include **Study Guide, Reinforcement,** and **Enrichment** worksheets. More strategies are suggested in **For Your Gifted Students** and **For Your Mainstreamed Students.**

Classes of Mollusks

Univalves

Meeting Different Ability Levels

For Section 14-1, use the following **Teacher Resource Masters** depending upon individual students' needs.

◆ **Study Guide Master** for all students.

● **Reinforcement Master** for students of average and above average ability levels.

▲ **Enrichment Master** for above average students.

Additional Teacher Resource Package masters are listed in the OPTIONS box throughout the section. The additional masters are appropriate for all students.

◆ STUDY GUIDE 55

cephalopod univalve bivalve

... predators by rapidly opening and closing their shells. As the water pours out, its force moves them rapidly in the opposite direction.

Cephalopods

The class of mollusks with the most specialized and complex members are the cephalopods. The word *cephalopod* means "head-footed" and describes the body structure of these invertebrates. Squid, octopuses, and the chambered nautilus belong to this class. Cephalopods have a large, well-developed head. Their "foot" is divided into many tentacles with strong suckers for capturing prey.

● REINFORCEMENT 55

▲ ENRICHMENT 55

TYING TO PREVIOUS KNOWLEDGE: Most students will be familiar with mollusks, such as oysters, clams, squid, garden snails, and octopuses. Use this knowledge to begin the discussion of mollusk characteristics.

OBJECTIVES AND SCIENCE WORDS: Have students review the objectives and science words to become familiar with this section.

2 TEACH

Key Concepts are highlighted.

CONCEPT DEVELOPMENT

▶ Use an overhead transparency of a snail to show the three main body parts shared by all mollusks—the muscular foot, the head, and the visceral mass.

▶ Explain that only mollusks have a radula.

▶ Ask: **How are clams, oysters, and scallops like sponges?** *They are filter feeders.*

▶ **Demonstration:** Insert a knife on the side of the hinge of a clam and cut away the muscles that hold the shell closed. Point out the mantle, siphons, gills, foot, sex organs, mouth, stomach, intestine, excretory organs, and heart.

CHECK FOR UNDERSTANDING

Ask students to describe how all mollusks are similar. They have the same basic body plan—a soft body covered by a hard shell, mantle, gills, muscular foot, and organs located in a visceral mass.

RETEACH

Give students a photocopied diagram of a snail and have them label the body parts.

EXTENSION

For students who have mastered this section, use the **Reinforcement** and **Enrichment** masters or other OPTIONS provided.

CHAPTER 14 325

Multicultural Awareness

Have students research the adaptations people have made to the environments in which they live. Students could research various aspects of adaptations, including food preferences, manner of dress, or architecture of homes in places as diverse as the desert, the tundra, or deciduous forests. Students could also contrast the adaptations made by native peoples to the same type of environment in different areas of the world.

Multicultural Awareness features highlight the contributions of individuals and societies from diverse cultures.

Numerous alternate assessment strategies are embedded in every chapter. Some of these are highlighted in the **Assessment Options** at the beginning of the chapter. Performance, portfolio, content, and group assessment strategies are included.

ASSESSMENT OPTIONS

PORTFOLIO
Refer to page 319 for suggested items that students might select for their portfolios.

PERFORMANCE ASSESSMENT
Skillbuilders, pp. 313, 318
MINI-Lab, p. 316
Activities, 13-1, p. 305; 13-2, p. 314
Using Lab Skills, p. 320

CONTENT ASSESSMENT
Assessment—Oral, pp. 310, 312
Skillbuilder, p. 304
Section Reviews, pp. 304, 307, 313, 318
Chapter Review, pp. 319-321
Mini Quizzes, pp. 303, 311, 317

GROUP ASSESSMENT
Opportunities for group assessment occur with Cooperative Learning Strategies and Flex Your Brain Activities.

A FULL RANGE OF SUPPORT MATERIALS TO MEET ALL YOUR NEEDS AND THOSE OF YOUR STUDENTS.

HANDS-ON LEARNING

Laboratory Manual contains at least two hands-on laboratory activities for each chapter. Students get more chances to acquire scientific knowledge while you get more opportunities to reinforce and apply chapter concepts.

Activity Worksheets include worksheets for every Mini-Lab and full-page Activity in the chapters.

Science Integration Activities are laboratory activities that relate physics, chemistry, and earth science to specific life science chapters.

REVIEW & REINFORCEMENT

Study Guide Worksheets are tailored to the needs of students who need a little extra help. They reinforce understanding of the topics and vocabulary found in each chapter.

Reinforcement Worksheets provide a variety of interesting activities to help students of average ability levels retain the important points in every chapter.

Concept Mapping masters reinforce learning by having students complete a concept map for each chapter.

Chapter Review Software presents chapter-end review questions in random order and provides feedback for incorrectly answered questions by noting the textbook page where the answer is found.

Chapter Review masters are two-page review worksheets consisting of 25 questions for each chapter. Ideal for test preparation, alternative tests, and vocabulary review.

ENRICHMENT AND APPLICATION

Enrichment Worksheets challenge your students of above average ability to design, interpret, and research scientific topics based on the text in each lesson.

Critical Thinking/Problem Solving helps your students develop important critical thinking and problem solving skills as they work through additional problems related to chapter topics.

Cross-Curriculum Connections are interdisciplinary worksheets. They emphasize learning by doing and provide valuable insight into the connection life science has with other disciplines.

Science And Society worksheets encourage further involvement with Science and Society lessons in the student text.

Technology masters explore recent developments in science and technology or explain how familiar machines, tools, or systems work.

ASSESSMENT

Chapter Test masters provide comprehensive tests for each chapter.

Computer Test Bank Package, available in Apple, IBM, and Macintosh versions, provides a convenient tool for creating your own chapter test. The software allows you to add your own problems or edit the existing questions.

Alternate Assessment in the Science Classroom explains the need for nontraditional methods of assessment such as performance assessments and portfolios. In addition to strategies it provides samples of questions, report forms, and a scoring rubric.

Performance Assessment provides 44 pages of specially designed assessments including eight unit summative performance tasks and 28 chapter-related skill assessments.

Performance Assessment in Middle School Science (PAMSS) includes the philosophy and strategies for assessing both the processes and products of science. The booklet contains practical performance assessment task lists and rubrics that you can use immediately to help you evaluate your students' performance.

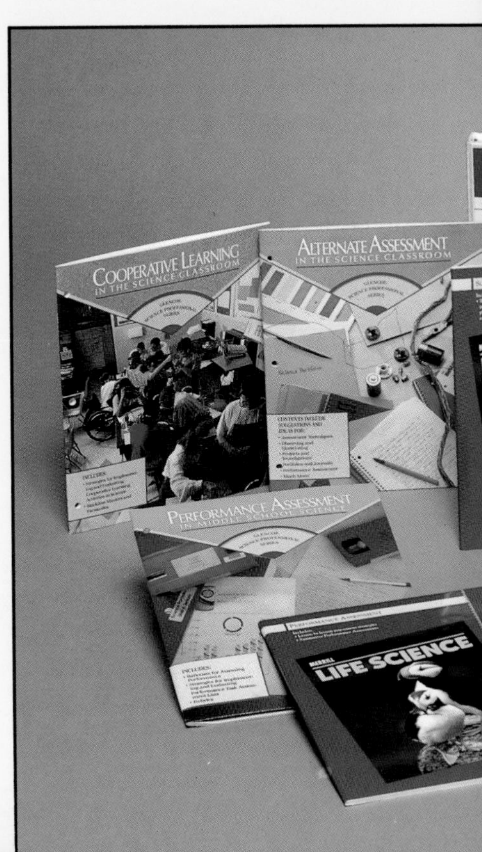

TEACHING AIDS

Color Transparency Package includes 50 full-color transparencies, some with overlays, in a three-ring binder with a resource book of blackline masters and student worksheets for each transparency. Excellent for direct instruction, reteaching, and review.

Transparency Masters include blackline reproductions and student worksheets for each of the program's full-color transparencies.

Spanish Resources provides Spanish translations of chapter objectives, glossary terms, and definitions. Ideal for bilingual classrooms.

Lesson Plan Book is a complete lesson planning resource for teachers. It is a correlation of lessons, objectives, features, and program resources.

Teacher Resource Guide contains a program planning guide, lab design and equipment, safety instruction, a media worksheet, and a Flex Your Brain worksheet in one convenient booklet.

English-Spanish Audiocassettes allow students with reading difficulties, students who absorb concepts better in an auditory way, or students for whom English is a second language to listen to chapters from the text in English and/or Spanish.

Cooperative Learning in the Science Classroom contains background information, strategies, and practical tips for using cooperative learning techniques whenever you do activities.

Lab and Safety Skills in the Science Classroom presents an overview of lab skills and safety skills related to lab activities. The booklet also contains lab and safety skills assessments for evaluating students' understanding of lab and safety skills.

TECHNOLOGY

Videodisc Correlation for Optical Data laserdisc program allows you to use this resource with the sweep of a light pen.

Science and Technology Videodisc Series, which features 7 discs containing over 280 videoreports produced and narrated by TV's "Mister Wizard," lets students catch a glimpse of real scientists exploring current problems and technological developments so they can discover the influence of science in day-to-day living. Videodisc Teacher Guides include bar code directories plus teaching strategies, research updates, and complete narration for each videoreport.

Lab Partner Software is a spreadsheet and graphing program that allows you and your students to record, collect, and graph data from laboratory activities simply and effectively. A **User Guide** is also available to assist students through the programs.

Infinite Voyage Video Series (videodiscs/VHS tapes) bring the amazing live action and animated sequences of this award-winning PBS series to your science classroom.

MAKE THE CHOICE... MERRILL LIFE SCIENCE
AND MAKE THE CONNECTION...
BETWEEN SCIENCE AND THE EVERYDAY WORLD

Student Edition	0-02-827027-4
Teacher Wraparound Edition	0-02-827028-2
Teacher Classroom Resources	0-02-827029-0
Teacher Resource Guide	0-02-827041-X
Study Guide—TE	0-02-827039-8
Reinforcement—TE	0-02-827037-1
Enrichment—TE	0-02-827036-3
Critical Thinking/Problem Solving	0-02-827034-7
Concept Mapping	0-02-827033-9
Cross-Curricular Connections	0-02-827035-5
Science Integration Activities	0-02-827064-9
Science and Society	0-02-827038-X
Transparency Masters	0-02-827043-6
Technology	0-02-827042-8
Activity Worksheets	0-02-827030-4
Laboratory Manual—TAE	0-02-827048-7
Spanish Resources	0-02-827050-9
Cooperative Learning in the Science Classroom	0-02-826430-4
Lab and Safety Skills in the Science Classroom	0-02-826853-9
Chapter Review	0-02-827032-0
Chapter and Unit Tests	0-02-827031-2
Performance Assessment in Middle School Science	0-02-827279-X
Performance Assessment	0-02-827063-0
Alternate Assessment in the Science Classroom	0-02-826429-0
Lesson Plans	0-02-827051-7
Color Transparency Package	0-02-827049-5
Study Guide—SE	0-02-827044-4
Reinforcement—SE	0-02-827045-2
Enrichment—SE	0-02-827046-0
Laboratory Manual—SE	0-02-827047-9
Chapter Review Software—Apple	0-02-827053-3
Chapter Review Software—IBM	0-02-827052-5
Chapter Review Software—Macintosh	0-02-827054-1
Computer Test Bank—Apple	0-02-827056-8
Computer Test Bank—IBM	0-02-827055-X
Computer Test Bank—Macintosh	0-02-827057-6
Lab Partner Software—Apple	0-675-92487-0
Lab Partner Software—IBM	0-675-17271-3
STVS Videodisc Correlation	0-02-827062-2
Optical Data Videodisc Correlation	0-02-827061-4
English Audio Tapes	0-02-826434-7
Spanish Audio Tapes	0-02-826431-2

Optical Data Videodiscs:
Order Direct from
Optical Data Corporation
30 Technology Drive
P.O. Box 4919
Warren, NJ 07060

For more information contact your nearest regional office or call 1-800-334-7344.

1. Northeast Region
GLENCOE
17 Riverside Drive
Nashua, NH 03062
603-880-4701
800-424-3451

2. Mid-Atlantic Region
GLENCOE
5 Terri Lane, Suite 5
Burlington, NJ 08016
609-386-7353
800-553-7515

3. Atlantic-Southeast Region
GLENCOE
Brookside Park
One Harbison Way, Suite 101
Columbia, SC 29212
803-732-2365

4. Southeast Region
GLENCOE
6510 Jimmy Carter Boulevard
Norcross, GA 30071
404-446-7493
800-982-3992

5. Mid-America Region
GLENCOE
4635 Hilton Corporate Drive
Columbus, OH 43232
614-759-6600
800-848-1567

6. Mid-Continent Region
GLENCOE
846 East Algonquin Road
Schaumburg, IL 60173
708-397-8448
800-762-4876

7. Southwest Region
GLENCOE
320 Westway Place, Suite 550
Arlington, TX 76018
817-784-2100
800-828-5096

8. Texas Region
GLENCOE
320 Westway Place, Suite 550
Arlington, TX 76018
817-784-2100
800-828-5096

9. Western Region
GLENCOE
610 East 42nd Street, #102
Boise, ID 83714
208-378-4002
800-425-6126
Includes Alaska

10. California Region
GLENCOE
15319 Chatsworth Street
P. O. Box 9609
Mission Hills, CA 91346
818-898-1391
800-423-9534
Includes Hawaii

Glencoe Catholic School Region
GLENCOE
25 Crescent Street, 1st Floor
Stamford, CT 06906
203-964-9109
800-551-8766

Canada
Maxwell Macmillan Canada
1200 Eglinton Avenue, East
Suite 200
Don Mills, Ontario M3C 3N1
Telephone: 416-449-6030
Telefax: 416-449-0068

Overseas
Macmillan/McGraw-Hill
International
10 Union Square East
New York, NY 10003
Telephone: 212-353-5700
Telefax: 212-353-5894

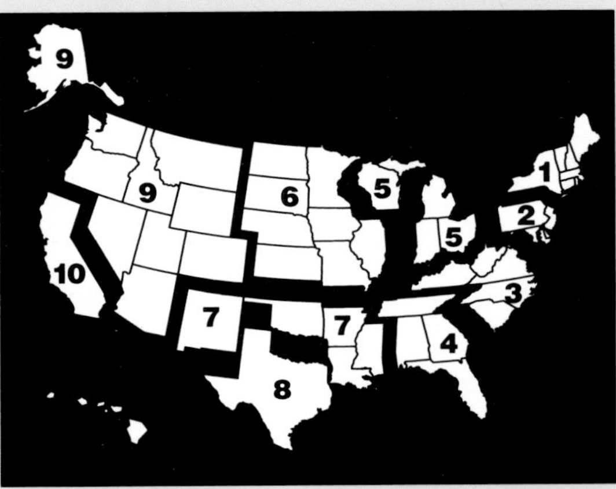

SC 90900-7

GLENCOE
Macmillan/McGraw-Hill
P.O. Box 508
Columbus, Ohio 43216

MERRILL

LIFE SCIENCE

TEACHER WRAPAROUND EDITION

Macmillan/McGraw-Hill

New York, New York   Columbus, Ohio   Mission Hills, California   Peoria, Illinois

MERRILL LIFE SCIENCE

Student Edition
Teacher Wraparound Edition
Teacher Classroom Resources
Study Guide, Student Edition
Study Guide, Teacher Edition
Reinforcement, Student Edition
Reinforcement, Teacher Edition
Enrichment, Student Edition
Enrichment, Teacher Edition
Activity Worksheets
Chapter and Unit Tests
Chapter Review
Concept Mapping
Critical Thinking/Problem Solving
Cross-Curricular Connections
Science and Society
Technology
Teacher Resource Guide
Transparency Package
Laboratory Manual
Laboratory Manual, Teacher Annotated Edition
Spanish Resources
Chapter Review Software (IBM, Apple, Macintosh)
Computer Test Bank (IBM, Apple, Macintosh)
Videodisc Correlation (Optical Data)
Science and Technology Videodisc Series (STVS)
Science and Technology Videodisc Series Teacher Guide
Lesson Plan Book
Science Integration Activities
Alternate Assessment in the Science Classroom
Performance Assessment
Cooperative Learning in the Science Classroom
Performance Assessment in Middle School Science (PAMSS)
Lab Partner Software (IBM, Apple)
Infinite Voyage Series (VHS, Videodiscs)

AUTHORS

Lucy Daniel
Edward Paul Ortleb
Alton Biggs

Send all inquiries to:
GLENCOE DIVISION
Macmillan/McGraw-Hill
936 Eastwind Drive
Westerville, Ohio 43081

Printed in the United States of America

ISBN 0-02-827027-4 (Student's Edition)

1 2 3 4 5 6 7 8 9–VH–02 01 00 99 98 97 96 95 94

ISBN 0-02-827028-2 (Teacher's Edition)

1 2 3 4 5 6 7 8 9–VH–02 01 00 99 98 97 96 95 94

Table of Contents

Teacher Guide

Student Edition

Merrill Life Science

Goals and Philosophy

*The goal of **Merrill Life Science** is to help students develop an understanding of the world around them through sound content and the practice of skills that will help them reach that understanding.*

Science educators daily face the task of helping students with different abilities and backgrounds to learn about the living world and to acquire skills that will increase their ability to evaluate that world. ***Merrill Life Science*** is dedicated and designed to help both students and teachers meet these goals.

In recent years, scientists and educators have collaborated in an effort to find a clear path toward better science education. Of all the reform-oriented proposals, **Science for All Americans,** a Project 2061 report on literacy goals in science, mathematics, and technology, sponsored by the American Association for the Advancement of Science, and **Scope, Sequence, and Coordination,** sponsored by the National Science Teachers Association, have attained national prominence. Both programs suggest major changes in what, when, and how science is taught. Both programs strongly suggest a more constructivist approach whereby the student, rather than the teacher, becomes the individual who identifies problems, proposes solutions, and then tests those proposals.

In response to the goals of science curriculum reform, ***Merrill Life Science*** promotes:

- solid, accurate content,
- a thematic orientation,
- hands-on activities that promote a constructivist approach through more student-planned activities,
- frequent practice of science process skills,
- integration of science concepts across the curriculum,
- numerous opportunities for authentic assessment, and
- decision making and problem solving.

Themes

Themes help students see how separate life science concepts are related.

Themes in science are broad, unifying ideas that integrate major concepts from many disciplines. Themes are an important part of any teaching strategy because they help students see the importance of truly understanding concepts rather than simply memorizing isolated facts.

In a life science course, many themes, such as energy, change over time, and stability, become evident as concepts are developed. These themes serve as conceptual frameworks for the course. They provide a rationale for the sequence of topics in the text. The three major themes in ***Merrill Life Science*** are:

- **Change over time:** From the outset, students are made aware that living things change and that adaptations are permanent evidence of change. Genetics and evolution are presented together to make students aware that evolution is the result of changes in the genes of populations. The evolution of plants and the origins of specific groups are presented.
- **Homeostasis and stability:** Traditional concepts of internal balance are highlighted throughout the human body and health units. But balance in the form of overall stability of all living systems is also emphasized in plant and animal chapters and in the ecology chapters, where the delicate nature of biomes is discussed.
- **Ecology:** The extensive and fragile network of relationships that pervades the living world is a theme that can be found in almost every chapter beginning at the outset with the characteristics and needs of living things in Chapter 1. A capsule scenario that emphasizes the importance that ***Merrill Life Science*** places on a knowledge of ecological relationships is found in the opening to Unit 8.

Constructivism in Science

Merrill Life Science provides a wide range of experiences that help students develop and apply thinking process skills.

Constructivism in Science

Strategies suggested in ***Merrill Life Science*** support a constructivist approach to science education. The role of the teacher is to provide an atmosphere where students design and direct activities. To develop the idea that science investigation is not made up of closed-end questions, the teacher should ask guiding questions and be prepared to help his or her students draw meaningful conclusions when their results do not match predictions. Through the numerous activities, cooperative learning opportunities, and a variety of critical-thinking exercises in ***Merrill Life Science,*** you can feel comfortable taking a constructivist approach to science in your classroom.

Activities

A constructivist approach to science is rooted in an activities-based plan. Students must be provided with sensory motor experiences as a base for developing abstract ideas. ***Merrill Life Science*** utilizes a variety of "learning by doing" opportunities. **Find Out** activities that open each chapter allow students to generate questions about the concepts to come, make observations, and share prior knowledge.

MINI-Labs and **Activities** involve students in learning and applying scientific methods to practice thinking skills and construct science concepts. They provide an engaging, diverse, active program. MINI-Labs require a minimum of equipment, and students may take responsibility for organization and execution.

Activities develop and reinforce or restructure concepts as well as develop the ability to use process skills. **Activity** formats are structured to guide students to make their own discoveries. Students collect real evidence and are encouraged through open-ended questions to reflect and reformulate their ideas based on this evidence.

In each chapter, there is one open-ended **Activity** called **"Designing an Experiment,"** that gives students a broad topic to explore and guides them with leading questions to the point where they can determine the direction of the investigation themselves. Students are asked to brainstorm hypotheses, make a decision to investigate one that can be tested, plan procedures, and in the end, think about why their hypotheses were supported or not.

Cooperative Learning

Cooperative Learning adds the element of social interaction to science learning. Group learning allows students to verbalize ideas, and encourages the reflection that leads to active construction of concepts. It allows students to recognize the inconsistencies in their own perspectives and the strengths of others. By presenting the idea that there is no one, "ready-made" answer, all students may gain the courage to try to find a viable solution. Cooperative Learning strategies appear in the *Teacher Wraparound Edition* margins whenever appropriate.

And More...

FLEX Your Brain, a self-directed critical-thinking matrix, is introduced in Chapter 1. This activity, referenced wherever appropriate in the *Teacher Wraparound Edition* margins, assists students in identifying what they already know about a subject, then in developing independent strategies to investigate further. **Revealing Misconceptions** suggests strategies the teacher may use to evaluate students' current perspectives.

Students are encouraged to discover the pleasure of solving a problem through a variety of features. **Apply** questions that require higher-level, divergent thinking appear in **Section Reviews.** The **Problem Solving, Technology,** and **Science and Society** features in each chapter invite students to confront real-life problems. **Think Critically** and **You Decide!** questions encourage students to reflect on issues related to technology and society. The **Skill Handbook** gives specific examples to guide students through the steps of acquiring thinking and process skills. As students complete each numbered section, a **Skill Builder** gives them a chance to assess and reinforce the concepts just learned through practice. **Think and Write Critically, Apply,** and **More Skill Builders** sections of the Chapter Review allow the teacher to assess and reward successful thinking skills.

Science, Technology, and Society

Merrill Life Science *provides a curriculum that helps prepare students for the relationships and responsibilities attendent to the interactions of science, technology, and society.*

Science involves the search for knowledge. Technology involves the application of this knowledge to human cultures. As society becomes more technologically complex, people increasingly rely upon it to satisfy their needs and solve their problems. Technology also increasingly impacts the environment and society, causing new challenges. Science literacy, then, is an important goal for all students so that they can use technology productively and be informed decision-makers. The ***Merrill Life Science*** curriculum incorporates strategies to develop critical and creative thinking for developing informed decision-making skills.

Critical thinking is characterized by:

- a search for clarity and accuracy,
- open-mindedness,
- taking and defending a position, and
- sensitivity to others' knowledge and feelings

Creative thinking is characterized by:

- engaging in tasks when answers are not apparent,
- pushing the limits of one's knowledge and abilities,
- generating and following one's own standard of evaluation, and
- generating new ways of viewing situations.

Merrill Life Science provides a curriculum that focuses on the interaction of science, technology, and society.

Student Edition

The Student Edition contains **Technology** features, **Science and Society** articles, and unit-ending **Global Connections** features.

- **Technology** features introduce students to applications of chapter concepts. *Think Critically* questions invite in-depth responses.
- **Science and Society** articles focus on the interaction of science and the impact of technology on culture. *You Decide!* questions call for informed decision-making.
- **Global Connections** features encourage awareness of science, technology, and society as a global responsibility.

Teacher Classroom Resources

As part of the *Teacher Classroom Resources,* there is a *Science and Society* booklet with readings and activities that relate science, technology, and society topics to chapter content. These one-page articles are different from the Science and Society features in the textbook. The *Science and Society* booklet articles present students with questions that are large societal life science issues to which there may be no easy answers, but about which students may face difficult decisions as they reach adulthood. There is one article per chapter. Teaching strategies are given.

In the *Technology* booklet, students are introduced to eight life-science related technological questions or advances. Based on overall unit topics, a teacher guide is provided with background information, teaching tips, and references.

Less controversial, but grounded in current scientific research, *Mr. Wizard's Science and Technology Videodisc Series* presents a series of full-motion videoreports designed to make the real world of research and technology real to students. The ***Merrill Life Science*** *Science and Technology Videodisc Teacher Guide* provides helpful extensions and alternate uses to help make students aware of their changing world and what roles they might eventually play in it.

Integrating Life Science Across the Curriculum

***Merrill Life Science** includes many opportunities to relate life science to other sciences and other disciplines across the curriculum.*

No subject exists in isolation. At appropriate points in ***Merrill Life Science,*** attention is called to other sciences and other disciplines across the curriculum in the *Student Edition,* the *Teacher Wraparound Edition,* and in the *Teacher Classroom Resources.*

Student Edition

- **Connect to ...**—margin features that relate basic questions in physics, chemistry, or earth science to life science
- **Global Connections**—a cross-curricular unit-end feature with a global perspective
- **Science and Literature/Science and Art**—a one-page feature on either art or literature related to the unit topic at the close of each unit

Teacher Wraparound Edition

- **Connect to ...**—suggested answers to the questions about physics, chemistry, and earth science that are asked in the *Student Edition* margin features.
- **Global Connections/Science and Literature/Science and Art**—provide teaching tips for these unit-end features
- **Cross Curriculum**—relates life science to health, history, language, math, and other disciplines

Teacher Classroom Resources

- **Science Integration Activities**—full-page activities related to chemistry, physics, or earth science
- **Cross-Curricular Connections**—relate life science to home economics, social studies, architecture, and other disciplines

Multicultural Awareness

"Multicultural education is an idea stating that all students, regardless of the groups to which they belong, such as those related to gender, ethnicity, race, culture, social class, religion or exceptionality, should experience education equality in the schools."—James Banks

American classrooms reflect the rich and diverse cultural heritages of the American people. Students come from different ethnic backgrounds and different cultural experiences into a common classroom that must assist all of them to learn. The diversity itself is an important focus of learning experience.

Diversity can be repressed, creating a hostile environment; ignored, creating an indifferent environment; or appreciated, creating a receptive and productive environment. Responding to diversity and approaching it as part of every curriculum is challenging to a teacher, experienced or not. The goal of science is understanding. The goal of multicultural education is to promote the understanding of how people from different cultures approach and solve the basic problems all humans have in living and learning. ***Merrill Life Science*** addresses this issue. In the Multicultural Awareness strategies in the *Teacher Wraparound Edition,* students are encouraged to become aware of how cultures around the world deal with topics related to life science. The intent is to build awareness and appreciation for the global community in which we all live.

Two books that provide additional valuable information on multicultural education are:

Banks, James A. (with Cherry A. Mcgee Banks). *Multicultural Education: Issues and Perspectives.* Boston: Allyn and Bacon, 1989.

Banks, James A. (with others). Curriculum Guidelines for Multiethnic Education. Washington, D.C.: National Council for the Social Studies, 1977.

Student Text Features

Chapter Organization

Chapter Introduction

Merrill Life Science chapter openers give students a hands-on way to explore a concept they'll learn more about in the chapter.

Each chapter of ***Merrill Life Science*** opens with a striking full-page photograph illustrating a concept introduced in the chapter. The chapter introductory paragraph establishes a relationship between the photo and the chapter.

FIND OUT!, an inquiry activity, is featured at each chapter's beginning. Designed to give students an opportunity to make observations and raise questions about upcoming chapter content, these brief activities require few materials. **FIND OUT!** provides a hands-on introduction to chapter content.

Gearing Up, an outline of the chapter sections, assists students in previewing the chapter and focusing on its most important ideas.

What's Next? leads students into the body of the chapter.

Benefits

- Students will confidently approach the chapter content with its friendly, real-life opening.
- Students are focused on and ready to learn the chapter concepts after the opening inquiry activity.

Lessons

Numbered lessons divide chapter content into amounts that are manageable for the students.

Each chapter is divided into three to five sections, with student objectives listed at the beginning of each. When appropriate, major sections are divided into subsections.

Objectives provide the framework for the review questions, chapter review, review worksheets (in the *Teacher Classroom Resources*), and chapter tests.

New Science Words, at the beginning of each lesson, are listed to help students preview the vocabulary words they will learn in that section.

Benefits

- Students can more easily understand the relationships between topics and the frameworks of different areas of life science.
- Students gain confidence as they see they can progress section by section to gain a full understanding of the chapter content.

Activities

Activities allow students to become fully engaged in the "doing" of science.

Each chapter of ***Merrill Life Science*** has two full-page laboratory activities that can be done in one laboratory period, along with shorter MINI-Labs, and Flex Your Brain activities.

Activities provide students with opportunities to learn by doing and to actively engage in learning concepts by using scientific methods. Many of the activities include a problem statement, materials list, step-by-step procedures. They also offer questions to help students review their observations, form hypotheses, design plans to test their hypotheses, and then analyze and draw conclusions about their investigations.

Designing an Experiment Activities are open-ended, student-directed activities. Students are guided to form hypotheses, design plans to test their hypotheses, and then analyze and draw conclusions about their investigations.

MINI-Labs, provide another opportunity to practice science process skills while using scientific methods. These hands-on activities can be completed in a short amount of time, and require simple, common materials.

Flex Your Brain is a self-directed activity students use while investigating content areas. It occurs as a one-page activity in Chapter 1, accompanying a text section discussing critical thinking and problem solving. The Teacher Resource Guide of the *Teacher Classroom Resources* provides a master with spaces for students to write their responses. See page 19T of this teacher guide for more information about **Flex Your Brain**.

Benefits

- The activities allow students to practice the thinking process skills in widely different contexts; this makes them better problem solvers for life.
- The different types of activities meet the diverse needs of students and teachers.

Section Review

The Section Review provides content assessment of chapter concepts.

Following each numbered section are three to five Section Review questions that may include an Apply question that demands a higher level of thinking to answer. A science integration question connects life science to physics, chemistry, or earth science.

Special Text Features

Special text features engage students as they make connections across the curriculum, and practice problem solving and decision making.

Connect to Other Sciences

Most numbered sections have a Connect to Physics, Connect to Earth Science, or Connect to Chemistry feature. These features integrate particular life science concepts to related concepts in other science disciplines.

Benefits

- Students see the natural interconnectedness of the sciences, giving them a broader and more in-depth understanding of science concepts.

Science and Society

Each chapter has a **Science and Society** section integrated into the chapters, featuring important science-related topics that affect society. At the conclusion of these sections is a **You Decide** feature that provides students an opportunity to practice critical-thinking and problem-solving skills as they formulate opinions on the issue discussed.

Benefits

- Students maintain high interest in science class as they see the relevance to society and their own lives.
- Students practice reasoning, critical thinking, decision making, and communicating their ideas after studying the issues.

Technology

Technology features in each chapter show "real-life" applications of the science concepts. A critical thinking question at the end of the feature gives students an opportunity to interact with the feature.

Benefits

- This feature makes science concepts interesting and relevant to students.
- Understanding technology will enable students to better handle their future challenges in terms of decisions about the environment and their careers.

Problem Solving

Problem Solving in each chapter feature stories about how people confront a problem linked to the chapter content. The feature's real-life application question engages students in the solving of the problem.

Benefits

- Students will benefit by learning problem-solving skills in a relevant and interesting way.

Margin Features

In Your Journal features encourage students to write creatively, record data, plan experiments, write letters, and state opinions on topics that relate to life science.

Science and Reading and **Science and Math** are features that provide problems or projects that require reading or math calculations to further explore science concepts. This reinforces and integrates reading and math skills in the science curriculum.

Periodic **Did You Know?** features present students with interesting facts related to the content being developed.

Eco Tips are simple for ways to have a positive effect on the environment. Eco Tips extend the environmental concepts of the book and increase students' sense that their actions have an effect in the world.

Periodic **Student notes** are blue margin questions about main ideas; they can serve as a study guide for students.

Chapter End

Chapter-end materials review, reinforce, and extend concepts.

The three-page chapter-end material begins with a Summary that concisely reviews the major concepts and principles of the chapter. Each summary statement is numbered to correspond to a section objective.

- **Key Science Words** list the chapter vocabulary terms in order of occurrence.
- **Understanding Vocabulary** is an exercise in matching definitions to key terms.
- **Checking Concepts** provides multiple-choice recall questions.
- **Using Lab Skills** provides assessment of various chapter activities.
- **Think and Write Critically** provides higher-level thinking questions.
- **Apply** questions require application of chapter concepts.
- **More Skill Builders** require students to use process skills as they answer content-related questions.
- **Projects** provide ideas for researching, creating, or investigating topics based on chapter concepts.

Answers to all chapter-end questions are provided in the margin of the *Teacher Wraparound Edition.*

Unit Introductions

What's Happening Here? photographs and text combine for an inquiry strategy to introduce each unit. The photos portray a puzzling situation, or an intriguing relationship related to the content of the upcoming unit.

End of Unit Features

Units close with four-page features that include:

Global Connections feature a two-page world map with interdisciplinary features from around the world. These features provide integration of sciences, strengthen geography knowledge and provide strong multicultural perspective.

Careers are features in which students use real-world applications of science knowledge acquired in the unit. Students can see a range of careers available to people with a variety of educational backgrounds. **Readings** provide several books or magazine articles related to the unit content.

Science and Literature or **Science and Art** features integrate science concepts to literature or art. Included may be fiction or nonfiction book excerpts, paintings, sculpture, photos, or music related to the unit.
Students are actively engaged at the close of the feature as they respond to application or critical-thinking questions.

Skill Reinforcement

Skills are reinforced throughout ***Merrill Life Science.*** Each section ends with a **Skill Builder** feature that challenges students to practice basic process skills on a specific science concept. The skill to be learned or practiced is explained in the **Skill Handbook,** a 14-page illustrated reference in the back of the student text. Specific examples are used to guide students through the steps of acquiring skills. **More Skill Builders** in the **Chapter Review** material also reference the **Skill Handbook.**

Appendices

There are four appendices that may be used to expand student learning or application of concepts. **Appendix A** shows care and use of the microscope and techniques for making a wet mount slide. **Appendix B** shows SI units and English/Metric conversions. It provides a table for quick reference to SI base units and derived units. **Appendix C** includes procedures students should practice to ensure lab safety. A chart of safety symbols that are used throughout the text activities alerts students to possible laboratory hazards. **Appendix D,** Classification, is available for reference to the members of the five kingdoms throughout the course.

Projects

Eight long-term projects are designed to be worked on during the course of individual units. Each project lends itself to having students work cooperatively in groups on topics that have students role-playing famous scientists or assessing people's real opinions about healthful food.

Glossaries and Index

The **English Glossary** and **Spanish Glossary** provide students with a quick reference to key terms and their pronunciations within the text. Page references are provided for all New Science Words from each chapter so students can easily locate the page on which a word is defined. Because it is complete and cross-referenced, the **Index** allows text material to be found quickly and easily.

Teacher Wraparound Edition

The ***Merrill Life Science*** *Teacher Wraparound Edition makes teaching life science EASY.*

Merrill Life Science *Teacher Wraparound Edition* has been designed and arranged to provide you with maximum support for maximum results. Support materials and strategies are there for activities, for curriculum integration, for alternate assessment, for planning, and for meeting the diverse needs of all your students.

Support for Activities

The Teacher Wraparound Edition provides you with information to make your science labs fun and productive.

In the margins of the *Teacher Wraparound Edition,* you will find:

- Materials and Teaching Tips for **Find Out!** activities;
- Objectives, Process Skills, Teaching Tips, and Answers for full-page activities;
- Objectives, Preparation, Materials, Teaching Tips, Guidance, and Answers for Designing an Experiment activities; and
- Materials and Suggested Outcomes for MINI-Labs.

Support for Integration

Cross-curricular connections in the *Teacher Wraparound Edition* save you research time. Look for **Cross Curriculum** teaching tips in the *Teacher Wraparound Edition* margins. Through the explicit integration of life science with other science and nonscience disciplines, students will come to appreciate that life science is connected to almost everything they do.

Global Connections features connect life science to other sciences as well as literature, art, history, and geography. Teacher background is provided for the teacher who is unfamiliar with that area. Help is provided in terms of lesson objectives, motivating ideas, teaching strategies, background, and extension strategies.

Teacher support for the **Connect to Earth Science, Connect to Physics,** and **Connect to Chemistry** features includes suggested answers to questions. **In Your Journal, Science and Math,** and **Science and Reading** features are supported with suggested responses to questions.

Support for Assessment

Use alternate assessments to confirm your students' grasp of content and skills.

Implementation of science reforms and the coming national science standards require that teachers use alternate forms of assessment to meet the changing curriculum. New assessment forms emphasize the processes of science as well as the products of science learning. Students are frequently challenged to solve complex problems or discover the many options of an open-ended activity. Assessment can be made as students create, produce, or perform; as they use critical-thinking and problem-solving skills; and as they perform tasks with real-world applications.

In the *Teacher Wraparound Edition,* every chapter opener lists assessment options for that chapter. **Assessment Options** are categorized in one of four areas:

- Portfolio suggestions,
- Performance-assessment items,
- Content-assessment features, and
- Group-assessment opportunities.

Portfolio suggestions are listed on the first page of the Chapter Review. Assessment in the form of modifications to activities are suggested in the teacher margin. Oral-assessment options are provided where appropriate so that teachers can quickly determine if students have internalized concepts and are ready to proceed. See an extended discussion of assessment on pages 22T through 24T in this guide.

Planning/Preparation/Background

All the resources and references you'll need to do your short-term planning are in one place—the *Teacher Wraparound Edition.* Use the two-page chapter planner at the beginning of each chapter to:

- Overview objectives,
- Plan content development,
- Select activities and materials, and
- Select from a wide assortment of program resources in the *Teacher Classroom Resources.*

For assistance in long-term planning, use the Planning Guide on page 16T. As the Planning Guide shows, the program has the flexibility to be used for full-year or semester courses with life science.

Meets Needs of ALL Students

The *Teacher Wraparound Edition* helps you meet the diverse needs of your students. ***Merrill Life Science*** *Teacher Wraparound Edition* provides you with a wide array of features and options designed to make science learning a successful experience for all your students.

To assist you in meeting the needs of all of your students, the *Teacher Wraparound Edition* provides:

- suggestions for using three types of ability-level worksheets for each section,
- strategies for gifted and mainstreamed students,
- guidelines for meeting the needs of challenged students,
- reteaching and extension strategies,
- cooperative learning strategies for activities, and
- multicultural awareness activities and topics for discussion.

Teacher Classroom Resources

An Effective Teaching Model

The ***Merrill Life Science*** *Teacher Wraparound Edition* delivers the collective teaching experience of its authors, consultants, and reviewers. By furnishing you with an effective teaching model, it saves you preparation time and energy. You, in turn, are free to spend that time and energy on your most important responsibility—your students.

As a professional, you will be pleased to find that this program provides you with readily available activities that will engage your students for the entire class period. Each major section of every chapter may be considered an individual lesson that includes a preparation section followed by a comprehensive three-part teaching cycle that, when utilized consistently, will result in better cognitive transfer for your students.

Theme Development begins each chapter, describing how one or all of the themes of the book are incorporated into the chapter.

Chapter Overview, following Theme Development, lists and decribes the material in each of the sections of the chapter.

Chapter Vocabulary is then listed for you in the order in which the words occur in the chapter.

FIND OUT! activities open the chapter. These activities help you focus the students' attention on the chapter and get them into the material immediately. Everything you will need to help the students complete the FIND OUT! is given alongside the student page. Any advanced preparation or materials are listed for you.

Cooperative Learning strategies are suggested and teaching tips are given to help you conduct the FIND OUT! activities.

The **OPTIONS** section of the chapter openers include activity-based strategies to help you with gifted or mainstreamed students you may have in your class.

PREPARATION

Preparation is an extremely important part of teaching any lesson. The **PREPARATION** section contains **SECTION BACKGROUND** that provides you with science content relevant to the section. Also provided for you in **PREPLANNING** is a list of things you may wish to do to prepare for the section in advance. The **PREPARATION** section will give you the foundation for keeping your lesson instructionally sound.

1 MOTIVATE

The first step to teaching any lesson is to motivate the students. A **Motivate** idea is provided for each section. The ideas may include demonstrations, cooperative learning techniques, audiovisuals, brainstorming, or other ideas that help motivate and focus the class so the lesson can begin. One unique way to get students interested in the section content is by using the FLEX YOUR BRAIN activity. This activity will give students confidence to begin learning section content by helping them find out what they already know and don't know about a certain topic.

Another important way to help motivate students is by connecting the current lesson with previous lessons or to common knowledge possessed by most students. The **Tying to Previous Knowledge** sections help you do this. Research has shown that connecting ideas provides for greater concept retention among students.

2 TEACH

The primary aim of the ***Merrill Life Science*** *Teacher Wraparound Edition* is to give you the tools to accomplish the task of getting concepts of life science across to your students. Many teaching suggestions are given to you under a series of clearly defined headings. Each section contains, under the **CONCEPT DEVELOPMENT** heading, ways for you to develop the content of the section. This may include a series of questions for you to ask students followed by possible student responses. These questions are a tool for you to use to develop the concepts in the section. Demonstration ideas also appear in Concept Development when they can help you develop student interest in the concepts to be taught.

To help you monitor and adjust your teaching to what students are learning, a **CHECK FOR UNDERSTANDING** idea is provided for you in each section at a point where it seems most appropriate. For students who are having trouble, a **RETEACH** suggestion immediately follows the CHECK FOR UNDERSTANDING hint. The RETEACH tip is a way to teach the same concepts or facts differently to adjust to students' individual learning styles. For those students in the class who do not need additional help understanding the lesson, there are suggestions for **ENRICHMENT** to allow these students to go on while others are reviewing the section. These ENRICHMENT ideas are provided for you in the **OPTIONS** boxes at the bottom of the *Teacher Wraparound Edition* pages.

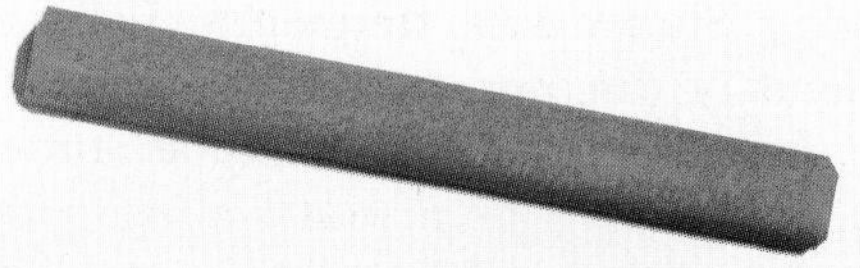

The components of the Teacher Classroom Resources *for* ***Merrill Life Science*** *provide background information and comprehensive teaching material to aid in the effective teaching of life science.*

The **TEACH** step of the *Teacher Wraparound Edition* provides many other strategies to help you teach the content of the section. A section may have a way to connect life science to another discipline in the **CROSS CURRICULUM** teaching tip. **MINI QUIZZES** assess students' mastery of the material. These can also be used as CHECK FOR UNDERSTANDING activities. An annotation key is provided with each MINI QUIZ to help you refer students to the page on which the answer is found.

The **REVEALING MISCONCEPTIONS** teaching tip suggests questions you may ask to elicit student misconceptions about section content or provide you with a possible misconception that is held by students and a method of correcting that misconception.

TEACHER F.Y.I. tips give you additional information to help you teach the section. These may be everyday applications or connections to other disciplines. Answers are given in the margin to any text questions to be answered by students. Also, for your convenience, we have highlighted the key concepts.

COOPERATIVE LEARNING suggestions are given in various locations of the *Teacher Wraparound Edition.* They are found in the FIND OUT activities, in the lab activities, and when appropriate within the Teach section. In each COOPERATIVE LEARNING tip, a specific cooperative grouping strategy is given to help you make the most of these activities.

You are also provided with information needed to teach each of the special features of the Student Edition such as: **MINI-Labs, Science and..., TECHNOLOGY, PROBLEM SOLVING, ACTIVITIES,** and **Connect to...** features.

And There's More

Our special *Teacher Wraparound Edition* has been designed to provide you with the most meaningful teacher information in the most convenient way.

The **OPTIONS** box, located at the bottom of most pages, contains a variety of items. The OPTIONS Box on the Chapter Opener pages provides you with ideas to help you teach the gifted students and the mainstreamed students in your class. Both of these groups pose a challenge to any teacher, and these tips are designed to help you teach the content of the chapter to these exceptional students.

In every classroom there are students with a variety of different ability levels. The OPTIONS box on the first two pages of each section shows you reduced copies of pages in the *Teacher Classroom Resources* to help you deal with all of the different ability levels. The Study Guide Master can be used by all students to review the basic content of the section. The Reinforcement Master will allow average students to go beyond the basic concepts of the chapter while reinforcing them. The Enrichment Master will provide those students who quickly master section content with an additional challenge to explore the ideas in the section more fully.

On the pages in the chapter, the OPTIONS Boxes provide you with Inquiry Questions and oral assessment questions to use with all students. These are suggested questions you may ask that require critical thinking on the part of the students. The questions relate specifically to the content of the page on which they are located.

With the materials present in the TEACH step of the teacher edition, your efficiency and productivity as a teacher will increase. The TEACH section brings together the major elements that form a sound teaching approach.

3 CLOSE

Closing the lesson is the last but one of the most important steps of teaching any lesson. This step is the complement in many ways to the MOTIVATE step. While the motivate step helps students become involved in the lesson, the Close step helps students bring things together in their minds to make sense of what went on in the lesson. The ***Merrill Life Science*** *Teacher Wraparound Edition* gives you a variety of ways to provide effective closure to a lesson. The Close options help summarize the section, bridge to the next lesson, or provide an application of the lesson.

As you review the ***Merrill Life Science*** *Teacher Wraparound Edition,* you will discover that you and your students are considered very important. With the enormous number of teaching strategies provided by the teacher edition, you should be able to accomplish the goals of your curriculum. The materials allow adaptability and flexibility so that student needs and curricular needs can be met.

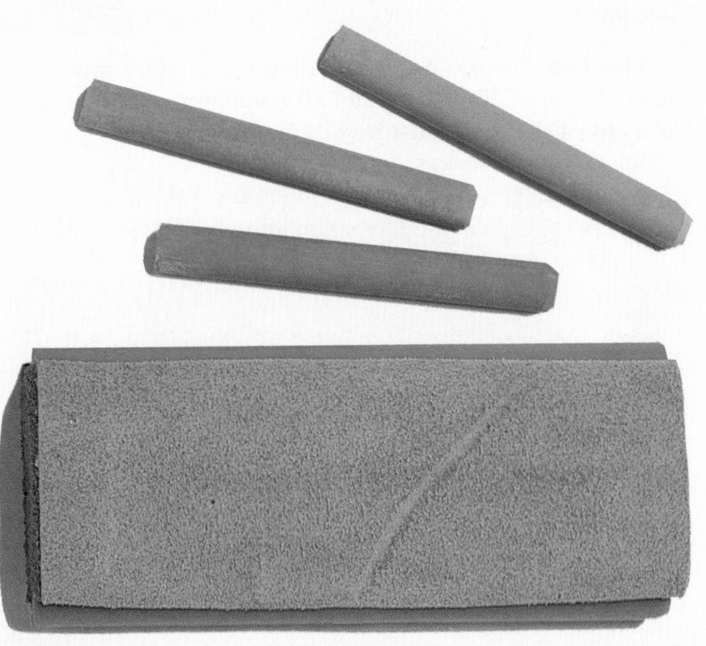

In addition to the wide array of instructional options provided in the Student and Teacher Wraparound Editions, ***Merrill Life Science*** also offers an extensive list of support materials and program resources. Some of these materials offer alternative ways of enriching or extending your life science program, others provide tools for reinforcing and assessing student learning, while still others will help you directly in delivering instruction.

For your convenience, appropriate resources are called to your attention in PROGRAM RESOURCES boxes and VideoDisc reference throughout the *Teacher Wraparound Edition.*

Hands-On Activities

Activity Masters provide a worksheet for every text Activity and MINI-Lab. This worksheet reproduces the complete text for the Activity or MINI-Lab, provides any needed table for data, and illustrates procedures.

The **Laboratory Manual** is a learning-through-doing program of activities. Based on the philosophy that scientific knowledge is acquired through individual activity and experimentation, the manual consists of many varied laboratory activities designed to reinforce concepts presented in ***Merrill Life Science.*** Laboratory investigations for each Student Edition chapter require students to work through a problem by observing, analyzing, and drawing conclusions. Some of the chemistry labs are microlabs, which make your lab program more efficient and cost effective by reducing the amounts of materials needed. The **Laboratory Manual, Teacher Annotated Edition**, consists of the Student Edition pages with teacher answers on reduced pages in the back. The reduced pages may include suggestions for alternate materials, teaching tips, and sample data, as appropriate, plus answers to all student questions.

The **Lab Partner Software Package** enables students to use an Apple or IBM computer to graph any quantitative data from any of the text Activities or laboratories.

Reinforcement Resources

The **Study Guide** worksheets are suitable for all students; they are closely tied to the student text and require recall of text content.

Reinforcement worksheets are for students of average and above-average ability; a variety of formats are used to reinforce each text lesson.

Concept Mapping masters challenge students to construct a visual representation of relationships among particular chapter concepts. This booklet is developmental in its approach; early concept maps are nearly complete; later ones provide only a skeleton and linking words.

Chapter Review masters are two-page review worksheets consisting of 25 questions for each chapter. They can be used to prepare for tests, as alternate tests, and as vocabulary review.

Chapter Review Software provides chapter-end review questions for student use. Based on the review questions found in the student text at the end of the chapter, the software presents questions in random order. Feedback for incorrectly answered questions provides the page in the student text where the answer is found.

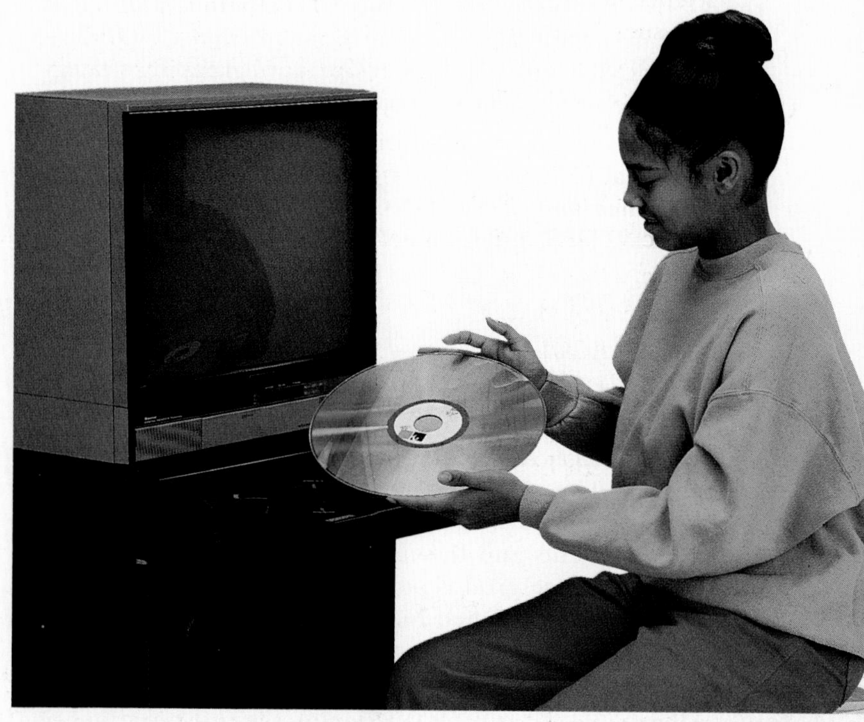

Enrichment Resources

Enrichment worksheets are tailored for students with above-average ability; a wide range of formats allows students to design, interpret, research, and create based on the text of each lesson.

Cross-Curricular Connections masters are interdisciplinary worksheets that relate science to other disciplines. There is one worksheet per chapter; the emphasis is on *doing* the related discipline whenever possible.

Science and Society worksheets show the impact of science on current societal issues and problems. There is one worksheet for each chapter; each ends with a question requiring students to draw conclusions and/or make decisions.

Critical Thinking/Problem Solving worksheets consist of a reading selection related to a chapter topic and questions that help to develop critical-thinking skills while applying concepts learned in the classroom to new situations.

Technology masters explain how something works and/or integrates the sciences. The topics are tied to the student text.

The seven-disc **Science and Technology Videodisc Series (STVS)** contains more than 280 full-motion videoreports on a broad spectrum of topics relating to current research in various science fields, innovations in technology, and science and society issues. In addition to reinforcing science concepts, the videoreports are ideal for illustrating science methods, laboratory techniques, and careers in science. Each disc is centered on a particular science topic.

- Disc 1: PHYSICS
- Disc 2: CHEMISTRY
- Disc 3: EARTH AND SPACE
- Disc 4: PLANTS AND SIMPLE ORGANISMS
- Disc 5: ANIMALS
- Disc 6: ECOLOGY
- Disc 7: HUMAN BIOLOGY

The **Videodisc Correlation** book contains the complete correlation with bar codes of OPTICAL DATA's Videodisc images to the content of *Merrill Life Science.*

Science Integration Activities broaden the students' understanding of life science by introducing masters with activities that relate to another science. There is one activity per chapter, each chosen for its relationship to the life science chapter content.

Assessment Resources

Alternate Assessment in the Science Classroom is a resource book that provides a rationale and strategies to use for authentically assessing student progress during your science course.

Performance Assessment contains eight unit-based performance-assessment activities that students can plan and perform as well as teaching strategies. It also contains one Skill Assessment for each chapter, each with teaching strategies.

Performance Assessment in Middle School Science (PAMSS) provides specific strategies for assessment, emphasizing use of the Performance Task Assessment Lists and rubrics that are included.

Chapter and Unit Tests includes one four-page test for each chapter and one two-page test for each unit. Questions range from simple recall to higher-order thinking process.

Computer Test Banks, available in Apple, IBM, and Macintosh versions, provide the ultimate flexibility in designing and creating your own test instruments. Select test items from two different levels of difficulty, or write and edit your own.

Additional Resources

The **Lesson Plan** booklet contains complete lesson plans for every lesson in the student text. Also included are references to all program components of *Merrill Life Science.*

The **Spanish Resources** book provides Spanish translations of Objectives, Summary statements, and Key Terms and their definitions for every chapter. Also included is a complete English/Spanish glossary to *Merrill Life Science.*

The **Color Transparency Package** contains 50 full-color transparencies, as well as the **Transparency Masters**, blackline versions of the transparencies. Also included are a student worksheet for each transparency and instructions for using the transparencies. These materials are conveniently packaged in a three-ring binder.

Planning for Emphasis

The purpose of a Planning Guide is to aid the teacher in developing a course that will offer the best possible program for students.

Merrill Life Science provides flexibility in the selection of topics and content, which allows teachers to adapt the text to the needs of individual students and classes. In this regard, the teacher is in the best position to decide what topics are to be presented, the pace at which the content is covered, and what material should be given the most emphasis. To assist the teacher in planning the course, a Planning Guide has been provided.

Two Semester Course

Merrill Life Science may be used in a full-year course of two semesters covering the text activities and chapter-end materials. It is assumed that a year-long course in life science will have 180 periods of approximately 45 minutes each. In the Planning Guide, each chapter is listed along with the number of class sessions recommended for teaching the chapter. Use the Planning Guide to gauge the amount of time you will spend on each topic.

One Semester Course

Through the selection of specific units, chapters, and sections, ***Merrill Life Science*** also may be used in a one-semester course that conforms to a local curriculum emphasis. Alternatives are presented in the Planning Guide for one-semester, 90-day courses in basic life science (LS), health (H), human body (HB), ecology (E), and science, technology, and society (STS). Columns in the Planning Guide list the number of class sessions suggested for the study of the chapter sections in each of the one-semester alternatives.

Flexibility

Please remember that the Planning Guide is provided as an aid in planning the best course for your students. You should use the Planning Guide in relation to your curriculum and the ability levels of the classes you teach, the materials available for activities, and the time allotted for teaching. You may decide to extend the scope and time devoted to certain topics through the use of enrichment activities and *Teacher Classroom Resources* materials. These materials are listed in the teacher margins and on the interleaf pages preceding each chapter. The Planning Guide will assist you in developing and following a schedule that will enable you to complete your goals for the school year or semester.

Planning Guide

CLASS SESSIONS

	Full-Year	One-Semester Options				
Chapter	**LS**	**LS**	**H**	**HB**	**E**	**STS**
1	5	3	3	3	4	4
2	6	5	4	5	—	—
3	7	4	4	4	—	—
4	7	4	4	4	—	—
5	7	5	4	5	—	8
6	6	4	—	4	4	8
7	6	3	3	3	4	—
8	5	5	4	—	5	4
9	5	4	4	—	5	4
10	6	4	—	—	5	4
11	6	5	—	—	5	4
12	6	5	—	—	4	4
13	6	5	—	—	5	—
14	8	5	—	—	5	—
15	6	5	—	—	5	—
16	6	5	—	—	5	—
17	5	4	—	—	4	—
18	7	—	7	7	—	1
19	7	—	7	5	—	1
20	7	—	7	6	—	1
21	7	—	7	6	—	1
22	7	—	7	6	—	1
23	7	—	7	7	—	5
24	7	—	7	5	—	5
25	7	—	7	5	—	5
26	7	5	—	5	10	10
27	7	5	—	5	10	10
28	7	5	4	5	10	10
TOTALS	**180**	**90**	**90**	**90**	**90**	**90**

(LS) Life Science; (H) Health; (HB) Human Body;
(E) Ecology; STS (Science, Technology, and Society)

Concept Maps

In science, concept maps make abstract information concrete and useful, improve retention of information, and show students that thought has shape.

Concept maps are visual representations or graphic organizers of relationships among particular concepts. Concept maps can be generated by individual students, small groups, or an entire class. ***Merrill Life Science*** develops and reinforces three types of concept maps—the **network tree, events chain,** and **cycle concept map**—that are most applicable to studying science. Examples of the three types and their applications are shown on this page.

Students can learn how to construct each of these types of concept maps by referring to pages 688 and 689 of the **Skill Handbook.** Throughout the course, students will have many opportunities to practice their concept mapping skills, as there is at least one concept mapping Skill Builder per chapter.

Network Tree

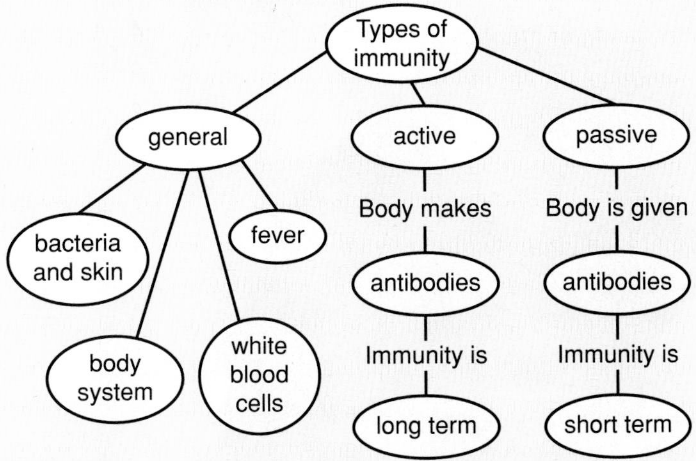

Applications:

- shows causal information
- a hierarchy
- branching procedures

Events Chain

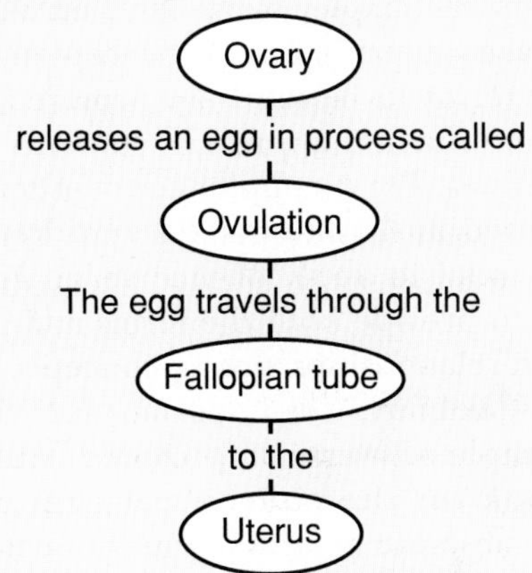

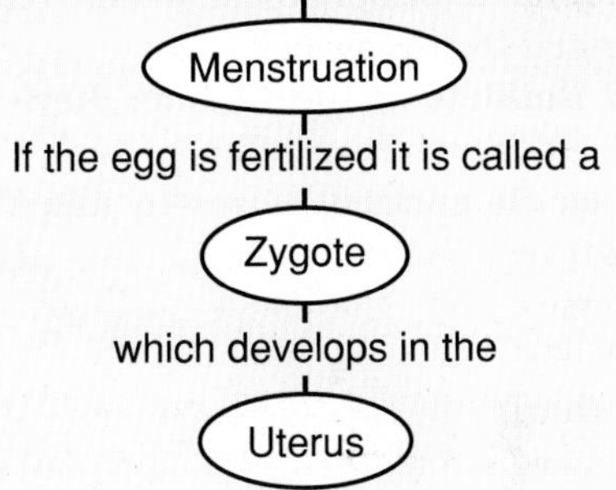

Applications:

- describes the stages of process
- the steps in a linear procedure
- a sequence of events

Cycle Concept Map

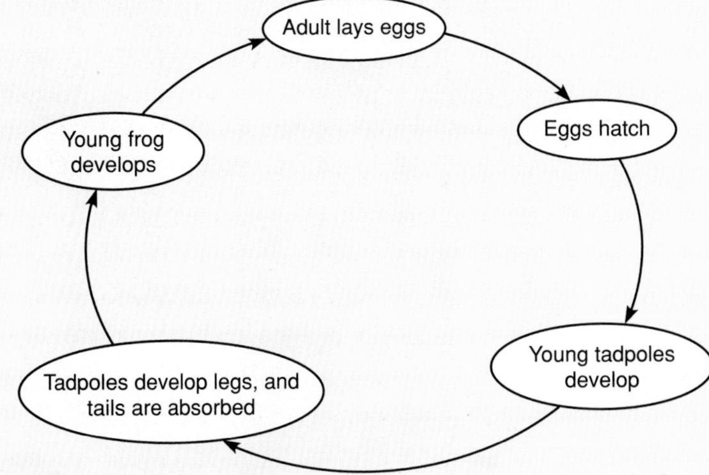

Application: shows how a series of events interact to produce a set of results again and again

Skill Builders/More Skill Builders

The section-ending **Skill Builders** and the **More Skill Builders** section of **Chapter Reviews** in early chapters direct students to make a specific type of concept map, and in most cases concept terms to be used are provided. In later chapters, in a developmental approach, students are given only general guidelines. For example, concept terms to be used may be provided and students are required to select the appropriate model to apply or vice versa. Finally, students may be asked to provide both the terms and type of concept map to explain relationships among concepts. When students are given this flexibility, it is important for you to recognize that, while sample answers are provided, student responses may vary. Look for the conceptual strength of student responses, not absolute accuracy. You'll notice that most network tree maps provide connecting words that explain the relationships among concepts. We recommend that you not require all students to supply these words, but many students may be challenged by this aspect.

More Skill Builders in the **Chapter Reviews** that ask students to make a concept map often provide the concept map format and specific concept terms to use. This will ensure more consistent student responses, and make assessment easier, than when students are asked to make their own concept map style.

Concept Mapping Booklet

The **Concept Mapping** book of the *Teacher Classroom Resources,* too, provides a developmental approach for students to practice concept mapping.

As a teaching strategy, generating concept maps can be used to preview a chapter's content by visually relating the concepts to be learned and allowing the students to read with purpose. Using concept maps for previewing is especially useful when there are many new key science terms for students to learn. As a review strategy, constructing concept maps reinforces main ideas and clarifies their relationships. Construction of concept maps using cooperative learning strategies as described in this Teacher Guide will allow students to practice both interpersonal and process skills.

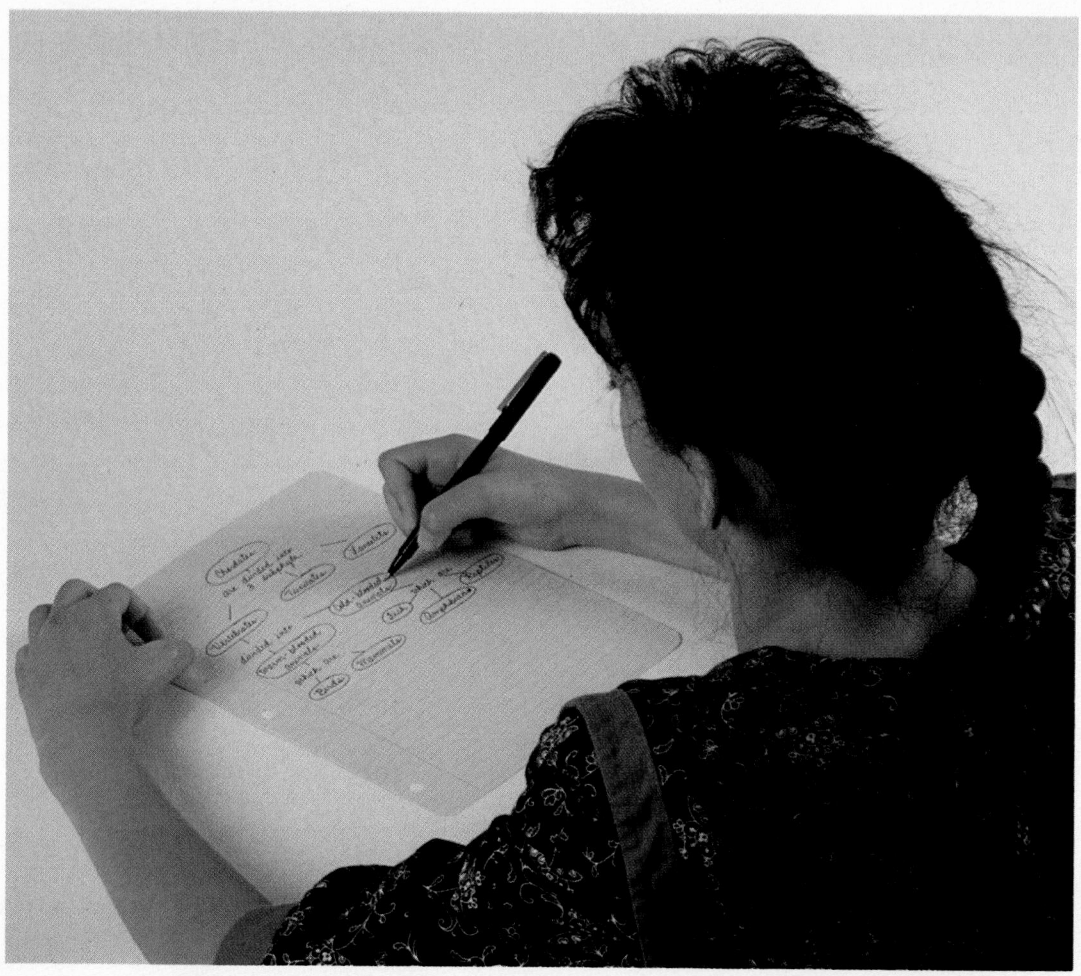

Flex Your Brain

What is FLEX YOUR BRAIN?

Flex Your Brain is a self-directed activity intended to assist students in developing critical-thinking skills while investigating content areas.

A key element in the coverage of problem-solving and critical-thinking skills in ***Merrill Life Science*** is a critical-thinking matrix called **Flex Your Brain.**

Flex Your Brain provides students with an opportunity to explore a topic in an organized, self-checking way, and then identify how they arrived at their responses during each step of their investigation. The activity incorporates many of the skills of critical thinking. It helps students to think about their own thinking and learn about thinking from their peers.

In a step-by-step approach, **Flex Your Brain** leads students to:

- **focus** on a single topic;
- **consider** their prior knowledge of the topic;
- **pose questions** regarding the topic;
- **hypothesize** answers to their questions;
- **evaluate** their **confidence** in their answer;
- **identify** ways to investigate the topic to prove, disprove, or amend their responses;
- **follow through** with their investigation;
- **evaluate** their **responses** to their questions, checking for their own misconceptions or lack of understanding;
- **reconsider** their question responses; incorporate new knowledge gained; and
- **review and share** the thinking processes that they used during the activity.

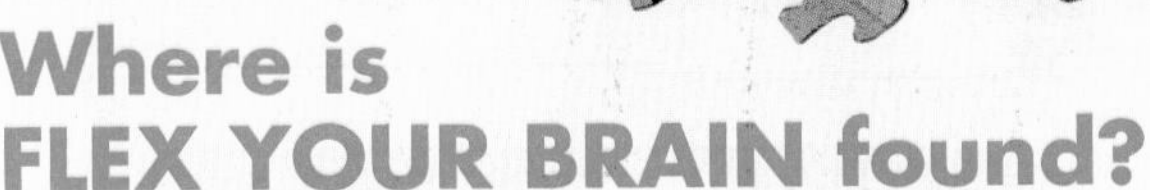

Where is FLEX YOUR BRAIN found?

In Chapter 1, on page 18 of the student text, is an introduction to the topics of critical thinking and problem solving. **Flex Your Brain** accompanies the text section as a one-page activity in Chapter 1. Brief student instructions are given, along with the matrix itself. A two-page version of **Flex Your Brain** appears as a worksheet on page 5 in the *Activity Worksheets* book of the *Teacher Classroom Resources.* This version provides spaces for students to write in their responses.

In the *Teacher Wraparound Edition,* suggested topics are given in each chapter for the use of **Flex Your Brain.** You can either refer students to Chapter 1 for the procedure or photocopy the worksheet master from the *Teacher Resource Guide.*

Use of **Flex Your Brain** is certainly not restricted to those topics suggested in the *Teacher Wraparound Edition.* Feel free to use it for practice in critical thinking about any concept or topic.

When to Use FLEX YOUR BRAIN

Flex Your Brain can be used as a whole-class activity or in cooperative groups, but is primarily designed to be used by individual students within the class. There are three basic steps.

1. Teachers assign a class topic to be investigated using **Flex Your Brain**.
2. Students use **Flex Your Brain** to guide them in their individual explorations of the topic.
3. After students have completed their explorations, teachers guide them in a discussion of their experiences with **Flex Your Brain,** bridging content and thinking processes.

Flex Your Brain can be used at many different points in the lesson plan.

▶**Introduction:** Ideal for introducing a topic, **Flex Your Brain** elicits students' prior knowledge and identifies misconceptions, enabling the teacher to formulate plans specific to student needs.

▶**Development:** **Flex Your Brain** leads students to find out more about a topic on their own, and develops their research skills while increasing their knowledge. Students actually pose their own questions to explore, making their investigations relevant to their personal interests and concerns.

▶**Review and Extension:** **Flex Your Brain** allows teachers to check student understanding while allowing students to explore aspects of the topic that go beyond the material presented in class.

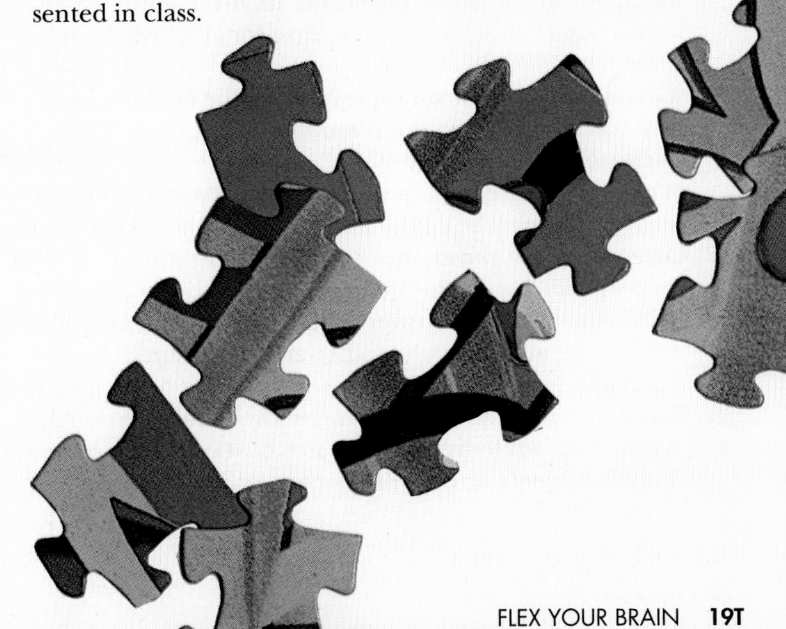

How to Use FLEX YOUR BRAIN

To assist teachers in using **Flex Your Brain,** an annotated version of the directions appearing on the student page is given below.

1. **Fill in the topic your teacher gives you.**
Focus students by providing a topic that you need to introduce, develop, or review. Topics should be fairly broad and stated in only a few words; for example, "frog life cycle" or "weathering." Later, when students are familiar with Flex Your Brain, you may want students to select the topic.

2. **Jot down what you already know about the topic. If you know more than five things, write them on another sheet of paper.**
As a class or cooperative group, this could be a brainstorming process. Otherwise, students should recall individually what they know about the topic. Set a reasonable time limit for this step.
As you track student progress through this step, look for evidence of misconceptions. Encourage students to draw knowledge from their own experiences as well as from academic sources.

3. **Using what you already know (Step 2), form a question about the topic. Are you unsure about one of the items listed? Do you want to know more about anything? Do you want to know what, how, or why? Write down your question.**
Students can pose their own questions about the topic and pursue their answers independently; or as a class or group, they can generate several questions. Groups may choose one of these questions to investigate, or the entire class can select one question for all to research.

4. **Guess an answer to your question. In the next few steps, you will be exploring the reasonableness of your answer. Write down your guess.**
The significance of this step is to get students to think, not just know the right answer. Based on what they already know about the topic, students should form hypotheses about the questions asked in Step 3. The correctness of their answers at this point is not as important as the means by which they arrive at them.
Guessing is not a poor means to arrive at an answer if students first consider the facts they know about the topic. Still, students may hesitate to write down answers that they aren't sure of. Encourage students to provide a "best guess."

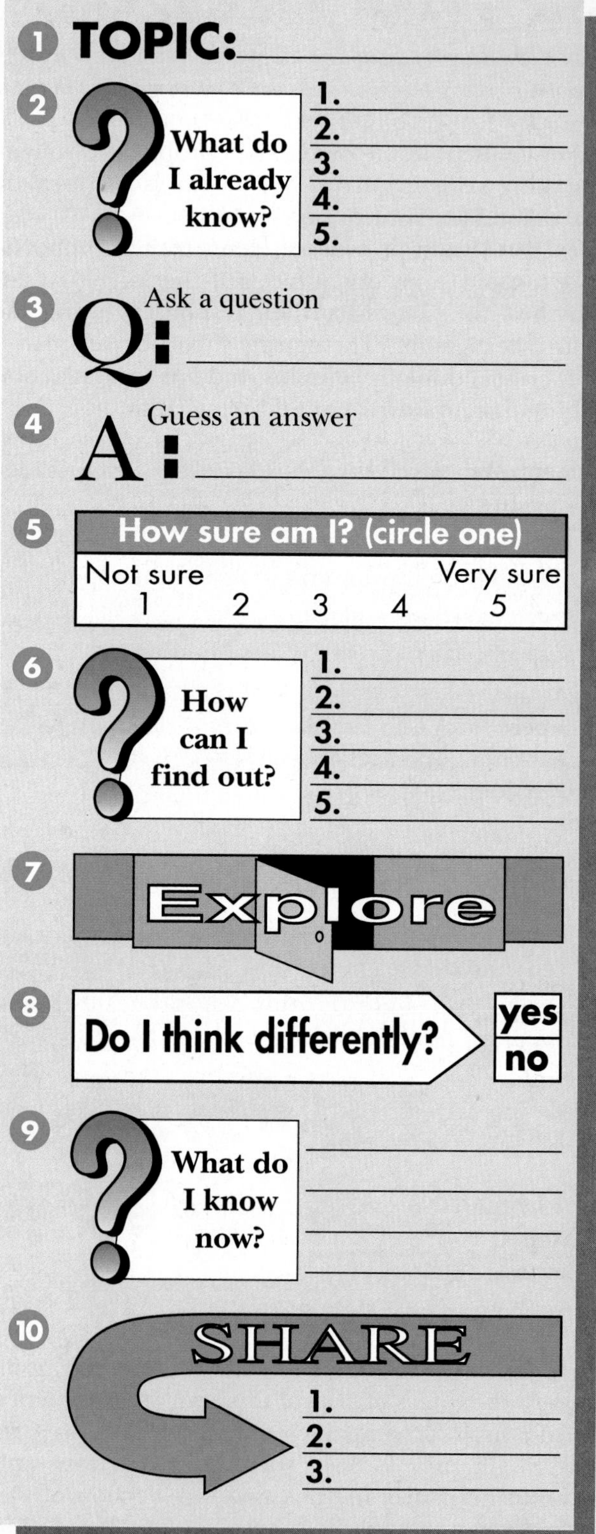

5 **Circle the number in the box that matches how sure you are of your answer in Step 4. This is your chance to rate your confidence in what you've done so far and, later, to see how your level of sureness affects your thinking.**
Students should evaluate their feelings at this point and begin to recognize the extent to which they are confident of their answers. Self-evaluation is critical to successful problem solving. This step becomes especially important when students review their thinking process in Step 10. Make sure students know that emotions and intuition are important aspects of problem solving.

6 **How can you find out more about your topic? You might want to read a book, ask an expert, or do an experiment. Write down ways you can find out more.**
The first time through this activity, you may want to have the class brainstorm ways to explore topics. Discuss the usefulness and appropriateness of the different means of investigation.
Evaluating the appropriateness of sources is an important skill to emphasize at this step. Encourage students to gain knowledge from the world at large, not just from expert sources. Be sure to bring out in your discussion that checking with a friend may be appropriate in some situations, but not in others.

7 **Make a plan to explore your answer. Use the resources you listed in Step 6. Then, carry out your plan.**
Students should use at least three resources before drawing conclusions. Encourage them to consider and use resources that might provide different perspectives on the topic in question.
Experimenting should be encouraged as a means of exploration. Review the scientific method when introducing this step and emphasize the science process skills.

8 **Now that you've explored, go back to your answer in Step 4. Would you answer differently? Mark one of the boxes.**
Students should consider whether they want to change their original answers. If their original answers require modifying in any way, they should mark the "yes" box.
At this step, students should begin to recognize their original misconceptions regarding the topic.

9 **Considering what you learned in your exploration, answer your question again. You may want to add new things you've learned or completely change your answer.**
If students' original answers were correct, request that they restate their answers incorporating new knowledge gained from their explorations. Otherwise, students should correct or modify their answers as needed.
This step leads students to draw conclusions. Help students to recognize how even a wrong answer in Step 4 contributed to the new answer by giving them a basis for exploration. Wrong answers are an important part of the scientific process.

10 **It's important to be able to talk about thinking. Choose three people to tell about how you arrived at your response in every step. For example, don't just read what you wrote in Step 2. Try to share how you thought of those things.**
The process of sharing ensures that students will reflect on what they have done. Encourage students to review each preceding step to determine how they arrived at their responses. Make sure students know that this is not an easy task and that they may struggle to put their process into words.
Verbalizing is important. When students put their own process into their own words, meaningful, internalized learning occurs. Discourage students from ridiculing other students during the sharing process.
The primary goal of **Flex Your Brain** is to improve students' ability to think effectively. Until their thought processes are somehow made clear to them, this will never occur. Step 10 achieves this by engaging students in metacognition, or the process of thinking about thinking.
Metacognition makes the process of thinking conscious and thus allows the process to be examined and improved. It is important to emphasize the importance of sharing and give adequate time to complete this step.

FLEX YOUR BRAIN is flexible.

Flex Your Brain was designed to be flexible in approaching critical thinking and problem solving. These directions are suggestions for one possible use of **Flex Your Brain,** but certainly not the only use. Teachers should feel free to adapt the activity to best suit their own needs and the needs of the diverse classes and students they teach.

Assessment

*What criteria do you use to assess your students as they progress through a course? Do you rely on formal tests and quizzes? To assess students' achievement in science, you need to measure not only their knowledge of the subject matter, but also their ability to handle apparatus, to organize, predict, record, and interpret data, to design experiments, and to communicate orally and in writing. The program presented by **Merrill Life Science** has been designed to provide you with a variety of assessment tools, both formal and informal, to help you develop a clearer picture of your students' progress.*

Merrill Life Science presents a wealth of opportunities for performance portfolio development. Each chapter in the student text contains projects, laboratory explorations, enrichment activities, skill builders, library research opportunities, and connections with life, society, and literature. Each of the student activities results in products. A mixture of these products can be used to document student growth during a grading period. At the beginning of each chapter in the ***Merrill Life Science*** *Teacher Wraparound Edition* you will find ASSESSMENT OPTIONS, a useful menu of assessment opportunities available in that particular chapter. These options are classified into performance, group, content, and portfolio suggestions.

PERFORMANCE ASSESSMENT

If we want students to be good problem solvers, then tests of problem-solving competency must logically assess performance *on problem-solving tasks. No paper-and-pencil test that grades an answer right or wrong can evaluate performance. Think about a musician, artist, basketball player, or writer. Their work is judged by a performance in a concert, a work of art, a game, or a book. These people do not take paper-and-pencil tests to demonstrate what they know—they perform!*

Performance assessments are becoming more common in today's schools. Science curriculums are being revised to prepare students to cope with change and with features that will depend on their abilities to think, learn, and solve problems. Although learning fundamental concepts will always be important in the science curriculum, the concepts alone are no longer sufficient in a student's scientific education. Performance assessments is a way of teaching and learning that involves both process and product. Teachers, as well as the students involved in these activities, will rate the performance and/or the products that result.

Performance Assessment in Middle School Science (PAMSS) and **Alternate Assessment in the Science Classroom**, two of Glencoe's professional handbooks, provide background information, specific examples, and details methods of evaluating performance assessments.

Merrill Life Science provides opportunities to observe student behavior both in informal and formal setting within the student text.

- Full-page activities contain suggestions for discussion or demonstration that will enable you to assess students' understanding of how the lab relates to the concepts presented in the text.
- MINI-Lab investigations present questions and other opportunities for students to demonstrate or practice skills related to content.
- Using Lab Skills, a section in the Chapter Review, allows students to practice skills by changing a variable in one of the Chapter Activities and MINI-Labs or by repeating the same exercise with entirely different materials.

Activity Worksheets booklet found in *Teacher Classroom Resources* provides formal assessment of student products. These sheets provide space for recording data and observations as students conduct Activities and MINI-Labs.

Merrill Life Science Teacher Wraparound Edition provides numerous alternate assessment ideas in the margins that will give you ideas for assessing mastery of skills and content Activities, MINI-Labs, and Skillbuilders.

***Merrill Life Performance Assessment* booklet** supplies another approach for assessing student mastery of concepts and skills in the laboratory. This booklet features twenty-eight one-page skill assessments and eight performance assessment tasks. These exercises give you additional opportunities to evaluate students' skills in handling laboratory equipment and students's knowledge of science processes. Rubrics and performance task assessment lists for evaluating these products can be found in Glencoe's **Performance Assessment in Middle School Science (PAMSS).**

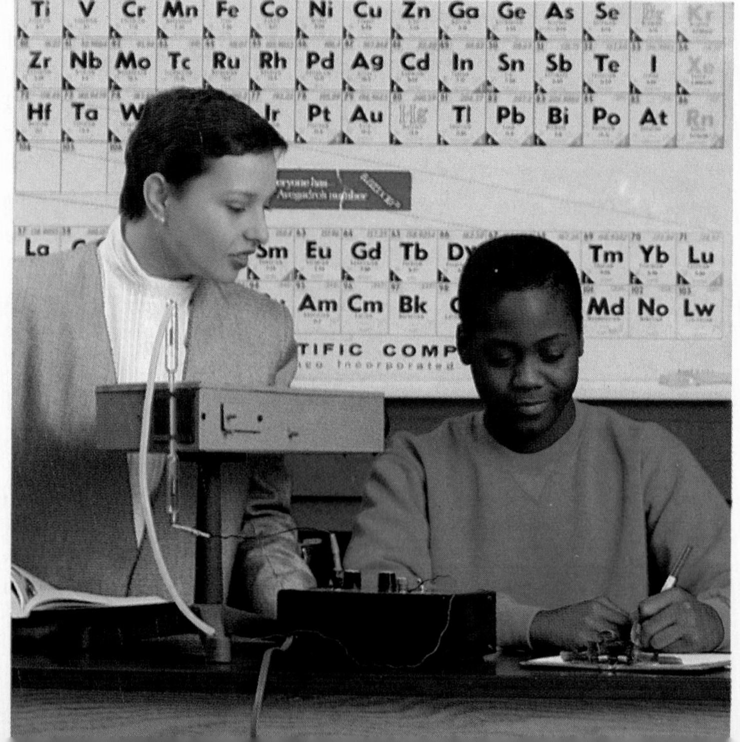

GROUP PERFORMANCE ASSESSMENT

The use of extended projects for the study of science and as an assessment tool to determine what students understand about science can be enhanced by having students work together in cooperative groups.

Recent research has shown that cooperative learning structures produce student learning outcomes for students of all ability levels. ***Merrill Life Science*** provides many opportunities for cooperative learning and, as a result, many opportunities to observe group work processes and products. Look for these opportunities in full-page Activities and periodically throughout the chapter.

Cooperative Group Assessment All members of the group contribute to the work process and the products it produces. For example, if a mixed ability, four-member laboratory work group conducts an activity, you can use a rating scale or checklist to assess the quality of both group interaction and work skills. An example, along with information about evaluating cooperative work, is provided in the booklet *Alternate Assessment in the Science Classroom.* All four members of the group are expected to review and agree on the data sheet produced by the group. You can require each member to certify the group results by signing the data sheet or lab report. Research shows that cooperative group assessment is as valid as individual assessment. The booklet *Performance Assessment in Middle School Science* deals with evaluation of group work and the individual.

STUDENTS' SELF-ASSESSMENT

An important life skill is the ability to self-assess and plan for improvement.

Students often complete their assignments expecting the teacher to grade and return it. Students should learn to thoughtfully study their own work and identify what they have done well and where they need to improve. The booklet *Performance Assessment in Middle School Science (PAMSS)* provides instructional materials to help students objectively assess their work.

ASSESSMENT RESOURCE BOOKS

Glencoe's ***Alternate Assessment in the Science Classroom***
Glencoe's ***Performance Assessment in Middle School Science (PAMSS)***
Merrill Life Science Performance Assessment

JOURNALS

Keeping a journal encourages a more thoughtful attitude toward written work and helps students learn more science.

Merrill Life Science strongly recommends the use of student journals. Journal entries are conducive to thinking about why something has been done. They can be used to note questions, insights, successes, and frustrations. The use of a journal can provide a history of a student's independent work that illuminates the growth of problem-solving or thinking skills. The importance of this is emphasized in that a special feature, **In Your Journal**, appears at least once in each chapter. Students may be asked to use their journals to write creatively or to answer specific questions, such as those in *Section Reviews* or *Chapter Reviews.*

Students are also encouraged to record data and observations, descriptions, and reflections in their journals. They may include diagrams and drawings. Excerpts from the student journals can be included in an individual or group portfolio.

PORTFOLIOS, PUTTING IT ALL TOGETHER

In portfolio assessment, the teacher must keep in mind that the pieces of work in the portfolio have been chosen by the students as representations of their best efforts.

The purpose of a student or cooperative group portfolio is to present examples of the individual or group's work in a "nontesting" environment. The performance portfolio is *not* intended to be a complete collection of all worksheets and other assignments for a grading period. At its best, the portfolio should include integrated performance products that show growth in concept attainment and skill development.

Merrill Life Science presents a wealth of opportunities for performance portfolio development. Each chapter in the student text contains projects, inquiry questions, enrichment activities, problem solving, skill builders, library research opportunities, and connections with other sciences, society, and literature. Each of the student activities results in a product. A mixture of these products can be used to document student growth during the grading period. The ***Merrill Life Science*** *Performance Assessments,* with skills and unit-related performance assessment exercises, provides additional materials that can be included in a portfolio submission. ***Merrill Life Science*** strongly suggests including excerpts from student journals in the individual or group portfolio. The booklet *Performance Assessment in Middle School Science (PAMSS)* provides additional background information and specific methods of handling portfolios within the classroom.

CONTENT ASSESSMENT

While new and exciting performance skills assessments are emerging, paper-and-pencil test are still a mainstay of student evaluation. Students must learn to conceptualize, process, and prepare for traditional content assessments. Presently and in the foreseeable future, students will still be required to pass pencil-and-paper tests to exit high school, and to enter college, trade schools, and other training programs.

Merrill Life Science contains numerous strategies and formative checkpoints for evaluating student progress toward mastery of science concepts. Traditional content assessment forms such as matching, multiple choice, and short essay items are effective in sampling contents.

- *Section Review* questions and application tasks are presented throughout the chapters in the student text. This spaced review process helps build learning bridges that allow all students to confidently progress from one lesson to the next.
- A *Summary* section is located on the first page of the three-page *Chapter Review.* After instruction for the chapter is complete, a summation of the major concepts is presented. The *Summary* enables students to check and reinforce their understanding of the content and concepts presented in the chapter.
- The *Chapter Review* also includes a formal review process for the written content assessment. Individual students or cooperative groups of three to five students can respond to *Chapter Review* questions to check their understanding of science terms, concepts, lab and critical-thinking skills, and problem-solving techniques. By evaluating the student responses to this extensive review, you can determine if any substantial reteaching is needed.

The ***Teacher Classroom Resources*** contains formal content assessment, including a two-page *Chapter Review,* a four-page *Chapter Test* for each chapter in the student text, and eight two-page unit tests.

- The *Review* can be used to help you determine the students' grasp of the concepts and supporting facts presented in the chapter. Using the review in a whole class session, you can correct any misperceptions and provide closure for a chapter.
- The *Chapter Test* enables students to document their mastery of the concepts developed in the chapter. The test includes multiple choice type items that test knowledge, skill questions, and numerous short answer and essay items that require students to apply and relate concepts.
- *Computer Test Banks* allow you to easily customize tests to fulfill individual assessment plans.

Cooperative Learning in Science

Cooperative Learning fosters academic, personal , and social success for all students.

What is Cooperative Learning?

In cooperative learning, students work together in small groups to learn academic material and interpersonal skills. Group members are responsible for the *group's* accomplishing an assigned task as well as for learning the material themselves. When compared to competitive or individual learning situations in which students work either against each other or alone, cooperative learning fosters academic, personal, and social success for all students. Research shows that successful cooperative learning results in:

- development of positive attitudes toward science;
- choosing to take more science courses as electives;
- positive attitudes toward science carrying over to positive attitudes toward school;
- lower drop-out rates for at-risk students;
- building respect for others regardless of race, ethnic origin, or sex;
- increased awareness of diverse perspectives;
- increased capability for problem solving in sciences;
- increased realization of potential for girls in science classes; and
- development of kindness, sensitivity, and tolerance.

The Teacher's Role

Decide Objectives

Before teaching the lesson, the teacher must decide the academic task and interpersonal skills students will learn or practice in groups. Students can learn most any academic objective in cooperative groups.

The teacher can also specify what interpersonal behaviors are necessary for a group to work cooperatively. When first starting out, it is wise to list and discuss with students basic interpersonal skills needed for people to work together. Basic interpersonal skills you might discuss include being responsible for your own actions, staying on task, listening while others are speaking, and respecting other people and their ideas. Students can learn and practice other interpersonal skills such as using quiet voices, encouraging other group members to participate, summarizing, checking for understanding, disagreeing constructively, reaching a group consensus, and criticizing ideas rather than people.

Select Groups

Cooperative groups usually contain from two to six students. If students are not experienced at working in groups, start with small groups. Consider grouping students in pairs and then joining pairs later to form groups of four or six.

Generally, it is best to assign students to heterogeneous groups. Be certain that each heterogeneous group contains a mixture of abilities, genders, and ethnicity. The use of heterogeneous groups exposes students to ideas different from their own and helps them learn how to work with persons different from themselves.

Change Groups

Initially, cooperative learning groups should work together for only a day or two. After the students are more experienced, they can do group work for longer periods of time. Some teachers change groups every week, while others keep groups of students together during the study of a unit or chapter. Keep groups together long enough for each group to experience success, and change groups often enough that students have the opportunity to work with others.

Structure Tasks

You can structure the learning task to promote the participation of each group member by the arrangement of your classroom and the provision of materials. Limiting the materials needed to accomplish the assigned task causes students to share them. Also, consider assigning students roles that contribute to accomplishing the group task. Student roles should be rotated so each group member has the opportunity to perform each role.

Group Assessment

Finally, decide on how you will assess the learning task and how well the students worked together. Because students are responsible for themselves and other group members learning the material, you can assess group performance during a lesson by frequently asking questions of group members picked at random. It is important that, although each student will be assessed as a group member, group grades are not given. Group grades may result in unequal sharing of responsibilities by group members. Individual learning can be assessed by traditional tests and quizzes.

To assess the learning of interpersonal skills, you can observe their use during the lesson. Groups can assess themselves by rating themselves on a scale from one to ten; they then list ways group members used interpersonal skills and ways to improve group performance.

Teaching the Lesson

Before the Group Task

- ▶**Academic Task:** Prepare students for the academic task by teaching any material they might need to know and by giving specific instructions for the task.
- ▶**Criteria for Success:** Instruct students that they are responsible for their own learning as well as the learning of other members of the group. Explain the sharing of materials and assuming of roles. Explain your criteria for evaluating group and individual learning.
- ▶**Interpersonal Skills:** Specify the interpersonal skills students will be working on, and list what behaviors look and sound like. Explain how you will assess interpersonal skills.
- ▶**Group Formation:** Divide the class into groups. Assign roles or divide materials.

During the Group Task

- ▶**Provide Assistance with the Academic Task:** Make certain each student can see and hear all other group members. When students are having trouble with the task, answer questions, clarify the assignment, reteach, or provide background as needed. When answering questions, make certain that no students in the group can answer the question before you reply.
- ▶**Monitor Student Behavior:** Spend most of your time monitoring the functioning of groups. Praise group cooperation and good use of interpersonal skills.
- ▶**Intervene to Teach Interpersonal Skills:** Whenever possible, allow groups to work out their own problems. When groups are having problems, ask group members to figure out how the group can function more effectively. Record your observations of the use of interpersonal skills; share your observations with the groups.

After the Group Task

- ▶**Provide Closure to the Lesson:** Reinforce student learning by having groups share their products or summarize the assignment. Answer any questions about the lesson.
- ▶**Assess Group and Individual Learning:** Use the criteria discussed before the lesson began to assess and give feedback on how well the academic task was mastered by the groups. Assess individual learning by your traditional methods.
- ▶**Assess How Well Groups Functioned:** Have students analyze how well their groups functioned and how well they used interpersonal skills. Groups can list what they did well and what they could do to improve. Have groups share their assessments with the class, and summarize the assessments of the whole class.

Using Cooperative Learning in *Merrill Life Science*

In ***Merrill Life Science*** *Teacher Wraparound Edition,* Activities, FIND OUTs, and Chapter Reviews suggest using Cooperative Learning. In addition, cooperative learning groups can be used with Section Reviews, Problem Solving, Concept Mapping, and Skill Builder features. The following cooperative learning strategies are referenced in the *Teacher Wraparound Edition.*

Paired Partners

Assign each student a partner, and ask a question or present a problem for them to solve. Each student composes an answer or solution; the pair then shares answers with each other. If partners disagree, they explain and discuss the issues until they agree. When both agree and can explain the answer, partners raise their hands to signify both agree and can explain the answer. After determining if groups have the correct answer, some teachers use thumbs up to indicate "correct" and thumbs down to indicate "incorrect." Paired Partners can be used for Problem Solving, Concept Mapping, and Skill Builder features.

Expert Teams

Give each group member a different part of an assignment to study and master. Send group members with the same part of the assignment from the different teams to work together to become experts on their parts. Bring the experts back to their original groups. Each group member teaches his or her part of the assignment to other members of the original group until everyone has mastered all the material. Expert Teams can be used for Section and Chapter Reviews.

Study Buddies

Study Buddies work together to help one another study for tests or to create concept maps. Group students in fours. After a chapter is completed, give students one class period to work in groups on the Chapter Review before a chapter test. If you wish, Study Buddies can be divided into Expert Teams to study and then teach the material. For concept maps, give each group member a different colored pen. Group members pass the concept map around the table adding to the map on each pass with their colored pens.

Numbered Heads Together

Form groups of three to five and have students number off. Then, ask a question or give an assignment. Have students in each group either agree on a group answer or, for higher- level thinking skills, name an example, make a prediction, or state an application. When students have agreed on an answer, call a number at random. Students with that number raise their hands and wait to be called on. Select one student to provide an answer. Determine if other students with that number have the correct answer by indicating thumbs up or down or by having them write their responses on index cards or on the chalkboard. Numbered Heads Together can be used for Section and Chapter Reviews, Problem Solving, Concept Mapping, and Skill Builders.

Problem-Solving Team

Form groups of four students, and assign roles. The reader reads the problem; the clarifier restates it; the solver suggests answers. If the group agrees on answers, the recorder writes the answers on a paper that all members sign. Review the answers and discuss the problem by calling on any group member. Use thumbs up or down, or write responses on response cards or the chalkboard to determine if all groups have the same answers. This strategy can be used for Problem Solving and Skill Builder features.

Science Investigation

Science Investigation group members work together to perform hands-on science investigations. The Science Investigation strategy is used for Activities, MINI-Labs, and FIND OUT features. In Science Investigation, each group member has a different role and duties to perform for the investigation. Some roles are working roles to accomplish the investigation, while others are interpersonal skill roles that help the group function effectively. Following are possible roles for Science Investigation groups.

Working Roles

Reader: reads any directions out loud

Materials Handler: obtains, dispenses, and returns all materials

Safety Officer: informs group of safety precautions; ensures group handles equipment safely

Recorder: records data collected during the activity; writes answers to questions; has all group members sign data collection and answer sheets

Reporter: reports data collected and answers to questions

Timekeeper: keeps group on task and manages the group's time

Calculator: performs calculations and measurements

Interpersonal Skill Roles

Monitor: ensures that each group member participates and encourages participation

Praiser: compliments group members on fulfilling their assigned tasks; compliments group members on use of interpersonal skills

Checker: checks on learning of group members; ensures that each group member can summarize the results of the activity and answer questions

Resources

Adams, D.M., and M.E. Hamm. *Cooperative Learning, Critical Thinking, and Collaboration Across the Curriculum.* Springfield, IL: Charles C. Thomas Publisher, 1990.

Association for Supervision and Curriculum Development. *Educational Leadership,* Volume 47, Number 4. December 1989–January 1990.

Foot, H.C., M.J. Morgan, and R.H. Shute. *Children Helping Children.* New York: John Wiley & Sons, 1990.

Johnson, D.W., and R.T. Johnson. *Learning Together and Alone: Cooperative, Competitive, and Individualistic Learning.* Englewood Cliffs, NJ: Prentice-Hall, 1987.

Johnson, D.W., and R.T. Johnson., E.J. Holubec, and P. Roy. *Circles of Learning: Cooperation in the Classroom.* Alexandria, VA: Association for Supervision and Curriculum Development, 1984.

Kagan, S. *Cooperative Learning: Resources for Teachers.* Riverside, CA: University of California, 1988.

Shlomo, S. *Cooperative Learning Theory and Research.* Westport, CT: Praeger, 1990.

Slavin, R. *Cooperative Learning Theory, Research, and Practice.* Englewood Cliffs, NJ: Prentice Hall, 1990.

Slavin, R. *Using Student Team Learning.* Baltimore, MD: The John Hopkins Team Learning Project, 1986.

Meeting Individual Needs

With careful planning, the needs of all students can be met in the science classroom.

	DESCRIPTION	SOURCES OF HELP/INFORMATION
Learning Disabled	All learning disabled students have a problem in one or more areas, such as academic learning, language, perception, social-emotional adjustment, memory, or attention.	*Journal of Learning Disabilities* *Learning Disability Quarterly*
Behaviorally Disordered	Children with behavior disorders deviate from standards or expectations of behavior and impair the functioning of others and themselves. These children may also be gifted or learning disabled.	*Exceptional Children* *Journal of Special Education*
Physically Challenged	Children who are physically disabled fall into two categories—those with orthopedic impairments and those with other health impairments. Orthopedically impaired children have the use of one or more limbs severely restricted, so the use of wheelchairs, crutches, or braces may be necessary. Children with other health impairments may require the use of respirators or have other medical equipment.	Batshaw, M.L. and M.Y. Perset. *Children with Handicaps: A Medical Primer.* Baltimore: Paul H. Brooks, 1981. Hale, G. (Ed.). *The Source Book for the Disabled.* NY: Holt, Rinehart & Winston, 1982. *Teaching Exceptional Children*
Visually Impaired	Children who are visually disabled have partial or total loss of sight. Individuals with visual impairments are not significantly different from their sighted peers in ability range or personality. However, blindness may affect cognitive, motor, and social development, especially if early intervention is lacking.	*Journal of Visual Impairment and Blindness* *Education of Visually Handicapped* American Foundation for the Blind
Hearing Impaired	Children who are hearing impaired have partial or total loss of hearing. Individuals with hearing impairments are not significantly different from their hearing peers in ability range or personality. However, the chronic condition of deafness may affect cognitive, motor, and social development if early intervention is lacking. Speech development also is often affected.	*American Annals of the Deaf* *Journal of Speech and Hearing Research* *Sign Language Studies*
Limited English Proficiency	Multicultural and/or bilingual children often speak English as a second language or not at all. Customs and behavior of people in the majority culture may be confusing for some of these students. Cultural values may inhibit some of these students from full participation.	*Teaching English as a Second Language Reporter* R.L. Jones, ed., *Mainstreaming and the Minority Child.* Reston, VA: Council for Exceptional Children, 1976.
Gifted	Although no formal definition exists, these students can be described as having above-average ability, task commitment, and creativity. Gifted students rank in the top 5% of their class. They usually finish work more quickly than other students, and are capable of divergent thinking.	*Journal for the Education of the Gifted* *Gifted Child Quarterly* *Gifted Creative/Talented*

TIPS FOR INSTRUCTION

1. Provide support and structure; clearly specify rules, assignments, and duties.
2. Establish situations that lead to success.
3. Practice skills frequently—use games and drills to help maintain student interest.
4. Allow students to record answers on tape and allow extra time to complete tests and assignments.
5. Provide outlines or tape lecture material.
6. Pair students with peer helpers, and provide class time for pair interaction.

1. Provide a clearly structured environment with regard to scheduling, rules, room arrangement, and safety.
2. Clearly outline objectives and how you will help students obtain objectives. Seek input from them about their strengths, weaknesses, and goals.
3. Reinforce appropriate behavior and model it for students.
4. Do not expect immediate success. Instead, work for long-term improvement.
5. Balance individual needs with group requirements.

1. Openly discuss with students any uncertainties you have about when to offer aid.
2. Ask parents or therapists and students what special devices or procedures are needed, and if any special safety precautions need to be taken.
3. Allow physically disabled students to do everything their peers do, including participating in field trips, special events, and projects.
4. Help nondisabled students and adults understand physically disabled students.

1. As with all students, help the student become independent. Some assignments may need to be modified.
2. Teach classmates how to serve as guides.
3. Limit unnecessary noise in the classroom.
4. Encourage students to use their sense of touch. Provide tactile models whenever possible.
5. Describe people and events as they occur in the classroom.
6. Provide taped lectures and reading assignments.
7. Team the student with a sighted peer for laboratory work.

1. Seat students where they can see your lip movements easily, and avoid visual distractions.
2. Avoid standing with your back to the window or light source.
3. Using an overhead projector allows you to maintain eye contact while writing.
4. Seat students where they can see speakers.
5. Write all assignments on the chalkboard, or hand out written instructions.
6. If the student has a manual interpreter, allow both student and interpreter to select the most favorable seating arrangements.

1. Remember that students' ability to speak English does not reflect their academic ability.
2. Try to incorporate the students' cultural experience into your instruction. The help of a bilingual aide may be effective.
3. Include information about different cultures in your curriculum to aid students' self-image—avoid cultural stereotypes.
4. Encourage students to share their cultures in the classroom.
5. Incorporate a variety of teaching strategies in your classroom to accommodate different learning styles.

1. Make arrangements for students to take selected subjects early and to work on independent projects.
2. Make public services available through a catalog of resources, such as agencies providing free and inexpensive materials, community services and programs, and people in the community with specific expertise.
3. Ask "what if" questions to develop high-level thinking skills; establish an environment safe for risk taking.
4. Emphasize concepts, theories, ideas, relationships, and generalizations.

Student Bibliography

GENERAL SCIENCE CONTENT

Barr, George. *Science Tricks and Magic for Young People.* New York: Dover Publications, Inc., 1987.

Cash, Terry. *175 More Science Experiments to Amuse and Amaze Your Friends: Experiments! Tricks! Things to Make!* New York: Random House, 1991.

Churchill, E. Richard. *Amazing Science Experiments with Everyday Materials.* New York: Sterling Publishing Co., Inc., 1991.

Gold, Carol. *Science Express,* "50 Scientific Stunts for the Ontario Science Centre." New York: Addison-Wesley, 1991.

Herbert, Don. *Mr. Wizard's Supermarket Science.* New York: Random House, 1980.

Lewis, James. *Hocus Pocus Stir and Cook, The Kitchen Science-Magic Book.* New York: Meadowbrook Press, Division of Simon and Shuster, Inc., 1991.

Mandell, Muriel, *Simple Science Experiments with Everyday Materials.* New York: Sterling Publishing Co., Inc., 1989.

Roberts, Royston. *Serendipity: Accidental Discoveries in Science.* New York: John Wiley and Sons, Inc., 1989.

Schultz, Robert F. *Selected Experiments and Projects.* Washington, DC: Thomas Alva Edison Foundation, 1988.

Strongin, Herb. *Science on a Shoestring.* Menlo Park, CA: Addison-Wesley Publishing Co., 1985.

Townsley, B.J. *Famous Scientists.* Los Angeles, CA: Enrich Education Division of Price Stern Sloan Inc., 1987.

PHYSICS

Aronson, Billy. "Water Ride Designers Are Making Waves." *3-2-1 Contact,* August, 1991, pp. 14-16.

Asimov, Isaac. *How Did We Find Out the Speed of Light?* New York: Walker, 1986.

Berger, Melvin. *Light, Lenses, and Lasers.* New York: Putnam, 1987.

Cash, Terry. *Sound.* New York: Warwick Press, 1989.

Catherall, Ed. *Exploring Sound.* Austin, TX: Steck-Vaughn Library, 1989.

Heiligman, Deborah. "There's a Lot More to Color Than Meets the Eye." *3-2-1 Contact,* November, 1991, pp. 16-20.

McGrath, Susan. *Fun with Physics.* Washington, DC: National Geographic Society, 1986.

Myles, Douglas. *The Great Waves.* New York: McGraw-Hill Book Company, 1985.

Taylor, Barbara. *Light and Color.* New York: Franklin Watts, 1990.

Taylor, Barbara. *Sound and Music.* New York: Warwick Press, 1990.

Ward, Allen. *Experimenting with Batteries, Bulbs, and Wires.* New York: Chelsea House, 1991.

Ward, Alan. *Experimenting with Sound.* New York: Chelsea Juniors, 1991.

Wood, Nicholas. *Listen . . . What Do You Hear?* Mahwah, NJ: Troll Associates, 1991.

CHEMISTRY

Barber, Jacqueline. *Of Cabbage and Chemistry.* Washington, DC: Lawrence Hall of Science, NSTA, 1989.

Barber, Jacqueline. *Chemical Reactions.* Washington, DC: Lawrence Hall of Science, NSTA, 1986.

Benrey, Ronald. *Alternative Energy Sources: Experiments You Can Do . . . from Edison.* Washington, DC: Thomas Alva Edison Foundation, Edison Electric Institute, 1988.

Cornell, John. *Experiments with Mixtures.* New York: John Wiley and Sons, Inc., 1990.

Matsubara, T. *The Structure and Properties of Matter.* New York: Springer-Verlag New York Inc., 1982.

Zubrewski, Bernie. *Messing Around with Baking Chemistry: A Children's Museum Activity Book.* Boston, MA: Little Brown and Co., 1981.

LIFE SCIENCE

Dewey, Jennifer Owings. *A Day and Night in the Desert.* Boston, MA: Little Brown, 1991.

Johnson, Cathy. *Local Wilderness.* New York: Prentice Hall, 1987.

Leslie, Clare Walker. *Nature All Year Long.* New York: Greenwillow, 1990.

McGrath, Susan. *The Amazing Things Animals Do.* Washington, DC: National Geographic Society, 1989.

Markmann, Erika. *Grow It! An Indoor/Outdoor Gardening Guide for Kids.* New York: Random House, 1991.

Children's Atlas of the Environment. Chicago, IL: Rand McNally, 1991.

Van Cleave, Janice Pratt. *Biology for Every Kid: 101 Easy Experiments that Really Work.* New York: Wiley, 1990.

EARTH SCIENCE

Ardley, Neil. *The Science Book of Air.* New York: Gulliver Books, Harcourt, Brace, Jovanovich, Publishers, 1991.

Ardley, Neil. *The Science Book of Water.* New York: Gulliver Books, Harcourt, Brace, Jovanovich, Publishers, 1991.

Barrow, Lloyd H. *Adventures with Rocks and Minerals: Geology Experiments for Young People.* Hillsdale, NJ: Enslow, 1991.

Booth, Basil. *Volcanoes and Earthquakes.* Englewood Cliffs, NJ: Silver Burdett Press, 1991.

Javna, John. *50 Simple Things Kids Can Do to Save the Earth.* Kansas City: The Earth Works Group, Andrews and McMeel, a Universal Press Syndicate Co., 1990.

Van Cleave, Janice. *Earth Science for Every Kid.* New York: John Wiley and Sons, Inc., 1991.

Wood, Robert W. *Science for Kids: 39 Easy Geology Activities.* Blue Ridge Summit, PA: Tab Books, 1992.

Teacher Bibliography

CURRICULUM

Aldridge, William G. "Scope, Sequence, and Coordination: A New Synthesis for Improving Science Education." *Journal of Science Education and Technology.*

Banks, James A. "Multicultural Education: For Freedom's Sake." *Educational Leadership,* December 1991/January 1992, pp. 32-35.

Beane, James A. "Middle School, The Natural Home of the Integrated Curriculum." *Educational Leadership,* October 1991, pp. 9-13.

Chemistry of Life: 1988 Curriculum Module, Princeton, NJ: Woodrow Wilson National Fellowship Foundation, 1988.

Driver, R. *The Children's Learning in Science Project, Monographs on Preconceptions in Science.* Center for Studies in Science and Mathematics Education, Department of Education, University of Leeds, 1984-1989.

Hazen, Robert M. and James Trefil. *Science Matter, Achieving Scientific Literacy.* New York: Doubleday, 1991.

Phillips, William. "Earth Science Misconceptions." *The Science Teacher,* October 1991, pp. 21-23.

Piaget, J. *To Understand Is to Invent: The Future of Education.* New York: Grossman Publishers, 1973.

Rutherford, E. James and Andrew Ahlgren. *Science for All Americans.* New York: Oxford University Press, 1990.

TEACHING METHODS

Altshular, Kenneth. "The Interdisciplinary Classroom." *The Physics Teacher,* October 1991, pp. 428-429.

Humphreys, David. *Demonstrating Chemistry.* Ontario, Canada: Chemistry Department, McMaster University Hamilton, 1983.

Johnson, David W., Roger T. Johnson, and Edythe Johnson Holubec. *Circles of Learning.* Edina, MN: Interaction Book Company, 1990.

Johnson, David W., Roger T. Johnson, and Edythe Johnson Holubec. *Cooperation in the Classroom.* Edina, MN: Interaction Book Company, 1991.

Johnson, David W., Roger T. Johnson. *Cooperative Learning: Warm-Ups, Grouping Strategies, and Group Activities.* Edina, MN: Interaction Book Company, 1985.

Novak, Joseph. "Clarify with Concept Maps." *The Science Teacher,* October 1991, pp. 44-49.

Penick, John. "Where's the Science?" *The Science Teacher,* May 1991, pp. 27-29.

CONTENT AREA BOOKS

Physics

Arons, A. B. *A Guide to Introductory Physics Teaching.* New York: John Wiley and Sons, 1990.

Berman, Paul. *Light and Sound.* New York: Marshall Cavendish, 1988.

Gardner, Robert. *Experimenting with Light.* New York: Franklin Watts, 1991.

Urone, Paul Peter. *Physics with Health Science Applications.* New York: Harper and Row Publishers, 1986.

Walpole, Brenda. *175 Science Experiments to Amuse and Amaze Your Friends.* New York: Random House, 1988.

Chemistry

Element of the Week. Batavia, IL: Flinn Scientific, Inc., 1990.

Ground to Grits: Scientific Concepts in Nutrition/Agriculture. Columbia, SC: South Carolina Department of Education, 1982.

Joesten, Melvin. *World of Chemistry.* Philadelphia, PA: Saunders College Publishing, 1991.

Mitchell, Sharon and Frederick Juergens. *Laboratory Solutions for the Science Classroom.* Batavia, IL: Flinn Scientific, Inc., 1991.

Solomon, Sally. "Qualitative Analysis of Eleven Household Compounds." *Journal of Chemical Education,* April 1991, pp. 328-329.

Life Science

Hancock, Judith M. *Variety of Life: A Biology Teacher's Sourcebook.* Portland, OR: J. Weston Walch, 1987.

Middleton, James I. "Student-Generated Analogies in Biology." *American Biology Teacher,* January 1991, pp. 42-46.

Vogel, Steven. *Life's Devices: The Physical World of Plants and Animals.* Princeton, NJ: Princeton University Press, 1989.

Earth Science

Callister, Jeffrey C., Lenny Coplestone, Gerald F. Consuegra, Sharon M. Stroud, and Warren E. Yasso. *Earthquakes.* Washington, DC: NSTA/FEMA, 1988.

Lasca, Norman P. "Build Me a River." *Earth,* January 1991, pp. 59-65.

Little, Jane Braxton. "California Town Unites to Save a Stream." *The Christian Science Monitor,* February 28, 1991.

Sae, Andy S. W. "Dynamic Demos." *The Science Teacher,* October 1991, pp. 23-25.

Laboratory Safety

Managing Activities

Hands-on activities provide another opportunity for students to learn to work together. Decisions and plans made by cooperative groups reflect the real world.

Preplanning and organization are important for successful use of hands-on activities in the classroom. Make copies of specific Activity worksheets ahead of time, from the **Activity Worksheets** booklet. Store materials for each activity in a box with a list of contents on the end. The boxes may be color-coded according to the unit of study. Place materials for each group of students in a plastic bag and label the bag with its contents. Students can be responsible for distributing the materials and checking the bag for its contents. Cleaning up and putting materials in their proper place is an essential part of the laboratory experience.

The materials used in ***Merrill Life Science*** are easily accessible. If the budget in your school is limited, have students bring in simple, inexpensive materials such as sugar, baking soda, and so on. Parents may also be willing to donate materials and their time to help organize the materials. Many parents genuinely want to help.

It is important to remember that there is no such thing as failure in science activities. If the experiment does not illustrate what you had intended, turn it into a question and hypothesis activity. Students evaluate their results and form hypotheses on how to alter the experiment. Students then test their hypotheses and draw conclusions. Activities can also end with a question or extension to encourage further exploration of concepts.

Laboratory Safety

Safety is of prime importance in every classroom. However, the need for safety is even greater when science is taught. Outlined on the next page are some considerations on laboratory safety.

The activities in ***Merrill Life Science*** are designed to minimize dangers in the laboratory. Even so, there are no guarantees against accidents. However, careful planning and preparation as well as being aware of hazards can keep accidents from happening. Numerous books and pamphlets are available on laboratory safety, with detailed instruction on preventing accidents. However, much of what they present can be summarized in the phrase: *Be prepared!* Know the rules and what common violations occur. Know the Safety Symbols used in this book (see p. 34T). Know where emergency equipment is stored and how to use it. Practice good laboratory housekeeping and management by observing these guidelines:

Classroom/Laboratory

1. Store chemicals properly. (see page 36T)
   - **a.** Separate chemicals by reaction type.
   - **b.** Label all chemical containers. Include purchase date, special precautions, and expiration date.
   - **c.** Discard chemicals when outdated, according to appropriate disposal methods.
   - **d.** Do not store chemicals above eye level.
   - **e.** Wood shelving is preferable to metal. All shelving should be firmly attached to walls. Anti-roll lips should be placed on all shelves.
   - **f.** Store only those chemicals that you plan to use.
   - **g.** Flammable and toxic chemicals require special storage containers.
2. Store equipment properly.
   - **a.** Clean and dry all equipment before storing.
   - **b.** Protect electronic equipment and microscopes from dust, humidity, and extreme temperatures.
   - **c.** Label and organize equipment so that it is accessible.
3. Provide adequate workspace.
4. Provide adequate room ventilation.
5. Post safety and evacuation guidelines.
6. Be sure safety equipment is accessible and works.
7. Provide containers for disposing of chemicals, waste products, and biological specimens. Disposal methods must meet local guidelines.
8. Use hot plates whenever possible as a heat source. If burners are used, a central shut-off valve for the gas supply should be available to the teacher. Never use open flames when a flammable solvent is in the same room.

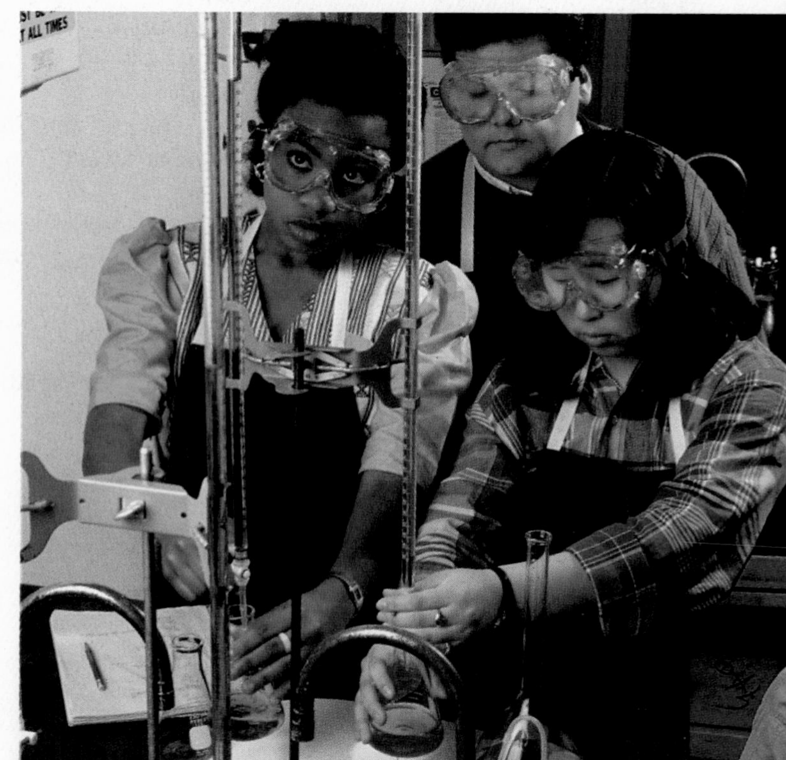

First Day of Class/Labs (with students)

1. Distribute and discuss safety rules, safety symbols, first aid guidelines, and safety contract found in the *Teacher Resource Guide.* Have students refer to Appendix C on pages 672-673, to review safety symbols and guidelines.
2. Review safe use of equipment and chemicals.
3. Review use and location of safety equipment.
4. Discuss safe disposal of materials and laboratory clean-up policy.
5. Discuss proper laboratory attitude and conduct.
6. Document students' understanding of above points.
   a. Have students sign the safety contract and return it.
   b. Administer the safety assessment found on page 39 in the *Teacher Resource Guide.* Reteach those points that students do not understand.

Before Each Investigation

1. Perform each investigation yourself before assigning it.
2. Arrange the lab in such a way that equipment and supplies are clearly labeled and easily accessible.
3. Have available only equipment and supplies needed to complete the assigned investigation.
4. Review the procedure with students, emphasizing any caution statements or safety symbols that appear.
5. Be sure all students know proper procedures to follow if an accident should occur.

During the Investigation

1. Make sure the lab is kept clean and free of clutter.
2. Insist that students wear safety goggles and aprons, if needed.
3. Never allow a student to work alone in the lab.
4. Never allow students to use a cutting device with more than one edge.
5. Students should not point the open end of a heated test tube toward anyone.
6. Remove broken glassware or frayed cords from use. Also clean up any spills immediately. Dilute solutions with water before removing.
7. Be sure all glassware that is to be heated is of a heat-treated type that will not shatter.
8. Remind students that hot glassware looks cool.
9. Prohibit eating and drinking in the lab.
10. Never stir with a thermometer.

After the Investigation

1. Be sure that the lab is clean.
2. Be certain that students have returned all equipment and disposed of broken glassware and chemicals properly.
3. Be sure all hot plates and electrical connections are off.
4. Insist that each student wash his or her hands when lab work is completed.

The ***Merrill Life Science*** program uses safety symbols to alert you and your students to possible laboratory dangers. These symbols are explained on the following page. Be sure your students understand each symbol before they begin any activity.

Safety Symbols

Safety symbols alert you and your students to cautions within activities.

DISPOSAL ALERT This symbol appears when care must be taken to dispose of materials properly.	**ANIMAL SAFETY** This symbol appears whenever live animals are studied and the safety of the animals and the students must be ensured.
BIOLOGICAL HAZARD This symbol appears when there is danger involving bacteria, fungi, or protists.	**RADIOACTIVE SAFETY** This symbol appears when radioactive materials are used.
OPEN FLAME ALERT This symbol appears when use of an open flame could cause a fire or an explosion.	**CLOTHING PROTECTION SAFETY** This symbol appears when substances used could stain or burn clothing.
THERMAL SAFETY This symbol appears as a reminder to use caution when handling hot objects.	**FIRE SAFETY** This symbol appears when care should be taken around open flames.
SHARP OBJECT SAFETY This symbol appears when a danger of cuts or punctures caused by the use of sharp objects exists.	**EXPLOSION SAFETY** This symbol appears when the misuse of chemicals could cause an explosion.
FUME SAFETY This symbol appears when chemicals or chemical reactions could cause dangerous fumes.	**EYE SAFETY** This symbol appears when a danger to the eyes exists. Safety goggles should be worn when this symbol appears.
ELECTRICAL SAFETY This symbol appears when care should be taken when using electrical equipment.	**POISON SAFETY** This symbol appears when poisonous substances are used.
PLANT SAFETY This symbol appears when poisonous plants or plants with thorns are handled.	**CHEMICAL SAFETY** This symbol appears when chemicals used can cause burns or are poisonous if absorbed through the skin.

Preparation of Solutions

The following text gives some general hints on solution preparation and some safety tips to keep in mind.

For best results, the preparation of each solution is tailored to the requirements of the Activity, MINI-Lab, or FIND OUT activity in which it is used. It is not recommended that solutions be made far in advance. Rather, they should be prepared fresh as needed.

Unless otherwise specified, solutions are prepared by adding the solid to a small amount of water and then diluting with water to the volume listed. Use distilled water for the preparation of solutions. For example, to make a 0.1*M* solution of aluminum sulfate, dissolve 34.2 g of $Al_2(SO_4)_3$ in a small amount of distilled water and dilute to a liter with water. If you use a hydrate that is different from the one specified in a particular preparation, you will need to adjust the amount of the hydrate to obtain the required concentration.

It is most important to use safe laboratory techniques when handling all chemicals. Many substances may appear harmless but are, in fact, toxic, corrosive, or very reactive. Always check the hazard information on the reagent bottle. If in doubt, check with the manufacturer or with Flinn Scientific Inc., (708) 879-6900. Chemicals should never be ingested. Be sure to use proper techniques to smell solutions or other reagents. Always wear safety goggles and an apron. The following general cautions should be used.

1. *Liquid and/or vapor poisonous/corrosive. Use in the fume hood.*
   acetic acid — hydrochloric acid
   ammonium hydroxide — nitric acid
2. *Poisonous and corrosive to eyes, lungs, and skin.*
   acids — limewater — iron(III) chloride
   bases — silver nitrate — potassium permanganate
   iodine
3. *Poisonous if swallowed, inhaled, or absorbed through the skin.*
   acetic acid, glacial — copper compounds
   barium chloride — lead compounds
   chromium compounds — lithium compounds
   cobalt(II) chloride — silver compounds
4. *Always add acids to water, never the reverse.*
5. *When either sulfuric acid or sodium hydroxide is added to water, a large amount of thermal energy is released. Sodium metal reacts violently with water. Use extra care if handling any of these substances.*

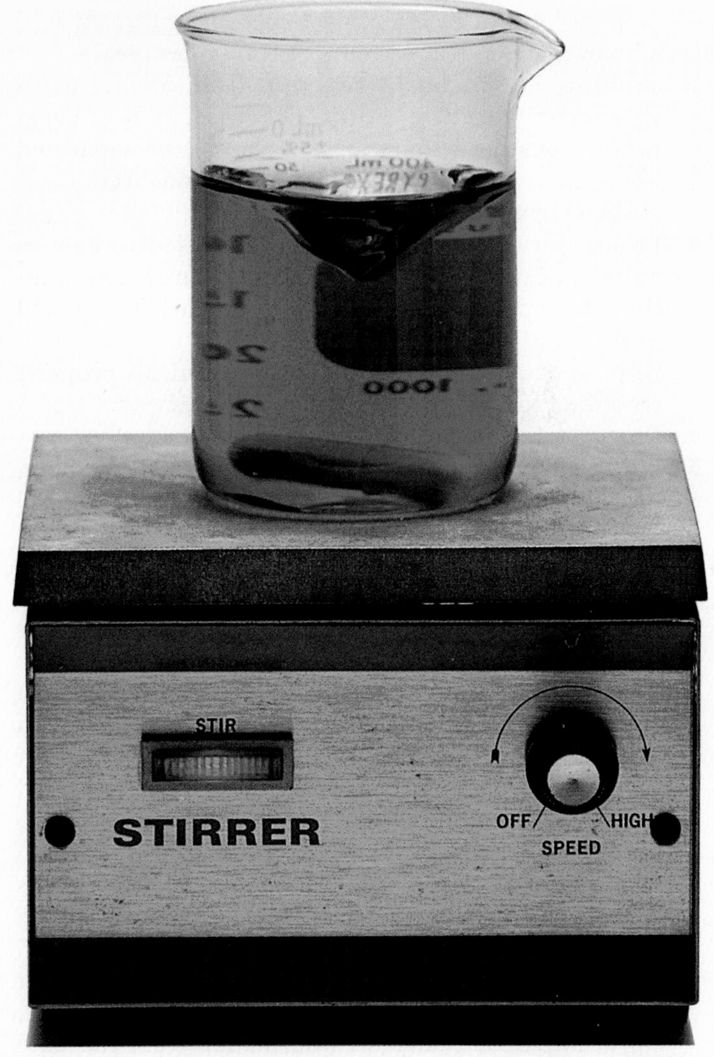

Chemical Storage and Disposal

General Guidelines

Be sure to store all chemicals properly. The following are guidelines commonly used. Your school, city, county, or state may have additional requirements for handling chemicals. It is the responsibility of each teacher to become informed as to what rules or guidelines are in effect in his or her area.

1. Separate chemicals by reaction type. Strong acids should be stored together. Likewise, strong bases should be stored together and should be separated from acids. Oxidants should be stored away from easily oxidized materials and so on.
2. Be sure all chemicals are stored in labeled containers indicating contents, concentration, source, date purchased (or prepared), any precautions for handling and storage, and expiration date.
3. Dispose of any outdated or waste chemicals properly according to accepted disposal procedures.
4. Do not store chemicals above eye level.
5. Wood shelving is preferable to metal. All shelving should be firmly attached to all walls and have anti-roll edges.
6. Store only those chemicals that you plan to use.
7. Hazardous chemicals require special storage containers and conditions. Be sure to know what those chemicals are and the accepted practices for your area.
8. When working with chemicals or preparing solutions, observe the same general safety precautions that you would expect from students. These include wearing an apron and goggles. Wear gloves and use the fume hood when necessary. Students will want to do as you do whether they admit it or not.
9. If you are a new teacher in a particular laboratory, it is your responsibility to survey the chemicals stored there and to be sure they are stored properly or disposed of. Consult the rules and laws in your area concerning what chemicals can be kept in your classroom. For disposal, consult up-to-date disposal information from the state and federal governments.

Disposal of Chemicals

Local, state, and federal laws regulate the proper disposal of chemicals. These laws should be consulted before chemical disposal is attempted. Although most substances encountered in school laboratories can be flushed down the drain with plenty of water, it is not safe to assume that is always true. It is recommended that teachers who use chemicals consult the following books from the National Research Council.

Prudent Practices for Handling Hazardous Chemicals in Laboratories. Washington, DC: National Academy Press, 1981.

Prudent Practices for Disposal of Chemicals from Laboratories. Washington, DC: National Academy Press, 1983.

These books are useful and still in print, although they are several years old. Current laws in your area would, of course, supersede the information in these books.

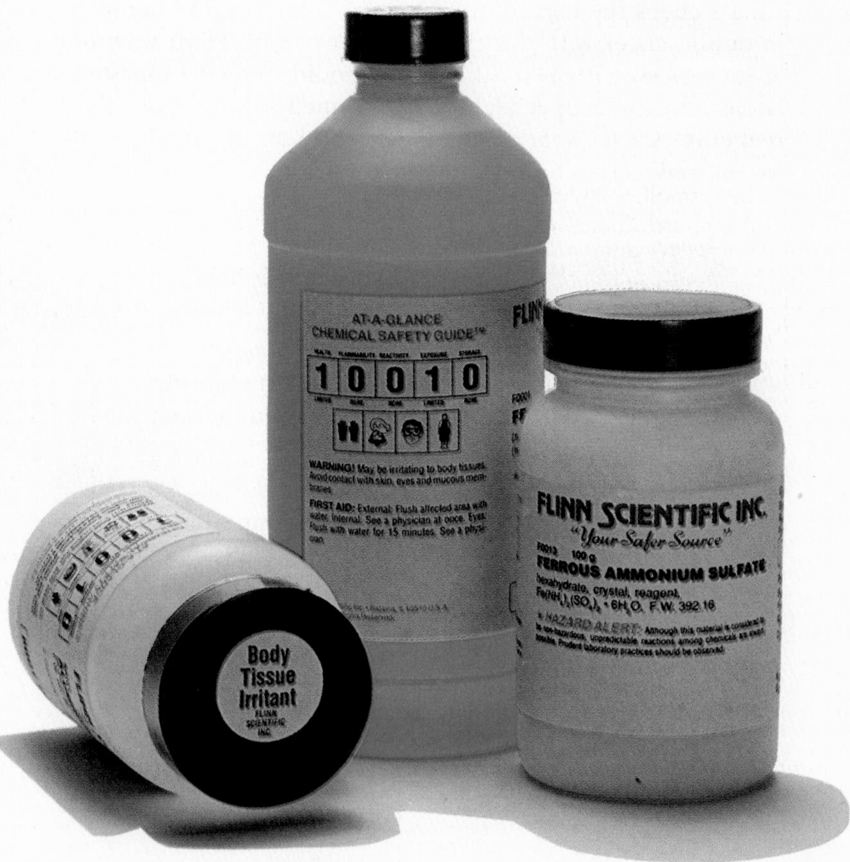

Life Science Materials List

NONCONSUMABLES			
ITEM	**ACTIVITY**	**MINI LAB**	**FIND OUT**
Aquarium with glass cover (1)	17-1		
or jar, 4-L (15)	15-2		
Aluminum foil, (1 roll)	28-1		
Bags, paper (15)	5-1		
plastic (15)			73
plastic, self-sealing (105)	24-1		275
Balance, pan (15)	3-2	21, 447	
Balloons (15)		353	
Beakers, (30)	14-1	59, 246 278	647
100-mL		38	
250-mL (45)	2-2, 8-2	447	
400-mL (30)	21-1		
600-mL (15)	15-1		
or glass (15)			5
Binoculars (15)	26-2		
Books, heavy (15)		156, 344	
Bowls, (15)		353	53, 647
glass (15)		31	
shallow (15)		375	
Box, cardboard, with lid (15)	6-1		
Brochures, pamphlets, veterinary (15)		184	
Calculators (15)			441
Cans, small, with lid (15)	17-2		
Cardboard, 5 cm × 20 cm, (120 strips)	28-2		
corrugated (15 pieces)		452	
Cards, index (2 pkgs)	22-2	115, 593	
Chalk (1 piece)		516	
Cheesecloth (15 pieces)		246	
Clay (8 boxes)		166	
Clocks		59	
or watches with second hands (15)	21-1	344	419, 463 487
Clothespins, snap-type (15)			419
Coffee filters			647
strips, 3 cm × 12 cm (45)			5
Containers,			
plastic milk (15)	28-1		
wide-mouthed, 0.5 L (45)	1-1		
500-mL, with covers (30)	3-1		
Cotton balls (1 pkg)	20-1, 24-1		557
Cotton swabs (15)	14-1		
Coverslips (75)	2-1, 2-2 9-1, 9-2 10-1, 11-2 12-1, 18-1 26-1		181
Culture dishes (15)	9-2	209	
Cups, measuring (15)		15	
paper (15)		137	

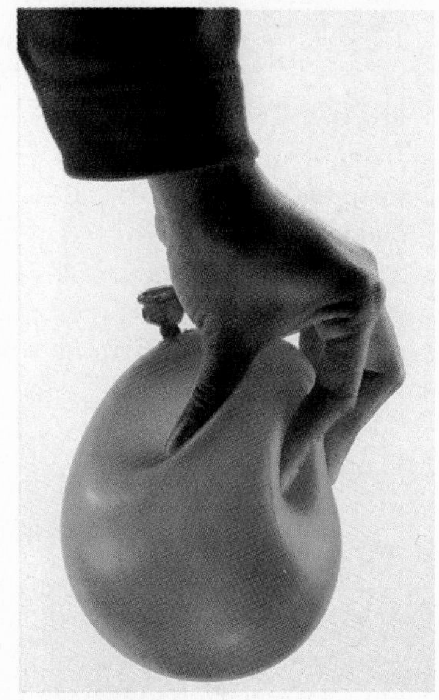

NONCONSUMABLES			
ITEM	ACTIVITY	MINI LAB	FIND OUT
Dishes, (15)	17-1		29
Disposable diapers, two brands (15 each)		15	
Dissecting needles (15)	10-1		
Dissecting pans, or cutting boards (15)	18-2		
Dissecting probes (15)	18-2		
Droppers, (15)	2-1, 2-2 9-1, 10-1 11-2, 13-2 18-1, 19-2 25-2, 26-1	280, 315 316, 380 435	
Dropping bottles (150)	19-1		
Dropping pipettes (15)			181
Feathers, contour (15)	16-1		
down (15)	16-1		
Field guides (15)	26-2		
Filter paper (60 sheets)	24-2		
Fishnets, small (15)	15-1, 15-2 17-1		
Flashlights (15)	14-1	12, 38	
Flowerpots, clay (30)	28-1		
Forceps (15)	2-1, 2-2 10-1, 11-1 12-1, 24-2		
Glasses, drinking (15)	19-2	31, 241	
Globe, or world map, (15)			627
biome (15)		633	
Glue (15 bottles)	22-2	78, 156	
Graduated cylinders (15)	3-1, 19-1 19-2, 21-1	246	
Gravel, (15 tsp.)			371
washed (1 bag)	15-2		
Hand lenses (15)	1-1, 7-1, 10-1 10-2, 11-1 11-2, 14-1 16-1, 17-2 18-1, 18-2 21-2, 26-2	5, 31, 241 263, 500	29, 73 201, 323
Hole punch (15)	28-2		
Ink pads (15)		115	
Jars, large,		358	
with screen wire cover (15)	17-2		
small (60)	24-2		
Knife (15)		241	
Labels (300)	1-1, 3-1 24-1, 24-2 27-2, 28-1		
Metersticks (15)	22-1		605
Microscopes, (15)	2-1, 2-2, 4-1 8-1, 8-2 9-1, 9-2 10-1, 10-2 11-2, 12-1 13-2, 18-1 18-2, 20-2 25-2, 26-1	380	181

NONCONSUMABLES

ITEM	ACTIVITY	MINI LAB	FIND OUT
Microscope, stereoscopic (15)	13-2, 15-2	315, 316	
Mitts, thermal (15)		447	
Mortar and pestle (15)		57	371
Newspaper, (30 sheets)		31, 156	
want ads (15 sheets)			129
Notebooks (15)	26-2		
Objects, small keys and shells (15)		137	
Oven (1)		447	
Over-the-counter products, different types (45)		584	
Paint brushes, small (15)	13-2		
Pans, shallow (15)	14-1		
Paper, (2 reams)	7-1, 7-2, 13-1 16-2, 23-1 25-1, 27-1	78, 106 115, 134 358, 508 537, 562, 584 611, 615 631, 659	105, 155 229, 301 393, 441 463, 487 579, 605
black	11-2, 17-2		
construction, black/white (15 sheets each)			129
construction, 6 colors (2 sheets each)	4-2		
graph (2 pkgs.)	5-2, 6-1, 6-2 23-2, 26-2	405	
heavy (15 sheets)	4-2		
typing, bond (15)	21-2		
waxed, 5 cm × 20 cm, (60 strips)	28-2		
Paper clips (15)		21	
Paper towels, (4 rolls)	9-2, 10-2 12-2, 14-1 24-1	214, 263 452	73 5
Pens, black, water soluble felt-tip			5
glass marking (15)	19-1, 19-2	78	
Pencils, (1 gross)	5-2, 6-1 6-2, 7-2, 10-1 13-1, 16-2 23-1, 24-1 24-2, 25-1 26-2, 27-1	106, 115 134, 358, 405 508, 537 562, 584, 611 615, 631, 659	105, 155 229, 301 441, 463 487, 579 605
red and blue (15 each)	23-2	466	
Pennies (1500)	6-1, 22-1		529
Petri dishes (60)	10-2 12-2, 28-1		
Pictures of beaks and feet of birds (15)	16-1		
Pictures of types of seeds including: coconuts, grasses, burrs, winged seeds (15)		265	
Pie pans, aluminum, (30)	28-1		
Planters (30)	27-2		
Plaster of paris, or clay (1 container)		137	
Plexiglass sheets 25 cm² (15)			393
Poster board, (30 sheets)		78, 156 185, 452	
Probes (15)		375	
Ring stands (15)	12-2		
Rocks (15)		21	

NONCONSUMABLES			
ITEM	**ACTIVITY**	**MINI LAB**	**FIND OUT**
Rulers, metric, (15)	5-2, 6-2 7-1, 17-1 22-2, 24-2	537	3, 29, 507
transparent (15)			323
Safety pins (225)		399	
Sand, or gravel (1 bag)	28-1		
washed, coarse (1 bag)	17-1		
Sandpaper (1 pkg.)	24-1		
Saucepans (15)		193	
Scale		650	
Scalpels (15)	11-1, 11-2 18-1, 18-2	500	
Scissors (15)	2-2, 4-2, 10-2 16-1, 24-2 28-1, 28-2	15	129
Slides, microscope (75)	2-1, 2-2, 8-2 9-1, 9-2 10-1, 11-2 12-1, 18-1 26-1	380	181
Soil, (3 bags)	14-1, 17-2,		
garden mixed with plant material (1 qt.)	17-2, 28-1		647
Spades, small or spoons (15)	27-2		
Spatulas, small (15)	8-2		
Splints, wooden (15)	1-1		
Sponge, small (15)	17-2		
Spoons	3-1		
metal (15)		21	
or trowel (15)	17-2		
Sphygmomanometers (15)	20-1		
Staplers with staples (15)	28-2		
Stethoscopes (15)	20-1	468	
Stirring rods (15)	3-2, 14-2 15-1		53, 647
Stopwatch, or watch with second hand (15)			463, 487
Straws, drinking (1 box)	21-1	278	
String (1 spool)	12-2, 28-2	246	
Tape,	4-2, 12-2		
transparent	24-2		
Test tubes, (150)	19-1, 19-2	57, 280, 560	
16 × 150 mL, with stoppers, (60)	3-2		
Test-tube racks (15)	3-2, 19-1 19-2	560	
Thermometers (15)	15-1, 17-1		
Thermos bottles (15)		193	
Thread (3 spools)		78	
Toothpicks (1 box)	22-2		
Trays, (5)		447	
cardboard or plastic (15)			529
Tuning fork (15)			349
Watch glass (15)	15-2	315, 316	
Yarn (2 pkg.)		78	

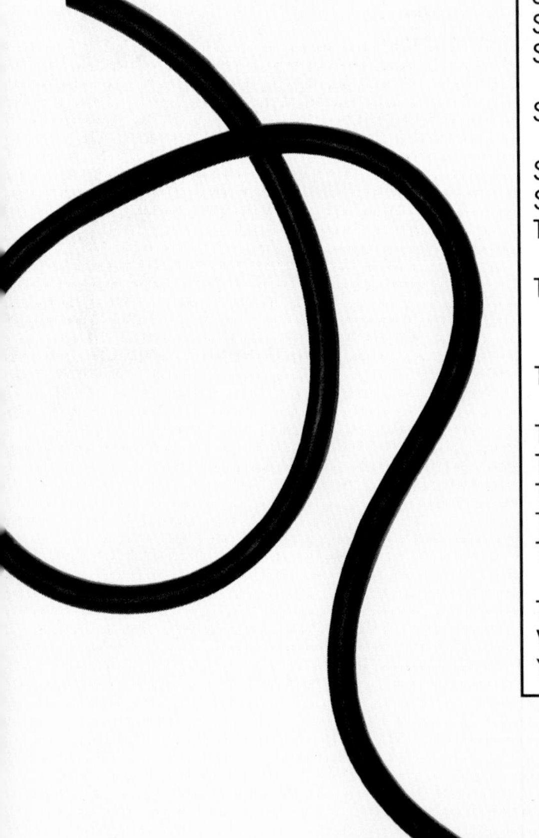

CONSUMABLES

ITEM	ACTIVITY	MINI LAB	FIND OUT
Apples, fresh (90)	24-1		
juice (1 can)	19-1		
peelings	28-1		
rotting (15)	24-1		
Beans, (830)	27-2		73
red (1500)	5-1		
white (1500)	5-1		
Beef, bones (15)	18-1		
ground, uncooked (5 Tbsp.)	14-2		
Bread crumbs (small box)	17-2		
Carrot juice (1 can)	19-1		
Celery, (1 bunch)		241	229
or carrots (15)		447	
Cranberry juice (1 bottle)	19-1		
Eggs, (15)	3-1	375	29
Fish food for guppies (small container)	17-1		
Food flavoring, peppermint or lemon (50 mL)			557
Gelatin, unflavored (30 pkgs.)	19-2	38	
Honey (small container)	17-2		
Lemon juice (1 bottle)	19-1		
Lettuce,			229
boiled (one head)	15-2		
in dish of water	12-1		
Lima beans (15)	11-1		
Lime juice (1 bottle)	19-1		
Liquid from sour cream (400 mL)		560	
Milk, (4 gal.)		57, 193	
fresh (400 mL)		560	
sour (400 mL)		560	
Mineral oil (30 mL)		560	
Mixed vegetable juice (1 can)	19-1		
Mushrooms (3 pkgs.)		214	201
Olives (1 jar)			229
Onions (3)	2-2		
plants with roots and leaves (15)			229
Orange juice (1 can)	19-1		
Peanuts, in shell (1 pkg.)			253
Pineapple juice (1 can)	19-1		
Potato slices, raw (30)			53
Seeds, (1500)	6-2		
corn (115)	11-1, 12-2		
sunflower (1 jar)			229
10 different (15 pkgs.)	7-1		
Sugar (small portion)	17-2		
Syrup, white table (250 mL)	3-1		
Tomatoes (3)			229
juice (1 bottle)	19-1		
Turkey leg, cooked	18-2		
Yogurt, (4 cartons)	8-2	560	181
starter (2 pkgs.)		193	

CHEMICAL SUPPLIES			
ITEM	ACTIVITY	MINI LAB	FIND OUT
Alcohol (1 bottle)	8-2, 20-1 24-1, 24-2		
Antacid tablets (30)		59	
Bromothymol blue, (150 mL)		280	
25 mL solution, (3 L)	21-1		
in dropping bottles (15)	3-2		
Cobalt chloride paper (15 pieces)		435	
Coffee, dilute solution of (50 mL)	25-2		
Cola, dilute solution of (50 mL)	25-2		
Cough medicine, dextromethorphen hydrobromide dilute solution (50 mL)	25-2		
Disinfectant (1 bottle)	24-2		
Ethyl alcohol, dilute solution of (50 mL)	25-2		
Food coloring (1 package, red)		241	
Hydrochloric acid, dilute (50 mL)	19-2		
Hydrogen peroxide (1 bottle)	24-2		
Indophenol solution (75 mL)	19-1		
Iodine solution	11-1, 21-2		
Methylene blue (350 drops)	8-2	560	181
Mouthwash (1 bottle)	24-2		
Pepsin powder (10 mL)	19-2		
Petroleum jelly (1 container)	28-2		
Phenol red (375 mL)		278	
Rennin tablets (15)		57	
Salt (225 g)			53
Salt solution,	12-1		
strong (500 mL)	1-1		
weak (500 mL)	1-1		
Soap, and water	24-1		
Sodium bicarbonate (15 g)	3-2		
Sugar solution (150 drops)		280	
Tobacco, dilute solution of (50 mL)	25-2		
Vinegar, (250 mL)	3-1, 14-1		
Water,	2-2, 9-2, 10-1 10-2, 11-2 12-2, 14-1 15-1, 17-2 18-1, 19-1 19-2, 27-2 28-1	15, 31, 38 59, 241, 246 259, 353 435, 452 560	5, 53
aquarium (5 gal.)	15-1		
distilled (1.5 L)	1-1, 26-1		
lake or pond (4 L)	15-2		
tap; aged three days	3-2,17-1 25-2		
Yeast suspension (300 drops)		280	

LIVING ORGANISMS

ITEM	ACTIVITY	MINI LAB	FIND OUT
Amoeba, culture of (15)	9-1		
Animals, or pictures of animals	16-2		
Ants (300)	17-2		
Clam shell (15)			323
Crayfish, in small aquarium (15)	14-2		
Daphnia culture	25-2		
Earthworm, live (30)	14-1		
Elodea plant (15)	2-1		
pieces (75)	3-2	278	
Euglena, culture of (15)	9-1		
Fern (15)	10-2		
Fern frond with spores (15)	10-2		
Fern spores	10-2		
Frog, egg masses (15)	15-2	358	
Fruit flies, cultures of (15)		337	
Gladiolus flower (30)	11-2		
Goldfish (2)	15-1		
Guppies (2)	17-1		
Gymnosperm cones, with seed (15)		263	
Leaf, (30)	28-1		
variety of types (200)		156	
Lichen, 3 different samples	9-2		
Liverworts (15)	10-1		
Micrasterias, culture of (15)	9-1		
Moss	10-1		
Mosses, dried reindeer (15)	9-2		
Nostoc and *Oscillatoria*, cultures of	8-1		
Paramecium, culture of (15)	9-1		
Planarians, culture of (15)	13-2	315, 316	
Plants, aquatic (15)	15-2, 17-1		
Seedling plants, in pots (15)			275
Seeds (15 Tbsp.)			371
Shrimp, brine, eggs	1-1		
Slime mold, culture		209	
Snails	17-1		
Spirogyra, culture of (15)	9-1		

PRESERVED SPECIMENS

ITEM	ACTIVITY	MINI LAB	FIND OUT
Blood, prepared slides of human and 2 other vertebrates, (fish, frog, reptile, bird) (15 each)	20-2		
Fern, prepared slides of (15)	10-2		
Gleocapsa and *Anabaena*, prepared slides of (15 each)	8-1		
Human cheek cell, prepared slide of (15)	2-1		
Kidney, of large animal (1)		500	
Liverwort gametophytes, and sporophytes, prepared slide of (15)	10-1		
Moss, prepared slide of (15)	10-1		
Muscles, prepared slides of smooth, skeletal, and cardiac (15)	18-2		
Nutrient Agar, sterile plates (75)	24-2		
Onion root tip, prepared slide of (30)	4-1		
Slime mold, prepared slide of (15)	9-1		
Sphagnum moss (55 tsp.)		246	
Tapeworm, prepared slide of (15)	13-2		
Vegetation, pond, dried (30)	26-1		
Whitefish, embryo, prepared slide (30)	4-1		

MERRILL LIFE SCIENCE

Supplier Addresses

Addresses and phone numbers throughout this book were accurate at the time of publication and are subject to change.

EQUIPMENT SUPPLIERS

Carolina Biological Supply Company
2700 York Road
Burlington, NC 27215

Central Scientific Company
11222 Melrose Avenue
Franklin Park, IL 60131

Edmund Scientific Company
101 Gloucester Pike
Barrington, NJ 08007

Fisher Scientific Company
4901 W. LeMoyne Avenue
Chicago, IL 60651

LaPine Scientific Company
13636 Western Avenue
Blue Island, IL 60406-0780

McKilligan Supply Corporation
435 Main Street
Johnson City, NY 13790

Nasco
901 Janesville Avenue
Fort Atkinson, WI 53538

Sargent-Welch Scientific Co.
911 Commerce Ct.
Buffalo Grove, IL 60089-2362

VWR of Canada, Ltd.
77 Enterprise N.
London, Ontario
Canada N6N 1A5

Science Kit and Boreal Labs
777 E. Park Drive
Tonawanda, NY 14150

Ward's Natural Science Establishment, Inc.
P.O. Box 92912
Henrietta, NY 14692

AUDIOVISUAL DISTRIBUTORS

Access Network
3720 76th Avenue
Edmonton, Alberta,
Canada T6B 2N9

Agency for Instructional Technology (AIT)
Box A
Bloomington, IN 47402

Aims Media
9710 Desoto Avenue
Chatsworth, CA 91311-4409

American Cancer Society
1599 Clifton Road NE
Atlanta, GA 30329

American Lung Association
1740 Broadway
New York, NY 10019

Benchmark Films, Inc.
569 N. State Road
Briarcliff Manor, NY 10510

Centre Communications, Inc.
1800 30 Street
Suite 207
Boulder, CO 80301

Churchill Films
12210 Nebraska Avenue
Los Angeles, CA 90025

Coronet/MTI Film and Video Distributors of LCA
108 Wilmot Road
Deerfield, IL 60015

CRM Films
2215 Faraday Avenue
Carlsbad, CA 92008

Educational Media International
Box 1288
175 Margaret Place
Elmhurst, IL 60126

Encyclopaedia Britannica Educational Corp. (EBEC)
310 S. Michigan Avenue
Chicago, IL 60604

Filmmakers Library, Inc.
124 E. 40th Street
Suite 901
New York, NY 10016

Films for the Humanities and Sciences, Inc.
(Current Affair)
P.O.Box 2053
Princeton, NJ 08543-2053

Films, Inc.
5547 N. Ravenswood Avenue
Chicago, IL 60640

Focus Media, Inc.
485 S. Broadway, Suite 12
Hicksville, NY 11801

Guidance Associates, Inc.
Box 1000
Communications Park
Mount Kisco, NY 10549

Hawkill Associates, Inc.
125 E. Gilman Street
Madison, WI 53703

Human Relations Media (HRM)
175 Tompkins Avenue
Pleasantville, NY 10570

Image Entertainment
9333 Oso Avenue
Chatsworth, CA 91311

Image Premastering Services
1781 Prior Avenue N.
Saint Paul, MN 55113

International Film Bureau, Inc. (IFB)
332 S. Michigan Avenue
Chicago, IL 60604

Lucerne Media
37 Ground Pine Road
Morris Plains, NJ 07950

Lumivision
1490 Lafayette
Suite 407
Denver, CO 80218

Marsh Film/Marsh Ware Enterprises, Inc.
P.O. Box 8082
Shawnee Missions, KS 66208

National Geographic Society
Educational Services
17th and "M" Streets, NW
Washington, DC 20036

Optical Data Corporation
P.O. Box 4919
30 Technology Drive
Warren, NJ 07059

PBS Video
1320 Braddock Place
Alexandria, VA 22314-1698

Phoenix Learning Group, Inc.
2349 Chaffee Drive
St. Louis, MO 63146-3306

Pyramid Film and Video
Box 1048
Santa Monica, CA 90406

Sunburst Communications
101 Castleton Street
Pleasantville, NY 10570

SysCon Corporation
2686 Dean Drive
Virginia Beach, VA 23452

Time-Life Videos
Time and Life Building
1271 Avenue of the Americas
New York, NY 10020

VideoDiscovery
1700 Westlake Avenue N.
Suite 600
Seattle, WA 98109-3012

SOFTWARE DISTRIBUTORS

Academic Hallmarks
P.O. Box 998
Durango, CO 81302

CONDUIT
The University of Iowa
Oakdale Campus
Iowa City, IA 52242

Cygnus Software
8002 E. Culver
Mesa, AZ 85207

D.C. Heath and Company
2700 N. Richardt Avenue
Indianapolis, IN 46219

Diversified Educational
Enterprises, Inc.
725 Main Street
Lafayette, IN 47901

Educational Activities, Inc.
P.O. Box 392
Freeport, NY 11520

Educational Materials and
Equipment Company (EME)
P.O. Box 2805
Danbury, CT 06813-2805

Encyclopaedia Britannica
Educational Corp. (EBEC)
310 S. Michigan Avenue
Chicago, IL 60604

Focus Media, Inc.
485 S. Broadway
Suite 12
Hicksville, NY 11801

IBM Educational Systems
Department PC
4111 Northside Parkway
Atlanta, GA 30327

Island Software, Inc.
P.O. Box 300
Lake Grove, NY 11755

J and S Software
14 Maple Street
Port Washington, NY 11050

Micro-ED, Inc.
P.O. Box 24750
Edina, MN 55424

Micro Power and Light Co.
8814 Sanshire Avenue
Dallas, TX 75231

National Geographic Society
Educational Services
17th and "M" Streets, NW
Washington, DC 20036

Opportunities for Learning, Inc.
941 Hickory Lane
P.O. Box 8103
Mansfield, OH 44901-8103

Queue, Inc.
338 Commerce
Fairfield, CT 06430

Scholastics Software
730 Broadway
New York, NY 10003

Sunburst Communications
101 Castleton Street
Pleasantville, NY 10570

Photo Credits

4T, Brent Turner/BLT Productions; 5T, Doug Martin; 6T, Ted Rice; 10T, Doug Martin; 12T, 13T, ©Ken Frick; 14T, 15T, KS Studios; 18T, Les Losego/Crown Studios; 19T, ©Ken Frick; 22T, Tim Courlas; 24T, ©Ken Frick; 27T, Cobalt Productions; 32T, Brent Turner/BLT Productions; 33T, Doug Martin; 34T, Les Losego/Crown Studios; 35T, 36T, Les Losego/Crown Studios; 37T, file photo; 38T-43T, Doug Martin.

We want your opinions!

We at Glencoe Publishing feel that with this edition of ***Merrill Life Science,*** we have produced a quality textbook program—but the final proof of that rests with you, the teachers who have had the opportunity to put our materials to use in your classrooms. That's why we would appreciate it if you would take the time to respond to any part of this questionnaire that is appropriate for you. In doing so, you will be letting us know how good a job we've done and where we can work to improve.

Please note: (1) you need not have used all of the program components to respond to this questionnaire; and (2) we encourage you to give us your honest and most candid opinions.

Student Text

Excellent				Poor	
5	4	3	2	1	Organization
5	4	3	2	1	Narrative style
5	4	3	2	1	Readability
5	4	3	2	1	Visual impact
5	4	3	2	1	Usable Table of Contents
5	4	3	2	1	Accuracy of content
5	4	3	2	1	Coverage of science principles
5	4	3	2	1	Reduced number of boldface terms
5	4	3	2	1	Skill builder questions
5	4	3	2	1	Skill Handbook
5	4	3	2	1	MINI-Labs/Activities
5	4	3	2	1	Margin Features
5	4	3	2	1	Using Lab Skills
5	4	3	2	1	Problem Solving features
5	4	3	2	1	Technology features
5	4	3	2	1	Science & Society Sections
5	4	3	2	1	Glossary and Index
5	4	3	2	1	Appendices
5	4	3	2	1	Projects
5	4	3	2	1	Global Connections features
5	4	3	2	1	Unit End features

Teacher Edition

Excellent				Poor	
5	4	3	2	1	Teachability
5	4	3	2	1	Planning charts
5	4	3	2	1	Organization of teaching cycle
5	4	3	2	1	Performance objectives
5	4	3	2	1	Assessment Options

Supplements

Excellent				Poor	
5	4	3	2	1	Teacher Classroom Resources
5	4	3	2	1	Laboratory Manual
5	4	3	2	1	Science Integration Activities
5	4	3	2	1	Color Transparency Package
5	4	3	2	1	Test Bank
5	4	3	2	1	Chapter Review Software
5	4	3	2	1	Science and Technology Videodisc Series
5	4	3	2	1	Alternate Assessment in the Science Classroom
5	4	3	2	1	Performance Assessment in Middle School Science
5	4	3	2	1	Performance Assessment

Comments (general or specific):

School Information

1. What is the grade level of the students you teach? 6 7 8 9
2. Total number of students in that grade? 1-50 51-100 101-200 200+
3. Average class size? 25 or fewer 26-30 31-40 41 or more
4. Total school enrollment? 1-200 201-500 501-1000 1000+
5. Ability level of your average class? Basic Average Advanced
6. How appropriate is this text for your class? Too easy On level Too difficult
7. How many years have you used this text? 1 2 3 4 5
8. What text were you using *before* you adopted this program?
   (Title/Publisher/copyright year) ____________________

Fold

Name ____________________ Date ____________________

School ____________________

Street ____________________

City ____________________ State ____________ Zip ____________

Fold

NO POSTAGE NECESSARY IF MAILED IN THE UNITED STATES

BUSINESS REPLY MAIL
FIRST CLASS PERMIT NO 284 COLUMBUS OHIO

POSTAGE WILL BE PAID BY ADDRESSEE
GLENCOE
SCIENCE PRODUCT MANAGER
PO BOX 508
COLUMBUS OHIO 43272-6174

MERRILL

LIFE SCIENCE

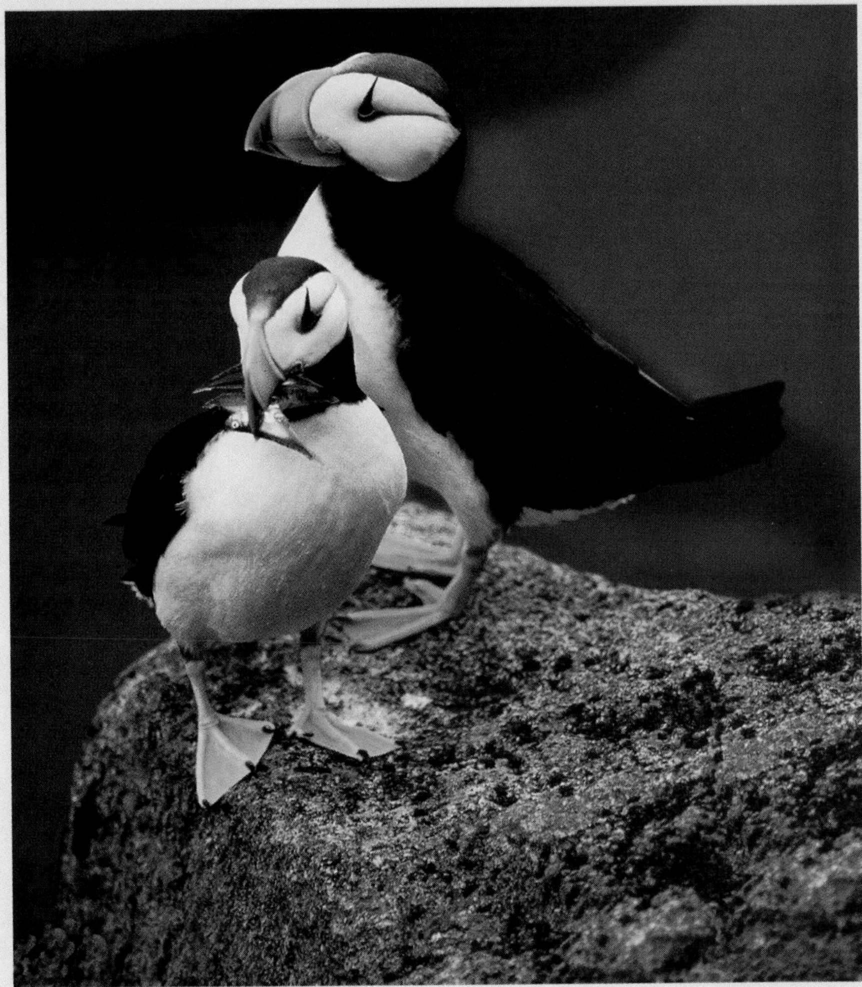

GLENCOE

Macmillan/McGraw-Hill

New York, New York   Columbus, Ohio   Mission Hills, California   Peoria, Illinois

A GLENCOE PROGRAM

MERRILL LIFE SCIENCE

Student Edition
Teacher Wraparound Edition
Teacher Resource Package
Study Guide, Student Edition
Reinforcement, Student Edition
Enrichment, Student Edition
Transparency Package
Laboratory Manual
Laboratory Manual, Teacher Annotated Edition
Spanish Resources
Chapter Review Software
Computer Test Bank
Videodisc Correlation
Science Integration Activities
Alternate Assessment in the Science Classroom
Performance Assessment
Science and Technology Videodisc Series
Science and Technology Videodisc Teacher Guide

REVIEWERS

Agnes B. Adamson
Rosarian Academy
West Palm Beach, Florida

Janice E. Barry
C.A. Gray Middle School
Moultrie, Georgia

James Chin
Day Junior High School
Newtonville, Massachusetts

Grant Critchfield
Smith Middle School
Killeen I.S.D.
Fort Hood, Texas

Gary E. Downes
Vista Verde Middle School
Irvine, California

Angeline Eliakopoulos
Chappell School
Chicago, Illinois

Donald E. Goldstein
Greenfield School
Phoenix, Arizona

Pilar Gonzalez de Killough
Silver Consolidated Schools
Silver City, New Mexico

Jeanne Lynn Helen Hatok
St. Charles Borromeo School
Albuquerque, New Mexico

George E. Judd
Elk Grove United School District
Elk Grove, California

Sherri Lynn Marchant
Austin Junior High
Irving, Texas

Amy L. Messinger
Guyan Valley High School
Branchland, West Virginia

Michael C. Meyers
United I.S.D.
Laredo, Texas

Theresa Lynn Parisi
Grapevine Middle School
Grapevine-Colleyville I.S.D.
Grapevine, Texas

Joe A. Starcher
Brooke County Schools
Follarsbee, West Virginia

Catherine R. Sullivan
Bellview Middle School
Pensacola, Florida

Sarah Wall
Clear Creek I.S.D.
League City, Texas

Cover Photograph: Horned puffins by Francis and Donna Caldwell

Send all inquiries to:
GLENCOE DIVISION
Macmillan/McGraw-Hill
936 Eastwind Drive
Westerville, OH 43081

ISBN 0-02-827027-4

Printed in the United States of America.

1 2 3 4 5 6 7 8 9-VH-02 01 00 99 98 97 96 95 94

AUTHORS

Lucy Daniel is a Science-helping Teacher for Rutherford County Schools, Spindale, North Carolina. She has thirty-five years of teaching experience in biology. Ms. Daniel holds a B.S. degree from the University of North Carolina at Greensboro and an M.A.S.E. from Western Carolina University at Cullowhee. She received the Presidential Award for Excellence in Science and Mathematics Teaching in 1984. She is a co-author of Merrill Publishing Company's *Biology: An Everyday Experience.*

Edward Paul Ortleb is the Science Supervisor for the St. Louis, Missouri Board of Education. He holds an A.B. in Education from Harris Teachers College, an M.A. in Education, and an Advanced Graduate Certificate in Science Education from Washington University, St. Louis. Mr. Ortleb is a lifetime member of NSTA, having served as its president in 1978-79. He is a contributing author for the Teacher Resource Books for Merrill Publishing Company's *Accent on Science* and *General Science* and is co-author of Merrill Publishing Company's *Science Connections.*

Alton Biggs is a biology instructor at Allen High School, Allen, Texas. Mr. Biggs received his B.S. in Natural Sciences and an M.S. in Biology from East Texas State University. He was a Resident in Science and Technology at Oak Ridge National Laboratory in 1986. Among the teaching awards he has received are Texas Outstanding Biology Teacher in 1982, Presidential Science Teacher Award Finalist in 1986, Teacher of the Year Award, Allen Independent School District in 1987, and Texas Teacher of the Year Finalist in 1988. Mr. Biggs is Past President of NABT. He is co-author of Merrill Publishing Company's *Biology: The Dynamics of Life.* Mr. Biggs is the founding president of TABT.

CONSULTANTS

Zoology:
Jerry Downhower, Ph.D.
Professor of Zoology
Department of Zoology
The Ohio State University
Columbus, Ohio

Genetics and Evolution:
Kathleen A. Fleiszar, Ph.D.
Professor of Biology
Department of Biology
Kennesaw State College
Marietta, Georgia

Human Body Systems:
Chris Teruo Hasegewa, Ph.D.
Associate Professor of Science Education
California State University, Sacramento
Sacramento, California

Cell Biology and Plants:
Eloy Rodriguez, Ph.D.
Professor of Biological Sciences
Department of Developmental and Cell Biology
University of California, Irvine
Irvine, California

Viruses, Immunity, and Drugs:
Melissa Sue Millam Stanley, Ph.D.
Professor of Biology
George Mason University
Fairfax, Virginia

Life Science and Ecology:
Richard D. Storey, Ph.D.
Associate Professor of Biology
Department of Biology
Colorado College
Colorado Springs, Colorado

Reading:
Barbara Pettegrew, Ph.D.
Director of Reading/Study Center
Assistant Professor of Education
Otterbein College
Westerville, Ohio

Gifted and Mainstreamed:
Barbara Murdock
Elementary Consultant For Instructions
Gahanna-Jefferson Public Schools
Gahanna, Ohio

Judy Ratzenberger
Middle School Science Instructor
Gahanna Middle School West
Gahanna, Ohio

Safety:
Robert Tatz, Ph.D.
Instructional Lab Supervisor
Department of Chemistry
The Ohio State University
Columbus, Ohio

Special Features:
Stephen C. Blume
Presidential Award for Excellence in Science and Mathematics, 1990
Elementary Science Specialist
St. Tammany Public School System
Slidell, Louisiana

Karen Muir, Ph.D.
Adjunct Professor
Social and Behavioral Sciences
Columbus State Community College
Columbus, Ohio

Mary Garvin
Managing Director, J.H. Barrow Biological Field Station
Hiram College
Hiram, Ohio

CONTENTS

v

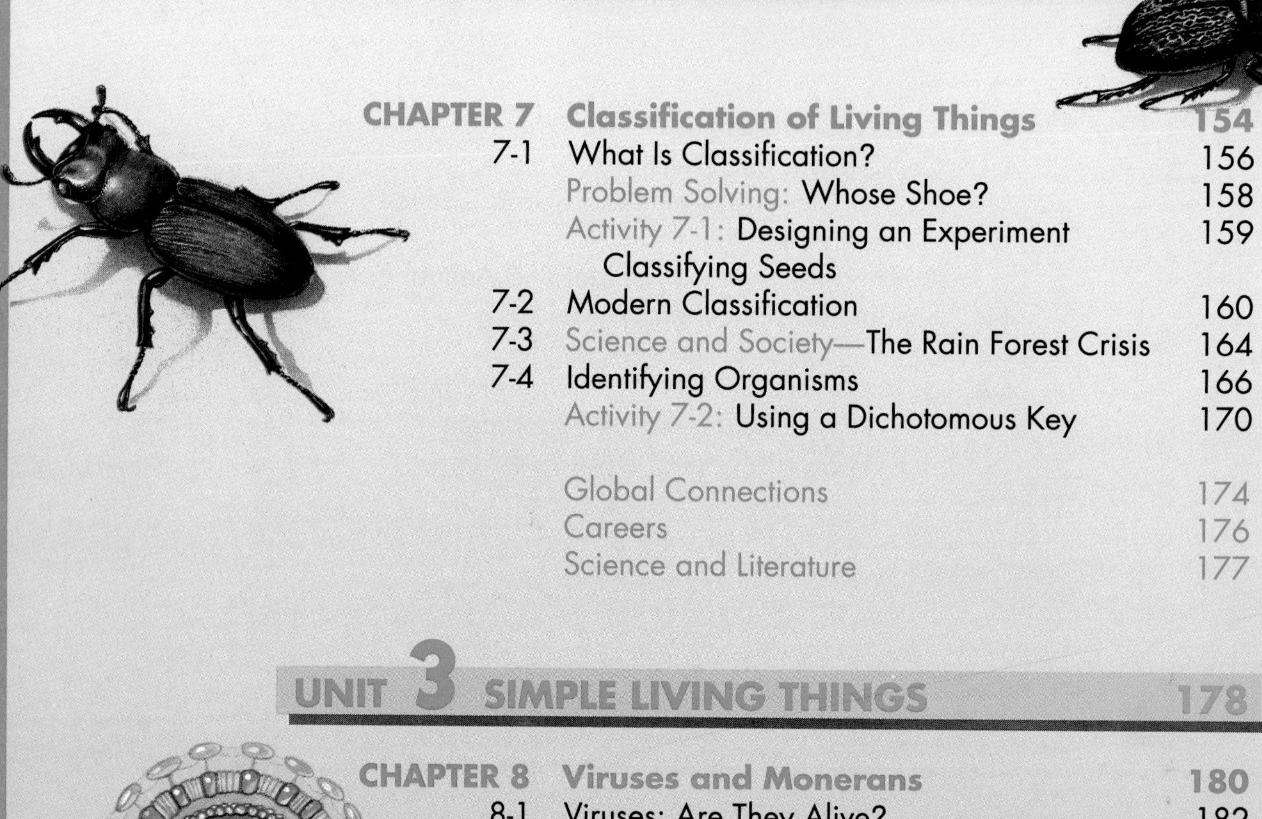

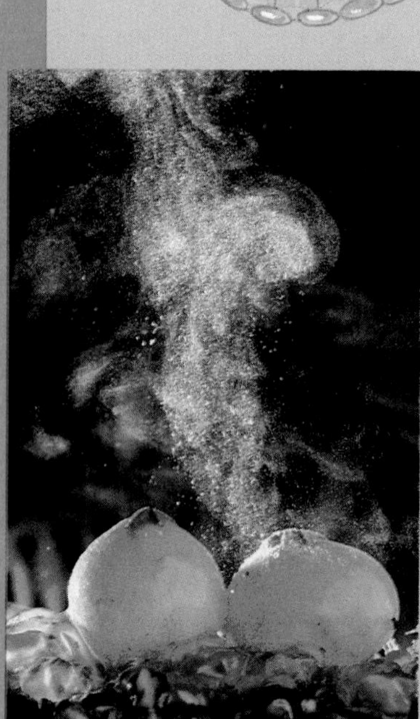

UNIT 4 PLANTS 226

UNIT 5 ANIMALS 298

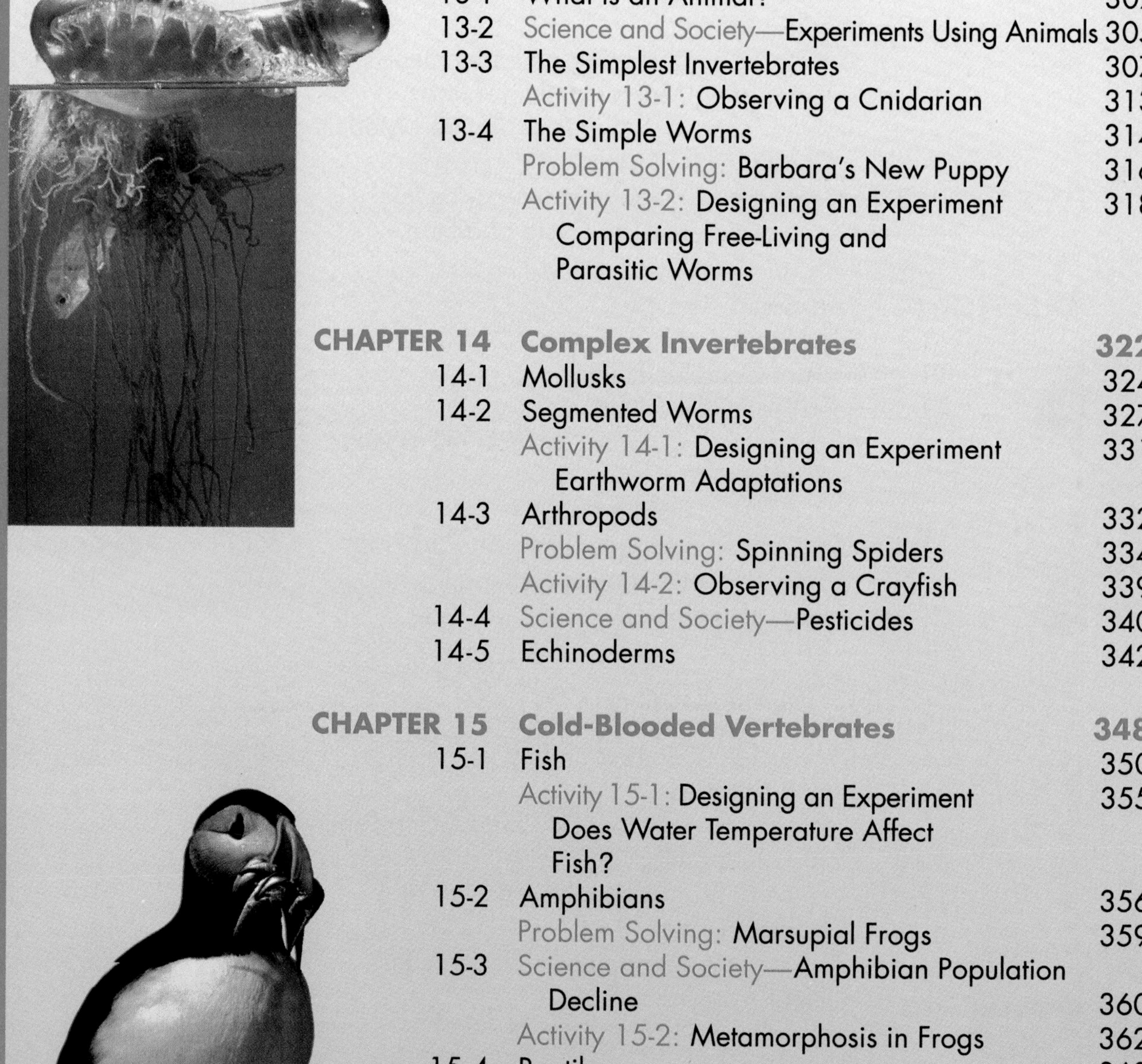

ACTIVITIES

MINI-Labs

PROBLEM SOLVING

TECHNOLOGY

SKILL BUILDERS

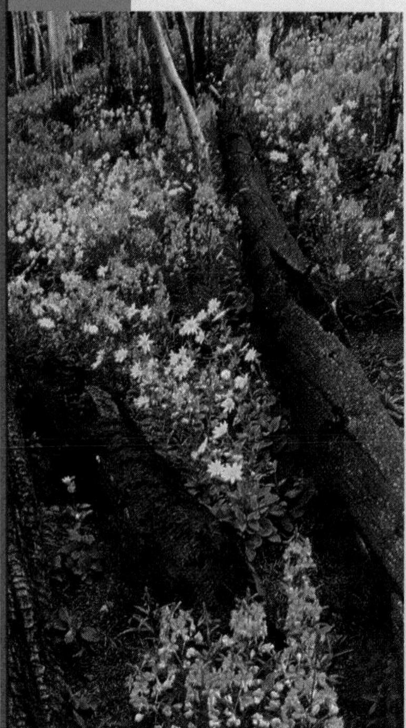

GLOBAL CONNECTIONS

CAREERS

SCIENCE AND LITERATURE/ART

USING MERRILL LIFE SCIENCE

Life Science is an everyday experience. It's a subject you're familiar with because every part of your day is based upon principles of life science... the simple act of walking involves your muscles to move your body, your nervous system to tell your muscles what to do and where you want to go, and your circulatory system to transport nutrients, energy, and oxygen to all parts of your body. Depending on where you walk, you may encounter a variety of insects, plants, and animals that are living their lives also based upon principles of life science. **Merrill Life Science** will help you understand life science principles and recognize how they affect you and the life around you everyday.

a quick tour of your textbook

What's happening here? Have you ever seen a raft setting on the tops of trees? Each unit begins with thought-provoking photographs that will make you wonder. The unit introduction then explains what is happening in the photographs and how the two relate to each other and to the content of the unit. What is a raft doing on the tops of trees? Read the opener to Unit 4 to find out.

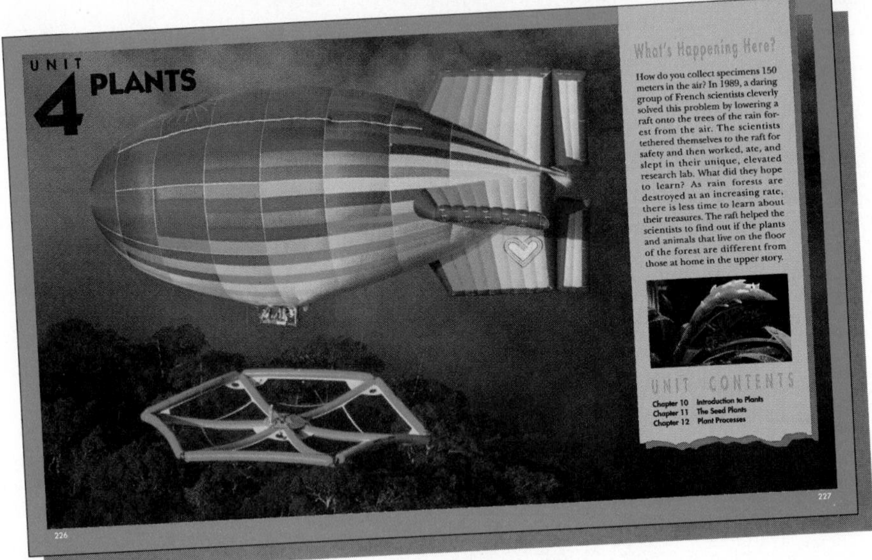

UNIT 4 PLANTS

What's Happening Here?

How do you collect specimens 150 meters in the air? In 1989, a daring group of French scientists cleverly solved this problem by lowering a raft onto the trees of the rain forest from the air. The scientists tethered themselves to the raft for safety and then worked, ate, and slept in their unique, elevated research lab. What did they hope to learn? As rain forests are destroyed at an increasing rate, there is less time to learn about their treasures. The raft helped the scientists to find out if the plants and animals that live on the floor of the forest are different from those at home in the upper story.

UNIT CONTENTS

Chapter 10 Introduction to Plants
Chapter 11 The Seed Plants
Chapter 12 Plant Processes

226 227

It's clearly organized to get you started and keep you going.

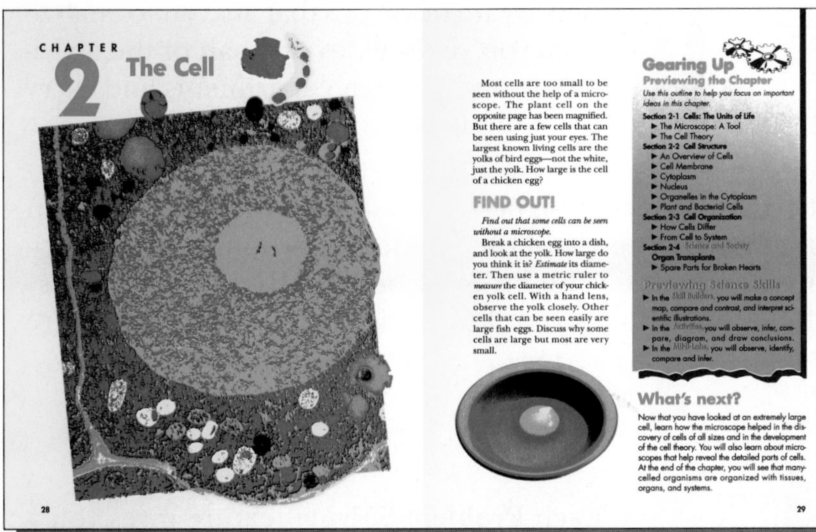

CHAPTER 2 The Cell

Most cells are too small to be seen without the help of a microscope. The plant cell on the opposite page has been magnified. But there are a few cells that can be seen using just your eyes. The largest known living cells are the yolks of bird eggs—not the white, just the yolk. How large is the cell of a chicken egg?

FIND OUT!

Find out that some cells can be seen without a microscope.

Break a chicken egg into a dish, and look at the yolk. How large do you think it is? *Estimate* its diameter. Then use a metric ruler to *measure* the diameter of your chicken yolk cell. With a hand lens, observe the yolk closely. Other cells that can be seen easily are large fish eggs. Discuss why some cells are large but most are very small.

Gearing Up

Previewing the Chapter

Use this outline to help you focus on important ideas in this chapter.

Section 2-1 Cells: The Units of Life
- The Microscope: A Tool
- The Cell Theory

Section 2-2 Cell Structure
- An Overview of Cells
- Cell Membrane
- Cytoplasm
- Nucleus
- Organelles in the Cytoplasm
- Plant and Bacterial Cells

Section 2-3 Cell Organization
- How Cells Differ
- From Cell to System

Section 2-4 Science and Society
Organ Transplants
- Spare Parts for Broken Hearts

Previewing Science Skills
- In the Skill Builders, you will make a concept map, compare and contrast, and interpret scientific illustrations.
- In the Activities, you will observe, infer, compare, diagram, and draw conclusions.
- In the MINI-Labs, you will observe, identify, compare and infer.

What's next?

Now that you have looked at an extremely large cell, learn how the microscope helped in the discovery of cells of all sizes and in the development of the cell theory. You will also learn about microscopes that help reveal the detailed parts of cells. At the end of the chapter, you will see that many-celled organisms are organized with tissues, organs, and systems.

28 29

As you begin each new chapter, use the **Gearing Up** to preview what topics are covered and how they are organized. You will also preview the skills you will use in this chapter.

After you've performed the **FIND OUT** activity and previewed the chapter, you're ready to further explore the topics ahead. Read **What's next** to see what's ahead.

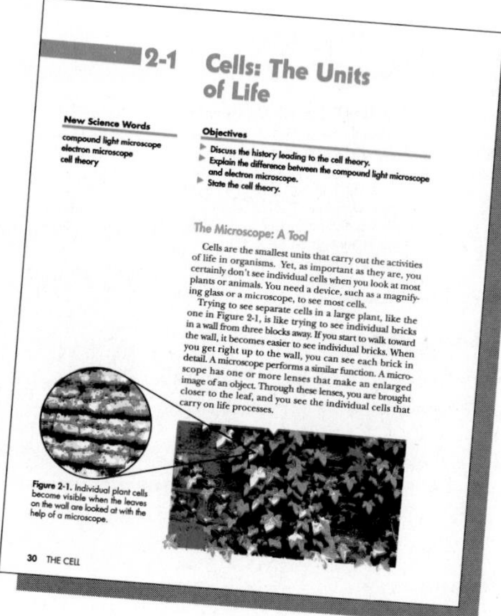

2-1 Cells: The Units of Life

New Science Words

compound light microscope
electron microscope
cell theory

Objectives
- Discuss the history leading to the cell theory.
- Explain the difference between the compound light microscope and electron microscope.
- State the cell theory.

The Microscope: A Tool

Cells are the smallest units that carry out the activities of life in organisms. Yet, as important as they are, you certainly don't see individual cells when you look at most plants or animals. You need a device, such as a magnifying glass or a microscope, to see most cells.

Trying to see separate cells in a large plant, like the one in Figure 2-1, is like trying to see individual bricks in a wall from three blocks away. If you start to walk toward the wall, it becomes easier to see individual bricks. When you get right up to the wall, you can see each brick in detail. A microscope performs a similar function. A microscope has one or more lenses that make an enlarged image of an object. Through these lenses, you are brought closer to the leaf, and you see the individual cells that carry on life processes.

Figure 2-1. Individual plant cells become visible when the leaves on the wall are looked at with the help of a microscope.

30 THE CELL

Chapters are organized into three to five numbered sections. The **Objectives** at the beginning of the numbered section tell you what major topics you'll be covering and what you should expect to learn about them. The **New Science Words** are also listed in the order in which they appear in the section.

Experience science by observing, experimenting, and asking questions.

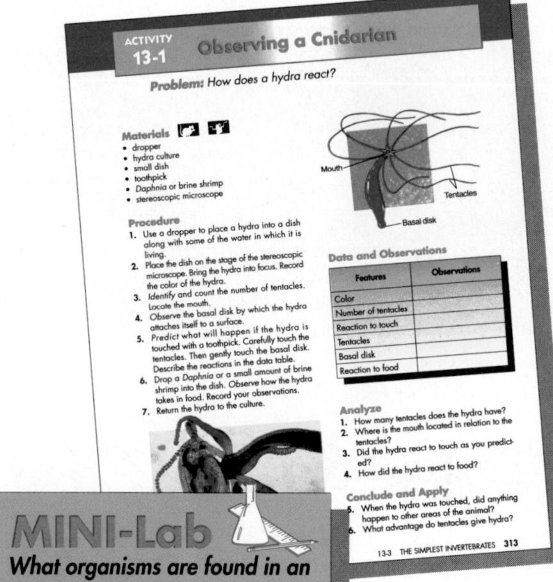

ACTIVITY 13-1

Observing a Cnidarian

Problem: How does a hydra react?

Materials

- dropper
- hydra culture
- small dish
- toothpick
- *Daphnia* or brine shrimp
- stereoscopic microscope

Procedure

1. Use a dropper to place a hydra into a dish along with some of the water in which it is living.
2. Place the dish on the stage of the stereoscopic microscope. Bring the hydra into focus. Record the color of the hydra.
3. *Identify* and count the number of tentacles. Locate the mouth.
4. *Observe* the basal disk by which the hydra attaches itself to a surface.
5. *Predict* what will happen if the hydra is touched with a toothpick. Carefully touch the tentacles. Then gently touch the basal disk. Describe the reactions in the data table.
6. Drop a *Daphnia* or a small amount of brine shrimp into the dish. *Observe* how the hydra takes in food. Record your observations.
7. Return the hydra to the culture.

Data and Observations

Features	Observations
Color	
Number of tentacles	
Reaction to touch	
Tentacles	
Basal disk	
Reaction to food	

Analyze

1. How many tentacles does the hydra have?
2. Where is the mouth located in relation to the tentacles?
3. Did the hydra react to touch as you predicted?
4. How did the hydra react to food?

Conclude and Apply

5. When the hydra was touched, did anything happen to other areas of the animal?
6. What advantage do tentacles give hydra?

13-3 THE SIMPLEST INVERTEBRATES 313

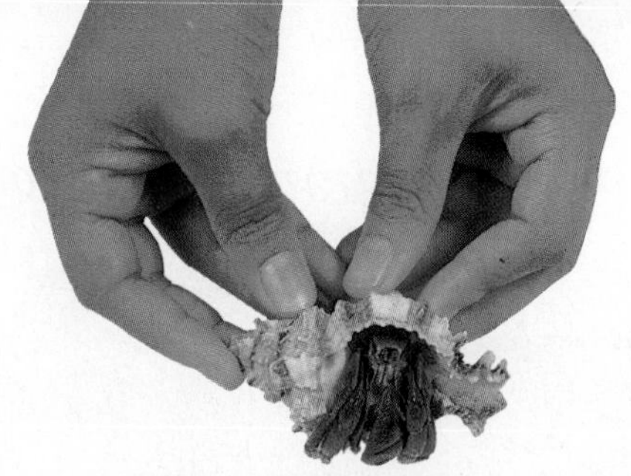

Science is more than words in a book. The Activities and MINI-Labs in each chapter give you the chance to explore and investigate the science topics hands on.

In the **Activities,** you'll use household items and laboratory equipment as you follow step-by-step procedures. At the end of each Activity are questions that ask you to analyze what you've done. In one-half of the activities, you will design experiments and carry them out to test your own hypotheses.

MINI-Lab

What organisms are found in an ecosystem?

Choose an ecosystem you are familiar with, such as a stream, garden plot, or pond, and identify the organisms found there. Make a list of all the populations you can see in the ecosystem. What is the niche of each species in the community?

Most **MINI-Labs** are designed so you can do them on your own or with friends outside of the science classroom using materials you find around the house. Doing a MINI-Lab is an easy and fun way to extend your knowledge about the topics you're studying.

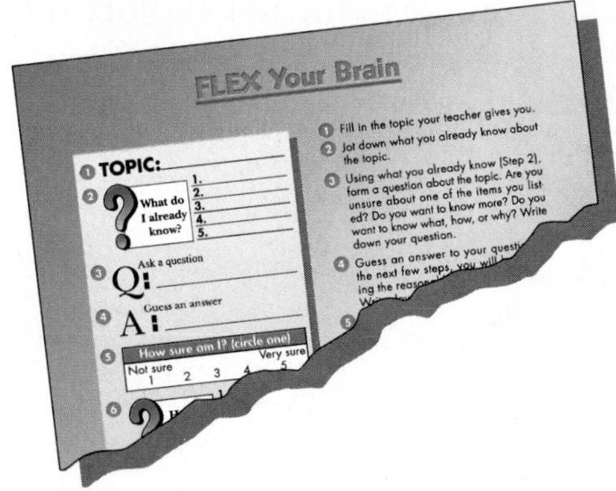

Flex Your Brain is a unique activity you can use to sharpen your critical thinking skills. Starting from what you already know about a science topic, you will apply a simple ten-step procedure to extend your knowledge about the topic from a perspective that interests you.

Each **Problem Solving** feature gives you a chance to solve a real life problem.

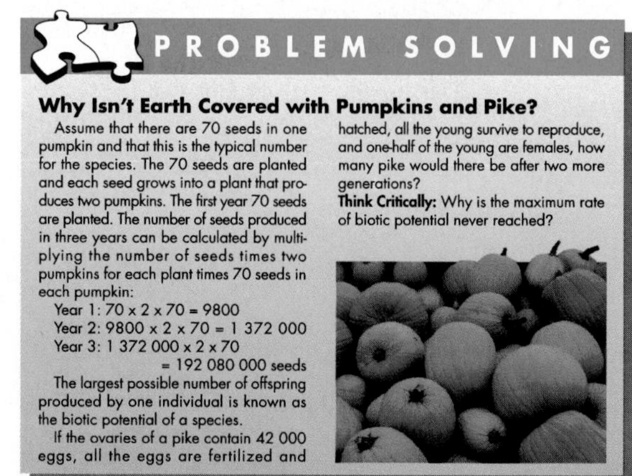

PROBLEM SOLVING

Why Isn't Earth Covered with Pumpkins and Pike?

Assume that there are 70 seeds in one pumpkin and that this is the typical number for the species. The 70 seeds are planted and each seed grows into a plant that produces two pumpkins. The first year 70 seeds are planted. The number of seeds produced in three years can be calculated by multiplying the number of seeds times two pumpkins for each plant times 70 seeds in each pumpkin:

Year 1: 70 x 2 x 70 = 9800
Year 2: 9800 x 2 x 70 = 1 372 000
Year 3: 1 372 000 x 2 x 70 = 192 080 000 seeds

The largest possible number of offspring produced by one individual is known as the biotic potential of a species.

If the ovaries of a pike contain 42 000 eggs, all the eggs are fertilized and hatched, all the young survive to reproduce, and one-half of the young are females, how many pike would there be after two more generations?

Think Critically: Why is the maximum rate of biotic potential never reached?

xx

Explore news-making issues, concerns about the environment, and how science shapes your world through technology.

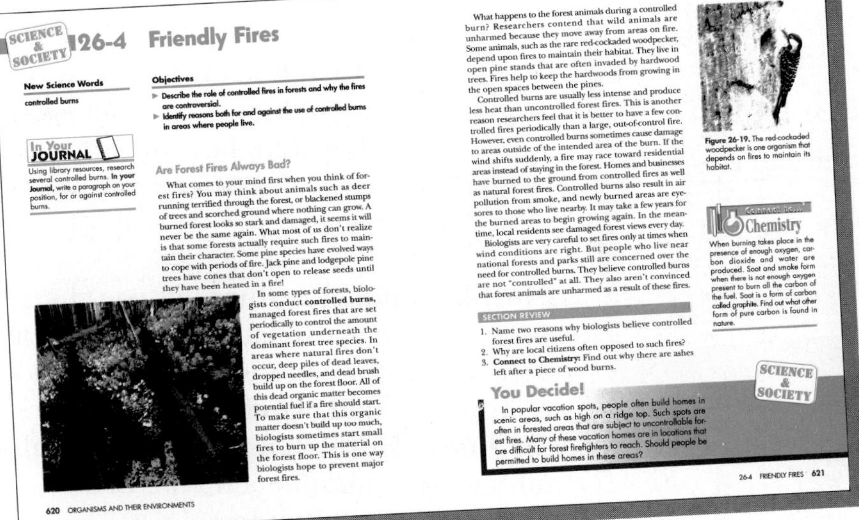

SCIENCE & SOCIETY **26-4 Friendly Fires**

New Science Words

controlled burns

Objectives

- Describe the role of controlled fires in forests and why the fires are controversial.
- Identify reasons both for and against the use of controlled burns in areas where people live.

In Your JOURNAL

Using library resources, research several controlled burns. **In your Journal,** write a paragraph on your position, for or against controlled burns.

Are Forest Fires Always Bad?

What comes to your mind first when you think of forest fires? You may think about animals such as deer running terrified through the forest, or blackened stumps of trees and scorched ground where nothing can grow. A burned forest looks so stark and damaged, it seems it will never be the same again. What most of us don't realize is that some forests actually require such fires to maintain their character. Some pine species have evolved ways to cope with periods of fire. Jack pine and lodgepole pine trees have cones that don't open to release seeds until they have been heated in a fire!

In some types of forests, biologists conduct **controlled burns,** managed forest fires that are set periodically to control the amount of vegetation underneath the dominant forest tree species. In areas where natural fires don't occur, deep piles of dead leaves, dropped needles, and dead brush build up on the forest floor. All of this dead organic matter becomes potential fuel if a fire should start. To make sure that this organic matter doesn't build up too much, biologists sometimes start small fires to burn up the material on the forest floor. This is one way biologists hope to prevent major forest fires.

620 ORGANISMS AND THEIR ENVIRONMENTS

What happens to the forest animals during a controlled burn? Researchers contend that wild animals are unharmed because they move away from areas on fire. Some animals, such as the rare red-cockaded woodpecker, depend upon fires to maintain their habitat. They live in open pine stands that are often invaded by hardwood trees. Fires help to keep the hardwoods from growing in the open spaces between the pines.

Controlled burns are usually less intense and produce less heat than uncontrolled forest fires. This is another reason researchers feel that it is better to have a few controlled fires periodically than a large, out-of-control fire. However, even controlled burns sometimes cause damage to areas outside of the intended area of the burn. If the wind shifts suddenly, a fire may race toward residential areas instead of staying in the forest. Homes and businesses have burned to the ground from controlled fires as well as natural forest fires. Controlled burns also result in air pollution from smoke, and newly burned areas are eyesores to those who live nearby. It may take a few years for the burned areas to begin growing again. In the meantime, local residents see damaged forest views every day.

Biologists are very careful to set fires only at times when wind conditions are right. But people who live near national forests and parks still are concerned over the need for controlled burns. They believe controlled burns are not "controlled" at all. They also aren't convinced that forest animals are unharmed as a result of these fires.

Figure 26-19. The red-cockaded woodpecker is one organism that depends on fires to maintain its habitat.

Connect to... Chemistry

When burning takes place in the presence of enough oxygen, carbon dioxide and water are produced. Soot and smoke form when there is not enough oxygen present to burn all the carbon of the fuel. Soot is a form of carbon called graphite. Find out what other form of pure carbon is found in nature.

SECTION REVIEW

1. Name two reasons why biologists believe controlled forest fires are useful.
2. Why are local citizens often opposed to such fires?
3. **Connect to Chemistry:** Find out why there are ashes left after a piece of wood burns.

You Decide!

In popular vacation spots, people often build homes in scenic areas, such as high on a ridge top. Such spots are often in forested areas that are subject to uncontrollable forest fires. Many of these vacation homes are in locations that are difficult for forest firefighters to reach. Should people be permitted to build homes in these areas?

SCIENCE & SOCIETY

26-4 FRIENDLY FIRES 621

The impact of science on society directly affects you. In the **Science and Society** section in each chapter, you'll learn about an issue that's affecting the world around you. The topics you'll read about are controversial, and you'll explore them from several sides. Then, you'll have a chance to express your opinion in the You Decide feature that follows.

In the **Technology** feature in each chapter, you'll read about recent discoveries, newly developed instruments, and applications of technology that have shaped our world and furthered our knowledge.

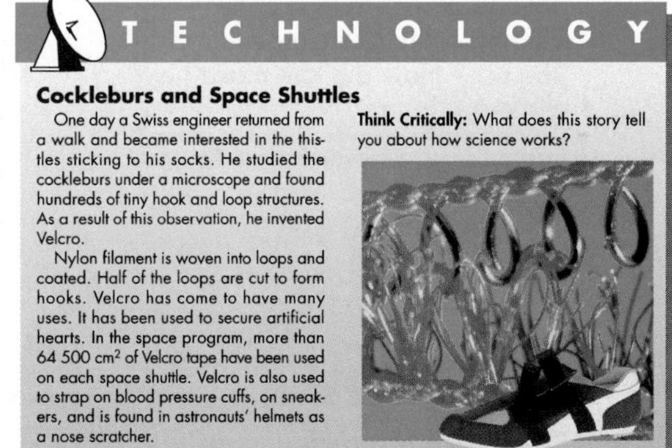

TECHNOLOGY

Cocklebursand Space Shuttles

One day a Swiss engineer returned from a walk and became interested in the thistles sticking to his socks. He studied the cockleburs under a microscope and found hundreds of tiny hook and loop structures. As a result of this observation, he invented Velcro.

Nylon filament is woven into loops and coated. Half of the loops are cut to form hooks. Velcro has come to have many uses. It has been used to secure artificial hearts. In the space program, more than 64 500 cm^2 of Velcro tape have been used on each space shuttle. Velcro is also used to strap on blood pressure cuffs, on sneakers, and is found in astronauts' helmets as a nose scratcher.

Think Critically: What does this story tell you about how science works?

Soil is a mixture of weathered rock and decaying organic matter (plant and animal). What roles do pioneer species such as lichens, mosses, and liverworts play in building soil?

All sciences are related. **Connect to...** features allow you to relate the life science topics you are studying to chemistry, physics, and earth science. Use a dictionary or basic library reference to find the answers to these questions.

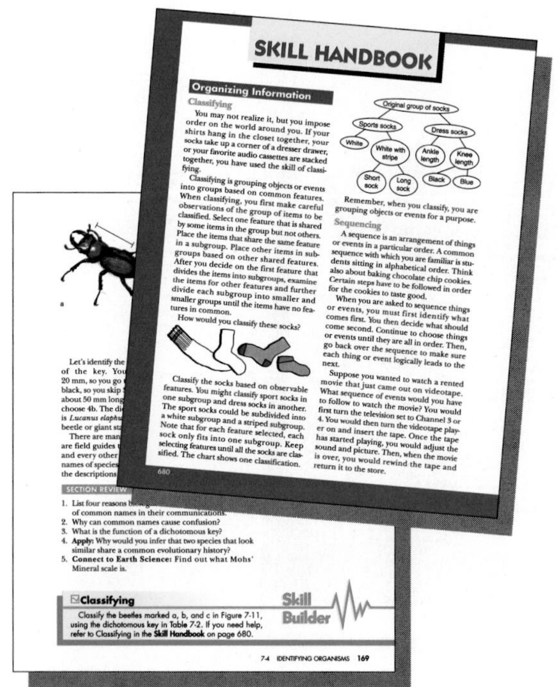

Discover that you can apply what you've learned as you answer questions and practice your science skills.

At the end of each section are several Section Review questions that help you test your knowledge. The last question challenges you to think critically and **Apply** what you've learned.

The **Skill Builder** feature lets you sharpen your science skills using only paper and pencil. If you need help with these skills, refer to the **Skill Handbook** at the back of the book. Here, you can find complete information about each type of skill covered in the Skill Builders.

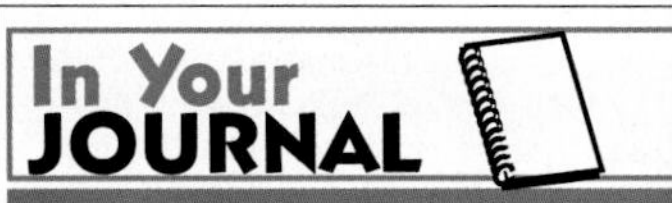

In your Journal, write about Barbara McClintock and her work with genes that change position on chromosomes. What organism did she use? How did she make her discovery?

In Your Journal features enable you to write expressively about life science topics you are studying. It includes writing creatively, giving reports, writing letters, and forming opinions. You may also keep lab notes and responses to Chapter Review questions in your Journal.

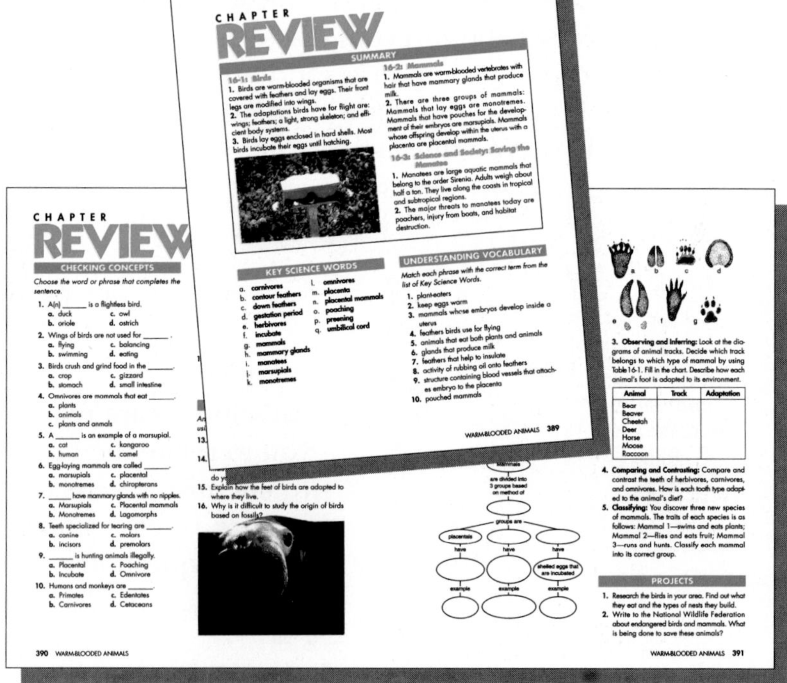

The **Chapter Review** starts with a summary so you can review the major concepts from each section. Then, you'll apply your knowledge and practice thinking skills as you answer the questions that follow. USING LAB SKILLS will help you practice your laboratory techniques.

Discover how life science topics relate to people and places all over the world.

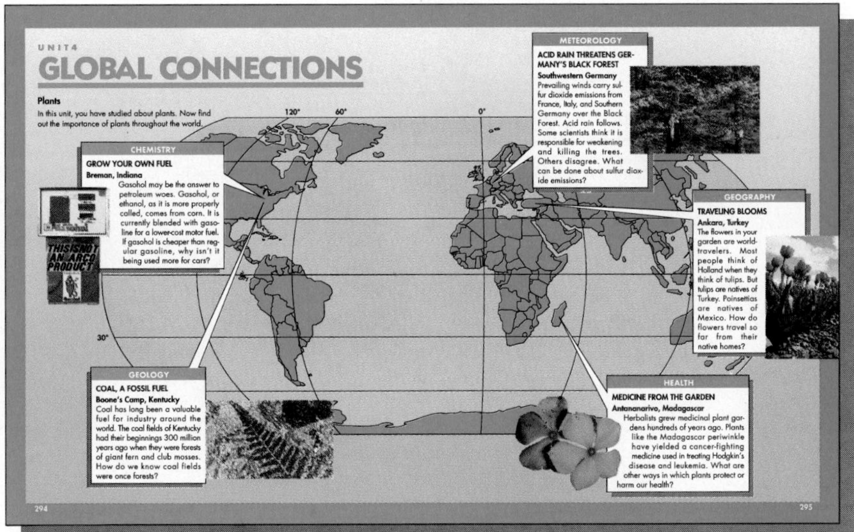
UNIT 4

GLOBAL CONNECTIONS

Plants

In this unit, you have studied about plants. Now find out the importance of plants throughout the world.

CHEMISTRY

GROW YOUR OWN FUEL

Breman, Indiana

Gasohol may be the answer to petroleum woes. Gasohol, or ethanol, as it is more properly called, comes from corn. It is currently blended with gasoline for a lower-cost motor fuel. If gasohol is cheaper than regular gasoline, why isn't it being used more for cars?

METEOROLOGY

ACID RAIN THREATENS GERMANY'S BLACK FOREST

Southwestern Germany

Prevailing winds carry sulfur dioxide emissions from France, Italy, and Southern Germany over the Black Forest. Acid rain follows. Some scientists think it is responsible for weakening and killing the trees. Others disagree. What can be done about sulfur dioxide emissions?

GEOGRAPHY

TRAVELING BLOOMS

Ankara, Turkey

The flowers in your garden are world-travelers. Most people think of Holland when they think of tulips. But tulips are natives of Turkey. Poinsettias are natives of Mexico. How do flowers travel so far from their native homes?

GEOLOGY

COAL, A FOSSIL FUEL

Boone's Camp, Kentucky

Coal has long been a valuable fuel for industry around the world. The coal fields of Kentucky had their beginnings 300 million years ago when they were forests of giant fern and club mosses. How do we know coal fields were once forests?

HEALTH

MEDICINE FROM THE GARDEN

Antananarivo, Madagascar

Herbalists grew medicinal plant gardens hundreds of years ago. Plants like the Madagascar periwinkle have yielded a cancer-fighting medicine used in treating Hodgkin's disease and leukemia. What are other ways in which plants protect or harm our health?

294 295

Global Connections help you to see how life science is related to other sciences as well as social studies, history and health.

Also at the end of each unit you will find two **Careers** that describe jobs that relate to the material in the unit you just read. What kind of jobs may be related to maintaining health and what do the people that have these jobs do? Read the careers at the end of Unit 7 to find out.

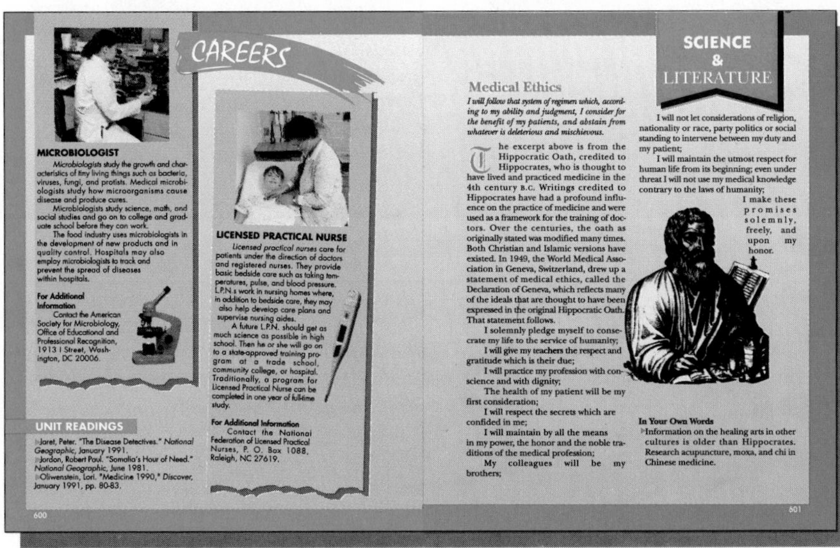
CAREERS

MICROBIOLOGIST

Microbiologists study the growth and characteristics of tiny living things such as bacteria, viruses, fungi, and protists. Medical microbiologists study how microorganisms cause disease and produce cures.

Microbiologists study science, math, and social studies and go on to college and graduate school before they can work.

The food industry uses microbiologists in the development of new products and in quality control. Hospitals may also employ microbiologists to track and prevent the spread of diseases within hospitals.

For Additional Information

Contact the American Society for Microbiology, Office of Educational and Professional Recognition, 1913 I Street, Washington, DC 20006.

LICENSED PRACTICAL NURSE

Licensed practical nurses care for patients under the direction of doctors and registered nurses. They provide basic bedside care such as taking temperatures, pulse, and blood pressure. L.P.N.s work in nursing homes where, in addition to bedside care, they may also help develop care plans and supervise nursing aides.

A future L.P.N. should get as much science as possible in high school. Then he or she will go on to a state-approved training program at a trade school, community college, or hospital. Traditionally, a program for Licensed Practical Nurse can be completed in one year of full-time study.

For Additional Information

Contact the National Federation of Licensed Practical Nurses, P. O. Box 1088, Raleigh, NC 27619.

UNIT READINGS

Jaret, Peter. "The Disease Detectives." *National Geographic*, January 1991.

Jordon, Robert Paul. "Somalia's Hour of Need." *National Geographic*, June 1981.

Oliwenstein, Lori. "Medicine 1990," *Discover*, January 1991, pp. 80-83.

600

SCIENCE & LITERATURE

Medical Ethics

I will follow that system of regimen which, according to my ability and judgment, I consider for the benefit of my patients, and abstain from whatever is deleterious and mischievous.

The excerpt above is from the Hippocratic Oath, credited to Hippocrates, who is thought to have lived and practiced medicine in the 4th century B.C. Writings credited to Hippocrates have had a profound influence on the practice of medicine and were used as a framework for the training of doctors. Over the centuries, the oath as originally stated was modified many times. Both Christian and Islamic versions have existed. In 1949, the World Medical Association in Geneva, Switzerland, drew up a statement of medical ethics, called the Declaration of Geneva, which reflects many of the ideals that are thought to have been expressed in the original Hippocratic Oath. That statement follows.

I solemnly pledge myself to consecrate my life to the service of humanity;

I will give my teachers the respect and gratitude which is their due;

I will practice my profession with conscience and with dignity;

The health of my patient will be my first consideration;

I will respect the secrets which are confided in me;

I will maintain by all the means in my power, the honor and the noble traditions of the medical profession;

My colleagues will be my brothers;

I will not let considerations of religion, nationality or race, party politics or social standing to intervene between my duty and my patient;

I will maintain the utmost respect for human life from its beginning; even under threat I will not use my medical knowledge contrary to the laws of humanity;

I make these promises solemnly, freely, and upon my honor.

In Your Own Words

Information on the healing arts in other cultures is older than Hippocrates. Research acupuncture, moxa, and chi in Chinese medicine.

601

What do life science and literature have in common? A lot, as you'll discover when you read the unit close to Unit 7. Each unit is closed with a reading from literature or an example of art that makes a connection with life science.

1

UNIT

1

LIFE

In Unit 1, students examine the features, needs, and origins of living things. Scientific methods, problem solving, and measurement are discussed.

CONTENTS

ADVANCE PREPARATION

Activities

▶ **Activity 1-1, page 24,** requires whole milk and vinegar.

▶ **Activity 2-1, page 43,** requires *Elodea* plants and prepared human cheek cell slides.

▶ **Activity 2-2, page 48,** requires hamburger, an onion, and iodine stain.

▶ **Activity 3-1, page 62,** calls for raw eggs, vinegar, distilled water, and a salt solution.

▶ **Activity 3-2, page 68,** requires sodium bicarbonate, bromothymol blue, and pieces of Elodea.

▶ **Activity 4-1, page 79,** calls for prepared slides of onion root tip and whitefish embryo.

▶ **Activity 4-2, page 94,** requires paper and crayons or markers.

2

OPTIONS

Cross Curriculum

▶ Have students find and report on news articles that discuss technological advances in life science.

PROJECT

During the course of this unit study, have students work cooperatively on PROJECT 1, *In Search of the Cell,* found on pages 694 and 695.

Science at Home

▶ Have students look for items in their home that are measured in metric units. Have them list each item with its measurement.

Cooperative Learning: Divide the class into groups of four. Have each group make a list of the characteristics of living things and give examples of each characteristic. Have each group decide what living things need to survive, how they obtain these things, and what happens when their needs cannot be met.

What's Happening Here?

A lion cub looks up from the feast he shares with a lioness. In his contented state, he is unaware that he is consuming energy, which will build muscle and bone and eventually bring him to the stage of development where he will reproduce. Living things require energy to grow, develop, and reproduce. The amoeba, the one-celled organism in the small photograph, is less dramatic than a lion, but it too requires energy, which it derives from other organisms. In this unit, you will learn about the characteristics that all living things share.

UNIT CONTENTS

3

INTRODUCING THE UNIT

What's Happening Here?

▶ Have students look at the photos and read the text. Ask them what's happening here. Explain that in this unit they will study the structure and function of the cell, the basic unit of life, and its relationship to the characteristics and needs of living things.

▶ **Background:** All living things, regardless of their size, need energy that is supplied by food. Green plants use the sun's energy, carbon dioxide, water, and minerals from the soil to make the food used by most organisms.

Previewing the Chapters

▶ In this first unit, students should be encouraged to look through the chapters to acquaint themselves with the features of the text. Use the To the Student Pages (pages XVIII through 1) prior to the Unit 1 opening pages as a guide. Have students find examples of these features in Chapters 1 through 4.

▶ Help students develop an overview of the material in the unit. The first chapter looks at life in general. With the help of Figures 1-2 and 1-5, begin to develop the concept of relationships between organisms. Help students to see that instruments such as the microscope in Chapter 2 are keys to learning about how organisms, including themselves, function. In Chapter 4, students will get an idea of how life is perpetuated.

Tying to Previous Knowledge

▶ Discuss with the students the building blocks of living and nonliving things. Elicit that in all living things, the basic unit of organization is the cell.

▶ Use the **inquiry questions** in the OPTIONS box below to investigate the characteristics of living things.

Multicultural Awareness

Different cultures have different ideas about how living forms appeared on Earth. Provide a variety of origin myths (stories) for students to read. Have students read some and select one to illustrate and share with the class. The Navajo traditions state that a deity named Changing Woman gave birth to the Hero Twins and that First Man and First Woman were formed from the ears of white and yellow corn. Traditions in Greece and India both teach that life resulted when a Cosmic Egg opened.

Inquiry Questions

Use the following questions to focus a discussion on how to determine if something is living or not.

▶ **Do all living things move?**

▶ **What are some of the ways in which they move?**

▶ **Why can movement alone not be used to classify living things?**

▶ **How do plants move?**

▶ **Can you say for sure that a rock is not a living thing?**

CHAPTER

1 Exploring Life

CHAPTER SECTION	OBJECTIVES	ACTIVITIES
1-1 Living Things (2 days)	1. **Identify** the features of living things in an organism. 2. **Recognize** the needs of living things and explain how these needs are provided. 3. **Determine** that living things and nonliving things interact in the environment.	
1-2 Where Does Life Come From? (1 day)	1. **Summarize** the results of Redi's and Spallanzani's experiments. 2. **Explain** how Pasteur's experiments disproved the theory of spontaneous generation. 3. **State** the theory of biogenesis.	
1-3 What Is Science? (2 days)	1. **Describe** methods scientists use to solve problems. 2. **Identify** and use the SI units of length, volume, mass, and temperature.	**MINI-Lab:** *What is an experiment?* p. 15 **MINI-Lab:** *How are things measured?* p. 21 **FLEX Your Brain:** p. 18
1-4 The Impact of Science On Your Life Science & Society (1 day)	1. **State** two ways in which life on Earth has been affected by technology in the life sciences.	**Activity 1-1:** *Designing an Experiment (Using a Scientific Method),* p. 24
Chapter Review		

ACTIVITY MATERIALS

FIND OUT	ACTIVITIES	MINI-LABS	
Page 5 2 or 3 coffee filter strips 3 cm × 12 cm 1 black water-soluble felt-tip pen 1 beaker or glass water	**1-1 Designing an Experiment, p. 24** 0.5-L wide–mouthed containers (3) brine shrimp eggs wooden splint distilled water (500 mL) weak salt solution (500 mL) strong salt solution (500 mL) 3 labels hand lens	**What is an experiment? p. 15** two brands disposable diapers scissors water measuring cups	**How are things measured? p. 21** paper clip small glass of water metal spoon rock pan balance

CHAPTER FEATURES	TEACHER RESOURCE PACKAGE	OTHER RESOURCES
Skill Builder: *Observing and Inferring,* p. 9	**Ability Level Worksheets** ◆ ***Study Guide,*** p. 5 ● ***Reinforcement,*** p. 5 ▲ ***Enrichment,*** p. 5 **Concept Mapping,** p. 7 **Cross-Curricular-Connections,** p. 5	**STVS:** Disc 4, Side 1
Skill Builder: *Observing and Inferring,* p. 12	**Ability Level Worksheets** ◆ ***Study Guide,*** p. 6 ● ***Reinforcement,*** p. 6 ▲ ***Enrichment,*** p. 6 **Critical Thinking/Problem Solving,** p. 5	**Lab Manual:** The Scientific Method, p. 1 **Lab Manual:** Using the Scientific Method, p. 3 **STVS:** Disc 6, Side 1
Problem Solving: *The Long Island Duck Problem,* p. 16 **Technology:** *Cockleburs and Space Shuttles,* p. 19 **Skill Builder:** *Measuring in SI,* p. 21	**Ability Level Worksheets** ◆ ***Study Guide,*** p. 7 ● ***Reinforcement,*** p. 7 ▲ ***Enrichment,*** p. 7 **Activity Worksheets,** pp. 11-12 **Technology,** pp. 5-6 **Transparency Masters,** pp. 1-4 **Activity Worksheets,** p. 5	**Color Transparency 1,** Flex Your Brain **Color Transparency 2,** SI/Metric to English Conversion **STVS:** Disc 4, Side 2 **Science Integration Activity 1**
You Decide! p. 23	**Ability Level Worksheets** ◆ ***Study Guide,*** p. 8 ● ***Reinforcement,*** p. 8 ▲ ***Enrichment,*** p. 8 **Activity Worksheets,** pp. 7-8 **Science and Society,** p. 5	**STVS:** Disc 1, Side 2
Summary Key Science Words Understanding Vocabulary Checking Concepts Using Lab Skills Think & Write Critically Apply More Skill Builders Projects	**ASSESSMENT RESOURCES** **Chapter Review,** pp. 5-6 **Chapter Test,** pp. 5-8 **Performance Assessment in Middle School Science (PAMSS)**	**Chapter Review Software** **Test Bank** **Alternate Assessment** **Performance Assessment**

◆ **Basic** ● **Average** ▲ **Advanced**

ADDITIONAL MATERIALS

SOFTWARE	AUDIOVISUAL	BOOKS/MAGAZINES
Ant Farm, Sunburst Communications (simulation) *Botanical Gardens,* Sunburst Communications (simulation) *Characteristics of a Scientist,* Cygnus Software. *Discover: A Science Experiment,* Sunburst Communications. *Discovery Lab,* MECC (simlulation)	*Biology: Exploring the Living World,* film, EBEC. *Inferring in Science,* film, AIT. *How Scientists Think and Work,* filmstrip, Hawkill. *Scientific Methods & Values,* filmstrip, Hawkill. *Conversations with Great Scientists Video Series—Francis Crick: Beyond the Double Helix,* video, Focus. *A Conversation with Stephen Jay Gould,* video, Focus. *What is Science?* film, Coronet/MTI.	Abbott, David, ed. *The Biographical Dictionary of Scientists: Biologists.* NY: Peter Bedrick Books, 1984. Bushman, Eva. M. *Biology.* Portland, OR: National Book Company, 1980. Mader, Sylvia, ed. *Inquiry into Life.* 6th ed. Dubuque, IA: William C. Brown Publishers, 1991.

CHAPTER

1

EXPLORING LIFE

THEME DEVELOPMENT: Three unifying themes occur throughout this book: evolution, homeostasis, and ecology. In this chapter, homeostasis and ecology are defined, and the discussion of adaptation provides the basis for the study of evolution.

CHAPTER OVERVIEW

▶**Section 1-1:** Section 1-1 identifies and gives examples of nine characteristics of living things. Energy and the materials needed for life are discussed.

▶**Section 1-2:** This section reviews the scientists and their experiments that led to the understanding of the concept of biogenesis. The work of Oparin and its importance to scientists of today are discussed.

▶**Section 1-3:** Section 1-3 investigates what science is, scientific methods, critical thinking, and theories and laws. The use of SI units and safety in science are explained.

▶**Section 1-4: Science and Society:** This section explores how the impact of technology affects life. The You Decide feature asks students to compare using polystyrene and paper, and to think about which might be more ecologically sound.

CHAPTER VOCABULARY

organisms
cells
stimulus
response
homeostasis
development
adaptation
life span
spontaneous generation
biogenesis
scientific methods
hypothesis
variable
control
theory
law
technology
ecology

CHAPTER 1

Exploring Life

4

OPTIONS

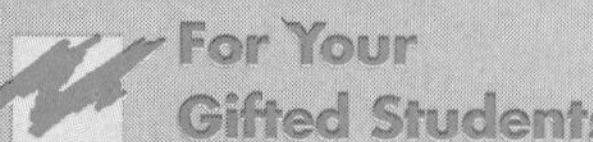

For Your Gifted Students

▶Have students prepare a commercial that would encourage choosing a career in the area of life science.

▶Ask students to write a rap or poem that will help them remember the safety rules.

▶Students can design a bulletin board for the room showing the steps in the scientific method.

For Your Mainstreamed Students

▶Have students choose and draw a scene on poster board or mural paper. They should label the objects in the pictures as living or nonliving.

▶Give students a label that names one of the steps in the scientific method. After all steps (labels) have been assigned, ask students to line up in the correct order. Teams may be formed, with the first team to line up correctly being the winner.

Many people think that science is the work of people in white coats who spend hours in a laboratory. But often the work of science begins when someone like you notices something new about an ordinary thing, and starts to ask questions.

FIND OUT!

Do this activity to find out about the first steps in scientific research.

Cut a strip of coffee filter paper 3 cm by 12 cm. Using a black felt tip pen, draw a line across the strip of filter about 2 cm from one end. Describe the color of the line. Put 1 cm of water in a small beaker or glass. Hang the filter strip over the beaker so that one end is in the water and the black line is above the water. *Predict* what will happen to the black line. Will you change your original description? If so, how?

Gearing Up

Previewing the Chapter

Use this outline to help you focus on important ideas in this chapter.

Section 1-1 Living Things
- ▶ Features of Life
- ▶ Needs of Living Things

Section 1-2 Where Does Life Come From?
- ▶ Life Comes from Life
- ▶ Origins

Section 1-3 What Is Science?
- ▶ The Work of Science
- ▶ Solving Problems
- ▶ Theories and Laws
- ▶ Critical Thinking
- ▶ Measuring in Science
- ▶ Safety First

Section 1-4 Science and Society
The Impact of Science on Your Life
- ▶ Times Have Changed
- ▶ Living with Technology

Previewing Science Skills

- ▶ In the Skill Builders, you will observe and infer, and measure in SI.
- ▶ In the Activities, you will Flex Your Brain, hypothesize and observe.
- ▶ In the MINI-Labs, you will design an experiment and measure in SI.

What's next?

In the Find Out activity, you have had a chance to use scientific methods to test an idea. Now find out about the features of living things and the methods that are used in science to study organisms.

INTRODUCING THE CHAPTER

Use the Find Out activity to introduce students to observing in science. Explain that they will be developing skills used in science by performing and designing experiments.

FIND OUT!

Preparation: Obtain coffee filter paper, beakers, and water soluble felt tip pens.

Materials: 2 or 3 filter strips 3 cm × 12 cm, 1 black water soluble felt tip pen, 1 beaker or glass, and water for every two students

Cooperative Learning: Use the Paired Partner strategy. Have one student gather the materials and one student record.

Teaching Tips

▶ Have students make sure the black line does not go under the water. If this happens, the colors will not separate as the water moves up the paper.

▶ When the water stops moving up the paper, have students remove the paper strips.

Gearing Up

Have students study the Gearing Up feature to familiarize themselves with the chapter. Discuss the relationships of the topics in the outline.

What's Next?

Before beginning the first section, make sure students understand the connection between the Find Out activity and the topics to follow.

ASSESSMENT OPTIONS

PORTFOLIO

Refer to page 25 for suggested items that students might select for their portfolios.

PERFORMANCE ASSESSMENT

See page 25 for additional Performance Assessment options.

Process

Skillbuilders, pp. 12, 21
MINI-Lab, pp. 15, 21
Activity, 1–1 p. 24
Using Lab Skills, p. 26

CONTENT ASSESSMENT

Assessment—Oral, pp. 8, 16
Skillbuilder, p. 9
Section Reviews, pp. 9, 12, 21, 23
Chapter Review, pp. 25-27
Mini Quizzes, p. 8, 20

GROUP ASSESSMENT

Opportunities for group assessment occur with Cooperative Learning Strategies and Flex Your Brain Activities.

SECTION 1-1

PREPARATION

SECTION BACKGROUND

▶All areas of science involve asking questions about nature and discovering facts that can help answer those questions.

▶All the characteristics of living things are needed for survival of the individual, except for reproduction. Reproduction is needed for the survival of the species.

▶Energy is the ability to do work. Work is the result of a force moving an object through a distance. The amount of work done in any action is calculated by the formula, $W = F \times d$, where F is the force exerted, and d is the distance an object is moved.

1 MOTIVATE

▶**Demonstration:** Place a petri dish half-filled with water on the overhead projector. Add two or three drops of Duco cement (cover the container so students will not see what it is). Have students observe. Add two or three small pieces of pencil shavings. Ask students what characteristics of living things they see. (Students should be able to observe movement and ingestion.) Discuss how nonliving things may appear to be living. To be classified as living, a thing must have all of the characteristics of life.

▶**Bulletin Board:** Arrange pictures of living things that show the characteristics of life. As the characteristics are studied, have students identify the pictures that represent each characteristic.

TYING TO PREVIOUS KNOWLEDGE: Students will have ideas about characteristics and needs of all living organisms. Ask what all living things have in common and write the answers on the chalkboard.

VideoDisc

STVS: Soil Crust in the Desert, Disc 4, Side 1 (**STVS** stands for *Science and Technology Videodisc Series.*)

1-1 Living Things

New Science Words

organisms
cells
stimulus
response
homeostasis
development
adaptation
life span

Objectives

- ▶ Identify the features of living things in an organism.
- ▶ Recognize the needs of living things and explain how these needs are provided.
- ▶ Determine that living things and nonliving things interact in the environment.

Features of Life

This book is about life—what it is and how it is maintained. Life is something we may take for granted. Have you ever tried to explain what it means to be alive? Think about life characteristics as you read the following.

Imagine that it's a hot summer day. You can't wait to get out of your hot, stuffy house or apartment. With your dog, you head off to a nearby stream. At first, you practice skipping stones across the water, while the dog runs along the bank barking. Then you sit down on some sand on the bank under a large tree. Dragonflies skim over the water, and the noise of the insects and birds is all you can hear. The dog settles down on the bank under a tree and appears to fall asleep. It sure is great to be alive. But what does this mean? Living things such as you and your dog and the tree are **organisms.** Organisms have certain features that rocks and water don't have. What makes you and your dog different from water and rocks? 1

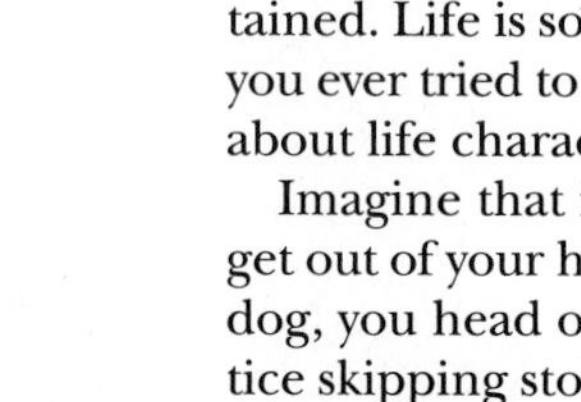
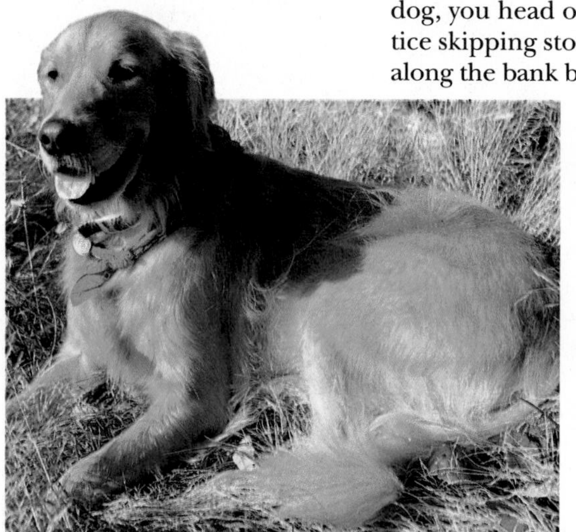

Figure 1-1. Organisms are made of one or more cells, use energy, and respond to their environment. The dog is a many-celled organism.

Organisms are made of one or more cells. You have some things in common with water and rocks in the stream because you are made up of chemicals as they are. However, if someone were to look at you or your dog or the tree under a microscope, it would be clear that each of you is made of units called cells. Water and rocks are not.
2 **Cells** are the smallest units of organisms that carry on the functions of life.

OPTIONS

Meeting Different Ability Levels

For Section 1-1, use the following **Teacher Resource Masters** depending upon individual students' needs.

◆ **Study Guide Master** for all students.

● **Reinforcement Master** for students of average and above average ability levels.

▲ **Enrichment Master** for above average students.

Additional Teacher Resource Package masters are listed in the OPTIONS box throughout the section. The additional masters are appropriate for all students.

◆ STUDY GUIDE 5

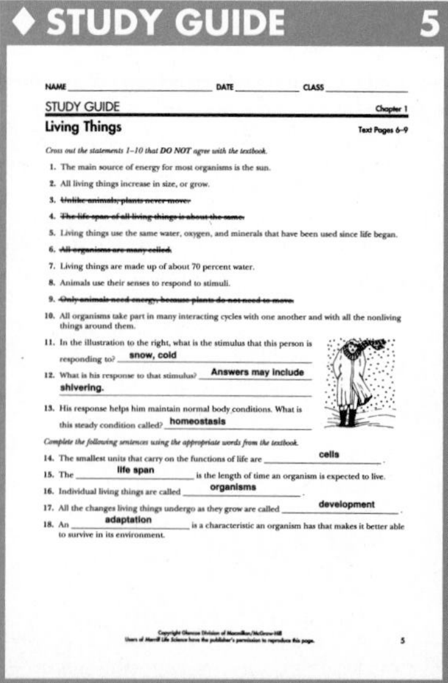

NAME ____________ DATE ________ CLASS ________

STUDY GUIDE — Chapter 1

Living Things — Text Pages 6–9

Cross out the statements 1–10 that ***DO NOT*** *agree with the textbook.*

1. The main source of energy for most organisms is the sun.
2. All living things increase in size, or grow.
3. ~~Unlike animals, plants never move.~~
4. ~~The life span of all living things is about the same.~~
5. Living things use the same water, oxygen, and minerals that have been used since life began.
6. ~~All organisms are many celled.~~
7. Living things are made up of about 70 percent water.
8. Animals use their senses to respond to stimuli.
9. ~~Only animals need energy, because plants do not need to move.~~
10. All organisms take part in many interacting cycles with one another and with all the nonliving things around them.
11. In the illustration to the right, what is the stimulus that this person is responding to? **snow, cold**
12. What is his response to that stimulus? **Answers may include shivering.**
13. His response helps him maintain normal body conditions. What is this steady condition called? **homeostasis**

Complete the following sentences using the appropriate words from the textbook.

14. The smallest units that carry on the functions of life are **cells**.
15. The **life span** is the length of time an organism is expected to live.
16. Individual living things are called **organisms**.
17. All the changes living things undergo as they grow are called **development**.
18. An **adaptation** is a characteristic an organism has that makes it better able to survive in its environment.

5

Figure 1-2. Organisms use energy that is supplied by food. For many animals, food is in the form of other animals.

Organisms move. If you ever watch a sleeping dog, it often looks lifeless. Then an ear twitches, or the dog may snore. What about plants? How do they move?

Organisms respond to changes in their environment. Does your dog come running every time it hears a can opener? Anything that an organism responds to, such as sound,
3 is a **stimulus.** The reaction of an organism to a stimulus, in this case probably tail wagging and barking, is a **response.** What other kinds of stimuli can you think of?

Organisms use energy. For organisms to respond to changes in their environment, they need energy. Energy is the ability to make things move or change. It takes energy for your dog to run and bark. It takes energy for you to carry out all the activities you do every day. In the presence of sunlight, plants produce food, which both plants and animals use for energy. Energy is then released from food in a process called respiration.

Organisms adjust to changes in their environment. Think about driving a car. The driver constantly makes small adjustments with the gas pedal to keep the car on the road at a steady rate of speed. The body systems of an organism do something similar. Maintaining steady conditions no matter what is going on inside or
4 outside the organism is called **homeostasis.** Controls in an organism work constantly to bring it back to "normal" after it has been stimulated.

Organisms reproduce. Your dog came from a litter of puppies. You might have looked at its parents to see what it would be like when full grown. Organisms produce new individuals that are usually very like the parent organisms. Oak trees produce acorns that become oak trees. Bluejays lay eggs that develop into young bluejays.

According to the Law of Conservation of Energy, energy can change form but cannot be created or destroyed under ordinary conditions. In many organisms, chemical energy is converted to mechanical energy. What is the source of chemical energy for animals? In what way do animals show the use of mechanical energy?

Figure 1-3. Organisms reproduce their own kind.

● REINFORCEMENT 5

NAME ________ DATE ________ CLASS ________

REINFORCEMENT — Chapter 1

Living Things — Text Pages 6–9

After each statement, write the feature of life that is illustrated.

1. "That boy shot up five inches in only one year." **Living things grow.**
2. "Our cat had a litter of kittens yesterday." **Living things reproduce.**
3. "To win at that sport, her muscle cells need to be worked hard every day." **Living things are made of one or more cells.**
4. "My dog has become much less clumsy now that he is a year old." **Living things develop.**
5. "Eat a good breakfast and you'll feel better through the morning." **Living things use energy.**
6. "When that car pulled in the driveway, my cat ran to hide under the porch." **Living things respond and adjust to changes in their environment.**
7. "The fish died after living in the aquarium for many years." **Living things have a life span.**
8. "The bird flew from tree to tree." **Living things move.**
9. "That owl's night vision allows it to see the movement of mice on even the darkest night." **Living things adapt.**

A B

10. In Figure A, the sun is what to the plant? **a stimulus**
11. In Figure B, the plant has shifted toward the sun. What is this reaction an example of? **a response**
12. Your body normally maintains a temperature of 98.6°F. This is an example of what? **homeostasis**
13. What are the smallest units that carry on the functions of life? **cells**
14. What adaptation does a giraffe have that allows it to eat from the high branches of a tree? **a long neck**

 5

▲ ENRICHMENT 5

NAME ________ DATE ________ CLASS ________

ENRICHMENT — Chapter 1

Living Things — Text Pages 6–9

THE FEATURES OF LIFE

Modern science has changed the way people understand their world. In the past, the people of many societies thought that mountains, rivers, and forests were "alive" in much the same way that animals and plants are alive. Today, most people recognize a difference between living and nonliving things. The features of life listed in your textbook help to make that distinction.

Fill out the table below. Use complete sentences to evaluate a dog and a fast-moving stream according to the features of life. Then answer the questions that follow.

Features of Life	Dog	Fast-moving stream
made of cells?	A dog is made of cells.	A stream is not made of cells.
uses energy?	A dog uses energy from food.	A stream does not use food energy.
moves?	A dog moves around.	A stream moves.
responds and adjusts to changes?	A dog responds to stimuli.	A stream does seem to respond to things, but not in the same way as animals.
reproduces?	A dog can reproduce.	A stream cannot reproduce.
grows?	A dog grows.	A stream can grow if there is more water.
develops?	A dog develops into an adult.	A stream might develop into a river if there is enough water.
adapts?	A dog has adaptations, such as sharp teeth for tearing food.	A stream does not have adaptations.
life span?	A dog has a definite life span.	A stream does not have a life span.

1. Did either the dog or the stream meet every condition of the features of life? **Yes, the dog did.**
2. Did the stream meet any of the conditions? **Yes, the stream moves and develops and even responds to things in certain ways.**
3. Why must all conditions be met to determine whether something is living or nonliving? **Nonliving things can sometimes meet some of the conditions that make up the features of life.**

 5

OBJECTIVES AND SCIENCE WORDS: Have students review the objectives and science words throughout the chapter to become familiar with each section.

2 TEACH

Key Concepts are highlighted.

CONCEPT DEVELOPMENT

▶ Discuss with students the ways in which they use energy.

▶ Explain that responses in plants are usually less obvious than in animals. Mimosa, however, shows rapid response to touch. Obtain a mimosa to demonstrate the rapid response.

REVEALING MISCONCEPTIONS

▶ Many students think respiration is the same as breathing. Breathing is taking in oxygen and giving off carbon dioxide. Respiration is a chemical reaction in cells of living things that releases energy.

Note: Connect to... features are intended to extend content. Students are not expected to find answers to these features in their text. Students should be able to find answers for most Connect to... features in a dictionary or basic library references or infer answers from text material.

Answer: Chemical energy: food
Mechanical energy: movement

CHECK FOR UNDERSTANDING

Use the Mini Quiz to check for understanding.

MINI QUIZ

Use the Mini Quiz to check students' recall of chapter content.

1. **A living thing is a(n) ______ .** *organism*
2. **The smallest units of organisms that carry on the functions of life are ______ .** *cells*
3. **Anything an organism responds to is a(n) ______ .** *stimulus*
4. **The ability to maintain steady conditions no matter what is going on inside or outside an organism is ______ .** *homeostasis*

RETEACH

Write the characteristics of life on 3 × 5 cards. Have students draw a card and give an example of the characteristic.

EXTENSION

For students who have mastered this section, use the **Reinforcement** and **Enrichment** masters or other OPTIONS provided.

In Your JOURNAL

Horse: 25-30 years; cow: 20+ years; dog: 20 years; cat (domestic): 13-17 years; shrew: 12-18 months; cockroach: oriental—40 days, American—200 days (no more than a year). Among these organisms, it appears that larger animals live longer.

What is an adaptation?

Organisms grow and develop. When a dog is born, it is very small. With care and feeding, it grows larger. Living things increase in size. They grow. All the changes organisms undergo as they grow are called **development.** Dogs can't see or walk when they are first born. But in eight or nine days, their eyes open, and their legs become strong enough to hold them up. A dog develops into an adult in about two years.

Figure 1-4. As organisms grow and develop, they usually change in appearance.

Organisms adapt. Any characteristic an organism has that makes it better able to survive in its environment is an **adaptation.** Your dog's coat is an adaptation. In summer, the dog sheds hair and his body is cooled. Dogs also pant to release excess body heat. If dogs didn't have these adaptations they would suffer in the summer heat. Plants have adaptations too, such as flower color or hairs. Adaptations are inherited. They are not merely responses the organism makes to an immediate need.

Organisms have life spans. The length of time an organism is expected to live is its **life span.** For some, the life span is very short. Millions of mayflies live only one day, but some bristlecone pine trees have been alive more than 4500 years. Your life span is about 80 years. What is the life span of your dog? How are the needs of living things met during this life span?

In Your JOURNAL

Use library references to find the life span of horses, cows, dogs, cats, shrews, and cockroaches. **In your Journal,** organize this information in a table. Is there any relationship between the size of an organism and its life span?

Needs of Living Things

Sitting on the bank of the stream, you and your dog are probably not aware that you interact with everything else in the environment. However, all organisms take part in many interacting cycles with each other and with all the nonliving things around them. When an organism is part of these relationships, its needs are met.

What are these needs? All organisms need energy and raw materials. The energy that you use comes from food. The main source of energy for living organisms is the sun. Green plants use the sun's energy, along with carbon dioxide, water, and minerals from the soil, to make food in the form of a sugar called glucose. The food is made in green leaves. This process also produces oxygen. Oxygen is used by most organisms to release energy from food.

OPTIONS

ASSESSMENT—ORAL

▶**How do plants move?** *bend toward the light, roots grow downward*

▶**Why do animals move?** *to find food, shelter, and mates*

▶**A thermostat responds to temperature change. How do you know it is not alive?** *It does not grow, develop, reproduce, adapt, move, or use energy.*

▶**What would happen to an organism that did not constantly expend energy to maintain itself?** *It could not live.*

▶**If you place a dried-out bath sponge in water, it becomes larger. Is the sponge growing? Explain your answer.** *No. It is just absorbing water that can be removed from the sponge.*

ENRICHMENT

▶Have students report on the adaptations that allow desert organisms, such as cacti, camels, and gerbils, to conserve and use water.

Raw materials that organisms use are the water, oxygen, and minerals that have been used since life began. Oxygen, water, carbon dioxide, and other chemicals are used, returned to the environment, and used over again.

Water is especially important. Your dog might live two or three weeks without food, but he would die in a few days without water. Living things are made up of about 70 percent water, and they need to maintain that level. Many substances in nature dissolve in water. Blood and the sap of trees are mostly water. Many organisms are born in or live in water all their lives.

Are you and your dog different from the water and rocks? To be considered alive, an organism must have certain features. It must be made of cells, use energy, be able to move, respond, adjust, reproduce, grow and develop, and adapt. Organisms need energy, water, and oxygen. Do rocks and water have these properties?

Figure 1-5. In the living world, energy is transferred from one organism to another. Grasshoppers and rabbits feed on plants that have trapped the sun's energy. A hawk obtains energy when it feeds on the rabbit.

What do organisms need for life?

SECTION REVIEW

1. What are the features of organisms?
2. What is the main source of energy used by most organisms?
3. What are the needs of organisms, and how are these needs supplied?
4. **Apply:** Explain how a tree shows the features of life.
5. **Connect to Physics:** Find out how radiant energy (sunlight) and chemical energy (sugar) are related in plants.

Observing and Inferring

Skill Builder

Virus crystals can remain in a jar on a shelf for years. Then once they are put into living cells, they reproduce. Are viruses alive? If you need help, refer to Observing and Inferring in the **Skill Handbook** on page 682.

CONCEPT DEVELOPMENT

▶Discuss the cycling of matter. Explain that every living thing is composed of matter and that life activities are based on chemical changes in matter.

3 CLOSE

▶Ask questions 1-3 and the **Apply** Question in the Section Review.

▶Set up a classroom aquarium with guppies, snails, and water plants. Discuss how the living things interact with the nonliving things. The aquarium can be used for the remainder of the year. Let students be responsible for its upkeep.

SECTION REVIEW ANSWERS

1. Features of an organism are: made of one or more cells, uses energy, moves, responds to changes in its environment, reproduces, grows and develops, adapts, has a life span, and adjusts to changes in its environment.

2. sun

3. They need energy and raw materials. Energy comes from food or the sun; raw materials from water, oxygen, and minerals.

4. Apply: A tree is made of cells, uses energy, grows and develops, moves by pushing roots down and stems outward, bears seeds, and eventually dies.

5. Connect to Physics: Plants use radiant energy to make chemical energy in the form of food.

PROGRAM RESOURCES

From the **Teacher Resource Package** use:

Concept Mapping, pages 7-8.

Cross-Curricular Connections, page 5, Photographing Nature.

Skill Builder

Although viruses exhibit some features of life, they do not exhibit all, and are not alive.

SkillBuilder

ASSESSMENT

Portfolio: Use this Skillbuilder to assess students' abilities to infer properties of viruses and infer their relation to living things.

SECTION 1-2

PREPARATION

SECTION BACKGROUND

▶ Redi believed that spontaneous generation was the only way to explain galls, swollen growths on plant stems and leaves. After Redi's death, one of his students showed that the gall was made by a wasp larva hatched from an egg inside the plant tissue.

▶ Most biologists now accept the idea of original spontaneous generation at the molecular level. This is considered to have happened under quite different environmental conditions from those that exist today.

1 MOTIVATE

▶ Obtain two beef bouillon cubes. Dissolve each cube in a separate flask of hot water. Boil the bouillon in one flask for several minutes. After it cools, seal the flask with a rubber stopper. Set the two flasks on the shelf in the classroom. Have the students observe the flasks each day and record what happens in their Journals.

VideoDisc

STVS: Desert in the Antarctic, Disc 6, Side 1

1-2 Where Does Life Come From?

New Science Words

spontaneous generation
biogenesis

Objectives

- ▶ **Summarize the results of Redi's and Spallanzani's experiments.**
- ▶ **Explain how Pasteur's experiments disproved the theory of spontaneous generation.**
- ▶ **State the theory of biogenesis.**

Life Comes from Life

Have you ever walked out after a thunderstorm and found earthworms all over the sidewalk? Earthworms have been found in large numbers after rainstorms for hundreds of years. It's no wonder that people used to think the earthworms had fallen from the sky when it rained. It was a logical conclusion based on repeated experience. But was it true? Jan Baptist van Helmont wrote a recipe for making mice by placing grain in a corner and covering it with rags. For much of history, people believed that living things came from nonliving matter, an idea called the theory of **spontaneous generation.**

People also believed that maggots came
from decaying meat. In 1668, Francesco 1
Redi, an Italian doctor, conducted one of the first controlled experiments in science. He showed that maggots hatch from eggs that flies had laid on meat, and not from the meat itself.

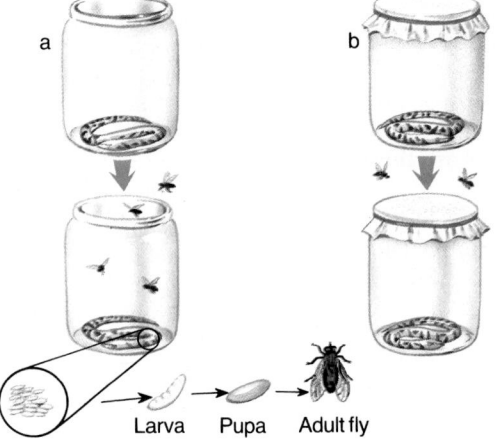

Figure 1-6. In Redi's experiment, maggots developed on snake meat that flies could reach (a), but not on meat that was protected from the flies (b).

a b Eggs Larva Pupa Adult fly

In the late 1700s, Lazzaro Spallanzani designed an experiment to show that tiny organisms came from other tiny organisms in the air. He boiled broth in two flasks, sealed one, and left the other one open to the air. The open flask became cloudy with organisms. The sealed flask developed no organisms. People believed Spallanzani had destroyed a "vital force" when he boiled the broth.

OPTIONS

Meeting Different Ability Levels

For Section 1-2, use the following **Teacher Resource Masters** depending upon individual students' needs.

◆ **Study Guide Master** for all students.

● **Reinforcement Master** for students of average and above average ability levels.

▲ **Enrichment Master** for above average students.

Additional Teacher Resource Package masters are listed in the OPTIONS box throughout the section. The additional masters are appropriate for all students.

◆ STUDY GUIDE 6

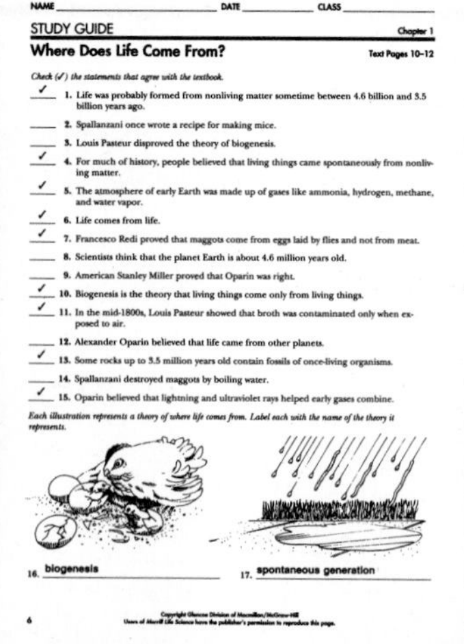
NAME ________ DATE ________ CLASS ________

STUDY GUIDE Chapter 1

Where Does Life Come From? Text Pages 10–12

Check (✓) the statements that agree with the textbook.

✓ 1. Life was probably formed from nonliving matter sometime between 4.6 billion and 3.5 billion years ago.
___ 2. Spallanzani once wrote a recipe for making mice.
___ 3. Louis Pasteur disproved the theory of biogenesis.
✓ 4. For much of history, people believed that living things came spontaneously from nonliving matter.
✓ 5. The atmosphere of early Earth was made up of gases like ammonia, hydrogen, methane, and water vapor.
✓ 6. Life comes from life.
✓ 7. Francesco Redi proved that maggots come from eggs laid by flies and not from meat.
___ 8. Scientists think that the planet Earth is about 4.6 million years old.
___ 9. American Stanley Miller proved that Oparin was right.
✓ 10. Biogenesis is the theory that living things come only from living things.
✓ 11. In the mid-1800s, Louis Pasteur showed that broth was contaminated only when exposed to air.
___ 12. Alexander Oparin believed that life came from other planets.
✓ 13. Some rocks up to 3.5 million years old contain fossils of once-living organisms.
___ 14. Spallanzani destroyed maggots by boiling water.
✓ 15. Oparin believed that lightning and ultraviolet rays helped early gases combine.

Each illustration represents a theory of where life comes from. Label each with the name of the theory it represents.

16. biogenesis   17. spontaneous generation

6

It was not until the mid-1800s that Louis Pasteur, a French chemist, showed conclusively that living things do not come from nonliving materials. In the experiment illustrated in Figure 1-7, Pasteur boiled broth in flasks with long, curved necks. The broth became contaminated only when dust that had collected in the curved neck of one flask was allowed to mix with the broth. The work of Redi, Spallanzani, Pasteur, and others provided enough evidence finally to disprove the theory of spontaneous generation. It was replaced with **biogenesis,** the theory that living things come only from other living things.

Broth is boiled.
Air and microbes driven out.
Microbes are trapped in the neck.
One year or more: broth remains without growth.
Flask is tilted.
Shortly, microbes develop in broth.

Figure 1-7. Pasteur's experiment disproved the theory of spontaneous generation.

Origins

If living things can come only from other living things, how then did life on Earth begin? Scientists hypothesize that about 5 billion years ago, the solar system was a whirling mass of gas and dust. The sun and planets formed from this mass. Our planet is thought to be about 4.6 billion years old. Rocks found in Australia that are more than 3.5 billion years old contain fossils of once-living organisms.

One hypothesis on the origin of life was proposed by Alexander I. Oparin, a Russian scientist. Oparin suggested that the atmosphere of early Earth was made up of gases similar to ammonia, hydrogen, methane, and water vapor. No free oxygen was present as it is today. Energy from lightning, and ultraviolet rays from the sun, helped these early gases to combine. The gases formed the chemical compounds of which living things are made. Oparin suggested that as the compounds were formed, they fell into hot seas. Over a period of time, the chemical compounds in the seas formed new and more complex compounds. Eventually, the complex compounds were able to copy themselves and make use of other chemicals for energy and food.

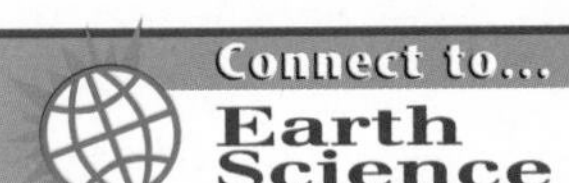

Scientists hypothesize that Earth's oceans originally formed when water vapor was released into the atmosphere from numerous volcanic eruptions. Once it cooled, rain fell to fill Earth's low areas. Find out what other materials are produced by volcanoes.

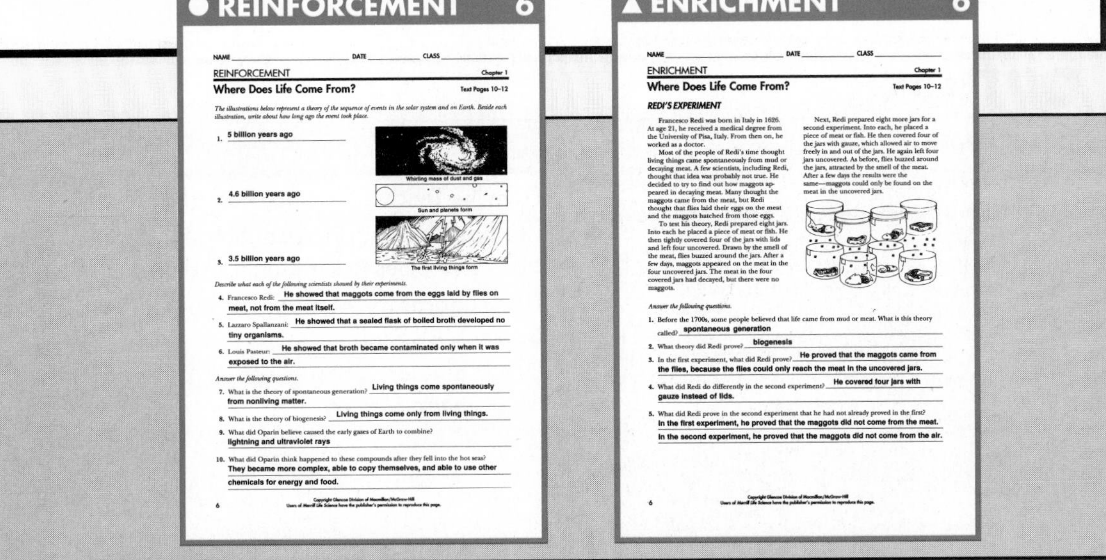

● REINFORCEMENT 6

NAME ______ DATE ______ CLASS ______

REINFORCEMENT Chapter 1

Where Does Life Come From? Text Pages 10–12

The illustrations below represent a theory of the sequence of events in the solar system and on Earth. Beside each illustration, write about how long ago the event took place.

1. 5 billion years ago — Whirling mass of dust and gas
2. 4.6 billion years ago — Sun and planets form
3. 3.5 billion years ago — The first living things form

Describe what each of the following scientists showed by their experiments.

4. Francesco Redi: He showed that maggots come from the eggs laid by flies on meat, not from the meat itself.
5. Lazzaro Spallanzani: He showed that a sealed flask of boiled broth developed no tiny organisms.
6. Louis Pasteur: He showed that broth became contaminated only when it was exposed to the air.

Answer the following questions.

7. What is the theory of spontaneous generation? Living things come spontaneously from nonliving matter.
8. What is the theory of biogenesis? Living things come only from living things.
9. What did Oparin believe caused the early gases of Earth to combine? lightning and ultraviolet rays
10. What did Oparin think happened to these compounds after they fell into the hot seas? They became more complex, able to copy themselves, and able to use other chemicals for energy and food.

6

▲ ENRICHMENT 6

NAME ______ DATE ______ CLASS ______

ENRICHMENT Chapter 1

Where Does Life Come From? Text Pages 10–12

REDI'S EXPERIMENT

Francesco Redi was born in Italy in 1626. At age 21, he received a medical degree from the University of Pisa, Italy. From then on, he worked as a doctor.

Most of the people of Redi's time thought living things came spontaneously from mud or decaying meat. A few scientists, including Redi, thought that idea was probably not true. He decided to try to find out how maggots appeared in decaying meat. Many thought the maggots came from the meat, but Redi thought that flies laid their eggs on the meat and the maggots hatched from those eggs.

To test his theory, Redi prepared eight jars. Into each he placed a piece of meat or fish. He then tightly covered four of the jars with lids and left four uncovered. Drawn by the smell of the meat, flies buzzed around the jars. After a few days, maggots appeared on the meat in the four uncovered jars. The meat in the four covered jars had decayed, but there were no maggots.

Next, Redi prepared eight more jars for a second experiment. Into each, he placed a piece of meat or fish. He then covered four of the jars with gauze, which allowed air to move freely in and out of the jars. He again left four jars uncovered. As before, flies buzzed around the jars, attracted by the smell of the meat. After a few days the results were the same—maggots could only be found on the meat in the uncovered jars.

Answer the following questions.

1. Before the 1700s, some people believed that life came from mud or meat. What is this theory called? spontaneous generation
2. What theory did Redi prove? biogenesis
3. In the first experiment, what did Redi prove? He proved that the maggots came from the flies, because the flies could only reach the meat in the uncovered jars.
4. What did Redi do differently in the second experiment? He covered four jars with gauze instead of lids.
5. What did Redi prove in the second experiment that he had not already proved in the first? In the first experiment, he proved that the maggots did not come from the meat. In the second experiment, he proved that the maggots did not come from the air.

6

TYING TO PREVIOUS KNOWLEDGE: Review the characteristics of life and the needs of living things from the previous section and connect those ideas with concepts of the origins of life in this section.

2 TEACH

Key Concepts are highlighted.

CONCEPT DEVELOPMENT

▶Emphasize the importance of repetition in experimentation. A hypothesis cannot be supported or disproved by one experiment.

Answer: Gases including carbon dioxide, plus compounds of silica, sulfur, and nitrogen.

CHECK FOR UNDERSTANDING

Have students explain the following misconceptions from the Middle Ages: (1) after spring rains, frogs appeared from the dried-out mud in ponds; (2) dead logs gave rise to all kinds of organisms; and (3) mice seemed to be made by throwing grains of wheat and a dirty shirt into a box.

RETEACH

Write a description of each of the experiments mentioned in this section on 3 × 5 cards. Have a student draw a card, read it, and describe the conclusions of the experiment.

EXTENSION

For students who have mastered this section, use the **Reinforcement** and **Enrichment** masters or other OPTIONS provided.

CROSS CURRICULUM

▶**Astronomy:** Have students use reference books to learn about planetary features such as temperature and relative distance from the sun, and relate these characteristics to the presence of life on Earth.

3 CLOSE

▶ Ask questions 1-2 and the **Apply** Question in the Section Review.

SECTION REVIEW ANSWERS

1. Spontaneous generation—living things from nonliving matter; Biogenesis—living things only from other living things of the same kind.

2. Redi conducted one of the first controlled experiments; Spallanzani's was an attempt at a controlled experiment. Both tried to demonstrate that life comes from life.

3. Apply: If sterilized, stoppered flasks remain sealed, they would still be uncontaminated.

4. Connect to Earth Science: Magma is molten rock below Earth's surface; lava is magma that reaches the surface.

Skill Builder

In a sense, he did. By boiling the broth, he destroyed the microorganisms that were in it at that time. People at that time did not understand that to grow and reproduce in the broth, the organisms had to be introduced.

Skillbuilder

ASSESSMENT

Performance: Assess students' ability to Observe and Infer by having them observe chicken broth, sealed and unsealed, for a week and infer what is happening from any changes that occur.

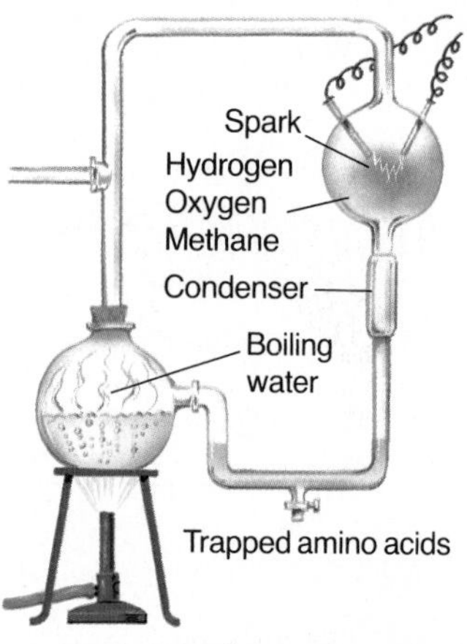

Figure 1-8. Miller's experiment was a model of what many scientists think Earth's early atmosphere was like. Volcanic eruptions like the one to the right remind us of what Earth was probably like while it was in its formative stages.

Ever since Oparin formed his hypothesis, other scientists have been testing it. In 1953, an American scientist, Stanley L. Miller, set up an experiment using the chemicals suggested in Oparin's hypothesis. Electrical sparks were sent through the mixture of chemicals. At the end of a week, new substances, similar to amino acids that are found in all living things, had formed. This showed that substances present in living things *could* come from nonliving materials in the environment. It did not prove that life was formed in this way.

Evidence suggests that life was formed from nonliving matter sometime between 4.6 billion and 3.5 billion years ago. However, scientists are still investigating where the first life came from.

SECTION REVIEW

1. Compare the theory of spontaneous generation with the theory of biogenesis.
2. What was most important about Redi's and Spallanzani's experiments?
3. **Apply:** Explain how it is possible for some of Pasteur's flasks to be uncontaminated after 100 years.
4. **Connect to Earth Science:** Find out the difference between magma and lava.

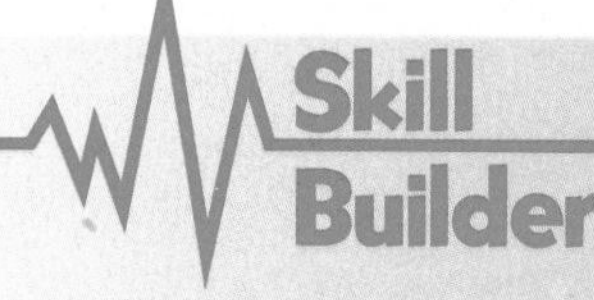

Observing and Inferring

People believed that Spallanzani had destroyed some "vital force" in the broth. Did he? Based on the experiment, what could they have inferred about where organisms come from? If you need help, refer to Observing and Inferring in the **Skill Handbook** on page 682.

OPTIONS

ENRICHMENT

▶ Have students report on the lives of John Needham, Francesco Redi, Lazzaro Spallanzani, and Louis Pasteur.

PROGRAM RESOURCES

From the **Teacher Resource Package** use:

Critical Thinking/Problem Solving, page 5, Fossil Clues to the Origin of Life.

Use **Laboratory Manual,** page 1, The Scientific Method.

Use **Laboratory Manual,** page 3, Using the Scientific Method.

What Is Science?

1-3

Objectives

- ▶ Describe methods scientists use to solve problems.
- ▶ Identify and use the SI units of length, volume, mass, and temperature.

New Science Words

scientific methods
hypothesis
variable
control
theory
law

The Work of Science

Where do the fleas on your dog come from? Have you ever tried to figure out how to get rid of them? Why do some trees lose their leaves in the fall and others in spring? If you have questions about the world in which you live, you think like a scientist. Scientists observe what goes on around them. They ask questions about their observations and may try to find answers to their questions by using tests called experiments.

Life scientists study living things. In this textbook, you'll be exposed to many branches of science as they touch on life science. For instance, you'll learn some basic chemistry, which is the study of the matter of which things are made. In the study of physics, you will learn that matter and energy are related. Earth science, the study of planet Earth, and life science use both chemistry and physics.

With millions of different kinds of organisms on Earth, it would be difficult to study every one. Many life scientists work with only one group of living organisms. Some life scientists are botanists who study plants. Others are zoologists who work with animals. More and more, scientists are seeing that none of these organisms exists alone. Ecologists are scientists who study how living things interact with each other and their environment on Earth. Genetics explains how traits of organisms are passed from generation to generation.

Figure 1-9. In life science, you will study plants and animals, what they are made of, and how they interact with each other.

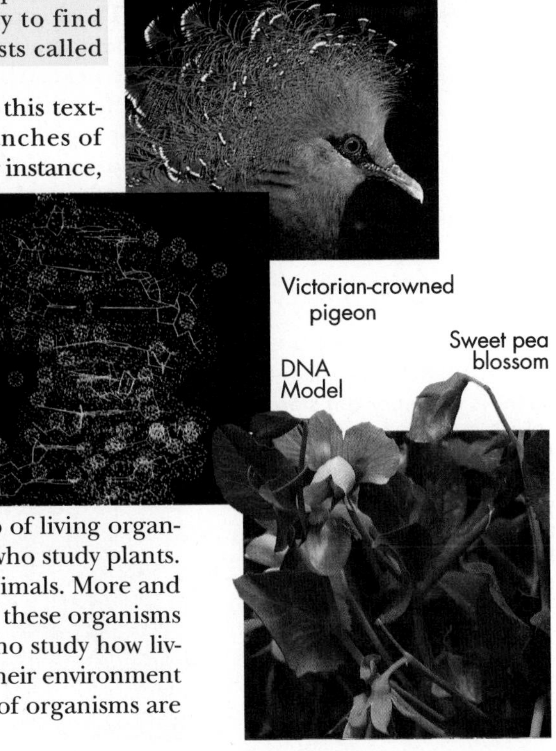

OPTIONS

PROGRAM RESOURCES

Use **Color Transparency** 1, Flex Your Brain.

Use **Color Transparency** 2, SI/Metric to English Conversions.

PROGRAM RESOURCES

From the **Teacher Resource Package** use:

Technology, pages 5-6, Nature's Glue.

Science Integration Activities 1, Measuring, Using SI Units.

Transparency Masters, pages 1-2, Flex Your Brain.

Transparency Masters, pages 3-4, SI/Metric Conversion.

SECTION 1-3

PREPARATION

SECTION BACKGROUND

▶ All scientific understanding is based on observation. Observing is using one or more of the five senses to become aware of objects or events.

▶ Scientists use many methods of investigation. They approach problems from a variety of perspectives and structure their research based on available data, tools, and time. Observations often generate new hypotheses that lead to additional testing.

▶ A scientist must be careful that his or her opinions and emotions do not influence observations. An opinion that influences an observation is a bias. A scientist's observations should be free of bias.

▶ The SI system was developed in France in the 1790s. SI stands for the French *Système International d'Unités.*

1 MOTIVATE

Cooperative Learning: Divide the class into groups of four. Ask the question: **What do you think life scientists do, and where do you think they work?** Have each group decide its own answer to the question. Have the groups present their conclusions to the class. When all groups have presented their ideas, discuss what most students perceive a life scientist to be. Dispel stereotyping.

▶ Invite scientists to visit the class and discuss their work with students.

VideoDisc

STVS: Detecting Climate Changes in Tree Rings, Disc 4, Side 2

TYING TO PREVIOUS KNOWLEDGE: Students will be familiar with everyday problem solving. Pose questions such as, **How did you decide what to wear to school today?** Have them identify problems they have solved. Relate solving everyday problems to scientific methods, theories, and laws.

2 TEACH

Key Concepts are highlighted.

CONCEPT DEVELOPMENT

▶**Demonstration:** Fill a 4-L glass jar with pond water. Place it in a lighted area. Have students hypothesize how the water will appear in ten days. Collect the written hypotheses, and at the end of ten days, have students check their hypotheses against the conditions of the jar.

▶Check scientific journals out of the library and display them. Explain that before scientists begin an experiment, they search for all the information that has already been written on a subject.

TEACHER F.Y.I.

▶The scientific method presented here is only one way to solve a problem.

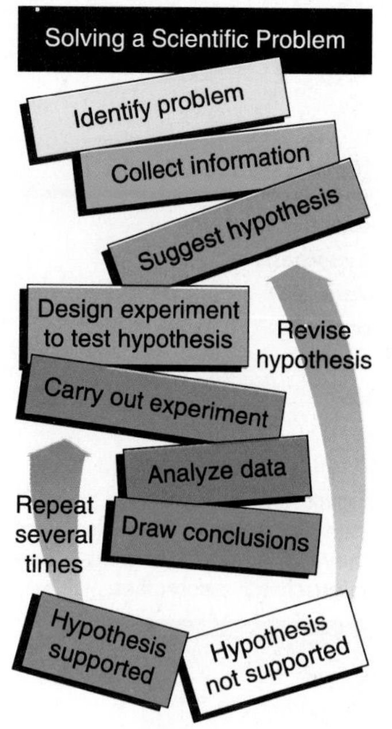

Solving Problems

Do scientists always find answers to their questions? No. Do they always do experiments in laboratories? Not always. There is no set method that all scientists use to solve problems. Each problem is different. Yet, solving any problem requires organization, doesn't it? In science, this organization often takes the form of a series of procedures or **scientific methods.**

The diagram to the left shows an order in which scientific methods might be used. You begin by observing something that you cannot explain. Suppose you awake one morning with a sore throat. It hurts all day, so you go to the doctor after school. The first step in a scientific method is to *state the problem.* A scientist can't begin to solve a problem until it's clearly stated. You tell the doctor that you have a sore throat.

The second step is to *gather information* about the problem. Tell the doctor everything you've observed about your sore throat—when it began hurting and how it feels now. The doctor makes her own observations. She takes your temperature and examines your throat.

What is a hypothesis?

The next step is to *form a hypothesis.* A **hypothesis** is a
1 prediction that can be tested. Based on her experience, the doctor hypothesizes that you have strep throat. She knows she can test her hypothesis right there in her laboratory. A hypothesis always has to be something you can test.

To test her hypothesis the doctor will *perform an experiment.* In an experiment, a series of steps is followed that tests a hypothesis using controlled conditions. The doctor

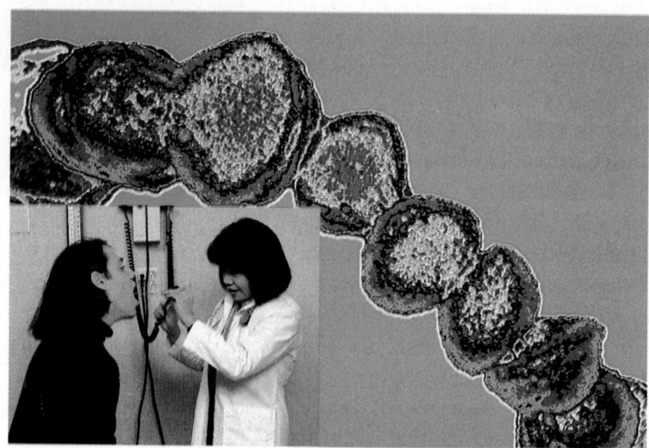

Figure 1-10. Performing an experiment is an important step in solving a problem scientifically. Here the doctor takes a throat culture to find out whether or not your throat infection is caused by the streptococcus bacterium. The larger photograph shows some streptococcus cells.

OPTIONS

Meeting Different Ability Levels

For Section 1-3, use the following **Teacher Resource Masters** depending upon individual students' needs.

◆ **Study Guide Master** for all students.

● **Reinforcement Master** for students of average and above average ability levels.

▲ **Enrichment Master** for above average students.

Additional Teacher Resource Package masters are listed in the OPTIONS box throughout the section. The additional masters are appropriate for all students.

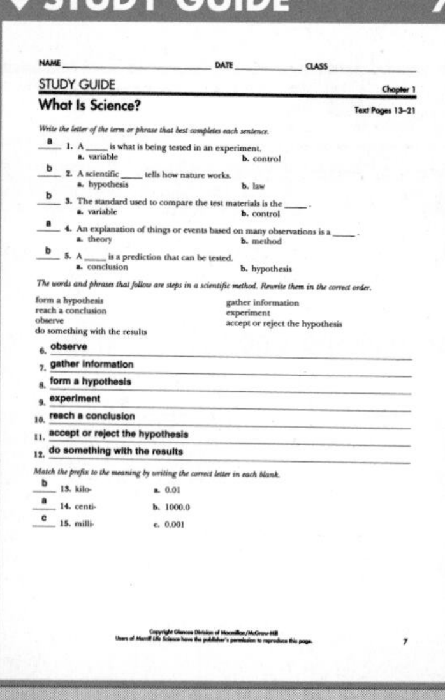

◆ STUDY GUIDE 7

NAME ______ DATE ______ CLASS ______

STUDY GUIDE — Chapter 1

What Is Science? — Text Pages 13–21

Write the letter of the term or phrase that best completes each sentence.

a 1. A _____ is what is being tested in an experiment.
a. variable b. control

b 2. A scientific _____ tells how nature works.
a. hypothesis b. law

b 3. The standard used to compare the test materials is the _____.
a. variable b. control

a 4. An explanation of things or events based on many observations is a _____.
a. theory b. method

b 5. A _____ is a prediction that can be tested.
a. conclusion b. hypothesis

The words and phrases that follow are steps in a scientific method. Rewrite them in the correct order.

form a hypothesis — gather information
reach a conclusion — experiment
observe — accept or reject the hypothesis
do something with the results

6. observe
7. gather information
8. form a hypothesis
9. experiment
10. reach a conclusion
11. accept or reject the hypothesis
12. do something with the results

Match the prefix to the meaning by writing the correct letter in each blank.

b 13. kilo- — a. 0.01
a 14. centi- — b. 1000.0
c 15. milli- — c. 0.001

 7

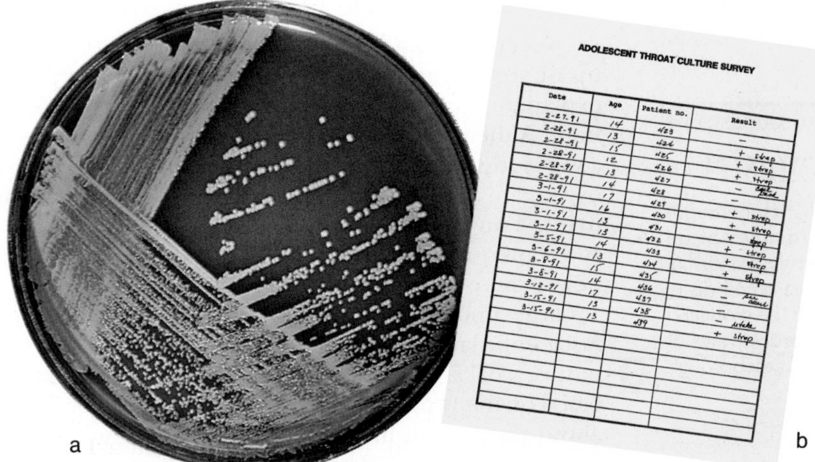

ADOLESCENT THROAT CULTURE SURVEY

Date	Age	Patient no.	Result
2-27-91	14	423	—
2-28-91	13	424	+ strep
2-28-91	15	425	+ strep
2-28-91	12	426	+ strep
2-28-91	13	427	— [illegible]
2-28-91	14	428	—
3-1-91	17	429	+ strep
3-1-91	16	430	+ strep
3-1-91	13	431	+ strep
3-1-91	13	432	+ strep
3-5-91	14	433	+ strep
3-6-91	13	434	+ strep
3-8-91	15	435	—
3-8-91	14	436	— [illegible]
3-12-91	17	437	—
3-15-91	13	438	— [illegible]
3-15-91	13	439	+ strep

a b

takes a throat culture from you. She has a petri dish with a growth medium, which she marks "experimental." She gently rubs the surface of the medium in the experimental dish with your throat culture. The dish marked experimental is the variable. A **variable** is the factor tested in an experiment. The doctor also has a sample of the organism 2
that is known to cause strep throat. She will use this sample as a control to compare with your culture. A **control** is the standard used to compare with the outcome of a test.

The doctor may give you a prescription for some capsules that will relieve your symptoms. She tells you that she will call you in 24 hours with the test results. The next day, a lab technician identifies the growth in the experimental petri dish. This *data is recorded* in your chart. The doctor uses the data to draw her conclusions. A *conclu-* 3
sion is a logical answer to a question based on data and observations of the test materials.

The next step in the search for an answer is to *accept or reject the hypothesis.* When you return, the doctor tells you that the growth in the experimental dish shows that your sore throat was caused by another kind of bacterium. You don't have strep throat. She has had to reject her original hypothesis.

The last step in solving a problem scientifically is to *do something with the results.* Your doctor may use the data in a paper she is preparing on different kinds of sore throats found in adolescents. She may have begun to notice some common factors in her data.

Figure 1-11. Research in life science may involve equipment such as a petri dish (a). Data is a record of observations (b). Data is recorded in a journal.

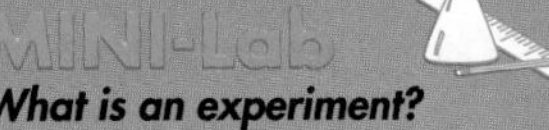

What is an experiment?

Using the information that you have just learned about scientific methods, **design an experiment** to find out how much water can be absorbed by different brands of disposable diapers. Work in groups to form a **hypothesis** and plan your experiment. Discuss your plan with your teacher. Plan to include as many scientific methods as you can, identify each method, and then perform the experiment. You will need to refer to Designing an Experiment in the **Skill Handbook** on pages 684–687.

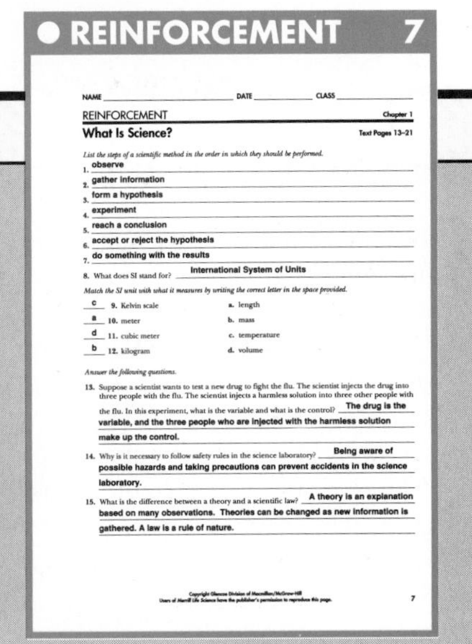

● REINFORCEMENT 7

NAME ____________ DATE ________ CLASS ________

REINFORCEMENT — Chapter 1

What Is Science? — Text Pages 13–21

List the steps of a scientific method in the order in which they should be performed.

1. observe
2. gather information
3. form a hypothesis
4. experiment
5. reach a conclusion
6. accept or reject the hypothesis
7. do something with the results
8. What does SI stand for? International System of Units

Match the SI unit with what it measures by writing the correct letter in the space provided.

c	9. Kelvin scale	a. length
a	10. meter	b. mass
d	11. cubic meter	c. temperature
b	12. kilogram	d. volume

Answer the following questions.

13. Suppose a scientist wants to test a new drug to fight the flu. The scientist injects the drug into three people with the flu. The scientist injects a harmless solution into three other people with the flu. In this experiment, what is the variable and what is the control? The drug is the variable, and the three people who are injected with the harmless solution make up the control.
14. Why is it necessary to follow safety rules in the science laboratory? Being aware of possible hazards and taking precautions can prevent accidents in the science laboratory.
15. What is the difference between a theory and a scientific law? A theory is an explanation based on many observations. Theories can be changed as new information is gathered. A law is a rule of nature.

7

▲ ENRICHMENT 7

NAME ____________ DATE ________ CLASS ________

ENRICHMENT — Chapter 1

What Is Science? — Text Pages 13–21

THE CASE OF THE POLLUTED STREAM

Scientists are often thought of as detectives who investigate natural things. Detectives are often thought of as scientists who investigate crime. The reason why the two are compared is because each goes about investigating in much the same way—both scientists and detectives use scientific methods.

These are the steps of a scientific method.

1. state the problem
2. gather information
3. form a hypothesis
4. experiment
5. reach a conclusion
6. accept or reject a hypothesis
7. do something with the results

Use these steps to write a detective story. The detective is you. In this story, you should solve the case by following the steps of a scientific method.

Everyone loves a secluded stream, with clear water and overhanging trees. That was what brought Jackie Shavers to Clear Creek on Thursday, July 2, at seven o'clock in the morning. He often fished in Clear Creek at that time of day. But on this day, Shavers was horrified to find the creek full of soap suds—his lovely stream had been polluted the night before. At least, that is what he thought. Shavers ran back to his house and called the police station. The sergeant put me on the case. The first thing I did was to head for Clear Creek.

Finish the story of the polluted stream on another sheet of paper. Use your imagination for the details. Be sure to use the steps of a scientific method to solve the case.

Student stories will vary.

7

CONCEPT DEVELOPMENT

▶ Explain that one of the most important characteristics of science is that conclusions have to be based on evidence known as data.

▶ Help students understand that inferring is going beyond what they actually observe.

MINI-Lab

Materials: two brands disposable diapers, scissors, water, measuring cups

Answer: Experiments can be simple comparisons of how much each brand can hold. Students should begin with the same size diaper, the same amount of water, and a valid way to measure and record data. Have them identify the variable. Plans may be rough, but discussion after can make for better experimental designs and give opportunity for discussing variables and controls.

MINI-Lab ASSESSMENT

Performance: To further test students' ability to design an experiment, have them design an experiment to test the absorbency of paper towels.

CHECK FOR UNDERSTANDING

Ask students to suggest solutions to a series of problems, such as: **How does a dog get fleas? Why do people get sore throats? How does a gardener get rid of weeds? How does a model ship get inside a bottle? How many ways can you divide a pizza?** List suggestions on the chalkboard.

RETEACH

Divide the class into groups of five. Have each group develop an investigation that would include collecting data from the entire school. Students should be able to explain how portions of their investigation reflect scientific problem-solving methods.

EXTENSION

For students who have mastered this section, use the **Reinforcement** and **Enrichment** masters or other OPTIONS provided.

CONCEPT DEVELOPMENT

▶Explain that a scientist may spend many years investigating one topic and collecting observations before drawing a conclusion and publishing the results.

▶This is a good time to have students begin science fair projects. You may want to have a brainstorming session on ideas for projects. In each project, the scientific method should be followed for both the actual project and the report of the project.

▶Carefully review and emphasize the relationship between *theory* and *hypothesis*.

▶Emphasize that hypotheses and theories may be changed or dropped as new information is obtained.

CROSS CURRICULUM

▶**Reading:** Have students research a major concept in life science and the person who made the discovery. Some scientists who have made notable discoveries are: Francesco Redi, William Harvey, Alexander Fleming, Barbara McClintock, and George Washington Carver.

PROBLEM SOLVING

Think Critically: Scientific research often takes many years of work to find answers. It often takes the scientist a long time to learn what the problem is before a plan can be laid for a solution. Dr. Price first had to find out what was causing the problem, then plan to solve it.

Connect to... Chemistry

In 1859, Robert Cheseborough, a chemist, took samples of an annoying, sticky, waxy material that no one wanted from an oil rig in Titusville, PA. From it, he produced Vaseline. What characteristics of a scientist did Mr. Cheseborough have?

Theories and Laws

Observations and conclusions in science either develop or support existing information. A **theory** is an explanation of things or events based on many observations. A theory is not someone's opinion, nor is a theory a vague idea. Hypotheses that have been tested over and over again and cannot be shown to be false support theories. You have already read about the theory of spontaneous generation. Theories can be changed as new data uncover new information.

Large amounts of data in science often show a trend. A scientific **law** based on these repeating data tells us how
4 nature works. A law is a reliable description of nature based on many observations. In life science, you will learn about the laws of heredity. Laws may change as more information becomes known.

Scientific methods help answer questions. Your questions may be as simple as "Where did I leave my house key?" or as complex as "What can we do about air pollution?" Will these methods guarantee that you will get an answer? Not always. Often they just lead you to more questions, but that is the work of science.

PROBLEM SOLVING

The Long Island Duck Problem

Dr. Jessie Price is a specialist in veterinary bacteriology, particularly diseases in birds. In the 1950s, while studying for her doctorate at Cornell University, Jessie Price was confronted with a major problem. Large flocks of White Pekin ducks on Long Island were dying from a mysterious illness. Long Island, New York was one of the largest producers of ducks for food in the United States. As many as 10% of these ducks were dying, costing farmers more than $250,000 a year.

It took several years of testing, experimenting, and comparing samples of blood from diseased ducks before Price identified the organism that infected the ducks with a bacterial disease. After six years of research, Price finally found a vaccine that protected new ducklings from infection.

Think Critically: How does Dr. Price's experience with scientific methods compare with the doctor on pages 14 and 15?

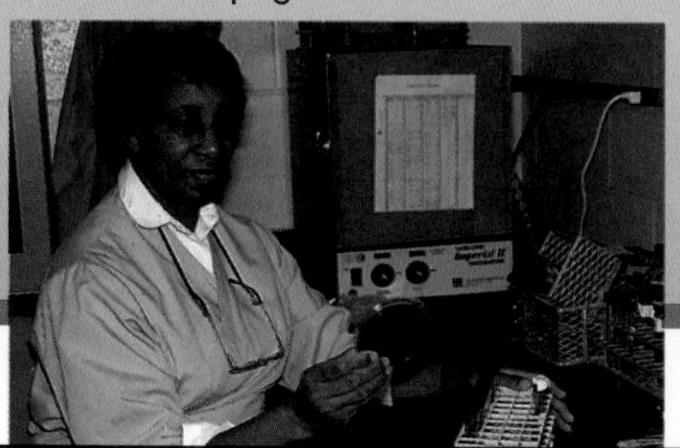

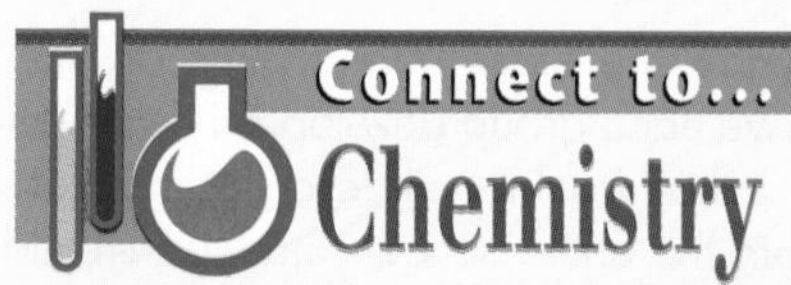

Answer: Mr. Cheseborough was observant and saw opportunity where others merely saw a problem.

OPTIONS

ASSESSMENT—ORAL

▶**Why should an experiment have only one variable?** *If an experiment had more than one variable, the scientist would not know which variable caused the result.*

▶**Why should the results of an experiment be published or shared?** *Other scientists will know about the results of the experiment and can verify them or use them to develop new hypotheses.*

▶**Is a control always necessary in an experiment? Explain your answer.** *Yes. If there is no control, the scientist does not have anything with which to compare the results obtained.*

▶**How is a hypothesis different from a theory?** *A theory is backed by more evidence than a hypothesis.*

▶**How is a theory different from a law?** *When a hypothesis explains why events occur, it becomes a theory. When a hypothesis explains how events occur, it becomes a law.*

Critical Thinking

Whether you become a scientist or not, you are going to solve problems all your life. Most of the problems you encounter will be solved in your head by a process of sorting through ideas to see what will or will not work. Suppose you make a batch of chili and it turns out badly. What went wrong with the chili? Even though you followed the usual recipe, it just didn't seem to have a zippy taste. After thinking for a bit, you realize that you left out one very important ingredient—chili powder! How did you figure out that lack of chili powder was the problem? Without being aware of it, you probably used a form of problem solving called critical thinking.

Critical thinking is a process that uses certain skills to solve problems. For example, you identified the problem by mentally comparing the not-so-great batch of chili with other batches of chili you've eaten. First you separated important information from unimportant information. For instance, you may have realized that temperature had little to do with the flavor of the chili. You may have examined your assumption that you followed the recipe correctly. After looking at the recipe again, you concluded that it was chili powder that was missing.

Finally, you probably went one step further and analyzed your conclusion. Would lack of chili powder have made the chili taste blah? If your answer was "yes," then you may have solved the problem.

"Flex Your Brain" is an activity that will help you think about and examine your way of thinking. "Flex Your Brain" is a way to keep your thinking on track when you are investigating a topic or problem. It takes you through steps of exploration from what you already know and believe, to new conclusions and awareness. Then, it encourages you to review and talk about the steps you took.

"Flex Your Brain" will help you improve your critical thinking skills. You'll become a better problem solver, and your next batch of chili will taste great. "Flex Your Brain" is found on the next page.

The Thinker
by Auguste Rodin

CONCEPT DEVELOPMENT

▶To acquaint students with this strategy, divide the class into small groups and have them explore a topic using the Flex Your Brain worksheet. Students should have some familiarity with the topic chosen. Possible topics might include cacti, volcanoes, robots, or topics from a list the class brainstorms. See pages 20T-21T for a more detailed description of the Flex Your Brain process.

▶Discuss inductive and deductive reasoning aspects in determining why the chili discussed in the text tasted bland. In the process of inductive reasoning, one generalizes from many experiences. After eating many bowls of chili, one can generalize on the attributes of chili, such as (a personal choice of) color, texture, and taste. After having made many bowls of chili, one can generalize about the techniques of making it, such as cooking time and temperature. Having reached a general idea of what "good" chili is *both* from making and eating a lot of it, one can evaluate a particular batch.

▶From generalizing about the cooking times and temperatures of many batches of chili, one can eliminate these as factors in making this batch bland, unless it was uncooked or burned. This elimination is an example of deductive reasoning; evaluating from a generalization.

▶**Why is an engineer called an applied scientist?** *An engineer applies scientific principles to situations rather than problems searching for new knowledge.*

ENRICHMENT

▶Have students find out about humane treatment of animals used in the laboratory for scientific testing. Pamphlets and books describing humane treatment may be obtained from the local humane society.

▶Discuss how life science has affected the quality of life. Begin with familiar examples—nutrition, exercise—and go to advances in medicine and agriculture.

FLEX Your Brain

Use the Flex Your Brain activity to have students explore PROBLEMS THAT LIFE SCIENTISTS ARE CURRENTLY TRYING TO SOLVE.

An example of how students might fill in the Flex Your Brain blanks:

1. Pollution
2. 1. Pollution harms wildlife
   2. Pollution affects the air, soil and water
3. Which country pollutes the most?
4. United States
5. 1 2 3 **4** 5
6. 1. Read magazine articles
   2. Write to environmental organizations
7. Students may work cooperatively.
8. Choice depends on Step 7.
9. Answers should reveal that students have learned something new.
10. Have students present their results.

▶Students should be encouraged to keep a record of their Flex Your Brain exercise in their Journals so that they can be submitted in their portfolios later.

FLEX Your Brain

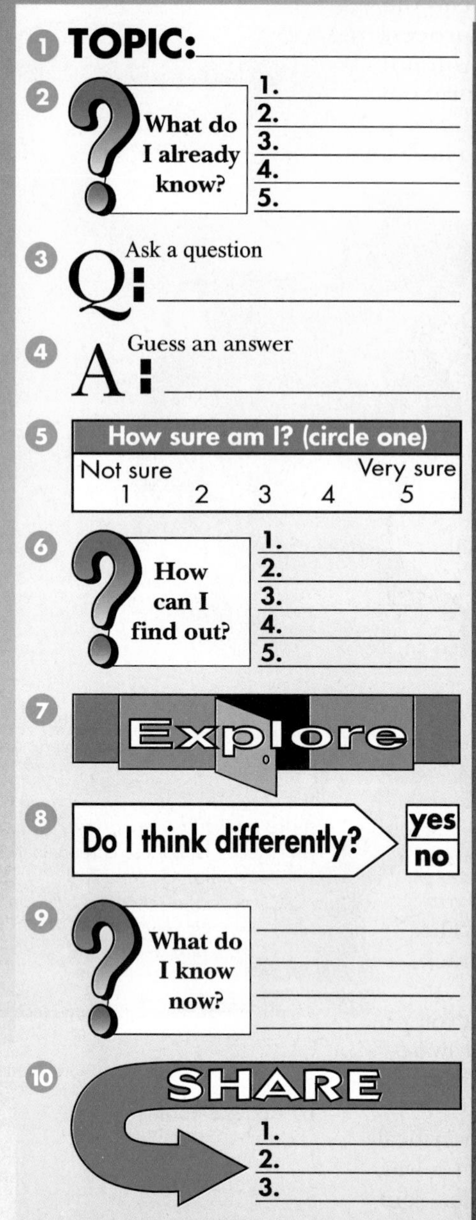

1 TOPIC:
2 What do I already know?
1. 2. 3. 4. 5.
3 Ask a question
Q:
4 Guess an answer
A:
5 How sure am I? (circle one)
Not sure Very sure
1 2 3 4 5
6 How can I find out?
1. 2. 3. 4. 5.
7 Explore
8 Do I think differently?
yes
no
9 What do I know now?
10 SHARE
1. 2. 3.

1. Fill in the topic your teacher gives you.
2. Jot down what you already know about the topic.
3. Using what you already know (Step 2), form a question about the topic. Are you unsure about one of the items you listed? Do you want to know more? Do you want to know what, how, or why? Write down your question.
4. Guess an answer to your question. In the next few steps, you will be exploring the reasonableness of your answer. Write down your guess.
5. Circle the number in the box that matches how sure you are of your answer in Step 4. This is your chance to rate your confidence in what you've done so far and, later, to see how your level of sureness affects your thinking.
6. How can you find out more about your topic? You might want to read a book, ask an expert, or do an experiment. Write down ways you can find out more.
7. Make a plan to explore your answer. Use the resources you listed in Step 6. Then, carry out your plan.
8. Now that you've explored, go back to your answer in Step 4. Would you answer differently? Mark one of the boxes.
9. Considering what you learned in your exploration, answer your question again, adding new things you've learned. You may completely change your answer.
10. It's important to be able to talk about thinking. Choose three people to tell about how you arrived at your response in every step. For example, don't just read what you wrote down in Step 2. Try to share how you thought of those things.

OPTIONS

PROGRAM RESOURCES

From the **Teacher Resource Package** use:

Activity Worksheets, page 5, Flex Your Brain.

Activity Worksheets, page 11, Mini-Lab: What is an experiment?

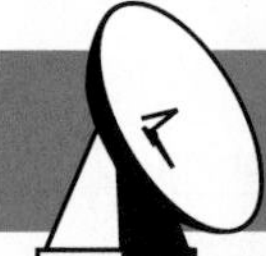

TECHNOLOGY

Cockleburs and Space Shuttles

One day a Swiss engineer returned from a walk and became interested in the cocklebur sticking to his socks. He studied the cocklebur under a microscope and found hundreds of tiny hook and loop structures. As a result of this observation, he invented Velcro.

Nylon filament is woven into loops and coated. Half of the loops are cut to form hooks. Velcro has come to have many uses. It has been used to secure artificial hearts. In the space program, more than 64 500 cm^2 of Velcro tape have been used on each space shuttle. Velcro is also used to strap on blood pressure cuffs, on sneakers, and is found in astronauts' helmets as a nose scratcher.

Think Critically: What does this story tell you about how science works?

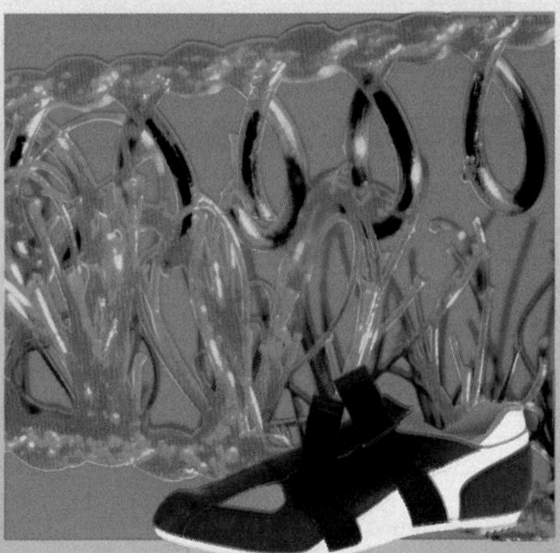

Measuring in Science

Think about how many things you use every day that measure or are measured. The thermostat keeps the air in your home at a certain temperature. Meters in your house measure water, electricity, and gas usage. Your food comes in pounds, ounces, and liters, and you step on a scale to check your weight.

Scientists use a system of measurement to make observations. Scientists around the world have agreed to use the International System of Units, or SI. SI is based on certain metric units. Using the same system gives scientists a common language. They can understand each other's research and compare results. Most of the units you will use in this textbook are shown in Table 1-1 on page 20. Because you are used to using the English system of pounds, ounces, and inches, a chart has been included in Appendix B on page 671 to help you convert these units to SI. 5

Science and READING

What do doctors, dentists, veterinarians, hairdressers, and school nurses have to learn about life science before they are allowed to do their jobs? Find these careers in the United States Government Guide to Careers and record the requirements for each **in your Journal.**

CONCEPT DEVELOPMENT

▶To introduce this section, show common items and point out the metric measurements—the mass of a bag of potato chips and other food items, the volume of drink bottles, the 35mm camera, and Celsius temperature readings in weather reports.

▶Display and discuss measuring instruments, such as a galvanometer, hourglass, barometer, pedometer, sundial, rain gauge, and an electric meter.

▶Demonstrate the use of a balance.

Science and READING

All of the professions listed require knowledge of biology. Doctors, dentists, veterinarians, and nurses all study anatomy, physiology, and pharmacology. Hairdressers have to know the bones and muscles of the head and face and the chemistry of skin and hair.

CROSS CURRICULUM

▶**History:** Have students research and report on early measurements and instruments such as the cubit, shadow clock, thermoscope, Galileo's thermometer, and the history of the metric system.

TECHNOLOGY

Extension: Ask students to come up with a new use for Velcro, draw a picture of the product, and make an ad to sell it. Share these as a class.

Reference: "Velcro: The Final Frontier" by Judith Stone, *Discover,* May 1988, pp. 82-84.

Think Critically: Some of the most interesting and useful discoveries in science result from accident or a new observation of a common item.

INQUIRY QUESTIONS

▶**Why is it important for scientists to measure things?** *It makes their observations and comparisons precise.*

▶**What makes using SI units difficult for people in the United States?** *They are accustomed to using pounds, ounces, inches, quarts, and so on.*

CONCEPT DEVELOPMENT

▶Discuss the fact that measurement is an integral part of our lives. Have students relate various measurements and measuring devices that are part of their daily lives from alarm clocks to clothing sizes and standard notebook paper.

▶Make sure that students understand the difference between weight and mass. Explain that the weight of an object on Earth is different from its weight on the moon but the mass of the object is the same in both places.

▶Avoid having students convert between metric and English systems. This is confusing and makes the metric system more complicated.

MINI QUIZ

Use the Mini Quiz to check students' recall of chapter content.

1. **A suggested solution to a problem based on known facts is a(n) ____.** *hypothesis*
2. **What is being tested in an experiment is the _____.** *variable*
3. **A logical answer to a problem based on observations is a(n) _____.** *conclusion*
4. **A rule of nature based on many observations is a(n) _____.** *law*
5. **The system of measurement used by all scientists is the _____.** *International System of Units*

Table 1-1

COMMON METRIC UNITS AND PREFIXES					
Length	**Mass**	**Volume**	**Prefix**	**Symbol**	**Meaning**
kilometer	kilogram	cubic meter	kilo-	k	1000.0
meter	gram	liter	centi-	c	0.01 (1/100)
centimeter		cubic centimeter	milli-	m	0.001 (1/1000)
millimeter	milligram	milliliter			

SI is based on units of ten. It is easy to use because calculations are made by multiplying or dividing by ten. Prefixes are used with units to change them to larger or smaller units.

The SI unit of length is the *meter.* A metric ruler or a meterstick is used to measure length. If you look at the table, you see that 1000 meters equal one kilometer. Large distances are measured in kilometers.

Mass is the amount of matter in an object. Mass is measured with a balance. The SI unit of mass is the *kilogram.* Smaller masses are measured in grams and milligrams.

The amount of space occupied by an object is its volume. Units of volume are based on units of length. Volume is found by multiplying the length times the width times the height. The SI unit of volume is the *cubic meter.* Cubic meters are too large to be of any use in the laboratory. Because of this, the cubic centimeter (cm^3) is used to measure volume. Liquid volumes are measured in liters (L). One liter has the same volume as 1000 cm^3. Milliliters are used to measure smaller volumes.

The *degree* is the unit for measuring temperature. The kelvin scale is the SI standard for measuring temperature. Scientists also often use the Celsius scale. On the Celsius scale, water freezes at 0°C and boils at 100°C at sea level.

Figure 1-12. A graduated cylinder (a) is used to measure volume. Metric rulers (b) are used to measure length. A pan balance (c) is used to measure mass.

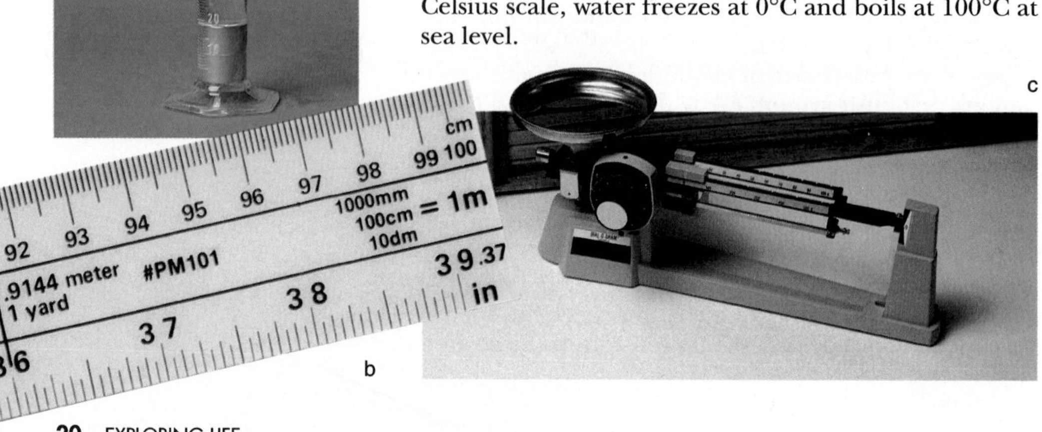

OPTIONS

INQUIRY QUESTIONS

▶**What makes SI units easy to use?** *They are based on multiples of ten.*

▶**How could you describe an object with measurements 10 cm long, 5 cm wide, and 10 cm high, in three different ways using SI units and prefixes?** *500 cubic centimeters, 500 milliliters, 0.5 liter*

▶**What is the difference between mass and weight?** *Mass is the measurement of the amount of matter, and weight is the measurement of the force of attraction between two objects caused by gravity.*

You will use these measurements to work in the laboratory this year. Working in the laboratory will help you understand the life science concepts in your textbook. You will observe and conduct experiments. Some of the observations you will make will be the same ones made by students and researchers for years and years. The important thing is that *you* will have seen these things yourself.

Safety First

Having the chance to do science is much more interesting than merely reading about it. Some of the scientific equipment that you will use is the same that scientists use in the field or in their laboratories. Of course, safety is of great importance in a laboratory. Most injuries are due to burns from heated objects or splatters and broken glass. The symbols shown below are used throughout your text to alert you to situations that require your special attention. A description of each symbol can be found on page 673 in the Safety Appendix. Following safety rules will protect you and others from injury and help you get more from your lab experiences.

SECTION REVIEW

1. What is a hypothesis?
2. How does a control differ from a variable?
3. What SI units are used to measure length, volume, and mass?
4. **Apply:** Identify the scientific methods that were used in the Find Out activity on page 5.
5. **Connect to Physics:** Find out the boiling point of water at sea level. What happens to this point at altitudes above sea level?

Skill Builder: Measuring in SI

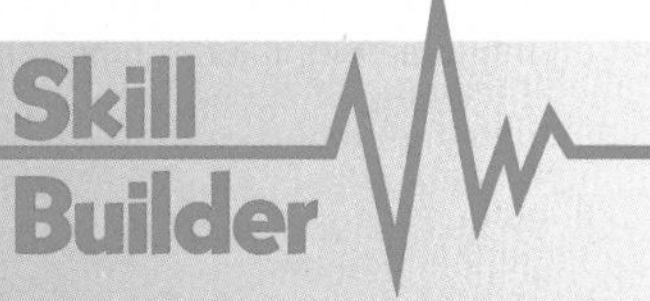

Measure the volume of space in your classroom in cubic meters. Then figure out how much of this volume each student occupies. If you need help, refer to Measuring in SI in the **Skill Handbook** on pages 684-685.

MINI-Lab

How are things measured?

Working in pairs, *measure* the mass of the following items: a small paper clip, small glass of water, a metal spoon, and a rock. Use a pan balance. *Record* your data in a table **in your Journal**. Are these items best measured in kilograms, milligrams, or grams? Why did you choose the unit you did?

MINI-Lab

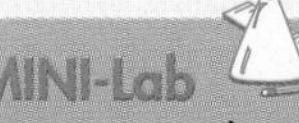

Answer: The rock could be measured in kilograms or grams; the spoon and paper clip would be measured in grams; water would be measured in milliliters or grams.

MINI-LAB ASSESSMENT

Performance: To further assess students' understanding of measurement, see USING LAB SKILLS, question 11, on p. 26.

3 CLOSE

▶Ask questions 1-3 and the **Apply** Question in the Section Review.

SECTION REVIEW ANSWERS

1. a prediction that can be tested

2. A variable is a factor tested; a control stays the same and is the factor against which the variable is measured.

3. meter, cubic meter, kilogram, respectively.

4. Apply: Students may have originally made a hypothesis that the black line was black. Then they performed an experiment and made observations. Because the line of ink separates into two or more colors, they may revise their hypothesis.

5. Connect to Physics: B.P. water is 100°C; as altitude increases, boiling point decreases.

Skill Builder

length × width × height = volume; volume divided by number of students = how much each student occupies

Skillbuilder ASSESSMENT

Performance: Assess students' abilities to Measure in SI by having them measure ten items at home.

ENRICHMENT

▶Have students identify different measuring devices (meterstick, balance, flask, thermometer, graduated cylinder), name the device, tell what it measures (volume, mass, length, temperature), and the unit of measurement (liter, gram, meter).

PROGRAM RESOURCES

From the **Teacher Resource Package** use:

Activity Worksheets, page 12, Mini-Lab: How are things measured?

SECTION 1-4

PREPARATION

SECTION BACKGROUND

▶Although much technology has been advantageous, some of it has been harmful to our environment.

1 MOTIVATE

▶Present pictures, a filmstrip, or a video of a third world country that has not been provided with technology.

TYING TO PREVIOUS KNOWLEDGE: Technology is used in many ways in the home. Ask students to name three ways.

2 TEACH

Key Concepts are highlighted.

CONCEPT DEVELOPMENT

▶Have students stand outside of the school building and make lists in their Journals of all technological advances they see. Have them divide the list into those that are positive and those that are negative for the environment.

VideoDisc

STVS: Computerized Apple, Disc 1, Side 2

PROGRAM RESOURCES

From the **Teacher Resource Package** use:

Science and Society, page 5, Is Fluoridated Water Harmless?

Activity Worksheets, page 5, Flex Your Brain.

1-4 The Impact of Science on Your Life

New Science Words

technology
ecology

Objectives

▶ **State two ways in which life on Earth has been affected by technology in the life sciences.**

Times Have Changed

Did you know that if you were born 100 years ago, you would probably not have lived past your first birthday? What a thought! When your parents were born, there were some places in the United States in which about one half of all infants died in their first year of life due to contaminated water and milk. This could still happen if your communities didn't have sanitation plants. Have you ever had a shot of penicillin to fight infection? Before the discovery of penicillin in 1940, infection was difficult to control, and many people died. Did you brush your teeth this morning? Before studies on why and how teeth decay, many people did not brush. Decayed teeth were pulled at an early age. Now most water systems are fluoridated, and tooth decay has decreased. Before modern techniques in food processing, such as irradiation of packaged foods, many diseases were acquired from contaminated foods. Sanitation, penicillin, fluoridated water, and irradiated food are examples of technologies that have extended lives. **Technology** is applied science, or the use of scientific knowledge to solve everyday problems or improve the quality of life.

Figure 1-13. Optical fibers are used to transmit voices and data. They are also used in new forms of surgery.

Living with Technology

Years of research in the life sciences have enabled people to benefit directly from science economically, nutritionally, medically, and environmentally. Advances in agriculture allow you to have fish, meat, bread, and milk regularly. Because of agricultural research, farmers

OPTIONS

Meeting Different Ability Levels

For Section 1-4, use the following **Teacher Resource Masters** depending upon individual students' needs.

◆ **Study Guide Master** for all students.

● **Reinforcement Master** for students of average and above average ability levels.

▲ **Enrichment Master** for above average students.

Additional Teacher Resource Package masters are listed in the OPTIONS box throughout the section. The additional masters are appropriate for all students.

◆ STUDY GUIDE 8

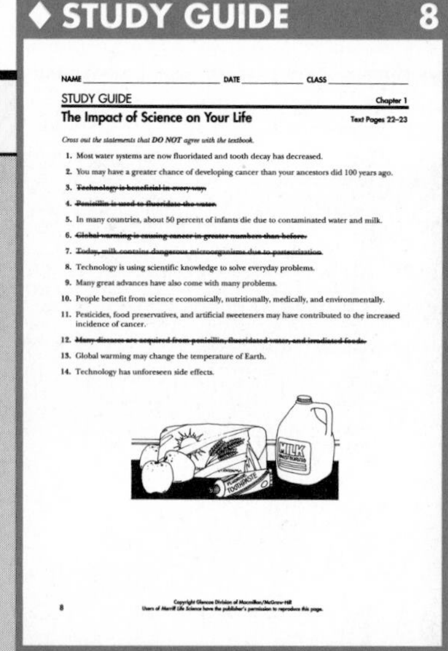

NAME ______ DATE ______ CLASS ______

STUDY GUIDE Chapter 1

The Impact of Science on Your Life Text Pages 22–23

*Cross out the statements that **DO NOT** agree with the textbook.*

1. Most water systems are now fluoridated and tooth decay has decreased.
2. You may have a greater chance of developing cancer than your ancestors did 100 years ago.
3. ~~Technology is beneficial in every way.~~
4. ~~Penicillin is used to fluoridate the water.~~
5. In many countries, about 50 percent of infants die due to contaminated water and milk.
6. ~~Global warming is causing cancer in greater numbers than before.~~
7. ~~Today, milk contains dangerous microorganisms due to pasteurization.~~
8. Technology is using scientific knowledge to solve everyday problems.
9. Many great advances have also come with many problems.
10. People benefit from science economically, nutritionally, medically, and environmentally.
11. Pesticides, food preservatives, and artificial sweeteners may have contributed to the increased incidence of cancer.
12. ~~Many diseases are acquired from penicillin, fluoridated water, and irradiated foods.~~
13. Global warming may change the temperature of Earth.
14. Technology has unforeseen side effects.

8

produce great quantities of healthful food at reasonable prices in many countries of the world.

Have all of the advances in the life sciences been beneficial? With the many great advances have also come many problems due to unforseen side effects. For example, you probably run a much greater chance of developing cancer than your ancestors of 100 years ago. The technology involved in the development of pesticides, food preservatives, and artificial sweeteners has contributed to increased incidence of cancer. Are the technologies the problem? Or is it how they are used that creates problems?

Now scientists are turning their energy to the environment. Many technologies seem to have a strong impact on the environment. **Ecology** is the study of how organisms interact with each other and the environment. Ecological problems such as global warming, the warming of Earth due to the buildup of carbon dioxide in the atmosphere, and deforestation are sources of concern being studied by scientists. Have these problems been caused by our own technology? Can we make use of all the knowledge we've gained through research to prevent these problems? Do we really know that technology is to blame for the problems of the environment? What is there in the framework of science that can help answer these questions?

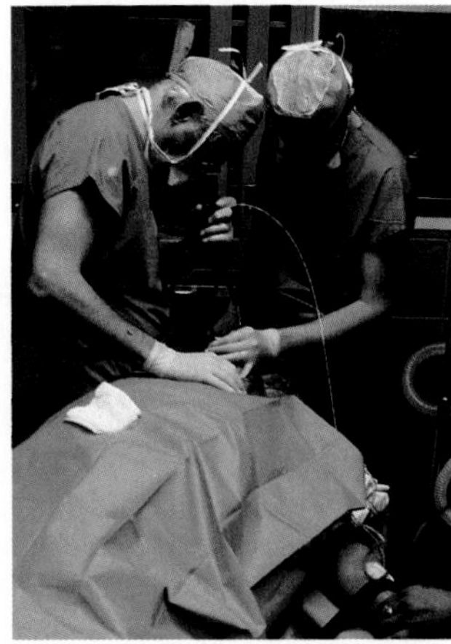

Figure 1-14. Laser surgery has reduced the time it takes to recover and has reduced infection.

SECTION REVIEW

1. Describe two ways in which people have benefitted from technology.
2. Describe two ways in which Earth may have been damaged as a result of technology.

You Decide!

How easy is it for a company to do what's best for the environment? Many fast-food companies are being urged to stop using styrofoam containers because the use of these products is thought to contribute to the destruction of the ozone layer. If companies go back to using paper products, however, where will these paper materials come from? If you were the company president, what would you have to do to make an informed decision?

● REINFORCEMENT 8

NAME ________ DATE ________ CLASS ________

REINFORCEMENT — Chapter 1

The Impact of Science on Your Life — Text Pages 22–23

Answer the following questions.

1. What is technology? **applied science, or the use of scientific knowledge, to solve everyday problems**
2. How has technology helped farmers? **Because of technology, farmers can produce great quantities of food at reasonable prices.**
3. What was an unforeseen side effect of food preservatives? **They increased the incidence of cancer.**
4. How is the milk you had for breakfast made safe to drink through technology? **It is pasteurized or heated until the dangerous microorganisms in it are killed.**
5. How has the development of penicillin helped save lives? **Penicillin controls infections that harm or kill people.**
6. Why might it be said that science is needed to protect us from our own scientific technology? **The same scientific technology that helps us also has unforeseen side effects that may harm us.**
7. What is global warming? **the warming of Earth due to the accumulation of carbon in the atmosphere**
8. Automobiles send carbon compounds into the atmosphere. How has the technology of the automobile benefited and also harmed us? **The automobile has been a benefit to our lives in terms of transportation, but it has also added to the pollution of the atmosphere. This pollution may cause global warming, which may change our climate.**

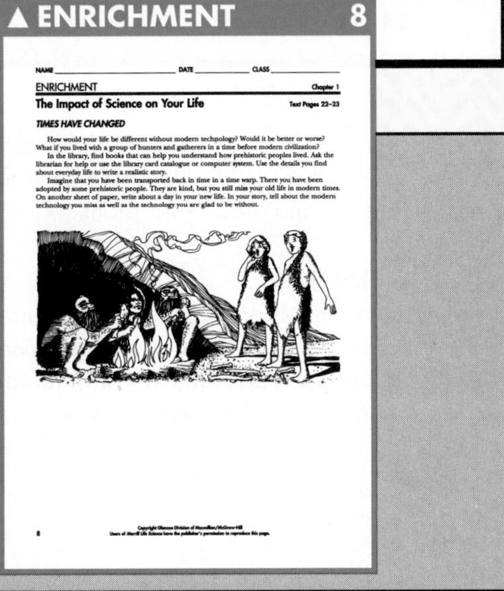

▲ ENRICHMENT 8

NAME ________ DATE ________ CLASS ________

ENRICHMENT — Chapter 1

The Impact of Science on Your Life — Text Pages 22–23

TIMES HAVE CHANGED

How would your life be different without modern technology? Would it be better or worse? What if you lived with a group of hunters and gatherers in a time before modern civilization?

In the library, find books that can help you understand how prehistoric peoples lived. Ask the librarian for help or use the library card catalogue or computer system. Use the details you find about everyday life to write a realistic story.

Imagine that you have been transported back in time in a time warp. There you have been adopted by some prehistoric people. They are kind, but you still miss your old life in modern times. On another sheet of paper, write about a day in your new life. In your story, tell about the modern technology you miss as well as the technology you are glad to be without.

CHECK FOR UNDERSTANDING

In a short paragraph, have students define *technology* and explain ten ways in which they have encountered technology since getting up.

RETEACH

Have students list the countries that they think are very technologically advanced and those they believe are not. First review the two lists, then discuss the effects of technology on the advanced countries and the advantages enjoyed by countries that have never developed such technologies.

EXTENSION

For students who have mastered this section, use the **Reinforcement** and **Enrichment** masters or other OPTIONS provided.

3 CLOSE

▶Ask questions 1-2 in the Section Review.

FLEX Your Brain

Use the Flex Your Brain activity to have students explore THE EFFECT OF TECHNOLOGY ON THEIR LIVES.

Flex Your Brain

ASSESSMENT

Portfolio: In Step 1, students may list cars/transportation, health, computers, entertainment, etc.

SECTION REVIEW ANSWERS

1. Some diseases are rare now; agricultural advances result in much higher yields of crops; medications, such as penicillin, fight or prevent infections; improved sanitation in handling food prevents transmission of diseases.

2. As new products are produced, their byproducts must be handled and disposed. Burning of some fuels changes air quality; pesticides; pollution from new technologies.

YOU DECIDE!

Students' answers may vary widely. Accept all reasonable answers that they can support.

ACTIVITY 1-1

OBJECTIVE: Design and carry out an experiment using a scientific method to infer why brine shrimp live in the ocean.
Time: One class period to brainstorm and set up the experiment, 5 minutes to observe for 3 days, and one-half class period to summarize on the last day.

PROCESS SKILLS applied in this activity are **observing, inferring, comparing and contrasting, recognizing cause and effect, interpreting data, hypothesizing,** and **using variables and controls.**

PREPARATION

- Dechlorinated water: Allow water to stand for 48 hours.
- Weak salt solution: Add 20 mL non-iodized salt to 4 L dechlorinated water. Stir until dissolved.
- Strong salt solution: Add 75 mL non-iodized salt to 4 L dechlorinated water. Stir until dissolved.
- Purchase brine shrimp from a pet store or a biological supply house.
- Pint mayonnaise jars may be used for the containers in this activity.

Cooperative Learning: Divide the class into Science Investigation Teams.

THINKING CRITICALLY

The steps are on pages 14 and 15. Students may suggest conducting an experiment with salt as the variable.

TEACHING THE ACTIVITY

*Refer to the **Activity Worksheets** for additional information and teaching strategies.*

- Do not place too many brine shrimp eggs in each container.
- Brine shrimp are orange-colored and swim with a jerky motion.
- To maintain the brine shrimp, add a small amount of yeast to the container two or three times a week.

SUMMING UP/ SHARING RESULTS

- The hypothesis that brine shrimp hatch best in strong salt solution was supported.
- Container 3 contained salt, water, oxygen, and received light.
- Students formed a hypothesis, performed an experiment with a variable and a control, recorded data, and drew conclusions.

GOING FURTHER!

They are near the ocean surface where there is light. The eggs did not hatch because there was not enough salt. Experimental designs should follow this activity's general format.

PROGRAM RESOURCES

From the **Teacher Resource Package** use:
Activity Worksheets, pages 7-8, Activity 1-1, Using a Scientific Method.

Activity ASSESSMENT

Performance: To further assess students' understanding of using scientific methods, see USING LAB SKILLS, Question 12, on page 26.

ACTIVITY 1-1 DESIGNING AN EXPERIMENT
Using a Scientific Method

Brine shrimp are tiny organisms that live in the ocean. Why do they live in the ocean? How can you find out why they live where they do?

Getting Started

You are to use a scientific method to determine how salt affects the growth of brine shrimp.

Thinking Critically

List the steps in problem solving using scientific methods. How can you use these steps to determine how salt affects the growth of brine shrimp?

Materials

Your cooperative group will use:

- 3 wide-mouthed 0.5-L containers
- brine shrimp eggs
- wooden splint
- 500 mL distilled water
- 500 mL weak salt solution
- 500 mL strong salt solution
- 3 labels
- hand lens

Try It!

1. Label the glass containers: 1-distilled water, 2 - weak salt solution, 3 - strong salt solution.
2. Pour the distilled water, weak salt solution, and strong salt solution into the designated jars.
3. Dip the end of the wooden splint into the brine shrimp eggs. Place the eggs that remain on the wooden splint into container 1. Repeat this step with containers 2 and 3.
4. Write a **hypothesis** to explain which container will have the most brine shrimp after 3 days.
5. For the next 3 days, use the hand lens to look for small animals swimming in jerking motion. *Observe* and *record* the number of the container in which you first see brine shrimp and the number of shrimp that hatch in each container each day.

Summing Up/Sharing Results

- Was your **hypothesis** supported by your data?
- What physical conditions in the container in which the brine shrimp hatched and grew best are similar to those in the ocean?
- How did you use a scientific method to solve this problem?

Going Further!

Where in the ocean do you think you would find brine shrimp? Why didn't the eggs hatch in container 1? **Design an experiment** to find out if there is a maximum amount of salt that can be added to water and still have brine shrimp grow.

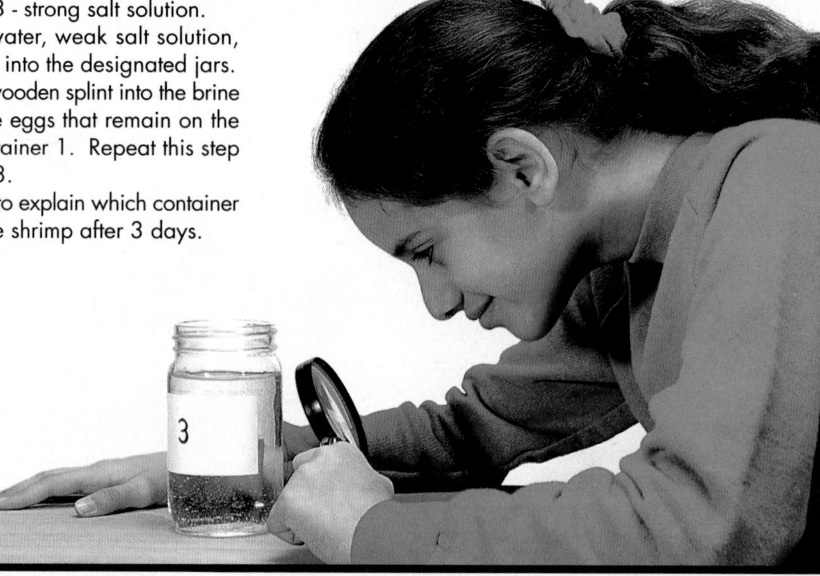

CHAPTER

REVIEW

SUMMARY

1-1: Living Things

1. Organisms are made up of cells, use energy, move, respond, adjust, reproduce, grow and develop, and adapt to their environment.
2. Organisms need energy and materials.
3. Organisms interact with other organisms and nonliving things in their environment.

1-2: Where Does Life Come From?

1. Early experiments of Redi and Spallanzani tried to disprove the idea that living things come from nonliving matter.
2. Pasteur's experiment proved the theory that life comes from life, biogenesis.
3. Scientists try to explain how life began on Earth. Oparin's hypothesis, which says that energy combined early gases to form the chemical compounds of life, has been tested.

1-3: What is Science?

1. Scientists investigate observations that are made about living and nonliving things with the help of problem-solving techniques.
2. Scientists use SI measurements to gather measurable data.
3. Safe laboratory practices help you to learn more about science.

1-4: Science and Society: The Impact of Science on Your Life

1. Technology has improved health but also produced some harmful side effects.

KEY SCIENCE WORDS

a. **adaptation**
b. **biogenesis**
c. **cells**
d. **control**
e. **development**
f. **ecology**
g. **homeostasis**
h. **hypothesis**
i. **law**
j. **life span**
k. **organisms**
l. **response**
m. **scientific methods**
n. **spontaneous generation**
o. **stimulus**
p. **technology**
q. **theory**
r. **variable**

UNDERSTANDING VOCABULARY

Match each phrase with the correct term from the list of Key Science Words.

1. the length of time an organism is expected to live
2. changes undergone during growth
3. a change in the environment that brings about a response
4. living things
5. the use of science for solving problems or improving life
6. theory that nonliving things produce living things
7. organized steps used to solve a problem
8. a prediction that can be tested
9. the units of life
10. a standard for comparing the result of an experiment

OPTIONS

ASSESSMENT

To assess student understanding of material in this chapter, use the resources listed in the Program Resources box to the right.

COOPERATIVE LEARNING

Consider using cooperative learning in the THINK AND WRITE CRITICALLY, APPLY, and MORE SKILL BUILDERS sections of the Chapter Review.

PROGRAM RESOURCES

From the **Teacher Resource Package** use:
Chapter Review, pages 5-6.
Chapter and Unit Tests, pages 5-8, Chapter Test.

CHAPTER

REVIEW

SUMMARY

Have students read the summary statements to review the major concepts of the chapter.

UNDERSTANDING VOCABULARY

1. j
2. e
3. o
4. k
5. p
6. n
7. m
8. h
9. c
10. d

ASSESSMENT
Portfolio
Encourage students to place in their portfolios one or two items of what they consider to be their best work. For each item, ask students to explain why that item was chosen and what they learned from it. Items might be selected from the following.

- MINI-Lab hypothesis and experimental design, p. 15
- Enrichment laboratory animal research, p. 17
- Activity 1-1 hypothesis, data, and conclusion, p. 24

Performance
Additional performance assessments may be found in *Performance Assessment* and *Science Integration Activities* that accompany **Merrill Life Science.** Performance Task Assessment Lists and rubrics for evaluating these activities and other products generated throughout the chapter can be found in Glencoe's *Performance Assessment in Middle School Science.*

CHAPTER

REVIEW

CHECKING CONCEPTS

1. b
2. c
3. a
4. d
5. c
6. d
7. a
8. b
9. b
10. b

USING LAB SKILLS

ASSESSMENT

Use these alternate lab excercises to assess students' understanding of skills used in this chapter.

11. Measurements will vary, but students will probably measure in meters and centimeters, giving the reason that in SI, length is measured in meters and centimeters.

12. Glass A contained the variable because it had all the same factors as Glass B except one, the addition of baking soda. Glass B was a control and was not subjected to baking soda. Therefore, the treatment in Glass A had something with which it could be compared.

THINK AND WRITE CRITICALLY

13. Growth is the process of becoming larger, such as becoming taller; development is a change in structure that brings forth new abilities, such as learning to walk.

14. Technology often enables scientists to see new problems to be solved.

15. Response is a reaction to a stimulus; an adaptation is an inherited trait that helps an organism to survive.

16. Energy is needed for all life functions: growth, movement, and so on. Without energy, these processes could not take place and organisms would eventually die.

CHAPTER

REVIEW

CHECKING CONCEPTS

Choose the word or phrase that completes the sentence.

1. An infant cutting teeth is an example of ______.
   a. growth c. respiration
   b. development d. adaptation
2. A bright light that causes you to shut your eyes is a ______.
   a. need c. stimulus
   b. response d. variable
3. The source of energy for life is ______.
   a. the sun c. chlorophyll
   b. oxygen d. carbon dioxide
4. Living things are made up of about 70 percent ______.
   a. oxygen c. minerals
   b. carbon dioxide d. water
5. ______ disproved the theory of spontaneous generation.
   a. Oparin c. Pasteur
   b. Spallanzani d. Redi
6. Scientists think that ______ was missing from Earth's early atmosphere.
   a. ammonia c. methane
   b. hydrogen d. oxygen
7. ______ are inherited characteristics that help an organism to survive.
   a. Adaptations c. Raw materials
   b. Stimuli d. Theories
8. The ______ is the part of the experiment that is tested.
   a. conclusion c. control
   b. variable d. data
9. A(n) ______ is a prediction that has to be testable.
   a. experiment c. theory
   b. hypothesis d. law
10. The SI unit used to measure liquids is ______.
    a. meter c. gram
    b. liter d. degree

USING LAB SKILLS

11. Use a meter stick to measure the length and width of your lab table. Record your results. Explain why you used the unit numbers that you used.
12. Label two glasses, one A and the other B. Add 100 mL of water and 5 mL of vinegar to each one. Then add 0.5 g of baking soda to A. Observe. Explain why one glass is the variable and why one is the control. Identify which is which.

THINK AND WRITE CRITICALLY

Answer the following questions in your Journal using complete sentences.

13. Explain the difference between growth and development. Give an example.
14. How does technology help science?
15. What is the difference between a response and an adaptation?
16. Explain why living things need energy.

APPLY

17. How does SI benefit scientists in different parts of the world?
18. Using a bird as an example, explain how it has all the characteristics of living things.
19. If a plant had no carbon dioxide available, would this affect animal life?
20. Show how Pasteur correctly used scientific methods to disprove the theory of spontaneous generation.
21. How are stimulus and response related to homeostasis?

MORE SKILL BUILDERS

If you need help, refer to the Skill Handbook.

1. **Designing an Experiment:** Design an experiment to test the effects of the plant food on growing plants. Identify the scientific methods in your experiment.
2. **Concept Mapping:** Use the following terms to complete a chain-of-events concept map showing the order in which you might use scientific methods: *collect data, perform an experiment, state your hypothesis.*

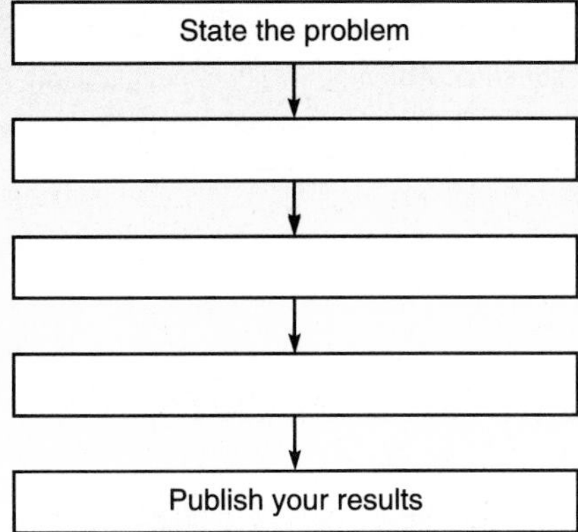

3. **Using Variables, Constants, and Controls:** You place one lima bean plant under a green light, another under a red light, and a third under a blue light. You measure their growth for 4 weeks to determine which color is most favorable to plant growth. What are the variables in this experiment? What is the hypothesis in this experiment? How could the experiment be improved?
4. **Observing and Inferring:** What evidence must scientists have to prove that cancer can be caused by cigarette smoking?
5. **Classifying:** Which type of measurement would you use for each of the following? Classify as meter, gram, cubic meter, degree, or liter.
   - **a.** distance run
   - **b.** how hot a pan of water is
   - **c.** how much juice is in a glass
   - **d.** your mass

PROJECTS

1. Interview people in your community whose jobs require a knowledge of life science. Make a Life Science Careers bulletin board. Summarize each person's job and what they had to study to prepare for that job.
2. Try to recreate Pasteur's experiment.

APPLY

17. Scientists can compare and repeat experiments; they have a common tool for measurement.

18. A bird is made up of cells; a bird uses energy to fly and breathe; a bird displays movement by flying or hopping; a bird jumps at a loud sound or responds to food; a bird maintains a constant body temperature; birds reproduce young birds; birds grow larger and develop the ability to fly; a bird is adapted with certain structures that enable it to fly; different species of birds live for one or more years.

19. No carbon dioxide would result in death of all life because it is needed for photosynthesis.

20. Pasteur's hypothesis was that air is necessary for an organism's survival and that air contained small particles of life. He had a control and a variable. The special flask allowed air in. Based on his results, he showed that no organisms grew in a curved-neck flask.

21. A stimulus momentarily causes an organism to respond. For example, you are thirsty (stimulus); you drink water to quench the thirst (response). As a result of responding to the stimulus by getting a drink of water, your body's homeostasis is maintained.

MORE SKILL BUILDERS

1. Designing an Experiment: Experiment should contain hypothesis, observations, variables, control, and explanation of how to interpret results.

2. Concept Mapping: Students' event chains should show the following sequence: state hypothesis, perform an experiment, collect data.

3. Using Variables, Constants, and Controls: The variables are the different colors of light and whatever effects they may produce. A hypothesis might be that green light will cause most favorable plant growth. This experiment could be improved by having fewer variables, more plants, and white light as a control.

4. Observing and Inferring: Scientists research the histories of cancer patients who have smoked; laboratory tests show tumors that have been caused by cigarette smoke.

5. Classifying: a. meter *(kilo-)*, b. degree, c. liter *(milli-)*, d. grams *(kilo-)*

CHAPTER

2 The Cell

CHAPTER SECTION	OBJECTIVES	ACTIVITIES
2-1 Cells: The Units of Life (2 days)	1. **Discuss** the history leading to the cell theory. 2. **Explain** the difference between the compound light microscope and electron microscope. 3. **State** the cell theory.	**MINI-Lab:** *How are objects magnified?* p. 31
2-2 Cell Structure (2 days)	1. **Diagram** a plant cell and an animal cell; **identify** the parts and the function of each part. 2. **Describe** the importance of the nucleus in the cell. 3. **Compare** and **contrast** prokaryotic and eukaryotic cells.	**MINI-Lab:** *What is cytoplasm like?* p. 38 **Activity 2-1:** *Comparing Plant and Animal Cells,* p. 43
2-3 Cell Organization (2 days)	1. **Recognize** that cells differ in size, shape, and function. 2. **Explain** how cells of one-celled organisms differ from cells of many-celled organisms. 3. **Explain** the differences among tissues, organs, and organ systems.	
2-4 Organ Transplants Science & Society (1 day)	1. **Relate** the importance of science in your life. 2. **Discuss** problems in organ transplants.	**Activity 2-2:** *Designing an Experiment (Comparing Plant and Animal Tissues),* p. 48
Chapter Review		

ACTIVITY MATERIALS

FIND OUT	ACTIVITIES		MINI-LABS	
Page 29 egg metric ruler hand lens	**2-1 Comparing Plant and Animal Cells, p. 43** microscope 1 microscope slide 1 coverslip forceps dropper *Elodea* plant prepared slide of human cheek cells	**2-2 Designing an Experiment, p. 48** onion plant microscope slides (3) coverslips (3) dropper water microscope 250-mL beaker forceps scissors apron	**How are objects magnified, p. 31** newspaper drinking glass glass bowl magnifying glass	**What is cytoplasm like? p. 38** 100-mL beaker unflavored gelatin flashlight

CHAPTER FEATURES	TEACHER RESOURCE PACKAGE	OTHER RESOURCES
Technology: *A Touch of Diamonds,* p. 33 **Skill Builder:** *Concept Mapping,* p. 35	**Ability Level Worksheets** ◆ ***Study Guide,*** p. 9 ● ***Reinforcement,*** p. 9 ▲ ***Enrichment,*** p. 9 **Activity Worksheets,** p. 20 **Concept Mapping,** pp. 9 **Transparency Masters,** pp. 5-6	**Color Transparency 3,** The Microscope **Lab Manual:** The Microscope, p. 7 **STVS:** Disc 2, Side 1 **Science Integration Activity 2**
Problem Solving: *A Tale of a Tail,* p. 40 **Skill Builder:** *Comparing and Contrasting,* p. 42	**Ability Level Worksheets** ◆ ***Study Guide,*** p. 10 ● ***Reinforcement,*** p. 10 ▲ ***Enrichment,*** p. 10 **Activity Worksheets,** pp. 14-15, 21 **Cross-Curricular Connections,** p. 6, 21 **Transparency Masters,** pp. 7, 8	**Color Transparency 4,** Animal and Plant Cells **Lab Manual:** Observing Cells, p. 11 **STVS:** Disc 4, Side 1
Skill Builder: *Interpreting Scientific Illustrations,* p. 45	**Ability Level Worksheets** ◆ ***Study Guide,*** p. 11 ● ***Reinforcement,*** p. 11 ▲ ***Enrichment,*** p. 11 **Critical Thinking/Problem Solving,** p. 6 **Science and Society,** p. 6	**STVS:** Disc 4, Side 2
You Decide! p. 47	**Ability Level Worksheets** ◆ ***Study Guide,*** p. 12 ● ***Reinforcement,*** p. 12 ▲ ***Enrichment,*** p. 12 **Activity Worksheets,** pp. 5, 16-17	**STVS:** Disc 7, Side 1
Summary Key Science Words Understanding Vocabulary Checking Concepts Using Lab Skills Think & Write Critically Apply More Skill Builders Projects	ASSESSMENT RESOURCES **Chapter Review,** pp. 7-8 **Chapter Test,** pp. 9-12 **Performance Assessment in Middle School Science (PAMSS)**	**Chapter Review Software** **Test Bank** **Alternate Assessment** **Performance Assessment**

◆ **Basic** ● **Average** ▲ **Advanced**

ADDITIONAL MATERIALS

SOFTWARE	AUDIOVISUAL	BOOKS/MAGAZINES
Cells, Educational Activities, Inc. *The Cell,* Queue.	*Microscopes and Modern Science,* filmstrip, National Geographic. *Cells: A First Film,* film, BFA. *The Cell: Structural Unit of LIfe,* film, Coronet/MTI. *The Living Cell: An Introduction,* film, EBEC.	Becker, Wayne M. *The World of the Cell.* Menlo, CA: Benjamin-Cummings Publishing Company, 1986. Brady, R.J. *Anatomy and Physiology: A Programmed Approach,* 15 bks. Englewood Cliffs, NJ: Prentice Hall, 1972. Fawcett, Don. *Cell.* 2nd ed. Philadelphia, PA: W.B. Saunders Company, 1981.

CHAPTER

2 THE CELL

THEME DEVELOPMENT: Evolution accounts for the number and variety of organisms. Students should relate that it is the cell that carries the information for evolution to occur. Homeostasis in organisms occurs when cells respond and adjust to their environment, whether the cell is a one-celled organism or part of a many-celled organism.

CHAPTER OVERVIEW

- **Section 2-1:** The importance of the development of microscopes in the discovery of cells is discussed. The cell theory is presented as one of the major theories in life science.
- **Section 2-2:** Prokaryotic and eukaryotic cells and the parts and functions common to all cells are discussed. Animal, plant, and bacterial cells are compared.
- **Section 2-3:** How cells are organized to form tissues, organs, systems, and organisms is presented.
- **Section 2-4: Science and Society:** Problems encountered with organ transplants are explored. You Decide asks students to consider the fate of body parts.

CHAPTER VOCABULARY

compound light microscope
electron microscope
cell theory
cell membrane
cytoplasm
organelles
nucleus
chromatin
endoplasmic reticulum
ribosomes
Golgi bodies
mitochondria
lysosomes
vacuoles
cell wall
chloroplast
tissues
organ
organ system
rejection

CHAPTER

2 The Cell

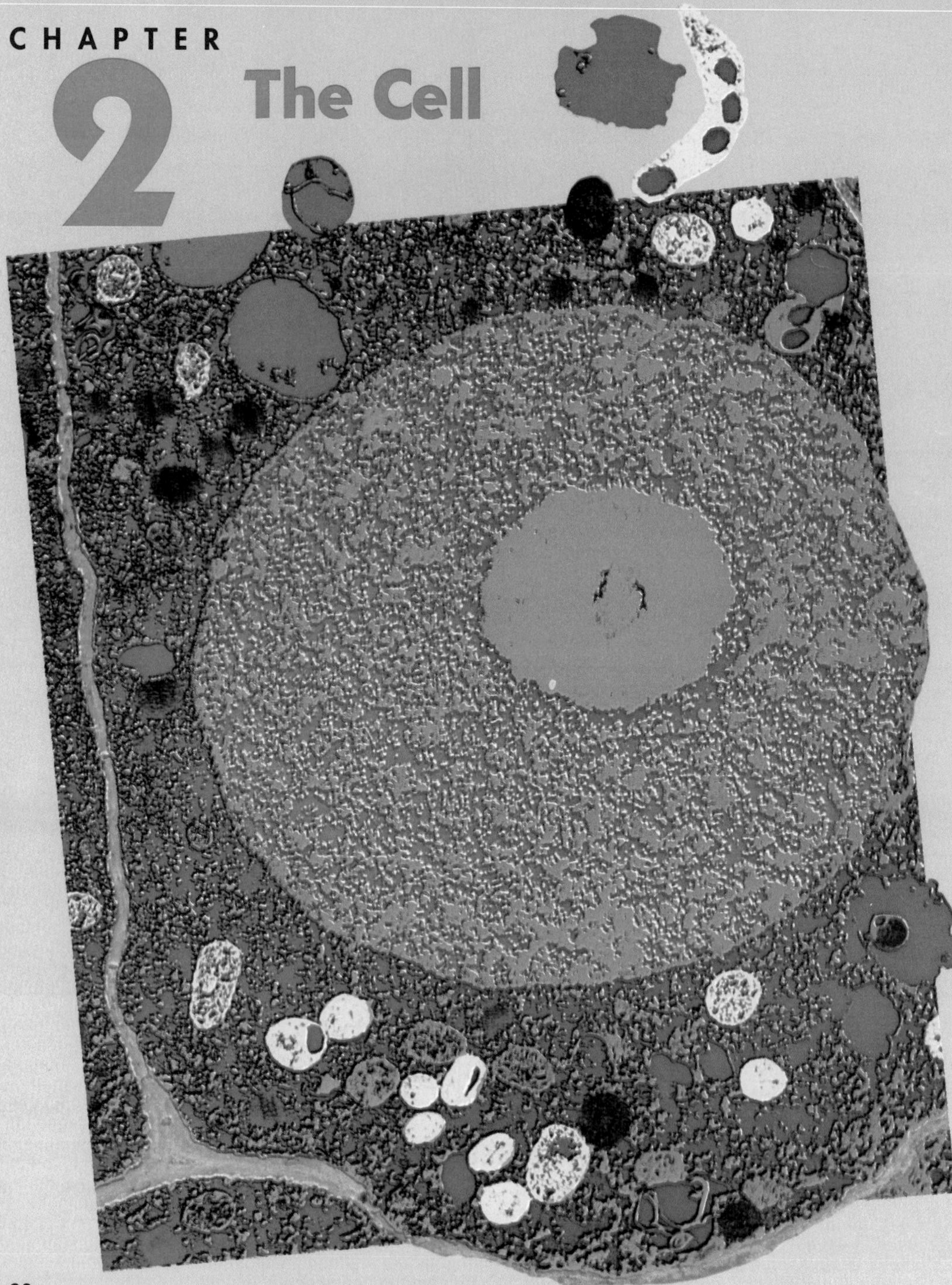

28

OPTIONS

For Your Gifted Students

After studying Section 2-3, have the students research the four different types of tissues. They should draw pictures of these tissues emphasizing the differences. They can make a board game using this information. Keeping within the cell context, have students draw a path for players to travel and devise game pieces and dice or a spinner. At certain locations on the board, students could be required to choose a picture card and name the tissue or its function or type.

For Your Mainstreamed Students

Students can observe plasmolysis using a microscope, slide, coverslip, *Elodea* plant, pipette, paper towel and a salt solution. Students will make a wet mount of the *Elodea* plant and draw the cell. Have them add a few drops of the salt solution and observe the cell and draw what they see. The cells will appear as small collapsed balloons within the cell wall.

Most cells are too small to be seen without the help of a microscope. The plant cell on the opposite page has been magnified. But there are a few cells that can be seen using just your eyes. The largest known living cells are the yolks of bird eggs—not the white, just the yolk. How large is the cell of a chicken egg?

FIND OUT!

Find out that some cells can be seen without a microscope.

Break a chicken egg into a dish, and look at the yolk. How large do you think it is? *Estimate* its diameter. Then use a metric ruler to *measure* the diameter of your chicken yolk cell. With a hand lens, observe the yolk closely. Other cells that can be seen easily are large fish eggs. Discuss why some cells are large but most are very small.

Gearing Up

Previewing the Chapter

Use this outline to help you focus on important ideas in this chapter.

Section 2-1 Cells: The Units of Life
- ▶ The Microscope: A Tool
- ▶ The Cell Theory

Section 2-2 Cell Structure
- ▶ An Overview of Cells
- ▶ Cell Membrane
- ▶ Cytoplasm
- ▶ Nucleus
- ▶ Organelles in the Cytoplasm
- ▶ Plant and Bacterial Cells

Section 2-3 Cell Organization
- ▶ How Cells Differ
- ▶ From Cell to System

Section 2-4 Science and Society
Organ Transplants
- ▶ Spare Parts for Broken Hearts

Previewing Science Skills

- ▶ In the Skill Builders, you will make a concept map, compare and contrast, and interpret scientific illustrations.
- ▶ In the Activities, you will observe, compare, diagram, and draw conclusions.
- ▶ In the MINI-Labs, you will observe, identify, compare and infer.

What's next?

Now that you have looked at an extremely large cell, learn how the microscope helped in the discovery of cells of all sizes and in the development of the cell theory. You will also learn about microscopes that help reveal the detailed parts of cells. At the end of the chapter, you will see that many-celled organisms are organized with tissues, organs, and systems.

INTRODUCING THE CHAPTER

Use the Find Out activity to introduce students to cells. Explain that they will be learning about the size of cells, cell parts and how cells were discovered.

FIND OUT!

Preparation: Provide one flat dish, one egg, one metric ruler, and one hand lens for every two students. The eggs can be used again in your next class.

Cooperative Learning: Use Paired Partners cooperative groups. Have each student measure the diameter of the yolk and look at the egg with the hand lens. Then have the students compare their results with their partner's and list reasons why different cells have different sizes.

Teaching Tips

▶ Have students measure in centimeters. Review the metric units of length from Chapter 1 and refer to Appendix C on p. 671.

▶ Students should wash their hands after they have completed the activity and return eggs to you.

Gearing Up

Have students study the Gearing Up feature to familiarize themselves with the chapter. Discuss the relationships of the topics in the outline.

What's Next?

Before beginning the first section, make sure students understand the connection between the Find Out activity and the topics to follow.

ASSESSMENT OPTIONS

PORTFOLIO

Refer to page 49 for suggested items that students might select for their portfolios.

PERFORMANCE ASSESSMENT

See page 49 for additional Performance Assessment options.

Process

Skillbuilders, pp. 35, 42, 45
MINI-Labs, pp. 31, 38
Activities, 2-1, p. 43; 2-2, p. 48
Using Lab Skills, p. 50

CONTENT ASSESSMENT

Assessment—Oral, pp. 32, 35, 38, 40
Section Reviews, pp. 35, 42, 45, 47
Chapter Review, pp. 49-51
Mini-Quizzes, pp. 33, 41

GROUP ASSESSMENT

Opportunities for group assessment occur with Cooperative Learning Strategies and Flex Your Brain Activities.

SECTION 2-1

PREPARATION

SECTION BACKGROUND

▶ About 150 years passed from the time that Hooke and Leeuwenhoek saw cells until scientists formulated the cell theory. In the 1830s, Robert Brown saw the nucleus, and Dujardin reported cells with a clear jelly-like material.

▶ A beam of electrons in an electron microscope is similar to the light used in a light microscope. Magnets in the electron microscope are like lenses in a light microscope. The magnets focus the electron beam on a fluorescent screen that produces an image of a specimen.

▶ Even though cells were discovered in the 17th century, most of the knowledge of cells has developed since 1940.

PREPLANNING

▶ Collect various magnifying devices, such as hand lenses, magnifying mirrors, light microscopes, and binoculars. Collect enough leaves for each student in the class.

1 MOTIVATE

▶ **PROJECT:** See PROJECT 1, In Search of Cells on pages 694 and 695, as you begin this chapter

▶ Display a light microscope and a stereomicroscope. Ask students to observe how they are different.

▶ Make a bulletin board using pictures of early microscopes and present-day microscopes.

Cooperative Learning: Assign Paired Partners cooperative groups to learn the parts and functions of the compound light microscope.

OBJECTIVES AND SCIENCE WORDS: Have students review the objectives and science words throughout the chapter as they study each section.

2-1 Cells: The Units of Life

New Science Words

compound light microscope
electron microscope
cell theory

Objectives

▶ **Discuss the history leading to the cell theory.**
▶ **Explain the difference between the compound light microscope and electron microscope.**
▶ **State the cell theory.**

The Microscope: A Tool

Cells are the smallest units that carry out the activities of life in organisms. Yet, as important as they are, you certainly don't see individual cells when you look at most plants or animals. You need a device, such as a magnifying glass or a microscope, to see most cells.

Trying to see separate cells in a large plant, like the one in Figure 2-1, is like trying to see individual bricks in a wall from three blocks away. If you start to walk toward the wall, it becomes easier to see individual bricks. When you get right up to the wall, you can see each brick in detail. A microscope performs a similar function. A microscope has one or more lenses that make an enlarged image of an object. Through these lenses, you are brought closer to the leaf, and you see the individual cells that carry on life processes.

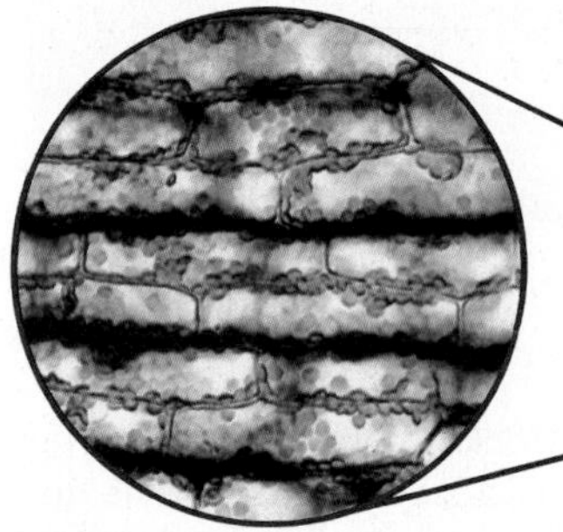

Figure 2-1. Individual plant cells become visible when the leaves on the wall are looked at with the help of a microscope.

OPTIONS

Meeting Different Ability Levels

For Section 2-1, use the following **Teacher Resource Masters** depending upon individual students' needs.

◆ **Study Guide Master** for all students.

● **Reinforcement Master** for students of average and above average ability levels.

▲ **Enrichment Master** for above average students.

Additional Teacher Resource Package masters are listed in the OPTIONS box throughout the section. The additional masters are appropriate for all students.

◆ STUDY GUIDE 9

NAME ______ DATE ______ CLASS ______

STUDY GUIDE — Chapter 2

Cells: The Units of Life — Text Pages 30–35

Complete the following sentences using appropriate terms from the textbook.

1. A light microscope having two lenses is called a **compound microscope**.
2. If the eyepiece lens of a microscope has a power of 10× and the objective lens has a power of 43×, then the **total magnification** is 430×.
3. Instead of a standard lens, the electron microscope uses a **magnetic field** to bend electrons.
4. The surfaces of whole objects can be examined with a **scanning** electron microscope.
5. If your class wanted to have a better look at an earthworm, a **stereoscopi.** light microscope could be used.
6. A **transmission electron microscope** is used to see what is inside of a cell.
7. The **cell theory** states that organisms are made of cells, cells are the basic units of structure and function in organisms, and cells come from preexisting cells.
8. A **light** microscope lets light pass through an object and then through two or more lenses.
9. The eyepiece lens usually has a power of **10x**.
10. SEM stands for **scanning electron microscope**.
11. The first person to use the term "cell" was looking through a microscope at a slice of **cork**.
12. Before the discovery that cells **divide** to form new cells, people thought life occurred spontaneously.
13. A magnifying glass works like a **simple** microscope because it has only one lens.
14. In a light microscope, the lenses **enlarge** the image and **bend** light toward your eye.

Match the statement on the left with the person on the right.

a 15. observed that every cell comes from a cell that existed before
c 16. made the first compound microscope
d 17. concluded that all animals are made up of cells
e 18. called boxlike structures in cork "cells"
b 19. made a simple microscope with a glass bead

a. Rudolph Virchow
b. Anton Van Leeuwenhoek
c. Zacharias Janssen
d. Theodor Schwann
e. Robert Hooke

 9

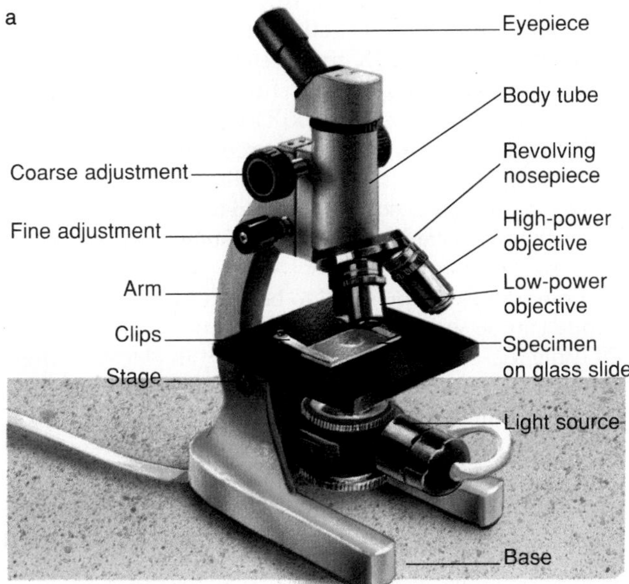

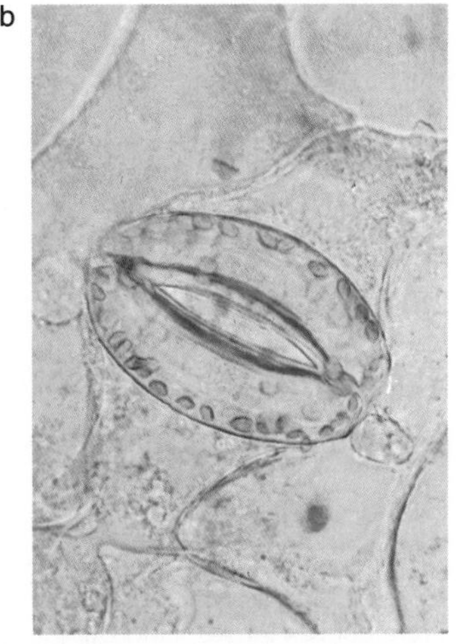

Figure 2-2. A compound light microscope (a) makes an enlarged image of something that you cannot see without help. Here (b) you see a small opening on the surface of a leaf.

Microscopes are simple or compound, depending on how many lenses they contain. A simple microscope is similar to a magnifying glass. It has only one lens. In 1590, a Dutch maker of reading glasses, Zacharias Janssen, put two magnifying glasses together in a tube. The result was the first crude compound microscope. By combining two lenses he got an image that was larger than an image made by only one lens. These early compound microscopes weren't satisfactory, however. The lenses would make a large image, but it wasn't always clear.

In the mid 1600s, Anton Van Leeuwenhoek, another Dutch scientist, made a simple microscope with a tiny glass bead for a lens. With it, he reported seeing things in pond water that no one had ever imagined before. His microscope could magnify up to 270 times. Another way to say this is that his microscope could make an image of an object 270 times larger than its actual size. Today we would say his lens had a power of 270X.

The microscope you will use in studying life science is a compound light microscope like the one in Figure 2-2. A **compound light microscope** lets light pass through an object and then through two or more lenses. The lenses enlarge the image and bend the light toward your eye. It has an eyepiece lens and an objective lens. An eyepiece

MINI-Lab

How are objects magnified?

Early scientists used various tools to help view objects. Try looking at a newspaper through the curved side of an empty glass, the flat bottom of an empty glass, a bowl filled with water, and a magnifying glass. *Compare* how well you can see through each. What did early scientists learn by using such tools?

● REINFORCEMENT 9

NAME ____________ DATE ________ CLASS ________

REINFORCEMENT — Chapter 2

Cells: The Units of Life — Text Pages 30–35

Part A

In the sentences below, a code letter has been substituted for each letter of the alphabet. To find out what the sentences say, use the following key to decode them. In the key, the code letters are shown directly below the alphabet letter they stand for. Write the correct letter above each code letter, then read the sentence aloud.

A B C D E F G H I J K L M N O P Q R S T U V W X Y Z
Z Y X W V U T S R Q P O N M L K J I H G F E D C B A

1. ALL ORGANISMS ARE MADE OF CELLS.
ZOO LITZMRHNH ZIV NZWV LU XVOOH.

2. A CELL IS THE BASIC UNIT OF STRUCTURE AND FUNCTION IN AN ORGANISM.
Z XVOO RH GSV YZHRX FMRG LU HGIFXGFIV ZMW UFMXGRLM RM ZM LITZMRHN.

3. EVERY CELL COMES FROM ANOTHER CELL.
VEVIB XVOO XLNVH UILN ZMLGSVI XVOO.

4. THIS IS CALLED THE CELL THEORY.
GSRH RH XZOOVW GSV XVOO GSVLIB.

Part B

Answer the following questions.

1. Who was the first person to see cells? **Robert Hooke**
2. In what material were cells first seen? **Cells were first seen in cork.**
3. Were the cells that were first seen living or nonliving? **The cells were nonliving.**
4. What did Schleiden and Schwann conclude about cells? **They were convinced that all living things are made of cells.**
5. What instrument that uses light and lenses is used to view cells? **a light microscope**

 9

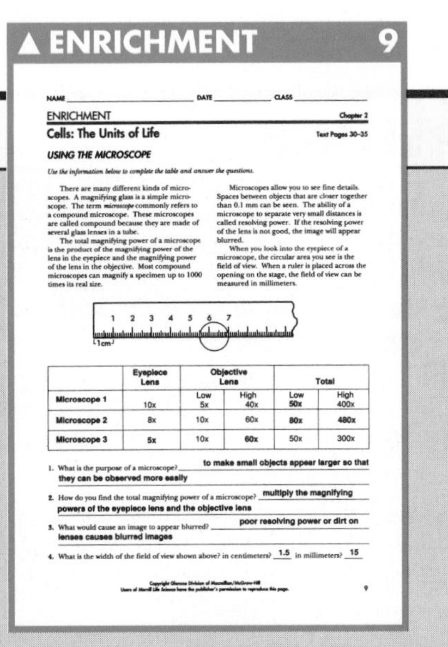

▲ ENRICHMENT 9

NAME ____________ DATE ________ CLASS ________

ENRICHMENT — Chapter 2

Cells: The Units of Life — Text Pages 30–35

USING THE MICROSCOPE

Use the information below to complete the table and answer the questions.

There are many different kinds of microscopes. A magnifying glass is a simple microscope. The term *microscope* commonly refers to a compound microscope. These microscopes are called compound because they are made of several glass lenses in a tube.

The total magnifying power of a microscope is the product of the magnifying power of the lens in the eyepiece and the magnifying power of the lens in the objective. Most compound microscopes can magnify a specimen up to 1000 times its real size.

Microscopes allow you to see fine details. Spaces between objects that are closer together than 0.1 mm can be seen. The ability of a microscope to separate very small distances is called resolving power. If the resolving power of the lens is not good, the image will appear blurred.

When you look into the eyepiece of a microscope, the circular area you see is the field of view. When a ruler is placed across the opening on the stage, the field of view can be measured in millimeters.

	Eyepiece Lens	Objective Lens		Total	
Microscope 1	10x	Low 5x	High 40x	Low **50x**	High 400x
Microscope 2	8x	10x	60x	**80x**	**480x**
Microscope 3	**5x**	10x	**60x**	50x	300x

1. What is the purpose of a microscope? **to make small objects appear larger so that they can be observed more easily**
2. How do you find the total magnifying power of a microscope? **multiply the magnifying powers of the eyepiece lens and the objective lens**
3. What would cause an image to appear blurred? **poor resolving power or dirt on lenses causes blurred images**
4. What is the width of the field of view shown above? in centimeters? **1.5** in millimeters? **15**

 9

TYING TO PREVIOUS KNOWLEDGE: Students may have used magnifying glasses and binoculars. Explain that a microscope magnifies in the same way.

2 TEACH

Key Concepts are highlighted.

CONCEPT DEVELOPMENT

▶ Have students look at leaves with hand lenses. You may wish to take them through the use of the microscope, Appendix A on page 670.

MINI-Lab

Materials: drinking glass, clear glass bowl, water, magnifying glass, newspaper pages

Teaching Tips

▶ Try this activity with the specific glasses to be used by students. Determine amount of water needed beforehand.

Answers: Early scientists learned that cells exist and began to infer that living things were made up of cells.

MINI-Lab ASSESSMENT

Performance: To further assess students' understanding of magnifying devices, see USING LAB SKILLS, question 11, on p. 50.

CONCEPT DEVELOPMENT

▶ Display photographs made through the compound light microscope, phase microscope, and electron microscope to show the magnifying power and resolving power of different kinds of microscopes.

▶ Have students study the photomicrographs at the bottom of the page.

REVEALING MISCONCEPTIONS

▶ Students may think that microscopes have only magnifying power. Explain that microscopes also have resolving power. Resolving power is the ability to see two objects as separate and distinct.

Science and MATH

The total magnification on high power is 400X.

Answer: A convex lens is thicker in the middle than at the edges. This causes the light rays to bend inward and meet at a point. Placing the object to be viewed a certain distance from the convex lens produces an enlarged image.

Science and MATH

Calculate the total magnification of your microscope if the eyepiece is 10X and the high power objective is 40X.

A magnifying glass is a convex lens. All microscopes use one or more convex lenses. Find out the shape of a convex lens and describe it. Explain how it magnifies.

lens usually has a power of 10X. An objective lens may have a power of 43X. Together, they have a total magnification of 430X (10X times 43X). Some compound microscopes have more powerful lenses that magnify an object up to 2000 times (2000X) its original size.

Your classroom may have a stereoscopic (stereo) light microscope that gives you a three-dimensional view of an
object. Stereo microscopes are used to look at thick struc- 3
tures that light can't pass through, such as whole insects or leaves, or your fingertips.

Things that are too small to be seen with a light microscope can be seen with an electron microscope. Instead of using lenses to bend beams of light, an **electron microscope** uses a magnetic field to bend beams of electrons. 4
Electron microscopes can magnify images more than 300 000 times. Figure 2-3 shows the kind of detail that can be seen with an electron microscope.

There are several kinds of electron microscopes. One is the transmission electron microscope (TEM), which is used to study parts inside a cell. The object has to be sliced very thin and placed in a vacuum. There is no air in a vacuum. As a result, only dead cells can be observed this way. A scanning electron microscope (SEM) is used to see the surfaces of whole objects. With an SEM, you can view and photograph living cells. From the time of Van Leeuwenhoek until the present, the microscope has been a valuable tool for studying cells. You will see how it was used to develop the cell theory.

Figure 2-3. The electron microscope shows details on these insects that can't be seen with a compound microscope.

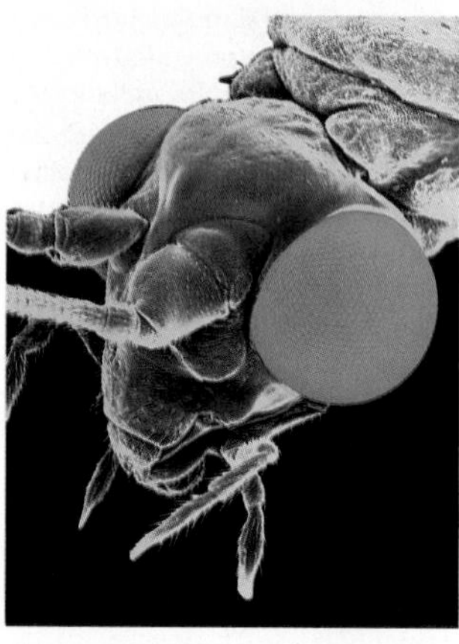

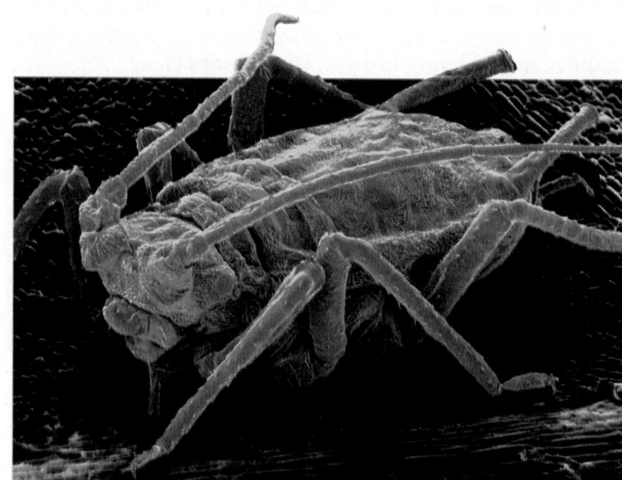

OPTIONS

ASSESSMENT—ORAL

▶ **Where do you think electron microscopes are used?** *laboratories, hospitals, colleges*

▶ **Why do you think electron microscopes are not used in classrooms?** *too expensive, persons have to be trained to use them* Explain that scientists use electron microscopes to find cell structures and other things about which little is known.

▶ **What kind of living things might early scientists have observed in rain water?** *one-celled organisms*

▶ **What words other than *cell* might Hooke have used to describe what he saw?** *blocks, boxes, rooms* Accept all reasonable answers.

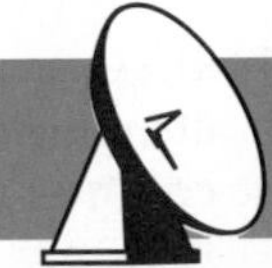

TECHNOLOGY

A Touch of Diamonds

One of the newest members of the microscope family is the atomic-force microscope. It makes use of a diamond-tipped probe that moves across the surface of a specimen. The tip moves up and down as it probes the hills and valleys of tissues. Just as a phonograph needle traces the grooves in a record, the diamond tip is moved up and down by the force of electrons on the surface of the atoms making up the tissue. As it does, it traces the shape of the surface of a cell. This tracing is then projected as a visual image on a screen. The diamond used is so small that human eyebrow hairs are used to move it when the probe is being assembled.

The atomic-force microscope "sees" individual living cells without damaging them. The first video made with an atomic-force microscope shows the clotting of blood. With this microscope, such events can now be watched as they take place.

Think Critically: How might an atomic-force microscope help medical research?

The Cell Theory

During the centuries when explorers like Columbus and Magellan set out to find new lands, scientists were busy observing everything they could about the smaller world around them. Curiosity made them look through their microscopes and lenses at mud from ponds and drops of rainwater. They examined blood and scrapings from their own teeth.

Cells weren't discovered until the microscope was improved. In 1665, Robert Hooke, an English scientist, made a very thin slice of cork and looked at it under his microscope. To Hooke, the cork seemed to be made up
1 of little empty boxes, which he called *cells*. The drawing of cork cells reproduced in Figure 2-4, was made by Robert Hooke more than 300 years ago. Actually, Hooke was not aware of the importance of what he was seeing.

In 1838, Matthias Schleiden, a German scientist, used a microscope to study plant parts. He concluded that all plants were made of cells. Just a year later, another

What did Robert Hooke see when he looked at cork?

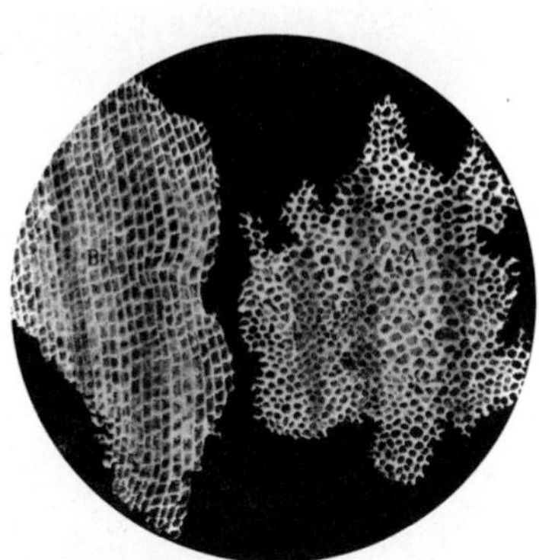

Figure 2-4.
Hooke's Cork Cells

CONCEPT DEVELOPMENT

▶ Review with students the definition of a theory presented in Chapter 1 and discuss the concept of theory.

▶ Explain that Hooke did not see cells; he saw the cell walls of dead cells from the bark of the cork oak tree. The cork tree is a live oak. It is green year-round and grows abundantly in Portugal and Spain.

TECHNOLOGY

For more information on different types of microscopes, see "Seeing Atoms" by James Trefil, *Discover*, June 1990, pp. 54-60.

Think Critically: It can be used to examine living tissues. Accept all reasonable answers.

CROSS CURRICULUM

▶ **Art:** Discuss the use of art before the camera was invented. Have students compare Hooke's drawing of cells with photographs throughout the chapter. Have students write their opinions of the advantages and disadvantages of using artwork and photography.

MINI QUIZ

Use the Mini Quiz to check students' recall of chapter content.

1. **____ was credited with discovering cells.** *Robert Hooke*
2. **A ______ allows light to pass through an object and two or more lenses.** *compound light microscope*
3. **A ______ is used for looking at things through which light cannot pass.** *stereo microscope*
4. **The ______ uses a magnetic field to bend electrons.** *electron microscope*

VideoDisc

STVS: Images of Atoms, Disc 2, Side 1

PROGRAM RESOURCES

From the **Teacher Resource Package** use:

Science Integration Activities, 3, The Magnifying Power of Lenses

PROGRAM RESOURCES

From the **Teacher Resource Package** use:

Concept Mapping, pages 9-10.

Transparency Masters, pages 5-6, The Microscope.

Activity Worksheets, page 20, Mini-Lab: How are objects magnified?

Use **Color Transparency** 3, The Microscope.

Use **Laboratory Manual,** page 7, The Microscope.

In Your JOURNAL

Historic scientific discoveries are often taken for granted. Dialogue in student responses should include what it must have been like to work with the equipment and information available at the time.

CONCEPT DEVELOPMENT

▶ Have students discuss the importance of the statement that all cells come from other cells and that everything that an organism is capable of doing is dependent on its cells.

CHECK FOR UNDERSTANDING

Review how the cell theory was developed. Quiz students about the parts of the cell theory.

RETEACH

Have students write an explanation of what each part of the cell theory means.

EXTENSION

For students who have mastered this section, use the **Reinforcement** and **Enrichment** masters or other OPTIONS provided.

In Your JOURNAL

Put yourself in the shoes of a Schleiden, Schwann, or Van Leeuwenhoek. Consider the times in which they lived. **In your Journal**, write a dialogue between the scientist and his son or daughter, as he tries to explain his discovery.

German scientist, Theodor Schwann, after observing many different animal cells, concluded that all animals were made up of cells. Together, they became convinced that all living things were made of cells.

About 15 years later, a German doctor, Rudolph Virchow, hypothesized that new cells don't form on their own. Instead, cells divide to form new cells. This was a startling idea. Remember that at that time, people thought earthworms fell from the sky when it rained. They thought that life came about spontaneously. What Virchow said was that every cell that is or has ever been came from a cell that already existed. The observations and conclusions of Schleiden, Schwann, Virchow, and other scientists became known as the cell theory. The major ideas of the **cell theory** are:

1. All organisms are made up of one or more cells.
2. Cells are the basic units of structure and function in all organisms.
3. All cells come from cells that already exist.

Figure 2-5. Cells divide to produce new cells. Pictured below are four early stages in the development of a frog.

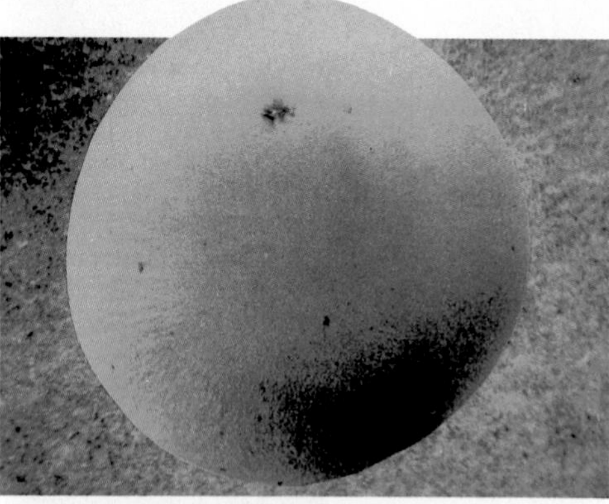

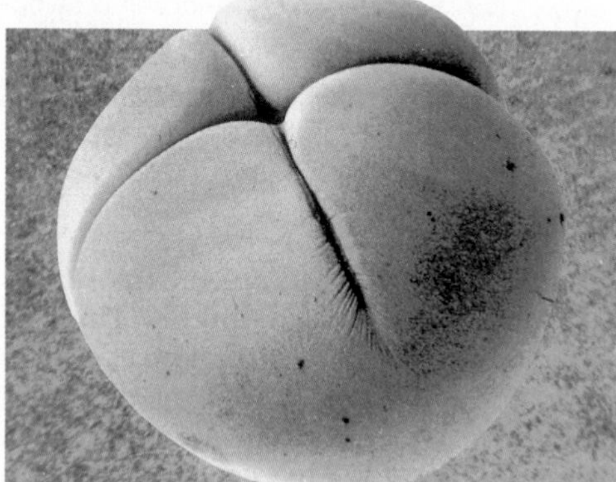

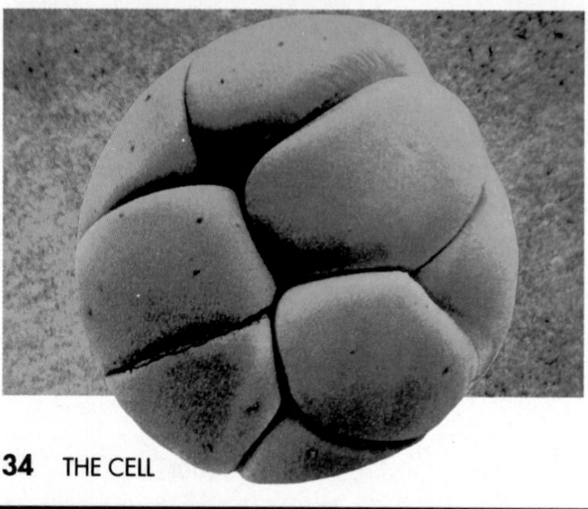

OPTIONS

ENRICHMENT

▶ Have students investigate and report on techniques for electron microscopy.

▶ Have students research the development of the compound light microscope, the phase microscope, and the electron microscope. To assist your students in researching this topic in an interesting way, see PROJECT 1, In Search of Cells, on pages 694 and 695.

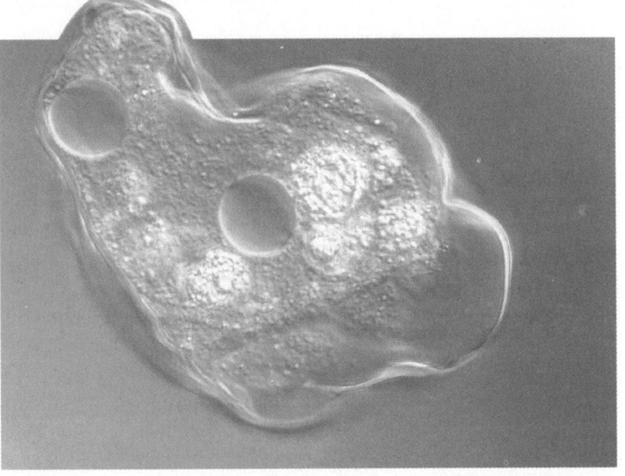

Figure 2-6. One-celled organisms like this amoeba may look simple, but they carry on many complex life processes.

Did You Know?

The most rapidly growing animal is the blue whale. It develops from an ovum (egg) with a mass of less than a milligram into a 26-ton adult in less than 2 years. All of this growth is the result of the division and metabolism of cells.

The cell theory is one of the major theories in science. It is not based on the hypotheses and observations of only one person, but is result of the discoveries of many scientists. Today it serves as the basis for scientists who study the parts of cells, how cells are organized, and how cells and organisms reproduce and change through time.

What are the three parts of the cell theory?

SECTION REVIEW

1. Explain why the invention of the microscope was important in the study of cells.
2. How is a compound light microscope different from an electron microscope?
3. Why is the cell theory important?
4. **Apply:** Why would it be better to be able to see the details of living cells than dead cells?
5. **Connect to Earth Science:** Find out about telescopes. How do they compare with microscopes?

Concept Mapping

Using a network tree concept map, show the differences between compound light microscopes and electron microscopes. If you need help, refer to Concept Mapping in the **Skill Handbook** on pages 688 and 689.

3 CLOSE

▶ Ask questions 1-3 and the **Apply** Question in the Section Review.

▶ Place a large drawing of a compound microscope on the bulletin board. Write the functions of each part on 3" × 5" cards. Have students select a card, name the part, and point out the part on the drawing.

SECTION REVIEW ANSWERS

1. Cells were too small to be seen without the microscope.
2. The light microscope uses lenses to bend beams of light. The electron microscope uses a magnetic field to bend beams of electrons.
3. It is the basis for scientists who study the parts of cells, how they are organized, and how they reproduce and change through time.
4. Apply: Accept all reasonable answers. Dead cells will show cell parts differently than will living cells.
5. Connect to Earth Science: Both have lenses and magnify objects invisible to the unaided eye. Telescopes draw people closer to stars and planets; microscopes to smaller objects.

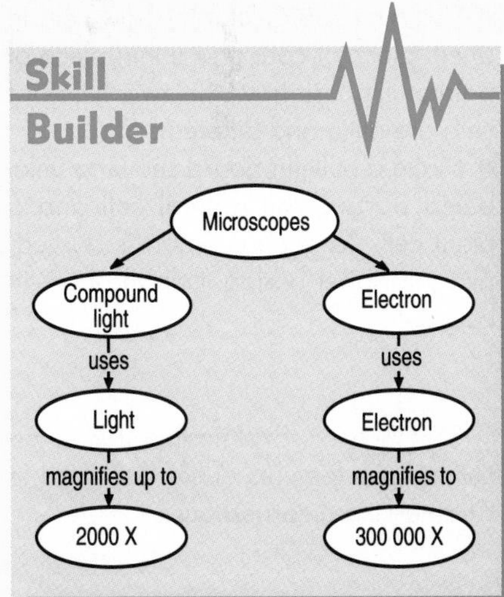

ASSESSMENT—ORAL

▶ **What is the relationship between technology and the development of the cell theory?** *Technology has produced better microscopes that have increased the ability to see smaller parts of cells and have increased our understanding of cells.*

▶ **What traits do you think Schleiden, Schwann, and Virchow must have had that helped them develop the cell theory?** *Curiosity, patience, keen observation skills, accurate record keeping, and any skills that a good scientist has.*

Skillbuilder ASSESSMENT

Performance: Use this Skillbuilder to assess students' Concept Mapping skills using the concept—Microscopes. Have students write a paragraph describing the limitations of each microscope as they understand them.

SECTION 2-2

PREPARATION

SECTION BACKGROUND

▶ The branch of biology that deals with the formation, structure, and function of cells is cytology.

▶ Cells range in size from 2 m long, as in the nerve cells that extend down a giraffe's leg, to 0.2 micrometer in diameter, as in some bacteria.

▶ Not all cells have nuclei. The red blood cells of mammals have nuclei when they are first formed in bone marrow. They lose their nuclei before entering circulation.

PREPLANNING

▶ Obtain *Elodea* and prepared cheek cell slides for Activity 2-1.

▶ Obtain materials for Mini-Lab on page 38.

1 MOTIVATE

▶ Use a microprojector to show cells in a thin piece of cork and a thin piece of potato. Discuss how the two types of cells are alike and different.

▶ Make a bulletin board showing unlabeled parts of an animal cell and a plant cell. As each cell part is studied, place its label beside it on the bulletin board.

TYING TO PREVIOUS KNOWLEDGE: Explain that it is in the cell that the features of life discussed in Chapter 1 are carried out.

VideoDisc

STVS: Bacterial Control of Mosquitoes, Disc 4, Side 1

2-2 Cell Structure

New Science Words

cell membrane
cytoplasm
organelles
nucleus
chromatin
endoplasmic reticulum
ribosomes
Golgi bodies
mitochondria
lysosomes
vacuoles
cell wall
chloroplasts

Objectives

- Diagram a plant cell and an animal cell; identify the parts and the function of each part.
- Describe the importance of the nucleus in the cell.
- Compare and contrast prokaryotic and eukaryotic cells.

An Overview of Cells

In contrast to the dry cork boxes that Hooke saw, living cells are dynamic and have several things in common. They all have a membrane and a gel-like material called cytoplasm inside the membrane. In addition, they all have something that controls the life of the cell. This control center is either a nucleus or nuclear material.

Scientists have found that there are two basic types of cells. Cells that have no membrane around their nuclear material are prokaryotic cells. Bacteria and cells that form pond scum are prokaryotic cells. A eukaryotic cell has a nucleus with a membrane around it. The animal and plant cells in this chapter are all eukaryotic cells.

How does a prokaryotic cell differ from a eukaryotic cell?

Figure 2-7. Pond scum is made up of prokaryotic cells.

OPTIONS

Meeting Different Ability Levels

For Section 2-2, use the following **Teacher Resource Masters** depending upon individual students' needs.

◆ **Study Guide Master** for all students.

● **Reinforcement Master** for students of average and above average ability levels.

▲ **Enrichment Master** for above average students.

Additional Teacher Resource Package masters are listed in the OPTIONS box throughout the section. The additional masters are appropriate for all students.

◆ STUDY GUIDE 10

NAME ________ DATE ________ CLASS ________

STUDY GUIDE — Chapter 2

Cell Structure — Text Pages 36–43

In the blank at the left, write the letter of the word or phrase that best completes the sentence.

c 1. Cells make their own protein on round structures called ____.
a. Golgi bodies b. mitochondria c. ribosomes d. lysosomes

c 2. Mitochondria supply the cell with ____.
a. oxygen b. nutrients c. energy d. enzymes

d 3. Plants can trap light energy in organelles called ____.
a. vacuoles b. cell walls c. mitochondria d. chloroplasts

a 4. Golgi bodies can be compared to ____ in a bakery.
a. loading docks b. power plants c. warehouses d. managers

b 5. Plant cells differ from animal cells because plant cells have ____.
a. cell membranes c. nuclear membranes
b. cell walls d. endoplasmic reticulum

d 6. Prokaryotic cells do not have any ____.
a. cytoplasm c. organelles
b. DNA d. membrane-covered organelles

d 7. Muscle cells have large numbers of ____.
a. chromosomes b. mitochondria c. chloroplasts d. cellulose fibers

a 8. Vacuoles may store ____.
a. waste products b. oxygen c. energy d. pigment

c 9. Plants can change light energy into chemical energy in the form of ____.
a. protein b. fat c. sugar d. fiber

c 10. Chromatin is made up of ____.
a. ATP b. chlorophyll c. DNA d. spores

In the blank at the left, write the letter of the term that best matches the phrase.

d 11. made of double layer of fats — a. cytoplasm
f 12. directs the cell's activities — b. endoplasmic reticulum
a 13. gel-like material inside cells — c. Golgi bodies
b 14. folded membranes that move materials in cells — d. cell membrane
c 15. package and move proteins — e. lysosome
e 16. digests wastes — f. nucleus
g 17. storage areas — g. vacuoles

10

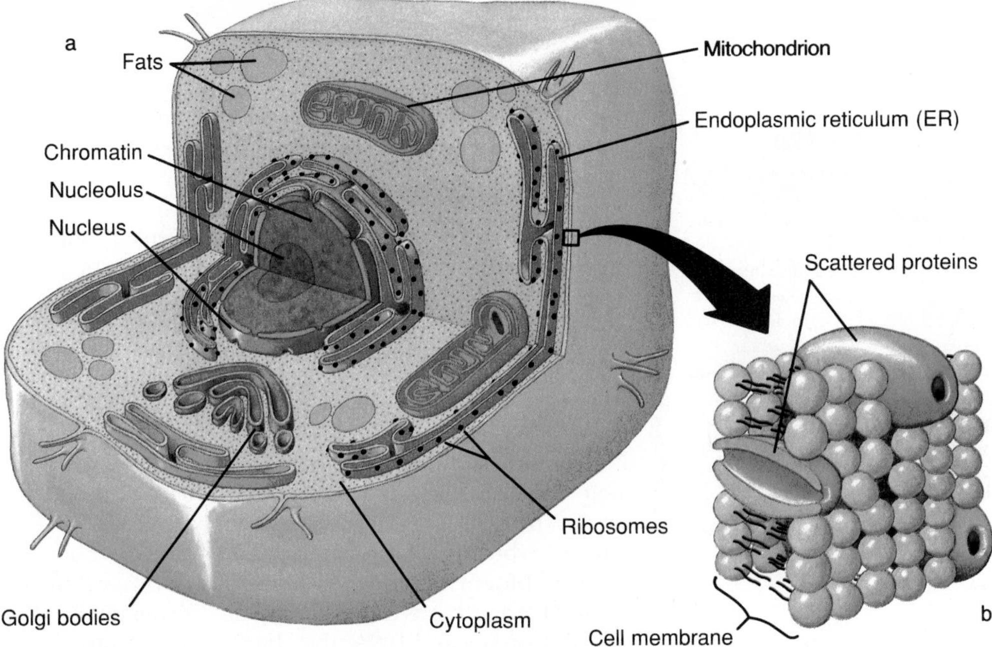

Figure 2-8. The parts of a typical animal cell are shown above (a). The cell membrane is made up of two layers (b).

Cell Membrane

Each cell in your body is constantly active and has a specific job to do. The activities in your cells might be compared to a business that operates 24 hours a day making dozens of different products. It operates inside a building. Only materials that are needed to make specific products are brought into the building. Finished products are then moved out. A cell is similar. It functions within a structure called the cell membrane.

1 The **cell membrane** is a structure that forms the outer boundary of the cell and allows only certain materials to move into and out of the cell. The membrane as shown in Figure 2–8, is flexible. It is made up of a double layer of fats with some proteins scattered throughout. The cell membrane helps to maintain a chemical balance between materials inside and outside the cell. Food and oxygen move into the cell through the membrane. Waste products also leave through the membrane. In Chapter 3, you will learn that different types of substances enter and leave the cell in different ways.

The cell membrane has been described as two layers of fats with large proteins here and there among the fat molecules. In some ways, it is similar to an ocean with icebergs. Icebergs are large pieces of ice that move randomly through the ocean. Explain how the iceberg analogy helps you visualize how the large protein molecules behave in a cell membrane.

2 TEACH

Key Concepts are highlighted.

CONCEPT DEVELOPMENT

Cooperative Learning: Assign Expert Teams to learn the functions of the parts of the cell.

▶ Use three-dimensional models of a plant cell and an animal cell to help students identify cell parts and learn their functions.

TEACHER F.Y.I.

▶ While fewer in number than lipid molecules, it is the scattered proteins that are responsible for most of the activities of the membrane.

▶ Plant cell membranes have no cholesterol.

CROSS CURRICULUM

▶ **Language Arts:** Greek and Latin words are used in naming cell parts. Help students pronounce the names of cell parts in the animal cell diagram. Have students make a list of cell parts, and use a dictionary to find the origins of the words and their meanings.

Answer: The large proteins move about through the fat layers.

● REINFORCEMENT 10

NAME ________ DATE ________ CLASS ________

REINFORCEMENT — Chapter 2

Cell Structure — Text Pages 36–43

Part A

Complete the following table with the correct cell part or function. Refer to your textbook for help.

Cell Part	Function
Cytoplasm	gel-like material inside cells
Cell membrane	**controls movement of materials into and out of cell**
Nucleus	directs cell activities
Endoplasmic reticulum	**moves materials within the cell**
Ribosome	makes protein
Mitochondria	release energy in the cell
Chromatin	**carries code that controls the cell**
Chloroplast	**changes light energy to chemical energy in plants**
Cell wall	**supports and protects the plant cell**
Vacuole	stores water, food, and waste products
Golgi bodies	**package and move proteins**
Lysosome	digests wastes

Part B

Label each cell part.

cell wall
cell membrane
cytoplasm
chromatin
nucleus
ribosomes
endoplasmic reticulum
mitochondria
chloroplasts
vacuole
Golgi body

10

▲ ENRICHMENT 10

NAME ________ DATE ________ CLASS ________

ENRICHMENT — Chapter 2

Cell Structure — Text Pages 36–43

BUILDING A CELL

Make a three-dimensional model of a cell using the following guidelines.

Materials

- glass cake pan or bowl
- light-colored gelatin
- various small items to make cell parts (may be food or nonfood items)

Procedure

1. Decide what items you will use to represent each of the cell parts listed below. Refer to the diagram of a cell in your textbook for ideas for the shapes and sizes of cell parts.
2. Prepare enough gelatin to fill your glass container about half to 3/4 full. Follow the instructions on the gelatin package.
3. When the gelatin becomes thick enough, add your cell part items to the gelatin.
4. Refrigerate your cell to finish setting.
5. Once you have built your cell, organize your information in the table below.

Data and Observations

Cell part	Item used	Function of actual cell part
cell membrane		**controls movement of materials into and out of cell**
nucleus		**directs cell activities**
chromatin	**Answers will vary.**	**controls the makeup of the cell's cytoplasm and cell activities**
endoplasmic reticulum		**moves materials around inside the cell**
ribosomes		**make protein**
mitochondria		**supply the cell with energy**
Golgi bodies		**package and move proteins**
vacuoles		**store water, food, or waste products**

10

REVEALING MISCONCEPTIONS

▶ Students often think that cells are flat. Use or make a three-dimensional model of a typical cell to help them visualize a cell and the arrangement of organelles within the cell. You may want to use the Mini-Lab to support the concept.

MINI-Lab

Materials: 100 mL beaker, unflavored gelatin, water, flashlight.

Answer to question: Particles represent organelles suspended in cytoplasm.

MINI-Lab ASSESSMENT

Performance: To further assess students' understanding of cytoplasm, see USING LAB SKILLS, question 12, p. 50.

CONCEPT DEVELOPMENT

▶ Emphasize that cytoplasm may appear clear in many living cells, but it is not watery. Cytoplasm is a gel, somewhat of the consistency of raw egg white.

▶ Ask students what color cells are. Living tissues are, for the most part, colorless. Photographs of cells have been color enhanced and prepared slides are generally stained. Ask students why stains are used.

▶ **Demonstration:** Use a microprojector to show the nuclei in prepared slides of frog blood and onion root tip.

▶ Explain that just as the cell is separated from the environment by the cell membrane, so are the cell's organelles separated from the cell environment by membranes.

MINI-Lab

What is cytoplasm like?

Cytoplasm of cells is a mixture of proteins, fats, carbohydrates, salts, water, and many other chemicals. The solid particles are suspended in the liquid part of this mixture. The cytoplasm can be gel-like. Make a *model* of cytoplasm. Fill a beaker with 100 mL of water. Add unflavored gelatin and stir. Shine a flashlight through the beaker. **In your Journal,** describe what you see. How does a model cell help you understand what a real one is like?

Cytoplasm

Cytoplasm is the gel-like material inside the cell membrane. Cytoplasm contains a large amount of water and many chemicals and structures that carry out the life processes in the cell. Unlike a gelatin dessert however, cytoplasm constantly moves or streams.

The structures within the cytoplasm of eukaryotic cells are **organelles.** Each one has a specific job. Some organelles break down food. Others move wastes to be expelled from the cell. Still others store materials. Most organelles are surrounded by membranes.

Nucleus

The largest organelle in the cytoplasm of a eukaryotic cell is the **nucleus,** a structure that directs all the activities of the cell. The nucleus is like a manager who directs everyday business for a company. The nucleus contains genetic blueprints for the operations of the cell in the form of long strands called **chromatin.** Chromatin is made up of proteins and DNA, the chemical that controls the activities of the cell. When a cell begins to divide, the strands of chromatin become thick and take on the form of chromosomes, which are easier to see. A structure called a nucleolus is also found in the nucleus.

A nucleus is separated from the cytoplasm by a nuclear membrane. Materials enter and leave through openings in the membrane.

Figure 2-9. A nucleus is surrounded by a membrane, or envelope. DNA in the nucleus controls the activities in the cell.

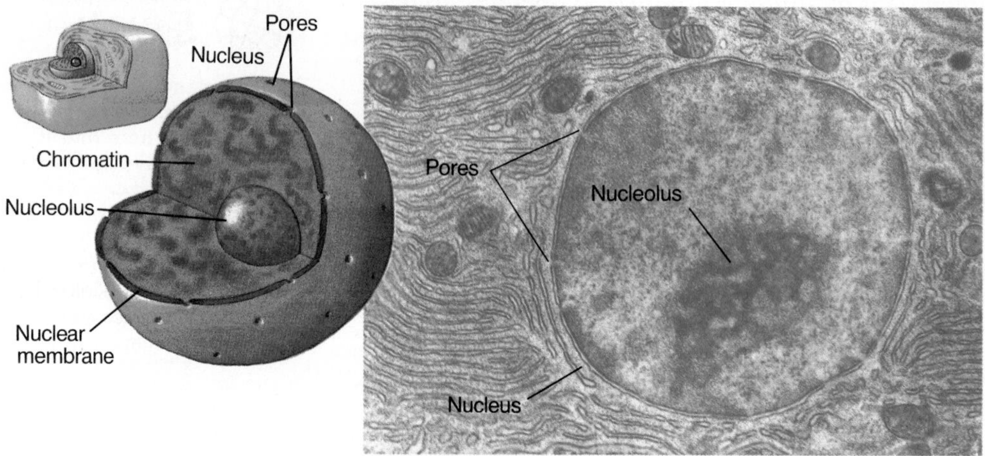

OPTIONS

ASSESSMENT—ORAL

▶ **What would happen if the nucleus of a cell became damaged in some way?** *The cell would no longer function correctly because the nucleus controls all the activities of the cell.*

▶ **Where do you think most of the materials that are needed by the cell are located?** *in the cytoplasm*

▶ **Why do you think cells are stained before they are viewed through a microscope?** *to make the organelles show up more clearly*

▶ **Why would a cell die if it had no ribosomes?** *Ribosomes make proteins needed for life.*

▶ **Why do you think that most ribosomes are located on the ER?** *The ER carries the proteins to where they are needed.*

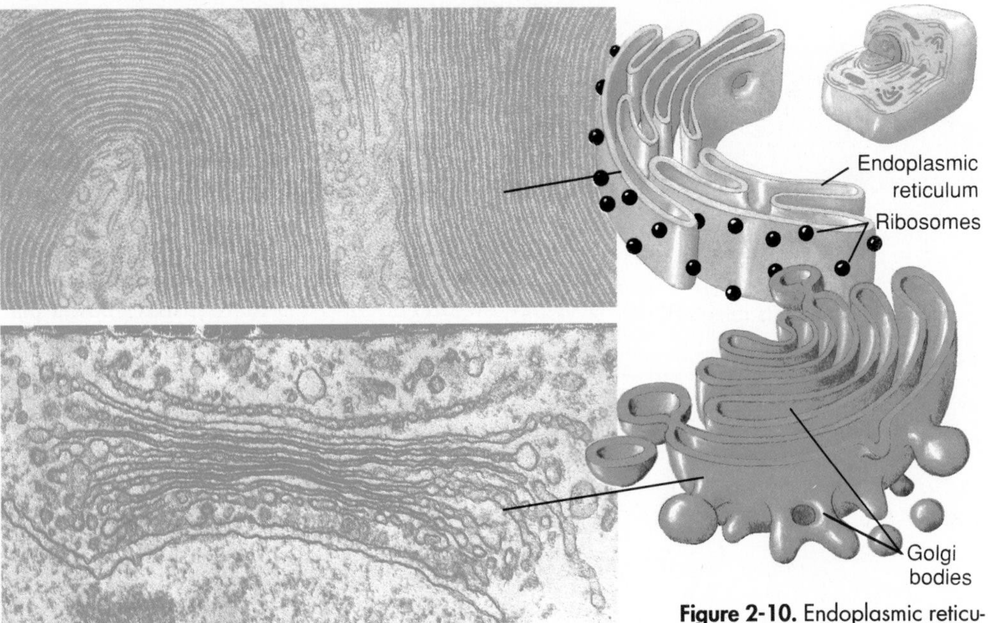

Figure 2-10. Endoplasmic reticulum is smooth or rough. Proteins are made on ribosomes. Golgi bodies may move materials to the outside of the cell.

Organelles in the Cytoplasm

The **endoplasmic reticulum (ER)** is a folded membrane that moves materials around in the cell. The ER extends from the nucleus to the cell membrane and takes up quite a lot of space in some cells. The ER is like a system of conveyor belts in a business that moves material from one place to another.

One chemical that takes part in nearly every cell activity is protein. Proteins are found in cell membranes. Other proteins are needed for chemical reactions that take place in the cytoplasm. Cells make their own proteins on small structures in the cytoplasm called **ribosomes.** Ribosomes receive directions from the nucleus on how to make specific proteins. Some ribosomes are scattered in the cytoplasm. Others are attached to the ER. Ribosome parts are made in the nucleolus.

Where are ribosomes found?

In a business, products are made, packaged, and moved to loading docks to be carried away. Structures called Golgi bodies are stacks of membrane-covered sacs that move proteins to the outside of the cell. **Golgi bodies** are the packaging and secreting organelles of the cell. When something is secreted, it is given off by the cell. 2

CONCEPT DEVELOPMENT

▶ Use an analogy for organelles, pointing out that just as different parts of a machine perform different functions in making a product, so each organelle performs a different function for the cell.

▶ Photomicrographs and separate drawings of cell parts have been provided on these pages to reinforce understanding of each cell part.

TYING TO PREVIOUS KNOWLEDGE: To help students visualize cytoplasm, point out that it is similar to the white of the raw egg in the Find Out activity.

TEACHER F.Y.I.

▶ Endoplasmic reticulum can be likened to a system of conveyor belts. Endoplasm is the inner portion of the cytoplasm, close to the nucleus. ER often appears continuous with the nuclear membrane.

MULTICULTURAL AWARENESS

▶ Have students research the work of Ernest E. Just and report on his contributions to the study of cell parts.

INQUIRY QUESTIONS

▶ **Looking at the photograph and drawing of the Golgi bodies, suggest how substances like proteins are moved out of the cell.** *Sections of the Golgi membrane pinch off in sacks that move to the cell membrane.*

PROGRAM RESOURCES

From the **Teacher Resource Package** use:

Transparency Masters, pages 7-8, Animal and Plant Cells.

Activity Worksheets, page 21, Mini-Lab: What is cytoplasm like?

Cross-Curricular Connections, page 6, The Cell in Disease.

Use **Laboratory Manual,** page 11, Observing Cells.

Use **Color Transparency** 4, Animal and Plant Cells.

CONCEPT DEVELOPMENT

▶ Emphasize that cell activities require energy, most of which is supplied through the functioning of mitochondria.

▶ Explain that mitochondria are composed of two membranes with the inner membrane folded into loops. Folding of the membrane increases the amount of surface area that can be contained within a small space.

PROBLEM SOLVING

Everything that goes on in the body of the organism is the result of the activity of cells.

Think Critically: Lysosomes; because lysosome chemicals can digest cell parts.

STUDENT TEXT QUESTION

▶ Page 40, paragraph 1: **Why would active cells have more mitochondria?** *to supply greater amounts of energy for the cell*

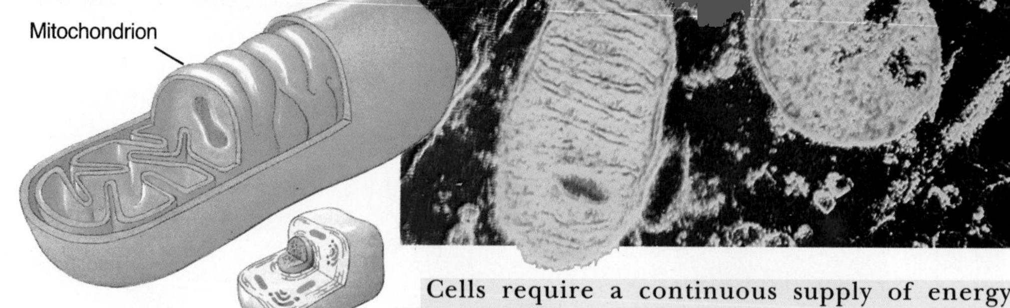

Figure 2-11. Energy is released in mitochondria.

Cells require a continuous supply of energy.
Mitochondria are organelles where food molecules are 3
broken down and energy is released. Just as a power plant supplies energy to a business, mitochondria generate energy for the cell. Some types of cells are more active than others. Muscle cells, which are always undergoing some type of movement, have large numbers of mitochondria. Why would active cells have more mitochondria?

An active cell also constantly produces waste products. In the cytoplasm, organelles called **lysosomes** contain chemicals that digest wastes and worn-out cell parts. When a cell dies, chemicals in the lysosomes act to quickly break down the cell. In a healthy cell, the membrane around the lysosome keeps it from breaking down the cell itself.

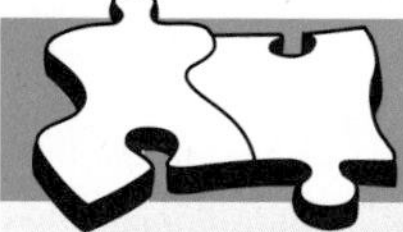

PROBLEM SOLVING

A Tale of a Tail

In September, Mrs. Tallman's class studied the parts of plant and animal cells.

In the spring, while on a field trip, the class captured tiny tadpoles in a local stream. Mrs. Tallman showed the students how to care for the tadpoles in the classroom. Gradually, as the animals grew, it was obvious that their bodies were changing shape. Back and front legs grew out, and the mouth expanded from a small hole to a large opening capable of swallowing large insects. In addition, the tails started to disappear. Mrs. Tallman told the whole class that there would be a bonus question about the tadpoles on the test on animals. For a study hint, she told them to look back at their notes on cells. What do changes in the body parts of the tadpoles have to do with cells?

Think Critically: Answer the bonus question: What cell part is making the tadpole tails disappear and why?

OPTIONS

ASSESSMENT—ORAL

▶ **Why might some muscle cells, which are always undergoing some movement, have larger numbers of mitochondria than other cells that are less active?** *They require more energy.*

▶ **Can cells make energy?** *No. Energy cannot be created or destroyed. It can be changed from one form to another.*

▶ **Where do cells get energy?** *Animals get it from the food they eat. Plants get it from the sun.*

▶ **Why are lysosomes critical to the health of an organism?** *Lysosomes get rid of wastes, dead cells, bacteria, and other materials not needed, to keep the cells of the organism healthy.*

▶ **Name the cell structures you have studied that are involved in making proteins.** *nucleus, nucleolus, ribosomes, chromosomes, and ER*

Many businesses have warehouses for storing products until they are sold. **Vacuoles** are storage areas in cells. They may store water, food, or waste products. Figure 2-12 shows that a vacuole may take up most of the space in a plant cell. In animal cells, vacuoles are small.

Plant and Bacterial Cells

How do plant and animal cells differ?

The major difference between an animal cell and a plant cell is that plant cells have cell walls. The **cell wall** is a rigid structure outside the cell membrane that supports and protects the plant cell. It is made of bundles of tough cellulose fibers and other materials made by the cell.

Plant cells also differ from animal cells because they can make their own food. **Chloroplasts** are organelles in plant cells in which light energy is changed into chemical energy in the form of a sugar called glucose. The chemical in chloroplasts that traps light energy is chlorophyll, the pigment that reflects green light.

Figure 2-12. The parts of a plant cell are shown below. Notice the thick cell wall and the chloroplasts.

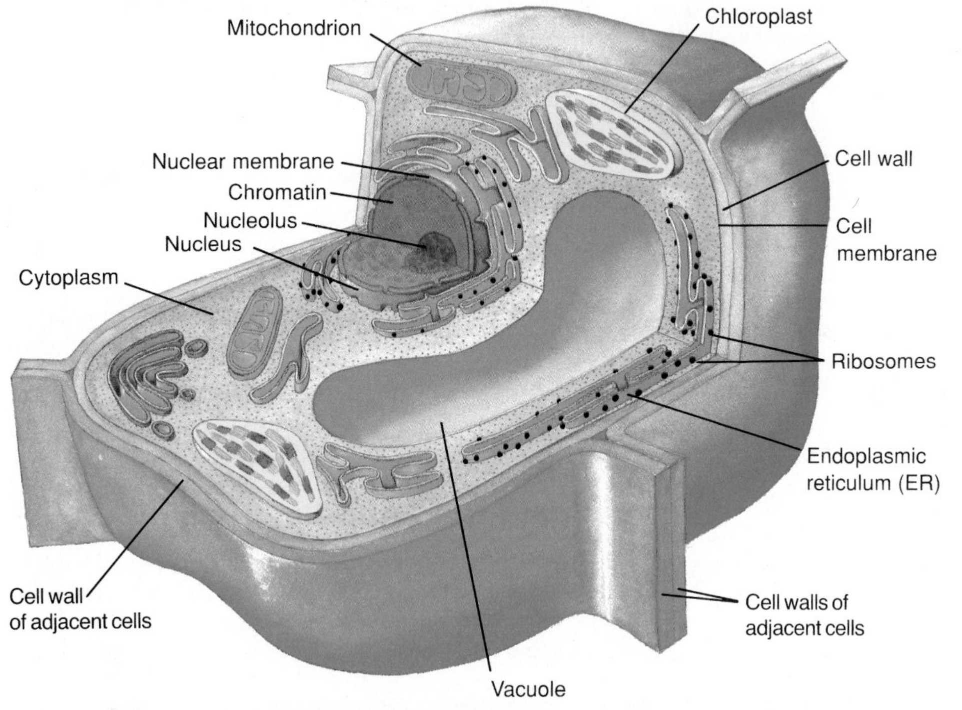

INQUIRY QUESTIONS

▶ **How do vacuoles in a plant cell differ from those in an animal cell?** *Plant cells usually have one large vacuole and animal cells may have several smaller ones or none that are at all noticeable.*

▶ **Where is the cell membrane located in a plant cell?** *inside the cell wall*

REVEALING MISCONCEPTIONS

▶ Some students may think the cell wall and the cell membrane are the same structure. Make a special effort to distinguish between the two structures and reinforce the differences as cells are studied.

▶ Recall the features of life from Chapter 1. Students may think that plant cells carry on photosynthesis, but not respiration. Make sure they understand that plant cells have mitochondria and carry on respiration as well as photosynthesis.

CONCEPT DEVELOPMENT

▶ Use the diagram of the plant cell to point out a vacuole, chloroplasts, and cell walls.

MINI QUIZ

Use the Mini Quiz to check students' recall of chapter content.

1. **The ____ is a structure that forms the outer boundary of the cell and allows only certain material to move in and out.** *cell membrane*
2. **____ are packaging and secreting organelles of the cell.** *Golgi bodies*
3. **____ are organelles where food molecules are broken down and energy is released.** *Mitochondria*

CHECK FOR UNDERSTANDING

Use a poster of a plant cell, animal cell, and bacterial cell to have students identify the parts and describe the functions of each part.

RETEACH

Have each student write an answer to the following question: **Based on what you know about cell structure, which do you think are more closely related, plant and bacterial cells, animal and plant cells, or animal and bacterial cells and why?** *Animal and plant cells because they contain more common parts.*

EXTENSION

For students who have mastered this section, use the **Reinforcement** and **Enrichment** masters or other OPTIONS provided.

3 CLOSE

▶ Ask the questions in the Section Review.

▶ Show the videocassette *The Cell: Structural Unit of Life.*

▶ Divide the class into two teams. Ask questions that have a cell part for the answer. Give points for each correct answer.

SECTION REVIEW ANSWERS

1. Student diagrams should reflect the diagram on page 37.

2. It directs the activities of the cell.

3. Bacterial cells do not have membrane-covered organelles.

4. Apply: The body is made up of eukaryotic cells, which contain organelles.

5. Connect to Chemistry: atom

Skillbuilder

ASSESSMENT

Performance: To further assess students' abilities to Compare and Contrast different types of cells, have them write a statement in their Journals comparing Figures 2-8 and 2-12.

To assess this product, refer to the Performance Task Assessment Lists in **Performance Assessment in Middle School Science**

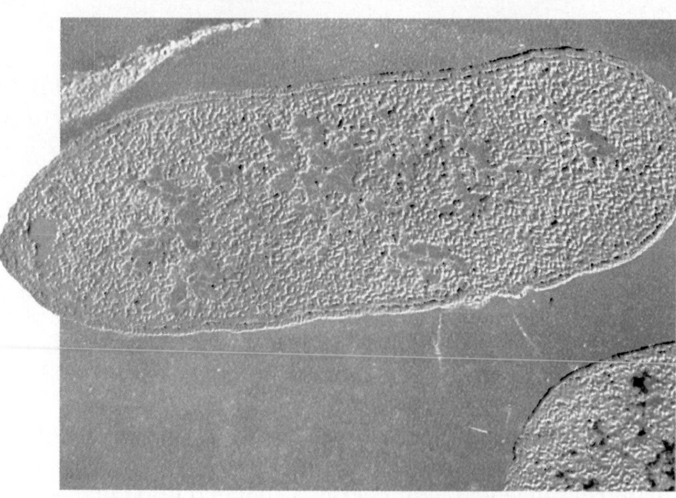

Figure 2-13. A bacterial cell is a one-celled prokaryotic organism. It has no membrane around its nuclear material. This type of bacterial cell lives in human intestines.

EcoTip

Cells need water to maintain health. Turning off the faucet every time you brush your teeth can conserve as much as five gallons of water that will benefit a living organism somewhere.

In contrast to plant and animal cells, bacterial cells are prokaryotic cells. Bacteria don't have membrane-covered organelles. In Figure 2-13, you can see the parts of a bacterial cell. It has a cell wall and cytoplasm, but it has only a single chromosome. There are no nuclei in bacteria, but they do contain ribosomes.

All cells have similar parts. Animal and plant cells have organelles. Bacterial cells have no membrane-covered structures.

SECTION REVIEW

1. Diagram an animal cell and label its organelles.
2. Explain the importance of the cell nucleus in the life of a cell.
3. How are bacterial cells different from other cells?
4. **Apply:** Explain why you could be called a eukaryotic organism.
5. **Connect to Chemistry:** The cell is the smallest unit of living matter. Find out what the smallest unit of nonliving matter is.

Skill Builder

Comparing and Contrasting

Organize information about cell organelles in a table. Use this information to compare and contrast animal cells, plant cells, and bacterial cells. If you need help, refer to Comparing and Contrasting in the **Skill Handbook** on page 683.

Skill Builder

There are several ways to answer this question. The table below is one option.

Feature	Plant cell	Animal cell	Bacterial cell
Prokaryotic/Eukaryotic	Eukaryotic	Eukaryotic	Prokaryotic
Cell membrane	yes	yes	yes
Cell wall	yes	no	yes
Nucleus	yes	yes	no
Chlorophyll	yes	no	Info not avail.
Vacuoles large or small	large	small	—

ACTIVITY 2-1

Comparing Plant and Animal Cells

Problem: *How do plant and animal cells differ?*

Materials   

- microscope
- 1 microscope slide
- 1 coverslip
- forceps
- dropper
- *Elodea* plant
- prepared slide of human cheek cells

Procedure

1. Follow the directions in Appendix A on page 670 for use of low and high power objectives on your microscope and for making a wet-mount slide.
2. With forceps, remove a young leaf from the tip of an *Elodea* plant. Place it on a microscope slide. Add a drop of water and a coverslip.
3. Place the slide on the microscope stage and observe the leaf on low power. Focus on the top layer of cells. Draw what you see.
4. Carefully focus down through the top layer of cells to observe the layers of cells.
5. Focus on one cell. Observe the movement of chloroplasts along the cell membrane.
6. Observe the cell nucleus. It looks like a clear ball. Look at the nucleus on high power.
7. Make a drawing of the *Elodea* cell. Label the cell wall, cytoplasm, chloroplasts, and nucleus. Return to low power and remove the slide.
8. Place a prepared slide of cheek cells on the microscope stage. Locate the cells under low power.
9. Switch to high power and observe the cell nucleus. Draw and label the cell membrane, cytoplasm, and nucleus.

Analyze

1. What parts of the *Elodea* cell were you able to identify?
2. How many cell layers could you see in the *Elodea* leaf?
3. Describe any movement you observed in the *Elodea* leaf.
4. What parts of the cheek cell were easy to identify?

Data and Observations

Sample Data

Cell Part	*Elodea*	Cheek
cytoplasm	X	X
nucleus	X	X
chloroplasts	X	
cell wall	X	
cell membrane	present, not visible	X

Conclude and Apply

5. *Compare and contrast* the shape of the cheek cell and the *Elodea* cell.
6. Identify the cell part that determines the shape of a plant cell. An animal cell.
7. What can you conclude about the differences between plant and animal cells?

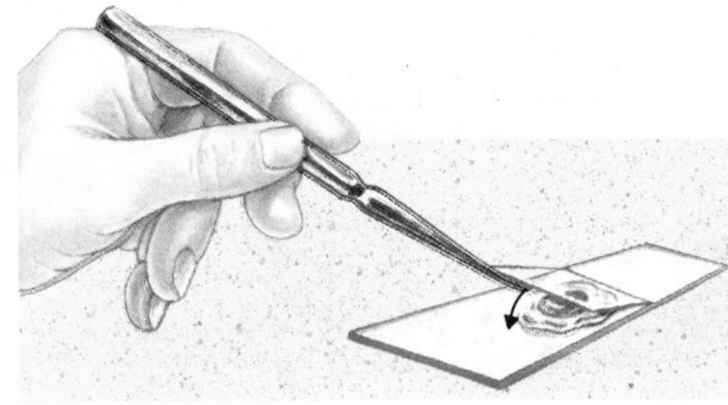

ACTIVITY 2-1

50 minutes

OBJECTIVE: Identify and **compare** the parts of a plant cell and an animal cell.

PROCESS SKILLS applied in this activity:
▶**Observing** in Procedure Steps 3, 5, 6, 8, and 9.
▶**Identifying** in Procedure Steps 5, 6, 7, and 9.
▶**Inferring** in Conclude and Apply Question 7.
▶**Diagramming** in Procedure Steps 7 and 9.

COOPERATIVE LEARNING
Use Science Investigation cooperative groups. Assign one student to obtain and set up the microscope and one student to prepare the Elodea wet mount and obtain the prepared cheek cell slide. Both students should observe the slides and record data.

TEACHING THE ACTIVITY

Alternate Materials: If *Elodea* is not available, plant cell parts can be seen in the epidermis of a lettuce leaf.

Troubleshooting: To see cells and movement of cytoplasm, use only leaves from the tips of *Elodea*. Help students focus up and down slowly so they will see cell layers in the leaves.

▶Demonstrate the technique for preparing a wet mount slide.

▶Caution students to clean slides and coverslips after use.

ANSWERS TO QUESTIONS

1. cell wall, chloroplasts, nucleus, and cytoplasm
2. usually 3 or 4
3. The cytoplasm flowed in a circular path.
4. cell membrane, nucleus, and cytoplasm
5. The *Elodea* cell is oblong and rectangular. The cheek cell is oval.
6. the cell wall; the cell membrane
7. Students will probably state that plant and animal cells differ by the presence or absence of a cell wall. Some may also say that animal cells have no chloroplasts.

PROGRAM RESOURCES

From the **Teacher Resource Package** use:
Activity Worksheets, pages 14-15, Activity 2-1, Comparing Plant and Animal Cells.

Activity
ASSESSMENT

Performance: To further assess students' ability to compare plant and animal cells, have them examine cells from lettuce leaves and other types of animal cells on prepared slides.

SECTION 2-3

PREPARATION

SECTION BACKGROUND

▶ Organization of parts, with each doing specific jobs, is known as division of labor.

PREPLANNING

▶ Obtain hamburger and onion and prepare iodine solution for Activity 2-2.

1 MOTIVATE

▶ Use the microprojector to show slides of different plant and animal tissues. Point out that each type of tissue is made of similar cells as is evidenced by the shape of the cells in the tissues.

2 TEACH

Key Concepts are highlighted.

CONCEPT DEVELOPMENT

▶ Name animal tissues and organs for students (muscle, bone, blood, nerve, kidneys, skin, heart, stomach).

VideoDisc

STVS: Salt-Resistant Crops, Disc 4, Side 2

2-3 Cell Organization

New Science Words

tissues
organ
organ system

Objectives

- ▶ Recognize that cells differ in size, shape, and function.
- ▶ Explain how cells of one-celled organisms differ from cells of many-celled organisms.
- ▶ Explain the differences among tissues, organs, and organ systems.

How Cells Differ

Cells come in a variety of sizes. A single nerve cell in your leg may be a meter in length. A human egg cell, on the other hand, is no bigger than a dot on this *i*. Going a step further, a human red blood cell is about one-tenth of the size of a human egg cell.

The shape of a cell may also tell you something about the job the cell does. The nerve cell in Figure 2-14 with its fine extensions sends impulses through your body. Look at its shape in contrast to the white blood cell, which can change shape. Some cells in plant stems are long and hollow with holes. They transport food and water through the plant. Human red blood cells, on the other hand, are disk-shaped and have to be small and flexible enough to move through tiny blood vessels.

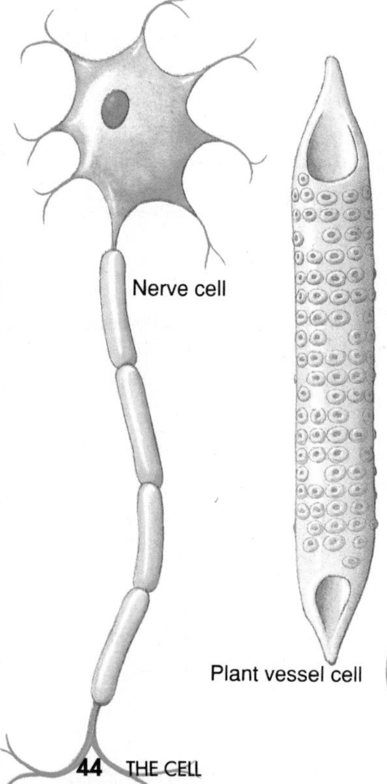

Human egg

Red blood cell

White blood cell

Figure 2-14. Often the shape of a cell tells you something about the job it performs.

From Cell to System

A one-celled organism performs all its life functions by itself. Cells in a many-celled organism, however, do not work alone. Instead, each cell depends on other cells in some way as the organism carries out its functions. This interaction helps the whole organism stay alive.

In Figure 2-15 you can see a single plant cell. In Figure 2-15 you also see a group of the same type of plant cells that together form a tissue on the outside of a plant leaf.

OPTIONS

Meeting Different Ability Levels

For Section 2-3, use the following **Teacher Resource Masters** depending upon individual students' needs.

◆ **Study Guide Master** for all students.

● **Reinforcement Master** for students of average and above average ability levels.

▲ **Enrichment Master** for above average students.

Additional Teacher Resource Package masters are listed in the OPTIONS box throughout the section. The additional masters are appropriate for all students.

◆ STUDY GUIDE 11

NAME ______ DATE ______ CLASS ______

STUDY GUIDE — Chapter 2

Cell Organization — Text Pages 44–45

Check (✔) the statements that agree with the textbook. Rewrite the statements that disagree.

✔ 1. A human egg cell is about ten times bigger than a human red blood cell.
✔ 2. Hollow cells in plant stems transport food and water.
___ 3. Human red blood cells have long extensions. **Human red blood cells are disk-shaped or nerve cells have long extensions.**
___ 4. The cells in a tissue do different sorts of work. **The cells in a tissue do the same sort of work.**
✔ 5. Different types of tissues in an organ work together.
___ 6. An organ system is a group of organisms working together. **An organ system is a group of organs working together.**
✔ 7. Your heart is an organ.
✔ 8. Some cells can change shape.
✔ 9. A whole plant is an organism.
✔ 10. Your heart and blood vessels work together as an organ system.
___ 11. In a many-celled organism, each cell works alone. **In a many-celled organism, each cell depends on other cells.**
___ 12. All cells are about the same size. **Cells come in different sizes.**
✔ 13. An organ is made up of different types of tissues.

Complete the following sentences using appropriate terms from the textbook.

14. Many organ systems working together make up an **organism**.
15. The leaf of a plant is an **organ**.
16. Groups of similar cells that do the same sort of work are called **tissues**.
17. **Nerve cells** have fine extensions.
18. Human red blood cells are **disk**-shaped.
19. The roots, stems, and leaves of a plant are called **organ systems**.
20. A single nerve cell can be as long as **a meter**.
21. Different types of **tissues** work together to make up an organ.
22. The **shape** of a cell tells you something about its function.
23. The leaf functions as an organ to make **food** for the plant.

11

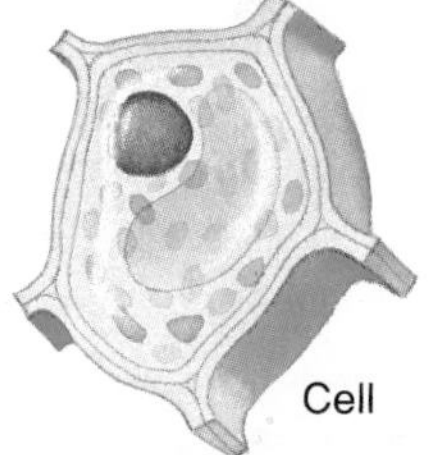

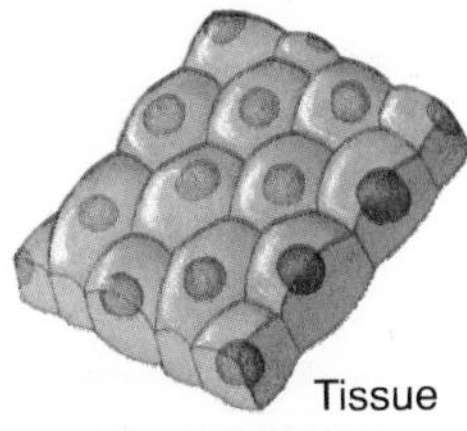

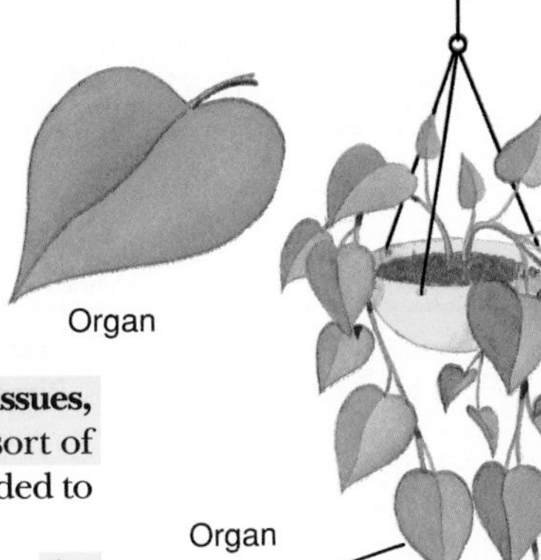

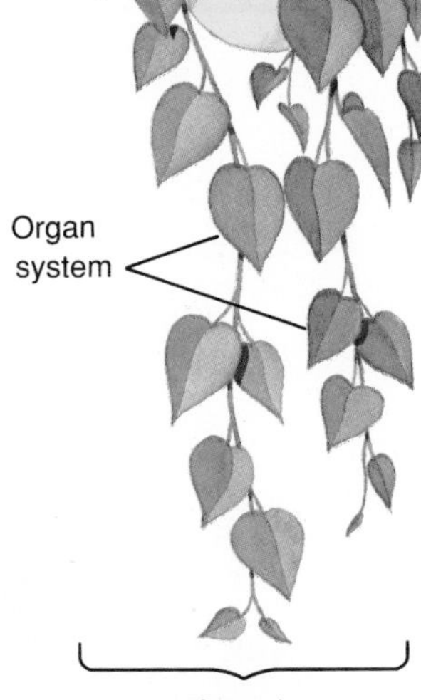

Figure 2-15. In a many-celled organism, different types of tissues are organized into organs and systems.

In many-celled organisms, cells are organized into **tissues,** which are groups of similar cells that do the same sort of work. Each tissue cell carries on all the functions needed to keep it alive. The cells in a tissue look alike.

Different tissues are further organized into organs. An **organ** is a structure made up of different types of tissues that work together to do a particular job. Your heart is an organ made up of muscle, nerve, and blood tissues. Several different tissues make up a plant leaf, an organ of the plant in which food is made.

A group of organs working together to do a certain job is an **organ system**. Your heart and blood vessels make up your cardiovascular system. What plant systems can you think of?

In a many-celled organism, several systems work together. Roots, stems, and leaves in a plant work together to keep the plant alive. Name three systems in your body.

Each cell in a many-celled organism carries on its own life functions. Although cells in an organism may differ in appearance and function, all the cells work together to keep the organism alive.

SECTION REVIEW

1. How is a tissue different from an organ system?
2. Give an example of an organ system in an animal and name the parts that make up the organ system.
3. **Apply:** How are cells of one-celled organisms different from those in many-celled organisms?
4. **Connect to Earth Science:** Systems are not just in living organisms. How does something like a solar system compare with a body of organ systems?

Interpreting Scientific Illustrations

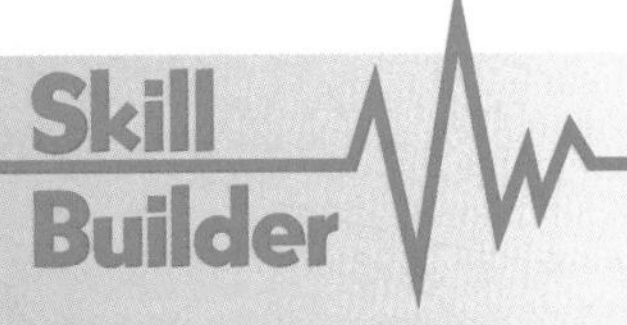

Use the diagram above to explain the levels of organization in living things. If you need help, refer to Interpreting Scientific Illustrations in the **Skill Handbook** on page 693.

● REINFORCEMENT 11

NAME ______ DATE ______ CLASS ______

REINFORCEMENT — Chapter 2

Cell Organization — Text Pages 44-45

Answer the following questions.

1. Why do cells differ in shape? **The shape of a cell is important for the special function of that particular cell type.**
2. Why are red blood cells so small? **So that they can fit through the smallest blood vessels.**
3. Why are some cells in plant stems hollow? **So that they can transport water and nutrients; together, they can act like a drinking straw.**
4. What makes up a many-celled organism? **A many-celled organism is made up of many cells, some of which may be organized into tissues, organs, or systems.**
5. Why do nerve cells have fine extensions? **The extensions are important for the function of sending impulses throughout the body.**
6. How would the cells of one-celled organisms differ from cells of many-celled organisms? **Cells in a one-celled organism work alone, carrying out all the necessary functions for life. Cells become specialized in many-celled organisms.**
7. Why is the heart considered an organ rather than a tissue? **The heart is made up of several different tissue types all working together to do a common job.**

Label the diagram of a plant using the following terms: tissue, organ, organ system, organism.

guard cells around pores on leaf's surface **tissue**
single leaf **organ**
stem **organ system**
all leaves **organ system**
roots **organ system**
all stems, roots, and leaves combined **organism**

11

▲ ENRICHMENT 11

NAME ______ DATE ______ CLASS ______

ENRICHMENT — Chapter 2

Cell Organization — Text Pages 44-45

The cells in complex, many-celled organisms do different jobs to keep the organisms alive. Cells become specialized. That means that they perform very specific jobs. For instance, in your body, a nerve cell performs a job that is very different from a skin cell. As animal or plant organisms become more complex, and as they develop, the number of different types of cells in the organisms increases. In complex organisms, groups of similar cells working together are called tissues.

A. Animal Tissues

Use other references to define the function of the following types of tissues and give their locations.

a. epithelial tissue **main function is protection; some form glands; covers all body surfaces and lines all cavities**
b. connective tissue **connects and holds body parts together; supports and protects organs; fat, bone, cartilage, blood are examples**
c. muscle tissue **moves body parts; attached to bones and lines organs**
d. nervous tissue **receives and sends messages; brain, spinal cord**

B. Plant Tissues

Plant tissue	Function	Where found?
Roots	1. anchor plants 2. absorb and transport nutrients and water	most underground
Leaves	1. trap light 2. make food for plant	attached to the stem of plants
Stem	1. support plant 2. transport water and food 3. may store food	most above ground

11

CHECK FOR UNDERSTANDING

Ask questions 1-2 and the **Apply** Question in the Section Review.

RETEACH

Ask students what organs work with the human nervous system to help a body respond to its environment.

EXTENSION

For students who have mastered this section, use the **Reinforcement** and **Enrichment** masters or other OPTIONS.

3 CLOSE

SECTION REVIEW ANSWERS

1. A tissue is made up of cells, and a system is made up of organs.
2. The circulatory system; heart, blood vessels, blood.
3. Apply: Cells of one-celled organisms work alone. In a many-celled organism, different cell types are interdependent.
4. Connect to Earth Science: A solar system is a star with a group of bodies bound to it by gravity; the sun and planets. They are interdependent. In organisms, systems are also interdependent.

Students should be able to relate that the level of organization goes from the cell on the left to the whole plant on the right, which represents several systems working together.

Skillbuilder

ASSESSMENT

Performance: Use this Skillbuilder to assess students' abilities to interpret how water might move through the vessel cell in Figure 2-14.

PROGRAM RESOURCES

From the **Teacher Resource Package** use for pages 46 and 47:

Critical Thinking/Problem Solving, page 6, Overcoming Rejection in Organ Transplants.

Science and Society, page 6, Selling Cells.

SECTION 2-4

PREPARATION

SECTION BACKGROUND

▶ Organ transplants are common but not always successful due to the action of the immune system that fights foreign substances.

1 MOTIVATE

▶ If possible, acquire a cow heart from a local butcher. Point out the many vessels that must be reconnected in a transplant.

2 TEACH

Key Concepts are highlighted.

CONCEPT DEVELOPMENT

▶ **Why are antirejection drugs important in organ transplants?** *They depress the body's immune system.* **Why is this dangerous for the patient?** *The body might become infected by bacteria or a virus.*

VideoDisc

STVS: Diagnosing Disease with Glowing Cells, Disc 7, Side 1

2-4 Organ Transplants

New Science Words

rejection

Objectives

▶ **Relate the importance of science in your life.**
▶ **Discuss problems in organ transplants.**

Spare Parts for Broken Hearts

Have you ever considered the thousands of things that take place in your body? Think of your body as an automobile. Mechanics are always busy replacing this belt, or that valve. The brakes, tires, battery, and spark plugs are all parts that often need to be changed. When an accident occurs, fenders, doors, or a hood may need to be replaced. Auto mechanics go to the local junkyard where usable parts can be recycled from cars that are no longer driveable. Who would have thought that a similar idea could someday be used to save lives?

Organ transplants have become quite common in today's medical world. You may have heard of the kidney transplant. This surgery is fairly common. Hearts and lungs have also been transplanted separately and together.

Unfortunately, organ transplants are not always successful. Just as patients who receive blood transfusions must be matched with the donor's blood type, organs also must match. The human immune system, the network in our body that fights infections, is an eternal watchdog that attacks foreign invaders. This is to your advantage when the invader is a virus or bacteria. However, **rejection** is a process whereby the immune system attacks the transplanted organ because it is foreign to the body. Care must be taken to match the donor organs with the tissues of the person receiving the organ.

OPTIONS

Meeting Different Ability Levels

For Section 2-4, use the following **Teacher Resource Masters** depending upon individual students' needs.

◆ **Study Guide Master** for all students.

● **Reinforcement Master** for students of average and above average ability levels.

▲ **Enrichment Master** for above average students.

Additional Teacher Resource Package masters are listed in the OPTIONS box throughout the section. The additional masters are appropriate for all students.

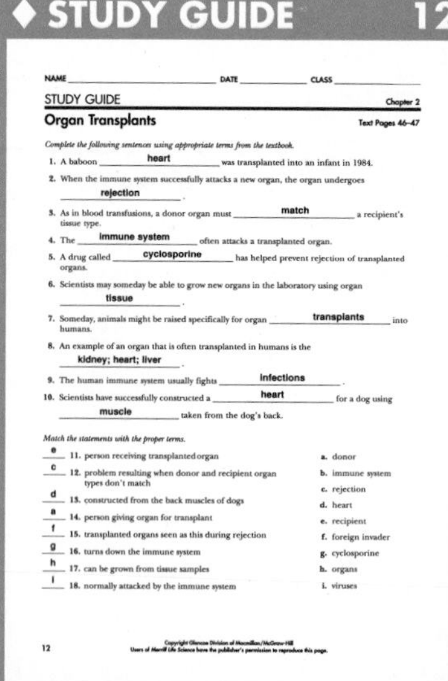
◆ STUDY GUIDE 12

NAME ________ DATE ________ CLASS ________

STUDY GUIDE Chapter 2

Organ Transplants Text Pages 46–47

Complete the following sentences using appropriate terms from the textbook.

1. A baboon **heart** was transplanted into an infant in 1984.
2. When the immune system successfully attacks a new organ, the organ undergoes **rejection**.
3. As in blood transfusions, a donor organ must **match** a recipient's tissue type.
4. The **immune system** often attacks a transplanted organ.
5. A drug called **cyclosporine** has helped prevent rejection of transplanted organs.
6. Scientists may someday be able to grow new organs in the laboratory using organ **tissue**.
7. Someday, animals might be raised specifically for organ **transplants** into humans.
8. An example of an organ that is often transplanted in humans is the **kidney; heart; liver**.
9. The human immune system usually fights **infections**.
10. Scientists have successfully constructed a **heart** for a dog using **muscle** taken from the dog's back.

Match the statements with the proper terms.

Answer	Statement	Term
e	11. person receiving transplanted organ	a. donor
c	12. problem resulting when donor and recipient organ types don't match	b. immune system
d	13. constructed from the back muscles of dogs	c. rejection
a	14. person giving organ for transplant	d. heart
f	15. transplanted organs seen as this during rejection	e. recipient
g	16. turns down the immune system	f. foreign invader
h	17. can be grown from tissue samples	g. cyclosporine
i	18. normally attacked by the immune system	h. organs
		i. viruses

12

Drugs are given to turn down the body's immune system and keep rejection from happening. Antirejection drugs have greatly increased the success of transplant surgeries by keeping the immune system from attacking the new organ. However, the patient has to take antirejection drugs every day of his or her life.

Today, hearts, kidneys, livers, and other organs are transplanted. Doctors believe that someday, with more research, certain animals could be raised specifically to supply organs for human transplants. Yet other exciting ideas are being considered. For example, can one part of the body be used to replace another? Scientists have already constructed a heart for a dog using the muscle taken from the dog's back.

Where will technology end? Someday scientists may be able to grow new organs in the laboratory using organ tissue. We may each have our own spare organs on the shelf ready to fix our broken parts!

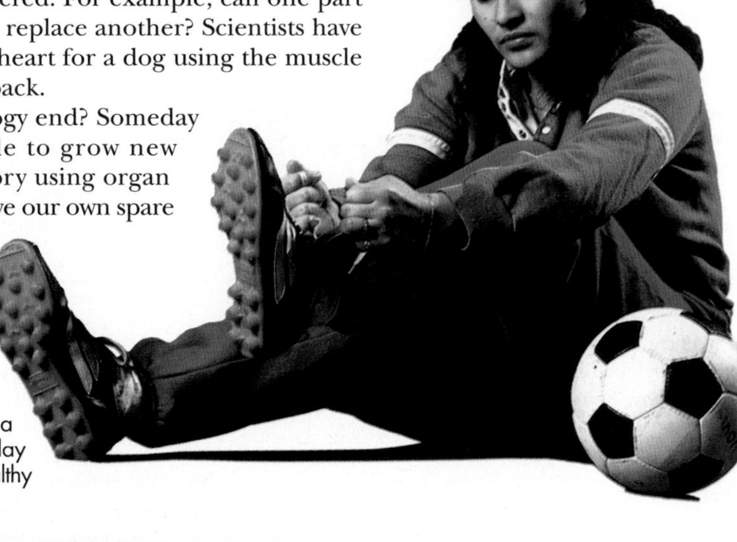

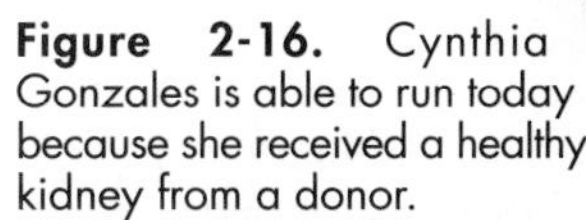

Figure 2-16. Cynthia Gonzales is able to run today because she received a healthy kidney from a donor.

SECTION REVIEW

1. Why have organ transplants become more successful?
2. Describe one problem with transplanting organs to humans.
3. **Connect to Physics:** Using what you know about cells, explain why it is beneficial for an organ that is being transplanted to be placed in cold storage until it is used.

You Decide!

Do you think that organs should be donated? Because there is such great demand for organs, but so few donors, some doctors are suggesting that donors be paid for their organs. Such an idea would make more organs available, but is it right to sell human body organs?

● REINFORCEMENT 12

NAME ______ DATE ______ CLASS ______

REINFORCEMENT — Chapter 2

Organ Transplants — Text Pages 46–47

Answer the following questions.

1. Why would someone need a kidney transplant? **If kidneys were diseased, injured, or otherwise failing, a person would need a kidney transplant.**
2. What problems can result from organ transplantation? **The body often rejects the donated organ.**
3. What precautions do doctors take to minimize complications in organ transplants? **Prior to organ transplants, donor and recipient tissue types are matched.**
4. Why does the body sometimes reject a donated organ? **The immune system sees the donated organ as a foreign invader and attacks the new tissue.**
5. Why is the drug, cyclosporine, given to organ recipients? **to prevent rejection**
   How does it work? **Cyclosporine turns down the recipient's immune system.**
6. What are two ways in which scientists might some day be able to make more organs available for transplantation? **by raising animals to serve as donors, and by growing organs from organ tissues in the laboratory**
7. What organs are typically transplanted into humans? **heart, kidneys, liver**

Indicate whether these procedures are routine, experimental, or untested with a check (✓).

Procedures	Now Done Routinely	Experimental Procedures	Untested Procedures
blood transfusions	✓		
growing new organs from organ tissue			✓
constructing new organs from tissue of same animal			✓
transplanting human kidneys into humans	✓		
transplanting organs from an animal into a human		✓	
use of cyclosporine	✓		
transplanting human livers into humans	✓		
raising animals specifically for human transplants			✓
matching donor and recipient organs	✓		
transplanting human hearts into humans	✓		

12

▲ ENRICHMENT 12

NAME ______ DATE ______ CLASS ______

ENRICHMENT — Chapter 2

Organ Transplants — Text Pages 46–47

DONOR BANKS

Banks aren't always places where money is kept. Special banks called donor banks, find, process, and transport organs to waiting recipients. Many donor banks only store information about patients needing transplants. Most donations of organs and tissues come from people who have recently died. Organs such as hearts and livers must be transplanted within hours of the donor's death. Corneas and kidneys can be kept for a few days. Bones and skin can be frozen and stored for years. Blood can be kept refrigerated for up to 40 days or it can be frozen for longer.

Live donors can give tissues and organs too. Because the body replaces lost blood, most healthy adults can donate a pint of blood every eight weeks. A kidney can be donated because it is possible to live with only one kidney. Bone marrow can also be donated by healthy adults because it is replaced over time.

Using the list below, write down the information you would collect from future donors for your donor bank and explain why it would be important. Then select information you would need to obtain from potential recipients. Use additional references when needed.

blood type	tissue type	which organ	medical records
next of kin	consent	cause of death	age

Donor Bank Data:

blood type: must match recipient's to prevent rejection
tissue type: must match recipient's to prevent rejection
which organ: need to know which organs will be available to recipients
next of kin: must sign a consent form in case there is any question about using tissue if the person is considered "brain dead"
consent: permission must be given to use organs
cause of death: some conditions may make it impossible to use donor organs
medical records: sometimes donor organs cannot be used because of diseases
age: important for size matching of organ—mainly for hearts and livers

Data from Recipient:

blood type: see above
tissue type: see above
which organ: see above
consent: the recipient must understand the risks
medical records: some conditions may make transplantation unwise
age: see above

12

CHECK FOR UNDERSTANDING

Use the You Decide! to check for understanding.

RETEACH

Have students write a one-paragraph recyclable parts analogy of organ transplants. The immune system and anti-rejection drugs must be portrayed in the analogy.

EXTENSION

For those students who have mastered this section, use the **Reinforcement** and **Enrichment** masters or other OPTIONS provided.

3 CLOSE

FLEX Your Brain

Use the Flex Your Brain activity to have students explore ORGAN TRANSPLANTS.

Flex Your Brain

ASSESSMENT

Portfolio: Use the Flex Your Brain activity to reinforce critical thinking and problem solving skills.

SECTION REVIEW ANSWERS

1. Antirejection drugs turn down the immune system.
2. Organs may not always be accepted due to the action of the immune system.
3. Connect to Physics: to slow down the activities of the cells in the organ

YOU DECIDE!

Accept all reasonable discussion. If students bring up artificial hearts or other artificial organs, mention that there has been a greater success rate in transplanting real hearts.

PROGRAM RESOURCES

From the **Teacher Resource Package** use:

Activity Worksheets, page 5, Flex Your Brain.

ACTIVITY 2-2

OBJECTIVE: Design and carry out an experiment to show the parts that make up an organism.
Time: One class period to brainstorm and complete the activity and one-fourth of a class period to summarize results.

PROCESS SKILLS applied in this activity are **observing, inferring, comparing and contrasting** and **predicting.**

PREPARATION

- Small green onions can be purchased in bunches at the grocery store.
- If the piece of onion bulb is placed in water, the onion skin tissue will be easier to place on the slide.

Cooperative Learning: Divide the class into Science Investigation Teams.

THINKING CRITICALLY

Student answers may include that the parts of an organism are cells, tissues, organs, and organ systems. Students may suggest that a one-celled organism performs all of its life functions by itself but the cells of many-celled organisms do not work alone.

TEACHING THE ACTIVITY

*Refer to the **Activity Worksheets** for additional information and teaching strategies.*

- Make sure that students record and analyze their observations accurately. Encourage them to draw what they see.
- Forceps can be used to remove thin layers from the roots, bulb, and leaves. You may wish to use a scalpel to cut a root in half for each student group.

SUMMING UP/SHARING RESULTS

Students will probably divide the onion into roots, bulbs, and leaves. These parts are organ systems. The thick layers observed are tissues that are made of cells. The onion plant has organs and each organ is composed of tissues made of cells that have the same function. The root cells absorb water, the stem cells (bulb) transport water and food, and the leaf cells make food for the onion plant.

GOING FURTHER!

Students will probably predict that similar structures would be found in any plant. They could check their predictions by making wet mount slides of tissues found in another plant and observing them under the microscope.

ACTIVITY 2-2 DESIGNING AN EXPERIMENT

Observing the Parts of an Organism

How are the parts of your body different? How are they alike? How are the parts of plants different? In Activity 2-1, you learned how plant and animal cells differ. Are all the cells in an organism alike?

Getting Started

In this activity, you will *observe* an organism, determine its parts, and *compare and contrast* the cells in each of the parts.

Thinking Critically

List the parts of an organism and explain how they work together. How are many-celled organisms different one-celled organisms?

Materials   

Your cooperative group will use:

- onion plant with roots and leaves
- microscope
- microscope slides (3)
- coverslips (3)
- 250-mL beaker with water
- dropper
- forceps
- scissors

Try It!

1. *Observe* the onion plant. Draw and label what you observe.
2. In your group, decide how the onion plant can be divided into its parts.
3. Use the scissors to cut the onion plant into parts.
4. Prepare a wet mount slide of a thin layer of each of the parts. *Observe* each slide under low power and then turn to high power. *Describe* what you see.

Summing Up/Sharing Results

- What parts did you divide the onion into and what are they called?
- What were the thin layers you observed under the microscope?
- What made up the thin layers?
- What are the parts of an onion plant and how do they work together?

Going Further!

Predict whether similar structures would be found in any plant. How would you check your predictions?

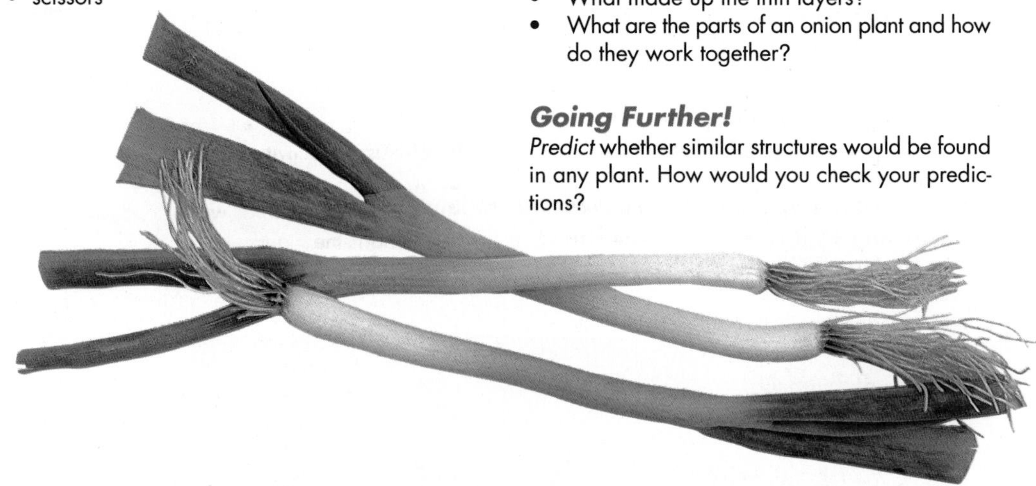

PROGRAM RESOURCES

From the **Teacher Resource Package** use:
Activity Worksheets, pages 16-17, Activity 2-2, Observing the Parts of an Organism.

Activity ASSESSMENT

Performance: To further assess students' understanding of polygenic inheritance, see USING LAB SKILLS, Question 12, on page 50.

CHAPTER

REVIEW

SUMMARY

2-1: Cells: The Units of Life

1. Janssen made the first compound microscope in 1590. Leeuwenhoek used fine lenses to make accurate observations. Hooke, Schleiden, and Schwann all drew their conclusions about cells with the help of the microscope.
2. Compound light microscopes use light and lenses to make images. Electron microscopes bend beams of electrons in a magnetic field.
3. According to the cell theory, the cell is the basic unit of life. Organisms are made of one or more cells and all cells come from other cells.

2-2: Cell Structure

1. There are differences among animal, plant, and bacterial cells. Plant cells have cell walls and chloroplasts. Bacteria have no membrane-covered organelles.
2. Cell functions are performed by organelles under control of DNA in the nucleus.
3. Prokaryotic cells do not have membrane-covered organelles; eukaryotic cells have membrane-bound organelles and a formed nucleus.

2-3: Cell Organization

1. Cells differ in size and shape. Shape is related to a cell's function.
2. One-celled organisms carry on all life activities within the single cell. Many-celled organisms may have tissues that perform separate functions.
3. Most many-celled organisms are organized into tissues, organs, and organ systems that perform specific jobs to keep an organism alive.

2-4: Science and Society: Organ Transplants

1. Through organ transplants, science works to provide ways to save lives.
2. Rejection and availability of matching organs are problems connected with organ transplants.

KEY SCIENCE WORDS

a. **cell membrane**
b. **cell theory**
c. **cell wall**
d. **chloroplasts**
e. **chromatin**
f. **compound light microscope**
g. **cytoplasm**
h. **electron microscope**
i. **endoplasmic reticulum**
j. **Golgi bodies**
k. **lysosomes**
l. **mitochondria**
m. **nucleus**
n. **organ**
o. **organ system**
p. **organelles**
q. **rejection**
r. **ribosomes**
s. **tissues**
t. **vacuoles**

UNDERSTANDING VOCABULARY

Match each phrase with the correct term from the list of Key Science Words.

1. hereditary material in the nucleus
2. directs cell activities
3. where energy is generated for the cell
4. store food, water, or waste products
5. supports the plant cell
6. contain all the same kind of cell
7. where proteins are made
8. gel-like substance inside cell membrane
9. trap light energy in plant cells
10. package and secrete substances

OPTIONS

ASSESSMENT

To assess student understanding of material in this chapter, use the resources listed in the Program Resources to the right.

COOPERATIVE LEARNING

Consider using cooperative learning strategies in the THINK AND WRITE CRITICALLY, APPLY, and MORE SKILL BUILDERS sections of the Chapter Review.

PROGRAM RESOURCES

From the **Teacher Resource Package** use:
Chapter Review, pages 7-8.
Chapter and Unit Tests, pages 9-12, Chapter Test.

CHAPTER

REVIEW

SUMMARY

Have students read the summary statements to review the major concepts of the chapter.

UNDERSTANDING VOCABULARY

1. e
2. m
3. l
4. t
5. c
6. s
7. r
8. g
9. d
10. j

ASSESSMENT

Portfolio

Encourage students to place in their portfolios one or two items of what they consider to be their best work. For each item, ask students to explain why that item was chosen and what they learned from it. Items might be selected from the following.

- Cross curriculum notes from discussion, p.33
- In Your Journal dialogue, p. 34
- Activity 2-2, data, answers, predictions, p. 48

Performance

Additional performance assessments may be found in *Performance Assessment* and *Science Integration Activities* that accompany **Merrill Life Science.** Performance Task Assessment Lists and rubrics for evaluating these activities and other products generated throughout the chapter can be found in Glencoe's *Performance Assessment in Middle School Science.*

CHAPTER

REVIEW

CHECKING CONCEPTS

1. a
2. c
3. a
4. c
5. d
6. b
7. b
8. c
9. a
10. a

USING LAB SKILLS

ASSESSMENT

Use these alternate lab exercises to assess students' understanding of skills used in this chapter.

11. The word appears magnified because the water drop is curved as if it were a lens.

12. Students' models should be logical, no matter what materials they choose.

THINK AND WRITE CRITICALLY

13. Microscopes enabled people to see cells for the first time. Improved lenses in compound microscopes enabled scientists to see individual parts of cells more clearly. Stereomicroscopes allow for three-dimensional viewing; smaller structures are seen and greater magnification is possible with electron microscopes.

14. Possible answers may include that lysosomes might only respond to old or worn out cell parts and substances foreign to the cell.

15. Accept all reasonable answers. Possible answers might include the observation that if something is made up of cells, it is an organism and that cells have to come from cells that already exist.

16. Accept all reasonable answers. Suggestions might be age of the donor organ, age of the two people who could receive the organ, overall health of these two people.

CHAPTER

REVIEW

CHECKING CONCEPTS

Choose the word or phrase that completes the sentence.

1. A microscope that uses lenses and objectives to magnify is the ______.
   - **a.** compound light microscope
   - **b.** scanning electron microscope
   - **c.** transmission electron microscope
   - **d.** atomic force microscope
2. A microscope that magnifies parts inside a cell 300 000 times or more is the ______.
   - **a.** compound light microscope
   - **b.** stereoscopic microscope
   - **c.** transmission electron microscope
   - **d.** scanning electron microscope
3. The scientist who gave the name *cells* to structures he viewed was ______.
   - **a.** Hooke **c.** Schleiden
   - **b.** Schwann **d.** Virchow
4. A ______ is an organelle that can destroy old cell parts.
   - **a.** chloroplast **c.** lysosome
   - **b.** vacuole **d.** cell wall
5. According to ______, cells are the basic units of function and structure of all living things and come from cells that already exist.
   - **a.** Janssen
   - **b.** Hooke
   - **c.** Leeuwenhoek
   - **d.** Schleiden, Schwann, Virchow, and other scientists
6. A structure that allows only certain things to pass in and out of the cell is the ______.
   - **a.** cytoplasm **c.** cell wall
   - **b.** cell membrane **d.** nuclear envelope
7. Structures in the cytoplasm of the eukaryotic cell are called ______.
   - **a.** organs **c.** organ system
   - **b.** organelles **d.** tissues
8. Materials can move around inside the cell through the folded membranes of ______.
   - **a.** chromatin **c.** endoplasmic reticulum
   - **b.** cytoplasm **d.** Golgi bodies
9. A bacterial cell has ______.
   - **a.** a cell wall **c.** mitochondria
   - **b.** lysosomes **d.** a nucleus
10. Groups of different tissues form an ______.
    - **a.** organ **c.** organ system
    - **b.** organelle **d.** organism

USING LAB SKILLS

11. In the Mini-Lab on page 31, you saw how well objects are magnified with ordinary materials. Place a sheet of waxed paper over a word in your book. Then place the drop of water on the waxed paper over the word. Explain what you see. Is the drop concave or convex?
12. In the Mini-Lab on page 38, you made a model of cytoplasm. Use materials that resemble a cell and cell organelles to make a model of either a plant cell or an animal cell. Make a key to the cell parts to explain your model.

THINK AND WRITE CRITICALLY

Answer the following questions in your Journal using complete sentences.

13. How has the development of different microscopes helped scientists study cells?
14. Suggest a reason why a lysosome can exist inside a living cell without destroying the cell.
15. Explain why the cell theory is important.
16. What are some things to consider when an organ becomes available and a doctor has to decide between two patients?

APPLY

17. What type of microscope would be best to use to look at a piece of moldy bread? Give reasons to support your choice.
18. What would happen to a plant cell that suddenly lost its chloroplasts?
19. What would happen to an animal cell if it didn't have ribosomes?
20. Choose two organelles of a plant cell and explain how they work together to help perform a life function.
21. How would you decide whether an unknown cell was an animal cell, a plant cell, or a bacterial cell?

MORE SKILL BUILDERS

If you need help, refer to the Skill Handbook.

1. **Interpreting Scientific Illustrations:** Use the illustrations of cells on page 44 to describe how the shape of each cell may be related to its function.
2. **Sequencing:** Sequence the major events and the scientists involved in the formation of the cell theory.
3. **Observing and Inferring:** Infer the effects of a damaged cell membrane on a cell.
4. **Making and Using Tables:** Make a table that lists the names of the cell structure(s) that are involved in each of the functions below. Structures may be listed in more than one column.

Function	Structures
Protection	Answers
Movement of materials	provided in
Breaking down substances	the teacher
Making substances	margin.
Storing substances	

5. **Concept Mapping:** Use the list of events below to make a series of events concept map to show, in correct order, the cell structures involved in each step of making a protein.
   - packages proteins for movement outside of the cell
   - gives instructions to make proteins
   - allows packaged proteins to leave the cell
   - receives instructions and is the place where proteins are made

PROJECTS

1. Using the directions for Activity 2-1 (page 43), observe cells in a flower petal or tomato skin. Write a report based on your observations. Illustrate your report.
2. As a class project, turn the classroom into a model of a plant cell so that students and desks are within the "cell." Begin to consider what activity is like inside a real cell.

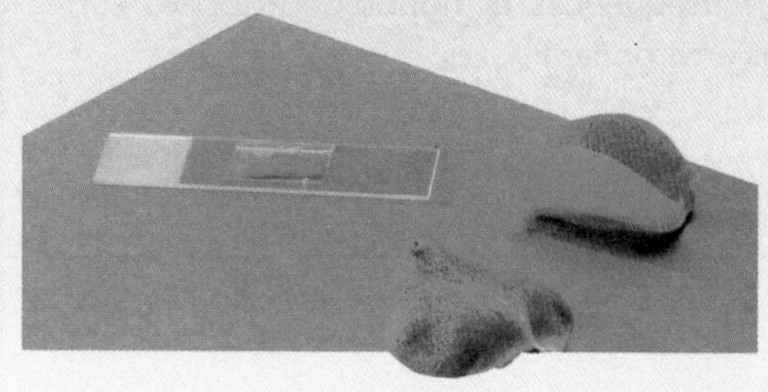

APPLY

17. A stereomicroscope would enable you to view a large part of the living organism as well as looking closely at the mold.
18. No light energy can be changed into food or chemical energy, and therefore the plant cell will die.
19. No proteins could be made and the animal cell would die.
20. For example, ribosomes make proteins and Golgi bodies send the proteins out of the cell.
21. Look for characteristic organelles. If there is a cell wall and/or chloroplasts, it is a plant cell; if there is a looped chromosome, it is a bacterial cell; if neither of the above fits, it is an animal cell.

MORE SKILL BUILDERS

1. Interpreting Scientific Illustrations: The long shape of nerve cells allows them to carry messages to other structures; hollow plant stem cells allow for transport of water; human blood cells are small enough to pass through vessels.
2. Sequencing: 1. Hooke identified boxlike structures in cork as "cells." 2. Schleiden determined that all plants are made of cells. 3. Schwann—that all animals are made of cells. 4. Virchow—all cells have come from other cells.
3. Observing and Inferring: There would be a lack of control of what could enter or leave the cell, causing damage and, eventually, death to the cell.
4. See table, below left.
5. Concept Mapping: Nucleus, ribosomes, Golgi bodies, cell membrane

4. Making and Using Tables:

Function	Structures
Protection	cell membrane; cell wall
Movement of materials	Golgi bodies; ER; cytoplasm
Breaking down substances	lysosomes; mitochondria
Making substances	ER; ribosomes; nucleus
Storing substances	Golgi bodies; ER; vacuoles

CHAPTER 3 Cell Processes

CHAPTER SECTION	OBJECTIVES	ACTIVITIES
3-1 Chemistry of Living Things (2 days)	1. **Describe** differences among atoms, elements, molecules, and compounds. 2. **Recognize** the relationship between chemistry and life science. 3. **Compare** inorganic and organic compounds.	**MINI-Lab:** *How do enzymes work?* p. 57
3-2 Cell Transport (2 days)	1. **Explain** the function of a selectively permeable membrane. 2. **Describe** the processes of diffusion and osmosis. 3. **Compare** and contrast passive transport and active transport and give examples of each.	**MINI-Lab:** *How does temperature affect the rate of diffusion of molecules?* p. 59 **Activity 3-1:** *Designing an Experiment (Observing Osmosis),* p. 62
3-3 Energy in Cells (2 days)	1. **Explain** the difference between producers and consumers. 2. **Compare** and **contrast** the processes of photosynthesis and respiration. 3. **Describe** how cells get energy from glucose through the process of fermentation.	
3-4 Nonbiodegradable Materials in Your Environment Science & Society (1 day)	1. **Explain** the consequences of nonbiodegradable substances in the environment.	**Activity 3-2:** *Photosynthesis and Respiration,* p. 68
Chapter Review		

ACTIVITY MATERIALS

FIND OUT	ACTIVITIES		MINI-LABS	
Page 53 2 small bowls salt raw potato	**3-1 Designing an Experiment, p. 62** 1 raw egg in shell 2 500-mL glass jars with covers 250 mL vinegar 250 mL white table syrup 250-mL graduated cylinder spoon 2 labels, A and B	**3-2 Photosynthesis and Respiration, p. 68** 4 test tubes 16 × 150 mL w/ stoppers 1 test-tube rack stirring rod balance sodium bicarbonate bromothymol blue solution in dropping bottle aged tap water 21 pieces *Elodea*	**How do enzymes work? p. 57** test tube milk rennin tablet mortar pestle	**How does temperature affect the rate of diffusion of molecules? p. 59** 2 beakers Alka-Seltzer tablet (2) clock cold water hot water

CHAPTER FEATURES	TEACHER RESOURCE PACKAGE	OTHER RESOURCES
Skill Builder: *Making and Using Tables,* p. 57	**Ability Level Worksheets** ◆ ***Study Guide,*** p. 13 ● ***Reinforcement,*** p. 13 ▲ ***Enrichment,*** p. 13 **Activity Worksheets,** p. 29	**Lab Manual:** Chemical Changes, p. 15 **STVS:** Disc 4, Side 2
Problem Solving: *What Happened to the Salad?* p. 61 **Skill Builder:** *Concept Mapping,* p. 61	**Ability Level Worksheets** ◆ ***Study Guide,*** p. 14 ● ***Reinforcement,*** p. 14 ▲ ***Enrichment,*** p. 14 **Activity Worksheets,** pp. 5, 23, 24 **Concept Mapping,** pp. 11	**Lab Manual:** Diffusion, p. 19
Technology: *Biodegradable Plastics,* p. 64 **Skill Builder:** *Comparing and Contrasting,* p. 65	**Ability Level Worksheets** ◆ ***Study Guide,*** p. 15 ● ***Reinforcement,*** p. 15 ▲ ***Enrichment,*** p. 15 **Critical Thinking/Problem Solving,** p. 7 **Cross-Curricular Connections,** p. 7	**Lab Manual:** Cell Respiration, p. 21 **STVS:** Disc 5, Side 1
You Decide! p. 67	**Ability Level Worksheets** ◆ ***Study Guide,*** p. 16 ● ***Reinforcement,*** p. 16 ▲ ***Enrichment,*** p. 16 **Science and Society,** p. 7 **Activity Worksheets,** pp. 25-26	**STVS:** Disc 6, Side 2 **Science Integration Activity 3**
Summary Key Science Words Understanding Vocabulary Checking Concepts Using Lab Skills Think & Write Critically Apply More Skill Builders Projects	ASSESSMENT RESOURCES **Chapter Review,** pp. 9-10 **Chapter Test,** pp. 13-16 **Performance Assessment in Middle School Science (PAMSS)**	**Chapter Review Software** **Test Bank** **Alternate Assessment** **Performance Assessment**

◆ **Basic** ● **Average** ▲ **Advanced**

ADDITIONAL MATERIALS

SOFTWARE	AUDIOVISUAL	BOOKS/MAGAZINES
Passive Transport, Classroom Consortia Media.	*Diffusion and Osmosis (2nd ed.),* film, EBEC. *Cell Biology: Life Functions,* film, Coronet/MTI. *Cell Biology: Motion and Function of the Living Cell,* laserdisc, VideoDiscovery. *Living Cells,* laserdisc, Aims Media.	Frienkel, Norbert, ed. *Contemporary Metabolism.* Vol 1. NY: Plenum Publishing Corporation, 1979. Karp, Gerald C. *Cell Biology.* 2nd ed. NY: McGraw-Hill Publishing, 1984. King, Barry, ed. *Cell Biology.* Cambridge, MA: Unwin Hyman, Inc., 1986.

CHAPTER

3

CELL PROCESSES

THEME DEVELOPMENT: Homeostasis is the major theme in this chapter. Living things function as a result of chemical reactions that take place in cells. The equilibrium maintained by cells results from their selectively permeable membranes. This is critical to the life of cells and the organism as a whole. Ecology is emphasized in the Science and Society section.

CHAPTER OVERVIEW

- **Section 3-1:** This section provides a basic explanation of the chemistry needed to understand life processes. Atoms make up elements. Elements are joined by chemical bonds to form compounds. Compounds are classified as organic and inorganic.
- **Section 3-2:** The processes of diffusion, osmosis, and active transport and how they move molecules in and out of cells are discussed.
- **Section 3-3:** The processes of photosynthesis, respiration, and fermentation and their roles in providing energy for cells of living organisms are described.
- **Section 3-4: Science and Society:** The use of biodegradable materials and recycling are explored. The You Decide feature asks students if all states should have mandatory recycling programs.

CHAPTER VOCABULARY

carbohydrates
lipids
proteins
enzymes
nucleic acids
passive transport
active transport
diffusion
equilibrium
osmosis
endocytosis
exocytosis
metabolism
producers
consumers
photosynthesis
cellular respiration
fermentation
biodegradable

CHAPTER

3 Cell Processes

52

OPTIONS

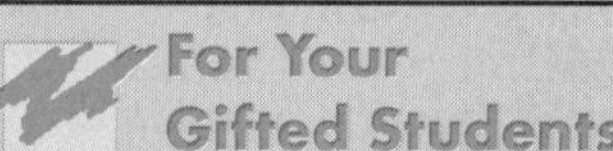

For Your Gifted Students

For one week, have students carry a trash bag with them during the day and place all their trash in that bag. At the end of the week, have students evaluate and classify the garbage. What is recyclable? Have a garage sale of those items they think may be used by others. Have them think of ways to reduce the amount of trash produced. Discuss whether it is possible to live without producing wastes. Help students to infer that even cells normally produce wastes.

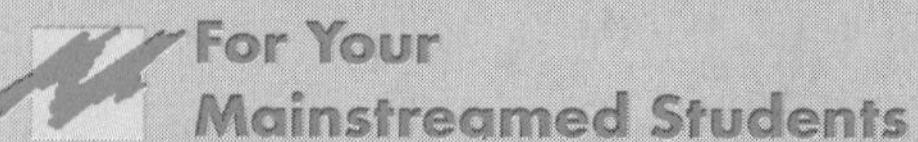

For Your Mainstreamed Students

Have students make a survey of biodegradable and nonbiodegradable materials used at school. They could start in their own classroom by measuring trash in the trash can. Have students record and chart results. They should look for any unnecessary waste. They can expand the survey to other classes. Contact the administration of the school to see if a recycling program can be established and student managed.

You may have washed dishes or stayed in a pool so long that your fingers became wrinkled. What caused your fingers to wrinkle? How long did it take your fingers to return to normal? Why does this happen?

FIND OUT!

Do this simple activity to find out why your fingers wrinkle.

Pour 250 mL of water into each of two small bowls. Stir 15 g of salt into one of the bowls of water. Label it salt water. Place slices of raw potato into each bowl. *Predict* how the potato slices will be affected. After 20 minutes, pick up the slices and examine them. Describe the slices. How does this compare with wrinkled fingers?

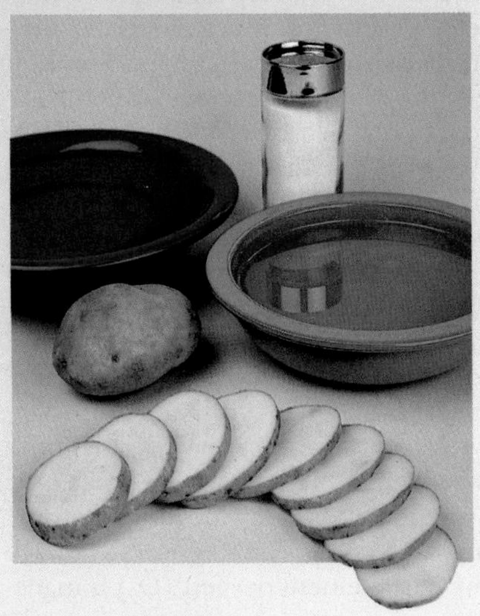

Gearing Up

Previewing the Chapter

Use this outline to help you focus on important ideas in this chapter.

Section 3-1 Chemistry of Living Things
- ▶ The Nature of Matter
- ▶ The Chemistry of Life

Section 3-2 Cell Transport
- ▶ Diffusion and Osmosis
- ▶ Other Kinds of Transport

Section 3-3 Energy in Cells
- ▶ Trapping Energy for Life
- ▶ Releasing Energy for Life

Section 3-4 Science and Society
Nonbiodegradable Materials in Your Environment
- ▶ Recycling and the Environment

Previewing Science Skills

- ▶ In the Skill Builders, you will make and use tables, make a concept map, and compare and contrast.
- ▶ In the Activities, you will measure, use tables, conclude, and infer.
- ▶ In the MINI-Labs, you will observe, infer, and predict.

What's next?

What you've just seen is the effect of water moving into and out of the potato cells. You will learn how atoms, molecules, and compounds relate to the movement of materials into and out of cells. You will also study which organisms produce food and how organisms use food as a source of energy.

INTRODUCING THE CHAPTER

Use the Find Out activity to show students that water moves into and out of potato cells. Explain that they will be learning about atoms, molecules, and compounds and that these are the materials that move into and out of cells.

FIND OUT!

Preparation: Prepare 0.5 cm thick potato slices, 2 bowls or baby food jars, and stirring rods.

Cooperative Learning: Use the Science Investigation strategy. Divide the class into groups of four. Assign one student to obtain the materials, one to record data, one to keep the group on task, and one to check for understanding.

Teaching Tips

▶ Have the water at room temperature so the salt will dissolve more readily.

▶ The amount of salt is not critical as long as the salt concentration is high.

Gearing Up

Have students study the Gearing Up feature to familiarize themselves with the chapter. Discuss the relationships of the topics in the outline.

What's Next?

Before beginning the first section, make sure students understand the connection between the Find Out activity and the topics to follow.

ASSESSMENT OPTIONS

PORTFOLIO

Refer to page 69 for suggested items that students might select for their portfolios.

PERFORMANCE ASSESSMENT

See page 69 for additional Performance Assessment options.

Process

Skillbuilders, pp. 61, 65
MINI-Lab, pp. 57, 59
Activities, 3-1, p. 62; 3-2, p. 68
Using Lab Skills, p. 70

CONTENT ASSESSMENT

Assessment—Oral, pp. 56, 60, 63
Skillbuilder, p. 57
Section Reviews, pp. 57, 61, 65, 67
Chapter Review, pp. 69-71
Mini Quizzes, pp. 60, 65

GROUP ASSESSMENT

Opportunities for group assessment occur with Cooperative Learning Strategies and Flex Your Brain Activities.

SECTION 3-1

PREPARATION

SECTION BACKGROUND

▶ Presently there are 109 known elements. Ninety occur naturally, and the others have been made in the laboratory. About 25 of the 92 elements are found in living matter. Eleven of these are listed in Table 3-1 on page 55. Carbon, hydrogen, oxygen, and nitrogen are the elements most abundant in living organisms.

▶ A diatomic molecule is made of two atoms. Hydrogen (H_2), oxygen (O_2), and nitrogen (N_2) are diatomic.

▶ It was once believed that organic compounds could form only within living organisms. However, scientists have been able to make organic compounds for many years. Plastics and synthetic fibers are examples.

PREPLANNING

▶ Obtain rennin tablets from a druggist for the Mini-Lab on page 57.

1 MOTIVATE

▶ Display a periodic table of elements and point out common elements. Ask students to give examples of objects that are made of elements such as aluminum, copper, and iron. Point out the different groups—metals, nonmetals, liquids, and gases.

VideoDisc

STVS: Straw as Feed, Disc 4, Side 2

TYING TO PREVIOUS KNOWLEDGE: Bring in labels from foods and cleaning products. Many foods and soaps contain sodium chloride (NaCl). Sodium hydroxide (lye, NaOH) is in drain cleaners. Have students use a copy of the periodic table and determine what elements are in the compounds.

3-1 Chemistry of Living Things

New Science Words

carbohydrates
lipids
proteins
enzymes
nucleic acids

Objectives

- ▶ Describe differences among atoms, elements, molecules, and compounds.
- ▶ Recognize the relationship between chemistry and life science.
- ▶ Compare inorganic and organic compounds.

The Nature of Matter

Do you know what makes up the universe? If you ask a scientist, he or she will probably say "matter and energy." Matter is anything that has mass and takes up space. How many things can you think of that fit that category?

Everything is made up of matter. You are made of matter and so are the fireflies on this page. Even the air that they fly through is made up of matter. Matter exists in the form of small units called atoms. Figure 3-1 shows a model of an oxygen atom. At the center of the atom is a nucleus, which contains two kinds of particles, protons and neutrons. A proton has a positive charge, and neutrons have no charge. Electrons are particles found outside the nucleus. They are negatively charged.

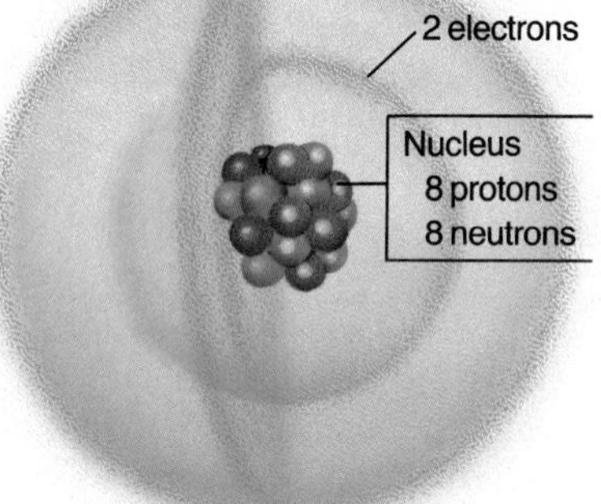

Figure 3-1. An Oxygen atom model shows the placement of electrons, protons, and neutrons.

What about the energy that the scientist mentioned? Energy in matter is locked in chemical bonds that hold atoms together. To release this energy, the atoms that are bound together have to be rearranged or broken apart in a chemical reaction. When fireflies blink and shine in the dark, energy is released in the form of light when chemical bonds are broken.

When something is made up of only one kind of atom, it is called an element. An element can't be broken down into any simpler form. The element oxygen (O_2) is made

OPTIONS

Meeting Different Ability Levels

For Section 3-1, use the following **Teacher Resource Masters** depending upon individual students' needs.

◆ **Study Guide Master** for all students.

● **Reinforcement Master** for students of average and above average ability levels.

▲ **Enrichment Master** for above average students.

Additional Teacher Resource Package masters are listed in the OPTIONS box throughout the section. The additional masters are appropriate for all students.

◆ STUDY GUIDE 13

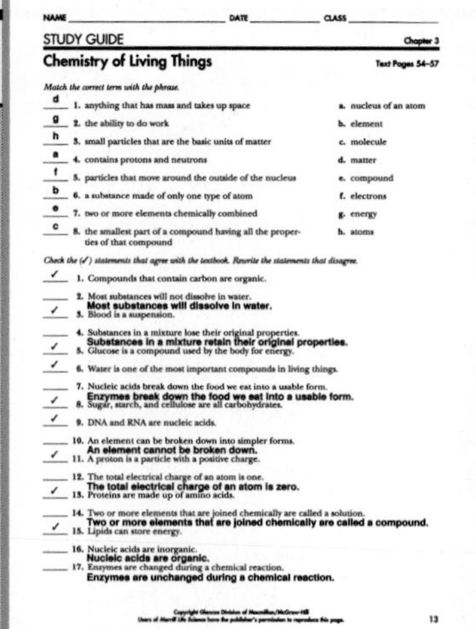

NAME ____________ DATE ________ CLASS ________

STUDY GUIDE — Chapter 3

Chemistry of Living Things — Text Pages 54–57

Match the correct term with the phrase.

Answer	Phrase		Term
d	1. anything that has mass and takes up space	a.	nucleus of an atom
g	2. the ability to do work	b.	element
h	3. small particles that are the basic units of matter	c.	molecule
a	4. contains protons and neutrons	d.	matter
f	5. particles that move around the outside of the nucleus	e.	compound
b	6. a substance made of only one type of atom	f.	electrons
e	7. two or more elements chemically combined	g.	energy
c	8. the smallest part of a compound having all the properties of that compound	h.	atoms

Check the (✓) statements that agree with the textbook. Rewrite the statements that disagree.

✓ 1. Compounds that contain carbon are organic.

___ 2. Most substances will not dissolve in water.
Most substances will dissolve in water.

✓ 3. Blood is a suspension.

___ 4. Substances in a mixture lose their original properties.
Substances in a mixture retain their original properties.

✓ 5. Glucose is a compound used by the body for energy.

✓ 6. Water is one of the most important compounds in living things.

___ 7. Nucleic acids break down the food we eat into a usable form.
Enzymes break down the food we eat into a usable form.

✓ 8. Sugar, starch, and cellulose are all carbohydrates.

✓ 9. DNA and RNA are nucleic acids.

___ 10. An element can be broken down into simpler forms.
An element cannot be broken down.

✓ 11. A proton is a particle with a positive charge.

___ 12. The total electrical charge of an atom is one.
The total electrical charge of an atom is zero.

✓ 13. Proteins are made up of amino acids.

___ 14. Two or more elements that are joined chemically are called a solution.
Two or more elements that are joined chemically are called a compound.

✓ 15. Lipids can store energy.

___ 16. Nucleic acids are inorganic.
Nucleic acids are organic.

___ 17. Enzymes are changed during a chemical reaction.
Enzymes are unchanged during a chemical reaction.

13

Table 3-1

ELEMENTS THAT COMPOSE THE HUMAN BODY		
Symbol	**Element**	**Percent**
O	Oxygen	65.0
C	Carbon	18.5
H	Hydrogen	9.5
N	Nitrogen	3.3
Ca	Calcium	1.5
P	Phosphorus	1.0
K	Potassium	0.4
S	Sulfur	0.3
Na	Sodium	0.2
Cl	Chlorine	0.2
Mg	Magnesium	0.1

up of only oxygen atoms, and hydrogen (H_2) is made up of only hydrogen atoms. Each element has its own symbol. More than 90 elements occur naturally on Earth. Everything, including you, is made of one or a combination of these elements. Table 3-1 lists elements as they occur in the human body. What are the two most common elements in your body?

When atoms of two or more elements are joined, or bonded together, a compound is formed. For example, water is made of the elements hydrogen and oxygen, as shown in Figure 3-2. In water, there are two atoms of hydrogen for every atom of oxygen. The chemical formula for water is H_2O. A chemical formula shows the kind and number of atoms that form a compound. The formula represents a single molecule of a compound. A molecule is the smallest part of a compound with all the properties of that compound. Glucose, the major form of sugar used by cells for energy, is a compound. It is made of the elements carbon, hydrogen, and oxygen, and its formula is $C_6H_{12}O_6$.

What is so important about molecules? In living organisms, cell membranes, cytoplasm, and other substances are all in the form of molecules. During its lifetime, a

Hydrogen
Hydrogen
Oxygen

Figure 3-2. The chemical formula for water is H_2O.

● REINFORCEMENT 13

NAME ______ DATE ______ CLASS ______

REINFORCEMENT — Chapter 3

Chemistry of Living Things — Text Pages 54–57

Answer the following questions.

1. What are atoms? **Atoms are the basic units of matter.**

Label the following diagram of an atom using the words: electron, proton, neutron.

electron
proton
neutron

2. Is water (H_2O) an element or a compound? **compound** Why? **It is made up of the elements hydrogen and oxygen, joined by chemical bonds.**

Study the following graph and answer the questions.

Elements That Make Up the Human Body
hydrogen
nitrogen
calcium
other
carbon
oxygen

1. Oxygen and carbon together make up about what percent of the elements in the human body? (Circle the correct answer.)

   a. 50% b. 80% (circled) c. 30% d. 10%

2. Why is there such a large percentage of carbon? **Organic compounds contain carbon. Carbohydrates, lipids, proteins, and enzymes are organic compounds.**
3. Why is there so much oxygen? **Respiration requires oxygen; oxygen is carried in the blood, and it is present in water, which makes up a large part of the human body.**
4. Which elements in the graph make up the compound glucose? **carbon, hydrogen, and oxygen**

 13

▲ ENRICHMENT 13

NAME ______ DATE ______ CLASS ______

ENRICHMENT — Chapter 3

Chemistry of Living Things — Text Pages 54–57

THE EFFECT OF TEMPERATURE ON SOLUBILITY

When a solid is dissolved in a liquid, the molecules of the solid mix with the molecules of the liquid. When you dissolve sugar in a cup of tea, you know it is there because of the taste, but you can't see the sugar.

In this experiment, you will see the effect of temperature on **solubility**. Solubility is the amount of a substance that dissolves in a solvent at a given temperature. The substance that is to be dissolved is called the **solute**, and the substance that it is dissolved in is called the **solvent**. In the example above, sugar is the solute and hot tea water is the solvent. When no more solute will dissolve at a given temperature, we say that the solution is **saturated**.

Materials
- table salt
- measuring spoons and measuring cups
- a shallow pan
- stirring rod
- thermometer

Procedure
1. Fill the measuring cup with 1/2 cup cold tap water. Measure the temperature. Add salt, 1 level teaspoon at a time while stirring. Repeat until the solution is saturated.
2. Record the number of teaspoons used and the temperature. Discard the solution.
3. Repeat the procedure using tap water that is about room temperature.
4. Repeat using very hot tap water. CAUTION: *Always be careful when handling hot objects.* After recording the data, pour the solution into the pan and allow the water to evaporate overnight.

Observations

Temperature	Teaspoons of salt (volume)

Plot your data on graph
temperature
volume of salt (tsp.)

Conclude and Apply
1. The salt seems to disappear as it goes into solution. How do we know it hasn't actually disappeared, other than taste? **The amount left in the pan after evaporation proves that the salt is present in the same quantity as was added to the solution.**
2. If you were to use a heat source to make the water even hotter, hypothesize how the solubility of salt would be affected. **More salt would dissolve at a higher temperature.**

 13

2 TEACH

Key Concepts are highlighted.

TEACHER F.Y.I.

▶ Elements are usually given names by their discoverers. The most common source for an element name is some property of an element. The name for chlorine, a greenish gas, comes from the Greek word *chloros,* which means "green." Some elements are named for the place where the element was discovered. Ytterbium is named for Ytterby, Sweden. Some elements are named to honor famous scientists. Einsteinium is in honor of Albert Einstein, fermium is named for Enrico Fermi, and curium is named for Marie and Pierre Curie.

CONCEPT DEVELOPMENT

▶ Life science deals with matter in all states. Discuss the three states of matter. Solids have a definite shape and occupy a definite volume. Liquids have a definite volume but change shape to fill their containers. Gases have no shape or volume but spread out to fill whatever space is available.

▶ Explain that the number of atoms of an element in a chemical formula is represented by a subscript. Ask how many atoms of oxygen are in a molecule of sugar, water, and other compounds containing oxygen.

OBJECTIVES AND SCIENCE WORDS: Have students review the objectives and science words throughout the chapter as they study each section.

CROSS CURRICULUM

▶ **Chemistry:** Have students read about the history of alchemy and write a report that includes when and where it was practiced and whether it helped in the development of chemistry.

CONCEPT DEVELOPMENT

▶ Explain that physical changes do not change the chemical properties of a substance; only the shape is changed (cutting paper, dissolving sugar in water, and boiling water). When a substance undergoes a change so that one or more new substances with different characteristics are formed, a chemical change has taken place.

REVEALING MISCONCEPTIONS

▶ Students may think that all fats are harmful. Explain that fats have many useful functions in living organisms, especially in cell membranes.

Both have the elements calcium, sodium, potassium, and magnesium.

CHECK FOR UNDERSTANDING

Use the Oral Assessments in the OPTIONS box to check understanding.

In Your JOURNAL

Answer: Percents will vary, but most will have high carbohydrate.

Figure 3-3. All organisms are dependent on water for life.

cell will put together and break apart many molecules to build new cell parts and supply itself with energy. Many of these molecules are found in solution in cytoplasm. A solution is a mixture in which one or more substances mix evenly with other substances. When you dissolve salt in water, you get a salt solution. You've probably noticed the taste of salt when you perspire. That's because your cells are bathed in a salt solution. Living organisms also have suspensions. A suspension is a mixture in which substances spread through a liquid or gas, then settle out over time. Your blood is a suspension. If a test tube of blood is left undisturbed, the red blood cells and white blood cells will gradually settle to the bottom. Of course, blood in your body is moved constantly by the pumping action of your heart. Therefore, the cells don't settle out.

Figure 3-4. Some salad dressings are suspensions. Large pieces separate and fall to the bottom after the dressing stands awhile.

In Earth's crust are silicon 27.7%, aluminum 8.1%, iron 5%, calcium 3.6%, sodium 2.8%, potassium 2.6%, and magnesium 2.1%. *Compare* these elements with those in the human body.

The Chemistry of Life

Compounds in living organisms are classified as organic and inorganic. Most compounds containing carbon are organic compounds. Organic compounds make up foods and cell membranes. Most inorganic compounds are made from elements other than carbon.

Water, an inorganic compound, makes up a large part of living matter. It is, therefore, one of the most important compounds in living things. In order for substances to be used in cells, they have to be dissolved in water. Nutrients and waste materials are carried throughout your body in solution form.

Four groups of organic compounds make up all living things. These are carbohydrates, lipids, proteins, and nucleic acids. **Carbohydrates** are organic compounds made up of carbon, hydrogen, and oxygen. Sugars, starch, and cellulose are carbohydrates. By breaking down

OPTIONS

ASSESSMENT—ORAL

▶ **Carbon monoxide is CO and carbon dioxide is CO_2. What do *mono* in carbon monoxide and *di* in carbon dioxide refer to?** *the number of oxygen atoms*

▶ **Both solutions and suspensions are mixtures. How can you distinguish between the two?** *Substances mix evenly in a solution and remain mixed. Substances spread through a suspension but settle out over time.*

▶ **Why is a study of chemicals important to the study of life science?** *Matter is made up of chemicals, and all living things are made of matter.*

▶ **What is the meaning of the word *organic* in organic gardening, organic fertilizer, and organic food?** *Only substances made from carbon are used in organic gardening, organic fertilizer, and in the production of organic food.*

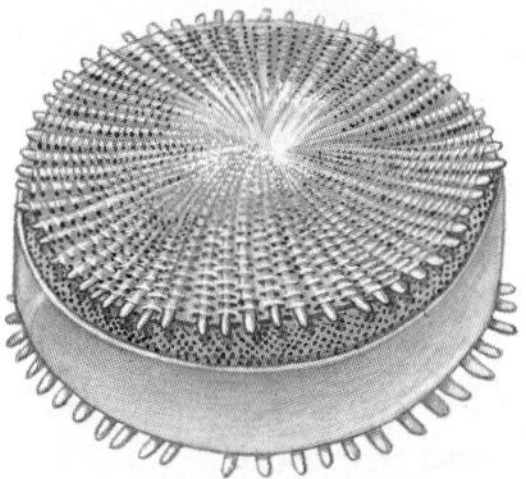
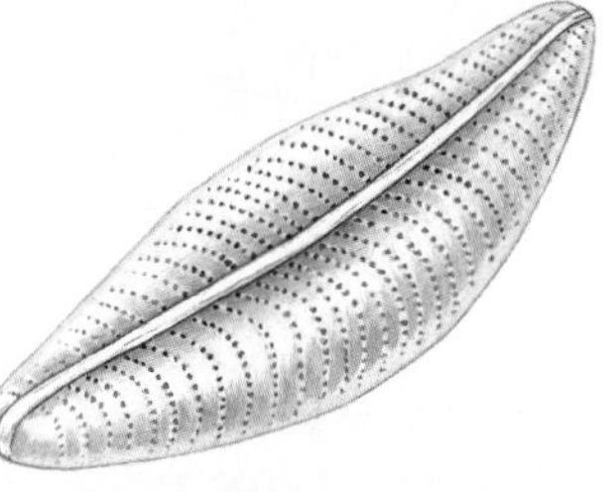

Figure 3-5. Some organisms, such as diatoms, store oils instead of carbohydrates as sources of energy.

these molecules, organisms release energy. The energy is then used to power cell processes.

Carbohydrates supply energy, but **lipids** are organic compounds that store and release even larger amounts of energy. In Chapter 2 you learned that cell membranes are made up of two layers of lipids. Fats, oils, and waxes are types of lipids found in different organisms.

Proteins, the third group of organic compounds, are used for building cell parts. Protein molecules are scattered throughout cell membranes. Proteins are made up of smaller molecules called amino acids. Certain proteins called **enzymes** (EN zimez) speed up chemical reactions in cells without being changed. Enzymes take part in nearly all the chemical reactions in cells.

Nucleic acids are large organic molecules that store important information in cells. One nucleic acid, DNA, is found in chromosomes, mitochondria, and chloroplasts. It carries information that directs each cell's activities. A second nucleic acid, RNA, carries information for making proteins and enzymes in cytoplasm.

MINI-Lab

How do enzymes work?

Fill a test tube with milk. Crush a rennin tablet and add the crushed tablet to the milk. Rennin is an enzyme. Let the milk stand during your class period. *Observe* what happens to the milk. *Infer* how the action of rennin on milk might be useful.

SECTION REVIEW

1. How are atoms different from molecules?
2. Why is water important to living organisms?
3. How do organic and inorganic compounds differ?
4. **Apply:** Diatoms are one-celled organisms that store oils instead of carbohydrates. How do they benefit from these oils?
5. **Connect to Chemistry:** Find out what the Periodic Table is and what it is used for.

In Your JOURNAL

In a table **in your Journal**, classify the foods you eat in a day as carbohydrates, lipids, or proteins. *Calculate* the percentage of your diet each represents for that day.

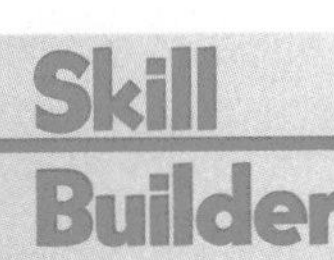

Skill Builder

Making and Using Tables

Use Table 3-1 on page 55 to analyze what elements make up more than 98 percent of the human body. If you need help, refer to Making and Using Tables in the **Skill Handbook** on page 690.

RETEACH

Display a variety of carbon compounds, such as soap, vinegar, alcohol, starch, and chalk. Ask why these are inorganic compounds.

EXTENSION

For students who have mastered this section, use the **Reinforcement** and **Enrichment** masters or other OPTIONS.

MINI-Lab

Materials: test tube, rennin tablet, milk

Answer: The milk will become curdled. The enzyme in the rennin tablet works to break down the protein in milk (casein).

MINI-Lab ASSESSMENT

Performance: To further assess students' understanding of enzymes, see USING LAB SKILLS, question 11, p. 70.

3 CLOSE

▶ Ask questions 1-3 and the **Apply** Question in the Section Review.

SECTION REVIEW ANSWERS

1. Atoms are the basic units of matter. Molecules are made up of atoms.
2. It makes up a large part of living matter and allows the transport of nutrients and waste materials in organisms.
3. Most compounds that contain carbon are organic compounds.
4. Apply: Their oils supply energy.
5. Connect to Chemistry: Elements are arranged in increasing atomic number; to classify elements.

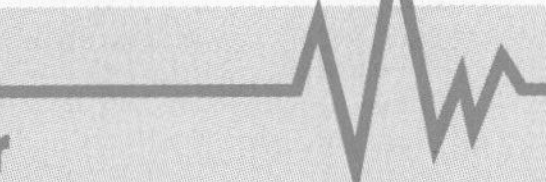

Skill Builder

oxygen (65 percent), carbon (18.5 percent), hydrogen (9.5 percent), nitrogen (3.3 percent), calcium (1.5 percent), phosphorus (1.0 percent)

Skillbuilder ASSESSMENT

Portfolio: Use this Skillbuilder to assess students' abilities to Use a Table.

▶ **If you were part of a space mission assigned to look for signs of life on another planet, what elements and compounds would you search for and why?** *Carbon, hydrogen, oxygen, nitrogen, sulfur, and phosphorus are found in all living things. Water makes up more than 70 percent of all living things. Carbohydrates, proteins, and lipids are used by all living things.*

PROGRAM RESOURCES

From the **Teacher Resource Package** use:

Activity Worksheets, page 29, Mini-Lab: How do enzymes work?

Use **Laboratory Manual,** page 15, Chemical Changes.

SECTION 3-2

PREPARATION

SECTION BACKGROUND

▶ Membrane permeability depends not only on size of molecules but on other properties, such as electrical charge and molecular shape.

▶ Molecular motion is directly proportional to temperature; therefore, passive transport is temperature dependent.

PREPLANNING

▶ Obtain vinegar, distilled water, salt, and eggs for Activity 3-1. Obtain antacid tablets for the Mini-Lab.

1 MOTIVATE

FLEX Your Brain

Use the Flex Your Brain activity to have students explore MEMBRANES.

Flex Your Brain
ASSESSMENT

Portfolio: Use the Flex Your Brain activity to reinforce critical thinking and problem solving skills. Students may need to review what they learned about cell membranes during this activity.

TYING TO PREVIOUS KNOWLEDGE: Review the parts of the cell in Chapter 2. The cell membrane maintains the balance between the cell and its environment.

3-2 Cell Transport

New Science Words

passive transport
active transport
diffusion
equilibrium
osmosis
endocytosis
exocytosis

Objectives

- ▶ Explain the function of a selectively permeable membrane.
- ▶ Describe the processes of diffusion and osmosis.
- ▶ Compare and contrast passive transport and active transport and give examples of each.

Diffusion and Osmosis

Cells obtain food, oxygen, and other substances from their environment. They also release waste materials. If a cell has a membrane around it, how do these things move into and out of the cell? How does the cell control what enters and leaves?

Have you ever seen marbles in a mesh bag? Although the mesh bag has holes in it, the marbles stay inside because they are larger than the holes. If you put sand in with the marbles, the sand will fall right through the holes because sand grains are smaller than the holes. The mesh bag is said to be selectively permeable because it allows some things to pass through it but not others. The marbles and sand are models for molecules. The cell membrane is selectively permeable. It allows some molecules to pass through but not others.

Figure 3-6. A cell membrane, like a mesh bag, will let some things through more easily than others.

If materials move through a cell membrane
without the help of energy, **passive transport** takes 1
place. If materials require energy to move through
a cell membrane, **active transport** takes place. 2

Passive transport depends on the fact that molecules in solids, liquids, and gases move about. Molecules move constantly. As they move, they move from places where they are crowded together into places where there are fewer of them. One type of passive transport is **diffusion,** the movement of molecules from an area where there are many to an area where there are few.

Molecules diffuse in liquids and in gases. You experience diffusion when someone opens a bottle of perfume in a closed room.

OPTIONS

Meeting Different Ability Levels

For Section 3-2, use the following **Teacher Resource Masters** depending upon individual students' needs.

◆ **Study Guide Master** for all students.

● **Reinforcement Master** for students of average and above average ability levels.

▲ **Enrichment Master** for above average students.

Additional Teacher Resource Package masters are listed in the OPTIONS box throughout the section. The additional masters are appropriate for all students.

◆ STUDY GUIDE 14

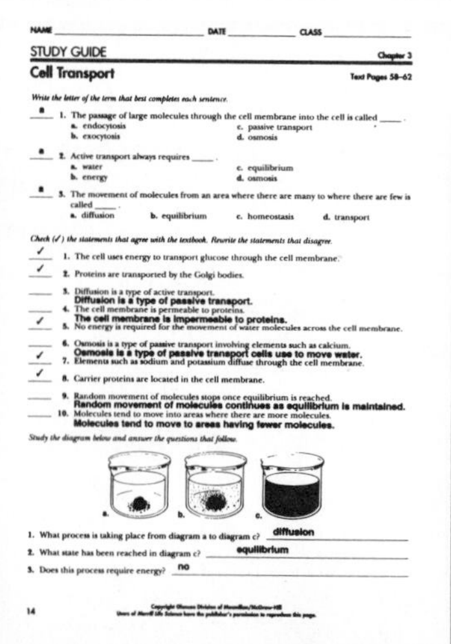

NAME ________ DATE ________ CLASS ________

STUDY GUIDE — Chapter 3

Cell Transport — Text Pages 58–62

Write the letter of the term that best completes each sentence.

a 1. The passage of large molecules through the cell membrane into the cell is called ____.
a. endocytosis c. passive transport
b. exocytosis d. osmosis

b 2. Active transport always requires ____.
a. water c. equilibrium
b. energy d. osmosis

a 3. The movement of molecules from an area where there are many to where there are few is called ____.
a. diffusion b. equilibrium c. homeostasis d. transport

Check (✓) the statements that agree with the textbook. Rewrite the statements that disagree.

✓ 1. The cell uses energy to transport glucose through the cell membrane.
✓ 2. Proteins are transported by the Golgi bodies.
___ 3. Diffusion is a type of active transport.
Diffusion is a type of passive transport.
___ 4. The cell membrane is permeable to proteins.
The cell membrane is impermeable to proteins.
✓ 5. No energy is required for the movement of water molecules across the cell membrane.
___ 6. Osmosis is a type of passive transport involving elements such as calcium.
Osmosis is a type of passive transport cells use to move water.
✓ 7. Elements such as sodium and potassium diffuse through the cell membrane.
✓ 8. Carrier proteins are located in the cell membrane.
___ 9. Random movement of molecules stops once equilibrium is reached.
Random movement of molecules continues as equilibrium is maintained.
___ 10. Molecules tend to move into areas where there are more molecules.
Molecules tend to move to areas having fewer molecules.

Study the diagram below and answer the questions that follow.

a. b. c.

1. What process is taking place from diagram a to diagram c? **diffusion**
2. What state has been reached in diagram c? **equilibrium**
3. Does this process require energy? **no**

14

Drop a sugar cube into a glass of water and taste the water. Then taste it again in three or four hours, the next day, and the next week. At first, the sugar molecules are concentrated near the sugar cube in the bottom of the glass. Then very slowly, the sugar molecules diffuse throughout the water until they are more evenly distributed. When the molecules of a substance are spread evenly throughout a space, a state called **equilibrium** occurs. But molecules don't stop moving when equilibrium is reached. They continue to move, and equilibrium is maintained.

You may remember that water makes up a large part of living matter. The passive transport of *water* through
3 a cell membrane by diffusion is called **osmosis.** Osmosis is important to cells because they are surrounded by water molecules and they contain water molecules. Water molecules move from where they are in large numbers to where they are in small numbers. When the number of water molecules inside and outside the cell is the same, a state of equilibrium is reached.

If you become extremely thirsty, you may say that you feel dehydrated. If cells don't have the water that they need, they dehydrate, or lose water. If you forget to water a plant, it will soon wilt. There is less water in the soil around the roots. Therefore, water tends to move out of the root cells. The cells in the rest of the plant don't get supplied with water from the roots. Cell membranes shrink away from cell walls. Then when the plant is watered, the water moves into the root cells and on up into the other parts of the plant. The cell membranes expand, and the plant becomes upright again.

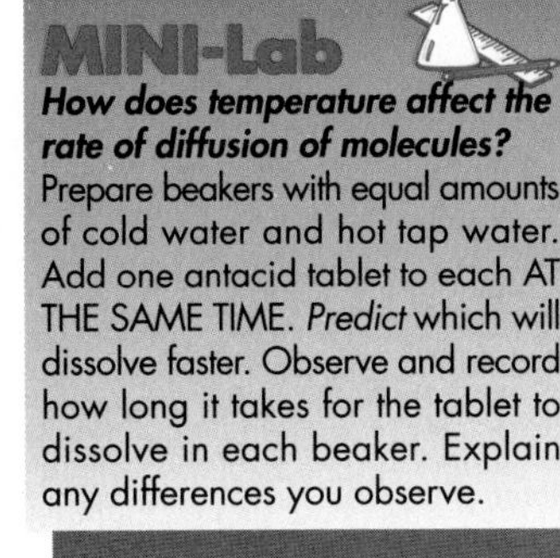

MINI-Lab

How does temperature affect the rate of diffusion of molecules?

Prepare beakers with equal amounts of cold water and hot tap water. Add one antacid tablet to each AT THE SAME TIME. *Predict* which will dissolve faster. Observe and record how long it takes for the tablet to dissolve in each beaker. Explain any differences you observe.

What is equilibrium?

Figure 3-7. Lack of water eventually causes a plant to wilt because more water diffuses from the cells than enters.

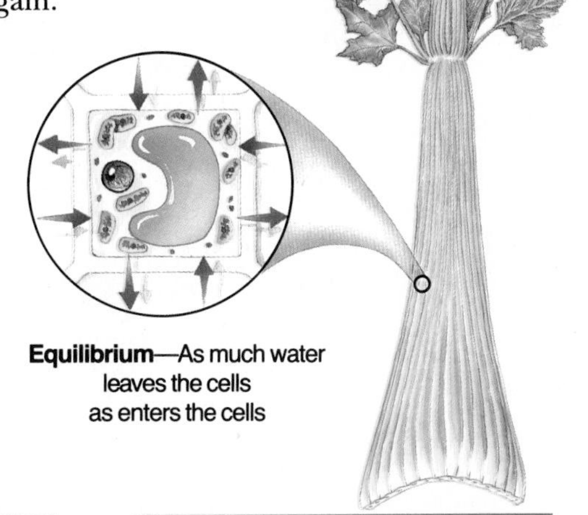

Wilting—More water leaves the cells than enters the cells

Equilibrium—As much water leaves the cells as enters the cells

2 TEACH

Key Concepts are highlighted.

CONCEPT DEVELOPMENT

▶ **Demonstration:** Place dried beans in a beaker until it is about one-third full. Fill the beaker with water and mark the water level on the outside of the beaker. Let it remain undisturbed overnight. Have students observe and explain what happened to the level of beans in the beakers. (Water diffuses into the beans and they increase in size, filling the beaker.)

▶ Discuss examples of osmosis in the human body, such as water moving from the blood into the cells, and water moving from the digestive tract into the blood.

▶ Early methods of preserving food included drying, smoking, salting, and sugar-curing. (Explain that these methods remove water from the foods by osmosis and prevent bacteria and molds from growing.)

MINI-Lab

Materials: 2 beakers, cold water, hot water, 2 antacid tablets, clock

Answer: The antacid tablet will dissolve faster in the hot tap water. Accept all reasonable answers. Students may predict, then report that hot water provides more energy for faster movement of molecules.

MINI-Lab

ASSESSMENT

Performance: To further assess students' understanding of the effect of temperature on solubility, have them repeat the activity using ice water instead of hot water.

● **REINFORCEMENT 14**

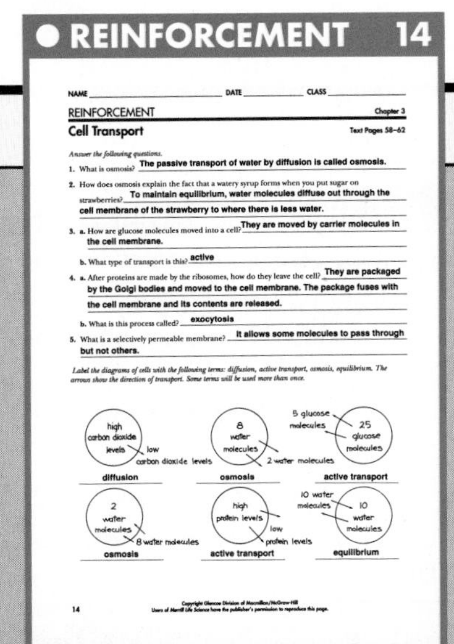

NAME ______ DATE ______ CLASS ______

REINFORCEMENT — Chapter 3

Cell Transport — Text Pages 58–62

Answer the following questions.

1. What is osmosis? **The passive transport of water by diffusion is called osmosis.**
2. How does osmosis explain the fact that a watery syrup forms when you put sugar on strawberries? **To maintain equilibrium, water molecules diffuse out through the cell membrane of the strawberry to where there is less water.**
3. a. How are glucose molecules moved into a cell? **They are moved by carrier molecules in the cell membrane.**
   b. What type of transport is this? **active**
4. a. After proteins are made by the ribosomes, how do they leave the cell? **They are packaged by the Golgi bodies and moved to the cell membrane. The package fuses with the cell membrane and its contents are released.**
   b. What is this process called? **exocytosis**
5. What is a selectively permeable membrane? **It allows some molecules to pass through but not others.**

Label the diagrams of cells with the following terms: diffusion, active transport, osmosis, equilibrium. The arrows show the direction of transport. Some terms will be used more than once.

14

▲ **ENRICHMENT 14**

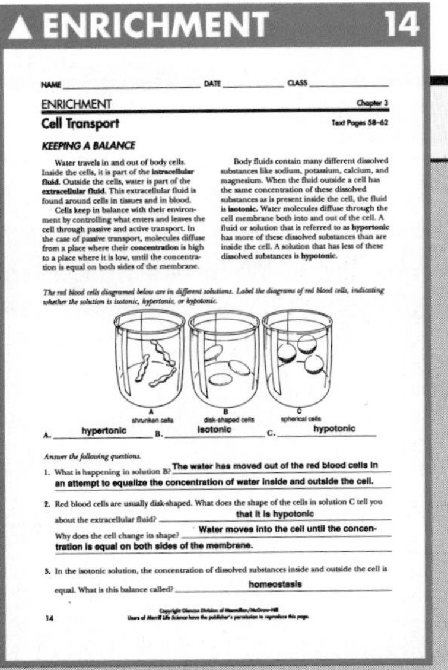

NAME ______ DATE ______ CLASS ______

ENRICHMENT — Chapter 3

Cell Transport — Text Pages 58–62

KEEPING A BALANCE

Water travels in and out of body cells. Inside the cells, it is part of the **intracellular fluid**. Outside the cells, water is part of the **extracellular fluid**. This extracellular fluid is found around cells in tissues and in blood.

Cells keep in balance with their environment by controlling what enters and leaves the cell through passive and active transport. In the case of passive transport, molecules diffuse from a place where their **concentration** is high to a place where it is low, until the concentration is equal on both sides of the membrane.

Body fluids contain many different dissolved substances like sodium, potassium, calcium, and magnesium. When the fluid outside a cell has the same concentration of these dissolved substances as is present inside the cell, the fluid is **isotonic.** Water molecules diffuse through the cell membrane both into and out of the cell. A fluid or solution that is referred to as **hypertonic** has more of these dissolved substances than are inside the cell. A solution that has less of these dissolved substances is **hypotonic**.

The red blood cells diagramed below are in different solutions. Label the diagrams of red blood cells, indicating whether the solution is isotonic, hypertonic, or hypotonic.

A. **hypertonic** B. **isotonic** C. **hypotonic**

Answer the following questions.

1. What is happening in solution B? **The water has moved out of the red blood cells in an attempt to equalize the concentration of water inside and outside the cell.**
2. Red blood cells are usually disk-shaped. What does the shape of the cells in solution C tell you about the extracellular fluid? **that it is hypotonic**
   Why does the cell change its shape? **Water moves into the cell until the concentration is equal on both sides of the membrane.**
3. In the isotonic solution, the concentration of dissolved substances inside and outside the cell is equal. What is this balance called? **homeostasis**

14

CONCEPT DEVELOPMENT

▶ Nerve cells use active transport to move sodium and potassium in and out during impulse transmission. Cells in the kidneys use active transport in separating useful materials from waste materials.

CHECK FOR UNDERSTANDING

Use the Mini Quiz to check for understanding.

MINI QUIZ

Use the Mini Quiz to check students' recall of chapter content.

1. **If the movement of materials through the cell membrane does not require the use of energy, it is called _____.** *passive transport*
2. **If the movement of materials through the cell membrane requires the use of energy, it is called _____.** *active transport*
3. **The passive transport of water through a membrane by diffusion is _____.** *osmosis*

RETEACH

Place several carrot strips in beakers of plain water, distilled water, and salt water. Ask students to observe and explain in terms of water movement.

EXTENSION

For students who have mastered this section, use the **Reinforcement** and **Enrichment** masters or other OPTIONS provided.

Answer: Water diffused out of the potato slice in the salt water.

PROBLEM SOLVING

Think Critically: The liquid in the bowl may have dripped off damp vegetables or diffused through the membranes of the vegetables' cells. The lettuce wilted because water diffused out of the cells to a place of lower concentration (due to the salt). Loss of water in lettuce cells caused the cells to shrink and lettuce to wilt.

Repeat the Find Out experiment on page 53. Using what you now know about osmosis and diffusion, *infer* what took place in the potato left in salt water.

Other Kinds of Transport

Cells take in a variety of substances. Some, like glucose molecules, are so large that they enter the cell only with the help of protein carrier molecules in the cell membrane. Each carrier molecule helps to move only one type of molecule through the cell membrane. Glucose molecules are too large to move quickly across the cell membrane, so carrier molecules help speed up their movement. Carrier proteins use energy when moving molecules into and out of a cell. Therefore, carrier molecules are involved in active transport.

Sometimes cells have to move substances from where there are small amounts to where there are larger amounts. This process is the opposite of diffusion. It requires energy and is a type of active transport. An example of this type of active transport can be seen when plant root cells take in minerals. These cells need minerals from the soil, and there are already more minerals in the root cells than in the water around the roots. The cells use energy to move additional minerals into the root cells. In your body, wastes are moved by active transport out of some kidney cells.

Some substances are too large to pass through the cell membrane by passive or active transport. Large protein molecules and bacteria enter the cell by becoming enclosed in a part of the membrane that folds in to form a sac. The sac pinches off and goes into the cytoplasm in a process called **endocytosis.** In the opposite way, wastes in vacuoles or proteins packaged by Golgi bodies move to the cell membrane. The package fuses with the cell membrane, and its contents are released from the cell in a process called **exocytosis.**

Figure 3-8. A white blood cell uses endocytosis to engulf a bacterial cell (a). In exocytosis, substances in small sacs are released at the cell membrane (b).

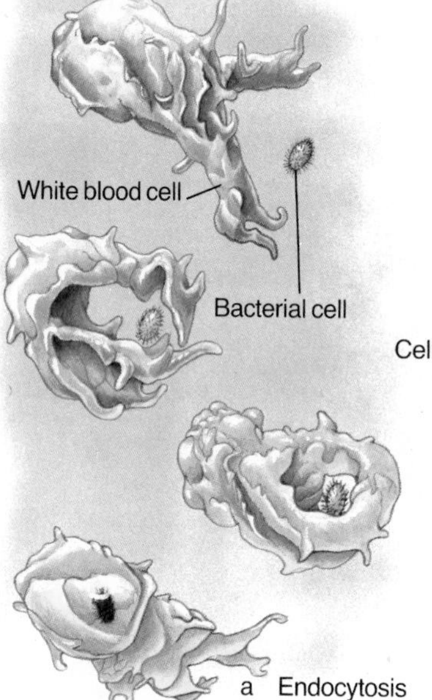

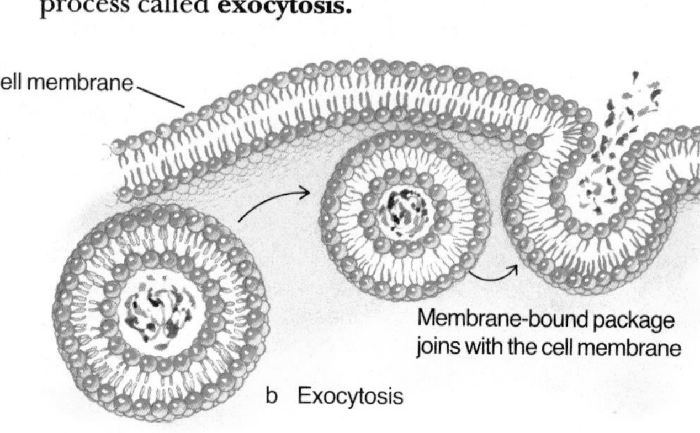

OPTIONS

ASSESSMENT—ORAL

▶ **People sometimes gargle with salt water to relieve a sore throat. What effect might salt have on bacteria that cause sore throat?** *The salt removes water from the bacteria and causes them to dehydrate.*

▶ **Would a cell placed in syrup lose or gain water? Explain your answer.** *It would lose water because water molecules move by diffusion from where there is a large concentration to where there is lesser concentration.*

PROGRAM RESOURCES

From the **Teacher Resource Package** use:

Activity Worksheets, page 30, Mini-Lab: How does temperature affect the rate of diffusion of molecules?

Activity Worksheets, page 5, Flex Your Brain.

Concept Mapping, pages 11-12.

Use **Laboratory Manual,** page 19, Diffusion.

PROBLEM SOLVING

What Happened to the Salad?

Lucy made a salad of lettuce, tomatoes, carrots, and cucumbers. She seasoned the salad with herbs, salt, and pepper. Then she placed it in the refrigerator for a couple of hours.

When Lucy returned, she took the salad from the refrigerator. The lettuce had wilted, and the other vegetables were limp. She noticed that there was liquid in the bottom of the bowl. Where did the liquid come from?

Think Critically: Why had the lettuce wilted?

Cells have to stay in balance with their environment. Cells keep this balance, called homeostasis, by controlling what enters and leaves the cells through passive and active transport. You can see that the selectively permeable cell membrane and its transport systems are important in keeping cells and organisms alive and healthy.

SECTION REVIEW

1. In what way are cell membranes selectively permeable?
2. Compare and contrast osmosis and diffusion.
3. How do active and passive transport differ?
4. How does a cell take in large bacterial cells?
5. **Apply:** Why are fresh fruits and vegetables sprinkled with water in produce markets?
6. **Connect to Physics:** Compare diffusion of perfume in a warm room and a cold room.

Concept Mapping

Skill Builder

Make a network tree concept map to use as a study guide to help you tell the difference between passive transport and active transport. Begin with the phrase *Transport through membranes*. If you need help, refer to Concept Mapping in the **Skill Handbook** on pages 688 and 689.

3 CLOSE

▶Ask questions 1-4 and the **Apply** Question in the Section Review.

▶Place two or three drops of vanilla in a balloon. Blow up the balloon and tie it. Have students observe the balloon. They will be able to smell the vanilla outside the balloon as diffusion takes place.

SECTION REVIEW ANSWERS

1. They allow some things to pass through but not others.

2. In both, molecules move from where there are many to where there are few. Osmosis is the diffusion of water through a selectively permeable membrane.

3. Active transport requires energy.

4. by the process of endocytosis

5. Apply: to keep water in the environment so it will diffuse into the fruits and vegetables and keep them from drying out

6. Connect to Physics: Diffusion is more rapid in warm room because molecules move faster.

Skill Builder

Transport through membrane
Active — requires — Energy — moves — Large molecules, Minerals
Passive — requires — No energy — examples — Diffusion, Osmosis

Skillbuilder
ASSESSMENT

Performance: Use this concept map to have students compare active and passive transport.

ACTIVITY 3-1

OBJECTIVE: Design and carry out an experiment to measure the amount of water that can move across a cell membrane.

Time: One class period to brainstorm and set up the activity, 5 minutes each day to observe, and 15 minutes to summarize results.

PROCESS SKILLS applied in this activity are **measuring, observing, inferring, interpreting data,** and **predicting.**

PREPARATION

• 500-mL jars with lids can be used for this activity.

• Thick white syrup gives better results than thin syrup.

Cooperative Learning: Divide the class into Science Investigation Teams.

THINKING CRITICALLY

Student answers will include that osmosis is the movement of water molecules from where they are greater in number to where they are fewer in number. Students may suggest measuring the amount of vinegar and syrup at the end of the activity.

TEACHING THE ACTIVITY

Refer to the ***Activity Worksheets*** *for additional information and teaching strategies.*

• If eggshells have not completely dissolved at the end of two days, leave them in the vinegar an extra day.

• Caution students to handle the eggs very carefully to avoid breaking the membrane.

SUMMING UP/SHARING RESULTS

• Student answers will vary as to the amount of water that moved into the egg. At the start, there was 250 mL and usually 210 to 220 mL will remain. It varies with the size of the egg. Student answers will vary as to the amount of water that moved out of the egg. At the start, there was 250 mL of syrup and at the end, there will be 280 to 290 mL of liquid.

• The movement of water in container B resulted from the egg having more water molecules in it than the syrup had, so the water molecules moved from the egg into the syrup.

• The part of the egg that controlled what moved into and out of the egg was the cell membrane.

GOING FURTHER!

Students should predict that the egg will increase in size when placed into water. The water molecules will move into the egg. There are more water molecules in the water than in the egg. Water molecules move from where there are more to where there are fewer molecules.

Activity ASSESSMENT

Performance: To further assess students' understanding of osmosis, see USING LAB SKILLS, Question 11, on page 70.

PROGRAM RESOURCES

From the **Teacher Resource Package** use:

Activity Worksheets, pages 23-24, Activity 3-1, Observing Osmosis.

ACTIVITY 3-1 DESIGNING AN EXPERIMENT
Observing Osmosis

It is difficult to see osmosis occurring in cells because of the small size of most cells. However you learned that there are a few cells that can be seen using just your eyes. Can you see osmosis in one of these cells?

Getting Started

You need to determine a way to measure the amount of water that can move through the membrane of an egg. Vinegar contains 5% acetic acid and 95% water. The acetic acid will dissolve the shell of an egg, leaving the membrane exposed.

Thinking Critically

What is osmosis? Brainstorm how you can use the following materials to determine how much water moves through the membrane of the egg.

Materials 

Your cooperative group will use:

- raw egg (1)
- 500-mL containers with covers (2)
- vinegar (250 mL)
- white table syrup (250 mL)
- 250-mL graduated cylinder
- spoon
- 2 labels, A and B

Try It!

1. Measure 250 mL of vinegar into container A. Add 250 mL of syrup to container B. Place the egg into container A. Cover both containers.
2. *Observe* the egg for 2 days. Record the appearance of the egg in a data table.
3. At the end of 2 days, use the spoon to remove the egg from container A. Rinse the egg and place it into container B.
4. *Observe* the egg on the next day. Record its appearance.
5. Measure the remaining liquid in containers A and B. Record the amounts.

Summing Up/Sharing Results

- Calculate the amount of water that moved from the vinegar into the egg and from the egg to the syrup.
- What caused the movement of water in container B?
- What part of the egg controlled what moved into and out of the egg?

Going Further!

Predict what would happen if you placed the egg you removed from the syrup into a beaker of water. Explain your answer.

Sample data

Day	0	1	2
Vinegar volume		300 mL	275 mL
Observations	Egg becomes larger.		
Table syrup volume		300 mL	325 mL
Observations	Egg becomes smaller and rubbery.		

Energy in Cells

3-3

Objectives

- Explain the difference between producers and consumers.
- Compare and contrast the processes of photosynthesis and respiration.
- Describe how cells get energy from glucose through the process of fermentation.

New Science Words

metabolism
producers
consumers
photosynthesis
cellular respiration
fermentation

Trapping Energy for Life

Think of all the energy used in a basketball game. Where do the players get all that energy? The simplest answer is, "from the food they eat". Cells take chemical energy stored in food and change it into other forms of
1 energy that can be used in metabolism (muh TAB uh lihz um). **Metabolism** is the total of all activities of an organism that enable it to stay alive, grow, and reproduce.

Living things are divided into two groups based on how they obtain their food energy. These two groups are pro- 2
ducers and consumers. Organisms that make their own food, such as plants, are called **producers.** Organisms that can't make their own food are **consumers.** Producers change light energy into chemical energy by a process called **photosynthesis** (foht oh SIHN thuh sus). During photosynthesis, the energy from sunlight is used to make glucose from carbon dioxide (CO_2) and water. During this process oxygen is also given off. Plants and other producers use a green pigment called chlorophyll found in chloroplasts. Therefore, photosynthesis in green plants takes place in the chloroplasts. Producers use some of the glucose they make during photosynthesis for energy. The rest is stored.

Do you eat vegetables? Have you ever watched sheep graze? Consumers eat producers such as vegetables and take in the stored energy. Consumers also eat other consumers in order to get energy. These relationships form a food chain. Producers are at the beginning of every food chain.

Figure 3-9. Green plants are producers because they use sunlight, carbon dioxide, and water to produce chemical energy in the form of glucose.

What is the difference between a producer and a consumer?

SECTION 3-3

PREPARATION

SECTION BACKGROUND

- In addition to glucose, all other organic materials originate from photosynthesis.
- Respiration releases 18 times more energy from a food molecule than fermentation does.

PREPLANNING

- Obtain *Elodea* for Activity 3-2.

1 MOTIVATE

- **Bulletin Board:** Display pictures of persons using energy—playing sports, gardening, and so on. Discuss the source of their energy (food from green plants which had captured the sun's energy).

CROSS CURRICULUM

- **Language Arts:** *Chlorophyll* comes from the Greek words *chloros* and *phyllon.* Ask students what they think these words mean (*chloros* = green, *phyllon* = leaf). *Photosynthesis* comes from the Greek words *photo* and *syntithenai* (*photo* = light, *syntithenai* = putting together).

OPTIONS

ASSESSMENT—ORAL

- **How are fermentation and respiration alike?** *Both processes release energy.*
- **How can fermentation cause bread to rise?** *Yeast fermentation produces bubbles of carbon dioxide that become trapped in the bread dough, causing it to rise.*
- **Why is sugar often added to bread dough?** *to provide food for the yeast*
- **From where do fossil fuels such as coal, oil, and gas derive their energy?** *from photosynthesis carried out by plants long ago*

PROGRAM RESOURCES

From the **Teacher Resource Package** use:

Critical Thinking/Problem Solving, page 7, Mysteries of the Mind.

Cross-Curricular Connections, page 7, Cholesterol in Your Diet.

Use **Laboratory Manual,** page 21, Cell Respiration.

2 TEACH

Key Concepts are highlighted.

CONCEPT DEVELOPMENT

▶ Emphasize that respiration takes place in all living things.

CHECK FOR UNDERSTANDING

Ask students to make a list of at least five producers and five consumers.

RETEACH

Have students identify those organisms in the classroom aquarium in which photosynthesis takes place and those in which respiration takes place.

EXTENSION

For students who have mastered this section, use the **Reinforcement** and **Enrichment** masters or other OPTIONS.

TECHNOLOGY

Think Critically: Bacterial plastics can biodegrade totally.

Reference: Fuller, R. Clinton, and Robert W. Lenz. "Natural Plastics." *Natural History*, May 1990, pp. 82-84.

VideoDisc

STVS Cockroach on a Treadmill, Disc 5, Side 1

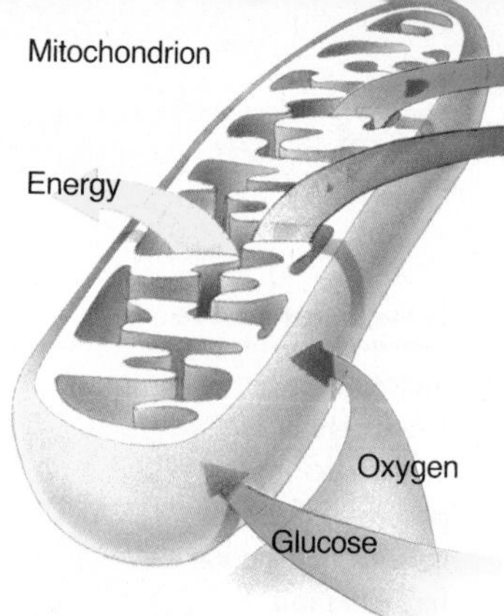

$$C_6H_{12}O_6 + 6\,O_2 \rightarrow 6\,CO_2 + 6\,H_2O + \text{Energy}$$

Glucose Oxygen Carbon dioxide Water

Figure 3-10. Respiration releases carbon dioxide, water, and energy. In what way do these products remind you of photosynthesis?

Releasing Energy for Life

Whether an organism is a producer or a consumer, it has to have some way to release energy from food. To do this, both producers and consumers break down food in their cells in processes called **cellular respiration.**

Inside most cells, glucose is the food that is broken down. The breakdown process takes place within mitochondria. This breakdown uses oxygen that you take in as you breathe.

During respiration, oxygen combines with glucose to release stored energy. The energy is released in a series of steps. Carbon dioxide and water are given off as waste

TECHNOLOGY

Biodegradable Plastics

Discarded plastics make up 32 percent of the garbage produced in the United States. Most of this plastic lasts forever. To reduce waste, scientists are working on plastics that can be broken down naturally.

One current method of breaking down plastic is to build molecules of cornstarch into the plastic. Cornstarch breaks down in nature, so the plastic is at least broken into smaller pieces. However, most of these plastic products are only 6 to 8 percent cornstarch, so the remaining plastic pieces are still large. Some scientists predict they will soon be able to produce plastics with a 50 percent cornstarch content.

A second method to help solve the plastic garbage problem involves the use of plastics produced by bacteria. Some soil bacteria are known to produce natural plastics and store them to use as energy sources. By changing factors such as temperature and food sources for the bacteria, scientists can adjust the characteristics of the plastic produced.

Think Critically: What is the advantage of bacterial plastics over plastic-cornstarch mixtures?

OPTIONS

Meeting Different Ability Levels

For Section 3-3, use the following **Teacher Resource Masters** depending upon individual students' needs.

◆ **Study Guide Master** for all students.

● **Reinforcement Master** for students of average and above average ability levels.

▲ **Enrichment Master** for above average students.

Additional Teacher Resource Package masters are listed in the OPTIONS box throughout the section. The additional masters are appropriate for all students.

◆ STUDY GUIDE 15

NAME ____ DATE ____ CLASS ____

STUDY GUIDE — Chapter 3

Energy in Cells — Text Pages 63-65

Complete the following sentences using appropriate terms from the textbook.

1. Plants get their energy from the **sun**.
2. The green pigment in plants is called **chlorophyll**.
3. Organisms that can make their own food are called **producers**.
4. A food chain always begins with a **producer**.
5. The process by which plants change light energy into chemical energy is called **photosynthesis**.
6. All the chemical changes that occur within the cells of an organism is called **metabolism**.
7. During fermentation, **carbon dioxide** and **alcohol** are produced.
8. Overworked muscles can still produce energy when oxygen levels are low by the process of **fermentation**.
9. The metabolism of glucose when oxygen is present is called **respiration**.
10. Consumers obtain energy by eating **producers** and other consumers.

Write the letter of the term that best matches each phrase.

Answer	Phrase	Term
k	1. energy is given off in the absence of oxygen	a. glucose
d	2. energy source necessary for photosynthesis	b. chlorophyll
l	3. causes muscles to tire	c. mitochondria
a	4. made and stored by plants	d. sunlight
f	5. obtain energy from producers	e. oxygen
b	6. traps radiant energy	f. consumers
j	7. energy can be lost in this form	g. alcohol
h	8. contain green pigment	h. chloroplasts
c	9. place where glucose is metabolized	i. water
e	10. combines with glucose during respiration	j. heat
g	11. fermentation product in yeast	k. fermentation
i	12. waste product in respiration	l. lactic acid

 15

products. Some of the energy produced in respiration is stored, and some of the energy is lost as heat.

Sometimes, however, during periods of strenuous activity, muscle cells run low on oxygen. The muscles begin to burn and sting. You begin to breathe harder and faster in an effort to supply the needed oxygen. But your body has another method to continue supplying small amounts of energy. When oxygen levels are low, the muscle cells begin to release energy from glucose by fermentation. **Fermentation** is a form of respiration that
3 releases energy from glucose when oxygen is insufficient. Lesser amounts of energy are produced by fermentation. In addition, carbon dioxide and an organic compound called lactic acid are formed. It is lactic acid that causes the muscles to burn and to be sore and stiff. Fermentation takes place in the cytoplasm of cells.

Yeast and some bacteria use fermentation to release energy. Yeast is an organism that breaks down the glucose in bread dough. Carbon dioxide and alcohol are released. The bubbles of carbon dioxide that are released cause the dough to rise. The alcohol is released into the air.

In summary, you can say that producers capture the sun's energy and store it in the form of food. Consumers eat producers. Cells use the glucose made by producers for energy.

Figure 3-11. During a sprint, oxygen can't be supplied to cells fast enough. Energy is supplied through fermentation and not respiration.

SECTION REVIEW

1. Explain the difference between producers and consumers.
2. How are chloroplasts important to photosynthesis?
3. Under what conditions do cells use fermentation?
4. Explain how the energy used by all living things on Earth can be traced back to sunlight.
5. **Apply:** Plants can use carbon monoxide, CO, instead of CO_2 for photosynthesis. CO is a major component of cigarette smoke. How can indoor plants help to purify the air in a room with cigarette smoke?
6. **Connect to Physics:** Find out how heat produced by respiration and body temperature are related.

Comparing and Contrasting — Skill Builder

Make a table that compares and contrasts respiration and fermentation. If you need help, refer to Comparing and Contrasting in the **Skill Handbook** on page 683.

● REINFORCEMENT 15

NAME ____________ DATE ________ CLASS ________

REINFORCEMENT — Chapter 3

Energy in Cells — Text Pages 63–65

Answer the following questions.

1. How do producers make their own food? **through photosynthesis**
2. Fill in the following equation for photosynthesis.
   sunlight + water + carbon dioxide + chlorophyll → **food** + **oxygen**
3. What are the end products of respiration? **carbon dioxide, water, and energy**
4. What is a food chain? **It is a process that begins as producers use energy from the sun to make food. Energy flows from producers to consumers when plant-eating consumers eat the producers. Consumers may eat other consumers.**
5. How does yeast cause bread to rise? **Through fermentation, yeast break down glucose and give off carbon dioxide and alcohol. The carbon dioxide gas causes the bread to rise.**
6. How do your muscles continue to get energy during high levels of activity when there is not enough oxygen? **Muscles can get energy from glucose through fermentation.**

Write the letter of the term that correctly answers each question or best completes each statement.

c 1. What process occurs in the mitochondria?
a. fermentation b. photosynthesis c. respiration d. metabolism

b 2. The green pigment in plants that traps radiant energy from the sun is called ____.
a. glucose b. chlorophyll c. oxygen d. water

c 3. During photosynthesis, plants produce glucose and give off ____.
a. carbon dioxide b. energy c. oxygen d. water

d 4. The energy used by all living things starts with ____.
a. producers b. consumers c. respiration d. sunlight

c 5. During respiration some energy is lost as ____.
a. chemical energy b. radiant energy c. heat d. carbon dioxide

a 6. Fermentation gives off energy without using ____.
a. oxygen b. glucose c. energy d. carbon dioxide

c 7. When muscles are overworked, soreness is caused by a buildup of ____.
a. glucose b. carbon dioxide c. lactic acid d. energy

15

▲ ENRICHMENT 15

NAME ____________ DATE ________ CLASS ________

ENRICHMENT — Chapter 3

Energy in Cells — Text Pages 63–65

During photosynthesis, radiant energy is used to make glucose from carbon dioxide and water. The glucose is stored in the leaves in the form of starch, another carbohydrate.

The following experiment will show whether starch is being made and stored when photosynthesis is interrupted. Iodine will be used as an **indicator** for the presence of starch. An indicator shows the presence of a certain substance.

Materials
- living green plant
- piece of black paper
- paper clips
- saucer with rubbing alcohol
- saucer with cornstarch solution
- medicine dropper
- tincture of iodine

Procedure
1. Attach a strip of black paper to the upper surface of 3 leaves. Cover about 1/3 of the leaf.
2. Place the plant in a sunny window for 2 or 3 days.
3. Cut the partly-covered leaves from the plant.
4. Remove the black paper and soak the leaves overnight in the saucer of alcohol.
5. Remove the leaves from the alcohol. With the medicine dropper, place a drop of iodine on parts of the leaves that were covered and on parts that were not covered.
6. Add a few drops of iodine to the cornstarch solution.

Observations

Leaf	Color		
	covered	uncovered	cornstarch
1	brown	blue-black	blue-black
2	brown	blue-black	
3	brown	blue-black	

Conclude and Apply
1. Iodine is used as an indicator solution in this experiment. What change do you see in the color of iodine when it is added to cornstarch? **It changes from brown to blue-black.**
2. Why do the uncovered parts of the leaves turn a different color than the covered parts? **No starch is stored in the covered parts because the chloroplasts in that part of the leaf are not receiving any radiant energy for photosynthesis.**

15

MINI QUIZ

Use the Mini Quiz to check students' recall of chapter content.

1. **The total of all the activities of an organism that enable it to live, grow, and reproduce is _____ .** *metabolism*
2. **Organisms that can make their own food are _____ .** *producers*
3. **The process in which energy is released without the use of oxygen is _____ .** *fermentation*

3 CLOSE

Cooperative Learning: Use Expert Teams to study photosynthesis, respiration, and fermentation.

▶ Ask questions 1-4 and the **Apply** Question in the Section Review.

SECTION REVIEW ANSWERS

1. Producers use sunlight to make food. Consumers obtain energy by eating producers or food producers make.
2. They contain chlorophyll and are where photosynthesis takes place.
3. when oxygen is depleted
4. Producers change light energy from the sun into chemical energy used by all living things.
5. Apply: Plants can remove carbon monoxide from the air by using it in photosynthesis and releasing oxygen.
6. Connect to Physics: heat—a by-product of respiration; temperature—a measure of body heat

Skill Builder

	Ferm.	Resp.
Start	Glucose	O_2, glucose
O_2?	No	Yes
Energy released	small amount	large amount

Skillbuilder ASSESSMENT

Portfolio: Use this Skillbuilder to assess students' abilities to Compare and Contrast photosynthesis and respiration.

SECTION 3-4

PREPARATION

SECTION BACKGROUND

▶ Recycling nonbiodegradable products could stop the accumulation of glass, plastic, aluminum, and motor oil in the environment.

1 MOTIVATE

▶ Display items acceptable for recycling in your community.

2 TEACH

Key Concepts are highlighted.

CONCEPT DEVELOPMENT

Cooperative Learning: Use Paired Partners cooperative strategy. Send each pair into the school yard to collect recyclable materials. Bring them back to the classroom and discuss the importance of removing these materials from the environment.

Connect to... Chemistry

Answer: It might not be necessary to mine bauxite; all aluminum could be recovered from recycling.

3-4 Nonbiodegradable Materials in Your Environment

New Science Words

biodegradable

Objectives

▶ **Explain the consequences of nonbiodegradable substances in the environment.**

Connect to... Chemistry

Bauxite is a claylike raw material that is the major source of aluminum compounds in the world. It takes 95% less energy to recycle aluminum products than it does to remove aluminum from bauxite. From this, what could you infer about the need to mine bauxite if everyone practiced recycling?

Recycling and the Environment

Do you ever drop candy or gum wrappers on the ground? It's likely to be a very long time before these wrappers disappear. Materials such as plastic wrappers and aluminum foil take years to break down in the environment.

Substances that do break down easily in the environment are **biodegradable.** A combination of sunlight, weathering, oxygen, bacteria and fungi, and moisture causes them to decompose, or break down, into the elements they are made of. For example, fruit and leaves are biodegradable. They are plant parts in a cycle that returns nutrients to the soil. Aluminum cans, glass bottles, and plastic bags, which are often seen along roadways, are nonbiodegradable. In fact, all of these take a very long time to decompose. A tin can needs 100 years to break down. An aluminum can requires more than 200 years to break down, and some types of plastics may require 450 years or more!

The situation becomes a problem when you consider the huge amounts of these materials that are dumped into the environment each year. The average United States cit-

OPTIONS

Meeting Different Ability Levels

For Section 3-4, use the following **Teacher Resource Masters** depending upon individual students' needs.

◆ **Study Guide Master** for all students.

● **Reinforcement Master** for students of average and above average ability levels.

▲ **Enrichment Master** for above average students.

Additional Teacher Resource Package masters are listed in the OPTIONS box throughout the section. The additional masters are appropriate for all students.

◆ **STUDY GUIDE 16**

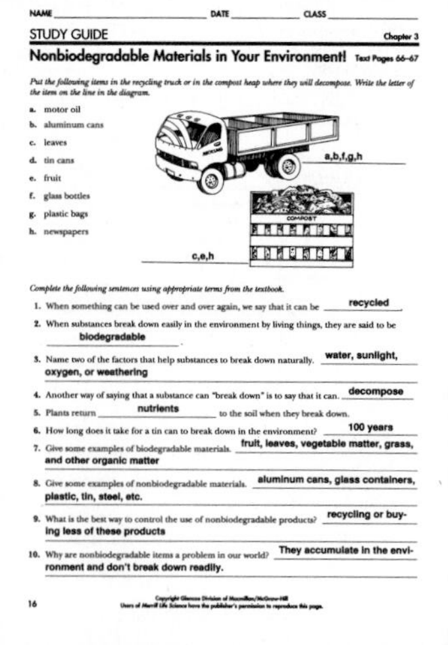

NAME ______ DATE ______ CLASS ______

STUDY GUIDE — Chapter 3

Nonbiodegradable Materials in Your Environment! Text Pages 66-67

Put the following items in the recycling truck or in the compost heap where they will decompose. Write the letter of the item on the line in the diagram.

a. motor oil
b. aluminum cans
c. leaves
d. tin cans
e. fruit
f. glass bottles
g. plastic bags
h. newspapers

Complete the following sentences using appropriate terms from the textbook.

1. When something can be used over and over again, we say that it can be **recycled**.
2. When substances break down easily in the environment by living things, they are said to be **biodegradable**.
3. Name two of the factors that help substances to break down naturally. **water, sunlight, oxygen, or weathering**
4. Another way of saying that a substance can "break down" is to say that it can. **decompose**
5. Plants return **nutrients** to the soil when they break down.
6. How long does it take for a tin can to break down in the environment? **100 years**
7. Give some examples of biodegradable materials. **fruit, leaves, vegetable matter, grass, and other organic matter**
8. Give some examples of nonbiodegradable materials. **aluminum cans, glass containers, plastic, tin, steel, etc.**
9. What is the best way to control the use of nonbiodegradable products? **recycling or buying less of these products**
10. Why are nonbiodegradable items a problem in our world? **They accumulate in the environment and don't break down readily.**

16

izen uses and throws away about 299 kilograms of nonbiodegradable material each year. At this rate, it probably wouldn't take very long for the trash produced by several people to fill up a space the size of a circus tent.

Are these products so bad that we should just stop producing them? That would be difficult. Nearly every day, you are in contact with or using some type of plastic, glass, or aluminum! There is no denying that many lives have been saved thanks to plastics in medical equipment and supplies.

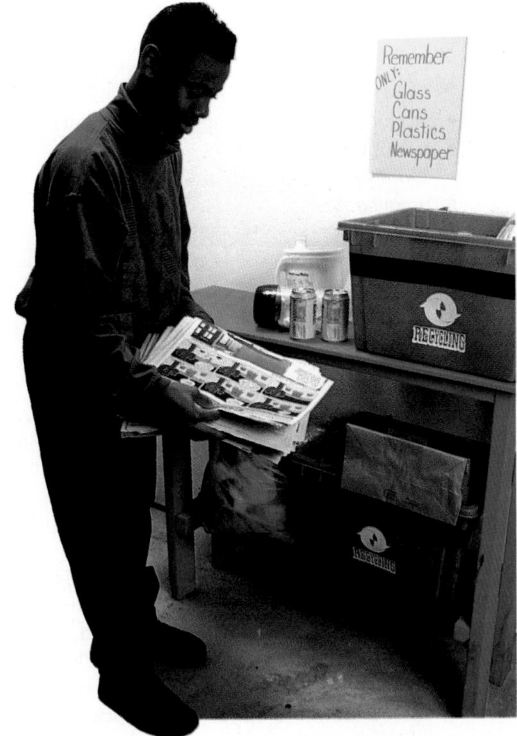

We aren't likely to quit using these products, and the best way to control their use may be to use them over and over again and change them into new products. We call this recycling. Many communities are very interested in finding new uses for nonbiodegradable materials and have begun recycling programs. Glass, plastic, aluminum, paper, and motor oil can all be recycled. Scientists are developing new forms of previously nonbiodegradable products, such as soap, paint, or trash bags, that decompose more easily. By buying such products and making an effort to recycle the others, we can become part of the effort to keep these materials from accumulating and continuing to cause environmental problems.

SECTION REVIEW

1. Give two ways in which you can help the current problem with nonbiodegradable materials.
2. Discuss how substances decompose naturally, and explain what happens to the materials that are released by decomposition.
3. **Connect to Chemistry:** Find out what becomes of used motor oil.

You Decide!

Some states have aluminum can deposit laws. Each empty aluminum can is worth five cents. This is one way to help motivate people to keep the roadsides free from litter. Should all states be required to develop recycling programs?

● REINFORCEMENT 16

NAME ______ DATE ______ CLASS ______

REINFORCEMENT — Chapter 3

Nonbiodegradable Materials in Your Environment! — Text Pages 66–67

Composting is one way of recycling organic materials. Gardeners often pile up leaves, dirt, and grass clippings and let them decompose. By the following season, the mixture has turned into soil. Other things can go into a compost mixture.

Circle the items you think could be added to a compost bin.

(eggshells)	(coffee grounds)	toothpaste
foil gum wrappers	plastic straws	motor oil
(weeds)	foam cups	(orange rinds)
(branches)	cellophane candy wrappers	glass bottles
candles	aluminum cans	(dirt)
(banana peels)	plastic gloves	(sawdust)
(wood ashes)	(paper shreds)	(manure)

Answer the following questions.

1. What do the items you circled have in common? **They are biodegradable.**
2. What happens when plants decompose? **Nutrients are returned to the soil as the plant material breaks down into its parts.**
3. What environmental factors help the process of decomposition? **sunlight, weathering, oxygen, and moisture**

   Knowing this, what is the best way to make your compost pile produce soil? **add water, stir or turn over with a shovel so oxygen gets to lower layers, and place the pile in sunlight**
4. Which of the listed items can be recycled? **glass, some plastics, aluminum cans, paper, and motor oil**
5. What are scientists doing about the problem of nonbiodegradable materials in the environment? **They are developing new products like soap, paint, or trash bags that will easily decompose.**
6. What steps can you take to help? **recycle, buy biodegradable products, buy fewer nonbiodegradable products, insist on less packaging, compost organic waste**

16

▲ ENRICHMENT 16

NAME ______ DATE ______ CLASS ______

ENRICHMENT — Chapter 3

Nonbiodegradable Materials in Your Environment! — Text Pages 66–67

RESOURCE RECOVERY

Take a look in the waste paper baskets at school. How much trash is generated by one classroom in a day? Where does it go after it leaves the school?

For a set period of time, measure how much waste paper accumulates in your school. A good way to measure paper is by mass. Measure the mass of a waste paper basket full of paper using a bathroom scale. Then measure the mass of the empty waste paper basket. Subtract the mass of the empty waste paper basket from the full basket. The result will be the mass of the waste paper.

Estimate how much waste paper is being thrown out every day from all the rooms at school. Then, find out its cash value. Look in the phone book to locate waste paper dealers in your area. They will give you prices for various kinds of paper. How much money is your accumulated paper worth each year? Maybe your school could start a recycling program.

Data

	Kilograms of waste paper	Cash value
day		
week		
year		

Make a map of your state showing where recycling of waste is done, and what kinds of wastes are being used. Include collecting sites for recyclable materials that are run by local businesses and organizations. Make a key indicating what materials are recycled in your state. The more information you can include on your map, the more helpful it will be to others. The Environmental Protection Agency (EPA) in your state is a good source of information.

Data for map

	Symbol on map	Recycler's name and address
paper		
glass		
aluminum		
plastic		
other		

Trace the outline of your state from an atlas. Use this drawing to make a large poster to promote recycling. Use drawings, advertisements, or photographs as illustrations on your map.

16

CHECK FOR UNDERSTANDING

Have the students write a paragraph on how they will contribute to a recycling effort.

RETEACH

Invite a speaker from a local recycling program to visit the class and explain the importance of recycling.

EXTENSION

For students who have mastered this section, use the **Reinforcement** and **Enrichment** masters or other OPTIONS.

3 CLOSE

▶ Ask questions 1-2 in the Section Review.

SECTION REVIEW ANSWERS

1. Convince someone to recycle, or begin recycling in your home. Try to purchase products in reusable containers.

2. Answers should state that a combination of moisture, sunlight, weathering, and oxygen, bacteria and fungi decompose substances and release nutrients to soil to be used again.

3. Connect to Chemistry: Used motor oil may be refined and reused.

YOU DECIDE!

Answers will vary. Students may agree that all states should have recycling programs but disagree on how these programs are handled.

VideoDisc

STVS: Garbage Science, Disc 6, Side 2

PROGRAM RESOURCES

From the **Teacher Resource Package** use:

Science and Society, page 7, Paper or Polystyrene Packaging?

Science Integration Activities, 3, Conservation-Recycling

ACTIVITY 3-2
45 minutes

OBJECTIVE: **Observe** plants carrying on both photosynthesis and respiration.

PROCESS SKILLS applied in this activity:
▶ **Measuring** in Procedure Steps 2, 3, and 4.
▶ **Observing** in Procedure Step 6 and Analyze Questions 2 and 3.
▶ **Using Tables** in Data and Observations.
▶ **Inferring** in Conclude and Apply Questions 5, 6, and 7.

COOPERATIVE LEARNING Use the Science Investigation strategy. In each group assign one person to gather materials, one person to fill the test tubes, one person to make the sodium bicarbonate solution, and one person to record data.

TEACHING THE ACTIVITY
Alternate Materials: Baby food jars with lids may be used for test tubes.
▶ *Elodea* should be kept in the dark for two days before the activity. Use sharp scissors to make a clean diagonal cut at the bottom of each stem.

Activity
ASSESSMENT
Performance: To further assess students' understanding of photosynthesis and respiration, have them now place test tube 1 in the dark for 30 minutes and explain the results.

ACTIVITY 3-2 Photosynthesis and Respiration

Problem: *When do plants carry on photosynthesis?*

Materials

- test tubes, 16 by 150 mm with stoppers (4)
- test-tube rack
- stirring rod
- balance
- sodium bicarbonate
- bromthymol blue solution in dropping bottle
- aged tap water
- pieces of *Elodea* (2)

Procedure
1. Label the test tubes 1 through 4. Fill each tube about three-fourths full with aged tap water.
2. Add 1 g of sodium bicarbonate to each test tube. Sodium bicarbonate releases carbon dioxide when mixed with water.
3. Add 5 drops of bromthymol blue solution to each test tube. Bromthymol blue turns green to yellow in the presence of an acid.
4. Cut two 10-cm pieces of *Elodea*. Place one piece of *Elodea* in test tube 1 and one piece in test tube 3. Stopper the test tubes. Record the color of the solution in each test tube.
5. Place test tubes 1 and 2 in bright light and test tubes 3 and 4 in the dark for 30 minutes.
6. Observe the test tubes and record the colors.

Analyze
1. What is indicated by the color of the water in all four tubes at the start of the activity?
2. What color were the liquids in tubes 1 and 3 after 30 minutes?
3. What color were the liquids in tubes 2 and 4 after 30 minutes?

Data and Observations Sample Data

Test tube	Color at start	Color after 30 minutes
1	All will	Blue
2	be	Yellow-green
3	yellow-	Yellow
4	green	Yellow-green

Conclude and Apply
4. Describe the purpose of tubes 2 and 4.
5. Explain the color of the liquid in tube 1 after 30 minutes.
6. Explain the color of the liquid in tube 3 after 30 minutes.
7. From your observations, *infer* when a green plant carries out photosynthesis.

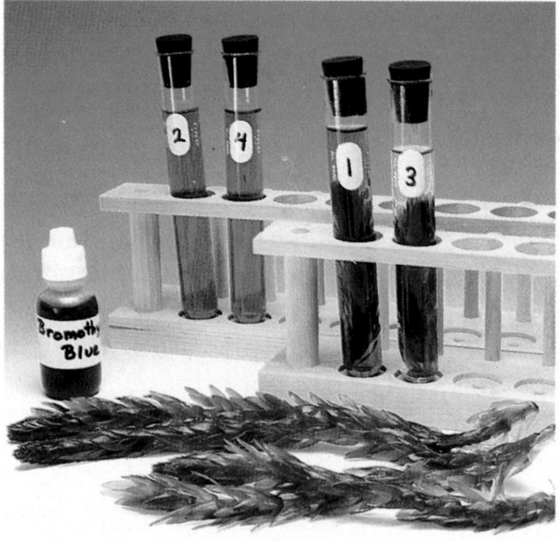

ANSWERS TO QUESTIONS
1. Carbon dioxide is present in the water.
2. One was blue and three was yellow-green.
3. They were both yellow-green.
4. controls to show how the carbon dioxide content of water changes over time
5. The change from yellow-green to blue shows that the plant is using carbon dioxide in photosynthesis.
6. The change to deep yellow shows that the plant is giving off carbon dioxide in respiration.
7. A green plant carries on photosynthesis in the presence of light.

PROGRAM RESOURCES
From the **Teacher Resource Package** use:
Activity Worksheets, pp. 25-26, Activity 3-2, Photosynthesis and Respiration.

CHAPTER

REVIEW

SUMMARY

3-1: Chemistry of Living Things

1. Everything is made of matter, which is composed of atoms arranged in elements, molecules and compounds.
2. Both inorganic and organic molecules are important to living things.
3. Carbohydrates, lipids, proteins, and nucleic acids are organic compounds. Most inorganic compounds do not contain carbon.

3-2: Cell Transport

1. The cell membrane controls what molecules can pass through it.
2. Molecules move by diffusion from areas of greater numbers to areas of lesser numbers. In osmosis, water diffuses through a selectively permeable membrane.
3. The cell expends energy to move molecules by active transport, but not by passive transport.

3-3: Energy in Cells

1. Producers make energy in the form of food, which consumers eat.
2. Green plants use light energy to make chemical energy in the form of glucose during photosynthesis. In respiration, glucose is broken down and energy is released.
3. Some yeasts, bacteria, and oxygen-deprived cells carry out fermentation to release small amounts of energy from glucose without using oxygen.

3-4: Science and Society: Nonbiodegradable Materials in Your Environment

1. Nonbiodegradable substances do not break down easily in the environment and accumulate as wastes.

KEY SCIENCE WORDS

a. **active transport**
b. **biodegradable**
c. **carbohydrates**
d. **cellular respiration**
e. **consumers**
f. **diffision**
g. **endocytosis**
h. **enzymes**
i. **equilibrium**
j. **exocytosis**
k. **fermentation**
l. **lipids**
m. **metabolism**
n. **nucleic acids**
o. **osmosis**
p. **passive transport**
q. **photosynthesis**
r. **producers**
s. **proteins**

UNDERSTANDING VOCABULARY

Match each phrase with the correct term from the list of Key Science Words.

1. organisms that make their own food
2. the movement of water molecules through a selectively permeable membrane
3. movement of molecules that requires energy
4. substance that is broken down in the environment by bacteria or fungi
5. capable of releasing more energy than carbohydrates
6. sugar, starch, and cellulose
7. made up of amino acids
8. use of light energy to make food
9. release of a substance from a small sac at the membrane
10. the breakdown of glucose without the presence of oxygen

CHAPTER

REVIEW

SUMMARY

Have students read the summary statements to review the major concepts of the chapter.

UNDERSTANDING VOCABULARY

1. r
2. o
3. a
4. b
5. l
6. c
7. s
8. q
9. j
10. k

ASSESSMENT

Portfolio

Encourage students to place in their portfolios one or two items of what they consider to be their best work. For each item, ask students to explain why that item was chosen and what they learned from it. Items might be selected from the following.

- In Your Journal food log, p. 57
- Skillbuilder concept map, p. 61
- Skillbuilder table, p. 65

Performance

Additional performance assessments may be found in *Performance Assessment and Science Integration Activities* that accompany **Merrill Life Science.** Performance Task Assessment Lists and rubrics for evaluating these activities and other products generated throughout the chapter can be found in Glencoe's *Performance Assessment in Middle School Science.*

OPTIONS

ASSESSMENT

To assess student understanding of material in this chapter, use the resources listed in the Program Resources box to the right.

COOPERATIVE LEARNING

Consider using cooperative learning in the THINK AND WRITE CRITICALLY, APPLY, and MORE SKILL BUILDERS sections of the Chapter Review.

PROGRAM RESOURCES

From the **Teacher Resource Package** use:

Chapter Review, pages 9-10.

Chapter and Unit Tests, pages 13-16, Chapter Test.

CHAPTER

REVIEW

CHECKING CONCEPTS

1. b	**6.** c
2. b	**7.** d
3. a	**8.** a
4. b	**9.** c
5. d	**10.** d

USING LAB SKILLS

ASSESSMENT

Use these alternate lab exercises to assess students' understanding of skills used in this chapter.

11. Raisins are dehydrated and water in the glass diffused into the raisins, causing them to become plump.

12. An enzyme in the fresh pineapple will keep pudding from solidifying; the enzyme in the canned pineapple has been destroyed.

THINK AND WRITE CRITICALLY

13. Diffusion is the movement of molecules from an area of higher concentration to one of lower concentration. Osmosis is the diffusion of water molecules through a selectively permeable membrane.

14. Accept all reasonable responses. One answer might be that respiration takes place in mitochondria and energy is released. Muscle cells utilize more energy than other cells.

15. Membranes are selectively permeable. Molecules pass through depending upon size.

16. Water molecules diffuse from outside to inside until there are 17 molecules inside the cell as well as outside.

17. At night, some starch is broken down by respiration and used to repair, maintain, or grow plant tissue; most remains stored.

CHAPTER

REVIEW

CHECKING CONCEPTS

Choose the word or phrase that completes the sentence.

1. Energy is used to move molecules by ____.
a. diffusion **c.** osmosis
b. active transport **d.** passive transport

2. Bacteria are taken into cells by ____.
a. osmosis **c.** exocytosis
b. endocytosis **d.** diffusion

3. An example of an organic compound is ____.
a. $C_6H_{12}O_6$ **c.** H_2O
b. NO_2 **d.** O_2

4. ____ are examples of carbohydrates.
a. Enzymes **c.** Waxes
b. Sugars **d.** Proteins

5. Organic compounds in the chromosomes are ____.
a. carbohydrates **c.** lipids
b. water molecules **d.** nucleic acids

6. The organic molecule that releases the greatest amount of energy is ____.
a. carbohydrate **c.** lipid
b. water **d.** nucleic acid

7. Salt is a(n) ____ molecule.
a. organic **c.** carbohydrate
b. lipid **d.** inorganic

8. ____ occurs when molecules are evenly distributed.
a. Equilibrium **c.** Fermentation
b. Metabolism **d.** Cellular respiration

9. ____ are organisms that can't make their own food.
a. Biodegradables **c.** Consumers
b. Producers **d.** Enzymes

10. Chlorophyll is needed for ____.
a. fermentation **c.** diffusion
b. cellular respiration **d.** photosynthesis

USING LAB SKILLS

11. In Activity 3-1 on page 62, you studied osmosis in an egg. Place 10 to 12 raisins in a glass of water and allow them to stand overnight. Explain your results.

12. The Mini-Lab on page 57 asked you to find out how enzymes work. Label two containers A and B. Make 1 cup of prepared instant pudding and divide it between the two containers. Add 1 teaspoon of fresh pineapple juice to container A and stir. Add 1 teaspoon of canned pineapple juice to container B and use a clean spoon to sir. What can you infer about differences between canned and fresh pineapple juice? What can you infer about your results in terms of enzyme activity?

THINK AND WRITE CRITICALLY

Answer the following questions in your Journal using complete sentences.

13. Compare and contrast diffusion and osmosis.

14. Why might there be many more mitochondria in muscle cells than in other types of cells?

15. Explain how some substances, but not others, can pass through the cell membrane.

16. Describe how a cell reaches equilibrium with its environment if there are ten water molecules inside and 24 molecules outside.

17. During the night, what happens to the starch the plant made and stored in its leaves?

APPLY

18. If you could place a single red blood cell in a glass of distilled water, what do you think you would see happen to the cell? Explain.

19. In snowy states, salt is used to melt ice on the roads. Explain what happens to many roadside plants as a result.

20. Explain why sugar dissolves faster in hot tea than in iced tea.

21. Explain what would happen to the consumers in a lake if all the producers died.

22. Meat tenderizers contain enzymes. How do these enzymes affect protein in meat?

MORE SKILL BUILDERS

If you need help, refer to the Skill Handbook.

1. **Interpreting Data:** In an experiment that tests the rate of photosynthesis in water plants, the water plants were placed at different distances from a light source. Bubbles of gas coming from the plants were counted to measure the rate. What can you say about how the rate is affected by the light?

Beaker number	Distance from light	Bubbles per minute
1	10 cm	45
2	30 cm	30
3	50 cm	19
4	70 cm	6
5	100 cm	1

2. **Making and Using Graphs:** Use the data from Question 1 to graph the relationship between photosynthesis rate and distance from light.
3. **Classifying:** Classify these common foods into a category of carbohydrates, lipids, or proteins: butter, bread, candy bar, cereals, cheese, cornstarch, fish, margarine, meats, pasta, peanut butter, shortening, sugar, vegetable oil.

4. **Concept Mapping:** Sequence the parts of matter from smallest to largest: element, atom, compound in an events chain concept map.
5. **Hypothesizing:** Make a hypothesis about what will happen to wilted celery when placed into a glass of plain water.

PROJECTS

1. Design an experiment to show that respiration takes place in growing lima bean seeds.
2. Make a list of elements found in ten substances in your home. Use the labels to determine the elements each product contains.

APPLY

18. The red blood cell would burst because so many water molecules would move into the cell. The water would move from an area of greater concentration (outside the cell) to an area of lesser concentration (inside the cell) in an effort to reach equilibrium.

19. Plants die as water molecules move out of the cells.

20. Sugar molecules move faster in hot water due to the increased energy in the heated water.

21. Consumers would also die; they depend on producers for food.

22. The enzymes break down the bonds of the proteins; this makes the meat more tender.

MORE SKILL BUILDERS

1. Interpreting Data: The closer a plant is to the light, the faster its rate of photosynthesis.

2. Making and Using Graphs:

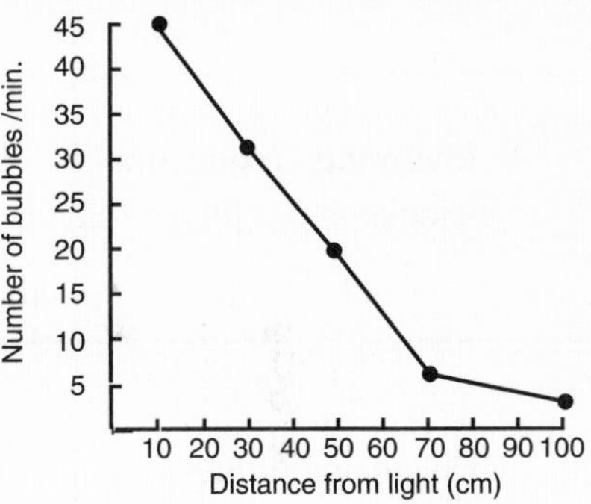

3. Classifying: Carbohydrates: bread, candy bar, cereals, cornstarch, pasta, sugar

Lipids: butter, margarine, peanut butter, shortening, vegetable oil

Proteins: cheese, fish, meats

4. Concept Mapping: Sequence in chain should be:

atom → element → compound

5. Hypothesizing: The wilted celery will become crisp as water molecules move by osmosis into its cells to reach an equilibrium.

CHAPTER

Cell Reproduction

CHAPTER SECTION	OBJECTIVES	ACTIVITIES
4-1 Cell Growth and Division (3 days)	1. **Describe** mitosis and explain its importance. 2. **Explain** differences between mitosis in plant and animal cells. 3. **Distinguish** between asexual and sexual reproduction and give two examples of asexual reproduction.	**MINI-Lab:** *How does one cell become two?* p. 78 **Activity 4-1:** *Mitosis in Plant and Animal Cells,* p. 79
4-2 Sexual Reproduction and Meiosis (1 day)	1. **Describe** the stages of meiosis and its end products. 2. **Name** the cells involved in fertilization and **explain** how fertilization occurs.	
4-3 DNA (2 days)	1. **Construct** and **identify** the parts of a model of a DNA molecule. 2. **Describe** how DNA copies itself.	**MINI-Lab:** *How is RNA made?* p. 90
4-4 Inventing Organisms Science & Society (1 day)	1. **Explain** the term *transgenic organism.* 2. **Explain** some advantages and disadvantages of patenting organisms.	**Activity 4-2:** *Designing an Experiment (Making a Model of DNA),* p. 94
Chapter Review		

ACTIVITY MATERIALS

FIND OUT	ACTIVITIES		MINI-LABS	
Page 73 5 soaked pinto bean seeds paper towels plastic bag hand lens	**4-1 Mitosis in Plant and Animal Cells, p. 79** prepared slide of an onion root tip prepared slide of a whitefish embryo microscope	**4-2 Designing an Exeriment, p. 94** 6 colors of construction paper, 2 pieces of each color scissors (4) heavy paper (1) tape	**How does one cell become two? p. 78** poster board yarn thread paper glue pens or markers	**How is RNA made? p. 90** none

CHAPTER FEATURES	TEACHER RESOURCE PACKAGE	OTHER RESOURCES
Problem Solving: *Cell Biology and Cancer Research,* p.78 **Skill Builder:** *Outlining,* p. 81	**Ability Level Worksheets** ◆ ***Study Guide,*** p. 17 ● ***Reinforcement,*** p. 17 ▲ ***Enrichment,*** p. 17 **Activity Worksheets,** pp. 32, 33, 38 **Concept Mapping,** p. 13 **Transparency Masters,** pp. 9-10	**Color Transparency 5,** Mitosis **Lab Manual:** Chromosomes, p. 25 **Lab Manual:** Mitosis, p. 23 **STVS:** Disc 4, Side 2
Skill Builder: *Making and Using Tables,* p.85	**Ability Level Worksheets** ◆ ***Study Guide,*** p. 18 ● ***Reinforcement,*** p. 18 ▲ ***Enrichment,*** p. 18 **Transparency Masters,** pp. 11-12	**Color Transparency 6,** Meiosis
Technology: *The Bacteria Factory,* p. 89 **Skill Builder:** *Concept Mapping,* p. 91	**Ability Level Worksheets** ◆ ***Study Guide,*** p. 19 ● ***Reinforcement,*** p. 19 ▲ ***Enrichment,*** p. 19 **Critical Thinking/Problem Solving,** p. 8 **Cross-Curricular Connections,** p. 8 **Activity Worksheets,** p. 39 **Transparency Masters,** pp. 13-14	**Color Transparency 7,** DNA Replication **Science Integration Activity 4**
You Decide! p. 93	**Ability Level Worksheets** ◆ ***Study Guide,*** p. 20 ● ***Reinforcement,*** p. 20 ▲ ***Enrichment,*** p. 20 **Activity Worksheets,** pp. 34-35 **Science and Society,** p. 8	**STVS:** Disc 4, Side 2
Summary Key Science Words Understanding Vocabulary Checking Concepts Using Lab Skills Think & Write Critically Apply More Skill Builders Projects	**ASSESSMENT RESOURCES** **Chapter Review,** pp. 11-12 **Chapter Test,** pp. 17-20 **Unit Test,** pp. 21-22 **Performance Assessment in Middle School Science (PAMSS)**	**Chapter Review Software** **Test Bank** **Alternate Assessment** **Performance Assessment**

◆ **Basic** ● **Average** ▲ **Advanced**

ADDITIONAL MATERIALS

SOFTWARE	AUDIOVISUAL	BOOKS/MAGAZINES
Cell Growth and Mitosis, Classroom Consortia Media. *Gene Machine,* Queue. Cell Functions: Growth and Mitosis, IBM. *The Fascinating Story of Cell Growth,* Queue.	*Cell Division and the Life Cycle,* filmstrip, Human Relations Media. *A Clone of Frogs,* film, EBEC. *The Living Cell: DNA,* film, EBEC. *Meiosis/Mitosis,* laserdisc, EBEC. *Reproduction in Organisms,* laserdisc, Aims Media.	Basera, Renato. *The Biology of Cell Production.* Cambridge, MA: Harvard University Press, 1985. Watson, James, D. and John Tooze. *Recombinant DNA: A Short Course.* NY: Freeman, W.H. and Company, 1983. Watson, James D. and Nancy A. Hopkins. *Molecular Biology of the Gene, Complete.* 4th ed. Redwood City, CA: Benjamin-Cummings Publishing Company, 1988.

CHAPTER

4

CELL REPRODUCTION

THEME DEVELOPMENT: The major themes in this chapter are evolution and homeostasis. DNA controls all cell activities by directing the production of proteins in living organisms. Changes in the DNA result in evolutionary changes that are inherited.

CHAPTER OVERVIEW

- **Section 4-1:** This section describes why cells divide, the cell cycle, and how body cells reproduce. Asexual and sexual reproduction are defined and discussed.
- **Section 4-2:** Sexual reproduction, the role of sex cells, and sex cell formation are presented in this section.
- **Section 4-3:** This section discusses the chemical composition of DNA, how DNA replicates, and how DNA directs the production of proteins. The section ends with an explanation of mutations and why they occur.
- **Section 4-4: Science and Society:** Transgenic organisms are discussed. The patenting of these organisms and some of the resulting questions are also explored.

CHAPTER VOCABULARY

mitosis	fertilization
chromosomes	zygote
asexual reproduction	DNA
sexual reproduction	gene
sperm	RNA
egg	mutation
meiosis	transgenic organisms

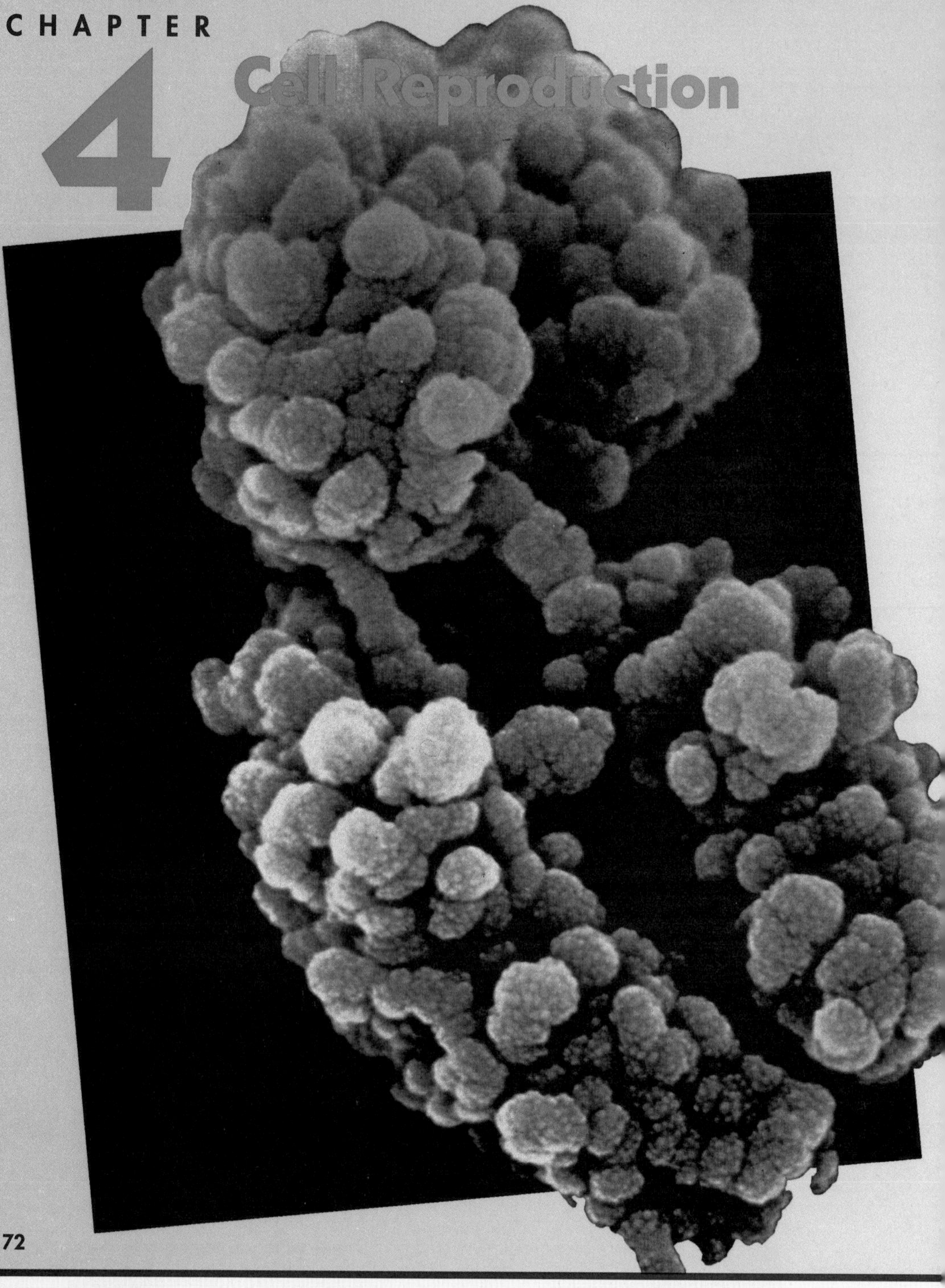

CHAPTER

4 Cell Reproduction

OPTIONS

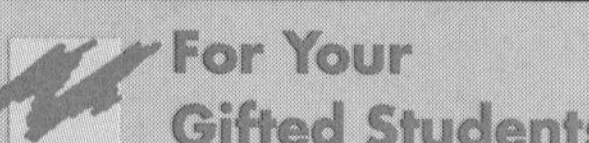

For Your Gifted Students

Have students interview a physician, nurse, or head of a local special education program to find out about Down syndrome. They should find out what the latest research shows about detection, causes, and care. Students will want to ask about the characteristics and capabilities of a person with Down syndrome.

For Your Mainstreamed Students

Collect planarians from a freshwater pond or stream. Have students place them in small dishes. Cut them in half across the middle to make head and tail sections. The specimens should be covered with pond water and kept in a cool, dark spot. Students should record their observations for a two-week period as regeneration occurs.

Do you know that your life hangs by a thread? Miles of thread-like DNA and protein condense to make up chromosomes in your cells, similar to the one to the left. When cells divide and organisms grow, chromosomes are the carriers of the information that shapes each new cell.

FIND OUT!

Do this activity to see what changes take place in a plant as it develops.

Carefully split open a pinto bean seed that has soaked in water overnight. Look at it with a hand lens. What do you think you find inside the top part of the seed? Place four other bean seeds in a moist paper towel in a plastic bag. *Observe* the seeds for a few days. What happened inside the seed? *Infer* how this happened.

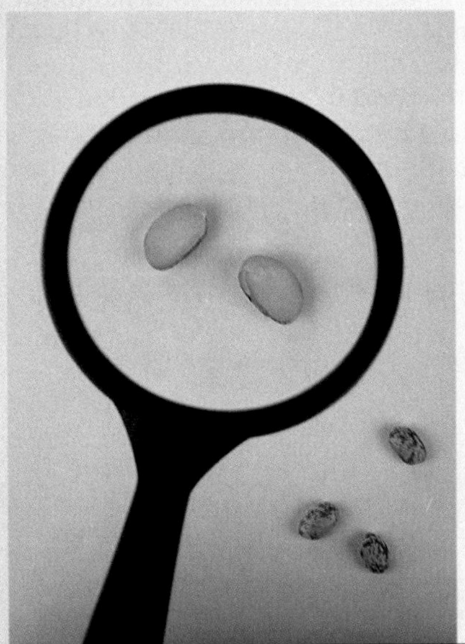

Gearing Up

Previewing the Chapter

Use this outline to help you focus on important ideas in this chapter.

Section 4-1 Cell Growth and Division
- What Happens When Cells Divide?
- The Cell Cycle
- Mitosis
- Types of Reproduction

Section 4-2 Sexual Reproduction and Meiosis
- Sexual Reproduction
- The Importance of Sex Cells
- Meiosis—Sex Cell Formation

Section 4-3 DNA
- What Is DNA?
- How DNA Copies Itself
- DNA and Genes
- Mutations

Section 4-4 Science and Society Inventing Organisms
- Transgenic Organisms
- Patenting Life

Previewing Science Skills

- In the **Skill Builders,** you will outline, make a table, and make a concept map.
- In the **Activities,** you will observe, compare, infer, and make a model.
- In the **MINI-Labs,** you will make a model and sequence events.

What's next?

In this chapter, you'll learn how cells divide and how organisms reproduce. You'll also learn about DNA—how it reproduces and controls proteins produced in the cells of an organism.

INTRODUCING THE CHAPTER

Use the Find Out activity to introduce students to growth. As they read the chapter they will understand that growth is the result of mitosis.

FIND OUT!

Preparation: Soak the pinto bean seeds for 24 hours so they will not be difficult to split open. Soak a few extra seeds for each class.

Materials: five soaked pinto bean seeds, paper towels, plastic bag, and hand lens for each group of two students

Cooperative Learning: Use the Paired Partners strategy. Have one student get the materials and read the activity. Have the other student do the activity and record the observations.

Teaching Tips

- Some students may have trouble splitting the seeds with their fingernails.

Gearing Up

Have students study the Gearing Up feature to familiarize themselves with the chapter. Discuss the relationships of the topics in the outline.

What's Next?

Before beginning the first section, make sure students understand the connection between the Find Out activity and the topics to follow.

ASSESSMENT OPTIONS

PORTFOLIO

Refer to page 95 for suggested items that students might select for their portfolios.

PERFORMANCE ASSESSMENT

See page 95 for additional Performance Assessment options.

Process

Skillbuilder, p. 85
MINI-Lab, pp. 78, 90
Activities, 4-1, p. 79; 4-2, p. 94
Using Lab Skills, p. 96

CONTENT ASSESSMENT

Assessment—Oral, pp. 76, 84, 90
Skillbuilders, pp. 81, 91
Section Reviews, pp. 81, 85, 91, 93
Chapter Review, pp. 95-97

GROUP ASSESSMENT

Opportunities for group assessment occur with Cooperative Learing Strategies and Flex Your Brain Activities.

PREPARATION

SECTION BACKGROUND

▶ Every species has a characteristic number of chromosomes in each cell. The number varies with the species. A cat has 32 chromosomes, whereas a potato and a chimpanzee each have 48 chromosomes. In all sexually reproducing organisms, chromosomes occur in pairs. The two chromosomes of each pair are called homologous chromosomes.

▶ The division of cytoplasm in a cell after mitosis is called cytokinesis. Cytokinesis also separates other structures distributed throughout the cytoplasm, such as mitochondria, ribosomes, and Golgi bodies. The two new cells formed are generally equal in size.

PREPLANNING

▶ Obtain prepared onion root tip and whitefish embryo slides for Activity 4-1.

1 MOTIVATE

▶ **Bulletin Board:** Have students bring one of their baby pictures and a first or second grade picture to class. Identify the pictures by number and place them on the bulletin board. Then have the students identify the person in each picture. When all the pictures have been identified, discuss how the students have grown and changed from the time they were babies until the present.

▶ Discuss injuries that students may have had. Help them realize that some cells in their bodies produce new cells in the healing of wounds and injuries. Areas that cannot produce new cells, fill in with scar tissue.

VideoDisc

STVS: Plant Clones, Disc 4, Side 2

4-1 Cell Growth and Division

New Science Words

mitosis
chromosomes
asexual reproduction
sexual reproduction

Objectives

▶ Describe mitosis and explain its importance.
▶ Explain differences between mitosis in plant and animal cells.
▶ Distinguish between asexual and sexual reproduction and give two examples of asexual reproduction.

What Happens When Cells Divide?

It's very likely that each time you go to the doctor, a nurse measures your height and mass. Over the years, similar data collected from thousands of people have given the medical profession an idea of how people grow. Much of the growth happens because the number of cells in your body increases as you develop. Other growth occurs because some cells become larger in size through metabolism.

The fact is that you are constantly changing. You aren't the same now as you were a year ago or even a few hours ago. At this very moment, as you read this page, groups of cells throughout your body are growing, dividing, and dying. Worn-out cells on the palms of your hands are being replaced. Cuts and bruises are healing. Red blood cells are being produced in your bones at a rate of two to three billion per second to replace those that wear out. Muscles that you exercise are getting larger. Other organisms undergo similar processes. A plant climbs a garden stake as the number of its cells increases. How does this happen?

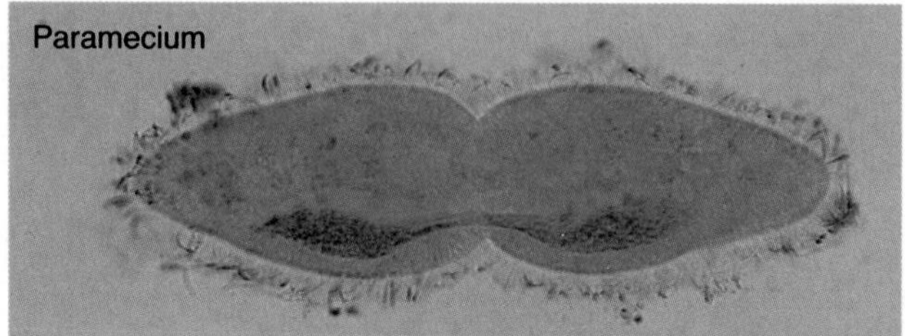

Figure 4-1. Many-celled organisms, such as the pole-bean plant above, grow by increasing numbers of cells. A one-celled organism reaches a certain size and then divides.

OPTIONS

Meeting Different Ability Levels

For Section 4-1, use the following **Teacher Resource Masters** depending upon individual students' needs.

◆ **Study Guide Master** for all students.
● **Reinforcement Master** for students of average and above average ability levels.
▲ **Enrichment Master** for above average students.

Additional Teacher Resource Package masters are listed in the OPTIONS box throughout the section. The additional masters are appropriate for all students.

◆ STUDY GUIDE 17

NAME ______ DATE ______ CLASS ______

STUDY GUIDE — Chapter 4

Cell Growth and Division — Text Pages 74–81

Write the name of the phase of the cell cycle next to each event described below.

anaphase 1. centromeres divide
prophase 2. centrioles move to opposite ends of the cell
telophase 3. nuclear membrane forms around each mass of chromosomes
anaphase 4. chromosome strands separate toward opposite ends of the cell
interphase 5. a copy of each chromosome is made
metaphase 6. centromeres attach to the spindle fibers
prophase 7. the nuclear membrane disappears
interphase 8. the material in the nucleus that appears grainy condenses to become visible as chromosomes
metaphase 9. double-stranded chromosomes line up in the center of the cell
interphase 10. chromatin condenses and becomes visible

Complete the following sentences using appropriate words from the textbook.

11. In animal cells, once the nucleus has divided, the cytoplasm pinches in to form two new cells.
12. Cell division resulting in two new nuclei having the same number of chromosomes as the original nucleus is called mitosis.
13. Eggs or sperm are sex cells.
14. Plant cells have no centrioles.
15. Plant and animal cells have spindle fibers during mitosis.
16. Bacteria reproduce asexually by means of a process called fission.
17. In plant cells, a structure called a cell plate forms between two new nuclei.
18. The process by which a new organism is produced when sex cells from two parents combine is called sexual reproduction.
19. Budding is a form of asexual reproduction.
20. A whole new organism can grow from just a piece of the parent in animals that have the ability to regenerate.

 17

The Cell Cycle

Organisms go through stages, or a life cycle, while they are alive. A simple life cycle is birth, growth and development, and death. Right now, you are in a stage in your life cycle called adolescence, a period of active growth and development. Cells also go through cycles. Most of the life of any cell is spent in a period of growth and development called *interphase.* Cells in your body that no longer divide, such as nerve and muscle cells, are always in interphase. Cells that actively divide, such as your skin cells, have a more complex cell cycle. The cell cycle, as shown in Figure 4-2, is a series of events that takes place in a cell from one division to the next.

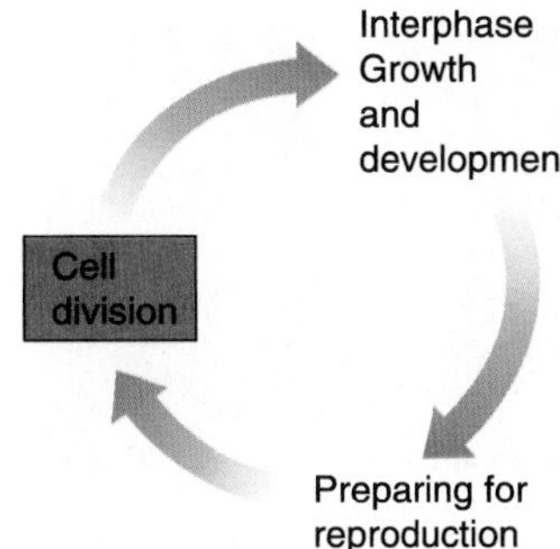

Figure 4-2. A Cell Cycle

Figure 4-3. An organism undergoes many changes as it develops. These photographs show some of the stages in the development of a pelican.

Mitosis

Cells divide in two steps. First the nucleus of the cell divides, and then the cytoplasm divides. **Mitosis** is the process in which a cell nucleus divides into two new nuclei, each of which contain the same number of chromosomes as the parent cell. The parent cell is the cell that undergoes division. Mitosis is described as a series of steps. These steps have been named prophase, metaphase, anaphase, and telophase.

TYING TO PREVIOUS KNOWLEDGE:

Review the structure of the nucleus and the function of each of its parts from Chapter 2. Tell students that mitosis involves only the nucleus of the cell.

OBJECTIVES AND SCIENCE WORDS:

Have students review the objectives and science words as they begin each section in the chapter.

2 TEACH

Key Concepts are highlighted.

CONCEPT DEVELOPMENT

▶ Make sure that students understand that during most of the cell cycle, the cell is in interphase.

▶ Explain that scientists have observed that size may be one factor that causes a cell to divide. As a cell grows, its surface area grows more slowly than its volume. After a while the cell membrane is not large enough to allow the necessary materials to enter the cell and wastes to leave.

▶ **Demonstration:** Show students two 10-cm cubes of plastic foam. Cut one cube of plastic foam into two equal halves. Then cut each half into five 1-cm cubes. Ask students which cube now has the most surface area, the 10-cm one or the one that has been cut.

● REINFORCEMENT 17

NAME ______ DATE ______ CLASS ______

REINFORCEMENT — Chapter 4

Cell Growth and Division — Text Pages 74–81

Label the following diagrams with the appropriate phase of mitosis.

1. anaphase
2. metaphase
3. prophase
4. telophase

Answer the following questions.

5. Why are skin cells undergoing mitosis continuously? **Skin cells need to be constantly replaced as they are injured or rubbed off.**
6. Once chromosomes have been copied during interphase, the cell is ready to begin what process? **mitosis**
7. During metaphase, the centromeres attach to what structures? **spindle fibers**
8. Why doesn't the cytoplasm pinch in following telophase in plant cells? **because the cell walls are rigid** What happens instead? **A cell plate forms between the two new nuclei and new cell walls form.**
9. How many chromosomes does each new cell contain after mitosis if the parent cell has 52 chromosomes? **52**
10. Why is mitosis a form of reproduction? **because the cell makes a copy of itself**
11. What are three types of asexual reproduction? **budding, fission, regeneration**
12. What types of cells in your body are no longer undergoing mitosis? **nerve and muscle cells**
13. What phase of the cell cycle are these cells in? **interphase**

 17

▲ ENRICHMENT 17

NAME ______ DATE ______ CLASS ______

ENRICHMENT — Chapter 4

Cell Growth and Division — Text Pages 74–81

THE EUKARYOTIC CELL CYCLE

As you learned in this chapter, the process of mitosis is one part of a continuous cycle of events in the life of a cell. The cell spends most of the time between divisions in interphase. Is the cell simply quiet during this phase?

Interphase is usually divided into two growth phases and a synthesis phase. Mitosis and interphase make up the cell cycle.

Synthesis
Growth 1
The Cell Cycle
Mitosis
Growth 2

Interphase

Interphase is the time between cell divisions when the cell is not going through mitosis. From the diagram, you can see that for cells that regularly divide, interphase is a period of growth and synthesis. During the synthesis phase the cell makes DNA, RNA, and proteins. DNA doubles and proteins that are needed to make spindles are made.

Mitosis (M)

As the chromosomes begin to coil into tightly compact bodies, the cell leaves the second growth phase of interphase and enters the first stage of mitosis, called prophase. The cell then continues through the remaining three stages of mitosis. Following mitosis, each new cell enters the cell cycle at the first growth phase, and the cycle repeats.

Answer the following questions about the cell cycle.

1. What are some cells that stay in interphase forever? **nerve and muscle cells**
2. Certain chemicals can be used experimentally to stop protein synthesis or DNA synthesis. If you used one of these on a cell during interphase, what would you expect to happen to the cell? **The cell would stop in interphase and would not go on to mitosis.**
3. According to the diagram, which stage of the cell cycle is the shortest? **mitosis**

 17

REVEALING MISCONCEPTIONS

▶ Many students may think that mitosis occurs in all the cells of an organism all the time. In some tissues, once the cells are formed, mitosis does not occur again. Red blood cells are formed, carry out their functions, and then are removed from the body. In plants, mitosis occurs in the growth regions and not throughout the whole plant.

CONCEPT DEVELOPMENT

▶ Mitosis and meiosis are difficult concepts for many students. Go over each stage carefully.

▶ Explain that to reproduce, the DNA in a cell's chromosomes must be copied exactly. The process of mitosis begins after a cell makes exact copies of its DNA and chromosomes. Students will learn more about DNA in Section 4-3.

▶ Make certain that students understand that most cells are not actively engaged in mitosis. Skin cells spend 15 to 20 days in interphase. Nerve cells spend 50 to 60 years.

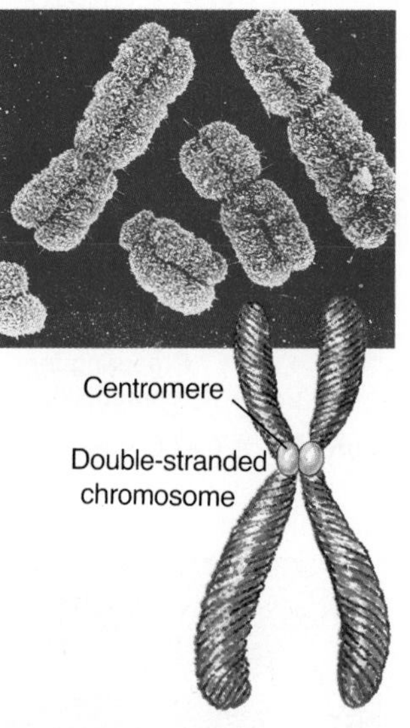

Figure 4-4. The steps of mitosis are shown in order on this page and the next.

Steps of Mitosis

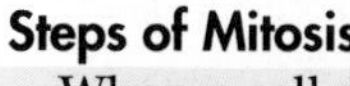

When a cell divides, the chromosomes in the nucleus play the most important part. **Chromosomes** are structures in the nucleus that contain DNA. During *interphase,* you can't see chromosomes but they are actively duplicating themselves. Then they condense and become visible. When the chromosomes appear, they are thick and doubled-stranded as in the illustrations on the left. Each double-stranded chromosome is held together at a region called a centromere. Once chromosomes are double-stranded, the cell is ready to begin the process of division. Follow the steps of mitosis in the illustrations in Figure 4-4 on these pages.

During *prophase,* chromosomes become fully visible. The nucleolus and the nuclear membrane fade and disappear. In animal cells, two small structures called centrioles move to opposite ends of the cell. Between the centrioles, threadlike spindle fibers begin to stretch across the cell, making a football-shaped network of fibers. Plant cells don't have centrioles.

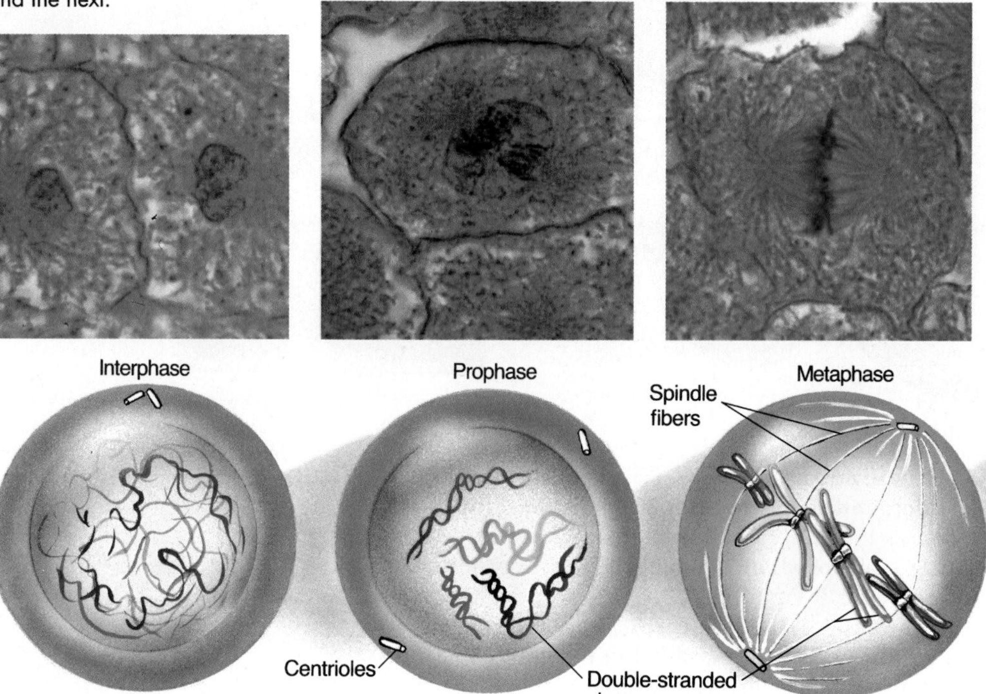

OPTIONS

ASSESSMENT—ORAL

▶ **Interphase used to be called the resting stage. Why is this not a good description of interphase?** *A cell in interphase is carrying out all of the life processes, and the chromosomes are doubling.*

▶ **Why do you think the nuclear membrane disappears during mitosis?** *to allow the chromosomes to move from the nucleus to the opposite ends of the cell*

▶ **How is mitosis different from cell division?** *Mitosis is the division of the nucleus, and cell division is the division of the cytoplasm and organelles.*

▶ **If the chromosomes did not separate during anaphase, what would the new nuclei be like?** *One would have double the number of chromosomes, and the other would not have any chromosomes.*

▶ **If a person has a spinal cord injury, it is often irreversible. Why do you think this is so?** *Nerve cells do not undergo mitosis and cannot form new cells.*

In the second step, *metaphase*, the double-stranded chromosomes line up around the center of the cell. Each centromere becomes attached to a spindle fiber. As the process enters *anaphase*, each centromere divides. The two strands of each chromosome separate. Then, the separate strands begin to move away from each other toward opposite ends of the cell.

In the final step of mitosis, *telophase*, centrioles and spindle fibers start to disappear. The chromosomes stretch out and become harder to see. A nuclear membrane forms around each mass of chromosomes, and a new nucleolus appears in each new nucleus.

In most organisms, once the nucleus has divided, the cytoplasm also separates and two whole new cells are formed. In animal cells, the cytoplasm pinches in to form the new cells. The new cells then begin a period of growth. They will take in water and other nutrients that they need to carry out cell processes.

Science and MATH

If a single cell undergoes mitosis every five minutes, how many cells will result from this single cell after one hour? *Calculate* the answer and record it **in your Journal**.

During what phase does a new nuclear membrane form?

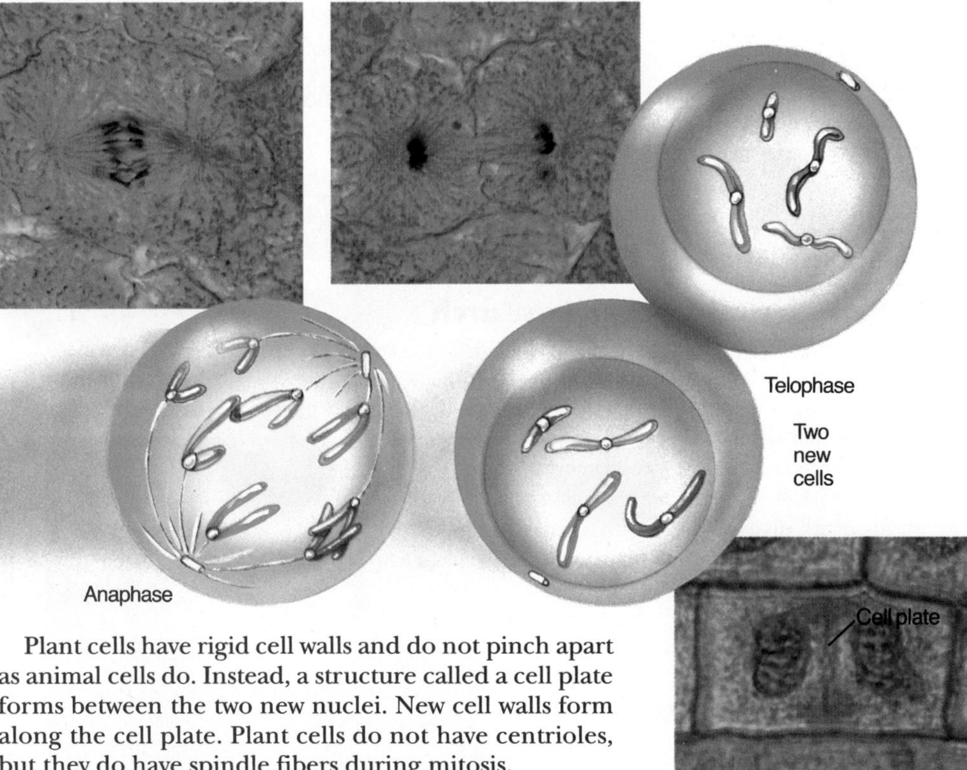

Plant cells have rigid cell walls and do not pinch apart as animal cells do. Instead, a structure called a cell plate forms between the two new nuclei. New cell walls form along the cell plate. Plant cells do not have centrioles, but they do have spindle fibers during mitosis.

Science and MATH

This is an example of exponential growth. One cell becomes two cells; two cells become four cells; four cells become eight cells; and so on. If cells divide every five minutes, there will be 12 divisions in one hour, for a total of 4096 cells ($2^{12} = 4096$ cells).

CONCEPT DEVELOPMENT

▶ **Demonstration:** To demonstrate chromosome thickening that occurs during interphase, take a telephone cord and stretch it out. It is long and thin. Then allow the cord to return to its usual position. It shortens and thickens. This is comparable to what happens to chromosomes during interphase and early prophase.

▶ Explain that the chromosomes are duplicated during a late stage of interphase.

▶ Have students find the meanings of the prefixes *inter-* (between), *pro-* (before), *meta-* (after), *ana-* (up), *telos-* (end).

CROSS CURRICULUM

▶ **Math:** It takes 15 minutes for certain embryo cells to divide. Have students calculate how many cells would be produced from one cell after 24 hours (8 388 608).

CHECK FOR UNDERSTANDING

Have students explain the meaning of Table 4-1 on page 78.

RETEACH

Provide students with a photocopied outline drawing of the cell cycle. Have them draw in the nucleus or chromosomes and describe what is occurring at each stage.

EXTENSION

For students who have mastered this section, use the **Reinforcement** and **Enrichment** masters or other OPTIONS provided.

ENRICHMENT

▶ Have some students research and report on cancer, others on aging. Then compare and contrast cancer and aging.

▶ Use the microprojector to show mitotic cell divisions in the growth of fertilized frog eggs. The increasing numbers of cells can be readily seen in the early stages of tadpole development.

MINI-Lab

Materials: posterboard, yarn, thread, paper, glue, markers, scissors

Answer: Students' posters should resemble the steps of mitosis as shown on pages 76 and 77.

MINI-Lab
ASSESSMENT

Performance: Assess students' understanding of the steps of mitosis by making flash cards of the stages and having students arrange them in proper order.

To assess this product, refer to the Performance Task Assessment Lists in **Performance Assessment in Middle School Science.**

PROBLEM SOLVING

Cells grown in a lab can be tested with different anti-cancer drugs without harm to the patient.

MINI-Lab

How does one cell become two?
Work in Paired Partners to *construct models* showing the different stages in mitosis. Use a cell with four chromosomes. Show each phase on a separate poster. Use yarn for chromosomes, sewing thread for spindle fibers, and nickel-sized paper dots for centrioles. Put the posters in correct sequence.

There are two important things to remember about mitosis. The first is that mitosis is the division of a nucleus. The second is that mitosis produces two new nuclei that have the same number of chromosomes as the original nucleus. Cells in your skin, like most of the cells in your body, each have 46 chromosomes. Each new skin cell produced by mitosis will also have 46 chromosomes. Cells in fruit flies have eight chromosomes. New fruit fly cells produced by mitosis will each have only eight chromosomes. Table 4-1 shows an additional example.

Table 4-1

CHROMOSOME NUMBERS		
Humans	**Fruit Flies**	**Carrots**
46	8	18
After mitosis	After mitosis	After mitosis
46, 46	8, 8	18, 18

PROBLEM SOLVING

Cell Biology and Cancer Research

Jewel Plummer Cobb first became fascinated with cells when she saw them through a microscope in high school. She has spent her life researching cells and is now one of the leading cancer researchers in the United States. Cancer is the result of uncontrolled cell division. Dr. Cobb's cancer research has been in two major areas. First, she wanted to find anti-cancer drugs that would not damage normal cells. Second, Dr. Cobb solved a major problem of cancer research by developing methods for growing human cancer cells in the laboratory.

Think Critically: What advantages are there in using cells grown in a lab rather than cells in a living person for research?

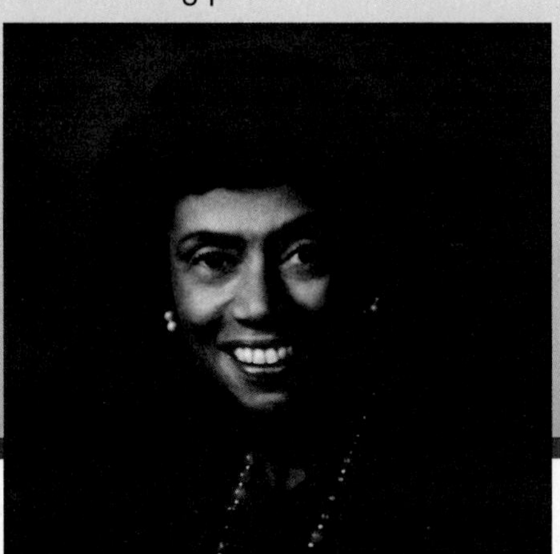

OPTIONS

INQUIRY QUESTIONS

▶**Explain why mitosis has to occur for an organism to grow.** *Cells can become only so large until there is not enough surface area to take in the materials needed for growth and to get rid of waste materials. Cells must divide if the organism is to grow.*

▶**Cancer is a disease in which there is abnormal, rapid growth in body cells. What happens to the rate of mitosis in cancer cells?** *Mitosis speeds up.*

PROGRAM RESOURCES

From the **Teacher Resource Package** use:

Activity Worksheets, page 38, Mini-Lab: How does one cell become two?

ACTIVITY 4-1

Mitosis in Plant and Animal Cells

Problem: *How is mitosis in a plant cell different from mitosis in an animal cell?*

Materials

- prepared slide of an onion root tip
- prepared slide of a whitefish embryo
- microscope

Procedure

1. Obtain prepared slides of onion root tip and whitefish embryo.
2. Set your microscope on low power and examine the onion root tip. Move the slide until you can see the area just behind the root tip. Turn the nosepiece to high power.
3. Use Figure 4-4 on pages 76-77 to help you find a cell in interphase. Draw and label the parts of the cell you observe.
4. Repeat Step 3 for prophase, metaphase, anaphase, and telophase.
5. Turn the microscope back to low power. Remove the onion root tip slide.
6. Place the whitefish embryo slide on the microscope stage under low power. Focus and find a region of dividing cells. Turn the microscope to high power.
7. Use Figure 4-4 to help you find a cell in interphase. In a data table, draw and label the parts of the cell you observe.
8. Repeat Step 7 for prophase, metaphase, anaphase, and telophase.
9. Return the nosepiece to low power. Remove the whitefish embryo slide from the microscope stage.

Analyze

1. How are the cells in the region behind the onion root tip different from those in the root tip?
2. What is the shape of the cells in the onion root tip?
3. In which stage do the chromosomes move to the center of the cell?
4. When do the chromosomes move to opposite ends of the cell?
5. What is the shape of the cells in the whitefish embryo?
6. When do the spindle fibers appear in both the onion root tip and the whitefish embryo?

Conclude and Apply

7. How do the whitefish embryo cells and the onion root tip cells *compare* in size?
8. How do the chromosomes of the whitefish embryo *compare* with those of an onion root tip?
9. What can you *conclude* about mitosis in the onion and in the whitefish embryo?

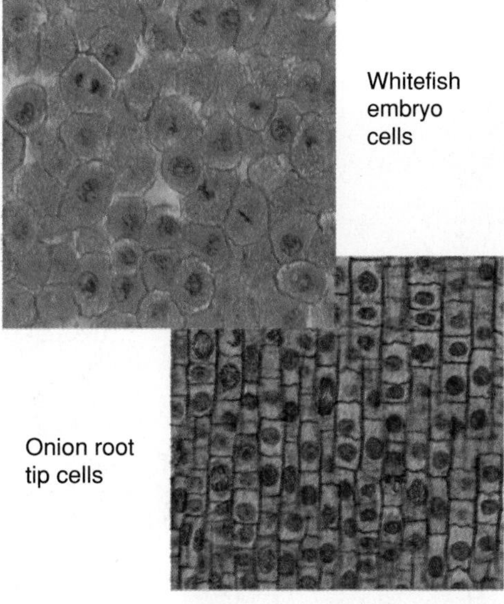

ACTIVITY 4-1
40 minutes

OBJECTIVE: Observe the stages of mitosis in plant cells and animal cells and **compare** the processes and results.

PROCESS SKILLS applied in this activity:
▶ **Observing** in Procedure Steps 3, 4, 6, 7, and 8.
▶ **Inferring** in Analyze Questions 1, 2, and Conclude and Apply questions 7, 8, and 9.

COOPERATIVE LEARNING
Use the Paired Partners strategy. Have students alternate using the microscope. Have both observe and record their observations.

TEACHING THE ACTIVITY

Alternate Materials: If microscopes are not available for all students, a microprojector can be used for the class to work together.

Troubleshooting: Students may have difficulty locating all phases in the onion root tip. You may want to place an onion root tip slide on the microprojector and point out the phases.

▶ Review the stages of mitosis before beginning this activity.

Activity ASSESSMENT

Performance: To further assess students' understanding of the stages of mitosis, give each one a stage on a slip of paper and have them describe what comes before and after that stage.

PROGRAM RESOURCES

From the **Teacher Resource Package** use:

Activity Worksheets, pages 32-33, Activity 4-1, Mitosis in Plant and Animal Cells.

ANSWERS TO QUESTIONS

1. Many cells are smaller. Chromosomes can be seen in many of the cells.
2. They are rectangular.
3. metaphase
4. anaphase
5. They are almost round.
6. prophase
7. The whitefish embryo cells are much larger.
8. The whitefish chromosomes are smaller.
9. Mitosis in whitefish and onion root cells is different. In whitefish embryo cells, there are centrioles at the poles of the spindle fibers during mitosis. The onion root tip cells do not have centrioles. In the whitefish embryo cells, the cell membrane pinches in or forms a furrow between the two new nuclei. In onion root tip cells, a cell plate forms in the center of the spindle between the two nuclei.

CONCEPT DEVELOPMENT

▶ Explain that asexually reproducing organisms produce offspring that are genetically identical to their parents. In one-celled organisms, new cells are produced by fission or mitosis. In many-celled organisms, new cells are formed by cloning or by budding. Sexually reproducing organisms reproduce offspring that are similar to their parents but not exactly like them.

▶ Explain that mitosis is responsible for asexual reproduction. All cells are identical to the parent cell.

Answer: Accept all reasonable responses. Living things grow larger by production of more cells or cells enlarge through metabolism. Nonliving things "grow" by addition of more of the same materials, but not by dividing.

▶ **Activity:** Have the students plant root cuttings of coleus, snake plant, and African violet leaves to show asexual reproduction.

▶ **Demonstration:** Place a sweet potato in a beaker of water. Have students check the potato weekly and note how long it takes for new plants to grow. Other plants may be used to illustrate asexual reproduction. Cut a 2-cm section from the top of a carrot, beet, or turnip, and remove any old leaves from it. Place the section so that the bottom of the cut part is resting in water. New leaves and stems should appear in a few days.

Types of Reproduction

Connect to... Earth Science

Many nonliving things, such as icicles, crystals, and sand dunes, appear to grow. Distinguish between the processes involved in the growth of living things and the "growth" of nonliving things.

Your body forms two types of cells—body cells and sex cells. Skin, liver, bones, kidneys, lungs, and muscles are made up of different types of body cells. By far, your body has many more body cells than sex cells. The only sex cells that you have are the eggs or sperm in your reproductive organs.

Reproduction is the process by which an organism produces others of the same kind. Among living organisms, there are two types of reproduction—asexual reproduction and sexual reproduction. In **asexual reproduction,** new organisms are produced from one parent. You've just seen examples of this in the process of mitosis. Offspring produced by asexual reproduction have DNA that is identical to the DNA of the parent organism.

How do asexual and sexual reproduction differ?

Several types of asexual reproduction are important in plants and animals. If you've ever grown a sweet potato in a jar of water, you've seen asexual reproduction take place. All the stems and leaves that grow out from the sweet potato have been produced by mitosis. New strawberry plants can be produced asexually from runners. Bacterial cells reproduce asexually by a process called fission. Fission is division of an organism into two equal parts.

Budding is also a type of asexual reproduction in which a new organism grows from the body of the parent organism. Hydra can reproduce this way. When the bud on the adult becomes large enough, it breaks away to live on its own.

A few organisms can repair damaged or lost body parts by regeneration. During regeneration, a whole organism may develop from a piece of the organism. Sponges, planaria, and sea stars regenerate. If the cells of a live sponge are separated, they collect together and form a whole

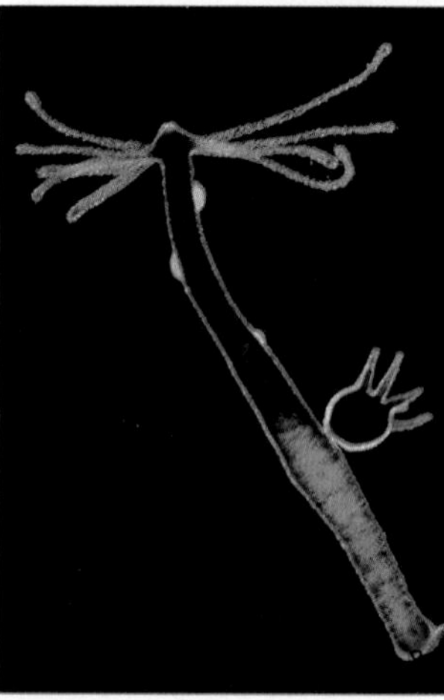

Figure 4-5. Many organisms reproduce asexually. Hydra, above, and strawberry runners, shown right, are examples.

OPTIONS

PROGRAM RESOURCES

From the **Teacher Resource Package** use:

Concept Mapping, pages 13-14.

Transparency Masters, pages 9-10, Mitosis.

Use **Color Transparency** 5, Mitosis.

Use **Laboratory Manual,** page 23, Mitosis.

Use **Laboratory Manual,** page 25, Chromosomes.

Figure 4-6. Some animals produce whole new body parts by regeneration. A sea star can develop a new ray when one is torn off.

new sponge body. If a sponge is cut into small pieces, a new sponge develops from each piece. How do you think sponge farmers increase their "crop"?

Through cell division, organisms grow, replace worn-out or damaged cells, or produce whole new organisms. Fission, budding, and regeneration are types of asexual reproduction that result from mitosis.

In **sexual reproduction,** a new organism is produced when sex cells from two parents combine. The DNA of the offspring is different from either parent. You will learn about sexual reproduction in Section 4-2.

Did You Know?

The lining of your digestive system is constantly worn away by the movement of food. Continuous cell division replaces this lining about every five days.

SECTION REVIEW

1. What is mitosis and how does it differ in plants and animals?
2. What are the stages of mitosis?
3. Distinguish between sexual and asexual reproduction.
4. Describe two types of asexual reproduction.
5. **Apply:** The body cells of a frog contain 26 chromosomes. If these cells undergo mitosis, how many chromosomes will be in each new cell produced?
6. **Connect to Chemistry:** Find out what the menstrual cycle is and what is produced during this cycle.

Outlining

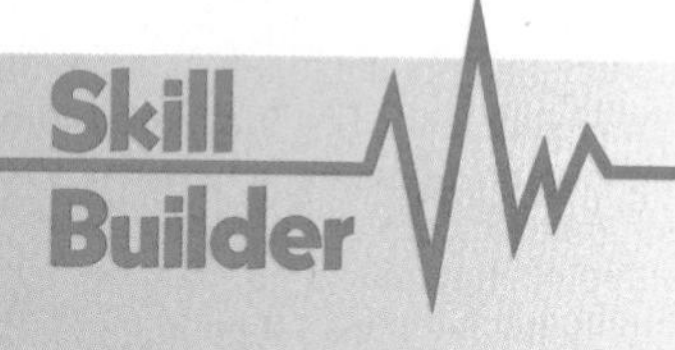

Outline the events in each stage of mitosis in animal cells using the stages of mitosis as heads. Begin with interphase. If you need help, refer to Outlining in the **Skill Handbook** on page 681.

3 CLOSE

▶ Ask questions 1-4 and the **Apply** Question in the Section Review.

▶ Show the film *Mitosis,* 2nd ed., Britannica, 14 min.

▶ Invite a doctor, nurse, or someone from the American Cancer Society to speak to the class about abnormal cell division.

SECTION REVIEW ANSWERS

1. Mitosis is the process in which a cell nucleus divides into two nuclei, each of which has the same number of chromosomes. Animal cells have centrioles. Plant cells do not. Plant cells also form a cell plate at telophase.

2. prophase, metaphase, anaphase, telophase

3. In sexual reproduction, a new organism is produced when cells from two parents combine. In asexual reproduction, a new organism is formed from one parent and is a duplicate of that parent genetically.

4. Fission is the process in which one cell divides in half to produce two cells. A new organism grows from a part of its parent in budding. A whole organism grows from a part of the parent organism in regeneration.

5. Apply: 26

6. Connect to Chemistry: A chemically controlled cycle in females that results in (haploid) sex cells or eggs.

Skillbuilder

ASSESSMENT

Portfolio: Use this Skillbuilder to assess students' ability to Outline the process of mitosis.

Skill Builder

I. Prophase
   A. Chromosomes become visible.
   B. Nucleolus and nuclear membrane fade, disappear.
   C. Centrioles move to opposite ends of the cell.
   D. Spindle fibers appear.

II. Metaphase
   A. Double-stranded chromosomes line up.
   B. Centromeres attach to spindle fibers.

III. Anaphase
   A. Centromeres divide.
   B. Two strands of chromosomes separate.
   C. Separated strands move toward opposite ends of cell.

IV. Telophase
   A. Centrioles and spindle fibers begin to disappear.
   B. Chromosomes become harder to see.
   C. Nuclear membrane forms in each cell.
   D. Nucleolus appears in each new nucleus.

SECTION 4-2

PREPARATION

SECTION BACKGROUND

▶ Sexual reproduction is the production of offspring through meiosis and the fusion of sex cells or gametes. Such offspring are genetically different from both parents because in meiosis, genes are combined in new ways through genetic recombination. Genetic recombination produces variation between parents and offspring and gives some offspring a better chance of surviving in a changing environment.

▶ In animals, meiosis results in haploid egg and sperm cells. In plants, meiosis results in haploid spores that later lead to the production of egg and sperm cells. In both plants and animals, the haploid cells fuse during fertilization to form a diploid zygote that develops into a new organism.

▶ With the 23 pairs of chromosomes in the human body, there are more than 8 million different combinations possible for every human cell formed by meiosis.

▶ Current research indicates that a human egg cell has not completed meiosis when released from the ovary. The cell is in meiosis II. The stimulus of a fertilizing sperm causes the cell to complete meiosis. If an egg is not fertilized, it does not complete meiosis.

1 MOTIVATE

▶ Show the film *Meiosis*, 2nd ed., Britannica, 15 min.

▶ Use a microprojector to project slides of the stages of meiosis in lily anthers. Point out the various changes that are taking place.

4-2 Sexual Reproduction and Meiosis

New Science Words

sperm
egg
meiosis
fertilization
zygote

Objectives

- ▶ **Describe the stages of meiosis and its end products.**
- ▶ **Name the cells involved in fertilization and explain how fertilization occurs.**

Sexual Reproduction

More than 2000 years ago, people debated how a human could grow from a single egg cell. Some thought the cell must contain a tiny, completely formed human. With the development of the microscope, scientists were able to disprove this idea. With the microscope, they were able to learn more about cells, marvel at reproduction, and watch growth take place.

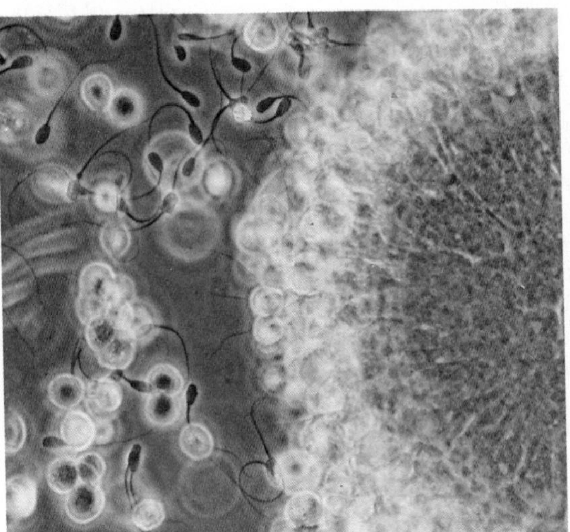

Figure 4-7. The large cell is a human egg cell, or ovum. The smaller cells are human sperm.

In the last section, you learned that asexual reproduction occurs with only one parent. In contrast, sexual reproduction requires two parents to produce offspring. During sexual reproduction two sex cells join to form a new individual. Each sex cell is produced by a different parent. The sex cell from the male parent is called the **sperm.** The sex cell from the female parent is called the **egg.** Sex cells like the ones in Figure 4-7 are usually different in size from one another. Eggs are usually large and contain food material. Sperm are small with whiplike tails. The sperm head is almost all nucleus.

A human body cell has 23 pairs of chromosomes, but human sex cells have only 23 chromosomes. How does the chromosome number become reduced? How do sex cells form? The process of nuclear division that produces sex cells is called **meiosis.** Meiosis takes place in cells of reproductive organs both in plants and animals.

OPTIONS

Meeting Different Ability Levels

For Section 4-2, use the following **Teacher Resource Masters** depending upon individual students' needs.

◆ **Study Guide Master** for all students.

● **Reinforcement Master** for students of average and above average ability levels.

▲ **Enrichment Master** for above average students.

Additional Teacher Resource Package masters are listed in the OPTIONS box throughout the section. The additional masters are appropriate for all students.

◆ STUDY GUIDE 18

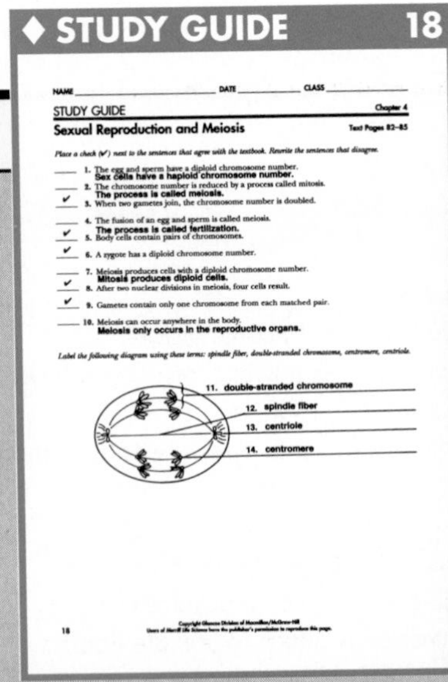

NAME ______ DATE ______ CLASS ______

STUDY GUIDE — Chapter 4

Sexual Reproduction and Meiosis — Text Pages 82–85

Place a check (✔) next to the sentences that agree with the textbook. Rewrite the sentences that disagree.

____ 1. The egg and sperm have a diploid chromosome number.
Sex cells have a haploid chromosome number.
____ 2. The chromosome number is reduced by a process called mitosis.
The process is called meiosis.
✔ 3. When two gametes join, the chromosome number is doubled.
____ 4. The fusion of an egg and sperm is called meiosis.
The process is called fertilization.
✔ 5. Body cells contain pairs of chromosomes.
✔ 6. A zygote has a diploid chromosome number.
____ 7. Meiosis produces cells with a diploid chromosome number.
Mitosis produces diploid cells.
✔ 8. After two nuclear divisions in meiosis, four cells result.
✔ 9. Gametes contain only one chromosome from each matched pair.
____ 10. Meiosis can occur anywhere in the body.
Meiosis only occurs in the reproductive organs.

Label the following diagram using these terms: spindle fiber, double-stranded chromosome, centromere, centriole.

18

The Importance of Sex Cells

In body cells, chromosomes are found in pairs. The 46 human chromosomes form 23 pairs of chromosomes. The pairs form because the chromosomes are alike. A cell that has two of every kind of chromosome is said to be diploid. Sex cells, on the other hand, contain only one chromosome from each matched pair. A sex cell that has just one chromosome of each pair is haploid. Haploid means "single form." A human sex cell has 23 chromosomes, not 23 pairs of chromosomes. Therefore, it is haploid. For corn, the diploid number is 20, and the haploid number is 10. Usually, the haploid number of chromosomes is found only in sex cells of an organism.

What is so important about sex cells? Sexual reproduction starts with the formation of sex cells and ends when one sex cell joins with another and a new organism is begun. The joining of an egg and a sperm is called **fertilization.** The cell that forms in fertilization is called a **zygote.** If an egg with 23 chromosomes joins with a sperm that has 23 chromosomes, a zygote forms that has 46 chromosomes, or the diploid chromosome number for that organism. A zygote then begins to undergo mitosis and the organism develops.

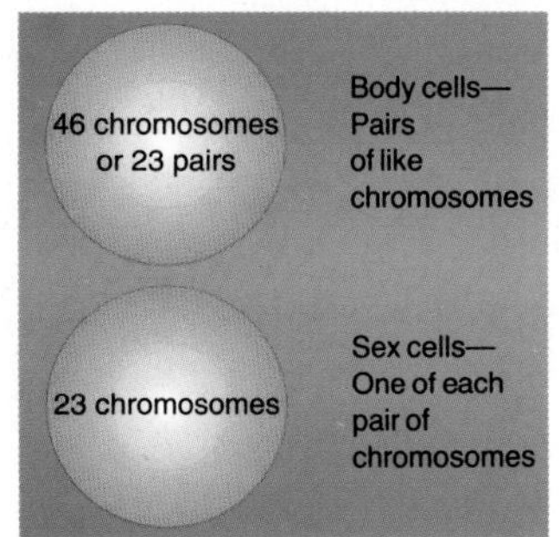

How many chromosomes will be in a goldfish zygote if the egg of the goldfish contains 94 chromosomes?

What is fertilization?

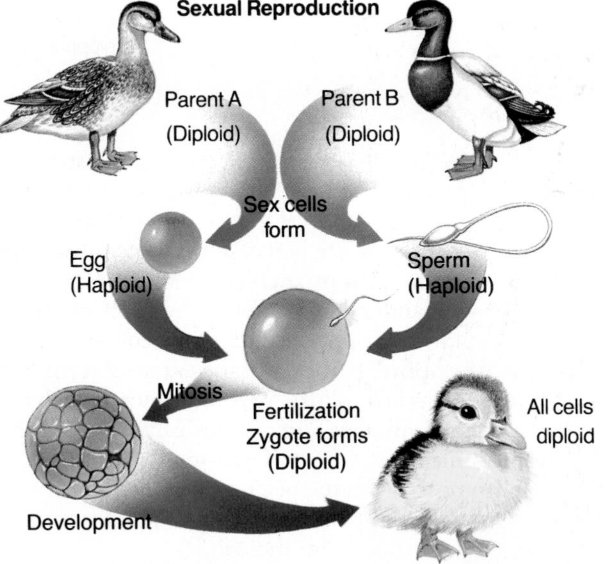
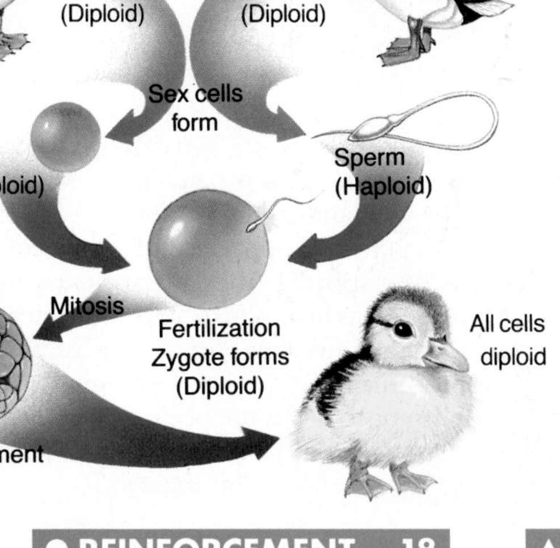

Figure 4-8. When sex cells join, a zygote forms. The zygote develops into a new individual.

TYING TO PREVIOUS KNOWLEDGE: Students will be familiar with cell division from having studied Section 4-1. Ask them to describe the stages of mitosis. Then tell the students that in meiosis the nucleus divides twice.

2 TEACH

Key Concepts are highlighted.

CONCEPT DEVELOPMENT

▶ Explain that sex cells in animals form in the testes of males and in the ovaries of females, so meiosis occurs only in the testes and ovaries.

▶ Explain that during fertilization, each sex cell contributes one chromosome to the resulting homologous pair. Therefore, each parent is contributing one homologous chromosome to its offspring. Meiosis makes it possible to produce sex cells with half the chromosome complement of the parent.

CROSS CURRICULUM

▶ **Reading:** Have students use library references to find the number of chromosomes in body cells of various plants and animals. Then have them find the number of chromosomes in the sex cells of each one.

● REINFORCEMENT 18

NAME ________ DATE ________ CLASS ________

REINFORCEMENT — Chapter 4

Sexual Reproduction and Meiosis — Text Pages 82–85

Label the following diagrams of the stages of meiosis.

1. anaphase I
2. anaphase II
3. metaphase II
4. prophase I

Answer the following questions.

5. In what way is meiosis II similar to mitosis? **Centromeres divide and double-stranded chromosomes separate.**
6. How are gametes different from other cells in the body? **They contain only half the number of chromosomes.**
7. What happens during fertilization? **Two gametes fuse to form a zygote.**
8. What is the total number of paired chromosomes in a body cell called? **diploid chromosome number**
9. Do centromeres divide at anaphase I or II? **anaphase II**
10. Starting with one cell, how many haploid sperm cells are formed after both phases of meiosis? **four**

18

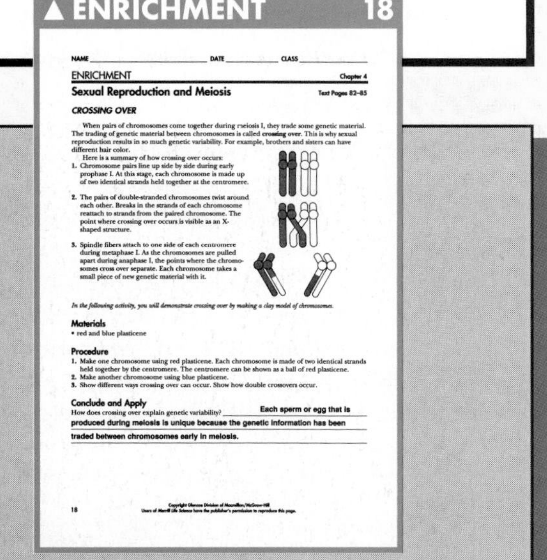

▲ ENRICHMENT 18

NAME ________ DATE ________ CLASS ________

ENRICHMENT — Chapter 4

Sexual Reproduction and Meiosis — Text Pages 82–85

CROSSING OVER

When pairs of chromosomes come together during meiosis I, they trade some genetic material. The trading of genetic material between chromosomes is called **crossing over**. This is why sexual reproduction results in so much genetic variability. For example, brothers and sisters can have different hair color.

Here is a summary of how crossing over occurs:

1. Chromosome pairs line up side by side during early prophase I. At this stage, each chromosome is made up of two identical strands held together at the centromere.
2. The pairs of double-stranded chromosomes twist around each other. Breaks in the strands of each chromosome reattach to strands from the paired chromosome. The point where crossing over occurs is visible as an X-shaped structure.
3. Spindle fibers attach to one side of each centromere during metaphase I. As the chromosomes are pulled apart during anaphase I, the points where the chromosomes cross over separate. Each chromosome takes a small piece of new genetic material with it.

In the following activity, you will demonstrate crossing over by making a clay model of chromosomes.

Materials

- red and blue plasticene

Procedure

1. Make one chromosome using red plasticene. Each chromosome is made of two identical strands held together by the centromere. The centromere can be shown as a ball of red plasticene.
2. Make another chromosome using blue plasticene.
3. Show different ways crossing over can occur. Show how double crossovers occur.

Conclude and Apply

How does crossing over explain genetic variability? **Each sperm or egg that is produced during meiosis is unique because the genetic information has been traded between chromosomes early in meiosis.**

18

CONCEPT DEVELOPMENT

Cooperative Learning: Use the Expert Teams strategy to have students study and master the stages of sex cell formation.

▶ Provide students with a photocopied outline drawing of cells. Have the students record the information on each stage of meiosis in their cell outlines.

▶ Explain that the major event of the first division of meiosis is the halving of the chromosome number. The first division is reduction-division. In meiosis, pairs of chromosomes separate.

▶ Mention that in meiosis, the stage of telophase I is very brief.

▶ Emphasize that meiosis has two major divisions and that double-stranded chromosomes are held together by a centromere. The centromere splits only in anaphase II.

Answer: The fact that other sperm are prevented from entering the egg.

Meiosis—Sex Cell Formation

In meiosis, there are two divisions of the nucleus: meiosis I and meiosis II. The different phases of each division have names like those in mitosis, so follow the steps of meiosis I carefully in Figure 4-9. In *prophase I,* double-stranded chromosomes and spindle fibers appear. The nuclear membrane and the nucleolus disappear. Like chromosomes come together in matching pairs.

In *metaphase I,* the pairs of chromosomes line up in the center of the cell. Their centromeres become attached to the spindle fibers. In *anaphase I,* each double-stranded chromosome separates from its matching chromosome. Each one is pulled to opposite ends of the cell. Then in *telophase I,* the cytoplasm divides and two cells form. Each chromosome is still double-stranded.

Figure 4-9. Meiosis begins with the pairing of like chromosomes. The steps of meiosis I are shown on this page.

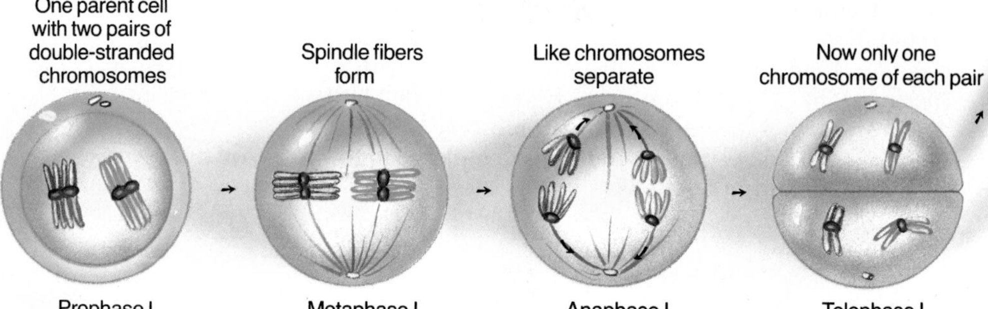

The human egg releases a chemical into the surrounding fluid that attracts sperm. Usually only one sperm fertilizes the egg, because after the sperm nucleus enters the egg, the cell membrane of the egg changes, preventing other sperm from entering. What adaptation in this process ensures that the zygote will be just diploid?

Meiosis II begins. In *prophase II,* the double-stranded chromosomes and spindle fibers reappear in each new cell. In *metaphase II,* the double-stranded chromosomes move to the center of the cell. There, the centromeres attach to spindle fibers. During *anaphase II,* the centromere divides, and the two strands of each chromosome separate and move to opposite ends of the cell. As *telophase II* begins, the spindle fibers disappear, and a nuclear membrane forms around the chromosomes at each end of the cell. Each nucleus contains only half the number of chromosomes that were in the original nucleus. A cell with 46 chromosomes at the beginning of meiosis I divides to produce cells that each have only 23 single-stranded chromosomes at the end of meiosis II.

OPTIONS

ASSESSMENT—ORAL

▶ **A sperm cell has 10 chromosomes. How many chromosomes were in the cell that produced it?** *20*

▶ **An egg cell with 10 chromosomes is fertilized by a sperm cell that has 10 chromosomes. What is the number of chromosomes in the resulting zygote?** *20*

▶ **Why do you think meiosis is sometimes referred to as reduction-division?** *The number of chromosomes is reduced and the cell divides.*

ENRICHMENT

▶ Have students use clay, pipe cleaners, pop beads, and yarn to make models of meiosis.

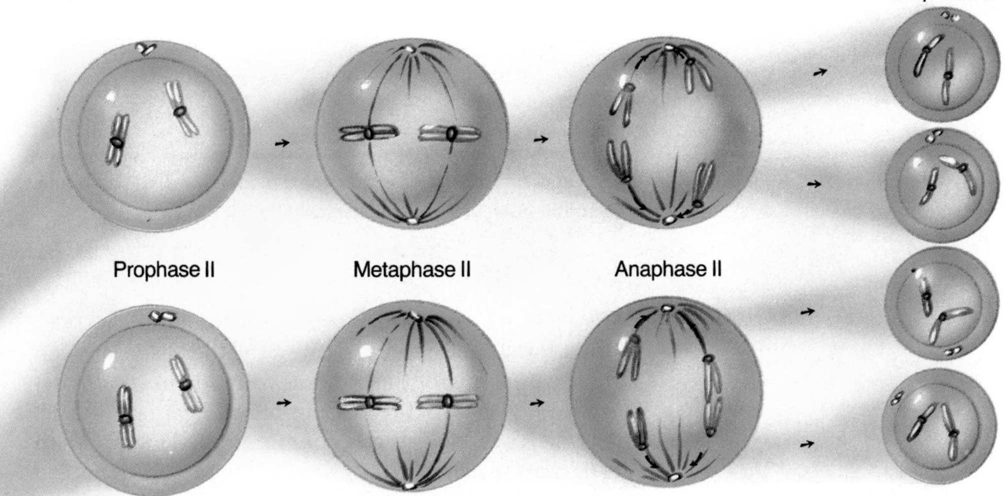

When meiosis II is finished, the cytoplasm divides. Meiosis I forms two cells. In meiosis II, both of these cells divide into two cells. The two nuclear divisions result in four cells. Each of these four cells is a sex cell. Each sex cell has one half the chromosomes of the original cell.

Figure 4-10. The steps of meiosis II are shown. Overall, meiosis results in four sex cells.

Are sex cells haploid or diploid?

SECTION REVIEW

1. Why is meiosis I sometimes called reduction division?
2. How many cells are there at the end of meiosis II?
3. What is a zygote, and how is it formed?
4. How do sex cells and body cells differ?
5. **Apply:** If body cells of a horse have 64 chromosomes, how many chromosomes are in a horse sperm cell?
6. **Connect to Physics:** Find out how sperm are stored in sperm banks.

Making and Using Tables

Skill Builder

Make a table to compare mitosis and meiosis in humans. Horizontal heads in your table should be: Feature, Mitosis, and Meiosis. Vertical heads under the Feature column should be: What type of cell (Body or Sex), Beginning cell (Haploid or Diploid), Number of cells produced, End-product cell (Haploid or Diploid), and Number of chromosomes in cells produced. If you need help, refer to Making and Using Tables in the **Skill Handbook** on page 690.

CHECK FOR UNDERSTANDING

Use the Skillbuilder to check for understanding.

RETEACH

Use an overhead transparency to show the stages of meiosis out of order. Have students identify each stage and place the stages in order.

EXTENSION

For students who have mastered this section, use the **Reinforcement** and **Enrichment** masters or other OPTIONS provided.

3 CLOSE

▶Ask questions 1-4 and the **Apply** Question in the Section Review.

▶Have students make a poster that compares the stages of mitosis and meiosis.

SECTION REVIEW ANSWERS

1. The chromosome number is reduced and the cell divides.

2. four

3. A zygote is a cell that forms in fertilization; it has a diploid number of chromosomes.

4. Body cells are diploid and sex cells are haploid.

5. Apply: 32

6. Connect to Physics: They are stored frozen in liquid nitrogen.

Skillbuilder

ASSESSMENT

Performance: Use this Skillbuilder to assess students' abilitites to Compare and Contrast meiosis and mitosis.

Skill Builder

COMPARING MITOSIS AND MEIOSIS

Feature	Mitosis	Meiosis
What type of cell (body or sex)	Body cell	Sex cell
Beginning cell (haploid or diploid)	Diploid	Diploid
Number of cells produced	Two	Four
End-product cell (haploid or diploid)	Diploid	Haploid
Number of chromosomes in cells produced	Same as parent—46	Half the parent number—23

PROGRAM RESOURCES

From the **Teacher Resource Package** use:

Transparency Masters, pages 11-12, Meiosis.

Use **Color Transparency** 6, Meiosis.

SECTION 4-3

PREPARATION

SECTION BACKGROUND

▶ DNA molecules make up the chromosomes of cells and contain the instructions for making proteins. In cell reproduction, the DNA molecules in a cell's chromosomes are copied exactly. A DNA molecule can make an exact copy of itself because the double-stranded structure separates and each half is copied exactly. The process of mitosis begins after a cell makes exact copies of its DNA and chromosomes.

▶ At one time, scientists hypothesized that the protein in the nucleus was the genetic material. By using the scientific method, data was collected that supported the theory that DNA, not the protein, was the source of genetic information in cells.

▶ Many individuals contributed to the discovery of the structure of DNA. The Watson and Crick model of DNA is an excellent example of how major scientific discoveries are usually based on the work of many scientists.

▶ This section contains complex material. You may wish to cover only a portion of the information, depending on the ability of your class.

PREPLANNING

▶ Obtain the paper, crayons, and scissors for Activity 4-2 on page 94.

1 MOTIVATE

▶ **Bulletin Board:** Display pictures of DNA molecules. Label the structure of a DNA molecule. Arrange articles that describe products being made or crops being improved through the technology of recombinant DNA.

▶ Show the film *The Living Cell: DNA*, Britannica, 20 min.

4-3 DNA

New Science Words

DNA
gene
RNA
mutation

Objectives

▶ **Construct and identify the parts of a model of a DNA molecule.**
▶ **Describe how DNA copies itself.**

What Is DNA?

Have you ever sent a message to someone using a code? In order for them to read your message, they had to understand the meaning of the symbols you used in your code. The chromosomes in the nucleus of a cell contain a code. This code is in the form of a chemical called DNA. When mitosis takes place, the DNA code in the nucleus is copied and passed to new cells. In this way, new cells receive the same coded information that was in the original cell. **DNA** controls the activities of cells with coded instructions. Every cell that has ever been formed in your body or in any plant or other animal contains DNA.

Since the mid-1800s, it has been known that the nuclei of cells contain chemicals called nucleic acids. What does DNA look like? Scientist Rosalind Franklin discovered that the DNA molecule was a strand of molecules in a spiral form. By using an X-ray technique, Dr. Franklin showed that the spiral was so large that it was probably made up of two spirals. As it turned out, the structure of DNA is similar to the handrails and steps of a spiral staircase. In 1953, using the work of Franklin and others, scientists James Watson and Francis Crick made a model of a DNA molecule. According to the Watson and Crick model, the sides ("the handrails") of the DNA molecule are made up of two twisted strands of sugar and phosphate molecules. The

Figure 4-11. James Watson and Francis Crick with their model of DNA.

OPTIONS

Meeting Different Ability Levels

For Section 4-3, use the following **Teacher Resource Masters** depending upon individual students' needs.

◆ **Study Guide Master** for all students.

● **Reinforcement Master** for students of average and above average ability levels.

▲ **Enrichment Master** for above average students.

Additional Teacher Resource Package masters are listed in the OPTIONS box throughout the section. The additional masters are appropriate for all students.

◆ STUDY GUIDE 19

NAME ______ DATE ______ CLASS ______

STUDY GUIDE — Chapter 4

DNA — Text Pages 86–91

Match the statements on the left with the terms on the right by writing the correct letter in the space provided.

c	1. units that make up proteins	a.	deoxyribose
d	2. used to demonstrate that the DNA molecule is a helix	b.	ribosomes
b	3. where messenger RNA attaches during protein construction	c.	amino acids
j	4. pairs with cytosine	d.	X rays
a	5. sugar molecules in DNA	e.	Watson and Crick
i	6. the part of DNA that directs the making of a specific protein	f.	Chargaff
f	7. hypothesized that nitrogen bases in DNA occur in pairs	g.	RNA
h	8. used to build cells and tissues	h.	protein
e	9. made a working model of the DNA molecule	i.	gene
g	10. contains uracil	j.	guanine

Complete the following sentences using the appropriate words from the textbook.

11. The process by which DNA copies itself is called **replication**.
12. Hair color and freckles are both **traits**.
13. During the making of a protein, amino acids are brought to the ribosome by **transfer RNA or tRNA**.
14. The "handrails" of each DNA strand are made up of **sugar or deoxyribose** and **phosphate groups**.
15. Any permanent change in the genetic material of a cell is called a **mutation**.
16. DNA strands held together by **nitrogen bases** are separated by an **enzyme** during replication.
17. Changes in the order of amino acids will change the **protein** produced.
18. **Messenger RNA or mRNA** carries the code for amino acids.
19. The basic units of inheritance are **genes**.
20. The shape of a DNA molecule is a **double helix or spiral**.

19

"stairs" that hold the two sugar-phosphate strands apart are made up of molecules called nitrogen bases. There are four kinds of nitrogen bases. These are adenine, guanine, cytosine, and thymine. In Figure 4-12, the bases are represented by the letters *A, G, C,* and *T.* An American biochemist named Edwin Chargaff had discovered in 1950 that the amount of cytosine in cells always equals the amount of guanine and that the amount of adenine always equals the amount of thymine. This led him to hypothesize that these bases occur in pairs in the molecule. The Watson and Crick model shows that adenine always pairs with thymine, and guanine always pairs with cytosine. Like interlocking pieces of a puzzle, each base pairs up only with its correct partner.

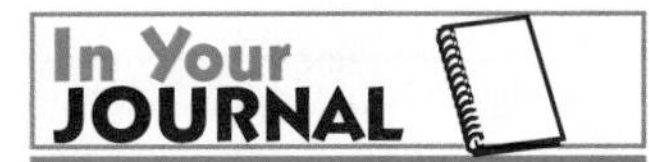

In Your JOURNAL

Read about the life and work of at least four people who contributed to the discovery of DNA. **In your Journal**, write a paragraph about the contribution of each one.

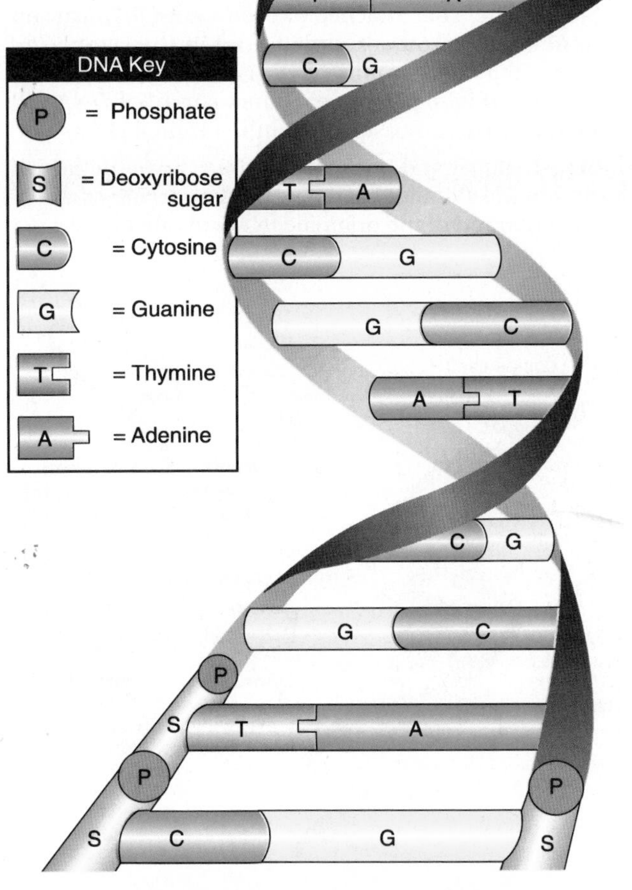

Figure 4-12. DNA stands for *de*oxyribo*n*ucleic *a*cid. Two models of the DNA molecule and its parts are shown.

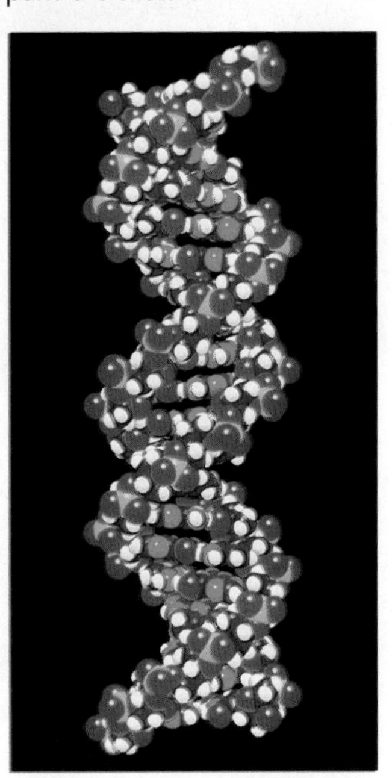

● REINFORCEMENT 19

NAME ______ DATE ______ CLASS ______

REINFORCEMENT — Chapter 4

DNA — Text Pages 86-91

Answer the following questions using information from the textbook.

1. Write the letter of the matching base of the following DNA strand.
   T: **A**
   G: **C**
   A: **T**
   T: **A**
   C: **G**
2. Write the names of the bases corresponding to the letters:
   A: **adenine**
   C: **cytosine**
   T: **thymine**
   G: **guanine**
3. What makes up the handrails of the helix of a DNA molecule? **deoxyribose (sugar) and phosphate groups**
4. What determines which proteins a cell makes? **A gene directs the making of specific proteins.**
5. What role does RNA play in protein production? **It carries the code for proteins from the nucleus to the ribosomes.**
6. How is RNA different from DNA? **It has uracil instead of thymine, and it is single-stranded.**
7. What are two kinds of RNA and what does each do? **Messenger RNA or mRNA carries the code from the nucleus to the cytoplasm where it attaches to ribosomes. Transfer RNA or tRNA picks up amino acids in the cytoplasm and brings them to the ribosomes.**
8. How does Down syndrome occur? **It occurs when a gamete forms with an extra number 21 chromosome.**
9. What can cause mutations? **X rays or toxic chemicals can result in errors in copying a gene or in a cell receiving an extra chromosome.**

19

▲ ENRICHMENT 19

NAME ______ DATE ______ CLASS ______

ENRICHMENT — Chapter 4

DNA — Text Pages 86-91

THE GENETIC CODE

How does the cell know which proteins to make? Each protein has a different amino acid sequence. The amino acids and the order they are arranged in are determined by the DNA code in the nucleus of the cell. The code for proteins leaves the cell nucleus on messenger RNA.

Scientists know from experiments and observation that the code on the mRNA for each amino acid is made of three bases. Each group of three bases is called a codon. The mRNA brings the codons to the ribosome, where proteins are assembled into a chain using amino acids from the cytoplasm. Because there are four bases in RNA, and each codon is made of three bases, there are 64 possible codons. Three of these give a signal to stop instead of coding for an amino acid. One of the codons, AUG (adenine-uracil-guanine), can either code for an amino acid called methionine (met) or it can be a signal to start.

Look at the genetic code dictionary below. You will notice that most amino acids are specified by more than one codon. Some codons like UUU and UUC code for the same amino acid, phenylalanine (phe). In the dictionary, abbreviations are given for all the amino acids.

Genetic Code Dictionary

First base in the codon	Second base in the codon				Third base in the codon
	U	C	A	G	
U	Phe Phe Leu Leu	Ser Ser Ser Ser	Tyr Tyr STOP STOP	Cys Cys STOP Trp	U C A G
C	Leu Leu Leu Leu	Pro Pro Pro Pro	His His Gln Gln	Arg Arg Arg Arg	U C A G
A	Ile Ile Ile (START) Met	Thr Thr Thr Thr	Asn Asn Lys Lys	Ser Ser Arg Arg	U C A G
G	Val Val Val Val	Ala Ala Ala Ala	Asp Asp Glu Glu	Gly Gly Gly Gly	U C A G

1. Using the genetic code dictionary, draw a diagram of mRNA showing the codons for an imaginary protein having three amino acids. Include start and stop signals.
   Answer will vary from example here. AUG;UUA;CUA;GAC;UAA start-leu-leu-asp-stop
2. Show how a change in one base will affect the protein you have designed.
   Answer will depend on previous answer. AUG;UUC;CUA;GAC;UAA start-phe-leu-leu-stop

19

In Your JOURNAL

Students may need help in locating information about scientists. Make sure they know how to research information in the school or local library.

TYING TO PREVIOUS KNOWLEDGE: Students are familiar with some template systems, such as keys and locks and peg and hole games. Ask for other examples. Use this knowledge to explain that RNA serves as a template for DNA in the nucleus.

2 TEACH

Key Concepts are highlighted.

CONCEPT DEVELOPMENT

▶ Draw a ladder on the chalkboard and identify the uprights of the ladder as well as the rungs to help students understand the structure of the DNA molecule.

CROSS CURRICULUM

▶ **Reading:** Have students research the history of DNA research and use poster board to draw and label a time line showing the events of DNA research since DNA was first removed from a cell nucleus in 1869.

CONCEPT DEVELOPMENT

▶ Students are probably familiar with Morse code. Morse code uses only two symbols, the dot and the dash, in varying combinations to represent numbers and the letters of the alphabet. Anything that can be expressed in language can be expressed in Morse code. The DNA code has four symbols. The order of nitrogen bases, rather than the sequence of dots and dashes, expresses the information needed for life processes.

▶ Explain that knowing the base-pairing rules allows you to predict the base sequence of a second strand of DNA if you know the sequence of the first strand.

▶ Make certain that students know that when they hear about the "code of life" they are hearing about the order of nitrogen bases in DNA.

▶ Ask students to predict the base sequence of the second strand of DNA if the sequence of one strand is T G A A C T C G A T A C. **Answer:** A C T T G A G C T A T G.

▶ **Activity:** Have students make a reproducible DNA model from common edible items such as marshmallows, gum drops, jelly beans, and so on. They can be joined together with toothpick halves.

Students' answers might include: to help doctors detect broken bones; to help dentists determine the position of teeth and cavities; to help determine structures of some molecules.

X rays, a form of electromagnetic radiation, have provided valuable information for research and health. Name three things that X rays can be used for.

How DNA Copies Itself

When chromosomes are doubled at the beginning of mitosis, the amount of DNA in the nucleus is also doubled. What is the process by which DNA copies itself? The Watson and Crick model can be used to show how this takes place. The two strands of DNA unwind and separate. Each strand then becomes a pattern on which a new strand is formed. Figure 4-13 follows a strand of DNA as it produces two new DNA strands identical to the original DNA. The events that take place while DNA copies itself are given below.

Step 1. DNA molecule before copying begins

Step 2. An enzyme breaks the bonds between the nitrogen bases. The two strands of DNA separate.

Step 3. The bases attached to each strand then pair up with new bases from a supply found in the cytoplasm. Adenine pairs with thymine. Cytosine pairs with guanine. The order of base pairs in each new strand of DNA will match the order of base pairs in the original DNA.

Step 4. Sugar and phosphate groups form the side of each new DNA strand. Each new DNA molecule now contains one strand of the original DNA and one new strand.

Figure 4-13. DNA uses itself as a pattern when it is copied.

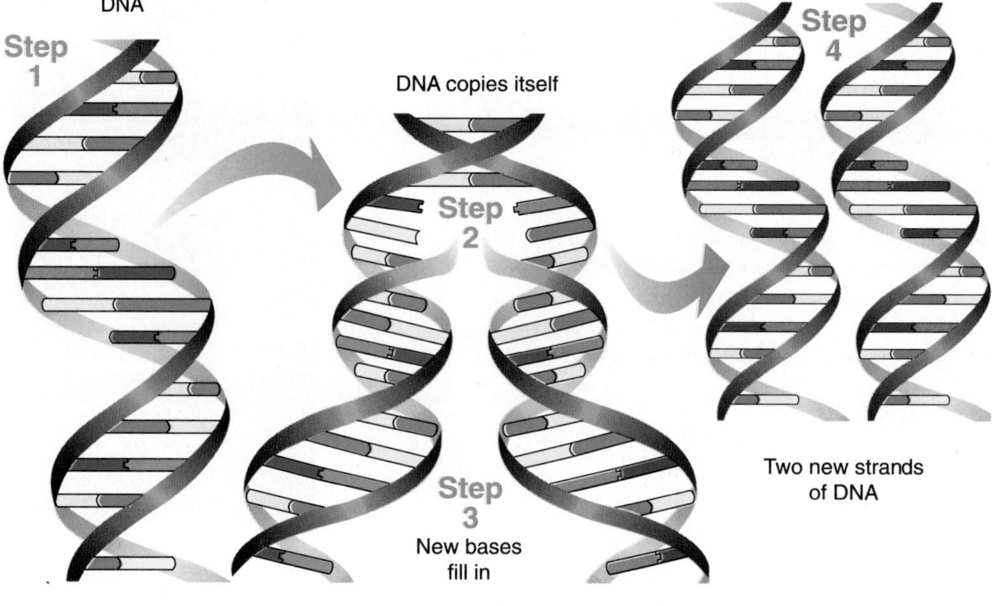

OPTIONS

PROGRAM RESOURCES

From the **Teacher Resource Package** use:

Transparency Masters, pages 13-14, DNA Replication.

Use **Color Transparency** 7, DNA Replication.

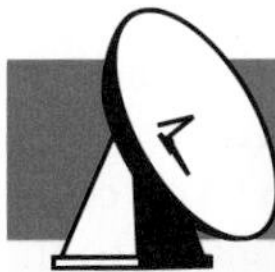

TECHNOLOGY

The Bacteria Factory

Bacteria can be turned into factories to produce important substances using DNA technology. Small circular molecules in some bacteria, called plasmids, are the key. A piece of DNA from another source is joined or spliced to the plasmid. This modified plasmid is mixed with bacterial cells and a few take in the new plasmid. Quickly, each cell grows into a colony of millions of cells—each producing the desired substance.

Insulin for diabetics and enzymes for making cheese have become common products of DNA technology. With the help of this technology, sugars in vegetable wastes have been converted into ethanol, a clean-burning fuel.

Think Critically: How would bacteria-produced ethanol be useful in the United States?

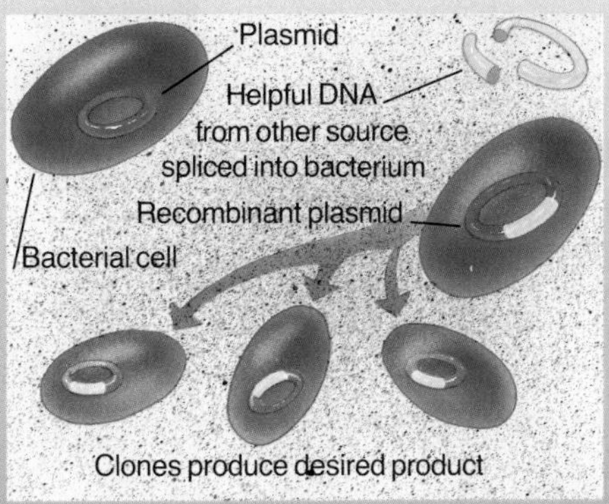

DNA and Genes

Why is DNA important? All of the characteristics that you have are affected by the DNA that you have in your cells. It controls the color of your eyes, the color of your hair, and whether or not you can digest milk. These characteristics are called traits. How traits appear in you depends on the kinds of proteins your cells make. DNA stores the blueprints for making proteins.

Proteins are made of units called amino acids that are linked together in a certain order. A protein may be made of hundreds or thousands of amino acids. Changing the order of the amino acids changes the kind of protein made. The section of DNA on a chromosome that directs the making of a specific protein is called a **gene.** Genes control proteins that build cells and tissues and work as enzymes. Think about what might happen if an important protein couldn't be made in your cells.

What is a gene?

How does a cell know which proteins to make? The gene gives the directions for the order in which amino acids will be arranged. This order results in a particular protein.

TECHNOLOGY

Extension: Have students find out if there are any ethanol plants or service stations operating in their state.

References: "Turning Plants into Fuel," *USA Today,* June 1990, p. 6. "Genetic Engineering," *World Book Encyclopedia,* Volume 8, 1990, pp. 85-87.

Think Critically: The process uses inedible vegetable matter, so it allows us to use otherwise useless materials. It might provide a cheaper substitute for petroleum-based fuels.

CONCEPT DEVELOPMENT

▶ Use the example of natural hair color to explain that the kind of protein you make shows up as your traits. Black hair color is due to a slightly different protein from that of blond hair color.

REVEALING MISCONCEPTIONS

▶ Students may not understand that while DNA is unique to every person, all the cells in an individual's body have the same kind of DNA.

▶ Explain to students that there are certain signals within the genetic code that direct the starting and stopping of various processes. The genetic code has punctuation just as a sentence begins with a capital letter and ends with a period. Certain start and stop signals have been recognized in the DNA code.

ENRICHMENT

▶ Have students read and report on the research of Freidrich Meischer, Robert Feulgen, Frank Griffith, Edwin Chargaff, Maurice Wilkins, Rosalind Franklin, James Watson, and Frances Crick.

▶ Have students research and report on Barbara McClintock's work on "jumping genes."

MINI-Lab

Answers: The order of bases in the RNA strand will be A G U C U A G C U. The order of bases in the tRNA strand will be U C A G A U C G A.

MINI-Lab

ASSESSMENT

Performance: To further assess students' understanding of base order in RNA, see USING LAB SKILLS, question 12, page 96.

CONCEPT DEVELOPMENT

▶ DNA produces RNA in a process called transcription. Messenger RNA carries the DNA code from the nucleus to the ribosome. On the ribosome, it directs the making of a protein. The process of making a protein from information in the mRNA is called translation.

CHECK FOR UNDERSTANDING

Ask students to explain how DNA, RNA, and ribosomes function in making proteins.

RETEACH

Have students make a drawing of DNA replication and protein synthesis.

EXTENSION

For students who have mastered this section, use the **Reinforcement** and **Enrichment** masters or other OPTIONS provided.

MINI-Lab

How is RNA made?

RNA is made in the nucleus on a pattern supplied by DNA. Look at Figures 4-12 and 4-13 to see the structure of DNA and how these strands separate to make new DNA. When RNA is made, the two DNA strands separate and one is used as a pattern for making a strand of RNA. Suppose a molecule of DNA untwists and separates and the left strand is used to make RNA. If the order of the bases in the DNA strand is T C A G A T C G A, what, in correct order, will be the bases in the RNA? If the new strand of RNA that you just made were a piece of mRNA, what tRNA bases would link up with it in the cytoplasm?

In Chapter 2, you learned that proteins are made on ribosomes in cytoplasm. How does the code in the nucleus reach the ribosomes out in the cytoplasm? The codes for making proteins are carried from the nucleus to the ribosomes by a second type of nucleic acid, called ribonucleic acid or **RNA.** RNA is different from DNA in that it is made up of only one strand, and it contains a nitrogen base called uracil (U) in place of thymine and the sugar ribose. RNA is made in the nucleus on a DNA pattern.

Two different kinds of RNA are made from DNA in the nucleus—messenger RNA (mRNA) and transfer RNA (tRNA). Protein assembly begins as mRNA moves out of the nucleus and attaches to ribosomes in the cytoplasm. Pieces of tRNA pick up amino acids in the cytoplasm and bring them to the ribosomes. There, tRNA temporarily matches with mRNA and the amino acids become arranged according to the code carried by mRNA. The amino acids become bonded together, and a protein molecule begins to form.

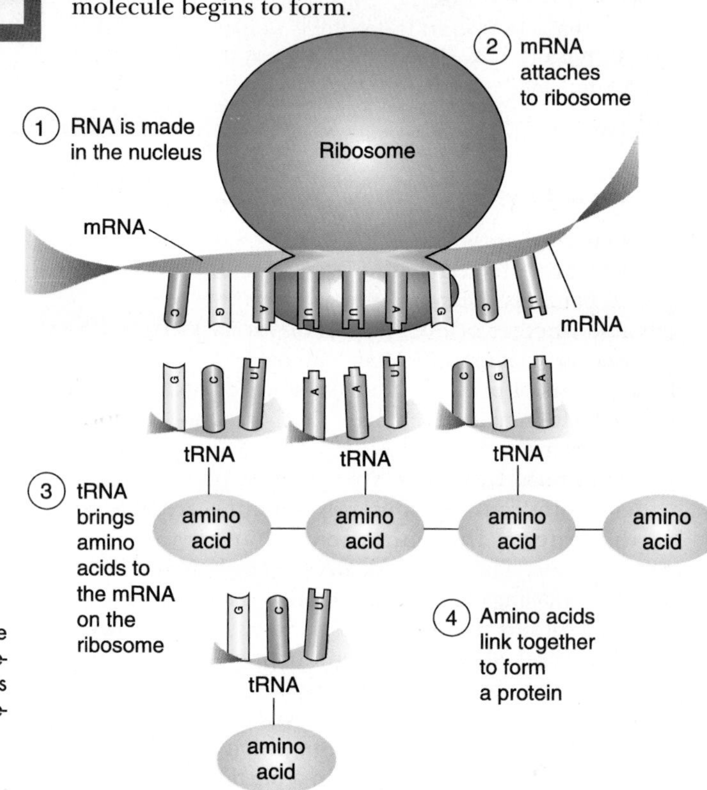

Figure 4-14. RNA carries the code for a protein from the nucleus to the ribosome. There, its message is translated into a specific protein.

OPTIONS

ASSESSMENT—ORAL

▶ **What would happen to a cell if it did not make copies of its chromosomes before it divided into two cells?** *The two new cells would have fewer chromosomes than the original cell and would not have the same instructions.*

▶ **Why is DNA replication necessary for life?** *Without DNA replication, new cells would not receive the instructions for the proteins needed for life.*

▶ **What is the significance of the fact that all life-forms have the same chemical responsible for regulating inherited traits?** *The common chemical makeup suggests common origin.*

PROGRAM RESOURCES

From the **Teacher Resource Package** use:

Activity Worksheets, page 39, Mini-Lab: How is RNA made?

Cross-Curricular Connections, page 8, Translation.

Mutations

Genes control the traits you inherit. Without correctly coded proteins, an organism can't grow, repair, or maintain itself. If a change occurs in a gene or chromosome, the traits of that organism are changed. Sometimes during replication an error is made in copying a gene. Occasionally, a cell receives an entire extra chromosome. Outside factors such as X rays and chemicals have been known to change or break chromosomes. Any permanent change in a gene or chromosome of a cell is called a **mutation.** If the mutation occurs in a body cell, it may or may not be life threatening to the organism. If, however, a mutation occurs in a sex cell, then all the cells that are formed from that sex cell will have that mutation. Down syndrome in humans occurs when a sex cell forms with an extra number 21 chromosome, so that the zygote that forms has three number 21 chromosomes. Many mutations are harmful to organisms. Some are not. Many times, an organism with a mutation doesn't survive. But mutations also add variety to a species.

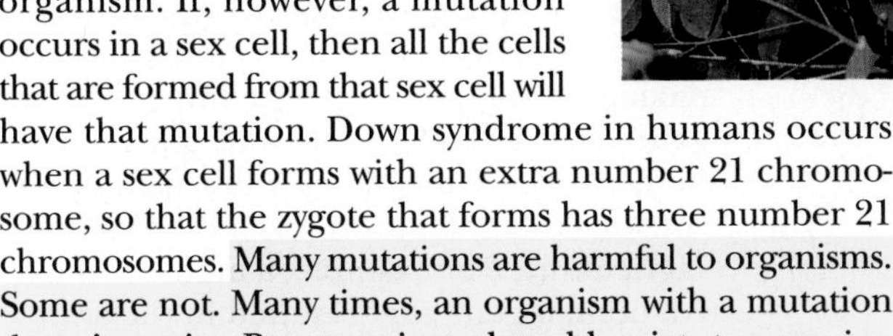

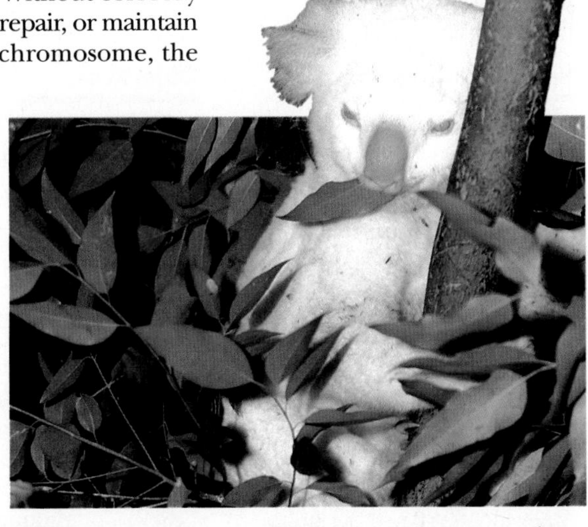

Figure 4-15. Albinism, resulting in the lack of pigments in skin, eyes, and hair, is an example of a mutation.

SECTION REVIEW

1. Which bases form pairs in a DNA molecule?
2. Why is protein production important in a cell?
3. How does DNA make a copy of itself?
4. **Apply:** A single strand of DNA has the bases AGTAAC. Using letters, show what bases would match up to form a matching DNA strand from this pattern.
5. **Connect to Chemistry:** Strands of DNA are held together by hydrogen bonds. What is the atomic number of hydrogen?

Concept Mapping — Skill Builder

Using a network tree concept map, show how DNA and RNA are alike and how they are different. If you need help, refer to Concept Mapping in the **Skill Handbook** on pages 688 and 689.

CONCEPT DEVELOPMENT

▶ Point out that not all mutations are harmful; some are helpful. Most, however, are probably neither helpful nor harmful.

3 CLOSE

▶ Ask questions 1-3 and the **Apply** Question in the Section Review.

FLEX Your Brain

Use the Flex Your Brain activity to have students explore ENVIRONMENTAL CAUSES OF MUTATIONS.

ASSESSMENT

Portfolio: Use the Flex Your Brain activity to reinforce critical thinking and problem solving skills. In Step 2, students may need to be led to common causes of cell mutation. Some examples are: radiation (including UV in sunlight), some chemicals, all carcinogens, and random chance.

SECTION REVIEW ANSWERS

1. adenine and thymine, guanine and cytosine
2. Proteins determine the traits an organism has.
3. The two strands of DNA separate and replicate identical strands. Each DNA molecule has one original strand and one new strand.
4. Apply: T C A T T G
5. Connect to Chemistry: one

PROGRAM RESOURCES

From the **Teacher Resource Package** use:

Activity Worksheets, page 5, Flex Your Brain.

Critical Thinking/Problem Solving, page 8, Genetic Testing and Missing Children.

Science Integration Activities 4, The Effects of Radiation on Seeds

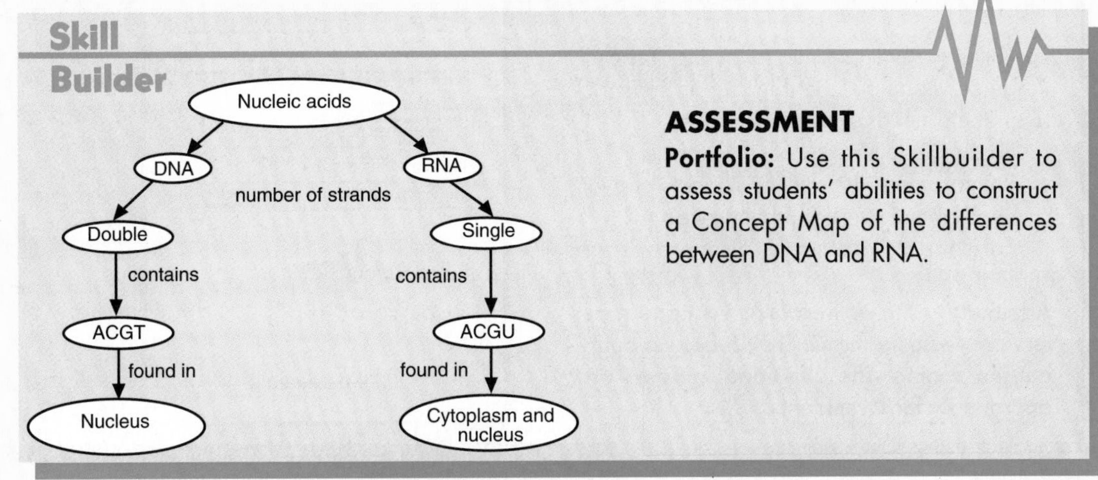

ASSESSMENT

Portfolio: Use this Skillbuilder to assess students' abilities to construct a Concept Map of the differences between DNA and RNA.

SECTION 4-4

PREPARATION

SECTION BACKGROUND

▶ Transgenic organisms are now being patented, an activity which is being opposed by individuals concerned with patenting life and preventing the sharing of knowledge with the scientific community.

1 MOTIVATE

▶ Obtain copies of patent applications and have students identify patent numbers on a variety of items.

2 TEACH

Key Concepts are highlighted.

CONCEPT DEVELOPMENT

▶ Ask the following question. **Why is it important that cells in experimental animals be genetically similar to human cells?** *They will react in similar ways to tested drugs.*

CHECK FOR UNDERSTANDING

Have students define the following terms in a short quiz: *transgenic, carcinogenic, patent.*

4-4 Inventing Organisms

New Science Words

transgenic organisms

Objectives

▶ Explain the term *transgenic organism.*
▶ Explain some advantages and disadvantages of patenting organisms.

Transgenic Organisms

Have you ever thought of actually being able to invent an organism? Today, scientists are doing just that. Through genetic engineering, they are able to produce **transgenic organisms,** organisms that contain genetic information from another species. Making a transgenic organism involves taking a gene from one organism and placing it in another. This technology provides many opportunities for research in medicine and agriculture. For example, a mouse received a human gene that resulted in the animal being very susceptible to human cancers. Because this mouse can develop certain types of tumors, it is a very valuable tool for cancer research. Using this organism, tests can be conducted to detect substances that cause cancer. Potential anti-cancer drugs can also be tested. In agriculture, a gene from bacteria can be transplanted into crops to prevent certain insect pests from eating them.

Patenting Life

Biologists are now asking for patents on genetically engineered organisms that they develop in the laboratory. A patent is a license issued by the government that gives the owner the legal right to control the manufacture and sale of an invention. In 1988, the United States government gave the first transgenic organism patent to Harvard University for its transgenic mouse.

OPTIONS

Meeting Different Ability Levels

For Section 4-4, use the following **Teacher Resource Masters** depending upon individual students' needs.

◆ **Study Guide Master** for all students.
● **Reinforcement Master** for students of average and above average ability levels.
▲ **Enrichment Master** for above average students.

Additional Teacher Resource Package masters are listed in the OPTIONS box throughout the section. The additional masters are appropriate for all students.

◆ STUDY GUIDE 20

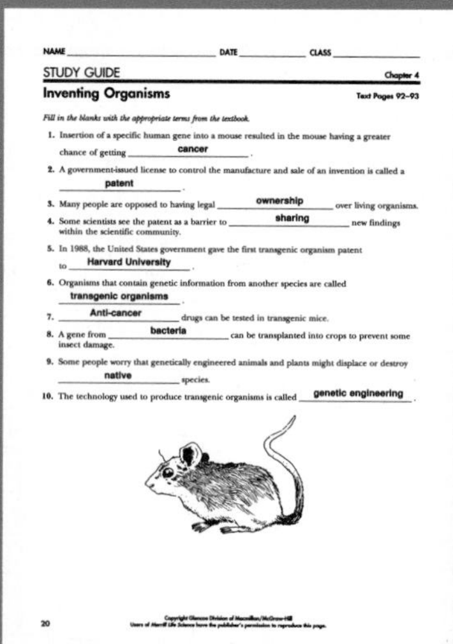

NAME ________ DATE ________ CLASS ________

STUDY GUIDE — Chapter 4

Inventing Organisms — Text Pages 92–93

Fill in the blanks with the appropriate terms from the textbook.

1. Insertion of a specific human gene into a mouse resulted in the mouse having a greater chance of getting **cancer**.
2. A government-issued license to control the manufacture and sale of an invention is called a **patent**.
3. Many people are opposed to having legal **ownership** over living organisms.
4. Some scientists see the patent as a barrier to **sharing** new findings within the scientific community.
5. In 1988, the United States government gave the first transgenic organism patent to **Harvard University**.
6. Organisms that contain genetic information from another species are called **transgenic organisms**.
7. **Anti-cancer** drugs can be tested in transgenic mice.
8. A gene from **bacteria** can be transplanted into crops to prevent some insect damage.
9. Some people worry that genetically engineered animals and plants might displace or destroy **native** species.
10. The technology used to produce transgenic organisms is called **genetic engineering**.

The development of transgenic organisms and the patenting of these organisms have met some opposition. People who are concerned about the consequences of such genetic technology worry that transgenic organisms may become pests or burdens to society. Some think that transgenic plants and animals might have the same effect on society as organisms introduced into the United States from other countries. Certain species of plants and insects have taken over or destroyed many native organisms. In addition, many people are opposed to the idea of having legal ownership of living organisms. Although the patents are designed to protect the biotechnology industry, some scientists see patenting of organisms as a barrier to the tradition of sharing information within the scientific community.

As technology continues to grow and play an important role in society, new developments may present new risks to the environment. People will have to compare these risks with the benefits that they may provide.

SECTION REVIEW

1. What is a transgenic organism?
2. List two risks of developing transgenic organisms.
3. What rights does a patent owner have?

You Decide!

After hospital patients have had surgery to remove tumors or tissues, these cells are sometimes saved. They are kept alive in the laboratory. Sometimes these cells, or products made from these cells, are sold for use in research. Patients are now claiming that they should have profits from products made from their own body tissues. Research companies that develop the products disagree. They argue that the cells all by themselves are useless without the time and skills of the researchers. Should patients be paid for products made from their cells?

● REINFORCEMENT 20

NAME ____________ DATE ________ CLASS ________

REINFORCEMENT — Chapter 4

Inventing Organisms — Text Pages 92-93

Answer the following questions using information from the textbook.

1. What are two reasons scientists want a mouse to be susceptible to human cancers? **to test whether substances cause cancer and to test anti-cancer drugs**
2. What is one way genetic engineering has been used in place of pesticides? **A bacterial gene transplanted into crops prevents insects from eating the crops.**
3. What is a patent? **It is a government-issued license giving the owner legal rights to control the manufacture and sale of an invention.**
4. Give two arguments against having patents for transgenic organisms. **Many people are opposed to having legal ownership of organisms. Some scientists see patents as a barrier to sharing new scientific findings within the scientific community.**
5. When was the first transgenic organism patented in the United States? **1988**
   What was this organism? **a transgenic mouse**
6. What is the technology called that has allowed scientists to transplant genes from one species to another? **genetic engineering**
7. Why do some people object to genetic engineering? **They worry that new organisms resulting from technology might displace or destroy native species.**

20

▲ ENRICHMENT 20

NAME ____________ DATE ________ CLASS ________

ENRICHMENT — Chapter 4

Inventing Organisms — Text Pages 92-93

GENETIC ENGINEERING

As scientists learned more about the coding of proteins, they realized that there might be an efficient way to use bacteria to produce large quantities of medically valuable proteins like insulin or interferon, a virus-fighting protein.

For example, the gene that codes for a protein such as human insulin can be inserted into the DNA of a bacterium. As the bacterium reproduces, the gene for insulin is copied. This process is called **gene splicing.** The resulting new DNA of the bacteria is called **recombinant DNA.**

Bacteria make good "factories" for producing many copies of an inserted gene. The inserted insulin gene tells the bacteria to make human insulin. Since bacteria are easy to grow and will multiply every 20 minutes, they can make a lot of human insulin quickly and inexpensively. And each new generation of bacteria contains copies of its own DNA and the genes for insulin.

Another reason why bacteria are used for gene splicing is that many strains of the bacteria contain free rings of DNA called plasmids. These plasmids are separate from the bacteria's chromosomes, but are copied and passed on with each division. The plasmids can be used like transport vehicles to get the new DNA into the bacteria.

Follow the diagram to see how gene splicing works.

1. Plasmid DNA rings are isolated from bacteria and cut open with enzymes that work at specific sites on the DNA.
2. Foreign DNA fragments are mixed with the cut plasmids and then fused together using an enzyme.
3. The new DNA, (now called recombinant DNA), is mixed with bacteria that have been treated so that they will absorb the plasmids.

plasmids + enzymes
enzyme
foreign genes
plasmid
chromosome

Answer the following questions.

1. Using gene-splicing techniques, how would you get bacteria to produce human insulin? (List the steps.) **1. Isolate human insulin gene and mix with cut plasmids. 2. Fuse DNA and put recombinant DNA into bacteria. 3. Grow bacteria and let them manufacture insulin.**
2. Using outside references, find out how the bacteria containing recombinant DNA is separated from the other bacteria. Give a basic explanation of how this is done. **A gene that gives resistance to a particular antibiotic can be attached to the foreign DNA. The bacteria are then grown in the presence of the antibiotic. Those bacteria that don't have the foreign DNA will be killed by the antibiotic.**

20

RETEACH

Present a video tape or filmstrip on animals in research or the development of disease-resistant crops.

EXTENSION

For students who have mastered this section, use the **Reinforcement** and **Enrichment** masters or other OPTIONS provided.

3 CLOSE

Use the Flex Your Brain activity to have students explore THE VALUE OF TRANSGENIC ORGANISMS.

Flex Your Brain

ASSESSMENT

Portfolio: Use the Flex Your Brain activity to reinforce critical thinking and problem solving skills.

▶ Ask questions 1-3 in the Section Review.

SECTION REVIEW ANSWERS

1. an organism that contains genetic information from another species

2. They may become pests or burdens to society. They may pose a barrier to the sharing of new scientific findings.

3. A patent holder has the legal right to control the manufacture and sale of an invention.

YOU DECIDE!

SCIENCE & SOCIETY

Students may disagree with each other and about the possible answers, but they should be able to support their answers.

VideoDisc

STVS: New Grains, Disc 4, Side 2

PROGRAM RESOURCES

From the **Teacher Resource Package** use:

Activity Worksheets, page 5, Flex Your Brain.

Science and Society, page 8, Glow-in-the-dark Genes.

ACTIVITY 4-2

OBJECTIVE: Design and construct a model of DNA and demonstrate how DNA copies itself.

Time: One class period to brainstorm and make patterns, one class period to complete the activity.

PROCESS SKILLS applied in this activity are **formulating a model, infering,** and **interpreting data.**

PREPARATION

- Obtain construction paper in six different colors.
- Obtain a three–dimensional model of DNA from a biological supply company or use pictures of a DNA molecule such as the one on page 87.
- Suggest that the groups use the same color for each of the molecules, for example: yellow for guanine, green for cytosine, purple for thymine, red for adenine, blue for sugar, and brown for phosphate.
- Students can model the sugars, phosphates, and bases from page 87 or pictures of other DNA molecules.

Cooperative Learning: Divide the class into Science Investigation Teams.

SAFETY

Remind students to use care with scissors.

THINKING CRITICALLY

The Watson and Crick model was like a spiral staircase, the side (handrails) are made up of twisted strands of sugar and phosphate molecules. The stairs that hold the two strands apart are made up of molecules called nitrogen bases that always occur in pairs.

TEACHING THE ACTIVITY

Refer to the ***Activity Worksheets*** *for additional information and teaching strategies.*

- Make sure that each group has 48 of each of the molecules that make up DNA. This will make a DNA molecule model that has 12 pairs that can be separated to copy itself.

ACTIVITY 4-2 DESIGNING AN EXPERIMENT
Making a Model of DNA

When you cut yourself, it heals through mitosis, where cells produce more cells. What ensures that each new cell is an exact copy of the original one?

Getting Started

DNA controls the activities of cells with coded instructions. What does a DNA molecule look like? In your group, you will design and construct a model of DNA and demonstrate with the model how DNA copies itself.

Thinking Critically

List the characteristics of the Watson and Crick model of the DNA molecule. How can these characteristics help you design and construct a DNA molecule model?

Materials

Your cooperative group will use:

- 6 colors of construction paper (8 1/2 x 11); 2 pieces of each color
- scissors (4 pairs)
- heavy paper for patterns (1 sheet)
- tape

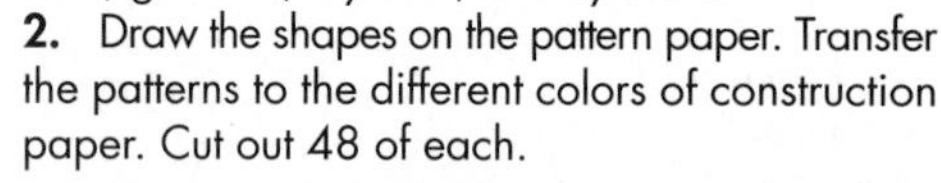

1. Decide what shapes you will use to represent each of the six molecules—sugar, phosphate, adenine, guanine, thymine, and cytosine.
2. Draw the shapes on the pattern paper. Transfer the patterns to the different colors of construction paper. Cut out 48 of each.
3. Join 12 sugar and 12 phosphate molecules in a chain and tape them. Repeat this step to make two sides of the DNA molecule.
4. Join the two sides of the DNA molecule using the four bases. Do not tape the bases together.
5. When you have observed the DNA molecule model, separate the bases and replicate the original molecule.
6. Use any combination of the DNA bases to create other molecule models.

Summing Up/Sharing Results

- Were the molecules your group created the same as those of other groups?
- Based on your observations of the DNA molecule model, *infer* why a DNA molecule seldom copies itself incorrectly.
- Explain why scientists use models.

Going Further!

How are RNA molecules different from DNA molecules? How are RNA molecules made?

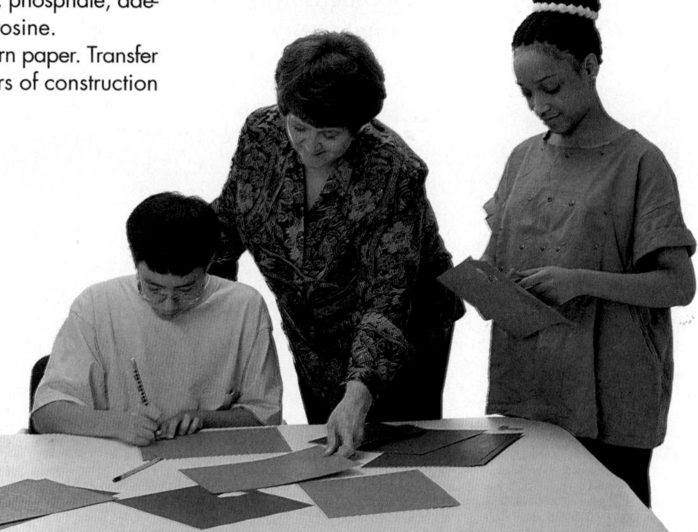

SUMMING UP/SHARING RESULTS

- The model of each of the groups was different from that of other groups.
- Only A and T pair and C and G pair so DNA seldom copies itself incorrectly.
- to help explain concept visually

GOING FURTHER!

Answers should include an RNA molecule is made of one strand and contains the nitrogen base uracil in place of thymine. RNA molecules are made on a DNA pattern.

PROGRAM RESOURCES

From the **Teacher Resource Package** use:

Activity Worksheets, pages 34-35, Activity 4-2, Making a Model of DNA.

Activity

ASSESSMENT

Performance: To further assess students' understanding of DNA Models, see USING LAB SKILLS, Question 10, on page 96.

CHAPTER

REVIEW

SUMMARY

4-1: Cell Growth and Division

1. Cells divide to form new cells by mitosis and divisions of cytoplasm. There are four stages of mitosis, during which chromosomes copy themselves, separate, and become part of new nuclei.
2. Animal cells have centrioles. New animal cells form when the cytoplasm pinches in. Plant cells lack centrioles; cytoplasm is separated by a cell plate following mitosis.
3. Asexual reproduction requires only one parent. Sexual reproduction requires two parents to produce a new individual. Fission, budding, and regeneration are all methods of asexual reproduction.

4-2: Sexual Reproduction and Meiosis

1. Sexual reproduction involves the production of sex cells through the process of meiosis.
2. Sex cells from two parents fuse during fertilization to produce a zygote, a cell with the diploid chromosome number.
3. Meiosis is a division of nuclei in reproductive cells that results in sex cells, four cells with the haploid chromosome number.

4-3: DNA

1. DNA is made up of bases that appear in specific sequences, and it directs all activities of a cell. DNA can copy itself and is the framework on which RNA is made.
2. RNA assembles proteins in the cytoplasm based on instructions from genes in the nucleus.
3. Mutations are changes in an organism's DNA. Most are harmful to the organism, but some are known to be beneficial.

4-4: Science and Society: Inventing Organisms

1. Transgenic organisms contain genes from an organism of another species.
2. Transgenic organisms can be used in research, but may become pests in nature.

KEY SCIENCE WORDS

a. **asexual reproduction**
b. **chromosomes**
c. **DNA**
d. **egg**
e. **fertilization**
f. **gene**
g. **meiosis**
h. **mitosis**
i. **mutation**
j. **RNA**
k. **sexual reproduction**
l. **sperm**
m. **transgenic organisms**
n. **zygote**

UNDERSTANDING VOCABULARY

Match each phrase with the correct term from the list of Key Science Words.

1. nuclear structures containing DNA
2. new individuals are identical to parent organism
3. segment of DNA that controls traits
4. deoxyribonucleic acid
5. formation of two new nuclei with identical chromosomes
6. fusion of an egg and a sperm
7. nuclear division that forms sex cells
8. a permanent change in a cell's DNA
9. organisms containing genetic information from another species
10. nucleic acid that carries information for making proteins from the nucleus

OPTIONS

ASSESSMENT

To assess student understanding of material in this chapter, use the resources listed.

COOPERATIVE LEARNING

Consider using cooperative learning in the THINK AND WRITE CRITICALLY, APPLY, and MORE SKILL BUILDERS sections of the Chapter Review.

PROGRAM RESOURCES

From the **Teacher Resource Package** use:

Chapter Review, pages 11-12.

Chapter and Unit Tests, pages 17-20, Chapter Test; pages 21, 22, Unit Test.

CHAPTER

REVIEW

SUMMARY

Have students read the summary statements to review the major concepts of the chapter.

UNDERSTANDING VOCABULARY

1. b	**6.** e
2. a	**7.** g
3. f	**8.** i
4. c	**9.** m
5. h	**10.** j

ASSESSMENT

Portfolio

Encourage students to place in their portfolios one or two items of what they consider to be their best work. For each item, ask students to explain why that item was chosen and what they learned from it. Items might be selected from the following.

- Enrichment research on cancer and aging, p. 77
- MINI-Lab mitosis models, p. 78
- Cross Curriculum poster on history of DNA reasearch, p. 87

Performance

Additional performance assessments may be found in *Performance Assessment* and *Science Integration Activities* that accompany **Merrill Life Science.** Performance Task Assessment Lists and rubrics for evaluating these activities and other products generated throughout the chapter can be found in Glencoe's *Performance Assessment in Middle School Science.*

CHAPTER

REVIEW

CHECKING CONCEPTS

1. d
2. d
3. b
4. c
5. a
6. a
7. d
8. d
9. c

USING LAB SKILLS

ASSESSMENT

Use these alternate lab excercises to assess students' understanding of skills used in this chapter.

10. The number of ways may seem endless.

11. Student models will be complex but should show mRNA and tRNA.

THINK AND WRITE CRITICALLY

12. A gene is a segment of DNA on a chromosome. Chromosomes are structures in the nucleus that contain DNA.

13. Offspring that are a result of asexual reproduction are identical genetically to the parent; offspring that are the result of sexual reproduction get half of their genetic material from each parent.

14. Spindle fibers pull chromatids to opposite poles.

15. Mitosis takes place in body cells; meiosis takes place in sex cells.

16. We notice the effects of harmful mutations more often, making us more aware of them.

CHAPTER

REVIEW

CHECKING CONCEPTS

Choose the word or phrase that completes the sentence.

1. ______ is a double spiral molecule with pairs of nitrogen bases.
a. RNA
b. An amino acid
c. A protein
d. DNA

2. RNA differs from DNA in that it contains ______.
a. thymine **c.** adenine
b. thyroid **d.** uracil

3. If a diploid tomato cell with 24 chromosomes undergoes meiosis, the sex cells produced will each have ______ chromosomes.
a. 6 **c.** 24
b. 12 **d.** 48

4. Chromosomes are doubled during ______.
a. anaphase **c.** interphase
b. metaphase **d.** telophase

5. During ______ in mitosis, double-stranded chromosomes separate.
a. anaphase **c.** metaphase
b. prophase **d.** telophase

6. The chromosome number in cells after mitosis is ______ the parent cell number.
a. the same as **c.** twice
b. half **d.** four times

7. Budding, fission, and regeneration are forms of ______.
a. mutations **c.** sexual reproduction
b. cell cycles **d.** asexual reproduction

8. ______ is any permanent change in a gene or a chromosome.
a. Fission **c.** Replication
b. Reproduction **d.** Mutation

9. Meiosis produces ______.
a. cells with the diploid chromosome number
b. cells with identical chromosomes
c. sex cells
d. a zygote

USING LAB SKILLS

10. In Activity 4-2 on page 94, you made a model of DNA using four bases. Design an alphabet with only four letters. How many words can you make using the four letters? The same letter may be used more than once. Using this exercise, what can you *infer* about the number of possible ways genes can combine?

11. Using the parts of the DNA model from Activity 4-2 and information in the MINI-Lab on page 90 about RNA, make a model showing how a strand of RNA is made from DNA.

THINK AND WRITE CRITICALLY

Answer the following questions in your Journal using complete sentences.

12. What is the difference between genes and chromosomes?

13. What are the differences between offspring formed by sexual reproduction and asexual reproduction?

14. What do spindle fibers do during mitosis and meiosis?

15. In what cells do mitosis and meiosis take place?

16. Why is more known about harmful mutations than beneficial ones?

APPLY

17. If one strand of DNA had bases ordered ATCCGTC, what would be the bases of its other strand?
18. A strand of RNA matching the DNA strand ATCCGTC would have what base sequence?
19. A mutation takes place in a human skin cell. Will this mutation be passed on to the person's offspring? Explain your answer.
20. What processes in mitosis provide both new cells with identical DNA?
21. How could a zygote end up with an extra chromosome?

MORE SKILL BUILDERS

If you need help, refer to the Skill Handbook.

1. **Making and Using Tables:** Compare and contrast DNA and RNA in a table.

Nucleic acid	DNA	RNA
Number of strands	2	1
Type of sugar	Deoxyribose	Ribose
Letter names of bases	ACTG	ACGU
Where found?	Nucleus	Nucleus and cytoplasm

Use a separate sheet of paper to complete your table.

2. **Hypothesizing:** Make a hypothesis about the effect of an incorrect mitotic division on the new cells produced.
3. **Comparing and Contrasting:** In a table, compare and contrast mitosis and meiosis as to: number of divisions, numbers of cells produced, numbers of chromosomes in parent cells and in sex cells.
4. **Concept Mapping:** Complete the events chain concept map of DNA synthesis.

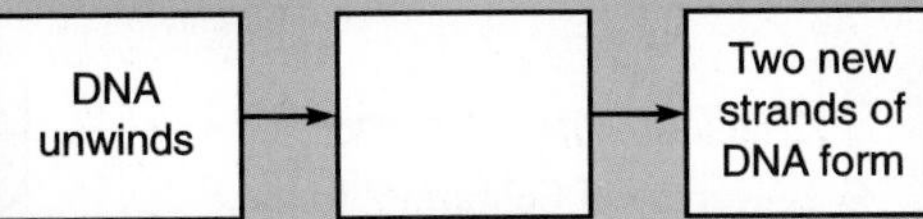
DNA unwinds
Two new strands of DNA form

5. **Sequencing:** Sequence the events that occur in sexual reproduction. Start with interphase in the parent cell and follow it through production of a sex cell by meiosis. End with the formation of the zygote. Tell whether the number of chromosomes present at each stage is haploid or diploid.

PROJECTS

1. Research the events involved in exploring DNA since it was first removed from a cell nucleus in 1869. Show your results on a time line.
2. Write a one-page paper about growth hormones called gibberellins and their use.

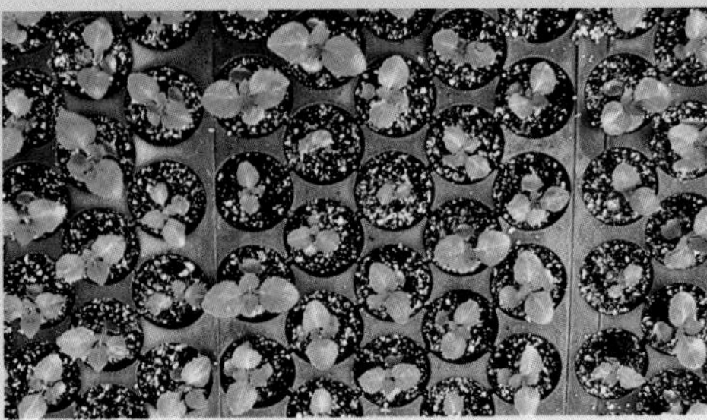

APPLY

17. A T C C G T C is matched by T A G G C A G.
18. U A G G C A G
19. No. In order for a mutation to be passed to offspring, the mutation must take place in a sex cell.
20. the copying of chromosomes in interphase; the separation of the copies at anaphase; the separation of two new cells at telophase
21. During meiosis, if one of the resulting eggs or sperm has both chromatids due to failure of the chromatids to separate, a zygote formed with that egg or sperm would have an extra chromosome.

MORE SKILL BUILDERS

1. Making and Using Tables: Answers are found printed in red in the table under question 1.
2. Hypothesizing: Incorrect division can result in an incorrect number of chromosomes, as in Down syndrome.
3. Comparing and Contrasting: Mitosis has one division, produces two cells, and has diploid number of chromosomes in both parent cells and cells produced. Meiosis has two divisions, produces four cells, and has diploid number of chromosomes in the parent cells and haploid in the cells produced, which are called sex cells.
4. Concept Mapping: Answer—new bases fill in.
5. Sequencing: Order of events given for meiosis should reflect information on pages 84 and 85; zygote information on page 83.

- Cell at the beginning of meiosis is diploid.
- Cells at end of meiosis are all haploid.

UNIT 1 GLOBAL CONNECTIONS

Objective

In this unit ending feature, the unit topic, "Life," is extended into other disciplines. Students will see the relationship between life science and other subjects.

Motivate

Cooperative Learning: Have students work in Expert Teams. Assign one Connection to each group of students. Have each group research the living things found in the geographic location of each Connection.

Teaching Tips

- Tell students to think about the characteristics of living things as they read this feature.
- After students have reviewed the Connections, ask them to form hypotheses about conditions that must be present for life to exist.

Wrap-Up

Have students make predictions about where in the world and universe life may exist. Have them back up their predictions using the characteristics of living things.

BIOLOGY

Background: How a cell "behaves" when certain components are missing helps to identify specific genes.

Discussion: Discuss the need for genetic research. Scientists are still learning about the functions of genes, where specific genes occur on the DNA ladder, and how the presence or absence of a specific gene affects the organism.

Answer to Question: By removing a single section of a chromosome, scientists can learn the location and function of a specific gene.

Extension: Have students read a description of how Watson and Crick discovered the structure of the DNA molecule.

UNIT 1

GLOBAL CONNECTIONS

Life

In this unit, you have studied life and the characteristics of living things. What is the relationship between life science and other subjects you will encounter in your lifetime?

120° 60°

BIOLOGY

CHROMOSOME SURGERY
University of California, Irvine, California

Chromosome surgery? Sounds impossible! Where would anyone find a scalpel small enough? Scientists at the University of California use a laser beam focused through a microscope to remove one part of a chromosome at a time. What can scientists learn from examining separate parts of chromosomes?

METEOROLOGY

LIFE FROM THE SKY?
Woods Hole, Massachusetts

Ships of Woods Hole Oceanographic Institute explore the sea, studying ocean life forms, charting currents and ocean depths, and studying the relationship between the ocean and rainfall. Why is rainfall important to living organisms?

98

METEOROLOGY

Background: Scientists have found that breaking ocean waves toss water droplets into the air. Specks of salt are released and travel to the clouds. Once embedded, they become nuclei for raindrops.

Discussion: Clouds are sometimes "salted" with silver compounds to produce rain. Would salt be an environmentally safer alternative?

Answer to Question: Rainfall furnishes water, which is necessary for life. Living things must have a source of water to exist.

Extension: Have students explore any temporary pools in your area at different times of the year. Be sure students notice the abundance of life when water is in the pool.

Multicultural Awareness

Use these Global Connections to have students increase their awareness of evidence of life forms as they are found around the world.

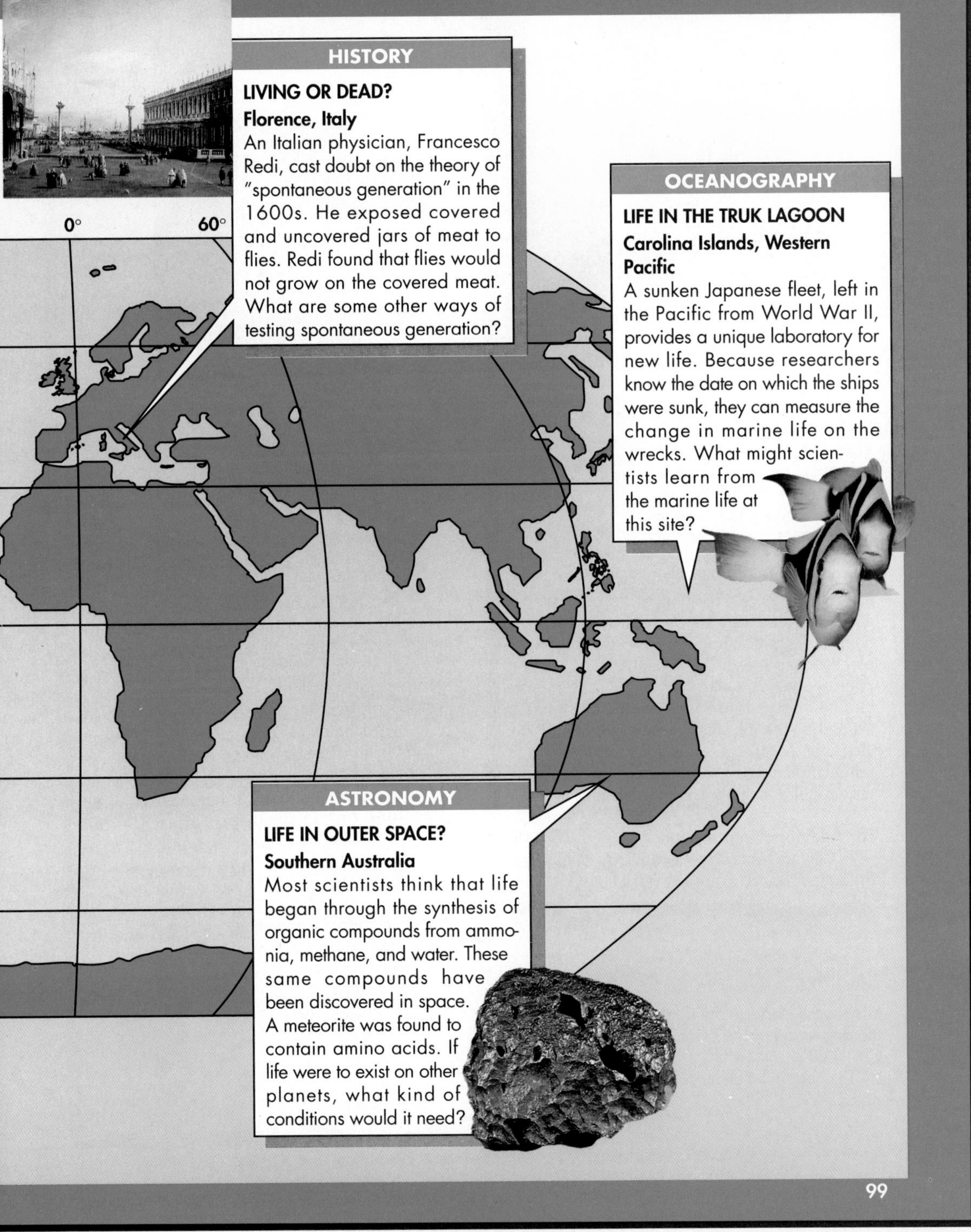

HISTORY

LIVING OR DEAD?

Florence, Italy

An Italian physician, Francesco Redi, cast doubt on the theory of "spontaneous generation" in the 1600s. He exposed covered and uncovered jars of meat to flies. Redi found that flies would not grow on the covered meat. What are some other ways of testing spontaneous generation?

OCEANOGRAPHY

LIFE IN THE TRUK LAGOON

Carolina Islands, Western Pacific

A sunken Japanese fleet, left in the Pacific from World War II, provides a unique laboratory for new life. Because researchers know the date on which the ships were sunk, they can measure the change in marine life on the wrecks. What might scientists learn from the marine life at this site?

ASTRONOMY

LIFE IN OUTER SPACE?

Southern Australia

Most scientists think that life began through the synthesis of organic compounds from ammonia, methane, and water. These same compounds have been discovered in space. A meteorite was found to contain amino acids. If life were to exist on other planets, what kind of conditions would it need?

HISTORY

Background: Spontaneous generation was a popular belief in the Middle Ages. Most scientists followed Aristotle, whose method was one of observation, thought, and logic, without experimentation.

Discussion: Ask students to solve the following question, "Do plants grow?" using Aristotle's method and then using scientific methods. Which method gives a more accurate answer?

Answer to Question: You might remove air from the meat as well as keeping it covered.

Extension: Have students investigate Stanley Miller's test of Alexander Oparin's hypothesis on the origin of life. Did Miller actually create life in a bottle?

OCEANOGRAPHY

Background: The sunken fleet in the Truk Lagoon leaks oil and chemicals into the surrounding ocean, but they seem to have little impact on sea life. Many unusual varieties of algae, jellyfish, and anemones flourish here.

Discussion: There is obviously some pollution occurring in the Lagoon. Discuss whether or not it should be "cleaned up" or left to nature.

Answer to Question: Oceanographers can tell how long it takes for certain kinds of marine life to establish in an area. They can also study rates of growth; relationships among plants, fish, and corals; and the effects of oil and chemicals on the environment.

Extension: Have students research other ship wrecks in both fresh and salt water to find out about the effects on the environment.

ASTRONOMY

Background: A meteorite that had recently fallen to Earth contained amino acids. Other meteorites contain cell-like fossils in such numbers that some scientists think life on other planets may have been much more prolific than on Earth.

Discussion: Discuss the planets of the solar system. Which might have the conditions necessary for life?

Answer to Question: Most of the life-forms we know need water, food, a temperate climate, and a breathable atmosphere.

Extension: Have students research other meteorites that have fallen to Earth. Have any others contained amino acids?

LABORATORY TECHNICIAN

Background: A laboratory technician may choose work in a variety of fields, including records, radiology, nuclear medicine, respiratory therapy, and cardiology. Preparation for this work can be obtained at a two-year college, in hospitals, medical centers, colleges, or trade schools.

Related Career	Education
Laboratory Technician	Jr. College
Radiology	Trade School
Ultrasound	Trade School

Career Issue: Every year types of technicians are required to be licensed. The proponents of this claim that it is necessary to protect the public. Others claim that it is government interference.

What do you think? Ask the students whether they think more technicians should be licensed.

BIOLOGY TEACHER

Background: A biology teacher works in the secondary education system, teaching junior and senior high school students about life science. Instruction usually includes class work, laboratory work and field trips. Biology teachers, in cooperation with other science instructors, may also provide opportunites for their students to become involved in science fairs. These fairs encourage individual and/or group research.

Related Career	Education
Teaching Assistant in High School	Jr. College
College Instructor or Professor	M.S. or Ph.D. degree

Career Issue: Biology teachers must be sensitive to controversies that may arise in their classrooms concerning such topics as animal rights.

What do you think? Can biology be taught well without having dissection labs?

CAREERS

LABORATORY TECHNICIAN

A *laboratory technician* provides daily care for laboratory experiments. Education requirements vary. Most laboratory technicians must have three years of work-related experience. Some laboratories may require one or two years of college-level courses in basic biology and English.

Other laboratory technicians assist scientists in setting up, maintaining, and monitoring specialized equipment for research projects in a university or government laboratory. For example, certain pieces of equipment such as electron microscopes and ultracentrifuges require specialized care. These technicians are usually trained on the job and by the equipment manufacturer. Most technician jobs require a high school diploma and mechanical aptitude. Some may require technical college training.

For Additional Information

Clinical Laboratory Technology, 818 Olive St. Suite 918, St. Louis, MO 63101

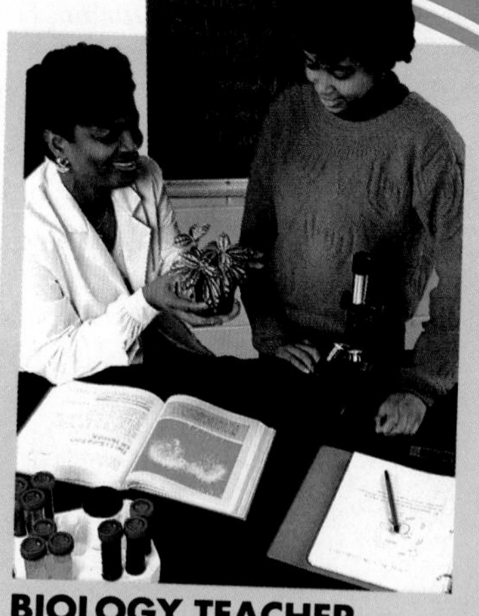

BIOLOGY TEACHER

A *biology teacher* is a science educator who works in schools helping students learn about living things.

Biology teachers teach about living things through lectures, demonstrations, and most importantly, having students involved in activities. At the middle/junior high school levels, the biology teacher and students may work in a laboratory doing experiments.

A bachelor's degree in biology is usually required by state departments of education to become certified as a biology teacher. Many teachers continue their professional development and obtain a master's degree. Often biology teachers take summer courses to experience field science and new teaching techniques.

For Additional Information

National Association of Biology Teachers, 11250 Roger Bacon Dr. #19, Reston, Virginia 22090.

UNIT READINGS

▶ Asimov, Isaac. *How Did We Find Out About the Beginning of Life?* Houston: Walker Publishing Co., 1982.

▶ "The Year in Science." *Discover,* January 1991, pp. 22-51.

100

UNIT READINGS

Background

▶ *How Did We Find Out About the Beginning of Life* is a comprehensive treatment of the historical search for the origin of life. It includes all the classics—Redi, Pasteur, Darwin, Oparin, Watson, etc.

▶ *The Atoms Within Us* is a well written tour of cell structure and other biological systems.

More Readings

1. Ward, Fred. "Florida's Reefs are Imperiled." *National Geographic,* July 1990.
2. Weaver, Robert F. "Beyond Supermouse: Changing Life's Genetic Blueprint." *National Geographic,* December 1984.
3. Gore, Rick. "The Awesome Worlds Within a Cell." *National Geographic,* September 1976.

Essays

by Sir Francis Bacon

Sir Francis Bacon wrote this essay in 1605 to illustrate the problems early scholars had developing scientific methods.

In the year of our Lord 1432, there arose a grievous quarrel among the brethren over the number of teeth in the mouth of a horse. For 13 days the disputation raged without ceasing. All the ancient books and chronicles were fetched out, and wonderful and ponderous erudition, such as was never before heard of in this region, was made manifest.

At the beginning of the 14th day, a youthful friar of goodly bearing asked his learned superiors for permission to add a word, and straightway, to the wonderment of the disputants, whose deep wisdom he sore vexed, he beseeched them to unbend in a manner coarse and unheard-of, and to look in the open mouth of a horse and find the answer to their questionings. At this, their dignity being grievously hurt, they waxed exceedingly wroth; and, joining in the mighty uproar, they flew upon him and smote him hip and thigh, and cast him out forthwith. For, said they, surely Satan hath tempted this bold neophyte to declare unholy and unheard-of ways of finding truth contrary to all the teachings of the fathers.

After many days more of grievous strife the dove of peace sat on the assembly, and they as one man, declaring the problem to be an everlasting mystery because of a grievous dearth of historical and theological evidence thereof, so ordered the same writ down."

In Your Own Words

▶ Write a report explaining what part of the scientific method the "youthful friar" wanted his superiors to use. Explain how this problem might be approached by researchers today.

101

Source: Sir Francis Bacon, *Essays*, written in 1597.

Biography: Sir Francis Bacon (1561-1626) was a British lawyer, philosopher, legislator, and author who, as was characteristic of "gentlemen" of that time, dabbled in many things. Of most interest to us is the fact that he was a leader in the struggle against the Aristotelian scholastic tradition and was the father of modern scientific thought.

TEACHING STRATEGY

If you have a flair for the dramatic, you may want to read this to your students. After the students have been exposed to the passage, have them respond to the discussion questions below.

Discussion Questions

1. **Have the students speculate on why Bacon wrote this passage.** *Bacon was a proponent of the experimental proof of scientific facts and theories. By holding the Aristotelians up to ridicule, he felt he could make his point.*
2. **Are there lessons in this passage for today?** *Note the reaction of the "established authorities" when their most basic beliefs are challenged. How might the young friar have made his point while avoiding becoming a social outcast? Set up cooperative learning teams to develop plans of action the friar might follow. Perhaps, one team could dramatize their plan.*

Classics

▶ Watson, James D. *The Double Helix.* New York: New American Library, 1969. Challenge, excitement, and competition in scientific research. Readable but considered gossipy by some reviewers.

▶ Darwin, Charles. *The Origin of the Species.* New York: New American Library, 1986. The watershed work in the evolution of "natural history" into biology.

UNIT

2

HEREDITY AND EVOLUTION

Unit 2 introduces students to how traits are inherited by living organisms. Evidence of evolution and how evolution takes place are discussed and used as a basis for classification of life on Earth.

CONTENTS

ADVANCE PREPARATION

Activities

▶ **Activity 5-1, page 113,** requires 100 red beans, 100 white beans, and paper bags.

▶ **Activity 5-2, page 124,** requires graph paper and rulers.

▶ **Activity 6-1, page 143,** calls for 100 pennies, cardboard boxes with lids, and graph paper.

▶ **Activity 6-2, page 150,** requires sunflower seeds and rulers.

▶ **Activity 7-1, page 159,** calls for 10 different kinds of seeds, hand lenses, and rulers.

Field Trips and Speakers

▶ Arrange a field trip to a museum that has fossil exhibits.

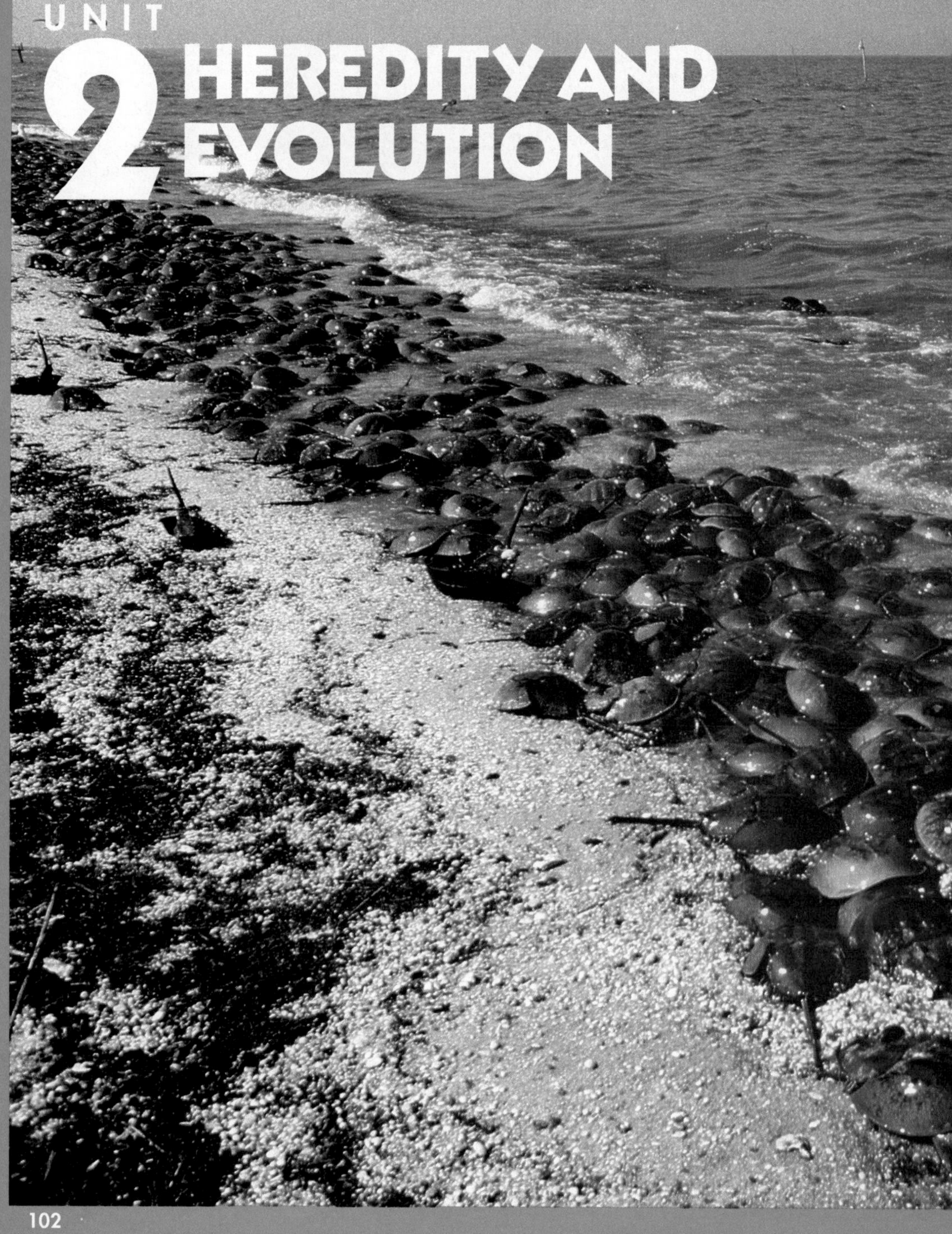

OPTIONS

Cross Curriculum

▶ Ask students to research how domesticating animals has affected human civilization.

▶ Have students use the dictionary to find the origin and meaning of the prefixes *paleo-*, *meso-*, *ceno-*, and the suffix *-zoic*.

PROJECT

During the course of this unit study, have students work cooperatively on PROJECT 2, *Keep Those Toes A-Tapping,* found on pages 696 and 697.

Science at Home

▶ Have students open 50 pea or bean pods. Count and keep a record of the number of seeds in each pod. Have them find the least number of seeds in a pod, the greatest number of seeds in a pod, and the average number of seeds in each pod.

Cooperative Learning: Have students work in groups of four to find out how fossils are dated and how this information has been used in constructing the geologic time table.

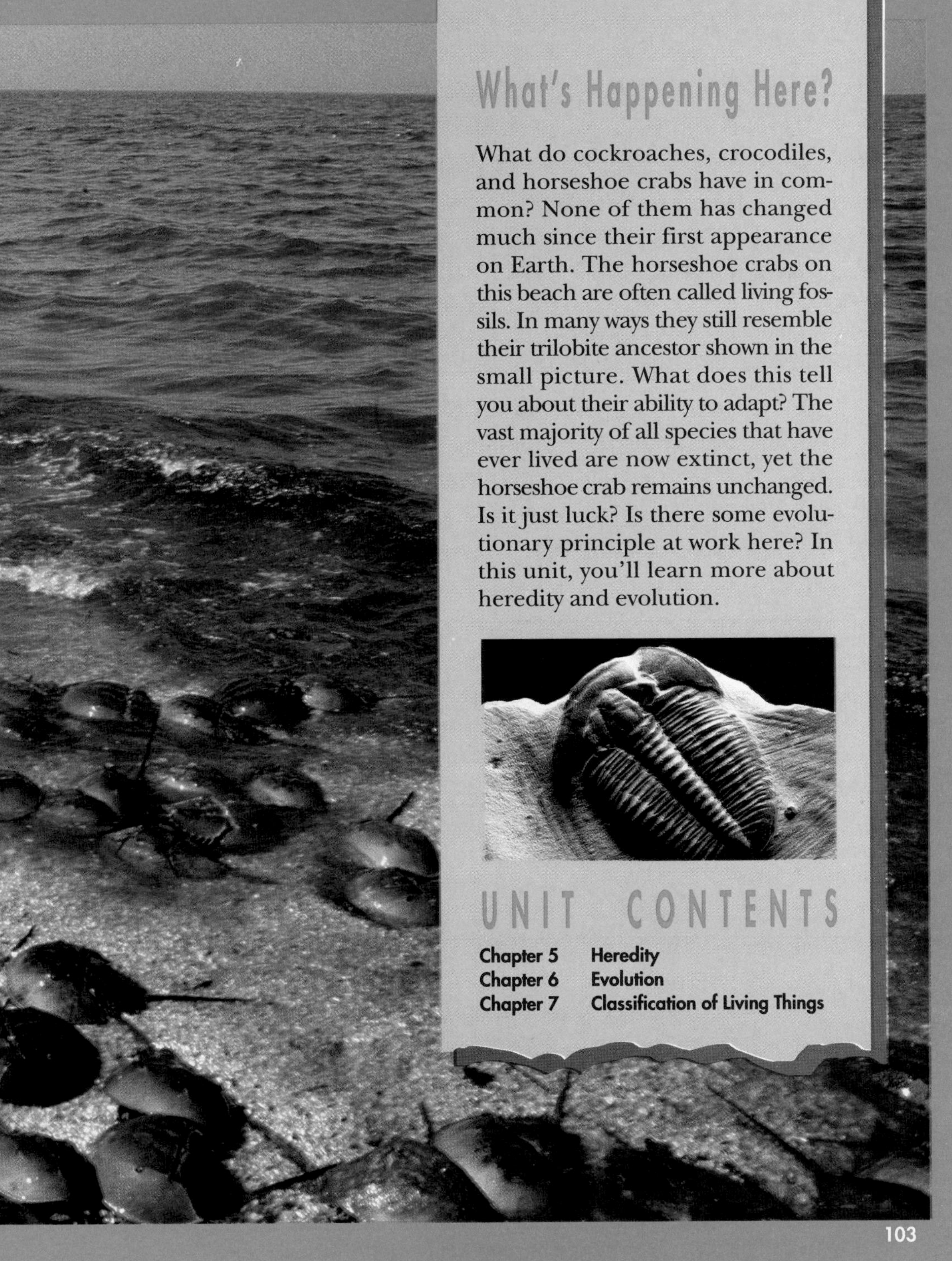

What's Happening Here?

What do cockroaches, crocodiles, and horseshoe crabs have in common? None of them has changed much since their first appearance on Earth. The horseshoe crabs on this beach are often called living fossils. In many ways they still resemble their trilobite ancestor shown in the small picture. What does this tell you about their ability to adapt? The vast majority of all species that have ever lived are now extinct, yet the horseshoe crab remains unchanged. Is it just luck? Is there some evolutionary principle at work here? In this unit, you'll learn more about heredity and evolution.

UNIT CONTENTS

INTRODUCING THE UNIT

What's Happening Here?

▶ Have students look at the photos and read the text. Ask them what's happening here. Point out that in this unit they will study the relationship between heredity and evolution and learn how evolutionary evidence is used to help classify organisms.

▶ **Background:** Horseshoe crab fossils show that today's examples are nearly identical to those that lived 190 million years ago. Modern cockroaches are similar to fossil cockroaches of the late Paleozoic, 320 million years ago. Neither crocodiles nor turtles have changed in appearance since the Triassic, 200 to 190 million years ago. Is there any kind of pattern in the survival of these disparate organisms?

Previewing the Chapters

▶ Students may be surprised that horseshoe crabs have such a long and successful evolutionary history. Have students research the life cycle and biology of horseshoe crabs to see how they are important to scientists today.

▶ Often animals that look alike are not classified similarly. Asian and African elephants actually belong to two different genuses, *Elephas* and *Loxodontia.*

Tying to Previous Knowledge

▶ Review the parts of the cell and how cells divide from Chapter 4. Stress the importance of meiosis to help students understand genetics and evolution.

▶ Display different kinds of fossils. Encourage students to find similarities and differences between fossil organisms and living organisms.

▶ Use the **inquiry questions** in the OPTIONS box below to investigate with students how organisms are classified.

Multicultural Awareness

Although humans in all cultures perceive the same range of the visible light spectrum, they do not have the same names or terms for all colors. Some cultures have many terms for color. The English language has 11 basic color terms. Some cultures, in New Guinea, for example, have only two color terms. These cultures group all colors into "black" or "white." Languages with three terms classify black, white, and red.

Inquiry Questions

Use the following questions to focus a discussion on how organisms are classified.

▶ **What is the first question you would ask to begin to classify a new organism? Why?**

▶ **What is the second question you would use? Why?**

▶ **How many questions would it take to be reasonably sure you were on the right track in classifying a new organism?**

CHAPTER 5 Heredity

CHAPTER SECTION	OBJECTIVES	ACTIVITIES
5-1 What Is Genetics? (3 days)	1. **Explain** how traits are inherited and Mendel's role in the history of genetics. 2. **Use** a Punnett square to predict the result of crosses. 3. **Explain** the difference between genotype and phenotype.	**MINI-Lab:** *What are some common traits?* p. 106 **Activity 5-1:** *Expected and Observed Results,* p. 113
5-2 Genetics Since Mendel (1 day)	1. **Explain** incomplete dominance. 2. **Compare** multiple allele and polygenic inheritance and give examples of both.	**MINI-Lab:** *What are fingerprints?* p. 115
5-3 Human Genetics (2 days)	1. **Describe** two human genetic disorders. 2. **Explain** inheritance of sex-linked traits. 3. **Explain** the importance of genetic engineering.	
5-4 The Human Genome Science & Society (1 day)	1. **Describe** the goal of the Human Genome Project. 2. **Explain** some advantages and disadvantages of genetic research.	**Activity 5-2:** *Designing an Experiment (Determining Polygenic Inheritance),* p. 124
Chapter Review		

ACTIVITY MATERIALS

FIND OUT	ACTIVITIES		MINI-LABS	
Page 105 pencil paper	**5-1 Expected and Observed Results, p. 113** 2 paper bags 100 red beans 100 white beans	**5-2 Designing an Experiment, p. 124** metric ruler graph paper pencil	**What are some common traits? p. 106** pencil paper	**What are fingerprints? p. 115** pencil paper

CHAPTER FEATURES	TEACHER RESOURCE PACKAGE	OTHER RESOURCES
Skill Builder: *Observing and Inferring,* p. 112	**Ability Level Worksheets** ◆ ***Study Guide,*** p. 21 ● ***Reinforcement,*** p. 21 ▲ ***Enrichment,*** p. 21 **Activity Worksheets,** pp. 5, 41-42, 47 **Concept Mapping,** p. 15 **Transparency Masters,** p. 15	**Lab Manual:** Genetic Traits, p. 29 **Lab Manual:** 50:50 Chances, p. 31 **Color Transparency 8,** Genetics **STVS:** Disc 4, Side 1
Skill Builder: *Comparing and Contrasting,* p. 116	**Ability Level Worksheets** ◆ ***Study Guide,*** p. 22 ● ***Reinforcement,*** p. 22 ▲ ***Enrichment,*** p. 22 **Activity Worksheets,** p. 48 **Cross-Curricular Connections,** p. 9	**Lab Manual:** Genes and Blood Clotting, p. 33 **STVS:** Disc 4, Side 1
Problem Solving: *Boy or Girl?* p. 118 **Technology:** *Karyotyping,* p. 120 **Skill Builder:** *Concept Mapping,* p. 121	**Ability Level Worksheets** ◆ ***Study Guide,*** p. 23 ● ***Reinforcement,*** p. 23 ▲ ***Enrichment,*** p. 23 **Critical Thinking/Problem Solving,** p. 9 **Science and Society,** p. 9 **Technology,** pp. 7-8	**STVS:** Disc 7, Side 2; Disc 4, Side 1 **Science Integration Activity 5**
You Decide! p. 123	**Ability Level Worksheets** ◆ ***Study Guide,*** p. 24 ● ***Reinforcement,*** p. 24 ▲ ***Enrichment,*** p. 24 **Activity Worksheets,** pp. 43-44	
Summary Key Science Words Understanding Vocabulary Checking Concepts Using Lab Skills Think & Write Critically Apply More Skill Builders Projects	ASSESSMENT RESOURCES **Chapter Review,** pp. 13-14 **Chapter Test,** pp. 32-35 **Performance Assessment in Middle School Science (PAMSS)**	**Chapter Review Software** **Test Bank** **Alternate Assessment** **Performance Assessment**

◆ **Basic** ● **Average** ▲ **Advanced**

ADDITIONAL MATERIALS

SOFTWARE	AUDIOVISUAL	BOOKS/MAGAZINES
Biology—Genetics, EBEC. *Cells & Genetics Picture File,* EBEC. *Modern Genetics: Chromosomes and Coding,* IBM. *Mendelian Genetics: The Science of Inheritance,* IBM. *Genetics,* Queue. *Heredity Dog,* Queue. *Mendel Bugs: A Genetics Simulation,* Focus.	*Genetics: Human Heredity,* film, Coronet/MTI. *Genetics and Plant Breeding,* film, BFA. *Genetics: Mendel's Laws,* film, Coronet/MTI. *Inheriting Your Physical Traits,* film, Coronet/MTI. *Mechanism of Inheritance,* film, McGraw-Hill.	Farnsworth, M.W. *Genetics.* 2nd ed. NY: Harper and Row Publishers, Inc., 1988. Fox, L. Raymond and Paul R. Elliot. *Heredity and You.* Dubuque, IA: Kendall-Hunt Publishing Company, 1983. Strickberger, Monroe, W., Jr. 3rd ed. *Genetics.* NY: Macmillan Publishing Company, Inc., 1985.

CHAPTER

5

HEREDITY

THEME DEVELOPMENT: The themes of evolution and homeostasis are further developed in this chapter. Homeostasis at the organism level is controlled by genes. This chapter introduces the basics of genetics and provides background for understanding evolution.

CHAPTER OVERVIEW

- **Section 5-1:** Genetics is defined within the context of the students' inheritance. Mendel's experiments and results are detailed.
- **Section 5-2:** Additional genetic concepts developed since Mendel's original work are introduced. Incomplete dominance is described. Multiple alleles are contrasted with polygenic inheritance.
- **Section 5-3:** Human genetics is explained in terms of Mendel's results. Some human genetic disorders and their causes are presented. Sex determination and sex-linked disorders are described.
- **Section 5-4: Science & Society:** The chapter ends with an overview of the goals of the human genome initiative. Students are asked to analyze the advantages and disadvantages of the project.

CHAPTER VOCABULARY

heredity
alleles
genetics
dominant
recessive
Punnett square
genotype
homozygous
heterozygous
phenotype
incomplete dominance
multiple alleles
polygenic inheritance
sex-linked gene
pedigree
genetic engineering
genome

CHAPTER

5 Heredity

OPTIONS

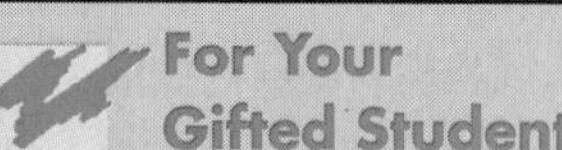

For Your Gifted Students

Have students research genetic engineering. They can use magazine and newspaper articles that discuss the topic. Have them discuss how genetic engineering is used and if it is beneficial to society. In a small group they can debate the pros and cons of the topic.

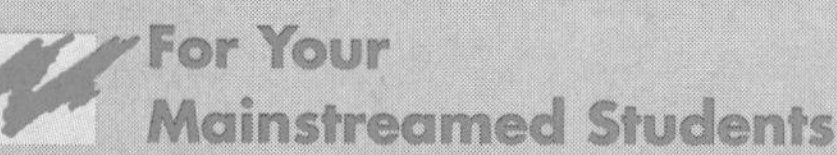

For Your Mainstreamed Students

Students can brainstorm a list of traits that are inherited (tongue rolling, eye color, handedness, etc.). Have them choose 1-3 traits and survey their classmates to see how many display the characteristics. Results can be graphed. The experiment can be expanded by surveying other classrooms. They can determine if the larger population has the same ratio of these traits as their classroom. You may wish to use the Mini-Lab on page 106 for this activity.

The tigers in the photo to the left have so many traits in common that they look almost identical. Most traits found in organisms have at least two different forms, such as free and attached earlobes in humans. How are these and other traits distributed among your classmates?

FIND OUT!

Do this activity to find out about the types of earlobes in your class.

Some people have earlobes that are attached, whereas others have earlobes that swing free. Count the number of your classmates who have attached earlobes. How many have free earlobes? *Record* the number of people in your class who have each type. What does your data tell you about the earlobe trait in your class? Do you have enough data to say which form of earlobe is more common in the human population?

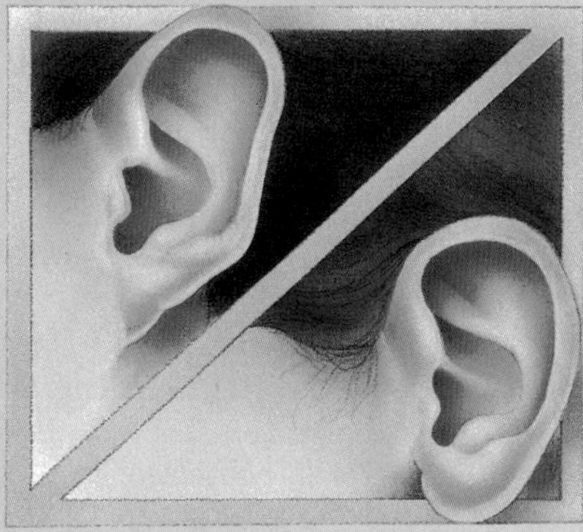

Gearing Up

Previewing the Chapter

Use this outline to help you focus on important ideas in this chapter.

Previewing Science Skills

- In the Skill Builders, you will observe and infer, compare and contrast, and use a concept map.
- In the Activities, you will predict, hypothesize, and collect and analyze data.
- In the MINI-Labs, you will observe, collect and analyze data.

What's next?

Now that you have seen that a single human trait can have more than one form, learn about some of the ways in which these traits are inherited, and learn about the importance of genetics in your life.

INTRODUCING THE CHAPTER

Use the Find Out activity to introduce students to inheritance. Inform students that they will be learning more about inheritance and genetics as they read the chapter.

FIND OUT!

Preparation: Make a table on the chalkboard or on an overhead transparency to complete for your classes. Continue to add data to the table throughout the day.

Answers: Results of individual classes will vary. The data from more than one class will not necessarily tell which is more common, because the sample is too small.

Cooperative Learning: Use Paired Partners or Science Investigation cooperative groups.

Teaching Tips

- Before the activity begins, you should clearly explain the difference between attached and free earlobes.
- Record data on the chalkboard for each class, so that students can compare their results with the whole.

Gearing Up

Have students study the Gearing Up feature to familiarize themselves with the chapter. Discuss the relationships of the topics in the outline.

What's Next?

Before beginning the first section, make sure students understand the connection between the Find Out activity and the topics to follow.

ASSESSMENT OPTIONS

PORTFOLIO

Refer to page 125 for suggested items that students might select for their portfolios.

PERFORMANCE ASSESSMENT

See page 125 for additional Performance Assessment options

Process

Skillbuilders, pp. 112, 116
MINI-Lab, p. 115
Activities, 5-1, 113; 5-2, 124
Using Lab Skills, p. 126

CONTENT ASSESSMENT

Assessment—Oral, pp. 110, 116, 117
Skillbuilder, p. 121
Section Reviews, pp. 112, 116, 121, 123
Chapter Review, pp. 125-7
Mini Quizzes, pp. 110, 119

GROUP ASSESSMENT

Opportunities for group assessment occur with Cooperative Learning Strategies and Flex Your Brain Activities.

SECTION 5-1

PREPARATION

SECTION BACKGROUND

▶ Mendel worked with mice before he worked with pea plants. The principles of genetics are the same for both plants and animals.

▶ Although the genotype may often be inferred from the phenotype, it is impossible to tell completely dominant homozygous and heterozygous genotypes apart by sight.

PREPLANNING

▶ To prepare for Activity 5-1, obtain 2 paper bags, 100 red beans, and 100 white beans for each pair of students.

1 MOTIVATE

FLEX Your Brain

Use the Flex Your Brain activity to have students explore the word INHERITANCE.

Flex Your Brain

ASSESSMENT

Performance: Use the Flex Your Brain activity to assess group critical thinking and problem solving skills.

▶ Make a collection of photographs that shows dominant and recessive traits in organisms. Examples might include a normally pigmented animal and an albino, a plant variety that has red flowers contrasted with the same species showing white flowers, etc. Have students try to explain how these traits are inherited.

TYING TO PREVIOUS KNOWLEDGE: Students will already be aware of many traits that are inherited in humans and other animals. Allow students to describe traits they have observed to be inherited in their family or in a pet.

5-1 What Is Genetics?

New Science Words

heredity
alleles
genetics
dominant
recessive
Punnett square
genotype
homozygous
heterozygous
phenotype

Objectives

- ▶ Explain how traits are inherited and explain Mendel's role in the history of genetics.
- ▶ Use a Punnett square to predict the results of crosses.
- ▶ Explain the difference between genotype and phenotype.

What Have You Inherited?

What's the first thing you think about when you hear the word *inheritance?* Money? A new home? People inherit all sorts of things. It might be a set of dishes or a set of tools. Of course, a lot of people will never inherit things like this. But there is a type of inheritance that all living things receive. For you, this inheritance is in the nucleus of each cell of your body in the form of sets of genes.

For centuries, people have been interested in why one generation looks like another. A new baby may look much like one of its parents. It may be the shape of its nose or its earlobes. In either case, a trait is being noticed. Every organism is a collection of traits, all inherited from its parents. **Heredity** (huh RED ut ee) is the passing of traits from parent to offspring. What controls these traits? As you will learn, traits are controlled by genes.

MINI-Lab

What are some common traits?

Survey 25 students in your class or school for the presence of freckles, dimples, tongue rolling, bent little fingers, and cleft chin. *Record* your results *in a table* **in your Journal.** Describe all the variations you notice.

In 1933, an American scientist, Thomas Hunt Morgan, stated that genes are found on chromosomes. Genes, as you learned in Chapter 4, are made up of DNA. They control all the traits that show up in an organism. When pairs of chromosomes separate into sex cells during meiosis, pairs of genes also separate from one another. As a result, each sex cell winds up with one form of a gene for each trait that an organism shows. If the trait is for earlobes, then the gene in one sex cell may control one form of the trait, such as attached earlobes. The gene for earlobes in the other sex cell may control the other form of the trait, such as free earlobes. The different forms a gene may have for a trait are its **alleles** (uh LEELZ). In life science, the study of how alleles affect generations of offspring is **genetics.**

What is an allele?

OPTIONS

Meeting Different Ability Levels

For Section 5-1, use the following **Teacher Resource Masters** depending upon individual students' needs.

◆ **Study Guide Master** for all students.

● **Reinforcement Master** for students of average and above average ability levels.

▲ **Enrichment Master** for above average students.

Additional Teacher Resource Package masters are listed in the OPTIONS box throughout the section. The additional masters are appropriate for all students.

◆ STUDY GUIDE 21

NAME ______ DATE ______ CLASS ______

STUDY GUIDE — Chapter 5

What Is Genetics? — Text Pages 106–113

For each item, cross out the phrase or phrases that do not accurately complete the statement.

1. Genes
   a. are made up of DNA.
   b. control the traits that show up in an organism.
   c. are found on chromosomes.
   d. ~~are never inherited by some people.~~
   e. have different forms called alleles.
2. Gregor Mendel
   a. is the Father of Genetics.
   b. lived in Europe in the 1800s.
   c. experimented with peas.
   d. ~~arrived at his conclusions by accident.~~
   e. determined the basic laws of genetics.
   f. ~~changed the thinking of the scientists of his day.~~
3. Alleles
   a. are forms of a gene.
   b. are the subject of genetics.
   c. can be dominant or recessive.
   d. ~~for a trait from both parents are found in one gamete.~~
4. A Punnett square
   a. is used to predict results in genetics.
   b. represents the genotypes of offspring that can result from the combination of alleles from two parents.
   c. ~~uses numbers to represent the offspring of two or more parents.~~
5. A genotype of an organism
   a. ~~is a physical trait.~~
   b. is its genetic makeup.
   c. determines its phenotype.
6. A recessive factor
   a. can seem to disappear in a generation of organisms.
   b. ~~is represented by a capital letter on a Punnett square.~~
   c. ~~covers, or dominates, a dominant factor.~~
7. An organism that is heterozygous for a trait
   a. has two different alleles for that trait.
   b. ~~is a purebred.~~
   c. ~~has the phenotype of the recessive allele.~~

21

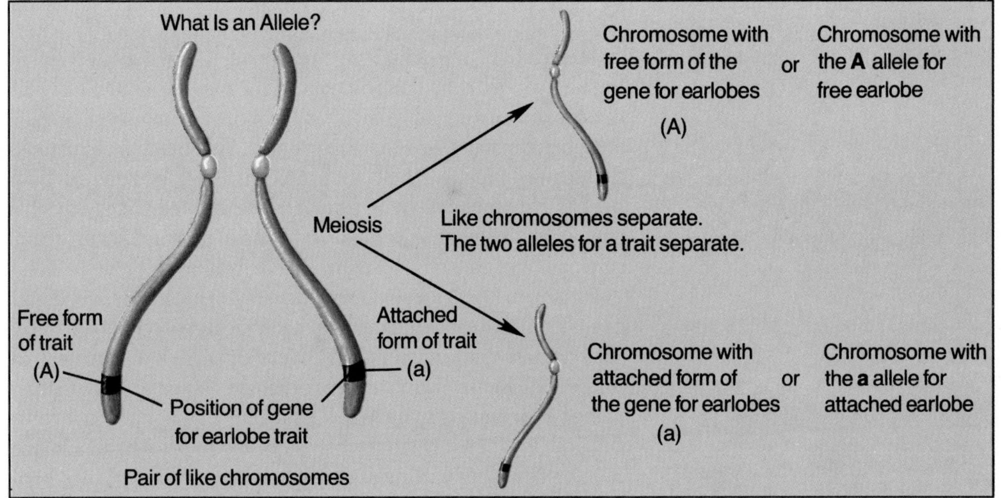

Figure 5-1. An allele is one form of a gene. Alleles separate into sex cells during meiosis.

The Father of Genetics

The first thorough scientific study of how traits pass from one generation to the next was done by a monk, Gregor Mendel. Mendel lived in the 1800s in a section of Eastern Europe that was part of Czechoslovakia. His father taught him about plants in the family's small orchard. Although everyone expected him to become a farmer, he became a priest. He studied science and math and eventually worked many years as a substitute teacher. While teaching, Mendel took over a garden plot at the monastery. There he experimented with plants. His observations of his father's orchard made him think that it was possible to predict the kinds of flowers and fruit a plant would produce. But something had to be known about the parents of the plant before a prediction could be made. Mendel made extremely careful use of scientific methods in his research. In 1866, after eight years of work on inheritance in plants, Mendel presented the results of his research to a group of scientists. Unfortunately, they didn't understand anything he was talking about. Mendel died in 1884, without knowing if his work would ever be understood.

In 1900, Mendel's work was rediscovered. By then, other scientists had come to the same conclusions that he had reached. Since then he has become known as the Father of Genetics.

Figure 5-2. Through experiments, Mendel discovered basic laws of inheritance.

2 TEACH

Key Concepts are highlighted.

CONCEPT DEVELOPMENT

▶ Use a model of a flower to show students how Mendel transferred pollen from the anther to the stigma. Explain how he prevented cross pollination by removing anthers and covering the female part of the flower with a bag.

MINI-Lab

Answer: Data in table should have horizontal heads: Trait and Number. Vertical heads should be: Freckles, Dimples, and so on.

MINI-Lab ASSESSMENT

Performance: To further assess students' understanding of inherited traits, have volunteers observe and tabulate several traits among children, parents, and grandparents.

PROGRAM RESOURCES

From the **Teacher Resource Package** use:

Activity Worksheets, page 47, Mini-Lab: What are some common traits?

Activity Worksheets, page 5, Flex Your Brain.

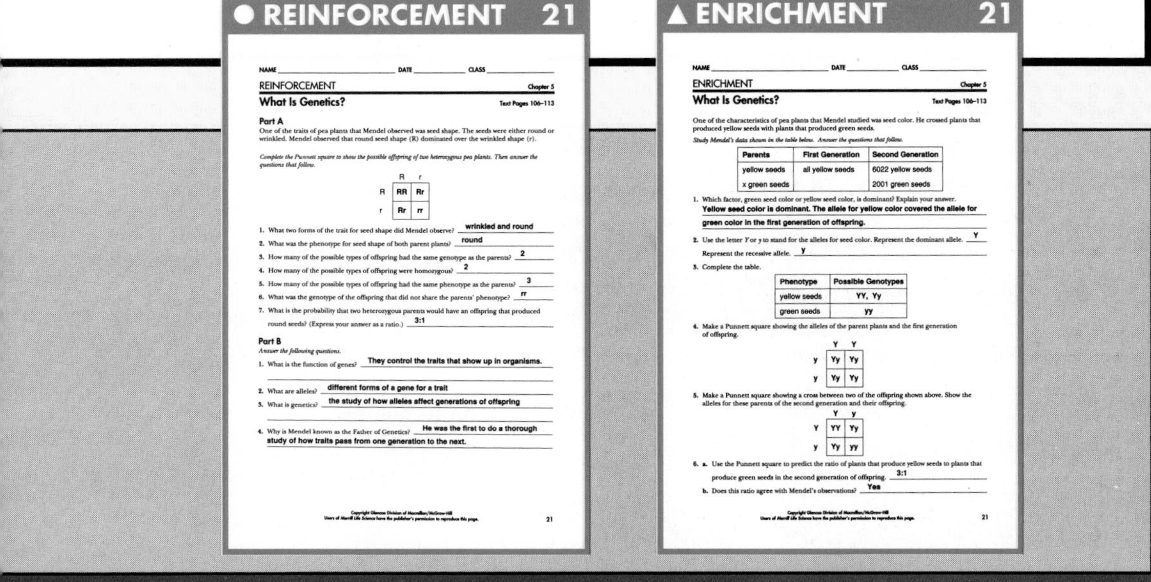

● REINFORCEMENT 21

NAME ______ DATE ______ CLASS ______

REINFORCEMENT — Chapter 5

What Is Genetics? — Text Pages 106–113

Part A

One of the traits of pea plants that Mendel observed was seed shape. The seeds were either round or wrinkled. Mendel observed that round seed shape (R) dominated over the wrinkled shape (r).

Complete the Punnett square to show the possible offspring of two heterozygous pea plants. Then answer the questions that follow.

	R	r
R	RR	Rr
r	Rr	rr

1. What two forms of the trait for seed shape did Mendel observe? **wrinkled and round**
2. What was the phenotype for seed shape of both parent plants? **round**
3. How many of the possible types of offspring had the same genotype as the parents? **2**
4. How many of the possible types of offspring were homozygous? **2**
5. How many of the possible types of offspring had the same phenotype as the parents? **3**
6. What was the genotype of the offspring that did not share the parents' phenotype? **rr**
7. What is the probability that two heterozygous parents would have an offspring that produced round seeds? (Express your answer as a ratio.) **3:1**

Part B

Answer the following questions.

1. What is the function of genes? **They control the traits that show up in organisms.**
2. What are alleles? **different forms of a gene for a trait**
3. What is genetics? **the study of how alleles affect generations of offspring**
4. Why is Mendel known as the Father of Genetics? **He was the first to do a thorough study of how traits pass from one generation to the next.**

 21

▲ ENRICHMENT 21

NAME ______ DATE ______ CLASS ______

ENRICHMENT — Chapter 5

What Is Genetics? — Text Pages 106–113

One of the characteristics of pea plants that Mendel studied was seed color. He crossed plants that produced yellow seeds with plants that produced green seeds.

Study Mendel's data shown in the table below. Answer the questions that follow.

Parents	First Generation	Second Generation
yellow seeds	all yellow seeds	6022 yellow seeds
x green seeds		2001 green seeds

1. Which factor, green seed color or yellow seed color, is dominant? Explain your answer. **Yellow seed color is dominant. The allele for yellow color covered the allele for green color in the first generation of offspring.**
2. Use the letter *Y* or *y* to stand for the alleles for seed color. Represent the dominant allele. **Y** Represent the recessive allele. **y**
3. Complete the table.

Phenotype	Possible Genotypes
yellow seeds	YY, Yy
green seeds	yy

4. Make a Punnett square showing the alleles of the parent plants and the first generation of offspring.

	Y	Y
y	Yy	Yy
y	Yy	Yy

5. Make a Punnett square showing a cross between two of the offspring shown above. Show the alleles for these parents of the second generation and their offspring.

	Y	y
Y	YY	Yy
y	Yy	yy

6. a. Use the Punnett square to predict the ratio of plants that produce yellow seeds to plants that produce green seeds in the second generation of offspring. **3:1**
   b. Does this ratio agree with Mendel's observations? **Yes**

 21

CONCEPT DEVELOPMENT

▶ Explain that Mendel worked with several thousand seeds in his tests.

▶ **How do Mendel's experiments illustrate the scientific method?** *Mendel (1) worked with a single trait, (2) made carefully controlled experiments, (3) collected and analyzed data, (4) reported his experiments so they could be repeated, and (5) worked with large samples.*

▶ Have students brainstorm organisms that they know are purebreds (cats, dogs, flowers, horses, and so on). Discuss why people are interested in organisms with pure traits.

▶ Have students find out the pros and cons of purebred breeding (i.e., some purebred dog breeds have fragile bone structures).

REVEALING MISCONCEPTIONS

▶ Ask students if there are usually more recessive or dominant alleles for a given trait. Most students will suggest that there are more dominant alleles. Explain that dominance does not mean that there will be more of a particular allele, and often the recessive allele occurs more often in a population.

OBJECTIVES AND SCIENCE WORDS: Have students review the objectives and science words in this chapter as they study each section.

VideoDisc

STVS: Selective Breeding in Cows, Disc 4, Side 1

STVS: Genetic Engineering in Barley, Disc 4, Side 1

STVS: Viruses and Plant Disease, Disc 4, Side 1

Figure 5-3. Many people breed dogs for specific traits. The dog above is a sheepdog, originally bred because it could be trained to tend and drive sheep.

Mendel's Experiments

Mendel chose ordinary green peas, like the ones you've eaten for dinner, for his experiments. Peas are easy to breed for pure traits. You've heard of purebred horses and purebred dogs. An organism that always produces the same traits in its offspring is pure, or purebred. Tall plants that always produce tall plants are pure for the trait of tall height. Altogether, Mendel studied seven traits of peas, which are shown in Table 5-1.

In one of his first experiments, Mendel crossbred tall plants with short plants. He took pollen from the male reproductive structures of flowers of pure tall plants and placed it on the female reproductive structures of flowers of pure short plants. This process is called cross-pollination. The results of this cross are shown in Figure 5-4. Notice that tall plants crossed with short plants produce all tall plants. It seemed as if whatever had caused the plants to be short had disappeared.

Mendel called the tall height form that appeared the
1 **dominant** factor, because it seemed to dominate or cover up the short height form. In this case, the tall height form was dominant. He called the form that seemed to disappear the **recessive** factor. But what had happened to the recessive form? He experimented to find out.

Mendel allowed the new tall plants to cross-pollinate. Then he collected the seeds from these tall plants and planted them. To his surprise, the plants that grew from these seeds were tall and short. The recessive form had reappeared. Mendel saw that for every three tall plants

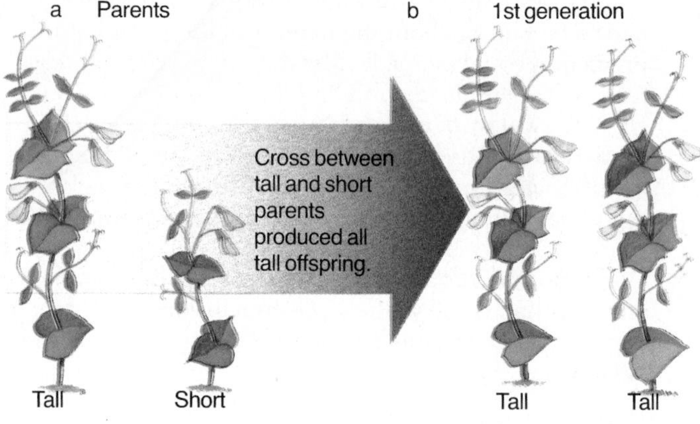

Figure 5-4. A cross between pure tall plants and pure short plants (a) produces a generation of all tall plants (b). When the first generation tall plants are crossed with each other, a second generation is produced with three tall plants for every one short plant (c).

OPTIONS

ENRICHMENT

▶ Have students make up a play that explains Mendel's life and work. They may present the play to other classes.

To assess this product, refer to the Performance Task Assessment Lists in **Performance Assessment in Middle School Science.**

INQUIRY QUESTIONS

▶ **How does heredity differ from genetics?** *Heredity is the passing of traits from parent to offspring. Genetics is the study of how genes or alleles affect offspring.*

▶ **Why do you think Mendel's work was successful?** *Because he used scientific methods and worked with many plants.*

▶ **Why are large numbers of tests important in scientific research?** *The larger the amount of data, the more accurate the results begin to become.*

Table 5-1

TRAITS COMPARED BY MENDEL						
Shape of seeds	**Color of seeds**	**Color of seed coats**	**Color of pods**	**Shape of pods**	**Plant height**	**Position of flowers**
Round	Yellow	Green	Green	Full	Tall	On side branches
Wrinkled	Green	White	Yellow	Flat or Constricted	Short	At tips of branches

Dominant traits appear in top row of the table.

there was one short plant, or a 3:1 ratio. He saw this 3:1 ratio often enough that he knew he could predict his results when he started a test. He knew that the *probability* was great that he would get that same outcome each time.

2 Probability is a science that helps you determine the chance that something will take place. Suppose your cat comes running every time you open a can of cat food. If the cat acts this way every day for a year, then you can say that the probability of the cat acting that way every day for the next year is very great. As a matter of fact, the probability is 100 percent. You can predict it.

Mendel also dealt with probabilities. One of the things that made his predictions accurate was that he worked with large numbers. He counted every plant and thousands of seeds. Now, this may not sound like fun, but by doing so, Mendel increased his chances of having accurate results. Scientific research is based on accurate, repeatable results.

Connect to...
Chemistry

If Mendel's pea plants produced plants with no chlorophyll, what is the probability of these plants surviving to produce flowers and seeds? Explain your answer.

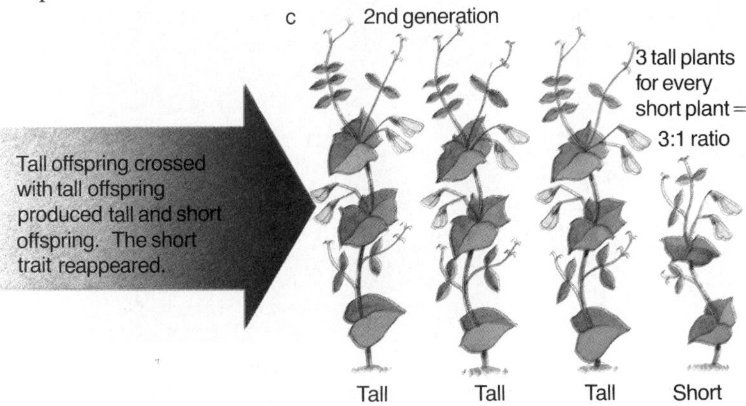
c
2nd generation
3 tall plants for every short plant = 3:1 ratio
Tall offspring crossed with tall offspring produced tall and short offspring. The short trait reappeared.
Tall
Tall
Tall
Short

CONCEPT DEVELOPMENT

▶ In Table 5-1, dominant traits are across the top, recessive on the bottom. Use Table 5-1 to reemphasize the difference between dominant and recessive. Ask them whether they eat more dominant or recessive peas.

▶ The diagram across pages 108 and 109 illustrates Mendel's basic F_1 and F_2 generation experiments. While these terms have not been used in the text, you may wish to introduce them to your students. Explain that *first* and *second filial* generations are terms understood by all geneticists.

▶ **Demonstration:** Flip a coin to demonstrate probability. Emphasize that the larger the number of tosses, the more accurate the probability.

CROSS CURRICULUM

▶ **Math:** Have students talk about examples of probability. Point out how probability is used in genetic crosses.

Connect to...
Chemistry

Answer: The probability would be 0 percent because without chlorophyll the pea plants would not produce food.

INQUIRY QUESTIONS

▶ **How does crossbreeding introduce variety into a species?** *Crossing organisms with two different forms for a trait may produce more varieties not seen in the traits of the two parents.*

PROGRAM RESOURCES

From the **Teacher Resource Package** use:

Concept Mapping, page 15.

Use **Laboratory Manual,** page 29, Genetic Traits, and page 31, 50:50 Chances.

CONCEPT DEVELOPMENT

▶It is usually easy for students to grasp the concepts of genotype and phenotype, homozygous and heterozygous. However, give them many examples for practice.

▶If this is the first time students have seen a Punnett square, make sure they understand the meaning of the caption for Figure 5-5.

Science and MATH

75 dozen tall × $20 per dozen = $1500

25 dozen short × $10 per dozen = $250

$1500 + $250 = $1750.

MINI QUIZ

Use the Mini Quiz to check students' recall of chapter content.

1. **What is the difference between a dominant factor and a recessive factor?** *Dominant factors seem to cover the recessive factors that seem to disappear when purebred organisms are crossed.*
2. **The science that helps you determine the chance something will take place is _____ .** *probability*
3. **Where are the letters representing the two alleles from each parent written on a Punnett square?** *One is written along the side and the other across the top.*

Science and MATH

Suppose Mendel's tall plants sell for $20 per dozen. The short ones sell for $10 per dozen. **In your Journal,** *calculate* how much money you would make if you sold all of 100 dozen offspring plants from a first generation cross.

Explain the difference between heterozygous and homozygous.

Using a Punnett Square

A handy tool used to predict results in genetics is the **Punnett square.** It uses your knowledge of alleles. Dominant and recessive alleles are represented by letters. A capital letter (*T*) stands for a dominant allele. A small letter (*t*) stands for a recessive allele. The letters are a form of shorthand. They show the genetic makeup, or **genotype,** of an organism. Once you understand what the letters mean, you can tell a lot about the inheritance of a trait in an organism.

Every cell in your body has two alleles for every trait. An organism with two alleles for a trait that are exactly the same is called **homozygous** (ho muh ZI gus). This would be written as *TT* or *tt*. An organism that has two different alleles for a trait is called **heterozygous** (het uh roh ZI gus). This condition would be written *Tt*. The purebred pea plants that Mendel used were homozygous for tall, or *TT*, and homozygous for short, or *tt*. The hybrid plants he produced were all heterozygous, or *Tt*.

The physical trait that shows as a result of a particular genotype is its **phenotype** (FEE nuh tipe). Red is the phenotype for red flowering plants. Short is the phenotype for short plants. If you have brown hair, then the phenotype for your hair color is brown.

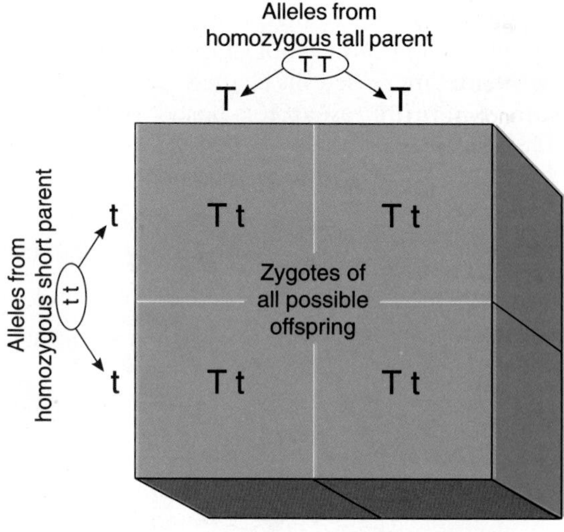

Figure 5-5. A Punnett square shows you all the ways in which alleles can combine. A Punnett square does *not* tell you how many offspring will be produced.

OPTIONS

ASSESSMENT—ORAL

▶**Why is Mendel called the Father of Genetics?** *Mendel was the first to conduct controlled experiments that produced results that eventually were accepted as the foundation for our present understanding of the science of genetics.*

▶**Why do you think Mendel used garden peas rather than horses or humans for his experiments?** *The life cycle of the garden pea is shorter than that of the horse or human. Garden peas require less space than experiments with horses or humans.*

▶**What would be the practicality of crossing a known purebred recessive organism with one of questionable lineage?** *This type of cross is known as a test cross and it can tell you if the questionable organism is a purebred or heterozygous individual.*

▶**With two alleles for a trait there are only four boxes in the Punnett square. How many boxes would there be if two traits were studied?** *Sixteen*

Look at the Punnett square in Figure 5-5. The letters representing the two alleles from one parent are written along the top of the square. Those of the second parent are placed along the side of the square. Each square in the middle is filled in much like a multiplication problem with one allele donated by each parent. The letters that you use to fill in each of the squares represent the genotypes of zygotes that the parents *could* produce. 3

The Punnett square in Figure 5-5 represents the first type of cross-pollination experiment by Mendel. You can see that each homozygous parent plant has two alleles for height. One parent is homozygous for tall (*TT*). The other parent is homozygous for short (*tt*). The alleles inside the squares are the genotypes of the possible offspring. All of them have the genotype *Tt*. They all have tall as a phenotype. Notice that you can't always figure out a genotype just by looking at the phenotype. The combination of *TT* or *Tt* both produce tall plants when *T* is dominant to *t*. Study the additional sample problems given below.

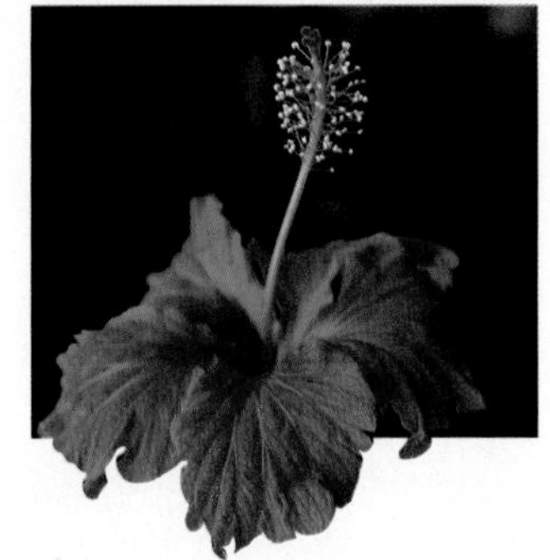

Figure 5-6. The color phenotype of this hibiscus flower is red.

EXAMPLE PROBLEMS: Using Punnett Squares

Problem 1: Color in Peas

Yellow color in peas is dominant to green peas. A homozygous yellow pea plant is crossed with a homozygous green pea plant. What will the genotypes of all the possible offspring be?

Outcome:

Genotypes of all possible offspring: All Yy

Phenotypes of offspring: All yellow

Yellow parent (YY)

Green parent (yy)	Y	Y
y	Yy	Yy
y	Yy	Yy

Problem 2: Wing Length in Fruit Flies

In fruit flies, long wings (L) are dominant to short wings (l). Two heterozygous long-winged fruit flies (both Ll) are crossed. What are the possible genotypes of their offspring? What are the phenotypes?

Outcome:

Genotypes of all possible offspring: LL, Ll, and ll

Phenotypes of all possible offspring:

LL and Ll = long wings

ll = short wings

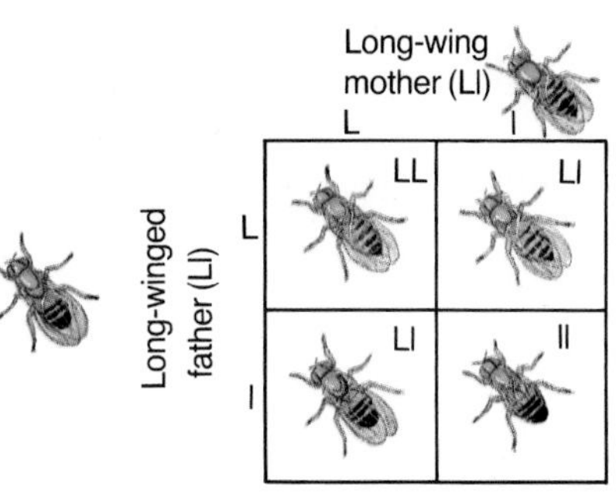

CONCEPT DEVELOPMENT

▶ Go over the two Example Problems on this page.

▶ Make sure students understand that in the inheritance of some traits, one phenotype may be the result of two different genotypes, as in tall being the result of TT or Tt in pea plants.

CHECK FOR UNDERSTANDING

▶ Work through the first and second generations of a sample genetics problem on the chalkboard or overhead projector.

▶ Ask questions 1-3 and the **Apply** Question in the Section Review on page 112.

RETEACH

Students usually have the most trouble determining the possible gene combinations. To demonstrate that the alleles separate, label ice cream sticks with letters to represent the alleles. Point out that the sex cell will have one kind or the other. Arrange the sticks in the same form as the Punnett square so that students can see the similarities.

EXTENSION

For students who have mastered this section, use the **Reinforcement** and **Enrichment** masters or other OPTIONS provided.

PROGRAM RESOURCES

From the **Teacher Resource Package** use:

Transparency Masters, pages 15-16, Genetics.

Use **Color Transparency** 8, Genetics.

3 CLOSE

▶ Ask students to summarize Mendel's experiments and results.

▶ Demonstrate the alleles along a chromosome by showing students two long strings of the large plastic beads used as a toddler's toy. Explain that each different shape and color represents a different trait. When the two strands are brought together you can demonstrate homozygous and heterozygous conditions.

SECTION REVIEW ANSWERS

1. Alleles are the forms of a gene that direct the formation of traits in an organism.

2. Genotype is the actual alleles an organism contains, and phenotype is the trait appearance of an organism. An example of a genotype is Tt for some plants. The phenotype is tall.

3. Mendel discovered that traits are controlled by "factors," (alleles), that may be dominant or recessive. He also found that an organism contains a pair of factors for each trait and these separate in sex cell formation.

4. Apply: The genotypes of the parents are BB (black) and bb (white). The genotypes of the offspring are all (Bb), and the phenotype of the puppies is black.

5. Connect to Earth Science: Weather forecasting; trends in populations

Skill Builder

Both parents would have to be heterozygous to produce children without the widow's peak trait.

Skillbuilder

ASSESSMENT

Performance: Use this Skillbuilder to assess students' abilities to Observe and Infer how the trait "hairline" is inherited.

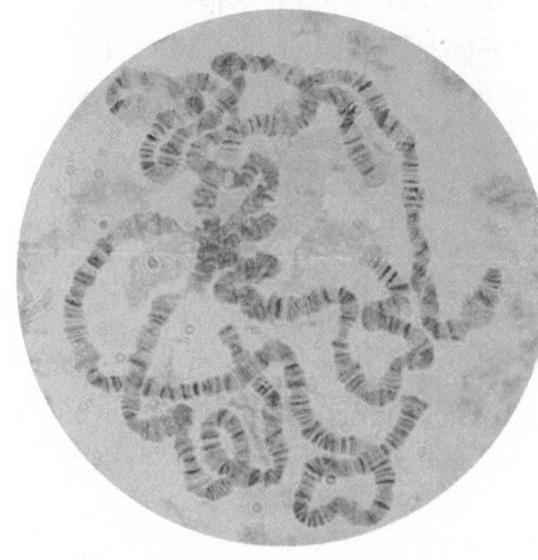

Figure 5-7. The markings on these chromosomes show where DNA is located.

Table 5-2

MENDELIAN INHERITANCE
1. Traits are controlled by alleles on chromosomes.
2. An allele may be dominant or recessive in form.
3. When a pair of chromosomes separates during meiosis, the different alleles for a trait move into separate sex cells.

Gregor Mendel didn't know anything about DNA, genes, or chromosomes. He did figure out that "factors" in the plants caused certain traits to appear. He also figured out that these factors separated when the plant reproduced. Mendel arrived at his conclusions by patient observation and careful analysis. His work is summed up briefly in Table 5-2.

SECTION REVIEW

1. How are alleles and traits related?
2. Differentiate between genotype and phenotype.
3. Explain Mendel's contribution to genetics.
4. **Apply:** Make a Punnett square showing a cross between two dogs. One dog is homozygous for black coat, and the other dog is homozygous for white coat. Black is dominant to white. Use *B* for the dominant allele and *b* for the recessive allele. What are the genotypes of the parents and the genotypes and phenotypes of the puppies?
5. **Connect to Earth Science:** Give an example of how probability is used in everyday life.

Skill Builder

Observing and Inferring

Hairline shape is an inherited trait in humans. A widow's peak is a V-shaped hairline in the middle of the forehead. People lacking the trait have a straight hairline. The widow's peak allele is dominant and the straight hairline allele is homozygous recessive. From your study of Mendel's experiments, infer how parents with widow's peaks could have a child without the trait. If you need help, refer to Observing and Inferring in the **Skill Handbook** on page 682.

OPTIONS

ENRICHMENT

▶ Have interested students form Punnett squares that show the results of 1st and 2nd generation crosses between organisms for two traits. Have them cross pure tall yellow peas with pure short green peas then cross their offspring. Show the gene combination and phenotypes. Have students prepare to answer this by studying the example problem on page 111 and discussing a plan with you. Pure tall yellow (TTYY) will produce sex cells (TY); pure short green plants (ttyy) will produce sex cells (ty). First generation will all be tall yellow (TtYy). Second generation will produce tall yellow, tall green, short yellow, and short green combinations.

ACTIVITY 5-1 Expected and Observed Results

Problem: *How does chance affect combinations of genes?*

Materials
- 2 paper bags
- 100 red beans
- 100 white beans

Procedure
1. Place 50 red beans and 50 white beans into a paper bag. Place 50 red beans and 50 white beans into a second bag. Each bean represents an allele for flower color.
2. Label one of the bags "female" for the female parent. Label the other bag "male" for the male parent.
3. Without looking inside the bags, remove one bean from each bag. The two beans represent the alleles that combine when sperm and egg join.
4. Use a Punnett square to predict how many red/red, red/white, white/white combinations will be selected.
5. Make a table like the one shown with room for 100 data entries. Record the color combination of the beans each time you remove two beans. Then return them to their original bags and shake the bags.
6. Repeat Step 5 *ninety-nine* times.
7. Count and record the total numbers of red/red, red/white, and white/white bean combinations in your data table.
8. Compile and record the class totals.

Data and Observations Sample Data

Beans	Red/Red	Red/White	White/White
Total	18	52	30
Class Total	376	751	373

Analyze
1. Which combination occurred most often?
2. If red is dominant and white is recessive, how many plants have heterozygous genes?
3. How did your *predicted* (expected) results compare with your *observed* (actual) results?
4. What was the ratio of red/red to red/white to white/white?

Conclude and Apply
5. What are the chances of selecting the same color in a pair of alleles each time?
6. Does chance affect allele combination? Explain.
7. How do the results of a small sample compare with the results of a large sample?
8. **Hypothesize** how you could get predicted results to be closer to actual results.

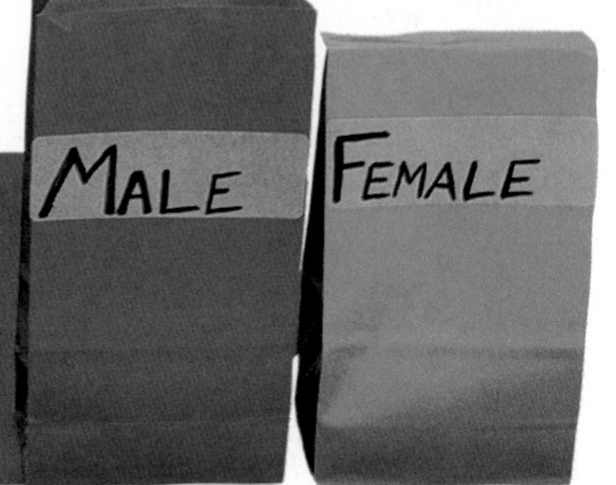

ANSWERS TO QUESTIONS
1. red/white
2. Answers will reflect the number of red/white combinations chosen by each student group.
3. Answers will vary according to the individual numbers of the various color combinations chosen, but should reflect a generally close following of the expected results.
4. Results should show a generally close ratio of 1:2:1.
5. one chance in two
6. Each time an allele is chosen, there is a 50/50 chance it or its corresponding other allele will be chosen.
7. Small samples can sometimes give distorted results. The larger the sample, the more accurate the results.
8. by increasing the size of the sample

ACTIVITY 5-1
one class period

OBJECTIVE: Students will **compare** expected results with observed results using Punnett squares and red and white beans, and will **investigate** how the principles of heredity are related to chance.

PROCESS SKILLS applied in this activity:
▶ **Predicting** in Procedure Step 4.
▶ **Observing** in Procedure Steps 5 and 6.
▶ **Recording data** in Procedure Steps 5 through 8.
▶ **Interpreting** in Analyze Questions 2 and 3 and Conclude and Apply Questions 5, 6, and 7.

COOPERATIVE LEARNING
Use Science Investigation cooperative groups. While one of the students places the beans in the bags, the other should make the Punnett square to predict the color ratios.

TEACHING THE ACTIVITY
▶ Emphasize the importance of completing all 100 trials. You might also challenge some students to do 1000 trials and have them compare the results of the two tests (100 vs. 1000).
Troubleshooting: Explain to students why beans need to be returned to the bag rather than kept out in order to keep the probabilities of choosing the color combinations the same throughout the activity.

Activity ASSESSMENT
Performance: To further assess students' understanding of probability, have them repeat the activity using three different kinds of beans.

PROGRAM RESOURCES
From the **Teacher Resource Package** use:
Activity Worksheets, pages 41-42, Activity 5-1, Expected and Observed Results.

SECTION 5-2

PREPARATION

SECTION BACKGROUND

▶ Many traits in humans are by polygenic inheritance.

PREPLANNING

▶ For the Mini-Lab you will need tape, index cards, hand lenses, and pads with washable ink.

1 MOTIVATE

▶ Point out the many hues of hair or eye color to be found within the class. Ask students how genetics can explain such variation.

▶ Discuss the importance of doctors knowing the ABO blood types of patients who need a transfusion.

Cooperative Learning: Using Expert Teams of three, have students research inheritance of coat color in palomino horses. The teams should then report, with visual aids, to the class.

In Your JOURNAL

Answer: Barbara McClintock won the Nobel Prize for Genetics with her discovery of "jumping genes"—genes that change position along a chromosome; corn; by years of patient observation

TYING TO PREVIOUS KNOWLEDGE: Explain that the concepts of incomplete dominance, multiple alleles, and polygenic inheritance do not displace Mendel's findings, but extend them.

VideoDisc

STVS: Breeding Fruit Flies, Disc 4, Side 1

STVS: Microinjecting Polygenes, Disc 4, Side 1

5-2 Genetics Since Mendel

New Science Words

incomplete dominance
multiple alleles
polygenic inheritance

Objectives

▶ **Explain incomplete dominance.**
▶ **Compare multiple allele and polygenic inheritance, and give examples of both.**

In Your JOURNAL

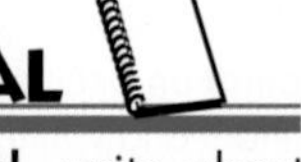

In your Journal, write about Barbara McClintock and her work with genes that change position on chromosomes. What organism did she use? How did she make her discovery?

Incomplete Dominance

When Mendel's work was rediscovered, scientists repeated his experiments. Mendel's results proved true for peas and other plants again and again. One scientist, Carl Correns, crossed pure red four o'clock plants with pure white four o'clocks. He expected to get all red flowers. But to his surprise, all the flowers were pink as in Figure 5-8a. Neither allele for flower color seemed dominant. Had the colors become blended like paint colors? He crossed the pink flowered plants with each other and red, pink, and white flowers were produced as in Figure 5-8b. The red and white alleles had not become "blended." When *both* alleles are expressed in offspring, the condition is called **incomplete dominance.** Coat color in cattle and horses and blood type in humans are examples of incomplete dominance.

Figure 5-8. The diagram below shows how color in four o'clock flowers is inherited by incomplete dominance. The first generation plants (a) are crossed with each other to obtain the second generation (b).

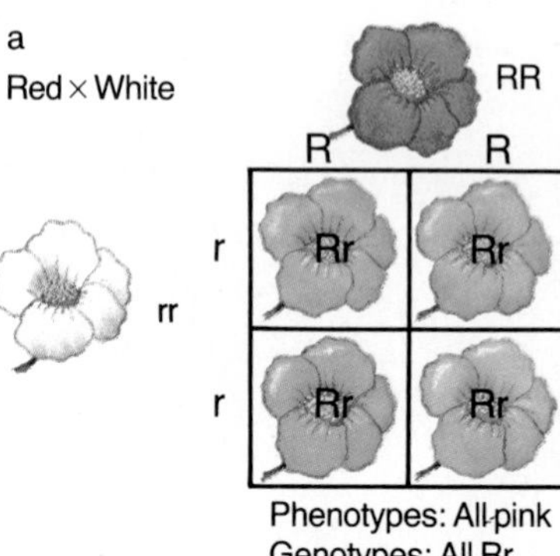

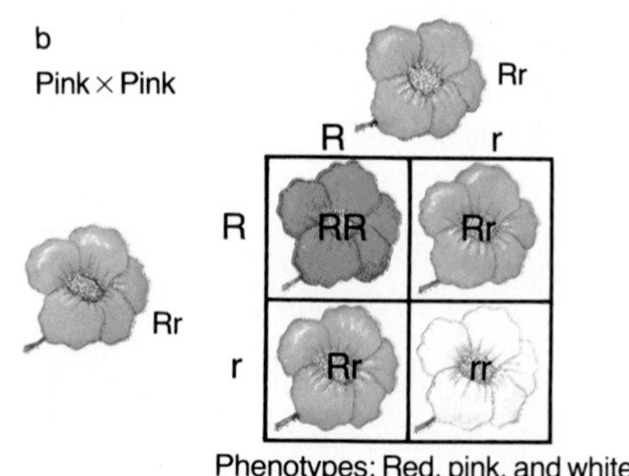

OPTIONS

Meeting Different Ability Levels

For Section 5-2, use the following **Teacher Resource Masters** depending upon individual students' needs.

◆ **Study Guide Master** for all students.

● **Reinforcement Master** for students of average and above average ability levels.

▲ **Enrichment Master** for above average students.

Additional Teacher Resource Package masters are listed in the OPTIONS box throughout the section. The additional masters are appropriate for all students.

◆ STUDY GUIDE 22

NAME ______ DATE ______ CLASS ______

STUDY GUIDE — Chapter 5

Genetics Since Mendel — Text Pages 114–116

Write the letter of the term or phrase that best completes the sentence.

d 1. When both alleles of a gene are expressed in the offspring, the condition is called ____.
a. heredity c. blending
b. mixing d. incomplete dominance

a 2. An example of incomplete dominance is ____.
a. a white allele and a red allele in a plant producing pink flowers
b. red flowers crossed with white flowers producing both red and white flowers
c. a red allele covering a white allele in red flowers
d. a dominant pink allele covering recessive red and white alleles

c 3. Because alleles A and B for blood type are inherited by incomplete dominance, a person with genotype AB would have the phenotype ____.
a. A c. AB
b. B d. O

d 4. Because the alleles A and B are both dominant and the O allele is recessive, a person with phenotype O would have genotype ____.
a. AO c. ABO
b. BO d. OO

d 5. A person with phenotype A blood could not be the parent of an offspring with phenotype ____ blood.
a. O c. AB
b. A d. B

For each item, identify the type of inheritance. Write "multiple allele" or "polygenic" in the blank.

polygenic 6. A group of genes acts together to produce a single trait.
multiple allele 7. One trait is controlled by more than two alleles of a gene.
multiple allele 8. There are three alleles for human blood type.
polygenic 9. Up to six gene pairs may control the color of human skin.
polygenic 10. The effect of a single allele may be small, but the combination of alleles from many genes produces a wide variety in a trait.
polygenic 11. Human traits such as eye and hair color, height, and weight are controlled by two or more gene pairs.

22

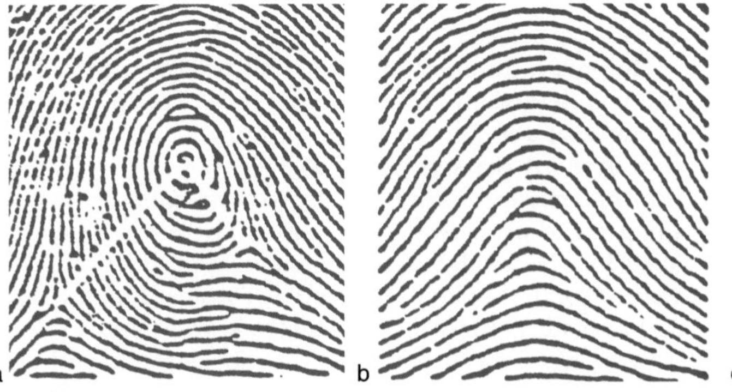

Figure 5-9. Fingerprints in your class will probably show varieties of the whorl (a), arch (b), and loop (c) patterns. There are many variations in these patterns.

Multiple Alleles

Mendel studied traits in peas that had two alleles controlling a trait. However, many traits have more than two alleles, or **multiple alleles,** that control them. In the human population, blood type is a trait that has three alleles.

In 1900, Dr. Karl Landsteiner found three blood types in the human population. He called them A, B, and O. A and B alleles are inherited by incomplete dominance. A person with AB blood type shows both alleles in his or her phenotype. However, both A and B alleles are dominant to the O allele, which is recessive.

Table 5-3 shows the ways in which the alleles for blood type can combine. Notice that a person with phenotype A blood has inherited either the genotype AA or AO. A person with phenotype B blood has inherited either genotype BB or BO. For a person to have type AB blood, an A allele is inherited from one parent and a B allele is inherited from the other parent. Finally, a person with phenotype O blood has inherited an O allele from each parent and has the genotype OO.

Table 5-3

HUMAN BLOOD TYPES	
Phenotype	**Genotype**
A	AA or AO
B	BB or BO
AB	AB
O	OO

MINI-Lab

What are fingerprints?

Fingerprints are formed before birth. The whorl pattern shown in Figure 5-9 is *LL*; the arch pattern is *ll*; the hybrid, or heterozygous, condition is *Ll* and shows a looped pattern. **In your Journal,** *collect data* to see which patterns appear in your class. Explain what type of inheritance is in action in fingerprints.

2 TEACH

Key Concepts are highlighted.

CONCEPT DEVELOPMENT

▶ Show the film *Inheritance in Man* (McGraw-Hill Films) that illustrates how human traits are governed by the laws of heredity.

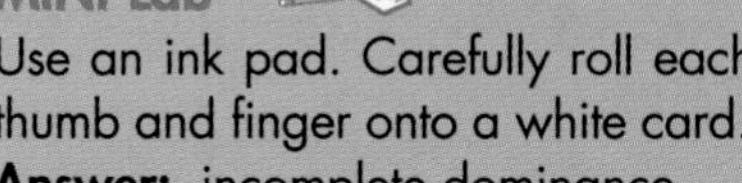

MINI-Lab

Use an ink pad. Carefully roll each thumb and finger onto a white card.
Answer: incomplete dominance

MINI-Lab ASSESSMENT

Performance: To further asess students' understanding of the inheritance of fingerprints, see USING LAB SKILLS, question 11, on p. 126.

PROGRAM RESOURCES

From the **Teacher Resource Package** use:

Activity Worksheets, page 48, Mini-Lab: What are fingerprints?

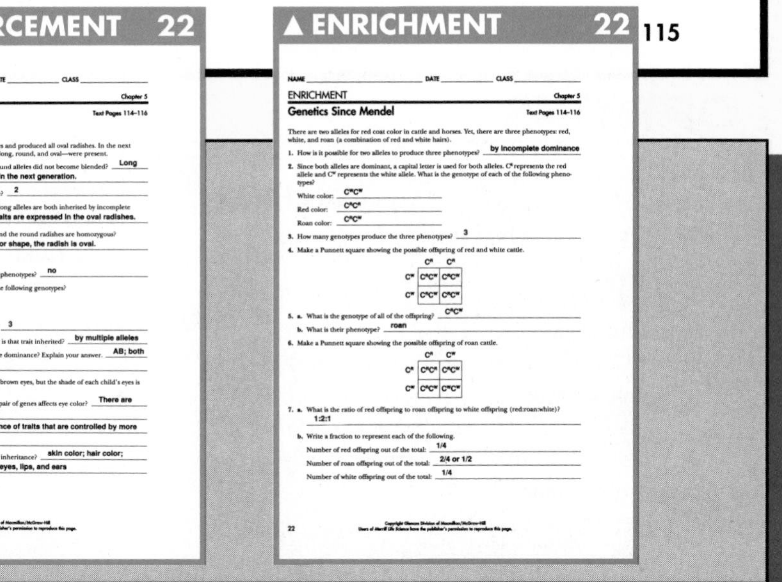

● REINFORCEMENT 22

NAME ________ DATE ________ CLASS ________

REINFORCEMENT — Chapter 5

Genetics Since Mendel — Text Pages 114–116

Answer the following questions.

1. Long radishes were crossed with round radishes and produced all oval radishes. In the next generation, all three phenotypes of radishes—long, round, and oval—were present.
   a. What evidence is there that the long and round alleles did not become blended? **Long and round radishes appeared again in the next generation.**
   b. How many alleles are there for radish shape? **2**
   c. What evidence is there that the round and long alleles are both inherited by incomplete dominance? **Both round and long traits are expressed in the oval radishes.**
   d. What evidence is there that both the long and the round radishes are homozygous? **When the radish is heterozygous for shape, the radish is oval.**
2. There are four phenotypes of human blood.
   a. Is it possible for two alleles to produce four phenotypes? **no**
   b. Which phenotype is produced by each of the following genotypes?
      AA **A** AO **A** OO **O**
      BB **B** BO **B** AB **AB**
   c. How many alleles are there for blood type? **3**
   d. When a trait has more than two alleles, how is that trait inherited? **by multiple alleles**
   e. Which blood type is inherited by incomplete dominance? Explain your answer. **AB; both alleles are visible in the phenotype.**
3. All of the children of brown-eyed parents have brown eyes, but the shade of each child's eyes is different.
   a. What evidence is there that more than one pair of genes affects eye color? **There are many shades of the same eye color.**
   b. What is polygenic inheritance? **Inheritance of traits that are controlled by more than one pair of genes**
   c. What other traits are examples of polygenic inheritance? **skin color; hair color; height; weight; body build; shape of eyes, lips, and ears**

22 Copyright Glencoe Division of Macmillan/McGraw-Hill
Users of Merrill Life Science have the publisher's permission to reproduce this page.

▲ ENRICHMENT 22

NAME ________ DATE ________ CLASS ________

ENRICHMENT — Chapter 5

Genetics Since Mendel — Text Pages 114–116

There are two alleles for red coat color in cattle and horses. Yet, there are three phenotypes: red, white, and roan (a combination of red and white hairs).

1. How is it possible for two alleles to produce three phenotypes? **by incomplete dominance**
2. Since both alleles are dominant, a capital letter is used for both alleles. C^R represents the red allele and C^W represents the white allele. What is the genotype of each of the following phenotypes?
   White color: **C^WC^W**
   Red color: **C^RC^R**
   Roan color: **C^RC^W**
3. How many genotypes produce the three phenotypes? **3**
4. Make a Punnett square showing the possible offspring of red and white cattle.

	C^R	C^R
C^W	C^RC^W	C^RC^W
C^W	C^RC^W	C^RC^W

5. a. What is the genotype of all of the offspring? **C^RC^W**
   b. What is their phenotype? **roan**
6. Make a Punnett square showing the possible offspring of roan cattle.

	C^R	C^W
C^R	C^RC^R	C^RC^W
C^W	C^RC^W	C^WC^W

7. a. What is the ratio of red offspring to roan offspring to white offspring (red:roan:white)? **1:2:1**
   b. Write a fraction to represent each of the following.
   Number of red offspring out of the total: **1/4**
   Number of roan offspring out of the total: **2/4 or 1/2**
   Number of white offspring out of the total: **1/4**

22 Copyright Glencoe Division of Macmillan/McGraw-Hill
Users of Merrill Life Science have the publisher's permission to reproduce this page.

CHECK FOR UNDERSTANDING

▶ Ask questions 1-2 and the **Apply** Question in the Section Review.

RETEACH

Have students compare the three types of inheritance in this section by making an outline, showing the number of chromosomes and the number of alleles that may be involved.

EXTENSION

For students who have mastered this section, use the **Reinforcement** and **Enrichment** masters or other OPTIONS provided.

3 CLOSE

▶ Have students answer Section Review questions.

SECTION REVIEW ANSWERS

1. Multiple allelic inheritance is governed by a single pair of genes that have more than two alleles. Polygenic inheritance is governed by multiple pairs of genes, each with two or more alleles.

2. A phenotype inherited by incomplete dominance appears because both alleles present express themselves.

3. Apply: Cross blue offspring among themselves. If all three phenotypes are produced, the trait is controlled by incomplete dominance.

4. Connect to Chemistry: deoxyribose

Skill Builder

Multiple alleles and multiple genes compare in that more than one gene is involved. They contrast in that in multiple alleles, more than two alleles exist that control a trait. In multiple genes, two or more gene pairs in an individual control a trait.

Skillbuilder
ASSESSMENT

Portfolio: Use this Skillbuilder to assess students' abilities to Compare and Contrast multiple alleles and multiple genes.

Figure 5-10. Skin color and eye color in humans are the results of the expression of more than one pair of genes.

Multiple Genes

How many different shades of blue or brown eye color can you detect among your classmates? Eye color is an example of a single trait that is produced by a combination of many pairs of genes. **Polygenic** (pahl ih JEHN ihk) **inheritance** occurs when a group of gene pairs act together to produce a single trait. The effect of each allele may be small, but the combination produces a wide variety. You'll probably have trouble classifying all the different shades of blue or brown eyes in your class.

Many human traits are controlled by polygenic inheritance. Height, weight, body build, and shape of eyes, lips, and ears are traits controlled by polygenic inheritance. It is estimated that skin color is controlled by three to six pairs of genes. Even more gene pairs may control the color of your hair and eyes.

Polygenic inheritance is, of course, not limited to human traits. Grain color in wheat, milk production in cows, and egg production in chickens are polygenic traits also.

The study of genes is no longer a simple look at a single trait controlled by one pair of alleles. Mendel would be astounded if he could see the amount of information that has come from his beginning work.

SECTION REVIEW

1. Compare multiple allele and polygenic inheritance.
2. Explain why a trait inherited by incomplete dominance is not a blend of two alleles.
3. **Apply:** A chicken that is purebred for black feathers is crossed with one that is purebred for white feathers. All the offspring produced have shiny blue feathers. Is this a case of incomplete dominance? Explain.
4. **Connect to Chemistry:** Genes are made of DNA. Find out what the sugar in DNA is called.

Skill Builder

Comparing and Contrasting

Using the information on pages 115 and 116, compare and contrast multiple alleles and multiple genes to explain differences between the two concepts. Include examples of each. If you need help, refer to Comparing and Contrasting in the **Skill Handbook** on page 683.

OPTIONS

ASSESSMENT—ORAL

▶ **Why might incomplete dominance also be called intermediate inheritance?** *It seems as if the trait shows up as an intermediate, or combination of the two genes.*

▶ **From your observations of humans, name a trait that you think is probably controlled by polygenic inheritance.** *Answers will vary, but can include almost any physical trait. Explain that height and (apparently) weight are polygenic in inheritance.*

PROGRAM RESOURCES

From the **Teacher Resource Package** use:

Cross-Curricular Connections, page 9, Eye Dominance and Heredity.

Use **Laboratory Manual,** page 33, Genes and Blood Clotting.

Human Genetics

5-3

Objectives

- ▶ **Describe two human genetic disorders.**
- ▶ **Explain inheritance of sex-linked traits.**
- ▶ **Explain the importance of genetic engineering.**

New Science Words

sex-linked gene
pedigree
genetic engineering

Genes and Health

Sometimes a gene undergoes mutation and results in an unwanted trait. If this happens in a sex cell, then all cells in future generations are affected. In Chapter 4, you learned that there are several types of mutations. Not all mutations are harmful, but some have resulted in genetic disorders among humans. Two disorders discussed here are the result of changes in DNA and are therefore gene mutations. They are sickle-cell anemia and cystic fibrosis.

Sickle-cell anemia (uh NEE mee uh) is a homozygous recessive disorder in which red blood cells are sickle-shaped instead of disc-shaped. Sickle cells can't deliver enough oxygen to the cells in the body. In addition, the misshapen cells don't move through blood vessels easily. Body tissues may be damaged due to insufficient oxygen. Sickle-cell anemia is found in tropical areas and in a small percentage of African Americans.

Cystic fibrosis is another homozygous recessive disorder. Thick mucus is produced
1 in the lungs and digestive system. Mucus in the lungs restricts oxygen intake. In the digestive system, enzymes can't reach food to break it down. Nutrients needed by the body aren't absorbed by body cells. However, with physical therapy exercises and improved medication, the lives of cystic fibrosis patients are being extended.

Did You Know?

In Africa, people who are heterozygous for sickle-cell condition appear to be better protected against malaria than people with normal disc-shaped red blood cells.

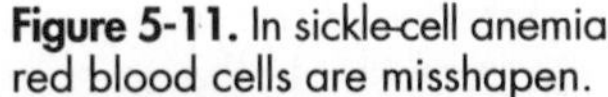

Figure 5-11. In sickle-cell anemia, red blood cells are misshapen.

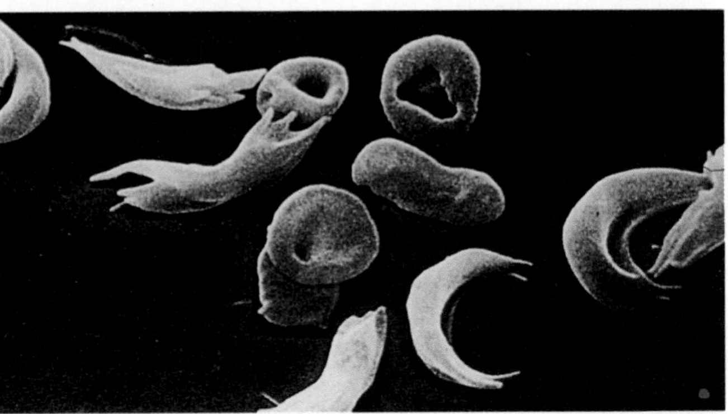

SECTION 5-3

PREPARATION

SECTION BACKGROUND

▶ About one thousand simple recessive human disorders are presently known.

▶ Most human genetic disorders are recessive.

▶ Screening for sickle-cell anemia, phenylketonuria, and other serious genetic disorders is common. Many states have laws that require doctors to screen for particular disorders.

▶ In 1990 the cystic fibrosis gene was "cured" in a test tube culture. This advance gives an indication that genetic engineering may make cures of genetic disorders possible within students' lifetimes.

1 MOTIVATE

▶ Ask students "if it were possible to cure a particular human genetic disorder by changing a person's genes, do you think scientists should? Why? Why not?"

MULTICULTURAL AWARENESS

Use the information on the relationship between sickle-cell anemia and malaria to increase student awareness of health in different cultures.

OPTIONS

ASSESSMENT—ORAL

▶ **Of what adaptive advantage might the sickle-cell allele be to a heterozygote?** *Fewer heterozygotes get malaria in populations where the disease is present.*

▶ **If the cystic fibrosis allele is found in four of every 100 Caucasians, what are the chances of two people producing a child with the disease?** *$(4/100 \times 1/2) \times (4/100 \times 1/2) = 1/2500$ or .004*

▶ **What are the chances of a color-blind man passing the trait on to his son?** *The chances are 0 because the man passes the Y chromosome to his sons and the gene for color-blindness is on the X chromosome.*

▶ **What are the chances that the hemophilia gene will be passed from a woman carrier to any of her sons?** *For each son the chances are 1 in 2 or 50 percent.*

TYING TO PREVIOUS KNOWLEDGE: Most students will be aware of at least one person with a genetic disorder. This section bridges the mechanisms of inheritance in the previous two sections with practical applications in the students' own lives.

2 TEACH

Key Concepts are highlighted.

MULTICULTURAL AWARENESS
Point out that certain traits are found more often in (but are not limited to) certain racial or ethnic groups. As examples, Tay-Sachs occurs more frequently among Ashkenazi Jews (of Central European descent), sickle-cell anemia among African Americans, and cystic fibrosis among Caucasians in the United States.

▶ You should use Punnett squares to demonstrate inheritance patterns discussed in this section.

ASSESSMENT
Performance: To assess students' understanding of genetics, assign a human genetic disorder to report on. Suggested topics are: albinism, sickle-cell anemia, phenyketonuria, polydactyly, hemophilia A, and Huntington's disease (chorea).

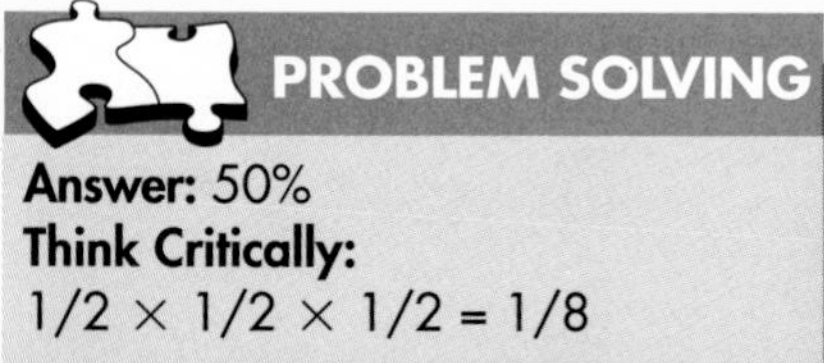

PROBLEM SOLVING

Answer: 50%
Think Critically:
$1/2 \times 1/2 \times 1/2 = 1/8$

VideoDisc

STVS: Detecting Cystic Fibrosis, Disc 7, Side 2

STVS: Selective Breeding in Micro-Pigs, Disc 4, Side 1

Figure 5-12. Sex in many organisms is determined by X (left) and Y (right) chromosomes. In the photograph above, you can see how X and Y chromosomes differ from one another in size.

Sex Determination

What determines the sex of an individual? Information on sex inheritance in many organisms, including humans, came from a study of fruit flies. Fruit flies have eight large chromosomes in their cells. They contain an X chromosome and a Y chromosome. By 1910, several scientists had concluded that the X and Y chromosomes contained genes that determine the sex of an individual. Females have two X chromosomes in their cells. Males have an X chromosome and a Y chromosome.

Females produce eggs that have only an X chromosome. Males, on the other hand, produce both X-containing sperm and Y-containing sperm. When an egg from a female is fertilized by an X sperm, the offspring is XX, a female. When an egg from a female is fertilized by a Y sperm, the zygote produced is XY, and the offspring is a male. What pair of sex chromosomes is in each of your cells?

PROBLEM SOLVING

Boy or Girl?

What is the probability of all children in a family being girls? Probability refers to the chance that an event will occur. For example, if you flip a coin, it will land in only one of two ways—heads or tails. The probability of the coin landing heads or tails is one out of two, or one-half, or 50 percent.

If you were to toss two coins at the same time, each coin still has only a one-half probability of landing with heads or tails up. Look at the combinations that could occur in a two-coin toss. The probability of each combination is the product of the probability of each. In the two-coin toss, the probability of both coins turning up heads is the product of the two, or:

$$1/2 \times 1/2 = 1/4$$

Jennifer's mother is going to have a baby. Jennifer already has a sister, so she really hopes this baby is a boy. What is the probability that the baby will be a boy?

Think Critically: What is the probability of a family having three female children in a row?

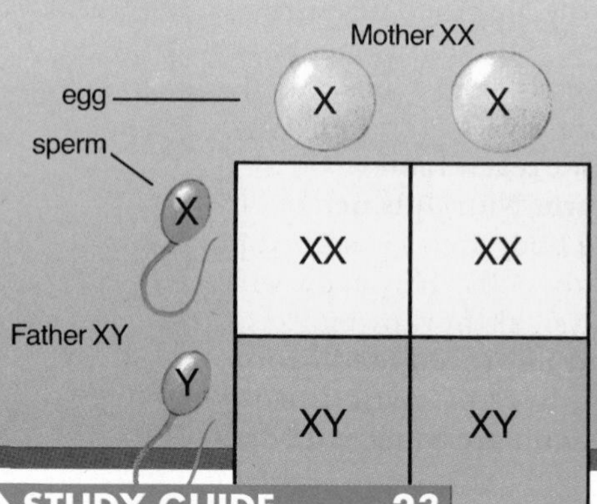

OPTIONS

Meeting Different Ability Levels

For Section 5-3, use the following **Teacher Resource Masters** depending upon individual students' needs.

◆ **Study Guide Master** for all students.

● **Reinforcement Master** for students of average and above average ability levels.

▲ **Enrichment Master** for above average students.

Additional Teacher Resource Package masters are listed in the OPTIONS box throughout the section. The additional masters are appropriate for all students.

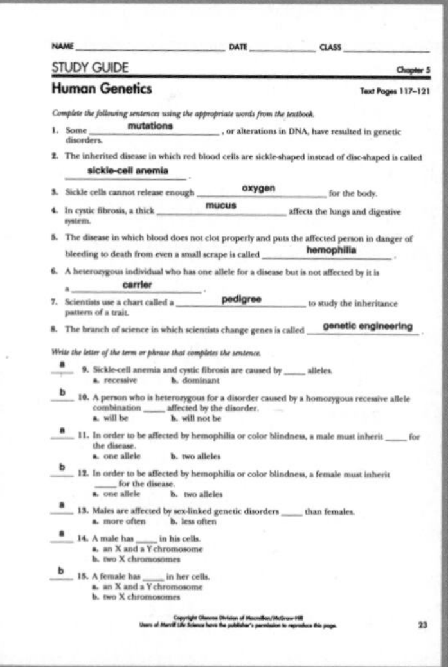

◆ STUDY GUIDE 23

NAME ____________ DATE ________ CLASS ________

STUDY GUIDE — Chapter 5

Human Genetics — Text Pages 117–121

Complete the following sentences using the appropriate words from the textbook.

1. Some **mutations**, or alterations in DNA, have resulted in genetic disorders.
2. The inherited disease in which red blood cells are sickle-shaped instead of disc-shaped is called **sickle-cell anemia**.
3. Sickle cells cannot release enough **oxygen** for the body.
4. In cystic fibrosis, a thick **mucus** affects the lungs and digestive system.
5. The disease in which blood does not clot properly and puts the affected person in danger of bleeding to death from even a small scrape is called **hemophilia**.
6. A heterozygous individual who has one allele for a disease but is not affected by it is a **carrier**.
7. Scientists use a chart called a **pedigree** to study the inheritance pattern of a trait.
8. The branch of science in which scientists change genes is called **genetic engineering**.

Write the letter of the term or phrase that completes the sentence.

a 9. Sickle-cell anemia and cystic fibrosis are caused by ____ alleles.
a. recessive b. dominant

b 10. A person who is heterozygous for a disorder caused by a homozygous recessive allele combination ____ affected by the disorder.
a. will be b. will not be

a 11. In order to be affected by hemophilia or color blindness, a male must inherit ____ for the disease.
a. one allele b. two alleles

b 12. In order to be affected by hemophilia or color blindness, a female must inherit ____ for the disease.
a. one allele b. two alleles

a 13. Males are affected by sex-linked genetic disorders ____ than females.
a. more often b. less often

a 14. A male has ____ in his cells.
a. an X and a Y chromosome
b. two X chromosomes

b 15. A female has ____ in her cells.
a. an X and a Y chromosome
b. two X chromosomes

23

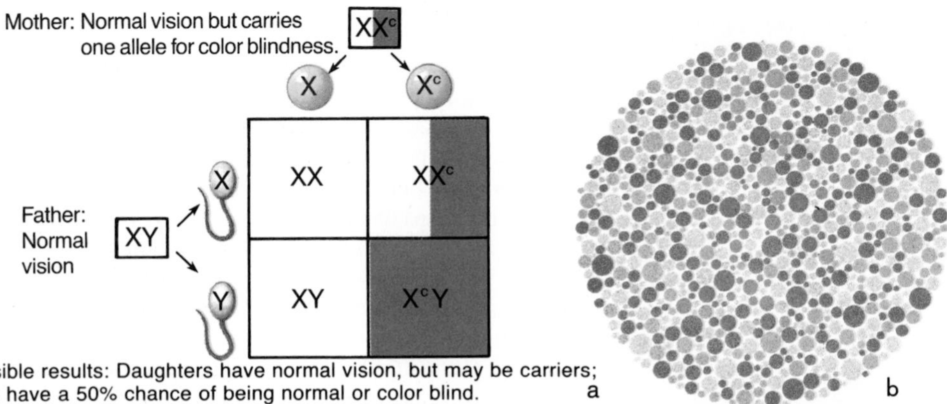
Mother: Normal vision but carries one allele for color blindness.
XX^c
X
X^c
Father: Normal vision
XY
X
Y
XX
XX^c
XY
X^cY
Possible results: Daughters have normal vision, but may be carriers; sons have a 50% chance of being normal or color blind.
a

b

Figure 5-13. Color blindness is a sex-linked trait. Females who are heterozygous see colors normally, but their sons may be color blind (a). In an eye examination, a color blind person will see a different number in the disc above (b) than a person with normal color vision.

Sex-linked Disorders

Some inherited conditions are closely linked with the X and Y chromosomes that determine the sex of an individual. A story is told that a young boy appeared to have normal intelligence, but couldn't be taught to pick ripe cherries on the family farm. His parents took him to a doctor. After observing him, the doctor concluded that the boy couldn't tell the difference between the colors red and green. Individuals who are red-green color-blind have inherited an allele on the X chromosome that prevents them from seeing these colors.

An allele inherited on a sex chromosome is a **sex-linked gene.** More males are color-blind than females. Can you figure out why? If a male inherits an X chromosome with the color-blind allele from his mother, he will be color-blind. Color-blind females are rare.

Another sex-linked gene disease is hemophilia (hee muh FIHL ee uh), a disorder in which blood does not clot properly. A scrape can be life threatening. Like color-blindness, males who inherit the X chromosome with the hemophilia allele will have the disorder. For a female to be a hemophiliac, she must inherit the defective allele on both X chromosomes. Hemophilia occurs in fewer than 1 in 7000 males. Females are usually just carriers.

How can you trace a trait through a family? A **pedigree** is a tool for tracing the occurence of a trait in a family. Males are represented by squares and females by circles. A solid circle or square shows that the person has the condition. Half colored circles or squares indicate carriers. A carrier has an allele for a trait, but does not show the trait.

2 **Figure 5-14.** Pedigrees show the occurrence of a trait in a family.

2
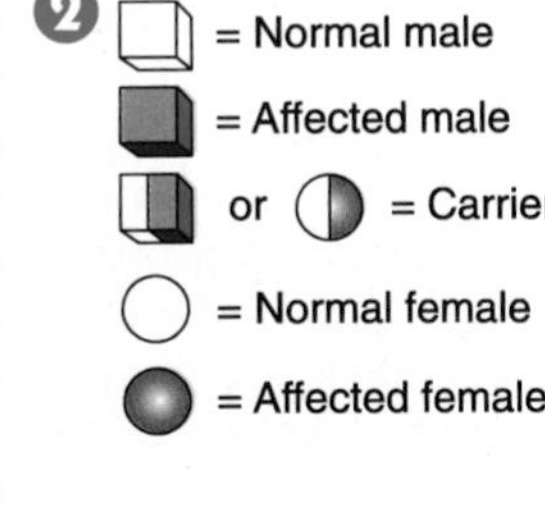
= Normal male
= Affected male
or
= Carrier
= Normal female
= Affected female

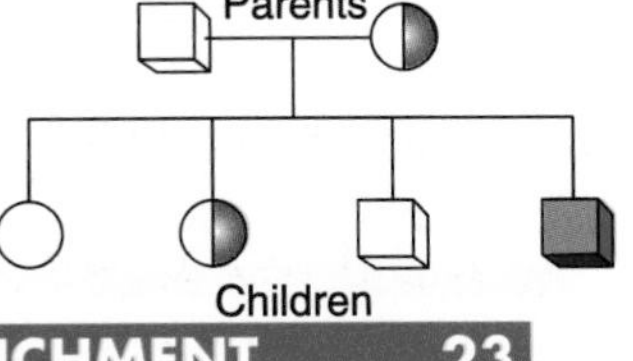
Parents
Children

REVEALING MISCONCEPTIONS

▶ Many students will have heard about hemophilia. Explain that hemophiliacs lack a clotting factor in their blood that can be replaced through transfusions. Because of these transfusions, hemophiliacs can lead a nearly normal life.

CROSS CURRICULUM

▶ **History:** The appearance of hemophilia in the royal families of Europe provides an example of how health impacts history. If possible, obtain a copy of the pedigree of Elizabeth I and discuss this with the students.

MINI QUIZ

Use the Mini Quiz to check students' recall of chapter content.

1. **What are the results of cystic fibrosis?** *Thick mucus is produced in the lungs and digestive system, restricting air and preventing digestion of food.*
2. **List two sex-linked disorders.** *colorblindness and hemophilia*
3. **The name given to the methods that may someday allow scientists to correct specific damaged or mutated genes is _____ .** *genetic engineering*

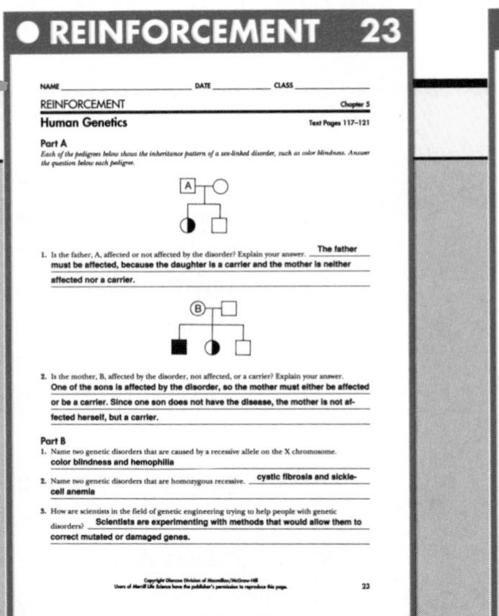

● REINFORCEMENT 23

NAME ________ DATE ________ CLASS ________

REINFORCEMENT — Chapter 5

Human Genetics — Text Pages 117–121

Part A

Each of the pedigrees below shows the inheritance pattern of a sex-linked disorder, such as color blindness. Answer the question below each pedigree.

A

1. Is the father, A, affected or not affected by the disorder? Explain your answer. **The father must be affected, because the daughter is a carrier and the mother is neither affected nor a carrier.**

B

2. Is the mother, B, affected by the disorder, not affected, or a carrier? Explain your answer. **One of the sons is affected by the disorder, so the mother must either be affected or be a carrier. Since one son does not have the disease, the mother is not affected herself, but a carrier.**

Part B

1. Name two genetic disorders that are caused by a recessive allele on the X chromosome. **color blindness and hemophilia**
2. Name two genetic disorders that are homozygous recessive. **cystic fibrosis and sickle-cell anemia**
3. How are scientists in the field of genetic engineering trying to help people with genetic disorders? **Scientists are experimenting with methods that would allow them to correct mutated or damaged genes.**

23

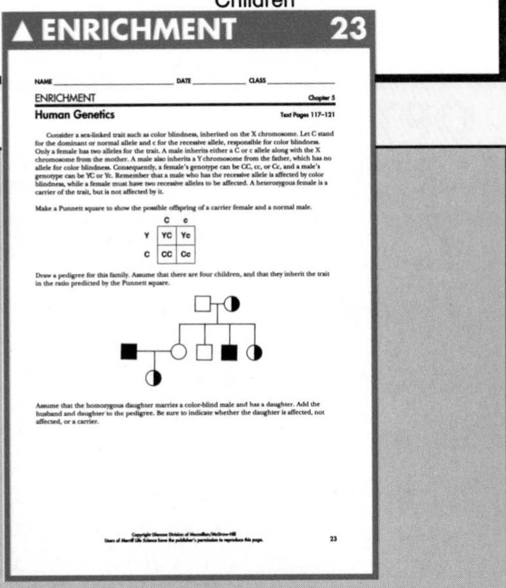

▲ ENRICHMENT 23

NAME ________ DATE ________ CLASS ________

ENRICHMENT — Chapter 5

Human Genetics — Text Pages 117–121

Consider a sex-linked trait such as color blindness, inherited on the X chromosome. Let C stand for the dominant or normal allele and c for the recessive allele, responsible for color blindness. Only a female has two alleles for the trait. A male inherits either a C or c allele along with the X chromosome from the mother. A male also inherits a Y chromosome from the father, which has no allele for color blindness. Consequently, a female's genotype can be CC, cc, or Cc, and a male's genotype can be YC or Yc. Remember that a male who has the recessive allele is affected by color blindness, while a female must have two recessive alleles to be affected. A heterozygous female is a carrier of the trait, but is not affected by it.

Make a Punnett square to show the possible offspring of a carrier female and a normal male.

	C	c
Y	YC	Yc
C	CC	Cc

Draw a pedigree for this family. Assume that there are four children, and that they inherit the trait in the ratio predicted by the Punnett square.

Assume that the homozygous daughter marries a color-blind male and has a daughter. Add the husband and daughter to the pedigree. Be sure to indicate whether the daughter is affected, not affected, or a carrier.

23

CHECK FOR UNDERSTANDING

Ask students why genetics is important. They should respond with information about genetic disorders and sex-linked inheritance.

RETEACH

Have students use a set of cut outs of X and Y chromosomes to model how a sex-linked disorder can be inherited.

EXTENSION

For students who have mastered this section, use the **Reinforcement** and **Enrichment** masters or other OPTIONS.

TECHNOLOGY

For more information on karyotying, see "Human Genetics" in *The New Encyclopaedia Brittanica*, Vol. 19, Chicago 1990, pp. 727–728.

Think Critically: Genes that control certain traits would be missing and the individual would not have those traits.

Answer: If ignored, persons with PKU suffer retardation. If detected early in urine or blood samples, diet can control and prevent symptoms.

Knowledge of genetic disorders has reached soft drink cans! Persons who have not inherited the enzyme to break down phenylalanine are warned about the presence of this chemical in certain artificially sweetened drinks. Find out what happens to people born with PKU.

Why Is Genetics Important?

If Mendel were to pick up a daily newspaper in most countries today, he probably wouldn't believe his eyes. There is hardly a day that goes by anymore that there isn't a news article about the latest information on genetic research. The word *gene* has become a household word. In this chapter, you have learned that every trait that you have inherited is the result of genes expressing themselves. The same laws that govern inheritance of traits in humans govern the inheritance of traits in watermelons, wheat, and mice as well.

In this section on human genetics, you've seen that there are inherited disorders that affect the human population. You may even know someone with one of these disorders. Many genetic disorders are controlled by diet and preventive measures. Genetics is no longer something that you can read about only in textbooks.

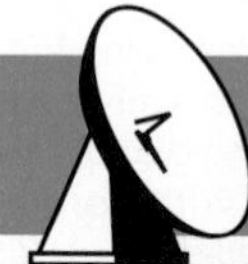

TECHNOLOGY

Karyotyping

Karyotyping is a process that allows scientists to study the chromosomes of an individual, sometimes even before birth.

Chromosomes are best seen during metaphase in mitosis when they are coiled. Skin cells divide frequently, so these cells are good to use for making a karyotype. The cells are fixed on microscope slides and dyed so that light and dark bands appear showing the position of DNA along the chromosomes. Then a photograph is taken of the chromosomes. The photograph is cut up into individual chromosomes. Scientists use the pattern of the stained bands, chromosome length, and the position of centromeres to identify pairs of matching chromosomes. Each pair is also numbered. The bands enable scientists to see if chromosomes have missing or duplicated parts.

Some genetic disorders can be identified just by looking at the karyotype of an individual. Sometimes this is done before birth when cells are taken from fluid around a fetus.

Think Critically: Why would a missing part of a chromosome affect the traits an individual would have?

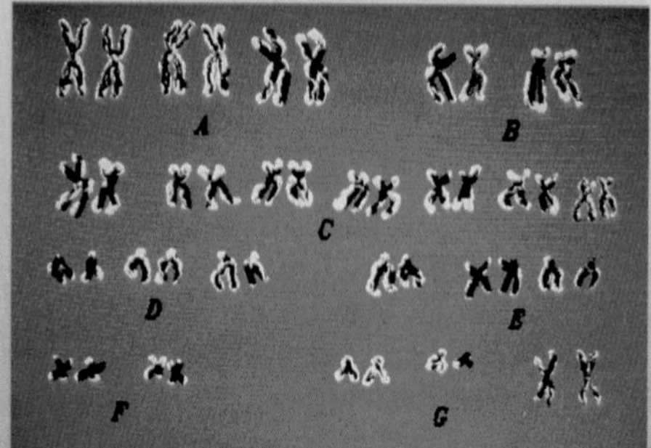

OPTIONS

ENRICHMENT

▶ Have students construct a pedigree of their family for one particular trait that they are interested in.

▶ Have students do research on different kinds of human genetic disorders. The March of Dimes can supply classroom quantities of some literature that may be useful.

PROGRAM RESOURCES

From the **Teacher Resource Package** use:

Critical Thinking/Problem Solving, page 9, Gene Therapy: Hope for Incurable Diseases.

Science and Society, page 9, Science in the Designer Genes Business.

Technology, pages 7-8, Fixing Genetic Material.

Science Integration Activities 5, The Law of Probability

Figure 5-15. Genetically engineered bacteria are being developed to clean up environmental problems such as oil spills.

Knowing how genes are inherited now causes some people to seek the advice of genetic counselors before having children. For those with children with genetic disorders, there are high hopes for cures through research.

3 Through **genetic engineering**, scientists are experimenting with methods that allow them to go into cells to change or correct specific damaged or mutated genes. Genetic engineering is already used to help produce large quantities of medicine, such as insulin, to meet the needs of people with the disease diabetes. In addition, genetic engineering research is being used to find new ways to provide people with more nutritious food.

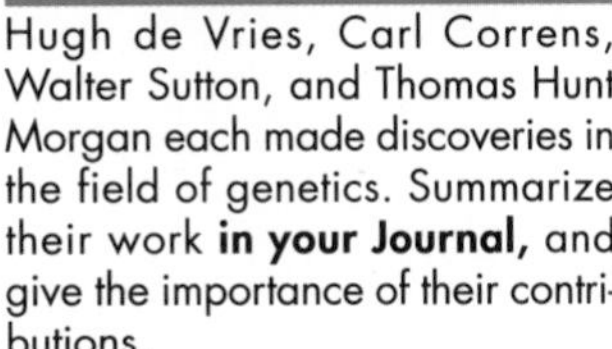

Hugh de Vries, Carl Correns, Walter Sutton, and Thomas Hunt Morgan each made discoveries in the field of genetics. Summarize their work **in your Journal,** and give the importance of their contributions.

What is genetic engineering?

SECTION REVIEW

1. Describe two genetic disorders.
2. Explain why males are affected more often than females by sex-linked genetic disorders?
3. Describe the importance of genetic engineering.
4. **Apply:** Use a Punnett square to explain how a woman who is a carrier for color blindness can have a daughter who is color-blind.
5. **Connect to Chemistry:** Find out how chemistry is used in genetic engineering.

Concept Mapping

Skill Builder

Use a network tree concept map to show how X and Y sex cells can combine to form zygotes. Begin with female sex cells each containing an X chromosome. Use two male sex cells. Indicate one with an X chromosome and one with a Y chromosome. If you need help, refer to Concept Mapping in the **Skill Handbook** on pages 688 and 689.

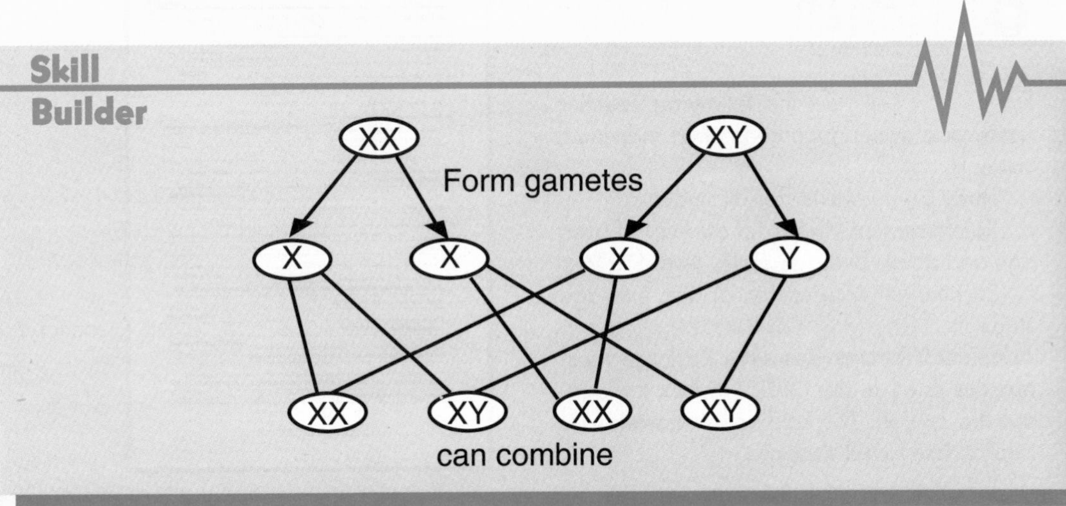

In Your JOURNAL

Reports can be two to three paragraphs for each scientist.

Walter Sutton: geneticist who first reported that chromosomes carry genes.

Thomas Hunt Morgan: established the chromosomal theory of inheritance; hypothesized the idea of sex-linked characters.

Hugo De Vries: one of several people who rediscovered Mendel's work; decided evolution could be studied by experiment; used the term *mutation* to describe variations that appeared suddenly.

Carl Correns: one of several who rediscovered and corroborated Mendel's work; researched incomplete dominance.

3 CLOSE

▶ Ask questions 1-3 and the **Apply** Question in the Section Review.

SECTION REVIEW ANSWERS

1. Answers will vary, but may include descriptions of cystic fibrosis and sickle-cell anemia.

2. Males inherit only one X chromosome, so a single recessive allele produces the same trait that the homozygous condition produces in females.

3. Genetic engineering produces large quantities of medicine, and may eventually allow scientists to correct human genetic disorders.

4. Apply: Answers will vary, but should show a woman carrier married to a man who is color-blind.

5. Connect to Chemistry: Chemicals are used to prepare cells for genetic transfer. Materials transferred are all chemical in nature.

Skillbuilder
ASSESSMENT

Performance: Use this Skillbuilder to assess students' abilities to form a Concept Map of how X and Y gametes combine to form zygotes.

SECTION 5-4

PREPARATION

SECTION BACKGROUND

▶ The human Genome Project was headed by Nobel Laureate, James Watson of Cold Spring Harbor Laboratory in New York from its founding in 1989 until 1992. Dr. Michael Gottesman, chief of the laboratory of cell biology at the National Cancer Institute, was appointed to replace Watson.

1 MOTIVATE

▶ Have students imagine a book that has all of the information about humans written in code. If the code were a single word for each gene, how long would the book have to be? The human genome consists of approximately 100 000 genes.

Answer: Computers are used to store data from almost every biological experiment. Also, computers do most of the calculations and illustrations in such experiments.

5-4 The Human Genome

New Science Words

genome

Objectives

- ▶ **Describe the goal of the Human Genome Project.**
- ▶ **Explain some advantages and disadvantages of genetic research.**

Tracking Human Genes

What is a genome?

Think of the DNA that makes up your chromosomes. It is somewhat like a long railroad track. Just as railroad engineers use maps showing the various towns along rail routes, genetic engineers also are now able to locate particular genes along each chromosome. The resulting chromosome map, or **genome,** is a chart that shows the location of individual genes on a chromosome. A project called the *Human Genome Project* is aimed at mapping all the genes on all the human chromosomes. In all, there are about 100 000 genes on the 46 chromosomes in each of the body cells of your body. Codes for part of about 5000 genes are known.

It is calculated that there are more than a billion base pairs in the DNA sequence of the human genome. Without integrated circuits and microchips developed from studies in physics, the computers needed to collect and store this data would not be possible. Besides storing DNA data, what other uses are there for computers in biology?

Much like a railroad engineer who stops at a specific station, scientists who study chromosomes are able to follow a piece of DNA and stop at the spot where a gene is located. Using chromosome maps, they can identify genes responsible for specific traits.

Why would such a map be useful? More than 3000 human disorders are known to be inherited, including cystic fibrosis, muscular dystrophy, and cancers caused by chemicals that change DNA. By examining a few drops of blood, scientists will be able to examine all chromosomes and find genes that code for certain disorders. After examining a person's genes, a doctor can tell if that person has a certain disorder. Doctors are then able to suggest measures to avoid development of the disorder. A person showing an inherited heart disease may begin a stress management program to reduce the chances of having a heart attack. Others might decide not to have children based on the information in their genome.

OPTIONS

Meeting Different Ability Levels

For Section 5-4, use the following **Teacher Resource Masters** depending upon individual students' needs.

◆ **Study Guide Master** for all students.

● **Reinforcement Master** for students of average and above average ability levels.

▲ **Enrichment Master** for above average students.

Additional Teacher Resource Package masters are listed in the OPTIONS box throughout the section. The additional masters are appropriate for all students.

◆ STUDY GUIDE 24

NAME ____________ DATE ________ CLASS ________

STUDY GUIDE — Chapter 5

The Human Genome — Text Pages 122–123

Answer the following questions using information from the textbook.

1. Where are genes located? **on chromosomes**
2. What does a chromosome map show? **the location of individual genes**
3. What is another name for the entire chromosome map of an organism? **genome**
4. Do scientists know the location of all the genes? **no**
5. How many chromosomes are there in each human body cell? **46**
6. About how many human genes are there? **100 000**
7. What is the human genome initiative? **a project aimed at mapping all the genes on all the human chromosomes**
8. What are genes made up of? **DNA**
9. What kind of diseases might scientists be able to detect by examining a person's genes? **genetic or inherited**
10. What benefit is there in knowing that a person may develop a disease? **A person may be able to take preventative measures to stop the development of the disease.**
11. Why might a person make a decision not to have children based on information in the genome? **There may be a great probability that an offspring would have a certain disease.**
12. What kind of discrimination might a person face whose genome indicates the possibility of the development of a certain disease? **A person may not be hired for a job if a certain disease could negatively affect performance on the job, even though the disease may not yet be developed.**
13. How might early detection of a physical or mental handicap benefit a child? **Early therapy could help improve the child's life.**
14. How might the human genome initiative benefit other sciences? **Technologies that can be used in other sciences will most likely be developed.**
15. What benefit is there in knowing the location and DNA code of disease-causing genes? **Scientists could use this information to detect disease, predict disease in offspring, or to possibly cure disease.**

24

This information may save lives, but there are other facets to consider with this new technology. For example, what if company officials wanted to examine the genome of anyone who applied to them for a job? This information might be able to identify inherited traits that would affect job performance. An airline company might be unwilling to hire a person who is genetically inclined to suffer from alcoholism. This technology may help prevent the development of many disorders, but is it fair to deny a job to someone who *may* develop a disorder?

What if a handicap or an inheritable disorder could be spotted while a child is still young? Early detection would allow therapy that could permit a much more enjoyable life.

Most scientists agree that the benefits of the Human Genome Project will far outweigh the costs. Such large research projects almost always result in other technologies that can be used in other sciences. Knowing the location and DNA code of some genes may even make it possible to cure human genetic disorders in the future.

Figure 5-16. Studies of parts of chromosomes can be used to produce chromosome maps that show where specific genes are located.

SECTION REVIEW

1. What is the purpose of the Human Genome Project?
2. List one possible advantage and one possible disadvantage of this type of technology.

You Decide!

Although the human genome initiative could mean good things for people with genetic disorders, the use of such technology may cause problems. Some genetic traits may hinder an individual's ability to perform a job safely. Do you feel that an employer should be given the right to use genetic information to judge an individual's ability to perform a job?

2 TEACH

Key Concepts are highlighted.

CONCEPT DEVELOPMENT

▶ The vast list of DNA sequences would be unmanageable without computer storage and retrieval capacities.

CHECK FOR UNDERSTANDING

Ask questions 1 and 2 in the Section Review.

RETEACH

Ask students how their lives might be affected by the human genome initiative.

EXTENSION

For students who have mastered this section, use the **Reinforcement** and **Enrichment** masters or other OPTIONS provided.

3 CLOSE

▶ Ask students how a greater knowledge of the human genome can improve lives of humans in the future.

SECTION REVIEW ANSWERS

1. The human genome initiative is aimed at mapping all the genes on all the human chromosomes.

2. By examining a few drops of blood, scientists will be able to examine all chromosomes and find genes that code for certain disorders. Companies might also exclude people from certain jobs if their gene maps were known.

YOU DECIDE!

Answers will vary. All answers that show clear and logical thinking should be accepted.

● REINFORCEMENT 24

NAME ____ DATE ____ CLASS ____

REINFORCEMENT — Chapter 5

The Human Genome — Text Pages 122–123

Fill in the table with information about the human genome initiative. Then answer the questions.

Name of project	**human genome initiative**
Purpose	**to map all genes on all human chromosomes**
Field of science	**human genetics**
Current progress	**about 1000 out of 100 000 genes mapped**
Possible benefits	**prevention of diseases, early detection of diseases, cures for diseases, development of new technologies**
Possible disadvantages	**misuse of genetic information, discrimination based on information from genome, high cost**

1. Why do scientists examine DNA in order to identify and locate genes? **Genes are on chromosomes which are made of DNA.**
2. What might be the advantage of early detection of an inherited disease or disorder? **It may be possible to take preventive measures to avoid the development of the disease. There may be therapies or programs to reduce the harmful effects of the disease. Some people may choose not to have children to avoid passing on a disease.**
3. What are some inherited diseases? **cystic fibrosis, muscular dystrophy, certain cancers, sickle-cell anemia, hemophilia, and Down Syndrome**
4. Give a specific example of misuse of genetic information. **Answers may vary and may include denial of a job based on the likelihood of the development of a disease.**

24

▲ ENRICHMENT 24

NAME ____ DATE ____ CLASS ____

ENRICHMENT — Chapter 5

The Human Genome — Text Pages 122–123

The human genome initiative project is expensive. The National Institute of Health, a government agency, provides much funding for the project. Many scientists agree that the benefits of the human genome initiative far outweigh the costs. It is also possible that money could be better spent on other areas of research. Or, the disadvantages, such as misuse of genetic information, might outweigh the benefits. Consider the opposing points of view. If possible, research recent magazines and newspapers for information about the project. Talk to people who have definite opinions about the issue. Decide which viewpoint you support. Choose one of the following.

A. Write a letter, as if to your congressional representative, urging him or her to **support** further spending for the human genome initiative project.

B. Write a letter, as if to your congressional representative, urging him or her to **oppose** further spending for the human genome initiative project.

24

ACTIVITY 5-2

OBJECTIVE: Design and carry out an experiment to determine polygenic inheritance.
Time: one class period

PROCESS SKILLS applied in this activity are **forming a hypothesis, measuring, recording data, calculating,** and **interpreting.**

PREPARATION

Collect materials needed for each group.

Cooperative Learning: Assign Science Investigation Teams.

HYPOTHESIZING

Student hypotheses will vary. A possible hypothesis might be: "The bar graph of class height data will look like a bell." or "The bar graph of class height data will have a slope from left to right."

TEACHING THE ACTIVITY

Refer to the ***Activity Worksheets*** *for additional information and teaching strategies.*

- Help students distinguish among range, median, mean, and mode.
- Results form a bell-shaped curve.

SUMMING UP/SHARING RESULTS

- The graph has a bell shape.
- Yes, because the trait shows a wide variety, not just one or two clearly different heights.

GOING FURTHER!

The data table of heights of girls may show the greater number of them with taller heights. This is normal for adolescents. Both graphs will still have a bell shape.

ACTIVITY 5-2 DESIGNING AN EXPERIMENT
Determining Polygenic Inheritance

How can the effect of polygenic inheritance be determined?

Getting Started

Devise a method of determining if height is controlled by polygenic inheritance.

Hypothesizing

Decide how data about students' heights in your classroom can be collected. Make a **hypothesis** about what a bar graph of the heights of students in your class will look like.

Materials

Your cooperative group will use:
- metric ruler
- graph paper
- pencil

Data and Observations Sample Data

Height in cm	Number of Students
A 101 - 110	0
B 111 - 120	1
C 121 - 130	2
D 131 - 140	9
E 141 - 150	8
F 151 - 160	4
G 161 - 170	1
H 171 - 180	0

Try It!

1. Make a data table like the one shown. Decide on the intervals of height to use (e.g. 131-140 cm).
2. *Measure and record* the height of every student in the class to the nearest centimeter.
3. Construct a bar graph from the data. (refer to sample graph)
4. The range of a set of data is the difference between the greatest measurement and the smallest measurement. The median is the middle number when the data are placed in order. The mean is the sum of all the data divided by the number of addends. The mode is the number that appears most often in the measurements. Calculate each of these numbers.

Summing Up/Sharing Results

- What does the bar graph look like?
- Can you *infer* from your data that height is controlled by more than two genes?

Going Further!

How would bar graphs of the heights of girls compare with bar graphs of the heights of boys? Try it and see if there is a difference.

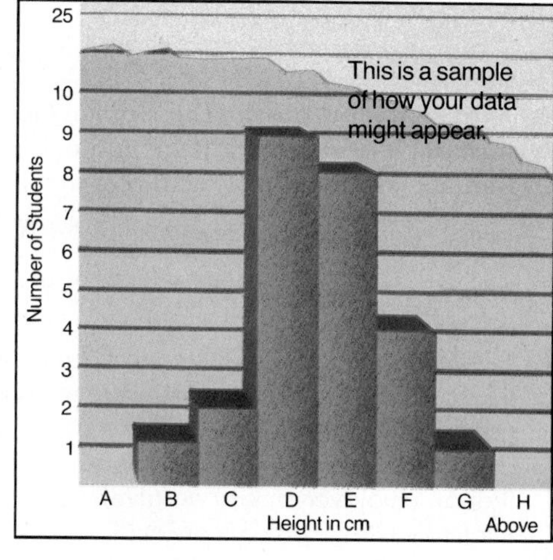

ANSWERS TO QUESTIONS

Questions 1-4 will have answers that depend on the data of the class.
5. The graph will look similar to sample on student page.
6. If there is wide variation in appearance of the trait with no real divisions among the variations, more than one gene is involved.
7. Yes, because the trait shows a wide variety, not just one or two clearly different heights.

PROGRAM RESOURCES

From the **Teacher Resource Package** use:
Activity Worksheets, pages 43-44, Activity 5-2, Determining Polygenic Inheritance.

Activity ASSESSMENT

Performance: To further assess students' understanding of polygenic inheritance, see USING LAB SKILLS, Question 12, on p. 126.

CHAPTER REVIEW

SUMMARY

5-1: What Is Genetics?

1. Traits are inherited through alleles passed from one generation to the next. Gregor Mendel determined the basic laws of genetics through controlled experiments.
2. The probability of types of offspring from parents can be predicted by using Punnett squares.
3. Genotype is the genetic makeup of an organism for a trait. Phenotype is the physical appearance that results from the genotype.

5-2: Genetics Since Mendel

1. When a trait is inherited by incomplete dominance, neither allele is completely dominant.
2. With multiple alleles, there are more than two alleles controlling a trait. In polygenic inheritance, more than one pair of alleles controls a trait. Eye and hair color are determined by polygenic inheritance.

5-3: Human Genetics

1. Hemophilia and cystic fibrosis are genetic disorders.
2. Color blindness and hemophilia are located on the X chromosomes. Males express sex-linked traits more than females. Pedigrees may show patterns of inheritance in a family.
3. Genetic engineering may be used to correct mutated genes, cure disease, and produce medicine and food.

5-4: Science and Society: The Human Genome

1. The Human Genome Project seeks to locate and identify all the genes on human chromosomes.
2. Genome information may help pinpoint specific genetic disorders.

KEY SCIENCE WORDS

a. **alleles**
b. **dominant**
c. **genetic engineering**
d. **genetics**
e. **genome**
f. **genotype**
g. **heredity**
h. **heterozygous**
i. **homozygous**
j. **incomplete dominance**
k. **multiple alleles**
l. **pedigree**
m. **phenotype**
n. **polygenic inheritance**
o. **Punnett square**
p. **recessive**
q. **sex-linked gene**

UNDERSTANDING VOCABULARY

Match each phrase with the correct term from the list of Key Science Words.

1. technology for changing a gene
2. gene located on the X chromosome
3. the allele that is hidden
4. different forms of the same gene
5. a tool for predicting possible offspring
6. an allele that does not show completely
7. the study of heredity
8. more than one pair of alleles controls a trait
9. shows the pattern of gene inheritance in a family
10. all the genes in a species

OPTIONS

ASSESSMENT

To assess student understanding of material in this chapter, use the resources listed.

COOPERATIVE LEARNING

Consider using cooperative learning in the THINK AND WRITE CRITICALLY, APPLY, and MORE SKILL BUILDERS sections of the Chapter Review.

PROGRAM RESOURCES

From the **Teacher Resource Package** use:

Chapter Review, pages 13-14.

Chapter and Unit Tests, pages 32-35, Chapter Test.

CHAPTER REVIEW

SUMMARY

Have students read the summary statements to review the major concepts of the chapter.

UNDERSTANDING VOCABULARY

1.	c	**6.**	j
2.	q	**7.**	d
3.	p	**8.**	n
4.	a	**9.**	l
5.	o	**10.**	e

ASSESSMENT

Portfolio Suggestions

Encourage students to place in their portfolios one or two items of what they consider to be their best work. For each item, ask students to explain why that item was chosen and what they learned from it. Items might be selected from the following.

- MINI-Lab trait survey and table, p. 106
- Enrichement Punnett squares, p. 112
- Enrichment family pedigree, p. 120

Performance

Additional performance assessments may be found in *Performance Assessment* and *Science Integration Activities* that accompany **Merrill Life Science.** Performance Task Assessment Lists and rubrics for evaluating these activities and other products generated throughout the chapter can be found in Glencoe's *Performance Assessment in Middle School Science.*

CHAPTER REVIEW

CHECKING CONCEPTS

1. a	**6.** b
2. c	**7.** d
3. a	**8.** a
4. c	**9.** a
5. d	**10.** b

USING LAB SKILLS

ASSESSMENT

Use these alternate lab excercises to assess students' understanding of skills used in this chapter.

11. Use caution in assigning this activity if you are aware of adoption situations in your classroom. Students could draw a pedigree with labeled fingerprints.

12. Have students measure from thumb to pinky finger with fingers spread wide. A bell-shaped curve indicates multiple alleles.

THINK AND WRITE CRITICALLY

13. Multiple alleles mean having more than 2 alleles, or types of genes, for a trait; the trait is controlled by one gene pair. Polygenic inheritance involves more than one gene pair controlling the trait. These gene pairs combine to produce the trait.

14. Mendel carried out his experiments carefully, keeping accurate records, and using large samples.

15. The blood types are A, B, AB, and O. A child with type O blood cannot have one or both parents with type AB blood; he can have parents with all other types as long as each has the O allele. A child with type AB blood cannot have one or both parents with type O; he cannot have both parents with type B or both parents with type A.

16. Homozygous means that both alleles for the trait are identical. Heterozygous means that the alleles for the trait are different.

17. Dominant means that the trait will show when there is a heterozygous genotype. It does not mean that it is the most common allele in a population. The most common may be a recessive allele.

CHAPTER REVIEW

CHECKING CONCEPTS

Choose the word or phrase that completes the sentence.

1. ______ are located on chromosomes.
   a. DNA codes **c.** Carbohydrates
   b. Pedigrees **d.** Zygotes

2. Color blindness results from an allele that is ______.
   a. dominant
   b. on the Y chromosome
   c. on the X chromosome
   d. present only in females

3. ______ results when two alleles both appear in the phenotype.
   a. Incomplete dominance
   b. Polygenic inheritance
   c. Multiple alleles
   d. Sex-linked genes

4. During meiosis, ______ for a trait separate.
   a. proteins **c.** alleles
   b. cells **d.** pedigrees

5. The major job of genes is to control ______.
   a. chromosomes **c.** cell membranes
   b. mitosis **d.** making proteins

6. Blood type is inherited through ______.
   a. polygenic inheritance
   b. multiple alleles
   c. incomplete dominance
   d. recessive genes

7. Sickle-cell anemia is inherited through ______.
   a. polygenic inheritance
   b. multiple alleles
   c. incomplete dominance
   d. recessive genes

8. Eye color is a trait inherited through ______.
   a. polygenic inheritance
   b. multiple alleles
   c. incomplete dominance
   d. recessive genes

9. A female is produced if an egg unites with a sperm containing ______.
   a. an X chromosome **c.** a Y chromosome
   b. XX chromosomes **d.** XY chromosomes

10. Punnett squares are used to ______ the outcome of crosses of traits.
   a. dominate **c.** assure
   b. predict **d.** number

USING LAB SKILLS

11. Using the information on fingerprint inheritance in the Mini-Lab on page 115, have at least ten classmates *volunteer* to see how patterns have been inherited in their families. Data in the form of fingerprints might be collected from family members and pedigrees drawn.

12. In the activity on page 124, you graphed a class trait controlled by polygenic inheritance. Perform a similar experiment testing hand spans. Find the mode, range, and median for your class. Record this data in your Journal and graph your results on a line graph.

THINK AND WRITE CRITICALLY

Answer the following questions in your Journal using complete sentences.

13. Compare and contrast multiple allele inheritance and polygenic inheritance.

14. Explain the importance of Mendel's experimental methods.

15. Using an example, explain how blood typing could be used to detect the correct parent of a newborn if there was a mix-up in a nursery.

16. What is the difference between homozygous and heterozygous?

17. Explain why dominant traits might not necessarily be the traits that show up most frequently.

APPLY

18. A pure black guinea pig is crossed with a pure white guinea. All the offspring are black. What are the genotypes of the parents and the offspring?
19. In breeding Andalusian variety chickens, farmers found that chickens with black feathers always had chicks with black feathers, and chickens with white feathers always had chicks with white feathers. But, mating a black with a white chicken produced some chickens with blue feathers. How can this trait be explained?
20. Explain the relationship between DNA, genes, alleles, and chromosomes.
21. How will finding the location of disease-causing genes on chromosomes help to cure genetic diseases?
22. Explain how an organism that has a genotype *Gg* could have the same phenotype as an organism with the genotype *GG*.

MORE SKILL BUILDERS

If you need help, refer to the Skill Handbook.

1. **Designing an Experiment:** Design an experiment to determine if a trait is transmitted by a dominant or recessive gene.
2. **Recognizing Cause and Effect:** A mutation in the gene for normal production of a plant's chlorophyll results in a plant with no chlorophyll. Explain the consequences.
3. **Classifying:** Classify the type of inheritance of each human trait by making a chart: blood type, color blindness, eye color, height, hemophilia, sickle-cell anemia, weight.
4. **Interpreting Scientific Illustrations:** What genotypes and phenotypes would be produced if you crossed a red offspring with a pink offspring from Figure 5-8b on page 114?
5. **Concept Mapping:** On a separate sheet of paper, use the following terms to complete the network tree concept map below: phenotype, recessive genes, dominant.

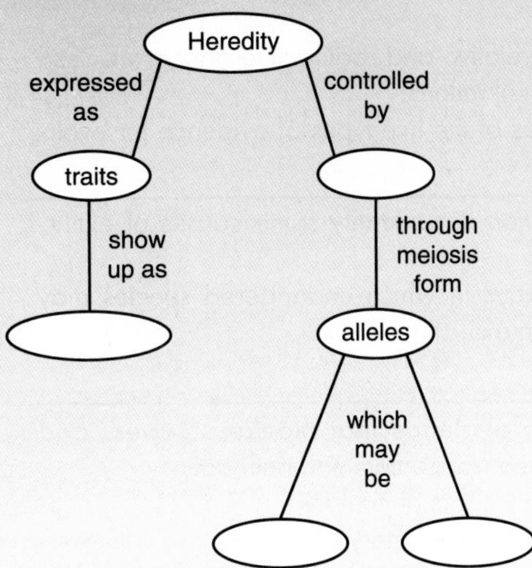

PROJECTS

1. Research a genetic disease such as cystic fibrosis. How is it inherited? Who does it affect? How can it be detected? What are its symptoms? How is it treated? Share your findings with the class in a written report.
2. Research sex determination in birds, butterflies, and frogs to find out how it differs from sex inheritance in humans.

APPLY

18. The parents are homozygous, one for black, one for white. All the offspring are heterozygous for black and white.
19. The blue feathers result from incomplete dominance of the black and white alleles, producing both black and white colors in feathers.
20. DNA is a chemical; a gene is a segment of DNA; an allele is a form of a gene specific for a trait; genes are located on chromosomes.
21. Finding where the genes are located may help predict the chances of having a child with the disease and eventually provide a cure.
22. In cases of simple dominant-recessive mechanism, the phenotype will show as dominant whether the genotype is homozygous or heterozygous dominant because the recessive gene does not show at all in the phenotype.

MORE SKILL BUILDERS

1. Designing an Experiment: The student needs to describe an experiment in which crosses can be made between pure-breeding organisms, similar to Mendel's experiments.
2. Recognizing Cause and Effect: The gene became defective and could not make the necessary plant pigment. Since there is no chlorophyll, the plant cannot change sunlight into chemical energy and therefore it will die. The mutation was harmful.
3. Classifying:

INCOMPLETE DOMINANCE
sickle-cell anemia

MULTIPLE ALLELES
blood type

POLYGENIC	SEX-LINKED
eye color	hemophilia
height	color-blindness
weight	

4. Interpreting Scientific Illustrations:
50 percent RR and red
50 percent Rr and pink

5. Concept Mapping:

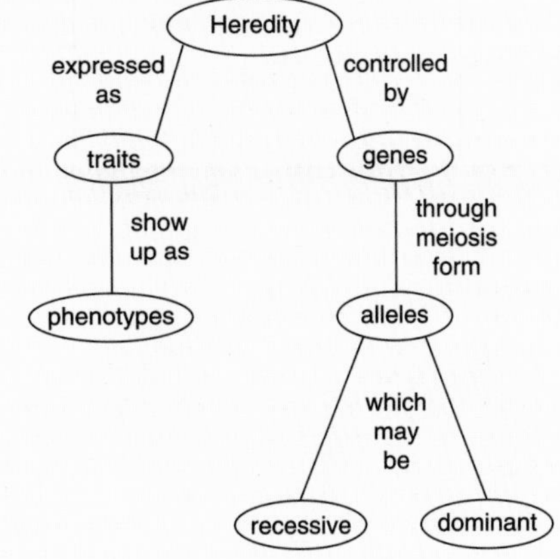

CHAPTER

6 Evolution

CHAPTER SECTION	OBJECTIVES	ACTIVITIES
6-1 Mechanisms of Evolution (2 days)	1. **Compare** and **contrast** Lamarck's theory of evolution with Darwin's. 2. **Explain** the importance of variations in organisms. 3. **Relate** how gradualism and punctuated equilibrium describe the rate of evolution.	**MINI-Lab:** *How does evolution occur?* p. 134
6-2 Evidence for Evolution (3 days)	1. **Describe** the importance of fossils as evidence of evolution. 2. **Explain** how relative and radioactive dating are used to date the fossil record. 3. **Give examples** of the five types of evidence for evolution.	**MINI-Lab:** *How are fossils made?* p. 137 **Activity 6-1:** *A Radioactive Dating Model*, p. 143
6-3 Plant and Animal Extinction Science & Society (1 day)	1. **Define** *extinction* and **identify** some causes of extinction. 2. **Name** two ways in which endangered species may be saved from extinction.	
6-4 Human Evolution (2 days)	1. **Describe** the evidence that monkeys, apes, and humans evolved from a common ancestor. 2. **Describe** the ancestors of humans. 3. **Trace** the evolutionary history of humans.	**Activity 6-2:** *Designing an Experiment (Recognizing Variations in a Population)*, p. 150
Chapter Review		

ACTIVITY MATERIALS

FIND OUT	ACTIVITIES		MINI-LABS	
Page 129 newspaper want ads black and white construction paper scissors	**6-1 A Radioactive Dating Model, p. 143** 100 pennies cardboard box with lid graph paper pencil	**6-2 Designing an Experiment, p. 150** 100 seeds metric ruler graph paper	**How does evolution occur? p. 134** paper pencil	**How are fossils made? p. 137** plaster of paris or clay

CHAPTER FEATURES	TEACHER RESOURCE PACKAGE	OTHER RESOURCES
Problem Solving: *Why Isn't Earth Covered with Pumpkins and Pike?* p. 133 **Skill Builder:** *Classifying,* p. 135	**Ability Level Worksheets** ◆ ***Study Guide,*** p. 25 ● ***Reinforcement,*** p. 25 ▲ ***Enrichment,*** p. 25 **Concept Mapping,** p. 17 **Activity Worksheets,** p. 56 **Transparency Masters,** pp. 17–18	**Color Transparency 9,** Horse Evolution **Lab Manual:** Overproduction, p. 37 **Lab Manual:** Variation, p. 41 **STVS:** Disc 6, Side 1; Side 2
Technology: *An Ostrich Egg Timer,* p. 140 **Skill Builder:** *Making and Using Tables,* p. 142	**Ability Level Worksheets** ◆ ***Study Guide,*** p. 26 ● ***Reinforcement,*** p. 26 ▲ ***Enrichment,*** p. 26 **Activity Worksheet,** p. 5, 50–51, 57 **Transparency Masters,** pp. 19–20	**Color Transparency 10,** Geologic Time Scale **STVS:** Disc 3, Side 2; Disc 2, Side 2 **Science Integration Activity 6**
You Decide! p. 145	**Ability Level Worksheets** ◆ ***Study Guide,*** p. 27 ● ***Reinforcement,*** p. 27 ▲ ***Enrichment,*** p. 27 **Critical Thinking/Problem Solving,** p. 10 **Science and Society,** p. 10	**STVS:** Disc 5, Side 1
Skill Builder: *Concept Mapping,* p. 149	**Ability Level Worksheets** ◆ ***Study Guide,*** p. 28 ● ***Reinforcement,*** p. 28 ▲ ***Enrichment,*** p. 28 **Cross-Curricular Connections,** p. 10 **Activity Worksheets,** pp. 52–53	
Summary Key Science Words Understanding Vocabulary Checking Concepts Using Lab Skills Think & Write Critically Apply More Skill Builders Projects	ASSESSMENT RESOURCES **Chapter Review,** pp. 15-16 **Chapter Test,** pp. 36-39 **Performance Assessment in Middle School Science (PAMSS)**	**Chapter Review Software** **Test Bank** **Alternate Assessment** **Performance Assessment**

◆ **Basic** ● **Average** ▲ **Advanced**

ADDITIONAL MATERIALS

SOFTWARE	AUDIOVISUAL	BOOKS/MAGAZINES
Evolution: Evolution and Natural Selection, CONDUIT. *Natural Selection,* EME. *Tracking Changes in Organisms Through Time,* Queue. *Evolutionary Trail,* Queue.	*Adaptations of Plants and Animals,* film, Coronet/MTI. *Darwin and the Theory of Natural Selection,* film, Coronet/MTI. *Fossils: Clues to Prehistoric Times,* film, Coronet/MTI. *Natural Selection,* film, Films for the Humanities and Science. *The Galapagos: Darwin's World Within Itself,* film, EBEC. *Australia's Improbable Animals,* laserdisc, Image Entertainment. *Insects: The Little Things That Run the World,* laserdisc, Lumivision. *Adaptations,* video, Coronet/MTI.	Lane, N. Gary. *Life of the Past.* Columbus, OH: Merrill Publishing Co., 1986. Mader, Sylvia S. *Biology: Evolution, Diversity, and the Environment.* Dubuque, IA: Wm. C. Brown Publishers, 1985. Mayr, E. *Populations, Species, and Evolution: An Abridgement of Animal Species and Evolution.* Cambridge, MA: Harvard University Press, 1970.

CHAPTER

6

EVOLUTION

THEME DEVELOPMENT: One of the central themes of biology, evolution, is discussed in this chapter. Theories of evolution and modifications are discussed along with indirect and direct evidence. The ecological theme is also developed in the third section. Plant and animal extinction rates are detailed.

CHAPTER OVERVIEW

- **Section 6-1:** Lamarck's theory and Darwin's theory of evolution are compared and contrasted in relation to observed variation. Gradualism and punctuated equilibrium are discussed as modifications of Darwin's theory.
- **Section 6-2:** The evidence for evolution by natural selection is presented. Relative and radioactive dating are compared.
- **Section 6-3: Science and Society:** Plant and animal extinctions are discussed in terms of the part humans play in the environment. Ways of saving endangered species are mentioned.
- **Section 6-4:** Evolution of primates from a common ancestor and the evolutionary history of humans is described.

CHAPTER VOCABULARY

species
evolution
natural selection
variation
population
gradualism
punctuated equilibrium
fossils
sedimentary rock
relative dating
radioactive elements
homologous
vestigial structure
embryology
extinction
endangered species
primates
hominids
Homo sapiens

CHAPTER 6

Evolution

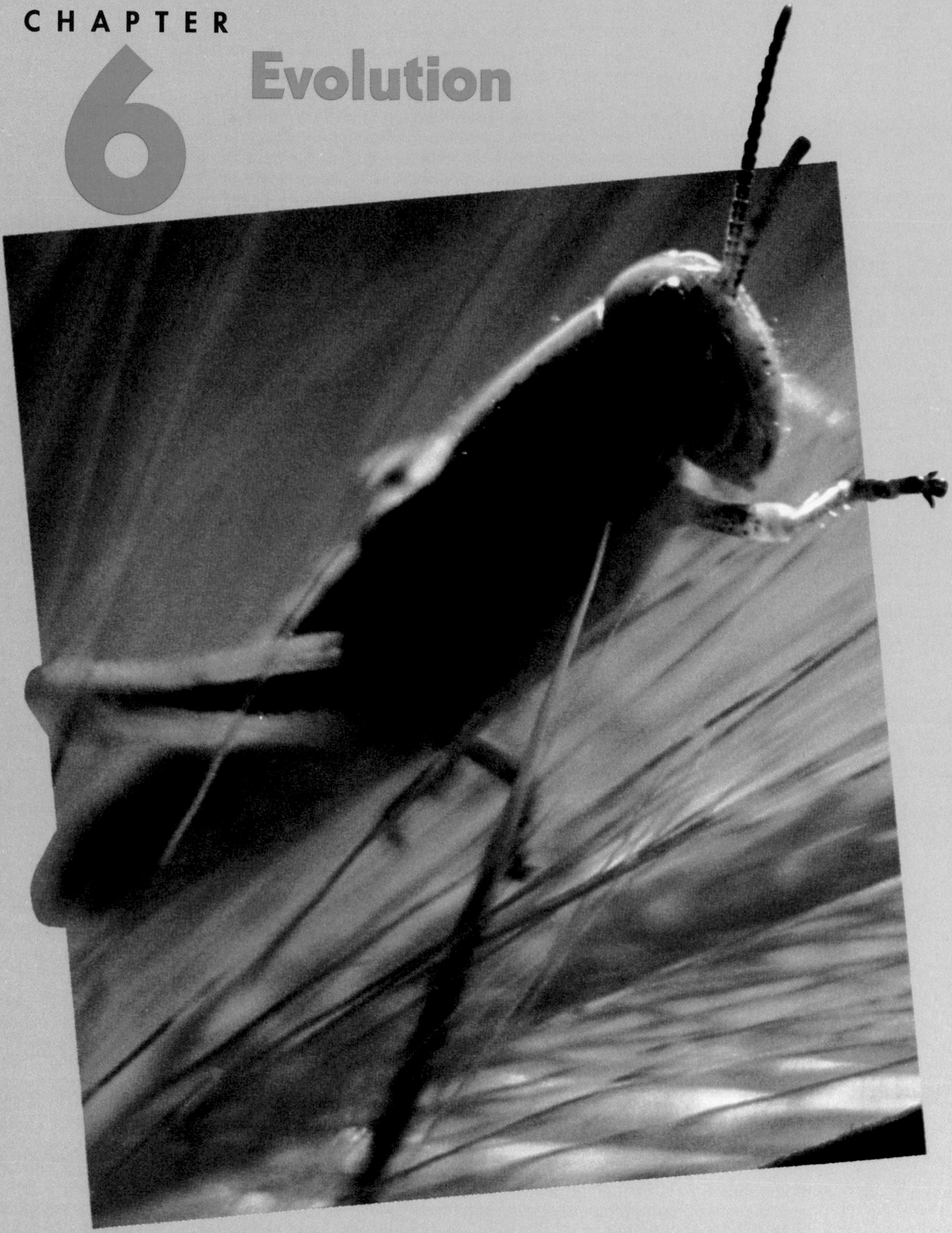

128

OPTIONS

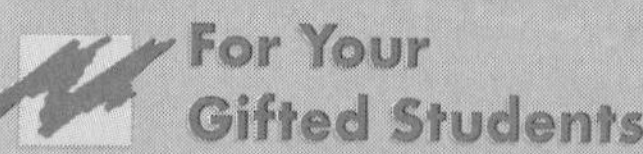

Students can create their own species (plant, animal, or other unknown). They can draw it, describe its habitat and diet (herbivore, carnivore, or other), and give the species a history. They should describe the changes that may have occurred within the species. Have students make a time line showing major physical changes and ecological events that coincided with changes in the species. The species may be from the past or the future.

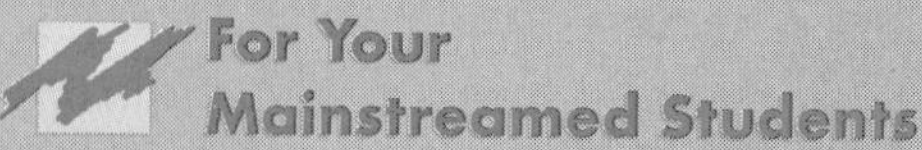

Have students make clay impressions (secretly) of various items of daily use. Plaster of Paris may be poured into the impression to form a fossil. Have the students analyze the objects as if they were anthropologists in the year 3000 A.D. who had just uncovered an ancient burial site. Who were these people? How did this culture live?

Have you ever wondered why some grasshoppers are green? Do you know why polar bears are white and grizzly bears brown?

FIND OUT!

Do this activity to find out why some organisms are the color they are. Record all data in your Journal.

Spread out a sheet of newspaper want ads on the floor. Cut ten 2 cm by 2 cm squares diagonally out of another sheet of want ads, a sheet of black paper, and a sheet of white paper. Scatter the 30 squares randomly across the sheet of newspaper on the floor. Have a partner time you for 10 seconds, and pick up squares from the newspaper. Count and record the number of squares of each color you picked up. Return the squares to the newspaper. Repeat this process three times. Average your results. If the squares represent organisms in an environment, *infer* which ones would most likely not be found. Why do you think you don't notice green grasshoppers easily?

Gearing Up

Previewing the Chapter

Use this outline to help you focus on important ideas in this chapter.

Section 6-1 Mechanisms of Evolution
- ▶ Early Theories of Evolution
- ▶ Evolution by Natural Selection
- ▶ Adaptation and Variation
- ▶ How Fast Does Evolution Occur?

Section 6-2 Evidence for Evolution
- ▶ Fossil Evidence
- ▶ The Fossil Record
- ▶ Other Evidence for Evolution

Section 6-3 Science and Society Plant and Animal Extinction
- ▶ Causes of Extinction
- ▶ Ways to Prevent Extinction

Section 6-4 Human Evolution
- ▶ Primates
- ▶ Human Ancestors
- ▶ Modern Humans

Previewing Science Skills

- ▶ In the Skill Builders, you will classify, use tables, and make a concept map.
- ▶ In the Activities, you will collect and organize data, interpret data, formulate models, hypothesize and experiment.
- ▶ In the MINI-Labs, you will make and use models and infer.

What's next?

In nature, organisms that are well suited to their environment have a better chance of surviving than those that aren't. In this chapter, you will learn why organisms live where they do and how organisms change. You will also learn about the process of evolution.

INTRODUCING THE CHAPTER

Use the Find Out activity to show students that organisms must have adaptations, such as cryptic coloration, for survival.

FIND OUT!

Preparation: For this activity, you will need more newspaper want ads than you think. Have the squares of want ads, black paper, and white paper cut before class.

Materials: newspaper want ads, black, and white construction paper, scissors

Cooperative Learning: Students should work as Paired Partners on the Find Out Activity.

Teaching Tips

▶ In general, students should pick up far more of the white and black squares than the newsprint squares.

▶ If you find that students are picking up most of the squares within the time limit, add more squares or use less time, or a combination.

Gearing Up

Have students study the Gearing Up feature to familiarize themselves with the chapter. Discuss the relationships of the topics in the outline.

What's Next?

Before beginning the first section, make sure students understand the connection between the Find Out activity and the topics to follow.

ASSESSMENT OPTIONS

PORTFOLIO

Refer to page 151 for suggested items that students might select for their portfolios.

PERFORMANCE ASSESSMENT

See page 151 for additional Performance Assessment options.

Process

Skillbuilders, pp. 135, 142
MINI-Lab, pp. 134, 137
Activities 6-1, p. 143; 6-2, p. 150
Using Lab Skills, p. 152

CONTENT ASSESSMENT

Assessment—Oral, pp. 134, 141, 148
Skillbuilder, p. 149
Mini Quizzes, pp. 133, 141, 148
Section Reviews, pp. 135, 142, 145, 149
Chapter Review, pp. 151-153

GROUP ASSESSMENT

Opportunities for group assessment occur with Cooperative Learning Strategies and Flex Your Brain Activities.

SECTION 6-1

PREPARATION

SECTION BACKGROUND

▶ Much attention has been given to Darwin's visit to the Galapagos, but he suspected evolution before he ever began the tour on the *Beagle*.

▶ Gradualism and punctuated equilibrium are not so much competing hypotheses as they are attempts to explain the fossil evidence. Both have a place in the history of evolution.

PREPLANNING

▶ You may want to cut out shapes of paper like those used in the Mini-Lab to use as an activity in class.

1 MOTIVATE

▶ **Brainstorming:** Have students list reasons why peacocks have such lavish colors while the peahens are a drab brown. Give other examples of dimorphism—mallard ducks for example where the male is colorful and the female drab.

▶ Have students look at Figure 6-1 and list all the differences and likenesses they observe in the horses. Discuss how these changes might have come about.

6-1 Mechanisms of Evolution

New Science Words

species
evolution
natural selection
variation
population
gradualism
punctuated equilibrium

Objectives

▶ Compare and contrast Lamarck's theory of evolution with Darwin's.
▶ Explain the importance of variations in organisms.
▶ Relate how gradualism and punctuated equilibrium describe the rate of evolution.

Figure 6-1. The evolution of the horse can be traced back in time for at least 55 million years. Notice the change from several toes to a single hoof.

Early Theories of Evolution

On Earth today, there are millions of different types of organisms. Included among these are different species of plants, animals, bacteria, fungi, and protozoa. A **species** is a group of organisms whose members look alike and successfully reproduce among themselves. Have any of these species of organisms changed since they first appeared on Earth? Are they still changing today? Evidence from observation and experimentation shows that living things have changed through time and are still changing. Change in the hereditary features of a species over time is **evolution.** Figure 6-1 shows how the horse has changed over time.

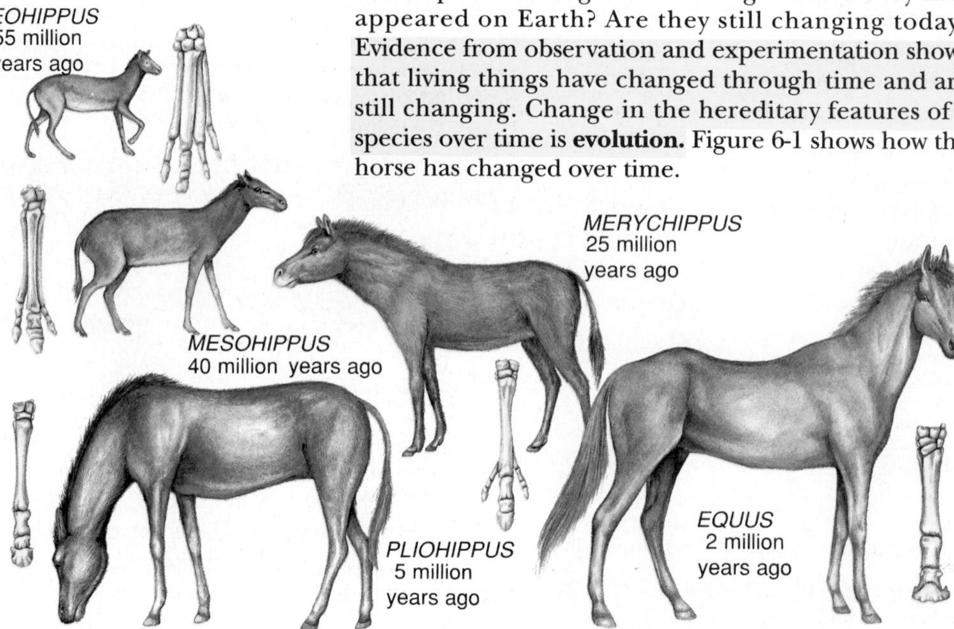

OPTIONS

Meeting Different Ability Levels

For Section 6-1, use the following **Teacher Resource Masters** depending upon individual students' needs.

◆ **Study Guide Master** for all students.

● **Reinforcement Master** for students of average and above average ability levels.

▲ **Enrichment Master** for above average students.

Additional Teacher Resource Package masters are listed in the OPTIONS box throughout the section. The additional masters are appropriate for all students.

◆ STUDY GUIDE 25

Figure 6-2. It takes years of practice and lessons for most people to acquire the ability to play an instrument well. However, the fact that you play well does not guarantee that the children you will have will play the piano well.

In 1809, Jean Baptiste de Lamarck, a French scientist, proposed one of the first theories to explain how species evolve or change. Lamarck hypothesized that species evolve by keeping traits that their parents developed during their life. Characteristics that were not used were lost from the species. According to Larmarck, if one of your parents was a bodybuilder and had large muscles, then you would be born with large muscles. Lamarck's theory of evolution is often called the theory of acquired characteristics.

When you study genetics, you learn that genes on chromosomes control the inheritance of traits. The traits you develop from your lifestyle, such as strong muscles from bodybuilding, are not inherited. After scientists collected large amounts of data on the inheritance of characteristics, Lamarck's theory wasn't accepted. The data showed that characteristics an organism develops during its life aren't passed on to its offspring.

2 **Figure 6-3.** According to Lamarck's theory, the necks of giraffes became longer when they stretched to reach high branches, and this acquired trait was passed on to their offspring. However, long necks on giraffes are an adaptation.

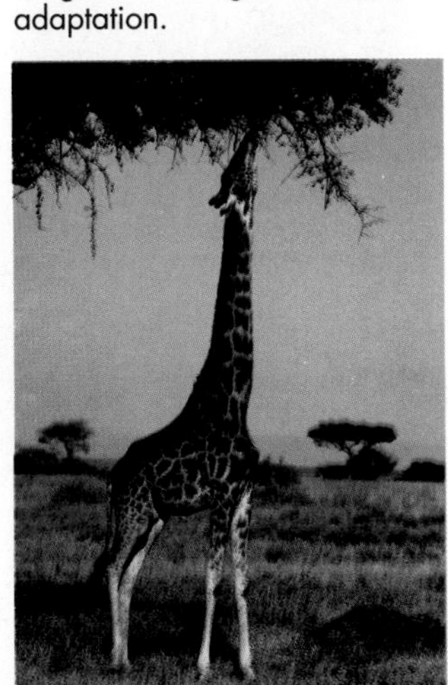

Evolution by Natural Selection

In the mid-1800s, Charles Darwin developed the theory of evolution still accepted today. At the age of 22, Darwin, after trying several other careers, became the ship's naturalist aboard the *HMS Beagle*. The *Beagle* was on a trip to survey the east and west coasts of South America. The ship sailed from England in December 1831. Darwin was responsible for recording information about all the plants and animals he observed during the journey.

● REINFORCEMENT 25

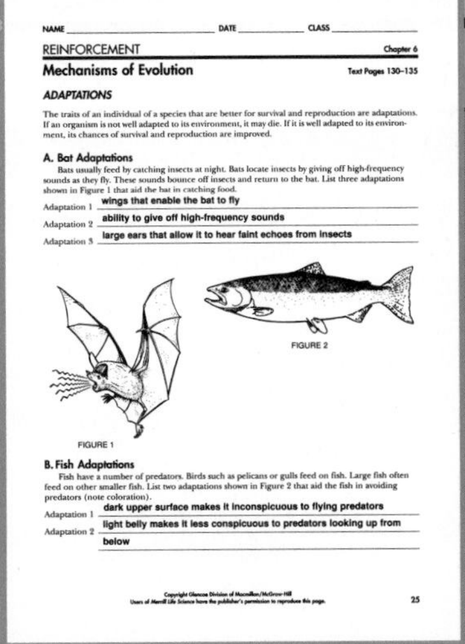

NAME ______ DATE ______ CLASS ______

REINFORCEMENT — Chapter 6

Mechanisms of Evolution — Text Pages 130–135

ADAPTATIONS

The traits of an individual of a species that are better for survival and reproduction are adaptations. If an organism is not well adapted to its environment, it may die. If it is well adapted to its environment, its chances of survival and reproduction are improved.

A. Bat Adaptations

Bats usually feed by catching insects at night. Bats locate insects by giving off high-frequency sounds as they fly. These sounds bounce off insects and return to the bat. List three adaptations shown in Figure 1 that aid the bat in catching food.

Adaptation 1 **wings that enable the bat to fly**

Adaptation 2 **ability to give off high-frequency sounds**

Adaptation 3 **large ears that allow it to hear faint echoes from insects**

FIGURE 1

FIGURE 2

B. Fish Adaptations

Fish have a number of predators. Birds such as pelicans or gulls feed on fish. Large fish often feed on other smaller fish. List two adaptations shown in Figure 2 that aid the fish in avoiding predators (note coloration).

Adaptation 1 **dark upper surface makes it inconspicuous to flying predators**

Adaptation 2 **light belly makes it less conspicuous to predators looking up from below**

25

▲ ENRICHMENT 25

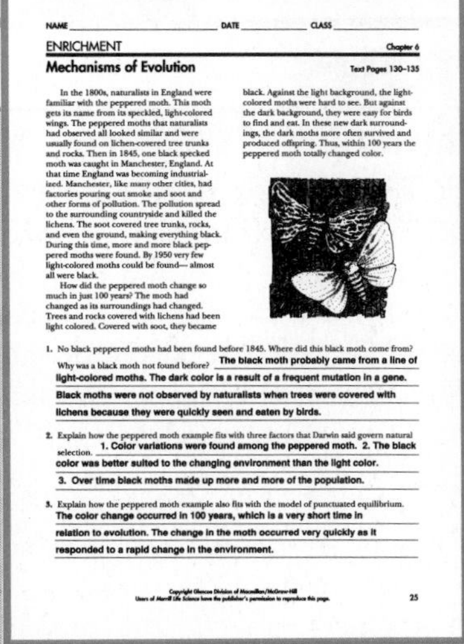

NAME ______ DATE ______ CLASS ______

ENRICHMENT — Chapter 6

Mechanisms of Evolution — Text Pages 130–135

In the 1800s, naturalists in England were familiar with the peppered moth. This moth gets its name from its speckled, light-colored wings. The peppered moths that naturalists had observed all looked similar and were usually found on lichen-covered tree trunks and rocks. Then in 1845, one black specked moth was caught in Manchester, England. At that time England was becoming industrialized. Manchester, like many other cities, had factories pouring out smoke and soot and other forms of pollution. The pollution spread to the surrounding countryside and killed the lichens. The soot covered tree trunks, rocks, and even the ground, making everything black. During this time, more and more black peppered moths were found. By 1950 very few light-colored moths could be found— almost all were black.

How did the peppered moth change so much in just 100 years? The moth had changed as its surroundings had changed. Trees and rocks covered with lichens had been light colored. Covered with soot, they became black. Against the light background, the light-colored moths were hard to see. But against the dark background, they were easy for birds to find and eat. In these new dark surroundings, the dark moths more often survived and produced offspring. Thus, within 100 years the peppered moth totally changed color.

1. No black peppered moths had been found before 1845. Where did this black moth come from? Why was a black moth not found before? **The black moth probably came from a line of light-colored moths. The dark color is a result of a frequent mutation in a gene. Black moths were not observed by naturalists when trees were covered with lichens because they were quickly seen and eaten by birds.**
2. Explain how the peppered moth example fits with three factors that Darwin said govern natural selection. **1. Color variations were found among the peppered moth. 2. The black color was better suited to the changing environment than the light color. 3. Over time black moths made up more and more of the population.**
3. Explain how the peppered moth example also fits with the model of punctuated equilibrium. **The color change occurred in 100 years, which is a very short time in relation to evolution. The change in the moth occurred very quickly as it responded to a rapid change in the environment.**

25

TYING TO PREVIOUS KNOWLEDGE: Reinforce the differences between the terms *hypothesis* and *theory*. Explain that a theory is much more than a hypothesis and is an accepted explanation based on a large number of tests.

OBJECTIVES AND SCIENCE WORDS: Have students review the objectives and science words throughout the chapter as they study each section.

2 TEACH

Key Concepts are highlighted.

CONCEPT DEVELOPMENT

▶ Emphasize that Darwin's theory of evolution is central to our present understanding of life science and provides many predictive insights into unsolved problems.

▶ It is interesting that Darwin had in his possession Mendel's report of his experiments with peas. If Darwin had understood Mendel's work, both men's lives might have been changed. Mendel would have certainly become more famous during his lifetime. Darwin would have had the evidence he needed to make his theory even more convincing.

▶ Discuss with students why evolution does not occur within individual organisms, but is a property of populations.

▶ **How are genetics and the theory of evolution related?** *Genetics provides the mechanism for evolution to work in a population. If traits were not inherited, there could be no evolution as we understand it.*

VideoDisc

STVS: Resistance to Pesticides in Cockroaches, Disc 6, Side 2

STVS: Aphid Mimics, Disc 6, Side 1

REVEALING MISCONCEPTIONS

▶ Ask students to list the inadequacies of Lamarck's theory. Students may agree subconsciously with Lamarck until they have critically examined the failures of his theory.

CHECK FOR UNDERSTANDING

Have students discuss the following question. **Why was Darwin impressed with the ability of humans to produce new breeds of plants and animals through artificial selection?** *He saw this as an analogy to the mechanism for evolution.*

RETEACH

Have students list the similarities between artificial selection and natural selection.

EXTENSION

For students who have mastered this section, use the **Reinforcement** and **Enrichment** masters or other OPTIONS provided.

Darwin collected many plants, animals, and fossils. He was amazed by the unique plants and animals he found in the Galapagos Islands. The Galapagos Islands are located off the coast of Ecuador in South America. He observed giant cactus trees, 14 species of very similar finches, and huge land tortoises. Darwin was very interested in the finches. He wondered how so many different, but closely related, species of finches could live in areas so near each other.

For 20 years after the voyage Darwin continued studying his collections. He thought about his observations and conducted further studies. He collected evidence of variations among species by breeding pigeons for racing. He also studied breeds of dogs, and other animals and varieties of flowers. Darwin knew that people used artificial selection in breeding plants and animals. They chose parents that had traits they wanted in the offspring.

Darwin concluded that individuals with traits most favorable for a specific environment survived and passed on these traits to their offspring. This is the theory of
3 evolution by **natural selection.** The factors Darwin identified that govern natural selection are:

1. Organisms produce more offspring than can survive.
2. Variations are found among individuals of a species.
3. Some variations enable members of a population to survive and reproduce better than others.
4. Over time, offspring of individuals with helpful variations make up more and more of a population.

Figure 6-4. A map of Darwin's voyage, and some of the unique organisms he found in the Galapagos Islands: (a) a tree cactus, (b) marine iguana, (c) the blue-footed booby, and (d) a Hood Island saddleback tortoise.

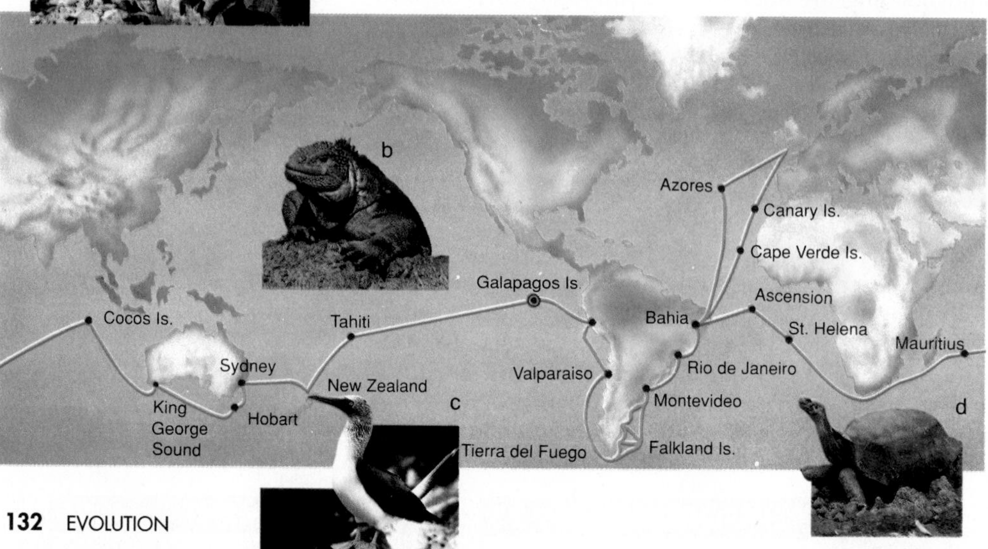

OPTIONS

ENRICHMENT

▶ Show the film *Evolution: The Four Billion Year Legacy* (Films Incorporated), which is narrated by Carl Sagan.

▶ Have students make a collection of butterflies or other insects. What variations do they see among members of the same species? Are these variations adaptive advantages?

PROGRAM RESOURCES

Use **Laboratory Manual,** page 37, Overproduction.

PROBLEM SOLVING

Why Isn't Earth Covered with Pumpkins and Pike?

Assume that there are 70 seeds in one pumpkin and that this is the typical number for the species. The 70 seeds are planted and each seed grows into a plant that produces two pumpkins. The first year 70 seeds are planted. The number of seeds produced in three years can be calculated by multiplying the number of seeds times two pumpkins for each plant times 70 seeds in each pumpkin:

Year 1: 70 x 2 x 70 = 9800
Year 2: 9800 x 2 x 70 = 1 372 000
Year 3: 1 372 000 x 2 x 70
= 192 080 000 seeds

The largest possible number of offspring produced by one individual is known as the biotic potential of a species.

If the ovaries of a pike contain 42 000 eggs, all the eggs are fertilized and hatched, all the young survive to reproduce, and one-half of the young are females, how many pike would there be after two more generations?

Think Critically: Why is the maximum rate of biotic potential never reached?

Natural selection results in organisms with traits best suited to their environments. This theory has also been called "survival of the fittest."

Darwin wrote a book describing his theory of evolution by natural selection. His book *On the Origin of Species by Means of Natural Selection* was published in 1859. Although minor changes have been made to Darwin's theory as new information has been gathered, his theory is one of the most important concepts in the study of life science.

In Your JOURNAL

In your Journal, describe how the work of Alfred Wallace interacted with that of Charles Darwin.

Adaptation and Variation

One of the points in Darwin's theory of evolution is that variations are found among individuals of a species. What are variations? A **variation** is an appearance of an inherited trait that makes an individual different from other members of the same species. Variations can be small, such as differences in the number of petals of a

What is a variation?

PROBLEM SOLVING

21 000 females × 42 000 eggs = 882 000 000 offspring in the first generation. 441 000 000 females × 42 000 eggs = 18 522 000 000 000 offspring in the second generation.

Think Critically: The maximum rate of biotic potential for the pike is never reached because all of the eggs do not hatch and many of the young are eaten by other organisms or die from environmental effects. All of the pumpkin seeds do not sprout. There is not enough food, water, space, and other necessities for the number that grow to survive.

CONCEPT DEVELOPMENT

▶ **Demonstration:** Show students photographs of different organisms. Let the students list the adaptations they think were selected.

In Your JOURNAL

Darwin and Wallace came to the same conclusion about natural selection. Wallace worked in South America amd Malaysia. Darwin presented the work of both in 1858.

MINI QUIZ

Use the Mini Quiz to check students' recall of chapter content.

1. **A group of organisms whose members look alike and successfully reproduce among themselves is a(n) _____ .** *species*
2. **Lamarck's theory is often called the theory of _____ .** *acquired characteristics*
3. **Darwin concluded that individuals with traits most favorable for a specific environment survive and pass these traits on to their offspring. This is the theory of _____ .** *natural selection*
4. **A(n) _____ is a group of organisms of one species that live in an area.** *population*

PROGRAM RESOURCES

From the **Teacher Resource Package** use:

Concept Mapping, pages 17-18.

Transparency Masters, pages 17-18, Horse Evolution.

Use **Color Transparency** 9, Horse Evolution.

Use **Laboratory Manual,** page 41, Variation.

MINI-Lab

Materials: paper and pencil

Answer: Accept any logical progression of ancestor shapes from the most simple to the most complex. There may be several branches involved in the progression. Students may begin with the vertical line as an ancestor or the plus sign. Using the plus will show more possible variations and progression.

MINI-Lab
ASSESSMENT

Performance: To further assess students' understanding of evolution, have them make up and explain another hypothetical evolutionary schema using whatever shapes or materials they choose.

CONCEPT DEVELOPMENT

▶ Use Figure 6-6 to help you explain the theories of gradualism and punctuated equilibrium to students.

▶ Explain that rapid evolution by punctuated equilibrium can occur several ways. It can occur if a small population becomes isolated from the main part of a population. Any genetic changes spread quickly through small populations. It can also occur when a group of organisms migrates to a new environment.

MINI-Lab

How does evolution occur?

Below are diagrams representing different species that came from the same ancestor. Determine which is the ancestor and then put the species in order, from simplest to most complex. Use one change in structure at a time. Each structural change is a variation and a new species forms. How does this help illustrate how evolution occurs?

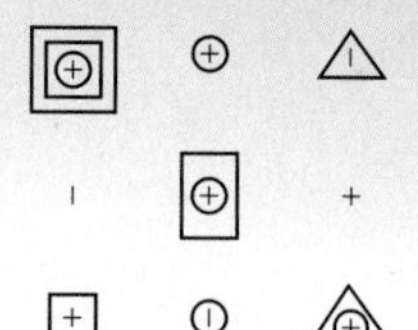

flower, or large, such as an albino deer or a fruit without seeds. Variations are important in populations of organisms. A **population** is a group of organisms of one species 4
that live in an area. If enough variations spread through a population as it produces new offspring, a new species may evolve from the existing species. Evolution of a new species probably takes many generations.

Some variations are more helpful than others. An adaptation is any variation that makes an organism better suited to its environment. The variations that result in adaptation can be in an organism's color, shape, behavior, or chemical makeup. Camouflage is a protective adaptation that lets an organism blend into its environment. An organism whose color or shape provides camouflage is more likely to survive and reproduce. Organisms that can't be easily seen are more likely to survive and reproduce. These types of variations result from mutations, changes in an organism's DNA. Mutations are the source of variation among organisms.

Several other factors bring about evolution. The movement of individuals into or out of a population brings in new genes and variations. Have you ever had an exchange student come to your school? The student probably brought new ideas, maybe a new style of dress, and even a new language. When new individuals come into an existing population, they can bring in new genes and variations in much the same way. The isolation of some individuals from others by geography and changes in climate can also result in evolutionary change. Each of these factors affects how fast evolution occurs.

Figure 6-5. Variation causes some organisms to blend into their environment. Variations among kittens often occur in the same litter.

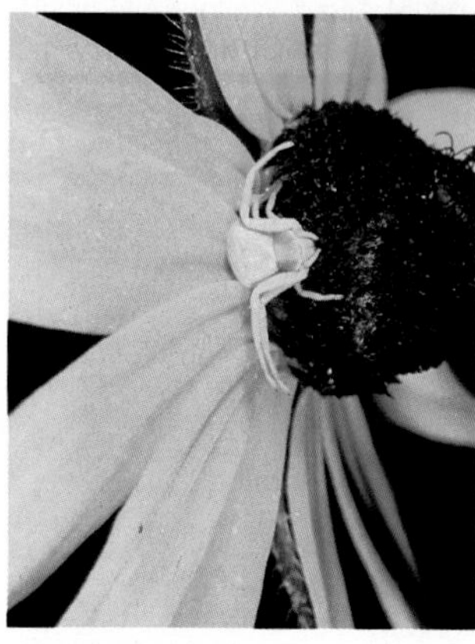

OPTIONS

ASSESSMENT—ORAL

▶ **Arctic foxes and hares have white coats in the winter and dark coats in the summer? How is this an advantage to both?** *Just as with the newspaper example, the coat colors protect the animals in their environments.*

▶ **Describe the major distinctions between the methods used by Lamarck and Darwin to make their theories.** *Lamarck made a hypothesis, but did little testing. His decision was made without much research. Darwin's hypotheses were tested for years by experiments and observations before he ever reached his decision about evolution.*

How Fast Does Evolution Occur?

How fast does evolution occur? Scientists are debating that question. Most scientists hypothesize that evolution occurs very slowly, perhaps taking tens or hundreds of millions of years. Other scientists hypothesize that evolution may occur very quickly, perhaps in a million years.

Darwin hypothesized that the rate of evolution was steady, slow, and continuous. The model that describes evolution as a slow change of one species to another new species is known as **gradualism.** In this theory there should be intermediate forms of all species.

Another model, the **punctuated equilibrium** model, shows that rapid evolution of species comes about by the mutation of a few genes. This is a steplike pattern of evolution. These mutations produce large changes in body form in a short period of time. The fossil record gives examples of this type of evolution.

What is the rate of evolution? As you study the evidence for evolution, you will see that evidence supports a combination of the two models.

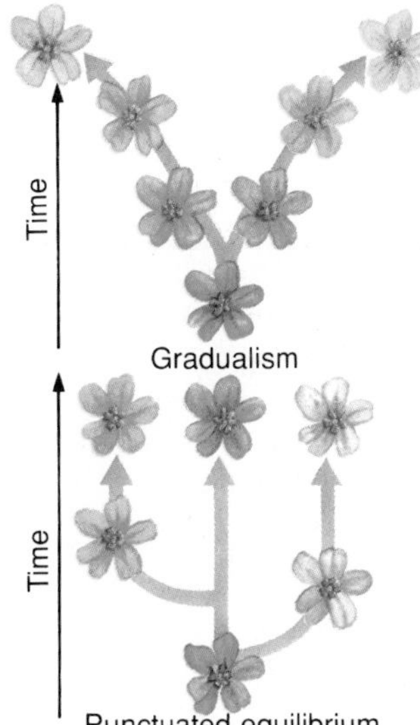

Figure 6-6. Evolution can occur very slowly, as in gradualism, or very rapidly, as in punctuated equilibrium.

SECTION REVIEW

1. Compare Lamarck's theory of evolution with Darwin's.
2. How are variations important in a population?
3. Explain how the gradualism model of evolution differs from the punctuated equilibrium model.
4. **Apply:** Which of the following variations would be beneficial to an animal living in the Arctic; thick fur, large size, large ears? Why?
5. **Connect to Earth Science:** What geologic feature do the Galapagos, Hawaiian and Canary Islands have in common?

Classifying

Skill Builder

Classify the following variations as being a part of an organism's shape, color, chemical makeup, or behavior.

1. Two species mate and don't produce offspring.
2. One species has large forelimbs and another species has short forelimbs.
3. Two species of birds build nests in different places.

If you need help, refer to Classifying in the **Skill Handbook** on page 680.

3 CLOSE

▶ Ask students to explain what is meant by the term *survival of the fittest.*
▶ Discuss how natural selection might occur in a population of water birds if there were a drying trend on Earth.
▶ Ask questions 1-3 and the **Apply** Question in the Section Review.

SECTION REVIEW ANSWERS

1. Lamarck's theory was similar to the one proposed by Darwin and Wallace in that organisms with the traits most suited to the environment survive. However, Lamarck thought traits acquired by adults were passed to offspring. Darwin and Wallace thought that only certain traits that were inherited were passed to offspring.
2. Some variations are more advantageous for survival and reproduction in the environment than others. Those variations that are best suited are the adaptations an organism makes to its environment.
3. Gradualism supports the hypothesis that an ancestral species slowly evolves to become another species. The punctuated equilibrium model explains that species tend to remain constant for millions of years and, suddenly, in the span of one or a few million years, become two or more species.
4. Apply: Thick fur would be more beneficial because of the cold winters that the animals must endure.
5. Connect to Earth Science: They are all volcanic in formation.

1. chemical make up
2. shape
3. behavioral

Skillbuilder
ASSESSMENT
Performance: To assess students' abilities to classify organisms by their variations, ask them to classify two species of birds that eat different types of food. (chemical)

▶ **How would the acrobatic ability of a child produced by two tight-rope walkers differ according to Lamarck's theory and Darwin's theory?** *According to Lamarck, the child should be better than others at tight-rope walking. According to Darwin, the child would only be better than others due to practice because there is no inherited factor for walking a tight-rope.*

PROGRAM RESOURCES

From the **Teacher Resource Package** use:
Activity Worksheets, page 56, Mini-Lab: How does evolution occur?

SECTION 6-2

PREPARATION

SECTION BACKGROUND

▶ Paleontology is the science of studying fossils.

▶ Usually, only the hard parts of a few organisms that were once alive are ever fossilized.

▶ The fossil record is incomplete. Not all organisms fossilize.

PREPLANNING

▶ Collect various fossils or photographs of fossils to show students during the discussion of this section.

1 MOTIVATE

▶ Compare the hand of a human, the wing of a bat, and the flipper of a sea lion. What relationship do these similar structures have?

FLEX Your Brain

Use the Flex Your Brain activity to have students explore FOSSILS.

Flex Your Brain

ASSESSMENT

Performance: Use the Flex Your Brain activity to reinforce critical thinking and problem solving skills. In Step 2 (see page 18), students may need to be led to what they know about dinosaur bones, shells, and plants.

6-2 Evidence for Evolution

New Science Words

fossils
sedimentary rock
relative dating
radioactive elements
homologous
vestigial structure
embryology

Objectives

- Describe the importance of fossils as evidence of evolution.
- Explain how relative and radioactive dating are used to date the fossil record.
- Give examples of five types of evidence for evolution.

Fossil Evidence

On a hot day in July 1975, in North Central Texas, a young couple was strolling along the breezy shores of Lake Lavon. The couple came across some odd-looking rocks projecting from the muddy shore. They noticed the rocks seemed different from the surrounding limestone rocks. They took a few of the rocks to Dr. Harold Laughlin, a local scientist specializing in reptiles and amphibians. Dr. Laughlin recognized the rocks as pieces of the skull of a fossil mosasaur (MA sah sawr), an extinct lizard that lived in salt water.

Figure 6-7. Uncovering fossils requires careful work.

OPTIONS

Meeting Different Ability Levels

For Section 6-2, use the following **Teacher Resource Masters** depending upon individual students' needs.

◆ **Study Guide Master** for all students.

● **Reinforcement Master** for students of average and above average ability levels.

▲ **Enrichment Master** for above average students.

Additional Teacher Resource Package masters are listed in the OPTIONS box throughout the section. The additional masters are appropriate for all students.

◆ STUDY GUIDE 26

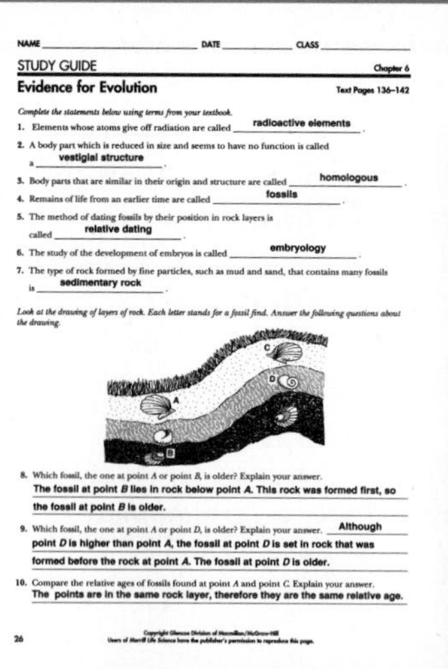

NAME ________ DATE ________ CLASS ________

STUDY GUIDE — Chapter 6

Evidence for Evolution — Text Pages 136–142

Complete the statements below using terms from your textbook.

1. Elements whose atoms give off radiation are called **radioactive elements**.
2. A body part which is reduced in size and seems to have no function is called a **vestigial structure**.
3. Body parts that are similar in their origin and structure are called **homologous**.
4. Remains of life from an earlier time are called **fossils**.
5. The method of dating fossils by their position in rock layers is called **relative dating**.
6. The study of the development of embryos is called **embryology**.
7. The type of rock formed by fine particles, such as mud and sand, that contains many fossils is **sedimentary rock**.

Look at the drawing of layers of rock. Each letter stands for a fossil find. Answer the following questions about the drawing.

8. Which fossil, the one at point *A* or point *B*, is older? Explain your answer. **The fossil at point *B* lies in rock below point *A*. This rock was formed first, so the fossil at point *B* is older.**
9. Which fossil, the one at point *A* or point *D*, is older? Explain your answer. **Although point *D* is higher than point *A*, the fossil at point *D* is set in rock that was formed before the rock at point *A*. The fossil at point *D* is older.**
10. Compare the relative ages of fossils found at point *A* and point *C*. Explain your answer. **The points are in the same rock layer, therefore they are the same relative age.**

26

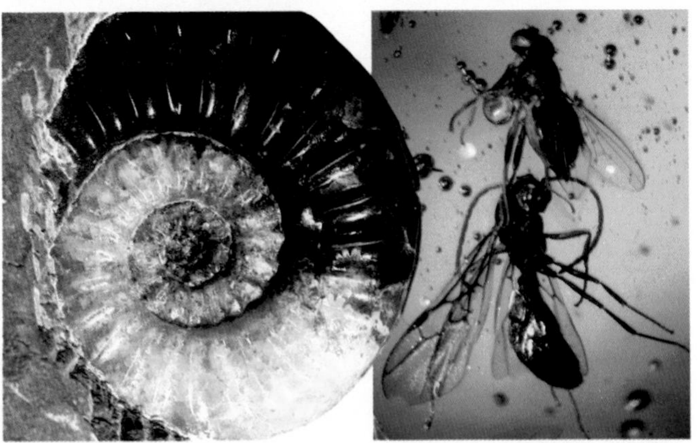

Figure 6-8. Three examples of fossils are shown: (a) an imprint made by a leaf, (b) a cast of an ancient mollusk, and (c) an insect caught in plant resin.

Dr. Laughlin contacted Professor Slaughter, a scientist who studies fossils, at Southern Methodist University. With the help of students, Professor Slaughter and Dr. Laughlin returned to the site and carefully dug up the rest of the fossil mosasaur. It took one of Slaughter's graduate students almost a year to study, preserve, and reassemble the 1.75-meter skull of the 15-meter-long fossil mosasaur. This find provides evidence that about 120 million years ago, the North Texas area—now more than 500 kilometers from the Gulf of Mexico—was covered by a shallow sea.

The most abundant evidence for evolution comes from fossils like those found on the shore of Lake Lavon in Texas. **Fossils** are any remains of life from an earlier time. ❶ Examples of fossils include:

1. the imprint of a leaf, animal, or feather in rock;
2. a cast made of minerals that filled in the hollows of an animal track or mollusk shell;
3. a piece of wood or bone replaced by minerals;
4. an animal or plant frozen in ice; and
5. an insect or reptile trapped in plant resin.

Sedimentary rock contains most fossils. **Sedimentary rock** is a rock type formed by mud, sand, or other fine particles that settle out of a liquid. ❷ Limestone, sandstone, and shale are all examples of sedimentary rock. Fossils are found more often in limestone than in any other kind of sedimentary rock.

MINI-Lab

How are fossils made?

Pour a small amount of plaster of Paris into a small paper cup or other small paper container. Press a small object such as a seashell, key, or leaf into the mixture. Carefully lift the object out of the container. Let the plaster dry for one to two days. Tear the paper away from the container. What type of fossil have you made? *Infer* whether a plant or an animal would be more likely to make this type of fossil. Explain your choice.

● REINFORCEMENT 26

NAME ______ DATE ______ CLASS ______

REINFORCEMENT — Chapter 6

Evidence for Evolution — Text Pages 136–142

Study the sketches below, and in the lines provided, tell how each fossil was formed.

a. wood replaced by minerals
b. imprint of a leaf in a rock
c. footprint left in sediment
d. insect in hardened plant resin
e. stone casts of mollusk shells
f. animal frozen in ice

Answer the following questions.

1. In which type of rock are most fossils found? **Most fossils are found in sedimentary rock.**
2. What two methods are used to determine the age of a rock or fossil? **The age of a rock or fossil can be measured by relative dating and radioactive dating.**
3. Why is the fossil record not complete? **The fossil record is incomplete because every living thing does not or cannot become fossilized.**
4. List other evidences of evolution.
   a. embryology
   b. vestigial structures
   c. homologous structures
   d. similarities in DNA

26

▲ ENRICHMENT 26

NAME ______ DATE ______ CLASS ______

ENRICHMENT — Chapter 6

Evidence for Evolution — Text Pages 136–142

Using household items or readily available supplies, create your own casts, molds, traces, and embedded fossils. Use your text and other resources to find out about each type of fossil. Use the table at the bottom of this page to record your methods and observations. Then bring the "fossils" you made to school to show your class.

TE NOTE: Some students' projects may include making a mold in damp sand or clay, filling the mold with plaster to make a cast, making a trace in damp sand or clay, and freezing an object in water.

Materials:
The following may be used as a suggestion of possible supplies. Access to a refrigerator and freezer may be helpful.

- sand
- clay
- plaster of paris
- water
- plastic spoons, bowls, and other containers
- objects to "fossilize" such as shells and bone

Type of fossil	Object "fossilized"	Materials used	Sketch of fossil
Mold		Answers will vary.	
Cast			
Trace			
Embedded fossil			

26

TYING TO PREVIOUS KNOWLEDGE: Review the structure and function of DNA with students while they study the evidence for evolution.

Field Trip: If your school is in a community that has a dry, rocky stream bed, students may be able to find rocks with fossil imprints there. Backyard gardens also yield rocks with imprints.

2 TEACH

Key Concepts are highlighted.

CONCEPT DEVELOPMENT

▶ Point out the difference between direct and indirect evidence. Compare the types of evidence with that presented in a court trial. "Enough" indirect evidence is sufficient to obtain a conviction in many cases.

▶ Industrial melanism, pesticide resistance in insects, antibiotic resistance in bacteria, and observed differences in salamanders and birds are all examples of direct evidence for evolution.

MINI-Lab

Time Allotment: 20 minutes

Alternate Materials: Clay can be substituted for the plaster of Paris.

Answer: An imprint fossil has been made. To make a cast fossil, students would need to fill the imprint with another material, then remove it to create a likeness, or cast, of the original object. Animal with hard body parts; plants, having few hard parts, have left fewer cast fossils.

MINI-Lab ASSESSMENT

Performance: To further assess students' understanding of fossils, see USING LAB SKILLS, question 11, on p. 152.

VideoDisc

STVS: Fossil CAT Scan, Disc 3, Side 2

STVS: Carbon-14 Dating, Disc 2, Side 2

CONCEPT DEVELOPMENT

▶ Ask students why the transitional forms of organisms in the fossil record are evidence of evolution. They show changes that occurred in the same kind of organism over a long period of time.

STUDENT TEXT QUESTION

▶ Page 138, paragraph 1: **How was that date obtained?** *The date of 120 million years was obtained by relative dating. Radioactive dating is difficult with limestone.*

CROSS CURRICULUM

▶ **Geology:** Ask students to identify some of the geological concepts that are used in this section. Examples include sedimentary rocks, deposition, geological time scale, fossils, and so on.

Cooperative Learning: Place students in Paired Partners or Science Investigation cooperative groups. Have each group make a model showing relative dating using clear plastic or glass containers and sand, dirt, salt, corn meal, or other materials. Each group should also place several "fossils" in the model and explain their relative ages. Have each group present their model to the class.

▶ Have students discuss what they observe in Figure 6-9 in terms of position of fossils and their apparent complexity.

The Fossil Record

You learned that the mosasaur fossil found in Texas was 120 million years old. How was that date obtained? Scientists have divided Earth's history into eras and periods. These divisions make up the geologic time scale as shown in Table 6-1. Unique rock layers and fossils give information about the geology, weather, and life forms of each time period. There are two basic methods for reading the record of past life. When these methods are used together, accurate estimates of the ages of certain rocks and fossils are made.

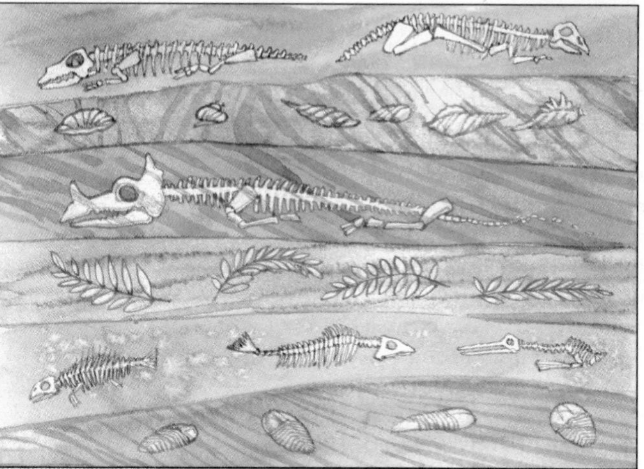

Figure 6-9. Fossils found in lower layers of sedimentary rock are usually older than fossils in upper layers. 3

One method often used to determine the approximate age of a rock layer, or fossils within the layer, is to look at where the particular rock layer is. In undisturbed areas, older rock layers lie below successively younger rock layers. Fossils found in the lower layers of rock are older than those in upper layers. This method of dating fossils is known as **relative dating.** Relative dating can only estimate the age of a fossil.

How are fossils dated using radioactive elements?

A method used to give a more accurate age to a rock layer or fossil is radioactive dating using radioactive elements. **Radioactive elements** are elements whose atoms give off radiation, a form of atomic energy. Uranium and a radioactive form of carbon are used in radioactive dating. Radioactive elements change to more stable products as they give off radiation. The radiation is given off at a constant rate, and the rate is different for each element. Scientists can measure how much of a radioactive element has changed. They can accurately determine the age of the rock by comparing the amount of stable product with the amount of radioactive element still present. For example, the radioactive element uranium changes to lead as it ages. Scientists can determine how old a fossil in a rock sample is by measuring the amounts of uranium and lead in the sample. The more lead there is, the older the fossil.

OPTIONS

ENRICHMENT

▶ Have students use the library to find out how dinosaurs evolved and what might have caused their extinction.

▶ **Field trip:** Visit a natural history museum to observe the fossil collections and dioramas on display.

PROGRAM RESOURCES

From the **Teacher Resource Package** use:

Activity Worksheets, page 57, Mini-Lab: How are fossils made?

Activity Worksheets, page 5, Flex Your Brain.

Science Integration Activities 6, Geologic Time

Science Integration Activities is found in Merrill Life Science Teacher Resource Manual

Table 6-1

GEOLOGIC TIME SCALE				
ERA	PERIOD	MILLION YEARS AGO	MAJOR EVOLUTIONARY EVENTS	REPRESENTATIVE ORGANISMS
Cenozoic	Quaternary		Humans evolve	
		5		
	Tertiary		First placental mammals	
		65		
Mesozoic	Cretaceous		Flowering plants dominant	
		144		
	Jurassic		First birds First mammals First flowering plants	
		213		
	Triassic		First dinosaurs	
		248		
Paleozoic	Permian		Cone-bearing plants dominant	
		286	First reptiles	
	Carboniferous	320	Great coal deposits form First seed plants	
		360		
	Devonian		First Amphibians	
		408	First land plants	
	Silurian		First jawed fish	
		438		
	Ordovician		Algae dominant First vertebrates	
		505		
	Cambrian		Simple invertebrates	
		544		
Precambrian			Life diversifies Eukaryotes Prokaryotes Life evolves	
		3500		

CHECK FOR UNDERSTANDING

Ask questions that allow students to differentiate between the types of fossils and evidence for evolution.

RETEACH

Demonstration: Present examples of as many kinds of fossils as possible. It should be fairly easy to show imprints, petrified wood, casts, and molds. You may be able to purchase amber with insects from a museum store or biological supply company. You may simulate by freezing insects in ice.

EXTENSION

For students who have mastered this section, use the **Reinforcement** and **Enrichment** masters or other OPTIONS provided.

CONCEPT DEVELOPMENT

▶ Throughout the plant and animal chapters of this book, geologic times are discussed in relation to the origin of organisms. Questions in the book do not require memorization of the eras. Some students may have difficulty comprehending the great spans of time. Use a time line clock to put the time in perspective. If a 12-hour period represents the total length of time since Earth was formed, humans appeared in the last fraction of a second. If a 4.5-meter-long time line is used, humans appeared in the last 0.1 cm.

INQUIRY QUESTIONS

▶ **Could relative dating be used to date fossils in rock layers that have been uplifted or disturbed in other ways?** *No, radioactive dating would have to be used to accurately date the fossil.*

PROGRAM RESOURCES

From the **Teacher Resource Package** use:

Transparency Masters, pages 19-20, The Geologic Time Scale.

Use **Color Transparency** 10, The Geologic Time Scale.

TECHNOLOGY

Extension: Have students brainstorm natural objects that, with minor changes, could have been used as utensils or tools by early humans (like eggshell water containers).

References: Bower, B. "Eggshells Help Date Ancient Sites" *Science News*, April 7, 1990, p. 215.
Marshall, Eliot. "Paleoanthropology Gets Physical." *Science*, Feb. 1990, pp. 198-201.

Think Critically: AAR applies to a period currently under debate. It may sort out Neanderthal and Cro-Magnon site dates and help resolve the evolutionary relationships among these forms and modern *Homo sapiens*.

REVEALING MISCONCEPTIONS

▶ Have students describe where they have observed fossils before. Answers may include in the stones on the sides of buildings, in museums, beside streams, or on the ground. Explain that fossils are not rare and are found in most places on Earth.

CONCEPT DEVELOPMENT

▶ Have students study The Geologic Time Scale to answer the following questions:

During which era did simple invertebrates appear? *the Cambrian Era*

How many million years ago did the first placental mammals appear? *65 million years ago*

During which period did the first amphibians evolve? *the Devonian period*

Ask other similar questions.

Science and MATH

3100 million years.

▶ Fossils, homologous structures, vestigial structures, embryology, and DNA studies are all examples of indirect evidence for evolution.

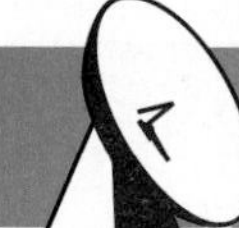

TECHNOLOGY

An Ostrich Egg Timer

A dating technique called amino acid racemization (AAR), used to date the age of fossil objects, may turn ordinary eggshells into geologic clocks.

AAR uses amino acids left in materials produced by living things to figure out how old the materials are. Over time, amino acids change into a slightly different form of their original form. By measuring the ratio of the original amino acids to the changed amino acids in an eggshell, the age of the shell can be determined. The technique works better in cold climates. In warm climates the rate of amino acid change is affected by temperature and by moisture.

Scientists are excited by AAR because it can be used to date objects that can't be dated by radioactive dating, such as eggshells, bits of rock, and tooth enamel. Ostrich eggshells are very common in many archeological sites. Early humans ate the eggs, and then used the shells as water containers. Using AAR to date the eggshells helps scientists determine the age of the site.

AAR is an important technique because it allows scientists to date back 200 000 years at warm-weather sites, and one million years at cold-weather sites. Current radioactive carbon dating methods can't date back this far.

Think Critically: How can AAR dating help us understand human evolution?

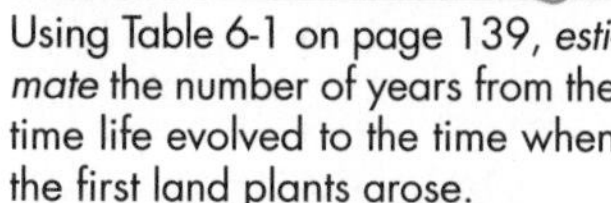

Science and MATH

Using Table 6-1 on page 139, *estimate* the number of years from the time life evolved to the time when the first land plants arose.

4 Fossils are a record of organisms that lived in the past. But the fossil record is incomplete, much like a book with some pages missing. Because every living thing doesn't or can't become fossilized, the record will never be complete. By looking at fossils, scientists have determined that many simpler forms of life existed earlier in Earth's history, and more complex forms of life appeared later. The oldest fossil bacteria appeared 3.8 billion years ago. Simple plants did not appear until the Ordovician period, about 440 million years ago. The first mammals and birds did not appear until the Jurassic period, about 190 million years ago. The fossil record gives scientists convincing evidence that living things evolved. There are still other types of evidence that support the theory of evolution.

OPTIONS

ASSESSMENT—ORAL

▶ **What is the evidence that North Texas was once covered by a shallow sea?** *Fossils that lived in salt water have been found in the rock layers of the area.*

▶ **How does sedimentary rock differ from ingenuous and metamorphic rock?** *Sedimentary rock is deposited without the great heat associated with igneous and metamorphic rock deposition.*

Other Evidence for Evolution

Besides fossils, what other evidence is there for evolution? Scientists have found more evidence by looking at similarities in the chemical makeup, the development, and the structure among organisms. You know that the functions of your arm, a dolphin's flipper, a bat's wing, and a bird's wing are all very different. Yet, as you can see in Figure 6-10, they are all made up of the same bones. Each has about the same number of bones, muscles, and blood vessels. Each of these limbs developed from similar tissues in the embryo. Body parts that are similar in origin and structure are called **homologous.** Homologous structures give evidence that two or more species share common ancestors. 5

Vestigial structures also give evidence for evolution. A **vestigial structure** is a body part that is reduced in size and doesn't seem to have a function. Examples of vestigial organs in humans are the appendix and the muscles that move the ear. Scientists think vestigial structures are parts that once functioned in an ancestor.

If two species of organisms have a similar type of protein, what are the chances that they share a common ancestor?

Figure 6-10. A bird wing, bat wing, dolphin flipper, and human arm are homologous. Each has about the same number of bones, muscles, and blood vessels.

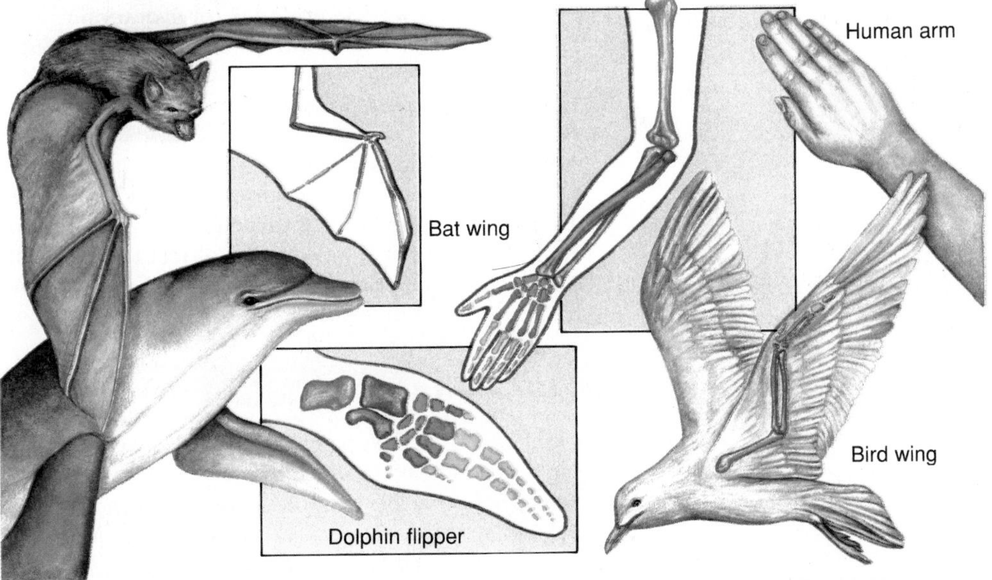

CONCEPT DEVELOPMENT

▶ Point out that transitional forms may be seen by studying homologous structures. An example is the ancestral horse which has slowly evolved from a dog-like form to the present horse.

MINI QUIZ

Use the Mini Quiz to check students' recall of chapter content.

1. **Any remains of life from an earlier time are called _____ .** *fossils*
2. **Most fossils are contained in _____ .** *sedimentary rock*
3. **Elements whose atoms give off radiation are _____ .** *radioactive elements*
4. **Fossils are a record of organisms that lived in the _____ .** *past*
5. **Body parts that are similar in origin and structure are called _____ .** *homologous*

Connect to... Chemistry

Note: Students should be able to find answers for most Connect to... features in basic library references or infer answers from text material.

Answer: Chances are good that they share a common ancestor.

ASSESSMENT—ORAL

▶ **Judging from what you know about conditions on Earth, which type of fossil is probably rarest?** *Animals and plants frozen in ice are rare because the conditions do not usually change abruptly from warm to cold.*

▶ **If a particular sedimentary rock layer were dated at 1.5 million years old, what statement could be made about the fossils in the layer?** *The fossils are also about the same age.*

3 CLOSE

▶Encourage students to bring fossils that they may have at home. Classify these according to type.
▶Ask questions 1-3 and the **Apply** Question in the Section Review.

SECTION REVIEW ANSWERS

1. Fossils of a wide variety of species appear to be related to one another in the fossil record.
2. Radioactive dating is used to give rocks and fossils dates by measuring the relative amounts of radioactive elements with their stable products.
3. fossil record; homologous structures; vestigial structures; genetics; embryology
4. Apply: shells, fish, leaves
5. Connect to Earth Science: sedimentary rock

Skill Builder

Jurassic; Cambrian (about 85 million years)

Skillbuilder
ASSESSMENT

Performance: Assess students' abilities to use tables by also having them examine humanity's place in the Geologic Time Scale table in relation to other organisms.

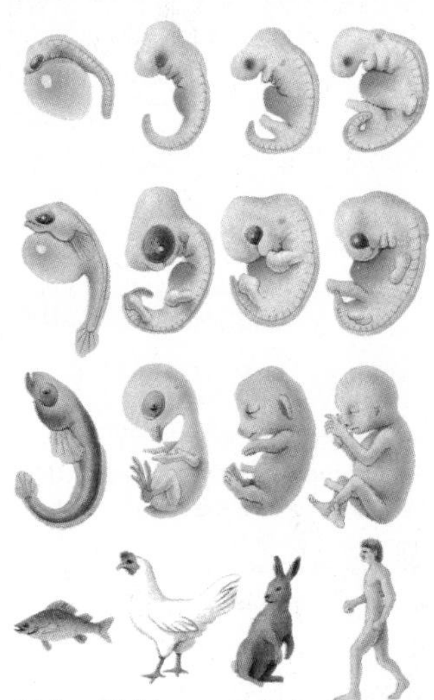

Figure 6-11. Similarities in the embryos of fish, chickens, rabbits, and humans show evidence of evolution.

The study of the development of embryos is called **embryology.** An embryo is an organism in its earliest stages of development. Compare the embryos of the organisms in Figure 6-11. Do you see similarities? In the early stages of development, the embryos of fish, reptiles, birds, and mammals have a tail and gills or gill slits. Fish keep their gills, but the other organisms lose them as their development continues. In humans, the tail disappears, but fish, birds, and lizards keep theirs. These similarities suggest an evolutionary relationship among all vertebrate species. This supports evidence from the fossil record that shows aquatic, gill-breathing organisms came before air-breathing land vertebrates.

DNA is the molecule that controls heredity. Scientists can also determine whether or not organisms are closely related by comparing their DNA. Organisms that are close relatives share more similarities in DNA with each other than with distant relatives. For example, by studying DNA, scientists have determined that dogs are the closest relatives of bears. You would probably not be surprised that gorillas and chimpanzees also have DNA that is very similar.

SECTION REVIEW

1. How do fossils provide evidence for evolution?
2. How is radioactive dating used to interpret the fossil record?
3. List five examples of evidence that support the theory of evolution.
4. **Apply:** Fossil leaves are found in the top layer, fossils of shells are found in the bottom layer, and fish bones are in the middle layer of three undisturbed beds of rock. List the fossil types from oldest to youngest.
5. **Connect to Earth Science:** What type of rock are the fossils in question 4 found in?

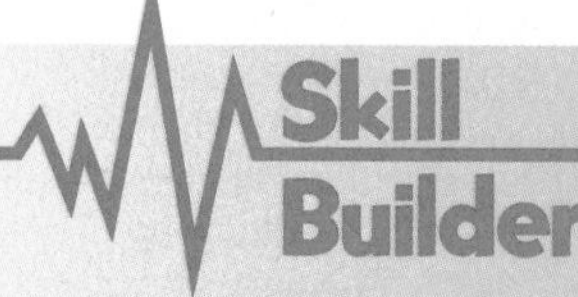

Skill Builder

Making and Using Tables

Use Table 6-1, The Geologic Time Scale, to answer the following questions. During which periods did the first mammals and flowering plants appear? Which was the longest period of the Paleozoic era? If you need help, refer to Making and Using Tables in the **Skill Handbook** on page 690.

OPTIONS

ENRICHMENT

▶ **Field Trip:** Visit a zoo to observe endangered species.
▶ **Guest Speaker:** Invite a zoo curator to speak to your class about efforts to save endangered species.

PROGRAM RESOURCES

From the **Teacher Resource Package** use:

Activity Worksheets, pages 50-51, 6-1 A Radioactive Dating Model.

ACTIVITY 6-1

A Radioactive Dating Model

Problem: *How can a radioactive element be used to determine a fossil's age?*

Materials

- 100 pennies
- cardboard box with lid
- graph paper
- pencil

Procedure

1. A rock or fossil may be dated by measuring the relative amount of a stable element with its radioactive parent element. As the rock ages, the amount of radioactive element become less and the amount of stable element increases. Examine the graph below to see the decrease of a radioactive element over time.
2. Make a data table like the one shown.
3. Put 100 pennies face up in the cardboard box and replace the lid.
4. Shake the coins in the box for 10 seconds.
5. Take off the lid and take out all coins that are face down.
6. Record the number of coins that you take out.
7. Repeat Steps 4 through 6 until all the coins have been removed.

Data and Observations

Sample Data

Trial #	0	1	2	3	4	5	6	7	8
Coins Left	100	57	30	18	12	7	5	2	1
# Removed	0	43	27	12	6	5	2	3	1

Analyze

1. What happens to the number of coins remaining after each trial?
2. Construct a graph of your results. Plot the number of coins remaining face up on the *y*-axis, and plot the trials on the *x*-axis. How does your graph compare with the graph below?
3. How does shaking the box represent the energy given off by radioactive elements when they become stable?

Conclude and Apply

4. How is this model similar to the decay of a radioactive element?
5. How is this model unlike the decay of a radioactive element?
6. Why do you think radioactive dating is considered more accurate than dates calculated from fossil beds?
7. Why are different radioactive elements used to date rocks and fossils?

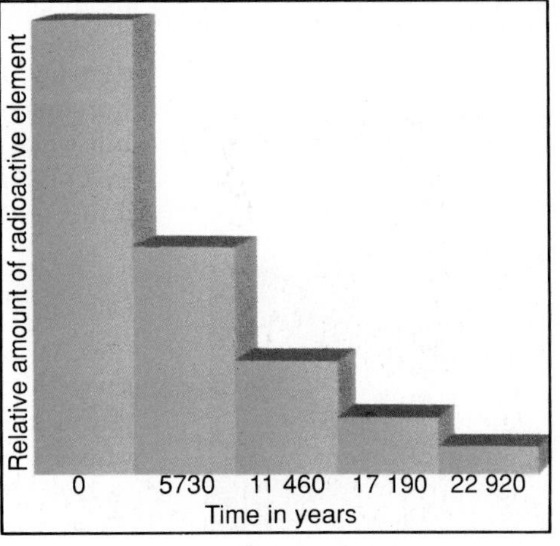

ACTIVITY 6-1

45 minutes

OBJECTIVE: Understand how radioactive dating works.

PROCESS SKILLS applied in this activity:
▶ **Measuring** in Procedure Step 2.
▶ **Formulating Models** in Procedure Steps 3 through 7.
▶ **Interpreting Data** in Analyze Questions 2 and 3.
▶ **Inferring** in Conclude and Apply Questions 4 through 7.

COOPERATIVE LEARNING Divide the class into Science Investigation cooperative groups.

TEACHING THE ACTIVITY

Troubleshooting: Students may need to be reminded how to make a graph. Have them refer to the Skill Handbook, page 691.

Activity ASSESSMENT

Performance: To further assess students' understanding of radioactive material, see USING LAB SKILLS, question 10, on p. 152.

ANSWERS TO QUESTIONS

1. The number decreases by about half.
2. This will be a graph similar to the one showing half-life in this activity.
3. Shaking the box requires energy and represents the energy given off when the elements change to become stable.
4. After a given period of time, about half the coins have changed because of the energy released.
5. The time for a given amount of a radioactive element to change to a stable element is exactly the same each time.
6. Radioactive dating is more accurate because the rate of decay is exact. Fossil beds may be formed at different rates in different time periods.
7. Not all rocks contain the same radioactive elements; also, not all radioactive elements have a decay period that can be used with every age of rock.

PREPARATION

SECTION BACKGROUND

▶ *The Endangered Species Handbook* is published by the Animal Welfare Institute, P.O. Box 3650, Washington, D.C., 20007.

1 MOTIVATE

▶ Ask students if a minnow should stop the building of a major dam that could supply water and electricity to thousands of people (snail darter story).

Connect to... Earth Science

Answer: Most dinosaurs became extinct during the Cretaceous period.

PROGRAM RESOURCES

From the **Teacher Resource Package** use:

Critical Thinking/Problem Solving, page 10, What is Happening to the Rain Forest?

Science and Society, page 10, Saving the Black Rhino.

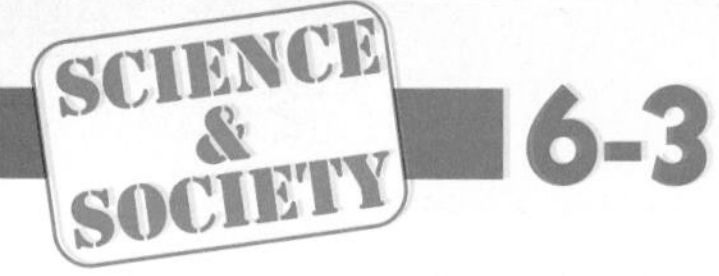

6-3 Plant and Animal Extinction

New Science Words

extinction
endangered species

Objectives

▶ Define extinction and identify some causes of extinction.
▶ Name two ways in which endangered species may be saved from extinction.

Connect to... Earth Science

Evidence that extinction is a natural event is found in the fossil record. Large scale extinction events, such as extinction of the dinosaurs, have taken place in the past. Find out when extinction of the dinosaurs took place.

Figure 6-12. Plant and animal habitats are being destroyed by development.

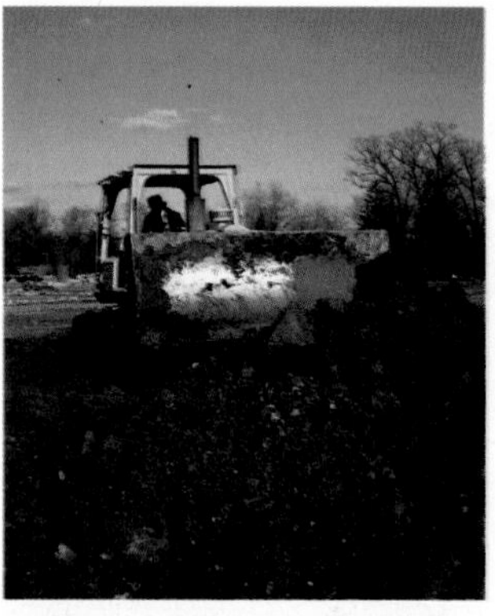

Causes of Extinction

By the time you finish reading this paragraph, one more animal or plant species may have joined the thousands that have become extinct. Some scientists estimate that one-fourth of all animal and plant species alive in the world today will be extinct in only 25 years! What's going on?

Extinction, the dying out of a species, is a natural event. Extinction occurs naturally because the environment is always changing. To survive, each species must adapt to it. In nature, only those species that can best adapt to the changing environment will escape extinction. Through the slow process of natural selection, those species that are genetically unable to change or adapt as the environment changes are naturally removed.

The rate of extinction has increased greatly due to destruction of natural habitats by humans. Not only have humans destroyed many species of plants and animals, but we also have destroyed habitats and food supplies that organisms depend on. With no home or food supply, a species will eventually die.

Ways to Prevent Extinction

How can extinction be prevented? One way is to preserve the remaining natural areas in which endangered plants and animals live. An **endangered species** is one whose numbers are so low that it is in danger of becoming extinct. By preserving areas that have not yet been destroyed, we can keep organisms in their natural homes

OPTIONS

Meeting Different Ability Levels

For Section 6-3, use the following **Teacher Resource Masters** depending upon individual students' needs.

◆ **Study Guide Master** for all students.

● **Reinforcement Master** for students of average and above average ability levels.

▲ **Enrichment Master** for above average students.

Additional Teacher Resource Package masters are listed in the OPTIONS box throughout the section. The additional masters are appropriate for all students.

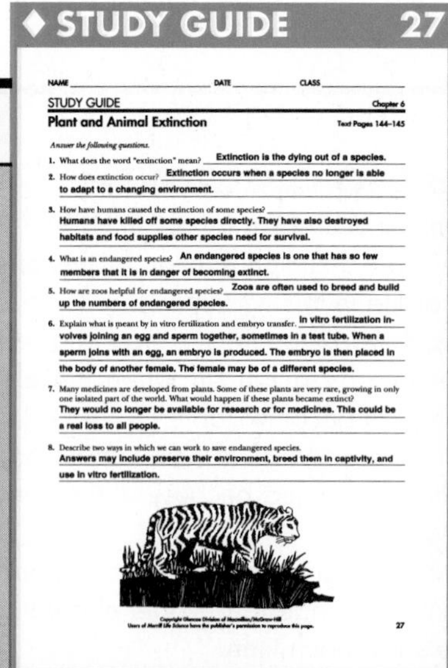

◆ STUDY GUIDE 27

NAME ________ DATE ________ CLASS ________

STUDY GUIDE — Chapter 6

Plant and Animal Extinction — Text Pages 144–145

Answer the following questions.

1. What does the word "extinction" mean? **Extinction is the dying out of a species.**
2. How does extinction occur? **Extinction occurs when a species no longer is able to adapt to a changing environment.**
3. How have humans caused the extinction of some species? **Humans have killed off some species directly. They have also destroyed habitats and food supplies other species need for survival.**
4. What is an endangered species? **An endangered species is one that has so few members that it is in danger of becoming extinct.**
5. How are zoos helpful for endangered species? **Zoos are often used to breed and build up the numbers of endangered species.**
6. Explain what is meant by in vitro fertilization and embryo transfer. **In vitro fertilization involves joining an egg and sperm together, sometimes in a test tube. When a sperm joins with an egg, an embryo is produced. The embryo is then placed in the body of another female. The female may be of a different species.**
7. Many medicines are developed from plants. Some of these plants are very rare, growing in only one isolated part of the world. What would happen if these plants became extinct? **They would no longer be available for research or for medicines. This could be a real loss to all people.**
8. Describe two ways in which we can work to save endangered species. **Answers may include preserve their environment, breed them in captivity, and use in vitro fertilization.**

27

and let them rebuild their populations. Organizations, such as the Nature Conservancy, are working hard to save what untouched natural areas remain.

What is an endangered species?

Another way to prevent extinction is to remove endangered animals from their natural habitats and breed them in captivity. After helping animals reproduce in a setting that provides food, shelter, and protection, researchers may then release them into the wild. Zoos have become very important in these types of programs.

Researchers are battling the extinction of Bengal tigers in this way. Only 3000 to 5000 of these tigers remain in the wild. This endangered species may be saved by a technique known as in vitro fertilization, joining an egg and sperm together in a test tube. An embryo results from the joining of an egg and sperm. Embryo transfer, the placing of the embryo in a female, then follows. If the procedure is successful, the embryo develops into a baby tiger. Zoos have been successful in breeding Bengal tigers by placing the embryo in a female Siberian tiger. Through surgery, the developed kittens were removed from the foster mother.

Figure 6-13. Bengal tigers are being bred in captivity to help save them from extinction.

It is hoped that in vitro fertilization can be used to help many other endangered animal species. By protecting the natural areas that are not yet destroyed and by breeding animals in captivity, preservationists are working to slow the current rate of extinction and preserve the species that remain.

SECTION REVIEW

1. Why is extinction not always a bad event?
2. Describe two ways to save endangered species.
3. **Connect to Earth Science:** Find out what hypotheses exist to explain extinction of the dinosaurs.

You Decide!

A lot of energy and money is put into programs that help us save endangered species from extinction. Are these species really worth our time and effort? Should we be concerned about saving an animal or plant that we'll never see?

● REINFORCEMENT 27

NAME ________ DATE ________ CLASS ________

REINFORCEMENT — Chapter 6

Plant and Animal Extinction — Text Pages 144–145

Did you know that almost all the organisms that ever lived on Earth are now extinct? Some scientists believe that fewer than one in 10 000 species that lived in the past are still alive today.

1. Does the above information fit with Darwin's theory of natural selection? Explain your answer. **Yes. Climates have changed often throughout the history of the world. In addition, local conditions have changed. Any organism that could not adapt, that was not naturally selected, would not be able to survive and reproduce. Any species that was unable to adapt would eventually become extinct.**
2. What natural events cause an animal or plant to become extinct? Other than nature, what else contributes to the extinction of plants and animals? **The environment changes and the species cannot adapt, so it dies out. Humans destroy or change environments for their own gain and cause species that cannot adapt to die out.**
3. What is an endangered species? **An endangered species is one whose numbers are so low that it is in danger of becoming extinct.**
4. Name two ways of preventing extinction. **preserving the remaining habitats in which plants and animals live and breeding in captivity**
5. How are zoos helping to save the Bengal tiger? **Zoos are breeding Bengal tigers in captivity through in vitro fertilization. They are implanting Bengal tiger embryos in female Siberian tigers.**

27

▲ ENRICHMENT 27

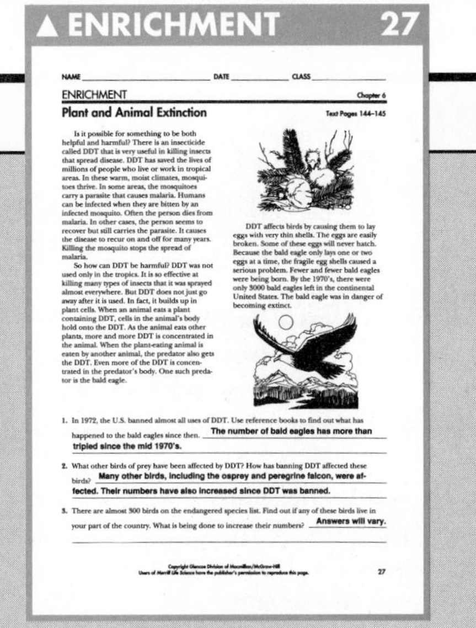

NAME ________ DATE ________ CLASS ________

ENRICHMENT — Chapter 6

Plant and Animal Extinction — Text Pages 144–145

Is it possible for something to be both helpful and harmful? There is an insecticide called DDT that is very useful in killing insects that spread disease. DDT has saved the lives of millions of people who live or work in tropical areas. In these warm, moist climates, mosquitoes thrive. In some areas, the mosquitoes carry a parasite that causes malaria. Humans can be infected when they are bitten by an infected mosquito. Often the person dies from malaria. In other cases, the person seems to recover but still carries the parasite. It causes the disease to recur on and off for many years. Killing the mosquito stops the spread of malaria.

So how can DDT be harmful? DDT was not used only in the tropics. It is so effective at killing many types of insects that it was sprayed almost everywhere. But DDT does not just go away after it is used. In fact, it builds up in plant cells. When an animal eats a plant containing DDT, cells in the animal's body hold onto the DDT. As the animal eats other plants, more and more DDT is concentrated in the animal. When the plant-eating animal is eaten by another animal, the predator also gets the DDT. Even more of the DDT is concentrated in the predator's body. One such predator is the bald eagle.

DDT affects birds by causing them to lay eggs with very thin shells. The eggs are easily broken. Some of these eggs will never hatch. Because the bald eagle only lays one or two eggs at a time, the fragile egg shells caused a serious problem. Fewer and fewer bald eagles were being born. By the 1970's, there were only 3000 bald eagles left in the continental United States. The bald eagle was in danger of becoming extinct.

1. In 1972, the U.S. banned almost all uses of DDT. Use reference books to find out what has happened to the bald eagles since then. **The number of bald eagles has more than tripled since the mid 1970's.**
2. What other birds of prey have been affected by DDT? How has banning DDT affected these birds? **Many other birds, including the osprey and peregrine falcon, were affected. Their numbers have also increased since DDT was banned.**
3. There are almost 300 birds on the endangered species list. Find out if any of these birds live in your part of the country. What is being done to increase their numbers? **Answers will vary.**

27

2 TEACH

Key Concepts are highlighted.

▶ Cheetahs have very little genetic variability, and this is the apparent reason they are endangered.

3 CLOSE

CHECK FOR UNDERSTANDING

Ask questions 1 and 2 in the Section Review.

RETEACH

Have students list the effects of human activity in rain forests, the Arctic tundra, and other fragile environments.

EXTENSION

For students who have mastered this section, use the **Reinforcement** and **Enrichment** masters or other OPTIONS provided.

SECTION REVIEW ANSWERS

1. Extinction is a natural event that can open habitats to new species.
2. Areas where endangered species live can be preserved. Endangered species can be removed and bred in captivity.
3. Connect to Earth Science: A widely held theory is that an asteroid hitting Earth caused major changes in temperature and the atmosphere.

YOU DECIDE!

Answers will vary. Some students may agree that the government should preserve all or some species. Others may think preservation should be left up to nature. Still other students may not think that extinction is important.

VideoDisc

STVS: Horseshoe Crab, Disc 5, Side 1

PREPARATION

SECTION BACKGROUND

▶ Primates evolved during the Eocene. There are presently about 200 species of primates alive.

▶ Genetic studies indicate that the chimpanzee and gorilla are more closely related to humans than to any other primate.

PREPLANNING

▶ Collect baseballs, batons, keys, and pencils for the Motivate activity.

1 MOTIVATE

▶ Have students look at Figure 6-14 and handle various implements to develop an appreciation for the opposable thumb. Have some students tape thumbs carefully to hand and see what can be accomplished without the thumb.

6-4 Human Evolution

New Science Words

primates
hominids
Homo sapiens

Objectives

- ▶ Describe the evidence that monkeys, apes, and humans evolved from a common ancestor.
- ▶ Describe the ancestors of humans.
- ▶ Trace the evolutionary history of humans.

Primates

1 Monkeys, apes, and humans belong to the group of mammals called **primates.** The primates share several characteristics that led scientists to think all primates evolved from a common ancestor. All primates have opposable thumbs that allow them to reach out and bring food to the mouth. Having an opposable thumb allows you to cross your thumb across your palm and touch your fingers. Think of the problems you might have if you didn't have this type of thumb!

Another important primate characteristic is binocular vision. Binocular vision permits a primate to judge depth or distance with its eyes. All primates have flexible shoulders and rotating forelimbs. These allow tree-dwelling primates to swing easily from branch to branch, and allow you to swing on a jungle gym. Each of these characteristics provides evidence that the primates have common ancestry.

Figure 6-14. An opposable thumb allows tree-dwelling primates to hold onto branches. It also allows you to use your hand in many ways.

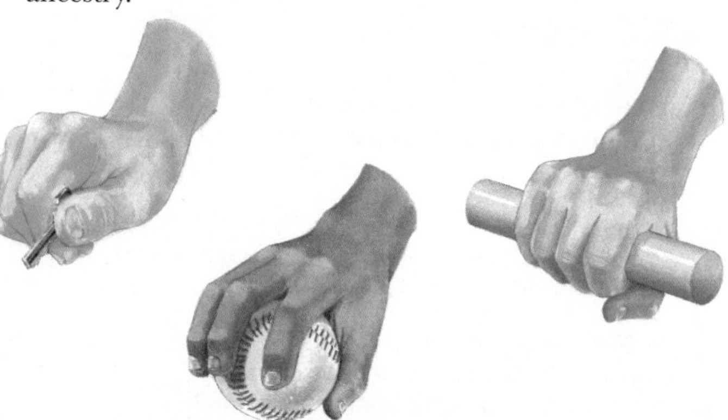

OPTIONS

Meeting Different Ability Levels

For Section 6-4, use the following **Teacher Resource Masters** depending upon individual students' needs.

◆ **Study Guide Master** for all students.

● **Reinforcement Master** for students of average and above average ability levels.

▲ **Enrichment Master** for above average students.

Additional Teacher Resource Package masters are listed in the OPTIONS box throughout the section. The additional masters are appropriate for all students.

◆ STUDY GUIDE 28

NAME ________ DATE ________ CLASS ________

STUDY GUIDE — Chapter 6

Human Evolution — Text Pages 146–149

Match the description in the first column with the item in the second column by writing the correct letter in the space provided.

d 1. humanlike primate — a. *Homo sapiens* Cro-Magnon

e 2. human ancestors that lived in South Africa, had a small brain, and walked upright — b. mammals

f 3. used stone tools, had a small chin and a large brain, died out about 35 000 years ago — c. *Homo sapiens*

h 4. animals with opposable thumbs — d. hominid

c 5. "wise human" — e. *Australopithecus*

a 6. lived in caves and looked like modern humans — f. *Homo sapiens* Neanderthal

g 7. used stone tools and lived about 1.5 to 2 million years ago — g. *Homo habilis*

h. primates

Fill in the blanks below with the appropriate terms from the textbook.

8. Monkeys, apes, and humans belong to the group of mammals called **primates**.
9. Primates have **opposable thumbs** that allow them to reach out and bring food to their mouths.
10. Primates have **binocular vision** that allows them to judge depth and distance with the eyes.
11. Primates have **flexible shoulders** and **rotating forelimbs** that allow them to swing on branches or jungle gyms easily.
12. The **DNA** of chimpanzees, gorillas, and humans is very similar.
13. The name ***Australopithecus*** means southern ape.
14. The name ***Homo habilis*** means handyman.
15. **Neanderthal** humans and **Cro-Magnon** humans were early *Homo sapiens*.
16. Neanderthal humans lived in **family** groups.
17. The oldest recorded art was painted on cave walls by **Cro-Magnon** humans.

28

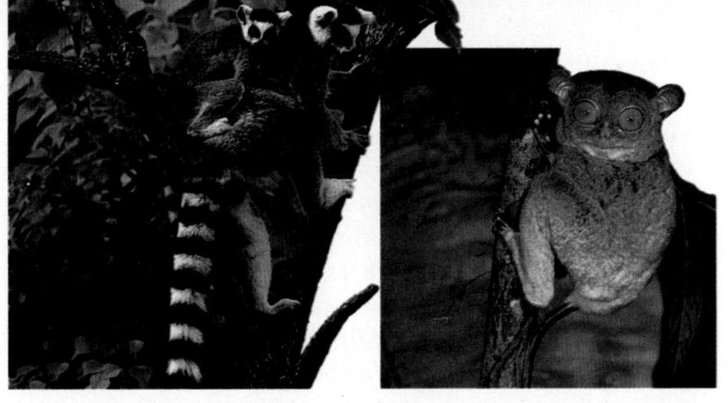

Figure 6-15. Lemurs (a) and tarsiers (b) belong to a subgroup of primates called the prosimians, which means "before apes."

Genetic evidence also supports the view that monkeys, apes, and humans all evolved from a common ancestor. The DNA of chimpanzees, gorillas, and humans is very similar. In addition, primates share many of the same proteins. Hemoglobin is a protein in red blood cells that carries oxygen. Many primates have hemoglobin that is almost identical.

Human Ancestors

Primates are divided into two major groups. The first group includes organisms such as lemurs and tarsiers. These animals have large eyes and are most active at night. The second group, the higher primates, includes monkeys, apes, and humans. About 4–6 million years ago the hominids, our earliest ancestors, branched off from the other higher primates. **Hominids** are humanlike primates that eat both meat and vegetables and walk upright on two feet. Hominids share some common characteristics with gorillas, orangutans, and chimpanzees. A larger brain size along with the characteristics mentioned above separated them from other great apes. ②

In the early 1920s, Raymond Dart, a South African scientist, discovered a fossil skull in a quarry in South Africa. The skull had a small brain cavity but humanlike jaw and teeth. Dart named his discovery *Australopithecus*. *Australopithecus,* one of the earliest hominid groups discovered, means "southern ape." In 1974, an almost complete skeleton of *Australopithecus* was discovered by an American scientist, Donald Johanson, and his colleagues. They named the fossil, estimated to be 2.9 to 3.4 million years old, Lucy. Lucy had a small brain but walked upright. Many scientists today think humans evolved in Africa from ancestors similar to Lucy.

Figure 6-16. The fossil remains of Lucy, a hominid, are estimated to be 2.9 to 3.4 million years old.

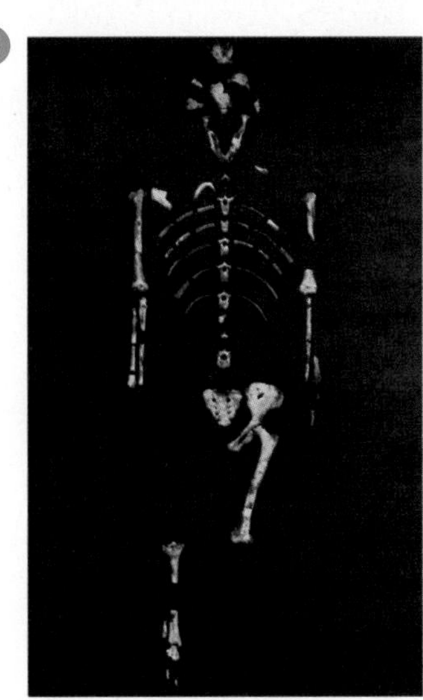

TYING TO PREVIOUS KNOWLEDGE: The concepts in this section are an extension of the concepts presented in the first two sections of this chapter.

2 TEACH

Key Concepts are highlighted.

CONCEPT DEVELOPMENT

▶ Have students find out about the Leakey family and their work in East Africa.

▶ You can demonstrate binocular vision by having students cover first one eye, and then the other. By having a portion of their visual field overlap, distance can be judged.

▶ Evolution of many human characteristics, including a two-hemisphere brain, larynx, and differences in height and weight of the sexes, appears to have been a gradual process rather than occurring all at once, as was previously believed.

▶ In describing Neanderthals and Cro-Magnon be certain to indicate that both of these were fully human. Their species, *Homo sapiens,* is the same as ours.

REVEALING MISCONCEPTIONS

▶ Ask students if humans evolved from apes according to the theory of evolution. Point out that the theory implies that apes and humans had a common, and as yet unknown, ancestor in the geological past.

MULTICULTURAL AWARENESS

▶ Use the information about fossil sites and researchers in Africa to increase students' awareness of African geography and their appreciation of the rich heritage found there.

● REINFORCEMENT 28

NAME ______ DATE ______ CLASS ______

REINFORCEMENT — Chapter 6

Human Evolution — Text Pages 146–149

In the table below list three physical characteristics that all primates share. Then describe how each of these characteristics functions or how it is adaptive.

Characteristic	Function/Adaption
opposable thumb	to grab or hold objects such as food or tools
binocular vision	to judge distances
rotating forelimbs	to swing from branches in a tree

Answer the following questions.

1. How do lemurs and tarsiers differ from the higher primates? **Lemurs and tarsiers have large eyes and are most active at night.**
2. How do hominids differ from monkeys and apes? **Hominids eat both meat and vegetables, walk upright, and have large brains.**
3. Would you expect the DNA of a human to be more similar to that of a gorilla or a lemur? Explain your answer. **Because humans are more closely related to gorillas than to lemurs, the DNA of a human would be more similar to that of a gorilla.**
4. In what ways do *Australopithecus* and *Homo habilis* differ? ***Homo habilis* had a larger brain than *Australopithecus*. *Homo habilis* was also more similar in form to a modern human.**
5. What traits did the early humans, Neanderthals and Cro-Magnons, share? **Both Neanderthal and Cro-Magnon humans were *Homo sapiens*, lived in family groups, and made and used stone tools.**
6. What social behaviors do we share with Cro-Magnon humans? **living in family groups, caring for the elderly, burying the dead, artistic expression through painting and carving**

28

▲ ENRICHMENT 28

NAME ______ DATE ______ CLASS ______

ENRICHMENT — Chapter 6

Human Evolution — Text Pages 146–149

Use your textbook and references to complete the table below with information concerning fossil evidence of early humans. You may wish to plot the location of these and other humanlike fossils that you may read about.

Example	Age	Description/Location
Australopithecus	3.4 to 2.9 million years old	small brain cavity; humanlike jaw and teeth; walked upright; found in South Africa
Homo habilis	2 to 1.5 million years old	used simple tools; larger brain cavity than *Australopithecus*; found in East Africa
Homo erectus	1 million to 250 000 years old	low forehead; heavy brow ridge; large jaw and teeth; used fire and tools; found in Java and China
Neanderthal	125 000 years old	a type of *Homo sapiens*; short heavy body with thick bones; small chin; heavy brow ridge; lived in family groups in caves; hunted large animals with stone tools; found in Africa and Europe
Cro-Magnon	40 000 to 10 000 years old	a type of *Homo sapiens*; almost identical to humans; lived in caves; painted people and animals on cave walls; made stone carvings; cared for their elderly; buried their dead; found in Europe, Asia, and Australia

28

In Your JOURNAL

Student ads should emphasize tools for hammering and cutting.

CHECK FOR UNDERSTANDING

Ask students to explain the common features of a lemur, monkey, and chimpanzee.

RETEACH

If students are unable to explain the common features, have them prepare a list of the characteristics common to humans and primates.

EXTENSION

For students who have mastered this section, use the **Reinforcement** and **Enrichment** masters or other OPTIONS provided.

MINI QUIZ

Use the Mini Quiz to check students' recall of chapter content.

1. **Monkeys, apes, and humans belong to the group of mammals called _____ .** *primates*
2. **About 6 million years ago, the _____ , our earliest ancestors, branched off from the other higher primates.** *hominids*
3. **Cro-Magnon people were almost identical to _____ .** *modern humans*

In Your JOURNAL

Using what you know of the characteristics of Neanderthals, write an advertisement **in your Journal** for *The Neanderthal News* for a modern-day product that you think humans might have been able to use then.

About 40 years after the discovery of *Australopithecus,* a discovery was made in East Africa by Louis, Mary, and Richard Leakey. The Leakeys discovered a fossil more similar to present-day humans than *Australopithecus.* They named this hominid *Homo habilis,* the "handy man," because they found simple stone tools near him. Scientists estimate *Homo habilis* to be 1.5 to 2 million years old. Homo habilis is thought to be a direct ancestor of humans because it had a larger brain and was more similar in form to humans than *Australopithecus.*

Modern Humans

Our species is named ***Homo sapiens,*** meaning "wise human." The fossil record shows that the human species, a very recent arrival on Earth, evolved about 300 000 years ago. By about 125 000 years ago, two early groups of *Homo sapiens,* Neanderthal human and Cro-Magnon human, probably lived at the same time in parts of Africa and Europe.

Neanderthal humans had short, heavy bodies with thick massive bones, small chins, and heavy brow ridges. The Neanderthal humans lived in family groups in caves and hunted mammoths, deer, and other large animals with well-made stone tools. For reasons that are not clear, Neanderthal humans disappeared from the fossil record about 35 000 years ago.

Cro-Magnon fossils have been found in Europe, Asia, and Australia. These fossils are dated from 40 000 to about
3 10 000 years ago. Cro-Magnon humans are thought to

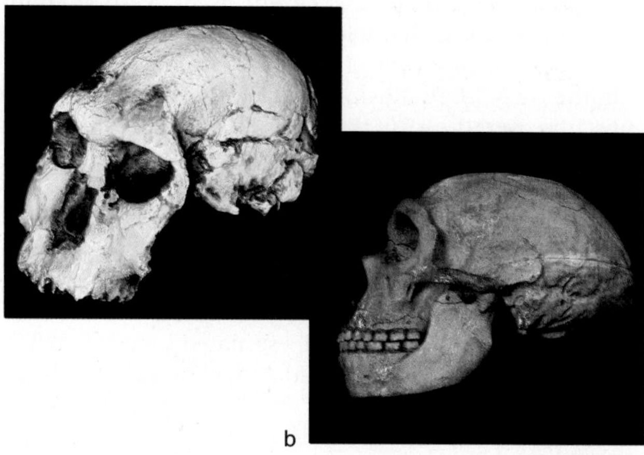

Figure 6-17. Fossil skulls of (a) *Homo habilis* and (b) Neanderthal are shown. Notice the larger brain area in the Neanderthal skull.

OPTIONS

ASSESSMENT—ORAL

▶ **What relationship is implied by the statement that monkeys, apes, and humans belong to the group of mammals called primates?** *The implication is that all of these organisms shared a common ancestor.*

▶ **Why is DNA considered evidence that monkeys, apes, and humans all evolved from a common ancestor?** *Because DNA is passed from organism to organism, shared ancestors should have similar, but not exactly the same, DNA.*

▶ **Why was *Homo habilis* called the handy man?** Homo habilis *made and used tools.*

▶ **Why can we say that Cro-Magnon humans were cultured?** *Cro-Magnon humans were the first artists. Art is considered by some people to be the hallmark of culture. Further, they made stone carvings, cared for their elderly, and buried their dead.*

b

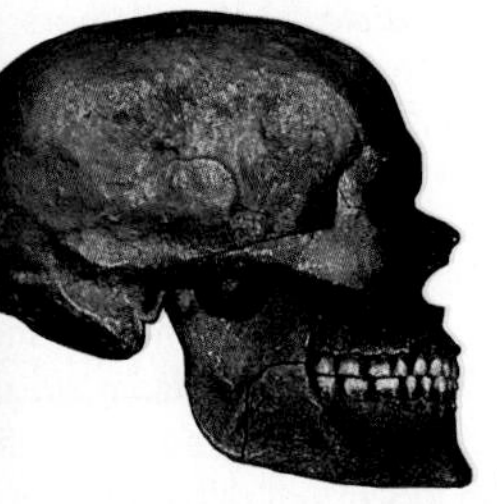

a

Figure 6-18. This figure shows a Cro-Magnon skull (a) thought to be 10 000 to 40 000 years old and a cave painting (b) done by Cro-Magnon man.

be the direct ancestors of modern humans. The oldest recorded art dates from the caves of France where Cro-Magnon humans first painted bison, horses, and spear-carrying people. Cro-Magnon humans lived in caves, made stone carvings, cared for their elderly, and buried their dead. Standing about 1.6–1.7 m tall, the physical appearance of Cro-Magnon people was almost identical to that of modern humans.

Many fossils from hominids and early humans have been found in the African rift valley. This area of Africa is where two tectonic plates of Earth's crust are moving past one another. Why might you expect to find many fossils here?

SECTION REVIEW

1. Describe at least three kinds of evidence that suggests all primates shared a common ancestor.
2. What is the importance of *Australopithecus*?
3. Describe the differences between Neanderthal humans, Cro-Magnon humans, and modern humans.
4. **Apply:** How was an opposable thumb important in human evolution?
5. **Connect to Earth Science:** Propose a hypothesis about why teeth represent the most abundant available fossils of hominids.

Concept Mapping

Using information is this section, make a concept map to show the sequence of human ancestors. Use the following terms: Neanderthal human, lemurs and tarsiers, *Homo habilis, Australopithecus,* modern *Homo sapiens*, Cro-Magnon human. If you need help, refer to Concept Mapping in the **Skill Handbook** on pages 688 and 689.

3 CLOSE

▶ Ask questions 1-3 and the **Apply** Question in the Section Review.

Answer: Fossils might be exposed in this area because of earthquakes and shifting of rock layers.

SECTION REVIEW ANSWERS

1. Homologous structures, similar genetic codes, and the fossil record suggest a common ancestry for all primates.

2. *Australopithecus* represents one of the earliest examples of a hominid that had a small brain case, but humanlike jaws and teeth.

3. Neanderthal humans had short, heavy bodies with thick, massive bones, small chins, and heavy brow ridges. Cro-Magnon humans stood about 1.8 meters tall, invented art, made stone carvings, cared for their elderly, and buried their dead. Cro-Magnon people looked almost the same as modern humans.

4. Apply: Opposable thumbs allowed humans to use weapons and tools that helped them survive.

5. Connect to Earth Science: Teeth are the hardest parts of the organism and therefore the most likely to be fossilized.

lemurs and tarsiers, *Australopithecus, Homo habilis,* Neanderthal human, Cro-Magnon human, and modern *Homo sapiens*

Skillbuilder
ASSESSMENT

Portfolio: Use this Skillbuilder to assess students' abilities to form a Concept Map of the sequence of human ancestors.

ENRICHMENT

▶ Have students write to The Paleontological Society, U.S.G.S., E/501 National Museum Building, Smithsonian Institution, Washington, DC 20006 for information about the work of a paleontologist.

▶ Have students write to The American Anthropological Association, 1703 New Hampshire Ave. NW, Washington, DC 20009 for information about the work of an anthropologist.

PROGRAM RESOURCES

From the **Teacher Resource Package** use:

Cross-Curricular Connections, page 10, Ice Age Cave Paintings.

Activity Worksheets, pp. 52-53, Activity 6-2, Designing an Experiment

ACTIVITY 6-2

OBJECTIVE: Design and carry out an experiment to show the variation in a population of seeds.
Time: One class to plan the investigation and a second class to complete the investigation.

PROCESS SKILLS applied in this activity are **forming a hypothesis, measuring, recording data, calculating,** and **interpreting.**

PREPARATION

Collect materials needed for each group.

Cooperative Learning: Assign Science Investigation Teams. Students can use the class data to calculate individually in Step 3.

HYPOTHESIZING

Student hypotheses will vary. Possible hypothesis might be: "A sample of sunflower seeds will exhibit variations." or "A sample of sunflower seeds will exhibit variations in length."

TEACHING THE ACTIVITY

*Refer to the **Activity Worksheets** for additional information and teaching strategies.*

- Guide students as they decide which trait to measure. Length, width, and circumference of seeds will be most easily measured with the materials suggested.
- For most students, larger seeds will be easier to work with. Sunflower, lima beans, pinto beans, and peas are large enough.
- For calculating range, mean, median, and mode, refer to Activity 5-2 on page 124.

SUMMING UP/SHARING RESULTS

- Answers will vary. Students will likely measure length or width of seeds.
- Accept any explanation of the hypothesis if its support is logcal and coherent.
- Range, mean, median, and mode should be similar between groups, but will not usually be exact. Similarities are explained by measuring the same trait.
- Student conclusions should accurately explain the results as expressed in the graph.

GOING FURTHER!

The results for another type of seed or clam shell should be similar to the variations found in the seeds studies. Sample size will influence results.

ACTIVITY 6-2 DESIGNING AN EXPERIMENT
Recognizing Variation in a Population

Have you ever noticed that when you first see a group of plants or animals of a species that they may all look alike? However, if you look closer you will notice that some are taller than others; some have slightly different colored parts. Variations must exist in a population for evolution to occur. Have you noticed variations in any plant or animal population?

Getting Started

You will need to determine the question you will investigate. You will recall from page 133 that variation is an appearance of an inherited trait that makes an individual different from other members of the species. Your task in this activity is to devise a method of determining if seeds exhibit variations.

Safety

CAUTION: *Do not put any seeds into your mouth.*

Hypothesizing

Make a **hypothesis** about whether seeds exhibit variation. What are your reasons for forming this **hypothesis**?

Materials

Your cooperative group will use:
- 100 seeds
- metric ruler
- graph paper

Try It!

1. Select one seed trait to measure.
2. *Design* a data table in which to record your results.
3. After collecting your data, *calculate* the range, mean, median, and mode of your sample of seeds.
4. *Graph* your data using a line graph.

Summing Up/Sharing Results

- What types of differences did you look for?
- Was your hypothesis supported by your data? Explain.
- Compare the range, mean, median, and mode of your data with results of other groups.
- Explain any similarities and differences.
- Write a conclusion based on your data.

Going Further!

Design an experiment that would reveal variations in another type of seed or in a collection of shells from a species of clam.

150 EVOLUTION

PROGRAM RESOURCES

From the **Teacher Resource Package** use:
Activity Worksheets, pages 52-53, Activity 6-2, Recognizing Variation in a Population.

Activity ASSESSMENT

Performance: To further assess students' understanding of variations in an inherited trait, have the class work cooperatively on Project 2, pages 696 and 697.

CHAPTER

REVIEW

SUMMARY

6-1: Mechanisms of Evolution

1. Hereditary features of a species of organisms evolve or change over time.
2. Lamarck developed the theory of acquired characteristics to explain evolution. His theory implied that characteristics parents developed during their lives were passed on to their offspring. Charles Darwin developed the natural selection theory of evolution. According to this theory, organisms best adapted to their environments survive and reproduce.
3. Variations are differences in inherited traits among members of the same species. Variations can allow an organism to be better suited to its environment.

6-2: Evidence for Evolution

1. Fossils are remains of life from an earlier time. They give important evidence of evolution.
2. The age of a fossil can be found by using relative dating or radioactive dating. In relative dating, the age of a fossil is found by looking where it is in the rock layers. Radioactive dating determines the age of a fossil by comparing amounts of radioactive and stable substances in a fossil.
3. Evidence for evolution is obtained by comparing embryology, homologous structures, chemical similarities, and vestigial structures.

6-3: Science and Society: Plant and Animal Extinction

1. Extinction is the dying out of a species due to changes in the environment.
2. Endangered species can be saved from extinction by preserving their environments and by breeding them in captivity.

6-4: Human Evolution

1. Primates share several common characteristics including opposable thumbs, binocular vision, and flexible shoulders and rotating forearms.
2. The earliest known hominid is *Australopithecus*. Hominids are humanlike primates. The hominid *Homo habilis* is thought to be the direct ancestor of humans.
3. Modern humans, *Homo sapiens,* evolved 300 000 years ago. They most likely evolved from Cro-Magnon humans.

KEY SCIENCE WORDS

a. **embryology**
b. **endangered species**
c. **evolution**
d. **extinction**
e. **fossils**
f. **gradualism**
g. **hominids**
h. **homologous**
i. ***Homo sapiens***
j. **natural selection**
k. **population**
l. **primates**
m. **punctuated equilibrium**
n. **radioactive elements**
o. **relative dating**
p. **sedimentary rock**
q. **species**
r. **variation**
s. **vestigial structure**

UNDERSTANDING VOCABULARY

Match each phrase with the correct term from the list of Key Science Words.

1. remains of once-living things
2. structure with no obvious use
3. similar organisms that successfully reproduce
4. body structures that are similar in origin
5. change in hereditary feature over time
6. a difference in an inherited trait
7. all the members of one species in an area
8. model of evolution showing slow change
9. the study of stages of an organism's early development
10. group containing monkeys, apes, and humans

CHAPTER

REVIEW

SUMMARY

Have students read the summary statements to review the major concepts of the chapter.

UNDERSTANDING VOCABULARY

1. e
2. s
3. q
4. h
5. c
6. r
7. k
8. f
9. a
10. l

ASSESSMENT

Portfolio

Encourage students to place in their portfolios one or two items of what they consider to be their best work. For each item, ask students to explain why that item was chosen and what they learned from it. Items might be selected from the following.

- MINI-Lab imprint fossils, p. 137
- In Your Journal advertisement, p. 148
- Enrichment copy of letter, p. 149

Performance

Additional performance assessments may be found in *Performance Assessment* and *Science Integration Activities* that accompany **Merrill Life Science.** Performance Task Assessment Lists and rrubrics for evaluating these activities and otheor products generated throughout the chapter can be found in Glencoe's *Performance Assessment in Middle School Science.*

OPTIONS

ASSESSMENT

To assess student understanding of material in this chapter, use the resources listed.

COOPERATIVE LEARNING

Consider using cooperative learning in the THINK AND WRITE CRITICALLY, APPLY, and MORE SKILL BUILDERS sections of the Chapter Review.

PROGRAM RESOURCES

From the **Teacher Resource Package** use:

Chapter Review, pages 15-16.

Chapter and Unit Tests, pages 36-39, Chapter Test.

CHAPTER
REVIEW

CHECKING CONCEPTS

1. b
2. b
3. d
4. c
5. d
6. c
7. c
8. a
9. b

USING LAB SKILLS

ASSESSMENT

Use these alternate lab exercises to assess students' understanding of skills used in this chapter.

10. Mass in milligrams with intervals from bottom to top of 25, 50, 75, 100, will be on the x-axis (the vertical axis). Time in millions of years on the y-axis (horizontal axis), with intervals from left to right of 1, 2, 3, 4, 5, 6, 7, 8, and so on as needed. Line graph will begin a 100 mg. at 1 million years, and decrease to less than 25 mg. at 8 years.

11. The cast made by a shell or footprint of an organism becomes filled with minerals; an imprint is made by an organism in a rock.

THINK AND WRITE CRITICALLY

12. Giraffes with longer necks (variation) survived because they outlived shorter-necked individuals since they could eat leaves higher up on trees. This variation was passed on to offspring because it is a genetic trait.

13. Variations result from differences in genes. The resulting adaptations allow the best organisms to survive, passing along these genes. The poorer traits are eliminated because these genes are not passed on to offspring.

14. The organisms could not interact. If the environments were different, different variations would be naturally selected and different species could evolve.

15. Gradualism explains evolution as slow, steady, and continuous. Punctuated equilibrium shows rapid change in DNA and rapid evolution.

16. Relative dating compares the age

CHAPTER
REVIEW

CHECKING CONCEPTS

Choose the word or phrase that completes the sentence or answers the questions.

1. _______ is the death of all members of a species.
a. Evolution
b. Extinction
c. Gradualism
d. Variation

2. The most accurate age of a fossil can be found using _______.
a. natural selection
b. radioactive elements
c. relative dating
d. camouflage

3. Homologous structures, vestigial structures, and fossils all provide evidence of _______.
a. extinction
b. species populations
c. food choice
d. evolution

4. A factor that controls natural selection is _______.
a. inheritance of acquired traits
b. unused traits become smaller
c. organisms produce more offspring than can survive
d. the size of an organism

5. A series of helpful variations in a species may eventually result in _______.
a. climate change
b. fossils
c. extinction
d. evolution into two species

6. Organisms adapted to their environment are _______.
a. extinct
b. not reproducing
c. surviving and reproducing
d. forming fossils

7. In order for evolution to occur, some change takes place in _______.
a. a single individual
b. an embryo
c. a population
d. the climate

8. Opposable thumbs and binocular vision are characteristics of _______.
a. all primates
b. hominids
c. humans only
d. monkeys and apes

9. The earliest paintings were done by _______.
a. *Australopithecus*
b. Cro-Magnon humans
c. *Homo sapiens*
d. Neanderthal humans

USING LAB SKILLS

10. Pretend that the pennies used in the Activity on page 143 each represent one milligram of radioactive material, and each time you shake the box represents one hundred thousand years. Make a line graph to show your results. Plot the mass on the x-axis and the time on the y-axis.

11. In the MINI-Lab on page 137, you made a model of an imprint fossil. Using clay, a stone, shell, or other small object of your choice, make a model of a cast fossil. How does a cast fossil differ from an imprint fossil?

THINK AND WRITE CRITICALLY

Answer the following questions in your Journal using complete sentences.

12. Use Darwin's theory of natural selection to explain how the giraffe got its long neck.

13. Describe what causes variations and how the resulting adaptations help explain evolution.

14. What could happen if members of a population became completely separated from each other?

15. Compare gradualism and punctuated equilibrium.
16. Compare relative dating to radioactive dating of fossils. Which do you think is more accurate and why?

APPLY

17. Explain how Lamarck and Darwin would have explained a duck's webbed feet.
18. Using an example, explain how a new species of organism could evolve. (HINT: isolation, change in climate)
19. How is the coloration of chameleons an adaptation to their environment?
20. Describe the process a scientist would use to figure out the age of a fossil.
21. Explain how an organism could become extinct. Give an example.

MORE SKILL BUILDERS

If you need help, refer to the Skill Handbook.

1. **Observing and Inferring:** Observe the birds' beaks pictured below. Describe each. Infer the types of food each would eat and explain why.

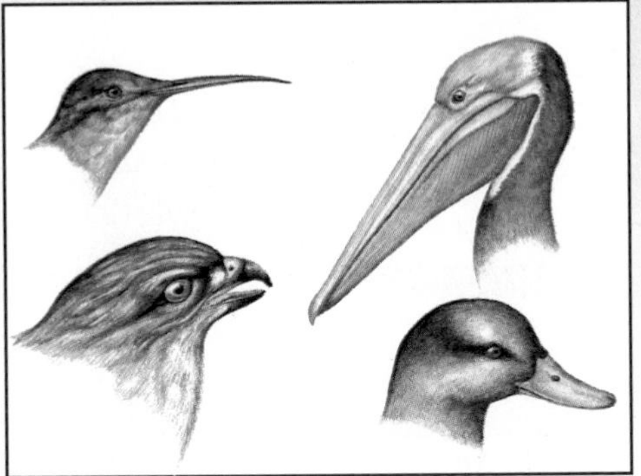

2. **Hypothesizing:** Frog eggs are common in ponds in spring. Make a hypothesis as to why ponds are not overpopulated by frogs in the summer. Use the ideas of natural selection to help you.
3. **Comparing and Contrasting:** Compare and contrast Cro-Magnon humans and Neanderthal humans.
4. **Outlining:** Outline the events that led Charles Darwin to his theory of natural selection.
5. **Interpreting Data:** Interpret the following data. The chemicals present in certain bacteria were studied. Listed below are the chemicals found in each type. Each letter represents a different chemical. Use this information to determine which of the bacteria are closely related.

Chemicals Present	
Bacteria 1	A, G, T, C, L, E, S, H
Bacteria 2	A, G, T, C, L, D, H
Bacteria 3	A, G, T, C, L, D, P, U, S, R, I, V
Bacteria 4	A, G, T, C, L, D, H

PROJECTS

1. Make a collection of fossils found in your area. Identify where your fossils were found. If you live in a city, visit a museum of natural history and report on the fossil history of the land the city is built on.
2. Research the work of Alfred Wallace. On a world map, show the sites of his work and that of Darwin. Compare and contrast the areas in which they worked.

of fossils found in different rock layers to each other. Radioactive dating gives a specific age to a fossil based on the actual amount of radioactive material present. This is more accurate because radioactivity can be measured.

APPLY

17. Lamarck: Ducks have webbed feet for swimming. The ducklings acquired the webbed feature.
Darwin: A variation in a duck's feet was an advantage to survival in the environment. This was a feature that was passed on to the offspring (because it was a genetic trait).
18. Geographical isolation due to a volcanic eruption can separate a population. Two species may form.
19. Chameleons blend into their environment. This ability to change color is an adaptation for survival.
20. The layer of rock where it was found would be noted. Radioactive dating would be done. Comparisons to other fossils would be made.
21. An organism can become extinct because its habitat is destroyed, its food supply disappears, it is killed off, or it is becoming less genetically stable. Rain forest animals, tigers, carrier pigeons, and cheetahs are examples.

MORE SKILL BUILDERS

1. Observing and Inferring: The hummingbird's beak is long and thin for sipping nectar. The bird of prey has a curved, sharp beak for catching and eating prey. The duck has a wide, flat beak for dabbling for food in the mud. The pelican has a long beak with storage for scooping up food.
2. Hypothesizing: Ponds are not overpopulated by frogs because not all eggs survive: some are eaten, the water dries up, and so on.
3. Comparing and Contrasting: Neanderthals had short, thick bodies, heavy bones, small chins, and heavy brow ridges. Cro-Magnon humans looked similar to modern humans. They both lived in caves, but the social structure of Cro-Magnon humans was more developed. Cro-Magnon made the first cave paintings. They also were the first to bury their dead.

4. Outlining:

I. Darwin took the position of ship's naturalist.
   A. Journeyed to South America and the Galapagos
   B. Observed and collected organisms
II. Darwin studied collections for 20 years.
   A. Observed variations in pigeons and flowers
   B. Observed artificial selection
   C. Developed theory of evolution by natural selection.

5. Interpreting Data: Bacteria 2 and 4 have the same chemicals and are, therefore, most closely related.

CHAPTER

7 Classification of Living Things

CHAPTER SECTION	OBJECTIVES	ACTIVITIES
7-1 What Is Classification? (2 days)	1. **Give** examples that show the need for classification systems. 2. **Describe** Aristotle's system of classification. 3. **Explain** Linnaeus's system of classification.	**Activity 7-1:** *Designing an Experiment (Classifying Seeds)*, p. 159 **MINI-Lab:** *How can leaves be classified?* p. 156
7-2 Modern Classification (1 day)	1. **Name** the five kingdoms of living things. 2. **Identify** characteristics and members of each kingdom. 3. **List** the groups within each kingdom.	
7-3 The Rain Forest Crisis Science & Society (1 day)	1. **Identify** two dangerous effects of tropical deforestation. 2. **Explain** the importance of the rain forest to us and to the people who live there.	
7-4 Identifying Organisms (2 days)	1. **List** reasons scientific names are more useful to scientists than common names. 2. **Identify** the functions of a dichotomous key. 3. **Demonstrate** how to use a dichotomous key.	**Activity 7-2:** *Using a Dichotomous Key*, p. 170 **MINI-Lab:** *How are organisms named?* p. 166
Chapter Review		

ACTIVITY MATERIALS

FIND OUT	ACTIVITIES		MINI-LABS	
Page 155 paper pencil	**7-1 Designing an Experiment, p. 159** packet of 10 different seeds hand lens metric ruler 2 sheets of paper	**7-2 Using a Dichotomous Key, p. 170** paper pencil	**How can leaves be classified? p. 156** variety of types of leaves newspaper heavy book poster board glue	**How are organisms named? p. 166** clay

CHAPTER FEATURES	TEACHER RESOURCE PACKAGE	OTHER RESOURCES
Problem Solving: *Whose shoe?,* p. 158 **Skill Builder:** *Observing and Inferring,* p. 158	**Ability Level Worksheets** ◆ ***Study Guide,*** p. 29 ● ***Reinforcement,*** p. 29 ▲ ***Enrichment,*** p. 29 **Concept Mapping,** p. 19 **Activity Worksheets,** pp. 5, 59-60, 65	**Lab Manual:** Classification, p. 43 **STVS:** Disc 5, Side 2
Technology: *Beyond Appearances,* p. 161 **Skill Builder:** *Concept Mapping,* p. 163	**Ability Level Worksheets** ◆ ***Study Guide,*** p. 30 ● ***Reinforcement,*** p. 30 ▲ ***Enrichment,*** p. 30 **Transparency Masters,** pp. 21-24	**Color Transparency 11,** Life's Five Kingdoms **Color Transparency 12,** Classification Pyramid
You Decide! p. 165	**Ability Level Worksheets** ◆ ***Study Guide,*** p. 31 ● ***Reinforcement,*** p. 31 ▲ ***Enrichment,*** p. 31 **Critical Thinking/Problem Solving,** p. 11 **Cross-Curricular Connections,** p. 11 **Science and Society,** p. 11	**Science Integration Activity 7**
Skill Builder: *Classifying,* p. 169	**Ability Level Worksheets** ◆ ***Study Guide,*** p. 32 ● ***Reinforcement,*** p. 32 ▲ ***Enrichment,*** p. 32 **Activity Worksheets,** pp. 61, 62, 66	**STVS:** Disc 5, Side 1
Summary Key Science Words Understanding Vocabulary Checking Concepts Using Lab Skills Think & Write Critically Apply More Skill Builders Projects	ASSESSMENT RESOURCES **Chapter Review,** pp. 17-18 **Chapter Test,** pp. 40-43 **Unit Test,** pp. 44-45 **Performance Assessment in Middle School Science (PAMSS)**	**Chapter Review Software** **Test Bank** **Alternate Assessment** **Performance Assessment**

◆ **Basic** ● **Average** ▲ **Advanced**

ADDITIONAL MATERIALS

SOFTWARE	AUDIOVISUAL	BOOKS/MAGAZINES
Classify—Classification Key Prog, Diversified Educ. Enterp. *Classification of Living Things,* Educated Activities, Inc. *Taxonomy: Classification and Organization,* IBM. *The Taxonomy Game,* Queue.	*Carolus Linnaeus,* film, EBEC. *Classifying Plants and Animals;* film, Coronet/MTI.	Ambrose, E.J. *The Nature and Origin of the Biological World.* Englewood Cliffs, NJ: Prentice Hall, 1982. Rose, Kenneth J. *Classification of the Animal Kingdom.* NY: David McKay Company, Inc., 1980. Starr, Cecie and Ralph Taggart. *Biology: Concepts and Applications.* Belmont, CA: Wadsworth Publishing Company, 1991.

CHAPTER

7

CLASSIFICATION OF LIVING THINGS

THEME DEVELOPMENT: The theme of evolution is used to explain how life scientists classify the millions of species of living things. The concept of evolutionary relatedness is emphasized. In addition, the themes of ecology and homeostasis are further developed in the section that discusses the dangers of tropical rain forest and other habitat destruction.

CHAPTER OVERVIEW

- **Section 7-1:** This section defines classification and explains its importance. The Linnaean system of binomial nomenclature is also explained.
- **Section 7-2:** Classification of organisms into the five kingdoms is discussed in this section. Taxonomic units from phylum to species are also illustrated.
- **Section 7-3: Science and Society:** The importance of organism diversity is presented using the tropical rain forest as an example. Dangers due to rain forest destruction are discussed.
- **Section 7-4:** The reasons scientific names are used are presented in this section. Students are given the opportunity to explore the use of a dichotomous key.

CHAPTER VOCABULARY

classify	phylum
taxonomy	division
kingdom	classes
binomial nomenclature	orders
species	families
genus	species diversity
phylogeny	dichotomous keys

CHAPTER 7

Classification of Living Things

154

OPTIONS

For Your Gifted Students

Students can research the rain forests past and present. They can draw a map of the world showing the existing rain forest and the area of the past 100 years. Using the information in the text, have them estimate how much forest will be left by the year 2000. They should indicate this on the map. Students can also use this information to equate the forest loss to the U.S. by the year 2000.

For Your Mainstreamed Students

Students can make classification mobiles. They should draw or find a picture of one member of each kingdom to hang as a mobile. Phyla and classes may be added.

You might have visited a zoo and discovered creatures you didn't even know existed. How did you decide if an animal was a type of monkey or a type of bear? Or if it was a bird or an insect? Think about how you made these distinctions.

FIND OUT!

Do this simple activity to find out how you might determine relationships between organisms.

Observe the organisms on the opposite page. Notice that all of the organisms are insects, all have wings, and all of them can fly. Now look closely at each of the organisms. Group them based on specific traits that are shared. For example, organisms that have feathery antenna can be grouped together. You might also group the organisms based on color, size, shape, or any other traits you observe. You might find organisms that can't be grouped with the others. Now try to divide your groups into smaller, more specific groups. How many different groups did you make?

Gearing Up

Previewing the Chapter

Use this outline to help you focus on important ideas in this chapter.

Previewing Science Skills

- ▶ In the Skill Builders, you will observe and infer, make a concept map, and classify.
- ▶ In the Activities, you will classify and observe.
- ▶ In the MINI-Labs, you will classify, make models, and apply knowledge.

What's next?

Life scientists look at many physical traits to categorize living things. In this chapter, you will learn how and why scientists categorize living things. You will learn about the history of this process and scientists who have developed widely-used classification methods. Finally, you will learn about the five kingdoms of living things and the groups within the kingdoms.

155

INTRODUCING THE CHAPTER

Use the Find Out activity to introduce students to classification. Inform students that in this chapter they will learn how organisms are classified.

FIND OUT!

Materials: pencil and paper

Cooperative Learning: Have students work as Paired Partners to complete the activity.

Teaching Tips

▶ Be sure students can identify the differences between moths and butterflies. The moths have feathered antennae and the butterflies have smooth.

▶ Students will probably first find three groups: the moths, the butterflies, and one dragonfly.

▶ Accept any logical classification groupings by students.

Gearing Up

Have students study the Gearing Up feature to familiarize themselves with the chapter. Discuss the relationships of the topics in the outline.

What's Next?

Before beginning the first section, make sure students understand the connection between the Find Out activity and the topics to follow.

OBJECTIVES AND SCIENCE WORDS: Have students review the objectives and science words throughout the chapter to become familiar with each section.

ASSESSMENT OPTIONS

PORTFOLIO

Refer to page 171 for suggested items that students might select for their portfolios.

PERFORMANCE ASSESSMENT

See page 171 for additional Performance Assessment options.

Process

Skillbuilders, pp. 158, 169
MINI-Labs, pp. 156, 166
Activities 7-1, p. 159; 7-2, p. 170
Using Lab Skills, p. 172

CONTENT ASSESSMENT

Assessment—Oral, pp., 158, 162, 168
Skillbuilder, p. 163
Section Reviews, pp. 158, 163, 165, 169
Chapter Review, pp. 171-173
Mini Quiz p. 162

GROUP ASSESSMENT

Opportunities for group assessment occur with Cooperative Learning Strategies and Flex Your Brain Activities.

SECTION 7-1

PREPARATION

▶ Linnaeus was able to classify only a few thousand organisms during his lifetime. Some scientists today conjecture there may be 10 million different species.

PREPLANNING

▶ You will need a variety of different household items for the Demonstration in the Motivate section.

MINI-Lab

▶ Have students collect 12 to 15 different types of leaves to classify.

▶ Caution students about poison ivy and poison oak.

▶ **Answer:** Students should group leaves according to leaf margins, number of lobes, or the shape of the leaves. Accept any logical classification scheme.

MINI-Lab

ASSESSMENT

Performance: To further assess students' understanding of classification, take them to the library to classify books.

1 MOTIVATE

FLEX Your Brain

Use the Flex Your Brain activity to have students explore what they already know about CLASSIFICATION.

Flex Your Brain

ASSESSMENT

Performance: Use the Flex Your Brain activity to reinforce critical thinking and problem solving skills.

VideoDisc

STVS: Insect Museum, Disc 5, Side 1
Naming Fish, Disc 5, Side 2

7-1 What Is Classification?

New Science Words

classify
taxonomy
kingdom
binomial nomenclature
species
genus

Objectives

- ▶ Give examples that show the need for classification systems.
- ▶ Describe Aristotle's system of classification.
- ▶ Explain Linnaeus's system of classification.

MINI-Lab

How can leaves be classified?
Collect a variety of types of leaves. Place the leaves between sheets of newspaper and press in a heavy book. *Classify* the leaves based on similarities. Glue the leaves onto poster board in their groups. **In your Journal,** explain how you classified the leaves.

Classifying

When you go into a grocery store do you usually go to one aisle to get milk, to another part of the store to get margarine, and to a third area to get yogurt? No, most grocery stores group similar items. The dairy products mentioned in the example above would be found together. When you place similar items together, you classify them. To **classify** means to group ideas, information, or objects based on similarities.

Classification is more a part of your life than you might think. Grocery stores, bookstores, and department stores group similar items together. In what other places is classification important?

Early History of Classification

More than 2000 years ago Aristotle, a Greek philosopher, developed a system to group living things. The science of grouping and naming organisms is called **taxonomy** (tak SAHN uh mee). Aristotle began his system of taxonomy by dividing organisms into two large kingdoms, the plant and animal kingdoms. A **kingdom** is the largest of the taxonomic categories. Aristotle then divided the animal kingdom into smaller groups based on where animals live. Animals that live on land were in one group, animals that live in water were in another group,

Figure 7-1. Why didn't frogs fit into Aristotle's system of classification?

OPTIONS

Meeting Different Ability Levels

For Section 7-1, use the following **Teacher Resource Masters** depending upon individual students' needs.

◆ **Study Guide Master** for all students.

● **Reinforcement Master** for students of average and above average ability levels.

▲ **Enrichment Master** for above average students.

Additional Teacher Resource Package masters are listed in the OPTIONS box throughout the section. The additional masters are appropriate for all students.

◆ STUDY GUIDE 29

NAME ______ DATE ______ CLASS ______

STUDY GUIDE — Chapter 7

What Is Classification? — Text Pages 156–158

Classifying

Fill in the blanks with the correct terms listed below.

study communicate similar kingdoms classify taxonomy groups

Classifying, or organizing living things into groups, is one way to **study** and **communicate** about organisms. To **classify** means to organize things into **groups**. The groups are based on the ways in which things are **similar**. The science of classifying living things is called **taxonomy**. Aristotle began his system of taxonomy by dividing organisms into two large **kingdoms**.

Scientific Naming

Fill in the blanks with the correct terms listed below. One term will be used twice.

species	scientific name	Panthera	genus
Carolus Linnaeus	Latin	names	bionomial nomenclature
organism	scientific name	two parts	leo

Organisms can have several common or popular **names**. It might be hard to identify a(n) **organism** if it has several names. To avoid this problem, scientists use a system that gives all organisms a(n) **scientific name**.

For example, the **scientific name** for a lion is *Panthera leo*. **Panthera** is the genus name for large cats, and lions belong to the species **leo**. The language used for naming organisms is **Latin**.

The system for giving organisms a scientific name was first developed by **Carolus Linnaeus**. His two-word naming system is called **binomial nomenclature**. The first part of the name is the **genus**; it is always capitalized. The last part of the name is the **species**; it starts with a small letter.

29

and animals that fly through the air were in a third group. The plant kingdom was also divided into three groups based on size and structure.

Eventually scientists began to criticize Aristotle's system because it had too many exceptions. Animals were classified according to where they lived, but what about frogs? Frogs spend part of their lives in water and part on land. There was a problem with Aristotle's classification of plants also. Trees aren't all related just because they're trees. For example, an apple tree is more closely related to a rose bush than to a maple tree. A more logical classification system was needed.

Scientific Naming

Like Aristotle, Carolus Linnaeus (luh NAY us), a Swedish physician and naturalist, created a system to classify organisms based on similarities in body structures and systems, size, shape, color, and methods of obtaining food.

Linnaeus's system gives a two-word name to every organism. The two-word naming system is called **binomial nomenclature.** *Binomial* means "two names." The two-word species name is commonly called the organism's scientific name or Latin name. The first word of an organism's scientific name is the genus, and the second is the specific name. **Species** is the smallest, most specific classification category. A **genus** is the next largest category. It is a group of different species that are similar. Different species that belong to the same genus have common ancestors.

An example of a two-word, or species, name is *Felis catus.* This is a common house cat. Notice that the first word, the genus name, always begins with a capital letter. The second word, the specific name, begins with a lowercase letter. Both words in a scientific name are written in italic or underlined. Linnaeus chose the Latin language for his naming system because the meanings for Latin words are the same around the world and the language is unchanging. Another important result of Linnaeus's system is that no two organisms have the same scientific name. Because of Linnaeus's system and the use of Latin, scientists around the world recognize *Felis catus* as a house cat and not a Bengal tiger, *Felis tigris.*

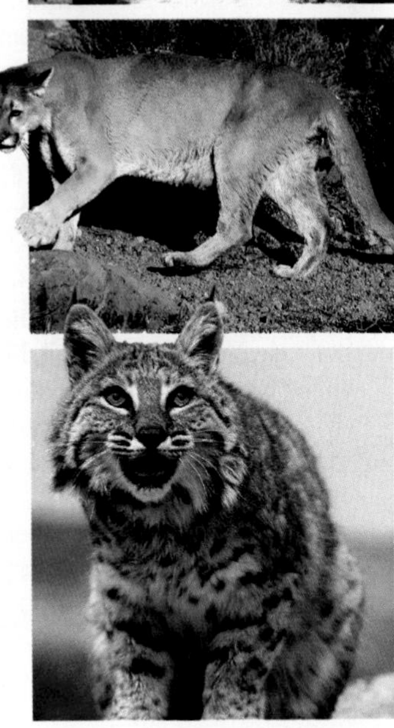

Figure 7-2. The top photo shows a housecat, *Felis catus.* Other members of the genus *Felis* are the cougar, *Felis concolor* (middle), and the bobcat, *Felis rufa* (bottom). Notice that there are three species, but that they all belong to the same genus.

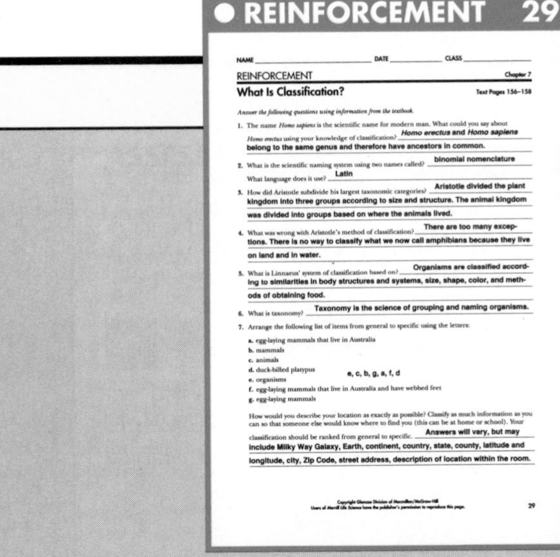

● REINFORCEMENT 29

NAME ______ DATE ______ CLASS ______

REINFORCEMENT — Chapter 7

What Is Classification? — Text Pages 156–158

Answer the following questions using information from the textbook.

1. The name *Homo sapiens* is the scientific name for modern man. What could you say about *Homo erectus* using your knowledge of classification? **Homo erectus and Homo sapiens belong to the same genus and therefore have ancestors in common.**
2. What is the scientific naming system using two names called? **binomial nomenclature** What language does it use? **Latin**
3. How did Aristotle subdivide his largest taxonomic categories? **Aristotle divided the plant kingdom into three groups according to size and structure. The animal kingdom was divided into groups based on where the animals lived.**
4. What was wrong with Aristotle's method of classification? **There are too many exceptions. There is no way to classify what we now call amphibians because they live on land and in water.**
5. What is Linnaeus' system of classification based on? **Organisms are classified according to similarities in body structures and systems, size, shape, color, and methods of obtaining food.**
6. What is taxonomy? **Taxonomy is the science of grouping and naming organisms.**
7. Arrange the following list of items from general to specific using the letters:
   a. egg-laying mammals that live in Australia
   b. mammals
   c. animals
   d. duck-billed platypus **e, c, b, g, a, f, d**
   e. organisms
   f. egg-laying mammals that live in Australia and have webbed feet
   g. egg-laying mammals

How would you describe your location as exactly as possible? Classify as much information as you can so that someone else would know where to find you (this can be at home or school). Your classification should be ranked from general to specific. **Answers will vary, but may include Milky Way Galaxy, Earth, continent, country, state, county, latitude and longitude, city, Zip Code, street address, description of location within the room.**

29

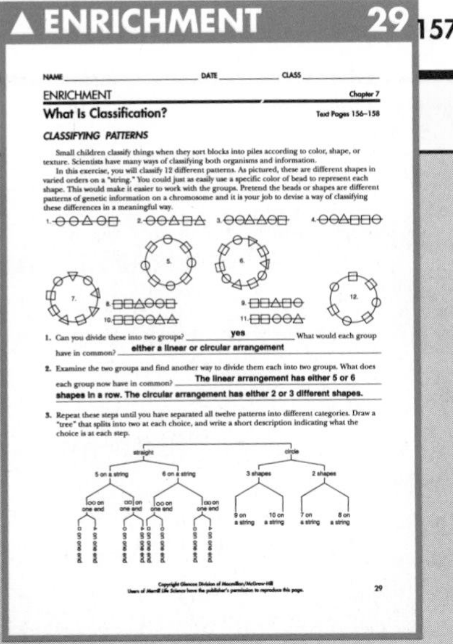

▲ ENRICHMENT 29

NAME ______ DATE ______ CLASS ______

ENRICHMENT — Chapter 7

What Is Classification? — Text Pages 156–158

CLASSIFYING PATTERNS

Small children classify things when they sort blocks into piles according to color, shape, or texture. Scientists have many ways of classifying both organisms and information.

In this exercise, you will classify 12 different patterns. As pictured, these are different shapes in varied orders on a "string." You could just as easily use a specific color of bead to represent each shape. This would make it easier to work with the groups. Pretend the beads or shapes are different patterns of genetic information on a chromosome and it is your job to devise a way of classifying these differences in a meaningful way.

1. Can you divide these into two groups? **yes** What would each group have in common? **either a linear or circular arrangement**
2. Examine the two groups and find another way to divide them each into two groups. What does each group now have in common? **The linear arrangement has either 5 or 6 shapes in a row. The circular arrangement has either 2 or 3 different shapes.**
3. Repeat these steps until you have separated all twelve patterns into different categories. Draw a "tree" that splits into two at each choice, and write a short description indicating what the choice is at each step.

29

2 TEACH

Key Concepts are highlighted.

CONCEPT DEVELOPMENT

▶ The second word in a species name is officially referred to as the specific epithet.

CROSS CURRICULUM

▶ **Math:** Ask students how numbers are classified. Answers may include real and imaginary, whole numbers and fractions, decimals and percents, rational and irrational.

CHECK FOR UNDERSTANDING

Use Figure 7-2 to check for understanding of scientific naming.

RETEACH

Have students tell something about an organism from the Latin derivation of the word. Examples that may be used include *Musca domestic* (housefly), *Equus zebra* (zebra), and *Camelus dromedarius* (dromedary camel).

EXTENSION

For students who have mastered this section, use the **Reinforcement** and **Enrichment** masters or other OPTIONS provided.

PROBLEM SOLVING

Think Critically: Students may want to use their own shoes, drawings, or pictures to help them solve the problem. The white shoe with laces and a leather sole belongs to Mike. The classification system developed by John's group is:

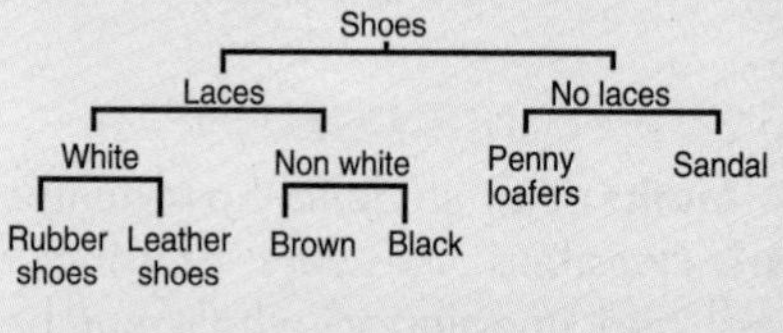

3 CLOSE

▶Ask questions 1-2 and the **Apply** Question in the Section Review.

SECTION REVIEW ANSWERS

1. The purpose is to arrange or group things according to similarities and differences.

2. Aristotle—2 Kingdoms; animals divided according to where they lived, plants divided according to size structure. Linneaus—groupings according to stuctures, systems, size, shape, color, food getting.

3. Apply: Answers will vary. Stamps, clothing, food at a grocery, and organisms are all acceptable answers.

4. Connect to Physics: solids, liquids, and gases

Skill Builder

You could infer that they all have similar ancestors and are closely related since they have the same genus name.

Skillbuilder

ASSESSMENT

Performance: Assess students' abilities to Observe and Infer the differences and similarities among maples; *Acer rubrum, Acer saccharum,* and *Acer pennsylvanicum.*

PROBLEM SOLVING

Whose Shoe?

After the bell had rung, Mrs. Spencer announced, "Yesterday, we discussed how scientists classify organisms. Today, I want you to report to your learning groups and develop a classification system for the shoes of the members of your group."

Dani's group divided the shoes into two kingdoms, one with shoelaces and one without. Then, the group divided the laces group into white and nonwhite shoes. Because Dani's white shoes had rubber soles and Mike's had leather soles, the type of sole was the characteristic the groups chose to subdivide the white shoe group. The group then divided the nonwhite shoes group into brown for Colleen's shoe and black for Albert's shoe.

The kingdom without laces was easy to divide. Maria's shoe was a penny loafer, and Lauren's shoe was a white sandal.

Whose shoe is a white shoe with laces and a leather sole?

Think Critically: Diagram the classification system developed by Dani's group.

SECTION REVIEW

1. What is the purpose of classification?
2. Compare and contrast Aristotle's and Linnaeus's classification systems.
3. **Apply:** List two examples of things that are classified based on their similarities.
4. **Connect to Physics:** Matter is classified into three states. What are these states?

Observing and Inferring

What could you infer about the following species of oaks: *Quercus alba, Quercus rubra, Quercus suber, Quercus macrocarpa.* If you need help, refer to Observing and Inferring in the **Skill Handbook** on page 682.

OPTIONS

ASSESSEMENT—ORAL

▶ **Why do people classify things?** *By classifying things, people can be more efficient and organized.*

▶ **What kinds of clothes might a clothing store owner group together?** *Clothing stores often have separate racks for cotton, polyester, and wool clothes. Further grouping occurs when tops are placed in one area and pants or skirts in another. Still further, clothes may be arranged by color or style.*

PROGRAM RESOURCES

From the **Teacher Resource Package** use:

Concept Mapping, pages 19-20.

Activity Worksheets, page 65, Mini-Lab: How can leaves be classified?

Activity Worksheets, page 5, Flex Your Brain.

Use **Laboratory Manual,** page 43, Classification.

ACTIVITY 7-1 DESIGNING AN EXPERIMENT
Classifying Seeds

Scientists have classification systems to show how things are related. In what ways can seeds be classified?

Getting Started
Find a system for classifying seeds.

Thinking Critically
List some ways the seeds can be classified. What characteristics will you use?

Materials
Your cooperative group will use:
- packet of 10 different seeds
- hand lens
- metric ruler
- 2 sheets of paper

1. Empty the packet of seeds on a sheet of paper. *Observe* each seed carefully.
2. Divide the 10 seeds into two groups, I and II. The seeds in each group must have at least one thing in common.
3. *Record* the common characteristic that Group I and Group II seeds have in a chart like the one shown.
4. Now, *classify* or divide the Group I seeds into two groups. Record the group characteristics.
5. Divide the seeds into two groups again. Record the common group characteristics.
6. Repeat Step 5 two more times.
7. Repeat Step 4 through 6 with Group II.
8. Give the seed packet and chart to another group. Ask that group to *identify* each seed using your classification system.

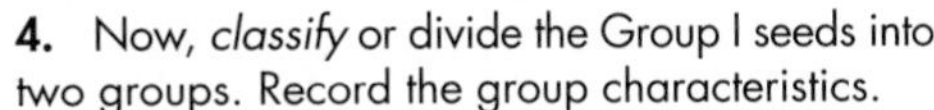

Summing Up/Sharing Results
- In what ways can groups of different types of seeds be classified?
- *Compare* your system with another groups'.
- Why is it an advantage for scientists to use a standardized system to classify organisms? What observations did your group make to support your answer?

Going Further!
How would you *classify* a collection of pine cones all from a single tree? Why would this grouping be harder to do than your packet of seeds?

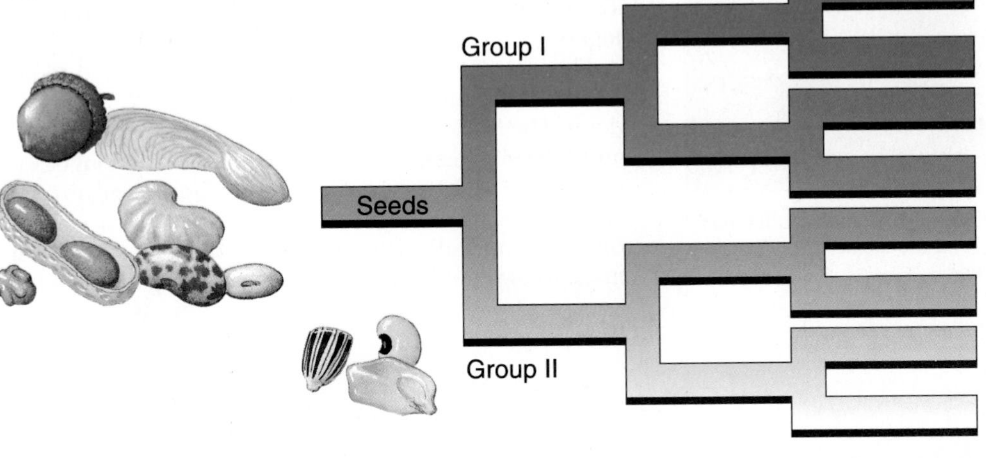

ACTIVITY 7-1

OBJECTIVE: Design and carry out an experiment to show the variation in a population of seeds.
Time: one class period

PROCESS SKILLS applied in this activity are **observing, classifying,** and **defining operationally.**

PREPARATION
Prepare packets of ten different seeds. Seeds that can easily be classified are lima beans, black-eyed peas, squash seeds, navy beans, peas, kidney beans, wrinkled corn seeds, popcorn, black beans, sunflower seeds, and pinto seeds.

Cooperative Learning: Have students worked as Paired Partners to complete this activity.

SAFETY
Caution students not to put any seeds into their mouths. Use only edible seeds, not treated seeds.

THINKING CRITICALLY
Students may consider such characteristics as size, color, shape, and texture. Students' answers will vary, but will use one of the above listed characteristics.

TEACHING THE ACTIVITY
*Refer to the **Activity Worksheets** for additional information and teaching strategies.*
- If students have difficulty dividing the seeds into groups, make sure they understand what *characteristics* are. Have them use a metric ruler and hand lens as tools to assist them.

SUMMING UP/SHARING RESULTS
- Size, color, shape, texture, how they are attached to the plant.
- Answers will vary. Comparing provides an opportunity to see how others have solved the problem.
- Different classification systems could result in confusion and differing interpretations of scientific observations. Answers will vary but should be based on the observation that not every student classified the seeds in the same way.

GOING FURTHER!
Classifying pine cones from a single tree would be more difficult than a collection of seeds. The pine cones would have much less variation but could still be classified by such variations as sizes and colors.

PROGRAM RESOURCES

From the **Teacher Resource Package** use:
Activity Worksheets, pages 59-60, Activity 7-1, Classifying Seeds.

Activity
ASSESSMENT
Performance: To further assess students' ability to classify, give them cutouts of ten different leaves and have them devise and describe a classification of the leaves.

SECTION 7-2

PREPARATION

SECTION BACKGROUND

▶ Many people today still use a two or three kingdom system of classification because that is what they learned when they were in school.

▶ Taxonomy is the science of classifying. Today, taxonomy is usually based on inferred evolutionary relationships.

▶ Explain to students that the classification of an organism is always subject to change, based on new evidence.

1 MOTIVATE

▶ Have students study Table 7-1 on page 162 and brainstorm the names of organisms (other than those pictured) that belong to each kingdom. Have them back up their responses by giving the characteristics they used to place the organisms in each kingdom.

TYING TO PREVIOUS KNOWLEDGE: Use the information in the previous section to guide students in their understanding of the modern classification system.

7-2 Modern Classification

New Science Words

phylogeny
phylum
division
classes
orders
families

Objectives

▶ Name the five kingdoms of living things.
▶ Identify characteristics and members of each kingdom.
▶ List the groups within each kingdom.

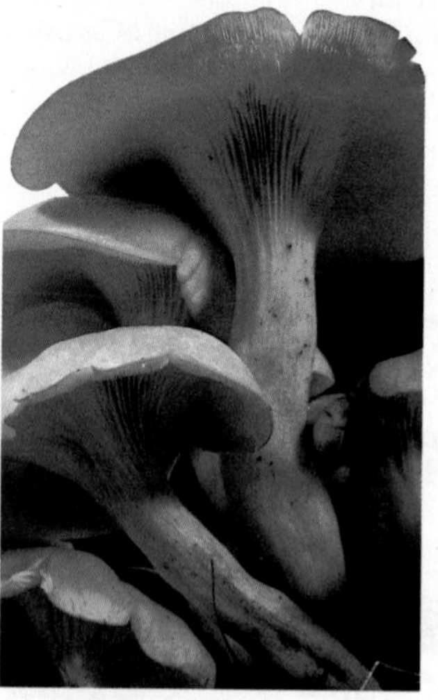

Figure 7-3. These colorful mushrooms are members of Kingdom Fungi.

Five-Kingdom System

How does the classification system used today by scientists differ from those of the past? Aristotle and Linnaeus developed their systems of classification using only characteristics of organisms that were easily observed. Besides easy to see characteristics, scientists today look at other types of traits to classify organisms. They look at the chemical makeup of the organism and its ancestors. Scientists find relationships among organisms by looking at similarities in genes and body structures. They also study fossils and the embryo of an organism as it develops. By studying all these things, scientists can determine an organism's phylogeny.
1 The **phylogeny** of an organism is its evolutionary history. Phylogeny tells scientists who the ancestors of an organism were and helps to classify it. Today classification of many organisms is based on phylogeny.

2 The classification system most commonly used today separates organisms into five major groups called kingdoms. The five kingdoms are Animal, Plant, Fungi, Protist, and Monera. Organisms are placed into a kingdom based on four characteristics. The first is whether or not the cells of the organism have a nucleus. Second, are there one or many cells present? Third, does the organism make its own food? And fourth, does the organism move?

Prokaryotes and Eukaryotes

Kingdom Monera is made up of one-celled organisms that are simple structures. Members of this kingdom include bacteria such as streptococcus, which causes strep

OPTIONS

Meeting Different Ability Levels

For Section 7-2, use the following **Teacher Resource Masters** depending upon individual students' needs.

◆ **Study Guide Master** for all students.
● **Reinforcement Master** for students of average and above average ability levels.
▲ **Enrichment Master** for above average students.

Additional Teacher Resource Package masters are listed in the OPTIONS box throughout the section. The additional masters are appropriate for all students.

◆ STUDY GUIDE 30

NAME ________ DATE ________ CLASS ________

STUDY GUIDE — Chapter 7

Modern Classification — Text Pages 160–163

Complete the following sentences using information from the textbook.

1. The kingdom **Monera** is made up of bacteria and cyanobacteria, single-celled plantlike organisms.
2. The **phylogeny** of an organism is its evolutionary history.
3. Monerans are called **prokaryotes** because they do not have an organized nucleus.
4. A characteristic of the members of the plant kingdom is **they can make their own food**.
5. Can there be more than one species in a genus? **yes**
6. The word eukaryote means **true nucleus**.
7. The first characteristic to consider when classifying an organism by kingdom, is whether or not the cells have a **nucleus**.
8. The first cells to form on Earth were in the kingdom **Monera**.

Use the clues below to complete the puzzle.

Across
1. an organism's evolutionary history
5. names a specific organism
6. divides the class Insecta into more than one part
8. classification that divides a phylum
9. created a system to classify organisms

Down
1. "before nucleus"
2. a group within a family
3. the two-word naming system, binomial ____
4. replaces phylum in the plant kingdom
7. classification that divides orders

30

TECHNOLOGY

Beyond Appearances

Besides looking at an animal's physical traits, scientists now check out an animal's DNA to classify it. The more similar the DNA between two animals is, the more closely they are related. By using DNA information, scientists can tell which similar animals are actually separate species. They can also study how geographic separation and other factors affect populations. The information scientists gather helps zoos and wildlife programs conserve many species of animals.

Genetic studies are already changing the way some organisms are classified. By studying DNA, scientists have found that pandas really are bears, and not relatives of the raccoon as they once thought. The "Asiatic" lions in U.S. zoos are not truly from Asia. They have African genes as well. White rhinos in North and South Africa are genetically the same in spite of being separated by 2000 miles. Some gray foxes living on islands off California's coast vary genetically from each other.

Think Critically: How do scientists use DNA to classify organisms?

throat, and cyanobacteria, a one-celled plantlike organism found in lakes and ponds. Unlike members of the other kingdoms, monerans don't have an organized nucleus surrounded by a membrane or membrane-bound organelles. Because monerans don't have an organized nucleus or membrane-bound organelles they are called prokaryotes. *Prokaryote* 3
means "before kernel." All organisms other than bacteria and cyanobacteria are eukaryotes. Eukaryotes are organisms 4
with a nucleus and organelles surrounded by membranes. *Eukaryote* means "true kernel." Figure 7-4 shows the differences between prokaryotes and eukaryotes. Table 7-1 on page 162 gives examples and basic information about each of the five kingdoms. You will learn more about each kingdom in coming chapters.

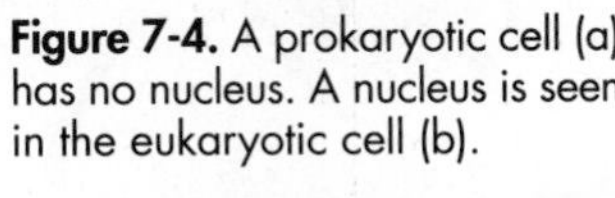

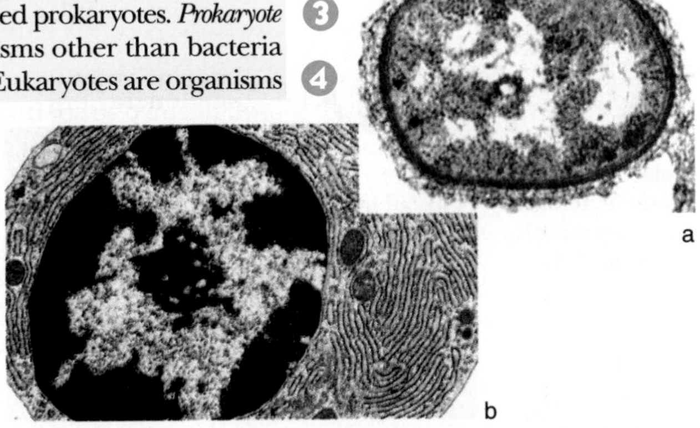

Figure 7-4. A prokaryotic cell (a) has no nucleus. A nucleus is seen in the eukaryotic cell (b).

● REINFORCEMENT 30

NAME ______ DATE ______ CLASS ______

REINFORCEMENT — Chapter 7

Modern Classification — Text Pages 160–163

Answer the following questions using information from the textbook.

1. What are the ways scientists determine an organism's phylogeny? **They look at external characteristics, chemical make-up, similarities in gene and body structure, study fossils, and study the embryology of the organism.**
2. What do the Animal and Fungi Kingdoms have in common? **They don't make their own food.**
3. How are members of the Animal Kingdom different from members of the Fungi Kingdom? **Animals can move.**
4. Organisms belonging to the Kingdoms Protist and Monera are unicellular. What is it that places them into different kingdoms? **Members of the Kingdom Protist have organized nuclei and monerans do not.**
5. How do we know that the first cells to form on Earth belonged to the Kingdom Monera? **Fossil evidence has indicated that moneran cells were the first on Earth.**
6. Put these Latin names in order from the most general to the most specific.

order—Diptera
class—Insecta
kingdom—Animal
phylum—Arthropoda
species—*melanogaster*
family—Tephritidae
genus—*Drosophila*

a. **Animal**
b. **Arthropoda**
c. **Insecta**
d. **Diptera**
e. **Tephritidae**
f. ***Drosophila***
g. ***melanogaster***

7. The organism you have classified is a fruit fly. What is its scientific name? ***Drosophila melanogaster***

30

▲ ENRICHMENT 30

NAME ______ DATE ______ CLASS ______

ENRICHMENT — Chapter 7

Modern Classification — Text Pages 160–163

Animals are many-celled organisms that must obtain food to stay alive. Most animals can move about in the environment. About 95 percent of the known species of animals do not have backbones or skeletons inside their bodies. In the animal kingdom, animals are classified into nine main groups called phyla.

Use reference books and your textbook to find information to complete the table below.

Animal Phyla Characteristics		
Phylum	Common Name	Characteristics
Porifera	sponges	thick sac of cells with pores, canals, and chambers; live in water; attached to one place
Cnidaria	cnidarians (sea anemone, hydra, jellyfish)	central cavity and mouth; most have tentacles; live in water; float or attached to one place
Platyhelminthes	flatworms	flattened body; freeliving in water or parasitic
Nematoda	roundworms	round body in cross section; free living in water or on land; some are parasitic
Annelida	segmented worms	body divided into segments that have bristles; freeliving in water or on land
Mollusca	mollusks (snails, octopus, squid, scallop)	soft bodies; most have shells or shell-like coverings; live in water or on land
Arthropoda	arthropods (insects, spiders, shrimp)	body divided into sections; jointed legs, have exoskeletons; live in water or on land
Echinodermata	echinoderms (sea cucumber, sea urchin, sea star)	spiny or leathery skin; radial symmetry; water vascular system; live in salt water
Chordata	chordates (vertebrates)	internal skeleton; specialized body systems; paired appendages; live in water and on land

30

TECHNOLOGY

For more information on using DNA to classify animals see "Genetics for Wildlife Conservation" by Jeffrey P. Cohn, *BioScience*, March 1990, pp. 167-171.

Think Critically: Genetic similarities are due to shared evolutionary history. They allow scientists to look beyond appearances.

TEACHER F.Y.I.

▶ Because the RNA in Archaebacteria is more like that in eukaryotes than prokaryotes, some taxonomists today suggest making a sixth kingdom positioned between the monerans and the protists.

2 TEACH

Key Concepts are highlighted.

CONCEPT DEVELOPMENT

▶ Emphasize to students that all living things can be traced back to the prokaryotes.

▶ Be sure your students understand the difference between taxonomy and phylogeny.

▶ Explain that fungi, plants, and animals evolved from the protists during the Precambrian.

▶ Tell students that not all of the kingdoms or other groups contain the same number of organisms. For example, some genera contain only a single species while others contain many. The animal kingdom accounts for 93 percent of all known organisms.

▶ Help students distinguish the characteristics of the prokaryotic cell and the eukaryotic cell in Figure 7-4.

MINI QUIZ

Use the Mini Quiz to check students' recall of chapter content.

1. **The _____ of an organism is its evolutionary history.** *phylogeny*
2. **The classification system most commonly used today separates organisms into five major groups called _____.** *kingdoms*
3. **Because monerans don't have an organized nucleus or organelles, they are called _____.** *prokaryotes*
4. **_____ are organisms with a nucleus and organelles surrounded by membranes.** *Eukaryotes*

Cooperative Learning: Arrange students in Paired Partners learning groups. Provide each group with illustrated guides to various groups of organisms or allow them to use an encyclopedia to find out the names for a list of organisms. Examples may include mockingbird, zinnia, lynx, cottontail, Monarch butterfly, blue crab, earthworm, and human. Each list should have 10 to 20 organisms, and every list should be different.

Answer: Igneous and metamorphic rocks have both been heated to melting. Metamorphic rock was a different kind of rock before being heated. It may have been igneous, sedimentary, or metamorphic.

Table 7-1

LIFE'S FIVE KINGDOMS

	Monera	Protist	Fungi	Plant	Animal
Type of cells	Prokaryotic	Eukaryotic	Eukaryotic	Eukaryotic	Eukaryotic
One-celled or Many-celled	One-celled	One- and many-celled	One- and many-celled	Many-celled	Many-celled
Movement	Some move	Some move	Don't move	Don't move	Move
Nutrition	Some members make their own food, others obtain it from other organisms.	Some members make their own food, others obtain it from other organisms.	All members obtain food from other organisms.	Members make their own food.	Members eat plants or other animals.
Examples					

Rocks are grouped into three main categories: igneous, sedimentary, and metamorphic. Find out the characteristics of igneous and metamorphic rocks.

Groups within Kingdoms

Imagine you received some money for your birthday. You decided to spend the money on the new tape you've been wanting, so you go to a music store at the mall. Now what do you do? Will you look through all the tapes in the store until you find the one you're looking for? You don't have to because the tapes are separated into categories of similar types of music such as rock, soul, classical, country, and jazz. Within each category, the tapes are divided by artists, and then into specific titles by the artist. Because of this classification system, you can easily find the tape you want.

OPTIONS

ASSESSMENT—ORAL

▶ **If a sixth kingdom were to be used, where would the organisms in it come from?** *They would be removed from an existing kingdom.*

▶ **Suppose a cell is so small that it has no mitochondria or chloroplasts. Into which kingdom would it be classed?** *Monera*

▶ **A slime mold is made up of single cells during most of its life. At some times, it looks like a fungus. Other times it appears to be an animal or plant. Which kingdom should it be classified in?** *Kingdom Protista*

PROGRAM RESOURCES

From the **Teacher Resource Package** use:

Transparency Masters, pages 21-22, Life's Five Kingdoms.

Transparency Masters, pages 23-24, Classification Pyramid.

Use **Color Transparency** 11, Life's Five Kingdoms.

Use **Color Transparency** 12, Classification Pyramid.

Scientists classify organisms into groups in just the same way. Organisms are placed into kingdoms. Each kingdom is separated into six groups, each smaller than the one before it. The largest group within a kingdom is a **phylum.** In the plant and fungi kingdoms, the word **division** is used in place of phylum. Each phylum or division is separated into **classes.** Classes are separated into **orders,** and orders are separated into **families.** A genus is a group within a family. A genus can have one or more species.

Scientists go through this series of categories to name a specific organism the same way you went through a series of categories to find your tape. To understand how an organism is classified, look at the classification of the Arctic wolf in Figure 7-5. Each grouping shows organisms that have more in common. Each grouping also contains fewer organisms. When you reach species, a specific organism has been named, the Arctic wolf.

Kingdom	Animal
Phylum	Chordata
Class	Mammalia
Order	Carnivora
Family	Canidae
Genus	*Canis*
Species	*Canis lupus*

Figure 7-5. The classification of the Arctic wolf shows its genus to be *Canis* and species to be *Canis lupus.*

SECTION REVIEW

1. Name and describe each kingdom. Identify a member of each.
2. Why are there smaller groups within each kingdom?
3. **Apply:** What kingdom does a single-celled, eukaryotic organism that makes its own food belong in?
4. **Connect to Chemistry:** Elements are classified into families in the periodic table. How many element families are there?

Concept Mapping

Use the following terms to make a network tree concept map. Provide your own linking words. *Cells, cell organelles, eukaryote, prokaryote, plants, animals, cyanobacteria, bacteria, no membrane-bound organelles, no nucleus, fungi, protists, organized nucleus.* If you need help, refer to Concept Mapping in the **Skill Handbook** on pages 688 and 689.

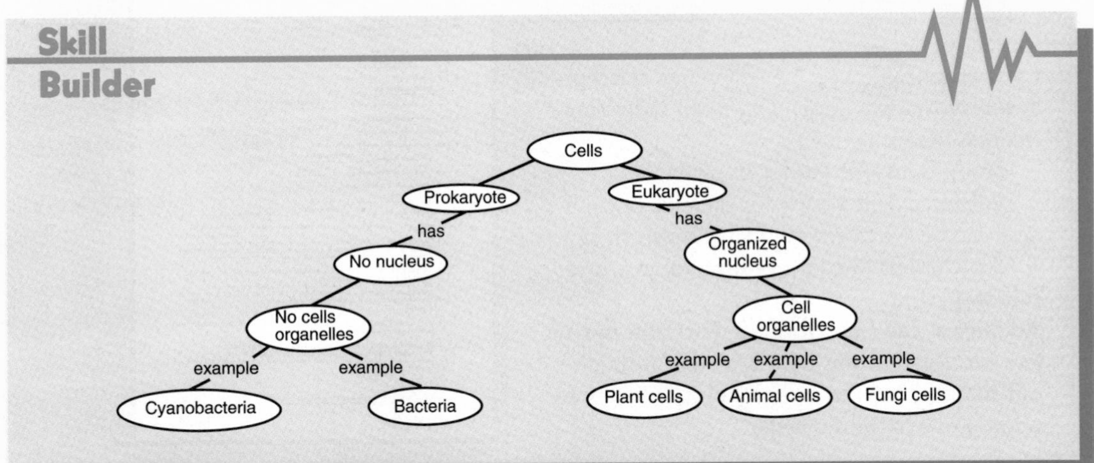

CHECK FOR UNDERSTANDING

Ask students to identify the classification group with the most members, with the least members. Ask them to explain.

RETEACH

Play a game in which the object is to decide the level of the group described. Example: **What level contains the bloodhound and poodle?** *species* **What level contains all the kinds of willow trees?** *genus* **What level contains all plants?** *kingdom*

EXTENSION

For students who have mastered this section, use the **Reinforcement** and **Enrichment** masters or other OPTIONS.

3 CLOSE

▶ Ask questions 1-2 and the **Apply** Question in the Section Review.

SECTION REVIEW ANSWERS

1. Monera: one-celled prokaryotics (streptococcus bacteria). Protista: one- and many-celled eukaryotes (paramecium). Fungi: one- and many-celled eukaryotes that can't move, obtain food from other organisms (mushroom). Plant: many-celled organisms that make their own food, can't move (oak tree). Animal: many-celled organisms that can move but can't make their own food (dog).
2. Scientists classify organisms into smaller groups with more characteristics in common. A genus and species identify a specific organism.
3. Apply: Protista
4. Connect to Chemistry: 18

Skillbuilder ASSESSMENT

Portfolio: Use this Skillbuilder to assess students' abilities to form a Concept Map of cell characteristics.

SECTION 7-3

PREPARATION

SECTION BACKGROUND

▶ Tropical rain forests occur near the Equator and may receive as much as 150-350 cm of rain per year.

▶ The Nature Conservancy and other similar groups are buying up parcels of the tropical rain forest in some countries to protect and preserve them.

PREPLANNING

▶ If you show the film listed in Motivate, you will need to order it.

1 MOTIVATE

▶ Show *The Tropical Rain Forest,* a short film available from The National Film Board of Canada.

TYING TO PREVIOUS KNOWLEDGE: Make reference to the changes that have occurred in your particular biome due to human activity.

2 TEACH

Key Concepts are highlighted.

CONCEPT DEVELOPMENT

▶ Ask students: **Why is the tropical rain forest soil poor in nutrients?** *Decomposers rapidly recycle organic materials into biomass.*

7-3 The Rain Forest Crisis

New Science Words

species diversity

Objectives

- ▶ **Identify two dangerous effects of tropical deforestation.**
- ▶ **Explain the importance of the rain forest to us and to the people who live there.**

Undiscovered Treasures

There are still a few places on Earth that have escaped human technology. Places such as dense, humid jungles hold thousands of undiscovered and unnamed species of plants and animals. The tropical rain forests of South America are just those places. Life there has gone on undisturbed for thousands of years, but this may all soon come to an end. There is a crisis occurring at this very moment.

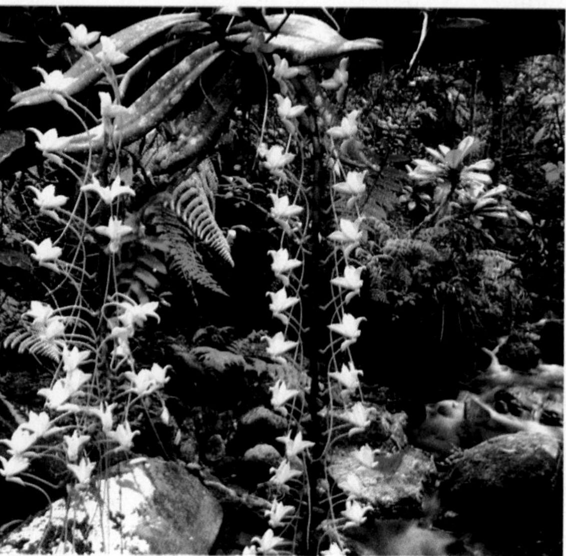

Figure 7-6. Many species of animals and plants will be lost if deforestation continues.

There is no time to waste. Every minute, more than 50 more acres of rain forest are cleared for farming, or cut for timber. Many organizations are hard at work to explore these mysterious rain forests and help stop deforestation, the destruction of forests. Each year, hundreds of field biologists hunt through the dense tropical forests to unlock their mysteries. They want to learn as much as possible about the plants and animals that live there before these organisms become extinct. Scientists think that only about 15 percent of all the species of plants and animals there have been identified. The rest may never be discovered. This great variety of plants and animals, or **species diversity,** in the rain forests of the world is being destroyed by humans at a rapid rate. In the next 30 years we may lose one-fourth of all the species alive today.

OPTIONS

Meeting Different Ability Levels

For Section 7-3, use the following **Teacher Resource Masters** depending upon individual students' needs.

◆ **Study Guide Master** for all students.

● **Reinforcement Master** for students of average and above average ability levels.

▲ **Enrichment Master** for above average students.

Additional Teacher Resource Package masters are listed in the OPTIONS box throughout the section. The additional masters are appropriate for all students.

◆ STUDY GUIDE 31

NAME ______ DATE ______ CLASS ______

STUDY GUIDE — Chapter 7

The Rain Forest Crisis — Text Pages 164–165

Fill in the blanks using the correct words below.

species diversity	thirty	extinction	farming	soils
desert	atmosphere	deforestation	rain forests	fifty
carbon dioxide	oxygen	logging		

1. Each minute we lose about **fifty** more acres of **rain forests**.
2. The great variety of plants and animals found in the rain forest is called **species diversity**.
3. The destruction of forests is called **deforestation**.
4. We may lose one-fourth of all species in the next **thirty** years.
5. South Americans need land for **farming** and **logging**.
6. Climatic changes could result from a damaged **atmosphere**.
7. Trees and plants remove **carbon dioxide** from the air.
8. Rain forests typically have poor **soils**.
9. With extensive loss of the rain forests, the southeastern United States may become a **desert**.
10. Deforestation may result in **extinction** of many species.

Place a check (✓) next to the statements that agree with the textbook. Rewrite the statements that disagree.

____ 1. There is no place on Earth that has not been touched by human technology. **The tropical rain forests of South America have escaped the effects of human technology.**

✓ 2. There are thousands of species of plants and animals yet to be discovered and named.

____ 3. The rain forests produce carbon dioxide, which is important for the air we breathe. **The rain forests produce oxygen.**

____ 4. Once a section of rain forest is cleared, it can be farmed over and over again. **It can only be farmed a few years before it is no longer fit to raise crops.**

✓ 5. Organisms that haven't even been discovered yet may become extinct if deforestation continues.

✓ 6. Scientists think that only 15 percent of all the species of plants and animals in the tropical rain forest have been identified.

____ 7. The evolution of species from a single ancestor is called species diversity. **The variety of species is called species diversity.**

✓ 8. Deforestation could add to the carbon dioxide in the atmosphere because plants and trees remove carbon dioxide.

 31

The Consequences

Will the destruction of the rain forest really affect you? Is the effort to save the forest really necessary? Scientists think so. Along with saving species of animals, saving the trees and plants is also very important. Our atmosphere is very dependent on them to remove carbon dioxide and produce oxygen. The air that we breath in the United States will be affected by the loss of tropical forests. The damaged atmosphere may cause changes in weather patterns and dangerous storms. The climate of areas of the United States could also change. The northeastern region could have the tropical climate of Florida. And the southeast may become a desert.

Wild animals and plants aren't the only ones dependent on the rain forest lands for survival. The native people depend on farming these areas for survival. Remember how the pioneers settled America? They were starving to death and many died before they cleared thousands of acres and mastered farming. The people of many tropical areas are in the same situation. They are dependent on farming or logging for their income and survival. Farming is very destructive to rain forests because they have poor soils. An area can be farmed for only a few years before it is no longer fit to raise crops. A new piece of forest must then be cleared to be used for farming. Every few years the people must cut a new area of forest to grow crops to survive.

Choose one of the rain forest plants, animals, or products listed here to research: toucans, squirrel monkeys, mahogany, kapok, curare, sloths, and bromeliads. Make a drawing of your choice **in your Journal**, and describe its location and its importance.

Figure 7-7. Rain forests are cleared for farming.

SECTION REVIEW

1. Why is tropical deforestation a concern to people in North America?
2. How are the native people dependent on rain forest lands?
3. **Connect to Earth Science:** Look at a world map and find out in what latitudes tropical rain forests are found.

You Decide!

Much of the research on the tropical rain forests is being conducted by scientists from countries outside the rain forest, such as the United States. Is it right for the United States to purchase and preserve land that other people depend upon for survival? Should another society walk into their homeland and demand that the rain forest destruction stop?

● REINFORCEMENT 31

NAME ______ DATE ______ CLASS ______

REINFORCEMENT — Chapter 7

The Rain Forest Crisis — Text Pages 164–165

Answer the following questions using information from the textbook.

1. At what rate is deforestation occurring in tropical rain forest areas? **Each minute 50 or more acres of rain forest are being lost.**
2. Why are rain forest lands a poor choice for agriculture? **The soils are so poor that the area can only be farmed for a few years.**
3. Without the large forest, there could be a buildup of **carbon dioxide gas** in the atmosphere.
4. Forests produce **oxygen** and remove carbon dioxide.
5. Why will species become extinct if deforestation continues unchecked? **Habitats for many species are being destroyed by deforestation.**
6. What is species diversity? **It is the great variety of plants and animals in a particular place, such as the rain forest.**
7. What percentage of all species of plants and animals do scientists think have been identified in the rain forests? **about 15%**
8. What is the hurry to identify and study new species? **They may soon become extinct, so it is important to find them now.**
9. What proportion of all species alive today might be lost in the next 30 years? **one-fourth**
10. What climatic changes might we expect to see in the future in the United States as a result of the loss of tropical forests? **Northeastern regions could have tropical climates like Florida; the Southeast may become a desert.**
11. Why do people cut down the trees in the rain forests? **People are dependent on the trees for timber and on the land for farming. Their income and survival depend on the forest resources.**
12. Where are some of the tropical rain forests located? **In South America**

 31

▲ ENRICHMENT 31

NAME ______ DATE ______ CLASS ______

ENRICHMENT — Chapter 7

The Rain Forest Crisis — Text Pages 164–165

DARWIN'S FINCHES

In this lesson, you learned about the high species diversity in the tropical rain forest. Why are there so many species, and what advantage might there be in having a high level of diversity?

Charles Darwin made an interesting observation when he first visited the Galapagos Islands in 1835. He found 14 different species of finches which had never before been identified. Each species had a different type of beak adapted for a specific diet and method of feeding. Darwin was the first to come up with a theory to explain **speciation**, the evolution of one species into others.

Use references to find out more about this topic and answer the following questions.

1. What is adaptive radiation? **It is the process by which new populations spread into different environments while changing in ways that allow individuals to take advantage of available niches.**
2. What is an ecological niche? **It is all the requirements for a species to live successfully and reproduce. It is a species' way of life.**
3. What usually happens if members of different species interbreed? **The offspring are sterile or do not live.**

Below is a drawing of some of the types of beaks belonging to finches that Darwin discovered.

4. Why are their beaks different? **Each species has a beak that is designed for a particular diet.**
5. What feature of their habitat helped the Galapagos finches develop different adaptations? **geographical isolation—separation of the volcanic islands**
6. When members of the species are able to take advantage of a niche by adapting to it, they usually survive and reproduce, passing their traits on to their offspring. What is this process called? **natural selection**

 31

CHECK FOR UNDERSTANDING

Ask students what they imagine life in a rain forest would have been like before and after human intrusion.

RETEACH

Cooperative Learning: Have students work as Expert Teams to make a script that describes rain forest destruction.

EXTENSION

For students who have mastered this section, use the **Reinforcement** and **Enrichment** masters or other OPTIONS.

MULTICULTURAL AWARENESS

Use Figures 7-6 and 7-7 to motivate students to learn how the destruction of a rain forest affects natives in that area.

3 CLOSE

▶ Ask questions 1-2 in the Section Review.

SECTION REVIEW ANSWERS

1. Deforestation decreases species diversity and may lead to climate change all over the world.

2. Native people depend on rain forests for work, shelter, food, and clothing.

3. Connect to Earth Science: Tropical rain forests occur in South America, Africa, Asia, and Australia between 30° north and 30° south latitude.

YOU DECIDE!

Answers will vary. All answers that show clear and logical thinking should be accepted.

PROGRAM RESOURCES

From the **Teacher Resource Package** use:

Critical Thinking/Problem Solving, page 11, The Role of Humans in the Extinction of Species.

Cross-Curricular Connections, page 11, Locating Tropical Rain Forests.

Science and Society, page 11, What Should Be Done to Protect the Rain Forests?

Science Integration Activities, 7, Human Impact on the Environment

SECTION 7-4

PREPARATION

SECTION BACKGROUND

▶ Explain that a key usually contains the most general statements first, and progresses to more specific characteristics until a species is identified.

▶ Extinct organisms known only from fossils can also be classified using specialized dichotomous keys.

PREPLANNING

▶ You will need to obtain various keys to organisms for use in the Motivate, Journal, and Close activities.

1 MOTIVATE

▶ Pass out examples of taxonomic keys or field guides to the class. Ask students to describe the function of these books and when they might be used. Challenge them to key out a particular bird or other organism they know about.

MINI-Lab

▶**Answer:** Students should give their organisms descriptive names. Be sure they are Latinized.

MINI-Lab ASSESSMENT

Performance: To further assess students' ability to name organisms, have them research the scientific names of three house pets.

7-4 Identifying Organisms

New Science Words

dichotomous keys

Objectives

- ▶ List reasons scientific names are more useful to scientists than common names.
- ▶ Identify the function of a dichotomous key.
- ▶ Demonstrate how to use a dichotomous key.

MINI-Lab

How are organisms named?

Make a fictitious organism out of clay. Give your organism a scientific name. Be sure the name is Latinized and *infers* information about its species. Present your organism and its name to the class.

Common Names and Scientific Names

Have you ever seen a bird like the one in Figure 7-8a? What do you call it? Have you heard anyone call this bird a *Turdus migratorius*? In much of the United States this bird is commonly called a robin, or a robin redbreast. However, people who live in England call the bird in Figure 7-8b a robin. In much of Europe the same bird is called a redbreast. If you lived in Australia, you'd call the bird in Figure 7-8c a robin, or yellow robin. Are these the same species of bird? No, these birds are obviously very different from one another. Yet they all have the same or a similar common name. If you were talking informally to someone in your area who knew the same birds you did, you could use common names. But the common names for the different birds could be confusing if you were in another country or in another area of the same country.

What would happen if life scientists used only common names when they tried to share information about the organisms they study? Many errors in understanding would probably result from the confusion. The system of binomial nomenclature developed by Linnaeus gives

Figure 7-8. These three robins have the same common name, yet they are three different species.

a

b

c

OPTIONS

Meeting Different Ability Levels

For Section 7-4, use the following **Teacher Resource Masters** depending upon individual students' needs.

◆ **Study Guide Master** for all students.

● **Reinforcement Master** for students of average and above average ability levels.

▲ **Enrichment Master** for above average students.

Additional Teacher Resource Package masters are listed in the OPTIONS box throughout the section. The additional masters are appropriate for all students.

◆ STUDY GUIDE 32

NAME ____________ DATE ________ CLASS ________

STUDY GUIDE — Chapter 7

Identifying Organisms — Text Pages 166–169

Place a check (✓) next to the statements that agree with the textbook. Rewrite the statements that disagree.

✓ 1. Animals with the same common name can actually be members of different species.

___ 2. Scientists use common names to avoid confusion and error.
Scientists use scientific or Latin names to avoid confusion.

___ 3. Organisms with similar evolutionary history are classified separately.
Organisms with similar evolutionary history are classified together.

✓ 4. The scientific name gives descriptive information about the species.

✓ 5. A dichotomous key is divided into steps, having two choices at each step.

___ 6. *Turdus migratorius* is the scientific name for the Australian robin.
It is the scientific name for the American robin.

Just for fun, see if you can identify yourself using a dichotomous key. Then identify a friend.

1a. If you are female go to step 2.
1b. If you are male go to step 5.
2a. If you have brown or black hair go to step 3.
2b. If you have blonde or red hair go to step 4.
3a. If you have blue, grey, or hazel eyes your name is: **first and last name**
3b. If you have brown eyes your name is: **first and last name**
4a. If you have blue, grey, or hazel eyes your name is: **first and last name**
4b. If you have brown eyes your name is: **first and last name**
5a. If you have brown or black hair go to step 6.
5b. If you have blonde or red hair go to step 7.
6a. If you have blue, grey, or hazel eyes your name is: **first and last name**
6b. If you have brown eyes your name is: **first and last name**
7a. If you have blue, grey, or hazel eyes your name is: **first and last name**
7b. If you have brown eyes your name is: **first and last name**

Answer the following questions.

1. Would many other people in your classroom fit into the same category as you? **yes**
2. List some characteristics that you might use to make a more complete dichotomous key.
Examples may include curly or straight hair, right or left handed, wear glasses or not, wear contact lenses or not, wear braces or not, freckles or not, taller than 5 feet or not.

32

each of these birds a unique scientific name. The scientific name for the bird in 7-8a is *Turdus migratorius.* The name for the bird in 7-8b is *Erithacus rubecula.* The name for the bird in 7-8c is *Eopsaltria australis.*

Scientific names serve four basic functions. First, scientific names help life scientists avoid errors in communication. A life scientist who studied yellow robins, *Eopsaltria australis,* would not be confused by information he or she read about *Turdus migratorius,* the American robin. Second, organisms with similar evolutionary histories are classified together. Because of this you know that organisms that share the same genus name are more related than those that don't. Third, the scientific name gives descriptive information about the species. What can you tell from the species name *Turdus migratorius*? It tells you that this bird must migrate from place to place. Fourth, the scientific names allow information about organisms to be organized and found easily and efficiently. Scientific names are used in guides to organisms called dichotomous (DI kaht uh mus) keys.

In Your JOURNAL

Select a field guide for flowers, trees, butterflies, birds, or mammals. **In your Journal**, write a book report describing the parts of the field guide. Select two organisms in the field guide that closely resemble each other. *Compare* them and explain how they differ, using labeled diagrams.

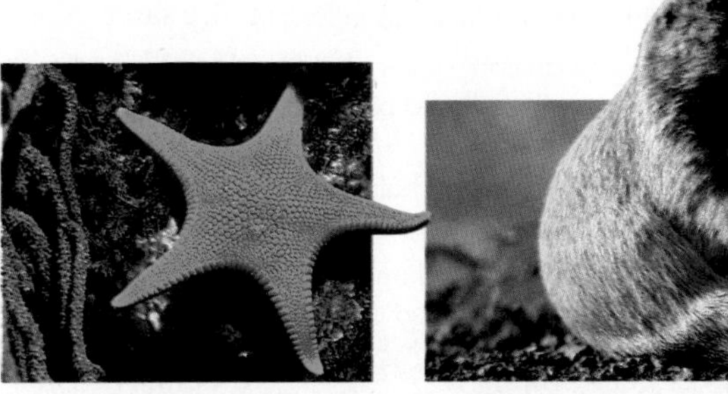

Figure 7-9. Common names can be misleading. Starfish are not fish and prairie dogs are more closely related to squirrels than to dogs.

Making and Using Dichotomous Keys

How would you go about finding out the name of an insect like the one pictured on page 168 in Table 7-2? The easiest way would be to ask an expert on insects, an entomologist at a college in your area. However, no one knows, or is even expected to know, all of the insects or any other major taxonomic group. The first thing the expert might tell you is that the insect is some kind of stag beetle. But, if you don't live near a college or natural history museum, what can you do? Should you give up? No, your next step should be to consult a field guide that contains descriptive guides as an aid to identification. Field guides can be found in a library.

● REINFORCEMENT 32

NAME ______ DATE ______ CLASS ______

REINFORCEMENT — Chapter 7

Identifying Organisms — Text Pages 166–169

Answer the following questions using information from the textbook.

1. Why don't scientists use common names to identify organisms? **Different organisms may go by the same common name in different places. Scientists would have difficulty sharing information.**
2. Why are scientific names important? Give four functions for scientific names.
   a. **to avoid errors in communication**
   b. **to classify organisms with similar evolutionary history together**
   c. **to give descriptive information about the species**
   d. **to organize and easily find information about organisms**
3. The system of **binomial nomenclature** gives each organism a unique scientific name.

Use the key to species of birch trees below to answer the questions that follow.

1a. bark dark, reddish-brown, yellowish-brown to black....go to 2
1b. bark creamy white, pinkish, or gray...go to 6
2a. bark and twigs with wintergreen fragrance when cut....go to 3
2b. bark and twigs without a fragrance when cut....go to 5
3a. leaves with 8-12 pairs of veins...go to 4
3b. leaves with 4-6 pairs of veins...*Betula uber*
4a. bark dark red to almost black; scales of cones smooth, 6-12 mm long...*Betula lenta*
4b. bark reddish-brown, peeling in loose, ragged sheets, scales of cone hairy, 5-7 mm....*Betula alleghaniensis*
5a. branchlets covered near tip with many small glands, Rocky Mountains or Western Canada...*Betula occidentalis*
5b. branchlets smooth, shiny, no glands present, eastern U.S....*Betula nigra*
6a. leaves hairy on lower surface...go to 7
6b. leaves smooth, hairless underside...go to 8
7a. leaves 5-13 cm long, pointed tip...*Betula papyrifer*
7b. leaves 3-7 cm long, pointed tip, winter buds shiny....*Betula pendula*
8a. bark dull gray to grayish-white, smooth and not peeling...*Betula populifolia*
8b. bark white to pinkish-white, peeling...go to 9
9a. leaves 6-10 cm, round base...*Betula caerulea*
9b. leaves 3-5 cm, squared base...*Betula pubescens.*

1. Are the leaves of *Betula populifolia* hairy or smooth on the lower surface? **smooth**
2. How many pairs of veins are on the leaves of *Betula lenta*? **8-12**
3. What is a characteristic of the bark of *Betula alleghaniensis*? **It peels off in loose, ragged sheets.**
4. When a twig of *Betula nigra* is broken, does it give off a wintergreen fragrance? **no**

32

▲ ENRICHMENT 32

NAME ______ DATE ______ CLASS ______

ENRICHMENT — Chapter 7

Identifying Organisms — Text Pages 166–169

A BUG OR A BEETLE?

Often we pick up an insect and casually refer to it as a bug or a beetle. Using scientific definitions the insect may not belong to either of these orders of insects. Bugs belong to the order Hemiptera and beetles belong to the order Coleoptera. Look carefully at these two pictures and use the key below to find out which order these insects belong to. The key is not complete for all orders of insects, but it should help you tell the difference between bugs and beetles. There are a few wingless bugs that are not covered by this key. Use references if needed.

A.

Head showing beak — Front wing

B.

1a. Wings transparent. Go to step 2.
1b. Front wings hard or leathery and covering hind wings. Go to step 2.
2a. Mouth parts chewing. Go to step 3.
2b. No chewing mouth parts. Go to step 4.
3a. Front wings with veins. Order: Orthoptera
3b. Front wings without veins. Antennae usually 11 segments, often 8–10. Order: Coleoptera
4a. Front wings nearly always thickened at the base; membranous at tip; beak comes from front or bottom of head; antennae has 4 or 5 segments. Order: Hemiptera
4b. Front wings of uniform texture; beak comes from hind part of head. Order: Homoptera

1. Which insect is a true bug and which is a true beetle? **Picture B is a bug, picture A is a beetle.**
2. What are some of the distinguishing characteristics of the insects shown? List them below.

	Bug	Beetle
Order:	**Hemiptera**	**Coleoptera**
Wings:	**transparent**	**have no veins**
Antennae:	**4–5 segments**	**11 segments**
Mouthparts:	**not chewing**	**chewing**

3. What is the common name of insect A? **sap beetle** insect B? **leaf bug**

32

TYING TO PREVIOUS KNOWLEDGE:

Bridge this section to the first two sections of this chapter where the classification system and groups within kingdoms are discussed.

2 TEACH

Key Concepts are highlighted.

CROSS CURRICULUM

▶ **Languages:** Latin and Greek names usually describe the appearance or behavior of an organism. Have students look up scientific names and suggest how such names reflect the characteristics of the organisms.

CONCEPT DEVELOPMENT

▶ Explain to students that taxonomic keys are based on the similarities and differences among organisms.

▶ Many field guides for flowers are arranged according to flower shape and color.

▶ Explain to students that many keys and field guides contain drawings or photographs to help identify an organism. When the differences between organisms are slight, a dichotomous identification key must be used.

▶ Emphasize that even expert taxonomists sometimes encounter difficulty in keying organisms.

In Your JOURNAL

Students reports should describe the different parts of the guide and give an importance for each part. Choice of organisms will vary.

VideoDisc

STVS: Insect Museum, Disc 5, Side 1

REVEALING MISCONCEPTIONS

▶ Many students will try to skip steps in a key to identify an organism. Point out that skipping steps leads to misidentification.

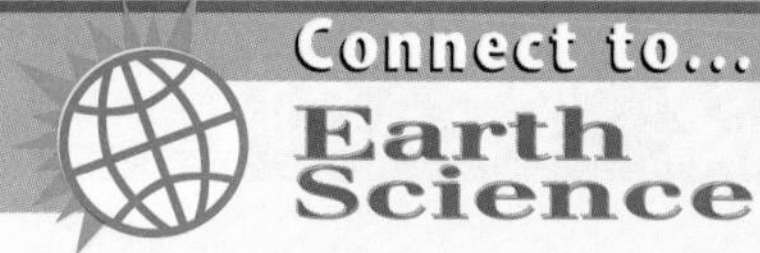

Answer: Several minerals may share the same characteristic, but a mineral can be indentified by a set of characteristics not shared by any other mineral.

CHECK FOR UNDERSTANDING

Have students use the key to identify the organism at the bottom. *(Pseudolucanus placidus)*

RETEACH

Demonstrate how to use a key by making a transparency of a key. Show students photographs of an organism that can be identified using the key. Have them orally go through the steps of identification.

EXTENSION

For students who have mastered this section, use the **Reinforcement** and **Enrichment** masters or other OPTIONS.

Keys can also be used to identify minerals. These keys are based on characteristics such as hardness, luster, color, streak, and cleavage. **Hypothesize** why you might need to know several characteristics to identify a mineral.

The keys found in most field guides are arranged in steps with two descriptive statements at each step. These keys are called **dichotomous keys.** Look at the dichotomous key for beetles in Table 7-2. Notice that the descriptions are labeled 1a and 1b, 2a and 2b, and so on. To use the key, you must always begin with a choice from the first pair of descriptions. Notice that the end of each description is either the name of a species or directions to go on to another step. If you use the dichotomous key properly, you will eventually wind up with the correct name for your species.

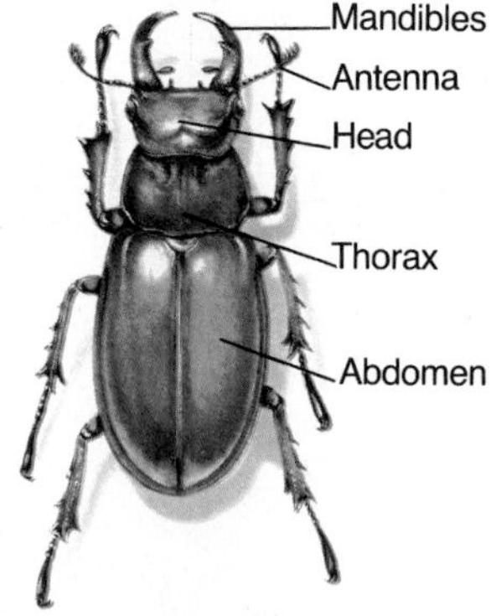

Figure 7-10. The external structures of a stag beetle

Table 7-2

SOME COMMON BEETLES OF NORTH AMERICA	
1a	If the beetle is longer than 20 mm, go to Step 2.
1b	If the beetle is less than 20 mm long, go to Step 5.
2a	If it is reddish brown, go to Step 3.
2b	If it is dark brown or black, go to Step 4.
3a	If the beetle has an oval patch of golden hairs on the front legs, it is commonly called a pinching bug, *Pseudolucanus capreolus.*
3b	If the beetle has no golden hairs on the front legs it is also called a pinching bug, *Pseudolucanus placidus.*
4a	If it is less than 30 mm long and the mandibles do not overlap, it is commonly called an antelope beetle, *Dorcus parallelus.*
4b	If it is more than 30 mm long with large overlapping mandibles, it is commonly called the elephant stag beetle or giant stag beetle, *Lucanus elaphus.*
5a	If the head is as wide as the thorax and it is reddish black, it is an oak stag beetle, *Platycerus agassizi.*
5b	If the head is much narrower than the thorax with a backward-directed horn, it is a rugose stag beetle, *Sinodendron rugosum.*

The actual size of this beetle is 50 mm.

OPTIONS

ASSESSMENT—ORAL

▶ **When using a dichotomous key can you begin with the second or third pair of descriptions?** *No, to accurately classify an organism you must begin with the first pair of descriptions.*

▶ **Can two different kinds of organisms be classified to be the same species?** *No. Two different organisms can not be the same species. If this happens an error has been made when classifying.*

PROGRAM RESOURCES

From the **Teacher Resource Package** use:

Activity Worksheets, page 66, Mini-Lab: How are organisms named?

Figure 7-11. Each of these beetles is a different species. The line next to each drawing is a scale that shows what the actual length of each beetle is. *Measure* each to determine its length in mm.

Let's identify the stag beetle in Table 7-2. Start at Step 1 of the key. Your beetle is much longer than 20 mm, so you go to Step 2. The beetle is dark brown to black, so you skip Step 3 and go to Step 4. Your beetle is about 50 mm long with overlapping mouth parts, so you choose 4b. The dichotomous key tells you that the insect is *Lucanus elaphus,* commonly called an elephant stag beetle or giant stag beetle.

There are many kinds of field guides available. There are field guides to plants, mushrooms, fish, butterflies, and every other kind of organism. You could find the names of species from all around the world by following the descriptions in the proper dichotomous key.

SECTION REVIEW

1. List four reasons biologists use scientific names instead of common names in their communications.
2. Why can common names cause confusion?
3. What is the function of a dichotomous key?
4. **Apply:** Why would you infer that two species that look similar share a common evolutionary history?
5. **Connect to Earth Science:** Find out what Mohs' Mineral scale is.

Classifying — Skill Builder

Classify the beetles marked a, b, and c in Figure 7-11, using the dichotomous key in Table 7-2. If you need help, refer to Classifying in the **Skill Handbook** on page 680.

CONCEPT DEVELOPMENT

Cooperative Learning: You may wish to have students work in Paired Partners or Numbered Heads Together groups for all activities that require using a dichotomous key.

3 CLOSE

▶ Ask questions 1-3 and the **Apply** Question in the Section Review.

SECTION REVIEW ANSWERS

1. Scientific names help avoid errors in communication. Organisms with similar evolutionary history are classified together. Scientific names give descriptive information about the species. Information about species may be organized and found easily and efficiently.

2. Two completely different organisms may share the same common name, causing confusion.

3. A dichotomous key allows you to identify an organism by choosing from pairs of descriptions of characteristics of the organism.

4. Apply: If two organisms have the same genus name, they share a common evolutionary history.

5. Connect to Earth Science: a reference scale of hardness that lists the hardness of 10 minerals with one (1) being the softest and ten (10) being the hardest

Skill Builder

Beetle *a* is an antelope beetle, *Dorcus parallelus.*

Beetle *b* is a pinching bug, *Pseudolucanus capreolus*

Beetle *c* is a rugose stag beetle, *Sinodendron rugosum.*

Skillbuilder

ASSESSMENT

Performance: Assess students' abilities to classify by placing organisms and keys to identify them in several locations in the room. Have students move to each station to identify the organisms.

EXTENSION

▶ Have students produce a dichotomous key to the fruits or vegetables available at the supermarket. Ask them to make up binomial names for their species.

▶ Have students examine field guides to see the similarities in members of organisms belonging to the same family.

ACTIVITY 7-2
30 minutes

OBJECTIVE: Use a dichotomous key to **identify** several organisms.

PROCESS SKILLS applied in this activity:
▶ **Observing** in Procedure Steps 1, 2, and 4.
▶ **Classifying** in Procedure Step 4.
▶ **Inferring** in Conclude and Apply Questions 3, 4, and 5.

COOPERATIVE LEARNING You may wish students to work in cooperative groups to complete this activity.

TEACHING THE ACTIVITY
▶Walk around the room as students are working. If they have the wrong answer, direct them to try again and to check their choices carefully.

Activity ASSESSMENT

Performance: To further assess students' understanding of dichotomous keys, see USING LAB SKILLS on p. 172.

ACTIVITY 7-2 Using a Dichotomous Key

Problem: *How can a dichotomous key be used to identify jays?*

Materials
- paper and pencil

Procedure
1. Look at the jays pictured below.
2. Begin with Step 1 of the Dichotomous Key to Jays of North America. Use the key to *identify* the bird below labeled A.
3. On your paper, make a data table like the one shown. Write the common name and scientific name for the jay.
4. Use the same procedure to *identify* the species of jay labeled B.

Data and Observations Sample Data

Jay	Scientific Name	Common Name
A	*Cyanocitta cristata*	blue jay
B	*Cyanocorax yncas*	green jay

Analyze
1. According to the key, how many species of jay are in North America?
2. How many genera (more than one genus) can be identified with this key?

Conclude and Apply
3. How do you know that this key doesn't contain all the species of jays in the world?
4. Why couldn't you be successful in identifying a robin using this key?
5. Why wouldn't it be a good idea to begin in the middle of a key, instead of with the first step?

KEY TO JAYS OF NORTH AMERICA

1a If the jay has a crest on the head, go to Step 2.
1b If the jay has no crest, go to Step 3.
2a If the jay's crest and upper body are mostly blue, it is a blue jay, *Cyanocitta cristata.*
2b If the jay's crest and upper body are brown or gray, it is a stellar's jay, *Cyanocitta stelleri.*
3a If the jay is mostly blue, go to Step 4.
3b If the jay has little or no blue, go to Step 6.
4a If the jay has a white throat outlined in blue, it is a scrub jay, *Aphelocoma coerulescens.*
4b If the throat is not white go to Step 5.
5a If the jay has a dark eye mask and gray breast, it is a gray-breasted jay, *Aphelocoma ultramarinus.*
5b If the jay has no eye mask and has a gray breast, it is a pinyon jay, *Gymnorhinus cyanocephalus.*
6a If the jay is mostly gray and has black and white head markings, it is a gray jay, *Perisoreus canadensis.*
6b If the jay is not gray, go to Step 7.
7a If the jay has a brilliant green body with some blue on the head, it is a green jay, *Cyanocorax yncas.*
7b If the jay has a plain brown body, it is a brown jay, *Cyanocorax moria.*

A B

ANSWERS TO QUESTIONS
1. 8 different species
2. 5 different genera
3. The title of the key specifies that it is a key to the jays of North America.
4. The key identifies jays, not robins.
5. The first pairs of descriptions are the most general. The following descriptions are more specific. The correct choice at the first pair of descriptions will lead to the correct identification.

PROGRAM RESOURCES

From the **Teacher Resource Package** use:

Activity Worksheets, pages 61-62, 7-2 Using a Dichotomous Key.

CHAPTER

REVIEW

SUMMARY

7-1: What Is Classification?

1. To *classify* means to group ideas, information, or objects based on similarities. Taxonomy is the study of grouping and naming organisms.
2. Aristotle was the first to develop a system to classify organisms.
3. Linnaeus developed a two-word system to name organisms.
4. Binomial nomenclature is a two-word naming system that gives every organism its own scientific name.

7-2: Modern Classification

1. The five kingdoms of living things are Monera, Protist, Fungi, Plant, and Animal.
2. Monerans, such as bacteria and cyanobacteria, are single-celled prokaryotes. The Kingdom Protist contains a variety of single-celled and many-celled eukaryotic organisms. Fungi are eukaryotic, many-celled organisms that can't make their own food. Plants and animals are both eukaryotic, many-celled organisms. Plants can make their own food, animals can't.
3. Organisms without a nucleus and membrane-bound cell organelles are prokaryotes. Eukaryotes have a nucleus and organelles.
4. Organisms are classified into seven categories. They are kingdom, phylum, class, order, family, genus, and species. The genus and species names a specific organism.

7-3: Science and Society: The Rain Forest Crisis

1. Many species of plants and animals are lost due to the deforestation of rain forests. The world's atmosphere may be damaged by the loss of the oxygen-producing trees.
2. Native people depend on rain forests for timber and farming to survive.

7-4: Identifying Organisms

1. Scientific names help scientists communicate. They also allow organisms with similar evolutionary histories to be classified together. Scientific names give descriptive information about the species and allow information about organisms to be organized.
2. Dichotomous keys are used to identify specific organisms.
3. Dichotomous keys are made up of steps of paired descriptions.

KEY SCIENCE WORDS

a. **binomial nomenclature**
b. **classes**
c. **classify**
d. **dichotomous keys**
e. **division**
f. **families**
g. **genus**
h. **kingdom**
i. **orders**
j. **phylogeny**
k. **phylum**
l. **species**
m. **species diversity**
n. **taxonomy**

UNDERSTANDING VOCABULARY

Match each phrase with the correct term from the list of Key Science Words.

1. to put similar organisms in groups
2. a group of similar species
3. the science of classification
4. system of naming organisms with two names
5. evolutionary history of organisms
6. tools for identifying organisms
7. subdivisions of a class
8. largest group within the Plant kingdom
9. variety of plants and animals
10. the second of an organism's scientific names

OPTIONS

ASSESSMENT

To assess student understanding of material in this chapter, use the resources listed.

COOPERATIVE LEARNING

Consider using cooperative learning in the THINK AND WRITE CRITICALLY, APPLY, and MORE SKILL BUILDERS sections of the Chapter Review.

PROGRAM RESOURCES

From the **Teacher Resource Package** use:

Chapter Review, pages 17-18.

Chapter and Unit Tests, pages 40-43, Chapter Test.

Chapter and Unit Tests, pages 44-45, Unit Test.

CHAPTER

REVIEW

SUMMARY

Have students read the summary statements to review the major concepts of the chapter.

UNDERSTANDING VOCABULARY

1. c
2. g
3. n
4. a
5. j
6. d
7. i
8. e
9. m
10. l

ASSESSMENT

Portfolio

Encourage students to place in their portfolios one or two items of what they consider to be their best work. For each item, ask students to explain why that item was chosen and what they learned from it. Items might be selected from the following.

- Skillbuilder concept map, p. 163
- Cooperative Learning script on rain forest destruction, p. 165
- Extension dichotomous key for fruits and vegetables, p. 169

Performance

Additional performance assessments may be found in *Performance Assessment* and *Science Integration Activities* that accompany **Merrill Life Science.** Performance Task Assessment Lists and rubrics for evaluating these activities and other products generated throughout the chapter can be found in Glencoe's *Performance Assessment in Middle School Science.*

CHAPTER

REVIEW

CHECKING CONCEPTS

1. b	**6.** c
2. d	**7.** c
3. a	**8.** d
4. d	**9.** b
5. d	**10.** d

USING LAB SKILLS

ASSESSMENT

Use these alternate lab exercises to assess students' understanding of skills used in this chapter.

11. Events should be listed as:

- Read the first two statements.
- Decide which statement is true.
- Proceed according to directions at the end of the statement.
- Contine choices until organism is identified.

12. Student keys could be completed as follows:

Key to the Kingdoms

1a If the organism has eukaryotic cells, go to step 2.

1b If the organism has prokaryotic cells, it is a member of Kingdom Monera.

2a If the organism has one cell or many cells, but no tissues, go to step 3.

2b If the organism has many cells arranged into tissues, go to step 4.

3a If the organism has no cell walls but makes its own food, it is a member of Kingdom Protista.

3b If the organism has cell walls, but does not make its own food, it is a member of Kingdom Fungi.

4a If the organism makes its own food, it is a member of the Plant Kingdom

4b If the organism eats plants or animals, it is a member of the Animal Kingdom

THINK AND WRITE CRITICALLY

13. The kingdom, phylum, class, and order are also the same.

14. Organisms that have similar or homologous structures and similar embryology are closely related and would be classified together.

15. Many species in the tropical rain forest have not yet been discovered and named. If we lose many more, the rain forest ecosystem may be upset, and we may be losing valuable sources of medicine and damaging our atmosphere.

16. The protist kingdom contains plantlike, animal-like and funguslike members. It seems as though organisms that did not quite fit into another kingdom were thrown into the protist kingdom.

CHAPTER

REVIEW

CHECKING CONCEPTS

Choose the word or phrase that completes the sentence.

1. The group with the most members is a ______.
   a. family c. genus
   b. kingdom d. order
2. Organisms that are the most similar belong in a ______.
   a. family c. genus
   b. class d. species
3. The most complex organisms are ______.
   a. animals c. plants
   b. monerans d. protists
4. The closest relative of *Canis lupus* is ______.
   a. *Quercus alba* c. *Felis tigris*
   b. *Equus zebra* d. *Canis familiaris*
5. The scientific name is written as ______.
   a. *Genus Species* c. genus Species
   b. genus species d. *Genus species*
6. The kingdom bacteria belong to is ______.
   a. Animal c. Monera
   b. Fungi d. Plant
7. The ______ do not have eukaryotic cells.
   a. animals c. monerans
   b. fungi d. plants
8. The simplest eukaryotes are ______.
   a. animals c. monerans
   b. fungi d. protists
9. Trees and flowers are ______.
   a. animals c. monerans
   b. plants d. protists
10. Cells without an organized nucleus are ______.
    a. eukaryotes c. species
    b. phylogeny d. prokaryotes

USING LAB SKILLS

11. Make an events chain to explain how the jays in the activity on page 170 were identified. For additional help, turn to Concept Mapping in the Skill Handbook on pages 688–689.
12. Using the information in Table 7-1 on page 162, make a dichotomous key to *identify* the kingdoms of all living things. Begin your key with:

    1a If the organism has eukaryotic cells, go to Step 2.

    1b If the organism has prokaryotic cells, it is a member of Kingdom Monera.

THINK AND WRITE CRITICALLY

Answer the following questions in your Journal using complete sentences.

13. Two organisms of a family also belong to what other same classification groups?
14. How do embryology and similar structures help classify organisms?
15. Why is it important to maintain the species diversity found in tropical rain forests?
16. Members of Kingdom Protist are often called "misfits." Can you explain why?

Gulfweed—A Protist

APPLY

17. Explain what binomial nomenclature is, and why it is important to scientists.
18. Name each of the five kingdoms, and *identify* a member of each kingdom.
19. Write a short dichotomous key to identify the members of your family. Use such things as sex, hair color, eye color, and weight in your key. Each individual's first and last name will make up the person's genus and species.
20. Discuss the relationship between tigers and lions, members of *Felis*.
21. Scientific names often describe a characteristic of the organism. What does *Lathyrus odoratus* tell you about a sweet pea?

MORE SKILL BUILDERS

If you need help, refer to the Skill Handbook.

1. **Comparing and Contrasting:** Compare the number and variety of organisms in a kingdom and genus.
2. **Classifying:** Use the Key to Jays of North America on page 170 to identify the birds below.
3. **Observing and Inferring:** You observe an organism you don't know. How would you go about determining its classification?
4. **Making and Using Graphs:** Make a pie graph to show the number of species in each kingdom listed in the table below.

Kingdom	Number of Species
Moneran	4 000
Protist	51 000
Fungi	100 000
Plant	285 000
Animal	1 million

5. **Concept Mapping:** Starting with Aristotle, use information in Sections 7-1 and 7-2 to make an events chain concept map to show events leading to modern classification.

PROJECTS

1. Make a poster of the five kingdoms. Draw organisms found in each kingdom, or make a collage using photographs.
2. As a class, make a local field guide for the school grounds or a nearby park. Draw maps and give the location of plants. Identify the plants and animals by common name and as many scientific names as you can.

a

b

c

APPLY

17. Binomial nomenclature is a two-word naming system for identifying specific organisms. The first is the organism's genus, the second, its species. Written in Latin, it is important because scientists can identify a specific organism. In the case of disease-causing organisms, it is especially important to correctly identify the right organism.

18. Accept all appropriate answers. Students can find answers in Table 7-1 and in the Classification Appendix, pages 674-679.

19. Accept all logical keys. Students should write dichotomous keys.

20. Since both tigers and lions belong in the same genus, they also have the same kingdom, phylum, class, order, and family. They differ in their species.

21. The name *odoratus* tells you that the sweet pea must have an odor.

MORE SKILL BUILDERS

1. Comparing and Contrasting: Kingdoms have a much larger number of organisms and therefore a greater variety of organisms than a genus.

2. Classifying: Jay *a* is a steller's jay, *Cyanocitta stelleri*. Jay *b* is a scrub jay, *Aphelocoma coeruliscens*. Jay *c* is a grey-breasted jay, *Aphelocoma ultramarinus*.

3. Observing: Looking at its physical characteristics is the first step in identifying an organism. The appropriate identification keys will allow further classification.

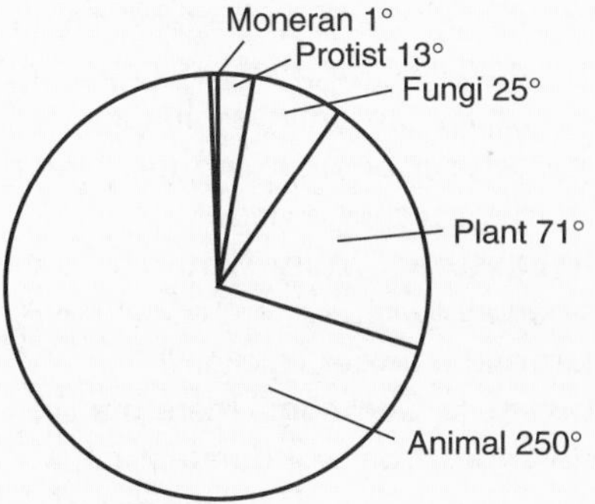

4. Making and Using Graphs: Students should find 69.4% of the pie graph, or 250°, represents the animals; 19.8% of the graph, or 71°, represents the plants; 6.9% of the graph, or 25°, represents the fungi; 3.5% of the graph, or 13°, represents the protists; and 0.28% of the graph, or 1°, represents the monerans.

5. Concept Mapping: Students' chain of events concept maps should begin with Aristotle. Linnaeus and his invention of binomial nomenclature should come next, followed by scientists studying phylogeny, cell type, similarities in genes and body structure, how embryos develop and fossils. The concept map should end with the development of a five kingdom system.

UNIT 2
GLOBAL CONNECTIONS

Objective

In this unit ending feature, the unit topic, "Heredity and Evolution," is extended into other disciplines. Students will see how heredity and evolution are connected.

Motivate

Cooperative Learning: Using the Expert Teams strategy, divide the class into two groups. Have one group research the work of Charles Darwin, and the second the work of Alfred Wallace. Have each group report on their findings.

Teaching Tips

- Have the students think about the connection between evolution and heredity in living things as they read this feature.
- Ask students to explain why Darwin is so famous, whereas Wallace is rarely mentioned.

Wrap-Up

Conclude this lesson by having students examine leaves from different tree species. Have them make hypotheses as to what hereditary traits led to tree evolution.

ASTRONOMY

Background: Astronomers would like to know how the universe began. By studying nearby stars and galaxies, astronomers hope to learn how the universe functions. The telescopes on Mauna Kea afford excellent viewing.

Discussion: Why would it be important to know more about how stars and galaxies function?

Answer to Question: There are six telescopes from six different countries. Each group of astronomers supports and aids the others.

Extension: Divide the class into two groups. Have one group study for a quiz together. In the other group, have each student study alone. Use quiz results to lead a discussion on sharing scientific information.

UNIT 2

GLOBAL CONNECTIONS

Heredity and Evolution

In this unit, you have studied the principles and importance of heredity and evolution. Now see how heredity and evolution influence people, places, and events throughout the world.

ASTRONOMY

LOOKING FOR THE BIG BANG
Mauna Kea, Hawaii

Fourteen thousand feet above the Pacific Ocean, at the Mauna Kea Observatory, six telescopes search the night skies to learn more about the origin of the universe. Why do you think there are six telescopes? What will they find?

BIOLOGY

MYSTERY OF THE JUMPING GENE
Cold Spring Harbor, New York

Geneticist Barbara McClintock puzzled over the different colors of Indian corn kernels. Through her studies, she learned that genes often change their positions on a single chromosome. She called them "jumping genes." Why are genes important?

174

BIOLOGY

Background: Barbara McClintock earned a Ph.D. in botany from Cornell in 1927. She published a paper on mobile genetic elements in 1953. Not until thirty years later did scientists realize the implications of her work. She received the Nobel Prize for physiology in 1983.

Discussion: Discuss the importance of publishing in scientific research.

Answer to Question: Scientists are trying to map the location of certain genes on the chromosome. Knowing that genes can change location helps in this work.

Extension: Invite a geneticist to speak to the class about the use of genetics in breeding better food crops.

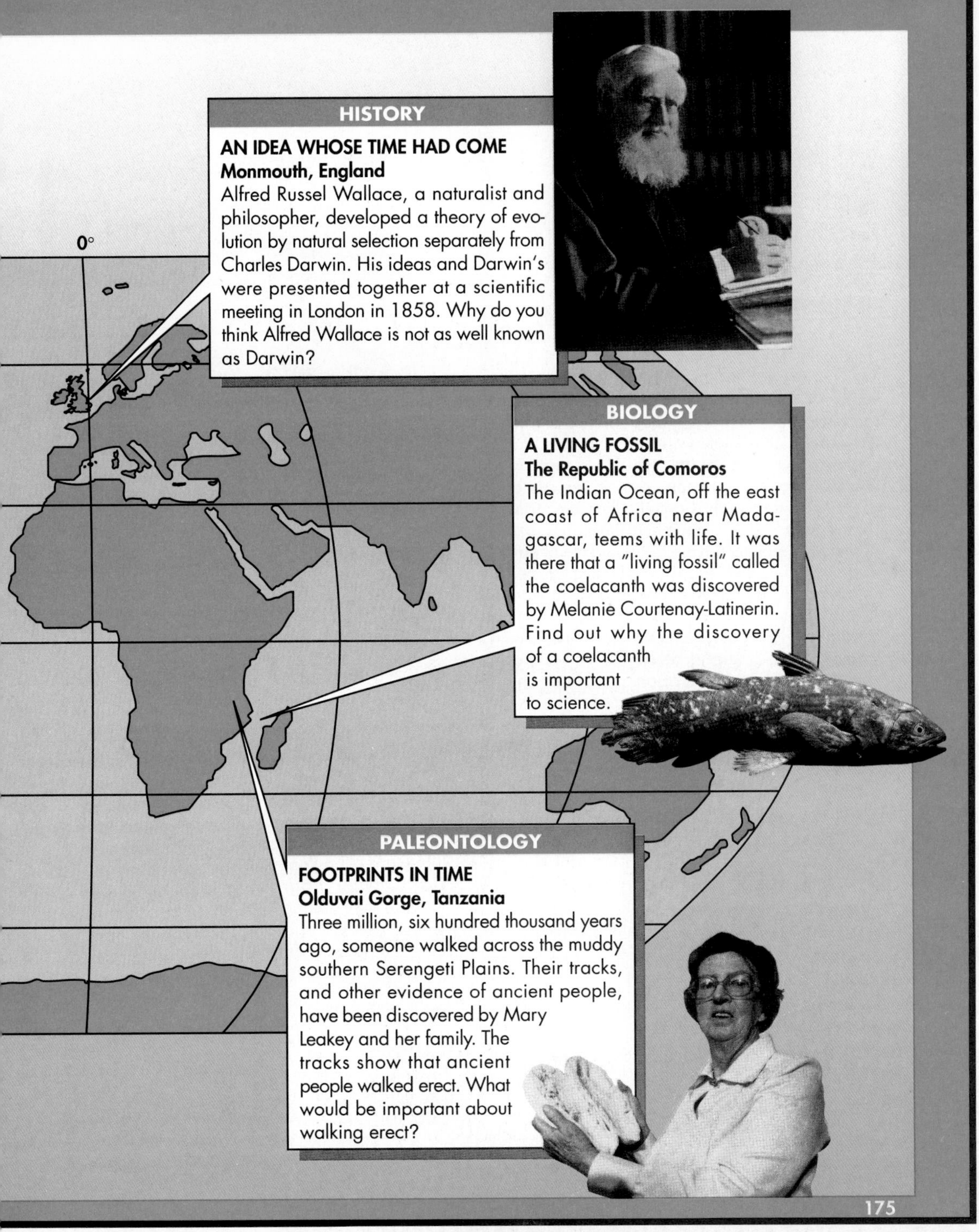

0°
HISTORY
AN IDEA WHOSE TIME HAD COME
Monmouth, England
Alfred Russel Wallace, a naturalist and philosopher, developed a theory of evolution by natural selection separately from Charles Darwin. His ideas and Darwin's were presented together at a scientific meeting in London in 1858. Why do you think Alfred Wallace is not as well known as Darwin?
BIOLOGY
A LIVING FOSSIL
The Republic of Comoros
The Indian Ocean, off the east coast of Africa near Madagascar, teems with life. It was there that a "living fossil" called the coelacanth was discovered by Melanie Courtenay-Latinerin. Find out why the discovery of a coelacanth is important to science.
PALEONTOLOGY
FOOTPRINTS IN TIME
Olduvai Gorge, Tanzania
Three million, six hundred thousand years ago, someone walked across the muddy southern Serengeti Plains. Their tracks, and other evidence of ancient people, have been discovered by Mary Leakey and her family. The tracks show that ancient people walked erect. What would be important about walking erect?
175

HISTORY

Background: Alfred Russel Wallace wrote a paper entitled "On the Law Which Has Regulated the Introduction of New Species" after he visited the Amazon and the Malay Archipelago. Wallace and Darwin presented their ideas together at the Royal Society of London in 1858.

Discussion: In science two or more people often arrive at the same conclusions independently.

Answer to Question: Darwin's "Theory of Evolution" upset religious leaders more than Wallace. Wallace believed in a spiritual interpretation of humanity and nature, even though he also supported the theory of evolution.

Extension: Ask students if any of them have ever had a brilliant idea, only to discover someone else had already thought of it. Have students make hypotheses to explain why this might occur.

BIOLOGY

Background: In 1938, the curator of a museum in South Africa received an unusual fish. She asked an ichthyologist, Dr. J.L.B. Smith, to identify it. The only thing Smith had seen that resembled it was a fossil perhaps 300 million years old. The coelacanth was an ancestor of modern vertebrates.

Discussion: Discuss the importance of finding a live specimen of an organism thought to be extinct for 300 million years.

Answer to Question: The discovery of the coelacanth was important because it allowed scientists to study the development of vertebrate animals.

Extension: Have students watch the movie, *Baby*, available on videocassette. Use the movie to discuss other "impossible" organisms such as the Loch Ness monster and Bigfoot.

PALEONTOLOGY

Background: A friend had sent Mary Leakey a hominid tooth which proved to be over two million years old. Working in the Laetoli plains, Mrs. Leakey found tracks of two hominids who had walked upright in the lava ash.

Discussion: Anthropologists estimated the height of the hominids from the length of their footprints. They were between 38.6 and 47.2 centimeters tall.

Answer to Question: Hominids who walked erect could use their hands for carrying tools, food, and children. They could shape pottery and tools.

Extension: Have students make casts of their hands or feet using plaster of paris. When casts are dry, mix them up and see if students can identify their own casts.

GENETIC COUNSELOR

Background: Many birth defects can occur only when both parents contribute a defective gene for the trait. Genetic testing can help couples make appropriate decisions about their chances of bearing a child with a hereditary problem.

Related Career	Education
Genetic counselor	M.S. degree
Geneticist	M.S. or Ph.D. degree
Biologist	B.S. degree

Career Issue: Some couples with a family history of genetic birth defects may decide to go ahead with conception. If amniocentesis reveals a birth defect, couples may decide to terminate the pregnancy.

What do you think? Lead students in a discussion about whether couples with a family history of genetic birth defects should try to have children or adopt.

FARMER

Background: Farmers today need more education to learn how to properly manage and run typical farms. Farmers use specialized equipment and computers to keep track of business expenses.

Related Career	Education
Farmer	high school, college
Extension agent	college degree
Pest control specialist	college degree

Career Issue: Many crops require huge amounts of chemical fertilizers and pesticides. Biological pest control methods include introducing predators of the pest species. But with biological controls there is always a certain amount of damage to the crop.

What do you think? Should farmers use chemical pesticides to kill pests? Should farmers accept some damage and use environmentally safe biological controls?

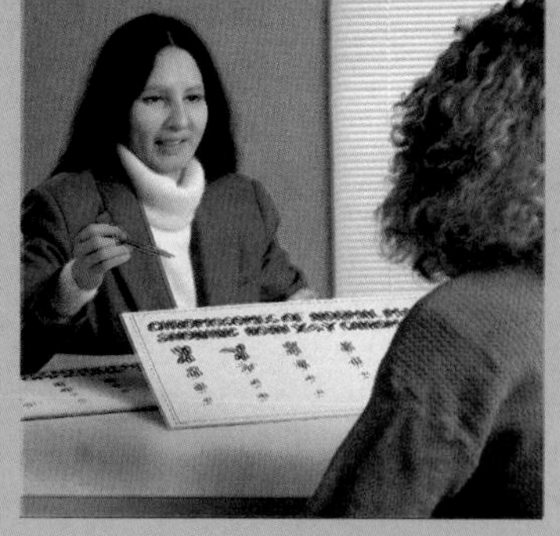

CAREERS

GENETIC COUNSELOR

Genetic counselors work in the field of medicine. They do genetic testing to help doctors understand the genetic background of a patient. Knowing a patient's genetic background can help doctors make a diagnosis or predict the risks of an inherited disease occurring. For instance, the risk for certain types of breast cancer can show in a person's genes. The risk of bearing a child with a genetic illness worries many people. With genetic testing, they can learn whether they have a good chance of having a healthy child. A person who is interested in becoming a genetic counselor should have a background in science, especially biology and genetics. A genetic counselor must have a master's degree and be certified by one of the health accrediting agencies. He or she should enjoy working with people.

For Additional Information

Contact the American Institute of Biological Sciences, Office of Career Service, 730 11th St. NW., Washington, DC 20001-4584

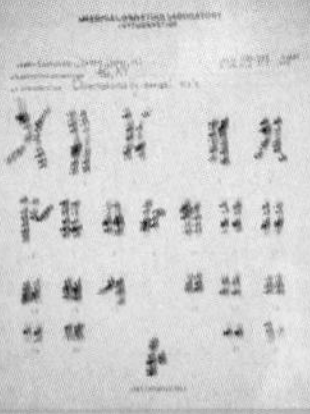

FARMER

Farmers who raise animals such as cattle or hogs or horses, keep records of the health and inheritance of these animals. By breeding the best of their animals, they improve their livestock. Most farmers are also aware of the benefits of genetically improved crop plants such as corn and soybeans. Many farmers learn farming through work on the family farm. Farmers spend much of their time working outdoors. They need to be in good physical health, because farm work involves hard physical labor. Because farming is changing, many farmers attend four-year college programs to learn advanced farming methods and business management. Organizations such as the National Farm Bureau and County Extension Agencies offer programs on livestock and crop production.

For Additional Information

Contact the National FFA Organization, Box 15160, 5632 Mt. Vernon Memorial Hwy., Alexandria, VA 22309; or the American Farm Bureau Federation, 225 Touhy Ave., Park Ridge, IL 60068.

UNIT READINGS

- "The Telltale Gene." *Consumer Reports,* July, 1990, pp. 483-488.
- "An Earlier Cancer Warning." *Newsweek,* Dec. 31, 1990, p. 68.

176

UNIT READINGS

Background

- Clymer, Eleanor. *Search for a Living Fossil.* Orlando, FL: Holt, Rinehart and Winston, 1963. A very readable adventure story that students will enjoy.
- "The Telltale Gene." Consumer Reports, July 1990. This article offers a fascinating view of the future of genetic screening in health care.

Classics

- Darwin, Charles. *Voyage of the Beagle.* NY: Penguin, 1989.
- Wallace, A.R. *The Maylay Archipelago.* NY: Dover, 1978.

The Voyage of the Beagle

by Charles Darwin

The following passage recounts Darwin's observations of tortoises on the Galapagos Islands in September 1835.

The tortoise is very fond of water, drinking large quantities, and wallowing in the mud. The larger islands alone possess springs. Near the springs it was a curious spectacle to behold many of these huge creatures, one set eagerly traveling onwards with their outstretched necks, and another set returning, after having drunk their fill. When the tortoise arrives at the spring, quite regardless of any spectator, he buries his head in the water above his eyes, and greedily swallows great mouthfulls, at the rate of about ten in a minute. The animal probably regulates them (its visits) according to the nature of the food on which it has lived. It is, however, certain, that tortoises can subsist even on these islands where there is no other water than what falls during a few rainy days in the year....

During the breeding season, when the male and female are together, the male utters a hoarse roar or bellowing, which, it is said, can be heard at the distance of more than a hundred yards. The female never uses her voice, and the male only at these times; so that when the people hear this noise, they know the two are together. They were at this time (October) laying their eggs.... The young tortoises, as soon as they are hatched, fall a prey in great numbers to the carrion-feeding buzzard. The old ones seem generally to die from accidents, as from falling down precipices: at least, several of the inhabitants told me, that they never found one dead without some evident cause.

The inhabitants believe that these animals are absolutely deaf; certainly do not overhear a person walking close behind them. I was always amused when overtaking one of these great monsters, as it was quietly pacing along, to see how suddenly, the instant I passed, it would draw in its head and legs, and uttering a deep hiss fall to the ground with a heavy sound, as if struck dead. I frequently got on their backs, and then giving a few raps on the hinder part of their shells, they would rise up and walk away, but I found it very difficult to keep my balance.

In Your Own Words

▶ In a brief essay, state the aspects of the tortoise's behavior that you may have found surprising.

Source: *The Voyage of the Beagle,* by Charles Darwin. NY: Penguin, 1989.

Biography: Charles Darwin was born February 12, 1809. He entered medical school at 16 and left after two years because he found it "intolerably dull." The only thing he had enjoyed was collecting marine specimens from tidal pools and listening to lectures about birds by the American visitor John James Audubon. His father sent him to Cambridge to become a minister. Charles managed to graduate, though his time in academic studies was, as he said, "wasted." He had no real training in any science or art. But at 22, he was recommended as a naturalist to sail aboard the *Beagle.*

TEACHING STRATEGY

Cooperative Learning: Have each group of students read Chapter XVII, Galapagos Archipelago, the paragraph on Ornithology, Curious Finches, from *Voyage of the Beagle* by Charles Darwin. Have them define the words, *gradation, diversity, intimately, fancy,* and *paucity.* What is Darwin saying in this paragraph?

▶ **Writing an Essay:** Could frogs have once existed on the islands? What might happen to frogs' eggs floating in sea water? In what other ways might frogs reach the island?

Other Works:

▶ Peterson, Roger Tory. "The Galapagos, Eerie Cradle of New Species." *National Geographic,* April, 1967.

Villiers, Alan. "In the Wake of Darwin's Beagle." *National Geographic,* October, 1969.

UNIT 3

SIMPLE LIVING THINGS

Unit 3 introduces students to the characteristics of viruses, monerans, protists, and fungi and discusses how these organisms affect other living things.

CONTENTS

ADVANCE PREPARATION

Activities

▶ **Activity 8-1, page 192,** requires live cultures of *Nostoc* and *Oscillatoria* and prepared slides of *Gloeocapsa* and *Anabaena*.

▶ **Activity 8-2, page 196,** requires cotton swabs, petri dishes, prepared slides of bacteria, forks or tongs, potatoes, hot plates, hot mitts, knives, and soap.

▶ **Activity 9-1, page 210,** requires cultures of *Paramecium, Amoeba, Euglena, Spirogyra,* and *Micrasterias;* prepared slides of slime molds; microscope slides and cover slips.

▶ **Activity 9-2, page 218,** requires stale bread, sealable plastic bags, prepared slides of molds, and materials to prepare wet mount slides.

Field Trips and Speakers

▶ Invite a scientist to discuss the study of viruses.

▶ Take students to a laboratory to see the equipment used to study microorganisms.

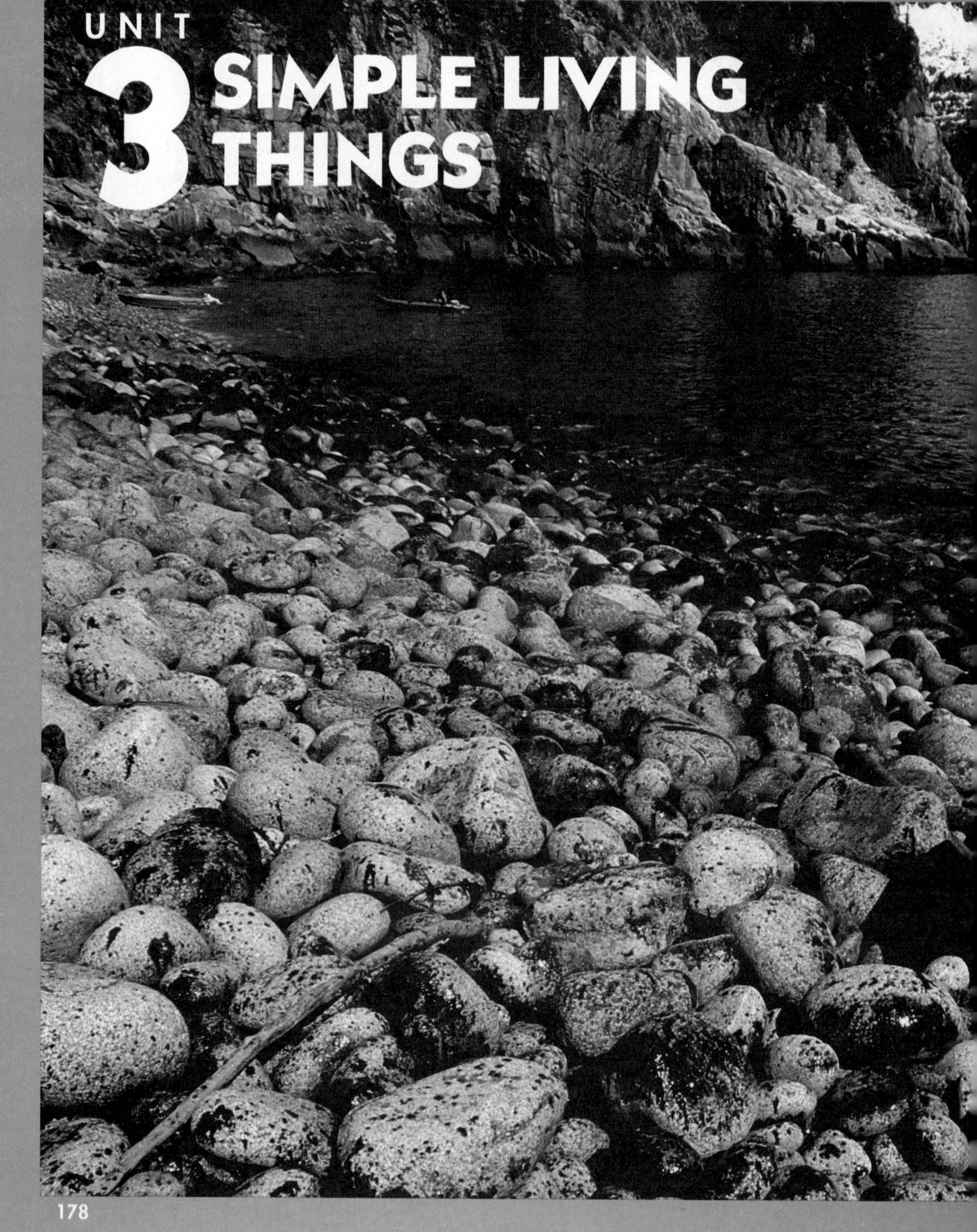

OPTIONS

Cross Curriculum

▶ Have students look for newspaper and magazine articles about how viruses, monerans, bacteria, and fungi affect the lives of humans. Display the articles on the bulletin board.

PROJECT

During the course of this unit study, have students work cooperatively on PROJECT 3, *Don't Toss It, Compost It!*, found on pages 698 and 699.

Science at Home

▶ Have students use yeast to bake bread and observe how the bread rises.

Cooperative Learning: Have students work in groups of four to research diseases caused by fungi. Have each group write a newspaper article on one disease. Have them include the symptoms of the diseases, how they may be prevented, and how they can be cured.

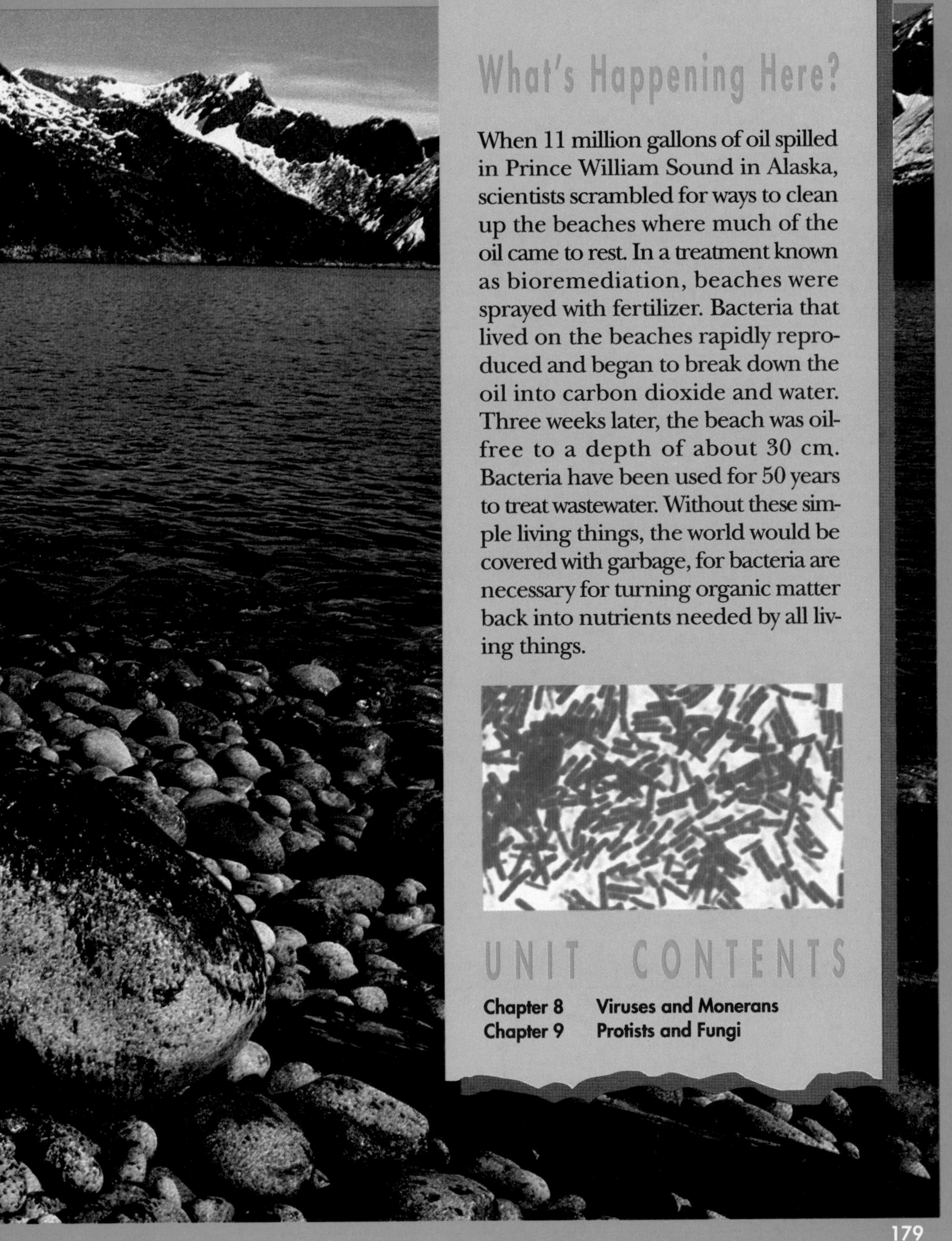

What's Happening Here?

When 11 million gallons of oil spilled in Prince William Sound in Alaska, scientists scrambled for ways to clean up the beaches where much of the oil came to rest. In a treatment known as bioremediation, beaches were sprayed with fertilizer. Bacteria that lived on the beaches rapidly reproduced and began to break down the oil into carbon dioxide and water. Three weeks later, the beach was oil-free to a depth of about 30 cm. Bacteria have been used for 50 years to treat wastewater. Without these simple living things, the world would be covered with garbage, for bacteria are necessary for turning organic matter back into nutrients needed by all living things.

UNIT CONTENTS

179

INTRODUCING THE UNIT

What's Happening Here?

▶ Have students look at the photos and read the text. Ask them what's happening here. Explain to students that in this unit they will study bacteria, viruses, protists, and fungi and how they affect other organisms and the environment.

▶ **Background:** Introduce students to the idea that bacteria and fungi are decomposers that turn dead organic material back into organic components that can be recycled. Explain that without bacteria and fungi, the world would be covered with dead material until all nutrients were gone—locked away in the dead organisms.

Previewing the Chapters

▶ Some students may not know that bacteria are used to clean up oil spills. Have students discuss how they think bacteria accomplish this task.

▶ Students should be encouraged to look through the chapters in this unit to become familiar with simple living things.

Tying to Previous Knowledge

▶ Some students will know that viruses are responsible for diseases such as the common cold and flu. Use this information to discuss what causes colds and flu, how the microorganisms enter the body, and the possible treatments. The discussion will help them see that there is no cure for colds. The effects of the virus are treated, but the virus is not killed.

▶ Use the **inquiry questions** in the OPTIONS box below to investigate with students the roles of simple living things in their lives.

Multicultural Awareness

All cultures must deal with diseases. People in tropical and subtropical areas particularly have problems because the warm, humid climate provides opportunities for bacteria and other disease-causing organisms to thrive. Malaria, for example, is caused by protozoa called plasmodium. Have students research the areas where malaria is a problem. Students can prepare a bulletin board with a map showing the countries affected, using tags and pictures to show the conditions that encourage the disease, and the effects malaria has on these cultures.

Inquiry Questions

Use the following questions to focus a discussion on the effects of simple living things on students' lives.

▶ **How is a virus different from a bacterium?**

▶ **Do all bacteria cause disease?**

▶ **How do you protect yourself from bacterial infections?**

▶ **Why is cooked food refrigerated?**

▶ **Why should bread be stored in a cool dry place?**

CHAPTER

8 Viruses and Monerans

CHAPTER SECTION	OBJECTIVES	ACTIVITIES
8-1 Viruses: Are They Alive? (2 days)	1. **Describe** the structure of a virus, and explain how viruses reproduce and cause disease. 2. **Explain** the value of vaccines. 3. **Describe** some helpful uses of viruses.	**MINI-Lab:** *Is your cat vaccinated?* p. 184
8-2 The Cost of Curing a Disease Science & Society (1 day)	1. **Define** AIDS and explain how it attacks the body. 2. **Estimate** the cost of curing AIDS.	
8-3 Kingdom Monera (2 days)	1. **Describe** the characteristics of moneran cells. 2. **Compare** aerobic and anaerobic organisms.	**Activity 8-1:** *Observing Cyanobacteria,* p. 192
8-4 Monerans in Your Life (1 day)	1. **Identify** some ways bacteria are helpful. 2. **Explain** the importance of nitrogen-fixing bacteria. 3. **Explain** how some monerans cause disease.	**MINI-Lab:** *How do you make yogurt?* p. 193 **Activity 8-2:** *Designing an Experiment (Determining What Gives Yogurt Its Taste),* p. 196
Chapter Review		

ACTIVITY MATERIALS

FIND OUT	ACTIVITIES		MINI-LABS	
Page 181 yogurt methylene blue dropping pipettes slides coverslips microscopes	**8-1 Observing Cyanobacteria, p. 192** microscope prepared slides of *Gleocapsa* and *Anabaena* living cultures of *Nostoc* and *Oscillatoria*	**8-2 Designing an Experiment, p. 196** plain yogurt small spatula 250-mL beaker of alcohol methylene blue 2 slides microscope	**Is your cat vaccinated? p. 184** brochures pamphlets	**How do you make yogurt? p. 193** milk saucepan yogurt starter thermos bottle fruit

CHAPTER FEATURES	TEACHER RESOURCE PACKAGE	OTHER RESOURCES
Skill Builder: *Hypothesizing,* p. 185	**Ability Level Worksheets** ◆ ***Study Guide,*** p. 33 ● ***Reinforcement,*** p. 33 ▲ ***Enrichment,*** p. 33 **Activity Worksheets,** pp. 5, 74 **Transparency Masters,** pp. 25-26	**Color Transparency 13,** Virus Reproduction **STVS:** Disc 4, Side 1
You Decide! p. 187	**Ability Level Worksheets** ◆ ***Study Guide,*** p. 34 ● ***Reinforcement,*** p. 34 ▲ ***Enrichment,*** p. 34 **Science and Society,** p. 12	
Problem Solving: *How are* E. Coli *bacteria helpful?* p. 190 **Skill Builder:** *Concept Mapping,* p. 191	**Ability Level Worksheets** ◆ ***Study Guide,*** p. 35 ● ***Reinforcement,*** p. 35 ▲ ***Enrichment,*** p. 35 **Concept Mapping,** pp. 21 **Activity Worksheets,** pp. 5, 68-69	**Lab Manual:** Monerans, p. 47 **Lab Manual:** Shapes of Bacteria, p. 49 **Lab Manual:** Bacterial Growth, p. 51 **STVS:** Disc 4, Side 1
Technology: *Hungry Bacteria,* p. 195 **Skill Builder:** *Measuring in SI,* p. 195	**Ability Level Worksheets** ◆ ***Study Guide,*** p. 36 ● ***Reinforcement,*** p. 36 ▲ ***Enrichment,*** p. 36 **Activity Worksheets,** pp. 70-71, 75 **Critical Thinking/Problem Solving,** p. 12 **Technology,** pp. 9-10 **Cross-Curricular Connections,** p. 12	**Science Integration Activity 8**
Summary Key Science Words Understanding Vocabulary Checking Concepts Using Lab Skills Think & Write Critically Apply More Skill Builders Projects	ASSESSMENT RESOURCES **Chapter Review,** pp. 19-20 **Chapter Test,** pp. 53-56 **Performance Assessment in Middle School Science (PAMSS)**	**Chapter Review Software** **Test Bank** **Alternate Assessment** **Performance Assessment**

◆ **Basic** ● **Average** ▲ **Advanced**

ADDITIONAL MATERIALS

SOFTWARE	AUDIOVISUAL	BOOKS/MAGAZINES
The Simplest Living Things, Collamore Educational Publishing. *Knowledge Master-Biology 3,* Academic Hallmarks.	*Bacteria—Friend and Foe,* film, EBEC. *Simple Organisms: Bacteria,* film, Coronet/MTI. *The Rotten World Around Us,* film, Films, Inc. *Viruses (2nd ed.),* film, Coronet/MTI. *Microbes: Bacteria and Fungi,* laserdisc, Aims Media. *Viruses: What They Are and How They Work/Bacteria,* laserdisc, EBEC.	Holt, John G. and N.R. Krieg, eds. *Bergey's Manual of Systematic Bacteriology,* Vol 1. Baltimore, MD: Williams and Wilkins, 1984. Ross, Fredrick C. *Introductory Microbiology. 2nd ed.* Glenview, IL: Scott Foresman and Company, 1986. Sleigh. *Microbes in the Sea.* Old Tappan, NJ: Prentice-Hall Press, 1987.

CHAPTER

8

VIRUSES AND MONERANS

THEME DEVELOPMENT: Homeostasis and ecology are the two major themes of this chapter. The role of monerans in maintaining a homeostatic balance is discussed. In addition to homeostasis, the ecological importance of bacteria is illustrated by using examples of niches filled by these organisms.

CHAPTER OVERVIEW

- **Section 8-1:** Viral structure and reproduction are described in this section. How vaccines and interferon are useful in preventing viral disease is discussed. The use of viruses in DNA technology is given as an example of a positive use of viruses.
- **Section 8-2: Science and Society:** The AIDS virus and epidemic are discussed in terms of how the disease is spread and numbers of people expected to get the disease.
- **Section 8-3:** Moneran characteristics are described in this section. Aerobic and anaerobic organisms are compared and contrasted.
- **Section 8-4:** The ways that bacteria are important to humans and the environment are emphasized in this section. Examples of helpful and pathogenic bacteria are given.

CHAPTER VOCABULARY

viruses	anaerobes
parasite	saprophyte
vaccine	nitrogen-fixing bacteria
AIDS	pathogen
flagellum	antibiotic
fission	toxins
aerobes	endospores

CHAPTER

8 Viruses and Monerans

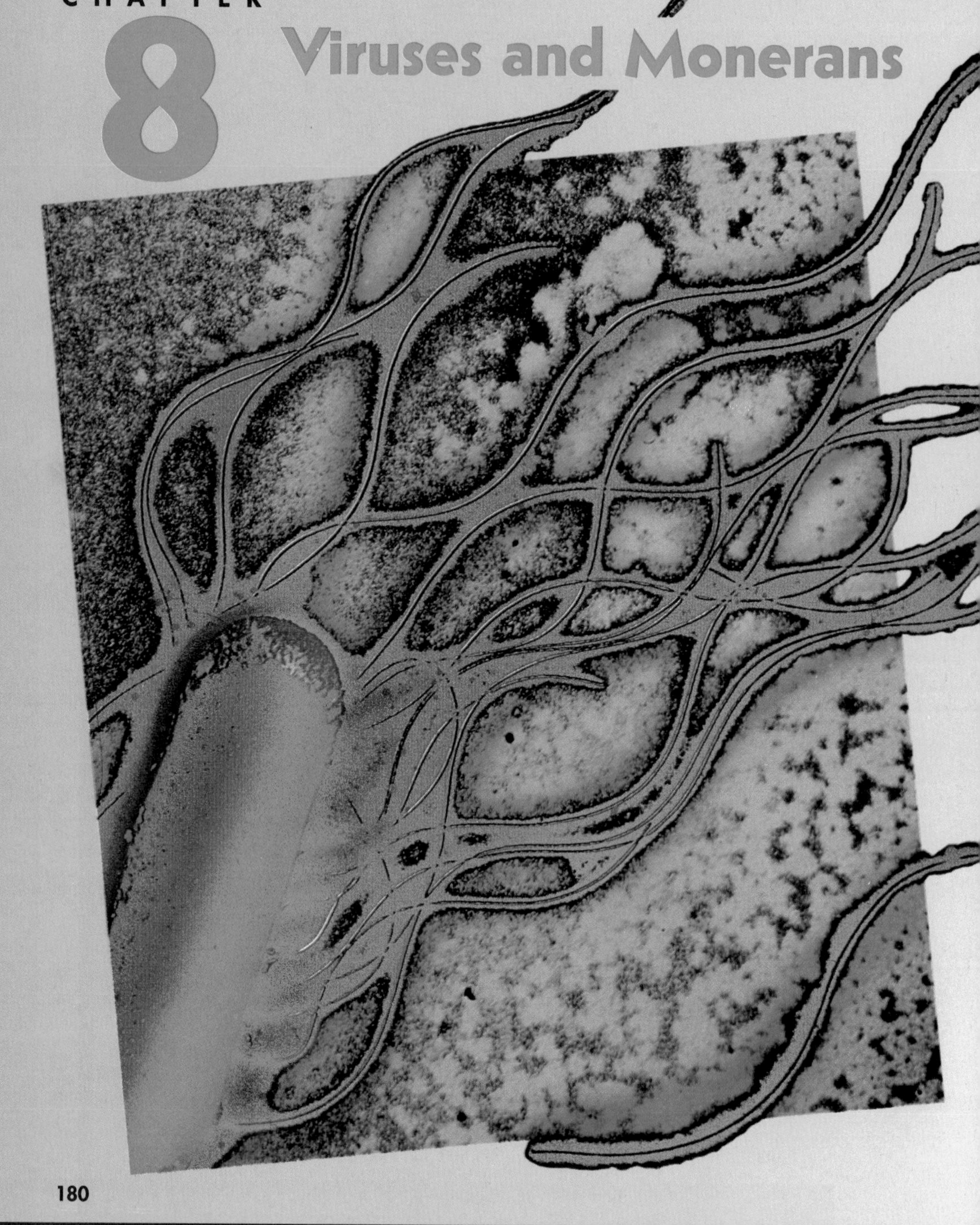

180

OPTIONS

For Your Gifted Students

Have students research AIDS. They can collect articles from magazines, newspapers, etc. They should find out as much as they can about the virus. As a group, ask them to problem solve and try to find a solution to help society cope socially with the disease.

For Your Mainstreamed Students

Students can research viruses to find out about their structure, life cycle, etc. They should write a storybook showing how a virus infects cells. Have them show that the transmission of viruses can be slowed or prevented.

Bacteria, like the one on the opposite page that causes food poisoning, are so small that you need a microscope to see them. Viruses are even smaller. Hundreds of virus particles can fit inside a single bacterium. Isn't it amazing that something so small can make you feel so bad? Are all bacteria and viruses harmful?

FIND OUT!

Do this activity to find out what some bacteria look like.

Bacteria are used to turn milk into yogurt. In a small dish, mix a small drop of yogurt with a drop of water. Now add a drop of methylene blue dye. Methylene blue stains the bacteria so that you can see them. Put a tiny drop of the mixture on a glass slide, set a coverslip on top, and *observe* the bacteria under low and then high magnification. What do these bacteria look like?

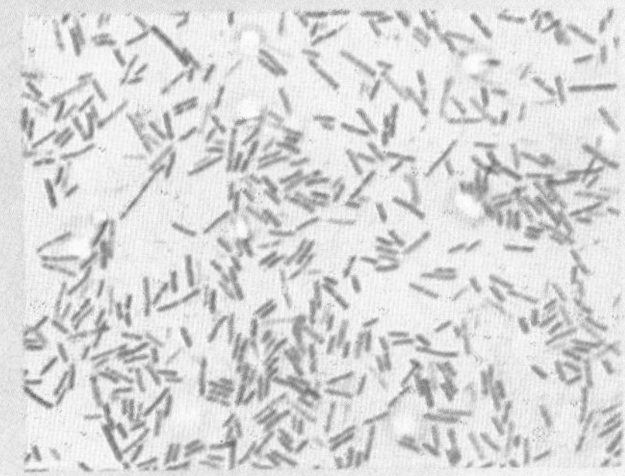

Gearing Up

Previewing the Chapter

Use this outline to help you focus on important ideas in this chapter.

Previewing Science Skills

- In the Skill Builders, you will hypothesize, make a concept map, and measure in SI.
- In the Activities, you will observe and collect and organize data.
- In the MINI-Labs, you will research and infer.

What's next?

You have just seen what some live bacteria look like. In this chapter, you will learn more about bacteria. But first you will learn about viruses, particles that are not thought of as living, but that affect you and most other organisms. Are all bacteria and viruses harmful? Is it possible that some of them are helpful? Find out.

INTRODUCING THE CHAPTER

Use the Find Out activity to introduce monerans. Bacteria are living organisms, as seen in the photograph, while life characteristics of viruses is debated. Explain that students will be learning about the characteristics of viruses, bacteria, and cyanobacteria (the monerans).

FIND OUT!

Preparation: Be sure to purchase fresh, plain yogurt that is made with live cultures.

Materials: yogurt, methylene blue, dropping pipettes, slides, coverslips, microscopes

Cooperative Learning: Use the Science Investigation strategy. Divide the class into groups of two. Have each group work together to prepare a single slide.

Teaching Tips

- You will probably want to check each microscope when students have the organism focused on high power to be certain that they are looking at the bacteria.
- Have the students make a drawing of what they see on the slide.

Gearing Up

Have students study the Gearing Up feature to familiarize themselves with the chapter. Discuss the relationships of the topics in the outline.

What's Next?

Before beginning the first section, make sure students understand the connection between the Find Out activity and the topics to follow.

ASSESSMENT OPTIONS

PORTFOLIO

Refer to page 197 for suggested items that students might select for their portfolios.

PERFORMANCE ASSESSMENT

See page 197 for additional Performance Assessment options.

Process

Skillbuilders, pp. 185, 191, 195
MINI-Lab, pp. 184, 193
Activities 8-1, p. 192; 8-2, p. 196
Using Lab Skills. p. 198

CONTENT ASSESSMENT

Assessment—Oral, pp. 185, 191, 193
Section Reviews, pp. 185, 187, 191, 195
Chapter Review, pp. 197-199
Mini Quizzes, pp. 185, 190, 194

GROUP ASSESSMENT

Opportunities for group assessment occur with Cooperative Learning Strategies and Flex Your Brain Activities.

SECTION 8-1

PREPARATION

SECTION BACKGROUND

▶ Viruses were first identified as the cause of disease in 1892.

▶ In the 1980s, the smallpox virus became the first infectious agent to be eliminated from the human population. This virus is held now only in certain laboratories around the world, including the Centers for Disease Control in Atlanta.

PREPLANNING

▶ You will need to obtain a Darwin tulip for the motivation activity.

1 MOTIVATE

▶ Obtain a Darwin tulip from your local florist. Ask students to explain how the flower obtained the streaks of colors. Tell them that a virus infected the plant to produce the streaks in the petals and sepals.

FLEX Your Brain

Use the Flex Your Brain activity to have students explore VIRUSES.

FLEX YOUR BRAIN ASSESSMENT

Portfolio: Use the Flex Your Brain activity to reinforce critical thinking and problem solving skills. In Step 2, students might list diseases caused by viruses, how viruses are transmitted, or symptoms of a viral disease.

OBJECTIVES AND SCIENCE WORDS: Have students review the objectives and science words in this chapter as they study each section.

PROGRAM RESOURCES

From the **Teacher Resource Package** use:

Activity Worksheets, page 5, Flex Your Brain

8-1 Viruses: Are They Alive?

New Science Words

viruses
parasite
vaccine

Objectives

- ▶ **Describe the structure of a virus, and explain how viruses reproduce and cause disease.**
- ▶ **Explain the value of vaccines.**
- ▶ **Describe some helpful uses of viruses.**

What Is a Virus?

Imagine something that doesn't grow, respond, or eat, yet it can reproduce itself. This something, which appears to be neither living nor nonliving, is a virus. **Viruses** are microscopic particles made up of either a DNA or RNA core covered by a protein coat. Viruses are so small that an electron microscope is needed to see them.

Viruses are not classified in any kingdom because they are not cells. They show almost none of the characteristics of living things. Some viruses can be made into crystals and stored in a jar on a shelf for years. Then, if they are put into an organism, presto, they reproduce and cause new infections.

The classification of viruses is based on the virus' shape, 1
the kind of nucleic acid it contains, and the kind of organism that the virus infects. The protein coat of a virus gives the particle its shape. Some viruses are many-sided and look somewhat like a soccer ball with 20 sides. Others look like rods. Some viruses have tails. Others, such as the AIDS virus, are spherical like a basketball.

Figure 8-1. Viruses have a variety of shapes. The photograph at the bottom shows virus particles that infect bacterial cells. You cannot see virus particles with a light microscope.

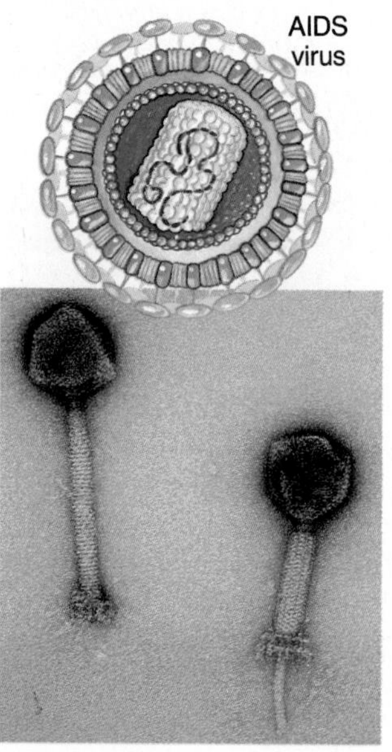

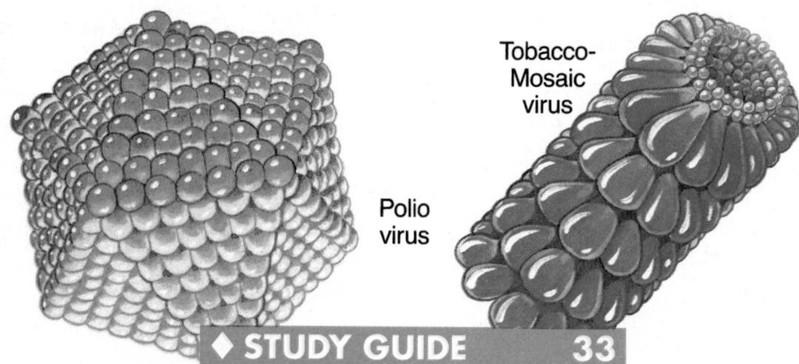

OPTIONS

Meeting Different Ability Levels

For Section 8-1, use the following **Teacher Resource Masters** depending upon individual students' needs.

◆ **Study Guide Master** for all students.

● **Reinforcement Master** for students of average and above average ability levels.

▲ **Enrichment Master** for above average students.

Additional Teacher Resource Package masters are listed in the OPTIONS box throughout the section. The additional masters are appropriate for all students.

◆ STUDY GUIDE 33

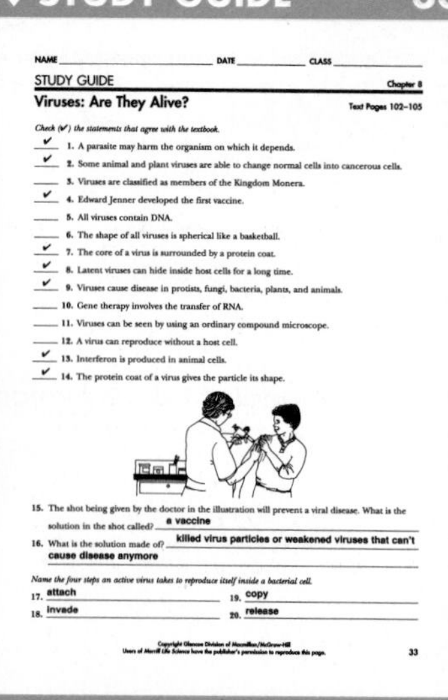

NAME ______ DATE ______ CLASS ______

STUDY GUIDE — Chapter 8

Viruses: Are They Alive? — Text Pages 102–105

Check (✔) the statements that agree with the textbook.

✔ 1. A parasite may harm the organism on which it depends.
✔ 2. Some animal and plant viruses are able to change normal cells into cancerous cells.
___ 3. Viruses are classified as members of the Kingdom Monera.
✔ 4. Edward Jenner developed the first vaccine.
___ 5. All viruses contain DNA.
___ 6. The shape of all viruses is spherical like a basketball.
✔ 7. The core of a virus is surrounded by a protein coat.
✔ 8. Latent viruses can hide inside host cells for a long time.
✔ 9. Viruses cause disease in protists, fungi, bacteria, plants, and animals.
___ 10. Gene therapy involves the transfer of RNA.
___ 11. Viruses can be seen by using an ordinary compound microscope.
___ 12. A virus can reproduce without a host cell.
✔ 13. Interferon is produced in animal cells.
✔ 14. The protein coat of a virus gives the particle its shape.

15. The shot being given by the doctor in the illustration will prevent a viral disease. What is the solution in the shot called? a vaccine

16. What is the solution made of? killed virus particles or weakened viruses that can't cause disease anymore

Name the four steps an active virus takes to reproduce itself inside a bacterial cell.

17. attach
18. invade
19. copy
20. release

 33

Making More Viruses

When most people hear the word *virus,* they relate it to a cold, AIDS, or a cold sore—anything but a pleasant experience. That's because viruses are generally destructive. They are a type of parasite. A **parasite** is anything that depends on an organism or cell to survive and may harm the thing on which it depends. As parasites, viruses depend on living cells. A virus has to be inside a cell in order to reproduce. The cell or organism that a virus
2 depends on is called a host. Once a virus is in a host cell, the virus can act in two ways. It can either be active, or it can become part of the cell for a while.

Active Viruses

If a virus enters a cell and becomes active right away, it causes the cell to make new viruses and destroys the host cell. Figure 8-2a shows the steps an active virus takes to reproduce itself inside a bacterial cell:

1. **Attach:** A specific virus attaches to the surface of a specific bacterial cell.
2. **Invade:** The nucleic acid of the virus injects itself into the cell.
3. **Copy:** The viral nucleic acid takes control of the cell, and the cell begins to make new virus particles.
4. **Release:** The cell bursts open, and hundreds of new virus particles are released from the cell. These new viruses go on to infect other cells.

Latent Viruses

Some viruses are called latent viruses. A latent virus enters a cell and becomes part of the cell's DNA without destroying the cell or making new viruses. The viral nucleic acid becomes part of the cell's own DNA as in Figure 8-2b. As the cell divides, the virus is reproduced right along with it. Latent viruses can hide inside host cells for a long time. Then, without warning, the virus becomes active. It forms new virus particles and destroys the cell. If you have ever had a cold sore, you've experienced a virus going from the latent phase into the active phase. The painful cold sore on your lip is a sign that the virus is active. It's destroying thousands of cells in your lip.
3 Stress factors, such as too much sun or a cold, may cause a virus to become active. When the cold sore disappears, the virus has become latent again. The virus is still in your body, you just don't realize it.

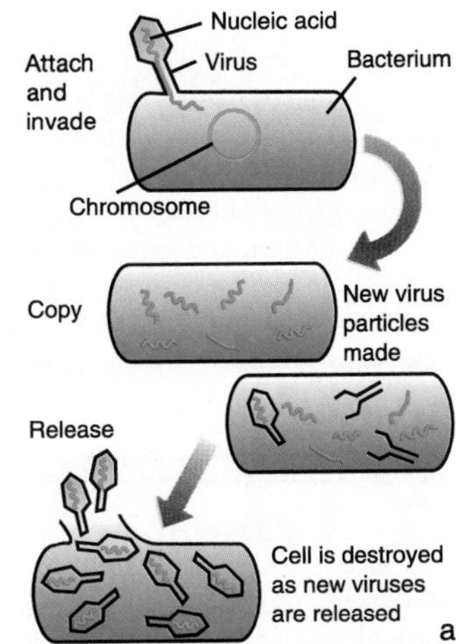

Figure 8-2. An active virus reproduces and destroys the cell (a). A latent virus, below, may not destroy a cell immediately (b).

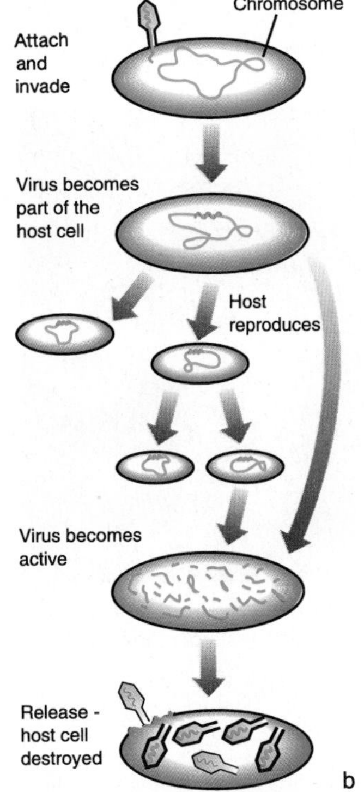

TYING TO PREVIOUS KNOWLEDGE: Use this section to review from Chapter 1 the characteristics of all living things.

2 TEACH

Key Concepts are highlighted.

CONCEPT DEVELOPMENT

▶ Explain that viruses infect members of every kingdom. The viruses that infect the bacteria are perhaps the best understood of all the viruses.

CHECK FOR UNDERSTANDING

Put the steps involved in virus replication, unordered, on an overhead transparency. Have students order and describe each step.

RETEACH

Allow students to build three-dimensional models of viruses. Tell them that only the nucleic acid core is used to make additional viruses once it reaches the interior of a cell.

EXTENSION

For students who have mastered this section, use the **Reinforcement** and **Enrichment** masters or other OPTIONS provided.

VideoDisc

STVS: Mapping a Virus, Disc 4, Side 1
STVS: Bacteriophage, Disc 4, Side 1

● REINFORCEMENT 33

NAME ______ DATE ______ CLASS ______

REINFORCEMENT Chapter 8

Viruses: Are They Alive? Text Pages 182–185

The pictures show a virus reproducing itself in a bacterial cell. Beside each picture, name the step and describe what is happening.

1. Step: Attach
   Description: The virus attaches to the surface of the cell.
2. Step: Invade
   Description: The nucleic acid of the virus is injected into the cell.
3. Step: Copy
   Description: The viral nucleic acid takes control of the cell and begins to make new virus particles.
4. Step: Release
   Description: The cell bursts open and hundreds of new virus particles are released from the cell.

Name three ways in which viruses are classified.

5. by their shape
6. by the kind of nucleic acid the virus contains
7. by the kind of organism the virus infects

Name the two parts of a virus particle.

8. DNA or RNA core
9. protein coat

Answer the following question.

10. What causes cold sores? Explain the process. A latent virus causes cold sores. It enters the body's cells without destroying them and reproduces along with the cells. Then, because of stress factors, it goes from a latent phase to an active phase, causing a cold sore on the lip. When a cold sore disappears, the virus has gone back into its latent phase.

 33

▲ ENRICHMENT 33

NAME ______ DATE ______ CLASS ______

ENRICHMENT Chapter 8

Viruses: Are They Alive? Text Pages 182–185

THE SIZE OF VIRUSES

Study the graph below. It shows the sizes of viruses and other things. Notice on the far right, the graph shows the size of the width of a pencil dot as 1 mm. Everything to the left of the 1 mm mark is smaller. In fact, each mark on the graph represents something that is 10 times smaller than the nearest mark to the right.

molecules
largest molecules
tiniest microscopic animals
atoms
pencil dot
viruses that attack bacteria
human blood cells
bacteria
viruses that attack plants
viruses that attack animals

After studying the graph, answer the following questions.

1. About how large is a pencil dot? 1 mm
2. How large are the smallest bacteria? 1/1 000 000 meter
3. What is the smallest thing shown on the graph? an atom
4. The smallest viruses attack what kinds of things? bacteria
5. The smallest viruses are about the size of what? the largest molecules
6. How large are the smallest viruses that attack animals? less than 1/10 000 000 meter
7. How large are human blood cells? about 1/10 000 meter
8. In general, which are larger, viruses or bacteria? bacteria

 33

CROSS CURRICULUM

▶ **History:** Have students research the contributions that Jenner, Pasteur, Ivanowski, and Beijerinck made to the understanding of viruses.

MINI-Lab

Materials: brochures, pamphlets

Answer: Information and diagrams are available from local veterinarians on the distribution of feline leukemia.

MINI-Lab

ASSESSMENT

Performance: To further assess students' knowledge of feline diseases, have them find out how rabies can be contracted by cats.

Answer: Viruses cannot be cells because they can pass through filters that cells cannot pass through.

Cooperative Learning: Have students work as Paired Partners to make an illustrated report on particular viruses. Each report should give the shape of the virus, organism infected, and symptoms of the disease. Possible topics include smallpox, chicken pox, herpes, rabies, mumps, yellow fever, hog cholera, foot-and-mouth disease, anthrax, wilt disease, and ring spot disease.

MINI-Lab

Is your cat vaccinated?

Research feline leukemia. Make a poster showing where the disease is found geographically, how a cat can be affected, and what can be done to prevent this disease. Make a presentation to your class.

Viruses are so small that they pass through fine porcelain filters that even the smallest cells cannot pass through. *Infer* from this information whether viruses are cells.

Figure 8-3. Edward Jenner developed the first vaccine in 1796.

Viruses Affect All Organisms

Viruses cause disease in plants, animals, fungi, bacteria, and protists. Tobacco mosaic virus infects tobacco plants and tomato plants. Animal viruses that you might know about are rabies, chicken pox, and mumps. If you have a cat, you might also know about the feline leukemia virus. Some RNA and DNA viruses can change healthy cells into cancerous tumor cells.

Preventing Diseases

There are no medications to *cure* viral diseases. But some viral diseases can be *prevented* by vaccines. A **vaccine** is made from damaged particles that can't cause disease anymore. Vaccines have been developed for measles and polio, and for bacterial diseases as well.

The first vaccine was developed in 1796 by Edward Jenner, an English doctor. Jenner developed a vaccine for smallpox, a disease that was greatly feared, even into the twentieth century. Jenner noticed that people who milked cows and came down with a disease called cowpox didn't get smallpox. He prepared a vaccine from the sores of milkmaids who had cowpox. When injected into healthy people, the cowpox vaccine seemed to protect them from smallpox. Did Jenner know he was fighting a virus? No. At that time, no one understood what caused disease or how the body fought disease.

One of the body's natural systems for fighting viral 4
infections is called interferon. Interferon is a protein produced in vertebrate cells that stops viruses from infecting other cells in that animal. Interferon "interferes" with the production of viruses and may have some antitumor features.

Table 8-1

VACCINES AGAINST DISEASE	
Caused by Viruses	**Caused by Bacteria**
Flu	Whooping cough
Measles	Diphtheria
German measles	Tuberculosis
Smallpox	Gangrene
Polio	Tetanus
Rabies	Typhoid fever
Mumps	Cholera

OPTIONS

ENRICHMENT

▶ Have students look up the terms *lysis, virion,* and *virology* and write a paragraph about the relevance of each to the study of viruses.

PROGRAM RESOURCES

From the **Teacher Resource Package** use:

Activity Worksheets, page 74, Mini-Lab: Is your cat vaccinated?

Transparency Masters, pages 25-26, Virus Reproduction.

Use **Color Transparency** 13, Virus Reproduction.

Are There Any Good Viruses?

Most of what you hear about viruses makes you think that viruses always act in a harmful way. However, there are some cases where, through research, scientists are discovering uses for viruses that may make them helpful.

One method, called gene therapy, involves substituting correctly coded DNA for a cell's incorrect DNA. Correctly coded DNA is enclosed in a virus. The virus then acts like an ambulance, taking the strand of DNA into defective cells to replace the incorrect DNA.

Using gene therapy, scientists hope to help people with genetic disorders. For example, some people have the genetic disorder sickle-cell anemia. Because of a defective gene, hemoglobin in their red blood cells does not release oxygen when it reaches body tissues. With the help of a virus, a repaired gene was allowed to "infect" blood cells in a mouse, and the mouse blood cells began to produce the correct substance. Researchers are hoping to use similar techniques for cancer patients.

Figure 8-4. The streaked color pattern in this tulip is the result of a viral infection. If the virus stops infecting the tulip, it will develop a more solid color.

SECTION REVIEW

1. Describe the structure of viruses and explain how viruses reproduce.
2. Explain how vaccines work.
3. Explain how viruses may be helpful in the process of gene therapy.
4. **Apply:** Explain why the doctor wouldn't give you a vaccine if you had a cold caused by a virus.
5. **Connect to Chemistry:** Find out what kind of nucleic acids (RNA or DNA) is in HIV and the virus that causes chicken pox.

Hypothesizing

Skill Builder

You are a researcher in a laboratory. The bacterial cells in an experiment you have been working on have reproduced in large numbers. Then, all of a sudden, most of the bacteria begin to die. You need to find out what caused the deaths of the bacteria. What hypothesis can you make about what is happening to the bacteria? If you need help, refer to Hypothesizing in the **Skill Handbook** on page 686.

MINI QUIZ

Use the Mini Quiz to check students' recall of chapter content.

1. **The classification of ____ is based on their shapes, the kind of nucleic acid they contain, and the kind of organisms infected.** *viruses*
2. **The cell or organism that a virus depends on is called a(n) ____.** *host*
3. **Stress factors, such as too much sun or a cold, may cause a virus to become ____.** *active*
4. **____ is produced in animal cells and stops viruses from infecting other cells.** *Interferon*

3 CLOSE

▶ Ask questions 1-3 and the **Apply** Question in the Section Review.

SECTION REVIEW ANSWERS

1. Student answers should reflect information on pages 182 and 183, including Figure 8-2(a) and (b).
2. Vaccines prevent viral diseases, they do not cure them.
3. They may be used to transfer correctly coded DNA into a cell.
4. Apply: There are no medicines to cure a viral infection.
5. Connect to Chemistry: HIV—RNA; chicken pox—DNA

Skill Builder

Because the bacteria grew through many generations and suddenly died, they may have been infected with a latent virus. Contamination by a new virus is also possible.

ASSESSMENT—ORAL

▶ **Why do scientists disagree as to whether viruses are alive or not alive?** *Viruses possess some, but not all, characteristics of living things.*

▶ **What factors do scientists use to classify viruses?** *shape, nucleic acid content, and the organism the virus infects*

▶ **How do viruses invade a cell?** *The virus injects its nucleic acid into the cell, which becomes infected.*

▶ **In what ways are latent viruses unlike active viruses?** *Latent viruses can go a long time inside the cell without causing the cell to produce new viruses or die.*

▶ **How can viral diseases be prevented?** *Vaccines made from virus particles or from weakened viruses can prevent certain viral diseases.*

Skillbuilder ASSESSMENT

Portfolio: Use this Skillbuilder to assess students' abilities to Hypothesize about an alternate cause of the change in the bacterial culture.

SCIENCE & SOCIETY

SECTION 8-2

PREPARATION

SECTION BACKGROUND

▶ The AIDS virus was first identified in 1983 at the Pasteur Institute in Paris after being observed for the first time in 1978.

1 MOTIVATE

▶ Ask students to explain why no one has ever contracted AIDS from donating blood, but some people did get the disease from receiving blood before the test for AIDS had been invented.

2 TEACH

Key Concepts are highlighted.

CONCEPT DEVELOPMENT

▶ Point out that the massive infusion of money and time into developing drugs and vaccines for AIDS has so far been only partially successful.

SCIENCE & SOCIETY

8-2 The Cost of Curing a Disease

New Science Words

AIDS

Objectives

- ▶ **Define AIDS and explain how it attacks the body.**
- ▶ **Estimate the cost of curing AIDS.**

AIDS

AIDS, acquired immune deficiency syndrome, is a group of diseases caused by an RNA virus called HIV. Within a decade of its discovery, AIDS became an epidemic, causing more than 300 000 deaths throughout the world. In 1992, more than one million people in the United States were thought to be infected with HIV. Of these, more than 20 percent have developed AIDS.

From an intensive study of HIV, scientists know that the virus is spread in body fluids. Most people get HIV through sexual contact. You can't get the virus through casual contacts such as hugging, kissing, or shaking hands. Many drug users get HIV by using contaminated syringes and needles. It is possible for a pregnant or nursing female with HIV to pass the virus to her baby. Before a test for HIV was developed, some people became infected through contaminated blood transfusions. Now the risk of infection from transfusion is low.

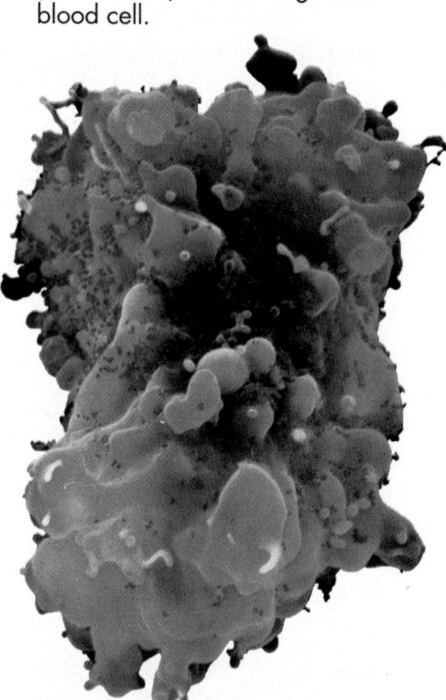

Figure 8-5. In this scanning electron micrograph, HIV particles (colored blue) attack a large white blood cell.

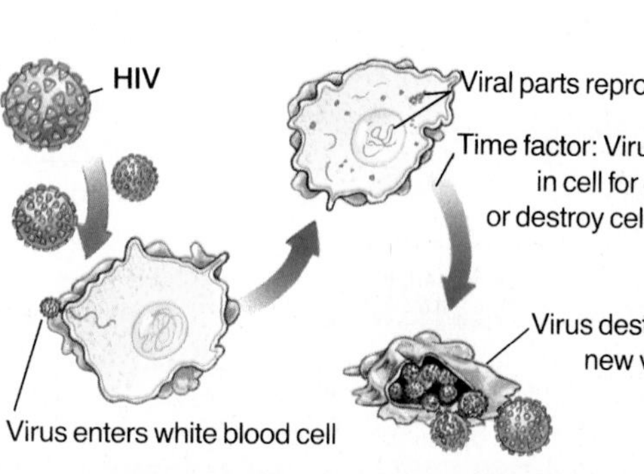
HIV
Viral parts reproduced in cells
Time factor: Virus may remain in cell for up to 10 years or destroy cell immediately.
Virus destroys cell and new virus particles are released.
Virus enters white blood cell

OPTIONS

Meeting Different Ability Levels

For Section 8-2, use the following **Teacher Resource Masters** depending upon individual students' needs.

◆ **Study Guide Master** for all students.

● **Reinforcement Master** for students of average and above average ability levels.

▲ **Enrichment Master** for above average students.

Additional Teacher Resource Package masters are listed in the OPTIONS box throughout the section. The additional masters are appropriate for all students.

◆ STUDY GUIDE 34

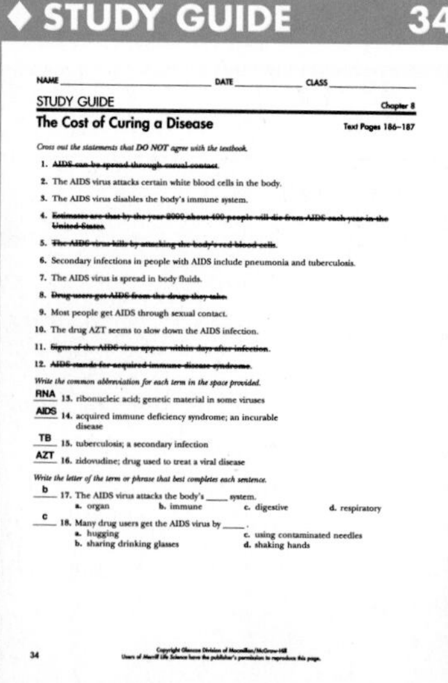

NAME ________ DATE ________ CLASS ________

STUDY GUIDE — Chapter 8

The Cost of Curing a Disease — Text Pages 186–187

Cross out the statements that DO NOT agree with the textbook.

1. ~~AIDS can be spread through casual contact.~~
2. The AIDS virus attacks certain white blood cells in the body.
3. The AIDS virus disables the body's immune system.
4. ~~Estimates are that by the year 2000 about 400 people will die from AIDS each year in the United States.~~
5. ~~The AIDS virus kills by attacking the body's red blood cells.~~
6. Secondary infections in people with AIDS include pneumonia and tuberculosis.
7. The AIDS virus is spread in body fluids.
8. ~~Drug users get AIDS from the drugs they take.~~
9. Most people get AIDS through sexual contact.
10. The drug AZT seems to slow down the AIDS infection.
11. ~~Signs of the AIDS virus appear within days after infection.~~
12. ~~AIDS stands for acquired immune disease syndrome.~~

Write the common abbreviation for each term in the space provided.

RNA 13. ribonucleic acid; genetic material in some viruses

AIDS 14. acquired immune deficiency syndrome; an incurable disease

TB 15. tuberculosis; a secondary infection

AZT 16. zidovudine; drug used to treat a viral disease

Write the letter of the term or phrase that best completes each sentence.

b 17. The AIDS virus attacks the body's ____ system.
a. organ b. immune c. digestive d. respiratory

c 18. Many drug users get the AIDS virus by ____.
a. hugging c. using contaminated needles
b. sharing drinking glasses d. shaking hands

34

HIV attacks certain white blood cells that normally protect the body from infection and disables the body's immune system. A person with HIV becomes unable to protect him- or herself from other types of viruses or bacteria that cause infections. People with HIV often develop pneumonia, tuberculosis, and certain forms of cancer. These diseases are called indicator illnesses. People with HIV die from these indicator illnesses rather than from HIV itself.

Scientists and medical doctors are working to combat AIDS by studying the virus and by using new drug therapies. No vaccine has yet been developed. Only a few drugs, AZT, ddI, and ddC have been approved for treatment. These drugs do not cure but only treat symptoms and, in some cases, prolong life. In order to make helpful drugs available as soon as possible, the Food and Drug Administration (FDA) has made exceptions to its own rules by approving drugs early for use in the fight against AIDS. AZT won approval for use in less than four months. Normally it takes an average of two years for a new drug to be approved for use.

AIDS is different in other respects. People disagree on the seriousness of the threat of HIV infection. They also disagree on how much money should be spent on finding a cure for AIDS and on educating people about HIV infection. Each year the government spends billions of dollars on research for diseases that are major causes of death in the United States. At the present time, more money is spent on AIDS research than on either cancer or heart disease, both of which cause more deaths than AIDS.

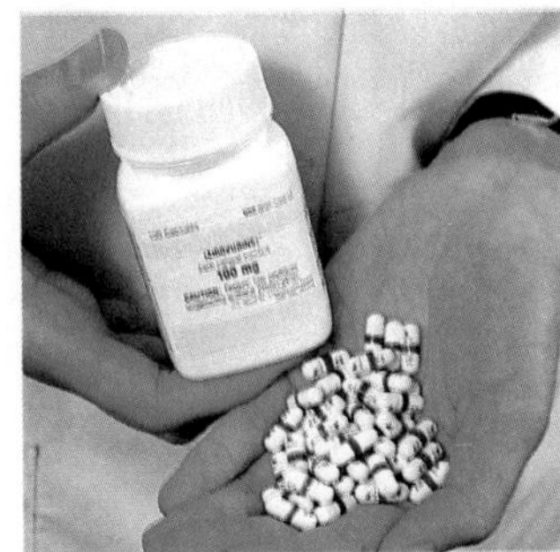

Figure 8-6. AZT is a drug used to combat AIDS infection.

Write a one-page paper **in your Journal** explaining why it is important to know the facts about AIDS.

SECTION REVIEW

1. What is AIDS and how is it spread?
2. How does HIV affect the immune system?
3. Why are people with AIDS more likely to die from indicator illnesses?
4. **Connect to Chemistry:** Find out how AZT, ddI, or ddC work in the body.

The cost of medical research and treatment is presently a topic of great concern to taxpayers. If you were in control of the federal budget, how would you determine how to spend money on research and treatment of AIDS and other diseases?

● REINFORCEMENT 34

NAME ______ DATE ______ CLASS ______

REINFORCEMENT — Chapter 8

The Cost of Curing a Disease — Text Pages 186–187

Complete the following table.

Ways the AIDS virus is spread	Ways the AIDS virus is NOT spread
1. blood transfusions	5. hugging
2. contaminated needles	
3. sexual contact	6. shaking hands
4. mothers nursing babies	

Answer the following questions.

7. How does the AIDS virus affect the body's immune system? **The AIDS virus attacks certain white blood cells that are part of the immune system and protect the body from infection.**
8. Why do people with AIDS develop secondary infections? **Because the AIDS virus disables the body's immune system, a person is not protected from bacteria and viruses that cause infection.**
9. What are some secondary infections that affect people with AIDS? **pneumonia, tuberculosis, and certain cancers**
10. Once a person is infected, how long does it take for AIDS symptoms to appear? **possibly years**
11. What makes AIDS a difficult disease to cure? **It is caused by a virus. No vaccine has been developed against this virus.**

34

▲ ENRICHMENT 34

NAME ______ DATE ______ CLASS ______

ENRICHMENT — Chapter 8

The Cost of Curing a Disease — Text Pages 186–187

AN AIDS SURVEY

Survey your parents and friends about their knowledge of AIDS.

Procedure

Make ten copies of this survey on a copier machine at the store or library. Then ask ten friends or relatives the questions in the survey, circling their response for each question. Inform the person that for each question, he or she may answer "yes," "no," or "not sure." Find the correct answers to questions 2-7 in your textbook or other reference. Combine your survey results with your classmates' to get a picture of your community's understanding of AIDS.

Survey Questions

1. Have you ever known anybody who has had AIDS?
   YES NO NOT SURE
2. Can you get AIDS through sexual contact?
   YES NO NOT SURE
3. Can you get AIDS through a mosquito bite?
   YES NO NOT SURE
4. Can you get AIDS by giving blood?
   YES NO NOT SURE
5. Can you get AIDS from a blood transfusion?
   YES NO NOT SURE
6. Can you get AIDS from shaking hands with a person with AIDS?
   YES NO NOT SURE
7. Can you get AIDS if you share a needle with someone else?
   YES NO NOT SURE
8. Would you feel safe living in the same house as an AIDS sufferer?
   YES NO NOT SURE
9. Would you feel safe sharing a meal with someone with AIDS?
   YES NO NOT SURE
10. Do you think there should be a law that prohibits discrimination against someone with AIDS?
    YES NO NOT SURE

Conclude and Apply

1. Did more of your friends or more adults have a better understanding of AIDS? **Answers will vary.**
2. Are your friends and relatives well-informed? **Answers will vary.** If not, what could you do to increase their knowledge of AIDS? **Students may share what they learned from the textbook.**
3. Questions 8–10 ask people about their attitudes toward people with AIDS. Did you find that your attitudes were similar to or different from the people you surveyed? Explain why. **Answers will vary. Differences of opinion may be from lack of information.**

34

3 CLOSE

In Your JOURNAL

Answers will vary but should include that knowing the facts about AIDS can save lives.

CHECK FOR UNDERSTANDING

Ask questions 1-3 in the Section Review.

RETEACH

Ask students to describe the part of the immune system that the AIDS virus attacks.

EXTENSION

For students who have mastered this section, use the **Reinforcement** and **Enrichment** masters or other OPTIONS.

SECTION REVIEW ANSWERS

1. AIDS is a disease caused by an RNA virus spread by close contact with body fluids containing the virus.
2. AIDS destroys the body's natural system of defense.
3. Because AIDS patients have disabled immune systems, organisms are able to invade the body and cause illness.
4. Connect to Chemistry: They block the mechanism by which HIV replicates.

YOU DECIDE!

Answers will vary. All reasonable answers should be accepted.

PROGRAM RESOURCES

From the **Teacher Resource Package** use:

Science and Society, page 12, Who Should Be Allowed to Manufacture AZT?

SECTION 8-3

PREPARATION

SECTION BACKGROUND

▶ Leewenhoek was probably not the first person to observe bacteria, but he was the first to keep convincingly accurate records of his observations.

▶ Prokaryotes contain very few cell organelles and none that are surrounded by membranes.

▶ The term *microbe* was introduced in 1881 by Sedillot, a French surgeon.

PREPLANNING

▶ If you prepare the materials for Activity 8-1 that require streaking bacteria onto agar media, be sure that you seal the petri plates so that students will not be able to come into contact with the innoculated media.

1 MOTIVATE

▶ Ask students whether they'd rather have $10 every 20 minutes for 13 hours or one penny doubled every 20 minutes for 13 hours. Point out that the growth rate of bacteria is like the penny ($400 by arithmetic growth or more than $5 million by exponential growth).

8-3 Kingdom Monera

New Science Words

flagellum
fission
aerobes
anaerobes

Objectives

▶ **Describe the characteristics of moneran cells.**
▶ **Compare aerobic and anaerobic organisms.**

What Is a Moneran?

In Chapter 2, you read about two types of cells—prokaryotic cells and eukaryotic cells. Prokaryotes are organisms whose cells have no membrane-bound organelles. Monerans are simple-looking, one-celled organisms. They are prokaryotes. Their nuclear material consists of a single circular chromosome. Monerans don't have organelles such as mitochondria or chloroplasts but they do contain ribosomes.

1

Monerans have a few plantlike characteristics. They have cell walls as plants do. Some monerans contain chlorophyll. This enables them to use carbon dioxide and sunlight to make their own food.

How do monerans obtain energy?

Most monerans, however, don't make their own food. That means they have to rely on other organisms to provide food. These monerans have to break down, or decompose, other living things to obtain energy. If you've ever helped clean out the refrigerator, you've probably run into some things that smelled bad. A moneran was at work decomposing a forgotten leftover. In nature, monerans and fungi keep the world free of wastes by breaking them down. In doing so, they also release nutrients into the soil for use again. There are two groups of monerans. One you know as bacteria. The other is cyanobacteria.

Figure 8-7. Bacteria are decomposers. They break down living tissues.

OPTIONS

Meeting Different Ability Levels

For Section 8-3, use the following **Teacher Resource Masters** depending upon individual students' needs.

◆ **Study Guide Master** for all students.

● **Reinforcement Master** for students of average and above average ability levels.

▲ **Enrichment Master** for above average students.

Additional Teacher Resource Package masters are listed in the OPTIONS box throughout the section. The additional masters are appropriate for all students.

◆ STUDY GUIDE 35

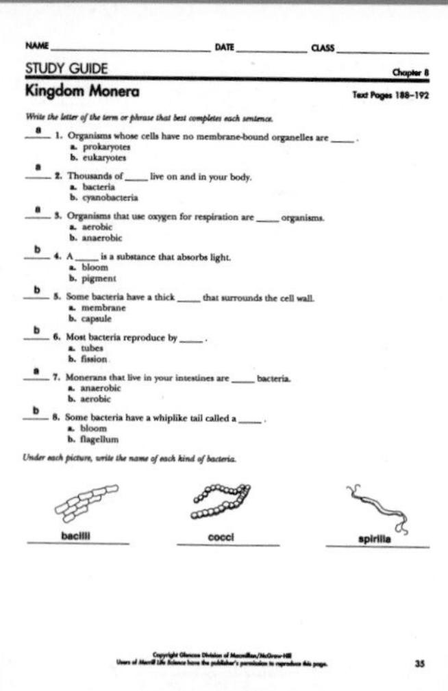
NAME ______ DATE ______ CLASS ______

STUDY GUIDE — Chapter 8

Kingdom Monera — Text Pages 188–192

Write the letter of the term or phrase that best completes each sentence.

a 1. Organisms whose cells have no membrane-bound organelles are ____.
a. prokaryotes
b. eukaryotes

a 2. Thousands of ____ live on and in your body.
a. bacteria
b. cyanobacteria

a 3. Organisms that use oxygen for respiration are ____ organisms.
a. aerobic
b. anaerobic

b 4. A ____ is a substance that absorbs light.
a. bloom
b. pigment

b 5. Some bacteria have a thick ____ that surrounds the cell wall.
a. membrane
b. capsule

b 6. Most bacteria reproduce by ____.
a. tubes
b. fission

a 7. Monerans that live in your intestines are ____ bacteria.
a. anaerobic
b. aerobic

b 8. Some bacteria have a whiplike tail called a ____.
a. bloom
b. flagellum

Under each picture, write the name of each kind of bacteria.

35

Bacteria

When most people hear the word *bacteria,* they probably associate it with sore throats or gum disease. However, very few bacteria cause illness. Most are important for other reasons. Bacteria are almost everywhere—in the air you breathe, the food you eat, and the water you drink. A shovelful of soil contains billions of them. Hundreds of thousands of bacteria live on and in your body. Most are beneficial to you.

The bacteria that normally inhabit your home and body have three basic shapes—spheres, rods, and spirals. Sphere-shaped bacteria are called *cocci,* rod-shaped bacteria are called *bacilli,* and spiral-shaped bacteria are called *spirilla.* The general characteristics of bacteria can be seen in the bacillus shown in Figure 8-9b. It contains cytoplasm, surrounded by a cell membrane and wall. The nuclear material is in the form of a single chromosome strand. Some bacteria have a thick gel-like capsule around the cell wall. The capsule helps the bacterium stick to surfaces. How would a capsule help a bacterium to survive?

Many bacteria float freely in the environment on air and water currents, your hands, your shoes, and the family dog or cat. Many that live in very moist conditions have a whiplike tail called a **flagellum** to help them move.

2 Most bacteria reproduce by fission, as shown in Figure 8-9a. **Fission** produces two cells with genetic material exactly like the parent cell's. Some bacteria also reproduce by a simple form of sexual reproduction. Two bacteria line up beside each other and exchange some DNA through a fine tube. This produces cells with different genetic material that may have an advantage in surviving in their environment.

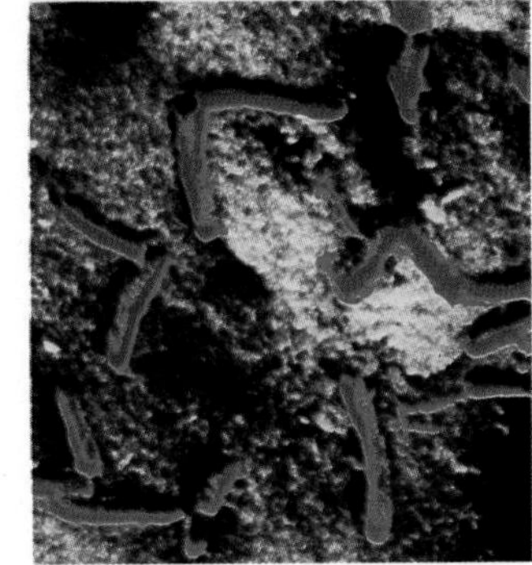

Figure 8-8. Bacteria that grow in plaque in your mouth produce acids that break down tooth enamel.

Figure 8-9. In the photograph, a bacterium is undergoing fission (a). The characteristics and common shapes of bacteria are also shown (b).

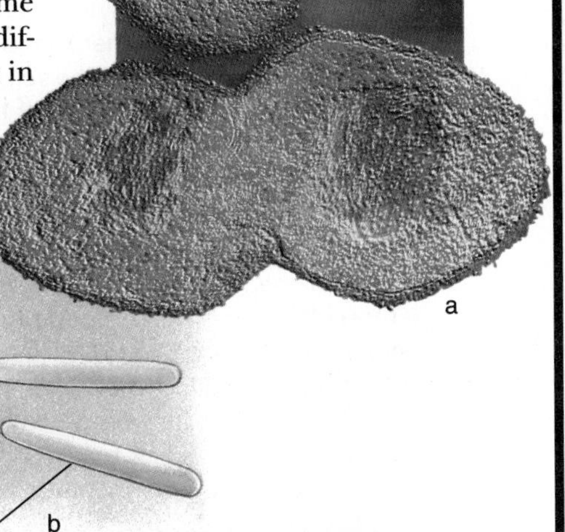

TYING TO PREVIOUS KNOWLEDGE: This section should be bridged with Chapters 1 and 2. The discussion of the characteristics of living things and parts of a cell should be reviewed thoroughly.

2 TEACH

Key Concepts are highlighted.

CONCEPT DEVELOPMENT

▶ **Demonstration:** Demonstrate the classification of bacteria through staining by allowing students to view commercially produced slides of gram positive and gram negative organisms.

▶ Be certain that students understand the three-dimensional nature of all cells.

▶ Scientists hypothesize that monerans resemble the first organisms on Earth.

STUDENT TEXT QUESTION

▶ Page 189, paragraph 2: **How would a capsule help a bacterium survive?** *It provides protection from adverse conditions in its surroundings.*

REVEALING MISCONCEPTIONS

▶ Point out that monerans only look simple. All living cells are, in fact, very complex.

FLEX Your Brain

Use the Flex Your Brain activity to have students explore ROLES OF BACTERIA.

FLEX YOUR BRAIN ASSESSMENT

Performance: Use the Flex Your Brain activity to reinforce critical thinking and problem solving skills. In Step 2, students might list foods changed by bacteria, decay, and other useful roles of bacteria.

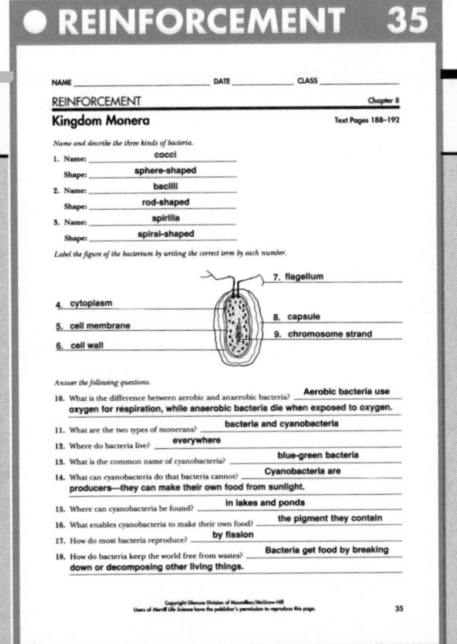

● REINFORCEMENT 35

NAME ________ DATE ________ CLASS ________

REINFORCEMENT — Chapter 8

Kingdom Monera — Text Pages 188–192

Name and describe the three kinds of bacteria.

1. Name: **cocci**
   Shape: **sphere-shaped**
2. Name: **bacilli**
   Shape: **rod-shaped**
3. Name: **spirilla**
   Shape: **spiral-shaped**

Label the figure of the bacterium by writing the correct term by each number.

4. **cytoplasm**
5. **cell membrane**
6. **cell wall**
7. **flagellum**
8. **capsule**
9. **chromosome strand**

Answer the following questions.

10. What is the difference between aerobic and anaerobic bacteria? **Aerobic bacteria use oxygen for respiration, while anaerobic bacteria die when exposed to oxygen.**
11. What are the two types of monerans? **bacteria and cyanobacteria**
12. Where do bacteria live? **everywhere**
13. What is the common name of cyanobacteria? **blue-green bacteria**
14. What can cyanobacteria do that bacteria cannot? **Cyanobacteria are producers—they can make their own food from sunlight.**
15. Where can cyanobacteria be found? **in lakes and ponds**
16. What enables cyanobacteria to make their own food? **the pigment they contain**
17. How do most bacteria reproduce? **by fission**
18. How do bacteria keep the world free from wastes? **Bacteria get food by breaking down or decomposing other living things.**

35

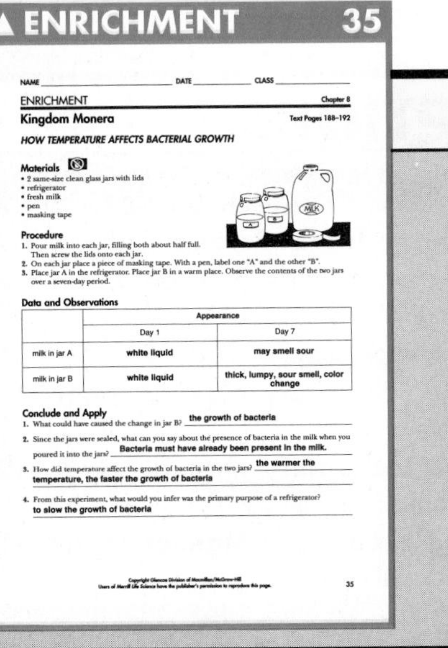

▲ ENRICHMENT 35

NAME ________ DATE ________ CLASS ________

ENRICHMENT — Chapter 8

Kingdom Monera — Text Pages 188–192

HOW TEMPERATURE AFFECTS BACTERIAL GROWTH

Materials
- 2 same-size clean glass jars with lids
- refrigerator
- fresh milk
- pen
- masking tape

Procedure
1. Pour milk into each jar, filling both about half full. Then screw the lids onto each jar.
2. On each jar place a piece of masking tape. With a pen, label one "A" and the other "B".
3. Place jar A in the refrigerator. Place jar B in a warm place. Observe the contents of the two jars over a seven-day period.

Data and Observations

	Appearance	
	Day 1	Day 7
milk in jar A	**white liquid**	**may smell sour**
milk in jar B	**white liquid**	**thick, lumpy, sour smell, color change**

Conclude and Apply
1. What could have caused the change in jar B? **the growth of bacteria**
2. Since the jars were sealed, what can you say about the presence of bacteria in the milk when you poured it into the jars? **Bacteria must have already been present in the milk.**
3. How did temperature affect the growth of bacteria in the two jars? **the warmer the temperature, the faster the growth of bacteria**
4. From this experiment, what would you infer was the primary purpose of a refrigerator? **to slow the growth of bacteria**

35

Connect to... Physics

Answer: Violet, indigo, blue, green, yellow, orange, and red are the colors in increasing wavelengths.

PROBLEM SOLVING

Think Critically: Boiling water kills *E. coli* and other disease-causing bacteria. The presence of *E. coli,* abundant in raw sewage, indicates that sewage containing human wastes has entered the drinking water-supply. Sewage contains many other disease-causing bacteria in addition to *E. coli.*

CHECK FOR UNDERSTANDING

Use the Mini Quiz to check for understanding.

MINI QUIZ

1. **The nuclear material of monerans consists of a single, circular _____.** *chromosome*
2. **______ produces two cells with genetic material exactly like the parent cell.** *Fission*
3. **Some organisms that die when they are exposed to oxygen are called ______.** *anaerobic*
4. **Cyanobacteria are monerans that are ______.** *producers*
5. **All species of cyanobacteria are one-celled organisms, but some of these live together in long chains, or _____.** *filaments*

RETEACH

To demonstrate decomposition, bring a piece of decomposing bread or fruit to class. Have students observe the process over a few days or a week.

EXTENSION

For students who have mastered this section, use the **Reinforcement** and **Enrichment** masters or other OPTIONS provided.

Connect to... Physics

One characteristic used to classify bacteria is the "wavelengths" of their flagella. In physics, each color in the visible spectrum has a wavelength. Find out the sequence of colors of light in the visible spectrum from the shortest to the longest wavelength.

Most bacteria live in places where there is a supply of oxygen. Organisms that use oxygen for respiration are called **aerobes.** You are an aerobic organism. In contrast,
some organisms, called **anaerobes,** can live without oxy- 3
gen. For a few types of bacteria, oxygen is even deadly.

Certain species of anaerobic bacteria are thought to have existed for millions of years. They are found in extreme conditions, such as hot springs, salty lakes, muddy swamps, the intestines of cattle, and near vents deep in the ocean where sunlight doesn't penetrate.

Figure 8-10. Bacteria that can survive in hot springs are similar to ancient forms.

PROBLEM SOLVING

How Are *E. coli* Bacteria Helpful?

Sarah's class was studying bacteria in life science. The teacher mentioned a common bacterium called *Escherichia coli,* which is found in the intestines of humans and other animals. The teacher explained that these bacteria help break down foods that otherwise would not be digested and produce materials such as vitamins.

While watching the news on TV, Sarah learned that people in a flooded area of her state were being told to boil their water. A test had shown high levels of *E. coli* bacteria in the water. She wondered why the water had been tested for *E. coli* and not for other disease-causing bacteria.

Think Critically: Why were the people in the flood area told to boil their water? Why are *E. coli* bacteria used as an indicator that water is harmful to drink?

OPTIONS

PROGRAM RESOURCES

From the **Teacher Resource Package** use:

Activity Worksheets, page 5, Flex Your Brain.

Concept Mapping, pages 21-22.

Use **Laboratory Manual,** page 47, Monerans.

Use **Laboratory Manual,** page 49, Shapes of Bacteria.

Use **Laboratory Manual,** page 51, Bacterial Growth.

Cyanobacteria

4 Cyanobacteria are monerans that are producers. They make their own food using carbon dioxide and sunlight. Cyanobacteria contain chlorophyll and another pigment that is blue. The pigment combination gives cyanobacteria their common name, blue-green bacteria. But, in fact, not all cyanobacteria are blue-green. Some are yellow or black or red. The Red Sea gets its name from red cyanobacterium.

All species of cyanobacteria are one-celled organisms. However, some of these organisms live together in long
5 chains or filaments. Many are covered with a gel-like substance. This allows them to live in globular groups called colonies. Individual cells reproduce by fission. They do not contain mitochondria, chloroplasts, or nuclei. Cyanobacteria are important for food production in lakes and ponds. Because cyanobacteria make food from sunlight, fish in a healthy pond can eat them and utilize the energy released from that food.

Have you ever seen a pond covered with smelly, green, bubbly slime? When large amounts of nutrients enter a pond, cyanobacteria increase in number and produce a matlike growth called a bloom. The cyanobacteria die. Bacteria feed on them and use up all the oxygen in the water. As a result, fish and other organisms die.

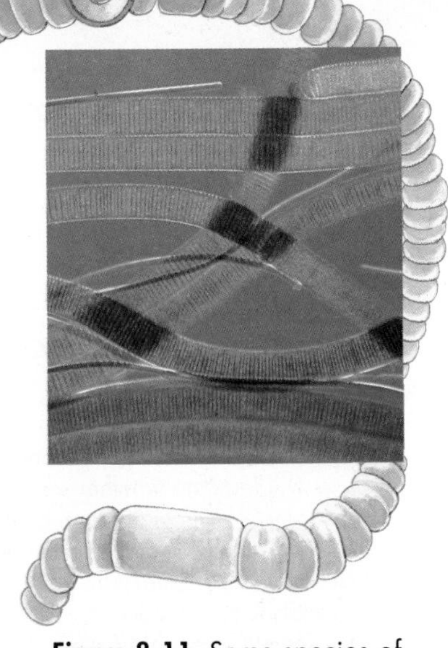

Figure 8-11. Some species of cyanobacteria exist as filaments. The photograph shows *Oscillatoria,* which grows on damp soil and can make stones slick in damp areas. The drawing is of a colony of *Anabaena*, commonly found in ponds.

SECTION REVIEW

1. What are the characteristics of monerans?
2. How can you infer that cyanobacteria undergo photosynthesis?
3. How do aerobic and anaerobic organisms differ?
4. **Apply:** A mat of cyanobacteria is found growing on a lake with dead fish floating along the edge. What has caused this to occur?
5. **Connect to Chemistry:** Find out how bacteria attack tooth enamel.

Concept Mapping

Use an events chain concept map to show what happens in a pond when a bloom of cyanobacteria dies. If you need help, refer to Concept Mapping in the **Skill Handbook** on pages 688 and 689.

Cooperative Learning: Using the Science Investigation strategy, divide the class into teams of three or four students. Using any materials available to them, have teams construct three-dimensional models of bacteria.

3 CLOSE

▶ Ask questions 1-3 and the **Apply** Question in the Section Review.

SECTION REVIEW ANSWERS

1. small cells without membrane-bound organelles
2. Cyanobacteria contain chlorophyll.
3. Aerobic bacteria require oxygen to live; anaerobic bacteria can live without oxygen.
4. **Apply:** The cyanobacteria probably removed too much of the oxygen and may have added poisonous materials to the water.
5. **Connect to Chemistry:** Food is decomposed by bacteria in the mouth. The bacteria form an acid that breaks down enamel.

The concept map should show the following events: the cyanobacteria bloom, then die; bacteria feed on them, using all the oxygen in the water; other organisms, such as fish, may die.

ASSESSMENT—ORAL

▶ **Why are monerans called prokaryotes?** *Monerans do not have a membrane surrounding their nuclear material.*

▶ **What are the plantlike characteristics of monerans?** *Monerans have cell walls, and some have a pigment similar to chlorophyll in plants.*

▶ **What are the major shapes of bacteria?** *spheres, rods, and spirals*

▶ **What name is given to the bacteria that are producers?** *cyanobacteria*

PROGRAM RESOURCES

From the **Teacher Resource Package** use:

Science Integration Activity 8, Water Puification

Skillbuilder
ASSESSMENT

Performance: Use this Skillbuilder to assess students' abilities to form a Concept Map of events stemming from the death of a cyanobacteria bloom.

VideoDisc

STVS: New Uses for Algae, Disc 4, Side 1

ACTIVITY 8-1
50 minutes

OBJECTIVE: Observe and **record** the characteristics of cyanobacteria.

PROCESS SKILLS applied in this activity:
▶ **Observing** in Procedure Steps 2 and 3.
▶ **Classifying** in Procedure Step 1 and Conclude and Apply Question 3.
▶ **Interpreting Data** in Conclude and Apply Question 4.

COOPERATIVE LEARNING Use the Paired Partners strategy. Assign one person to observe slides and the other person to record data. Have students switch roles.

TEACHING THE ACTIVITY

Troubleshooting: If students have trouble seeing the jellylike layer, have them reduce the amount of light coming through the diaphragm of the microscope.

ANSWERS TO QUESTIONS

1. Cyanobacteria are bluer than leaves but contain chlorophyll. Therefore, they can undergo photosynthesis.

2. It keeps the cells of the colony or filament together.

3. Cyanobacteria do not have visible nuclei and are very small cells.

4. Cyanobacteria appear blue-green in color and may form slimy colonies in water.

Activity ASSESSMENT

Performance: To further assess students' ability to Observe cyanobacteria, set up a practical to have them identify organisms and/or answer questions about them.

PROGRAM RESOURCES

From the **Teacher Resource Package** use:

Activity Worksheets, pages 68-69, Activity 8-1, Observing Cyanobacteria.

ACTIVITY 8-1 Observing Cyanobacteria

Problem: *What do cyanobacteria look like?*

Materials  

- microscope
- prepared slides of *Gloeocapsa* and *Anabaena*
- living cultures of *Nostoc* and *Oscillatoria*

Procedure

1. Make a data table like the one shown. In it, you will indicate whether each cyanobacteria sample is in colony form or filament form. Write a yes or no for the presence or absence of each characteristic in each type of cyanobacteria observed.
2. *Observe* the prepared slides of *Gloeocapsa* and *Anabaena* under the low and high power of the microscope. Notice the difference in the arrangement of the cells. The large cells in the *Anabaena* filaments fix nitrogen. The jelly-like capsules around *Gloeocapsa* cells help them to stick together in a group. Draw and label a few cells of each cyanobacterium.
3. Make a wet mount of each living culture. Observe under low and high power of the microscope. Watch for larger nitrogen-fixing cells in *Nostoc*. **In your Journal,** draw and label a few cells of each cyanobacterium.
4. *Return all slides with living material to your teacher for disposal.*
5. Wash your hands before continuing.

Analyze

1. How does the color of cyanobacteria compare with the color of leaves on trees? What can you *infer* from this?
2. What is the purpose of the jelly-like layers around some cyanobacteria?

Conclude and Apply

3. How can you tell by observing them through a microscope that cyanobacteria belong to Kingdom Monera?
4. Describe the activities and general appearance of live cyanobacteria.

Data and Observations

Structure	Anabaena	Gloeocapsa	Nostoc	Oscillatoria
Filament or colony				
Nucleus				
Chlorophyll				
Jelly-like layer				
Nitrogen-fixing cells				

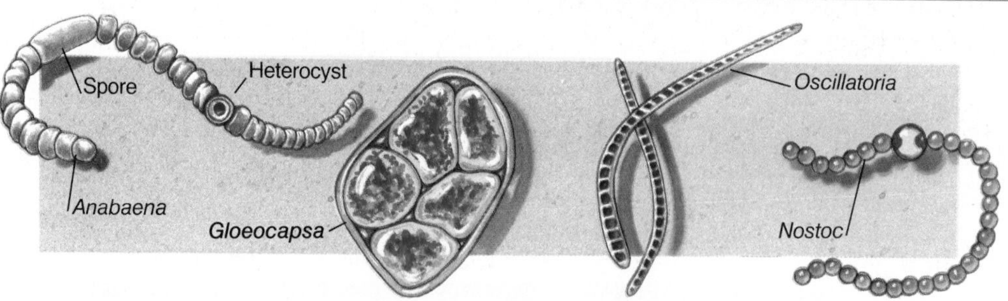

Monerans in Your Life

8-4

Objectives

- ▶ Identify some ways bacteria are helpful.
- ▶ Explain the importance of nitrogen-fixing bacteria.
- ▶ Explain how some monerans cause disease.

New Science Words

saprophyte
nitrogen-fixing bacteria
pathogen
antibiotic
toxins
endospores

Helpful Monerans

Have you had any monerans for lunch lately? Anytime you eat cheese or yogurt, you eat some bacteria. Bacteria break down substances in milk to make these everyday products. If you have eaten sauerkraut, you ate a product made with cabbage and a bacterial culture. Vinegar is also made by a bacterium.

1 Many industries rely on bacteria. Biotechnology has put bacteria to use in making medicines, enzymes, cleansers, adhesives, and other products. Bacteria have been extremely important in helping to clean up the extensive oil spills in Alaska, California, and Texas.

2 Monerans called saprophytes (SAP ruh fitz) help maintain nature's balance. A **saprophyte** is any organism that uses dead material as a food and energy source. Saprophytes digest dead organisms and recycle nutrients so that they are available for use by other organisms. Without saprophytic monerans, there would be layers of dead material deeper than you are tall spread over all of Earth.

MINI-Lab

How do you make yogurt?

Bring a quart of milk almost to a boil in a saucepan. **CAUTION:** *Always be careful when using a stove or hot plate.* Remove the pan from the burner. Cool the milk to lukewarm. Add one or two heaping tablespoons of yogurt starter and stir. Pour the mixture into a clean thermos bottle and put on the lid. Let stand for six hours and then refrigerate. *Infer* why you let the milk cool before adding the starter.

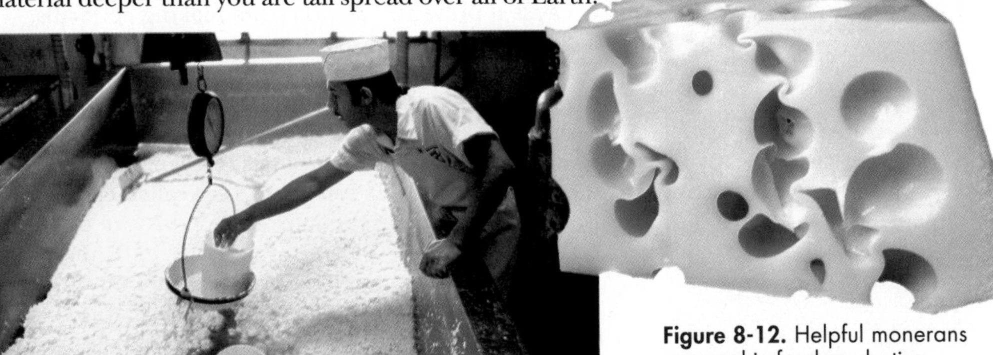

Figure 8-12. Helpful monerans are used in food production.

OPTIONS

ASSESSMENT—ORAL

▶ **Why are streptococci and staphylococci called pathogens?** *They cause disease in humans.*

▶ **Why should you be careful when handling any bacteria?** *Even normally safe bacteria can cause disease if they get into the bloodstream through a cut.*

▶ **Why is pasteurization an important industrial process?** *Pasteurization kills harmful bacteria that may be in such foods as milk and fruit juices.*

PROGRAM RESOURCES

From the **Teacher Resource Package** use:

Activity Worksheets, page 75, Mini-Lab: How do you make yogurt?

Critical Thinking/Problem Solving, page 12, Bacteria Clean Up!

Cross-Curricular Connections, page 12, Preserving Food.

Technology, pages 9-10, A Beneficial Toxin.

SECTION 8-4

PREPARATION

SECTION BACKGROUND

▶ Bacteria are necessary to many organisms because they break down food, and they are necessary to the environment because they break down dead material and release the nutrients for recycling.

PREPLANNING

▶ For the Mini-Lab, obtain milk, yogurt starter, and thermos bottle. Obtain potatoes, petri dishes, and nutrient agar for Activity 8-2.

1 MOTIVATE

▶ Discuss bacteria as simbionts. (Bacteria in humans supply some vitamins and are useful in the digestion of foods high in fiber.)

MINI-Lab

Materials: Milk, saucepan, yogurt starter, thermos bottle

Answer: Answers will vary. Yogurt culture is live and may be killed by excess heat. At lukewarm, culture may grow well.

MINI-Lab ASSESSMENT

Performance: To further assess students' understanding of how bacteria make foods, see USING LAB SKILLS, question number 12, on p. 198.

VideoDisc

STVS: Bacterial Waste Treatment, Disc 4, Side 1

STVS: Using Bacteria to Detect Carcinogens, Disc 4, Side 1

STVS: Detecting Salmonella, Disc 4, Side 1

2 TEACH

Key Concepts are highlighted.

CONCEPT DEVELOPMENT

▶ Remind students that only a few bacterial species are associated with disease. Most bacteria are helpful.

MINI QUIZ

Use the Mini Quiz to check students' recall of chapter content.

1. **Many industries rely on ______ .** *bacteria*
2. **Monerans called ______ also help maintain nature's balance.** *saprophytes*
3. **A(n) ______ is a drug used to kill bacteria.** *antibiotic*
4. **Some pathogens produce poisons called ______ .** *toxins*

CHECK FOR UNDERSTANDING

Have students draw and label pictures that illustrate the benefits of bacteria.

RETEACH

Show students photographs of processes carried out by bacteria. Have students identify each process and its importance.

EXTENSION

For students who have mastered this section, use the **Reinforcement** and **Enrichment** masters or other OPTIONS.

The roots of plants such as peanuts and peas contain **nitrogen-fixing bacteria** in growths called nodules. These bacteria change nitrogen from the air into forms useful for plants and animals. Plants benefit by using the nitrogen, and animals benefit because they obtain nitrogen for making needed proteins. It is estimated that nitrogen-fixing bacteria save United States farmers several million dollars in fertilizer costs every year.

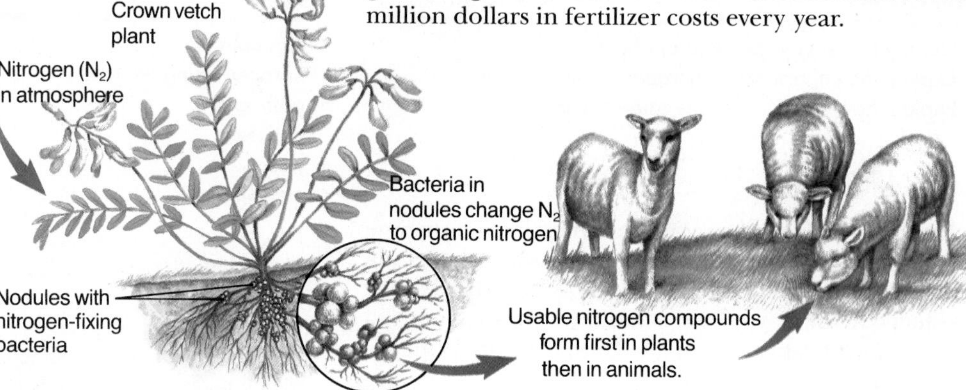

Figure 8-13. Nitrogen-fixing bacteria live in nodules on roots.

Harmful Monerans

Some monerans are pathogens. A **pathogen** is any organism that produces disease. If you have ever had strep throat, you have had firsthand experience with a bacterial pathogen. Other pathogenic bacteria are anthrax in cattle, and diphtheria, tetanus, and whooping cough in humans. Bacterial diseases in animals are treated effectively with antibiotics. An **antibiotic** is a substance
produced by one organism that inhibits or kills another 3
organism. Penicillin, a well-known antibiotic, works because it prevents bacteria from making cell walls.

Some pathogens produce poisons called **toxins.** 4
Botulism, a type of food poisoning, is caused by a toxin that can cause paralysis and death. Many bacteria that produce toxins are able to produce thick walls around themselves and make themselves resistant to heat or drying. These thick-walled structures are called **endospores.** Botulism endospores require long exposure to heat to be destroyed. Once the endospores are in canned food, the bacteria can change back to regular cells and start producing toxins. Botulism bacteria are able to grow inside cans because they are anaerobes and do not need oxygen to live. Botulism toxins can be destroyed by heat.

Figure 8-14. Bacterial endospores are heat-resistant structures.

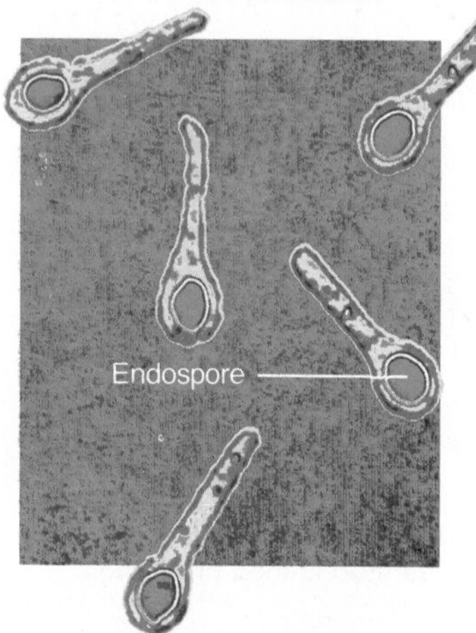

OPTIONS

Meeting Different Ability Levels

For Section 8-4, use the following **Teacher Resource Masters** depending upon individual students' needs.

◆ **Study Guide Master** for all students.

● **Reinforcement Master** for students of average and above average ability levels.

▲ **Enrichment Master** for above average students.

Additional Teacher Resource Package masters are listed in the OPTIONS box throughout the section. The additional masters are appropriate for all students.

◆ STUDY GUIDE 36

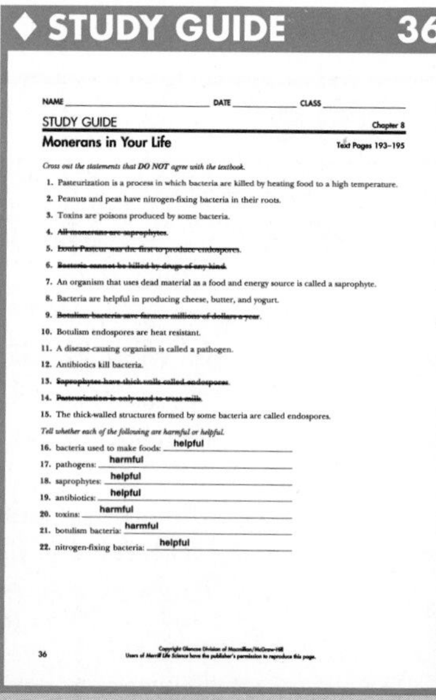

NAME ______ DATE ______ CLASS ______

STUDY GUIDE — Chapter 8

Monerans in Your Life — Text Pages 193–195

Cross out the statements that DO NOT agree with the textbook.

1. Pasteurization is a process in which bacteria are killed by heating food to a high temperature.
2. Peanuts and peas have nitrogen-fixing bacteria in their roots.
3. Toxins are poisons produced by some bacteria.
4. ~~All monerans are saprophytes.~~
5. ~~Louis Pasteur was the first to produce endospores.~~
6. ~~Bacteria cannot be killed by drugs of any kind.~~
7. An organism that uses dead material as a food and energy source is called a saprophyte.
8. Bacteria are helpful in producing cheese, butter, and yogurt.
9. ~~Botulism bacteria save farmers millions of dollars a year.~~
10. Botulism endospores are heat resistant.
11. A disease-causing organism is called a pathogen.
12. Antibiotics kill bacteria.
13. ~~Saprophytes have thick walls called endospores.~~
14. ~~Pasteurization is only used to treat milk.~~
15. The thick-walled structures formed by some bacteria are called endospores.

Tell whether each of the following are harmful or helpful.

16. bacteria used to make foods: helpful
17. pathogens: harmful
18. saprophytes: helpful
19. antibiotics: helpful
20. toxins: harmful
21. botulism bacteria: harmful
22. nitrogen-fixing bacteria: helpful

36

TECHNOLOGY

Hungry Bacteria

Scientists are exploring a new way to remove poisons from soil and water. They are using bacteria to eat away the problem in a technique called *bioremediation.* The name refers to the use of living organisms to break down pollutants.

To clean a toxic site, large populations of resistant bacteria are brought to a problem area. The bacteria reproduce rapidly and eat away the pollutant over several years. When the pollutant is gone, so is the bacteria's food source, and they die off. Bioremediation has been used to clean up oil, gasoline, PCBs, and pesticides.

Think Critically: What are some disadvantages of bioremediation?

Pasteurization, a process of heating food to a temperature that kills harmful bacteria, is used in the food industry. You are most familiar with it in pasteurized milk, but beer and fruit juices are also pasteurized. The term comes from Louis Pasteur, who first formulated the process for the wine industry in the 19th century in France.

Many foods are "vacuum packed." What does this mean? Find out how a vacuum helps prevent food spoilage.

SECTION REVIEW

1. Why are saprophytes helpful and necessary?
2. What is a pathogen?
3. Why are nitrogen-fixing bacteria important?
4. **Apply:** Why is botulism associated with canned foods and not fresh foods?
5. **Connect to Chemistry:** Find out why you have to take penicillin for ten days.

Measuring in SI

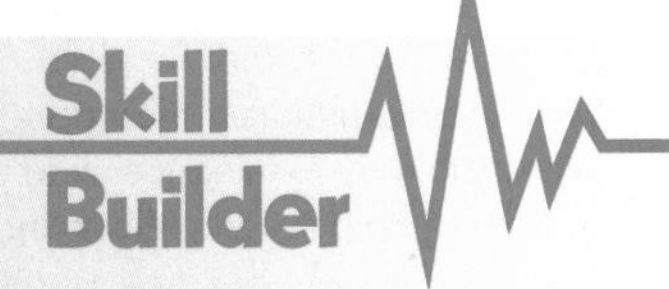

Air may have more than 3500 bacteria per cubic meter. How many bacteria might be in your classroom? If you need help, refer to Measuring in SI in the **Skill Handbook** on pages 684 and 685.

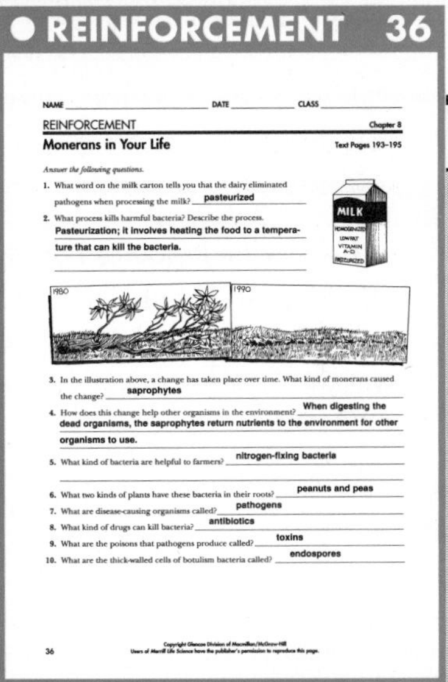

● REINFORCEMENT 36

NAME ______ DATE ______ CLASS ______

REINFORCEMENT — Chapter 8

Monerans in Your Life — Text Pages 193–195

Answer the following questions.

1. What word on the milk carton tells you that the dairy eliminated pathogens when processing the milk? **pasteurized**
2. What process kills harmful bacteria? Describe the process. **Pasteurization; it involves heating the food to a temperature that can kill the bacteria.**

MILK HOMOGENIZED LOWFAT VITAMIN A-D PASTEURIZED

1980 1990

3. In the illustration above, a change has taken place over time. What kind of monerans caused the change? **saprophytes**
4. How does this change help other organisms in the environment? **When digesting the dead organisms, the saprophytes return nutrients to the environment for other organisms to use.**
5. What kind of bacteria are helpful to farmers? **nitrogen-fixing bacteria**
6. What two kinds of plants have these bacteria in their roots? **peanuts and peas**
7. What are disease-causing organisms called? **pathogens**
8. What kind of drugs can kill bacteria? **antibiotics**
9. What are the poisons that pathogens produce called? **toxins**
10. What are the thick-walled cells of botulism bacteria called? **endospores**

36

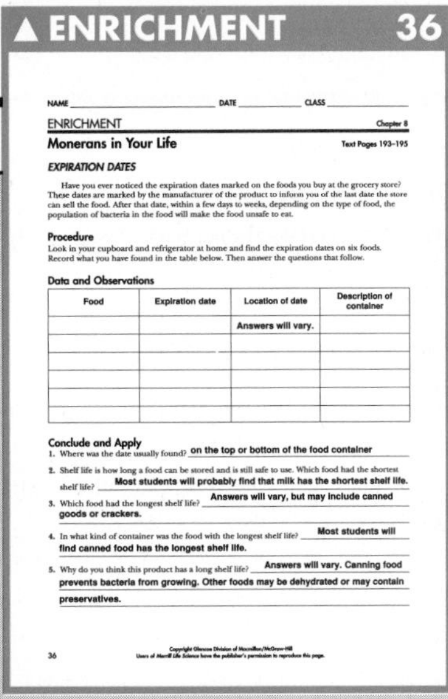

▲ ENRICHMENT 36

NAME ______ DATE ______ CLASS ______

ENRICHMENT — Chapter 8

Monerans in Your Life — Text Pages 193–195

EXPIRATION DATES

Have you ever noticed the expiration dates marked on the foods you buy at the grocery store? These dates are marked by the manufacturer of the product to inform you of the last date the store can sell the food. After that date, within a few days to weeks, depending on the type of food, the population of bacteria in the food will make the food unsafe to eat.

Procedure

Look in your cupboard and refrigerator at home and find the expiration dates on six foods. Record what you have found in the table below. Then answer the questions that follow.

Data and Observations

Food	Expiration date	Location of date	Description of container
		Answers will vary.	

Conclude and Apply

1. Where was the date usually found? **on the top or bottom of the food container**
2. Shelf life is how long a food can be stored and is still safe to use. Which food had the shortest shelf life? **Most students will probably find that milk has the shortest shelf life.**
3. Which food had the longest shelf life? **Answers will vary, but may include canned goods or crackers.**
4. In what kind of container was the food with the longest shelf life? **Most students will find canned food has the longest shelf life.**
5. Why do you think this product has a long shelf life? **Answers will vary. Canning food prevents bacteria from growing. Other foods may be dehydrated or may contain preservatives.**

36

TECHNOLOGY

▶ For more information see "The Tiniest Toxic Avengers" by Robert D. Hof, *Business Week*, June 4, 1990, p. 96.

Think Critically: It takes a long time. Also, if more than one pollutant is present, it may be difficult to use different microbes together.

Answer: High heat and fast sealing are used to drive air from food in closed containers; cutting off oxygen makes most bacteria unable to reproduce.

3 CLOSE

▶ Ask questions 1-3 and the **Apply** Question in the Section Review.

SECTION REVIEW ANSWERS

1. They maintain homeostasis by recycling nutrients.

2. A pathogen is an organism that causes a disease.

3. They change nitrogen from the atmosphere into a form that can be used by plants and animals.

4. Apply: The bacteria that cause botulism are anaerobes.

5. Connect to Chemistry: Penicillin breaks down bacterial cell walls.

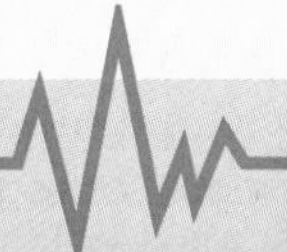

Skill Builder

Have students measure the classroom to find the length, width, and height in meters. Then have them multiply these measurements to obtian the volume in cubic meters. Multiply the volume by 3500 to obtain the number of bacteria per cubic meter.

Skillbuilder ASSESSMENT

Performance: Assess students' abilities to Measure in SI Units by having them calculate the same situation for a room at home.

ACTIVITY 8-2

OBJECTIVE: Design and carry out an experiment to determine what gives yogurt its distinctive taste.
Time: one class period

PROCESS SKILLS applied in this activity are **observing, inferring,** and **recognizing cause** and **effect.**

PREPARATION

• Cultured buttermilk or sour cream can be used instead of plain yogurt.
• You may want to obtain a microbiology book for students to see pictures of the bacteria.
• Methylene blue stain: dissolve 1.5 g methylene blue in 100 mL ethyl alcohol. Dilute by adding 10 mL solution to 90 mL water.

Cooperative Learning: Divide the class into Science Investigation Teams.

THINKING CRITICALLY

Milk has been pasteurized to kill bacteria that cause disease and the cold temperature slows the growth of other bacteria. The bacteria that were not destroyed when the milk was pasteurized remain in the milk so even at a lower temperature, the milk will eventually sour.

TEACHING THE ACTIVITY

Refer to the ***Activity Worksheets*** *for additional information and teaching strategies.*
• Caution students to use a small amount of yogurt.
• Alcohol is used to remove fat particles and causes bacteria to stick to the slide. A coverslip is not needed.
• Streptococcus lactic bacteria change lactose (milk sugar) into several acids, including lactic acid, which causes the taste in yogurt.

SUMMING UP/SHARING RESULTS

• The cells were small without a nucleus and other cell parts. Most will be spherical. Some may be paired but most will be in chains.
• It makes the bacteria more visible.
• The cells produce substances during metabolism and respiration that cause the distinctive flavor of yogurt.

GOING FURTHER!

• Buttermilk, cottage cheese, and sour cream are other foods made from sour milk. Bacteria give each of these foods their distinctive flavor.

PROGRAM RESOURCES

From the **Teacher Resource Package** use:
Activity Worksheets, pages 70-71, Activity 8-2, Determining What Gives Yogurt Its Taste?

Activity ASSESSMENT

Performance: To further assess students' understanding of the bacteria involved in the production of yogurt, have them use Steps 3-7 and substitute buttermilk.

ACTIVITY 8-2

DESIGNING AN EXPERIMENT
What Gives Yogurt Its Taste?

Have you tasted yogurt? How did it taste? Bacteria are microscopic organisms that are found everywhere. Can bacteria get the materials they need to live—oxygen and water—from yogurt?

Getting Started

You need to find out what causes the taste found in yogurt. Make a **hypothesis** as to what you think may cause the taste.

Thinking Critically

Why does the milk you buy keep for two to three weeks in the refrigerator? What causes milk in the refrigerator to finally sour or spoil?

Materials

Your cooperative group will use:
• plain yogurt
• small spatula
• 250-mL beaker of alcohol
• methylene blue stain
• 2 slides
• microscope

Try It!

1. Place a small amount of yogurt on a clean slide. Spread out the yogurt with the edge of another clean slide so that it forms a very thin layer. Allow the yogurt to dry for 5 to 10 minutes.
2. After 10 minutes, remove the slide and *examine* it under low power of the microscope and then turn to high power. Record what you observe.
3. Place two or three drops of methylene blue on the slide. **CAUTION: Methylene blue stains.** Be careful not to get it on you or the tabletop. Wait 2 or 3 minutes, then carefully rinse off the excess stain by running the slide under a very gentle flow of cold water. Allow the slide to dry.
4. *Examine* the slide under low power and then turn to high power. Draw any cells that you observe.

Summing Up/Sharing Results

• Describe the cells you observed. How were they grouped?
• Why did you use methylene blue stain?
• Why do the cells cause yogurt to have a distinctive taste?

Going Further!

What other foods are made from sour milk? What gives each of these foods a distinctive flavor?

CHAPTER
REVIEW

SUMMARY

8-1: Viruses: Are They Alive?

1. A virus is a structure containing nucleic acid surrounded by a protein coat. A virus can reproduce only inside a living cell. It may destroy a cell immediately or be latent.
2. Vaccines prevent viral infections. Interferon blocks viruses from reproducing.
3. Viruses are used in gene therapy to replace defective DNA with correctly coded DNA.

8-2: Science and Society: The Cost of Curing a Disease

1. AIDS is a deadly human viral disease affecting the immune system.
2. AIDS detroys the body's ability to combat disease, leaving a person with AIDS defenseless. Death results from other infections.
3. The cost of treating AIDS is more than that spent on diseases that affect many more people, such as heart disease, diabetes, and cancer.

8-3: Kingdom Monera

1. Monerans are cells that contain DNA, ribosomes, and cytoplasm; they lack organelles. Most reproduce by fission.
2. Cyanobacteria are monerans that make their own food.
3. The oldest types of monerans are anaerobic, whereas more complex cells need oxygen and are called aerobes.

8-4: Monerans in Your Life

1. Monerans are helpful through recycling nutrients, nitrogen fixation, and use in food production.
2. Some bacteria are harmful because they cause disease. Some diseases are anthrax, tetanus, and botulism.
3. A process like pasteurization keeps food and the environment free of harmful bacteria.

KEY SCIENCE WORDS

a. **aerobes**
b. **AIDS**
c. **anaerobes**
d. **antibiotic**
e. **endospores**
f. **fission**
g. **flagellum**
h. **nitrogen-fixing bacteria**
i. **parasite**
j. **pathogen**
k. **saprophyte**
l. **toxins**
m. **vaccine**
n. **viruses**

UNDERSTANDING VOCABULARY

Match each phrase with the correct term from the list of Key Science Words.

1. organisms that decompose dead organisms
2. structure by which some organisms move
3. heat-resistant structures in bacteria
4. substance that can prevent, not cure, a disease
5. something that does harm to its host
6. incurable disease caused by an RNA virus
7. organisms that can survive without oxygen
8. harmful chemicals produced by some pathogens
9. bacteria in roots that make nitrogen available to organisms
10. microscopic particles that are not alive

OPTIONS

ASSESSMENT

To assess student understanding of material in this chapter, use the resources listed.

COOPERATIVE LEARNING

Consider using cooperative learning in the THINK AND WRITE CRITICALLY, APPLY, and MORE SKILL BUILDERS sections of the Chapter Review.

PROGRAM RESOURCES

From the **Teacher Resource Package** use:
Chapter Review, pages 19-20.
Chapter and Unit Tests, pages 53-56, Chapter Test.

CHAPTER
REVIEW

SUMMARY

Have students read the summary statements to review the major concepts of the chapter.

UNDERSTANDING VOCABULARY

1. k
2. g
3. e
4. m
5. i
6. b
7. c
8. l
9. h
10. n

ASSESSMENT

Portfolio

Encourage students to place in their portfolios one or two items of what they consider to be their best work. For each item, ask students to explain why that item was chosen and what they learned from it. Items might be selected from the following.

- Cooperative Learning illustrated report on viruses, p. 180
- Cooperative Learning three-dimensional model of bacteria, p. 191
- Check for Understanding pictures, p. 194

Performance

Additional performance assessments may be found in *Performance Assessment* and *Science Integration Activities* that accompany **Merrill Life Science.** Performance Task Assessment Lists and rubrics for evaluating these activities and other products generated throughout the chapter can be found in Glencoe's *Performance Assessment in Middle School Science.*

CHAPTER
REVIEW

CHECKING CONCEPTS

1. c **6.** a
2. b **7.** b
3. a **8.** c
4. d **9.** d
5. a **10.** a

USING LAB SKILLS

ASSESSMENT

Use these alternate lab exercises to assess students' understanding of skills used in this chapter.

11. $50\ 000 \times 30 \times 250$, or 375 000 000 bacteria

12. Student experiments might use the regular method for making yogurt as a control. Variables could be—raising the temperature higher before adding starter, or adding starter to cold milk.

THINK AND WRITE CRITICALLY

13. Most will say no, because viruses show limited signs of "life." Students who say viruses should be in a kingdom should justify their answers with a reasonable explanation.

14. It has been contaminated by exposure to the air.

15. The AIDS virus leaves the body unable to fight infections. As a result, people with AIDS get many secondary infections, one of which is eventually the cause of death.

16. Prokaryotes, of which monerans are examples, have no nuclei or organelles.

CHAPTER
REVIEW

CHECKING CONCEPTS

Choose the word or phrase that completes the sentence.

1. ______ is an example of a viral disease.
a. Tuberculosis **c.** Smallpox
b. Anthrax **d.** Tetanus

2. Moneran cells contain ______.
a. nuclei **c.** mitochondria
b. DNA **d.** no chromosomes

3. Monerans that make their own food have ______.
a. chlorophyll **c.** Golgi bodies
b. lysosomes **d.** mitochondria

4. Most monerans are ______.
a. anaerobic **c.** many-celled
b. pathogens **d.** beneficial

5. Bacteria that are rod shaped are ______.
a. bacilli **c.** spirilli
b. cocci **d.** colonies

6. The structure that allows a bacterium to stick to surfaces is the ______.
a. capsule **c.** chromosome
b. flagella **d.** cell wall

7. Blooms in ponds are caused by ______.
a. archaebacteria **c.** cocci
b. cyanobacteria **d.** viruses

8. Nutrients and carbon dioxide are returned to the environment by ______.
a. producers **c.** saprophytes
b. flagella **d.** pathogens

9. ______ is caused by a pathogenic moneran.
a. Nitrogen-fixation **c.** Cheese
b. Antibiotics **d.** Strep throat

10. Organisms that may find oxygen deadly are ______.
a. anaerobes **c.** producers
b. aerobes **d.** viruses

USING LAB SKILLS

11. The Find Out! on page 181 asked you to look at bacteria in yogurt. Assume that there are 50 000 bacteria in every drop of yogurt and there are thirty drops of yogurt in a milliliter. *Calculate* the number of bacteria in a 250-milliliter serving of yogurt.

12. Look at the MINI-Lab on page 193 again. *Design an experiment* that would test the effect of temperature on the final product—homemade yogurt. Carry out your experiment.

THINK AND WRITE CRITICALLY

Answer the following questions in your Journal, using complete sentences.

13. Would your classify viruses in a kingdom? Explain your answer.

14. Milk produced by cows is free of bacteria, yet several hours later, it needs to be pasteurized. Why?

15. Why do most AIDS victims die from secondary infections?

16. Explain why monerans are prokaryotes.

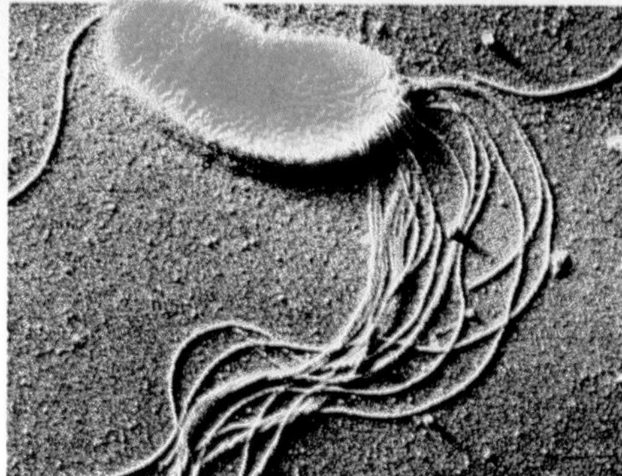
Bacterial cell with flagella

APPLY

17. What would happen if nitrogen-fixing bacteria could no longer live on the roots of plants?
18. Why are bacteria capable of surviving in all environments of the world?
19. Why is it difficult to get rid of viruses?
20. The organism that causes bacterial pneumonia is called *Pneumococcus*. What is its shape?
21. What precautions can be taken to prevent food poisoning?

MORE SKILL BUILDERS

If you need help, refer to the Skill Handbook.

1. **Sequencing:** Sequence the events of a virus actively infecting a cell.
2. **Making and Using Graphs:** Graph the data of viruses reproducing at different temperatures in a human body.

Virus Reproduction

Body Temperature °C	Millions of Viruses
36.9	1.0
37.2	1.0
37.5	0.5
37.8	0.25
38.3	0.10
38.9	0.05

3. **Interpreting Data:** What can you conclude about the effect of rising temperature on the viruses in Question 2?

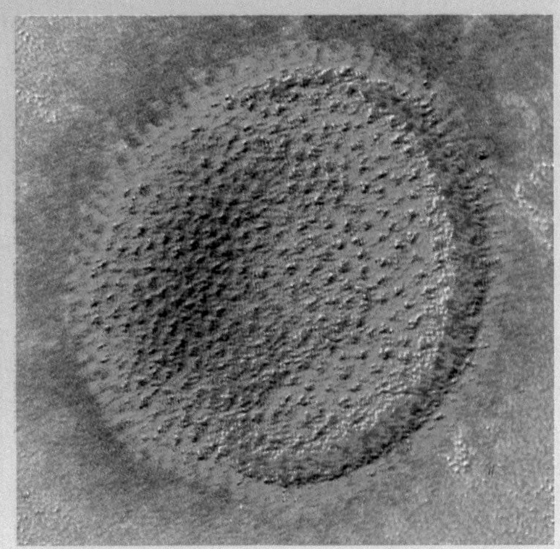

4. **Concept Mapping:** Make an events chain concept map to show what happens when a latent virus becomes active.
5. **Comparing and Contrasting:** In a table, compare and contrast a virus, a bacterial cell, and a eukaryote cell according to the structures they contain.

PROJECTS

1. **In your Journal,** draw and label a timeline illustrating the history of vaccination. Look up the term *variola* in a library reference and read about early attempts to protect people from disease, especially smallpox.
2. Research the components of a successful compost heap. As a class project, set up a model composting project at your school, indicating the role of bacteria in the process.

APPLY

17. Nitrogen would no longer be available in the form that plants could use; therefore the plants would die unless fertilizer were added.
18. Bacteria can reproduce quickly, have means of moving, and can form endospores to survive extreme conditions.
19. Once a virus infects a cell, it uses the cell to produce more viruses. The immune system has to work to get rid of viruses because there are no drugs that can kill viruses.
20. round
21. Using fresh foods and keeping them refrigerated, washing hands and all surfaces and utensils, and properly cooking foods help to prevent food poisoning.

MORE SKILL BUILDERS

1. Sequencing: attaches to cell, injects nucleic acid, cell is taken over by viral nucleic acids; the cell begins to make new virus particles, the cell bursts, releasing more viruses
2. Making and Using Graphs: See below. Body temperature is in °C.

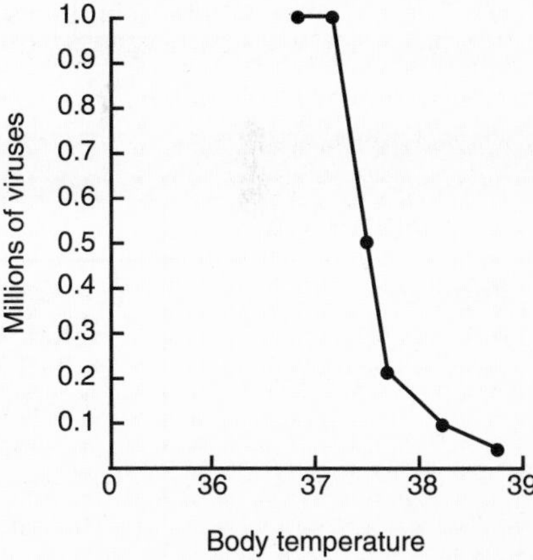

3. Interpreting Data: Viruses cannot survive as well at higher temperatures and therefore a fever helps to kill them.
4. Concept Mapping: Map should show the following order: latent virus enters cell; virus becomes part of cell's DNA; cell divides; virus reproduced as part of cell division; virus becomes active; virus forms new virus particles; cell destroyed.

5. Comparing and Contrasting:

Virus or cell type	Structures	Nucleus
Virus	Nucleic acid and protein coat	No
Bacteria	Nucleic acid, cell wall, cytoplasm, ribosomes	No
Eukaryote cell	cell membrane, cytoplasm, organelles, DNA in a nucleus	Yes

CHAPTER

9 Protists and Fungi

CHAPTER SECTION	OBJECTIVES	ACTIVITIES
9-1 Kingdom Protista (3 days)	1. **Identify** the characteristics shared by all protists. 2. **Describe** the three groups of protists. 3. **Compare** and **contrast** the protist groups.	**MINI-Lab:** *What do slime molds look like?* p. 209 **Activity 9-1:** *Comparing Algae and Protozoa,* p. 210
9-2 Kingdom Fungi (2 days)	1. **Identify** the characteristics shared by all fungi. 2. **Classify** fungi into groups based on their methods of reproduction. 3. **Describe** the difference between the imperfect fungi and all other fungi.	**MINI-Lab:** *How are spore prints made?* p. 214
9-3 Fungus—Can't Live Without It Science & Society (1 day)	1. **State** two types of drugs that are produced from fungi. 2. **Appreciate** the role of fungi in medicine.	**Activity 9-2:** *Designing an Experiment (Investigating Lichens),* p. 218
Chapter Review		

ACTIVITY MATERIALS

FIND OUT	ACTIVITIES		MINI-LABS	
Page 201	**9-1 Comparing Algae and Protozoa, p. 210**	**9-2 Designing an Experiment, p. 218**	**What do slime molds look like? p. 209**	**How are spore prints made? p. 214**
mushroom magnifying glass	Cultures of *Paramecium, Amoeba, Euglena, Spirogyra,* and *Micrasterias* prepared slide of slime mold 5 coverslips microscope dropper 5 microscope slides	3 lichen samples dried reindeer moss microscope slide coverslip microscope culture dish paper towels water	slime mold culture culture dishes	mushrooms (3) white paper towel

CHAPTER FEATURES	TEACHER RESOURCE PACKAGE	OTHER RESOURCES
Problem Solving: *Puzzled about Slime,* p. 208 **Skill Builder:** *Making and Using Tables,* p. 209	**Ability Level Worksheets** ◆ ***Study Guide,*** p. 37 ● ***Reinforcement,*** p. 37 ▲ ***Enrichment,*** p. 37 **Activity Worksheets,** pp. 77-78, 83 **Cross-Curricular Connections,** p. 13 **Transparency Masters,** pp. 27-28	**Color Transparency 14,** Protists **Lab Manual:** Life in Pond Water, p. 53
Technology: *A Yeast Library,* p. 213 **Skill Builder:** *Using Variables, Constants, and Controls,* p. 215	**Ability Level Worksheets** ◆ ***Study Guide,*** p. 38 ● ***Reinforcement,*** p. 38 ▲ ***Enrichment,*** p. 38 **Activity Worksheets,** p. 84 **Critical Thinking/Problem Solving,** p. 13 **Concept Mapping,** p. 23	**Lab Manual:** Molds, p. 55 **Lab Manual:** Yeasts, p. 57 **Lab Manual:** Lichens, p. 61 **Science Integration Activity 9** **STVS:** Disc 4, Side 1
You Decide! p. 217	**Ability Level Worksheets** ◆ ***Study Guide,*** p. 39 ● ***Reinforcement,*** p. 39 ▲ ***Enrichment,*** p. 39 **Activity Worksheets,** pp. 79-80 **Science and Society,** p. 13	**STVS:** Disc 4, Side 1
Summary Key Science Words Understanding Vocabulary Checking Concepts Using Lab Skills Think & Write Critically Apply More Skill Builders Projects	**ASSESSMENT RESOURCES** **Chapter Review,** pp. 21-22 **Chapter Test,** pp. 57-60 **Unit Test,** pp. 61-62 **Performance Assessment in in Middle School Science (PAMSS)**	**Chapter Review Software** **Test Bank** **Alternate Assessment** **Performance Assessment**

◆ **Basic** ● **Average** ▲ **Advanced**

ADDITIONAL MATERIALS

SOFTWARE	AUDIOVISUAL	BOOKS/MAGAZINES
Protozoa, Ventura Educational Systems. *The Microorganism Simulator,* Career Aids, Inc. *Organizing Protists and Fungi,* Queue. *The Microorganism Simulator,* Focus.	*Amoeba: One-Celled Organism,* film, International Film Bureau. *Fungi: The One Hundred Thousand,* film, Coronet/MTI. *Fungi and Man,* film, Benchmark. *Life of Molds,* film, McGraw-Hill. *Living Things in a Drop of Water,* film, EBEC. *Protozoa: Structures and Life Functions,* film, Coronet/MTI. *The Protist Kingdom,* film, BFA.	John, Theodore L. *How to Know the Protozoa.* 2nd ed. Dubuque, IA: William C. Brown Publishers, 1978. Sleigh, Michael. *Protozoa and Other Protists.* NY: Routledge, Chapman, and Hall, Inc., 1989. Weber, Nancy S. and Alexander H. Smith. *A Field Guide to Southern Mushrooms.* East Lansing, MI: Michigan State University Press, 1985.

CHAPTER

9

PROTISTS AND FUNGI

THEME DEVELOPMENT: The evolution and ecology themes are emphasized throughout this chapter. Evolutionary and ecological relationships of the protist and fungi are discussed.

CHAPTER OVERVIEW

- **Section 9-1:** Characteristics of all protists are first discussed in this section. Then the three groups of protists are described, compared, and contrasted with each other.
- **Section 9-2:** In this section, fungi characteristics are identified. Fungi are then classified based on their methods of reproduction.
- **Section 9-3: Science and Society:** The role of fungi in drug production is discussed in this section.

CHAPTER VOCABULARY

protists
algae
protozoa
pseudopods
cilia
hyphae
chitin
spores
sporangia
asci
budding
basidium
lichen
symbiosis
mutualistic

CHAPTER

9 Protists and Fungi

200

OPTIONS

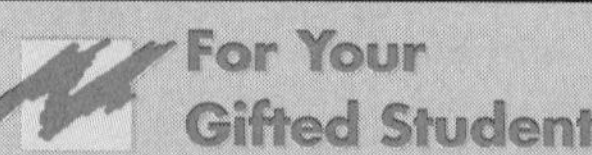

For Your Gifted Students

Students can make a flip book for each type of protist (flagellates, ciliates, sarcodines, etc.). On each page of the flip book they should show one stage of the animal's movement. When the pages are placed together and flipped through, the protist should appear to move. The more pages in the book showing a small segment of the movement, the more interesting the book will be.

For Your Mainstreamed Students

Have students make and display three-dimensional papier mâché or clay models of the major protists. Include flagellates, ciliates, and sarcodines. Students should label some of the structures that allow the protists to move.

Have you ever seen mushrooms growing in your yard or in a field? You may even like to eat mushrooms in salads, casseroles or on pizza. But what do you really know about mushrooms?

FIND OUT!

Do this simple activity to find out more about one of the members of Kingdom Fungi.

Examine a mushroom from the produce section of a grocery store. Carefully pull the cap off the stalk and lay it aside. Use your fingers to pull the stalk apart lengthwise. Continue to pull the stalk apart until the pieces are as small as you can get them. What do you see? Before the mushroom was picked, the stalk was connected to an underground structure by thin strands of tubelike cells. The strands of cells grew up from the ground to form the stalk.

Now look at the underside of the cap. *Observe* the many thin membranes. The brown color comes from the spores along the edges of each membrane. Why are the spores on the membranes?

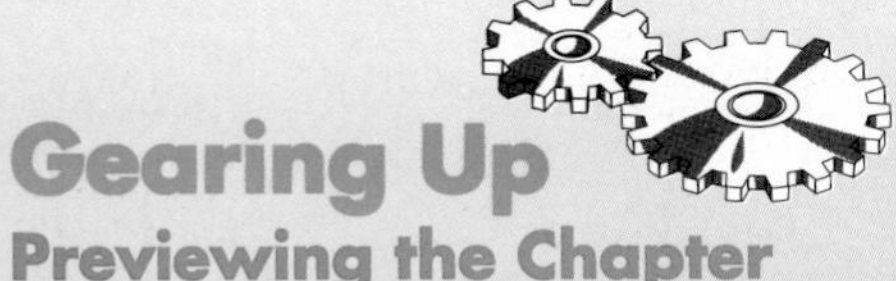

Gearing Up

Previewing the Chapter

Use this outline to help you focus on important ideas in this chapter.

Previewing Science Skills

- In the Skill Builders, you will make and use tables and use variables, constants, and controls.
- In the Activities, you will observe, compare, and design an experiment.
- In the MINI-Labs, you will make observations, predict, experiment, and estimate.

What's next?

Mushrooms are only one type of fungi. In this chapter you will learn more about the organisms in the Kingdom Fungi and their importance to the living world. But first you will learn about the different types of organisms in Kingdom Protista.

INTRODUCING THE CHAPTER

Use the Find Out activity to introduce the students to fungi. Inform students that they will be learning more about both protists and fungi as they read the chapter.

FIND OUT!

Preparation: Buy enough mushrooms so that each student can have one.

Materials: mushroom and magnifying glass for each student

Teaching Tips

- The best mushrooms to use will be those that have the membrane of the cap just separated from the stalk.
- Encourage students to examine the mushroom gills thoroughly using a magnifying glass.

Gearing Up

Have students study the Gearing Up feature to familiarize themselves with the chapter. Discuss the relationships of the topics in the outline.

What's Next?

Before beginning the first section, make sure students understand the connection between the Find Out activity and the topics to follow.

ASSESSMENT OPTIONS

PORTFOLIO

Refer to page 219 for suggested items that students might select for their portfolios.

PERFORMANCE ASSESSMENT

See page 219 for additional Performance Assessment options.

Process

Skillbuilders, pp. 209, 215
MINI-Labs, pp. 209, 214
Activities, 9-1, 210; 9-2, 218
Using Lab Skills, p. 220

CONTENT ASSESSMENT

Assessment—Oral, pp. 204, 206, 215
Section Reviews, pp. 209, 215, 217
Chapter Review, pp. 219-221
Mini Quizzes, pp. 207, 213

GROUP ASSESSMENT

Opportunities for group assessment occur with Cooperative Learning Strategies and Flex Your Brain Activities.

SECTION 9-1

PREPARATION

SECTION BACKGROUND

▶ Algae absorb much of the carbon dioxide produced by burning fossil fuels and release large amounts of oxygen to the atmosphere. Algae are the basis for most food chains in freshwater and marine habitats, and they are used to make industrial products.

▶ Iodine can be obtained from eating certain kelps.

▶ The population of diatoms may be so high in some ocean environments that a cubic meter of water may contain more than 35 000 000 of them.

▶ Red algae are a source of agar, a material on which bacteria and fungi are grown in laboratories.

PREPLANNING

▶ Obtain 200-300 buttons of different colors and shapes for the Motivate activity.

▶ Order cultures of protists needed for Activity 9-1.

▶ You will need plant and animal cell models for the demonstration on page 206.

▶ Make a culture of slime mold for the Check for Understanding.

▶ Innoculate oranges or bread with mold for the Motivate in Section 9-3.

1 MOTIVATE

▶ Have students bring in pond water. Have them use a microscope and draw the protists observed.

Cooperative Learning: Divide students into cooperative groups. Provide each group with about 40 buttons to arrange in a classification system. Relate this activity to the variety of protists and the difficulty involved in classifying them.

9-1 Kingdom Protista

New Science Words

protists
algae
protozoa
pseudopods
cilia

Objectives

▶ Identify the characteristics shared by all protists.
▶ Describe the three groups of protists.
▶ Compare and contrast the protist groups.

Figure 9-1. Kingdom Protista is made up of a diverse group of organisms.

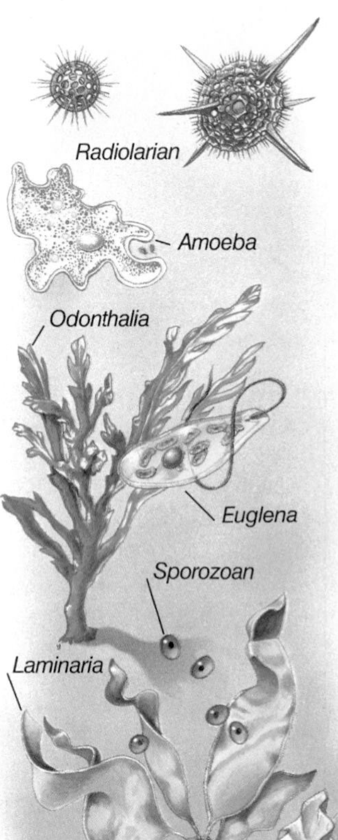

What Is a Protist?

Look at the organisms in Figure 9-1. Do you see any similarities among them? As different as they appear, all these organisms belong to the protist kingdom. Kingdom Protista (pruh TIHS tuh) contains many different types of organisms that share more characteristics with each other than with members of any other kingdom. All of the **protists** have a nucleus and therefore are eukaryotic. Some protists are one-celled, and others are many-celled. None share the complex organization found in plants and animals. Some protists contain chlorophyll and make their own food and others don't. Protists are plantlike, animal-like, and funguslike.

1 Most scientists think that protists evolved from the monerans because protists, although simple, are more complex in structure than monerans. Bacteria are thought to be the ancestors of animal-like and funguslike protists because these organisms can't make their own food. Plantlike protists probably evolved from cyanobacteria because both produce their own food.

Scientists hypothesize that protists are ancestors of the fungi, plant, and animal kingdoms. Plantlike protists, such as green algae, are the most probable ancestors of plants. Animal-like protists, the protozoa, are hypothesized to be the ancestors of animals because they have so many animal characteristics. Knowledge of the evolution of protists is incomplete because many lack hard parts and, as a result, there aren't very many fossils of these organisms.

OPTIONS

Meeting Different Ability Levels

For Section 9-1, use the following **Teacher Resource Masters** depending upon individual students' needs.

◆ **Study Guide Master** for all students.

● **Reinforcement Master** for students of average and above average ability levels.

▲ **Enrichment Master** for above average students.

Additional Teacher Resource Package masters are listed in the OPTIONS box throughout the section. The additional masters are appropriate for all students.

◆ STUDY GUIDE 37

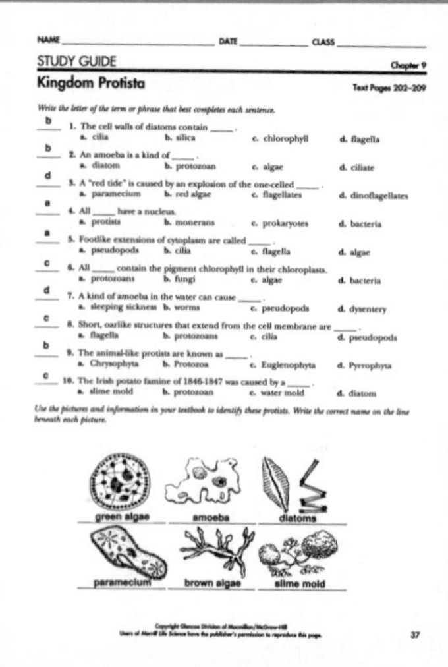

NAME ________ DATE ________ CLASS ________

STUDY GUIDE — Chapter 9

Kingdom Protista — Text Pages 202–209

Write the letter of the term or phrase that best completes each sentence.

b 1. The cell walls of diatoms contain ____.
a. cilia b. silica c. chlorophyll d. flagella

b 2. An amoeba is a kind of ____.
a. diatom b. protozoan c. algae d. ciliate

d 3. A "red tide" is caused by an explosion of the one-celled ____.
a. paramecium b. red algae c. flagellates d. dinoflagellates

a 4. All ____ have a nucleus.
a. protists b. monerans c. prokaryotes d. bacteria

a 5. Footlike extensions of cytoplasm are called ____.
a. pseudopods b. cilia c. flagella d. algae

c 6. All ____ contain the pigment chlorophyll in their chloroplasts.
a. protozoans b. fungi c. algae d. bacteria

d 7. A kind of amoeba in the water can cause ____.
a. sleeping sickness b. worms c. pseudopods d. dysentery

c 8. Short, oarlike structures that extend from the cell membrane are ____.
a. flagella b. protozoans c. cilia d. pseudopods

b 9. The animal-like protists are known as ____.
a. Chrysophyta b. Protozoa c. Euglenophyta d. Pyrrophyta

c 10. The Irish potato famine of 1846-1847 was caused by a ____.
a. slime mold b. protozoan c. water mold d. diatom

Use the pictures and information in your textbook to identify these protists. Write the correct name on the line beneath each picture.

37

Plantlike Protists

Plantlike protists are known as **algae.** Some species of algae are one-celled and others are many-celled. All algae can make their own food because they contain the pigment chlorophyll in their chloroplasts. Remember pigments are colored materials that absorb certain wavelengths of light. Chlorophyll is a green pigment. Even though all algae have chlorophyll, not all of them look green. Many have other pigments that cover up their chlorophyll. Species of algae are grouped into phyla mainly according to their pigments and the form in which they store food. There are six main phyla of algae. Each phylum has its own unique characteristics.

Euglenas

Algae that belong to the phylum Euglenophyta have characteristics of both plants and animals. A typical euglena is the bright green *Euglena gracilis,* shown in Figure 9-2. Like plants, these one-celled algae have chloroplasts and produce carbohydrate as food. When light is not present, euglenas require other food sources. Although euglenas have no cell walls, they do have a strong, flexible layer inside the cell membrane that helps them move and change shape. Many move by using flagella. Another animal-like characteristic of euglenas is that they have an adaptation called an eyespot that responds to light. The reddish eyespot helps the euglena locate light it needs to produce food by photosynthesis.

Eyespot
Cell membrane
Flagellum
Nucleus
Chloroplast

Figure 9-2. Euglenas are protists that have both plantlike and animal-like characteristics.

Diatoms

2 The most numerous of all the algae are the diatoms. They belong to the phylum Chrysophyta (kruh SAHF uh tuh). *Chryso* means "golden brown." Diatoms are photosynthetic one-celled algae that store food in the form of oil. They have a golden-brown pigment that masks the green chlorophyll. The shells of diatoms contain silica, the main element in glass. The body of a diatom is like a small box with a lid. One half of the shell fits inside the other half. Diatom shells are covered with markings and pits that form patterns.

You can find diatoms living in both freshwater and saltwater habitats. They are an important food source for many aquatic organisms.

Figure 9-3. The glassy cell walls of diatoms are covered with many beautiful markings.

TYING TO PREVIOUS KNOWLEDGE: Bridge this section with what students learned in Chapter 7 about classification.

OBJECTIVES AND SCIENCE WORDS: Have students review the objectives and science words in this chapter as they begin each section.

2 TEACH

Key Concepts are highlighted.

CONCEPT DEVELOPMENT

▶ **How are algae like plants?** *Algae contain chloroplasts and cell walls. Many algae are also many-celled.*

TEACHER F.Y.I.

▶ The *Euglena* shares characteristics of both algae and protozoans. When placed in darkness, euglenas can lose their chloroplasts and become even more like a protozoan.

FLEX Your Brain

Use the Flex Your Brain activity to have students explore PROTISTS.

Flex Your Brain
ASSESSMENT
Portfolio: Use the Flex Your Brain activity to reinforce critical thinking and problem solving skills. In Step 2, students might list size of protists, habitats, and names.

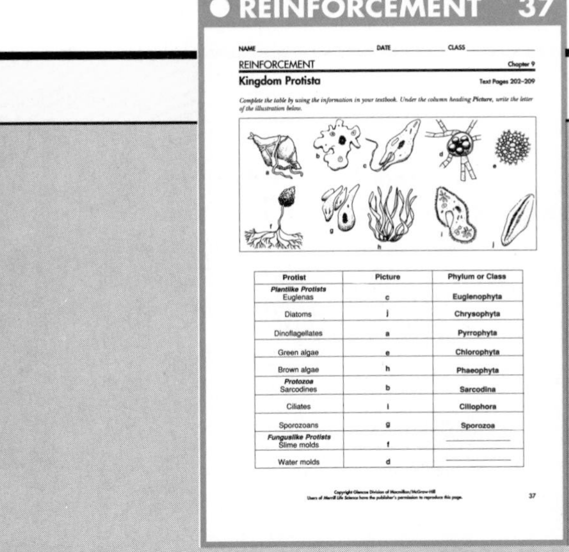

● REINFORCEMENT 37

NAME ____ DATE ____ CLASS ____

REINFORCEMENT Chapter 9

Kingdom Protista Text Pages 202–209

Complete the table by using the information in your textbook. Under the column heading ***Picture****, write the letter of the illustration below.*

Protist	Picture	Phylum or Class
Plantlike Protists Euglenas	c	Euglenophyta
Diatoms	j	Chrysophyta
Dinoflagellates	a	Pyrrophyta
Green algae	e	Chlorophyta
Brown algae	h	Phaeophyta
Protozoa Sarcodines	b	Sarcodina
Ciliates	i	Ciliophora
Sporozoans	g	Sporozoa
Funguslike Protists Slime molds	f	
Water molds	d	

37

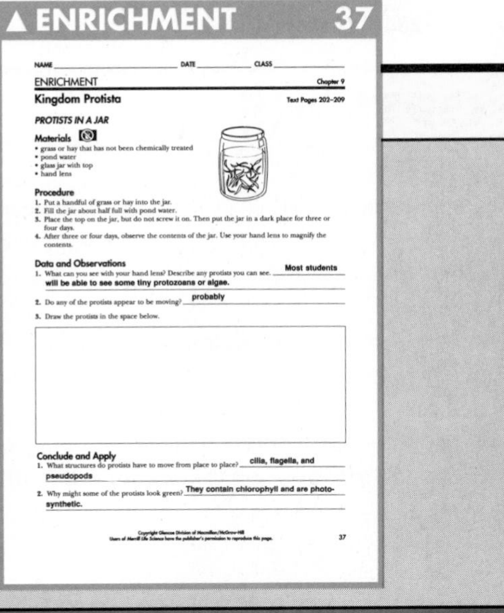

▲ ENRICHMENT 37

NAME ____ DATE ____ CLASS ____

ENRICHMENT Chapter 9

Kingdom Protista Text Pages 202–209

PROTISTS IN A JAR

Materials

- grass or hay that has not been chemically treated
- pond water
- glass jar with top
- hand lens

Procedure

1. Put a handful of grass or hay into the jar.
2. Fill the jar about half full with pond water.
3. Place the top on the jar, but do not screw it on. Then put the jar in a dark place for three or four days.
4. After three or four days, observe the contents of the jar. Use your hand lens to magnify the contents.

Data and Observations

1. What can you see with your hand lens? Describe any protists you can see. **Most students will be able to see some tiny protozoans or algae.**
2. Do any of the protists appear to be moving? **probably**
3. Draw the protists in the space below.

Conclude and Apply

1. What structures do protists have to move from place to place? **cilia, flagella, and pseudopods**
2. Why might some of the protists look green? **They contain chlorophyll and are photosynthetic.**

37

CONCEPT DEVELOPMENT

▶ Overlapping protein fibers in the pellicles of many protists allow them to alter their shape considerably.
▶ Concentrate on structure and function of the organisms in this section rather than the scientific names.
▶ Each year, more than 200 000 metric tons of diatomaceous earth is mined in a quarry in California. Ask students what this means in terms of the number of diatoms it took to accumulate such a deposit. Also ask them what this means in terms of where the ocean was located at one time.

Diatoms reproduce in unbelievably large numbers. When the organisms die, their small shells sink to the floor of the body of water and collect in very deep layers. Deposits made of these diatoms are mined with power shovels and used in insulation, filters, and road paint. Diatom shells produce the sparkle that makes road lines visible at night.

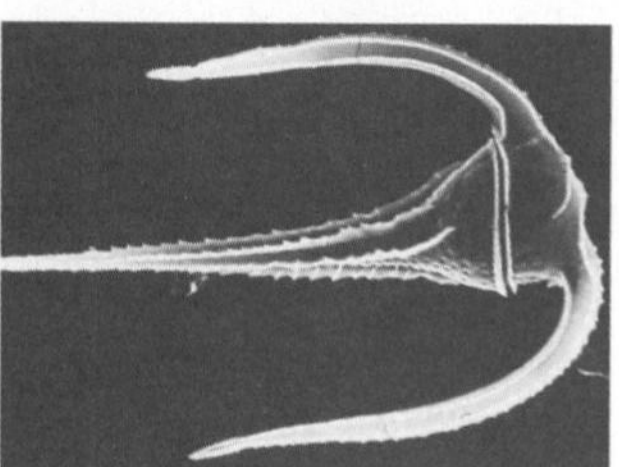

Figure 9-4. A scanning electron micrograph of a dinoflagellate.

Dinoflagellates

Phylum Pyrrophyta (puh RAHF uh tuh), the "fire" algae, contains species of one-celled algae called dinoflagellates that have red pigments. The name *dinoflagellate* means "spinning flagellates." One of the flagella moves the cell, and the other circles the cell, causing it to spin with a motion similar to a top. Dinoflagellates store food in the form of starch and oils.

Almost all dinoflagellates live in salt water. They are important food sources for many saltwater organisms. For unknown reasons, every once in awhile the population of certain species of dinoflagellates increases very rapidly. Red pigments in these algae cause the water to look red. This explosion of algae is known as a "red tide."
3 Some dinoflagellates that cause red tides contain a nerve poison that can kill fish. People who eat fish or shellfish that have absorbed the poisons produced by these algae can also become ill and sometimes die.

Figure 9-5. There are many different shapes among the species of green algae. Phytoplankton (a) and *Spirogyra* (b) are shown here.

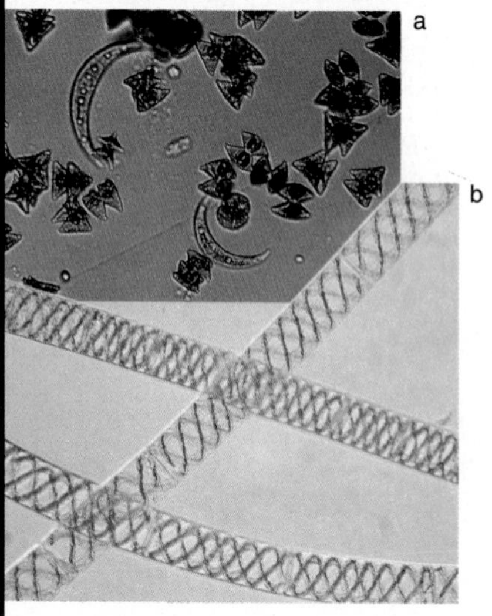

Green Algae

Species of green algae form the phylum Chlorophyta (kloh RAHF uh tuh). The presence of chlorophyll in green algae tells you that they undergo photosynthesis and produce food in the form of starch. Figure 9-5 shows different forms of green algae. Although most green algae live in water, others can live in many other environments, including trunks of trees and even on other organisms! Green algae can be one-celled or many-celled. *Chlamydomonas* is an example of a one-celled green alga. *Ulva,* also called sea lettuce, is a many-celled saltwater alga that grows in thin sheets. Many-celled species of green algae can be either chainlike or form colonies. *Spirogyra* is a chainlike freshwater alga that has spiral-shaped chloroplasts. *Volvox* is a freshwater form that occurs in ball-shaped colonies. The colony rolls through the water using its flagella.

OPTIONS

ASSESSMENT—ORAL

▶ **Why are protists in a separate kingdom?** *Protists have more traits in common with each other than with monerans, fungi, plants, or animals.*
▶ **Why are dinoflagellates important?** *They are important food sources for many saltwater organisms.*
▶ **How do algae obtain their energy?** *Algae are able to carry on photosynthesis like plants.*
▶ **How can populations of dinoflagellates have economic impact on a fishing industry?** *They can have great economic impact when they cause red tides because fish that eat them die, making fish unavailable as a source of food for people.*

Red Algae

The red algae belong to the phylum Rhodophyta (roh DAHF uh tuh). *Rhodo* means "red" and describes the color of members of this phylum. If you've ever eaten pudding, or used toothpaste, you have eaten something made with red algae! A carbohydrate called carrageenan 4 found in red algae is used to give toothpaste and pudding their smooth, creamy textures. Most red algae are many-celled. Some species of red algae can live up to 175 meters deep in the ocean. Their red pigment allows them to absorb the limited amount of light that penetrates to those depths and enables them to produce a type of starch.

Brown Algae

Brown algae make up the phylum Phaeophyta (fee AHF uh tuh). Members of this phylum are many-celled and vary greatly in size. Large brown algae called kelps may be as much as 100 meters long. Many kelps are important food sources for fish and invertebrates. They form a dense mat of stalks and leaflike blades where many small fish and other animals live.

People in many parts of the world eat brown algae. The thick texture of food such as ice cream and marshmallows is produced by a carbohydrate called algin found in this alga. Brown algae are also used to make fertilizer.

Choose one species of plantlike protist and imagine that you become that protist for a day. **In your Journal**, after researching the protist, describe what your day as a protist might be like.

Table 9-1

THE PLANTLIKE PROTISTS

Phylum	Example	Pigments	Other Characteristics
Euglenophyta Euglenas		Chlorophyll	One-celled algae that move with a flagellum; has eyespot to detect light.
Chrysophyta Diatoms		Golden Brown	One-celled algae with body made of two halves. Cell walls contain silica.
Pyrrophyta Dinoflagellates		Red	One-celled algae with two flagella. Flagella cause cell to spin. Some species cause red tide.
Chlorophyta Green Algae		Chlorophyll	One- and many-celled species. Most live in water, some live out of water, in or on other organisms.
Rhodophyta Red Algae		Red	Many-celled algae; carbohydrate in red algae is used to give some foods a creamy texture.
Phaeophyta Brown Algae		Brown	Many-celled algae, most live in salt water; important food source in aquatic environments.

In Your JOURNAL

Students should choose a protist and find out about its habitat, what it feeds on, and so on. Students' descriptions of their day as that organism should reflect this information.

To assess this product, refer to the Performance Task Assessment Lists in **Performance Assessment in Middle School Science.**

REVEALING MISCONCEPTIONS

▶ Point out that even though some of the brown algae, such as kelps, may grow 100 meters long, they are not plants because they do not develop the complex structures of plants.

CONCEPT DEVELOPMENT

▶ Have students use Table 9-1 as a way to summarize the characteristics of plantlike protists. Ask questions based on the table and text.

PROGRAM RESOURCES

From the **Teacher Resource Package** use:

Cross-Curricular Connections, page 13, Dysentery and Malaria.

Activity Worksheets, page 5, Flex Your Brain.

Use **Laboratory Manual** page 53, Life in Pond Water.

CONCEPT DEVELOPMENT

▶The White Cliffs of Dover in England are made up primarily of the shells of sarcodines.

▶Of the 100 000 or so protozoans, only a few cause disease, but the illnesses they do cause can be severe or even fatal.

▶Malaria, sleeping sickness, and dysentery are diseases produced by species of protozoans.

▶**How are protozoans like animals?** *Protozoans do not have cell walls, and most have some means of movement. Protozoans also cannot make their own food.*

Cooperative Learning: Interested students can work in Paired Partners to find out more about particular parasitic protozoans. Have them report to the class on their findings.

▶**Demonstration:** Use model plant and animal cells to illustrate the organelles in common between algae and plants and between protozoans and animals.

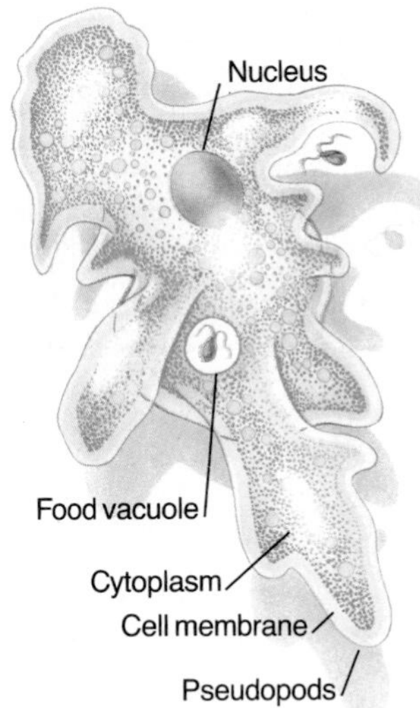

Figure 9-6. An amoeba constantly changes shape as it forms pseudopods.

Animal-like Protists

5 One-celled animal-like protists are known as **protozoa** (proht uh ZOH uh). These complex organisms live in water, soil, and dead organisms. Many types of protozoans are parasites. Remember that a parasite is an organism that lives in or on another organism. The parasite harms the other organism. In addition to the usual cell organelles, protozoans contain special vacuoles for digesting food and getting rid of excess water. Four species of protozoan are classified based on method of movement.

Sarcodines

The first protozoans were probably similar to members of the phylum Sarcodina (sar kuh DI nuh). The *Amoeba* shown in Figure 9-6 is a typical species of this phylum. Sarcodines move about and feed using temporary bulges of their cytoplasm called **pseudopods** (SEWD uh pahdz). The word *pseudopod* means "false foot." Pseudopods are footlike extensions of cytoplasm the organism uses to move and to trap food. An amoeba extends the cytoplasm of a pseudopod on either side of a bit of food such as a bacterium. The two ends of the pseudopod close like two fingers. As the fingers of cytoplasm close, the bacterium is trapped. A vacuole forms around the food and it is digested. Sarcodines are found in freshwater or saltwater environments and in other animals as parasites. You may have heard that it isn't wise to drink water in certain countries. This is because many areas of the world have a species of amoeba in the water that can cause dysentery. Dysentery can induce a severe form of diarrhea.

Figure 9-7. Many of the saltwater sarcodines have skeletons made of a chalklike material called calcium carbonate. The White Cliffs of Dover in England are made almost entirely of the shells of billions of these organisms. The photo at the right shows the shells of some sarcodines.

OPTIONS

ASSESSMENT—ORAL

▶**Why is it improper to say that mosquitoes cause malaria?** *Malaria is caused by a sporozoan type of protozoa. The sporozoan is spread from one organism to another through the mosquito intermediate.*

▶**Are protists simple or complex? Defend your answer.** *Protists have simpler cells than plants or animals, but all cells are complex enough to carry on all life processes.*

▶**Describe the three stages in the life of a slime mold.** *There is a single-celled flagellate stage, a slimy amoeboid stage, and a funguslike reproductive stage.*

Flagellates

Protozoans that move using flagella are called flagellates (FLAJ uh layts) and belong to the phylum Mastigophora (mas tuh GAHF uh ruh). All of the flagellates have one or many long flagella that whip through a watery environment, moving the organism along. Many species of flagellates live in fresh water, but some are parasites.

Trypanosoma is a flagellate that causes African sleeping sickness in humans and other animals. It is spread by the tsetse fly in Africa. The disease causes fever, swollen glands, and extreme sleepiness. Another flagellate lives in the digestive system of termites. These flagellates are beneficial to the termites, because they produce enzymes that digest the wood the termites eat. The termites can't digest the wood without the flagellates.

Ciliates

The most complex protozoans belong to the phylum Ciliophora. Members of this phylum move by using cilia. **Cilia** are short, threadlike structures that extend from the cell membrane. Ciliates may be covered with cilia, or have cilia grouped in special areas of the cell. The beating of the cilia is organized so the organism can move swiftly in any direction.

A typical ciliate is *Paramecium* in Figure 9-8. In the *Paramecium* you can see another characteristic of the ciliates: they have two nuclei, a macronucleus, and a micronucleus. The large macronucleus controls the everyday functions of the cell. The smaller micronucleus functions in reproduction. Paramecia usually feed on bacteria swept into the oral groove. Once the food is inside the cell, a food vacuole forms and the food is digested. The contractile vacuole removes excess water from the cell. Wastes are removed through the anal pore.

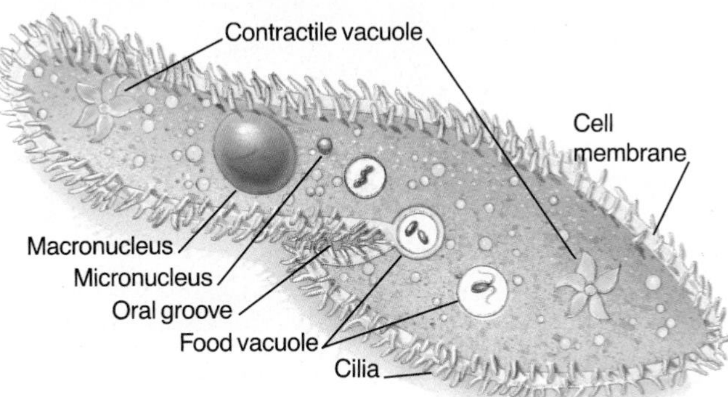

Some termites feed on aboveground buildings or structures made from wood and trees. These termites are completely dependent on their intestinal protozoans for survival. If these protozoans were all removed from a termite, it would continue to feed, but would eventually starve to death. What carbohydrate in wood do the protozoans break down?

Figure 9-8. *Paramecium* is a typical ciliate.

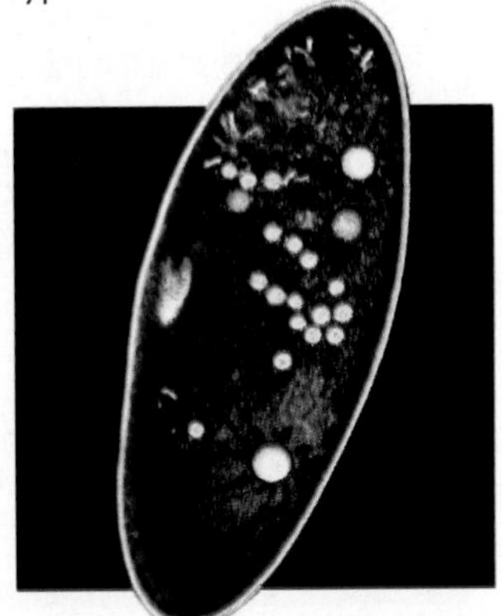

CROSS CURRICULUM

▶**English:** Ask students to imagine that they are a particular protist. Have them write a short story about the environment in which they live.

MINI QUIZ

Use the Mini Quiz to check students' recall of chapter content.

1. **Most scientists think that protists evolved from ____ because protists are more complex.** *monerans*
2. **The most numerous of all the algae are the ____ .** *diatoms*
3. **Some _____ that cause red tides contain a nerve poison that can kill fish.** *dinoflagellates*
4. **A carbohydrate found in _____ is used to give toothpaste and pudding their smooth, creamy texture.** *red algae*
5. **One-celled animal-like protists are known as ____ .** *protozoa*
6. **Slime molds and water molds are ____ protists.** *funguslike*

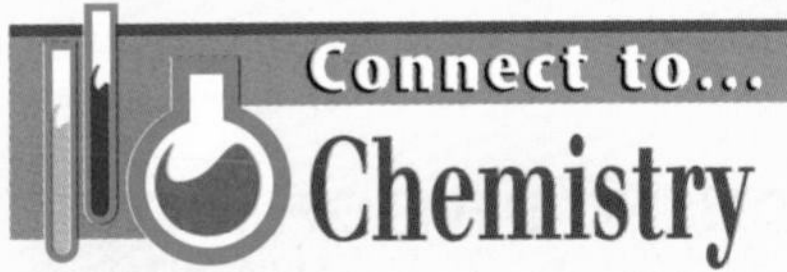

Answer: cellulose

CONCEPT DEVELOPMENT

▶To help students envision what the beating of cilia is like, show them a photograph of a rowing crew. All members move together so that their oars are all in the same position at the same time. This propels the boat through water as cilia propel a paramecium or other ciliate through its watery medium.

PROGRAM RESOURCES

From the **Teacher Resource Package** use:

Activity Worksheets, page 83, Mini-Lab: What do slime molds look like?

Transparency Masters, pages 27-28, Protists.

Use **Color Transparency** 14, Protists.

PROBLEM SOLVING

Think Critically: Slime molds have characteristics of organisms in both protist and fungus kingdoms. Slime molds resemble animal-like protists in their methods of locomotion and ingesting food. Slime molds resemble fungi in their methods of spore formation.

CHECK FOR UNDERSTANDING

Use a culture of slime mold to illustrate the phases of its life cycle. It has a flagellate stage, a plasmodium stage, and a funguslike stage.

RETEACH

Show students photographs or illustrations of various protists. Have students identify the life cycle phase of each.

EXTENSION

For students who have mastered this section, use the **Reinforcement** and **Enrichment** masters or other OPTIONS provided.

CONCEPT DEVELOPMENT

▶ **How are the slime molds and water molds like fungi?** *They are sometimes found on decaying materials like fungi, and one part of their life cycle resembles a fungus.*

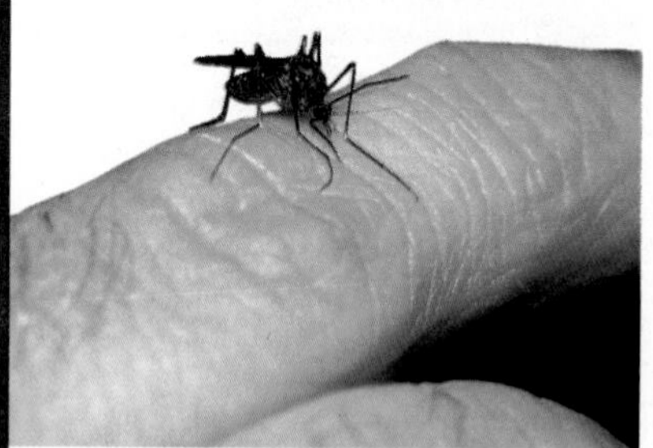

Figure 9-9. Female *Anopheles* mosquitoes spread the sporozoan that causes malaria.

Sporozoans

The phylum Sporozoa contains only small, parasitic protozoans. Sporozoans have no way of moving on their own. All are parasites that live in and feed on the blood of humans, and other animals.

Figure 9-9 shows a mosquito that carries *Plasmodium,* a sporozoan common in tropical climates. This sporozoan causes malaria in humans. Malaria is spread when an infected mosquito bites a human. There is no vaccine for malaria. It causes many deaths each year.

Funguslike Protists

Funguslike protists include several small phyla of protists that have features of both protists and fungi. Slime
6 molds and water molds are funguslike protists. Slime molds are much more attractive than their name sounds. Many are very brightly colored. They form a delicate weblike structure on the surface of their food supply. They obtain energy by decomposing organic materials.

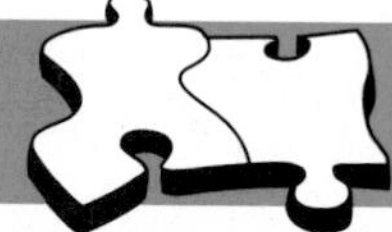

PROBLEM SOLVING

Puzzled about Slime

Stan and his cousin, Aaron, were walking through the woods near Stan's home when Stan discovered an orange mass on a fallen log. Stan told Aaron it was slime mold. His class had been studying protists, and he had seen some slime mold at school. He had read that slime molds were classified in the protist kingdom because they are similar to amoebas, using pseudopods to move and ingest food.

Aaron said, "You're wrong. My textbook classifies slime molds in the fungus kingdom. Slime molds have reproductive structures just like fungi."

The two boys were puzzled as to why slime molds were placed in the protist kingdom in one book and in the fungus kingdom in another. What evidence is there that slime molds are protists?

Think Critically: Why do some scientists classify slime molds as protists and others classify them as fungi?

OPTIONS

ENRICHMENT

▶ Have students analyze the following story by making a graph. A person who contracted malaria had the following temperatures at the given times. When will the person next have chills? DAY 1: 1:00 PM - 103° (chills); 6:00 PM - 101° (sweating); DAY 2: 1:00 AM - 97°; 6:00 AM - 98°; 1:00 PM - 97°; 6:00 PM - 99°; DAY 3: 1:00 AM - 101° (chills); 6:00 AM - 103°; 1:00 PM - 103°; 6:00 PM - 101° (sweating) (Answer: DAY 5 at 1:00 AM)

Slime Molds

Slime molds like the one pictured in Problem Solving have some protozoan characteristics. During part of their life cycle, the cells of slime molds move by means of pseudopods and behave like amoebas. Slime molds reproduce with spores the way fungi do. You will learn about reproduction in fungi in the next section.

Although most slime molds live on decaying logs or dead leaves in moist, cool, shady woods, one common slime mold is sometimes found crawling across lawns and mulch. It creeps along feeding on bacteria and decayed plants and animals. When conditions become less favorable, reproductive structures form on stalks, and spores are produced.

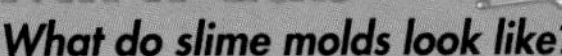

MINI-Lab

What do slime molds look like?

Obtain materials for growing the slime mold *Physarum* from your teacher. Follow the instructions for growing the mold. **In your Journal,** make daily drawings and observations of the mold as it grows. *Predict* the conditions under which the slime mold would change from the amoeboid form.

Water Molds and Mildew

Water molds, downy mildews, and white rusts make up another phylum of funguslike protists. The most well-known member of this phylum is the water mold that caused the Irish potato famine in the 1840s. Potatoes were Ireland's main crop and the main food source for its people. Nearly one million people died in the famine.

Water molds have cell walls as do fungi, but their relatively simple cells are more like those of protozoans. Figure 9-10 shows a parasitic water mold that grows on decaying fish. If you have an aquarium, you may see water molds attack a fish and cause its death.

Figure 9-10. A parasitic water mold growing on a fish.

SECTION REVIEW

1. What are the main characteristics of all protists?
2. Describe the characteristics of the three groups of protists.
3. **Apply:** Why aren't there many fossils of the different groups of protists?
4. **Connect to Earth Science:** Find out what plankton is, where it is located, and what protists are in plankton.

Making and Using Tables

Skill Builder

Make a table that compares the characteristics of the four phyla of protozoans. Include phylum, example species, mode of locomotion, and other characteristics. If you need help, refer to Making and Using Tables in the **Skill Handbook** on page 690.

Skill Builder

Phylum	Example	Locomotion	Other Characteristics
Sarcodina	Amoeba	Pseudopod	Some have hard shells
Mastigophora	Trypanosome	Flagella	Many parasites
Ciliophora	Paramecium	Cilia	At least two nuclei
Sporozoans	Plasmodium	None	Parasitic

MINI-Lab

Materials: slime mold culture, culture dishes

Teaching Tips

▶Obtain kit for growing slime mold from a biological supply company. *Physarum polycephalum* produces good results.

Answer: Students should observe a slimy mass called plasmodium. They should also observe the mold move across the culture dish. The second stage of slime mold growth involves the development of sporangia and spores.

MINI-Lab ASSESSMENT

Performance: To assess students' understanding of slime molds, have them try to grow slime mold cultures on different surfaces with different nutrients.

3 CLOSE

▶Ask questions 1-3 and the **Apply** Question in the Section Review.

SECTION REVIEW ANSWERS

1. Protists are simple organisms, mostly single-celled, that possess membrane-bound nuclei, but don't have tissues or organs. Protists may be plantlike, animal-like, or funguslike.

2. Algae are plantlike protists with chloroplasts and cell walls. Protozoans are single-celled animal-like protists that have no cell walls and cannot make their own food. The funguslike protists are all consumers that produce structures similar to the fungi.

3. Apply: Protists are small and generally do not have any hard parts to be fossilized. The protists decay rapidly, leaving no trace of their existence.

4. Connect to Earth Science: Plankton are plantlike and animal-like organisms that float in the uppermost layer of water. They are found in fresh and salt water worldwide. Protists in plankton are diatoms and dinoflagellates.

Skillbuilder ASSESSMENT

Performance: Assess students' abilities to Make and Use Tables by having them write a statement in their journals comparing movement among protozoans.

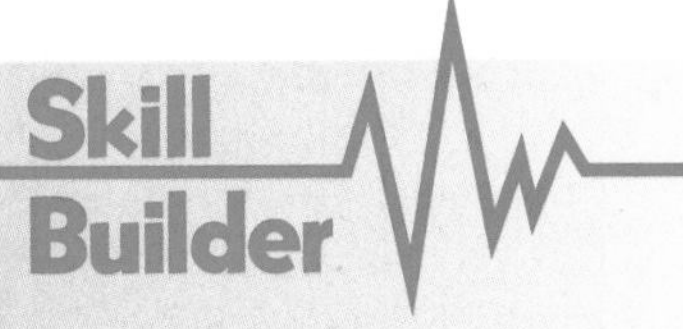

ACTIVITY 9-1

50 minutes

OBJECTIVE: Observe the differences between algae and protozoa.

PROCESS SKILLS applied in this activity:
▶ **Observing** in Procedure Steps 3-5.
▶ **Inferring** in Analyze Question 1.
▶ **Defining Operationally** in Conclude and Apply Questions 5-6.

COOPERATIVE LEARNING Use the Science Investigation Team strategy. One member of each group should make a data table. Students should take turns recording observations.

TEACHING THE ACTIVITY

Alternate Materials: Other protozoans and algae may be used. It is best to use local materials from the same phyla if available. Prepared slides may be used if living material is not available.

Troubleshooting: Maintain separate cultures for each organism. Prevent students from using the same dropper in more than one culture.

PROGRAM RESOURCES

From the **Teacher Resource Package** use:

Activity Worksheets, pages 77-78, Activity 9-1, Comparing Algae and Protozoa.

ACTIVITY 9-1 Comparing Algae and Protozoa

Problem: *What are the differences between algae and protozoans?*

Materials

- cultures of *Paramecium, Amoeba, Euglena, Spirogyra,* and *Micrasterias*
- prepared slide of slime mold
- 5 coverslips
- microscope
- dropper
- 5 microscope slides

Procedure

1. Make a data table **in your Journal** for your drawings and observations.
2. Make a wet mount of the *Paramecium* culture.
3. Observe the wet mount first under low and then under high power. Draw and label the organism.
4. Repeat Steps 2 and 3 with the other cultures. *Return all preparations to your teacher and wash your hands.*
5. *Observe* the slide of slime mold under low and high power. *Record* your observations.

Analyze

1. For each organism that could move, label the structure that enabled the movement.
2. Which protists could not move? Why?
3. Describe whether each protist had one cell or many cells.

Conclude and Apply

4. Which protists make their own food? *Explain* how you know that they make their own food.
5. Which protists had animal characteristics?
6. Which protists had fungus characteristics?

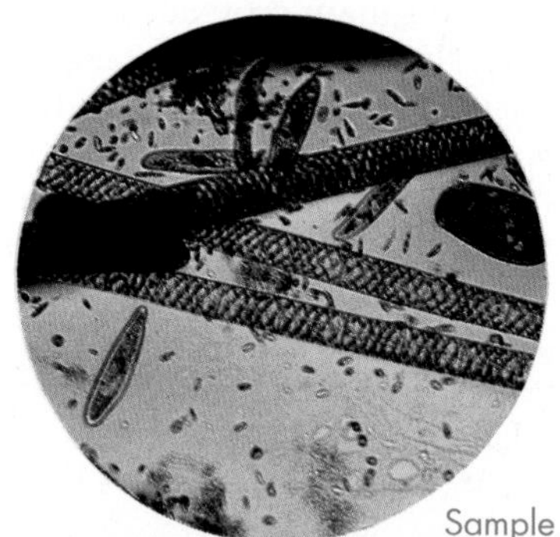

Data and Observations

Sample Data

	Paramecium	*Amoeba*	*Euglena*	*Spirogyra*	*Micrasterias*
Drawing	Students' drawings should resemble *Paramecium* on page 207.	Students' drawings should resemble *Amoeba* on page 206.	Students' drawings should resemble *Euglena* on page 203.	Students' drawings should resemble *Spirogyra* on page 204.	Students' drawings should show a single-celled green alga with a flowerlike shape.

ANSWERS TO QUESTIONS

1. *Paramecium*—cilia; *Amoeba*—pseudopoda; *Euglena*—flagella
2. *Spyrogyra* and *Micrasterias* have no structures to enable them to move.
3. Only *Spyrogyra* is many-celled; the rest of the protists are one-celled.
4. *Spyrogyra, Micrasterias,* and *Euglena* make their own food. We know they make their own food because they contain choloroplasts and therefore, undergo photosynthesis. Organisms that undergo photosynthesis are producers, capable of making their own food.
5. *Paramecium, Amoeba,* and *Euglena* are like animals because they obtain their food from other sources.
6. None had fungus characteristics.

Activity ASSESSMENT

Performance: To further assess students' abilities to identify algae and protozoans, have students make and view scrapings from fish tanks, making sure they wash their hands thoroughly after.

Kingdom Fungi

9-2

Objectives

- ▶ Identify the characteristics shared by all fungi.
- ▶ Classify fungi into groups based on their methods of reproduction.
- ▶ Describe the difference between the imperfect fungi and all other fungi.

New Science Words

hyphae
chitin
spores
sporangia
asci
budding
basidium
lichen
symbiosis
mutualistic

What Are Fungi?

Do you believe you can find members of Kingdom Fungi in a quick trip around your house or apartment? Well, you probably can. You can find fungi in your kitchen if you have canned mushrooms, mushroom soup, or fresh mushrooms. You may also find mold, a type of fungus, growing on an old loaf of bread, or mildew, another fungus, growing on your shower curtain. Yeasts are a type of fungi used to make bread, cheese, beer, and wine.

As important as fungi seem in the production of different foods, they are most important in their role as organisms that decompose or break down organic materials. Food scraps, clothing, dead plants, and animals are all made of organic material. Fungi work to decompose, or break down, all these materials and return them to the soil. The materials returned to the soil are then reused by plants. Fungi, along with bacteria, are nature's recyclers. They keep Earth from becoming buried under mountains of waste materials.

Fungi were once classified as plants. But, unlike plants, fungi do not make their own food or have the specialized tissues and organs of plants. You will learn about plant structures in the next unit. Most species of fungi are many-celled. The body of a fungus is usually a mass of many-celled, threadlike tubes called **hyphae** (HI fee). 1
The mat of hyphae is called a mycelium (mi SEE lee um). The cell walls of hyphae are made of cellulose or of **chitin,** a strong, flexible carbohydrate that is also found in the body covering and wings of insects.

Figure 9-11. A fungus is made up of many threadlike tubes called hyphae.

OPTIONS

PROGRAM RESOURCES

From the **Teacher Resource Package** use:

Activity Worksheets, page 5, Flex Your Brain.

Critical Thinking/Problem Solving, page 13, Fungi: A Source of Indoor Air Pollution.

Concept Mapping, pages 23-24.

Use **Laboratory Manual** page 55, Molds.

SECTION 9-2

PREPARATION

SECTION BACKGROUND

▶ Mycology is the study of fungi.

▶ The poison of the *Amanita* group of mushrooms acts like the venom of a rattlesnake.

▶ Many people are allergic to the spores produced by fungi.

PREPLANNING

▶ Make a yeast suspension for the Motivate activity and demonstration.

▶ Obtain a loaf of stale bread for Activity 9-2.

▶ Make a collection of photographs of fungi for use in Reteach.

1 MOTIVATE

▶ Mix some yeast suspension with some flour and water and put in a warm place. Ask students to predict what will happen over the course of the day. (The dough will rise as carbon dioxide is made by the yeast.)

FLEX Your Brain

Use the Flex Your Brain activity to have students explore FUNGI.

Flex Your Brain
ASSESSMENT

Portfolio: Use the Flex Your Brain activity to reinforce critical thinking and problem solving skills. In Step 2, students might list different fungi and their habitats.

▶ Ask students if they have ever eaten a mushroom. Ask what it tasted like. Point out that mushrooms are only one group of fungi. Other fungi are used to make bread, wine, beer, cheese, and even medicines.

VideoDisc

STVS: Fungal Collection for Research, Disc 4, Side1

TYING TO PREVIOUS KNOWLEDGE: Compare the role bacteria play in decomposition with the same role of fungi.

2 TEACH

Key Concepts are highlighted.

CONCEPT DEVELOPMENT

▶ **What is an ascus?** *An ascus is a sac that contains spores of the Ascomycotes.*

▶ **What is a sporangium?** *A sporangium is a spherical structure that produces spores in Zygomycotes.*

▶ Caution students not to pick fungi because some may be poisonous.

CONCEPT DEVELOPMENT

▶ **Demonstration:** Fill a test tube with some water mixed with white syrup in a ratio of about 3:1. Add a few milliliters of yeast suspension. Place your thumb over the top of the full tube and turn it upside down in a beaker containing the same water and syrup mixture. Observe the production of carbon dioxide bubbles.

CAUTION

▶ Contact the school nurse if a student shows signs of an allergic reaction to fungi used in demonstrations during this section. It is rare, but some students may experience breathing difficulties in the presence of spores.

Figure 9-12. Fungi reproduce by spores. This puffball mushroom releases many spores at once.

Fungi don't contain chlorophyll and can't make their own food. Most fungi feed on dead organisms. Organisms that obtain food in this way are called saprophytes. A fungus secretes enzymes to digest food outside of itself. Then the fungus cells absorb the digested food. Fungi that cause athlete's foot and ringworm are parasites. They obtain their food directly from living things.

Fungi grow best in warm, humid areas. Moist places such as tropical regions of the world or the spaces between your toes are homes for many types of fungi.

Look at the mushroom in Figure 9-12. The puff of "smoke" above it is actually made of the structures fungi use for reproduction, spores. (2) **Spores** are reproductive cells that form new organisms without fertilization. The structures in which fungi produce spores are used to classify fungi into one of four divisions.

Divisions of Fungi

How are fungi classified?

Zygote Fungi

The fuzzy black mold that you sometimes find growing on an old loaf of bread is a type of zygote fungus. Fungi that belong to this division, the division Zygomycota (zi goh mi KOH tuh), produce spores in round spore cases called **sporangia** on the tips of upright hyphae. The sporangia turn black as they mature. The black fuzz you see on the bread is actually a mass of mature spore cases. When each sporangium splits open, hundreds of spores are released into the air. Each spore will grow into more mold if it lands where there is enough moisture, a warm temperature, and a food supply.

Figure 9-13. In zygote fungi, spores are produced in round black spore cases on the tips of upright hyphae. The photo shows the zygote fungus *Rhizopus* growing on bread.

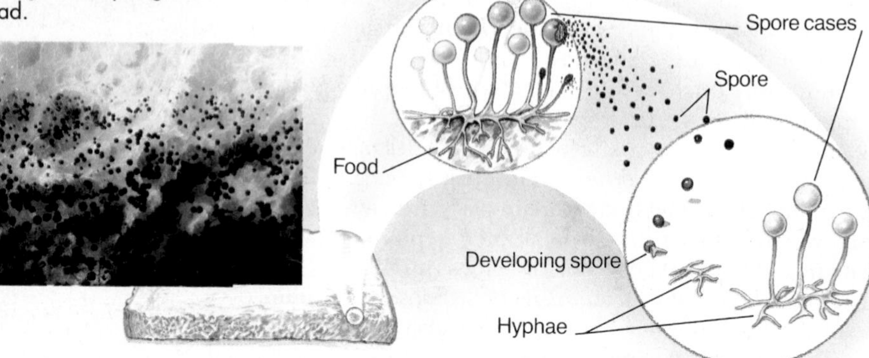

212 PROTISTS AND FUNGI

OPTIONS

Meeting Different Ability Levels

For Section 9-2, use the following **Teacher Resource Masters** depending upon individual students' needs.

◆ **Study Guide Master** for all students.

● **Reinforcement Master** for students of average and above average ability levels.

▲ **Enrichment Master** for above average students.

Additional Teacher Resource Package masters are listed in the OPTIONS box throughout the section. The additional masters are appropriate for all students.

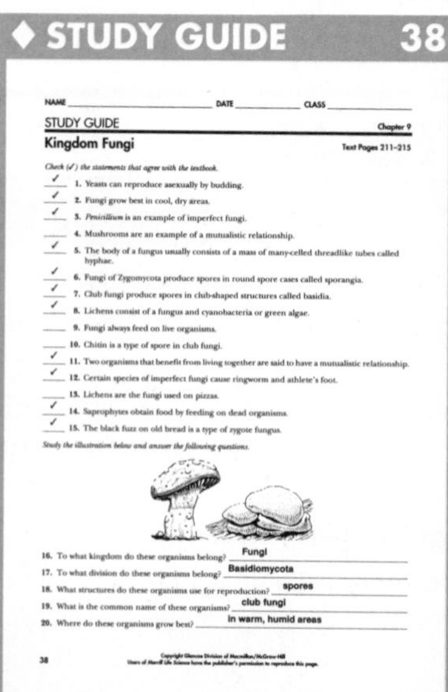

◆ STUDY GUIDE 38

NAME ____ DATE ____ CLASS ____

STUDY GUIDE — Chapter 9

Kingdom Fungi — Text Pages 211–215

Check (✓) the statements that agree with the textbook.

✓ 1. Yeasts can reproduce asexually by budding.
✓ 2. Fungi grow best in cool, dry areas.
✓ 3. *Penicillium* is an example of imperfect fungi.
___ 4. Mushrooms are an example of a mutualistic relationship.
✓ 5. The body of a fungus usually consists of a mass of many-celled threadlike tubes called hyphae.
✓ 6. Fungi of Zygomycota produce spores in round spore cases called sporangia.
✓ 7. Club fungi produce spores in club-shaped structures called basidia.
✓ 8. Lichens consist of a fungus and cyanobacteria or green algae.
___ 9. Fungi always feed on live organisms.
___ 10. Chitin is a type of spore in club fungi.
✓ 11. Two organisms that benefit from living together are said to have a mutualistic relationship.
✓ 12. Certain species of imperfect fungi cause ringworm and athlete's foot.
___ 13. Lichens are the fungi used on pizzas.
✓ 14. Saprophytes obtain food by feeding on dead organisms.
✓ 15. The black fuzz on old bread is a type of zygote fungus.

Study the illustration below and answer the following questions.

16. To what kingdom do these organisms belong? **Fungi**
17. To what division do these organisms belong? **Basidiomycota**
18. What structures do these organisms use for reproduction? **spores**
19. What is the common name of these organisms? **club fungi**
20. Where do these organisms grow best? **in warm, humid areas**

38

Copyright Glencoe Division of Macmillan/McGraw-Hill
Users of Merrill Life Science have the publisher's permission to reproduce this page.

Sac Fungi

Yeasts, molds, morels, and truffles are all examples of sac fungi. The spores of these fungi are produced in little sacs called **asci.** The division Ascomycota is named for these sacs. The ascospores are released when the tip of an ascus breaks open. ③

Many sac fungi are well known by farmers because they destroy plant crops. Diseases caused by sac fungi are Dutch elm disease, apple scab, and ergot disease of rye.

Yeast is an economically important sac fungus. Yeasts don't always reproduce by forming spores. They also reproduce asexually by budding. **Budding** is a form of asexual reproduction in which a new organism grows off the side of the parent. Yeasts are used in the baking industry. As yeasts grow, they use sugar for energy and produce alcohol and carbon dioxide as waste products. The carbon dioxide causes bread to rise. All the little holes you see in a slice of bread were formed from little bubbles of carbon dioxide produced by yeast.

Figure 9-14. Yeasts can reproduce by forming buds off the side of the parents. The bud pinches off and forms an identical cell.

TECHNOLOGY

A Yeast Library

Scientists have a goal to map the entire human DNA pattern, or genome, within fifteen years. Surprisingly, yeast may be an important tool. The genome maps the order of all the bases in human DNA and allows scientists to locate a particular gene on a specific piece of DNA. The gene can then be analyzed to find the exact bases involved and their order. Yeast artificial chromosomes (YAC) are providing a library for storing human DNA segments.

Human DNA is cut into pieces, and these pieces are inserted into yeast chromosomes. Scientists then try to reorder the DNA as it would appear on a human chromosome and map the nucleotide bases present. YACs are useful because one YAC holds more than ten times as many bases as other artificial chromosomes. The longer the sequence of bases in a segment, the easier it is to piece segments together. The most complete YAC collection of human DNA contains 60 000 YACs.

Think Critically: How would knowing the human genome help us?

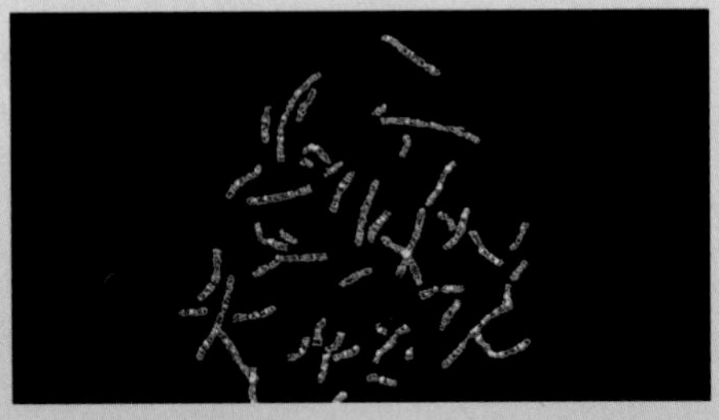

● REINFORCEMENT 38

NAME ____ DATE ____ CLASS ____

REINFORCEMENT — Chapter 9

Kingdom Fungi — Text Pages 211–215

On the line below each diagram, write the name of the fungi division that is identified with the diagram.

1. sac fungi
2. zygote fungi
3. imperfect fungi
4. club fungi
5. lichen
6. Why are saprophytes good for the environment? **Saprophytes break down and return dead organisms to the soil. The materials they break down are used by plants.**
7. In what conditions do fungi grow best? **in warm, humid areas**
8. If you find a mushroom in the wild, why shouldn't you eat it? **Many mushrooms are poisonous, and a mistake in identification could be fatal.**
9. What makes up a lichen? **A lichen consists of a fungus and cyanobacteria or green algae.**
10. What are asci? **Asci are little sacs in which the spores of sac fungi are produced.**
11. What is basidium? **A basidium is a club-shaped structure in which the spores of club fungi are produced.**
12. What are sporangia? **Sporangia are the round spore cases in which the spores of zygote fungi are produced.**
13. How does yeast cause bread to rise? **As yeast grows, it produces alcohol and carbon dioxide as waste products. The carbon dioxide causes the bread to rise.**
14. What kind of fungi causes athlete's foot? **imperfect fungi**

38

▲ ENRICHMENT 38

NAME ____ DATE ____ CLASS ____

ENRICHMENT — Chapter 9

Kingdom Fungi — Text Pages 211–215

WHAT MAKES DOUGH RISE?

Materials
- 1 cup flour
- 1 tablespoon of sugar
- 1 package of dried baker's yeast
- measuring cup
- glass jar
- marker
- clock
- warm tap water
- spoon

Procedure
1. Put the flour and sugar into the glass jar.
2. Put 1/4 cup warm tap water into the measuring cup. Add the package of yeast, stirring with a spoon.
3. Slowly stir the yeast water into the flour and sugar mixture. Stir until dough forms. Push the dough down to the bottom of the jar. Mark the height of the dough on the jar.
4. Put the jar in a warm place. You could put it by a heat outlet or place the jar in a pan of warm water.
5. Watch the clock. Every ten minutes, mark the height of the dough on the outside of the jar. Continue this for an hour.

Conclude and Apply
1. What happened to the dough during the hour? **The dough increased in size, rising in the jar.**
2. What kind of fungus is the yeast you added? **sac fungi**
3. What do yeasts use for energy? **sugar**
4. What did you add as food for the yeast? **1 tablespoon of sugar**
5. What are the waste products when yeasts consume food? **alcohol and carbon dioxide**
6. After an hour, could you see any bubbles in the dough? **yes**
7. What could you conclude was the gas in these bubbles? **carbon dioxide**
8. Then what made the dough rise? **The consumption of the sugar by the yeast produced carbon dioxide, which made the dough rise.**

38

TECHNOLOGY

For more information on mapping the human genome, see "The Genome Initiative: How to Spell 'Human,'" by Norton Zinder, *Scientific American,* July 1990, p. 128.

Teaching Tip: The Human Genome project is being funded by NIH and research is being shared. Discuss why many scientists/companies do not want to share research. Should they have to share? What are the benefits of sharing information?

Think Critically: It could help us to determine the causes of genetic problems, and ultimately, to repair them. It would give a better understanding of recessive genes in populations and how polygenic traits work.

CHECK FOR UNDERSTANDING

Use the Mini Quiz to check for understanding.

MINI QUIZ

Use the Mini Quiz to check students' recall of chapter content.

1. **The body of a fungus is usually a mass of many-celled, threadlike tubes called _____ .** *hyphae*
2. **_____ are reproductive cells that form new organisms without fertilization.** *Spores*
3. **The spores of sac fungi are produced in a little sac called a(n) _____ .** *ascus*
4. **The spores of Basidiomycota are produced in a club-shaped structure called a(n) _____ .** *basidium*
5. **Phylum Deuteromycota, the imperfect fungi, is composed of species of fungi in which the _____ has never been observed.** *sexual stage*
6. **Many lichens are easily affected by _____ .** *air pollution*

RETEACH

Show students various illustrations of fungi. Ask them to identify the parts of each fungus.

EXTENSION

For students who have mastered this section, use the **Reinforcement** and **Enrichment** masters or other OPTIONS provided.

CROSS CURRICULUM

▶ **Math:** The fairy ring toadstool advances outward from the center at the rate of about 30 centimeters per year. One ring is estimated to be 150 years old. How many meters in diameter is the ring? (150 years × 30 cm/year × 2 radii = 900 cm = 9 meters)

MINI-Lab

Materials: mushrooms, paper towels

Teaching Tips

Be sure students do not disturb the mushrooms while spore prints are being made.

Answers

Students should see prints made from the spores of the mushrooms. Count the number of spores by sampling and estimating.

MINI-Lab

ASSESSMENT

Performance: To further assess students' understanding of mushrooms and fungi in general, see USING LAB SKILLS, question 12, on page 220.

Figure 9-15. A mushroom is the spore-producing structure of a club fungus. Spores are contained in many club-shaped structures that line the gills of a mushroom cap.

MINI-Lab

How are spore prints made?

Obtain several mushrooms from the grocery store. Let them age several days until the undersides look very brown. Remove the stems and arrange the mushrooms caps down on a white paper towel. Be sure the gills are against the paper towel. Let them sit undisturbed overnight. Lift the mushrooms from the paper towel. What do you see? What made the marks? How could you *estimate* the number of new mushrooms that could come from this one cap?

Club Fungi

The mushroom shown in Figure 9-15 is a member of the division Basidiomycota (buh SIHD ee uh mi koht uh). These fungi are commonly known as club fungi. 4 The spores of these fungi are produced in a club-shaped structure called a **basidium.** Under the cap of a mushroom, the reproductive structure of the fungus, are thin sheets of tissue called gills. The tiny club-shaped structures line the gills. The spores you observed on the gills of the mushroom in the Find Out activity at the beginning of this chapter were produced in the basidia. Puffball mushrooms, mushrooms you eat on pizza, and bracket, or shelf, fungi that grow on the trunks of trees are all club fungi.

Many of the club fungi are economically important. Rusts and smuts damage billions of dollars of food crops each year. Mushrooms are an important food crop, but you should never eat a wild mushroom because many are poisonous. Even experts at fungi identification sometimes find it difficult to tell the difference between edible and poisonous mushrooms. Mistakes in identification can be fatal.

Imperfect Fungi

5 The imperfect fungi, division Deuteromycota, is composed of species of fungi in which a sexual stage has never been observed. When a sexual stage of one of these fungi is observed, the species is immediately classified as one of the other three phyla. *Penicillium* is one example from this group. Penicillin, an antibiotic, is an important product of this fungus. Other examples of imperfect fungi are species that cause ringworm and athlete's foot.

OPTIONS

ENRICHMENT

▶ Allow students to use field guides to identify fungi they find in the wild. Caution students to observe but not pick or eat the fungi because some may be poisonous.

PROGRAM RESOURCES

From the **Teacher Resource Package** use:

Activity Worksheets, page 84, Mini-Lab: How are spore prints made?

Science Integration Activities, 9, Preparation of Carbon Dioxide

Use **Laboratory Manual** page 57, Yeasts.

Use **Laboratory Manual** page 61, Lichens.

Lichens

The colorful organisms in Figure 9-16 are lichens. A **lichen** is an organism that is made of a fungus and green alga or a cyanobacterium. When two organisms live together, they often have a symbiotic relationship. **Symbiosis,** a close relationship between two organisms, can have several results. The fungus and cyanobacterium, or green alga, both benefit from living together, and they are said to have a **mutualistic** relationship.

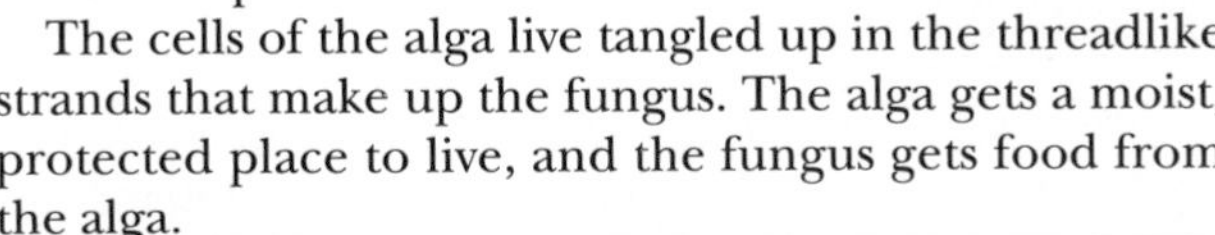

The cells of the alga live tangled up in the threadlike strands that make up the fungus. The alga gets a moist, protected place to live, and the fungus gets food from the alga.

Lichens are an important food source for many animals. Reindeer and caribou feed on reindeer moss, a lichen that
grows in arctic regions. Many lichens are easily affected by 6
air pollution. Lichen growth on tree trunks and stone buildings can be used to monitor pollution levels.

Figure 9-16. Lichens can grow upright, like the British soldier lichen (left), or appear leafy (right). Some other forms grow flat, much like a crust.

Connect to... Earth Science

Lichens are often the first organisms to colonize bare rock. For this reason, lichens are called pioneer species. Pioneer species release acids as a part of their metabolism. **Hypothesize** the effect of pioneer species on rock.

SECTION REVIEW

1. How do fungi obtain food?
2. Why are Ascomycota called sac fungi?
3. In a lichen, how do both the fungus and cyanobacterium or alga benefit from living together?
4. **Apply:** If an imperfect fungus were found to produce basidia under some circumstances, how would the fungus be reclassified?
5. **Connect to Physics:** How do air currents help in the spread of fungi?

Using Variables, Constants, and Controls

Skill Builder

In an experiment, yeasts were cultured at different temperatures in a particular concentration of syrup and water. Different amounts of carbon dioxide were produced at each temperature. Identify the constant and variables in this experiment. If you need help, refer to Using Variables, Constants, and Controls in the **Skill Handbook** on pages 686-687.

ASSESSMENT—ORAL

▶ **Explain how fungi obtain food.** *Fungi secrete enzymes that digest materials outside their cells. The fungi then absorb the digested materials.*

▶ **Why are decomposers like fungi important?** *Were it not for the decomposition processes of fungi, organic material would remain locked up in dead bodies and eventually all life would die.*

▶ **Why is it beneficial for fungi to produce both sexually and asexually?** *Asexual reproduction allows for fungi to take full and rapid advantage of a food source. Sexual reproduction provides variety for surviving in different environmental conditions.*

▶ **Why are the Deuteromycotes called imperfect fungi?** *They have never been observed to reproduce sexually.*

CONCEPT DEVELOPMENT

Cooperative Learning: Have students work in fours to interview a local physician about the role of fungi in medicine.

3 CLOSE

▶ Ask questions 1-3 and the **Apply** Question in the Section Review.

▶ Draw a mushroom on the chalkboard. Have students tell what the parts of the mushroom are and how it reproduces.

Connect to... Earth Science

Answer: Acid released from lichens and other pioneer species begin to break rock into smaller pieces.

SECTION REVIEW ANSWERS

1. Some fungi get their food from decaying organisms, and others are parasites on living organisms.
2. Ascomycota produce spores in a saclike structure called an ascus.
3. The alga or cyanobacterium gets a protected, moist environment, and the fungus gets food from the alga.
4. **Apply:** It would be reclassified as a club fungus belonging to phylum Basidiomycota.
5. **Connect to Physics:** Air currents carry spores to new locations.

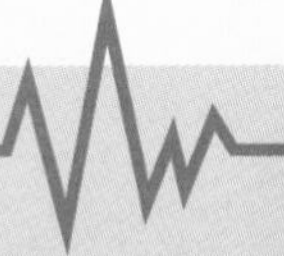

Skill Builder

The constant was the concentration of syrup and water, and the variables were temperature and amount of carbon dioxide produced.

Skillbuilder ASSESSMENT

Performance: Have students design an experiment based on the Skillbuilder and carry it out.

SECTION 9-3

PREPARATION

SECTION BACKGROUND

▶ There is currently a debate over whether more research money should be spent to cure bacterial and viral diseases or on ways to prevent them.

▶ Some people believe antibiotic use should be limited only to particular bacteria so that resistant strains are not developed as rapidly.

▶ Aureomycin, erythromycin, and penicillin are three antibiotics produced from fungi.

1 MOTIVATE

▶ Have students make wet mount slides of mold from an orange and examine the mold.

PROGRAM RESOURCES

From the **Teacher Resource Package** use:

Science and Society, page 13, Choosing the Right Medicine.

VideoDisc

STVS: Disease-Resistant Tomatoes, Disc 4, Side 1

9-3 Fungus—Can't Live Without It

Objectives

- **State two types of drugs that are produced from fungi.**
- **Appreciate the role of fungi in medicine.**

Beneficial Fungi

Did you know that fungus has probably saved your life? In fact, this may have happened several times! Before the introduction of antibiotic drugs during the 1940s, many deaths occurred due to diseases caused by bacteria. It was common at that time for people to die of pneumonia, tuberculosis, and other infections. Today, however, antibiotics, chemicals that have a harmful effect on bacteria, are used to treat these infections. Many antibiotics are produced from fungi that grow in the soil. You have probably heard of penicillin, the most famous antibiotic. It is produced by a fungus, *Penicillium chrysogenum.* You may have seen this fungus growing on a moldy orange.

Figure 9-17. *Penicillium* grows easily on oranges.

Antibiotics prevent bacterial diseases by preventing bacteria from reproducing and growing. Most bacterial infections can be cured with antibiotics. Unfortunately, some people cannot be treated with certain antibiotics because of allergic reactions.

Now scientists face a new problem: bacteria are able to develop new strains that are not affected by antibiotics. For example, at one time, penicillin was effective against many types of bacteria. Today, it has no effect on most of these bacteria. Other antibiotics that have this same problem are streptomycin, tetracycline, and erythromycin. This keeps scientists working to develop new antibiotics to work against these harmful types of bacteria.

Antiviral drugs are used to treat infections caused by viruses. However, they were much more difficult to

OPTIONS

Meeting Different Ability Levels

For Section 9-3, use the following **Teacher Resource Masters** depending upon individual students' needs.

◆ **Study Guide Master** for all students.

● **Reinforcement Master** for students of average and above average ability levels.

▲ **Enrichment Master** for above average students.

Additional Teacher Resource Package masters are listed in the OPTIONS box throughout the section. The additional masters are appropriate for all students.

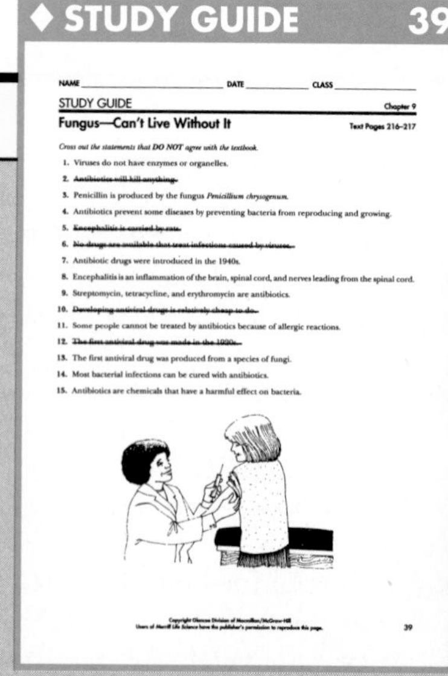

◆ STUDY GUIDE 39

NAME ______ DATE ______ CLASS ______

STUDY GUIDE — Chapter 9

Fungus—Can't Live Without It — Text Pages 216–217

Cross out the statements that DO NOT agree with the textbook.

1. Viruses do not have enzymes or organelles.
2. ~~Antibiotics will kill anything.~~
3. Penicillin is produced by the fungus *Penicillium chrysogenum.*
4. Antibiotics prevent some diseases by preventing bacteria from reproducing and growing.
5. ~~Encephalitis is carried by rats.~~
6. ~~No drugs are available that treat infections caused by viruses.~~
7. Antibiotic drugs were introduced in the 1940s.
8. Encephalitis is an inflammation of the brain, spinal cord, and nerves leading from the spinal cord.
9. Streptomycin, tetracycline, and erythromycin are antibiotics.
10. ~~Developing antiviral drugs is relatively cheap to do.~~
11. Some people cannot be treated by antibiotics because of allergic reactions.
12. ~~The first antiviral drug was made in the 1990s.~~
13. The first antiviral drug was produced from a species of fungi.
14. Most bacterial infections can be cured with antibiotics.
15. Antibiotics are chemicals that have a harmful effect on bacteria.

develop. This is because viruses do not have enzymes or organelles to destroy. Drugs developed to treat viral infections often resulted in being harmful to the patient as well.

Encephalitis is an inflammation of the brain, spinal cord, and the nerves leading from the spinal cord. It is carried by mosquitoes and can be deadly. In 1978, an antiviral drug, vidarabine, was finally approved by the United States Food and Drug Administration for treatment of encephalitis. This was the very first antiviral drug. It was produced from a species of fungi.

In 1983 another drug produced from fungi was approved for use. The drug cyclosporine was approved for use in organ transplant patients. Cyclosporine is made from an imperfect fungus. The drug helps prevent the rejection of transplanted organs. Because of cyclosporine, organ transplants have become more common. One problem, however, is that the patient must take cyclosporine for life, and it can become expensive. It can also have some bad side effects, such as kidney failure.

Due to the great expense of research programs that were needed to develop these drugs, some people argue that more effort should be made to prevent these diseases rather than cure them. Encephalitis, for example, could be prevented by controlling the mosquitoes that transmit it. Society must work with science if such medical problems are to be prevented.

Figure 9-18. Anti-rejection drugs suppress the immune system, and must be taken daily.

SECTION REVIEW

1. State two types of drugs that are products of fungi.
2. Why have fungi become so important to medicine?
3. **Connect to Earth Science:** Find out where most fungi that produce antibiotics are usually found.

You Decide!

In the fall of 1990, there was an outbreak of encephalitis spread by mosquitoes in central and southern Florida. Health officials urged people to limit evening activities outdoors because mosquitoes are most active and feeding then. By keeping people away from the mosquitoes, they hoped to limit the spread of encephalitis. Why do you think it was important for people to stay away from the mosquitoes even though there is a drug to fight the disease? Would you have taken the advice of the health officials?

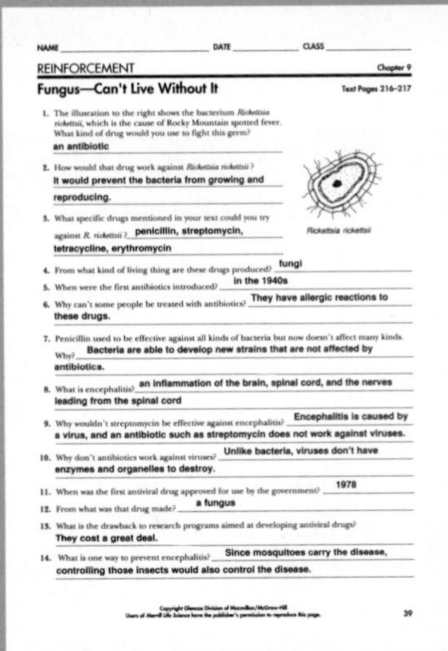

● REINFORCEMENT 39

NAME ______ DATE ______ CLASS ______

REINFORCEMENT — Chapter 9

Fungus—Can't Live Without It — Text Pages 216–217

1. The illustration to the right shows the bacterium *Rickettsia rickettsii*, which is the cause of Rocky Mountain spotted fever. What kind of drug would you use to fight this germ? **an antibiotic**
2. How would that drug work against *Rickettsia rickettsii*? **It would prevent the bacteria from growing and reproducing.**
3. What specific drugs mentioned in your text could you try against *R. rickettsii*? **penicillin, streptomycin, tetracycline, erythromycin**

Rickettsia rickettsii

4. From what kind of living thing are these drugs produced? **fungi**
5. When were the first antibiotics introduced? **in the 1940s**
6. Why can't some people be treated with antibiotics? **They have allergic reactions to these drugs.**
7. Penicillin used to be effective against all kinds of bacteria but now doesn't affect many kinds. Why? **Bacteria are able to develop new strains that are not affected by antibiotics.**
8. What is encephalitis? **an inflammation of the brain, spinal cord, and the nerves leading from the spinal cord**
9. Why wouldn't streptomycin be effective against encephalitis? **Encephalitis is caused by a virus, and an antibiotic such as streptomycin does not work against viruses.**
10. Why don't antibiotics work against viruses? **Unlike bacteria, viruses don't have enzymes and organelles to destroy.**
11. When was the first antiviral drug approved for use by the government? **1978**
12. From what was that drug made? **a fungus**
13. What is the drawback to research programs aimed at developing antiviral drugs? **They cost a great deal.**
14. What is one way to prevent encephalitis? **Since mosquitoes carry the disease, controlling those insects would also control the disease.**

 39

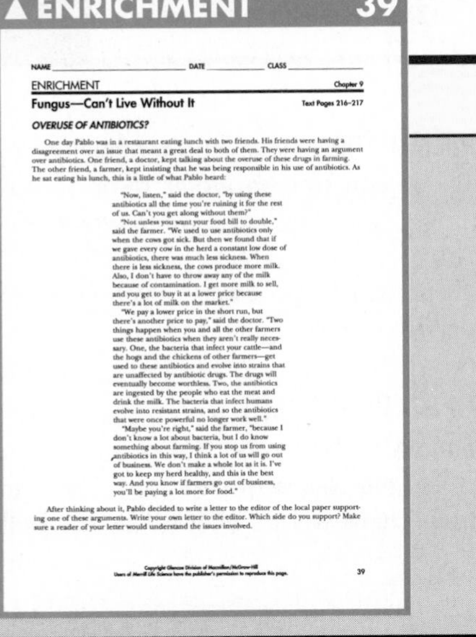

▲ ENRICHMENT 39

NAME ______ DATE ______ CLASS ______

ENRICHMENT — Chapter 9

Fungus—Can't Live Without It — Text Pages 216–217

OVERUSE OF ANTIBIOTICS?

One day Pablo was in a restaurant eating lunch with two friends. His friends were having a disagreement over an issue that meant a great deal to both of them. They were having an argument over antibiotics. One friend, a doctor, kept talking about the overuse of these drugs in farming. The other friend, a farmer, kept insisting that he was being responsible in his use of antibiotics. As he sat eating his lunch, this is a little of what Pablo heard:

"Now, listen," said the doctor, "by using these antibiotics all the time you're ruining it for the rest of us. Can't you get along without them?"

"Not unless you want your food bill to double," said the farmer. "We used to use antibiotics only when the cows got sick. But then we found that if we gave every cow in the herd a constant low dose of antibiotics, there was much less sickness. When there is less sickness, the cows produce more milk. Also, I don't have to throw away any of the milk because of contamination. I get more milk to sell, and you get to buy it at a lower price because there's a lot of milk on the market."

"We pay a lower price in the short run, but there's another price to pay," said the doctor. "Two things happen when you and all the other farmers use these antibiotics when they aren't really necessary. One, the bacteria that infect your cattle—and the hogs and the chickens of other farmers—get used to these antibiotics and evolve into strains that are unaffected by antibiotic drugs. The drugs will eventually become worthless. Two, the antibiotics are ingested by the people who eat the meat and drink the milk. The bacteria that infect humans evolve into resistant strains, and so the antibiotics that were once powerful no longer work well."

"Maybe you're right," said the farmer, "because I don't know a lot about bacteria, but I do know something about farming. If you stop us from using antibiotics in this way, I think a lot of us will go out of business. We don't make a whole lot as it is. I've got to keep my herd healthy, and this is the best way. And you know if farmers go out of business, you'll be paying a lot more for food."

After thinking about it, Pablo decided to write a letter to the editor of the local paper supporting one of these arguments. Write your own letter to the editor. Which side do you support? Make sure a reader of your letter would understand the issues involved.

 39

2 TEACH

Key Concepts are highlighted.

CONCEPT DEVELOPMENT

▶ Point out that without antibiotics, many parasitic bacteria would severely weaken or kill a human who contracted the bacterial disease.

▶ Emphasize that antibiotics have no effect on viral diseases.

CHECK FOR UNDERSTANDING

Ask students to compare and contrast treatments of bacterial and viral diseases and their relationship to antibiotics.

RETEACH

Show a filmstrip on the development of antibiotics.

EXTENSION

For students who have mastered this section, use the **Reinforcement** and **Enrichment** masters or other OPTIONS provided.

3 CLOSE

▶ Ask questions 1-2 in the Section Review.

▶ Ask students to explain why it is important not to use antibiotics unless prescribed by a physician. (If antibiotics are taken randomly, bacteria are more likely to develop resistant strains.)

SECTION REVIEW ANSWERS

1. antibiotics and antiviral drugs

2. They produce a wide variety of compounds that are useful in treating bacterial and viral diseases.

3. Connect to Earth Science: Most fungi that are used to produce antibiotics originate in soil.

YOU DECIDE!

Answers will vary. All answers that demonstrate clear and logical thinking should be accepted.

ACTIVITY 9-2

OBJECTIVE: Examine and **classify** lichen samples and determine how lichens are adapted for mutualism.
Time: one class period

PROCESS SKILLS applied in this activity are **observing, classifying, inferring,** and **interpreting data.**

PREPARATION

- Lichens may be collected and occasionally misted with water or stored in a terrarium.
- Dry samples of reindeer moss (Clasonia) for several days prior to the activity.
- Place numbers on the lichens specimens to be used in the activity. The reindeer moss used for the microscope part of the activity should not be dried.

Cooperative Learning: Divide the class into Science Investigation Teams.

THINKING CRITICALLY

Both lichens and mosses are small and greenish in color. Lichens are the first organisms to inhabit rocky surfaces where they release acids and break down rocks into soil.

TEACHING THE ACTIVITY

*Refer to the **Activity Worksheets** for additional information and teaching strategies.*

- Crustlike lichen samples may have different color and textures. Leaflike lichens will have a thin leathery feeling, and shrublike will have a powdery texture.
- The reindeer moss that is soaked in water will be darker green and feel like a wet sponge.
- Lichens are sometimes called an index species because the number of lichens, the density of lichens, and the variety of lichens that can grow in an area can indicate how clean or polluted the air is.

SUMMING UP/SHARING RESULTS

- The algae and cyanobacteria are green in color due to the presence of chlorophyll. They make food for the lichen.
- Lichens can absorb water during heavy rains and store it for times when water is not available.
- Lichens absorb pollutants which can build up in the tissues killing the lichens.

GOING FURTHER!

If there are no lichens on the school campus, try to find the source of pollution. The transplanted lichens will be affected by the pollution.

PROGRAM RESOURCES

From the **Teacher Resource Package** use:
Activity Worksheets, pages 79-80, Activity 9-2, Investigating Lichens.

Activity ASSESSMENT

Performance: To further assess students' understanding of the kinds of lichens, provide other samples of them to classify.

ACTIVITY 9-2

DESIGNING AN EXPERIMENT
Investigating Lichens

Lichens grow on tree trunks, brick buildings, and bare rocks. What characteristics do lichens have that enable them to live in so many different places?

Getting Started

You will examine and classify samples of lichens and determine what characteristics they have and how they are adapted for a mutualistic relationship.

Thinking Critically

Why are lichens sometimes mistaken for mosses? Lichens are referred to as pioneer plants. How can this be explained?

Materials

Your cooperative group will use:
- 3 lichen samples
- dried reindeer moss
- microscope slide
- coverslip
- microscope
- culture dish half full of water
- paper towels

Try It!

1. Obtain 3 lichen samples from your teacher. Examine each one with a hand lens. Answer the following questions for each sample:
- Is the lichen crustlike, leaflike, or shrublike?
- Describe its color and texture.

Make a data table to record the information.

2. Obtain a piece of dried reindeer moss. Describe the general texture and color. Place the lichen in a culture dish that is half full of water. After 30 minutes, remove the lichen from the culture dish. Describe its color and texture.

3. Make a wet mount slide of a lichen sample provided by your teacher. Crush the lichen with the eraser of your pencil. Add a few drops of water. Cover the mashed lichen with a coverslip. *Observe* the slide under low power and then under high power. Draw and label what you observe.

Summing Up/Sharing Results

- What color are the algae and cyanobacteria that make up lichens? What do you think is the role of this part of the lichen?
- Describe a lichen's water-holding capacity. How is this an adaptation for survival in many climates?
- Why do you think lichens are good indicators of air pollution?

Going Further!

If your school campus does not have lichens, transplant healthy lichens and their substrates to the campus. *Observe* the lichens for the effects of air pollution.

CHAPTER

REVIEW

SUMMARY

9-1: Kingdom Protista

1. Protists are eukaryotic organisms. They can be one- or many-celled but do not have the complex organization found in plants and animals.
2. The Protist Kingdom is made up of a variety of organisms. Some are plantlike, others are animal-like, and others are funguslike.
3. Algae are photosynthetic and classified according to their pigments. Other protists, like flagellates and ciliates, are animal-like and have special means of locomotion. Funguslike protists include slime molds and water molds.

9-2: Kingdom Fungi

1. Fungi are saprophytes. They reproduce by spores. The structures in which a fungus produces spores are used to classify fungi into one of four divisions.
2. Zygote fungi produce spores in small round spore cases on the tips of upright hyphae. Bread mold is a zygote fungus. Sac fungi produce spores in small sacs called asci. Yeast is an example of a sac fungus. Mushrooms are club fungi. Club fungi produce spores in a club-shaped structure called a basidium. These structures line the gills of a mushroom cap.
3. Imperfect fungi are those for which no sexual stage has been observed. Penicillin is an important product of these fungi.

9-3: Science and Society: Fungus—Can't Live Without It

1. Fungi are important sources of drugs such as penicillin, streptomycin, vidarabine, and cyclosporine.
2. Medicines produced from fungus products have saved the lives of millions of people with diseases such as pneumonia and tuberculosis, as well as those who have had organ transplants.

KEY SCIENCE WORDS

a. **algae**
b. **asci**
c. **basidium**
d. **budding**
e. **chitin**
f. **cilia**
g. **hyphae**
h. **lichen**
i. **mutualistic**
j. **protists**
k. **protozoa**
l. **pseudopods**
m. **sporangia**
n. **spores**
o. **symbiosis**

UNDERSTANDING VOCABULARY

Match each phrase with the correct term from the list of Key Science Words.

1. footlike cytoplasmic extensions
2. reproductive cells of fungi
3. eukaryotic organisms that are animal-like, plantlike, or funguslike
4. animal-like protists
5. plantlike protists
6. threadlike structures used for movement
7. a close relationship between two organisms
8. threadlike strings of a fungus
9. contain spores in zygote fungi
10. strong carbohydrate in fungus' cell walls

OPTIONS

ASSESSMENT

To assess student understanding of material in this chapter, use the resources listed.

COOPERATIVE LEARNING

Consider using cooperative learning in the THINK AND WRITE CRITICALLY, APPLY, and MORE SKILL BUILDERS sections of the Chapter Review.

PROGRAM RESOURCES

From the **Teacher Resource Package** use:

Chapter Review, pages 21-22.

Chapter and Unit Tests, pages 57-60, Chapter Test.

Chapter and Unit Tests, pages 61-62, Unit Test.

CHAPTER

REVIEW

SUMMARY

Have students read the summary statements to review the major concepts of the chapter.

UNDERSTANDING VOCABULARY

1.	l	**6.**	f
2.	n	**7.**	o
3.	j	**8.**	g
4.	k	**9.**	m
5.	a	**10.**	e

ASSESSMENT

Portfolio

Encourage students to place in their portfolios one or two items of what they consider to be their best work. For each item, ask students to explain why that item was chosen and what they learned from it. Items might be selected from the following.

- In Your Journal description of a protist's day, p. 205
- Enrichment graph of a malaria patient's temperatures, p. 208
- MINI-Lab spore prints, p. 214

Performance

Additional performance assessments may be found in *Performance Assessment* and *Science Integration Activities* that accompany **Merrill Life Science.** Performance Task Assessment Lists and rubrics for evaluating these activities and other products generated throughout the chapter can be found in Glencoe's *Performance Assessment in Middle School Science.*

CHAPTER

REVIEW

CHECKING CONCEPTS

1. d	**6.** a
2. c	**7.** d
3. d	**8.** b
4. c	**9.** b
5. b	**10.** c

USING LAB SKILLS

ASSESSMENT

Use these alternate lab exercises to assess students' abilities to use the skills in this chapter.

11. Students may test apples, lemons, bananas, or any other fruits. Apples seem to take longer to succumb to mold formation.

12. Students should find that both are made up of hyphae.

THINK AND WRITE CRITICALLY

13. Algae are plantlike in that they have chloroplasts and are able to make their own food.

14 Fungi are important to the environment because they help break down or decompose organic material and return nutrients to the soil.

15. Protozoans are like animals in that they cannot make their own food and many have some method of movement.

16. Red tide is caused by a population explosion of dinoflagellates. The millions of these organisms with red pigments cause the water to look red. Some species of dinoflagellates have a nerve toxin that can kill fish and be absorbed by shellfish. Humans who eat fish or shellfish poisoned by this toxin can become ill.

APPLY

17. *Spirogyra* are named for their spiral-shaped chloroplasts.

CHAPTER

REVIEW

CHECKING CONCEPTS

Choose the word or phrase that completes the sentence.

1. ______ (is) are examples of one-celled algae.
   a. Paramecia **c.** Amoebas
   b. *Plasmodium* **d.** Dinoflagellates

2. Members of phylum Chrysophyta are ______ in color.
   a. green **c.** golden-brown
   b. red **d.** brown

3. Large numbers of ______ cause red tides.
   a. *Euglena* **c.** *Ulva*
   b. diatoms **d.** dinoflagellates

4. Brown algae belong to phylum ______.
   a. Rhodophyta **c.** Phaeophyta
   b. Chrysophyta **d.** Pyrrophyta

5. ______ moves by using cilia.
   a. *Amoeba* **c.** *Plasmodium*
   b. *Paramecium* **d.** *Euglena*

6. Funguslike protists can live well ______.
   a. on decaying logs **c.** on dry surfaces
   b. in bright light **d.** on metal surfaces

7. Decomposition is an important role of ______.
   a. protozoans **c.** plants
   b. algae **d.** fungi

8. All mutualistic relationships ______.
   a. are harmful **c.** form lichens
   b. are helpful **d.** pollute

9. ______ produce the spores in mushrooms.
   a. Sporangia **c.** Asci
   b. Basidia **d.** Hyphae

10. An example of an imperfect fungi is ______.
   a. mushroom **c.** *Penicillium*
   b. yeast **d.** lichen

USING LAB SKILLS

11. **In your Journal,** make a hypothesis and design an experiment to test what fruits allow *Penicillium* mold to grow. Conduct your experiment. Collect data and state a conclusion.

12. In the MINI-Lab on page 214, you made a spore print of the gills of a mushroom cap. Now carefully tease apart the top of the cap and make a wet mount slide of an *extremely* small piece. Do the same with the stalk. **In your Journal**, draw and describe what you see, and compare the two.

THINK AND WRITE CRITICALLY

Answer the following questions in your Journal using complete sentences.

13. Explain why algae are plantlike.

14. How are fungi important to the environment?

15. How are protozoans like animals?

16. What are the causes and effects of a red tide like the one shown below?

APPLY

17. Look at Figure 9-5b again. Why is *Spirogyra* a good name for this green algae?
18. Compare and contrast one-celled, colonial, chain, and many-celled algae.
19. Discuss why scientists find it difficult to trace the origin of fungi. Why are fossils of fungi rare?
20. Explain the adaptations of fungi that enable them to get food.
21. What kind of environment is needed to prevent fungal growth?

MORE SKILL BUILDERS

If you need help, refer to the Skill Handbook.

1. **Observing and Inferring:** Match the prefix of each alga: *Chloro-, Chryso-, Phaeo-, Pyrro-, Rhodo,* with the correct color: brown, green, fire, gold, red.
2. **Concept Mapping:** Complete the following concept map on a separate sheet of paper.

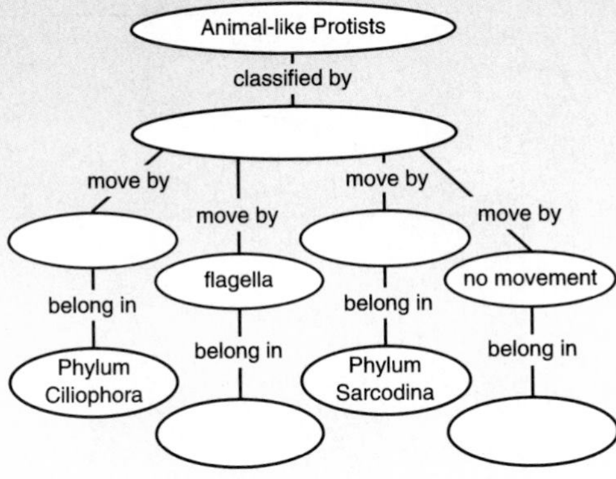

3. **Comparing and Contrasting:** Make a chart comparing and contrasting sac fungi, zygote fungi, and club fungi.
4. **Classifying:** Classify the organisms based on their method of movement. *Euglena, Amoeba,* dinoflagellates, *Paramecium, Plasmodium,* slime molds, *Trypanosoma, Volvox.*
5. **Observing and Inferring:** The figure below shows some of the structures and specialized adaptations of kelp, including the stipe, blade, holdfast, and gas floats. What purposes do these adaptations serve for kelp?

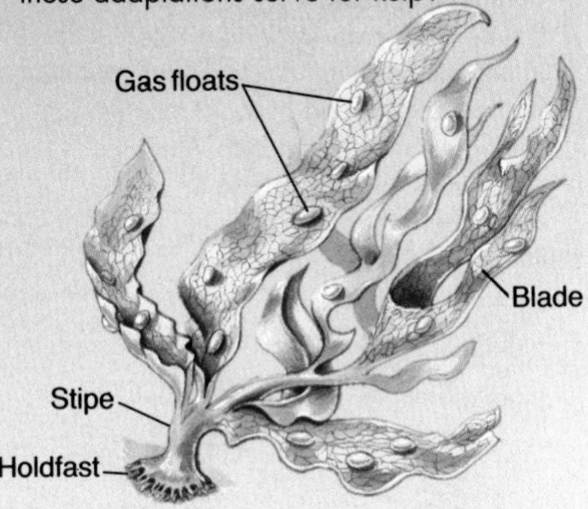

PROJECTS

1. Bring in labels of foods that contain carrageenans. Find out what protist these are from.
2. Use samples from an aquarium, trees, and any rocks with a slick green growth from a damp area near school. Make wet mount slides and draw and label any algal growths that you find. Indentify the parts of each that take part in producing food.

18. unicellular—one-celled; colonial—many cells arranged in a group; chains—many cells arranged in a long filament; multicellular—organism has many different cells and is large

19. Fungi are made of soft parts that do not form fossils easily. They also live in warm, moist, humid areas that would allow them to decompose rapidly. Because fungi have characteristics of plants but also their own unique characteristics, their origin is difficult to trace.

20. Fungi make enzymes that digest food outside of the organism. Then the nutrients are absorbed by the cells.

21. No warmth, low humidity, and low moisture all inhibit fungal growth.

MORE SKILL BUILDERS

1. Observing and Inferring: *chloro*—green, *chryso*—gold, *pyrro*—fire, *rhodo*—red, *phaeo*—brown

2. Concept Mapping:

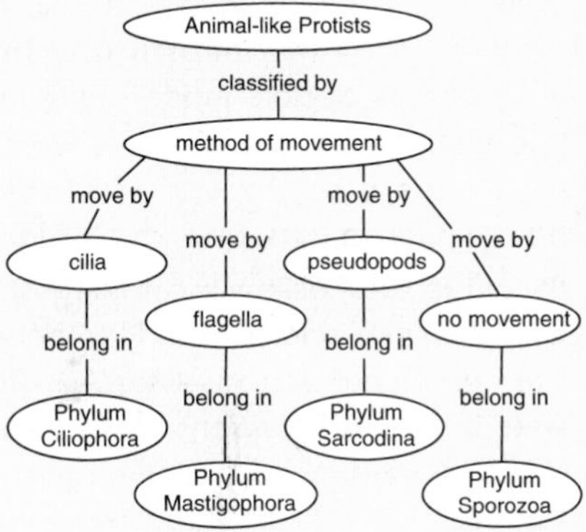

3. Comparing and Contrasting: Students' tables should contain the following information. Ascomycota reproduce by budding and making spores, example yeast; Zygomycota reproduce by producing spores in sporangia, example bread mold; Basidiomycota reproduce by making spores in basidia, example mushrooms.

4. Classifying: Move by means of flagella—*Euglena,* dinoflagellates, *Volvox, Trypanosoma.* Move by means of pseudopods—*Amoeba,* slime molds. Move by means of cilia—*Paramecium.* No movement—*Plasmodium.*

5. Observing and Inferring: The holdfast keeps the alga attached to one place. The gas floats keep the alga upright and floating near the surface to receive plenty of light. The stalk connects the blades.

UNIT 3
GLOBAL CONNECTIONS

Objective

In this unit ending feature, the unit topic, "Simple Living Things," is extended into other disciplines. Students will see how simple living things affect human health and history.

Motivate

Cooperative Learning: Using the Expert Teams strategy, divide the class into several groups. Assign one Connection to each group of students. Have each group research the geographic location of the Connection to find out more about its climate, culture, and ecology.

Teaching Tips

- ▶ Tell the students to keep in mind the connection between simple living things and more complex forms of life.
- ▶ Have students investigate the hypotheses doctors made about the spread of malaria and yellow fever.

Wrap-Up

Conclude this lesson by having students design a scientific study to determine what causes a disease. Ask students to list the questions researchers would ask to discover how the disease is spread.

GEOLOGY

Background: Some kinds of bacteria get energy from sulfides and release sulphur. They also get energy from sunlight, just as plants do. These bacteria may be among the oldest life-forms on Earth.

Discussion: Here is another kind of "helpful" bacteria. Review the nitrogen cycle with students. How do the nitrogen bacteria aid growing plants?

Answer to Question: Bacteria have probably changed very little since they first appeared on Earth.

Extension: Have students make a list of areas in their homes where bacteria may be found, then compare lists. Why do students think disinfectants are used in kitchens and bathrooms?

UNIT 3
GLOBAL CONNECTIONS

The Simplest Living Things

In this unit, you have studied viruses and simple organisms. Now find out the effects of viruses, bacteria, protists, and fungi around the world.

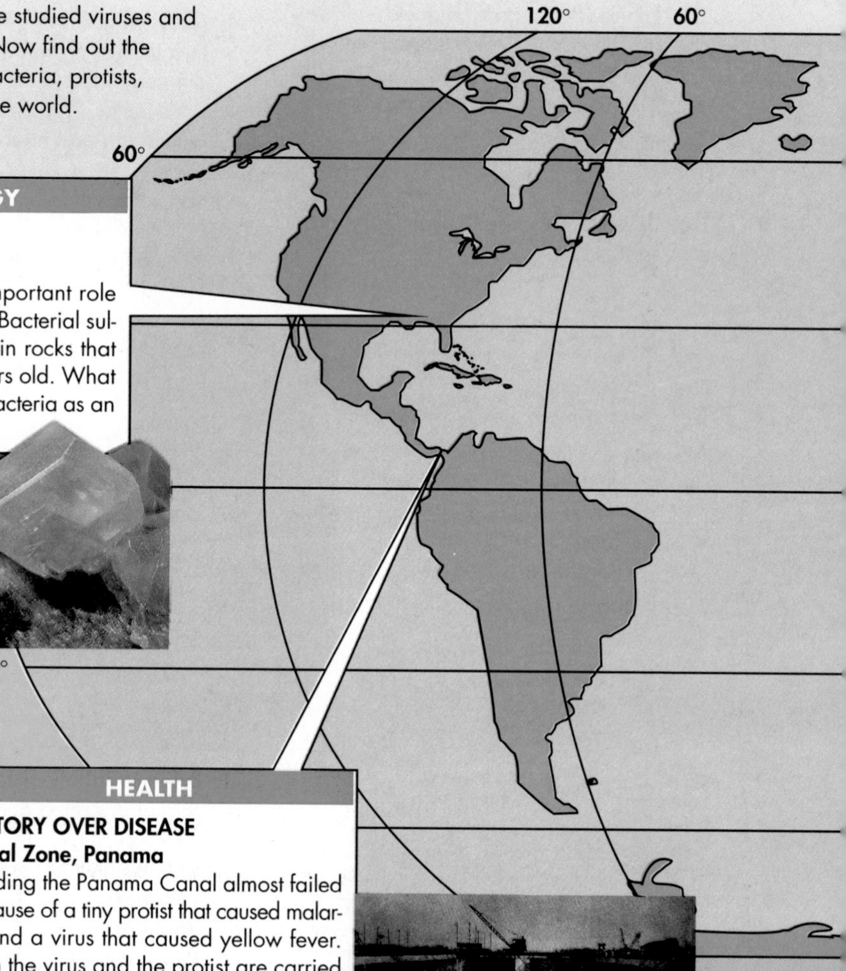

120° 60° 60° 30°

GEOLOGY

SULFUR BACTERIA
Sulphur, Louisiana
Bacteria play an important role in producing sulfur. Bacterial sulfur has been found in rocks that are 800 million years old. What can you tell about bacteria as an early life form?

HEALTH

VICTORY OVER DISEASE
Canal Zone, Panama
Building the Panama Canal almost failed because of a tiny protist that caused malaria and a virus that caused yellow fever. Both the virus and the protist are carried by mosquitoes. How could these diseases be controlled?

222

HEALTH

Background: Panama has an extremely high annual rainfall, which encourages the mosquito population. The French government tried in the 19th century to build a canal, but too many workers died due to yellow fever and malaria. By the start of the Spanish-American War in 1898, U.S. doctors had learned that mosquitoes spread these diseases.

Discussion: Even today mosquitoes are a problem in the Canal Zone. Water plants such as hydrilla and water hyacinths protect mosquito larva from oil spray. Discuss with students the advantages of biological controls over chemical pesticides.

Answer to Question: Yellow fever and malaria can be controlled by controlling mosquitoes, which are the vectors for these diseases.

Extension: Have students read the local newspaper for one week and cut out any articles dealing with the spread of disease. Arrange the articles on a bulletin board under headings that reflect articles found, such as "water-borne" or "vector-borne" diseases.

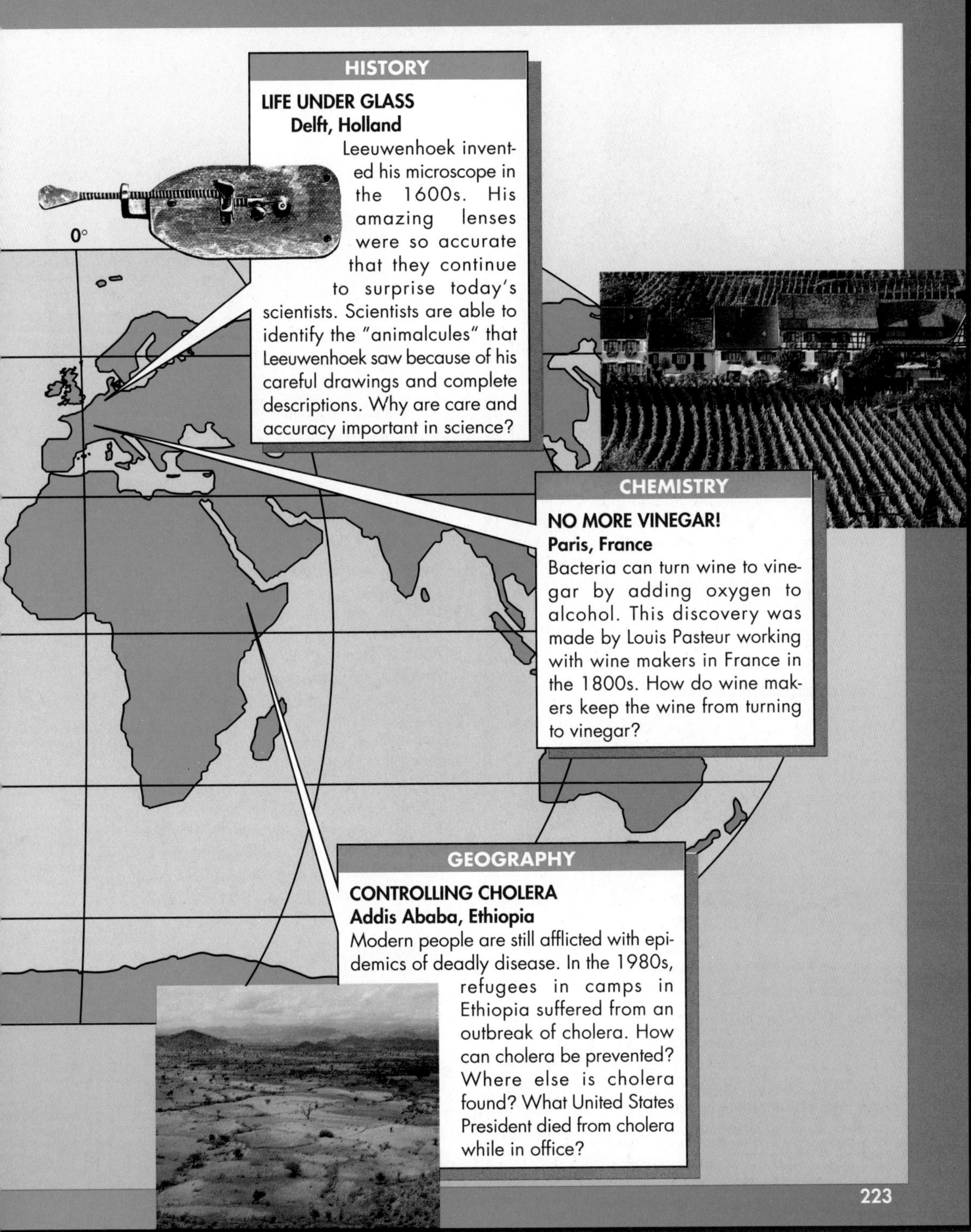

HISTORY

LIFE UNDER GLASS
Delft, Holland

Leeuwenhoek invented his microscope in the 1600s. His amazing lenses were so accurate that they continue to surprise today's scientists. Scientists are able to identify the "animalcules" that Leeuwenhoek saw because of his careful drawings and complete descriptions. Why are care and accuracy important in science?

CHEMISTRY

NO MORE VINEGAR!
Paris, France

Bacteria can turn wine to vinegar by adding oxygen to alcohol. This discovery was made by Louis Pasteur working with wine makers in France in the 1800s. How do wine makers keep the wine from turning to vinegar?

GEOGRAPHY

CONTROLLING CHOLERA
Addis Ababa, Ethiopia

Modern people are still afflicted with epidemics of deadly disease. In the 1980s, refugees in camps in Ethiopia suffered from an outbreak of cholera. How can cholera be prevented? Where else is cholera found? What United States President died from cholera while in office?

223

HISTORY

Background: Leeuwenhoek sold linen and woolen cloth. In his business, he used a small magnifying glass called a "linen counter," which enabled him to grade the cloth. He began to grind lenses as a hobby.

Discussion: Hobbies often add to knowledge. Encourage students to discuss their hobbies and the things they have learned from them.

Answer to Question: Care and accuracy are important in science because an accurate hypothesis is based on accurate observations.

Extension: Have students fill glass containers of various heights and thicknesses with water. Using a penny taped to a pencil, have students compare the magnification powers of the containers, then make hypotheses about the shape and thickness of lenses needed to magnify organisms so they can be seen easily.

CHEMISTRY

Background: Pasteur had already solved the age-old question of "spontaneous generation." Now he put the information to practical use by solving the dilemma of the wine makers. He suggested that the wine be heated to 140–160 degrees Fahrenheit in order to kill the bacteria that soured the wine.

Discussion: Pasteur's discoveries were put to use in practical ways. Ask the students if they know of other ways in which Pasteur's discoveries were used, i.e., pasteurization of milk. How would Pasteur's work affect food preparation and preservation? Ask students how Pasteur's work probably changed people's view of science.

Answer to Question: Wine could be protected from air or heated to kill bacteria which make vinegar.

Extension: Have students choose a piece of bread or fruit and place them into sealable plastic bags. Have students observe what happens when the plastic bags are kept sealed on a counter for two weeks. Is this a demonstration of spontaneous generation? Why?

GEOGRAPHY

Background: Cholera is a bacterial disease that is spread through contaminated water sources. The danger of cholera appears when a supply of clean water is affected by natural disasters or war or whenever sanitation systems are disrupted and fecal material gets into public water supplies. Severe diarrhea occurs and the extreme loss of body fluids can result in death within 24 hours. It can be treated by restoring the body's electrolyte and glucose supply.

Discussion: Discuss the importance of sanitation systems and relate the role of bacteria to the spread of disease.

Answer to Question: Cholera has been found in every major country of the world throughout history. In the 1980s and 1990s, it has been a particular problem in Africa, South America, and in Bangladesh. Zachary Taylor, 12th President of the United States, died of cholera, July 9, 1850, after eating contaminated food.

Extension: Have students visit a local water treatment plant. If not possible, then have students research and devise a flow chart that shows water treatment in their community.

EPIDEMIOLOGIST

Background: The work of the epidemiologist is closely allied with medicine. The epidemiologist studies diseases which spread rapidly, such as certain strains of flu, AIDS, malaria, and yellow fever. Many disease epidemics occur in developing nations. Many epidemiologists work outside their own countries.

Related Career	Education
Epidemiologist	M.D.
Microbiologist	M.S.
Public Health Nurse	B.S.N.

Career Issue: Recently, the drug AZT has been found to help AIDS patients. It is manufactured by only one drug company and is so expensive that many patients cannot afford to buy it. The government is considering allowing other companies to produce the drug in order to gain lower prices.

What do you think? Should one company control the production of a special drug or should the government determine which companies produce a drug and what they may charge for it?

MUSHROOM FARMER

Background: Mushroom farmers grow mushrooms for produce markets and restaurants. Because of interest in ethnic food and gourmet cooking, the kinds of mushrooms being grown are expanding. Mushroom growers need to be aware of exotic new varieties of mushrooms and their culture.

Related Career	Education
Mushroom Farmer	on the job
Farmer	on the job
Apiculturalist	Tech. School

Career Issue: Mushroom growers probably find the best markets near large cities. Since fresh mushrooms do not ship well, mushroom farmers need to be near their markets.

What do you think? Have students discuss the relationship between products and their market. What factors determine where a product is marketed?

CAREERS

EPIDEMIOLOGIST

An *epidemiologist* studies the causes and spread of infectious diseases that affect large numbers of people. He or she may work in a laboratory, or at the site of an outbreak of a disease. A student interested in becoming an epidemiologist should study biology and chemistry in high school.

An epidemiologist will need four years of college plus graduate work in the fields of microbiology, biochemistry, physiology, and biotechnology. Many epidemiologists are medical doctors or people with doctorates who work in pathology laboratories. They may work in a national disease control center or in a university laboratory.

For Additional Information

Contact the American Society for Microbiology, Office of Education and Professional Recognition, 1913 I Street, Washington, DC 20006.

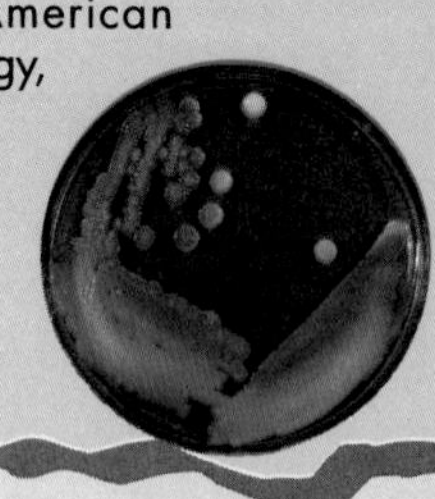

MUSHROOM FARMER

A *mushroom farmer* must have a knowledge of soils, as well as the growing conditions that mushrooms require. He or she will also need a knowledge of the varieties of mushrooms. Many new varieties of mushrooms are being cultivated because of the increase in ethnic cooking and specialty restaurants.

Most of the training a mushroom farmer needs can be learned on-the-job. To advance, these farmers should have some training in small business management. Such training can be obtained at trade schools or through community college programs. Usually, small business training programs take two years.

For Additional Information

Contact the American Farm Bureau Federation, 225 Touhy Ave., Park Ridge, IL 60068.

UNIT READINGS

▶Lee, Douglas. "Slime Mold: The Fungus That Walks." *National Geographic,* July 1981.

▶de Kruif, Paul. *Microbe Hunters.* Orlando, FL: Harcourt, Brace, 1926. This classic details the lives of pioneers in microbiology and their search for the cause of disease.

224

UNIT READINGS

Background

▶ Weaver, Kenneth F. "Electronic Voyage Through an Invisible World." *National Geographic,* February 1977. Minimal reading, but lots of excellent pictures and fine explanation of scanning electron microscope and field ion microscope.

Classics

▶ de Kruif, Paul. *Microbe Hunters.* New York: Harcourt, Brace, 1926. This book details the history of early microbe pathologists.

More Readings

1. Schwartz, George I. *Life in a Drop of Water.* Garden City, NY: Natural History Press, 1970. A delightful and very readable book accompanied by excellent photos by the author.

2. Anderson, M.D. *Through the Microscope.* Garden City, NY: Natural History Press, 1965. A fascinating and highly readable book which highlights the development of the microscope and allied sciences.

The Sea Around Us

by Rachel Carson

The following passage is taken from a chapter titled The Changing Year, which describes life in the sea as the seasons change.

In the sea, as on land, spring is a time for renewal of life. During the long months of winter in the temperate zones the surface waters have been absorbing cold. Now the heavy water begins to sink, slipping down and displacing the warmer layers below. Rich stores of minerals have been accumulating on the floor of the continental shelf—some freighted down by the rivers from the lands; some derived from sea creatures that have died and whose remains have drifted down to the bottom; some from the shells that once encased a diatom, the streaming protoplasm of a radiolarian, or the transparent tissues of a pteropod. Nothing is wasted in the sea; every particle of material is used over and over again, first by one creature, then by another. And when in spring the waters are deeply stirred, the warm bottom water brings to the surface a rich supply of minerals ready for use by new forms of life.

Just as land plants depend on minerals in the soil for their growth, every marine plant, even the smallest, is dependent upon the nutrient salts or minerals in the sea water. Diatoms must have silica, the element of which their fragile cells are fashioned. For these and all other microplants, phosphorus is an indispensable mineral. Some of these elements are in short supply and in winter may be reduced below the minimum necessary for growth. The diatom population must tide itself over this season as best it can. It faces a stark problem of survival, with no opportunity to increase, a problem of keeping alive the spark of life by forming tough protective spores against the stringency of winter, a matter of existing in a dormant state in which no demands shall be made on an environment that already withholds all but the most meager necessities of life. So the diatoms hold their place in the winter sea, like seeds of wheat in a field under snow and ice, the seeds from which the spring growth will come.

In Your Own Words

▶ Many of the simplest living things survive unfavorable living conditions as spores. Describe how this allows these life forms to survive. Write a brief essay describing the characteristics of spores. Tell what problems in the environment may cause an organism to form a spore. Also tell what must happen for the spore to resume normal growth and development.

Source: *The Sea Around Us,* by Rachel Carson. New York: Oxford University Press, 1951.

Biography: Rachel Carson, an American scientist and writer, was the author of many articles and books that brought biology and the nature of the living world to public attention in the 1950s and 1960s. Rachel Carson was born in 1907. She graduated from Pennsylvania College for Women in 1929 and obtained a master's degree from Johns Hopkins University in 1932. She went to work at Woods Hole Oceanographic Institute and the U.S. Fish and Wildlife Service. *The Sea Around Us,* published in 1951, paints a picture of everything from the power of the ocean, to the changes that occur there from season to season, to a profile of the rich life-forms found in a single drop of seawater. In 1962, *Silent Spring* aroused the public to the dangers of indiscriminant use of pesticides and to environmental concerns in general. Carson died in 1964.

TEACHING STRATEGY

Obtain drawings of the ocean floor so that students can see the continental shelf and develop an appreciation of the different depths of the ocean.

Cooperative Learning: Have students work in Paired Partners to find out the relationship between diatoms and oil deposits and deposits of diatomaceous earth.

Writing an Essay: How is Rachel Carson's work on environment relevant to today?

Other Works:

▶ Aronson, Richard B. "An Octopus's Garden." *Natural History,* February 1991, pp. 30-37.

▶ Martin, Glen. "Otter Madness." *Discover,* July 1990, pp. 36-39.

UNIT

4

PLANTS

In Unit 4, students are introduced to the world of plants. The unit is organized along evolutionary lines, from the simplest plants, to more complex plants, to plant processes and relationships.

CONTENTS

ADVANCE PREPARATION

Activities

▶ **Activity 10-1, page 240,** requires live specimens and prepared slides of mosses and liverworts.

▶ **Activity 10-2, page 248,** requires live specimens of ferns, fronds with spores, spores, and prepared slides of ferns.

▶ **Activity 11-1, page 267,** requires lima beans, corn, and iodine solution.

▶ **Activity 11-2, page 270,** requires live specimens of gladiolus flowers and black pepper.

▶ **Activity 12-1, page 281,** requires lettuce leaves and salt solution.

▶ **Activity 12-2, page 290,** requires corn seeds.

Field Trips and Speakers

▶ Arrange to take your class to a conservatory to identify plants with different adaptations.

OPTIONS

Cross Curriculum

▶ Ask students to keep track of how often plants are mentioned in other classes.

PROJECT

During the course of this unit study, have students work cooperatively on PROJECT 4, *Animal Engineers,* found on pages 700 and 701.

Science at Home

▶ Have students grow plants in various places inside their homes—by a sunny window, in a dark closet, or by a window that does not get direct sunlight. Have students explain what happened to the plants.

Cooperative Learning: Assign students to one of four or five groups. Assign each group a plant part such as root, stem, leaf, seed, or flower. Have each group research to determine what common food items are derived from their plant part. Have each group report on their findings.

What's Happening Here?

How do you collect specimens 150 meters in the air? In 1989, a daring group of French scientists cleverly solved this problem by lowering a raft onto the trees of the rain forest from the air. The scientists tethered themselves to the raft for safety and then worked, ate, and slept in their unique, elevated research lab. What did they hope to learn? As rain forests are destroyed at an increasing rate, there is less time to learn about their treasures. The raft helped the scientists to find out if the plants and animals that live on the floor of the forest are different from those at home in the upper story.

UNIT CONTENTS

INTRODUCING THE UNIT

What's Happening Here?

▶ Have students look at the photos and read the text. Ask them what's happening here. Point out to students that in this unit they will study the characteristics of plants and plant processes, and the relationships of plants to other organisms.

▶ **Background:** Most of the organisms in tropical rain forests live in the trees, especially in the top layer or canopy. Working from the forest floor made it difficult for scientists to observe the organisms that never come down out of the trees. By working from this raft, these scientists are able to observe and collect organisms never before seen.

Previewing the Chapters

▶ Many students will be surprised that algae are not classified as plants. Be sure to discuss the differences and similarities between algae and the bryophytes.

▶ Ask students to create their own classification scheme to order protists, monerans, and plants. Ask what are the most important characteristics to include in such a system.

▶ Students should be encouraged to look through the unit chapters to see what organisms are classified as plants.

Tying to Previous Knowledge

▶ Lead students in a discussion of their own knowledge of plants. Many will already be able to identify common plants by leaf, seed, flower, or fruit.

▶ Have students brainstorm a list of products they use that come from plants. Have them identify nonfood items such as clothing or building materials.

▶ Use the **inquiry questions** in the OPTIONS box below to investigate with students the common characteristics of plants.

Multicultural Awareness

People in all cultures rely a great deal on various parts of plants for food. Most cultures rely on a specific source of carbohydrate, derived from a specific grain indigenous to their part of the world. Students can prepare posters showing different plant parts that are used for food. More familiar foods like peas, cinnamon, and tapioca can be presented as well as grains, dandelion greens, prickly pears, pine nuts, banana flowers, the Hawaiian dish poi, and bamboo shoots.

Inquiry Questions

Use the following questions to focus a discussion of what characteristics define a plant.

▶ **What color are plants?**

▶ **Can plants be any other color?**

▶ **Can plants move?**

▶ **How do plants move?**

▶ **How are plants important to humans? to other animals?**

CHAPTER 10 Introduction to Plants

CHAPTER SECTION	OBJECTIVES	ACTIVITIES
10-1 Characteristics of Plants (2 days)	1. **List** the characteristics of plants. 2. **Describe** adaptations of plants that made it possible for them to survive on land. 3. **Compare** vascular and nonvascular plants.	
10-2 Seedless Plants (3 days)	1. **Describe** the life cycles of mosses and ferns. 2. **Compare** simple nonvascular plants with simple vascular plants. 3. **State** the importance of simple nonvascular and vascular plants.	**Activity 10-1:** *Designing an Experiment (Comparing Mosses and Liverworts),* p. 240 **MINI-Lab:** *Where does water move in a plant?* p. 241
10-3 Peat Moss as Fuel Science & Society (1 day)	1. **State** two advantages of using peat as fuel. 2. **Understand** the environmental cost of using peat as fuel.	**MINI-Lab:** *Does moss hold much water?* p. 246 **Activity 10-2:** *The Life Cycle of a Fern,* p. 248
Chapter Review		

ACTIVITY MATERIALS

FIND OUT	ACTIVITIES		MINI-LABS	
Page 229	**10-1 Designing an Experiment, p. 240**	**10-2 The Life Cycle of a Fern, p. 248**	**Where does water move in a plant? p. 241**	**Does moss hold much water? p. 246**
paper pencil	hand lens forceps dropper 2 microscope slides microscope dissecting needle prepared slides of moss and liverwort gametophytes and sporophytes mosses liverworts water 2 coverslips	fern fern frond with spores fern spores hand lens petri dish paper towel scissors water prepared slides of fern microscope	celery, glass of water, red food coloring	*Sphagnum* moss (3 teaspoons) cheesecloth rubber band water beaker glass dish

CHAPTER FEATURES	TEACHER RESOURCE PACKAGE	OTHER RESOURCES
Technology: *Oil from Desert Plants*, p. 235 **Skill Builder:** *Hypothesizing*, p. 235	**Ability Level Worksheets** ◆ ***Study Guide***, p. 40 ● ***Reinforcement***, p. 40 ▲ ***Enrichment***, p. 40 **Critical Thinking/Problem Solving,** p. 14 **Concept Mapping,** p. 25 **Transparency Masters,** pp. 29-30	**Color Transparency 15,** Plant Phylogeny **Lab Manual:** Vascular and Nonvascular Plants, p. 63 **STVS:** Disc 4, Side 2
Problem Solving: *What is in Nature's Medicine Chest?*, p. 245 **Skill Builder:** *Concept Mapping*, p. 245	**Ability Level Worksheets** ◆ ***Study Guide***, p. 41 ● ***Reinforcement***, p. 41 ▲ ***Enrichment***, p. 41 **Activity Worksheets,** pp. 5, 86-87, 92 **Transparency Masters,** pp. 31-34	**Color Transparency 16,** Moss Life Cycle **Color Transparency 17,** Fern Life Cycle **Science Integration Activity 10**
You Decide! p. 247	**Ability Level Worksheets** ◆ ***Study Guide***, p. 42 ● ***Reinforcement***, p. 42 ▲ ***Enrichment***, p. 42 **Activity Worksheets,** pp. 88-89, 93 **Cross-Curricular Connections,** p. 14 **Science and Society,** p. 14	**STVS:** Disc 4, Side 2
Summary Key Science Words Understanding Vocabulary Checking Concepts Using Lab Skills Think & Write Critically Apply More Skill Builders Projects	ASSESSMENT RESOURCES **Chapter Review,** pp. 23-24 **Chapter Test,** pp. 68-71 **Performance Assessment in Middle School Science (PAMSS)**	**Chapter Review Software** **Test Bank** **Alternate Assessment** **Performance Assessment**

◆ **Basic** ● **Average** ▲ **Advanced**

ADDITIONAL MATERIALS

SOFTWARE	AUDIOVISUAL	BOOKS/MAGAZINES
The Green Machine, Micro-ED. *Organizing Plants*, Queue.	*Mosses, Liverworts, and Ferns*, film, Coronet/MTI. *Life of a Plant*, film, EBEC. *How Plants Are Classified*, filmstrip, EBEC. *Kingdom of Plants*, filmstrip, National Geographic. *The Bio Libe Encyclopedia: A Photographic Record of the Earth's Fauna and Flora*, laserdisc, Image Premastering Services.	Mickel, John T. *How to Know the Ferns and Fern Allies.* Dubuque, IA: Wm. C. Brown Company, 1979. Tryon, Rolla M., and Alice F. Tryon. *Ferns and Allied Plants: With Special Reference to Tropical America.* New York: Springer-Verlag New York, Inc., 1982.

CHAPTER

10 INTRODUCTION TO PLANTS

THEME DEVELOPMENT: The evolution of plants from plantlike protists is detailed in this chapter. Emphasis is placed on the adaptations of plants for life on land. The ecology of simple plants is also explained.

CHAPTER OVERVIEW

- **Section 10-1:** Characteristics of plants are listed in this section. The adaptations of plants for survival on land are described. Differences between vascular and nonvascular plants are given.
- **Section 10-2:** The life cycles of mosses and ferns are described. A comparison between nonvascular and vascular plants is made. Ecological importance of nonvascular and vascular plants is discussed.
- **Section 10-3: Science and Society:** The advantages of using peat as a fuel are compared with the environmental cost of its use.

CHAPTER VOCABULARY

cellulose	alternation of generations
cuticle	pioneer species
vascular plants	epiphyte
nonvascular plants	rhizome
rhizoids	frond
sporophyte	sori
gametophyte	prothallus
	bogs

CHAPTER

10 Introduction to Plants

228

OPTIONS

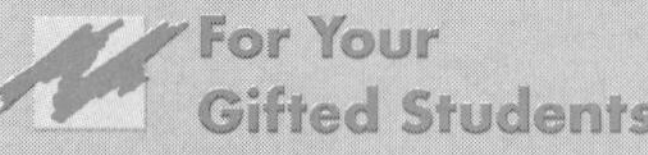

For Your Gifted Students

- Students can research the types of plants that grow in different climate regions and altitudes. They should decide how the plants are suited to the environment. Students may wish to make a picture or mural to help explain their research.
- Have students discuss the effects of plants on the home and work environment. They can brainstorm a list of all the benefits.

For Your Mainstreamed Students

Students can make a terrarium or dish garden of their own or for the class. The garden should reflect a particular environment. Students should be encouraged to learn as much as possible about the environment before planning and planting begins. Students may be interested in finding out the origin of bottle gardens.

Do you eat salads? A salad can be made up of almost any food you can think of. What plants would you choose for a salad? Do you know what plant parts you would be eating?

FIND OUT!

Do this simple activity to find out what plant parts you eat.

Make a list of five plants that you might find in a garden salad. *Compare* your list with a classmate's list. Did you choose different things? Now *decide* what part of the plant each salad item is. Did you choose carrots? Carrots are roots. If you chose lettuce, you would be eating leaves. What about some sprigs of parsley? They are stems. Bean sprouts, pineapple fruit, and sunflower seeds are other parts of plants you might find in a salad.

Gearing Up

Previewing the Chapter

Use this outline to help you focus on important ideas in this chapter.

Previewing Science Skills

- In the **Skill Builders**, you will hypothesize and make a concept map.
- In the **Activities**, you will compare, observe, and collect data.
- In the **MINI-Labs**, you will experiment, observe, measure, and calculate.

What's next?

So far, you have identified some plant parts that are common food sources. In this chapter you will learn more about the characteristics of plants and how plants are adapted to life on land.

229

INTRODUCING THE CHAPTER

Use the Find Out activity to introduce students to plant parts. Inform students that they will be learning more about simple plants and their parts in this chapter.

FIND OUT!

Materials: Ingredients for a salad might include lettuce, carrots, olives, tomatoes, and sunflower seeds.

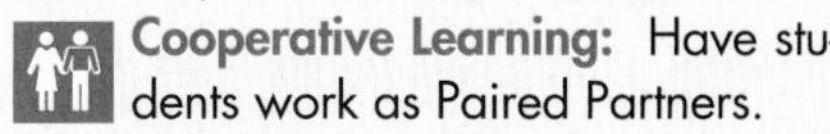

Cooperative Learning: Have students work as Paired Partners.

Teaching Tips

- After making a garden salad for the students, have them identify the plant parts as they eat the salad.
- You may wish to offer suggestions of additional vegetables if students do not produce a sufficiently diverse list.

Gearing Up

Have students study the Gearing Up feature to familiarize themselves with the chapter. Discuss the relationships of the topics in the outline.

What's Next?

Before beginning the first section, make sure students understand the connection between the Find Out activity and the topics to follow.

ASSESSMENT OPTIONS

PORTFOLIO

Refer to page 249 for suggested items that students might select for their portfolios.

PERFORMANCE ASSESSMENT

See page 249 for additional Performance Assessment options.

Process

Skillbuilder, p. 235
MINI-Lab, pp. 241, 246
Activities, 10-1, p. 240; 10-2, p. 248
Using Lab Skills, p. 250

CONTENT ASSESSMENT

Assessment—Oral, pp. 234, 242, 243
Skillbuilder, p. 245
Section Reviews, pp. 235, 245, 247
Chapter Review, pp. 249-251
Mini Quizzes, pp. 234, 242

GROUP ASSESSMENT

Opportunities for group assessment occur with Cooperative Learning Strategies and Flex Your Brain Activities.

SECTION 10-1

PREPARATION

SECTION BACKGROUND

▶ The characteristic tissue formation of plants sets them apart from the plantlike protists.

▶ Fossils of tropical plants have been found in Antarctica, indicating that the continent was once located nearer the equator.

▶ Reproduction without the requirement of water was a major adaptation of land plants.

▶ Emphasize the fact that bryophytes do not contain roots, leaves, or stems because bryophytes have no vascular tissue.

PREPLANNING

▶ Obtain living specimens of mosses and liverworts for Activity 10-1.

1 MOTIVATE

▶ Ask students what features can be found in most plants.

▶ Use the Find Out activity on page 229 as a focus for the lesson.

TYING TO PREVIOUS KNOWLEDGE: Discuss the basic relationship between plants and animals. Review the concepts of producer and consumer with students.

OBJECTIVES AND SCIENCE WORDS: Have students review the objectives and science words throughout the chapter as they study each section.

VideoDisc

STVS: Rubber from Guayule, Disc 4, Side 2

STVS: Water Hyacinth, Disc 4, Side 2

10-1 Characteristics of Plants

New Science Words

cellulose
cuticle
vascular plants
nonvascular plants

Objectives

▶ List the characteristics of plants.
▶ Describe adaptations of plants that made it possible for them to survive on land.
▶ Compare vascular and nonvascular plants.

What Is a Plant?

Have you ever walked in a park? Maybe you've taken off your shoes and wriggled your toes in the cool grass. Perhaps you climbed a tree and looked at the sky through green leaves. Or you may have walked down a nature trail like the one shown in Figure 10-1 to a low, damp place where ferns grew. Grass, trees, and ferns all are members of the Plant Kingdom.

Now look at Figure 10-2. These organisms, mosses and liverworts, have characteristics that identify them as plants, too. What do they have in common with grass, trees, and ferns? What makes a plant a plant?

1 All plants are many celled and most contain the green pigment chlorophyll. Cell walls surround plant cells and give them structure. Most plants have roots or rootlike

Figure 10-1. All plants are many-celled and most contain chlorophyll. Grass, trees, and ferns all are members of the Plant Kingdom.

OPTIONS

Meeting Different Ability Levels

For Section 10-1, use the following **Teacher Resource Masters** depending upon individual students' needs.

◆ **Study Guide Master** for all students.

● **Reinforcement Master** for students of average and above average ability levels.

▲ **Enrichment Master** for above average students.

Additional Teacher Resource Package masters are listed in the OPTIONS box throughout the section. The additional masters are appropriate for all students.

◆ STUDY GUIDE 40

NAME ______ DATE ______ CLASS ______

STUDY GUIDE — Chapter 10

Characteristics of Plants — Text Pages 230–235

Part A

Use the clues to fill in the boxes and to complete the word puzzle. Then answer the riddle below.

What goes around a room and around plant cells? **walls**

1. W A T E R
2. C A R O T E N O I D S
3. P L A N T
4. C U T I C L E
5. N O N V A S C U L A R

Clues

1. needed by all plants to carry nutrients to cells and wastes away from cells
2. red, yellow, or orange pigments found within chloroplasts
3. a many-celled organism that can make its own food and that usually contains chlorophyll
4. waxy, protective layer coating some stems and leaves
5. Bryophytes are often referred to as _____ plants.

Part B

Answer the following questions.

1. What is the organic compound that makes cell walls rigid? **cellulose**
2. The oldest plant fossils are from the Devonian period. How old are those fossils? **about 400 million years old**
3. What is the name of the plant division that includes the mosses and liverworts? **Bryophyta**
4. What is the pigment found in most plant cells? **chlorophyll**
5. What type of plant has tubes that transport materials from place to place within the plant? **vascular**

40

structures that hold them in the ground, so plants usually do not move around. Plants are eukaryotes that have adapted successfully to life on land. In fact, they have adapted so well that they are found in nearly every environment on Earth. From the frigid, ice-bound Antarctica to the hot, dry deserts of Africa, plants have evolved to survive even the most extreme conditions.

Most plants live on land, but many live in or near water.
2 Plants range in size from tiny water ferns that are so small they require a hand lens to see, to the giant sequoia trees of the western United States, some of which are over a thousand years old and 100 m in height.

About 285 000 plant species have been identified, and scientists think there are many more still to be found, mainly in tropical rain forests. If you were asked to make a list of all the plants you could name, you probably would name vegetables, fruits, and field crops like wheat, rice, or corn. These plants are important food sources to humans and other consumers. Without green plants, life on Earth as we know it would not be possible.

Figure 10-2. Simple plants include mosses and liverworts like these.

Why are green plants important to consumers?

Origin and Evolution of Plants

Where did the first plants come from? Like all life, early plants probably came from the sea, evolving from plantlike protists. What evidence is there that this occurred? Both plants and green algae have the same
3 types of chlorophyll as well as carotenoids (kuh RAT uh noydz) in their cells. Carotenoids are red, yellow, or orange pigments found in chloroplasts and in all cyanobacteria. Carrots are orange in color because of these pigments.

One way of understanding the evolution of plants is to look at the fossil record. Unfortunately, the fossil record for plants is not as good as that for animals. Plants usually decay before they form fossils. The oldest fossil plants are from the Devonian period and are about 400 million years old. Available fossils show that early plants were very similar to the plantlike protists. Fossils of *Rhynia major* represent the earliest land plants known. This fossil plant is illustrated in Figure 10-3. These plants had no leaves, and their stems grew underground. Scientists hypothesize that these kinds of plants evolved into the simple plants of today.

Figure 10-3. *Rhynia major* is one of the earliest land plants known.

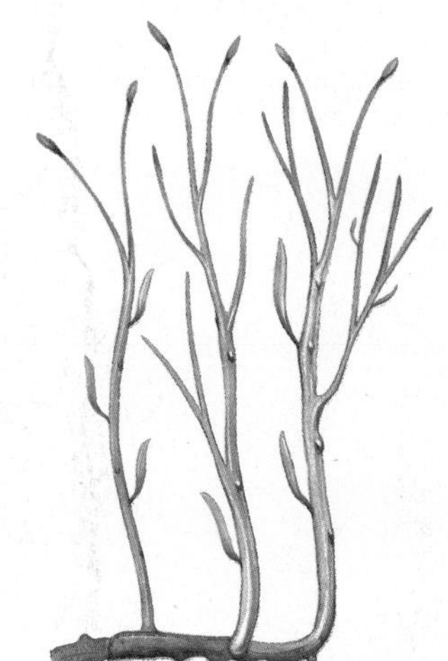

REVEALING MISCONCEPTIONS

▶ Most students will be unaware that some plants do not have true roots, stems, or leaves. Point out that they will be learning about these simpler plants in this section.

TEACHER F.Y.I.

▶ Fossils of *Rhynia major* show characteristics that make them distinctly plants and not protists. Their internal tissues were arranged the same way as the tissues in the roots of modern plants; that is, the xylem and phloem formed a solid cylinder with the phloem on the outside.

2 TEACH

Key Concepts are highlighted.

CONCEPT DEVELOPMENT

▶ Prepare a bulletin board that shows many different environments in which plants live.
▶ Have students list the parts of plants and the function of each part.
▶ Students might bring in fossils of plants that they may have found. But there are fewer fossilized plants because plant tissues were more easily destroyed. They also have few if any, harder structures to fossilize.

● REINFORCEMENT 40

NAME ____ DATE ____ CLASS ____

REINFORCEMENT — Chapter 10

Characteristics of Plants — Text Pages 230–235

Part A

Answer the following questions.

1. Name two things that all plants have in common.
They are many-celled and their cells have walls.
2. What are two characteristics shared by plants and plantlike protists?
Both have two types of chlorophyll and carotenoids.
3. How do cell walls and cuticles help plants adapt to life on land?
Both slow down water loss. Cell walls also support the plant, replacing the buoyant support of water.
4. Name one benefit to a plant of living on land compared to living in water. **sunlight**
5. In what division do the simplest plants belong? **Bryophyta**
6. What is the difference between vascular plants and nonvascular plants?
Vascular plants have vessels that transport materials through the plant; nonvascular plants do not.

Part B

Classify the following plants as vascular or nonvascular.

vascular **vascular** **nonvascular**

vascular **nonvascular** **vascular**

40

▲ ENRICHMENT 40

NAME ____ DATE ____ CLASS ____

ENRICHMENT — Chapter 10

Characteristics of Plants — Text Pages 230–235

SWEET CELERY

The following activity will show you how a vascular plant moves nutrients to all its parts.

Materials

- 2 fresh stalks of celery with leaves (stalks from the middle of the bunch work best)
- 2 tall glasses
- sugar
- measuring spoons
- masking tape
- pen

SWEET
WATER

Procedure

1. Fill each glass half full with water.
2. Add 4 tablespoons of sugar to one of the glasses. Label the cup "sweet" with masking tape.
3. Label the other cup "water."
4. Put a celery stalk in each glass and leave them for 48 hours.
5. Taste the leaves from each stalk. **CAUTION:** *Never taste anything in a lab setting. This activity is safe because it uses only celery, sugar, and water.*

Data and Observations

1. How did the celery leaf from the sugar water taste? **sweet**
2. How did the celery leaf from the plain water taste? **not sweet; like a plain celery leaf**

Conclude and Apply

1. How did the sugar get from the water to the celery leaf?
The sugar dissolved in the water and moved through vascular tubes to the leaf.
2. How do plants get nutrients from the soil to their leaves?
Nutrients in the soil dissolve in water and move through the vascular tubes to the leaves.
3. What advantages do plants with this adaptation have?
These plants could grow very tall and still get nutrients to their leaves. They do not have to live in damp environments.

40

CONCEPT DEVELOPMENT

▶ Remind students of the processes of osmosis and diffusion. It would also be useful to remind students about what they learned about algae in Chapter 9.

▶ Discuss with students the controversy over where green algae are placed taxonomically—with protists or with plants.

▶ Ask a florist or horticulturist to discuss the characteristics of plants that make them attractive and useful.

▶ Set up stations around the classroom that contain different plant parts. Have students write down the name of each plant part as they travel from station to station.

CROSS CURRICULUM

▶ **Earth Science:** The fossils of plants found in such places as Antarctica and high on mountains provide evidence of continental drift and mountain building.

Adapting to Land

Imagine life for a one-celled green alga floating in a shallow pool. The water in the pool surrounds and supports the alga cell. It can make its own food through the process of photosynthesis, and materials move freely into and out of the cell membrane by osmosis and diffusion. The alga has everything it needs to survive.

Now imagine a summer drought. The pool begins to dry up. Soon the alga lies on the damp mud. What will happen to the alga? It can still make its own food, so it won't starve. As long as the ground stays damp, the alga can move materials in and out through the cell membrane. But it is no longer supported by the pool's water. What will happen to the alga if the land continues to dry up? The alga will lose water through the cell membrane by osmosis. Remember that in osmosis, water molecules diffuse through a membrane from an area of higher concentration to one of lower concentration. Water will diffuse out of the alga to the drier pond bottom. Without water, the alga will die.

What adaptations would make it possible for plants to survive on land? What would help the plant conserve water? Loss of water is a major problem plants face. Plant cells have cell membranes like the protists do, but they also have rigid cell walls outside the membrane. Cell walls are made of **cellulose,** an organic compound made up

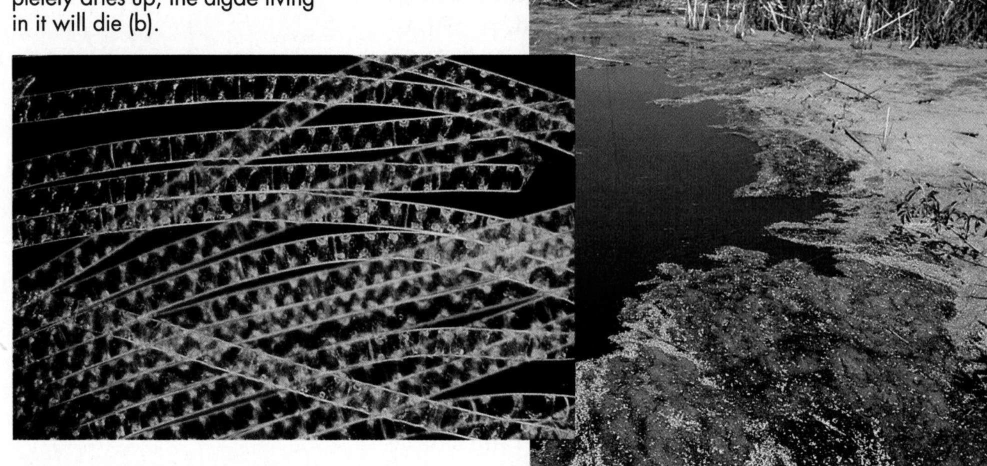

Figure 10-4. Green algae (a) produce their own food and move materials in and out through the cell membrane. Algae must have water to survive. If a pond completely dries up, the algae living in it will die (b).

OPTIONS

INQUIRY QUESTIONS

▶ **How can you tell the difference between a plantlike protist and a plant?** *Plantlike protists do not have tissues as true plants do.*

▶ **Why would you expect there to be few fossils of the earliest plants?** *Such plants would have probably been small, with few hard tissues for preservation.*

Figure 10-5. Cell walls and waxy cuticles are adaptations that enable plants to survive on land.

of long chains of sugar molecules. Wood is about 50 percent cellulose. Cell walls help prevent plant cells from drying out and provide structure and support. Most land plants also have a waxy, protective layer on stems and leaves called a **cuticle.** The next time it rains, go outside and see how raindrops bead up on the surface of most plant leaves or flower petals. The rain rolls right off because of the waxy cuticle. The cuticle is an adaptation that helped plants to keep from drying out on land.

How does a cuticle help a plant adapt to life on land?

Another problem plants faced in the move onto land was reproduction. Plants evolved from organisms that reproduced in water. As a matter of fact, these species were completely dependent on water for reproduction and survival. Simple plants still require water to reproduce. You will learn more about reproduction in simple plants in Section 10-2. More complex plants evolved ways to reproduce that do not require water. Reproduction in these plants is discussed in Chapter 11.

There are some advantages to life on land for plants. For one thing, there's a lot more sunlight available for photosynthesis on land than in water. Another advantage of life on land is the availability of carbon dioxide. There is much more carbon dioxide in the air than in the water. Carbon dioxide is a gas that plants use to produce food. All plants still need water to carry nutrients into and waste products out of their cells. They also need structures to support their weight on land. These supports eventually came in the form of stems and roots.

Cellulose, found in plant cell walls, is an organic compound composed of many glucose units strung together. More than half the carbon in plants is in the form of cellulose. Raw cotton is over 90 percent cellulose. Which physical property do you think makes cellulose ideal for helping plants survive on land?

CONCEPT DEVELOPMENT

▶ **Demonstration:** Use a spray mister to mist water onto a plant leaf that has a thick cuticle. Allow students to observe how the water beads up and runs off.

▶ **Demonstration:** Place a stalk of celery in a glass of water to which some food coloring has been added. Once the colored liquid has reached the leaves, pull apart the stalk to show the parallel vascular tubes in the celery.

▶ Water pressure in plant cells provides some rigidity and support to land plants.

▶ Discuss how leaves of plants are modified to conserve water or to be exposed to more sun. Cacti leaves are reduced to spines to conserve water in dry environments. Broadleaf deciduous trees have flat, thin leaves that spread out to provide more surface area for absorption of sunlight.

▶ One other adaptation to conserve water is the shedding of leaves, or leaf abscission. Deciduous trees such as oaks, hickories, elms, and maples lose leaves each fall to prevent water loss in cold weather through transpiration.

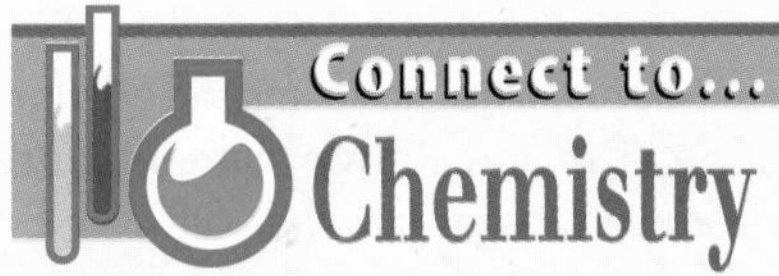

Note: Students should be able to find answers for most Connect to... features in basic library references or infer answers from text material.

Answer: Cellulose is a tough material that enables plants to stand upright. By standing upright, plant leaves are able to catch more sunlight and make more food.

ENRICHMENT

▶ Show the film, *Evolution and the Vascular Plants*, available from Encyclopaedia Britannica Educational Corporation.

MINI QUIZ

Use the Mini Quiz to check students' recall of chapter content.

1. **All plants are many-celled and most contain the green pigment _____.** *chlorophyll*
2. **Plants range in size from tiny _____ to the giant _____.** *water ferns, sequoias*
3. **_____ are red, yellow, or orange pigments found in chloroplasts.** *Carotenoids*
4. **Plants with vessels that transport materials to other cells are called _____.** *vascular plants*

CHECK FOR UNDERSTANDING

Have students make a list of everything they know about plants. Review their lists to see that students reflect an understanding of the interrelationship between vascular and nonvascular, and producer and consumer.

RETEACH

Ask students to describe what the results for animals would have been if plants had not adapted to live on land.

EXTENSION

For students who have mastered this section, use the **Reinforcement** and **Enrichment** masters or other OPTIONS provided.

3 CLOSE

▶ Ask questions 1-3 and the **Apply** Question in the Section Review.

Classification of Plants

Today, plants usually are classified into major groups called divisions. A division is the same level as the phylum you've studied in the Protist Kingdom. The simplest plants—mosses and liverworts—are placed in the division Bryophyta (bri uh FITE uh).

Bryophytes are small plants found in damp environments like the forest floor, the edges of ponds and streams, and near the ocean. Bryophytes are usually just a few cells thick, so they are able to absorb water directly through their cell walls. Plants in the more complex divisions of the Plant Kingdom have systems in which water and nutrients are transported through the plant in tubelike vessels. In a sense, these vessels are similar to blood vessels. Vessels transport materials to all plant cells. Plants with vessels (4) are more than a few cells thick. Plants that have vessels are **vascular plants.** Most of the plants you know, such as trees, bushes, and flowers, are vascular plants. Plants without vessels, the bryophytes, are **nonvascular plants.** Why is it an advantage for nonvascular plants to be located in damp areas?

What are plants with and without vessels called?

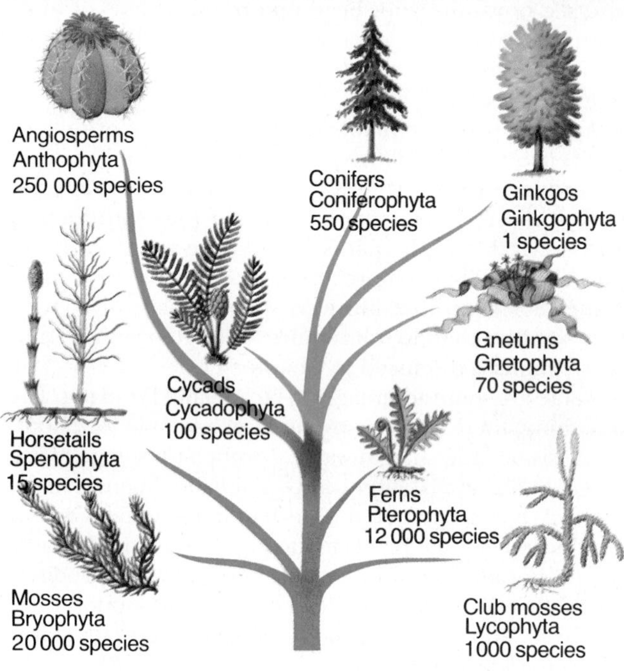

Figure 10-6. The Plant Kingdom

OPTIONS

ASSESSMENT—ORAL

▶ **Suppose you find a new species of plant that grows a few feet tall. Would it be a vascular plant or a nonvascular plant. Why?** *The plant would be a vascular plant because plants need vascular tissue if they grow more than a few inches in height.*

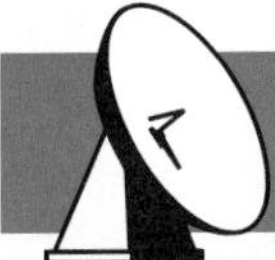

TECHNOLOGY

Oil from Desert Plants

The jojoba (ho HO bah), a desert plant, produces brown, peanut-sized seeds that contain an oil very similar to sperm whale oil. Trade in sperm whale oil was banned in the United States in 1970. Sperm whale oil was used in many products, from face creams to transmission fluids. Faced with a ban on sperm whale oil, industries tried new oils, including jojoba oil.

Jojoba seeds contain 50 percent oil by weight. This oil has a very high boiling point (398°C) and is valuable for lubricating machinery. Jojoba oil has no cholesterol and can be a healthy substitute for cooking oils. Used also in soap and cosmetics, jojoba has a promising future. However, it is expensive because, until recently, most oil was extracted from seeds collected in the desert. Only a few thousand hectares have been planted commercially so far.

Think Critically: How does the availability of jojoba oil affect its usefulness to industry?

SECTION REVIEW

1. List the characteristics of plants.
2. What adaptations make life on land possible for plants? What are some advantages of life on land for plants?
3. Compare vascular and nonvascular plants.
4. **Apply:** If you left a board lying on the grass for a few days, what would happen to the grass underneath the board? Why?
5. **Connect to Chemistry:** To which class of organic compounds (proteins, carbohydrates, or lipids) does the material that makes plant cuticles belong?

Skill Builder

Hypothesizing

From what you have learned about adaptations necessary for life on land, make a **hypothesis** as to what types of adaptations land plants might have if they had to survive submerged in water. If you need help, refer to Hypothesizing in the **Skill Handbook** on page 686.

PROGRAM RESOURCES

From the **Teacher Resource Package** use:

Critical Thinking/Problem Solving, page 14, Adaptations of Desert Plants.

Concept Mapping, pages 25-26.

Transparency Masters, pages 29-30, Plant Phylogeny.

Use **Color Transparency** 15, Plant Phylogeny.

Use **Laboratory Manual,** page 63, Vascular and Nonvascular Plants.

TECHNOLOGY

▶ **Extension:** Research plants used in cosmetics. Make a class chart showing the plants, products in which they are used, and what characteristics of the plant contribute to the product.

▶ **Reference:** Villena-Denton, Vicky. "A Whale of a Substitute," *The Oil Daily,* May 9, 1990, p. B2.

▶ **Think Critically:** If more jojoba oil is produced, its price is likely to fall. Lower cost would make it a more attractive alternative to other oils and encourage companies to develop it as a resource.

SECTION REVIEW ANSWERS

1. All plants are many-celled and most contain chlorophyll and carotenoids. Most plants have roots or rootlike fibers that hold them in the ground.

2. Vascular tissue, a cuticle layer on leaves, and a way to reproduce without water are all critical adaptations plants have for living on land. Plants that live on land have a greater supply of light and carbon dioxide.

3. Vascular plants have tubes that carry water and food. Nonvascular plants get water and food by diffusion.

4. Apply: The grass would turn yellow. Light is needed to keep up production of chlorophyll.

5. Connect to Chemistry: lipids

Skill Builder

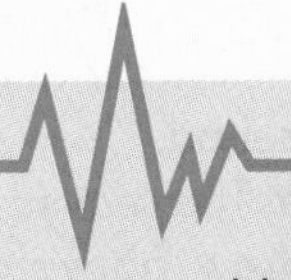

Plants submerged in water would need more efficient chloroplasts because of less light. In addition, they would need ways to get additional carbon dioxide and ways to keep a balance between water inside cells and outside cells.

Skillbuilder

ASSESSMENT

Performance: Assess students' abilities to Hypothesize about the adaptations necessary for land plants to live in water by having them devise a model for one of their suggested adaptations.

SECTION 10-2

PREPARATION

SECTION BACKGROUND

▶ There are more species in the bryophyte division than in any other, except anthophytes, the flowering plants.

▶ Some mosses do have water-conducting cells, but they are not organized as vascular tissue.

▶ Mosses, like amphibians, evolved from water-dwelling organisms and still require water for fertilization.

▶ Rhizoids hold the plant to the ground, but do not conduct water to other cells in the plant as roots do.

▶ There are few differences between fossil mosses and liverworts and those that are alive today.

▶ As plants become more complex, the size of the gametophyte decreases.

▶ Most of the club mosses and horsetail species became extinct during the Mesozoic era. Only a few species survive today.

▶ The ferns are the most diverse group of simple plants. They range in size from the tiny water fern to the tree fern, and in form from the maidenhair fern to the staghorn fern.

▶ Tree ferns can be thought of as living fossils. The species of tree ferns alive today live mostly in tropical rain forests. No tree ferns grow as tall as those that lived during the Devonian period.

▶ The leaves of ferns, called fronds, grow above ground from the buried stems (or rhizomes). Rhizomes are underground structures.

10-2 Seedless Plants

New Science Words

rhizoids
sporophyte
gametophyte
alternation of generations
pioneer species
epiphyte
rhizome
frond
sori
prothallus

Objectives

▶ Describe the life cycles of mosses and ferns.
▶ Compare simple nonvascular plants with simple vascular plants.
▶ State the importance of simple nonvascular and vascular plants.

Seedless Nonvascular Plants

If you were asked to name the parts of a plant, you probably would list roots, stems, leaves, and perhaps flowers. You may also know that many plants grow from seeds. But did you know that some simple plants have none of these parts? Nonvascular plants, the bryophytes, do not have roots, stems, or leaves. They do have rootlike fibers, stalks that look like stems, and leaflike green growths. Instead of growing from seeds, bryophytes grow from spores. The nonvascular plants include the mosses and liverworts. Figure 10-7 shows some common types of nonvascular plants.

What is a moss?

A moss is a simple rootless plant with leaflike growths in a spiral around a stalk. Moss plants are held in place by rootlike filaments or threads made up of only a few long
1 cells called **rhizoids.** A liverwort is a simple, rootless plant that has a flattened, leaflike body. Liverworts get their

Figure 10-7. The nonvascular plants include the mosses (a) and the liverworts (b).

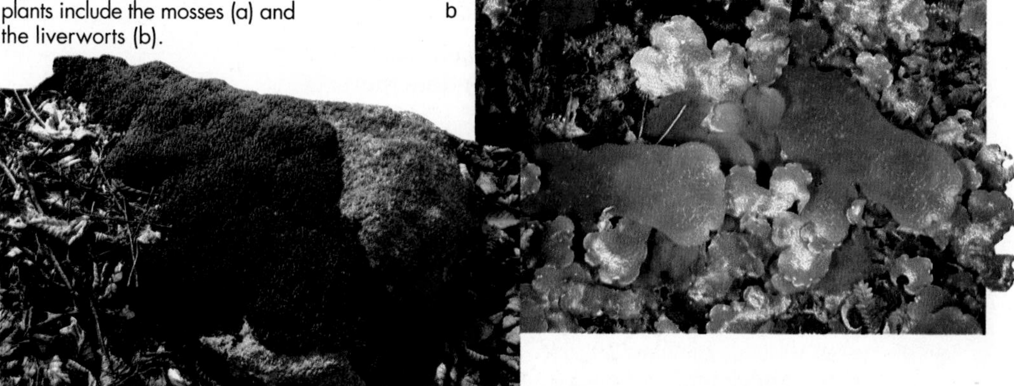

OPTIONS

Meeting Different Ability Levels

For Section 10-2, use the following **Teacher Resource Masters** depending upon individual students' needs.

◆ **Study Guide Master** for all students.

● **Reinforcement Master** for students of average and above average ability levels.

▲ **Enrichment Master** for above average students.

Additional Teacher Resource Package masters are listed in the OPTIONS box throughout the section. The additional masters are appropriate for all students.

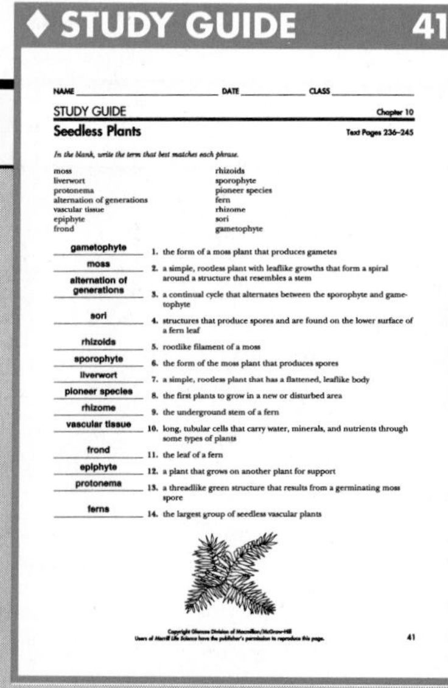
◆ STUDY GUIDE 41

NAME ______ DATE ______ CLASS ______

STUDY GUIDE — Chapter 10

Seedless Plants — Text Pages 236–245

In the blank, write the term that best matches each phrase.

moss	rhizoids
liverwort	sporophyte
protonema	pioneer species
alternation of generations	fern
vascular tissue	rhizome
epiphyte	sori
frond	gametophyte

gametophyte 1. the form of a moss plant that produces gametes
moss 2. a simple, rootless plant with leaflike growths that form a spiral around a structure that resembles a stem
alternation of generations 3. a continual cycle that alternates between the sporophyte and gametophyte
sori 4. structures that produce spores and are found on the lower surface of a fern leaf
rhizoids 5. rootlike filament of a moss
sporophyte 6. the form of the moss plant that produces spores
liverwort 7. a simple, rootless plant that has a flattened, leaflike body
pioneer species 8. the first plants to grow in a new or disturbed area
rhizome 9. the underground stem of a fern
vascular tissue 10. long, tubular cells that carry water, minerals, and nutrients through some types of plants
frond 11. the leaf of a fern
epiphyte 12. a plant that grows on another plant for support
protonema 13. a threadlike green structure that results from a germinating moss spore
ferns 14. the largest group of seedless vascular plants

41

name because, to some people, they look like a liver. In the ninth century, liverworts were thought to be useful in treating diseases of the liver. The ending, *-wort,* means "herb," so the word *liverwort* means "herb for the liver." In liverworts, each rhizoid is made up of just one cell. Both mosses and liverworts are small plants that grow in damp areas. They range in size from 2 to 5 centimeters in height.

Of approximately 20 000 species of nonvascular plants, most are classified as mosses. Have you ever seen mosses growing on tree trunks, rocks, or the ground in damp or humid areas? If conditions are right, mosses grow right on the sides of wood houses.

The Moss Life Cycle

A typical growth of moss plants looks like a soft, green carpet on the forest floor, as shown in Figure 10-7a. Now look at the closeup of the moss plants in Figure 10-8. Can you see the structures sticking up out of the leaflike plants? These are the moss sporophytes. A **sporophyte** includes a stalk and a capsule in which numerous haploid spores are produced. Eventually, the capsule breaks open and the spores are scattered. When a haploid spore lands on wet soil or rocks, it germinates into a thin, threadlike green structure. Within a few days, small gametophyte moss plants begin to grow here and there along the thread. A **gametophyte** is the form of a moss plant that produces sex cells. Sometimes a moss gametophyte produces only male or female sex cells, but often both types are produced. During a heavy dew or rain, the male sperm get splashed onto the female gametophyte. They swim to the female eggs. When the male and female sex cells unite, a diploid zygote forms. The zygote divides by mitosis to form an embryo, which in turn develops into a sporophyte, and the cycle begins again. In plants, this continual cycle, which alternates between the spore-producing phase and

Figure 10-8. The leafy green gametophytes of moss support the stalks and capsules of the moss sporophytes.

What is a gametophyte?

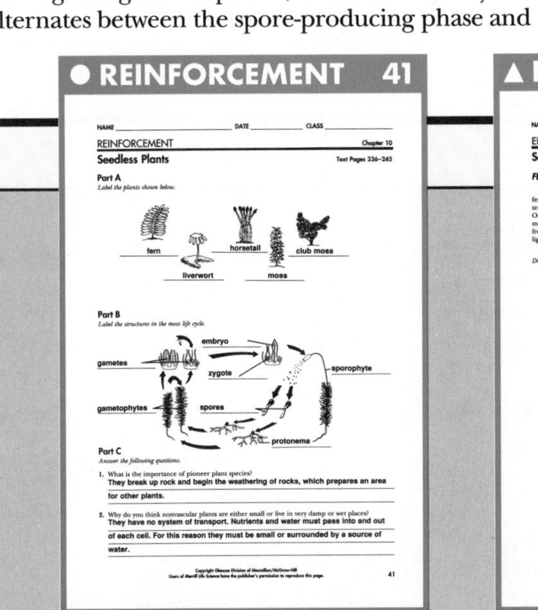

● REINFORCEMENT 41

NAME ______ DATE ______ CLASS ______

REINFORCEMENT — Chapter 10

Seedless Plants — Text Pages 236–245

Part A

Label the plants shown below.

Part B

Label the structures in the moss life cycle.

Part C

Answer the following questions.

1. What is the importance of pioneer plant species?
   They break up rock and begin the weathering of rocks, which prepares an area for other plants.

2. Why do you think nonvascular plants are either small or live in very damp or wet places?
   They have no system of transport. Nutrients and water must pass into and out of each cell. For this reason they must be small or surrounded by a source of water.

 41

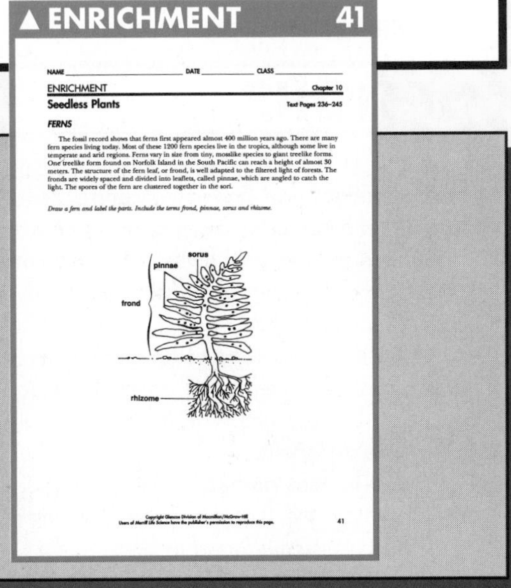

▲ ENRICHMENT 41

NAME ______ DATE ______ CLASS ______

ENRICHMENT — Chapter 10

Seedless Plants — Text Pages 236–245

FERNS

The fossil record shows that ferns first appeared almost 400 million years ago. There are many fern species living today. Most of these 1200 fern species live in the tropics, although some live in temperate and arid regions. Ferns vary in size from tiny, mosslike species to giant treelike forms. One treelike form found on Norfolk Island in the South Pacific can reach a height of almost 30 meters. The structure of the fern leaf, or frond, is well adapted to the filtered light of forests. The fronds are widely spaced and divided into leaflets, called pinnae, which are angled to catch the light. The spores of the fern are clustered together in the sori.

Draw a fern and label the parts. Include the terms frond, pinnae, sorus and rhizome.

 41

PREPLANNING

▶ Collect different kinds of ferns to bring to class to display.

▶ If possible, obtain moss sporophytes and leaves of ferns with mature sori.

1 MOTIVATE

▶ The protonema looks like a filamentous green alga.

▶ Display different species of moss plants. If possible, have a few with sporophytes.

▶ Have students measure moss plants to see how small they are.

▶ Ask students to collect mosses from areas near their homes. Mosses are often easier to see in fall and winter when other plants die back.

TYING TO PREVIOUS KNOWLEDGE: Point out that students may know club mosses as ground pines from holiday decorations, and they know ferns from houseplants.

REVEALING MISCONCEPTIONS

▶ Some students may not think the nonvascular plants have any importance. Emphasize the importance of all species and point out the important roles nonvascular plants have played in the production of fossil fuels.

In Your JOURNAL

Student responses will vary. Mnemonics might play off the fact that many terms in the life cycle begin with the letter s.

TEACHER F.Y.I.

▶ Elizabeth Britton, born in 1858 in New York City, became America's foremost bryologist. She helped to establish the New York Botanical Garden.

2 TEACH

Key Concepts are highlighted.

CONCEPT DEVELOPMENT

▶ Some plants are called mosses, but really are not bryophytes. Spanish moss is a flowering plant, and reindeer moss is a lichen.
▶ A single capsule of a moss sporophyte can release 50 million spores.
▶ Bryophyte species in temperate climates take 6 to 18 months to mature.
▶ Mosses are important in preventing soil erosion in damp areas.
▶ Mosses are eaten by birds and mammals.
▶ The flattened, leaflike body of a liverwort is called a *thallus.*
▶ Liverworts also reproduce asexually when lobes of the thallus elongate. When the main part of the thallus dies, each lobe may become a new plant.
▶ The asexual cuplike structures on liverworts are *gemmae.* Gemmae are produced by mitosis and will mature into plants that are identical to the parent plant.

FLEX Your Brain

Use the Flex Your Brain activity to have students explore LIFE CYCLE.

Flex Your Brain
ASSESSMENT
Portfolio: Use the Flex Your Brain activity to reinforce critical thinking and problem solving skills.

What is alternation of generations?

sex cell-producing phase is called **alternation of generations**. In bryophytes, the most visible plant is the green, leaflike gametophyte. The life cycle of a moss, including the alternation of generations, is shown in Figure 10-9. Liverworts have a life cycle similar to that of the mosses.

In addition to the alternation of generations, both mosses and liverworts reproduce asexually. New moss plants can develop when a small piece of the parent plant breaks off. Liverworts develop small balls of cells within cuplike structures on the surface of the leaflike body. These balls of cells can be carried by water to new areas where they grow into new plants.

All plants have a life cycle in which the sporophyte and gametophyte alternate. In other words, all plants go through a diploid and a haploid stage. In nonvascular plants like mosses and liverworts, the sporophytes depend on the gametophytes for water and nutrients. In the more complex vascular plants, like tulips and oak trees, the sporophyte doesn't depend on the gametophyte for these things.

In your Journal, write a poem or a memory code to help you remember the life cycle of mosses.

Figure 10-9. The Moss Life Cycle

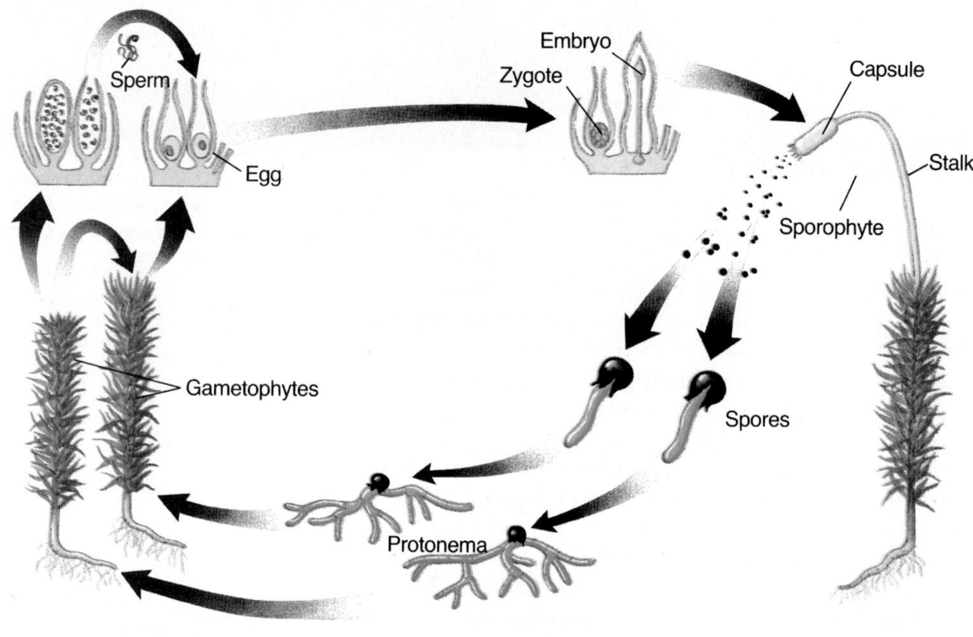

OPTIONS

INQUIRY QUESTIONS

▶ **Where might you find a concentration of mosses and liverworts?** *near cool, wet areas*
▶ **How are mosses and liverworts different?** *Mosses grow upright, and liverworts form a flattened, leaflike body.*
▶ **What is the advantage of an alternation of generations?** *By alternating generations and producing spores, simple nonvascular plants can survive dry spells.*
▶ **How do mosses and ferns differ?** *Mosses do not contain the transport tubes that make up the vascular tissue found in ferns.*

Figure 10-10. Mosses and liverworts are pioneer plants that often are among the first living things to inhabit a new environment like this lava field.

Importance of Mosses and Liverworts

Mosses and liverworts are important in the ecology of many areas. They are often among the first plants to grow in new environments, such as lava fields, or disturbed environments, such as forests destroyed by fire. When a volcano erupts, the lava covers the land and destroys the plants living there. After the lava cools, the spores of mosses and liverworts are carried to the new rocks by the wind and begin to grow wherever there is enough water. When they grow on rocks, their rhizoids can actually begin to penetrate tiny cracks in the rocks' surfaces. Weathering of rocks often begins when mosses release chemicals that begin to break rocks down. Organisms that are the first 3 to grow in new or disturbed areas like these are called **pioneer species.** Pioneer plant species grow and die and begin to build up decaying plant material that provides nutrients for other, less hardy, plants. That is, these pioneer plants change the conditions in the environment so that other plants can grow there, too. Eventually, larger vascular plants can begin to grow, sending their roots down through the decaying plants and into the rocks underneath. As the roots grow, they crack the rocks apart even more, and the weathering process slowly turns the rocks into new soil.

Soil is a mixture of weathered rock and decaying organic matter (plant and animal). What roles do pioneer species such as lichens, mosses, and liverworts play in building soil?

CONCEPT DEVELOPMENT

▶ Point out the stages in the life cycle of mosses using the diagrams in the text.

▶ Emphasize that spores are not seeds.

▶ Explain that many of the nonvascular plants and ferns became extinct at the same time as the dinosaurs during the Mesozoic era.

▶ **Demonstration:** The differences between mosses and liverworts can be explained by showing these plants to students.

Cooperative Learning: Use Paired Partners cooperative groups. Provide pairs of students with a variety of plants to care for during this study of plants. Have them conduct library research about their plant's habitat requirements. Allow students to present a two- to three-minute report on their findings.

Answer: Lichens break down rock with acids; decaying mosses and liverworts provide organic matter that mixes with rock particles. In time, enough of this mixture gives larger plants a place to anchor themselves.

CROSS CURRICULUM

▶ **History:** Pioneers sometimes used *Sphagnum* moss as dressing on wounds. The antibiotic properties of the moss could sometimes prevent infections.

PROGRAM RESOURCES

From the **Teacher Resource Package** use:

Activity Worksheets, page 5, Flex Your Brain.

Transparency Masters, pages 31-32, Moss Life Cycle.

Science Integration Activities 10, Carbon Impressions

Use **Color Transparency** 16, Moss Life Cycle.

ACTIVITY 10-1

OBJECTIVE: Design and carry out an experiment to compare and contrast the structure of mosses and liverworts.
Time: 30 minutes

PROCESS SKILLS applied in this activity are **observing, classifying,** and **inferring.**

PREPARATION

Collect materials needed for each group. Microscope slides may need to be borrowed.

Cooperative Learning: Assign Science Investigation Teams. Students may work in groups of two to four in preparing the materials for this investigation. If possible, students should work in pairs at the microscope.

THINKING CRITICALLY

Characteristics that can be compared are the appearances of the rhizoids, gametophytes, and sporophytes and whether spores are formed.

TEACHING THE ACTIVITY

*Refer to the **Activity Worksheets** for additional information and teaching strategies.*

- The species of mosses and liverworts used in this exercise do not matter. If you cannot obtain these locally, near the school in a wooded lot, they can be purchased from a biological supply company.
- If the moss spores are not liberated using water only, you may add an equal volume of glycerin to the water. This mixture will alow the spores to come out of the capsule more freely.

ACTIVITY 10-1 DESIGNING AN EXPERIMENT
Comparing Mosses and Liverworts

How are mosses and liverworts similar? How are they different?

Getting Started
Your task in this activity is to compare and contrast two seedless, nonvascular plants.

Thinking Critically
List some characteristics that can be compared. What characteristics will you use?

Materials 
Your cooperative group will use:
- hand lens
- forceps
- dropper
- 2 microscope slides
- microscope
- dissecting needle
- prepared slides of moss and liverwort gametophytes and sporophytes
- pencil with eraser on end

Try It!

1. Obtain a moss. The leafy part of the moss is the gametophyte. With a hand lens, *observe* the gametophyte. Locate and *observe* the rhizoids.
2. *Examine* prepared slides of sections of the sporophyte and gametophyte moss plants under high and low power. Draw what you see.
3. On the moss gametophyte, locate the capsule at the tip of the stalk. Remove the capsule and place it in a drop of water on a slide. Place a coverslip on the capsule. Using the eraser end of your pencil, gently crush the capsule to release the spores. **CAUTION:** *Do not break the coverslip. Observe* the spores under low and high power. Draw and label the spores.
4. Obtain a liverwort. Repeat Steps 1-3.
5. Record information collected from your observations in a data table.

Structure	Liverwort	Moss
Sporophyte		
Gametophyte		
Spores		

Summing Up/Sharing Results
- What function do the rhizoids have?
- Where are spores formed in a moss? In a liverwort?
- *Compare* the gametophytes and sporophytes of the mosses and liverworts.

Going Further!
Remove a piece of the leafy portion of the moss and of the liverwort with your forceps and make a wet mount. *Examine* the cell layers under low power. How many cell layers were you able to observe in the moss? In the liverwort?

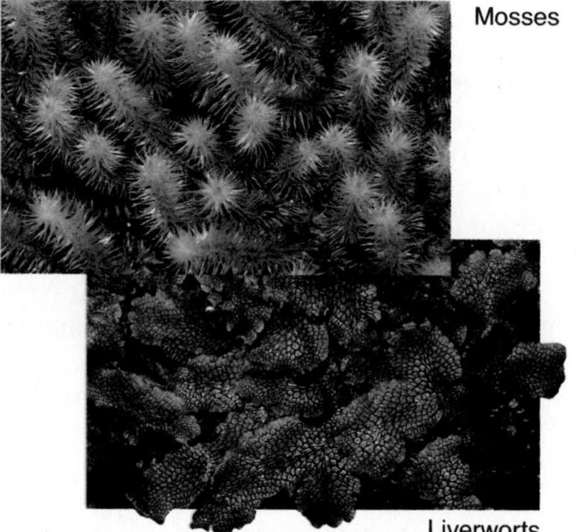

SUMMING UP/SHARING RESULTS

- Rhizoids hold the plants to the ground.
- in the capsule at the top of the stalk of the moss and under the top of the umbrella-like structure at the top of the liverwort stalk.
- The gametophyte of the moss is a low-growing, leafy structure in a whorl around a stalk, that of the liverwort is a flat, leaflike form. Sporophytes of mosses form a stalk with a capsule containing spores at the top. In liverworts, the sporophytes form umbrella-like structures with spores in cases underneath.

GOING FURTHER!

Mosses have one or two cell layers; liverworts have one to three.

PROGRAM RESOURCES

From the **Teacher Resource Package** use:
Activity Worksheets, pages 86-87, Activity 10-1, Comparing Mosses and Liverworts.

Activity ASSESSMENT

Performance: Assess students' understanding of the parts of mosses and liverworts by having them identify the parts of living specimens and describe a function for each.

Seedless Vascular Plants

A second and larger group of simple plants includes the seedless vascular plants. These plants differ from mosses mainly because they have vascular tissue. They differ from other vascular plants because they produce spores rather than seeds. The vascular tissue in the seedless vascular plants is made up of long, tube-like cells. These cells carry water, minerals, and nutrients to all the cells throughout the plant. Why is having cells like these an advantage to a plant? Remember that bryophytes are only a few cells thick. Each cell absorbs water directly from its environment. As a result, these plants do not grow very big. Plants with vascular tissue, on the other hand, can grow bigger and thicker because each cell gets water and nutrients through the vascular tissue.

Seedless vascular plants include the club mosses, spike mosses, horsetails, and ferns. Today, there are about 1000 species of club mosses, spike mosses, and horsetails. Ferns are more abundant, with at least 12 000 species known. During the Paleozoic era, there were many species that we now know of only from fossils. In the warm, moist Paleozoic era, some species of horsetails grew 15 meters tall. In contrast, simple vascular plant species today reach no more than 1 or 2 meters in height.

What is vascular tissue?

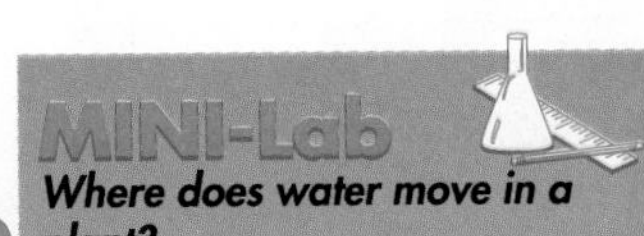

4 **MINI-Lab**

Where does water move in a plant?

Make a clean cut across the bottom of a celery stalk that has leaves. Put the cut end in a glass of water to which red food coloring has been added. Let the stalk stand overnight. *Observe* to see where the colored water has traveled in the plant. **In your Journal,** *relate* the movement of the water through the stalk to vascular tissue and describe where vascular tissue is located.

Figure 10-11. The seedless vascular plants include club mosses, spike mosses, horsetails, and ferns.

MINI-Lab

Materials: water, glass, food coloring, celery stalks, sharp knife, hand lens

Answers: Students may see that the color appears in the veins of the leaves. Journal entries should indicate that the colored water has moved up through the vascular tissues of the plant and that vessels are located in the stalk and throughout the leaves.

MINI-Lab
ASSESSMENT

Performance: Assess students' ability to recognize vascular tissue in other plants by having them identify it in carnations and carrots.

CONCEPT DEVELOPMENT

- Ask what adaptations ferns had that allowed them to surpass the other seedless plants in the variety of forms and habitats.
- Prepare a simple key that can be used to identify the groups of seedless plants. Make copies of the key and have students key seedless plants available in the classroom. Refer to Ch. 7 for information on keys.

OPTIONS

ENRICHMENT

- Have students read natural history books to find out why ferns have such common names as Boston fern, maidenhair fern, staghorn fern, cinnamon fern, hayscented fern, and bracken fern.

PROGRAM RESOURCES

From the **Teacher Resource Package** use:

Activity Worksheets, page 92, Mini-Lab: Where do the seedless vascular plants live?

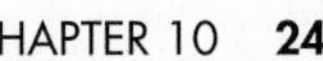

CONCEPT DEVELOPMENT

▶ *Equisetum* is the scientific genus name for the horsetails, or scouring rushes.

▶ If specimens of club moss and horsetails are available, show students the cones that carry spores. Cones in many species are modified leaves.

▶ *Lycopodium* is used in holiday decorations such as wreaths.

▶ Club mosses are evergreen; grow in moist, wooded areas; and generally are less than 30 cm tall.

▶ Modern species of horsetails all belong to one genus, *Equisetum*.

▶ Horsetails are hollow inside.

▶ Horsetails reproduce sexually by alternation of generations. They reproduce asexually by sending up new shoots from underground stems.

▶ There are only 15 species of *Equisetum* today.

MINI QUIZ

Use the Mini Quiz to check students' recall of chapter content.

1. **The rootlike filaments that hold moss plants in place are called _____.** *rhizoids*
2. **The continual cycle that alternates between the sporophyte and gametophyte of the moss plant is called _____.** *alternation of generations*
3. **Plants that are the first to grow in new or disturbed areas are called _____.** *pioneer species*
4. **Club mosses, spike mosses, horsetails, and ferns are all included in the _____ vascular plants.** *seedless*
5. **The leaf of a fern is called a(n) _____.** *frond*

Club Mosses and Spike Mosses

Look at the photographs of club mosses and spike mosses in Figure 10-12. Club mosses produce spores at the end of the stems in structures that look like tiny pine cones. The upright stems of club mosses have needlelike leaves. One kind of club moss, *Lycopodium,* is called the ground pine because it looks like a miniature pine tree. *Lycopodium* is found from arctic regions to the tropics, but never in large numbers. In some areas *Lycopodium* is endangered because of overcollection. Most tropical species of *Lycopodium* are epiphytes. An **epiphyte** is a plant that grows on other plants for support.

Figure 10-12. Club mosses (a) and spike mosses (b) produce spores at the end of stems in structures similar to pine cones. Early photographers often used the dry, flammable spores of club moss as flash powder to provide light to take photographs.

Spike mosses look similar to club mosses. One species of spike moss, the resurrection plant, is adapted to desert conditions. When water isn't available, the plant curls up and looks dead. But when water falls in the desert, the resurrection plant unfurls its green leaves and begins making food again. The plant can go through this process many times if necessary.

Horsetails

Horsetails have a stem structure unique among the vascular plants. Their stems are jointed and have a hollow center surrounded by a ring of vascular tissue. Leaves grow in a spiral around each joint. In Figure 10-13 you can see these joints easily. If you pull on a horsetail stem it will pop apart in sections. Spores grow in a structure at the tip of the stems, as in the club mosses. The stems of the horsetails contain silica, a gritty substance found in sand. In the days when

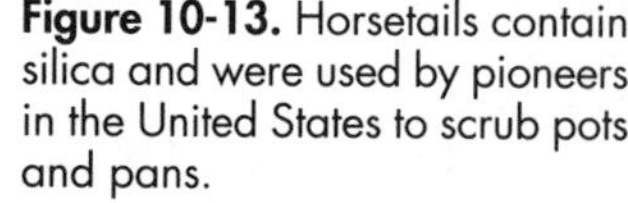

Figure 10-13. Horsetails contain silica and were used by pioneers in the United States to scrub pots and pans.

OPTIONS

ASSESSMENT—ORAL

▶ **What is the advantage to a plant of being an epiphyte?** *Epiphytes depend on other plants or objects for structure and support. Because they grow in a wide range of conditions they are not dependent on any one kind of plant for support.*

▶ **How can you tell the difference between mosses and club mosses?** *Mosses produce spores in capsules on stalks. Club mosses produce spores at the ends of stems in structures that look like tiny pine cones.*

pioneers traveled westward across the United States, horsetail stems were used to scour pots and pans. The common name of *Equisetum* today is still "scouring rush."

Ferns

The largest group of seedless vascular plants is the division Pterophyta (teh ruh FITE uh), the ferns. There are about 12 000 living species of ferns, but like the other divisions of vascular plants, many more species are known only as fossils.

Figure 10-14. The unusual cinnamon fern bears its spores on separate stalks.

During the Carboniferous period of the Paleozoic era, most ferns grew much larger than any fern species alive at present. Species of tree ferns found today in tropical areas may reach 3 to 5 meters in height, but ancient tree ferns living in the swamps of the Carboniferous period grew as high as 25 meters. When they died, tree ferns and other plant species became submerged in water and mud before they could decompose. This plant material built up, became compacted and compressed, and eventually became coal. Today, a similar process is taking place in peat bogs. Coal and peat are two types of fuel used in many parts of the world today for energy. Coal, peat, petroleum, and natural gas are all known as fossil fuels because they were formed from the plants and other organisms of the Carboniferous period.

Peat is an organic fuel found in Earth at the surface and to depths less than a hundred feet. Coal is found at a depth of a few hundred to a thousand feet. Oil is found from a thousand feet to seven thousand feet. Gas is found in the same depths as oil, but also to a depth of ten thousand feet. *Make a bar graph* to illustrate the depths at which each of these organic fuels is found.

Figure 10-15. The Carboniferous swamp forest contained many more species of club mosses, spike mosses, horsetails, and ferns than are alive today.

TEACHER F.Y.I.

▶ At least 12 000 species of ferns have been identified. Ferns may be as small as a tiny water fern or large as trees.

Answer: The bar graph will have the names of the fuels across the x-axis and the depths along the y-axis.

To assess this product, refer to the Performance Task Assessment Lists in **Performance Assessment in Middle School Science.**

CONCEPT DEVELOPMENT

▶ Cross sections of fern stems show a ring of xylem tissue, with phloem on the inside and outside of the xylem.
▶ The asparagus fern is not a fern, nor is it an asparagus!
▶ Fern fronds have fewer layers of tissue than leaves of flowering plants. They do not have stomata, so water can evaporate from the whole surface of the fern frond.
▶ Fronds consist of leaflets called *pinnae*.
▶ The dry, fibrous parts of some ferns have been used as mattress stuffing.
▶ Orchids are sometimes grown on a base of ferns. Some orchids are epiphytes.
▶ Young fern fronds called fiddleheads are used for food in Asia and elsewhere.
▶ Fern spores usually are shed in the autumn, but do not germinate into sporophytes until the warmer weather of spring.
▶ Display different species of ferns. Have students discuss the differences between ferns and liverworts and the plantlike protists.

ASSESSMENT—ORAL

▶ **What are the four fossil fuels used for energy worldwide?** *coal, peat, natural gas, petroleum*
▶ **Why isn't wood a fossil fuel?** *Woody plants did not exist during the Carboniferous period when fossil fuels were formed.*

CHECK FOR UNDERSTANDING

Provide each student with a list of terms that can be used in an explanation of moss and fern life cycles. Have students write paragraphs that compare and contrast these two life cycles.

RETEACH

Give pairs of students drawings of the stages of moss and fern life cycles on 3" × 5" notecards. Have students arrange the cards and label the drawings.

EXTENSION

For students who have mastered this section, use the **Reinforcement** and **Enrichment** masters or other OPTIONS provided.

3 CLOSE

▶ Ask questions 1-3 and the **Apply** Question in the Section Review.

▶ Ask students: **Why are ferns classified as vascular rather than nonvascular plants?** *because they have vascular tissue*

The Fern Life Cycle

In the life cycle of a fern, the gametophyte and sporophyte do not depend on each other. This means that both forms of the fern plant can produce their own food. The fern plants that you may have seen growing in the woods or in an arrangement from a florist are the sporophytes. A fern has leaves that grow above the ground from an underground stem called a **rhizome.** Roots grow from the rhizome to anchor the plant in the soil. The leaf of a fern is called a **frond.** Young fronds are often called fiddleheads because their curled shape looks similar to the tops of violins, or fiddles. Ferns produce spores in structures called **sori** (*sing.* sorus) on the lower sides of the mature fronds. The sori usually look like crusty rust, brown, or blackish-colored bumps. When a sorus opens, spores are released at incredible speeds. They land on damp soil or rocks and germinate into small, green, heart-shaped gametophyte plants. The gametophyte is called a **prothallus.** The prothallus produces sex cells that unite to form the zygote. The zygote develops into the embryo, then sporophyte, as in the mosses. But in ferns, the gametophyte also produces new rhizomes, which can grow into the separate sporophyte fern plants you are familiar with. You can see the life cycle of a fern in Figure 10-16.

What is a frond?

5

Ferns are interesting to scientists because they have characteristics of both nonvascular and vascular plants. Like the bryophytes, ferns produce spores, yet they have

Figure 10-16. The Fern Life Cycle. Compare the position of the spore-producing structures on this fern with the cinnamon fern on page 243.

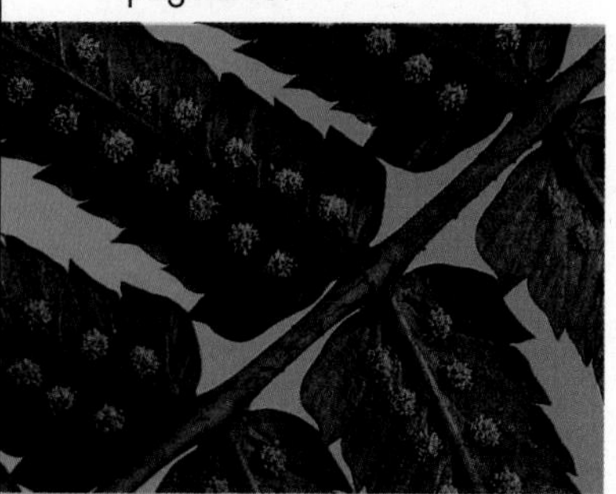

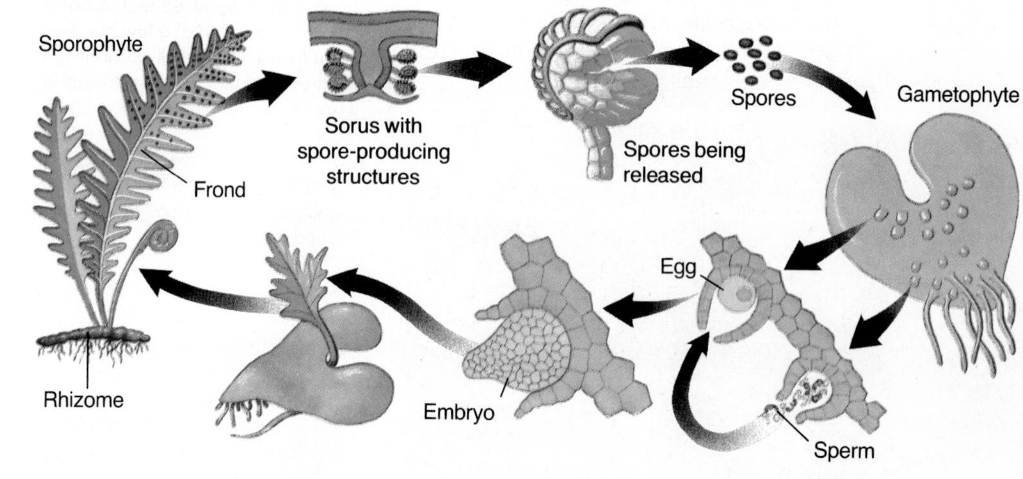

OPTIONS

ENRICHMENT

▶ Have students research the conditions under which sori break open to release spores.

PROGRAM RESOURCES

From the **Teacher Resource Package** use:

Transparency Masters, pages 33-34, Fern Life Cycle.

Use **Color Transparency** 17, Fern Life Cycle.

PROBLEM SOLVING

What is in Nature's Medicine Chest?

People in all cultures have found ways to treat illnesses, and many of these cures involve plants. Various native American cultures used willow bark to cure headaches. Heart problems were treated with foxglove in England, sea onions in Egypt, and seeds in Africa. In Peru, the bark of the cinchona tree was used to treat malaria. Scientists have found that many native cures are medically sound. Willow bark contains salicylates, the main ingredient in aspirin. Foxglove, sea onions, and the African seeds all contain digitalis or similar drugs. Cinchona bark contains quinine, an antimalarial drug.

Think Critically: How might the destruction of rain forests affect research for new drugs from plants?

vascular tissue like seed-bearing plants. Ferns have leaves (the fronds), stems (the rhizomes), and roots. If you were a scientist, how would you classify ferns?

SECTION REVIEW

1. Describe the life cycle of mosses.
2. How are club and spike mosses similar to mosses?
3. Explain the stages in the life cycle of a fern.
4. **Apply:** List ways simple plants affect your life each day. (HINT: Where do electricity and heat for homes come from?)
5. **Connect to Earth Science:** Plants are made up of materials supplied from soil. Horsetails contain large amounts of silica. What can you infer about the source of silica in horsetails?

Skill Builder

Concept Mapping

Make a concept map showing how nonvascular and vascular plants are related. Include these terms in the concept map: *plant kingdom, bryophytes, liverworts, mosses, seedless plants, nonvascular plants, seedless vascular plants, club mosses, ferns, horsetails,* and *spike mosses.* If you need help, refer to Concept Mapping in the **Skill Handbook** on pages 688 and 689.

PROBLEM SOLVING

Think Critically: Plants containing substances that might benefit health might be destroyed and the species lost.

SECTION REVIEW ANSWERS

1. Spores land on the ground and produce the upright, leafy gametophytes. Gametes are produced at the tops of the gametophytes. Sperm swim in water to reach the female gametes. When united, the gametes produce a zygote. The zygote grows into a spore-producing sporophyte that is dependent on the gametophyte.

2. Both club mosses and spike mosses produce spores and reproduce through alternation of generations.

3. A spore produces a heart-shaped gametophyte. The gametophyte produces gametes that unite to form a zygote. A zygote then produces an embryo that grows into a sporophyte.

4. Apply: Simple plants such as mosses and ferns formed fossil fuels. These fuels are burned to produce electricity for homes, to power automobiles, and to provide heat.

5. Connect to Earth Science: Silica is found in soil.

Skillbuilder

ASSESSMENT

Portfolio: Use this Skillbuilder to assess students' abilities to form a Concept Map of the plant kingdom.

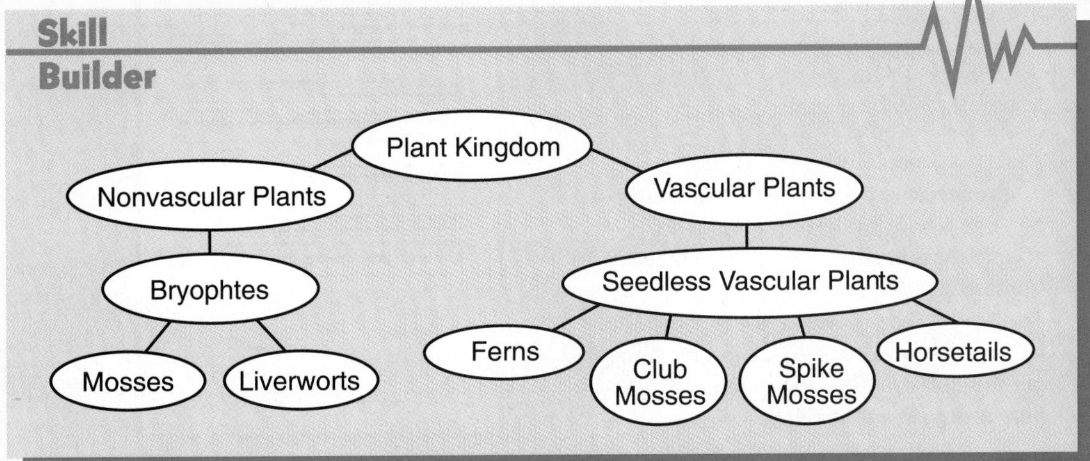

SECTION 10-3

PREPARATION

SECTION BACKGROUND

▶ Quaking bogs form a dense mat of material above water. The mat may be thick enough to walk on in some places, but not in others.

▶ Peat moss is valuable as a water-absorbing material in which horticulturists plant orchids and other plants that need lots of water.

1 MOTIVATE

▶ Ask students what kinds of materials people use for fuel. Emphasize that peat is a source of fuel similar in many ways to coal, oil, and gasoline.

MINI-Lab

Materials: *Sphagnum* moss, cheesecloth, beakers, water, graduated cylinders, balance

Answer: *Sphagnum* moss will soak up about 100 mL of water.

MINI-Lab ASSESSMENT

Performance: To further assess students' understanding of the absorption capability of *Sphagnum* moss, see USING LAB SKILLS, question 12, on p. 250.

PROGRAM RESOURCES

From the **Teacher Resource Package** use:

Activity Worksheets, page 93, MINI-Lab: Does moss hold much water?

Science and Society, page 14, Should Forest Fires Be Prevented in Our National Parks?

Cross-Curricular Connections, page 14, The First Disposable Diaper.

SCIENCE & SOCIETY 10-3 Peat Moss as Fuel

New Science Words

bogs

Objectives

- ▶ State two advantages of using peat as fuel.
- ▶ Understand the environmental cost of using peat as fuel.

What's In a Bog?

What is a bog?

Bogs are very mysterious places. Human bodies 7000 years old have been recovered from these wet areas all over the world. Sound like the beginning of a horror movie? What exactly is a bog? **Bogs** are poorly drained areas with spongy, wet ground that is composed mainly of dead and decaying plants. The principal plants in bogs are nonvascular plants that belong to the genus *Sphagnum. Sphagnum* moss forms large mats on top of the water surface in bogs. The water in a bog is very acidic with little oxygen. These conditions slow down or stop the growth of the bacteria that would normally decompose dead organic matter. As a result, moss and other plants that die in bogs don't decay. Dead moss builds up in the bog and is compressed by the weight of the other plants and water above. Gradually the moss becomes peat, a combination of partially decaying sphagnum moss and other organisms. Peat has been used as a low-cost fuel for centuries. Blocks of black peat can be cut out of the bog, dried in the sun, then burned in a stove or furnace to produce heat. Peat is actually the first step in the geological process of turning plant material into coal. Lignite coal is formed when peat has been compressed and heated by the geologic force of Earth.

MINI-Lab

Does moss hold much water?

Place a few teaspoons of *Sphagnum* moss on a piece of cheesecloth. Twist the cheesecloth to form a ball and tie it securely. Weigh the ball. Put 200 mL of water in a beaker. Add the ball. *Predict* how much water the ball will absorb. Wait 15 minutes; remove the ball, draining off excess water. Weigh the ball and *measure* the amount of water left in the beaker. *Calculate* how much water the *Sphagnum* moss soaked up.

Peat has been used as a fuel for many years in Northern Europe, England, and Ireland. It is very cheap and burns

OPTIONS

Meeting Different Ability Levels

For Section 10-3, use the following **Teacher Resource Masters** depending upon individual students' needs.

◆ **Study Guide Master** for all students.

● **Reinforcement Master** for students of average and above average ability levels.

▲ **Enrichment Master** for above average students.

Additional Teacher Resource Package masters are listed in the OPTIONS box throughout the section. The additional masters are appropriate for all students.

◆ STUDY GUIDE 42

NAME ______ DATE ______ CLASS ______

STUDY GUIDE — Chapter 10

Peat Moss as Fuel — Text Pages 246–247

Answer the following questions.

1. What are bogs? **Bogs are low-lying areas with spongy, wet ground, composed mainly of dead and decaying plants.**
2. What is peat? **Peat is a combination of partially decaying *Sphagnum* moss and other plants.**
3. How is coal formed? **Coal is formed when partially decayed vegetation, such as peat, is compressed for millions of years.**
4. Why might peat be a good source of fuel for New England? **It might be a good source because there is a large quantity in Maine and because coal, oil, and gas are not found in that area.**
5. What are two consequences to the environment if peat is burned as fuel? **Burning peat emits carbon dioxide and sulfur dioxide. These contribute to air pollution.**
6. How would using peat as fuel in the United States affect the bogs? **Using peat will deplete the bogs. Bogs serve as a home for certain plants and animals that can live nowhere else. Because they hold large amounts of water, bogs also help control flooding. They also filter impurities out of water before it reaches lakes and streams. Losing the bogs would take away all of these advantages.**

42

slowly for long periods of time. After it is dried, the most decomposed peat burns with greater fuel efficiency than wood. Developing countries such as Burundi and Rwanda are looking into the use of peat as an alternative to importing expensive petroleum products. In the United States, geologists estimate that Maine has 300 000 hectares of peat available to use for fuel, whereas Minnesota has about three million hectares. In Maine, America's first peat-fueled power plant is already supplying energy to Boston. This is very economical for the New England area because oil, gas, and coal resources are all very far away.

Have we found the perfect source of fuel? Environmentalists don't think so. They fear that the bogs of the United States are headed toward the same fate as those in Ireland. Today, Ireland generates 21 percent of its electric power with peat. However, in the last 40 years, the Irish have destroyed 95 percent of their bogs. These bogs took nature 10 000 years to build! Why save bogs? They are home to animals and plants that can live no other place in the world. Bogs help control flooding by holding huge amounts of water. Peat moss can hold water in amounts up to 20 times its own dry weight. Bogs also help keep water resources such as rivers and lakes clean by filtering out silt and chemicals. Although peat is a cheap source of energy, it is a nonrenewable fossil fuel, just like coal, petroleum, and natural gas. When it burns, it releases chemicals such as carbon dioxide, sulfur dioxide, and nitrogen oxides, all sources of air pollution.

Figure 10-17. Unique plants of the peat bog include: (a) arethusa, (b) bog rosemary, (c)leather leaf, (d) pitcher plant, (e) false Solomon's seal, (f) sundew, (g) dewthread, (h) cranberry, (i) Labrador tea, and (j) *Sphagnum* moss.

SECTION REVIEW

1. Why is peat considered a good alternative fuel?
2. Why are environmentalists concerned with its use?
3. **Connect to Earth Science:** Do you think *Sphagnum* in a bog has any potential anti-infection properties?

You Decide!

The state of Maine has certain environmental standards for peat harvesters. Maine peat has little sulfur and emits little sulfur dioxide. Excess water is captured in ponds to prevent runoff, and vegetation is to be restored to the area after the harvest. Are these sufficient trade-offs for losing bogs that took thousands of years to form?

2 TEACH

Key Concepts are highlighted.

CONCEPT DEVELOPMENT

MULTICULTURAL AWARENESS

Have students consider the use and availability of fossil fuels to different cultures and our own.

▶ **Demonstration:** Use a world map to illustrate the locations of major bogs in the world.

CHECK FOR UNDERSTANDING

Ask students to give their opinions about the use of peat as a fuel source.

RETEACH

Have students make comparative drawings as they imagine a peat bog before and after harvesting of peat.

EXTENSION

For students who have mastered this section, use the **Reinforcement** and **Enrichment** masters or other OPTIONS provided.

3 CLOSE

▶ Ask questions 1-2 in the Section Review.

SECTION REVIEW ANSWERS

1. Peat is cheap and burns slowly and efficiently.

2. Destruction of bogs may destroy habitats for plants and animals, as well as pollute water and air resources.

3. Connect to Earth Science: Possible anti-bacterial qualities; if it doesn't decay easily or at all, then bacteria can't grow.

YOU DECIDE!

Answers will vary. All answers should be supported.

VideoDisc

STVS: Oil from Wood, Disc 4, Side 2

● REINFORCEMENT 42

NAME ______ DATE ______ CLASS ______

REINFORCEMENT — Chapter 10

Peat Moss as Fuel — Text Pages 246–247

Part A

Number the following steps to show the sequence of the formation of coal. You may use your textbook and other resources.

4 a. mass of layers of dead vegetation and rock causes pressure
5 b. formation of lignite
3 c. formation of peat
2 d. high acidity and low oxygen level prevent growth of bacteria
1 e. growth of *Sphagnum* moss and other plants

Part B

1. Why are well-preserved fossils often found in bogs? **The high acid content and low oxygen level prevent growth of large amounts of bacteria which normally break down dead organisms.**
2. Although peat forms over a period of years, it takes from 1 to 400 million years for coal to form. Do you think it might be better to use peat rather than coal as a fuel? Explain your answer. **Answers will vary. Although coal is a nonrenewable resource and burning coal gives off pollution, there are also problems with the use of peat. The main problems are the destruction of bogs and air pollution.**
3. What are some of the reasons for saving bogs? **Bogs are the home for animals and plants that can live nowhere else. They help control flooding and they keep rivers clean by filtering out silt and chemicals.**

42

▲ ENRICHMENT 42

NAME ______ DATE ______ CLASS ______

ENRICHMENT — Chapter 10

Peat Moss as Fuel — Text Pages 246–247

ACID RAIN

Sulfur dioxide is one gas given off by burning peat. It can have harmful effects on people and the environment. Released into the air we breathe, sulfur dioxide can irritate the eyes and respiratory system. It may also dissolve in water droplets to form acid rain, which can damage buildings and harm or even kill wildlife.

You can see for yourself the effects of acid rain on plants.

Materials

- 3 1-quart jars
- measuring cups
- 3 small houseplants or 3 cuttings
- vinegar or lemon juice
- masking tape
- marking pen

Procedure

1. Make 3 labels with the masking tape. Label one plant "tap water," label the second plant "diluted acid," and the third "concentrated acid."
2. Measure 1/4 cup of vinegar or lemon juice (both are acidic) into one jar. Fill the rest of the jar with tap water. Label this jar "diluted acid." Water the plant labeled "diluted acid" with this solution.
3. Measure 1 cup of vinegar or lemon juice into another jar. Label this jar "concentrated acid." Water the plant labeled "concentrated acid" with this solution.
4. Fill the third jar with tap water. Water the plant labeled "tap water" with this.
5. Provide your plants with the same amount of sunlight, and water them every 2-4 days.
6. Record your observations in the table below.

Data and Observations

Plant	Appearance			
	Day 1	Day 3	Day 6	Day 9
diluted acid				
concentrated acid				
tap water				

Conclude and Apply

1. How did the appearance of the plants watered with the acid solutions change? **The color faded and the leaves wilted.**
2. How did the amount of acid in the plant water relate to the growth of the plant? **The more acid in the plant water, the sooner the plant died.**

42

ACTIVITY 10-2
30 minutes

OBJECTIVE: Students will **observe** the stages in the life cycle of a fern.

PROCESS SKILLS applied in this activity:
▶ **Observing** in Procedure Steps 3-6.
▶ **Classifying** in Analyze Questions 3 and 4.
▶ **Compare and contrast** in Conclude and Apply Questions 5 and 6.

TEACHING THE ACTIVITY

Alternate Materials: Grow fern gametophytes by sprinkling the spores on a clay pot turned bottom up. First soak the pot overnight. Keep the pot in a plastic bag with water in the bottom to maintain a high humidity. Materials may be purchased from a biological supply house for little cost.

Troubleshooting: Sometimes mold will overtake the fern gametophyte stages. Cultures should be discarded. Have additional specimens ready.

ANSWERS TO QUESTIONS

1. sporophyte
2. gametophyte
3. fronds
4. rhizome
5. Rhizomes are underground stems with true roots attached. Rhizoids are rootlike structures that lack vascular tissue.
6. The sporophyte is large and has true roots, stems, and leaves. The gametophyte is small and lacks roots or stems. Both are green and independent.
7. Powerful projectiles send spores in all directions, increasing their chances of starting new plants.

Activity
ASSESSMENT

Performance: To further assess students' abilities to observe features of a fern life cycle, have them point out individual parts that you ask for.

PROGRAM RESOURCES

From the **Teacher Resource Package** use:

Activity Worksheets, pages 88-89, 10-2 The Life Cycle of a Fern.

ACTIVITY 10-2 The Life Cycle of a Fern

Problem: *How can stages in the life cycle of a fern be observed?*

Materials  

- fern
- fern frond with spores
- fern spores
- hand lens
- petri dish
- paper towel
- scissors
- water
- prepared slides of fern
- microscope

Procedure

1. Cut a piece of paper towel into a circle that will just fit into the bottom of the petri dish.
2. Soak the paper towel circle with water. Use your fingernail to gently scrape away the protective membrane of the sori on a fern frond. Tap the sori to dislodge the spores.
3. Sprinkle a few fern spores on the paper towel and cover the petri dish. *Observe* the spores in the petri dish with a hand lens once each week for 6 weeks. Add water if necessary.
4. Make a data table like the one shown. Draw and label what you see in the petri dish each week. Write your observations in the table.
5. *Observe* the fronds of the fern plant for evidence that it is a sporophyte.
6. Use the microscope to look at prepared slides of the fern life cycle under low and high power.

Analyze

1. In which stage in the life cycle of a fern are spores formed?
2. In which stage are sex organs formed?
3. What is the name given to fern leaves?
4. What is the name of the underground stem?

Data and Observations Sample Data

Week	Fern Gametophyte	Fern Sporophyte
1		
2	Student drawings will vary but	
3	should resemble fern life cycle	
4	stages shown in Figure 10–16.	
5		
6		

Conclude and Apply

5. *Compare and contrast* rhizoids and rhizomes.
6. *Compare* the sporophyte with the gametophyte of a fern.
7. What advantage is there to having fern spores shot out of the sori?

CHAPTER
REVIEW

SUMMARY

10-1: Characteristics of Plants

1. Plants are many-celled eukaryotes adapted to life on land.

2. Plants evolved from the plantlike protists.

3. Simple plants include nonvascular plants, the bryophytes, and vascular plants with vessels that conduct materials to plant cells.

10-2: Seedless Plants

1. Bryophytes—mosses and liverworts—reproduce through the process of alternation of generations, as do all plants.

2. Ferns, club mosses, and horsetails are seedless vascular plants with vascular tissue.

3. Nonvascular plants are pioneer species that help to break down rock into soil for other plants. Simple vascular plants have characteristics of both nonvascular and vascular plants.

10-3: Science and Society: Peat Moss as Fuel

1. Peat is a cheap fuel that burns slowly for long periods of time. It is available in places that are far away from oil, gas, or coal resources.

2. Bogs are important homes for unique organisms, hold water to prevent flooding, and help filter out silt and chemicals in water.

KEY SCIENCE WORDS

a. **alternation of generations**
b. **bogs**
c. **cellulose**
d. **cuticle**
e. **epiphyte**
f. **frond**
g. **gametophyte**
h. **nonvascular plants**
i. **pioneer species**
j. **prothallus**
k. **rhizoids**
l. **rhizome**
m. **sori**
n. **sporophyte**
o. **vascular plants**

UNDERSTANDING VOCABULARY

Match each phrase with the correct term from the list of Key Science Words.

1. plant that produces sex cells
2. wet ground made of dead, non-decaying plants
3. plant that produces spores
4. rootlike filaments
5. underground fern stem
6. spore-producing capsules
7. plants containing tube-like conducting cells
8. plant group including mosses and liverworts
9. the first plants to grow in new environments
10. waxy layer on stems and leaves

OPTIONS

ASSESSMENT

To assess student understanding of material in this chapter, use the resources listed.

Cooperative Learning

Consider using cooperative learning in the THINK AND WRITE CRITICALLY, APPLY, and MORE SKILL BUILDERS sections of the Chapter Review.

PROGRAM RESOURCES

From the **Teacher Resource Package** use:

Chapter Review, pages 23-24.

Chapter and Unit Tests, pages 68-71, Chapter Test.

CHAPTER
REVIEW

SUMMARY

Have students read the summary statements to review the major concepts of the chapter.

UNDERSTANDING VOCABULARY

1. g
2. b
3. n
4. k
5. l
6. m
7. o
8. h
9. i
10. d

ASSESSMENT

Portfolio

Encourage students to place in their portfolios one or two items of what they consider to be their best work. For each item, ask students to explain why that item was chosen and what they learned from it. Items might be selected from the following.

- Skillbuilder hypothesis on plant adaptations, p. 235
- In Your Journal poem or memory code for remembering the life cycles of mosses, p. 238
- You Decide! issue controversy about bogs in Maine, p. 247

Performance

Additional performance assessments may be found in *Performance Assessment* and *Science Integration Activities* that accompany **Merrill Life Science.** Performance Task Assessment Lists and rubrics for evaluating these activities and other products generated throughout the chapter can be found in Glencoe's *Performance Assessment in Middle School Science.*

CHAPTER
REVIEW

CHECKING CONCEPTS

1. d
2. b
3. a
4. d
5. b
6. d
7. a
8. c
9. b
10. d

USING LAB SKILLS

ASSESSMENT

Use these alternate lab exercises to assess students' understanding of skills used in this chapter.

11. Student plans will vary considerably. Student records should show a plan for the experiment, materials, and how they would time the activity. Students will need to set time intervals as part of their expermental plan. One suggestion is that they put at least three celery stalks in colored water all at the same time. Then, at a given interval, one stalk can be removed and cut at a particular distance from the bottom of the stalk to see if the color has moved that far up the stalk. One stalk should be left and the time checked when color first appears in the leaves.

12. Data would be best illustrated with a bar graph.

THINK AND WRITE CRITICALLY

13. Alternation of generations appears to be an adaptation, assures production of new plants.

14. The sporophyte depends on the food produced by the gametophyte.

15. Without a cuticle, land plant cells lose too much water and die.

16. Nonvascular plant cells absorb water directly from their surroundings. Vascular plant cells get water through vascular tissue.

CHAPTER
REVIEW

CHECKING CONCEPTS

Choose the word or phrase that completes the sentence.

1. Roots, cell walls of cellulose, and having many cells are examples of _______ characteristics.
   a. moneran c. animal
   b. protist d. plant
2. Plants probably evolved from _______.
   a. animals c. fungi
   b. protists d. monerans
3. _______ are plants only a few cells thick.
   a. Bryophytes c. Horsetails
   b. Ferns d. Vascular
4. _______ plants are plants with vessels.
   a. Bryophyte c. Nonvascular
   b. Moss d. Vascular
5. Nonvascular plants do not have _______.
   a. zygotes c. rhizoids
   b. rhizomes d. stems
6. Plants without vascular tissues are _______.
   a. ferns c. moss
   b. liverworts d. nonvascular
7. Ferns reproduce by forming _______.
   a. spores c. seeds
   b. vascular tissue d. flowers
8. _______ are seedless vascular plants.
   a. Mosses c. Horsetails
   b. Liverworts d. Oaks
9. Of these, ferns do not have _______.
   a. fronds c. rhizomes
   b. rhizoids d. spores
10. All plants have a life cycle with _______.
    a. seeds c. flowers
    b. fruits d. alternation of generations

USING LAB SKILLS

11. Design an experiment using celery stalks with leaves, a sharp knife, water with food coloring, and a timer to determine the rate at which water moves up through vessels. Describe and record your experiment in your Journal.
12. Graph the data from the MINI-Lab on page 246 to compare the weight of the mass of Sphagnum before and after water is absorbed, or the amount of water in the beaker or jar before and after.

THINK AND WRITE CRITICALLY

Answer the following questions in your Journal using complete sentences.

13. Why is it an advantage for simple plants to alternate generations?
14. Why are sporophytes dependent on the gametophytes in bryophytes?
15. What would happen if a land plant's waxy cuticle was destroyed?
16. Compare how the cells of nonvascular and vascular plants obtain water.

APPLY

17. Human remains sometimes found in peat bogs are very well-preserved. Explain why this occurs.
18. Discuss the importance of water in reproduction of bryophytes and ferns.
19. Why is *Sphagnum* moss added to garden soil?
20. Explain why mosses are found on moist forest floors.
21. How do mosses change rocky areas so that other plants can take root?

MORE SKILL BUILDERS

If you need help, refer to the Skill Handbook.

1. **Concept Mapping:** Fill in the map showing the divisions of the Plant Kingdom.

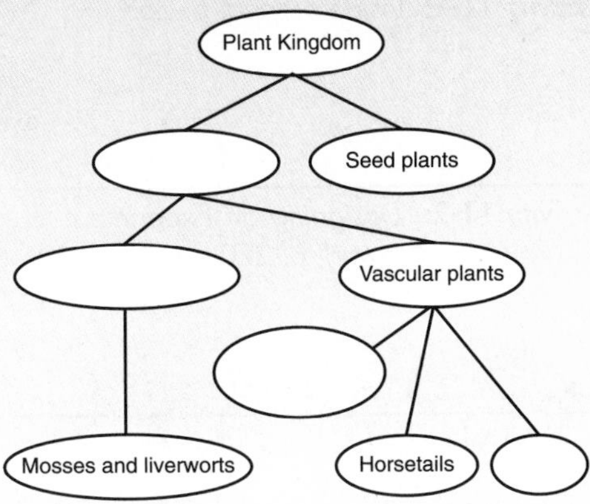

2. **Classifying:** Classify these plants in the correct group, bryophytes or seedless vascular: 1. club moss, 2. fern, 3. liverwort, 4. *Equisetum*, 5. horsetail, 6. *Lycopodium*, 7. moss, 8. spikemoss, 9. *Sphagnum*, 10. peat moss
3. **Sequencing:** Put the steps of the moss life cycle in sequence starting with the sporophyte stage.
4. **Making and Using Graphs:** Make a pie graph using the data below.

Simple Plants	Number of Species
Bryophytes	20 000
Club mosses, spike mosses, and horsetails	1000
Ferns	12 000
Total	33 000

PROJECTS

1. Develop a model of how coal is formed. Include a time line and a map of the parts of the U.S. where coal mines are located. Obtain samples of "soft" and "hard" coal and describe differences in their composition, origin, and formation.
2. Go for a walk in the woods and see how many different kinds of mosses there are. List the locations and conditions under which each is found growing.

APPLY

17. Bogs are acidic; this kills the bacteria that normally cause decay.
18. Male gametes need water because they swim to female gametes during reproduction in simple plants.
19. *Sphagnum* moss retains water, thus allowing soil to hold water for plants growing in the garden soil.
20. Mosses absorb water directly into cells, so they must grow where it is wet.
21. Mosses begin to break down rocks, releasing the minerals that other plants need. These plants can then begin to grow on this new soil.

MORE SKILL BUILDERS

1. **Concept Mapping:** See below for possible solution.
2. **Classifying:** Bryophytes: 3, 7, 9, 10; Seedless vascular: 1, 2, 4, 5, 6, 8
3. **Sequencing:** Sporophyte produces haploid spores.
Spores germinate into threadlike green structures.
Gametophyte grows.
Sex cells are produced.
Sex cells unite.
Zygote forms.
Embryo develops.
Embryo develops into sporophyte.
4. **Making and Using Graphs:** See below.

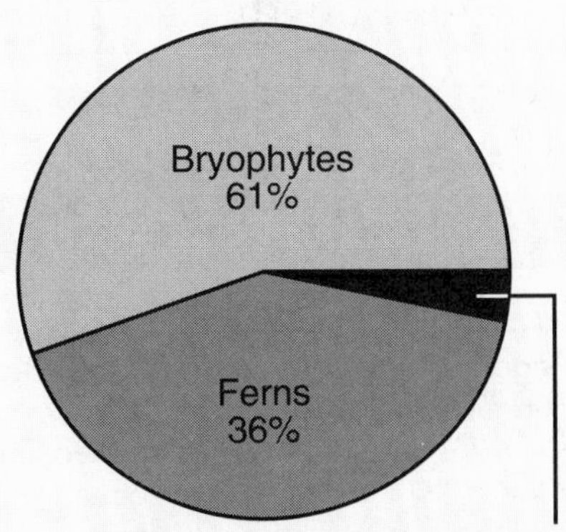

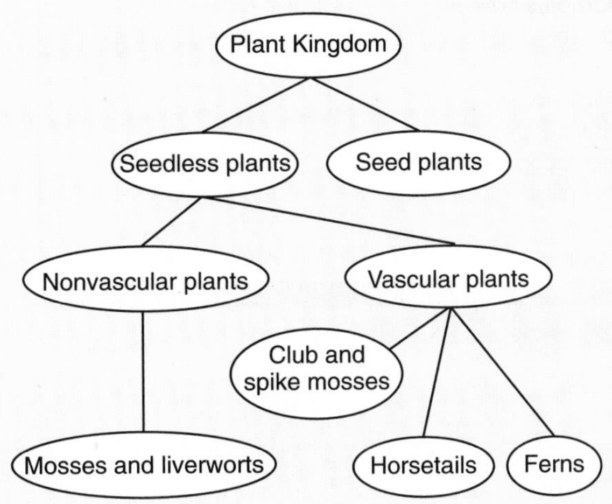

CHAPTER

11 The Seed Plants

CHAPTER SECTION	OBJECTIVES	ACTIVITIES
11-1 The Seed Plants (2 days)	1. **List** the main characteristics of seed plants. 2. **Describe** the main characteristics of gymnosperms and angiosperms and their importance. 3. **Compare** the evolution of gymnosperms and angiosperms.	
11-2 Parts of Complex Plants (1 day)	1. **Describe** the structures of roots, stems, and leaves. 2. **Describe** the functions of roots, stems, and leaves. 3. **Explain** some unique adaptations of roots, stems, and leaves.	
11-3 Seed Plant Reproduction (2 days)	1. **Sequence** the stages in the life cycles of typical gymnosperms and angiosperms. 2. **Describe** the structure and function of the flower. 3. **Describe** methods of seed dispersal in seed plants.	**MINI-Lab:** *What's in a pine cone?* p. 263 **MINI-Lab:** *How do plants disperse seeds?* p. 265 **Activity 11-1:** *Inside a Seed,* p. 267
11-4 Effects of Acid Rain Science & Society (1 day)	1. **State** the primary causes of acid rain. 2. **Explain** why the solution to this problem requires international cooperation.	**Activity 11-2:** *Designing an Experiment (Parts of a Flower),* p. 270
Chapter Review		

ACTIVITY MATERIALS

FIND OUT	ACTIVITIES		MINI-LABS	
Page 253 peanut in its shell	**11-1 Inside a Seed, p. 267** soaked lima bean soaked seed corn forceps scalpel iodine solution hand lens	**11-2 Designing an Experiment, p. 270** gladiolus flower scalpel hand lens black paper microscope microscope slide coverslip dropper water	**What's in a pine cone? p. 263** hand lens gymnosperm cone towel	**How do plants disperse seeds? p. 265** picture of types of seeds including: coconuts, grasses, burrs, winged seeds

CHAPTER FEATURES	TEACHER RESOURCE PACKAGE	OTHER RESOURCES
Skill Builder: *Comparing and Contrasting,* p. 258	**Ability Level Worksheets** ◆ ***Study Guide,*** p. 43 ● ***Reinforcement,*** p. 43 ▲ ***Enrichment,*** p. 43 **Transparency Masters,** pp. 35-36	**Color Transparency 18,** Monocots and Dicots **STVS:** Disc 4, Side 2
Technology: *Plants in Space?* p. 260 **Skill Builder:** *Classifying,* p. 261	**Ability Level Worksheets** ◆ ***Study Guide,*** p. 44 ● ***Reinforcement,*** p. 44 ▲ ***Enrichment,*** p. 44	**STVS:** Disc 4, Side 2
Problem Solving: *How Can You Tell if Seeds Are Living?* p. 265 **Skill Builder:** *Hypothesizing,* p. 266	**Ability Level Worksheets** ◆ ***Study Guide,*** p. 45 ● ***Reinforcement,*** p. 45 ▲ ***Enrichment,*** p. 45 **Activity Worksheets,** pp. 5, 95-96, 101, 102 **Cross-Curricular Connections,** p. 15 **Concept Mapping,** p. 27 **Critical Thinking/Problem Solving,** p. 15 **Transparency Masters,** pp. 37-38	**Color Transparency 19,** Parts of a Flower **STVS:** Disc 4, Side 2
You Decide! p. 269	**Ability Level Worksheets** ◆ ***Study Guide,*** p. 46 ● ***Reinforcement,*** p. 46 ▲ ***Enrichment,*** p. 46 **Activity Worksheets,** pp. 97-98 **Science and Society,** p. 15	**STVS:** Disc 6, Side 2 **Science Integration Activity 11**
Summary Key Science Words Understanding Vocabulary Checking Concepts Using Lab Skills Think & Write Critically Apply More Skill Builders Projects	ASSESSMENT RESOURCES **Chapter Review,** pp. 25-26 **Chapter Test,** pp. 72-75 **Performance Assessment in Middle School Science (PAMSS)**	**Chapter Review Software** **Test Bank** **Alternate Assessment** **Performance Assessment**

◆ **Basic** ● **Average** ▲ **Advanced**

ADDITIONAL MATERIALS

SOFTWARE	AUDIOVISUAL	BOOKS/MAGAZINES
Plant Growth, Classroom Consortia Media. *Reproduction in Plants,* J & S Software. *Leaf: Structure and Function,* Classroom Consortia Media. *Leaf: Structure and Physiology,* IBM. *Pollination and Fertilization: Seeds, Fruits, and Embryos,* IBM.	*Flowering Plants and Their Parts,* film, EBEC. *Flowers at Work,* (2nd ed.), film, EBEC. *Flowers: Structure and Function,* film, Coronet/MTI. *Growth of Flowers* (2nd ed.), film, Coronet/MTI. *How Seeds Are Made,* film, International Film Bureau. *Plants That Grow From Leaves, Stems, and Roots* (revised), film, Coronet/MTI. *Pollination,* film, National Geographic. *Seeds: How They Germinate,* film, Coronet/MTI.	Dowden, Anne O. *From Flower to Fruit.* NY: Thomas Y. Crowell Company, 1984. Fahn, A. *Plant Anatomy.* 3rd ed. Elmsford, NY: Pergamon Printers Inc., 1982. Hamilton, David. *Flowers.* NY: Arcade Publishing, Inc., 1990.

CHAPTER

11

THE SEED PLANTS

THEME DEVELOPMENT: This chapter develops the ecology theme by emphasizing the roles gymnosperms and angiosperms play in the environment. In addition, the evolutionary theme is used in a discussion of the phylogeny of these plants.

CHAPTER OVERVIEW

- **Section 11-1:** The characteristics of seed plants, gymnosperms, and angiosperms are listed in this section. The importance of the gymnosperms and angiosperms is discussed. A comparison of evolution of gymnosperms and angiosperms is given.
- **Section 11-2:** This section describes the structure and function of roots, stems, and leaves. Some of the unique adaptations of these plant parts are explained.
- **Section 11-3:** The life cycle of gymnosperms and angiosperms is sequenced in this section. The structure and function of the flower is described, and methods of seed dispersal are given.
- **Section 11-4: Science and Society:** The causes of acid precipitation are discussed. International cooperation to find a solution to the problem is explored.

CHAPTER VOCABULARY

gymnosperms
angiosperms
monocots
dicots
xylem
phloem
cambium
stomata
guard cells
ovules
pollen grains
stamen
pistil
ovary
pollination
acid rain

CHAPTER

11 The Seed Plants

252

OPTIONS

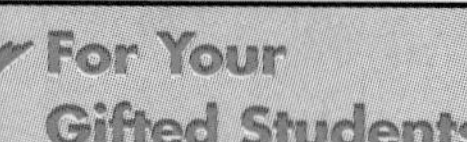

For Your Gifted Students

- Students can record the development of a plant by taking video shots or photos each day.
- Have students research the cactus plant to find out about its unique water storage system.
- Using various types of pine cones, students can create original art designs or pictures. Students can dry flowers in silica gel to add to their artwork.

For Your Mainstreamed Students

- Have students dissect different parts of the seed to see what can be missing, and still allow the seed to germinate. For example, remove the seed coat and try to germinate.
- Students can make a poster to compare the monocots and dicots from seed to flower.

What's inside a seed? Are there green leaves, stems, and roots in a seed? Is a seed anything like a full-grown plant?

FIND OUT!

Do this simple activity to find out what's in a seed.

Obtain a raw peanut that is still in its shell. Peanuts are seeds, and shells are dried fruit. Open the shell and *observe* the seeds. Take the reddish-brown papery covering off one of the peanuts. Carefully pull apart the halves of the peanut. Look for a small bump on one half. This is the embryo plant. *Identify* parts that you think will become leaves, a stem, or roots.

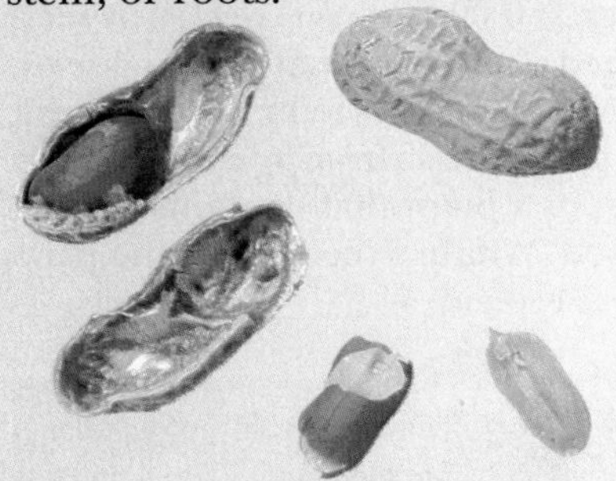

Gearing Up

Previewing the Chapter

Use this outline to help you focus on important ideas in this chapter.

Previewing Science Skills

- In the Skill Builders, you will compare and contrast, classify, and hypothesize.
- In the Activities, you will observe, analyze, collect and interpret data, and compare.
- In the MINI-Labs, you will observe and identify.

What's next?

Now you know that a seed contains an embryo plant. In this chapter you'll learn about the kinds of plants that make seeds, how seed plants are important in the environment, and how the evolutionary history of seed plants is related to the evolution of other organisms.

ASSESSMENT OPTIONS

PORTFOLIO

Refer to page 271 for suggested items that students might select for their portfolios.

PERFORMANCE ASSESSMENT

See page 271 for additional Performance Assessment options.

Process

Find Out, p. 253
Skillbuilders, pp. 258, 261, 266
MINI-Lab, p. 263
Activities, 11-1, p. 267; 11-2, p. 270
Using Lab Skills, p. 272

CONTENT ASSESSMENT

Assessment—Oral, pp. 256, 258, 259, 266
Section Reviews, pp. 258, 261, 266, 269
Chapter Review, pp. 271-273

GROUP ASSESSMENT

Opportunities for group assessment occur with Cooperative Learning Strategies and Flex Your Brain Activities.

INTRODUCING THE CHAPTER

Use the Find Out activity to introduce students to the parts of seeds. Inform students that they will learn more about seeds and the plants that make seeds in this chapter.

FIND OUT!

Preparation: Purchase raw peanuts in the shell from a grocery produce section.

Materials: Each student will need one seed from a peanut.

Cooperative Learning: Have students work in Science Investigation or Paired Partners cooperative groups to complete this activity.

Teaching Tips

- As you teach the parts of the peanut, use the scientific names for the parts. The shell is a dried ovary; peanuts are seeds; seed leaves are cotyledons, etc.

Find Out

ASSESSMENT

Performance: To further assess students' understanding of seed parts, see USING LAB SKILLS, question 11, p. 272.

Gearing Up

Have students study the Gearing Up feature to familiarize themselves with the chapter. Discuss the relationships of the topics in the outline.

What's Next?

Before beginning the first section, make sure students understand the connection between the Find Out activity and the topics to follow.

OBJECTIVES AND SCIENCE WORDS: Have students review the objectives and science words throughout the chapter as they study each section.

SECTION 11-1

PREPARATION

SECTION BACKGROUND

▶ There are fewer than 1000 species of gymnosperms. Of these, about 550 are conifers including pines, firs, junipers, and cedars.

▶ A single giant redwood may produce as much as 480 000 board feet of lumber.

▶ The economic value of seed plants to humans is practically incalculable. They are sources of food, shelter, and nourishment.

PREPLANNING

▶ Obtain several boxes of cereal for the Motivation.

▶ Obtain a collection of plants to show students at different points throughout this chapter.

▶ Cut out photographs of plants to use during class discussions.

REVEALING MISCONCEPTIONS

▶ Ask students where most species of seed plants are found. Most of the 260 000 living species survive in tropical rain forests.

FLEX Your Brain

Use the Flex Your Brain activity sheet to analyze what information students already have about SEED PLANTS.

Flex Your Brain
ASSESSMENT

Portfolio: Use the Flex Your Brain activity to reinforce critical thinking and problem solving skills. In Step 2, students might list fruits they've eaten that have seeds and what those plants are like; or describe other seeds they've seen.

VideoDisc

STVS: Seed Banks, Disc 4, Side 2

11-1 Seed Plants

New Science Words

gymnosperms
angiosperms
monocots
dicots

Objectives

- ▶ List the main characteristics of seed plants.
- ▶ Describe the main characteristics of gymnosperms and angiosperms and their importance.
- ▶ Compare the evolution of gymnosperms and angiosperms.

What Is a Seed Plant?

Have you ever eaten Chinese vegetables like those shown in Figure 11-1? These vegetables include bamboo shoots, water chestnuts, and snow peas. In the Philippines and Malaysia, coconut milk and banana flowers are used in many dishes. Corn and beans are staple foods in Mexico. All of these foods come from seed plants. What fruits and vegetables have you eaten today? If you had an apple, a peanut butter and jelly sandwich, and a glass of orange juice for lunch, you ate foods that came from seed plants.

How are seed plants different from simple plants?

1 Nearly all the plants you are familiar with are seed plants. There are over 250 000 known species of seed plants in the world. Seed plants have roots, stems, leaves, and vascular tissue. What makes a seed plant different from simple plants is that it grows from a seed. A seed is the reproductive part of a plant that contains a plant embryo and stored food. Within a seed are all the parts needed to produce a new plant.

Figure 11-1. Plants like these provide food for humans and most other consumers on Earth.

OPTIONS

Meeting Different Ability Levels

For Section 11-1, use the following **Teacher Resource Masters** depending upon individual students' needs.

◆ **Study Guide Master** for all students.

● **Reinforcement Master** for students of average and above average ability levels.

▲ **Enrichment Master** for above average students.

Additional Teacher Resource Package masters are listed in the OPTIONS box throughout the section. The additional masters are appropriate for all students.

◆ STUDY GUIDE 43

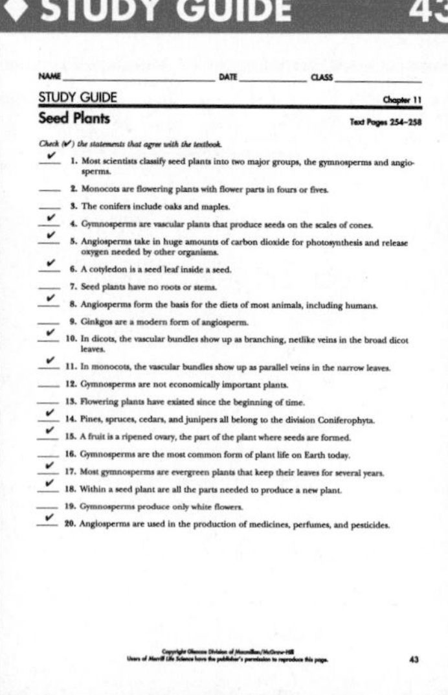

NAME ______ DATE ______ CLASS ______

STUDY GUIDE — Chapter 11

Seed Plants — Text Pages 254–258

Check (✔) the statements that agree with the textbook.

✔ 1. Most scientists classify seed plants into two major groups, the gymnosperms and angiosperms.
___ 2. Monocots are flowering plants with flower parts in fours or fives.
___ 3. The conifers include oaks and maples.
✔ 4. Gymnosperms are vascular plants that produce seeds on the scales of cones.
✔ 5. Angiosperms take in huge amounts of carbon dioxide for photosynthesis and release oxygen needed by other organisms.
✔ 6. A cotyledon is a seed leaf inside a seed.
___ 7. Seed plants have no roots or stems.
✔ 8. Angiosperms form the basis for the diets of most animals, including humans.
___ 9. Ginkgos are a modern form of angiosperm.
✔ 10. In dicots, the vascular bundles show up as branching, netlike veins in the broad dicot leaves.
✔ 11. In monocots, the vascular bundles show up as parallel veins in the narrow leaves.
___ 12. Gymnosperms are not economically important plants.
___ 13. Flowering plants have existed since the beginning of time.
✔ 14. Pines, spruces, cedars, and junipers all belong to the division Coniferophyta.
✔ 15. A fruit is a ripened ovary, the part of the plant where seeds are formed.
___ 16. Gymnosperms are the most common form of plant life on Earth today.
✔ 17. Most gymnosperms are evergreen plants that keep their leaves for several years.
✔ 18. Within a seed plant are all the parts needed to produce a new plant.
___ 19. Gymnosperms produce only white flowers.
✔ 20. Angiosperms are used in the production of medicines, perfumes, and pesticides.

43

Most scientists classify seed plants into two major groups, the gymnosperms and angiosperms. Both of these groups produce seeds, but you will see that they have some very important differences.

Gymnosperms

The oldest trees alive today are gymnosperms (JIHM nuh spurmz)—in fact, there's one bristlecone pine tree in the White Mountains of eastern California that is 4900 years old. **Gymnosperms** are vascular plants that produce seeds on the scales of female cones. The word *gymnosperm* comes from the Greek language and means "naked seed." Seeds of gymnosperms are not protected by a fruit. Gymnosperms do not produce flowers. Leaves of most gymnosperms are needlelike or scalelike. Most gymnosperms are evergreen plants that keep their leaves for several years. 2

What are gymnosperms?

Gymnosperms include four divisions of plants—the conifers, cycads, ginkgos, and gnetophytes. Of these, you are probably most familiar with the pines, firs, spruces, cedars, and junipers in the division Coniferophyta (KAHN uh fur AHF uh tuh). This division contains the greatest number of species of gymnosperms. Some conifers produce male and female cones on separate trees, but other species produce both types of cones on the same tree. Figure 11-2 shows examples of the four divisions of gymnosperms.

Figure 11-2. The gymnosperms include conifers (a), cycads (b), ginkgos (c), and gnetophytes (d).

a

b

c

d

● REINFORCEMENT 43

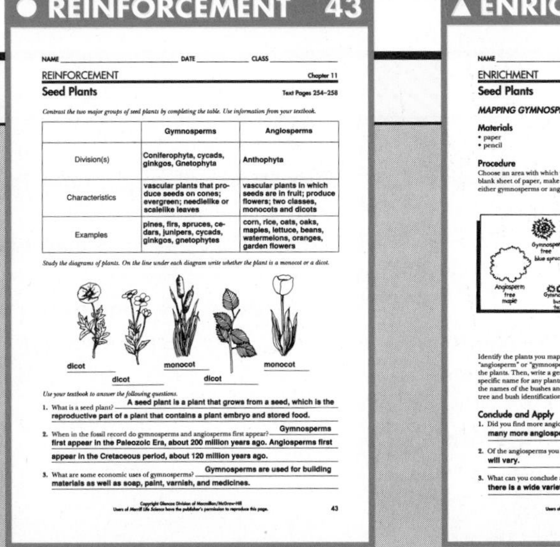

NAME ______ DATE ______ CLASS ______

REINFORCEMENT Chapter 11

Seed Plants Text Pages 254–258

Contrast the two major groups of seed plants by completing the table. Use information from your textbook.

	Gymnosperms	Angiosperms
Division(s)	Coniferophyta, cycads, ginkgos, Gnetophyta	Anthophyta
Characteristics	vascular plants that produce seeds on cones; evergreen; needlelike or scalelike leaves	vascular plants in which seeds are in fruit; produce flowers; two classes, monocots and dicots
Examples	pines, firs, spruces, cedars, junipers, cycads, ginkgos, gnetophytes	corn, rice, oats, oaks, maples, lettuce, beans, watermelons, oranges, garden flowers

Study the diagrams of plants. On the line under each diagram write whether the plant is a monocot or a dicot.

dicot dicot monocot dicot monocot

Use your textbook to answer the following questions.

1. What is a seed plant? **A seed plant is a plant that grows from a seed, which is the reproductive part of a plant that contains a plant embryo and stored food.**
2. When in the fossil record do gymnosperms and angiosperms first appear? **Gymnosperms first appear in the Paleozoic Era, about 200 million years ago. Angiosperms first appear in the Cretaceous period, about 120 million years ago.**
3. What are some economic uses of gymnosperms? **Gymnosperms are used for building materials as well as soap, paint, varnish, and medicines.**

 43

▲ ENRICHMENT 43

NAME ______ DATE ______ CLASS ______

ENRICHMENT Chapter 11

Seed Plants Text Pages 254–258

MAPPING GYMNOSPERMS AND ANGIOSPERMS

Materials
- paper
- pencil

Procedure
Choose an area with which you are familiar—a yard or a small park or the school grounds. Using a blank sheet of paper, make a map of the area that shows the major plants and identifies them as either gymnosperms or angiosperms. Make your map similar to the one below.

Angiosperm grass
Gymnosperm tree blue spruce
Angiosperm bushes forsythia
Gymnosperm tree white pine
Angiosperm tree maple
Angiosperm tree maple
Gymnosperm bushes taxis
Angiosperm flowers tulips
Gymnosperm bushes taxis
Angiosperm tree oak

Identify the plants you map as specifically as you can. As shown on the example map, first write "angiosperm" or "gymnosperm." You may use the information in your textbook to help you identify the plants. Then, write a general term such as "tree," "bush," "flower," or "grass." Finally, write the specific name for any plants that you know. Ask adults for help. Many adults will be glad to tell you the names of the bushes and trees they have planted in their yards. Also, you can find books on tree and bush identification in almost any library.

Conclude and Apply
1. Did you find more angiosperms or gymnosperms? **Most students will have found many more angiosperms.**
2. Of the angiosperms you identified, which are monocots and which are dicots? **Answers will vary.**
3. What can you conclude about the variety of plants in an area? **Most students will say there is a wide variety of plants in an area.**

 43

1 MOTIVATE

▶ Show students a pine cone, a rose, and a walnut. Ask students to explain what the plants that produce these structures all have in common.

▶ Line up several boxes of cereal. Have students identify the plants that produced the contents.

TYING TO PREVIOUS KNOWLEDGE: Review the characteristics of plants from the previous chapter. Point out that seed plants share some characteristics with mosses, liverworts, and ferns, but are also different in some important ways.

2 TEACH

Key Concepts are highlighted.

CONCEPT DEVELOPMENT

▶ Point out the largest group of gymnosperms in the division Coniferophyta.

▶ **Demonstration:** Use a variety of monocots and dicots to demonstrate the characteristics of these two groups of plants.

▶ Cut out photographs of various gymnosperms and angiosperms and use them to make a bulletin board.

▶ Begin one of the ideas from CLOSE on page 258 at this point. Students will need time to do the grocery survey or time to collect products made from seed plants. Most will bring in cereals, but oils and coffee are also derived from seeds.

CHECK FOR UNDERSTANDING

Ask students to compare the monocots and dicots according to their characteristics.

RETEACH

Emphasize the important characteristics of the monocots and dicots again. Place several monocots and dicots around the room. Write a hint on a 3″ × 5″ card for each one. (Example: Notice the parallel veins in the leaves of this plant.) Have students identify them by using the hints.

EXTENSION

For students who have mastered this section, use the **Reinforcement** and **Enrichment** masters or other OPTIONS provided.

In Your JOURNAL

The students might also take a poll to see how many households bag their grass. Once they know the facts, they might decide to try other means of informing the public.

To assess this product, refer to the Performance Task Assessment Lists in **Performance Assessment in Middle School Science**

CROSS CURRICULUM

▶ **Art:** Have students produce watercolor paintings or colored pencil drawings that are representative of particular plants. You may choose the style of the work or leave it up to the students.

Angiosperms

What are angiosperms?

When you bite into a pear or an apple, you are eating part of an angiosperm (AN jee uh spurm). **Angiosperms**
3 are vascular plants in which the seed is enclosed inside a fruit. A fruit is a ripened ovary, the part of the plant where seeds are formed. If you eat most of your pear or apple, you'll find seeds inside. All angiosperms produce flowers and are classified in the division Anthophyta (an THAWF uh tuh). More than half of all known plant species are angiosperms.

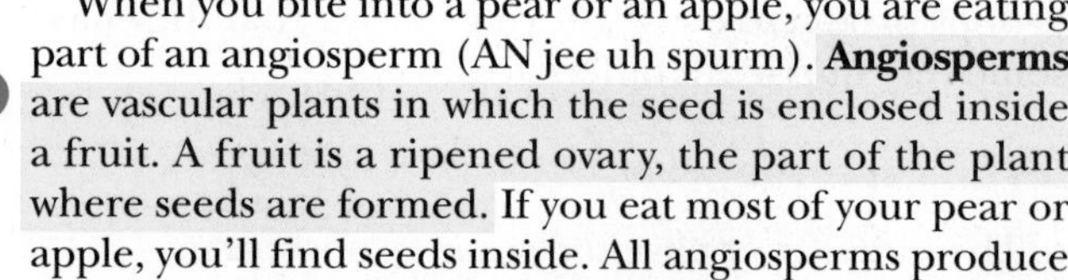

There are two classes of angiosperms, or flowering plants: the monocots (MAHN uh kahts) and the dicots (DI kahts). The terms *monocot* and *dicot* are shortened forms of the words monocotyledon and dicotyledon. A cotyledon is a seed leaf inside a seed. **Monocots** have one seed leaf inside their seeds; **dicots** have two. Monocots are flowering plants with flower parts in threes and vascular bundles throughout the stems. The scattered bundles show up as parallel veins in their narrow leaves. If you had cereal for breakfast this morning, you ate a bowlful of parts of monocot seeds! Cereal grains such as corn, rice, oats, and wheat are all monocots.

In Your JOURNAL

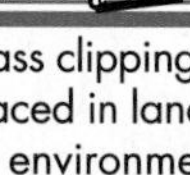

The collection of grass clippings as part of garbage placed in landfills has become a big environmental issue. Study the issue. Then, **in your Journal,** write a rough draft of a letter to the editor of your local paper supporting one side or the other. Make a clean copy and send it to the paper.

Dicots are flowering plants with flower parts in fours or
4 fives and vascular bundles in rings inside the stems. The bundles show up as branching, netlike veins in the broad dicot leaves. Examples of dicots include trees like oaks and maples, vegetables like lettuce and beans, fruits like watermelons and oranges, and many garden flowers.

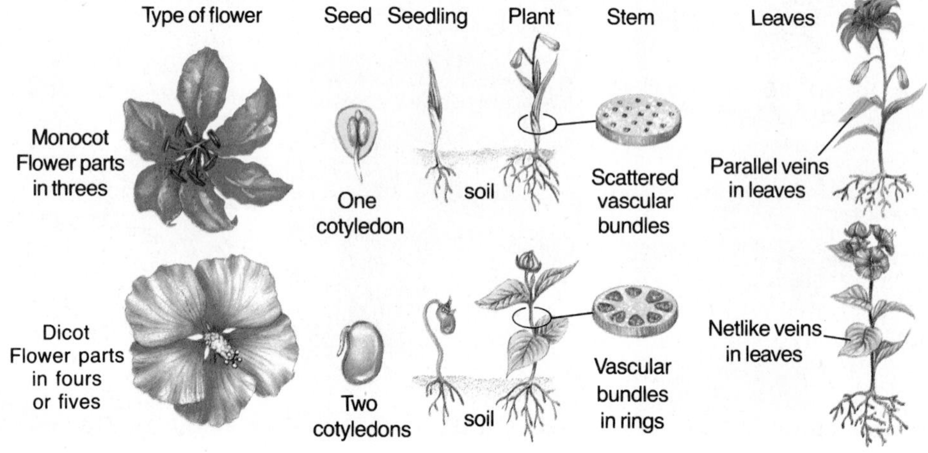

Figure 11-3. The Characteristics of Monocots and Dicots

OPTIONS

ASSESSMENT—ORAL

▶ **Explain why scientists agree that seed plants evolved after mosses and ferns.** *Seed plants are more complex. The fossil record contains fossils of mosses and ferns that are much older than any seed plant.*

▶ **Gymnosperms can live in cold climates. What adaptations might help gymnosperms live in these environments?** *Leaves that are evergreen; leaves with reduced surface area; and thick, resinous sap that does not easily freeze are all adaptations for living in cold environments.*

▶ **Which group of seed plants is most fine furniture made from?** *Angiosperms (particularly hardwoods) because they have denser wood than gymnosperms.*

▶ **Which groups of gymnosperms are not conifers?** *cycads, ginkgos, and gnetophytes*

Figure 11-4. Rice is a staple food crop for humans worldwide. Ninety percent of all rice is grown in Asia, but, in the United States, Texas and California are major producers.

Origin and Evolution of Seed Plants

Seed plants have some different needs from those of simple vascular plants like ferns. For example, ferns depend on water for reproduction, but seed plants do not. A seed is an adaptation that gives plants the ability to survive on land away from bodies of water. Gymnosperms and angiosperms today are found in tropical, temperate, and even extremely dry or cold environments.

Gymnosperms probably evolved from a group of plants that grew about 350 million years ago. Cycads and conifers have been found in the fossil record since the Paleozoic era, 200 million years ago. Flowering plants did not exist until the Cretaceous period, about 120 million years ago. Gaps in the fossil record prevent scientists from tracing the exact origins of flowering plants. After their appearance in the Cretaceous period, angiosperms evolved into many different forms. Today they are the most common form of plant life on Earth.

Importance of Seed Plants

Imagine that your class is having a picnic in the park. You cover the picnic table with a red-checked tablecloth and pass out paper cups and plates. You toast hot dogs and buns, eat potato chips, and drink apple cider. Perhaps you collect leaves or flowers for a science project. Later you clean up and put leftovers in paper bags.

Throughout history, different cultures have depended on grains as the basis of their diets. Find out what complex carbohydrate is common to many grains such as corn, beans, and rice.

MINI QUIZ

Use the Mini Quiz to check students' recall of chapter content.

1. **A seed is the reproductive part of a plant that contains a plant ____ and stored food.** *embryo*
2. **____ are vascular plants that produce seeds on the scales of cones.** *Gymnosperms*
3. **____ are vascular plants in which the seed is enclosed inside a fruit.** *Angiosperms*
4. **____ are flowering plants with flower parts in fours or fives, and vascular bundles in rings inside the stems.** *Dicots*
5. **____ are used in the production of medicines, rubber, oils, perfumes, pesticides, and some industrial chemicals.** *Angiosperms*

CONCEPT DEVELOPMENT

▶ The viability of seeds can be affected by abiotic factors such as water, temperature, and common chemicals.

▶ Dormant seeds can survive during unfavorable conditions such as intense heat or extreme cold. A species that has seeds able to survive by undergoing dormancy has an advantage over species in which seeds do not have a dormant period.

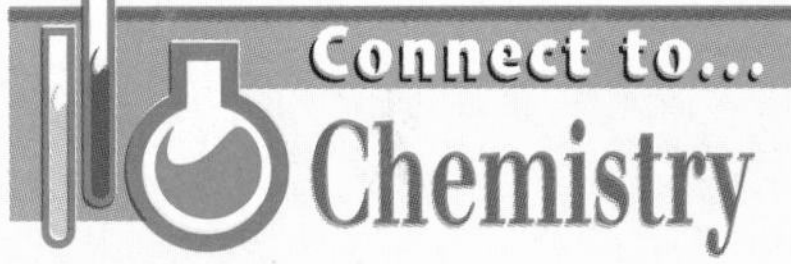

Answer: starch

MULTICULTURAL AWARENESS

Use the Connect ot Chemistry feature to broaden students' awareness of the fact that many cultures depend on a specific grain. Discuss some examples. **ASK: What grains does the United States depend on?** *(corn and wheat) Have students find out about millet, sorghum and amaranth.*

ENRICHMENT

Bring in several different kinds of seeds. Mix all the seeds together. Allow students to choose a seed and grow the plant. Have them identify the plant as a monocot, dicot, or gymnosperm as soon as the characteristics become visible.

PROGRAM RESOURCES

From the **Teacher Resource Package** use:

Activity Worksheets, page 5, Flex Your Brain.

Transparency Masters, pages 35-36, Monocots and Dicots.

Use **Color Transparency** 18, Monocots and Dicots.

3 CLOSE

▶ Ask questions 1-3 and the **Apply** Question in the Section Review.

▶ Have students make a survey of the kinds of plants they can find at a supermarket. Discuss the list in class. Are more gymnosperms, monocots, or dicots on the list?

▶ Allow students to bring in products of seed plants. Discuss the plant that produced the product and where the plant grows. Try to classify the plant as a gymnosperm, monocot, or dicot.

SECTION REVIEW ANSWERS

1. Seed plants have roots, stems, leaves, and vascular tissue.

2.

	Gymnosperms	Angiosperms
Flowers	No	Yes
Cones	Yes	No
Seeds	Yes	Yes
Fruits	No	Yes

3. Seed plants consume carbon dioxide and produce oxygen, form the basis for the diets of animals, and can be used in many industries.

4. Apply: Male trees produce only male flowers, and do not produce fruits.

5. Connect to Earth Science: Carboniferous period

Skill Builder

Monocots have one seed leaf inside their seeds, while dicots have two. Monocots have parallel veins in their leaves, and dicots have netlike veination. Monocots have flower parts in threes, and dicots have their flowering parts in fours or fives.

Skillbuilder

ASSESSMENT

Performance: To assess students' abilities to Compare and Contrast monocots and dicots, have them tie the CLOSE survey of grocery products (above) made from seed plants to this Skillbuilder by giving examples of monocots and dicots from the CLOSE activities.

Figure 11-5. Linen, a type of cloth, is made from fibers of the flax plant, *Linum usitatissimum*.

Now let's imagine this scene if there were no seed plants on Earth. There would be no picnic table and no pulp to make paper products such as cups, plates, and bags. Bread for buns, apples for cider, and potatoes for chips all come from plants. The tablecloth is made of cotton, a plant. Without seed plants, there would be no picnic!

On a global scale, conifers are the most economically important gymnosperms. Most of the wood used for building construction and for paper production comes from conifers such as pines and spruces. Resin, a waxy substance secreted by conifers, is used to make chemicals found in soap, paint, varnish, and some medicines.

The most common plants on Earth are the angiosperms. They are important to all life because they form the basis for the diets of most animals. Grains such as barley and wheat and legumes such as peas and lentils were the first plants ever grown by humans. Angiosperms take in huge amounts of carbon dioxide for photosynthesis and release oxygen needed by other organisms. Many of the fibers used in clothing such as flax and cotton fibers come from angiosperms. Angiosperms are used in the production of medicines, rubber, oils, perfumes, pesticides, and some industrial chemicals. 5

SECTION REVIEW

1. What are the main characteristics of a seed plant?
2. Make a chart to compare the characteristics of gymnosperms and angiosperms.
3. Give three reasons why seed plants are important.
4. **Apply:** The fruit of the ginkgo has an unpleasant smell, so only male trees are planted near homes. Why?
5. **Connect to Earth Science:** Use the geologic time scale on page 139 to find out during what period, the first seed plants evolved.

Skill Builder

Comparing and Contrasting

Monocots and dicots have some similarities and differences. Compare and contrast these two classes of plants. If you need help, refer to Comparing and Contrasting in the **Skill Handbook** on page 683.

OPTIONS

ASSESSMENT—ORAL

▶ **Why are gymnosperms economically important?** *They produce wood for construction, and chemicals found in soap, paint, varnish, and some medicines.*

▶ **Why are angiosperms economically important?** *They produce fibers for clothing, grains for food, and may be used to make medicines, rubber, oils, perfumes, pesticides, and some chemicals.*

Parts of Complex Plants

11-2

Objectives

- ▶ Describe the structures of roots, stems, and leaves.
- ▶ Describe the functions of roots, stems, and leaves.
- ▶ Explain some unique adaptations of roots, stems, and leaves.

New Science Words

xylem
phloem
cambium
stomata
guard cells

Roots

Imagine a large tree growing alone on top of a hill. What is the largest part? Chances are you named the trunk or the branches. But did you consider the roots? The root systems of most plants are as large or larger than the aboveground stems and leaves. Why do you think root systems are so large?

Why are roots so important to plants? All the water and
1 minerals used by a plant enter by way of its roots. Roots have vascular tissue to move water and minerals from the ground up through the stems to the leaves. Roots also anchor plants in soil. If they didn't, plants could be blown away by wind, or washed away by water. Each root system must support the plant parts that are above the ground—the stem, branches, and leaves of a tree, for example. In Figure 11-6, you can see the root system of a dandelion as compared with the rest of the plant.

Roots also store food. When you eat carrots or beets you eat roots swollen with stored food. Root tissues may also perform special functions such as absorption of oxygen and the process of photosynthesis.

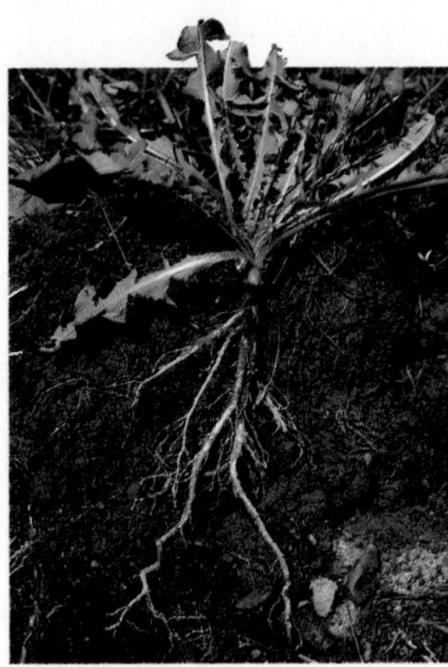

Figure 11-6. The Root System of a Dandelion

Stems

Did you know that the trunk of a tree is really its stem?
2 The main functions of a stem are to support the aboveground parts of the plant and to allow movement of materials between leaves and roots. Some stems also have adaptations that allow them to store food. Potatoes and onions are underground stems with stored food. Sugarcane has an aboveground stem that stores large

SECTION 11-2

PREPARATION

SECTION BACKGROUND

▶ The sweet potato is a fleshy root that contains stored food, but the Irish potato is a fleshy underground stem that contains stored food.

▶ Cactus spines are modified leaves that prevent water loss. Another modification of leaves is food storage as seen in celery and rhubarb.

PREPLANNING

▶ Obtain some sod for the TEACH on page 260.

▶ Obtain carrots, asparagus, and lettuce for the Motivation activity.

VideoDisc

STVS: Sounds of Thirsty Plants, Disc 4, Side 2

STVS: Growing Plants in Space, Disc 4, Side 2

OPTIONS

ASSESSMENT—ORAL

▶ **How could you prove that roots transport water and minerals through a plant?** *Answers will vary, but some type of experiment that traces water and minerals should be accepted.*

▶ **What would happen if you cut a ring through the bark of a tree all the way around the trunk?** *The tree would die because food could not get down to the roots through the phloem.*

▶ **Why do plants have leaves that are shaped differently?** *Leaf shape is an adaptation for light absorption. Generally large, wide leaves are found in plants that grow in shady areas, and narrow thicker leaves are found in plants that grow in full sun.*

▶ **Why is one layer of plant leaves called spongy?** *The spongy layer is so named because of the many air pockets contained within the cells that allow carbon dioxide to be transported to all the cells that photosynthesize. (Save the questions on leaf structure for page 261.)*

1 MOTIVATE

▶ Give each student a piece of carrot, asparagus, and lettuce. Ask what part of a plant each comes from and what they have in common. Let the students eat their plant parts to see that each part stores food.

TECHNOLOGY

For more information see: Whittingham-Barnes, Donna, et al. "Growing Gardens in Space," *Black Enterprise*, Feb. 1990, p. 116.

Think Critically: Food can be grown in areas where temperature, insects, amount of water, and other factors prohibit traditional farming methods. Reducing pesticide use benefits the environment as well.

2 TEACH

Key Concepts are highlighted.

CONCEPT DEVELOPMENT

▶ Root systems are so large because they must collect all the water and minerals a plant needs.

▶ Obtain a piece of healthy sod and have students pull small pieces apart to see how extensive and intertwined the roots are.

Phloem – transports sugars throughout plant
Cambium – produces additional xylem and phloem
Xylem – transports minerals and water throughout plant
Bark
Wood

Figure 11-7. The vascular tissue of seed plants includes xylem, phloem, and cambium.

quantities of food. Stems of cacti are adapted to carry on photosynthesis and make food for the rest of the plant.

Three main tissues make up the vascular system in the roots, stems, and leaves. **Xylem** (ZI lum) tissue is made up of tubular vessels that transport water and minerals up from the roots throughout the plant. **Phloem** (FLOH em) is a plant tissue made up of tubular cells that move food from leaves and stems, where it is made, to other parts of the plant for direct use or storage. Between xylem and phloem is the cambium (KAM bee um). **Cambium** is a tissue that produces new xylem and phloem cells. All three tissues are shown in Figure 11-7.

Plant stems are either herbaceous (hur BAY shus) or woody. Herbaceous stems are soft and green. Examples are the stems of peppers, corn, and tulips.

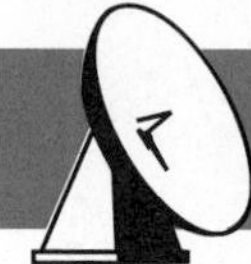

TECHNOLOGY

Plants in Space?

Growing plants without soil is not new, but scientists are moving beyond traditional hydroponics, or the growing of plants in a nutrient solution.

New hydroculture techniques are being developed for use by astronauts on space missions. These techniques allow use of highly nutritious root crops. Instead of immersing roots in nutrient solutions, nutrients are sprayed directly on plant roots. This allows the plants better access to oxygen in the atmosphere, shortening growing time. The environment is carefully controlled, so the vegetables are virtually bacteria-free and pesticides are not needed. Growing trays can be tilted so less room is needed.

These techniques produce good quality vegetables without the need for farmers to worry about seasons, weather, insects, or any other environmental factors.

Think Critically: How might hydroculture be used to help the world food supply?

OPTIONS

Meeting Different Ability Levels

For Section 11-2, use the following **Teacher Resource Masters** depending upon individual students' needs.

◆ **Study Guide Master** for all students.

● **Reinforcement Master** for students of average and above average ability levels.

▲ **Enrichment Master** for above average students.

Additional Teacher Resource Package masters are listed in the OPTIONS box throughout the section. The additional masters are appropriate for all students.

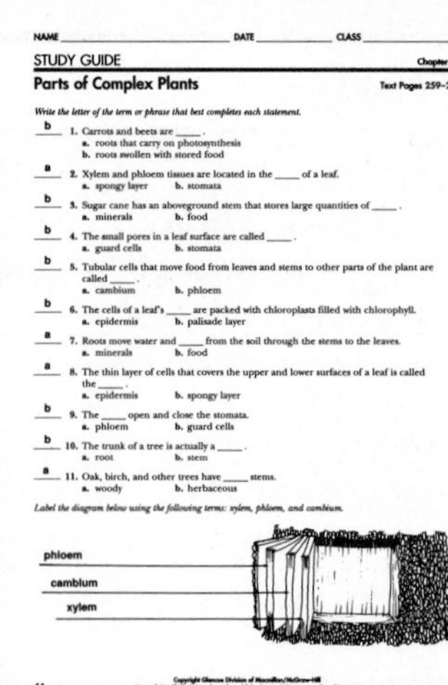

◆ STUDY GUIDE 44

NAME ______ DATE ______ CLASS ______

STUDY GUIDE Chapter 11

Parts of Complex Plants Text Pages 259–261

Write the letter of the term or phrase that best completes each statement.

__b__ 1. Carrots and beets are _____.
a. roots that carry on photosynthesis
b. roots swollen with stored food

__a__ 2. Xylem and phloem tissues are located in the _____ of a leaf.
a. spongy layer b. stomata

__b__ 3. Sugar cane has an aboveground stem that stores large quantities of _____.
a. minerals b. food

__b__ 4. The small pores in a leaf surface are called _____.
a. guard cells b. stomata

__b__ 5. Tubular cells that move food from leaves and stems to other parts of the plant are called _____.
a. cambium b. phloem

__b__ 6. The cells of a leaf's _____ are packed with chloroplasts filled with chlorophyll.
a. epidermis b. palisade layer

__a__ 7. Roots move water and _____ from the soil through the stems to the leaves.
a. minerals b. food

__a__ 8. The thin layer of cells that covers the upper and lower surfaces of a leaf is called the _____.
a. epidermis b. spongy layer

__b__ 9. The _____ open and close the stomata.
a. phloem b. guard cells

__b__ 10. The trunk of a tree is actually a _____.
a. root b. stem

__a__ 11. Oak, birch, and other trees have _____ stems.
a. woody b. herbaceous

Label the diagram below using the following terms: xylem, phloem, and cambium.

phloem
cambium
xylem

44

Oak, birch, and other trees and shrubs have woody stems. Woody stems are hard and rigid.

Leaves

Have you ever rested in the shade of a tree's leaves on a hot, summer day? Leaves are the plant organs that usually trap light and make food for the plant through the process of photosynthesis.

Look at the structure of the leaf shown in Figure 11-8. The epidermis is a thin layer of cells that covers and protects both the upper and lower surfaces of a leaf. A waxy cuticle that protects the plant from wilting or drying out covers the epidermis of many leaves. Another adaptation of leaves are small pores in the leaf surfaces called **stomata** (STOH mut uh). Stomata allow carbon dioxide, water, and oxygen to enter and leave a leaf. The stomata are surrounded by **guard cells** that open and close the pores. The cuticle, stomata, and guard cells all are adaptations that enabled plants to survive on land by conserving water.

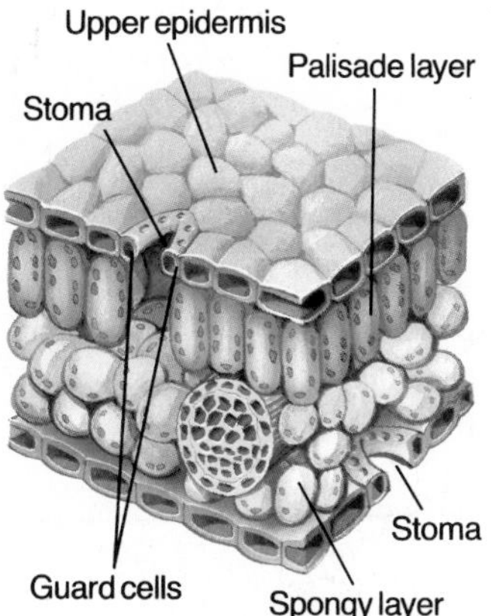

Figure 11-8. The Layers of a Leaf

Inside a typical leaf are different layers of cells. One layer, the palisade layer, has rows of closely packed cells just below the upper epidermis. Palisade cells are packed with chloroplasts filled with chlorophyll. Most of the food made by leaves is made here. Between the palisade cells and the lower epidermis is a layer of loosely arranged cells, the spongy layer, and many air spaces. Xylem and phloem vessels are located in this layer. 3

Where is most plant food made?

SECTION REVIEW

1. What are the functions of roots, stems, and leaves?
2. What is the advantage of having most of the chloroplasts in cells toward the top of a leaf?
3. **Apply:** The cuticle and epidermis are transparent. How is this an advantage for a plant?
4. **Connect to Physics:** Why is most of the food made in leaves produced in the layer under the upper epidermis?

Classifying

Skill Builder

From what you have learned in this section, classify the following vegetables as roots, stems, or leaves: turnip, lettuce, asparagus, potato, onion, and carrot. If you need help, refer to Classifying in the **Skill Handbook** on page 680.

● REINFORCEMENT 44

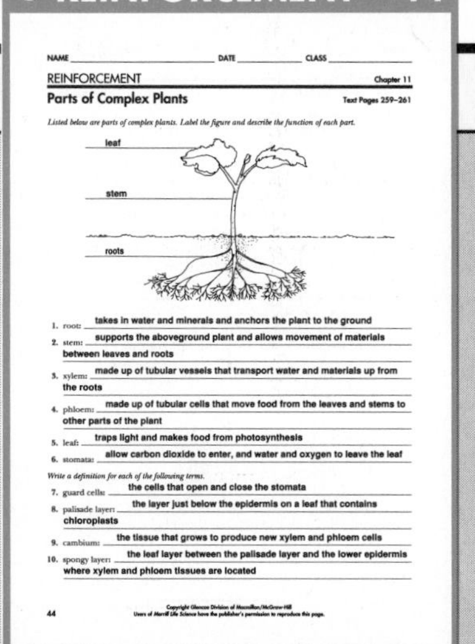

NAME ______ DATE ______ CLASS ______

REINFORCEMENT — Chapter 11

Parts of Complex Plants — Text Pages 259–261

Listed below are parts of complex plants. Label the figure and describe the function of each part.

1. root: **takes in water and minerals and anchors the plant to the ground**
2. stem: **supports the aboveground plant and allows movement of materials between leaves and roots**
3. xylem: **made up of tubular vessels that transport water and materials up from the roots**
4. phloem: **made up of tubular cells that move food from the leaves and stems to other parts of the plant**
5. leaf: **traps light and makes food from photosynthesis**
6. stomata: **allow carbon dioxide to enter, and water and oxygen to leave the leaf**

Write a definition for each of the following terms.

7. guard cells: **the cells that open and close the stomata**
8. palisade layer: **the layer just below the epidermis on a leaf that contains chloroplasts**
9. cambium: **the tissue that grows to produce new xylem and phloem cells**
10. spongy layer: **the leaf layer between the palisade layer and the lower epidermis where xylem and phloem tissues are located**

44

▲ ENRICHMENT 44

NAME ______ DATE ______ CLASS ______

ENRICHMENT — Chapter 11

Parts of Complex Plants — Text Pages 259–261

MODIFIED LEAVES

Some plants grow in acid bogs where nutrients are easily washed away. These plants have structures to help supplement their nutrition. These carnivorous plants trap insects with modified leaves. The leaves contain nectar glands that attract insects. Leaf hairs in or on the leaves also aid in capturing prey. Once an insect is trapped, the leaves begin to produce digestive enzymes. Nutrients from the digested insects are absorbed by the leaves. Carnivorous plants also make carbohydrates through photosynthesis as do other green plants, but because their roots are small, they are not efficient at absorbing nutrients.

Carnivorous plants have several types of traps. The Venus's flytrap is an example of an active trap. When an insect lands on the sensitive hairs on the leaf, the leaf folds to keep the insect from escaping.

The largest of the carnivorous plants are the pitcher plants, which have pitfall traps. The leaves of these plants form a slender tube. Nectar glands on the lip of the tube attract insects. The insects land on a slick area of the tube and fall into a pool of digestive juices at the bottom. Hairs inside the plant prevent the insect from crawling out.

The sundew is a flypaper trap. The leaves of this plant are covered with hairs that secrete sticky droplets. The insect then becomes entangled in the hairs on the leaf, where enzymes digest the soft parts of the insect.

Venus's flytrap — Pitcher plant — Sundew

1. Why do carnivorous plants trap and digest insects? **They trap insects to supplement their nutritional needs.**
2. List the three types of trap structures and name a plant that uses each structure.
   a. **active trap — Venus's flytrap**
   b. **pitfall trap — pitcher plant**
   c. **flypaper trap — sundew**
3. How are carnivorous plants similar to other plants? How are they different? **Carnivorous plants are similar in that they make food by photosynthesis. They also have leaves and roots like other plants. They are different in that they have modified leaves that act as insect traps, and they supplement their nutrition by digesting insects.**

44

CHECK FOR UNDERSTANDING

Use the Mini Quiz to check for understanding.

MINI QUIZ

Use the Mini Quiz to check students' recall of chapter content.

1. **The ____ have vascular tissue to move water from the ground up to the leaves.** *roots*
2. **A ____ supports the above-ground parts of the plant.** *stem*
3. **Most of the food made by leaves is made in the ____ .** *palisade layer*

RETEACH

Observe prepared slides of leaf cross sections under the microscope. Have students draw and label the parts. Ask students to explain the function of each leaf part.

EXTENSION

For students who have mastered this section, use the **Reinforcement** and **Enrichment** masters or other OPTIONS.

3 CLOSE

▶ Ask questions 1-2 and the **Apply** Question in the Section Review.

SECTION REVIEW ANSWERS

1. Roots must carry water and minerals to other plant parts.
2. Top is where most sunlight hits, and therefore, where most photosynthesis takes place.
3. Apply: Cuticles are transparent to allow light to reach the cells underneath for photosynthesis.
4. Connect to Physics: This is where most of the chloroplasts are found.

Skill Builder

turnip—underground stem; lettuce—leaves; asparagus—stem; potato—underground stem; onion—stem; carrot—root

Skillbuilder

ASSESSMENT

Performance: Assess students' abilities to Classify vegetables by structure by giving them actual vegetables to classify.

SECTION 11-3

PREPARATION

SECTION BACKGROUND

▶ Flowers that contain both stamens and pistils are perfect. Imperfect flowers may be either male if they have only stamens or female if they have only pistils.

PREPLANNING

▶ Obtain a model of a flower for the Teach activity.

REVEALING MISCONCEPTIONS

▶ Emphasize that not all flowers are colorful. Many students will not recognize that trees such as oaks, pecans, and elms have flowers.

FLEX Your Brain

Use the Flex Your Brain activity sheet to identify what students already know about FLOWERS.

Flex Your Brain

ASSESSMENT

Portfolio: Use the Flex Your Brain activity to reinforce critical thinking and problem solving skills.

VideoDisc

STVS: Super Trees, Disc 4, Side 2

11-3 Seed Plant Reproduction

New Science Words

ovules
pollen grains
stamen
pistil
ovary
pollination

Objectives

- ▶ Sequence the stages in the life cycles of typical gymnosperms and angiosperms.
- ▶ Describe the structure and function of the flower.
- ▶ Describe methods of seed dispersal in seed plants.

Gymnosperm Reproduction

Have you ever collected pine cones in the fall to use in making decorations? If you have, you've noticed that pine cones come in many different shapes and sizes. Some people can identify a pine tree just by looking at its cones.

Figure 11-9. Male cones release clouds of tiny pollen grains.

Pine trees are typical gymnosperms. Each pine tree produces male cones and female cones on the sporophyte tree. Female cones consist of a spiral of woody scales on a short stem. Two **ovules** (OHV
yewls) are produced on top of each scale. 1
As an ovule develops, it produces an egg cell, nutrient tissue, and a sticky fluid.
Pollen grains containing sperm develop 2
on the smaller male cone. Wind carries pollen grains to the female gymnosperm cones. Millions of pollen grains may be released in a great cloud from a single male cone. You can see this in Figure 11-9. Most of this pollen never reaches the female cones because the wind blows it onto other plants, water, and the ground. When a pollen grain does get blown between the scales of a female cone of the same species, it gets caught in the sticky fluid secreted by the ovule. A pollen tube grows from the pollen grain to the ovule. A sperm swims down the pollen tube and fertilizes the egg cell. As a result, a zygote forms that develops into an embryo.

OPTIONS

Meeting Different Ability Levels

For Section 11-3, use the following **Teacher Resource Masters** depending upon individual students' needs.

◆ **Study Guide Master** for all students.

● **Reinforcement Master** for students of average and above average ability levels.

▲ **Enrichment Master** for above average students.

Additional Teacher Resource Package masters are listed in the OPTIONS box throughout the section. The additional masters are appropriate for all students.

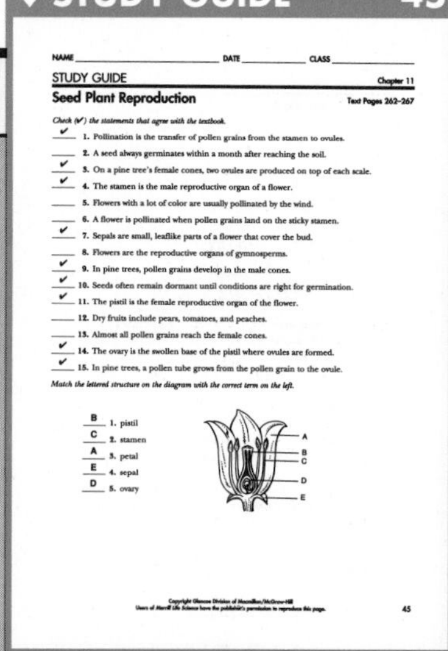

◆ STUDY GUIDE 45

NAME ________ DATE ________ CLASS ________

STUDY GUIDE — Chapter 11

Seed Plant Reproduction — Text Pages 262-267

Check (✔) the statements that agree with the textbook.

✔ 1. Pollination is the transfer of pollen grains from the stamen to ovules.
___ 2. A seed always germinates within a month after reaching the soil.
✔ 3. On a pine tree's female cones, two ovules are produced on top of each scale.
✔ 4. The stamen is the male reproductive organ of a flower.
___ 5. Flowers with a lot of color are usually pollinated by the wind.
___ 6. A flower is pollinated when pollen grains land on the sticky stamen.
✔ 7. Sepals are small, leaflike parts of a flower that cover the bud.
___ 8. Flowers are the reproductive organs of gymnosperms.
✔ 9. In pine trees, pollen grains develop in the male cones.
✔ 10. Seeds often remain dormant until conditions are right for germination.
✔ 11. The pistil is the female reproductive organ of the flower.
___ 12. Dry fruits include pears, tomatoes, and peaches.
___ 13. Almost all pollen grains reach the female cones.
✔ 14. The ovary is the swollen base of the pistil where ovules are formed.
✔ 15. In pine trees, a pollen tube grows from the pollen grain to the ovule.

Match the lettered structure on the diagram with the correct term on the left.

B 1. pistil
C 2. stamen
A 3. petal
E 4. sepal
D 5. ovary

45

Female cones of pine trees mature, open, and release their seeds during the fall or winter months. It may take a long time for seeds to be released from a pine cone. From the moment a pollen grain falls on the female cone until the seeds are released may take as long as two years. Once seeds are released from a pine cone, they are carried away, eaten, or buried by animals. The buried seeds may germinate and eventually grow into new pine trees.

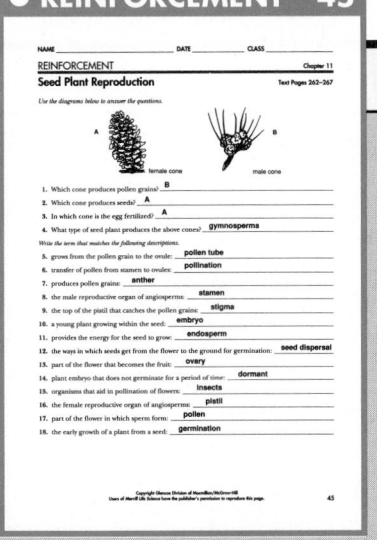

MINI-Lab

What's in a pine cone?

Using a hand lens, look at the parts that make up a gymnosperm cone. Examine the parts and draw some **in your Journal**. On a large paper towel, open the cone. Where are the seeds located? Draw some of the seeds. *Predict* how this location is an advantage for the tree species.

Angiosperm Reproduction

3 Angiosperms all produce flowers. Flowers are important because they are the reproductive organs of angiosperms. Some flowers are not brightly colored, and many don't smell at all. Have you ever looked at the flowers of wheat, rice, or grass? Why do you think there is such variety among flowers?

What is a flower?

The Flower

The shape, size, or color of flowers tells you something about the life of the plant. Large flowers with brightly colored petals often attract insects. In turn, these insects help to pollinate the flowers. Flowers that aren't colored often depend on wind for pollination. Their petals may be small, or the flowers may have no petals at all! Flowers that open at night often have very strong scents.

4 Usually the colored parts of a flower are the petals. Outside the petals are some small, leaflike parts called sepals. Sepals are easy to see when a flower is still a bud because they cover the bud. In some flowers, the sepals are as colorful as or even larger than petals.

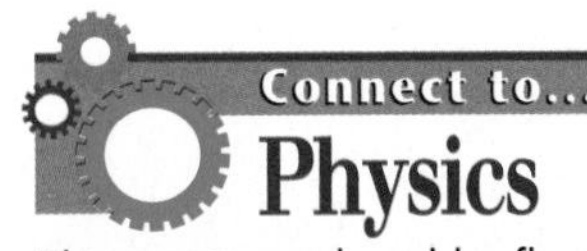

The scents produced by flowers are volatile compounds. Such compounds diffuse into the air as the kinetic energy of air molecules strike the scent molecules. Why would you be able to smell a group of flowers better on a windless day than a windy one?

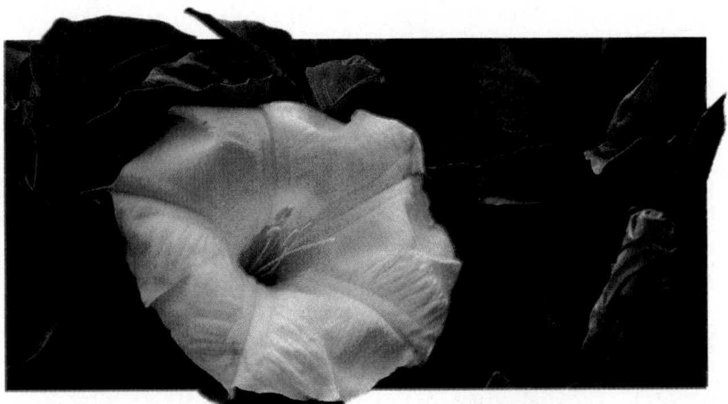

Figure 11-10. Some flowers bloom at night, depending on pollinators that are active at night.

● REINFORCEMENT 45

NAME ______ DATE ______ CLASS ______

REINFORCEMENT — Chapter 11

Seed Plant Reproduction — Text Pages 262–267

Use the diagrams below to answer the questions.

A — female cone; B — male cone

1. Which cone produces pollen grains? **B**
2. Which cone produces seeds? **A**
3. In which cone is the egg fertilized? **A**
4. What type of seed plant produces the above cones? **gymnosperms**

Write the term that matches the following descriptions.

5. grows from the pollen grain to the ovule: **pollen tube**
6. transfer of pollen from stamen to ovules: **pollination**
7. produces pollen grains: **anther**
8. the male reproductive organ of angiosperms: **stamen**
9. the top of the pistil that catches the pollen grains: **stigma**
10. a young plant growing within the seed: **embryo**
11. provides the energy for the seed to grow: **endosperm**
12. the ways in which seeds get from the flower to the ground for germination: **seed dispersal**
13. part of the flower that becomes the fruit: **ovary**
14. plant embryo that does not germinate for a period of time: **dormant**
15. organisms that aid in pollination of flowers: **insects**
16. the female reproductive organ of angiosperms: **pistil**
17. part of the flower in which sperm form: **pollen**
18. the early growth of a plant from a seed: **germination**

45

▲ ENRICHMENT 45

NAME ______ DATE ______ CLASS ______

ENRICHMENT — Chapter 11

Seed Plant Reproduction — Text Pages 262–267

HOW MUCH POLLEN DOES A FLOWER MAKE?

Materials

- 3 sheets of dark construction paper
- ruler
- hand lens
- pencil
- 3 flowers

Procedure

1. Use your ruler to divide each sheet into eight equal parts.
2. Obtain three flowers from a local flower shop or from your yard.
3. Take one flower and one sheet of divided construction paper. Place the paper on a table. Turn the flower upside down and shake the pollen from the flower over the entire paper.
4. With the hand lens, count the number of grains of pollen in one section of the sheet. Then multiply that number by eight to get the estimated total number of grains that fell from the flower.
5. Repeat this process for each of the other two flowers.
6. Add the three totals together and then divide by three to obtain the average number of pollen grains that fell from each flower.

Conclude and Apply

1. What was the average number of pollen grains that fell from each flower? **Answers will vary widely. A flower can produce millions of pollen grains.**
2. From what part of the flower are these pollen grains falling? **from the stamen**
3. Why would a flower make so many pollen grains? **The pollen grains are carried by insects or wind to other flowers of the same kind to pollinate those flowers. There must be a great number of pollen grains because that process is not very efficient.**

45

1 MOTIVATE

▶ Show several different kinds of seeds to the students. Have the students explain how each seed is dispersed from the parent plant.

TYING TO PREVIOUS KNOWLEDGE: Anyone who has blown the fluff off a dandelion stem has acted as a seed dispersal mechanism. Ask students to recall any other times they have helped in the dispersal of seeds. One answer may be—eating a watermelon.

2 TEACH

Key Concepts are highlighted.

Answer: Scent molecules would remain closer together and closer to the source on a windless day.

MINI-Lab

Materials: cones with seeds, paper toweling

Answer: Students will find seeds between the woody scales. Students may predict that protection of the seeds insures that more seeds will survive.

MINI-Lab ASSESSMENT

Performance: To further assess students' understanding of gymnosperm cones, have students place several cones in the refrigerator overnight, then examine them on removal and several hours later. Have students state in their journals, how cone activity might be an advantage to gymnosperms.

CONCEPT DEVELOPMENT

▶ Have students identify whether a plant is a monocot or dicot by observing whether flower parts are in groups of three, four, or five.

▶ **Demonstration:** Use a model of a perfect flower to show students the structures. Emphasize the functions for each structure.

Cooperative Learning: Have students work in Expert Teams to write a short story about a particular kind of flower and its pollination mechanism. The story should include a day in the life of the flower.

CHECK FOR UNDERSTANDING

Use an overhead transparency to explain the sequence of gymnosperm and angiosperm life cycles.

RETEACH

Bring in plants, flowers, and seeds to demonstrate the parts of the life cycles of these plants. Have students place the plants and parts in the proper sequence or dissect the flower with a hand lens. Have students write a paragraph that details the sequence according to their placement.

EXTENSION

For students who have mastered this section, use the **Reinforcement** and **Enrichment** masters or other OPTIONS.

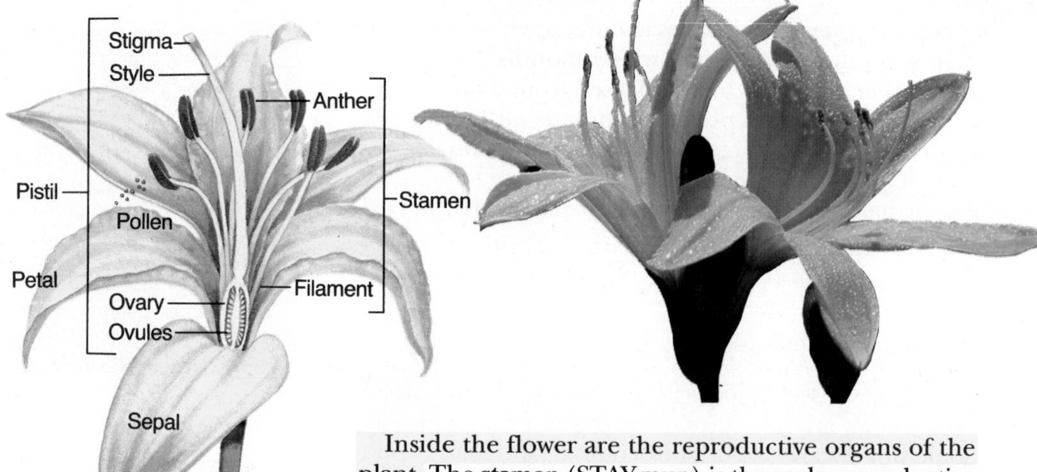

Figure 11-11. Parts of a Flower

What is a pistil?

Inside the flower are the reproductive organs of the plant. The **stamen** (STAY mun) is the male reproductive organ of the flower. Stamens consist of a filament and an anther where pollen grains form. The sperm develop in the pollen grain.

The **pistil** (PIHS tul) is the female reproductive organ 5
of a flower. Pistils consist of a sticky stigma where the pollen grains land, a long stalklike style, and an ovary. In angiosperms, the **ovary** is the swollen base of the pistil where ovules are formed. Eggs are produced as the ovule develops. You can see the parts of a typical flower in Figure 11-11.

Development of a Seed

How does a seed develop? Pollen grains reach the ovule in a variety of ways. Pollen is carried by the wind or by animals such as insects, birds, and mammals. A flower is pollinated when pollen grains land on the sticky stigma. The pollen tube grows from the pollen grain down through the style and into the ovary until it reaches the ovules. The sperm then travels down the pollen tube and unites with, or fertilizes, the egg. This zygote develops into the plant embryo part of the seed. The transfer of pollen grains from the stamen to ovules is the process of **pollination.**

Figure 11-12. The Parts of a Seed

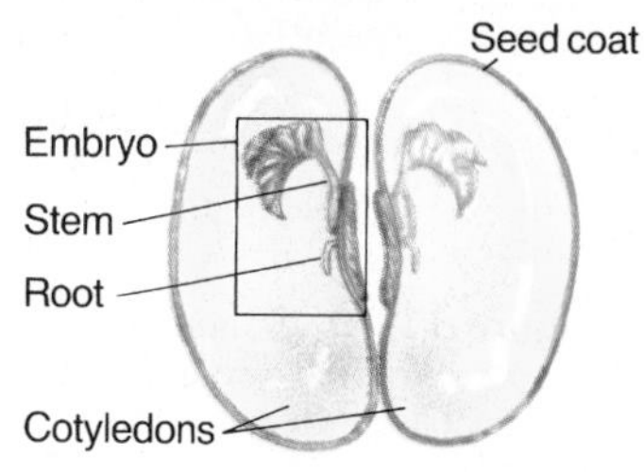

An embryo plant consists of cotyledons, stem, and root. Protected by a seed coat, the embryo also has a supply of stored food called the endosperm. After the seed germinates, the embryo will use the endosperm until it can begin to produce food on its own. You can see the parts of a seed in Figure 11-12.

OPTIONS

PROGRAM RESOURCES

From the **Teacher Resource Package** use:

Activity Worksheets, page 5, Flex Your Brain.

Cross-Curricular Connections, page 15, Planning a Vegetarian Meal.

Transparency Masters, pages 37-38, Parts of a Flower.

Use **Color Transparency** 19, Parts of a Flower.

Seed Dispersal and Germination

How do seeds get from the flower to the ground for germination? You know that the seeds grow inside the ovary. As an ovary grows larger, it becomes a fruit. Fruits may be either fleshy, like peaches or apples, or dry, like a walnut. As fruits are eaten by animals, the seeds are moved to new locations where they may be able to germinate. When you eat watermelon and spit out the seeds, you move the watermelon seeds to a new place! Both dry and fleshy fruits protect the seeds. Some seeds cannot germinate until the fruit has been digested by an animal, and the seeds are released in the animal's feces. Seeds are also dispersed by the wind or water or carried along on fur or clothing.

MINI-Lab

How do plants disperse seeds?

Look at the seeds pictured in Figure 11-13. How are the seeds of each plant dispersed—by wind, water, insects, birds, or mammals? *Identify* features that tell you how each is dispersed?

a

b

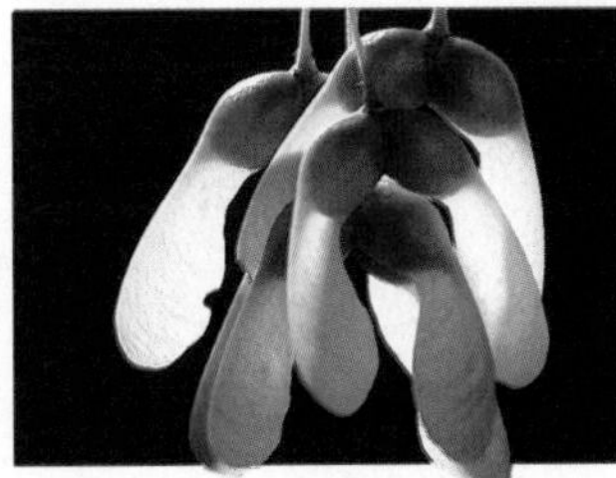

c

Figure 11-13. Seeds can be dispersed by floating (a); attaching to fur, feathers, or clothing (b); or wind (c).

PROBLEM SOLVING

How Can You Tell If Seeds Are Living?

Aurora visited her grandfather, who lived in a nursing home. They went to buy flower seeds to plant in the courtyard of the nursing home. When examining a package of petunia seeds, Aurora noticed that the package said the seeds were 95 percent viable. Her grandfather explained that *viable* meant living. He said that 95 out of 100 seeds would germinate and grow.

They purchased the petunia seeds and several other packages of flower seeds and headed back to the nursing home to plant them. How could Aurora and her grandfather determine if the seeds they planted in the courtyard were viable?

Think Critically: How could Aurora determine if the seed company's claim on the seed package were true?

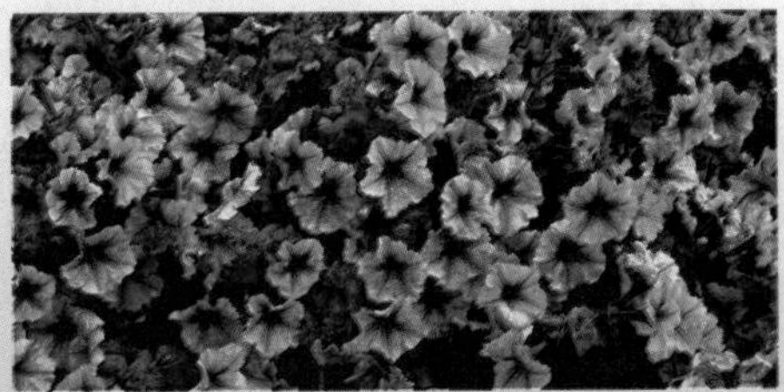

MINI QUIZ

Use the Mini Quiz to check students' recall of chapter content.

1. **Two ____ are produced on top of each scale.** *ovules*
2. **_____ containing the sperm, or male gametophytes, develop on the male cone.** *Pollen grains*
3. **Angiosperms all bear _____ .** *flowers*
4. **Flowers that open at _____ have very strong scents.** *night*
5. **The _____ is the female reproductive organ of the flower.** *pistil*
6. **Usually the _____ begins to elongate and grow out of the seed first.** *root*

MINI-Lab

Materials: pictures of types of seeds including: coconuts, grasses, burrs, and winged seeds

Answer: Coconuts and sedges are dispersed by water. They are buoyant and float from place to place. Cockleburs catch onto fur, feathers, or clothing with tiny "hooks." They fall off or are picked off in a different place. Maple and ash seeds are carried on winged samaras and are dispersed by the wind.

PROBLEM SOLVING

Think Critically: If the seeds Aurora and her grandfather planted germinate and grow, the seeds are viable. Aurora could soak 100 seeds and count the number that germinate. If 95 or more of the seeds germinate, the claims of the seed company are true.

PROBLEM SOLVING

ASSESSMENT

Performance: To further assess students' understanding of viability, see USING LAB SKILLS, Question 12, p. 272.

ENRICHMENT

Visit an art museum to observe how many pieces of art contain flowers or other plant parts.

PROGRAM RESOURCES

From the **Teacher Resource Package** use:

Concept Mapping, pages 27-28.

Critical Thinking/Problem Solving, page 15, Saving Endangered Plants.

Activity Worksheets, page 102, Mini-Lab: How do plants disperse seeds?

3 CLOSE

▶ Ask questions 1-3 and the **Apply** Question in the Section Review.

SECTION REVIEW ANSWERS

1. The life cycles of angiosperms and gymnosperms are similar except that the angiosperms produce their gametophytes in flowers and gymnosperms produce gametes, or sex cells, in cones.
2. Student diagrams may resemble those in Figure 11-11.
3. wind, eaten, carried in fur
4. Apply: This arrangement prevents the tree's flowers from fertilizing themselves. Because the wind blows the pollen, it will tend to fall onto the pistils of nearby plants and so will not land on the flowers above it.
5. Connect to Physics: The force is generated by water absorbed by the roots.

Skill Builder

The traits listed are all common to wind-pollinated plants.

Skillbuilder
ASSESSMENT

Performance: Assess students' abilities to Observe and Infer how a flower is pollinated by giving each student a flower and having them identify how this particular specimen would be pollinated.

Figure 11-14. In dicots, the cotyledons raise the seed above the soil. As the endosperm is used the cotyledons shrivel and fall off.

What is germination?

When a seed reaches the soil, it may not germinate right away. Seeds often remain dormant until conditions are right for germination. Some seeds can remain dormant for hundreds of years, yet still be able to grow when environmental conditions become favorable. In 1982 seeds of the East Indian lotus germinated after 466 years!

Germination, or the development of a seed into a new plant, begins when water is taken into the seed tissues. Water signals the seed to begin growth. The endosperm provides the energy for the seed to grow. Usually the root begins to elon- 6
gate and grow out of the seed first. If the seed is a monocot, the seed containing the endosperm remains below the soil. In a dicot seed, such as a bean, the cotyledons contain most of the endosperm of the seed.They also raise the seed above the surface. Eventually, as the food is used, the cotyledons shrivel and fall off dicot plants. Then the plant takes over and makes all of its own food by photosynthesis.

SECTION REVIEW

1. Compare life cycles of angiosperms and gymnosperms.
2. Diagram the parts of a flower and label their functions.
3. List three methods of seed dispersal in plants.
4. **Apply:** Some conifers bear female cones on the top half of the tree and male cones on the bottom half. What do you think this arrangement of male and female cones on trees helps to prevent or promote?
5. **Connect to Physics:** A watered seed will soon germinate. What force enables the seedling to push up through soil?

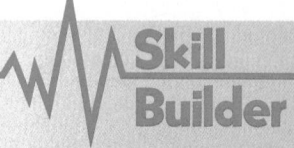

Skill Builder — **Hypothesizing**

A corn plant produces thousands of pollen grains on top of the plant in flowers that have no odor or color. The pistils grow from the cob lower down on the plant. Explain how a corn plant is probably pollinated. If you need help, refer to Hypothesizing in the **Skill Handbook** on page 686.

OPTIONS

ASSESSMENT—ORAL

▶ **What is the pollination agent in pine trees?** *Pine trees are pollinated by the wind.*

▶ **If a flower is large and brightly colored, what is its likely pollinator?** *Large, brightly colored flowers are usually pollinated by insects.*

▶ **What is the gametophyte generation of an angiosperm?** *The pollen contains the male gametophyte, and the ovule contains the female gametophyte of flowering plants.*

▶ **Explain at least three ways in which animals can be effective in dispersing seeds.** *Animals may ingest the seed, carry the seed in fur or feathers, or bury the seed in another location.*

ACTIVITY
11-1

Inside a Seed

Problem: *What can you see inside a seed?*

Materials  

- soaked lima bean
- soaked seed corn
- forceps
- scalpel
- iodine solution
- hand lens

Procedure

1. Make a data table in your Journal.
2. Examine the soaked corn seed. Place the seed flat on the table and, using a scalpel, cut through the center of the light-colored area on the front of the seed. **CAUTION:** *Always be careful when handling a scalpel.*
3. **Hypothesize** which part or parts contain starch for food. Place a drop of iodine solution on different parts of the seed to test for starch. A blue-black color indicates the presence of starch. **CAUTION:** *Iodine is poisonous and may stain and burn the skin.*
4. Locate and *observe* the embryo corn plant with its single cotyledon. Find the embryo stem and the embryo root. Draw and label the seed **in your Journal.**
5. Using the hand lens, *observe* the lima bean.
6. Locate the two thick cotyledons and the embryo plant. Separate the cotyledons. Draw and label the seed **in your Journal.**
7. Test the lima bean for starch. Which part or parts contain starch?

Data and Observations Sample Data

	Number of Leaves	Size of Endosperm
Monocot	1	smaller
Dicot	2	larger

Analyze

1. What is the function of the outer covering of the seed?
2. Which parts tested positive for starch?
3. What is the function of the endosperm?
4. What parts are found in both monocot and dicot seeds?

Conclude and Apply

5. Which seed had the larger cotyledon? Is there an adaptive advantage for a seed to have a larger cotyledon? Explain.
6. How can you tell the difference between a monocot and a dicot by dissecting the seed?
7. **Design an experiment** to show which parts of the embryo become roots, stems, or leaves.

ACTIVITY 11-1

30 minutes

OBJECTIVE: Students will **observe** the parts of monocot and dicot seeds and **test** for the presence of starch in the endosperm.

PROCESS SKILLS applied in this activity:
▶**Observing** in Procedure Steps 3 through 5.
▶**Inferring** in Analyze Question 3, and Conclude and Apply Question 6.

COOPERATIVE LEARNING Have students work in Science Investigation cooperative groups. Each student should be responsible for the dissection of one of the seeds.

TEACHING THE ACTIVITY

Alternate Materials: Any monocot or dicot seeds can be used. However, the larger the seeds, the easier it will be for students to identify the various parts.

Troubleshooting: Be sure the seeds have been soaked in water for at least 24 hours. Some students will find it difficult to orient the seeds properly for cutting. You should demonstrate where to make the cuts.

PROGRAM RESOURCES

From the **Teacher Resource Package** use:

Activity Worksheets, pages 95-96, Activity 11-1, Inside a Seed.

ANSWERS TO QUESTIONS

1. protects the embryo
2. root
3. The endosperm contains stored food in the form of starch.
4. cotyledons, stem, root, seed coat, embryo
5. The dicot has the largest cotyledon. A larger cotyledon contains more food for the embryo. It can grow bigger before it must make its own food.
6. The dicot has two cotyledons, and the monocot has only one.
7. Student designs will vary greatly. A large lima bean might be marked with different colors of permanent marker and allowed to germinate. Students would have to watch development carefully on a daily basis to track changes.

Activity
ASSESSMENT

Performance: Use germinated seeds to have students answer questions placed on index cards about plant parts and presence or absence of starch.

PREPARATION

SECTION BACKGROUND

▶ The first reports of acidification of lakes came from Sweden and Norway, several hundred kilometers from the source of the pollution in Europe.

REVEALING MISCONCEPTIONS

▶ Point out that acid precipitation is a weak acid, usually much weaker than the vinegar in a pickle. When students think of acids they think only of concentrated hydrochloric or similar acids.

1 MOTIVATE

▶ Squeeze a lemon into a beaker, and dip a piece of pH paper into the juice. Point out that the most acidic precipitation was only slightly less than that of lemon juice.

2 TEACH

Key Concepts are highlighted.

VideoDisc

STVS: Acid Rain and Plants, Disc 6, Side 2

STVS: Fish and Acid Rain, Disc 6, Side 2

11-4 Effects of Acid Rain

New Science Words

acid rain

Objectives

- ▶ State the primary causes of acid rain.
- ▶ Explain why the solution to this problem requires international cooperation.

Acid Rain: An International Problem

Suppose you and your friends go on a camping trip. As you are walking into the forest you are shocked by the death around you. The trees are brown. There are no plants on the forest floor. You can hear no birds or animals. You go fishing in a small lake, but there are no fish there to catch.

It is hypothesized that during the 1980s, southeastern Canada, northeastern and northwestern United States, and much of Europe began to experience the effects of acid rain. **Acid rain** is precipitation that falls to Earth after combining with sulfur dioxide and nitrogen oxide in the air. Sulfur dioxide and nitrogen oxide are gases released when fossil fuels such as coal and oil are burned. Sulfur dioxide typically enters the air in emissions from coal-burning power plants, whereas nitrogen oxide is released in the exhuast of automobiles. In the air, sulfur dioxide and nitrogen oxide combine with water vapor to form sulfuric acid and nitric acid. When it rains, snows, or fog covers the ground, these acids land on water or soil. Acid rain is particularly harmful to organisms that live in water.

Figure 11-15. Forests affected by acid rain often die from the top down.

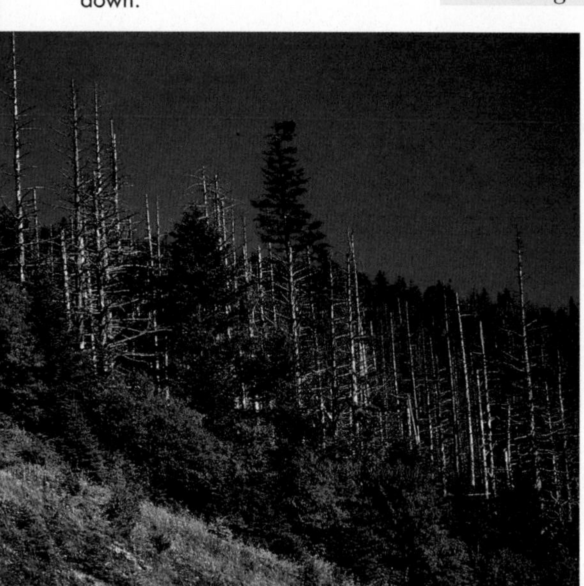

Hundreds of thousands of lakes in the United States and Canada have been damaged due to acid rain. Forests, buildings, outdoor artwork, and groundwater are also affected by this problem. It is thought that the decline in maple syrup production in the United States and Canada over the last

OPTIONS

Meeting Different Ability Levels

For Section 11-4, use the following **Teacher Resource Masters** depending upon individual students' needs.

◆ **Study Guide Master** for all students.

● **Reinforcement Master** for students of average and above average ability levels.

▲ **Enrichment Master** for above average students.

Additional Teacher Resource Package masters are listed in the OPTIONS box throughout the section. The additional masters are appropriate for all students.

◆ STUDY GUIDE 46

NAME ______ DATE ______ CLASS ______

STUDY GUIDE Chapter 11

Effects of Acid Rain Text Pages 268–269

Cross out the statements that DO NOT agree with the textbook.

1. ~~European countries have no acid rain problems.~~
2. ~~All of the acid rain in Canada results from pollution produced in Canada.~~
3. A scrubber blows a fine mist of water through the gases released from burning fuel.
4. Acid rain contains sulfur dioxide and nitrogen oxide that were released into the air as pollution.
5. Acid rain causes the chemistry of soil to change.
6. ~~Acid rain does not include snow or fog.~~
7. ~~Only a few lakes in Canada and the United States have been affected by acid rain.~~
8. About 75 percent of the acid rain in some parts of Canada is the result of pollution from the United States.
9. Outdoor artwork can be harmed by acid rain.
10. Acid rain is sometimes a thousand times more acidic than normal rain.
11. ~~Scrubbers are cheap to use and install.~~
12. Nitrogen oxides are released by automobile exhausts.
13. Many forests in Europe have been badly damaged by acid rain.
14. The maple syrup industry in the United States and Canada has been harmed by acid rain.
15. Studies show that decreasing the amount of chemicals released into the air decreases the harm done to Earth.
16. Acid rain can be in the form of rain, snow, or fog.
17. ~~Buildings are too solid to be affected by acid rain.~~
18. ~~Fish cannot be harmed by acid rain.~~
19. A source of sulfur dioxide pollution is power generating stations that burn coal.
20. ~~Sulfur dioxide and nitrogen oxide combine with water vapor in the air to form hydrochloric acid and nitric acid.~~

46

ten years may be due to acid rain. Many forests in Europe are badly damaged. Acid rain causes the chemistry of the soil to change. Some minerals such as calcium and iron are washed away, while unusual amounts of other minerals such as aluminum are released from the soil and taken up by plants.

Much of the acid rain that falls in the eastern parts of the United States and Canada was produced in the Midwest. Scientists in Canada and the United States are working together to try and solve this vast problem. In some parts of Canada, up to 75 percent of the acid rain is due to pollution from the United States. Most of the emissions that result in acid rain in the United States originated here. Studies show that by decreasing the amount of these chemicals released into the air, the damage they cause can be greatly decreased. Many industries in the United States and Canada use scrubbers to remove some of these chemicals. A scrubber is a machine that blows a fine mist of water through gases leaving a power plant's smokestacks. Sulfur dioxide and nitrogen oxide combine with the water in the mist instead of water vapor in the air. The scrubbers are very expensive, yet they prevent much of the sulfur dioxide and nitrogen oxide from being released.

Other means of reducing the emission of the chemicals that cause acid rain are also being developed. It is hoped that each country will develop its own acid rain control program to allow all countries to benefit from an undamaged environment.

Figure 11-16. Installing scrubbers in smokestacks can reduce emissions that result in acid rain.

Acid rain is carried from its sources by winds. Because of this, acid rain is a global problem. Find out what trade winds, prevailing westerlies, polar easterlies, and jet streams are.

SECTION REVIEW

1. What chemicals are responsible for acid rain?
2. **Connect to Earth Science:** Why is acid rain an international issue?

You Decide!

Some people believe the damaging effects of acid rain are not important enough to discontinue the use of coal and other fossil fuels. Others feel that the use of fossil fuels should be minimized unless expensive scrubbers are used. How do you think the problem should be solved?

● REINFORCEMENT 46

NAME ______ DATE ______ CLASS ______

REINFORCEMENT — Chapter 11

Effects of Acid Rain — Text Pages 268–269

Write the pollutants released in the space provided next to each picture.

1. nitrogen oxide
2. sulfur dioxide
3. Why might a Canadian be upset with automobile drivers in the United States? **The automobile of a driver in the U.S. releases nitrogen oxide in the exhaust. Wind takes the polluted air into Canada, where it falls as acid rain.**
4. What forms can acid rain take? **rain, snow, or fog**
5. How does acid rain affect soil? **It changes the chemistry of soil. Some minerals, such as calcium and iron, are washed away, while unusual amounts of other minerals are released and taken up by plants.**
6. What has been damaged by acid rain? **lakes, forests, soil, groundwater, buildings, outdoor artwork, plants, animals, industries**
7. How does a scrubber work? **The scrubber blows a fine mist of water through the gases released from burning fuel.**
8. What does a scrubber prevent from being released into the air? **sulfur dioxide and nitrogen oxide**
9. Why would industries not want to install scrubbers on their smokestacks? **Scrubbers are very expensive.**

46

▲ ENRICHMENT 46

NAME ______ DATE ______ CLASS ______

ENRICHMENT — Chapter 11

Effects of Acid Rain — Text Pages 268–269

ACIDITY

Chemists classify substances by saying a substance is either an acid or a base. But some acids are more acidic than others. Think of lightness and darkness. Pitch black darkness is one extreme. Bright light is the other extreme. In between are various shades of lightness and darkness. In chemistry, "acid" is at one end of the scale, while "base" is at the other. Midway between is called "neutral." By studying a substance, chemists can place it somewhere on the scale. Chemists call this scale the pH Scale. The pH Scale measures the amount of hydrogen ions present in a substance. The scale ranges from 0 to 14, with 7 being neutral.

Study this pH Scale. It shows where on the scale some common substances fall. Using the scale, answer the questions that follow.

0 Acid — Acid Rain — 7 Neutral — 14 Base
battery acid, lemon juice, vinegar, coffee, tomato, normal rain, milk, distilled water, lard, baking soda, milk of magnesia, ammonia, lime (calcium oxide)

1. What is the most acidic substance shown on this scale? **battery acid**
2. What is the most basic (or base) substance shown on this scale? **lime**
3. What substance is neutral on the pH scale? **distilled water**
4. Which is more acidic, coffee or baking soda? **coffee**
5. Which is more basic, vinegar or ammonia? **ammonia**
6. Some people take milk of magnesia to "neutralize their stomach acid." What property of milk of magnesia is being used by those people? **Milk of magnesia is a base.**
7. Would you say that normal rain is slightly acidic or slightly basic? **slightly acidic**
8. The worst acid rain is almost as acidic as what substance? **battery acid**

46

Answer: *Trade wind*—constant easterly winds across tropics and subtropics; *prevailing westerlies*—winds moving toward the west; *polar easterlies*—wind blowing toward the east; *jet stream*—strong, westerly, narrow band in upper troposphere

3 CLOSE

CHECK FOR UNDERSTANDING

▶ Ask questions 1 and 2 in the Section Review.

RETEACH

Put distilled water in a bottle. Test it with pH paper to see that the water has a neutral pH. Put a lid on the bottle, and lower a burning piece of sulfur into it. When the sulfur fire goes out, remove the sulfur. Shake the bottle and retest the pH. Explain that the acidity increased when the sulfur dioxide combined chemically with the water.

EXTENSION

For students who have mastered this section, use the **Reinforcement** and **Enrichment** masters or other OPTIONS.

SECTION REVIEW ANSWERS

1. sulfur dioxide and nitrogen oxide
2. **Connect to Earth Science:** The countries producing the emissions that cause acid deposition are not the countries where the deposition occurs.

YOU DECIDE!

Answers will vary. All acceptable answers will be supported with clear and logical thinking.

PROGRAM RESOURCES

From the **Teacher Resource Package** use:

Science and Society, page 15, The Acid Rain Problem.

Science Integration Activities, 11, Acid Rain

ACTIVITY 11-2

OBJECTIVE: Design and carry out an experiment to identify the parts of a flower
Time: one class period

PROCESS SKILLS applied in this activity are **observing, inferring,** and **comparing.**

PREPARATION

Collect materials needed for each group.

Cooperative Learning: Have students work in Science Investigation Teams.

SAFETY

Caution students to be careful using sharp instruments such as the scalpel

THINKING CRITICALLY

Students will list such structures as the sepals, petals, stamens, and pistils. To locate these parts and some of the subparts will require a hand lens, microscope, and a cutting tool such as a scalpel.

TEACHING THE ACTIVITY

Refer to the ***Activity Worksheets*** *for additional information and teaching strategies.*

- Other flowers that may be used are tulips or lilies.
- Free or inexpensive flowers may be obtained from florists or the flower department in large supermarkets.

SUMMING UP/SHARING RESULTS

- There are three stamens, three petals, and three sepals.
- The anther has four pollen sacs; the stigma has three parts.
- three
- The petals are colorful and might attract insects.
- Pollen may fall onto the stigma or be carried by insects, birds, water, or wind.
- It is sticky.

ACTIVITY 11-2 DESIGNING AN EXPERIMENT
Parts of a Flower

In order to understand how seeds are produced, you will need to know the parts of a flower.

Getting Started

You need to determine what type of flower to dissect and how to identify its parts.

Thinking Critically

List some structures of the flower that you will want to identify. What will you need to do to locate them?

Materials

Your cooperative group will use:

- gladiolus flower
- scalpel
- hand lens
- black paper
- microscope
- microscope slide
- coverslip
- dropper
- water

Try It!

1. Record your observations in a data table.
2. *Examine* a gladiolus flower using Figure 11-11 on page 264 as a guide.
3. Remove the outer row of sepals. The leaflike structures inside the sepals are the petals. Remove the petals.
4. Locate and remove the stamens. *Examine* the stamens with a hand lens. *Observe* the anther and the filament.
5. Tap the anther on black paper to knock out the pollen grains. *Observe* under a microscope.
6. The structure that remains is the pistil. Identify the stigma, the style, and the pistil.
7. Use a scalpel to make a cross section of the ovary. **CAUTION:** *Always be careful with sharp instruments.* Use a hand lens to look at the ovules. Make a drawing in your table.

Summing Up/Sharing Results

- How do the numbers of stamens, petals, and sepals compare?
- Describe the appearance of the anther and the stigma.
- How many compartments are in the cross section of the ovary?
- What functions might the petals have?
- How might pollen travel to the stigma?
- What adaptations does the stigma have for trapping pollen grains.

Going Further!

Observe other types of flowers and their parts. *Compare* monocot and dicot flowers. Are there any differences in the number of flower parts? Is there a pattern to the number of parts?

GOING FURTHER!

There are differences in the number of flower parts between monocots and dicots. A pattern that can be seen is that monocots will have flower parts in multiples of three, while dicots will have flower parts in multiples of four or five.

Activity ASSESSMENT

Performance: To further assess students' understanding of flower parts, see MORE SKILL BUILDERS, Question 15, on page 273.

PROGRAM RESOURCES

From the **Teacher Resource Package** use:
Activity Worksheets, pages 97-98, Activity 11-2, Parts of a Flower.

CHAPTER

REVIEW

SUMMARY

11-1: Seed Plants

1. Seed plants have stems, roots, and leaves, and grow from seeds.
2. Gymnosperms produce seeds on cones; angiosperms produce seeds in flowers.
3. Seed plants are sources of oxygen, food, clothing, furniture, and other products.

11-2: Parts of Complex Plants

1. Roots absorb water and minerals, store food, and anchor the plant.
2. Stems transport food and water between roots and leaves and support the plant.
3. The palisade layer of leaves carries out photosynthesis. Xylem and phloem tissues are located in the spongy layer of the leaf.

11-3: Seed Plant Reproduction

1. Wind carries gymnosperm pollen to female cones where fertilization occurs.
2. In angiosperms, pollen reaches the stigma and travels to the ovary to fertilize the ovules.
3. Flowers contain male and female structures. The female pistils consist of the stigma, style, ovary, and ovules. The male stamens consist of the filament and anther, where pollen grains are formed.

11-4: Science and Society: Effects of Acid Rain

1. Acid rain results when sulfur dioxide and nitrogen oxide combine with water vapor in the air to form sulfuric acid and nitric acid.
2. Most acid rain results from emissions from the burning of fossil fuels such as coal and oil.
3. Acid rain may travel far from its source and cause problems in other states or countries. International cooperation is needed to solve this problem.

KEY SCIENCE WORDS

a. **acid rain**
b. **angiosperms**
c. **cambium**
d. **dicots**
e. **guard cells**
f. **gymnosperms**
g. **monocots**
h. **ovary**
i. **ovules**
j. **phloem**
k. **pistil**
l. **pollen grains**
m. **pollination**
n. **stamen**
o. **stomata**
p. **xylem**

UNDERSTANDING VOCABULARY

Match each phrase with the correct term from the list of Key Science Words.

1. transfer of pollen grains to ovules
2. cone-bearing plants
3. male part of a flower
4. small pores in leaf surfaces
5. water containing sulfur dioxide
6. flower-producing plants
7. female part of a flower
8. vessels that transport water and minerals
9. plants with one embryo leaf
10. moves food from leaves and stems

OPTIONS

ASSESSMENT

To assess student understanding of material in this chapter, use the resources listed.

COOPERATIVE LEARNING

Consider using cooperative learning in the THINK AND WRITE CRITICALLY, APPLY, and MORE SKILL BUILDERS sections of the Chapter Review.

PROGRAM RESOURCES

From the **Teacher Resource Package** use:

Chapter Review, pages 25-26.

Chapter and Unit Tests, pages 72-75, Chapter Test.

CHAPTER

REVIEW

SUMMARY

Have students read the summary statements to review the major concepts of the chapter.

UNDERSTANDING VOCABULARY

1. m		**6.** b	
2. f		**7.** k	
3. n		**8.** p	
4. o		**9.** g	
5. a		**10.** j	

ASSESSMENT

Portfolio Suggestions

Encourage students to place in their portfolios one or two items of what they consider to be their best work. For each item, ask students to explain why that item was chosen and what they learned from it. Items might be selected from the following

- In Your Journal letter to the editor, p 256
- Cross Curriculum paintings and drawings of plants, p. 256
- Cooperative Learning short story about flower pollination, p. 264

Performance

Additional performance assessments may be found in *Performance Assessment* and *Science Integration Activities* that accompany **Merrill Life Science.** Performance Task Assessment Lists and rubrics for evaluating these activities and other products generated throughout the chapter can be found in Glencoe's *Performance Assessment in Middle School Science.*

CHAPTER

REVIEW

CHECKING CONCEPTS

1. d	6. b
2. b	7. d
3. a	8. b
4. d	9. b
5. a	10. c

USING LAB SKILLS

ASSESSMENT

Use these alternate exercises to assess students' understanding of skills used in this chapter.

11. Students may need a hand lens and several peanuts to complete their drawings. You may also want to use other seeds.

12. Students should plant 100 seeds and derive the viability percentage. Accuracy would increase with a larger sample, i.e. 1000 seeds.

THINK AND WRITE CRITICALLY

13. Pollen sticks to the stigma on top of the style through which a tube forms allowing the pollen to move down to the ovary.

14. One acceptable hypothesis is that natural selection favors colored petals since pollinators may be attracted by the colors. Other acceptable hypotheses may deal with chemical (scent) attractants.

15. A waxy cuticle protects the epidermis from water loss. Guard cells control the leaf's openings so water is not continually lost.

APPLY

16. Resin prevents fungi and bacterial growth on the tree; insects cannot invade as easily.

17. Succulents can be found in areas that are hot or dry where plants must store water for long periods of time.

18. Water loss is prevented in conditions that are especially dry.

19. Students can describe any brightly colored large flower that they know: goldenrod, lilies, dandelions, and daisies are good examples.

20. Fleshy: apple, cherry, orange
Dry: acorn, almond, peanut

CHAPTER

REVIEW

CHECKING CONCEPTS

Choose the word or phrase that completes the sentence.

1. Gymnosperms include ______.
   - **a.** oaks
   - **b.** flowers
   - **c.** gladiolus
   - **d.** pines
2. Angiosperms produce ______.
   - **a.** cones
   - **b.** flowers
   - **c.** needles
   - **d.** woody scales
3. Colorful flowers are usually pollinated by______.
   - **a.** insects
   - **b.** wind
   - **c.** clothing
   - **d.** birds
4. A ______ is the sticky part of a flower.
   - **a.** sepal
   - **b.** ovary
   - **c.** stamen
   - **d.** stigma
5. The ______ contains food for the embryo.
   - **a.** endosperm
   - **b.** pollen grain
   - **c.** stigma
   - **d.** root
6. The seed leaves of an embryo are ______.
   - **a.** root hairs
   - **b.** cotyledons
   - **c.** guard cells
   - **d.** stomata
7. ______ anchor plants in soil.
   - **a.** Leaves
   - **b.** Stems
   - **c.** Stomata
   - **d.** Roots
8. ______ produces new xylem and phloem.
   - **a.** Epidermis
   - **b.** Cambium
   - **c.** Pollen
   - **d.** An ovule
9. Most photosynthesis occurs in the ______.
   - **a.** guard cells
   - **b.** palisade layer
   - **c.** epidermis
   - **d.** cuticle
10. A ______ provides waxy protection for the leaf.
    - **a.** guard cell
    - **b.** epidermis
    - **c.** cuticle
    - **d.** loose layer

USING LAB SKILLS

11. In the Find Out! on page 253, you looked at the parts of a peanut seed. Now that you know about the different parts of a seed, draw and label the embryo, stem, roots, and endosperm of the peanut.
12. **Design an experiment** to test the viability of bean seeds as in the Problem Solving on page 265. Carry out your experiment. What percentage of viability did you find? How could you increase the accuracy of the result?

THINK AND WRITE CRITICALLY

Answer the following questions in your Journal using complete sentences.

13. What adaptations of flowers help pollination?
14. Both petals and sepals of the gladiolus flower are brightly colored rather than just green. Make a hypothesis about why natural selection might have favored colored petals and sepals over plain green ones.
15. What prevents water loss in leaves?

APPLY

16. What function does resin serve for the gymnosperm that secretes it?
17. Plants called succulents store large amounts of water in their leaves. In what kind of environment would you find succulents?
18. Why would cacti benefit from thick cuticles?
19. Use an example to describe a flower that is pollinated by insects.
20. Classify each fruit as dry or fleshy: acorn, almond, apple, cherry, orange, peanut.

MORE SKILL BUILDERS

If you need help, refer to the Skill Handbook.

1. **Comparing and Contrasting:** Compare the functions of roots, stems, leaves, and seeds in seed plants.
2. **Comparing and Contrasting:** Compare and contrast these structures of monocots and dicots: number of seed leaves, bundles arranged in stems, veins in leaves, flower parts.
3. **Concept Mapping:** Fill in the following concept map to show the sequence of pollination and fertilization.

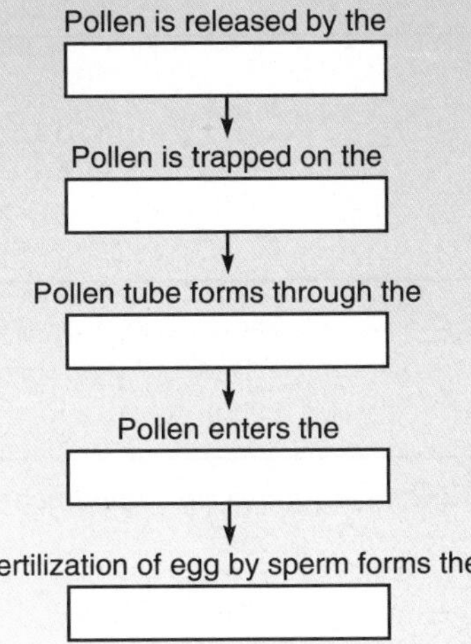

Pollen is released by the
Pollen is trapped on the
Pollen tube forms through the
Pollen enters the
Fertilization of egg by sperm forms the

4. **Interpreting Data:** Interpret the data in the table below concerning stomata of these leaf types. What do these data tell you about where gas exchange occurs in each type?

Stomata (per mm^2)	Upper Surface	Lower Surface
Pine	50	71
Bean	40	281
Fir	0	228
Tomato	12	13

5. **Observing and Inferring:** Observe pictures or specimens of flowers and infer how each is pollinated. What features does each flower have to enable the pollination you suggest?

PROJECTS

1. Identify the types of trees in your yard or at your school. You can collect leaves and press them between papers placed inside a phone book. Identify leaves by using a guide book from the library. Bring your collection to school.
2. Research how hydroponics is used to grow plants. Using this technique, try to grow a plant.

MORE SKILL BUILDERS

1. Comparing and Contrasting: Roots—anchor in soil, absorb water and nutrients
Stems—hold plant upright, transport water, food, and nutrients
Leaves—make food for plant, transpire, move food to other plant parts
Seeds—reproductive part of plant, produce new plant embryos

2. Comparing and Contrasting: Monocots have 1 seed leaf, scattered vascular bundles, parallel veins in leaves, and flower parts in 3s. Dicots have two seed leaves, vascular bundles in rings, net veins in leaves, and flower parts in 4s and 5s.

3. Concept Mapping:

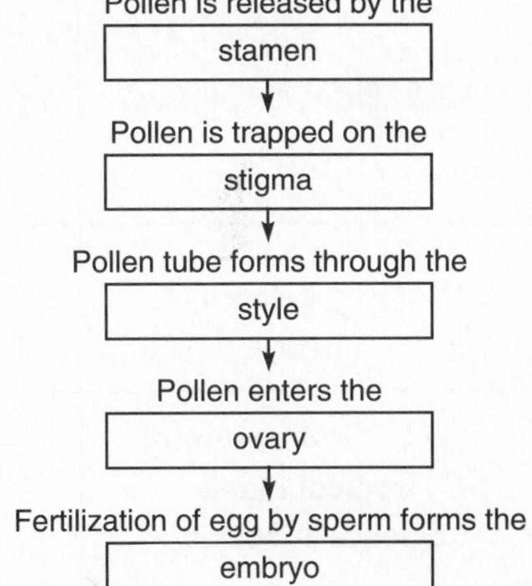

Pollen is released by the
stamen
Pollen is trapped on the
stigma
Pollen tube forms through the
style
Pollen enters the
ovary
Fertilization of egg by sperm forms the
embryo

4. Interpreting Data: Pine and tomato have about equal numbers of stomata on both surfaces indicating equal exchange on each leaf surface. Beans must use the lower stomata more; fir have no stomata on the upper surface and therefore all gas exchange occurs on the bottom.

5. Observing and Inferring: Good pictures of specimens include: trumpet creeper flower pollinated by hummingbirds; goldenrod, clover, and dandelions pollinated by bees; wind-pollinated include ragweed, willow, corn, grasses.

CHAPTER

12 Plant Processes

CHAPTER SECTION	OBJECTIVES	ACTIVITIES
12-1 Photosynthesis and Respiration (3 days)	1. **Describe** the process of gas exchange in plants. 2. **Explain** the process and importance of photosynthesis. 3. **Describe** the process and importance of respiration and its relationship to photosynthesis.	**MINI-Lab:** *How can the use of carbon dioxide by plants be shown?* p. 278 **MINI-Lab:** *How can respiration in yeast be shown?* p. 280 **Activity 12-1:** *Stomata in Leaves,* p. 281
12-2 Plant Responses (2 days)	1. **Explain** the relationship between stimuli and tropisms in plants. 2. **Differentiate** between long-day and short-day plants. 3. **Explain** the relationship between plant hormones and responses.	
12-3 Plant Relationships (1 day)	1. **Define** and give examples of parasitism. 2. **Define** and give examples of mutualism. 3. **Describe** an example of coevolution.	
12-4 The Treasure of Tropical Plants Science & Society (1 day)	1. **Identify** the role of plants in medicine. 2. **Describe** how destruction of tropical forests may destroy plant and animal species that could have beneficial uses in modern medicine.	**Activity 12-2:** *Designing an Experiment (Plant Tropisms),* p. 290
Chapter Review		

ACTIVITY MATERIALS

FIND OUT	ACTIVITIES		MINI-LABS	
Page 275 plastic bags rubber bands seedling plants in pots	**12-1 Stomata in Leaves, p. 281** lettuce in dish of water coverslip microscope microscope slide salt solution forceps	**12-2 Designing an Experiment, p. 290** petri dish tape string corn seeds paper towel water ring stand	**How can the use of carbon dioxide by plants be shown? p. 278** 25 mL bromthymol blue 25 mL of phenol red beakers (2) straws (2) pieces of *Elodea*	**How can respiration in yeast be shown? p. 280** 10 mL of bromthymol blue 20 drops of yeast suspension 10 drops of sugar solution test tube

CHAPTER FEATURES	TEACHER RESOURCE PACKAGE	OTHER RESOURCES
Technology: *Designer Plants,* p. 278 **Skill Builder:** *Comparing and Contrasting,* p. 280	**Ability Level Worksheets** ◆ ***Study Guide,*** p. 47 ● ***Reinforcement,*** p. 47 ▲ ***Enrichment,*** p. 47 **Activity Worksheets,** pp. 5, 104, 105, 110, 111 **Concept Mapping,** pp. 29 **Transparency Masters,** pp. 39-40 **Technology,** pp. 11-12	**Lab Manual:** Transpiration, p. 73 **Lab Manual:** Oxygen and Photosynthesis, p. 75 **Lab Manual:** Carbon Dioxide & Photosynthesis, p. 79 **Lab Manual:** Plant Respiration, p. 83 **Science Integration Activity 12** **Color Transparency 20,** Gas Exchange in a Leaf
Problem Solving: *How Do Plants Climb Fences?* p. 283 **Skill Builder:** *Comparing and Contrasting,* p. 284	**Ability Level Worksheets** ◆ ***Study Guide,*** p. 48 ● ***Reinforcement,*** p. 48 ▲ ***Enrichment,*** p. 48 **Critical Thinking/Problem Solving,** p. 16 **Cross-Curricular Connections,** p. 16	**STVS:** Disc 4, Side 2 **Lab Manual:** Tropisms, p. 85
Skill Builder: *Hypothesizing,* p. 287	**Ability Level Worksheets** ◆ ***Study Guide,*** p. 49 ● ***Reinforcement,*** p. 49 ▲ ***Enrichment,*** p. 49	**STVS:** Disc 4, Side 2
You Decide! p. 289	**Ability Level Worksheets** ◆ ***Study Guide,*** p. 50 ● ***Reinforcement,*** p. 50 ▲ ***Enrichment,*** p. 50 **Activity Worksheet,** pp. 106-107 **Science and Society,** p. 16	**STVS:** Disc 4, Side 2
Summary Key Science Words Understanding Vocabulary Checking Concepts Using Lab Skills Think & Write Critically Apply More Skill Builders Projects	**ASSESSMENT RESOURCES** **Chapter Review,** pp. 27-28 **Chapter Test,** pp. 76-79 **Unit Test,** pp. 80-81 **Performance Assessment in Middle School Science (PAMSS)**	**Chapter Review Software** **Test Bank** **Alternate Assessment** **Performance Assessment**

◆ **Basic** ● **Average** ▲ **Advanced**

ADDITIONAL MATERIALS

SOFTWARE	AUDIOVISUAL	BOOKS/MAGAZINES
Green Plants, Educational Activities, Inc. *Photosynthesis and Light Energy,* Classroom Consortia Media. *Light, Plants, and Photosynthesis: Energy in Conversion,* IBM. *Photosynthesis: Unlocking the Power of the Sun,* Queue. *Computer Investigations: Plant Growth, and the Plant Growth Simulator,* Focus.	*Photosynthesis (2nd ed.),* film, EBEC. *Discovering the Forest,* film, EBEC. *Green Plants and Sunlight,* film, EBEC. *The Photosynthesis and Respiration Cycle,* film, Churchill. *Death Trap,* laserdisc, Image Entertainment. *Rain Forest,* laserdisc, Image Entertainment.	Devlin, Robert M. and Francis H. Witham. *Plant Physiology. 4th ed.* Belmont, CA: Wadsworth Publishing Company, 1983. Kaufman, Peter B. *Plants: Their Biology and Importance.* NY: Harper and Row Publishers, Inc., 1989. Raven, Peter H. *Biology of Plants.* NY: Worth Publishers, Inc., 1986.

CHAPTER

12

PLANT PROCESSES

THEME DEVELOPMENT: This chapter continues to develop the homeostasis and ecology themes. Particular emphasis is placed on photosynthesis and respiration as homeostatic mechanisms and the ecological role of symbiotic relationships between plants and other organisms.

CHAPTER OVERVIEW

- **Section 12-1:** The process and importance of photosynthesis and aerobic respiration are discussed in this section. Gas exchange is explained as a process that is needed in plants and animals.
- **Section 12-2:** In this section the relationship between stimuli and tropisms in plants is explained. The flowering response of plants to specific periods of darkness is explained and examples are given. Plant hormones are discussed in terms of the responses they produce in plants.
- **Section 12-3:** Symbiotic relationships of parasitism and mutualism are defined, and examples are given. The process of coevolution is explained.
- **Section 12-4: Science and Society:** The role of plants in medicine and the connection between this role and tropical forests is discussed.

CHAPTER VOCABULARY

transpiration
photosynthesis
respiration
stimulus
tropism
auxin
photoperiodism
long-day plants
short-day plants
day-neutral plants
coevolution
ethnobotany

CHAPTER

12 Plant Processes

274

OPTIONS

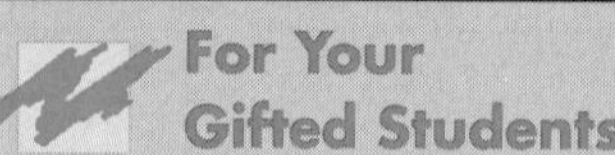

For Your Gifted Students

Have students look into the process of aerobic respiration during aerobic exercise. Aerobic exercise refers to any exercise that requires delivery of a large amount of oxygen to muscles over a certain period of time. Have students make a hypothesis as to whether the heart, lungs, and blood vessels work harder during aerobic exercise than during rest. What happens in the body?

For Your Mainstreamed Students

Have students hypothesize why all animals depend on plants for their survival. Have them list the foods they eat in a normal day, then trace the foods back to their plant origins. Remind them that plants produce their own food through photosynthesis.

Have you ever forgotten to water a potted plant? What happened to the plant after awhile?

FIND OUT!

Do the following activity to find out how water enters and leaves a plant.

Obtain a resealable plastic bag and a small potted plant from your teacher. Water the plant, then cover the plant and its pot with the plastic bag. Seal the bag. Place the bag in a sunny window and observe. What collects on the inside of the bag? Where does this substance come from? Some materials move in and out of leaves through tiny openings on leaf surfaces. Carbon dioxide is one of the substances that enters leaves through these openings. If enough water is lost by a plant and not replaced, what do you *predict* will happen to the plant?

Gearing Up

Previewing the Chapter

Use this outline to help you focus on important ideas in this chapter.

Section 12-1 Photosynthesis and Respiration
- Gas Exchange
- Photosynthesis
- Respiration

Section 12-2 Plant Responses
- What Are Plant Responses?
- Plant Responses and Plant Hormones
- Photoperiods

Section 12-3 Plant Relationships
- Parasitic Relationships
- Mutual Relationships
- Coevolution

Section 12-4 Science and Society The Treasure of Tropical Plants
- Medical Miracles from the Tropics

Previewing Science Skills

- In the **Skill Builders,** you will compare and contrast and hypothesize.
- In the **Activities,** you will observe, compare, interpret data, and make and use tables.
- In the **MINI-Labs,** you will relate, experiment, and compare.

What's next?

In this chapter you will learn about the process of photosynthesis, how plants respond to changes in chemicals, light, and gravity. You will also learn about the relationship of plants to other living things.

INTRODUCING THE CHAPTER

Use the Find Out activity to introduce students to plant processes. Inform students that they will be learning more about plant processes as they read the chapter.

FIND OUT!

Preparation: Obtain the materials needed for the Find Out activity at least two weeks in advance. The seedlings should be started in time for the plants to be large enough to show transpiration.

Materials: resealable bags and seedling plants in pots

Cooperative Learning: Have students work as Paired Partners. Have them make hypotheses and observations together.

Teaching Tips

- Have students record their hypotheses about what will happen when the plastic bags are placed over the seedlings.

Gearing Up

Have students study the Gearing Up feature to familiarize themselves with the chapter. Discuss the relationships of the topics in the outline.

What's Next?

Before beginning the first section, make sure students understand the connection between the Find Out activity and the topics to follow.

ASSESSMENT OPTIONS

PORTFOLIO

Refer to page 291 for suggested items that students might select for their portfolios.

PERFORMANCE ASSESSMENT

See page 291 for additional Performance Assessment options.

Process

Skillbuilders, pp. 284, 287
MINI-Labs, pp. 278, 280
Activities 12-1, p. 281; 12-2, p. 290
Using Lab Skills, p. 292

CONTENT ASSESSMENT

Assessment—Oral, pp. 278, 284, 285
Skillbuilder, p. 280
Section Reviews, pp. 280, 284, 287, 289
Chapter Review, pp. 291-293
Mini Quizzes, pp. 279, 283

GROUP ASSESSMENT

Opportunities for group assessment occur with Cooperative Learning Strategies and Flex Your Brain Actitivities.

SECTION 12-1

PREPARATION

SECTION BACKGROUND

▶ Photosynthesis accounts for more than 99 percent of the energy used by living organisms on Earth. Chemosynthesis by bacteria accounts for the rest.

▶ The chlorophyll is contained in stacks of membranes within the chloroplasts.

▶ Guard cells are usually the only cells in the leaf epidermis that contain chloroplasts.

PREPLANNING

▶ Obtain balloons and cellophane tape for the Motivation demonstration.

REVEALING MISCONCEPTIONS

▶ Ask students which cells of plants are able to make their own food through photosynthesis. Point out that only cells that contain chloroplasts are able to make food. Most cells in a typical plant do not carry out photosynthesis.

FLEX Your Brain

Use the Flex Your Brain activity sheet to analyze what students already know about PHOTOSYNTHESIS.

Flex Your Brain ASSESSMENT

Portfolio: Use the Flex Your Brain activity to reinforce critical thinking and problem solving skills.

OBJECTIVES AND SCIENCE WORDS: Have students review the objectives and science words throughout the chapter as they study each section.

12-1 Photosynthesis and Respiration

New Science Words

transpiration
photosynthesis
respiration

Objectives

- ▶ **Describe the process of gas exchange in plants.**
- ▶ **Explain the process and importance of photosynthesis.**
- ▶ **Describe the process and importance of respiration and its relationship to photosynthesis.**

Gas Exchange

When you breathe in, your lungs take in a mixture of gases called air. Air is made up of nitrogen, oxygen, and a small amount of carbon dioxide. On exhaling, you release a mixture of gases, now with slightly more carbon dioxide. Gas exchange is one of the ways living cells obtain raw materials and get rid of waste products.

Plants need water and carbon dioxide to survive. Plants absorb water and minerals through their roots. Water travels up from the roots through the stem to the leaves. Water evaporates from leaf tissues and is released through stomata as water vapor. Carbon dioxide, a gas, gets into plants through stomata on leaf surfaces. Because leaves usually have more stomata on the lower surface, more carbon dioxide enters the spaces of the spongy layer, as shown in Figure 12-1. Water vapor is also found in the air spaces of the spongy layer. Carbon dioxide and water vapor are exchanged by diffusion 1
through the stomata on leaf surfaces.

How does carbon dioxide enter a leaf? Each

Figure 12-1. Carbon dioxide is taken into leaves through stomata. Water vapor is lost to the air when stomata open.

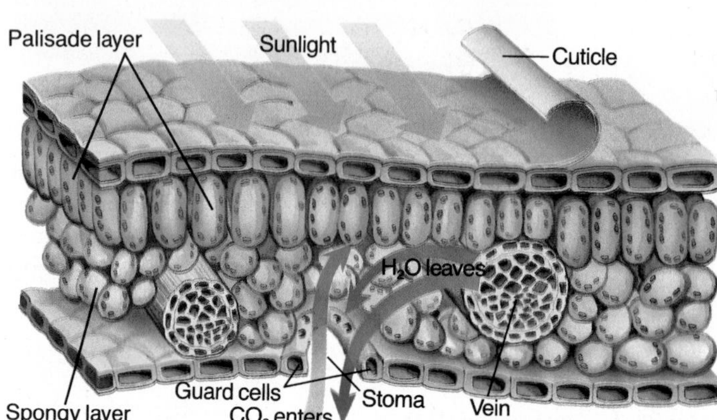

OPTIONS

Meeting Different Ability Levels

For Section 12-1, use the following **Teacher Resource Masters** depending upon individual students' needs.

◆ **Study Guide Master** for all students.

● **Reinforcement Master** for students of average and above average ability levels.

▲ **Enrichment Master** for above average students.

Additional Teacher Resource Package masters are listed in the OPTIONS box throughout the section. The additional masters are appropriate for all students.

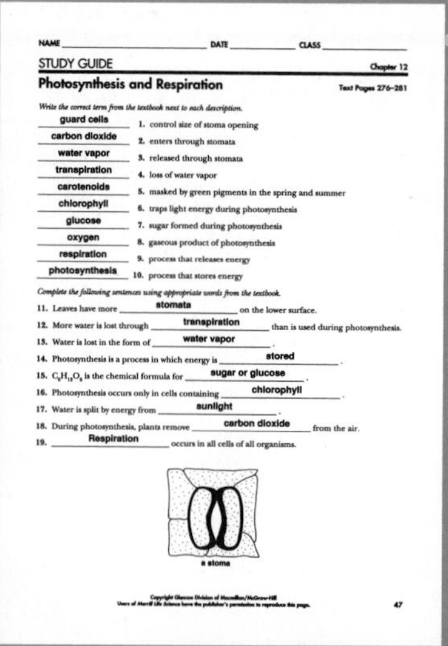
◆ STUDY GUIDE 47

NAME ____ DATE ____ CLASS ____

STUDY GUIDE — Chapter 12

Photosynthesis and Respiration — Text Pages 276-281

Write the correct term from the textbook next to each description.

guard cells 1. control size of stoma opening
carbon dioxide 2. enters through stomata
water vapor 3. released through stomata
transpiration 4. loss of water vapor
carotenoids 5. masked by green pigments in the spring and summer
chlorophyll 6. traps light energy during photosynthesis
glucose 7. sugar formed during photosynthesis
oxygen 8. gaseous product of photosynthesis
respiration 9. process that releases energy
photosynthesis 10. process that stores energy

Complete the following sentences using appropriate words from the textbook.

11. Leaves have more **stomata** on the lower surface.
12. More water is lost through **transpiration** than is used during photosynthesis.
13. Water is lost in the form of **water vapor**.
14. Photosynthesis is a process in which energy is **stored**.
15. $C_6H_{12}O_6$ is the chemical formula for **sugar or glucose**.
16. Photosynthesis occurs only in cells containing **chlorophyll**.
17. Water is split by energy from **sunlight**.
18. During photosynthesis, plants remove **carbon dioxide** from the air.
19. **Respiration** occurs in all cells of all organisms.

a stoma

47

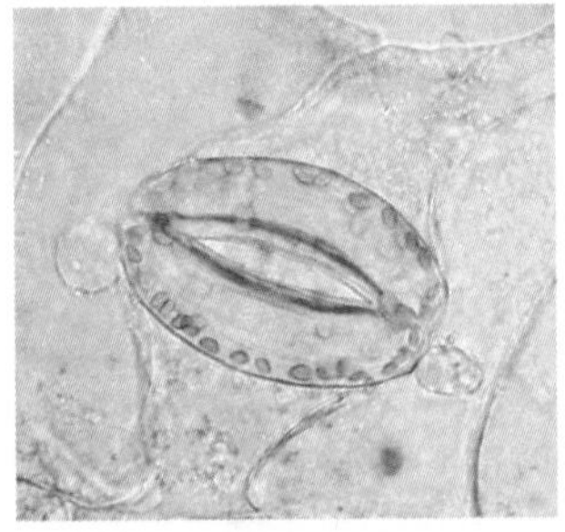

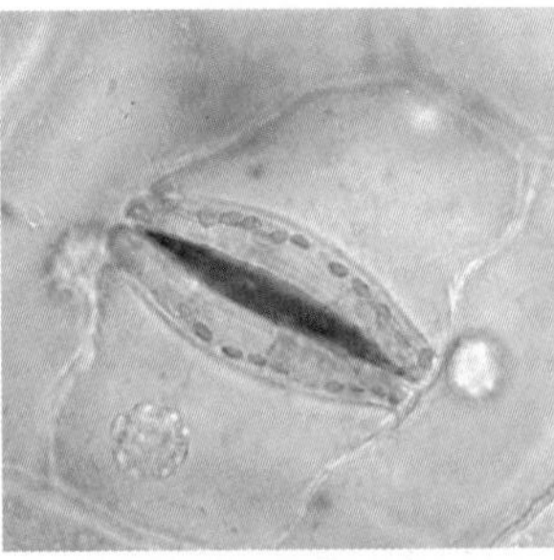

Figure 12-2. Stomata open when guard cells absorb water and close when water is lost.

stoma is surrounded by two guard cells that control the size of the opening. Water moves into and out of guard cells by osmosis. When guard cells absorb water, they swell and the stoma opens, letting carbon dioxide in. Water vapor escapes during the process. When guard cells lose water, they relax and the stoma closes. Figure 12-2 shows open and closed stomata.

Light, water, and carbon dioxide all affect the opening and closing of stomata. Stomata usually are open during the day when photosynthesis is going on. When photosynthesis stops at night, the stomata are usually closed. Less carbon dioxide enters and less water vapor escapes from the leaf when stomata are closed.

If you did the Find Out activity on page 275, you saw that water vapor condensed on the inside of the plastic bag. Loss of water vapor through stomata of a leaf is called
2 **transpiration.** Far more water is lost through transpiration than is used during photosynthesis.

Water lost through transpiration

Water absorbed by roots

Figure 12-3. Every plant loses water through transpiration.

Photosynthesis

Why aren't the leaves of the tree in Figure 12-4 green? If you live in a place that has changing seasons, you know that this photograph was taken in the fall. Many trees and bushes change color as the days get shorter and the weather grows colder. Plants may change from green to red, brown, yellow, or orange. Some show all of these colors at the same time. These colors are the results of pigments like the carotenoids that are present in the green leaves. In the spring and summer, the green pigment chlorophyll is so abundant in the leaves that it covers up, or masks, all other pigments. In the fall, plant cells stop producing chlorophyll, and the other pigments become visible. Chlorophyll is a pigment that traps light

Figure 12-4. Deciduous trees like this red maple change color in the fall, then drop their leaves.

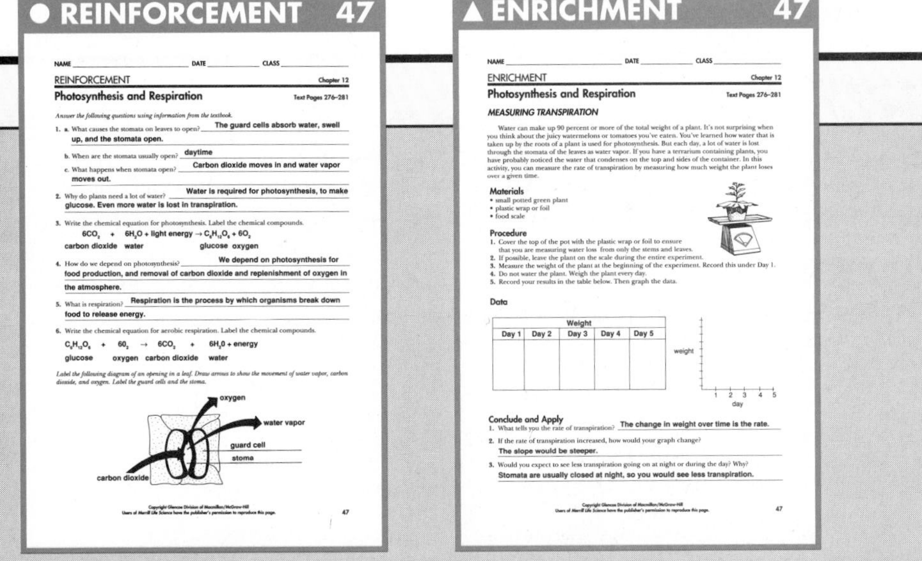

● REINFORCEMENT 47

NAME ______ DATE ______ CLASS ______

REINFORCEMENT — Chapter 12

Photosynthesis and Respiration — Text Pages 276–281

Answer the following questions using information from the textbook.

1. a. What causes the stomata on leaves to open? **The guard cells absorb water, swell up, and the stomata open.**
   b. When are the stomata usually open? **daytime**
   c. What happens when stomata open? **Carbon dioxide moves in and water vapor moves out.**
2. Why do plants need a lot of water? **Water is required for photosynthesis, to make glucose. Even more water is lost in transpiration.**
3. Write the chemical equation for photosynthesis. Label the chemical compounds.
   $6CO_2 + 6H_2O + \text{light energy} \rightarrow C_6H_{12}O_6 + 6O_2$
   carbon dioxide, water, glucose, oxygen
4. How do we depend on photosynthesis? **We depend on photosynthesis for food production, and removal of carbon dioxide and replenishment of oxygen in the atmosphere.**
5. What is respiration? **Respiration is the process by which organisms break down food to release energy.**
6. Write the chemical equation for aerobic respiration. Label the chemical compounds.
   $C_6H_{12}O_6 + 6O_2 \rightarrow 6CO_2 + 6H_2O + \text{energy}$
   glucose, oxygen, carbon dioxide, water

Label the following diagram of an opening in a leaf. Draw arrows to show the movement of water vapor, carbon dioxide, and oxygen. Label the guard cells and the stoma.

47

▲ ENRICHMENT 47

NAME ______ DATE ______ CLASS ______

ENRICHMENT — Chapter 12

Photosynthesis and Respiration — Text Pages 276–281

MEASURING TRANSPIRATION

Water can make up 90 percent or more of the total weight of a plant. It's not surprising when you think about the juicy watermelons or tomatoes you've eaten. You've learned how water that is taken up by the roots of a plant is used for photosynthesis. But each day, a lot of water is lost through the stomata of the leaves as water vapor. If you have a terrarium containing plants, you have probably noticed the water that condenses on the top and sides of the container. In this activity, you can measure the rate of transpiration by measuring how much weight the plant loses over a given time.

Materials
- small potted green plant
- plastic wrap or foil
- food scale

Procedure
1. Cover the top of the pot with the plastic wrap or foil to ensure that you are measuring water loss from only the stems and leaves.
2. If possible, leave the plant on the scale during the entire experiment.
3. Measure the weight of the plant at the beginning of the experiment. Record this under Day 1.
4. Do not water the plant. Weigh the plant every day.
5. Record your results in the table below. Then graph the data.

Data

Weight				
Day 1	Day 2	Day 3	Day 4	Day 5

Conclude and Apply
1. What tells you the rate of transpiration? **The change in weight over time is the rate.**
2. If the rate of transpiration increased, how would your graph change? **The slope would be steeper.**
3. Would you expect to see less transpiration going on at night or during the day? Why? **Stomata are usually closed at night, so you would see less transpiration.**

47

1 MOTIVATE

VideoDisc

STVS: City Trees, Disc 4, Side 2
STVS: Farming Indoors, Disc 4, Side 2

▶ **Demonstration:** You can model guard cells of a stoma by using the following demonstration. Blow up two balloons. Show students that when the balloons are blown up they remain straight. Now place cellophane tape along the side of two balloons as you blow them up. The balloons will bow outward. When put together, they will form an opening resembling a stoma.

TYING TO PREVIOUS KNOWLEDGE: This section should be tied to Chapter 2 where the structure and function of chloroplasts and mitochondria are explained.

2 TEACH

Key Concepts are highlighted.

CONCEPT DEVELOPMENT

▶ **Demonstration:** Place a prism in a bright light source. As the colors of white light are separated, point out that objects show only the colors that are reflected. All other colors are absorbed. Chlorophyll absorbs every color but green. Green light waves are reflected and thus are not used in photosynthesis. Show a series of objects of different colors and have students tell what colors are absorbed and reflected.

▶ Relate absorbed colors to energy absorbed.

MINI-Lab

Materials: 25 mL of phenol red, two beakers, two straws, pieces of *Elodea*

Teaching Tips

▶ Test this before class to be sure color changes are apparent.

Answers to Questions

▶ Phenol red changes to orange when carbon dioxide is added.

▶ When *Elodea* is added and light is provided, the plants begin photosynthesis and remove the added carbon dioxide, causing a change from orange to red in the beakers.

MINI-Lab
ASSESSMENT

Performance: To further assess students' understanding of laboratory methods used to detect carbon dioxide, see USING LAB SKILLS, question 12, on p. 292.

TECHNOLOGY

Think Critically: Designer plants would be resistant to disease and grow faster and better.

MINI-Lab

How can the use of carbon dioxide by plants be shown?

Pour 25 mL of phenol red solution into a beaker. Note the original color. Blow carefully through a straw. **CAUTION:** *Do not inhale.* Note the color change. What have you added to the solution? Repeat this procedure in a second beaker. Add an *Elodea* sprig to one beaker. Place under a bright light for 15 minutes, observing every 5 minutes. What do you see? *Relate* any changes you see to photosynthesis.

energy for plants to use in making food. **Photosynthesis** is the process in which plants use light energy to produce food. 3

What do plants need besides light to make food? During photosynthesis, a plant makes food from light energy, carbon dioxide, and water in a series of chemical reactions. Some of the light energy trapped in the chlorophyll is used to split water molecules into hydrogen and oxygen. Light energy is then used to join hydrogen and carbon dioxide together to form a new molecule of sugar. The sugar formed is glucose, the food a plant uses for maintenance and growth. The process is illustrated in the following equation:

$$\underset{\text{carbon dioxide}}{6CO_2} + \underset{\text{water}}{6H_2O} + \text{light energy} \xrightarrow[\text{chlorophyll}]{} \underset{\text{sugar}}{C_6H_{12}O_6} + \underset{\text{oxygen}}{6O_2}$$

It takes six molecules of carbon dioxide and six molecules of water to form one molecule of sugar, plus six molecules of oxygen. That is, the products of photosynthesis are sugar and oxygen.

TECHNOLOGY

Designer Plants

Using a gun that blasts tiny DNA-coated metal spheres into plant cells, scientists are trying to change the genetic codes of food crops such as wheat, rice, and corn.

Microbeads are a technique for introducing new genetic information into monocot plants such as grains. Cell transplants and gene transfers to plant embryos work well on dicots such as soybeans, tomatoes, and potatoes. The tomato mosaic virus costs the food industry $50 million each year, but resistant tomato plants can now be made available. In other tomato plants, a transfer gene provides DNA that blocks production of a chemical that triggers decay. As a result, the tomatoes have a much longer shelf life.

Think Critically: How could these designer plants benefit both farmers and consumers?

OPTIONS

ASSESSMENT—ORAL

▶ **Where does the energy for photosynthesis come from?** *Light absorbed by the chlorophyll is the source of energy to drive the photosynthetic reactions.*

▶ **What are the end products of photosynthesis?** *sugar and oxygen*

ENRICHMENT

▶ Students can learn the colors of the visible light spectrum by memorizing a mnemonic like ROY G BIV.

What happens to these products in a plant? Plant cells make use of the products of photosynthesis in many ways. Some of the oxygen is used by the plant cells for respiration. The rest of the oxygen is released and leaves the leaf through stomata. Sugar produced by plants is used for plant life processes such as growth, or it can be stored in the roots or stems of the plant as sugar, starch, or other products. When you eat beets, carrots, potatoes, or onions, you are eating stored food.

Why is photosynthesis important to living things? The most important role of photosynthesis is food production. Photosynthetic organisms are producers that provide food for nearly all the consumers on Earth. Secondly, through photosynthesis, plants remove carbon dioxide from the air and produce the oxygen that most organisms need to stay alive. As much as 90 percent of the oxygen entering our atmosphere today probably comes as a result of photosynthesis.

Respiration

Look at the photographs in Figure 12-5. Do these organisms have anything in common? All of these organisms are similar in that they break down food to produce energy. **Respiration** is the process by which organisms break down food to release energy. Respiration that uses oxygen to break food down chemically is called aerobic respiration. Aerobic respiration occurs in the mitochondria of all eukaryotic cells. The overall chemical equation for aerobic respiration is: 4

$$\underset{\text{sugar}}{C_6H_{12}O_6} + \underset{\text{oxygen}}{6O_2} \rightarrow \underset{\text{carbon dioxide}}{6CO_2} + \underset{\text{water}}{6H_2O} + \text{energy}$$

Connect to... Physics

Light energy can be described by waves or particles. Particles of light energy are called photons. One photon of light contains enough energy to excite a chlorophyll molecule. What color of light is least used by a chlorophyll molecule? (HINT: What color do most leaves reflect in mid-summer?)

In Your JOURNAL

In most parts of the world, trees lose their leaves in fall and winter. What continues to produce oxygen during these seasons? **In your Journal,** after doing some researching, discuss what you think keeps the world supplied with oxygen.

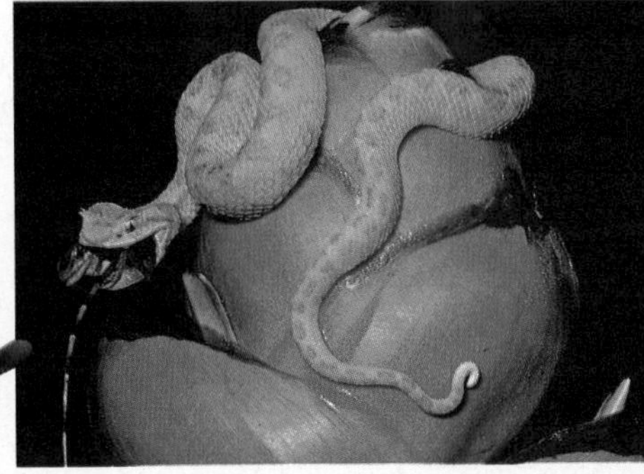

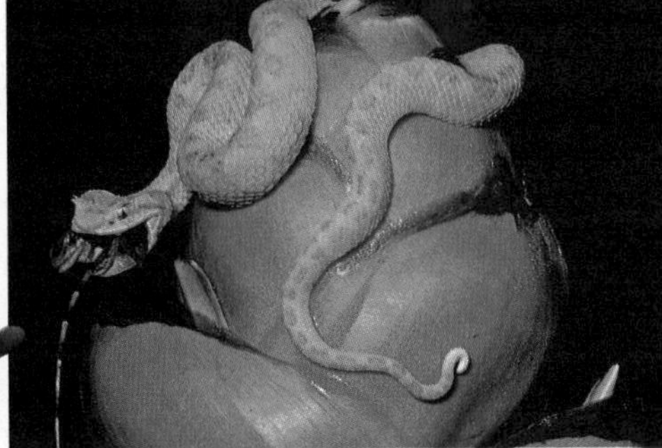

Figure 12-5. Respiration enables organisms to release energy that is found in the chemical bonds of food. Respiration occurs in all cells of all organisms.

Connect to... Physics

Answer: Chlorophyll does not use green light and it is reflected.

CHECK FOR UNDERSTANDING

▶ Ask why plants must be placed in sufficient light.

▶ Use the Mini Quiz below to check for understanding.

RETEACH

Place one plant in the light and another in a dark area. Keep all other variables constant. Observe the changes in the two plants for a week.

EXTENSION

For students who have mastered this section, use the **Reinforcement** and **Enrichment** masters or other OPTIONS provided.

MINI QUIZ

Use the Mini Quiz to check students' recall of chapter content.

1. **Carbon dioxide and water vapor are gases that are exchanged by _____ through the stomata on leaf surfaces.** *diffusion*
2. **Loss of water vapor through stomata of a leaf is called _____.** *transpiration*
3. **_____ is the process in which plants use energy from light to produce food.** *Photosynthesis*
4. **Aerobic respiration occurs in the _____ of cells.** *mitochondria*

PROGRAM RESOURCES

From the **Teacher Resource Package** use:

Science Integration Activities, 12, Capillary Action

PROGRAM RESOURCES

From the **Teacher Resource Package** use:

Transparency Masters, pages 39-40, Gas Exchange in a Leaf.

Use **Color Transparency** 20, Gas Exchange in a Leaf.

Use **Laboratory Manual** page 73, Transpiration; page 75, Oxygen and Photosynthesis; page 79, Carbon Dioxide and Photosynthesis; page 83, Plant Respiration.

In Your JOURNAL

Most of the world's oxygen is produced by phytoplankton (all the plant-like protists).

Materials: 10 mL of bromothymol blue, 20 drops of yeast suspension, 10 drops of sugar solution, test tube

Teaching Tips

►Try the activity several times to gauge the amount of color change after 5, 10, and 15 minutes.

Answers to Questions

►Yeast carries on respiration, adding carbon dioxide to the solutions. The color gradually changes from blue to green over the 15-minute period.

MINI-Lab

ASSESSMENT

Performance: to further assess students' understanding of bromothymol blue as an indicator, see question 12, USING LAB SKILLS, on p. 292.

3 CLOSE

► Ask questions 1-3 and the **Apply** Question in the Section Review.

► Have students write their own definitions of *photosynthesis* and *respiration.*

SECTION REVIEW ANSWERS

1. Carbon dioxide and water vapor are exchanged by diffusion through the stomata on leaf surfaces.

2. In photosynthesis, water splits into hydrogen and oxygen; oxygen escapes; and hydrogen is added to carbon dioxide, forming sugar. In respiration, oxygen is combined with sugar, and carbon dioxide, water vapor, and energy are released.

3. Photosynthesis and respiration form a cycle of energy capture and use.

4. Apply: Humidity is high in tropical rain forests because of the tremendous amount of water vapor transpired.

5. Connect to Earth Science: oxygen

Skill Builder

See Table 12-1 at the top of student page 280.

Skillbuilder

ASSESSMENT

Portfolio: Use this Skillbuilder to assess students' abilities to Compare and Contrast photosynthesis and respiration.

Table 12-1

COMPARING PHOTOSYNTHESIS AND RESPIRATION				
	Energy	**Raw Materials**	**End Products**	**In What Cells**
Photosynthesis	stored	light energy, water, carbon dioxide	sugar, oxygen	cells with chlorophyll
Respiration	released	sugar, oxygen	water, carbon dioxide, energy	all cells

Is the equation for respiration familiar? How does it relate to the chemical reaction for photosynthesis? During photosynthesis, energy is stored in food. Photosynthesis occurs only in cells that contain chlorophyll, such as those in the leaves of plants. *Respiration occurs in all cells of all organisms.* Respiration releases energy from food. Table 12-1 compares the two processes of photosynthesis and respiration.

Why do plant cells need energy? Plants use energy to build and repair their cells, tissues, and organs. Sugar is used to make cellulose to build cell walls. Plants use energy from the food they make during photosynthesis for growth and reproduction.

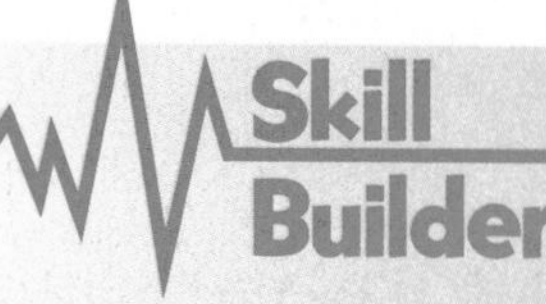

How can respiration in yeast be shown?

Pour 10 mL of bromothymol blue into a test tube. Add 20 drops of yeast suspension and 10 drops of sugar solution. What color change is seen after 5 minutes? 10 minutes? 15 minutes? What has caused this color change? *Compare* the results of the MINI-Lab on page 278 with the results of this MINI-Lab.

SECTION REVIEW

1. Explain how carbon dioxide and water vapor are exchanged in a leaf.
2. Describe the chemical reactions that take place in photosynthesis and respiration.
3. Why are photosynthesis and respiration important?
4. **Apply:** Humidity is caused by water vapor in the air. Why is humidity high in lush tropical rain forests?
5. **Connect to Earth Science:** What gas is put into the atmosphere as a result of photosynthesis?

Skill Builder

Comparing and Contrasting

Compare the raw materials and end products of photosynthesis and respiration and what happens to the energy involved in each process. If you need help, refer to Comparing and Contrasting in the **Skill Handbook** on page 683.

OPTIONS

PROGRAM RESOURCES

From the **Teacher Resource Package** use:

Concept Mapping, pages 29-30.

Technology, pages 11-12, Custom Crops.

Activity Worksheets, page 5, Flex Your Brain.

Activity Worksheets, page 110, Mini-Lab: How can the use of carbon dioxide by plants be shown?; page 111, Mini-Lab: How can respiration in yeast be shown?

ACTIVITY 12-1

Stomata in Leaves

Problem: *Where are stomata in lettuce leaves?*

Materials  

- lettuce in dish of water
- coverslip
- microscope
- microscope slide
- salt solution
- forceps

Procedure

1. Remove a piece of an outer green leaf of the lettuce. Choose a leaf that is turgid (stiff) from absorbing water in the dish.
2. Bend the leaf back and use the forceps to strip off some of the transparent tissue covering the leaf. This is the epidermis. Prepare a wet mount of a small piece of this tissue.
3. *Examine* the preparation under low and then high power on the microscope. Draw and label the leaf section in your data table.
4. Note the location and spacing of the stomata. Count the number of stomata that are open.
5. Make a second wet mount of the lettuce leaf epidermis. This time place a few drops of salt solution on the leaf. *Predict* what will happen to the stomata.
6. *Examine* the preparation under low and then high power on the microscope. Draw and label the leaf section in your data table.
7. Note the location and spacing of the stomata. Count the number of stomata that are open.

Data and Observations Sample Data

Water Mount	Salt Solution Mount
Number of stomata open	Number of stomata open
16 out of 17 open	13 out of 13 closed

Analyze

1. Describe the guard cells around a stoma.
2. How many stomata did you see in each leaf preparation?
3. Are guard cells different from the other cells of the leaf epidermis? Explain.
4. **Hypothesize** why a lettuce leaf becomes stiff in water.

Conclude and Apply

5. Were more stomata open or closed in the salt solution? Explain.
6. What can you *infer* about the function of stomata in a leaf?
7. How can you *conclude* that guard cells take part in photosynthesis?

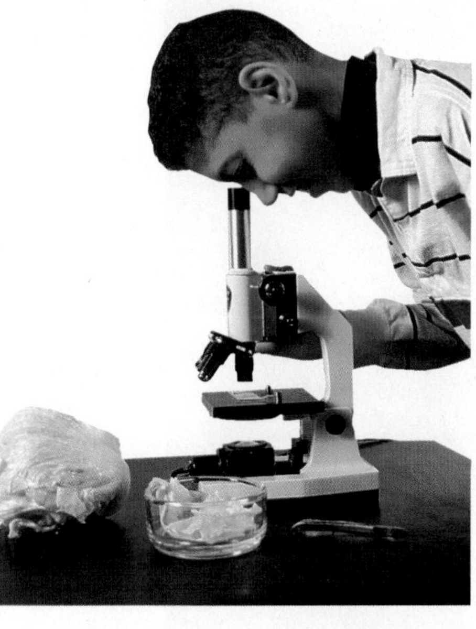

ACTIVITY 12-1

30 minutes

OBJECTIVE: Observe the activity of stomata in green plants.

PROCESS SKILLS applied in this activity:
▶ **Observing** in Procedure Steps 4 and 7.
▶ **Comparing** in Analyze Questions 2 and 3.
▶ **Inferring** in Analyze Question 4.

TEACHING THE ACTIVITY

Alternate Materials: Other leaf tissues that may be substituted for lettuce include onion skin and celery epidermis.

Troubleshooting: If an alternate leaf tissue is used, try the activity beforehand to make sure observations will be the same

▶ Remind students that osmosis is diffusion of water across a semipermeable membrane.

Activity ASSESSMENT

Performance: To further assess students' understanding of stomata, have them use one of the materials from the alternate materials list and describe the outcome.

ANSWERS TO QUESTIONS

1. Guard cells look like two sausages with the ends touching; contain chloroplasts.
2. Answers will vary depending on the surface chosen.
3. The guard cells are the only cells in the epidermis that have chloroplasts.
4. Water entered the cells by osmosis.
5. Suggested hypothesis: More were open in salt solution.
6. Stomata provide a passage into the leaf for carbon dioxide from the atmosphere.
7. Guard cells contain chloroplasts.

PROGRAM RESOURCES

From the **Teacher Resource Package** use:

Activity Worksheets, pages 104-105, Activity 12-1, Stomata in Leaves.

SECTION 12-2

PREPARATION

SECTION BACKGROUND

▶ Japanese scientists Yabuta and Sumiki discovered growth hormones called gibberellins during a study of rice plants that grew extremely tall because of "foolish rice disease," later found to be an overproduction of the hormone.

▶ Cytokinins are plant hormones that cause plants to produce an undifferentiated mass that can later be separated to produce many plant clones.

▶ Poinsettias can be made to flower by controlling the amount of darkness they receive several weeks prior to the time flowers are desired.

PREPLANNING

▶ You will need various plants and photographs for the demonstrations of tropisms in this section.

1 MOTIVATE

▶ If possible, obtain a *Mimosa pudica* plant from a biological supply house. Touch one of the leaflets with a finger. Ask students to explain what they see.

TYING TO PREVIOUS KNOWLEDGE: Review plant parts such as roots, stems, and leaves with students. Have them give the functions of each part.

12-2 Plant Responses

New Science Words

stimulus
tropism
auxin
photoperiodism
long-day plants
short-day plants
day-neutral plants

Objectives

- ▶ **Explain the relationship between stimuli and tropisms in plants.**
- ▶ **Differentiate between long-day and short-day plants.**
- ▶ **Explain the relationship between plant hormones and responses.**

Figure 12-6. The *Mimosa* plant's leaflets (a) respond to touch by folding up (b). The response to touch is called *thigmotropism.*

a

b

What Are Plant Responses?

If you touch the leaflets of the *Mimosa pudica* plant in Figure 12-6a, its leaflets will close up, as shown in Figure 12-6b. Why does this happen? The leaflets fold up in response to the stimulus of being touched. A **stimulus** is ❶
anything in the environment that causes a change in the behavior of an organism. The response of a plant to a stimulus is a **tropism.** Most plants respond to stimuli by growing. Tropisms can be positive or negative, that is, result in growth toward or away from the stimulus. Plants respond to stimuli by changing their behavior.

Touch is one stimulus that results in a change in a plant's behavior. Plants also respond to the stimuli of light, gravity, temperature, and amount of water. Some of the responses of plants are the result of plant hormones or chemical reactions in plant cells.

Did you ever see a plant that appeared to be leaning toward a window? Light is an important stimulus to plants. Plants respond to sunlight by growing toward it. The growth response of a plant to light is called *phototropism.* When the plant responds to light by growing toward it, the response is a positive phototropism. An example of ❷
negative phototropism occurs when roots of a plant respond to light by growing away from it.

The growth response of an organism to gravity is called *gravitropism.* Plant roots tend to grow downward and the stems to grow upward. The downward growth of a plant's roots in response to the force of gravity is a positive gravitropism. Upward growth of the plant's stem away from the force of gravity is called negative gravitropism.

OPTIONS

Meeting Different Ability Levels

For Section 12-2, use the following **Teacher Resource Masters** depending upon individual students' needs.

◆ **Study Guide Master** for all students.

● **Reinforcement Master** for students of average and above average ability levels.

▲ **Enrichment Master** for above average students.

Additional Teacher Resource Package masters are listed in the OPTIONS box throughout the section. The additional masters are appropriate for all students.

◆ STUDY GUIDE 48

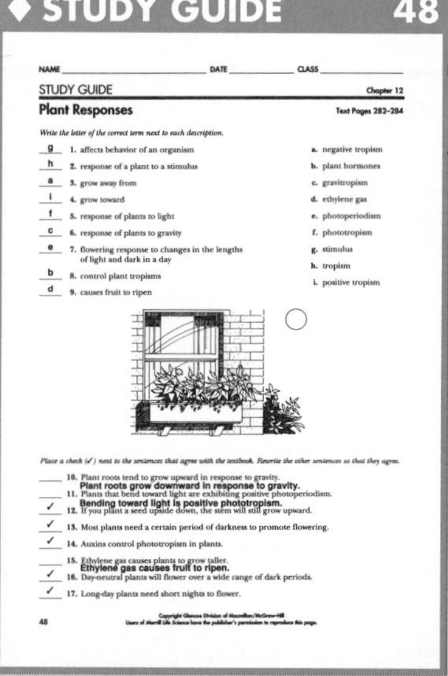

NAME ______ DATE ______ CLASS ______

STUDY GUIDE — Chapter 12

Plant Responses — Text Pages 282-284

Write the letter of the correct term next to each description.

Answer	Description	Term
g	1. affects behavior of an organism	a. negative tropism
h	2. response of a plant to a stimulus	b. plant hormones
a	3. grow away from	c. gravitropism
i	4. grow toward	d. ethylene gas
f	5. response of plants to light	e. photoperiodism
c	6. response of plants to gravity	f. phototropism
e	7. flowering response to changes in the lengths of light and dark in a day	g. stimulus
b	8. control plant tropisms	h. tropism
d	9. causes fruit to ripen	i. positive tropism

Place a check (✓) next to the sentences that agree with the textbook. Rewrite the other sentences so that they agree.

____ 10. Plant roots tend to grow upward in response to gravity.
Plant roots grow downward in response to gravity.
____ 11. Plants that bend toward light are exhibiting positive photoperiodism.
Bending toward light is positive phototropism.
✓ 12. If you plant a seed upside down, the stem will still grow upward.
✓ 13. Most plants need a certain period of darkness to promote flowering.
✓ 14. Auxins control phototropism in plants.
____ 15. Ethylene gas causes plants to grow taller.
Ethylene gas causes fruit to ripen.
✓ 16. Day-neutral plants will flower over a wide range of dark periods.
✓ 17. Long-day plants need short nights to flower.

48

PROBLEM SOLVING

How Do Plants Climb Fences?

While emptying the garbage at the back of his family's restaurant, Hazizi noticed some small plants growing near the fence. Over the next month, Hazizi stopped to look at these plants whenever he helped at the restaurant. As the plants grew, Hazizi noticed little twisted leaves growing out of the sides of the stems. These twisted leaves looked like the metal springs underneath his mattress.

One day Hazizi noticed that a plant seemed to be caught on the fence. He very carefully untwisted the curly leaves that were wrapped around the metal wires of the fence. But the next afternoon, two of the twisted leaves again were wrapped around the wires of the fence. This time Hazizi did not try to untangle them.

A week later Hazizi saw that every plant growing near the fence was climbing slowly up the fence by using the special twisted leaves. How were the plants able to climb the fence?

Think Critically: What caused the plants to twist their leaves around the fence wires?

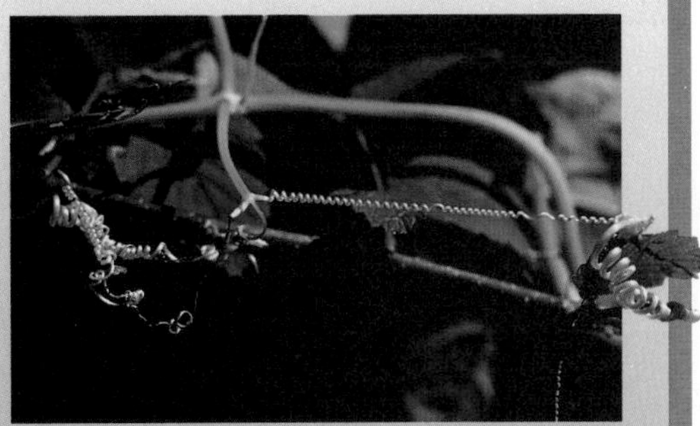

Plant Responses and Plant Hormones

In garden centers and greenhouses pots in which seedlings are grown are turned every few days. This is done because the seedlings lean toward the sunlight. As you have learned, this is an example of positive phototropism. But why do the seedlings respond to light in this way?

Plant tropisms are controlled by plant hormones, which are chemical substances that affect growth. An **auxin** is a type of plant hormone. Auxins cause plant stems and leaves to exhibit positive phototropism. If light shines on a plant from one side, the auxin moves to the shaded side of the stem. The auxin causes cells on the shaded side of the stem to grow longer than cells on the lighted side. This causes the stem to curve toward the light.

Ethylene gas, a simple chemical compound of carbon and hydrogen, is another plant hormone. Many plants produce ethylene gas, which causes fruits to ripen.

Connect to... Earth Science

The yearly motion of Earth around the sun is called its revolution. Earth is tilted during orbit, and as a result, seasons are produced. Seasons are times when the photoperiods change. Describe the four seasons in terms of photoperiods, explaining which would produce long-day and short-day plants.

● REINFORCEMENT 48

NAME ______ DATE ______ CLASS ______

REINFORCEMENT — Chapter 12

Plant Responses — Text Pages 282–284

Write the letter of the term that best completes each sentence.

b 1. A plant's response to gravity is called ____.
a. phototropism b. gravitropism c. photosynthesis d. photoperiodism

d 2. The flowering of a plant in response to change of light or dark is called ____.
a. photosynthesis b. phototropism c. gravitropism d. photoperiodism

a 3. Anything in the environment that affects the behavior of an organism is called a ____.
a. stimulus b. positive tropism c. response d. hormone

c 4. Plants that require long nights to flower are called ____.
a. long-day plants b. long-night plants c. short-day plants d. day-neutral plants

d 5. Auxins cause cells to grow ____.
a. longer on the sunny side of the stem c. shorter on the shaded side of the stem
b. shorter on the sunny side of the stem d. longer on the shaded side of the stem

a 6. Ethylene gas is ____.
a. a plant hormone that affects ripening of fruit
b. used by growers to cause stems to lengthen
c. a stimulus
d. a plant hormone that affects seed germination

b 7. Because of the effect of auxins on cell growth, plant stems grow ____.
a. away from gravity b. toward light c. toward touch d. straight

d 8. The response of roots growing downward is an example of ____.
a. negative phototropism
b. negative gravitropism
c. negative thigmotropism
d. both negative phototropism and positive gravitropism

Label the responses of the stem and roots in the following diagram. Figure 1 shows a plant's response after being tipped on its side for a few days. Figure 2 shows a plant's response to sunlight.

FIGURE 1
A. stem: negative gravitropism
B. roots: positive gravitropism

FIGURE 2
A. stem: positive phototropism
B. roots: negative phototropism

48

▲ ENRICHMENT 48

NAME ______ DATE ______ CLASS ______

ENRICHMENT — Chapter 12

Plant Responses — Text Pages 282–284

COMMERCIAL USE OF PLANT HORMONES

One way growers make sure that the produce you buy in the store looks perfect is to treat it with natural or synthetic plant hormones. Some methods involve the use of chemicals that cause the plant to produce the hormone. Many of the foods we eat have been treated before they get to the grocery shelves.

Use references to find out more about the use of these substances and fill in the table below.

Hormone	Effect
gibberellin	produces larger grapes; stalk on each cluster lengthens
auxin	promotes growth of plants from root cuttings
ethylene gas	ripens bananas
auxin	prevents fruit from falling off trees before harvest
auxin	can be used as weedkiller, promotes uncontrolled growth

Answer the following questions.

1. Why are hormones used? They are used to improve quality, uniformity of size or color, and to control ripening time.
2. What is an easy way to speed up the ripening of store-bought fruit? Place it in a bag with a piece of ripe fruit. Ripe fruit produces ethylene gas.

48

2 TEACH

Key Concepts are highlighted.

CONCEPT DEVELOPMENT

▶ Use rapid cycling *Brassica* available from biological supply houses to demonstrate plant responses to hormones.

▶ **Demonstration:** Cut two stems of coleus. Apply a rooting hormone on one stem. Have students observe both stems for several days. Ask students to explain the results.

Cooperative Learning: Have students work as Paired Partners to design and carry out their own experiment on a particular plant tropism.

PROBLEM SOLVING

Think Critically: The curly leaves are tendrils. Tendrils demonstrate thigmotropism. They respond to touch by curling around things. This helps plants climb towards the sunlight.

Connect to... Earth Science

Answer: Summer has the longest photoperiod. Spring and autumn have photoperiods longer than winter's shortest photoperiod.

MINI QUIZ

Use the Mini Quiz to check students' recall of chapter content.

1. **A(n) _____ is anything in the environment that affects the behavior of an organism.** *stimulus*
2. **When the plant responds to light by growing toward it, the response is called _____ phototropism.** *positive*
3. **_____ is the flowering response of plants to the change in the lengths of darkness.** *Photoperiodism*

CHECK FOR UNDERSTANDING

Ask students to give examples of different plant tropisms.

RETEACH

Use different plants to demonstrate various tropisms. As an example, you might use a Venus's flytrap to demonstrate thigmotropism or a plant that leans toward the light to demonstrate phototropism.

EXTENSION

For students who have mastered this section, use the **Reinforcement** and **Enrichment** masters or other OPTIONS provided.

3 CLOSE

▶ Ask questions 1-3 and the **Apply** Question in the Section Review.

SECTION REVIEW ANSWERS

1. Phototropism: movement in response to light. Gravitropism: response to gravity. Thigmotropism: response to touch.
2. Long-day plants require short nights to produce flowers. Short-day plants require long nights to produce flowers. Day-neutral plants may produce flowers during a range of night lengths.
3. Auxins cause cells to lengthen on the shady side of the plant causing the plant to bend toward the light.
4. **Apply:** Plant hormones are responsible for some tropisms.
5. **Connect to Physics:** When an insect touches hairs on a leaf, cells in the leaf rib change, closing the leaf on the insect.

Skill Builder

Roots tend to grow away from light, while leaves and stems tend to grow toward light.

Skillbuilder
ASSESSMENT

Performance: Use this Skillbuilder to assess students' abilities to Compare and Contrast the responses of roots, stems, and leaves to light by having them explain their results to you.

Figure 12-7. Long-day plants such as dill and short-day plants such as primroses flower in response to specific periods of darkness.

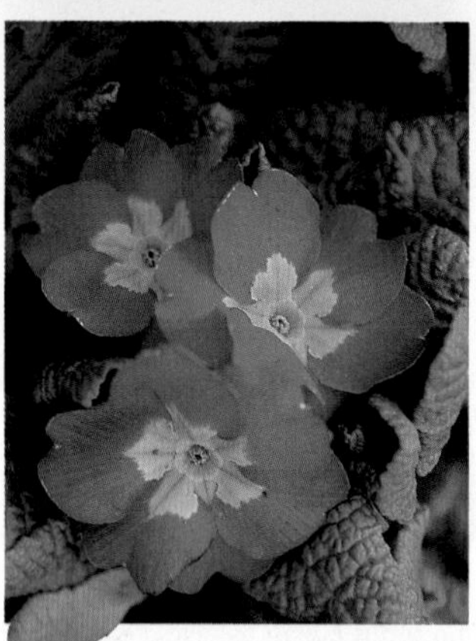

When fruits such as oranges, tomatoes, grapes, and bananas are picked before they have ripened, they stay green because their cells won't produce ethylene. Growers and shippers of fruit found out that it doesn't matter to the fruit where ethylene comes from for ripening. Growers now pick fruit before it is ripe. Shippers treat the picked fruit with ethylene gas during shipping. By the time fruit arrives at the store, it has ripened.

Photoperiods

Sunflowers bloom in the summer and cherry trees flower in the spring. Different plant species produce flowers
at different times during the year. **Photoperiodism** is the 3
flowering response of a plant to the change in the lengths
of light and dark in a day.

Most plants require a specific period of darkness to flower. Plants that require short nights to flower are called **long-day plants.** You may be familiar with long-day plants such as spinach, lettuce, and potatoes. Similarly, plants that require long nights to flower are called **short-day plants.** Some short-day plants are poinsettias, strawberries, and ragweed. Other plants, such as dandelions, corn, and tomatoes, flower over a wide range of night lengths. Plants like these that aren't very sensitive to the number of hours of darkness have a range of flowering times and are called **day-neutral plants.**

SECTION REVIEW

1. Describe the three major forms of tropisms.
2. Compare three different photoperiod responses.
3. Explain how auxins cause positive phototropism.
4. **Apply:** What is the relationship between plant hormones and tropisms?
5. **Connect to Physics:** Find out how a Venus's-flytrap closes up when an insect gets too close.

Skill Builder — Comparing and Contrasting

Different plant parts exhibit positive and negative tropisms. Compare and contrast the responses of roots, stems, and leaves to light. If you need help, refer to Comparing and Contrasting in the **Skill Handbook** on page 683.

OPTIONS

ASSESSMENT—ORAL

▶ **In what ways are tropisms different from and similar to animal behaviors?** *Tropisms represent plant behavior. Both are reactions to stimuli, but animal behaviors generally occur much faster.*

▶ **What adaptive advantages might plants obtain from tropisms?** *Tropisms allow plants to survive under different conditions in their environments.*

▶ **What gas can be used by grocers to speed the ripening of fruit?** *ethylene gas*

PROGRAM RESOURCES

From the **Teacher Resource Package** use:

Critical Thinking/Problem Solving, page 16, Night-Blooming Plants.

Cross-Curricular Connections, page 16, Symmetry.

Use **Laboratory Manual,** page 85, Tropisms.

Plant Relationships 12-3

SECTION 12-3

Objectives

- ▶ Define and give examples of parasitism.
- ▶ Define and give examples of mutualism.
- ▶ Describe an example of coevolution.

New Science Words

coevolution

Parasitic Relationships

How does a plant that has no chlorophyll get its food? The dodder plant is a parasite. A parasite is an organism
1 that lives on or in an organism of a different species and gets its nutrients from that organism. Although most plants make their own food by photosynthesis, parasitic plants get nutrients from other plants or animals. Parasitism can occur between any two species of organisms. It is one of the relationships plants have with other organisms.

The Indian Pipe in Figure 12-8 is a ghostly white parasitic plant often mistaken for a fungus. Because it has no chlorophyll, it must get its food from other sources, such as the rotting leaves and stems of other plants.

Several kinds of animals are parasites of plants. Scale insects and aphids suck out the sap from plants. If the scale insect or aphid populations increase to very high levels, plants will eventually die from loss of food and
2 water. Most parasites do not kill their hosts. What would probably happen if they did?

Connect to... Chemistry

Insect-eating plants, such as pitcher plants and sundews, often grow in areas where heavy rains wash nitrogen from the soil. Few other plants can grow there. *Infer* how these plants get the nitrogen they need if nitrogen is not available from soil.

a

b

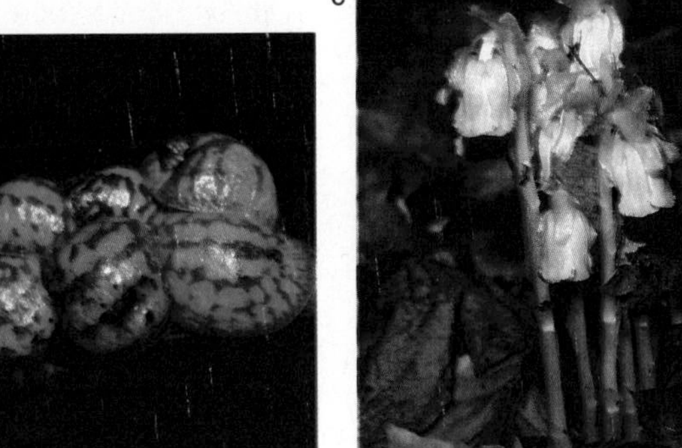

c

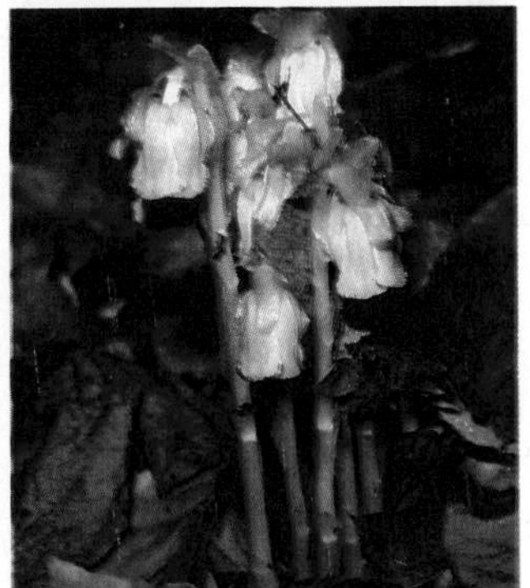

Figure 12-8. Organisms like orange dodder plants (a), scale insects (b), and Indian Pipes (c) are parasites.

PREPARATION

SECTION BACKGROUND

▶ The starflower of South America smells like rotting meat. It has a mutualistic relationship with the carrion flies that pollinate it.

▶ Bright colors, sweet odors, and nectar are only a few of the most obvious adaptations to attract insects.

▶ Coralroot is a native American orchid that lives off decaying material on the forest floor.

▶ Mistletoe is a parasitic plant that has roots that grow into the wood of host trees.

Connect to... Chemistry

Answer: They get nitrogen from the bodies of insects they trap and eat.

1 MOTIVATE

▶ Ask students to explain the relationship between bees and plants. Point out the mutualistic nature of these species.

OPTIONS

ASSESSMENT—ORAL

▶ **Why are Indian Pipe and dodder plants parasites?** *The plants have no chloroplasts with which to produce their own food.*

▶ **Why might you be able to find certain species of butterflies in close association with particular flowers?** *Some butterflies have a mutualistic relationship with plants that they pollinate in return for nectar.*

▶ **Does coevolution produce changes in one or both species?** *Coevolution occurs when each species evolves in response to the other.*

ENRICHMENT

▶ Many films have been made about the Galapagos Islands. Choose one that emphasizes the coevolution of cacti and tortoises to show to students.

TYING TO PREVIOUS KNOWLEDGE: Refer students to the discussion of parasitic and mutualistic relationships of some fungi in Chapter 9.

2 TEACH

Key Concepts are highlighted.

CONCEPT DEVELOPMENT

► Have students make a table that shows the kinds of interrelationships between organisms, the characteristics of the relationships, and examples.

CHECK FOR UNDERSTANDING

Ask students how their relationships with their parents are mutualistic in some ways and parasitic in some ways.

RETEACH

Show students photographs of different plants that show parasitic and mutualistic relationships.

EXTENSION

For students who have mastered this section, use the **Reinforcement** and **Enrichment** masters or other OPTIONS provided.

VideoDisc

STVS: Bats as Pollinators, Disc 4, Side 2

Mutual Relationships

What kind of relationship can you infer between the bat and the flower in Figure 12-9? The bat visits the flower to eat the nectar it produces. When the bat puts its head into the flower, some pollen is rubbed off onto the animal's fur. The bat travels from one flower to the next, pollinating flowers at each stop. Both flowers and bat benefit from their relationship. A relationship between two
3 species that benefits both is called mutualism. Mutualism often occurs between plants and their pollinators. Most angiosperms pollinated by insects, birds, or mammals have mutualistic relationships with their pollinators.

Figure 12-9. The bat and the bird and the flowers they pollinate have mutualistic relationships. Relationships like these may be a result of coevolution.

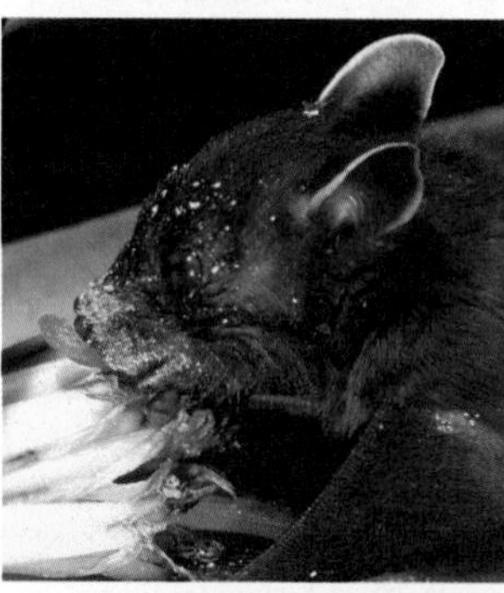

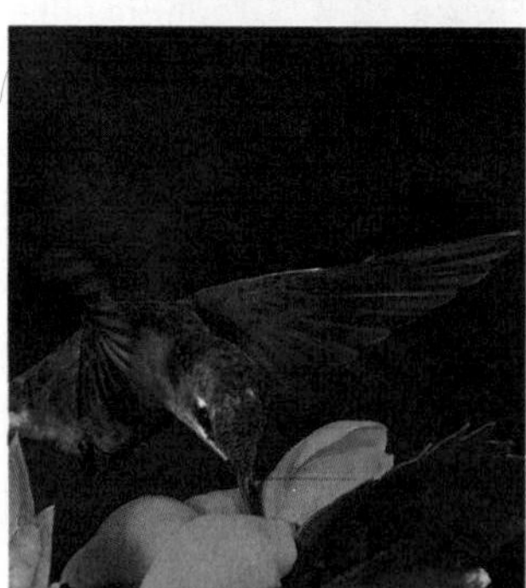

Other mutualistic relationships that do not involve pollination occur between plants and insects. One such relationship has evolved between the thorny acacia trees of Central America and Africa and a species of ant. The ants hollow out and occupy the inside of the acacia tree's thorns. The acacias produce a sweet sap that the ants eat. The tree provides both shelter and food for the ants. The ants attack and bite large insects and herbivores that browse on the acacias. As a result, these animals stay away. Acacia trees without a mutualistic relationship with ants grow more slowly and have a higher death rate.

Coevolution

An evolutionary relationship that sometimes occurs between specific plants and animals is coevolution. **Coevolution** occurs when two species of organisms evolve structures and behaviors in response to changes in each other over a long period of time. The pollination of flowers by animals is the most common example of coevolution.

Insects and birds already existed before flowering plants appeared. When dinosaurs died out, so did many gymnosperms, leaving room for new species to emerge. Insects, birds, mammals, and angiosperms evolved together and interacted in ways that improved each species' chances for survival. Today, flowers that depend on bees for pollination often are colored blue, yellow, or ultraviolet. Bees see these colors best and so are attracted to flowers with these colors. Flowers pollinated by moths are usually white, with a strong scent. Moths generally fly at night and may be directed to flowers by scent as well as color. Birds can

OPTIONS

Meeting Different Ability Levels

For Section 12-3, use the following **Teacher Resource Masters** depending upon individual students' needs.

◆ **Study Guide Master** for all students.

● **Reinforcement Master** for students of average and above average ability levels.

▲ **Enrichment Master** for above average students.

Additional Teacher Resource Package masters are listed in the OPTIONS box throughout the section. The additional masters are appropriate for all students.

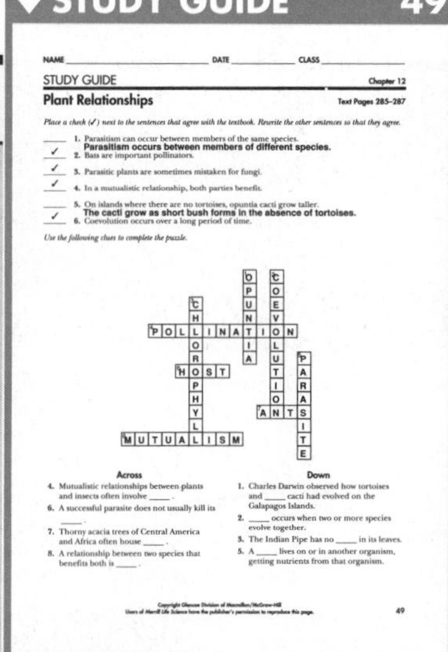

◆ STUDY GUIDE 49

NAME ______ DATE ______ CLASS ______

STUDY GUIDE — Chapter 12

Plant Relationships — Text Pages 285–287

Place a check (✓) next to the sentences that agree with the textbook. Rewrite the other sentences so that they agree.

____ 1. Parasitism can occur between members of the same species.
Parasitism occurs between members of different species.

✓ 2. Bats are important pollinators.

✓ 3. Parasitic plants are sometimes mistaken for fungi.

✓ 4. In a mutualistic relationship, both parties benefit.

____ 5. On islands where there are no tortoises, opuntia cacti grow taller.
The cacti grow as short bush forms in the absence of tortoises.

✓ 6. Coevolution occurs over a long period of time.

Use the following clues to complete the puzzle.

Across

4. Mutualistic relationships between plants and insects often involve ____ .
6. A successful parasite does not usually kill its ____ .
7. Thorny acacia trees of Central America and Africa often house ____ .
8. A relationship between two species that benefits both is ____ .

Down

1. Charles Darwin observed how tortoises and ____ cacti had evolved on the Galapagos Islands.
2. ____ occurs when two or more species evolve together.
3. The Indian Pipe has no ____ in its leaves.
5. A ____ lives on or in another organism, getting nutrients from that organism.

49

see red and orange very well. They pollinate flowers such as the red and orange bird-of-paradise flower.

Another example of coevolution can be found in the Galapagos Islands once visited by Charles Darwin. Opuntia cacti are found on several of these islands. On some islands, the cacti grow as short bushy forms near ground level. On other islands, this same species of cactus grows as taller, upright tree forms. Galapagos tortoises eat opuntia cactus, but there are no tortoises on islands that have short bushy cacti. Galapagos tortoises are found on islands where the cacti grow as trees. Tortoises on these islands have shells that arch up high above their heads. These tortoises can stretch their necks up high in order to feed. Grazing by the tortoises on these islands apparently eliminated the short opuntias that grew close to the ground. The taller forms survived to reproduce, resulting in the evolution of tall, upright tree forms. As the opuntias evolved into upright forms, the tortoises that had higher shells were able to graze higher in the opuntia trees on their islands. The tortoises with higher shells continued to find food and reproduce. Scientists hypothesize that these tortoises favored the evolution of taller opuntia cacti, and the taller cacti favored the evolution of tortoises with higher shells. These two species favored the evolution of some adaptations in one another.

Figure 12-10. Short opuntia cacti (a) grow on islands without tortoises, whereas taller tree forms of cacti (b) are only found where tortoises live (c).

a

b

c

SECTION REVIEW

1. What is a parasitic relationship?
2. Describe the features of a mutualistic relationship and give an example.
3. What is coevolution?
4. **Apply:** Some flowers are pollinated by flies that lay eggs on dead or decaying animals. What do you think these flowers smell like? How is this adaptation helpful?
5. **Connect to Earth Science:** How are evolution of living things and evolution of Earth similar?

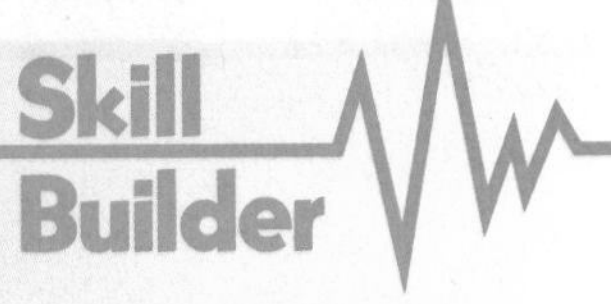

Hypothesizing

In a particular relationship between a kind of orchid and a bee, the orchid flower looks and smells like the female bee. The male bee is attracted to the flower and helps in the pollination. Make a hypothesis about the evolutionary history of these species. If you need help, refer to Hypothesizing in the **Skill Handbook** on page 686.

● REINFORCEMENT 49

NAME ________ DATE ________ CLASS ________

REINFORCEMENT — Chapter 12

Plant Relationships — Text Pages 285–287

Answer the following questions.

1. Describe the relationship between pollinators and plants. **It is a mutual relationship in which both organisms benefit. Pollinators usually get nectar and the plant is able to spread its pollen, increasing its chances for successful reproduction.**
2. What is a parasite? **An organism that lives on or in an organism of a different species and gets its nutrients from that organism is considered a parasite.**
3. What happens if the population of scale insects on a plant increases drastically? **The plant will die from loss of food and water.**
4. a. How does the Indian Pipe get its nutrients? **from rotting leaves and stems of other plants**
   b. What type of relationship is this? **parasitic**
5. a. What is one example of a mutualistic relationship between plants and insects that does not involve pollination? **ants and acacia trees**
   b. Explain how each organism benefits from this relationship. **The ant gets shelter and sweet sap; the acacia gets protection from large insects and browsing herbivores.**
6. What happens to acacia trees that do not have mutualistic relationships with ants? **They grow more slowly and have a higher death rate.**
7. Why do opuntia cacti grow as trees rather than bushes on islands where tortoises live? **Grazing by tortoises has eliminated the short opuntias.**
8. What is the relationship between opuntia cacti and tortoises an example of? **The relationship is an example of coevolution.**

 49

▲ ENRICHMENT 49

NAME ________ DATE ________ CLASS ________

ENRICHMENT — Chapter 12

Plant Relationships — Text Pages 285–287

MUTUALISM

Relationships between two different species are called mutualistic when both partners benefit from the relationship. Reproduction in most plants requires a mutualistic relationship with pollinators. These pollinators include insects, bats, and hummingbirds. More flowers are pollinated by bees than by any other insect. Flowers have evolved sweet odors and bright colors to attract bees and other pollinators. The pollen is carried by the pollinator from flower to flower as it is collected. The flowers are pollinated and the pollinator uses the pollen for food. Both species benefit from the relationship. In some cases, the species in a mutualistic relationship rely on each other so much that one lives inside of the other.

Investigate a relationship of this type between plants called legumes and a type of bacteria called nitrogen-fixing bacteria. Use reference books in the library to find out more about mutualism of this type. Write a report below describing the relationship.

These soil bacteria infect the roots of legumes such as soybeans, alfalfa, peas, and clover. As the bacteria divide, the host plants respond by increasing the number and size of the root cells. The roots become swollen into nodules where the bacteria live. The bacteria make nitrogen in the soil available to the plant in a usable form. In return, the host plant provides nutrients for the bacteria. Legumes, as a result of this relationship with nitrogen-fixing bacteria, are extremely important crops because they don't deplete the soil of nutrients. Growers don't have to use large amounts of fertilizer. Legumes are important links in the cycling of nitrogen.

 49

MINI QUIZ

Use the Mini Quiz to check students' recall of chapter content.

1. **A(n) _____ is an organism that lives on or in an organism of a different species and gets its nourishment from that organism.** *parasite*
2. **Most parasites do not kill their _____ .** *hosts*
3. **A relationship between two species that benefits both is called _____ .** *mutualism*

CONCEPT DEVELOPMENT

Cooperative Learning: Have groups of students research relationships plants share with other plants or animals. Each group should present a report including photographs or drawings that show the relationships.

3 CLOSE

▶ Ask questions 1-3 and the **Apply** Question in the Section Review.

SECTION REVIEW ANSWERS

1. One member of the relationship benefits at the expense of the other.
2. Both members of the relationship benefit, as in bees pollinating flowers.
3. Coevolution occurs when two or more species live, interact, and evolve together over time.
4. Apply: dead or decaying animals. The smell is attractive to the flies which may not (otherwise) visit the flowers.
5. Connect to Earth Science: Both living things and Earth have changed over long periods of time. The DNA of living things change, while the crust of Earth erodes.

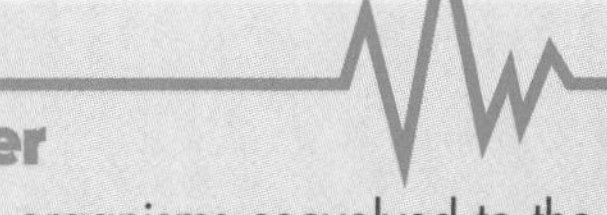

The two organisms coevolved to the point that they began a mutualistic relationship.

Skillbuilder
ASSESSMENT

Performance: Assess students' understanding of coevolution by having them explain one of the relationships researched in Concept Development above.

SECTION 12-4

PREPARATION

SECTION BACKGROUND

▶ The poisonous seeds of the castor bean can be made into a laxative called caster oil.

PREPLANNING

▶ Bring an aloe vera plant to class to use in the Teach section.

1 MOTIVATE

▶ Ask students what medicine they might take for a headache. When they suggest aspirin, point out that it can be obtained from willow bark.

MULTICULTURAL AWARENESS

Have students investigate the plant and cultural origin of a drug, such as aspirin or quinine.

PROGRAM RESOURCES

From the **Teacher Resource Package** use:

Science and Society, page 16, A Controversial Clearing.

12-4 The Treasure of Tropical Plants

New Science Words

ethnobotany

Objectives

- ▶ **Identify the role of plants in medicine.**
- ▶ **Describe how destruction of tropical forests may destroy plant and animal species that could have beneficial uses in modern medicine.**

Medical Miracles from the Tropics

Have you ever imagined yourself as the discoverer of a new drug that could be used to treat what was once an incurable disease? Surely, years and years of research are required to develop such miracle drugs! In fact, many of the drugs and medications we use today came from common plants. The knowledge of these plants was gained from ancient cultures in tropical regions of the world. Rosy periwinkle, for example, is a plant that grows on the island of Madagascar off the coast of Africa. A total of 75 chemicals are derived from this plant, including two that are used to successfully treat childhood leukemia and Hodgkin's disease.

Figure 12-11. Rosy periwinkle is the source of numerous chemicals used to treat disease.

Ethnobotany is the study of how people use plants. Many plants that are used in drugs and other medications were first discovered through such studies. Today, researchers document the knowledge of traditional healers in tribes that inhabit tropical environments. These healers have been taught about the qualities of local plants and how they can be used for medical treatment. In many instances, this information has never been written down. Rather, it has been handed down by word of mouth over thousands of years. Some

OPTIONS

Meeting Different Ability Levels

For Section 12-4, use the following **Teacher Resource Masters** depending upon individual students' needs.

◆ **Study Guide Master** for all students.

● **Reinforcement Master** for students of average and above average ability levels.

▲ **Enrichment Master** for above average students.

Additional Teacher Resource Package masters are listed in the OPTIONS box throughout the section. The additional masters are appropriate for all students.

◆ STUDY GUIDE 50

NAME ____________ DATE ________ CLASS ________

STUDY GUIDE — Chapter 12

The Treasure of Tropical Plants — Text Pages 288–290

Complete the following sentences using words from the following list.

drugs	National Cancer Institute	word-of-mouth	deforestation
document	tumors	healers	ethnobotany

1. The study of people and their use of plants is called **ethnobotany**.
2. Cultures that have traditional healers rely on **word-of-mouth** to hand down their knowledge of medicinal plants.
3. Funding for collecting and testing tropical plants has come from the **National Cancer Institute**.
4. Many **drugs** we use today are derived from common plants.
5. Traditional **healers** know how local plants can be used for medical treatment.
6. More and more species of plants are being lost because of tropical **deforestation**.
7. Researchers must carefully **document** the knowledge of traditional healers.
8. Scientists are testing tropical plants for their ability to slow or stop the growth of **tumors**.

Place a check (✓) next to the following sentences that agree with the textbook. Rewrite the other sentences so that they agree.

___ 9. Deforestation is mainly a problem that affects humans. **Loss of the tropical forests also affects many species of plants and animals that may become extinct.**

✓ 10. Many of the drugs and medications we use today are derived from common plants.

___ 11. Traditional healers have kept careful written records for thousands of years. **Traditional healers have passed on knowledge by word-of-mouth.**

✓ 12. Thousands of hectares of tropical forests are cut or burned every day.

___ 13. The National Cancer Institute is funding investigations to stop deforestation. **It is funding investigations of tropical plants that might slow or stop tumor growth.**

✓ 14. Ethnobotanists have only investigated a few of the remaining tribes in the tropics.

✓ 15. Rosy periwinkle grows on the island of Madagascar.

___ 16. Acacia trees contain chemicals used to treat leukemia and Hodgkin's disease. **Rosy periwinkle contains chemicals used to treat leukemia and Hodgkin's disease.**

50

ethnobotanists live with tribal groups for a few months every year, collecting specimens, seeds, and flowers, as well as writing down information about how each plant is used. Today, cultures that have used this kind of information are threatened by the destruction of the tropical rain forests where they live.

Thousands of hectares of tropical forests are cut down or burned each day, and as they are, more and more species of animals and plants are lost. Few of the remaining tribes in the tropics have been thoroughly investigated and studied by ethnobotanists. Without documentation of the knowledge of those remaining tribes, the world may suffer great scientific and economic losses.

The National Cancer Institute has seen the urgency of this situation and is taking action. The Institute has awarded millions of dollars to researchers to collect and test species of tropical plants for their ability to stop or slow the growth of tumors. By taking immediate action, researchers may be able to help find the source of such drugs before they are lost forever due to tropical deforestation.

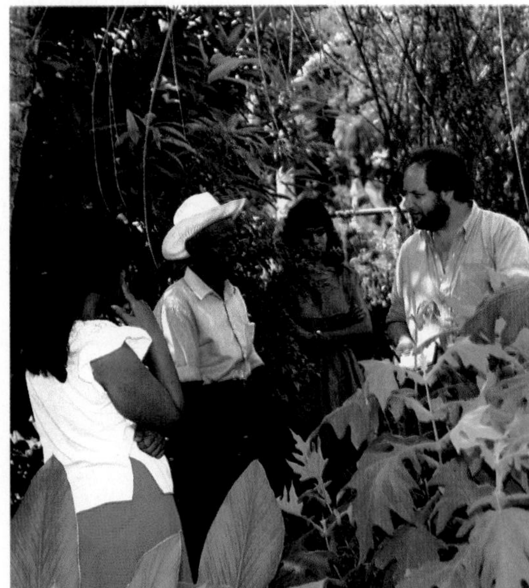

Figure 12-12. Ethnobotanists work with local healers to identify plants with medicinal value.

SECTION REVIEW

1. What are two diseases that can be treated with drugs derived from plants?
2. What work do ethnobotanists do?
3. **Connect to Earth Science:** Use a world map to locate the island of Madagascar off the east coast of Africa. Describe the island and its products.

You Decide!

Some people believe that the people of developing countries could take advantage of the plants that have medicinal value. Factories could be developed to extract the chemicals from these plants. This would make jobs and help the economies of these countries. However, such a life-style change could also change the cultures of tribal groups forever. Should countries become involved in developing the resources of plants with medicinal value in this manner?

● REINFORCEMENT 50

NAME ______ DATE ______ CLASS ______

REINFORCEMENT — Chapter 12

The Treasure of Tropical Plants — Text Pages 288–290

Answer the following questions using information from the textbook.

1. Why do researchers consult traditional healers about tropical plants? **Traditional healers have knowledge of the qualities of local plants and how they can be used for medical treatment.**
2. How has information been passed down in tribes about medically valuable plants? **Traditional healers in tribes have passed this information down for thousands of years by word-of-mouth.**
3. What is an example of a disease that can be treated using a drug from a tropical plant? What is the name of the plant? **leukemia or Hodgkin's disease; rosy periwinkle**
4. Other than effects on climate, why should we care about the loss of tropical rain forests? **A great number of plant and animal species would be lost if the rain forests disappeared. Many of these plants may be useful as medicines.**
5. What are the researchers who are being funded by the National Cancer Institute trying to do? **They are collecting and testing species of tropical plants to see if any of them will slow or stop tumor growth.**
6. What is an ethnobotanist? **An ethnobotanist studies people and their use of plants.**
7. What is happening to the tribal cultures that have the knowledge of medically useful plants? **These cultures are being displaced by development of the tropical rain forests where they live.**
8. What are conservation groups doing about this problem? **They are working to slow the rate of tropical deforestation.**

50

▲ ENRICHMENT 50

NAME ______ DATE ______ CLASS ______

ENRICHMENT — Chapter 12

The Treasure of Tropical Plants — Text Pages 288–290

STARTING A LIBRARY

Libraries aren't always full of books. A library is a place where you can look something up. Information is cataloged and can be retrieved in a library. An herbarium is really a library of plants. It is used as a reference. Whenever a new plant species is identified, the specimen is carefully pressed, dried, and labeled. Then it is stored at one of several herbariums. Today many species are becoming extinct as tropical rain forests are cut down or burned. There are herbarium specimens that are the only representatives of a particular species of plant. As the rain forests disappear, botanists are hurrying to gather plants of all kinds. Whenever a new species is identified, it is dried on location and shipped to one of many herbariums where the identity is checked. The plant specimen is then labeled and stored for future reference.

Develop your own library of identified plant specimens. Start in your own backyard or neighborhood. Be sure to get permission before you collect on someone else's property. **CAUTION:** *Do not collect poisonous plants.*

The best time for identification is when plants are flowering. Cut a section of plant that includes stem, leaves, and flower or fruit if present. For trees, include several leaves as well as flower or seed pods. Collect only a few specimens at one time. Try to identify them using a good identification book or key. Record the date and location for each plant. Dry your specimens using the following procedure.

Materials
- 2 rectangular or square pieces of wood
- heavyweight acid-free paper of the same size as the wood
- newspapers
- 2 nylon straps that can be tightened so that they will stay tight

Specimen Ready to Press

boards

Plant Press

specimens

Procedure
1. As soon as you have identified the plant, spread it out so that it is as flat as possible on a piece of heavyweight paper.
2. Repeat this for as many plants as you have collected.
3. Sandwich the plants on the heavyweight paper between several layers of newspaper.
4. Place the stack of layers between the boards being careful not to disturb the layers.
5. Tighten the nylon straps around the "sandwich" to provide equal pressure at each end.
6. Press the plants for several days until they are thoroughly dry. Store the specimens flat.
7. When you have dried a few specimens, see if you can use your library as a reference. Have a friend use it to identify an "unknown" plant from your backyard.

50

2 TEACH

Key Concepts are highlighted.

CONCEPT DEVELOPMENT

▶ **Demonstration:** Bring an aloe vera plant to class. If possible, obtain products that advertise aloe vera extract as an ingredient. Break off a leaf to show students the jelly-like sap of the plant.

CHECK FOR UNDERSTANDING

Ask students where most medicines come from.

RETEACH

Demonstration: Bring in empty medicine bottles. Emphasize that the medicine in each bottle probably was derived directly or indirectly from chemicals isolated from plants.

EXTENSION

For students who have mastered this section, use the **Reinforcement** and **Enrichment** masters or other OPTIONS.

3 CLOSE

▶ Ask questions 1-2 in the Section Review.

SECTION REVIEW ANSWERS

1. Answers will vary, but may include childhood leukemia and Hodgkin's disease.

2. Ethnobotanists study how people use plants.

3. Connect to Earth Science: Location—off the east coast of Africa; a large island; little remaining rain forest; produces rice, coffee, sugarcane

YOU DECIDE!

Answers will vary. All answers that indicate clear and logical reasoning should be accepted.

VideoDisc

STVS: Managing a Forest, Disc 4, Side 2

ACTIVITY 12-2

OBJECTIVE: Design and carry out an experiment to observe plant tropisms in plant seedlings.
Time: 30 minutes.

PROCESS SKILLS applied in this activity are **forming a hypothesis, observing,** and **inferring.**

PREPARATION

Obtain materials needed for each group. One petri dish can be prepared beforehand as a model.

Cooperative Learning: Use Science Investigation Teams.

HYPOTHESIZING

Student hypotheses will vary. Possible hypotheses for this activity are: "Corn seedling roots will show positive gravitropism;" "Corn seedlings stems will show negative gravitropism," and "Corn seedlings will not show any tropisms."

TEACHING THE ACTIVITY

*Refer to the **Activity Worksheets** for additional information and teaching strategies.*

• Make sure the petri dish is handed down from the ring stand vertically rather than horizontally.

• Make sure students understand that negative phototropism also affects the downward growth of roots; positive phototropism affects the upward growth of stems and leaves.

SUMMING UP/SHARING RESULTS

• Answers will vary. Two to four days are typical answers.

• The roots of each seed grow downward. The stems and leaves grow out from the seed and upward.

• positive gravitropism; negative gravitropism

ACTIVITY 12-2 DESIGNING AN EXPERIMENT
Plant Tropisms

A plant's response to a stimulus is called a tropism. Tropisms can be positive or negative. What kind of tropisms will plants respond to?

Getting Started

Your task in this activity is to devise a method to observe plant seedling response to a stimulus.

Hypothesizing

Make a **hypothesis** about how a plant seedling will respond to a stimulus. What are your reasons for forming this hypothesis?

Materials

Your cooperative group will use:

- petri dish
- tape
- string
- corn seeds
- paper towel
- water
- ring stand

Try It!

1. Place the bottom of the petri dish on a paper towel, draw around the dish, then cut the circle to fit the dish.
2. Tape the paper in the bottom of the dish.
3. Arrange ten corn seeds at equal distances around the outer edge of the petri dish with all the points of the seeds pointing to the middle of the dish.
4. Tape the seeds to the paper towel and wet the towel. Place another wet towel in the petri dish. Put the cover on the dish and tape it closed.
5. Tape a piece of string to the back of the dish and hand the dish from a ring stand. *Predict* the direction in which the roots will grow.
6. After 24 hours, unwrap and open the petri dish to check the seeds. Add water if the seeds or paper are dry. Reseal the dish. Repeat this step to check the seeds daily for the next five days. Draw your observations each day.

Summing Up/Sharing Results

- When did the seeds first germinate?
- In what direction did the roots appear to grow? In what direction did the stems and leaves grow?
- What name would you give to the movement of the roots? To the movement of stems and leaves?

Going Further!

Design an experiment to prove that gravitropism was involved in this investigation and not phototropism.

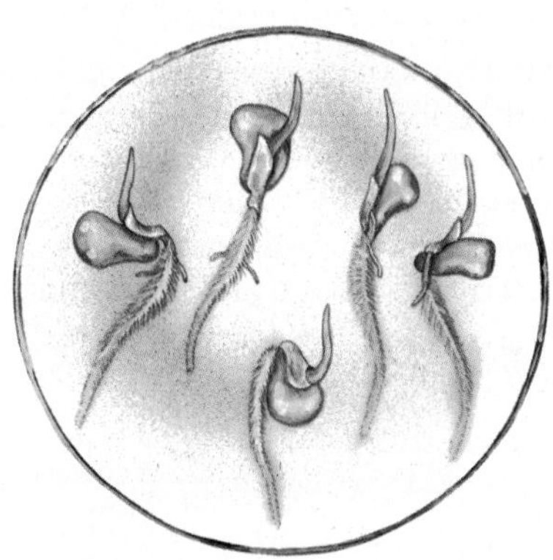

290 PLANT PROCESSES

GOING FURTHER!

An experiment to prove that phototropism is not involved would include leaving some of the seeds uncovered. Perform the same experiment again and observe any differences. If differences are due to light, growth of seeds in the two dishes will differ.

Activity ASSESSMENT

Performance: To further assess students' understanding of plant tropisms, have them design an activity to demonstrate other tropisms described in this chapter.

PROGRAM RESOURCES

From the **Teacher Resource Package** use:
Activity Worksheets, pages 106-107, Activity 12-2, Plant Tropisms.

CHAPTER

REVIEW

SUMMARY

12-1: Photosynthesis and Respiration

1. Carbon dioxide needed for photosynthesis enters leaves through stomata. Water vapor escapes during this process.

2. Chlorophyll in plants captures sunlight. Photosynthesis splits water and makes glucose and oxygen, which are needed by all organisms.

3. Through respiration, energy stored in glucose becomes available for cell activities.

12-2: Plant Responses

1. Plants respond to stimuli by growing toward or away from light, gravity, or water.

2. Plants depend on a critical period of darkness in order to flower.

3. Many tropisms are controlled by hormones produced by plants.

12-3: Plant Relationships

1. Some plants live as parasites, deriving food from other plants.

2. Some plants have mutualistic relationships with other organisms in which both benefit.

3. Plants and animals have coevolved, developing unique relationships with each other.

12-4: Science and Society: The Treasure of Tropical Plants

1. Many medicines used today are derived from common plants.

2. Destroying tropical forests may destroy unknown medicinal plants as well.

KEY SCIENCE WORDS

a. **auxin**
b. **coevolution**
c. **day-neutral plants**
d. **ethnobotany**
e. **long-day plants**
f. **photoperiodism**
g. **photosynthesis**
h. **respiration**
i. **short-day plants**
j. **stimulus**
k. **transpiration**
l. **tropism**

UNDERSTANDING VOCABULARY

Match each phrase with the correct term from the list of Key Science Words.

1. anything in the environment that affects behavior of an organism
2. a plant hormone
3. flowering response of plant to light
4. using light to make glucose and oxygen
5. positive and negative plant responses
6. loss of water through stomata
7. plants that require short nights to flower
8. plants that are not sensitive to hours of darkness
9. releasing energy from food
10. when species evolve together

OPTIONS

ASSESSMENT

To assess student understanding of material in this chapter, use the resources listed.

COOPERATIVE LEARNING

Consider using cooperative learning in the THINK AND WRITE CRITICALLY, APPLY, and MORE SKILL BUILDERS sections of the Chapter Review.

PROGRAM RESOURCES

From the **Teacher Resource Package** use:

Chapter Review, pages 27-28.

Chapter and Unit Tests, pages 76-79, Chapter Test.

Chapter and Unit Tests, pages 80-81, Unit Test.

CHAPTER

REVIEW

SUMMARY

Have students read the summary statements to review the major concepts of the chapter.

UNDERSTANDING VOCABULARY

1.	j	**6.**	k
2.	a	**7.**	e
3.	f	**8.**	c
4.	g	**9.**	h
5.	l	**10.**	b

ASSESSMENT

Portfolio

Encourage students to place in their portfolios one or two items of what they consider to be their best work. For each item, ask students to explain why that item was chosen and what they learned from it. Items might be selected from the following.

- Flex Your Brain, p. 276
- Cooperative Learning report on plant relationships with other plants or animals, p. 287
- Skillbuilder hypothesis about evolutionary history of bees and orchids, p. 287

Performance

Additional performance assessments may be found in *Performance Assessment* and *Science Integration Activities* that accompany **Merrill Life Science.** Performance Task Assessment Lists and rubrics for evaluating these activities and other products generated throughout the chapter can be found in Glencoe's *Performance Assessment in Middle School Science.*

CHAPTER

REVIEW

CHECKING CONCEPTS

1. c	**6.** b
2. b	**7.** d
3. d	**8.** c
4. c	**9.** d
5. d	**10.** a

USING LAB SKILLS

ASSESSMENT

Use these alternate lab exercises to assess students' understanding of skills used in this chapter.

11. Students will have to select fruit that is very green. Bananas and peaches often arrive at groceries very underripe. Controls would be a green banana or peach in separate resealable plastic bags. Hypotheses might include: A whole, unpeeled apple will ripen fruit twice as fast as if the fruit is left to ripen itself.

12. Students can use Elodea sprigs and indicator tests for color changes that indicate the production of carbon dioxide or oxygen under different light conditions (lack of light to simulate night). They can also experiment with keeping plants in darkness for different lengths of time to see if this influences respiration.

THINK AND WRITE CRITICALLY

13. Plants use energy from sunlight in photosynthesis.

14. If it did, it would die also.

15. Indian Pipe cannot produce its own food, so it must take nourishment from a producer.

CHAPTER

REVIEW

CHECKING CONCEPTS

Choose the word or phrase that completes the sentence.

1. Stomata open to allow _______ the plant.
a. sugar into **c.** carbon dioxide into
b. sugar out of **d.** light into

2. A product of photosynthesis is _______.
a. CO_2 **c.** H_2O
b. O_2 **d.** H_2

3. Water, carbon dioxide, and energy are all products of _______.
a. cell division **c.** growth
b. photosynthesis **d.** respiration

4. A _______ plant needs short nights to flower.
a. day-neutral **c.** long-day
b. short-day **d.** nonvascular

5. Light, touch, and gravity are all examples of _______.
a. tropisms **c.** responses
b. growth behaviors **d.** stimuli

6. _______ is a plant's response to gravity.
a. Phototropism **c.** Thigmotropism
b. Gravitropism **d.** Hydrotropism

7. Stems, roots, and fruits are all plant parts that may respond to _______.
a. tropisms **c.** germination
b. sugar **d.** hormones

8. The _______ is an example of a parasitic plant.
a. oak **c.** dodder
b. acacia **d.** dandelion

9. The relationship that occurs between ants and acacias is _______.
a. parasitism **c.** photoperiodism
b. independence **d.** mutualism

10. Adaptations of cacti and tortoises over time are examples of _______.
a. coevolution **c.** mutualism
b. parasitism **d.** tropism

USING LAB SKILLS

11. **Design an experiment** to determine if apples can be used to promote ripening in fruit. State a **hypothesis** clearly. Be sure to identify a control. Then perform your experiment.

12. Using some of the techniques you learned in the MINI-Labs on pages 278 and 280, **design an experiment** to test a **hypothesis** that states: Respiration takes place in plants only at night.

THINK AND WRITE CRITICALLY

Answer the following questions in your Journal using complete sentences.

13. Why is it important that plants grow toward the light as in the photograph below?

14. Why doesn't the dodder plant kill its host?

15. What advantage does the Indian Pipe have living off a plant?

APPLY

16. Growers of bananas pick green bananas, then treat them with ethylene during shipping. Why?
17. Identify the stimulus and response as positive or negative.
    —stem grows up
    —tendrils grow up
    —roots grow down
    —plant grows toward light
18. Scientists who study sedimentary rocks and fossils suggest that oxygen did not occur on Earth until plantlike protists appeared. Why?
19. Explain why crab apple trees bloom in the spring, and sometimes in the fall, but not in the summer.
20. Some tropical flowers have fruity odors and are light in color. Bats that are nocturnal eaters have long tongues and fur on their faces. Explain how these two organisms are examples of coevolution.

MORE SKILL BUILDERS

If you need help, refer to the Skill Handbook.

1. **Hypothesizing:** Make a hypothesis as to the action of leaf guard cells in desert plants.
2. **Observing and Inferring:** Based on your knowledge of plants, infer the amount and location of stomata in land and water plants.
3. **Classifying:** Classify each as short-day, long-day, or day-neutral plants by making a chart: corn, chrysanthemum, dandelions, dill, iris, poinsettia, red clover, spinach, sunflower, tomato.
4. **Comparing and Contrasting:** Compare and contrast the action of each chemical on a plant: auxin, ethylene.
5. **Designing an Experiment:** Design an experiment concerning the effect of day length on flowering.

PROJECTS

1. **Design an experiment** with a control to test how well plant cuttings root with and without rooting hormone.
2. **Design an experiment** to show the rate of yeast respiration at three different temperatures.
3. Research tropical plants studied for their medicinal value. Illustrate two and describe the culture from which they derive.

APPLY

16. If bananas were picked when ripe, they may be over ripe by the time they arrived at a store.
17. negative gravitropism or positive phototropism (stem and tendrils up); positive gravitropism (roots down); positive phototropism (plant toward light)
18. Oxygen is a by-product of photosynthesis. Until organisms that used carbon dioxide for photosynthesis existed, little oxygen was available.
19. Crab apple trees bloom in the spring, and sometimes in the fall, when the days are short and nights long. During these periods, the increased dark period may contribute to the trees producing flowers.
20. Light colors and fruity odors attract bats. Bats have evolved furry faces that pick up pollen when they visit flowers to sip nectar. Both bats and flowers benefit from this mutualistic relationship that has resulted in coevolution of bats and night-blooming flowers.

MORE SKILL BUILDERS

1. Hypothesizing: When water is limited and plants lose water through transpiration, guard cells close stomata to prevent further water loss.
2. Observing and Inferring: Water plants that float on water have stomata on upper leaf surfaces, whereas most land plants have more stomata on lower leaf surfaces than on upper leaf surfaces.
3. Classifying: Short-day plants include iris, chrysanthemum, and poinsettia. Day-neutral plants include corn, tomatoes, and dandelions. Long-day plants include dill, red clover, spinach, and sunflower.
4. Comparing and Contrasting: auxin—stem grows toward light; ethylene—fruits ripen
5. Designing an Experiment: Experiments may include growing flowers in pots and keeping them in darkness for various periods of time each day, then recording which ones flower first. Make sure there is a control in each experiment.

UNIT 4
GLOBAL CONNECTIONS

Objective

In this unit ending feature, the unit topic, "Plants," is extended into other disciplines. Students will see the effects of plants on themselves and other organisms, and the importance of plants to Earth.

Motivate

Cooperative Learning: Have students work in Expert Teams. Assign one Connection to each group of students. Have each group research the geographic region of the Connection to find out more about the kinds of plants found and the special adaptations that enable them to live in such surroundings.

Teaching Tips

- Have students state a hypothesis about the basic conditions plants need to live and grow.
- Tell students to think about the connection between plants and other living things as they read this feature.

Wrap-Up

Conclude this lesson by having students predict what would happen to life on Earth if sudden darkness prevented plants from producing food through photosynthesis.

CHEMISTRY

Background: Dependency on foreign oil has long been a problem. Ethanol is a possible substitute. It can be added to gasoline or used by itself.
Discussion: The U.S. has an annual surplus of corn. Would it be in the best interest of the country to use some of it in the production of ethanol?
Answer to Question: The production of ethanol is less costly, plus it is a renewable energy source.
Extension: Obtain small amounts of gasohol and regular lead-free gasoline. Have students compare their physical properties.

UNIT 4
GLOBAL CONNECTIONS

Plants

In this unit, you have studied about plants. Now find out the importance of plants throughout the world.

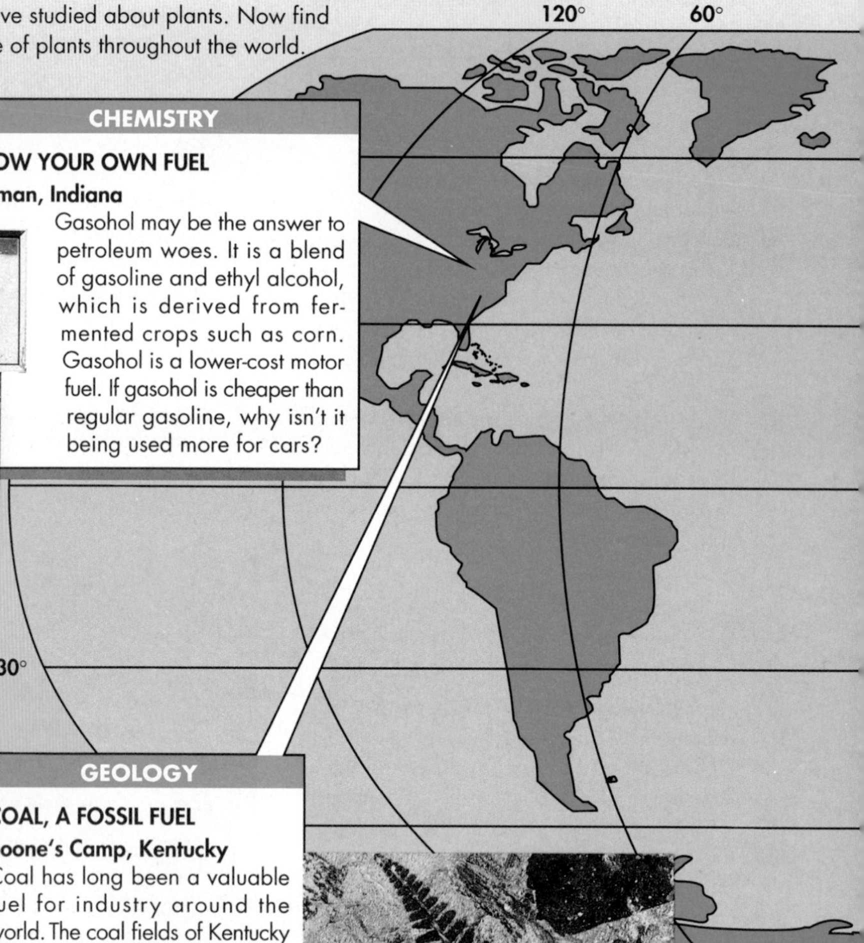

120° 60° 30°

CHEMISTRY

GROW YOUR OWN FUEL
Breman, Indiana 
Gasohol may be the answer to petroleum woes. It is a blend of gasoline and ethyl alcohol, which is derived from fermented crops such as corn. Gasohol is a lower-cost motor fuel. If gasohol is cheaper than regular gasoline, why isn't it being used more for cars?

THIS IS NOT AN ARCO PRODUCT
Gasohol

GEOLOGY

COAL, A FOSSIL FUEL
Boone's Camp, Kentucky
Coal has long been a valuable fuel for industry around the world. The coal fields of Kentucky had their beginnings 300 million years ago when they were forests of giant fern and club mosses. How do we know coal fields were once forests?

294

GEOLOGY

Background: Students should remember that plants use water, carbon dioxide, and light to produce new plant cells. These cells contain stored chemical energy. In the case of coal, it has been stored for millions of years.
Discussion: Have students review the process of photosynthesis. Discuss the importance of this process to all other living things.
Answer to Question: Parts of fossilized trees have been found in layers of coal.
Extension: Have students examine samples of bituminous and anthracite coal. Ask students to try to determine which coal type is harder and how each is constructed.

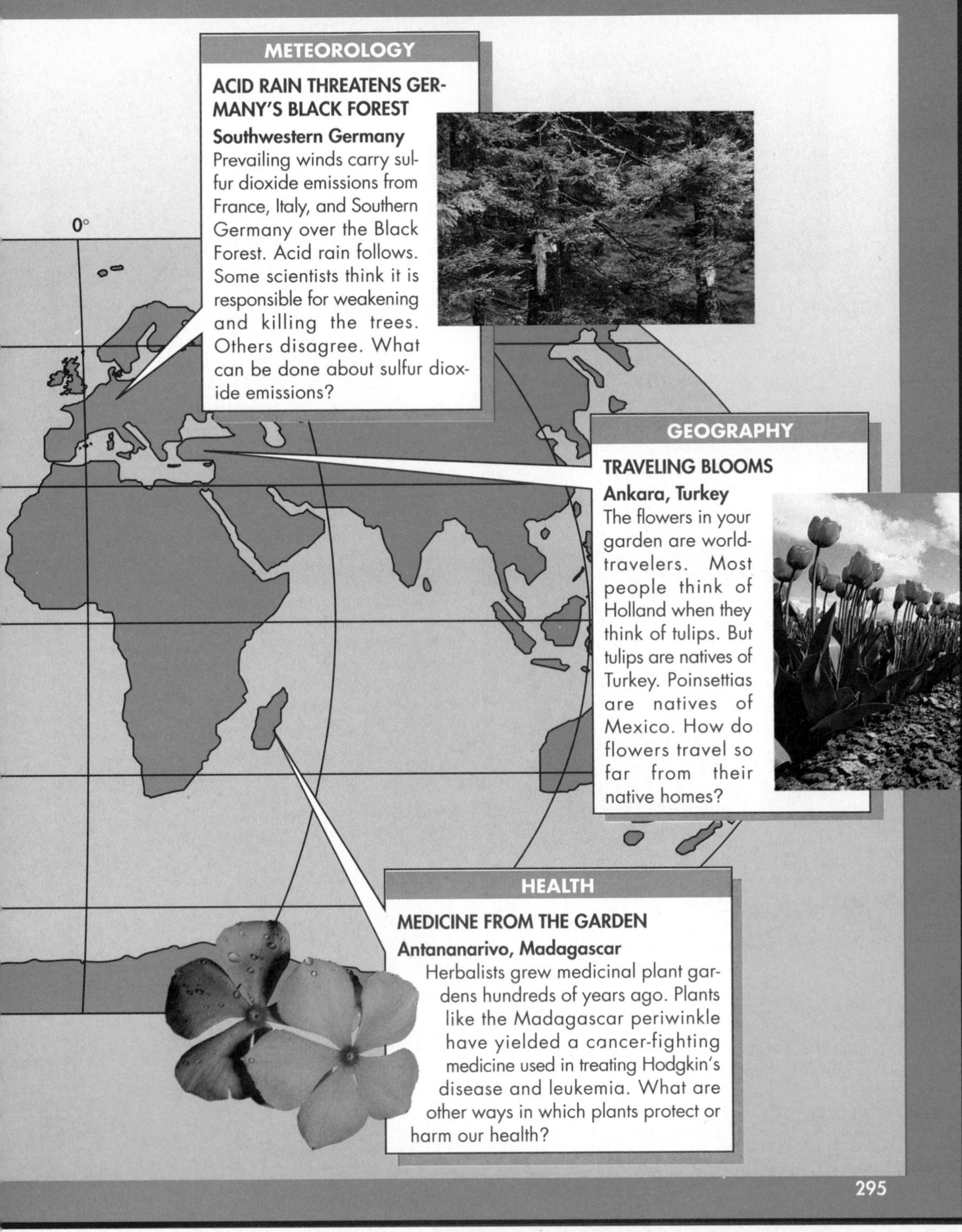

0°

METEOROLOGY

ACID RAIN THREATENS GERMANY'S BLACK FOREST

Southwestern Germany

Prevailing winds carry sulfur dioxide emissions from France, Italy, and Southern Germany over the Black Forest. Acid rain follows. Some scientists think it is responsible for weakening and killing the trees. Others disagree. What can be done about sulfur dioxide emissions?

GEOGRAPHY

TRAVELING BLOOMS

Ankara, Turkey

The flowers in your garden are world-travelers. Most people think of Holland when they think of tulips. But tulips are natives of Turkey. Poinsettias are natives of Mexico. How do flowers travel so far from their native homes?

HEALTH

MEDICINE FROM THE GARDEN

Antananarivo, Madagascar

Herbalists grew medicinal plant gardens hundreds of years ago. Plants like the Madagascar periwinkle have yielded a cancer-fighting medicine used in treating Hodgkin's disease and leukemia. What are other ways in which plants protect or harm our health?

295

METEOROLOGY

Background: Acid rain threatens forests throughout Europe. The amount of damage from acid rain depends on the original acidity of the soil.

Discussion: Acid rain causes damage to forests as well as marble sculptures and limestone buildings. Ask students to explain how acid rain affects marble and limestone.

Answer to Question: Factories can install "scrubbers" to reduce the amount of sulfur emissions into the air; less use of fossil fuels such as coal and oil; or more efficient use of these fuels.

Extension: Set up two separate fish tanks. Place the same type and number of plants in each. Every week, spray plants in one tank with plain water; spray those in the second tank with a weak solution of vinegar and water. Have students record responses over three weeks.

GEOGRAPHY

Background: Explorers and conquerors have carried flowers and plants from country to country. The tulip, so connected with Holland, came from Turkey. French marigolds originated in South America.

Discussion: Ask students to suggest some reasons why plants and flowers have traveled so far from their original countries.

Answer to Question: Many world travelers and explorers carried roots or seeds of flowers back to their home countries.

Extension: Provide different types of seeds to examine, such as maple, petunia, and milkweed. Have students find out which seed travels farthest and how seeds are carried from place to place.

HEALTH

Background: Herbs such as foxglove, willow, and valerian have long provided medicines. The world's oldest botanical garden at Oxford University, England, is still a laboratory for plant science.

Discussion: Botany was once the most advanced science. Ask students why they think this was so.

Answer to Question: Answers will vary. Plants provide us with fresh vegetables and fruits, but they also cause reactions like poison ivy rash or produce fruit that is poisonous.

Extension: Have students examine some common products that contain plant products (cooking oils, shoe polishes, and shampoos) or are derived from plants.

PLANT PATHOLOGIST

Background: Plant pathologists study plant diseases and research methods of disease control. A college degree in biological science is needed for a career in plant pathology, plus graduate work in the area of the specialty.

Related Career	Education
Plant Pathologist	MS degree
Botanist	BS degree
Horticulturist	2 yr. college

Career Issue: Genetic engineering may be used to develop disease-resistant plants. Such plants might alleviate world hunger. However, some people are concerned about genetic manipulation of existing species.

What do you think? What might be the benefits of disease-resistant plants?

PROFESSIONAL GARDENER

Background: Professional gardeners learn most of the basics of their trade on the job. Two years at a trade school or community college, however, are beneficial in terms of knowledge, skill, and business management.

Related Career	Education
Gardener	On the job
Landscape Designer	College
Landscape Architect	BS degree

Career Issue: Some of the so-called "safe" pesticides cause allergies, nausea, and headaches. Some weed killers are highly toxic to pets and small children.

What do you think? Have students discuss safe methods of insect and weed control that professional gardeners might use.

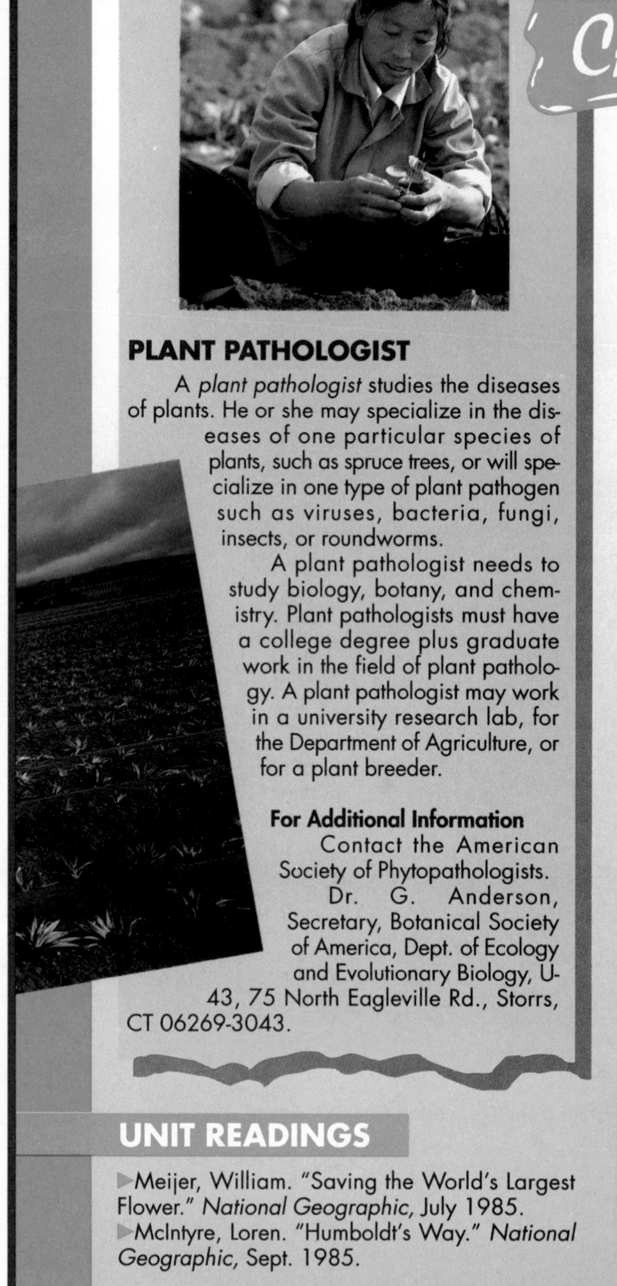

CAREERS

PLANT PATHOLOGIST

A *plant pathologist* studies the diseases of plants. He or she may specialize in the diseases of one particular species of plants, such as spruce trees, or will specialize in one type of plant pathogen such as viruses, bacteria, fungi, insects, or roundworms.

A plant pathologist needs to study biology, botany, and chemistry. Plant pathologists must have a college degree plus graduate work in the field of plant pathology. A plant pathologist may work in a university research lab, for the Department of Agriculture, or for a plant breeder.

For Additional Information

Contact the American Society of Phytopathologists. Dr. G. Anderson, Secretary, Botanical Society of America, Dept. of Ecology and Evolutionary Biology, U-43, 75 North Eagleville Rd., Storrs, CT 06269-3043.

PROFESSIONAL GARDENER

A *professional gardener* should have an interest in plants, landscaping, and the conditions plants need for growing. Professional gardeners need to have a knowledge of the kinds of plants to use in a particular setting and what kinds of grass and trees will grow well in a particular soil. He or she should enjoy working out-of-doors.

Professional gardeners must have a knowledge of gardening tools such as mechanical tree planters, pruning devices, and tillers. A professional gardener can learn much through on-the-job training. Programs are also offered through trade schools, community colleges, or county agricultural extension agencies.

For Additional Information

For additional information, contact the Professional Grounds Management Society, 12 Galloway Ave., Suite 1E, Cockneysville, MD 21030.

UNIT READINGS

▶Meijer, William. "Saving the World's Largest Flower." *National Geographic,* July 1985.
▶McIntyre, Loren. "Humboldt's Way." *National Geographic,* Sept. 1985.

296

UNIT READINGS

Background

▶Farney, Dennis. "The Tallgrass Prairie—Can It Be Saved?" *National Geographic,* January 1980. This article details attempts to locate and preserve undisturbed tallgrass prairie.

▶Meijer, William. "Saving the World's Largest Flower." *National Geographic,* July 1985. Students who enjoy the *Guiness Book of World Records* will enjoy reading about *Rafflesia arnoldii,* a flower that weighs 6 kilograms and measures almost a meter across.

▶McIntyre, Loren. "Humboldt's Way." *National Geographic,* September 1985, tells of the expedition of Alexander Von Humboldt to the Americas, early in the 19th century. He made an intensive study and collection of plants from the rain forests.

Where the Sky Began

by John Madson

The following is an excerpt from Where the Sky Began, *by John Madson.*

Grass and sky would be enough. With only those, the summer prairie would be a smiling, running spread of cloud shadow and wind pattern.

But the tall prairie goes beyond that. From the first greening of spring to the full ripening of autumn, it is spangled by a vivid progression of flowers—a rainbow host that first enamels the burned slopes of early spring and ends months later with great nodding blooms that rise above a man's head. Sometimes as secret and solitary as jewels, but often in broad painted fields, the prairie flowers come on—lavender, indigo, creamy white, pink, coral, gold, magenta, crimson, orange, and palest yellow and blue, their flowers tending from ice to flame....

The prairie flowers come on in waves, each in its own time, some blooming very briefly and others persisting for weeks. Except for a short period early in the growing season, the flowers must compete with a rising tide of grasses. All may begin growing at about the same time; some just mature much later than others, needing months of growth if they are to compete with the towering August bluestem. Spring or fall, prairie flowers are as tall as they need to be.

Most of these earliest prairie flowers appear before mid-May and are never much more than six inches tall. The best known is the pasqueflower, with tulip-like blooms ranging from white to pale lavender, its stem and leaves wearing a dense covering of fine silken hairs. It gives the impression of a small flower trying to keep warm and having a tough time doing so, for pasqueflowers may bloom on bare, exposed crests of old glacial moraines while there are still patches of snow on the sheltered slopes behind them. Brave little flowers, often braver than I. More than once, the lowering clouds and sharp gray winds of late March have hustled me off the prairie before I'd finished photographing the first pasqueflowers. But no matter—at such times their blooms are usually closed anyway.

In Your Own Words

▶ Choose a wildflower, native to your area, and write a brief, descriptive report on its habitat, time of bloom, and any special features of the flower. (Hint: dandelions are not native wildflowers!)

Source: John Madson. *Where the Sky Began*. Boston: Houghton Mifflin, 1983.

Biography: John Madson lives in Godfrey, a small town in southwestern Illinois. He is a naturalist, poet, and scholar with a deep enthusiasm and concern for the tallgrass prairie.

TEACHING STRATEGY

Have students read through the passage by John Madson. Then have them discuss the questions below.

Discussion Questions

1. **Why do the smallest and most delicate flowers appear in the early spring?** *When small flowers bloom early, they do not have to compete for sunshine and nutrients with the tall summer grasses.*
2. **What is the feeling that you get from Madson's description of the earliest flowers?** *Madson describes the pasqueflower as a "brave little flower" which seems to be "trying to keep warm." The "lowering clouds and sharp gray winds" give a strong impression of the prairie in early spring.*
3. **What do you think is Madson's reason for writing about the tallgrass prairie and its wild flowers?** *Much of the tallgrass prairie has disappeared. Madson writes to concern us with the need to preserve what little remains and to let us "see" how remarkable the prairie once was.*

More Readings

1. Aikman, Lonnelle. "Herbs For All Seasons." *National Geographic,* March 1983, describes plants that are used in cooking, as medicine, and as insect repellents.

2. Hapgood, Fred. "The Prodigious Soybean." *National Geographic,* July 1987. Why are soybeans important to world economy? This article details some of their uses in common consumer products.

Classics

▶ Leopold, Aldo. *Sand County Almanac.* New York: Ballantine, 1986. The effect of the changing seasons and years is lyrically described in this book.

▶ Madson, John. *Where the Sky Began.* Boston: Houghton Mifflin, 1983. The land of the tallgrass prairie as it was before the white settlers came.

Other Works

▶ Rolvaag, O.E. *Giants in the Earth.* New York: Harper & Brothers, 1927. Tells the story of early settlers on the prairie.

UNIT

5

ANIMALS

In Unit 5, students look into characteristics of animals and what distinguishes invertebrates from vertebrates.

CONTENTS

ADVANCE PREPARATION

Activities

▶ **Activity 13-2, page 314,** requires live cultures of hydra, *Daphnia,* or brine shrimp, and stereoscopic microscopes.

▶ **Activity 14-1, page 331,** requires live earthworms, soil, vinegar, cotton swabs, and hand lenses.

▶ **Activity 14-2, page 339,** calls for live crayfish and uncooked ground beef.

▶ **Activity 15-1, page 355,** requires goldfish or guppies, aquarium water, small nets, beakers, rods, thermometers, and ice water.

298

OPTIONS

Cross Curriculum

▶ Have students research and report on the Great Barrier Reef near Australia.

▶ Insects are often used in the art and architecture of other countries. Have students look through magazines to find examples of items that have insects on them.

PROJECT

During the course of this unit study, have students work cooperatively on PROJECT 5, *Was It a Reptile? Was It a Bird? Was It Really Warm?*, found on pp. 702 and 703.

Science at Home

▶ Have students observe the behavior of a pet at home. Have them record how the pet reacts to people, food, and other animals and share their observations with the class.

Cooperative Learning: Divide the class into groups of four. Assign each group one of the invertebrate phyla. Have them make posters to illustrate the characteristics of the assigned phylum.

What's Happening Here?

These ants aren't preparing to feast on the tiny aphids, but they do get food from them. Ants "farm" aphids for the sweet nectar they produce. The ants keep the aphids in "pens" in underground burrows. In another type of relationship, a border collie protects its sheep from predators such as wolves by keeping the sheep together. Animals have many different types of relationships with one another. Many of these relationships work for the benefit of all the animals involved. You'll learn more about animals and about their relationships in this unit.

UNIT CONTENTS

299

▶ **Activity 15-2, page 362,** requires live frog eggs, lake or pond water, glass jars or aquaria, nets, aquatic plants, gravel, lettuce, watch glasses, and stereoscopic microscopes.

▶ **Activity 16-1, page 378,** requires down and contour feathers, hand lenses, scissors, and balances.

▶ **Activity 17-1, page 400,** requires aquaria, covers, thermometers, sand, aged tap water, rulers, guppies, snails, water plants, guppy food, and nets.

▶ **Activity 17-2, page 408,** requires ants, soil, jars, wire covers, black paper, sponges, honey, sugar, bread crumbs, small cans with lids, spoons or trowels, and hand lenses.

Field Trips and Speakers

▶ Invite an entomologist to speak to the class on the adaptations of insects.

INTRODUCING THE UNIT

What's Happening Here?

▶ After students look at the photos and read the text, ask them what's happening here. Explain that in this unit they will study the relationships between animals and other organisms, and the characteristics of each major animal group.

▶ **Background:** The relationship between ants and aphids is a mutualistic one because they both benefit from their association.

Previewing the Chapters

▶ Some students may not think of invertebrates as animals. Have students review animal characteristics to see how invertebrates match up.

Tying to Previous Knowledge

▶ Use the **inquiry questions** in the OPTIONS box below to investigate whether humans are animals.

Multicultural Awareness

Have students research the rules various cultures have about eating vertebrates. What are the rules, and how do the cultures explain these rules? For example, many Middle Eastern cultures forbid eating pork. In India, many people are not allowed to eat meat, although cows are numerous. In some cultures dogs are domesticated for use as food, but in the United States, where dogs and cats are treated as companions, eating these animals would be frowned upon and prohibited.

Inquiry Questions

Use the following questions to focus a discussion of whether humans are animals.

▶ **Are humans alive?**

▶ **How do you know?**

▶ **Do humans fit into one of the chapters in this unit?**

▶ **Which group includes humans?**

▶ **Do humans have the characteristics of living things?**

▶ **If humans aren't animals, what are they?**

CHAPTER

13 Introduction to Animals

CHAPTER SECTION	OBJECTIVES	ACTIVITIES
13-1 What Is an Animal? (2 days)	1. **Identify** the characteristics of animals. 2. **Determine** how the body plans of animals differ. 3. **Distinguish** between invertebrates and vertebrates.	**Activity 13-1:** *Determining Symmetry,* p. 313
13-2 Experiments Using Animals Science & Society (1 day)	1. **Acknowledge** the advancements made due to experimentation with animals. 2. **Develop** an idea for a compromise between progress in medicine and the welfare of laboratory animals.	
13-3 The Simplest Invertebrates (3 days)	1. **Identify** the structures that make up sponges and cnidarians. 2. **Describe** how sponges and cnidarians obtain food and how they reproduce.	**Activity 13-2:** *Designing an Experiment (Comparing Flatworms and Roundworms),* p. 318
13-4 The Simple Worms (2 days)	1. **Compare** the body plans of flatworms and roundworms. 2. **Distinguish** between free-living and parasitic organisms. 3. **Identify** disease-causing flatworms and roundworms.	**MINI-Lab:** *How do planarians move?* p. 314 **MINI-Lab:** *How do planarians respond to light?* p. 315
Chapter Review		

ACTIVITY MATERIALS

FIND OUT	ACTIVITIES		MINI-LABS	
Page 301 paper pencil	**13-1 Determining Symmetry, p. 313** paper pencil	**13-2 Designing an Experiment, p. 318** culture of planarians prepared slide of tapeworm dropper small piece of liver small paintbrush compound microscope stereoscopic microscope	**How do planarians move? p. 314** culture of planarians medicine dropper watch glass stereomicroscope	**How do planarians respond to light? p. 315** culture of planarians medicine dropper watch glass stereomicroscope

CHAPTER FEATURES	TEACHER RESOURCE PACKAGE	OTHER RESOURCES
Skill Builder: *Concept Mapping,* p. 304	**Ability Level Worksheets** ◆ ***Study Guide,*** p. 51 ● ***Reinforcement,*** p. 51 ▲ ***Enrichment,*** p. 51 **Cross-Curricular Connections,** p. 17 **Concept Mapping,** p. 31 **Activity Worksheets,** pp. 5, 113-114 **Transparency Masters,** pp. 41-42	**Color Transparency 21,** Animal Phylogeny **STVS:** Disc 5, Side 1
You Decide! p. 306	**Ability Level Worksheets** ◆ ***Study Guide,*** p. 52 ● ***Reinforcement,*** p. 52 ▲ ***Enrichment,*** p. 52 **Science and Society,** p. 17	**STVS:** Disc 5, Side 2
Technology: *Sea Pharmacy,* p. 309 **Skill Builder:** *Comparing and Contrasting,* p. 312	**Ability Level Worksheets** ◆ ***Study Guide,*** p. 53 ● ***Reinforcement,*** p. 53 ▲ ***Enrichment,*** p. 53 **Activity Worksheets,** pp. 115-116 **Critical Thinking/Problem Solving,** p. 17 **Transparency Masters,** pp. 43-44	**Color Transparency 22,** Cnidarians **STVS:** Disc 5, Side 1 **Science Integration Activity 13**
Problem Solving: *Barbara's New Puppy,* p. 316 **Skill Builder:** *Outlining,* p. 317	**Ability Level Worksheets** ◆ ***Study Guide,*** p. 54 ● ***Reinforcement,*** p. 54 ▲ ***Enrichment,*** p. 54 **Activity Worksheets,** p. 119, 120 **Transparency Masters,** pp. 45-46	**Color Transparency 23,** Planaria **STVS:** Disc 5, Side 1
Summary Key Science Words Understanding Vocabulary Checking Concepts Using Lab Skills Think & Write Critically Apply More Skill Builders Projects	ASSESSMENT RESOURCES **Chapter Review,** pp. 29-30 **Chapter Test,** pp. 89-92 **Performance Assessment in Middle School Science (PAMSS)**	**Chapter Review Software** **Test Bank** **Alternate Assessment** **Performance Assessment**

◆ Basic ● Average ▲ Advanced

ADDITIONAL MATERIALS

SOFTWARE	AUDIOVISUAL	BOOKS/MAGAZINES
Animal, Compuware. *The Worm,* Ventura Educational Systems. *Organizing Animals,* Queue.	*How Animals Are Classified,* filmstrip, EBEC. *Animals Without Backbones,* film, Coronet/MTI. *All Things Animal,* film, Barr Productions, Inc. *The Bio Libe Encyclopedia: A Photographic Record of the Earth's Fauna and Flora,* laserdisc, Image Premastering Services.	Alexander, R.M. *Animals.* NY: Cambridge University Press., 1990. Buchsbaum, Ralph. *Animals Without Backbones.* 3rd ed. Chicago, IL: University of Chicago Press, 1987. Ferry, Georgina. ed. *The Understanding of Animals.* NY: Basil Blackwell, 1984.

CHAPTER

13

INTRODUCTION TO ANIMALS

THEME DEVELOPMENT: Evolution can be used to show students the diversity in the Animal Kingdom. Sponges appeared on Earth about 600 million years ago. There is evidence that they evolved from a type of protozoan. The discussion of animals includes the themes homeostasis and ecology.

CHAPTER OVERVIEW

- **Section 13-1:** This section describes the characteristics that all animals have in common. The classification of animals as invertebrates or vertebrates and symmetry are discussed.
- **Section 13-2: Science and Society:** This section examines the use of animals in experiments. The You Decide feature asks if students would buy products that had been tested on animals.
- **Section 13-3:** This section introduces the animal phyla beginning with Porifera and cnidaria. Their structure, methods of feeding, methods of reproduction, origins, and importance are described.
- **Section 13-4:** The flatworms and roundworms are described in detail. The body plan and way of life of each group is considered.

CHAPTER VOCABULARY

vertebrates
invertebrates
radial symmetry
bilateral symmetry
sessile
Porifera
filter feeders
collar cells
regeneration
hermaphrodite
larva
cnidarians
tentacles
polyp
medusa
free-living
anus
cyst

CHAPTER

13 Introduction to Animals

300

OPTIONS

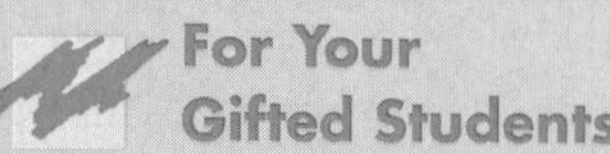

For Your Gifted Students

Have students look into the issue of using animals in scientific research. They can form two groups, with one group trying to find information that gives the point of view of the animal rights groups as the other group searches for the opinion of the researchers. After the information has been gathered, the students can conduct a debate on the issue.

For Your Mainstreamed Students

Students can make chalk or pencil rubbings of the items in the room or school that show bilateral symmetry. They can make and display a collage of the rubbings. Have them follow the same procedure for items that show radial symmetry and those that show no symmetry.

Close your eyes and picture an animal. What comes to your mind? Did you imagine something with four legs and fur, or did you see an insect, a worm, or a person?

FIND OUT!

Do this simple activity to find out how different members of the Animal Kingdom are.

Look at the large photograph at the left. On a sheet of paper, make a *list* of all the animals you recognize in the photo. How many animals are on your list? Did you list five or six? Of course the human and the fish are animals, but so are the corals, sponges, and fine, branching organisms. **In your Journal,** write a one sentence definition of the word "animal."

Gearing Up

Previewing the Chapter

Use this outline to help you focus on important ideas in this chapter.

Section 13-1 What Is an Animal?
- ▶ Animal Characteristics
- ▶ Animal Classification

Section 13-2 Science and Society
Experiments Using Animals
- ▶ To Test or Not To Test

Section 13-3 The Simplest Invertebrates
- ▶ Sponges
- ▶ Origin and Importance of Sponges
- ▶ Cnidarians
- ▶ Origin and Importance of Cnidarians

Section 13-4 The Simple Worms
- ▶ Flatworms
- ▶ Roundworms

Previewing Science Skills

- ▶ In the Skill Builders, you will make a concept map, compare and contrast, and outline.
- ▶ In the Activities, you will observe, hypothesize, and collect data.
- ▶ In the MINI-Labs, you will observe, describe, experiment, and predict.

What's next?

Now that you know that there are many different types of animals, you will learn what makes them different and how they are classified. You will also learn about some invertebrates—sponges, cnidarians, and simple worms.

INTRODUCING THE CHAPTER

Use the Find Out activity to introduce the diversity of animals. Animal diversity can be seen in the chapter opening photograph. Explain to students that they will be learning more about the characteristics of animals.

FIND OUT!

Materials: pencil and paper

Cooperative Learning: Use the Paired Partners strategy for students to discuss what animal each pictured when their eyes were closed.

Teaching Tips

▶ Ask students what animals are the most familiar to them. They probably will suggest dogs, cats, hamsters, fish, and other pets.

▶ Some students may not be able to identify the corals and sponges in the photograph.

▶ Write on the chalkboard the animals they pictured with their eyes closed.

Gearing Up

Have students study the Gearing Up feature to familiarize themselves with the chapter. Discuss the relationships of the topics in the outline.

What's Next?

Before beginning the first section, make sure students understand the connection between the Find Out activity and the topics to follow.

OBJECTIVES AND SCIENCE WORDS: Have students review the objectives and science words in this chapter as they study each section.

ASSESSMENT OPTIONS

PORTFOLIO

Refer to page 319 for suggested items that students might select for their portfolios.

PERFORMANCE ASSESSMENT

See page 319 for additional Performance Assessment options.

Process

Skillbuilder, p. 318
MINI-Lab, p. 316
Activities, 13-1, p. 305; 13-2, p. 314
Using Lab Skills, p. 320

CONTENT ASSESSMENT

Assessment—Oral, pp. 310, 312
Skillbuilder, p. 304
Section Reviews, pp. 304, 307, 313, 318
Chapter Review, pp. 319-321
Mini Quizzes, pp. 303, 311, 317

GROUP ASSESSMENT

Opportunities for group assessment occur with Cooperative Learning Strategies and Flex Your Brain Activities.

SECTION 13-1

PREPARATION

SECTION BACKGROUND

▶ The branch of biology that deals with the study of animals is called zoology. Those scientists who study animals are zoologists. Zoologists group the animal kingdom into phyla. The nine largest phyla will be studied in this text. These phyla contain the majority of animal species.

PREPLANNING

▶ Collect pictures of invertebrates and vertebrates for the bulletin board.

1 MOTIVATE

FLEX Your Brain

Use the Flex Your Brain activity to have students explore ANIMAL CHARACTERISTICS.

FLEX YOUR BRAIN ASSESSMENT

Performance: Use the Flex Your Brain activity to reinforce critical-thinking and problem-solving skills. Students are likely to brainstorm familiar animals.

PROGRAM RESOURCES

From the **Teacher Resource Package** use:

Activity Worksheets, page 5, Flex Your Brain.

13-1 What Is an Animal?

New Science Words

vertebrates
invertebrates
radial symmetry
bilateral symmetry

Objectives

- ▶ Identify the characteristics of animals.
- ▶ Determine how the body plans of animals differ.
- ▶ Distinguish between invertebrates and vertebrates.

Figure 13-1. Animals depend on other living things for their food.

Animal Characteristics

If someone asked you if it was important to protect giant pandas to keep them from becoming extinct, you would probably say yes. But would you feel the same if you were asked about an endangered worm or beetle or spider? How are animals' relationships to each other and to us important? Let's look at the characteristics that all animals have in common.

1. Animals cannot make their own food. They depend on other living things in the environment for food. Some eat plants, some eat other animals, and some eat both plants and animals.
2. Animals digest their food. They can't use the proteins, fats, and carbohydrates in foods directly. Instead, food must be broken down into molecules small enough for their bodies to use.
3. Many animals move from place to place. Moving around lets them find food, escape from enemies, find a better place to live, and find mates. Animals that move very slowly or not at all, have adaptations that let them take care of these needs.
4. Animals have many cells. Different cells carry out different functions such as digesting food, reproduction, and getting rid of wastes.
5. Animal cells are eukaryotic. They have a nucleus surrounded by a membrane and organelles surrounded by membranes.

An animal is a many-celled eukaryotic organism that must find and digest its food. How does this description compare with the one you wrote for the activity on page 301?

OPTIONS

Meeting Different Ability Levels

For Section 13-1, use the following **Teacher Resource Masters** depending upon individual students' needs.

◆ **Study Guide Master** for all students.

● **Reinforcement Master** for students of average and above average ability levels.

▲ **Enrichment Master** for above average students.

Additional Teacher Resource Package masters are listed in the OPTIONS box throughout the section. The additional masters are appropriate for all students.

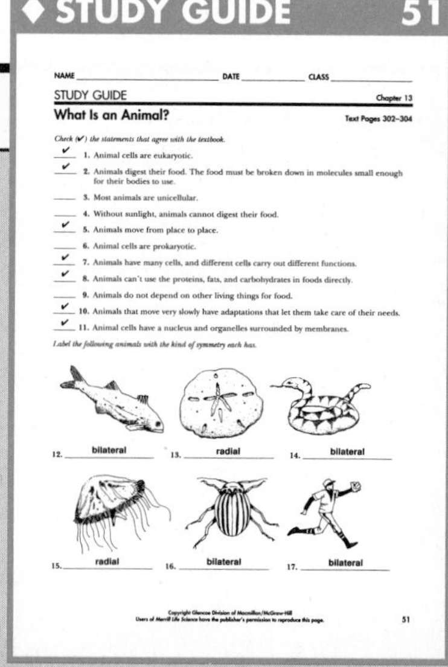

◆ STUDY GUIDE 51

NAME ______ DATE ______ CLASS ______

STUDY GUIDE — Chapter 13

What Is an Animal? — Text Pages 302–304

Check (✔) the statements that agree with the textbook.

✔ 1. Animal cells are eukaryotic.
✔ 2. Animals digest their food. The food must be broken down in molecules small enough for their bodies to use.
___ 3. Most animals are unicellular.
___ 4. Without sunlight, animals cannot digest their food.
✔ 5. Animals move from place to place.
___ 6. Animal cells are prokaryotic.
✔ 7. Animals have many cells, and different cells carry out different functions.
✔ 8. Animals can't use the proteins, fats, and carbohydrates in foods directly.
___ 9. Animals do not depend on other living things for food.
✔ 10. Animals that move very slowly have adaptations that let them take care of their needs.
✔ 11. Animal cells have a nucleus and organelles surrounded by membranes.

Label the following animals with the kind of symmetry each has.

12. bilateral 13. radial 14. bilateral
15. radial 16. bilateral 17. bilateral

 51

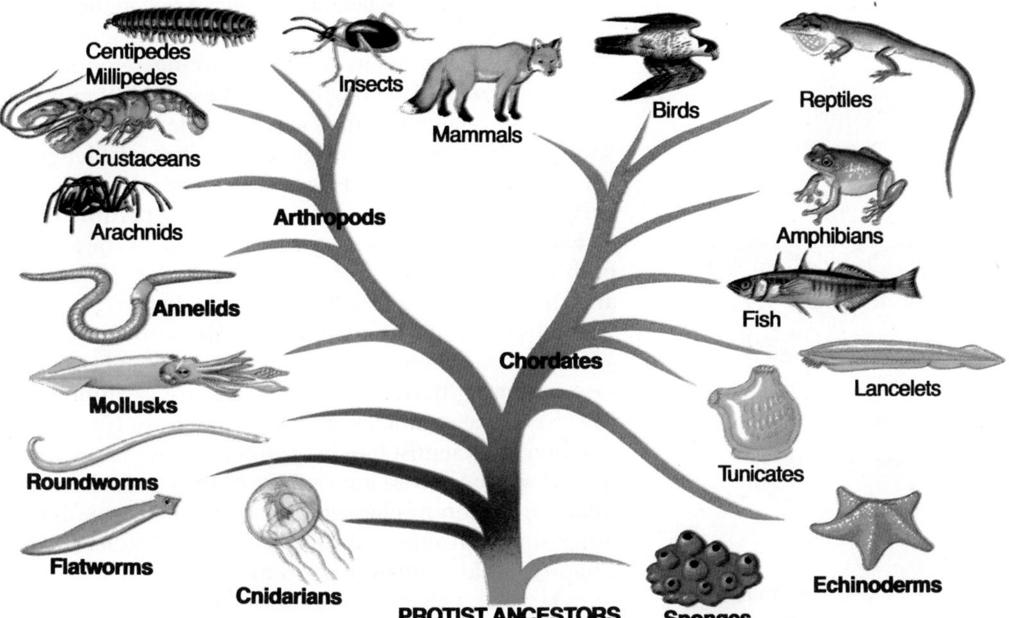

Figure 13-2. This diagram shows the relationships between different groups in the animal kingdom.

Animal Classification

Scientists have identified and named over one million species of animals. Some estimate that there are three to thirty million more to identify and name. There are nine major phyla in the animal kingdom. The phyla are shown in Figure 13-2. All animals have the characteristics listed on page 302, but when a scientist comes across a new animal, how does he or she begin to classify it?

1 The first thing the scientist will do is look to see if the animal has a backbone. Animals with backbones are called **vertebrates.** Some vertebrate animals are fish, humans, whales, and snakes. About 97 percent of all other animal 2 species are invertebrates. **Invertebrates** are animals that don't have a backbone. Sponges, jellyfish, worms, insects, and clams are all invertebrates.

The next thing the scientist will look at is the arrangement of the animal's body parts. This is called the animal's symmetry. Some animals have body parts arranged in a circle around a central point, the way spokes are arranged around the hub on a bicycle wheel. These animals have **radial symmetry.** Hydra and sea anemones have radial symmetry.

Among vertebrates, there are several groups of fish including cartilaginous fish and bony fish. Find out what elements the skeleton of a bony fish contains that cartilaginous skeletons don't have.

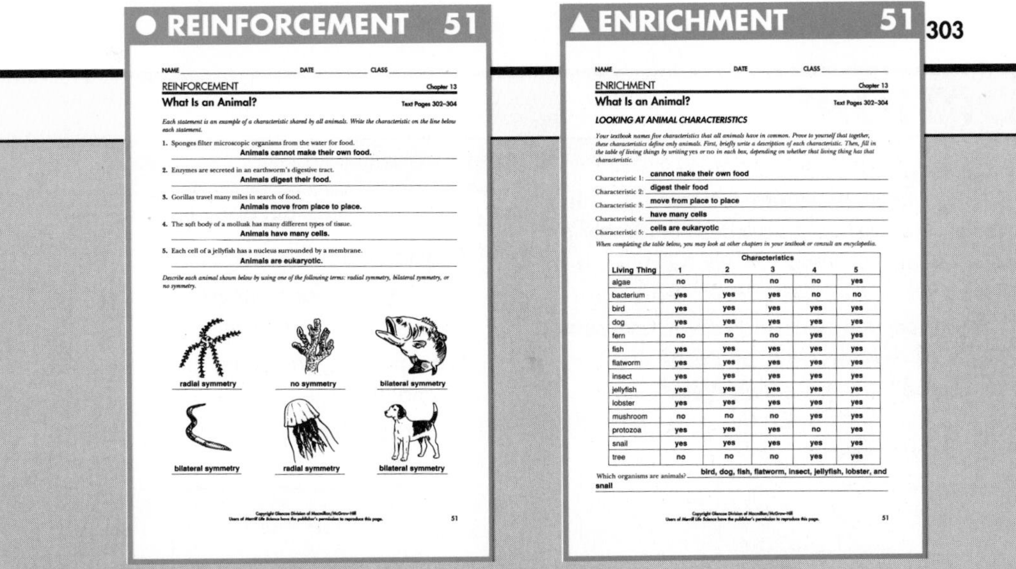

● REINFORCEMENT 51

NAME ______ DATE ______ CLASS ______

REINFORCEMENT Chapter 13

What Is an Animal? Text Pages 302–304

Each statement is an example of a characteristic shared by all animals. Write the characteristic on the line below each statement.

1. Sponges filter microscopic organisms from the water for food.
   Animals cannot make their own food.
2. Enzymes are secreted in an earthworm's digestive tract.
   Animals digest their food.
3. Gorillas travel many miles in search of food.
   Animals move from place to place.
4. The soft body of a mollusk has many different types of tissue.
   Animals have many cells.
5. Each cell of a jellyfish has a nucleus surrounded by a membrane.
   Animals are eukaryotic.

Describe each animal shown below by using one of the following terms: radial symmetry, bilateral symmetry, or no symmetry.

radial symmetry — no symmetry — bilateral symmetry

bilateral symmetry — radial symmetry — bilateral symmetry

 51

▲ ENRICHMENT 51

NAME ______ DATE ______ CLASS ______

ENRICHMENT Chapter 13

What Is an Animal? Text Pages 302–304

LOOKING AT ANIMAL CHARACTERISTICS

Your textbook names five characteristics that all animals have in common. Prove to yourself that together, these characteristics define only animals. First, briefly write a description of each characteristic. Then, fill in the table of living things by writing yes or no in each box, depending on whether that living thing has that characteristic.

Characteristic 1: **cannot make their own food**
Characteristic 2: **digest their food**
Characteristic 3: **move from place to place**
Characteristic 4: **have many cells**
Characteristic 5: **cells are eukaryotic**

When completing the table below, you may look at other chapters in your textbook or consult an encyclopedia.

	Characteristics				
Living Thing	1	2	3	4	5
algae	no	no	no	no	yes
bacterium	yes	yes	yes	no	no
bird	yes	yes	yes	yes	yes
dog	yes	yes	yes	yes	yes
fern	no	no	no	yes	yes
fish	yes	yes	yes	yes	yes
flatworm	yes	yes	yes	yes	yes
insect	yes	yes	yes	yes	yes
jellyfish	yes	yes	yes	yes	yes
lobster	yes	yes	yes	yes	yes
mushroom	no	no	no	yes	yes
protozoa	yes	yes	yes	no	yes
snail	yes	yes	yes	yes	yes
tree	no	no	no	yes	yes

Which organisms are animals? **bird, dog, fish, flatworm, insect, jellyfish, lobster, and snail**

 51

TYING TO PREVIOUS KNOWLEDGE: Review the characteristics of animals in Chapter 7. Make certain that students understand that animals are heterotrophic, have eukaryotic cells, and are many-celled.

2 TEACH

Key Concepts are highlighted.

CONCEPT DEVELOPMENT

▶ **Demonstration:** Compare symmetrical and asymmetrical objects. Use rocks, chalk, and other classroom objects.

REVEALING MISCONCEPTIONS

▶ Many students do not think of invertebrates as animals. Elicit student help in writing on the chalkboard a list of familiar invertebrates. This will allow you to correct misconceptions. Emphasize that the animals on the list are related by the absence of a structural characteristic—a backbone—rather than the presence of a characteristic.

Connect to... Chemistry

Answer: calcium and phosphorus

MINI QUIZ

Use the Mini Quiz to check students' recall of chapter content.

1. **When a scientist finds a new animal, how does he or she begin to classify it?** *by looking to see if it has a backbone*
2. **Are most animals invertebrates or vertebrates?** *invertebrates*
3. **What type of symmetry do most animals have?** *bilateral*

VideoDisc

STVS: Zooplankton, Disc 5, Side 1

CHECK FOR UNDERSTANDING

▶ Ask questions 1-3 and the **Apply** Question in the Section Review.

RETEACH

Have students make diagrams of a fish, a butterfly, and a dog and label the dorsal and ventral sides and the anterior and posterior ends.

EXTENSION

For students who have mastered this section, use the **Reinforcement** and **Enrichment** masters or other OPTIONS.

3 CLOSE

▶ Show the filmstrip *Animals without Backbones* from Encyclopaedia Britannica.

SECTION REVIEW ANSWERS

1. cannot make their own food, digest their food, most move, have many cells, and have eukaryotic cells

2. radial symmetry—hydra; bilateral symmetry—dog, person; asymmetry—sponges

3. invertebrates do not have a backbone; vertebrates have backbones

4. Apply: dorsal—back, ventral—lower, anterior—front, posterior—back

5. Connect to Earth Science: sedimentary rock; formed when sediments become pressed or cemented together

Skill Builder

Students' events chain concept maps should begin with scientists determining if the animal is a vertebrate or invertebrate. The next step should be the scientist determining the animal's symmetry. The concept map should end with scientists comparing the organism's characteristics with those of other organisms.

Skillbuilder

ASSESSMENT

Portfolio: Use this Skillbuilder to assess students' abilities to create a chain of events Concept Map of the steps scientists use to classify a new animal.

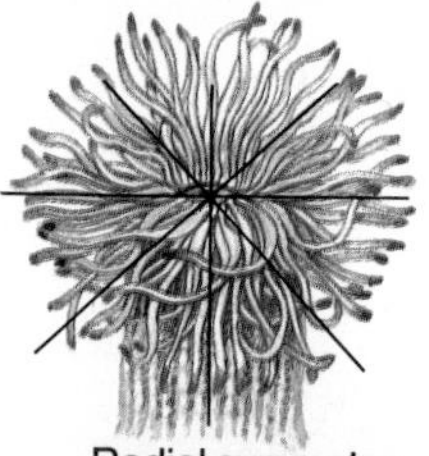

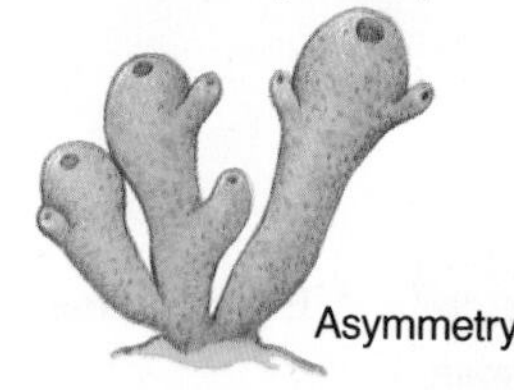

Figure 13-3. Sea anemones have radial symmetry, butterflies have bilateral symmetry, and sponges are asymmetrical.

Most animals have bilateral symmetry. Look in the mirror. 3 Does your body look the same on both sides? An animal with **bilateral symmetry** has its body parts arranged in the same way on both sides of its body. In Latin the word *bilateral* means "two sides." Bilateral animals can be divided into right and left halves by drawing an imaginary line down the length of the body. Each half is a mirror image of the other half.

Animals with bilateral symmetry have a definite front, or *anterior*, end. They also have a definite back, or *posterior*, end. The upper side of the animal is called the *dorsal* side, and the lower side is called the *ventral* side.

Some organisms have no definite shape and are called asymmetrical. There is no way their bodies can be divided into matching halves. Many sponges are asymmetrical. Three types of symmetry are shown in Figure 13-3.

When the scientist has determined an animal's symmetry and whether it's an invertebrate or vertebrate, he or she will have to identify characteristics it has in common with other animal groups in order to classify it. You learned in Chapter 6 that animals in a group have similar characteristics because they descended from a common ancestor. The evolutionary history and relationships among animal phyla are represented by the diagram in Figure 13-2. It shows what are thought to be some of the major developments in animal evolution that led to the diversity we see today.

SECTION REVIEW

1. What are five characteristics of animals?
2. What are the types of symmetry? Give examples.
3. How are invertebrates different from vertebrates?
4. **Apply:** Identify your dorsal and ventral sides and your anterior and posterior ends.
5. **Connect to Earth Science:** Find out what kind of rock most fossil animals have been found in. How is this rock formed?

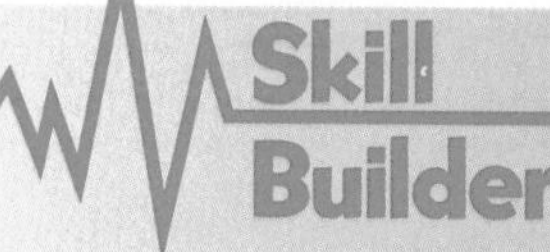

Skill Builder — Concept Mapping

Using the information on pages 303 and 304, make an events chain concept map showing the steps a scientist might use to classify a new animal. If you need help, refer to Concept Mapping in the **Skill Handbook** on pages 688 and 689.

OPTIONS

INQUIRY QUESTIONS

▶ **Which form of symmetry do you think allows an animal to move more efficiently on land?** *bilateral* **Why?** *It allows animals to organize their movement in a straight line.*

▶ **Where do you think the nerves and sense organs of animals with bilateral symmetry are located?** *in the anterior end* **Why?** *It allows animals to sense the environment as they move forward.*

PROGRAM RESOURCES

From the **Teacher Resource Package** use:

Concept Mapping, pages 31-32.

Cross-Curricular Connections, page 17, How Many Invertebrates?

Transparency Masters, pages 41-42, Animal Phylogeny.

Use **Color Transparency** 21, Animal Phylogeny.

Experiments Using Animals

13-2 

Objectives

- ▶ Acknowledge the advancements made due to experimentation with animals.
- ▶ Develop an idea for a compromise between progress in medicine and the welfare of laboratory animals.

To Test or Not To Test

Did you know that you're probably alive today because of laboratory animals? You may have received a shot of penicillin or a flu vaccine. Such vaccines have wiped out diseases such as polio and smallpox in the United States. Malaria, caused by a microscopic parasite, has killed more people in the world than any other disease. In order to develop an anti-malarial medicine, canaries were infected with malaria to allow scientists to learn more about the parasite. In fact, every surgical technique and drug used today was first tried in animals. But what about the animals? Is it truly necessary to experiment on animals?

It is estimated that each year about 100 million animals are used for research, education, or product testing. Cats are used to learn about the dangerous effects of drug addiction. Digestive enzymes are taken from pig digestive systems and used to help people with cystic fibrosis digest food. Antibiotics are first tested on rats before they are tried on humans. Organ transplants are possible because surgeons first practice on cows and other animals. Dogs and rabbits were used to understand diabetes and in the discovery of insulin. Treatments for cancer, alcoholism, and heart disease have all been greatly improved through experimentation using animals. Due to the advancements in medicine resulting from experiments with animals, a baby born today has a much greater chance of living to a very old age than ever before in the history of the world.

Figure 13-4. Jonas Salk developed the vaccine for polio that underwent much testing before it was made available for people.

OPTIONS

Meeting Different Ability Levels

For Section 13-2, use the following **Teacher Resource Masters** depending upon individual students' needs.

◆ **Study Guide Master** for all students.

● **Reinforcement Master** for students of average and above average ability levels.

▲ **Enrichment Master** for above average students.

Additional Teacher Resource Package masters are listed in the OPTIONS box throughout the section. The additional masters are appropriate for all students.

◆ STUDY GUIDE 52

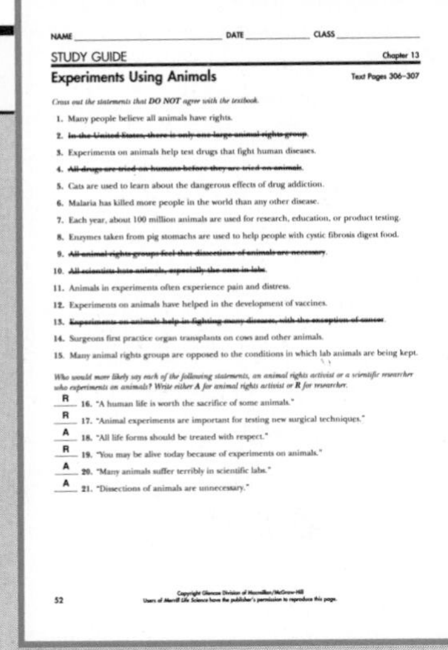

NAME ______ DATE ______ CLASS ______

STUDY GUIDE — Chapter 13

Experiments Using Animals — Text Pages 306–307

Cross out the statements that DO NOT agree with the textbook.

1. Many people believe all animals have rights.
2. ~~In the United States, there is only one large animal rights group.~~
3. Experiments on animals help test drugs that fight human diseases.
4. ~~All drugs are tried on humans before they are tried on animals.~~
5. Cats are used to learn about the dangerous effects of drug addiction.
6. Malaria has killed more people in the world than any other disease.
7. Each year, about 100 million animals are used for research, education, or product testing.
8. Enzymes taken from pig stomachs are used to help people with cystic fibrosis digest food.
9. ~~All animal rights groups feel that dissections of animals are necessary.~~
10. ~~All scientists hate animals, especially the ones in labs.~~
11. Animals in experiments often experience pain and distress.
12. Experiments on animals have helped in the development of vaccines.
13. ~~Experiments on animals help in fighting many diseases, with the exception of cancer.~~
14. Surgeons first practice organ transplants on cows and other animals.
15. Many animal rights groups are opposed to the conditions in which lab animals are being kept.

*Who would more likely say each of the following statements, an animal rights activist or a scientific researcher who experiments on animals? Write either **A** for animal rights activist or **R** for researcher.*

R 16. "A human life is worth the sacrifice of some animals."
R 17. "Animal experiments are important for testing new surgical techniques."
A 18. "All life forms should be treated with respect."
R 19. "You may be alive today because of experiments on animals."
A 20. "Many animals suffer terribly in scientific labs."
A 21. "Dissections of animals are unnecessary."

52

PREPARATION

SECTION BACKGROUND

▶ Through experimentation on laboratory animals, scientists have developed many lifesaving drugs. Animal rights groups are opposed to such experimentation. They are concerned about humane treatment of animals.

1 MOTIVATE

▶ Rabbits and rats are often used for medical research. Arrange to have a rabbit and a rat brought to class. Discuss with students the importance of both animals in developing and testing drugs.

2 TEACH

Key Concepts are highlighted.

CONCEPT DEVELOPMENT

▶ Invite a technician from a local hospital or a college biology professor to speak to the class about how animals are used in research.

Cooperative Learning: Use Problem Solving team cooperative groups. Assign them the task of testing drugs without the use of experimental animals.

3 CLOSE

CHECK FOR UNDERSTANDING

▶ Ask questions 1 and 2 in the Section Review.

RETEACH

Give your students several days to prepare a chart of five different types of experimental animals and the diseases that research on these animals has helped treat.

EXTENSION

For students who have mastered this section, use the **Reinforcement** and **Enrichment** masters or other OPTIONS provided.

FLEX Your Brain

Use the Flex Your Brain activity to have students explore ANIMAL RESEARCH.

FLEX YOUR BRAIN ASSESSMENT

Portfolio: Use the Flex Your Brain activity to reinforce critical-thinking and problem-solving skills.

SECTION REVIEW ANSWERS

1. cancer, heart disease, alcoholism

2. Animals deserve respect and should not be used unnecessarily for research or student dissection.

Answers may vary, but students should be able to support their answers.

PROGRAM RESOURCES

From the **Teacher Resource Package** use:

Activity Worksheets, page 5, Flex Your Brain.

Science and Society, page 17, Animal Rights: Unfair to Humans?

VideoDisc

STVS: Animal Models of Human Physiology, Disc 5, Side 2

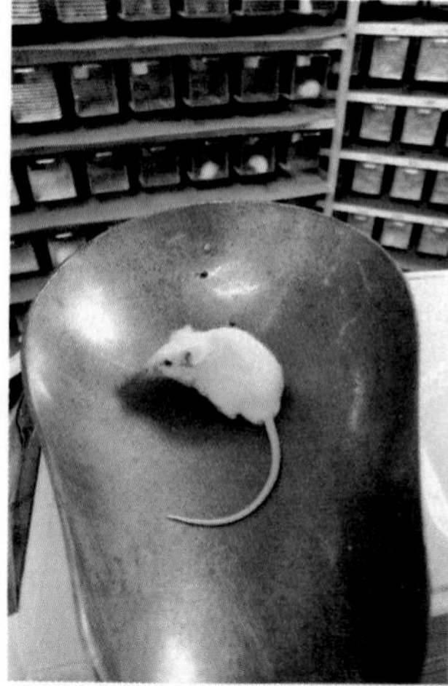

Figure 13-5. Many animals are kept in laboratories and used for research.

Today animal rights activists are fighting for the rights that they say animals are entitled to. There are about 10 million people in the United States who belong to animal rights groups. It's not just the deaths of animals that they are concerned with. Many are opposed to the conditions in which the lab animals are kept. Others feel that dissections of cats, frogs, pigs, and other animals by school-age students are unnecessary.

Some of the tests performed on the animals are very painful. In one experiment, dogs were fed pesticides until they were so sick they could not move. Some reforms are in progress. One of the most common tests for irritancy is the Draize test. In this test, high concentrations of the substance being tested are squirted into the eyes of rabbits. Many companies are looking for alternatives to this test and are reducing the concentrations of test substances. Researchers think that saving human lives through research is more important than the animal lives that are sacrificed. They feel that reforms and regulations may threaten the future of science.

The controversy will surely go on for years. Perhaps a compromise of some type will be reached. Activists often acknowledge the importance of these experiments and argue that only those that are necessary should be allowed. But where can we draw the line? Often basic research that appears unnecessary is later shown to be very important in solving a new medical problem. Will a solution ever be reached? Although animal research has saved many lives, pain and distress for the animals often results in their death.

SECTION REVIEW

1. Name three diseases for which treatments have been greatly improved due to research using animals.
2. What is the basic belief of animal rights groups?

You Decide!

You have your heart set on a brand new pair of acid washed jeans. Before you go to buy them, you hear on the news that the company that makes your favorite brand uses animals in experiments. To determine if it would irritate humans, the company tests the bleach solution used to lighten the jeans on the eyes and skin of animals. Knowing this, would you still buy the jeans?

● REINFORCEMENT 52

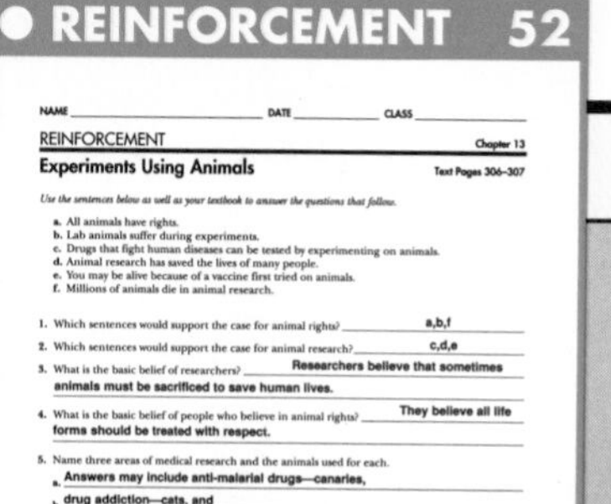

NAME ______ DATE ______ CLASS ______

REINFORCEMENT — Chapter 13

Experiments Using Animals — Text Pages 306–307

Use the sentences below as well as your textbook to answer the questions that follow.

a. All animals have rights.
b. Lab animals suffer during experiments.
c. Drugs that fight human diseases can be tested by experimenting on animals.
d. Animal research has saved the lives of many people.
e. You may be alive because of a vaccine first tried on animals.
f. Millions of animals die in animal research.

1. Which sentences would support the case for animal rights? **a,b,f**
2. Which sentences would support the case for animal research? **c,d,e**
3. What is the basic belief of researchers? **Researchers believe that sometimes animals must be sacrificed to save human lives.**
4. What is the basic belief of people who believe in animal rights? **They believe all life forms should be treated with respect.**
5. Name three areas of medical research and the animals used for each.
   a. **Answers may include anti-malarial drugs—canaries,**
   b. **drug addiction—cats, and**
   c. **antibiotics—rats.**
6. What might be a good compromise between animal rights and the needs of researchers? **Answers will vary. Compromises could include making labs as humane for animals as possible and doing animal research only when necessary.**

▲ ENRICHMENT 52

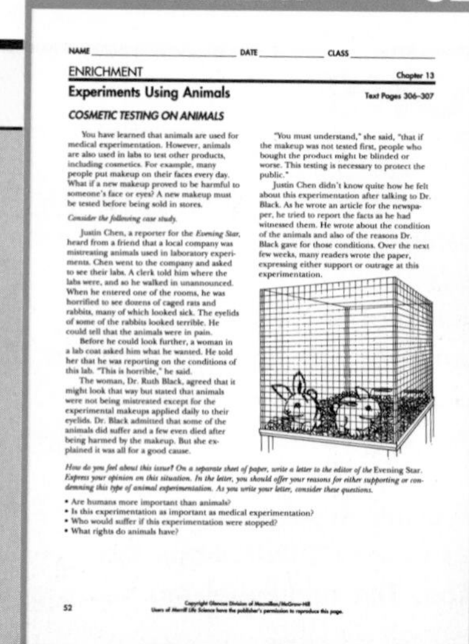

NAME ______ DATE ______ CLASS ______

ENRICHMENT — Chapter 13

Experiments Using Animals — Text Pages 306–307

COSMETIC TESTING ON ANIMALS

You have learned that animals are used for medical experimentation. However, animals are also used in labs to test other products, including cosmetics. For example, many people put makeup on their faces every day. What if a new makeup proved to be harmful to someone's face or eyes? A new makeup must be tested before being sold in stores.

Consider the following case study.

Justin Chen, a reporter for the *Evening Star*, heard from a friend that a local company was mistreating animals used in laboratory experiments. Chen went to the company and asked to see their labs. A clerk told him where the labs were, and so he walked in unannounced. When he entered one of the rooms, he was horrified to see dozens of caged rats and rabbits, many of which looked sick. The eyelids of some of the rabbits looked terrible. He could tell that the animals were in pain.

Before he could look further, a woman in a lab coat asked him what he wanted. He told her that he was reporting on the conditions of this lab. "This is horrible," he said.

The woman, Dr. Ruth Black, agreed that it might look that way but stated that animals were not being mistreated except for the experimental makeups applied daily to their eyelids. Dr. Black admitted that some of the animals did suffer and a few even died after being harmed by the makeup. But she explained it was all for a good cause.

"You must understand," she said, "that if the makeup was not tested first, people who bought the product might be blinded or worse. This testing is necessary to protect the public."

Justin Chen didn't know quite how he felt about this experimentation after talking to Dr. Black. As he wrote an article for the newspaper, he tried to report the facts as he had witnessed them. He wrote about the condition of the animals and also of the reasons Dr. Black gave for those conditions. Over the next few weeks, many readers wrote the paper, expressing either support or outrage at this experimentation.

How do you feel about this issue? On a separate sheet of paper, write a letter to the editor of the Evening Star. *Express your opinion on this situation. In the letter, you should offer your reasons for either supporting or condemning this type of animal experimentation. As you write your letter, consider these questions.*

- Are humans more important than animals?
- Is this experimentation as important as medical experimentation?
- Who would suffer if this experimentation were stopped?
- What rights do animals have?

52

The Simplest Invertebrates

13-3

Objectives

- Identify the structures that make up sponges and cnidarians.
- Describe how sponges and cnidarians obtain food and how they reproduce.

New Science Words

sessile
Porifera
filter feeders
collar cells
regeneration
hermaphrodite
larva
cnidarians
tentacles
polyp
medusa

Sponges

Let's take a closer look at the different animal phyla. The "simplest" of the animal groups are sponges, cnidarians (ni DAIR ee uhnz), and simple worms.

Does the sponge in Figure 13-6 look like an animal to you? Well, it is. Years ago, early scientists thought sponges were plants. However, sponges make up the group of animals considered to be the simplest in the animal kingdom. They have simple body plans and no body tissues, organs, or organ systems.

All sponges live in water. Most are found in warm, shallow salt water near the coast, although some are found at ocean depths of 8500 meters or more. A few species live in freshwater rivers, lakes, and streams. Sponges grow in many shapes, sizes, and colors. Some have radial symmetry, but most are asymmetrical. They may be smaller than a marble or larger than a compact car!

Adult sponges live attached to one place. They are often found with other sponges in colonies that never move, unless they are washed away by a strong wave. Organisms that remain attached to one place during their lifetime are **sessile.** Early scientists classified sponges as 1
plants because they didn't move. As microscopes were improved, scientists observed that sponges couldn't make their own food, and so they reclassified them as animals.

The body of a sponge is covered with many small openings called pores. It is from the pores that sponges get their phylum name, **Porifera.** *Porifera* comes from a Latin word meaning "pore-bearing."

Figure 13-6. Some species of sponge can grow to be larger than humans.

SECTION 13-3

PREPARATION

SECTION BACKGROUND

- Most sponges live in salt water. Less than 200 of all species live in fresh water.
- Nearly 80 percent of all life on Earth inhabits the oceans. Sea organisms make some of the most lethal poisons known.
- During periods of cold or dry weather, some freshwater sponges form specialized buds called gemmules. A gemmule is a cluster of sponge cells with a food supply surrounded by a protective covering. When conditions become favorable, the gemmule opens and the cells grow into a new sponge.

PREPLANNING

- Order hydra and *Daphnia* for Activity 13-2.

1 MOTIVATE

- Many students will know of sponges only through the synthetic ones that have now replaced the natural ones for household use. Show students a natural sponge. Ask how many would even guess that it was an animal. Have students compare the natural sponge to an artificial sponge. Make certain that each student observes and touches both sponges. Students should see that both sponges are porous and hold large amounts of water.

Cooperative Learning: Use Problem Solving Team strategy. Divide the class into groups of four. Give each group equal sized pieces of a natural sponge and an artificial sponge. Have them determine which sponge will absorb more water. *The natural sponge will.*

VideoDisc

STVS: Jellyfish, Disc 5, Side 1

OPTIONS

Meeting Different Ability Levels

For Section 13-3, use the following **Teacher Resource Masters** depending upon individual students' needs.

- ◆ **Study Guide Master** for all students.
- ● **Reinforcement Master** for students of average and above average ability levels.
- ▲ **Enrichment Master** for above average students.

Additional Teacher Resource Package masters are listed in the OPTIONS box throughout the section. The additional masters are appropriate for all students.

◆ STUDY GUIDE 53

NAME ____________ DATE ________ CLASS ________

STUDY GUIDE — Chapter 13

The Simplest Invertebrates — Text Pages 308–313

Write the letter of the term or phrase that best completes each statement.

b 1. A sponge is covered with many small openings called ____.
a. larva b. pores c. polyps d. filters

a 2. The ability of an organism to replace body parts is called ____.
a. regeneration b. reproduction c. restitution d. reintroduction

b 3. All cnidarians have ____.
a. reverse symmetry b. radial symmetry c. bilateral symmetry d. radical symmetry

a 4. Cells that line the inside of a sponge are called ____.
a. collar cells b. polyps c. filter feeders d. medusa

c 5. Organisms that are attached to one place throughout their life span are ____.
a. cnidarians b. filter feeders c. sessile d. medusa

c 6. The phylum name of sponges is ____.
a. Precambrium b. Cnidarian c. Porifera d. Spongerium

Use the diagram to answer the following questions.

A B C D

7. This diagram illustrates reproduction of what form of cnidarians? medusa
8. What does the free-swimming medusa release into the water? eggs and sperm
9. Is that release part of sexual reproduction or asexual reproduction? sexual reproduction
10. What do the fertilized eggs grow into? larvae Which drawing represents that? C
11. In the next step, has the cnidarian become a medusa or a polyp? a polyp
12. What buds off a polyp to complete the cycle? a medusa
13. Is the budding a form of sexual or asexual reproduction? asexual reproduction

 53

TYING TO PREVIOUS KNOWLEDGE: Review cell structure and function and levels of tissue organization in Chapter 2 to help students understand differences in the body structures of sponges and cnidarians.

2 TEACH

Key Concepts are highlighted.

CONCEPT DEVELOPMENT

▶ Sponges do not have specialized tissues. Each type of cell is independent of other cells. Collar cells are unique to sponges. Cnidarians are more advanced than sponges because their cells are organized into tissues.

▶ Explain that the bodies of sponges would collapse without some type of supporting structure.

▶ **Demonstration:** Mix soil and plant materials in a beaker of water. Pour the water mixture through a coffee filter into a container. Let students examine the contents of the filter and the container to see how the cells of a sponge might filter food particles from the water.

Connect to... Chemistry

Answer: Ocean water contains calcium and silica, which must be where sponges get their supplies of these elements for the formation of calcium carbonate and silica spicules.

How do filter feeders obtain food?

Ocean water contains many dissolved elements. Spicules of "glass" sponges are composed of silica. Other sponges have spicules of calcium carbonate. *Relate* the composition of spicules to the composition of seawater.

Figure 13-7 shows the body plan of a sponge. A sponge's body is a hollow tube closed at the bottom but with an opening in the top. The body wall has two cell layers made up of several different types of cells. There are cells that help a sponge get food, cells that digest food, cells that carry nutrients to all parts of the sponge, and cells that enable water to flow into the sponge.

Sponges obtain food from water that is pulled in through their pores. They filter bacteria, algae, protozoans, and other materials out of the water. Organisms that obtain food this way are called **filter feeders.** Cells that line the inside of the sponge, called **collar cells,** help water move through the sponge. You can see in Figure 13-7 that collar cells have flagella. The beating of the flagella in these cells brings water into the sponge. The water moving through the sponge also brings oxygen to the cells and carries away wastes.

The bodies of many sponges contain sharp, pointed structures called spicules. The soft-bodied sponges people use to take baths or to wash their cars have a skeleton of a fibrous material called spongin. Other sponges are supported by both spicules and spongin. Scientists classify sponges based on the kinds of materials that make up their skeletons.

Sponges reproduce both asexually and sexually. Sponges can reproduce asexually by forming buds. New sponges can also form from small pieces that break off the parent sponge. Sponge growers actually cut sponges into pieces, attach weights to them, and put them back into the ocean so the sponges can regenerate. **Regen-**
2 **eration** is the ability of an organism to replace body parts.

Figure 13-7. Sponges are made of several different types of specialized cells. The spicules provide structure for the sponge.

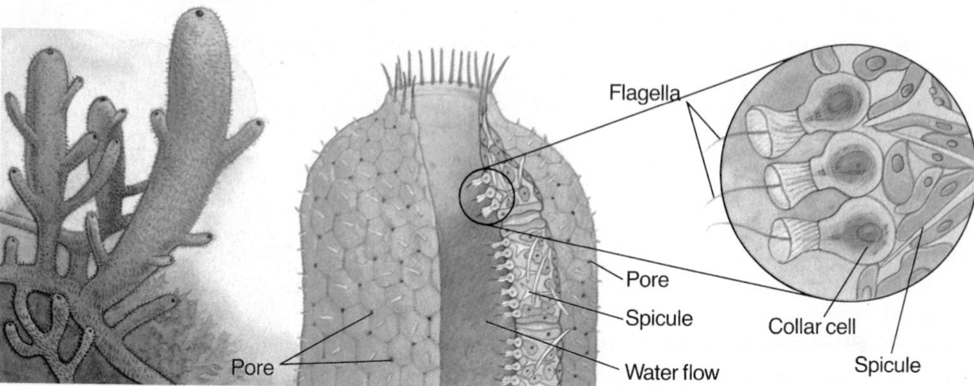

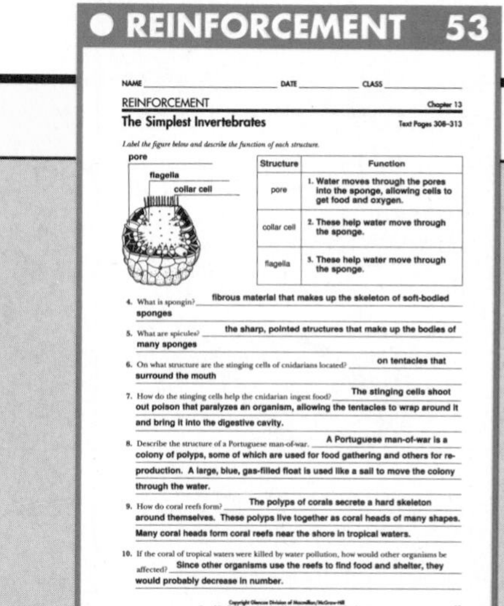

● REINFORCEMENT 53

NAME ____ DATE ____ CLASS ____

REINFORCEMENT — Chapter 13

The Simplest Invertebrates — Text Pages 308–313

Label the figure below and describe the function of each structure.

pore
flagella
collar cell

Structure	Function
pore	1. Water moves through the pores into the sponge, allowing cells to get food and oxygen.
collar cell	2. These help water move through the sponge.
flagella	3. These help water move through the sponge.

4. What is spongin? **fibrous material that makes up the skeleton of soft-bodied sponges**
5. What are spicules? **the sharp, pointed structures that make up the bodies of many sponges**
6. On what structure are the stinging cells of cnidarians located? **on tentacles that surround the mouth**
7. How do the stinging cells help the cnidarian ingest food? **The stinging cells shoot out poison that paralyzes an organism, allowing the tentacles to wrap around it and bring it into the digestive cavity.**
8. Describe the structure of a Portuguese man-of-war. **A Portuguese man-of-war is a colony of polyps, some of which are used for food gathering and others for reproduction. A large, blue, gas-filled float is used like a sail to move the colony through the water.**
9. How do coral reefs form? **The polyps of corals secrete a hard skeleton around themselves. These polyps live together as coral heads of many shapes. Many coral heads form coral reefs near the shore in tropical waters.**
10. If the coral of tropical waters were killed by water pollution, how would other organisms be affected? **Since other organisms use the reefs to find food and shelter, they would probably decrease in number.**

 53

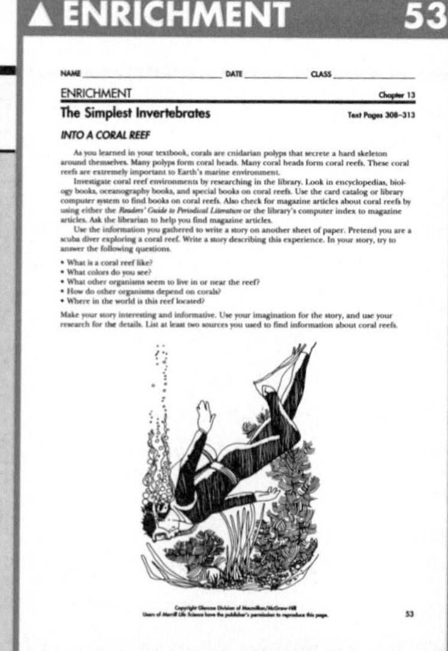

▲ ENRICHMENT 53

NAME ____ DATE ____ CLASS ____

ENRICHMENT — Chapter 13

The Simplest Invertebrates — Text Pages 308–313

INTO A CORAL REEF

As you learned in your textbook, corals are cnidarian polyps that secrete a hard skeleton around themselves. Many polyps form coral heads. Many coral heads form coral reefs. These coral reefs are extremely important to Earth's marine environment.

Investigate coral reef environments by researching in the library. Look in encyclopedias, biology books, oceanography books, and special books on coral reefs. Use the card catalog or library computer system to find books on coral reefs. Also check for magazine articles about coral reefs by using either the *Readers' Guide to Periodical Literature* or the library's computer index to magazine articles. Ask the librarian to help you find magazine articles.

Use the information you gathered to write a story on another sheet of paper. Pretend you are a scuba diver exploring a coral reef. Write a story describing this experience. In your story, try to answer the following questions.

- What is a coral reef like?
- What colors do you see?
- What other organisms seem to live in or near the reef?
- How do other organisms depend on corals?
- Where in the world is this reef located?

Make your story interesting and informative. Use your imagination for the story, and use your research for the details. List at least two sources you used to find information about coral reefs.

 53

Sponges can reproduce sexually by egg and sperm. Some species of sponges have separate sexes, but most are hermaphrodites. A **hermaphrodite** (hur MAF ruh dite) is an animal that produces both sperm and eggs. 3 Sperm are released in the water and carried by currents to other sponges where they fertilize the eggs. The fertilized egg develops into a young organism called a **larva.** 4 Larvae usually look very different from adults. Sponge larvae have cilia that allow it to swim about in the water. After a short time, the larva settles down on an object. There it will remain and grow into an adult sponge.

Origin and Importance of Sponges

Sponges appeared on Earth about 600 million years ago in the Cambrian period. Their collar cells are similar to a type of colonial protozoan that is thought to be the ancestor of sponges. No other animal species is known to have evolved from sponges.

TECHNOLOGY

Sea Pharmacy

Eighty percent of all life on Earth lives in the sea. Scientists hope that the 400 000 or more species of ocean plants and animals will provide chemicals to produce new medicines. But with 400 000 plus species, where do scientists start looking? One place is in organisms that use chemicals for protection such as sponges, soft corals, and sea squirts. These chemicals may one day be useful to humans. Sea specimens are also tested for the ability to stimulate the human immune system or to limit the growth of bacteria, viruses, or tumors.

There are many promising leads. The blue blood of horseshoe crabs is used extensively to test for the presence of bacteria that cause infectious diseases. One rare sponge called *Luffariella variabilis* contains a chemical that blocks pain and inflammation, and may be used for patients with arthritis and muscular dystrophy. Algae, sponges, and sea squirts have provided potential anticancer drugs. Poison from puffer fish may one day be used as an anesthetic.

Think Critically: Why are drugs from the sea just beginning to be thoroughly researched?

CONCEPT DEVELOPMENT

▶ Self-fertilization rarely, if ever, occurs in hermaphroditic species. Hermaphroditism is common in many invertebrates that are sessile, move slowly, or live in low density populations.

▶ Explain that budding, regeneration, and sexual reproduction may occur simultaneously in sponges.

REVEALING MISCONCEPTIONS

▶ Many students do not think of sponges as animals because they are sessile. Explain that they carry on all the processes of life just as other animals do.

CROSS CURRICULUM

▶ **Math:** A 1-cm^3 sponge can filter 1 L of water in a 24-hour period. How many liters of water can a 2-m^3 sponge filter in a 48-hour period? (2 m^3 = 2 000 000, 2 000 000 L of water in 24 hours, 4 000 000 L of water in 48 hours)

TECHNOLOGY

Extension: Research the kinds of sea organisms that live in or near a coral reef.

References: "Drugs from the Sea" by Ricki Lewis, *Discover*, May 1988, pp. 61-69. "The Pharmacy of the Seas" by Sharon Begley, *Newsweek*, May 21, 1990, pp. 77-78.

Think Critically: We are just developing the technology to dive to the depths required to find some organisms. We are also developing better techniques to sample and test specimens.

OPTIONS

ASSESSMENT—ORAL

▶ **The sponge growing industry has decreased in the last 50 years. Suggest why this has happened.** *Synthetic sponges are less expensive and have replaced natural sponges.*

▶ **The presence of freshwater sponges indicates a clean environment free of pollution. Suggest a reason for this.** *Sponges are filter feeders and take in food from the water around them. In polluted waters, they would take in food that was polluted.*

▶ **Why would natural sponges be better for polishing than synthetic ones?** *The synthetic ones do not have spongin and spicules which would aid in polishing.*

▶ **Since sponges are sessile animals, how is it possible that a population of sponges can spread into a larger area?** *The larvae have cilia and are able to swim to new locations before attaching.*

CONCEPT DEVELOPMENT

▶ Display a piece of coral. Discuss the ecological value of coral. Many fish, live, and feed around coral reefs.

▶ Explain that even though cnidarians have different body forms, they all have stinging cells, tentacles, two cell layers arranged into tissues, a digestive cavity, and radial symmetry.

Field Trip: Arrange a trip to a freshwater pond to collect hydra.

TEACHER F.Y.I.

▶ The phylum Cnidaria is also known as Coelenterata. The prefix *coel-* comes from the Greek word meaning "hollow." *Enteron* is a Greek word that means "intestine." The name implies that these animals have a coelom, but they do not. Cnidaria is less confusing.

MINI QUIZ

Use the Mini Quiz to check students' recall of chapter content.

1. **Organisms that remain attached to one place during their lifetime are _____.** *sessile*
2. **The ability of an organism to replace parts is _____.** *regeneration*
3. **An animal that produces both sperm and eggs is a(n) _____.** *hermaphrodite*
4. **A fertilized egg develops into a young organism called a(n) _____.** *larva*

Sponges provide food for many fish, snails, and starfish. They are shelter for smaller organisms that live in their bodies. Natural sponges are used in pottery, painting, in other arts and crafts, and as bath sponges. Researchers are collecting and testing sponges as possible sources of medicines.

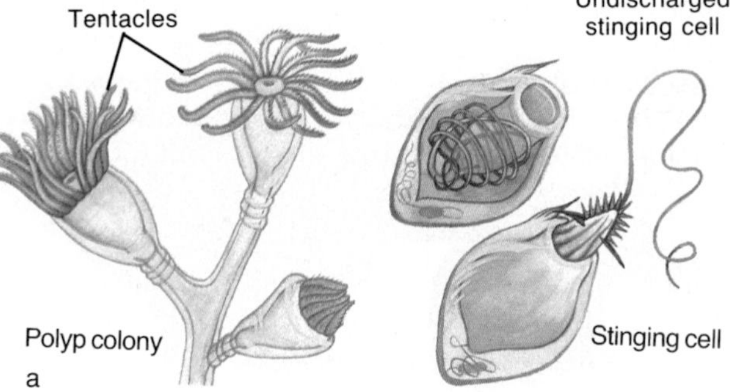

Figure 13-8. The polyp colony (a), Portuguese man-of-war (b), and sea anemone (c) all contain stinging cells characteristic of the phylum Cnidaria.

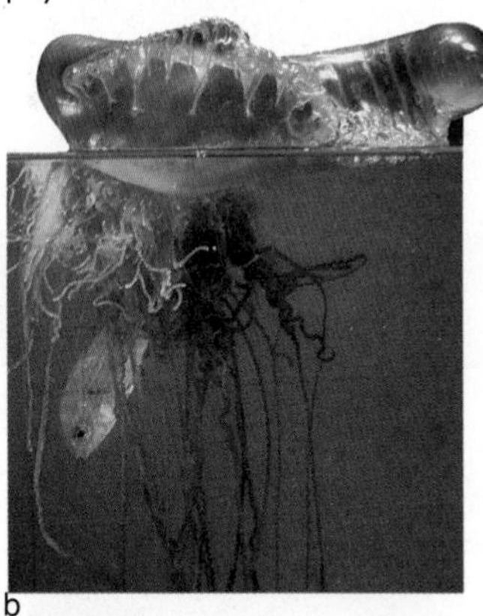
b

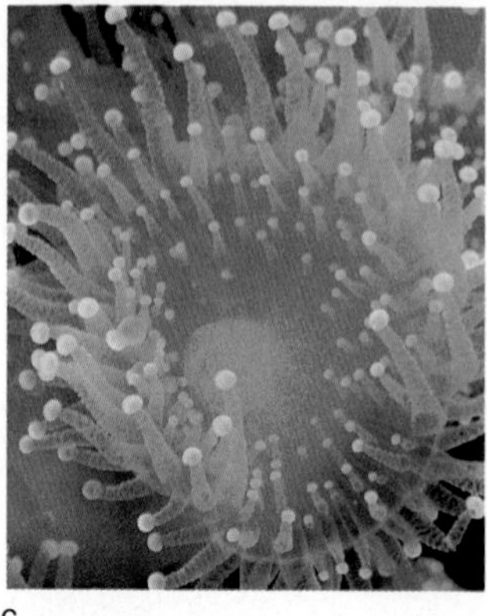
c

Cnidarians

The iridescent Portuguese man-of-war, colorful sea anemone, and hydra you see in Figure 13-8 belong to a group of animals called **cnidarians.** This name describes the stinging cells that all members of this phylum have. The word *cnidaria* is Latin for "stinging cells."

Although many types of hydra live in fresh water, most cnidarians live in salt water. Cnidarians live as single organisms and in colonies. All cnidarians have radial symmetry and bodies more complex than those of sponges. They have two cell layers that are arranged into tissues. Cnidarians have a digestive cavity where food is broken down. Most cnidarians are adapted with armlike structures called **tentacles** that surround the mouth. The tentacles are armed with the stinging cells that help the organism capture food.

The stinging cell of a cnidarian is actually a capsule or sac that contains a coiled harpoonlike thread and poison. Each capsule has a hair that works like a trigger. When a small organism bumps into the hair, the capsule explodes and shoots out its thread and poison. The poison paralyzes the organism until the tentacles wrap around it to pull it into the mouth. In most species, food is digested by enzymes in the digestive cavity. Undigested materials leave through the mouth.

ENRICHMENT

▶ Research and report on *Xestospongia muta*, a barrel sponge that grows like a giant vase off the Cayman Islands. Sponges of this type are being tested for potential medicines.

▶ Prepare a report on the commercial harvest of sponges.

PROGRAM RESOURCES

From the **Teacher Resource Package** use:

Critical Thinking/Problem Solving, page 17, Protecting the Coral Reefs.

Transparency Masters, page 43-44, Cnidarians.

Science Integration Activities 13, Salt Concentration in Ocean Water

Use **Color Transparency** 22, Cnidarians.

Cnidarians have two different body plans. Figure 13-9 shows the two body plans. The **polyp** is shaped like a tube or vase and is usually sessile. The **medusa** is bell-shaped and free swimming. Some cnidarians go through both polyp and medusa stages during their life cycles.

The Portuguese man-of-war on page 311 is really a colony of polyps. Some of these polyps gather food, and others are used for reproduction. The man-of-war is adapted with a large, blue, gas-filled float that is used like a sail and moves the colony through the water.

Cnidarians don't have complex nervous systems. They have a system of nerve cells called a nerve net. The nerve net carries impulses and connects all parts of the organism. This makes cnidarians capable of some simple responses and movements. For example, a hydra can somersault away from a threatening situation.

Cnidarians reproduce asexually and sexually. Polyps reproduce asexually by producing buds that eventually fall off the parent and grow into new polyps. Polyps can also reproduce sexually by producing either an egg or sperm. Sperm are released into the water and fertilize the eggs.

Medusa forms of cnidarians have both an asexual stage of reproduction and a sexual one. You can see the cycling between these two stages in Figure 13-10. Free-swimming medusae produce eggs and sperm and release them into the water. The eggs are fertilized and grow into larvae. The larvae eventually settle down and grow into polyps. Young medusae bud off the polyp and the cycle begins again.

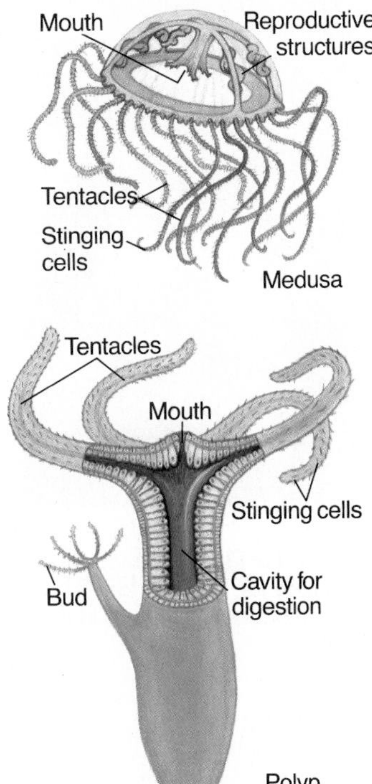

Figure 13-9. The polyp form and medusa form are the two body plans of cnidarians.

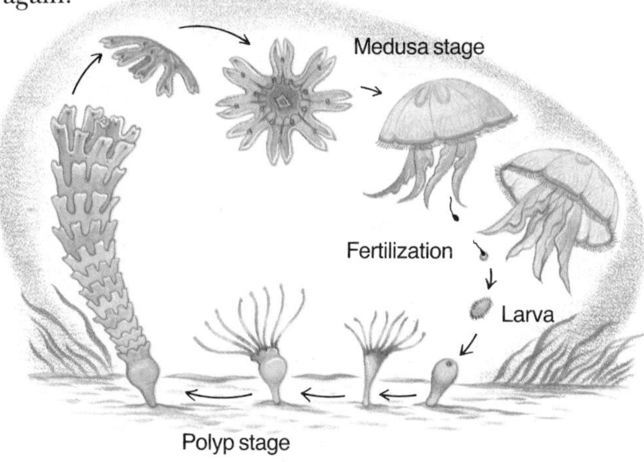

Figure 13-10. Some cnidarians alternate between polyp and medusa forms in their life cycles.

CONCEPT DEVELOPMENT

▶ Point out that common names are often confusing. The jellyfish is a cnidarian and not a fish.

▶ Explain that corals live in colonies. Dead corals' skeletons form reefs.

▶ *Polyp* is derived from the Greek word meaning "many feet." Medusa was the mythological Greek figure with writhing snakes instead of hair.

▶ **Demonstration:** Dissect a preserved specimen of a sea anemone for students to see the tentacles, gastrovascular cavity, and basal disc.

In Your JOURNAL

Although they are not very mobile, jellyfish can be quite large and their tentacles can extend great distances. Student handbook entries should say that jellyfish remain poisonous even after they have washed ashore and should be avoided.

OPTIONS

ASSESSMENT—ORAL

▶ **Sea stars, sea urchins, and sand dollars have pentaradial symmetry. What does *penta-* mean?** *five-part symmetry*

▶ **Why do you think that algae are only found inside the bodies of corals that live in shallow waters?** *Algae must have light to carry on photosynthesis.*

▶ **How do the tentacles of a jellyfish and a sea anemone differ?** *The tentacles of the jellyfish point downward and the tentacles of the sea anemone point upward.* **How is this difference an advantage to each animal?** *The mouth of the sea anemone is on the dorsal side and the mouth of the jellyfish is on the ventral side.*

▶ **The hydra was named for a giant mythical water monster with nine heads. Hercules, a Greek hero, was supposed to slay the monster. But each time he cut off one head, Hydra grew two more to take its place. How is the real hydra like the mythical one?** *A real hydra can be cut into pieces and each piece will regenerate a new hydra.*

3 CLOSE

CHECK FOR UNDERSTANDING

▶ Ask questions 1-2 and the **Apply** Question in the Section Review.

RETEACH

Give students a photocopied diagram of a sponge and a hydra and have them label the structures.

EXTENSION

For students who have mastered this section, use the **Reinforcement** and **Enrichment** masters or other OPTIONS provided.

SECTION REVIEW

1. Cidarians trap and stun prey with their tentacles and stinging cells. Sponges filter bacteria, protozoans, and other materials from water.
2. Polyps reproduce asexually by budding, sexually by egg or sperm; medusas alternate between sexual and asexual reproduction.
3. Apply: Corals secrete a hard skeleton around themselves which fossilizes easily. Fossil sponges might be rare except for fossilized spicules.
4. Connect to Earth Science: A barrier reef is separated from shore by a lagoon; an atoll is a ring of coral reef surrounding a lagoon.

Skill Builder

Sponges filter bacteria, algae, protozoans, and other materials out of the water. Collar cells help water move through the sponge. Cnidarians use the stinging cells on the tentacles to paralyze an organism, then the tentacles pull that organism into the mouth.

Skillbuilder ASSESSMENT

Portfolio: Use this Skillbuilder to assess students' abilities to Compare and Contrast the food gathering methods of sponges and cnidarians. State which is active and which is passive.

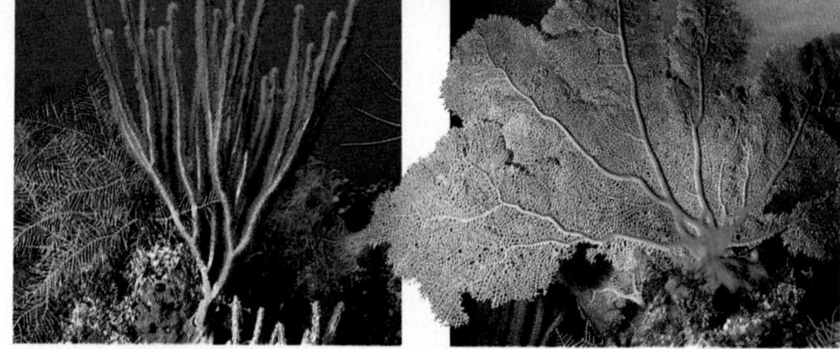

a b c

Figure 13-11. Elkhorn coral (a), sea rods (b), and sea fans (c) are all cnidarians that help build coral reefs.

In Your JOURNAL

Imagine your class is planning a field study trip to the ocean. **In your Journal,** write a section for a student handbook to explain how to deal with jellyfish.

Origin and Importance of Cnidarians

Cnidarians were present on Earth during the Precambrian Era over 600 million years ago. Scientists think that the first form of cnidarian was the medusa. Polyps may have formed from larvae of medusae that became permanently attached to a surface. Most of the cnidarian fossils found are fossils of corals.

Corals are cnidarian polyp colonies that secrete a hard skeleton around themselves. Many of these polyps living together form coral heads of many different shapes. Some are round, and others look like antlers of elk or deer. Many coral heads together form large coral reefs near shores in warm tropical waters. A large variety of fish and other organisms live on coral reefs. Food and shelter are provided for these organisms in the reef.

Coral reefs protect beaches and shorelines from being washed away by ocean waves. The reefs are also areas of recreation for many people. The beautiful corals and large variety of sea life provide scuba divers and snorklers with a wonderful view of life-forms in a reef.

SECTION REVIEW

1. Compare how cnidarians and sponges obtain food.
2. Describe the two body plans of cnidarians and their forms of reproduction.
3. **Apply:** Why are most fossils of cnidarians coral fossils? Would you ever find a fossil sponge? Explain.
4. **Connect to Earth Science:** Find out what a coral barrier reef is, and what an atoll is.

Skill Builder — Comparing and Contrasting

Compare and contrast the methods of getting food in sponges and cnidarians. If you need help, refer to Comparing and Contrasting in the **Skill Handbook** on page 683.

ENRICHMENT

▶ Have students research the Great Barrier Reef off the coast of Australia.

ACTIVITY 13-1

Observing a Cnidarian

Problem: *How does a hydra react?*

Materials  

- dropper
- hydra culture
- small dish
- toothpick
- *Daphnia* or brine shrimp
- stereoscopic microscope

Procedure

1. Use a dropper to place a hydra into a dish along with some of the water in which it is living.
2. Place the dish on the stage of the stereoscopic microscope. Bring the hydra into focus. Record the color of the hydra.
3. *Identify* and count the number of tentacles. Locate the mouth.
4. *Observe* the basal disk by which the hydra attaches itself to a surface.
5. *Predict* what will happen if the hydra is touched with a toothpick. Carefully touch the tentacles. Then gently touch the basal disk. Describe the reactions in the data table.
6. Drop a *Daphnia* or a small amount of brine shrimp into the dish. *Observe* how the hydra takes in food. Record your observations.
7. Return the hydra to the culture.

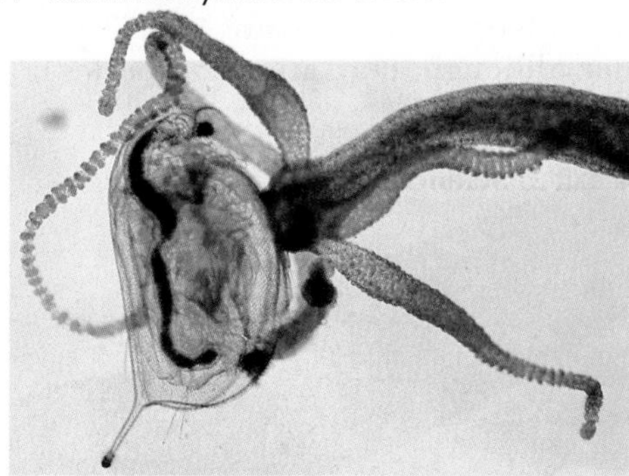

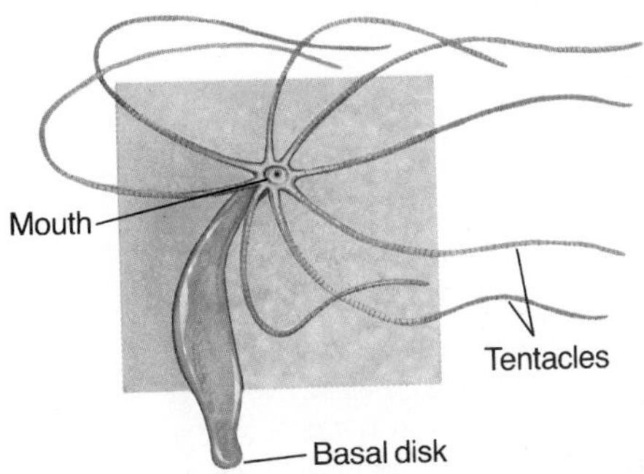

Data and Observations Sample Data

Features	Observations
Color	varies with species
Number of tentacles	varies with species
Reaction to touch	hydra should move away
Tentacles	varies with species
Basal disk	hydra should move
Reaction to food	hydra should move

Analyze

1. How many tentacles does the hydra have?
2. Where is the mouth located in relation to the tentacles?
3. Did the hydra react to touch as you predicted?
4. How did the hydra react to food?

Conclude and Apply

5. When the hydra was touched, did anything happen to other areas of the animal?
6. What advantage do tentacles give hydra?

ACTIVITY 13-1

40 minutes

OBJECTIVE: Observe the structure and method of feeding in a cnidarian.

PROCESS SKILLS applied in this activity:
▶ **Observing** in Procedure Steps 2, 3, 4, and 6.
▶ **Identifying** and **counting** in Procedure Step 3.
▶ **Hypothesizing** in Procedure Step 5.
▶ **Interpreting Data** in Analyze Questions 1, 2, 3, and 4.
▶ **Inferring** in Conclude and Apply Question 6.

COOPERATIVE LEARNING
Use the Paired Partners strategy. Assign one person to place the hydra in the dish under the stereoscopic microscope and the other one to record data.

TEACHING THE ACTIVITY

Troubleshooting: Place several coverslips into the jar the day before the hydra are needed, because they will probably cling to the coverslips. When you need hydra, remove the coverslips with forceps and place the coverslips directly into the dish.

Activity
ASSESSMENT
Performance: To further assess students' ability to Observe the structure and method of feeding in a cnidarian, have them view a short video on cnidarians and describe their activity.

ANSWERS TO QUESTIONS

1. Hydra usually have 5 to 7 tentacles.
2. The mouth is at the center of the tentacles.
3. Students who hypothesized that the hydra would react or contract should answer yes.
4. The tentacles stretched out to get the food.
5. When certain parts were touched, other parts may have responded.
6. The tentacles wrap around food and bring it into the mouth. Stinging cells in tentacles also stun prey.

PROGRAM RESOURCES

From the **Teacher Resource Package** use:

Activity Worksheets, pages 113-114, Activity13-1, Observing a Cnidarian.

SECTION 13-4

PREPARATION

SECTION BACKGROUND

▶ One of the world's most serious health problems is caused by three species of blood fluke. Two to three hundred million people, mostly in Asia, Africa, and South America are affected by these parasites.

PREPLANNING

▶ Obtain planaria for the Mini-Lab.

1 MOTIVATE

▶ Obtain specimens of tapeworms and Ascaris from a local veterinarian or a biological supply company. Have students observe and list the characteristics of each.

MINI-Lab

Materials: culture of planarians, dropper, watch glass or clean jar lid, stereomicroscope

Troubleshooting: If the planarians don't move, students may have to gently prod them with a small brush.

Answer: Students should describe planarians as creeping or swimming. Some students may infer that movement is by means of muscular undulations.

MINI-Lab

ASSESSMENT

Performance: To further assess students' understanding of planarians, have students feed them pin-head sized pieces of liver twice a week and describe behavior.

VideoDisc

STVS: Nematodes, Disc 5, Side 1

STVS: Sheep Parasite, Disc 5, Side 1

13-4 The Simple Worms

New Science Words

free-living
anus
cyst

Objectives

- ▶ Compare the body plans of flatworms and roundworms.
- ▶ Distinguish between free-living and parasitic organisms.
- ▶ Identify disease-causing flatworms and roundworms.

MINI-Lab

How do planarians move?

Use a dropper to transfer a planarian to a watch glass. Add enough water so the worm can move freely. Place the glass under a stereomicroscope and *observe*. **In your Journal,** describe how the organism moves in the dish. Draw the organism.

Flatworms

The animal you most likely think of when you hear the word *worm* is the earthworm—the worm that crawls across pavement after a rain, or the worm used to bait a fishing hook. You probably wouldn't think immediately of tapeworms or any of the other many types of worms in the world. Just what is a worm? Worms are invertebrates with soft bodies and bilateral symmetry. They have three layers of tissues, organs, and organ systems. Worms live in many different environments. Some are very beautiful; others are not so attractive. There are flatworms, roundworms, and worms with segments. In this chapter, you will learn about flatworms and roundworms.

As their name implies, flatworms have flattened bodies. They are members of the phylum Platyhelminthes (plat ih hel MIHN theez). Members of this phylum include planarians and tapeworms. Some flatworms are free-living but most are parasites. Remember a parasite depends on another organism for food and a place to live. Unlike parasites, **free-living** organisms don't depend on one particular organism for food or a place to live. Most flatworms live in salt water, although there are a few species that live in fresh water.

Figure 13-12. Planarians have a simple body structure.

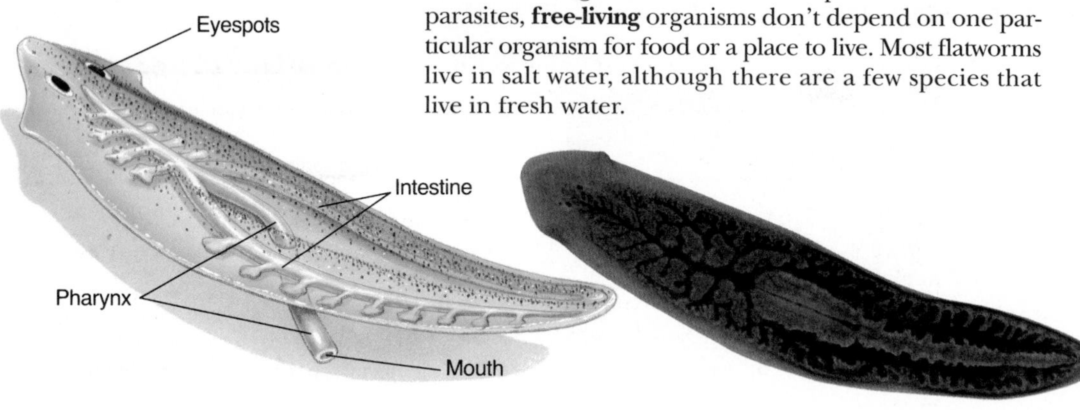

OPTIONS

ENRICHMENT

▶ Have students use the library to research parasitic roundworms such as pinworms, whipworms, guinea worms, and filaria worms.

MULTICULTURAL AWARENESS

▶ Have students research the economic effects of disease-causing roundworms in developing countries.

PROGRAM RESOURCES

From the **Teacher Resource Package** use:

Activity Worksheets, page 119, Mini-Lab: How do planarians move?

Activity Worksheets, page 120, Mini-Lab: How do planarians respond to light?

Transparency Masters, pages 45-46, Planaria.

Use **Color Transparency** 23, Planaria.

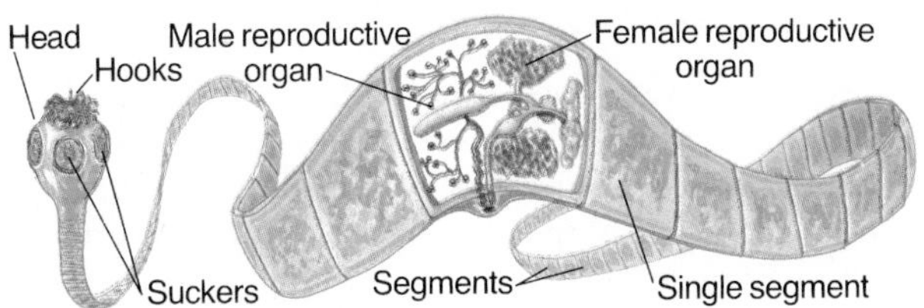

1 A planarian, like the one in Figure 13-12, is an example of a free-living class of flatworms. It has a triangle-shaped head with two eyespots. There is one body opening, a mouth, on the ventral side of the body. A muscular tube called the pharynx connects the mouth and the digestive tract. A planarian feeds on small organisms and dead bodies of larger organisms. Most planarians live in fresh water, under rocks, or on plant material. Some live in moist places on land. The body of a planarian is usually about one centimeter long and is covered with cilia. The cilia move the worm along in a slimy mucus track secreted by the underside of the planarian.

Planarians reproduce asexually or sexually. They reproduce asexually simply by dividing in two. Planarians also have the ability to regenerate. A planarian can be cut in two, and each piece will grow into a new worm. Planarians reproduce sexually by egg and sperm. Most are hermaphrodites, and they exchange sperm with one another. They lay fertilized eggs that hatch in a few weeks.

2 Tapeworms are parasitic members of the phylum Platyhelminthes. These worms use hooks and suckers to attach themselves to the intestine of a host organism. Dogs, cats, humans, and other animals are hosts for tapeworms. A tapeworm doesn't have a mouth or a digestive system. Food that has already been digested by the host is absorbed from the intestine of the host by the worm. A tapeworm grows by producing new body segments behind the head. Its ribbonlike body may grow to be 12 meters long.

Each body segment of the tapeworm has both male and female reproductive organs. The segments produce eggs and sperm, and the eggs are fertilized in the segment. Once a segment is filled with fertilized eggs, it breaks off and passes out of the host's body. If a fertilized egg is eaten by another host, the egg hatches and develops into a new worm.

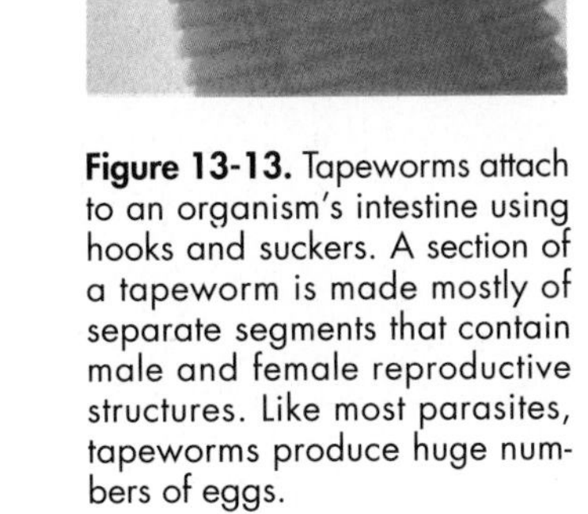

Figure 13-13. Tapeworms attach to an organism's intestine using hooks and suckers. A section of a tapeworm is made mostly of separate segments that contain male and female reproductive structures. Like most parasites, tapeworms produce huge numbers of eggs.

MINI-Lab

How do planarians respond to light?

Use a dropper to transfer a planarian to a watch glass. Add enough water so the worm can move freely. Place a piece of foil to cover half of the watch glass; place under a stereomicroscope and observe. *Predict* the direction the worm will move in—toward or away from the light.

TYING TO PREVIOUS KNOWLEDGE: Review the definitions of *tissue, organs,* and *organ systems* from Chapter 2. Review *bilateral symmetry* from Section 13-1. Discuss the presence or absence of tissues of organisms in this chapter—sponges, cnidarians, and now flatworms.

2 TEACH

Key Concepts are highlighted.

CONCEPT DEVELOPMENT

▶ Obtain posters of the life cycle of parasitic worms from the health department or a biological supply company. Display them on the bulletin board.

Cooperative Learning: Use Problem Solving Teams or Expert Teams cooperative groups. Have each group diagram and label the life cycle of a different flatworm not illustrated in the text. Display the diagrams.

CROSS CURRICULUM

▶ **Math:** Tapeworms as long as nine meters have been found in humans. Have students use a meterstick and measure nine meters to have an understanding of just how long tapeworms are.

MINI-Lab

Materials: culture of planarians, dropper, watch glass, stereomicroscope

Answer: Most wil predict that the worm will move toward light. However, planaria use their eyespots to detect and move away from light. The organism should spend most of its time under the covered part of the watch glass.

MINI-Lab

ASSESSMENT

Performance: To further assess students' ability to Predict how planarians respond to light, see USING LAB SKILLS, Question 12 on p. 320.

OPTIONS

Meeting Different Ability Levels

For Section 13-4, use the following **Teacher Resource Masters** depending upon individual students' needs.

◆ **Study Guide Master** for all students.

● **Reinforcement Master** for students of average and above average ability levels.

▲ **Enrichment Master** for above average students.

Additional Teacher Resource Package masters are listed in the OPTIONS box throughout the section. The additional masters are appropriate for all students.

◆ STUDY GUIDE 54

NAME ______ DATE ______ CLASS ______

STUDY GUIDE — Chapter 13

The Simple Worms — Text Pages 315–318

Complete the following sentences using the appropriate terms from the textbook.

1. The worms called **tapeworms** use hooks and suckers to attach themselves to the intestine of a host organism.
2. The roundworms have a digestive system with a(n) **mouth** and an anus.
3. You can get sick by eating undercooked **pork** that contains ***Trichinella*** cysts.
4. Worms with flattened bodies are called **flatworms**.
5. Worms have **bilateral** symmetry.
6. Humans can get **hookworm** by walking barefoot over soil.
7. An example of a free-living class of flatworms are the **planarians**.
8. Masses of ***Ascaris*** worms can block the intestines of animals and cause death.
9. A tapeworm can grow to be **12 meters** long.
10. Dogs can get **heartworm** through a mosquito bite.
11. Parasites live off other organisms, while **free-living** organisms don't depend on others for food or a place to live.
12. The body of a planarian is usually covered with **cilia**.
13. Roundworms make up the phylum called **Nematoda**.
14. Flatworms are members of the phylum called **Platyhelminthes**.
15. A **cyst** is a young worm with a protective covering.

Tapeworm

54

CONCEPT DEVELOPMENT

▶ Explain that mosquitoes ingest microfilaria along with blood when they bite an infected animal. Microfilaria develop into larvae in the mosquito. When a mosquito bites a healthy dog, larvae are transmitted to the dog. The larvae mature into worms in the dog's heart. In 6 or 7 months, the worms produce microfilaria which travel through the dog's blood vessels. Heartworm disease is prevented by medication that kills larvae before they mature.

PROBLEM SOLVING

Think Critically: For the medication to control heartworm disease effectively, a minimum dosage must be in the bloodstream; therefore, the dosage must be increased as the puppy gains weight.

MINI QUIZ

Use the Mini Quiz to check students' recall of chapter content.

1. **An example of a free-living flatworm is the _____ .** *planarian*
2. **Tapeworms attach themselves to their host with _____ .** *hooks and suckers*
3. **The opening at the end of the digestive tract where wastes leave is the _____ .** *anus*

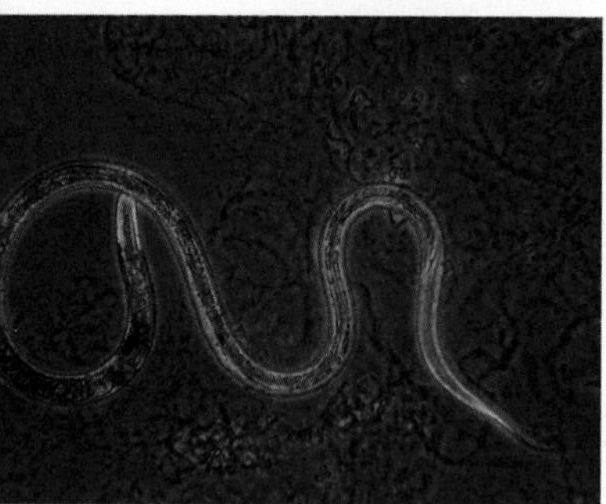

Figure 13-14. This free-living species of roundworm lives among cyanobacteria.

Roundworms

If you own a dog, you've probably had to get medicine from your vet to protect your dog from heartworms. Heartworm disease in dogs is caused by roundworms. Roundworms make up the largest phylum of worms, the phylum Nematoda. It is estimated that there are more than a half million species of roundworms in the phylum Nematoda. Nematodes are found in soil, in animals, and in fresh water and salt water. Many are parasitic, but most are free-living.

Roundworms are slender and tapered at both ends. The body is a tube within a tube, with fluid in between. Unlike the organisms you have studied so far, Nematodes 3 have two body openings, a mouth and an anus. The **anus** is an opening at the end of the digestive tract through which wastes leave the body.

Some roundworms are parasites of humans. The most common roundworm parasites of humans are *Ascaris*, hookworm, and *Trichinella*. Humans get hookworms by

PROBLEM SOLVING

Barbara's New Puppy

Barbara received a new puppy for her birthday. Barbara and her parents took the puppy to a veterinarian for a checkup. After the examination, the veterinarian gave Barbara some medication to give her puppy monthly to prevent heartworm disease. He explained how she should increase the dosage as the puppy gains weight.

The vet told her that it was important to give the puppy the medication monthly. Heartworms are spread from dog to dog by mosquitoes. When a mosquito bites a dog infected with heartworms, it can pick up and carry them to a dog not infected. Dogs can get heartworms at anytime during their life if they don't receive medication.

How can heartworm disease be controlled? **Think Critically:** Why should the dosage of heartworm medication be increased as the puppy gains weight?

● REINFORCEMENT 54

NAME ____________ DATE ________ CLASS ________

REINFORCEMENT — Chapter 13

The Simple Worms — Text Pages 315–318

The table below lists six worms you learned about in the textbook. Complete the table by referring to your textbook. Some answers have already been filled in to help you.

Worm	Kind	Phylum	Reproduction	Way of life
Ascaris	**roundworm**	**Nematoda**	sexual	parasite
heartworm	roundworm	**Nematoda**	**sexual**	**parasite**
hookworm	**roundworm**	**Nematoda**	**sexual**	**parasite**
planarian	**flatworm**	**Platyhelminthes**	sexual or asexual	**free-living**
tapeworm	**flatworm**	**Platyhelminthes**	**sexual**	**parasite**
Trichinella	**roundworm**	Nematoda	**sexual**	**parasite**

Answer the following questions by referring to your textbook.

1. How can you prevent getting trichinosis? **by eating only thoroughly cooked pork**
2. How can you prevent getting hookworm? **by wearing shoes when outdoors**
3. *Ascaris* eggs enter the host's body through contaminated food or water. What might you do to prevent getting sick from this parasite? **You could cook the food and boil the water.**
4. Where are the reproductive organs on a tapeworm? **Each body segment has both male and female reproductive organs.**
5. What characteristics do all worms have? **They are invertebrates with soft bodies and bilateral symmetry. Worms have three layers of tissues, organs, and organ systems.**

54

▲ ENRICHMENT 54

NAME ____________ DATE ________ CLASS ________

ENRICHMENT — Chapter 13

The Simple Worms — Text Pages 315–318

PLANARIAN REGENERATION

Materials

- large wide-mouth jar
- hand lens
- pond water
- spring water at room temperature
- cooked egg yolks or raw meat
- magnifying glass
- art knife

Procedure

1. Planarians can be found near the shoreline of ponds that have decaying leaves and other debris. Fill your wide-mouth jar with this shoreline material and a few inches of pond water.
2. Let the jar sit overnight. In the morning, planarians will likely be clinging to the sides of the jar. Planarians look like small, dark, jelly-like blobs. Remove much of the shoreline material, and add some more pond water.
3. Once a day, feed the planarians some egg yolk or raw meat. After about an hour, remove the remaining egg yolk or meat, drain as much of the water away as possible, and replace with fresh pond water or bottled spring water.
4. Using the magnifying glass and art knife, carefully cut two planarians in half as shown above. **CAUTION:** *The art knife is very sharp. Be careful when using sharp objects.*
5. Observe the planarian parts for a week and then return them to a pond.

Data and Observations

What do your planarian parts look like after a week? **Answers will vary, although most planarian parts will have regenerated into whole planarians.**

Conclude and Apply

1. Why did the planarian parts change? **Answers will vary. Some students may answer that the planarian parts died because of inadequate food or water temperature or because cutting the worms in half killed them. Some may answer that planarians can regenerate damaged or lost parts.**
2. How would regeneration affect the planarian population? Explain your answer. **Regeneration increases the planarian popluation. Planarians don't die when cut in half, but grow into two worms. When planarians are injured, instead of losing a member of the population, regeneration adds a new member to the population.**

54

walking barefoot over dirt or through fields. Hookworm eggs hatch in warm, moist soil. Wearing shoes is the best protection against hookworms.

Ascaris is found in the intestines of pigs, horses, and humans. Eggs enter the host's body in contaminated food or water. They travel to the intestines where they mature and mate. Masses of worms can block the intestines and cause death if left untreated.

Trichinella worms cause the disease trichinosis (trihk uh NOH sus). Humans become infected when they eat undercooked pork that has *Trichinella* cysts. A **cyst** is a young worm with a protective covering. Trichinosis can be prevented by thoroughly cooking pork.

The heartworms you treat your dog for enter the blood of a dog through a mosquito bite. They move to the heart, where they grow and reproduce. They can block the valves of the heart that lead to and from the lungs if the infection is left untreated. Medicine can be given to dogs to prevent heartworm disease.

The flatworms and roundworms are more complex than sponges and cnidarians. They have bilateral symmetry, three well-developed tissue layers, and organ systems. The roundworms have digestive systems with a mouth and an anus. Like sponges and cnidarians, these invertebrates probably evolved in the sea. They still live in moist environments.

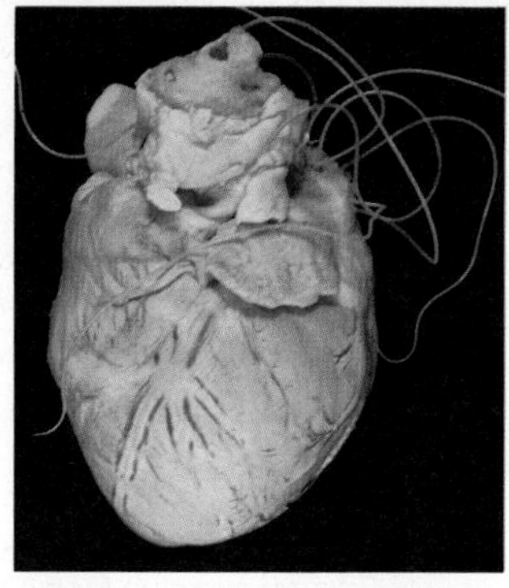

Figure 13-15. This heart of a dog is infected with heartworms.

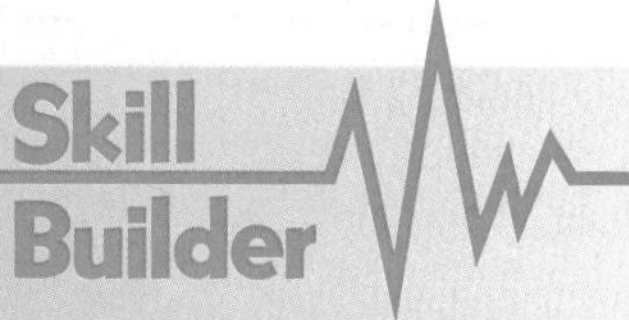

In your Journal, write a public service announcement for your local television station informing the community about heartworm disease in dogs. A veterinarian can probably provide much useful information.

SECTION REVIEW

1. Compare body plans of flatworms and roundworms.
2. Give an example of a free-living flatworm and a parasitic flatworm.
3. What are three roundworms that cause diseases in humans? How can humans prevent infection?
4. **Apply:** Which organism is more complex, a cnidarian or a flatworm? Explain.
5. **Connect to Chemistry:** Suggest a reason why tapeworms are not digested by their hosts.

Skill Builder: Outlining

Make an outline of the concepts found in Section 13-4. If you need help, refer to Outlining in the **Skill Handbook** on page 681.

CHECK FOR UNDERSTANDING

Ask questions 1-3 and the **Apply** Question in the Section Review.

RETEACH

Have each student write one short-answer question relating to flatworms and roundworms. Read the questions aloud. Ask students to respond as a group.

EXTENSION

For students who have mastered this section, use the **Reinforcement** and **Enrichment** masters or other OPTIONS provided.

3 CLOSE

SECTION REVIEW ANSWERS

1. Flatworms have flattened bodies with one body opening, the mouth. Roundworms' bodies have a tube within a tube with fluid in between, with two body openings—a mouth and an anus.

2. free-living—planarian; parasitic—tapeworm

3. Hookworm—wear shoes; Ascaris—don't eat contaminated food or drink contaminated water; Trichinosis—cook pork thoroughly

4. Apply: A flatworm. It has three layers of tissues, organs, and organ systems.

5. Connect to Chemistry: Some parasites are protected by a thick mucous; hosts may not have enzymes to break down the mucous.

Skill Builder

Accept all logical outlines. Students' outlines may be similar to:

I The Simple Worms
- **A.** Flatworms
  - **1.** flattened bodies
  - **2.** phylum Platyhelminthes
  - **3.** mostly parasites
- **B.** Roundworms
  - **1.** round, slender bodies; tapered at ends
  - **2.** phylum Nematoda
  - **3.** 2 body openings; mouth and anus
  - **4.** most free-living

Skillbuilder

ASSESSMENT

Performance: Assess students' abilities to Outline by having them outline information on reproduction of each animal group in the chapter.

ACTIVITY 13-2

OBJECTIVE: Compare and contrast free-living and parasitic worms.
Time: One class period to complete the activity and one-half class period to summarize.

PROCESS SKILLS applied in this activity are **comparing and contrasting, observing**, and **recognizing cause and effect.**

PREPARATION

- Planaria can be found in most streams, rivers, and lakes, underneath rocks or leaves.
- Use prepared whole mount slides of tapeworms with heads and proglottides.
- Use a small brush to transfer the planaria from the culture to the culture dishes.

Cooperative Learning: Divide the class into Science Investigation Teams.

THINKING CRITICALLY

Student responses will vary. Tapeworms and flukes are parasitic flatworms. Parasitic worms have body parts for attaching to a host and free-living worms have body parts for the environment in which they live.

TEACHING THE ACTIVITY

*Refer to the **Activity Worksheets** for additional information and teaching strategies.*

- Make sure the fed planaria are returned to a labeled dish so they will not be used again the same day.
- Student observations of the planaria may include the following: Planaria move with a gliding motion. The most sensitive body parts are those closestto the head. The tube-like structure on the underneath side pulls the food into the body. It has a flat body. The mouth is on the underneath side of the body.

SUMMING UP/SHARING RESULTS

- Parasitic worms have body parts adapted for attaching to a host.
- Body systems that are better developed in free-living worms are the nervous systems and digestive systems.
- In parasitic worms, the reproductive system is more complex.

GOING FURTHER!

Answer may include that parasitic worms do not have to move to find food once they have attached to a host. Two things that may ensure the success of parasitic worms are the ability to reproduce and to live off the host without killing the host.

PROGRAM RESOURCES

From the **Teacher Resource Package** use:

Activity Worksheets, pages 115-116, Activity 13-2, Comparing Free-Living and Parasitic Worms.

Activity ASSESSMENT

Performance: To further assess students' understanding of the differences between free-living and parasitic worms, assign different cooperative groups and have them use their drawings and observations to explain the differences. Assess explanations for accuracy.

ACTIVITY 13-2 DESIGNING AN EXPERIMENT

Comparing Free-Living and Parasitic Worms

Flatworms include species that are free-living and parasitic. In parasitic worms, certain systems are reduced or lost.

Getting Started

You need to observe flatworms to determine how parasitic worms make up for the systems that have been lost. You will also find out how free-living worms are adapted to their lifestyle.

Thinking Critically

Make a list of flatworms that are parasites. Make a list of flatworms that are free-living. How do you think they are different?

Materials  

Your cooperative group will use:

- culture dish with a planarian
- prepared slide of a tapeworm
- dropper
- small piece of liver
- small paintbrush
- compound microscope
- stereoscopic microscope

Try It!

1. Obtain a planarian in a culture dish from your teacher. Examine and draw the planarian under the stereoscopic microscope. *Observe* how it moves. Use the paintbrush togently touch the body of the planarian. Record your observations.
2. Add a small piece of liver to the culture dish and *observe* how the planarian feeds.
3. Use low power of the compound microscope to observe and draw the prepared slide of the tapeworm. Locate the head with its hooks and suckers. The segments behind the head produce fertilized eggs. The tapeworm grows by adding new segments.

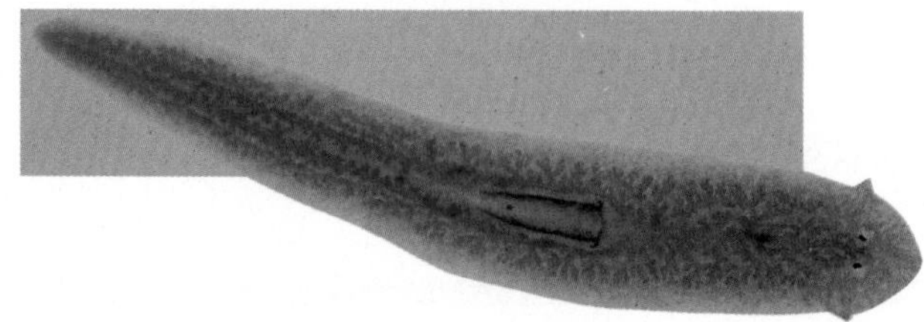

Summing Up/Sharing Results

- *Compare and contrast* parasitic and free-living worms.
- Which body systems are better developed in free-living worms?
- Which body system is more complex in parasitic worms?

Going Further!

Suggest a reason for the lack of locomotion in parasitic worms. What are two things that ensure the success of parasitic worms.

CHAPTER

REVIEW

SUMMARY

13-1: What Is an Animal?

1. Animals are many-celled eukaryotic organisms that must find and digest their food.
2. Invertebrates are animals that don't have backbones. Animals with backbones are vertebrates.
3. Animals that have body parts arranged the same way on both sides of the body have bilateral symmetry. Animals with body parts in a circle around a central point are radially symmetrical.

13-2: Science and Society: Experiments Using Animals

1. Many lifesaving drugs and vaccines have been developed using animals for testing.
2. Many people feel animals have rights and should be treated humanely and with respect.

13-3: The Simplest Invertebrates

1. Sponges are considered the simplest animals in the animal kingdom. They are sessile and obtain food and oxygen by filtering water through pores. Sponges reproduce by egg and sperm and also by budding or regeneration.
2. Cnidarians are hollow-bodied animals with radial symmetry. Most have tentacles with stinging cells to obtain food. Digestion takes place in a central cavity. Jellyfish, hydras, and corals are cnidarians.

13-4: The Simple Worms

1. Flatworms belong to the phylum Platyhelminthes. They have bilateral symmetry. There are both free-living and parasite forms. Planarians have one body opening. Tapeworms are parasitic forms of flatworms that have hooks and suckers to attach to a host. Roundworms belong to the phylum Nematoda. They have a tube within a tube body plan and bilateral symmetry. Roundworms have two body openings, a mouth and anus.
2. Free-living organisms don't depend on one particular organism for food or a place to live. Parasites depend on other organisms for food and a place to live.
3. Some parasitic roundworms that infect humans are hookworms, *Ascaris*, and *Trichinella*. Heartworms are roundworms that infect dogs.

KEY SCIENCE WORDS

a. **anus**
b. **bilateral symmetry**
c. **cnidarians**
d. **collar cells**
e. **cyst**
f. **filter feeders**
g. **free-living**
h. **hermaphrodite**
i. **invertebrates**
j. **larva**
k. **medusa**
l. **polyp**
m. **Porifera**
n. **radial symmetry**
o. **regeneration**
p. **sessile**
q. **tentacles**
r. **vertebrates**

UNDERSTANDING VOCABULARY

Match each phrase with the correct term from the list of Key Science Words.

1. body parts the same on each side
2. a young worm enclosed in a covering
3. a young organism different from the adult
4. opening for digestive wastes
5. attached, nonmoving animal
6. organisms with no backbones
7. animals that have backbones
8. cnidarian shaped like a tube or vase
9. an organism that makes both egg and sperm
10. used by cnidarians to capture food

OPTIONS

ASSESSMENT

To assess student understanding of material in this chapter, use the resources listed.

COOPERATIVE LEARNING

Consider using cooperative learning in the THINK AND WRITE CRITICALLY, APPLY, and MORE SKILL BUILDERS sections of the Chapter Review.

PROGRAM RESOURCES

From the **Teacher Resource Package** use:
Chapter Review, pages 29-30.
Chapter and Unit Tests, pages 89-92, Chapter Test.

CHAPTER

REVIEW

SUMMARY

Have students read the summary statements to review the major concepts of the chapter.

UNDERSTANDING VOCABULARY

1. b
2. e
3. j
4. a
5. p
6. i
7. r
8. l
9. h
10. q

ASSESSMENT

Portfolio

Encourage students to place in their portfolios one or two items of what they consider to be their best work. For each item, ask students to explain why that item was chosen and what they learned from it. Items might be selected from the following.

- For Your Mainstreamed Students symmetry rubbings, p. 300
- Skillbuilder concept map of animal classification, p. 304
- In your Journal nonfiction writing on jellyfish, p. 313

Performance

Additional performance assessments may be found in *Performance Assessment* and *Science Integration Activities* that accompany **Merrill Life Science.** Performance Task Assessment Lists and rubrics for evaluating these activities and other products generated throughout the chapter can be found in Glencoe's *Performance Assessment in Middle School Science.*

In Your JOURNAL

Students' announcements should reflect accurate information and creativity. (See p. 318, student text.)

CHAPTER
REVIEW

CHECKING CONCEPTS

1. d **6.** c
2. a **7.** b
3. b **8.** c
4. a **9.** c
5. c **10.** a

USING LAB SKILLS

ASSESSMENT
Use these alternate lab exercises to assess students' understanding of skills used in this chapter.

11. Both flatworms and roundworms are invertebrates, have bilateral symmetry and include parasitic and free-living worms. Flatworms have a flat body shape and lack the tube-within-a-tube body plan of the roundworms.

12. Student hypotheses need to be testable. Experimental designs should include no less than five trials. Data should be collected and a conclusion statement made.

THINK AND WRITE CRITICALLY

13. An organism of both should be used. Cnidarians for radial; any of the worms for bilateral.

14. Budding involves cells or a new individual developing on and then breaking off the parent. Regeneration involves replacing lost parts. Both involve mitosis.

15. Research may be halted or slowed down due to regulations. The progress made on some studies may be jeopardized.

16. The tapeworm has hooks and suckers to hold onto the intestinal cells of its host; it has no digestive system because it gets already-digested food from its host.

CHAPTER
REVIEW

CHECKING CONCEPTS

Choose the word or phrase that completes the sentence.

1. Animal characteristics include all of the following except ______.
   a. movement **c.** eukaryotic cells
   b. digestion **d.** prokaryotic cells
2. All of these belong to the same group except ______.
   a. fish **c.** jellyfish
   b. clam **d.** sponge
3. ______ is the opposite of dorsal.
   a. Anterior **c.** Radial
   b. Ventral **d.** Bilateral
4. Scientists classify sponges based on ______.
   a. the material that makes up their skeletons
   b. reproduction
   c. method of obtaining food
   d. symmetry
5. Sponges are members of Phylum ______.
   a. Cnidaria **c.** Porifera
   b. Nematoda **d.** Platyhelminthes
6. Sponges reproduce asexually by ______.
   a. medusas **c.** budding
   b. polyps **d.** egg and sperm
7. The body plans of cnidarians are polyp and ______.
   a. larva **c.** ventral
   b. medusa **d.** buds
8. All are examples of cnidarians except ______.
   a. coral **c.** planarian
   b. hydra **d.** jellyfish
9. An example of a parasite is a ______.
   a. sponge **c.** tapeworm
   b. planarian **d.** jellyfish
10. Separate sexes are found in ______.
   a. *Ascaris* **c.** sponges
   b. planarian **d.** cnidarians

USING LAB SKILLS

11. The Activity on page 318 asked you to compare different flatworms. Using similar methods, compare and contrast the flatworms observed with roundworms, such as Trichinella and vinegar eels.
12. Use your prediction about planaria response in the MINI-Lab on page 315, to **hypothesize** how the planarian would react to light. Then **design an experiment** to test your hypothesis.

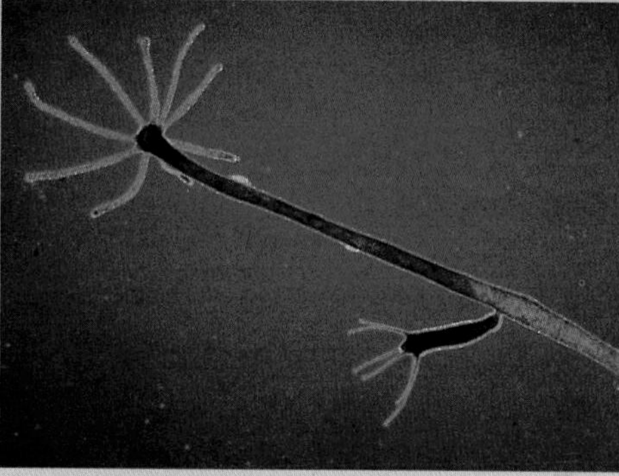

THINK AND WRITE CRITICALLY

Answer the following questions in your Journal using complete sentences.

13. Using an example, explain the difference between bilateral and radial symmetry.
14. Explain the difference between budding and regeneration.
15. Why do scientists feel that reforms and regulations in animal research may threaten the future of science?
16. Explain how the structure of a tapeworm and its life-style are related.

APPLY

17. Compare the body organization of a sponge to that of a simple worm.
18. What is the advantage of having more than one means of reproduction in simple animals?
19. List the types of food sponges, hydras, and planarians eat. Explain why the size of the food particles is different in each organism.
20. Compare and contrast the medusa and polyp body forms of cnidarians.
21. What are some reasons why scientists think the medusa stage was the first stage of the cnidarians?

MORE SKILL BUILDERS

If you need help, refer to the Skill Handbook.

1. **Sequencing:** Sequence the order of heartworm infection in a dog.
2. **Using Variables, Constants, and Controls:** Design an experiment to test the sense of touch in a planarian.
3. **Concept Mapping:** Complete the concept map of classification in the animal kingdom.

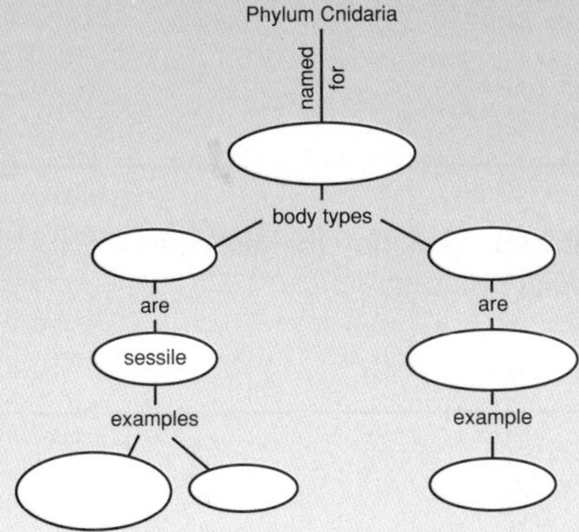

4. **Hypothesizing:** Make a hypothesis about cooking pork at high temperatures to prevent worms from developing, if present in the uncooked meat.
5. **Observing and Inferring:** Use Figure 13-2 on page 303 to determine to which phylum each of the following animals belong: jellyfish, crayfish, sponge, spider, starfish, oyster, snail, sea anemone, and sea urchin.

PROJECTS

1. Make a map of the areas of the world where parasitic worms are problems for humans. What conditions cause the high rate of infection in people in these areas?
2. With your teacher, look in a stream on and under rocks for flatworms and roundworms. Observe them under the microscope.

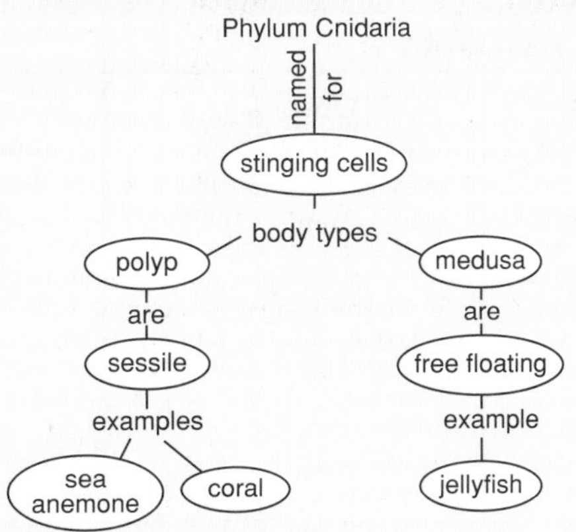

APPLY

17. Sponges have two cell layers; simple worms have three. Sponges have little cell organization; simple worms have tissues, organs and organ systems. Sponges are asymmetrical; simple worms have bilateral symmetry.

18. Simple animals can reproduce in a variety of ways which allows continuation of the species. These organisms would otherwise not be able to survive as well.

19. Sponges are filter-feeding omnivores: they eat bacteria, algae, and protozoans. Hydra are carnivores that eat small animals they capture with their tentacles, such as *Daphnia*. Planarians are carnivores that eat smaller animals.

20. The medusa stage is bell-shaped and free-swimming. The polyp stage is tube-shaped and sessile.

21. The medusa is a moving form. The polyp stage develops from the medusa. More polyp fossils exist, possibly indicating later existence.

MORE SKILL BUILDERS

1. Sequencing: Infected mosquito bites dog—blood infected—to heart where the heartworms grow and reproduce—valves are blocked—death if untreated.

2. Using Variables, Constants, and Controls: From their observations of Planarians, students should be able to suggest an experiment that involves simply tapping the watch glass and observing the movement or touching the worm with a blunt instrument.

3. Concept Mapping: See below, left, for possible solution.

4. Hypothesizing: Cysts are destroyed by heat—the worm has no protection and is killed.

5. Observing and Inferring:

jellyfish - cnidaria
crayfish - arthropoda
sponge - porifera
spider - arthropoda
starfish - echinodermata
oyster - mollusca
snail - mollusca
sea anemone - cnidaria
sea urchin - echinodermata

CHAPTER

14 Complex Invertebrates

CHAPTER SECTION	OBJECTIVES	ACTIVITIES
14-1 Mollusks (1 day)	1. **Identify** the features of mollusks. 2. **Name** three classes of mollusks and **identify** a member of each.	
14-2 Segmented Worms (2 days)	1. **Describe** the traits of segmented worms. 2. **Describe** the structures and digestive process of an earthworm. 3. **Identify** the evolutionary relationship between segmented worms and mollusks.	**Activity 14-1:** *Designing an Experiment (Determining Earthworm Adaptations)*, p. 331
14-3 Arthropods (3 days)	1. **Identify** features used to classify arthropods. 2. **Relate** the structure of the exoskeleton to its function. 3. **Distinguish** between complete and incomplete metamorphosis.	**MINI-Lab:** *What type of metamorphosis do fruit flies undergo?* p. 337 **Activity 14-2:** Observing a Crayfish, p. 339
14-4 Pesticides Science & Society (1 day)	1. **Describe** the importance of pesticides in agriculture. 2. **Identify** the impacts of pesticides on the environment.	
14-5 Echinoderms (1 day)	1. **Identify** the features of echinoderms. 2. **Describe** how sea stars get and digest food.	**MINI-Lab:** *How do tube feet open clam shells?* p. 344
Chapter Review		

ACTIVITY MATERIALS

FIND OUT	ACTIVITIES		MINI-LABS	
Page 323 1 clam shell hand lens transparent metric ruler	**14-1 Designing an Experiment, p. 331** live earthworm moist paper towels shallow pan beaker of water hand lens loose soil cotton swab vinegar flashlight	**14-2 Observing a Crayfish, p. 339** crayfish in a small aquarium stirrer uncooked ground beef	**What type of metamorphosis do fruit flies undergo? p. 337** cultures of fruit flies	**How do tube feet open clam shells? p. 344** heavy books clock with second hand

CHAPTER FEATURES	TEACHER RESOURCE PACKAGE	OTHER RESOURCES
Skill Builder: *Comparing and Contrasting,* p. 326	**Ability Level Worksheets** ◆ ***Study Guide,*** p. 55 ● ***Reinforcement,*** p. 55 ▲ ***Enrichment,*** p. 55	**STVS:** Disc 5, Side 1 **Color Transparency 24,** Mollusk Features **Science Integration Activity 14**
Technology: *Leeches to the Rescue,* p. 329 **Skill Builder:** *Interpreting Scientific Illustrations,* p. 330	**Ability Level Worksheets** ◆ ***Study Guide,*** p. 56 ● ***Reinforcement,*** p. 56 ▲ ***Enrichment,*** p. 56 **Critical Thinking/Problem Solving,** p. 18 **Cross-Curricular Connections,** p. 18 **Activity Worksheets,** pp. 122-123 **Technology,** pp. 13-14 **Transparency Masters,** pp. 47-50	**Color Transparency 25,** Parts of the Earthworm **Lab Manual:** Earthworm Anatomy, p. 91 **STVS:** Disc 5, Side 1
Problem Solving: *Spinning Spiders,* p. 334 **Skill Builder:** *Concept Mapping,* p. 338	**Ability Level Worksheets** ◆ ***Study Guide,*** p. 57 ● ***Reinforcement,*** p. 57 ▲ ***Enrichment,*** p. 57 **Activity Worksheets,** pp. 124, 125, 128 **Transparency Masters,** pp. 51-52	**Color Transparency 26,** Metamorphosis **Lab Manual:** Collecting Insects, p. 95; Grasshopper Anatomy, p. 97; Metamorphosis, p. 101 **STVS:** Disc 5, Side 1
You Decide! p. 341	**Ability Level Worksheets** ◆ ***Study Guide,*** p. 58 ● ***Reinforcement,*** p. 58 ▲ ***Enrichment,*** p. 58 **Science and Society,** p. 18	**STVS:** Disc 4, Side 2; Disc 6, Side 1
Skill Builder: *Observing and Inferring,* p. 344	**Ability Level Worksheets** ◆ ***Study Guide,,*** p. 59 ● ***Reinforcement,*** p. 59 ▲ ***Enrichment,*** p. 59 **Concept Mapping,** pp. 33 **Activity Worksheets,** p. 129	**STVS:** Disc 5, Side 1
Summary Key Science Words Understanding Vocabulary Checking Concepts Using Lab Skills Think & Write Critically Apply More Skill Builders Projects	ASSESSMENT RESOURCES **Chapter Review,** pp. 31-32 **Chapter Test,** pp. 93-96 **Performance Assessment in Middle School Science (PAMSS)**	**Chapter Review Software** **Test Bank** **Alternate Assessment** **Performance Assessment**

◆ **Basic** ● **Average** ▲ **Advanced**

ADDITIONAL MATERIALS

SOFTWARE	AUDIOVISUAL	BOOKS/MAGAZINES
Classifying Animals Without Backbones, D.C. Heath and Company. *The Insect World: A Science Discovery Unit,* Ventura Educational Systems. *The Insect World,* Queue. *The World of Insects,* Focus.	*Amazing Ants,* film, Coronet/MTI. *Insect Life Cycles,* film, Centron. *Insect Metamorphasia,* film, BFA. *Insects,* film, EBEC. *Crustaceans,* film, EBEC. *Wasps: Paper Makers of the Summer,* video, LCA.	Shepherd, Elizabeth. *No Bones: A Key to Bugs and Slugs, Worms and Ticks, Spiders and Centipedes, and Other Creepy Crawlies.* New York: Macmillan Publishing Co., 1988. Stokes, Donald W. *A Guide to Observing Insect Lives.* Boston, MA: Little, Brown and Company, 1983.

CHAPTER

14 COMPLEX INVERTEBRATES

THEME DEVELOPMENT: The theme of evolution is shown in the comparisons of the adaptations that have evolved in the mollusks, segmented worms, and arthropods. Ecology is addressed in the ways the organisms of each phylum live and affect their environment.

CHAPTER OVERVIEW

- **Section 14-1:** This section describes the basic body plan of all mollusks before the characteristics of univalves, bivalves, and cephalopods are discussed. The importance of mollusks is also discussed.
- **Section 14-2:** A description of features of segmented worms is followed by a discussion of earthworms and leeches. The section ends with information about the evolution of segmented worms and mollusks.
- **Section 14-3:** The features used to classify arthropods are described. Complete and incomplete metamorphosis are discussed.
- **Section 14-4: Science and Society:** The importance of pesticides in agriculture, the impact of pesticides on the environment, and the need for more efficient application of pesticides are explored. The You Decide feature asks students if we can live without pesticides.
- **Section 14-5:** The features of echinoderms are presented in this section. How sea stars obtain and digest food is described.

CHAPTER

14 Complex Invertebrates

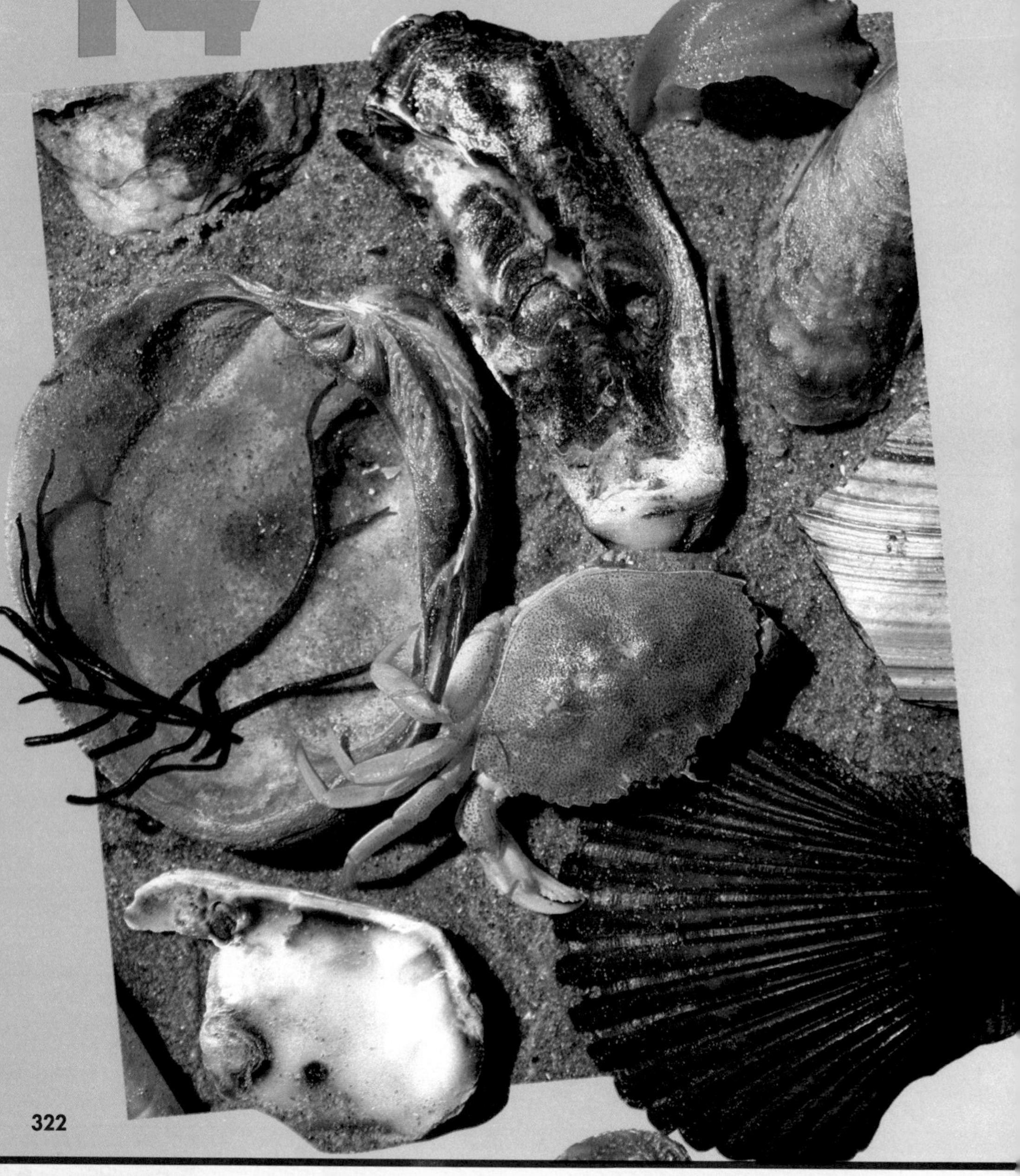

322

OPTIONS

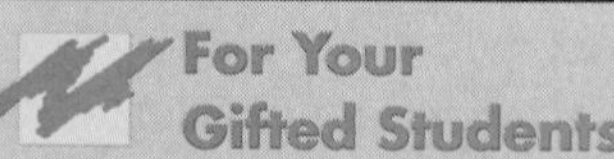

For Your Gifted Students

Have students research the major complex invertebrates in the chapter. They should make identification cards of the ones in which they are most interested. On one side of the card they can draw a picture of the invertebrate and on the other give the name and major characteristics. A game can be made by keeping track of the number of organisms students can identify.

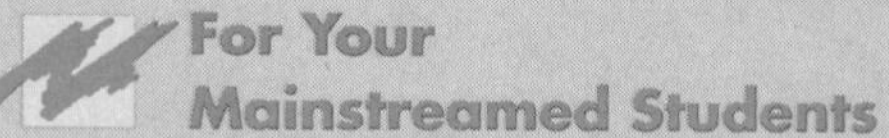

For Your Mainstreamed Students

Have students make papier mâché models of all the major complex invertebrates mentioned in the chapter. Name cards and a list of reasons why they are important creatures in the ecosystem should be attached to each.

If you've ever walked along a beach, you've probably seen lots of seashells. Seashells come in many different colors, shapes, and sizes. Each shell is made of many rings or bands. Can you learn anything about the shell and the organism that made it by looking at the bands?

FIND OUT!

Do this simple activity to find out how many bands are on a clam shell and what they mean.

Use a hand lens to observe a clam shell. Count the number of rings or bands on the shell. Count the large top point called the crown as one ring. Are all of the bands the same width? *Predict* what the width of the bands tells you about a year in the life of the clam. *Compare* your results with those of your classmates. Do all of the clams have the same number of bands? *Infer* what the bands represent. Why do you think some are wider than others?

Gearing Up

Previewing the Chapter

Use this outline to help you focus on important ideas in this chapter.

Previewing Science Skills

- ▶ In the Skill Builders, you will compare and contrast, interpret scientific illustrations, make a concept map, and observe and infer.
- ▶ In the Activities, you will observe, experiment, and collect and record data.
- ▶ In the MINI-Labs, you will observe and record data.

What's next?

In this chapter you will study mollusks, segmented worms, arthropods, and echinoderms. You will learn about the lives of these organisms and about how these organisms are important to you and the environment.

323

CHAPTER VOCABULARY

mollusks	appendages
mantle	exoskeleton
gills	molting
open circulatory system	spiracles
radula	metamorphosis
closed circulatory system	pesticide
setae	echinoderms
crop	water-vascular system
gizzard	tube feet
Arthropoda	

INTRODUCING THE CHAPTER

Use the Find Out activity to introduce students to mollusks. Explain that they will be learning about mollusks and other complex invertebrates in the chapter.

FIND OUT!

Preparation: Obtain one clam shell, a hand lens, and a transparent metric ruler for each pair of students.

Cooperative Learning: Use the Paired Partners strategy. Have each pair of students count the bands and answer the questions.

Teaching Tips

▶ As a clam grows, so does its shell. A new ring or band shows on the shell each year. Count the bands to tell how old a clam is. The crown is one year's growth. Food supply, water temperature, oxygen content, pollutants, and the calcium carbonate in the water affect the width of the bands.

Gearing Up

Have students study the Gearing Up feature to familiarize themselves with the chapter. Discuss the relationships of the topics in the outline.

What's Next?

Make sure students understand the connection between the Find Out activity and the topics to follow.

ASSESSMENT OPTIONS

PORTFOLIO

Refer to page 345 for suggested items that students might select for their portfolios.

PERFORMANCE ASSESSMENT

See page 345 for additional Performance Assessment options.

Process

Skillbuilders, pp. 326, 330
MINI-Labs, pp. 337, 344
Activities, 14-1, p. 331; 14-2, p. 339
Using Lab Skills, p. 346

CONTENT ASSESSMENT

Assessment—Oral, pp. 327, 334, 336
Skillbuilder, pp. 338, 344
Section Reviews, pp. 326, 330, 338, 341, 344
Chapter Review, pp. 345-347
Mini Quizzes, pp. 329, 336, 343

GROUP ASSESSMENT

Opportunities for group assessment occur with Cooperative Learning Strategies and Flex Your Brain Activities.

SECTION 14-1

PREPARATION

SECTION BACKGROUND

▶ The phylum Mollusca contains several classes. The three main classes are Gastropoda, the stomach-footed mollusks; Bivalvia, the two-shelled mollusks; and Cephalopoda, the head-footed mollusks. The bodies of all these mollusks are soft and unsegmented.

▶ Scientists have discovered mollusks living in hydrothermal vents in the ocean floor. They get their food from bacteria that live in their bodies. The bacteria use sulfide compounds given off by the vents and produce sugars that the mollusks use for food.

PREPLANNING

▶ Obtain snails for the Motivate activity.

▶ Purchase clams for the demonstration on page 325.

1 MOTIVATE

▶ **Activity:** Collect snails from a local pond or purchase them from a pet shop. Have students observe how the snails move. Have them examine their eyes with a hand lens. Have them feed lettuce to the snails and observe their feeding method with the hand lens.

TYING TO PREVIOUS KNOWLEDGE: Most students will be familiar with mollusks such as oysters, clams, squid, garden snails, and octopuses.

PROGRAM RESOURCES

From the **Teacher Resource Package** use:

Technology, pages 13-14, Building with Barnacles.

Transparency Master, pages 47-48, Mollusk Features.

Use **Color Transparency** 24, Mollusk Features.

Science Integration Activities, 14, Ocean Life

14-1 Mollusks

New Science Words

mollusks
mantle
gills
open circulatory system
radula
closed circulatory system

Objectives

▶ **Identify the features of mollusks.**
▶ **Name three classes of mollusks and identify a member of each.**

Figure 14-1. Most mollusks have the same basic body plan: a shell, foot, mantle, and visceral mass.

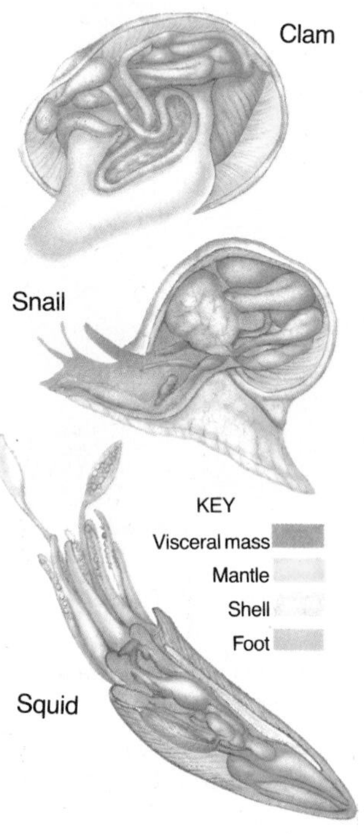

Features of Mollusks

If you've collected shells on the beach, watched a snail crawl, or eaten oysters or squid, you are familiar with mollusks. The word *mollusk* comes from the Latin word meaning "soft," and describes the bodies of the organisms in the phylum Mollusca (mah LUS kuh). **Mollusks** are soft-bodied invertebrates that usually have shells. They are found on land, in fresh water, and in salt water. Like the simple worms, mollusks have bilateral symmetry. Unlike the simple worms, sponges, and cnidarians, mollusks have a fluid-filled body cavity that provides space for the body organs.

Mollusks come in many different sizes, shapes, and colors. They vary in size from the tiny aquarium snail to the giant squid that can reach a length of 18 m. Different as they are, all mollusks have the same basic body plan.

All mollusks have a soft body usually covered by a hard shell. Covering the soft body is the mantle. The **mantle** is a thin layer of tissue that secretes the shell or protects the body if the mollusk does not have a shell. Between the soft body and the mantle is a space called the mantle cavity. In it are the **gills,** organs that exchange oxygen and carbon dioxide with the water. The body organs of mollusks are located together in an area called the visceral mass. The mantle covers the visceral mass. Finally, all mollusks have a muscular foot used for movement.

The circulatory system of most mollusks is an **open circulatory system.** In this type of system blood isn't completely contained in vessels the way your blood is. Blood bathes a mollusk's organs directly in some areas of their bodies.

The classification of mollusks is based on whether or not a shell is present; if a shell is present, the kind of shell it is; and the kind of foot. In this section, you will learn about three classes of mollusks.

OPTIONS

Meeting Different Ability Levels

For Section 14-1, use the following **Teacher Resource Masters** depending upon individual students' needs.

◆ **Study Guide Master** for all students.

● **Reinforcement Master** for students of average and above average ability levels.

▲ **Enrichment Master** for above average students.

Additional Teacher Resource Package masters are listed in the OPTIONS box throughout the section. The additional masters are appropriate for all students.

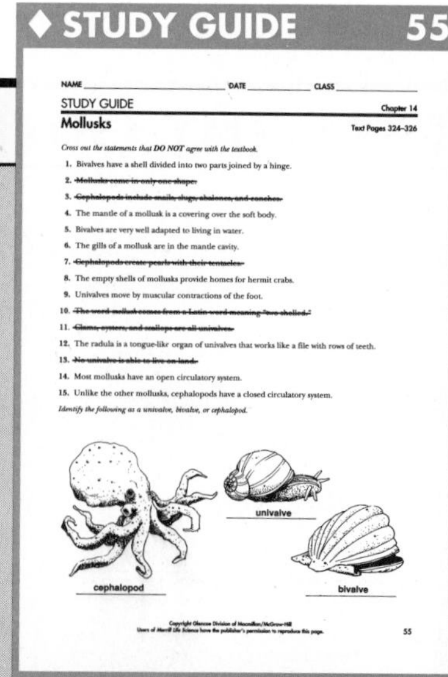
◆ STUDY GUIDE 55

NAME ______ DATE ______ CLASS ______

STUDY GUIDE — Chapter 14

Mollusks — Text Pages 324–326

Cross out the statements that ***DO NOT*** *agree with the textbook.*

1. Bivalves have a shell divided into two parts joined by a hinge.
2. ~~Mollusks come in only one shape.~~
3. ~~Cephalopods include snails, slugs, abalones, and conches.~~
4. The mantle of a mollusk is a covering over the soft body.
5. Bivalves are very well adapted to living in water.
6. The gills of a mollusk are in the mantle cavity.
7. ~~Cephalopods create pearls with their tentacles.~~
8. The empty shells of mollusks provide homes for hermit crabs.
9. Univalves move by muscular contractions of the foot.
10. ~~The word mollusk comes from a Latin word meaning "two shelled."~~
11. ~~Clams, oysters, and scallops are all univalves.~~
12. The radula is a tongue-likr organ of univalves that works like a file with rows of teeth.
13. ~~No univalve is able to live on land.~~
14. Most mollusks have an open circulatory system.
15. Unlike the other mollusks, cephalopods have a closed circulatory system.

Identify the following as a univalve, bivalve, or cephalopod.

cephalopod

univalve

bivalve

55

Science Integration Activities is found in **Merrill Life Science Teacher Resource Manual.**

Classes of Mollusks

Univalves

The largest class of mollusks includes snails, slugs, abalones, whelks, sea slugs, and conches. Look at the organism in the top photograph of Figure 14-2. Can you see why these organisms are called one-shelled mollusks, or univalves? Except for slugs, each member of this class has a single shell. Many have a pair of tentacles with eyes at the tips. Food is obtained with a radula. The **radula** is a tongue-like organ with rows of teeth that work like a file. It is an adaptation used for scraping and tearing algae and other food materials.

Univalves are adapted for different ways of life. Slugs and many snails are adapted to life on land. They move by muscular contractions of the foot. Glands in the foot secrete a layer of mucus for the foot to slide along in. Slugs live where it is moist. Slugs do not have shells, but they are protected by a layer of mucus.

Bivalves

Bivalves are two-shelled mollusks that have a two-part shell joined by a hinge. *Bivalve* means "two-shelled." Clams, oysters, and scallops are bivalve mollusks. These animals pull their shells closed with powerful muscles. To open their shells, they relax these muscles. The shell is made up of several layers made by the mantle. The smoothest layer is on the inside and protects the soft body.

Bivalves are very well adapted to living in water. For protection, clams burrow deep into the sand with their muscular foot. Mussels and oysters cement themselves to a solid surface or attach themselves with a strong thread. This keeps strong waves from washing them away. Scallops escape predators by rapidly opening and closing their shells. As the water pours out, its force moves them rapidly in the opposite direction.

Cephalopods

The class of mollusks with the most specialized and complex members are the cephalopods. The word *cephalopod* means "head-footed," and describes the body structure of these invertebrates. Squid, octopus, and the chambered nautilus belong to this class. Cephalopods have a large, well-developed head. Their "foot" is divided into many tentacles with strong suckers for capturing

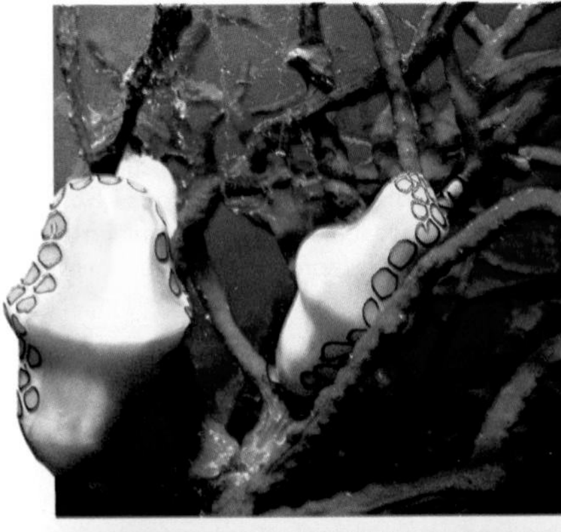

Figure 14-2. The organism in the top photo is a representative of the univalves. The scallop in the middle photo is a bivalve. The octopus in the bottom photo is a cephalopod.

● REINFORCEMENT 55

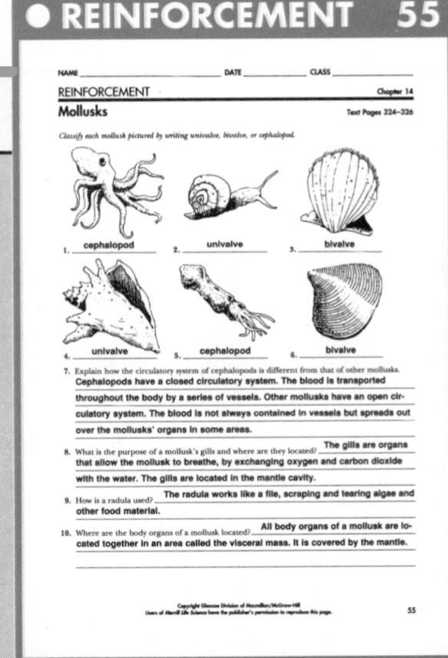

NAME ____ DATE ____ CLASS ____

REINFORCEMENT — Chapter 14

Mollusks — Text Pages 324–326

Classify each mollusk pictured by writing univalve, bivalve, or cephalopod.

1. cephalopod
2. univalve
3. bivalve
4. univalve
5. cephalopod
6. bivalve
7. Explain how the circulatory system of cephalopods is different from that of other mollusks. **Cephalopods have a closed circulatory system. The blood is transported throughout the body by a series of vessels. Other mollusks have an open circulatory system. The blood is not always contained in vessels but spreads out over the mollusks' organs in some areas.**
8. What is the purpose of a mollusk's gills and where are they located? **The gills are organs that allow the mollusk to breathe, by exchanging oxygen and carbon dioxide with the water. The gills are located in the mantle cavity.**
9. How is a radula used? **The radula works like a file, scraping and tearing algae and other food material.**
10. Where are the body organs of a mollusk located? **All body organs of a mollusk are located together in an area called the visceral mass. It is covered by the mantle.**

55

▲ ENRICHMENT 55

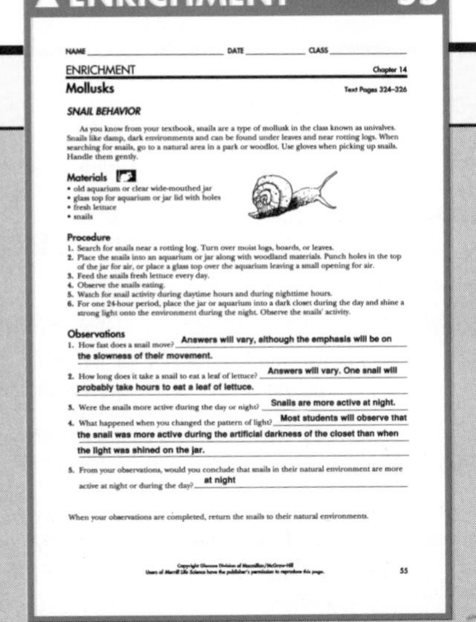

NAME ____ DATE ____ CLASS ____

ENRICHMENT — Chapter 14

Mollusks — Text Pages 324–326

SNAIL BEHAVIOR

As you know from your textbook, snails are a type of mollusk in the class known as univalves. Snails like damp, dark environments and can be found under leaves and near rotting logs. When searching for snails, go to a natural area in a park or woodlot. Use gloves when picking up snails. Handle them gently.

Materials

- old aquarium or clear wide-mouthed jar
- glass top for aquarium or jar lid with holes
- fresh lettuce
- snails

Procedure

1. Search for snails near a rotting log. Turn over moist logs, boards, or leaves.
2. Place the snails into an aquarium or jar along with woodland materials. Punch holes in the top of the jar for air, or place a glass top over the aquarium leaving a small opening for air.
3. Feed the snails fresh lettuce every day.
4. Observe the snails eating.
5. Watch for snail activity during daytime hours and during nighttime hours.
6. For one 24-hour period, place the jar or aquarium into a dark closet during the day and shine a strong light onto the environment during the night. Observe the snails' activity.

Observations

1. How fast does a snail move? **Answers will vary, although the emphasis will be on the slowness of their movement.**
2. How long does it take a snail to eat a leaf of lettuce? **Answers will vary. One snail will probably take hours to eat a leaf of lettuce.**
3. Were the snails more active during the day or night? **Snails are more active at night.**
4. What happened when you changed the pattern of light? **Most students will observe that the snail was more active during the artificial darkness of the closet than when the light was shined on the jar.**
5. From your observations, would you conclude that snails in their natural environment are more active at night or during the day? **at night**

When your observations are completed, return the snails to their natural environments.

55

OBJECTIVES AND SCIENCE WORDS: Have students review the objectives and science words throughout the chapter to become familiar with each section.

2 TEACH

Key Concepts are highlighted.

CONCEPT DEVELOPMENT

▶ Use an overhead transparency of a snail to show the three main body parts shared by all mollusks—the muscular foot, the head, and the visceral mass.

▶ Ask: **How are clams, oysters, and scallops like sponges?** *They are filter feeders.*

▶ **Demonstration:** Obtain some lively clams from a seafood market. Place them in water. Have students observe bubbling as the clams siphon water through. Some may also extend the ventral foot.

CHECK FOR UNDERSTANDING

Ask students to describe how all mollusks are similar: a soft body usually covered by a hard shell, mantle, gills, muscular foot, and organs located in a visceral mass.

RETEACH

Give students a photocopied diagram of a snail and have them label the body parts.

EXTENSION

For students who have mastered this section, use the **Reinforcement** and **Enrichment** masters or other OPTIONS.

VideoDisc

STVS: Conch Farming, Disc 5, Side 1

CROSS CURRICULUM

▶**Reading:** Read the poem "The Chambered Nautilus" by Oliver Wendell Holmes. Provide a nautilus shell to examine. Have them draw the shell and write a report in their journals about how the shell forms.

To assess this product, refer to the Performance Task Assessment Lists in **Performance Assessment in Middle School Science.**

Answer: Muscles exert force on water under the mantle. Water being forced out exerts force back, resulting in movement backward.

3 CLOSE

▶Use the Assessment—Oral questions on the bottom of page 327.

▶Ask questions 1-2 and the **Apply** Question in the Section Review.

SECTION REVIEW ANSWERS

1. kind of foot; presence or absence of shell; if present, kind of shell

2. univalves—snails, slugs, abalones, whelks, sea slugs, and conches; bivalves—clams, oysters, and scallops; cephalopods—squid, octopuses, and chambered nautilus

3. Apply: The radula is the adaptation that a snail uses to scrape algae from the aquarium glass.

4. Connect to Physics: By closing its shell sharply, water is forced out, causing the scallop to shoot backward.

Skillbuilder

ASSESSMENT

Performance: Assess students' abilities to Make and Use Tables by having them write a paragraph in their journals to compare the ways in which the three types of mollusks feed.

According to Newton's Third Law of Motion, when one object exerts a force on a second object, the second object exerts an equal but opposite force on the first. *Infer* how this law is represented in the movement of a squid.

Figure 14-3. Pearls are formed in many bivalve mollusks. Smooth mother of pearl is secreted by the mantle in layers around a grain of sand or other particle trapped between the mantle and the shell.

prey. Squid and octopuses have a well-developed nervous system and large eyes similar to human eyes. Unlike the other mollusks, in head-footed mollusks, blood containing food and oxygen is contained and transported in a series of vessels, in a **closed circulatory system.**

All cephalopods live in oceans and have bodies adapted for swimming. They move quickly by jet propulsion. When the space under the mantle is filled with water, they contract their mantle muscles and force water from an opening near the head. The jet of water sends them backward. A squid can swim at more than 60 m per second this way. Although octopuses can swim by jet propulsion like squid can, they usually use their tentacles to creep more slowly over the ocean floor.

Importance of Mollusks

Mollusks provide food for fish, sea stars, and birds. Their empty shells provide homes for invertebrates such as hermit crabs. Clams, oysters, snails, and scallops are used for food. Pearls are produced by many species of mollusks, but most are made by pearl oysters. Scientists are studying the nervous systems of squid and octopus to understand how learning takes place and how memory works. Mollusks can also cause problems. Snails and slugs feed on plants and damage crops. Certain species of snails are hosts of parasites that infect humans.

SECTION REVIEW

1. What features are used to classify mollusks?
2. Name the three classes of mollusks and identify a member from each class.
3. **Apply:** What adaptation makes a snail useful to have in an aquarium?
4. **Connect to Physics:** Find out what a scallop does to escape a predator.

Comparing and Contrasting

Make a table that compares and contrasts the features of one-shelled mollusks, two-shelled mollusks, and head-footed mollusks. If you need help, refer to Comparing and Contrasting in the **Skill Handbook** on page 683.

This table represents one way students might answer the Skillbuilder question.

Skill Builder

Type of Mollusk	Nutrition	Movement	Circulation	Habitat
one-shelled mollusk	feed on plant material using radula	muscular contractions of foot, mucus layer	open circulatory system	land and water
two-shelled mollusk	use siphons fo filter feed	sessile or use foot to crawl, open and close shell	open circulatory system	water
head-footed mollusk	use tentacles with suckers to capture prey	jet propulsion or use tentacles	closed circulatory system	water

Segmented Worms

14-2

Objectives

- ▶ Describe the traits of segmented worms.
- ▶ Describe the structures and digestive process of an earthworm.
- ▶ Identify the evolutionary relationship between segmented worms and mollusks.

New Science Words

setae
crop
gizzard

Features of Segmented Worms

The worms you see crawling across sidewalks and driveways after a hard rain and the night crawlers you dig up for fishing belong to the phylum Annelida (uh NEL ud uh). The phylum Annelida is made up of segmented worms. Their tube-shaped bodies are divided into many little sections, or segments. The word *annelid* means "little rings" and describes the bodies of these worms. Segmented worms are found in fresh water, salt water, and moist soil. Earthworms, leeches, and beautiful fan worms belong to this phylum.

Besides segments, members of this phylum have several other characteristics in common. Like mollusks, all segmented worms have a body cavity that holds their organs. On the outside of each segment, are bristle-like structures called **setae** (SEE tee) that help the worms move. Have you ever watched a bird try or tried yourself to pull an earthworm out of the ground? It's not that easy, is it? Segmented worms use their setae to hold on to the soil. Let's look more closely at two classes of segmented worms—earthworms and leeches. 1

Types of Segmented Worms

Earthworms

The tube-like body of an earthworm is divided into more than 100 segments. The segments aren't only on the outside; the body cavity is divided also. Each segment is identical except for those near the front and hind ends. Each body segment, except for the first and last, has four pairs of setae. Earthworms move by using their setae and two sets of muscles in the body wall.

Figure 14-4. Featherduster worms (top) and fireworms (bottom) are two types of segmented worms that live in the ocean.

SECTION 14-2

PREPARATION

SECTION BACKGROUND

▶ Earthworms are hermaphroditic. Fertilization occurs when two earthworms align their bodies next to each other and exchange sperm. The clitellum secretes a ring of mucus into which the egg and sperm are deposited as the clitellum moves forward. As the mucus passes over the head of the worm, it closes into a cocoon from which the eggs later hatch into young.

PREPLANNING

▶ Obtain live earthworms for Activity 14-1.

1 MOTIVATE

▶ Display specimens (or pictures on a bulletin board) of annelids (clam worms, tube worms, fan worms, leeches, Tubifex worms, redworms, and earthworms). Ask students how they are similar.

VideoDisc

STVS: Leeches, Disc 5, Side1

OPTIONS

ASSESSMENT—ORAL

▶ **How is a squid adapted for being a predator?** *It has tentacles with strong suckers for capturing prey, a well developed nervous system and large eyes, and jet propulsion for rapid movement.*

▶ **Suggest ways in which the shell of a mollusk is an adaptation.** *protection, conserves water, and camouflage*

▶ **Slugs can live in places where the soil is low in calcium but snails cannot. How does this fact relate to the two different organisms?** *Snails use calcium for their shells. Slugs do not have shells.*

TYING TO PREVIOUS KNOWLEDGE: Students have probably seen earthworms burrowing in the soil or crawling on sidewalks after a hard rain. Use their experiences with earthworms to discuss their observations about worms, where they live, and what they eat. Review flatworms and roundworms from Section 13-4.

2 TEACH

Key Concepts are highlighted.

CONCEPT DEVELOPMENT

▶ Invite a farmer or someone from the agricultural extension service to discuss the importance of earthworms and how they are raised commercially.

▶ **Demonstration:** Dissect a preserved earthworm and have students examine its digestive, excretory, circulatory, reproductive, and nervous systems.

▶ Explain that earthworm aeration of the soil is so beneficial that their presence in hundreds or thousands per acre is a characteristic of a productive farm. The earthworm, *Lumbricus terrestris,* ingests its own weight in soil and decaying matter every 24 hours.

▶ Explain that the skin of an earthworm is coated with mucus that keeps the skin moist so oxygen from the air can diffuse into the blood. Explain that earthworms survive in thin films of water in slightly moist soil, but if the soil becomes flooded, the earthworm can drown.

To get food, earthworms eat soil! While they burrow through the soil they are actually eating it. Earthworms get the energy they need to live from the bits of leaves and other plant and animal materials found in the soil. The digestive system of an earthworm is made up of a crop, gizzard, and intestine. The soil eaten by an earthworm moves to the **crop,** which is a sac used for storage. Behind the crop is a muscular structure, the **gizzard,** that grinds the soil. In the intestine, which follows, food is broken down and absorbed by the blood. Undigested soil and waste materials leave the worm through the anus.

Look at the body structure of an earthworm in Figure 14-5. Earthworms have two blood vessels along the sides of the body that meet in the anterior, or front, end. The vessels meet here and form five pairs of structures, called aortic arches, that pump blood through the body. Smaller vessels go into each body segment. Because an earthworm's blood is all contained in tubes and vessels, it has a closed circulatory system. Wastes are removed from earthworms by small, coiled tubes found in each segment. Earthworms have small, simple brains in the anterior segment. Nerves in each segment join to form a main nerve cord that connects to the brain. They respond to light, temperature, and moisture. Can you think of any times when they respond to light, temperature, and moisture?

2 An earthworm doesn't have gills or lungs. It lives in a thin film of water where it exchanges oxygen and carbon dioxide by diffusion through its skin. Each worm has both male and female reproductive structures, but an individual worm can't fertilize its own eggs. It has to exchange sperm to reproduce.

Figure 14-5. Segmented worms have circulatory, respiratory, excretory, digestive, muscular, and reproductive systems.

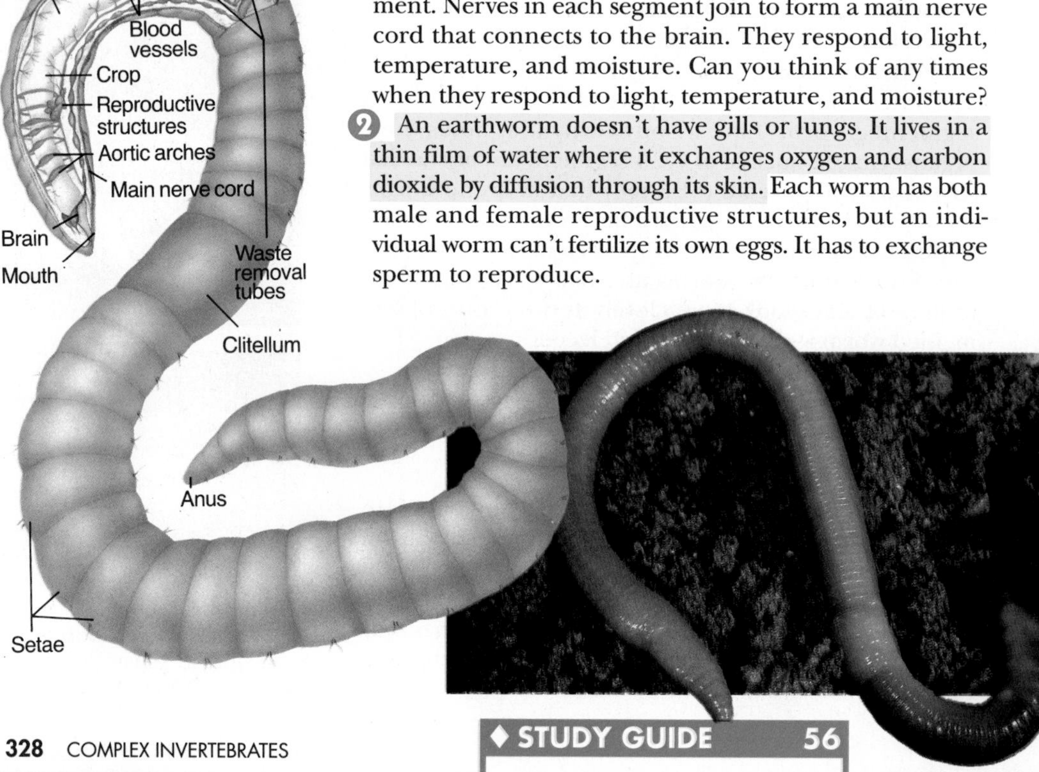

OPTIONS

Meeting Different Ability Levels

For Section 14-2, use the following **Teacher Resource Masters** depending upon individual students' needs.

◆ **Study Guide Master** for all students.

● **Reinforcement Master** for students of average and above average ability levels.

▲ **Enrichment Master** for above average students.

Additional Teacher Resource Package masters are listed in the OPTIONS box throughout the section. The additional masters are appropriate for all students.

◆ STUDY GUIDE 56

NAME ____ DATE ____ CLASS ____

STUDY GUIDE — Chapter 14

Segmented Worms — Text Pages 327–330

Check (✓) the statements that agree with the textbook.

____ 1. The lungs of worms are directly related to the gills of mollusks.
✓ 2. Setae are the bristle-like structures on the outside of an earthworm's body.
✓ 3. Worms and leeches are in the phylum Annelida.
✓ 4. Five pairs of aortic arches pump blood through an earthworm's body.
____ 5. Leeches only feed on the blood of humans.
____ 6. The earthworm nerve cord connects the crop to the intestine.
✓ 7. The gizzard is a muscular structure that grinds soil.
____ 8. Earthworms have an open circulatory system.
✓ 9. The crop of an earthworm is a sac used for storage.
✓ 10. Mollusks and segmented worms probably share a common ancestor.
✓ 11. Earthworms have two blood vessels that run along the sides of the body.
____ 12. The earthworm intestine is called a setae.
✓ 13. Leeches secrete a substance to keep blood from clotting.
____ 14. Every segment of an earthworm is different from all the other segments.
✓ 15. The first stage of development in both the mollusks and the segmented worms is a structure called a larva.

Complete the following sentences using appropriate words from the textbook.

16. The earthworm uses its **gizzard** to grind the soil it eats.
17. The body of an earthworm is divided into more than 100 **segments**.
18. Leeches use two **suckers** to attach to an animal.
19. An earthworm stores soil in its **crop**.
20. The earthworm's two long blood vessels meet at the organism's **anterior**, or front, end.
21. An earthworm can cling to the soil with its **setae**.
22. The **larva** is the best evidence that mollusks and segmented worms share a common ancestor.

56

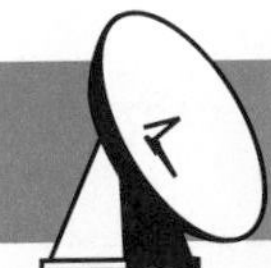

TECHNOLOGY

Leeches to the Rescue

No one likes to meet a leech by accident, but some hospital pharmacies now stock leeches, in various sizes, for use in delicate microsurgery. In fact, a major leech farm ships 25 000 leeches each year to two dozen countries.

When doctors perform microsurgery to reattach small body parts like ears, fingers, and toes, they may have problems due to the tiny veins involved. These tend to become blocked and cut off the blood flow to the reattached tissue. Without a good blood flow, the tissue dies. To prevent this, surgeons attach leeches to the microsurgery site. The leeches suck some of the congested blood, and chemicals in their saliva keep the blood flowing for several hours.

Other chemicals in leeches' saliva may be beneficial. Scientists have found in leech saliva an anticlotting substance, an anesthetic, and a chemical that increases the size of veins.

Think Critically: How could these chemicals be useful for patients with circulatory problems?

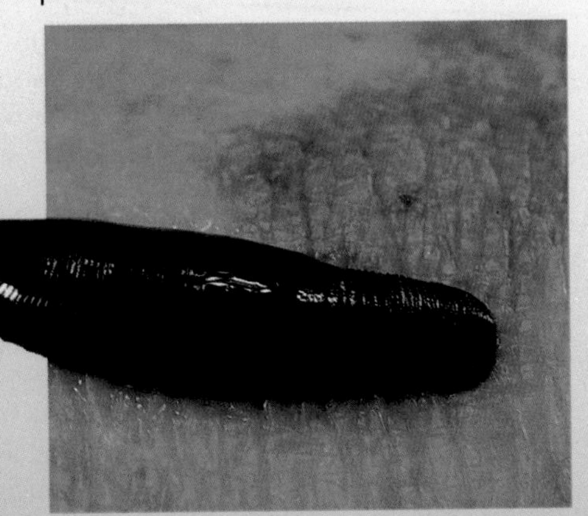

Leeches

You or someone you know may have had to remove a leech from your body while swimming in a pond, lake, or river. Leeches belong to another class of segmented worms and are adapted to a life-style very different from earthworms. Their bodies are not as round or long as earthworms', and they don't have setae. Leeches feed on the blood of ducks, fish, and even humans. Two suckers, one at each end of the body, are used to attach to an animal. After the leech has attached itself, it cuts open a wound and sucks two to ten times its weight in blood. If you have had a leech attached to you, you know that when you pulled it off, the wound didn't stop bleeding for quite a long time. This is because the leech secretes a substance to keep your blood from clotting so it can feed more easily.

Connect to... Chemistry

When a leech feeds on an organism, it may take in and store enough blood to keep itself alive for four months. A leech also produces an anticoagulant while feeding and during the months of digestion. Find out what an anticoagulant is and *infer* how it benefits the leech.

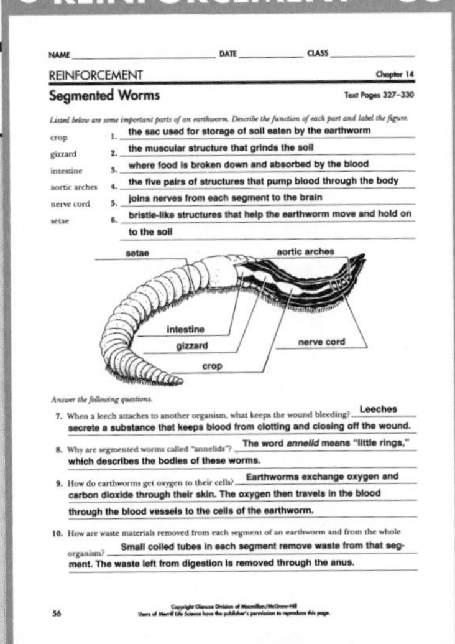

● REINFORCEMENT 56

NAME ______ DATE ______ CLASS ______

REINFORCEMENT — Chapter 14

Segmented Worms — Text Pages 327–330

Listed below are some important parts of an earthworm. Describe the function of each part and label the figure.

crop 1. the sac used for storage of soil eaten by the earthworm

gizzard 2. the muscular structure that grinds the soil

intestine 3. where food is broken down and absorbed by the blood

aortic arches 4. the five pairs of structures that pump blood through the body

nerve cord 5. joins nerves from each segment to the brain

setae 6. bristle-like structures that help the earthworm move and hold on to the soil

Answer the following questions.

7. When a leech attaches to another organism, what keeps the wound bleeding? Leeches secrete a substance that keeps blood from clotting and closing off the wound.
8. Why are segmented worms called "annelids"? The word *annelid* means "little rings," which describes the bodies of these worms.
9. How do earthworms get oxygen to their cells? Earthworms exchange oxygen and carbon dioxide through their skin. The oxygen then travels in the blood through the blood vessels to the cells of the earthworm.
10. How are waste materials removed from each segment of an earthworm and from the whole organism? Small coiled tubes in each segment remove waste from that segment. The waste left from digestion is removed through the anus.

56

▲ ENRICHMENT 56

NAME ______ DATE ______ CLASS ______

ENRICHMENT — Chapter 14

Segmented Worms — Text Pages 327–330

EARTHWORM COMPOST

Farmers and gardeners have long known the value of earthworms. By their natural actions of burrowing and feeding, earthworms turn and enrich the soil. According to one estimate, an acre of earthworms will produce 12 to 18 tons of topsoil a year.

Today, earthworms are also beginning to provide a service to city dwellers who produce unwanted garbage. By following the directions here, you can make an indoor compost pile for home or school. The earthworms in the compost pile will turn wasted food into new soil that can be added to gardens in the spring.

Materials

- wooden box
- newspapers
- peat moss
- bucket
- water
- screening
- earthworms
- tape

Procedure

1. Place the wooden box in an out-of-the-way place such as the basement, a closet, or underneath some shelves.
2. Fill the bottom of the box with shredded newspaper. About half the box should contain this material.
3. In the bucket, mix about a gallon of water with a few handfuls of peat moss. Make the mixture thick with the peat moss, but keep the mixture mostly liquid.
4. Pour the water-peat moss mixture into the box.
5. Add a box of earthworms purchased at a bait shop.
6. Place the screening over the box and tape it into place.
7. After about a week, begin adding food wastes and other organic garbage. Lift the screen and bury the waste under the old compost. Do this daily. Do not add bones or meat.

Observations

1. What happened to the shredded newspaper after a few weeks? The newspaper is no longer evident.
2. What do you think happened to the newspaper? It was eaten by the earthworms.
3. What took the newspaper's place? earthworm waste
4. Did your earthworm compost have a bad smell? No. The earthworms take care of the organic material so quickly that the compost just smells like dirt.

56

TECHNOLOGY

► For more information on leeches, see "The Return of the Bloodsuckers" by Matt Clark, *Newsweek*, Feb. 2, 1987, p. 58, or "The Little Suckers Have Made a Comeback" by Richard Conniff, *Discover*, August, 1987, pp. 84-86.

Think Critically: These chemicals could be used to increase blood to the heart muscle after a heart attack, or to help circulation problems due to clots.

CHECK FOR UNDERSTANDING

Have students summarize the characteristics that make the segmented worms more advanced evolutionarily than flatworms and mollusks.

RETEACH

Have students make a chart that compares the structures, environment, and way of life (free-living or parasitic) of flatworms, mollusks, and segmented worms.

EXTENSION

For students who have mastered this section, use the **Reinforcement** and **Enrichment** masters or other OPTIONS provided.

MINI QUIZ

Use the Mini Quiz to check students' recall of chapter content.

1. **What are three characteristics of segmented worms?** *body segments, a body cavity, and setae*
2. **How do earthworms exchange oxygen and carbon dioxide?** *through their skin*
3. **What is the best evidence that segmented worms and mollusks share a common ancestor?** *the larvae*

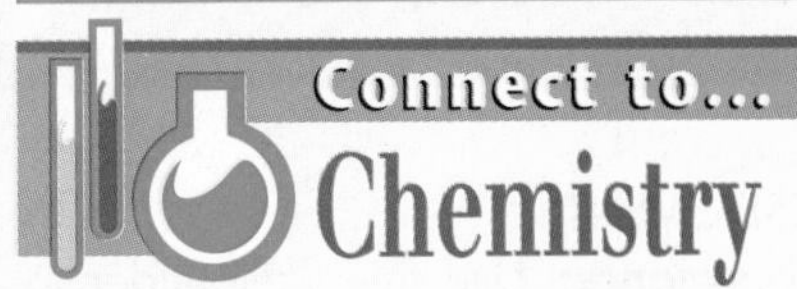

Connect to... Chemistry

Answer: An anticoagulant keeps blood from clotting. This works while the leech feeds and keeps stored blood from clotting while it is being digested.

CROSS CURRICULUM

▶ **Reading:** Have students research and report on the use of leeches for medicinal bloodletting throughout the middle ages and up until the early 1900s.

3 CLOSE

▶ Ask questions 1-4 and the **Apply** Question in the Section Review.

SECTION REVIEW ANSWERS

1. body segments
2. It takes in soil which moves through the digestive tract. Food (plant and animal material in the soil) is removed from the soil and digested. Waste materials leave through the anus.
3. They both have a body that has space for body organs, and the larvae of both are similar.
4. Leeches don't have setae. They have suckers on each end. They feed on blood instead of organic material in the soil.
5. **Apply:** Earthworms move deeper into soil where there is the necessary amount of moisture. They can't live in dry soil.
6. **Connect to Earth Science:** Earthworms break up hard, packed soil as they burrow through it feeding. This allows oxygen and water to penetrate the soil more easily to reach the roots of plants.

Skill Builder

nervous system: brain, main nerve cord
reproductive system: reproductive structures, clitellum
circulatory system: aortic arches, blood vessels
digestive system: mouth, crop, gizzard, intestine, anus
excretory system: waste removal tubes

Skillbuilder
ASSESSMENT

Performance: Use this Skillbuilder to assess students' abilities to Interpret the Scientific Illustration by pointing out external features in the photograph in Figure 14-5.

Figure 14-6. Leeches were once used to try to cure sick people.

Before the 1900s, physicians used leeches to drain blood from sick people. They thought that bleeding by leeches would drain away sickness and make the person well. The treatment usually caused patients to become weaker. Sometimes it even killed them.

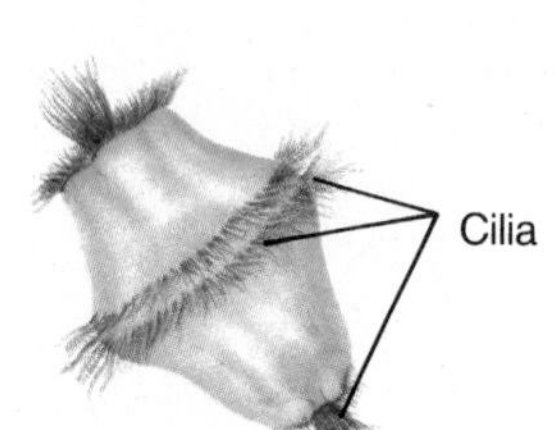

Figure 14-7. Mollusks and segmented worms have the same type of larva.

Evolution of Segmented Worms and Mollusks

Scientists infer that mollusks and segmented worms share a common ancestor. They were the first of the animal groups to have a body with space for the body organs. In both groups, the first stage of development is a structure that looks like a spinning top with cilia. This
3 structure, a larva, is shown in Figure 14-7. This larva is the best evidence that mollusks and segmented worms share a common ancestor.

SECTION REVIEW

1. What is the most noticeable feature of annelids?
2. Describe how an earthworm feeds and digests its food.
3. How are segmented worms related to mollusks?
4. How are leeches different from earthworms?
5. **Apply:** During dry periods, where are earthworms found in soil? Explain.
6. **Connect to Earth Science:** Why do farmers depend on earthworms that live in the soil of their crop fields?

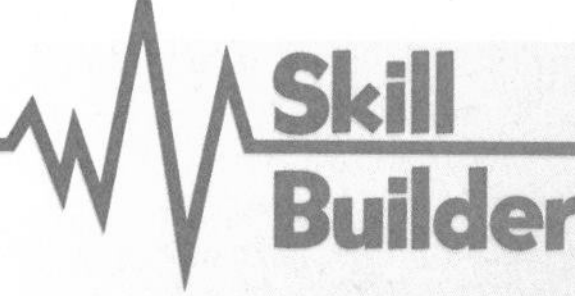

Skill Builder

☑ Interpreting Scientific Illustrations

Use the diagram in Figure 14-5 to list the body systems of an earthworm and organs in each system. If you need help, refer to Interpreting Scientific Illustrations in the **Skill Handbook** on page 693.

OPTIONS

PROGRAM RESOURCES

From the **Teacher Resource Package** use:

Critical Thinking/Problem Solving, page 18, Putting Worms to Work.

Cross-Curricular Connections, page 18, Leeches and Medicine.

Transparency Master, pages 49-50, Parts of the Earthworm.

Use **Color Transparency** 25, Parts of the Earthworm.

Use **Laboratory Manual,** page 91, Earthworm Anatomy.

ACTIVITY 14-1

DESIGNING AN EXPERIMENT
Earthworm Adaptations

Why are earthworms on top of the soil after a lot of rain? Any characteristic an organism has that makes it better able to survive in an environment is an adaptation.

Getting Started

You will examine a living earthworm to determine how its body is arranged and adapted to its environment

Thinking Critically

Make a list of the characteristics of earthworms that you know about. *Predict* why earthworms come to the top of the soil after a rain.

Materials  

Your cooperative group will use:

- live earthworms in a covered container
- moist paper towel
- shallow pan
- beaker of water
- hand lens
- loose soil
- 2 cotton swabs
- 10 mL of vinegar
- flashlight

1. **Hypothesize** how the earthworms will respond to light, being held with the fingers, touching its ends, and vinegar. Make a table to record all of your predictions and the earthworm's responses.
2. Be sure some of the earthworms are on top of the soil. Shine the flashlight on the worms.
3. Keep your hands wet while working with the earthworms. Remove an earthworm from the container your teacher has provided. **CAUTION:** *Use care when working with live animals.*
4. Use the hand lens to *observe* the structure of the earthworm. Locate the mouth, anus, and clitellum. Estimate the number of body segments the earthworm has. Hold the worm gently between your thumb and forefinger. *Observe* its movements. Use the hand lens to observe the setae on the underneath side.
5. Touch the anterior and posterior body areas with a moist cotton swab.
6. Use a cotton swab to draw a "line" of vinegar in front of the earthworm.
7. Place the earthworm on the surface of a thick layer of loose soil in a pan.
8. Return the earthworm to the place your teacher has designated. Wash your hands.

Summing Up/Sharing Results

- How is the earthworm's response to light an adaptation for living in soil?
- How are the setae and body shape useful for living in soil?
- How is the earthworm's structure related to its way of life?

Going Further!

The body of a live earthworm is wet, moist, and slimy. If the earthworm's body surface dries out, it will die. Why does this happen?

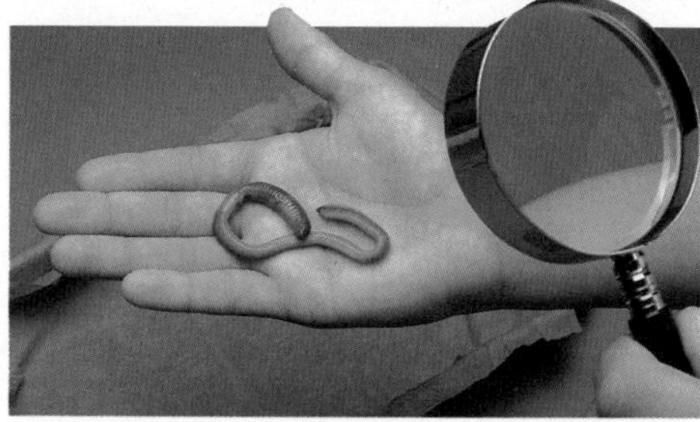

ACTIVITY 14-1

OBJECTIVE: **Observe** an earthworm and **carry out experiments** to determine how the structure of an earthworm is related to its way of life.
Time: one class period

PROCESS SKILLS applied in this activity are **observing, inferring, recognizing cause and effect, forming a hypothesis,** and **experimenting.**

PREPARATION

• Earthworms may be collected from the soil from late spring until early fall.
• Distribute materials before distributing the earthworms.
• Caution students to keep their hands wet and to handle the worms gently.
Cooperative Learning: Divide the class into Science Investigation Teams.

THINKING CRITICALLY

Student answers will vary. The earthworm exchanges oxygen and carbon dioxide through its moist skin. If there is too much water, the earthworm cannot get oxygen and give off carbon dioxide.

TEACHING THE ACTIVITY

*Refer to the **Activity Worksheets** for additional information and teaching strategies.*
• If the soil in Step 7 is not moist and at least 6 cm deep, the earthworms will not burrow.
• If the worms are not on top of the soil, they can be placed in the dark for a few minutes.
• Worm responses may include: tries to move under moist paper towel; tries to pry itself away from finger grasp; when touched in its anterior and posterior areas, it shortens its body and moves away from touch; moves away from vinegar; and burrows into soil.

SUMMING UP/ SHARING RESULTS

• It allows the earthworm to stay buried in the soil where it will not dry out.
• Setae anchor the earthworm, help in movement, and prevent it from being pulled easily from the soil. The front end of the earthworm is pointed so that it can move through the soil. The long body keeps the burrow open and stretches to become thin in tight places.
• Its structure allows food and oxygen to be taken in from the soil.

GOING FURTHER!

The surface of the earthworm must remain moist to allow for the diffusion of oxygen and carbon dioxide during respiration.

Activity
ASSESSMENT

Performance: To further assess students' observations of segmented worms, have them test the earthworm's response to temperature.

PROGRAM RESOURCES

From the **Teacher Resource Package** use:
Activity Worksheets, pages 122-123, Activity 14-1, Earthworm Adaptations.

SECTION 14-3

PREPARATION

SECTION BACKGROUND

▶ Animals with arthropod characteristics evolved more than 600 million years ago. In the course of evolution the bodies of arthropods came to have fewer segments. These segments have become fixed in number and more specialized in function.

▶ Millipedes and centipedes are often grouped as myriapods. *Myriapod* is Greek for "many feet."

▶ Centipedes can detach legs when attacked. The detached legs continue to wiggle while the centipede escapes.

PREPLANNING

▶ Obtain crayfish for Activity 14-2.

1 MOTIVATE

▶ **Bulletin Board:** Prepare a bulletin board with pictures of a wide variety of arthropods. As each arthropod class is studied, separate arthropods from that group and label them arachnids, centipedes, millipedes, crustaceans, and insects.

Cooperative Learning: Use Paired Partners cooperative groups. Give each group a dissecting tray containing a preserved millipede, centipede, crayfish, grasshopper, and spider. Have the students list the characteristics of the animals and determine how the five organisms are similar and how they are different.

14-3 Arthropods

New Science Words

Arthropoda
appendages
exoskeleton
molting
spiracles
metamorphosis

Objectives

- ▶ **Identify features used to classify arthropods.**
- ▶ **Relate the structure of the exoskeleton to its function.**
- ▶ **Distinguish between complete and incomplete metamorphosis.**

Features of Arthropods

Have you ever swatted at a fly, been bitten by a mosquito, or been stung by a bee? If you have, you've been bugged by an arthropod (AR thruh pahd). The arthropods make up the largest phylum of animals in the animal kingdom. They are adapted to almost every environment on Earth. Insects, shrimp, spiders, and centipedes are all members of the phylum **Arthropoda.**

What are arthropods?

Arthropoda means "jointed foot" and describes the jointed appendages of arthropods. **Appendages** are structures that grow from the body. Your arms and legs are appendages. The jointed appendages of arthropods include legs, antennae, claws, and pinchers.

The bodies of arthropods are divided into segments like those of segmented worms. Because of this, scientists hypothesize that arthropods and segmented worms have a common ancestor. The bodies of some arthropods have many segments, whereas others have segments that are fused together to form body regions. Figure 14-8 shows three regions formed by fused segments: the head, the thorax, and the abdomen.

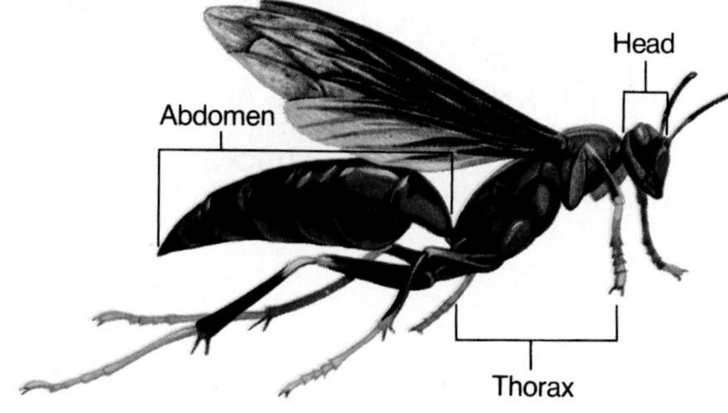

Figure 14-8. The segments of arthropods are fused together to form three regions: the head, the thorax, and the abdomen.

OPTIONS

Meeting Different Ability Levels

For Section 14-3, use the following **Teacher Resource Masters** depending upon individual students' needs.

◆ **Study Guide Master** for all students.

● **Reinforcement Master** for students of average and above average ability levels.

▲ **Enrichment Master** for above average students.

Additional Teacher Resource Package masters are listed in the OPTIONS box throughout the section. The additional masters are appropriate for all students.

◆ STUDY GUIDE 57

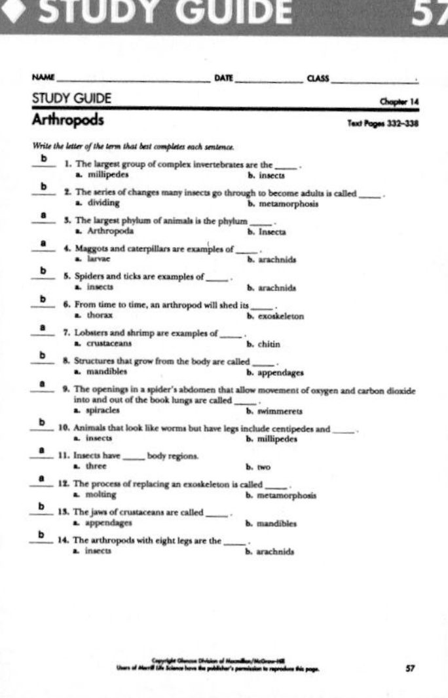

NAME ____________ DATE ________ CLASS ________

STUDY GUIDE — Chapter 14

Arthropods — Text Pages 332-338

Write the letter of the term that best completes each sentence.

b 1. The largest group of complex invertebrates are the _____.
a. millipedes b. insects

b 2. The series of changes many insects go through to become adults is called _____.
a. dividing b. metamorphosis

a 3. The largest phylum of animals is the phylum _____.
a. Arthropoda b. Insecta

a 4. Maggots and caterpillars are examples of _____.
a. larvae b. arachnids

b 5. Spiders and ticks are examples of _____.
a. insects b. arachnids

b 6. From time to time, an arthropod will shed its _____.
a. thorax b. exoskeleton

a 7. Lobsters and shrimp are examples of _____.
a. crustaceans b. chitin

b 8. Structures that grow from the body are called _____.
a. mandibles b. appendages

a 9. The openings in a spider's abdomen that allow movement of oxygen and carbon dioxide into and out of the book lungs are called _____.
a. spiracles b. swimmerets

b 10. Animals that look like worms but have legs include centipedes and _____.
a. insects b. millipedes

a 11. Insects have _____ body regions.
a. three b. two

a 12. The process of replacing an exoskeleton is called _____.
a. molting b. metamorphosis

b 13. The jaws of crustaceans are called _____.
a. appendages b. mandibles

b 14. The arthropods with eight legs are the _____.
a. insects b. arachnids

57

In addition to jointed appendages and segmented bodies, all arthropods have an external covering called the **exoskeleton.** The exoskeleton covers, supports, and protects the body. The lightweight exoskeleton is made of protein and a carbohydrate called chitin (KITE un). This covering also keeps the animal's body from drying out.

The exoskeleton is made of nonliving material, so it can't grow as the animal grows. From time to time, the old exoskeleton is shed and replaced by a new one in a

1 process called **molting.** The new exoskeleton is soft and takes a while to harden. During this time the animal is not well protected from its predators. Many people enjoy eating soft-shelled crabs. The crab has just molted and its new shell hasn't hardened yet.

Arthropods have a body cavity and a digestive system with two openings, a mouth and an anus. They have a nervous system similar to that of annelids but with a larger brain. You will learn about five classes of arthropods in this section.

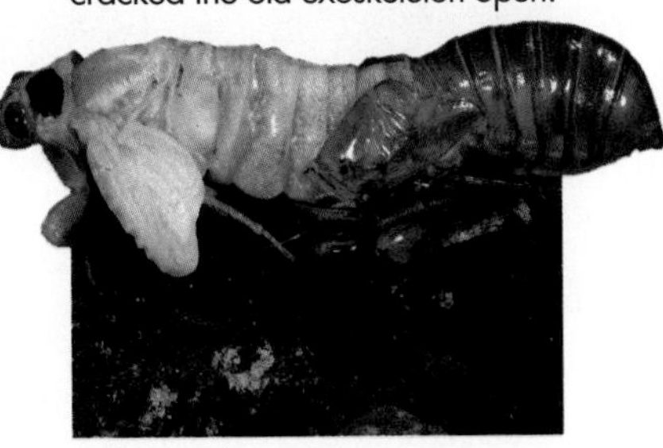

Figure 14-9. Arthropods molt several times as they grow. In order to molt, this cicada took a large amount of air into its digestive and respiratory systems, causing the body to expand. Expansion cracked the old exoskeleton open.

Classes of Arthropods

Arachnids

Spiders, scorpions, mites, and ticks are arachnids (uh RAK nudz). Arachnids are arthropods with two body regions, a head-chest region called the cephalothorax and an abdomen, no antennae, and four pairs of legs. You can tell an arachnid from an insect because the arachnid has eight legs. Arachnids are adapted to kill prey with poison glands, stingers, or fangs. Appendages near the mouth are used to hold food.

The spider in Figure 14-10 is a common arachnid. Spiders have eight simple eyes they use to sense light and darkness. A spider cannot chew food. Using a pair of fangs, a spider injects poison into its prey; the poison paralyzes the prey and turns it into a liquid. The spider then sucks up the liquid. Oxygen and carbon dioxide are exchanged in structures called book lungs. They are called book lungs because they look like the pages of a book. Openings in

2 the abdomen called **spiracles** allow oxygen and carbon dioxide to move into and out of the book lungs.

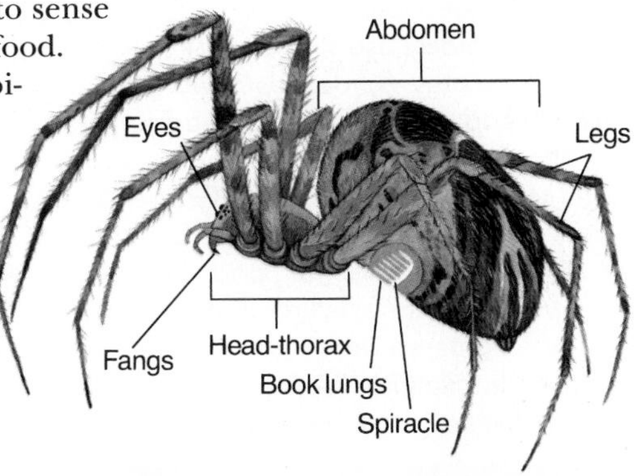

Figure 14-10. The external structures and book lungs of a common spider

● REINFORCEMENT 57

NAME ____ DATE ____ CLASS ____

REINFORCEMENT — Chapter 14

Arthropods — Text Pages 332–338

This drawing of an arthropod shows three distinct body regions. Label them with the correct terms.

head
thorax
abdomen

Answer the following questions.

4. Why do scientists think arthropods and segmented worms have a common ancestor? **The bodies of both are divided into segments.**
5. Why is molting necessary for arthropod growth? **An arthropod's exoskeleton is made of nonliving material, so it can't grow with the organism. By molting, the arthropod can increase in size before the new exoskeleton hardens.**
6. If spiders cannot chew, how can they eat? **A spider injects poison into its prey, paralyzing the prey and turning it into liquid. Then the spider sucks up the liquid material.**
7. What are two diseases spread by ticks? **Rocky Mountain spotted fever and Lyme disease**
8. What is the main difference between centipedes and millipedes? **Centipedes have one pair of joined legs per segment, while millipedes have two pairs per segment.**
9. What is attached to an insect's thorax? **three pairs of jointed legs and, in many species, one or two pairs of wings**
10. What organ systems do insects have? **systems for digestion, reproduction, and waste removal**
11. What are the four stages of development in complete metamorphosis? **egg, larva, pupa, adult**
12. What is a nymph? **the middle stage in incomplete metamorphosis, between the egg and the adult**

 57

▲ ENRICHMENT 57

NAME ____ DATE ____ CLASS ____

ENRICHMENT — Chapter 14

Arthropods — Text Pages 332–338

CRICKETS

A common arthropod of the insect class is the cricket, a small black bug with antennae, wings, and powerful hind legs. The word *cricket* comes from a French word meaning "to click or creak." Anyone who has had a cricket in the house has observed their noise-making abilities. Crickets can easily be found from spring through the fall in fields and vacant lots. They can also be purchased at bait stores that sell fishing supplies.

Materials

- wide-mouthed clear glass jar
- rubber band
- old nylon stocking
- scissors
- thermometer
- clock with second hand
- cricket

Procedure

1. Catch a cricket in a field or buy one from a bait store.
2. Place the cricket in the jar. Cut a top for the jar from the old nylon stocking. Secure the cover with the rubber band.
3. Observe the cricket's body parts and make a sketch of the arthropod. Label your sketch with these terms: head, thorax, abdomen.
4. Wait until the cricket is chirping regularly. While watching the second hand of the clock, count how many chirps the cricket makes in 15 seconds. Add 40 to the number of chirps you counted. Compare that sum to the temperature of the room.
5. Move the jar to a warmer place (near a sunny window, for example). Make your count again.

Observations

Draw your sketch here.

thorax
head
abdomen

Conclude and Apply

1. How did the sum of the number of chirps compare with the temperature reading? **Answers will vary, but the numbers should be close.**
2. What happened to the number of chirps when you moved the jar to a warmer place? **The number increased.**
3. How does temperature affect cricket activity? **Crickets become more active with a rise in temperature.**

 57

TYING TO PREVIOUS KNOWLEDGE: Students will be familiar with many species in the Phylum Arthropoda. Many have eaten shrimp and lobsters. All will have some knowledge of insects and spiders. Use their knowledge of arthropods in identifying common characteristics.

2 TEACH

Key Concepts are highlighted.

CONCEPT DEVELOPMENT

▶ **Analogy:** Compare the limitations of the arthropod exoskeleton with a person wearing a suit of armor.

▶ Explain that the chitin of the exoskeleton is coated with a thin, waxy layer that makes the exoskeleton waterproof. In addition to protecting the body and preventing it from drying out, the exoskeleton provides a place for the attachment of muscles.

▶ Most arthropods undergo four to seven molts before reaching adult size. After shedding the old exoskeleton, the body is quite soft. The animal swells by taking in extra water or air while the new exoskeleton hardens. During this time, the animal is not well protected from its enemies.

CHECK FOR UNDERSTANDING

Have students describe the arthropod exoskeleton and its limitations.

RETEACH

Have students explain why arthropods molt several times during their lifetime.

EXTENSION

For students who have mastered this section, use the **Reinforcement** and **Enrichment** masters or other OPTIONS provided.

VideoDisc

STVS: Tick Research, Disc 5, Side 1

STVS: March of the Spiny Lobster, Disc 5, Side 1

CONCEPT DEVELOPMENT

▶ Remind students that most animals are not harmful to people unless provoked or threatened.

▶ It may interest students that arachnids range in size from 0.25 mm long to a species of African scorpion that is 18 cm long. Mites are the most widespread of the arachnids.

▶ A tarantula provides an interesting classroom animal. It can be kept in a terrarium covered with screen wire with a 5 cm layer of gravel at the bottom. It will eat mealworms and crickets. A water dish should be kept filled at all times. Students can observe behavior, feeding, and molting.

REVEALING MISCONCEPTIONS

▶ Many students may think that all spiders are harmful. In the United States, only a few species are reported to cause reactions, and only two, the black widow and the brown recluse, can make persons extremely ill. While some deaths have been reported from spider bites, they usually result from complications rather than the bite itself.

PROBLEM SOLVING

Think Critically: The spider can walk to the prey because the "spoke" threads on which the spider walks are not sticky. The circular threads are the sticky ones.

Figure 14-11. Scorpions (a) and housedust mites (b) are both arachnids.

Ticks and mites are arachnids that are parasites. Mites are very tiny. Housedust mites feed on dead skin cells found in the dust on floors and in bedding. There is even a type of mite that lives in the follicles of human eyelashes. Ticks attach to the skin of a host and feed on its blood. Some ticks spread diseases such as Rocky Mountain spotted fever and Lyme disease.

Scorpions have a sharp stinger at the end of the abdomen that contains poison. Scorpions grab their prey with their pinchers and inject venom to paralyze them. The sting of a scorpion is very painful to humans, but it usually is not fatal.

PROBLEM SOLVING

Spinning Spiders

Have you ever looked closely at a spider web? Spiders make these interesting structures using three pairs of openings on the underside of the abdomen called spinnerets. A fluid protein is squeezed out of the spinnerets and hardens into a silk thread when exposed to air.

Some spiders use silk to construct webs to capture prey. Spiders spin two types of silk thread. One type of thread is very strong, whereas the other type is elastic and contains sticky droplets. Garden spiders spin the sticky thread along a spiral line that forms the trapping part of the web.

Spiders use the silk thread for other purposes. Some spiders use silk thread to change habitat. Maybe you've seen a spider glide through the air or lower itself from a tree using a thread. Also, silk threads are used to make cocoons for fertilized eggs or a nest in some species.

Think Critically: Imagine that an insect flies into a garden spider's web. Explain how the insect is captured by the silk, but the spider is able to walk to the prey.

OPTIONS

ASSESSMENT—ORAL

▶ **During molting, an arthropod that lives in water absorbs water and swells. What is the advantage of this behavior?** *Swelling cracks the exoskeleton thereby assisting molting.*

▶ **Why are arthropods more vulnerable during molting?** *They are not well protected by their exoskeletons.*

▶ **How do the feeding habits of spiders benefit humans?** *They feed on insects that damage crops that humans use for food.*

▶ **How are arachnids different from other arthropods?** *Arachnids have four pairs of legs.*

▶ **Why are the prefixes *centi-* and *milli-* used to name centipedes and millipedes?** *Centipedes appear to have a hundred legs and millipedes appear to have a thousand legs.*

▶ **Why do you think crustaceans such as crayfish, lobster, and shrimp are mistakenly called "shellfish"?** *They have a hard exoskeleton, and they live in water and breathe with gills; they are not fish however, and do not have shells.*

Centipedes and Millipedes

Even though centipedes and millipedes look like worms, you know they are not because worms don't have legs. Centipedes and millipedes make up two classes of arthropods. They have long bodies with many segments, an exoskeleton, jointed legs, antennae, and simple eyes. They live on land in moist environments. Both reproduce sexually and make nests for their eggs. They stay with the eggs until they hatch.

Compare the centipede and millipede below. The centipede has one pair of jointed legs per segment, whereas the millipede has two pairs of legs per segment. Centipedes hunt for their food and have a pair of poison claws used to inject venom into their prey. Centipedes feed on snails, slugs, and worms. Their bites are very painful to humans. Millipedes don't move as quickly as centipedes and feed on plants.

In your Journal, develop a plan for a new business involving one of the many animals in this chapter. Of course, your plan is to be an ethical and environmentally sound venture. Explain your business plan to your future investors—the class.

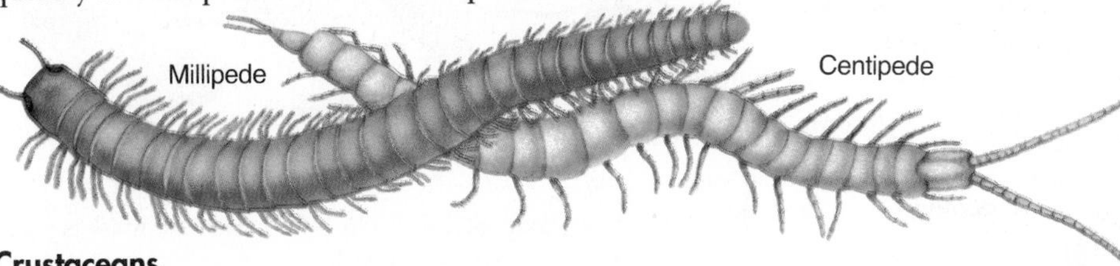

Crustaceans

Crabs, crayfish, lobsters, shrimp, pill bugs, and water fleas all belong to the Class Crustacea. Crustaceans are arthropods that have one or two antennae and jaws called mandibles used for crushing food. Most crustaceans live in water, but some, like pill bugs, live on land in moist environments. If you turn over a board or look under some leaves in your yard, you may find a pill bug.

The crayfish in Figure 14-12 is a typical crustacean. Although some crustaceans have three body regions, the crayfish has a head and thorax joined to form one region.

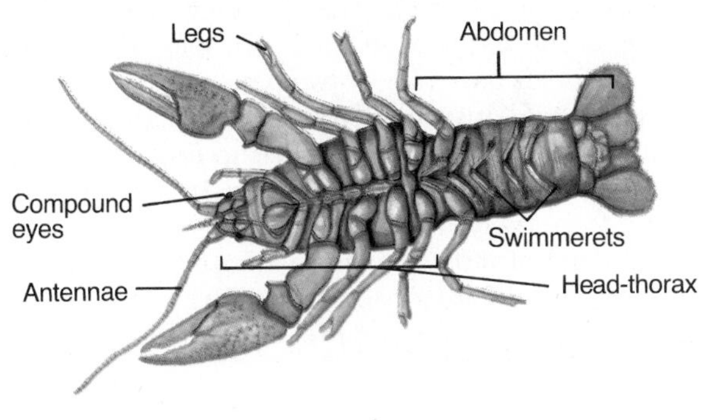

Figure 14-12. All crustaceans have body structures similar to those of a crayfish. Here you can see the features of the ventral surface of the crayfish.

CONCEPT DEVELOPMENT

▶ Show students the differences in centipedes and millipedes by using living or preserved specimens or photographs. Point out the number of legs per segment and the poisonous claws of the centipede. Explain that compared to centipedes, millipedes are less active and move more slowly.

▶ Point out that when disturbed, both millipedes and centipedes roll into a ball.

▶ Explain that even though all centipedes have poison glands, their bite is not lethal to humans. The poison of tropical species that reach a length of 30 cm is not dangerous. The bite can be painful but is no more lethal than a hornet sting.

▶ It may interest students that millipedes range in length from 3 mm to 28 cm. They are found throughout the world except in polar regions.

▶ **Demonstration:** Use the microprojector to show students a Daphnia. Explain that most crustaceans are small.

▶ Explain that if a crayfish loses an appendage in a fight or an accident, it is replaced in the next molt.

In Your JOURNAL

This is a good place to discuss the ethics of humans using other species for their own ends. Obviously, many ideas will be both ethical and profitable.

REVEALING MISCONCEPTIONS

▶ Using the word *shellfish* to refer to crustaceans is misleading. Crustaceans are not fish, nor do they have shells. They have an exoskeleton that is very different from the shells of clams and snails.

ENRICHMENT

▶ Have students report on Rocky Mountain spotted fever and Lyme disease.

▶ Have students prepare an illustrated report on the various hunting methods of spiders. Include the wolf spider and trap door spider.

▶ Have each student choose an arthropod and report on its life cycle, natural history, distribution, and way of getting food.

PROGRAM RESOURCES

Use **Laboratory Manual,** page 95, Collecting Insects.

Use **Laboratory Manual,** page 97, Grasshopper Anatomy.

Science and READING 

Students should find out that many species of bats are insectivorous and consume large quantities of insects each year. For example, Mexican free-tail bats feed on as much as 20 000 pounds of insects each year. The residents of Austin, Texas, are especially appreciative of these bats for controlling mosquito populations.

CONCEPT DEVELOPMENT

▶ Have students look under rocks and logs for isopods. Isopods are land crustaceans. There are two easily distinguishable types. When disturbed, the pill bug curls into a ball. A sow bug does not curl into a ball. Isopods can be maintained in the classroom in a jar of moist soil with a slice of potato in it.

▶ Explain that some insects, such as ants, bees, termites, and wasps, live in highly organized societies. Each member has a specific job necessary for the survival of the colony.

▶ Explain that beetles are by far the largest and most successful insect group. More than one-third of all insects are beetles.

▶ Use overhead transparencies to compare a spider and a typical insect such as a grasshopper.

▶ Invite a person from the agriculture extension service to talk about the beneficial and harmful effects of insects.

CROSS CURRICULUM

▶ **Home Economics:** Brainstorm with students to make a list of those crustaceans used for food.

MINI QUIZ

Use the Mini Quiz to check students' recall of chapter content.

1. **The replacing of the old exoskeleton by a new one is called _____ .** *molting*
2. **Spiders exchange oxygen and carbon dioxide through openings in the abdomen called _____ .** *spiracles*
3. **The series of changes that insects and other animals that hatch from eggs go through is _____ .** *metamorphosis*

Science and READING 

Insects are an important food source for many species of bats. In turn, the bats help control the populations of several types of insects, such as mosquitoes. Research the types of insects bats feed on. Find out the areas of the country where bats are important in controlling insect populations.

Crustaceans have five pairs of legs. The first pair of legs are claws that catch and hold food. The other four pairs are walking legs. The five pairs of appendages on the abdomen are swimmerets. They help the crustacean move and are used in reproduction. The swimmerets also force water over the feathery gills that crustaceans use for breathing. If a crustacean happens to lose an appendage, it can grow back, or regenerate, the lost part.

Barnacles are crustaceans very different in form from the others. They have a very hard, stony shell and spend their lives cemented to boats, piers, or animals such as whales.

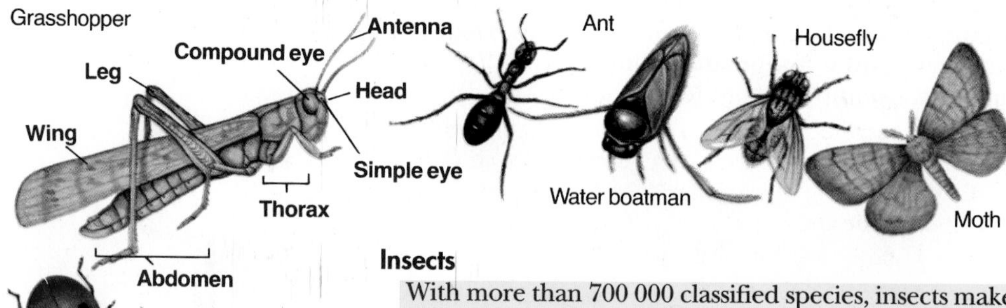

Figure 14-13. All insects have six legs, a head, a thorax, and an abdomen. They also have one or two pairs of wings, one pair of antennae, and a pair of compound eyes.

Insects

With more than 700 000 classified species, insects make up the largest group of complex invertebrates. Scientists describe thousands more each year. Actually insects are the most successful group of animals. They are the only invertebrates that are able to fly. Let's look at the features that distinguish insects from other arthropods and find out why they are so successful.

Insects have three body regions: a head, a thorax, and an abdomen. With some insects, it is almost impossible to see where one region stops and the next one begins. The head has a pair of antennae, eyes, and a mouth. The thorax has three pairs of jointed legs and, in many species, one or two pairs of wings. The abdomen is divided into 11 segments and has neither wings nor legs attached to it.

Most insects have one or two pairs of wings. Flies have one pair. Some insects such as silverfish and fleas don't have wings. Flying enables insects to find new places to live, new food sources, and mates. It also helps them escape from their enemies.

The head of most insects has simple and compound eyes and a pair of antennae. Simple eyes are used to detect light and dark. Compound eyes contain many lenses and

OPTIONS

ASSESSMENT—ORAL

▶ **How is the growth and development of a puppy different from growth and development of a butterfly?** *The puppy looks like the adult and does not go through a larval or pupal stage.*

▶ **How are incomplete and complete metamorphosis similar?** *Both are a series of changes.*

▶ **How are incomplete and complete metamorphosis different?** In complete metamorphosis, the young look very different from the adults.

▶ **Insects are often described as the most successful group of animals. Can you suggest reasons that could account for this?** *abundance, small size, eat a variety of foods, short life cycles*

▶ **What are examples of social behavior in insects that you have observed?** *ants in a line, crickets chirping, bees in a hive, wasps in a nest*

detect some colors and movement. The antennae are used for touch and smell.

Insects have open circulatory systems. The open circulatory system carries digested food to cells and removes wastes. Insect blood does not carry oxygen. As in arachnids, insects have spiracles on the abdomen and thorax through which air enters and waste gases are expelled. You can see that insects have systems for digestion, reproduction, and removing wastes.

Insects reproduce sexually. Females lay thousands of eggs but only a fraction of the eggs develop into adults. Think about how overproduction ensures that the species will continue.

Many species of insects and other animals that hatch from fertilized eggs go through a series of changes in body form to become adults. This series of changes is called **metamorphosis.** Metamorphosis is controlled by chemicals secreted by the body of the animal. There are two kinds of metamorphosis—complete and incomplete. 3

Most insects, including butterflies, beetles, ants, bees, moths, and flies, develop through complete metamorphosis. The four stages of development are egg, larva, pupa, and adult. A fertilized egg hatches into a wormlike larva stage. Maggots and caterpillars are larvae. The larva spends its time eating and growing before forming a shelter called a cocoon or chrysalis and going into a resting stage called the pupa. Inside, the larva changes and develops into an adult in a process that is little understood. The cocoon eventually opens and the adult comes out.

MINI-Lab

What type of metamorphosis do fruit flies undergo?

Obtain from your teacher a vial containing food and fruit flies in various stages of metamorphosis. *Identify* and draw all the stages of metamorphosis you see. Observe the vial every day for two weeks and record your observations. What type of metamorphosis do fruit flies undergo? About how long does each stage last? In what stages are the flies the most active?

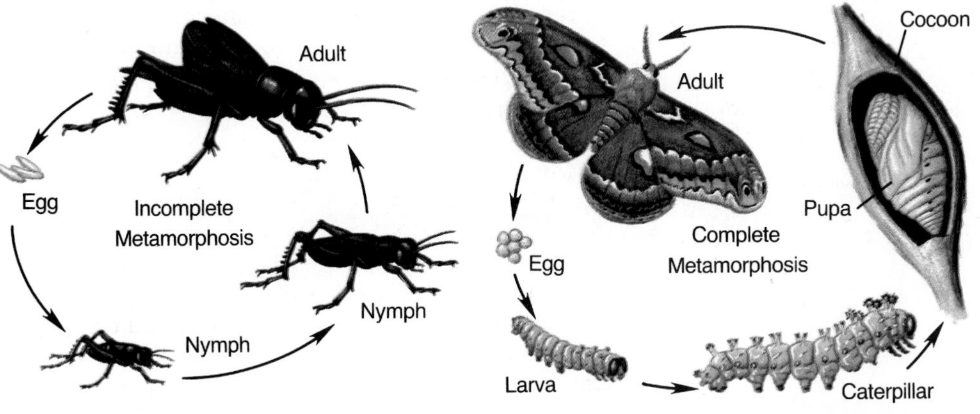

Figure 14-14. Butterflies and moths undergo complete metamorphosis. A moth forms a cocoon. A butterfly forms a chrysalis. Crickets undergo incomplete metamorphosis.

▶Caterpillars can be collected during the summer and early fall. They may be kept in a widemouth glass jar with a two-piece lid. A piece of screen wire can be substituted for the lid. Feed the caterpillars a supply of fresh leaves from the plant where they were collected. When they begin to pupate, place bottle caps of water in the jar.

CROSS CURRICULUM

▶**Art:** Provide construction paper, modeling clay, pipe cleaners, cardboard tubes, and other materials for students to construct a model of an arachnid, centipede, millipede, crustacean, or insect.

MINI-Lab

Materials: cultures of fruit flies

Troubleshooting: Be sure vials contain flies at all stages of metamorphosis. Students may have a difficult time seeing the egg stage. Be sure to point it out to them.

Answers: Fruit flies undergo complete metamorphosis. Depending on temperature, eggs hatch within 24 to 48 hours. Larva pupate in four days. Adults emerge from the pupa in about ten days. The flies are the most active in the larva and adult stages.

MINI-Lab

ASSESSMENT

Performance: Assess students' understanding of metamorphosis by having them perform USING LAB SKILLS, question 11, on page 346.

ENRICHMENT

▶Have interested students collect and identify insects. Field study guides provide information on the collection and preservation of insects.

▶Have students research and report on the uses of chitin extracted from crab, shrimp, and lobster shells. Reference: "How an Old Crab Could Keep You in Stitches." *Discover,* February, 1987, p. 16.

PROGRAM RESOURCES

From the **Teacher Resource Package** use:

Activity Worksheets, page 128, Mini-Lab: What type of metamorphosis do fruit flies undergo?

Transparency Master, pages 51-52, Metamorphosis: Complete and Incomplete.

Use **Color Transparency** 26, Metamorphosis: Complete and Incomplete.

Use **Laboratory Manual,** page 101, Metamorphosis.

Answer: sugar

3 CLOSE

▶Ask questions 1-3 and the **Apply** Question in the Section Review.

SECTION REVIEW ANSWERS

1. jointed legs, body segments, exoskeleton

2. Advantages: provides covering to prevent dehydration, supports body, protects body; Disadvantages: body outgrows exoskeleton, animal vulnerable while new exoskeleton hardens

3. Complete: egg, larva, pupa, adult. Egg hatches into larva, which spends its time eating and growing before pupating and going into resting stage; larva changes inside cocoon or chrysalis before emerging as an adult. Incomplete: egg, nymph, adult. Egg hatches into nymph, which looks like small adult with no wings; nymph molts several times to become adult.

4. Apply: Students may choose any insect to describe. For example, a grasshopper is green or brown to blend in with grasses.

5. Connect to Physics: They rub a scraper on one forewing against a vein on the other forewing to produce chirping sounds.

Skill Builder

Students should make a cycle concept map for each type of metamorphosis. Complete metamorphosis should show egg → larva → pupa → adult. The map of incomplete metamorphosis should show egg → nymph → adult. Each map should be made in a circular fashion to show the cycle continues.

Skillbuilder

ASSESSMENT

Portfolio: Use this Skillbuilder to assess students' abilities to form a Concept Map of the events in complete and incomplete insect metamorphosis.

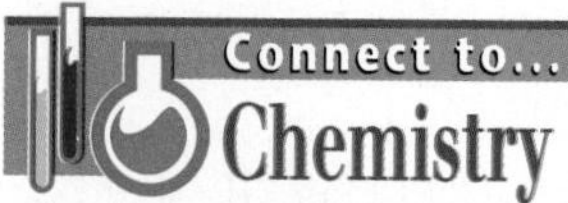

Honeybees make honey by sucking up nectar from flowers and storing it in their stomachs where other chemicals are added to it. When the stomach is filled, the honeybee returns to the hive and puts the nectar into an empty cell. In the honeycomb, the water in the nectar evaporates and the chemicals turn the nectar into honey. Find out what chemical makes up the major part of honey.

Grasshoppers, silverfish, lice, and crickets develop through incomplete metamorphosis. The three stages of development are egg, nymph, and adult. The fertilized egg hatches into a nymph that looks like a small adult, but without wings. A nymph molts several times before reaching the adult stage.

Adaptations such as an exoskeleton with wings and jointed appendages allow insects to live on land and to fly. Insects adapt to new environments rapidly because they have short life spans. Natural selection takes place more quickly in insect populations than in organisms that take longer to reproduce. Small size allows them to live in a wide range of environments, use a small amount of food, and hide from their enemies. Many species of insects can live in the same area and not compete with one another for food.

Many people think the world would be better off without insects. They cause problems for people by eating crops, clothing, and wood in buildings. Some insects spread diseases. But it would be a serious mistake to think that we could live without them, for they are also helpful. They pollinate flowering plants that produce fruits and vegetables. They feed on decaying wood, and produce honey and other useful products.

SECTION REVIEW

1. What are three features of all arthropods?
2. What are the advantages and disadvantages of an exoskeleton?
3. Describe the stages in complete metamorphosis and the stages in incomplete metamorphosis.
4. **Apply:** Choose an insect you are familiar with and explain how it's adapted to its environment.
5. **Connect to Physics:** Find out how male crickets use sound to attract females.

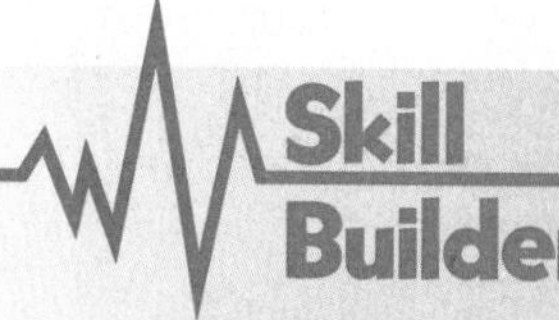

Concept Mapping

Make a concept map showing the events in the complete metamorphosis of an insect. Make another concept map showing the events in incomplete metamorphosis. If you need help, refer to Concept Mapping in the **Skill Handbook** on pages 688 and 689.

OPTIONS

ENRICHMENT

▶Set up numbered specimens of the five arthropod classes around the room. Provide students with the following key and have them use the key to determine the class of each organism.

1. a. Walking legs, more than 5 pairs
   go to 2
   b. Walking legs, 5 or fewer pairs go to 3
2. a. 1 pair of legs on each segment
   Class Chilopoda
   b. 2 pairs of legs on each segment
   Class Diplopoda
3. a. Antennae present
   go to 4
   b. Antennae absent
   Class Arachnida
4. a. 1 pair of antennae
   Class Insecta
   b. More than 1 pair of antennae
   Class Crustacea

ACTIVITY
14-2

Observing a Crayfish

Problem: *How does a crayfish use its appendages?*

Materials  

- crayfish in a small aquarium
- stirrer
- uncooked ground beef

Procedure

1. Your teacher will provide you with a crayfish in an aquarium. **CAUTION:** *Use care when working with live animals. Leave the crayfish in the aquarium while you do the activity.*
2. Touch the crayfish with the stirrer. How does the body feel?
3. *Observe* the crayfish moving in the water. How many body regions does it have? Name the regions. Use the diagram on page 335 to help you. How many appendages does it have on each body region? What is the function of the appendages? Record your answers in a data table like the one shown.
4. Observe the compound eyes. On which body region are they located?
5. Drop a small piece of ground beef into the aquarium and observe how the crayfish eats.
6. Return the crayfish and aquarium to the place your teacher has assigned. Wash your hands.

Analyze

1. Describe the texture of the body of the crayfish.
2. How many body regions does it have?
3. How many appendages are located on each body region?
4. What structures does the crayfish use to get food?

Conclude and Apply

5. *Infer* how the location of the eyes is an advantage for the crayfish.
6. How does the structure of the claws aid in getting food?
7. What can you *infer* about the exoskeleton and protection?

Data and Observations

Sample Data

Body Region	Appendages	Function
Head	5 pairs including antennae and mouth parts	The appendages on this region are used to sense the environment and hold and chew food.
Thorax	5 pairs including claws (chelipeds) and walking legs	These appendages are used to capture food and for defense. Walking legs allow the crayfish to walk.
Abdomen	5 pairs including swimmerets and the tail	These appendages create water currents to help crayfish move. They also function in reproduction.

ACTIVITY 14-2

35 minutes

OBJECTIVE: Observe the structure of a crayfish and **determine** how it uses its appendages.

PROCESS SKILLS applied in this activity:
▶**Observing** in Procedure Steps 2-5.
▶**Interpreting Data** in Analyze Questions 1-4.
▶**Inferring** in Conclude and Apply Questions 5-7.

COOPERATIVE LEARNING Use the Paired Partner strategy with two students working together to do the activity.

TEACHING THE ACTIVITY

Alternate Materials: Preserved crayfish may be used.

Troubleshooting: Students may not be able to count the appendages in the aquarium. Use a transparency of a crayfish on the overhead projector to help them.

▶Crayfish may be obtained from a local stream or a biological supply company. Return them to a local stream when the activity is completed.

PROGRAM RESOURCES

From the **Teacher Resource Package** use:

Activity Worksheets, pages 124-125, Activity 14-2, Observing a Crayfish.

Activity ASSESSMENT

Performance: To further assess students' understanding of crayfish, have them put a plastic bag containing ice cubes at one end of a long aquarium and a bag of warm water at the other end to test crayfish response to different temperatures. Then have students infer how this response is reflected in nature.

ANSWERS TO QUESTIONS

1. It is very hard.
2. two
3. The first body region has 2 pairs of antennae, 1 pair of claws, and 4 pairs of walking legs. The second body region has 5 pairs of swimmerets.
4. the claws
5. The eyes at the end of stalks allow the crayfish to look in different directions at the same time.
6. The serrated edges help hold prey.
7. The hard structure protects the body organs.

SECTION 14-4

PREPARATION

SECTION BACKGROUND

▶ The process in which levels of chemicals increase in animals higher on the food chain is called biological magnification.

1 MOTIVATE

FLEX Your Brain

Use the Flex Your Brain activity to have students explore PESTICIDES.

Flex Your Brain
ASSESSMENT
Portfolio: Use the Flex Your Brain activity to reinforce critical thinking and problem solving skills.

2 TEACH

Key Concepts are highlighted.

CONCEPT DEVELOPMENT

▶ Ask students if they would be willing to eat less-than-"perfect" fruits and vegetables that have not been treated with pesticides.

VideoDisc

STVS: Insecticides from Desert Plants, Disc 4, Side 2
STVS: Soldier Bugs, Disc 6, Side 1

SCIENCE & SOCIETY 14-4 Pesticides

New Science Words

pesticides

Objectives

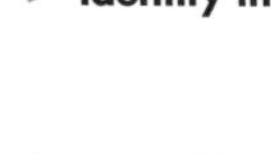

▶ Describe the importance of pesticides in agriculture.
▶ Identify the impact of pesticides on the environment.

Insect Pests?

What's your first reaction when you see insects? Are you inclined to avoid them? Many people are! We typically view insects as pests that are better dead than walking around in our homes. Most insects are harmless. In fact, many, such as the praying mantis, are beneficial. These insects eat harmful insects. On the other hand, farmers and gardeners often find it necessary to destroy insects that damage crops and ornamental plants. Many farmers would not have successful crop yields without the use of pesticides. **Pesticides** are chemicals that kill undesirable plants and insects. It is estimated that for every $3 million spent on pesticides, about $12 million is returned in additional crops. In addition, the price of pesticides has increased much more slowly than the price of other farm expenses. This makes pesticides a very smart investment for farmers.

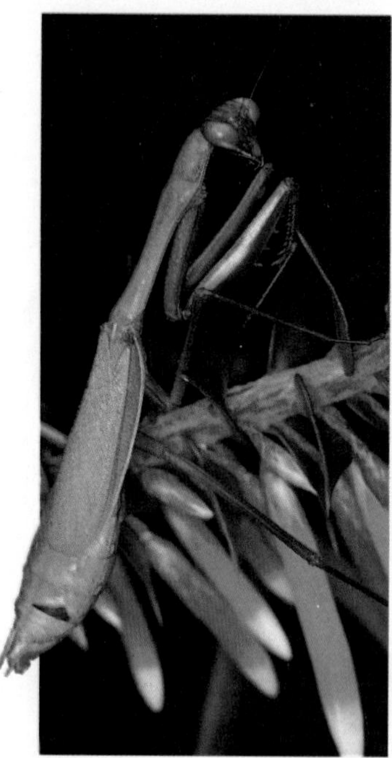

Figure 14-15. The praying mantis is a beautiful and helpful insect.

Although pesticides are very beneficial to farmers, the environmental damage caused as a result of pesticides is thought to be very great. Studies show that very little of the pesticide applied to crops actually reaches the target organism. Instead, about 99 percent of the pesticide winds up in the soil, water, and air. Here, it is not only the harmful insects that are killed, but most of the nation's 200 000 species of plants and animals are also affected. For example, animals that feed on the plant sap or tissue may come in contact with the pesticides. Some larger animals, including birds and mammals, may either come in direct contact with the pesticide or feed on plants and animals that have been contaminated. One study showed that the pesticide DDT accumulates in the environment and can contaminate organisms. DDT is a colorless, odorless pesticide that is toxic to many animals. DDT is no longer in use in the United States. The study showed that in soil containing

OPTIONS

Meeting Different Ability Levels

For Section 14-4, use the following **Teacher Resource Masters** depending upon individual students' needs.

◆ **Study Guide Master** for all students.
● **Reinforcement Master** for students of average and above average ability levels.
▲ **Enrichment Master** for above average students.

Additional Teacher Resource Package masters are listed in the OPTIONS box throughout the section. The additional masters are appropriate for all students.

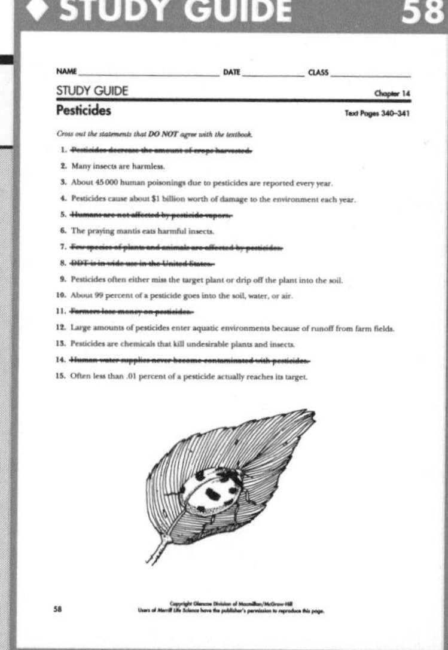

◆ STUDY GUIDE 58

NAME ______ DATE ______ CLASS ______

STUDY GUIDE Chapter 14

Pesticides Text Pages 340–341

*Cross out the statements that **DO NOT** agree with the textbook.*

1. ~~Pesticides decrease the amount of crops harvested.~~
2. Many insects are harmless.
3. About 45 000 human poisonings due to pesticides are reported every year.
4. Pesticides cause about $1 billion worth of damage to the environment each year.
5. ~~Humans are not affected by pesticide vapors.~~
6. The praying mantis eats harmful insects.
7. ~~Few species of plants and animals are affected by pesticides.~~
8. ~~DDT is in wide use in the United States.~~
9. Pesticides often either miss the target plant or drip off the plant into the soil.
10. About 99 percent of a pesticide goes into the soil, water, or air.
11. ~~Farmers lose money on pesticides.~~
12. Large amounts of pesticides enter aquatic environments because of runoff from farm fields.
13. Pesticides are chemicals that kill undesirable plants and insects.
14. ~~Human water supplies never become contaminated with pesticides.~~
15. Often less than .01 percent of a pesticide actually reaches its target.

58

10 parts per million of DDT, earthworms contained 141 parts per million of DDT, and robins that fed on those worms contained 444 parts per million in their brain tissues. It is estimated that there are about 45 000 human poisonings due to pesticides every year.

Pesticides that are sprayed over great distances, especially with airplanes, cause much pollution in the air. These vapors are easily inhaled by humans. The soil is the primary place for toxins to accumulate because pesticides often miss the target plant or drip off the plant into the soil. Although very little pesticide is applied directly to the water system, large amounts of pesticides ultimately reach rivers, lakes, and streams, due to the runoff of pesticides in drainage from agricultural fields. Human water supplies are very likely to be contaminated from such runoff.

EcoTip

Flea collars, sprays, dips, and powders often put poisons in the environment. Instead of using these items, comb your pet often with a flea comb and bathe it regularly with a nontoxic shampoo.

Figure 14-16. Planes are used to spray crops with pesticides.

SECTION REVIEW

1. Why are pesticides so important to our economy?
2. What are the current problems with pesticide application methods?

You Decide!

Our society has gained much from pesticides, but are we paying for them with our health? Can some alternative solution be reached? Improved application methods may prevent some of the problem. Use of chemicals that break down easily with the help of sunlight, water, and biological action may also be an option. Even if these solutions are able to cut the current problem in half, should we still continue to use pesticides? Can we live without them?

● REINFORCEMENT 58

NAME ______ DATE ______ CLASS ______

REINFORCEMENT — Chapter 14

Pesticides — Text Pages 340–341

Complete the following sentences using the appropriate terms from the textbook.

1. Farmers use **pesticides** to eliminate undesirable plants and animals.
2. Pesticides sprayed from airplanes cause air **pollution**.
3. A dangerous pesticide no longer in use in the United States is called **DDT**.
4. Pesticides reach water supplies because of **runoff** from agricultural fields.

Answer the following questions.

5. To a farmer, why is the use of pesticides a good business decision? **Compared to other farm expenses, pesticides are cheap. And the return in crop production is much more than the cost of the pesticides.**
6. What happens to most of the pesticide applied to a field? **Most of it goes into the soil, water, and air.**
7. Pesticides are often aimed at insects but end up affecting larger animals. Explain. **Larger animals may feed on the crops that have been sprayed or may feed on smaller animals that have become contaminated with the pesticide.**
8. A study showed earthworms contained 141 parts per million of DDT while robins contained 444 parts per million. How can you account for this difference? **The robins ate the worms contaminated with DDT. The DDT in the worms became concentrated in each robin.**
9. How can humans be affected by pesticide use? **Spraying pesticides causes air pollution. Runoff from fields causes water pollution. Both forms of pollution can affect humans.**
10. Food production might suffer without pesticides. Yet pesticides cause pollution and harm many species. How might we cut back on their use? **Answers will vary. Students may suggest using improved application methods or using chemicals that break down easily in the environment.**

58

▲ ENRICHMENT 58

NAME ______ DATE ______ CLASS ______

ENRICHMENT — Chapter 14

Pesticides — Text Pages 340–341

Most people keep some kind of pesticide around the home. These may include ant and roach sprays, flea sprays and dips, weed killers, and mosquito sprays and repellents. These chemicals are often kept under the kitchen sink, out in the garage or tool shed, or down in the basement. Find out what pesticides are in and around your home by looking around and by asking the adults in your home. Read the warning label or cautionary statement on each box or can. Consider how they are used. Use the table below to record what you find.

Observations

Product	Use	Caution or warning on product
1.	**Answers will vary.**	
2.		
3.		
4.		

Conclude and Apply

1. Where were most of the pesticide products in your home kept? **Answers will vary.**
2. How dangerous do you think these products are? **Most students will say that the products are dangerous.**
3. Do you and others in your home sometimes use these products carelessly? **Most students will answer yes.**
4. Ninety-nine percent of agricultural pesticides do not reach the intended target organism. How effective are the products in your home in reaching their targets? Give an example. **Most students will say that most of the spray from these products does not hit the target. Examples may include flea spray or ant and roach spray.**
5. Could any of these products be causing pollution in your home? How could this be prevented? **Most students will say yes. Students may suggest discontinuing use of the product or using natural alternatives.**

58

CHECK FOR UNDERSTANDING

Ask questions 1-2 in the Section Review.

RETEACH

Have students make a cause and effect chart describing the causes of pesticide use and the effects of its use. These should be both negative and positive.

EXTENSION

For students who have mastered this section, use the **Reinforcement** and **Enrichment** masters or other OPTIONS provided.

3 CLOSE

▶ Show students an example of a food chain and have them explain how a pesticide affects organisms higher in the chain.

SECTION REVIEW ANSWERS

1. They limit crop damage at very low price to farmers and increase crop yield.

2. Current application methods allow pesticides to enter the soil, air, and water causing environmental damage.

YOU DECIDE!

Have students explore alternate methods to control pests including genetic controls such as releasing sterile males, environmental controls such as altering planting time and crop diversity, and chemical controls such as pheromones and insect hormones. Students' answers will vary on the use of pesticides. Acceptable answers should be reasonable and have logical rationale.

PROGRAM RESOURCES

From the **Teacher Resource Package** use:

Activity Worksheets, page 5, Flex Your Brain.

Science and Society, page 18, Better Ways to Battle Bugs.

SECTION 14-5

PREPARATION

SECTION BACKGROUND

▶ Phylum Echinodermata is made up of eleven classes, six of which are extinct and known only from fossils.

▶ The fossil record indicates that echinoderms date back to the Cambrian period over 500 million years ago. The fossil record also indicates that early adult forms were probably sessile and not free-living as are most modern forms.

▶ Echinoderms show an important evolutionary relationship to invertebrates in that they may share a common ancestor with the lower chordates.

1 MOTIVATE

▶ Display preserved specimens of sea stars, sand dollars, sea cucumbers, and sea urchins. Point out the characteristics—spiny skin, endoskeleton, tube feet, and radial symmetry. Discuss why their characteristics enable them to live in a water habitat.

▶ Use a bicycle pump to show how pressure changes can create suction. Compare this to the thousands of tiny pumps in a sea star's water vascular system.

Answer: A large number of echinoderm fossils exist because their hard outer calcium plates fossilize easily.

TYING TO PREVIOUS KNOWLEDGE: Even if students have not seen living echinoderms, they have probably seen dried sea stars and sand dollar skeletons. Use their knowledge of echinoderms to introduce this lesson.

VideoDisc

STVS: Sea Urchins and Power Plants, Disc 5, Side 1

14-5 Echinoderms

New Science Words

echinoderms
water-vascular system
tube feet

Objectives

▶ Identify the features of echinoderms.
▶ Describe how sea stars get and digest food.

Connect to... Earth Science

Scientists have found many echinoderm fossils from the Cambrian period but few fossils of other species that lived during the Cambrian period. *Infer* what adaptation might explain the large number of fossilized echinoderms.

Features of Echinoderms

Unless you live near the ocean, you may not have seen an echinoderm, but most of you know what a sea star is. Sea stars, sea urchins, sand dollars, and sea cucumbers are all echinoderms (ih KI nuh durmz). **Echinoderms** are spiny-skinned invertebrates that live on the ocean bottom. The name *echinoderm* means "spiny-skin." The spiny part refers to the spines that cover the outside of these animals. Their bodies are supported and protected by an internal skeleton made of calcium plates. The plates are covered by the thin, spiny skin.

A unique characteristic shared by all echinoderms is a water-vascular system. The **water-vascular system** is a network of water-filled canals. Thousands of tube feet are connected to this system. **Tube feet** act like suction cups and help the animal move and feed. Look at the tube feet shown in Figure 14-18. Another very obvious trait of echinoderms is their symmetry. The bodies of many echinoderms can be divided into five sections all located around a central point. Let's look more closely at one of the most well known echinoderms, a sea star.

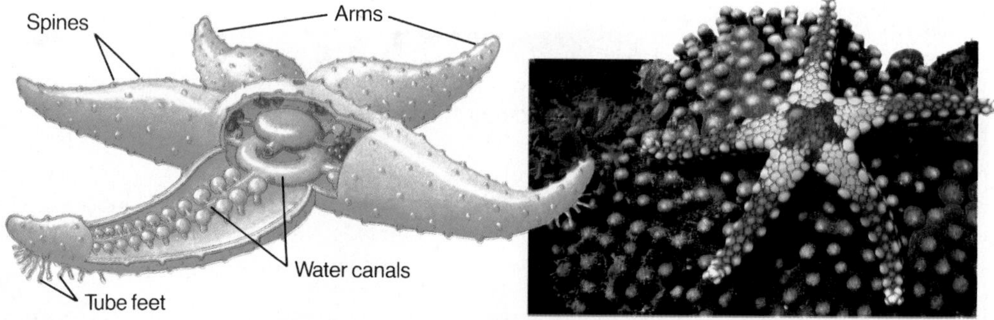

Figure 14-17. Sea stars and all other echinoderms have a water-vascular system that allows them to move, eat, get oxygen, and get rid of waste.

OPTIONS

Meeting Different Ability Levels

For Section 14-5, use the following **Teacher Resource Masters** depending upon individual students' needs.

◆ **Study Guide Master** for all students.

● **Reinforcement Master** for students of average and above average ability levels.

▲ **Enrichment Master** for above average students.

Additional Teacher Resource Package masters are listed in the OPTIONS box throughout the section. The additional masters are appropriate for all students.

◆ STUDY GUIDE 59

NAME ______ DATE ______ CLASS ______

STUDY GUIDE — Chapter 14

Echinoderms — Text Pages 342–344

Check (✓) the statements that agree with the textbook.

____ 1. The echinoderms are the most primitive form of invertebrates.
✓ 2. Sea stars feed on mollusks by using the suction power of their tube feet.
✓ 3. Echinoderms live on the ocean bottom.
✓ 4. A characteristic of echinoderms is radial symmetry.
____ 5. The name *echinoderm* means "scaly-skin."
____ 6. Oyster farmers love to see echinoderms in their oyster beds.
✓ 7. The water vascular system is a network of water-filled canals.
✓ 8. Spines cover the outside of echinoderms.
____ 9. The bodies of echinoderms are divided into seven sections located around a central point.
✓ 10. The tube feet of echinoderms are connected to the water vascular system.
____ 11. Sand dollars are not echinoderms but a form of mollusk.
____ 12. A sea star uses its spines to attack clams.
✓ 13. A sea star can regrow lost arms.
✓ 14. To eat a clam, a sea star turns its stomach inside out.

Match the echinoderm with the illustrations by writing the correct letter next to the organism's name.

d 15. sea star
a 16. sea urchin
c 17. sand dollar
b 18. brittle star
e 19. sea cucumber

a b c d e

Sea Stars

If you have ever pried open the shell of an oyster or a clam, you know it's not an easy job. Sea stars feed on these mollusks. Sea stars have five or more arms arranged around a central point. These arms are lined with thousands of tube feet. The sea star wraps its arms around a mollusk and uses the suction power of its tube feet to open the shell.

Figure 14-19 on page 344 shows a sea star using its tube feet to open a clam. Once the shell is open slightly, the sea star turns its stomach inside out, pushes it through its mouth, and surrounds the soft body of the mollusk. The stomach secretes enzymes that digest the animal. Then when the meal is over, the sea star takes the stomach out of the clam shell and pulls it back inside its own body.

1 Sea stars reproduce sexually by releasing eggs and sperm into the water. Females can produce 200 000 eggs in one season. They can also repair themselves by regeneration. If a sea star loses an arm, it can grow a new one. If enough of the arm is lost, the arm itself can grow into a whole new sea star.

Figure 14-18. All echinoderms have tube feet. This photograph shows the tube feet on the underside of a sea urchin.

Table 14-1

OTHER ECHINODERMS

Brittle Stars

Brittle stars are very secretive echinoderms. They live hidden under rocks or litter on the ocean floor. They move much more quickly than sea stars and break off their arms as a defense when disturbed. The broken parts are quickly regrown.

Sea Urchins and Sand Dollars

Sea urchins and sand dollars make up another class of echinoderms. Both have a skeleton made of calcium carbonate plates, and both are covered with spines. Sand dollars are very flat and covered with small, fine spines, whereas sea urchins are round and covered with longer spines.

Sea Cucumbers

Sea cucumbers are soft-bodied echinoderms with a leathery covering. They have tentacles around the mouth and rows of tube feet on both the dorsal and ventral sides.

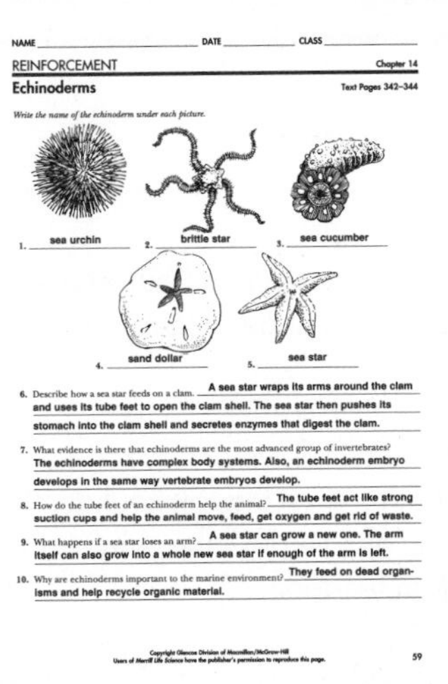

● REINFORCEMENT 59

NAME ______ DATE ______ CLASS ______

REINFORCEMENT — Chapter 14

Echinoderms — Text Pages 342–344

Write the name of the echinoderm under each picture.

1. sea urchin
2. brittle star
3. sea cucumber
4. sand dollar
5. sea star
6. Describe how a sea star feeds on a clam. **A sea star wraps its arms around the clam and uses its tube feet to open the clam shell. The sea star then pushes its stomach into the clam shell and secretes enzymes that digest the clam.**
7. What evidence is there that echinoderms are the most advanced group of invertebrates? **The echinoderms have complex body systems. Also, an echinoderm embryo develops in the same way vertebrate embryos develop.**
8. How do the tube feet of an echinoderm help the animal? **The tube feet act like strong suction cups and help the animal move, feed, get oxygen and get rid of waste.**
9. What happens if a sea star loses an arm? **A sea star can grow a new one. The arm itself can also grow into a whole new sea star if enough of the arm is left.**
10. Why are echinoderms important to the marine environment? **They feed on dead organisms and help recycle organic material.**

59

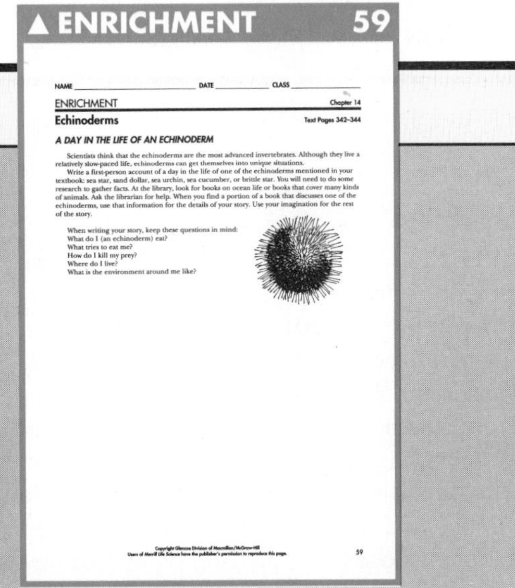

▲ ENRICHMENT 59

NAME ______ DATE ______ CLASS ______

ENRICHMENT — Chapter 14

Echinoderms — Text Pages 342–344

A DAY IN THE LIFE OF AN ECHINODERM

Scientists think that the echinoderms are the most advanced invertebrates. Although they live a relatively slow-paced life, echinoderms can get themselves into unique situations.

Write a first-person account of a day in the life of one of the echinoderms mentioned in your textbook: sea star, sand dollar, sea urchin, sea cucumber, or brittle star. You will need to do some research to gather facts. At the library, look for books on ocean life or books that cover many kinds of animals. Ask the librarian for help. When you find a portion of a book that discusses one of the echinoderms, use that information for the details of your story. Use your imagination for the rest of the story.

When writing your story, keep these questions in mind:
What do I (an echinoderm) eat?
What tries to eat me?
How do I kill my prey?
Where do I live?
What is the environment around me like?

59

2 TEACH

Key Concepts are highlighted.

CONCEPT DEVELOPMENT

▶ Ask: **What are the characteristics of echinoderms?** *radial symmetry, spiny skin, endoskeleton, marine, no head region, and no segmentation* **Where is the sea star's mouth?** *in the middle on the ventral side* **What are the functions of a sea star's tube feet?** *movement and feeding*

MINI QUIZ

Use the Mini Quiz to check students' recall of chapter content.

1. **How do sea stars reproduce?** *by eggs and sperm and regeneration*
2. **How are echinoderms important to the marine environment?** *They feed on dead organisms and help recycle materials.*
3. **Why do scientists think echinoderms are the most advanced group of invertebrates?** *They have complex body systems and the embryos develop in the same way that the embryos of chordates do.*

PROGRAM RESOURCES

From the **Teacher Resource Package** use:

Activity Worksheets, page 129, Mini-Lab: How do tube feet open clam shells?

Concept Mapping, pages 33-34.

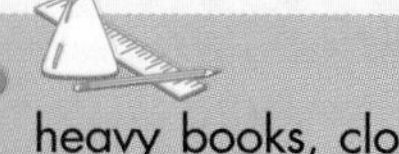

MINI-Lab

Materials: heavy books, clock with a second hand

Teaching Tips

▶ Make sure the book students use is heavy but not impossibly heavy.

Answer: Students should respond that the book is putting pressure on their arm muscles causing them to tire and drop the book. Sea stars put pressure on the muscles of a clam by using the suction power of their tube feet. Eventually the muscles of the clam tire, and the shell opens.

MINI-Lab
ASSESSMENT

Performance: To further assess students' understanding of sea stars, see USING LAB SKILLS, question12, p. 346.

3 CLOSE

CHECK FOR UNDERSTANDING

▶ Ask questions 1-3 and the **Apply** Question in the Section Review.

RETEACH

Have students explain why echinoderms have few predators.

EXTENSION

For students who have mastered this section, use the **Reinforcement** and **Enrichment** masters or other OPTIONS.

SECTION REVIEW ANSWERS

1. Spiny skin refers to the spines that cover the outside of the body.

2. spiny skin, endoskeleton made of calcium plates, water vascular system, tube feet, radial symmetry

3. Tube feet act like strong suction cups and help them move and get food.

4. Apply: Sea stars graze on organisms that live among the coral.

5. Connect to Physics: The water vascular system must have water in it for the tube feet to have the suction that enables them to move.

MINI-Lab

How do tube feet open clam shells?

Carry out this modeling activity to show how sea stars can open clam shells that are held closed by muscles. Hold your arm straight out, palm up. Place a heavy book on your hand. Have your partner time how long you can hold your arm up with the book on it. *Describe* how your arm feels after awhile. If the book represents the sea star and your arm represents the clam, *infer* how the sea star successfully overcomes the clam to obtain food.

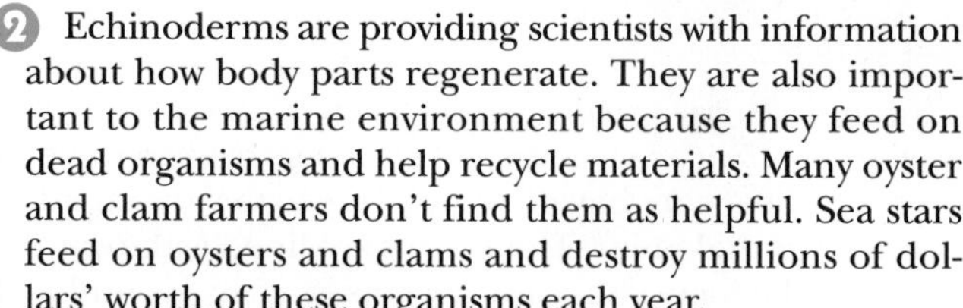

2 Echinoderms are providing scientists with information about how body parts regenerate. They are also important to the marine environment because they feed on dead organisms and help recycle materials. Many oyster and clam farmers don't find them as helpful. Sea stars feed on oysters and clams and destroy millions of dollars' worth of these organisms each year.

3 Scientists think that echinoderms are the most advanced group of invertebrates. They have radial symmetry like some of the less complex invertebrates, but they also have complex body systems. Another reason scientists view them as advanced is the way an echinoderm embryo develops. It develops the same way the embryos of chordates do, and not like the embryos of less complex invertebrates. Humans belong to the chordate phylum which you will study in the next chapter.

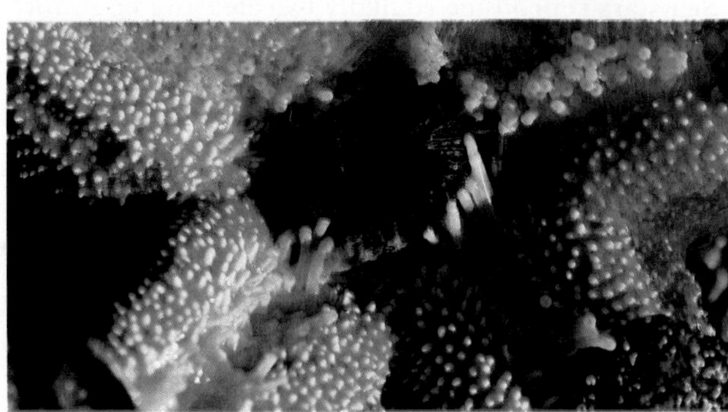

Figure 14-19. A sea star using its tube feet to open an oyster.

SECTION REVIEW

1. Why are echinoderms called spiny-skinned animals?
2. What features do all echinoderms have in common?
3. How do echinoderms move and get their food?
4. **Apply:** Why are sea stars common in coral reefs?
5. **Connect to Physics:** Sea stars that wash onto the beach cannot crawl back into the water. Why?

Skill Builder

Observing and Inferring

Observe the echinoderms pictured on page 343 and infer why they are slow moving and live on the ocean floor. If you need help, refer to Observing and Inferring in the **Skill Handbook** on page 682.

Skill Builder

With the exception of the brittle star, the echinoderms pictured have no means of movement except for their tube feet. Therefore they are not capable of much rapid movement. Because they are not able to run from predators, each has its own unique way of protecting itself for life on the ocean floor. The sea biscuit is camouflaged, the sea urchin has long pointed spines, the sea cucumber looks unappetizing as well as having other defenses. The brittle star is capable of rapid movement, but spends most of its time hidden under debris.

Skillbuilder
ASSESSMENT

Portfolio: Use this Skillbuilder to assess students' abilities to Infer why echinoderms are bottom feeders.

CHAPTER
REVIEW

SUMMARY

14-1: Mollusks

1. Mollusks are soft-bodied invertebrates usually covered by a hard shell. Mollusks move by means of a muscular foot. A mollusk's body organs are located together in the visceral mass.
2. Mollusks with one shell, such as snails, are univalves. Bivalves are mollusks with two shells hinged together. Clams, and oysters are bivalves. Squid are cephalopods. A cephalopod has a foot divided into tentacles, and it has an internal shell.

14-2: Segmented Worms

1. Annelids are animals with segmented bodies. Annelids have a body cavity to hold internal organs. They have bristle-like structures called setae to help them move.
2. Earthworms have a digestive system made of a mouth, crop, gizzard, intestine, and anus. Earthworms burrow through soil, eating it as they go.
3. Both mollusks and annelid worms have a body cavity to hold body organs. They both develop from the same type of larva.

14-3: Arthropods

1. Arthropods are classified by number of body segments and appendages.
2. Arthropods are covered by an exoskeleton. The exoskeleton covers, protects, and supports the body.
3. Metamorphosis is a series of developmental stages. Complete metamorphosis has four stages: egg, larva, pupa, and adult. Incomplete metamorphosis has three stages: egg, nymph, and adult.

14-4: Science and Society: Pesticides

1. Pesticides control populations of insects that damage crops and ornamental plants. They allow farmers to produce greater amounts of crops.
2. Pesticides can build up in, and poison, animals that come in contact with them. Pesticides that are sprayed can be inhaled by humans. They can also accumulate in the soil and run off into water supplies and contaminate them.

14-5: Echinoderms

1. Echinoderms are spiny-skinned invertebrates. They all have a water-vascular system to help the animal move and feed.
2. Sea stars use their tube feet to open the shells of mollusks. They push their stomach into the mollusk and use enzymes to digest the food.

KEY SCIENCE WORDS

a. **appendages**
b. **Arthropoda**
c. **closed circulatory system**
d. **crop**
e. **echinoderms**
f. **exoskeleton**
g. **gills**
h. **gizzard**
i. **mantle**
j. **metamorphosis**
k. **mollusks**
l. **molting**
m. **open circulatory system**
n. **pesticides**
o. **radula**
p. **setae**
q. **spiracles**
r. **tube feet**
s. **water-vascular system**

UNDERSTANDING VOCABULARY

Match each phrase with the correct term from the list of Key Science Words.

1. temporary food storage site
2. tongue-like organ in univalves
3. bristles earthworms use to hold on to soil
4. structure in earthworms that grinds soil
5. arms, legs, and antennae
6. outer covering made of chitin
7. shedding of the exoskeleton
8. spiny-skinned invertebrates
9. openings for air in some arthropods
10. structure that secretes the shell of mollusks

OPTIONS

ASSESSMENT

To assess student understanding of material in this chapter, use the resources listed.

COOPERATIVE LEARNING

Consider using cooperative learning in the THINK AND WRITE CRITICALLY, APPLY, and MORE SKILL BUILDERS sections of the Chapter Review.

PROGRAM RESOURCES

From the **Teacher Resource Package** use:

Chapter Review, pages 31-32.

Chapter and Unit Tests, pages 93-96, Chapter Test.

CHAPTER
REVIEW

SUMMARY

Have students read the summary statements to review the major concepts of the chapter.

UNDERSTANDING VOCABULARY

1. d
2. o
3. p
4. h
5. a
6. f
7. l
8. e
9. q
10. i

ASSESSMENT

Portfolio

Encourage students to place in their portfolios one or two items of what they consider to be their best work. For each item, ask students to explain why that item was chosen and what they learned from it. Items might be selected from the following.

- For Your Mainstreamed Students models of invertebrates, p. 322
- Skillbuilder table comparing and contrasting mollusks, p. 326
- You Decide! issue controversy on the use of pesticides, p. 341

Performance

Additional performance assessments may be found in *Performance Assessment* and *Science Integration Activities* that accompany **Merrill Life Science.** Performance Task Assessment Lists and rubrics for evaluating these activities and other products generated throughout the chapter can be found in Glencoe's *Performance Assessment in Middle School Science.*

CHAPTER REVIEW

CHECKING CONCEPTS

1. d
2. d
3. a
4. a
5. c
6. d
7. c
8. d
9. a
10. b

USING LAB SKILLS

ASSESSMENT

Use these alternate lab exercises to assess students' understanding of skills used in this chapter.

11. length of time it takes for each stage of development

12. Yes. They act like suction cups.

THINK AND WRITE CRITICALLY

13. Univalues use their radula to tear plant materials and scrape algae; cephalopods use tentacles with strong suction cups to capture organisms and pry open shells; two-shelled mollusks are filter feeders that get food by the action of sweeping cilia.

14. Clams burrow into the sand; oysters attach themselves by stong sticky threads to rocks; scallops eject water and move by rapidly opening and closing their shells; a squid ejects water and escapes by jet propulsion.

15. Grit in the earthworm gizzard grinds food much the way teeth break up and grind food in the mouth.

16. A space in the body to hold organs is present in both groups. The larva of both groups is similar.

CHAPTER REVIEW

CHECKING CONCEPTS

Choose the word or phrase that completes the sentence.

1. The ______ hold(s) the organs of mollusks.
a. gills **c.** mantle
b. foot **d.** visceral mass

2. An example of an annelid is a(n) ______.
a. snail **c.** octopus
b. slug **d.** earthworm

3. The organism with a closed circulatory system is a(n) ______.
a. earthworm **c.** oyster
b. snail **d.** slug

4. The largest phylum of animals are the ______.
a. arthropods **c.** annelids
b. mollusks **d.** echinoderms

5. Organisms with two body regions are ______.
a. insects **c.** arachnids
b. mollusks **d.** annelids

6. An example of an arthropod is a(n) ______.
a. snail **c.** slug
b. earthworm **d.** scorpion

7. ______ are organisms with radial symmetry.
a. Annelids **c.** Echinoderms
b. Mollusks **d.** Arthropods

8. The structures echinoderms use to move and open shells are ______.
a. calcium plates **c.** spines
b. arms **d.** tube feet

9. ______ body parts can be replaced by regeneration.
a. Sea star **c.** Octopus
b. Snail **d.** Slug

10. The most DDT would be found in ______.
a. soil **c.** a worm
b. a bird **d.** water

USING LAB SKILLS

11. In the Mini-Lab on page 337, you observed metamorphosis of fruit flies. Obtain a larva of a moth or butterfly, an insect cage, and leaves for the larva to eat. Place the larva in the insect cage and observe it until it becomes a moth or butterfly. How is metamorphosis of a moth or butterfly different from fruit fly metamorphosis?

12. In the Mini-Lab on page 344, you modeled how some echinoderms open a clam. Wet a suction cup and press it against a variety of surfaces. Can you pick up objects by holding the suction cup? Compare this to tube feet.

THINK AND WRITE CRITICALLY

Answer the following questions in your Journal using complete sentences.

13. Describe how each class of mollusks obtains food.

14. Compare the ability of clams, oysters, scallops, and squid to protect themselves.

15. Compare an earthworm gizzard with teeth.

16. What evidence do scientists have that mollusks and annelids share a common ancestor?

APPLY

17. Describe the adaptations of the spinnerets and swimmerets.
18. Why is it unwise to cut apart sea stars and throw the parts back into a water area which is already overpopulated with them?
19. What problems do arthropods have immediately after molting?
20. Based on your observations of the structure of a segmented worm in Activity 14-1, explain why you see many earthworms on the surface of the soil after a heavy rain.
21. What structures do insects have that enable them to pollinate flowers?

MORE SKILL BUILDERS

If you need help, refer to the Skill Handbook.

1. **Classifying:** Group the following animals into arthropod classes: spider, pill bug, crayfish, grasshopper, crab, silverfish, cricket, sow bug, tick, scorpion, shrimp, barnacle, butterfly.
2. **Concept Mapping:** Copy and complete the map below that describes the characteristics of arthropods.
3. **Observing and Inferring:** Of what advantage is it to barnacles to live on a whale?
4. **Classifying:** The suffix *-ptera* means wings. Find out the meaning of the prefix of each insect class and give an example of a member of each group.
DIPTERA HOMOPTERA
ORTHOPTERA HEMIPTERA
COLEOPTERA
5. **Making and Using Graphs:** Make a pie graph of the species of arthropods.

Class of Arthropod	# of species
Arachnids	100 000
Crustaceans	25 000
Insects (known)	700 000

PROJECTS

1. Find out about the dance bees use to communicate to each other. Explain and diagram the dance to your class.
2. Begin an insect collection of your own. There are many good books at your local library to guide you in how to collect and identify insects.

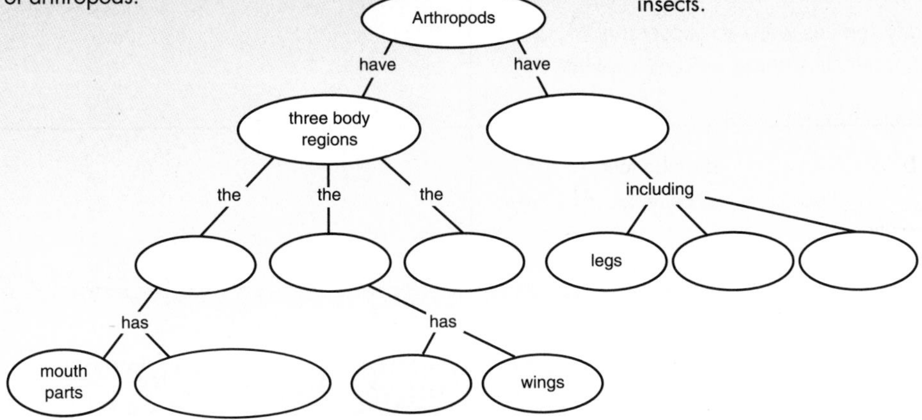

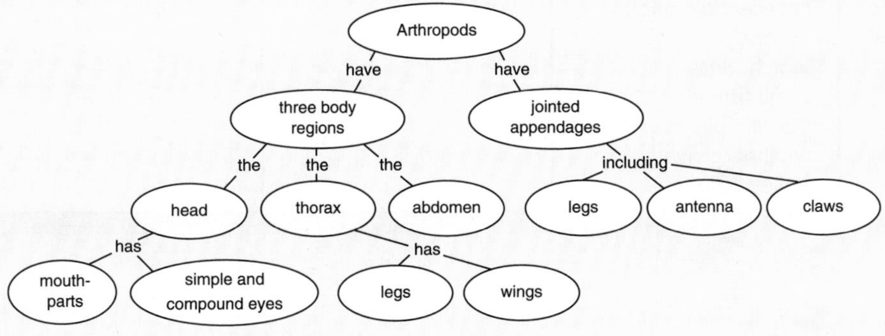

APPLY

17. Spinnerets are specialized appendages of arachnids used to release silk and spin webs; swimmerets are specialized appendages that help crustaceans move and are used in reproduction.

18. Sea stars regenerate. Cutting them apart causes the growth of more sea stars.

19. The exoskeleton is not fully hardened right after molting. Organisms usually hide at this time, until their exoskeleton is hard enough to protect them from predators.

20. An earthworm cannot take in oxygen from water-laden soil and must come to the surface for air.

21. Insects have wings, allowing them to go from place to place, to hover above a flower; hairs on their bodies pick up pollen.

MORE SKILL BUILDERS

1. Classifying: Arachnids: spider, tick, scorpion
Crustaceans: pill bug, crayfish, crab, sow bug, shrimp, barnacle
Insects: grasshopper, silverfish, cricket, butterfly

2. Concept Mapping: See Below

3. Observing and Inferring: Whales move, providing new food sources to attached barnacles.

4. Classifying: Diptera = 2 wings; flies
Orthoptera = straight wings; grasshoppers
Coleoptera = shielded wings; beetles
Homoptera = whole wings; leaf hoppers
Hemiptera = half wings; water bugs

5. Making and Using Graphs: Students should find 12% of the pie graph or 43° represents the arachnids, 3% of the graph or 11° represents the crustaceans, and 85% or 306° represents the insects.

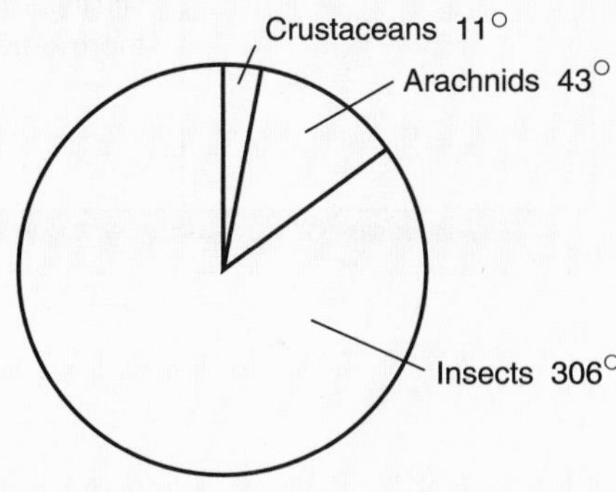

CHAPTER

15 Cold-Blooded Vertebrates

CHAPTER SECTION	OBJECTIVES	ACTIVITIES
15-1 Fish (3 days)	1. **Identify** the major characteristics of chordates. 2. **Explain** the differences between cold-blooded animals and warm-blooded animals. 3. **Describe** the characteristics that identify the three classes of fish.	**MINI-Lab:** *How does a fish adjust to depths?* p. 353 **Activity 15-1:** *Designing an Experiment (Does Water Temperature Affect Fish?),* p. 355
15-2 Amphibians (2 days)	1. **Describe** the adaptations that amphibians have for living in water and living on land. 2. **Identify** the three kinds of amphibians and describe the characteristics of each. 3. **Describe** the metamorphosis of a frog.	**MINI-Lab:** *How do frogs behave?* p. 358
15-3 Amphibian Population Decline Science & Society (1 day)	1. **State** three possible reasons for the recent decline in amphibian populations. 2. **Explain** why amphibians are likely to be affected by pollution.	**Activity 15-2:** *Metamorphosis in Frogs,* p. 362
15-4 Reptiles (1 day)	1. **Identify** the adaptations that enable reptiles to live on land. 2. **Infer** why the early reptiles were so successful. 3. **Describe** the characteristics of the modern reptiles.	
Chapter Review		

ACTIVITY MATERIALS

FIND OUT	ACTIVITIES		MINI-LABS	
Page 349 tuning fork Also see **Science Integration Activity 15**	**15-1 Designing an Experiment, p. 355** goldfish aquarium water small fishnet 600-mL beaker container of ice water stirring rod thermometer	**15-2 Metamorphosis in Frogs, p. 362** 4-L aquarium or jar frog egg masses lake or pond water stereoscopic microscope watch glass small fishnet aquatic plants washed gravel boiled lettuce	**How does a fish adjust to depths? p. 353** balloons bowls water	**How do frogs behave? p. 358** paper pencil

CHAPTER FEATURES	TEACHER RESOURCE PACKAGE	OTHER RESOURCES
Skill Builder: *Outlining,* p. 354	**Ability Level Worksheets** ◆ ***Study Guide,*** p. 60 ● ***Reinforcement,*** p. 60 ▲ ***Enrichment,*** p. 60 **Activity Worksheets,** pp. 131-132, 137 **Cross-Curricular Connections,** p. 19 **Transparency Masters,** pp. 53-54	**Lab Manual:** Vertebrates, p. 105 **Color Transparency 27,** Bony Fish **STVS:** Disc 5, Side 2 **Science Integration Activity 15**
Problem Solving: *Marsupial Frogs,* p. 359 **Skill Builder:** *Sequencing,* p. 359	**Ability Level Worksheets** ◆ ***Study Guide,*** p. 61 ● ***Reinforcement,*** p. 61 ▲ ***Enrichment,*** p. 61 **Activity Worksheets,** p. 138 **Transparency Masters,** pp. 55-56	**Lab Manual:** Frog Anatomy, p. 107 **Color Transparency 28,** Frog Metamorphosis
You Decide! p. 361	**Ability Level Worksheets** ◆ ***Study Guide,*** p. 62 ● ***Reinforcement,*** p. 62 ▲ ***Enrichment,*** p. 62 **Activity Worksheets,** pp. 5, 133-134 **Science and Society,** p. 19	
Skill Builder: *Concept Mapping,* p. 366 **Technology:** *Snake Oil Medicines,* p. 366	**Ability Level Worksheets** ◆ ***Study Guide,*** p. 63 ● ***Reinforcement,*** p. 63 ▲ ***Enrichment,*** p. 63 **Critical Thinking/Problem Solving,** p. 19 **Concept Mapping,** p. 35	**STVS:** Disc 5, Side 2
Summary Key Science Words Understanding Vocabulary Checking Concepts Using Lab Skills Think & Write Critically Apply More Skill Builders Projects	ASSESSMENT RESOURCES **Chapter Review,** pp. 33-34 **Chapter Test,** pp. 97-100 **Performance Assessment in Middle School Science (PAMSS)**	**Chapter Review Software** **Test Bank** **Alternate Assessment** **Performance Assessment**

◆ **Basic** ● **Average** ▲ **Advanced**

ADDITIONAL MATERIALS

SOFTWARE	AUDIOVISUAL	BOOKS/MAGAZINES
Operation Frog, Scholastics, Inc. *Visifrog: Vertebrate Anatomy,* Queue. *Anatomy of a Fish,* Queue.	*Amphibians,* film, Coronet/MTI. *Reptiles and Their Characteristics,* film, Coronet/MTI. *Crocodile,* film, Centre Productions, Inc. *Frogs and Toads—Watch Them Sing!,* film, International Film Bureau, Inc. *Reptiles,* film, EBEC. *Cities of Coral,* (NOVA) laserdisc, Image Entertainment. *Dragons of Galapagos (Cousteau),* laserdisc, Image Entertainment. *Encyclopedia of Animals: Reptiles and Amphibians,* laserdisc, Optical Data Corp. *Snakes, Scorpions, and Spiders,* video, LCA.	Blum, V., tr by A.C. Whittle. *Vertebrate Reproduction.* NY: Springer-Verlag, 1986. Ricciuti, Edward. "Quacking Ducks That Are Never There (Because They're Frogs)." *Audubon,* March 1988, pp. 54-55. Young, John Z. *The Life of Vertebrates. 3rd ed.* NY: Oxford, University Press, 1981.

CHAPTER

15

COLD-BLOODED VERTEBRATES

THEME DEVELOPMENT: The theme of evolution is used to introduce chordates and vertebrates. The theme of homeostasis is developed through the discussion of adaptations that fish, amphibians, and reptiles have for living in their environments. The ecological importance of fish, amphibians, and reptiles is presented.

CHAPTER OVERVIEW

- **Section 15-1:** This section identifies the major characteristics of chordates. The structure of fish and three classes of fish are described. Examples of why fish are important are given.
- **Section 15-2:** Amphibian adaptations for living in water and on land are examined. The characteristics of frogs, toads, and salamanders are described.
- **Section 15-3: Science and Society:** The reasons for the decline in amphibian populations are discussed. The You Decide feature asks students if saving amphibians is more important than work for people in the logging industry.
- **Section 15-4:** This section identifies the adaptations that enable reptiles to live on land and discusses why early reptiles were so successful. Modern reptiles are presented.

CHAPTER VOCABULARY

chordates
notochord
dorsal nerve cord
gill slits
endoskeleton
cold-blooded
warm-blooded
fish
fins
scales
cartilage
predator
prey
amphibians
hibernation
estivation
biological indicators
reptile
amniote egg

CHAPTER

15 Cold-Blooded Vertebrates

348

OPTIONS

For Your Gifted Students

Students can make a flip book of the stages of growth of the frog from egg to adult. Or, a video camera or 35mm camera can be used to record the frog's development. Have students take a few seconds of film or a still picture daily until the frog is fully developed.

For Your Mainstreamed Students

Have students make a mobile of the different types of fish mentioned in this chapter (jawless, cartilaginous, and bony). Beneath each type, they should write the fact they find most interesting about it. The same thing can be done with amphibians and reptiles.

Exactly how much do you know about reptiles? For example, do snakes sting with their tongues? How can a snake swallow something as big as a deer? And, if snakes have no ears, how do they hear?

FIND OUT!

Do this simple activity.

Hold a tuning fork by the stem and tap it on a hard piece of rubber. Then hold it next to your ear. What happens? Now tap it again and this time, press the base of the stem hard against your chin. What do you observe? Snakes receive similar vibrations from the environment. How do you think a snake detects vibrations? *Predict* how different animals make use of vibrations.

Gearing Up

Previewing the Chapter

Use this outline to help you focus on important ideas in this chapter.

Section 15-1 Fish
- ▶ Vertebrate Characteristics
- ▶ What Is a Fish?
- ▶ Jawless Fish
- ▶ Cartilaginous Fish
- ▶ Bony Fish

Section 15-2 Amphibians
- ▶ What Is an Amphibian?
- ▶ Origin of Amphibians
- ▶ Frogs, Toads, and Salamanders

Section 15-3 Science and Society
Amphibian Population Decline
- ▶ Where Have All the Frogs Gone?

Section 15-4 Reptiles
- ▶ What Is a Reptile?
- ▶ Origin of Reptiles
- ▶ Modern Reptiles

Previewing Science Skills

- ▶ In the Skill Builders, you will outline, sequence, and make a concept map.
- ▶ In the Activities, you will hypothesize, observe, collect data, measure, and experiment.
- ▶ In the MINI-Labs, you will compare and observe.

What's next?

In this chapter, you will learn about cold-blooded vertebrates. First you will learn about fish and amphibians. Then you will study about reptiles, what they are and how they live.

INTRODUCING THE CHAPTER

Use the Find Out activity to introduce students to reptiles, one of the cold-blooded vertebrate classes they will study in this chapter.

OBJECTIVES AND SCIENCE WORDS: Have students review the objectives and science words as they begin each section.

FIND OUT!

Materials: one tuning fork for every two students

Cooperative Learning: Use the Paired Partner strategy. Ask students to share what they observe with their partner.

Teaching Tips

▶ Advise students not to touch the prongs of the tuning fork after hitting it on the piece of rubber.

▶ Sound was heard because vibrations traveled through the bones in the jaw and skull to the fluid in the inner ear. Here, the vibrations were translated into nerve impulses and interpreted as sound by the brain. In many snakes and salamanders, the jawbone is sensitive to ground vibrations.

Gearing Up

Have students study the Gearing Up feature to familiarize themselves with the chapter. Discuss the relationships of the topics in the outline.

What's Next?

Before beginning the first section, make sure students understand the connection between the Find Out activity and the topics to follow.

ASSESSMENT OPTIONS

PORTFOLIO

Refer to page 367 for suggested items that students might select for their portfolios.

PERFORMANCE ASSESSMENT

See page 367 for additional Performance Assessment options.

Process

Skillbuilders, pp. 354, 359, 366
MINI-Labs, pp. 353, 358
Activities 15-1, p. 355, 15-2, p. 362
Using Lab Skills, p. 368

CONTENT ASSESSMENT

Assessment—Oral, p. 352
Section Reviews, pp. 354, 359, 361, 366
Chapter Review, pp. 367-369
Mini Quizzes, pp. 353, 358, 365

GROUP ASSESSMENT

Opportunities for group assessment occur with Cooperative Learning Strategies and Flex Your Brain Activities.

SECTION 15-1

PREPARATION

SECTION BACKGROUND

▶ Adaptations in fish are the result of millions of years of evolution by natural selection. The first known vertebrates were small, jawless, fishlike organisms called ostracoderms that appeared on Earth more than 500 million years ago. Scientists hypothesize that all other vertebrates can be traced back to the ostracoderms. The first modern fish appeared about 225 million years ago during the Triassic Period.

PREPLANNING

▶ Obtain goldfish or guppies for Activity 15-1.
▶ Obtain balloons for the MINI-Lab.

1 MOTIVATE

▶ Display preserved specimens of a tunicate, an amphioxus, and a frog so that students can observe characteristics of the three chordate subphyla.
▶ Provide a skeleton of a fish (cold-blooded vertebrate) and a cat (warm-blooded vertebrate). Have students observe the skeletons and determine how they are similar and how they are different.

TYING TO PREVIOUS KNOWLEDGE: Vertebrates are animals that are familiar to students. Ask them to name as many kinds of vertebrates as possible. Then ask what characteristics these animals have in common.

15-1 Fish

New Science Words

chordates
notochord
dorsal nerve cord
gill slits
endoskeleton
cold-blooded
warm-blooded
fish
fins
scales
cartilage
predator
prey

Objectives

- ▶ Identify the major characteristics of chordates.
- ▶ Explain the differences between cold-blooded animals and warm-blooded animals.
- ▶ Describe the characteristics that identify the three classes of fish.

Figure 15-1. Tunicates, top, and lancelets, bottom, are both chordates.

Vertebrate Characteristics

Most of the groups of animals you've studied so far live in water. They are all invertebrates, and their bodies are well adapted to aquatic life. Animals with backbones, the vertebrates, can be found in almost every environment—oceans, fresh water, land, and air. How are these organisms adapted to all these types of environments? To answer that question, let's look at some of the traits of vertebrates.

Vertebrate animals belong to the phylum Chordata (kor DAHT uh). The chordates are grouped into three subphyla, the largest being the vertebrates. The two small subphyla are made up of tunicates and lancelets, pictured in Figure 15-1. Tunicates, also called sea squirts, are sessile. They live a life very much like that of a clam or sponge. They are filter-feeders and live attached to objects in the sea. Lancelets also live in salt water and look a lot like fish. They can swim freely, but spend most of their time buried in the sand with only their heads sticking out. Lancelets are also filter-feeders.

What do sea squirts and lancelets have in common with vertebrates such as whales, bears, fish, and humans? Actually, three things: at sometime during their lives all **chordates** have a notochord, a dorsal hollow nerve cord, and gill slits. The **notochord** is a flexible, rodlike structure along the dorsal side, or back, of an animal. In vertebrates, the notochord is eventually replaced by bones that make up a backbone in the adult. A **dorsal nerve cord** is a bundle of nerves that lies above the notochord. In most vertebrates, it develops into a spinal cord

OPTIONS

Meeting Different Ability Levels

For Section 15-1, use the following **Teacher Resource Masters** depending upon individual students' needs.

- ◆ **Study Guide Master** for all students.
- ● **Reinforcement Master** for students of average and above average ability levels.
- ▲ **Enrichment Master** for above average students.

Additional Teacher Resource Package masters are listed in the OPTIONS box throughout the section. The additional masters are appropriate for all students.

◆ STUDY GUIDE 60

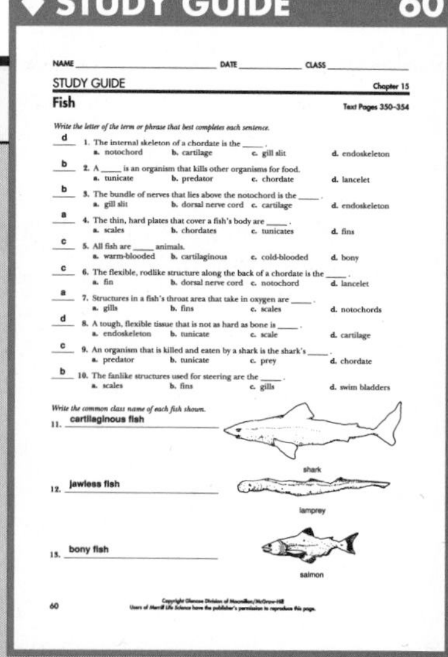

NAME ______ DATE ______ CLASS ______

STUDY GUIDE — Chapter 15

Fish — Text Pages 350-354

Write the letter of the term or phrase that best completes each sentence.

d 1. The internal skeleton of a chordate is the _____.
a. notochord b. cartilage c. gill slit d. endoskeleton

b 2. A _____ is an organism that kills other organisms for food.
a. tunicate b. predator c. chordate d. lancelet

b 3. The bundle of nerves that lies above the notochord is the _____.
a. gill slit b. dorsal nerve cord c. cartilage d. endoskeleton

a 4. The thin, hard plates that cover a fish's body are _____.
a. scales b. chordates c. tunicates d. fins

c 5. All fish are _____ animals.
a. warm-blooded b. cartilaginous c. cold-blooded d. bony

c 6. The flexible, rodlike structure along the back of a chordate is the _____.
a. fin b. dorsal nerve cord c. notochord d. lancelet

a 7. Structures in a fish's throat area that take in oxygen are _____.
a. gills b. fins c. scales d. notochords

d 8. A tough, flexible tissue that is not as hard as bone is _____.
a. endoskeleton b. tunicate c. scale d. cartilage

c 9. An organism that is killed and eaten by a shark is the shark's _____.
a. predator b. tunicate c. prey d. chordate

b 10. The fanlike structures used for steering are the _____.
a. scales b. fins c. gills d. swim bladders

Write the common class name of each fish shown.

11. cartilaginous fish — shark

12. jawless fish — lamprey

13. bony fish — salmon

60

with a brain at the front end. **Gill slits** are paired openings located in the throat behind the mouth. Gill slits develop into gills in fish. Traces of gills can be seen in human embryos.

Vertebrates are named for the column of bones called vertebrae that encloses the dorsal nerve cord. The vertebrae, skull, and the rest of the internal skeleton is called the **endoskeleton.** *Endo-* means "within." An endoskeleton supports and protects internal organs and is the place where muscles are attached.

There are seven classes of vertebrates. Members of five of the seven classes of vertebrates are cold-blooded. The body temperature of a **cold-blooded** animal changes with the temperature around it. **Warm-blooded** animals such as birds and mammals have a body temperature that stays the same in any environment. You are warm-blooded. In this chapter, you will study cold-blooded vertebrates—fish, amphibians, and reptiles.

In a fluid, forces are transmitted equally in every direction. Some organisms have hydrostatic skeletons, fluid-filled bodies that support body shape and organs. Which organisms you have studied thus far do you think have hydrostatic skeletons?

What Is a Fish?

When you hear the word *fish,* the first thing that probably comes to mind is a pet goldfish, or a can of tuna in the kitchen cabinet. Actually, there are more than 30 000 different species of fish—more than all other species of vertebrates. What is a fish, and how is it adapted to its environment?

Fish are cold-blooded vertebrates that have three adaptations that allow them to live in water. These adaptations are gills, fins, and scales. All fish have gills for breathing.
1 Located in the throat area, gills are structures that take oxygen from water as it passes over them. At the same time, carbon dioxide is given off to the water surrounding
2 the gills. Second, most fish have **fins,** fanlike structures used for steering, balancing, and moving. The large tail fin moves back and forth to propel the fish through the water.
3 Third, most fish have **scales,** hard, thin, overlapping plates that cover and protect a fish's body. Each scale grows larger as the fish grows. You can see the yearly growth rings in Figure 15-2. Like the age of some trees, the age of some fish is estimated by counting growth rings.

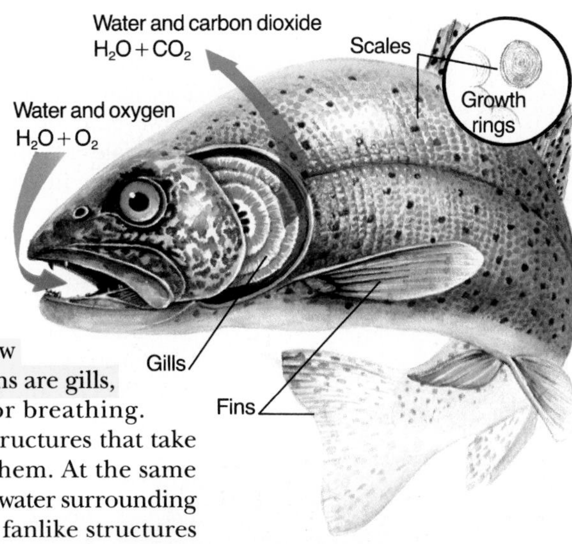

Figure 15-2. Fish have gills for taking in oxygen and scales for protection. Fins are used for steering, balancing, and moving.

● REINFORCEMENT 60

NAME ______ DATE ______ CLASS ______

REINFORCEMENT — Chapter 15

Fish — Text Pages 350–354

For each of the following fish, write the common name of the class to which it belongs.

1. shark: **cartilaginous fish**
2. lungfish: **bony fish**
3. perch: **bony fish**
4. lamprey: **jawless fish**
5. ray: **cartilaginous fish**
6. To which class do 95 percent of all fish belong? **bony fish**

Name and describe the three adaptations that allow fish to live in water. Then label the figure with those terms.

7. **Gills are the structures in the throat area that take in oxygen as water passes over them.**
8. **Fins are the fanlike structures used for steering, balancing, and propelling a fish.**
9. **Scales are the overlapping, thin, hard plates that cover and protect a fish's body.**

gills — fins — scales

Name and describe the three structures that all chordates have in common.

10. **the notochord; a flexible, rodlike structure along the back of an animal**
11. **the dorsal nerve cord; a bundle of nerves that lies above the notochord, usually turning into the spinal cord and brain**
12. **the gill slits; paired openings located in the throat area**

60

▲ ENRICHMENT 60

NAME ______ DATE ______ CLASS ______

ENRICHMENT — Chapter 15

Fish — Text Pages 350–354

CHORDATE HEARTS

The hearts of fish, amphibians, and reptiles are all different. This is because each type of animal has different needs and different body functions.

Fish have hearts with only two chambers, an atrium and a ventricle. Their blood flows in a simple path from the heart to the gills, where it is oxygenated, to body tissues, and back to the heart. This two-chambered heart supports the life of the fish because fish don't need as high a level of energy for body functions as land vertebrates. Fish are supported and buoyed up by the water they live in.

Amphibians and reptiles have a more complex circulatory system than fish. This better enables them to deal with the environmental problems of living on land. Amphibians and reptiles have a third heart chamber that helps the heart to keep oxygenated and deoxygenated blood separate.

Amphibians carry on respiration through their skin as well as lungs. Unlike fish, they have two atria instead of one. One atrium receives oxygenated blood from the lungs, and the other receives deoxygenated blood from body tissues. Both atria empty into a single ventricle. The ventricle lets blood bypass the lungs when the animal is under water.

Reptiles don't have skin respiration. Reptile hearts have two atria and a ventricle. But unlike the amphibian's, the reptile heart has a partially divided ventricle. This wall further decreases the mixing of oxygenated and deoxygenated blood and allows oxygenated blood to be pumped continually to the body.

Decide if the following are hearts of fish, amphibians, or reptiles. Label the diagrams, then answer the questions that follow.

atrium, ventricle — **fish**; atrium, ventricle — **amphibian**; atrium, ventricle — **reptile**

Conclude and Apply

1. Summarize how the number of chambers in the hearts of fish, amphibians, and reptiles relates to their lifestyles. **The more energy an organism requires, the more complex the heart. Generally, land-living organisms require more energy.**
2. Why do you think humans and birds have a more advanced type of heart than fish, amphibians, and reptiles? **Answers may vary, but might include that warm-blooded animals such as humans and birds lead a far more active life. Their hearts have four chambers to meet higher oxygen needs. The mammal heart efficiently prevents mixing of oxygenated and deoxygenated blood.**

60

2 TEACH

Key Concepts are highlighted.

CONCEPT DEVELOPMENT

▶ Point out that although there are only about 200 species of cephalochordates known, they are numerous in certain parts of the world. The lancelet is used for food in China, Japan, and several other countries.

▶ Explain that there are important similarities between gills and lungs. Both provide a large surface area on which gas exchange may take place. Although lungs are found in air-breathing animals, the exchange still takes place through moist membranes.

▶ Tell students that fish have many adaptations for living in different environments, getting food, and protection. Ice fish in Antarctica have a special kind of protein in their blood that keeps the blood from freezing, just as antifreeze keeps water in a car's radiator from freezing. A lionfish has needle-sharp spines loaded with poison. When a porcupine fish is disturbed, it gulps down water and blows up like a balloon covered with prickly spines.

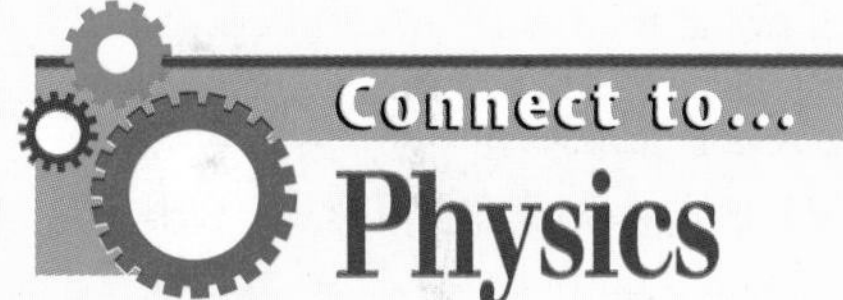

Answer: Planaria, nematodes, earthworms

CHECK FOR UNDERSTANDING

Ask students to list three characteristics of chordates.

RETEACH

Have students write a paragraph explaining why vertebrates are classified as chordates, but not all chordates are classified as vertebrates. Remind students that humans are vertebrates. Have them point out vertebrate characteristics in themselves.

EXTENSION

For students who have mastered this section, use the **Reinforcement** and **Enrichment** masters or other OPTIONS provided.

CONCEPT DEVELOPMENT

▶ Give examples of other things made of cartilage, such as the floppy ears of a dog and parts of the kneecap.

▶ Point out that young lampreys do not look like the adults and are not parasites. The bottom-dwelling larvae hatch and live for three to seven years by filtering nutrients from the water.

▶ Tell students that a shark's mouth has 6 to 20 rows of backward-pointing teeth. When one tooth breaks or wears down, a replacement moves forward. One shark may use more than 20 000 teeth in its lifetime.

Cooperative Learning: Divide the class into Expert Teams cooperative groups. Assign each group a well-known shark—great white, bull, hammerhead, or nurse. Have each group research characteristics, feeding habits, and behavior of the sharks and report to the class.

REVEALING MISCONCEPTIONS

▶ Many students may think that all sharks attack people. Fewer than ten percent of shark species are known to attack humans unprovoked. Have students find out which sharks attack people and what safety rules are used to prevent shark attacks.

VideoDisc

STVS: Hump Back Fish, Disc 5, Side2
STVS: Studying Sharks, Disc 5, Side 2
STVS: Raising Super Fish, Disc 5 Side 2

Fish are grouped into three classes, jawless fish, cartilaginous (kart uhl AJ uh nuhs) fish, and bony fish.

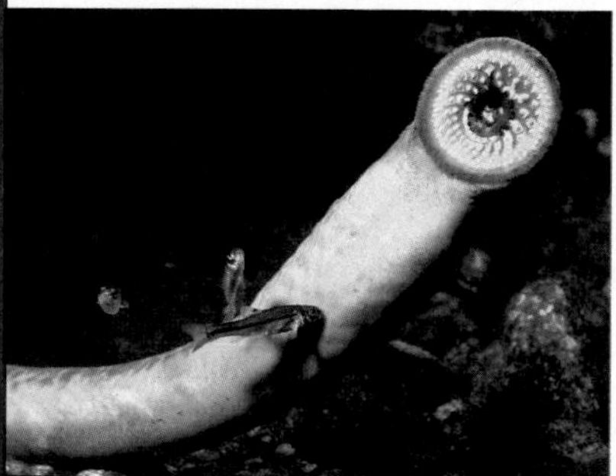

Figure 15-3. Lampreys are jawless fish that feed on the blood and body fluids of other fish.

Jawless Fish

The lamprey in Figure 15-3 belongs to the class Agnatha. *Agnatha* is a Greek word that means "jawless." These jawless fish have round mouths and long tubelike bodies covered with slimy skin with no scales. Fish in this
4 class have very flexible bodies made of cartilage. **Cartilage** is a tough, flexible tissue that is not as hard as bone. Feel your ears and the tip of your nose. These are also made of cartilage.

Hagfish live only in salt water, but many lampreys live in fresh water. Lampreys use their round mouths to attach to other fish by suction. They cut into the fish with toothlike structures and feed on its blood and body fluids. You can see the mouth of a lamprey in Figure 15-3.

Cartilaginous Fish

Sharks, skates, and rays are members of the class Chondrichthyes (kahn DRIHK thee eez). They have skeletons made of cartilage like jawless fish. Unlike jawless fish however, chondrichthyes have movable jaws and scales. The scales resemble vertebrate teeth and cause their skin to feel like fine sandpaper.

The fact that sharks are predators gives them a repu-
5 tation for being killers. A **predator** is an organism that kills and eats another for food. The organism that is killed and eaten is the **prey.** Unless they are provoked, few sharks will attack humans.

Shark eggs are fertilized inside the female's body. In most species, the fertilized eggs develop in the body and young are born alive.

Figure 15-4. Rays (left) and sharks (right) are members of the class Chondrichthyes.

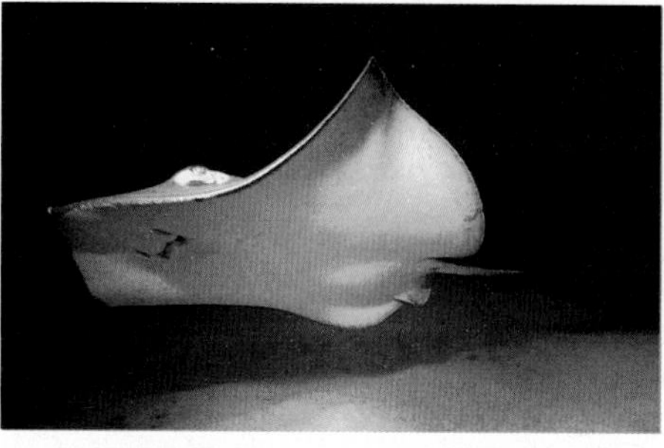

OPTIONS

ASSESSMENT—ORAL

▶ **Cold-blooded animals do not perspire or pant. How do you think they control body temperature?** *The temperature of a cold-blooded animal changes with its surroundings. If it becomes too warm, it makes contact with cooler surroundings. If it becomes too cold, it makes contact with warmer surroundings to change its body temperature.*

▶ **Rays have flat bodies and fins that look like flat wings. Where in the ocean do these adaptations enable the ray to live? Explain.** *on the ocean bottom where they can blend in with the sand*

▶ **Do you think lampreys would ever be a threat to swimmers? Why or why not?** *Yes; if they became too numerous and there were not enough fish, they might attack swimmers.*

▶ **Why do you think unprovoked sharks attack people?** *They could mistake swimmers or persons on surfboards for a food source, such as seals.*

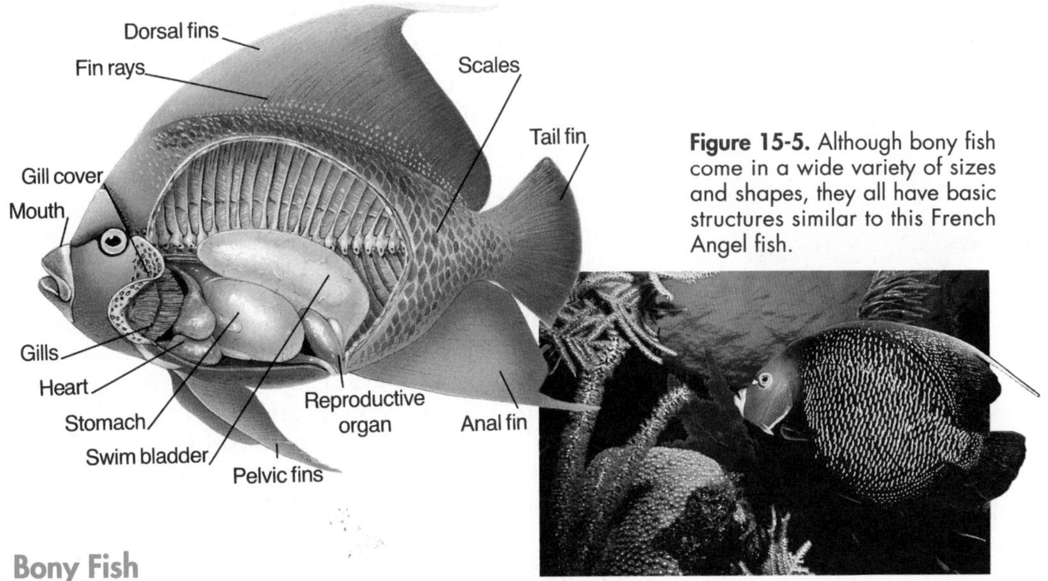

Figure 15-5. Although bony fish come in a wide variety of sizes and shapes, they all have basic structures similar to this French Angel fish.

Bony Fish

Bony fish make up the class Osteichthyes (ahs tee IHK thee eez). The word *osteon* is Greek for bone, and *ichthyes* is Greek for fish. Fish that belong to this class have skeletons made of bone. About 95 percent of all species of fish belong to this class.

Floating is an important adaptation for fish. Most bony fish have adaptations for floating. Some float by trapping gases in a balloonlike structure called the swim bladder. By adjusting the amount of gas in the swim bladder, fish can stay at a particular depth in the water with little effort. Find the swim bladder in Figure 15-5.

The body structure of bony fish, a French angel fish, is shown in Figure 15-5. The gills are covered and protected by a bony flap called the gill cover. Gill covers aid the fish in breathing. They open and close, moving the water taken in through the mouth over the gills.

Most kinds of bony fish have separate males and females. The females release large numbers of eggs into the water in a behavior called spawning. Males then swim over the eggs and release sperm. The joining of egg and sperm cells outside the female body is called external fertilization.

Scientists who specialize in studying fish have organized bony fish into three groups: lobe-finned fish, lungfish, and ray-finned fish. Most fish are ray-finned fish.

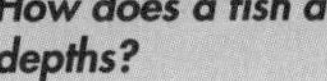

MINI-Lab

How does a fish adjust to depths?

Fill a balloon with air. Place it in a bowl of water. What happens? What structure in a fish does the balloon model? *Compare* where in the water (surface, deep) a fish would be as this structure is being filled with air.

CONCEPT DEVELOPMENT

▶ **Demonstration:** Fish scales show annual rings of growth. Set up microscopes with slides of fish scales (or use the microprojector) and have students count the growth rings and note the differences in thickness.

▶ Find a picture of a coelacanth, the only species of the lobe-finned fish alive today, and show it to the class.

MINI-Lab

Materials: balloons, water, bowls

Answer: The balloon will float; it is like the swim bladder. The actual action of the swim bladder cannot be shown because the balloon, no matter how filled, will float. The more the swim bladder is filled, however, the closer to the surface the fish will come; the less air, the deeper the fish will be in the water.

MINI-Lab ASSESSMENT

Performance: To further assess students' understanding of swim bladders in fish, have them view goldfish in an aquarium and describe how they rise and fall in the water.

CROSS CURRICULUM

▶ **Health:** Many people include fish in their diet as a source of protein and to reduce their fat consumption. Have students visit a grocery and make a list of fish used for food. Have them find out which fish contain the most fat.

MINI QUIZ

Use the Mini Quiz to check students' recall of chapter content.

1. **What are gills?** *structures in the throat area that take oxygen from the water and give off carbon dioxide*
2. **What are fins?** *fanlike structures used by fish in steering, balancing, and moving*
3. **What are scales?** *overlapping, thin, hard plates that cover and protect a fish's body*
4. **What is cartilage?** *a tough, flexible tissue that is not as hard as bone*
5. **What is a predator?** *an organism that kills and eats another for food*

ENRICHMENT

▶ Fish farmers raise and breed catfish, salmon, trout, bass, and other types of fish. Have students research and report on fish farming.

PROGRAM RESOURCES

From the **Teacher Resource Package** use:

Science Integration Activities, 15, Sound Waves and Pitch

PROGRAM RESOURCES

From the **Teacher Resource Package** use:

Cross-Curricular Connections, page 19, Finding Information.

Transparency Masters, pages 53-54, Bony Fish.

Activity Worksheets, page 137, Mini-Lab: How does a fish adjust to depths?

Use **Color Transparency** 27, Bony Fish.

Use **Laboratory Manual,** page 105, Vertebrates.

Science Integration Activities is found in **Merrill Life Science Teacher Resource Manual.**

Science and READING

Have students study fish native to waters in your area and consider which fish to offer in their store.

3 CLOSE

▶Ask questions 1-4 and the **Apply** Question in the Section Review.

SECTION REVIEW ANSWERS

1. notochord, a dorsal hollow nerve cord, and gill slits

2. supports and protects the internal organs and provides a place for muscles to attach

3. The body temperatures of cold-blooded animals change with their surroundings. The body temperatures of warm-blooded animals stay the same in any surroundings.

4. Student answers should reflect information on pages 352 and 353.

5. Apply: Many eggs are not fertilized, and some eggs and young fish are eaten by predators.

6. Connect to Physics: They are adapted to living and feeding on the bottom and do not need to float to different depths.

Figure 15-6. Fish are food for many animals, including this Puffin.

Science and READING

Imagine you have just opened a fish market in your hometown. What different species will you offer, where will you get them, and what facts do you need to know about them?

Ray-finned fish have fins made of long, thin bones covered with skin. Yellow perch, tuna, salmon, swordfish, and eels are all ray-finned fish.

Lungfish have both gills and lungs for breathing. This adaptation allows many to live in shallow waters that dry up in summer. When the water evaporates, lungfish burrow into the mud and cover themselves with mucus until water returns. Lungfish have been found along the coast of South America, Australia, and Africa.

Lobe-finned fish have fins that are lobe-like and fleshy. It was thought that these organisms had been extinct for more than 70 million years. But in 1938, one was caught in a net by some South African fishermen. Several living lobe-finned fish have been studied since. Scientists observed these fish using their lobed fins to swim in a way similar to the way land vertebrates walk. Lobe-finned fish are important because scientists think these fish are the ancestors of the first land vertebrates.

Fish are an important food source for animals such as raccoons, bears, and sea birds. They are also an important food source for humans. Some researchers hypothesize fish oils protect blood vessels from fat deposits that lead to heart disease. Because fish feed on insect larvae such as those of mosquitoes, they help to control insect populations that are pesty or dangerous. Some species of fish feed on plants that clog waterways. This helps keep them clear for boat traffic.

SECTION REVIEW

1. What are three characteristics of chordates?
2. What is the function of the endoskeleton in vertebrates?
3. Explain the differences between cold-blooded animals and warm-blooded animals.
4. What are the characteristics of the three fish classes?
5. **Apply:** Female fish lay thousands of eggs. Why are the waters not overcrowded with fish?
6. **Connect to Physics:** Suggest a reason why bottom-dwelling fish usually don't have swim bladders.

Skill Builder

Outlining

Outline the characteristics of chordates, vertebrates, and fish. If you need help, refer to Outlining in the **Skill Handbook** on page 681.

Skill Builder

A. Chordates
   1. have notochord
   2. dorsal hollow nerve cord
   3. gill slits

B. Vertebrates
   1. are chordates
   2. bones called vertebrae enclose nerve cord

C. Fish
   1. are chordates
   2. are vertebrates
   3. have gills, fins, and scales
   4. live in water

Skillbuilder ASSESSMENT

Performance: Assess students' abilities to Outline by having them outline characteristics of jawless, cartilaginous, and bony fish.

ACTIVITY 15-1 DESIGNING AN EXPERIMENT
Does Water Temperature Affect Fish?

Sometimes warmer water is dumped into bodies of water in which fish live. What happens to a fish if the water temperature in which it lives changes?

Getting Started

Determine a way to measure the effects of water temperature on a fish. What adaptations do fish have for breathing? How can you count how many times a fish breathes in one minute?

Thinking Critically

Where does the oxygen that fish breathe come from? How do the gills function in breathing?

Materials 

Your cooperative group will use:

- goldfish
- aquarium water
- 1 small fishnet
- 600-mL beaker
- container of ice water
- stirring rod
- thermometer

Try It!

1. Fill the beaker about one-half full of aquarium water. With the fishnet, transfer one fish from the aquarium to the beaker.
2. Use the thermometer to measure the temperature of water in the beaker and record the temperature.
3. *Observe* the movement of the fish and the movements of the gill covers. Count and record the number of times the fish opens its gill covers in one minute. Repeat this step three more times.
4. **Hypothesize** how the breathing rate and movements of the fish will be affected by cooling the temperature of the water.
5. Place the beaker in a container of ice water. Allow the water in the beaker to cool to approximately 14°C. Use the stirring rod and gently stir the water in the beaker. **CAUTION:** *Do not injure the fish.* Record the water temperature.
6. Repeat Step 3 and record your results as trials 1, 2, 3, and 4.
7. Return the fish to the aquarium.

Summing Up/Sharing Results

- What were you measuring when you counted the gill cover openings?
- How does lowering the water temperature affect fish? *Predict* what would happen if the water temperature increased?
- Could your results be used to determine the kind of environment in which the fish can live?

Going Further!

Fish may live in water that is totally covered by ice. How is this possible? What would happen to a fish if the water were to become very warm?

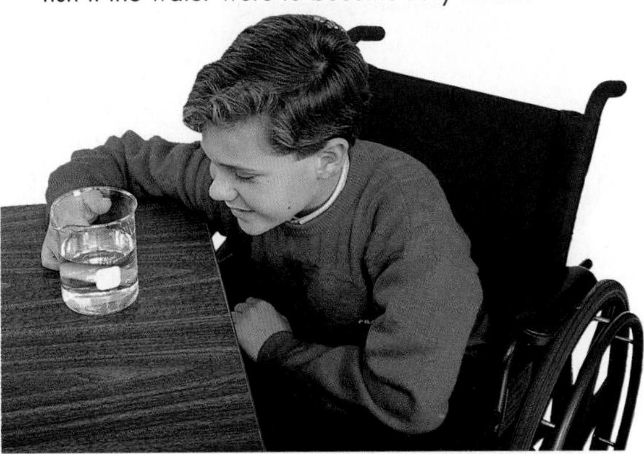

ACTIVITY 15-1

OBJECTIVE: Design and carry out an experiment to measure the effect of water temperature on fish.
Time: one class period

PROCESS SKILLS applied in this activity are **observing, measuring,** and **interpreting data.**

PREPARATION

- Pet shops stock feeder fish that are not expensive. If guppies are used, do not lower the temperature below 18°C.

Cooperative Learning: Divide the class into Science Investigation Teams.

THINKING CRITICALLY

Oxygen is dissolved in the water. The gills take oxygen from water as it passes over them.

TEACHING THE ACTIVITY

*Refer to the **Activity Worksheets** for additional information and teaching strategies.*

- Students should handle fish with extreme care.
- The starting temperature of the water is the temperature of the water in the aquarium or beaker.
- Students may notice that the number of times a fish's mouth opens is directly related to the openings of the gill coverings.

SUMMING UP/SHARING RESULTS

- The rate of breathing was measured by the gill cover openings.
- As water temperature decreases, breathing rate decreases, and movements become slower. If the water temperature increased, the gill cover openings would increase and the movements would increase.
- Student answers will vary but should include that fish need an environment in which water is not too warm.

GOING FURTHER!

Cold water contains much dissolved oxygen. Warm water has less oxygen and the fish must breathe faster using energy faster than it can take in oxygen and will die.

PROGRAM RESOURCES

From the **Teacher Resource Package** use:
Activity Worksheets, pages 131-132, Activity 15-1, Does Water Temperature Affect Fish?

Activity ASSESSMENT

Performance: To further assess students' understanding of the effect of water temperature on fish, see Question 11 USING LAB SKILLS, on page 368.

SECTION 15-2

PREPARATION

SECTION BACKGROUND

▶ Biologists divide amphibians into three orders: Urodela, the salamanders; Anura, the frogs and toads; and Apoda, the wormlike caecilians. There are only about 250 species of salamanders in the world. Most are found in North America.

▶ Hell benders and mud puppies are two of the largest salamanders found in the United States. They reach lengths of 43 cm. Mud puppies have red external gills. Hell benders have skin that looks wrinkled and too large for their bodies.

▶ Caecilians are tropical amphibians that are not well known. The legless, wormlike creatures look and, to some extent, act like earthworms. They average 30 cm in length, have small eyes, and are often blind. They burrow in underground tunnels and hunt and eat worms and other invertebrates. Many species of caecilians have bony scales, like those of fish, embedded in their skin.

PREPLANNING

▶ Obtain a frog or toad in a terrarium for the Motivate activity. Obtain recordings of frog calls.

1 MOTIVATE

▶ Place a frog or toad in a terrarium. In many places, one may be obtained from a local pond and later returned. Have the students observe its characteristics and behavior.

▶ Obtain a recording of frog calls from the public library or a biological supply company. Ask students to make a hypothesis about frog calls. (Reproductive strategy)

15-2 Amphibians

New Science Words

amphibians
hibernation
estivation

Objectives

- ▶ Describe the adaptations that amphibians have for living in water and living on land.
- ▶ Identify the three kinds of amphibians and describe the characteristics of each.
- ▶ Describe the metamorphosis of a frog.

What Is an Amphibian?

The word *amphibian* comes from the Greek word *amphibios*, which means "double life." They are well named for **amphibians** are cold-blooded vertebrates that spend part of their lives in water and part on land. Frogs, toads, and salamanders are amphibians. What adaptations do these animals have that allow them to live both on land and in water? Amphibians have moist, smooth, thin skin without scales. They can breathe through their skin. Oxygen and carbon dioxide are exchanged through the skin and the lining of the mouth. Amphibians also have very small, simple, saclike lungs in the chest cavity to use for breathing.

Because amphibians are cold-blooded, their body temperatures change when the temperature of their surroundings changes. Did you ever wonder how frogs and toads survive cold winters in the northern states? Amphibians have adaptations that protect them from very cold and very warm temperatures. During the cold winter months they become inactive. They bury themselves in mud or leaves until the temperature warms up. The period of inactivity in the winter is called **hibernation.** Inactivity 1
during the hot, dry summer months is called **estivation.**

Another adaptation amphibians have to life on land is a strong skeleton. Organisms living in water are supported by the water. However, on land a stronger structure is needed to support the body.

Because their eggs lack shells and must be kept moist, most species of amphibians return to water to lay their eggs. The eggs hatch into larvae that live in water until they become adults and move to land.

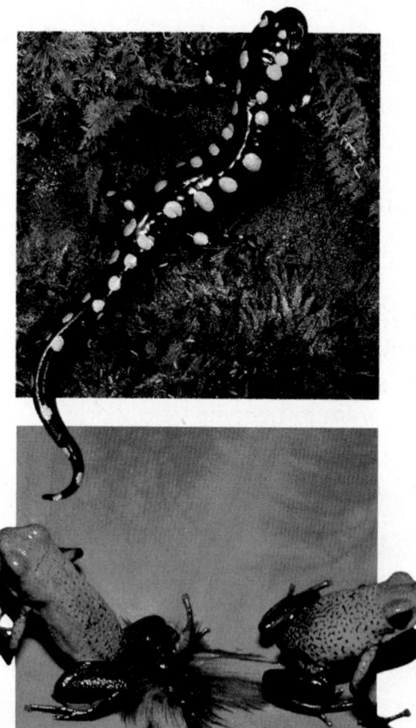

Figure 15-7. Salamanders, top, and frogs, bottom, are both types of amphibians.

OPTIONS

Meeting Different Ability Levels

For Section 15-2, use the following **Teacher Resource Masters** depending upon individual students' needs.

◆ **Study Guide Master** for all students.
● **Reinforcement Master** for students of average and above average ability levels.
▲ **Enrichment Master** for above average students.

Additional Teacher Resource Package masters are listed in the OPTIONS box throughout the section. The additional masters are appropriate for all students.

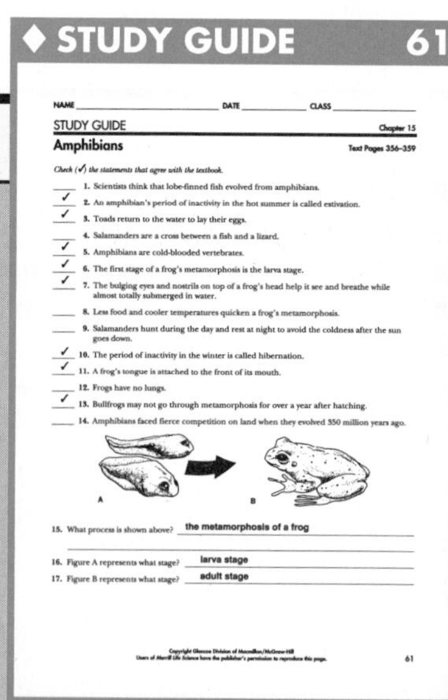

◆ STUDY GUIDE 61

NAME ______ DATE ______ CLASS ______

STUDY GUIDE Chapter 15

Amphibians Text Pages 356–359

Check (✓) the statements that agree with the textbook.

____ 1. Scientists think that lobe-finned fish evolved from amphibians.
✓ 2. An amphibian's period of inactivity in the hot summer is called estivation.
✓ 3. Toads return to the water to lay their eggs.
____ 4. Salamanders are a cross between a fish and a lizard.
✓ 5. Amphibians are cold-blooded vertebrates.
✓ 6. The first stage of a frog's metamorphosis is the larva stage.
✓ 7. The bulging eyes and nostrils on top of a frog's head help it see and breathe while almost totally submerged in water.
____ 8. Less food and cooler temperatures quicken a frog's metamorphosis.
____ 9. Salamanders hunt during the day and rest at night to avoid the coldness after the sun goes down.
✓ 10. The period of inactivity in the winter is called hibernation.
✓ 11. A frog's tongue is attached to the front of its mouth.
____ 12. Frogs have no lungs.
✓ 13. Bullfrogs may not go through metamorphosis for over a year after hatching.
____ 14. Amphibians faced fierce competition on land when they evolved 350 million years ago.

A B

15. What process is shown above? the metamorphosis of a frog
16. Figure A represents what stage? larva stage
17. Figure B represents what stage? adult stage

 61

Origin of Amphibians

The fossil record shows that fish were the first vertebrates on Earth. For about 150 million years they were the only vertebrates. Then as competition for food and space increased, the lobe-finned fish may have easily traveled across land searching for deeper water. The lobe-finned fish had lungs and bony fins that could have supported their weight on land. Amphibians are thought to have evolved from the lobe-finned fish about 350 million years ago.

Evolution favored the development of amphibians because there wasn't much competition on land. Almost free of predators, land provided a great supply of food in the form of insects, spiders, and other invertebrates. Amphibians were able to reproduce in large numbers, and many new species evolved. For 100 million years or more, amphibians were the dominant land animals.

For organisms adapted to life in water, why is land a hostile environment?

Frogs, Toads, and Salamanders

Frogs and toads are amphibians that as adults have short, broad bodies with no neck, no tail, and four legs. The strong hind legs are longer than the front and are used for swimming and jumping. Bulging eyes and nostrils on top of the head let them see and breathe while almost totally submerged in water. On spring nights, they make their presence known with loud, distinctive sounds.

Figure 15-8 shows the external body structures of a
2 frog. A frog's tongue is attached to the front of its mouth. When a frog sees an insect, it flips the loose end of its tongue out and the insect gets stuck in the sticky saliva on the tongue. Then the frog flips its catch back into its mouth. For hearing, frogs have round membranes called tympanic membranes on each side of the head just behind the eyes.

Toads live in drier environments than frogs. They have thick, rough skin that keeps them from drying out. Both frogs and toads return to the water to lay eggs.

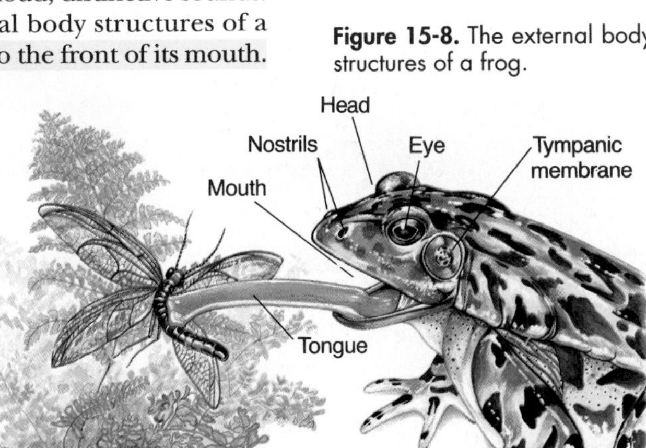

Figure 15-8. The external body structures of a frog.

● REINFORCEMENT 61

NAME ____________ DATE ________ CLASS ________

REINFORCEMENT Chapter 15

Amphibians Text Pages 356–359

Listed below are some of the parts of a frog. Describe the structure and function of each part and label the figure.

1. skin: **moist, smooth, and thin without scales; through which oxygen and carbon dioxide are exchanged**
2. tongue: **attached to front of mouth; covered with sticky saliva; used to catch insects by flipping it out and then pulling the food back in**
3. nostrils: **openings on top of head; allow the frog to breathe while almost totally submerged in water**
4. eyes: **bulging, located on top of head; allow frog to see while almost totally submerged in water**
5. back legs: **stronger and longer than the front legs; used for swimming and jumping**

nostril
tongue
eye
skin
back leg

Answer the following questions.

6. Compare and contrast hibernation and estivation. **Both describe periods of inactivity. Hibernation is inactivity in the cold winter months, while estivation is inactivity in the hot summer months.**
7. What are three kinds of amphibians? **frogs, toads, and salamanders**
8. What is metamorphosis? **a series of changes that a larva goes through to become an adult**

 61

▲ ENRICHMENT 61

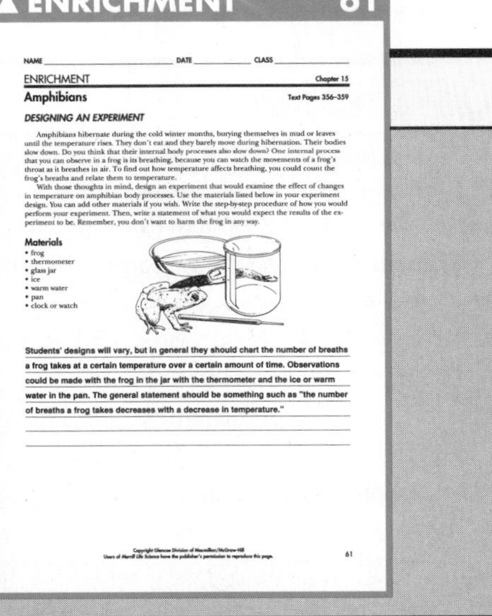

NAME ____________ DATE ________ CLASS ________

ENRICHMENT Chapter 15

Amphibians Text Pages 356–359

DESIGNING AN EXPERIMENT

Amphibians hibernate during the cold winter months, burying themselves in mud or leaves until the temperature rises. They don't eat and they barely move during hibernation. Their bodies slow down. Do you think that their internal body processes also slow down? One internal process that you can observe in a frog is its breathing, because you can watch the movements of a frog's throat as it breathes in air. To find out how temperature affects breathing, you could count the frog's breaths and relate them to temperature.

With those thoughts in mind, design an experiment that would examine the effect of changes in temperature on amphibian body processes. Use the materials listed below in your experiment design. You can add other materials if you wish. Write the step-by-step procedure of how you would perform your experiment. Then, write a statement of what you would expect the results of the experiment to be. Remember, you don't want to harm the frog in any way.

Materials
- frog
- thermometer
- glass jar
- ice
- warm water
- pan
- clock or watch

Students' designs will vary, but in general they should chart the number of breaths a frog takes at a certain temperature over a certain amount of time. Observations could be made with the frog in the jar with the thermometer and the ice or warm water in the pan. The general statement should be something such as "the number of breaths a frog takes decreases with a decrease in temperature."

 61

TYING TO PREVIOUS KNOWLEDGE: Review the information in the previous section about lobe-finned fish. Ask students what adaptations would be necessary for a fish to survive on land.

2 TEACH

Key Concepts are highlighted.

CONCEPT DEVELOPMENT

▶ Explain that many amphibians have poison glands in the skin. In most cases, the poison is a bad-tasting irritant that discourages predators. But some species produce powerful poisons that can kill enemies, including people.

▶ Some amphibians change color. They have pigment-containing cells called chromatophores that are responsible for the color changes.

Answer: A land environment is drier, creating respiratory and reproductive challenges for organisms.

CROSS CURRICULUM

▶ **Language Arts:** Have students find the meaning of the prefix *amphi-* and explain why it is used to describe amphibians.

TEACHER F.Y.I.

▶ Frogs and toads have vocal cords. Most male frogs have inflatable throat pouches that increase the volume of the sounds they make.

MINI-Lab
Teaching Tips

▶ Place any good references on amphibians on reserve in your school library.

▶ Life in water—some webbing of feet, thin skin, strong hind legs. Life on land—strong skeleton.

MINI-Lab
ASSESSMENT

Performance: To further assess students' observation skills, have them distinguish between a toad and a frog.

CHECK FOR UNDERSTANDING

MINI QUIZ

Use the Mini Quiz to check students' recall of chapter content.

1 **How is hibernation different from estivation?** *Hibernation is inactivity in the winter. Estivation is inactivity during hot, dry, summer months.*

2 **What is a frog adaptation for catching insects?** *the tongue is attached to the front of the mouth*

3 **Why are salamanders often mistaken for lizards?** *because of their long, slender bodies*

RETEACH

Show pictures or slides of amphibians and have students identify each as a frog, toad, salamander, or caecilian.

EXTENSION

For students who have mastered this section, use the **Reinforcement** and **Enrichment** masters or other OPTIONS.

MINI-Lab

How do frogs behave?

Obtain a live frog in a large empty jar. Carefully *observe* the frog. Notice the position of its legs as it sits. Observe the position of its mouth, eyes, nostrils, and ears. Observe the color of its dorsal and ventral sides. Return the frog to your teacher to place in the aquarium. Observe how the frog swims. **In your Journal,** describe the adaptations the frog has for living in water. What adaptations does it have for living on land?

Recall that metamorphosis is a series of changes that a larva goes through to become an adult. Most frogs and toads have a two-stage life cycle, a larva that lives in water and an adult that lives on land. Figure 15-9 shows the metamorphosis of a leopard frog.

The rate at which amphibians grow and the length of time they spend as larvae depend on the species, the water temperature, and available food. The less food and the cooler the temperature, the longer it takes for metamorphosis. Bullfrogs may not go through metamorphosis for a year or more after hatching, whereas spadefoot toads change into adults in about two weeks.

Salamanders are amphibians often mistaken for lizards because of their long, slender bodies. They have smooth, 3
moist skin and short legs. They rest under dead leaves and rocks during the day to avoid the drying heat of the sun. At night they come out and use their well-developed senses of smell and vision to feed on worms, crustaceans, and insects.

Amphibians are unique animals in many ways. Their adaptations allow them to live very successfully both on land and in water. The complete transition to life on land did not occur until the reptiles appeared. You will learn about reptiles in the next section.

Figure 15-9. Frogs undergo a series of changes from egg to adult.

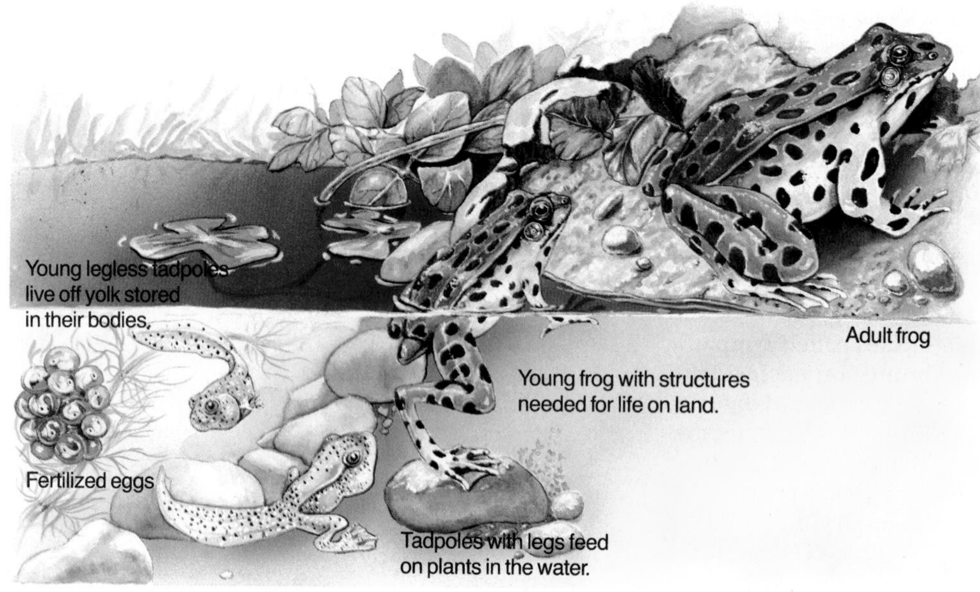
Young legless tadpoles live off yolk stored in their bodies.
Adult frog
Young frog with structures needed for life on land.
Fertilized eggs
Tadpoles with legs feed on plants in the water.

OPTIONS

PROGRAM RESOURCES

From the **Teacher Resource Package** use:

Transparency Masters, pages 55-56, Frog Metamorphosis.

Activity Worksheets, page 138, Mini-Lab: How do frogs behave?

Use **Color Transparency** 28, Frog Metamorphosis.

Use **Laboratory Manual,** page 107, Frog Anatomy.

ENRICHMENT

▶ Have students research and report on protective coloration in amphibians.

▶ Have students research and report on one of the following early amphibians. Ichthyostega, one of the earliest amphibians, was about 1 m long, had four strong legs, a strong tail, and ate insects and other small animals. Diplocaulus was a water-dwelling amphibian with a strange, horned head. Mastodonsaurus was about 4 m long and spent most of its time in swamps.

PROBLEM SOLVING

Marsupial Frogs

Most frogs deposit their eggs in water where the eggs hatch into fishlike tadpoles. The tadpoles exchange gases and get nutrients in the water environment. Over a period of time, the tadpoles develop into adult frogs.

Researchers studying tree frogs in tropical rain forests have found that some species reproduce using a special pouch. Because of the pouch, they call these marsupial frogs.

The female marsupial frog lays eggs, and the male frog maneuvers the eggs one-by-one into a pouch on her back. The eggs are fertilized by sperm the male frog has deposited on the female's back.

The eggs are enclosed in a fluid-filled pouch lined with many blood vessels. In the pouch, the eggs undergo metamorphosis. The researchers think the tadpoles exchange gases, fluids, nutrients, and wastes with the female but get most of the nutrients they need from the yolk of the egg.

After development is complete, the female uses her toes to open the pouch and release the small frogs.

Think Critically: How is reproduction in marsupial frogs similar to that of other amphibians? How is it different?

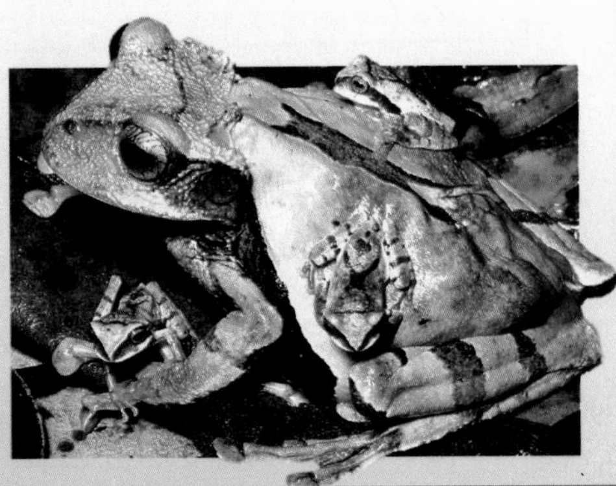

SECTION REVIEW

1. List three characteristics of amphibians for living in water and three characteristics for living on land.
2. Describe the three kinds of amphibians.
3. Describe how a tadpole is different from a frog.
4. **Apply:** How would you explain the appearance of frogs and toads after a rain?
5. **Connect to Earth Science:** Find out some advantages to life on land.

Sequencing

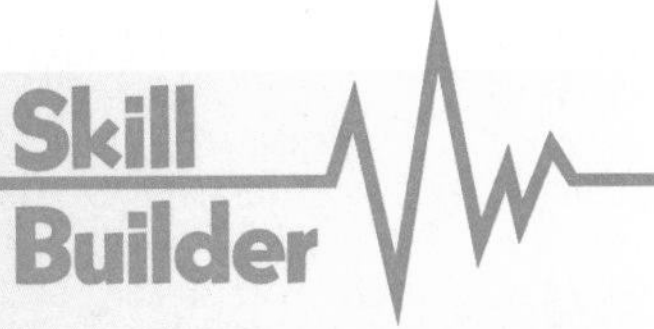

Sequence and describe each stage of frog metamorphosis. If you need help, refer to Sequencing in the **Skill Handbook** on page 680.

SECTION REVIEW ANSWERS

1. water—moist, smooth skin; bulging eyes and nostrils on top of the head; strong hind legs for swimming; land—lungs; strong skeleton; strong hind legs for jumping; moist, smooth skin

2. frogs have short, broad bodies with no neck, no tail, and four legs; toads have thick, rough skin that keeps them from drying out; salamanders have smooth, moist skin and legs.

3. Tadpoles live in water, breathe through gills, feed on plant matter, and have a tail. Frogs live on land and in water, feed on insects, breathe through skin and small lungs, and have no tail.

4. Apply: They come out to hunt food when the air is moist.

5. Connect to Earth Science: Little competition; abundance of food; few predators, so reproductive potential is great.

PROBLEM SOLVING

Think Critically: It is similar because young are hatched from fertilized eggs. It is different because marsupial frogs carry developing young in pouches.

CONCEPT DEVELOPMENT

Cooperative Learning: Divide the class into Expert Team cooperative groups. Have each group research and report on the habitats and breeding behaviors of one of the following: Suriname Toad, Midwife Toad, Green-and-Black Dart-Poison Frog, Smith Frog, Glass Frog, or Darwin's Frog.

REVEALING MISCONCEPTIONS

▶ Some people describe snakes and other reptiles as slimy, but it is the amphibian that has this trait. The slime, or mucus, comes from glands in the amphibian's skin, and its function is to keep the skin moist.

3 CLOSE

▶ Ask questions 1-3 and the **Apply** Question in the Section Review.

▶ Have students write a poem about an amphibian.

Skill Builder

Students should sequence the steps of frog development in this order: egg—legless tadpole—tadpole with legs—young frog with developing legs—adult frog on land. Descriptions should be similar to those in Figure 15-9 on page 358.

Skillbuilder
ASSESSMENT

Performance: Assess students' understanding of metamorphosis by having them compare frog metamorphosis with insect metamorphosis.

SECTION 15-3

PREPARATION

SECTION BACKGROUND

▶ Clear-cutting is fast and very economical. It creates new habitats for grazing animals and adds to surface run-off, creating larger water supplies downstream.

▶ Clear-cutting also causes erosion and flooding. Debris can clog rivers and streams. It also destroys the habitats of many organisms.

1 MOTIVATE

▶ Have students list all the things they come in contact with each day that are made of wood. Have them guess how many trees it takes to make those things.

TYING TO PREVIOUS KNOWLEDGE: Have students recall from the previous section the characteristics of amphibians. They should recall habitat, food source, and reproduction.

PROGRAM RESOURCES

From the **Teacher Resource Package** use:

Science and Society, page 19, Something's Fishy.

Activity Worksheets, page 5, Flex Your Brain.

15-3 Amphibian Population Decline

New Science Words

biological indicators

Objectives

- ▶ State three possible reasons for the recent decline in amphibian populations.
- ▶ Explain why amphibians are likely to be affected by pollution.

Where Have All the Frogs Gone?

Where have all the amphibians gone? That's what scientists all over the world are wondering. They have noticed a decline in the populations of toads, frogs, and salamanders in places like the United States, Costa Rica, Japan, and Australia. Yosemite toads, leopard frogs, and chorus frogs are a few of the many species of amphibians now in danger of becoming extinct.

What's causing this decline of amphibians? In 1990, a meeting of scientists was called by the National Academy of Science to discuss the problem. A variety of possible factors was suggested.

Figure 15-10. Many amphibian habitats are being lost due to the clear-cutting of forests.

Some scientists think that the changes people have made in our environment are responsible for this disappearing act. Destruction of native habitats is likely to be a contributing factor. Amphibians typically require moist habitats to breed. With the continual expansion of urban areas, drainage of wetlands, and with deforestation, suitable breeding areas for many types of amphibians are becoming very limited. In the northwestern United States where thousands of acres of forest have been clear-cut, or completely cleared of trees, scientists report noticeably fewer frogs than in unlogged areas.

OPTIONS

Meeting Different Ability Levels

For Section 15-3, use the following **Teacher Resource Masters** depending upon individual students' needs.

◆ **Study Guide Master** for all students.

● **Reinforcement Master** for students of average and above average ability levels.

▲ **Enrichment Master** for above average students.

Additional Teacher Resource Package masters are listed in the OPTIONS box throughout the section. The additional masters are appropriate for all students.

◆ STUDY GUIDE 62

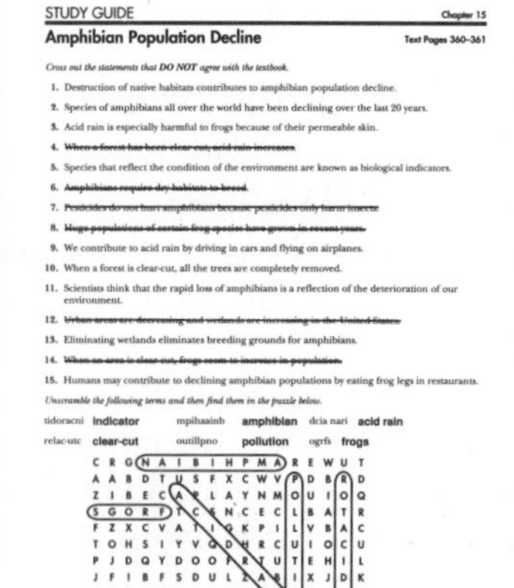

NAME ______ DATE ______ CLASS ______

STUDY GUIDE — Chapter 15

Amphibian Population Decline — Text Pages 360–361

*Cross out the statements that **DO NOT** agree with the textbook.*

1. Destruction of native habitats contributes to amphibian population decline.
2. Species of amphibians all over the world have been declining over the last 20 years.
3. Acid rain is especially harmful to frogs because of their permeable skin.
4. ~~When a forest has been clear-cut, acid rain increases.~~
5. Species that reflect the condition of the environment are known as biological indicators.
6. ~~Amphibians require dry habitats to breed.~~
7. ~~Pesticides do not hurt amphibians because pesticides only harm insects.~~
8. ~~Huge populations of certain frog species have grown in recent years.~~
9. We contribute to acid rain by driving in cars and flying on airplanes.
10. When a forest is clear-cut, all the trees are completely removed.
11. Scientists think that the rapid loss of amphibians is a reflection of the deterioration of our environment.
12. ~~Urban areas are decreasing and wetlands are increasing in the United States.~~
13. Eliminating wetlands eliminates breeding grounds for amphibians.
14. ~~When an area is clear-cut, frogs seem to increase in population.~~
15. Humans may contribute to declining amphibian populations by eating frog legs in restaurants.

Unscramble the following terms and then find them in the puzzle below.

tidoracni **indicator** mpihaainb **amphibian** dcia nari **acid rain**

relac-utc **clear-cut** outillpno **pollution** ogrfs **frogs**

C R G N A I B I H P M A R E W U T
A A B D T U S F X C W V P D B R D
Z I B E C A R L A Y N M O U I O Q
S G O R F T C S N C E C L B A T R
F Z X C V A T I G K P I L V B A C
T O H S I Y V Q D H R C U I O C U
P J D Q Y D O O F R T U T E H I L
J F I B F S D U L Z A B I X J D K
O I U U Y T R E W C R I O M O N W
P C L E A R C U T D F B N L P I K
J O I N B H U Y G V C F T K R S O

62

Industry and automobiles contribute to acid rain and snow, which can kill salamander eggs as well as forests . Other possible causes of amphibian extinction include pesticides and the introduction of amphibian-eating game fish to new locations. In addition, countries like Indonesia, India, and Malaysia provide huge amounts of frog legs to satisfy the appetites of many people. Although drought may be a natural cause of this loss of amphibians, people may be playing a major role.

Because of the way amphibians live, it makes sense that they would be very sensitive to pollutants in the environment. They live mostly in water or in damp soil. They have no scales or shells for protection. And because their skin absorbs oxygen, poisonous gases and chemicals can also be absorbed. For this reason amphibians are considered **biological indicators**—species that reflect the condition of the environment.

What can be done? Amphibians are the source of food for many birds, mammals, fish, and reptiles. They also feed on many insects and other invertebrates. Will enough species be able to survive to prevent a major disturbance in the world's food chain?

Figure 15-11. Many species of frogs and salamanders are endangered by human activities.

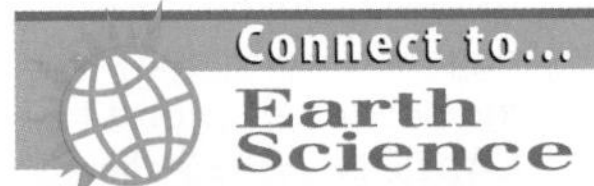

Find out what took place during the Devonian and Carboniferous periods when amphibians became successful on Earth.

SECTION REVIEW

1. What two current environmental problems may be contributing to the decline in amphibian populations?
2. Why are amphibians more likely to be affected by acid rain than reptiles, which also often live in the water?

You Decide!

Many environmentalists think that the clear-cutting of forests in the northwestern United States is too destructive to be continued. Some scientists have found half the number of amphibians in logged areas as compared to unlogged areas. However, many people depend on logging to support their families. Without the logging industry, the economy of many areas would be ruined. What do you think the answer to this problem is?

● REINFORCEMENT 62

NAME ______ DATE ______ CLASS ______

REINFORCEMENT — Chapter 15

Amphibian Population Decline — Text Pages 360–361

Beside each picture, explain how the activity shown contributes to amphibian population decline.

1. **Clear-cutting forests destroys native habitats of amphibians and eliminates suitable breeding grounds.**
2. **Automobile exhaust contributes to acid rain, a form of pollution. Amphibians are especially at risk because of their permeable skin. Pollution kills amphibians.**
3. **Eating frogs has an impact on their populations.**

Answer the following questions.

4. What is a biological indicator? **a species that reflects the condition of the environment**
5. What does the rapid loss of amphibians say about our environment? **Amphibians are biological indicators that reflect the deterioration of the environment for all living things, including humans.**
6. How does the use of pesticides affect amphibians? **Pesticides kill insects, thus eliminating the main source of food for amphibians. Also, pesticides may directly harm amphibians. Using pesticides therefore causes population decline among amphibians.**
7. How are amphibians affected when a swamp is filled in to build houses? **Amphibians need moist areas to breed. When a swamp is filled in, a habitat is lost and amphibian populations decline.**

62

▲ ENRICHMENT 62

NAME ______ DATE ______ CLASS ______

ENRICHMENT — Chapter 15

Amphibian Population Decline — Text Pages 360–361

AN ECOLOGICAL FAIRY TALE

Once upon a time there lived a prince who had everything. He lived with his father, the king, in a castle. The king was a wise man, but he had long since retired, leaving the family business—ruling the kingdom—to his son. The prince loved to build things, and he often directed his subjects to fill in the swamps of the kingdom so he could build new tennis courts and swimming pools.

One day the prince was playing with his favorite ball when it accidentally rolled into the swamp—the only remaining swamp in the kingdom.

"Oh, no," said the prince, "I've lost my favorite ball."

"But you haven't," said a croaking voice. "I'll get it for you if you grant my request."

The prince saw that the voice came from an old female bullfrog, who was sticking her head out from the waters of the swamp.

"If you can get the ball, I'll grant you any request," replied the prince.

"My request," said the frog, "is that you leave this swamp alone so that my babies can live out their lives."

The prince thought only a moment before agreeing to her request, whereupon she dove into the depths of the swamp and retrieved his ball.

"Ha!" cried the prince when he had the ball in his hand. "You'll get nothing from me, you old frog. Tomorrow I drain this swamp, and I'll never lose my ball again."

That night the king and the prince were sitting at their long table eating their ten-course meal when suddenly the king gagged on his food.

"What in heaven's name is happening?" he asked when he had finally cleared his throat. For before him, in a puddle of slimy swamp water, squatted one old female frog. Entering the room behind her were her many babies and dozens of other frogs.

"I am the Frog Queen of the swamp," she announced grandly, "and I have come to tell the king of his son's promise."

Finish writing this fairy tale, using information from your textbook. What will the Frog Queen say to the king? How can she convince him to honor his son's promise not to fill in the swamp?

Answers will vary.

62

2 TEACH

Key Concepts are highlighted.

CONCEPT DEVELOPMENT

▶ Be sure students realize clear-cutting can be made environmentally acceptable by cutting smaller areas, blending cuts with the terrain, and by replanting with grasses and trees to control erosion and ensure regrowth. Waste products can be used to make paper or other wood products.

CHECK FOR UNDERSTANDING

Ask questions 1 and 2 in the Section Review.

RETEACH

Show students more areas that have been clear-cut. Have them list all the problems that may be associated with clearing trees in this manner.

EXTENSION

For students who have mastered this section, use the **Reinforcement** and **Enrichment** masters or other OPTIONS provided.

3 CLOSE

Use the Flex Your Brain activity to have students explore AMPHIBIAN POPULATION DECLINE.

Flex Your Brain

ASSESSMENT

Portfolio: Use the Flex Your Brain activity to reinforce critical thinking and problem solving skills.

SECTION REVIEW ANSWERS

1. destruction of native habitats, pollution

2. Their skin is permeable and in direct contact with the polluted water.

YOU DECIDE!

Answers will vary. Students should be able to support their answers.

ACTIVITY 15-2
25 min. on day 1

OBJECTIVE: Observe tadpole development and use observations to study the life cycle of a frog.

PROCESS SKILLS applied in this activity:
▶ **Observing** in Procedure Steps 4-5.
▶ **Recording** in Procedure Steps 4-6.
▶ **Identifying** in Procedure Step 6.
▶ **Inferring** in Conclude and Apply Questions 5 and 7.

COOPERATIVE LEARNING Use the Science Investigation strategy. Divide the class into groups of four.

TEACHING THE ACTIVITY

Troubleshooting: The eggs must be handled carefully if they are to hatch. Make sure the temperature of the water remains about 20°C.

▶ Field-collected eggs can be shipped in January, February, and March. Peeper or toad eggs develop rapidly.

▶ Boil the lettuce leaves for ten minutes and then cool them.

▶ Aged tap water can be used instead of pond water.

PROGRAM RESOURCES

From the **Teacher Resource Package** use:

Activity Worksheets, pages 133-134, 15-2 Metamorphosis in Frogs.

ACTIVITY 15-2 Metamorphosis in Frogs

Problem: *How long does it take for a tadpole to develop into a frog?*

Materials  

- 4-L aquarium or jar
- frog egg masses
- lake or pond water
- stereoscopic microscope
- watch glass
- small fishnet
- aquatic plants
- washed gravel
- boiled lettuce

Procedure

1. Collect a mass of frog eggs and 4 L of water from a lake or pond.
2. Prepare an aquarium with gravel, the pond water, and five or six aquatic plants.
3. Place the egg mass in the aquarium. Use the fishnet to obtain a few eggs and place the eggs in a watch glass. The eggs should have the dark side up and the white yolk down. **CAUTION:** *Handle the eggs with care.*
4. Observe the eggs and record your observations in the data table.
5. *Observe* the eggs twice each week and *record* any changes that you observe.
6. When the tadpoles hatch, observe the stages of development twice a week. Identify the fin on the back, mouth, eyes, gill cover, gills, nostrils, hind and front legs. Feed the tadpoles boiled lettuce. Observe how they feed.

Analyze

1. How long does it take for the eggs to hatch?
2. How long does it take for a tadpole to develop legs?
3. What organs do tadpoles have for breathing?
4. Which pair of legs appears first?

Conclude and Apply

5. What is the advantage of the jellylike coating around the eggs?
6. *Compare* the eyes of the young tadpoles with the eyes of the older ones.
7. Why are plants needed in the aquarium?
8. Are tadpoles plant-eaters or meat-eaters?
9. How long does it take for a tadpole to develop into a frog?

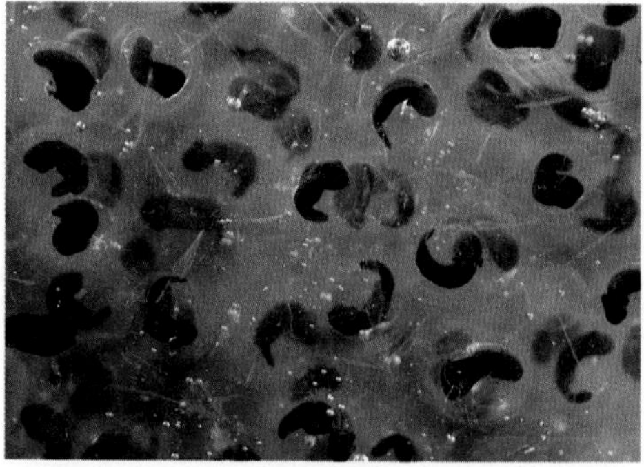

Data and Observations

Sample Data

Date	Observations
	This will vary according to the kind of frog eggs used. Students should observe fish-like appearance of tadpoles as well as the disappearance of the tail and the development of legs.

ANSWERS TO QUESTIONS

1. answers will vary; 8 to 20 days
2. approximately four weeks, depending on the species
3. gills
4. the hind legs
5. It protects the eggs and keeps them from drying out.
6. The eyes of young are on each side of the head, as in fish. In older ones, they are nearer the top of the head.
7. to add oxygen to the water and remove carbon dioxide
8. plant-eaters
9. Approximately two to four months, depending on the species.

Activity ASSESSMENT

Performance: To further assess students' ability to Observe tadpole development, see Question 12, USING LAB SKILLS, on page 368.

Reptiles

15-4

Objectives

- Identify the adaptations that enable reptiles to live on land.
- Infer why the early reptiles were so successful.
- Describe the characteristics of the modern reptiles.

New Science Words

reptile
amniote egg

What Is a Reptile?

The class Reptilia includes lizards, snakes, turtles, crocodiles, and alligators. They have many characteristics that are adaptations for life on land. A thick, dry, 1 waterproof skin covered with scales prevents drying out and injury. With the exception of snakes, reptiles have four legs with claws that hold the body off the ground for moving quickly. The claws are used to dig, climb, and run.

Remember from the last section that even though amphibians live on land, they still need water for respiration and reproduction. Scientists hypothesize that, over time, some amphibians became less dependent on water and became the ancestors of reptiles. A **reptile** is a cold-blooded vertebrate with dry, scaly skin. Reptiles lay eggs covered with leathery shells.

Reptiles vary greatly in size, shape, and color. There are giant pythons, longer than 10 meters, that can swallow crocodiles whole. Some sea turtles have a mass of more than 0.9 metric tons and can swim faster than you can run. Three-horned lizards have movable eye sockets and tongues as long as their bodies. Reptiles live on every continent except Antarctica and in all the oceans except those in the polar regions.

2 Reptiles breathe with lungs. Even turtles that live in water must come to the surface to breathe air. A reptile's
3 heart has three chambers. The lower chamber is partially divided. Oxygen-rich blood coming from the lungs is kept separated from blood that contains carbon dioxide returning from the body. This type of circulatory system provides a lot of oxygen to all parts of the body. Remember that oxygen is needed to release energy by cell respiration.

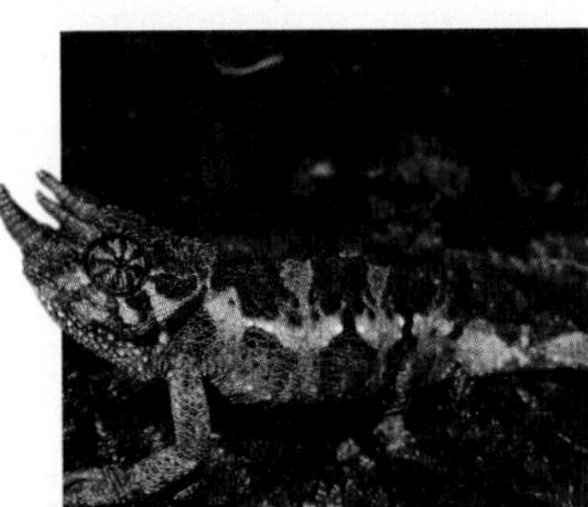

Figure 15-12. A three-horned lizard, top, and an Eastern box turtle, bottom, are both reptiles.

SECTION 15-4

PREPARATION

SECTION BACKGROUND

- Most lizards have four legs with five clawed toes on each foot, long tails, movable eyelids, and external ear openings. A few species, such as glass snakes and slow worms, have no legs. Lizards feed on anything they can catch and swallow.
- Snakes don't have legs, eyelids, outside ear openings, or bladders. They have a very flexible spine made up of 100 to 400 vertebrae, each attached to a pair of thin ribs. All snakes are meat-eaters.
- Turtles are the only reptiles with a shell: the carapace, which covers the back, the plastron, which covers all or part of the underneath side, and the bridge, which connects the two. Turtles feed on insects, worms, and fish, fruit and other plant material.
- The amniote egg allows reptiles to live on land. It has four specialized membranes: the amnion, yolk sac, allantois, and chorion. It is named for the amnion, which surrounds the fluid in which the embryo floats. The yolk sac encloses the yolk, which is the food supply for the developing embryo. The allantois stores wastes produced by the embryo until the egg hatches. The chorion lines the outer shell enclosing the embryo and all other membranes.

1 MOTIVATE

- Invite a herpetologist to bring specimens of snakes to show the class.
- Obtain a small turtle and place it in a terrarium for students to observe. Have the students use references to find out about care and feeding of turtles.

VideoDisc

STVS: Rattlesnakes, Disc 5, Side 2

OPTIONS

ENRICHMENT

- Have students research and report on reptiles common to your area. Students should include habitat, type of food, and unusual habits of the reptiles in their report. Reports may be presented to the class.

PROGRAM RESOURCES

From the **Teacher Resource Package** use:

Critical Thinking/Problem Solving, page 19, Sneaky Snakes.

Concept Mapping, pages 35-36.

TYING TO PREVIOUS KNOWLEDGE: Most students are familiar with dinosaurs. Explain that dinosaurs and modern alligators and crocodiles are all reptiles.

OBJECTIVES AND SCIENCE WORDS: Have students review the objectives and science words to become familiar with this section.

2 TEACH

Key Concepts are highlighted.

CONCEPT DEVELOPMENT

▶ Have students brainstorm a list of as many reptiles as they know or have seen.

▶ Explain that the amniote egg gets its name from the amnion, the thin membrane that encloses the fluid in which the embryo floats.

In Your JOURNAL

Several theories attempt to explain the disappearance of dinosaurs. Students should choose one theory and support it.

CROSS CURRICULUM

▶ **Reading:** Have students research and report on a species of endangered sea turtle.

Cooperative Learning: Divide the class into Expert Teams cooperative groups and have them research the characteristics, habitats, and behaviors of reptiles that lived long ago, such as *Pteranodon, Elasmosaurus, Archelon, Tyrannosaurus,* and *Deinosuchus.*

VideoDisc

STVS: Reptilian Sex Change, Disc 5, Side 2

STVS: Sea Turtle Mystery, Disc 5, Side 2

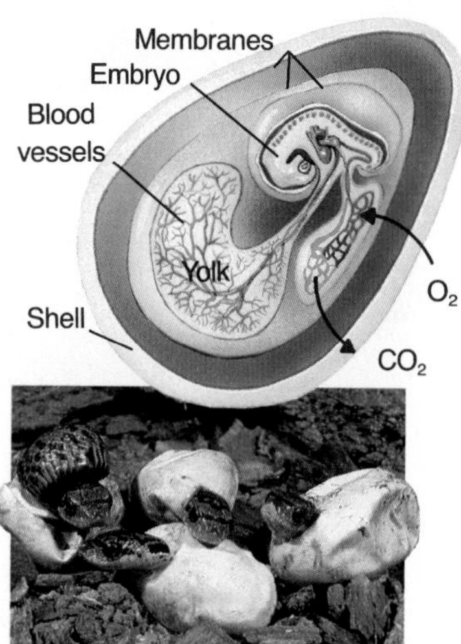

Figure 15-13. The amniote egg, top, is one of the adaptations reptiles have for living on land. Above, young reptiles hatch from their eggs fully developed.

One of the most important adaptations of reptiles for living on land is the way they reproduce. Unlike the eggs of fish and amphibians, eggs of reptiles are fertilized internally, inside the body of the female. After fertilization, the female secretes a leathery shell around each egg, and then lays the eggs on land.

Figure 15-13 shows the structures in a reptile egg. The egg provides a complete environment for the embryo to develop in. This type of egg, called an **amniote egg,** contains membranes to protect and cushion the embryo, and to help it get rid of wastes. It also contains a large food supply, the yolk, for the embryo to use as it develops. Small holes in the shell, called pores, allow oxygen and carbon dioxide to be exchanged. By the time it hatches, a young reptile looks like a small adult.

Have you ever been hiking and seen the dry outer layer of skin a snake has left behind? Many snakes and other reptiles molt from time to time, depending on how fast they grow. The skin is a good source of protein so they often eat it after molting.

Origin of Reptiles

4 Reptiles evolved slowly over a period of millions of years. During the Mesozoic Era, they came to dominate Earth. The Mesozoic Era, which lasted over 160 million years, is known as the Age of the Reptiles. Many species emerged from the earliest reptiles called cotylosaurs (KAHD uh luh sahrs). Their numbers increased. Some returned to the water to live. Others became adapted with large, heavy, scale-like plates and speed. At least two species of flying reptiles evolved.

The ancient reptiles were very successful. Why, then, did they die off? Scientists have developed many hypotheses to explain this disappearance. One is that the climate cooled toward the end of the Cretaceous Period, and it became too cold for the dinosaurs. They were too large to hibernate and had no body covering for protection. Another theory is that the explosion of a nearby star gave off dangerous radiation and caused cold, unfavorable weather for thousands of years. There are several other hypotheses to explain the disappearance of dinosaurs. Many scientists suggest that no one theory completely explains why the dinosaurs died.

In Your JOURNAL

There is no one convincing answer that explains the disappearance of the dinosaurs from Earth. Research this subject and choose the theory you most agree with. **In your Journal,** write a report explaining the theory and why you chose it.

OPTIONS

Meeting Different Ability Levels

For Section 15-4, use the following **Teacher Resource Masters** depending upon individual students' needs.

◆ **Study Guide Master** for all students.

● **Reinforcement Master** for students of average and above average ability levels.

▲ **Enrichment Master** for above average students.

Additional Teacher Resource Package masters are listed in the OPTIONS box throughout the section. The additional masters are appropriate for all students.

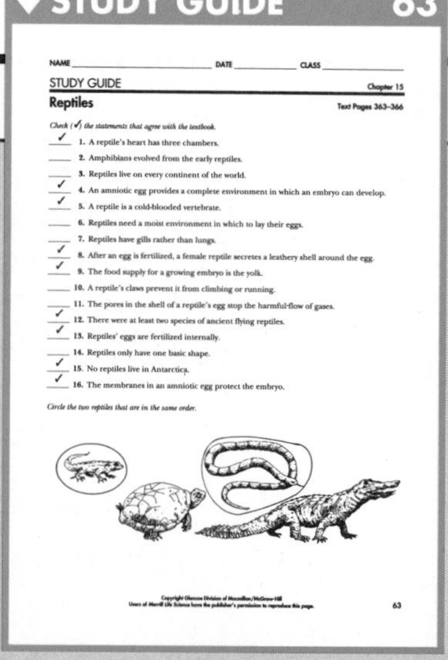

◆ STUDY GUIDE 63

NAME ________ DATE ________ CLASS ________

STUDY GUIDE — Chapter 15

Reptiles — Text Pages 363-366

Check (✓) the statements that agree with the textbook.

✓ 1. A reptile's heart has three chambers.
___ 2. Amphibians evolved from the early reptiles.
___ 3. Reptiles live on every continent of the world.
✓ 4. An amniotic egg provides a complete environment in which an embryo can develop.
✓ 5. A reptile is a cold-blooded vertebrate.
___ 6. Reptiles need a moist environment in which to lay their eggs.
___ 7. Reptiles have gills rather than lungs.
✓ 8. After an egg is fertilized, a female reptile secretes a leathery shell around the egg.
✓ 9. The food supply for a growing embryo is the yolk.
___ 10. A reptile's claws prevent it from climbing or running.
___ 11. The pores in the shell of a reptile's egg stop the harmful flow of gases.
✓ 12. There were at least two species of ancient flying reptiles.
✓ 13. Reptiles' eggs are fertilized internally.
___ 14. Reptiles only have one basic shape.
✓ 15. No reptiles live in Antarctica.
✓ 16. The membranes in an amniotic egg protect the embryo.

Circle the two reptiles that are in the same order.

Modern Reptiles

Modern reptiles probably descended from similar reptilian ancestors. Today there are three orders of reptiles: turtles; crocodiles and alligators; lizards and snakes.

Table 15-1

ORDERS OF MODERN REPTILES

Turtles
Turtles make up a very successful order of animals. They can be found on almost every continent and in most of the world's oceans. The body of a turtle is covered by a hard shell on both top and bottom. Most turtles can withdraw into their shell for protection. Turtles have no teeth and use their beaks to feed on insects, worms, fish, and plants.

Painted Turtle

Green Turtle

Loggerhead Turtle

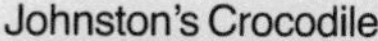

Johnston's Crocodile

American Alligator

Speckled Caiman

Crocodiles and Alligators
Crocodiles and alligators are among the world's largest living reptiles. They make up the order Crocodilia. Members of this order can be found in or near water in tropical climates. Crocodiles have long slender snouts and are very aggressive. They can attack animals as large as cattle. Alligators are less aggressive. They have broad snouts and feed on fish, turtles, and waterbirds.

Lizards and Snakes
Lizards and snakes make up the largest group of reptiles. Lizards have moveable eyelids, external ears, and legs with clawed toes on each foot. They feed on reptiles, insects, spiders, worms, and mammals. Snakes don't have legs, eyelids, or external ears. Snakes are meat-eaters. Some snakes wrap around and constrict their prey. Others inject their prey with poison venom. Many snakes feed on rodents and help control their populations.

Rough Green Snake

Land Iguana

Madagascar Day Gecko

● REINFORCEMENT 63

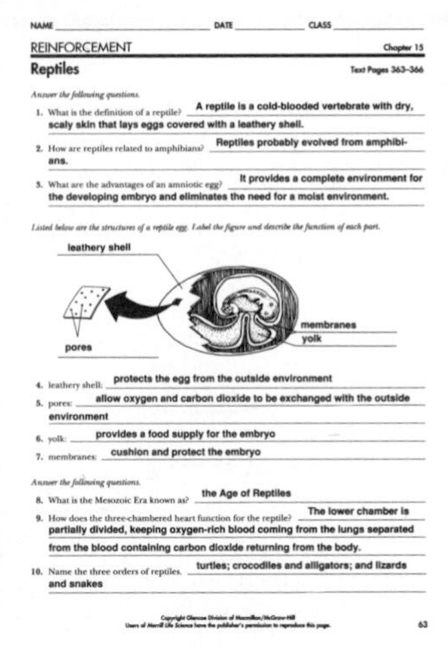

NAME ______ DATE ______ CLASS ______

REINFORCEMENT — Chapter 15

Reptiles — Text Pages 363–366

Answer the following questions.

1. What is the definition of a reptile? **A reptile is a cold-blooded vertebrate with dry, scaly skin that lays eggs covered with a leathery shell.**
2. How are reptiles related to amphibians? **Reptiles probably evolved from amphibians.**
3. What are the advantages of an amniotic egg? **It provides a complete environment for the developing embryo and eliminates the need for a moist environment.**

Listed below are the structures of a reptile egg. Label the figure and describe the function of each part.

4. leathery shell: **protects the egg from the outside environment**
5. pores: **allow oxygen and carbon dioxide to be exchanged with the outside environment**
6. yolk: **provides a food supply for the embryo**
7. membranes: **cushion and protect the embryo**

Answer the following questions.

8. What is the Mesozoic Era known as? **the Age of Reptiles**
9. How does the three-chambered heart function for the reptile? **The lower chamber is partially divided, keeping oxygen-rich blood coming from the lungs separated from the blood containing carbon dioxide returning from the body.**
10. Name the three orders of reptiles. **turtles; crocodiles and alligators; and lizards and snakes**

63

▲ ENRICHMENT 63

NAME ______ DATE ______ CLASS ______

ENRICHMENT — Chapter 15

Reptiles — Text Pages 363–366

WHERE DID ALL THE DINOSAURS GO?

Dinosaurs lived on Earth during the Mesozoic Era, about 225 million to 65 million years ago. Some were as long as a couple of school buses and others were as small as chickens. Although dinosaurs are classified as reptiles, they may be more closely related to our modern-day birds than to today's reptiles.

The climate during most of the Mesozoic Era was almost tropical. Earth had one large land mass that was beginning to break up into the continents, and dinosaurs could move easily from place to place over these land connections. The most common plants were cone-bearing trees, ferns, cycads, and mosses. Crocodiles, frogs, lizards, insects, and a few small mammals lived on the land. Some reptiles could fly, but there were no birds.

By the end of the Mesozoic Era, the climate had become cooler and drier. Trees, such as oaks, maples, poplars, and redwoods, grew in forests and flowering plants were common. Birds and snakes were present, and modern fish and huge turtles lived in the sea. And the dinosaurs disappeared from the fossil record.

What happened to the dinosaurs? Did the change in climate and plant and animal life have anything to do with their extinction? Scientists have several theories that are based on climatic changes.

One theory is that a large star exploded and the radiation given off caused Earth to experience cold weather for thousands of years. The dinosaurs simply couldn't survive the cold conditions. Another theory is that new plants that developed during the latter part of the Mesozoic Era couldn't be eaten by the plant-eating dinosaurs. The plant-eaters starved and therefore, the meat-eaters had no animals to eat. Other scientists think a giant asteroid may have collided with Earth and the dust cloud that resulted blocked the sunlight for several months. With little sunlight, most of the plants died, then the plant-eaters died. Finally the meat-eaters died when the number of prey was greatly reduced.

No one really knows what happened to the dinosaurs. The explanation may be a combination of many theories.

Answer the following questions.

1. Think of your own theory to explain the disappearance of dinosaurs. **Answers will vary, but may include the emergence of a predator or a disease that wiped out the dinosaur population.**
2. How do you think scientists can find out more about why dinosaurs died out? **Answers will vary, but may include study of dinosaur fossils, fossils of plants and animals of the same period, climatic changes of the period and their possible causes.**

63

REVEALING MISCONCEPTIONS

▶ Many people think snakes sting with their tongues. Actually, snakes use their tongues to detect chemicals.

CONCEPT DEVELOPMENT

▶ Reptiles, being cold-blooded, turn their bodies to get the maximum or minimum amount of heat from the sun to maintain an almost constant body temperature.

CHECK FOR UNDERSTANDING

Have students make a list of differences between reptiles and amphibians.

RETEACH

Prepare a quiz game with photographs or descriptions of lizards, turtles, alligators, crocodiles, snakes, frogs, toads, and salamanders. Let students classify the animals as amphibians or as reptiles and name a characteristic of each.

EXTENSION

For students who have mastered this section, use the **Reinforcement** and **Enrichment** masters or other OPTIONS provided.

MINI QUIZ

Use the Mini Quiz to check students' recall of chapter content.

1. **Three characteristics of reptiles are _____.** *cold-blooded vertebrate; dry, scaly skin; lay eggs covered with a leathery shell*
2. **Reptiles breathe with _____.** *lungs*
3. **The reptile heart has _____ chambers.** *three*
4. **The Age of Reptiles was the _____ Era.** *Mesozoic*

TECHNOLOGY

For more information, see "Viper Venom for Stroke" in *Prevention,* May, 1989.
Think Critically: as soon as possible to dissolve the clot and restore blood supply to affected areas

3 CLOSE

▶ Ask questions 1-4 and the **Apply** Question in the Section Review.

SECTION REVIEW ANSWERS

1. a thick, dry, waterproof skin covered with scales; four legs with claws to hold the body off the ground (with the exception of snakes); three-chambered heart; lungs
2. The earliest reptiles were called cotylosaurs. Many species evolved from them. Some returned to the water; others developed large, heavy, scale-like plates, and speed. There were at least two species of flying reptiles.
3. Turtles are covered with a hard shell on both top and bottom. Most can withdraw into this shell for protection.
4. Snakes don't have legs, eyelids, or external ears. Lizards have external ears, movable eyelids, and legs with clawed toes.
5. Apply: Predators think they are poisonous snakes and leave them alone.
6. Connect to Chemistry: red blood cells and the nervous system

Skillbuilder
ASSESSMENT
Performance: Use this Concept Map to have students Compare and Contrast crocodiles and alligators. Have students identify these two groups from photographs.

To assess this product, refer to the Performance Task Assessment Lists in **Performance Assessment in Middle School Science.**

TECHNOLOGY

Snake Oil Medicines

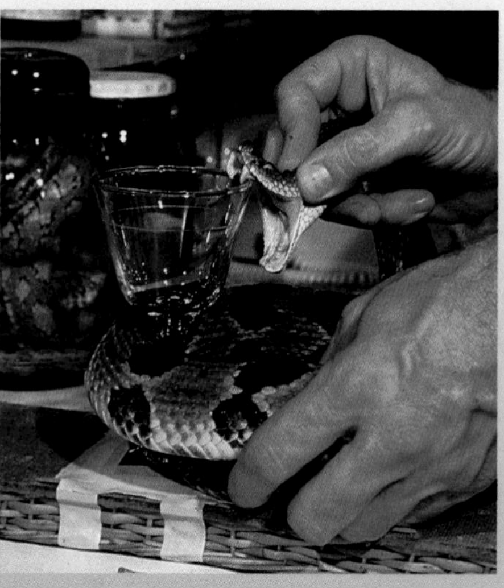

Scientists are finding potential medicines in unusual places. Ancrod, an ingredient in the venom of Malayan pit vipers, may help human stroke victims. Ancrod is a substance that stops blood from clotting. The victim of the pit viper dies from the large dose of ancrod it receives. In the correct doses, ancrod can prevent the damage of strokes caused by blood clots that block blood vessels in the brain of a human.

The snake venom is collected from the snakes by pressing their upper jaw against the top of a container. This causes the venom to drip from the snake's fangs.
Think Critically: About how long after a stroke would this treatment need to be tried to be successful?

SECTION REVIEW

1. What adaptations do reptiles have for living on land?
2. Describe the ancient reptiles.
3. How do turtles differ from other groups of reptiles?
4. Explain how snakes are different from lizards.
5. **Apply:** Poisonous coral snakes and some harmless snakes have bright red, yellow, and black colors. How is this an advantage to nonpoisonous snakes?
6. **Connect to Chemistry:** Snake venom contains poisonous proteins or toxins. What do you think hemotoxins and neurotoxins attack?

Skill Builder

Concept Mapping

Make a concept map showing the major characteristics and orders of the reptile class. If you need help, refer to Concept Mapping in the **Skill Handbook** on pages 688 and 689.

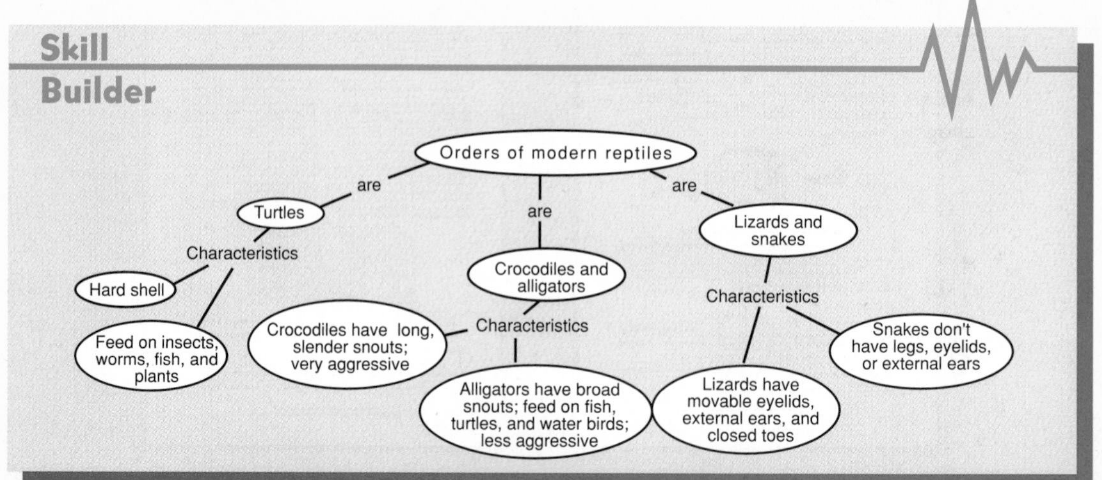

CHAPTER
REVIEW

SUMMARY

15-1: Fish

1. The phylum Chordata includes lancelets, tunicates, and vertebrates: all have a notochord, dorsal nerve cord, and gill slits.
2. Warm-blooded animals maintain a constant body temperature. The body temperature of a cold-blooded animal changes with its environment.
3. Classes of fish include jawless fish, cartilaginous fish, and bony fish.

15-2: Amphibians

1. Amphibians are cold-blooded vertebrates that live both in water and on land. The adaptations of amphibians include moist skin, mucus glands, and lungs.
2. Frogs live in wet areas, have jumping legs, and sensory organs. Toads live in drier areas and have hopping legs. Salamanders have smooth, moist skin and short legs.
3. Frogs go through metamorphosis from egg, to tadpole, to adult. Legs develop, lungs replace gills, and the tail is lost.

15-3: Science and Society: Amphibian Population Decline

1. Amphibian populations are declining due to habitat destruction, acid rain, pesticides, and predation.
2. Amphibians are sensitive to pollution because they have no scales or shells for protection. They also absorb materials through their skin, including pollutants.

15-4: Reptiles

1. Reptiles are true land animals having dry, scaly skin. They lay amniote eggs with leathery shells.
2. Early reptiles were successful because of their adaptations to living on land.
3. Reptile groups include turtles with tough shells, meat-eating crocodiles and alligators, and the largest group including snakes and lizards.

KEY SCIENCE WORDS

a. **amniote egg**
b. **amphibians**
c. **biological indicators**
d. **cartilage**
e. **chordates**
f. **cold-blooded**
g. **dorsal nerve cord**
h. **endoskeleton**
i. **estivation**
j. **fins**
k. **fish**
l. **gill slits**
m. **hibernation**
n. **notochord**
o. **predator**
p. **prey**
q. **reptile**
r. **scales**
s. **warm-blooded**

UNDERSTANDING VOCABULARY

Match each phrase with the correct term from the list of Key Science Words.

1. dorsal structure that becomes the backbone
2. cold-blooded land and water animals
3. the internal skeleton
4. tough, flexible tissue
5. organism that kills others for food
6. inactivity in winter
7. inactivity in summer
8. cold-blooded organisms with dry skin
9. fanlike steering structures in fish
10. species that reflect the condition of the environment.

OPTIONS

ASSESSMENT

To assess student understanding of material in this chapter, use the resources listed.

COOPERATIVE LEARNING

Consider using cooperative learning in the THINK AND WRITE CRITICALLY, APPLY, and MORE SKILL BUILDERS sections of the Chapter Review.

PROGRAM RESOURCES

From the **Teacher Resource Package** use:

Chapter Review, pages 33-34.

Chapter and Unit Tests, pages 97-100, Chapter Test.

CHAPTER
REVIEW

SUMMARY

Have students read the summary statements to review the major concepts of the chapter.

UNDERSTANDING VOCABULARY

1. n
2. b
3. h
4. d
5. o
6. m
7. i
8. q
9. j
10. c

ASSESSMENT

Portfolio

Encourage students to place in their portfolios one or two items of what they consider to be their best work. For each item, ask students to explain why that item was chosen and what they learned from it. Items might be selected from the following.

- Cooperative Learning report on sharks, p. 352
- Skillbuilder sequencing of frog metamorphosis, p. 359
- You Decide! issue controversy on clear-cutting of forests, p. 361

Performance

Additional performance assessments may be found in *Performance Assessment* and *Science Integration Activities* that accompany **Merrill Life Science.** Performance Task Assessment Lists and rubrics for evaluating these activities and other products generated throughout the chapter can be found in Glencoe's *Performance Assessment in Middle School Science.*

CHAPTER

REVIEW

CHECKING CONCEPTS

1. d	**6.** a
2. b	**7.** d
3. d	**8.** a
4. a	**9.** a
5. c	**10.** d

USING LAB SKILLS

ASSESSMENT

Use these alternate lab exercises to assess students' understanding of skills used in this chapter.

11. from under the gill flaps; a few minutes

12. Frogs have lungs to breathe air and strong legs for mobility on land. Tadpoles breathe with gills and have tails with which to swim in water.

THINK AND WRITE CRITICALLY

13. Fish have gills, a streamlined body, and fins.

14. Lampreys feed on other fish.

15. Amphibians have moist skin with mucous glands; they respire through their skin and their lungs. Reptiles have dry, scaly skin to prevent water loss, and they lay amniote eggs.

16. Frogs undergo metamorphosis. The eggs laid in water hatch into tadpoles. The tadpoles feed on plants while developing legs. Eventually metamorphosis is complete, and the adult moves onto land. The cycle then repeats itself.

CHAPTER

REVIEW

CHECKING CONCEPTS

Choose the word or phrase that completes the sentence.

1. Animals that have fins, scales, and gills are ______.
   a. amphibians **c.** reptiles
   b. crocodiles **d.** fish
2. Jawless fish do not have ______.
   a. cartilage **c.** slimy skin
   b. scales **d.** long tubelike bodies
3. An example of a cartilaginous fish is a ______.
   a. hagfish **c.** perch
   b. lamprey **d.** shark
4. Most fish species belong to the ______ group.
   a. osteichthyes **c.** cartilaginous
   b. jawless **d.** chondrichthyes
5. A fish with both gills and lungs is a ______.
   a. shark **c.** lungfish
   b. ray **d.** perch
6. The group of vertebrates with lungs and moist skin is ______.
   a. amphibians **c.** reptiles
   b. fish **d.** lizards
7. A ______ is *not* a reptile.
   a. lizard **c.** snake
   b. turtle **d.** salamander
8. The largest group of reptiles includes lizards and ______.
   a. snakes **c.** crocodiles
   b. turtles **d.** alligators
9. Reptiles lay ______ eggs.
   a. amniote **c.** jelly-like
   b. brown **d.** hard-shelled
10. At some time in their lives, all ______ have a notochord, dorsal nerve cord, and gill slits.
   a. echinoderms **c.** arthropods
   b. mollusks **d.** chordates

USING LAB SKILLS

11. In Activity 15-1, you determined how water temperature affects fish. Place a small drop of slightly diluted food coloring in front of the mouth of a goldfish or guppy. Where does the colored water come out? How long does it take to appear?
12. In Activity 15-2, you observed metamorphosis in frogs. Based on your observations, what evidence do you have that frogs are adapted for life on land and tadpoles are adapted for living in water?

THINK AND WRITE CRITICALLY

Answer the following questions in your Journal using complete sentences.

13. Describe the structural adaptations that allow fish to live in water.
14. Explain why a lamprey is a predator.
15. Compare and contrast the adaptations of amphibians and reptiles to life on land.
16. Describe the life cycle of a frog.

APPLY

17. Accept all logical answers with appropriate support.

18. The amniote egg provides the developing embryo protection from drying out and from being eaten. An adequate supply of food is also provided. Frogs lay eggs in large quantities in the water. The eggs have no shells but do have a food supply. These eggs are in danger of being eaten or drying up. The large quantities laid help ensure the survival of some of the eggs.

19. The skin of amphibians absorbs oxygen, as well as any poisonous gases or chemicals in the area. Because of this, amphibians can be considered biological indicators—species that reflect the condition of the environment.

20. Tunicates and sponges are both sessile filter feeders.

21. All vertebrates have, at some time in their life, a notochord that becomes the vertebrae in vertebrates, a dorsal nerve cord, and gill slits. They also all have an endoskeleton.

APPLY

17. Why do you think there are fewer species of amphibians on Earth than any other type of vertebrate?
18. What is the advantage of the amniotic egg over the eggs laid by amphibians?
19. Why are amphibians considered biological indicators?
20. In what ways are tunicates similar to sponges?
21. What features are common to all vertebrates?

MORE SKILL BUILDERS

If you need help, refer to the Skill Handbook.

1. **Sequencing:** Sequence the order in which these structures appeared in evolutionary history and explain what type of organism had this adaptation and why: skin has mucus glands; skin has scales; fins used for movement; dry, scaly skin
2. **Concept Mapping:** Complete the concept map below describing the chordates.

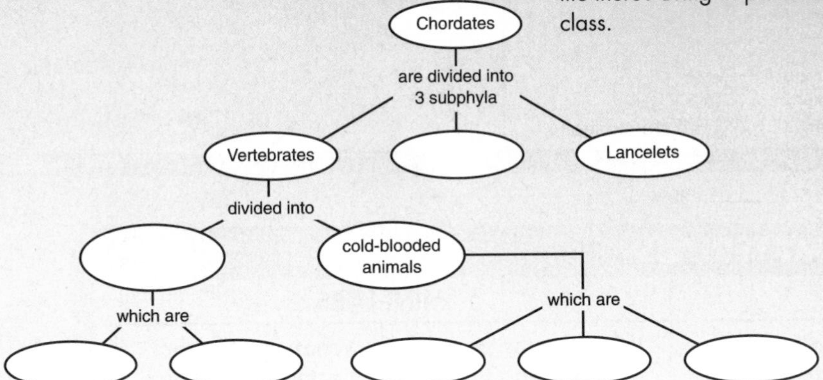

3. **Comparing and Contrasting:** Make a chart comparing the features of fish, amphibians, and reptiles.
4. **Designing an Experiment:** Design an experiment to find out the effect of water temperature on frog egg development.
5. **Making and Using Graphs:** Make a pie graph of the species of fish in the table below.

Classes of Fish	# of Species
Jawless Fish	45 species
Chondrichthyes	275 species
Osteichthyes	25 000 species

PROJECTS

1. Check the local library or nature center and find out what kinds of amphibians and reptiles are common in your area. Arrange for a local naturalist to visit class with representative animals from your area.
2. Study the tropical rain forests' amphibians. What special adaptations do they have for life there? Bring in pictures to share with the class.

MORE SKILL BUILDERS

1. Sequencing: Scales and fins are present in fish, allowing them to move in water; skin with mucous glands is present in amphibians, which began to live on dry land; dry, scaly skin developed in reptiles that lived their life on land.

2. Concept Mapping: See below, left.

3. Comparing and Contrasting: Students' charts should list fish, amphibians, and reptiles down the left side of the chart. Across the top they should have the following categories: habitat, food, body features, and reproduction. Characteristics can be found in text. Accept all logical charts.

4. Designing an Experiment: Students should design experiments with a control—a set temperature—and groups with higher or lower temperatures—the variables.

5. Making and Using Graphs: Students should find that 90% of the graph or 324° represents osteichthyes, 9% of the graph or 32° represents amphibians, about 1% or 4° represents chondrichthyes, and about 0.16% or about 0.6° represents jawless fish.

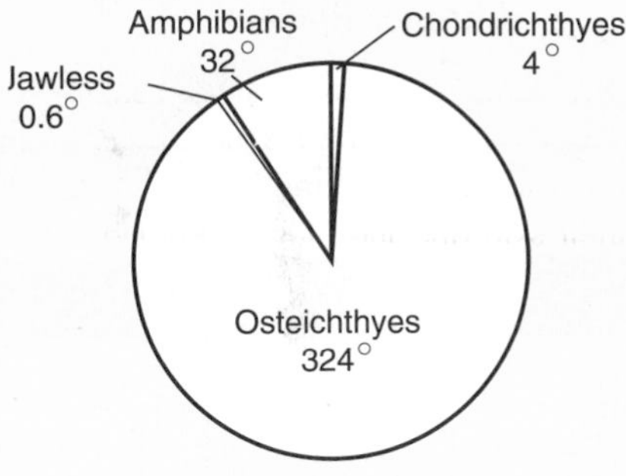

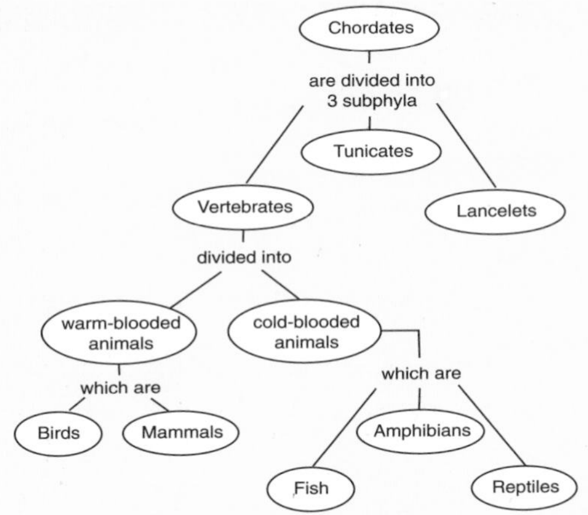

CHAPTER

16 Warm-Blooded Animals

CHAPTER SECTION	OBJECTIVES	ACTIVITIES
16-1 Birds (3 days)	1. **Identify** the characteristics of birds. 2. **Identify** the adaptations birds have for flight. 3. **Explain** how birds reproduce and develop.	**MINI-Lab:** *What are the parts of a bird's egg?* p. 375 **Activity 16-1:** *Designing an Experiment (Investigating Adaptations of Birds),* p. 378
16-2 Mammals (3 days)	1. **Identify** the characteristics of mammals and **explain** how they enable mammals to adapt to different environments. 2. **Distinguish** among monotremes, marsupials, and placental mammals. 3. **Compare** reproduction and development in the three kinds of mammals.	**MINI-Lab:** *What are the characteristics of hair?* p. 380
16-3 Saving the Manatee Science & Society (1 day)	1. **Identify** the characteristics of manatees. 2. **Explain** the major threats to manatees today.	**Activity 16-2:** *Classifying Vertebrates,* p. 388
Chapter Review		

ACTIVITY MATERIALS

FIND OUT	ACTIVITIES		MINI-LABS	
Page 371 1 mortar and pestle 1 tablespoon of seeds 1 teaspoon of gravel	**16-1 Designing an Experiment, p. 378** contour feather down feather hand lens scissors pictures of bird beaks pictures of bird feet	**16-2 Classifying Vertebrates, p. 388** paper pencil animals or pictures of animals	**What are the parts of a bird's egg? p. 375** chicken eggs shallow bowls probes	**What are the characteristics of hair? p. 380** slides droppers microscopes

CHAPTER FEATURES	TEACHER RESOURCE PACKAGE	OTHER RESOURCES
Technology: *Healthier Eggs,* p. 376 **Skill Builder:** *Comparing and Contrasting,* p. 377	**Ability Level Worksheets** ◆ ***Study Guide,*** p. 64 ● ***Reinforcement,*** p. 64 ▲ ***Enrichment,*** p. 64 **Activity Worksheets,** pp. 140-141, 146 **Cross-Curricular Connections,** p. 20 **Science and Society,** p. 20 **Transparency Masters,** pp. 57-60	**Color Transparency 29,** Feathers **Color Transparency 30,** Bird Anatomy **Lab Manual:** Owl Pellets, pp. 113-114 **Science Integration Activity 16** **STVS:** Disc 5, Side 2
Problem Solving: *What Colors Can Spot See?* p. 384 **Skill Builder:** *Classifying,* p. 385	**Ability Level Worksheets** ◆ ***Study Guide,*** p. 65 ● ***Reinforcement,*** p. 65 ▲ ***Enrichment,*** p. 65 **Activity Worksheets,** p. 147 **Critical Thinking/Problem Solving,** p. 20 **Concept Mapping,** p. 37	**STVS:** Disc 5, Side 2
You Decide! p. 387	**Ability Level Worksheets** ◆ ***Study Guide,*** p. 66 ● ***Reinforcement,*** p. 66 ▲ ***Enrichment,*** p. 66 **Activity Worksheets,** pp. 142-143	
Summary Key Science Words Understanding Vocabulary Checking Concepts Using Lab Skills Think & Write Critically Apply More Skill Builders Projects	ASSESSMENT RESOURCES **Chapter Review,** pp. 35-36 **Chapter Test,** pp. 101-104 **Performance Assessment in Middle School Science (PAMSS)**	**Chapter Review Software** **Test Bank** **Alternate Assessment** **Performance Assessment**

◆ **Basic** ● **Average** ▲ **Advanced**

ADDITIONAL MATERIALS

SOFTWARE	AUDIOVISUAL	BOOKS/MAGAZINES
Animal Reproduction, J&S Software.	*In the Company of Birds,* film, Information Service of India. *Marine Life Preview,* film, Marineland of Florida. *Birds and Their Characteristics,* film, Coronet/MTI. *Mammals and Their Characteristics,* film, Coronet/MTI. *Animal Olympians (NOVA),* laserdisc, Image Entertainment. *Animal Series (4),* laserdisc, EBEC. *Desert Whales (Cousteau),* laserdisc, Image Entertainment. *Encyclopedia of Animals: Carnivores and Sea Mammals; Ostriches, Penguins, Birds of Prey, Cranes; Herbivores; Primates,* laserdisc, Optical Data Corporation.	Dorst, Jean. *The Life of Birds.* New York: Columbia University Press, 1974. Young, John. *The Life of Vertebrates.* 3rd ed. New York: Oxford University Press, 1981.

CHAPTER

16

WARM-BLOODED ANIMALS

THEME DEVELOPMENT: A discussion of adaptations of birds to life in the air and on land emphasizes the steps in the evolution of species that have moved from water to land environments. The theme of evolution is continued in the discussion of mammal characteristics and origins. Both birds and mammals have developed homeostatic mechanisms for living on land.

CHAPTER OVERVIEW

- **Section 16-1:** This section describes the internal and external characteristics of birds and how these characteristics show adaptations for flight. Kinds of birds, the origin of birds, and their importance are also discussed.
- **Section 16-2:** This section continues to build on the concept of diversity through adaptation and classification. The orders of mammals are listed, and the section concludes with the origin and importance of mammals.
- **Section 16-3: Science and Society:** Reasons why manatees were at one time heavily hunted are discussed, and the major threats to manatees today are explored. The You Decide feature asks students what they think about spending large amounts of money on saving injured manatees.

CHAPTER VOCABULARY

incubate	placental mammals
contour feathers	gestation period
down feathers	placenta
preening	umbilical cord
mammals	marsupials
mammary glands	monotremes
herbivores	manatees
carnivores	poaching
omnivores	

CHAPTER

16 Warm-Blooded Animals

OPTIONS

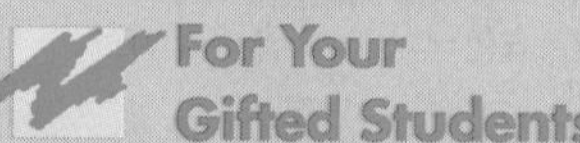

For Your Gifted Students

- Ask students to contact a local birding chapter to see if they can find a member to talk to the class about starting a bird watching club.
- Ask students to draw or take pictures of the different birds found around school or their home. They should identify the birds and start a bird diary. Have them note the time of day, location, and any other interesting observations.

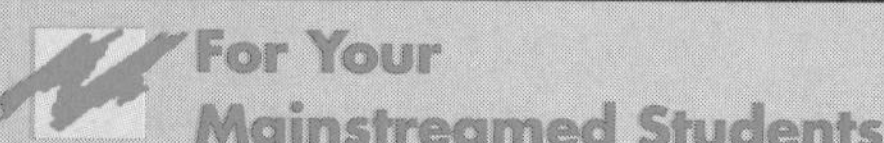

For Your Mainstreamed Students

Have students make or find a wall map of the world. On that map, they can place drawings and magazine pictures of some of the mammals that occupy the different areas of the world. Students can visit a local zoo to take pictures of animals to add to the map.

Have you ever watched birds eat from a feeder? If you have, you know that they eat one seed after another without stopping. Birds don't chew their food before swallowing, as you do. They don't have teeth, but they do have a muscular grinding sac called a gizzard. How does the gizzard work?

FIND OUT!

Do this simple activity to find out how a bird's gizzard works.

Place some cracked corn, birdseed, sunflower seeds, nuts, and some gravel in a mortar. Use a pestle to grind the seeds. What do the seeds look like now? A bird's gizzard works in much the same way. The gizzard has no teeth, but contains gravel to crush the food. *Infer* how a bird's gizzard helps in digestion.

Gearing Up

Previewing the Chapter

Use this outline to help you focus on important ideas in this chapter.

Section 16-1 Birds
- Characteristics of Birds
- Kinds of Birds
- Origin and Importance of Birds

Section 16-2 Mammals
- Characteristics of Mammals
- Kinds of Mammals
- Origin and Importance of Mammals

Section 16-3 Science and Society
Saving the Manatee
- "Sea Cows"?!
- The Future of Manatees

Previewing Science Skills

- In the Skill Builders, you will compare and contrast, and classify.
- In the Activities, you will observe, predict, infer, and classify.
- In the MINI-Labs, you will observe, infer, and describe.

What's next?

In this chapter, you'll learn more about birds. You'll learn how they are adapted for flight and how they live. You'll also learn about mammals, the class to which you belong. You will learn how mammals are classified and how they live.

INTRODUCING THE CHAPTER

Use the Find Out activity to introduce students to characteristics of birds. Explain that they will be learning more about the external and internal characteristics of both birds and mammals in this chapter.

FIND OUT!

Preparation: Purchase packaged bird seed and mix with some corn and sunflower seeds. Collect gravel or pebbles.

Materials: one mortar and pestle, approximately one tablespoon of seeds, and one teaspoon of gravel for every three students

Cooperative Learning: Group students into Science Investigation groups. Assign one student to be the reader, one to get the materials and grind the seeds, and one to be the recorder.

Teaching Tips
- You may want to demonstrate how to use a mortar and pestle if students have not used one before.
- Have the recorder time how long it takes to grind the seeds into small pieces.
- Provide a container to collect the ground seeds. They can be placed in bird feeders.

Gearing Up

Have students study the Gearing Up feature to familiarize themselves with the chapter. Discuss the relationships of the topics in the outline.

What's Next?

Before beginning the first section, make sure students understand the connection between the Find Out activity and the topics to follow.

ASSESSMENT OPTIONS

PORTFOLIO

Refer to page 389 for suggested items that students might select for their protfolios.

PERFORMANCE ASSESSMENT

See page 389 for additional Performance Assessment options.

Process
Skillbuilders, pp. 377, 385
MINI-Lab, p. 380
Activities, 16-1, p. 378; 16-2, p. 388
Using Lab Skills, p. 390

CONTENT ASSESSMENT

Assessment—Oral, pp. 374, 379
Section Reviews, pp. 377, 385, 387
Chapter Review, pp. 389-91

GROUP ASSESSMENT

Opportunities for group assessment occur with Cooperative Learning Strategies and Flex Your Brain Activities.

SECTION 16-1

PREPARATION

SECTION BACKGROUND

- Birds and mammals are warm-blooded, or endothermic, vertebrates. They produce heat to stay warm and lose heat to stay cool. Endothermic vertebrates have higher metabolic rates than cold-blooded vertebrates. They have developed adaptations such as feathers, fur, body hair, and layers of fat for retaining heat. They sweat and pant to release body heat.
- Only birds and mammals have complete four-chambered hearts. A four-chambered heart keeps oxygenated blood completely separate from deoxygenated blood.

PREPLANNING

- Make plans to obtain a live bird for the Motivate activity.
- Obtain feathers for Activity 16-1.

In Your JOURNAL

Bird species will vary with geographic location. Feeders may be used or students might plant specific plant species that attract specific birds.

1 MOTIVATE

- If someone in the class has a pet bird (canary, parakeet, parrot), ask that student to bring it to class so the students can observe its characteristics and behavior.

16-1 Birds

New Science Words

incubate
contour feathers
down feathers
preening

Objectives

- **Identify the characteristics of birds.**
- **Identify the adaptations birds have for flight.**
- **Explain how birds reproduce and develop.**

Figure 16-1. Birds are the only organisms with feathers. Most birds incubate their eggs until they hatch.

Characteristics of Birds

There probably isn't a day that goes by in which you don't see a bird. Birds are found on rooftops, backyards, city streets, and in farmers' fields. What are the characteristics that make a bird a bird? First, unlike cold-blooded reptiles and amphibians, birds are warm-blooded. Warm-blooded organisms maintain a constant body temperature. A bird's body temperature is about 40°C. You are warm-blooded also and maintain a body temperature of 37°C.

Second, birds have feathers and scales. They are the only members of the animal kingdom that have feathers. A third characteristic shared by all birds is they lay eggs enclosed in a hard shell. The parents keep the eggs warm, or **incubate** them, until they hatch. Fourth, the front legs of all birds are modified into wings. The two hind legs support the body and usually have claws.

People have always been fascinated by the ability of birds to fly. Now that you know the major characteristics of birds, let's look at the adaptations birds have for flight.

In Your JOURNAL

Form a group in your class to develop a bird-feeding plan for your area of the country. *Determine* which species you should be feeding. **In your Journal,** draw the kinds of feeders you will use. List the locations and kinds of food these birds will need.

Adaptations for Flight

Flight requires a light, yet strong skeleton, wings, and feathers to help lift the bird off the ground. Keen senses, especially eyesight, and tremendous amounts of energy are also needed for flight.

A bird's body is covered with two types of feathers, contour feathers and down feathers. Strong, lightweight **contour feathers** give birds their coloring and smooth, sleek shape. These are the feathers birds use to fly. The long contour feathers on the wings and tail help the bird steer and keep from tipping over. Have you ever watched ducks swimming in a pond on a freezing cold day, and

OPTIONS

Meeting Different Ability Levels

For Section 16-1, use the following **Teacher Resource Masters** depending upon individual students' needs.

- ◆ **Study Guide Master** for all students.
- ● **Reinforcement Master** for students of average and above average ability levels.
- ▲ **Enrichment Master** for above average students.

Additional Teacher Resource Package masters are listed in the OPTIONS box throughout the section. The additional masters are appropriate for all students.

◆ STUDY GUIDE 64

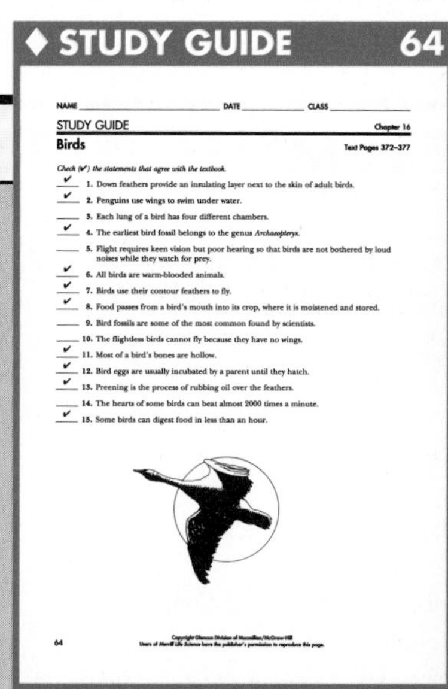

NAME ______ DATE ______ CLASS ______

STUDY GUIDE — Chapter 16

Birds — Text Pages 372–377

Check (✔) the statements that agree with the textbook.

✔ 1. Down feathers provide an insulating layer next to the skin of adult birds.
✔ 2. Penguins use wings to swim under water.
___ 3. Each lung of a bird has four different chambers.
✔ 4. The earliest bird fossil belongs to the genus *Archaeopteryx*.
___ 5. Flight requires keen vision but poor hearing so that birds are not bothered by loud noises while they watch for prey.
✔ 6. All birds are warm-blooded animals.
✔ 7. Birds use their contour feathers to fly.
✔ 8. Food passes from a bird's mouth into its crop, where it is moistened and stored.
___ 9. Bird fossils are some of the most common found by scientists.
___ 10. The flightless birds cannot fly because they have no wings.
✔ 11. Most of a bird's bones are hollow.
✔ 12. Bird eggs are usually incubated by a parent until they hatch.
✔ 13. Preening is the process of rubbing oil over the feathers.
___ 14. The hearts of some birds can beat almost 2000 times a minute.
✔ 15. Some birds can digest food in less than an hour.

64

wondered how they keep warm? Soft, fluffy **down feathers** provide an insulating layer next to the skin of adult birds and cover the bodies of young birds. Feathers help birds maintain their constant body temperature. They grow in much the same way as your hair grows. The base of each feather shaft is rooted in a tiny follicle and receives nutrients from the blood. When it is fully grown, the feather either falls off or is pushed out by a new feather. As you can see in Figure 16-2, the shaft of the feather has two vanes, each with many branches called barbs. Each barb has many cross braches that give the feather strength.

To care for its feathers, a bird has an oil gland located just above the base of its tail. Using its beak, the bird rubs oil from the gland over its feathers in a process called **preening.** The oil conditions the feathers and helps make them water repellent.

A bird sheds its feathers just as a reptile sheds its skin. Whereas a reptile sheds its skin all at once, a bird sheds and replaces its feathers a few at a time. The shedding of old feathers accompanied by the growth of new ones is called molting.

Flight requires that a bird have very sharp vision and hearing. Keen vision is necessary for taking off, landing, and hunting. For example, at night an owl can see a mouse on the forest floor from high in the branches of a tree. Acute hearing helps some species of bird locate prey, and it allows all birds to sense danger. Some of the adaptations birds have for flight can't be seen, because they are internal. One of these is a bird's skeleton. Many of the bones in a bird's skeleton are fused for extra strength and more stability for flight.To make a bird lighter, most of its bones

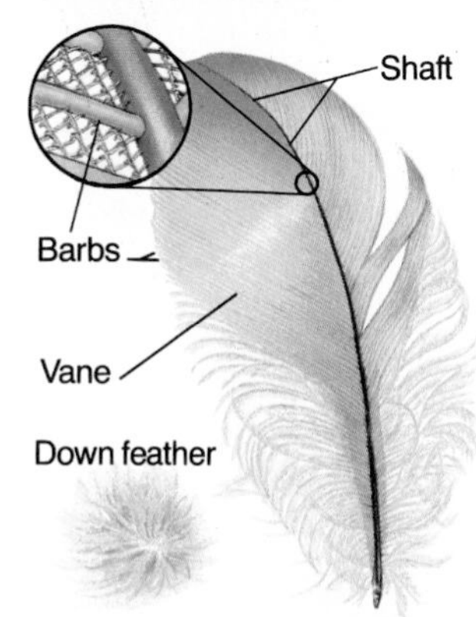

Figure 16-2. Preening conditions feathers and helps make them waterproof.

Figure 16-3. Keen vision allows some birds to locate prey from a long distance.

TYING TO PREVIOUS KNOWLEDGE: Compare reptile characteristics from Chapter 15 with the characteristics of birds. Point out the traits of reptiles that are modified in birds.

2 TEACH

Key Concepts are highlighted.

CONCEPT DEVELOPMENT

▶ **Demonstration:** Divide the class into pairs. Have one student keep time and count while the other holds his or her arms straight out and flaps. See how many times a person can flap in one minute and how long it takes for the person to tire. Explain that just as our leg muscles are well-developed for walking, the chest muscles of most birds are well-developed for flying.

▶ Explain that birds belong to phylum Chordata and class Aves. *Aves* is the Latin word for bird and it is the root of bird words such as *avian* and *aviary.*

▶ Explain that birds preen their feathers to clean them and remove parasites. Preening also helps zip together separated barbs. Many birds spread oil on their feathers when they preen. The oil helps to keep birds' feathers water-repellant.

▶ **Demonstration:** To show how oil helps waterproof feathers, cut two 6-cm squares of construction paper. Cover one with petroleum jelly and leave the other one plain. Drop 6 drops of water on each square. Have students write a sentence or two about what they observe.

OBJECTIVES AND SCIENCE WORDS: Have students review the objectives and science words throughout the chapter as they study each section.

VideoDisc

STVS: Peregrine Falcon, Disc 5, Side 2

STVS: Sexing Birds, Disc 5, Side 2

● REINFORCEMENT 64

NAME ______ DATE ______ CLASS ______

REINFORCEMENT — Chapter 16

Birds — Text Pages 372-377

Describe the four characteristics of birds.

1. **Birds are warm-blooded. They keep a constant body temperature of about 40°C.**
2. **Birds are covered with feathers.**
3. **Birds lay eggs enclosed in a hard shell.**
4. **Birds have two front legs that have been modified into wings.**

Label the drawing of a contour feather with the names of its parts.

5. **vane**
6. **barb**
7. **shaft**

Write the name of the organ responsible for each step in the digestive process of a bird.

crop 8. where food is moistened and stored
stomach 9. where food is partially digested
gizzard 10. where grain, seeds, and other foods are ground and crushed
small intestine 11. where the rest of digestion occurs

Answer the following questions.

12. How do the lungs of birds help make the body lighter? **Balloonlike air sacs are attached to each lung. The air sacs spread to various parts of the body, including the hollow bones.**
13. What is an amniotic egg? **It is an egg that provides a complete environment for the developing embryo.**
14. What is the purpose of preening? **Preening spreads oil over the feathers, making them water repellent.**
15. Compare and contrast contour feathers and down feathers. **Contour feathers are long and sleek, while down feathers are soft and fluffy. Contour feathers are used in flight, while down feathers are used for insulation.**

64

▲ ENRICHMENT 64

NAME ______ DATE ______ CLASS ______

ENRICHMENT — Chapter 16

Birds — Text Pages 372–377

DETERGENTS THAT HARM BIRDS

You will observe the effects of detergent on oil in the following experiment.

Materials

- large clear, glass bowl
- mineral oil
- powdered detergent
- teaspoon

Procedure

1. Pour about a cup of clear water into the bowl.
2. Add one teaspoon of mineral oil. Observe what happens to the oil.
3. Sprinkle two teaspoons of detergent over the liquid. Then gently stir the mixture.
4. Observe what happens to the oil.

Conclude and Apply

1. What happened to the oil when you added it to the water? **The oil stayed on the surface.**
2. What happened to the oil when you added the detergent? **Some of the oil sank to the bottom of the bowl and some of it stayed at the surface in bubbles.**
3. What do birds do when they preen? **They use their beaks to rub oil over their feathers.**
4. What does a bird's oil do for its feathers? **The oil waterproofs the feathers.**
5. What would happen if a bird got detergent on its feathers? **The oil would ball up and the feathers would no longer be waterproof.**
6. When the detergent we use reaches streams, rivers, and lakes, how are some birds probably harmed? **The detergent pollution robs the birds of their natural waterproofing.**

64

CONCEPT DEVELOPMENT

▶ Explain that airplanes are designed to use an airfoil similar to those of flying birds. Lift and streamlining enable both birds and airplanes to fly.

Answer: Cold-blooded animals maintain body temperature from sunlight or warm objects. Warm-blooded animals generate body heat as a result of respiration. **Note:** Students must also be aware that cold-blooded animals do undergo respiration.

TEACHER F.Y.I.

▶ Large eyes located on the sides of the head in most birds allow birds to see much better than other animals, including people. The eyes focus independently of each other so a bird can see two different images at the same time. This is called monocular vision. Many predatory birds such as owls and eagles can focus straight ahead with both eyes focusing on the same image (like humans). Both eyes focusing on the same image is called binocular vision. This allows the bird to judge depth and distance and helps the bird capture prey. Most birds can see in color, which helps them find food.

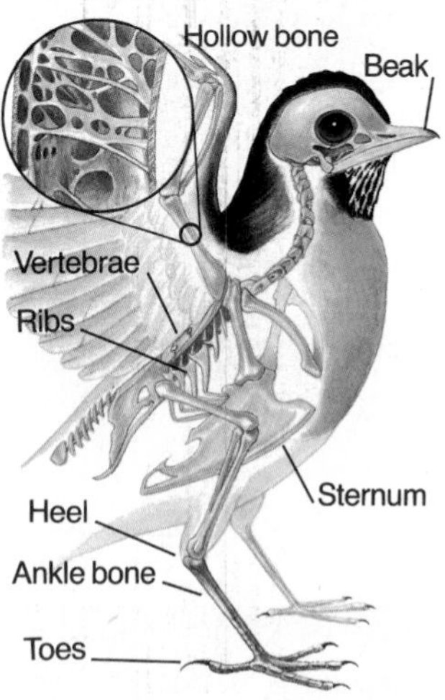

Figure 16-4. The hollow bones of birds are an adaptation for flight.

Compare how cold-blooded animals and warm-blooded animals maintain body temperatures.

are hollow with thin cross braces of bone inside for strength and support. The hollow spaces are filled with air. Why do you think it's important for a bird to have a light skeleton?

You can see the skeleton of a bird in Figure 16-4. Notice that the sternum, or breastbone, supports the large chest muscles. The last bones of the spine support the tail feathers, which play an important part in steering and balancing during flight and landing.

Of course, one of the most important adaptations birds have for flight is wings. All birds have wings, even birds that don't fly such as ostriches and penguins. In flying birds, wings are attached to powerful chest muscles. Wings are curved on top and flat or slightly curved on the bottom. The shape gives the bird lift to get off the ground.

Wings also serve important functions for birds that don't fly. Penguins use their wings to swim under water. While running or walking, ostriches and rheas balance with their wings.

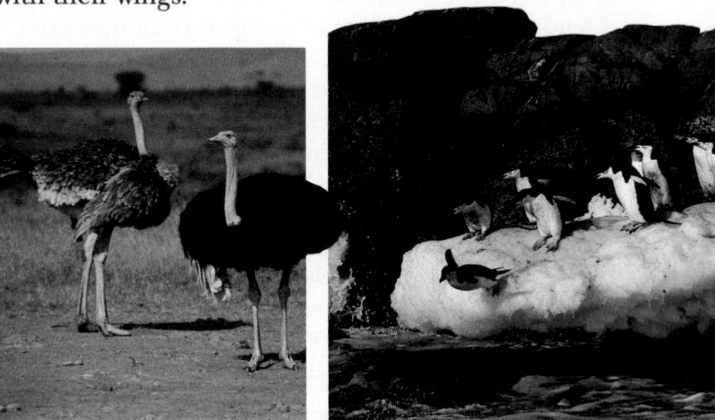

Figure 16-5. Flightless birds use their wings for other purposes. Ostriches (left) use their wings for balance, and penguins (right) use them for swimming.

Digestion and Respiration

Birds get the energy to fly from the food they eat. Because flying requires large amounts of energy, birds need to eat large amounts of food. An efficient digestive system breaks down the food quickly to supply birds with the energy they need. Whereas it takes us as long as a day to digest the food we eat, some birds can digest food in less than an hour.

OPTIONS

ASSESSMENT—ORAL

▶ **How do people benefit from wearing down jackets?** *The down insulates and provides warmth.*

▶ **Where do you think the expressions "free as a bird," "light as a feather," and "birds of a feather" come from?** *Accept any logical answers; freedom in flight, feathers have little mass, and birds of the same kind have the same kind of feathers.*

▶ **Why do most insect-eating birds migrate?** *Certain insects become scarce in areas that become cooler during winter.*

▶ **How are birds' eggs better protected than the eggs of cold-blooded vertebrates?** *Birds' eggs are protected by a hard shell, and the parents incubate them.*

▶ **Why do you think the sense of smell is not as important to birds as the senses of sight and hearing?** *Birds hunt food from the air, where scents do not linger.*

▶ **What is a disadvantage of being warm-blooded?** *It takes more food to maintain body temperature.*

You learned about part of a bird's digestive process in the Find Out activity at the beginning of this chapter. Food first passes from the mouth into the crop, where it is moistened and stored. From there, it moves into the first part of the stomach, where it is partially digested. Then it moves into the gizzard. Small stones and grit in the gizzard grind and crush seeds and other foods. Food then passes into the small intestine, where the rest of digestion occurs. Finally, nutrients are absorbed by the bloodstream.

Birds also need energy to maintain body temperature. Body heat is generated from the energy in food. Oxygen is needed to convert food to energy. To obtain oxygen, birds have an efficient respiratory system. As you can see in Figure 16-6, birds have two lungs with balloonlike air sacs attached to each one. The air sacs spread into different parts of the body, including the hollow bones. These air sacs increase the amount of oxygen a bird can take in. Having these air sacs spread throughout the body also makes a bird lighter.

Birds have a heart with four chambers. Blood with oxygen is kept separate from blood without oxygen. Blood is circulated very quickly because birds have a rapid heartbeat rate. The hearts of some birds beat almost 1000 times in a minute.

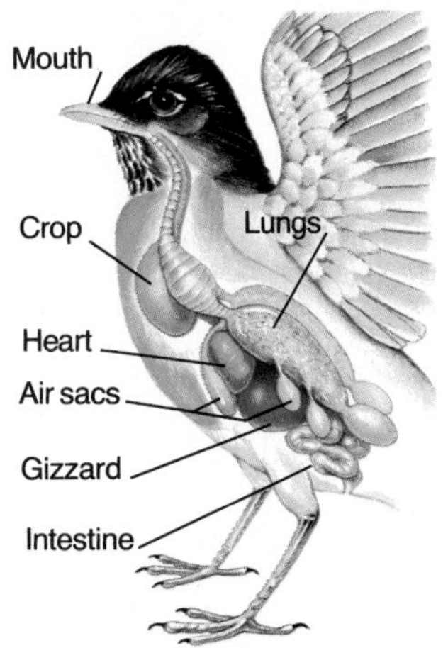

Figure 16-6. A bird's digestive system processes large amounts of food needed for energy. A bird's respiratory system allows a bird to take in large quantities of oxygen.

Reproduction and Development

If you've ever cooked scrambled eggs, you are familiar with a bird's egg. Like reptiles, birds reproduce by laying an amniote egg with a shell. Remember that an amniote egg provides a complete environment for a developing embryo. Fertilization is internal, and the shell is formed around the fertilized egg. Bird eggs have the same membranes to protect and nourish the embryo as reptile eggs. Unlike a reptile's egg though, a bird's egg has a hard shell made of calcium carbonate, the same chemical that makes up seashells, limestone, and marble. Birds also usually stay with their eggs to incubate them, and they care for their young.

The female bird usually lays eggs in a nest. Most birds lay from two to eight eggs at one time. All the eggs laid at one time are called a clutch. The length of time the eggs must be incubated varies with the species and the size of the species.

MINI-Lab

What are the parts of a bird's egg?

Open a chicken egg into a shallow bowl. *Observe* the shell and the contents of the egg. *Identify* the parts of the egg from the diagram below. What do you *infer* about the function of each part?

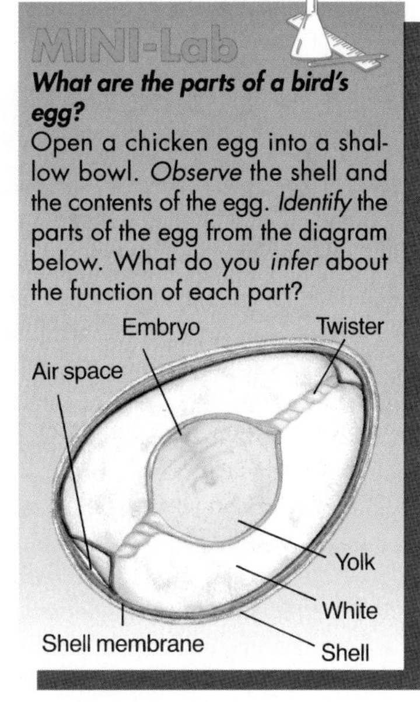

CHECK FOR UNDERSTANDING

▶ Ask questions 1-3 and the **Apply** Question in the Section Review.

▶ Use diagrams to show the internal and external structures of a bird. Point out the digestive tract, oil gland, and air sacs. Have students make a list of the bird's internal and external characteristics.

RETEACH

Cooperative Learning: Form Numbered Heads Together cooperative groups to have students describe how the area near the school can be made into a more favorable habitat for birds.

EXTENSION

For students who have mastered this section, use the **Reinforcement** and **Enrichment** masters or other OPTIONS provided.

MINI-Lab

Materials: chicken eggs, shallow bowls, probes

Teaching Tips

▶ Be sure students wash their hands thoroughly after handling the eggs. Raw eggs carry the bacteria *Salmonella.*

▶ Students should not break the yolk while examining the egg.

Answers: Students should be able to identify the yolk, white, twisters, shell, embryo, shell membrane, and air space.

MINI-Lab

ASSESSMENT

Performance: To assess students' understanding of parts of eggs, have students compare bird eggs with what they know of reptilian eggs.

CONCEPT DEVELOPMENT

▶ Have students research bird statistics, naming the birds with the following characteristics: largest wingspan (Andean condor, albatross, marabou stork), tallest and heaviest (ostrich), earliest known bird (Archaeopteryx), smallest (hummingbird of Cuba), fastest (swift), fastest diver (peregrine), largest eyes (ostrich), largest eggs (ostrich), smallest eggs (hummingbird), fastest swimmer (penguin).

ENRICHMENT

▶ About 80 species of birds have become extinct since the 1600s. Have students research and report on the moa, the dodo, the great auk, the Labrador duck, the heath hen, and the passenger pigeon.

▶ Have students research and report on the life and work of Roger Tory Peterson.

▶ Have students research and report on how birds communicate—song, calls, and drumming noises.

PROGRAM RESOURCES

From the **Teacher Resource Package** use:

Activity Worksheets, page 146, Mini-Lab: What are the parts of a bird's egg?

Transparency Masters, pages 57-58, Feathers.

Science Integration Activities 16, Thermal Energy from Foods

Use **Color Transparency** 29, Feathers.

In Your JOURNAL

The *Exxon Valdez* spill in Alaska in March 1989 is a good example for students to see the effects of an oil spill on birds. Students can review the literature from that incident. Oil economics versus environmental risks should make for a good class debate. See student page 377 for this feature.

TECHNOLOGY

▶ Eggs are a popular food all by themselves, but they are also part of many foods we eat. As a class, list as many different foods that contain eggs as you can.

▶ For more information on low cholesterol eggs see: "Reinventing the Egg" by Michelle Stacey. *New Yorker*, October 1, 1990, pp. 73-88.

Think Critically: Many people are concerned about their cholesterol level. Whatever the chicken eats may be reflected in the egg. Lowering dietary cholesterol for chickens may produce eggs lower in cholesterol.

Kinds of Birds

Almost 9000 species of birds belong to the Class Aves. Within this class, birds are classified into orders based on characteristic beaks, feet, feathers, and other physical features. The four most common orders of birds are: the flightless birds, water birds, birds of prey, and perching birds.

Figure 16-7. The four most common orders of birds are represented by the emu (a), the wood duck (b), the barred owl (c), and the Altamira oriole (d).

TECHNOLOGY

Healthier Eggs

Scientists and farmers are working to lower the amount of cholesterol in eggs. Two major tactics involve changes in feed and in the hen's environment. One way is to feed hens a diet high in fiber and natural vitamins and minerals. Another way is to use iodine-enriched feed, not to lower egg cholesterol, but to help people who eat the egg to process its cholesterol more quickly. Environmental changes have included removing dust from the air, providing artificial light similar to sunlight, treating water with ultraviolet rays, and reducing electromagnetic radiation from machine motors.

Think Critically: How might changing a chicken's diet lower cholesterol in eggs?

OPTIONS

INQUIRY QUESTIONS

▶ **What natural conditions may have favored flightlessness in large birds such as ostriches?** *The lack of predators or few predators.*

▶ **What are the advantages of both parents sharing incubation of eggs?** *The eggs are cared for and kept warm at all times even while one parent has to be away from the nest.*

PROGRAM RESOURCES

From the **Teacher Resource Package** use:

Cross-Curricular Connections, page 20, Bird Migration.

Science and Society, page 20, Saving the Birds of Guam.

Transparency Masters, pages 59-60, Bird Anatomy.

Use **Color Transparency** 30, Bird Anatomy.

Use **Laboratory Manual,** pages 113-114, Owl Pellets.

Origin and Importance of Birds

Scientists have learned about the origins of birds by studying their fossils. Feathers and hollow bones decompose easily. As a result, there are few fossils to study. Those that do exist have convinced scientists that birds developed from reptiles millions of years ago. Today, birds still have some characteristics of reptiles, including reptile-like scales on their feet and legs.

The earliest bird fossils belong to the genus *Archaeopteryx* (ar kee AHP tuh rihks). Fossils of *Archaeopteryx* are about 150 million years old. The fossil image in Figure 16-8 shows that almost all the body had feathers. *Archaeopteryx* also had teeth, a reptile-like tail, solid bones, and three claws on each wing. It may have used its claws to climb trees. No one has been able to figure out if it could fly.

Birds play many very important roles in the environment. Birds eat insects and help control insect populations. Birds of prey control populations of rats and mice. Some birds, such as turkey vultures, feed on dead and decaying materials and in return help clean the environment. Hummingbirds pollinate nectar-producing flowers as they feed on the nectar.

Our endless demand for oil has led to an increased probability of oil spills. **In your Journal,** answer these questions: How do oil spills affect birds? What is being done to prevent these spills?

Figure 16-8. *Archaeopteryx* is considered a link between reptiles and birds.

SECTION REVIEW

1. List four characteristics shared by all birds.
2. How are a bird's feathers, air sacs, and skeleton adaptations for flight?
3. Describe reproduction in birds.
4. **Apply:** DDT, a pesticide, was once used extensively to control populations of insect pests such as mosquitoes. Contact with this pesticide caused eagles and other birds to lay eggs with very thin, brittle shells. What do you think was the effect on the populations of these birds? Explain.
5. **Connect to Physics:** Find out how the shape of a bird's wings enables the bird to fly.

Skill Builder: Comparing and Contrasting

Using the photographs in this chapter, compare and contrast the beaks and feet of flightless birds, water birds, birds of prey, and perching birds. If you need help, refer to Comparing and Contrasting in the **Skill Handbook** on page 683.

INQUIRY QUESTIONS

▶ **What changes would there be in the environment if there were no birds?** *There would be more insects, rats, mice, and decaying materials. Not as many flowers would be pollinated.*

▶ **Why do you think there are so many people who enjoy bird watching?** *People find birds interesting, enjoy their beauty, have a desire to know more about birds. Accept any answer.*

3 CLOSE

STUDENT TEXT QUESTION

▶ Page 377, paragraph 1: **Today, birds still have some characteristics of reptiles. What are some of these characteristics?** *Both lay eggs, have partially hollow bones, have similar types of skulls and earbones, and have scales covering parts of their bodies.*

SECTION REVIEW ANSWERS

1. warm-blooded, feathers, lay eggs with a hard shell, incubate eggs
2. Contour feathers steer and keep the bird upright. The hollow bones in a bird's skeleton makes the bird lighter, as do inflated air sacs.
3. Birds have internal fertilization and lay shelled eggs that are incubated until hatched.
4. **Apply:** The populations decreased because the thin-shelled eggs could not hold up.
5. **Connect to Physics:** A bird's wing is curved on top, only slightly curved on bottom. Air flows faster over the top surface than over the flatter bottom surface. Faster airflow reduces air pressure on top of the wing. Air beneath pushes the wing upward, causing lift. Lift enables the bird to overcome gravity and rise into the air.

Skill Builder

Water birds have beaks modified for dabbling in the mud for food or for catching prey. Their feet are webbed for swimming or have long toes to walk on mud. Perching birds have beaks adapted to eating seeds and insects. Their feet are adapted for perching in trees. Birds of prey have beaks and feet adapted for catching prey. Flightless birds have feet adapted to walking on land and beaks adapted for grazing, eating insects, or hunting.

Skillbuilder ASSESSMENT

Performance: Assess students' abilities to Compare and Contrast by having them examine beaks and feet of birds indigenous to your area.

ACTIVITY 16-1

OBJECTIVE: Design and carry out an experiment to determine how the structure of feathers, beaks, and feet adapt birds for a particular way of life.
Time: one class period

PROCESS SKILLS applied in this activity are **observing, inferring,** and **interpreting data.**

PREPARATION

- Obtain feathers from a biological supply company or a craft store. If obtained elsewhere, be certain they are clean. Most feathers can be rinsed in warm water (no soap) and when dry, placed in a microwave oven for one to two minutes.
- Collect pictures (beaks and feet) for students to use.

Cooperative Learning: Divide the class into Science Investigation Teams.

THINKING CRITICALLY

Feathers are used to fly, maintain body temperature, and to attract mates. A beak's primary function is to get food. Feet are designed for perching, climbing, swimming, and capturing prey.

TEACHING THE ACTIVITY

*Refer to the **Activity Worksheets** for additional information and teaching strategies.*

- For students with allergies, feather can be kept in clear plastic bags and observed.

SUMMING UP/ SHARING RESULTS

- Contour feathers are larger with a more rigid shaft and the barbs are hooked together. Down feathers are smaller, fluffier, and softer. The hairlike filaments are not hooked together, and the shaft is very flexible.
- A contour feather is lightweight, flat, and spread out to push the air. The locked barbs give it strength. The down feather traps air that is warmed by the bird's body. It provides a layer of insulation.
- cardinal — eats seeds, perches; brown pelican — eats fish, swims; American redstart — eats insects, perches; snowy egret — eats fish, wades; mallard — strains food from water, swims; pileated woodpecker — eats insects, climbs; screech owl — eats animals, preys; hummingbird — probes for food, perches; nuthatch — eats insects, climbs; peregrine falcon —eats animals, preys.

GOING FURTHER!

The more streamlined the bird's body, the less air resistance the bird will meet and the less effort it will take to fly. Cardinals should live in wooded areas, snowy egrets near marshes, shall ponds, or streams, pileated woodpeckers in wooded areas, and mallards near lakes, ponds, or streams.

PROGRAM RESOURCES

From the **Teacher Resource Package** use:

Activity Worksheets, pages 140-141, Activity 16-1, Investigating Adaptations of Birds.

Activity ASSESSMENT

Performance: To further assess students' understanding of bird's feathers, see USING LAB SKILLS, Question 11, on page 390.

ACTIVITY 16-1

DESIGNING AN EXPERIMENT
Investigating Adaptations of Birds

What are the obvious characteristics of birds? How are these structures adaptations for a bird's unique way of life?

Getting Started

In this activity, you will examine some of the structures of birds and determine how each is designed for a certain function.

Thinking Critically

Make a list of the characteristics of birds and how each functions in a bird's unique way of life.

Materials 

Your cooperative group will use:

- contour feather
- down feather
- hand lens
- scissors
- pictures of the beaks of birds
- pictures of the feet of birds

Try It!

1. Use a hand lens to examine a contour feather. Find the shaft, vanes, and barbs. Hold the shaft by the thicker end and carefully bend the opposite end. *Observe* what happens when you release the bent end. Record all of your observations. With the scissors, cut about 2 cm from the end of the shaft. *Examine* the cut end. Separate the barbs near the center of the vanes. Use your fingers and gently rub the feather where you separated it. *Observe* what happens.
2. Use a hand lens to observe a down feather. *Observe* how it is different from a contour feather.
3. Look at the pictures of the beaks. Record the kind of food you think each bird would eat.
4. Study the feet of each bird in the pictures. *Infer* the function of each kind of feet.
5. Wash your hands when you have completed the activity.

Summing Up/Sharing Results

- *Compare* the shape, filaments (barbs), and shaft of the down feather with those of the contour feather.
- Based on your observations in Steps 1 and 2, *infer* how the structure of a contour feather is an adaptation for flight and how a down feather is an adaptation for maintaining body heat.
- Based on your observations in Steps 5 and 6, describe the way of life for which each is adapted.

Going Further!

Why is it important for a bird to have a smooth, streamlined body? Based on the feet and beaks of birds such as cardinals, snowy egrets, woodpeckers, and mallards, in what habitat would you expect them to live?

Mammals

16-2

Objectives

- Identify the characteristics of mammals and explain how they enable mammals to adapt to different environments.
- Distinguish among monotremes, marsupials, and placental mammals.
- Compare reproduction and development in the three kinds of mammals.

New Science Words

mammals
mammary glands
herbivores
carnivores
omnivores
placental mammals
gestation period
placenta
umbilical cord
marsupials
monotremes

Characteristics of Mammals

Cats, whales, moles, bats, horses, and people are all mammals. Mammals live almost everywhere, from tropical and arctic regions, to deserts, woodlands, and oceans. You may be wondering what you, a mole, and a whale have in common. Each species of mammal is adapted to its way of life, but all mammals share some characteristics. **Mammals** are warm-blooded vertebrates that have hair and produce milk to feed their young.

Nearly all mammals have hair on their bodies at some time during their lives. Many have thick fur that covers all or parts of their bodies to keep them warm. Others, such as whales and dolphins, have little hair. However, they do have a thick layer of fat that helps keep them warm. Elephants and rhinoceroses have little hair, but they live in warm climates. Porcupine quills and the spines of a hedgehog are modified hairs that protect them from their enemies. Sensory hairs, called whiskers, around the mouth of many mammals, such as cats, mice, and sea otters, help them keep in touch with their environment.

Skin covers and protects the bodies of all mammals. Hair, horns, claws, nails, and hooves are produced by the skin. The skin contains many different kinds of glands. **Mammary glands,** which are characteristic of all mammals, produce the milk female mammals use to feed their young. Oil glands produce oil to lubricate and condition the hair and skin, and sweat glands help mammals stay cool. Many mammals have scent glands that are used for self-defense and to recognize members of their own species.

Figure 16-9. Mammals care for their young after they are born. Mammary glands in the female secrete milk to feed the young.

SECTION 16-2

PREPARATION

SECTION BACKGROUND

- The monotremes and marsupials make up only 5 percent of all mammals. Three species of monotremes, the duck-billed platypus and two species of echidnas, exist today. Monotremes live in Australia and on nearby islands in a wide range of habitats.
- The duck-billed platypus gets its name from the broad, flat, hairless snout that resembles the bill of a duck. It uses the bill to dig burrows in stream banks and to scoop up worms and other invertebrates from the bottoms of streams. The platypus reaches a length of about 46 cm and has thick brown fur, a broad flat tail, and webbed feet. The feet have claws, and the hind legs have sharp spines connected to poison glands used in defense.
- Most of the 250 species of marsupials live in Australia. Only one, the opossum, lives in North America.

PREPLANNING

- Collect preserved specimens or photographs for Activity 16-2.
- Make arrangements to obtain a small mammal for class observation.

1 MOTIVATE

- Provide a hamster, gerbil, or guinea pig for students to observe mammal characteristics and behavior.

VideoDisc

STVS: Kit Fox, Disc 5, Side 2
STVS: Temperature Regulation in Dogs, Disc 5, Side 2
STVS: How Bats Hear, Disc 5, Side 2

OPTIONS

ASSESSMENT—ORAL

- **Which group of mammals is more similar to birds?** *monotremes* **Why?** *They lay eggs.*
- **What disadvantages do the developing young of monotremes have when compared with the developing young of marsupials?** *The developing young of marsupials are protected in their mother's pouch. The young of monotremes are left alone while the parent(s) leave(s) to find food, making the young more vulnerable to predators.*
- **Many mammals, such as antelopes, deer, elephants, and whales, make their homes out in the open. Do you think their young would be well-developed or poorly developed at birth?** *well-developed* **Why?** *There is no den or nest for protection.*
- **Do you think there is a connection between the size of mammals and their gestation periods?** *Generally the gestation period of larger species is longer than that of smaller species.*

2 TEACH

Key Concepts are highlighted.

MINI-Lab

Materials: slides, droppers, microscopes

Teaching Tips

▶ Be sure students obtain hairs with the root still attached.

▶ You may have to help students focus on the hairs on high power.

Answers

Students should see that the inside of the hair is made of parallel layers of material while the outside is covered with overlapping layers of cells.

To assess this product, refer to the Performance Task Assessment Lists in Performance Assessment in Middle School Science.

MINI-Lab

ASSESSMENT

Performance: To further assess students' ability to evaluate mammal characterestics, see USING LAB SKILLS on page 390.

▶ Explain that the way mammals live has a lot to do with the food they eat. Kangaroos and bison are grazers, eating mainly grasses. Deer, giraffes, and koalas feed on leaves, stems, twigs, and barks of shrubs and trees and are called browsers. Lions, weasels, and wolves usually hunt and kill other animals. Jackals and hyenas hunt and they also eat the remains of dead animals. Moles, shrews, pangolins, and many bats eat insects almost exclusively. Bears, opossums, and raccoons are omnivores.

▶ Ask students: **Why do omnivores have an advantage over herbivores and carnivores?** *If one food becomes scarce, they can easily shift to another.*

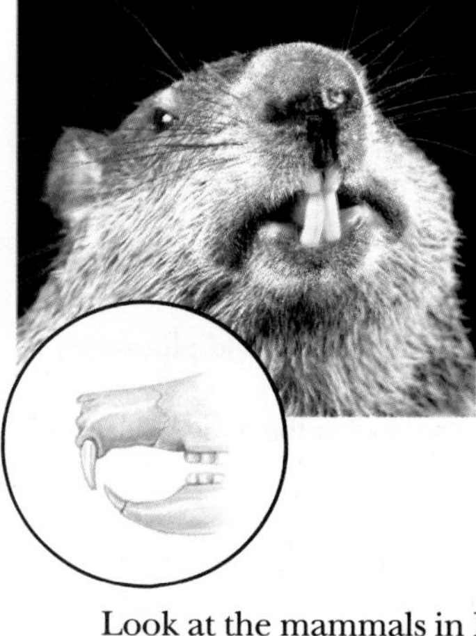

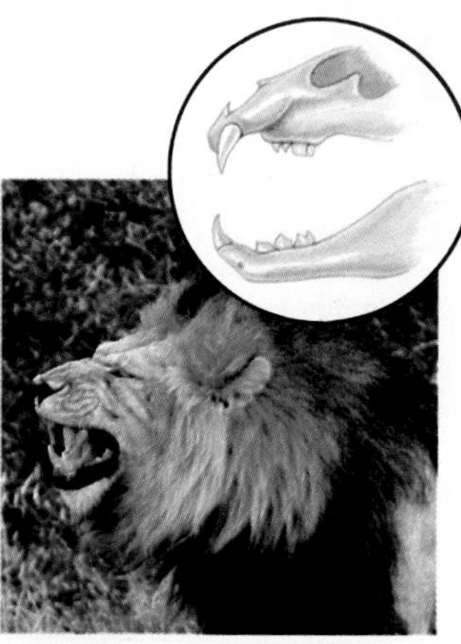

Figure 16-10. Mammals have teeth specialized for the food they eat.

Look at the mammals in Figure 16-10. Can you tell anything about what these animals eat by looking at their teeth? Almost all mammals have specialized teeth. Front teeth, called incisors, bite and cut. Next to the incisors, pointed canine teeth grip and tear. Premolars and molars shred, grind, and crush. Rabbits use their large incisors
1 to cut off blades of grass. Animals like the rabbit that eat plants are called **herbivores.** Many mammals are herbivores. **Carnivores,** such as tigers, are flesh-eating animals. Some animals, like yourself, eat both plants and animals and are called **omnivores.**

MINI-Lab

What are the characteristics of hair?

Brush or comb your hair to remove a few loose hairs. Take two hairs from your brush that look like they still have the root attached. Make a wet mount slide of the two hairs being sure to include the root. Focus on the hairs with the low power objective. Draw what you see. Switch to the high power objective and focus on the hairs. Draw what you see. *Describe* the characteristics of the hair and root.

Mammals live very active lives. Think of all the things you do each day! The body systems of mammals allow them to be very active and to adapt to many environments. All mammals have a four-chambered heart and a network of blood vessels. This type of heart allows blood coming from the lungs, full of oxygen, to be pumped directly to the body. Mammals have well-developed lungs made up of millions of microscopic sacs that increase the surface area for exchange of carbon dioxide and oxygen.

The digestive system of mammals varies according to the kind of food the animal eats. Carnivores have a short digestive system compared to that of herbivores. Meat is more easily digested than plant material. Herbivores need a long digestive system to help them break down the carbohydrate called cellulose found in plants.

The nervous system is made up of a brain, spinal cord, and nerves. The brain in most mammals is larger than those of other animals the same size. Mammals are able to learn and remember more than other animals.

OPTIONS

Meeting Different Ability Levels

For Section 16-2, use the following **Teacher Resource Masters** depending upon individual students' needs.

◆ **Study Guide Master** for all students.

● **Reinforcement Master** for students of average and above average ability levels.

▲ **Enrichment Master** for above average students.

Additional Teacher Resource Package masters are listed in the OPTIONS box throughout the section. The additional masters are appropriate for all students.

◆ **STUDY GUIDE 65**

Reproduction and Development

All mammals reproduce sexually. Most mammals give birth to live young after a period of development inside a uterus. One or both parents care for the young after they are born.

Mammals are classified into three groups based on how the young develop. The three groups are: placental mammals, marsupials, and monotremes.

In **placental mammals,** embryos develop inside the 2
female in an organ called the uterus. The time during which the embryo develops in the uterus is called the **gestation period.** Gestation periods range from 16 days in 3
hamsters to 650 days in elephants. Placental mammals are named for the **placenta,** an organ developed by the growing embryo that attaches to the uterus. An **umbilical** 4
cord attaches the embryo to the placenta. Several blood vessels make up the umbilical cord. The placenta absorbs oxygen and food from the mother's blood. The umbilical cord transports the food and oxygen to the embryo. The mother's blood never mixes with the blood of the embryo.

The kangaroo in Figure 16-12 is an example of a marsupial. **Marsupials** are pouched mammals that give birth to tiny, immature offspring. Young kangaroos are born only a few days after fertilization takes place and are about the size of a honeybee at birth. Immediately after birth, the young kangaroo crawls into the pouch on the female's abdomen and attaches to a nipple. It remains protected in the pouch while it develops more completely. After a few months, the young kangaroo, called a joey, is better developed and crawls out of the pouch. The joey returns to its mother's pouch for protection.

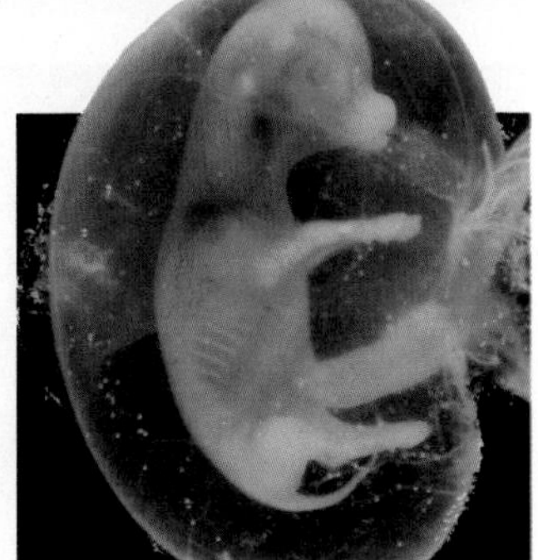

Figure 16-11. A mammal embryo develops inside a sac of fluid inside the uterus of a female.

Figure 16-12. Marsupials carry their developing young in a pouch outside their bodies. A kangaroo and baby (right) and an opossum with babies (left) are marsupials.

CHECK FOR UNDERSTANDING

Have students list the characteristics of mammals.

RETEACH

Have students group the following characteristics in three columns under the headings Mammals, Birds, Both Mammals and Birds: have the most fully developed brains, warm-blooded, hollow bones, have hair, have feathers, lay shelled eggs, nurse their young, give birth to live young, give extended parental care, do not have teeth, have a diaphragm, have outer ears, have a four-chambered heart, belong to the phylum Chordata.

EXTENSION

For students who have mastered this section, use the **Reinforcement** and **Enrichment** masters or other OPTIONS.

CONCEPT DEVELOPMENT

▶ Explain that even though no direct connection exists between the blood vessels of the mother and the embryo, the capillaries of each run close together. Therefore, almost anything in the mother's blood can have an effect on the embryo.

▶ Explain that a very inefficient placenta exists for a very brief time in marsupials, and that they don't have an umbilical cord. Nutrients and oxygen from the mother slowly pass through the marsupial placenta directly into the fetus.

▶ **Field Trip:** Visit a local dairy to see how cows are milked and how the milk is collected and stored.

● REINFORCEMENT 65

NAME ________ DATE ________ CLASS ________

REINFORCEMENT — Chapter 16

Mammals — Text Pages 379–385

Classify each of these animals according to what they eat.

1. carnivore 2. herbivore 3. omnivore

Cross out the animal name that does not belong in each row. Then write the name of the order to which the other three belong. Use Table 16-2 in your textbook if needed.

4. camels	cows	~~seals~~	deer	**Artiodactyls**
5. ~~bats~~	rabbits	pikas	hares	**Lagomorphs**
6. monkeys	humans	apes	~~dogs~~	**Primates**
7. ~~manatees~~	whales	dolphins	porpoises	**Cetaceans**
8. ~~walruses~~	raccoons	moles	bears	**Carnivores**
9. horses	zebras	~~cows~~	rhinoceroses	**Perissodactyls**
10. beavers	mice	squirrels	~~rabbits~~	**Rodents**

Use Table 16-2 to find the order that these characteristics describe. Write the name of the order in the space provided.

Proboscideans 11. elongated nose forms a trunk; enlarged incisors form tusks
Pinnipedes 12. hind limbs modified as flippers; teeth adapted for catching fish
Chiropterans 13. front limbs adapted for flying
Primates 14. long arms; opposable thumbs; eyes face forward

Describe the three groups of mammals based on how the young develop.

15. placental mammals: **The embryos develop inside the female in the uterus; the embryo is attached to the placenta by the umbilical cord.**
16. marsupials: **The young are born only a few days after fertilization; after birth, the young spend months in the female's pouch.**
17. monotremes: **Female lays leathery eggs; after hatching, the young lick milk from the skin and hair surrounding the female's mammary glands.**

65

▲ ENRICHMENT 65

NAME ________ DATE ________ CLASS ________

ENRICHMENT — Chapter 16

Mammals — Text Pages 379–385

OBSERVATIONS OF A MAMMAL

You will observe characteristics of a mammal in the following activity.

Procedure

Choose a family pet or some pet you know well in the neighborhood. Make sure the pet is a mammal. Pets might include a dog, a cat, a rabbit, a gerbil, a horse, or a mouse. Observe the mammal, keeping in mind the characteristics that you learned about in your textbook. Add to your observations by looking in an encyclopedia or some other reference book that might include information about your mammal. Complete the table below with your observations and the information you gained by reading about this mammal.

Data and Observations

Characteristic	Observation/Information
Animal type	**Answers will vary depending on mammal observed.**
Scientific classification (kingdom, class, order)	
Outer covering	
Teeth	
Food	
Reproduction (type, mammal classification, gestation period)	
Other characteristics	

Conclude and Apply

1. After observing the mammal's teeth and food choices, would you classify it as an herbivore, a carnivore, or an omnivore? **Answers will vary depending on mammal observed.**
2. What characteristics of the mammal you observed seem to make it well suited to living in your home or neighborhood? **Answers will vary.**

65

CONCEPT DEVELOPMENT

▶ Tell students that monotreme eggs are flexible and leathery like reptile eggs, not hard and smooth like bird eggs. But, unlike reptiles and like most birds, female monotremes incubate their eggs. Platypuses often curl up around their eggs. Echidnas carry their eggs around with them in a pouch that develops only during breeding season.

Answer: Bats send out sounds that bounce off objects, giving the bat an idea of where potential food is located and whether it is moving.

CROSS CURRICULUM

▶ **Language Arts:** Ask students to make a list of expressions that draw an analogy between the behavior of a mammal and the behavior of a person. Some examples are: blind as a bat, eats like a pig, quiet as a mouse, busy as a beaver, sly as a fox, slothful.

▶ Explain that homes for mammals such as bears, foxes, and chipmunks are underground burrows called dens. Squirrels and chimpanzees make leafy nests in treetops. Home for newborn rabbits is a fur-lined nest in a tangle of grasses or in the ground.

MINI QUIZ

Use the Mini Quiz to check students' recall of chapter content.

1. **Animals that eat plants are ______ .** herbivores
2. **Placental mammals develop inside the female in an organ called the ______.** *uterus*
3. **The time during which an embryo develops is called the ______ .** *gestation period*
4. **The ______ attaches the embryo to the placenta.** *umbilical cord*
5. **Mammals that incubate their eggs are ______ .** *monotremes*

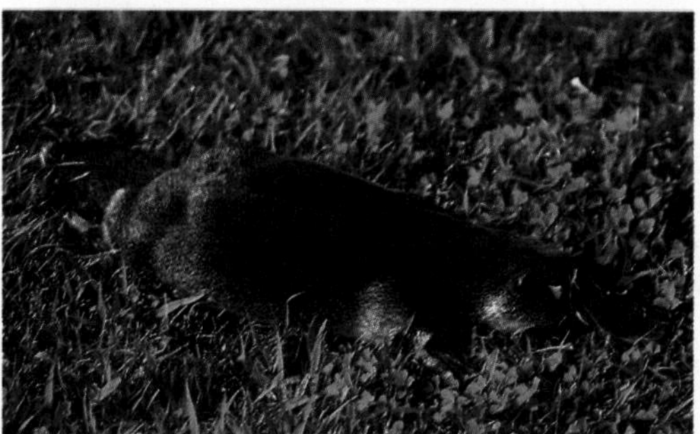

Figure 16-13. A duckbilled platypus is a monotreme. Monotremes are mammals that lay eggs.

Echolocation is used in sonar and by some animals. High frequency sound waves are given off and bounce off objects. Some bats are capable of echolocation. In what way is this behavior an advantage for the bat in obtaining food?

The marsupials you are probably the most familiar with are kangaroos and opossums. Opossums are the only marsupials native to North America. Other marsupials are the koala, Tasmanian devil, and wallaby. Marsupials live primarily in Australia, Tasmania, and New Guinea.

Monotremes do not give birth to live young as placental mammals and marsupials do. Duckbilled platypuses are the most well-known monotreme. **Monotremes** are mammals that lay eggs with tough, leathery shells. 5 The female incubates the eggs. Mammary glands of the monotremes do not have nipples. When the young hatch, they nurse by licking milk from the skin and hair surrounding the female's mammary glands.

Most placental mammals, as well as all marsupials and monotremes, are nearly helpless, and sometimes even blind, when they are born. They can't do much for themselves the first several days or even months. They just eat, sleep, and develop.

Mammal parents are very protective. They make homes in which their young can grow protected from predators and the weather. For some, such as bears and foxes, home is an underground burrow called a den. For chimpanzees and squirrels, home is a nest in the top of a tree. The young of some mammals, such as antelope, deer, elephants, whales, and dolphins, must be well developed at birth to be able to move with their constantly moving parents. Those that live on land can usually stand by the time they are a few minutes old. Marine mammals can swim as soon as they are born.

During the nursing period, young mammals learn skills they need to survive. Among most kinds of mammals, the mother raises the young alone. In some species, males help with nest building, finding food, and protecting the young.

Kinds of Mammals

There are 18 orders of modern mammals. Monotremes and marsupials make up one order each. Placental mammals make up the other orders. Each order has adaptations that help identify it. Characteristics and examples of several of the most common orders and examples are given in Table 16-1.

OPTIONS

ENRICHMENT

▶ Have students research and report on gestation periods of mammals and the average number of offspring born to different species. Ask students to evaluate the data by comparing the size or activity of these organisms to the length of gestation. Some gestation periods are:

opossum	12 days
hamster	16 days
mouse	20 days
rabbit	31 days
dog	61 days
cat	63 days
guinea pig	68 days
cow	281 days
horse	336 days
camel	406 days
giraffe	442 days
whale	450 days
elephant	650 days

Table 16-1

MAJOR ORDERS OF PLACENTAL MAMMALS
Chiropterans Bats Front limbs adapted for flying; feed on fruit, insects or blood; active at night
Rodents Mice, rats, squirrels and beavers Pair of greatly enlarged incisors on upper and lower jaws; specialization for movements such as gliding, swimming, leaping, hopping, and running
Lagomorphs Rabbits, hares, and pikas Two pairs of upper incisors; bodies covered with soft fur; long legs adapted to jumping and running
Cetaceans Whales, dolphins, and porpoises Forelimbs are modified into flippers; breathe through one or two blowholes on top of their heads; little or no hair; largest animals on Earth
Perissodactyls Horses, zebras, and rhinoceroses Hooves made of a hard protein; odd number of toes; flat, grinding molars; large skeletons adapted for running
Artiodactyls Deer, moose, camels, pigs, and cows Hooves made of a protein called keratin; even number of toes; large, flat molars; complex stomachs and intestines
Carnivores Cats, dogs, bears, and racoons Long, sharp canine teeth; long, sharp claws; many hunt and kill prey
Proboscideans Elephants Elongated nose forms a trunk to eat and drink; pair of enlarged incisors forms tusks; leathery skin
Primates Humans, apes, and monkeys Long arms with grasping hands and five fingers; opposable thumb; eyes face forward; large brain

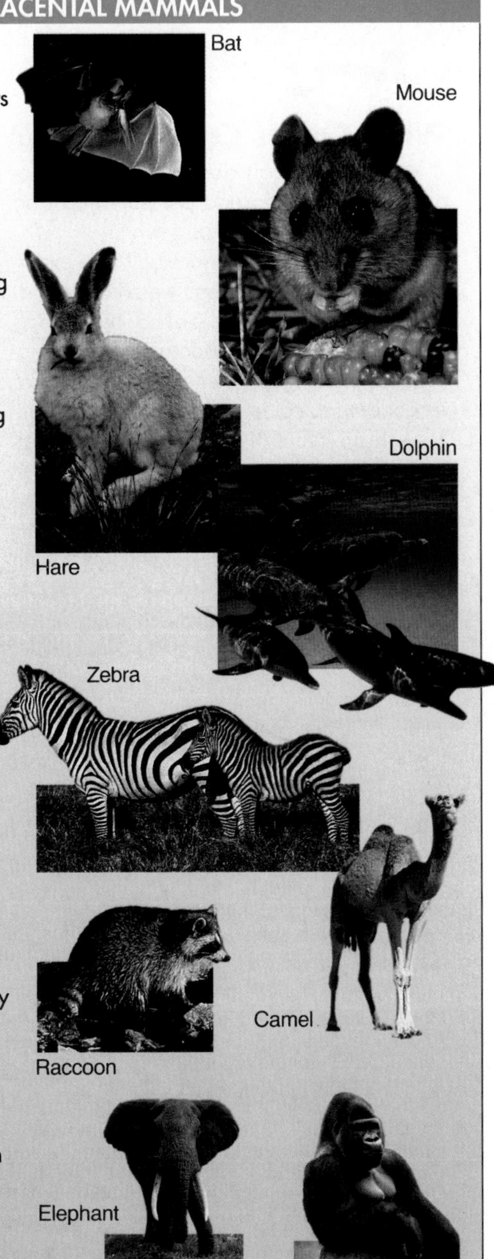

REVEALING MISCONCEPTIONS

▶ Many people use the words *monkey* and *ape* interchangeably, but the words refer to different kinds of animals. The main differences between the two have to do with the structure of the skeleton, especially the skull, and the fact that most monkeys have tails.

▶ Camels do not store water in their humps. The humps contain fat, an energy-conserving adaptation that enables them to survive in the desert where food and water are scarce.

▶ Bats are not blind. All bats can see and a few even have better vision than people do. Most do not have a keen sense of sight and rely on echolocation to help them navigate.

CONCEPT DEVELOPMENT

▶ Point out that today there are over 4000 species of mammals divided into about 20 orders on the basis of similar anatomy, physiology, ancestry, and other characteristics.

▶ Have students research and report on the migration patterns of one of the following: the gray whale, the caribou, the Arctic wolf, the Bighorn sheep.

▶ Have students research and report on digestion in even-toed hoofed animals such as giraffes and cattle.

INQUIRY QUESTIONS

▶ **In what ways do retractable claws enable lions and other cats to survive?** *They use their claws to climb trees and hold prey. They can be retracted, enabling the animals to walk and run faster.*

▶ **Some animals such as elephants and whales have little body hair. Suggest reasons why this adaptation exists in these mammals.** *Elephants have thick skin that insulates, and whales have blubber or fat that insulates. Wet hair would not be a good insulator in whales.*

PROGRAM RESOURCES

From the **Teacher Resource Package** use:

Activity Worksheets, page 147, Mini-Lab: What are the characteristics of hair?

Concept Mapping, pages 37-38.

Critical Thinking/Problem Solving, page 20, Playing Cat and Mouse.

PROBLEM SOLVING

Scientists think dogs can see colors at opposite ends of the spectrum. They can see blue, but greens, yellows, oranges, and reds are seen as one color.

Think Critically: Both experiments indicated that dogs responded to color; however, in the recent investigation scientists controlled differences in brightness.

CONCEPT DEVELOPMENT

▶ Explain that in terms of evolutionary history, mammals are some of the youngest animals around.

MULTICULTURAL AWARENESS

▶ Use the caption for Figure 16-14 and information on elephant ivory on page 385 to increase students' awareness of organisms indigenous to other countries.

FLEX Your Brain

Use the Flex Your Brain activity to have students explore the PROTECTION OF WILD ANIMALS.

Flex Your Brain

ASSESSMENT

Portfolio: Use the Flex Your Brain activity to reinforce critical thinking and problem solving skills.

PROBLEM SOLVING

What Colors Can Spot See?

What colors can dogs see? Most people believe that dogs are color-blind and see the world in black and white.

In 1969, scientists performed some experiments to determine if dogs were color-blind. The experiments indicated that dogs respond to color. But, because the experiments did not test whether the dogs were responding to color or differences in brightness, many scientists still believed dogs were color-blind.

A recent study indicates that dogs are not color-blind. The dogs in the experiments could tell the difference between white light and colored light and between colors at the opposite ends of the spectrum. The scientists were careful to control differences in brightness. Based on their investigation, the scientists speculate that dogs can see blues, but they see greens, yellows, oranges, and reds as one shade.

Think Critically: Compare and contrast the 1969 experiment with the more recent investigation.

Origin and Importance of Mammals

About 65 million years ago, dinosaurs became extinct. This opened up new habitats for mammals, and they began to branch out into many different species. Eventually, they became the most numerous animals on Earth. Some species died out, but others gave rise to modern mammals. Today, more than 4000 species of mammals have evolved from tiny, shrewlike creatures that lived 200 million years ago.

Figure 16-14. Large carnivores such as this cheetah help control populations of grazing animals in Africa, India, and parts of Asia.

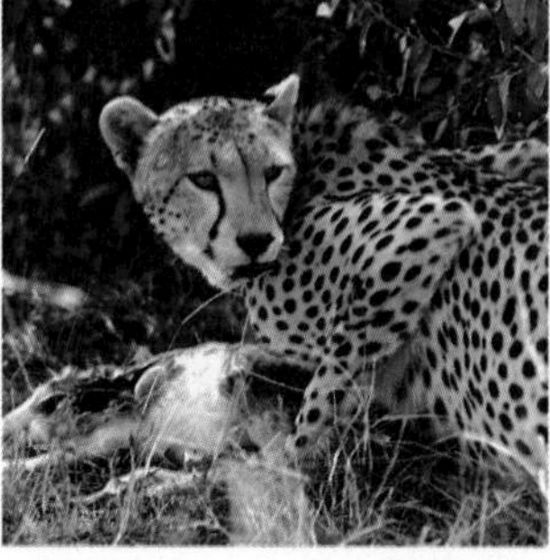

Mammals are very important in maintaining a balance in the environment. Large carnivores such as lions help control populations of grazing animals. Bats help pollinate flowers, and some pick up plant seeds in their fur and distribute them.

Mammals are in trouble today. As millions of acres of wildlife habitat are developed for shopping centers, recreational areas, and housing, many mammals are left without food, shelter, and space to survive.

OPTIONS

INQUIRY QUESTIONS

▶ **How were mammals important to our early ancestors?** *They supplied food, hides for clothing and shelter, tusks for tools and weapons, and inspiration for art and religious ceremonies.*

▶ **What mammals have been used for transportation?** *horse, camel, llama, donkey, ox, elephant, sled-dog*

▶ **Why do you think the first domestic animals were dogs?** *They were used to hunt other animals.*

▶ **Why do you think our ancestors began keeping herds of domestic animals?** *They no longer had to hunt each time they needed food or products that came from these animals.*

▶ **Why are mammals important to people today?** *We get meat, milk, hides, and fur from them. We make pets of mammals, use them for recreation, and use them in scientific research. They are a critical part of the food chain.*

Overhunting has threatened many mammal populations in the past, but most countries now have strict hunting regulations. However, these regulations don't keep people from killing and capturing wild mammals to sell. People use poisons to kill many mammals such as rats, prairie dogs, and rodents. Insect-eating mammals are affected by insecticides. Because of these problems, many people are working hard to protect mammals. You will learn more about measures being taken to protect endangered mammals and other species in Chapter 28.

Figure 16-15. Many elephants have been killed for their ivory tusks.

SECTION REVIEW

1. Describe five characteristics of all mammals.
2. Differentiate among placental mammals, monotremes, and marsupials.
3. Give examples of six orders of mammals, and briefly list characteristics of each.
4. **Apply:** Suggest a reason why placental mammals are far more numerous than monotremes and marsupials.
5. **Connect to Physics:** Find out what evaporation is and how it benefits warm-blooded animals.

Skill Builder

Classifying

Using Table 16-1, classify the following mammals into three groups: whales, echidnas, koalas, horses, elephants, opossums, kangaroos, rabbits, bats, bears, platypuses, monkeys. Compare and contrast their characteristics. If you need help, refer to Classifying in the **Skill Handbook** on page 680.

ENRICHMENT

▶ Have students research and report on the work of Dian Fossey with mountain gorillas in Africa and Jane Goodall's work with chimpanzees.

PROGRAM RESOURCES

From the **Teacher Resource Package** use:

Activity Worksheets, page 5, Flex Your Brain.

3 CLOSE

▶ Ask questions 1-3 and the **Apply** Question in the Section Review.

▶ Take students on a field trip to a zoo. Visit the zoo before the trip and find out which animals have young, the feeding times of the animals, and other information. Prepare a sheet for the students to record their observations.

SECTION REVIEW ANSWERS

1. warm-blooded, have hair, produce milk to feed their young, have teeth, have a four-chambered heart, have well-developed lungs, have a large brain, reproduce sexually

2. In placental mammals, embryos develop inside the female uterus. Marsupials give birth to tiny, immature offspring that develop in their pouches. Monotremes lay eggs with a tough, leathery shell, and the parent(s) incubate(s) the eggs.

3. Use any six from page 383.

4. Apply: They are fully developed at birth and have a better chance of surviving and reproducing.

5. Connect to Physics: Evaporation occurs when water molecules absorb thermal energy and escape from the surface. Where this happens on an animal, the skin temperature drops and the organism is cooled.

Skill Builder

Students should classify the organisms into three groups: placentals—horse, elephant, rabbits, bats, bears, monkeys; marsupials—koalas, opposum, kangaroos; monotremes—echidnas, platypuses. They should compare and contrast characteristics such as habitat, reproduction, nutrition, and locomotion. A chart or table would be the most useful method for students to use.

Skillbuilder
ASSESSMENT

Performance: Assess students' abilities to Classify another selection of mammals from Table 16-1.

SECTION 16-3

PREPARATION

SECTION BACKGROUND

▶ Manatees, "sea cows," are large marine mammals that live in warm climates. In the United States, they are found off the coast of Florida.

PREPLANNING

▶ Bring in pictures of manatees and pictures of the developed Florida coast along which they live.

1 MOTIVATE

▶ Have students discuss what is going on in the photograph in Figure 16-16. It connotes interaction with manatees and humans and also tells something about the life of the animal.

2 TEACH

Key Concepts are highlighted.

CONCEPT DEVELOPMENT

▶ Ask students the following question: **Do manatees play a very important role in the food chain in the big picture?** *No, there are very few manatees left now. If we lost the remaining animals, we would not notice a change in the environment.*

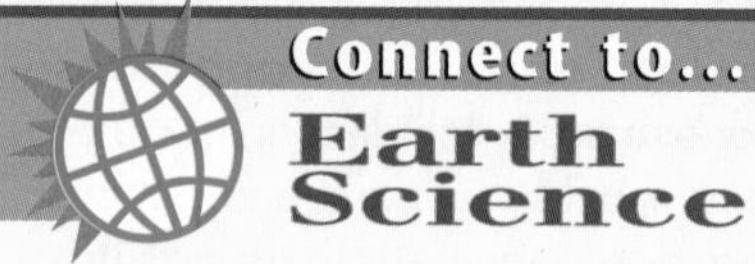

Connect to... Earth Science

Answer: Student opinions will vary but should be accompanied by reasons based on facts.

16-3 Saving the Manatee

New Science Words

manatees
poaching

Objectives

▶ Identify the characteristics of manatees.
▶ Explain the major threats to manatees today.

"Sea Cows"?!

Have you ever thought about running into a cow in the water? Unless you live in Florida, you probably didn't realize that "sea cows" exist. Sea cows, or **manatees,** are large mammals belonging to the order Sirenia that live in salt waters. Adults weigh about half a ton. They are known to live in the rivers, estuaries, and coasts in the tropical and subtropical regions. They cannot tolerate cold temperatures.

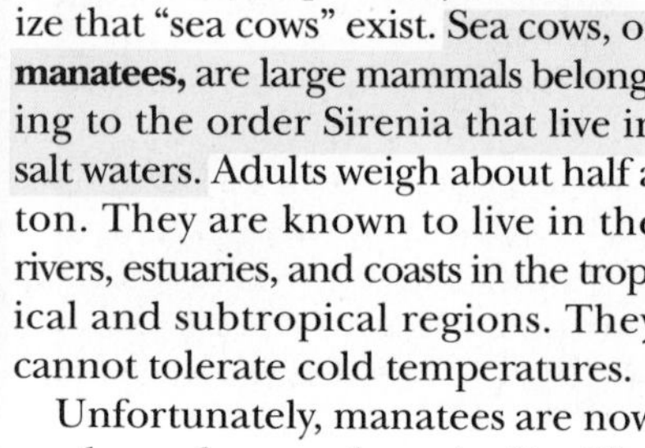

Unfortunately, manatees are now on the endangered species list. The slow, curious "sea cows" are not afraid to approach and investigate visiting humans, making them very easy to hunt. In the past, they were heavily hunted for their meat, oil, and hides.

Today, manatees are still dying. Manatees are in danger from barges and motorboats. They often collide with these boats and are injured or cut by propellers. In fact, collisions are so common with manatees that researchers are able to identify more than 900 individual manatees just by using their distinctive scars caused by these collisions. Manatees are herbivores and feed on aquatic plants that clog waterways. Many manatees have been poisoned by herbicides used to kill these plants.

Connect to... Earth Science

Dinosaurs became extinct by the Cenozoic Era, when mammals and flowering plants began to flourish. **In your Journal,** write an opinion about whether strong measures should be taken to prevent extinction of species when it is known that they are endangered.

The Future of Manatees

Researchers are very concerned about the future of these huge, quiet mammals. Although the United States protected them by law in 1983, the state of Florida cannot employ

OPTIONS

Meeting Different Ability Levels

For Section 16-3, use the following **Teacher Resource Masters** depending upon individual students' needs.

◆ **Study Guide Master** for all students.

● **Reinforcement Master** for students of average and above average ability levels.

▲ **Enrichment Master** for above average students.

Additional Teacher Resource Package masters are listed in the OPTIONS box throughout the section. The additional masters are appropriate for all students.

◆ STUDY GUIDE 66

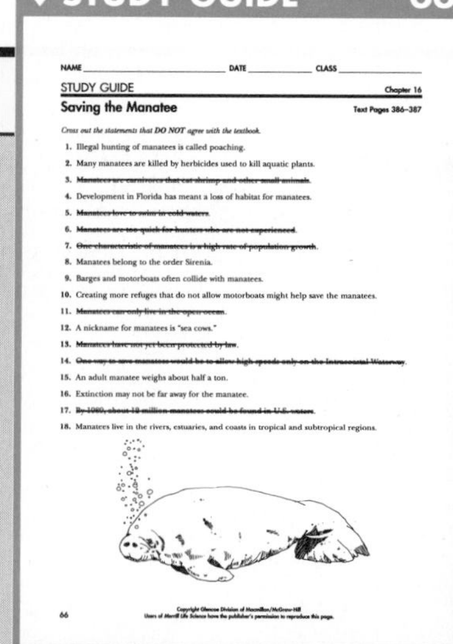

NAME ______ DATE ______ CLASS ______

STUDY GUIDE — Chapter 16

Saving the Manatee — Text Pages 386–387

Cross out the statements that DO NOT agree with the textbook.

1. Illegal hunting of manatees is called poaching.
2. Many manatees are killed by herbicides used to kill aquatic plants.
3. ~~Manatees are carnivores that eat shrimp and other small animals.~~
4. Development in Florida has meant a loss of habitat for manatees.
5. ~~Manatees love to swim in cold waters.~~
6. ~~Manatees are too quick for hunters who are not experienced.~~
7. ~~One characteristic of manatees is a high rate of population growth.~~
8. Manatees belong to the order Sirenia.
9. Barges and motorboats often collide with manatees.
10. Creating more refuges that do not allow motorboats might help save the manatees.
11. ~~Manatees can only live in the open ocean.~~
12. A nickname for manatees is "sea cows."
13. ~~Manatees have not yet been protected by law.~~
14. ~~One way to save manatees would be to allow high speeds only on the Intracoastal Waterway.~~
15. An adult manatee weighs about half a ton.
16. Extinction may not be far away for the manatee.
17. ~~By 1980, about 10 million manatees could be found in U.S. waters.~~
18. Manatees live in the rivers, estuaries, and coasts in tropical and subtropical regions.

66

enough law enforcers to prevent illegal hunting called **poaching.** In 1988, 133 manatees were killed in Florida. By 1989 only about 1200 manatees were believed to be living in United States waters.

Manatees have a very slow rate of reproduction. Manatees that are killed by accident are not quickly replaced. Combine this steady decrease in population with the loss of habitat due to development (nearly 1000 people move to Florida each day) and extinction does not look to be very far away.

What can be done to save the manatees? Some suggest developing more refuges where boats would not be allowed to travel. Because Florida real estate is very valuable, this is not likely to be popular with coastal landowners. Another option is to enforce a slower speed limit on the boats that travel in the Intracoastal Waterway where many manatees are injured.

Whatever the solution, manatees need more protection soon if we hope to save them from their current fate of extinction.

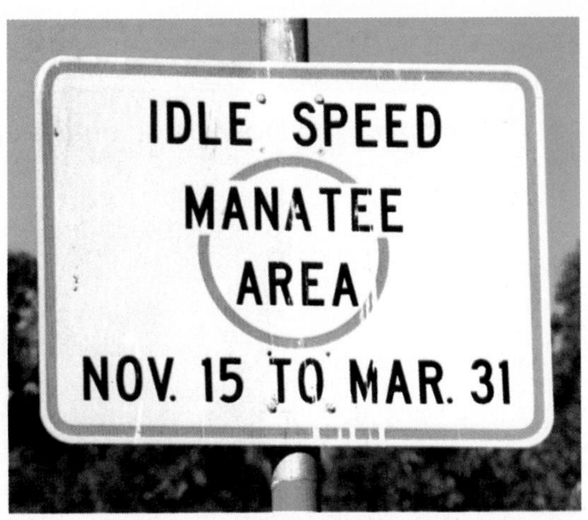

Figure 16-16. Warning signs are being placed in areas where manatees are found to help protect them from being injured.

SECTION REVIEW

1. What are manatees and where do they live?
2. What two major factors account for the gradual decrease in the population?
3. **Connect to Earth Science:** Find out when large mammals such as the wooly mammoth and saber-tooth tiger became extinct. Hypothesize what could have caused their extinction.

You Decide!

In Miami, Florida, injured manatees are taken by a special ambulance to Miami's Seaquarium where they can be treated. Scientists are hoping to release the treated manatees and help the population become more stable. The estimated cost of treatment for one injured manatee is about $18 000. When returned to its environment, there is no guarantee that it will escape another, perhaps fatal, injury. Some people believe that it's not right to spend this much money on saving the manatees. What do you think?

CHECK FOR UNDERSTANDING

▶ Ask questions 1-2 in the Section Review.

RETEACH

Present a nature video on marine mammals to the class.

EXTENSION

For students who have mastered this section, use the **Reinforcement** and **Enrichment** masters or other OPTIONS.

3 CLOSE

▶ On the chalkboard, have students list the consequences of losing the manatees forever. Then have students list the importance of continued development to the state of Florida.

SECTION REVIEW ANSWERS

1. Manatees are large marine mammals that belong to the order *Sirenia*. They live in tropical rivers and estuaries.
2. poaching and colliding with boats and barges
3. Connect to Earth Science: They became extinct during the last ice age. They could have been trapped in the glaciers at that time.

YOU DECIDE!

Answers will vary. All answers that show clear and logical thinking should be accepted.

● REINFORCEMENT 66

NAME ______ DATE ______ CLASS ______

REINFORCEMENT — Chapter 16

Saving the Manatee — Text Pages 386–387

Answer the following questions.

1. What are manatees also called? **sea cows**
2. How are manatees classified according to their food? **They are herbivores.**
3. How can their actions be described? **calm and slow-moving**
4. Where can manatees be found in the United States? **Florida**
5. What kind of conditions do manatees like? **Manatees like the hot tropical and subtropical regions because they cannot stand cold temperatures.**
6. Why are poachers able to hunt manatees so easily? **Manatees are slow-moving and curious.**
7. What are three causes of death for the manatee?
   a. **poaching**
   b. **herbicides used to kill aquatic plants**
   c. **collisions with barges and motorboats**
8. How does the movement of people to Florida affect the manatee? **Florida's population growth causes a loss of habitat for the manatee.**
9. Why is the manatee's slow rate of reproduction significant? **The manatees that are killed by poaching, herbicides, and boating accidents cannot be replaced quickly.**
10. What are two ways that the manatees might be saved?
    a. **create more refuges for manatees where no motorboats are allowed**
    b. **enforce a speed limit on the Intracoastal Waterway**

This sculpture reminds boaters in Florida to watch for manatees.

66

▲ ENRICHMENT 66

NAME ______ DATE ______ CLASS ______

ENRICHMENT — Chapter 16

Saving the Manatee — Text Pages 386–387

Imagine that by some miracle manatees could suddenly speak English and express themselves on the issues of the day. To protest their potential extinction, they organize a rally at the town council meeting in a small southern Florida town. They appear with protest signs, and one manatee rises from the water to address the citizens of the town. From the council members and townspeople come counter arguments. Some feel they shouldn't be expected to change their lives just to protect some slow sea cows.

You are a reporter for the *Southern Florida News and Sun* sent by your editor to cover the town council meeting. Write an article reporting what you saw and heard. Reread Section 16-3 so that you have the facts of the manatee situation in mind. As you write the article, answer the following questions in the speeches of the manatee and the townspeople.

- How has the manatee population declined in recent years?
- What is killing the manatee?
- What do you think would be the attitude of Florida residents if forced to change their lives to protect manatees?
- What laws could the town council pass to protect the manatees while still pleasing the people?

Be creative in your description of this meeting and rally. Use the details from your text to make it true to life. Yet make the people—and the manatees—believable to your readers.

SAVE the MANATEE

66

ACTIVITY 16-2
35 minutes

OBJECTIVE: Identify and **classify** selected vertebrates to class.

PROCESS SKILLS applied in this activity:
▶ **Classifying** in Procedure Steps 3-5.
▶ **Inferring** in Conclude and Apply Question 3.

COOPERATIVE LEARNING Use the Paired Partners strategy, but have both students record their observations.

TEACHING THE ACTIVITY
Alternate Materials: Pictures of vertebrates can be used.
Troubleshooting: Students may have difficulty with Chondrichthyes and Osteichthyes. Make certain they know which fish have a skeleton of bone and which have a skeleton of cartilage.
▶ The characteristics of hair, feathers, jaws, paired fins, cartilage skeleton, bony skeleton, and scales should be visible on the animals used. If pictures are used, characteristics can be written on the pictures.

Activity
ASSESSMENT
Performance: To further assess students' ability to Classify vertebrates, have them move among numbered stations where there are living or preserved animals set up. The students can identify organisms or answer a question placed at each station.

ACTIVITY 16-2 Classifying Vertebrates

Problem: *How do scientists classify vertebrates?*

Materials
- paper
- pencil
- animals or pictures of animals

Procedure
1. Copy the four-column chart shown under Data and Observations.
2. Write the name of each animal you are studying in the first column.
3. *Predict* the class of the animal and write the predicted class in the second column.
4. Write the characteristics of the animal that helped you decide its class.
5. Use the key to find the class of each animal. Write the class name in the last column.

Data and Observations

Animal	Predicted class	Characteristics	Actual class

Analyze
1. How many of your predictions were correct?
2. What characteristics of each vertebrate class are used to *identify* the animals?

Conclude and Apply
3. How do scientists classify vertebrates?

1. Does the animal have a backbone?
YES Go to 2.
NO Invertebrate

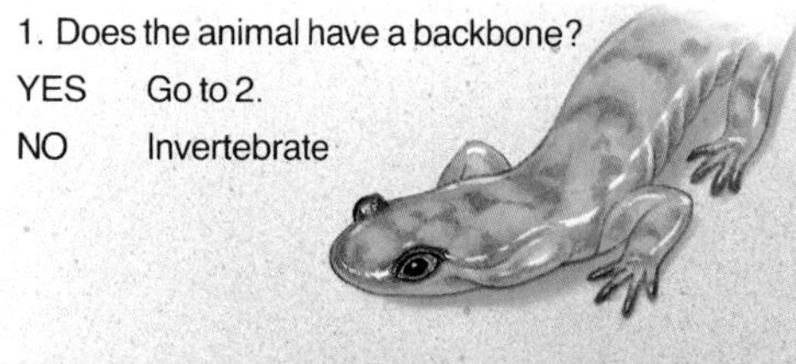

2. Does the animal have feathers?
YES Class Aves
NO Go to 3.

3. Does the animal have jaws?
YES Go to 4.
NO Class Agnatha

4. Does the animal have paired fins?
YES Go to 5.
NO Go to 6.

5. Does the animal have a bony skeleton?
YES Osteichthyes
NO Chondrichthyes

6. Does the animal have hair?
YES Mammal
NO Go to 7.

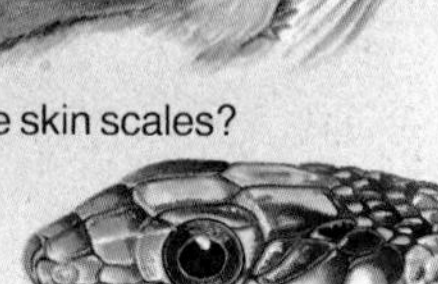

7. Does the animal have skin scales?
YES Reptilia
NO Amphibia

ANSWERS TO QUESTIONS
1. Answers will vary depending on students' predictions.
2. Mammalia—hair; Aves—feathers; Reptilia—scales; Amphibia—skin; Agnatha—no jaws; Chondrichthyes—fins and cartilage; Osteichthyes—fins and bony skeleton
3. Scientists use a key with characteristics that will identify each class of vertebrates.

PROGRAM RESOURCES
From the **Teacher Resource Package** use:
Activity Worksheets, pages 142-143, Activity 16-2, Classifying Verte-brates.

CHAPTER

REVIEW

SUMMARY

16-1: Birds

1. Birds are warm-blooded organisms that are covered with feathers and lay eggs. Their front legs are modified into wings.
2. The adaptations birds have for flight are: wings; feathers; a light, strong skeleton; and efficient body systems.
3. Birds lay eggs enclosed in hard shells. Most birds incubate their eggs until hatching.

16-2: Mammals

1. Mammals are warm-blooded vertebrates with hair. Female mammals have mammary glands that produce milk.
2. There are three groups of mammals. Mammals that lay eggs are monotremes. Mammals that have pouches for the development of their embryos are marsupials. Mammals whose offspring develop within the uterus with a placenta are placental mammals.

16-3: Science and Society: Saving the Manatee

1. Manatees are large aquatic mammals that belong to the order Sirenia. Adults weigh about half a ton. They live along the coasts in tropical and subtropical regions.
2. The major threats to manatees today are poachers, injury from boats, and habitat destruction.

KEY SCIENCE WORDS

a. **carnivores**
b. **contour feathers**
c. **down feathers**
d. **gestation period**
e. **herbivores**
f. **incubate**
g. **mammals**
h. **mammary glands**
i. **manatees**
j. **marsupials**
k. **monotremes**
l. **omnivores**
m. **placenta**
n. **placental mammals**
o. **poaching**
p. **preening**
q. **umbilical cord**

UNDERSTANDING VOCABULARY

Match each phrase with the correct term from the list of Key Science Words.

1. plant-eaters
2. keep eggs warm
3. mammals whose embryos develop inside a uterus
4. feathers birds use for flying
5. animals that eat both plants and animals
6. glands that produce milk
7. feathers that help to insulate
8. activity of rubbing oil onto feathers
9. structure containing blood vessels that attaches embryo to the placenta
10. pouched mammals

OPTIONS

ASSESSMENT

To assess student understanding of material in this chapter, use the resources listed.

COOPERATIVE LEARNING

Consider using cooperative learning in the THINK AND WRITE CRITICALLY, APPLY, and MORE SKILL BUILDERS sections of the Chapter Review.

PROGRAM RESOURCES

From the **Teacher Resource Package** use:

Chapter Review, pages 35-36.

Chapter and Unit Tests, pages 101-104, Chapter Test.

CHAPTER

REVIEW

SUMMARY

Have students read the summary statements to review the major concepts of the chapter.

UNDERSTANDING VOCABULARY

1. e
2. f
3. n
4. b
5. l
6. h
7. c
8. p
9. q
10. j

ASSESSMENT

Portfolio

Encourage students to place in their portfolios one or two items of what they consider to be their best work. For each item, ask students to explain why that item was chosen and what they learned from it. Items might be selected from the following.

- For Your Gifted Students bird diary, p. 370
- Skillbuilder classification of mammals, p. 385
- Connect to Earth Science opinion paper on preventing extinction of endangered species, p. 386

Performance

Additional performance assessments may be found in *Performance Assessment* and *Science Integration Activities* that accompany **Merrill Life Science.** Performance Task Assessment Lists and rubrics for evaluating these activities and other products generated throughout the chapter can be found in Glencoe's *Performance Assessment in Middle School Science.*

CHAPTER
REVIEW

CHECKING CONCEPTS

1. d	**6.** b
2. d	**7.** b
3. c	**8.** a
4. c	**9.** c
5. c	**10.** a

USING LAB SKILLS

ASSESSMENT

Use these alternate lab exercises to assess students' understanding of skills used in this chapter.

11. Down feathers are soft, fluffy, and have air spaces that insulate. Some synthetic coat materials are also soft, fluffy and insulate. Synthetic fibers are somewhat spongy.

12. Cat and dog hair is shorter, usually tapered, and may show several different color patterns. Horse hair is extremely coarse.

THINK AND WRITE CRITICALLY

13. To have enough energy to maintain body temperature and to fly, birds consume large quantities of food.

14. Placental mammals have a long period of development inside the mother. This provides food and protection for the embryo and a greater chance for survival.

15. Water birds have webbed feet for swimming. Shore birds have feet with long toes for wading through mud. Perching birds have claws for gripping onto branches. Birds of prey have clawed toes for catching prey.

16. Because the hollow bones and feathers of birds decompose easily, there are not many fossils to study.

CHAPTER
REVIEW

CHECKING CONCEPTS

Choose the word or phrase that completes the sentence.

1. A(n) ______ is a flightless bird.
a. duck **c.** owl
b. oriole **d.** ostrich

2. Wings of birds are not used for ______ .
a. flying **c.** balancing
b. swimming **d.** eating

3. Birds crush and grind food in the ______.
a. crop **c.** gizzard
b. stomach **d.** small intestine

4. Omnivores are mammals that eat ______.
a. plants
b. animals
c. plants and anmals

5. A ______ is an example of a marsupial.
a. cat **c.** kangaroo
b. human **d.** camel

6. Egg-laying mammals are called ______.
a. marsupials **c.** placental
b. monotremes **d.** chiropterans

7. ______ have mammary glands with no nipples.
a. Marsupials **c.** Placental mammals
b. Monotremes **d.** Lagomorphs

8. Teeth specialized for tearing are ______.
a. canine **c.** molars
b. incisors **d.** premolars

9. ______ is hunting animals illegally.
a. Placental **c.** Poaching
b. Incubate **d.** Omnivore

10. Humans and monkeys are ______.
a. primates **c.** edentates
b. carnivores **d.** cetaceans

USING LAB SKILLS

11. In Activity 16-1 on page 378, you observed the structure of feathers. Based on your observations, explain why down feathers of birds are used to make warm coats for humans. Examine some synthetic materials used in making warm coats and compare them with down feathers. How are they alike? How are they different?

12. In the MINI-Lab on page 380, you looked at some of your own hairs. Examine cat, dog, or horse hair to compare them with human hair.

THINK AND WRITE CRITICALLY

Answer the following questions in your Journal using complete sentences.

13. Why is it incorrect to tell someone who eats very little that he or she "eats like a bird"?

14. There are far more species of placental mammals than of marsupials or monotremes. Why do you think this is so?

15. Explain how the feet of birds are adapted to where they live.

16. Why is it difficult to study the origin of birds based on fossils?

APPLY

17. Discuss the differences in reproduction between birds and reptiles.
18. Name a mammal group that probably doesn't have sweat glands and explain why it doesn't.
19. Which type of bird would have lighter wing bones—ducks or turkeys? Explain.
20. What features of birds allow them to be fully adapted to life on land?
21. Which mammal group has the best adaptations for the survival of its embryo? Why?

MORE SKILL BUILDERS

If you need help, refer to the Skill Handbook.

1. **Sequencing:** Sequence the order in which food passes through a bird's digestive system.
2. **Concept Mapping:** Complete the concept map describing the groups of mammals.

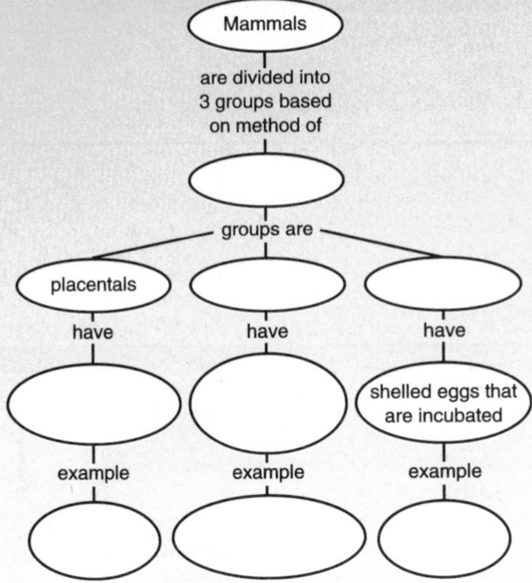

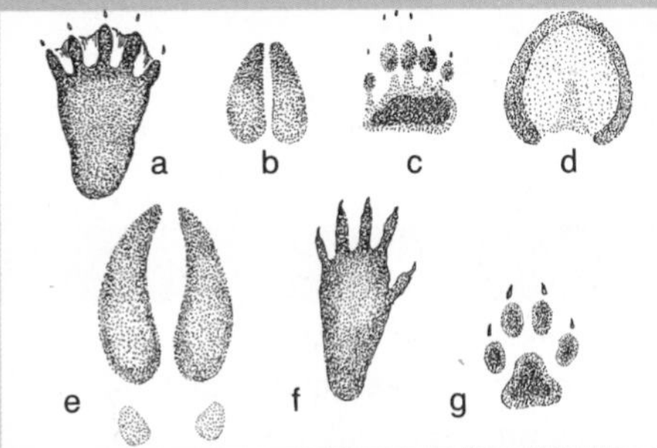

3. **Observing and Inferring:** Look at the diagrams of animal tracks. Decide which track belongs to which type of mammal by using Table 16-1. Fill in the chart. Describe how each animal's foot is adapted to its environment.

Animal	Track	Adaptation
Bear		
Beaver		
Cheetah		
Deer		
Horse		
Moose		
Raccoon		

4. **Comparing and Contrasting:** Compare and contrast the teeth of herbivores, carnivores, and omnivores. How is each tooth type adapted to the animal's diet?
5. **Classifying:** You discover three new species of mammals. The traits of each species is as follows: Mammal 1—swims and eats plants; Mammal 2—flies and eats fruit; Mammal 3—runs and hunts. Classify each mammal into its correct group.

PROJECTS

1. Research the birds in your area. Find out what they eat and the types of nests they build.
2. Write to the National Wildlife Federation about endangered birds and mammals. What is being done to save these animals?

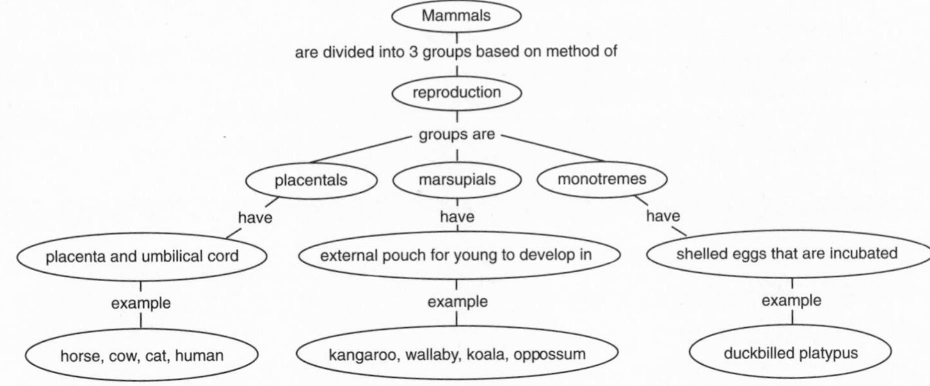

APPLY

17. Both birds and reptiles lay eggs. Reptiles lay large quantities of leathery-shelled eggs and do not incubate them. Birds lay fewer eggs at a time. The shells of bird eggs are hard, and the parents incubate them.
18. The cetaceans probably do not have sweat glands because they live in water.
19. Ducks have lighter wing bones than turkeys. Turkeys do not fly long distances and would be less likely to need this trait.
20. The amniote egg, specialized beaks, feet, and wings allow land adaptation.
21. The placental mammals. Placental mammals' embryos are well protected inside the female and have easier access to nutrients through the umbilical cord.

MORE SKILL BUILDERS

1. Sequencing: mouth–crop–stomach–gizzard–small intestine
2. Concept Mapping: See below, left.
3. Observing and Inferring: a. beaver, b. deer, c. bear, d. horse, e. moose, f. racoon, g. cheetah
Beavers' feet are adapted to living in water. The cheetah, racoon, and bear have adapted for support and catching food. The deer and moose have feet adapted for support and running. The horse's hoof is adapted for running.
4. Comparing and Contrasting: Herbivores have large incisors for cutting off blades of grass. Carnivores have small incisors and large canine teeth to grip and tear food. Omnivores have canines and incisors to feed on both plants and animals.
5. Classifying: Mammal 1 belongs to the order Sirenia. Mammal 2 belongs to the order Chiroptera. Mammal 3 belongs to the order Carnivora.

CHAPTER

17 Animal Behavior

CHAPTER SECTION	OBJECTIVES	ACTIVITIES
17-1 Types of Behavior (2 days)	1. **Distinguish** between innate and learned behavior. 2. **Recognize** reflex and instinctive actions and explain how they help organisms live. 3. **Describe** and give examples of imprinting, trial and error, conditioning, and insight.	**MINI-Lab:** *How does insight help you solve problems?* p. 399 **Activity 17-1:** *Designing an Experiment (Investigating Conditioning),* p. 400
17-2 Behavioral Adaptations (2 days)	1. **Recognize** the importance of behavioral adaptations. 2. **Explain** how courtship behavior increases the chances of reproductive success. 3. **Evaluate** the importance of social behavior and cyclic behavior.	**MINI-Lab:** *How does an animal enter into hibernation?* p. 405
17-3 Rehabilitation of Wild Animals Science & Society (1 day)	1. **Explain** some of the effects of human development on wildlife. 2. **Describe** the disadvantages of releasing rehabilitated animals into the wild.	**Activity 17-2:** *Observing Social Behavior in Ants,* p. 408
Chapter Review		

ACTIVITY MATERIALS

FIND OUT	ACTIVITIES		MINI-LABS	
Page 393 Plexiglas sheet paper	**17-1 Designing an Experiment, p. 400** aquarium glass cover for aquarium thermometer washed coarse sand tap water aged three days metric ruler dish food for guppies fish net guppies snails water plants	**17-2 Observing Social Behavior in Ants, p. 408** 20 ants large jar with screen wire cover soil black paper small moist sponge water food—honey, sugar, bread crumbs small can with lid spoon or trowel hand lens	**How does insight help you solve problems? p. 399** safety pins	**How does an animal enter into hibernation? p. 405** graph paper

CHAPTER FEATURES	TEACHER RESOURCE PACKAGE	OTHER RESOURCES
Problem Solving: *The Disappearing Lizards,* p. 398 **Skill Builder:** *Making and Using Tables,* p. 399	**Ability Level Worksheets** ◆ ***Study Guide,*** p. 67 ● ***Reinforcement,*** p. 67 ▲ ***Enrichment,*** p. 67 **Activity Worksheets,** pp. 149-150, 155 **Critical Thinking/Problem Solving,** p. 21 **Cross-Curricular Connections,** p. 21 **Transparency Masters,** pp. 61-62	**Color Transparency 31,** Mating Behavior of Stickleback Fish **Lab Manual:** Environment and Behavior, p. 115 **Science Integration Activity 17** **STVS:** Disc 5, Side 2
Technology: *Looking for a Sign,* p. 404 **Skill Builder:** *Using Variables, Constants, and Controls,* p. 405	**Ability Level Worksheets** ◆ ***Study Guide,*** p. 68 ● ***Reinforcement,*** p. 68 ▲ ***Enrichment,*** p. 68 **Activity Worksheets,** p. 156 **Concept Mapping,** pp. 39-40	**STVS:** Disc 5, Side 1; Side 2
You Decide! p. 407	**Ability Level Worksheets** ◆ ***Study Guide,*** p. 69 ● ***Reinforcement,*** p. 69 ▲ ***Enrichment,*** p. 69 **Activity Worksheets,** pp. 151-152 **Science and Society,** p. 21	**STVS:** Disc 6, Side 1
Summary Key Science Words Understanding Vocabulary Checking Concepts Using Lab Skills Think & Write Critically Apply More Skill Builders Projects	ASSESSMENT RESOURCES **Chapter Review,** pp. 37-38 **Chapter Test,** pp. 105-108 **Unit Test,** pp. 109-110 **Performance Assessment in Middle School Science (PAMSS)**	**Chapter Review Software** **Test Bank** **Alternate Assessment** **Performance Assessment**

◆ **Basic** ● **Average** ▲ **Advanced**

ADDITIONAL MATERIALS

SOFTWARE	AUDIOVISUAL	BOOKS/MAGAZINES
Describing the Behavior of Organisms, Queue. *The Honey Factory,* CBS Software.	*Beyond Words: Animal Communication,* film, National Geographic. *Biological Clocks,* filmstrip, National Geographic. *African Wildlife,* laserdisc, Image Entertainment. *Among the Wild Chimpanzees,* laserdisc, Image Entertainment. *Gorilla,* laserdisc, Image Entertainment. *Great Ape,* laserdisc, Image Entertainment. *Signs of the Apes/Songs of the Whales* (NOVA), laserdisc, Image Entertainment.	Aronson, Elliot. *The Social Animal.* 4th ed. NY: W.H. Freeman, 1984. Berdecio, Susana and Leanne T. Nash. *Chimpanzee Visual Communication.* Tucson, AZ: University of Arizona Press, 1981. Brenner, Barbara. *Gorilla Signs Love.* NY: Lothrop, Lee & Shepard Books, 1984.

CHAPTER

17

ANIMAL BEHAVIOR

THEME DEVELOPMENT: Help students see that innate behavior has developed in the course of evolution. Another major focus should be that reflexes are homeostatic mechanisms. Ecological conditions also influence behaviors.

CHAPTER OVERVIEW

- **Section 17-1:** This section discusses reflexes and instincts as innate behaviors. Learning, imprinting, trial and error, conditioning, insight, and examples of each are presented for students to study learned behavior.
- **Section 17-2:** This section describes territorial behavior, courtship behavior, social behavior, and cyclic behavior.
- **Section 17-3: Science and Society:** How animals are injured and how injured animals are rehabilitated are explored. The You Decide feature asks students if it would be wrong to let nature take its course when a young bird falls from a nest.

CHAPTER VOCABULARY

behavior
innate behavior
reflex
instinct
learning
imprinting
trial and error
motivation
conditioning
insight
territory
aggression
courtship behavior
social behavior
society
communication
cyclic behaviors
circadian rhythm
migration
carrying capacity

OBJECTIVES AND SCIENCE WORDS: Have students review the objectives and science words in this chapter as they study each section.

CHAPTER

17 Animal Behavior

392

OPTIONS

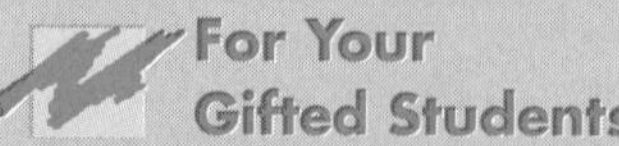

For Your Gifted Students

Ask students to choose an animal mentioned in the chapter to research. They should write a "want ad" for it, including the behaviors that would identify the animal (cyclic, innate, territorial, courtship).

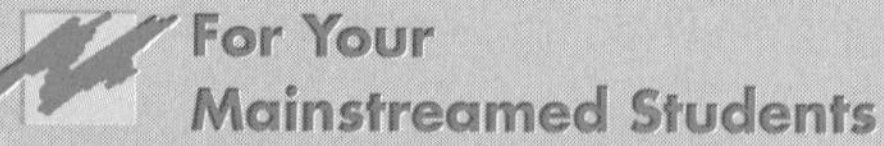

For Your Mainstreamed Students

Allow students to collect crickets in an aquarium containing moist soil and lettuce or grass. They can observe them through a hand lens to note how their chirping sound is produced.

Owls fluff up their feathers and spread their wings to appear larger when threatened. You quickly pull your hand away from a hot stove or iron, and blink your eyes when you see something rushing toward your face. Do organisms learn these actions, or do they occur naturally?

FIND OUT!

Do this simple activity to find out if some simple responses of organisms are learned or natural.

Obtain a piece of clear plexiglass and a wadded up sheet of paper from your teacher. Have a partner hold the plastic a few inches from his or her face. Stand three feet away and gently toss the wadded up paper toward your partner's face. *Observe* what happened to his or her eyes. Toss the paper ball again. This time tell your partner not to blink. Was he or she successful? Do you think this behavior is learned, or does it occur naturally? Do you think it is good for some behaviors to occur naturally?

Gearing Up

Previewing the Chapter

Use this outline to help you focus on important ideas in this chapter.

Previewing Science Skills

- In the Skill Builders, you will make and use tables and use variables, constants and controls.
- In the Activities, you will observe and collect and record data.
- In the MINI-Labs, you will hypothesize, experiment, and graph data.

What's next?

The Find Out activity demonstrated some of the things you do naturally and are not learned. In this chapter you will learn more about these types of behaviors and about behaviors that are learned. You will also learn how animals behave to be able to survive in their environments.

INTRODUCING THE CHAPTER

Use the Find Out activity to introduce the students to innate behavior. Tell them that they will be learning about both innate and learned behaviors.

FIND OUT!

Preparation: Before you do the activity, have 25 cm^2 Plexiglass sheets cut.

Materials: one Plexiglass sheet and one sheet of wadded paper for every three students

Cooperative Learning: Use the Science Investigation strategy. Divide the class into groups of three. Have one student hold the Plexiglass in front of the face, one throw the wadded paper, and one observe the eye movement of the student holding the Plexiglass.

Teaching Tips

- If Plexiglass is not available, sheets of plastic wrap and cotton balls may be used. Do not use wadded paper with plastic wrap.

Gearing Up

Have students study the Gearing Up feature to familiarize themselves with the chapter. Discuss the relationships of the topics in the outline.

What's Next?

Before beginning the first section, make sure students understand the connection between the Find Out activity and the topics to follow.

ASSESSMENT OPTIONS

PORTFOLIO

Refer to page 409 for suggested items that students might select for their portfolios.

PERFORMANCE ASSESSMENT

See page 409 for additional Performance Assessment options.

Process

Skillbuilders, pp. 399, 405
MINI-Labs, pp. 399, 405
Activities, 17-1, p. 400; 17-2, p. 408
Using Lab Skills, p. 410

CONTENT ASSESSMENT

Assessment—Oral, pp. 397, 404
Section Reviews, pp. 399, 405, 407
Chapter Review, pp. 409-411
Mini Quizzes, pp. 398, 403

GROUP ASSESSMENT

Opportunities for group assessment occur with Cooperative Learning Strategies and Flex Your Brain Activities.

SECTION 17-1

PREPARATION

SECTION BACKGROUND

▶ The scientific study of animal behavior as it occurs in the organism's natural environment is called ethology.

▶ Konrad Z. Lorenz, an Austrian naturalist and a founder of ethology, during the 1930s studied the behavior of birds and developed a theory of animal behavior that stressed its inherited aspects. Lorenz and two other ethologists, Karl von Frisch of Austria and Nikolaas Tinbergen from The Netherlands, received the 1973 Nobel Prize for their work.

▶ Explain that scientists must understand the perceptual world of their animal subject before interpreting behavioral data. The interpretation of what is not human in terms of human characteristics is anthropomorphism.

PREPLANNING

▶ Collect pictures of animal behaviors for the bulletin board.

▶ Obtain one 4-L aquarium for each group to conduct Activity 17-1. Purchase sand, water plants *(Elodea)*, guppies, and snails for each group.

1 MOTIVATE

▶ **Bulletin Board:** Arrange pictures of animal behaviors on the bulletin board. Use pictures of behaviors associated with food getting, protection, and reproduction. Have students list behaviors that come under the three categories.

VideoDisc

STVS: Development of Walking in Chicks, Disc 5, Side 2

17-1 Types of Behavior

New Science Words

- behavior
- innate behavior
- reflex
- instinct
- learning
- imprinting
- trial and error
- motivation
- conditioning
- insight

Objectives

- ▶ Distinguish between innate and learned behavior.
- ▶ Recognize reflex and instinctive actions and explain how they help organisms survive.
- ▶ Describe and give examples of imprinting, trial and error, conditioning, and insight.

Innate Behavior

When you come home from school, does your dog run to meet you, barking and wagging its tail? After you play with him a while, does he sit at your feet and watch every move you make? Why does he do these things? Dogs are pack animals that generally follow a leader. They have been living with people for about 50 000 years, so it's easy for a dog to adopt a human as its leader.

If you have a cat, it doesn't jump up and lick your face when you come home, does it? Cats are domestic animals too. Even though cats are happy living with people, they keep some traits of wild cats. Living with people doesn't change some of the behaviors of cats and dogs.

1 **Behavior** is the way an organism acts toward its environment. Anything in the environment to which an organism reacts is called a stimulus. You are the stimulus that causes your dog to bark and wag its tail. Your dog's reaction to you is a response.

Dogs and cats are quite different from one another in their behavior. They were born with certain behaviors, and they have learned others. A behavior that an organism is born with is an **innate behavior.** Such behaviors are inherited and they do not have to be learned.

Innate behavior patterns are usually correct the first time an animal responds to a stimulus. Kittiwakes are sea birds that nest on narrow ledges. The chicks stand still as soon as they hatch. The chicks of a related bird, the herring gull, which nests on the ground, move around as soon as they can stand. A kittiwake chick can't do this because one step could mean instant death. They don't have time to learn.

Figure 17-1. Some behaviors an animal is born with are necessary for survival. Kittiwakes, shown here, are adapted for nesting on narrow ledges.

OPTIONS

Meeting Different Ability Levels

For Section 17-1, use the following **Teacher Resource Masters** depending upon individual students' needs.

◆ **Study Guide Master** for all students.

● **Reinforcement Master** for students of average and above average ability levels.

▲ **Enrichment Master** for above average students.

Additional Teacher Resource Package masters are listed in the OPTIONS box throughout the section. The additional masters are appropriate for all students.

◆ STUDY GUIDE 67

NAME ______ DATE ______ CLASS ______

STUDY GUIDE — Chapter 17

Types of Behavior — Text Pages 394-399

Use the clues to complete the puzzle.

Across

3. behavior that is modified by experience is called trial and ____ (ERROR)
5. something inside an animal that causes it to act; necessary for learning to occur (MOTIVATION)
7. complex pattern of innate behavior, often having several parts (INSTINCT)
9. behavior that an organism is born with and which doesn't have to be learned (INNATE)
10. behavior that develops through experience (LEARNED)

Down

1. a type of learning in which an animal forms a bond to another organism shortly after birth or hatching (IMPRINTING)
2. the way an organism acts toward its environment (BEHAVIOR)
4. behavior that is modified so that a response becomes associated with another stimulus (CONDITIONING)
6. a form of reasoning that allows animals to use past experiences to solve new problems (INSIGHT)
8. an automatic response that does not involve the brain (REFLEX)

Complete the following sentences using the terms below.

trial and error, recognition, conditioning, motivation, imprinting, stimulus, reflex, instinct, learn

1. Mating behavior of the male stickleback fish is an example of **instinct**.
2. Shivering is a **reflex** behavior.
3. Anything in the environment to which an organism responds is called a **stimulus**.
4. Older quail will not crouch if a leaf falls because of their ability to **learn**.
5. Imprinting is a learning pattern involving **recognition**.
6. Pavlov's dogs secreted saliva when a bell was rung even when no food was present because of **conditioning**.
7. A hungry rat will be able to learn a maze if it receives food at the end. This is an example of **motivation**.
8. Riding a skateboard is learned by **trial and error**.
9. Wolfgang Kohler's experiment showed that chimpanzees used **insight** when they piled up boxes in order to reach bananas.

67

The lives of most insects are too short for the young to learn from the parents. In many cases, the parents have died by the time the young hatch. And yet, every insect reacts automatically to its environment. A moth will fly toward a light, and a cockroach will run away from it. Fleas are attracted to warm-blooded animals by the heat their bodies give off, and fruit flies are attracted to the odor of overripe fruit. They do not have to spend time learning what to do.

Figure 17-2. Spiders must automatically respond to their environment. There is no time for learning in their short life cycle.

The simplest innate behaviors are reflex
2 actions. A **reflex** is an automatic response that does not involve the brain. Sneezing, shivering, yawning, jerking your hand away from a hot surface, and blinking your eyes when something is thrown toward you are all reflex actions. All animals have reflexes. The fur on your cat's back stands on end when she is frightened. An octopus changes colors when it senses danger.

During a reflex, a message passes from a sense organ along the nerve to the spinal cord and back to the muscles. The message does not go to the brain. You are aware of the reaction only after it has happened. A reflex is not the result of conscious thinking.

An **instinct** is a complex pattern of innate behavior. Have you ever watched a spider spin a web? Spinning a web is very complicated, and yet spiders spin webs correctly on the first try. Unlike reflexes, instinctive behaviors may have several parts and take weeks to complete. Instinctive behavior begins when the animal recognizes a stimulus and continues until all parts of the behavior have been performed.

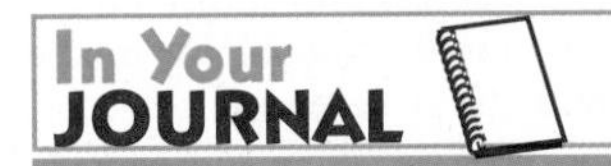

In your Journal, make a list of some of the problems of using animals (or humans) in behavioral studies.

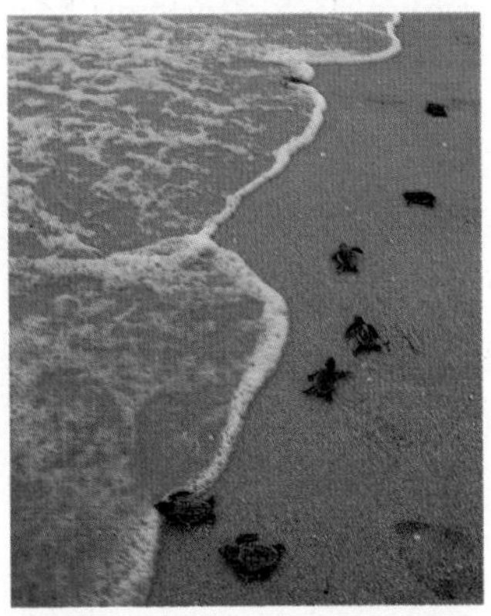

Figure 17-3. When sea turtles hatch, they instinctively run to the ocean.

TYING TO PREVIOUS KNOWLEDGE: Students are familiar with pet behaviors. Ask them to relate instances of feeding behavior and evidence of communication with other animals and with humans. Review the definitions of the terms *stimulus* and *response*.

2 TEACH

Key Concepts are highlighted.

CONCEPT DEVELOPMENT

▶ Explain that innate behavior is also called inborn behavior.

▶ An example of instinctive behavior is found in the digger wasp. After the digger wasp comes from her underground pupa, she lives only a few weeks. In this time, she digs a hole for a nest, hunts prey to nourish her young, and lays her eggs. Her parents died several months before she emerged, so she was not taught these behaviors.

In Your JOURNAL

Accept all reasonable ideas. Animals other than humans cannot communicate with behavioral researchers in words. Humans in studies can give erroneous answers to questions. Equating nonverbal animal responses to those of humans may not always work.

● REINFORCEMENT 67

NAME ______ DATE ______ CLASS ______

REINFORCEMENT — Chapter 17

Types of Behavior — Text Pages 394–399

Tell which of the following examples are innate behavior and which are learned behavior. Write the word innate or learned in the space provided.

innate 1. Sowbugs move away from the light.
learned 2. A dog lies down when given a command to do so.
innate 3. A male stickleback fish protects its young.
innate 4. The pupil of a cat's eye increases in size in dim light.
learned 5. A seeing-eye dog leads a blind person.
learned 6. A dog barks twice for a bone.
innate 7. Spiders spin webs correctly on their first try.
innate 8. Birds fly south for the winter.
learned 9. A person rides a bicycle.
innate 10. A person shivers when cold.

Examples of types of behavior are given below. Identify each behavior and tell how it is helpful.

Behavior	Example	How it is helpful
conditioning	Fish will move to the place where they are fed when you tap on the tank.	useful in training animals
trial and error	A person tries the keys on a key ring until one works.	helps avoid making similar mistakes
imprinting	A duckling becomes socially attached to its mother.	parental care helps animals survive
reflex	A person sneezes.	helps body protect itself
insight	A chimp stacks boxes to reach a banana hung out of its reach.	helps solve problems

 67

▲ ENRICHMENT 67

NAME ______ DATE ______ CLASS ______

ENRICHMENT — Chapter 17

Types of Behavior — Text Pages 394–399

ANIMAL BEHAVIOR RESEARCH

The Austrian scientist Konrad Lorenz and the Dutch zoologist Niko Tinbergen were the founders of an approach to behavior called ethology. These two men shared the 1973 Nobel Prize in medicine and physiology with the German biologist Karl von Frisch. **Ethology** is the study of animal behavior in nature, and is concerned with innate behavior patterns. Instincts and reflexes are innate behaviors.

Learned behavior is acquired as the result of experience. There are several kinds of learned behavior. **Habituation** is a form of learning in which an animal learns not to respond to a stimulus. Animals may learn not to respond to stimuli that are repeated frequently and that are unimportant to them. Conditioning is another type of learned behavior. There are two kinds of conditioning. Ivan Pavlov, a Russian psychologist, knew that dogs respond to the smell of food by salivating. This response is innate. Pavlov wanted to know if the salivation response could be caused by a less likely stimulus. Pavlov rang a bell at the same time the dog was given food. Eventually, the sound of the bell was enough to cause the dog to salivate. This type of conditioning is known as **classical conditioning**.

The American psychologist B.F. Skinner used pigeons to study another type of conditioning. He realized that if a certain behavior were followed by a reinforcement or reward, the behavior was more likely to be repeated in the future. To study this phenomenon, he invented the Skinner box. The box contained a typewriter-like key connected to a food supply. If the pigeon pecked at the key, it would be rewarded with food. This type of learning is called **operant conditioning**. If the response is reinforced with a reward, the behavior tends to be repeated.

Determine if the following behaviors are innate or learned. If the behavior is innate, determine if the behavior is an instinct or a reflex. If the behavior is learned, determine if the behavior is habituation, classical conditioning, or operant conditioning.

1. The first time a squirrel sees a nut, it tries to bury it. — innate — instinct
2. Jane touches a hot curling iron and jerks her hand away. — innate — reflex
3. John is startled by a loud noise outside his window. The same noise is repeated several times, but John no longer responds. — learned — habituation
4. Cindy is teaching her dog tricks. When the dog rolls over at her command, she rewards it with praise. — learned — operant conditioning
5. A planarian moves away from bright light. It curls up when electrically shocked. When a bright light is closely followed by an electric shock, the planarian curls up. Eventually, the light alone causes the planarian to curl. — learned — classical conditioning
6. Kittens stalk imaginary prey. — innate — instinct

 67

CONCEPT DEVELOPMENT

▶ **Activity:** Have students make bird feeders from milk cartons and hang them outside the classroom window. Place birdseed in the feeders and watch for birds to come to the feeders. Have them record how different birds act when they feed.

To assess this product, refer to the Performance Task Assessment Lists in **Performance Assessment in Middle School Science.**

▶ Ask students to give examples of other innate behaviors. Some are: yawning, coughing, sneezing, blinking, crying, growling or snapping in a dog, and grooming in a cat.

▶ Explain that in animals that develop without parental care, behaviors are more likely to be innate.

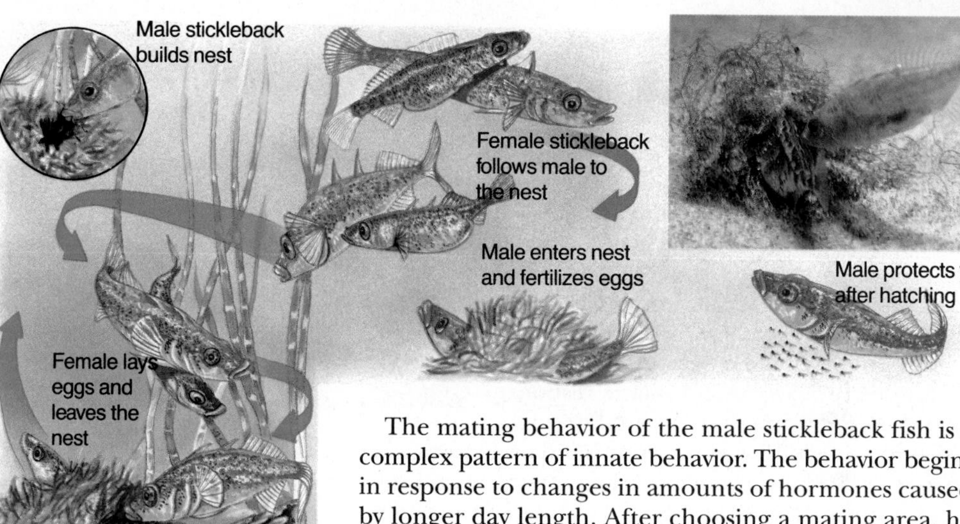
Male stickleback builds nest
Female stickleback follows male to the nest
Male enters nest and fertilizes eggs
Male protects young after hatching
Female lays eggs and leaves the nest

Figure 17-4. The male stickleback fish does not learn mating behavior. It is a complex innate behavior pattern.

The mating behavior of the male stickleback fish is a complex pattern of innate behavior. The behavior begins in response to changes in amounts of hormones caused by longer day length. After choosing a mating area, he drives other fish from it. Then he collects plants and makes them into a small mound in which he creates a tunnel. The tunnel becomes his nest. Meanwhile, his normally dull, brown-colored body changes to bright red on the underside and bluish-white on top. When a female enters the mating area, he does a zigzag dance to lead her into the nest. She lays her eggs in the nest and swims away. The male fertilizes the eggs and stays with them. When the young hatch, he protects them until they are well developed. All male sticklebacks are born with this pattern of behavior. Even if young sticklebacks are not raised with other sticklebacks, they follow the same behavior pattern.

Figure 17-5. Mammals have many more learned behaviors than other animals.

Learned Behavior

All animals have both innate and learned behaviors. **Learning** is a behavior that develops through experience. Instinct almost completely determines the behavior of insects, spiders, and other arthropods. These animals are able to learn very little, so their survival depends on innate behavior.

Fish, reptiles, amphibians, birds, and mammals can learn more than invertebrates. Fish behave more by instinct than birds. Birds show more instincts than mammals. In humans, babies suckle, smile, and grasp instinctively, but as they grow older, most of their behavior is learned.

OPTIONS

ENRICHMENT

▶ Have students use the library to report on the work of Konrad Lorenz.

▶ Have students report on the work of Nikolaas Tinbergen.

PROGRAM RESOURCES

From the **Teacher Resource Package** use:

Critical Thinking/Problem Solving, page 21, Sign Stimuli.

Transparency Master, pages 61-62, Mating Behavior of Stickleback Fish.

Cross-Curricular Connections, page 21, Animals that Help People.

Use **Color Transparency** 31, Mating Behavior of Stickleback Fish.

Use **Laboratory Manual,** page 115, Environment and Behavior.

Instincts can be modified by learning. Grouse and quail chicks leave their nest the day they hatch. They can run and find food, but they can't fly. When something moves above them, they crouch down and keep perfectly still until the danger is past. They will crouch without moving even if the falling object is only a leaf. Older birds have learned that leaves will not harm them, but they too freeze when a hawk moves overhead.

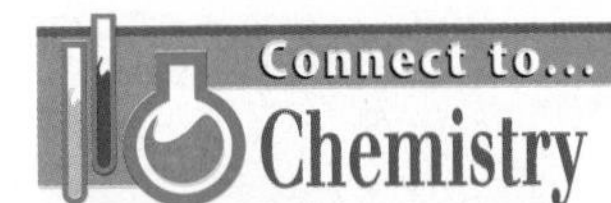

When a salmon hatches, the chemical makeup of the water around it is imprinted in its memory. As an adult, the salmon follows streams with a chemical makeup that most closely corresponds to this memory. Why is this imprinting important for salmon survival?

Imprinting

You have probably seen young ducks following their mother. This is an important behavior because the adult bird has had more experience in finding food, escaping predators, and getting along in the world. **Imprinting** is a type of learning in which an animal forms a social attachment to another organism within a specific time period after birth or hatching.

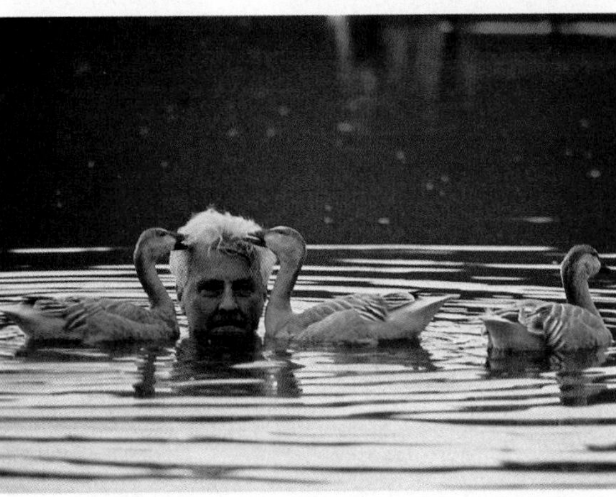

Figure 17-6. Young goslings may imprint on any object or organism present during hatching. Konrad Lorenz is shown here swimming with imprinted goslings.

Konrad Lorenz, an Austrian naturalist, developed the concept of imprinting. Working with geese, he discovered that a gosling follows the first moving object it sees after hatching. It recognizes the moving object as its parent. It later recognizes similar objects as members of its own species. This behavior works well when the first moving object a young goose sees is an adult female goose. But, goslings hatched in an incubator may see a human first and may imprint on him or her. Animals that become imprinted toward animals of another species never learn to recognize members of their own species.

Trial and Error

Can you remember when you learned to ride a bicycle? You probably fell many times before you learned to balance on the bicycle. But after a while you could ride without having to think about it. You have many skills that you have learned through trial and error. Skating and riding a bicycle are just a few.

Behavior that is modified by experience is called **trial and error** learning. Both invertebrates and vertebrates learn by trial and error. When baby chicks first learn to

CONCEPT DEVELOPMENT

▶ Explain that imprinting is an example of a behavior that is both innate and learned. Following and becoming attached to a moving object after birth or hatching is innate behavior. Recognizing that object is learned behavior.

▶ **Activity:** Assemble a large puzzle and then separate it into 12 or 14 sections (one section for every two students). Take the pieces of each section apart, place them in paper bags, and give one to each group. After each group has put its section of the puzzle together, have students fit their sections together. Discuss what type of learning this is. *(Learned)*

REVEALING MISCONCEPTIONS

▶ Many people attribute human feelings and attitudes to animals. This can lead to a misinterpretation of an animal's behavior. Animals are not recognized to have any conscious motive for their behavior other than being genetically programmed to behave as they do.

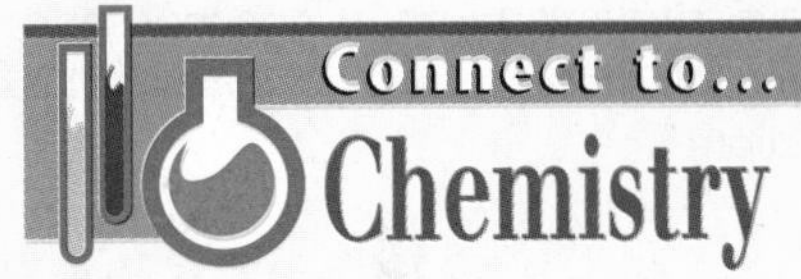

Answer: Imprinting allows the salmon to return to a place where it can reproduce. Therefore, the population survives.

ASSESSMENT—ORAL

▶ **Why do you think innate behaviors are inherited?** *They are critical to survival.*

▶ **Why are reflexes important?** *They provide protection by allowing an animal to respond rapidly to a sudden change.*

▶ **How is innate behavior an advantage to an animal that hatches from an egg after its parents die?** *It will use its instincts to find food and shelter.*

▶ **How does imprinting combine instinct and learning?** *Following and becoming attached to a moving object after birth or hatching is innate. Recognizing that object is learning.*

▶ **If a hungry rat eats a small amount of poisoned food that makes it ill, it will never eat poisoned food again. How can you explain this?** *The rat's behavior changes because of the experience.*

PROBLEM SOLVING

The lizards' behavior was innate.
Think Critically: Cold-blooded animals have little control over their body temperatures. The lizards were sunning themselves to warm up after a cool night. When they became too warm, they burrowed underground. At sunset, when it became cooler, they dug out to look for food.

CHECK FOR UNDERSTANDING

Use the Mini Quiz to check for understanding.

MINI QUIZ

Use the Mini Quiz to check students' recall of chapter content.

1. **The way an organism acts toward its environment is _____.** *behavior*
2. **A reflex is automatic because it does not involve the _____.** *brain*
3. **_____ is something inside an animal that causes an animal to act.** *Motivation*

RETEACH

Have students make a concept map using all the vocabulary words in this section.

EXTENSION

For students who have mastered this section, use the **Reinforcement** and **Enrichment** masters or other OPTIONS.

Figure 17-7. Many pets become conditioned to be fed at certain times of the day.

feed themselves, they peck at many spots before they get any food. As a result of trial and error, they soon learn to peck only at grain.

A particular stimulus may cause a different response in the same animal at different times. When your cat sees food, it will eat it if it's hungry, but if it has just eaten, it will ignore the food. For a hungry rat, the motive to learn a maze may be the food at the end of the maze. **Motivation** is something inside an animal that causes the animal to act. It is necessary for learning to take place.

Conditioning

Animals often learn new behaviors by conditioning. In **conditioning,** behavior is modified so that a response previously associated with one stimulus becomes associated with another. Russian scientist Ivan P. Pavlov was the first person to study conditioning. He knew that the sight and smell of food made hungry dogs secrete saliva. Pavlov

PROBLEM SOLVING

The Disappearing Lizards

Julio's science class is studying vertebrate animals. His teacher asked each member of the class to choose a vertebrate animal, observe it for several days, and prepare a report.

Shortly after sunup, Julio went to a hot desert area near his home to observe lizards. Julio saw three lizards lying motionless on some rocks. He watched for a while, then went home for a cool drink. When he returned, the lizards seemed to have disappeared. The same thing happened on the next day.

One morning Julio stayed a little longer. One lizard left its rock and burrowed into the sandy soil. Soon, the other two lizards did the same thing. Then, on Saturday, Julio made more frequent visits to the rock. The lizards stayed underground until sunset when they dug out and began hunting for food.

Was the lizards' behavior innate or learned?

Think Critically: Use what you learned about cold-blooded animals to explain the behavior of the lizards.

OPTIONS

INQUIRY QUESTIONS

▶ **How could a dog trainer motivate a dog to shake hands or roll over?** *by giving it a dog treat after it has completed the behavior*

▶ **Why do you think the term "mouth watering" is used to describe appetizing foods?** *The response to the stimulus of food is production of saliva. Upon conditioning, the sight or smell of food can cause the production of saliva.*

▶ **What value does learning have for animals with long life spans?** *Learning enables them to adapt to changes in the environment.*

ENRICHMENT

▶ Have students research the work of B. F. Skinner and operant conditioning.

PROGRAM RESOURCES

From the **Teacher Resource Package** use:

Activity Worksheets, page 155, Mini-Lab: How does insight help you solve problems?

Science Integration Activities, 17, Relationships

added another stimulus. He rang a bell when he gave the dogs food. The dogs began to connect the sound of the bell with food. Then Pavlov rang the bell without giving the dogs food. The dogs secreted saliva when the bell was rung even though he did not show them food. The dogs were conditioned to respond to the bell.

The American psychologist John B. Watson demonstrated that responses of humans can also be conditioned. In one experiment, he struck a metal object each time an infant touched a furry animal. The loud noise frightened the child. In time, the child became frightened by the furry animal when no sound was made.

Insight

When you solve a math problem, do you use what you have previously learned in math to solve the problem? **Insight** is a form of reasoning that enables animals to use past experiences to solve new problems. In Wolfgang Kohler's experiments with chimpanzees, a bunch of bananas was placed too high for the chimpanzees to reach. Chimpanzees piled up boxes found in the room, climbed up on them, and reached the bananas. Much of adult human learning is based on insight. Younger children learn by trial and error.

MINI-Lab

How does insight help you solve problems?

Obtain a chain of 15 safety pins. Count the safety pins on the chain to be sure there are 15. Make a **hypothesis** as to the *fewest* number of times you can open and close pins to break the chain into the following:

5 pins attached
4 pins attached
3 pins attached
2 pins attached, and
1 alone

Each open-and-close counts as **one.** Test your hypothesis and compare with those of your classmates. (All pins must be closed at the end!)

SECTION REVIEW

1. Compare innate behavior with learned behavior.
2. Compare a reflex with an instinct.
3. Describe the four types of learned behavior.
4. How are trial and error and insight behaviors important to humans?
5. **Apply:** A family moves into a new home near an airport. They are awakened by the planes during the first few nights. Explain why after a week or two, they stop waking up even though the planes continue to fly.
6. **Connect to Chemistry:** Write a paragraph about how some drugs are used in a beneficial way to modify behavior.

Skill Builder

Making and Using Tables

Make a table that shows the kinds of innate and learned behaviors in this section. Include examples of each. If you need help, refer to Making and Using Tables in the **Skill Handbook** on page 690.

Skill Builder

INNATE BEHAVIORS		LEARNED BEHAVIORS	
Type	**Example**	**Type**	**Example**
reflex	sneezing, blinking eyes	imprinting	ducklings following mother
instinct	spider spinning web, turtles running to sea	trial and error	riding a bike
		conditioning	a specific response to a specific stimulus

MINI-Lab

Materials: safety pins

Teaching Tips

▶ Have pins linked together in sets of 15 before the start of the activity.

Answer: The individual sets of pins can be made in three open and close moves. This can be accomplished by leaving the sixth pin open and later closing it.

MINI-Lab
ASSESSMENT

Performance: To further assess students' understanding of insight, have them describe examples of when they used insight to solve problems.

3 CLOSE

▶ Ask questions 1-4 and the **Apply** Question in the Section Review.

SECTION REVIEW ANSWERS

1. An innate behavior is a behavior an animal is born with. Learned behavior is a behavior that develops through experience.

2. A reflex is an automatic response that does not involve the brain. An instinct is a complex pattern of innate behavior.

3. Student responses should reflect information on pages 396-399.

4. They both allow humans to learn through experience.

5. Apply: The family becomes conditioned to sleep through the noise of the planes.

6. Connect to Chemistry: Accept all reasonable answers. Drugs have been used beneficially to modify behavior of persons under stress or with nervous conditions they could not control unaided.

Skillbuilder
ASSESSMENT

Performance: Use this Skillbuilder to assess students' abilities to Make and Use Tables. Ask students to identify a knee jerk and piano playing as innate or learned behaviors.

ACTIVITY 17-1

OBJECTIVE: Design and carry out an experiment to demonstrate conditioning in fish.

Time: One class period to set up aquarium, 10 minutes each day for 2 to 3 weeks to experiment and collect data.

PROCESS SKILLS applied in this activity are **observing, experimenting, interpreting data,** and **recognizing cause and effect.**

PREPARATION

- Aged tap water: allow to stand open for 3 days.
- Small aquariums may be set up using 4-L glass jars or 3 L drink bottles.

Cooperative Learning: Divide the class into Science Investigation Teams.

THINKING CRITICALLY

Fish use their fins for movement. The gills open and close as they breathe. They open their mouths and take in food. Student answers may include examples of teaching an animal to do tricks for food, and noises that animals and people become conditioned to.

TEACHING THE ACTIVITY

Refer to the ***Activity Worksheets*** *for additional information and teaching strategies.*

- Possible stimuli for students to use are tapping on the tank, turning on a light, ringing a bell, and playing music.
- Caution students not to overfeed the fish to use extreme care when working with live animals.

SUMMING UP/SHARING RESULTS

- Answers will vary. The fish may be conditioned in two to three weeks.
- The fish were trained to come to the top of the aquarium in response to whatever stimulus the group used instead of the presence of food. Normally the response would be produced by the presence of food.

GOING FURTHER!

Students may design an experiment using different kinds of music, different color of lights, etc.

ACTIVITY 17-1 DESIGNING AN EXPERIMENT
Investigating Conditioning

In conditioning, behavior is modified so that a response previously associated with one stimulus becomes associated with another.

Getting Started

You need to determine a way to find out if fish can be conditioned.

Thinking Critically

How do fish move, breathe, and get their food? Make a list of behaviors that are examples of conditioning.

Materials 

Your cooperative group will use:

- aquarium
- glass cover for aquarium
- thermometer
- washed coarse sand
- tap water aged three days
- metric ruler
- dish
- food for guppies
- fish net
- guppies
- snails
- water plants

Try It!

1. Wash and rinse the aquarium thoroughly. Place it on a flat surface where it will receive indirect light or use and aquarium light.

2. Place 3 to 4 cm of sand on the bottom of the aquarium. Place a dish on the sand and pour aged tap water on the dish until the water level is about 5 cm above the sand. Anchor plants near the back of the aquarium about 5 cm apart. Add aged tap water slowly to fill the aquarium. Allow it to stand for one day.

3. Add one guppy and one snail for each four liters of water. Cover the aquarium. Maintain a water temperature of about 24°C.

4. Decide what stimulus you will use to condition the fish. **Hypothesize** how long it will take to condition the fish.

5. Feed the fish at the same time each day. Feed the same amount and kind of food. When you feed the fish, apply the stimulus. Do this for two weeks and record your observations.

6. After two weeks, at the normal feeding time, apply the stimulus where the food is usually placed. Do not put any food into the water. *Observe* the fish. Feed the fish when you have finished observing.

Summing Up/Sharing Results

- What stimulus did you use to condition the fish?
- How long did it take the fish to become conditioned? Was your **hypothesis** supported by your data?
- Write a brief conclusion based on your data.

Going Further!

Based on your data and observations, **design another experiment** using another stimulus to condition fish to food.

Activity ASSESSMENT

Performance: To further assess students' understanding of conditioning, see USING LAB SKILLS, Question 11, on page 410.

PROGRAM RESOURCES

From the **Teacher Resource Package** use:

Activity Worksheets, pages 149-150, Activity 17-1, Investigating Conditioning.

Behavioral Adaptations 17-2

Objectives

- Recognize the importance of behavioral adaptations.
- Explain how courtship behavior increases the chances of reproductive success.
- Evaluate the importance of social behavior and cyclic behavior.

New Science Words

territory
aggression
courtship behavior
social behavior
society
communication
cyclic behaviors
circadian rhythm
migration

Territorial Behavior

Many animals set up territories for feeding, mating, and raising young. A **territory** is an area that an animal defends from other members of the same species. Ownership of a territory is set up in different ways. Songbirds sing to set up territories. Sea lions bellow and squirrels chatter. Other animals leave scent marks. Some patrol the area to warn intruders.

Pet dogs and cats have territories. A dog often barks and nips at other dogs around its home. Cats and dogs mark their territories with urine. A cat out hunting will sniff for scent marks of other cats. If the scent marks are fresh, the cat will take another path.

The size of a territory varies with the species. Song sparrows have territories that are as large as 3000 square meters where they feed, mate, and nest. Gulls, penguins, and other water birds nest in large groups, but each male and his mate have their own territory. Male sea lions defend small territories that are used only for mating. Squirrels have large territories for feeding. Female lizards defend feeding territories, and male lizards defend mating areas.

Figure 17-8. These puffins are defending their nesting territory.

Have you ever watched a dog approach another dog eating a bone? What happens to the appearance of the dog with the bone? The hair stands up on its back, the lips curl, and the dog makes growling noises. This behavior is aggression. **Aggression** is a forceful act used to dominate or control another animal. Fighting and threatening are aggressive behaviors animals use to defend their territories, protect their young, or to get food. 1

Aggressive behaviors seen in birds include letting the wings droop below the tail feathers, taking another's

OPTIONS

PROGRAM RESOURCES

From the **Teacher Resource Package** use:
Concept Mapping, pages 39-40.

ENRICHMENT

- Have students select a specific animal and research how that animal communicates.

SECTION 17-2

PREPARATION

SECTION BACKGROUND

- In territorial species, only individuals with territories mate. This helps ensure an adequate food supply for offspring. It also passes on the genes of the best adapted individuals to the next generation.
- Animals use a variety of navigational devices to find their way. Honeybees use polarized light. Birds use the sun much as humans use a compass and map. Many birds migrate at night by observing the stars.
- One of the main characteristics of a society is the division of labor among its members. Specific tasks are carried out by specific groups or individuals of the society. In some cases, the individuals that perform a certain task have specialized body structures that the other individuals do not have.
- Some animal societies are organized into different levels of authority, such as a pecking order within a society of chickens.

PREPLANNING

- Obtain jars and soil for Activity 17-2.

1 MOTIVATE

- Invite a beekeeper to visit the class and discuss bee behavior. If the beekeeper has a demonstration hive, ask that it be displayed.

Cooperative Learning: Have students work in Paired Partners cooperative groups. Have each group select a team sport or a group activity, then diagram the organization of that group and explain the importance of each member to the group.

VideoDisc

STVS: Tagging Ants, Disc 5, Side 1
STVS: Alligator Courtship, Disc 5, Side 2
STVS: Migration of Killer Bees, Disc 5, Side 1

TYING TO PREVIOUS KNOWLEDGE: Students will be familiar with aggressive and territorial behaviors of pets and other animals. Discuss familiar examples of territorial displays and threatening behaviors.

2 TEACH

Key Concepts are highlighted.

CONCEPT DEVELOPMENT

▶ Have students suggest reasons that territorial displays and aggressive behavior may actually reduce the number of fights and injured animals. Other animals may be reluctant to challenge stronger animals; fights may end as soon as dominance is demonstrated.

▶ Explain that courtship patterns are examples of complex innate behavior.

▶ Point out that the ranges of the golden-fronted woodpecker and the red-bellied woodpecker overlap in a narrow range of Texas. Despite strong similarities in appearances, vocalization, and behavior, they do not interbreed. They obviously use different clues in recognition of mates.

Connect to... Chemistry

Answer: The ability to glow, bioluminesence, is due to a chemical reaction in which molecules of the compound luciferin become excited and, as a result, emit photons of light. This behavior enables females to locate males for the purpose of reproduction.

perch, and thrusting the head forward in a pecking motion. These behaviors are intended to avoid physical contact. Fighting wastes valuable energy. And a missing feather or two can greatly reduce a bird's ability to fly.

Animals seldom fight to the death. They rarely use their teeth, beaks, claws, or horns to fight members of their own species. These structures are used for killing prey or for defense against members of another species. To avoid being killed, a defeated animal shows submission by lowering its posture or retreating.

Connect to... Chemistry

There are many kinds of courtship behaviors in nature. Find out what causes the flashes of light produced by male fireflies. How does this behavior help in firefly survival?

Courtship Behavior

You have probably seen a male peacock spread the beautiful feathers on his lower back. The male frigate bird in Figure 17-9 has a bright red pouch on his throat that takes about 25 minutes to blow up. A male sage grouse fans his tail, fluffs his feathers, and blows up his two air sacs. These are examples of a behavior that both males and females of a species perform before mating. This (2) type of behavior is called **courtship behavior.** Courtship behaviors allow male and female members of a species to recognize each other. They also allow males and females to be ready to mate at the same time. This helps provide reproductive success.

In most species, the males perform courtship displays to attract a mate. Songbirds sing. Other birds depend on bright colors and attention-getting movements or postures. Many species of birds move the head, wings, or other body parts. The response of the female may closely resemble the male's display, and so the two look as if they are dancing.

Some courtship behaviors allow males and females to find each other across distances. Male fireflies produce different

Figure 17-9. The male peacock (left) and the male frigate bird (right) are both trying to attract females with their displays.

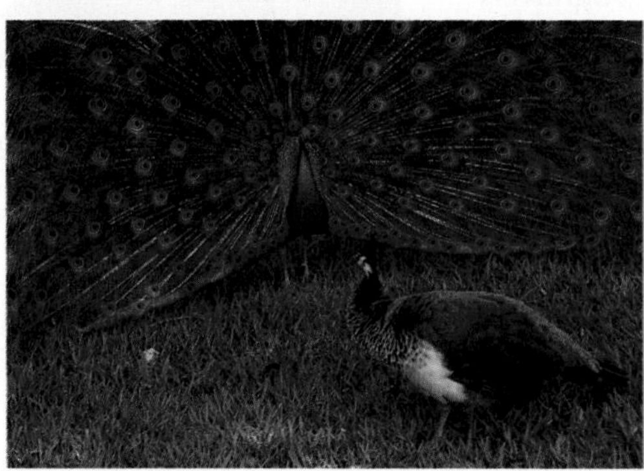

OPTIONS

Meeting Different Ability Levels

For Section 17-2, use the following **Teacher Resource Masters** depending upon individual students' needs.

◆ **Study Guide Master** for all students.

● **Reinforcement Master** for students of average and above average ability levels.

▲ **Enrichment Master** for above average students.

Additional Teacher Resource Package masters are listed in the OPTIONS box throughout the section. The additional masters are appropriate for all students.

◆ **STUDY GUIDE 68**

NAME ______ DATE ______ CLASS ______

STUDY GUIDE — Chapter 17

Behavioral Adaptations — Text Pages 401–405

Match the definitions on the left with the terms on the right by writing the correct letter in the space provided.

d 1. a forceful act used to dominate or control another animal
h 2. an exchange of information
b 3. an instinctive seasonal movement of animals
c 4. an area that an animal defends from other members of its species
i 5. animals that are active during the night
a 6. a group of animals of the same species that live and work together in an organized way
f 7. attracts male moths from several kilometers away
e 8. a cyclic response to cold temperatures and limited food supplies
j 9. used to mark territories by cats and dogs
g 10. allows males and females to be ready to mate at the same time

a. society
b. migration
c. territory
d. aggression
e. hibernation
f. pheromones
g. courtship behavior
h. communication
i. nocturnal
j. urine

Place a check (✓) next to the sentences that agree with the textbook. Rewrite the others so that they agree.

____ 1. Cyclic behaviors are learned by trial and error.
Cyclic behaviors are innate behaviors.
✓ 2. An estivating animal becomes inactive and its metabolism slows down.
✓ 3. Territories may be designated for one purpose such as feeding.
____ 4. Animals usually fight members of their own species to the death.
Animals of the same species avoid killing one another.
✓ 5. Bright colors in birds are often designed to attract a mate.
____ 6. The dominant male controls the mating of females in a wolf pack.
The dominant female controls the mating of other females in a wolf pack.
✓ 7. Pheromones are a form of communication.
____ 8. Crickets make sounds to warn others away from their territory.
Crickets make sounds to call to possible mates.
____ 9. Animals that are inactive in the daytime are hibernating.
Animals that are inactive during the day are nocturnal.
____ 10. Migration occurs in groups as a learned behavior.
Migration is an instinctive behavior.

68

patterns in their flashes of light. A female of the same species recognizes the pattern and flashes back. Female gypsy moths release chemical messengers called pheromones that attract males. Very small amounts of pheromone can attract male moths several kilometers away. Sound is another way males and females find each other. Insects such as crickets, grasshoppers, and locusts call to each other by sounds made by rubbing their legs or wings together. Frogs make sounds that can be heard hundreds of meters away.

Social Behavior

Do you know why geese fly in flocks and some fish swim in schools? Animals live together in groups for several reasons. One reason is that there is safety in large numbers. A wolf pack is less likely to attack a herd of musk oxen than an individual musk oxen. In some groups, large numbers of animals help keep each other warm. Penguins in Antarctica huddle together against the cold winds. Migrating animals in large groups are less likely to get lost than if they traveled alone.

Interactions among organisms of the same species are examples of **social behavior.** Social behaviors include courtship and mating, caring for the young, claiming territories, protecting each other, and getting food. These behaviors provide advantages for survival of the species.

Insects such as ants, bees, and termites live together in societies. A **society** is a group of animals of the same species living and working together in an organized way. Each member has a certain job. Usually there is a female that lays eggs, a male that fertilizes the eggs, and workers that do all the other jobs in the society. 3

Some societies are organized by dominance. Wolves usually live together in packs. In a wolf pack, there is a dominant female. The top female controls the mating of the other females. If there is plenty of food, she mates and allows the others to do so. If food is scarce, she allows less mating, and usually she is the only one to mate.

In all social behavior, communication is important. 4
Communication is an exchange of information. How do you communicate with the people around you? Animals in a group communicate with sounds and actions. Alarm calls, pheromones, speech, courtship behavior, and aggression are all forms of communication.

Imagine you are a scientist who studies behavior. **In your Journal,** write a proposal to study the behavior of an animal that interests you. Be sure to be specific about what behavior you want to study and why you think the study is important.

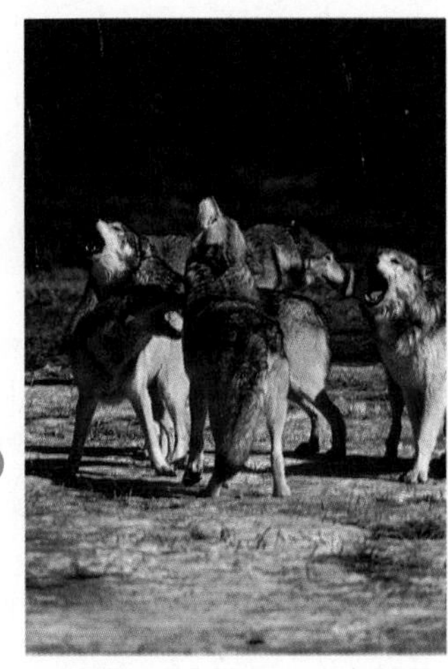

Figure 17-10. Wolves live together socially in packs. They hunt and raise their young together. Strong bonds are formed among members of the pack.

● REINFORCEMENT 68

NAME ______ DATE ______ CLASS ______

REINFORCEMENT — Chapter 17

Behavioral Adaptations — Text Pages 401–405

Answer the following questions using information from your textbook.

1. What purpose does aggression serve? **It is used by animals to defend territories, protect young, or to get food.**
2. Why do some animals live in groups? **There is safety in numbers, warmth, and less chance that an individual will get lost.**
3. What is a circadian rhythm? **It is a cyclic behavior that is based on a 24-hour cycle.**
4. What is the purpose of courtship displays? **They allow species to recognize each other and are used to attract a mate at the correct time.**
5. What are some examples of social behavior? **courtship and mating, caring for young, claiming territories, protecting each other, getting food**
6. What happens to an animal when it hibernates? **Its body temperature drops to near that of its surroundings and breathing rate is reduced.**
7. What signals some animals to hibernate? **cold temperatures and limited food supplies**
8. What courtship behaviors help males and females find each other across a distance? **light, fireflies use flashing signals; chemical messengers or pheromones; sound**
9. Why do some animals migrate? **It is a cyclic behavior in response to seasonal changes.**
10. What is communication? **It is an exchange of information.**

68

▲ ENRICHMENT 68

NAME ______ DATE ______ CLASS ______

ENRICHMENT — Chapter 17

Behavioral Adaptations — Text Pages 401–405

ADAPTATIONS FOR NIGHT-LIVING

Many animals sleep during the day and become active at night. These animals are nocturnal. When we think of all the things an animal must do to survive, it is apparent that night-living requires some special sensory adaptations. An owl has such good hearing that it can pick up the smallest sounds of a rodent stepping over dry leaves or twigs 23 meters away!

Procedure

Use the library to find out more about nocturnal animals such as the owl. Fill in the table below with information gathered from your library research.

Data and Observations

Sensory Adaptations for the owl	
Vision	
Position of eyes:	The eyes of an owl are positioned to look straight ahead. Because the owl sees an object with both eyes, it can focus in a wider range.
Size of eyes:	An owl's eyes are quite large. Because of their size, they can catch most of the available light.
Hearing	
Size of ear openings:	Instead of small holes in the skull, the holes are huge, allowing more sound to enter.
Directing sound into ear openings:	The owl has two facial discs around each eye that funnel the sound into the ear openings.
Other adaptations for hunting and foraging at night:	The feathers have a soft edge. This results in noiseless flight. The owl hunts by listening and watching while flying very close to the ground. It can hear better and fly close to its prey without being heard.

Conclude and Apply

1. What is another animal you have read about that is nocturnal? **Answers will vary, but may include bats, raccoons, or opossums.**
2. What adaptations does it have for night-living? **Answers will vary.**

68

In Your JOURNAL

Be sure students' proposals include logical scientific methods of investigation. Student proposals should state how the behavior of their animal choices is important.

CONCEPT DEVELOPMENT

▶ Ask students to define the word *social* (living together in a group or community).

▶ Discuss the fact that many animals live together in groups and display cooperation. This offers protection to the animals and helps with food getting. A single wolf cannot easily kill a moose, but a pack of wolves can. Have students speculate how this behavior helps maintain balance in nature.

CROSS CURRICULUM

▶ **Math:** The distance between New York and San Francisco is 4800 km. A list of animals and the one-way distances they migrate each year is given below. Calculate how many times these animals could have gone from one coast of the United States to the other.

		Answer
Indigo bunting	3200 km	0.67
Arctic tern	17 200 km	3.6
Loggerhead turtle	2300 km	0.48

MINI QUIZ

Use the Mini Quiz to check students' recall of chapter content.

1. **A forceful act used to dominate or control another animal is _____ .** *aggression*
2. **Courtship behaviors allow male and female members of a species to _____ .** *recognize each other*
3. **A group of animals of the same species living and working together in an organized way is a _____ .** *society*
4. **The exchange of information is _____ .** *communication*
5. **Animals that are active at night are said to be _____ .** *nocturnal*

TECHNOLOGY

For more information, see "Loulis, the Talking Chimp," by Mike Toner. *National Wildlife*, Feb./Mar. 1986, p. 24.

Think Critically: It indicates that the apes find language useful, because they are teaching it for its own value, not for reward from humans. It also allows scientists to see how the apes modify the language and its use.

CHECK FOR UNDERSTANDING

Have students give examples of animal behaviors, along with the stimulus and response for each behavior.

RETEACH

Have students make a chart with the headings Territorial Behavior, Courtship Behavior, Social Behavior, and Cyclic Behavior with examples of each behavior.

EXTENSION

For students who have mastered this section, use the **Reinforcement** and **Enrichment** masters or other OPTIONS.

Cyclic Behavior

What determines when an owl sleeps and when it wakes up? Animals show regularly repeated behaviors such as feeding in the day and sleeping at night or the opposite. Many reproduce every spring and migrate every spring and fall.

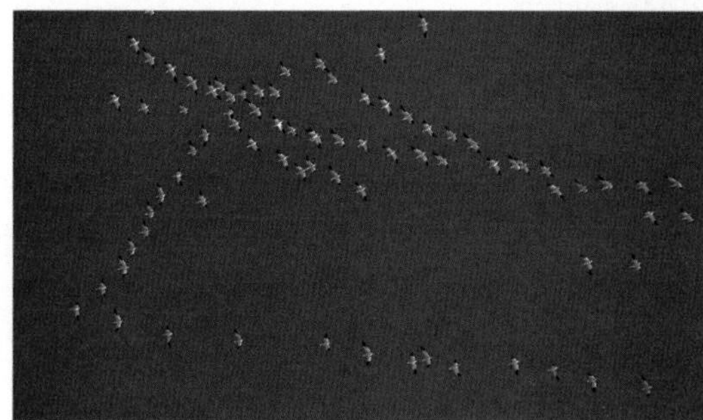

Figure 17-11. Migration is a cyclic behavior.

Cyclic behaviors are innate behaviors that occur in a repeating pattern. They are often repeated in response to changes in the environment. Behavior that is based on a 24-hour cycle is called a **circadian rhythm.** Animals that are active during the day are *diurnal.* Animals that are active at night are *nocturnal.* 5

Cyclic behaviors also occur over long periods of time. Hibernation is a cyclic response to cold temperatures and limited food supplies. During hiber-

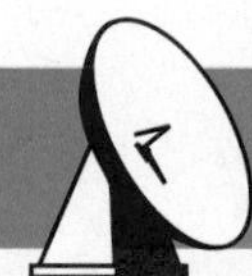

TECHNOLOGY

Looking for a Sign

Young Loulis is learning language from his friends. This hardly seems unusual, except the language is American Sign Language (AMESLAN), and his friends are laboratory chimps.

Scientists have used many different techniques to study the language-learning abilities of nonhuman primates. Most have involved humans teaching apes language, but now some involve letting apes teach other apes. AMESLAN was originally used for studies because apes cannot produce certain human word sounds. AMESLAN is a complete, nonverbal language. Some scientists have made up new languages and have taught apes these.

Apes are using language, creating new words, and teaching these words to other apes. But ape language studies affect humans, too. Yerkish, a symbol language developed for apes, has allowed severely handicapped humans to communicate with others for the first time.

Think Critically: What is the significance of apes, not humans, teaching other apes language?

OPTIONS

ASSESSMENT—ORAL

▶ **In what ways does communication help an animal species survive?** *It allows animals to find food, defend against enemies, and locate mates.*

▶ **How do fish benefit from forming schools?** *Predators are less able to attack a single fish while it is in a school. It is more vulnerable when it swims alone.*

▶ **What do you think is the greatest benefit for animals to exist in groups?** *defense against predators*

▶ **Why is it important for a species to have behaviors that protect members from predators?** *to ensure that members of the species can reproduce*

▶ **What kind of communication is: a blind person's ability to read Braille?** *touch;* **the cry of a baby?** *sound;* **an insect using pheromones?** *smell*

nation, an animal's body temperature drops to near that of its surroundings, and its breathing rate is greatly reduced. An animal in hibernation survives on stored body fat. The animal remains inactive until the weather becomes warm in the spring. Some mammals and many amphibians and reptiles hibernate.

Have you ever seen large flocks of birds flying overhead in the fall or spring? Many birds and mammals move to new locations when the seasons change instead of hibernating. This instinctive seasonal movement of animals is called **migration.** Many species of birds fly hours or days without stopping. The blackpoll warbler flies nonstop a distance of more than 4000 km from North America to its winter home in South America. The trip takes nearly 90 hours. Arctic terns fly about 17 700 km from their breeding grounds in the Arctic to their winter home in the Antarctic. They return a few months later, traveling a distance of more than 35 000 km in less than a year. Gray whales swim from the cold Arctic waters to the waters off the coast of California. When the young are born, they make the return trip.

Animals have many different behaviors. Some behaviors are innate and some are learned. Many are a combination of innate and learned behaviors. Appropriate behaviors help animals survive, reproduce, and maintain the species.

MINI-Lab

How does an animal enter into hibernation?

Some animals respond to their cold environment by lowering the thermostat of their body. Use the data of a ground squirrel over a period of hours to *graph* this reaction.

Temperature	Hours
37°C	1
25°C	2
30°C	3
15°C	4
14°C	5
13°C	6
10°C	7

Body temperature is maintained between 5°C and 10°C. If the temperature drops below this level, the animal becomes active. How is this helpful?

SECTION REVIEW

1. How do animals communicate?
2. What are some examples of courtship behavior?
3. Give three reasons why animals live in groups.
4. How are cyclic behaviors a response to stimuli in the environment?
5. **Apply:** What behaviors do pet dogs have that indicate they are related to wolves?
6. **Connect to Earth Science:** Find out about a flyway of a migratory bird.

Using Variables, Constants, and Controls — Skill Builder

Design an experiment to show that ants leave chemical trails to show other ants where food can be found. If you need help, refer to Using Variables, Constants, and Controls in the **Skill Handbook** on page 686.

MINI-Lab

Materials: graph paper

Teaching Tips

▶ Students can make bar graphs or line graphs. Hours (elapsed) should be on the x-axis and temperature on the y-axis.

Answer: The animal becomes active to prevent itself from freezing to death.

MINI-Lab

ASSESSMENT

Performance: To further assess students' understanding of cyclic behavior, have them identify species that practice estivation.

3 CLOSE

▶ Ask questions 1-4 and the **Apply** Question in the Section Review.

SECTION REVIEW ANSWERS

1. fighting, threatening, courtship displays, chemicals, sound

2. the bright red pouch on the throat of the male frigate bird, pheromones in gypsy moths

3. safety, warmth, migrating animals are less likely to get lost

4. They are responses to changes in the environment.

5. Apply: aggressive and threatening behaviors such as growling, hair standing up on their backs, lips curling.

6. Connect to Earth Science: Choices will vary. Some birds have flyways or routes that are thousands of miles long and cross two continents using stars as their guide.

Skill Builder

Accept all logical experiments. Students should identify a variable, constant, and control.

Skillbuilder

ASSESSMENT

Performance: Assess students' abilities to Use Variables, Constants, and Controls by having them carry out their experiments.

INQUIRY QUESTIONS

▶ **Why is hibernation a benefit to all animals, not just those that hibernate?** *There is less competition for food.*

▶ **What do you think annual rhythms are?** *those that occur once every year*

▶ **What causes jet lag?** *Changing time zones affect the circadian rhythm pattern.*

ENRICHMENT

▶ Have students describe ways they have seen animals communicate.

▶ Have students read "Sleeping Beauties" on pp. 62-65 of *Discover,* April 1989. Discuss the article with the students.

SECTION 17-3

PREPARATION

SECTION BACKGROUND

▶Because of human interference, millions of animals are killed each year.

1 MOTIVATE

▶Many local zoos have rehabilitation programs that may be visited or outreach programs that bring rehabilitated animals to the classroom.

TYING TO PREVIOUS KNOWLEDGE: Natural selection works to keep only the healthiest animals in a population. Humans disrupt natural selection by rehabilitating and returning inferior animals to the wild.

2 TEACH

Key Concepts are highlighted.

VideoDisc

STVS: Red Wolf, Disc 6, Side 1

PROGRAM RESOURCES

From the **Teacher Resource Package** use:

Science and Society, page 21, Helping Sea Turtles.

17-3 Rehabilitation of Wild Animals

New Science Words

carrying capacity

Objectives

- ▶ Explain some of the effects of human development on wildlife.
- ▶ Describe the disadvantages of releasing rehabilitated animals into the wild.

Saving Injured Animals

Each year, millions of animals are killed due to the movement of humans into animal territories. Thousands of birds are killed each year after they fly into electrical towers during their night migration. You might have seen wild animals lying along the road after being hit by a car. Thousands of large birds like hawks, owls, and eagles, which prey on other animals, are electrocuted each year while nesting on telephone poles or electric wires.

Figure 17-12. A human caretaker interacts with a young bird using a bird-like puppet to encourage near-natural imprinting.

Many volunteers around the country are working to save animals that have been injured due to human presence. Nursing wild animals back to health is becoming very popular in the United States.

These wild animal rehabilitation programs can be very difficult to conduct. Wild animals are often very uncomfortable out of their natural environment. They require special conditions in order for them to heal and to prevent them from becoming too familiar with their human caretakers. In many rehabilitation programs for birds of prey, the young owls, hawks, or eagles never see their human caretakers. Instead, they are fed by puppets that look very much like their natural parents. By using such puppets to feed them their natural food source, these young have a better chance of surviving on their own later in the wild.

OPTIONS

Meeting Different Ability Levels

For Section 17-3, use the following **Teacher Resource Masters** depending upon individual students' needs.

◆ **Study Guide Master** for all students.

● **Reinforcement Master** for students of average and above average ability levels.

▲ **Enrichment Master** for above average students.

Additional Teacher Resource Package masters are listed in the OPTIONS box throughout the section. The additional masters are appropriate for all students.

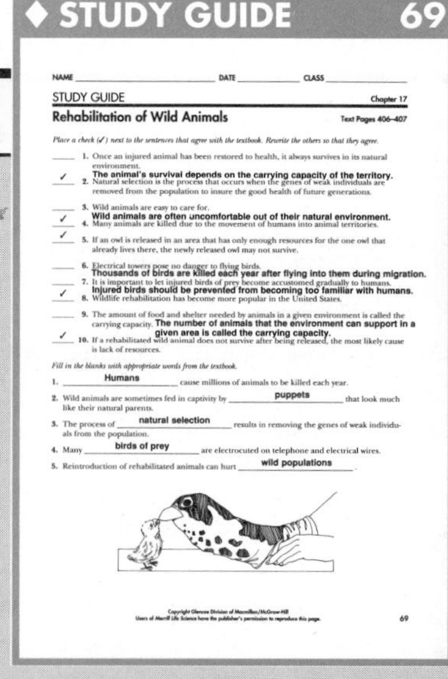

◆ STUDY GUIDE 69

NAME ______ DATE ______ CLASS ______

STUDY GUIDE — Chapter 17

Rehabilitation of Wild Animals — Text Pages 406–407

Place a check (✓) next to the sentences that agree with the textbook. Rewrite the others so that they agree.

____ 1. Once an injured animal has been restored to health, it always survives in its natural environment. **The animal's survival depends on the carrying capacity of the territory.**

✓ 2. Natural selection is the process that occurs when the genes of weak individuals are removed from the population to insure the good health of future generations.

____ 3. Wild animals are easy to care for. **Wild animals are often uncomfortable out of their natural environment.**

✓ 4. Many animals are killed due to the movement of humans into animal territories.

✓ 5. If an owl is released in an area that has only enough resources for the one owl that already lives there, the newly released owl may not survive.

____ 6. Electrical towers pose no danger to flying birds. **Thousands of birds are killed each year after flying into them during migration.**

____ 7. It is important to let injured birds of prey become accustomed gradually to humans. **Injured birds should be prevented from becoming too familiar with humans.**

✓ 8. Wildlife rehabilitation has become more popular in the United States.

____ 9. The amount of food and shelter needed by animals in a given environment is called the carrying capacity. **The number of animals that the environment can support in a given area is called the carrying capacity.**

✓ 10. If a rehabilitated wild animal does not survive after being released, the most likely cause is lack of resources.

Fill in the blanks with appropriate words from the textbook.

1. **Humans** cause millions of animals to be killed each year.
2. Wild animals are sometimes fed in captivity by **puppets** that look much like their natural parents.
3. The process of **natural selection** results in removing the genes of weak individuals from the population.
4. Many **birds of prey** are electrocuted on telephone and electrical wires.
5. Reintroduction of rehabilitated animals can hurt **wild populations**.

69

Return to the Wild

Although injured wild animals are often nursed back to good health, some scientists argue that while the individual animal is saved, wild populations do not benefit from releasing rehabilitated animals. We may actually be hurting wild populations by returning these animals to the wild. Through natural selection, the genes of weak individuals are removed from the population to ensure the good health of future generations. Although saving these animals appears to be a very humane venture, the consequences of rehabilitation must be carefully considered. A great deal of research is needed to study the effects of a released animal on its new environment and the effects of the new environment on the animal. Many animals are very territorial. Each territory is designed to hold a certain number of organisms. The number of animals that the environment can support with food and shelter in a given area is called the **carrying capacity.** For example, if an owl was released in a forest that provided only enough resources for the one owl that already lived there, the newly released owl would not survive. Similarly, if the new owl established itself in the forest, the established owl may be forced to leave. In each case, one animal is likely to die due to a lack of resources.

Although rehabilitation appears to be very kind and noble, we must think carefully about its consequences. We must try to reach a balance between our humane instincts and the reality of the natural world.

Earth's natural resources, including water, soil, minerals, and living space, are limited. Because these resources are limited, population size of humans and other organisms is limited. How can you help preserve Earth's natural resources?

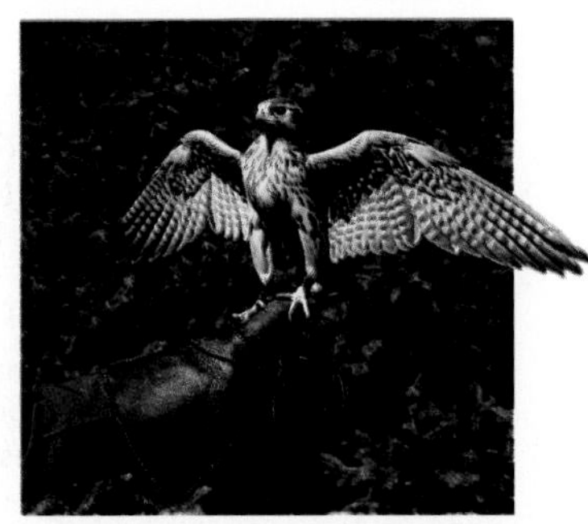

Figure 17-13. A rehabilitated falcon is being released into its natural habitat.

SECTION REVIEW

1. Name three ways in which humans cause injury or death to wild animals.
2. Explain some problems of releasing wild animals into new environments.
3. **Connect to Physics:** Find out why birds are normally not harmed when landing on electric wires.

You Decide!

Often during the spring months newly hatched birds are found on the ground after falling out of their nests. Our first tendency may be to help the young bird by bringing it inside and trying to nurse it back to health. Some people might argue that the young bird will serve as food for another animal or for insects. Is it wrong to let nature take its course?

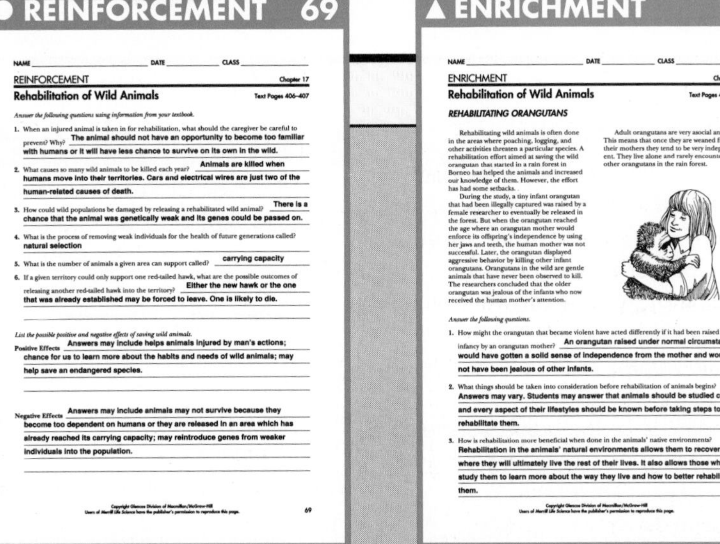

● REINFORCEMENT 69

NAME ________ DATE ________ CLASS ________

REINFORCEMENT — Chapter 17

Rehabilitation of Wild Animals — Text Pages 406–407

Answer the following questions using information from your textbook.

1. When an injured animal is taken in for rehabilitation, what should the caregiver be careful to prevent? Why? **The animal should not have an opportunity to become too familiar with humans or it will have less chance to survive on its own in the wild.**
2. What causes so many wild animals to be killed each year? **Animals are killed when humans move into their territories. Cars and electrical wires are just two of the human-related causes of death.**
3. How could wild populations be damaged by releasing a rehabilitated wild animal? **There is a chance that the animal was genetically weak and its genes could be passed on.**
4. What is the process of removing weak individuals for the health of future generations called? **natural selection**
5. What is the number of animals a given area can support called? **carrying capacity**
6. If a given territory could only support one red-tailed hawk, what are the possible outcomes of releasing another red-tailed hawk into the territory? **Either the new hawk or the one that was already established may be forced to leave. One is likely to die.**

List the possible positive and negative effects of saving wild animals.

Positive Effects **Answers may include helps animals injured by man's actions; chance for us to learn more about the habits and needs of wild animals; may help save an endangered species.**

Negative Effects **Answers may include animals may not survive because they become too dependent on humans or they are released in an area which has already reached its carrying capacity; may reintroduce genes from weaker individuals into the population.**

 69

▲ ENRICHMENT 69

NAME ________ DATE ________ CLASS ________

ENRICHMENT — Chapter 17

Rehabilitation of Wild Animals — Text Pages 406–407

REHABILITATING ORANGUTANS

Rehabilitating wild animals is often done in the areas where poaching, logging, and other activities threaten a particular species. A rehabilitation effort aimed at saving the wild orangutan that started in a rain forest in Borneo has helped the animals and increased our knowledge of them. However, the effort has had some setbacks.

During the study, a tiny infant orangutan that had been illegally captured was raised by a female researcher to eventually be released in the forest. But when the orangutan reached the age where an orangutan mother would enforce its offspring's independence by using her jaws and teeth, the human mother was not successful. Later, the orangutan displayed aggressive behavior by killing other infant orangutans. Orangutans in the wild are gentle animals that have never been observed to kill. The researchers concluded that the older orangutan was jealous of the infants who now received the human mother's attention.

Adult orangutans are very asocial animals. This means that once they are weaned from their mothers they tend to be very independent. They live alone and rarely encounter other orangutans in the rain forest.

Answer the following questions.

1. How might the orangutan that became violent have acted differently if it had been raised from infancy by an orangutan mother? **An orangutan raised under normal circumstances would have gotten a solid sense of independence from the mother and would not have been jealous of other infants.**
2. What things should be taken into consideration before rehabilitation of animals begins? **Answers may vary. Students may answer that animals should be studied closely and every aspect of their lifestyles should be known before taking steps to rehabilitate them.**
3. How is rehabilitation more beneficial when done in the animals' native environments? **Rehabilitation in the animals' natural environments allows them to recover where they will ultimately live the rest of their lives. It also allows those who study them to learn more about the way they live and how to better rehabilitate them.**

 69

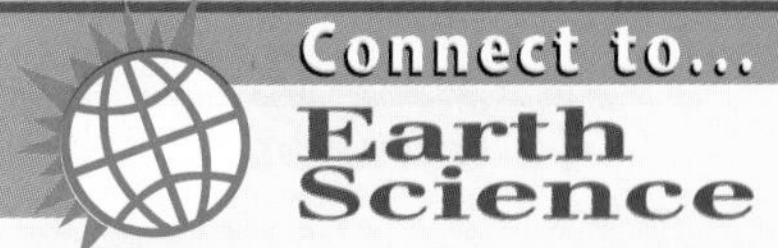

Answer: Accept all reasonable responses. Answers may include recycling and limiting water use.

CHECK FOR UNDERSTANDING

Have students write a paragraph about the benefits and problems of the rehabilitation of animals.

RETEACH

Create a scenario concerning an animal that has been injured due to human interference. Follow the animal through the injury, rehabilitation, and release into the wild.

EXTENSION

For students who have mastered this section, use the **Reinforcement** and **Enrichment** masters or other OPTIONS.

3 CLOSE

▶ Ask questions 1-2 in the Section Review.

SECTION REVIEW ANSWERS

1. hit by car, flying into man-made objects, electrocuted by perching on telephone pole or wire

2. The new environment may already be taken by another individual; the territory may not contain enough food for more animals.

3. Connect to Physics: Wires are insulated.

YOU DECIDE!

Answers that show clear and logical thinking should be accepted.

ACTIVITY 17-2
45 minutes

OBJECTIVE: Observe the interaction of ants in a colony.

PROCESS SKILLS applied in this activity:
▶ **Observing** in Procedure Steps 5 and 6.
▶ **Recording** in Procedure Step 6.
▶ **Inferring** in Analyze Questions 2-4.

COOPERATIVE LEARNING Use the Science Investigation strategy. Divide the class into groups of four. Have one person get the materials, one person record, one person read the activity aloud, and one person check the time.

TEACHING THE ACTIVITY
Alternate Materials: Any glass container may be used.
Troubleshooting: To prevent ants from crawling out of the jars, rub vegetable oil around the top inside.
▶ Collect the soil near the anthill. The ants can be collected at the same time. Try to find a queen for each colony.
▶ Have students discuss what they have observed about anthills. (a cone of excavated material with a hole)

ACTIVITY 17-2 Observing Social Behavior in Ants

Problem: *How do ants live?*

Materials
- 20 ants
- large jar with screen wire cover
- soil
- black paper
- small moist sponge
- water
- food—honey, sugar, bread crumbs
- small can with lid
- spoon or trowel
- hand lens

Procedure
1. Fill the jar 3/4 full with moistened loose soil. Place the damp sponge on top of the soil.
2. Find an anthill. Use the spoon or trowel to place loose soil from the nest into the small can. Look for ants, small white eggs, cocoons, and larvae. Try to find a queen. The queen will be larger than the other ants. Cover. Place the ants in the refrigerator for a few minutes to slow their movements.
3. Gently place the ants in the jar. Add a small amount of each food in different areas. Place the screen wire cover on the jar.
4. Tape the black paper around the jar. Leave about 2 cm at the top uncovered.
5. Take the paper off for a short period of time each day and *observe* the ants at work. Use a hand lens to observe individual ants.
6. Keep the sponge moistened and add food each day. *Record* your observations each day.

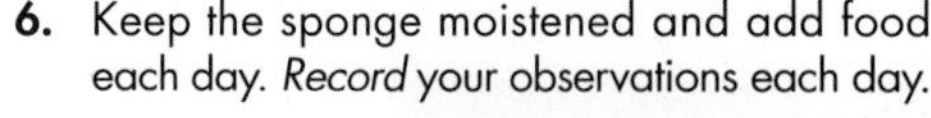

Data and Observations Sample Data

Date	Number of ants visible	Other observations
		See teacher margin.

Analyze
1. How long did it take for the ants to build tunnels after they were placed in the jar?
2. What did you observe that indicated the ants were working together?
3. What kinds of foods do ants prefer?
4. How do the ants in the jar differ?

Conclude and Apply
5. What is a society?
6. What evidence do you have that ants are social insects?

ANSWERS TO QUESTIONS
1. Answers will vary. Usually, they take from one day to five days.
2. Two or more ants may have been observed moving food or building a tunnel.
3. usually something sweet— answers will vary
4. The queen will appear different from the workers.
5. a group of animals of the same species living and working together in an organized way
6. They live in colonies.

PROGRAM RESOURCES

From the **Teacher Resource Package** use:
Activity Worksheets, pages 151-152, Activity 17-2, Observing Social Behavior in Ants.

Activity ASSESSMENT
Performance: To further assess students' ability to Observe food preferences, see question 12, USING LAB SKILLS, on page 410.

CHAPTER

REVIEW

SUMMARY

17-1: Types of Behavior

1. Behavior that an animal is born with is innate behavior. Some animal behaviors are learned. These behaviors are formed through experience.

2. Reflexes are the simplest innate behaviors. An instinct is a complex pattern of innate behavior. A spider spins a web by instinct. Birds build their nests by instinct.

3. Imprinting is a learned behavior in which an animal forms a social attachment to another organism soon after birth. Behavior that is modified by experience is learning by trial and error. Conditioning occurs when a response previously associated with one stimulus becomes associated with another. Insight is a form of reasoning that uses past experiences to solve new problems.

17-2: Behavioral Adaptations

1. Behavioral adaptations such as defense of territory, courtship behavior, and social behavior help species of animals survive and reproduce.

2. Courtship behaviors allow males and females to recognize each other. Courtship behaviors help provide for reproductive success.

3. Interactions among members of the same species are social behaviors. Cyclic behaviors are behaviors that occur in repeating patterns. Cyclic behaviors such as hibernation and migration help animals survive harsh environmental conditions.

17-3: Science and Society: Rehabilitation of Wild Animals

1. Millions of animals are killed or injured each year due to the movement of humans into animal territories. Many animals are hit by cars. Some animals fly into electrical towers, and others are electrocuted while perching on telephone poles or on electric wires.

2. Rehabilitated animals released back into the wild may not be able to find food on their own. The carrying capacity of the environment may also be disturbed by adding these additional animals.

KEY SCIENCE WORDS

a. **aggression**
b. **behavior**
c. **carrying capacity**
d. **circadian rhythm**
e. **communication**
f. **conditioning**
g. **courtship behavior**
h. **cyclic behaviors**
i. **imprinting**
j. **innate behavior**
k. **insight**
l. **instinct**
m. **learning**
n. **migration**
o. **motivation**
p. **reflex**
q. **social behavior**
r. **society**
s. **territory**
t. **trial and error**

UNDERSTANDING VOCABULARY

Match each phrase with the correct term from the list of Key Science Words.

1. an inherited, not learned behavior
2. automatic response to a stimulus
3. forming a social bond to another organism after birth
4. the way an organism behaves toward its environment
5. change in behavioral response to a given stimulus
6. a defended area
7. using reasoning to solve a problem
8. an organized group of animals in a species
9. behaviors before mating
10. instinctive seasonal movement

OPTIONS

ASSESSMENT

To assess student understanding of material in this chapter, use the resources listed.

COOPERATIVE LEARNING

Consider using cooperative learning in the THINK AND WRITE CRITICALLY, APPLY, and MORE SKILL BUILDERS sections of the Chapter Review.

PROGRAM RESOURCES

From the **Teacher Resource Package** use:

Chapter Review, pages 37-38.

Chapter and Unit Tests, pages 105-108, Chapter Test.

Chapter and Unit Tests, pages 109-110, Unit Test.

CHAPTER

REVIEW

SUMMARY

Have students read the summary statements to review the major concepts of the chapter.

UNDERSTANDING VOCABULARY

1. j
2. p
3. i
4. b
5. f
6. s
7. k
8. r
9. g
10. n

ASSESSMENT

Portfolio

Encourage students to place in their portfolios one or two items of what they consider to be their best work. For each item, ask students to explain why that item was chosen and what they learned from it. Items might be selected from the following.

- Cooperative Learning diagram of the organization of a team sport or group activity, p. 401
- MINI-Lab graph of a squirrel's body temperature response to a cold environment, p. 405
- Skillbuilder experimental design, p. 405

Performance

Additional performance assessments may be found in *Performance Assessment* and *Science Integration Activities* that accompany **Merrill Life Science.** Performance Task Assessment Lists and rubrics for evaluating these activities and other products generated throughout the chapter can be found in Glencoe's *Performance Assessment in Middle School Science.*

CHAPTER

REVIEW

CHECKING CONCEPTS

1. c	**6.** d
2. a	**7.** a
3. d	**8.** b
4. c	**9.** c
5. a	**10.** a

USING LAB SKILLS

ASSESSMENT

Use these alternate lab exercises to assess students' understanding of skills used in this chapter.

11. Student designs will vary. In all cases, students should be cautioned to treat their pet kindly. Students who do not have a dog can work with any pet they may have, or work with another student who has a pet. Students should be ready to discuss varying difficulties and successes in conditioning.

12. This suggestion is untested. Students may wish to use animals other than ants. Mealworms are a possibility. However, they must make comparisons based on data to draw valid conclusions.

THINK AND WRITE CRITICALLY

13. Living in a group allows for sharing of food, water, and space; protection from invaders; and division of jobs.

14. Reflex behaviors are automatic responses, so no learning, thinking, or insight is necessary.

15. Any example of trial-and-error is fine—tying of shoes, writing, riding a bike, and so on.

16. Territories are expressed by marking with urine, chasing intruders away, making sounds (bird calls, howls, and so on), patrolling, or leaving scent markers.

17. Conditioning: learned response associated with stimuli; Pavlov's dog bell with food. Trial-and-error: learning by repeating; tying shoes. Imprinting: attachment to another right after birth or hatching; goslings and mother goose.

CHAPTER

REVIEW

CHECKING CONCEPTS

Choose the word or phrase that completes the sentence.

1. An example of a reflex is ______.
a. writing **c.** sneezing
b. talking **d.** riding a bicycle

2. An instinct is an example of ______.
a. innate behavior **c.** imprinting
b. learned behavior **d.** conditioning

3. Nest building is an example of ______.
a. conditioning **c.** learned behavior
b. imprinting **d.** an instinct

4. The animals that depend least on instinct and more on learning are ______.
a. birds **c.** mammals
b. fish **d.** amphibians

5. ______ involves using reasoning to solve problems.
a. Insight **c.** Conditioning
b. Imprinting **d.** Trial and error

6. Something inside an animal that causes the animal to act is ______.
a. conditioning **c.** insight
b. imprinting **d.** motivation

7. Teeth, beaks, and claws are used for ______.
a. killing prey **c.** hurting other birds
b. fighting **d.** sending signals

8. All are examples of courtship behavior except ______.
a. fluffing feathers
b. taking over a perch
c. singing songs
d. releasing pheromones

9. An organized group of animals working specific jobs is a ______.
a. community **c.** society
b. territory **d.** circadian rhythm

10. The response of inactivity and slowed metabolism that occurs during cold conditions is ______.
a. hibernation **c.** migration
b. imprinting **d.** circadian rhythm

USING LAB SKILLS

11. In Activity 17-1, you learned how to condition fish. Based on your observations, design an experiment to show how you could condition your dog to return a ball or frisbee to you. Carry out your experiment.

12. Repeat Activity 17-2 using different food types: aspartame (a sugar-substitute), apple pieces or molasses. Compare your observations with those made in your original laboratory activity.

THINK AND WRITE CRITICALLY

Answer the following questions in your Journal using complete sentences.

13. What is the advantage of living in a group?

14. What are the advantages of reflex behaviors?

15. Give an example of something you learned by trial and error.

16. Describe the ways territoriality is expressed to other members of a species.

17. Compare conditioning, trial and error, and imprinting using examples of each.

APPLY

18. Explain the type of behavior involved when the bell rings at the end of class.
19. Discuss the advantages and disadvantages of migration as a means of survival.
20. Explain how a habit, such as tying your shoes, is different from a reflex.
21. Use one example to explain how behavior increases an animal's chance for survival.
22. Hens lay more eggs in the spring when day length increases. How can farmers use this knowledge of behavior to their advantage?

MORE SKILL BUILDERS

If you need help, refer to the Skill Handbook.

1. **Classifying:** Make a list of 25 things that you do regularly. Classify each as an innate or learned behavior. Which behaviors do you have more of?
2. **Concept Mapping:** Complete the following concept map outlining the types of behavior.

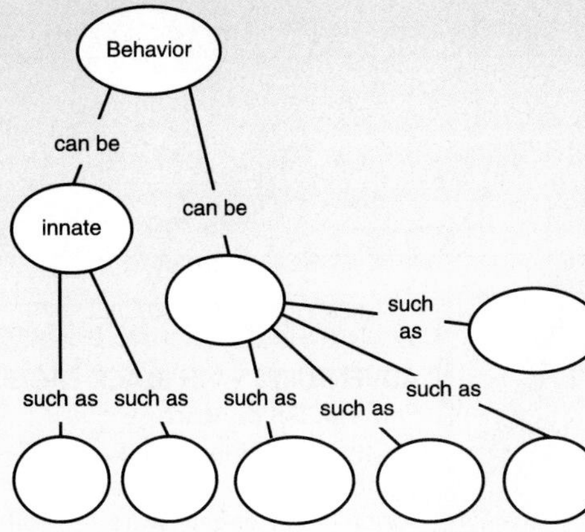

3. **Observing and Inferring:** Make observations of a dog, cat, or bird for a week. Record what you see. How did the animal communicate with other animals and with you?
4. **Using Variables, Constants, and Controls:** Design an experiment to get a specific response to a stimulus by an animal.
5. **Hypothesizing:** Make a hypothesis about how frogs commmunicate with each other. How could you test your hypothesis?

PROJECTS

1. View the film *Never Cry Wolf* (Disney). Write a summary of the behaviors of the wolves.
2. Draw a map showing the migration route of caribou, the gray whale, or arctic wolf.

APPLY

18. Leaving the room when the bell rings is a learned conditioned response.

19. Migration allows organisms to survive changes in weather. The journey in many cases is long and stressful.

20. A habit is a learned behavior that has become automatic. A reflex is innate (not learned) behavior.

21. Behaviors that help get food, are protective, or are defensive help organisms to survive.

22. A farmer can artificially lengthen the amount of 'daylight' hens are exposed to by using lights, thus stimulating hens to lay more eggs.

MORE SKILL BUILDERS

1. Classifying: Most behaviors will be learned, for example, brushing teeth or combing hair. Innate behaviors include sneezing or yawning and other similar actions.

2. Concept Mapping:

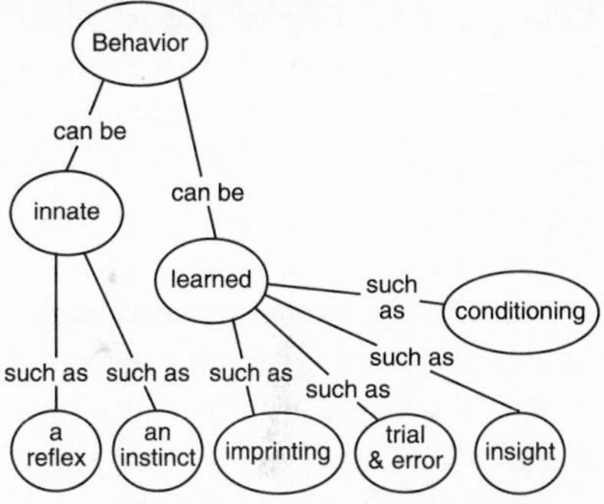

3. Observing and Inferring: Observations will vary, but communication should be noted.

4. Using Variables, Constants, and Controls: Any controlled experiment should be accepted. Be sure students identify the variable, constant, and control.

5. Hypothesizing: Students will probably hypothesize that frogs communicate by using vocalizations. Accept all logical ways to test this hypothesis.

UNIT 5
GLOBAL CONNECTIONS

Objective
In this unit ending feature, the unit topic, "Animals," is extended into other disciplines. Students will see how changes in the environment and in the relationship of one organism to another affect animal species and their numbers on Earth.

Motivate
Cooperative Learning: Assign one Connection to each group of students. Using the Expert Teams strategy, have each group research to find out more about the geographic location of the Connection—its climate, culture, plants, animals, and ecological issues.

Wrap-Up
Conclude this lesson by contacting a local zoo to see if it has an Adopt-an-Animal Program.

OCEANOGRAPHY
Background: Zebra mollusks first appeared in Lake St. Clair near Detroit in 1988. They allegedly came from a tanker that was expelling ballast water. Since then, these tiny invaders have spread rapidly through the Great Lakes. With few predators to keep them in check, the mollusks have clogged municipal water systems and damaged dock pilings. Adult zebra mollusks can produce up to 30,000 larvae annually. The larvae are so small that they can infiltrate boat motors and grow inside the engine.
Discussion: How were the zebra mollusks able to take such control in U.S. waters? How could zebra mollusks be controlled? What happens when populations go unchecked?
Answer to Question: Possible controls for zebra mollusks might be a disease that would affect only the mollusk, making them infertile, or changing the environment in some way.
Extension: Have students set up a simulation that shows what takes place when an organism has an unlimited food supply and no predators, a limited food supply, and an unlimited food supply and predators.

UNIT 5
GLOBAL CONNECTIONS

Animals
In this unit, you have studied a variety of animals. Now find out how animals are connected to other subjects in different parts of the world.

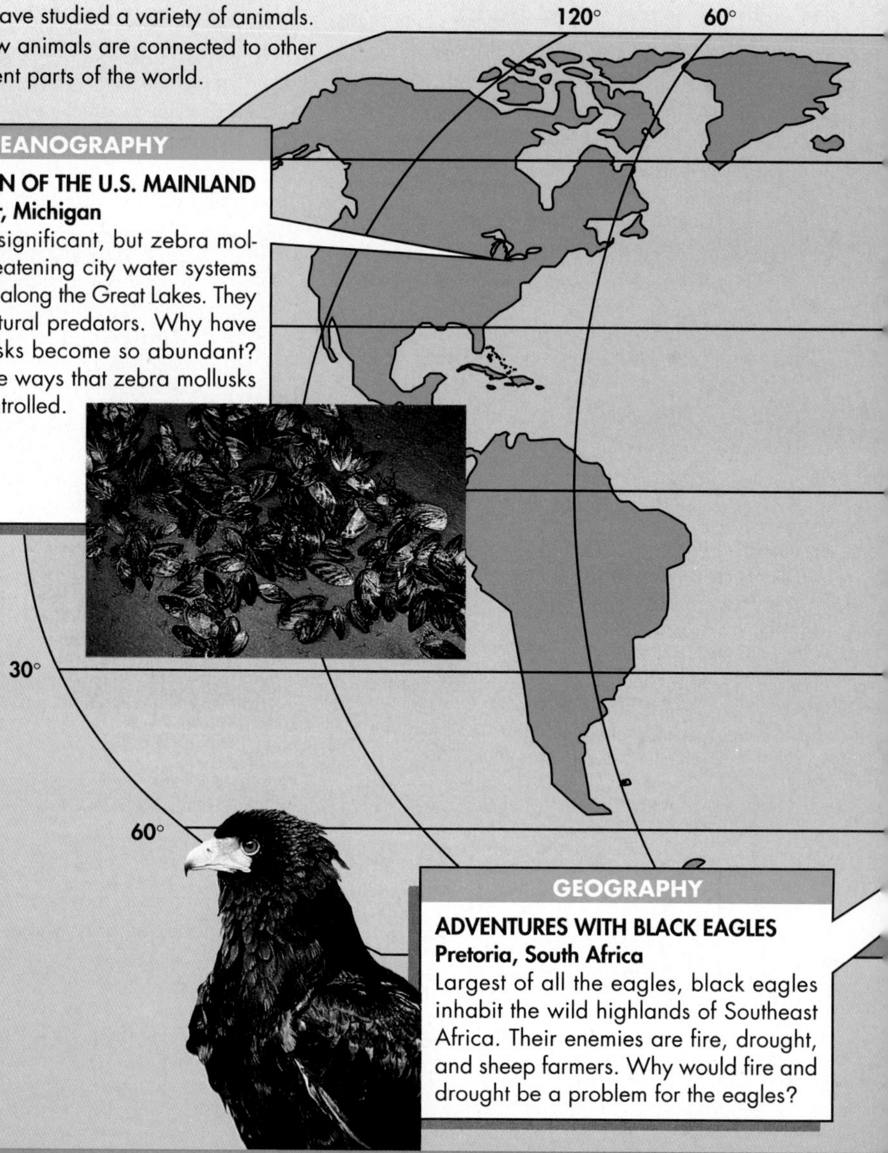

120° 60° 30° 60°

OCEANOGRAPHY
AN INVASION OF THE U.S. MAINLAND
Lake St. Clair, Michigan
They look insignificant, but zebra mollusks are threatening city water systems and beaches along the Great Lakes. They have few natural predators. Why have zebra mollusks become so abundant? Suggest some ways that zebra mollusks might be controlled.

GEOGRAPHY
ADVENTURES WITH BLACK EAGLES
Pretoria, South Africa
Largest of all the eagles, black eagles inhabit the wild highlands of Southeast Africa. Their enemies are fire, drought, and sheep farmers. Why would fire and drought be a problem for the eagles?

412

GEOGRAPHY
Background: Black eagles, with a seven-foot wing span, are one of the largest eagles in the world. Their range is from the tip of South Africa to Egypt. Ornithologists study them in order to learn about their habits and relationship to other eagles in the world. Like most eagles, they prefer remote areas where game is plentiful.
Discussion: What function do the students think black eagles perform in the ecology of the area? What can scientists learn by studying eagles?
Answer to Question: Both fire and drought could reduce the populations on which the eagles feed.
Extension: Have students investigate the affects and aftermath of the fires in Yellowstone National Park in 1988.

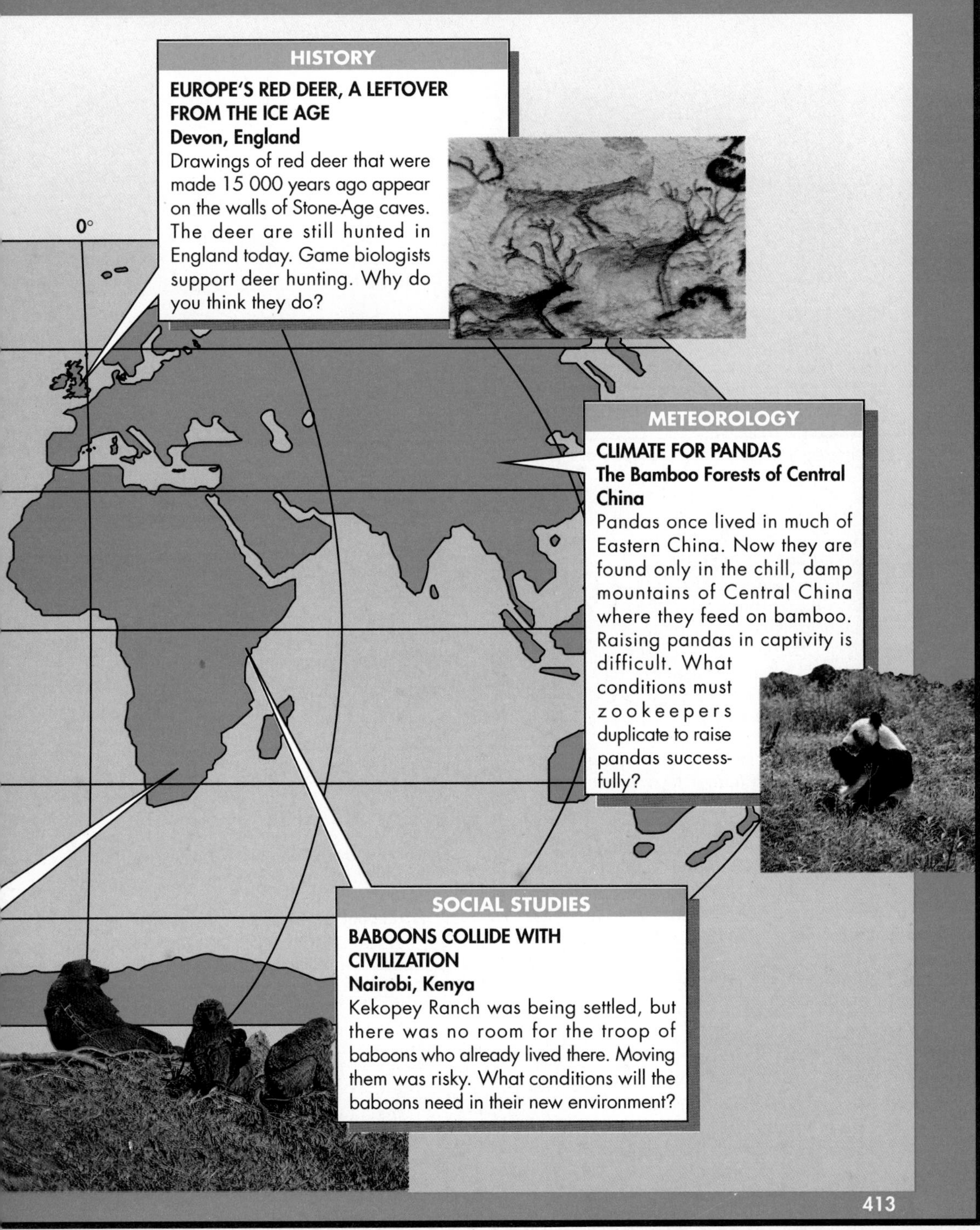

HISTORY
EUROPE'S RED DEER, A LEFTOVER FROM THE ICE AGE
Devon, England
Drawings of red deer that were made 15 000 years ago appear on the walls of Stone-Age caves. The deer are still hunted in England today. Game biologists support deer hunting. Why do you think they do?
0°
METEOROLOGY
CLIMATE FOR PANDAS
The Bamboo Forests of Central China
Pandas once lived in much of Eastern China. Now they are found only in the chill, damp mountains of Central China where they feed on bamboo. Raising pandas in captivity is difficult. What conditions must zookeepers duplicate to raise pandas successfully?
SOCIAL STUDIES
BABOONS COLLIDE WITH CIVILIZATION
Nairobi, Kenya
Kekopey Ranch was being settled, but there was no room for the troop of baboons who already lived there. Moving them was risky. What conditions will the baboons need in their new environment?
413

METEOROLOGY

Background: Pandas are gentle animals who feed mostly on bamboo, although they are known to be carnivorous. Fossils indicate they once roamed much of eastern China. Now they are found only in the chill, damp mountains of central China. The Chinese government has studied the habits of pandas thoroughly in an effort to protect them.

Discussion: What do students think may be the cause of the decline of pandas in eastern Chinas? Why would it be important to be able to raise pandas in captivity? What would zoologists need to know about pandas to raise them successfully?

Answer to Question: Zookeepers would need to duplicate the climate, kinds of food, and family structure that pandas have in the wild. They would need to know whether pandas are more active at night or during the day and how much space a panda needs to feel comfortable.

Extension: Obtain some bamboo from a wholesale florist. Have students examine it to find out something about panda diets. Post a map of China and have students study its climate and geography. If you are near a zoo that houses pandas, arrange a visit to it.

HISTORY

Background: Most deer are raised on hunting preserves where they roam freely. Once a year, the deer are hunted. Deer hunting to many people seems very cruel. But if deer herds were not culled, many deer would die of starvation.

Discussion: If a population of animals grows too large for a source of food, some will starve or become weak and diseased and eventually die. In the past, natural predators kept deer herds from growing too large. Would it be better to encourage natural predators to control the deer herds?

Answer to Question: Most game biologists are concerned about large deer populations. Starving deer suffer cruelly. Large herds do damage in the forests, eat the food supplies of other species, and are nuisances to farmers and gardeners.

Extension: Have students make a bulletin board display of five organisms in the U.S. that are under protection. Have extensions from the map show the areas where the organism is protected, what it looks like, what its food sources are and what steps have been taken to protect them.

SOCIAL STUDIES

Background: As in many parts of the world, land in East Africa is being developed for farming. Many native animals are a problem for farmers and settlers. They dig up and eat garbage, spoil crops, and bother homeowners. To combat such a problem, three troops of baboons were moved to a remote area by truck.

Discussion: What problems do the students think might be encountered in attempting to move 130 wild baboons several hundred miles? They should remember that other baboons live in the new area.

Answer to Question: The baboons will need a source of food and water. They may need to learn about new foods. They will have to adjust to other baboon tribes.

Extension: Have students discuss what it is like to have to move into a new place.

ZOOLOGIST

Background: A zoologist may choose among several careers. With a Ph.D., there is opportunity to do research, teach in a college, or conduct independent study. A Master's degree offers many occupations.

Related Career	Education
Zoologist	M.S. degree
Game Biologist	B.S. degree
Agricul. scientist	B.S. degree

Career Issue: Some zoologists must use animals in their research. Many animal rights people are concerned about this.

What do you think? How can animals be protected without jeopardizing valuable research?

VETERINARIAN'S TECHNICIAN

Background: Most veterinarian's technicians receive their training on the job.

Related Career	Education
Veterinary Assistant	On-the-job
Agricul. Technician	Tech. School
Lab. Animal Assistant	On-the-job

Career Issue: In some states, veterinarian's technicians need to be licensed. To become licensed, they must take a test.

What do you think? Why license veterinarian's technicians?

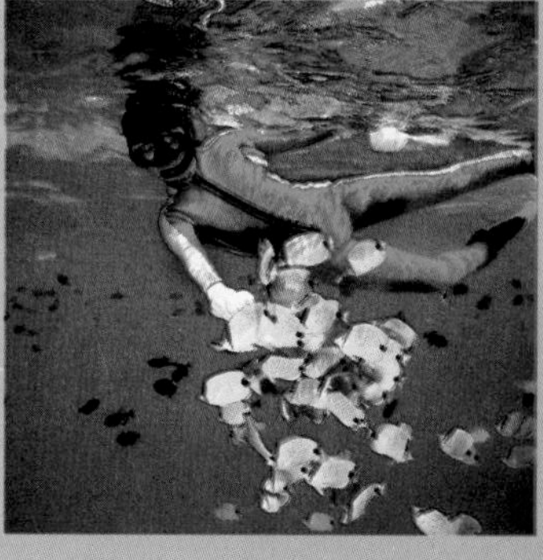

ZOOLOGIST

A *zoologist* is a scientist who studies animals. These scientists are interested in the animal's habitat, behavior, diseases, and life processes. Zoologists who study birds are called ornithologists. Those who study reptiles are called herpetologists. Mammalogists study mammals and ichthyologists study fish.

A person who wants to become a zoologist should like working with animals and be willing to spend long and irregular hours with them. Zoologists need a background in science, particularly biology, zoology, and animal behavior. A trained zoologist will need to have a bachelor of science degree for non-research work. If they plan to do research, they will need to have a master's degree.

For Additional Information

Contact the American Society of Zoologists, 104 Sirius Circle, Thousand Oaks, CA 91360.

CAREERS

VETERINARIAN'S TECHNICIAN

A *veterinarian's technician* likes animals and enjoys working with them. He or she should be calm and reliable, as well as willing to work hard. Some of the work that veterinarian's technicians do is not easy. Cages have to be cleaned, animals groomed, and excited animals must be calmed. Most veterinary technicians are trained on the job. It is possible to start as a part-time animal attendant and advance to technician. Experienced technicians may also find work in kennels, small zoos, pet stores, as assistant laboratory animal technicians, or with the Humane Society.

Some veterinarian technicians want to open their own kennels to raise, groom, or board small animals.

For Additional Information

Contact Animal Caretakers Information, The Humane Society of the U.S., Companion Animals Division, Suite 100, 5430 Grosvenor Lane, Bethesda, MD 20814.

UNIT READINGS

▶"Zebras Are Coming." *Popular Mechanics,* February 1991.
▶Schaller, George G. "Pandas in the Wild." *National Geographic,* December 1981.
▶Clutton-Brock, T.H., "Red Deer and Man." *National Geographic,* October 1986.

414

UNIT READINGS

Background

▶ "Zebras are Coming." *Popular Mechanics,* February 1991. This article explains some of the dilemmas in controlling these mollusks.

▶ Schaller, George B. "Pandas in the Wild." *National Geographic,* December 1981. An interesting account of studies conducted by the U.S. and China on wild pandas.

▶ Clutton-Brock, T.H. "Red Deer and Man." *National Geographic,* October 1986. Discusses how deer have survived the pressures of civilization.

More Readings

1. Strum, Shirley C. "The Pumphouse Gang Moves to a Strange New Land." *National Geographic,* November 1987. Anthropologist Strum has written entertainingly of her experiences with a troop of baboons.

2. Cowden, Jeanne. "Adventures With South Africa's Black Eagles." *National Geographic,* October 1969. When an eaglet fell over a precipice, Cowden knew she had to rescue him—but dealing with the furious parents made it nearly impossible.

SCIENCE & ART

John James Audubon
The American Bison

Many people know of Audubon's beautiful bird paintings. In fact, Audubon's birds are what he is best known for today. But Audubon painted other animals, or quadrupeds, as he called them. This painting of the American bison was done in 1845 during a trip Audubon made to the Dakota Prairies.

Audubon did this painting of a male bison. As you study it, you will notice that the painting is so carefully done, it seems almost like a photograph. All attempts were made to be true to nature. Audubon had a dead bison measured and weighed, and made careful notes of the measurements. The painting was one seventh the original size of the animal.

Drawing birds and wild animals was the only thing that made Audubon really happy. Born in Haiti, he spent his early years in France before going to America at 18. His father, a wealthy French planter, tried to set him up in business, but Audubon failed miserably because he was, as a business partner said, continually in the forest. To develop detailed paintings of birds, he used dead birds in their natural settings as his models. In 1826, he began to publish volumes of his works on North American birds. By 1845, Audubon had sold a series of subscriptions for his bird paintings throughout Europe and America. The work had taken him over twenty years to produce.

Now he wanted to complete a series of quadrupeds as well. But he was 53 years old and losing his memory. Only 150 prints of animals were ever completed. Audubon was deeply saddened by the killing of bison and wrote of a hunt he had seen, "What a terrible destruction of life, as it were for nothing, or next to nothing....."

In Your Own Words

▶ Investigate an extinct or endangered species of North American animal. Describe the animal and how it became endangered or extinct. Illustrate the animal in some way.

Source: John James Audubon, *The American Bison.* This painting was part of a project to illustrate all of the animals of North America. Audubon completed his 150 illustrations before his death.

Biography: Audubon was born in Haiti, on April 26, 1785. His father, Jean Audubon, was a wealthy planter as well as a lieutenant in the French navy. He owned estates in the West Indies, Pennsylvania, and France. Audubon was raised in France, but came to Mill Grove, Pennsylvania, as a young man to manage his father's interest in a lead mine. He was, however, much more interested in the birds and animals of the Pennsylvania countryside. Though his father tried to set him up in business, Audubon found such a life intolerable. He decided instead to paint all the birds of North America in their natural surroundings. Audubon used dead birds for models. The paintings would then be copied on copper plates, printed and colored, and sold in complete sets to subscribers, $1000 per subscription. The project took twenty years.

TEACHING STRATEGY

You may want to display other examples of Audubon's paintings so that students can see the detail. Point out that Audubon was not a well-known artist at the time he did his works. All that he knew about birds and animals he had learned from seeing them in the wild. However, Audubon's paintings are today considered among the best in the world.

Cooperative Learning: Have Paired Partner's choose one particular Audubon painting and research the animal or bird that is illustrated. Have them also report on the status of the animal today. Is the animal or bird still in existence? Is it endangered? What is its natural habitat? Is the habitat endangered?

Classics

▶ Seton, Ernest Thompson. *Wild Animals I Have Known.* This is an old book and may be difficult to find. The Grolier Society reprinted some of the stories in a paperback in 1967. The stories relate some of the attitudes and thinking of early settlers towards wolves and bears.

▶ Audubon, John James. *Audubon's America.* Edited by Donald Culross Peattie, Houghton Mifflin Co., 1940. In addition to his well-known paintings, Audubon was a fine writer. This book was taken from the diaries he kept as he traveled across early America.

Other Works

See Audubon's *Birds of America.* Audubon, John James, *Audubon's America,* edited by Donald Culross Peattie, Houghton Mifflin Co., 1940.

UNIT

6

THE HUMAN BODY

Unit 6 is organized into chapters that deal with body systems.

CONTENTS

ADVANCE PREPARATION

Activities

▶ **Activity 18-1, page 425,** requires fresh beef bones.

▶ **Activity 18-2, page 436,** requires cooked turkey legs and prepared slides of muscles.

▶ **Activity 19-1, page 450,** requires juice samples and indophenol solution.

▶ **Activity 19-2, page 458,** requires gelatin, pepsin powder, and dilute hydrochloric acid.

▶ **Activity 20-1, page 470,** requires a sphygmomanometer, a stethoscope, alcohol, and cotton balls.

▶ **Activity 20-2, page 477,** calls for prepared slides of vertebrate blood cells.

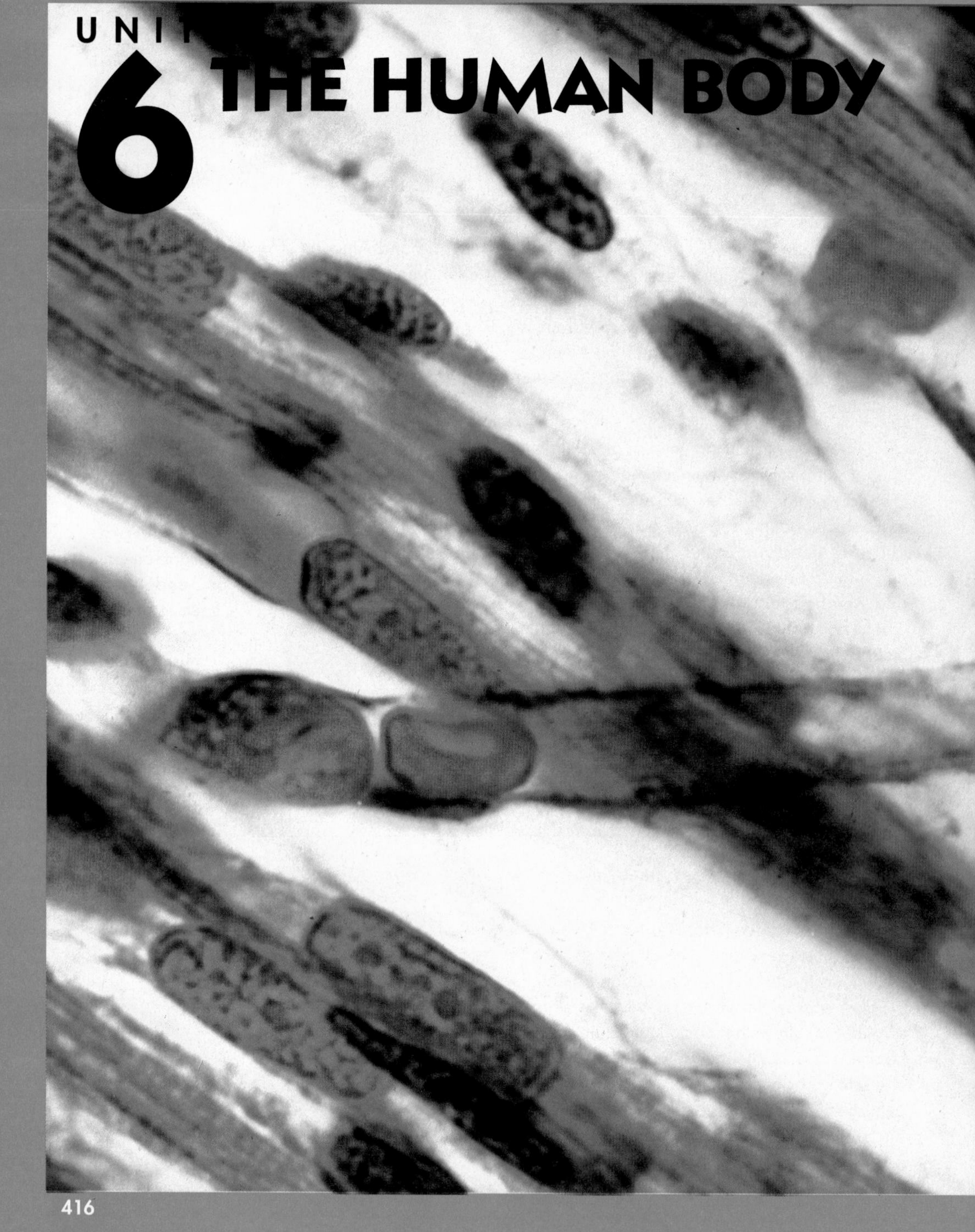

OPTIONS

Cross Curriculum

▶ Ask students to write short essays detailing what they know about one of the body systems in the unit. When the unit is completed, return essays to students for editing.

PROJECT

During the course of this unit study, have students work cooperatively on PROJECT 6, *On Your Mark! Get Set! Go!,* found on pages 704 and 705.

Science at Home

▶ Ask students to find out how much they weighed and how tall (long) they were at birth. Have students weigh and measure themselves to compare with birth statistics.

Cooperative Learning: Have students make a list of all the food items they consume in one day, including junk food. Working in small groups, students should place each item on the list into one food group. Students should look up nutritional information on each food, and identify the most nutritious items in each group listed.

What's Happening Here?

Blood surges through your blood vessels with every beat of your heart. Red blood cells carrying oxygen bend and contort as they are pushed through narrow capillaries. White blood cells tumble past in the river of plasma, on their way to the day-to-day work of keeping you healthy. A ride through a water slide may give you an idea of what it is like to be a blood cell on this journey. In this unit, you will learn about your circulatory system. You will also learn about the other body systems on which your life depends.

UNIT CONTENTS

▶ **Activity 21-1, page 495,** calls for straws and bromothymol blue solution.
▶ **Activity 21-2, page 502,** requires iodine solution and paper.
▶ **Activity 22-1, page 514,** requires pennies and metersticks.
▶ **Activity 22-2, page 519,** requires index cards, toothpicks, and glue.
▶ **Activity 23-2, page 546,** requires graph paper and red and blue pencils.

Field Trips and Speakers
▶ Arrange for speakers from health organizations to visit your classroom.

INTRODUCING THE UNIT

What's Happening Here?
▶ Have students look at the photos and read the text. Explain that in this unit they will learn all about the human body and how body systems function.
▶ **Background:** Point out that body systems operate together and influence each other. The circulatory system, for example, brings oxygen-laden blood to all body cells, but the blood gets the oxygen in the lungs, part of the respiratory system.

Previewing the Chapters
▶ Encourage students to look at the diagrams of body systems found in this unit.

Tying to Previous Knowledge
▶ Have students brainstorm what the functions of bones, muscles, and skin are. Make sure they see that skin is a part of their immune system, too.
▶ Use the **inquiry questions** in the OPTIONS box below to investigate with students the relationship between digestion and excretion.

Multicultural Awareness
Have students research the activities associated with "coming-of-age" ceremonies in various cultures. In some cultures, girls are not considered women until they begin menstruation. Boys in many cultures must undergo some kind of ceremony to prove their "manliness" before they become men. Have students speculate on what the "coming-of-age" rituals are in their own families and cultures, and compare them to such ceremonies in other cultures.

Inquiry Questions
Use the following questions to focus a discussion on the relationship between digestion and excretion.
▶ **How do you get energy?**
▶ **What do you need energy for?**
▶ **Do you use all the energy in the food you eat?**
▶ **How are digestion and excretion related?**
▶ **What would be the result if humans had no excretory system?**

CHAPTER

18 Bones, Muscles, and Skin

CHAPTER SECTION	OBJECTIVES	ACTIVITIES
18-1 The Skeletal System (3 days)	1. **Identify** the five major functions of the skeletal system. 2. **Describe** how a fracture heals. 3. **Compare** and **contrast** movable and immovable joints.	**Activity 18-1:** *Designing an Experiment (Observing Bones)*, p. 425
18-2 The Muscular System (2 days)	1. **Describe** the major function of muscles. 2. **Compare** and **contrast** three types of muscles. 3. **Explain** how muscle action results in movement of body parts.	**MINI-Lab:** *How do muscle pairs work?* p. 428
18-3 Drugs for Fitness? Science & Society (1 day)	1. **List** side effects of anabolic steroid use. 2. **Discuss** the use of anabolic steroids among athletes.	
18-4 Skin (1 day)	1. **Compare** and **contrast** the epidermis and dermis of the skin. 2. **List** the functions of the skin. 3. **Discuss** how skin protects the body from disease and how it heals itself.	**MINI-Lab:** *Is there water in sweat?* p. 435 **Activity 18-2:** *Observing Muscle*, p. 436
Chapter Review		

ACTIVITY MATERIALS

FIND OUT	ACTIVITIES		MINI-LABS	
Page 419 snap-type clothespin clock or watch	**18-1 Designing an Experiment, p. 425** beef bone hand lens scalpel microscope slide microscope coverslip dropper water	**18-2 Observing Muscle, p. 436** prepared slides of smooth, skeletal, and cardiac muscles microscope cooked turkey leg dissecting pan or cutting board dissecting probes (2) hand lens	**How do muscle pairs work?, p. 428** none	**Is there water in sweat? p. 435** cobalt chloride paper water

CHAPTER FEATURES	TEACHER RESOURCE PACKAGE	OTHER RESOURCES
Skill Builder: *Outlining,* p. 424	**Ability Level Worksheets** ◆ ***Study Guide,*** p. 70 ● ***Reinforcement,*** p. 70 ▲ ***Enrichment,*** p. 70 **Activity Worksheets,** pp. 5, 158-159 **Transparency Masters,** pp. 63-64 **Critical Thinking/Problem Solving,** p. 22 **Cross-Curricular Connections,** p. 22	**Lab Manual:** Analyzing Bones, p. 119 **Color Transparency 32,** Human Skeletal System **Science Integration Activity 18** **STVS:** Disc 7, Side 2
Problem Solving: *High Altitude Bones,* p. 429 **Skill Builder:** *Sequencing,* p. 429	**Ability Level Worksheets** ◆ ***Study Guide,*** p. 71 ● ***Reinforcement,*** p. 71 ▲ ***Enrichment,*** p. 71 **Activity Worksheets,** p. 164 **Transparency Masters,** pp. 65-66	**Lab Manual:** Muscle Action, p. 123 **Color Transparency 33,** Human Muscle System **STVS:** Disc 7, Side 1
You Decide! p. 431	**Ability Level Worksheets** ◆ ***Study Guide,*** p. 72 ● ***Reinforcement,*** p. 72 ▲ ***Enrichment,*** p. 72 **Science and Society,** p. 22	
Technology: *Robot Skin,* p. 434 **Skill Builder:** *Concept Mapping,* p. 435	**Ability Level Worksheets** ◆ ***Study Guide,*** p. 73 ● ***Reinforcement,*** p. 73 ▲ ***Enrichment,*** p. 73 **Concept Mapping,** p. 41 **Activity Worksheets,** pp. 160, 161, 165	**STVS:** Disc 7, Side 1
Summary Key Science Words Understanding Vocabulary Checking Concepts Using Lab Skills Think & Write Critically Apply More Skill Builders Projects	ASSESSMENT RESOURCES **Chapter Review,** pp. 39-40 **Chapter Test,** pp. 122-125 **Performance Assessment in Middle School Science (PAMSS)**	**Chapter Review Software** **Test Bank** **Alternate Assessment** **Performance Assessment**

◆ **Basic** ● **Average** ▲ **Advanced**

ADDITIONAL MATERIALS

SOFTWARE	AUDIOVISUAL	BOOKS/MAGAZINES
The Skeletal System, Queue. *The Heart,* Queue. *Human Anatomy,* EBEC. *Anatomy Challenge,* Island Software.	*Bones and Joints,* film, BFA. *Human Body Skeleton,* film, Coronet/MTI. *Muscles and Movement,* Mechanics of Life Series, film, BFA. *The Skeleton,* film, EBEC. *How the Body Moves,* film, Time-Life. *Human Body: Muscular System,* film, Coronet/MTI. *Work of the Heart/Muscles: Structure and Function,* laserdisc, EBEC. *I Am Joe's Skin,* film, Pyramid Film.	Berger, Gilda. *Human Body*. NY: Doubleday and Company, Inc., 1989. Cohn. *Anatomy*. 8th ed. NY: Elsevier Science Publishing Company, Inc., 1986.

CHAPTER

18

BONES, MUSCLES, AND SKIN

THEME DEVELOPMENT: One of the major themes emphasized in this text is energy. The skeletal and muscular systems work in an integrated manner to produce movement. Energy is used, and work is done. The food we eat is processed by the digestive system to convert it into substances that the body can use. The chemical energy in glucose is used by muscles and transformed into mechanical energy.

CHAPTER OVERVIEW

▶ **Section 18-1:** This section discusses the five major functions of the skeletal system. In addition, there are details of the characteristics, composition, and development of bone, and healing of fractures. Bone joints are also discussed.

▶ **Section 18-2:** The major functions and types of muscles are discussed in this section. An explanation of how muscle action produces movement is given.

▶ **Section 18-3: Science and Society:** The possible side effects of anabolic steroid use are explored. The You Decide feature asks students to consider the issue of the use of anabolic steroids by athletes.

▶ **Section 18-4:** This section discusses the tissue layers and structures of the skin. The five major functions of the skin are reviewed.

CHAPTER VOCABULARY

skeletal system	involuntary muscle
periosteum	skeletal muscles
marrow	tendons
cartilage	smooth muscles
fracture	cardiac muscle
joint	anabolic steroids
ligament	epidermis
immovable joint	

CHAPTER

18 Bones, Muscles, and Skin

418

OPTIONS

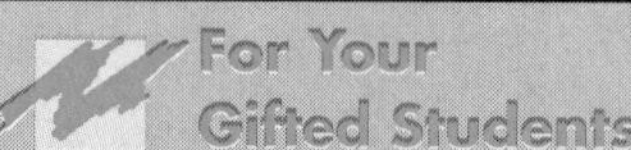

For Your Gifted Students

Students can make a life-sized model of the human skeleton. Have them use tag board or other lightweight cardboard and draw the major bones found in the human skeletal system. They can connect the bones with brads to form the skeleton.

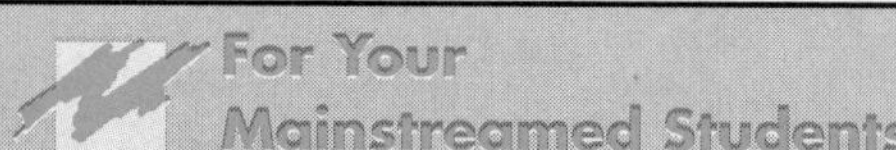

For Your Mainstreamed Students

Have students find a diagram of the human skull. Let students make clay or papier-mâché models of the skull. Students can search for diagrams of different animals and make similar models. Challenge students to make the lower jaw articulate.

Like the speed skater on the left, you might have experienced tired muscles after gym class or a soccer game. Although large muscles are the ones that you notice when they tire, your body has many small muscles that can tire after just a little bit of work.

FIND OUT!

Do this activity to find out about some of your small muscles.

Hold a snap-type clothespin between the thumb and forefinger of your writing hand. Rapidly pinch the clothespin open and shut. Count the number of times you can pinch the clothespin before the fingers in your hand become tired. *Record* the number. Use the thumb and forefinger of your other hand. How do they compare? Do small muscles that are exercised more (your writing hand) tire faster or slower than those in the hand that gets less exercise?

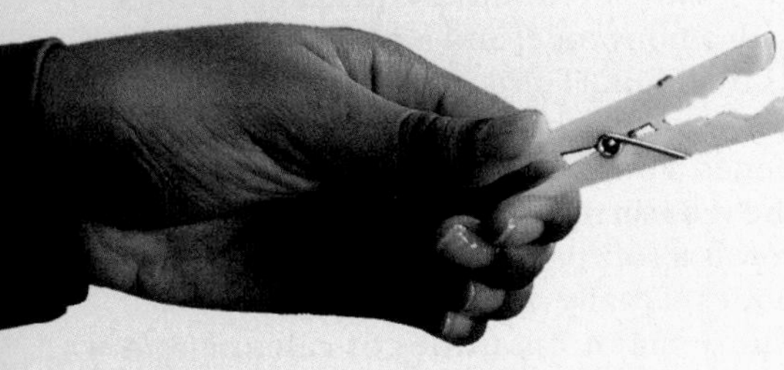

Gearing Up

Previewing the Chapter

Use this outline to help you focus on important ideas in this chapter.

Section 18-1 The Skeletal System
- What Is a Skeletal System?
- Bone
- Bone Development
- Fractures
- Joints

Section 18-2 The Muscular System
- The Muscular System
- Three Types of Muscle Tissue
- Muscles at Work

Section 18-3 Science and Society
Drugs for Fitness?
- Do Steroids Give a "Winning Edge"?

Section 18-4 Skin
- The Body's Largest Organ
- Skin at Work

Previewing Science Skills

- In the Skill Builders, you will outline, sequence, and make a concept map.
- In the Activities, you will observe and collect data.
- In the MINI-Labs, you will observe and experiment.

What's next?

Throughout the day, hundreds of muscles, both large and small, work to move your body. In this chapter, you'll learn how muscles and bones provide the body with a working structure. You'll also learn about skin, your body's largest organ, and its many functions.

419

movable joint
muscle
voluntary muscle
melanin
dermis

INTRODUCING THE CHAPTER

Use the Find Out activity to introduce students to muscles. Inform students they will learn more about the muscular and skeletal systems as they read the chapter.

FIND OUT!

Preparation: Before you begin this chapter, have students bring in wooden snap-type clothespins. Provide extras.

Cooperative Learning: Use Science Investigation cooperative groups. While one student pinches the clothespin, another can count the number of pinches. The third can record. Then have them rotate.

Teaching Tips

- Tell students to count only until the first sign of tiredness, not until exhaustion or cramping. Remind students that this is not a competition.

Gearing Up

Have students study the Gearing Up feature to familiarize themselves with the chapter. Discuss the relationships of the topics in the outline.

What's Next?

Before beginning the first section, make sure students understand the connection between the Find Out activity and the topics to follow.

ASSESSMENT OPTIONS

PORTFOLIO

Refer to page 437 for suggested items that students might select for their portfolios.

PERFORMANCE ASSESSMENT

See page 437 for additional Performance Assessment options.

Process

Skillbuilders, pp. 424, 429
MINI-Lab, p. 435
Activities 18-1, p. 425; 18-2, p. 436
Using Lab Skills, p. 438

CONTENT ASSESSMENT

Assessment—Oral, pp. 423, 429, 434
Section Reviews, pp. 424, 429, 431, 435
Chapter Review, pp. 437-439
Mini Quizzes, pp. 423, 431

GROUP ASSESSMENT

Opportunities for group assessment occur with Cooperative Learning Strategies and Flex Your Brain Activities.

SECTION 18-1

PREPARATION

SECTION BACKGROUND

▶ Bone tissue is composed chiefly of collagen and the mineral hydroxyapatite $Ca_{10}(PO_4)_6(OH)_2$.

▶ In an embryo, the skeleton begins as a cartilaginous-membrane system. In the last months of the pregnancy, the cartilage and the membranes are replaced by bone.

▶ Bone tissue is constantly being reabsorbed and reformed. Calcium and vitamin D are necessary for the deposition of new bone material.

PREPLANNING

▶ To prepare for Activity 18-1, obtain a long beef bone. Boil the bone and have it cut in half along the length and width by a butcher.

1 MOTIVATE

▶ Discuss how the skeleton gives general shape to the body. Ask students if there are any bones on their skeleton that they can actually feel. Students may indicate ribs, wrist, knuckles, elbows, ankles, and vertebrae.

▶ Place a chicken bone in a jar of vinegar for several days. Remove the bone and examine it. Have students note how the mineral salts have dissolved, leaving only elastic tissue.

Cooperative Learning: Assign Problem Solving teams to find the inaccuracies in cardboard or model skeletons displayed at Halloween.

OBJECTIVES AND SCIENCE WORDS: Have students review the objectives and science words throughout the chapter to become familiar with each section.

VideoDisc

STVS: Measuring Calcium Deficiency, Disc 7, Side 2

18-1 The Skeletal System

New Science Words

skeletal system
periosteum
marrow
cartilage
fracture
joint
ligament
immovable joint
movable joint

Objectives

▶ **Identify the five major functions of the skeletal system.**
▶ **Describe how a fracture heals.**
▶ **Compare and contrast movable and immovable joints.**

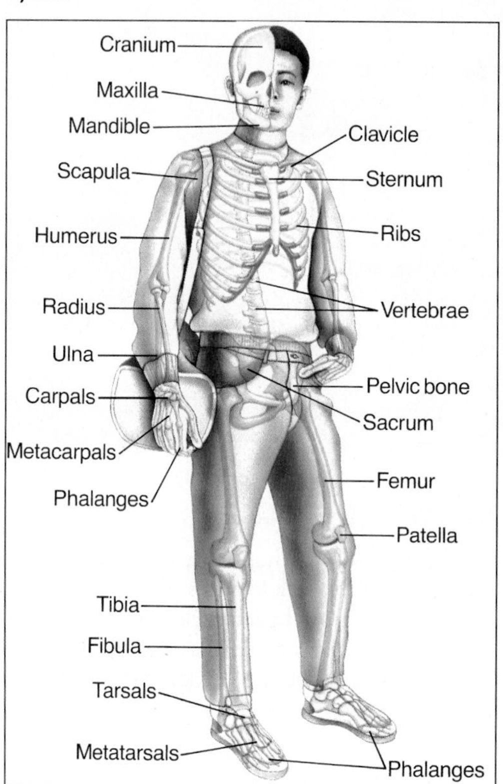

Figure 18-1. The Human Skeletal System

What Is a Skeletal System?

Does the doctor you go to have a human skeleton in the office? If so, you might have the impression that all bones are dead structures made of dry, rocklike material. It's true that the skeleton's bones are no longer living, but the bones in your body are very much alive. Each bone within your body is a living organ made of several different tissues. Cells in these bones take in food and expend energy. They have the same requirements as other cells in your body.

Because you have a skeleton, you stand, walk, run, and breathe. All the bones together make up your **skeletal system,** which is the framework of your body. The human skeletal system has five major functions. First, it gives shape and support to your body, much like the framework of a building. Second, bones protect your internal organs. In the skeleton shown in Figure 1
18-1, ribs surround the heart and lungs with a bony cage, and a hard skull encloses the brain. Third, major muscles of the body are attached to bone. Muscles move bones. Fourth, blood cells are formed in the red marrow of some bones. Bone marrow is a soft tissue in the center of many bones. Finally, the skeleton is the place where major quantities of calcium and phosphorus compounds are stored for later use in the body. Calcium and phosphorus make bone hard.

OPTIONS

Meeting Different Ability Levels

For Section 18-1, use the following **Teacher Resource Masters** depending upon individual students' needs.

◆ **Study Guide Master** for all students.

● **Reinforcement Master** for students of average and above average ability levels.

▲ **Enrichment Master** for above average students.

Additional Teacher Resource Package masters are listed in the OPTIONS box throughout the section. The additional masters are appropriate for all students.

◆ STUDY GUIDE 70

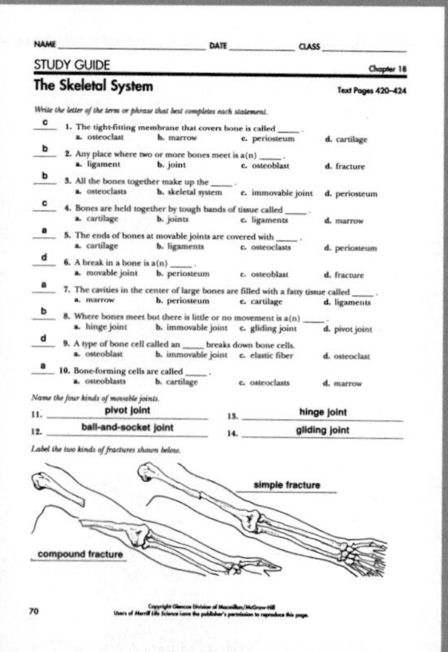
NAME ____ DATE ____ CLASS ____

STUDY GUIDE — Chapter 18

The Skeletal System — Text Pages 420–424

Write the letter of the term or phrase that best completes each statement.

c 1. The tight-fitting membrane that covers bone is called ____.
a. osteoclast b. marrow c. periosteum d. cartilage

b 2. Any place where two or more bones meet is a(n) ____.
a. ligament b. joint c. osteoblast d. fracture

b 3. All the bones together make up the ____.
a. osteoclasts b. skeletal system c. immovable joint d. periosteum

c 4. Bones are held together by tough bands of tissue called ____.
a. cartilage b. joints c. ligaments d. marrow

a 5. The ends of bones at movable joints are covered with ____.
a. cartilage b. ligaments c. osteoclasts d. periosteum

d 6. A break in a bone is a(n) ____.
a. movable joint b. periosteum c. osteoblast d. fracture

a 7. The cavities in the center of large bones are filled with a fatty tissue called ____.
a. marrow b. periosteum c. cartilage d. ligaments

b 8. Where bones meet but there is little or no movement is a(n) ____.
a. hinge joint b. immovable joint c. gliding joint d. pivot joint

d 9. A type of bone cell called an ____ breaks down bone cells.
a. osteoblast b. immovable joint c. elastic fiber d. osteoclast

a 10. Bone-forming cells are called ____.
a. osteoblasts b. cartilage c. osteoclasts d. marrow

Name the four kinds of movable joints.

11. pivot joint
12. ball-and-socket joint
13. hinge joint
14. gliding joint

Label the two kinds of fractures shown below.

simple fracture

compound fracture

70

Bone

Pick up a bone and you'll find that it isn't all smooth. There are bumps, edges, round ends, rough spots, and many pits and holes. Each of these marks has a purpose. Muscles attach to some of the bumps and pits. Blood vessels and nerves enter and leave through the holes.

Your skeleton has 206 bones of various sizes and shapes. Bones are frequently classified according to shape. Figure 18-2 shows bones that are long, short, flat, or irregular. Their shapes are genetically controlled and modified by the work of muscles attached to them.

Many internal and external characteristics of bone are easily seen in the humerus shown in Figure 18-3. The humerus is a long bone in your upper arm. The surface of the long portion of the bone is covered with a tough, tight-fitting membrane called **periosteum** (per ee AHS tee um). Small blood vessels in the periosteum carry nutrients into the bone. This membrane is also important in the growth and repair of bone. Under the periosteum is compact bone, a hard, strong layer of bone. Compact bone contains bone cells, blood vessels, the minerals calcium and phosphorus, and elastic fibers. Rickets and osteoporosis are two diseases that result from lack of minerals in bone. Elastic fibers keep bone from being too rigid and brittle or easily broken.

Spongy bone in a long bone is found toward the ends of the bone. Spongy bone is much less compact and has many small open spaces that make the bone lightweight. If all your bones were completely solid, you'd have a much greater mass. Long bones have large openings, or cavities. The cavities in the center of long bones and the

2

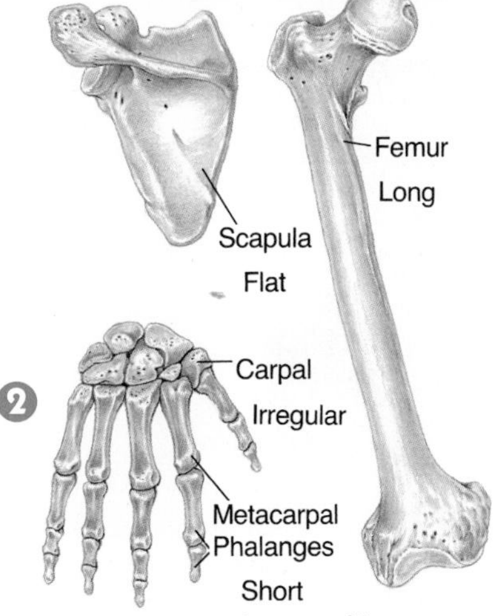

Figure 18-2. Shapes of human bones show something about their functions.

One of the minerals found in bone is calcium carbonate. This compound makes up 3.6 percent of Earth's crust. Find out the relationship between calcium carbonate and limestone. What is limestone formed from?

Figure 18-3. Bone is made up of living tissue.

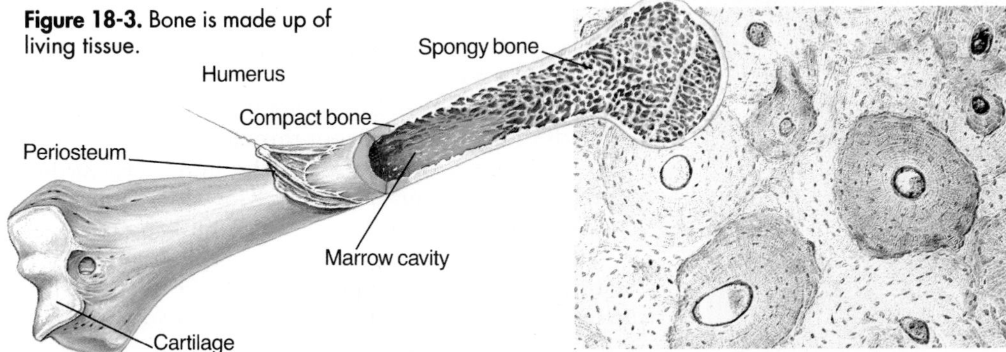

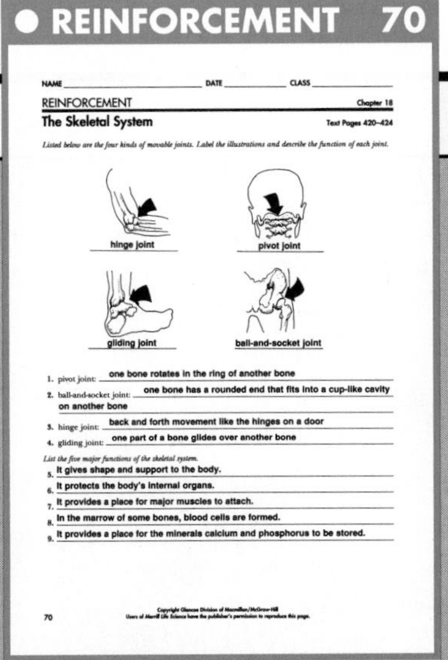

● REINFORCEMENT 70

NAME ______ DATE ______ CLASS ______

REINFORCEMENT — Chapter 18

The Skeletal System — Text Pages 420–424

Listed below are the four kinds of movable joints. Label the illustrations and describe the function of each joint.

hinge joint — pivot joint — gliding joint — ball-and-socket joint

1. pivot joint: one bone rotates in the ring of another bone
2. ball-and-socket joint: one bone has a rounded end that fits into a cup-like cavity on another bone
3. hinge joint: back and forth movement like the hinges on a door
4. gliding joint: one part of a bone glides over another bone

List the five major functions of the skeletal system.

5. It gives shape and support to the body.
6. It protects the body's internal organs.
7. It provides a place for major muscles to attach.
8. In the marrow of some bones, blood cells are formed.
9. It provides a place for the minerals calcium and phosphorus to be stored.

70

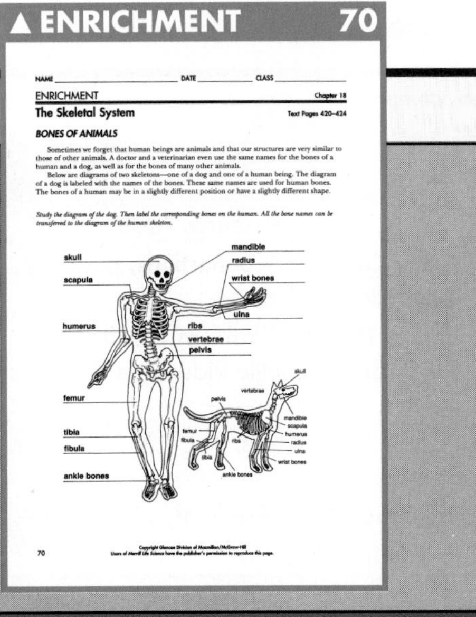

▲ ENRICHMENT 70

NAME ______ DATE ______ CLASS ______

ENRICHMENT — Chapter 18

The Skeletal System — Text Pages 420–424

BONES OF ANIMALS

Sometimes we forget that human beings are animals and that our structures are very similar to those of other animals. A doctor and a veterinarian even use the same names for the bones of a human and a dog, as well as for the bones of many other animals.

Below are diagrams of two skeletons—one of a dog and one of a human being. The diagram of a dog is labeled with the names of the bones. These same names are used for human bones. The bones of a human may be in a slightly different position or have a slightly different shape.

Study the diagram of the dog. Then label the corresponding bones on the human. All the bone names can be transferred to the diagram of the human skeleton.

70

TYING TO PREVIOUS KNOWLEDGE: Recall the role of the skeletal system in classifying vertebrate animals. Remind students of the function of the skeletal system in providing internal support and framework for the vertebrate body.

Answer: Limestone is calcium carbonate. Most limestone is the result of decayed skeletons of marine invertebrates.

2 TEACH

Key Concepts are highlighted.

CONCEPT DEVELOPMENT

▶ **Demonstration:** Use 4 × 6 cards and tape to make one rectangular column and one cylindrical tube. Stand the structures on a flat surface and balance textbooks on the top of them. Note which structure collapses first. Relate this strength to the long, round bones of our body.

FLEX Your Brain

Use the Flex Your Brain activity to have students explore BONE.

Flex Your Brain

ASSESSMENT

Portfolio: Use the Flex Your Brain activity to reinforce critical thinking and problem solving skills. In Step 2, students might list what they know about bone function, location, or names.

REVEALING MISCONCEPTIONS

▶ Students often believe that the cartilage of the embryo and fetus is *changed* into bone. Have students note that the cartilage breaks down and is *replaced* by bone.

CROSS CURRICULUM

▶ **Language Arts:** Have students look up *periosteum* and *osteoblast* in the dictionary. Have volunteers explain how the roots of the words reflect their meanings.

TEACHER F.Y.I.

▶ The bones of the foot are arranged to form a transverse and a longitudinal arch. Arches form a very strong structure to support the weight of the body.

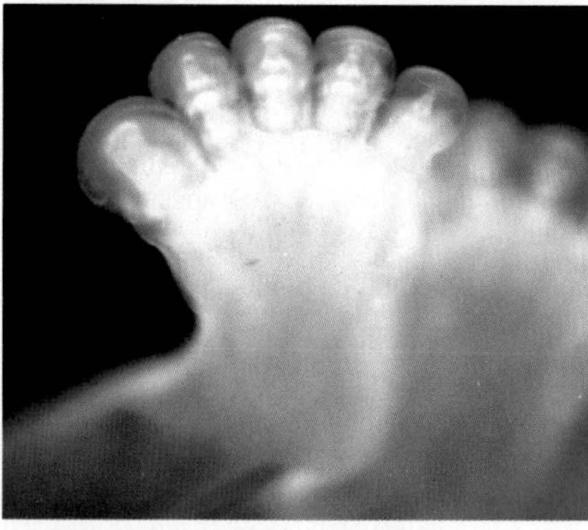

Figure 18-4. During development before birth, the skeleton is first made up of cartilage, which is then gradually replaced by bone.

spaces in spongy bone are filled with a fatty tissue called **marrow.** In a healthy person, marrow produces red blood cells at an incredible rate of between two and three million cells per second. White blood cells are also produced in bone marrow, but in lesser amounts.

Unless you have a condition called arthritis, you move about fairly comfortably. This is because the ends of your bones are covered with a thick, smooth layer of tissue called **cartilage.** Cartilage does not contain blood vessels or minerals. It is flexible and is important at joints, where it absorbs shock and makes movement easier. 3

Bone Development

Before you were born, your skeleton was first made in the form of cartilage. Gradually the cartilage was broken down and replaced by bone-forming cells called osteoblasts (AHS tee oh blastz). These cells deposit calcium and phosphorus that make bone tissue hard. At birth, your skeleton was made up of more than the 206 bones you now have. As you developed, some bones fused, or grew together.

How is bone formed and reformed?

Healthy bone tissue is dynamic. It is always being formed and reformed by the cells it contains. A second type of bone cell, an osteoclast, breaks down bone tissue. This is a normal process in a healthy person. Osteoclasts break down bone and release calcium and phosphorus. This process keeps the calcium and phosphorus in your bloodstream at about the same level.

Figure 18-5. For a fracture to heal, the broken ends of the bone have to be brought together.

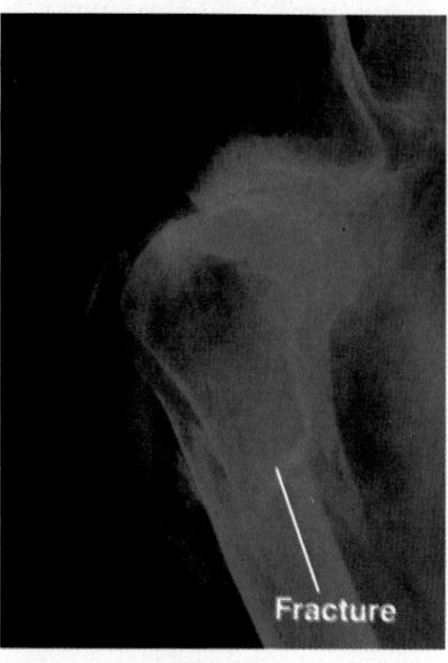

Fractures

Although bones are strong and flexible, they can break. A break in a bone is a **fracture.** A simple fracture occurs when a bone has broken but the broken ends do not break through the skin. In a compound fracture, the broken ends of the bone stick out through the skin. 4

To heal, the broken ends of the bone have to be brought in contact with each other. The periosteum starts to make new bone cells. A thick band of new bone forms around the break, acting like a built-in splint. Over time, the thickened band disappears as bone is reshaped with the help of osteoclasts. Breaks are usually held immobile with a cast.

OPTIONS

PROGRAM RESOURCES

From the **Teacher Resource Package** use:

Cross-Curricular Connections, page 22, Leonardo da Vinci.

Critical Thinking/Problem Solving, page 22, A Space Shuttle Ride for a Phantom Head.

Activity Worksheets, page 5, Flex Your Brain.

Use **Laboratory Manual,** page 119, Analyzing Bones.

Joints

Think of all the different actions you performed this morning getting ready for school. You opened your mouth to yawn, chewed your breakfast, reached for a toothbrush, and stretched out your arm to turn the doorknob as you walked out the door. All these motions were possible because your skeleton has joints.

Any place where two or more bones meet is a **joint.** A model of a joint can be made from two cardboard tubes and some rubber bands. The tubes represent bones, and the rubber bands represent ligaments. A **ligament** is a tough band of tissue that holds bones together at joints. Many joints, such as your knee, are held together by more than one ligament.

Joints are classified as immovable or movable depending on how much movement can take place. An **immovable joint** allows little or no movement. The joints of the bones of your skull and pelvis are classified as immovable. A **movable joint** allows the body to make a wide range of movements. Shooting baskets or working the controls of a video game require movable joints. There are several types of movable joints: pivot, ball-and-socket, hinge, and gliding

In Your JOURNAL

You may have heard of an athlete who has had arthroscopic surgery on a knee. Find out about this procedure. **In your Journal,** describe how this surgery is different from other types of surgery.

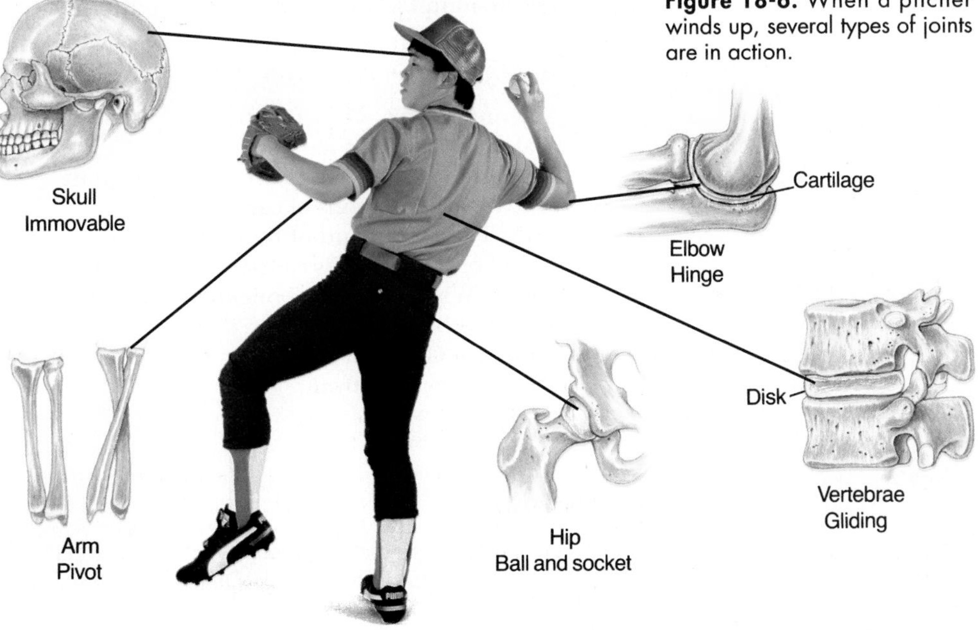

Figure 18-6. When a pitcher winds up, several types of joints are in action.

MINI QUIZ

Use the Mini Quiz to check students' recall of chapter content.

1. **Give an example of how the skeleton protects organs.** *Ribs surround the heart and lungs; skull encloses the brain.*
2. **What are the four shapes of your bones?** *long, short, flat, and irregular*
3. **Why is cartilage important at joints?** *It absorbs shock and makes movement easier.*
4. **What is a fracture?** *a break in a bone*
5. **How does a hinge joint work?** *It has a back and forth movement like the hinge on a door.*

In Your JOURNAL

Rather than making a long incision as in regular surgery, a surgeon makes a very small incision in arthroscopic surgery. An arthroscope is then inserted through that incision and the surgeon can look around inside the joint. Repair of the joint is also done through the small incision by means of special instruments designed to work in conjunction with the arthroscope.

CHECK FOR UNDERSTANDING

▶ **What would happen if there was no cartilage covering on the ends of the bones at movable joints?** *The movement would be stiff and uncomfortable.*

▶ Ask questions 1-4 and the **Apply** Question in the Section Review on page 424.

▶ Have students infer the role of cartilage between the vertebrae.

RETEACH

Bring in the leg and wing bones of cooked chicken to examine the ends of the bones for cartilage.

EXTENSION

For students who have mastered this section, use the **Reinforcement** and **Enrichment** masters or other OPTIONS provided.

ASSESSMENT—ORAL

▶ **Infer the function of a fluid-filled cavity between the two cartilage surfaces.** *The fluid serves to lubricate the cartilage and also acts as a shock absorber.*

▶ **Some joints have small sacs called bursae, under the ligaments or muscle tissue. Infer the function of the bursae.** *The bursae serve to lessen the friction between the bone and ligaments or muscle and to absorb shock.*

Science Integration Activities is found in **Merrill Life Science Teacher Resource Manual**

PROGRAM RESOURCES

From the **Teacher Resource Package** use:

Transparency Masters, pages 63-64, Human Skeletal System.

Use **Color Transparency** 32, Human Skeletal System.

Science Integration Activities 18, Differences in a Species.

3 CLOSE

▶ Have an athletic trainer, sports medicine physician, or sports physical therapist come to the class to discuss athletic injuries common to adolescents and adults.

SECTION REVIEW ANSWERS

1. gives shape and support; protects internal organs; provides place for muscle attachment, place for blood cell production, place for storing minerals

2. contains small blood vessels that carry nutrients to the bone; important in growth and repair of bone

3. Possible answers include: pivot—base of skull; ball and socket—shoulder, hip; hinge—elbow, knee, fingers; gliding—wrists, ankles.

4. helps body move more easily by providing padding; also provides shape for ears and nose

5. Apply: osteoclasts

6. Connect to Chemistry: titanium and ceramic

Skillbuilder
ASSESSMENT

Performance: To further assess students' abilities to Outline, have them outline the internal and external features of bone.

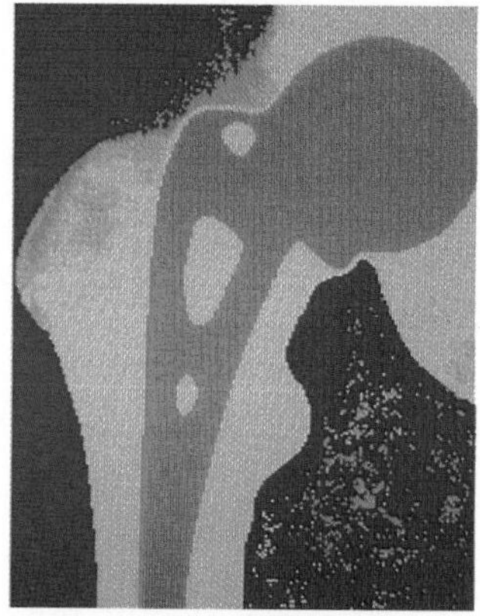

Figure 18-7. Certain joints, especially those in the hip, knee, and elbow, have been successfully replaced by artificial joints.

joints. In a pivot joint, one bone rotates in a ring of another bone. Turning your head is an example of a pivot movement. In a ball-and-socket joint, one bone has a rounded end that fits into a cuplike cavity on another bone. Hips and shoulders are examples.

5 A third type of joint is a hinge joint. This joint has a back and forth movement like hinges on a door. Elbows, knees, and fingers have hinge joints. A fourth type of joint is a gliding joint where one part of a bone glides over another bone. Gliding joints are found in your wrists and ankles and between vertebrae. Gliding joints are the most frequently used joints in your body. You can't write a word, pick up a sock, or take a step without using a gliding joint.

Cartilage helps make joint movements easier. It covers the ends of bones in movable joints. Your outer ear and the end of your nose are made of cartilage. Pads of cartilage called disks are also found between the vertebrae. Here, cartilage also acts as a cushion and prevents injury to your spinal cord.

Your skeleton is a living framework in the form of bones. Bones not only support the body but also supply it with minerals and blood cells. Joints are places between bones that enable the framework to be flexible and to be more than just a storehouse for minerals.

SECTION REVIEW

1. What are the five major functions of a skeleton?
2. What are two jobs of the periosteum?
3. Name and give an example of a movable joint.
4. What are the functions of cartilage?
5. **Apply:** The thick band of bone that forms around a healing broken bone is called a callus. In time it disappears. What is there in bone that enables this extra amount of bone to disappear?
6. **Connect to Chemistry:** Find out what materials many implants are made from.

Outlining

Outline the major functions of bone and give an example of each. If you need help, refer to Outlining in the **Skill Handbook** on page 681.

Skill Builder

This is an example of how this material may be outlined.

I. Functions of Bone
  A. Shape and support
    1. carpals
    2. vertebrae
  B. Protect organs
    1. ribs around heart and lungs
    2. skull around brain
  C. Muscle attachment
    1. skeletal muscles attached to bones
  D. Blood cell production
    1. bone marrow
  E. Stores minerals
    1. calcium
    2. phosphorus

ACTIVITY 18-1 DESIGNING AN EXPERIMENT
Observing Bones

Bones are living organs and are made of several different kinds of tissue. What are some of the major tissues of bone? What do they look like?

Getting Started
You need to determine a way to examine a bone. Your task in this activity is to devise a method to identify and describe the parts of a beef bone.

Thinking Critically
List some ways a bone can be examined. How will you identify the various tissues?

Materials 
Your cooperative group will use:
- beef bone
- hand lens
- scalpel
- microscope slide
- microscope
- coverslip
- dropper
- water

Try It!

1. Make a data table to record the parts of the bone examined and their descriptions.
2. Obtain a beef bone that has been cut in half along the length and width.
3. *Observe* the bone with a hand lens. Identify the periosteum, compact bone, spongy bone, and any marrow that is present.
4. Draw a diagram of the bone and label the parts that you have been able to recognize. In the table, describe each part.
5. Use a scalpel to carefully remove a very small piece of red bone marrow. **CAUTION:** Always use extreme care when using sharp instruments.
6. Make a temporary wet mount slide of the marrow. *Examine* it under low power. Describe and draw what you see.

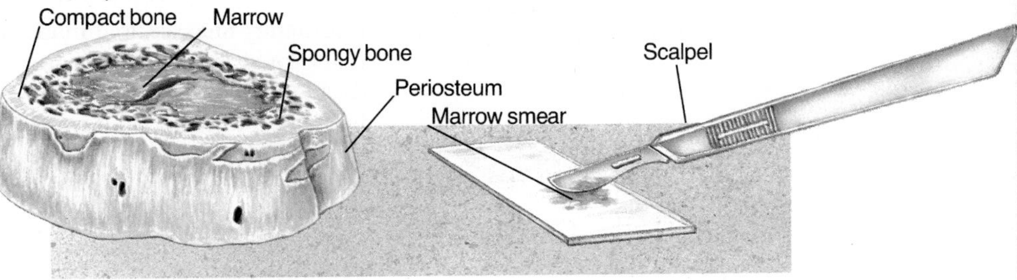

Summing Up/Sharing Results
- What parts of the bone were you able to identify?
- Describe the portions of the bone that contain marrow.
- What differences did you observe in the overall shape of the bone?
- How might the function of the spongy bone area and the compact bone areas relate to the shape of the areas in which they were found?
- From your observations, where would you *infer* that most calcium is stored in a long bone?

Going Further!
Compare the beef bone with a turkey leg bone. What are some similarities and differences? Can you observe any adaptations for flight in the turkey bone?

GOING FURTHER!
The turkey bone is similar in shape to the beef bone, but it is slimmer, less compact, and has a hollow center. The light weight of the bones is an adaptation to flight.

PROGRAM RESOURCES

From the **Teacher Resource Package** use:
Activity Worksheets, pages 158-159, Activity 18-1, Observing Bones.

Activity ASSESSMENT
Performance: To further assess students' understanding of bones, see USING LAB SKILLS, Question 11, on page 438.

ACTIVITY 18-1

OBJECTIVE: Design and carry out an experiment to identify and describe the parts of a beef bone
Time: one class period

PROCESS SKILLS applied in this activity are **observing** and **inferring.**

PREPARATION
Obtain the materials for each group. Have turkey leg bones available for the Going Further activity.
Cooperative Learning: Divide the class into Science Investigation Teams.

SAFETY
Caution students to use extreme care when using sharp instruments such as a scalpel.

THINKING CRITICALLY
Answers will vary but students will likely list the periosteum, compact bone, spongy bone, and marrow. Identification of bone parts can be done using Figure 18-3 on page 421.

TEACHING THE ACTIVITY
*Refer to the **Activity Worksheets** for additional information and teaching strategies.*
- Sharp kitchen knives or artists' knives may be used in place of scalpels but with the same caution.
- Do not attempt to use chicken bones for this activity. They are less compact, making it difficult to observe the various parts.

SUMMING UP/SHARING RESULTS
- Answers will vary, but may include the periosteum, compact bone, spongy bone, and red and yellow marrow.
- the long portion of the bone
- The shaft is narrow, while the ends are rounded and expanded.
- The long part of the bone supports weight and therefore must be compact.
- The ends of the bone can be spongy, since they distribute mass over their areas.
- in the shaft in the compact bone

SECTION 18-2

PREPARATION

SECTION BACKGROUND

▶ All the movements of the body are the result of a unique specialization of muscle tissue, contractility. The shortening of the muscles pulls on the body parts to which they are attached and movement occurs.

▶ A nerve impulse traveling to a muscle causes chemicals called neurotransmitters to be released at the junction of the nerve and muscle. The chemicals cause the muscle to contract.

▶ Muscle cells respond to any stimulus by contracting to their maximum; there is no partial contraction. This phenomenon is known as the all-or-none principle. If a muscle only partially contracts, it is because only some of its cells are contracting.

▶ Muscle activity produces heat. It is this heat that the circulatory system moves to other body tissues to help maintain the body temperature.

1 MOTIVATE

▶ Obtain a fresh chicken leg. Remove the skin and have students observe the muscle tissue and tendons.

FLEX Your Brain

Use the Flex Your Brain activity to have students explore MUSCLE.

Flex Your Brain

ASSESSMENT

Portfolio: Use the Flex Your Brain activity to reinforce critical thinking and problem solving skills. In Step 2, students might list strength training, muscle location, and muscle function.

▶ Discuss how muscle tissue on the skeletal system gives shape to the body. Emphasize the need for activity to keep muscles strong and firm.

VideoDisc

STVS: Electric Heart, Disc 7, Side 1

18-2 The Muscular System

New Science Words

muscle
voluntary muscles
involuntary muscles
skeletal muscles
tendons
smooth muscles
cardiac muscle

Objectives

- ▶ **Describe the major function of muscles.**
- ▶ **Compare and contrast three types of muscles.**
- ▶ **Explain how muscle action results in movement of body parts.**

Figure 18-8. Muscles of the Human Body

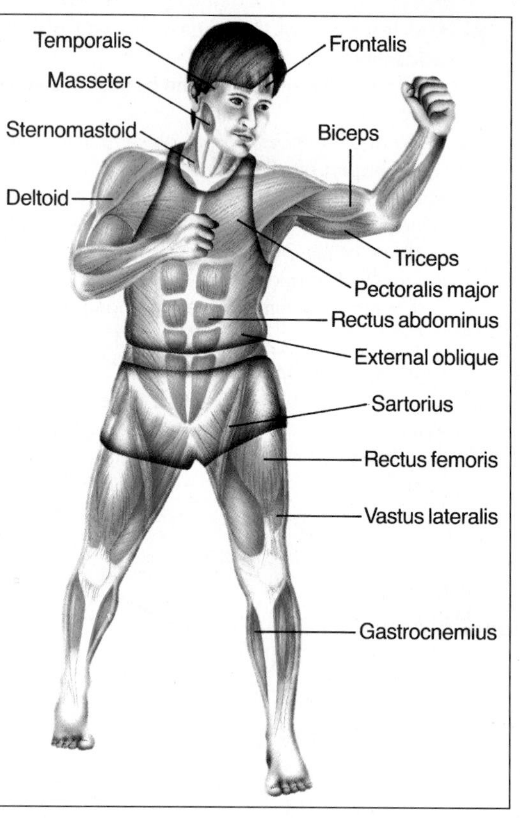

The Muscular System

Bones and joints together are somewhat like simple machines, but on their own they have no power to move. Muscles are the motors that move body parts. There are more than 600 muscles in your body. This means that nearly 35 to 40 percent of your body mass is muscle tissue. No matter how still you might try to be, there is always movement taking place in your body. Many systems in your body contain some type of muscle tissue as a part of that system. As a result, most parts of your body move at some time during your life, if not every day. A **muscle** is an organ that contracts and gets shorter. As a result, body parts move. Energy is used, and work is done.

Voluntary muscles are muscles that you control. Your arm and leg muscles are voluntary. So are the muscles of your hands and face. You can choose to move them or not to move them. In contrast, **involuntary muscles** are muscles you can't consciously control. You don't have to decide to make these muscles work. They just go on working all day long, all your life. Blood gets pumped through blood vessels, and food is moved through your digestive system by the action of involuntary muscles. You can sleep at night without having to think about how to keep these muscles working.

OPTIONS

Meeting Different Ability Levels

For Section 18-2, use the following **Teacher Resource Masters** depending upon individual students' needs.

◆ **Study Guide Master** for all students.

● **Reinforcement Master** for students of average and above average ability levels.

▲ **Enrichment Master** for above average students.

Additional Teacher Resource Package masters are listed in the OPTIONS box throughout the section. The additional masters are appropriate for all students.

◆ STUDY GUIDE 71

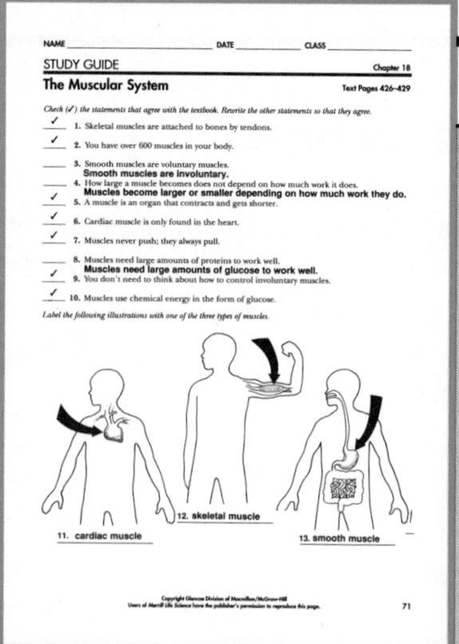

NAME ______ DATE ______ CLASS ______

STUDY GUIDE — Chapter 18

The Muscular System — Text Pages 426–429

Check (✓) the statements that agree with the textbook. Rewrite the other statements so that they agree.

✓ 1. Skeletal muscles are attached to bones by tendons.
✓ 2. You have over 600 muscles in your body.
___ 3. Smooth muscles are voluntary muscles.
Smooth muscles are involuntary.
___ 4. How large a muscle becomes does not depend on how much work it does.
Muscles become larger or smaller depending on how much work they do.
✓ 5. A muscle is an organ that contracts and gets shorter.
✓ 6. Cardiac muscle is only found in the heart.
✓ 7. Muscles never push; they always pull.
___ 8. Muscles need large amounts of proteins to work well.
Muscles need large amounts of glucose to work well.
✓ 9. You don't need to think about how to control involuntary muscles.
✓ 10. Muscles use chemical energy in the form of glucose.

Label the following illustrations with one of the three types of muscles.

11. cardiac muscle
12. skeletal muscle
13. smooth muscle

71

Table 18-1

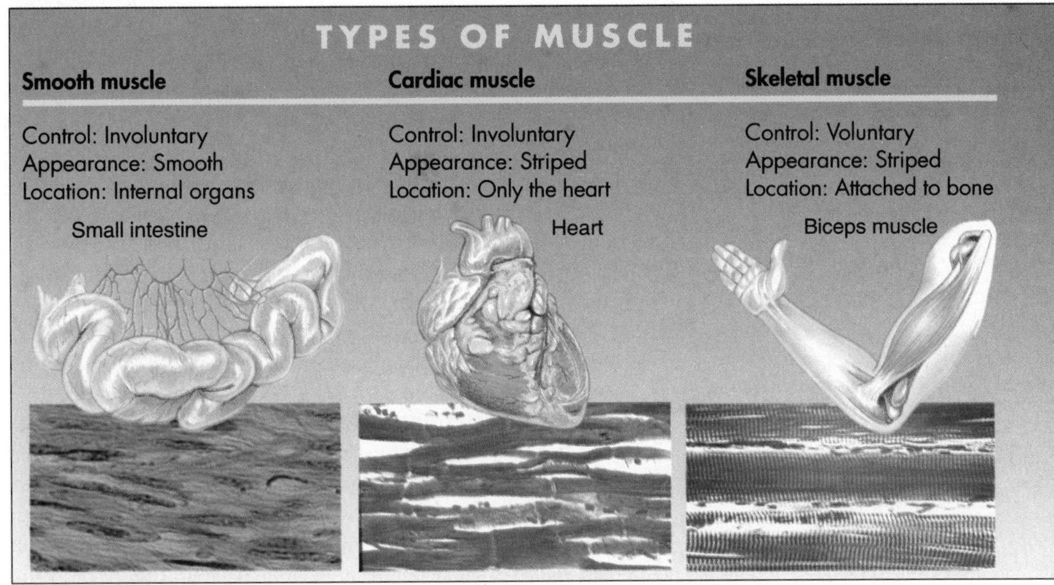

TYPES OF MUSCLE

Smooth muscle	Cardiac muscle	Skeletal muscle
Control: Involuntary	Control: Involuntary	Control: Voluntary
Appearance: Smooth	Appearance: Striped	Appearance: Striped
Location: Internal organs	Location: Only the heart	Location: Attached to bone

Three Types of Muscle Tissue

There are three types of muscle tissue in your body: skeletal, smooth, and cardiac. Skeletal muscles are the most numerous muscles in the body. Under a microscope, skeletal muscle cells look striped, or striated. **Skeletal muscles** are attached to bones by tendons. **Tendons** are thick bands that pull on the bone as the muscle contracts. Skeletal muscles are voluntary muscles. You can control their use. You choose when to walk or not to walk. Skeletal muscles tend to contract quickly and tire easily.

Smooth muscles are involuntary. The walls of the stomach, intestine, uterus, bladder, penis, and blood vessels are some of the places where one or more layers of smooth muscle are found. Smooth muscles contract and relax slowly. They have no striations.

The third type of muscle, **cardiac muscle,** is found only in the heart. Cardiac muscle is also involuntary. As you can see from Table 18-1, cardiac muscle has striations like skeletal muscle. However, cardiac muscle is like smooth muscle in that it is involuntary. Cardiac muscle contracts about 70 times per minute every day of your life. You know each contraction as a heartbeat, but it isn't something you can control.

Science and MATH

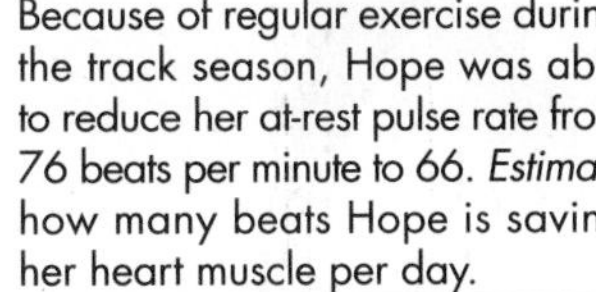

Because of regular exercise during the track season, Hope was able to reduce her at-rest pulse rate from 76 beats per minute to 66. *Estimate* how many beats Hope is saving her heart muscle per day.

● REINFORCEMENT 71

NAME ______ DATE ______ CLASS ______

REINFORCEMENT — Chapter 18

The Muscular System — Text Pages 426–429

Think of the muscle at each of the following body parts. In the space provided, write the name of the muscle associated with that body part. Also tell whether that muscle is voluntary or involuntary.

1. thigh: **skeletal, voluntary**
2. upper arm: **skeletal, voluntary**
3. intestine: **smooth, involuntary**
4. heart: **cardiac, involuntary**
5. calf: **skeletal, voluntary**
6. stomach: **smooth, involuntary**
7. hand: **skeletal, voluntary**
8. blood vessels: **smooth, involuntary**
9. uterus: **smooth, involuntary**
10. neck: **skeletal, voluntary**
11. The two illustrations show an ankle bending. Label the second illustration, showing which muscle contracts and which muscle relaxes.

contracts / relaxes — relaxes / contracts

12. How many muscles are in the body? **over 600**
13. How do muscles make mechanical energy? **When muscles contract, the bonds in glucose break down. Chemical energy is changed to mechanical energy.**
14. What happens when glucose is not available to muscles? **The muscles have no energy supply so they tire and need rest.**

 71

▲ ENRICHMENT 71

NAME ______ DATE ______ CLASS ______

ENRICHMENT — Chapter 18

The Muscular System — Text Pages 426–429

MUSCLES AND TENDONS

Materials
- frozen chicken leg and thigh
- sharp kitchen knife

Procedure
1. Defrost the chicken leg and thigh until they are almost completely thawed.
2. Use the knife to cut the skin away. Try not to cut any of the meat. **CAUTION:** *Always use extreme care when using sharp instruments.*
3. Closely examine the bundles of muscle under the skin. Then examine the ends of the muscles at the knee joint.
4. Carefully cut away all the flesh from around the knee joint. Try to count how many different bundles there are in the chicken leg.

Data and Observations
1. The word *striated* means "made up of elongated fibers." Are the muscles of a chicken's leg striated or smooth? **striated**
2. How are the muscles of the chicken leg attached to the bones? **by tendons**
3. As you cut the muscles away from the joint, how many bundles did you count? **Most students will count four to six.**

Conclude and Apply
1. The chicken leg is the same as what part of our leg? **the calf**
2. What part of the chicken do we call meat? **the muscles**

 71

TYING TO PREVIOUS KNOWLEDGE: Everyone has had sore muscles from overworking them. Have students recall that the soreness often appears in muscles that are not properly used or are infrequently exercised.

2 TEACH

Key Concepts are highlighted.

CONCEPT DEVELOPMENT

▶ If possible, obtain an anatomy chart of the muscular system. Use this to illustrate the large percent of the body mass that is muscle tissue.

▶ Ask students: **Why aren't all muscles voluntary?** *It would be impossible for a person to think about all the body functions that require muscle action—breathing, heart beating, stomach churning, etc.*

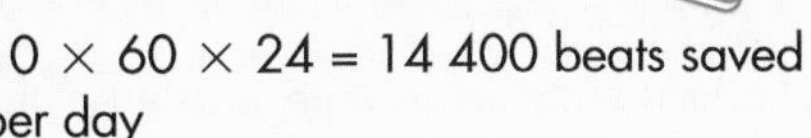

Science and MATH

$10 \times 60 \times 24 = 14\,400$ beats saved per day

TEACHER F.Y.I.

▶ At movable joints the place of muscle attachment to the less movable bone is called the origin; the place of muscle attachment to the bone of greater movement is the insertion.

MINI-Lab
Answer: The *bicep* muscle flexes, the *tricep* extends, the arm.

MINI-Lab
ASSESSMENT
Performance: To further assess students' understanding of muscle pairs, have them repeat the lab using the jaw muscles.

CHECK FOR UNDERSTANDING
Use the oral assessment questions at the bottom of Teacher page 429 to check for understanding.

RETEACH
Cooperative Learning: Using the Paired Partner strategy, have students make flash cards with names of body organs that require muscle action. On the reverse side of each card, identify the type of muscle (skeletal, smooth, or cardiac) involved.

EXTENSION
For students who have mastered this section, use the **Reinforcement** and **Enrichment** masters or other OPTIONS.

Muscles at Work

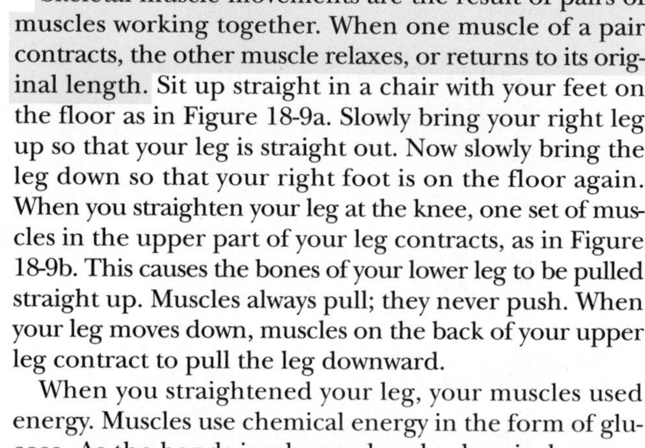

Skeletal muscle movements are the result of pairs of muscles working together. When one muscle of a pair contracts, the other muscle relaxes, or returns to its original length. Sit up straight in a chair with your feet on the floor as in Figure 18-9a. Slowly bring your right leg up so that your leg is straight out. Now slowly bring the leg down so that your right foot is on the floor again. When you straighten your leg at the knee, one set of muscles in the upper part of your leg contracts, as in Figure 18-9b. This causes the bones of your lower leg to be pulled straight up. Muscles always pull; they never push. When your leg moves down, muscles on the back of your upper leg contract to pull the leg downward.

When you straightened your leg, your muscles used energy. Muscles use chemical energy in the form of glucose. As the bonds in glucose break, chemical energy changes to mechanical energy and the muscle contracts. When the supply of glucose in a muscle gets used up, it becomes tired and needs to rest. Muscles also produce thermal energy when they contract. The heat produced by muscle contraction helps to keep your body temperature constant.

Over a period of time, muscles can become larger or smaller, depending on how much work they do. Skeletal muscles that do a lot of work, such as those in your writing hand or in the arms of a brick layer, become large and strong. In contrast, if you just sit and watch TV all day, your muscles will become soft and flabby, and will lack strength. Muscles that aren't exercised become smaller in size.

MINI-Lab
How do muscle pairs work?
Work with a partner. Stretch your arm out straight. Then bring your hand to your shoulder, then down again. Using a muscle chart, have your partner *determine* which skeletal muscles in your upper arm enable you to perform this action. **In your Journal,** draw and label the changes that take place in the arm.

Figure 18-9. Skeletal muscles work in pairs. When one muscle in a pair contracts, the other is relaxed.

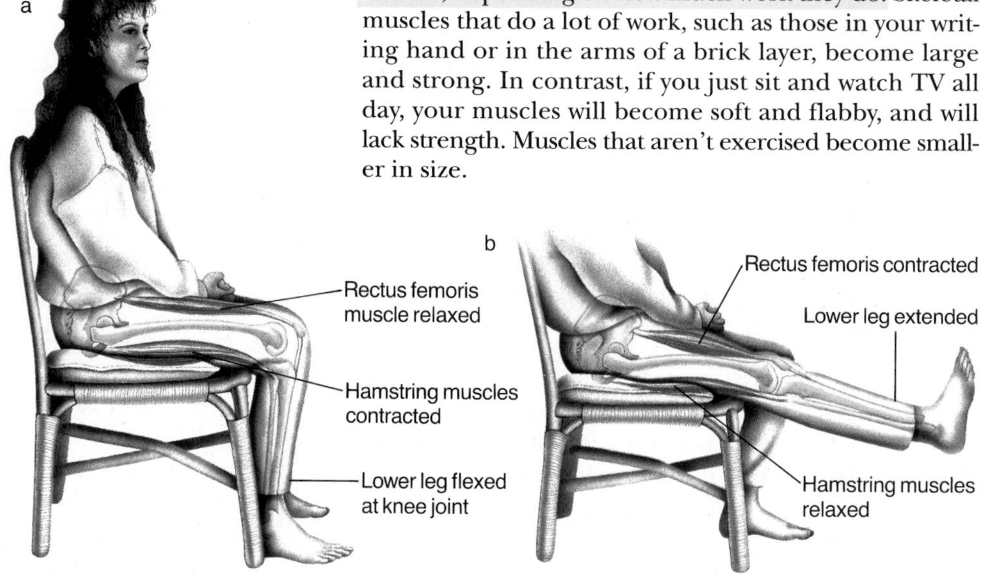

OPTIONS

ENRICHMENT
▶ Have students investigate what happens when muscles become fatigued.

PROGRAM RESOURCES

From the **Teacher Resource Package** use:

Activity Worksheets, page 164, Mini-Lab: How do muscles work in pairs?

Transparency Masters, pages 65-66, Human Muscle System.

Activity Worksheets, page 5, Flex Your Brain.

Use **Color Transparency** 33, Human Muscle System.

Use **Laboratory Manual,** page 123, Muscle Action.

PROBLEM SOLVING

High Altitude Bones

People living at high altitudes have a number of environmental stresses different from those faced by people at lower altitudes. Oxygen pressure is low and soil is poor. There are strong winds and high solar radiation. Many people live at high altitudes, but the two areas studied the most are the Andes in Peru and the Himalayas in Tibet. Individuals in both of these places have bones that grow more slowly than the bones of people living at sea level. Because their skeletal growth is slower, people living at high altitudes tend to be relatively short. They also reach maturity as much as two to four years later than people living at sea level.

Think Critically: Which two stresses mentioned here contribute to slowed skeletal growth?

Could you lift and lower your leg all day? What would it feel like after one minute? After five minutes? Your leg would probably feel tired and achy after five minutes of flexing, don't you think? Large and small muscles need to rest from time to time to be resupplied with glucose.

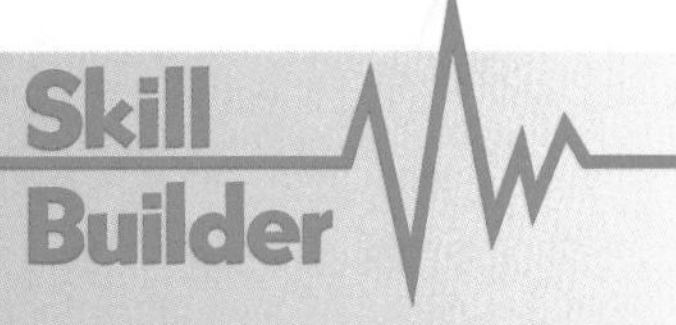

Muscle cells contain a protein called myoglobin that is similar to hemoglobin in red blood cells. Hemoglobin is a red pigment that carries oxygen. From this information, what can you *infer* about myoglobin?

SECTION REVIEW

1. What is the function of the muscular system?
2. Compare and contrast the three types of muscle.
3. What attaches a muscle to a bone?
4. **Apply:** What happens to your muscles when you bend your arm at the elbow?
5. **Connect to Physics:** What three forms of energy are involved in a muscle contraction?

Sequencing — Skill Builder

Sequence the activities that take place when you bend your leg at the knee. If you need help, refer to Sequencing in the **Skill Handbook** on page 680.

ASSESSMENT—ORAL

▶ **What is the advantage of the biceps muscle of the upper arm having one of its insertions farther away from the elbow than its other insertion?** *It provides greater mechanical advantage for the lever action.*

▶ **Why is the arrangement of smooth muscle in the walls of blood vessels circular or spiral rather than longitudinal?** *The function of the smooth muscles is to control the diameter of the vessel openings and not the length of the vessel.*

▶ **Why must the contracting activities of cardiac muscle be coordinated?** *If the contractions were at random, there would be no coordinated pumping action of the atria and ventricles of the heart.*

PROBLEM SOLVING

Think Critically: Lack of oxygen will have some effect. People in these parts of the world also experience larger hearts and lungs to compensate for lower oxygen pressure. Poor soil contributes to poor nutrition, which also can account for smaller stature.

Connect to... Chemistry

Answer: Myoglobin is a red pigment that can carry oxygen.

3 CLOSE

▶Ask questions 1-3 and the **Apply** Question in the Section Review.

SECTION REVIEW ANSWERS

1. movement
2. Skeletal and cardiac are similar because they have striations, but different because the first is voluntary and the second is involuntary. Smooth and cardiac are alike because they are involuntary but different because smooth muscle is not striated and cardiac is.
3. tendons
4. **Apply:** Muscles on the inside of the arm contract while muscles on the outside of the arm relax.
5. **Connect to Physics:** chemical, mechanical, thermal

The muscles on the underside of your thigh contract and the foot is drawn toward the thigh. At the same time the muscles on the top of the thigh relax and lengthen.

Skillbuilder

ASSESSMENT

Performance: Assess students' abilites to Sequence by having them order the activities that take place when the ankle is flexed.

SCIENCE & SOCIETY SECTION 18-3

PREPARATION

SECTION BACKGROUND

▶ Regular exercise helps enlarge muscles. It does not increase the number of muscle fibers, but rather increases the size and strength of the muscle fibers already there.

▶ An additional benefit of muscle-building activities is that they prevent atrophy, the condition that is characterized by decrease in size and strength of muscle tissues.

1 MOTIVATE

▶ Have students discuss the relative values of the "body beautiful" versus the healthy body. Analyze teen magazine articles and advertisements aimed at external beauty.

▶ Have students collect current newspaper and magazine articles about steroid use by athletes.

TYING TO PREVIOUS KNOWLEDGE: Have students recall that during adolescence physical changes are caused by hormones, one of which is testosterone.

PROGRAM RESOURCES

From the **Teacher Resource Package** use:

Science and Society, page 22, Should Doctors Prescribe Anabolic Steroids for Athletes?

18-3 Drugs for Fitness?

New Science Words

anabolic steroids

Objectives

- ▶ List side effects of anabolic steroid use.
- ▶ Discuss the use of anabolic steroids among athletes.

Do Steroids Give a "Winning Edge"?

Have you stood in front of the mirror flexing your muscles, wishing you looked like the bodybuilders you've seen on TV or in magazines? In our society, a well-built body is often admired. Athletes from junior high school students to the top professionals feel the desire to be bigger, better, or faster. They spend hours each week training and building their muscles. But some athletes feel that they aren't getting big enough fast enough. Some even turn to misuse of drugs called anabolic steroids.

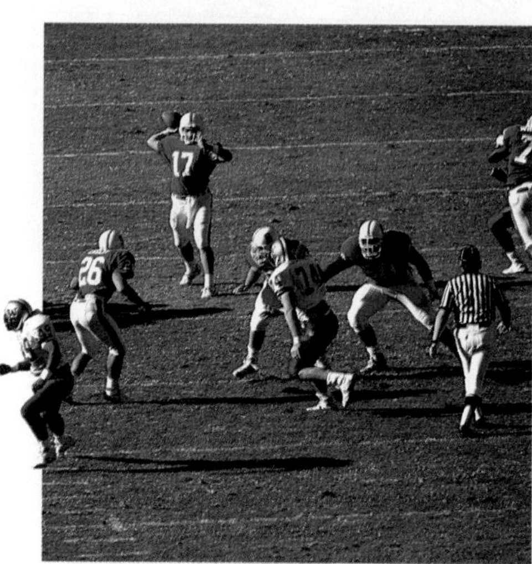

Figure 18-10. Abuse of anabolic steroids among high school athletes causes health problems.

Anabolic steroids are drugs that contain a variation of the hormone testosterone. 1 Testosterone occurs naturally in both men and women, but it reaches much higher levels in males. It produces male characteristics such as a deep voice. Anabolic steroids were originally developed in the 1930s to help sick people build muscle tissue.

Many athletes feel that if normal amounts of testosterone are good for muscle building, then increased amounts are even better. But is this true?

Athletes who use anabolic steroids may endanger their health. When women use steroids, they may grow excessive facial hair, their voices may deepen, and their breasts may shrink. In males, steroid use may cause testicles to shrink and small lumps of breast tissue to develop. Both males and females may develop hypertension and experience emotional swings. When taken in abusive amounts, some athletes have become extremely aggressive, retain excess water, and show excessive acne. When adolescents use steroids, their growth can be stunted, and that growth can't be made up in later life.

OPTIONS

Meeting Different Ability Levels

For Section 18-3, use the following **Teacher Resource Masters** depending upon individual students' needs.

◆ **Study Guide Master** for all students.

● **Reinforcement Master** for students of average and above average ability levels.

▲ **Enrichment Master** for above average students.

Additional Teacher Resource Package masters are listed in the OPTIONS box throughout the section. The additional masters are appropriate for all students.

◆ STUDY GUIDE 72

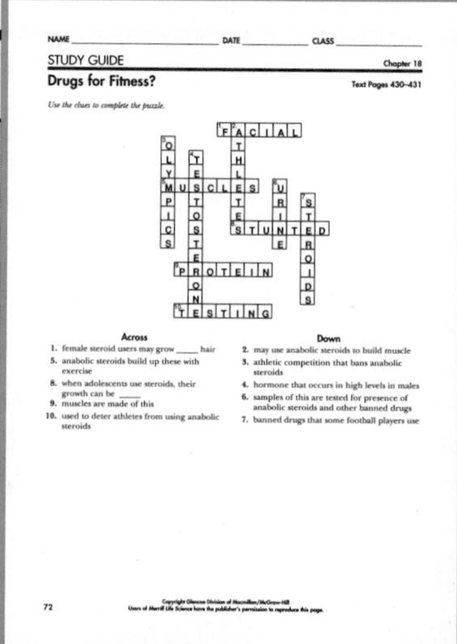

NAME ______ DATE ______ CLASS ______

STUDY GUIDE — Chapter 18

Drugs for Fitness? — Text Pages 430-431

Use the clues to complete the puzzle.

Across

1. female steroid users may grow _____ hair
5. anabolic steroids build up these with exercise
8. when adolescents use steroids, their growth can be _____
9. muscles are made of this
10. used to deter athletes from using anabolic steroids

Down

2. may use anabolic steroids to build muscle
3. athletic competition that bans anabolic steroids
4. hormone that occurs in high levels in males
6. samples of this are tested for presence of anabolic steroids and other banned drugs
7. banned drugs that some football players use

72

Is steroid use worth it? Does it give an athlete the "winning edge"?

In 1988, Canadian athlete Ben Johnson won a gold medal at the Olympic games in Seoul, Korea. He ran the 100-meter race in 9.79 seconds, a world record. Three days later Ben Johnson's gold medal and world record were taken from him. Olympic officials had found anabolic
2 steroids in his urine sample. Like most major amateur sporting events, the Olympics ban the use of steroids and other drugs. Johnson has since said that he had used steroids for about seven years.

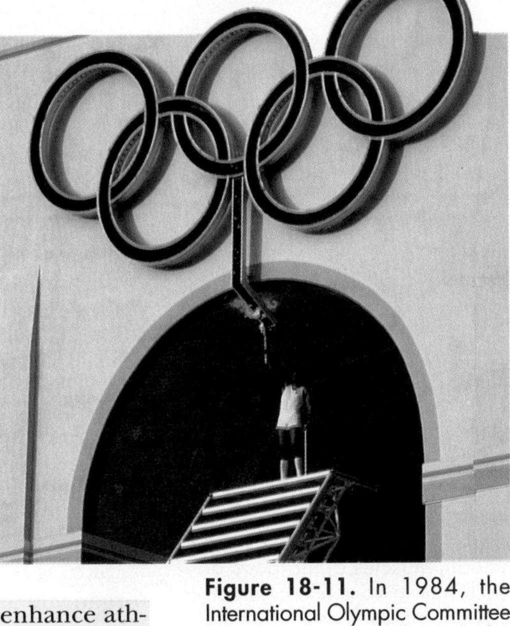

Figure 18-11. In 1984, the International Olympic Committee banned the use of anabolic steroids to maintain fairness in the events.

Some college and professional athletes are known to use anabolic steroids. Some feel strongly that when taken in prescribed doses, these drugs are safe, promote healing of damaged tissues, and enhance athletic ability. The greater concern comes from the use of these drugs by developing adolescents who may think more is better and misuse the drugs.

An additional problem becomes clear. If athletes use drugs such as anabolic steroids, is real athletic ability being demonstrated? How fair is the high school track meet if some athletes use the drugs? And what price do students pay? One researcher has warned that the drugs are too powerful not to have some damaging effects if used over a long period.

SECTION REVIEW

1. Why were anabolic steroids originally developed?
2. What are some of the side effects of anabolic steroid use?

You Decide!

Tests have been developed to detect the presence of anabolic steroids in blood and urine. Should professional and college athletes be allowed to use these drugs? What about younger athletes? Steroid abuse has been found in junior high and high school students. Should school athletes be allowed to use these drugs, even if prescribed?

● REINFORCEMENT 72

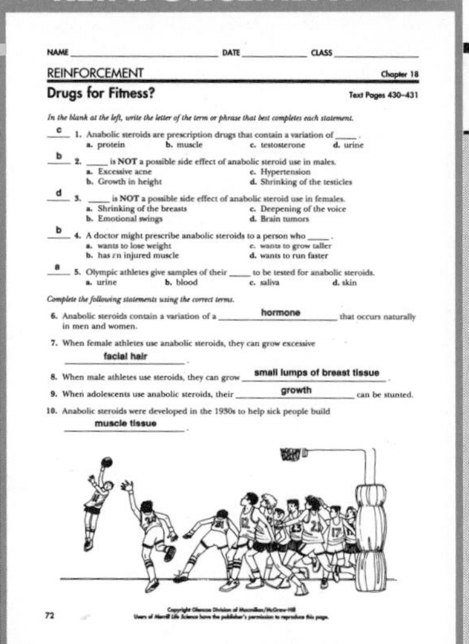

NAME ______ DATE ______ CLASS ______

REINFORCEMENT Chapter 18

Drugs for Fitness? Text Pages 430–431

In the blank at the left, write the letter of the term or phrase that best completes each statement.

c 1. Anabolic steroids are prescription drugs that contain a variation of ____.
a. protein b. muscle c. testosterone d. urine

b 2. ____ is NOT a possible side effect of anabolic steroid use in males.
a. Excessive acne c. Hypertension
b. Growth in height d. Shrinking of the testicles

d 3. ____ is NOT a possible side effect of anabolic steroid use in females.
a. Shrinking of the breasts c. Deepening of the voice
b. Emotional swings d. Brain tumors

b 4. A doctor might prescribe anabolic steroids to a person who ____.
a. wants to lose weight c. wants to grow taller
b. has an injured muscle d. wants to run faster

a 5. Olympic athletes give samples of their ____ to be tested for anabolic steroids.
a. urine b. blood c. saliva d. skin

Complete the following statements using the correct terms.

6. Anabolic steroids contain a variation of a **hormone** that occurs naturally in men and women.
7. When female athletes use anabolic steroids, they can grow excessive **facial hair**.
8. When male athletes use steroids, they can grow **small lumps of breast tissue**.
9. When adolescents use anabolic steroids, their **growth** can be stunted.
10. Anabolic steroids were developed in the 1930s to help sick people build **muscle tissue**.

72

▲ ENRICHMENT 72

NAME ______ DATE ______ CLASS ______

ENRICHMENT Chapter 18

Drugs for Fitness? Text Pages 430–431

DIFFERENT KINDS OF STEROIDS

In this lesson, you learned about anabolic steroids. Anabolic steroids enhance appetite, promote muscle growth, and help tissue repair itself. They also help relieve bone pain from osteoporosis, a disease common in older women. Doctors prescribe anabolic steroids to elderly patients and to patients who have had surgery. However, there are several other kinds of steroids.

All sex hormones are steroids. Your body produces the hormones estrogen, progesterone, and testosterone. Women have higher levels of estrogen and progesterone in their bodies. Men have higher levels of testosterone in their bodies. Both men and women produce all three hormones, but different levels of each kind produce female or male characteristics.

Cortisone and hydrocortisone are steroids often prescribed by doctors for arthritis, joint injuries, or skin rashes. These steroids reduce swelling and itching, but they also suppress your immune system and can cause kidney damage if too much is taken for too long. Hydrocortisone cream relieves itching of the skin and can be bought over the counter at your local drugstore. On the other hand, oral cortisone steroids must be taken by prescription, and injections of cortisones must be given by a doctor or nurse.

Steroids come in many different forms. You or a member of your family may have used a prescription steroid drug without even knowing. Steroids can be helpful, but most should be used with a doctor's supervision.

Answer the following questions.

1. What type of steroid might be used to treat a knee injury? **cortisone**
2. What type of steroid might be used to treat an elderly patient who had lost weight after surgery? **anabolic steroid**
3. What might a doctor prescribe to a woman with a hormone imbalance? **estrogen or progesterone**
4. What might be prescribed to a man with a hormone imbalance? **testosterone**

72

2 TEACH

Key Concepts are highlighted.

CONCEPT DEVELOPMENT

▶ Explain to students that some steroid effects may be immediate and irreversible.

CHECK FOR UNDERSTANDING

Use the Mini Quiz to check for understanding.

MINI QUIZ

Use the Mini Quiz to check students' recall of chapter content.

1. **______ are hormone drugs.** *Anabolic steroids*
2. **A(n) ______ test is often used to detect steroids in the body.** *urine*

RETEACH

Compare and contrast exercise and weight training with "quick-result" steroid use.

EXTENSION

For students who have mastered this section, use the **Reinforcement** and **Enrichment** masters or other OPTIONS provided.

3 CLOSE

▶ Ask questions 1 and 2 in the Section Review.

SECTION REVIEW QUESTIONS

1. to help sick people build muscle tissue

2. Growth can be stunted, voices deepen, hypertension, and mood swings are some effects.

YOU DECIDE!

Answers in the discussion will vary. Many will be against use. There is much controversy over their use. Young people in middle and high school are undergoing bursts of development and the question remains as to whether these substances are harmful, especially at these times.

SECTION 18-4

PREPARATION

SECTION BACKGROUND

▶ The epidermis of skin is thickest in areas where there is persistent friction, such as on the palms of the hands and the soles of the feet.

▶ Fingerprints are ridges on the surface of the skin due to the ridges in the upper surface of the dermis.

▶ Hair develops from cells in the base of the hair follicle. As the cells move upward, they die and become keratinized.

PREPLANNING

▶ To prepare for Activity 18-2, obtain the slides for skeletal, smooth, and cardiac muscles.

1 MOTIVATE

▶ Discuss with students what they believe is the largest organ of the body. Few will consider the skin. An average-sized person has 1.6 to 1.8 square meters of skin.

▶ Have students discuss the common skin disorder of acne and how proper skin care can alleviate the problem.

TYING TO PREVIOUS KNOWLEDGE: Recall the characteristics of mammals from Chapter 16. One of those characteristics is a hair-covered body. Humans lack the extensive hair covering of many other animals, but have the most hair on top of their heads.

VideoDisc

STVS: Through-the-Skin Heart Drug, Disc 7, Side 1

18-4 Skin

New Science Words

epidermis
melanin
dermis

Objectives

- ▶ **Compare and contrast the epidermis and dermis of the skin.**
- ▶ **List the functions of the skin.**
- ▶ **Discuss how skin protects the body from disease and how it heals itself.**

Did You Know?

Goose bumps result when small smooth muscles at the base of each hair in your skin contract. Each hair stands on end as the muscle pulls on the hair.

The Body's Largest Organ

Your skin, hair, nails, and millions of sweat and oil glands are part of your body's largest system, the integumentary (ihn teg yuh MENT uh ree) system.

Skin is the largest organ of your body. It is the one organ you are most familiar with, yet you may know less about your skin than about your muscles. Much of the information you receive about your environment comes through your skin. Only a few millimeters thick in most places, skin is made up of two layers of tissue, the epidermis and the dermis. The **epidermis** is the surface layer of your skin. The cells on the top of the epidermis are dead. Thousands of these cells rub off every time you take a shower, shake hands, blow your nose, or scratch your elbow. New cells are constantly produced at the bot-

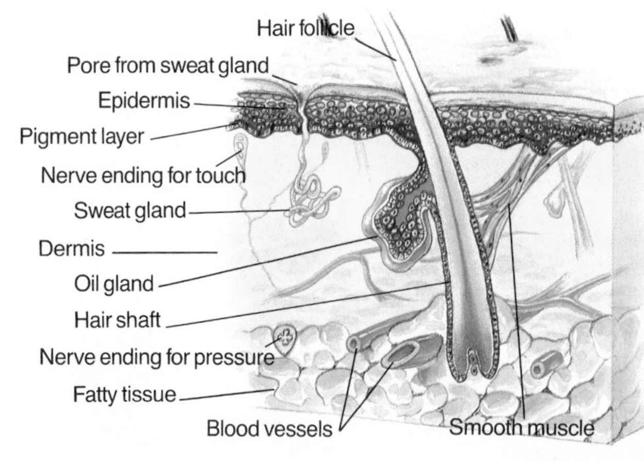

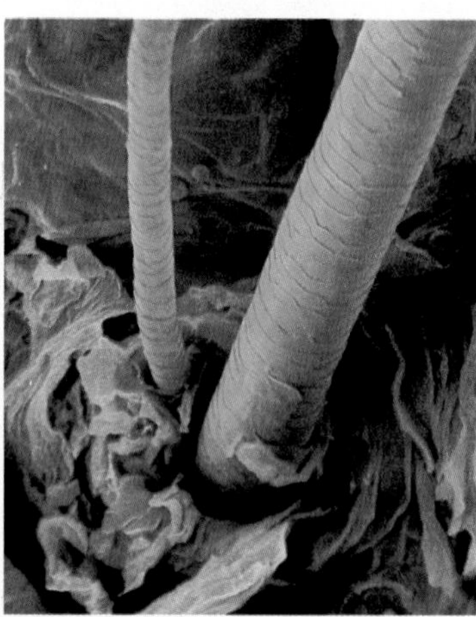

Figure 18-12. Skin is made up of two layers, the epidermis and the dermis. The dermis contains blood vessels, glands, and nerves.

OPTIONS

Meeting Different Ability Levels

For Section 18-4, use the following **Teacher Resource Masters** depending upon individual students' needs.

◆ **Study Guide Master** for all students.

● **Reinforcement Master** for students of average and above average ability levels.

▲ **Enrichment Master** for above average students.

Additional Teacher Resource Package masters are listed in the OPTIONS box throughout the section. The additional masters are appropriate for all students.

◆ STUDY GUIDE 73

NAME ______ DATE ______ CLASS ______

STUDY GUIDE — Chapter 18

Skin — Text Pages 432-435

*Cross out the statements that **DO NOT** agree with the textbook.*

1. ~~Pigment cells are in the dermis layer of the skin.~~
2. ~~There are only three sweat glands in the body.~~
3. Skin is the largest organ of the body.
4. The dermis is the layer of tissue under the epidermis.
5. The skin regulates body temperature.
6. Small amounts of vitamin D are produced in the epidermis in the presence of sunlight.
7. ~~Melanin functions as a kind of sweat gland.~~
8. ~~Skin is often 1 meter thick.~~
9. The epidermis is the surface layer of skin.
10. Thousands of your skin cells rub off every time you take a shower.
11. The more melanin you produce, the darker the color of your skin.
12. ~~No nerve endings are in the skin.~~
13. ~~The integumentary system includes the eyes and ears.~~
14. Skin excretes wastes from the body.
15. Skin serves as a protective covering for the body.

Label the two layers of skin in the diagram below.

16. epidermis
17. dermis

73

tom of the epidermis to replace the ones that are rubbed off. Cells in the epidermis produce the chemical melanin (MEL uh nun). **Melanin** is a pigment that gives your skin color. The more melanin, the darker the color of the skin. Melanin increases when your skin is exposed to the ultraviolet rays of the sun. If someone has very few melanin-producing cells, very little color gets deposited. These people have less protection from the sun. They burn more easily and may develop skin cancer more easily.

The **dermis** is the layer of tissue under the epidermis. This layer is thicker than the epidermis and contains many blood vessels, nerves, and oil and sweat glands. Notice in Figure 18-12 that under the dermis there are fat cells. This fatty tissue insulates the body. When a person gains too much weight, this is also where much of the extra fat is deposited.

The epidermis may be very thin, but injury to large areas of the skin can cause death. If the epidermis is burned away in a third degree burn, there are no cells left that can divide to replace this lost layer. As a result, nerve endings and blood vessels in the dermis are exposed. Water is lost rapidly from the dermis and muscle tissues. Body tissues are exposed to bacteria and to potential infection, shock, and death.

If injury to the epidermis is slight, a scab forms as in Figure 18-14. In scab formation, cells of the deepest layer of the epidermis can reproduce to cover the injured dermis. Later, the scab falls off on its own.

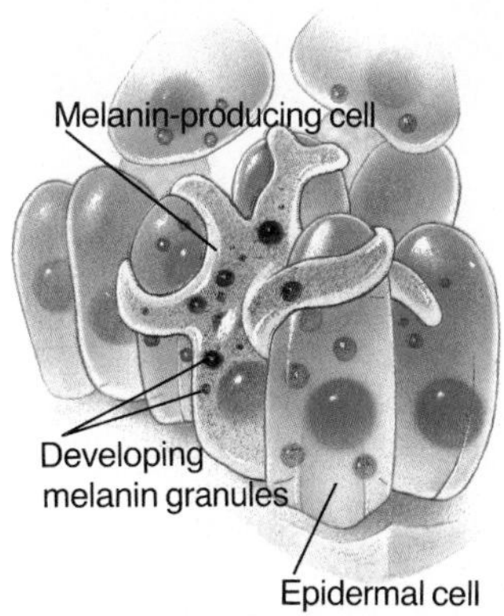

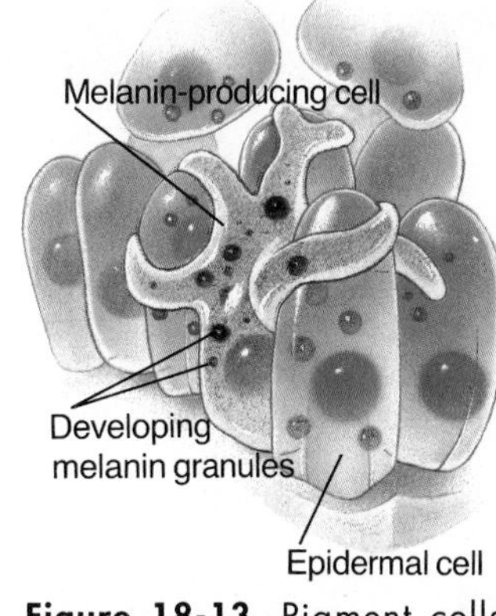
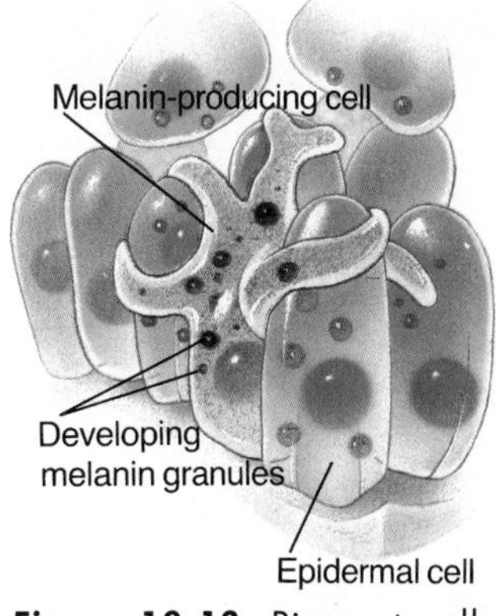

Figure 18-13. Pigment cells release melanin, which moves into other cells in the skin.

Figure 18-14. Within minutes of an injury, a loose blood clot forms (a). In a few days, new cells from the epidermis grow under the hardening clot (b). In a week, the hardened scab is ready to be pushed off (c).

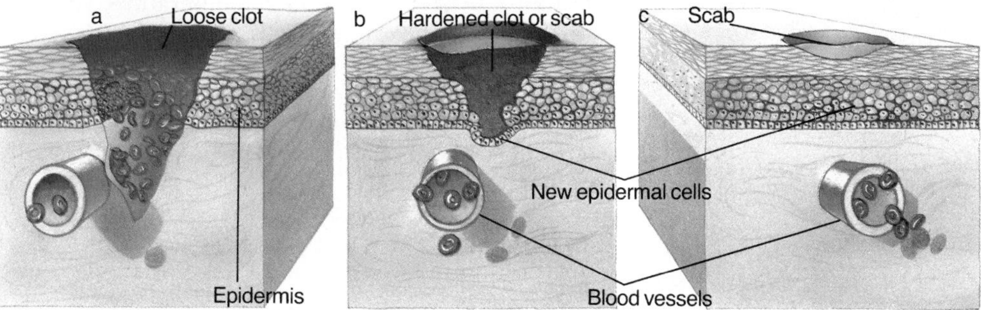

2 TEACH

Key Concepts are highlighted.

CONCEPT DEVELOPMENT

▶Refer to the diagram of the features of skin in Figure 18-12.

▶Have students use a metric ruler to see three millimeters, the approximate thickness of the skin.

▶**Demonstration:** Have students experience the cooling effect of evaporation of water from the skin. Each student should pat a wet cotton ball onto the wrist. Then have them gently blow the wet surface. The evaporation removes body heat. Relate this experience to the evaporation of perspiration from the skin during hot weather.

TEACHER F.Y.I.

▶To move from the lowest layer of the epidermis to the surface of the skin takes cells approximately 14 days. Cells on the surface remain for about thirty days before being shed.

CROSS CURRICULUM

▶**Health:** Have students find out about the formation of vitamin D in the skin and its transformation in the kidneys and liver before it is usable by the body.

● REINFORCEMENT 73

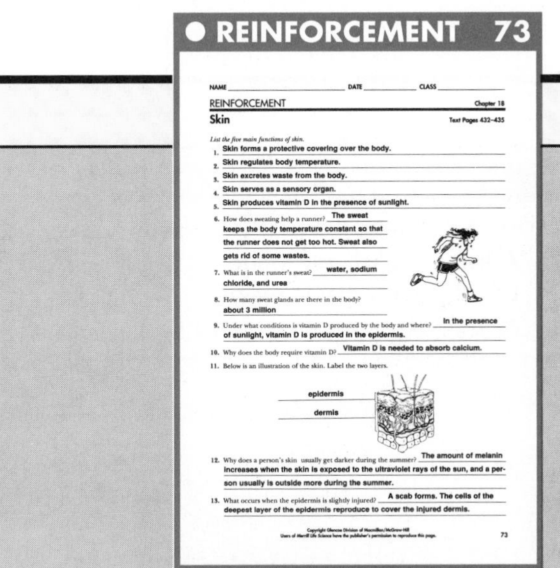

NAME ________ DATE ________ CLASS ________

REINFORCEMENT — Chapter 18

Skin — Text Pages 432–435

List the five main functions of skin.

1. **Skin forms a protective covering over the body.**
2. **Skin regulates body temperature.**
3. **Skin excretes waste from the body.**
4. **Skin serves as a sensory organ.**
5. **Skin produces vitamin D in the presence of sunlight.**
6. How does sweating help a runner? **The sweat keeps the body temperature constant so that the runner does not get too hot. Sweat also gets rid of some wastes.**
7. What is in the runner's sweat? **water, sodium chloride, and urea**
8. How many sweat glands are there in the body? **about 3 million**
9. Under what conditions is vitamin D produced by the body and where? **In the presence of sunlight, vitamin D is produced in the epidermis.**
10. Why does the body require vitamin D? **Vitamin D is needed to absorb calcium.**
11. Below is an illustration of the skin. Label the two layers.

epidermis
dermis

12. Why does a person's skin usually get darker during the summer? **The amount of melanin increases when the skin is exposed to the ultraviolet rays of the sun, and a person usually is outside more during the summer.**
13. What occurs when the epidermis is slightly injured? **A scab forms. The cells of the deepest layer of the epidermis reproduce to cover the injured dermis.**

73

▲ ENRICHMENT 73

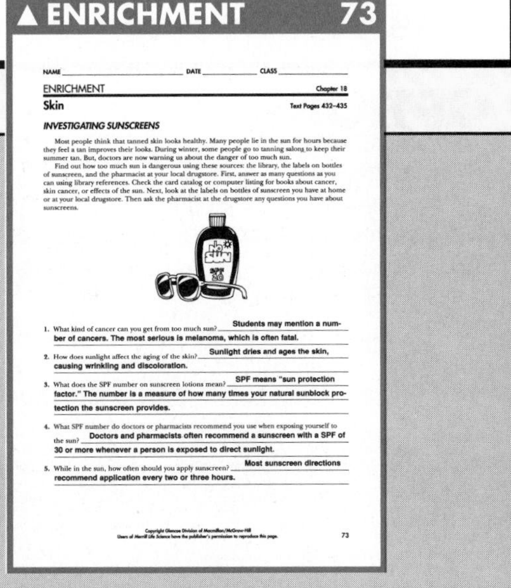

NAME ________ DATE ________ CLASS ________

ENRICHMENT — Chapter 18

Skin — Text Pages 432–435

INVESTIGATING SUNSCREENS

Most people think that tanned skin looks healthy. Many people lie in the sun for hours because they feel a tan improves their looks. During winter, some people go to tanning salons to keep their summer tan. But, doctors are now warning us about the danger of too much sun.

Find out how too much sun is dangerous using these sources: the library, the labels on bottles of sunscreen, and the pharmacist at your local drugstore. First, answer as many questions as you can using library references. Check the card catalog or computer listing for books about cancer, skin cancer, or effects of the sun. Next, look at the labels on bottles of sunscreen you have at home or at your local drugstore. Then ask the pharmacist at the drugstore any questions you have about sunscreens.

1. What kind of cancer can you get from too much sun? **Students may mention a number of cancers. The most serious is melanoma, which is often fatal.**
2. How does sunlight affect the aging of the skin? **Sunlight dries and ages the skin, causing wrinkling and discoloration.**
3. What does the SPF number on sunscreen lotions mean? **SPF means "sun protection factor." The number is a measure of how many times your natural sunblock protection the sunscreen provides.**
4. What SPF number do doctors or pharmacists recommend you use when exposing yourself to the sun? **Doctors and pharmacists often recommend a sunscreen with a SPF of 30 or more whenever a person is exposed to direct sunlight.**
5. While in the sun, how often should you apply sunscreen? **Most sunscreen directions recommend application every two or three hours.**

73

TECHNOLOGY

Think Critically: Answers will vary. The skin cannot be too heavy and must look natural.

Reference: Vogel, Shawna. "Smart Skin," *Discover*, April 1990, page 26.

CHECK FOR UNDERSTANDING

Have students infer how humans would receive information about the environment if they did not have skin.

RETEACH

Have students prepare an outline of the skin functions correlated with the skin layers (protection—epidermis and dermis; vitamin D production—epidermis, and so on).

EXTENSION

For students who have mastered this section, use the **Reinforcement** and **Enrichment** masters or other OPTIONS.

Connect to... Earth Science

Answer: Heat is not what burns skin at these altitudes. Rather, a thinner atmosphere can't block the sun's ultraviolet rays, and so the skin is exposed to more harmful radiation.

TECHNOLOGY

Robot Skin

Industrial robots are programmed to use a specific force to pick up an object. So long as they are built to handle a specific type of object, this works well. But, what if a robot had sense organs? Then it could adjust itself to doing different tasks.

Scientists are developing an artificial skin for robots that can sense the difference between a rock and a tomato and handle each correctly. The artificial skin is modeled after human skin. The outer layer is made of electrode sheets with gel between them. This layer reads the amount of pressure placed on an object. The inner layer consists of a layer of sensors between two rubber sheets. These sensors can detect very fine textures and friction.

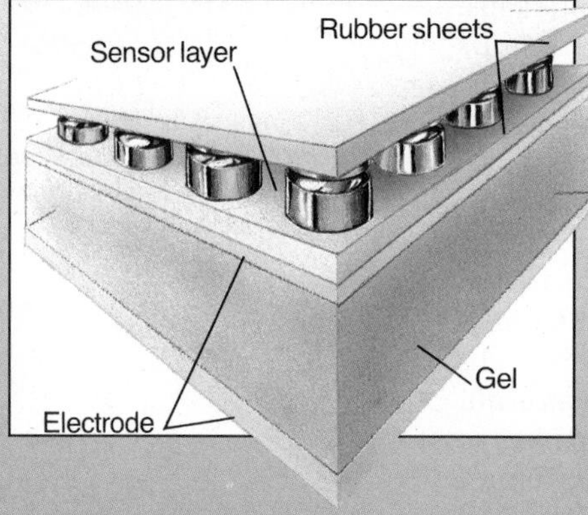

Right now, scientists are looking for a way to make the artificial skin without making it too thick.

Think Critically: Explain why artificial skin should not be too thick.

Connect to... Earth Science

Find out about the effects of ultraviolet radiation on skin. Why do mountain climbers risk becoming severely sunburned even in freezing temperatures?

Skin at Work

Your skin has five major functions. First, it forms a protective covering over the body. As a covering, it prevents injury and disease. A large bacterial population that lives on the surface works to destroy invading pathogens. Skin prevents excess water loss from body tissues.

Second, skin has an important homeostasis function because it helps regulate body temperature. Humans, unlike most fur-bearing animals, have very little hair to help them regulate body temperature. Hair is an adaptation that usually helps control body temperature. Instead, sweating cools the body. Blood vessels in the skin can help hold or release heat.

There are about three million sweat glands in the dermis. These glands help rid the body of wastes. They are also part of the body's method of keeping body temperature constant. When the body gets hot, blood vessels expand or dilate. Pores leading to sweat glands in the

OPTIONS

ASSESSMENT—ORAL

▶ **What problems arise when a large portion of a person's skin is burned?** *bacterial infection and the loss of moisture from the underlying tissues*

▶ **What causes the skin to produce "goose pimples"?** *When chilled, a small muscle in the skin pulls on the hair follicle, causing the hair to stand on end and an indentation surrounding the hair. This produces a small bump.*

PROGRAM RESOURCES

From the **Teacher Resource Package** use:

Activity Worksheets, page 165, MINI-Lab: Is there water in sweat?

Concept Mapping, pages 41-42.

Figure 18-15. The epidermis of many mammals produces much more hair than in humans.

skin open. Perspiration, or sweat, moves out onto the skin. The body then cools as sweat evaporates. This system balances heat produced by muscle contractions.

A third function of the skin is to excrete wastes. Sweat glands release water, sodium chloride, and a protein product called urea. Fourth, skin serves as a sensory organ. Nerve endings in skin pick up information about temperature, touch, and pressure. The fifth function is the formation of vitamin D. Small amounts of this vitamin are produced in the epidermis in the presence of sunlight. Vitamin D is needed to absorb calcium.

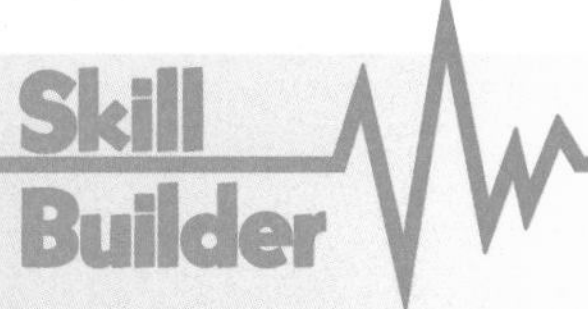

MINI-Lab

Is there water in sweat?

Note the color of a piece of cobalt chloride paper; record. With a medicine dropper, add a drop of water. *Observe* and *record* the color now. Do vigorous jumping jacks for three minutes, being sure to clap your hands. Then hold the cobalt chloride paper in your hand. What color is it now? What can you *conclude* about the distribution of sweat glands on your hands? Is there an advantage to having sweat glands on the palms of your hands?

SECTION REVIEW

1. Compare and contrast the epidermis and dermis.
2. List the five functions of skin.
3. How does skin help prevent disease in the body?
4. Explain how a scab forms.
5. **Apply:** Why is a person who has been severely burned in danger of death from loss of water?
6. **Connect to Physics:** Find out what insulation is, and why hair and fur are insulators.

Concept Mapping

Skill Builder

Make an events chain concept map to show how skin helps keep body temperature constant. If you need help, refer to Concept Mapping in the **Skill Handbook** on pages 688 and 689.

ENRICHMENT

- Have students research the function of melanocytes in the epidermis in response to ultraviolet radiation.
- Have students find out about the secretions of the apocrine glands and their relationship to body odors.

MINI-Lab

Answer: Cobalt chloride turns from blue to pink in the presence of water. Sweat contains water, thus the paper should turn pink. Sweat glands on the palms of hands increase the surface for evaporation and cooling.

MINI-Lab

ASSESSMENT

Performance: To further assess students' understanding of sweat, see USING LAB SKILLS, question 12, on p. 438.

3 CLOSE

- Ask questions 1-4 and the **Apply** Question in the Section Review.

SECTION REVIEW ANSWERS

1. The epidermis is the thin outer layer of skin. It contains melanin. The dermis is under the epidermis. It is thicker and contains many blood vessels, nerves, and sweat glands.

2. protection, regulates body temperature, excretes waste, serves as sensory organ, formation of vitamin D

3. As long as it is unbroken, skin prevents pathogens from entering the body. Bacteria on skin destroy harmful bacteria. Skin prevents water loss.

4. Loose blood clot forms; cells in epidermis grow under hardening clot; hardened scab gets pushed off.

5. Apply: because they have lost the protection for body tissues and also from loss of water

6. Connect to Physics: Insulating materials do not conduct heat. Hair and fur do not conduct heat, so body heat gets trapped between hairs and skin.

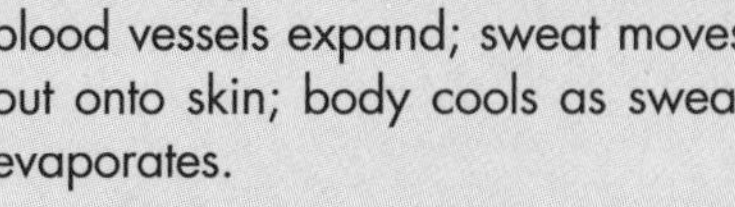

Skill Builder

Students' sequence should approximate: Body becomes hot or warm; blood vessels expand; sweat moves out onto skin; body cools as sweat evaporates.

ACTIVITY 18-2

1 class period

OBJECTIVE: Observe and **identify** parts of skeletal muscle and **describe** how muscle is attached to bone.

PROCESS SKILLS applied in this activity:
▶ **Observing** in Procedure Steps A1 and B4.
▶ **Identifying** in Procedure Steps A1, B4, and in Analyze Questions 1 and 3.
▶ **Measuring** in Procedure Step 5 and Analyze Question 2.
▶ **Inferring** in Analyze Question 4 and Conclude and Apply Questions 5 and 6.

COOPERATIVE LEARNING: Have students work as Paired Partners to complete this activity.

TEACHING THE ACTIVITY

Troubleshooting: Caution students to use probes carefully when teasing the turkey tissue. They should spread out the tissue as finely as possible and notice the tapered ends of the muscle.

ACTIVITY 18-2 Observing Muscle

Problem: *What do different types of muscle look like?*

Materials   

- prepared slides of smooth, skeletal, and cardiac muscles
- microscope
- cooked turkey leg
- dissecting pan or cutting board
- dissecting probes (2)
- hand lens

Procedure

Part A

1. Using the microscope, first on low power, and then on high power, *observe* three different types of muscle on prepared slides.
2. On a sheet of paper, draw and label each type of muscle that you observe.

Part B

3. Muscle tissue is made up of groups of cells held together in fibers, usually by a transparent covering called connective tissue. Obtain a piece of cooked turkey leg from your teacher. On the cutting board, use the two probes to tease the muscle fibers apart.
4. *Use a hand lens* to examine the muscle fibers and any connective tissue you see from the turkey leg.
5. *Draw* and *measure* five turkey leg fibers and describe the shape of these muscle fibers.

Data and Observations Sample Data

Skeletal	Cardiac			Smooth	
Student drawings should resemble photos on p 427.					
Length of fibers	Lengths will vary. 1 cm	2.5 cm	1.5 cm	0.75 cm	1 cm
Fibers are long, tapered at ends.					

Analyze

1. Which muscles on the slides have stripes, or striations?
2. How long are the muscle fibers in the turkey leg?
3. What type of muscles are there in a turkey leg?
4. How are muscle fibers arranged in the prepared slides and in the turkey leg?

Conclude and Apply

5. *Predict* how the shape of a muscle fiber *relates* to its function.
6. Can you *conclude* that striations have anything to do with whether a muscle is voluntary or involuntary? Explain.

ANSWERS TO QUESTIONS

1. skeletal and cardiac muscles
2. Lengths will vary from 1 to 5 cm or more.
3. skeletal muscles
4. Accept all reasonable descriptions. Most will reply that the fibers lay side by side along the length of the bone.
5. The fiber is long, thicker in the center than it is at the ends. When a muscle contracts, it becomes even thicker in the center, but the ends stay about the same.
6. No. Both cardiac and skeletal are striated, but cardiac is involuntary and skeletal is voluntary.

PROGRAM RESOURCES

From the **Teacher Resource Package** use:
Activity Worksheets, pages 160-161, Activity 18-2, Observing Muscle.

Activity ASSESSMENT

Performance: To further assess students' understanding of muscle, have them examine a piece of turkey heart.

CHAPTER
REVIEW

SUMMARY

18-1: The Skeletal System

1. Bones are living structures that protect, support, make blood cells, store minerals, and provide for muscle attachment.
2. Broken bones heal when the broken ends contact and a thick band of cells produced by the periosteum forms.
3. Movable joints move freely and are pivot, hinge, ball-and-socket, or gliding in type. The skull and pelvic joints in adults do not move and are classified as immovable.

18-2: The Muscular System

1. Muscle contracts to move bones and body parts.
2. Skeletal muscle is voluntary and moves bones. Smooth muscle is involuntary and controls movement in internal organs. Cardiac muscle is involuntary and located only in the heart.
3. Muscles contract to move body parts. Muscles pull, never push, body parts. Skeletal muscles work in pairs. When one contracts, the other relaxes.

18-3: Science and Society: Drugs for Fitness?

1. Anabolic steroids build muscle, but the effects of abuse may be threatening to health.
2. Use of anabolic steroids is controversial.

18-4: Skin

1. The epidermis is the outer layer with dead cells on the surface. Bottom layers of epidermis give rise to new skin cells. Melanin is found in the inner layers. The dermis is the inner layer with hair follicles, nails, nerves, sweat and oil glands, and blood vessels.
2. Protection, water retention, formation of vitamin D, and helping with body temperature are the skin's jobs.
3. Undamaged epidermis destroys bacteria and can form protective scabs.

KEY SCIENCE WORDS

a. **anabolic steroids**
b. **cardiac muscle**
c. **cartilage**
d. **dermis**
e. **epidermis**
f. **fracture**
g. **immovable joint**
h. **involuntary muscles**
i. **joint**
j. **ligament**
k. **marrow**
l. **melanin**
m. **movable joint**
n. **muscle**
o. **periosteum**
p. **skeletal muscles**
q. **skeletal system**
r. **smooth muscles**
s. **tendons**
t. **voluntary muscles**

UNDERSTANDING VOCABULARY

Match each phrase with the correct term from the list of Key Science Words.

1. tough outer covering of bone
2. internal body framework
3. tissue or organ that contracts
4. a broken bone
5. voluntary muscles
6. involuntary heart muscle
7. outer layer of skin
8. skin pigment
9. holds bones together
10. attach muscles to bones

OPTIONS

ASSESSMENT

To assess student understanding of material in this chapter, use the resources listed.

COOPERATIVE LEARNING

Consider using cooperative learning in the THINK AND WRITE CRITICALLY, APPLY, and MORE SKILL BUILDERS sections of the Chapter Review.

PROGRAM RESOURCES

From the **Teacher Resource Package** use:
Chapter Review, pages 39-40.
Chapter and Unit Tests, pages 122-125, Chapter Test.

CHAPTER
REVIEW

SUMMARY

Have students read the summary statements to review the major concepts of the chapter.

UNDERSTANDING VOCABULARY

1. o
2. q
3. n
4. f
5. p
6. b
7. e
8. l
9. j
10. s

ASSESSMENT

Portfolio

Encourage students to place in their portfolios one or two items of what they consider to be their best work. For each item, ask students to explain why that item was chosen and what they learned from it. Items might be selected from the following.

- For Your Gifted Students model of the human skeleton, p. 418
- In Your Journal report on orthroscopic surgery, p. 423
- Skillbuilder outline on bones, p. 424

Performance

Additional performance assessments may be found in *Performance Assessment* and *Science Integration Activities* that accompany **Merrill Life Science.** Performance Task Assessment Lists and rubrics for evaluating these activities and other products generated throughout the chapter can be found in Glencoe's *Performance Assessment in Middle School Science.*

CHAPTER

REVIEW

CHECKING CONCEPTS

1. a
2. d
3. a
4. a
5. d
6. b
7. c
8. c
9. d
10. c

USING LAB SKILLS

ASSESSMENT

Use these alternate lab exercises to assess students' understanding of skills used in this chapter.

11. Bones in flying birds are more hollow and lighter in weight.

12. You may wish to have only volunteers do this alternate activity. They can report their results to the class. Do not have students apply the cobalt chloride paper for very long because the number of sweat glands on the soles of the feet is large (about 3000 per square inch) and the result will be hard to detect.

THINK AND WRITE CRITICALLY

13. Ligaments connect bones to bones; tendons connect muscles to bones.

14. Blood supply to bone might be cut off; broken bones would have a difficult time healing or would not heal at all.

15. Cartilage allows a joint to operate smoothly. When the cartilage is damaged or destroyed, the harder inner parts of the bones rub together, the movement is rough, and irritation of that joint results in pain and inflammation.

16. Accept all reasonable responses. Use of anabolic steroids can cause an athlete to mature earlier, or build muscle mass, thus being more of an asset for the team. All coaches want a great team. Coaches also know, however, that use of those steroids can cause lasting problems for athletes, thus shortening their careers.

17. Failure to sweat when the body is overheated can result in heatstroke, which can lead to death.

CHAPTER

REVIEW

CHECKING CONCEPTS

Choose the word or phrase that completes the sentence.

1. ______ bone is the most solid form of bone.
- **a.** Compact **c.** Spongy
- **b.** Periosteum **d.** Marrow

2. Blood cells are made in the ______.
- **a.** compact bone **c.** cartilage
- **b.** periosteum **d.** marrow

3. Minerals are stored in ______.
- **a.** bone **c.** muscle
- **b.** skin **d.** blood

4. ______ covers the ends of bones.
- **a.** Cartilage **c.** Ligaments
- **b.** Tendons **d.** Muscle

5. Immovable joints are found ______.
- **a.** at the elbow **c.** in the wrist
- **b.** at the neck **d.** in the skull

6. The knees and fingers are examples of a ______ joint.
- **a.** pivot **c.** gliding
- **b.** hinge **d.** ball-and-joint

7. Vitamin ______ is made in the skin.
- **a.** A **c.** D
- **b.** B **d.** K

8. Dead cells are found on the ______.
- **a.** dermis **c.** epidermis
- **b.** marrow **d.** periosteum

9. Anabolic steroids are used to promote ______ growth.
- **a.** blood **c.** nerve
- **b.** bone **d.** muscle

10. ______ helps retain fluids in the body.
- **a.** Bone **c.** Skin
- **b.** Muscle **d.** A steroid

USING LAB SKILLS

11. In the Activity on page 425 you observed the parts of a long bone from a mammal. Obtain a long bone from a flying bird such as a turkey, duck, or goose and compare it to the beef bone. What noticeable modifications for flight are present in the bird bone?

12. In the MINI-Lab on page 435 you noted the water in secretions from the sweat glands. Use the cobalt chloride paper to determine the distribution of sweat glands on the soles of your feet.

THINK AND WRITE CRITICALLY

Answer the following questions in your Journal using complete sentences.

13. Distinguish between the functions of ligaments and tendons.

14. What would result if a person had a disease of the periosteum?

15. In arthritis, cartilage is frequently damaged. Why does movement then become painful?

16. From a coach's point of view, why are there advantages and disadvantages to use of anabolic steroids?

17. Predict what would happen if a person's sweat glands didn't produce sweat.

APPLY

18. When might skin not be able to produce enough vitamin D?
19. What effects do sunblocks have on melanin?
20. What would lack of calcium do to bones?
21. Using a microscope, how could you distinguish among the three muscle types?
22. What function of skin in your lower lip changes when a dentist gives you novocaine for a filling in your bottom teeth? Why?

MORE SKILL BUILDERS

If you need help, refer to the Skill Handbook.

1. **Hypothesizing:** Make a hypothesis to explain why a person who "slips a disc" is in pain.
2. **Designing an Experiment:** Design an experiment to compare the heartbeat of athletes and nonathletes in your class.
3. **Hypothesizing:** Make a hypothesis about the distribution of sweat glands throughout the body. Are they evenly distributed?
4. **Concept Mapping:** Complete the events chain concept map to describe scab formation.

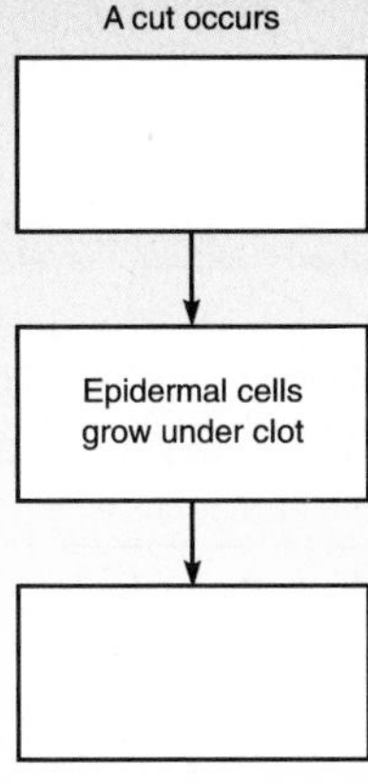

A cut occurs
Epidermal cells grow under clot

5. **Observing and Inferring:** The joints in the skull of a newborn baby are soft, whereas those of a 17-year-old are tightly grown together. Infer why the infant's skull joints are soft.

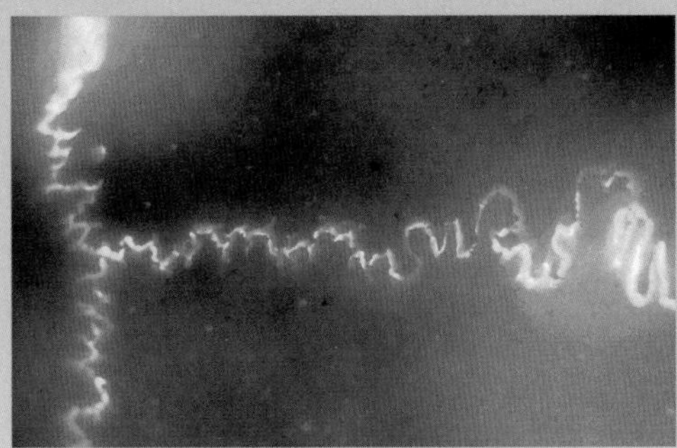

PROJECTS

1. Find out the differences among first, second, and third degree burns. A local hospital's burn unit or fire department are sources of information on burns. **In your Journal,** draw and describe the differences.
2. How do tanning booths give you a suntan? As a class project, research the safety of this equipment. Present class results and assemble a safety pamphlet that reflects concepts from this chapter.

APPLY

18. If there is not enough exposure to sunlight, not enough vitamin D would be produced.
19. Sunblocks "mask" the skin cells and prevent full penetration of the sun's rays to the melanin of these cells.
20. Bones lacking calcium become too flexible and misshapen.
21. Smooth muscle is made of individual, spindle-shaped cells in layers; skeletal muscle is made of striped fibers; cardiac muscle is a branching, fiber-like structure.
22. Novocaine numbs the nerves and your skin can no longer sense stimuli. Novacaine numbs nerves in the dermis.

MORE SKILL BUILDERS

1. Hypothesizing: The disc prevents the vertebrae from rubbing against each other; if it is out of place, there is friction of bone-on-bone and nerves are also stimulated.
2. Designing an Experiment: The experiment should involve comparing both resting and exercising heart rates.
3. Hypothesizing: Sweat glands are not evenly distributed on the skin. This is known because, when exercising, sweat is found in certain areas (face, underarms) and not others.
4. Concept Mapping:

A cut occurs

Loose blood clot forms
↓
Epidermal cells grow under clot
↓
Scab ready to be pushed off

5. Observing and Inferring: Accept all reasonable answers. Softer joints are generally thought to be helpful during birth.

CHAPTER 19 Nutrients and Digestion

CHAPTER SECTION	OBJECTIVES	ACTIVITIES
19-1 Nutrition (3 days)	1. **List** the six classes of nutrients. 2. **Describe** the importance of each type of nutrient. 3. **Explain** the relationship between diet and health.	**MINI-Lab:** *How much water?* p. 447 **Activity 19-1:** *Identifying Vitamin C Content,* p. 450
19-2 Your Digestive System (3 days)	1. **Distinguish** between mechanical and chemical digestion. 2. **Name** the organs of the digestive system and **describe** what takes place in each. 3. **Explain** how homeostasis is maintained in digestion.	**MINI-Lab:** *What is the advantage of a rough inner lining?* p. 452
19-3 Eating Disorders Science & Society (1 day)	1. **Name** and **describe** two types of eating disorders. 2. **Infer** the consequences of improper diets associated with eating disorders.	**Activity 19-2:** *Designing an Experiment (Protein Digestion),* p. 458
Chapter Review		

ACTIVITY MATERIALS

FIND OUT	ACTIVITIES		MINI-LABS	
Page 441 paper pencil	**19-1 Identifying Vitamin C Content, p. 450** indophenol solution graduated cylinder glass marking pencil 10 test tubes test-tube rack 10 dropping bottles containing water, orange, pineapple, apple, lemon, tomato, cranberry, carrot, lime, and mixed vegetable juices	**19-2 Designing an Experiment, p. 458** unflavored gelatin dropper 2 test tubes in rack drinking glass pepsin powder cold water graduated cylinder marking pen dilute hydrochloric acid	**How much water? p. 447** pan balance 200-mL beaker celery or carrots tray oven mitt oven	**What is the advantage of a rough inner lining? p. 452** smooth poster board corrugated cardboard water

CHAPTER FEATURES	TEACHER RESOURCE PACKAGE	OTHER RESOURCES
Technology: *Fake Fat,* p. 445 **Problem Solving:** *The Big Race,* p. 449 **Skill Builder:** *Making and Using Tables,* p. 449	**Ability Level Worksheets** ◆ ***Study Guide,*** p. 74 ● ***Reinforcement,*** p. 74 ▲ ***Enrichment,*** p. 74 **Critical Thinking/Problem Solving,** p. 23 **Cross-Curricular Connections,** p. 23 **Activity Worksheets,** pp. 167-168, 173	**Lab Manual:** Testing for Carbohydrates, p. 127; Testing for Proteins, p. 131; Digestion of Fats, p. 133 **STVS:** Disc 7, Side 2
Skill Builder: *Observing and Inferring,* p. 455	**Ability Level Worksheets** ◆ ***Study Guide,*** p. 75 ● ***Reinforcement,*** p. 75 ▲ ***Enrichment,*** p. 75 **Concept Mapping,** p. 43 **Activity Worksheets,** p. 174 **Transparency Masters,** pp. 67-70	**Color Transparency 34,** Human Digestive System **Color Transparency 35,** Digestive Organs **Science Integration Activity 19**
You Decide! p. 457	**Ability Level Worksheets** ◆ ***Study Guide,*** p. 76 ● ***Reinforcement,*** p. 76 ▲ ***Enrichment,*** p. 76 **Science and Society,** p. 23 **Activity Worksheet,** pp. 169-170	**STVS:** Disc 7, Side 2
Summary Key Science Words Understanding Vocabulary Checking Concepts Using Lab Skills Think & Write Critically Apply More Skill Builders Projects	**ASSESSMENT RESOURCES** **Chapter Review,** pp. 41-42 **Chapter Test,** pp. 126-129 **Performance Assessment in Middle School Science (PAMSS)**	**Chapter Review Software** **Test Bank** **Alternate Assessment** **Performance Assessment**

◆ **Basic** ● **Average** ▲ **Advanced**

ADDITIONAL MATERIALS

SOFTWARE	AUDIOVISUAL	BOOKS/MAGAZINES
Digestion, J&S Software. *Explorer's Digest,* Micro-Ed, Inc. *Let's Eat,* Focus. *Easy Search: Diet Detective,* Focus. *The Digestion Simulator,* Focus.	*Biology: Nutrition Research For a Healthier Life,* film, Science Screen Report. *Diet for All Reasons,* film, Churchill. *Digestion: Chemical Changes,* film, Lucerne. *Human Body: Nutrition and Metabolism,* film, Coronet/MTI. *The Digestive System,* film, EBEC. *Nutrition: Foods, Fads, Frauds, Facts,* filmstrip, Guidance Associates.	Fekete, Irene and Peter D. Ward. *Your Body. World of Science Series.* New York: Facts On File, Inc.

CHAPTER

19 NUTRIENTS AND DIGESTION

THEME DEVELOPMENT: Energy is a major theme in this chapter. Nutrients contain chemical energy. Complex food compounds become simpler compounds that the body cells can use for metabolism. Chemical energy is transformed into heat and mechanical energy.

CHAPTER OVERVIEW

- **Section 19-1:** This section focuses on the nutrients required by the body to carry on the activities of life. The six classes of nutrients and their functions within the body are also described.
- **Section 19-2:** The process of digestion is introduced in this section. Each of the major organs of the digestive system is described and its specific function is detailed.
- **Section 19-3: Science and Society:** Two major types of eating disorders are defined and the consequences of improper diets are explored. The You Decide feature asks students to consider how to help a friend who they think has an eating disorder.

CHAPTER VOCABULARY

nutrients
carbohydrates
proteins
amino acids
fats
vitamins
minerals
food group
digestion
mechanical digestion
chemical digestion
saliva
peristalsis
chyme
villi
anorexia nervosa
bulimia

CHAPTER

19 Nutrients and Digestion

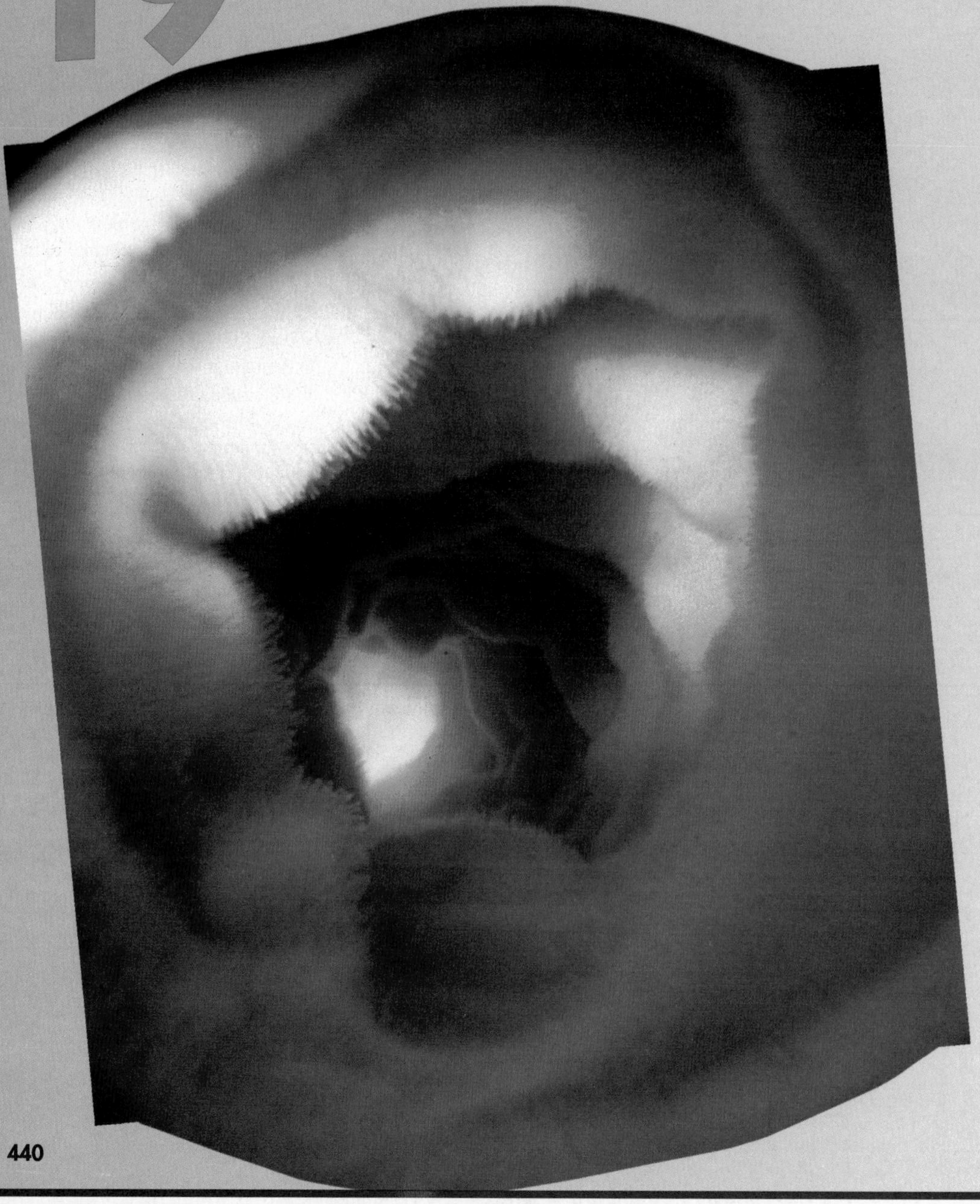

440

OPTIONS

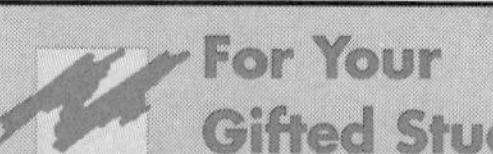

For Your Gifted Students

Have students write a story from the point of view of "the engineer in charge of the digestive system." Have this "engineer" describe "a typical day at work," describing what happens when food or drink enters the mouth. The "engineer" will discuss the role of all the digestive system "employees" and how they work together to be successful. The "engineer" may also be able to relate how eating disorders affect the digestive system.

For Your Mainstreamed Students

Students can discover how the lack of a sense of smell or sight affects the sense of taste. Have students wear a blindfold as they taste bits of apples, pears, and potatoes (things with similar texture). They should report what they taste. Students may try the same experiment, this time using a nose plug.

Food that you eat is broken down and absorbed in the soft tissues of your small intestine as shown in the photograph on page 440. Energy from this food is measured in kilocalories. You know these simply as calories. How many calories do you use each day?

FIND OUT!

Calculate the rate at which your body uses energy.

Males use about 1.0 calorie per kilogram (kg) of body mass per hour. Females burn 0.9 calorie per kg of body mass per hour.

1. Convert your mass in pounds to kilograms.
   Pounds ÷ 2.2 lbs/kg = kg
2. Multiply your mass by either 1.0 or 0.9 based on your sex. Your answer is how much energy (cal) you use per hour.
3. Multiply the calories used in one hour by 24 (hours per day). Your answer is the number of calories you burn per day.

Gearing Up

Previewing the Chapter

Use this outline to help you focus on important ideas in this chapter.

Previewing Science Skills

- ▶ In the Skill Builders, you will make and use tables and observe and infer.
- ▶ In the Activities, you will predict, experiment, and collect data.
- ▶ In the MINI-Labs, you will measure in SI and make models.

What's next?

Food that you eat each day gives you the energy you need to live. Find out how your body processes food that supplies you with this energy.

441

INTRODUCING THE CHAPTER

Use the Find Out activity to introduce students to the digestive system. Inform students that they will be learning about nutrients and the digestive system as they read the chapter. Explain that the chapter-opening photograph is a view of the inside of the small intestine. The soft appearance is the result of thousands of fingerlike structures through which nutrients get absorbed.

FIND OUT!

Materials: paper and pencil, or calculator

Teaching Tips

▶ Students may need to convert their mass in pounds to kilograms. Use this formula: Divide the number of pounds by 2.2 to get the number of kilograms.

▶ Have students calculate the percent of their daily caloric needs that a meal of a hamburger, french fries, and soft drink (approximately 650 calories) provides.

▶ Explain that in the United States, most balances in doctor's offices and fitness centers still register mass in pounds (the English system) rather than in kilograms (metric).

Gearing Up

Have students study the Gearing Up feature to familiarize themselves with the chapter. Discuss the relationships of the topics in the outline.

What's Next?

Before beginning the first section, make sure students understand the connection between the Find Out activity and the topics to follow.

ASSESSMENT OPTIONS

PORTFOLIO

Refer to page 459 for suggested items that students might select for their portfolios.

PERFORMANCE ASSESSMENT

See page 459 for Additional Performance Assessment options.

Process

Skillbuilders, p. 449
MINI-Lab, pp. 447, 452
Activities, 19-1, p. 450; 19-2, p. 458
Using Lab Skills, p. 460

CONTENT ASSESSMENT

Assessment—Oral, pp. 444, 454
Section Reviews, pp. 449, 455, 457
Chapter Review, pp. 459-461
Mini Quizzes, pp. 448, 454

GROUP ASSESSMENT

Opportunities for group assessment occur with Cooperative Learning Strategies, Flex Your Brain Activities, and the Science and Society article.

SECTION 19-1

PREPARATION

SECTION BACKGROUND

▶ Nutrients not immediately required for body activities are stored in cells in the form of glycogen or fat.

▶ The unique sequence of amino acids in a protein determines the shape of the molecule, and the shape of the protein molecule determines its function.

▶ Vitamins A, D, E, and K are fat-soluble and are absorbed with the fats into cells.

PREPLANNING

▶ To prepare for Activity 19-1, obtain the indophenol solution and the fruit and vegetable juices needed.

1 MOTIVATE

▶ Have students read the labels of canned and packaged foods. Discuss the major sources of various nutrients in the foods they eat each day.

▶ Have the school lunchroom manager or a hospital dietician discuss with students how a balanced diet can be achieved with a variety of foods.

Cooperative Learning: Assign Problem Solving Teams to find food to which minerals and vitamins have been added to make them more healthful.

TYING TO PREVIOUS KNOWLEDGE: Have students discuss what is meant by a "balanced diet."

MULTICULTURAL AWARENESS

Amaranth, pictured in Figure 19-2, is a source of protein in South America and Africa. Discuss different grains that are staples for various cultures.

VideoDisc

STVS: Measuring Body Fat, Disc 7, Side 2

19-1 Nutrition

New Science Words

nutrients
carbohydrates
proteins
amino acids
fats
vitamins
minerals
food group

Objectives

- ▶ **List the six classes of nutrients.**
- ▶ **Describe the importance of each type of nutrient.**
- ▶ **Explain the relationship between diet and health.**

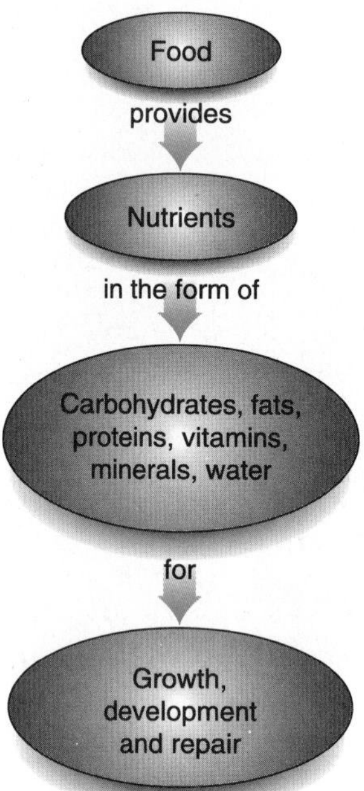

Figure 19-1. Food from a variety of sources provides your body with six types of nutrients.

Why Do You Eat?

While eating cereal for breakfast, you've probably read all the ads and offers on the box. You may also have noticed the list of nutrients on the side panel. By law, the amount of each nutrient in the cereal has to be listed for you on the label. You think you're eating toasted corn with raisins, but you're really taking in nutrients. **Nutrients** 1 are substances in foods that provide energy and materials for cell development, growth, and repair.

There are six kinds of nutrients available in food: carbohydrates, proteins, fats, vitamins, minerals, and water. Carbohydrates, proteins, vitamins, and fats are all organic nutrients. In contrast, minerals and water are inorganic. They do not contain carbon. Foods containing carbohydrates, fats, and proteins are usually too complex to be absorbed right away by your body. These substances need to be broken down into simpler molecules before the body can make use of them. In contrast, minerals and water can be absorbed directly into your bloodstream. They don't require digestion or breakdown.

OPTIONS

Meeting Different Ability Levels

For Section 19-1, use the following **Teacher Resource Masters** depending upon individual students' needs.

◆ **Study Guide Master** for all students.

● **Reinforcement Master** for students of average and above average ability levels.

▲ **Enrichment Master** for above average students.

Additional Teacher Resource Package masters are listed in the OPTIONS box throughout the section. The additional masters are appropriate for all students.

◆ **STUDY GUIDE 74**

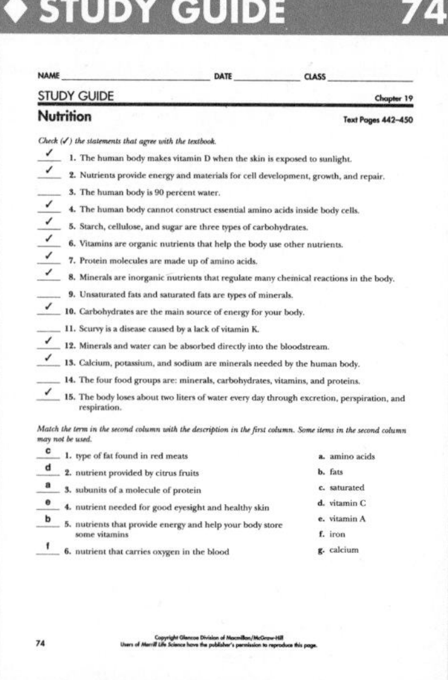

NAME ______ DATE ______ CLASS ______

STUDY GUIDE — Chapter 19

Nutrition — Text Pages 442–450

Check (✓) the statements that agree with the textbook.

✓ 1. The human body makes vitamin D when the skin is exposed to sunlight.
✓ 2. Nutrients provide energy and materials for cell development, growth, and repair.
___ 3. The human body is 90 percent water.
✓ 4. The human body cannot construct essential amino acids inside body cells.
✓ 5. Starch, cellulose, and sugar are three types of carbohydrates.
✓ 6. Vitamins are organic nutrients that help the body use other nutrients.
✓ 7. Protein molecules are made up of amino acids.
✓ 8. Minerals are inorganic nutrients that regulate many chemical reactions in the body.
___ 9. Unsaturated fats and saturated fats are types of minerals.
✓ 10. Carbohydrates are the main source of energy for your body.
___ 11. Scurvy is a disease caused by a lack of vitamin K.
✓ 12. Minerals and water can be absorbed directly into the bloodstream.
✓ 13. Calcium, potassium, and sodium are minerals needed by the human body.
___ 14. The four food groups are: minerals, carbohydrates, vitamins, and proteins.
✓ 15. The body loses about two liters of water every day through excretion, perspiration, and respiration.

Match the term in the second column with the description in the first column. Some items in the second column may not be used.

c	1. type of fat found in red meats	a. amino acids
d	2. nutrient provided by citrus fruits	b. fats
a	3. subunits of a molecule of protein	c. saturated
e	4. nutrient needed for good eyesight and healthy skin	d. vitamin C
b	5. nutrients that provide energy and help your body store some vitamins	e. vitamin A
f	6. nutrient that carries oxygen in the blood	f. iron
		g. calcium

74

Carbohydrates

If you look at the panels on several boxes of cereal, you'll notice that carbohydrates are frequently listed first. That means that most of the nutrient in that particular cereal is in carbohydrate form. **Carbohydrates** are the main sources of energy for your body. They contain carbon, hydrogen, and oxygen atoms. During respiration, energy is released when molecules of carbohydrate break down in your cells. Starch, cellulose, and sugar are three
2 types of carbohydrates. Starch and cellulose are complex carbohydrates. Starch is in foods such as potatoes and those made from grains such as pasta. Cellulose occurs in plant cell walls. There are many types of sugars. You're probably most familiar with one called table sugar. Table sugar is an example of a simple carbohydrate. Fruits, honey, and milk are sources of sugar. Your cells use sugar in the form of glucose.

There are many kinds of organic compounds, but only carbohydrates have a ratio of two hydrogen atoms to one oxygen atom. What would be the chemical formula for the most simple carbohydrate?

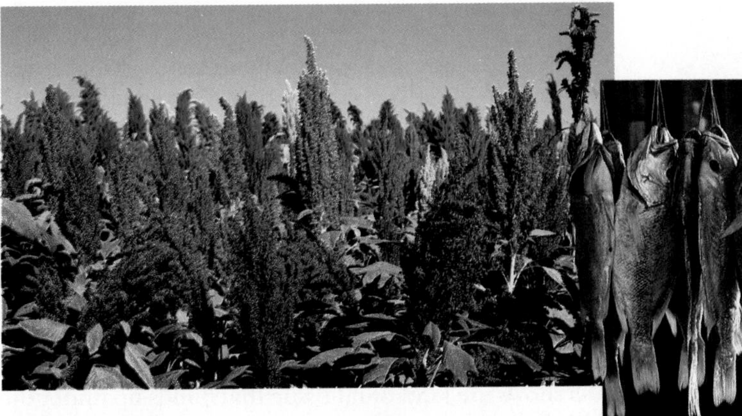

Figure 19-2. Grains and fish are good sources of protein. The grain shown here is amaranth, grown in the tropics.

Proteins

Proteins are nutrients needed throughout your body for growth, as enzymes, and in the replacement and repair of body cells. Proteins are large molecules that contain carbon, hydrogen, oxygen, and nitrogen. A molecule of protein is made up of a large number of subunits or building blocks called **amino acids.** In Chapter 4 you learned that proteins are made according to directions supplied by genes that you inherit. Your body needs 20 different amino acids to be able to construct the proteins needed in your cells. Most of these amino acids can be made in your cells. The eight remaining amino acids are called *essential* amino acids. Your body doesn't have genetic instructions to construct

● REINFORCEMENT 74

NAME ______ DATE ______ CLASS ______

REINFORCEMENT — Chapter 19

Nutrition — Text Pages 442–450

Name the six nutrients available in food and briefly describe each.

1. **carbohydrates** — **These nutrients contain C, H, and O atoms. Carbohydrates are the main source of energy.**
2. **proteins** — **Proteins are made up of amino acids. Proteins are used for growth, replacement, and repair.**
3. **fats** — **Fats provide energy and help the body store some vitamins.**
4. **vitamins** — **Vitamins are essential, organic nutrients that help the body use other nutrients.**
5. **minerals** — **Minerals are inorganic nutrients that regulate many chemical reactions in the body.**
6. **water** — **Water enables the chemical reactions to take place in the body and removes waste products from the cells.**

Answer the following questions.

7. What foods supply saturated fats and unsaturated fats? **Red meats supply saturated fats and plant foods supply unsaturated fats.**
8. How much of a human body is water? **about 60 percent**
9. What are the four food groups? **the milk group, the fruit and vegetable group, the grain group, and the meat group**
10. Which food group is represented by pizza? **Pizza is an example of a food group combination.**

Use Tables 19:1, 19:2, and 19:3 in your textbook to answer the following questions.

11. Which minerals help make strong bones? **calcium and phosphorus**
12. Which vitamin is essential for good eyesight? **vitamin A**
13. In what way do we lose the most water? **through the excretory system**
14. Where is most of the body's water located? **around cells in tissues**
15. What food sources provide vitamin C? **citrus fruits, tomatoes, green leafy vegetables**
16. How does a person get iodine? **by eating seafood and iodized salt**

74

▲ ENRICHMENT 74

NAME ______ DATE ______ CLASS ______

ENRICHMENT — Chapter 19

Nutrition — Text Pages 442–450

DO YOU EAT TOO MUCH FAT?

Carbohydrates, proteins, fats, vitamins, and minerals are nutrients that are important to good health and growth. To maintain health, you need to eat lots of carbohydrates. Proteins and fats are needed in smaller amounts.

Most people don't realize that fats are important to your health. Doctors recommend that 30 percent of the calories you eat come from fat. However, many people eat too many fatty foods. In most Americans' diets, 40 percent of the calories come from fat. Many people like the way high-fat foods taste. One gram of fat contains 9 calories. On the other hand, 1 gram of carbohydrate or protein contains only 4 calories. So, you can eat less fat than carbohydrate, but the total calories from fat could be higher.

Are you eating too much fat? Find out how many calories come from fat in five of your favorite foods. Nutritional information can be found on the labels of packaged foods or in nutrition or diet books at the library.

Procedure

1. Look at the nutritional information on the package. The amount of fat is usually listed in grams per serving. Fat content may be divided into saturated fat and unsaturated fat. To find the total amount of fat in a serving, add these two types of fat together.
2. Find the number of calories from fat per serving. Multiply total grams of fat by 9 calories.
3. Find the total number of calories per serving on the label. Divide the number of calories from fat by the total number of calories. Multiply this number by 100 to get the percentage of calories from fat.
4. Record your answers in the table below.

	Fat (grams) per serving	Calories per serving	Percentage of calories from fat
Food 1			
Food 2			
Food 3			
Food 4			
Food 5			

Conclude and Apply

1. Which of your favorite foods were high in fat?
Answers will vary, but may include potato chips or cheese.
2. How can you make sure you aren't eating too much fat?
Choose low-fat snacks or eat small amounts of high-fat foods and eat plenty of foods high in carbohydrates.

74

OBJECTIVES AND SCIENCE WORDS: Have students review the objectives and science words throughout the chapter as they study each section.

2 TEACH

Key Concepts are highlighted.

CONCEPT DEVELOPMENT

▶ Have students consider the meaning of the expression, "You are what you eat," before studying this section. Reconsider the meaning of the expression after completing this section.

▶ Depending on the class ability level, you may want to introduce the terms *monosaccharide* and *polysaccharide, dipeptide* and *polypeptide.*

▶ Students should be informed that table sugar is not glucose, but a larger molecule called sucrose.

TEACHER F.Y.I.

▶ Approximately 100 grams of body proteins are broken down into amino acids each day and resynthesized into new protein molecules.

Connect to...
Chemistry

Note: For all Connect to ... questions in this chapter, students should be able to determine answers from logic, from a dictionary, or from basic library references. **Answer:** The most basic carbohydrate would have the formula CH_2O (formaldehyde).

CONCEPT DEVELOPMENT

▶ **Does everyone need the same amount of food each day?** *No. Adolescents need more food during their growing stage. Older adults need less food. Even within age groups, calorie requirements vary from person to person.*

▶ **Is there any one food item that supplies the body with all of the nutrients it needs?** *No. However, milk does supply many of the nutrients.*

▶ Emphasize that fat is a greater source of energy; that a certain amount of fat is needed for health to be maintained; that too much fat can cause weight and other health problems.

CROSS CURRICULUM

▶ **Mathematics:** Have students check the nutrient label of a food item for the grams of fat contained. Multiply this number by 9, the number of calories used when one gram of fat is oxidized. Divide this number by the total number of calories in the food item. The calculation will give the percent of calories from fat. Many dieticians recommend that no more than 30 percent of your calorie intake come from fat.

Did You Know?

People in cultures that drink blood as a part of their diet supply some of their nutritional needs for sodium, protein, and iron.

them in your cells. Therefore, they have to be supplied through food in your diet. Eggs, milk, and cheese contain all the essential amino acids. Beef, pork, fish, chicken, and nuts supply only some of them. You might be surprised to know that whole grains such as wheat, rice, and soybeans
3 supply many needed amino acids in addition to carbohydrates.

Fats

Fats are nutrients that provide energy and help your body store some vitamins. For good health, your diet should be no more than 30 percent fat. Fats are stored in your body in the form of fat tissue that cushions your internal organs. Carbohydrates supply most people with most of the energy they need, because people eat a lot of carbohydrates. However, a molecule of fat releases more energy than a molecule of carbohydrate. A single molecule of fat breaks down into smaller molecules called fatty acids and glycerol.

There are two types of fats: unsaturated and saturated fats. Plants supply unsaturated fats. Corn, safflower, and soybean oils are all unsaturated fats. Some unsaturated fats are also found in poultry, fish, and nuts. Saturated fats are found in red meats. Saturated fats have been associated with high levels of blood cholesterol that contribute to heart disease. A lot has been written about the bad effects of cholesterol. However, cholesterol does occur normally in all your cell membranes. But, too much cholesterol in your diet causes fat deposits to form on the walls of blood vessels, resulting in a cutoff of blood supply to organs and an increase in blood pressure. Figure 19-3 shows the kind of fat tissue that builds up under the skin.

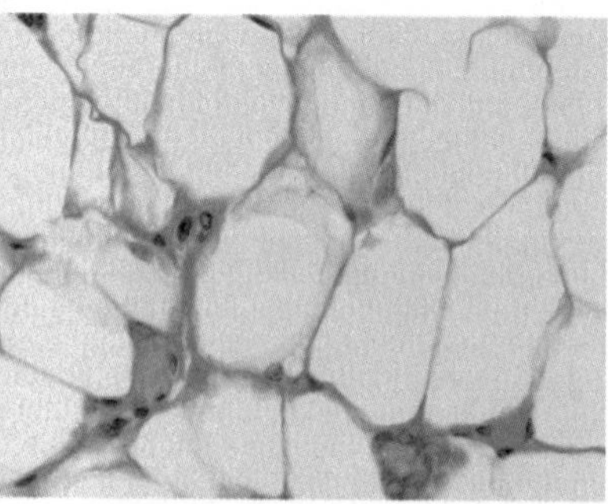

Figure 19-3. Certain cells in your body become filled with fat. The cytoplasm and nucleus get pushed to the edge of the cell.

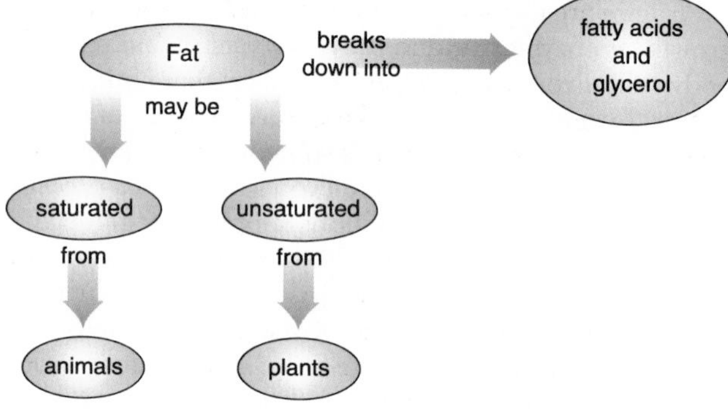

Figure 19-4. In your digestive system, a single molecule of fat breaks down into fatty acids and glycerol.

OPTIONS

ASSESSMENT—ORAL

▶ **If you eat a balanced diet, is it necessary to take vitamin and mineral pills?** *Most people get their daily requirements of minerals and vitamins if they eat a balanced diet. Some people may need supplemental vitamins and minerals due to their diet or some body disorder.*

▶ **How do foods supplemented with vitamins and minerals add to the health of many people?** *Foods supplemented with vitamins and minerals allow people to take in the recommended amounts of these supplements without having to pay much attention to what they eat.*

▶ **What are the most commonly consumed vitamin supplemented foods that the greatest number of people consume?** *vitamin-enriched, commercially prepared cereals*

TECHNOLOGY

Fake Fat

Americans obtain an average of 37 percent of their daily calorie intake from fat. Fats and sodium often make the food taste great. Various companies are experimenting with fat substitutes so that people can eat rich-tasting foods, without gaining weight.

Foods using fat substitutes or fake fats have to be lower in calories than real fat and still try to maintain the taste of fat-rich foods. Some fake fats are useful in frozen products, but can't be used to cook with because they break down when heated. Many fail taste tests.

One fat substitute, however, is waiting approval by the Food and Drug Administration. It is composed of table sugar molecules attached to fatty acid molecules. It looks like fat and tastes like it, too. This product passes through the digestive system without being absorbed. In addition, it can be used in cooking and has no calories.

Think Critically: What are the possible disadvantages of using fake fat in your diet?

Vitamins

Vitamins are essential, organic nutrients needed in small quantities to help your body use other nutrients. For instance, vitamin D is needed for bone cells to use the mineral calcium. In general, vitamins promote growth and regulate many body functions.

Most foods supply some vitamins, but no one food has them all. Eating a variety of foods usually supplies all the vitamins the body needs. Some vitamins dissolve easily in water and are called water-soluble vitamins. Others dissolve only in fat and are called fat-soluble vitamins. Although some people feel that taking extra vitamins is helpful, normally eating a balanced diet is sufficient to give your body all the vitamins it needs. Vitamins are found in a variety of meats and vegetables. Your body makes vitamin D when your skin is exposed to sunlight. Some vitamin K is made with the help of bacteria that live in your large intestine. Table 19-1, on the next page, lists major vitamins, their effects on your health, and some of the foods that provide them.

In Your JOURNAL

What might be the disadvantage of substituting vitamin pills for a proper diet? Find out which vitamins are water soluble and which are fat soluble. **In your Journal,** state why vitamin pills are not a substitute for a balanced diet.

What are two types of vitamins?

TECHNOLOGY

Extension: As a class, collect ads for food substitutes and make a bulletin board showing each "real" food and its substitute.

References: "A Heavyweight Fuss Over the New 'Fake Fat'" by Barbara Kantrowitz, *Newsweek,* March 5, 1990, p. 41.

"A New Fat" by Bonnie Tandy Leblang, *American Health,* May 1990, pp. 86-88.

Think Critically: Thinking that foods like cake and sweets have few calories might lead one to consume a great deal of these foods instead of having a balanced diet including fruits and vegetables and sources of protein.

In Your JOURNAL

Other nutrients besides vitamins are needed for maintaining health. A well-balanced diet supplies the range of needed vitamins. Vitamins A, D, E, and K are fat-soluble. B-complex and vitamin C are water soluble.

CONCEPT DEVELOPMENT

▶ Emphasize the difference between fat-soluble and water-soluble vitamins. Relate this idea to intake of excess vitamins. Ask students to suggest what might happen to excess fat-soluble vitamins. Fat-soluble vitamins accumulate in fat cells. It is not entirely certain how this harms the body.

▶ Explain how vitamins were discovered—by evidence of the absence of certain foods.

REVEALING MISCONCEPTIONS

▶ Students may have the mistaken notion that if vitamins are good for you, "more is better." Some vitamins should not be taken in high doses. Excess of vitamin A can cause liver, skin, and muscle problems. Too much vitamin D can cause an imbalance of blood calcium resulting in digestive and nervous system disorders.

ENRICHMENT

▶ Have students research a particular vitamin deficiency. If possible, they should include information on how it was discovered.

PROGRAM RESOURCES

From the **Teacher Resource Package** use:

Critical Thinking/Problem Solving, page 23, Sweeteners, Not Sweets, Spoil Your Appetite.

Cross-Curricular Connections, page 23, Should Cosmetic Use of Retin-A be Allowed?

Use **Laboratory Manual** page 127, Testing for Carbohydrates, page 131, Testing for Proteins, and page 133, Digestion of Fats.

CONCEPT DEVELOPMENT

▶ Ask students to study the photo of the hurdlers and discuss how certain minerals enable athletes to perform this activity. Have students refer to specific minerals from Table 19-2 on page 447 in their discussion.

CHECK FOR UNDERSTANDING

Have students review the vitamin table on this page to obtain a general concept of sources of each type of vitamin. Then have students infer the vitamins contained in some similar but unlisted fruits and vegetables.

RETEACH

Cooperative Learning: Using the Paired Partner strategy, make flashcards with names of fruits or vegetables. On the reverse side identify the major vitamin content of each food. Partners can test each other using the flashcards.

EXTENSION

For students who have mastered this section, use the **Reinforcement** and **Enrichment** masters or other OPTIONS provided.

Table 19-1

VITAMINS		
Vitamin	**Health Effect**	**Food Sources**
A	growth, good eyesight, healthy skin	green/yellow vegetables, liver and fish liver oils, milk, yellow fruit
B (thiamine, riboflavin, niacin, B_6, B_{12})	growth, healthy nervous system, use of carbohydrates, red blood cell production	meat, eggs, milk, cereal grains, green vegetables
C	growth, healthy bones and teeth, wound healing	citrus fruits, tomatoes, green leafy vegetables
D	absorption of calcium and phosphorus by bones and teeth	milk, eggs, fish
E	formation of cell membranes	vegetable oils, eggs, grains
K	blood clotting, wound healing	green leafy vegetables, egg yolks, tomatoes

Minerals

Minerals are inorganic nutrients that regulate many chemical reactions in your body. Minerals are chemical elements such as phosphorus. About 14 elements are used by your body for building cells for chemical reactions in cells, for sending nerve impulses throughout your body, and for carrying oxygen to body cells. Minerals used in the largest amounts in your body are given in Table 19-2. These include calcium, phosphorus, potassium, and sodium. Some minerals, called trace minerals, are required in only very small amounts. Trace minerals include iron, copper, and iodine.

Figure 19-5. Minerals are needed for healthy bones and muscle contraction.

OPTIONS

INQUIRY QUESTIONS

▶ **In addition to the saturated fats found in red meats, what are some other food sources of saturated fats?** *milk, butter, and cheese*

▶ **What might cause fatty deposits to form at certain places on artery walls?** *Areas of accumulation usually occur where the arteries are injured or where they branch.*

Water

You don't have to be lost in a desert to know how important water is for your body. Next to oxygen, water is the most vital factor for survival. You could live a few weeks without food, but only a few days without water. Most of the nutrients you have studied in this chapter can't be used by your body unless they are carried in a solution. This means that they have to be dissolved in water. Water enables chemical reactions to take place in cells.

Where is water found in the body?

Your body is about 60 percent water by weight. This water is found in and around cells and in plasma and lymph. Water removes waste products from cells. Wastes dissolved in water leave your body as urine or perspiration. To balance water lost each day, you need to drink about two liters of liquids. But don't think that you have to drink just water to keep your cells supplied. Most foods have more water in them than you realize. An apple is about 80 percent water and many meats are as much as 90 percent water.

Table 19-2

MINERALS		
Mineral	**Health Effect**	**Food Sources**
Calcium	strong bones and teeth, blood clotting, muscle and nerve activity	milk, eggs, green leafy vegetables
Phosphorus	strong bones and teeth, muscle contraction, stores energy	cheese, meat, cereal
Potassium	balance of water in cells, nerve impulse conduction	bananas, potatoes, nuts, meat
Sodium	fluid balance in tissues, nerve impulse conduction	meat, milk, cheese, salt, beets, carrots
Iron	carries oxygen in hemoglobin in red blood cells	raisins, beans, spinach, eggs
Iodine	thyroid activity, stimulates metabolism	seafood, iodized salt

MINI-Lab

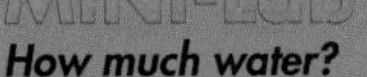

How much water?

Using a pan balance, find the mass of an empty 250-mL beaker. Fill the beaker with sliced celery or sliced carrots and find the mass of the filled beaker. *Estimate* the amount of water you think is in the vegetable. Obtain permission to dry the vegetable pieces overnight in an oven on a flat tray at very low heat. Use an oven mitt to remove the tray. Allow the tray and vegetables to cool. Then *determine* the mass of the dried vegetables. How much water was in the fresh vegetables?

MINI-Lab

Materials: pan, balances, cut up vegetables, beakers, oven, oven mitt

Troubleshooting: CAUTION: *Remind students to be careful when handling hot materials.*

Answer: The water content will vary with the freshness of the vegetables used, but students should find that the vegetables used are mostly water, some as much as 85 percent water. Be sure to dry vegetables completely and to allow them to return to room temperature before weighing.

MINI-Lab
ASSESSMENT

Performance: To further assess students' ability to determine how much water is in fresh fruits, refer to USING LAB SKILLS, question 11, on page 460.

CONCEPT DEVELOPMENT

▶ Discuss water in vegetables such as turnips and squashes. Many have the impression that drinking water is the only source of water.

▶ Point out that water is necessary for chemical reactions in the body and that water is a by-product of the breakdown of all molecules in the process of respiration.

▶ Have students comment on the statement that where there is water, there is life.

ENRICHMENT

▶ Have students keep track of all the foods and liquids they consume in one day. Estimate how much water was taken in and compare with the two liters or so of water lost daily.

PROGRAM RESOURCES

From the **Teacher Resource Package** use:

Activity Worksheets, page 173, Mini-Lab: How much water?

MINI QUIZ

Use the Mini Quiz to check students' recall of chapter content.

1. **____ provide energy and materials for development, growth, and repair.** *nutrients*
2. **What are three types of carbohydrates?** *starch, cellulose, and sugar*
3. **Many needed amino acids are supplied by ____.** *whole grains*
4. **Why does the body need vitamins?** *to help the body use other nutrients*
5. **What percent by weight of the human body is water?** *60%*

CONCEPT DEVELOPMENT

▶ Explain that fiber is an indigestible food residue made up mostly of cellulose. Explain that fiber resists digestion in humans because we do not have the enzyme to break it down.

▶ Ask students which food groups on this page provide fiber.

PROBLEM SOLVING

Jerry should choose menu b because it contains more complex carbohydrates and less fat and protein than the other menus.

Think Critically: Jerry wants to load his liver and muscles with glycogen to use as energy during the race. Even with carbohydrate loading, Jerry will probably deplete his carbohydrate supply after running about 30-33 kilometers, at which time his body will begin to use stored fat.

Table 19-3

WATER LOSS	
Through	Amount (mL/day)
Exhaled air	350
Feces	150
Skin (mostly as sweat)	500
Excretory system	1800

Your body also loses about two liters of water every day through excretion, perspiration, and respiration. Table 19-3 shows how water is lost from the body. The body is equipped to maintain its fluid content, however. When your body needs water, it sends messages to your brain. A feeling of thirst develops. Drinking a glass of water usually restores the body's homeostasis, and the signal to the brain stops.

Food Groups

Because no one food has every nutrient, you need to eat a variety of foods. Nutritionists have developed a simple system to help people plan meals that include all the nutrients required for good health.

Foods that contain the same nutrients belong to a **food group.** Figure 19-6 shows a food pyramid with the basic food groups with the serving suggestions for maintaining health. Eating a certain amount from each food group will supply your body with the nutrients it needs for energy and growth. Of course, most people eat foods in combined forms. Combinations of food contain ingredients from more than one food group and supply the same nutrients as the foods they contain. Examples of

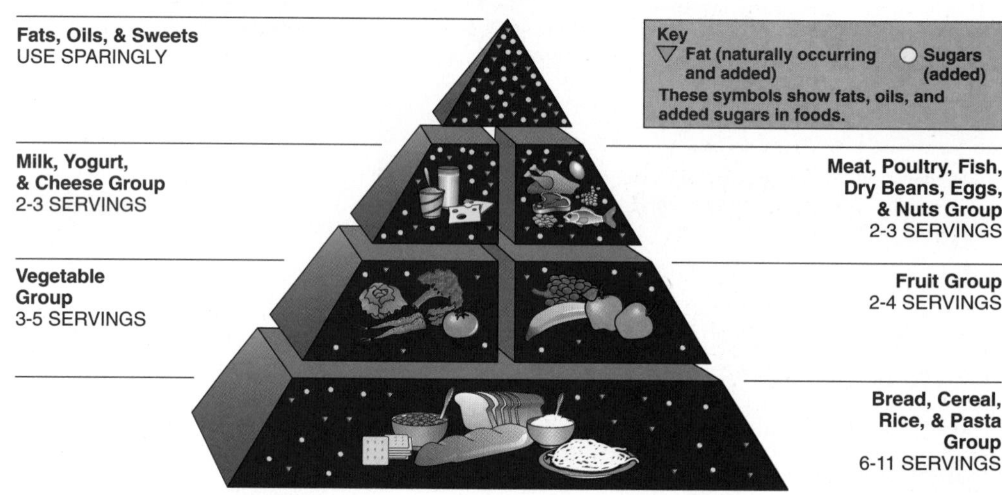

Figure 19-6. The pyramid shape reminds you that you should consume more servings from the bread and cereal group than from the meat and milk group. The least number of servings should come from the fats and oils group.

OPTIONS

INQUIRY QUESTIONS

▶ **Why do fruits and vegetables contain fiber?** *They are plants and plant cell walls containing cellulose.*

▶ **What are people's attitudes about food? Why can't people be satisfied with a "nutrient shake" for meeting their daily requirements?** *Discussion will vary. People are equipped with taste buds and enjoy variety in taste, plus color and texture.*

PROBLEM SOLVING

The Big Race

Jerry is a long-distance runner. He plans to run the Boston Marathon on Patriots' Day. This world-famous race is more than 42 kilometers. His mother asked him to choose one of the following menus for dinner the night before the big race.

a) hamburger, french fries, chocolate cake, a large glass of cola
b) spaghetti with tomato sauce, fresh vegetable salad, whole wheat roll, fresh fruit, a large glass of water
c) fried chicken, mashed potatoes and gravy, broccoli with cheese sauce, roll and butter, and ice cream.

Jerry wants a meal with a lot of complex carbohydrates. Which menu should Jerry choose for dinner?

Think Critically: Why does Jerry want to eat carbohydrates before the big race?

food group combinations include chili and macaroni and cheese.

SECTION REVIEW

1. List six classes of nutrients and give one example of a food source for each.
2. Describe a major function of each class of nutrients.
3. Discuss the relationship between your diet and your health.
4. Explain the importance of water in the body.
5. **Apply:** What foods from each food group would provide a balanced breakfast? Explain your choices.
6. **Connect to Earth Science:** Find out about the source of trace minerals essential for the normal functions of the human body.

EcoTip

Eat lower on the food chain by consuming more vegetables and fruits.

Making and Using Tables

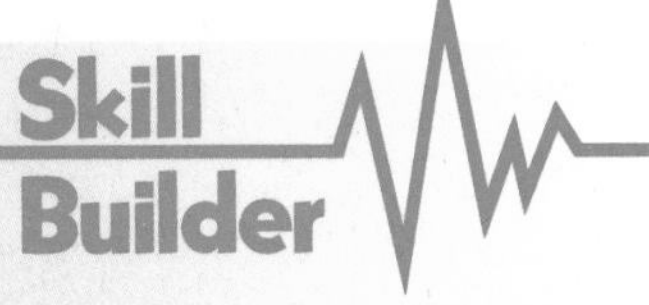

Use the information in Table 19-1 and Table 19-2 to determine the vitamins and minerals needed for healthy bones. If you need help, refer to Making and Using Tables in the **Skill Handbook** on page 690.

3 CLOSE

▶ Ask questions 1-4 and the **Apply** Question in the Section Review.

SECTION REVIEW ANSWERS

1. carbohydrates; proteins; fats; vitamins; minerals; water (examples of each will vary but might include, respectively, bread; fish, meat; cooking oil; fruit, milk; fruits and vegetables; water.)

2. carbohydrates: source of energy; proteins: growth of cells; fats: source of energy; vitamins: growth; minerals: nerve activity; water: removes wastes from cells

3. Answers will vary. Students should recognize the relationship between eating nutritionally balanced diets and getting adequate amounts of substances needed for growth and good health.

4. dissolves and carries nutrients in the body and removes wastes

5. Apply: Answers will vary. Students should be able to support their answers with information in Figure 19-6.

6. Connect to Earth Science: Trace elements are found in soils and water and become part of the food we eat.

Skill Builder

vitamins C and D, calcium, and phosphorus

Skillbuilder
ASSESSMENT

Performance: Assess students' abilities to Make and Use Tables by having them use Table 19-3 to find out how much water is lost through skin.

EXTENSION

▶ Have students find out about the American Heart Association diet and how a heart-healthy diet compares to the recommended servings of food groups.

ACTIVITY 19-1

50 minutes

OBJECTIVE: Identify and **compare** juices that contain vitamin C.

PROCESS SKILLS applied in this activity:
▶ **Observing** in Procedure Step 5.
▶ **Classifying** in Analyze Questions 2 and 3.
▶ **Measuring** in Procedure Steps 4, 5, and 6.
▶ **Predicting** in Procedure Step 3.
▶ **Using Numbers** in Procedure Steps 1 and 2.
▶ **Interpreting Data** in questions 2 and 3 of Analyze, and Conclude and Apply Question 5.
▶ **Experimenting** in Procedure Steps 4 to 8.

COOPERATIVE LEARNING

Divide the class into Science Investigation Teams of three. One student should make the data table and record all data. Another should set up and label the test tubes and the third should add the indophenol and juices to the tubes. All should observe the experiment in process.

TEACHING THE ACTIVITY

Alternate Materials: If indophenol is not available, make a starch solution according to the directions given in the Preparation of Solutions in the Teacher Guide. Add 3 or 4 drops of iodine solution until the starch solution turns a very light blue. This may be used in place of indophenol.

Troubleshooting: Caution students not to mix up the droppers from the test solution bottles.

Activity ASSESSMENT

Performance: To further assess students' ability to Identify vitamin C content, have them test for vitamin C in water from shredded lettuce or shredded cabbage cores that have sat overnight in a small amount of water.

PROGRAM RESOURCES

From the **Teacher Resource Package** use:

Activity Worksheets, pages 167-168, Activity 19-1, Your Digestive System.

ACTIVITY 19-1 Identifying Vitamin C Content

***Problem:** Which juices contain vitamin C?*

Materials

- indophenol solution
- graduated cylinder
- glass marking pencil
- 10 test tubes
- test-tube rack
- 10 dropping bottles containing water, orange juice, pineapple juice, apple juice, lemon juice, tomato juice, cranberry juice, carrot juice, lime juice, mixed vegetable juice

Procedure

1. Make a data table like the one shown to record your observations.
2. Label the test tubes 1 through 10.
3. Predict which juices contain vitamin C. Record your predictions in your table.
4. Measure 5 mL of indophenol into each of the 10 test tubes. **CAUTION:** *Wear your goggles and apron.* Indophenol is a blue liquid that turns colorless when vitamin C is present. The more vitamin C in a juice, the less juice it takes to turn indophenol colorless.
5. Add 20 drops of water to test tube 1. Record your observations.
6. Begin adding orange juice, one drop at a time, to test tube 2.
7. Record the number of drops needed to turn indophenol colorless.
8. Use Steps 6 and 7 to test the other juices.

Data and Observations

Sample Data

Test tube	Juice	Prediction (yes or no)	Number of drops
1	water	Student	20
2	orange	prediction	5-6
3	pineapple	will	12
4	apple	vary	20
5	lemon		5-6
6	tomato		7-8
7	cranberry		20
8	carrot		20
9	lime		5-6
10	vegetable		20

Analyze

1. What was the purpose of test tube 1?
2. Which juice did not contain vitamin C?
3. Does the amount of vitamin C vary in fruit juices?

Conclude and Apply

4. What is the best way to take in vitamins?
5. Suggest why scurvy, a deficiency of vitamin C, is uncommon today.

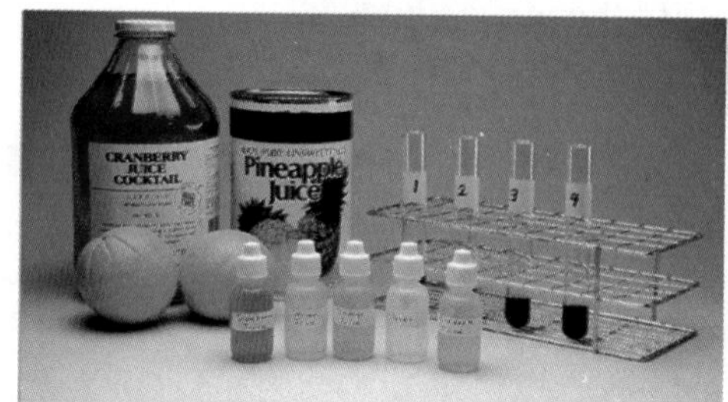

ANSWERS TO QUESTIONS

1. control
2. apple, cranberry, carrot, and vegetable. Answers may vary if juices fortified with vitamins are used.
3. Yes, the amount of vitamin C is influenced by whether C has been added, water has been added, or if the juice is fresh, frozen, or canned.
4. by eating a balanced diet
5. Fruits and vegetables containing vitamin C are included in diets. People consume a substantial variety of vitamin-fortified foods.

Your Digestive System 19-2

Objectives

- Distinguish between mechanical and chemical digestion.
- Name the organs of the digestive system and describe what takes place in each.
- Explain how homeostasis is maintained in digestion.

New Science Words

digestion
mechanical digestion
chemical digestion
saliva
peristalsis
chyme
villi

Processing Food

Like the other animals you have studied so far, you are a consumer. The energy you need for life comes from food sources outside yourself. To keep the cells in your body alive, you take in food every day. Food is processed in your body in four phases: ingestion, digestion, absorption, and elimination. Whether it is a fast-food burger or a home-cooked meal, all the food you eat is treated to the same processes in your body. As soon as it enters your mouth, or is ingested, food begins to be broken down. **Digestion** is the process that breaks food down into small molecules so they can be absorbed, or taken into the cells of your body. Food molecules then move into cells during the process of absorption. Molecules that aren't absorbed are eliminated and pass out of your body as wastes.

Figure 19-7 shows the major organs of your digestive tract: mouth, esophagus (i SAH fuh guhs), stomach, small intestine, large intestine, rectum, and anus. Food passes *through* all of these organs. However, food *doesn't* pass through your liver, pancreas, or gallbladder. These three organs produce or store enzymes and chemicals that help break down food as it passes through the digestive tract.

Digestion is both mechanical and chemical. **Mechanical digestion** takes place when food is chewed and mixed in the mouth, churned in your stomach, and acted on by bile. **Chemical digestion** breaks down large molecules of food into smaller molecules that can be absorbed by cells. Chemical digestion takes place in your mouth, stomach, and small intestine.

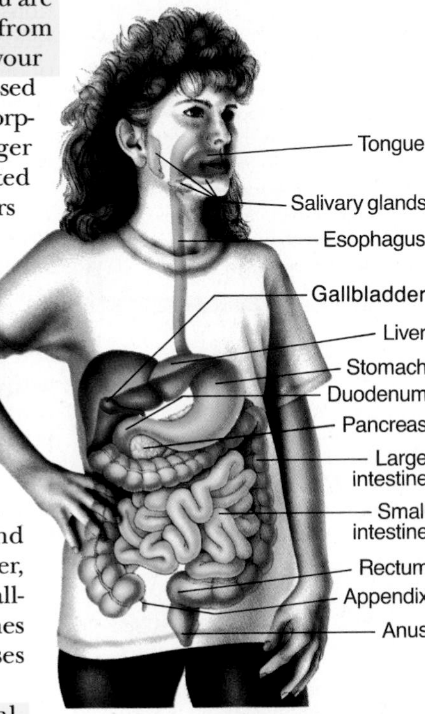

Figure 19-7. The Major Organs for Digestion

SECTION 19-2

PREPARATION

SECTION BACKGROUND

- Near the junction of the small and large intestines is a small tube-like pouch called the appendix. It does not appear to have any function in the digestive process. The organ can become inflamed and require removal.

PREPLANNING

- To prepare for Activity 19-2, obtain the dilute hydrochloric acid (1%) solution and the pepsin powder.

1 MOTIVATE

- Discuss how the teeth of carnivores and herbivores are specialized for their diet. Compare with the teeth of humans.

TYING TO PREVIOUS KNOWLEDGE: Recall from Chapter 1 that one feature of living things is that they use energy. Discuss with students that food is the source of energy for cell activities, but that it must be processed into usable molecules. This is the function of the digestive system.

OPTIONS

PROGRAM RESOURCES

From the **Teacher Resource Package** use:

Concept Mapping, pages 43-44.

Activity Worksheets, page 174, Mini-Lab: What is the advantage of a rough inner lining?

Transparency Masters, pages 67-68, Human Digestive System.

Use **Color Transparency** 34, Human Digestive System.

2 TEACH

Key Concepts are highlighted.

CONCEPT DEVELOPMENT

▶ Use the diagrams of the digestive system throughout this section for teaching this chapter.

▶ **Demonstration:** To illustrate the need for mechanically breaking up food into smaller pieces, do this demonstration. Into one small container with water at room temperature, drop a piece of hard candy. Into another similar container, also with water at room temperature, drop a hard candy that has been crushed into small pieces. Record the time it takes each candy to dissolve completely. Discuss with students the relationship of particle size and rate of dissolving.

MINI-Lab

Materials: posterboard, corrugated cardboard

▶ **Teaching Tips:** Place the cardboard on paper towels before adding water; limit the amount of water given to each group.

▶ Corrugated paper can be found in packaging, or old bulletin board borders work well, too.

Answer: The corrugated paper represents the small intestine (stomach too). It has greater surface area. The posterboard tube is smooth like the esophagus, and the corrugated cardboard folds in and out like the villi of the small intestine.

MINI-Lab ASSESSMENT

Performance: To further assess students' understanding of the importance of surface area, have students measure the areas of both the dry cardboard and the wet expanded cardboard and make a comparative statement.

PROGRAM RESOURCES

From the **Teacher Resource Package** use:

Activity Worksheets, page 5, Flex Your Brain.

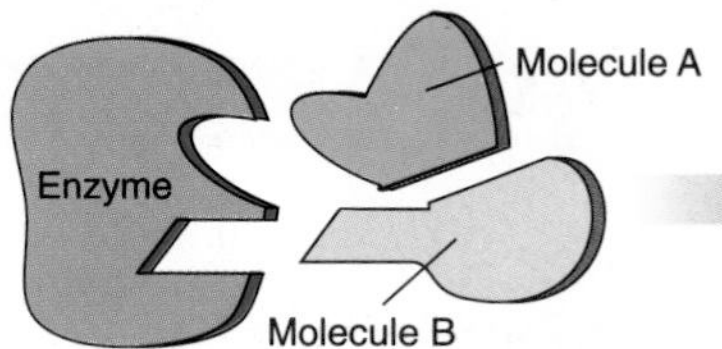

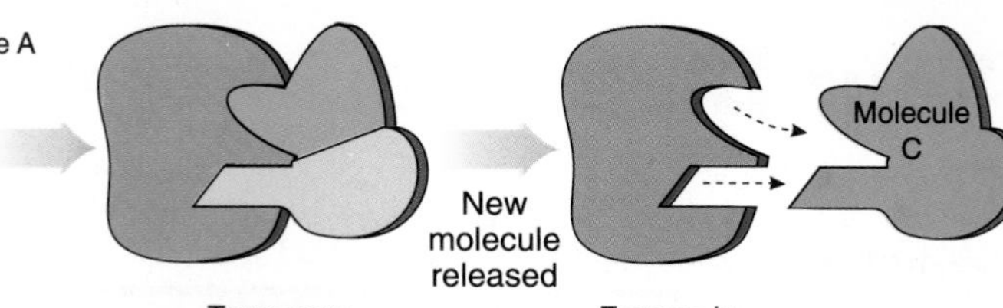

Figure 19-8. Enzymes are unchanged at the end of a chemical reaction.

Enzymes are proteins that are vital in chemical digestion. An enzyme enters a reaction and helps to join or break up two substances. The enzyme then leaves the reaction without being changed itself. You could say that enzymes are regulators. Some enzymes speed up reac- 1
tions. Others enable a large molecule to break up without releasing a lot of excess heat energy in the cell. Thousands of enzyme-assisted reactions are taking place in your cells right now. But you aren't at all aware of this activity.

MINI-Lab

What is the advantage of a rough inner lining?

Cut equal pieces of smooth posterboard and corrugated cardboard. Roll each into a tube. Unroll the tubes. Wet both pieces with water. Which paper has the greater surface area? *Infer* which tube might represent the esophagus, and which the small intestine. How do the two tubes show the differences between the two organs?

In Your Mouth

The process of digestion begins in your mouth. There, your tongue and teeth mechanically break food up into small pieces. Humans are adapted with several kinds of teeth for cutting, grinding, tearing, and crushing.

Some chemical digestion also starts in your mouth. As you chew, your tongue moves food around and mixes it with a watery substance called **saliva.** Saliva is produced by three sets of glands in your mouth that are shown in Figure 19-7. Saliva is made up mostly of water, but it also contains mucus and an enzyme called amylase. Amylase starts the breakdown of starch to sugar in your mouth. Food that is mixed with saliva becomes a soft mass. The food mass is moved to the back of your tongue where it is swallowed and passes into your esophagus. Now the process of ingestion is complete, and the process of digestion has begun.

Your esophagus is a muscular tube about 25 cm long. Through it, food passes to your stomach in about four to eight seconds. No digestion takes place there. Smooth muscles in the walls of the esophagus move food downward by a squeezing action. These waves or contractions, called **peristalsis** (pe ruh STAHL sis), move food along throughout the digestive system. Figure 19-10 shows how peristalsis works.

Figure 19-9. Taste buds help you to figure out whether you like the taste of a particular food.

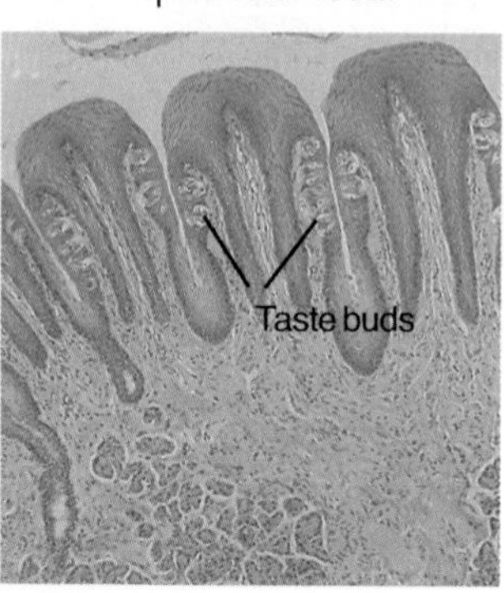

OPTIONS

Meeting Different Ability Levels

For Section 19-2, use the following **Teacher Resource Masters** depending upon individual students' needs.

◆ **Study Guide Master** for all students.

● **Reinforcement Master** for students of average and above average ability levels.

▲ **Enrichment Master** for above average students.

Additional Teacher Resource Package masters are listed in the OPTIONS box throughout the section. The additional masters are appropriate for all students.

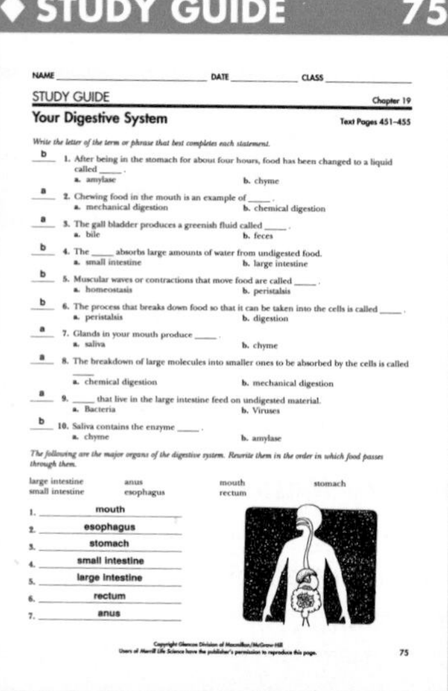

◆ STUDY GUIDE 75

NAME ________ DATE ________ CLASS ________

STUDY GUIDE Chapter 19

Your Digestive System Text Pages 451–455

Write the letter of the term or phrase that best completes each statement.

b 1. After being in the stomach for about four hours, food has been changed to a liquid called ____. a. amylase b. chyme

a 2. Chewing food in the mouth is an example of ____. a. mechanical digestion b. chemical digestion

a 3. The gall bladder produces a greenish fluid called ____. a. bile b. feces

b 4. The ____ absorbs large amounts of water from undigested food. a. small intestine b. large intestine

b 5. Muscular waves or contractions that move food are called ____. a. homeostasis b. peristalsis

b 6. The process that breaks down food so that it can be taken into the cells is called ____. a. peristalsis b. digestion

a 7. Glands in your mouth produce ____. a. saliva b. chyme

a 8. The breakdown of large molecules into smaller ones to be absorbed by the cells is called ____. a. chemical digestion b. mechanical digestion

a 9. ____ that live in the large intestine feed on undigested material. a. Bacteria b. Viruses

b 10. Saliva contains the enzyme ____. a. chyme b. amylase

The following are the major organs of the digestive system. Rewrite them in the order in which food passes through them.

large intestine, small intestine, anus, esophagus, mouth, rectum, stomach

1. mouth
2. esophagus
3. stomach
4. small intestine
5. large intestine
6. rectum
7. anus

 75

In Your Stomach

Your stomach is a muscular bag. When it is empty, it is somewhat sausage-shaped with folds on the inside. As food enters from the esophagus, the stomach expands and the folds smooth out. Both mechanical and chemical digestion take place in the stomach. Mechanically,
2 food is mixed by the muscular walls of the stomach and by peristalsis. Food is also mixed with strong digestive juices. Hydrochloric acid and enzymes are made by cells in the walls of the stomach. The stomach also produces mucus that lubricates the food, making it more slick. Mucus also protects the stomach itself from the strong digestive juices. Food moves through your stomach in about four hours. At the end of this time, it doesn't look like food anymore, but has been changed to a thin, watery liquid called **chyme**. Little by little, chyme moves out of your stomach and into your small intestine.

Muscles contract
Food mass
Muscles relax
Food moves down

Figure 19-10. During peristalsis, muscles behind the food contract and push the food forward. Muscles in front of the food relax.

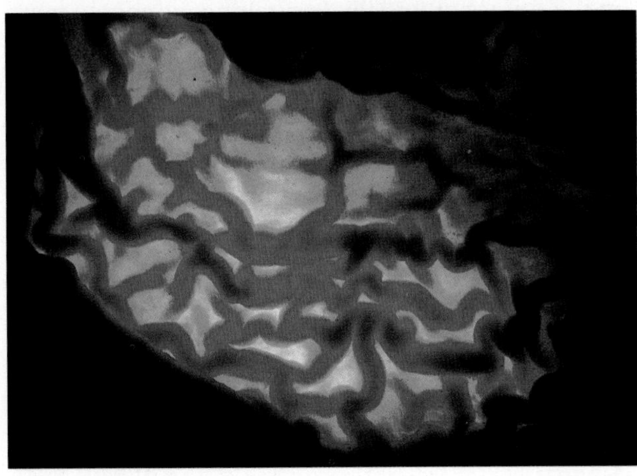

Figure 19-11. Wrinkles in the stomach wall smooth out as it fills with food.

In Your Small Intestine

Your small intestine is small in diameter, but it is nearly seven meters in length. As chyme leaves your stomach, it enters the first part of your small intestine, called the duodenum. The major portion of all digestion takes place in your duodenum. A lot of different things take place here at the same time. Here, digestive juices from the liver and pancreas are added to the mixture. Your liver produces a greenish fluid called bile, which it stores in a

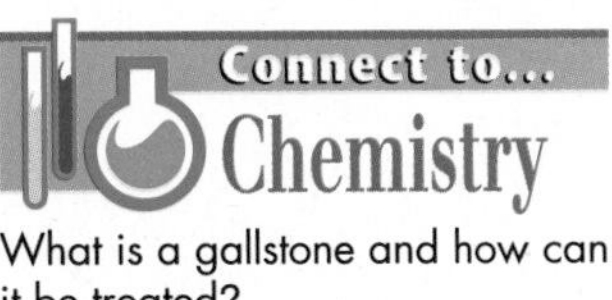

What is a gallstone and how can it be treated?

CONCEPT DEVELOPMENT

▶ Have students identify the mechanical and chemical digestion processes that take place in the mouth, stomach, small intestine, and large intestine.

▶ **Demonstration:** Carefully drip several drops of vegetable oil into a bowl of water. Have students note the one large ring the oil forms on the surface of the water. Drip several drops of dishwashing detergent into the middle of the ring. Observe the breakup of the ring into smaller globules. This is similar to the action of bile on fat particles in the small intestine.

Connect to... Chemistry

Answer: crystallized cholesterol; can be dissolved, crushed ultrasonically, or removed surgically

FLEX Your Brain

Use the Flex Your Brain activity to have students explore DIGESTION.

Flex Your Brain

ASSESSMENT

Portfolio: Use the Flex Your Brain activity to reinforce critical thinking and problem solving skills.

● REINFORCEMENT 75

NAME ____________ DATE ________ CLASS ________

REINFORCEMENT — Chapter 19

Your Digestive System — Text Pages 451–455

Listed below are organs that aid in the digestion of food. Describe the function of each organ and label the figure.

1. mouth: begins both chemical and mechanical digestion; glands in mouth produce saliva, contain amylase, which breaks down starch
2. esophagus: connects the throat to the stomach; this muscular tube moves food downward by a squeezing action called peristalsis
3. stomach: a muscular bag where chemical and mechanical digestion continue; food stays here for about four hours, during which it is changed to chyme
4. small intestine: a tube nearly seven meters long where digestive juices from the liver and pancreas are added; villi here absorb molecules from the chyme
5. pancreas: a small organ that produces substances that stop the action of stomach acid, and enzymes that break down carbohydrates, fats, and proteins
6. large intestine: absorbs water from undigested food; where the unabsorbed materials become more solid
7. liver: produces bile, which is stored in the gall bladder; bile physically breaks up large particles of fats into smaller particles
8. gall bladder: a small sac that stores bile produced by the liver
9. rectum: where muscles control the release of wastes from the body

mouth
esophagus
liver
stomach
gall bladder
pancreas
small intestine
large intestine
rectum

 75

▲ ENRICHMENT 75

NAME ____________ DATE ________ CLASS ________

ENRICHMENT — Chapter 19

Your Digestive System — Text Pages 451–455

DOES DIGESTION BEGIN IN THE MOUTH?

The starch you eat is changed to sugar through the process of digestion. Where does this process begin? You can find out by testing for the presence of starch in an unchewed cracker and a chewed cracker.

Materials
- 2 small clear glasses
- tincture of iodine
- saltine crackers
- spoon
- eyedropper

IODINE

Procedure
1. Crumble half a cracker into one of the glasses. Add 2 spoonfuls of water and mix well.
2. Use the eyedropper to add 3 drops of iodine to the mixture. Iodine changes the color of any starchy substance from pale purple to dark purple. **CAUTION:** *Iodine is a poison. Handle with care.* Record the color of this mixture.
3. Put the other half of the cracker into your mouth and chew it for about a minute. Spit the chewed cracker into the other glass. Add 2 spoonfuls of water and mix well.
4. Use the eyedropper to add 3 drops of iodine to the mixture in the second glass. Observe the color of the mixture. Record your observations.

Data and Observations

Mixture	Color	Starch present?
unchewed cracker		
chewed cracker		

Conclude and Apply
1. Was there a difference in color in the mixtures in the two glasses? Explain. Yes, the mixture in the first glass was darker than the mixture in the second.
2. Since you know that iodine tests for starch, which cracker had more starch in it, the unchewed cracker or the chewed cracker? the unchewed cracker
3. What enzyme produced by glands in the mouth could have changed the starch in the chewed cracker? amylase
4. Is the change in color evidence of chemical digestion or mechanical digestion? chemical digestion

 75

MINI QUIZ

Use the Mini Quiz to check students' recall of chapter content.

1. **Proteins that speed up digestive reactions are called ______ .** *enzymes*
2. **What causes the mechanical mixing of food in the stomach?** *peristalsis*
3. **What structures increase the surface area of the small intestine?** *villi*

CHECK FOR UNDERSTANDING

Have students infer what would happen if chyme did not remain in the large intestine for several days, as in the case of diarrhea.

RETEACH

Have students prepare a chart with these parts of the digestive system listed in the first column: mouth, stomach, small intestine, and large intestine. In the second column, across from the part, give a brief description of the digestive process associated with it.

EXTENSION

For students who have mastered this section, use the **Reinforcement** and **Enrichment** masters or other OPTIONS.

Figure 19-12. The liver, gallbladder, and pancreas are at the beginning of the small intestine.

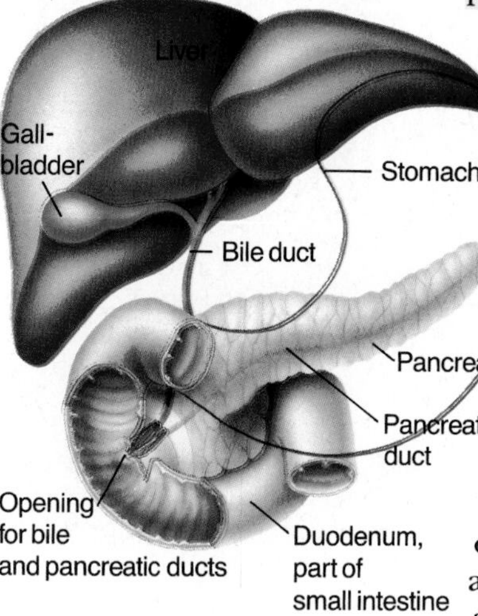

Where are nutrients absorbed?

small sac called the gallbladder. The acid from the stomach makes large fat particles float to the top of the liquid. Bile physically breaks up these particles into smaller pieces, the way detergent acts on grease on dishes.

Your pancreas makes insulin and substances that stop the action of stomach acid. It also produces enzymes that break down carbohydrates, fats, and proteins still further.

The walls of your small intestine are not smooth like the inside of a garden hose. Rather, they have many ridges and folds. These folds are covered with tiny, fingerlike projections called **villi.** Villi make the surface of the small intestine larger so that there are more places for food to be absorbed. Villi are shown in Figure 19-13. 3

By this time, chyme has become a soup of molecules that is ready to be absorbed through the cells on the surface of the villi. Peristalsis continues to move and mix the chyme. In addition, the villi themselves move and are bathed in the soupy liquid. Molecules of nutrients pass by diffusion, osmosis, or active transport into blood vessels in each villus. From there, blood transports the nutrients to all the cells of the body. Peristalsis continues to slowly force the remaining materials that are not absorbed along into the large intestine.

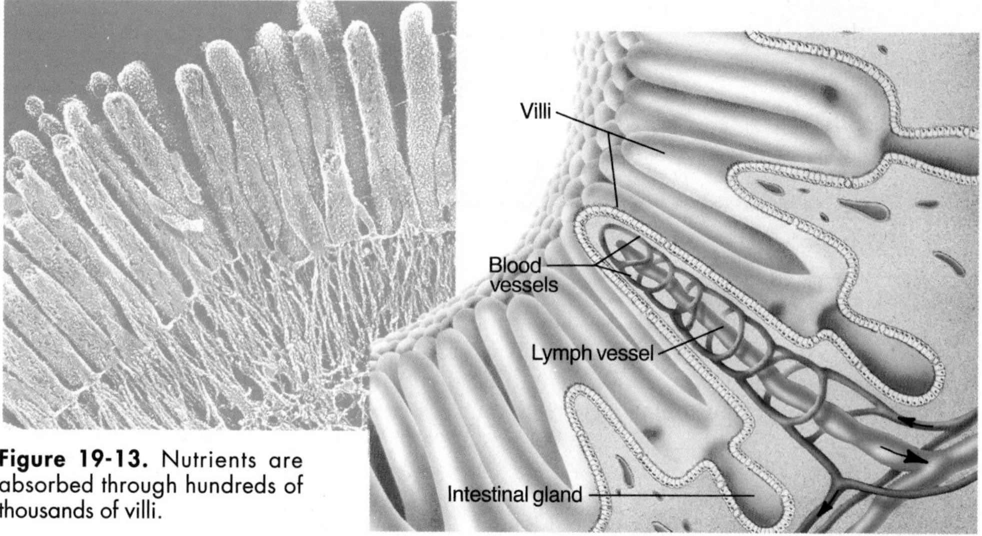

Figure 19-13. Nutrients are absorbed through hundreds of thousands of villi.

OPTIONS

ASSESSMENT—ORAL

▶ **Why is it necessary for acid-neutralizing secretions to be added to the food entering the small intestine?** *to reduce the acidity in order for other enzymes to do their work*

▶ **How is the liver involved in the digestion process?** *It produces the bile that is stored in the gall bladder and released into the small intestine.*

▶ **Sometimes bile becomes crystallized in the gall bladder and forms rock-like particles known as gallstones. What problems could these gallstones cause?** *The stones could block the ducts leading to the small intestine and produce jaundice or pass through the ducts and cause pain.*

In Your Large Intestine

As chyme enters the large intestine, it is still a thin, watery mixture. The main job of the large intestine is to absorb this water from the undigested mass. In doing so, large amounts of water are returned to the body and homeostasis is maintained. Peristalsis slows down somewhat in the large intestine. As a result, chyme may stay in the large intestine as much as three days. During this time, excess water and sodium are absorbed back into the bloodstream. The remaining materials, consisting of undigested cellulose and bacteria, become more solid. Bacteria that live in your large intestine feed on undigested materials like cellulose. This is a symbiotic relationship. The bacteria feed on cellulose and in return they produce several vitamins that you need. Muscles in the rectum and anus control the release of solidified wastes from the body in the form of feces.

Food is processed in your digestive system for the purpose of supplying your body with raw materials for metabolism. These raw materials are in the form of nutrients. Those that are not digested and absorbed are eliminated.

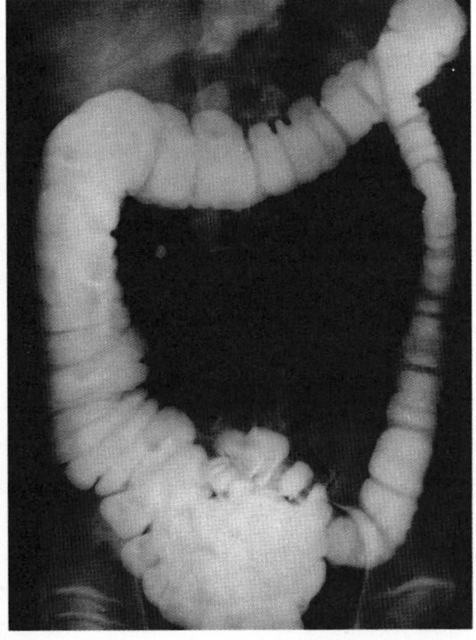

Figure 19-14. The large white area in this X ray is the large intestine.

SECTION REVIEW

1. Name in order the organs through which food passes as it moves through the digestive system.
2. What is the difference between mechanical and chemical digestion?
3. How are the liver, pancreas, and gallbladder related to digestion?
4. How do activities in the large intestine help maintain homeostasis?
5. **Apply:** Crackers contain starch. Explain why a soda cracker held in your mouth for five minutes might begin to taste sweet.
6. **Connect to Physics:** Talk with a dentist to find out about the number of pounds of pressure that can be exerted by the teeth.

Observing and Inferring — Skill Builder

What would happen to food if the pancreas did not secrete its juices into the small intestine? If you need help, refer to Observing and Inferring in the **Skill Handbook** on page 682.

3 CLOSE

▶ Ask questions 1-4 and the **Apply** Question in the Section Review.

SECTION REVIEW ANSWERS

1. mouth, esophagus, stomach, small intestine, large intestine, rectum, anus

2. Mechanical digestion involves cutting, grinding, tearing, and crushing by the teeth and tongue in the mouth and peristalsis by the walls of the stomach. Chemical digestion occurs in parts of the digestive system through enzyme action.

3. The liver produces bile, stored in the gall bladder, which helps break down fat. The pancreas produces substances that stop the action of stomach acid and enzymes that break down protein, carbohydrates, and fats.

4. Water is removed and returned to the body.

5. Apply: Complex carbohydrates are broken into simple sugar.

6. Connect to Physics: The molars can exert a force of about 200 pounds; the incisors exert only about 25 or 30 pounds.

Skill Builder

The food would be undigested and continue to be very acidic from the stomach, possibly causing ulcers.

Skillbuilder
ASSESSMENT

Portfolio: Assess students' abilities to Infer by having them suggest what happens to digestion and nutrition in people who have lost their teeth.

ENRICHMENT

▶ Have students research how strong emotions affect the nervous system, causing interference with the digestive process.

▶ Have students find out about the action of microorganisms in the small and large intestines that cause the breakdown of proteins and complex carbohydrates.

PROGRAM RESOURCES

From the **Teacher Resource Package** use:

Transparency Masters, pages 69-70, Digestive Organs.

Science Integration Activities 19, The Breakdown of Starch

Use **Color Transparency** 35, Digestive Organs.

SECTION 19-3

PREPARATION

SECTION BACKGROUND

▶ Anorexia nervosa claims the life of about 15 percent of those who develop it. Death is due to starvation, complications such as dehydration and infections, or suicide by some individuals.

1 MOTIVATE

▶ Bring in advertisements for over-the-counter weight loss or diet pills. Analyze them for cost and actual effectiveness. Compare with diets that only require a calorie-limited, balanced diet and moderate exercise.

Connect to... Chemistry

Answer: The loss of potassium and sodium salts results in reduced efficiency of nerve transmission and weakening of skeletal muscle activity. Lack of fluids causes dehydration of cells and tissues.

In Your JOURNAL

Upon investigation, students should understand that by not meeting diet goals, health is impaired.

VideoDisc

STVS: Bulimia, Disc 7, Side 2

PROGRAM RESOURCES

From the **Teacher Resource Package** use:

Science and Society, page 23, Advertising and Anorexia.

19-3 Eating Disorders

New Science Words

anorexia nervosa
bulimia

Objectives

- ▶ Name and describe two types of eating disorders.
- ▶ Infer the consequences of improper diets associated with eating disorders.

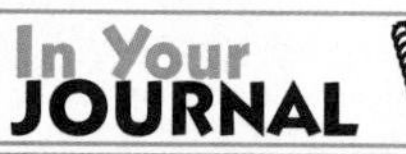

In Your JOURNAL

The United States Department of Agriculture has composed a list of seven diet goals for improving and maintaining health. Find out what the seven diet goals are. **In your Journal,** write about how people with eating disorders fall short of these goals.

Connect to... Chemistry

Excessive vomiting can result in the loss of potassium, sodium, and fluids. What are some symptoms of the excessive loss of salts and fluids?

Are Looks That Important?

Have you ever tried on a bathing suit or a pair of jeans in the store only to be disappointed by the fact that they just didn't fit? Do you feel bad when someone comments on your weight? American culture is so fashion conscious that a person's self-worth sometimes becomes dependent on looks. This can be quite a problem because people often rely on food. Sometimes the thought of looking fit and trim is so important that people become obsessed with loosing extra mass as quickly as possible. But often people begin a diet or exercise program, and then return to old eating habits after just a few days.

Eating Disorders

Anorexia nervosa and bulimia are two eating disorders that are very dangerous. Although these disorders are more common in adolescent girls, a small percentage of boys have been known to have problems as well.

Anorexia nervosa is an eating disorder that involves extreme weight loss to the point of damaging vital organs and even death. It is characterized by preoccupation with eating as little food as possible due to the fear of gaining mass. The goal is to become more attractive, but the opposite effect often results. The extreme lack of food results in starvation-like symptoms, including dry skin and hair, cold hands and feet, and general weakness. There is also a tendency toward increased infections, stress fractures in bones, and weakness of the heart muscle that has resulted in death in some cases.

Bulimia is an eating disorder that involves binging, or eating huge amounts of food in a short period of time, fol-

OPTIONS

Meeting Different Ability Levels

For Section 19-3, use the following **Teacher Resource Masters** depending upon individual students' needs.

◆ **Study Guide Master** for all students.

● **Reinforcement Master** for students of average and above average ability levels.

▲ **Enrichment Master** for above average students.

Additional Teacher Resource Package masters are listed in the OPTIONS box throughout the section. The additional masters are appropriate for all students.

◆ STUDY GUIDE 76

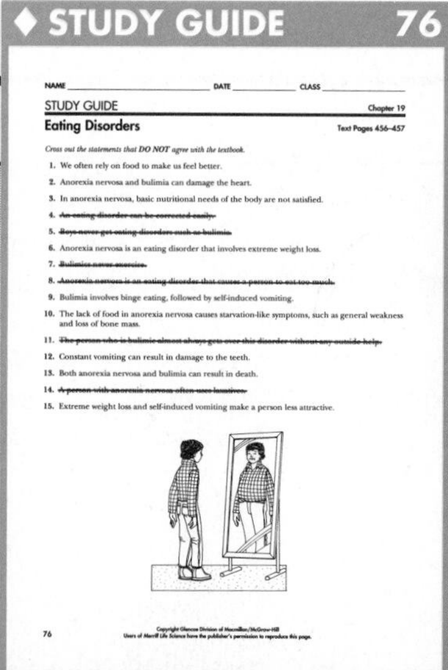

NAME ________ DATE ________ CLASS ________

STUDY GUIDE — Chapter 19

Eating Disorders — Text Pages 456–457

*Cross out the statements that **DO NOT** agree with the textbook.*

1. We often rely on food to make us feel better.
2. Anorexia nervosa and bulimia can damage the heart.
3. In anorexia nervosa, basic nutritional needs of the body are not satisfied.
4. ~~An eating disorder can be corrected easily.~~
5. ~~Boys never get eating disorders such as bulimia.~~
6. Anorexia nervosa is an eating disorder that involves extreme weight loss.
7. ~~Bulimics never exercise.~~
8. ~~Anorexia nervosa is an eating disorder that causes a person to eat too much.~~
9. Bulimia involves binge eating, followed by self-induced vomiting.
10. The lack of food in anorexia nervosa causes starvation-like symptoms, such as general weakness and loss of bone mass.
11. ~~The person who is bulimic almost always gets over this disorder without any outside help.~~
12. Constant vomiting can result in damage to the teeth.
13. Both anorexia nervosa and bulimia can result in death.
14. ~~A person with anorexia nervosa often uses laxatives.~~
15. Extreme weight loss and self-induced vomiting make a person less attractive.

76

lowed by self-induced vomiting, for the purpose of not adding mass to the body. Bulimics also fast and abuse laxatives to move food through their intestines quickly. As a result, the small intestine does not absorb vital nutrients. They sometimes exercise fanatically. Severe dental problems result from exposing the teeth to stomach enzymes and acid during vomiting. Painful ulcers may develop in the mouth and stomach and muscle weakness (the heart is a muscle too) may also develop. In both cases, basic nutritional needs of the body are not satisfied. Over a period of time, the body is starved to death.

Some people may turn and look in admiration at a very thin person. Everyone likes to receive positive comments on his or her appearance. However, the consequences of extreme weight loss or self-induced vomiting make a person look much less attractive. By talking and learning about the problem, victims of eating disorders will come to feel better about themselves and restore themselves to health.

Figure 19-15. Persons with eating disorders may have difficulties with their self-esteem.

What Can You Do?

Do you or does someone that you know have an eating disorder? If so, what can you do? It is often impossible to deal with an eating disorder alone. However, many people are willing to help. Through treatment, the problem can be controlled. The school guidance counselor and nurse are specially trained to help with these problems. Programs have been specifically developed to control these disorders.

SECTION REVIEW

1. Describe two types of eating disorders.
2. What are the effects of eating disorders?
3. **Connect to Chemistry:** What is the nutritional relationship between self-induced vomiting over a period of time and starvation?

You Decide!

Individuals with eating disorders are usually very unwilling to talk about their problems with friends. Maybe they're embarrassed about the condition, or afraid that their families may eventually find out. If you think a friend has an eating disorder, would it be right for you to mention this to the school nurse or guidance counselor? Would you be a caring friend if you chose not to seek help for that person?

● REINFORCEMENT 76

NAME ______ DATE ______ CLASS ______

REINFORCEMENT — Chapter 19

Eating Disorders — Text Pages 456–457

List two types of eating disorders and briefly describe each.

1. **anorexia nervosa; This eating disorder involves extreme weight loss. The person eats very little food for fear of gaining weight.**
2. **bulimia; This eating disorder involves binge eating followed by vomiting. Bulimics often use laxatives and exercise fanatically in order to lose weight.**

Answer the following questions.

3. What are some symptoms of anorexia nervosa? **dry skin and hair, cold hands and feet, stress fractures in bones, general weakness, increased infections, weakness of the heart muscle, extreme thinness**
4. How does American culture contribute to the cause of eating disorders? **Americans are so fashion conscious that people become overly concerned with how they look rather than how healthy they are. People become obsessed with losing weight.**
5. Why do bulimics take laxatives? **They want to move food through the digestive system quickly so that it is not absorbed by the body.**
6. Why is an eating disorder difficult to deal with alone? **Eating disorders are often used to cope with emotions, such as anger, sorrow, and fear. These emotions must be dealt with along with the disorder.**
7. Where can a person with an eating disorder get help? **school guidance counselor, school nurse, special programs**

76

▲ ENRICHMENT 76

NAME ______ DATE ______ CLASS ______

ENRICHMENT — Chapter 19

Eating Disorders — Text Pages 456–457

Eating disorders are characterized by obsessive concern with appearance and weight. Many pressures in our society cause people to feel overweight. Experts say that almost all Americans think they are overweight, even when they are not.

Take a survey of your family, friends, and people in your neighborhood. Ask the questions below of five males and five females. Record their answers in the table below.

Questions to ask:

1. Do you feel that you generally eat too little, just the right amount, or too much?
2. Do you consider yourself underweight, just the right weight, or overweight?
3. Have you ever gone on a diet to lose weight?

	Answer 1	Answer 2	Answer 3	Your assessment of person's weight
Male 1				
Male 2				
Male 3				
Male 4				
Male 5				
Female 1				
Female 2				
Female 3				
Female 4				
Female 5				

Conclude and Apply

1. How did most people feel about their weight? **Most will probably have said "overweight."**
2. Can you conclude from your survey that more women or more men generally feel overweight? **Most women think they are overweight, and the survey will probably reflect that.**
3. Have most people dieted at some time? **Answers will vary, although dieting is very common.**
4. Did your own assessment of a person's weight always match the person's self-assessment? **Students' judgments will sometimes differ.**

76

2 TEACH

Key Concepts are highlighted.

CONCEPT DEVELOPMENT

▶ **Why do crash diets not work?** *Weight loss requires a change in eating and exercise habits. A quick change in diet stresses the body, and vital nutrients may be eliminated.*

CHECK FOR UNDERSTANDING

Have students explain the difference between anorexia and bulimia.

RETEACH

Cooperative Learning: Use the Numbered Heads Together strategy to investigate why the greater percent of adolescents with eating disorder problems are females.

EXTENSION

For students who have mastered this section, use the **Reinforcement** and **Enrichment** masters or other OPTIONS.

3 CLOSE

▶ Ask questions 1-2 in the Section Review.

SECTION REVIEW ANSWERS

1. Anorexia nervosa involves extreme weight loss due to eating as little as possible. Bulimia involves binging and purging in order not to gain weight.
2. dry skin and hair, cold hands and feet, increased infections, stress fractures in bones, weakening of heart, tooth decay, malnutrition
3. Connect to Chemistry: Self-induced vomiting over a period of time results in starvation because nutrients are thrown up before they have a chance to reach the small intestine where they normally would be absorbed.

YOU DECIDE!

Answers will vary, reflecting varying views of friendship. Accept all answers that students can support.

ACTIVITY 19-2

OBJECTIVE: **Design and carry out an experiment** to determine what happens to proteins in the digestive system.
Time: one class period

PROCESS SKILLS applied in this activity are **observing, measuring, interpreting data, experimenting,** and **inferring.**

PREPARATION

Obtain materials needed for each group. Have an alkaline solution such as baking soda in water available for use to test the inference in the Going Further activity.

Cooperative Learning: Divide the class into Science Investigation Teams.

SAFETY

Caution student to use care when handling any acids. Wash hands thoroughly after working with acid.

THINKING CRITICALLY

Enzymes responsible for protein digestion are pepsins, trypsin, carboxypeptidase, and aminopeptidase. Pepsin requires an acidic condition such as found in the stomach to initiate protein digestion.

TEACHING THE ACTIVITY

*Refer to the **Activity Worksheets** for additional information and teaching strategies.*

- **CAUTION:** Hydrochloric acid (HCl) is an irritant and corrosive. Vapors are harmful. Any spillage or contact with skin should be flushed with water. If a baking soda solution is used in the Going Further activity to neutralize the HCl, foaming will take place. Do not let the foam spill over the top of the test tube since this will obscure the results.

SUMMING UP/SHARING RESULTS

- The test tubes with acid had more liquid than those without acid. Yes, more liquid in the test tube with acid.
- to keep them at a cool temperature to prevent the gelatin from melting
- gelatin and pepsin
- Pepsin is active only in an acid medium.
- Acid is added to the food material from the walls of the stomach. Secretion added by the pancreas into the small intestine stop the action of the acid.

GOING FURTHER!

The pepsin activity stops in the small intestine because something neutralizes the acid. The addition of an alkaline will stop the action of the acid.

ACTIVITY 19-2

DESIGNING AN EXPERIMENT
Protein Digestion

What causes large protein molcules to break down for easy digestion? What are the conditions needed for protein digestion?

Getting Started

Your task in this activity is to devise a method to observe the conditions necessary for protein digestion.

Thinking Critically

What enzymes are responsible for protein digestion? Are there special conditions in the digestive tract that are required for enzyme activity to take place?

Materials 

Your cooperative group will use:

- unflavored gelatin
- dropper
- 2 test tubes in a rack
- drinking glass
- pepsin powder
- cold water
- graduated cylinder
- marking pen
- dilute hydrochloric acid

Try It!

1. Copy the data table shown.
2. Prepare the gelatin according to the directions on the package. Pour 10 mL of the gelatin into each of the 2 test tubes. Place the test tubes in a rack and do not disturb them while the gelatin sets.
3. Label one test tube "pepsin with acid," and the other one "pepsin without acid."
4. When the gelatin has set, add 10 drops of the dilute acid to the test tube labeled "pepsin with acid." **CAUTION:** *Always use care when working with acid. Wash your hands thoroughly after pouring the acid.*
5. Sprinkle enough pepsin powder to cover the surface of the gelatin in both test tubes. Place the 2 test tubes upright into a drinking glass containing cold water.
6. *Observe* the gelatin every 10 minutes, and record your results in the data table after each observation.

Summing Up/Sharing Results

- What did you *observe* in the test tubes? Was there a difference?
- Why were the test tubes kept in cold water?
- What were the constants in this experiment?
- Did the acid have any effect on the activity of the pepsin? How does this relate to the activity of this enzyme in the stomach?

Going Further!

Infer why the activity of the pepsin stops in the small intestine. Test your inference and see if the enzyme activity stops.

Data and Observations Sample Data

Time minutes	Pepsin with acid	Pepsin without acid
10	liquid from acid only	no liquid
20	slightly more liquid	no liquid
30	some increase in amount of liquid	very slight amount of liquid
40	same as above	same as above
50	some increase in liquid	same as above

PROGRAM RESOURCES

From the **Teacher Resource Package** use:
Activity Worksheets, pages 169-170, Activity 19-2, Protein Digestion.

Activity ASSESSMENT

Performance: Assess students' understanding of protein digestion see USING LAB SKILLS, Question 12, on page 460.

CHAPTER

REVIEW

SUMMARY

19-1: Nutrition

1. Carbohydrates, fats, proteins, minerals, vitamins, and water are the six nutrients found in food.
2. Carbohydrates provide energy; proteins are needed for growth and repair; fats store energy and cushion organs. Vitamins and minerals regulate functions. Water makes up about 60 percent of body mass and is used for a variety of homeostatic functions.
3. Health is affected by the combination of foods that make up a diet.

19-2: Your Digestive System

1. The digestive system breaks food down mechanically by chewing and churning, and chemically, with the help of enzymes, into substances that cells can absorb and use.
2. Food passes through the mouth, esophagus, stomach, small intestine, large intestine, rectum, and out the anus.
3. Homeostasis is maintained by absorption of water in the large intestine.

19-3: Science and Society: Eating Disorders

1. Anorexia nervosa and bulimia are two eating disorders that involve the use of starvation and extreme behaviors to control body mass.
2. Eating disorders result in damage to organs and even death.

KEY SCIENCE WORDS

a. **amino acids**
b. **anorexia nervosa**
c. **bulimia**
d. **carbohydrates**
e. **chemical digestion**
f. **chyme**
g. **digestion**
h. **fats**
i. **food group**
j. **mechanical digestion**
k. **minerals**
l. **nutrients**
m. **peristalsis**
n. **proteins**
o. **saliva**
p. **villi**
q. **vitamins**

UNDERSTANDING VOCABULARY

Match each phrase with the correct term from the list of Key Science Words.

1. process of breaking down food
2. muscular contractions that move food
3. enzyme-containing fluid in the mouth
4. fingerlike projections in small intestine
5. starch and sugars
6. nutrients that form a cushioning tissue
7. physical breakdown of food
8. inorganic nutrients
9. subunits of proteins
10. eating disorder with induced vomiting

OPTIONS

ASSESSMENTS

To assess student understanding of material in this chapter, use the resources listed.

COOPERATIVE LEARNING

Consider using cooperative learning in the THINK AND WRITE CRITICALLY, APPLY, and MORE SKILL BUILDERS sections of the Chapter Review.

PROGRAM RESOURCES

From the **Teacher Resource Package** use:

Chapter Review, pages 41-42.

Chapter and Unit Tests, pages 126-129, Chapter Test.

CHAPTER

REVIEW

SUMMARY

Have students read the summary statements to review the major concepts of the chapter.

UNDERSTANDING VOCABULARY

1. g	**6.** h
2. m	**7.** j
3. o	**8.** k
4. p	**9.** a
5. d	**10.** c

ASSESSMENT

Portfolio

Encourage students to place in their portfolios one or two items of what they consider to be their best work. For each item, ask students to explain why that item was chosen and what they learned from it. Items might be selected from the following.

- For Your Gifted Students story about the digestive "engineer,"p. 440
- Flex Your Brain, p. 453
- Skillbuilder opinion on dealing with friends who have eating disorders, p. 457

Performance

Additional performance assessments may be found in *Performance Assessment* and *Science Integration Activities* that accompany **Merrill Life Science.** Performance Task Assessment Lists and rubrics for evaluating these activities and other products generated throughout the chapter can be found in Glencoe's *Performance Assessment in Middle School Science.*

CHAPTER

REVIEW

CHECKING CONCEPTS

1. a	**6.** d
2. b	**7.** b
3. d	**8.** b
4. d	**9.** a
5. a	**10.** b

USING LAB SKILLS

ASSESSMENT

Use these alternate lab exercises to assess students' understanding of skills used in this chapter.

11. Answers may vary depending on variety of fruit used.

Apple = 85%
Pear = 83%

12. The baking soda solution is alkaline and does not support pepsin activity.

THINK AND WRITE CRITICALLY

13. Surface area is increased, making it easier for the enzymes to contact the substrates.

14. Ingestion is eating; digestion is breaking down of food; absorption is the taking in of nutrients by the blood and cells; elimination is ridding the body of wastes.

15. Using elaborate recipes does not add vitamins or minerals to food. It does, however, make food appear nicer sometimes and more appetizing. People prefer to eat food that looks, smells, and tastes good.

16. The pancreas and liver add enzymes to the small intestines.

17. A correct diet gives the correct nutrients for all functions and reactions of the body.

APPLY

18. Fat, because the gall bladder stores bile, which aids in the digestion of fats.

19. The stomach, because acids are produced there. Antacids are used to counteract an excess of acid.

20. Bile breaks fats down into smaller droplets, which have an increased surface area, so digestion occurs more quickly and easily. This is comparable to the action of soap on grease when washing dishes.

21. The body would retain those vitamins that are fat-soluble. They are stored in fat deposits. The water-soluble vitamins cannot be stored in fat deposits.

22. The body uses for its growth, maintenance, and energy those substances that we eat. If we eat good, nutritious food, the body will have good building blocks to use. If we eat an unbalanced or poor diet, the body will not have good building blocks to use.

CHAPTER

REVIEW

CHECKING CONCEPTS

Choose the word or phrase that completes the sentence.

1. Most digestion occurs in the______.
a. duodenum **c.** liver
b. stomach **d.** large intestine

2. The ______ makes bile, which acts on fats.
a. gallbladder **c.** stomach
b. liver **d.** small intestine

3. Water is absorbed in the ______.
a. liver **c.** esophagus
b. small intestine **d.** large intestine

4. Food does not pass through the ______.
a. mouth **c.** small intestine
b. stomach **d.** liver

5. Nutrients that store vitamins are _______.
a. fats **c.** carbohydrates
b. proteins **d.** minerals

6. The vitamin not used for growth is ______.
a. A **c.** C
b. B **d.** K

7. In the ______, hydrochloric acid is added to the food mass.
a. mouth **c.** small intestine
b. stomach **d.** large intestine

8. The ______ produces enzymes that digest proteins, fats, and carbohydrates.
a. mouth **c.** large intestine
b. pancreas **d.** gallbladder

9. The food group containing yogurt and cheese is ______.
a. milk **c.** meat
b. grain **d.** fruit

10. Carbohydrates are best obtained from ______.
a. milk **c.** meat
b. grains **d.** eggs

USING LAB SKILLS

11. The MINI-Lab on page 447 investigates the amount of water in vegetables. Repeat this activity to determine the amount of water in an apple or a pear. What percent of water do these fruits have?

12. In the Activity on page 458 you were asked to determine the conditions necessary for protein digestion by pepsin. Repeat the experiment using a baking soda solution instead of dilute hydrochloric acid. What effect does the baking soda solution have on pepsin activity?

THINK AND WRITE CRITICALLY

Answer the following questions in your Journal using complete sentences.

13. What do chewing and the action of bile have in common in the digestive system?

14. Describe the differences among ingestion, digestion, absorption, and elimination.

15. Explain why it isn't really necessary to have food prepared using elaborate recipes. Why do people go to the trouble of cooking meals?

16. List the structures that add enzymes to chyme in the small intestine.

17. Why is a balanced diet important to maintaining homeostasis?

APPLY

18. What types of food would a person whose gall bladder has been removed need to limit in his or her diet? Explain your choice.

19. In what part of the digestive system do antacids work? Explain your choice.

20. Bile's action is similar to soap. Use this information to explain bile working on fats.
21. Vitamins are in two groups: water- or fat-soluble. Which of these might your body retain? Explain your answer.
22. Based on your knowledge of food groups and nutrients, discuss the meaning of the familiar statement: "You are what you eat."

MORE SKILL BUILDERS

If you need help, refer to the Skill Handbook.

1. **Making and Using Graphs:** Recommended Dietary Allowances (RDA) are made for the amounts of nutrients people should take in to maintain health. Prepare a bar graph of the percent of RDA of each nutrient from the breakfast product information listed below.

Nutrient	Percent US RDA
Protein	2
Vitamin A	20
Vitamin C	25
Vitamin D	15
Calcium (Ca)	less than 2
Iron (Fe)	25
Zinc (Zn)	15

Which nutrients are given the greatest percent of the Recommended Dietary Allowance? Could a person on a fat-restricted diet eat this product? Explain.

2. **Sequencing:** In a table, sequence the order of organs through which food passes in the digestive system. Indicate whether ingestion, digestion, absorption, or elimination takes place in the individual organs.
3. **Comparing and Contrasting:** Compare and contrast the location, size and functions of the esophagus, stomach, small intestine, and large intestine.
4. **Sequencing:** Sequence the digestion of a starch, a fat, and a protein, listing where it occurs, what happens, the enzymes used, and the end product of this digestion.
5. **Interpreting Data:** Use a label from a food, list the nutrients and food groups to which each belongs. How many calories are provided by each type?

PROJECTS

1. Research the ingredients of products used in diarrhea and laxatives. Where in the digestive system do these products act and how?
2. Collect labels from breakfast cereals. Compare the amounts of vitamins and minerals. Are the vitamins and minerals a natural part of the ingredients? Were any vitamins or minerals added to make the cereal more healthful?

MORE SKILL BUILDERS

1. **Making and Using Graphs:** Vitamin C and iron are given the greatest percent of RDA. A person on a fat-restricted diet could not eat it because the fat content is not listed and therefore is not known. *See graph at bottom of page.*
2. **Sequencing:** Mouth carries out chemical digestion of carbohydrates and mechanical digestion of foods. When food is swallowed, it is ingested. Peristalsis occurs in the esophagus. The stomach begins chemical digestion of proteins and mechanical digestion through peristalsis. The small intestine finishes chemical digestion of all food types, mechanically digests through peristalsis, absorbs final end products through the villi, passes undigestible wastes to large intestine. The large intestine absorbs water and passes undigestible wastes to the rectum, then anus for elimination.
3. **Comparing and Contrasting:** See #2 for functions. Esophagus is a tube between the mouth and stomach, 25 cm. Stomach is a sac located between the esophagus and small intestine. The small intestine is located between the stomach and large intestine, 7 m. The large intestine, 1.5 m, is located between the small intestine and rectum.
4. **Sequencing:** Digestion of carbohydrates begins in the mouth by amylase to a simple sugar; complex sugars are broken down in the SI to simple sugars. Proteins are broken down in the stomach to simple proteins and into amino acids in the SI. Fats are broken apart mechanically by bile in the duodenum and further digested in the SI, forming fatty acids and glycerol.
5. **Interpreting Data:** Answers will vary, depending on the label.

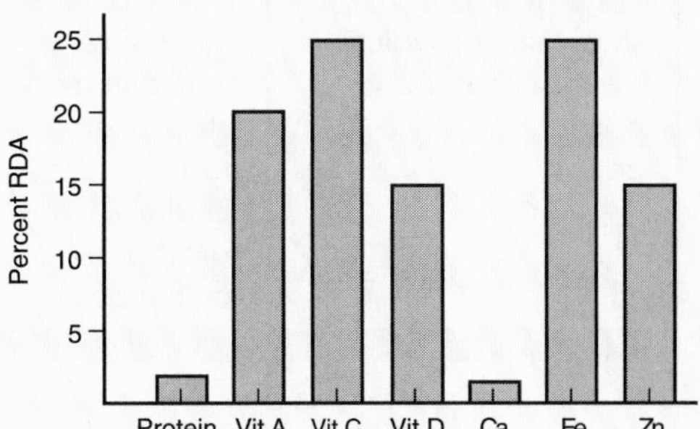

CHAPTER

20 Your Circulatory System

CHAPTER SECTION	OBJECTIVES	ACTIVITIES
20-1 Circulation (3 days)	1. **Compare** arteries, veins, and capillaries. 2. **Trace** the pathway of blood through the chambers of the heart. 3. **Describe** pulmonary and systemic circulation.	**MINI-Lab:** *What are the parts of the heart?* p. 466 **MINI-Lab:** *How does a stethoscope work?* p. 468 **Activity 20-1:** Taking Blood Pressure, p. 470
20-2 Blood (2 days)	1. **Describe** the characteristics and functions of the parts of blood. 2. **Explain** the importance of checking blood types before a transfusion is given. 3. **Describe** a disease and a disorder of blood.	**Activity 20-2:** *Designing an Experiment (Comparing Blood Cells)*, p. 477
20-3 Autologous Blood Transfusions Science & Society (1 day)	1. **Explain** the advantages of an autologous blood transfusion. 2. **Name** one problem with saving one's own blood.	
20-4 Your Lymphatic System (2 days)	1. **Describe** the functions of the lymphatic system. 2. **Explain** where lymph comes from. 3. **Explain** the role of lymph organs in fighting infections.	
Chapter Review		

ACTIVITY MATERIALS

FIND OUT	ACTIVITIES		MINI-LABS	
Page 463 stopwatch or watch with second hand paper pencil	**20-1 Taking Blood Pressure, p. 470** sphygmomanometer stethoscope table chairs (2) alcohol cotton balls	**20-2 Designing an Experiment, p. 477** prepared slides of human blood and blood of two other vertebrates (fish, frog, reptile, bird) microscope	**What are the parts of the heart? p. 466** Figure 20-3	**How does a stethoscope work? p. 468** stethoscope

CHAPTER FEATURES	TEACHER RESOURCE PACKAGE	OTHER RESOURCES
Technology: *An Assist for the Heart,* p. 469 **Skill Builder:** *Concept Mapping,* p. 469	**Ability Level Worksheets** ◆ ***Study Guide,*** p. 77 ● ***Reinforcement,*** p. 77 ▲ ***Enrichment,*** p. 77 **Activity Worksheets,** pp. 5, 176-177, 182-183 **MINI-Lab Worksheet,** pp. 182, 183 **Transparency Masters,** pp. 71-74 **Concept Mapping,** p. 45	**Color Transparency 36,** Pulmonary Circulation **Color Transparency 37,** Human Circulatory System **Lab Manual:** Heart Structure, p. 137 **Lab Manual:** Measuring Heartbeat, p. 141 **Lab Manual:** Blood Pressure, p. 145 **STVS:** Disc 7, Side 1
Problem Solving: *The Blood Type Mystery,* p. 475 **Skill Builder:** *Making and Using Tables,* p. 476	**Ability Level Worksheets** ◆ ***Study Guide,*** p. 78 ● ***Reinforcement,*** p. 78 ▲ ***Enrichment,*** p. 78 **Activity Worksheets,** pp. 178-179 **Critical Thinking/Problem Solving,** p. 24 **Cross-Curricular Connections,** p. 24	
You Decide! p. 479	**Ability Level Worksheets** ◆ ***Study Guide,*** p. 79 ● ***Reinforcement,*** p. 79 ▲ ***Enrichment,*** p. 79 **Science and Society,** p. 24	**Science Integration Activity 20**
Skill Builder: *Comparing and Contrasting,* p. 482	**Ability Level Worksheets** ◆ ***Study Guide,*** p. 80 ● ***Reinforcement,*** p. 80 ▲ ***Enrichment,*** p. 80 **Transparency Masters,** pp. 75-76	**Color Transparency 38,** Lymphatic System
Summary Key Science Words Understanding Vocabulary Checking Concepts Using Lab Skills Think & Write Critically Apply More Skill Builders Projects	ASSESSMENT RESOURCES **Chapter Review,** pp. 43-44 **Chapter Test,** pp. 130-133 **Performance Assessment in Middle School Science (PAMSS)**	**Chapter Review Software** **Test Bank** **Alternate Assessment** **Performance Assessment**

◆ **Basic** ● **Average** ▲ **Advanced**

ADDITIONAL MATERIALS

SOFTWARE	AUDIOVISUAL	BOOKS/MAGAZINES
Circulation—Organs, Micro Power & Light Co. *Heart Lab,* Educational Activities, Inc. *All About Circulation,* Queue. *The Heart Simulator,* Focus. *Heart Abnormalities and EKGs,* Focus.	*Heart, Lungs, and Circulation,* film, Coronet/MTI. *The Life of a Red Blood Cell,* film, Educational Media International. *The Blood,* film, EBEC. *The Heart and Circulatory System,* film, EBEC. *The Life of a Red Blood Cell,* laserdisc, AIMS Media.	Anthony, Catherine P. and Gary A. Thibodeaux. *Structure and Function of the Body.* 6th ed. St Louis, MO: C.V. Mosley Company, 1983. Avraham, Regina. *Circulatory System.* NY: Chelsea House Publishers, 1989. Clemente, Carmine D. Anatomy: *A Regional Atlas of the Human Body.* 3rd ed. Baltimore, MD: Urban and Schwarzenberg, 1987.

CHAPTER

20

YOUR CIRCULATORY SYSTEM

THEME DEVELOPMENT: One of the major themes in this text is homeostasis. The circulatory system is an excellent example of a body system that functions to maintain the stability of life processes in the human body.

CHAPTER OVERVIEW

- **Section 20-1:** This section provides an overview of the major structures and functions of the circulatory system. The importance of the system is highlighted with a discussion of the diseases and disorders of the heart.

- **Section 20-2:** This section focuses on blood, the flowing tissue of the circulatory system. The liquid and solid components of blood and their role in transporting essential materials are discussed.

- **Section 20-3: Science and Society:** The advantages of an autologous blood transfusion are explored. The You Decide feature asks students to consider the AIDS notification policy of the American Red Cross.

- **Section 20-4:** The lymphatic system is introduced, and its important role in fighting infections is detailed.

CHAPTER VOCABULARY

atria
ventricles
pulmonary circulation
arteries
veins
capillaries
systemic circulation
coronary circulation
blood pressure
atherosclerosis
hypertension
plasma
hemoglobin
platelets
homologous blood transfer
autologous blood transfer
lymphatic system
lymph
lymphocytes
lymph nodes

CHAPTER

20 Your Circulatory System

462

OPTIONS

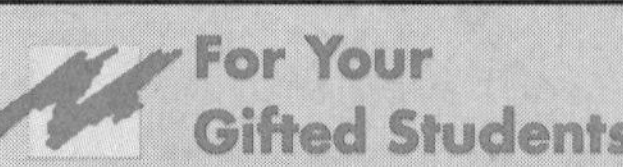

For Your Gifted Students

Have students investigate different types of hearts in vertebrates. Students should diagram the hearts of a fish, an amphibian, and a bird. Diagrams should be labeled with parts of each heart and indicate whether the blood in each chamber is rich in oxygen, poor in oxygen, or mixed. Students should be able to explain how these hearts contrast with the human heart.

For Your Mainstreamed Students

Label index cards with the names of each part of the heart, the blood vessels between the heart and lungs, and each type of blood vessel in the body. (See Figures 20-2 and 20-3.) Shuffle the cards and pass them out to students. Then have students form a human model of the cardiovascular system. Once in correct position, beginning with the right atrium, have students tell what happens to blood in that part of the system.

Strenuous swimming could tire you out, but because your heart constantly pumps blood throughout your body you can continue to swim and play. How do you know that your heart is pumping? Why does it pump faster at certain times?

FIND OUT!

Do this activity to find out about the pumping activity of your heart.

Place your middle and index fingers over one of the carotid arteries in your neck. These arteries are on either side of your trachea. Don't press too hard. You should feel movement as blood is pumped through this artery. This movement is a pulse. Count your carotid pulse rate for 15 seconds, then multiply it by four. This number is your resting heartbeat rate. Now, jog in place for one minute. Take your pulse again. Is there a difference from your resting heart rate? *Hypothesize* why exercise causes a difference in pulse rate.

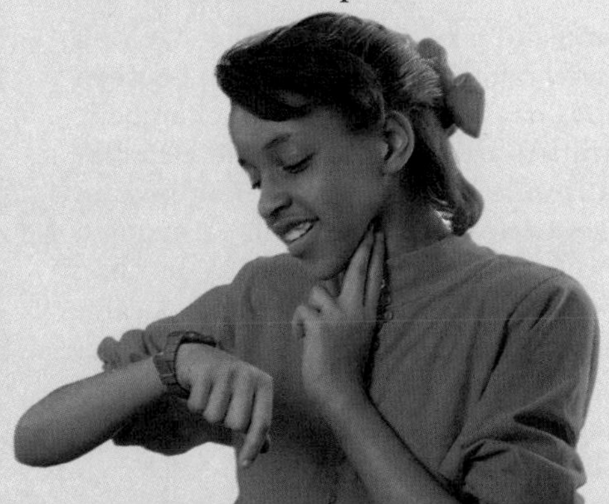

Gearing Up

Previewing the Chapter

Use this outline to help you focus on important ideas in this chapter.

Section 20-1 Circulation
- Your Cardiovascular System
- Your Heart
- Pathways of Circulation
- Blood Pressure
- Diseases and Disorders of the Heart

Section 20-2 Blood
- Functions of Blood
- Parts of Blood
- Blood Clotting
- Blood Types
- Diseases and Disorders of Blood

Section 20-3 Science and Society
Autologous Blood Transfusions
- Saving Your Own Blood

Section 20-4 Your Lymphatic System
- Functions of Your Lymphatic System
- Lymphatic Organs
- A Disease of the Lymphatic System

Previewing Science Skills

- In the Skill Builders, you will make a concept map, interpret data, and compare and contrast.
- In the Activities, you will observe, experiment, and collect data.
- In the MINI-Labs, you will diagram and observe and interpret.

What's next?

You've just learned that your pulse can tell you something about your heartbeat rate. In this chapter, you will learn about your heart and your blood and your lymphatic system.

INTRODUCING THE CHAPTER

Use the Find Out activity to introduce the role of the heart in keeping the blood flowing in the circulatory system.

FIND OUT!

Preparation: Model the proper way to take a pulse on the neck. Have available stopwatches, watches, or a clock with a second hand.

Cooperative Learning: Use Paired Partners or Science Investigation cooperative groups. As one student takes his or her pulse, another monitors the time. The third student records the data and calculates the rate for one minute. Then have students rotate roles.

Teaching Tips
- Any student unable to jog can be timekeeper or recorder for a group.
- Students should count pulse beats silently to themselves and then report the results to the recorder.
- Explain the reason for multiplying by four to obtain the pulse rate for one minute.

Gearing Up

Have students study the Gearing Up feature to familiarize themselves with the chapter. Discuss the relationships of the topics in the outline.

What's Next?

Before beginning the first section, make sure students understand the connection between the Find Out activity and the topics to follow.

ASSESSMENT OPTIONS

PORTFOLIO

Refer to page 483 for suggested items that students might select for their portfolios.

PERFORMANCE ASSESSMENT

See page 483 for additional Performance Assessment options.

Process
Skillbuilders, pp. 469, 476
MINI-Labs, pp. 466, 468
Activities, 20-1, p. 470; 20-2, p. 477
Using Lab Skills, p. 484

CONTENT ASSESSMENT

Assessment—Oral, pp. 468, 474, 482
Skillbuilder, p. 482
Section Reviews, pp. 469, 476, 479, 482
Chapter Review, pp. 483-485
Mini Quizzes, pp. 466, 475, 481

GROUP ASSESSMENT

Opportunities for group assessment occur with Cooperative Learning Strategies and Flex Your Brain Activities.

SECTION 20-1

PREPARATION

SECTION BACKGROUND

▶ The heart and the blood vessels make up the cardiovascular system of the circulatory system. The action of the heart drives the blood through a closed system of tubes.

▶ Arteries have greater elasticity than veins do in order to withstand the forceful movement of blood through them.

▶ Inhibitory and excitatory impulses to the heart are regulated by the cardiac center of the medulla.

▶ Blood distributes materials for cell activities and functions in maintaining homeostasis within the cell.

PREPLANNING

▶ To prepare for Activity 20-1, obtain the necessary sphygmomanometers and stethoscopes.

1 MOTIVATE

▶ Discuss with students any evidence they have that there is a circulatory system within each of their bodies. Students may indicate they can see veins, feel and hear their heart beating, and see blood flow from a cut.

▶ **Demonstration:** Obtain a beef heart to examine. Draw students' attention to the thick, muscular walls of the ventricles, which are necessary to supply a lifetime of pumping action.

VideoDisc

STVS: Modeling Blood Flow, Disc 7, Side 1

STVS: Measuring Blood Pressure, Disc 7, Side 1

STVS: Testing Heart Valves, Disc 7, Side1

20-1 Circulation

New Science Words

atria
ventricles
pulmonary circulation
arteries
veins
capillaries
systemic circulation
coronary circulation
blood pressure
atherosclerosis
hypertension

Objectives

- ▶ **Compare arteries, veins, and capillaries.**
- ▶ **Trace the pathway of blood through the chambers of the heart.**
- ▶ **Describe pulmonary and systemic circulation.**

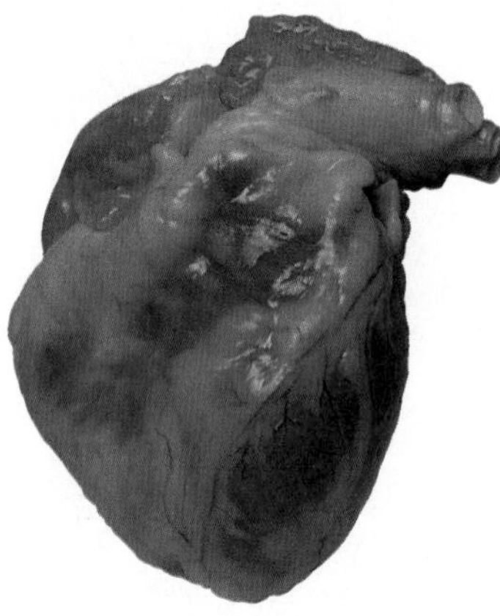

Figure 20-1. A Human Heart

Your Cardiovascular System

With a body made up of trillions of cells, you're quite different from a one-celled amoeba living in a puddle of water. But are you really that different? Even though your body is larger and made up of complex systems, the cells in your body have the same needs as the smaller life-form. Both of you need a continuous supply of oxygen and nutrients and a way to remove cell wastes.

An amoeba takes oxygen directly from its watery environment. Nutrients are distributed throughout its single cell by cytoplasmic streaming. In your body, a cardiovascular system moves oxygen and nutrients to cells and removes carbon dioxide and other wastes. *Cardio-* means heart and *vascular* means vessel. This system includes your heart, blood, and kilometers of vessels that carry blood to every part of your body. It is a closed system because blood moves within vessels. How do the parts of this system work?

Your Heart

Your heart is a muscular organ about the size of a closed fist. It is located behind your sternum and between your lungs. Your heart has four cavities called chambers. The two upper chambers are the right and left **atria** (AY tree uh) (*singular* atrium). The two lower chambers are the right and left **ventricles** (VEN trih kulz). During a single heartbeat, both atria contract at the same time. Then both ventricles contract at the same time. A valve separates each atrium from the ventricle below it, so that blood flows only from an atrium to a ventricle. 1

OPTIONS

Meeting Different Ability Levels

For Section 20-1, use the following **Teacher Resource Masters** depending upon individual students' needs.

◆ **Study Guide Master** for all students.

● **Reinforcement Master** for students of average and above average ability levels.

▲ **Enrichment Master** for above average students.

Additional Teacher Resource Package masters are listed in the OPTIONS box throughout the section. The additional masters are appropriate for all students.

◆ STUDY GUIDE 77

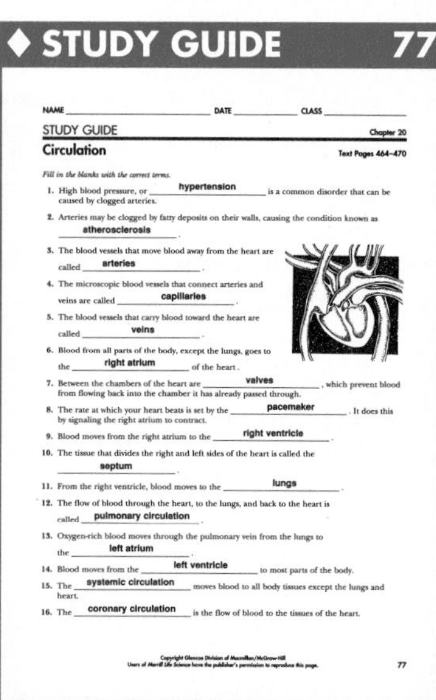

NAME ____ DATE ____ CLASS ____

STUDY GUIDE Chapter 20

Circulation Text Pages 464–470

Fill in the blanks with the correct terms.

1. High blood pressure, or **hypertension** is a common disorder that can be caused by clogged arteries.
2. Arteries may be clogged by fatty deposits on their walls, causing the condition known as **atherosclerosis**.
3. The blood vessels that move blood away from the heart are called **arteries**.
4. The microscopic blood vessels that connect arteries and veins are called **capillaries**.
5. The blood vessels that carry blood toward the heart are called **veins**.
6. Blood from all parts of the body, except the lungs, goes to the **right atrium** of the heart.
7. Between the chambers of the heart are **valves**, which prevent blood from flowing back into the chamber it has already passed through.
8. The rate at which your heart beats is set by the **pacemaker**. It does this by signaling the right atrium to contract.
9. Blood moves from the right atrium to the **right ventricle**.
10. The tissue that divides the right and left sides of the heart is called the **septum**.
11. From the right ventricle, blood moves to the **lungs**.
12. The flow of blood through the heart, to the lungs, and back to the heart is called **pulmonary circulation**.
13. Oxygen-rich blood moves through the pulmonary vein from the lungs to the **left atrium**.
14. Blood moves from the **left ventricle** to most parts of the body.
15. The **systemic circulation** moves blood to all body tissues except the lungs and heart.
16. The **coronary circulation** is the flow of blood to the tissues of the heart.

 77

Pathways of Circulation

Three major pathways for blood movement through your circulatory system are called pulmonary (PUL mo ner ee), systemic (sihs TEM ihk), and coronary circulations.

Pulmonary Circulation

Pulmonary circulation is the flow of blood through the heart, to the lungs, and back to the heart. Blood from body cells is high in carbon dioxide and enters the right atrium of the heart through large veins called the vena cavae. When the right atrium contracts, this blood is forced into the right ventricle. The right ventricle then contracts, and blood leaves the heart through the pulmonary arteries to go to the lungs. As blood circulates through the lungs, carbon dioxide is exchanged for oxygen. Oxygen-rich blood then returns to the heart through the pulmonary vein and fills up the left atrium. These are the only veins in the body that carry oxygen-rich blood. When the left atrium is full of oxygen-rich blood, it contracts and forces blood into the left ventricle. The final step of this path through the heart occurs when the left ventricle contracts and forces blood up and out of the heart into the largest artery of the body, called the aorta. The aorta carries blood away from the heart to many branching arteries that distribute it to all 2

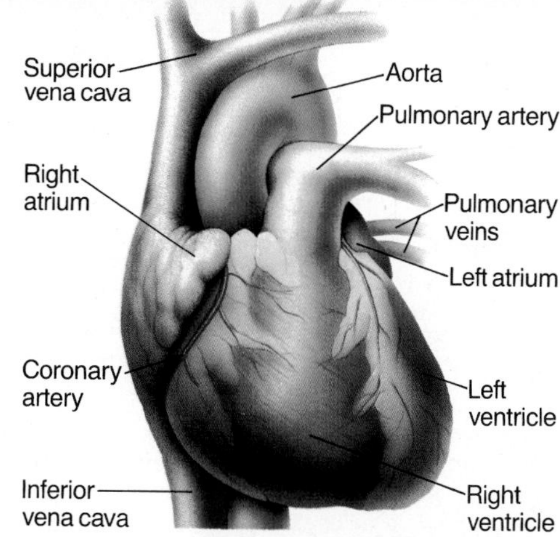

Figure 20-2. In mammals, the heart is a two-pump organ. The right ventricle pumps oxygen-poor blood to the lungs, and the left ventricle pumps oxygen-rich blood to body tissues.

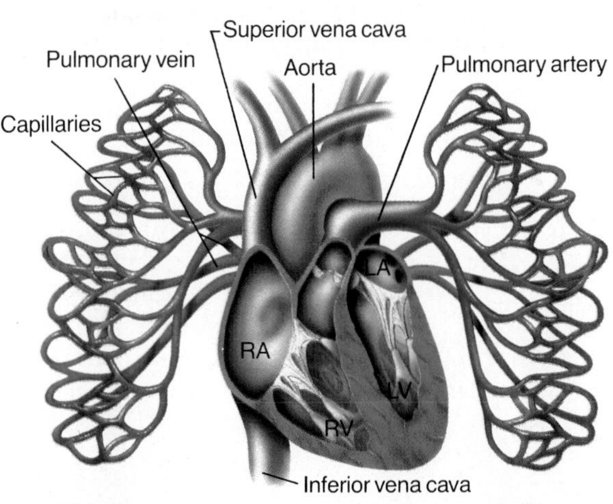

Figure 20-3. Pulmonary circulation moves blood between the heart and lungs.

● REINFORCEMENT 77

NAME ______ DATE ______ CLASS ______

REINFORCEMENT — Chapter 20

Circulation — Text Pages 464–470

Label the diagram of the heart. Include the following terms on your diagram: right atrium, left atrium, right ventricle, left ventricle, superior vena cava, inferior vena cava, pulmonary artery, aorta. Use a pencil to draw arrows showing the path of oxygen-rich blood. Use a pen to show the path of oxygen-poor blood.

superior vena cava
aorta
pulmonary artery
right atrium
to lungs
from lungs
left atrium
to smaller arteries & capillaries
pen
pencil
inferior vena cava
right ventricle
left ventricle

Fill in the blanks with the correct terms for questions 1-4.

1. The **pulmonary vein** is the only vein in the body that carries oxygen-rich blood. This vein moves blood from the **lungs** to the **heart**.
2. Valves inside the veins prevent blood from flowing **away from** the heart.
3. Blood in **veins** carries wastes away from the cells of the body back to the heart.
4. A measure of the flow of blood in arteries is called **blood pressure**.
5. What is the function of pulmonary circulation? **Pulmonary circulation takes blood high in carbon dioxide and other cell wastes to the lungs. In the lungs, the blood exchanges carbon dioxide for oxygen and returns to the heart. The heart then pumps the blood to body tissues.**

77

▲ ENRICHMENT 77

NAME ______ DATE ______ CLASS ______

ENRICHMENT — Chapter 20

Circulation — Text Pages 464–470

EXERCISE AND CIRCULATION

Design an experiment to test the circulation of your friends and family members. Choose an exercise that will increase circulation, such as running in place or jumping jacks. Take a pulse reading or a blood pressure reading for each person before and after exercising. Make sure the exercise is not too strenuous for anyone you are testing.

Fill in the following table as you conduct your experiment. The rate in the table refers to the pulse rate or blood pressure reading. If you need more space, use a separate sheet of paper.

Person						
Age of person						
Rate before exercise						
Type of exercise						
Length of time exercising						
Rate after exercising						
Rate after resting						

Write a conclusion for your experiment. Include answers to the following questions.
1. Did the heart rate differ with age?
2. Did the heart rate go up by the same amount in all the people tested?
3. How long did each person rest after exercising? Had his or her heart rate returned to the first measurement? Test yourself to see how long it takes for your heart rate to return to normal.

Conclusion:

77

TYING TO PREVIOUS KNOWLEDGE: Recall information in Unit 5 about cold-blooded and warm-blooded vertebrates. Relate the role of the circulatory system in maintaining a relatively constant temperature within our bodies.

OBJECTIVES AND SCIENCE WORDS: Have students review the objectives and science words throughout the chapter to become familiar with each section.

2 TEACH

Key Concepts are highlighted.

CONCEPT DEVELOPMENT

▶ **What does the word *circulate* mean?** The word *circulate* refers to movement or flow in a circle. Discuss with students the concept of fluid movement within a closed system.

▶ **What is required to move a fluid in a closed system?** *a pumping mechanism*

▶ **Demonstration:** Have each student make a closed fist with the right hand and hold it on the chest "to the left of the sternum." (Students might think that the heart is under the sternum, directly in the middle of the chest.) This will give them an idea of the size and position of the heart.

MINI-Lab

Teaching Tip

A diagram of the heart can be reproduced and handed out to students. Diagrams should reflect information in Figure 20-3.

Answer: The septum separates oxygen-rich blood in the left side of the heart from oxygen-poor blood in the right side of the heart. A "blue baby" has a congenital heart defect whereby some blood goes from the right side to the left side of the heart through a hole in the septum.

MINI-Lab

ASSESSMENT

Performance: To further assess students' understanding of blood flow through the heart, have them add direction arrows and color the veins and arteries.

CHECK FOR UNDERSTANDING

Check students' understanding of the direction of blood flow in arteries (from the heart) and in veins (to the heart) and how the blood gets from the arteries to the veins (capillaries). Divide the class into groups of two students. Hang a large paper on the wall for each pair. One student will draw an outline of the body and the major vessels. The other student will label the vessels.

RETEACH

On a chart of the circulatory system, have students trace with their fingers the flow of blood from the heart to a major artery and back to the heart.

EXTENSION

For students who have mastered this section, use the **Reinforcement** and **Enrichment** masters or other OPTIONS.

MINI QUIZ

Use the Mini Quiz to check students' recall of chapter content.

1. **The lower chambers of the heart are the _____ .** *ventricles*
2. **What is the largest artery?** *aorta*
3. **What is coronary circulation?** *flow of blood to tissues of the heart*
4. **Another name for high blood pressure is _____ .** *hypertension*

MINI-Lab

What are the parts of the heart?
Draw and label a diagram of the heart. Color the oxygen-rich parts red and the carbon-dioxide parts blue. *Interpret* the purpose served by the thick septum or wall of muscle between the left and right sides of the heart. How is the heart of a "blue baby" different?

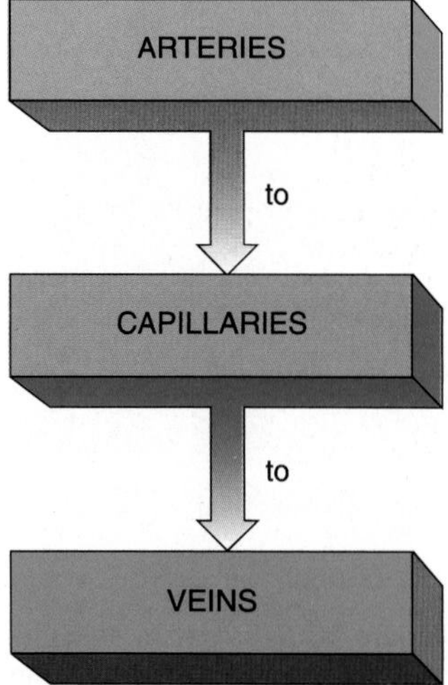

Figure 20-4. Blood flows from arteries to capillaries to veins. Valves in veins and muscles to the right help blood flow back toward the heart.

body parts. Use Figure 20-3 to trace the pathway of blood through the heart, to the lungs, and back to the heart again.

Blood Vessels

Until the work of William Harvey was published in 1628, many people believed that blood moved back and forth in the body like an ocean tide. His work showed that blood circulates only in one direction and that it is moved by the pumping action of the heart. He found that blood flows from arteries to veins. But he couldn't figure out how blood got from arteries to veins. It wasn't until the invention of the microscope that capillaries, the tiniest vessels of the cardiovascular system, were seen.

When blood is pushed out of your heart, it begins a journey through your arteries, capillaries, and veins. **Arteries** are blood vessels that move blood *away* from your heart. Arteries have thick elastic walls that are lined with smooth muscle. Each ventricle of the heart is connected to an artery, so with each contraction blood is moved from the heart into arteries.

Veins are blood vessels that move blood *to* the heart. Veins have one-way valves to keep blood moving toward the heart. If there is a backward movement of the blood, the pressure of the blood closes the valves. The greatest number of valves are in veins in the legs. Why do you think this is so? Veins that are near skeletal muscles are squeezed when these muscles contract. This action also helps blood move toward the heart. Blood in veins carries waste materials from cells and is therefore low in oxygen.

Capillaries are microscopic blood vessels that connect arteries and veins. The walls of capillaries are only one

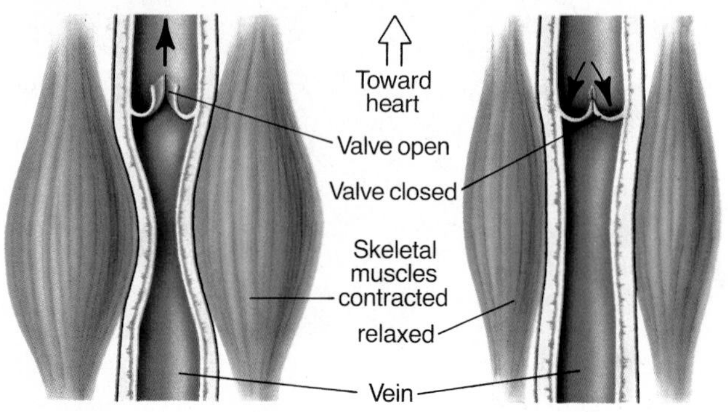

OPTIONS

ENRICHMENT

- Have students research how food and oxygen move from capillaries into surrounding cells.
- Have students research how artificial pacemakers stimulate heart contractions.
- Have students investigate the relationship between intake of dietary cholesterol and fatty deposits on artery walls.

PROGRAM RESOURCES

From the **Teacher Resource Package** use:

Transparency Masters, pages 71-72, Pulmonary Circulation.

Transparency Masters, pages 73-74, Human Circulatory System.

Use **Color Transparency** 36, Pulmonary Circulation.

Use **Color Transparency** 37, Human Circulatory System.

cell thick. Food and oxygen diffuse to body cells through the thin capillary walls. Materials move from body cells into the capillaries to be carried back to the heart.

Systemic circulation moves blood to all body tissues except the lungs and heart. This is the longest of the three pathways. It carries oxygen-rich blood from the left ventricle through the aorta to arteries and capillaries in all the organs and tissues of the body. Nutrients and oxygen are exchanged for carbon dioxide and wastes. Blood returns to the heart in veins from the head and neck through the superior vena cava. Blood returns from your abdomen and the lower parts of your body through the inferior vena cava to the right atrium. Then, oxygen-poor blood is sent to the lungs.

Your heart has its own blood vessels that supply it with nutrients and oxygen and remove wastes. These blood vessels are the coronary arteries and veins. It is on these blood vessels that coronary by-pass surgery is performed. **Coronary circulation** is the
3 flow of blood to the tissues of the heart.

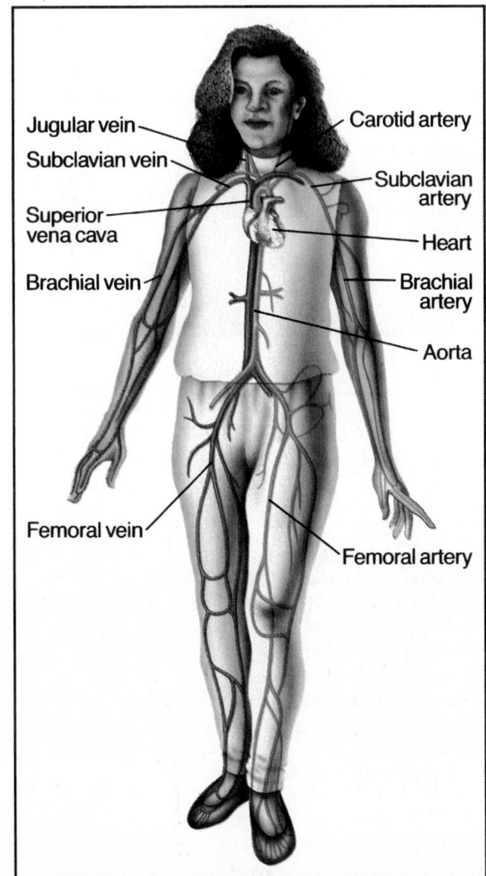

Figure 20-5. Systemic Circulation

Blood Pressure

When you pump up a bicycle tire, you can feel the pressure of the air on the walls of the tire. In the same way, when the heart pumps blood through the cardiovascular system, blood exerts a force called **blood pressure** on the walls of the vessels. This pressure is highest in arteries.

As the wave of pressure rises and falls in your arteries, it is felt throughout your body as a pulse. Normal pulse rates are between 65 and 80 beats per minute. There is less pressure in capillaries and even less in veins.

Blood pressure is measured in large arteries and is expressed by two numbers, such as 120 over 80. The first number is a measure of the pressure caused when the ventricles contract and blood is pushed out of the heart. Then blood pressure suddenly drops as the ventricles relax. The lower number is a measure of the pressure when the ventricles are filling up, just before they contract again.

According to a law of physics, when pressure is exerted on a fluid in a closed container, the pressure is transmitted through the liquid in all directions. Your circulatory system is like a closed container and each heart beat exerts pressure. Where do you think that pressure is transmitted?

CROSS CURRICULUM

▶ **Mathematics:** Have students calculate how many times a heart beats in one hour, day, month, year, and in 70 years. Use a pulse rate of 70 beats per minute.

REVEALING MISCONCEPTIONS

▶ There is a general misconception that veins never carry oxygen-rich blood. Students will learn that the pulmonary vein carries oxygen-rich blood from the lungs to the left atrium.

CONCEPT DEVELOPMENT

▶ **Why does blood pressure drop when the ventricles relax?** *The pressure drops because the blood is no longer being forced into the arteries.*

▶ A normal range of blood pressure for young adults is 110 to 140 for the systolic pressure and 65 to 90 for the diastolic pressure.

Answer: through blood to the walls of blood vessels

FLEX Your Brain

Use the Flex Your Brain activity to have students explore BLOOD VESSELS.

Flex Your Brain

ASSESSMENT

Portfolio: Use the Flex Your Brain activity to find out what students know about blood vessels. At Step 2, students may list hardening of arteries or coronary arteries.

PROGRAM RESOURCES

From the **Teacher Resource Package** use:

Concept Mapping, pages 45-46.

Activity Worksheets, page 182, MiniLab: What are the parts of the heart?

Activity Worksheets, page 5, Flex Your Brain.

Science Integration Activities 20, Preparation of Oxygen

MINI-Lab

Teaching Tips

▶ Show students the proper use of a stethoscope.

▶ Have students find out if the heart sounds are as easy to hear on the surface as from the ventral surface.

Troubleshooting: Have students work in same-sex pairs.

MINI-Lab

ASSESSMENT

Performance: Assess students' understanding of heart sounds by referring to USING LAB SKILLS, question 11, on p. 484.

CONCEPT DEVELOPMENT

▶ Explain to students that heart disease is the major cause of death in our country. Nearly one-third of all deaths are due to heart disease.

▶ The peak pressure of blood flow caused by the contraction of the ventricles is known as systolic pressure. The low pressure of blood flow produced when the heart relaxes is known as diastolic pressure.

In Your JOURNAL

Student reports will vary in length and detail, but should include that bypass surgery enables blood to flow past blocked coronary arterties through grafted vessels.

Diseases and Disorders of the Heart

Any disease or disorder that affects the cardiovascular system can be serious to a person's health. The term *heart disease* is often used to describe any of the health problems that affect the heart. Heart disease is the major cause of death in the United States.

One leading cause of heart disease is **atherosclerosis** (ath uh roh skluh ROH sus), a condition of fatty deposits on arterial walls. Eating too many foods that contain cholesterol can cause these deposits. The fat can build up and form a hard mass that clogs the inside of the vessel. As a result, less blood flows through the artery. If the artery is clogged completely, blood is not able to flow through. When this occurs in a coronary artery, it can cause a heart attack. If an artery in the brain is clogged, a stroke can occur.

MINI-Lab

How does a stethoscope work? Listen to your heart through a stethoscope. What is causing the sounds you hear? *Describe* the two sounds.

4 Another common disorder of the cardiovascular system is high blood pressure, or **hypertension** (hi pur TEN chun). One cause of hypertension is the condition described above. A clogged artery can cause the pressure within the vessel to increase. This can cause the walls to lose their ability to contract and dilate. There also is extra strain on the heart to work harder to keep blood flowing. Being overweight as well as eating foods with too much salt and fat may contribute to hypertension. Smoking and stress also can increase blood pressure.

In Your JOURNAL

What is a bypass operation? Write to the local office of the American Heart Association and find out about bypass operations. Write a report **in your Journal** about how this operation aids victims of atherosclerosis.

Your cardiovascular system efficiently provides your body cells with nutrients and oxygen and removes cell wastes. A well-functioning cardiovascular system is important for good health. You can promote a healthy heart! Choose to eat a healthful diet and exercise regularly.

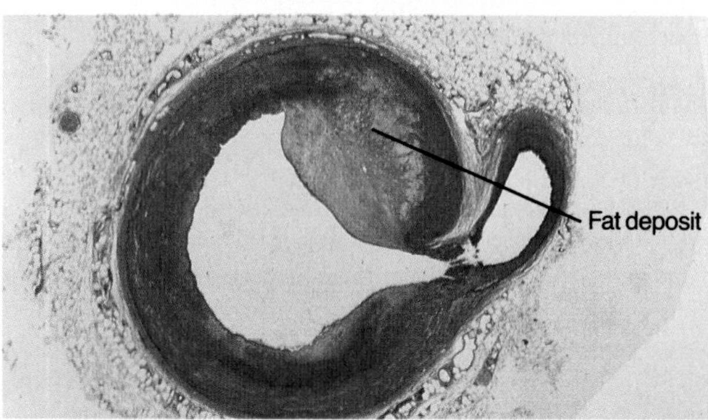

Figure 20-6. Atherosclerosis interferes with blood flow by blocking blood vessels with fatty substances.

OPTIONS

ASSESSMENT—ORAL

▶ **How do the elastic walls of arteries help move blood?** *When blood pulses through an artery, the arterial walls expand and then contract to push the blood forward.*

▶ **What would happen if the valves in a major vein would not function?** *Blood flow would be sluggish, and cellular waste products would not be removed.*

▶ **When you receive a cut on the surface of the skin, what type of blood vessels are most likely damaged?** *capillaries*

PROGRAM RESOURCES

Activity Worksheets, p. 183, MINI-Lab: How does a stethoscope work?

Use **Laboratory Manual,** page 137, Heart Structure.

Use **Laboratory Manual,** page 141, Measuring Heartbeat.

Use **Laboratory Manual,** page 145, Blood Pressure.

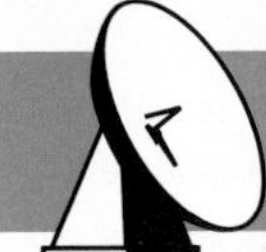

TECHNOLOGY

An Assist for the Heart

Some patients need temporary help until a damaged or transplanted heart is able to function properly. Miniature heart pumps, like the Hemopump, may be the answer.

The Hemopump is the size of the eraser of a number 2 pencil. When a transplanted heart shows signs of being rejected, the Hemopump can take over, pumping blood for several days until antirejection drugs allow the patient's new heart to function.

The Hemopump is inserted into an artery in the upper part of the leg. From here it is carefully pushed up through the blood vessels into the left ventricle. A tiny blade, making 25 000 revolutions per minute, draws the blood through the tube and spills it into the aorta. The pump is connected to a motor outside the patient's body by a thin cable. The operation to insert the pump takes 20 minutes.

Think Critically: How could the Hemopump be used to help victims of heart attack?

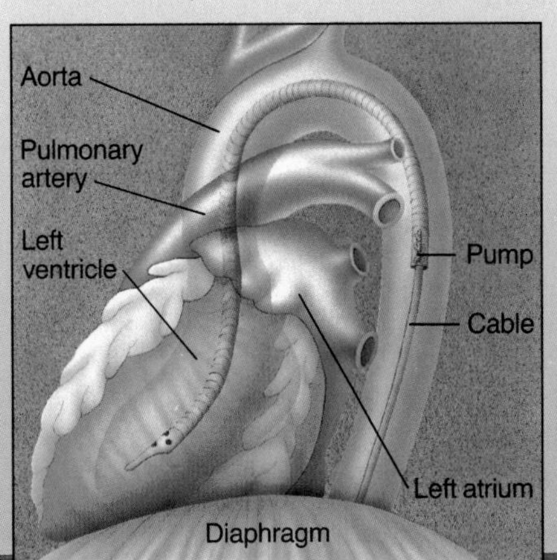

SECTION REVIEW

1. Compare and contrast the three types of blood vessels.
2. Explain the pathway of blood through the four chambers of the heart.
3. Contrast pulmonary and systemic circulations.
4. **Apply:** Your heart, like any muscle, needs exercise. What exercises should you do to give your heart a workout?
5. **Connect to Chemistry:** What respiration waste product builds up in blood and cells when the heart is unable to pump blood efficiently?

Concept Mapping

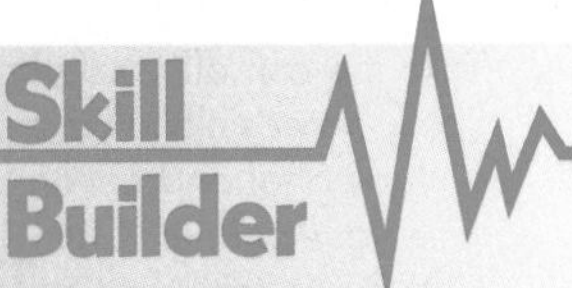

Make an events chain concept map to show pulmonary circulation beginning at the right atrium and ending at the aorta. If you need help, refer to Concept Mapping in the **Skill Handbook** on pages 688 and 689.

INQUIRY QUESTIONS

▶ **In what vessel is the oxygen content very high?** *pulmonary vein*

▶ **What chamber of the heart receives blood high in carbon dioxide?** *right atrium*

▶ **What prevents blood from leaking from one chamber of the heart into another?** *valves*

TECHNOLOGY

Research the development of the Jarvik-7 artificial heart and discuss why heart pumps might be more beneficial than artificial hearts.

Think Critically: It could help keep a patient alive until the heart can work normally again.

3 CLOSE

▶ Ask questions 1-3 and the **Apply** Question in the Section Review.

SECTION REVIEW ANSWERS

1. Students' answers should reflect information on pages 466 and 467.
2. Blood enters the right atrium and then moves into the right ventricle, which pumps blood to the lungs. From the lungs, blood enters the left atrium and then moves into the left ventricle, and out to the body through the aorta.
3. Students' answers should reflect information on page 465 and page 467.
4. Apply: Exercises, such as jogging, walking, swimming, or biking, when sustained for at least 20 minutes will provide exercise for the heart.
5. Connect to Chemistry: CO_2

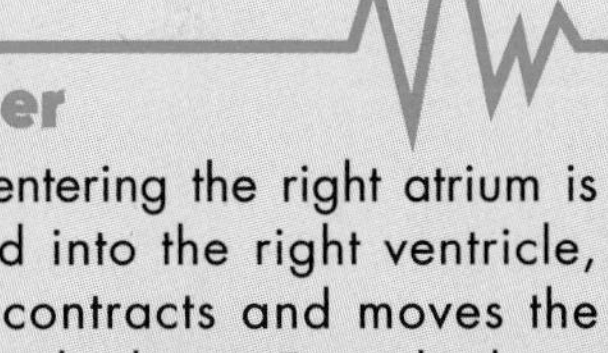

Blood entering the right atrium is pumped into the right ventricle, which contracts and moves the blood to the lungs. From the lungs the blood enters the left atrium, which contracts and forces the blood into the left ventricle, which then contracts to move the blood to the body through the aorta.

Skillbuilder
ASSESSMENT

Performance: Assess students' abilities to form a Concept Map of the flow of blood through the three types of vessels.

ACTIVITY 20-1
30 minutes

OBJECTIVE: **Measure** arterial pressure and interpret results on the data.

PROCESS SKILLS applied in this activity: ▶ **Interpreting Data** in Conclude and Apply Question 3.

COOPERATIVE LEARNING Divide the class into Science Investigation cooperative groups. Students will take turns measuring arterial blood pressure on a partner. Then students will rotate roles.

TEACHING THE ACTIVITY
Troubleshooting: Remind students to close the valve on the pump. The pump should not be squeezed so the needle exceeds 160 on the gauge. The valve of the pump should be opened slowly.

ACTIVITY 20-1 Taking Blood Pressure

Problem: *How can you measure arterial pressure?*

Materials 
- sphygmomanometer
- stethoscope
- table
- chairs (2)
- alcohol
- cotton balls

Procedure
1. Your partner should sit down on a chair with his or her arm on the table. One hand should face up as in the photo.
2. Sit and face your partner. Place the blood pressure cuff on your partner's arm above the elbow. Align the arrows on the cuff. Close the cuff snugly.
3. Close the valve on the pump. Place the ear pieces of the stethoscope in your ears. Place the round stethoscope head on your partner's brachial artery as shown in photograph. Squeeze the pump until the needle reaches 160 on the gauge. No blood will flow through this constricted area.
4. *Slowly* open the valve of the pump. *Do not open the valve quickly. Observe* the gauge. The needle will drop at a constant rate. Watch for the number on the gauge when you first hear a beat. This is the top, or systolic, number.
5. Keep listening for a heartbeat. *Observe* the needle on the gauge. Note the number on the gauge when you no longer hear a beat. This is the bottom, or diastolic, number. At this point, blood pressure has dropped; blood flow has returned to normal.
6. Take the cuff off your partner's arm. Record your partner's blood pressure. Clean ear pieces with alcohol and cotton balls. Trade places with your partner. **CAUTION:** *Do not do this activity on the same person more than one time per class period.*
7. *Make* and fill in *a table* like the one shown.

Analyze
1. Describe what happened in your brachial artery during this activity.
2. As you listened to your partner's blood pressure, what did you notice about the needle on the gauge when you first heard the heartbeat rate and later when it stopped?

Conclude and Apply
3. How was your blood pressure different from your partner's?
4. What can you *infer* about a person who has low blood pressure?

Data and Observations

Blood Pressure	Example	Yours	Partner's
Systolic	120		
Diastolic	80		

ANSWERS TO QUESTIONS
1. The pressure in the brachial artery is temporarily closed off, then released.
2. The needle dropped on the gauge.
3. Pressures will probably be about the same unless other factors such as stress or health complications exist.
4. Accept all reasonable conclusions. Some will answer that this person will be tired, lacking in energy, and may appear pale.

PROGRAM RESOURCES

From the **Teacher Resource Package** use:
Activity Worksheets, pages 176-177, Activity 1, 20-1 Taking Blood Pressure.

Activity
ASSESSMENT
Performance: To further assess students' understanding of blood pressure, have students graph the class data to find the average range of pressures.

Blood

20-2

Objectives

- Describe the characteristics and functions of the parts of blood.
- Explain the importance of checking blood types before a transfusion is given.
- Describe a disease and a disorder of blood.

New Science Words

plasma
hemoglobin
platelets

Functions of Blood

Blood makes up about eight percent of your body's total mass. If you weigh 45 kg, you have about 3.6 kg of blood moving through your body. The amount of blood in an adult would fill five one-liter bottles. If this volume falls, the body is sent into shock because blood pressure falls rapidly.

Blood is a tissue consisting of cells, cell fragments, and liquid. Blood has many important functions. It plays a part in every major activity of your body. First, blood carries oxygen from your lungs to all body cells. It also removes carbon dioxide from your body cells and carries it to the lungs to be exhaled. Second, it carries waste products of cell activity to your kidneys to be removed. Third, blood transports nutrients from the digestive system to body cells. Fourth, materials in blood fight infections and help heal wounds. Anything that disrupts or changes any of these functions affects all the tissues of the body.

1 What does blood consist of?

Parts of Blood

If you've ever taken a ride in a water slide at an amusement park, you have some idea of the twisting and turning travels of a blood cell inside a blood vessel. Surrounded by water, you travel rapidly through a narrow watery passageway, much like a red blood cell moves in the plasma of blood.

Figure 20-7. Donated blood is measured in units. The units pictured show how much blood is in your body.

OPTIONS

INQUIRY QUESTIONS

- **What is the approximate ratio of white blood cells to red blood cells in a millimeter of blood?** *1/800*
- **Why are red blood cells formed at such high rates?** *The life span of red blood cells is only 120 days. A large number wear out and must be replaced.*
- **How are bacteria in tissues captured?** *White blood cells move out of the capillaries and into the tissues to ingest them.*

SECTION 20-2

PREPARATION

SECTION BACKGROUND

- Cell and cell fragments make up the formed elements in blood. The percent of formed elements in total blood volume is about 46 percent.
- White blood cells can be classified into five groups—neutrophils, eosinophils, basophils, monocytes, and lymphocytes.
- Blood platelets are fragments of large cells called megakaryocytes that are formed in red bone marrow.
- A normal component of plasma is cholesterol, a lipid substance, necessary for many cell activities. Excess dietary cholesterol can cause fatty deposits in blood vessels.

PREPLANNING

- To prepare for Activity 20-2, obtain the necessary microscope slides.

1 MOTIVATE

- **Demonstration:** Fill two and one-half 2-liter plastic soft drink bottles full of water with red food coloring. The water represents the 5 liters of blood in an adult.
- To demonstrate the small size of red blood cells, have students examine the tip of a pencil or head of a straight pin and tell them that either space could hold more than five million red blood cells.

VideoDisc

STVS: White Cells in Action, Disc 7, Side 1
STVS: Natural Time-Release Capsules, Disc 7, Side 1

TYING TO PREVIOUS KNOWLEDGE: Recall from Chapter 18 that red blood cells form in bone marrow. Relate the necessity of this when discussing the life span of red blood cells.

In Your JOURNAL

CPR stands for cardiopulmonary resuscitation. CPR is an emergency procedure that is used with mouth-to-mouth resuscitation when the heart has stopped beating. Call the local American Red Cross for more information.

TEACHER F.Y.I.

▶ About one-third of a red blood cell is hemoglobin. When combined with oxygen, it is bright red; it is darker red when oxygen has been released.

2 TEACH

Key Concepts are highlighted.

CONCEPT DEVELOPMENT

▶ Stress the *four* major functions of blood. Students generally only think of blood as carrying oxygen and nutrients to cells.

▶ **What chemical in red blood cells carries the oxygen and carbon dioxide gases?** *hemoglobin*

▶ **How do blood platelets function in maintaining homeostasis when the skin is cut?** *Platelets form blood clots to help prevent further loss of blood.*

▶ **Demonstration:** Have students view a film clip of amoeba activity to draw an analogy to white blood cell movements and ingestation.

In Your JOURNAL

CPR is a life-saving technique everyone should know. **In your Journal,** write what the letters stand for. Analyze the parts of the words to find out what they mean. Find out where you can learn CPR.

If you examine blood closely, you would see that is it not just a red-colored liquid. Blood is a tissue made of red and white blood cells, platelets, and plasma. **Plasma** is the liquid part of blood and consists mostly of water. Plasma makes up more than half the volume of blood. Nutrients, minerals, and oxygen are dissolved in plasma.

A cubic millimeter of blood has more than five million red blood cells. These disk-shaped blood cells contain **hemoglobin,** a chemical that can carry oxygen and carbon dioxide. Hemoglobin carries oxygen from your lungs to body cells. Red blood cells also carry carbon dioxide from body cells to your lungs. Red blood cells have a life span of about 120 days. They are formed in the marrow of long bones at a rate of two to three million per second and contain no nuclei. About an equal number of old ones wear out and are destroyed in the same time period.

2 In contrast to red blood cells, there are only about five to ten thousand white blood cells in a cubic millimeter of blood. White blood cells fight bacteria, viruses, and other foreign substances that constantly try to invade your body. Your body reacts to infection by increasing its number of white blood cells. White blood cells slip between the cells of capillary walls and out around the tissues that have been invaded. Here, they ingest foreign substances and dead cells. The life span of white blood cells varies from a few days to many months.

Platelets are irregularly shaped cell fragments that help clot blood. A cubic millimeter of blood may contain as many as 400 thousand platelets. Platelets have a life span of five to nine days. Table 20-1 summarizes the solid parts of blood and their functions.

Figure 20-8. A cut vessel below shows an escaping red blood cell. Blood as shown on the right, contains disc-shaped red blood cells, a white cell (with a wrinkled appearance), and platelets, small irregularly shaped fragments.

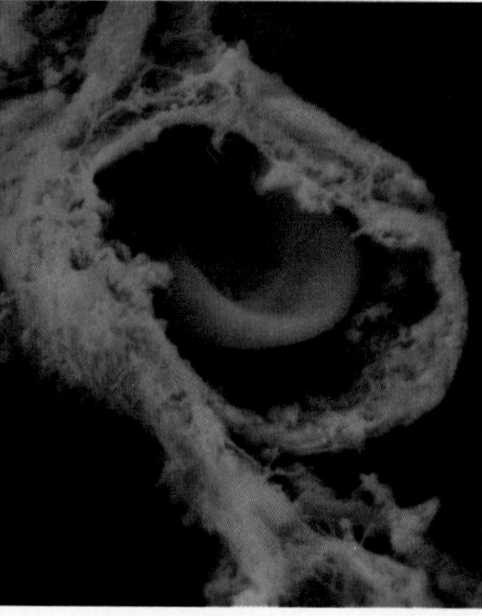

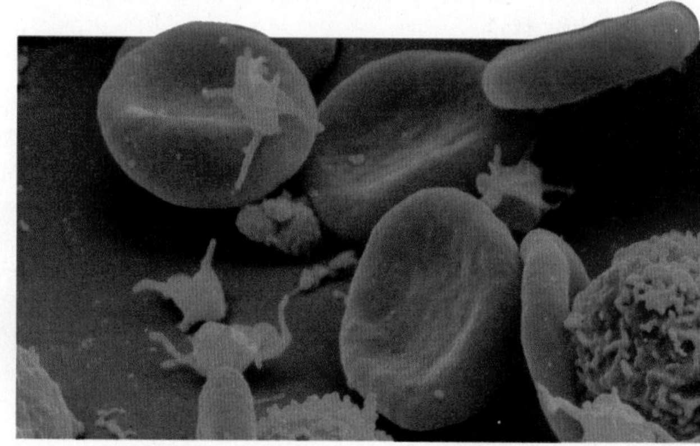

OPTIONS

Meeting Different Ability Levels

For Section 20-2, use the following **Teacher Resource Masters** depending upon individual students' needs.

◆ **Study Guide Master** for all students.

● **Reinforcement Master** for students of average and above average ability levels.

▲ **Enrichment Master** for above average students.

Additional Teacher Resource Package masters are listed in the OPTIONS box throughout the section. The additional masters are appropriate for all students.

◆ STUDY GUIDE 78

NAME ________ DATE ________ CLASS ________

STUDY GUIDE — Chapter 20

Blood — Text Pages 471–477

Answer the following questions.

1. What percentage of your body mass is blood? **about 8%**
2. What is blood made of? **red and white blood cells, platelets, and plasma**
3. What is plasma made of? **nutrients, minerals, and oxygen dissolved in water**
4. Which gas does blood carry to cells of the body? What else does it carry to the cells? **oxygen, nutrients**
5. Which gas does blood carry away from cells of the body? What else does it carry away from the cells? **carbon dioxide, wastes**
6. Which type of blood cells are more numerous in the blood? **red blood cells**
7. What type of blood cells fight infection? **white blood cells**
8. What blood type contains both substance A and substance B? **type AB**
9. What blood type contains neither substance A nor substance B? **type O**
10. How long do red blood cells live? **about 120 days** How long do white blood cells live? **about 30 days**

Use terms from your textbook to complete the sentences below.

11. Red blood cells carry **hemoglobin**, a chemical that can pick up oxygen and carbon dioxide.
12. A person receives blood or blood parts during a **blood transfusion**.
13. A disease in which one or more types of white blood cells are produced in greater number is called **leukemia**.
14. A disorder in which there are few red blood cells or too little hemoglobin in the blood is called **anemia**.
15. The liquid part of blood that is mostly made of water is called **plasma**.
16. The odd-shaped cell fragments that help clot blood are called **platelets**.
17. **White blood cells** fight bacteria, viruses, and other foreign substances in the body.
18. Blood groups and **Rh factor** are checked before transfusions are given.
19. Your body reacts to infection by **increasing** its white blood count.

78

Table 20-1

TYPES OF BLOOD CELLS			
Name	**Nucleus**	**Function**	**Where Formed**
Red blood cells	no	carry oxygen and carbon dioxide	red bone marrow
White blood cells	yes	destroy bacteria, foreign substances	red bone marrow, some move to lymph tissue to produce more white blood cells
Platelets	no	help form blood clots	red bone marrow (fragments of another type of blood cell)

Blood Clotting

It would be serious if you got a paper cut that would not stop bleeding. Platelets in your blood help prevent loss of blood by making a blood clot. A blood clot is somewhat like a bandage. When you cut yourself, a series of chemical reactions cause threadlike fibers called fibrin to form a sticky net that traps escaping blood cells and plasma. This forms a clot and helps prevent further loss of blood. Blood clots are important for homeostasis. Figure 20-9 shows blood clotting after a cut.

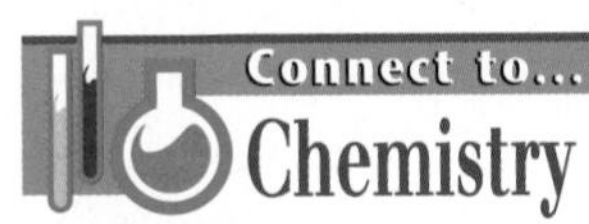

When you have a cut, substances released by broken platelets start chemical reactions that cause fibers to form. These fibers trap cells and a clot forms. Find out about the role of calcium and vitamin K in blood clotting.

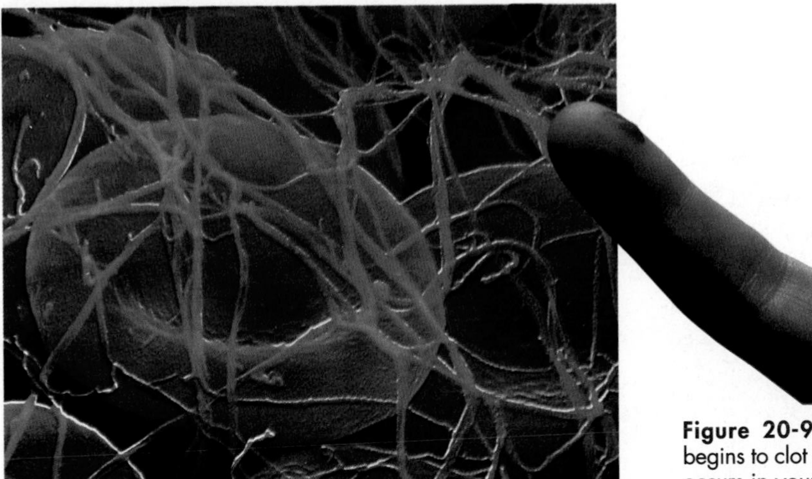

Figure 20-9. Healthy blood begins to clot as soon as a break occurs in your skin.

● REINFORCEMENT 78

NAME ______ DATE ______ CLASS ______

REINFORCEMENT — Chapter 20

Blood — Text Pages 470–477

Fill in the missing words to describe the four functions of blood.

1. carries **oxygen** to and removes **carbon dioxide** from all body cells
2. carries **waste products** of cell activity to **kidneys** to be removed
3. transports **nutrients** from the digestive system to all cells
4. carries materials that fight **infections** and heal wounds

The parts of human blood are shown below. Label each part, and underneath, list the phrase that describes the part: help clot blood; transport oxygen; contains nutrients and minerals; help fight infection.

plasma

white blood cells	red blood cells	platelets	contains nutrients
help fight infection	transport oxygen	help clot blood	and minerals

Complete the following table. Possible blood types of the donor are listed horizontally. The possible blood types of the receiver are listed vertically. Make a check (✓) in the box if the receiver can receive blood from the donor directly above. Write no in the box if the receiver cannot receive blood from the donor directly above. Then answer the questions below the table.

Donor (can give blood to); Receiver (can receive blood from)

Blood Type	O	A	B	AB
O	✓	no	no	no
A	✓	✓	no	no
B	✓	no	✓	no
AB	✓	✓	✓	✓

1. Who can receive any type blood? **people with type AB blood**
2. Who can receive only type O blood? **people with type O blood**

78

▲ ENRICHMENT 78

NAME ______ DATE ______ CLASS ______

ENRICHMENT — Chapter 20

Blood — Text Pages 471–477

BLOOD TYPING

Blood may be typed as A, B, AB, or O. To find the blood type, a drop of blood is mixed with special liquids called serums. Depending on the blood type, clumps will form in the drop of blood or it will be unchanged. For example, when a drop of type A blood is mixed with anti-A serum, clumps form. Clumps form because antibodies in the serum attach to antigens on the red blood cells. Antibodies are substances that provide immunity. Antigens are substances that signal the immune system to react. The antigens and antibodies stick together in clumps. When type A blood is mixed with anti-B serum, no clumps form and the drop of blood appears unchanged.

In the same way, type B blood clumps when mixed with anti-B serum but does not clump when mixed with anti-A serum. Type AB blood clumps with both anti-A and anti-B serum. Type O blood doesn't clump when mixed with any serum.

Look at the microscope slides below. Samples of blood have been mixed with anti-A and anti-B serums. Find out the type of each blood sample and complete the chart.

Sample			Type
Sample 1	anti-A serum	anti-B serum	B
Sample 2	anti-A serum	anti-B serum	AB
Sample 3	anti-A serum	anti-B serum	O
Sample 4	anti-A serum	anti-B serum	A

78

CONCEPT DEVELOPMENT

▶ **Demonstration:** Put a thin layer of cotton fibers on the bottom of a wire sieve. Pour in some white glue. Have students notice how the glue is prevented from oozing through by the network of fibers. Relate this to a blood clot. The dried glue and fibers simulate a scab.

CROSS CURRICULUM

▶ **History:** Have students research how blood transfusion experiments started in the early 1800s.

TEACHER F.Y.I.

▶ Treatment for the Rh problem, *erythroblastosis fetalis,* includes blood transfusion and exposure to fluorescent light.

Answer: Both calcium and vitamin K must be present in order for the chemical reactions in blood clotting to take place.

CHECK FOR UNDERSTANDING

Provide students with a table similar to Table 20-2. Only identify the blood types in the first column. Leave the other two columns blank. Have students complete the table to determine if they understand how the blood groups react.

RETEACH

Cooperative Learning: Use Problem Solving cooperative groups. Have each student pick a different blood group. Let students decide who can give blood to whom or from whom they can receive blood.

EXTENSION

For students who have mastered this section, use the **Reinforcement** and **Enrichment** masters or other OPTIONS provided.

PROBLEM SOLVING

Think Critically: The robber didn't realize that even though he had received type O blood, which can be given to almost anyone, his own type A could not be hidden.

Blood Types

Sometimes, after an accident or during surgery, a person loses a lot of blood. This person may receive blood from another person through a blood transfusion. During a blood transfusion, a person receives blood or parts of blood. In the past, transfusions could cause red blood cells to clump together. Clots would form in blood vessels and result in death. Then, in the early 1900s, Dr. Karl Landsteiner, an Austrian-American scientist, discovered that there are four different types of blood. He called the four types A, B, AB, and O. Landsteiner found that each type has a chemical identification tag called an antigen on its red blood cells. Type A blood has A antigens. Type B has B antigens. Type AB has both A and B antigens on each blood cell. Type O has no A or
3 B antigens.

Each type also has specific antibodies in the plasma of that blood. Antibodies are proteins that destroy or neutralize foreign substances, such as pathogens, in your body. The antibodies prevent certain different blood types from mixing. Type A blood has antibodies against type B. If you mix type A blood with type B blood, type A red blood cells react to type B blood as if it were a foreign substance. The antibodies in type A respond by clumping the type B blood. Type B blood has antibodies against type A. Type AB has no antibodies, so it can receive blood from A, B, and AB. Type O has both A and B antibodies. Table 20-2 lists the four blood types, what they can receive, and what types they can donate to.

Table 20-2

BLOOD TRANSFUSION POSSIBILITIES

Type	Can receive	Can donate to
A	O, A	A, AB
B	O, B	B, AB
AB	all	AB
O	O	all

OPTIONS

ASSESSMENT—ORAL

- **Name the four types of blood groups.** *A, B, AB, O*
- **What blood type may be given to all blood groups?** *O*
- **Name another substance in blood that must be checked before making a transfusion.** *Rh factor*

PROBLEM SOLVING

The Blood Type Mystery

Detective Johnson was on to something. He had been assigned to a confusing robbery case. He had received conflicting testimony from the people involved. Now he had the last piece of evidence he needed to solve the crime.

A person had come home to find someone he knew robbing his apartment. The victim struggled with the robber. He broke a vase over the robber's head. The victim later identified the robber in a police lineup, but the robber said he had an alibi.

While investigating, Detective Johnson found blood on the vase and carpet. He sent the blood to a lab to be analyzed. The blood of the victim and robber also were analyzed, even though the robber offered proof that he had just received a blood transfusion with type O blood.

When Detective Johnson received the blood analysis report, the blood on the vase and carpet was found to be type A. The victim's blood type was B. The robber's blood type was A. The robber was found guilty. The case was solved.

Think Critically: Explain what the robber didn't know about the blood he received.

If you have already studied the chapter on genetics in this textbook, then you know that blood type is inherited. You cannot change your blood type.

The Rh factor is another inherited substance in blood. If present, the person has Rh-positive (Rh+) blood. If it is not present, the person is said to be Rh-negative (Rh−). An Rh− person receiving blood from an Rh+ person will produce antibodies against the Rh+ factor.

A problem also occurs when an Rh− mother carries an Rh+ baby. Close to the time when the baby is about to be born, antibodies from the mother can cross the placenta and destroy the baby's red blood cells. If this happens, the baby can receive a blood transfusion before or right after birth. At 28 weeks and immediately after the birth, the mother can receive an injection that destroys any Rh+ antibodies she has made. Thanks to the work of Dr. Landsteiner, blood groups and Rh factor now are checked before transfusions and during pregnancies.

Did You Know?

Rh, or Rhesus, factor was first discovered in the rhesus monkey in 1940. From 85 to 90 percent of the people in the United States have this antigen, meaning that they are Rh-positive.

PROGRAM RESOURCES

From the **Teacher Resource Package** use:

Critical Thinking/Problem Solving, page 24, Too Much of a Good Thing?

Cross-Curricular Connections, page 24, The Red Cross.

Activity Worksheets, page 5, Flex Your Brain.

MINI QUIZ

Use the Mini Quiz to check students' recall of chapter content.

1. **What is blood?** *Blood is a tissue consisting of cells, cell fragments, and liquid.*
2. **Are there more red blood cells or white blood cells in the body?** *red blood cells*
3. **What type of blood group does a person have if the blood has no A or B antigens?** *O*
4. **What is the name of the disease in which white blood cells are produced in increased numbers?** *leukemia*

FLEX Your Brain

Use the Flex Your Brain activity to have students explore BLOOD.

Flex Your Brain

ASSESSMENT

Portfolio: Use the Flex Your Brain activity to reinforce critical thinking and problem solving skills.

MULTICULTURAL AWARENESS

Some cultures consider blood as a needed part of their diet. Have students find out what necessary nutrients are supplied with this characteristic. (Usually sodium and iron).

3 CLOSE

▶ Ask questions 1-4 and the **Apply** Question in the Section Review.

SECTION REVIEW ANSWERS

1. carries oxygen from the lungs to all body cells and removes carbon dioxide from body cells to the lungs, carries wastes to kidneys, transports nutrients from the digestive system to body cells, has materials to fight infections and help heal wounds
2. red blood cells—transport oxygen and carbon dioxide; white blood cells—fight infections; plasma—liquid part of blood that carries dissolved nutrients, minerals, oxygen; platelets—help in blood clotting
3. to prevent blood cells from clumping
4. Leukemia is a disease in which one or more types of white blood cells are produced in increased numbers; anemia is a disorder in which there is too little hemoglobin in the red blood cells.
5. Apply: Respiration could not take place. Cells in tissues would not have oxygen necessary for releasing energy from food. In time, tissues would die.
6. Connect to Physics: The sickle shaped cells can get caught in capillaries and cause blockage. This results in damage to organs when blood flow is cut off, and oxygen is not delivered properly.

Skill Builder

A can receive from A and O
B can receive from B and O
AB can receive from A, B, AB, O
O can receive from O

Skillbuilder
ASSESSMENT
Performance: Use this Skillbuilder to assess students' abilities to Use Table 20-2 to determine acceptable donors for someone with type AB blood.

Figure 20-10. A sample of blood from a person with leukemia shows red blood cells and immature white blood cells.

Diseases and Disorders of Blood

Blood, like any other tissue in your body, is subject to illness. Because blood circulates to all parts of the body and performs so many vital functions, any illness related to blood or circulation is cause for concern. Anemia is a disorder in which there are too few red blood cells, or too little hemoglobin in the red blood cells. Because of this, body tissues do not receive enough oxygen. They are unable to carry on their usual activities. Sometimes the loss of great amounts of blood or improper diet will cause anemia. Anemia can also result from the effects of a disease or side effects of treatment for a disease.

Leukemia is a disease in which one or more types of white 4
blood cells are produced in increased numbers. However, these cells are immature and do not effectively fight infections. Blood transfusions and bone marrow transplants are used to treat this disease, but they are not always successful.

Blood has many functions. It transports oxygen and nutrients to body cells. Blood takes wastes from these cells to organs for removal. Cells in blood help fight infection and heal wounds. You can understand why blood is sometimes called the tissue of life.

SECTION REVIEW

1. What are the four functions of blood in the body?
2. Compare blood cells, plasma, and platelets.
3. Why is blood type checked before a transfusion?
4. Describe a disease and a disorder of blood.
5. **Apply:** Think about the main job of your red blood cells. If red blood cells couldn't pick up carbon dioxide and wastes from your cells, what would be the condition of your tissues?
6. **Connect to Physics:** In sickle cell anemia abnormal hemoglobin causes some red blood cells to have a sickle shape. How can these odd shaped cells cause problems in the body?

Skill Builder — Making and Using Tables

Look at the data in Table 20-2 on page 474 about blood group interactions. To which group(s) can type AB donate blood? If you need help, refer to Making and Using Tables in the **Skill Handbook** on page 690.

OPTIONS

ENRICHMENT

▶ Have students research the role of vitamin K in blood clotting.
▶ Have students research the relationship of a red blood cell's shape and its ability to transport oxygen and carbon dioxide.

ACTIVITY 20-2

DESIGNING AN EXPERIMENT

Comparing Blood Cells

Blood is an important tissue for all vertebrates. How do human blood cells compare with those of other vertebrates?

Getting Started

Your task in this activity is to devise a method to *compare and contrast* blood cells.

Thinking Critically

List some general characteristics of red and white blood cells and of platelets. What are some ways blood cells of other vertebrates could be alike or different?

Materials

Your cooperative group will use:

- prepared slides of human blood and blood of two other vertebrates (fish, frog, reptile, bird)
- microscope

1. Under low power, *observe* the prepared slide of human blood. Locate the red blood cells. *Examine* the red blood cells under high power. In a data table like the one shown, *draw, count,* and *describe* the red blood cells.

2. Move the slide to another position. Find one or two white blood cells. They will be stained blue or purple. *Draw, count,* and *describe* the white blood cells in the space provided in the table.

3. Still using high power, *examine* the slide for very small blue fragments. These are platelets. *Draw, count,* and *describe* the platelets.

4. Follow Steps 1 to 7 for each of the two other slides of vertebrate blood.

Summing Up/Sharing Results

- Which type of blood cells are present in the greatest number?
- How do red blood cells, white blood cells, and platelets compare among vertebrates?
- Does each of the vertebrates studied have all three cell types?
- Describe the function of each blood cell.

Going Further!

Describe how technology might design a new type of red blood cell to carry more oxygen.

Data and Observations

Sample Data

Vertebrate type	Blood cell type	Description	Number in field	Drawing
Human	Red	Round biconcave; no nucleus	200	Red
	White	Round or oval with nucleus	0 – 4	White
	Platelets	Cell fragments	10 – 12	Platelets
Bird	Red	Oval; nucleus	Numbers	Red
	White	Round-oval; nucleus	will	White
	Platelets	Spindle-shaped; nucleus	vary.	Platelets
Frog	Red	Oval; nucleus	Numbers	Red
	White	Round-oval; nucleus	will	White
	Platelets	Spindle-shaped; nucleus	vary.	Platelets

ACTIVITY 20-2

OBJECTIVE: Design and carry out an experiment to observe and compare the characteristics of blood cells of selected vertebrates.

Time: one class period

PROCESS SKILLS applied in this activity are **observing, classifying, communicating,** and **inferring.**

PREPARATION

Obtain the necessary microscope slides for comparisons. Review procedures for working with prepared microscope slides to prevent damage to them.

Cooperative Learning: Divide the class into Science Investigation Teams.

THINKING CRITICALLY

Answers will vary, but might include such characteristics as size and shape of the cells and of the nucleus. Cells could be similar or different according to the characteristics mentioned above. Red cells might all be similar in shape. White blood cells might all have differently shaped nuclei. Platelets might all be fragments.

TEACHING THE ACTIVITY

*Refer to the **Activity Worksheets** for additional information and teaching strategies.*

- Remind students of the differences between red and white blood cells.

SUMMING UP/SHARING RESULTS

- red blood cells
- The red blood cells of the bird and frog are oval and have nuclei; human blood cells are round without nuclei. All of the white blood cells are similar. The bird and frog platelets are alike in that they are spindle-shaped and have nuclei.
- yes
- Red blood cells carry oxygen and carbon dioxide. White blood cells ingest foreign substances and dead cells. Platelets help clot blood.

GOING FURTHER!

Answers will vary. A new red blood cell might have more hemoglobin in it or another substance that could carry more oxygen to cells.

PROGRAM RESOURCES

From the **Teacher Resource Package** use:
Activity Worksheets, pages 178-179, Activity 20-2, Comparing Blood Cells.

Activity

ASSESSMENT

Performance: To further assess students' understanding of blood cells, see USING LAB SKILLS, Question 12, on page 484.

SECTION 20-3

PREPARATION

SECTION BACKGROUND

▶ A serious concern about transfusions has been the possibility of contracting hepatitis or HIV. Mandatory screening and testing for these conditions have helped relieve fears of donors and recipients.

1 MOTIVATE

▶ Discuss the importance of blood donations to the success of surgery and the recovery of the patient.

2 TEACH

Key Concepts are highlighted.

CROSS CURRICULUM

▶ **Physics:** Have students research how a laser generates its intense beam of light.

PROGRAM RESOURCES

From the **Teacher Resource Package** use:

Science and Society, page 24, Should Health-Care Workers Be Tested for AIDS?

20-3 Autologous Blood Transfusions

New Science Words

homologous blood transfer
autologous blood transfer

Objectives

- ▶ **Explain the advantages of an autologous blood transfusion.**
- ▶ **Name one problem with saving one's own blood.**

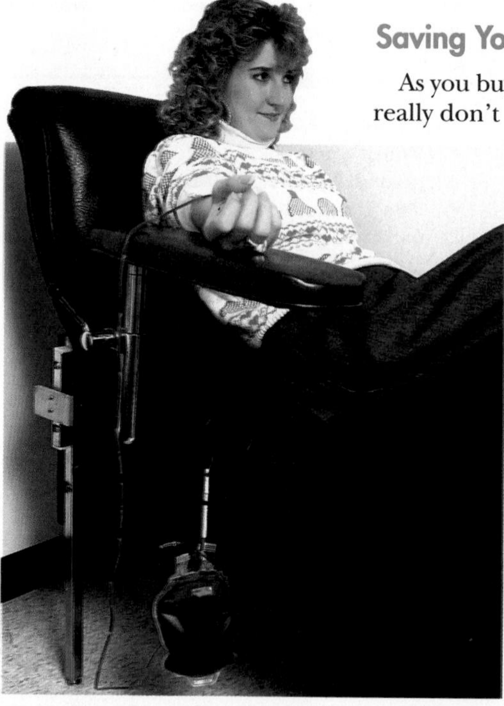

Figure 20-11. Autologous blood transfer involves giving and storing your own blood for your own future use.

Saving Your Own Blood

As you buy a snack, new T-shirt, or magazine (that you really don't need), you might feel a bit guilty about not saving that money away somewhere so it will be safe until you really need it. Today, doctors are suggesting that money isn't the only thing that people should be putting in the bank. Have you ever read about someone storing his or her own blood just in case it is needed?

Because new blood cells are constantly being produced in the body, lost blood is replaced quickly in a healthy person. You probably know that the Red Cross Blood Mobile collects units of blood from individuals willing to donate to others who might need blood. A **homologous blood transfer** is blood taken from one person and given to another. Some people store their own blood if they are anticipating surgery. **Autologous** (aw TAHL uh gus) **blood transfer** is the use of one's own blood for transfusion. This blood is collected prior to surgery.

Unfortunately, there are several drawbacks to this technology. The biggest problem lies in the expense of storing blood. Doctors estimate that it costs about $200 per year to safely store blood. In addition, the patient may be hundreds or thousands of miles away from his or her blood bank when the blood is needed.

However, in response to people who are interested in this procedure, doctors at Northwestern Memorial

OPTIONS

Meeting Different Ability Levels

For Section 20-3, use the following **Teacher Resource Masters** depending upon individual students' needs.

◆ **Study Guide Master** for all students.

● **Reinforcement Master** for students of average and above average ability levels.

▲ **Enrichment Master** for above average students.

Additional Teacher Resource Package masters are listed in the OPTIONS box throughout the section. The additional masters are appropriate for all students.

◆ **STUDY GUIDE 79**

NAME ____ DATE ____ CLASS ____

STUDY GUIDE — Chapter 20

Autologous Blood Transfusions — Text Pages 478–479

Answer the following questions.

1. What is the difference between an autologous and a homologous blood transfer? **In an autologous blood transfer, blood is drawn from a person for his or her own use later. In a homologous blood transfer, blood is taken from one person and given to another.**
2. When might a person choose to have blood taken for an autologous blood transfusion? **before a scheduled surgery**
3. Give two advantages of an autologous blood transfusion over a homologous blood transfusion. **1. Homologous blood transfusions can spread diseases of the donor—this risk does not exist with autologous blood transfusions. 2. There would be no problem matching blood types.**
4. Why is an autologous blood transfusion like "money in the bank"? **Like money in the bank, having blood drawn for an autologous blood transfusion is saving something valuable for a time when it might be needed.**
5. What are two disadvantages for storing your blood for an autologous blood transfusion some time in the future? **1. Blood is expensive to store for long periods of time. 2. At the time it is needed, the person may be far away from the place where the blood is stored.**
6. Describe how blood "lost" in surgery can be reused. **Blood lost during surgery can be collected by suction devices and sponges. The blood must then be treated with salt solution and then may be injected into the patient.**
7. When might the use of artificial hemoglobin be helpful? **Artificial hemoglobin could be used when it is not possible to collect a patient's blood. This could happen if a person was bleeding before getting to the hospital.**
8. In addition to artificial hemoglobin, what other discoveries may be helpful in avoiding homologous blood transfers? Explain how they would be helpful. **A hormone being researched would force the body to produce blood cells more quickly. Using lasers instead of scalpels would reduce the amount of blood lost.**

79

Hospital in Chicago have developed two forms of autologous blood transfusion. Blood lost during surgery is collected with suction devices. This blood may be put back into the patient's body after it has been cleansed. Lost blood also can be collected with sponges and squeezed out into a bowl of saline, a type of salt solution. Within 15 minutes, it is processed and put back into the patient's circulation. By using both methods, doctors believe that they can recover up to 90 percent of blood that would otherwise be lost.

Still, in some situations, it is very difficult to collect blood from a patient if much of it is lost before medical help is available. To solve this problem, doctors are working on ways to develop artificial hemoglobin that would temporarily transport oxygen and carbon dioxide throughout the body. Another idea being researched is the reproduction of a hormone that causes the body to produce blood cells much more quickly than it normally does. This would allow the body to replace much of its own blood rather than having to receive new blood. One last idea includes doctors using lasers, rather than scalpals, to perform surgery. Lasers cause much less damage to body tissue, thus preventing blood loss due to the use of a scalpal.

Researchers are working hard to develop new techniques to protect people from blood contaminated by disease. Saving one's own blood for later use appears to be a very good idea. What do you think?

Artificial blood substances have been developed to use in blood transfusions. They can carry oxygen and carbon dioxide. What other properties must they have to be safe?

SECTION REVIEW

1. Compare autologous and homologous blood transfers.
2. What are two problems with autologous blood banks?
3. **Connect to Chemistry:** Find out what jaundice is. How is it related to blood?

You Decide!

When the American Red Cross collects blood from volunteer donors, the blood is screened for the AIDS virus. If a donor is found to have the virus, it is the policy of the Red Cross that that individual is notified. Why would the Red Cross have such a policy? Do you think this policy is a good one?

● REINFORCEMENT 79

NAME ____________ DATE ________ CLASS ________

REINFORCEMENT — Chapter 20

Autologous Blood Transfusions — Text Pages 478–479

Answer the following questions.

1. What is an autologous blood transfer and how does it differ from a homologous blood transfer? **In an autologous blood transfer, blood is drawn from a person for his or her own use later. In a homologous blood transfer, blood is taken from one person and given to another.**
2. You have learned about typing blood. Would this testing be necessary if there were only autologous blood transfusions? **If there came a time when there were no homologous blood transfusions, typing of blood would probably be unnecessary. However, it is unlikely that there will ever be a time with no homologous blood transfusions.**
3. Would tests for Rh factors be necessary if there were only autologous blood transfusions? **Yes, testing for Rh factors would still be important. It will always be necessary to learn whether a pregnant woman has Rh blood factors.**
4. What techniques and developments may reduce the need for homologous blood transfers? **Reusing blood after surgery could help reduce the need for homologous blood transfers. Using lasers for surgery, artificial hemoglobin, and chemicals that cause the body to make more blood would also reduce the need for homologous blood transfers.**
5. What are two drawbacks of autologous blood transfers? **Blood is expensive to store for long periods of time. At the time blood is needed the person may be far away from the place where the blood is stored.**

79

▲ ENRICHMENT 79

NAME ____________ DATE ________ CLASS ________

ENRICHMENT — Chapter 20

Autologous Blood Transfusions — Text Pages 478–479

TESTING BLOOD

Many people are now concerned about getting AIDS from a blood transfusion. Donated blood has been screened for antibodies to the AIDS virus since 1985. Antibodies are substances which form when you are exposed to anything new to your body, such as bacteria or viruses.

Did you know that there are other diseases that may be carried in blood products? One such disease is hepatitis. Hepatitis is a disease that prevents the liver from functioning properly. One form of the disease, hepatitis B, can be deadly. Until the 1970s, it was possible to get hepatitis B from blood transfusions. Then, a test was invented to detect the disease-causing virus. In the last several years, another dangerous form of the disease, hepatitis C, has occasionally been carried in blood transfusions. After much hard work, a test for antibodies to hepatitis C was found in 1989.

A new method may soon be used to prevent any viruses or bacteria from spreading through the blood supply. Blood can be treated with special drugs that attach to bacteria and viruses. When certain types of light are then shined on the treated blood, the drugs release a form of oxygen that kills the viruses and bacteria.

Find out all you can about current methods of testing and treating blood to prevent the spread of diseases by transfusions. Use the space below and the back of this page to summarize what you have learned. Use reference materials and contact your local Red Cross or a blood bank for more information. Find out how many people use autologous blood transfusions. Include in your summary an explanation of why testing for antibodies may not guarantee absence of a disease-causing bacteria or virus.

Summaries will vary. If blood is collected from an infected person before antibodies have formed, the blood will not test positive for the disease.

79

Connect to... Chemistry

Answer: Artificial blood must not cause clumping or be diseased.

CHECK FOR UNDERSTANDING

Use the You Decide discussion to check for understanding.

RETEACH

Have students infer why it might be necessary to store one's blood until needed. Use this discussion to review autologous blood transfusions.

EXTENSION

For students who have mastered this section, use the **Reinforcement** and **Enrichment** masters or other OPTIONS provided.

3 CLOSE

▶Ask questions 1-2 in the Section Review.

▶Discuss with students that researchers are looking for blood substitutes that could be used during surgery and recovery until the body produced its own replacement.

SECTION REVIEW ANSWERS

1. Autologous blood transfer is the use of a person's own blood for transfusion; homologous blood transfer is a transfusion from one person to another.

2. the expense of storing blood and the need to be near the blood bank

3. Connect to Chemistry: Jaundice is a yellowing of the skin and eyes caused by bilirubin, a product formed when excessive numbers of red blood cells break down.

YOU DECIDE!

The person may not be aware that he or she has AIDS. Opinions will vary.

SECTION 20-4

PREPARATION

SECTION BACKGROUND

▶ Lymph nodes are approximately 3 cm in length and usually occur in groups. Major groups of nodes are located in the neck, under the arms, and in the groin.

▶ Macrophages are large, connective tissue cells found in the linings of blood vessels and in such organs as the liver and spleen. They also are found in lymph nodes and serve to remove foreign particles in the lymph.

▶ The AIDS virus destroys a particular kind of lymphocyte, causing the body to lose its ability to fight disease pathogens.

1 MOTIVATE

▶ Discuss the lymphatic system as an often overlooked component of the circulatory system. Because lymph is essentially a colorless, watery fluid, students may not realize that a cut on the skin also causes the loss of lymph as well as blood. Stress the importance of the system as one component in maintaining homeostasis of body fluids and as the major defense against pathogens.

▶ Collect newspaper and magazine articles about the latest research about HIV. Discuss how the lymphatic system is involved in HIV infection. Ask students why HIV itself does not kill the victim.

TYING TO PREVIOUS KNOWLEDGE: Ask students to recall when they may have had an infection and experienced swollen lymph nodes in the neck or under the arms. What do they think caused this?

20-4 Your Lymphatic System

New Science Words

lymphatic system
lymph
lymphocytes
lymph nodes

Objectives

▶ Describe the functions of the lymphatic system.
▶ Explain where lymph comes from.
▶ Explain the role of lymph organs in fighting infections.

Functions of Your Lymphatic System

What is the function of your lymphatic system?

You have learned that blood carries nutrients and oxygen to cells. Molecules of these substances pass through capillary walls to be absorbed by nearby cells. Some of the water and dissolved substances around cells becomes part of tissue fluid between cells. What would happen if this fluid kept collecting in the spaces? The tissue would swell and eventually burst. Obviously, this does not happen. Your **lymphatic** (lihm FAT ihk) **system** collects fluid from body tissue spaces and returns it to the blood through lymph capillaries and larger lymph vessels. This system also contains cells that help your body defend itself against pathogens.

Lymphatic Organs

The lymphatic system carries fluid away from tissues into the lymphatic capillaries found in most tissues. This fluid is known as **lymph** (LIHMF). Lymph enters the cap-

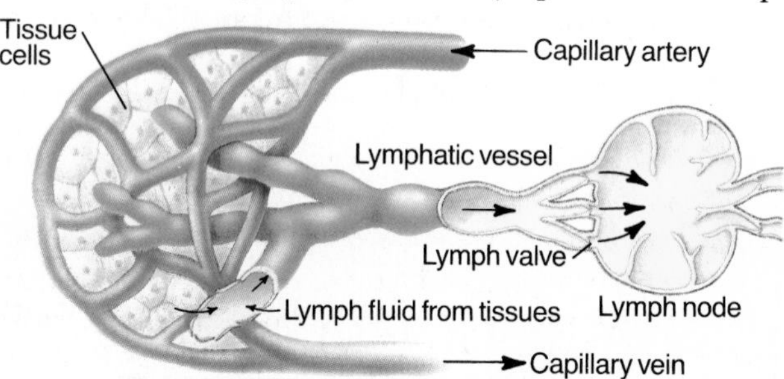

Figure 20-12. Lymph is fluid that has moved from around cells into lymph vessels.

OPTIONS

Meeting Different Ability Levels

For Section 20-4, use the following **Teacher Resource Masters** depending upon individual students' needs.

◆ **Study Guide Master** for all students.

● **Reinforcement Master** for students of average and above average ability levels.

▲ **Enrichment Master** for above average students.

Additional Teacher Resource Package masters are listed in the OPTIONS box throughout the section. The additional masters are appropriate for all students.

◆ STUDY GUIDE 80

NAME ________ DATE ________ CLASS ________

STUDY GUIDE — Chapter 20

Your Lymphatic System — Text Pages 480–482

Use terms from the textbook to fill in the blanks below.

1. A soft mass of tissue that produces lymphocytes is the **thymus**.
2. Microorganisms and foreign materials are filtered out of the lymph in the **lymph nodes**.
3. The system that collects fluid from body tissue spaces and returns the fluid to the blood is called the **lymphatic system**.
4. Fluid found outside body cells is called **lymph**.
5. The organ that filters bacteria, foreign particles, and damaged red blood cells out of the blood is the **spleen**.
6. White blood cells that fight infection are called **lymphocytes**.

Mark each of the following statements true or false. If a statement is false, change the underlined term to make the statement true.

7. A person with AIDS is able to fight infections of the body due to the number of T cells present. **False, unable**
8. Bacteria and other foreign materials are filtered out of the blood by the lymph nodes. **False, spleen**
9. Like arteries, the lymphatic vessels have valves that prevent the backward flow of fluid. **False, veins**
10. If lymph were not carried away from tissues of the body by the lymphatic system, the tissues would swell and eventually burst. **True**
11. The thymus protects the mouth and nose from infection. **False, tonsils**
12. The lymph is made up of water, dissolved substances, and lymphocytes. **True**
13. After lymph enters blood, it is filtered in structures called lymph nodes. **False, before**
14. The thymus produces red blood cells. **False, lymphocytes, a type of white blood cells**
15. Unlike the circulatory system, the lymphatic system has no structure to pump the fluid through the body. **True**
16. AIDS is a virus that destroys all lymphocytes. **False, Helper T lymphocytes**
17. The circulatory system helps the body fight infection. **False, lymphatic**

80

1 illaries by absorption and diffusion. Lymph consists mostly of water, dissolved substances, and **lymphocytes** (LIHM fuh sites), a type of white blood cell. The lymphatic capillaries join with larger vessels that eventually drain the lymph into large veins near the heart. There is no heartlike structure to pump the lymph through the lymphatic system. The movement of lymph is due to contraction of skeletal muscles and the smooth muscles in lymph vessels. Like veins,
2 the lymphatic vessels have valves that prevent the backward flow of lymph.

Before lymph enters blood, it passes through bean-shaped structures throughout the body known as **lymph nodes.** Lymph nodes filter out microorganisms and foreign materials that have been engulfed by lymphocytes. Sometimes an infection takes over a lymph node. It becomes inflamed and tender to the touch. You have felt an enlarged lymph node in your neck when you've had a cold.

Larger groups of lymph nodes are the tonsils, thymus, and spleen. Tonsils are in the back of the throat. They provide protection to the mouth and nose against pathogens. The thymus is a soft mass of tissue located behind the sternum. The thymus produces lymphocytes that travel to other lymph organs. The spleen is the largest organ of the lymphatic system and is located behind the upper left part of the stomach. Blood flowing through the spleen gets filtered. Here, old, worn out, and damaged red blood cells are broken down. Large, specialized cells in the spleen engulf and destroy bacteria and other foreign substances.

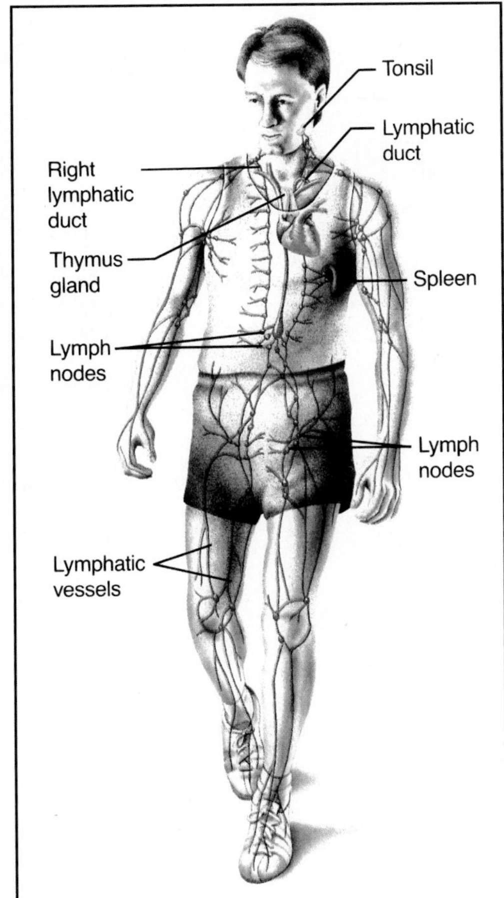

Figure 20-13. The Human Lymphatic System

What is the function of a lymph node?

A Disease of the Lymphatic System

As you read in Chapter 8, the HIV is a deadly virus. When the HIV enters a person's body, it attacks and destroys a particular kind of lymphocyte called helper T cells. Normally, helper T cells help produce antibodies to fight infections. With fewer numbers of helper  3

2 TEACH

Key Concepts are highlighted.

CONCEPT DEVELOPMENT

▶ If possible, obtain a chart of the lymphatic system in order to illustrate the extensive system of lymphatic vessels.

CHECK FOR UNDERSTANDING

Use the Mini Quiz to check for understanding.

MINI QUIZ

Use the Mini Quiz to check students' recall of section content.

1. **How does tissue fluid get into lymphatic capillaries?** *absorption and diffusion*
2. **____ are structures in lymph vessels that prevent the backward flow of lymph.** *Valves*
3. **How does HIV cause the body to be less able to fight pathogens?** *It destroys Helper T cells, which help produce antibodies to fight infections.*

RETEACH

Have students construct a table to illustrate the functions of lymph nodes, the thymus, and the spleen.

EXTENSION

For students who have mastered this section, use the **Reinforcement** and **Enrichment** masters or other OPTIONS provided.

● REINFORCEMENT 80

NAME ____ DATE ____ CLASS ____

REINFORCEMENT Chapter 20

Your Lymphatic System Text Pages 480-482

Label the drawing below of the lymphatic system. Include the terms lymph node, thymus, tonsils, and spleen.

tonsil
thymus
spleen
lymph node

Answer the questions below.

1. What are three main functions of the lymphatic system? **The lymphatic system drains fluids from around tissue cells, filters these fluids, and fights infection.**
2. What do lymph nodes do? Why do they sometimes become large and tender? **They filter the lymph. They become large and tender when they are infected.**
3. What is the role of the thymus? **It produces lymphocytes.**
4. How is the spleen like a lymph node? How do their functions differ? **They both filter impurities from the body. Lymph nodes filter lymph; the spleen filters blood.**

In the space below, write a paragraph describing how the AIDS virus affects the body.

The AIDS virus destroys Helper T cells and generally weakens the immune system. A person with AIDS loses the ability to fight infection and becomes sick from common pathogens in the environment. Eventually the infections overwhelm the person.

80

▲ ENRICHMENT 80

NAME ____ DATE ____ CLASS ____

ENRICHMENT Chapter 20

Your Lymphatic System Text Pages 480-482

DESIGNING A SYSTEM

The lymphatic system is sometimes compared to the circulatory system. However, the lymphatic system has no pump, like the heart, to keep lymph circulating through the body. In addition, the circulatory system is a continuous system—the lymphatic system is not. In the circulatory system, arteries leaving the heart connect to smaller and smaller arteries that connect to capillaries. The capillaries connect to small veins that connect to larger and larger veins. The large veins lead back to the heart, where the circuit is complete. In the lymphatic system, although lymph capillaries lead to larger lymph vessels, the lymph capillaries begin "blindly" in the tissues. The lymph does not go in a complete circuit through the body.

Part A

Compared to the circulatory system, do you think that the lymphatic system is well-adapted to carrying material through the body? Explain your answer. **Students may respond that valves in the lymphatic system prevent the backward flow of materials. If they look up additional materials they may also learn that recent studies show that lymph vessels contract rhythmically. These contractions, along with contractions of the body muscles, propel lymph through the body.**

Part B

Design a lymphatic system for an imaginary species of the future. Draw a picture of the system and explain how it compares to the human lymphatic system.

Drawings and answers will vary.

80

Answer: Gravity is a force that tends to hold or pull a body toward Earth.

3 CLOSE

▶ Ask questions 1-3 and the **Apply** Question in the Section Review.

▶ Review with students the sequence of events from the initial entrance of a microorganism into the body and its destruction within a lymph node.

SECTION REVIEW ANSWERS

1. Your lymphatic system collects fluid from body tissue spaces and returns it to the blood. This system also contains cells that help your body defend itself against pathogens.

2. Lymph comes from tissue fluids and gets into lymphatic capillaries by absorption and diffusion.

3. It attacks and destroys a particular kind of lymphocyte called Helper T cells.

4. Apply: Accept reasonable answers. Lymphatic vessels collect fluid from tissues and move it back toward veins. They contain valves. Arteries contain blood and fluids that are moving toward tissues.

5. Connect to Physics: The fluid will be forced out of the spaces and eventually passes into lymph vessels.

Skill Builder

alike—transport fluids with cells, have vessels, some lymph vessels, like veins, have valves; unlike—type of fluid transported, different kinds of vessels, fluid not pumped by a heart

Skillbuilder

ASSESSMENT

Portfolio: Use this Skillbuilder to assess students' abilities to Compare and Contrast the lymphatic and cardiovascular systems.

Lymph from the head and neck region of the body flows into veins due to the force of gravity. Find out what gravity is.

Figure 20-14. Ryan White was infected with the HIV when receiving blood transfusions before donated blood was tested for the HIV. Here Ryan (right) is shown with Lukas Haas, who played Ryan in a movie about his life.

T cells, a person is less able to fight pathogens. A person infected with the HIV develops infections that a healthy person would be able to fight. These infections become difficult to treat and often are the cause of death.

Your lymphatic system collects extra fluid from body tissue spaces. It also produces lymphocytes that fight infections from microorganisms and foreign material that enter the body. This system works to keep your body healthy.

SECTION REVIEW

1. Describe the role of your lymphatic system.
2. Where does lymph come from and how does it get into the lymphatic capillaries?
3. What happens when the HIV enters the body?
4. **Apply:** Compare lymphatic vessels and arteries.
5. **Connect to Physics:** When the amount of fluid in the spaces between cells increases, so does the pressure in these spaces. What do you *infer* will happen?

Skill Builder

Comparing and Contrasting

Compare and contrast your lymphatic system and your cardiovascular system. If you need help, refer to Comparing and Contrasting in the **Skill Handbook** on page 683.

OPTIONS

ASSESSMENT—ORAL

▶ **How is lymph returned to the circulatory system?** *Lymph vessels empty into veins near the heart.*

▶ **What is an inflamed lymph node a sign of?** *an infection by microorganisms*

PROGRAM RESOURCES

From the **Teacher Resource Package** use:

Transparency Masters, pages 75-76, Lymphatic System.

Use **Color Transparency** 38, Lymphatic System.

CHAPTER REVIEW

SUMMARY

20-1: Circulation

1. Arteries carry blood from the heart; capillaries exchange food and oxygen and wastes in cells; veins bring blood back to the heart.
2. Blood enters the right atrium of the heart through veins, moves to the right ventricle, to the lungs by the pulmonary artery. Blood rich in oxygen returns to the left atrium, moves into the left ventricle, and then out through the aorta to the body.
3. Pulmonary circulation is the path of blood to and from the heart and lungs. Circulation through the rest of the body is systematic circulation.

20-2: Blood

1. Red blood cells carry oxygen; platelets form clots; white blood cells fight infection.
2. A, B, AB, and O blood types are determined by the presence or absence of antigens. Correct type transfusion is necessary for survival.
3. In anemia, too little oxygen reaches tissues because of too few red blood cells or too little hemoglobin. In leukemia, larger numbers of white blood cells are produced, none of which can fight disease.

20-3: Science and Society: Autologous Blood Transfusions

1. Autologous blood transfusion is the use of one's own blood for transfusion.
2. Cost of storage and having the blood easily available are problems.

20-4: Your Lymphatic System

1. Lymphatic vessels return fluid to the circulatory system and fight disease.
2. Lymph is fluid between cells.
3. Lymph structures filter blood, produce certain white blood cells that destroy bacteria, and destroy worn-out red and white blood cells.

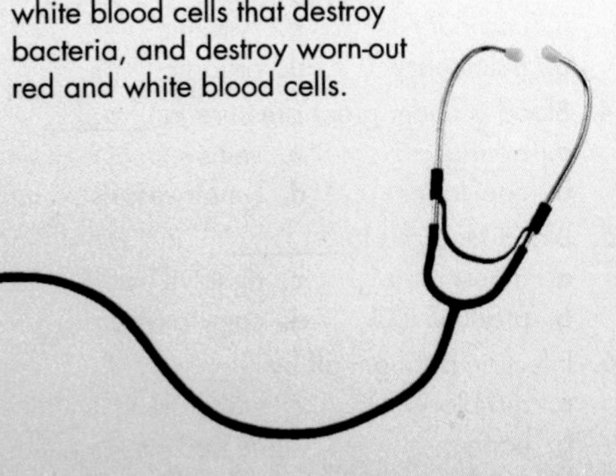

KEY SCIENCE WORDS

a. **arteries**
b. **atherosclerosis**
c. **atria**
d. **autologous blood transfer**
e. **blood pressure**
f. **capillaries**
g. **coronary circulation**
h. **hemoglobin**
i. **homologous blood transfer**
j. **hypertension**
k. **lymph**
l. **lymphatic system**
m. **lymph nodes**
n. **lymphocytes**
o. **plasma**
p. **platelets**
q. **pulmonary circulation**
r. **systemic circulation**
s. **veins**
t. **ventricles**

UNDERSTANDING VOCABULARY

Match each phrase with the correct term from the list of Key Science Words.

1. filter microorganisms
2. upper heart chambers
3. vessels connected to the heart ventricles
4. path of blood between heart and lungs
5. path of blood to the heart tissue
6. fatty deposit on artery walls
7. high blood pressure
8. liquid portion of the blood
9. blood vessels that connect arteries to veins
10. active in blood clot formation

CHAPTER REVIEW

SUMMARY

Have students read the summary statements to review the major concepts of the chapter.

UNDERSTANDING VOCABULARY

1. m
2. c
3. a
4. q
5. g
6. b
7. j
8. o
9. f
10. p

ASSESSMENT

Portfolio

Encourage students to place in their portfolios one or two items of what they consider to be their best work. For each item, ask students to explain why that item was chosen and what they learned from it. Items might be selected from the following.

- MINI-Lab diagram of the heart, p. 466
- Skillbuilder concept map of pulmonary circulation, p. 469
- Cross Curriculum–Physics research on lasers, p. 478

Performance

Additional performance assessments may be found in *Performance Assessment* and *Science Integration Activities* that accompany **Merrill Life Science.** Performance Task Assessment Lists and rubrics for evaluating these activities and other products generated throughout the chapter can be found in Glencoe's *Performance Assessment in Middle School Science.*

OPTIONS

ASSESSMENT

To assess student understanding of material in this chapter, use the resources listed.

COOPERATIVE LEARNING

Consider using cooperative learning in the THINK AND WRITE CRITICALLY, APPLY, and MORE SKILL BUILDERS sections of the Chapter Review.

PROGRAM RESOURCES

From the **Teacher Resource Package** use:

Chapter Review, pages 43-44.

Chapter and Unit Tests, pages 130-133, Chapter Test.

CHAPTER

REVIEW

CHECKING CONCEPTS

1. b	**6.** c
2. b	**7.** a
3. c	**8.** c
4. a	**9.** d
5. d	**10.** a

USING LAB SKILLS

ASSESSMENT

Use these alternate lab exercises to assess students understanding of skills used in this chapter.

11. Sound is different; it sounds like air moving through a tube.

12. Insects have two basic types of blood cells but with a great diversity of shapes and average about 50 000 per cubic millimeter.

THINK AND WRITE CRITICALLY

13. The red blood cell may hold more hemoglobin.

14. Valves help keep blood moving against gravity back to the heart.

15. With anemia, a person's body tissues do not receive enough oxygen. This can result in a lack of energy and shortness of breath.

CHAPTER

REVIEW

CHECKING CONCEPTS

Choose the word or phrase that completes the sentence.

1. Exchange of food, oxygen, and wastes occurs through the _______.
a. arteries **c.** veins
b. capillaries **d.** lymph vessels

2. Oxygen-rich blood first enters the _______.
a. right atrium **c.** left ventricle
b. left atrium **d.** right ventricle

3. Circulation to all body organs is _______.
a. coronary **c.** systemic
b. pulmonary **d.** organic

4. Blood is under great pressure in _______.
a. arteries **c.** veins
b. capillaries **d.** lymph vessels

5. Blood functions to _______.
a. digest food **c.** dissolve bone
b. produce CO_2 **d.** carry oxygen

6. Infection is fought off by _______ cells.
a. red blood **c.** white blood
b. bone **d.** nerve

7. In blood, oxygen is carried by _______.
a. red blood cells **c.** white blood cells
b. platelets **d.** lymph

8. Clotting of blood requires _______.
a. plasma **c.** platelets
b. oxygen **d.** carbon dioxide

9. Type O blood has _______ antigen(s).
a. A **c.** A and B
b. B **d.** no

10. The largest filtering lymph organ is (are) the _______.
a. spleen **c.** tonsils
b. thymus **d.** node

USING LAB SKILLS

11. The MINI-Lab on page 468 deals with the use of a stethoscope to hear your heart sounds. Use the stethoscope placed on a person's back to listen to the sounds of air moving in and out of the lungs. How does this sound compare with the heart sound?

12. In the Activity on page 477 you compared human blood cells with other vertebrates. Obtain a slide of insect blood cells. How do these blood cells compare with those of humans?.

THINK AND WRITE CRITICALLY

Answer the following questions in your Journal using complete sentences.

13. A mature human red blood cell has no nucleus. How might this be an advantage?

14. What is the relationship between valves in veins and lymphatic capillaries and gravity?

15. Why would a person with anemia be tired?

APPLY

16. Identify the following as having oxygen-rich or carbon-dioxide full blood: aorta, coronary arteries, coronary veins, inferior vena cava, left atrium, left ventricle, right atrium, right ventricle, superior vena cava.
17. Discuss some benefits of autologous blood transfer.
18. Explain how the lymphatic system works with the cardiovascular system.
19. Why is cancer of the blood or lymph hard to control?
20. Pulse is usually taken at the neck or wrist. Why do you think this is so even though there are many arteries?

MORE SKILL BUILDERS

If you need help, refer to the Skill Handbook.

1. **Sequencing:** Put the path of blood in sequence from the heart, to the lungs, and out to the body.
2. **Comparing and Contrasting:** Compare the life span of the different types of blood cells.
3. **Interpreting Data:** Interpret the data obtained in a lab. Find the average heartbeat rate of four males and four females and compare the two averages.
   Males: 72, 64, 65, 72
   Females: 67, 84, 74, 67
4. **Designing an Experiment:** Design an experiment to compare the heartbeat rate at rest and after exercising.
5. **Hypothesizing:** Make a hypothesis to suggest the effects of smoking on heartbeat rate.

PROJECTS

1. With supervision, prepare a heart healthy recipe. Check for recipes through the American Heart Association cookbooks. Bring a sample in to share with the class.
2. Write a report on heart transplants, their success, and what a heart-transplant patient has to do to remain healthy.

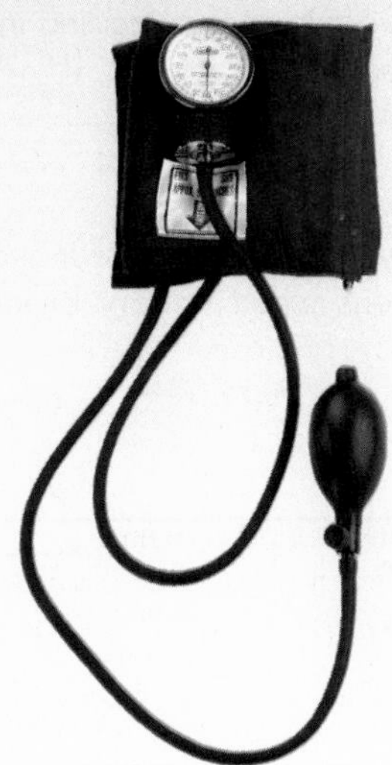

APPLY

16. oxygen-rich blood: aorta, coronary arteries, left atrium, left ventricle; carbon-dioxide full blood: coronary veins, inferior vena cava, right atrium, right ventricle, superior vena cava
17. Benefits include guarantee of a good blood match and no waiting for blood.
18. The lymphatic system, like your veins, carries fluid away from body tissues and returns it to your circulatory system by way of your veins.
19. Blood and lymph move throughout the entire body.
20. These arteries are closest to the skin and pulse can be more easily felt.

MORE SKILL BUILDERS

1. Sequencing: superior vena cava—right atrium—right ventricle—pulmonary arteries—lungs—pulmonary vein—left atrium—left ventricle—aorta—body
2. Comparing and Contrasting:
red blood cells: 120 days
white blood cells: a few days to several months
platelets: 5 to 9 days
3. Interpreting Data: Males = 68. Females = 73. In general, males' heartbeat rates are lower than females'.
4. Designing an Experiment: An experiment should include trials for both resting and exercising heartbeat rates with different subjects.
5. Hypothesizing: Cigarettes contain nicotine, which is a stimulant. Heartbeat rate will increase.

CHAPTER

21 Respiration and Excretion

CHAPTER SECTION	OBJECTIVES	ACTIVITIES
21-1 Your Respiratory System (3 days)	1. **State** three functions of the respiratory system. 2. **Explain** how oxygen and carbon dioxide are exchanged in the lungs and in tissues. 3. **Trace** the pathway of air in and out of the lungs. 4. **Name** three effects of smoking on the respiratory system.	**MINI-Lab:** *What is percussing?* p. 494 **Activity 21-1:** *Designing an Experiment (The Effects of Respiration),* p. 495
21-2 Dangerous Breathing Science & Society (1 day)	1. **State** two reasons why exercise is important to good health. 2. **Become aware** of the dangers of exercising in environments that are polluted.	
21-3 Your Urinary System (3 days)	1. **List** three functions of the urinary system. 2. **Describe** how your kidneys work. 3. **Describe** the excretory functions of the skin and lungs. 4. **Explain** what happens when urinary organs don't work.	**MINI-Lab:** *What do kidneys look like?* p. 500 **Activity 21-2:** *Designing an Experiment (Sweat Glands in the Skin),* p. 502
Chapter Review		

ACTIVITY MATERIALS

FIND OUT	ACTIVITIES		MINI-LABS	
Page 487 stopwatch or watch-with second hand paper pencil	**21-1 Designing an Experiment, p. 495** clock or watch with second hand drinking straws 200 mL bromothymol blue solution 400-mL beakers (2) graduated cylinder	**21-2 Designing an Experiment, p. 502** iodine solution bond typing paper hand lens	**What is percussing? p. 494** no materials	**What do kidneys look like? p. 500** large animal kidneys scapels magnifying glass

CHAPTER FEATURES	TEACHER RESOURCE PACKAGE	OTHER RESOURCES
Skill Builder: *Sequencing,* p. 494	**Ability Level Worksheets** ◆ ***Study Guide,*** p. 81 ● ***Reinforcement,*** p. 81 ▲ ***Enrichment,*** p. 81 **Activity Worksheets,** pp. 185-186, 191 **Critical Thinking/Problem Solving,** p. 25 **Science and Society,** p. 25 **Transparency Masters,** pp. 77-78	**Color Transparency 39,** The Respiratory System **Lab Manual:** Lung Capacity, p. 149 **Science Integration Activity 21**
You Decide! p. 497	**Ability Level Worksheets** ◆ ***Study Guide,*** p. 82 ● ***Reinforcement,*** p. 82 ▲ ***Enrichment,*** p. 82	**STVS:** Disc 7, Side 2
Problem Solving: *Frederick's First Baseball Game,* p. 499 **Technology**: *Kidney Transplants,* p. 501 **Skill Builder:** *Concept Mapping,* p. 501	**Ability Level Worksheets** ◆ ***Study Guide,*** p. 83 ● ***Reinforcement,*** p. 83 ▲ ***Enrichment,*** p. 83 **Activity Worksheets,** pp. 187, 188, 192 **Concept Mapping,** p. 47 **Cross-Curricular Connections,** p. 25 **Transparency Masters,** pp. 79-80	**Color Transparency 40,** Urinary System **STVS:** Disc 7, Side 2
Summary Key Science Words Understanding Vocabulary Checking Concepts Using Lab Skills Think & Write Critically Apply More Skill Builders Projects	ASSESSMENT RESOURCES **Chapter Review,** pp. 45-46 **Chapter Test,** pp. 134-137 **Performance Assessment in Middle School Science (PAMSS)**	**Chapter Review Software** **Test Bank** **Alternate Assesssment** **Performance Assessment**

◆ Basic ● Average ▲ Advanced

ADDITIONAL MATERIALS

SOFTWARE	AUDIOVISUAL	BOOKS/MAGAZINES
Second Wind, Micro-ED, Inc. *The Human Systems Series 3,* Focus. *Human Life Processes II: Systems Level,* IBM.	*I Am Joe's Kidney,* film, Pyramid Film. *Breathing Easy,* film, American Lung Association. *Human Body: Excretory System,* film, Coronet/MTI. *Human Body: Respiratory System,* film, Coronet/MTI.	Edelman, Norman H. and Theodoro Santiago. *Breathing Disorders of Sleep.* New York: Churchill Livingstone, Inc., 1986. Haas, Francois and Sheila S. Haas. *The Essential Asthma Book: A Manual for Asthmatics of All Ages.* New York: Ivy Books, 1988. Torrens, M.J. and J.F. Morrison. *The Physiology of the Lower Urinary Tract.* New York: Springer-Verlag New York, Inc., 1987. Ward, Brian. *The Lungs and Breathing.* The Human Body Series. New York: Watts, Franklin, Inc., 1982.

CHAPTER

21

RESPIRATION AND EXCRETION

THEME DEVELOPMENT: Energy transformation is a central theme in this text. Cells require oxygen in order to utilize nutrients. The cell respiration process provides the energy for all cellular activities. Chemical energy is transformed into heat and mechanical energy. Another theme developed in this chapter is homeostasis. The urinary system has an important function in maintaining the balance of liquids and various dissolved elements and compounds.

CHAPTER OVERVIEW

- **Section 21-1:** This section details the basic anatomy and physiology of the respiratory system. The mechanics of breathing are given. A brief description of diseases and disorders of the respiratory system concludes this section.
- **Section 21-2: Science and Society:** The dangers of breathing polluted air are explained. The You Decide feature asks students to consider the issue of smoking in public places.
- **Section 21-3:** The organs of the urinary system and other excretory organs are introduced, and their role in maintaining homeostasis is explained.

CHAPTER VOCABULARY

pharynx	nitrogen dioxide
larynx	urinary system
trachea	kidneys
bronchi	nephrons
alveoli	urine
diaphragm	ureters
emphysema	bladder
chronic bronchitis	urethra
asthma	

CHAPTER

21 Respiration and Excretion

486

OPTIONS

For Your Gifted Students

Try to determine the profile of a smoker versus a nonsmoker. Students can create a survey that individuals can take anonymously. Ask questions regarding age, sex, and if parents were smokers or nonsmokers. Do friends smoke? Do they participate in sports? Students should brainstorm as many questions as possible. The results can be tallied and analyzed for trends. Results can be reported.

For Your Mainstreamed Students

Students can look through magazines for advertisements related to tobacco (pipes, chewing tobacco, cigarettes, and cigars). All the advertisements can be made into a collage. Students should write all the things the advertisements "say" about smoking. How many of the ideas are true about tobacco? Some students may want to make their own advertisement showing facts related to nontobacco use and health.

Have you ever run so fast that it felt like your lungs would burst? What did you do to ease your pain and get more air to your lungs? How long did it take your breathing rate to return to normal?

FIND OUT!

Do this activity to find out about your breathing rate.

Put your hand on your chest. Notice your breathing. You can feel your chest move up and down slightly. Take a deep breath. Notice how your rib cage moves out and upward when you inhale. *Count* your breathing rate for 15 seconds. Multiply by four to figure your breathing rate for one minute. Jog in place for one minute and count your breathing rate again. How long does it take for your breathing rate to return to normal? How is your breathing rate *related* to activity?

Gearing Up

Previewing the Chapter

Use this outline to help you focus on important ideas in this chapter.

Section 21-1 Your Respiratory System
- Functions of Your Respiratory System
- Organs of Your Respiratory System
- How You Breathe
- Diseases and Disorders of the Respiratory System

Section 21-2 Science and Society Dangerous Breathing
- Watch Where You Exercise

Section 21-3 Your Urinary System
- Functions of Your Urinary System
- Organs of Your Urinary System
- Other Excretory Organs
- Diseases and Disorders of the Urinary System

Previewing Science Skills

- In the Skill Builders, you will sequence and make a concept map.
- In the Activities, you will observe, compare, and predict.
- In the MINI-Labs, you will observe, diagram, and describe.

What's next?

Now that you know that exercise and breathing rate are related, find out how your body uses the air you breathe. Also learn how your body gets rid of liquid wastes through your lungs, skin, and kidneys.

487

INTRODUCING THE CHAPTER

Use the Find Out activity to discover how the respiratory system responds to physical activity.

FIND OUT!

Preparation: Have one stopwatch—or a clock or watch with a second hand—for every three students. Check with a school nurse to determine if any student should not jog or engage in activities that stress the heart and respiratory system.

Cooperative Learning: Use Science Investigation cooperative groups. Group students into threes. As one student breathes, a partner counts his or her respiration rate. The third student monitors the time, records the data,and calculates the rate for one minute. Then have students rotate roles.

Teaching Tips

- Any student not able to participate in this activity can be assigned as a timekeeper or recorder for a group without doing the jogging activity.
- Review the reason for multiplying by four to obtain the pulse rate for one minute.

Gearing Up

Have students study the Gearing Up feature to familiarize themselves with the chapter. Discuss the relationships of the topics in the outline.

What's Next?

Before beginning the first section, make sure students understand the connection between the Find Out activity and the topics to follow.

ASSESSMENT OPTIONS

PORTFOLIO

Refer to page 503 for suggested items that students might select for their portfolios.

PERFORMANCE ASSESSMENT

See page 503 for additional Performance Assessment options.

Process

Skillbuilder, p. 494
MINI-Labs, pp. 494, 501
Activities, 21-1, p. 495; 21-2, p. 502
Using Lab Skills, p. 504

CONTENT ASSESSMENT

Assessment—Oral, pp. 490, 500
Skillbuilder, p. 501
Section Reviews, pp. 494, 497, 501
Chapter Review, pp. 503-505
Mini Quizzes, pp. 490, 500

GROUP ASSESSMENT

Opportunities for group assessment occur with Cooperative Learning Strategies and Flex Your Brain Activities.

SECTION 21-1

PREPARATION

SECTION BACKGROUND

▶ The term *breathing* refers to the process of inhaling and exhaling air. *Respiration* refers to the exchange of oxygen and carbon dioxide in the lungs (external respiration) or between the blood and body cells (internal respiration).

▶ The left lung is slightly smaller than the right lung in order to accommodate the heart.

▶ During one minute while the body is at rest, approximately 200 mL of oxygen are used by body cells; an equal amount of carbon dioxide is produced.

PREPLANNING

▶ To prepare for Activity 21-1, obtain the bromothymol blue solution.

1 MOTIVATE

▶ Demonstrate the oxidation of sugar by burning a sugar cube or small pile of crystals on a piece of aluminum foil. To facilitate the start of burning, add a sprinkle of charcoal to the corner of the cube or top of the pile before igniting with a match. Use caution because the sugar will melt and boil. Extinguish the flame. Lead a discussion of the necessity of oxygen for the oxidation of sugar within the cells. The respiratory system provides oxygen for this chemical activity that releases energy for cell functions.

▶ Discuss with students ways singers practice their breathing in order to have greater breathing control while singing.

OBJECTIVES AND SCIENCE WORDS: Have students review the objectives and science words throughout the chapter as they study each section.

21-1 Your Respiratory System

New Science Words

pharynx
larynx
trachea
bronchi
alveoli
diaphragm
emphysema
chronic bronchitis
asthma

Objectives

- ▶ State three functions of the respiratory system.
- ▶ Explain how oxygen and carbon dioxide are exchanged in the lungs and in tissues.
- ▶ Trace the pathway of air in and out of the lungs.
- ▶ Name three effects of smoking on the respiratory system.

Functions of Your Respiratory System

People have always known that air and food are needed for life. However, no one knew what it was about air that made it so vital. In 1774, a British chemist, Joseph Priestley, published the results of some experiments with air. He discovered that a mouse couldn't live in a container in which a candle had previously been burned. He reasoned that a gas in the air had been destroyed when the candle burned. He also discovered that if he put a plant into the container, the gas necessary for life returned in eight or nine days. Then a mouse again could live in this container. Think about photosynthesis. What do you think went on when the plant was in the container? The gas for life was later named oxygen.

Respiration is the process in which energy is released from glucose in cells. People often get the terms *breathing* and *respiration* mixed up. Breathing is the process whereby air moves into and out of lungs. Air contains oxygen, which is carried to cells by your circulatory system. At the same time, your digestive system has prepared a supply of glucose in your cells. Now oxygen, which has made this long journey, goes to work and plays a key role in releasing energy from glucose. At the end of this process, carbon dioxide wastes get carried back to your lungs in your blood. There, it is expelled from your body. Now think about why the first mouse was not able to stay alive in the container. 1

Did You Know?

Joseph Priestley was also an ordained minister. In the early 1790s, his home in England was burned because he was sympathetic to the French Revolution. In 1794, he emigrated to the United States and settled in Pennsylvania. He died there in 1804.

OPTIONS

Meeting Different Ability Levels

For Section 21-1, use the following **Teacher Resource Masters** depending upon individual students' needs.

◆ **Study Guide Master** for all students.

● **Reinforcement Master** for students of average and above average ability levels.

▲ **Enrichment Master** for above average students.

Additional Teacher Resource Package masters are listed in the OPTIONS box throughout the section. The additional masters are appropriate for all students.

◆ STUDY GUIDE 81

NAME ____ DATE ____ CLASS ____

STUDY GUIDE — Chapter 21

Your Respiratory System — Text Pages 488–495

Study the diagram below. Then label each of the numbered structures.

1. nasal cavity
2. pharynx
3. trachea
4. alveoli
5. lung
6. bronchi
7. diaphragm

Fill in the blank with the term that best completes each sentence.

8. The smallest tubes in the lungs are the **bronchioles**.
9. The **epiglottis** prevents food from entering the trachea.
10. The pharynx is a passageway for **food** and **air**.
11. The alveoli are surrounded by **capillaries**.
12. Air entering your body is first moistened and warmed in the **nasal cavity**.
13. The trachea is kept open by rings made of **cartilage**.

Place a check (✓) beside the sentences that agree with the textbook. Rewrite the others so that they agree.

____ 14. Emphysema can result in the blood being low in hemoglobin. **Emphysema can result in the blood being low in oxygen.**
✓ 15. The greatest contributing factor to lung cancer is inhaling the tar in cigarette smoke.
____ 16. People who produce too much mucus in the bronchial tubes have a disease called emphysema. **People who produce too much mucus in the bronchial tubes have chronic bronchitis.**
✓ 17. A lung disorder often associated with allergies is called asthma.
✓ 18. The diaphragm is a muscle beneath the lungs.
____ 19. The mucus lining the trachea moves foreign particles to the esophagus. **The cilia move the foreign particles.**
____ 20. When you inhale, your diaphragm relaxes. **The diaphragm contracts on inhalation.**

81

Organs of Your Respiratory System

Your respiratory system is made of body parts that
1 help you breathe by taking oxygen into your body and removing carbon dioxide. The major organs of your respiratory system are shown in Figure 21-1. These organs include your nasal cavity, pharynx (FER ingks), larynx, trachea, bronchi, and lungs. Air enters your body through two openings in your nose called nostrils. Hair inside your nostrils traps dust from the air. Your nostrils lead to
2 your nasal cavity, where air gets moistened and warmed. Glands that produce sticky mucus line the nasal cavity. The mucus traps dust, pollen, and other materials. This helps filter the air you breathe. Tiny hairlike structures, called cilia, move mucus and trapped material to the back of the throat where it can be swallowed.

How do cilia in your respiratory system help you?

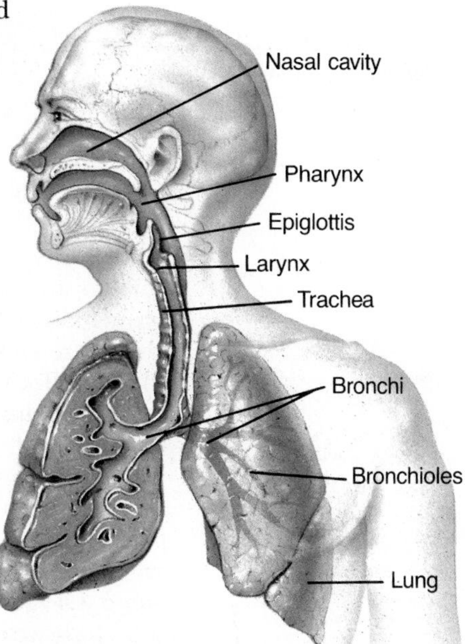

Figure 21-1. The Structures of the Human Respiratory System

Warm, moist air now moves to the **pharynx,** a tubelike passageway for both food and air. The pharynx is located between your nasal cavity and your esophagus. At the lower end of the pharynx is a flap of tissue called the epiglottis (ep uh GLAHT us). When you swallow, the epiglottis closes over your larynx. By doing this, food or liquid is prevented from entering your larynx by accident. The food goes to your esophagus instead. What do you think could happen if you talk or laugh while eating?

Between the pharynx and the trachea is a structure called the **larynx,** to which your vocal cords are attached. When you speak, muscles tighten or loosen your vocal cords. Sound is produced when air moves past, causing them to vibrate. The vocal cords of males are longer than those of females.

How is sound produced?

Below the larynx is the **trachea,** a tube about 12 cm
3 in length. C-shaped rings of cartilage keep the trachea open and prevent it from collapsing. The trachea is lined with mucous membranes and cilia to trap dust, bacteria, and pollen. Why is it necessary for the trachea to stay open all the time?

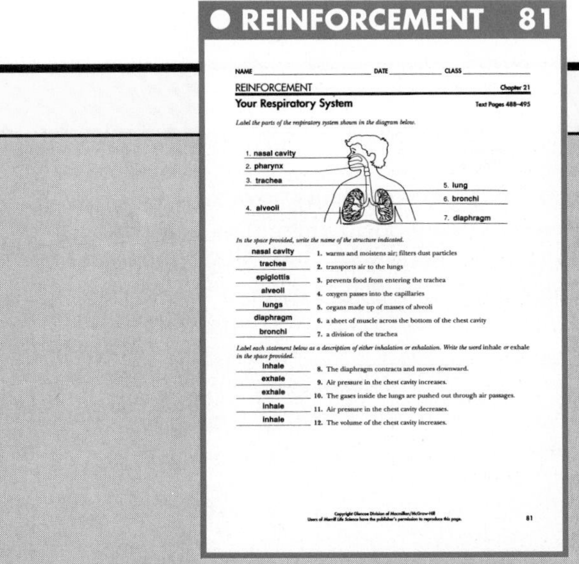

● REINFORCEMENT 81

NAME ______ DATE ______ CLASS ______

REINFORCEMENT — Chapter 21

Your Respiratory System — Text Pages 488–495

Label the parts of the respiratory system shown in the diagram below.

1. nasal cavity
2. pharynx
3. trachea
4. alveoli
5. lung
6. bronchi
7. diaphragm

In the space provided, write the name of the structure indicated.

nasal cavity 1. warms and moistens air; filters dust particles
trachea 2. transports air to the lungs
epiglottis 3. prevents food from entering the trachea
alveoli 4. oxygen passes into the capillaries
lungs 5. organs made up of masses of alveoli
diaphragm 6. a sheet of muscle across the bottom of the chest cavity
bronchi 7. a division of the trachea

Label each statement below as a description of either inhalation or exhalation. Write the word inhale *or* exhale *in the space provided.*

inhale 8. The diaphragm contracts and moves downward.
exhale 9. Air pressure in the chest cavity increases.
exhale 10. The gases inside the lungs are pushed out through air passages.
inhale 11. Air pressure in the chest cavity decreases.
inhale 12. The volume of the chest cavity increases.

 81

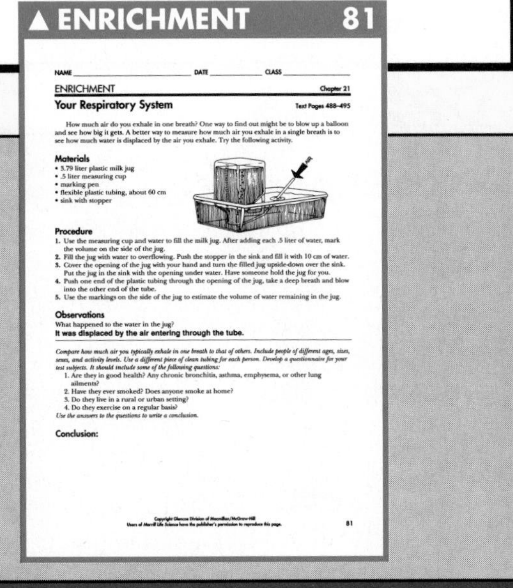

▲ ENRICHMENT 81

NAME ______ DATE ______ CLASS ______

ENRICHMENT — Chapter 21

Your Respiratory System — Text Pages 488–495

How much air do you exhale in one breath? One way to find out might be to blow up a balloon and see how big it gets. A better way to measure how much air you exhale in a single breath is to see how much water is displaced by the air you exhale. Try the following activity.

Materials
- 3.79 liter plastic milk jug
- .5 liter measuring cup
- marking pen
- flexible plastic tubing, about 60 cm
- sink with stopper

Procedure
1. Use the measuring cup and water to fill the milk jug. After adding each .5 liter of water, mark the volume on the side of the jug.
2. Fill the jug with water to overflowing. Push the stopper in the sink and fill it with 10 cm of water.
3. Cover the opening of the jug with your hand and turn the filled jug upside-down over the sink. Put the jug in the sink with the opening under water. Have someone hold the jug for you.
4. Push one end of the plastic tubing through the opening of the jug, take a deep breath and blow into the other end of the tube.
5. Use the markings on the side of the jug to estimate the volume of water remaining in the jug.

Observations
What happened to the water in the jug?
It was displaced by the air entering through the tube.

Compare how much air you typically exhale in one breath to that of others. Include people of different ages, sizes, sexes, and activity levels. Use a different piece of clean tubing for each person. Develop a questionnaire for your test subjects. It should include some of the following questions:
1. Are they in good health? Any chronic bronchitis, asthma, emphysema, or other lung ailments?
2. Have they ever smoked? Does anyone smoke at home?
3. Do they live in a rural or urban setting?
4. Do they exercise on a regular basis?

Use the answers to the questions to write a conclusion.

Conclusion:

 81

TYING TO PREVIOUS KNOWLEDGE: Recall from Chapter 20 the relationship of the circulatory system to the lungs. Remind students of the role of blood in providing oxygen to body cells.

TEACHER F.Y.I.

▶ Sinuses are air-filled cavities in the cranial bones that open into the nasal cavity. The cavities act as resonant chambers and affect the sound of our speech.

2 TEACH

Key Concepts are highlighted.

CONCEPT DEVELOPMENT

▶ If possible, obtain pig lungs from a meat packing house or biological supply company for students to examine. Students are often surprised at the light weight of the organs. Lead a discussion on why lungs aren't heavy.

▶ **Demonstration:** For comparison, obtain microscopic slides or microphotographs of normal lung tissue and diseased lung tissue due to smoking.

FLEX Your Brain

Use the Flex Your Brain activity to have students explore RESPIRATION.

Flex Your Brain
ASSESSMENT

Portfolio: Use the Flex Your Brain activity to reinforce critical thinking and problem solving skills. In Step 2, students might list how breathing feels, its function, what effects breathing rate, or things that interfere with breathing.

REVEALING MISCONCEPTIONS

▶ When a person chokes on a piece of food, he or she often comments that "it went down the wrong tube." The pharynx serves as both the passageway of food into the esophagus and air into the trachea. The epiglottis is a small flap of tissue that normally closes over the tracheal opening when food is swallowed. Sometimes food or liquid gets past the epiglottis and goes into the trachea, triggering the choking reflex.

CONCEPT DEVELOPMENT

▶ Ask students to verbally explain the pathway oxygen takes from air into the alveoli.

▶ Explain that air is very dry when it enters the nostrils, but by the time it reaches the alveoli, enough moisture has been added to it so that the humidity is about 100 percent.

MINI QUIZ

Use the Mini Quiz to check students' recall of chapter content.

1. **Taking air into your lungs and removing carbon dioxide is known as _____ .** *breathing*
2. **Where is inhaled air moistened and warmed?** *nasal cavity*
3. **What keeps the trachea tube from collapsing?** *rings of cartilage*
4. **Masses of _____ are described as grape-like clusters.** *alveoli*
5. **When relaxed, the diaphragm has a _____ shape.** *dome*

Increased lung surface area allows the human body to take in greater amounts of oxygen. Oxygen is absorbed through the walls of the alveoli.

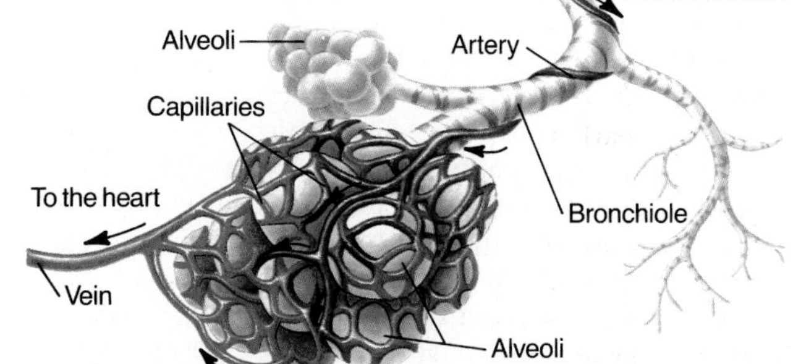

Figure 21-2. In the lungs, alveoli are surrounded by capillaries.

At the lower end of the trachea are two short branches, called **bronchi,** that carry air into the lungs (*singular:* bronchus). Your lungs take up most of the space in your chest cavity. Within the lungs, the bronchi branch into smaller and smaller tubes. The smallest tubes are the bronchioles (BRAHN kee ohlz). At the end of each bronchiole are clusters of tiny thin-walled sacs called **alveoli** (al VE uh li). Your lungs are masses of alveoli arranged in grape-like clusters. Capillaries surround the alveoli. The exchange of oxygen and carbon dioxide takes place between the alveoli and capillaries. This happens easily as the walls of the alveoli and capillaries are only one cell thick. Oxygen diffuses through the walls of the alveoli and then through the walls of the capillaries into the blood. Oxygen is picked up by hemoglobin in red blood cells and carried to all body cells. As this happens, carbon dioxide coming back from body cells diffuses through the walls of the capillaries and through the walls of the alveoli. Carbon dioxide leaves your body when you breathe out, or exhale. 4

The surface area of your lungs is about 20 times as great as that of your skin. What does this mean about how well adapted the body is to take in oxygen?

Figure 21-3. Air under pressure moves out of a squeezed bottle.

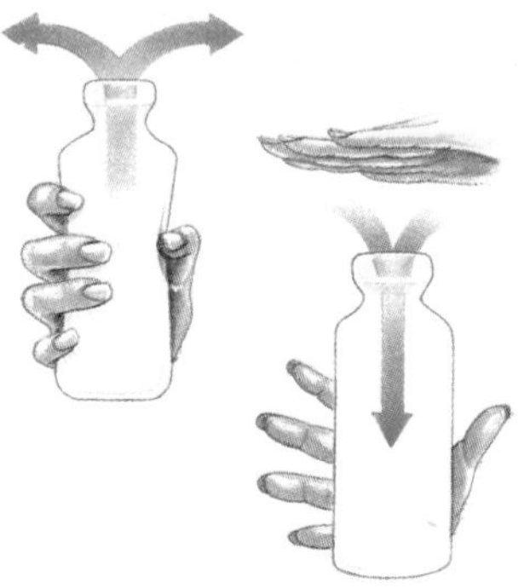

How You Breathe

Breathing is partly the result of changes in air pressure. Under normal conditions, a gas moves from an area of high pressure to an area of low pressure. When you squeeze an empty plastic bottle, air rushes out. This happens because pressure outside the top of the bottle is less than inside the bottle. As you release your grip on the bottle, the pressure inside the bottle becomes less than outside the bottle. Air rushes back in.

Your lungs work in a similar way to the squeezed bottle. Your **diaphragm** (DI uh fram) is a muscle beneath

OPTIONS

ASSESSMENT—ORAL

▶ **Why is chronic bronchitis due to smoking often said to be a "vicious circle" disease?** *Coughing destroys the cilia of the air passages; excess mucus is produced leading to more coughing and more damage to cilia. Finally, the damaged air passages become coated with tars that cannot be moved away because of insufficient cilia, and more coughing results.*

▶ **Why does your body take in greater amounts of air for awhile after exercise? Why does breathing slow after awhile?** *Large amounts of air are taken in to meet the body's increased need for oxygen because cells are undergoing respiration at a faster rate and CO_2 is building at a faster rate. Once the production of CO_2 slows down, breathing slows.*

your lungs that helps move air in and out of your body. Your diaphragm contracts and relaxes when you breathe. Like your hands on the plastic bottle, the diaphragm exerts pressure or relieves pressure on your lungs. Remember earlier when you felt your chest move up and down? When you inhale, your diaphragm contracts and moves down. The upward movement of your rib cage and the downward movement of your diaphragm cause the volume of your chest cavity to increase. Air pressure is reduced in your chest cavity, and your lungs fill with air. Air under pressure outside the body pushes into your air passageways and lungs. Your lungs are somewhat elastic and expand as the air rushes into them.

5 When you exhale, your diaphragm relaxes and returns to its dome shape. Your rib cage moves downward. These two actions reduce the size of your chest cavity. Your lungs also return to their original position. Pressure on your lungs is increased. The gases inside your lungs are pushed out through air passages.

Even when you exhale forcefully, a little bit of air is always left in your lungs. During times of quiet activity, such as reading or doing homework, your lungs inhale and exhale about 500 mL of air with every breath. When you exercise vigorously, you may inhale and exhale as much as 2000 mL of air per breath.

What causes the movement of your diaphragm?

Science and MATH

At the beginning of this chapter you figured out your breathing rate for one minute. Use this number to *calculate* how many breaths you take in a day. If every two breaths filled a liter bottle, how many bottles would you fill in a day?

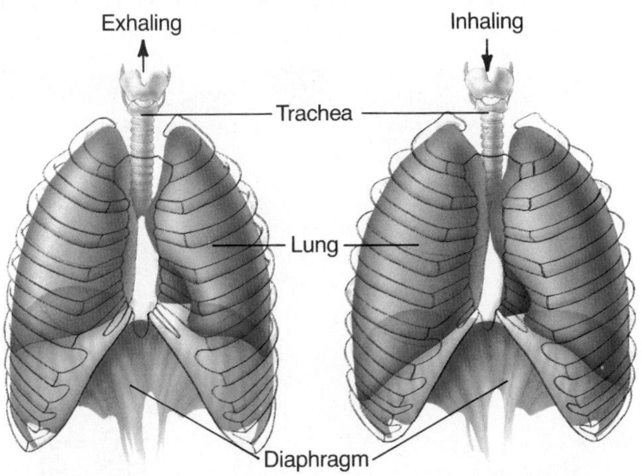

Figure 21-4. When you exhale, your diaphragm relaxes and pushes up on your lungs. When you inhale, your diaphragm contracts and lowers.

CONCEPT DEVELOPMENT

▶ Use the diagrams in Figure 21-4 to understand the position of the diaphragm during the processes of inhaling and exhaling.

Science and MATH

An adult inhales about 12 times each minute at rest. In one day a person exhales and inhales about 17 280 times. If a person exercises, this number increases; 8640 L bottles.

CHECK FOR UNDERSTANDING

Have students write a paragraph describing why air moves into the lungs when the ribs move upward and the diaphragm moves downward.

RETEACH

Demonstration: Use a lung demonstration apparatus to illustrate how the downward movement of the diaphragm causes reduced air pressure within the chest cavity.

EXTENSION

For students who have mastered this section, use the **Reinforcement** and **Enrichment** masters or other OPTIONS provided.

ENRICHMENT

▶ Have students research what takes place physically and physiologically in the condition of conjestive heart failure.

PROGRAM RESOURCES

From the **Teacher Resource Package** use:

Transparency Masters, pages 77-78, The Respiratory System.

Activity Worksheets, page 5, Flex Your Brain.

Science Integration Activities, 21, Air

Use **Color Transparency** 39, The Respiratory System.

CONCEPT DEVELOPMENT

Cooperative Learning: Have students work in Paired Partners to investigate one of the respiratory diseases or conditions described on these pages. Others that may be added are cystic fibrosis and tuberculosis. After both students have investigated the problem, an oral report should be presented. One student should present the disease and its causes. The second student can present treatment and discuss any equipment used to help persons with the particular condition.

Answer: Black lung is caused by inhalation of mineral dusts such as silica, asbestos and coal dust. Fibrous tissue develops around the particles, replacing lung tissue. Less O_2-CO_2 exchange takes place.

Black lung is a disease common among coal miners and sand blasters. Find out what causes black lung and how it affects lung tissue.

Diseases and Disorders of the Respiratory System

If you were asked to list some of the things that can harm your respiratory system, you would probably put smoking at the top. Many serious diseases are related to smoking. Being around others who smoke also can harm your respiratory system. Smoking, polluted air, and coal dust have been related to respiratory problems such as emphysema, bronchitis, asthma, and cancer.

What happens when a person has emphysema?

Emphysema (em fuh SEE muh) is a disease in which the alveoli in the lungs lose their ability to expand and contract. Most cases of emphysema result from smoking. When a person has emphysema, smoke becomes trapped in the alveoli in the lungs. This eventually causes the alveoli to stretch and lose their elasticity. As a result, alveoli can't push air out of the lungs. Less oxygen moves into the bloodstream from the alveoli. Blood becomes low in oxygen and high in carbon dioxide. This condition results in shortness of breath. Some people affected with emphysema can't blow out a match or walk up a flight of stairs. Because the heart works harder to supply oxygen to body cells, people who have emphysema often develop heart problems as well.

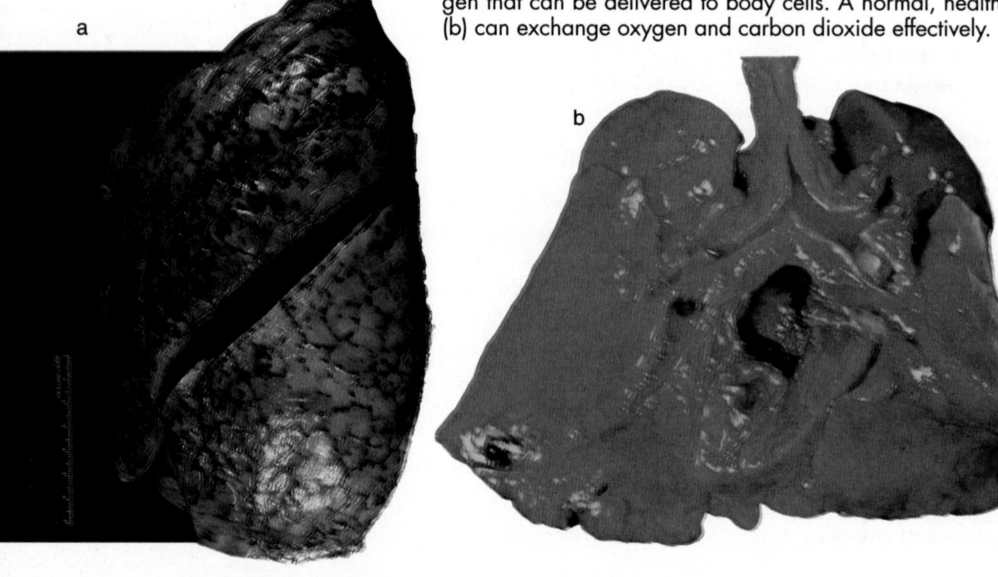

Figure 21-5. A diseased lung (a) cuts down on the amount of oxygen that can be delivered to body cells. A normal, healthy lung (b) can exchange oxygen and carbon dioxide effectively.

OPTIONS

PROGRAM RESOURCES

From the **Teacher Resource Package** use:

Critical Thinking/Problem Solving, page 25, Additional Hazards of Cigarette Smoking.

Science and Society, page 25, Should Asbestos Be Banned?

Use **Laboratory Manual,** page 149, Lung Capacity.

Figure 21-6. Coughing is a reflex that moves unwanted matter from respiratory passages. Cilia help trap and move this foreign matter. When cilia are damaged, the lungs lose a defense against disease. The photograph to the right shows hair-like cilia around cancer cells in the respiratory passage.

Chronic bronchitis is a disease in which too much mucus is produced in the bronchial tubes. Most cases of this disease result from smoking. People who have chronic bronchitis cough often to try to clear the mucus from the airway. However, the more a person coughs, the more the cilia and bronchial tubes can be harmed. When cilia are damaged, their ability to move mucus, bacteria, and dirt out of the lungs is impaired. When this happens, harmful substances, such as sticky tar from burning tobacco, build up in the airways.

Asthma is a disorder of the lungs in which there may be shortness of breath, wheezing, or coughing. When a person has an asthma attack, the bronchial tubes contract quickly. Asthma is often an allergic reaction. An asthma attack can result from a reaction to breathing certain substances, such as plant pollen. Eating certain foods or stress have also been related to onset of asthma attacks.

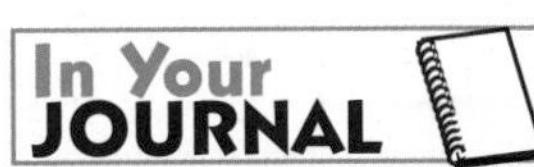

Write a letter to the local office of the American Lung Association. Find out more about a particular respiratory disease. **In your Journal,** write about the causes of the disease, its symptoms, and treatment. Include information on how many people have the disease in the United States.

In Your JOURNAL

Have students collect materials from the American Lung Association to learn about respiratory diseases and disorders.

To assess this product, refer to the Performance Task Assessment Lists in **Performance Assessment in Middle School Science.**

TEACHER F.Y.I.

▶ Hyperventilation is taking rapid and deep breaths of air. People may do this before swimming underwater, incorrectly thinking it will enable them to stay under longer. The decrease in carbon dioxide in the blood can cause a decrease in blood pressure and result in losing consciousness underwater.

CONCEPT DEVELOPMENT

▶ Have students study the damaged and healthy lungs in the photographs in Figure 21-5. Make sure students understand that if lung tissue is damaged, then that much less oxygen can be picked up in the alveoli.

3 CLOSE

▶ Discuss how the lungs function as excretory organs by exhaling carbon dioxide and water vapor. This will be studied in Section 21-3 in this chapter.

▶ Ask questions 1-4 and the **Apply** Question in the Section Review on page 494.

INQUIRY QUESTIONS

▶ **How is the amount of healthy lung tissue related to the amount of oxygen that can be supplied to a person's body tissues?** *the healthier the lungs, the more oxygen that will be delivered to tissues.*

▶ **How would diseased or damaged lung tissue affect a person's heart?** *Increased amounts of carbon dioxide would eventually damage the heart as it works harder to clear blood of this waste and supply oxygen to tissues.*

MINI-Lab

Troubleshooting: Have students work in same-sex pairs.

▶Caution students to be moderate in this exercise.

Answer: the sound produced will indicate whether there is air or fluid there.

MINI-Lab
ASSESSMENT

Performance: Have students develop models of clear and fluid filled lungs using balloons and 2L drink bottles and explain differences in sounds made when tapped.

SECTION REVIEW ANSWERS

1. supplies oxygen to the blood and removes carbon dioxide

2. Oxygen diffuses through the walls of the alveoli and then through the walls of the capillaries into the blood. Carbon dioxide diffuses in the opposite direction.

3. movement of the diaphragm

4. alveoli stretch and lose their elasticity

5. Apply: Digestive system provides food for respiration in cells. The circulatory system transports oxygen to break down food and carries respiration waste products to the lungs to be expelled.

6. Connect to Chemistry: Anemia is a deficiency of red blood cells. Anemia results in less oxygen being delivered to body cells.

Skill Builder

atmosphere—nasal cavity—pharynx—larynx—trachea—bronchi—bronchioles—alveoli—capillaries—blood—to all body cells—carbon dioxide comes back from body cells—capillaries—alveoli—and reverse of above

SMOKING POLLUTES
YOU AND EVERYTHING ELSE
American Cancer Society

Figure 21-7. Everyone should be aware of the effects of smoking on lungs.

Lung cancer is the leading cause of cancer deaths in men and women in the United States. Inhaling the tar in cigarette smoke is the greatest contributing factor to lung cancer. Smoking also is believed to be a factor in the development of cancer of the mouth, esophagus, larynx, and pancreas.

Except for certain bacteria, all living things would die without oxygen. Your respiratory system takes in oxygen and gets rid of carbon dioxide. This system also helps get rid of some pathogens. You can help keep your respiratory system healthy. Avoid smoking and polluted air. Regular exercise helps increase your body's ability to use oxygen.

MINI-Lab

What is percussing?

Have you ever gone to the doctor and had him or her thump you on the back? This is called percussing. A doctor can tell whether a body part is solid or air filled by percussing. Put one hand flat on a partner's back and tap the third finger with the third finger of your other hand. Lungs should make a clear, hollow sound. Pneumonia patients have fluid around their lungs. *Explain* how percussing can help a doctor diagnose the condition of a patient's lungs.

SECTION REVIEW

1. What is the main function of the respiratory system?
2. What happens when oxygen and carbon dioxide are exchanged in the lungs?
3. What causes air to move in and out of the lungs?
4. How does emphysema affect a person's alveoli?
5. **Apply:** How is the work of the digestive and circulatory systems related to the respiratory system?
6. **Connect to Chemistry:** Find out what anemia is. What is its relationship to respiration in cells?

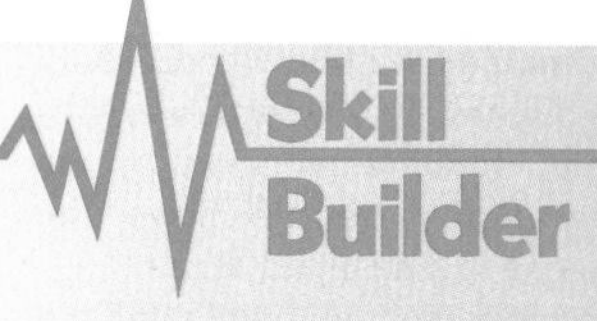

Skill Builder

Sequencing

Sequence the pathway of air through the respiratory organs from the atmosphere to the blood and back to the atmosphere. If you need help, refer to Sequencing in the **Skill Handbook** on page 680.

OPTIONS

Skillbuilder
ASSESSMENT

Performance: Use this Skillbuilder to assess students' abilities to Sequence the path of air through the respiratory system. Have students hold 3 × 5 index cards, each identified with a respiratory structure or process. Then have students assemble themselves in correct sequence and explain their individual step in the sequence.

PROGRAM RESOURCES

From the **Teacher Resource Package** use:

Activity Worksheets, page 191, Mini-Lab: What is percussing?

ACTIVITY 21-1 DESIGNING AN EXPERIMENT
The Effects of Respiration

Breathing rate increased with the increase in physical activity. How will increase in physical activity affect the carbon dioxide exhaled?

Getting Started
Your task in this activity is to *compare* the amount of carbon dioxide in exhaled air at rest and after activity. *Design an experiment* to test your **hypothesis.** Show your design to your teacher before conducting it.

Hypothesizing
Make a **hypothesis** about how exercise will affect the amount of carbon dioxide exhaled by the lungs.

Materials 
Your cooperative group will use:
- clock or watch with second hand
- drinking straws
- bromthymol blue solution (200 mL)
- 400-mL beakers (2)
- graduated cylinder

Try It!
1. To test for the amount of carbon dioxide exhaled from your lungs, label beaker A and beaker B. Use beaker A for your resting test and beaker B for your activity test. Pour 100 mL bromothymol blue solution into each of the beakers.
2. Test for exhaled carbon dioxide: Look at the clock and begin timing. Exhale through the straw into the bromthymol blue solution in your beaker. **CAUTION:** *Do not inhale through the straw.* Continue exhaling for 15 seconds or until the bromthymol blue solution changes color. Note the color. Record the time it takes for the color change to change color. Carbon dioxide causes this color change to occur.
3. Complete your experiment. Remember to exhale using the same force for both your resting and your active tests.
4. *Compare* your data with the class data.

Summing Up/Sharing Results
- What caused the bromthymol blue in the beaker to change color? What color was it at the conclusion of each test?
- *Compare* the time it took the bromthymol blue solution to change color before exercise and after exercise. Explain any difference.
- What was the control in this experiment?
- Was your **hypothesis** supported?

Going Further!
How could you determine the presence of any other gas exhaled from the lungs? Devise a test to see if water vapor is exhaled.

ACTIVITY 21-1

OBJECTIVE: Design and carry out an experiment to show how exercise affects the amount of carbon dioxide exhaled by the lungs.
Time: one class period

PROCESS SKILLS applied in this activity are **communicating, hypothesizing, designing an experiment, inferring, interpreting data,** and **comparing.**

PREPARATION
Obtain materials needed for cooperative groups.
 Cooperative Learning: Students may work in Paired Partners.

SAFETY
Caution students *not inhale through the straw.* Do not let students share the same straw for this activity. Before the students exercise, make sure no student has a medical problem that the exercise would complicate.

HYPOTHESIZING
A possible hypothesis may be: "If you exercise, your lungs will exhale more carbon dioxide."

TEACHING THE ACTIVITY
*Refer to the **Activity Worksheets** for additional information and teaching strategies.*
- Experimental designs may vary. One possibility: breathe into Beaker A before exercising and then breathe into Beaker B after three minutes of exercise, such as running in place.
- The carbon dioxide bubbled into the water forms carbonic acid that causes the color change in the bromthymol blue solution.
- Remind students to blow quickly into the bromthymol blue solution after exercising.

SUMMING UP/SHARING RESULTS
- the presence of carbon dioxide; yellow
- Bromthymol blue takes less time to change color after exercise. More carbon dioxide was produced as a result of exercise.
- The control was breathing into beaker A, which contained the bromthymol blue solution, before exercising.
- Student answers may vary. Refer to original hypothesis suggested.

GOING FURTHER!
Breathe on a cool, shiny metal or glass surface. Minute water droplets can be seen and felt. Paper treated with cobalt chloride will turn from blue to pink when in contact with moisture.

Activity ASSESSMENT
Performance: To further assess students' ability to analyze the effects of respiration, see USING LAB SKILLS, Question 11, on page 504.

PROGRAM RESOURCES

From the **Teacher Resource Package** use:

Activity Worksheets, pages 185-186, 21-1 The Effects of Respiration.

SECTION 21-2

PREPARATION

SECTION BACKGROUND

▶ Nitrogen comprises about 78 percent of the atmosphere. Because it is an inert gas, it is not utilized by the body. High concentrations of the gas can produce an anesthetic effect on the central nervous system.

1 MOTIVATE

▶ Discuss with students how many people engage in jogging and aerobics to increase their physical well-being. Exercise causes the respiratory rate to increase. Any pollutants in the air are, therefore, taken up more rapidly. This section will deal with the effects of one such pollutant, nitrogen dioxide.

Answer: Smog is gray or brown air resulting from industrial use of fossil fuels or the result of sunlight reacting with nitrogen dioxide in the air.

2 TEACH

Key Concepts are highlighted.

CONCEPT DEVELOPMENT

Cooperative Learning: Use Problem Solving Team cooperative groups. Present a community problem of smoking in a certain public place. One group defends the rights of smokers. Another group defends the rights of nonsmokers. A third group must present a diplomatic resolution to the problem.

21-2 Dangerous Breathing

New Science Words

nitrogen dioxide

Objectives

- ▶ State two reasons why exercise is important to good health.
- ▶ Become aware of the dangers of exercising in environments that are polluted.

Watch Where You Exercise

You probably realize that exercise is important for good health. When you participate in regular exercise, you feel good, your muscle tone improves, and fat is lost. With regular exercise, lungs become capable of holding greater amounts of air.

What happens to your lungs with regular exercise?

Many people participate in aerobics or go jogging during lunch or after school or work. Unfortunately, busy city streets and parks where people exercise can be polluted with exhaust fumes. Although most people realize that it is not healthful to breathe polluted air, it actually may be more dangerous to do so while exercising.

Two scientists at the Los Alamos National Laboratory in New Mexico discovered that exercising in polluted environments can be dangerous to your health. **Nitrogen dioxide,** found in cigarette smoke and automobile exhaust, is a chemical that is harmful if inhaled. In a research project involving rats exposed to nitrogen dioxide, researchers found that rats that exercised after being exposed to nitrogen dioxide suffered five times more lung damage than those rats that did not exercise after being exposed to the toxins. The rats were exposed to only half of the nitrogen dioxide that a typical smoker breathes in during one puff of a cigarette! This means that exercising in a polluted environment is more harmful than not exercising at all. What does this mean for our society when so many people now exercise by running? Persons who smoke while doing strenuous work or exercises should reconsider their need to smoke. Joggers should avoid areas of heavy traffic.

Smog has become a feature of many large cities and even the Grand Canyon. Find out what smog is and what causes it to form.

Nitrogen dioxide also is found in coal mines and in farm silos, where it is given off by decaying plant matter.

OPTIONS

Meeting Different Ability Levels

For Section 21-2, use the following **Teacher Resource Masters** depending upon individual students' needs.

◆ **Study Guide Master** for all students.

● **Reinforcement Master** for students of average and above average ability levels.

▲ **Enrichment Master** for above average students.

Additional Teacher Resource Package masters are listed in the OPTIONS box throughout the section. The additional masters are appropriate for all students.

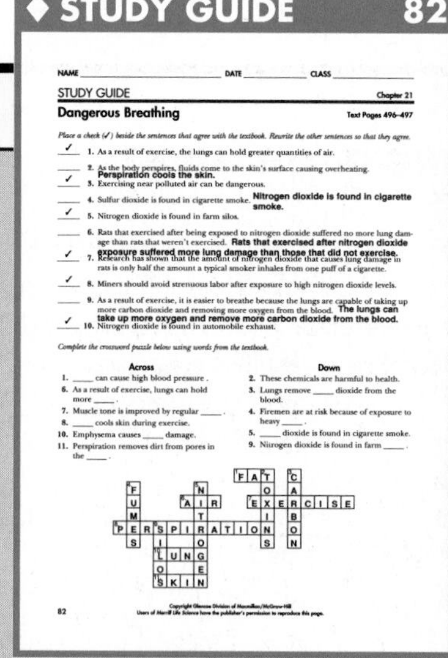

◆ STUDY GUIDE 82

NAME ______ DATE ______ CLASS ______

STUDY GUIDE — Chapter 21

Dangerous Breathing — Text Pages 496–497

Place a check (✓) beside the sentences that agree with the textbook. Rewrite the other sentences so that they agree.

✓ 1. As a result of exercise, the lungs can hold greater quantities of air.

____ 2. As the body perspires, fluids come to the skin's surface causing overheating. **Perspiration cools the skin.**

✓ 3. Exercising near polluted air can be dangerous.

____ 4. Sulfur dioxide is found in cigarette smoke. **Nitrogen dioxide is found in cigarette smoke.**

✓ 5. Nitrogen dioxide is found in farm silos.

____ 6. Rats that exercised after being exposed to nitrogen dioxide suffered no more lung damage than rats that weren't exercised. **Rats that exercised after nitrogen dioxide exposure suffered more lung damage than those that did not exercise.**

✓ 7. Research has shown that the amount of nitrogen dioxide that causes lung damage in rats is only half the amount a typical smoker inhales from one puff of a cigarette.

✓ 8. Miners should avoid strenuous labor after exposure to high nitrogen dioxide levels.

____ 9. As a result of exercise, it is easier to breathe because the lungs are capable of taking up more carbon dioxide and removing more oxygen from the blood. **The lungs can take up more oxygen and remove more carbon dioxide from the blood.**

✓ 10. Nitrogen dioxide is found in automobile exhaust.

Complete the crossword puzzle below using words from the textbook.

Across

1. ____ can cause high blood pressure .
6. As a result of exercise, lungs can hold more ____ .
7. Muscle tone is improved by regular ____ .
8. ____ cools skin during exercise.
10. Emphysema causes ____ damage.
11. Perspiration removes dirt from pores in the ____ .

Down

2. These chemicals are harmful to health.
3. Lungs remove ____ dioxide from the blood.
4. Firemen are at risk because of exposure to heavy ____ .
5. ____ dioxide is found in cigarette smoke.
9. Nitrogen dioxide is found in farm ____ .

82

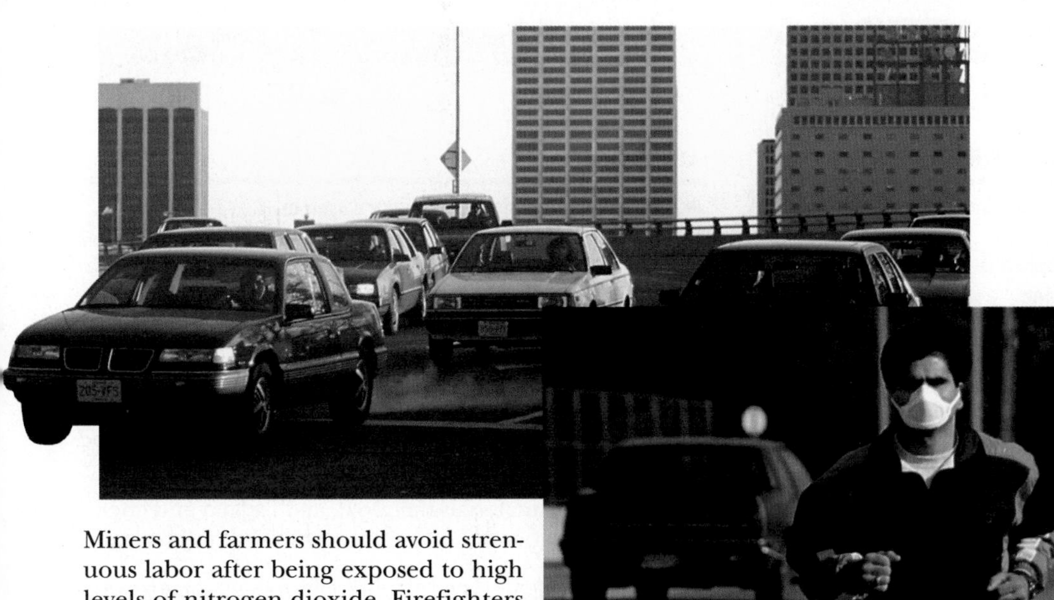

Miners and farmers should avoid strenuous labor after being exposed to high levels of nitrogen dioxide. Firefighters have also been found to be at high risk due to their exposure to heavy fumes while working hard to extinguish fires.

Although research in this area is just beginning, there appears to be a great deal of potential harm associated with exercising after and during exposure to nitrogen dioxide. It is hoped that there will be more information about this subject in the future. Until then, be careful about the air you breathe during exercise. Remember, you only have one set of lungs!

SECTION REVIEW

1. Name two benefits of regular exercise.
2. Name three places where nitrogen dioxide is found.

You Decide!

Most restaurants and other public places have designated smoking areas. Some people say smoking should be banned from all public places. How do you feel about this statement? Where do you think people should have the right to smoke?

● REINFORCEMENT 82

NAME ________ DATE ________ CLASS ________

REINFORCEMENT — Chapter 21

Dangerous Breathing — Text Pages 496–497

Write the letter of the item that best completes each sentence.

__c__ 1. Regular exercise has been shown to have all of the following benefits **EXCEPT** ____.
a. increased lung capacity
b. improved muscle tone
c. decreased life span
d. loss of fat

__c__ 2. ____ is a known toxin.
a. carbon dioxide b. oxygen c. nitrogen dioxide d. nitrogen

__a__ 3. In experiments, rats that developed the most lung damage ____.
a. exercised after exposure to nitrogen dioxide
b. exercised before exposure to nitrogen dioxide
c. did not exercise after exposure to nitrogen dioxide
d. did not exercise before exposure to nitrogen dioxide

__b__ 4. All of the following contain potentially harmful levels of nitrogen dioxide **EXCEPT** ____.
a. automobile exhaust
b. acid rain
c. coal mines
d. cigarette smoke

__c__ 5. One way to avoid lung damage in a polluted environment is to ____.
a. wear a mask
b. slow the pace of exercise
c. exercise indoors
d. smoke cigarettes with filters

Answer the following questions using information from the textbook.

6. What effect does regular exercise have on lung function? **Lungs are more able to take up oxygen and remove carbon dioxide from the blood; the lungs can hold more air.**
7. What happens when the body perspires? **The skin is cooled by sweat and dirt is removed from the pores. Small amounts of salt are lost, as well.**
8. What are some common pollutants in cities? **automobile exhaust, industrial wastes, and cigarette smoke**
9. Where would it be safest to exercise in a large city? **indoors; in a health club, gymnasium, or spa**
10. What places should joggers avoid? **areas of heavy traffic**
11. Describe what scientists have observed in experiments studying the effects of exposure to nitrogen dioxide after exercise. **Rats exercised after exposure to nitrogen dioxide had more lung damage than rats that did not exercise after exposure.**

82

▲ ENRICHMENT 82

NAME ________ DATE ________ CLASS ________

ENRICHMENT — Chapter 21

Dangerous Breathing — Text Pages 496–497

SMOKING AGAINST YOUR WILL

Today, we are becoming more concerned about the effects of secondary or passive smoking. Families of smokers are more likely to develop serious lung ailments, even though they do not smoke themselves. Reports by several surgeon generals have stressed that the health effects of passive smoking can be even more serious than direct smoking. Passive smoking can cause lung cancer, emphysema, chronic bronchitis, and asthma. National surveys show attitudes toward smoking have changed, and Americans are increasingly aware of the negative health effects of passive smoking. Many cities now have regulations governing smoking in certain places.

Find out more about what is going on in your city. Most information can be gathered over the telephone. When you call, be sure to introduce yourself and state why you are calling. Write out your questions so that you can read them easily. Record your data in the table below.

Places to Survey

restaurants, schools, colleges or universities, museums
malls, grocery stores, city parks, bus stations
hospitals, libraries, health clubs, movie theaters

Questions to Ask

Is smoking permitted in the establishment? If so, is a nonsmoking section available? You might pursue additional questions if time allows and if the person you are interviewing is receptive.

Data

Places surveyed	Smoking allowed?	Nonsmoking section?	Other questions

Conclude and Apply

1. Where do nonsmokers encounter the most risk from passive smoking (not including nonsmokers who live with people who smoke)? **Answers will vary but may include bus stations and some restaurants.**
2. How do you think nonsmokers can be better protected? **more stringent smoking regulations in public places**

82

CHECK FOR UNDERSTANDING

Have students list ten ways that they could change their daily routine to avoid being exposed to nitrogen dioxide.

RETEACH

Use the photographs on this page to underscore the causes of poor air quality and the need for taking precautions when exercising in these conditions.

EXTENSION

For students who have mastered this section, use the **Reinforcement** and **Enrichment** masters or other OPTIONS provided.

CROSS CURRICULUM

▶ **Art:** Have students develop a poster that explains the dangers of exercising in polluted environment.

3 CLOSE

▶ Ask questions 1-2 in the Section Review.

VideoDisc

STVS: Children and Smog, Disc 7, Side 2

SECTION REVIEW ANSWERS

1. increased muscle tone, fat loss, increased lung capacity

2. cigarette smoke, automobile exhaust, coal mines, farm silos

YOU DECIDE!

Review current literature about the dangers of passive smoke. What legislation has been passed regarding passive smoke?

SECTION 21-3

PREPARATION

SECTION BACKGROUND

▶ Kidneys produce hormones that can bring about changes to increase blood pressure or increase formation of red blood cells.

PREPLANNING

▶ To prepare for Activity 21-2, obtain the iodine solution. Test the bond paper to be used for reaction to the iodine. A small drop of iodine should cause a purple/black spot to appear.

1 MOTIVATE

▶ Discuss how your community's water is made safe to drink.

Cooperative Learning: Assign Problem Solving Teams to find ways to filter a sample of muddy water. Suggested filters to be used include filter paper, cotton balls, sand, or cotton cloth.

TYING TO PREVIOUS KNOWLEDGE: Recall the function of a filter in a coffee-making machine. Only liquid is allowed to flow through. Have students compare this to the function of the urinary system.

VideoDisc

STVS: Laser Treatment of Bladder Cancer, Disc 7, Side 2

21-3 Your Urinary System

New Science Words

urinary system
kidneys
nephrons
urine
ureters
bladder
urethra

Objectives

- ▶ List three functions of the urinary system.
- ▶ Describe how your kidneys work.
- ▶ Describe the excretory functions of the skin and lungs.
- ▶ Explain what happens when urinary organs don't work.

Functions of Your Urinary System

Waste water in your community is treated after it leaves your home. Filters strain out large materials. Settling basins remove materials that float, and oxygen and chemicals kill bacteria and other organisms that might cause disease.

Your excretory system works in a similar way to the equipment that purifies water. Your excretory organs are
1 your kidneys, lungs, and skin. These organs help your body get rid of wastes. If they didn't do this, you could become very sick from a buildup of toxic substances in cells. Organs could be damaged.

Why is the amount of water in blood important?

The organs of your urinary system are excretory organs. Your **urinary system** is made up of organs that rid your blood of wastes and control blood volume by removing excess water produced by cells. The amount of water in blood is important to maintain normal blood pressure, the movement of gases, and excretion of solid wastes. Your urinary system also balances certain salts and water that must be present in specific concentrations for cell activities to take place.

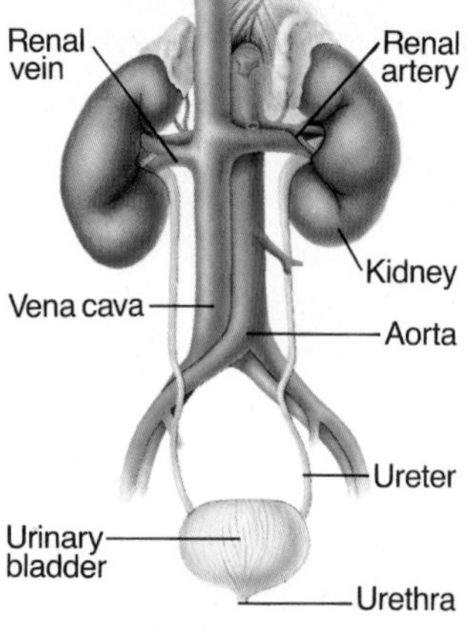

Figure 21-8. Organs of the Urinary System

Organs of Your Urinary System

The major organs of your urinary system are two bean-shaped kidneys. Kidneys are located on the back wall of the abdomen at about waist level. The **kidneys** filter blood that has collected wastes from cells. All of your blood passes through your kidneys many times a day. Figures 21-8 and 21-9 show the structures of the kidneys. Here you can see that blood enters the kidneys through a large artery and leaves through a large vein.

OPTIONS

Meeting Different Ability Levels

For Section 21-3, use the following **Teacher Resource Masters** depending upon individual students' needs.

◆ **Study Guide Master** for all students.

● **Reinforcement Master** for students of average and above average ability levels.

▲ **Enrichment Master** for above average students.

Additional Teacher Resource Package masters are listed in the OPTIONS box throughout the section. The additional masters are appropriate for all students.

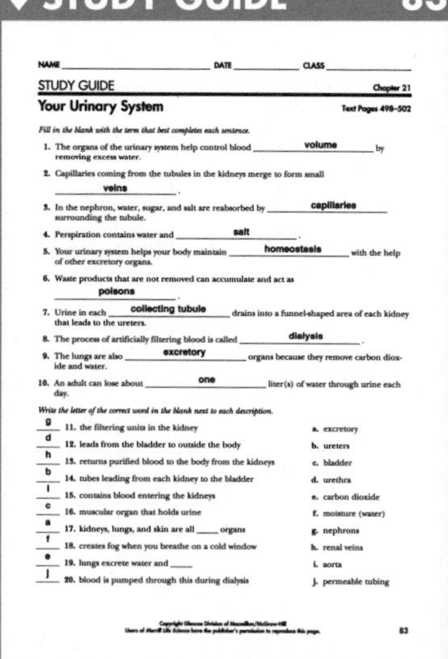

◆ STUDY GUIDE 83

NAME ________ DATE ________ CLASS ________

STUDY GUIDE — Chapter 21

Your Urinary System — Text Pages 498–502

Fill in the blank with the term that best completes each sentence.

1. The organs of the urinary system help control blood **volume** by removing excess water.
2. Capillaries coming from the tubules in the kidneys merge to form small **veins**.
3. In the nephron, water, sugar, and salt are reabsorbed by **capillaries** surrounding the tubule.
4. Perspiration contains water and **salt**.
5. Your urinary system helps your body maintain **homeostasis** with the help of other excretory organs.
6. Waste products that are not removed can accumulate and act as **poisons**.
7. Urine in each **collecting tubule** drains into a funnel-shaped area of each kidney that leads to the ureters.
8. The process of artificially filtering blood is called **dialysis**.
9. The lungs are also **excretory** organs because they remove carbon dioxide and water.
10. An adult can lose about **one** liter(s) of water through urine each day.

Write the letter of the correct word in the blank next to each description.

g	11. the filtering units in the kidney	a. excretory
d	12. leads from the bladder to outside the body	b. ureters
h	13. returns purified blood to the body from the kidneys	c. bladder
b	14. tubes leading from each kidney to the bladder	d. urethra
i	15. contains blood entering the kidneys	e. carbon dioxide
c	16. muscular organ that holds urine	f. moisture (water)
a	17. kidneys, lungs, and skin are all ____ organs	g. nephrons
f	18. creates fog when you breathe on a cold window	h. renal veins
e	19. lungs excrete water and ____	i. aorta
j	20. blood is pumped through this during dialysis	j. permeable tubing

83

2 Each kidney is made up of about one million **nephrons** (NEF rahnz), the tiny filtering units of the kidney. Each nephron has a cuplike structure and a duct. Water, sugar, salt, and wastes from your blood pass into the cuplike structure. There, liquid is squeezed into a narrow tubule. Capillaries that surround the tubule reabsorb most of the water, sugar, and salt and return it to the blood. These capillaries merge to form small veins. The small veins merge to form the renal veins, which return purified blood to be circulated throughout your body. The liquid left behind flows into collecting tubules in each kidney. This waste liquid, **urine,** contains excess water, salts, and other wastes. The average adult produces about 1 L of urine per day.

The urine in each collecting tubule drains into a funnel-shaped area of each kidney that leads to the ureters (YOOR ut urz). **Ureters** are tubes that lead from each kidney to the bladder. The **bladder** is a muscular organ that holds urine until it leaves the body. A tube called the **urethra** (yoo REE thruh) carries urine from the bladder to the outside of the body. The amount you lose depends on the amount of fluid you drink.

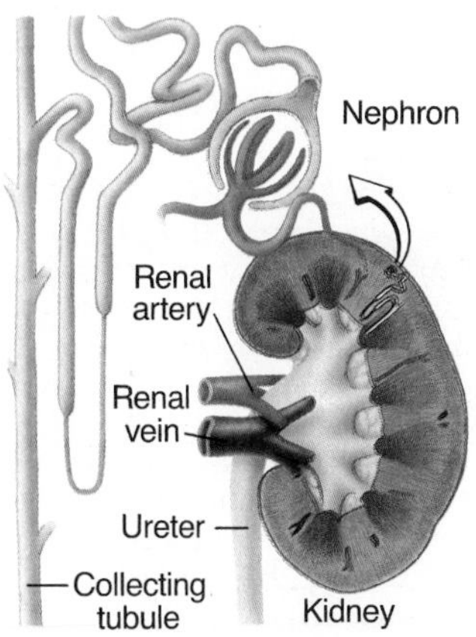

Figure 21-9. The kidneys are made up of many nephrons. A
3 single nephron is shown in detail by the arrow.

PROBLEM SOLVING

Frederick's First Baseball Game

Last summer, Frederick played baseball for the first time in his life. Frederick said, "I had a great time, but it was hard for me to play at times."

Frederick was born without sweat glands. He finds it difficult to stay outdoors for any length of time, especially in the summer. Frederick is not usually able to enjoy a day at the beach or to participate in sports.

However, a local restaurant raised money to develop a "cool suit" for Frederick. The suit resembles a space suit and fits under his clothing. Frederick uses a thermostat to control the pumping of a cool solution through the suit. The suit enables him to spend more time outside.

Think Critically: Explain how the circulating solution keeps Frederick cool.

● REINFORCEMENT 83

NAME ________ DATE ________ CLASS ________

REINFORCEMENT — Chapter 21

Your Urinary System — Text Pages 498–502

Complete the following sentences using information from the textbook.

1. What functions do the organs of the urinary system perform? **They rid the blood of wastes, help control blood volume, and get rid of excess salts.**
2. What happens when the kidneys don't work? **Water accumulates, causing swelling. Wastes accumulate, acting as poisons. An imbalance of salts prevents normal cell functioning. Kidney failure can eventually lead to death.**
3. What are the excretory organs of the body? **kidneys, lungs, and skin**
4. Why do you see your breath on a cold day? **You exhale water vapor and carbon dioxide. The water vapor forms a cloud of small droplets in the cold air.**
5. Why is skin considered an excretory organ? **Some of the body's wastes are removed through the skin.**

Number the following events to place them in the correct order.

7 Urine flows from ureters to the bladder.
2 Blood enters the nephrons.
1 Blood enters the kidneys through the aorta.
6 Urine drains from each collecting tubule into funnel-shaped areas of the kidney.
4 Capillaries reabsorb most of the water, sugar, and salt.
3 Water, sugar, salt, and wastes in the blood enter a cuplike structure.
5 The renal vein returns purified blood to be circulated.
8 Urine flows from the bladder through the urethra and out of the body.

Label the parts of the urinary system shown in the diagram below using the following terms: urethra, ureter, bladder, kidney, aorta, renal vein. (Arrows show the direction of blood flow.)

aorta
renal vein
kidney
ureter
bladder
urethra

 83

▲ ENRICHMENT 83

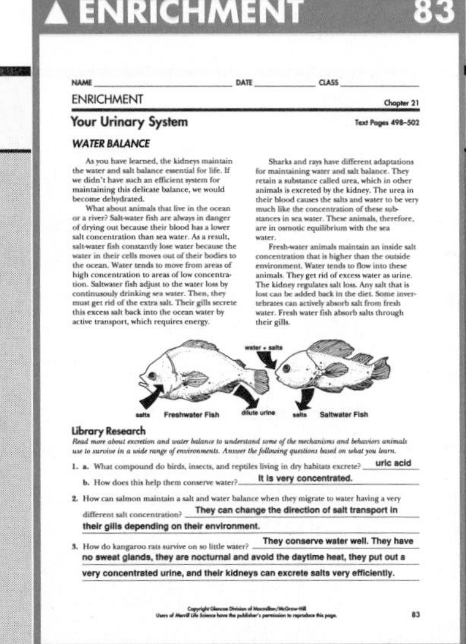

NAME ________ DATE ________ CLASS ________

ENRICHMENT — Chapter 21

Your Urinary System — Text Pages 498–502

WATER BALANCE

As you have learned, the kidneys maintain the water and salt balance essential for life. If we didn't have such an efficient system for maintaining this delicate balance, we would become dehydrated.

What about animals that live in the ocean or a river? Salt-water fish are always in danger of drying out because their blood has a lower salt concentration than sea water. As a result, salt-water fish constantly lose water because the water in their cells moves out of their bodies to the ocean. Water tends to move from areas of high concentration to areas of low concentration. Saltwater fish adjust to the water loss by continusouly drinking sea water. Then, they must get rid of the extra salt. Their gills secrete this excess salt back into the ocean water by active transport, which requires energy.

Sharks and rays have different adaptations for maintaining water and salt balance. They retain a substance called urea, which in other animals is excreted by the kidney. The urea in their blood causes the salts and water to be very much like the concentration of these substances in sea water. These animals, therefore, are in osmotic equilibrium with the sea water.

Fresh-water animals maintain an inside salt concentration that is higher than the outside environment. Water tends to flow into these animals. They get rid of excess water as urine. The kidney regulates salt loss. Any salt that is lost can be added back in the diet. Some invertebrates can actively absorb salt from fresh water. Fresh water fish absorb salts through their gills.

water + salts
salts — Freshwater Fish — dilute urine
salts — Saltwater Fish

Library Research

Read more about excretion and water balance to understand some of the mechanisms and behaviors animals use to survive in a wide range of environments. Answer the following questions based on what you learn.

1. a. What compound do birds, insects, and reptiles living in dry habitats excrete? **uric acid**
   b. How does this help them conserve water? **It is very concentrated.**
2. How can salmon maintain a salt and water balance when they migrate to water having a very different salt concentration? **They can change the direction of salt transport in their gills depending on their environment.**
3. How do kangaroo rats survive on so little water? **They conserve water well. They have no sweat glands, they are nocturnal and avoid the daytime heat, they put out a very concentrated urine, and their kidneys can excrete salts very efficiently.**

 83

2 TEACH

Key Concepts are highlighted.

CONCEPT DEVELOPMENT

▶ Use an anatomical model to demonstrate the location of the kidneys and how well they are protected from being easily injured.

▶ **Demonstration:** Set up a diffusion experiment using a cellophane bag with a sugar solution placed in a container of water.

▶ Explain to students that approximately 180 liters of plasma are filtered by the kidneys every 24 hours.

TEACHER F.Y.I.

▶ All the uncoiled nephrons in a kidney stretched out and laid end to end would reach 120 kilometers.

REVEALING MISCONCEPTIONS

▶ Students often do not recognize the need for drinking adequate amounts of water each day. Water is necessary for the proper functioning of the kidneys. Water leaves our bodies through perspiration from the skin, through water vapor in exhaled breath, and in urine.

CROSS CURRICULUM

▶ **Mathematics:** Calculate the ratio of body weight to amount of urine excreted. If an 11-kg child excretes 600 mL, what would you expect a 45-kg person to excrete? *(2455 mL)* However, the amount is actually only around 1500 mL. The amount excreted by children is greater in proportion to their weight.

PROBLEM SOLVING

Think Critically: The circulating solution carries heat produced by Frederick away from his body much as evaporation of sweat would normally.

MINI-Lab

Materials: large animal kidneys, scalpels, magnifying glasses

CAUTION: *Caution students about the use of sharp objects.*

Answer: Students' drawings should reflect Figure 21-9 on page 499. Kidneys are red because they filter blood.

CHECK FOR UNDERSTANDING

Use the Mini Quiz to check for understanding.

MINI QUIZ

Use the Mini Quiz to check students' recall of chapter content.

1. **What are your main excretory organs?** *kidneys, lungs, skin*
2. **The filtering unit of the kidneys is the _____.** *nephron*
3. **The _____ are the tubes that carry urine from the kidneys to the bladder.** *ureters*
4. **What is dialysis?** *when blood from a person's artery is pumped through tubing that is bathed in a salt solution*

RETEACH

Have students prepare a basic diagram of the sequence of organs involved in the processing of liquid wastes in the body.

EXTENSION

For students who have mastered this section, use the **Reinforcement** and **Enrichment** masters or other OPTIONS provided.

Connect to... Physics

Answer: water

MINI-Lab

What do kidneys look like?

Examine a model of a kidney or some kidney that your teacher has obtained from a grocery. Carefully cut the tissue in half around the outline of the kidney. Use a magnifying glass to *observe* the internal features of the kidneys or view the features in the model. **In your Journal,** draw and describe what you see. Explain why the kidney is red in color.

Other Excretory Organs

Your urinary system helps your body maintain homeostasis with the help of other excretory organs. In Chapter 18, you read that skin releases perspiration to cool your body. Perspiration contains water and small amounts of salt. An adult normally loses up to 0.5 L of water in a day.

Your lungs are excretory organs, too. They remove carbon dioxide produced by cell activity in your body. You also lose water each time you breathe out. The air you exhale has some moisture in it. This is evident when you see your breath on a cold day or breathe on a cold window pane and notice a cloud or fog. Each day about 350 mL of water is removed from your body through your respiratory system.

Diseases and Disorders of the Urinary System

What happens when someone's urinary organs don't work properly? Waste products that are not removed build up and act as poisons in body cells. Water that would normally be removed from body tissues accumulates and causes swelling of the ankles and feet. Fluids can also build up around the heart. The heart works harder to move less to the lungs. Without elimination, there also could be an imbalance of salts. The body responds by trying to restore this balance. If the balance is not restored, the kidneys and other organs can be damaged.

Persons who have extremely defective kidneys may have to have their blood filtered by an artificial kidney machine
in a process called dialysis. During dialysis, blood from 4
an artery is pumped through tubing that is bathed in a salt solution similar to blood plasma. Waste materials diffuse from the tube containing blood and are washed away by the salt solution. The cleaned blood left behind is returned to a vein. A person who has only one kidney can still function normally. One healthy kidney can do the job of two.

The urinary system is a purifying unit for the circulatory system. Wastes are filtered from blood as it passes through the kidneys. Some water and salts are reabsorbed to maintain homeostasis. Waste materials, dissolved in water, are eliminated from the body. This system helps to maintain the health of cells and, therefore, the entire body.

Connect to... Physics

Changes in pressure in blood vessels and nephrons cause fluids to filter from the blood into the kidney or from the kidney into the blood. What substance in blood is excreted and reabsorbed the most?

OPTIONS

ASSESSMENT—ORAL

▶ **What happens to the blood after it leaves the kidneys?** *It returns to the heart via veins.*

▶ **Which urinary system organ has walls that allow it to expand?** *the bladder*

▶ **What evidence is there that the two kidneys in our body are more than enough to take care of the excretory functions?** *People have led normal lives after the removal of one kidney.*

PROGRAM RESOURCES

From the **Teacher Resource Package** use:

Activity Worksheets, page 192, Mini-Lab: What do kidneys look like?

Concept Mapping, pages 47-48.

Cross-Curricular Connections, page 25.

Transparency Masters, pages 79-80, The Urinary System.

Use **Color Transparency** 40, The Urinary System.

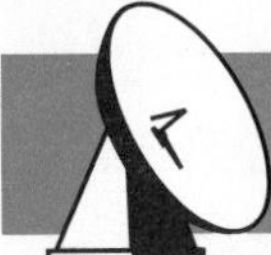

TECHNOLOGY

Kidney Transplants

A person can have just one kidney and still function normally. However, when both kidneys are lost, due to an accident or disease, a kidney transplant is needed. Kidneys from close relatives are preferred for transplants. This is because the relative may have a genetically similar immune system. If this is the case, the patient will be less likely to reject the new kidney.

The donor's kidney is removed and flushed with a sterile solution so that no trace of the donor's blood remains. An incision is made in the patient's abdomen, and a new kidney is inserted in the abdomen. The new kidney's vessels are stitched to the patient's blood vessels, and the urethra is inserted into the bladder. The operation is simple, but the patient's body will mobilize its immune system against the donor kidney just as it would against an infection. The patient must be given drugs to stop this immune system response.

Think Critically: Why would many transplant patients first suffer more from infection than from organ rejection?

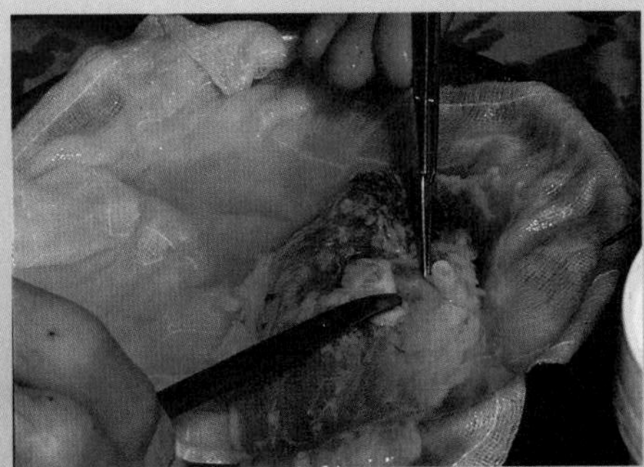

SECTION REVIEW

1. Describe three functions of the urinary system.
2. Explain how the kidneys remove wastes and keep fluids and salts in balance.
3. Why are skin and lungs excretory organs?
4. What happens when urinary organs don't work?
5. **Apply:** Explain why reabsorption of certain materials in the kidneys is important.
6. **Connect to Chemistry:** Find out what urea is. Find out two places in the body where urea is excreted.

Skill Builder

Concept Mapping

Using a network tree concept map, compare the excretory functions of the kidneys and the lungs. If you need help, refer to Concept Mapping in the **Skill Handbook** on pages 688 and 689.

TECHNOLOGY

Have students research the history of kidney transplants.

Think Critically: The immunosuppressive drugs make the patient vulnerable to infection.

3 CLOSE

▶ Ask questions 1-4 and the **Apply** Question in the Section Review.

▶ Small stones composed of calcium, uric acid, or magnesium compounds may form in the kidney, become lodged in the ureter, and cause severe pain. A sonic device can be used to break up the kidney stones instead of surgical removal.

SECTION REVIEW ANSWERS

1. rids the body of wastes, controls blood volume by removing excess water, balances salts and water

2. Kidneys filter the blood to remove wastes, water, and salt. Necessary amounts of water, sugar, and salt are returned to the blood.

3. They help your body get rid of water and other wastes.

4. Waste products accumulate and act as poisons to the body.

5. Apply: Many of the substances that are reabsorbed are needed by the body.

6. Connect to Chemistry: Urea is an end product when amino acids are broken down. Urea is excreted mainly through the kidneys but some through the skin.

Skill Builder

Waste in blood — are carried to → Lungs, Kidneys

Lungs — expel → Water and CO_2 — from → Cells

Kidneys — expel → Water and salts — from → Cells

Skillbuilder

ASSESSMENT

Portfolio: Use this Skillbuilder to assess students' abilities to form a Concept Map of the excretory functions of the lungs and kidneys.

MINI-Lab (p. 500)

ASSESSMENT

Performance: Have students identify parts of the kidney from a classroom model.

ACTIVITY 21-2

OBJECTIVE: Design and carry out an experiment to determine the number and position of sweat glands.
Time: one class period

PROCESS SKILLS applied in this activity are **experimenting, observing, measuring,** and **analyzing.**

PREPARATION

Obtain 2% tincture of iodine. Test the bond typing paper to ensure the starch/iodine reaction takes place.

 **Cooperative Learning:** Form Paired Partners.

SAFETY

Students should pull up sleeves during this experiment. Iodine is a poison and should not be put in the mouth or eyes.

THINKING CRITICALLY

Students will likely list "sweaty" areas such as the forehead, hands, underarms, and the feet. Suggestions of areas having few sweat glands might include back of hand, forearm, and calf of leg. Students may suggest a variety of means including visual examination with hand lenses. The small size of the glands requires some method to make them visible. A dye might be suggested.

TEACHING THE ACTIVITY

*Refer to the **Activity Worksheets** for additional information and teaching strategies.*

- Use typewriter bond paper.

SUMMING UP/SHARING RESULTS

- Answers will vary due to individual differences. One test indicated 18 dots per centimeter.
- The iodine reacted with the starch to form a purple/black compound wherever moisture from sweat glands touched the paper.
- Sweat glands have greater distribution on the palm of the hand than on the back of the hand. Results will vary on different parts of the body.

ACTIVITY 21-2 DESIGNING AN EXPERIMENT
Sweat Glands in the Skin

Sweat glands are special structures in the skin that excrete waste.

Getting Started

Your task in this activity is to devise a method of determining how sweat glands are distributed on the body.

Thinking Critically

List places on the body where your group thinks there are many sweat glands and where there are few sweat glands. How can individual sweat glands be detected?

Materials    

Your cooperative group will use:

- iodine solution
- bond typing paper
- hand lens

Try It!

1. Place a 1 cm x 1 cm square on the skin of a palm of one hand with two percent tincture of iodine. **CAUTION:** *Iodine is a poison; do not get it in your mouth or eyes. Iodine can stain clothing.* Allow the iodine solution to dry on your skin for three to four minutes.
2. Using the thumb from the other hand, hold a 2 cm x 2 cm square piece of bond typing paper on the iodine spot for one minute.
3. List up the paper and examine it with a hand lens. You should see some purple/black dots. Each dot represents a sweat gland. Count the number of dots in 1 cm^2 of the paper.
4. Select another spot, such as the back of the hand or forearm, to test for sweat gland distribution. Repeat steps 1 to 3.

Summing Up/Sharing Results

- How many dots per centimeter did you *observe* on your palm? On the second spot tested?
- Bond paper has starch in it. What caused the purple/black dots to appear on the paper?
- From your observations, *compare* the distribution of sweat glands in different body areas.

Going Further!

If this test were done on the palm of your hand after exercise, *predict* what you would expect to see. Try it and see if there is any effect on the number of dots that become visible.

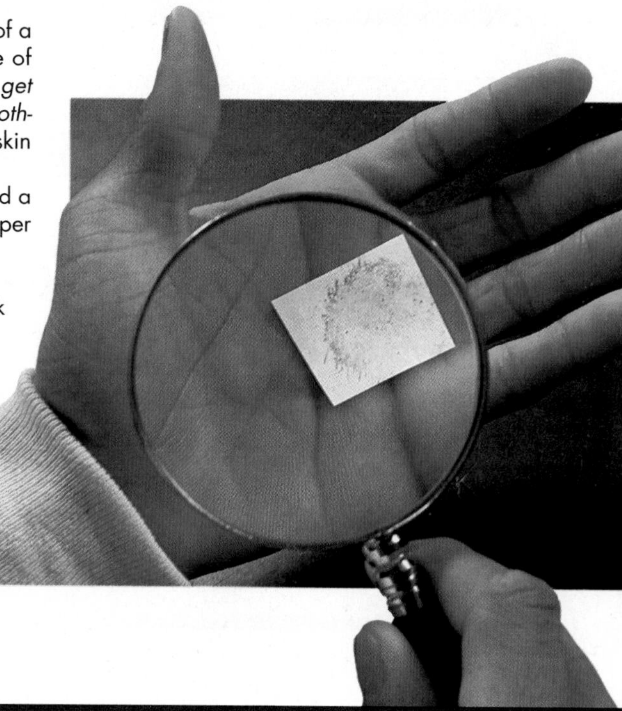

GOING FURTHER!

More dots would be seen because more sweat glands would be active and secrete moisture.

Activity ASSESSMENT

Performance: To further assess students' ability to use skills that located the position of sweat glands, see USING LAB SKILLS, Question 12, on page 504.

PROGRAM RESOURCES

From the **Teacher Resource Package** use:
Activity Worksheets, pages 187-188, Activity 21-2, Sweat Glands in the Skin.

CHAPTER

REVIEW

SUMMARY

21-1: Your Respiratory System

1. Your respiratory system helps you take oxygen into your lungs and body cells and helps you remove carbon dioxide.
2. Inhaled air passes through the nasal cavity, pharynx, larynx, trachea, bronchi, bronchioles, and into the alveoli of the lungs.

3. Breathing results in part from the diaphragm's movement, which changes the pressure within the lungs.
4. Smoking causes many problems throughout the respiratory system.

21-2: Science and Society: Dangerous Breathing

1. Regular exercise helps your body feel and work well.
2. Exercising in polluted environments can be dangerous to your health.

21-3: Your Urinary System

1. Kidneys filter blood to remove wastes and keep sodium and other chemicals in balance.
2. The kidneys are the major organs of the urinary system; they filter wastes from all of the blood in your body.
3. The skin and lungs are also excretory organs.
4. When kidneys fail to work, dialysis may be used.

KEY SCIENCE WORDS

a. **alveoli**
b. **asthma**
c. **bladder**
d. **bronchi**
e. **chronic bronchitis**
f. **diaphragm**
g. **emphysema**
h. **kidneys**
i. **larynx**
j. **nephrons**
k. **nitrogen dioxide**
l. **pharynx**
m. **trachea**
n. **ureters**
o. **urethra**
p. **urinary system**
q. **urine**

UNDERSTANDING VOCABULARY

Match each phrase with the correct term from the list of Key Science Words.

1. harmful chemical in auto exhaust
2. clusters of air sacs
3. where vocal cords are attached
4. branches of the trachea
5. muscle involved in breathing
6. disease of alveoli that have lost their elasticity
7. a cartilage-reinforced tube through which air moves to the bronchi
8. major urinary organs
9. tubes from kidney to bladder
10. fluid waste

OPTIONS

ASSESSMENT

To assess student understanding of material in this chapter, use the Program Resources listed.

COOPERATIVE LEARNING

Consider using cooperative learning in the THINK AND WRITE CRITICALLY, APPLY, and MORE SKILL BUILDERS sections of the Chapter Review.

PROGRAM RESOURCES

From the **Teacher Resource Package** use:

Chapter Review, pages 45-46.

Chapter and Unit Tests, pages 134-137, Chapter Test.

CHAPTER

REVIEW

SUMMARY

Have students read the summary statements to review the major concepts of the chapter.

UNDERSTANDING VOCABULARY

1. k
2. a
3. i
4. d
5. f
6. g
7. m
8. h
9. n
10. q

ASSESSMENT

Portfolio

Encourage students to place in their portfolios one or two items of what they consider to be their best work. For each item, ask students to explain why that item was chosen and what they learned from it. Items might be selected from the following.

- Skillbuilder sequence of respiratory pathway, p. 494
- Check for Understanding list of how to avoid nitrogen dioxide exposure, p. 497
- Cross Curriculum–Art poster on dangers of exercising in polluted environments, p. 497

Performance

Additional performance assessments may be found in *Performance Assessment* and *Science Integration Activities* that accompany **Merrill Life Science.** Performance Task Assessment Lists and rubrics for evaluating these activities and other products generated throughout the chapter can be found in Glencoe's *Performance Assessment in Middle School Science.*

CHAPTER
REVIEW

CHECKING CONCEPTS

1. b
2. c
3. a
4. b
5. a
6. d
7. a
8. b
9. c
10. c

USING LAB SKILLS

ASSESSMENT

Use these alternate lab exercises to assess students' understanding of skills used in this chapter.

11. Accept all reasonable designs. Students should be cautioned to breathe normally and should be encouraged to estimate the amount of water vapor.

12. There are more sweat glands on the palms of the hands than on the backs of the hands. An advantage is in the ability to grip.

THINK AND WRITE CRITICALLY

13. Air is warmed, filtered, and moistened in the nasal cavity.

14. The trachea must remain open for air to enter through it; the esophagus is soft and does not need to be "open" except when food is moving through.

15. The kidneys filter water, sugar, salt, and wastes from blood. Then most water, sugar, and salt is reabsorbed. Nitrogen waste is eliminated.

16. Both systems rid the body of wastes, carbon dioxide, and urea, keeping the blood chemicals balanced and free of toxic materials.

CHAPTER
REVIEW

CHECKING CONCEPTS

Choose the word or phrase that completes the sentence.

1. When you inhale, your ______ contract(s) and move(s) down.
a. bronchioles **c.** nephrons
b. diaphragm **d.** kidneys

2. Air is moistened, filtered, and warmed in the ______.
a. larynx **c.** nasal cavity
b. pharynx **d.** trachea

3. Exchange of gases occurs between the ______ and capillaries.
a. alveoli **c.** bronchioles
b. bronchi **d.** trachea

4. The rib cage ______ when you exhale.
a. moves up **c.** moves out
b. moves down **d.** stays the same

5. ______ is a lung disorder that may occur as an allergic reaction.
a. Asthma **c.** Emphysema
b. Chronic bronchitis **d.** Cancer

6. A condition worsened by smoking is ______.
a. arthritis **c.** excretion
b. respiration **d.** emphysema

7. ______ are filtering units of the kidney.
a. Nephrons **c.** Neurons
b. Ureters **d.** Alveoli

8. Urine is temporarily held in the ______.
a. kidneys **c.** ureter
b. bladder **d.** urethra

9. About 1 liter of water is lost per day through ______.
a. sweat **c.** urine
b. lungs **d.** none of these

10. All except ________ is(are) reabsorbed by blood after passing through the kidneys.
a. salt **c.** wastes
b. sugar **d.** water

USING LAB SKILLS

11. The Activity on page 495 relates to carbon dioxide exhaled by the lungs. Design an experiment to detect the water vapor exhaled in an hour; in a day.

12. In the Activity on page 502, you determined the distribution of sweat glands on the palms of your hand. Repeat the experiment to determine distribution of sweat glands on the backs of your hands. If there is a difference, suggest an advantage to having more sweat glands on one surface than another.

THINK AND WRITE CRITICALLY

Answer the following questions in your Journal using complete sentences.

13. Why is it better to breathe through your nose?
14. Why does the trachea have cartilage but the esophagus does not?
15. Explain how kidneys maintain homeostasis.
16. Describe how the respiratory and excretory systems maintain homeostasis.

APPLY

17. Compare air pressure in the lungs during inhalation and exhalation.
18. What is the advantage of the lungs having many air sacs instead of being just two large sacs, like balloons?
19. Explain the damage smoking does to cilia, alveoli, and lungs.
20. What would happen to the blood if the kidneys stopped working?

MORE SKILL BUILDERS

If you need help, refer to the Skill Handbook.

1. **Making and Using Graphs:** Make a pie graph of total lung capacity.
   - Tidal volume (inhaled or exhaled during a normal breath) = 500 mL
   - Inspiratory reserve volume (air that can be forcefully inhaled after a normal inhalation) = 3000 mL
   - Expiratory reserve volume (air that can be forcefully exhaled after a normal expiration) = 1100 mL
   - Residual volume (air left in the lungs after forceful exhalation) = 1200 mL
2. **Interpreting Data:** Interpret the data below. How much of each substance is reabsorbed into the blood in the kidneys? What substance is totally excreted in the urine?

Substance	Amount moving through kidney to be filtered	Amount excreted in urine
water	125 liters	1 liter
salt	350 grams	10 grams
urea	1 gram	1 gram
glucose	50 grams	0 grams

3. **Recognizing Cause and Effect:** Discuss how lack of oxygen is related to lack of energy.
4. **Hypothesizing:** Hypothesize the number of breaths you would expect a person would take per minute in each situation and give a reason for each hypothesis.
   - while sleeping
   - while exercising
   - while on top of Mount Everest
5. **Concept Mapping:** Make an events chain concept map showing what happens when urine forms in the kidneys. Begin with the phrase, *In the nephron...*

PROJECTS

1. Find out what the Heimlich maneuver is and how it is used in saving lives. How is its success based on the presence of residual air, the small amount of air that is always left in the lungs? Make poster showing the correct procedure and demonstrate the procedure for the class.
2. Contact your local Kidney Foundation for information on dialysis machines and kidney transplants. Ask your teacher if a guest speaker can talk to the class.

APPLY

17. When you inhale, the volume of your chest cavity increases. Air pressure is reduced in your chest cavity, and your lungs fill with air. When you exhale, the size of your chest cavity decreases. Pressure in your lungs increases and you breathe out.

18. Having many air sacs increases the surface area for gas exchange over one large surface.

19. Cilia are destroyed and therefore cannot move particles out of the lungs; alveoli lose elasticity, decreasing surface area for gas exchange.

20. If the kidneys stopped working, blood would accumulate the waste products of the body's organs. It would become poisonous, and the person would need immediate medical care.

MORE SKILL BUILDERS

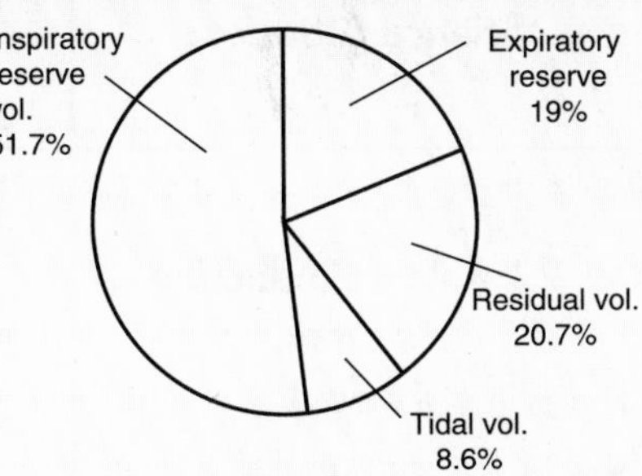

1. Making and Using Graphs:

2. Interpreting Data: water—124 liters; salt—340 grams; urea—all is excreted; glucose—50 grams

3. Recognizing Cause and Effect: Insufficient amounts of oxygen to cells result in a lack of energy because cells need oxygen to break down food to release energy. If oxygen isn't supplied, energy is not released.

4. Hypothesizing: Number of breaths while sleeping is low because less oxygen is needed for rest. While exercising, breaths per minute increase to get oxygen to the muscles. On top of Mt. Everest there is less oxygen in the air (high altitude), and therefore more breaths per minute are taken.

5. Concept Mapping: Events should be in the following order in a chain of boxes: (1) In the nephron, water, sugar, salt, and wastes are filtered. (2) Capillaries reabsorb most of these substances and return them to the blood. (3) Waste-free blood is circulated throughout the body. (4) Excess waste is eliminated from the body.

CHAPTER

22 Body Regulation

CHAPTER SECTION	OBJECTIVES	ACTIVITIES
22-1 Your Nervous System (3 days)	1. **Describe** the basic structure of a neuron and how an impulse moves from one neuron to the next. 2. **Compare** the central and peripheral nervous systems and **identify** their parts and functions. 3. **Interpret** the pathway of a reflex.	**MINI-Lab:** *Who has a better sense of smell?* p. 508 **Activity 22-1:** *Reaction Time,* p. 514
22-2 The Senses (2 days)	1. **List** the sensory receptors in each sense organ. 2. **Explain** what type of stimulus each sense organ responds to and how. 3. **Explain** the need for healthy senses.	**MINI-Lab:** *How is balance maintained?* p. 516 **Activity 22-2:** *Designing an Experiment (Investigating Skin Sensitivity),* p. 519
22-3 Alzheimer's Disease Science & Society (1 day)	1. **Describe** Alzheimer's disease. 2. **State** some of the results of Alzheimer's disease on society. 3. **Explain** two possible causes of Alzheimer's.	
22-4 Your Endocrine System (1 day)	1. **Explain** the function of hormones. 2. **Name** three endocrine glands and explain the effects of their hormones. 3. **Explain** how a feedback system works.	
Chapter Review		

ACTIVITY MATERIALS

FIND OUT	ACTIVITIES		MINI-LABS	
Page 507 metric ruler	**22-1 Reaction Time, p. 514** penny meterstick	**22-2 Designing an Experiment, p. 519** index card (3" × 5") 8 toothpicks glue metric ruler	**Who has a better sense of smell? p. 508** paper pencil	**How is balance maintained? p. 516** chalk chalkboard

CHAPTER FEATURES	TEACHER RESOURCE PACKAGE	OTHER RESOURCES
Technology: *Watching the Brain at Work,* p. 511 **Skill Builder,** *Concept Mapping,* p. 513	**Ability Level Worksheets** ◆ ***Study Guide,*** p. 84 ● ***Reinforcement,*** p. 84 ▲ ***Enrichment,*** p. 84 **Activity Worksheets,** pp. 5, 194, 195, 200 **Transparency Masters,** pp. 81-84	**Color Transparency 41,** The Synapse **Color Transparency 42,** The Reflex **Lab Manual:** Reaction Time, p. 151 **STVS:** Disc 7, Side 1; Disc 5, Side 2
Skill Builder, *Making and Using Tables,* p. 518	**Ability Level Worksheets** ◆ ***Study Guide,*** p. 85 ● ***Reinforcement,*** p. 85 ▲ ***Enrichment,*** p. 85 **Activity Worksheets,** pp. 196, 197, 201 **Cross-Curricular Connections,** p. 26 **Technology,** pp. 15-16	**Lab Manual:** Parts of the Eye, p. 155 **STVS:** Disc 7, Side 1; Disc 7, Side 2 **Science Integration Activity 22**
You Decide! p. 521	**Ability Level Worksheets** ◆ ***Study Guide,*** p. 86 ● ***Reinforcement,*** p. 86 ▲ ***Enrichment,*** p. 86	
Problem Solving: *Why am I so Tired?* p. 524 **Skill Builder:** *Comparing and Contrasting,* p. 524	**Ability Level Worksheets** ◆ ***Study Guide,*** p. 87 ● ***Reinforcement,*** p. 87 ▲ ***Enrichment,*** p. 87 **Critical Thinking/Problem Solving,** p. 26 **Concept Mapping,** p. 49 **Science and Society,** p. 26 **Transparency Masters,** pp. 85-86	**Color Transparency, 43** Endocrine Glands **STVS:** Disc 7, Side 2
Summary Key Science Words Understanding Vocabulary Checking Concepts Using Lab Skills Think & Write Critically Apply More Skill Builders Projects	ASSESSMENT RESOURCES **Chapter Review,** pp. 47-48 **Chapter Test,** pp. 138-141 **Performance Assessment in Middle School Science (PAMSS)**	**Chapter Review Software** **Test Bank** **Alternate Assessment** **Performance Assessment**

◆ **Basic** ● **Average** ▲ **Advanced**

ADDITIONAL MATERIALS

SOFTWARE	AUDIOVISUAL	BOOKS/MAGAZINES
Senses, Ventura Educational System. *Regulation and Homeostasis: Systems in Balance,* IBM. *The Ear,* Queue. *The Eye,* Queue.	*The Endocrine System (2nd ed.),* film, EBEC. *The Human Brain,* film, EBEC. *The Hidden Universe: The Brain,* film, McGraw-Hill. *The Human Body: The Nervous System,* film, Coronet/MTI. *Ears and Hearing/Eyes and Seeing,* laserdisc, EBEC. *Human Brain/Nervous System,* laserdisc, EBEC.	Chapouthier, G. and J.J. Matras. *Nervous System and How It Functions.* Cambridge, MA: Abacus Press, 1986. Facklam, Margery and Howard Facklam. *Brain: Magnificent Mind Machine.* San Diego, CA: Harcourt Brace Jovanovich, Inc., 1982. Little, Marjorie. *Endocrine System.* NY: Chelsea House Publishers, 1989. Talbot, Mary. *Senses.* NY: Chelsea House Publishers, 1989.

CHAPTER

22

BODY REGULATION

THEME DEVELOPMENT: A major concept of science is homeostasis. This chapter on body regulation emphasizes the theme of homeostasis. The structures and functions of the nervous and endocrine systems are described. Their integrated efforts bring about the stable environment necessary for the healthy functioning of the body.

CHAPTER OVERVIEW

▶**Section 22-1:** This section centers on the functioning of the nervous system and the movement of impulses along neurons. Also described are the two major divisions of the nervous system.

▶**Section 22-2:** This section describes the sense organs that enable the body to distinguish changes in its internal and external environments. The mechanisms of reception and response are discussed.

▶**Section 22-3: Science and Society:** The symptoms and causes of Alzheimer's disease are explored. The You Decide feature asks students to consider the advantages and disadvantages of keeping a family member with Alzheimer's at home.

▶**Section 22-4:** The endocrine system is studied in this section. Its effect on tissues via its secretions, hormones, is described.

CHAPTER VOCABULARY

neuron
dendrites
axon
sensory neurons
interneurons
motor neurons
synapse
central nervous system
peripheral nervous system
cerebrum
cerebellum
brainstem
reflex
retina
cochlea
olfactory cells
taste buds
Alzheimer's disease
hormones
target tissues

CHAPTER

22 Body Regulation

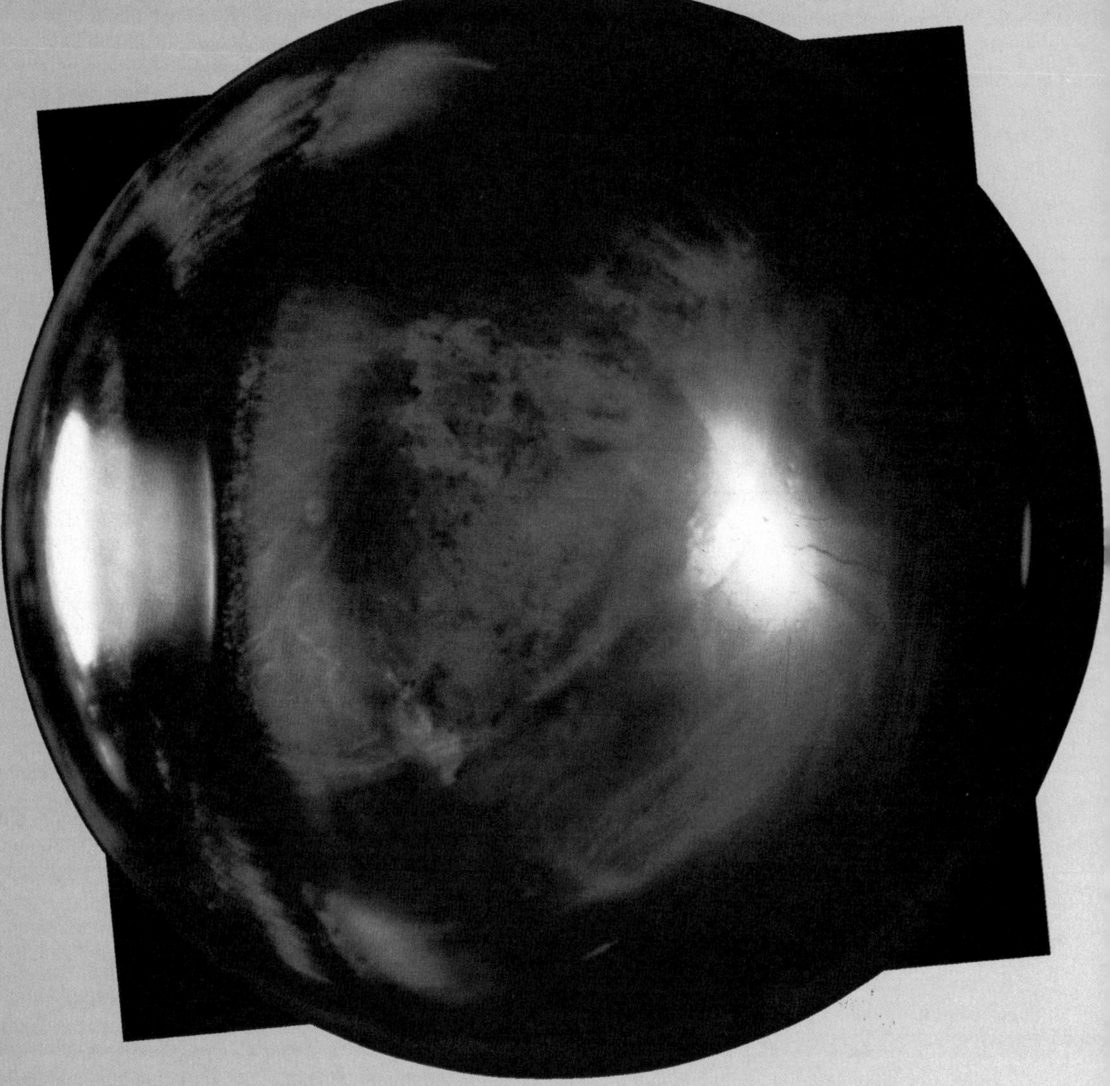

506

OPTIONS

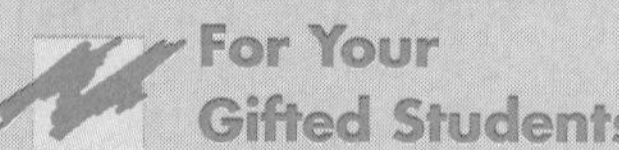

For Your Gifted Students

Students can research the senses of animals. Have them find out how they compare with human senses. Have them find out how the animals sense their surroundings (an insect's sight, an owl's ability to see at night). Ask students to write about the most unusual animals that they find. Have them try to create a situation that might enable the class to simulate a particular animal's sense.

For Your Mainstreamed Students

As you progress through the chapter, have students make a poster display showing how the nervous system works. They should show the structures and divisions of the nervous system. Have them show how the brain receives and sends messages.

The eye on the page to your left is a complex organ that is your key to the world of vision. Sight is one of the senses that helps you keep in touch with what is going on around you. Do your eyes ever play tricks on you? Have your senses ever fooled you?

FIND OUT!

Do this activity to find out if your senses ever mislead you.

Look at the figure at the bottom of the page. *Estimate* the size difference between the left outer circle and the right inner circle. Now use a metric ruler to measure the sizes of the two circles to determine if their diameters differ. What did you find out?

Gearing Up

Previewing the Chapter

Use this outline to help you focus on important ideas in this chapter.

Previewing Science Skills

- ▶ In the **Skill Builders,** you will make a concept map, make and use tables, and compare and contrast.
- ▶ In the **Activities,** you will experiment, predict, measure, and collect data.
- ▶ In the **MINI-Labs,** you will experiment and collect and interpret data.

What's next?

Your senses may have misled you in the Find Out activity, but your nervous system generally works to keep you aware and responsive to your environment. In this chapter, you will also learn about chemical control systems that maintain the homeostasis of all other body systems.

INTRODUCING THE CHAPTER

Use the Find Out activity to introduce students to one aspect of the nervous system, vision. Inform students they will be learning more about the nervous system and also about the endocrine system as they read the chapter.

FIND OUT!

Preparation: Collect some additional illustrations of optical illusions, such as inkblot tests, to extend the learning of this activity.

Cooperative Learning: Use the Paired Partners strategy and have each student estimate the size difference. Then have one student place a nickel within the upper left circle. Now make a revised estimate about the difference in size of the two circles.

Teaching Tips

▶Have students devise their own optical illusions in the form of inkblot tests.

Gearing Up

Have students study the Gearing Up feature to familiarize themselves with the chapter. Discuss the relationships of the topics in the outline.

What's Next?

Before beginning the first section, make sure students understand the connection between the Find Out activity and the topics to follow.

ASSESSMENT OPTIONS

PORTFOLIO

Refer to page 525 for suggested items that students might select for their portfolios.

PERFORMANCE ASSESSMENT

See page 525 for additional Performance Assessment options.

Process

Skillbuilders, pp. 518, 524
MINI-Labs, pp. 508, 516
Activities, 22-1, p. 514; 22-2, p. 519
Using Lab Skills, p. 526

CONTENT ASSESSMENT

Assessment—Oral, pp. 510, 513, 518
Skillbuilder, p. 513
Section Reviews, pp. 513, 518, 521, 524
Chapter Review, pp. 525-527
Mini Quizzes, pp. 510, 517, 521, 523

GROUP ASSESSMENT

Opportunities for group assessment occur with Cooperative Learning Strategies and Flex Your Brain Activities.

SECTION 22-1

PREPARATION

SECTION BACKGROUND

▶ Inside a neuron at rest, the concentration of potassium ions is greater than that of sodium ions. A stimulus causes a brief reversal of the concentration of the ions. This produces a small electrical charge that travels from the excited area to the adjacent resting area. The entire action causes the impulse to travel along the nerve fiber.

PREPLANNING

▶ To prepare for Activity 22-1, obtain pennies and a meterstick for each set of partners.

1 MOTIVATE

▶ In a school yard or athletic field, position two students 100 meters apart. Use this visual representation to illustrate the distance that some nerve impulses travel in one second.

MINI-Lab

Materials: various foods, colognes, household products, paper, pencils

Answer: A 1988 survey of several thousand people by National Geographic indicates that females are more acutely aware of odors than males.

MINI-Lab
ASSESSMENT

Performance: To further assess students' understanding of the sense of smell, see USING LAB SKILLS, question 11, on page 526.

PROGRAM RESOURCES

From the **Teacher Resource Package** use:

Activity Worksheets, page 200, Who has a better sense of smell?

22-1 Your Nervous System

New Science Words

neuron
dendrites
axon
sensory neurons
interneurons
motor neurons
synapse
central nervous system
peripheral nervous system
cerebrum
cerebellum
brainstem
reflex

Objectives

- ▶ Describe the basic structure of a neuron and how an impulse moves from one neuron to the next.
- ▶ Compare the central and peripheral nervous systems and identify their parts and functions.
- ▶ Interpret the pathway of a reflex.

The Nervous System at Work

It's your night to do the dishes. To make the job easier, you decide to plug in a favorite tape. About halfway through the chore, you turn to put a plate in the dish drainer and come face to face with your brother wearing an ugly rubber Halloween mask. You scream, and the dish clatters into the sink. Instantly you are out of breath and your hands begin to shake. Your knees feel wobbly, and your heartbeat speeds up. But then, after a few minutes, your breathing returns to normal and your heartbeat is back to its regular rate. What's going on here?

The scene just pictured is an example of how your body responds to changes in its environment and adjusts itself. Your body makes these adjustments with the help of your nervous system. Any change inside or outside your body that brings about a response is called a stimulus. On an average day, you're bombarded by thousands of stimuli. Noise, light, the smell of food in the cafeteria, and the feel of a cold metal doorknob in your hand are all stimuli from outside your body. A growling stomach is an example of an internal stimulus.

How can your body handle all these stimuli at the same time and yet appear so calm? How is it possible for all your systems to work in a coordinated way? Your body has hundreds of internal control systems that maintain homeostasis. Breathing rate, heartbeat rate, and digestion are just a few of the activities in your body that are constantly checked and regulated. Your nervous system and the endocrine system, a chemical control described later in this chapter, are the main mechanisms by which homeostasis is maintained in your body.

MINI-Lab

Who has a better sense of smell?

Design an experiment to test classmates' ability to smell different foods, colognes, and household products. Record their responses in a data table according to sex. *Interpret* your data. Are there differences between males and females in the ability to detect odor?

OPTIONS

Meeting Different Ability Levels

For Section 22-1, use the following **Teacher Resource Masters** depending upon individual students' needs.

◆ **Study Guide Master** for all students.

● **Reinforcement Master** for students of average and above average ability levels.

▲ **Enrichment Master** for above average students.

Additional Teacher Resource Package masters are listed in the OPTIONS box throughout the section. The additional masters are appropriate for all students.

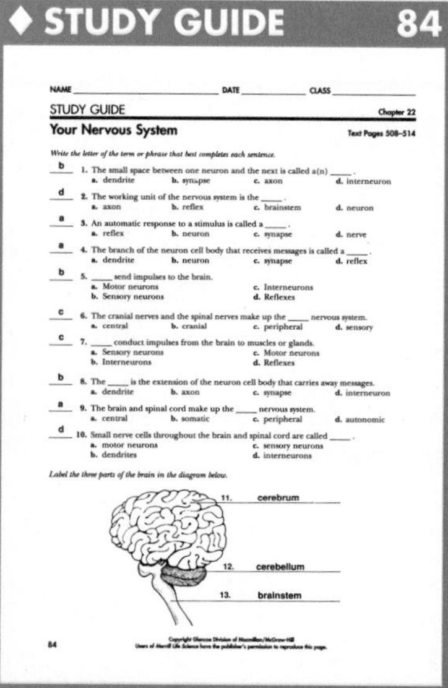

◆ STUDY GUIDE 84

NAME ____________ DATE ________ CLASS ________

STUDY GUIDE — Chapter 22

Your Nervous System — Text Pages 508–514

Write the letter of the term or phrase that best completes each sentence.

b 1. The small space between one neuron and the next is called a(n) _____.
a. dendrite b. synapse c. axon d. interneuron

d 2. The working unit of the nervous system is the _____.
a. axon b. reflex c. brainstem d. neuron

a 3. An automatic response to a stimulus is called a _____.
a. reflex b. neuron c. synapse d. nerve

a 4. The branch of the neuron cell body that receives messages is called a _____.
a. dendrite b. neuron c. synapse d. reflex

b 5. _____ send impulses to the brain.
a. Motor neurons b. Sensory neurons c. Interneurons d. Reflexes

c 6. The cranial nerves and the spinal nerves make up the _____ nervous system.
a. central b. cranial c. peripheral d. sensory

c 7. _____ conduct impulses from the brain to muscles or glands.
a. Sensory neurons b. Interneurons c. Motor neurons d. Reflexes

b 8. The _____ is the extension of the neuron cell body that carries away messages.
a. dendrite b. axon c. synapse d. interneuron

a 9. The brain and spinal cord make up the _____ nervous system.
a. central b. somatic c. peripheral d. autonomic

d 10. Small nerve cells throughout the brain and spinal cord are called _____.
a. motor neurons b. dendrites c. sensory neurons d. interneurons

Label the three parts of the brain in the diagram below.

84

Neurons

The working unit of the nervous system is the nerve cell, or **neuron.** The neuron in Figure 22-2 is made up of a cell body and branches called dendrites and axons.
1 **Dendrites** receive messages and send them to the cell body. An **axon** carries messages away from the neuron cell body. Any message carried by a neuron is called an impulse.

There are three types of neurons in your body: sensory neurons, motor neurons, and interneurons. **Sensory neurons** receive information and send impulses *to* the spinal cord or brain. Your skin and other sense organs are equipped with structures called receptors that respond to stimuli. These stimuli may be changes in temperature, pain, pressure, odor, sound waves, and chemicals around you. This information is picked up by nearby sensory neurons and sent in the form of impulses to your brain or spinal cord. There, a second type of neuron, an interneuron, passes the impulses to motor neurons. *Inter*
2 means "between." **Interneurons** are nerve cells throughout the brain and spinal cord that relay impulses from sensory neurons to motor neurons. There are more interneurons in your body than the other two types. **Motor neurons** conduct impulses *from* the brain or spinal cord to muscles or glands throughout the body. For instance, when you saw the Halloween mask, sensory receptors in your eyes were stimulated. A message was sent to your brain by way of sensory neurons. Your brain responded by sending back impulses along motor neurons to your muscles. Your heart immediately started to pound and your breathing rate increased. Together, these three types of neurons act like a relay team, moving impulses through your body from stimulus to response.

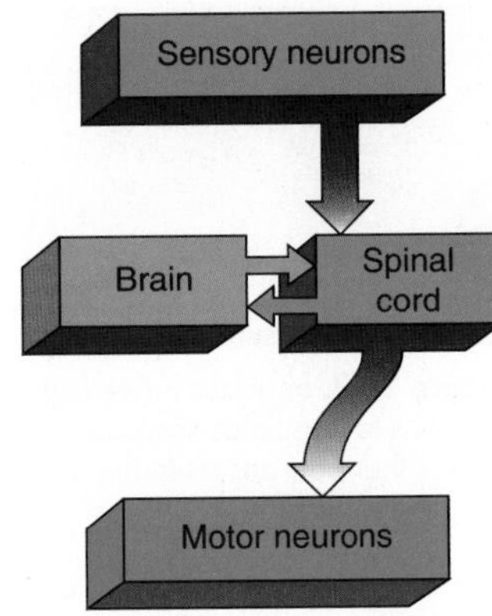

Figure 22-1. In your nervous system, impulses travel a pathway that takes them from sensory neurons to interneurons in your brain and spinal cord to motor neurons.

In what direction do sensory neurons move impulses?

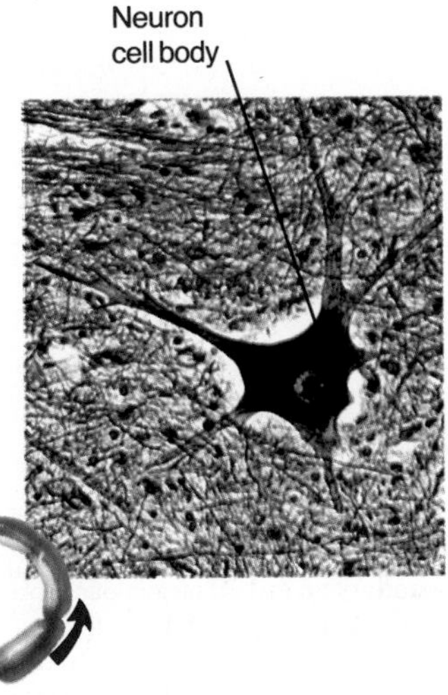

Dendrites
Nucleus
Cell body
Axon
Direction of impulse

Figure 22-2. A neuron is made up of a cell body, dendrites, and an axon.

OBJECTIVES AND SCIENCE WORDS: Have students review the objectives and science words for the chapter as they study each section.

2 TEACH

Key Concepts are highlighted.

CONCEPT DEVELOPMENT

▶Obtain a model of the human brain from a local high school to use in teaching this chapter.

▶Ask the question: **What would happen if all your sensory neurons stopped working?** *Your brain would stop receiving stimuli from inside and outside your body, making it impossible to maintain homeostasis.*

▶**Demonstration:** Have three students stand side by side. Their arms and fingers should be outstretched at their sides. Starting at one end of the line, have the students pass a metric ruler from person to person. Students' fingers should not touch. Use this to show students how an impulse travels from a sensory neuron through an interneuron to a motor neuron. Have students identify the axon and dendrites of each "nerve cell."

● REINFORCEMENT 84

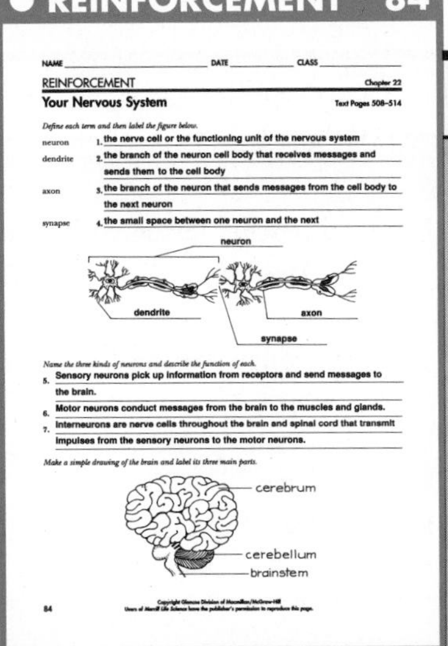

NAME ____ DATE ____ CLASS ____

REINFORCEMENT — Chapter 22

Your Nervous System — Text Pages 508–514

Define each term and then label the figure below.

neuron 1. **the nerve cell or the functioning unit of the nervous system**

dendrite 2. **the branch of the neuron cell body that receives messages and sends them to the cell body**

axon 3. **the branch of the neuron that sends messages from the cell body to the next neuron**

synapse 4. **the small space between one neuron and the next**

neuron — dendrite — axon — synapse

Name the three kinds of neurons and describe the function of each.

5. **Sensory neurons pick up information from receptors and send messages to the brain.**
6. **Motor neurons conduct messages from the brain to the muscles and glands.**
7. **Interneurons are nerve cells throughout the brain and spinal cord that transmit impulses from the sensory neurons to the motor neurons.**

Make a simple drawing of the brain and label its three main parts.

cerebrum
cerebellum
brainstem

84

▲ ENRICHMENT 84

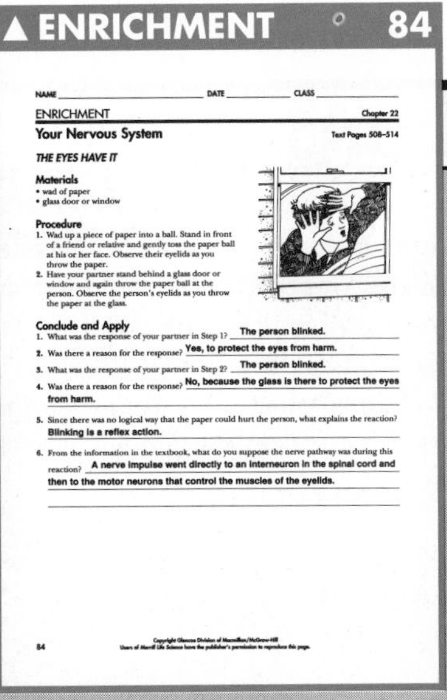

NAME ____ DATE ____ CLASS ____

ENRICHMENT — Chapter 22

Your Nervous System — Text Pages 508–514

THE EYES HAVE IT

Materials
- wad of paper
- glass door or window

Procedure
1. Wad up a piece of paper into a ball. Stand in front of a friend or relative and gently toss the paper ball at his or her face. Observe their eyelids as you throw the paper.
2. Have your partner stand behind a glass door or window and again throw the paper ball at the person. Observe the person's eyelids as you throw the paper at the glass.

Conclude and Apply
1. What was the response of your partner in Step 1? **The person blinked.**
2. Was there a reason for the response? **Yes, to protect the eyes from harm.**
3. What was the response of your partner in Step 2? **The person blinked.**
4. Was there a reason for the response? **No, because the glass is there to protect the eyes from harm.**
5. Since there was no logical way that the paper could hurt the person, what explains the reaction? **Blinking is a reflex action.**
6. From the information in the textbook, what do you suppose the nerve pathway was during this reaction? **A nerve impulse went directly to an interneuron in the spinal cord and then to the motor neurons that control the muscles of the eyelids.**

84

MINI QUIZ

Use the Mini Quiz to check students' recall of chapter content.

1. **What is the function of the dendrite?** *to receive the messages and send them to the cell body*
2. **_____ are nerve cells that relay impulses from sensory to motor neurons.** *Interneurons*
3. **The peripheral nervous system consists of _____ and _____ nerves.** *cranial, spinal*
4. **The _____ connects the brain to the spinal cord.** *brainstem*

CROSS CURRICULUM

▶**Language Arts:** Have students look up *synapse* in the dictionary. Students will find that it comes from the Greek roots *sun* (together) and *haptein* (unite). Ask students to think about how the roots of the word reflect its meaning.

Answer: 2500 times faster

TEACHER F.Y.I.

▶Some neurons in the brain may have in excess of 100 000 synaptic connections.

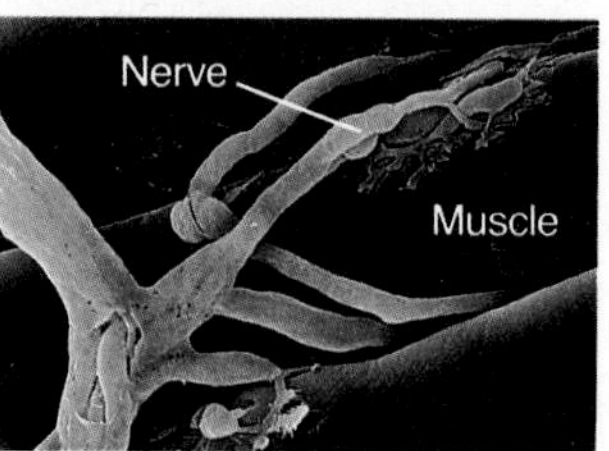

Figure 22-3. Impulses move from neurons to muscle as seen above, or to other neurons as in the illustration to the right.

Synapse

Neurons don't touch each other. How does an impulse move from one neuron to another? To get from one neuron to the next, an impulse jumps across a small space called a **synapse.** In Figure 22-3, you can see that when an impulse reaches the end of an axon, a nerve transmitting chemical is released by the axon. This chemical diffuses across the synapse and starts an impulse in the next neuron. In this way, an impulse moves from one neuron to another.

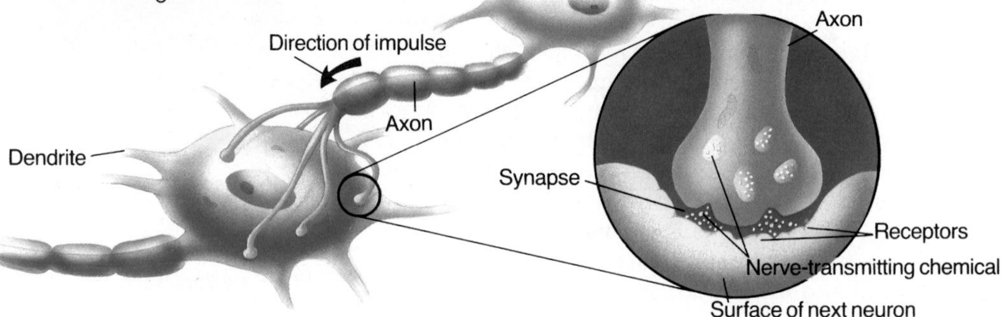

Divisions of the Nervous System

Organs of the nervous system are grouped into two major divisions: the central nervous system (CNS) and the peripheral nervous system (PNS). The **central nervous system** is made up of the brain and spinal cord. The
3 **peripheral nervous system** is made up of all the nerves outside the CNS, including cranial nerves and spinal nerves. These nerves connect the brain and spinal cord to other body parts.

The Central Nervous System

Because you have a central nervous system with a brain, your body activities are coordinated. If someone pokes you in the ribs, your whole body is aware. Your neurons are adapted in such a way that impulses move in only one direction. Impulses in sensory neurons move from a receptor to the brain or spinal cord.

The brain is made up of approximately 100 billion neurons. Figure 22-4 shows that the brain is divided into three major parts: cerebrum, cerebellum, and brainstem. The largest part of the brain, the **cerebrum,** is divided into two large sections called hemispheres. Here, impuls-

Connect to... Physics

Nerve impulses can travel at speeds of up to 120 meters per second in the body. Compare this to the speed of electricity, which moves at about 300 000 kilometers per second. How many times faster is an electric impulse than a nerve impulse?

OPTIONS

ASSESSMENT—ORAL

▶**What structures are involved in a reflex action pathway?** *receptors, sensory neurons, interneurons, motor neurons, effectors*

▶**Why might an anesthesiologist test for a reflex?** *to determine how the drug (anesthetic) being used is affecting the nervous system*

PROGRAM RESOURCES

From the **Teacher Resource Package** use:

Activity Worksheets, page 5, Flex Your Brain.

Transparency Masters, pages 81-82, The Synapse.

Use **Laboratory Manual,** page 151, Reaction Time.

Use **Color Transparency** 41, The Synapse.

es from the senses are interpreted, memory is stored, and the work of voluntary muscles is controlled. The outer layer of the cerebrum, the cortex, is marked by many ridges and grooves. The diagram also shows some of the tasks that sections of the cortex control.

A second part of the brain, the **cerebellum,** is behind and under the cerebrum. It coordinates voluntary muscle movements and maintains balance and muscle tone.

4 The **brainstem** extends from the cerebrum and connects the brain to the spinal cord. It is made up of the midbrain, the pons, and the medulla. The brainstem controls homeostasis of heartbeat, breathing, and blood pressure. It also coordinates involuntary muscle movements.

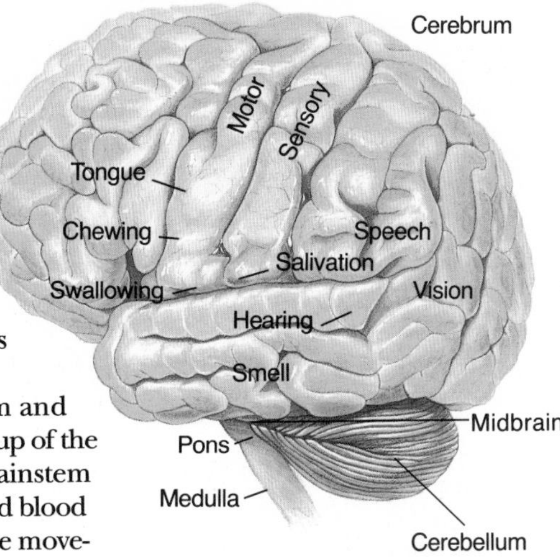

Figure 22-4. Different areas of the brain control specific body activities.

TECHNOLOGY

Watching the Brain at Work

Many medical centers in the United States are now using positron emission tomography (PET) for research and diagnosis of the brain. The simple sugar, glucose, is tagged with a radioactive tracer and fed to the patient. The tracer moves through the circulatory system of the brain and gives off particles called positrons. Positrons collide with electrons from the body and release a form of energy. The path of the energy appears as an image on a color monitor. The image shows where energy is used in different areas of the brain as these areas become stimulated.

Scientists can use PET to go beyond looking at just the physical makeup of the brain. By comparing PET images formed when people perform different tasks, researchers have pinpointed the areas of the brain used for seeing, reading, hearing, speaking, and thinking.

Think Critically: How can PET help scientists understand what happens when someone has an epileptic seizure or when the brain processes language?

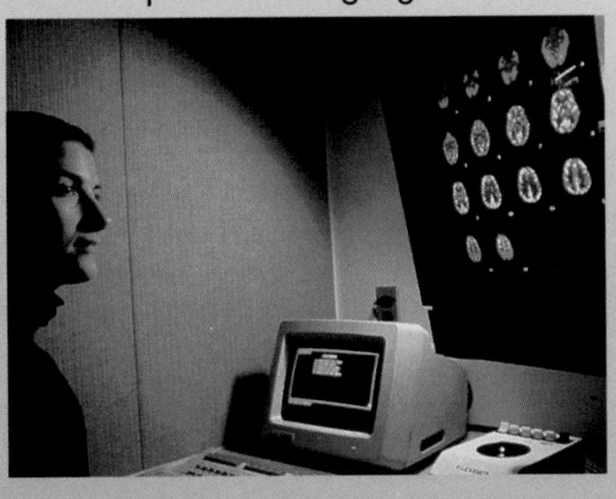

CHECK FOR UNDERSTANDING

▶ Ask the question: **What might be the purpose of the extensive folding of the surface of the cerebrum?** *The folding provides a greater surface area to be contained within the enclosure of the skull.*

▶ Ask the question: **Why might a hard blow to the back of the head cause a vision disorder?** *The primary visual area of the brain is located at the lower rear side of the cerebrum.*

RETEACH

Have students make a papier-mâché or modeling clay model of the brain and its major parts. Use the model to review the parts and their functions.

EXTENSION

For students who have mastered this section, use the **Reinforcement** and **Enrichment** masters or other OPTIONS provided.

TECHNOLOGY

Extension: Discuss how knowledge of the functions of different brain locations can help scientists cure disorders throughout the body.

References: "The Restless Brain" by Geoffrey Montgomery, *Discover,* March 1989, pp. 57-68. "Epileptic PET Probes" by Bruce Bower, *Science News,* April 30, 1990, pp. 280-281.

Think Critically: The technology enables scientists to see what different parts of the brain look like as the activity takes place.

PROGRAM RESOURCES

From the **Teacher Resource Package** use:

Transparency Masters, pages 83-84, The Reflex.

Use **Color Transparency** 42, The Reflex.

VideoDisc

STVS: PETT Scanner, Disc, 7, Side 1

STVS: Panic Disorder and Brain Activity, Disc 7 Side 1

STVS: Nerve Regeneration in Garfish, Disc 5, Side 2

CONCEPT DEVELOPMENT

Cooperative Learning: Use the Expert Teams strategy to study the various functions of the parts of the central and peripheral nervous systems.

TEACHER F.Y.I.

►An unusual reflex is the consenual reflex of the iris. Light shone into one eye causes the iris of the other eye to constrict as well.

3 CLOSE

►Ask questions 1-4 and the **Apply** Question in the Section Review.

►Have students, under teacher supervision, test the knee-jerk reflex by tapping the patellar ligament just below the knee cap.

►Have students discuss the information received by the complex and specialized sensory organs that will be studied in Section 22-2.

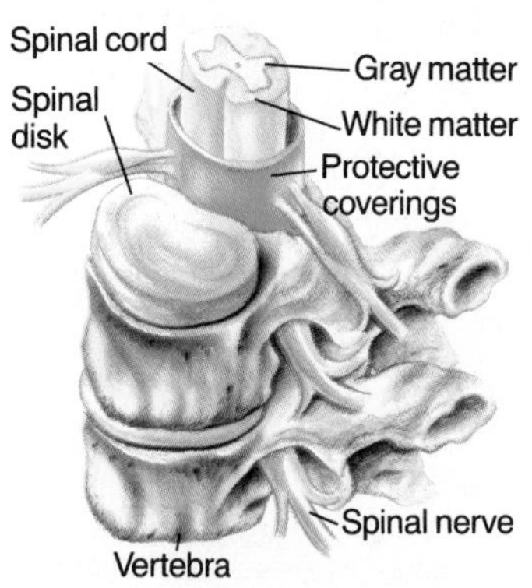

Figure 22-5. Impulses travel to and from the brain by way of the spinal cord.

Your spinal cord is an extension of the brainstem. It is made up of bundles of neurons that carry impulses from all parts of the body to the brain and from the brain to all parts of your body. The spinal cord is about as big around as an adult thumb and 43 cm long. The CNS is protected by a bony cap called the skull, by vertebrae, and by three layers of membranes. Between some of these membranes is a fluid called cerebrospinal fluid. What purpose might this fluid serve?

The Peripheral Nervous System

Your brain and spinal cord are connected to the rest of your body by the peripheral nervous system. The PNS is made up of 12 pairs of cranial nerves from your brain and 31 pairs of spinal nerves from your spinal cord. These nerves link your central nervous system with all parts of your body. Spinal nerves are made up of bundles of sensory and motor neurons. For this reason, a single spinal nerve may have impulses going to and from the brain at the same time.

There are two divisions of the peripheral nervous system. The *somatic system* consists of the cranial and spinal nerves that go from the central nervous system to your skeletal muscles. The second division, the *autonomic system*, controls your heartbeat rate, breathing, digestion, and gland functions. When your salivary glands release saliva, your autonomic system is at work. Use Figure 22-6 to help you remember these two divisions.

Figure 22-6. The Divisions of the Peripheral Nervous System

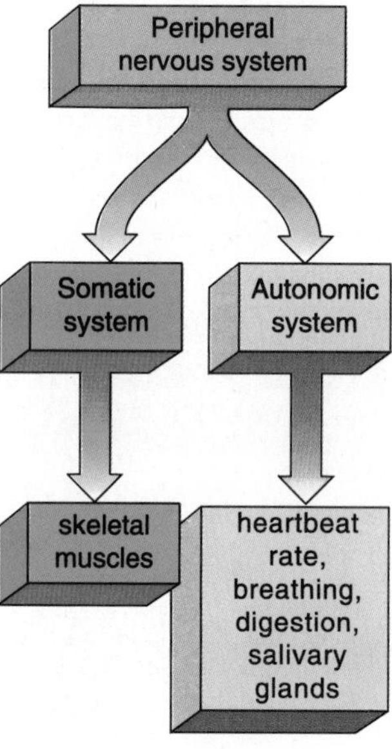

Reflexes

Have you ever jumped back from something hot or sharp? Then you've experienced a reflex. A **reflex** is an involuntary and automatic response to a stimulus. You can't control reflexes, but many times they have saved lives. A reflex involves a simple nerve pathway called a reflex arc. Figure 22-7 shows a reflex arc. As someone hands you a piece of pizza, some very hot cheese falls on your finger. Sensory receptors in your finger respond to the hot cheese, and an impulse is sent to the spinal cord. The impulse passes to an interneuron in the spinal cord that immediately relays the impulse to motor neurons. Motor neurons transmit the impulse to muscles in your arm. Instantly, without thinking, you pull your arm back in response to the burning food. This is a withdrawal

OPTIONS

ENRICHMENT

►Have students find out about the significance of the gray and white matter of the cerebrum.

►Have students find out about right and left brain activities.

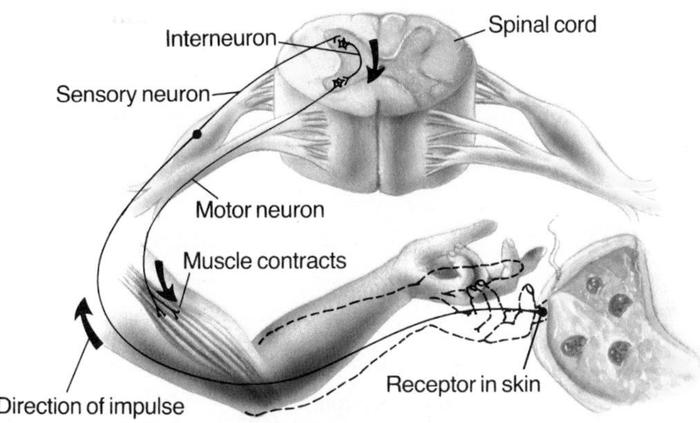

Figure 22-7. Your response in a reflex is controlled in your spinal cord, not in your brain.

reflex. A reflex allows the body to respond without having to think about what action to take. Reflex responses are controlled in your spinal cord, not in your brain. Your brain acts after the reflex to help you figure out what to do to make the pain stop.

Where are reflexes controlled?

Remember the rubber mask scare? What would have happened if your breathing and heartbeat rate didn't calm down within a few minutes? Your body system can't be kept in a state of continual excitement. The organs of your nervous system control and coordinate responses to maintain homeostasis within your body.

SECTION REVIEW

1. Draw and label the parts of a neuron.
2. Compare the central and peripheral nervous systems.
3. Explain what happens to an impulse at a synapse.
4. Compare sensory and motor neurons.
5. **Apply:** Describe the reflex pathway of nerve impulses when you step accidentally on a sharp object.
6. **Connect to Chemistry:** Find out what two chemical elements play roles in moving an impulse along an axon.

Concept Mapping

Skill Builder

Prepare a chain of events concept map of the different kinds of neurons an impulse moves along from the stimulus to a response. If you need help, refer to Concept Mapping in the **Skill Handbook** on pages 688 and 689.

SECTION REVIEW ANSWERS

1. See Figure 22-2.
2. The central nervous system is made up of the brain and spinal cord. The peripheral nervous system is made up of cranial nerves and spinal nerves.
3. An impulse jumps across a synapse.
4. Sensory neurons receive information and send impulses to the spinal cord or brain. Motor neurons conduct impulses from the brain or spinal cord to muscles or glands throughout the body.
5. Apply: Sensory receptors in your foot respond, and an impulse is sent to the spinal cord. An interneuron relays the impulse to motor neurons, which transmit the impulse to your leg. You withdraw from the sharp object.
6. Connect to Chemistry: K-potassium, Na-sodium

Skill Builder

The concept map should include the following: stimulus→sensory neurons→interneurons→motor neurons→response

Skillbuilder
ASSESSMENT
Portfolio: Use this Skillbuilder to assess students' abilities to form a concept map of the chain of neurons involved between a stimulus and a response.

To assess this product, refer to the Performance Task Assessment Lists in **Performance Assessment in Middle School Science.**

ASSESSMENT—ORAL

▶**Why is it good that the reflex arc does not include the brain?** *It would take much longer to get a response.*
▶**What are some common reflexes?** *blinking, pulling away from hot or cold, knee jerk, gagging*

ACTIVITY 22-1
30 minutes

OBJECTIVE: Observe reaction time using three different activities.

PROCESS SKILLS applied in this activity:
▶**Observing, Communicating,** and **Using Numbers** in Procedure Step 4 in Part A, Step 4 in Part B, and Steps 4 and 5 in Part C.
▶**Recognizing and Using Spatial Relationships** in Procedure Step 2 in Part A, Step 2 in Part B, and Step 2 in Part C.
▶**Measuring** in Procedure Step 4 in Part C and Conclude and Apply Question 6.
▶**Interpreting Data** in Conclude and Apply Questions 4-6.

COOPERATIVE LEARNING
Use the Paired Partners strategy for Procedure Parts B and C.

TEACHING THE ACTIVITY
Troubleshooting: Students should hold the meterstick in a vertical position.

PROGRAM RESOURCES
From the **Teacher Resource Package: Activity Worksheets,** pages 194-195, Activity 22-1, Reaction Time.

ACTIVITY 22-1 Reaction Time

Problem: *How can reaction time be measured?*

Materials
- penny
- meterstick

Procedure
Part A
1. Hold your right arm out with the palm down. Put a penny on the center of the back of your hand.
2. Tilt your hand so that the penny slides off. Try to catch the penny with your right hand.
3. Repeat Step 2 nine more times. In a table like the one shown, record how many times you catch the penny and how many times you drop the penny.
4. Repeat Steps 2 and 3 with your left hand. *Record* your results.

Part B
1. Have a partner hold a penny about 0.5 m above the palm of your hand.
2. When your partner drops the penny, move your hand before the penny hits it.
3. Repeat Step 2 for different distances. Record your results in the table.
4. Repeat Steps 2 and 3 with your other hand. *Record* your results.

Part C
1. Have a partner hold a meterstick at one end so that the other end of the ruler is between your thumb and index finger.
2. Keep your eyes on the bottom of the ruler as your partner releases the ruler. Try to catch the ruler.
3. Repeat Step 2 nine more times.
4. Record the distance the ruler has fallen each time you catch it.
5. Repeat Steps 2 and 3 with your other hand. *Record* your results.

Data and Observations Sample Data

Part A

Trial	Right Hand Caught	Right Hand Not caught	Left Hand Caught	Left Hand Not caught
1	7	3	6	4
2	8	2	7	3

Part B

Trial	Distance above hand	Hit	Not hit
1	0.5 m	1	9
2	0.4 m	0	10

Part C

Trial	Distance fallen: Right hand	Distance fallen: Left hand
1	17 cm	19 cm
2	19 cm	21 cm

Analyze
1. What was the stimulus in each activity?
2. What was the response in each activity?
3. What was the variable in each activity?

Conclude and Apply
4. Did you catch the penny, avoid the penny, and catch the ruler faster with your writing hand?
5. How did your response time improve?
6. How can reaction time be measured?
7. *Draw a conclusion* about how practice *relates* to stimulus-response time.

ANSWERS TO QUESTIONS
1. A—the penny sliding off the hand; B—dropping the penny; C—dropping the ruler
2. A—catching the penny; B—avoiding the penny; C—catching the ruler
3. A—left hand if you are right-handed; B—distance; C—other hand
4. Answers will vary. The student will probably catch faster with the writing hand.
5. Reaction time will probably show improvement with practice.
6. Reaction time can be measured by recording the length of time it takes to respond to a stimulus each time.
7. With practice, stimulus-response time will probably get better.

Activity ASSESSMENT
Performance: To further assess students' abilities to measure and improve Reaction Time, have them analyze how long it would take the non-writing hand to be trained to respond as the writing hand responds.

The Senses

22-2

Objectives

- ▶ List the sensory receptors in each sense organ.
- ▶ Explain what type of stimulus each sense organ responds to and how.
- ▶ Explain the need for healthy senses.

New Science Words

retina
cochlea
olfactory cells
taste buds

In Touch with Your Environment

Many stories by science fiction writers talk about energy force fields around spaceships. When some form of energy tries to enter the ship's force field, the ship is put on alert. Your body has an alerting system as well, in the form of sense organs. Your senses enable you to see, hear, smell, taste, touch, and feel whatever comes into your personal territory. The energy that stimulates your sense organs may be in the form of light rays, heat, sound waves, chemicals, or pressure. Sense organs are adapted for capturing and transmitting these different forms of energy.

Lenses in eyes are convex, but their shape can change over time. Find out what happens to the lens in a condition called *presbyopia.* What is the result?

Vision

Think about the different kinds of things you look at every day. It's amazing that, all in one view, you can see words on a page, photographs in color, and a cat as it walks across the floor.

Light travels in a straight line unless something bends it. Your eyes are equipped with structures that bend light.
1 Light rays are first bent by the cornea and then a lens. The lens directs the rays onto the retina (RET nuh). The **retina** is a tissue at the back of the eye that is sensitive to light energy. Rods and cones are cells in the retina. Cones respond to bright light and color. Rods respond to dim light. Light energy stimulates impulses in these cells. The impulses pass to the optic nerve, which carries them to the brain, where the impulses are interpreted and you "see" what you are looking at.

Figure 22-8. The Parts of the Human Eye

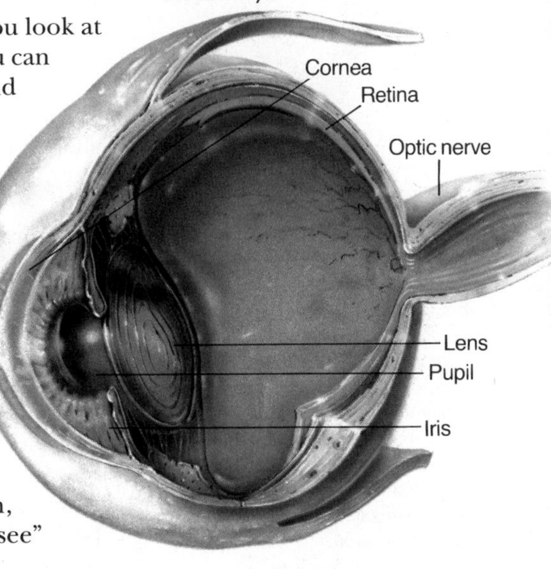

OPTIONS

ENRICHMENT

▶ Have students research the method used to correct vision defects so that an image is focused directly on the retina.

▶ Have students find out about otoliths—"stones" in the inner ear that are important in providing information about the position of the head in relation to gravity.

Science Integration Activities is found in **Merrill Life Science Teacher Resource Manual.**

PROGRAM RESOURCES

From the **Teacher Resource Package** use:

Cross-Curricular Connections, page 26, Reading About Personal Experiences.

Technology, pages 15-16, Laser Eye Surgery.

Activity Worksheets, page 201, Mini-Lab: How is balance maintained?

Science Integration Activities, 22, Conduction of Heat

Use **Laboratory Manual,** page 155, Parts of the Eye.

SECTION 22-2

PREPARATION

SECTION BACKGROUND

▶ Sense organs receive stimuli that are conveyed by nerve impulses to the brain, where they are interpreted.

▶ The senses of sight, hearing, balance, smell, and taste have highly specialized structures located in specific areas of the body.

▶ Receptors of sensations of touch, pressure, pain, and temperature have less specialized structures and are more generally located in the body.

PREPLANNING

▶ To prepare for Activity 22-2, obtain meter rulers, toothpicks, glue, and index cards.

1 MOTIVATE

▶ Discuss how information received from sense organs not only results in muscular movement, but may be stored in the brain for later recall. Ask students how they recognize someone or something they have seen before.

▶ **Demonstration:** Obtain a double convex lens to illustrate the refraction and focusing of light rays similar to that which occurs in the eye.

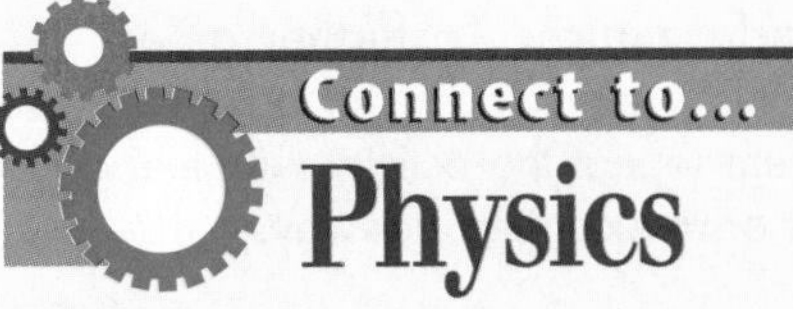

Answer: In middle age, the lens loses its elasticity and becomes slightly elongated. This usually results in farsightedness.

VideoDisc

STVS: Vision Diagnosis, Disc 7, Side 1
STVS: Nearsightedness Surgery, Disc 7, Side 2
STVS: Ear Implants, Disc 7, Side 2
STVS: Sounds Made by the Ear, Disc 7, Side 2

TYING TO PREVIOUS KNOWLEDGE: Have students compare the sensory system of information processing with that of a telephone system. Have them brainstorm similarities and differences.

2 TEACH

Key Concepts are highlighted.

CONCEPT DEVELOPMENT

▶Obtain models or charts of the eye and ear to use in teaching this section.

▶Explain why humans can't hear all sounds. *The nerve cells in the auditory pathway respond only to frequencies between 100 and 24 000 Hz (cycles per second). Anything above or below that range is not heard.*

Science and READING

Students may enjoy seeing a movie about Helen Keller, titled *The Miracle Worker*. Discuss it after viewing.

MINI-Lab

Materials: chalkboard, chalk

Answer: Students will sway slightly when trying to stand still. With the eyes closed, there should be even more movement. With eyes open, a person can focus on a point to help the body remain still.

MINI-Lab ASSESSMENT

Performance: To further assess students' understanding of balance, have them repeat the activity with feet apart or arms extended sideways.

Science and READING

Find out how Helen Keller was able to learn to communicate even though she was deprived of sight and hearing at an early age.

Hearing

Sound energy is to hearing as light energy is to vision. When an object vibrates, it causes the air around it to also vibrate, thus producing energy in the form of sound waves. When sound waves reach your ears, they stimulate nerve cells deep in your ear. Impulses are sent to the brain. The brain responds, and you hear a sound.

Figure 22-9 shows that your ear is divided into three sections: the outer, middle, and inner ear. Your outer ear traps sound waves and funnels them down the ear canal to the middle ear. Once there, the sound waves cause the eardrum to vibrate much like the membrane on a drum. These vibrations then move through three little bones
in your middle ear called the hammer, anvil, and stirrup. 2
The stirrup bone rests against a second membrane on an opening to the inner ear.

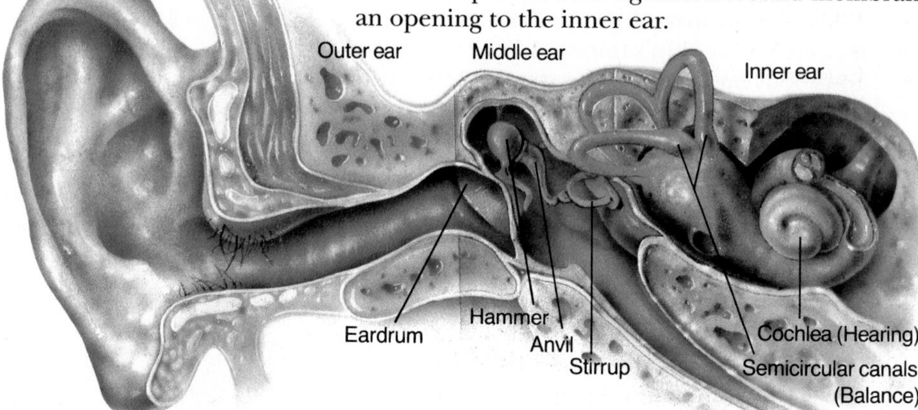

Figure 22-9. Your ear responds to sound waves and to changes in the position of your head.

The **cochlea** (KOH klee uh) is a fluid-filled structure shaped like a snail's shell in the inner ear. When the stirrup vibrates, fluids in the cochlea also begin to vibrate. These vibrations stimulate nerve endings in the cochlea, and impulses are sent to the brain by the auditory nerve. Depending on how the nerve endings are stimulated, you hear a different type of sound. High-pitched sounds make the endings move differently from lower, deeper sounds. Sound produced by jet engines or rock instruments is greater than your ear is adapted to handle and may produce pain and hearing loss.

Balance is also controlled in the inner ear. Special structures and fluids in the inner ear constantly adjust to the position of your head. This stimulates impulses to the brain, which interprets the impulses and helps you make the necessary adjustments to maintain your balance.

MINI-Lab

How is balance maintained?

Draw two parallel vertical lines on the chalkboard and have a student stand between them, as still and straight as possible for three minutes. *Observe* how well balance is maintained. Next, have the student close his or her eyes and repeat standing within the lines for three more minutes. When was balance more difficult to maintain? Suggest an explanation.

OPTIONS

Meeting Different Ability Levels

For Section 22-2, use the following **Teacher Resource Masters** depending upon individual students' needs.

◆ **Study Guide Master** for all students.

● **Reinforcement Master** for students of average and above average ability levels.

▲ **Enrichment Master** for above average students.

Additional Teacher Resource Package masters are listed in the OPTIONS box throughout the section. The additional masters are appropriate for all students.

◆ STUDY GUIDE 85

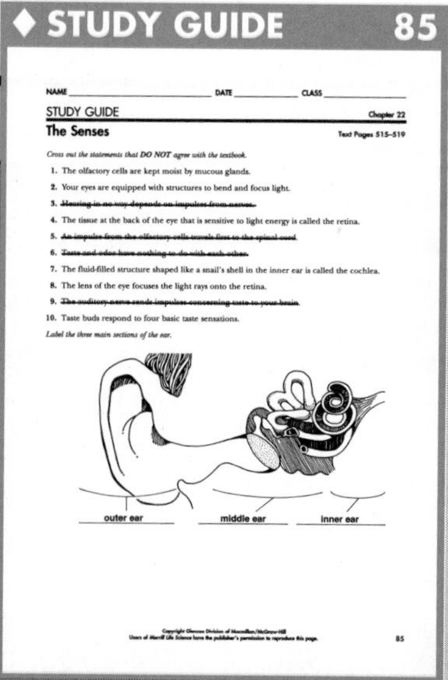

NAME ________ DATE ________ CLASS ________

STUDY GUIDE — Chapter 22

The Senses — Text Pages 515–519

Cross out the statements that DO NOT agree with the textbook.

1. The olfactory cells are kept moist by mucous glands.
2. Your eyes are equipped with structures to bend and focus light.
3. ~~Hearing in no way depends on impulses from nerves.~~
4. The tissue at the back of the eye that is sensitive to light energy is called the retina.
5. ~~An impulse from the olfactory cells travels first to the spinal cord.~~
6. ~~Taste and odor have nothing to do with each other.~~
7. The fluid-filled structure shaped like a snail's shell in the inner ear is called the cochlea.
8. The lens of the eye focuses the light rays onto the retina.
9. ~~The auditory nerve sends impulses concerning taste to your brain.~~
10. Taste buds respond to four basic taste sensations.

Label the three main sections of the ear.

outer ear — middle ear — inner ear

85

Smell

A bloodhound is able to track a particular scent through fields and forest. Our ability to detect odors is not as sharp, but our sense of smell is important for survival and enjoyment.

You can smell food because it gives off molecules into the air. Nasal passages contain sensitive nerve cells called
3 **olfactory cells** that are stimulated by gas molecules. The cells are kept moist by mucous glands. When gas molecules in the air dissolve in this moisture, the cells become stimulated. If enough gas molecules are present, an impulse starts in these cells and travels to the brain.

The brain interprets the stimulus. If it is recognized from previous experience, you can identify the odor. If you can't recognize a particular odor, it is remembered and can be identified the next time, especially if it's a bad one. Certain odors even let you recall events from your childhood.

Taste

Have you ever tasted a new food or medicine with the tip of your tongue and found that it tasted sweet? Then you may have been unpleasantly surprised to find that when you swallowed it, it tasted bitter. **Taste buds** on your
4 tongue are the major sensory receptors for taste. About ten thousand taste buds are found all over your tongue, enabling you to tell one taste from another.

Taste buds respond to chemical stimuli. When you think of food, your mouth begins to water with saliva. This adaptation is helpful because in order to taste something, it has to be dissolved in water. Saliva in your mouth dissolves food. The solution washes over the taste buds, and an impulse is started that is sent to your brain. The brain interprets the impulse and enables you to tell what you are tasting.

Taste buds respond to more than one taste sensation. However, there are areas of the tongue that seem to respond to one taste more easily than another. There are four basic taste sensations: sweet, salty, sour, and bitter. Figure 22-10 shows where these tastes are commonly stimulated on your tongue.

Figure 22-10. Taste buds contain nerve endings that react to chemicals in food.

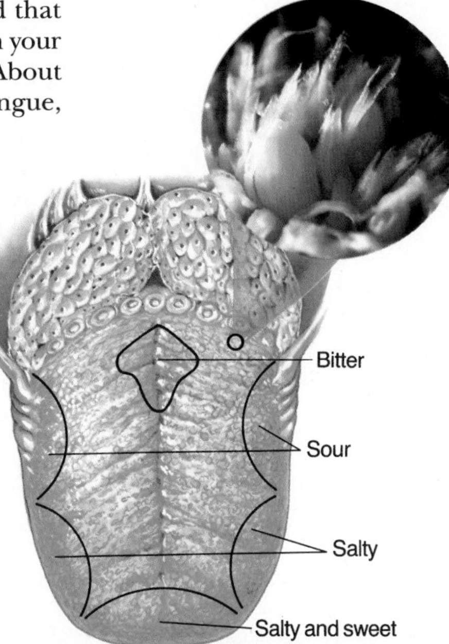

● REINFORCEMENT 85

NAME ______ DATE ______ CLASS ______

REINFORCEMENT — Chapter 22

The Senses — Text Pages 515–519

Use the following terms to label the figure below.

rods and cones retina cornea visual cortex optic nerve lens

cornea
visual cortex
optic nerve
rods and cones
retina
lens

Answer the following questions.

1. Using the terms with which you labeled the figure, describe how light entering your eye becomes an image you see. **Light rays first pass through the cornea and onto the lens. The lens focuses the light onto the retina. Rods and cones on the retina respond to the light, sending impulses to the optic nerve. The optic nerve carries the impulses to the brain. The brain interprets the impulses, allowing you to see an image.**
2. What are the three main sections of the ear? **the outer ear, the middle ear, and the inner ear**
3. What is the cochlea and what does it do? **The cochlea is a fluid-filled structure shaped like a snail's shell in the inner ear. It picks up vibrations in the fluid and sends impulses to the auditory nerve.**
4. In what cells does a smell impulse begin? **olfactory cells**
5. What four basic taste sensations do the taste buds have? **sweet, salty, sour, and bitter**
6. How do your senses help you maintain homeostasis? **The senses pick up any changes in the environment and transmit this information to the brain and spinal cord. The body responds to that information to protect itself or maintain homeostasis.**

 85

▲ ENRICHMENT 85

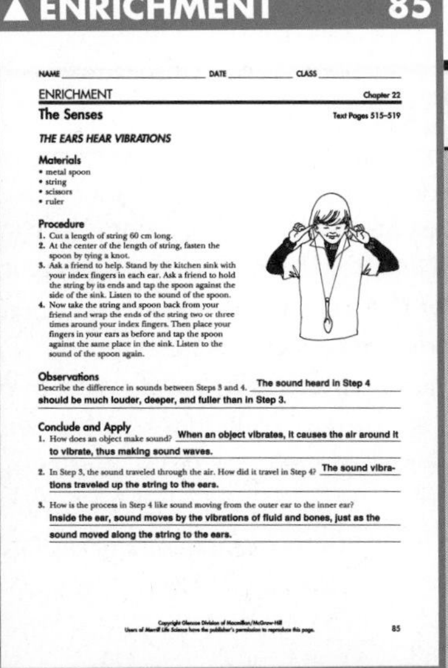

NAME ______ DATE ______ CLASS ______

ENRICHMENT — Chapter 22

The Senses — Text Pages 515–519

THE EARS HEAR VIBRATIONS

Materials
- metal spoon
- string
- scissors
- ruler

Procedure
1. Cut a length of string 60 cm long.
2. At the center of the length of string, fasten the spoon by tying a knot.
3. Ask a friend to help. Stand by the kitchen sink with your index fingers in each ear. Ask a friend to hold the string by its ends and tap the spoon against the side of the sink. Listen to the sound of the spoon.
4. Now take the string and spoon back from your friend and wrap the ends of the string two or three times around your index fingers. Then place your fingers in your ears as before and tap the spoon against the same place in the sink. Listen to the sound of the spoon again.

Observations

Describe the difference in sounds between Steps 3 and 4. **The sound heard in Step 4 should be much louder, deeper, and fuller than in Step 3.**

Conclude and Apply
1. How does an object make sound? **When an object vibrates, it causes the air around it to vibrate, thus making sound waves.**
2. In Step 3, the sound traveled through the air. How did it travel in Step 4? **The sound vibrations traveled up the string to the ears.**
3. How is the process in Step 4 like sound moving from the outer ear to the inner ear? **Inside the ear, sound moves by the vibrations of fluid and bones, just as the sound moved along the string to the ears.**

 85

CONCEPT DEVELOPMENT

▶Discuss why a cold or the flu can cause difficulty in detecting odors or tastes. The mucous layer becomes so congested that the sensory cells are unable to function properly.

CROSS CURRICULUM

▶**Physical Science:** Use a tuning fork to demonstrate how the vibrations produce sound waves that move through the air and reach our ears.

REVEALING MISCONCEPTIONS

▶**CAUTION:** During times of solar eclipses, students and adults often want to view the sun through sunglasses, smoked glass, or black film negatives. *Explain that none of these procedures is safe!* The only safe way is to indirectly view an image of the sun projected onto a piece of paper. Direct viewing can cause severe eye damage and result in blindness.

MINI QUIZ

Use the Mini Quiz to check students' recall of chapter content.

1 **What is the function of the lens?** *to focus light rays onto the retina*

2 **What are the three small bones in the middle ear?** *hammer, anvil, stirrup*

3 **What are the sensitive nerve cells in your nasal passages called?** *olfactory cells*

4 **The major sensory receptors for taste are found on your _____.** *tongue*

TEACHER F.Y.I.

▶The eustachian tube allows for equalization of air pressure on both sides of the eardrum.

CHECK FOR UNDERSTANDING

Ask questions 1-4 and the **Apply** Question in the Section Review on page 518.

RETEACH

Cooperative Learning: Using the Study Buddies strategy, have students make concept maps of the sensory organs.

EXTENSION

For students who have mastered this section, use the **Reinforcement** and **Enrichment** masters or other OPTIONS.

3 CLOSE

▶ Discuss with students various types of visual and auditory pollution.

SECTION REVIEW ANSWERS

1. It is a tissue at the back of the eye sensitive to light energy, which stimulates impulses in the rods and cones.

2. outer ear—traps sound waves and funnels them to the middle ear; eardrum, hammer, anvil, and stirrup—vibrate; nerve endings in cochlea—stimulated

3. Food is often smelled when molecules come in contact with the membranes in your nasal passages.

4. The body responds to protect itself or maintain homeostasis.

5. Apply: sensory receptors in your internal organs as well as your skin

6. Connect to Physics: The crystals, acted on by gravity, help maintain balance.

Skill Builder

Make sure the table lists sense organs for vision, hearing, smell, taste, touch, pressure, pain, and temperature. The energy listed should include light, heat, sound waves, chemicals, and pressure.

Skillbuilder

ASSESSMENT

Performance: Assess students' abilities to Make and Use Tables by having them write a statement comparing the different types of energy that stimulate the senses.

Did You Know?

Referred pain is pain felt in a region of the body that is not the source of the pain. For example, pain from an inflamed gallbladder may be felt on the right side of the neck, under the right shoulder blade, and at the diaphragm.

Smell and taste are related. When you have a head cold with a stuffy nose, food seems tasteless because it is blocked from contacting the moist membranes in your nasal passages. On the other hand, hot food is often smelled when odor molecules from food in your mouth come in contact with the membranes.

Touch, Pressure, Pain, and Temperature

How important is it to be able to feel pain inside your body? Sensory receptors in your internal organs, as well as throughout your skin, respond to touch, pressure, pain, and temperature. These receptors pick up changes in touch, pressure, and temperature and transmit impulses to the brain or spinal cord. The body responds in such a way to protect itself or maintain homeostasis.

Your fingertips have many different types of receptors for touch. As a result, you can tell whether an object is rough or smooth, hot or cold, light or heavy. Your lips are sensitive to heat and prevent you from drinking something so hot that it would burn you. Pressure sensitive cells in the dermis give warning of danger to a body part and enable you to move to avoid injury.

Your senses are adaptations that help you enjoy or avoid things around you. Through your senses you constantly make small and large adjustments to your environment.

SECTION REVIEW

1. What is the retina and how is it stimulated?
2. How do sound waves affect different parts of the ear?
3. How are smell and taste related?
4. Why is it important to have receptors for pain and pressure in your internal organs as well as in your skin?
5. **Apply:** What sensory receptors are involved when you accidentally close your hand in a door?
6. **Connect to Physics:** Find out the function of tiny calcium carbonate stones called otoliths in your inner ears.

Skill Builder — Making and Using Tables

Organize the information on senses in a table that names the sense organs and gives the type of energy to which they respond. If you need help, refer to Making and Using Tables in the **Skill Handbook** on page 690.

OPTIONS

ASSESSMENT—ORAL

▶ **How does the brain "know" where a stimulus is coming from?** *The brain interprets the pathway of the stimulus and then recognizes the location of the source of the stimulus.*

▶ **From your own experience, do you think there are ranges of sensitivity to hot and cold?** *Thermoreceptors are more sensitive to and respond to warm/hot temperatures above 25°C and to cold/cool temperatures below 10°C and 20°C.*

ACTIVITY 22-2 DESIGNING AN EXPERIMENT

Investigating Skin Sensitivity

Your body responds to touch, pressure, and temperature. These sensitivities can be for protection or to maintain homeostasis. What areas of your skin are more sensitive than others?

Getting Started

Devise a method of determining what areas of the skin are more sensitive than others.

Hypothesizing

Make a **hypothesis** as to what areas of the skin are more sensitive. Explain your reasoning.

Materials 

Your cooperative group will use:

- 3″ x 5″ index cards
- toothpicks
- glue
- metric ruler

Try It!

1. *Predict* what skin areas are more sensitive to touch. Rank the areas from 5 (the most sensitive) to 1 (the least sensitive) in a data table.
2. Glue toothpicks onto the card as shown in the figure to the right so they are 1 mm, 3 mm, 5 mm, and 10 mm apart. Label the card.
3. With your partner's eyes closed, use the part of the card with toothpicks 1 mm apart and carefully touch the skin surface. Touch the fingertip, palm of the hand, back of the hand, forearm, and back of the neck. **CAUTION:** *Do not apply heavy pressure.*
4. If your partner feels two points, record a plus (+) in the table. If your partner cannot feel both points, record a minus (-).
5. Again, with your partner's eyes closed, use the toothpicks 3 mm apart to touch each areas listed in the table. Record whether your partner can feel one or two points.
6. Repeat Steps 5 and 6 with 5 mm and 10 mm sections of the card.

Summing Up/Sharing Results

- Which part of the body tested can distinguish between the closest stimuli?
- Which body part is least sensitive?
- Rank the body parts tested from most sensitive to least sensitive. How did your test results compare with your predictions?
- What is located in the dermis that provides a sense of touch?
- What areas tested were sensitive to touch?
- What do your answers to the first two questions indicate about the distribution of touch receptors in the skin?

Going Further!

Plan an investigation to determine if there are areas of the skin that are more sensitive to cold. Try it and see if your inference was correct.

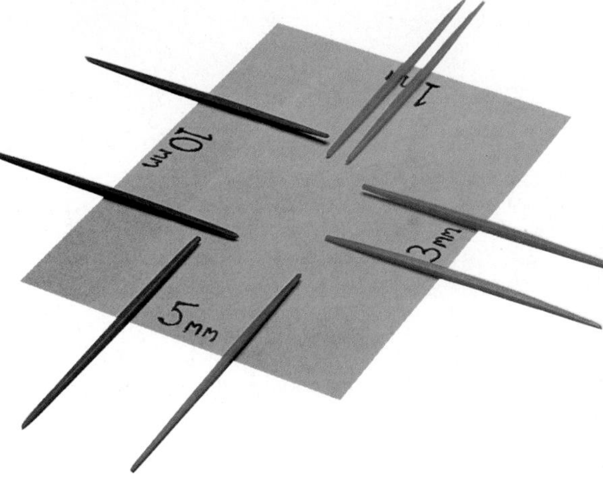

ACTIVITY 22-2

OBJECTIVE: Design and carry out an experiment to determine the sensitivity of skin on various parts of the body by testing for the location of receptors in the skin.

Time: One class period to plan and hypothesize; one period to collect and analyze data.

PROCESS SKILLS applied in this activity are **observing, forming a hypothesis, experimenting,** and **analyzing data.**

PREPARATION

Obtain materials for cooperative groups. Prepare the test cards the day before the activity to allow glue to dry.

Cooperative Learning: Form Paired Partners.

HYPOTHESIZING

Student hypotheses will vary. Possible hypotheses are: "The whole body is equally sensitive," "Nerve endings are various distances apart," and "The most sensitive parts of the body are the fingertips."

TEACHING THE ACTIVITY

Refer to the ***Activity Worksheets*** *for additional information and teaching strategies.*

- Discuss the fact that sensory receptors are actual structures in the dermis.

SUMMING UP/SHARING RESULTS

- Most students should indicate the fingertips.
- Most students should indicate the back of the hand or forearm.
- Answers will vary. Students should answer honestly if predictions do not match outcomes.
- sensory or touch receptors
- Fingertips and palms are more sensitive to touch.
- Touch receptors are closer together in the skin of the fingertips and further apart on the back of the hand and forearm. Receptors in the palm and the back of the neck vary.

GOING FURTHER!

Answers will vary, but may include inferences about the cheek, neck, and wrist.

PROGRAM RESOURCES

From the **Teacher Resource Package** use:

Activity Worksheets, pages 196-197, Activity 22-2, Investigating Skin Sensitivity

Activity ASSESSMENT

Performance: To further assess students' understanding of skin sensitivity, see USING LAB SKILLS, Question 12, on page 526.

SCIENCE & SOCIETY SECTION 22-3

PREPARATION

SECTION BACKGROUND

▶Not all cases of confusion and in older people are the result of Alzheimer's disease. An estimated 10 to 20 percent of older persons with these symptoms are treatable.

1 MOTIVATE

▶Have students collect newspaper and magazine articles about new research dealing with the disease.

2 TEACH

Key Concepts are highlighted.

CONCEPT DEVELOPMENT

▶Ask the question: **What might be one of the first indications that a person has Alzheimer's disease?** *forgetting where items were placed or not recognizing family members*

TYING TO PREVIOUS KNOWLEDGE: In Alzheimer's disease, the transmission of impulses from one neuron to the next is prevented.

22-3 Alzheimer's Disease

New Science Words

Alzheimer's disease

Objectives

- ▶ Describe Alzheimer's disease.
- ▶ State some of the results of Alzheimer's disease on society.
- ▶ Explain two possible causes of Alzheimer's.

What Is Alzheimer's Disease?

Alzheimer's disease is the failure of nerve cells in the brain to communicate with each other. Researchers are attempting to determine what causes this breakdown. They have found that Alzheimer's patients have unusually low amounts of the enzyme that signals the cell to
1 produce a nerve-transmitting chemical called acetylcholine. Without this chemical, nerve impulses aren't carried from one neuron to the next. It has also been found that the brains of Alzheimer's patients use far less oxygen than normal.
Autopsies of Alzheimer's patients show that 2
many neurons have died. The destruction of brain cells caused by Alzheimer's disease results in such severe memory loss that, over a period of years, the victim becomes a very different person. At first, patients forget simple things. Items are misplaced around the house. Later, they may not recognize family members or even know who they are. Eventually, after losing physical functioning, they die. What causes this condition?

Researchers are still working to determine the cause of Alzheimer's. One idea is that the disease may be inherited. Some researchers believe that a defect exists on chromosome 21. Children of Alzheimer's victims show a greater chance of developing the disease than others. Other evidence, however, shows that a protein called amyloid B, which helps cell growth in young brain cells, has the opposite effect in older people. There is also evidence that young people who have experimented

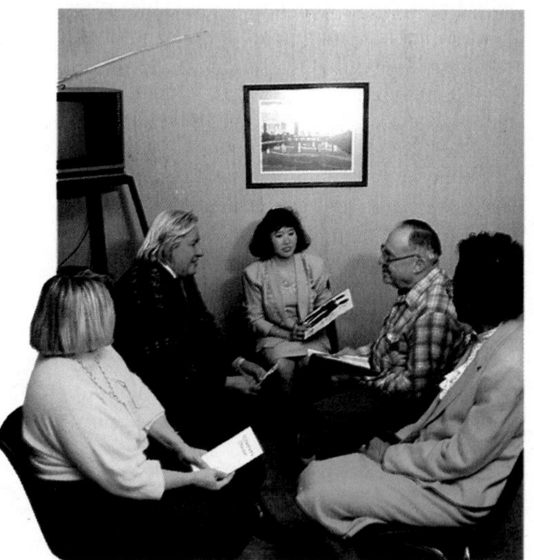

Figure 22-11. Support groups help people who have the responsibility for caring for people with Alzheimer's disease.

OPTIONS

Meeting Different Ability Levels

For Section 22-3, use the following **Teacher Resource Masters** depending upon individual students' needs.

◆ **Study Guide Master** for all students.

● **Reinforcement Master** for students of average and above average ability levels.

▲ **Enrichment Master** for above average students.

Additional Teacher Resource Package masters are listed in the OPTIONS box throughout the section. The additional masters are appropriate for all students.

◆ STUDY GUIDE 86

NAME ________ DATE ________ CLASS ________

STUDY GUIDE — Chapter 22

Alzheimer's Disease — Text Pages 520–521

Cross out the statements that DO NOT agree with the textbook.

1. Children of Alzheimer's victims have a 50 percent chance of getting the disease later in life.
2. Without acetylcholine, nerve impulses are not carried from one neuron to the next.
3. The cost of keeping someone in a nursing home averages about $25 000 a year.
4. ~~Alzheimer's disease is harmful but no one dies from it.~~
5. ~~The cure for Alzheimer's disease has just been found.~~
6. Alzheimer's disease is a failure of nerve cells in the brain to communicate with one another.
7. Some researchers think that a defect in chromosome 19 can cause Alzheimer's disease.
8. ~~Alzheimer's victims lose muscle control but retain good memory.~~
9. ~~Alzheimer's disease definitely has no genetic cause.~~
10. Alzheimer's disease causes severe memory loss.
11. Caregivers of Alzheimer's victims often suffer from the stress of the situation.
12. ~~The drugs that produce more acetylcholine have no side effects.~~
13. ~~Only 1000 people die from Alzheimer's every year.~~
14. Alzheimer's destroys brain cells.
15. Some scientists think that Alzheimer's may be caused by a virus.
16. ~~In patients with Alzheimer's, the chemical that crosses the synapse from one nerve cell to another is present.~~
17. Over a period of years, victims of Alzheimer's become very different people.
18. The protein, amyloid B, may work differently in older people.

with so-called designer drugs have developed a form of Alzheimer's disease.

The Cost of Disease

It is believed that about 100 000 people die with Alzheimer's disease in the United States every year. The cost of this disease to families and society is very high. Home health care for a patient ranges from $18 000 to $20 000 a year. In a skilled care nursing home, the cost averages about $25 000 per year. In 1985, it was estimated that Alzheimer's disease and similar disorders involving loss of brain function cost the United States about $88 billion. None of these people can be cured.

There are other costs as well. The stress involved in caring for someone with Alzheimer's disease can be damaging in itself. Studies show that the extreme stress harms caregivers' immune systems and can make them more susceptible to illness.

Is a cure for Alzheimer's in sight? Because intensive research has been conducted for only about a decade, little progress has been made. Drugs that help produce more acetylcholine have been developed and tried. However, these were found to have severe side effects. Until scientists are able to find the cause of Alzheimer's disease, it will be difficult to find a cure.

In Your JOURNAL

Find out what Medicare is. **In your Journal,** write a one-page description of this health plan. Include in your report how a person qualifies for Medicare benefits.

What is Alzheimer's disease?

SECTION REVIEW

1. What happens to the brain of an Alzheimer's patient that causes memory loss?
2. What is the cost of this disease for society? How might society benefit from what is learned about Alzheimer's disease?
3. What are two possible causes of Alzheimer's?

You Decide!

It can be very difficult to care for a family member who has Alzheimer's disease. Due to loss of memory, the patient needs constant care. Would you be able to help take care of such a family member in your home? What might be the advantages and disadvantages of letting the person stay in a nursing care facility instead?

● REINFORCEMENT 86

NAME ______ DATE ______ CLASS ______

REINFORCEMENT — Chapter 22

Alzheimer's Disease — Text Pages 520–521

List four possible causes of Alzheimer's disease.

1. **a virus**
2. **the protein amyloid B**
3. **genetic factors**
4. **use of designer drugs**

Answer the following questions.

5. How does Alzheimer's disease affect nerve impulses in the brain? **The disease causes less production of the chemical acetylcholine, which is necessary for nerve impulses to be carried from one neuron to the next. Some neurons also die.**
6. What symptom does the destruction of brain cells cause in a person with Alzheimer's? **severe memory loss**
7. Are the children of an Alzheimer's victim at risk of getting the disease? **Yes; they have a 50 percent chance of getting Alzheimer's.**
8. What chromosome is thought to contribute to Alzheimer's disease? **chromosome 19**
9. How many people die of Alzheimer's every year in the United States? **about 100 000**
10. How much progress has been made in curing Alzheimer's? **Very little progress has been made because the cause has not yet been found.**

86

▲ ENRICHMENT 86

NAME ______ DATE ______ CLASS ______

ENRICHMENT — Chapter 22

Alzheimer's Disease — Text Pages 520–521

NO GUILT IN ALZHEIMER'S

Imagine that you are an adult and have moved to another state. One day you go to the store and run into an old school friend, Lisa. After greeting each other and talking about old times, she mentions that her grandmother has become quite ill.

"You remember my grandmother," Lisa says. "Well, she's come down with something called Alzheimer's disease," Lisa adds, "and she seems as crazy as you can get—always forgetting things and sometimes saying the most ridiculous stuff you'd ever hear."

"I know something about Alzheimer's" you respond, "and from what I know—"

"Yeah, well I know something about it, too," she says, interrupting you, "and from what I know, my grandfather is the guilty one here. He never made enough money, so my grandmother was always worried. And they fought a lot, too. I think he drove her nuts. Actually, I think I'd go out of my mind if I had to live with him these days. He used to be kind of nice, but now he's always depressed. You'd think he'd try to stay upbeat about things so my grandmother could get better. But he's always down. I'll bet he feels guilty about driving her crazy."

As you can see, Lisa is misinformed about Alzheimer's disease. In the space below, continue this conversation. What can you tell Lisa to let her know that her grandfather is not at fault for her grandmother's illness and that it is normal for him to feel down now? Use the information from your textbook about the causes and symptoms of Alzheimer's. Be creative in writing the rest of this dialogue, as if you were writing a screenplay to be made into a television movie. Make the words that you and Lisa say seem real. Don't just write a speech. Remember that conversations generally move back and forth between people.

86

CHECK FOR UNDERSTANDING

Use the Mini Quiz to check for understanding.

MINI QUIZ

Use the Mini Quiz to check students' recall of chapter content.

1. **What nerve-transmitting chemical is involved in Alzheimer's disease?** *acetylcholine*
2. **What happens to the neurons in the brain of an Alzheimer's patient?** *They die.*

RETEACH

Have a nursing home or care facility health specialist come to the class to discuss the care required for Alzheimer's patients.

EXTENSION

For students who have mastered this section, use the **Reinforcement** and **Enrichment** masters or other OPTIONS.

In Your JOURNAL

Medicare is a government sponsored and funded medical care program. Information may be obtained from a local library.

3 CLOSE

▶ Ask questions 1-3 in the Section Review.

▶ Review recent information on latest research on causes and cures for Alzheimer's.

SECTION REVIEW ANSWERS

1. Nerve impulses aren't carried from one neuron to the next.

2. thousands of dollars in home and nursing home care and the cost of stress for the care giver; scientists may learn more about brain functions

3. a defect on chromosome 19; the effect of a protein called amyloid B in older people

YOU DECIDE!

Answers will vary. Advantages include lower stress for family members. Worry by family members is one disadvantage. Expense is a large consideration.

SECTION 22-4

PREPARATION

SECTION BACKGROUND

▶Glands that secrete substances such as oils, sweat, or mucus through ducts are *exocrine* glands.

▶The hormones of the endocrine system regulate many of the body's metabolic functions, usually causing the rate of these functions to speed up or slow down.

PREPLANNING

▶Prepare for Cooperative Learning by having material ready that can be used for flash cards.

1 MOTIVATE

▶Discuss with students that certain hormones are released into the bloodstream in response to a nerve impulse.

Connect to... Chemistry

Answer: In the kidney and bladder, the pituitary gland controls when fluids are retained and released; it also controls growth in young animals and metabolism in adults.

22-4 Your Endocrine System

New Science Words

hormones
target tissues

Objectives

- ▶ Explain the function of hormones.
- ▶ Name three endocrine glands and explain the effects of their hormones.
- ▶ Explain how a feedback system works.

Connect to... Chemistry

The pituitary gland has been called the "master gland" because it controls so many other endocrine glands in the body. Find out how the pituitary affects the kidneys and bladder, and growth.

Endocrine System

"The tallest man in the world!" and "the shortest woman in the land!" were commonly seen entertainers in circuses of the past. These people were ordinary persons except for their extraordinary height or lack of height. In most cases, their sizes were the result of a malfunction in their endocrine (EN duh krun) systems.

The endocrine system consists of ductless glands throughout the body. Your salivary glands and your liver
1 are glands that produce substances that flow through ducts. Endocrine glands, however, have no ducts. Their secretions move directly from the cells of the gland into your bloodstream. Endocrine secretions, called **hormones,** usually control activities in parts of the body other than right around the gland. Hormones affect specific tissues called **target tissues.** Target tissues are frequently in another part of the body at a distance from the gland that affects them. Table 22-1 shows the position of eight endocrine glands, the hormones they produce, and their target tissues.

In Chapter 19, the pancreas was described for its role in producing a digestive enzyme. However, other groups of cells in the pancreas also secrete hormones. One of
these hormones, insulin (IN suh lin), enables cells to 2
take in glucose. Glucose is the main source of energy for respiration in cells. Normally, insulin enables glucose to pass from the bloodstream through cell membranes. Persons who can't make insulin are diabetic. That means that insulin isn't there to enable glucose to get into cells.

Figure 22-12. Some athletes are extremely tall as a result of excess growth hormone.

OPTIONS

Meeting Different Ability Levels

For Section 22-4, use the following **Teacher Resource Masters** depending upon individual students' needs.

◆ **Study Guide Master** for all students.

● **Reinforcement Master** for students of average and above average ability levels.

▲ **Enrichment Master** for above average students.

Additional Teacher Resource Package masters are listed in the OPTIONS box throughout the section. The additional masters are appropriate for all students.

◆ STUDY GUIDE 87

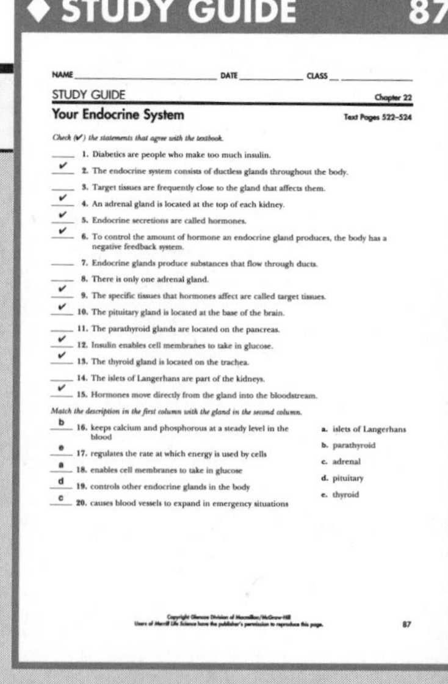

NAME ______ DATE ______ CLASS ______

STUDY GUIDE — Chapter 22

Your Endocrine System — Text Pages 522-524

Check (✔) the statements that agree with the textbook.

___ 1. Diabetics are people who make too much insulin.
✔ 2. The endocrine system consists of ductless glands throughout the body.
___ 3. Target tissues are frequently close to the gland that affects them.
✔ 4. An adrenal gland is located at the top of each kidney.
✔ 5. Endocrine secretions are called hormones.
✔ 6. To control the amount of hormone an endocrine gland produces, the body has a negative feedback system.
___ 7. Endocrine glands produce substances that flow through ducts.
___ 8. There is only one adrenal gland.
✔ 9. The specific tissues that hormones affect are called target tissues.
✔ 10. The pituitary gland is located at the base of the brain.
___ 11. The parathyroid glands are located on the pancreas.
✔ 12. Insulin enables cell membranes to take in glucose.
✔ 13. The thyroid gland is located on the trachea.
___ 14. The islets of Langerhans are part of the kidneys.
✔ 15. Hormones move directly from the gland into the bloodstream.

Match the description in the first column with the gland in the second column.

b 16. keeps calcium and phosphorous at a steady level in the blood
e 17. regulates the rate at which energy is used by cells
a 18. enables cell membranes to take in glucose
d 19. controls other endocrine glands in the body
c 20. causes blood vessels to expand in emergency situations

a. islets of Langerhans
b. parathyroid
c. adrenal
d. pituitary
e. thyroid

87

Table 22-1

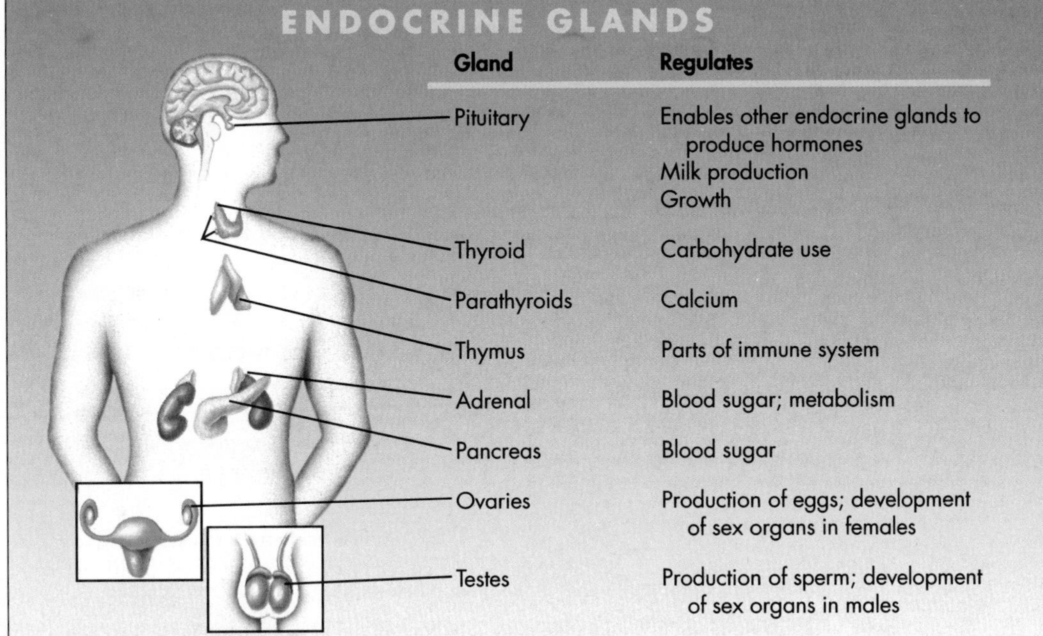

ENDOCRINE GLANDS

Gland	Regulates
Pituitary	Enables other endocrine glands to produce hormones Milk production Growth
Thyroid	Carbohydrate use
Parathyroids	Calcium
Thymus	Parts of immune system
Adrenal	Blood sugar; metabolism
Pancreas	Blood sugar
Ovaries	Production of eggs; development of sex organs in females
Testes	Production of sperm; development of sex organs in males

The reproductive glands also produce hormones. The functions of the hormones of the ovaries and testes will be discussed in Chapter 23.

A Negative Feedback System

To control the amount of hormone an endocrine gland produces, the endocrine system sends chemical information back and forth to itself. This is a negative feedback system. It works much the way a thermostat works. When the temperature in a room drops below a certain level, a thermostat signals the furnace to turn on. Once the furnace has raised the temperature to the level set on the thermostat, the furnace shuts off. It will stay off until the thermostat signals again. Once a target tissue responds to its hormone, the tissue sends a chemical signal back to the gland. This signal causes the gland to stop or slow down production. When the level of the hormone in the
3 bloodstream drops again, the endocrine gland is signaled to start the secretion of the hormone again. In this way, the concentration of the hormone in the bloodstream is kept at the needed level, and homeostasis is maintained.

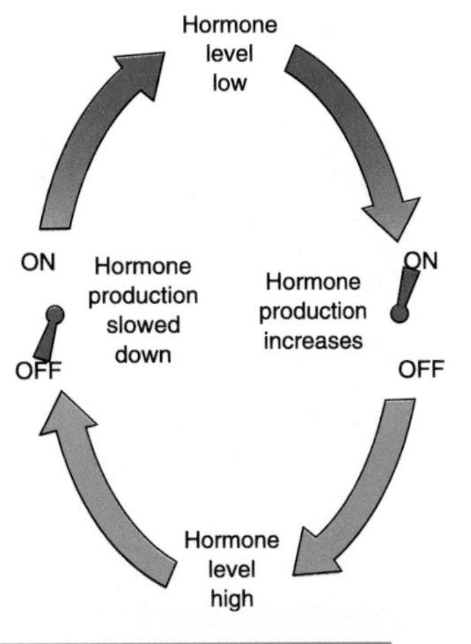

● REINFORCEMENT 87

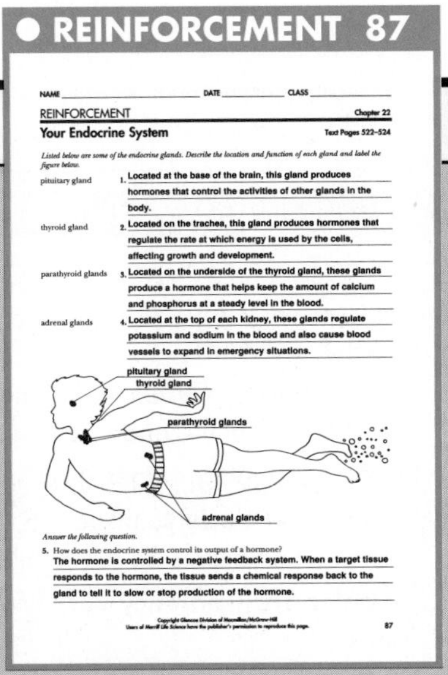

NAME ____ DATE ____ CLASS ____

REINFORCEMENT — Chapter 22

Your Endocrine System — Text Pages 522–524

Listed below are some of the endocrine glands. Describe the location and function of each gland and label the figure below.

pituitary gland — 1. **Located at the base of the brain, this gland produces hormones that control the activities of other glands in the body.**

thyroid gland — 2. **Located on the trachea, this gland produces hormones that regulate the rate at which energy is used by the cells, affecting growth and development.**

parathyroid glands — 3. **Located on the underside of the thyroid gland, these glands produce a hormone that helps keep the amount of calcium and phosphorus at a steady level in the blood.**

adrenal glands — 4. **Located at the top of each kidney, these glands regulate potassium and sodium in the blood and also cause blood vessels to expand in emergency situations.**

pituitary gland
thyroid gland
parathyroid glands
adrenal glands

Answer the following question.

5. How does the endocrine system control its output of a hormone? **The hormone is controlled by a negative feedback system. When a target tissue responds to the hormone, the tissue sends a chemical response back to the gland to tell it to slow or stop production of the hormone.**

87

▲ ENRICHMENT 87

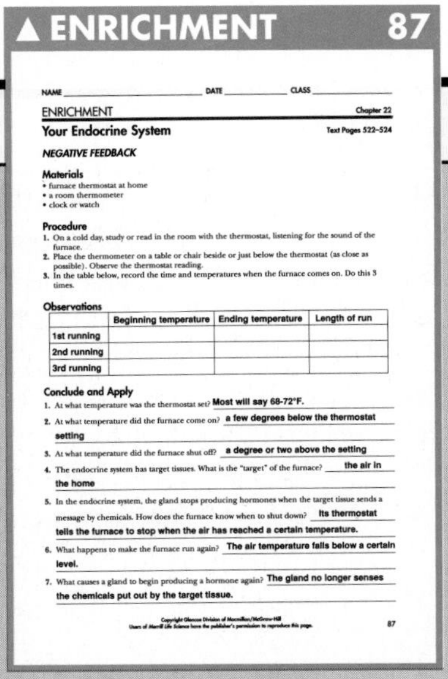

NAME ____ DATE ____ CLASS ____

ENRICHMENT — Chapter 22

Your Endocrine System — Text Pages 522–524

NEGATIVE FEEDBACK

Materials
- furnace thermostat at home
- a room thermometer
- clock or watch

Procedure
1. On a cold day, study or read in the room with the thermostat, listening for the sound of the furnace.
2. Place the thermometer on a table or chair beside or just below the thermostat (as close as possible). Observe the thermostat reading.
3. In the table below, record the time and temperatures when the furnace comes on. Do this 3 times.

Observations

	Beginning temperature	Ending temperature	Length of run
1st running			
2nd running			
3rd running			

Conclude and Apply
1. At what temperature was the thermostat set? **Most will say 68-72°F.**
2. At what temperature did the furnace come on? **a few degrees below the thermostat setting**
3. At what temperature did the furnace shut off? **a degree or two above the setting**
4. The endocrine system has target tissues. What is the "target" of the furnace? **the air in the home**
5. In the endocrine system, the gland stops producing hormones when the target tissue sends a message by chemicals. How does the furnace know when to shut down? **Its thermostat tells the furnace to stop when the air has reached a certain temperature.**
6. What happens to make the furnace run again? **The air temperature falls below a certain level.**
7. What causes a gland to begin producing a hormone again? **The gland no longer senses the chemicals put out by the target tissue.**

87

2 TEACH

Key Concepts are highlighted.

CONCEPT DEVELOPMENT

Cooperative Learning: Use the Paired Partners strategy to have students make flash cards with names of endocrine glands listed in Table 22-1. On the reverse side, identify the part of the body where the gland is found and its major functions. Partners can test each other using the flash cards.

▶Ask the question: **If there is a problem with the absorption of calcium by the bone, what endocrine gland might be involved?** *parathyroid*

▶Ask the question: **What happens to excess hormones in the bloodstream?** *The hormones are removed by the kidneys and the liver.*

CHECK FOR UNDERSTANDING

Use the Mini Quiz to check understanding.

MINI QUIZ

Use the Mini Quiz to check students' recall of chapter content.

1. **Endocrine glands do not have ___.** *ducts*
2. **What is one of the hormones produced by the pancreas?** *insulin*
3. **What signals a gland to start secreting its hormone again?** *when the level of hormone in the bloodstream drops*

RETEACH

Using a chart of the circulatory system, have students trace the pathway of a hormone to its target tissue.

EXTENSION

For students who have mastered this section, use the **Reinforcement** and **Enrichment** masters or other OPTIONS provided.

VideoDisc

STVS: Insulin Pills, Disc 7, Side 2
STVS: Insulin Pump, Disc 7, Side 2

PROBLEM SOLVING

The pancreas is involved in diabetes.

Thinking Critically: Carrie's body cells are not able to use the glucose in her blood. Thus, her body cells have no source of energy.

3 CLOSE

►Ask questions 1-3 and the **Apply** Question in the Section Review.

SECTION REVIEW ANSWERS

1. They are endocrine secretions that control activities in parts of the body not near to the gland.
2. This system allows the concentration of the hormone in the bloodstream to be kept at the needed level.
3. pancreas—secretes insulin, refer to Table 22-1 for other answers
4. Apply: Insulin enables glucose to pass from the bloodstream into cells. Cells need glucose for respiration. Without insulin, cells can't respire.
5. Connect to Chemistry: Iodine is found in the thyroid hormone. If iodine is lacking, the hormone cannot be produced. Supplementary iodine is provided in the diet in the form of iodized salt.

Skill Builder

Both systems send and receive messages and help maintain homeostasis.

PROBLEM SOLVING

Why am I so tired?

Carrie hasn't been feeling well lately. She complains about feeling tired all the time.

"She just doesn't have any get-up-and-go," her mother told the doctor. The doctor ordered some blood tests and questioned Carrie about her diet. After school one afternoon, Carrie went to the doctor's office to hear the results of the tests. The tests showed that her blood sugar was much too high. The doctor gave her strict instructions to follow detailing what she was allowed to eat every day. He told Carrie that she showed signs of diabetes. What endocrine gland is involved in diabetes?

Think Critically: How could Carrie be tired all the time and yet have a high blood sugar reading?

Hormones secreted by endocrine glands go directly into the bloodstream and affect target tissues. The level of the hormone is controlled by a negative feedback system. In this way, many chemicals in the blood and many body functions are controlled.

SECTION REVIEW

1. What is the function of a hormone?
2. What is a negative feedback system?
3. Explain how three hormones work in the body.
4. **Apply:** How would a lack of insulin affect respiration in a person's cells?
5. **Connect to Chemistry:** Find out why people in most areas need to use iodized salt.

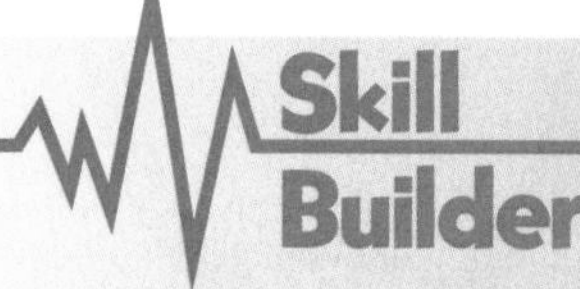

Skill Builder

☑ Comparing and Contrasting

In what ways are the nervous system and endocrine system alike? If you need help, refer to Comparing and Contrasting in the **Skill Handbook** on page 683.

OPTIONS

ENRICHMENT

►Have students research the function of the chemical messengers called prostaglandins.

Skillbuilder

ASSESSMENT

Performance: Use this Skillbuilder to assess students' abilities to Compare and Contrast the nervous and endocrine systems.

PROGRAM RESOURCES

From the **Teacher Resource Package** use:

Transparency Masters, pages 85-86, Endocrine Glands.

Critical Thinking/Problem Solving, page 26, Diseases of the Endocrine System.

Concept Mapping, pages 49-50.

Science and Society, page 26, Should Alzheimer's Patients Be Hospitalized?

Use **Color Transparency** 43, Endocrine Glands.

CHAPTER

REVIEW

SUMMARY

22-1: Your Nervous System

1. A neuron is the basic unit of structure and function of the nervous system.
2. A stimulus is detected by sensory neurons. Impulses are carried to the interneurons where the proper response is determined and transmitted to the motor neurons. A response is made automatically; this is a reflex.
3. The central nervous system contains the brain and spinal cord. The peripheral nervous system is composed of cranial and spinal nerves.

22-2: The Senses

1. The eyes respond to light energy, and the ears respond to sound waves. Interpretation of the stimuli is done by the cerebrum.
2. Olfactory cells of the nose and taste buds of the tongue are stimulated by chemicals. This stimulation produces impulses that are interpreted as different tastes.

22-3: Science and Society: Alzheimer's Disease

1. Neural transmitters carry impulses across synapses between neurons.
2. In Alzheimer's disease, neural transmission is impaired, resulting in memory loss and physical disfunctioning.
3. Scientists think a virus, chemical imbalance, or heredity are possible causes of Alzheimer's disease.

22-4: Your Endocrine System

1. Endocrine glands secrete hormones directly into the bloodstream.
2. Hormones affect specific tissues throughout the body.
3. A change in the body causes a gland to function. Once homeostasis is reached, the gland receives a signal to slow or stop its secretion.

KEY SCIENCE WORDS

a. **Alzheimer's disease**
b. **axon**
c. **brainstem**
d. **central nervous system**
e. **cerebellum**
f. **cerebrum**
g. **cochlea**
h. **dendrites**
i. **hormones**
j. **interneurons**
k. **motor neurons**
l. **neuron**
m. **olfactory cells**
n. **peripheral nervous system**
o. **reflex**
p. **retina**
q. **sensory neurons**
r. **synapse**
s. **target tissues**
t. **taste buds**

UNDERSTANDING VOCABULARY

Match each phrase with the correct term from the list of Key Science Words.

1. neurons carrying impulses from the brain
2. basic unit of nervous system
3. nerve cells between sensory and motor neurons
4. gap between neurons
5. division containing brain and spinal cord
6. an automatic response to stimuli
7. light-sensitive tissue of the eye
8. center for coordination of voluntary muscle action
9. inner ear structure containing fluid
10. nerve cells that respond to gas molecules

OPTIONS

ASSESSMENT

To assess student understanding of material in this chapter, use the resources listed.

COOPERATIVE LEARNING

Consider using cooperative learning in the THINK AND WRITE CRITICALLY, APPLY, and MORE SKILL BUILDERS sections of the Chapter Review.

PROGRAM RESOURCES

From the **Teacher Resource Package** use:
Chapter Review, pages 47-48.
Chapter and Unit Tests, pages 138-141, Chapter Test.

CHAPTER

REVIEW

SUMMARY

Have students read the summary statements to review the major concepts of the chapter.

UNDERSTANDING VOCABULARY

1. k
2. l
3. j
4. r
5. d
6. o
7. p
8. e
9. g
10. m

ASSESSMENT

Portfolio

Encourage students to place in their portfolios one or two items of what they consider to be their best work. For each item, ask students to explain why that item was chosen and what they learned from it. Items might be selected from the following.

- For Your Mainstreamed Students poster on the nervous system, p. 506
- Reteach model of the brain, p. 511
- In Your Journal report on Medicare, p. 521

Performance

Additional performance assessments may be found in *Performance Assessment* and *Science Integration Activities* that accompany **Merrill Life Science.** Performance Task Assessment Lists and rubrics for evaluating these activities and other products generated throughout the chapter can be found in Glencoe's *Performance Assessment in Middle School Science.*

CHAPTER
REVIEW

CHECKING CONCEPTS

1. d
2. b
3. c
4. c
5. a
6. c
7. a
8. d
9. c
10. c

USING LAB SKILLS

ASSESSMENT

Use these alternate lab exercises to assess students' understanding of skills used in this chapter.

11. Encourage students to test two widely separated age groups, such as children between 11 and 13 and adults between 45 and 54. Results will vary, but generally children will show more sensitivity to different odors.

12. Have students clearly delineate areas of the leg—*i.e.* knee, front of leg between knee and ankle, back of knee, back of calf, top of foot, sole. Responses will vary but the back of the knee and sole of the foot will probably show more responsiveness.

THINK AND WRITE CRITICALLY

13. Taste is remembered. If something tastes bad and results in sickness, you remember it and do not taste the food again. Smell provides the same protection.

14. If a synapse were blocked, the chemicals that transmit nerve impulses would not be able to do this; therefore, these nerves could not be stimulated.

15. breathing and heartbeat rates increase; more glucose put into your blood for energy; tense and ready muscles; and increased blood supply to muscles, heart, and lungs

CHAPTER
REVIEW

CHECKING CONCEPTS

Choose the word or phrase that completes the sentence.

1. Impulses cross synapses by way of ______.
   a. osmosis c. a cell body
   b. interneurons d. a chemical
2. The neuron structures that carry impulses to the cell body are ______.
   a. axons c. synapses
   b. dendrites d. nucleus
3. Neurons detecting stimuli in the skin and eyes are ______.
   a. interneurons c. sensory neurons
   b. motor neurons d. synapses
4. The ______ is the largest part of the brain.
   a. cerebellum c. cerebrum
   b. brainstem d. pons
5. The ______ controls voluntary muscle.
   a. cerebellum c. cerebrum
   b. brainstem d. pons
6. The ______ is the part of the brain that is divided into two hemispheres.
   a. cerebellum c. cerebrum
   b. brainstem d. spinal cord
7. The ______ is (are) controlled by the somatic division of the PNS.
   a. skeletal muscles c. glands
   b. heart d. salivary glands
8. The inner ear contains the ______.
   a. anvil c. eardrum
   b. hammer d. cochlea
9. The ______ gland controls many other endocrine glands throughout the body.
   a. adrenal c. pituitary
   b. thyroid d. pancreas
10. Ductless glands produce ______.
    a. enzymes c. hormones
    b. target tissues d. saliva

USING LAB SKILLS

11. The MINI-Lab on page 508 compared the sense of smell in males and females. Repeat this experiment to test differences in ability to smell between children and adults. Are there differences in their ability to detect odors?
12. The Activity on page 519 focused on skin sensitivity on the upper body. Try this activity on the lower part of the leg and on the foot. What did you find out about the distribution there?

THINK AND WRITE CRITICALLY

Answer the following problems in your Journal using complete sentences.

13. Discuss why taste and smell are protective.
14. What would happen if a chemical was blocking many synapses?
15. List the effects adrenal gland hormones would have on your body as you prepare to run a race.

APPLY

16. Why is it helpful to have impulses move in only one direction in a neuron?
17. How are reflexes protective?
18. How is your endocrine system like the thermostat in a house?
19. Describe an example of a problem that results from improper gland functioning.
20. If a fly were to land on your face and on your back, which might you feel first? Explain how you would test your choice.

MORE SKILL BUILDERS

If you need help, refer to the Skill Handbook.

1. **Classifying:** Classify the types of neurons as to their location and direction of impulse.
2. **Comparing and Contrasting:** Compare and contrast the structures and functions of the cerebrum, cerebellum, and brainstem. Include in your discussion the following functions: balance, homeostasis, involuntary muscles, muscle tone, memory, voluntary muscles, thinking, senses.
3. **Sequencing:** Sequence the structures through which light passes in the eye.
4. **Interpreting Scientific Illustrations:** Using the diagram of the synapse on page 510, explain how an impulse moves from one neuron to another.
5. **Observing and Inferring:** If an impulse traveled down one neuron, but failed to move on to the next neuron, what might you infer about the first neuron?

PROJECTS

1. Work in Paired Partners to find out how Annie Sullivan communicated with Helen Keller by drawing letters in her hand. Obtain Braille books and Braille writing tools. Learn how these instruments are used.
2. Find out how hearing aids work. Explain why not all hearing-impaired persons can use hearing aids.

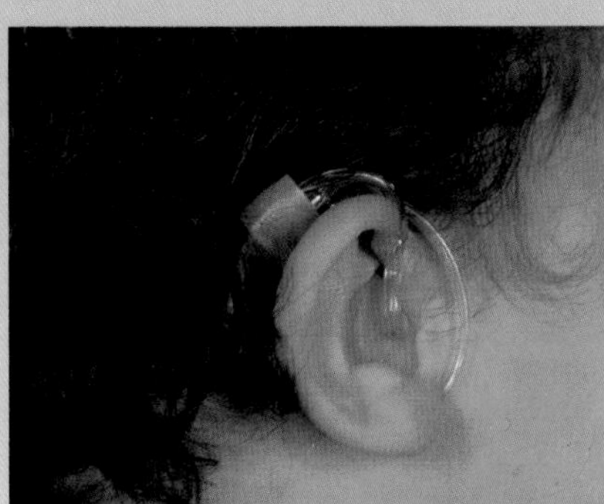

APPLY

16. Accept all reasonable responses; so messages do not get mixed; so that responses can be coordinated

17. Reflexes are automatic acts that occur without our thinking about them. Therefore, they happen very quickly and can sheild our bodies from danger, such as sharp or hot objects.

18. According to the level of hormone in the blood, target tissue sends a chemical signal back to the gland to stop or start hormone secretion.

19. Examples include giantism, dwarfism, hyperthyroidism, and Cushing's syndrome.

20. The skin on your face has more neurons or more closely arranged neurons than the skin on your back. The sense of touch is more pronounced on the face. You could check this out with a series of tests using something very lightweight, such as a feather or paint brush hair to test sensitivity on the face and neck.

MORE SKILL BUILDERS

1. Classifying:
Sensory neurons—Location: sense organs and spinal cord; Impulse Direction: carry impulses to brain
Interneurons—Location: in CNS; Impulse Direction: carry impulses from CNS to motor neurons
Motor neurons—Location: muscles and glands; Impulse Direction: carry impulses from brain to muscles and glands

2. Comparing and Contrasting: cerebrum—memory, senses, thinking; cerebellum—voluntary muscles, balance, muscle tone; brainstem—involuntary muscles, homeostasis of heartbeat, breathing, and blood pressure

3. Sequencing: cornea → lens → retina → optic nerve → brain

4. Interpreting Scientific Illustrations: A nerve-transmitting chemical is released from the axon of one neuron, diffuses across the synapse, and starts an impulse in the next neuron.

5. Observing and Inferring: It may be lacking the nerve-transmitting chemical released by the axon.

CHAPTER 23 Reproduction and Growth

CHAPTER SECTION	OBJECTIVES	ACTIVITIES
23-1 Human Reproduction (3 days)	1. **Explain** the function of the reproductive system. 2. **Identify** the major structures of the male reproductive system. 3. **Describe** the functions of the major female reproductive organs. 4. **Explain** the stages of the menstrual cycle.	**Activity 23-1:** *Interpreting Diagrams,* p. 534
23-2 Fertilization to Birth (2 days)	1. **Describe** how an egg becomes fertilized. 2. **Identify** the major events in the stages of development of an embryo and fetus. 3. **Differentiate** between fraternal and identical twins.	**MINI-Lab:** *How long is an embryo?* p. 537
23-3 Development after Birth (2 days)	1. **State** the sequence of events of childbirth. 2. **Compare** the stages of infancy and childhood. 3. **Relate** adolescence to preparation for adulthood.	**MINI-Lab:** *What is the immunization schedule for babies and young children?* p. 541
23-4 Aging Science & Society (1 day)	1. **Discuss** concerns in society for the elderly. 2. **Explain** the chemical process of aging.	**Activity 23-2:** *Average Growth Rate in Humans,* p. 546
Chapter Review		

ACTIVITY MATERIALS

FIND OUT	ACTIVITIES		MINI-LABS	
Page 529 penny cardboard or plastic tray	**23-1 Interpreting Diagrams, p. 534** paper pencil	**23-2 Average Growth Rate in Humans, p. 546** graph paper red and blue pencils	**How long is an embryo? p. 537** paper pencil metric ruler	**What is the immunization schedule for babies and young children? p. 541** none

CHAPTER FEATURES	TEACHER RESOURCE PACKAGE	OTHER RESOURCES
Skill Builder: *Sequencing,* p. 533	**Ability Level Worksheets** ◆ ***Study Guide,*** p. 88 ● ***Reinforcement,*** p. 88 ▲ ***Enrichment,*** p. 88 **Activity Worksheet,** pp. 203-204 **Transparency Masters,** pp. 87-90	**Color Transparency 44,** Male and Female Reproductive Systems **Color Transparency 45,** The Menstrual Cycle
Problem Solving: *When Is the Baby Due?* p. 536 **Technology:** *Operating in the Womb,* p. 538 **Skill Builder:** *Making and Using Graphs,* p. 539	**Ability Level Worksheets** ◆ ***Study Guide,*** p. 89 ● ***Reinforcement,*** p. 89 ▲ ***Enrichment,*** p. 89 **Activity Worksheets,** p. 209 **Critical Thinking/Problem Solving,** p. 27 **Transparency Masters,** pp. 91-92	**Lab Manual:** Fetal Development, p. 159 **Color Transparency 46,** An Embryo in the Uterus **Science Integration Activity 23**
Skill Builder: *Outlining,* p. 543	**Ability Level Worksheets** ◆ ***Study Guide,*** p. 90 ● ***Reinforcement,*** p. 90 ▲ ***Enrichment,*** p. 90 **Concept Mapping,** p. 51 **Cross-Curricular Connections,** p. 27	**STVS:** Disc 7, Side 2
You Decide! p. 545	**Ability Level Worksheets** ◆ ***Study Guide,*** p. 91 ● ***Reinforcement,*** p. 91 ▲ ***Enrichment,*** p. 91 **Activity Worksheets,** pp. 205, 206, 210 **Science and Society,** p. 27	**STVS:** Disc 7, Side 2
Summary Key Science Words Understanding Vocabulary Checking Concepts Using Lab Skills Think & Write Critically Apply More Skill Builders Projects	ASSESSMENT RESOURCES **Chapter Review,** pp. 49, 50 **Chapter Test,** pp. 142-145 **Unit Test,** pp. 146-147 **Performance Assessment in Middle School Science (PAMSS)**	**Chapter Review Software** **Test Bank** **Alternate Assessment** **Performance Assessment**

◆ **Basic** ● **Average** ▲ **Advanced**

ADDITIONAL MATERIALS

SOFTWARE	AUDIOVISUAL	BOOKS/MAGAZINES
Reproduction Organs, Micro Power & Light Co. *Reproduction Process,* Micro Power & Light Co. *Human Life Processes III:* Development and Differentiation, IBM. *All About Reproduction,* Queue.	*Human Body: Reproductive System,* film, Coronet/MTI. *The Miracle of Life,* film, Time-Life. *Birth: How Life Begins,* film, EBEC. *Life Cycles,* laserdisc, Videodiscovery. *Lifetimes of Change: Development and Growth,* laserdisc, Aims Media.	Anselmo, Sandra. *Early Childhood Development: Prenatal Through Age Eight.* Columbus, OH: Merrill Publishing, 1987. Madar, Sylvia S. *Human Reproductive Biology.* Dubuque, IA: William C. Brown Publishers, 1980. Villee. *Human Reproduction.* St. Louis, MO: C.V. Mosby, 1991.

CHAPTER

23 REPRODUCTION AND GROWTH

THEME DEVELOPMENT: A central theme in this text is that of energy. Energy drives many biological systems and is involved in the reproduction, growth, and development of organisms. The unique functioning of the human reproductive system is dependent on energy for cell division and the subsequent growth and differentiation of cells.

CHAPTER OVERVIEW

- **Section 23-1:** In this section the structure and function of the major organs of the female and male reproductive systems are introduced. Special attention is given to the menstrual cycle.
- **Section 23-2:** This section follows the development of a fertilized egg to the full-term fetus. The causes for multiple births are explored.
- **Section 23-3:** This section opens with the details of childbirth. The development of a human is traced from infancy through adulthood. Characteristics of each stage are given.
- **Section 23-4: Science and Society:** This section deals with the concerns in society for the elderly. The problems elderly people encounter in society today are discussed, along with research into halting the process of aging.

CHAPTER VOCABULARY

testes	pregnancy
sperm	embryo
semen	amniotic sac
ovaries	fetus
ovulation	infancy
uterus	childhood
vagina	adolescence
menstrual cycle	adulthood
menstruation	catalase
menopause	

CHAPTER

23 Reproduction and Growth

528

OPTIONS

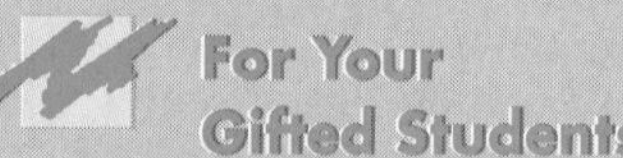

For Your Gifted Students

Allow students to visit an elementary school to interview teachers at grades K-4. Have the students prepare a set of questions that will help determine the physical and mental capabilities of the students at each level. Students can present their information on the developmental progression of childhood. Have them check their findings with books on the topic of childhood development.

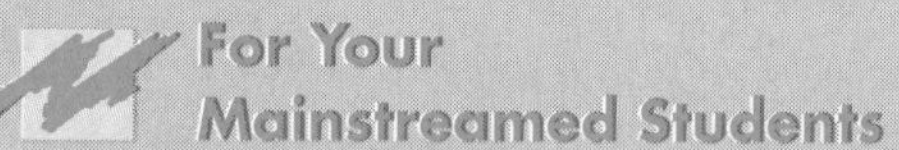

For Your Mainstreamed Students

Have students investigate the reproductive cycles of other animals, such as dogs, fish, elephants, and so on.

You may have heard a newspaper or television story about a large family that has all girls or all boys. What do you think are the odds of a family having all boys or all girls? What do you think is the proportion of boys to girls among children born in the general population?

FIND OUT!

Do this simple activity to find out what proportion of newborns are boys and what proportion are girls.

Take a penny and toss it in the air. There is an equal chance that it will land heads or tails. Toss the penny a hundred times and keep a record of the times it falls heads up and the number of times it falls tails up. Keep a record of the order in which the penny landed. What did your record show? Did you sometimes have five or more heads before a tails fell? What application can you make to families who have five girls or five boys? *Infer* the chances that a girl would be born if there were a sixth child.

Gearing Up

Previewing the Chapter

Use this outline to help you focus on important ideas in this chapter.

Previewing Science Skills

- ▶ In the Skill Builders, you will sequence, graph, and outline data.
- ▶ In the Activities, you will interpret and graph data.
- ▶ In the MINI-Labs, you will interpret data and do research.

What's next?

In this chapter, you will learn about the structures and function of the male and female reproductive systems. You will also learn about the development of a human from fertilization to old age.

529

INTRODUCING THE CHAPTER

Use the Find Out activity to introduce students to the reproductive system. Inform students that they will be learning more about the functions of the reproductive system and the development of a human.

FIND OUT!

Preparation: Collect small cardboard or foam trays to deaden sound and catch the tossed pennies.

Cooperative Learning: Place students into Science Investigation teams. One student tosses the penny, and the second student observes whether it lands heads or tails and reports this to the third student, who serves as the recorder. When the hundred trials are completed, the students rotate roles.

Teaching Tips

▶ Have the student who observes the side of the coin that lands face up convey the information quietly to a recorder. This will expedite the process and reduce the noise of the activity.

▶ Share the class data to see any long sequence of heads or tails. Combine all the totals to determine how close to a 50:50 ratio was established.

Gearing Up

Have students study the Gearing Up feature to familiarize themselves with the chapter. Discuss the relationships of the topics in the outline.

What's Next?

Before beginning the first section, make sure students understand the connection between the Find Out activity and the topics to follow.

ASSESSMENT OPTIONS

PORTFOLIO

Refer to page 547 for suggested items that students might select for their portfolios.

PERFORMANCE ASSESSMENT

See page 547 for additional Performance Assessment options.

Process

Skillbuilders, pp. 533, 539
MINI-Lab, pp. 537, 541
Activities 23-1, p. 534; 23-2, p. 546
Using Lab Skills, p. 548

CONTENT ASSESSMENT

Assessment—Oral, pp. 532, 539, 542
Section Reviews, pp. 533, 539, 543, 545
Chapter Review, pp. 547-549
Mini Quizzes, pp. 532, 538

GROUP ASSESSMENT

Opportunities for group assessment occur with Cooperative Learning Strategies and Flex Your Brain Activities.

SECTION 23-1

PREPARATION

SECTION BACKGROUND

▶ As in all male mammals, the urethra carries both urine and semen outside the body. The release of urine through the urethra is controlled by sphincter muscles located above the opening of the vas deferens into the urethra. This system prevents urine and semen from mixing in the urethra.

▶ Occasionally, an egg from the ovary does not enter the opening of the oviduct and enters the abdominal cavity and becomes fertilized. An ectopic pregnancy occurs but does not continue because the tissues cannot sustain the embryo.

Answer: The streamlined head of a sperm reduces friction as it moves through liquids. The active tail propels the sperm in a direction that can be against gravity.

1 MOTIVATE

▶ Discuss with students how cells (blood, skin, hair) are constantly being replaced in the body by cell division. However, the development of a human requires more than cell division. The union of a male sex cell with that of a female sex cell initiates human growth.

23-1 Human Reproduction

New Science Words

testes
sperm
semen
ovaries
ovulation
uterus
vagina
menstrual cycle
menstruation
menopause

Shapes of objects are often related to their function. *Infer* why sperm have a streamlined head and an active tail.

Objectives

- ▶ Explain the function of the reproductive system.
- ▶ Identify the major structures of the male reproductive system.
- ▶ Describe the functions of the major female reproductive organs.
- ▶ Explain the stages of the menstrual cycle.

The Reproductive System

Reproduction is the process that continues life on Earth. Human and all other sexually reproducing organisms form eggs and sperm to carry genetic information from one generation to the next. If you have ever babysat, you know that babies and young children require nearly constant attention. Mammals, including humans, produce few offspring at one time compared to invertebrate animals. Some tropical termite queens can lay 80 000 eggs a day for nearly 30 years! Can you imagine babysitting for that many offspring? Unlike many other animals, mammals must provide care for their offspring to ensure their survival.

In previous chapters, you have read that most human body systems are alike in males and females. This is not the case for the reproductive system. Males and females each have structures specialized for their role in reproduction.

The Male Reproductive System

Figure 23-1 shows the structures of the male reproductive system. The external organs of the male reproductive system are the penis and scrotum (SKROH tum). The penis is the male organ for reproduction and urination. Behind the penis is a saclike pouch called the scrotum that holds the testes. During puberty, the two
1 **testes** begin to produce **sperm,** the male reproductive cells. Sperm are single cells with a head and tail. The tail moves the sperm, and the head contains genetic information. The scrotum helps regulate temperature for sperm production. It holds the testes outside the male's body. As a result, the testes have a lower temperature. This cooler temperature is necessary to produce sperm.

OPTIONS

Meeting Different Ability Levels

For Section 23-1, use the following **Teacher Resource Masters** depending upon individual students' needs.

◆ **Study Guide Master** for all students.

● **Reinforcement Master** for students of average and above average ability levels.

▲ **Enrichment Master** for above average students.

Additional Teacher Resource Package masters are listed in the OPTIONS box throughout the section. The additional masters are appropriate for all students.

◆ STUDY GUIDE 88

NAME ________ DATE ________ CLASS ________

STUDY GUIDE — Chapter 23

Human Reproduction — Text Pages 530-533

Complete the following sentences using appropriate terms from the textbook.

1. The **scrotum** helps regulate body temperature for sperm production.
2. The production of eggs and sperm begins during **puberty**.
3. Sperm is produced in the **testes**.
4. The urethra is a passageway for urine as well as for **semen**.
5. Egg cells are produced in the **ovaries**.
6. Fertilization usually takes place in the **oviduct**.
7. The menstrual cycle is controlled by **hormones** from the pituitary gland.
8. There is a reduction in **ovulation** and **menstruation** as menopause approaches.
9. The head of a sperm contains **genetic information**.
10. The gland that produces a nourishing fluid for sperm is called the **seminal vesicle**.

Complete the crossword puzzle below using terms from the textbook.

MENSTRUATION, MENSTRUAL, OVA, SPERM, SEMEN, MENOPAUSE, UTERUS, TESTES, OVARIES, VAGINA

Across

1. monthly discharge of uterine lining and blood
4. monthly cycle in female that is controlled by hormones produced in the pituitary gland and the ovaries
6. female reproductive cells
8. male reproductive cells
9. mixture of sperm and fluid

Down

1. menstrual cycle becomes irregular and stops
2. pear-shaped muscular organ in which a baby develops
3. produce sperm
5. female organs that produce eggs
7. passageway from the uterus to the outside of a female's body

88

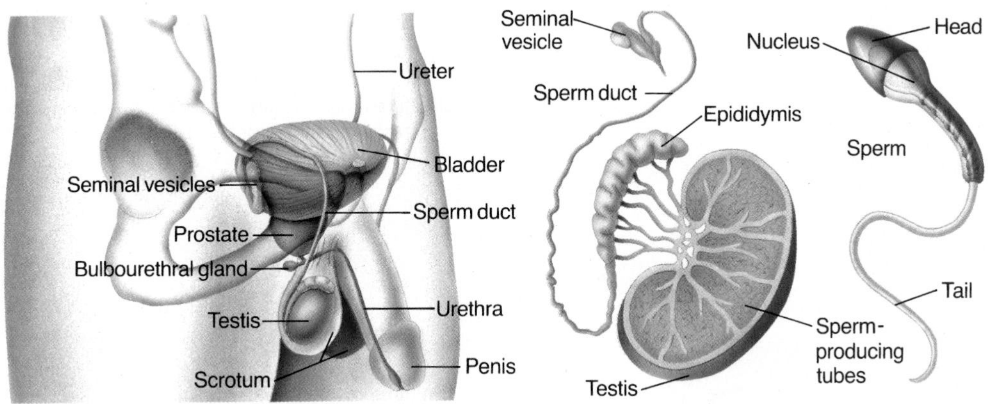

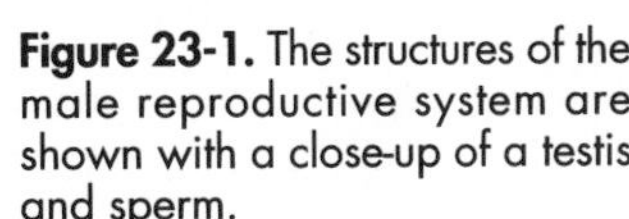

Figure 23-1. The structures of the male reproductive system are shown with a close-up of a testis and sperm.

Many organs help in the production, transport, and storage of sperm inside the male body. After sperm are produced, they travel from the testes through tubes that circle the bladder. Behind the bladder, a gland called the seminal vesicle provides sperm with a fluid that gives them energy and helps them move. This mixture of sperm and fluid is called **semen.** Semen leaves the body 2
through the urethra, the same tube that carries urine from the body. Semen and urine never mix. A muscle at the back of the bladder contracts to prevent urine from entering the urethra as sperm are ejected from the body.

The Female Reproductive System

When a female begins puberty, eggs start to develop in **ovaries**, the female sex organs. Unlike the male, most of the reproductive organs of the female are internal. The ovaries are located in the lower part of the body cavity. Each of the two ovaries is about the size and shape of an almond. About once a month an egg is released from an ovary. This process is called **ovulation** (ahv yuh LAY shun). The two ovaries take turns releasing the eggs. One month, the first ovary releases an egg; next month, the other ovary releases an egg and so on. When the egg is released, it enters the oviduct. If the egg is fertilized by a 3
sperm, it will occur in an oviduct. Cilia help sweep the egg through the oviduct to the uterus. The **uterus** is a hollow, pear-shaped, muscular organ with thick walls in which a fertilized egg develops. The lower end of the uterus is connected to the outside of the body by a muscular tube called the **vagina.** The vagina is also called the birth canal because a baby passes through this passage-

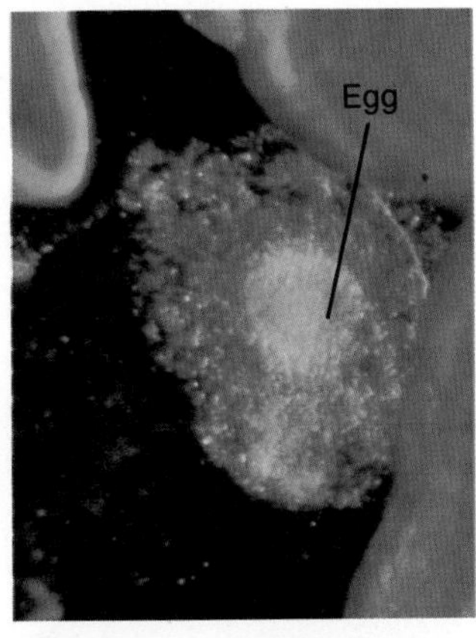

Figure 23-2. In the center of the photograph is an egg just released from an ovary. It will begin its journey down the oviduct.

● REINFORCEMENT 88

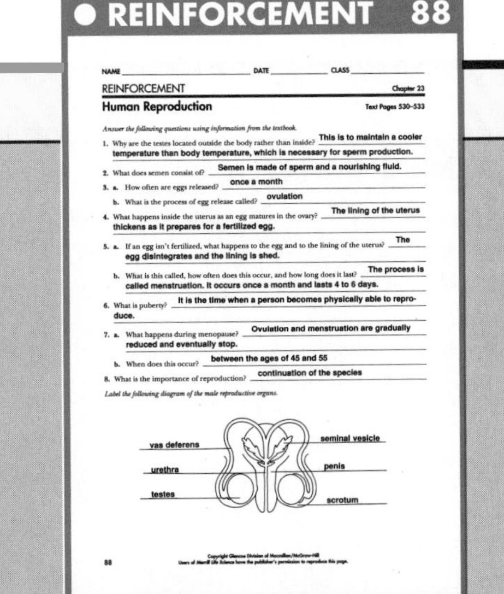

NAME ______ DATE ______ CLASS ______

REINFORCEMENT Chapter 23

Human Reproduction Text Pages 530–533

Answer the following questions using information from the textbook.

1. Why are the testes located outside the body rather than inside? **This is to maintain a cooler temperature than body temperature, which is necessary for sperm production.**
2. What does semen consist of? **Semen is made of sperm and a nourishing fluid.**
3. a. How often are eggs released? **once a month**
   b. What is the process of egg release called? **ovulation**
4. What happens inside the uterus as an egg matures in the ovary? **The lining of the uterus thickens as it prepares for a fertilized egg.**
5. a. If an egg isn't fertilized, what happens to the egg and to the lining of the uterus? **The egg disintegrates and the lining is shed.**
   b. What is this called, how often does this occur, and how long does it last? **The process is called menstruation. It occurs once a month and lasts 4 to 6 days.**
6. What is puberty? **It is the time when a person becomes physically able to reproduce.**
7. a. What happens during menopause? **Ovulation and menstruation are gradually reduced and eventually stop.**
   b. When does this occur? **between the ages of 45 and 55**
8. What is the importance of reproduction? **continuation of the species**

Label the following diagram of the male reproductive organs.

88

▲ ENRICHMENT 88

NAME ______ DATE ______ CLASS ______

ENRICHMENT Chapter 23

Human Reproduction Text Pages 530–533

HORMONAL CONTROL OF THE MENSTRUAL CYCLE

Using some references, find out more about the hormonal control of the menstrual cycle. Use information obtained from your research to answer the following questions.

1. What is menarche? **This is the first menstrual period.**
2. What is the average length of the menstrual cycle? **28 days**
3. What is the endometrium? **This is the uterine lining.**
4. How does FSH influence the ovarian follicle? **It causes the follicle to mature.**
5. What happens when estrogen levels begin to increase? **The uterine lining thickens.**
6. a. When is an ovum released? **in the middle of the cycle**
   b. What hormone is in the highest amount at the time of ovulation? **LH**
   c. What ovarian hormone is produced in response? **progesterone**
7. In the absence of pregnancy, what happens to the levels of estrogen and progesterone? **These hormones gradually decrease.**
8. Which hormones are necessary to maintain the uterine lining? **estrogen and progesterone**
9. a. What does the hormone called human chorionic gonadotropin (HCG) do? **It maintains high levels of estrogen and progesterone during pregnancy.**
   b. Where is HCG made? **It is produced by the placenta.**
10. How do birth control pills work? **They contain estrogen and progesterone. These suppress the production of FSH and LH and so ovulation does not occur and a woman cannot become pregnant.**
11. FSH and LH are abbreviations. What are the names of these hormones? **follicle-stimulating hormone and luteinizing hormone**

88

TYING TO PREVIOUS KNOWLEDGE: Recall from Chapter 4 cell division and the importance of meiosis to produce cells with 23 chromosomes in order to have a fertilized egg with the full complement of 46 chromosomes.

OBJECTIVES AND SCIENCE WORDS: Have students review the objectives and science words throughout the chapter to become familiar with each section.

2 TEACH

Key Concepts are highlighted.

CONCEPT DEVELOPMENT

▶ If possible, use an anatomical model in teaching this section.

▶ During the discussion of the function of the reproductive system, ask this question: **Why is it not necessary for a human female to produce thousands of eggs at one time?** *Eggs of humans are better protected and have a better chance of being fertilized.*

▶ Use a question box for questions students may not want to ask in class. Use the questions to develop content of lessons in the text.

TEACHER F.Y.I.

▶ Sperm production is a continuous process from puberty throughout the life of the male. Oocyte production in females begins before birth. There is no production after birth. Each ovary may have nearly a million primary oocytes, cells from which eggs form.

CROSS CURRICULUM

▶ **Language Arts:** Have students look up the words *sperm* and *ovary* and find out why the Latin roots are appropriate to their meaning.

CONCEPT DEVELOPMENT

▶ Use Figure 23-4 to develop the concept of the thickening of the lining of the uterus wall and its consequent shedding.

▶ **What initiates the formation of the hormones of the menstrual cycle?** *hormones from the pituitary gland*

MINI QUIZ

Use the Mini Quiz to check students' recall of chapter content.

1. **Males produce ____ and females produce eggs.** *sperm*
2. **The passage for both sperm and urine out of the male body is the ____.** *urethra*
3. **Where is the egg usually fertilized?** *in the upper third of the oviduct*
4. **What is the normal length of the menstrual period?** *4 to 6 days*

REVEALING MISCONCEPTIONS

▶ Menstruation is often thought of as a "disorder" of the female reproductive system. Although there may be discomfort associated with the menstrual period of some persons, the process is a normal one and all students, male and female, should have this understanding after studying this section.

CHECK FOR UNDERSTANDING

Have students diagram the movement of a mature egg from the ovary to the uterus.

way when being born. Figure 23-3 shows the structures of the female reproductive system.

Figure 23-3. The structures of the female reproductive system are shown from the front (right) and the side (left).

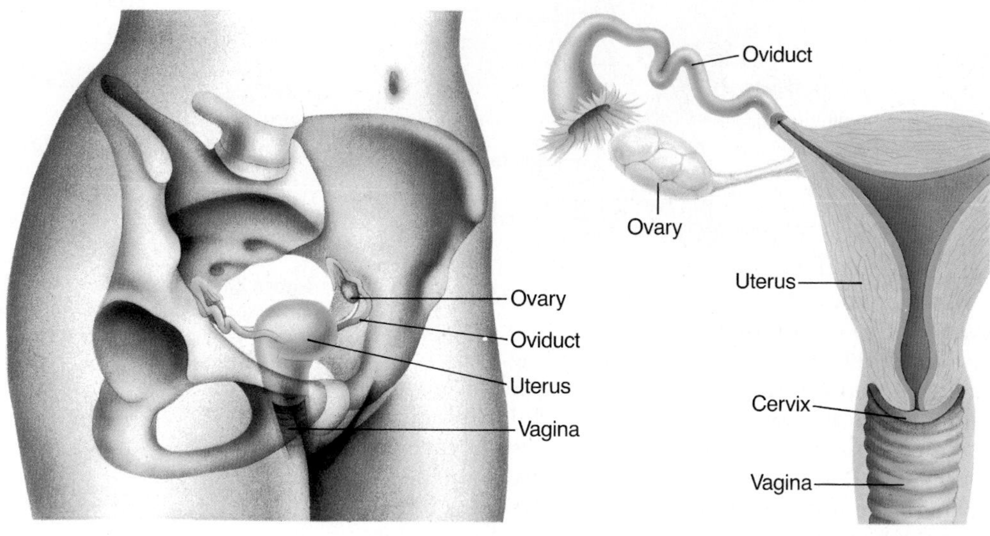

The Menstrual Cycle

Before and after an egg is released from an ovary, the uterus undergoes certain changes. The **menstrual cycle** is the monthly cycle of changes in the female reproductive system. This cycle of a human female averages 28 days. However, the cycle can vary in some individuals from 20 to 40 days. The changes include the maturing of an egg and the preparation of the uterus to receive a fertilized egg, as shown in Figure 23-4. When an egg inside an ovary matures, the lining of the uterus thickens as it prepares for a fertilized egg.

When does the menstrual cycle begin?

On approximately day 14 of a 28-day menstrual cycle, an egg is released from an ovary. It remains alive for about 24 to 48 hours. During this time, it travels along the oviduct. A female can become pregnant at this time in her menstrual cycle. The egg can be fertilized if live sperm are present in the oviduct 48 hours before or after ovulation. If an egg is not fertilized in the oviduct, it disintegrates. The thickened lining of the uterus is shed and bleeding occurs. This monthly discharge of the lining of the uterus and blood through the vagina is called

OPTIONS

ASSESSMENT—ORAL

▶ **How are ovulation and menstruation related?** *After ovulation, if there is no implantation of the egg, the uterus wall is shed and the material is discharged.*

▶ **In a normal year, without a pregnancy, how many eggs would be released by one ovary?** *Six. Another six would be released by the other ovary.*

ENRICHMENT

▶ Have students do research to compare the size of a uterus before pregnancy and during pregnancy.

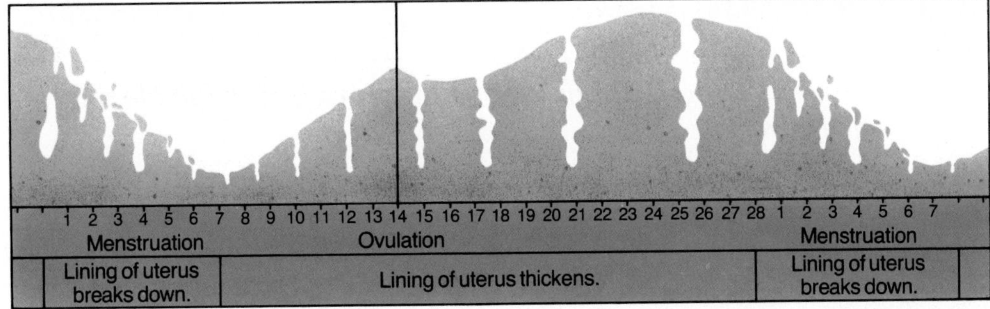

menstruation (men STRAY shun). The menstrual period usually lasts from four to six days and is a normal function of the female reproductive system.

4 **Figure 23-4.** The menstrual cycle begins on day one of menstruation. Ovulation occurs near day 14 of a regular 28-day cycle.

When menstruation begins, another egg begins maturing. The uterus again prepares for the egg. Hormones from the pituitary gland and the ovaries control the events of the menstrual cycle.

Menstruation begins when a girl's reproductive organs have matured. For most females, the first menstrual period happens between ages 8 and 13 and continues until age 45 to 55. Then, there is a gradual reduction of ovulation and menstruation. **Menopause** occurs when the menstrual cycle becomes irregular and eventually stops.

When the reproductive systems of males and females become mature, sperm and eggs are produced. The reproductive process allows for the species to continue.

SECTION REVIEW

1. What is the major function of a reproductive system?
2. Trace and label the route of sperm movement through the male reproductive system.
3. List the organs of the female reproductive system and describe their functions.
4. Explain the cause of menstrual flow.
5. **Apply:** What happens to the lining of the uterus after a women goes through menopause?
6. **Connect to Chemistry:** Find out why adolescent females often require additional amounts of iron in their diet.

Sequencing

Skill Builder

Sequence the movement of an egg through the female reproductive system. If you need help, refer to Sequencing in the **Skill Handbook** on page 680.

PROGRAM RESOURCES

From the **Teacher Resource Package** use:

Transparency Masters, pages 87-88, Male and Female Reproductive Systems; pages 89-90, The Menstrual Cycle.

Use **Color Transparency** 44, Male and Female Reproductive Systems; 45, The Menstrual Cycle.

RETEACH

Use diagrams to illustrate the pathway of the egg as it moves from the ovary through the Fallopian tube into the uterus and becomes embedded in the uterus lining.

EXTENSION

For students who have mastered this section, use the **Reinforcement** and **Enrichment** masters or other OPTIONS provided.

3 CLOSE

▶ Ask questions 1-4 and the **Apply** Question in the Section Review.

SECTION REVIEW ANSWERS

1. Its function is to reproduce the organism through eggs and sperm that carry genetic information from one generation to the next.

2. Sperm are produced in the testes, travel through tubes that encircle the bladder, are mixed with seminal fluid from the seminal vesicle, and leave the body via the urethra.

3. The ovaries produce the eggs; oviducts carry the egg to the uterus, where it will develop if fertilized; the vagina connects the uterus to the outside of the body and serves as the birth canal.

4. If the egg is not fertilized and does not implant in the uterus, the uterine lining is shed.

5. Apply: It does not build up because the hormones necessary for its buildup are not present.

6. Connect to Chemistry: If large amounts of blood are lost during the menstrual period, the body may have temporary iron-deficiency because iron is a component of hemoglobin lost in menstrual blood.

Skill Builder

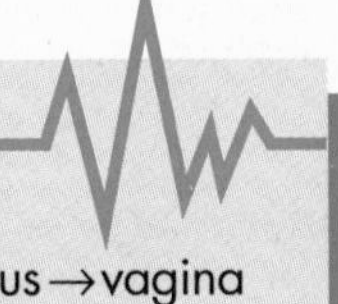

ovary→oviduct→uterus→vagina

Skillbuilder
ASSESSMENT

Performance: Assess students' abilities to Sequence by having them order the thicknesses of the uterus during a menstrual cycle using Figure 23-4.

ACTIVITY 23-1

40 minutes

OBJECTIVE: Examine and **interpret** diagrams of the menstrual cycle.

PROCESS SKILLS applied in this activity:
▶ **Observing** in Procedure Steps 2 and 3.
▶ **Interpreting Data** in Analyze Questions 1-4 and Conclude and Apply Questions 5-12.

COOPERATIVE LEARNING Use the Paired Partners strategy. Have students discuss the possible answers to the questions before recording their answers.

TEACHING THE ACTIVITY

Troubleshooting: Stress the importance of examining labels and captions when interpreting diagrams.

PROGRAM RESOURCES

From the **Teacher Resource Package** use:

Activity Worksheets, pages 203-204, Activity 23-1, Interpreting Diagrams.

Activity
ASSESSMENT

Performance: To further assess students' ability to interpret diagrams, see USING LAB SKILLS, Question 12, on p. 548.

ACTIVITY 23-1 Interpreting Diagrams

***Problem:** What happens to the uterus during a female's monthly cycle?*

Materials

- paper and pencil

Procedure

1. The diagrams below show what is explained in the previous section on the menstrual cycle.
2. Study the diagrams and their labels.
3. Use the information in Section 23-1 and the diagrams below to complete a table like the one shown.

Data and Observations Sample Data

Days	Condition of uterus	What happens
1 – 6	breakdown of lining	menstruation
7 – 12	lining begins to thicken	egg matures in the ovary
13 – 14	lining is thicker	ovulation
15 – 28	lining thickens	egg moves to uterus

Analyze

1. What do the diagrams represent?
2. What do the pink- or red-colored parts of the diagrams represent?
3. How does the figure caption relate to the diagram in Figure 23-4?
4. At what stage in the menstrual cycle is the uterine lining the most thickened?

Conclude and Apply

5. How long is the average menstrual cycle?
6. How many days does menstruation usually last?
7. How are the diagrams different?
8. On approximately what day in a 28-day cycle is the egg released from the ovary?
9. On what days does the lining of the uterus build up?
10. Why is this process called a cycle?
11. How many days before menstruation does ovulation usually occur?
12. *Interpret* the diagram to explain the menstrual cycle.

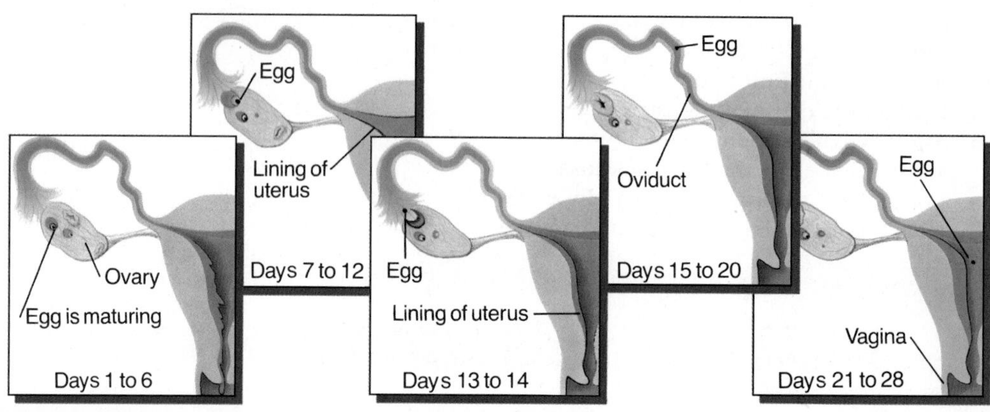

ANSWERS TO QUESTIONS

1. the menstrual cycle in human females
2. the lining of the uterus wall and the time of menstruation
3. It explains what is shown in the diagram.
4. between days 20 and 28
5. 28 days
6. four to six days
7. Each shows the condition of the uterus at a different time during the menstrual cycle.
8. day 14
9. days 7 through 28
10. If fertilization does not occur, menstruation takes place, another egg is formed, and the process is repeated.
11. about 14
12. You can see what is happening to the lining of the uterus each day of the cycle. The labels show the sequence of things taking place.

Fertilization to Birth

23-2

Objectives

- Describe how an egg becomes fertilized.
- Identify the major events in the stages of development of an embryo and fetus.
- Differentiate between fraternal and identical twins.

New Science Words

pregnancy
embryo
amniotic sac
fetus

Fertilization

Before the invention of powerful microscopes, some people imagined a sperm to be a miniature person that grew in the uterus of a female. Others thought the egg contained a miniature individual that started to grow when stimulated by semen. In the latter part of the 1700s an Italian naturalist, Lazzaro Spallanzani, showed with experiments using amphibians that contact between an egg and sperm is necessary for life to begin development. With the formulation of the cell theory in 1839, scientists recognized that a human develops from a single egg that has been fertilized by a sperm. The uniting of a sperm with an egg is known as fertilization.

Figure 23-5. Only one sperm will fertilize an egg.

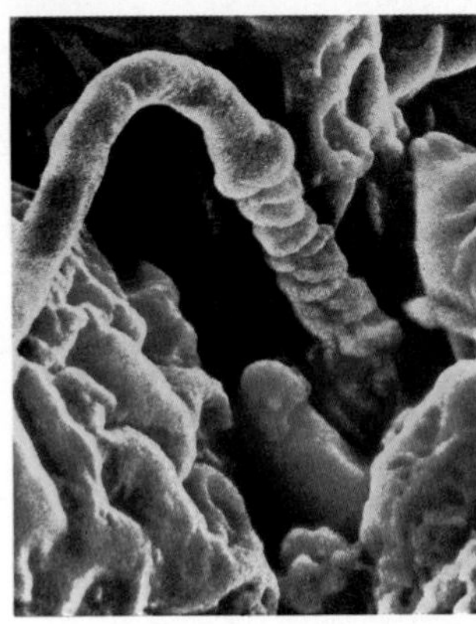

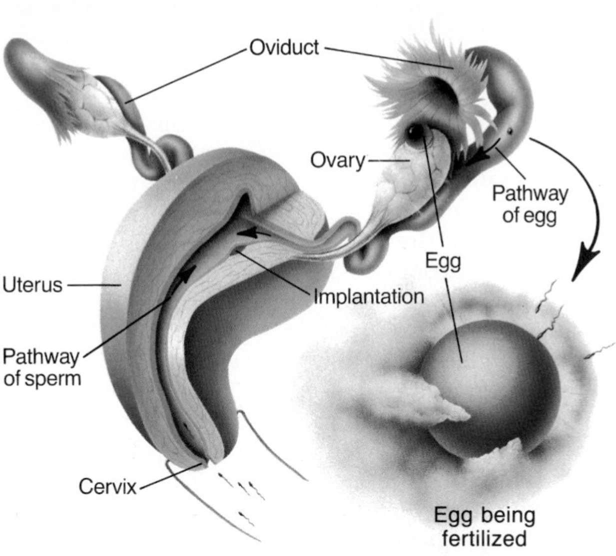

Figure 23-6. An egg is usually fertilized in an oviduct and is implanted in the uterus. The egg being fertilized has been enlarged for purposes of illustration.

SECTION 23-2

PREPARATION

SECTION BACKGROUND

- At the tip of a head of a sperm is a small sac filled with the enzyme hyaluronidase. This enzyme allows the sperm to penetrate the cell membrane of the egg.
- When fully developed, the placenta is about 2 cm in thickness and has a diameter of about 20 cm.
- The enlargement of the fetus and uterus eventually causes them to outgrow the space in the pelvis area. They then grow upward and displace such organs as the intestines and stomach.

1 MOTIVATE

- Discuss with students the remarkable process that transforms a single fertilized egg into an organism having billions of differentiated cells.
- Discuss with students the health research that indicates the importance of good nutrition during pregnancy and the necessity of refraining from drugs, alcohol, and tobacco to help ensure a healthy baby.

TYING TO PREVIOUS KNOWLEDGE: Have students recall from Chapter 19 the general need for good nutrition for the body and relate it also to the needs of the developing fetus. Draw attention to the government warnings on alcoholic beverages that indicate the dangers of alcohol consumption during pregnancy.

OPTIONS

PROGRAM RESOURCES

From the **Teacher Resource Package** use:

Critical Thinking/Problem Solving, page 27, Identical Twins.

Activity Worksheets, page 209, Mini-Lab: How long is an embryo?

Transparency Masters, pages 91-92, An Embryo in the Uterus.

Use **Color Transparency** 46, An Embryo in the Uterus.

Science Integration Activities 23, Chemical Reactions

Science Integration Activities is found in **Merrill Life Science Teacher Resource Manual.**

2 TEACH

Key Concepts are highlighted.

CONCEPT DEVELOPMENT

▶ **Demonstration:** Use Styrofoam balls to construct four- and eight-cell models to illustrate the earliest stages of growth of a zygote. Have students infer how these early stages develop into a hollow ball of cells.

▶ **What is the difference between an embryo and a fetus?** *During the first two months of pregnancy the unborn child is called an embryo; thereafter, it's called a fetus.*

▶ Have students describe how any identical or fraternal twins they know are alike.

PROBLEM SOLVING

Doctors are concerned about the duration of gestation because the risk of infection and possibility of cesarean section increases when labor is induced for an overterm baby.

Think Critically: The researchers should continue their research and measure the period of gestation of a larger number of women.

CROSS CURRICULUM

▶ **History:** Have students look up early misconceptions of the role of sperm and eggs in the process of fertilization and development.

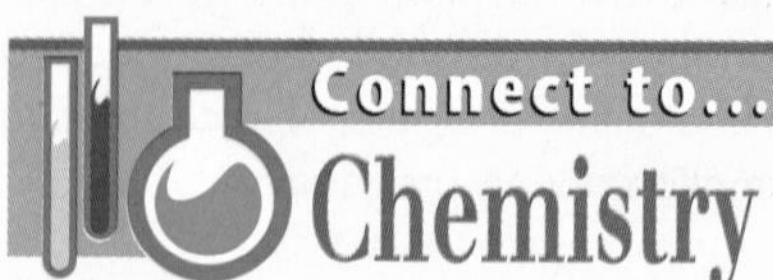

Connect to... Chemistry

Answer: Answers will vary and may include that an egg fertilized by two sperms would have 69 chromosomes, producing a zygote with an abnormal number of chromosomes. It would not develop properly.

PROBLEM SOLVING

When Is the Baby Due?

One of the first questions expectant mothers ask is "When is the baby due?" The due date is important to the family of the baby. What does a family do to prepare for a baby? The date also is important for the obstetrician, a doctor who specializes in childbirth. This doctor keeps track of the growth of the fetus. The obstetrician is concerned about the health of the baby and the mother.

Doctors generally estimate the period of human gestation at nine and one-half menstrual cycles, or about 266 days from conception. This date is estimated from the beginning of the mother's last menstrual period. Conception is thought to have occurred two weeks later.

Researchers have recently measured the periods of gestation among a small group of women with normal pregnancies. They found that the median gestation period was longer than 266 days and think that the normal period of gestation might be longer. Why are doctors interested in the duration of gestation?

Think Critically: Identify the part of scientific investigation researchers need to do to determine a normal period of gestation?

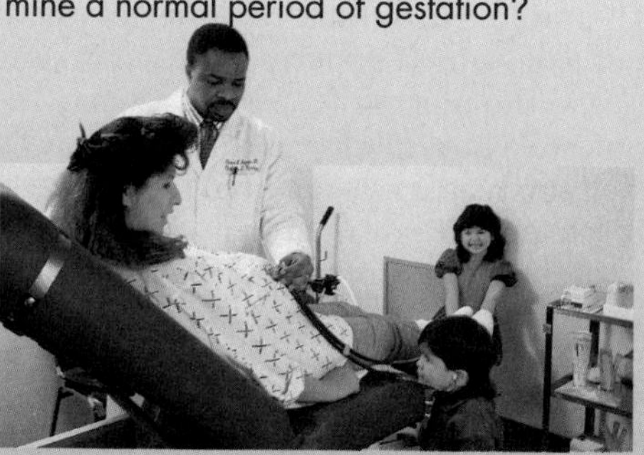

Connect to... Chemistry

Enzymes released from the head of a sperm break down the cell membrane of the egg and allow the sperm to enter the egg. *Relate* this to chromosomes and mitosis in the fertilized egg. *Infer* why no more than one sperm fertilizes an egg.

Sperm deposited into the vagina move through the uterus into the oviducts. Whereas only one egg is usually present, nearly 200 to 300 million sperm are deposited.
1 Only one sperm will fertilize the egg. Once the sperm penetrates the outer surface of the egg, the egg sets up a chemical barrier to keep out the other millions of sperm. The nucleus of the sperm and the nucleus of the egg have 23 chromosomes each. When the egg and sperm unite, a zygote with 46 chromosomes is formed. Most fertilization occurs in an oviduct.

The zygote moves along the oviduct to the uterus. During this time, the zygote is dividing. After about seven days, the ball of cells is implanted in the wall of the uterus. The uterine wall has been thickening in preparation to receive a fertilized egg. Here the egg will develop for nine months until the birth of the baby. This period of time is known as **pregnancy.**

OPTIONS

Meeting Different Ability Levels

For Section 23-2, use the following **Teacher Resource Masters** depending upon individual students' needs.

◆ **Study Guide Master** for all students.

● **Reinforcement Master** for students of average and above average ability levels.

▲ **Enrichment Master** for above average students.

Additional Teacher Resource Package masters are listed in the OPTIONS box throughout the section. The additional masters are appropriate for all students.

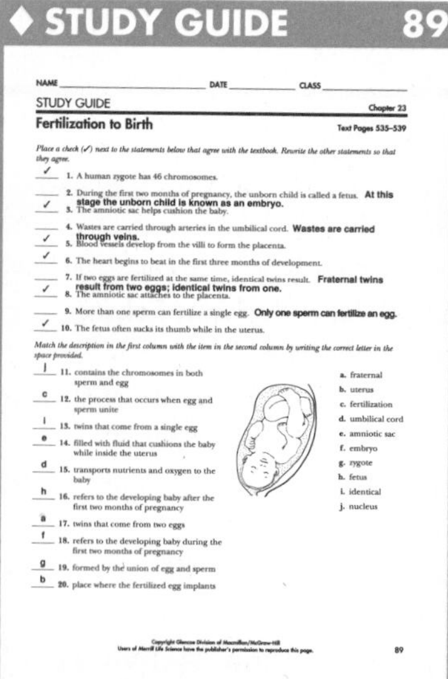

◆ STUDY GUIDE 89

NAME ________ DATE ________ CLASS ________

STUDY GUIDE — Chapter 23

Fertilization to Birth — Text Pages 535–539

Place a check (✓) next to the statements below that agree with the textbook. Rewrite the other statements so that they agree.

✓ 1. A human zygote has 46 chromosomes.

___ 2. During the first two months of pregnancy, the unborn child is called a fetus. **At this stage the unborn child is known as an embryo.**

✓ 3. The amniotic sac helps cushion the baby.

___ 4. Wastes are carried through arteries in the umbilical cord. **Wastes are carried through veins.**

✓ 5. Blood vessels develop from the villi to form the placenta.

✓ 6. The heart begins to beat in the first three months of development.

___ 7. If two eggs are fertilized at the same time, identical twins result. **Fraternal twins result from two eggs; identical twins from one.**

✓ 8. The amniotic sac attaches to the placenta.

___ 9. More than one sperm can fertilize a single egg. **Only one sperm can fertilize an egg.**

✓ 10. The fetus often sucks its thumb while in the uterus.

Match the description in the first column with the item in the second column by writing the correct letter in the space provided.

j 11. contains the chromosomes in both sperm and egg
c 12. the process that occurs when egg and sperm unite
i 13. twins that come from a single egg
e 14. filled with fluid that cushions the baby while inside the uterus
d 15. transports nutrients and oxygen to the baby
h 16. refers to the developing baby after the first two months of pregnancy
a 17. twins that come from two eggs
f 18. refers to the developing baby during the first two months of pregnancy
g 19. formed by the union of egg and sperm
b 20. place where the fertilized egg implants

a. fraternal
b. uterus
c. fertilization
d. umbilical cord
e. amniotic sac
f. embryo
g. zygote
h. fetus
i. identical
j. nucleus

 89

The Stages of the Embryo and Fetus

The outer cells of a fertilized egg attach to the wall of the uterus. During the first two months of pregnancy, the unborn child is known as an **embryo.** Nutrients from the wall of the uterus are received by the embryo through villi. Blood vessels develop from the villi and form the
2 placenta (pluh SENT uh). The umbilical (um BIHL ih kul) cord attaches at the embryo's navel with the placenta. The umbilical cord transports nutrients and oxygen from the mother to the baby through arteries. Carbon dioxide and other wastes are carried through veins in the umbilical cord back to the mother's blood. Other substances in the mother's blood can pass to the embryo. These include drugs, toxins, and disease organisms. These substances can cause great harm to the embryo. For this reason, a mother should avoid harmful drugs, alcohol, and tobacco during pregnancy.

During the third week of pregnancy, a thin membrane begins to form around the embryo. This is called the amnion or the amniotic sac and is filled with a clear liquid called amniotic fluid. The amniotic fluid in the **amniotic sac** helps cushion the embryo against blows and can store nutrients and wastes. This sac attaches to the placenta.

During the first three months of development, all of
an embryo's major organs form. A heart structure begins 3
to beat and move blood through the embryo's blood ves-

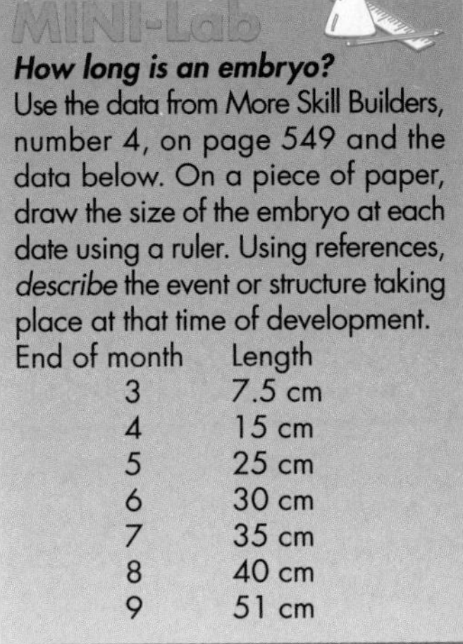

MINI-Lab

How long is an embryo?

Use the data from More Skill Builders, number 4, on page 549 and the data below. On a piece of paper, draw the size of the embryo at each date using a ruler. Using references, *describe* the event or structure taking place at that time of development.

End of month	Length
3	7.5 cm
4	15 cm
5	25 cm
6	30 cm
7	35 cm
8	40 cm
9	51 cm

How is the embryo cushioned against blows?

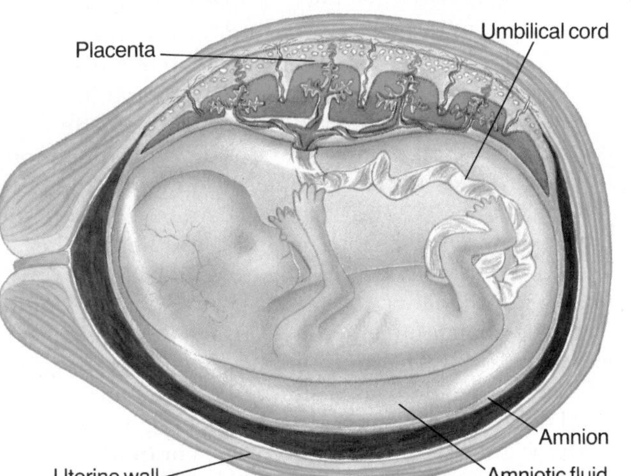

Figure 23-7. The developing embryo is surrounded by protective membranes.

CHECK FOR UNDERSTANDING

▶ Have students describe how substances are exchanged between the embryo or fetus and the mother. Determine if they understand that there is also no direct exchange of materials. Recall that there is also diffusion of substances other than oxygen, carbon dioxide, and nutrients.

RETEACH

Demonstration: Conduct a diffusion experiment to illustrate the process and reinforce the concept of the diffusion through the placenta.

EXTENSION

For students who have mastered this section, use the **Reinforcement** and **Enrichment** masters or other OPTIONS provided.

MINI-Lab

Students should see that the greatest advances in growth, in comparison to the previous month's size, occurs in the early months and that growth is steadier in later months.

MINI-Lab
ASSESSMENT

Performance: To further assess students' understanding of embryo size, see USING LAB SKILLS, question 11, on p. 548.

● REINFORCEMENT 89

NAME ____________ DATE ________ CLASS ________

REINFORCEMENT — Chapter 23

Fertilization to Birth — Text Pages 535–539

Answer the following questions using information from the textbook.

1. Why can only one sperm fertilize an egg? **The egg sets up a chemical barrier once fertilization has occurred.**
2. a. How many chromosomes are in the sperm cell nucleus? **23**
   b. How many chromosomes are in the zygote? **46**
3. Where does fertilization usually occur? **It usually occurs in an oviduct.**
4. What happens to the zygote after about seven days? **The zygote divides, becoming a ball of cells. This ball of cells is implanted in the wall of the uterus.**
5. Describe how the embryo receives oxygen and nutrients and how it gets rid of wastes. **Oxygen and nutrients are carried through the arteries of the umbilical cord. Wastes are removed through the umbilical veins.**
6. While developing in the uterus, at what point does an unborn baby have recognizable facial features? **at five weeks**
7. At the time body organs are present, what is the developing baby called? **a fetus**
8. a. Explain how fraternal twins develop. **Two eggs are released and fertilized by two sperm.**
   b. Why do identical twins look exactly alike? **They develop from a single egg, which splits after fertilization, therefore they have the same genes.**

Place the following events on the time line below. Write the letter of each event on the appropriate space above the time line.

a. recognizable eyes, nose, mouth form
b. mother feels movements of the fetus
c. fertilization
d. fatty tissue develops
e. fingers and toes develop
f. fetus shifts to head-down position

c a e b d f
0 1 2 3 4 5 6 7 8 9
(months)

 89

▲ ENRICHMENT 89

NAME ____________ DATE ________ CLASS ________

ENRICHMENT — Chapter 23

Fertilization to Birth — Text Pages 535–539

BIRTH "TRAUMA"

After nine months in the uterus, a baby is suddenly exposed to a very different environment. For the baby, we can only imagine that this comes as a complete shock.

Think about the changes a baby faces, then fill in the table below. You may need to use references.

Conditions	In the Uterus	After Birth
temperature	body temperature	about 14°C colder than in uterus
light	no light	bright, blinding lights
sound	sounds of mother's heartbeat, digestive sounds, muffled voices	loud, unfamiliar noises
oxygen	oxygen intake through umbilical arteries	first breath requires five times more effort than normal breath as air sacs are expanded; lungs must be cleared of mucus
food	nourishment through fetal circulation	breastmilk or formula; first sensation of hunger
wastes	wastes removed through umbilical vein	gas pains during digestion; dirty diapers; spitting up undigested milk
protection	cushioned within amniotic sac	left alone in crib for intervals
touch	contact with uterine walls through amniotic sac so that kicking results in feeling resistance	not in contact with living warmth except when being held; in contact with hard surfaces; weighted down instead of floating free

Use the library to find additional information on birth, then answer the following questions.

1. What sort of trauma would the baby experience during the process of birth? **The baby is being pushed and squeezed through a narrow tunnel with tremendous force.**
2. What procedures are performed on the baby in the delivery room or within hours after birth? **Answers will vary, but may include the cord is cut, the nose and throat are often suctioned to remove mucus; the baby is washed, weighed, and measured; a solution of silver nitrate is put in its eyes; vitamin K is injected to promote clotting; the baby is foot-printed, wiped, changed, handled, and may have blood tests. If the baby is a boy, it may be circumcised.**

 89

TECHNOLOGY

For more information see: "The Tiniest Patients" by Jean Seligmann, *Newsweek,* June 11, 1990, pp. 56-57.

Think Critically: This could be used to treat blockages of organs and to remove tumors.

MINI QUIZ

Use the Mini Quiz to check students' recall of chapter content.

1. **Of the millions of sperms deposited, how many will fertilize the egg?** *one*
2. **The _____ attaches the embryo to the placenta.** *umbilical cord*
3. **When does the embryo heart begin to beat?** *early in the first three months of development*
4. **The fetus usually has shifted to a head-down position by the _____ month.** *ninth*
5. **Why are some twins identical?** *Each egg has the same set of genes.*

FLEX Your Brain

Use the Flex Your Brain activity to have students explore MULTIPLE BIRTHS.

Flex Your Brain

ASSESSMENT

Portfolio: Use the Flex Your Brain activity to reinforce critical thinking and problem solving skills.

TECHNOLOGY

Operating in the Womb

Doctors performing fetal surgery may have patients that are only 24 weeks beyond conception and only inches in size.

Using ultrasound, doctors can see an image of the developing fetus in the womb and can detect fetal defects that frequently mean death soon after birth. One of these involves a hole in the diaphragm that allows the intestines, stomach, spleen, and liver to move into the chest cavity. When this happens, the growth of the fetus' lungs is stunted, and the baby is unable to breathe when born. The only hope has been to operate on the baby immediately following birth, but this is successful only 20 percent of the time. Now, doctors are able to operate on such a fetus while it is still in the uterus. To do this, a small incision is made in the mother's uterus. The fetus's chest is opened and abdominal organs are returned to their proper place. Finally, a patch is placed over the hole in the diaphragm. The baby now has a good chance to survive after birth.

Think Critically: What other prenatal conditions might be corrected using this method?

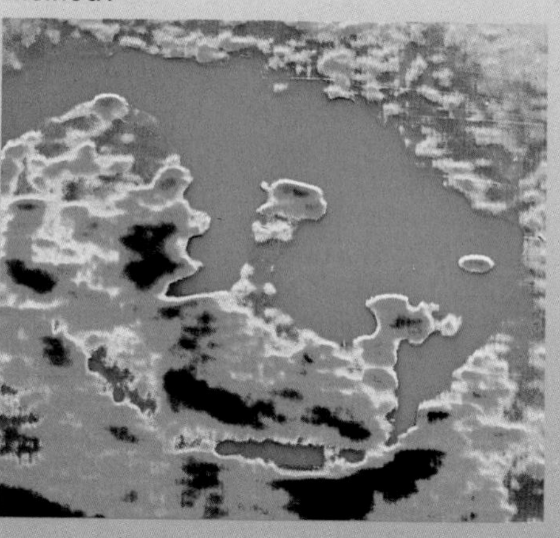

sels. At five weeks, the embryo is only as large as a single grain of rice, but there is a head with recognizable eye, nose, and mouth features. During the sixth and seventh weeks, tiny arms and legs develop fingers and toes.

After the first two months of pregnancy, the developing baby is called a **fetus** (FEE tus). At this time, body organs are present. Around the third month, the fetus is 8-10 cm long, the heart can be heard beating, and the mother may feel the baby's movements within her uterus. The fetus sucks its thumb. By the end of the seventh month of pregnancy, the fetus is 30 to 38 cm in length. Fatty tissue builds up under the skin, and the fetus appears
4 less wrinkled. By the ninth month, the fetus usually has shifted to a head-down position within the uterus. The head usually is in contact with the opening of the uterus to the vagina. The fetus is about 50 cm in length and weighs from 2.5 to 3.5 kg. The baby is ready for birth.

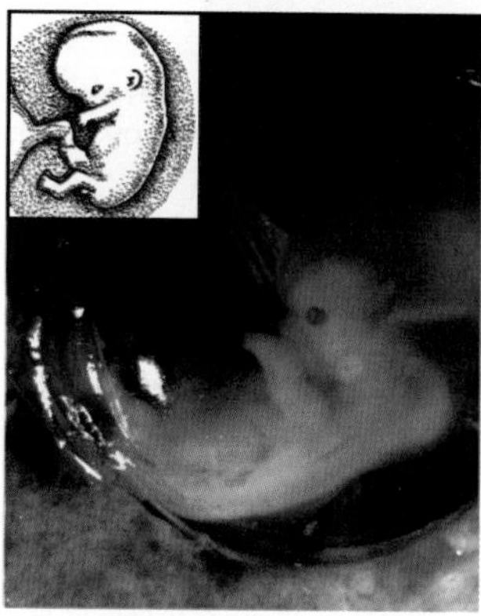

Figure 23-8. A two-month embryo shown enlarged in the photo and actual size in the drawing.

OPTIONS

ENRICHMENT

- Have students research how suspending an object in water helps cushion the object from blows and sudden movements.
- Have students find out how ultrasound is used to diagnose certain disorders while the fetus is still in the uterus.
- Research methods that can be used to evaluate the fetus before birth (for example, amniocentesis, chorionic villi sampling, and so on).
- Have a gynecologist, pediatrician, or pediatric nurse visit with the class to explain the physical and psychological problems of teenage pregnancies.

Multiple Births

In some cases, two eggs leave the ovary at the same time. If both eggs are fertilized and both develop, twins are born. When two different eggs were fertilized by two different sperm, the two babies are called fraternal twins. Fraternal twins may be two girls, two boys, or a boy and a girl. Fraternal twins do not always look alike. In other cases of twins, a single egg may split apart shortly after it is fertilized. Each part of the egg then develops and forms an embryo. Because both children developed from the same egg and sperm, they are identical twins. Each has the
5 same set of genes. Identical twins must be either two girls or two boys. These twins look exactly alike.

You may know some twins. In what ways are they alike or different? Are they fraternal or identical twins? Triplets and other multiple births may occur when either three or more eggs are produced at one time or an egg splits into three or more parts.

A remarkable series of events changes a single fertilized egg into a baby with billions of cells. The mother's body prepares for the baby. Special tissues nourish and protect the developing embryo and fetus. After a nine-month period, the fetus is ready to live outside the mother's body.

Figure 23-9. Pick out the fraternal and identical twins.

SECTION REVIEW

1. What happens when an egg is fertilized?
2. What is one major event that occurs during the embryo and fetal stages?
3. Explain how twins are fraternal or identical.
4. **Apply:** Why can't identical twins be a girl and boy?
5. **Connect to Chemistry:** How are nutrients, oxygen, and wastes exchanged between the fetus and the mother?

Making and Using Graphs — Skill Builder

Make a graph of the embryo and fetus sizes and the months of development using this data: 1 month = 0.5 cm, 2 months = 3 cm, 3 months = 7.5 cm, 4 months = 15 cm, 5 months = 25 cm, 6 months = 30 cm, 7 months = 35 cm, 8 months = 40 cm, 9 months = 50 cm. If you need help, refer to Making and Using Graphs in the **Skill Handbook** on page 691.

3 CLOSE

▶ Ask questions 1-3 and the **Apply** Question in the Section Review.

SECTION REVIEW ANSWERS

1. Sperm and egg combine, producing a zygote with 46 chromosomes.
2. During the embryo stage, the amniotic sac and placenta develop and all the embryo's major organs form. During the fetal stage, the fetus continues to grow and develop.
3. Identical twins result when one fertilized egg splits and develops into two individuals. Fraternal twins are the result of two fertilized eggs.
4. Apply: In order to get a boy and a girl, two eggs would be needed. The genetic material would need to be different.
5. Connect to Chemistry: Since there is no direct connection between the fetus and the placenta of the mother, the exhange is due to diffusion between the capillaries of the umbilical cord and the placenta.

Skill Builder

Students should make line or bar graphs. Time should be plotted on the y-axis and size on the x-axis.

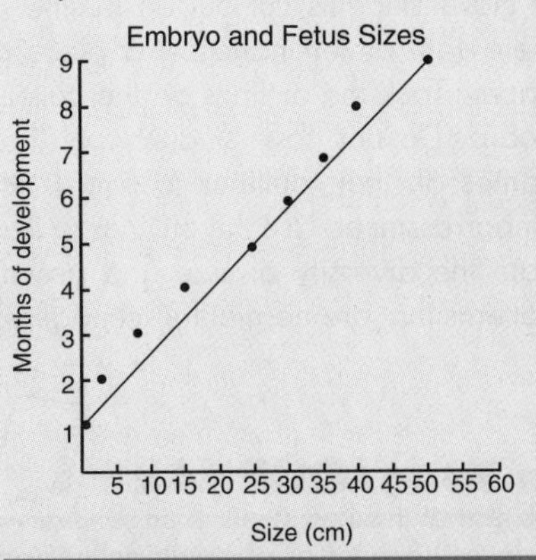

ASSESSMENT—ORAL

▶ **Why might fraternal twins not look like each other even if they are the same sex?** *The fraternal twins come from two different eggs, each having its own set of genes.*

▶ **Infer how fraternal quadruplets are produced.** *Four eggs are produced at one time and then are all fertilized at the same time.*

PROGRAM RESOURCES

From the **Teacher Resource Package** use:

Activity Worksheets, page 5, Flex Your Brain.

Use **Laboratory Manual,** page 159, Fetal Development.

Skillbuilder ASSESSMENT

Performance: Assess students' abilities to Make and Use Graphs by having them make a statement about how the growth rate of a fetus compares to the growth rate of an embryo.

SECTION 23-3

PREPARATION

SECTION BACKGROUND

▶ The pressure and stretching of the uterine wall and an increase in progesterone cause contractions of the uterus, stretching of the uterine walls.

▶ One of the most dramatic events for the newborn is the first breath. Air must be forced into the lungs because they are collapsed and the moist membranes are held together by surface tension.

▶ During the adolescent growth sprint, the hands and feet of teenagers may grow faster than the rest of the body. Eventually the arms, legs, and chest also grow, bringing the body into proportion.

PREPLANNING

▶ To prepare for Activity 23-2, obtain graph paper and red and blue pencils.

1 MOTIVATE

▶ Discuss how the newborn infant must adjust to the new environment into which it has been born. Have students compare the conditions of the two environments.

▶ Have students cut out an outline of their right or left hand on a piece of paper. Tack the outlines on the bulletin board. Do not have students put their names on their outlines to avoid any embarrassment. Use the outlines to illustrate the diversity of size and growth patterns that are normal for teenagers.

In Your JOURNAL

A local hospital or obstetrician or book on childbirth preparation can provide information.

VideoDisc

STVS: Obesity and Heredity, Disc 7, Side 2

23-3 Development after Birth

New Science Words

infancy
childhood
adolescence
adulthood

Objectives

▶ State the sequence of events of childbirth.
▶ Compare the stages of infancy and childhood.
▶ Relate adolescence to preparation for adulthood.

In Your JOURNAL

The Lamaze method of childbirth has become a popular technique. **In your Journal,** define what is it and tell why is it so popular?

Childbirth

After nine months of developing within the mother, the baby is ready to be born. Like all newborn living things, the baby will find itself suddenly pushed out into the world. Within the uterus and amniotic sac the baby was in a warm, watery, dark, and protected environment. In contrast, the new environment is cooler, drier, brighter, and not as well protected.

The process of childbirth begins with labor, the muscular contractions of the uterus. As the contractions increase in strength and frequency, the amniotic sac usually breaks and releases its fluid. This is known as "breaking water." Usually, over a period of hours, the contractions cause the opening of the uterus to widen to allow the baby to pass through. More powerful and frequent contractions push the baby out through the vagina into its new environment. Sometimes, the mother's pelvis is too small for the baby to fit through or the baby is in the wrong position for birth. In cases like this, the baby is delivered through an incision in the mother's uterus and abdomen. This surgery is called cesarean section.

What is cesarean section?

At birth, the baby is still attached to the umbilical cord and placenta. The person assisting with the birth of the baby ties the cord and then cuts it. The scar that later forms where the cord was attached is the navel. Soon after the baby's delivery, contractions expel the placenta from the mother's body. The baby takes its first breath of oxygen without the umbilical cord. The baby may cry. Crying forces air into its lungs.

OPTIONS

Meeting Different Ability Levels

For Section 23-3, use the following **Teacher Resource Masters** depending upon individual students' needs.

◆ **Study Guide Master** for all students.
● **Reinforcement Master** for students of average and above average ability levels.
▲ **Enrichment Master** for above average students.

Additional Teacher Resource Package masters are listed in the OPTIONS box throughout the section. The additional masters are appropriate for all students.

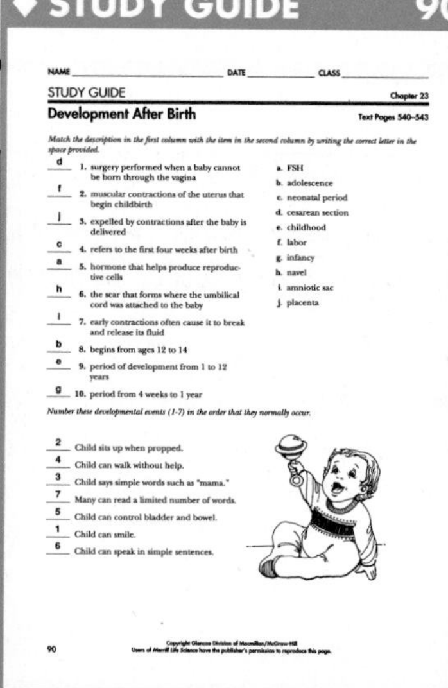

◆ STUDY GUIDE 90

NAME ______ DATE ______ CLASS ______

STUDY GUIDE Chapter 23

Development After Birth Text Pages 540-543

Match the description in the first column with the item in the second column by writing the correct letter in the space provided.

d 1. surgery performed when a baby cannot be born through the vagina
f 2. muscular contractions of the uterus that begin childbirth
j 3. expelled by contractions after the baby is delivered
c 4. refers to the first four weeks after birth
a 5. hormone that helps produce reproductive cells
h 6. the scar that forms where the umbilical cord was attached to the baby
i 7. early contractions often cause it to break and release its fluid
b 8. begins from ages 12 to 14
e 9. period of development from 1 to 12 years
g 10. period from 4 weeks to 1 year

a. FSH
b. adolescence
c. neonatal period
d. cesarean section
e. childhood
f. labor
g. infancy
h. navel
i. amniotic sac
j. placenta

Number these developmental events (1-7) in the order that they normally occur.

2 Child sits up when propped.
4 Child can walk without help.
3 Child says simple words such as "mama."
7 Many can read a limited number of words.
5 Child can control bladder and bowel.
1 Child can smile.
6 Child can speak in simple sentences.

90

Infancy and Childhood

The first four weeks after birth are known as the neonatal (ne o NA tal) period. *Neonatal* means "new born." During this time, the baby adjusts to life outside of the uterus. Body functions such as respiration, digestion, and excretion are now performed by the baby rather than through the placenta. Unlike some other living things, the human baby must depend on others to survive. A newborn colt begins walking a few hours after its birth. The human baby is fed and has its diaper changed. It is not able to take care of itself.

The next stage of development is **infancy,** the period from neonatal to one year. It is a period of rapid growth and development for both mental and physical skills. An early skill is the ability to smile that usually begins around six weeks of age. At four months, most babies can laugh, sit up when propped, and recognize their mothers' faces. At eight months, the infant is usually able to say a few simple words, such as "mama" and "kitty." One of the major events within the first year is the ability to stand unsupported for a few seconds.

After infancy is the **childhood** stage, which lasts until age 12. The physical growth rate for height and weight is not as rapid as in infancy. However, there is development of muscular coordination and mental abilities. By 18 months, the child is able to walk without help. Between two and three years, the child controls his or her bladder and bowel. At age three, the child can speak in simple sentences. By age five, many children can read a limited number of words. Throughout this stage, children develop their ability to speak, read, write, and reason. At the same time, children also mature emotionally and learn how to get along with other people. Find out how old you were when you began to talk. What were your first words?

Science and MATH

Cotton diapers can be reused up to 100 times and decompose in one to six months! Disposable diapers take up to 500 years to decompose. **In your Journal,** *calculate* the cost of disposable diapers for a year if an average of eight are used per day.

MINI-Lab

What is the immunization schedule for babies and young children?

Find out when and what vaccines are given to protect babies and young children. *Compare* your own immunization record. Record your own record **in your Journal.** What boosters were you given? What immunizations does your school require?

a

b

Figure 23-10. Compare the different abilities of an infant (a) and a six-year old (b).

● REINFORCEMENT 90

NAME ______ DATE ______ CLASS ______

REINFORCEMENT Chapter 23

Development After Birth Text Pages 540–543

Answer the following questions using information from the textbook.

1. What causes secondary sex characteristics to develop? **They are caused by the hormones FHS and LH, which are produced by the pituitary gland.**
2. a. When does the adolescent growth spurt begin and end for girls? **It begins between 11 and 13 and ends between ages 15 and 16.**
   b. For boys? **It begins between 13 and 15 and ends between ages 17 and 18.**
3. a. What physical changes might you expect during the period from age 30 into your 50s? **a decrease in physical strength, less efficient circulation and respiration, bones more brittle, and skin less elastic**
   b. What is this stage called? **middle adulthood**
4. What stage includes puberty? **adolescence**
5. How is a baby delivered if the mother's pelvis is too small for the baby to fit through? **through an incision in the abdomen and uterus called a cesarean section**
6. What is labor? **muscular contractions of the uterus**
7. Describe the events that lead up to childbirth. **Contractions usually cause the amniotic sac to break. Over several hours the baby moves out of the uterus into the vagina. More powerful contractions push the baby out.**
8. What are some secondary sex characteristics in males? **facial, pubic, and underarm hair; deeper voice; increased muscle size; sexual attraction to other sex**
9. What are some secondary sex characteristics in females? **breast development, pubic and underarm hair; fatty tissue in buttocks and thighs; sexual attraction to other sex**

90

▲ ENRICHMENT 90

NAME ______ DATE ______ CLASS ______

ENRICHMENT Chapter 23

Development After Birth Text Pages 540–543

SURVEY OF HEIGHT

This exercise will require you to collect some data. It may work best if you work in teams. Once the data is collected, you can work individually to plot the data.

Procedure

1. Find out the height of all the students in your class. Use a tape measure.
2. Organize your data separately for males and females.
3. Record your data in the table below.

Data and Observations

FEMALES	MALES
Height	Height

Conclude and Apply

1. In the space below, make a bar graph showing the heights of the people in your class. An example is given. Draw the data for males and females in different colors.

Number of Students; 0, 2, 4, 6, 8, 10; 120-130, 130-140, 140-150; Height (cm); Female; Male

Graphs will vary.

2. What can you tell about the height of people in your class from the bar graph? **A few people are very tall, a few are very short, and most are somewhere in the middle.**
3. Who are generally taller, males or females? **Answers will vary depending on the sample, but females are often taller at age 13.**

90

Science and MATH

Students should find the cost of disposable diapers. Calculate the cost per year by multiplying the cost per diaper by 8 diapers per day by 365 days in a year.

2 TEACH

Key Concepts are highlighted.

CONCEPT DEVELOPMENT

▶ If possible, obtain X rays of the bones of infants or young children and compare with X rays of adults to show the difference in cartilage and bone ratios.

MINI-Lab

▶ The school nurse can provide information concerning immunization schedules and those required for school admission. Parents will generally have their children's records.

Answer: Student immunization records should be similar to Table 24-3 on page 569. They should interview the school nurse to find out the immunizations your school requires.

MINI-Lab ASSESSMENT

Performance: To further assess students' understanding of immunization, have them check with grandparents or older doctors to find what immunizations were used in earlier years.

CROSS CURRICULUM

▶ **Health:** Have students find out about changes in the bones of older adults and the condition of kyphosis.

CHECK FOR UNDERSTANDING

▶ Ask questions 1-4 and the **Apply** Question in the Section Review.

RETEACH

Have students research old photographs of students. Compare the physical development of adolescents 100 years ago to that of today's adolescents. Infer what factors might have caused puberty to begin later than is common today.

EXTENSION

For students who have mastered this section, use the **Reinforcement** and **Enrichment** masters or other OPTIONS provided.

FLEX Your Brain

Use the Flex Your Brain activity to have students explore ADOLESCENCE.

Flex Your Brain

ASSESSMENT

Portfolio: Use the Flex Your Brain activity to reinforce critical thinking and problem solving skills.

Answer: Differential growth causes the head to grow less rapidly after birth and the length of the head is approximately one-eighth of the body length.

Adolescence

The next stage of development is adolescence; you are in this stage. **Adolescence** begins from ages 12 to 14. A part of adolescence is puberty. As you read earlier, puberty is the time of development when a person becomes physically able to reproduce. For girls, puberty occurs between ages 8 and 13. For boys, puberty occurs between ages 13 and 15.

Figure 23-11. Adolescence is a time for many changes.

During puberty, hormones are produced by the pituitary gland that cause changes in the body. The hormone FSH helps produce reproductive cells. LH helps with the production of other hormones. As a result, secondary sex characteristics result. In females, the breasts develop, pubic and underarm hair appears, and fatty tissue is added to the buttocks and thighs. In males, the hormones cause the growth of facial, pubic, and underarm hair; a deepened voice; and an increase in muscle size. Many begin to experience feelings of sexual attraction.

Adolescence also is a time of a growth spurt. Are you shorter or taller than your classmates? Because of differences in the time hormones begin functioning among individuals and between males and females, there are differences in boys' and girls' growth rates. Girls often begin their increase in growth between the ages of 11 and 13 and end at ages 15 to 16. Boys usually start their growth spurt at ages 13 to 15 and end at 17 to 18 years of age. Hormonal changes also cause underarm sweating and acne, requiring extra cleanliness and care. All of these physical changes can cause you to feel different or uncomfortable. This is normal. As you move through the period of adolescence, you will find that you will become more coordinated, be better able to handle problems, and gain improved reasoning abilities. You'll find that you enjoy spending more time with your friends.

At birth, the head of the newborn is about one-quarter of the body length. *Measure* classmates and **in your Journal,** *calculate* the proportion of the head length to body length for adolescents.

Adulthood

The final stage of development is that of **adulthood.** It begins with the end of adolescence and extends to old age. There are several stages of adulthood. The early years of adulthood occur when people are in their 20s. Many of these adults are completing an education, finding employment, and possibly marrying and beginning a family. The growth of the muscular and skeletal system stops.

OPTIONS

ASSESSMENT—ORAL

▶ **Compare the neonatal period with the infancy stage of development.** *The neonatal period is only one month in length; the child is adjusting to life outside the uterus. Infancy is about 11 months in length; the child is growing and developing rapidly and becoming more independent.*

▶ **Do all children learn to walk and talk at the same time during childhood?** *No; each individual is different and has his or her own rate of development.*

PROGRAM RESOURCES

From the **Teacher Resource Package** use:

Concept Mapping, pages 51-52.

Cross-Curricular Connections, page 27, Babies—a Natural for Art.

Activity Worksheets, page 5, Flex Your Brain.

Activity Worksheets, page 210, Mini-Lab: What is the immunization schedule for babies and young children?

People in their 30s to 50s make up the stage of middle adulthood. During these years, physical strength begins to decline. As a person ages, blood circulation and respiration are less efficient. Bones become more brittle, and the skin's elastic tissues are lost, causing the skin to become wrinkled. This group is busy with family and work commitments. People in this group often care for aging parents as well as children.

Think about someone you know over age 65. How is that person alike or different from you? The young-old age group includes people from 65 to 74 years old. Around this age, many people retire. To fill their time, they may take up hobbies, travel, or volunteer at hospitals, schools, and community organizations. Many older people are an active part of society. The old-old age group is made up of people 75 years and older. Many people in this group are frail and need assistance in meeting their needs.

After birth, the stages of development of the human body begin with a baby making adjustments to a new environment. From infancy to adolescence, the body's systems mature, enabling the person to be physically and mentally ready for adulthood. During the next 50 years or longer, people live their lives making contributions to family, community, and society. Throughout the life cycle, people who care for their health enjoy a higher quality of life.

Figure 23-12. Many people in the 65- to 74-year age group continue to work.

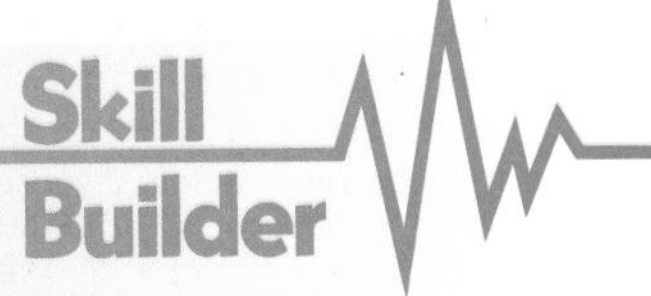

Interview someone who's over 60 years of age. Find out what his or her life was like when he or she was your age. What does the person like to do now? Write a report **in your Journal** to share with the class.

SECTION REVIEW

1. What are major events during childbirth?
2. Compare infancy with childhood.
3. How does the period of adolescence prepare you for the stage of adulthood?
4. Define the stages of adulthood.
5. **Apply:** Why is it hard to compare the growth and development of different teenagers?
6. **Connect to Physics:** Compare the growth of boys and girls during childhood, ages 18 months through 13 years.

Outlining — Skill Builder

Prepare an outline of the various life stages of human development from neonatal to adulthood. If you need help, refer to Outlining in the **Skill Handbook** on page 681.

Skill Builder

A. Neonatal
   1. first four weeks after birth
   2. baby adjusts to life outside the uterus
B. Infancy
   1. neonatal to one year
   2. period of rapid growth, development
C. Childhood
   1. ages 1 to 12
   2. development of muscular coordination and mental abilities
D. Adolescence
   1. ages 12 to 14
   2. time of development when a person becomes physically able to reproduce
E. Adulthood
   1. end of adolescence to old age
   2. several stages
      a. early adulthood—twenties
      b. middle adulthood—30s to 50s
      c. young-old—65 to 74 years
      d. old-old—75 years +

In Your JOURNAL

Students should interview someone with whom they are close, such as a grandparent, to get a good exchange of ideas and communication.

3 CLOSE

▶ Have students discuss the extension of the life span of humans due to improved health care. Specifically, discuss the control of disease as a bridge to concepts introduced in Chapter 24.

MULTICULTURAL AWARENESS

▶ Have students find out about ceremonies in other societies, past or present, that marked the transition of young people to adults.

SECTION REVIEW ANSWERS

1. Events include contractions, rupture of the amniotic sac, widening of the opening of the uterus to allow the baby to pass through that opening and the birth canal, and expulsion of the placenta.
2. Infancy is from age one month to one year and involves rapid growth and development of both mental and physical skills. Childhood is from age one year to age 12. There is development of muscular coordination and mental abilities.
3. Adolescence is a transition time between childhood and adulthood. It is a time of physical growth to adult size and sexual maturity.
4. Early adulthood is the 20s. Middle adulthood is the 30s to 50s. Young-old is from 65 to 74. Old-old is 75 and older.
5. Apply: Each person has his or her own rate and schedule of development. Some are early bloomers, others are later bloomers.
6. Connect to Physics: Student answers should reflect basic information about development as on pages 541 and 542.

Skillbuilder ASSESSMENT

Portfolio: Use this Skillbuilder to assess students' abilities to Outline the stages of human development.

SECTION 23-4

PREPARATION

SECTION BACKGROUND

▶ Have students take a survey of parents with young children. They should ask them questions such as: At what age did your child get its first tooth? walk? talk? The class can then compile and discuss their results.

1 MOTIVATE

▶ Ask a parent with a child, age 1 to 2, to bring the child to class and discuss the major changes that have occurred in his or her life since birth.

Answer: A diet lacking in calcium can cause osteoporosis since calcium is necessary for normal bone cell replacement.

PROGRAM RESOURCES

From the **Teacher Resource Package** use:

Science and Society, page 27, Living Longer.

23-4 Aging

New Science Words

catalase

Objectives

▶ Discuss concerns in society for the elderly.
▶ Explain the chemical process of aging.

Prolonging Life

Take a look in a mirror and try to picture yourself at 65 years of age. Most junior high school students probably don't give too much thought to aging. In fact, you are probably eager to be older than you are. You may view people older than yourself as having freedom, money, and many privileges. However, if you think about some elderly persons that you know, you may be reminded of some disadvantages of growing older. There are many chronic diseases associated with people who are in their 60s, 70s, and 80s. Conditions associated with aging include Alzheimer's disease, arthritis, osteoporosis, diabetes, cancer, and heart disease. An older person also may have less energy than when he or she was younger and have memory loss. Hearing and sight may decline.

You might be aware of the rising costs of health care. This especially affects the elderly. Many older persons cannot afford adequate health care. This problem contributes to a general decline in health maintenance. It is estimated that within 60 years, almost a fourth of the population will be age 65 and older. With a growing older population, adequate health care for older persons is a real concern in our country.

Congress has set aside millions of dollars for research and programs on aging. Many persons are concerned that living longer does not mean living healthfully. No one wants to live 20 years longer if its means enduring a painful, chronic disease. However, the goal of many researchers is to increase years of health, not just lengthen life. In the past several years, researchers and medical professionals have found that a person can do much to promote his or her health. This includes eating right, exercising regularly, monitoring stress, and not smoking.

Osteoporosis is a disease characterized by the deterioriation of bone. *Predict* why the diet of some people can result in this disease.

OPTIONS

Meeting Different Ability Levels

For Section 23-4, use the following **Teacher Resource Masters** depending upon individual students' needs.

◆ **Study Guide Master** for all students.

● **Reinforcement Master** for students of average and above average ability levels.

▲ **Enrichment Master** for above average students.

Additional Teacher Resource Package masters are listed in the OPTIONS box throughout the section. The additional masters are appropriate for all students.

◆ STUDY GUIDE 91

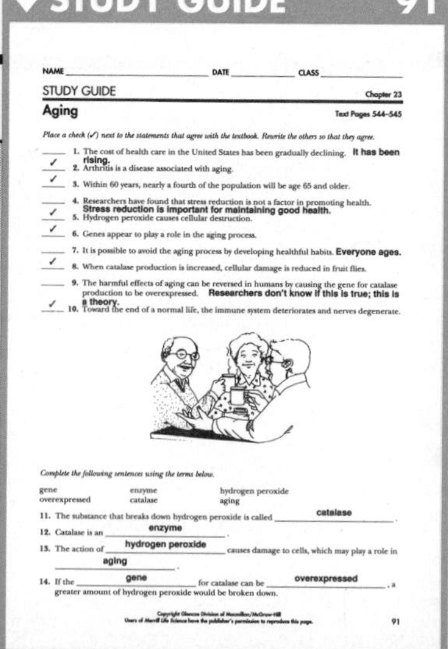

NAME ______ DATE ______ CLASS ______

STUDY GUIDE — Chapter 23

Aging — Text Pages 544–545

Place a check (✓) next to the statements that agree with the textbook. Rewrite the others so that they agree.

____ 1. The cost of health care in the United States has been gradually declining. **It has been rising.**
✓ 2. Arthritis is a disease associated with aging.
✓ 3. Within 60 years, nearly a fourth of the population will be age 65 and older.
____ 4. Researchers have found that stress reduction is not a factor in promoting health. **Stress reduction is important for maintaining good health.**
✓ 5. Hydrogen peroxide causes cellular destruction.
✓ 6. Genes appear to play a role in the aging process.
____ 7. It is possible to avoid the aging process by developing healthful habits. **Everyone ages.**
✓ 8. When catalase production is increased, cellular damage is reduced in fruit flies.
____ 9. The harmful effects of aging can be reversed in humans by causing the gene for catalase production to be overexpressed. **Researchers don't know if this is true; this is a theory.**
✓ 10. Toward the end of a normal life, the immune system deteriorates and nerves degenerate.

Complete the following sentences using the terms below.

gene	enzyme	hydrogen peroxide
overexpressed	catalase	aging

11. The substance that breaks down hydrogen peroxide is called **catalase**.
12. Catalase is an **enzyme**.
13. The action of **hydrogen peroxide** causes damage to cells, which may play a role in **aging**.
14. If the **gene** for catalase can be **overexpressed**, a greater amount of hydrogen peroxide would be broken down.

91

By choosing healthful habits, some people will have much healthier lives than others. However, everyone does age. The body breaks down as nerves degenerate, the immune system deteriorates, hormones change, and death approaches.

Researchers believe that certain genes are responsible for the aging process. It is believed that the chemical destruction of cells by hydrogen peroxide plays a large role in the aging process. Hydrogen peroxide is produced by many cellular reactions. Dr. Glenn Bewley at North Carolina State University found that he could isolate in fruit flies the gene that produces **catalase,** the enzyme that breaks down hydrogen peroxide. When he did this, less catalase was produced and the life span of the fruit flies was shortened. If the gene for catalase production can be overexpressed, meaning producing more catalase, a greater amount of hydrogen peroxide would be broken down. Fewer body cells would be damaged. If researchers developed a way to overexpress the gene for catalase, would it even be useful?

SECTION REVIEW

1. What can you do now to increase your years of health?
2. What role is catalase believed to play in the aging process?
3. **Connect to Chemistry:** Find out how wrinkles form in the skin of adults.

You Decide!

The life spans of many people are increasing due to the many new treatments for illness and disease doctors and scientists are discovering. In 1960, the average life expectancy for women was 73 years and 67 years for men. In 1988, it increased to 78 years for women and 72 years for men. Although people are living longer, many can't live on their own. They must depend on relatives and friends to help them with basic needs. Many also spend the rest of their lives dependent on medicine and medical technology. Do you think life should be extended when the person can no longer maintain the quality of life they once had?

● REINFORCEMENT 91

NAME ____ DATE ____ CLASS ____

REINFORCEMENT — Chapter 23

Aging — Text Pages 544–545

Answer the following questions using information from the textbook.

1. Why are so many teens in a hurry to grow up? **Adults seem to have freedom, money, and privileges.**
2. Why can't many older people afford good health care? **They are retired and have a fixed income.**
3. What are some of the factors researchers have found to promote health? **eating right, exercise, monitoring stress, and not smoking**
4. What does hydrogen peroxide do to cells? **It causes cell damage.**
5. What does this have to do with aging? **Cell damage may be an important part of the aging process as cells break down.**
6. Name some of the chronic diseases associated with aging. **Alzheimer's disease, arthritis, osteoporosis, diabetes, cancer, and heart disease**
7. Approximately what percent of the population will be age 65 and older within 60 years? **25%**
8. Why are researchers interested in the gene for catalase production? **Catalase breaks down hydrogen peroxide, so it may be able to reduce cell damage.**
9. Why are fruit flies a part of catalase research? **because the gene for catalase has been isolated in fruit flies**
10. What happens if the gene for catalase is overexpressed? **If the gene is overexpressed, more catalase will be produced and a greater amount of hydrogen peroxide will be broken down.**

91

▲ ENRICHMENT 91

NAME ____ DATE ____ CLASS ____

ENRICHMENT — Chapter 23

Aging — Text Pages 544–545

HOW WILL YOU AGE?

Scientists have proposed many theories about aging. The actual cause is probably a combination of many things. Some age-related diseases appear to run in families. You can find out what kinds of habits prevent or control some of these conditions.

We can have some control over certain factors associated with aging. Wrinkled skin, for example, can be lessened by using a good sunscreen or avoiding the sun. Wearing sunscreen is also important for preventing skin cancer. Ultraviolet rays from the sun are thought to play a role in cataract formation, so it is a good idea to wear protective sunglasses.

Other habits to develop now are to get regular exercise, eat a diet low in fat with plenty of fresh fruits and vegetables, eat calcium-rich foods, avoid overly salty foods, maintain a desirable weight, don't start to smoke, and always wear a seat belt when riding in an automobile. The habits you form now will not guarantee you a longer life, but your older years should be much healthier and enjoyable.

Procedure

1. Find out more about your family's medical history. With the help of your parents or other relatives, try to find out how long your ancestors lived and how they died.
2. Find out about your living relatives' health. There may be a particular health condition that surfaces again and again on one side of the family. If so, there is likely to be a genetic factor involved.
3. Record your family medical history in the table on the right.

Data

	Medical history	Age at death
mother		
her siblings		
her parents		
other relatives		
your father		
his siblings		
his parents		
other relatives		
your medical history		
your siblings		

Conclude and Apply

1. While researching your family medical history, you may have discovered heart problems, high blood pressure, or stroke in members of your family. What sort of habits would help prevent these conditions? **eat a balanced diet, exercise, and do not smoke**
2. Make a list of good health habits you practice now, and habits you intend to change or adopt.
   Good health habits: **Answers will vary, but may include not smoking or always wearing a seat belt.**
   Things to change: **Answers will vary, but may include eating better or getting more exercise.**

91

2 TEACH

Key Concepts are highlighted.

CONCEPT DEVELOPMENT

Students may not be familiar with the effects of diseases such as Alzheimer's disease, arthritis, osteoporosis, or diabetes. A brief discussion of each of these will help students.

CHECK FOR UNDERSTANDING

Have students brainstorm some common daily problems of the aging.

RETEACH

Have students make a list of things they can do to live a longer life.

EXTENSION

For students who have mastered this section, use the **Reinforcement** and **Enrichment** masters or other OPTIONS provided.

3 CLOSE

▶ Ask questions 1-2 in the Section Review.

SECTION REVIEW ANSWERS

1. Don't abuse drugs or alcohol or tobacco; choose healthful habits, foods, and a healthful life-style.

2. It breaks down hydrogen peroxide, which damages cells. When some cells are damaged, they are not replaced, resulting in aging.

3. Connect to Chemistry: Breakdown of collagen and elastic fibers in the dermis causes the skin to become wrinkled and to sag.

YOU DECIDE!

This question may elicit responses that are passionate. Be prepared to be a mediator and allow all views to be expressed.

VideoDisc

STVS: Orthopedic Implants, Disc 7 Side 2

ACTIVITY 23-2

30 minutes

OBJECTIVE: Graph the average growth rate in males and females, and **determine** if there are differences in the average growth rates of males and females.

PROCESS SKILLS applied in this activity:
▶ **Graphing Information** in Procedure Steps 1-5.
▶ **Interpreting Data** in Analyze Questions 1 and 2 and Conclude and Apply Questions 3-7.

TEACHING THE ACTIVITY

▶ Have students refer to Making and Using Graphs in the Skill Handbook for help in making a line graph.

Activity ASSESSMENT

Performance: To further assess students' understanding of growth, have them correlate the range of heights in their class to the statistical average.

PROGRAM RESOURCES

From the **Teacher Resource Package: Activity Worksheets,** pages 205-206, Activity 23-2, Average Growth Rate in Humans.

ACTIVITY 23-2 Average Growth Rate in Humans

Problem: *Is average growth rate the same in males and females?*

Materials

- graph paper
- red and blue pencils

Procedure

1. Construct a graph similar to graph A below. Plot mass on the vertical axis and age on the horizontal axis.
2. Plot the data given under Data and Observations for the average female growth in mass from ages 8 to 18. Connect the points with a red line.
3. On the same graph, plot the data for the average male growth in mass from ages 8 to 18. Connect the points with a blue line.
4. Construct a separate graph similar to graph B below. Plot height on the vertical axis and age on the horizontal axis.
5. Plot the data for the average female growth in height from ages 8 to 18. Connect the points with a red line. Plot the data for the average male growth in height from ages 8 to 18. Connect the points with a blue line.

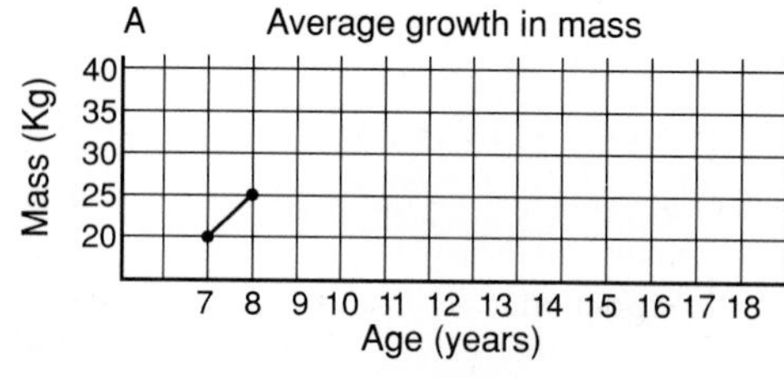

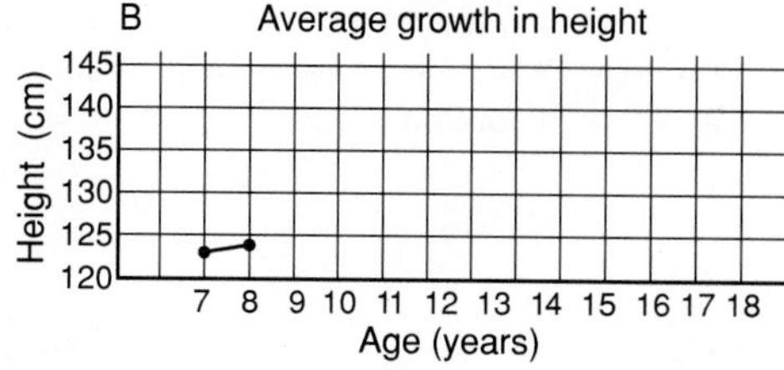

Data and Observations

Averages for Growth in Humans				
Age	Mass (kg)		Height (cm)	
	Female	Male	Female	Male
8	25	25	123	124
9	28	28	129	130
10	31	31	135	135
11	35	37	140	140
12	40	38	147	145
13	47	43	155	152
14	50	50	159	161
15	54	57	160	167
16	57	62	163	172
17	58	65	163	174
18	58	68	163	178

Analyze

1. Up to what age is average growth in mass similar in males and females?
2. Up to what age is average growth in height similar in males and females?

Conclude and Apply

3. Between what ages do females increase most in height?
4. During what ages do males increase most in height?
5. When does the mass of females change most?
6. How can you explain the differences in growth between males and females?
7. *Interpret* the data to find if the average growth rate is the same in males and females.

ANSWERS TO QUESTIONS

1. up to about age 11
2. up to about age 12
3. Females increase the most in height between the ages of 11 and 13.
4. Males increase the most in height at ages 13, 14, and 15.
5. Females increase the most in mass between the ages of 11 and 13.
6. Answers may vary slightly. Most students will say that girls may enter adolescence at an earlier age than boys.
7. Average growth is the same until puberty. At puberty, girls have an earlier growth spurt than boys, but on the average boys grow taller than girls.

CHAPTER

REVIEW

SUMMARY

23-1: Human Reproduction

1. The reproductive system allows new organisms to be formed.
2. The testes produce sperm that exit the male through the penis.
3. The female ovary produces an egg that can be fertilized to produce a zygote that develops within the uterus.
4. When an egg is not fertilized, it disintegrates; the lining of the uterus is shed.

23-2: Fertilization to Birth

1. After fertilization, the zygote undergoes developmental changes to become an embryo surrounded by amniotic fluid, then a fetus.
2. The uterine muscles contract to push the baby out of the mother's uterus and vagina.
3. Twins occur when two eggs are fertilized or when a single egg splits after fertilization.

23-3: Development after Birth

1. Birth begins with labor, muscular contractions of the uterus. The amniotic sac breaks, then usually after several hours, the contractions force the baby out of the body.
2. Infancy is the stage of development from neonatal to one year. It is a period of rapid growth of mental and physical skills. Childhood, which lasts to age 12 is marked by development of muscular coordination and mental abilities.
3. Adolescence is the stage of development when a person becomes physically able to reproduce. The final stage of development, adulthood, consists of many stages. All physical growth is complete.

23-4: Science and Society: Aging

1. Many conditions are associated with aging, including cancer, heart disease, and diabetes. Due to rising health care costs, many elderly are unable to get proper medical care.
2. Researchers believe the chemical hydrogen peroxide produced in cells plays a large role in the aging process.

KEY SCIENCE WORDS

a. **adolescence**
b. **adulthood**
c. **amniotic sac**
d. **catalase**
e. **childhood**
f. **embryo**
g. **fetus**
h. **infancy**
i. **menopause**
j. **menstrual cycle**
k. **menstruation**
l. **ovaries**
m. **ovulation**
n. **pregnancy**
o. **semen**
p. **sperm**
q. **testes**
r. **uterus**
s. **vagina**

UNDERSTANDING VOCABULARY

Match each phrase with the correct term from the list of Key Science Words.

1. birth canal
2. male sex cells
3. egg-producing organs
4. release of egg from the ovary
5. nourishing fluid for sperm
6. ending of the menstrual cycle with age
7. sperm-producing organs
8. place where a fertilized egg develops into a baby
9. stores nutrients for the unborn baby
10. name for the unborn child the first two months

OPTIONS

ASSESSMENT

To assess student understanding of material in this chapter, use the resources listed.

COOPERATIVE LEARNING

Consider using cooperative learning in the THINK AND WRITE CRITICALLY, APPLY, and MORE SKILL BUILDERS sections of the Chapter Review.

PROGRAM RESOURCES

From the **Teacher Resource Package** use:

Chapter Review, pages 49-50.

Chapter and Unit Tests, pages 142-145, Chapter Test.

Chapter and Unit Tests, pages 146-147, Unit Test.

CHAPTER

REVIEW

SUMMARY

Have students read the summary statements to review the major concepts of the chapter.

UNDERSTANDING VOCABULARY

1. s
2. p
3. l
4. m
5. o
6. i
7. q
8. r
9. c
10. f

ASSESSMENT

Portfolio

Encourage students to place in their portfolios one or two items of what they consider to be their best work. For each item, ask students to explain why that item was chosen and what they learned from it. Items might be selected from the following.

- Enrichment research on suspension in water as protection, p. 538
- Skillbuilder graph on embryo and fetus sizes, p. 539
- In Your Journal interview with someone over 60 years of age, p. 543

Performance

Additional performance assessments may be found in *Performance Assessment* and *Science Integration Activities* that accompany **Merrill Life Science.** Performance Task Assessment Lists and rubrics for evaluating these activities and other products generated throughout the chapter can be found in Glencoe's *Performance Assessment in Middle School Science.*

CHAPTER

REVIEW

CHECKING CONCEPTS

1. c	**6.** c
2. b	**7.** d
3. a	**8.** c
4. a	**9.** b
5. b	**10.** a

USING LAB SKILLS

ASSESSMENT

Use these alternate lab exercises to assess students' understanding of skills used in this chapter.

11. Answers will vary. Possible objects may include:

3 mos. = width of 3 × 5 card
4 mos. = length of dollar bill
5 mos. = diameter of dinner plate
6 mos. = 12 inch ruler
7 mos. = length of two pencils laid end to end
8 mos. = width of wall calendar
9 mos. = ½ length of meterstick

12. Student answers should reflect information from the Activity diagrams and content on multiple births on page 539.

THINK AND WRITE CRITICALLY

13. Sperm need to travel a great distance to fertilize the egg; therefore, many are needed so that one might make it to the egg.

14. Semen protects, nourishes, and helps the sperm move.

15. The lining provides a space for the zygote to attach.

16. Identical twins = one egg fertilized by one sperm: zygote splits after fertilization. Fraternal twins = two eggs fertilized by two different sperm.

17. The woman's body has a thickened uterine wall; the placenta develops to transport food and wastes through the umbilical cord; amniotic fluid cushions the embryo.

CHAPTER

REVIEW

CHECKING CONCEPTS

Choose the word or phrase that completes the sentence.

1. The embryo develops in the ______.
a. oviduct **c.** uterus
b. ovary **d.** vagina

2. The monthly process of egg release is called ______.
a. fertilization **c.** menstruation
b. ovulation **d.** puberty

3. The union of an egg and sperm is ______.
a. fertilization **c.** menstruation
b. ovulation **d.** puberty

4. The egg is fertilized in the ______.
a. oviduct **c.** vagina
b. uterus **d.** ovary

5. Mental and physical skills rapidly develop during ______.
a. neonatal period **c.** childhood
b. infancy **d.** adolescence

6. Puberty occurs during ______.
a. childhood **c.** adolescence
b. adulthood **d.** infancy

7. Sex characteristics common to males and females include ______.
a. breasts **c.** increased fat
b. increased muscles **d.** pubic hair

8. Growth of the skeleton and muscles stops during ______.
a. childhood **c.** adulthood
b. adolescence **d.** infancy

9. The period of development with three stages is ______.
a. infancy **c.** adolescence
b. adulthood **d.** childhood

10. The ability to reproduce begins at ______.
a. adolescence **c.** childhood
b. adulthood **d.** infancy

USING LAB SKILLS

11. The MINI-Lab on page 537 concerns the size of an embryo/fetus during development. Find and measure common household items that correspond to the fetus sizes. Example 6 months = 12 inch ruler. Set up a display of these items.

12. In the Activity on page 534, you learned how to interpret diagrams of the menstrual cycle. Use the diagram to explain how fraternal twins form.

THINK AND WRITE CRITICALLY

Answer the following questions in your Journal using complete sentences.

13. Why are so many sperm released if only one is needed to fertilize an egg?

14. Why is semen necessary for sperm survival?

15. What is the purpose of the thickened uterine lining?

16. Explain the difference between identical and fraternal twins.

17. What features of a pregnant woman protect the developing child?

APPLY

18. Explain the similar functions of the ovaries and testes.
19. Identify the structure in which each process occurs: meiosis, ovulation, fertilization, and implantation.
20. Describe the structural differences between embryo and fetus.
21. What kind of cell division occurs as the zygote develops?
22. Describe one major change in each stage of human development.

MORE SKILL BUILDERS

If you need help, refer to the Skill Handbook.

1. **Classifying:** Classify each structure as female or male and internal or external: ovary, penis, scrotum, testes, uterus, vagina.
2. **Concept Mapping:** Fill in the concept map of egg release.

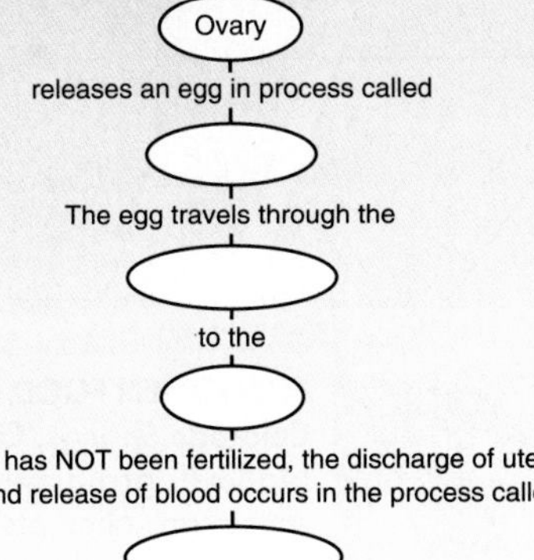

Ovary
releases an egg in process called
The egg travels through the
to the
If the egg has NOT been fertilized, the discharge of uterine lining and release of blood occurs in the process called

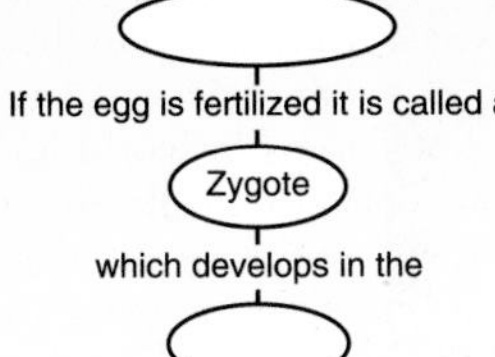

If the egg is fertilized it is called a
Zygote
which develops in the

3. **Hypothesizing:** Make a hypothesis about the effects of raising identical twins apart from each other.
4. **Making and Using Graphs:** Use the data to make a graph showing the day of development versus size of the embryo. When is the fastest period of growth?

Week after Fertilization	Size
3	3 mm
4	6 mm
6	12 mm
7	2 cm
8	4 cm
9	5 cm

5. **Sequencing:** Sequence the steps involved in the birth process.

PROJECTS

1. Find newspaper articles on the effects of smoking on the health of the developing embryo and newborns. Discuss this with your class.
2. Put a soft fruit or vegetable (plum, peach, tomato) into a self-sealing plastic bag filled with water. Use this model to observe how the fetus is protected from shock by the amniotic fluid.

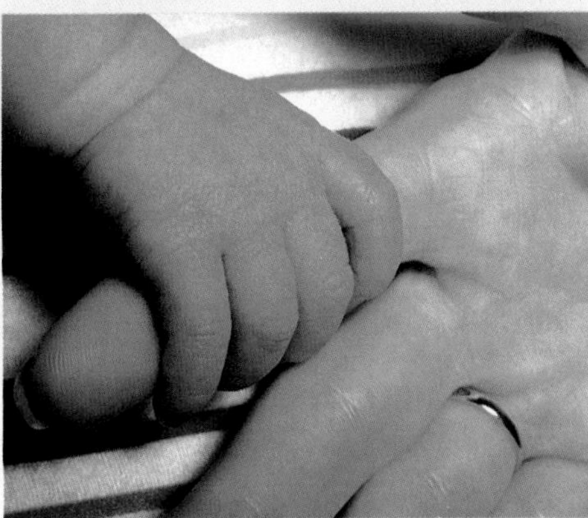

APPLY

18. Both are paired organs and produce sex cells.

19. Meiosis occurs in gametes; ovulation occurs in the ovaries; fertilization occurs in the oviduct; implantation occurs in the uterus.

20. Embryo is the stage of development for the first two months; it is small and undergoes beginning development of all structures. The fetus is the developmental stage from two months until birth; it is more developed and its movement can be felt.

21. Mitosis continues to produce new cells.

22. Neonatal: needs to be taken care of

Infancy: rapid growth and development of mental and physical skill: speak, stand

Childhood: development of mental and muscle coordination: walk, talk, read, reason

Adolescence: puberty—ability to reproduce; secondary sex characteristics and growth; increased reasoning

Adulthood: three stages; growth of skeleton and muscles stops

MORE SKILL BUILDERS

1. Classifying: Female internal: ovary, uterus, vagina

Male external: penis, scrotum, testes

2. Concept Mapping:

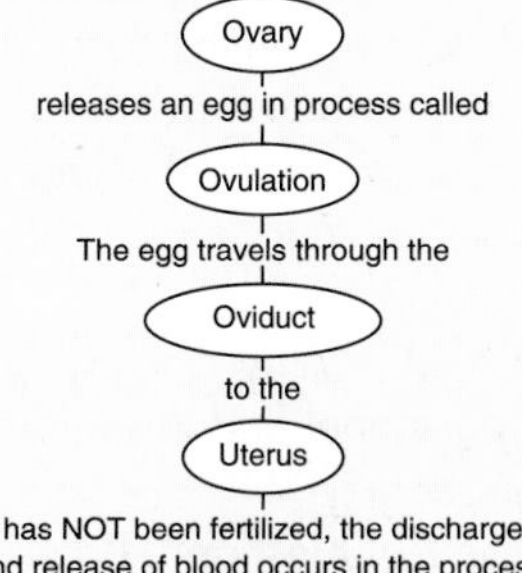

Ovary
releases an egg in process called
Ovulation
The egg travels through the
Oviduct
to the
Uterus
If the egg has NOT been fertilized, the discharge of uterine lining and release of blood occurs in the process called
Menstruation
If the egg is fertilized it is called a
Zygote
which develops in the
Uterus

3. Hypothesizing: Students might hypothesize that they will be the same because of their genes or they might say the environment will have an effect, making them different from each other. Research now indicates the genes rule.

4. Making and Using Graphs: Students should make line or bar graphs with Week After Fertilization on the x-axis and size on the y-axis. The most rapid period of growth occurs between weeks 7 and 8.

5. Sequencing: Labor—amniotic sac breaks—baby moves out of uterus and vagina.

UNIT 6
GLOBAL CONNECTIONS

Objective

In this unit ending feature, the unit topic, "The Human Body," is extended into other disciplines. Students will see how knowledge of the human body can help people stay healthy.

Motivate

Cooperative Learning: Have students work in Expert Teams. Assign one Connection to each group of students. Have each group research the topic in the assigned Connection and report to the class.

Teaching Tips

- Ask students to consider how the environment affects them physically.
- Excretion, digestion, respiration, and a sensory system are attributes of many other organisms besides humans.

Wrap-Up

Conclude this lesson by having students respond to the question, "What if you had no skin?" Students should answer in a short paragraph.

SOCIAL STUDIES

Background: Physical anthropologists can tell much about people from looking at their bones. Wear on bones and teeth, scars, loss of calcium, and hundreds of other characteristics furnish clues about the age, sex, and medical history of the skeleton.

Discussion: Why do law enforcement personnel find this information so important?

Answer to Question: Bones show the marks of stress. A man's bones are often heavier than a woman's. The ridges above the eyes on a man's skull are different.

Extension: Have students measure their wrists. Record all the wrist measurements. Can the size of the wrist be used to tell the sex of the person?

UNIT 6
GLOBAL CONNECTIONS

The Human Body

In this unit, you have studied the human body. Find out how humans influence and have been influenced by their world.

120° 60° 60° 60°

SOCIAL STUDIES

WRITTEN IN BONE
University of Pennsylvania

Bones can tell anthropologists much about an individual. For instance, bones reveal the age, sex, race, and medical history of a person. How is it possible for anthropologists to detect this information from bones?

HISTORY

FORGOTTEN FOOD CROPS
Colorado Springs, Colorado

Natural grains that native Americans once ate are being rediscovered. These food plants offer science some exciting solutions to world food problems. Why could these "new" old plants be important?

550

HISTORY

Background: Many of the wild plants once grown by Native Americans are being rediscovered. Guayule (gwy OO lee) produces rubber. Some plants that grow well in arid conditions show promise as new food crops.

Discussion: Knowledge of plants used by Native Americans may help the world in many ways. How do the students think these plants might be used?

Answer to Question: These plants might furnish substitutes for petroleum products, boost the economy of developing nations, or become a food source for people.

Extension: Have students take a trip to a grocery store's fresh produce section to make a list of food products that are new or unknown to them.

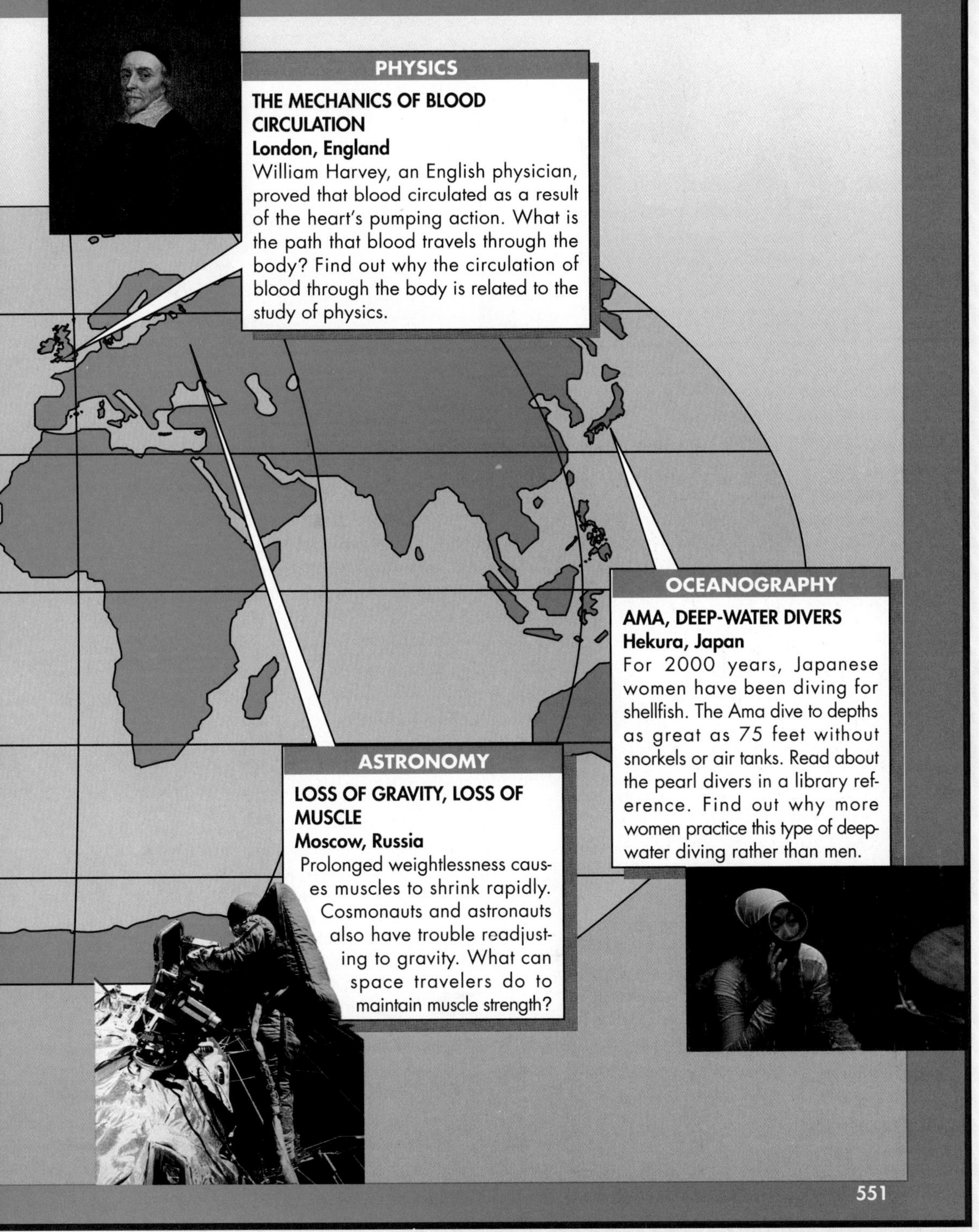

PHYSICS
THE MECHANICS OF BLOOD CIRCULATION
London, England
William Harvey, an English physician, proved that blood circulated as a result of the heart's pumping action. What is the path that blood travels through the body? Find out why the circulation of blood through the body is related to the study of physics.
OCEANOGRAPHY
AMA, DEEP-WATER DIVERS
Hekura, Japan
For 2000 years, Japanese women have been diving for shellfish. The Ama dive to depths as great as 75 feet without snorkels or air tanks. Read about the pearl divers in a library reference. Find out why more women practice this type of deep-water diving rather than men.
ASTRONOMY
LOSS OF GRAVITY, LOSS OF MUSCLE
Moscow, Russia
Prolonged weightlessness causes muscles to shrink rapidly. Cosmonauts and astronauts also have trouble readjusting to gravity. What can space travelers do to maintain muscle strength?
551

PHYSICS

Background: Three thousand years ago, a Chinese medical manual stated, "The heart regulates all the blood of the body; the blood flows...in a circle and never stops." This wisdom was lost until William Harvey determined that blood circulated because of the action of the heart and followed a circular path.

Discussion: Discuss why the ancient Chinese knowledge of circulation was unknown in the western world.

Answer to Question: Refer to Chapter 20 for information on blood circulation. Blood is forced through the body under pressure. It is a liquid that responds to pressure according to physical laws.

Extension: Have students place fingers on their wrists or neck to feel their pulses. Have them count their pulses.

OCEANOGRAPHY

Background: The Ama of Japan have been deep-water diving for centuries without any kind of breathing apparatus. They may dive to depths of 75 feet one hundred times a day. The Ama dive for shellfish, especially abalone.

Discussion: Though the Ama have been diving for more than two thousand years, the tradition is dying. Why do students think this is happening?

Answer to Question: Ocean water is very cold. Women have an extra layer of fat and can dive for three to four hours before they become too chilled.

Extension: Have students hypothesize why women's bodies have a layer of fat that men's do not.

ASTRONOMY

Background: Soviet cosmonauts have collected twelve years of data on the effect of weightlessness on the human body. Despite two hours of daily exercise, cosmonauts still lose muscle. After 4-5 months in space, cosmonauts are too weak to walk.

Discussion: Weightlessness is a problem in performing many tasks on board space ships. How do students think the problems of weightlessness might be solved?

Answer to Question: Weightlifting is not possible, so cosmonauts exercise on a treadmill and stationary bicycle.

Extension: Have students try to lift weights while they are standing in a swimming pool or sitting in a bathtub.

CAREERS

PHYSICAL THERAPIST

Background: Physical therapists may work in sports therapy, in pediatric therapy, or with all kinds of patients. A physical therapist must have at least a Bachelor of Science degree. Many physical therapists pursue a Master's degree in a particular subject area.

Related Career	Education
Physical Therapist	BS degree
Occupational Therapist	BS degree
Speech Pathologist	BS degree

Career Issue: Some sports injuries are painful and take time to heal. Should athletes who are not fully healed be given pain-killing drugs and allowed to play?

What do you think? Ask how students feel about injured athletes playing before they have healed.

HOME HEALTHCARE AIDE

Background: The role of the home healthcare aide is an important one. With an aging population, there will be much more demand for this profession. Healthcare aides provide home management services, personal care, and light housekeeping.

Related Career	Education
Healthcare aide	High school
Nursing Aide	High school
Childcare attendant	High school

Career Issue: Only a small number of men are employed in home healthcare. Many more are needed. Unfortunately, this type of work does not pay well.

What do you think? What kinds of changes are needed to make home healthcare more attractive as a career to both men and women?

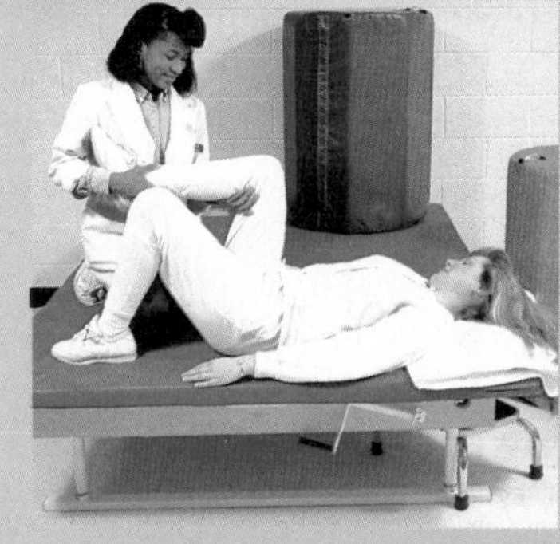

CAREERS

PHYSICAL THERAPIST

A *physical therapist* works with people who have had accidents, handicaps, arthritis, and heart disease. The patients range from newborn to elderly. Some physical therapists work with a wide range of problems. Others specialize, working in pediatrics, sports therapy, or neurology, as well as other specialties. Physical therapists need to be supportive of their patients, flexible, and physically strong. The work of the physical therapist is emotionally demanding but also rewarding. A physical therapist must have at least a bachelor of science degree. He or she will need to have courses in basic science, anatomy and physiology, as well as kinesiology, and a background in psychology.

For Additional Information

Contact the American Physical Therapy Association, 1111 N. Fairfax St., Alexandria, VA 22314.

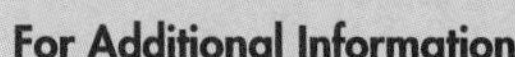

HOME HEALTHCARE AIDE

Home healthcare aides work with the elderly, disabled, or ill. The job of the healthcare aide varies from patient to patient. Schedules also vary so that a healthcare aide needs to be very adaptable. Many people need help with personal care, such as bathing and shampooing, either for a short time after a hospital stay, or for longer periods in the case of chronic illnesses. Aides check a patient's pulse and respiration, help with prescribed exercises, assist with medicines, and even prepare meals.

Home healthcare aides should be able to read and write and have a high school diploma, plus a deep interest in helping people. New laws require 75 hours of training with a combination of 16 hours of classroom study and 16 hours of supervised practical training. Training programs may be offered by the employing agency, the Red Cross, or a community college.

For Additional Information

Contact The National Association for Homecare, 519 C St. NE, Washington, DC 20002.

UNIT READINGS

▶Golden, Frederic. "Clever Kanzi." *Discover,* March 1991, p. 20.

▶Minsky, Marvin. "The Intelligence Transplant." *Discover,* Oct. 1989, p. 52.

552

UNIT READINGS

Background

▶ Canby, Thomas Y. "Are the Soviets Ahead in Space?" *National Geographic,* October 1986. This is a fairly comprehensive article on the Soviet space program.

▶ Vietmeyer, Noel D. "Rediscovering America's Forgotten Crops." *National Geographic,* May 1981. This article reveals some interesting ideas for old crops.

▶ Marden, Luis. "Ama, Sea Nymphs of Japan." *National Geographic,* July 1971. After thousands of years, scientists are interested in the amazing Ama divers of Japan.

More Readings

1. Sochurek, Howard. "Medicine's New Vision." *National Geographic,* January 1987.
2. Nourse, Alan E. *The Body.* New York: Life Science Library, Time, Inc. 1964.

Sula

by Toni Morrison

The following passage tells of two girlhood friends, Nel and Sula, who have been reunited after many years.

Nel alone noticed the peculiar quality of the May that followed the leaving of the birds. It had a sheen, a glimmering as of green, rain-soaked Saturday nights (lit by the excitement of newly installed street lights); of lemon-yellow afternoons bright with iced drinks and splashes of daffodils. It showed in the damp faces of her children and the river-smoothness of their voices. Even her own body was not immune to the magic. She would sit on the floor to sew as she had done as a girl, fold her legs up under her or do a little dance that fitted some tune in her head. There were easy sun-washed days and purple dusks in which Tar Baby sang "Abide With Me" at prayer meetings, his lashes darkened by tears, his silhouette limp with regret against the whitewashed walls of Greater Saint Matthew's. Nel listened and was moved to smile. To smile at the sheer loveliness that pressed in from the windows and touched his grief, making it a pleasure to behold.

Author Toni Morrison

Although it was she alone who saw this magic, she did not wonder at it. She knew it was all due to Sula's return to the Bottom. It was like getting the use of an eye back, having a cataract removed. Her old friend had come home. Sula. Who made her laugh, who made her see old things with new eyes, in whose presence she felt clever, gentle and a little raunchy. Sula, whose past she had lived through and with whom the present was a constant sharing of perceptions. Talking to Sula had always been a conversation with herself. Was there anyone else before whom she could never be foolish? In whose view inadequacy was a mere idiosyncrasy, a character trait rather than a deficiency? Sula never competed; she simply helped others define themselves. Other people seemed to turn their volume on and up when Sula was in the room. More than any other thing, humor returned. She could listen to the crunch of sugar underfoot that the children had spilled without reaching for the switch; and she forgot the tear in the living-room window shade.

In Your Own Words

▶ Scientists and doctors often notice the effect of mental attitude on the human body. In this passage, you can see the effect Sula's return had on her friend, Nel. Think of an event in your own life that has changed your outlook either for better or worse. How did you feel after it? What physical changes did you feel? Write a short story telling of this event and how it affected you.

553

SCIENCE & LITERATURE

Source: Toni Morrison. *Sula.* NY: Alfred A. Knopf, 1974.

Biography: Toni Morrison was born in Lorain, Ohio. She graduated from Howard University and received her Master's degree from Cornell. She taught English and humanities at Texas Southern University and at Howard for nine years before she published her first novel, *The Bluest Eye.*

TEACHING STRATEGY

Have students read through the passage by Toni Morrison. Then have them respond to the discussion questions below.

Discussion Questions

1. **What do you think the "magic" that Nel noticed was?** *The answers to this question may vary, but most students will probably mention the return of Sula and the happiness seeing her brought to Nel.*
2. **What, more than any other thing, changed Nel's attitude toward her house and her children?** *Humor allowed Nel to ignore the torn shade and to overlook her children's shortcomings.*

Cooperative Learning: Divide the class into small groups. Using the Expert Teams strategy, have each group of students reread the passage. Then let them discuss what they think Nel was like before her friend Sula returned. How did she change? Let each group choose sentences from the passage to illustrate the change. Ask each group to list ways in which joy or sorrow affect the way people feel physically.

Classics

▶ Asimov, Isaac. *The Human Body.* Boston: Houghton, Mifflin, 1963. This book is written with the attention to detail characteristic of Asimov.

Other Works:

▶ Huston, Larry. "The Beat Goes On." *Discover,* March 1991. pp. 26-27.

UNIT

7

STAYING HEALTHY

In Unit 7, students are introduced to the facts about staying healthy and drug-free. The unit is organized into two chapters which deal with the immune system and with how drugs affect the human body.

CONTENTS

ADVANCE PREPARATION

Activities

▶ **Activity 24-1, page 567,** requires fresh and rotten apples.

▶ **Activity 24-2, page 574,** requires agar plates, filter paper, disinfectant, hydrogen peroxide, mouthwash, and alcohol.

▶ **Activity 25-2, page 594,** requires *Daphnia,* tap water, and dilute solutions of coffee, cola, ethyl alcohol, tobacco, and cough medicine (dextromethorphen hydrobromide).

Field Trips and Speakers

▶ Arrange for a pharmacist to visit the class to discuss controlled substances and over-the-counter medications.

▶ Arrange for a police officer working in narcotics to speak to the class on illegal substances and the penalties for their use.

▶ Arrange for a representative from the American Lung Association to bring a display on the hazards of smoking to the class.

OPTIONS

Cross Curriculum

▶ Have students research smallpox and find out how this disease was eradicated worldwide. Have students write papers hypothesizing how other diseases could be similarly eradicated.

PROJECT

During the course of this unit study, have students work cooperatively on PROJECT 7, *To Your Health!,* found on pages 706 and 707.

Science at Home

▶ Have students debate the pros and cons of using paper cups rather than drinking glasses in family bathrooms.

Cooperative Learning: Have students monitor newspapers and television news coverage for one week to see how often public health information is discussed. At the end of the week, divide the class into two groups and have each group list the public health issues that were discussed in each medium. Compare the lists.

What's Happening Here?

The mighty macrophage puts forth tendril-like extensions as it reaches out to snag bacterial cells that daily attack the fortress of your body. Constantly on guard, these large white blood cells patrol the body, slipping between cells in capillary walls to sweep up invaders that threaten your health. Your body is equipped with a variety of defenses that maintain your health even though you remain unconscious of their activity. The child in the smaller photograph is very conscious of his defense system. Lacking a natural defense system, he depends on the walls of his artificial environment to protect him from assault by disease-causing agents.

UNIT CONTENTS

555

INTRODUCING THE UNIT

What's Happening Here?

▶ Have students look at the photos and read the text. Ask them what's happening here. Point out to students that in this unit they will study their immune system and how it keeps them healthy, and the effects that drugs have on their bodies and their health.

▶ **Background:** Children are born with some immunities from their mother, but these wear off after a few months. Born without an effective immune system, this child cannot live in the outside world.

Previewing the Chapters

▶ Encourage students to look through the chapters in this unit to find answers to questions they may have about diseases. For questions that are beyond the scope of this book, refer students to their family doctors, school nurses, or public health facilities.

Tying to Previous Knowledge

▶ Lead students in a discussion of their experiences with diseases. Talk about immunizations they have had. Encourage students to recognize how much they already know about diseases.

▶ Use the **inquiry questions** in the OPTIONS box below to investigate with students how tobacco and alcohol affect the human body.

Multicultural Awareness

Students can research various activities and their effects on the transmission of disease. International movement of people and products sometimes results in movement of diseases. Clearing land for farming or highways in Latin America provides new breeding areas for mosquitos and other disease carriers. Have students research an historical example of disease transmission such as typhoid Mary, smallpox among native Americans, or a current example.

Inquiry Questions

Use the following questions to focus a discussion of how use of tobacco and alcohol affects the human body.

▶ **Do you know someone who is using tobacco and/or alcohol?**

▶ **How did that person feel the first time he or she used one of these drugs?**

▶ **Why did they feel sick?**

▶ **If they felt sick after trying one of these drugs, why would they use one again?**

▶ **What are some of the reasons people use tobacco and alcohol?**

CHAPTER

24 Immunity

CHAPTER SECTION	OBJECTIVES	ACTIVITIES
24-1 The Nature of Disease (2 days)	1. **Describe** the work of Pasteur, Koch, and Lister in the discovery and prevention of disease. 2. **List** diseases caused by viruses and bacteria. 3. **Discuss** sexually transmitted diseases (STDs), their causes and treatments.	**MINI-Lab:** *Are bacteria present in food?* p. 560 **MINI-Lab:** *How fast do bacteria reproduce?* p. 562
24-2 Your Immune System (3 days)	1. **Explain** the natural defenses your body has against disease. 2. **Describe** differences between active and passive immunity. 3. **Explain** how HIV affects the immune system.	**Activity 24-1:** *Microorganisms and Disease,* p. 567
24-3 Preventing Disease Science & Society (1 day)	1. **Explain** how vaccination prevents certain diseases. 2. **Explain** how disease can be prevented from spreading.	
24-4 Noncommunicable Disease (2 days)	1. **List** two noncommunicable diseases. 2. **Describe** the basic characteristics of cancer. 3. **Name** two chronic diseases of the immune system.	**Activity 24-2:** *Designing an Experiment (Preventing Microorganism Growth),* p. 574
Chapter Review		

ACTIVITY MATERIALS

FIND OUT	ACTIVITIES		MINI-LABS	
FIND OUT Page 557 peppermint or lemon food flavoring cotton ball	**24-1 Microorganisms and Disease, p. 567** fresh apples (6) rotting apple (1) alcohol self-sealing plastic bags (6) labels and pencil sandpaper cotton ball (1) soap and water paper towels	**24-2 Designing an Experiment, p. 574** transparent tape sterile nutrient agar plates (5) filter paper disinfectant hydrogen peroxide pencil and labels scissors mouthwash metric ruler forceps alcohol small jars (4)	**Are bacteria present in food? p. 560** fresh milk sour milk yogurt liquid from sour cream methylene blue mineral oil test tubes labels	**How fast do bacteria reproduce? p. 562** pencil paper

CHAPTER FEATURES	TEACHER RESOURCE PACKAGE	OTHER RESOURCES
Skill Builder: *Recognizing Cause and Effect,* p. 562	**Ability Level Worksheets** ◆ ***Study Guide,*** p. 92 ● ***Reinforcement,*** p. 92 ▲ ***Enrichment,*** p. 92 **Activity Worksheets,** p. 218, 219 **Cross-Curricular Connections,** p. 28	**Lab Manual:** Communicable and Noncommunicable Disease, p. 161 **Lab Manual:** Sexually Transmitted Diseases, p. 163
Technology: *Super Sleuth!* p. 565 **Skill Builder:** *Comparing and Contrasting,* p. 566	**Ability Level Worksheets** ◆ ***Study Guide,*** p. 93 ● ***Reinforcement,*** p. 93 ▲ ***Enrichment,*** p. 93 **Activity Worksheets,** pp. 212-213 **Concept Mapping,** p. 53	**Science Integration Activity 24** **STVS:** Disc 7, Side 1
You Decide! p. 569	**Ability Level Worksheets** ◆ ***Study Guide,*** p. 94 ● ***Reinforcement,*** p. 94 ▲ ***Enrichment,*** p. 94 **Critical Thinking/Problem Solving,** p. 28 **Science and Society,** p. 28	
Problem Solving: *Fighting TB,* p. 573 **Skill Builder:** Making and Using Tables, p. 573	**Ability Level Worksheets** ◆ ***Study Guide,*** p. 95 ● ***Reinforcement,*** p. 95 ▲ ***Enrichment,*** p. 95 **Activity Worksheets,** pp. 214-215	**STVS:** Disc 7, Side 1; Side 2
Summary Key Science Words Understanding Vocabulary Checking Concepts Using Lab Skills Think & Write Critically Apply More Skill Builders Projects	ASSESSMENT RESOURCES **Chapter Review,** pp. 51-52 **Chapter Test,** pp. 161-164 **Performance Assessment in Middle School Science (PAMSS)**	**Chapter Review Software** **Test Bank** **Alternate Assessment** **Performance Assessment**

◆ **Basic** ● **Average** ▲ **Advanced**

ADDITIONAL MATERIALS

SOFTWARE	AUDIOVISUAL	BOOKS/MAGAZINES
Malana, Compuware. *Sexually Transmitted Diseases,* HRM. *Pathology: Diseases and Defense,* IBM. *Understanding the Human Fight to Stay Healthy,* Queue.	*Diabetes: Beating the Needle,* film, Filmmakers Library. *Infection and Immunity,* film, I.F.B. *Microorganisms That Cause Disease,* film, Coronet/MTI. *Herpes: Facing the Realities,* filmstrip, Sunburst. *Infectious Diseases and Man-made Defenses,* film, Coronet/MTI. *The Discoverers—Searching for the Cure,* film, American Cancer Society.	Benenson, Abram S., ed. *Control of Communicable Diseases in Man.* Washington, DC: American Public Health Association, 1985. Conte, John E. and Steven L. Barriere. *Manual of Antibiotics and Infectious Diseases.* 6th ed. Philadelphia, PA: Lea and Febinger, 1988. Silverstein, Alvin and Virginia B. Silverstein. *Diabetes: The Sugar Disease.* NY: Harper and Row Junior Books Group, 1980.

CHAPTER

24 IMMUNITY

THEME DEVELOPMENT: One of the major themes of science is interaction among systems. In this chapter, the focus will be on interactions within the body. Intricate natural defenses in the body aid in preventing disease. Vaccines also provide active immunity to the body systems.

CHAPTER OVERVIEW

- **Section 24-1:** This section describes the discovery of the causes of disease. There is a special focus on sexually transmitted diseases.
- **Section 24-2:** The mechanisms of active and passive immunity are introduced. The action of the AIDS virus on the immune system is highlighted.
- **Section 24-3: Science and Society:** The role of vaccination in disease prevention is discussed. The You Decide feature asks students to consider whether or not the last of the smallpox virus should be destroyed.
- **Section 24-4:** This section explores characteristics of noncommunicable diseases. There is a special emphasis on cancer.

CHAPTER VOCABULARY

pasteurization	passive immunity
disinfectant	lymphocytes
antiseptic	vaccination
communicable disease	noncommunicable diseases
sexually transmitted diseases (STDs)	chronic diseases
immune system	cancer
antigens	tumor
antibody	chemotherapy
active immunity	allergy
	allergens

CHAPTER

24 Immunity

556

OPTIONS

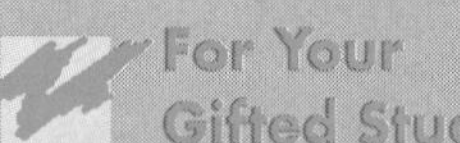

For Your Gifted Students

Have students research the immunizations that are required when visiting other countries. A map and comparison chart can be made to show the results.

For Your Mainstreamed Students

- Ask students to contact the school nurse to find out what immunizations their school district requires before a student can be enrolled. Have them find out if these requirements have changed during the past 15 years and why.
- Students can conduct a survey of their classmates to determine the most common allergies. Results can be graphed.

Aaahhh-choooo! Does pollen make you sneeze? Have you noticed that when one person in your class gets a cold, others soon have it too? Allergies and colds are examples of your body's response to substances and organisms in the environment.

FIND OUT!

Do this activity to see how disease-causing organisms can be spread.

Work with a partner. Place a drop of peppermint or lemon food flavoring on a cotton ball. Pretend that the flavoring is a mass of cold viruses. Next, rub the cotton ball in the shape of an X over the palm of your right hand and let it dry. Can you smell the flavoring? Now, shake hands with a classmate. **Hypothesize** how some diseases can be spread. *Design an experiment* to determine how many persons could be infected by the peppermint "virus."

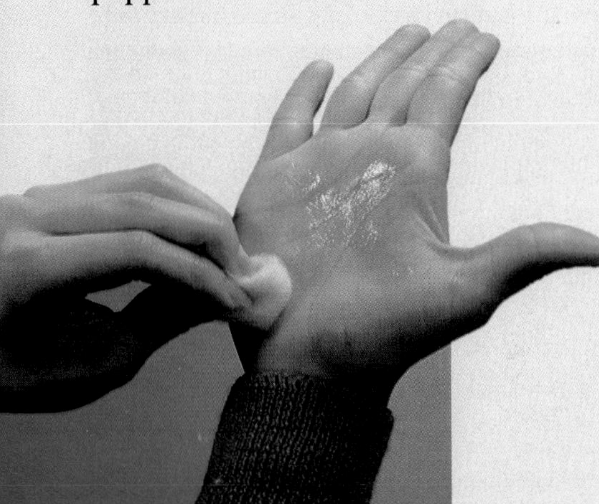

Gearing Up

Previewing the Chapter

Use this outline to help you focus on important ideas in this chapter.

Previewing Science Skills

- ▶ In the Skill Builders, you will recognize cause and effect, compare and contrast, and make and use tables.
- ▶ In the Activities, you will experiment, hypothesize, and analyze data.
- ▶ In the MINI-Labs, you will test and observe, and calculate and graph results.

What's next?

You have just seen how disease can be spread from person to person. Discover ways that disease can be spread by bacteria and viruses and what steps you and your body take to prevent this from happening. You will also learn about diseases and disorders that are caused by genetic, metabolic, and life-style factors.

557

INTRODUCING THE CHAPTER

Use the Find Out activity to introduce students to how disease can be spread. Inform students that they will be learning more about diseases and the immune system as they read the chapter.

FIND OUT!

Preparation: Before beginning this chapter, obtain the food flavoring and cotton balls. Other flavors that can be used are orange, almond, or banana.

Materials: peppermint or lemon food flavorings, one cotton ball for each pair

Teaching Tips

▶ Have students discuss how the mass of cold virus could have gotten on the hands in the first place.

▶ Have students infer if the infection could be passed around on drinking glasses and handkerchiefs.

▶ **Caution:** *At the end of the activity, have all students wash their hands thoroughly.*

Gearing Up

Have students study the Gearing Up feature to familiarize themselves with the chapter. Discuss the relationships of the topics in the outline.

What's Next?

Before beginning the first section, make sure students understand the connection between the Find Out activity and the topics to follow.

ASSESSMENT OPTIONS

PORTFOLIO

Refer to page 575 for suggested items that students might select for their portfolios.

PERFORMANCE ASSESSMENT

See page 575 for additional Performance Assessment options.

Process

Skillbuilders, pp. 562, 573
MINI-Labs, pp. 560, 562
Activities, 24-1, p. 567; 24-2, p. 574
Using Lab Skills, p. 576

CONTENT ASSESSMENT

Assessment—Oral, pp. 560, 566, 572
Section Reviews, pp. 562, 566, 569, 573
Skillbuilder, p. 566
Chapter Review, pp. 575-577
Mini Quizzes, pp. 560, 565, 571

GROUP ASSESSMENT

Opportunities for group assessment occur with Cooperative Learning Strategies, Flex Your Brain Activities, and the Science and Society article.

PREPARATION

SECTION BACKGROUND

▶ Even after the work of Pasteur and Koch became known, the idea of diseases being caused by microorganisms was not accepted and was challenged many times.

▶ AZT is a drug that helps slow the multiplication of HIV. DDI, a drug similar to AZT, is also undergoing research.

PREPLANNING

▶ To prepare for the Demonstration under Motivate, have a disposable petri dish with agar ready for each student.

1 MOTIVATE

▶ Have students compare the population figures of their town with the 25 000 000 people who died of the plague. This is a graphic way of illustrating the effect disease has on history.

▶ **Demonstration:** Have students wash their hands and then lightly press their fingertips onto the surface of a sterile agar plate. Student petri dishes can be coded. Culture in a dark place, and have students observe the extent of the microorganism growth.

In Your JOURNAL

Student journal reports should include that building the canal was held up for years until mosquitos causing malaria and yellow fever were brought under control. The canal was completed in 1914.

24-1 The Nature of Disease

New Science Words

pasteurization
disinfectant
antiseptic
communicable disease
sexually transmitted diseases (STDs)

Objectives

- ▶ **Describe the work of Pasteur, Koch, and Lister in the discovery and prevention of disease.**
- ▶ **List diseases caused by viruses and bacteria.**
- ▶ **Discuss sexually transmitted diseases (STDs), their causes and treatments.**

In Your JOURNAL

What is the relationship between yellow fever and the building of the Panama Canal? Write a report **in your Journal** about the effect of yellow fever on the construction of the Panama Canal.

Discovering Disease

"Ring around the rosie, A pocket full of posies,
Ashes, Ashes, We all fall down."

Do you know that this rhyme is more than 600 years old? The rhyme is thought to be about a disease called the Black Death or the Plague. In the thirteenth century, the Plague killed one-fourth of the people in Europe. "Ring around the rosie" was a symptom of the disease—a ring around a red spot on the skin. People carried a pocket full of flower petals ("posies") and spices to keep away the stench of dead bodies. But still, 25 000 000 people "all fell down," dead from this terrible disease.

1 Today we know that diseases are caused by viruses and by harmful bacteria known as pathogens, and not by some fault in a person's behavior as was once believed.

Louis Pasteur, a French chemist, was the first to discover that harmful bacteria could cause disease. He

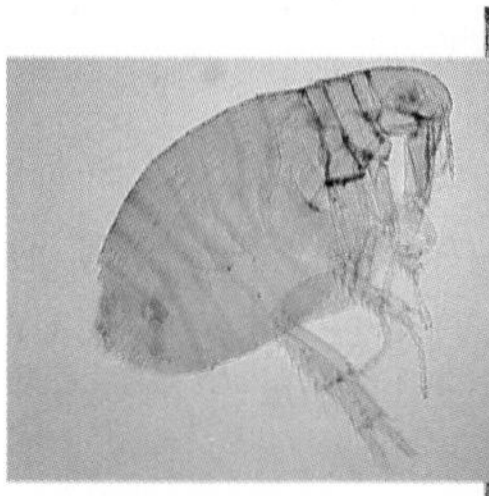

Figure 24-1. Plague is caused by a bacterium that reproduces in fleas that live on rats.

OPTIONS

Meeting Different Ability Levels

For Section 24-1, use the following **Teacher Resource Masters** depending upon individual students' needs.

◆ **Study Guide Master** for all students.

● **Reinforcement Master** for students of average and above average ability levels.

▲ **Enrichment Master** for above average students.

Additional Teacher Resource Package masters are listed in the OPTIONS box throughout the section. The additional masters are appropriate for all students.

◆ STUDY GUIDE 92

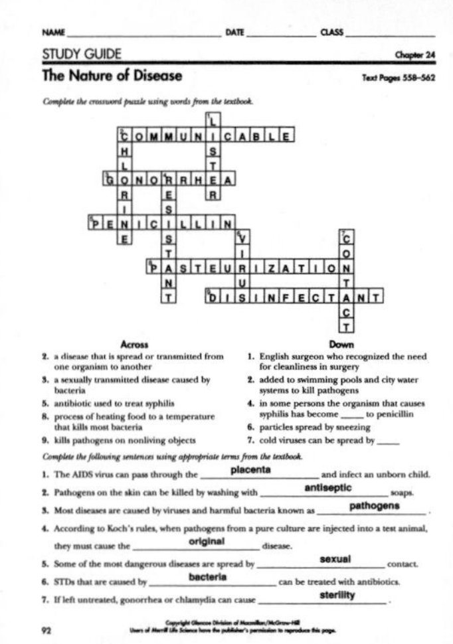

NAME ____________ DATE ________ CLASS ________

STUDY GUIDE — Chapter 24

The Nature of Disease — Text Pages 558–562

Complete the crossword puzzle using words from the textbook.

Across

2. a disease that is spread or transmitted from one organism to another
3. a sexually transmitted disease caused by bacteria
5. antibiotic used to treat syphilis
8. process of heating food to a temperature that kills most bacteria
9. kills pathogens on nonliving objects

Down

1. English surgeon who recognized the need for cleanliness in surgery
2. added to swimming pools and city water systems to kill pathogens
4. in some persons the organism that causes syphilis has become _____ to penicillin
6. particles spread by sneezing
7. cold viruses can be spread by _____

Complete the following sentences using appropriate terms from the textbook.

1. The AIDS virus can pass through the placenta and infect an unborn child.
2. Pathogens on the skin can be killed by washing with antiseptic soaps.
3. Most diseases are caused by viruses and harmful bacteria known as pathogens.
4. According to Koch's rules, when pathogens from a pure culture are injected into a test animal, they must cause the original disease.
5. Some of the most dangerous diseases are spread by sexual contact.
6. STDs that are caused by bacteria can be treated with antibiotics.
7. If left untreated, gonorrhea or chlamydia can cause sterility.

92

developed a method of using heat to kill pathogens. **Pasteurization,** named for him, is the process of heating food to a temperature that kills most bacteria. Pasteur's work began the science of bacteriology.

Being able to tell *which* organism caused a disease was a problem. It was a German doctor, Robert Koch, who first developed a way to isolate and grow, or culture, just one type of bacteria at a time. Koch developed a set of rules to be used for figuring out which organism caused a particular disease. These rules are:

1. In every case of a particular disease, the organism thought to cause the disease must be present.
2. The organism has to be separated from all other organisms and grown in a pure culture.
3. When the organism from the pure culture is injected into a test animal, it must cause the original disease.
4. Finally, when the suspect organism is removed from the test animal and cultured again, it must be compared with the original organism to see if they are the same. Only when they match can you say that that organism is the pathogen that causes that disease.

Keeping Clean

Do you make it a habit to wash your hands and clean your fingernails? Into the late 1800s, doctors regularly operated in their street clothes and with their bare hands. More patients died than survived as a result of surgery. Joseph Lister, an English surgeon, was horrified. He recognized the relationship between the infection rate and cleanliness in surgery. Lister dramatically reduced deaths among his surgical patients by washing their skin and his hands before surgery. Lister used a disinfectant, but today antiseptics are used on people. A **disinfectant** kills pathogens on objects such as instruments, floors, toilets, and bathtubs. An **antiseptic** kills pathogens on skin and prevents them from growing there for sometime after. Today doctors wash their hands often with antiseptic soaps to keep from spreading pathogens from person to person. However, even with the use of disinfectants, some diseases could not be controlled. In the late 1800s and early 1900s, viruses were discovered.

Because of the work of these early scientists, and many others who followed, more is understood about the cause and control of disease.

Did You Know?

Researchers at an institute in England spent forty years trying to find a cure for the common cold. The institute is now closed due to a lack of success.

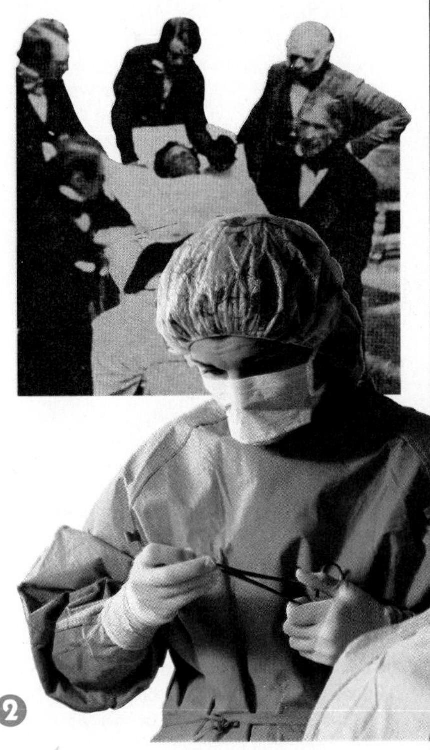

Figure 24-2. Antiseptics and strict surgical methods have made operations safer than they once were.

Cooperative Learning: The Demonstration activity above can be done using the Paired Partners strategy by having one partner wash his or her hands and clean under his or her fingernails. At the conclusion of the activity, the class can discuss the value of cleanliness and determine if regular hand washing would be adequate for a surgeon.

OBJECTIVES AND SCIENCE WORDS: Have students review the objectives and science words in this chapter as they study each section.

2 TEACH

Key Concepts are highlighted.

CONCEPT DEVELOPMENT

- If possible, obtain microscope slides of stained bacteria to give students an idea of their size and number.
- Have students take a survey of products at home that are either disinfectants or antiseptics.
- Ignaz Semmelweis, a Hungarian doctor, preceded Joseph Lister in making a connection between disease and cleanliness. Semmelweis published his findings in 1861, but they were not well received.

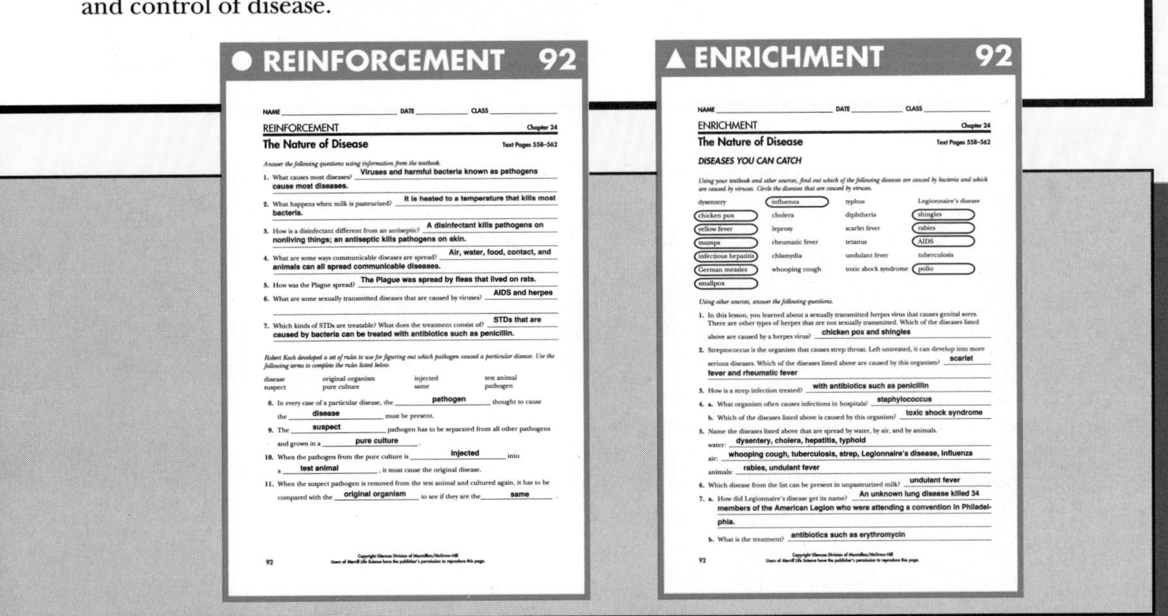

● REINFORCEMENT 92

NAME ______ DATE ______ CLASS ______

REINFORCEMENT — Chapter 24

The Nature of Disease — Text Pages 558–562

Answer the following questions using information from the textbook.

1. What causes most diseases? **Viruses and harmful bacteria known as pathogens cause most diseases.**
2. What happens when milk is pasteurized? **It is heated to a temperature that kills most bacteria.**
3. How is a disinfectant different from an antiseptic? **A disinfectant kills pathogens on nonliving things; an antiseptic kills pathogens on skin.**
4. What are some ways communicable diseases are spread? **Air, water, food, contact, and animals can all spread communicable diseases.**
5. How was the Plague spread? **The Plague was spread by fleas that lived on rats.**
6. What are some sexually transmitted diseases that are caused by viruses? **AIDS and herpes**
7. Which kinds of STDs are treatable? What does the treatment consist of? **STDs that are caused by bacteria can be treated with antibiotics such as penicillin.**

Robert Koch developed a set of rules to use for figuring out which pathogen caused a particular disease. Use the following terms to complete the rules listed below.

disease	original organism	injected	test animal
suspect	pure culture	same	pathogen

8. In every case of a particular disease, the **pathogen** thought to cause the **disease** must be present.
9. The **suspect** pathogen has to be separated from all other pathogens and grown in a **pure culture**.
10. When the pathogen from the pure culture is **injected** into a **test animal**, it must cause the original disease.
11. When the suspect pathogen is removed from the test animal and cultured again, it has to be compared with the **original organism** to see if they are the **same**.

92

▲ ENRICHMENT 92

NAME ______ DATE ______ CLASS ______

ENRICHMENT — Chapter 24

The Nature of Disease — Text Pages 558–562

DISEASES YOU CAN CATCH

Using your textbook and other sources, find out which of the following diseases are caused by bacteria and which are caused by viruses. Circle the diseases that are caused by viruses.

dysentery	(influenza)	typhus	Legionnaire's disease
(chicken pox)	cholera	diphtheria	(shingles)
(yellow fever)	leprosy	scarlet fever	(rabies)
(mumps)	rheumatic fever	tetanus	(AIDS)
(infectious hepatitis)	chlamydia	undulant fever	tuberculosis
(German measles)	whooping cough	toxic shock syndrome	(polio)
(smallpox)			

Using other sources, answer the following questions.

1. In this lesson, you learned about a sexually transmitted herpes virus that causes genital sores. There are other types of herpes that are not sexually transmitted. Which of the diseases listed above are caused by a herpes virus? **chicken pox and shingles**
2. Streptococcus is the organism that causes strep throat. Left untreated, it can develop into more serious diseases. Which of the diseases listed above are caused by this organism? **scarlet fever and rheumatic fever**
3. How is a strep infection treated? **with antibiotics such as penicillin**
4. a. What organism often causes infections in hospitals? **staphylococcus**
   b. Which of the diseases listed above is caused by this organism? **toxic shock syndrome**
5. Name the diseases listed above that are spread by water, by air, and by animals.
   water: **dysentery, cholera, hepatitis, typhoid**
   air: **whooping cough, tuberculosis, strep, Legionnaire's disease, influenza**
   animals: **rabies, undulant fever**
6. Which disease from the list can be present in unpasteurized milk? **undulant fever**
7. a. How did Legionnaire's disease get its name? **An unknown lung disease killed 34 members of the American Legion who were attending a convention in Philadelphia.**
   b. What is the treatment? **antibiotics such as erythromycin**

92

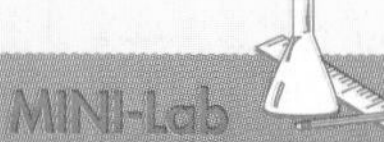

Teaching the Activity: Try using dairy products, such as fresh milk, sour milk, yogurt, and liquid from sour cream that is older than the sell-by date.

Answer: Length of time will vary with age of product used. By eating and drinking fresh foods, the contact with potentially harmful bacteria will be reduced.

MINI-Lab

ASSESSMENT

Performance: To further assess students' understanding of bacteria on food, see USING LAB SKILLS, question 11, on page 576.

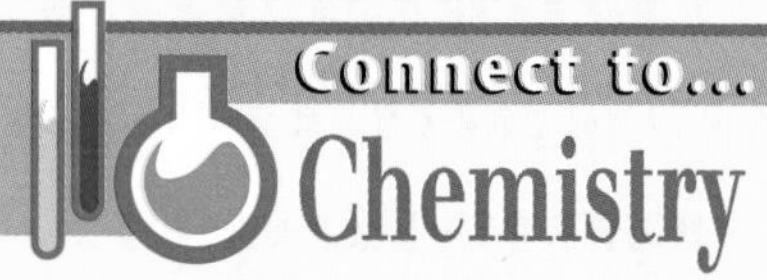

Answer: Surface tension is an elastic-like force on the surface of a body, usually a liquid. Answers will vary but may include concentration, time, temperature, pH, type and number of organisms.

CHECK FOR UNDERSTANDING

▶ Have students explain the difference between disinfectants and antiseptics.

▶ Have students infer what a germicide is.

RETEACH

Review the concepts of microorganisms as causes of infections and the need to keep clean.

EXTENSION

For students who have mastered this section, use the **Reinforcement** and **Enrichment** masters or other OPTIONS.

MINI QUIZ

Use the Mini Quiz to check students' recall of chapter content.

1. **Harmful bacteria and viruses are called _____.** *pathogens*
2. **What kind of substances are used to kill microbes on people?** *antiseptics*
3. **STDs are caused by _____.** *viruses, bacteria*

MINI-Lab

Are bacteria present in food?

Methylene blue is a dye used to detect bacteria. The faster the color fades when we add the dye to a substance, the more bacteria the substance contains. Use the samples provided by your teacher to compare the amount of bacteria in each. Label four test tubes 1, 2, 3, and 4. Fill the first three half full with the substance to be tested. Fill the fourth with water. Add 20 drops of methylene blue to each test tube. Add 2 drops of mineral oil to each tube. Place the tubes into a warm water bath. Write down the time, and observe. Record and *compare* how long it takes each tube to lose its color. *Conclude* why it is important to eat and drink only the freshest foods.

Communicable Diseases

Have you ever shared a soft drink with a friend? Did you both use the same straw? Sharing eating or drinking implements is one way that many diseases are spread from person to person. A disease that is spread, or transmitted, from one organism to another is a **communicable disease.**

What causes a communicable disease? Communicable diseases are caused by agents such as viruses, pathogenic bacteria, protists, and fungi. Some diseases caused by these agents are given in Table 24-1.

Table 24-1

DISEASES AND THEIR AGENTS			
Bacteria	**Protists**	**Fungi**	**Viruses**
Tetanus Tuberculosis Typhoid fever Strep throat Pink eye Bacterial pneumonia Plague	Malaria Sleeping sickness Mosquito	Athlete's foot Ringworm Virus	Colds Influenza AIDS Measles Mumps Yellow fever Polio Smallpox Viral pneumonia

Communicable diseases are spread through water and air, on food, by contact with contaminated objects, and by vectors. City water systems and swimming pools add chlorine to prevent disease. Many diseases are spread through air. When you have a cold and sneeze, you hurl thousands of virus particles through the air. Each time you turn a doorknob, or press the button on a water fountain at school, your skin comes in contact with bacteria and viruses. Finally, some of the most dangerous diseases are transmitted by sexual contact.

Examples of vectors that spread disease are rats, mice, flies, fleas, mosquitoes, birds, and cats and dogs. Six hundred years ago, the bacterium that causes Plague was spread by fleas that lived on rats.

The spread of communicable diseases is closely watched by agencies in every country. In the United States, the Centers for Disease Control (CDC) monitors the spread of diseases throughout the country. The CDC also watches for diseases brought into the country.

Many chemical disinfectants lower the surface tension of bacteria, allowing the chemical to be more easily absorbed. Find out what surface tension is. What makes a disinfectant effective?

OPTIONS

ASSESSMENT—ORAL

▶ **How might a cold virus be passed to all members of a family?** *Through contact with one another's eating and drinking utensils and food, contaminated handkerchiefs, and coughs and sneezes.*

▶ **Infer how disease-carrying fleas were transported from one area to another during the Plague of 1665 in London.** *The fleas were on rats and on people.*

▶ **Why might a disease brought in from outside the country be especially dangerous?** *There may not be proper medicines, vaccines, or methods of treatment available here. Some students will also say that people would not have resistance to the disease.*

Sexually Transmitted Diseases

Diseases transmitted from person to person during sexual contact are called **sexually transmitted diseases (STDs).** You can become infected with an STD by engaging in sexual activity with an infected person. STDs are caused by both viruses and bacteria. 3

In Chapter 8, you learned that HIV is a latent virus that can exist in blood and body fluids. You can contract AIDS by having sex with an HIV-infected person, or by sharing an HIV-contaminated needle used to inject drugs. AIDS is also transmitted by contaminated blood transfusions, and a pregnant female with AIDS may infect her child when the virus passes through the placenta or when nursing after birth. There is no vaccine to prevent AIDS and no medication to cure it. In the next section, more information will be given on how AIDS breaks down the immune system.

Genital herpes causes painful blisters on the sex organs. Herpes can be transmitted by an infected mother to her child during birth. The herpes virus, also a latent virus, hides in the body for long periods and then reappears suddenly. There is no cure for herpes, and there is no vaccine to prevent it.

Gonorrhea and chlamydia are STDs caused by bacteria that may not produce any symptoms. When symptoms do appear, they may include painful urination, genital discharge, and sores. Penicillin and other antibiotics are used to treat these diseases. If left untreated, either disease can cause sterility, the inability to reproduce.

Table 24-2

STDs	
Agent	**Disease**
Bacteria	Gonorrhea Chlamydia Syphilis
Viruses	Genital Herpes AIDS Genital warts

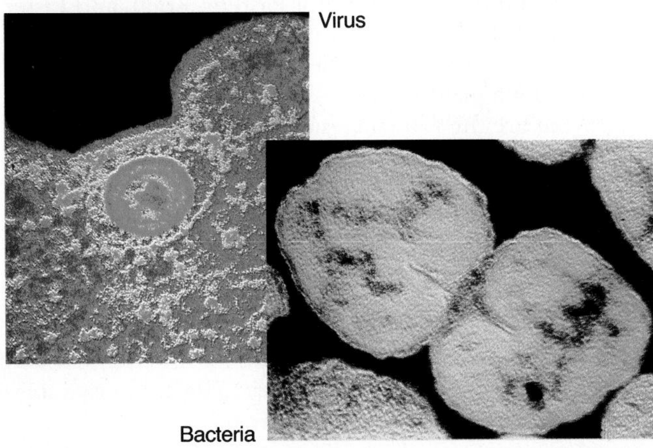

Figure 24-3. Sexually transmitted diseases are caused by certain viruses and bacteria.

CONCEPT DEVELOPMENT

▶ **What steps are taken to purify drinking water taken from a reservoir or river?** *The steps of water purification include sedimentation, coagulation, filtration, and chlorination.*

▶ Discuss the derivation of the term *venereal disease* formerly used to label STDs.

3 CLOSE

FLEX Your Brain

Use the Flex Your Brain activity to have students explore DISEASE.

Flex Your Brain

ASSESSMENT

Portfolio: Use the Flex Your Brain activity to reinforce critical thinking and problem solving skills.

▶ Have students discuss AIDS in terms of its transmission by sexual contact and the sharing of needles used in drug injection. Drug abuse is a topic of Chapter 25.

▶ Ask questions 1-3 and the **Apply** Question in the Section Review.

PROGRAM RESOURCES

From the **Teacher Resource Package** use:

Activity Worksheets, page 218, Mini-Lab: Are bacteria present in food?

Activity Worksheets, page 5, Flex Your Brain.

Use **Laboratory Manual,** page 161, Communicable and Noncommunicable Diseases.

MINI-Lab

Answer: At the five-hour point, there will be 32 768 bacteria. To draw the graph, students should label the vertical axis *thousands of bacteria* and the horizontal axis *time in hours.* Students should learn that bacteria divide very quickly and can take over the body with amazing speed. Tell students that it is very important to take *all* of any antibiotic that is prescribed and not to quit when feeling better.

MINI-Lab
ASSESSMENT

Performance: To further assess students' understanding of bacterial multiplication, have them multiply their totals by 10, by 100, and by 1000.

SECTION REVIEW ANSWERS

1. Pasteur—discovered bacteria could cause diseases; Koch—developed a way to isolate and grow one type of bacteria at a time; Lister—related infection and cleanliness

2. Answers will vary. Possible answers are virus—viral pneumonia; bacterium—strep throat; protist—sleeping sickness; fungus—athlete's foot.

3. (1) a sore throat for 10-14 days, (2) rash, fever, or swollen glands, (3) symptoms may disappear, (4) cardiovascular and nervous systems infected

4. Apply: Answers will vary. Koch used strict isolation techniques and retested the suspected pathogen. He used controlled techniques.

5. Connect to Physics: Ultraviolet light kills germs on the surface of objects and substances.

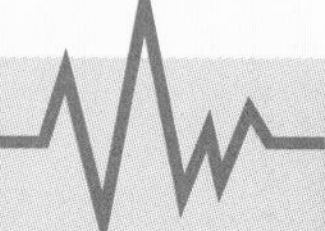

Skill Builder

You can infer that not washing hands means that many viruses and bacteria accumulate on the skin and will be spread to other people and surfaces, increasing the chance of spreading disease.

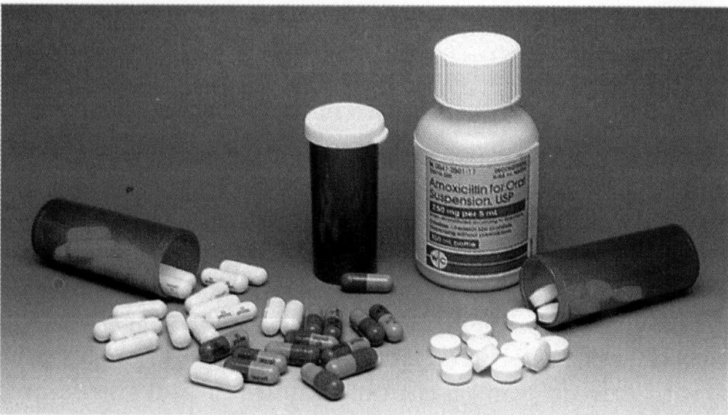

Figure 24-4. Penicillin is used to treat many diseases including some STDs.

MINI-Lab

How fast do bacteria reproduce? Bacteria divide every 20 minutes. If your body did not defend itself, it would be invaded by thousands of bacteria in just a few hours. You can't believe that? Then make a chart like the one below. Complete it to the fifth hour. How many bacteria will there be after five hours? *Graph* your data. Can you see why it is important to take antibiotics promptly if you have an infection?

Time	Number of Bacteria
0 hours 0 minutes	1
20 minutes	2
40 minutes	4
1 hour 0 minutes	8
20 minutes	
40 minutes	

Syphilis has several stages. In stage 1, a sore apears on the mouth or genitals that lasts 10 to 14 days. Stage 2 may involve a rash, a fever, and swollen lymph glands. During stage 3, these symptoms then may also disappear. The victim often believes that the disease has gone away, but it hasn't. In Stage 4, syphilis may infect the cardiovascular and nervous systems. At this point, it is too late to treat syphilis. Nerve damage and death may result. Syphilis can be treated and cured with antibiotics only in the early stages.

Sexually transmitted diseases are difficult to treat. For years, penicillin was used to treat syphilis. However, the organism that causes syphilis has become resistant to the antibiotic in some persons. In 1989, an artificial form of penicillin was made in the laboratory that may be effective against the disease.

SECTION REVIEW

1. How did the discoveries of Pasteur, Koch, and Lister help in the battle against disease?
2. List a communicable disease caused by a virus, a bacterium, a protist, and a fungus.
3. What are the four stages of syphilis?
4. **Apply:** In what ways does Koch's procedure follow the scientific method?
5. **Connect to Physics:** Find out how ultraviolet light is used to disinfect.

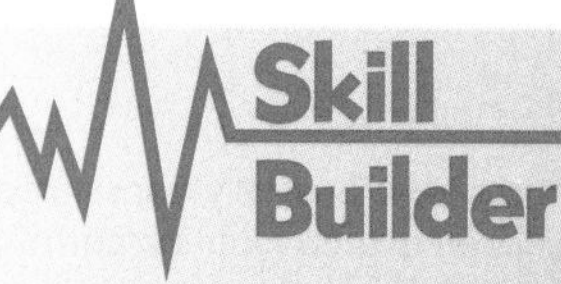

Skill Builder

Recognizing Cause and Effect

How is not washing your hands related to the spread of disease? If you need help, refer to Recognizing Cause and Effect in the Skill Handbook on page 683.

OPTIONS

PROGRAM RESOURCES

From the **Teacher Resource Package** use:

Activity Worksheets, page 219, Mini-Lab: How fast do bacteria reproduce?

Cross-Curricular Connections, page 28, The History of Penicillin.

Use **Laboratory Manual,** page 163, Sexually Transmitted Diseases.

Skillbuilder
ASSESSMENT

Performance: Assess students' abilities to Recognize Cause and Effect by having them describe the effect of sneezing on the spread of microorganisms.

Your Immune System 24-2

Objectives

- Explain the natural defenses your body has against disease.
- Describe differences between active and passive immunity.
- Explain how HIV affects the immune system.

New Science Words

immune system
antigens
antibody
active immunity
passive immunity
lymphocytes

Natural Defenses

A healthy body is like a well-equipped fortress. Your **immune system** is a complex group of defenses that your body has to fight disease. It is made up of cells, tissues, organs, and body systems that fight bacteria, viruses, harmful chemicals, and cancer cells.

In Section 24-1, harmful bacteria were discussed for their disease-causing properties. However, most bacteria do not cause disease. Millions of helpful bacteria live on your skin and give you your first line of defense by killing many harmful types of bacteria. Disease-causing bacteria enter your body through breaks in your skin. 1 Even then, your body is not defenseless. It mobilizes a series of defenses against disease-causing intruders.

Several other body systems maintain health. Your circulatory system contains white blood cells that engulf and digest foreign organisms and chemicals. These white blood cells constantly patrol the body, sweeping up and digesting bacteria that manage to get into the body. They slip between cells in the walls of capillaries to destroy bacteria around cells. When the white cells cannot destroy the bacteria fast enough, a fever may develop. But fever generally also helps to fight the pathogen.

Your respiratory system contains cilia and mucus to trap pathogens. When you cough, you expel trapped bacteria. In the digestive system, enzymes in the stomach, pancreas, and liver destroy pathogens. Hydrochloric acid 2 in your stomach kills bacteria that enter your body on food that you eat.

All of these processes are general defenses that work to keep you disease-free. But if you do get sick, your body has another line of defense in the form of active and passive immunity.

Figure 24-5. White blood cells leave capillaries and engulf harmful bacteria in surrounding tissues.

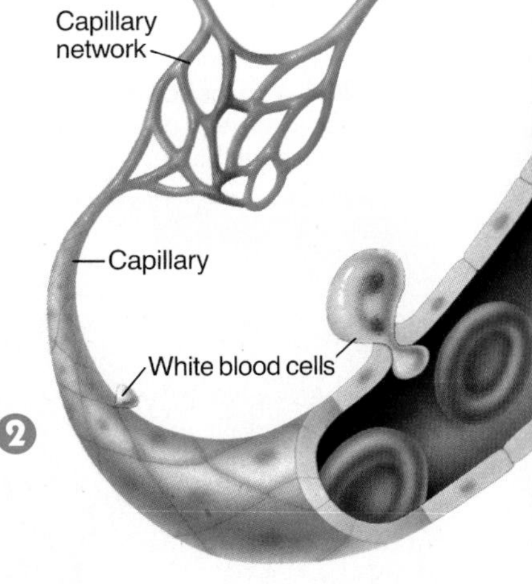

PREPARATION

SECTION BACKGROUND

- Antibodies from a mother are passed in the breast milk fed to her infant.
- A difficulty in finding a vaccine against HIV is due to the fact that HIV has the ability to evade detection by the immune system. HIV can be inactive for a long time.

PREPLANNING

- To prepare for Activity 24-1, obtain a rotting apple or prepare one by scratching a fresh apple and rubbing dirt into the wounds. Allow this prepared apple to rot several days before use.

1 MOTIVATE

- Have students infer how the body might prevent disease-causing bacteria from entering.
- Have students discuss any diseases they had and how they were treated. Also have them recall any diseases for which they received a vaccination.

ENRICHMENT

- Have students define chemotherapy and find out how chemotherapy is used in the control and treatment of disease. Their definitions and use of the term should encompass more than just cancer.
- Have students investigate the actual major causes of death of AIDS victims.

TYING TO PREVIOUS KNOWLEDGE: Recall the body systems studied so far and the various tissues, organs, and structures in the circulatory, respiratory, digestive, and integumentary systems that have a role in defending against microbes.

2 TEACH

Key Concepts are highlighted.

CONCEPT DEVELOPMENT

▶ Have students discuss how their bodies responded to a minor cut and describe the healing process. Contrast the healing process for cuts that were adequately cleaned with the process for cuts that were not cleaned.

Cooperative Learning: Use the Expert Teams strategy to study active and passive immunity. Expert Teams can be used for review of these concepts.

What is an antigen?

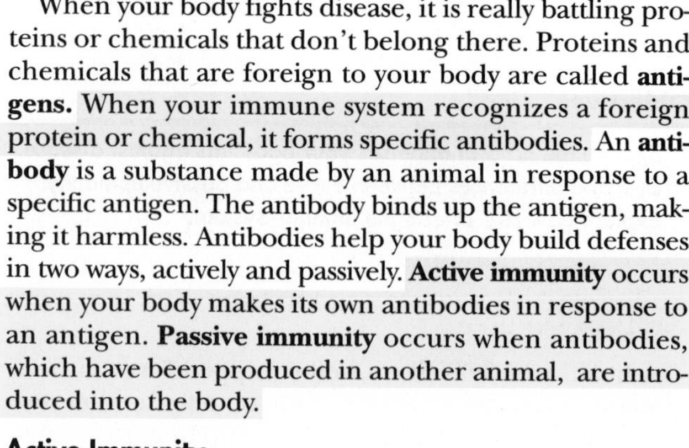

Specific Defenses

When your body fights disease, it is really battling proteins or chemicals that don't belong there. Proteins and chemicals that are foreign to your body are called **antigens.** When your immune system recognizes a foreign protein or chemical, it forms specific antibodies. An **antibody** is a substance made by an animal in response to a specific antigen. The antibody binds up the antigen, making it harmless. Antibodies help your body build defenses in two ways, actively and passively. **Active immunity** occurs when your body makes its own antibodies in response to an antigen. **Passive immunity** occurs when antibodies, which have been produced in another animal, are introduced into the body.

3

EcoTip

Reduce the use of pesticides in your home and around your garden. Many cause harmful effects on the immune systems of people.

Active Immunity

If your body is invaded by a pathogen, it immediately starts to make antibodies to inactivate the antigen. Once enough antibodies form, you get better. These antibodies stay in your blood on duty, becoming active when you again encounter the disease. Under these conditions, your body has active immunity to a particular disease.

Another way to develop active immunity to a particular disease is to be inoculated with a vaccine. A vaccine gives you active immunity against a disease without you having to get the disease first. For example, suppose a vaccine for measles is injected into your body. Your body to forms antibodies against the measle antigen. If you later encounter the virus, antibodies necessary to fight that virus are already in your bloodstream ready to destroy the pathogen. Antibodies that immunize you against one virus may not guard against a different virus. As you grow older, you will be exposed to many more types of viruses and will build a separate immunity to each one.

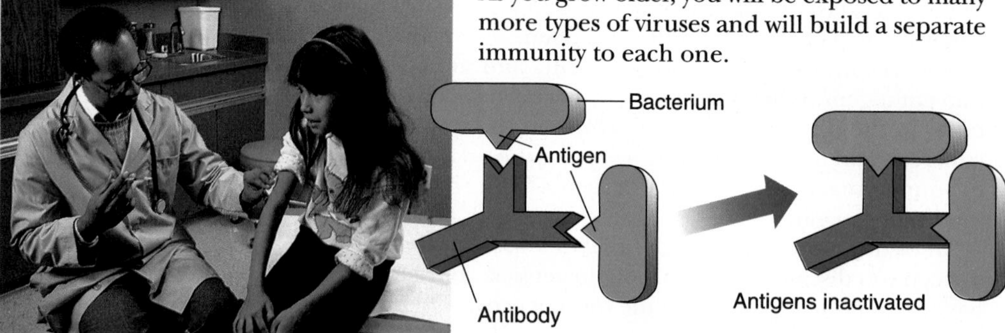

Figure 24-6. Vaccines containing weakened antigens cause the body to make antibodies.

OPTIONS

Meeting Different Ability Levels

For Section 24-2, use the following **Teacher Resource Masters** depending upon individual students' needs.

◆ **Study Guide Master** for all students.

● **Reinforcement Master** for students of average and above average ability levels.

▲ **Enrichment Master** for above average students.

Additional Teacher Resource Package masters are listed in the OPTIONS box throughout the section. The additional masters are appropriate for all students.

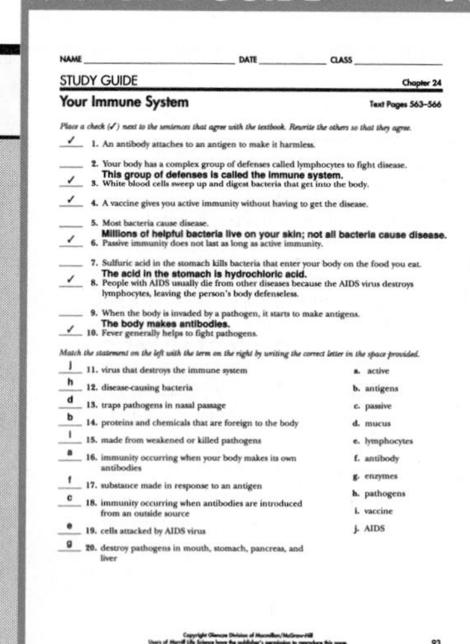

◆ STUDY GUIDE 93

NAME ______ DATE ______ CLASS ______

STUDY GUIDE — Chapter 24

Your Immune System — Text Pages 563–566

Place a check (✓) next to the sentences that agree with the textbook. Rewrite the others so that they agree.

✓ 1. An antibody attaches to an antigen to make it harmless.

___ 2. Your body has a complex group of defenses called lymphocytes to fight disease.
This group of defenses is called the immune system.

✓ 3. White blood cells sweep up and digest bacteria that get into the body.

✓ 4. A vaccine gives you active immunity without having to get the disease.

___ 5. Most bacteria cause disease.
Millions of helpful bacteria live on your skin; not all bacteria cause disease.

✓ 6. Passive immunity does not last as long as active immunity.

___ 7. Sulfuric acid in the stomach kills bacteria that enter your body on the food you eat.
The acid in the stomach is hydrochloric acid.

✓ 8. People with AIDS usually die from other diseases because the AIDS virus destroys lymphocytes, leaving the person's body defenseless.

___ 9. When the body is invaded by a pathogen, it starts to make antigens.
The body makes antibodies.

✓ 10. Fever generally helps to fight pathogens.

Match the statement on the left with the term on the right by writing the correct letter in the space provided.

j	11. virus that destroys the immune system	a. active
h	12. disease-causing bacteria	b. antigens
d	13. traps pathogens in nasal passage	c. passive
b	14. proteins and chemicals that are foreign to the body	d. mucus
i	15. made from weakened or killed pathogens	e. lymphocytes
a	16. immunity occurring when your body makes its own antibodies	f. antibody
f	17. substance made in response to an antigen	g. enzymes
c	18. immunity occurring when antibodies are introduced from an outside source	h. pathogens
e	19. cells attacked by AIDS virus	i. vaccine
g	20. destroy pathogens in mouth, stomach, pancreas, and liver	j. AIDS

93

Passive Immunity

How is passive immunity different from active immunity? Passive immunity comes from several sources. As a newborn, you were a bundle of passive immunity. You were born with all the antibodies that your mother had in her blood. However, these antibodies stayed with you only a few months. Passive immunity does not last as long as active immunity. Newborn babies loose their passive immunity in a few months. Then they are vaccinated to develop their own immunity. 

Tetanus is an example. Tetanus toxin is produced by a bacterium in soil. The toxin paralyzes muscles. Death can occur by suffocation when the muscles that control breathing become paralyzed. Tetanus antitoxin is produced in horses when they respond to injections of the tetanus antigen. Samples of antibodies made by horses are injected into humans. These antibodies provide limited immunity. Booster shots are needed throughout life, especially in the case of puncture wounds.

Select a disease from this list: diphtheria, pertussis, tetanus. Using library references, write **in your Journal** about what causes the disease; what its symptoms are; and how technological advances have developed treatment or prevention of the disease.

TECHNOLOGY

Super Sleuth!

Doctors have several tools for diagnosing diseases. One of the newest tools is magnetic resonance imaging (MRI).

Here's how it works. A person with a condition that needs to be diagnosed is placed in an MRI machine. The machine generates a powerful magnetic field in the body. In the body, the nuclei inside the atoms line up parallel to the magnetic field. Then, the nuclei are jolted with very fast radio waves. The nuclei resonate, or vibrate, like tiny tuning forks, producing a faint radio signal. This signal creates a picture of the body in great detail.

Using MRI, scientists have discovered changes in the brain that may be used to diagnose Alzheimer's disease. Doctors hope that MRI can be used to evaluate medicines for heart problems and cancer.

Think Critically: What advantages does MRI have over other methods that are used to diagnose disease?

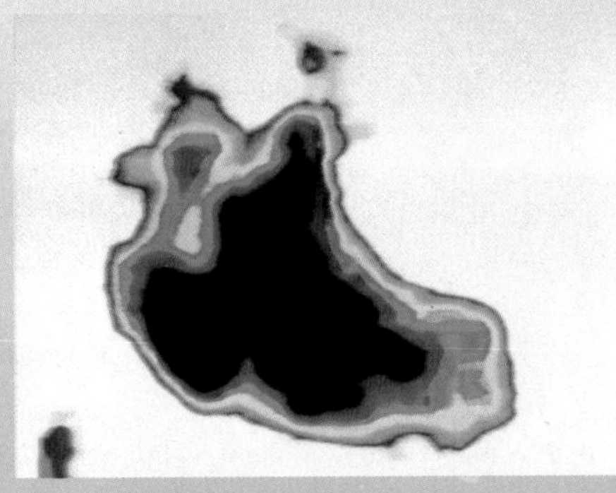

● REINFORCEMENT 93

NAME ______ DATE ______ CLASS ______

REINFORCEMENT — Chapter 24

Your Immune System — Text Pages 563–566

Answer the following questions using information from the textbook.

1. What is an antibody and how does it function? **An antibody is a substance made by an animal in response to an antigen. It attaches to the antigen, making it harmless.**
2. What is one source of passive immunity? **The mother gives her newborn baby passive immunity. Some vaccines, such as the one for tetanus, provide passive immunity.**
3. What does the AIDS virus do to the immune system? **It attacks lymphocytes that normally fight antigens and chemicals. The body can't fight off invading antigens as a result.**
4. What happens if disease-causing bacteria get through a break in the skin and enter the circulatory system? **White blood cells sweep them up and digest them.**
5. What are vaccines made from? **Vaccines are made from weakened or killed pathogens.**
6. How are pathogens trapped by and expelled from the respiratory system? **Mucus and cilia trap the pathogens; they are then expelled by coughing.**
7. What usually causes a person with AIDS to die? **diseases such as pneumonia, cancer, or tuberculosis**
8. a. How do vaccines work? **They cause the body to form its own antibodies.**
   b. What is this type of immunity called? **active immunity**
   c. What is another way to get this type of immunity? **By actually getting the disease, the body makes antibodies that will stay in the blood after the disease is gone.**
9. What defenses does the digestive system have against pathogens? **Enzymes in the mouth, stomach, pancreas, and liver destroy pathogens. Hydrochloric acid in the stomach kills bacteria entering the body on food.**
10. What is an antigen? **Proteins and chemicals that are foreign to your body are antigens.**

93

▲ ENRICHMENT 93

NAME ______ DATE ______ CLASS ______

ENRICHMENT — Chapter 24

Your Immune System — Text Pages 563–566

HUMAN IMMUNODEFICIENCY VIRUS (HIV)

By now, you've heard a lot about Acquired Immune Deficiency Syndrome, or AIDS. People who are infected with the virus that causes AIDS, the human immunodeficiency virus (HIV), are called HIV positive.

The HIV virus attaches itself to a cell and then it injects its RNA. HIV attacks specific cells in the immune system called T4 lymphocytes or helper T cells. The virus enters the T cells and takes over, turning the T cells into virus factories that produce many copies of the virus. As new virus particles are released into the bloodstream, the original T cell dies. The T cells that fight infection are gradually killed off.

Normally, healthy people have around 1000 T cells per cubic millimeter of blood. People infected with HIV usually show no outward sign of illness until their T cell counts fall to between 400 and 200 cells per cubic millimeter. At this point, their bodies lose the ability to fight infections. At this stage, the disease is called AIDS.

The first signs of a weakened immune system are chronically swollen lymph nodes and infections of the skin and mouth. People with AIDS have unexplained fevers, night sweats, diarrhea, and weight loss. Pathogens that are normally present in their bodies now become a problem. Because their immune systems aren't working well, disease-causing organisms take advantage of an opportunity to become established. These infections are called opportunistic infections. Patients with AIDS may have opportunistic infections such as Pneumocystic carinii pneumonia (PCP), cryptococcal meningitis, which is caused by a fungus, and toxoplasmosis, a parasitic infection of the brain. These, along with a cancer called Kaposi's sarcoma (KS), are the major killers of people with AIDS.

Read more about AIDS and answer the following questions.

1. What treatments for AIDS are currently available? **Answers will vary. The antiviral drug AZT is available. Other antiviral drugs, such as DDI, are still being tested. Drugs to prevent or combat opportunistic infections are in use and being tested. Pentamidine and bactrim are used to prevent and treat PCP. Immune modulators such as interferon are experimental, but are being used to treat people with KS.**
2. Is there a vaccine for AIDS? **Answers will vary. A vaccine has been developed. At the time of publication, it's still experimental. For more information, see "AIDS." *The World Book Health and Medical Annual.* 1991 ed.**
3. How are new AIDS drugs tested? **Clinical trials are run at several sites throughout the U.S. Physicians refer their patients to researchers if they wish to participate in a study.**

93

In Your JOURNAL 

Students should be directed to a current encyclopedia or microbiology reference. Emphasize that all three diseases are thought of as old and no longer common in the United States, but are still possible if preventive measures were to be stopped.

To assess this product, refer to the Performance Task Assessment Lists in **Performance Assessment in Middle School Science.**

MINI QUIZ

Use the Mini Quiz to check students' recall of chapter content.

1. **How do disease-causing bacteria usually enter the body?** *through a break in the skin*
2. **What destroys pathogens in your digestive system?** *enzymes, also hydrochloric acid in the stomach*
3. **Substances made in response to antigens are called _____ .** *antibodies*
4. **What type of immunity do newborn babies have?** *passive*
5. **HIV attacks cells called _____ .** *lymphocytes*

TECHNOLOGY

For more information on how the brain works, see "The Anatomy of Memory Loss." *Science News,* March 24, 1990, p. 204.

Think Critically: When perfected, MRI may be able to give instantaneous diagnosis. It does not involve the pain and possible problems of diagnostic surgery.

VideoDisc

STVS: Detecting the Body's Magnetic Fields, Disc 7, Side 1

STVS: Combined Immune Deficiency, Disc 7, Side 1

CHECK FOR UNDERSTANDING

▶Ask questions 1-3 and the **Apply** Question in the Section Review.

RETEACH

Use an anatomy chart or model to illustrate the various body systems involved in defense against microbes.

EXTENSION

For students who have mastered this section, use the **Reinforcement** and **Enrichment** masters or other OPTIONS.

Answer: Many valuable materials may be lost to the world if rain forests are destroyed before we even know what has been lost.

3 CLOSE

▶Have students discuss what health practices they can follow to prevent infection from disease organisms.

SECTION REVIEW ANSWERS

1. white blood cells, cilia, mucus, coughing, enzymes in the digestive system, hydrochloric acid, active immunity
2. Passive immunity does not last as long as active immunity.
3. lymphocytes
4. Apply: Some antibody injections do not last as long as others.
5. Connect to Chemistry: Through genetic engineering, scientists put genes into plants and animals. These genes produce chemicals that will kill specific disease organisms.

Skill Builder

Compare: both confer immunity; contrast: in active immunity, the body makes its own antibodies, and in passive immunity, antibodies are introduced.

Skillbuilder
ASSESSMENT
Portfolio: Use this Skillbuilder to assess students' abilities to Compare and Contrast active and passive immunity.

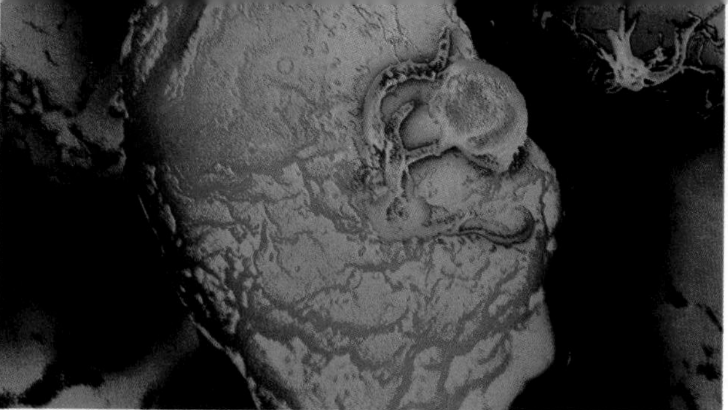

Figure 24-7. Infection by HIV can result in AIDS. HIV destroys lymphocytes, cells in the body that are vital to fighting disease.

HIV and Your Immune System

5 HIV is different from other viruses. It attacks cells in the immune system called lymphocytes. **Lymphocytes** are white blood cells throughout the lymphatic system that chemically recognize antigens. They then produce antibodies and destroy invading antigens.

Because HIV destroys lymphocytes, the body is left with no way to fight invading antigens. The whole immune system breaks down. The body is unable to fight HIV, or any other pathogen. For this reason, people with AIDS die from other diseases such as pneumonia, cancer, or tuberculosis. The victim's body becomes defenseless.

When a microbe attacks your body, it must get past all of your natural defenses. If it gets past your skin or other defenses, it encounters your immune system—your last line of defense. But if HIV has destroyed the immune system, there is no defense left.

In laboratory tests, a chemical from a tree in the rain forest of a Samoan island has been found to shield cells from HIV. How does this discovery stress the need to prevent the destruction of rain forests?

SECTION REVIEW

1. List natural defenses your body has against disease.
2. Why does passive immunity need to be renewed throughout life?
3. What cells does HIV attack?
4. **Apply:** Why might someone have to receive booster injections of antibodies?
5. **Connect to Chemistry:** Find out how genetic engineers are creating plants and animals that are pest resistant.

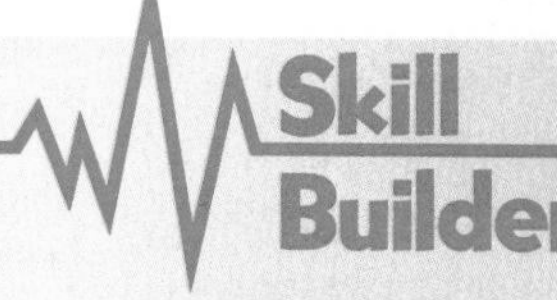

Comparing and Contrasting

Compare and contrast active and passive immunity. If you need help, refer to Comparing and Contrasting in the **Skill Handbook** on page 683.

OPTIONS

ASSESSMENT—ORAL

▶**What initiates the manufacturing of antibodies by your body?** *the invasion of a pathogen*

▶**Why is someone with HIV defenseless against other pathogens?.** *The lymphocytes that normally would fight invading antigens have been attacked and destroyed.*

Science Integration Activities is found in **Merrill Life Science Teacher Resource Manual.**

PROGRAM RESOURCES

From the **Teacher Resource Package** use:
Concept Mapping, pages 53-54.
Science Integration Activities, 24, Acids, Bases and Indicators

ACTIVITY 24-1

Microorganisms and Disease

Problem: *How do microorganisms cause infection?*

Materials  

- fresh apples (6)
- rotting apple
- alcohol
- self-sealing plastic bags (6)
- labels and pencil
- paper towels
- sandpaper
- cotton ball
- soap and water

Procedure

1. Label the plastic bags 1 through 6.
2. Put a fresh apple in bag 1 and seal the bag.
3. Rub the rotting apple over the entire surface of the remaining 5 apples. The rotting apple is your source of microorganisms. **CAUTION:** *Always wash your hands after handling microorganisms.* Take one of the apples and put it in bag number 2.
4. Drop one apple to the floor from a height of about two meters. Put this apple into bag number 3.
5. Rub one of the remaining apples with sandpaper. Place this apple in bag number 4.
6. Wash one of the last two apples with soap and water. Dry the apple well with a paper towel. Put this apple in bag number 5.
7. Use a cotton ball to spread alcohol over the last apple. Let the apple air-dry for a short time. Watch it and as soon as it is dry place the apple in bag number 6.
8. Place all of the apples in a dark place for one week. Then wash your hands.
9. Write a **hypothesis** to explain what you think will happen to each apple.
10. At the end of the week, compare all the apples. Record your observations. **CAUTION:** *Give all apples to your teacher for proper disposal.*

Data and Observations Sample Data

Apple	Observations
1	No change
2	Many brown spots, decay
3	Brown spots at breaks
4	Some brown, soft areas
5	Little or no decay
6	No change

Analyze

1. What was the purpose of apple number 1?
2. Did you *observe* any changes in apple number 2? Explain.
3. What happened on apples 3 and 4?
4. What effect did the soap and water have?

Conclude and Apply

5. Did you *observe* changes in apple number 6?
6. Why is it important to wash your hands before eating?
7. Why is it important to clean a wound?
8. Were your **hypotheses** supported?
9. *Relate* microorganisms and infection.

ACTIVITY 24-1

20 minutes

OBJECTIVE: Predict and **explain** how microbes spread.

PROCESS SKILLS applied in this activity:
▶**Observing** and **Communicating** in Procedure Step 10.
▶**Inferring** in Analyze Questions 3-5.
▶**Interpreting Data** in Analyze Questions 2-5.
▶**Forming Hypotheses** in Procedure Step 9.

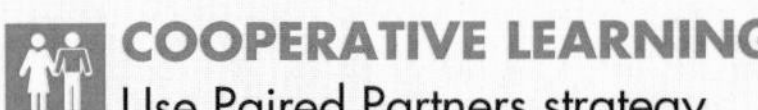

COOPERATIVE LEARNING Use Paired Partners strategy.

TEACHING THE ACTIVITY

▶Point out that brown areas found below the skin of apples (bruises) may be caused by bacteria.

PROGRAM RESOURCES

From the **Teacher Resource Package** use:

Activity Worksheets, pages 212-213, Activity 24-1, Microorganisms and Disease.

Activity ASSESSMENT

Performance: To further assess students' understanding of microorganisms and disease, see USING LAB SKILLS, Question 12, on page 576.

ANSWERS TO QUESTIONS

1. Number 1 is a control and a standard for comparison.
2. It decayed.
3. The bruise showed some decay.
4. Apple 5 decayed less than apples 2, 3, and 4.
5. Apple 6 did not decay. The skin was not broken, so pathogens could not enter the apple. Alcohol could have killed some.
6. Washing hands with soap and water removes pathogens that might be harmful if taken into the mouth.
7. Cleaning a wound removes pathogens, and infection is prevented.
8. Answers will vary with the hypotheses.
9. Pathogens can cause infection by entering damaged or cut surfaces of an organism.

SECTION 24-3

PREPARATION

SECTION BACKGROUND

▶ Some vaccination serums are made from toxins excreted as wastes by certain bacteria, such as those that produce diphtheria and tetanus. The toxins are known as exotoxins.

1 MOTIVATE

▶ Explain the nature of polio as a disease in the 1940s and 1950s. Explain what the function of an iron lung was. Emphasize that the vaccines are *not* cures, but preventive measures.

2 TEACH

Key Concepts are highlighted.

▶ Review Jenner's work with the development of a smallpox vaccine as described on page 184.

Answer: D = toxoid from bacteria that cause diptheria; **P** = vaccines from bacteria that causes pertussis; **T** = toxoid from bacteria that causes tetanus

24-3 Preventing Disease

New Science Words

vaccination

Objectives

- ▶ Explain how vaccination prevents certain diseases.
- ▶ Explain how disease can be prevented from spreading.

Inactivated substances from infectious organisms that are used for vaccinations are called toxoids. Find out what toxoids are in DPT vaccine.

Guarding Against Disease

Vaccination is the process of giving a vaccine by injection or by mouth. In Chapter 8, you learned that the first vaccine was invented by an English doctor, Edward Jenner. An epidemic of smallpox was killing many people in Europe. In 1980, the World Health Organization promoted a successful vaccination program to eliminate smallpox from the world.

In the 1950s, polio vaccines were developed by Jonas Salk and Albert Sabin. Because children are vaccinated against polio, your chances of getting this disease are near zero. Other diseases that you can be vaccinated against include measles, mumps, and diphtheria. Early in life, most of you received vaccines to protect you from these diseases.

Some of these diseases have begun to reappear. In the United States, more cases of measles were reported in 1989 than in 1988. College campuses are reporting cases of mumps in young adults. About 20 000 new cases of tuberculosis (TB) are reported each year.

CASES OF TUBERCULOSIS IN LARGE CITIES OF THE UNITED STATES

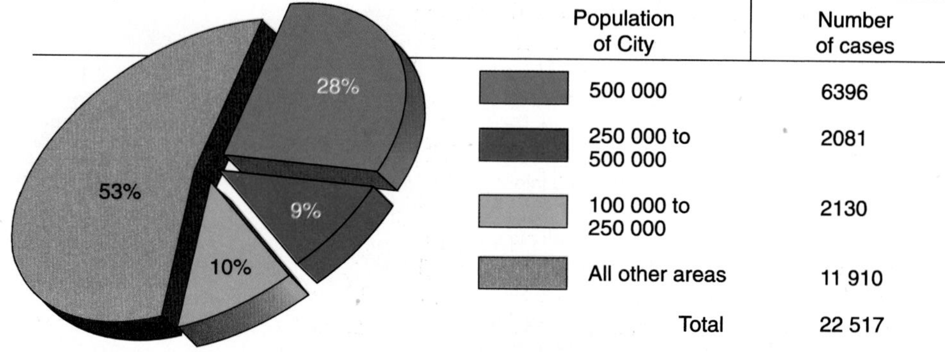

Population of City	Number of cases
500 000	6396
250 000 to 500 000	2081
100 000 to 250 000	2130
All other areas	11 910
Total	22 517

OPTIONS

Meeting Different Ability Levels

For Section 24-3, use the following **Teacher Resource Masters** depending upon individual students' needs.

◆ **Study Guide Master** for all students.

● **Reinforcement Master** for students of average and above average ability levels.

▲ **Enrichment Master** for above average students.

Additional Teacher Resource Package masters are listed in the OPTIONS box throughout the section. The additional masters are appropriate for all students.

◆ STUDY GUIDE 94

NAME ______ DATE ______ CLASS ______

STUDY GUIDE — Chapter 24

Preventing Disease — Text Pages 568–569

Match the statement on the left with the term on the right by writing the correct letter in the space provided.

Answer	Statement	Term
f	1. needed when one dose of vaccine doesn't provide long-lasting protection	a. tuberculosis
c	2. developed by Salk and Sabin	b. vaccine
a	3. often developed by people with AIDS	c. polio vaccine
i	4. time when vaccines against mumps, measles, tetanus, whooping cough, and polio should be given	d. Edward Jenner
j	5. time when a second vaccination is recommended	e. smallpox
g	6. virus that damages the immune system	f. booster shot
d	7. developed the first vaccine	g. AIDS
b	8. a weakened virus that allows you to develop immunity against a disease	h. medical records
e	9. disease eliminated by Jenner's vaccine	i. 15 months
h	10. show whether you have been vaccinated	j. 6th or 7th grade

Place a check (✓) next to the sentences that agree with the textbook. Rewrite the others so that they agree.

____ 11. Jonas Salk and Carl Sabin invented the first vaccine.
Edward Jenner invented the first vaccine.

✓ 12. Measles and mumps are beginning to reappear in the United States.

✓ 13. Adults must remember to get booster shots for tetanus.

____ 14. All babies are currently being vaccinated against mumps, measles, tetanus, whooping cough, and polio.
Many babies are not being vaccinated.

✓ 15. Cleanliness, eating right, and getting plenty of rest and exercise are some of the ways you can prevent disease.

____ 16. Over 20 000 new cases of whooping cough are being reported each year.
Over 20 000 new tuberculosis cases are being reported each year.

✓ 17. A vaccine may be injected or swallowed.

____ 18. Those who work or live with people infected with tuberculosis stand the chance of developing AIDS. **They stand the chance of developing tuberculosis from those who are infected.**

✓ 19. Tuberculosis has returned in part because people with AIDS are unable to fight off the infection, and often develop the disease.

____ 20. Doctors do not recommend getting a second vaccination.
Doctors recommend a second vaccination for some diseases such as measles and tetanus.

94

According to the Centers for Disease Control, there are two reasons for the return of these diseases. First, many babies are not being vaccinated against mumps, measles, tetanus, whooping cough, and polio. While some children have shown reactions to the whooping cough vaccine, most children receive vaccines for these diseases by the age of 15 months. Second, just one dose of vaccine does not always provide enough protection. Many adults forget that they need booster shots for tetanus and possibly measles.

Tuberculosis is returning due in part to the damaged immune systems of AIDS victims. Unable to fight off the tuberculosis bacterium, AIDS patients often develop that disease. Those who work or live with such individuals are then exposed to tuberculosis and have a greater chance of developing the disease.

What can you do to prevent becoming infected with these diseases? Make sure you have been vaccinated. Check your medical records to be sure you are vaccinated against all of these diseases. Also, doctors recommend a second vaccination for measles, especially among college students, to rebuild their immunity to the virus. Keeping clean, eating a balanced diet, getting rest and exercise, and getting medical care when you need it are ways to maintain health.

Table 24-3

VACCINATION SCHEDULE

Vaccine	Age
Oral polio	2 months 4 months 18 months 5-6 years
Diphtheria pertussis, tetanus (DPT)	2 months 4 months 6 months 18 months 5-6 years
Tetanus-diphtheria (Td)	14-16 years, every 10 years
Measles, Mumps, Rubella (MMR)	15 months
Measles	17 years
Tuberculosis (TB)	Test in U.S.
	Some countries give a vaccine

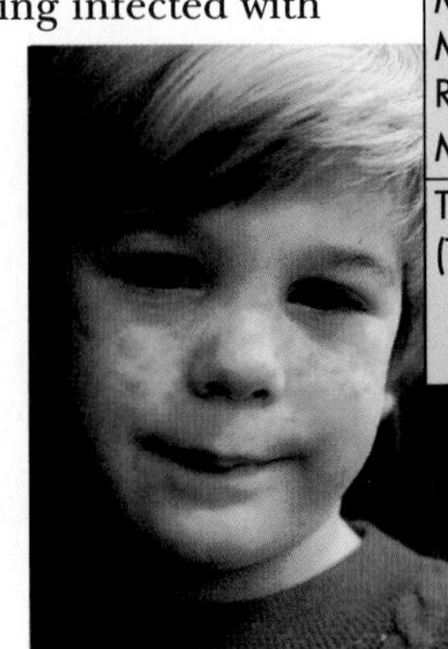

Figure 24-8. This child shows symptoms of the measles virus.

SECTION REVIEW

1. How does vaccination help prevent disease?
2. Why is vaccination important for the population of a country?
3. **Connect to Chemistry:** Find out about the use of gamma globulin as an innoculation for certain diseases.

You Decide!

The last case of smallpox was in 1980 in Somalia in Africa. Now the only places where the smallpox virus exists is in the Centers for Disease Control in Atlanta, GA, and in a similar facility in the Soviet Union. Should these countries destroy the remaining virus samples? Should they keep them? Can you think of any benefit for keeping the virus samples?

569

● REINFORCEMENT 94

NAME ______ DATE ______ CLASS ______

REINFORCEMENT Chapter 24

Preventing Disease Text Pages 568–569

Answer the following questions using information from the textbook.

1. Which diseases have started making a comeback in numbers that cause concern in the United States? **Measles, mumps, and tuberculosis have reappeared.**
2. What led to the invention of the first vaccine? **The smallpox epidemic was killing many people in Europe and led to the invention of the smallpox vaccine.**
3. What should you do to keep from being infected with a communicable disease? **Make sure you've been vaccinated, get booster vaccinations, keep clean, eat a balanced diet, get plenty of rest and exercise, don't smoke, don't use alcohol or drugs, and see a doctor when you need to.**
4. What is a vaccine? **It is a weakened virus that is injected or swallowed to develop immunity against a disease.**
5. a. When is a booster shot necessary? **When just one dose of vaccine doesn't provide enough protection, you need a booster shot.**
   b. Which diseases require booster shots? **tetanus and maybe measles**
6. Why do people with AIDS often become infected with tuberculosis? **They cannot fight off the tuberculosis infection because their own immune systems are damaged.**
7. According to the Centers for Disease Control, what are two reasons for the reappearance of some diseases? **1. Many babies are not being vaccinated; 2. One dose of the vaccine does not always provide enough protection.**
8. If you work or live with someone with tuberculosis, what would you expect to happen? **You would have a greater chance of developing tuberculosis.**
9. How can you find out whether or not you've had the right vaccinations at the appropriate times? **Check your medical records.**

94

▲ ENRICHMENT 94

NAME ______ DATE ______ CLASS ______

ENRICHMENT Chapter 24

Preventing Disease Text Pages 568–569

YOUR MEDICAL RECORDS

Perhaps you've heard your parents talk about having the measles or mumps as a child. Your grandparents may have known someone who was quarantined for whooping cough or friends who had polio. We often take our good health for granted because we get active immunity from vaccines rather than the disease itself.

Question your family members or close friends to find out what childhood diseases they've had and what diseases they've been vaccinated against. By interviewing people from different generations, you should be able to draw some conclusions about medical progress.

Make three tables like the one below: one for yourself, one for a parent or person from your parents' generation, and one for a grandparent or person from your grandparents' generation. When you fill in the table, note that a combination vaccine called DPT is now given for diphtheria, pertussis (whooping cough), and tetanus. Similarly, a combination vaccine called MMR is given for measles, mumps, and rubella (German measles).

	vaccinations (date or age)	diseases (yes or no)
smallpox		
polio		
mumps		
measles		
rubella		
diphtheria		
pertussis		
tetanus		
influenza		

Answer the following questions using the data you gathered and other sources.

1. What vaccinations did the older generations get that are no longer required? Why? **smallpox; No cases of this disease have been diagnosed for many years.**
2. Did anyone you interviewed get a disease despite having been vaccinated against that disease? If so, what could be the explanation? **They were probably vaccinated too early or needed a booster. In some instances, the vaccine may not have been effective.**
3. Even today, some people in the United States are not vaccinated. Why? **religious objections, allergy to eggs in which some live viruses are cultured, fear of adverse reaction to the vaccine, lack of knowledge**

94

CHECK FOR UNDERSTANDING

Ask questions 1-2 in the Section Review.

RETEACH

Conduct a school survey to find how many students or members of their families have had measles, chickenpox, mumps, or whooping cough.

EXTENSION

For students who have mastered this section, use the **Reinforcement** and **Enrichment** masters or other OPTIONS.

3 CLOSE

▶ Have students locate areas of the world where yellow fever is a problem. Infer why it has not been eradicated.

SECTION REVIEW ANSWERS

1. A vaccination causes the body to form antibodies against a specific antigen.

2. to prevent an epidemic or widespread outbreak of a disease

3. Connect to Chemistry: Gamma globulin from blood serum is collected from persons who have immunity to specific diseases. It is administered to persons who need antibodies against those specific diseases.

YOU DECIDE!

You might divide students into two groups and assign an opinion to each. Those in favor of destroying will probably cite getting rid of the disease for good. Those in favor of saving may say samples may be needed at some time to fight a similar disease.

PROGRAM RESOURCES

From the **Teacher Resource Package** use:

Science and Society, page 28, Immunization.

Critical Thinking/Problem Solving, page 28, A New Weapon Against the Flu.

SECTION 24-4

PREPARATION

SECTION BACKGROUND

▶ Diseases not caused by pathogens do not respond to antibodies. Chronic diseases cause more deaths than communicable diseases.

▶ For some forms of cancer, less radical operations are being performed. Combined radiation and chemotherapy are used as postoperative treatment to eliminate small masses of cancer cells in the circulatory system.

PREPLANNING

▶ To prepare for Activity 24-2, prepare the sterile nutrient agar plates. Disinfect the plates before disposing of them at the end of the activity.

1 MOTIVATE

▶ Conduct an anonymous-response class survey to determine the number of people students have known who have had heart disease or cancer. Compile the statistics to emphasize the great number of people who have noncommunicable diseases compared to those with communicable diseases.

▶ Have a representative from the American Cancer Society come to class to discuss the progress made in the detection and treatment of cancer.

TYING TO PREVIOUS KNOWLEDGE: Recall the information in Chapter 20 concerning diseases and disorders of the heart. Review the role of diet and exercise in preventing heart disease.

VideoDisc

STVS: Patient Simulator, Disc 7, Side 1
STVS: Arthritis Research, Disc 7, Side 2

24-4 Noncommunicable Disease

New Science Words

noncommunicable diseases
chronic diseases
cancer
tumor
chemotherapy
allergy
allergens

Objectives

- ▶ List two noncommunicable diseases.
- ▶ Describe the basic characteristics of cancer.
- ▶ Name two chronic diseases of the immune system.

Chronic Disease

Diseases and disorders such as diabetes, allergies, asthma, cancer, and heart disease are not caused by pathogens. These diseases are called **noncommunicable diseases** because they are not spread from one person to another. You can't "catch" them. Allergies, genetic disorders, life-style diseases, or chemical imbalances such as diabetes, are not spread by sneezes or handshakes.

Some noncommunicable diseases are called **chronic diseases** because they last a long time. Some chronic dis- 1
eases can be cured. Others cannot. Chronic diseases may result from improperly functioning organs, contact with harmful chemicals, or an unhealthy life-style. For example, your pancreas produces the hormone insulin. Diabetes is a chronic disease in which the pancreas cannot produce the amount of insulin the body needs.

Arthritis is a chronic disease that results from a faulty immune system. The immune system begins to treat the body's normal proteins as if they were antigens. The faulty immune system forms antibodies against the normal proteins in joints. Movement becomes difficult and painful.

Some chronic diseases are caused by chemicals.
Household cleaners, car 2
exhaust, and cigarette smoke are examples. Cigarette smoke has been linked with lung cancer, other lung diseases, and heart disease.

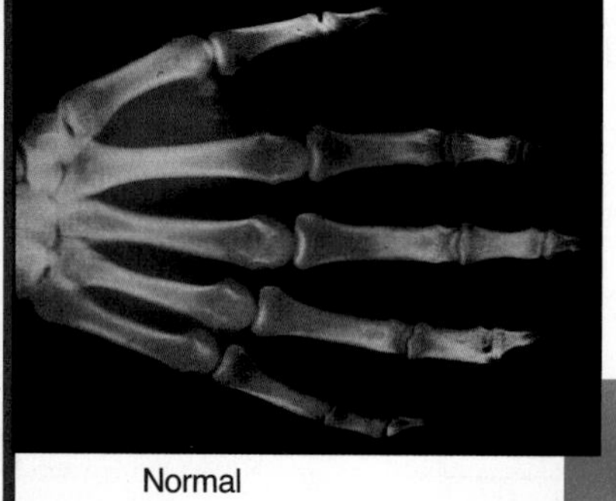

Figure 24-9. Arthritis turns the body's immune system against itself. Joints may be severely deformed.

OPTIONS

Meeting Different Ability Levels

For Section 24-4, use the following **Teacher Resource Masters** depending upon individual students' needs.

◆ **Study Guide Master** for all students.

● **Reinforcement Master** for students of average and above average ability levels.

▲ **Enrichment Master** for above average students.

Additional Teacher Resource Package masters are listed in the OPTIONS box throughout the section. The additional masters are appropriate for all students.

◆ STUDY GUIDE 95

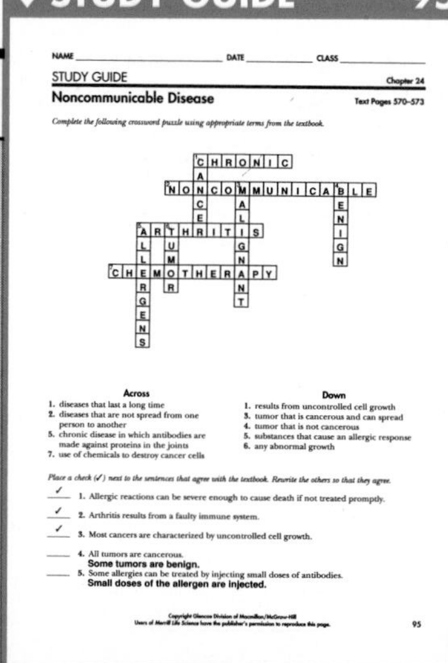

NAME ______ DATE ______ CLASS ______

STUDY GUIDE — Chapter 24

Noncommunicable Disease — Text Pages 570-573

Complete the following crossword puzzle using appropriate terms from the textbook.

Across
1. diseases that last a long time
2. diseases that are not spread from one person to another
5. chronic disease in which antibodies are made against proteins in the joints
7. use of chemicals to destroy cancer cells

Down
1. results from uncontrolled cell growth
3. tumor that is cancerous and can spread
4. tumor that is not cancerous
5. substances that cause an allergic response
6. any abnormal growth

Place a check (✓) next to the sentences that agree with the textbook. Rewrite the others so that they agree.

✓ 1. Allergic reactions can be severe enough to cause death if not treated promptly.
✓ 2. Arthritis results from a faulty immune system.
✓ 3. Most cancers are characterized by uncontrolled cell growth.
___ 4. All tumors are cancerous.
Some tumors are benign.
___ 5. Some allergies can be treated by injecting small doses of antibodies.
Small doses of the allergen are injected.

95

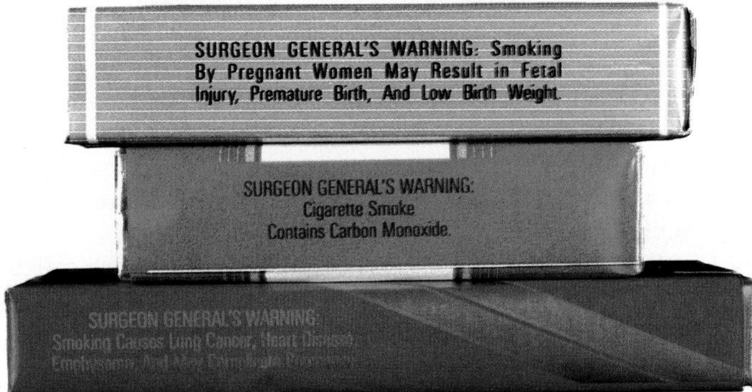

Figure 24-10. Tobacco products have been directly linked to lung cancer.

Cancer

Cancer is a major chronic disease. **Cancer** results from uncontrolled cell growth. There are many different types of cancer, but most of them have these characteristics:

1. Uncontrolled cell growth results in large numbers of cells.
2. The large number of cells do not function as a part of the body.
3. The cells take up space and interfere with normal bodily functions.
4. The cells do not remain in one place, but travel throughout the body. In this way, cancer spreads and grows in many areas of the body.

A **tumor** is an abnormal growth anywhere in the body. If a tumor is near the surface of the body, you may feel it as a lump. But if it is deep inside the body, it may go undetected for years. There are two kinds of tumors, benign 3
and malignant (muh LIHG nuhnt). Benign tumors are not cancerous. Malignant tumors are cancerous and can spread.

Treatment for cancer includes surgery to remove cancer tissue, radiation with X rays to kill cancer cells, and chemotherapy. **Chemotherapy** is the use of chemicals to destroy cancer cells.

Cancers are complicated, and no one fully understands how they form. Some scientists hypothesize that cancer cells form regularly in the body. However, the immune system destroys them. Only if the immune system fails or becomes overwhelmed, does the cancer begin to expand. Smoking, poor diet, and exposure to harmful chemicals encourage some cancers to form.

Science and MATH

Aside from the health reasons for not smoking, look at an economic one. **In your Journal,** *calculate* the amount of money a person will save in a week and in a year by not smoking. Assume that the person smokes a pack a day and the cost is $2.00 per pack.

Table 24-4

CAUSES OF CANCER	
Carcinogens	
Substances that cause cancer:	
smoking	air pollution
asbestos dust	high fat diet
ultraviolet light	aflatoxins
radiation	
Oncogenes	
Genes that cause a normal cell to become cancerous	

● REINFORCEMENT 95

NAME ______ DATE ______ CLASS ______

REINFORCEMENT — Chapter 24

Noncommunicable Disease — Text Pages 570–573

Write the letter of the term or phrase that best completes each sentence.

c 1. Diseases that are not caused by pathogens are ____.
a. communicable c. noncommunicable
b. chronic d. viral

a 2. A tumor is usually located ____.
a. anywhere in the body c. near the surface of the body
b. deep inside the body d. on the skin

b 3. Proven cancer treatments may involve ____.
a. injections of megadoses of vitamins c. physical therapy
b. radiation d. gene therapy

d 4. A chronic disease involving a faulty immune system is ____.
a. diabetes b. cancer c. heart disease d. arthritis

b 5. A tumor that can spread is said to be ____.
a. chronic b. malignant c. pathogenic d. benign

c 6. When you come in contact with an allergen, your immune system forms ____.
a. hormones b. lymphocytes c. antibodies d. antigens

a 7. An allergy is a strong reaction of the ____.
a. immune system c. respiratory system
b. circulatory system d. digestive system

d 8. Some noncommunicable diseases are called chronic diseases because ____.
a. they are present at birth c. they are short-lived
b. they are inherited d. they last a long time

d 9. Cigarette smoke has been linked with all of the following **EXCEPT** ____.
a. lung cancer b. lung diseases c. heart disease d. arthritis

c 10. Diabetes is a chronic disease resulting from ____.
a. contact with harmful chemicals c. an improperly functioning organ
b. an unhealthy lifestyle d. cigarette smoking

Complete the following statements by selecting the appropriate terms from the list.

normal function uncontrolled large travel

Most cancers are characterized by:

1. **Uncontrolled** cell growth results in **large** numbers of cells.
2. The large number of cells do not **function** as part of the body.
3. The cells take up space and interfere with **normal** bodily functions.
4. The cells do not remain in one place, but **travel** all over the body.

95

▲ ENRICHMENT 95

NAME ______ DATE ______ CLASS ______

ENRICHMENT — Chapter 24

Noncommunicable Disease — Text Pages 570–573

ASTHMA: A CHRONIC DISEASE

Over 9 million Americans have asthma, and about a third to one-half of them are children. Asthma causes children to miss more days of school, have more hospital stays, and have more restriction in their activities than does any other disease.

It can be frightening to see someone having an asthma attack—wheezing, coughing, breathing rapidly, or even gasping for air. It is important to understand that although this disease is chronic, it can be controlled. Asthma is not contagious.

Anyone can develop asthma. In children, the symptoms usually appear before age three. Some children seem to outgrow the disease, but often it reappears. In most cases, people learn to prevent asthma attacks and how to treat them when they do occur. Asthma patients must learn what triggers their asthma attacks so that they can prevent them.

During an asthma attack, the muscles around the bronchial tubes tighten and go into a spasm. Extra mucus is produced and interferes with breathing. The lining of the bronchial tubes swells, making the airway smaller. The person has trouble passing air in and out and begins to wheeze or cough.

Asthma attacks can be caused by many things. These causes are usually specific for each person and are often a combination of things. The following list includes most of the better-known "triggers" for asthma.

- allergies—to grasses, trees, weeds, pollens, mold, dust, animal fur, food, and drugs
- respiratory infections—colds, sore throats
- irritants—perfume, smoke, air pollution
- physical exercise
- sudden change in air temperature or humidity
- emotional stress

Several medications are available. Bronchodilators relax the bronchial muscles in the lungs. Cortisone reduces the swelling. Antibiotics fight infections. Allergy shots can reduce the sensitivity to allergens. Newer medications help reduce the allergic response. Also, a person can avoid things he or she is allergic to. This may mean getting rid of pets or avoiding certain foods. It is also wise to stay away from cigarette smoke.

Answer the following questions.

1. Do you know anyone who has asthma? Find out how they deal with their chronic disease. Are they able to participate in most activities? **Answers will vary, but most asthmatics can participate in many activities with a doctor's supervision and medication.**
2. Think about ways your school could make it easier for people with asthma to be just like everyone else. Keep in mind that many schools do not let children carry inhalers (for medicine) with them. Why do you think this is a good or bad policy? In addition, many classrooms have furry animals or birds. What options are there for children whose allergies to these animals could trigger an asthma attack? **Answers will vary, but may include having some "no pets" classrooms and training for teachers and students on what to do when someone has an asthma attack.**

95

2 TEACH

Key Concepts are highlighted.

CONCEPT DEVELOPMENT

▶ Obtain pamphlets from health agencies such as the American Heart Association, American Cancer Society, American Diabetes Association, and Arthritis Foundation to learn about new treatment methods for these chronic diseases.

▶ Have students collect newspaper and magazine articles about the health effects of inhaling passive smoke from tobacco products.

▶ Have students distinguish between an antigen and an allergen. Note that an allergen is an antigen that causes an allergic response in the body.

▶ Have a student volunteer who has an allergy explain to the class how the doctor determined the exact cause of the allergy and how he or she is being treated for it.

MINI QUIZ

Use the Mini Quiz to check students' recall of chapter content.

1. **Can some chronic diseases be cured?** *yes*
2. **What are some chemicals that can cause chronic disease?** *household cleaners, car exhaust, cigarette smoke*
3. **The two kinds of tumors are ____.** *benign, malignant*
4. **Substances that cause allergic responses are called ______.** *allergens*

Science and MATH

At the rate of $2.00 per pack per day, a smoker would save $14.00 a week and $728 a year by not smoking.

Connect to... Chemistry

Answer: Antihistamines may cause drowsiness, dry mouth, and blurred vision. For this reason users are cautioned against driving or operating heavy machinery.

REVEALING MISCONCEPTIONS

▶Many persons are diagnosed as being allergic to dust and feathers. In many cases the persons are reacting to the feces of a very tiny mite that lives on pillows, sheets, and mattresses, feeding on shed epidermal cells.

CHECK FOR UNDERSTANDING

Have students give a general definition of *cancer* and briefly summarize the four major characteristics of cancer as related to a specific type of cancer (such as a brain tumor, leukemia, skin cancer, breast cancer, or colon cancer).

RETEACH

Cooperative Learning: Using the Expert Teams strategy, review the information about cancer. Check to see that students understand that there are some ways individuals can reduce the risk of developing cancer.

EXTENSION

For students who have mastered this section, use the **Reinforcement** and **Enrichment** masters or other OPTIONS.

Allergies

Have you ever broken out in an itchy rash after eating a favorite food? An **allergy** is an overly strong reaction of the immune system to a foreign substance. Many people have allergic reactions to cosmetics, shrimp, strawberries, and bee stings. Allergic reactions to some things such as antibiotics can even be fatal.

Antihistamines are chemicals that are used to counteract the histamines released in response to allergens. What side effects may be associated with antihistamines?

4 Substances that cause the allergic response are called **allergens.** These are substances that the body would normally respond to as a mild antigen. Chemicals, dust, grass pollen, food, pollen, molds, and some antibiotics are allergens for some sensitive people. When you come in contact with an allergen, your immune system forms antibodies, and your body may react in many ways. When the body responds to an allergen, chemicals called histamines are released. Histamines promote red, swollen tissues. Allergic reactions are sometimes treated with antihistamines. Pollen is an allergen that causes a stuffy nose, breathing difficulties, watery eyes, and a tired feeling in some people. Some foods cause blotchy rashes such as hives or, stomach cramps and diarrhea. Most allergic reactions are minor. But severe allergic reactions can occur, causing shock and even death if not treated promptly. Some severe allergies are treated with repeated injections of small doses of the allergen, which allows the body to become less sensitive to the allergen.

Noncommunicable diseases aren't spread from person to person. So, you don't have to worry about catching

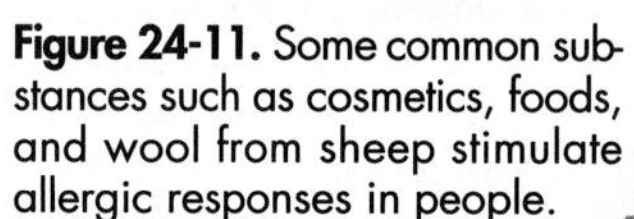

Figure 24-11. Some common substances such as cosmetics, foods, and wool from sheep stimulate allergic responses in people.

OPTIONS

ASSESSMENT—ORAL

▶**The symptoms of an allergic reaction include a runny nose, watery eyes, and sneezing. These conditions are caused by chemicals called histamines. Can you infer what the effect of antihistamine medicines is on the body?** *The antihistamines reduce the effects of the histamines but do not cure the allergy.*

PROGRAM RESOURCES

From the **Teacher Resource Package** use:

Activity Worksheets, page 5, Flex Your Brain.

PROBLEM SOLVING

Fighting TB

Annie Dodge Wauneka, the daughter of a Navajo chief, has spent much of her life fighting tuberculosis (TB). She was the first woman elected to the Tribal Council, and was appointed chairperson of the health committee. TB killed many people on the reservation each year, so Wauneka decided to learn everything she could to fight it. She wrote a Navajo-English medical dictionary and got Navajo and non-Navajo doctors to work together on the disease. She also encouraged Navajo patients to trust non-Navajo doctors and to get early diagnosis. Her war against TB was so successful that she was awarded the Freedom Award, and the Navajo gave her the name "Warrior Who Scouts the Enemy."

Think Critically: Annie Dodge Wauneka was not a doctor. How was she able to help fight a disease without medical knowledge?

diabetes, cancer, or allergies from someone else. But, left untreated, certain noncommunicable diseases can be as deadly as AIDS or untreated syphilis. After all, most people die from chronic heart disease and cancer, and not from a virus or bacterium.

SECTION REVIEW

1. What are two noncommunicable diseases?
2. What are the basic characteristics of cancer?
3. What are two chronic diseases of the immune system?
4. **Apply:** Joel has an ear infection. The doctor prescribes an antibiotic. After taking the antibiotic, Joel breaks out in a rash and has difficulty breathing. What is happening to him? What should he do immediately?
5. **Connect to Chemistry:** Find out about diseases associated with a diet that is deficient in iron.

Making and Using Tables

Make a table that lists some chronic diseases and their treatments. Use the information you have read in this section. If you need help, refer to Making and Using Tables in the **Skill Handbook** on page 690.

Skill Builder

Chronic disease	Treatment
Cancer	Radiation Chemotherapy
Allergies	Injections of allergens
Diabetes	Insulin

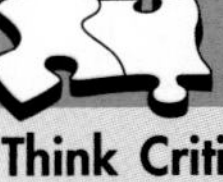

PROBLEM SOLVING

Think Critically: Wauneka helped with the social aspects of medical treatment. She brought together medical "experts" from both cultural traditions. She gave them ways to communicate with each other. She helped overcome the conflicts and distrust between native healers, patients, and non-Navajo doctors.

3 CLOSE

FLEX Your Brain

Use the Flex Your Brain activity to have students explore ALLERGIES.

Flex Your Brain

ASSESSMENT

Portfolio: Use the Flex Your Brain activity to reinforce critical thinking and problem solving skills. Students might begin in Step 2 by listing what they know of hayfever.

▶ Ask questions 1-3 and the **Apply** Question in the Section Review.

▶ Have students find out about noncommunicable diseases of the nervous system, such as cerebral palsy, epilepsy, and multiple sclerosis.

SECTION REVIEW ANSWERS

1. Answers may be: diabetes, allergies, asthma, cancer, heart disease.

2. uncontrolled cell growth, large number of cells that do not function as part of the body, cells interfere with normal bodily functions, cells travel throughout the body

3. allergies and cancer

4. Apply: Joel's immune system seems to be reacting to the antibiotic. Joel should seek immediate medical care.

5. Connect to Chemistry: Insufficient iron in the diet can cause anemia.

Skillbuilder

ASSESSMENT

Performance: Use this Skillbuilder to assess students' abilities to Make and Use Tables by having them classify other diseases and name the treatments given.

ACTIVITY 24-2

OBJECTIVE: Design and carry out an experiment to determine the chemicals that prevent growth of microorganisms.
Time: One class period to hypothesize, plan procedures, and collect materials. Fifteen minutes to inoculate plates. One-half hour to read plates and draw conclusions.

PROCESS SKILLS applied in this activity are **observing, hypothesizing, experimenting, predicting,** and **analyzing data.**

PREPARATION

Prepare nutrient agar plates, five for each cooperative group. Collect small jars, four for each group.

Cooperative Learning: Divide class into Science Investigation Teams.

SAFETY

Caution students to handle forceps carefully. Do not allow students to open plates. Plates should be given to the teacher for proper disposal. Spray the cultures with a strong household disinfectant. Reseal dishes before disposal.

HYPOTHESIZING

Student hypotheses will vary. Possible hypotheses are: "The chemicals will be equally effective in preventing microorganism growth." or "Hydrogen peroxide is most effective in preventing microorganism growth."

TEACHING THE ACTIVITY

*Refer to the **Activity Worksheets** for additional information and teaching strategies.*

• Explain what nutrient agar is and how agar in petri dishes can help them with their investigation.
• Demonstrate the correct way to inoculate agar plates.

SUMMING UP/SHARING RESULTS

• Answers will vary depending on the substances used. Household disinfectants and antiseptics (e.g. hydrogen peroxide) will have an effect. Some microorganisms are more susceptible to chemicals than others.
• Answers will vary but there should be considerable microorganism growth.
• Disinfectants and antiseptics limit the growth of microorganisms.

GOING FURTHER!

Sunlight does have a negative effect on many microorganism. Results may vary depending on the amount of time the dish was exposed to sunlight.

PROGRAM RESOURCES

From the **Teacher Resource Package** use:
Activity Worksheets, pages 214-215, Activity 24-2, Preventing Microorganism Growth.

Activity ASSESSMENT

Performance: To further assess students' understanding of microorganism growth, have them repeat their experiment using different mouthwashes, antiseptics, and disinfectants.

ACTIVITY 24-2 DESIGNING AN EXPERIMENT
Preventing Microorganism Growth

Without cleanliness, there is a good chance of getting an infection when you scrape an arm or knee. Infections are caused by microorganisms. Disinfectants are chemicals that kill or remove disease organisms from objects. Antiseptics are chemicals that kill or prevent growth of disease organisms on living tissues. What methods can be used to prevent microorganism growth?

Getting Started

You need to determine what microorganisms need to grow. Devise a method to prevent the growth of microorganisms.

Hypothesizing

Make a **hypothesis** about what method will prevent the growth of microorganisms.

Materials    

Your cooperative group will use:
- transparent tape
- sterile nutrient agar plates (5)
- filter paper
- disinfectant
- hydrogen peroxide
- pencil and labels
- scissors
- mouthwash
- metric ruler
- forceps
- alcohol
- small jars (4)

Try It!

1. Label five nutrient agar petri plates 1-5. Remove the covers from the five petri dishes and rub your finger over the agar in each plate. Put the covers back on and seal dish 5 with tape.
2. Label four jars and four 2-cm squares of filter paper: D, H, M, and A.
3. Pour 50 mL of each of the following solutions into the jars: disinfectant, 3% hydrogen peroxide, mouthwash, and alcohol. Drop each square into its corresponding solution: D=disinfectant, H=hydrogen peroxide, M=mouthwash, and A= alcohol.
4. Using forceps, remove each piece of paper and place one on the agar in each of the four plates. Cover the dishes and seal with tape. **CAUTION:** *Handle the forceps carefully.* Store the dishes in a warm, dark place for 48 hours.
5. *Predict* what will happen in each dish. After 2 days, examine the dishes. *Compare* the growth beneath and around each square.
6. Record your observations in a data table. **CAUTION:** *Give your plates to your teacher for proper disposal.*

Summing Up/Sharing Results

- What substances appeared to be most effective in controlling microorganisms? To be least effective?
- What happened to the control?
- How do disinfectants and antiseptics affect microorganisms growth?

Going Further!

Design an experiment to determine if sunlight has an effect on microorganism growth. Try it and see if there is an effect. *Compare* to the effect of antiseptics and disinfectants.

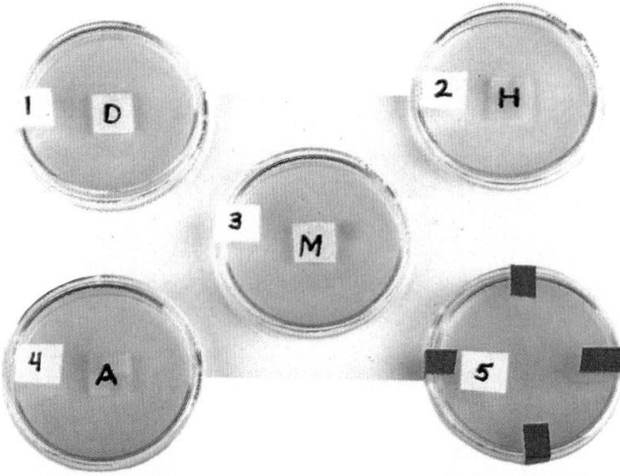

CHAPTER
REVIEW

SUMMARY

24-1: The Nature of Disease

1. Pasteur and Koch discovered that diseases are caused by microbes. Lister used disinfectants to help control microbes.

2. Communicable diseases caused by pathogenic bacteria, viruses, fungi, and protists can be passed from one person to another by air, water, food, and animal contact.

3. Sexually transmitted diseases (STDs) are passed between persons during sexual contact. They include genital herpes, gonorrhea, chlamydia, and syphilis. AIDS is classified as a communicable disease and as an STD.

24-2: Your Immune System

1. The body is protected against pathogens by skin, cilia and mucus in the respiratory system, white blood cells in the circulatory system, and digestive enzymes. The purpose of the immune system is to fight disease.

2. Active immunity is long-lasting; passive immunity does not last.

3. AIDS is a communicable disease that may be transmitted by sexual contact, by using a needle contaminated by someone who already has the disease, by transfusion with contaminated blood, and through the placenta to a developing fetus. AIDS damages the body's immune system so that it cannot fight any disease.

24-3: Science and Society: Preventing Disease

1. A vaccine is a weakened virus delivered by injection or swallowed to develop immune protection against a disease.

2. Disease can be prevented by vaccination and good health habits.

24-4: Noncommunicable Disease

1. Causes of noncommunicable disease include genetics, chemicals, poor diet, and uncontrolled cell growth.

2. Chronic noncommunicable diseases include diabetes, cancer, arthritis, and allergies.

3. Most noncommunicable diseases can be medically treated.

KEY SCIENCE WORDS

a. **active immunity**
b. **allergens**
c. **allergy**
d. **antibody**
e. **antigens**
f. **antiseptic**
g. **cancer**
h. **chemotherapy**
i. **chronic diseases**
j. **communicable disease**
k. **disinfectant**
l. **immune system**
m. **lymphocytes**
n. **noncommunicable diseases**
o. **passive immunity**
p. **pasteurization**
q. **sexually transmitted diseases (STDs)**
r. **tumor**
s. **vaccination**

UNDERSTANDING VOCABULARY

Match each phrase with the correct term from the list of Key Science Words.

1. cause allergic reactions
2. disease spread through air, water, or contact
3. chemical that prevents pathogen growth on the skin
4. foreign proteins attacked by the body
5. introduces antibodies for short-term immunity
6. white blood cells that fight pathogen
7. long-lasting, noncommunicable diseases
8. uncontrolled cell division
9. use of chemicals to destroy cancer cells
10. introduces antigens for long-term immunity

OPTIONS

ASSESSMENT

To assess student understanding of material in this chapter, use the resources listed.

COOPERATIVE LEARNING

Consider using cooperative learning in the THINK AND WRITE CRITICALLY, APPLY, and MORE SKILL BUILDERS sections of the Chapter Review.

PROGRAM RESOURCES

From the **Teacher Resource Package** use:

Chapter Review, pages 51-52.

Chapter and Unit Tests, pages 161-164, Chapter Test.

CHAPTER
REVIEW

SUMMARY

Have students read the summary statements to review the major concepts of the chapter.

UNDERSTANDING VOCABULARY

1.	b	**6.**	m
2.	j	**7.**	i
3.	f	**8.**	g
4.	e	**9.**	h
5.	o	**10.**	s

ASSESSMENT

Portfolio

Encourage students to place in their portfolios one or two items of what they consider to be their best work. For each item, ask students to explain why that item was chosen and what they learned from it. Items might be selected from the following.

- For Your Mainstreamed students notes on immunizations required by the school, p. 556
- In Your Journal report on the effect of yellow fever on the construction of the Panama Canal, p. 558
- You Decide! opinion about keeping smallpox virus samples, p. 569

Performance

Additional performance assessments may be found in *Performance Assessment* and *Science Integration Activities* that accompany **Merrill Life Science.** Performance Task Assessment Lists and rubrics for evaluating these activities and other products generated throughout the chapter can be found in Glencoe's *Performance Assessment in Middle School Science.*

CHAPTER

REVIEW

CHECKING CONCEPTS

1. a	**6.** b
2. d	**7.** a
3. c	**8.** b
4. a	**9.** a
5. d	**10.** d

USING LAB SKILLS

ASSESSMENT

Use these alternate lab exercises to assess students' understanding of skills used in this chapter.

11. Raw hamburger most likely will contain more bacteria.

12. It is an effective disinfectant for many kinds of bacteria.

THINK AND WRITE CRITICALLY

13. His experiments helped make food safer to eat.

14. With arthritis, the immune system begins to treat the body's normal proteins as if they were antigens.

15. All of these activities put pathogens on your hands. If any of these pathogens gets inside your body, you could get a disease.

16. Antibiotics kill bacterial STDs. Early treatment can prevent damage to body systems.

17. With active immunity, your body produces its own long-lasting antibodies when you get a disease or receive a vaccine. Passive immunity comes from several sources as the body doesn't produce antibodies to particular antigens.

CHAPTER

REVIEW

CHECKING CONCEPTS

Choose the word or phrase that completes the sentence.

1. A pathogen causes a specific disease if it is _______.
   - **a.** present in all cases of the disease
   - **b.** does not infect other animals
   - **c.** causes other diseases
   - **d.** treated with heat
2. Communicable diseases can be caused by _______.
   - **a.** heredity **c.** chemicals
   - **b.** allergies **d.** organisms
3. _______ is an example of an STD.
   - **a.** Anthrax **c.** AIDS
   - **b.** Malaria **d.** Pneumonia
4. All of these diseases are caused by bacteria except _______.
   - **a.** AIDS **c.** chlamydia
   - **b.** gonorrhea **d.** syphilis
5. Your body's defenses against pathogens include all of the following except _______.
   - **a.** stomach enzymes **c.** white blood cells
   - **b.** skin **d.** hormones
6. All of these are noncommunicable diseases except _______.
   - **a.** allergies **c.** asthma
   - **b.** syphilis **d.** diabetes
7. Lymphocytes are attacked by the virus that causes _______.
   - **a.** AIDS **c.** flu
   - **b.** chlamydia **d.** polio
8. _______ is a chronic joint disease.
   - **a.** Asthma **c.** Muscular dystrophy
   - **b.** Arthritis **d.** Diabetes
9. Cancer cells are destroyed by _______.
   - **a.** chemotherapy **c.** vaccines
   - **b.** antigens **d.** viruses
10. _______ are formed by the blood to fight invading antigens.
    - **a.** Hormones **c.** Pathogens
    - **b.** Allergens **d.** Antibodies

USING LAB SKILLS

11. In the MINI-Lab on page 560 a method to detect bacteria in foods is described. Use this method to compare the amount of bacteria in raw versus cooked hamburger. What did you discover?
12. The Activity on page 567 asked you to find out how microorganisms cause infections. Try using hydrogen peroxide (3%) to clean an infected apple. Determine if it stops bacterial growth.

THINK AND WRITE CRITICALLY

Answer the following questions in your Journal using complete sentences.

13. How did Pasteur's experiments help society?
14. What makes arthritis different from other immune disorders?
15. Why is it very important to wash your hands after using a restroom, petting a dog, handling a pet gerbil or parakeet, and before eating?
16. If a person gets a bacterial STD, why must it be treated promptly with antibiotics?
17. What's the difference between active and passive immunity?

APPLY

18. Which is better—to vaccinate people or to wait until they build their own immunity?
19. What advantage might a breast-fed baby have compared to a bottle-fed baby?
20. How does your body protect itself from antigens?
21. How do lymphocytes eliminate antigens?
22. Describe the differences among antibodies, antigens, and antibiotics.

MORE SKILL BUILDERS

If you need help, refer to the Skill Handbook.

1. **Making and Using Tables:** Make a chart comparing the following diseases and their prevention: cancer, diabetes, tetanus, measles.
2. **Concept Mapping:** Make network tree concept map, comparing the defenses your body has against disease. Compare general defenses, active immunity, and passive immunity.
3. **Classifying:** Classify the following diseases as communicable or noncommunicable: diabetes, gonorrhea, herpes, strep throat, syphilis, cancer, flu.
4. **Recognizing Cause and Effect:** Use a library reference to identify the cause of each disease as bacteria, virus, fungus, or protist: athlete's foot, AIDS, cold, dysentery, flu, pinkeye, pneumonia, strep throat, ringworm.
5. **Making and Using Graphs:** Interpret the graph below showing the rate of polio cases. Explain the rate of cases between 1950 and 1965. What conclusions can you draw about the effectiveness of the polio vaccines?

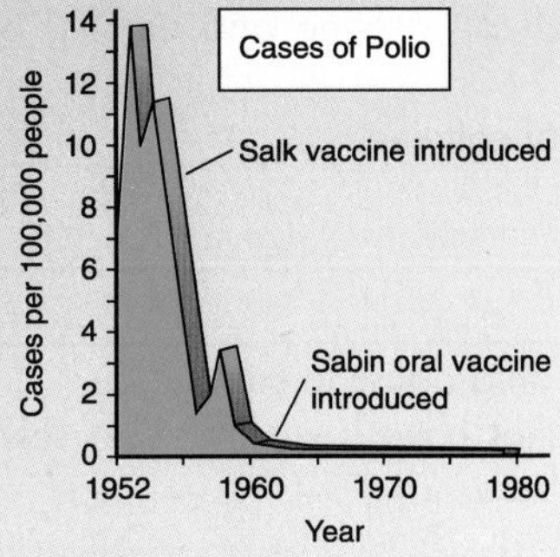

PROJECTS

1. Design a flow chart (diagram) to show how a person with a cold could spread the disease to family members and to others.
2. Write a report on Lyme disease, its causes, symptoms, and treatment.

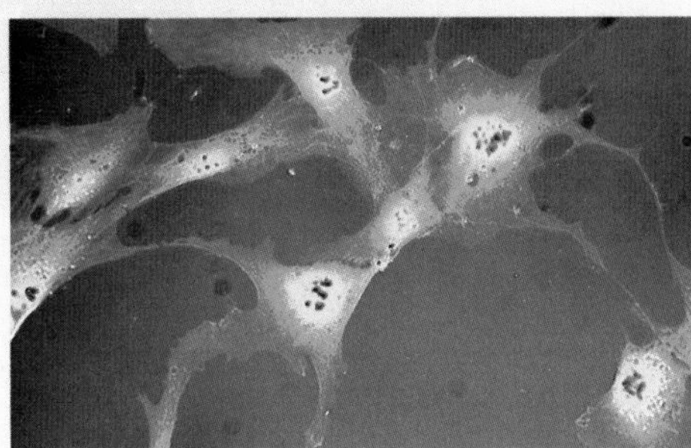

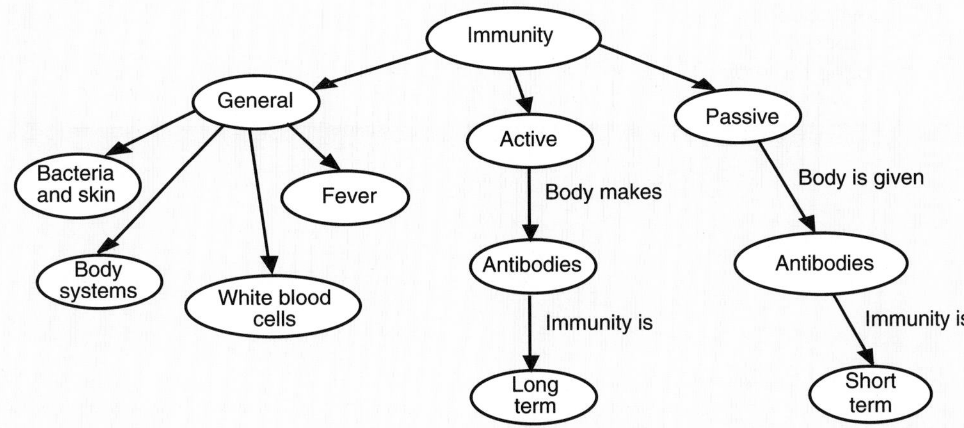

APPLY

18. In some cases, it is better to vaccinate. Some people may not produce antibodies to certain antigens and become very ill.
19. Breast-fed babies receive immunity antibodies in their mothers' milk. These antibodies help fight diseases.
20. Your skin, respiratory, circulatory, and digestive systems all help defend your body.
21. Lymphocytes produce antibodies and destroy invading antigens.
22. Antigens are foreign proteins and chemicals that invade your body. Antibodies are formed by your body's immune system to destroy and remove the antigens. Antibiotics are drugs used to kill antigens when your own antibodies can't do the job.

MORE SKILL BUILDERS

1. Making and Using Tables:

Disease	Prevention
cancer	avoid smoking, alcohol; eat a balanced diet
diabetes	diet and/or insulin
tetanus	vaccination for tetanus; revaccinate if vaccination is over three years old
measles	prevent with vaccination

2. Concept Mapping: See below.
3. Classifying: communicable: gonorrhea, herpes, strep throat, syphilis, flu; noncommunicable: diabetes, cancer
4. Recognizing Cause and Effect: *Students will need to consult references other than their texts.* Bacteria: strep throat, pneumonia, pinkeye. Virus: AIDS, cold, flu, pneumonia, pinkeye. Fungus: ringworm, athlete's foot. Protist: dysentery.
5. Making and Using Graphs: The rate of cases dropped right after the polio vaccine was introduced. The dramatic decline in polio cases was due to the newly discovered vaccines.

CHAPTER

25 Facts about Drugs

CHAPTER SECTION	OBJECTIVES	ACTIVITIES
25-1 Drugs and Health (2 days)	1. **Explain** how drug misuse differs from drug abuse. 2. **Describe** the effects of nicotine and alcohol on the body. 3. **Explain** the effects of caffeine.	**Activity 25-1:** *Interpreting Drug Label Information,* p. 582 **MINI-Lab:** *A Childproof Package?* p. 584
25-2 Drugs in Society **Science & Society** (1 day)	1. **Explain** the cost of drug abuse to society. 2. **List** two attempts that have been made to solve the problem of drug abuse.	
25-3 Problems with Illegal Drugs (2 days)	1. **Compare** the effects of marijuana and tobacco use. 2. **Describe** the effects of stimulants and depressants. 3. **Describe** the effects of hallucinogens. 4. **Explain** problems of using opiates.	**MINI-Lab:** *How are drugs classified?* p. 593 **Activity 25-2:** *Designing an Experiment (The Effect of Drugs on Heartbeat Rate),* p. 594
Chapter Review		

ACTIVITY MATERIALS

FIND OUT	ACTIVITIES		MINI-LABS	
Page 579 paper pencil	**25-1 Interpreting Drug Label Information, p. 582** paper pencil	**25-2 Designing an Experiment, p. 594** dilute solutions of coffee, cola, ethyl alcohol, tobacco, and cough medicine (dextromethorphen hydrobromide) *Daphnia* culture microscope dropper aged tap water	**A Childproof Package? p. 584** different types of over-the-counter products	**How are drugs classified? p. 593** cardboard

CHAPTER FEATURES	TEACHER RESOURCE PACKAGE	OTHER RESOURCES
Technology: *Taking Your Medicine,* p. 581 **Problem Solving,** *Passive Smoke,* p. 583 **Skill Builder:** *Concept Mapping,* p. 585	**Ability Level Worksheets** ◆ ***Study Guide,*** p. 96 ● ***Reinforcement,*** p. 96 ▲ ***Enrichment,*** p. 96 **Activity Worksheets,** pp. 221-222, 227 **Critical Thinking/Problem Solving,** p. 29 **Technology,** pp. 17-18	**Lab Manual:** Effects of Nicotine, p. 165 **Lab Manual:** Effects of Caffeine, p. 169 **STVS:** Disc 7, Side 2; Disc 5, Side 2
You Decide! p. 587	**Ability Level Worksheets** ◆ ***Study Guide,*** p. 97 ● ***Reinforcement,*** p. 97 ▲ ***Enrichment,*** p. 97 **Science and Society,** p. 29	**Science Integration Activity 25**
Skill Builder: *Outlining,* p. 593	**Ability Level Worksheets** ◆ ***Study Guide,*** p. 98 ● ***Reinforcement,*** p. 98 ▲ ***Enrichment,*** p. 98 **Activity Worksheets,** pp. 223, 224, 228 **Concept Mapping,** p. 55 **Cross-Curricular Connections,** p. 29	
Summary Key Science Words Understanding Vocabulary Checking Concepts Using Lab Skills Think & Write Critically Apply More Skill Builders Projects	ASSESSMENT RESOURCES **Chapter Review,** pp. 53-54 **Chapter Test,** pp. 165-168 **Unit Test,** pp. 169-170 **Performance Assessment in Middle School Science (PAMSS)**	**Chapter Review Software** **Test Bank** **Alternate Assessment** **Performance Assessment**

◆ **Basic** ● **Average** ▲ **Advanced**

ADDITIONAL MATERIALS

SOFTWARE	AUDIOVISUAL	BOOKS/MAGAZINES
Alcohol—An Educational Simulation, Marshware. *Drugs,* Intellectual Software. *Alcohol: Making the Choice,* Focus. *The Great Knowledge Race: Substance Abuse,* Focus.	*Alcohol, Tobacco, and Drugs vs. Physical Fitness,* film, AIMS. *Drugs and the Nervous System, 2nd ed.,* film, Churchill. *Straight Talk About Drugs: Stimulants and Narcotics,* filmstrip, Guidance Associates. *Alcoholism: Who Pays the Tab?* filmstrip, Current Affairs.	Auraham, Regina. *Substance Abuse: Prevention and Treatment.* Edgemont, PA: Chelsea House Publications, 1988. Berger, Gilda. Ed. Jennie Rakos. *Drug Abuse: The Impact on Society.* NY: Franklin Watts, Inc., 1988. Coulson, C.J. *Molecular Mechanisms of Drug Action.* NY: Taylor and Francis, Inc., 1988.

CHAPTER

25

FACTS ABOUT DRUGS

THEME DEVELOPMENT: Homeostasis is one of the major themes emphasized in this text. Illness can cause the body to lose its homeostasis. Drugs are often used to bring the body back into equilibrium and self-regulation. Drug abuse is a purposeful change of body homeostasis for short durations at the risk of deterioration of mental and physical health that sometimes leads to death.

CHAPTER OVERVIEW

▶ **Section 25-1:** This section explains the difference between drug misuse and drug abuse. There is a special focus on the drugs alcohol, nicotine, and caffeine.

▶ **Section 25-2: Science and Society:** The costs to society of drug abuse are discussed. The You Decide feature asks students to consider the amount and type of drug-related publicity found in advertising.

▶ **Section 25-3:** The problems associated with the use of illegal drugs are the focus of this section. The effects of stimulants, depressants, hallucinogens, and opiates are explained.

CHAPTER VOCABULARY

drug
drug misuse
drug abuse
psychological dependence
physical dependence
tolerance
withdrawal
stimulant
fermentation
depressant
methamphetamine
marijuana
hashish
cocaine
crack
hallucinogens
heroin

CHAPTER

25 Facts about Drugs

578

OPTIONS

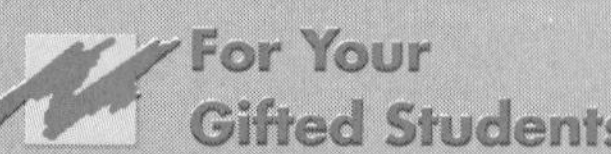

For Your Gifted Students

Have students compose a song that emphasizes being drug free and talks about activities a person can do instead of using drugs.

For Your Mainstreamed Students

Have students design and put up a bulletin board that emphasizes being drug free. Two topics that might be featured are drugs' effects on the body and community organizations that could use teenage volunteer help, such as a senior citizen home, the local library, and local parks or recreation areas.

Drugs are chemical substances that are used to help lessen pain, to cure disease, or control a chemical imbalance in the body. Drugs have important roles in overcoming illness and maintaining a healthy life.

FIND OUT!

Do this activity to find out which foods and drinks contain caffeine.

Read the labels of food and medicine products to see if the stimulant drug caffeine is listed as an ingredient. Design a chart to show the amount of caffeine per serving or dose in ten of these items. *Compare* and rank the products from those containing the most caffeine to those containing the least. What products surprised you because of the large amount of caffeine they contain?

Gearing Up

Previewing the Chapter

Use this outline to help you focus on important ideas in this chapter.

Previewing Science Skills

- ▶ In the Skill Builders, you will make a concept map and outline.
- ▶ In the Activities, you will analyze, observe, and collect data.
- ▶ In the MINI-Labs, you will make a model and classify.

What's next?

Now you know that many foods, medicines, and drinks contain the drug caffeine. In this chapter, you will find out what effects caffeine and other drugs have on the body.

INTRODUCING THE CHAPTER

Use the Find Out activity to introduce students to drugs and health. Inform students that they will be learning more about drugs and their effects on the body as they read the chapter.

FIND OUT!

Preparation: Remind students that on their assignment at the grocery store they should take along pad and pencil to record the information about caffeine.

Materials: pad and pencil

Teaching Tips

Cooperative Learning: Use the Study Buddies strategy to have students make a chart showing the amount of caffeine per serving or dose in each item.

▶ Some products may not list the specific amount of caffeine they contain, but only list it as an ingredient. The ingredients are listed in order of the percent of total weight/volume from highest to lowest. Estimates may have to be made as to the amount of caffeine present.

▶ Suggested medicines to check are over-the-counter drugs for staying awake and diet pills.

Gearing Up

Have students study the Gearing Up feature to familiarize themselves with the chapter. Discuss the relationships of the topics in the outline.

What's Next?

Before beginning the first section, make sure students understand the connection between the Find Out activity and the topics to follow.

ASSESSMENT OPTIONS

PORTFOLIO

Refer to page 595 for suggested items that students might select for their portfolios.

PERFORMANCE ASSESSMENT

See page 595 for additional Performance Assessment options.

Process

Skillbuilders, pp. 585, 593
MINI-Lab, p. 584
Activities 25-1, p. 582; 25-2, p. 594
Using Lab Skills, p. 596

CONTENT ASSESSMENT

Assessment—Oral, pp. 583, 590
Section Reviews, pp. 585, 587, 593
Chapter Review, pp. 595-597
Mini Quizzes, pp. 581, 590

GROUP ASSESSMENT

Opportunities for group assessment occur with Cooperative Learning Strategies and Flex Your Brain Activities.

SECTION 25-1

PREPARATION

SECTION BACKGROUND

▶ Drug abuse among adolescents is often related to the tendency of teenagers to consider themselves immune to health problems.

▶ Caffeine is a compound found only in plants. Coffee beans have about 1 percent caffeine, cocoa beans about 0.5 percent, and tea leaves from 2 to 5 percent caffeine.

PREPLANNING

▶ Obtain additional labels from over-the-counter drugs to use as a follow-up to Activity 25-1. Have students make comparisons of the various labels.

1 MOTIVATE

Cooperative Learning: Form Numbered Heads Together groups to agree whether given examples of substances are stimulants or depressants or to name examples of stimulants and depressants. Some examples to use: espresso coffee, jasmine tea, beer, cigars, vodka, chewing tobacco, wine cooler, cola, and hot chocolate.

TYING TO PREVIOUS KNOWLEDGE: Recall from Chapters 20 and 21 the effects of smoking on the circulatory and respiratory systems.

25-1 Drugs and Health

New Science Words

drug
drug misuse
drug abuse
psychological dependence
physical dependence
tolerance
withdrawal
stimulant
fermentation
depressant

Objectives

- ▶ Explain how drug misuse differs from drug abuse.
- ▶ Describe the effects of nicotine and alcohol on the body.
- ▶ Explain the effects of caffeine.

How Drugs Affect the Body

Make a list of five things that you think include drugs. Does your list include foods, drinks, tobacco, medicines, or illegal drugs? A **drug** is any chemical substance that changes the way a person thinks, feels, or acts. Drugs provide no nutrition for the body. Different drugs can act to slow down or speed up the nervous system or to change the functions of some cells. Most drugs are derived from plants, fungi, and bacteria.

When used properly, drugs benefit health. When used improperly, they can harm the user. For example, taking too much of some kinds of blood pressure medicine can lower blood pressure so much that a person feels faint. Improper use also can cause behavior that harms the health and safety of others. Driving after drinking can result in injury to the driver and others. Because of these concerns, federal laws were passed to control the manufacture, distribution, and possession of all drugs.

Have you ever been really sick with a cold and cough? Your doctor may have prescribed medicine with a cough suppressant to keep you from coughing so hard and to let you rest. Suppose you decided to take an extra spoonful before bedtime to make sure you were able to sleep all night. This behavior is a form of drug misuse. **Drug misuse** is use of a drug for the purpose for which it was made, but taking it improperly. Taking too much or too little medicine according to directions is drug misuse. Using someone else's medicine is also drug misuse. Drug mis-
1 use is harmful to health. **Drug abuse** is the deliberate use of a drug for other than its intended purpose. Using illegal drugs is drug abuse. Drug abuse damages a person's health.

Medicines
Drugs that help the body fight and prevent some diseases.

Prescription drugs
Medicine that only can be obtained with a doctor's prescription.

Over-the-counter drugs
Drugs purchased without a prescription at drugstores and supermarkets.

Figure 25-1. Many kinds of drugs are made for helping people regain health.

OPTIONS

Meeting Different Ability Levels

For Section 25-1, use the following **Teacher Resource Masters** depending upon individual students' needs.

- ◆ **Study Guide Master** for all students.
- ● **Reinforcement Master** for students of average and above average ability levels.
- ▲ **Enrichment Master** for above average students.

Additional Teacher Resource Package masters are listed in the OPTIONS box throughout the section. The additional masters are appropriate for all students.

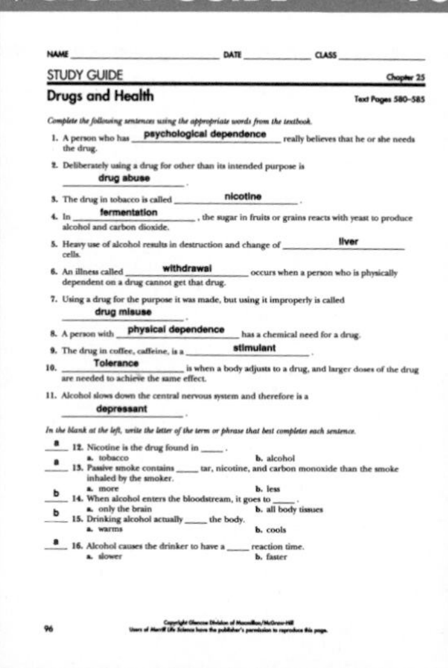

◆ STUDY GUIDE 96

NAME ______ DATE ______ CLASS ______

STUDY GUIDE — Chapter 25

Drugs and Health — Text Pages 580-585

Complete the following sentences using the appropriate words from the textbook.

1. A person who has **psychological dependence** really believes that he or she needs the drug.
2. Deliberately using a drug for other than its intended purpose is **drug abuse**.
3. The drug in tobacco is called **nicotine**.
4. In **fermentation**, the sugar in fruits or grains reacts with yeast to produce alcohol and carbon dioxide.
5. Heavy use of alcohol results in destruction and change of **liver** cells.
6. An illness called **withdrawal** occurs when a person who is physically dependent on a drug cannot get that drug.
7. Using a drug for the purpose it was made, but using it improperly is called **drug misuse**.
8. A person with **physical dependence** has a chemical need for a drug.
9. The drug in coffee, caffeine, is a **stimulant**.
10. **Tolerance** is when a body adjusts to a drug, and larger doses of the drug are needed to achieve the same effect.
11. Alcohol slows down the central nervous system and therefore is a **depressant**.

In the blank at the left, write the letter of the term or phrase that best completes each sentence.

a 12. Nicotine is the drug found in ____.
a. tobacco b. alcohol

a 13. Passive smoke contains ____ tar, nicotine, and carbon monoxide than the smoke inhaled by the smoker.
a. more b. less

b 14. When alcohol enters the bloodstream, it goes to ____.
a. only the brain b. all body tissues

b 15. Drinking alcohol actually ____ the body.
a. warms b. cools

a 16. Alcohol causes the drinker to have a ____ reaction time.
a. slower b. faster

96

When drugs are misused or abused, serious health problems can result. A person can develop chemical dependence. With this condition, a person has a psychological and/or physical need for a drug. A person who has **psychological dependence** really believes that he or she needs the drug. The person may believe that use of the drug can be controlled. A person who has **physical dependence** has a chemical need for a drug. This also is called addiction. The person becomes so used to the effects of the drug that he or she needs it to function. There are two parts to physical dependence—tolerance and withdrawal. **Tolerance** occurs when the body adjusts to a drug and needs increasingly larger doses to produce the desired effect. **Withdrawal** is an illness
2 that occurs when the drug a person is physically dependent on is removed. Withdrawal symptoms are real and can be very painful. They include nausea, vomiting, headaches, and chills.

What are the two parts to physical dependence?

TECHNOLOGY

Taking Your Medicine

Some people need medicine or pain killers regularly. People with diabetes or cancer seem to respond best to frequent, low doses of medication. Taking medication by mouth can cause high blood levels of medicine as the medicine takes effect, followed by low blood concentrations as it wears off. This takes the patient on a roller coaster ride of pain or problems. Various techniques are being explored that allow a steady release medication system, flexible enough to allow a higher dosage on demand.

A drug pump may be implanted under skin and deliver medication as directed by a desktop programmer. A syringe-type pump is frequently used for newborns and for cancer patients. Other systems use biodegradable implants that slowly release medication as the implant breaks down. Ultrasound waves, magnetism, heat, and enzyme triggers are all being studied as ways of allowing patients to increase the dosage released by an implant.

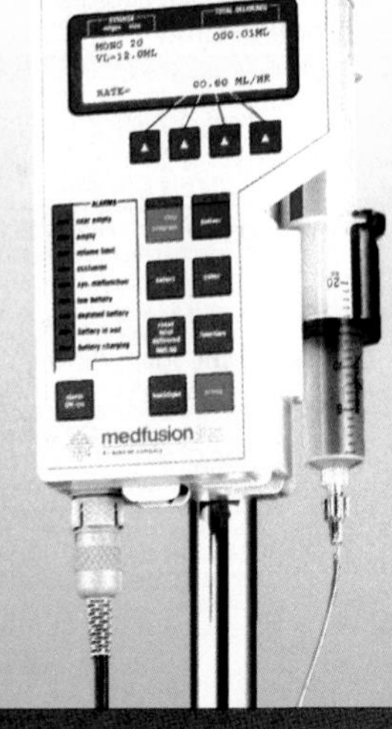

Think Critically: What kinds of illness might this system be most useful for?

● REINFORCEMENT 96

NAME ______ DATE ______ CLASS ______

REINFORCEMENT Chapter 25

Drugs and Health Text Pages 580–585

Answer the following questions using the information in your textbook.

1. What is a drug? **A drug is any chemical substance that changes the way a person thinks, feels, or acts.**
2. What is drug misuse? **Drug misuse is the use of a drug for the purpose it was made, but taking it improperly.**
3. What is drug abuse? **Drug abuse is the deliberate use of a drug for other than its intended purpose.**
4. What is psychological dependence? **when a person really believes he or she needs a drug**
5. What effect does alcohol have on blood vessels? **Alcohol causes the blood vessels to enlarge.**
6. How is alcohol the opposite of caffeine? **Alcohol is a depressant of the central nervous system, while caffeine is a stimulant of the central nervous system.**
7. What are some symptoms of withdrawal? **nausea, vomiting, headaches, and chills**
8. What effects does nicotine have on the body? **Nicotine speeds up the heart, raises blood pressure, constricts blood vessels, increases the chances for heart attacks and strokes, and causes psychological and physical dependence.**
9. Can a person be harmed if someone else is smoking nearby? **Yes; the smoke from a burning cigar, cigarette, or pipe—known as passive smoke—can enter the respiratory system of a person nearby and cause harm.**
10. What is the stimulant in coffee and in what else can it be found? **caffeine; found in tea, cola soft drinks, chocolate, and cocoa**

96

▲ ENRICHMENT 96

NAME ______ DATE ______ CLASS ______

ENRICHMENT Chapter 25

Drugs and Health Text Pages 580–585

CIGARETTE SURVEY

Although less than half of adults smoke, many adults who don't smoke now did use cigarettes in the past. Often they started smoking in high school or college.

Take a survey of five adults who smoke now or who smoked at some earlier age. Use a copier at the library to make four more copies of this survey form or write out the survey form on four sheets of paper.

Age: ______

Sex: male or female (circle one)

At what age did you smoke your first cigarette? ______

Why did you smoke the first one? ______

Did you like the first one? ______

How long did it take you to get hooked on cigarettes? ______

How many cigarettes a day do (did) you smoke? ______

Do (Did) you enjoy every single cigarette? If not, why do (did) you smoke the ones you do (did) not enjoy? ______

Have you ever tried to quit smoking? ______

When you quit or tried to quit, what were your withdrawal symptoms? ______

What do (did) you enjoy about smoking? ______

What do (did) you dislike about smoking? ______

Do you regret having smoked cigarettes? ______

Do you have any lasting effects or diseases from smoking? ______

Do you fear any future effects from smoking? ______

After you have taken your survey of five adults, compare your findings with others in the class.

96

OBJECTIVES AND SCIENCE WORDS: Have students review the objectives and science words as they begin each section of this chapter.

2 TEACH

Key Concepts are highlighted.

CONCEPT DEVELOPMENT

FLEX Your Brain

Use the Flex Your Brain activity to have students explore the term *DRUG*.

Flex Your Brain
ASSESSMENT
Portfolio: Use the Flex Your Brain activity to reinforce critical thinking and problem solving skills.

▶ Have students discuss examples of potential drug misuse situations that they might know of.

MINI QUIZ

Use the Mini Quiz to check students' recall of chapter content.

1. **What is drug abuse?** *the deliberate use of a drug for other than its intended purpose*
2. **The illness that occurs when the drug a person is physically dependent upon is removed is called _____.** *withdrawal*
3. **Chocolate and cocoa contain the stimulant _____.** *caffeine*

TECHNOLOGY

▶ For more information on drug pump implants, see "Implanting Anesthetics," *USA Today*, February 1990, p. 12.

Think Critically: It would be most useful for treating diabetes, asthma, and high blood pressure. It could be used for birth control and medications given after operations.

ACTIVITY 25-1

30 minutes

OBJECTIVE: Interpret data from the label of an over-the-counter drug.

PROCESS SKILLS applied in this activity:
▶ **Communicating** in Procedure Step 2.
▶ **Analyzing Data** in Analyze Questions 1-6 and Conclude and Apply Questions 8-10.
▶ **Comparing** in Conclude and Apply Question 7.

TEACHING THE ACTIVITY

Alternate Materials: Provide additional over-the-counter labels for students to study. Enlarge the labels and photocopy them for student use.

PROGRAM RESOURCES

From the **Teacher Resource Package** use:

Activity Worksheets, pages 221-222, Activity 25-1, Interpreting Drug Label Information.

Activity ASSESSMENT

Performance: To further assess students' understanding of drug label information, see USING LAB SKILLS, Question 11, on page 596.

ACTIVITY 25-1 Interpreting Drug Label Information

Problem: *What information is on the label of an over-the-counter drug?*

Materials

- paper
- pencil

Procedure

1. The photograph below shows a label from an over-the-counter drug. Read it carefully.
2. Make a data table like the one shown. Use the information on the label to complete the data table.

Data and Observations Sample Data

Information	Drug Label
Product	Information
Number of pills	will reflect
Ingredients in the medicine	label below
Directions for use	or one of
Warnings	your
Possible side effects	choosing.
Expiration date	
Storage of drug	

Arthritis Strength

PROVIDE FAST PAIN RELIEF AND HELP PROTECT AGAINST ASPIRIN STOMACH UPSET.
• Provides effective temporary relief of the minor aches and pains, stiffness, swelling and inflammation of arthritis.
* Buffered formulation which helps prevent the stomach upset that plain aspirin can cause.
• Coated caplets for easy swallowing.
DOSAGE: Adults—2 caplets with water every 4 hours as needed. Do not exceed 8 caplets in 24 hours, or give to children under 12 unless directed by physician. CAUTION: If pain persists for more than 10 days or redness is present, or in arthritic or rheumatic conditions affecting children under 12, consult a physician immediately. Do not take without consulting a physician if under medical care. WARNING: **Children and teenagers should not use this medicine for chicken pox or flu symptoms before a doctor is consulted about Reye syndrome, a rare but serious illness reported to be associated with aspirin. KEEP THIS AND ALL MEDICINES OUT OF CHILDREN'S REACH. IN CASE OF ACCIDENTAL OVERDOSE, CONTACT A PHYSICIAN OR POISON CONTROL CENTER IMMEDIATELY. If dizziness, impaired hearing or ringing in the ears occurs, discontinue use. As with any drug, if you are pregnant or nursing a baby, seek the advice of a health professional before using this product.** ACTIVE INGREDIENT (PER CAPLET): Aspirin (500 mg) in a formulation buffered with Calcium Carbonate, Magnesium Oxide and Magnesium Carbonate. **Other Ingredients:** Benzoic Acid, Carnauba Wax, Citric Acid, Corn Starch, FD&C Blue No. 1, Hydroxypropyl Methylcellulose, Mineral Oil, Polysorbate 20, Povidone, Propylene Glycol, Simethicone Emulsion, Sodium Phosphate, Sorbitan Monolaurate, Titanium Dioxide. May also contain: Glyceryl Behenate, Magnesium Stearate, Sodium Lauryl Sulfate, Sodium Stearyl Fumarate, Stearic Acid, Zinc Stearate. Remove cotton and recap bottle. Store at room temperature.

Analyze

1. What is an over-the-counter drug?
2. What is a side effect? What side effects are caused by this drug?
3. What information is given on the label of this over-the-counter drug?
4. For whom does the label specify the drug is intended?
5. Who should not take this drug without consulting a physician?
6. What is the maximum number of doses that should be taken in 24 hours?

Conclude and Apply

7. Explain how over-the-counter drugs differ from prescription drugs.
8. Why should a person never take any kind of drug without reading the label?
9. *Infer* why the age of a person would determine drug dose.
10. Why should a person never take more than the recommended amount of an over-the-counter drug?

ANSWERS TO QUESTIONS

1. a drug that you can buy legally without a prescription
2. an unexpected result of a drug; dizziness, impaired hearing, or ringing
3. use; how much to take; how often to take; how long to take; warnings; cautions; interactions; ingredients
4. adults and children over 12 years of age
5. children under 12
6. 2 every 4 hours as needed; not to exceed 8 caplets in 24 hours
7. Prescription drugs are ordered by a physician; over-the-counter drugs can be purchased by anyone without a prescription.
8. The drug could cause a drug interaction, or the person might have a disease that the drug would affect adversely.
9. Accept all reasonable answers. Young children usually receive smaller doses of a drug because of less mature body systems and lower weight.
10. Higher doses may cause undesirable effects.

Tobacco

When a person first begins to smoke, he or she may feel sick. It takes time for the body to get used to nicotine, the drug in tobacco. Many people do not think of the nicotine in tobacco as a drug. However, it is a **stimulant,** a drug that speeds up the nervous system. Nicotine speeds up the heartbeat rate and raises blood pressure. It constricts blood vessels, reducing blood flow. Nicotine also contributes to the buildup of fatty substances in blood vessels. The chance of developing blood clots that cause heart attacks and strokes increases. Nicotine causes physical and psychological dependence.

Tobacco smoke brings other substances into the body. Carbon monoxide is a harmful gas found in tobacco smoke. Carbon monoxide becomes attached to hemoglobin in red blood cells more easily than does oxygen. The hemoglobin cannot carry oxygen. As a result, cells do not get enough oxygen. The respiratory and cardiovascular systems have to work harder.

In Chapter 21, you read that tar is a thick, sticky fluid produced when tobacco burns. Tar sticks to the respiratory tract. It damages cilia and contains cancer-causing chemicals.

In Your JOURNAL

More than 300 000 people die each year in the United States due to smoking. In spite of the fact that many people have stopped smoking, more cases of lung cancer are being reported. It takes 20 to 30 years for lung cancer to develop. Many people who stopped smoking in the 1960s were diagnosed with cancer only in the 1980s. Why smoke cigarettes? *Analyze* tobacco advertisements. **In your Journal,** describe techniques used by cigarette manufacturers to suggest smoking a particular brand.

PROBLEM SOLVING

Passive Smoke

Tim and his family were eating in the nonsmoking section of a restaurant. However, when Tim began to eat, smoke from a burning cigarette in the smoking section wafted under his nose. It affected the taste of Tim's food. He was upset that his dinner was being disturbed by this smoke. Tim knows that nonsmokers who are exposed to tobacco smoke from cigarettes or smoke exhaled by a smoker have an increased risk of developing the same diseases as those who smoke. What would you have done if you were Tim's family?

Think Critically: Tim thinks smoking shouldn't be allowed in public places. What do you think about smoking in public places?

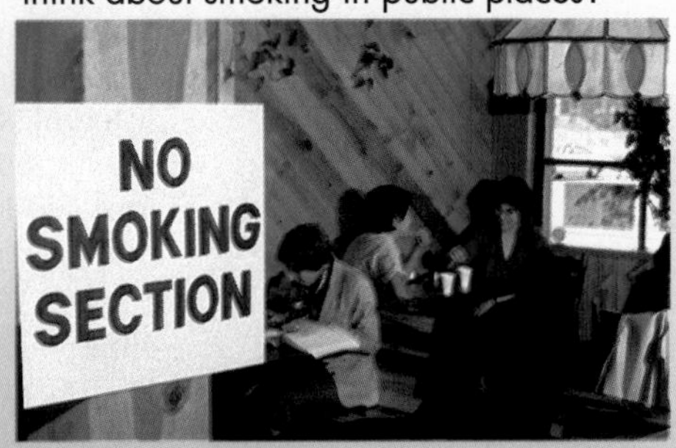

CONCEPT DEVELOPMENT

▶ **In addition to the physiological effects of smoking tobacco, what are some physical conditions resulting from smoking?** *The fingers and teeth become stained; the breath has an odor; and clothes smell of smoke.*

▶ **What are three substances brought into the body by smoking tobacco?** *nicotine, carbon monoxide, tar*

In Your JOURNAL

People in cigarette ads often appear young, glamorous, successful, and active—all things youg people are drawn to identify with.

PROBLEM SOLVING

▶ Discuss whether smoking should be banned in all public places.

Think Critically: Most students will agree with Tim because passive smoke puts nonsmokers at risk for smoking-related diseases.

TEACHER F.Y.I.

▶ Nicotine has been used as an insecticide since 1690. It was a common fumigant in greenhouses until the mid-1900s.

CHECK FOR UNDERSTANDING

Have students discuss how a person might have psychological and not physical dependence on a drug.

RETEACH

Have students outline the difference between psychological and physical dependence. Include the two parts of physical dependence.

EXTENSION

For students who have mastered this section, use the **Reinforcement** and **Enrichment** masters or other OPTIONS.

VideoDisc

STVS: Nicotine and the Lungs, Disc 7, Side 2

STVS: Drugs from Snake Venom, Disc 5, Side 2

OPTIONS

ASSESSMENT—ORAL

▶ **One of the most commonly abused drugs is aspirin. What are some of the side-effects of aspirin abuse?** *stomach ulcers, hemorrhagic stroke, ringing of ears, loss of hearing, nausea*

▶ **What might be the role of a pharmacist in preventing drug abuse?** *The pharmacist can check a prescription to see that it is not forged or changed by a person seeking to illegally obtain drugs.*

PROGRAM RESOURCES

From the **Teacher Resource Package** use:

Activity Worksheets, page 5, Flex Your Brain.

Technology, pages 17-18, Breaking Through the Blood-Brain Barrier.

Science Integration Activities, 25, Solutions

CONCEPT DEVELOPMENT

▶**Demonstration:** Add a small amount of dry yeast to 25 mL of sweetened grape juice and place in a warm place. Observe the fermentation of the fruit sugar as it produces bubbles of carbon dioxide gas. Alcohol is also produced, but in small amounts.

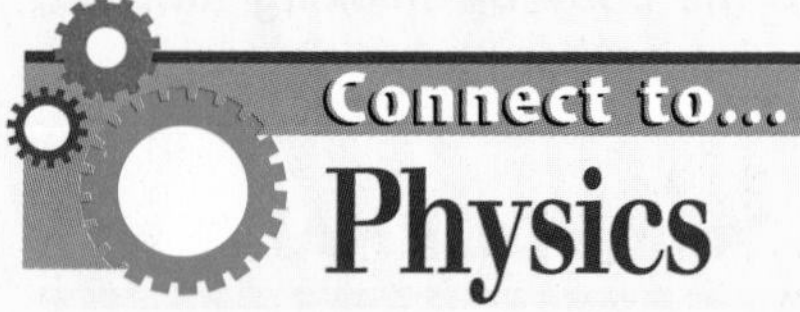

Answer: In the presence of enough depressant, the substance that normally diffuses across the synapse is prevented from doing so.

MINI-Lab

Materials: different types of over-the-counter products

Answer: Answers will vary with design. Students should be encouraged to help solve this problem. Worthy examples could be sent to pharmaceutical companies.

MINI-Lab
ASSESSMENT

Performance: To further assess students' understanding of problems with packaging, have students interview druggists to find out how many people request alternate packaging.

3 CLOSE

▶Ask questions 1-4 and the **Apply** Question in the Section Review.

MINI-Lab
A Childproof Package?
Survey different types of packaging used in over-the-counter drugs to make them tamperproof and childproof. *Design* a childproof package for a product. Could your package be opened easily by a person with arthritis?

Connect to... Physics

Depressants cause changes in the nervous system. In some extreme cases, the heart stops beating. What happens to the principle of diffusion at a synapse when a depressant drug is present?

Unfortunately, you cannot escape the danger of tobacco smoke by just not smoking. There are real health risks from just being in a room with tobacco smoke. If you are in a room with someone who is smoking, harmful substances from smoke will enter your respiratory system. The smoke from a burning cigarette, cigar, or pipe is known as passive smoke. Passive smoke contains more tar, nicotine, carbon monoxide, and other chemicals than that inhaled by the smoker.

Alcohol

Alcohol is also a drug. Alcoholic drinks have been made and used in most cultures since ancient times. Yeast utilizes the sugar in fruits to produce alcohol and carbon dioxide in a process known as **fermentation.**

Most alcohol is absorbed through the walls of the stomach and small intestine. It enters the circulatory system undigested. Carried by the blood, all body tissues become exposed to it directly. Alcohol is often viewed as a stimulant, but it is a **depressant.** As such, alcohol changes the function of cell membranes, slows down the central nervous system, and affects all functions of the brain. Judgment, reasoning, memory, and concentration are impaired. Muscle functions are affected. Even though alcohol is a depressant, it slightly increases heartbeat and pulse rate. It causes blood vessels to enlarge. As a result, a feeling of warmth passes through the body. In reality, however, the body is losing heat from the capillaries near the surface of the skin. Heavy use of alcohol results in destruction of brain and liver cells.

Large amounts of alcohol in the bloodstream dull the senses and cause the drinker to have a slower reaction time, lose coordination, and have slurred speech. Excessive amounts of alcohol can result in unconsciousness and death.

Most states have laws that define when a person is legally drunk, based on levels of alcohol in the person's blood. The amount of alcohol that causes abnormal behavior and loss of judgment varies considerably from person to person. Some people react after one drink.

Sometimes, people become physically dependent on alcohol. Alcoholism is an incurable disease characterized by the abuse of alcohol. The person with this disease cannot control his or her drinking. Continued use of alcohol causes permanent damage to all body cells. Alcoholism

OPTIONS

ENRICHMENT

▶Have students research the ingredients of common OTC cough medicines, pain relievers, or diet pills to find the active ingredients and classify them as stimulants or depressants. Use the *Physicians' Desk Reference* for OTCs to obtain the information.

▶Have students investigate local and national organizations such as Students Against Drunk Driving (SADD) that focus on the issues of drunk driving.

PROGRAM RESOURCES

From the **Teacher Resource Package** use:

Activity Worksheets, page 227, Mini-Lab: A Childproof Package?

Critical Thinking/Problem Solving, page 29, Alcoholism as a Disease.

Use **Laboratory Manual,** page 165, Effects of Nicotine.

Use **Laboratory Manual,** page 169, Effects of Caffeine.

also causes personality and behavioral changes. These problems affect the alcoholic and others around him or her. Approaches used to treat alcoholism include drugs, therapy, and self-help programs.

Caffeine

What do coffee, tea, and cola soft drinks have in common? You probably know that it is the drug caffeine. 3 Caffeine is a stimulant, a type of drug that stimulates your central nervous system and speeds up other body systems. You may hear people say they need a cup of coffee to get started in the morning or a cola soft drink for energy in the afternoon. They may be dependent on the caffeine in the drink to feel more energetic. Caffeine also is found in chocolate and cocoa. It is probably the most used stimulant in the world because it is so readily available.

Too much caffeine can increase heartbeat rate and cause a person to be restless, have tremors, and be unable to sleep. It can also stimulate the kidneys to produce more urine. Caffeine can cause physical dependence. When people who take in a lot of caffeine stop, they can have headaches and nausea. In spite of these known effects, a study released in 1990 showed that no damage could be found from the use of caffeine products. The study followed the caffeine habits of more than 10 000 people for a period of more than five years.

SECTION REVIEW

1. How does drug misuse differ from drug abuse?
2. How does nicotine affect the body?
3. How does alcohol affect the brain?
4. What happens when a person consumes too much caffeine?
5. **Apply:** Why does constant alcohol use cause permanent damage to blood vessels?
6. **Connect to Chemistry:** What does *transdermal* mean? Find out how transdermal patches work.

In Your JOURNAL

Find out about the programs of Alcoholics Anonymous, Al-Anon, and Alateen. Describe the work of these groups **in your Journal** and share the information with the class.

Skill Builder — Concept Mapping

Using a network tree concept map, show some differences between depressants and stimulants discussed in this section and give examples. If you need help, refer to Concept Mapping in the **Skill Handbook** on pages 688 and 689.

In Your JOURNAL

Alcoholics Anonymous—organization of persons who help each other deal with alcoholism, founded in 1935; people meet in groups; 12 steps—for alcohol-free living.

Al-Anon—worldwide organization of families and friends of alcoholics who help each other deal with the effects of living with alcoholism.

Alateen—Teen division of Al-Anon.

CONCEPT DEVELOPMENT

▶ Have students count the number of decaffeinated food and beverage items they can find in the grocery store.

SECTION REVIEW ANSWERS

1. Drug misuse is use of a drug for the purpose for which it was made, but taking it improperly. Drug abuse is the deliberate use of a drug for other than its intended purpose.

2. Nicotine speeds up the heartbeat rate, raises blood pressure, constricts blood vessels, and contributes to plaque in blood vessels. The chances of blood clots increase.

3. Alcohol slows down the central nervous system, affecting judgment, reasoning, memory, and concentration.

4. Too much caffeine can increase heartbeat rate and cause a person to be restless, have tremors, and be unable to sleep. It can stimulate the kidneys to produce more urine.

5. Apply: Alcohol causes blood vessels to enlarge. Constant alcohol use reduces the ability of blood vessels to constrict.

6. Connect to Chemistry: Transdermal means *across the dermis* or *skin.* Mild doses of medication are continuously administered when a patch is applied to the skin.

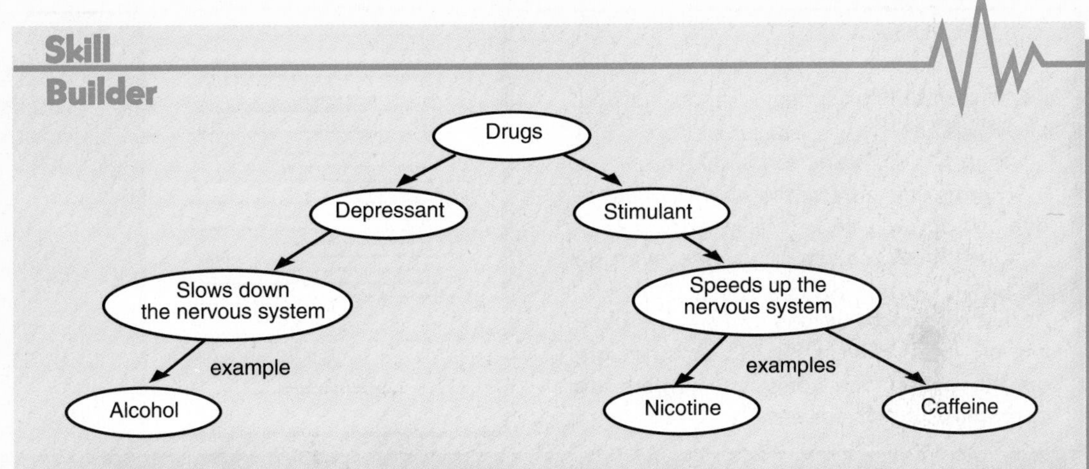

Skillbuilder ASSESSMENT

Performance: Use this Skillbuilder to assess students' abilities to compare stimulants and depressants. Have them write a statement in their journals that compares the effects of the two on the nervous system.

SECTION 25-2

PREPARATION

SECTION BACKGROUND

▶ Treatment programs can cost more than $10 000 per addict per month.

1 MOTIVATE

▶ Discuss why drug use on the job might be dangerous to drug users, coworkers, or the general public.

2 TEACH

Key Concepts are highlighted.

CONCEPT DEVELOPMENT

▶ Discuss the success or lack of success of drug education programs.

Connect to... Chemistry

Answer: prescription: amphetamines, barbituates, sedatives, morphine, tranquilizers

legal, non-prescription: alcohol, nicotine, inhalant fumes from cleaning fluids, gasoline, model airplane glue

CROSS CURRICULUM

▶ **Mathematics:** Have students calculate the cost per person in the United States (250 million population) for the $160,000,000,000 business loss each year due to drug abuse.

25-2 Drugs in Society

New Science Words

methamphetamine

Objectives

- ▶ **Explain the cost of drug abuse to society.**
- ▶ **List two attempts that have been made to solve the problem of drug abuse.**

Costs of Drug Abuse

For some time, drug abuse has been America's greatest concern. Many people suffer because of the effects of illegal drugs. In 1989, 2500 people in the United States died as a result of drug overdoses, drug-related crimes, or drug-related accidents. Babies born to mothers that abuse drugs may be addicted and suffer effects of the drugs. In the business world, of 1000 corporate executives surveyed, 67 percent agreed that drug abuse is the nation's number one labor problem. Loss of productivity, industrial accidents, higher health care cost, absenteeism, and lateness due to employee drug abuse results in $160 billion in business loss each year. The General Motors Corporation reports that it loses $1 billion each year to employee drug abuse. Indirectly, this loss is then added to the price of new cars.

Figure 25-2. Campaigns to stop drug abuse are visible in many places.

How can these problems be solved? In 1982, a war on drugs was declared in the United States. It was designed to attack the problem of drug abuse by educating the general public about the effects of drugs. Law enforcement was strengthened through increased crackdowns on dealers, expanded courts and prisons, and even military forces to keep drugs out of the country. The program appears to have had some impact on the problem. But despite attempts to prevent illegal drugs from entering the United States, drugs continue to arrive. The 60 to 75 percent drop in the prices of some drug on the streets shows how easily available drugs are. What is the real source of the drug problem? Is it lack of education?

Connect to... Chemistry

Find examples of the following types of drugs that are used often in an abusive manner: prescription drugs, and legal, non-prescription drugs.

OPTIONS

Meeting Different Ability Levels

For Section 25-2, use the following **Teacher Resource Masters** depending upon individual students' needs.

◆ **Study Guide Master** for all students.

● **Reinforcement Master** for students of average and above average ability levels.

▲ **Enrichment Master** for above average students.

Additional Teacher Resource Package masters are listed in the OPTIONS box throughout the section. The additional masters are appropriate for all students.

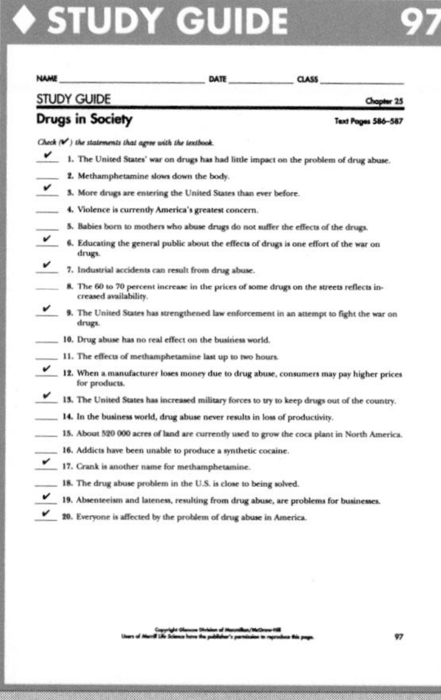

◆ STUDY GUIDE 97

NAME ________ DATE ________ CLASS ________

STUDY GUIDE — Chapter 25

Drugs in Society — Text Pages 586–587

Check (✔) the statements that agree with the textbook.

✔ 1. The United States' war on drugs has had little impact on the problem of drug abuse.
___ 2. Methamphetamine slows down the body.
✔ 3. More drugs are entering the United States than ever before.
___ 4. Violence is currently America's greatest concern.
___ 5. Babies born to mothers who abuse drugs do not suffer the effects of the drugs.
✔ 6. Educating the general public about the effects of drugs is one effort of the war on drugs.
✔ 7. Industrial accidents can result from drug abuse.
___ 8. The 60 to 70 percent increase in the prices of some drugs on the streets reflects increased availability.
✔ 9. The United States has strengthened law enforcement in an attempt to fight the war on drugs.
___ 10. Drug abuse has no real effect on the business world.
___ 11. The effects of methamphetamine last up to two hours.
✔ 12. When a manufacturer loses money due to drug abuse, consumers may pay higher prices for products.
✔ 13. The United States has increased military forces to try to keep drugs out of the country.
___ 14. In the business world, drug abuse never results in loss of productivity.
___ 15. About 520 000 acres of land are currently used to grow the coca plant in North America.
___ 16. Addicts have been unable to produce a synthetic cocaine.
✔ 17. Crank is another name for methamphetamine.
___ 18. The drug abuse problem in the U.S. is close to being solved.
✔ 19. Absenteeism and lateness, resulting from drug abuse, are problems for businesses.
✔ 20. Everyone is affected by the problem of drug abuse in America.

97

In South America, it is estimated that millions of hectares of land are used to grow the coca plant from which cocaine is produced. Even if this crop were destroyed, experts predict that addicts would produce a synthetic cocaine to satisfy their needs. This has already happened in the United States. A very powerful drug called **methamphetamine** (meth am FET uh men), crank, or ice speeds up body functions. Its effects last up to eight hours. Do drug suppliers provide their product because they know they will always have customers?

Some people argue that there is no solution to controlling the drug abuse problem in the United States. Drug education programs and increased law enforcement have proven effective for some, but not for all. It may be a while before the positive effects of drug education can be seen. Some experts suggest that drugs should be legalized. They suggest that the war on drugs is not practical because millions of Americans are paying for failed attempts to capture and reform hard-core drug producers. After all, alcohol and caffeine are legal. As law enforcement gets tougher, dealers may only make the drugs stronger, easier to hide, and more dangerous. What would happen if these drugs were made legal? How do you think the rate of abuse would change?

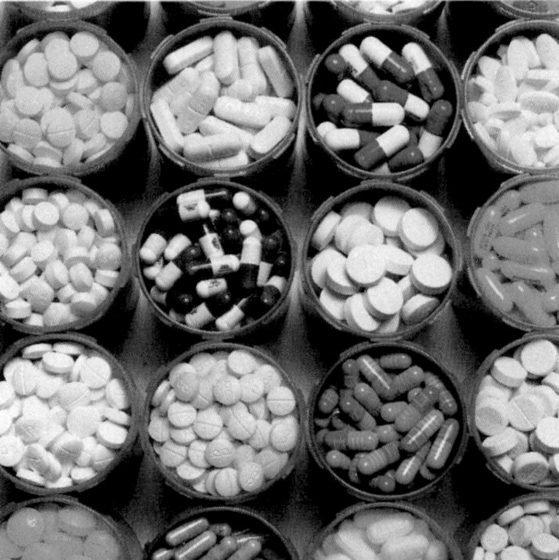

Figure 25-3. Drugs exist to help, not to be abused.

SECTION REVIEW

1. Why does drug abuse cause such a great loss to businesses each year?
2. List ways in which the government attempted to declare war on drugs to help solve the problem of drug abuse.

You Decide!

How much publicity do drugs receive? Many magazines, movies, and television programs still make drug use appear glamorous. Even tobacco and alcohol ads make it appear that people who use these products are young, well-dressed, and always having a great time. Does advertising give the wrong impression? What can be done to make people understand the truth about drugs?

● REINFORCEMENT 97

NAME ______ DATE ______ CLASS ______

REINFORCEMENT — Chapter 25

Drugs in Society — Text Pages 586–587

In the blank at the left, write the letter of the word or phrase that best completes each sentence.

b 1. Drug abuse is America's _____ concern.
a. useless b. greatest c. smallest d. least important

c 2. In 1982, a war on drugs was designed to attack drug abuse in all of the following ways EXCEPT _____.
a. educating the public about drugs
b. using the military to keep drugs out of the country
c. making prisons less enjoyable
d. increasing our police forces

d 3. Some prices of illegal drugs have dropped 60 to 75 percent because _____.
a. drug dealers want to sell more
b. drug users couldn't afford the higher prices
c. no one wants these drugs
d. there is more of the drug available

a 4. Methamphetamine is also called _____.
a. crank b. pot c. crack d. snow

b 5. _____ is an illegal drug.
a. Tobacco b. Cocaine c. Caffeine d. Alcohol

c 6. Babies born to mothers who abuse drugs _____.
a. always weigh more than other babies
b. are always born after the mothers' due dates
c. may be addicted to drugs
d. often have stronger hearts

d 7. Some experts suggest that _____.
a. drugs should be free
b. caffeine should be illegal
c. drug abusers always recover
d. drugs should be legalized

a 8. Methamphetamine _____.
a. speeds up the body
b. keeps you awake for days
c. slows down the body
d. makes you hallucinate

Answer the following questions.

9. What problems does drug abuse cause for employers? **loss of productivity, industrial accidents, higher health care cost, absenteeism, and lateness**
10. How much money does General Motors lose each year to drug abuse? **one billion dollars**
11. In what countries is the coca plant grown that is used to make cocaine? **Peru, Bolivia, and Columbia**
12. What might happen if all the coca fields were destroyed? **Those who used and sold cocaine might make a synthetic version of the drug.**

 97

▲ ENRICHMENT 97

NAME ______ DATE ______ CLASS ______

ENRICHMENT — Chapter 25

Drugs in Society — Text Pages 586–587

HELP FOR DRUG ABUSERS

What would you do if you discovered that your best friend had a drug problem? Most likely, you would want to help your friend get better. There are many programs for people with drug problems and their families.

Look in the yellow pages to find out who your friend could call for help. Try looking under "Drug Abuse and Addiction," "Alcoholism Information and Treatment," or "Drug Treatment." List some of the programs you find below. Find two programs of each kind listed. Some programs may fit into more than one category.

1. Programs for Adolescents
Answers will vary.
2. Programs for Alcoholism
Answers will vary, but may include Alcoholics Anonymous or Salvation Army.
3. Programs for Family Members of Drug Abusers
Answers will vary, but may include Al-Anon.
4. Residential Programs
Answers will vary.
5. Support Groups
Answers will vary, but may include Narcotics Anonymous.

Which program would be best for these people?

6. A 13 year old with a cocaine addiction: **Answers should include programs for adolescents or for cocaine abusers.**
7. The brother of an alcoholic: **Answers may include Al-Anon or other programs for family members of drug abusers.**
8. A drug abuser who feels that he or she needs treatment, counseling, and daily support: **Answers should include residential programs.**
9. An adult with a drinking problem: **Answers may include Alcoholics Anonymous.**
10. A recovering drug abuser who is tempted to try drugs again: **Answers may include support groups such as Narcotics Anonymous.**

 97

CHECK FOR UNDERSTANDING

▶ Ask questions 1-3 in the Section Review.

RETEACH

Cooperative Learning: Use the Paired Partners strategy to investigate how smoking and drinking could affect a person's ability to work in an assembly line plant.

EXTENSION

For students who have mastered this section, use the **Reinforcement** and **Enrichment** masters or other OPTIONS.

3 CLOSE

▶ Have students suggest new ideas of what would make a good drug education program.

SECTION REVIEW ANSWERS

1. Drug abuse results in loss of productivity, industrial accidents, higher health care costs, and absenteeism.
2. By educating the general public about the effects of drugs, strengthening law enforcement, and expanding courts, prisons, and military forces.

YOU DECIDE!

Encourage students to think outside of their usual responses. Provide magazine ads that display a glamourous image of tobacco and alcohol use. You might mention that tobacco ads on TV have been banned for over ten years.

PROGRAM RESOURCES

From the **Teacher Resource Package** use:

Science and Society, page 29, Drug Abuse: The Human Cost.

SECTION 25-3

PREPARATION

SECTION BACKGROUND

▶ The active ingredient in marijuana is delta-9-tetrahydrocannabinal (THC). THC can act as an analgesic, hallucinogen, sedative, or stimulant depending on the dose and psychological state of the drug user.

▶ Stimulants are drugs that affect the central nervous system by causing an excitatory action on certain smooth muscles and the heart. This increases the rate of glycogen conversion to sugars and stimulates the respiratory system.

▶ Depressants can be classified into two major groups—sedatives and tranquilizers.

▶ Certain cacti, peyote and psilocybe, have chemicals that cause hallucinations. They have been used by Native Americans and the ancient Aztecs in religious ceremonies to induce a trance-like state.

▶ A major factor in the spread of opiate addiction in the mid-1800s to early 1900s was the invention of the hypodermic needle.

PREPLANNING

▶ To prepare for Activity 25-2, obtain a culture of *Daphnia*. Allow the culture to remain undisturbed for at least one day before starting the activity.

1 MOTIVATE

Cooperative Learning: Use the Numbered Heads Together strategy to prepare a three-column paper that lists 1) the name of an illegal drug, 2) what students think its medical uses are, and 3) what students think it does to the body. Use the information to illustrate the general misinformation most people have about drugs.

▶ Discuss how the body makes its own drugs, endorphin and enkephalin, that provide relief from pain.

25-3 Problems with Illegal Drugs

New Science Words

marijuana
hashish
cocaine
crack
hallucinogens
heroin

Figure 25-4. Marijuana is made from the parts of the hemp plant.

Objectives

- ▶ Compare the effects of marijuana and tobacco use.
- ▶ Describe the effects of stimulants and depressants.
- ▶ Describe the effects of hallucinogens.
- ▶ Explain problems of using opiates.

Marijuana

Indian hemp is a weed plant found worldwide. An illegal drug called **marijuana** or pot is made from the stems, leaves, flowers, and seeds of this hemp plant. These plant parts are crushed up and made into a cigarette (joint) for smoking. The sticky tarlike substance from the flowering top of the marijuana plant is known as **hashish.** This is 1
more concentrated and powerful than marijuana. Hashish is smoked, chewed, or put in a drink to get its effect.

Marijuana is considered to be more dangerous than cigarettes. Many harmful chemicals have been identified in marijuana. The most powerful is THC, which produces a mind-altering effect. The effects of marijuana use depend a lot on the user's surroundings and feelings at the time it is smoked. It can both stimulate and depress. The person who uses marijuana may feel calm, relaxed, and have a feeling of well-being. Then the user may become sensitive to sights and sounds. Marijuana use impairs reaction time and coordination. It harms a person's ability to think, learn, and remember. The effects of smoking marijuana last from two to four hours. Regular users can suffer long-term effects, including losing interest in life. They substitute marijuana for other goals. They also experience feelings of anxiety, panic, and periods of depression.

Marijuana use has physical effects as well. Smoking the drug increases heartbeat rate and therefore is dangerous for persons with heart conditions. Its effects on the lungs are similar to those of tobacco use. Lung tissue can be damaged. The risks of bronchitis, emphysema, and lung cancer increase.

OPTIONS

Meeting Different Ability Levels

For Section 25-3, use the following **Teacher Resource Masters** depending upon individual students' needs.

◆ **Study Guide Master** for all students.

● **Reinforcement Master** for students of average and above average ability levels.

▲ **Enrichment Master** for above average students.

Additional Teacher Resource Package masters are listed in the OPTIONS box throughout the section. The additional masters are appropriate for all students.

◆ STUDY GUIDE 98

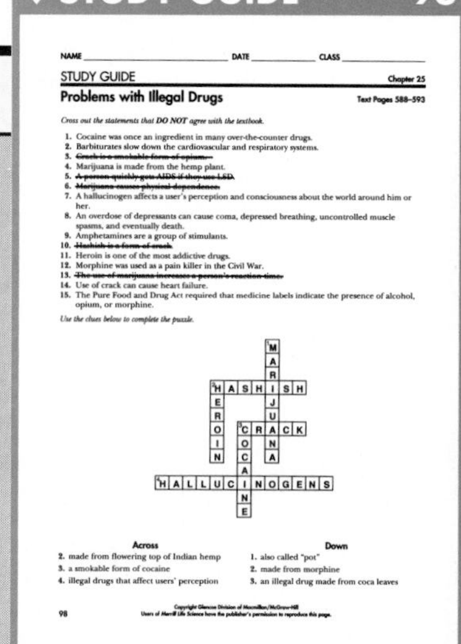

NAME ____________ DATE ________ CLASS ________

STUDY GUIDE — Chapter 25

Problems with Illegal Drugs — Text Pages 588–593

*Cross out the statements that **DO NOT** agree with the textbook.*

1. Cocaine was once an ingredient in many over-the-counter drugs.
2. Barbiturates slow down the cardiovascular and respiratory systems.
3. ~~Crack is a smokable form of opium.~~
4. Marijuana is made from the hemp plant.
5. ~~A person quickly gets AIDS if they use LSD.~~
6. ~~Marijuana causes physical dependence.~~
7. A hallucinogen affects a user's perception and consciousness about the world around him or her.
8. An overdose of depressants can cause coma, depressed breathing, uncontrolled muscle spasms, and eventually death.
9. Amphetamines are a group of stimulants.
10. ~~Hashish is a form of crack.~~
11. Heroin is one of the most addictive drugs.
12. Morphine was used as a pain killer in the Civil War.
13. ~~The use of marijuana increases a person's reaction time.~~
14. Use of crack can cause heart failure.
15. The Pure Food and Drug Act required that medicine labels indicate the presence of alcohol, opium, or morphine.

Use the clues below to complete the puzzle.

Across
2. made from flowering top of Indian hemp
3. a smokable form of cocaine
4. illegal drugs that affect users' perception

Down
1. also called "pot"
2. made from morphine
3. an illegal drug made from coca leaves

98

Even though marijuana does not cause physical dependence, users want the good feeling to continue. They develop a psychological dependence on the drug.

Stimulants

As you read earlier, nicotine and caffeine are stimulants. However, there are much stronger stimulants than those. One group of stimulants is amphetamines (am FET uh meenz). In the past, amphetamines were prescribed to increase a person's energy and decrease appetite to control weight. This is rarely done now.

What effect do stimulants have on the body?

2 Illegal amphetamines are known as pep pills and uppers. Another form of illegal amphetamines is known as speed or crank. Crank is also called ice. These drugs suppress hunger and cause the user to experience increased energy. Their use increases heartbeat rate, blood pressure, and breathing. Used over a long period of time, these drugs can lead to anxiety and feelings that someone will cause the person harm. Physical and psychological dependence can result. Withdrawal from use of these drugs can cause depression.

Another illegal stimulant is **cocaine,** a white powder made from the leaves of the South American coca plant. It is thought that the leaves of the coca plant are chewed by workers in the Andes Mountains to stimulate their bodies in this low-oxygen environment. Cocaine was an ingredient in many nonprescription drugs in the 1800s. A popular cola soft drink originally had cocaine in it and was advertised as a stimulant.

Cocaine is inhaled through the nose, swallowed, or injected. It is absorbed by membranes and enters the bloodstream quickly. Cocaine is a drug that users can become dependent on in a matter of days. The drug has a short-lived "high," or feeling of well-being. Anxiety soon replaces the good feeling. To get the same feeling, a user must take increasingly larger doses of the drug. Cocaine increases heartbeat rate and blood pressure and has been associated with sudden death due to a heart attack.

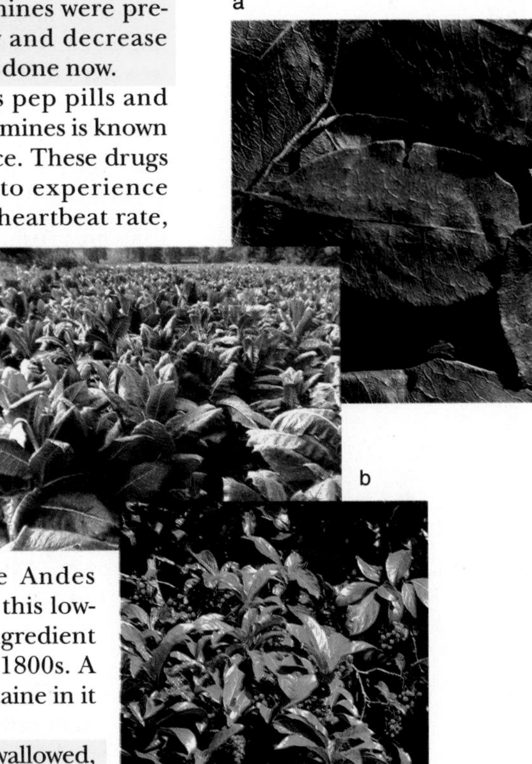

Figure 25-5. Coca leaves (a), tobacco (b), and coffee (c) contain substances that stimulate the body to different degrees. The stimulants produced from coca leaves are controlled, illegal substances.

TYING TO PREVIOUS KNOWLEDGE: Recall the function of the nervous system in maintaining homeostasis within the body. Have students discuss how drug abuse can bring about a disequilibrium of body functions.

2 TEACH

Key Concepts are highlighted.

CONCEPT DEVELOPMENT

- If possible, have a speaker from a drug rehabilitation center discuss with students reasons given by drug abusers for starting drug use. Use this as a springboard to consider alternatives to drug use.
- Explain to students that smoking marijuana can cause some of the same health problems as smoking tobacco even though there is no nicotine. Ask students why this might be true. *Marijuana has other cancer-causing chemicals, and the smoke is inhaled more deeply.*
- Amphetamines produce stimulation by causing the release of stored energy from mitochondria in the cells.
- Discuss with students the situation in the late 1970s of overprescribing drugs such as tranquilizers and Valium. This has led to physicians being more conservative in prescribing them.

● REINFORCEMENT 98

NAME ______ DATE ______ CLASS ______

REINFORCEMENT — Chapter 25

Problems with Illegal Drugs — Text Pages 588–593

Using the information in your textbook, classify the following drugs by writing stimulant, depressant, hallucinogen, *or* opiate.

stimulant 1. cocaine
hallucinogen 2. LSD
stimulant 3. amphetamine
opiate 4. heroin
hallucinogen 5. PCP
depressant 6. alcohol
depressant 7. barbiturate
stimulant 8. caffeine
opiate 9. morphine
stimulant 10. crack

Match the drug with its effect by writing the correct letter in the space provided.

c 11. constricted blood vessels, irregular heartbeat, heart failure — a. heroin
a 12. dreamlike state, drowsiness, nausea, physical addiction — b. pot
d 13. hallucinations, dramatic changes in personality — c. crack
f 14. increased heart rate, anxiety, fear — d. LSD
e 15. depressed breathing, confusion, coma — e. downer
b 16. relaxation, sensitive to sights and sounds, loss of interest in life — f. speed

Answer the following questions.

17. How does a pregnant woman who smokes crack affect her unborn baby? **The baby can feel the effects of the drug. The baby can also become physically dependent on the drug. After birth, the baby may have to undergo withdrawal from crack.**
18. How might a user of opiates get AIDS? **AIDS is spread among heroin users when they share needles to use the drug.**

98

▲ ENRICHMENT 98

NAME ______ DATE ______ CLASS ______

ENRICHMENT — Chapter 25

Problems with Illegal Drugs — Text Pages 588–593

FINDING ARTICLES ABOUT DRUGS

Almost every day a newspaper will carry articles on problems associated with illegal drugs. Some articles focus on medical issues, others focus on the violence associated with drug abuse, and still others focus on what drugs are doing to society.

Each day, Monday through Friday, buy a copy of your local newspaper. You may want to work with other students to share the cost of buying these papers through the week. Or bring your home-delivered copy to class.

Search through the paper each day for articles, columns, editorials, or pictures that have something to do with drugs. Cut out the articles and tape or paste them onto poster board for display at school. To make the posters, you may want to wait until the end of the week, when you have collected many articles. Then group articles together, according to their focus. For example, all articles about drug killings should be grouped together.

98

REVEALING MISCONCEPTIONS

▶ Young people often have the idea that because they are young and healthy, "doing" a little drugs will not do any harm. Emphasize that each individual reacts differently to drugs and even small amounts can have adverse health consequences. In addition, most street drugs are mixed with other substances that can cause reactions, including death.

MINI QUIZ

Use the Mini Quiz to check students' recall of chapter content.

1. **What powerful drug is made from the resin of the marijuana plant?** *hashish*
2. **What are some street names for illegal amphetamines?** *pep pills, uppers (speed, crank, ice)*
3. **A stronger form of cocaine that is often smoked is called _______ .** *crack*
4. **What two systems are slowed down by depressants?** *cardiovascular and respiratory systems*

CROSS CURRICULUM

▶ **Language Arts:** Have students look up *morphine, hashish,* and *caffeine* in the dictionary. Students should try to find how their roots reveal something about the origin of the drug or its effects on the body.

How does crack affect the body?

3 A stronger form of cocaine is called **crack.** Crack is a smokable form of cocaine. Since the introduction of crack around 1986, it has rapidly become a much abused drug. Unfortunately, many people did not realize the rapid addictive powers of cocaine or its dangerous effects on the body. The effects of crack are more intense and short-lived than cocaine. As with all stimulants, the drug affects the central nervous system. Crack use causes constricted blood vessels, causing the heart to work harder to move blood throughout the body. As a result, blood pressure increases, and an irregular heartbeat can develop. Heart failure can occur. The initial high of the drug is followed by a period of depression in which the user may feel anxious and irritable and have a strong craving for more of the drug. Pregnant females using the drug pass the drug through the placenta to the developing fetus. These babies can develop physically dependent. After birth, these babies go through withdrawal.

Depressants

You read earlier about depressants, drugs that slow down the central nervous system. Alcohol, sleeping pills, tranquilizers, and a class of drugs called barbiturates (bar BIH chuh ruts) are all depressants. A slang word for depressant pills is "downers." Depressants slow down or prevent impulses from moving across a synapse from one neuron to another. How would this affect your heart and your respiratory system? 4 Blood vessels dilate and blood pressure lowers. Breathing rates slow. A person under the influence of depressants may be confused, lack coordination, appear to be intoxicated, and have slow reflexes. This person is prone to accidents.

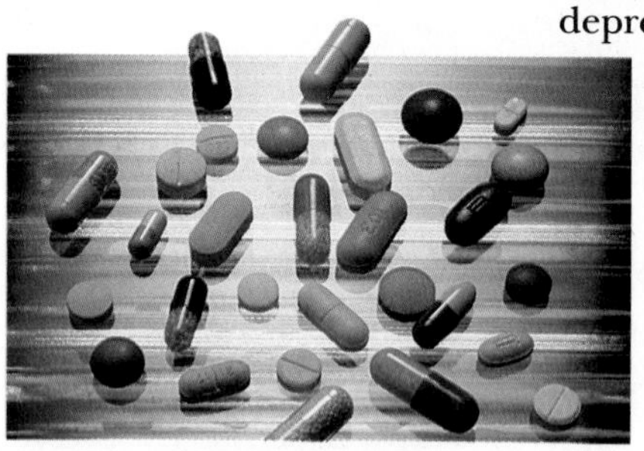

Figure 25-6. Taking combinations of depressants can be deadly.

An overdose of depressants can cause coma, depressed breathing, uncontrolled muscle spasms, and eventually death. Depressants taken in combination with other depressants, such as sleeping pills or alcohol, increase the dangerous effects. Withdrawal from depressants is difficult.

Some persons take a combination of stimulants and depressants to try to keep in a constant state of drug enjoyment. The physical and psychological damage to the body is severe and can lead to death.

OPTIONS

ASSESSMENT—ORAL

▶ **What is the stimulant most widely used in the world today?** *caffeine in coffee, soft drinks, and medicines*

▶ **Infer the dangers of taking amphetamines to stay awake for long periods of time.** *The lack of sleep may eventually cause the person to be exhausted and fall into a deep sleep. If the person is driving a car or engaged in other activities that require alertness, accidents can result, causing injury to self or other persons.*

Hallucinogens

Hallucinogens (huh LEWS un uh junz) are a group of drugs that affect users' perceptions and consciousness of the world around them. Under the influence of hallucinogens, the user often experiences an extreme sensitivity to light, sound, taste, and odor. This causes confusion, and the person cannot tell the difference between reality and fantasy.

In addition to psychological dependence, the use of hallucinogens can cause a person to have unpredictable, violent behavior. Coma, convulsions, and death can occur with the use of these drugs.

Two of the more powerful hallucinogens are LSD and PCP, also called angel dust. These drugs can produce dramatic effects from taking even small amounts. The senses become highly stimulated, and users talk about "seeing" sounds, "tasting" odors, and "hearing" colors. The effects may last for several hours. The hallucinations of a bad experience with these drugs can cause anxiety and panic and lead to suicide. LSD is also known to cause flashbacks to hallucinations that occurred years before.

The use of hallucinogens can bring about dramatic changes in personality. Persons who use these drugs find they can't control themselves after using even a small amount. Users may react violently toward themselves or others. In some cases, they may slip into a long, semiconscious state. The results of using these drugs are unpredictable.

Connect to... Chemistry

Ergot is a parasitic fungus that grows on rye and some other plants that are classified as grasses. It contains LSD. What condition does it stimulate? What are its symptoms?

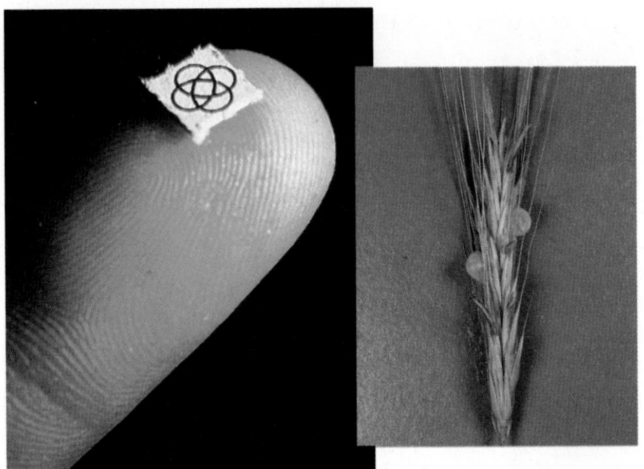

Figure 25-7. Even small amounts of LSD, such as those found in the small patch to the left, can cause hallucinogenic responses. The photograph on the right shows the fungus ergot on a head of rye grain. Ergot contains the drug LSD.

CONCEPT DEVELOPMENT

▶ Explain to students that hallucinogenic drugs such as LSD and PCP are so powerful that only very small amounts (micrograms) are needed to bring about the hallucinogenic effect.

Connect to... Chemistry

Answer: Ergot causes ergotism in humans and cattle. Humans get this disease from eating bread made from infected rye grain. Gangrene and convulsions result. Ergotism is now uncommon due to improved methods of cleaning grain.

CHECK FOR UNDERSTANDING

▶ Ask questions 1-4 and the **Apply** Question in the Section Review.

▶ Have students compare the abuse of aspirin to the abuse of amphetamines.

▶ Have students find out about the effects of combining drugs that can bring about a reaction that is much more than the arithmetic sum of their effects. Compare arithmetic progression with geometric progression.

RETEACH

FLEX Your Brain

Use the Flex Your Brain activity to study ILLEGAL DRUGS.

Flex Your Brain

ASSESSMENT

Portfolio: Use the Flex Your Brain activity to reinforce critical thinking and problem solving skills and to uncover basic information students may have missed.

EXTENSION

For students who have mastered this section, use the **Reinforcement** and **Enrichment** masters or other OPTIONS provided.

ENRICHMENT

▶ Have students research the drug contents of patent medicines available in the latter part of the 1800s.

▶ Have students investigate the use of narcotic antagonists in the treatment of heroin addiction.

PROGRAM RESOURCES

From the **Teacher Resource Package** use:

Activity Worksheets, page 5, Flex Your Brain.

CONCEPT DEVELOPOMENT

▶ Discuss with students another morphine derivative, codeine, that is used in cough medicines. Before being regulated there was some abuse in the use of codeine-based medicines.

▶ **What pain-killing drug was used extensively during the Civil War?** *morphine*

3 CLOSE

▶ Have students discuss the legal stimulants and depressants often abused and contrast with the abuse of illegal stimulants and depressants.

▶ Collect articles about drug use among athletes and use the information to discuss how drug use has ruined careers of some sports figures.

In Your JOURNAL

For some people, aspirin contains ingredients (salicylates) that upset the stomach.

Opiates

Using plant substances to relieve pain has occurred often throughout history. During the third century B.C., the Greeks used a substance from the white poppy to relieve suffering from pain. This same substance was introduced to the Orient in the 1800s. However, the drug soon was used less for medical purposes than for putting oneself into a dreamlike state. Physical dependence on the drug became widespread. This drug was opium, made from the juices of the seed capsule of the white poppy. Opium and its active chemical, morphine, were used as pain killers during the Civil War. Many wounded soldiers became dependent on morphine. Because there were no laws regulating nonprescription drugs, many of them also contained opium or morphine. People became physically dependent on these medicines. There was such a protest against the use of these drugs, that Congress passed the Pure Food and Drug Act in 1906. This law required that labels on medicine show whether they contain alcohol, opium, or morphine.

Because morphine was a good pain killer but could cause physical dependence, researchers tried to make other pain-relieving drugs from opium. The powerful drug **heroin** was made from morphine. It was not supposed to cause physical dependence. However, this was not true. A much stronger drug than morphine, heroin is one of the most addictive drugs and is widely abused in the United States.

In Your JOURNAL

An analgesic is a drug that relieves pain. Aspirin and morphine are both analgesics. So is acetaminophen. **In your Journal,** explain why some people cannot take aspirin.

What does the Pure Food and Drug Act require?

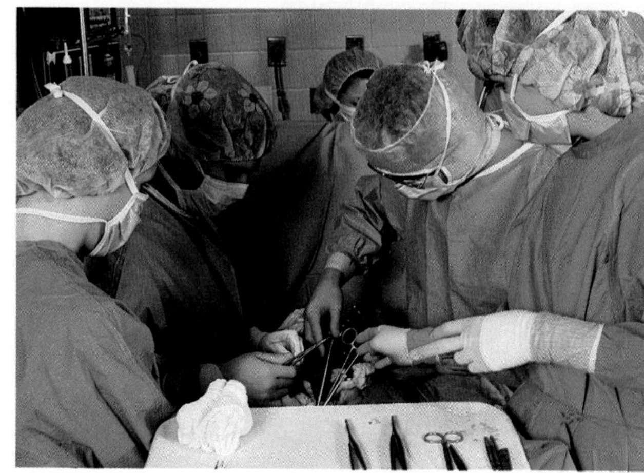

Figure 25-8. Morphine is used medically as a pain reliever after some surgeries and for some terminal cancer patients. Its use is strictly regulated.

OPTIONS

PROGRAM RESOURCES

From the **Teacher Resource Package** use:

Concept Mapping, pages 55-56.

Cross-Curricular Connections, page 29, Debating an Issue.

Activity Worksheets, page 228, Mini-Lab: How are drugs classified?

Figure 25-9. Opium is made from the juices of the seed capsule of the white poppy.

Opiates, also called narcotics, cause a brief, dreamlike state, drowsiness, constricted pupils, and nausea. Because these drugs are often injected and the needles shared, users can develop infections from contaminated needles. Tetanus; hepatitis (hep uh TITE us), a disease of the liver; and HIV are spread among heroin users when sharing needles to use the drug.

Drug abuse has become one of the greatest problems in our country today. The waste of human resources and lives due to drug abuse has affected every community. Sometimes, people feel the only way to escape their problems is to use drugs. Unfortunately, by doing this they exchange one set of problems for another. Problems of drug dependence lead to mental and physical illness and sometimes death. There is no future in drug abuse.

MINI-Lab

How are drugs classified?

Make flash cards for the following drugs. *Classify* them into categories of stimulants, depressants, hallucinogens, and opiates. Drugs: alcohol, amphetamines, barbiturates, crank/ice, caffeine, cocaine, crack, LSD, heroin, morphine, PCP, sleeping pills, speed, tranquilizers.

SECTION REVIEW

1. How is marijuana use similar to tobacco use?
2. Why are stimulants and depressants dangerous?
3. What are the dangers to a person's health if hallucinogens are used?
4. What are additional risks of using narcotics besides the effect of the drugs?
5. **Apply:** Why is it dangerous for a person who has been using marijuana to drive?
6. **Connect to Chemistry:** Find out what flashbacks are and why they can occur after use of a hallucinogen.

Outlining

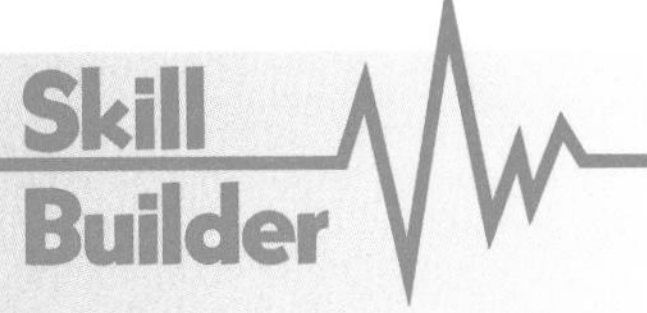

Outline the effects of two drugs discussed in Section 25-3. Title your outline, Problems with Illegal Drugs. If you need help, refer to Outlining in the **Skill Handbook** on page 681.

Skill Builder

Choices of drugs will vary.

25-3 Problems with Illegal Drugs

I. Effects of Marijuana
  A. may depress or stimulate
  B. depend on surroundings and feelings
  C. psychological effects
    1. calm, well-being, anxiety, panic, depression
    2. harms ability to think, learn, remember
  D. physical effects
    1. increases heartbeat rate
    2. increases risk of bronchitis, emphysema, lung cancer

II. Effects of Hallucinogens
  A. psychological effects
    1. extreme sensitivity to light, sound, taste, and odor
    2. confusion between reality and fantasy
    3. flashbacks may occur years later
  B. physical effects
    1. unpredictable, violent behavior
    2. semiconsciousness at times

MINI-Lab

Answers: *Stimulants*—amphetamines, crank/ice, caffeine, cocaine, crack, speed; *Depressants*—alcohol, barbiturates, sleeping pills, tranquilizers; *Hallucinogens*—LSD, PCP; *Opiates*—morphine, heroin, opium

SECTION REVIEW ANSWERS

1. Smoking marijuana increases heartbeat rate and can damage lung tissue. The risks of bronchitis, emphysema, and lung cancer increase.

2. Stimulants increase heartbeat rate, blood pressure, and breathing. These drugs can lead to anxiety and feelings that someone will cause the person harm. Depressants cause blood vessels to dilate and blood pressure to lower. Breathing rates slow. A person who uses depressants may be confused, lack coordination, appear to be intoxicated, and have slow reflexes.

3. Using hallucinogens can cause confusion, and the person cannot tell the difference between reality and fantasy.

4. Narcotics are often injected. When the needles are shared, users can develop infections. Tetanus, hepatitis, and the AIDS virus are spread when sharing needles to use drugs.

5. Apply: The THC in marijuana produces a mind-altering effect. Marijuana use also impairs reaction time and coordination. It harms a person's ability to think, learn, and remember. Driving a car under these conditions would be dangerous.

6. Connect to Chemistry: A flashback is the recurrence of negative or frightening sensory experiences some time after a drug has been taken and without taking the drug again. It is not clear why these occur.

Skillbuilder

ASSESSMENT

Performance: Use this Skillbuilder to have students Compare and Contrast the psychological effects of marijuana and hallucinogens.

ACTIVITY 25-2

OBJECTIVE: Design and carry out an experiment to observe and communicate how drugs affect heartbeat rate.
Time: One class period to plan the activity. One class to do the experiment.

PROCESS SKILLS applied in this activity are **observing, experimenting, predicting, interpreting data,** and **communicating.**

PREPARATION

- Let water stand overnight.
- Preparation of Solutions: ethyl alcohol — add 2 mL ethyl alcohol to 98 mL distilled water; nicotine — soak a nonfiltered cigarette in 100 mL distilled water for one hour and strain the solution; prepare weak coffee; dilute cola 1 part water to 1 part cola; cough medicine — add 2 mL cough medicine to 98 mL distilled water.
- Students may need to use more than one *Daphnia* as the animal may not survive more than two drug doses.

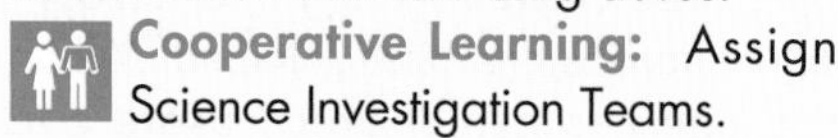
Cooperative Learning: Assign Science Investigation Teams.

THINKING CRITICALLY

Students should consider a number of invertebrate organisms. The heartbeat rate is determined by counting the number of beats per minute. Careful handling, using aged tap water, and using a new organism for each step will help prevent injury.

TEACHING THE ACTIVITY

*Refer to the **Activity Worksheets** for additional information and teaching strategies.*
Sample drug solutions and associated heartbeat rates per minute are: no drug-240; coffee-280; cola-260; ethyl alcohol-210; tobacco-295-300; cough medicine-200-220.

SUMMING UP/SHARING RESULTS

- The control is when no drug is used. The variables are each of the drug solutions.
- See students' tables.
- Coffee and tobacco
- Coffee, cola, and tobacco are stimulants. Ethyl alcohol is a depressant. Cough medicine may be listed as either.
- Students should compare between predictions and results.
- Stimulants increase heartbeat rate; depressants decrease heartbeat rate.

GOING FURTHER!

Drugs have similar effects on people and can alter the normal functioning of their central nervous system.

Activity
ASSESSMENT
Performance: To further assess students' understanding of the effects of drugs on heartbeat rate, have them compare four over-the-counter cough medicines. Have students write a conclusion about how they think these medications would affect heartbeat rate.

PROGRAM RESOURCES
From the **Teacher Resource Package** use:
Activity Worksheets, pages 223-224, Activity 25-2, The Effects of Drugs on Heartbeat Rate.

ACTIVITY 25-2 DESIGNING AN EXPERIMENT
The Effects of Drugs on Heartbeat Rate

Stimulants and depressants have a direct effect on the central nervous system. The CNS controls the homeostasis of heartbeat.

Getting Started

You need to determine what chemicals can be safely used to test for heartbeat effects. Devise a method to show how drugs affect the heartbeat rate of an animal.

Thinking Critically

List some invertebrates that might be used for this activity. How will the heartbeat rate be determined? What precautions must be used to prevent injury to the organism?

Materials   

Your cooperative group will use:
- dilute solutions of coffee, cola, ethyl alcohol, tobacco, and cough medicine (dextromethorphen hydrobromide)
- *Daphnia* culture
- microscope
- microscope culture slide
- dropper
- aged tap water

Try It!

1. Use a dropper to place a single *Daphnia* crustacean on a culture slide.
2. Place the slide under the microscope. Use low power and count the number of times the heart of the *Daphnia* beats in one minute. (See photo for location of the animal's heart.) Record all of your observations in a data table.
3. *Predict* whether the drugs are stimulants or depressants.
4. Add two drops of one of the drug solutions to the *Daphnia* in the culture slide and count the heartbeat rate again.
5. Use a dropper to remove the solution. Flush the slide and *Daphnia* carefully with aged tap water. **CAUTION:** *Flush the used* Daphnia *into a beaker of aged tap water provided by your teacher.*
6. Repeat Steps 1 through 6 for each of the drugs solutions provided. **CAUTION:** *Use a new* Daphnia *for each step.* Record the effects of each drug on heartbeat rate.

Summing Up/Sharing Results

- What was the control in this experiment? What were the variables?
- How did the drugs affect *Daphnia*?
- Which drugs caused the greatest change in the number of heartbeats per minute?
- Which drugs were stimulants? Which drugs were depressants?
- *Compare* your predicted results with the experimental results.
- How do drugs affect the rate of heartbeat of an animal?

Going Further!

How does the data collected in this experiment help you understand how drugs affect people who use them?

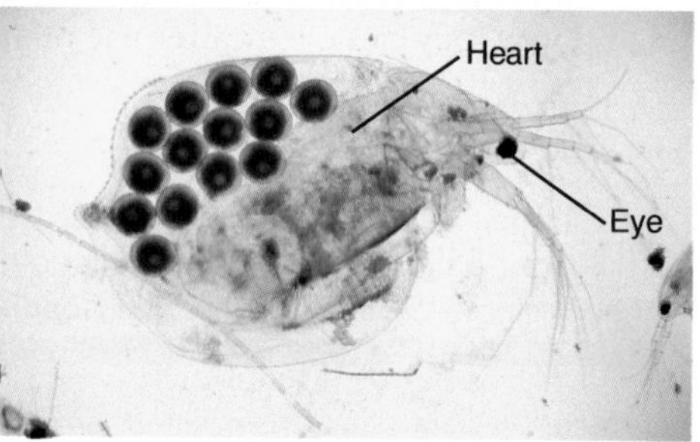

CHAPTER

REVIEW

SUMMARY

25-1: Drugs and Health

1. Drug misuse is using a drug for the purpose it was made, but taking it improperly. Drug abuse is use of a drug for other than its intended purpose. Both are harmful to health and can cause chemical dependence.
2. Nicotine in tobacco speeds up the heartbeat rate; alcohol slows down body processes.
3. Caffeine is a stimulant.

25-2: Science and Society: Drugs in Society

1. Drug abuse costs can be measured in increased health care costs, on-the-job losses, and the cost of educating people on why they should avoid usage.
2. Education, increased law enforcement, and cooperation of countries from which drugs come are attempts to solve the drug problem.

25-3: Problems with Illegal Drugs

1. Marijuana is more dangerous than cigarettes; marijuana produces a mind-altering effect.
2. Stimulants, including amphetamines, cocaine, and crack, speed up body functions; depressants, including alcohol, tranquilizers, and barbiturates, slow down body functions.
3. Hallucinogens affect perception and consciousness; users can react violently.
4. Opiates, such as morphine and heroin are narcotics and are very addictive.

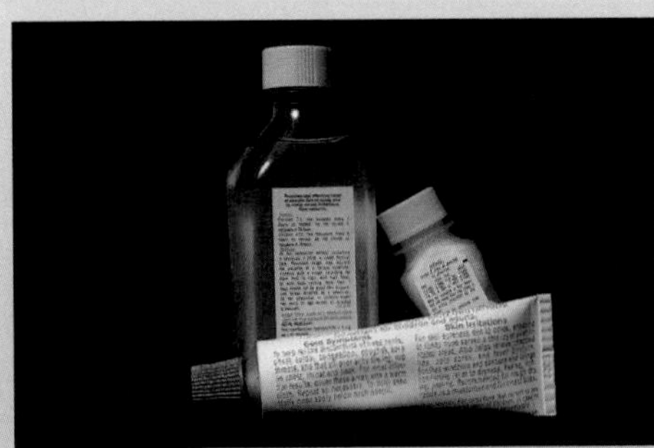

KEY SCIENCE WORDS

a. **cocaine**
b. **crack**
c. **depressant**
d. **drug**
e. **drug abuse**
f. **drug misuse**
g. **fermentation**
h. **hallucinogens**
i. **hashish**
j. **heroin**
k. **marijuana**
l. **methamphetamine**
m. **physical dependence**
n. **psychological dependence**
o. **stimulant**
p. **tolerance**
q. **withdrawal**

UNDERSTANDING VOCABULARY

Match each phrase with the correct term from the list of Key Science Words.

1. larger doses are needed to produce the effects of the drug
2. stronger form of cocaine
3. taking too much medicine
4. made from sticky residue of marijuana plants
5. using illegal drugs
6. process of making alcohol
7. drug that slows the body down
8. affect perception and consciousness
9. narcotic made from morphine
10. any drug that increases heartbeat rate and breathing

OPTIONS

ASSESSMENT

To assess student understanding of material in this chapter, use the resources listed.

COOPERATIVE LEARNING

Consider using cooperative learning in the THINK AND WRITE CRITICALLY, APPLY, and MORE SKILL BUILDERS sections of the Chapter Review.

PROGRAM RESOURCES

From the **Teacher Resource Package** use:

Chapter Review, pages 53-54.

Chapter and Unit Tests, pages 165-168, Chapter Test.

Chapter and Unit Tests, pages 169-170, Unit Test.

CHAPTER

REVIEW

SUMMARY

Have students read the summary statements to review the major concepts of the chapter.

UNDERSTANDING VOCABULARY

1. p
2. b
3. f
4. i
5. e
6. g
7. c
8. h
9. j
10. o

ASSESSMENT

Portfolio

Encourage students to place in their portfolios one or two items of what they consider to be their best work. For each item, ask students to explain why that item was chosen and what they learned from it. Items might be selected from the following.

- For Your Gifted Students song about being drug-free, p. 578
- Cooperative Learning chart on caffeine, p. 579
- MINI-Lab childproof package design, p. 584

Performance

Additional performance assessments may be found in *Performance Assessment* and *Science Integration Activities* that accompany **Merrill Life Science.** Performance Task Assessment Lists and rubrics for evaluating these activities and other products generated throughout the chapter can be found in Glencoe's *Performance Assessment in Middle School Science.*

CHAPTER
REVIEW

CHECKING CONCEPTS

1. c
2. b
3. b
4. d
5. a
6. d
7. d
8. c
9. d
10. a

USING LAB SKILLS

ASSESSMENT

Use these alternate lab exercises to assess students' understanding of science skills used in this chapter.

11. Cautions usually include informtion that use of the prouduct may cause headaches, nervousness, sleeplessness, and/or palpitations.

12. Responses will vary but may include some special way of aligning the cap to allow easy opening or a code to read that allows easy opening.

THINK AND WRITE CRITICALLY

13. Psychological dependence—a person really believes that he or she needs the drug; physical dependence—a person has a chemical need for a drug.

14. The effects of the drug can be passed to the developing fetus, and the baby will undergo withdrawal after birth.

15. Drugs that are used improperly can harm health. Federal law, controlling the manufacture, distribution, and possession of all drugs, attempts to ensure that drugs are used properly and safely.

CHAPTER
REVIEW

CHECKING CONCEPTS

Choose the word or phrase that completes the sentence.

1. Methamphetamine is also called ______.
a. marijuana **c.** ice
b. heroin **d.** crack

2. Alcohol is absorbed directly into the ______.
a. heart **c.** large intestine
b. stomach **d.** liver

3. The drug ______ is found in tobacco.
a. tar **c.** carbon monoxide
b. nicotine **d.** hashish

4. Smoking affects all of these body parts except ______.
a. heart **c.** lungs
b. blood vessels **d.** bones

5. ______ is an example of a depressant.
a. Alcohol **c.** Caffeine
b. Nicotine **d.** Amphetamine

6. An example of a stimulant is ______.
a. LSD **c.** marijuana
b. alcohol **d.** amphetamines

7. Symptoms of depressants include all of the following except ______.
a. slow reflexes
b. lowered blood pressure
c. dilated blood vessels
d. increased breathing rate

8. ______ is obtained from the white poppy.
a. Caffeine **c.** Opium
b. Marijuana **d.** LSD

9. An example of a narcotic is ______.
a. alcohol **c.** PCP
b. caffeine **d.** opium

10. A majority of drugs are obtained from ______.
a. plants **c.** soil
b. animals **d.** minerals

USING LAB SKILLS

11. In the Activity on page 582 the label of an over-the-counter aspirin was analyzed for information. Obtain the label of over-the-counter diet pills. What cautions are given about their use? Can they be dangerous?

12. In the MINI-Lab on page 584 you were asked to design a childproof package for a product. How would you design a childproof cap for a pill bottle that can be opened by persons with arthritis in their hands?

THINK AND WRITE CRITICALLY

Answer the following questions in your Journal using complete sentences.

13. What is the difference between psychological and physical dependence?

14. What can happen to her baby if a pregnant female uses crack?

15. Explain why drug use is strictly controlled by the government.

APPLY

16. What are some effects on the body that are common to all stimulants?
17. What are some effects on the body that are common to all depressants?
18. Explain how industry is affected by drug abuse.
19. What differences would the way a drug is taken have on its effects?
20. Explain the physical effects of one drug discussed in this chapter.

MORE SKILL BUILDERS

If you need help, refer to the Skill Handbook.

1. **Comparing and Contrasting:** In a chart, compare and contrast the effects of stimulants and depressants.
2. **Making and Using Graphs:** Use the data from Activity 25-2 (page 594) to make a bar graph of substances and their effect on heartbeat rate in *Daphnia*.
3. **Hypothesizing:** Make a hypothesis as to what effects might be felt by a person who is smoking a cigarette and drinking coffee at the same time.
4. **Interpreting Scientific Illustrations:** Compare adult's dosage with child's on the label in Activity 25-1 (page 582). Why do you think a child's dose is different from an adult's?
5. **Concept Mapping:** Choose a brand name headache remedy and a generic brand. Compare the ingredients, cost, and amount of medication in each container. Make a network tree concept map showing your results.

PROJECTS

1. For one week, collect newspaper articles that have drug-related information in them. Bring these articles to school to share and discuss with your class.
2. Research the pattern of smoking habits among people in the United States over the past ten years. Who smokes more, men or women; what age group; how much do people smoke?

APPLY

16. Increase in heartbeat rate, blood pressure, and breathing.
17. Decrease in blood pressure and breathing; confusion, lack of coordination, slow reflexes.
18. Loss of productivity, industrial accidents, higher health care costs, absenteeism, and lateness.
19. The faster a drug is able to get into the blood, the faster its effects will occur. Injections and snorting move the drug quickly to the blood.
20. Student answers will vary.

MORE SKILL BUILDERS

1. Comparing and Contrasting: Accept all reasonable answers. Stimulants and depressants are alike because they generally affect the nervous and circulatory systems. Stimulants and depressants are different because stimulants increase the rate at which body systems function and depressants slow down the rate at which body systems function.

2. Making and Using Graphs:

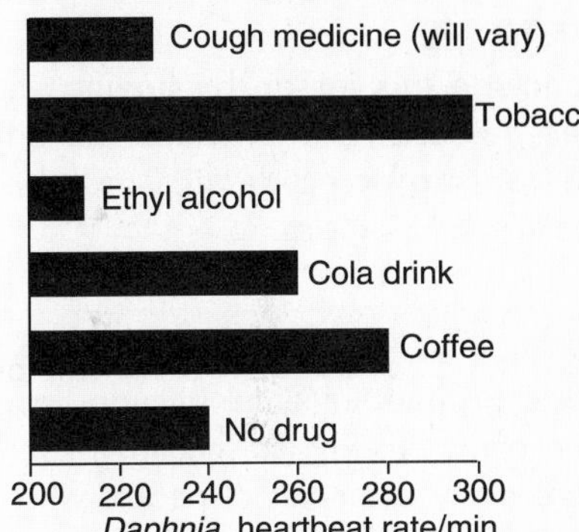

3. Hypothesizing: Accept all reasonable hypotheses. A person who smokes a cigarette and drinks coffee at the same time will show an increase in each drug's usual stimulating effects.

4. Interpreting Scientific Illustrations: Adult medicine is taken in a greater dosage and more often compared to children's medicine.

5. Concept Mapping:

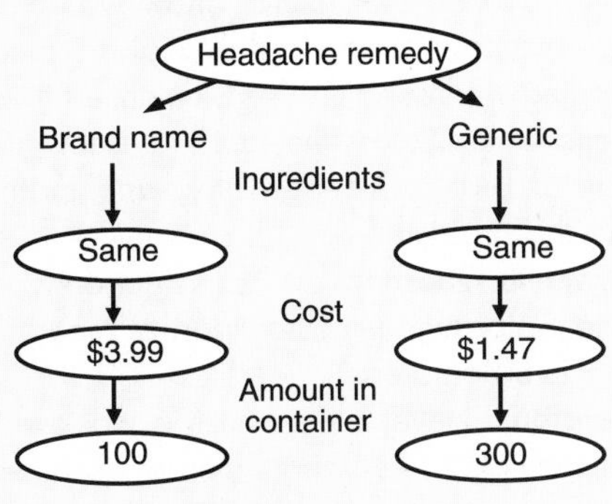

UNIT 7
GLOBAL CONNECTIONS

Objective
In this unit ending feature, the unit topic, "Staying Healthy," is extended into other disciplines. Students will see how important good health is to their own lives.

Motivate
Cooperative Learning: Using the Expert Teams strategy, divide the class into groups. Assign each group a different phase of health, including personal, family, community, national, and international health. Have groups investigate how actions in each area affect the others.

Teaching Tips
- Tell students to think about the connection between personal health and community health while they are reading this feature. Have students consider how each Connection affects their lives.

Wrap-Up
Conclude this lesson by having students research the newspaper for articles on major epidemics or other health-related articles.

HEALTH

Background: Mental attitude has an effect on physical health. There are many examples of people who are "fighters" successfully fending off diseases that killed others in similar circumstances.
Discussion: Ask the students if they know anyone whose attitude has helped him or her fight off diseases or recover from serious injuries.
Answer to Question: Mental attitude can have a profound effect on physical health. A person's attitude affects the way he or she feels, as well as how he or she behaves.
Extension: Ask if any student has an elderly relative who is ill or has been injured. Have students relate any incidents in which they think mental attitude played a role in the ability of their relatives to recover.

UNIT 7
GLOBAL CONNECTIONS

Staying Healthy
In this unit, you have studied health, disease, and drugs. Why is health important to all people? What influence does health have worldwide?

HEALTH

ZAPPING GERMS MAY HELP CURE DISEASE
Houston, Texas
A positive mental attitude may help cure diseases of the immune system. Patients at the M.D. Anderson Hospital can use a video game to zap germs. Fun may be an important part of the cure. How can mental health influence physical health?

HISTORY

BAD WATER EQUALS BAD HEALTH
London, England
In 1854, more than 500 people from the same neighborhood died of cholera in less than a week. Why? Dr. John Snow found that they all got their water from the same well. When the well was closed, the epidemic stopped. Why is it important to have a source of clean water?

HISTORY

Background: In the nineteenth century, millions of people were killed by cholera. Snow was one of the first to apply demographics to the analysis of epidemics. By mapping where victims lived, he linked them to one well.
Discussion: Discuss the fact that Snow used his knowledge of biology, geography, and math (statistics) to solve this problem.
Answer to Question: Water is not only a basic need, it is the major means of transport of many diseases, especially in urban areas.
Extension: Have students visit a wastewater or sewage treatment plant. Ask what types of disease-causing organisms are normally found in wastewater.

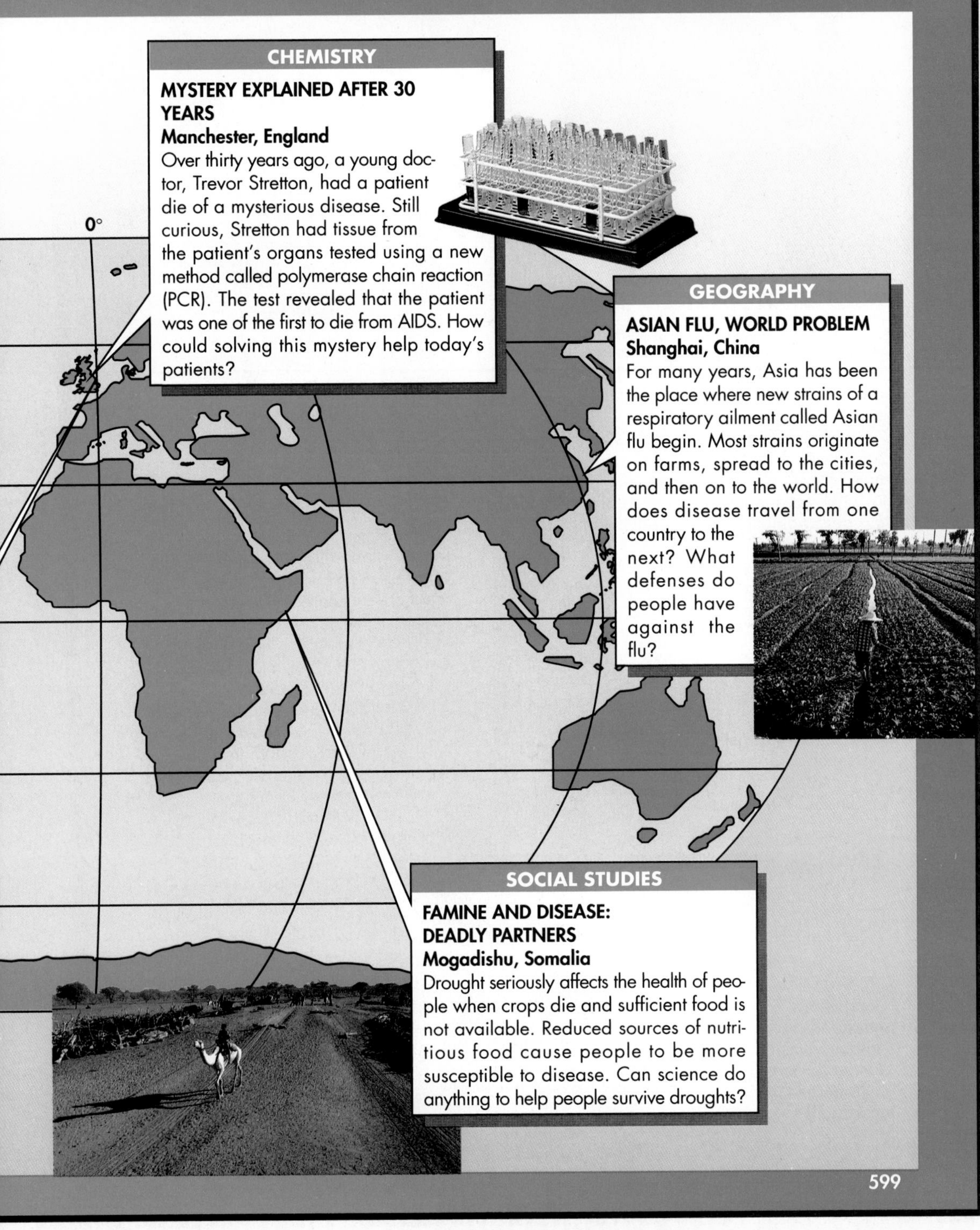

0°
CHEMISTRY
MYSTERY EXPLAINED AFTER 30 YEARS
Manchester, England
Over thirty years ago, a young doctor, Trevor Stretton, had a patient die of a mysterious disease. Still curious, Stretton had tissue from the patient's organs tested using a new method called polymerase chain reaction (PCR). The test revealed that the patient was one of the first to die from AIDS. How could solving this mystery help today's patients?
GEOGRAPHY
ASIAN FLU, WORLD PROBLEM
Shanghai, China
For many years, Asia has been the place where new strains of a respiratory ailment called Asian flu begin. Most strains originate on farms, spread to the cities, and then on to the world. How does disease travel from one country to the next? What defenses do people have against the flu?
SOCIAL STUDIES
FAMINE AND DISEASE: DEADLY PARTNERS
Mogadishu, Somalia
Drought seriously affects the health of people when crops die and sufficient food is not available. Reduced sources of nutritious food cause people to be more susceptible to disease. Can science do anything to help people survive droughts?
599

GEOGRAPHY

Background: The influenza virus easily mutates. This accounts for its ability to overcome the human immune system. The combination of ducks, pigs, and people on Chinese farms creates a situation that produces new strains of virus.

Discussion: Travelers have always been a source of disease. Europeans brought diseases such as measles that decimated certain American Indian tribes. What can be done to protect such populations?

Answer to Question: Diseases often are carried from one country to the next by infected human travelers. Prevention and immunization forestall the spread of diseases.

Extension: Have students find out from the public health department how their favorite restaurants are rated.

SOCIAL STUDIES

Background: Drought and social unrest have subjected Somalia's people to more than a decade of famine and disease.

Discussion: What are the basic needs of humans? How can they be best met over the long term? Many of Somalia's health problems are a result of over-population and political instability.

Answer to Question: Modern technology is being used to search for new underground aquifers in Sudan, Somalia, and Ethiopia. Scientists are also developing vaccines and medicines to combat the diseases that affect people in these areas.

Extension: Show students a photograph of people bathing in the Ganges River in India. Would students bathe in water that cattle are drinking? What if it is the only water available?

CHEMISTRY

Background: In 1959, a young sailor appeared at the Manchester Royal Infirmary. He died and tissue from several organs was preserved in blocks of paraffin. Recently, Manchester scientists identified the AIDS virus in four of the six specimens.

Discussion: Discuss how the knowledge of many disciplines can make a person a better scientist.

Answer to Question: The chemical test confirmed the disease. This case becomes part of a data base that could show patterns in the spread of the disease.

Extension: Divide the class into small groups. Give each group a list of symptoms of a certain disease and medical guides to identify the disease.

MICROBIOLOGIST

Background: Microbiologists spend most of their time in laboratories. They investigate the growth and characteristics of microscopic organisms such as bacteria, viruses, and fungi. A Ph.D. is required for most basic research.

Related Career	Education
Food Inspector	B.S. degree
Pathologist	M.D.
Physiologist	Graduate school

Career Issue: Recently, chemical and biological weapons have been developed by scientists. Some scientists feel that they should refuse to do this sort of work.

What do you think? Do scientists have a moral responsibility for the manner in which their work is used?

LICENSED PRACTICAL NURSE

Background: L.P.N.s work under the direction of doctors and registered nurses. In addition to basic bedside care, some L.P.N.s assist in the delivery, care, and feeding of infants. Some experienced L.P.N.s supervise nursing assistants and aids.

Related Career	Education
Emergency Medical Technician	High school+
Medical Records Technician	Community College+, exam
Nurse's Aid	Community college

Career Issue: L.P.N.s can face considerable danger from AIDS and other infectious diseases. In some situations, proper protection is not available or not provided.

What do you think? If proper safeguards are not provided, does a nurse have the right to refuse to treat a patient?

MICROBIOLOGIST

Microbiologists study the growth and characteristics of tiny living things such as bacteria, viruses, fungi, and protists. Medical microbiologists study how microorganisms cause disease and produce cures.

Microbiologists study science, math, and social studies and go on to college and graduate school before they can work.

The food industry uses microbiologists in the development of new products and in quality control. Hospitals may also employ microbiologists to track and prevent the spread of diseases within hospitals.

For Additional Information

Contact the American Society for Microbiology, Office of Educational and Professional Recognition, 1913 I Street, Washington, DC 20006.

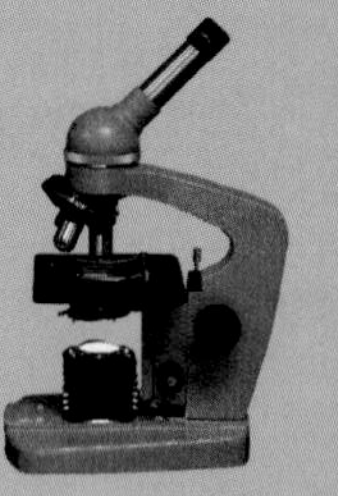

CAREERS

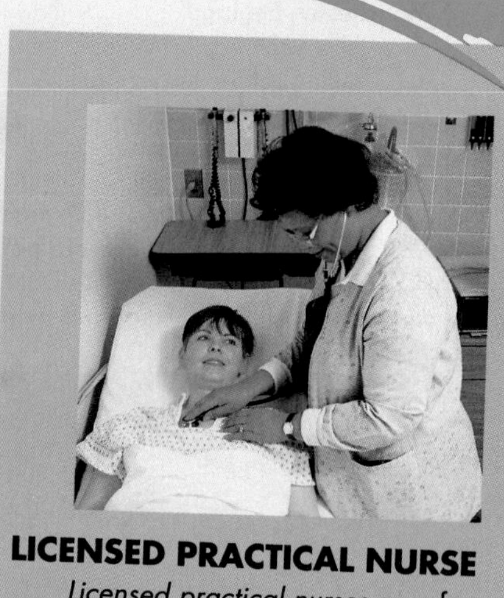

LICENSED PRACTICAL NURSE

Licensed practical nurses care for patients under the direction of doctors and registered nurses. They provide basic bedside care such as taking temperatures, pulse, and blood pressure. L.P.N.s work in nursing homes where, in addition to bedside care, they may also help develop care plans and supervise nursing aides.

A future L.P.N. should get as much science as possible in high school. Then he or she will go on to a state-approved training program at a trade school, community college, or hospital. Traditionally, a program for Licensed Practical Nurse can be completed in one year of full-time study.

For Additional Information

Contact the National Federation of Licensed Practical Nurses, P. O. Box 1088, Raleigh, NC 27619.

UNIT READINGS

- Jaret, Peter. "The Disease Detectives." *National Geographic,* January 1991.
- Jordon, Robert Paul. "Somalia's Hour of Need." *National Geographic,* June 1981.
- Oliwenstein, Lori. "Medicine 1990," *Discover,* January 1991, pp. 80-83.

600

UNIT READINGS

Background

▶ *The Disease Detectives* is a beautifully written and illustrated history of epidemiology. It includes especially comprehensive treatments of AIDS and Lyme disease.

▶ "Our Immune System: The Wars Within" is a comprehensive and clear explanation of how the immune system works. The article contains especially stunning artwork and photographs.

▶ "Somalia's Hour of Grief" is a case book study of what happens when a natural disaster (drought) and political unrest occur together. While the article is a bit old, the same conditions persist today.

More Readings

1. Duplaix, Nicole. "Fleas: The Lethal Leapers." *National Geographic,* May 1988.
2. Dubos, René and Maya Pines. "Health and Disease." *Time, Inc.,* 1965.
3. Dubos, René. *Mirage of Health.* New York: Doubleday, 1965.

Medical Ethics

I will follow that system of regimen which, according to my ability and judgment, I consider for the benefit of my patients, and abstain from whatever is deleterious and mischievous.

The excerpt above is from the Hippocratic Oath, credited to Hippocrates, who is thought to have lived and practiced medicine in the 4th century B.C. Writings credited to Hippocrates have had a profound influence on the practice of medicine and were used as a framework for the training of doctors. Over the centuries, the oath as originally stated was modified many times. Both Christian and Islamic versions have existed. In 1949, the World Medical Association in Geneva, Switzerland, drew up a statement of medical ethics, called the Declaration of Geneva, which reflects many of the ideals that are thought to have been expressed in the original Hippocratic Oath. That statement follows.

I solemnly pledge myself to consecrate my life to the service of humanity;

I will give my teachers the respect and gratitude which is their due;

I will practice my profession with conscience and with dignity;

The health of my patient will be my first consideration;

I will respect the secrets which are confided in me;

I will maintain by all the means in my power, the honor and the noble traditions of the medical profession;

My colleagues will be my brothers;

I will not let considerations of religion, nationality or race, party politics or social standing to intervene between my duty and my patient;

I will maintain the utmost respect for human life from its beginning; even under threat I will not use my medical knowledge contrary to the laws of humanity;

I make these promises solemnly, freely, and upon my honor.

In Your Own Words

▶ Information on the healing arts in other cultures is older than Hippocrates. Research acupuncture, moxa, and chi in Chinese medicine.

Other Works

▶ Dubos, René. *Man Adapting.* New Haven, CT: Yale University Press, 1965.

Classics

▶ Bettman, Otto L. *A Pictorial History of Medicine.* Springfield, IL: C.C. Thomas, 1956.

▶ Dubos, René and Jean Dubos. *The White Plague.* Boston: Little Brown, 1952.

SCIENCE & LITERATURE

Sources: Hutchins, Robert M., ed. *Great Books of the Western World,* Vol 10. *Hippocratic Writings/On the Natural Faculties by Galen.* Chicago, IL: Encyclopaedia Britannica, Inc., 1952.

Cameron, Nigel M. de S. *The New Medicine—Life and Death After Hippocrates.* Wheaton, IL: Crossway Books, 1991.

Biography: Hippocrates, a Greek physician, was born in 460 B.C. He is thought to have founded a school for physicians at Cos. In the century following his death, writings, eventually called the Hippocratic Collection, were attributed to him. Among these were discourses on surgery, plagues, fractures and the so-called Hippocratic oath, a code of ethics that physicians of that school were thought to have sworn to. The writings in the collection reveal that Hippocrates' lasting legacy was his dependence on facts derived from observation of the whole patient and his or her condition, including the climate, in making a diagnosis.

The wording of the Hippocratic Oath has undergone modification many times over the centuries, but the intent remained the same. The World Medical Association was formed in 1948 in Geneva. In 1949, the Association formulated the Declaration, which comes somewhat close to a modern version of the oath. Even that oath has undergone change.

TEACHING STRATEGY

Discuss the term *ethics* with students. Does every group have a code of ethics? Have students discuss why various groups have written or unwritten codes for their professions.

Cooperative Learning: Divide the class into as many sections as there are sections of the Declaration. Have each group of students analyze a different section and propose how it applies to contemporary life. Have students imagine that they are physicians. How might the Declaration affect how they would practice medicine?

UNIT

8

ECOLOGY

In Unit 8, students are introduced to the fundamentals of ecology, including the relationships among organisms and between organisms and their environments.

CONTENTS

ADVANCE PREPARATION

Activities

▶ **Activity 26-1, page 609,** requires permanent markers and clear plastic sheets.
▶ **Activity 26-2, page 622,** requires field guides, binoculars, and graph paper.
▶ **Activity 27-2, page 642,** requires pinto bean seeds, planters, and soil.
▶ **Activity 28-1, page 652,** requires soil, flower pots, leaves, aluminum foil, plastic milk cartons, aluminum pie plates, sand or gravel, newspaper, apples, and plastic foam.
▶ **Activity 28-2, page 662,** requires cardboard, waxed paper, string, petroleum jelly, and a hole punch.

Field Trips and Speakers

▶ Arrange to take your class on a trip to a recycling center.
▶ Arrange for presentations in the classroom by environmental and conservation organizations.

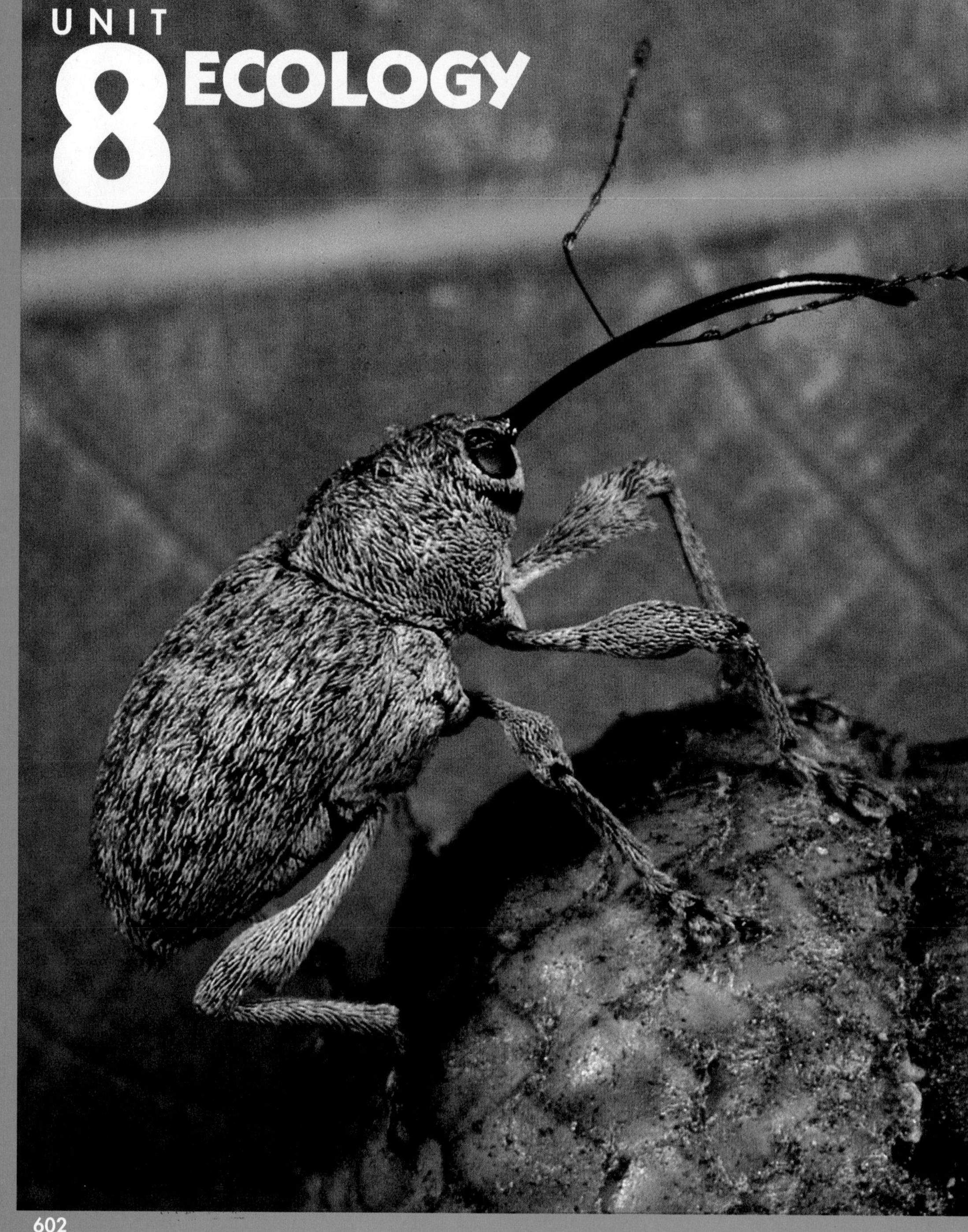

602

OPTIONS

Cross Curriculum

▶ Ask students to identify all of the natural resources that they use and write a poem about recycling renewable resources.
▶ Ask students to collect items normally thrown away and create a piece of art.

PROJECT

During the course of this unit study, have students work cooperatively on PROJECT 8, *Where in the World?*, found on pages 708 and 709.

Science at Home

▶ As an extension to Activity 28-2, have students try to identify air pollution sites in their own homes. Places to put collector strips may include heating vents, bathrooms, kitchens, basements, and garages.

Cooperative Learning: Have students monitor references to the environment or to other ecological concepts in the newspaper or on television for one week. Have each student compile a list of topics covered to compare with classmates' lists.

What's Happening Here?

Mighty oak trees grow from acorns like these, but not all acorns become trees. As acorns ripen and fall off the tree, many become food for bears, deer, and mice. Squirrels, magpies, and jays bury acorns as winter approaches. Some acorns are attacked by acorn weevils that drill holes through the hulls to feed and lay their eggs. Once the acorn is damaged, other organisms move in. Filbert worm moths use the holes made by the weevils to lay their own eggs. When their larvae hatch, they eat their way out. Green fungi invade and decay accelerates. Maggots and springtails arrive to eat the fungi. What's in a nutshell? Life!

UNIT CONTENTS

603

INTRODUCING THE UNIT

What's Happening Here?

▶ Have students look at the photos and read the text. Ask them what's happening here. Point out to students that in this unit they will study the relationships among organisms and their interactions with their environments.

▶ **Background:** Ecosystems exist at all levels. Ecologists identify organisms and describe the nonliving components of ecosystems, then study the interactions involved. Even a square meter of garden soil is filled with information on the living world.

Previewing the Chapters

▶ Some students may have trouble with the ecological meanings of population and community. Although population of a country is easy for them to understand, make sure they can see that every kind of organism lives in a population in nature.

▶ Students should be encouraged to look through the chapters in this unit and study the diagrams dealing with ecological principles such as succession and energy pyramids.

Tying to Previous Knowledge

▶ Have students brainstorm interactions among organisms. Information from previous chapters can be integrated in this unit so that students can see beyond body systems of organisms to how the organisms affect and are affected by their environments.

▶ Lead students in a discussion of ecosystems they may be familiar with, such as water ecosystems, or parks and wilderness areas. Encourage students to recognize how much they already know about ecology.

▶ Use the **inquiry questions** in the OPTIONS box below to investigate with students the use and misuse of natural resources.

Multicultural Awareness

Some people living in desert areas wear loose-fitting, light-weight, light-colored garments that reflect sun and heat. They live in tents that enable them to take advantage of the smallest breezes. A tundra, rainforest, or deciduous forest would call for different cultural responses. Have students construct a physical model to show one culture's modifications in food, housing, and clothing to meet the environment in which it exists.

Inquiry Questions

Use the following questions to focus a discussion on natural resources.

▶ **What are natural resources?**

▶ **Where do natural resources come from?**

▶ **Can natural resources be used up? Why?**

▶ **Why is it important to reuse some resources?**

▶ **What can you do to use fewer natural resources every day?**

CHAPTER

26 Organisms and Their Environments

CHAPTER SECTION	OBJECTIVES	ACTIVITIES
26-1 Organisms and Their Environments (3 days)	1. **Identify** the biotic and abiotic factors in the biosphere. 2. **Describe** the characteristics of populations. 3. **Compare** a species' habitat and its niche within a community.	**Activity 26-1:** *Designing an Experiment (Determining How Populations Change),* p. 609 **MINI-Lab:** *What organisms are found in an ecosystem?* p. 611
26-2 Biotic Relationships (2 days)	1. **Identify** kinds of relationships organisms have with each other. 2. **Describe** food chains and food webs in a community. 3. **Explain** how energy is transferred through a community.	**MINI-Lab:** *What are the requirements of ecosystems?* p. 615
26-3 Abiotic Factors in the Biosphere (1 day)	1. **Describe** how materials in the biosphere are reused in a continuous cycle. 2. **Diagram** the cycles of water, carbon, oxygen, and nitrogen in the biosphere. 3. **Discuss** the importance of recycling these materials in the biosphere.	
26-4 Friendly Fires Science & Society (1 day)	1. **Describe** the role of controlled fires in forests and why the fires are controversial. 2. **Identify** reasons both for and against the use of controlled burns in areas where people live.	**Activity 26-2:** *Studying an Ecosystem,* p. 622
Chapter Review		

ACTIVITY MATERIALS

FIND OUT	ACTIVITIES		MINI-LABS	
Page 605 meterstick	**26-1 Designing an Experiment, p. 609** 1-L glass container with lid dried pond vegetation 1-L distilled water microscope 3 microscope slides 3 coverslips 3 droppers	**26-2 Studying an Ecosystem, p. 622** graph paper notebook pencil binoculars field guides hand lens	**What organisms are found in an ecosystem? p. 611** paper pencil	**What are the requirements of ecosystems? p. 615** pencil paper

CHAPTER FEATURES	TEACHER RESOURCE PACKAGE	OTHER RESOURCES
Problem Solving: *The Milk Carton Garden,* p. 607 **Skill Builder:** *Observing and Inferring,* p. 611	**Ability Level Worksheets** ◆ ***Study Guide,*** p. 99 ● ***Reinforcement,*** p. 99 ▲ ***Enrichment,*** p. 99 **Activity Worksheets,** pp. 230, 231, 236 **Concept Mapping,** p. 57 **Cross-Curricular Connections,** p. 30	**STVS:** Disc 6, Side 1
Skill Builder: *Interpreting Scientific Illustrations,* p. 616	**Ability Level Worksheets** ◆ ***Study Guide,*** p. 100 ● ***Reinforcement,*** p. 100 ▲ ***Enrichment,*** p. 100 **Activity Worksheets,** p. 237 **Critical Thinking/Problem Solving,** p. 30 **Transparency Masters,** pp. 93-94	**Lab Manual:** Communities, p. 171; Living Things in Soil, p. 175 **Color Transparency 47,** Food Web, Energy Pyramid **STVS:** Disc 6, Side 1
Technology: *Monitoring Mayflies,* p. 618 **Skill Builder:** *Sequencing,* p. 619	**Ability Level Worksheets** ◆ ***Study Guide,*** p. 101 ● ***Reinforcement,*** p. 101 ▲ ***Enrichment,*** p. 101 **Transparency Masters,** pp. 95-96	**Color Transparency 48,** Cycles **STVS:** Disc 6, Side 2 **Science Integration Activity 26**
You Decide! p. 621	**Ability Level Worksheets** ◆ ***Study Guide,*** p. 102 ● ***Reinforcement,*** p. 102 ▲ ***Enrichment,*** p. 102 **Activity Worksheets,** pp. 232-233 **Science and Society,** p. 30	
Summary Key Science Words Understanding Vocabulary Checking Concepts Using Lab Skills Think & Write Critically Apply More Skill Builders Projects	ASSESSMENT RESOURCES **Chapter Review,** pp. 55, 56 **Chapter Test,** pp. 176-179 **Performance Assessment in Middle School Science (PAMSS)**	**Chapter Review Software** **Test Bank** **Alternate Assessment** **Performance Assessment**

◆ **Basic** ● **Average** ▲ **Advanced**

ADDITIONAL MATERIALS

SOFTWARE	AUDIOVISUAL	BOOKS/MAGAZINES
Balance—Predator-Prey Simulation, Diversified Educational Enterprises. *Food Webs,* Diversified Educational Enterprises. *The Environment I and II,* IBM. *Odell Lake,* Queue.	*Food Cycle and Food Chains,* film, Coronet/MTI. *Introducing Ecology,* film, Coronet/MTI. *The Ecosystem, Network of Life,* film, BFA. *Ecology: Populations, Communities and Biomes,* filmstrip, Science and Mankind. *Relationships,* laserdisc, Syscon Corp. *The Marshes,* video, Coronet/MTI.	Allaby, M. *Dictionary of the Environment.* 3rd ed. NY: New York University Press, 1989. Odum, Eugene P. *Ecology and Our Endangered Life-Support Systems.* Sunderland, MA: Sinauer Associates, Inc., 1989. Purdom, P. Walton and Stanley H. Anderson. *Environmental Science.* Columbus, OH: Merrill Publishing, 1983.

CHAPTER

26

ORGANISMS AND THEIR ENVIRONMENTS

THEME DEVELOPMENT: This chapter ties directly to the ecology theme of the book. In learning about relationships among living and nonliving things in populations, communities, and ecosystems, students will see that all organisms interact with and adapt to their environments. Adaptation, which results from evolution by natural selection, is a factor that unifies the structure and behavior of all organisms throughout the biosphere.

CHAPTER OVERVIEW

- **Section 26-1:** The ecological units that make up the biosphere and the ways in which biotic and abiotic factors within the biosphere interact are discussed in this section.
- **Section 26-2:** This section focuses on how energy flows through an ecosystem. All organisms need energy to carry on life processes. Organisms store only part of the energy they consume.
- **Section 26-3:** This section discusses the cycling of materials—water, nitrogen, oxygen, and carbon—through the biosphere.
- **Section 26-4: Science and Society:** This section explores the role of controlled fires in forests and why they are controversial.

CHAPTER VOCABULARY

biosphere	camouflage
ecology	mimicry
biotic factors	food chain
abiotic factors	food web
population	energy pyramid
population density	water cycle
community	carbon-oxygen cycle
habitat	nitrogen cycle
niche	controlled burns
ecosystem	
competition	

CHAPTER

26 Organisms and Their Environments

OPTIONS

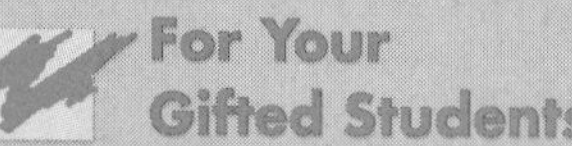

For Your Gifted Students

Have students find and read articles about forest fires in library magazines. Students can create and write a story about an animal or human that is involved in a forest fire. The character should describe what started the fire, how it spread, how it affected the environment, etc.

For Your Mainstreamed Students

Let students act out the following terms as they relate to the chapter: *competition, camouflage, mimicry.* Students can also act out a food chain in a community. The participants can be labeled to show whether they are producers, herbivores, carnivores, or omnivores.

Animals such as butterflies and ladybird beetles often live in large groups. They share space, food, and nesting sites. How does the number of individuals in a group affect each organism? You share your science classroom with other individuals. How much space does each person have in your science classroom?

FIND OUT!

Do this simple activity to find out how much space each person has in your classroom.

Use a meterstick to *measure* the length and width of the classroom. Multiply the length times the width to find the area in square meters. *Count* the number of individuals in your class. Divide the number of square meters in the classroom by the number of individuals. How much space does each person have? *Predict* the amount of space each person would have if your class size doubled.

Gearing Up

Previewing the Chapter

Use this outline to help you focus on important ideas in this chapter.

Section 26-1 Organisms and Their Environments
- What Is the Biosphere?
- Populations
- Communities
- Ecosystems

Section 26-2 Biotic Relationships
- Feeding Relationships
- Symbiotic Relationships in a Community
- The Transfer of Energy in a Community

Section 26-3 Abiotic Factors in the Biosphere
- Cycles of Matter
- The Water Cycle
- The Carbon-Oxygen Cycle
- The Nitrogen Cycle

Section 26-4 Science and Society Friendly Fires
- Are Forest Fires Always Harmful?

Previewing Science Skills

- In the Skill Builders, you will observe and infer, interpret scientific illustrations, and sequence.
- In the Activities, you will predict, count, sample, observe, record, and use field guides.
- In the MINI-Labs, you will list, identify, and classify.

What's next?

Now you know that the number of individuals in a population determines the amount of space each has. In this chapter you'll learn about populations and how organisms and environments interact.

605

INTRODUCING THE CHAPTER

Use the Find Out activity to give students firsthand experience in determining the density of a population. Tell students they will learn more about populations and how populations interact with other populations and nonliving things in the environment.

FIND OUT!

Preparation: Obtain metersticks for the activity.

Materials: one meterstick for each pair of students

Cooperative Learning: Use Paired Partner strategy. Have one student measure and the other student record. Then they can find the answers to the questions together.

Teaching Tips

► If the student population is 30 and the classroom size is 240 m^2, then each student would have 8 m^2. Explain that if the population density of the classroom doubled, each person would have only 4 m^2. However, to calculate population density you divide the number of individuals in the population by the area to get individuals per unit area. In this case the population density would be

$$\frac{30 \text{ students}}{240 \text{ m}^2} = \frac{3 \text{ students}}{24 \text{ m}^2}$$

$$= \frac{1}{8} = 0.125 \text{ students/m}^2$$

Gearing Up

Have students study the Gearing Up feature to familiarize themselves with the chapter. Discuss the relationships of the topics in the outline.

What's Next?

Before beginning the first section, make sure students understand the connection between the Find Out activity and the topics to follow.

ASSESSMENT OPTIONS

PORTFOLIO

Refer to page 623 for suggested items that students might select for their portfolios.

PERFORMANCE ASSESSMENT

See page 623 for additional Performance Assessment options.

Process

Skillbuilders, pp. 611, 619
MINI-Labs, pp. 611, 615
Activities, 26-1, p. 609; 26-2, p. 622
Using Lab Skills, p. 624

CONTENT ASSESSMENT

Assessment—Oral, pp. 610, 615
Section Reviews, pp. 611, 616, 619, 621
Skillbuilder, p. 616
Chapter Review, pp. 623-625
Mini Quiz, p. 610

GROUP ASSESSMENT

Opportunities for group assessment occur with Cooperative Learning Strategies.

SECTION 26-1

PREPARATION

SECTION BACKGROUND

▶No living thing lives alone. Every living thing depends somehow upon certain other living and nonliving things.

▶The word *ecology* comes from the Greek word *oikos* meaning "house" or place to live and *-ology* meaning "the study of."

▶Ecologists study how organisms act together and how they are adapted to their environments.

▶Ecologists learn about the biosphere by studying smaller and simpler ecological units within it.

PREPLANNING

▶Obtain permanent markers and clear plastic sheets for Activity 26-1.

1 MOTIVATE

▶Discuss and compare the habitats of different peoples on Earth. Show pictures of different cultures in distant parts of the world, such as Australia, Japan, Nigeria, the United Kingdom, India, Turkey, and others. Have students research the habitats of other peoples.

Connect to... Chemistry

Answer: In photosynthesis, the energy from sunlight is used by green plants to make glucose from carbon dioxide and water. In chemosynthesis, bacteria produce food and oxygen by using dissolved sulfur compounds that escape from lava.

TYING TO PREVIOUS KNOWLEDGE: Review producers and consumers from Chapter 3 and classification and species from Chapter 7. Students will be familiar with organisms from their own environment. Ask how these organisms are interacting in their environment.

26-1 Organisms and Their Environments

New Science Words

biosphere
ecology
biotic factors
abiotic factors
population
population density
community
habitat
niche
ecosystem

Objectives

- ▶ **Identify the biotic and abiotic factors in the biosphere.**
- ▶ **Describe the characteristics of populations.**
- ▶ **Compare a species' habitat and its niche within a community.**

What Is the Biosphere?

Deep in the Pacific Ocean is a region where the continental plates are moving apart. As these plates pull apart, volcanic rocks and lava seep into the resulting cracks, or rifts, from beneath Earth's crust. Water near this rift zone heats up, rises, cools, and falls, creating small areas of nutrient-rich warmth in the pitch black, freezing cold waters. Few organisms are found in such waters 2.5 kilometers below the sea's surface. But in these waters the crew of a tiny submarine called *ALVIN* made a discovery that astounded scientists worldwide. An entire community of organisms was found living in darkness near the rift. The community included clams, crabs, mussels, and bacteria. With no sunlight available for photosynthesis, these bacteria produce food by chemical synthesis. About 20 of these deep ocean rift communities have now been identified in this newly discovered part of the living world.

Figure 26-1. *ALVIN'S* crew photographed organisms living without light in the rift zone deep in the Pacific Ocean.

The processes of photosynthesis and chemosynthesis both produce food for organisms. Define chemosynthesis. *Compare* these two processes.

Deep ocean rift communities and all the parts of Earth where life is found make up the **biosphere** (BI uh sfihr). 1
The biosphere extends from the deepest oceans to the upper atmosphere and includes all the air, land, and water where life exists. Within the biosphere, all living things depend upon and interact with each other and with the nonliving things in their environment. The study of interactions between organisms and their environments is called **ecology** (ih KAHL uh jee). Living things such as plants and animals in the environment are called **biotic** (bi AHT ihk) **factors.** Nonliving things in the envi-

OPTIONS

Meeting Different Ability Levels

For Section 26-1, use the following **Teacher Resource Masters** depending upon individual students' needs.

◆ **Study Guide Master** for all students.

● **Reinforcement Master** for students of average and above average ability levels.

▲ **Enrichment Master** for above average students.

Additional Teacher Resource Package masters are listed in the OPTIONS box throughout the section. The additional masters are appropriate for all students.

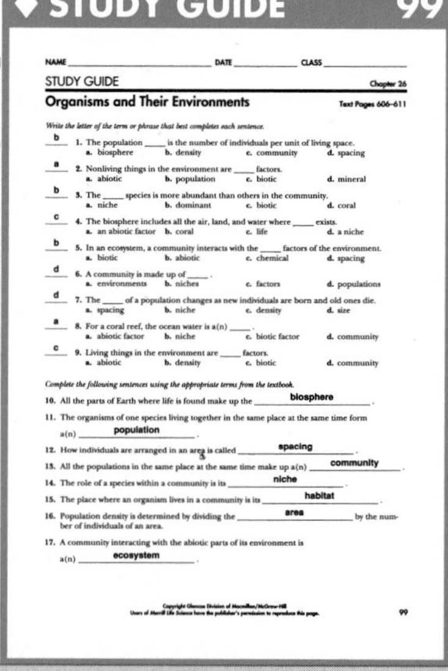

◆ STUDY GUIDE 99

NAME ________ DATE ________ CLASS ________

STUDY GUIDE — Chapter 26

Organisms and Their Environments — Text Pages 606–611

Write the letter of the term or phrase that best completes each sentence.

b 1. The population ____ is the number of individuals per unit of living space.
a. biosphere b. density c. community d. spacing

a 2. Nonliving things in the environment are ____ factors.
a. abiotic b. population c. biotic d. mineral

b 3. The ____ species is more abundant than others in the community.
a. niche b. dominant c. biotic d. coral

c 4. The biosphere includes all the air, land, and water where ____ exists.
a. an abiotic factor b. coral c. life d. a niche

b 5. In an ecosystem, a community interacts with the ____ factors of the environment.
a. biotic b. abiotic c. chemical d. spacing

d 6. A community is made up of ____.
a. environments b. niches c. factors d. populations

d 7. The ____ of a population changes as new individuals are born and old ones die.
a. spacing b. niche c. density d. size

a 8. For a coral reef, the ocean water is a(n) ____.
a. abiotic factor b. niche c. biotic factor d. community

c 9. Living things in the environment are ____ factors.
a. abiotic b. density c. biotic d. community

Complete the following sentences using the appropriate terms from the textbook.

10. All the parts of Earth where life is found make up the **biosphere**.
11. The organisms of one species living together in the same place at the same time form a(n) **population**.
12. How individuals are arranged in an area is called **spacing**.
13. All the populations in the same place at the same time make up a(n) **community**.
14. The role of a species within a community is its **niche**.
15. The place where an organism lives in a community is its **habitat**.
16. Population density is determined by dividing the **area** by the number of individuals of an area.
17. A community interacting with the abiotic parts of its environment is a(n) **ecosystem**.

99

ronment, such as soil, water, temperature, air, light, wind, and minerals, are **abiotic** (ay bi AHT ihk) **factors.**

Look at the pond in Figure 26-2. The abiotic factors include the pond's water, the soil at the pond bottom, and the temperatures of the air and water. The amount of light available for photosynthesis and the minerals dissolved in the water also are abiotic factors. What organisms make up the biotic factors in the pond? Monerans, protists, fungi, plants, and animals are the biotic factors.

Each species in the pond makes up a population. All of the populations in the pond make up the pond community. Together, the pond community and the abiotic factors make up an ecological system, or ecosystem. You'll learn more about populations, communities, and ecosystems in the following sections.

Figure 26-2. Biotic and abiotic factors work together in ecological systems like this pond.

PROBLEM SOLVING

The Milk Carton Garden

Paulo was disappointed. He was eager to start his spring vegetable garden, but the weather forecaster predicted cold temperatures for at least another week.

Because he couldn't plant outside due to the cold temperatures, he decided to plant some bean seeds inside and transplant the seedlings when the weather warmed up.

Paulo cut the top off a 0.5-liter milk carton and filled it with nutrient-rich soil. He planted 25 seeds in the carton, placed it on a window ledge in a sunny area, and kept the soil moistened. After ten days, the bean plants had sprouted and seemed to be growing well. But, after three weeks, the plants had tall, thin stems and yellowish leaves. Several plants had withered and died. What should Paulo have done to take care of the plants?

Think Critically: Explain why the plants were healthy when they sprouted but were unhealthy after three weeks.

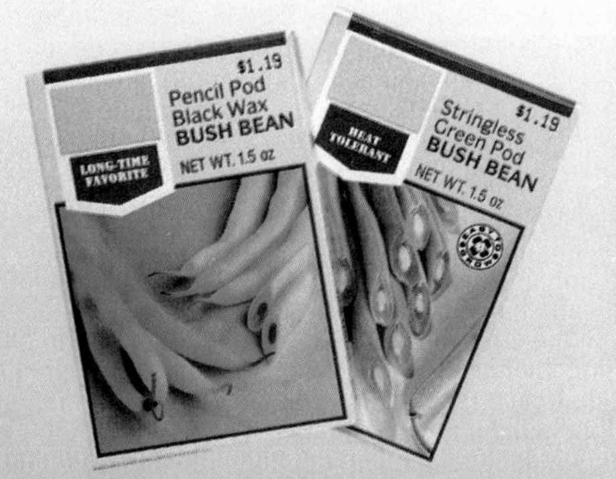

● REINFORCEMENT 99

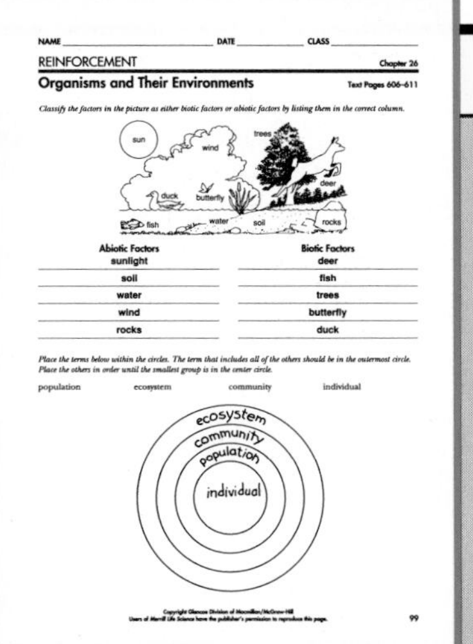

NAME ______ DATE ______ CLASS ______

REINFORCEMENT Chapter 26

Organisms and Their Environments Text Pages 606–611

Classify the factors in the picture as either biotic factors or abiotic factors by listing them in the correct column.

Abiotic Factors	Biotic Factors
sunlight	deer
soil	fish
water	trees
wind	butterfly
rocks	duck

Place the terms below within the circles. The term that includes all of the others should be in the outermost circle. Place the others in order until the smallest group is in the center circle.

population ecosystem community individual

99

▲ ENRICHMENT 99

NAME ______ DATE ______ CLASS ______

ENRICHMENT Chapter 26

Organisms and Their Environments Text Pages 606–611

FINDING NICHES

Choose two different species of birds that you have seen near your home. Your choices will vary widely, depending on where you live. The birds you choose are a part of the biological community. If you don't know the name of the birds you have seen, ask an adult in your home or neighborhood.

Investigate the birds' niches. Their niches will not be exactly the same. Every species has its own niche. Use your observations to complete the table below. You may find additional information on your birds at the library. A field identification guide or books on the specific bird you chose are good sources.

	Bird 1	Bird 2
Kind (species)		
Food		
Home/Nest		
Habits		
Description		

99

2 TEACH

Key Concepts are highlighted.

CONCEPT DEVELOPMENT

▶**Demonstration:** Obtain six 1-pint milk cartons. Cut the side out of each, and fill each with potting soil. Soak bean seeds overnight and plant two seeds in the first carton, four seeds in the second, six seeds in the third, eight in the fourth, ten in the fifth, and twelve in the sixth. Label each carton with the number of seeds. Water the seeds with the same amount of water and keep the cartons in a place where they can receive the same amount of light and where the temperature will be the same. Have students observe the plants. Ask students to account for any differences in appearance of the plants as they grow.

▶Ask students to list the biotic and abiotic factors that would affect the population growth of frogs near the pond.

PROBLEM SOLVING

Paulo should have made sure the plants had nutrient-rich soil, sunlight, and water.

Think Critically: When the plants sprouted, they had sufficient nutrients, sunlight, and water. After three weeks of growing, the plants had used up the nutrients in the soil and thus lack of nutrients became a limiting factor to their growth.

CONCEPT DEVELOPMENT

▶Remind students of the Find Out activity when you discuss population density. The number of organisms on land is measured in square units (square meters/kilometers), and aquatic organisms are measured by cubic units (cubic meters).

▶Have students list the populations of living things that might be in the area where they live.

CROSS CURRICULUM

▶**Geography:** Have students use an atlas to find the population and size of New York City to calculate its population density.

Science and MATH

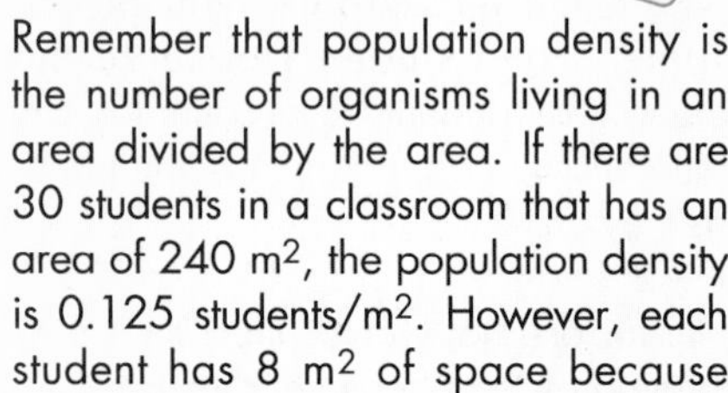

Remember that population density is the number of organisms living in an area divided by the area. If there are 30 students in a classroom that has an area of 240 m^2, the population density is 0.125 students/m^2. However, each student has 8 m^2 of space because

$$\frac{240\ m^2}{30\ students} = 8\ m^2/student.$$

VideoDisc

STVS: Preserving Duck Habitats, Disc 6, Side 1

STVS: Fish Survey, Disc 6, Side 1

Populations

Have you ever gone for a walk at dusk and noticed birds lined up on a telephone wire as shown in Figure 26-4? These birds all live together in the same place at the same time; they make up a population. All the people living in your town, the flocks of geese on a pond, and the worms in the soil under a baseball diamond each represent a population. A **population** consists of organisms of one species living together in the same place at the same time.

Populations are described according to certain characteristics. The number of individuals in a given area 2
is called **population density.** In the chapter opening activity, you found how much space each student had in your classroom. To find the population density, you must divide the area by the number of individuals in the area (density = individuals/unit area).

Another characteristic of populations is spacing. Spacing is how individuals are arranged in an area. They may be arranged evenly, like the birds on the wire, unevenly, or in random clumps. Figure 26-4 shows the types of spacing found in populations.

Size is an important characteristic of populations. The size of a population is constantly changing as individuals move into and out of a community. Populations also change when new individuals are born and when older individuals die. The total size of a population can be determined by studying these changes over time. The United States Census Bureau uses information such as this to determine the size of the human population in the United States every ten years. Figure 26-3 shows how population size changes over time.

Figure 26-3. The world's population size changes over time.

Science and MATH

Calculate the population density of your city or county and compare it to that of New York City.

Figure 26-4. Spacing in populations varies from even (a), to uneven (b), to random clumps (c).

OPTIONS

INQUIRY QUESTIONS

▶**What advantages does a population that feeds on several kinds of organisms have over a population that eats only one kind of plant or animal?** *Feeding on several kinds of organisms assures the population of a food supply if one food source becomes unavailable.*

▶**Which would be easier, finding the size of a plant population or finding the size of an animal population? Explain.** *It would be easier to find the size of a plant population because plants don't move.*

PROGRAM RESOURCES

From the **Teacher Resource Package** use:

Concept Mapping, pages 57-58.

Cross-Curricular Connections, page 30, Analyzing a Computer Model of the Global Ecosystem.

ACTIVITY 26-1

DESIGNING AN EXPERIMENT

Determining How Population Changes

Factors such as food, space, and competition from other organisms limit the growth rate of a population. Can you observe population changes in a closed ecosystem?

Getting Started

You will establish an ecosystem and determine if the populations in the ecosystems change over time.

Thinking Critically

Make a list of abiotic factors. *Predict* how the abiotic factors may affect an ecosystem.

Materials

Your cooperative group will use:

- 1 1-L glass container with lid
- dried pond vegetation
- 1 L distilled water
- microscope
- 3 microscope slides
- 3 coverslips
- 3 droppers

Try It!

1. Place some dried plants from a pond into a clean glass container. Fill the container with distilled water. Place a lid on the container loosely.
2. Let the container stand in a lighted area that has a constant temperature of at least 21°C.
3. *Predict* what populations you may expect to observe in the water ecosystem.
4. *Observe* the container each day. Record any changes. If any of the water evaporates, refill the jar with distilled water.
5. At the end of three days, use the dropper and from the top of the container, take a sample of water. Make a wet mount slide of the water. Examine the slide under the microscope on low power and then switch to high. Record the number of different organisms that you observe.
6. Use a clean dropper and repeat Step 5 taking water from the middle of the container.
7. Use a clean dropper and repeat Step 5 taking water from the bottom of the container.
8. Repeat Steps 5 through 7 every two days for eight more days.

Summing Up/Sharing Results

- What were the biotic factors in the ecosystem? The abiotic factors?
- How many populations did you observe the first time you examined the water?
- Were the populations the same at the top, middle, and bottom of the container?
- *Compare* the number of populations of organisms in the container the first time you observed with the number at the end of the activity.
- Where did the populations come from?

Going Further!

How can you explain the changes you observed? Would you expect other samples of pond water to have the same populations?

ACTIVITY 26-1

OBJECTIVE: Design and carry out an experiment to determine if populations in a closed ecosystem changes over time.
Time: One class period to set up and 15 minutes every other day to observe water samples.

PROCESS SKILLS applied in this activity are **observing, inferring, comparing and contrasting,** and **interpreting data.**

PREPARATION

Collect plant material from an established pond. Spread the plant material on waxed paper and let it dry. Have students being in 1 L glass jars with lids to use in this activity.

Cooperative Learning: Divide class into Science Investigation Teams.

THINKING CRITICALLY

Abiotic factors may include soil, water, temperature, air, light, wind, and minerals. The living organisms are the biotic factors. Cold temperatures, the lack of water, strong winds, and the lack of minerals in soil and water environments could cause populations to die or adapt.

TEACHING THE ACTIVITY

Refer to the ***Activity Worksheets*** *for additional information and teaching strategies.*

- You may obtain reference books with pictures of algae, protozoans, and other one-celled organisms for students to use when observing their slides.
- You may wish to set up a biotic glass container at the same time to be used in studying succession in Chapter 27.

SUMMING UP/ SHARING RESULTS

- The living things came from the dried pond vegetation. The abiotic factors were light, temperature, and air.
- Student answers will vary based upon the vegetation placed in the container.
- The populations may be the same at the beginning and change over the observation period.
- At the beginning, the only population may have been small flagellates. At the end, many populations may have been observed including algae, Vorticella, amoeba, rotifers, and others.
- The populations came from the dried pond vegetation containing eggs, spores, and other dormant stages of organisms.

PROGRAM RESOURCES

From the **Teacher Resource Package** use:

Activity Worksheets, pages 43-44, Activity 26-1, Determining How Populations Change.

GOING FURTHER!

The smaller organisms may have been used for food by the larger ones. It may have taken the eggs of some longer to hatch. The populations could be different because of the wide number of populations found in pond water.

Activity ASSESSMENT

Performance: Have students repeat the activity using plants from a stream or creek. The kinds and number of populations they start with may be different from those first observed in this activity.

CONCEPT DEVELOPMENT

▶Ask students to estimate how many populations there might be in the community in which they live.

▶Provide students with information about the history of your area for them to determine the original nature of their hometown (at one time it may have been a forest or a field).

▶Make certain that students understand that the community and ecosystem occupy the same area of land or water. The difference is that the community interacting with abiotic factors makes up the ecosystem.

CHECK FOR UNDERSTANDING

Use the Mini Quiz to check for understanding.

MINI QUIZ

Use the Mini Quiz to check students' recall of chapter content.

1. **All the parts of Earth where life is found is the ____.** *biosphere*
2. **The number of individuals per unit of living space is the ______.** *population density*
3. **What a species eats, how it gets its food, and how it interacts within a community is its ____.** *niche*
4. **The place where an organism lives is its ____.** *habitat*

RETEACH

Explain the relationships between population, community, and ecosystem. Point out that populations interact to make a community. Stress that communities and nonliving things make up the ecosystem.

EXTENSION

For students who have mastered this section, use the **Reinforcement** and **Enrichment** masters or other OPTIONS.

Communities

What is a community?

Hundreds of populations of organisms make up the oak-hickory forest community shown in Figure 26-6. A **community** is made up of all the populations of different species that live in the same place at the same time. In a community, a few species are usually more abundant than others. These species are called dominant species.

Figure 26-5. Although all three warblers feed in the same spruce tree, they occupy very different niches.

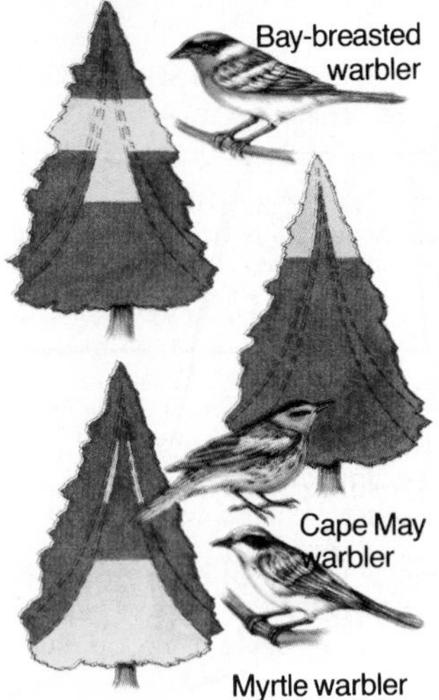

The oak and hickory trees are the dominant species in the community shown in Figure 26-6. When their leaves open in the spring, they create so much shade that plants underneath the trees cannot grow. In the autumn, their leaves fall and provide food for bacteria, fungi, and worms on the forest floor. Many other organisms eat the acorns and nuts produced by the trees. The tree trunks and branches are homes for insects, birds, squirrels, lichens, and other organisms. All the populations in this forest community depend on one another for survival.

The place where an organism lives in a community is
4 its **habitat.** Squirrels live in the trees in the forest. You would probably find crayfish, minnows, and other organisms living in the streams in the forest. Water is the habitat of these organisms. What is your habitat?

What is a niche?

No two species have exactly the same needs or role in a community. The role of a species within a community
3 is its **niche** (NIHCH). What a species eats, how it gets its food, and how it interacts with other organisms are all part of its niche. You may have seen many birds at a bird feeder in the park. Sparrows, doves, and cardinals all feed on birdseed—but not necessarily the same kinds of seeds. Two different species can occupy the same habitat, but they usually cannot occupy the same niche. Figure 26-5 shows how three bird species can live in one tree and still occupy very different niches.

Figure 26-6. In an oak-hickory forest, oaks and hickories are the dominant species.

To compare a community, a habitat, and a niche, we can look at the floor of the oak-hickory forest community. When a termite lives on and eats the wood of a fallen tree, the tree is the termite's habitat. Breaking down the parts of the dead tree is the termite's niche within the community.

OPTIONS

ASSESSMENT—ORAL

▶**How is the habitat of a squirrel different from its niche?** *A squirrel lives in trees in a forest. Its niche is gathering and storing nuts to eat.*

▶**What kinds of organisms live in a human habitat?** *pets, mice, rodents, spiders, birds, etc.*

▶**What might happen if two populations occupied the exact same niche?** *Competition could lead to extinction, migration, or adaptation of one of the populations.*

▶**Where might you expect to find communities with only a few populations?** *in places with unfavorable abiotic conditions such as a cave, high on top of a mountain, or where it is extremely cold*

ENRICHMENT

▶Have students research and report on the artificial reefs that have been built to provide a habitat for sea plants and animals.

Ecosystems

An oak-hickory forest is one type of ecosystem. A coral reef such as the one shown in Figure 26-7 is another type of ecosystem. A coral reef is composed of corals, sponges, sea stars, sea anemones, clown fish, trigger fish, groupers, algae, clams, mussels, oysters, and many other organisms. The ocean water, with its currents, temperature, and salinity, affects all of the organisms living on or near the coral reef. The amount of sunlight reaching the reef helps determine how much food the plantlike protists produce, and it affects those organisms that eat the protists. Together, these biotic and abiotic factors interact and function as an ecosystem. An **ecosystem** is a community interacting with the abiotic parts of its environment. Ecosystems may be as small as the acorn you studied in the unit opener, or as large as the Pacific Ocean. There are some very large natural ecosystems, such as forests, on Earth. You will learn about these ecosystems in Chapter 27.

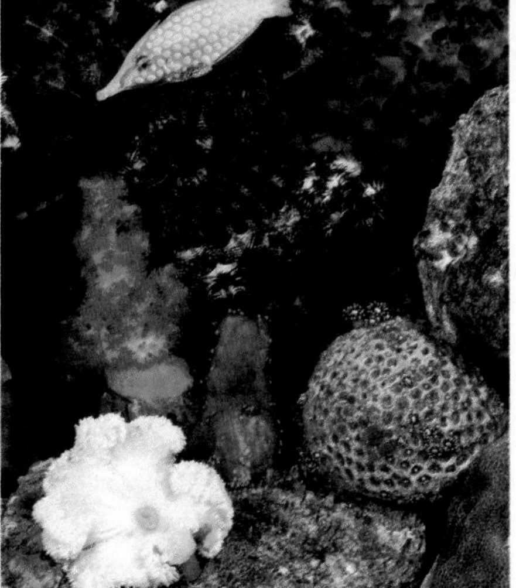

Figure 26-7. All of the organisms living on or near this coral reef are part of the community that interacts with the ocean water to form this ecosystem.

What is an ecosystem?

SECTION REVIEW

1. Name five biotic and five abiotic factors in the biosphere.
2. Describe the difference between a population and a community.
3. How are niches and habitats different ?
4. **Apply:** Ecosystems such as coral reefs are in a delicate balance. Hypothesize what would happen to that balance if one abiotic factor such as amount of sunlight suddenly changed.
5. **Connect to Earth Science:** Find out how the eruption of Mt. St. Helen in Washington state in 1980 affected the population densities of area organisms.

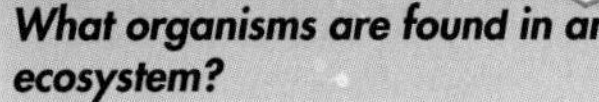

MINI-Lab

What organisms are found in an ecosystem?

Choosing an ecosystem you are familiar with, such as a stream, garden plot, or empty lot, *identify* the organisms found there. Make a list of all the populations you can see in the ecosystem. *Infer* the niche of each species in the community.

Observing and Inferring — Skill Builder

Each person lives in a population as part of a community. Describe your population, community, habitat, and niche. (HINT: You may belong to more than one of each!) If you need help, refer to Observing and Inferring in the **Skill Handbook** on page 682.

MINI-Lab

Materials: pencil and paper

Teaching Tips

▶ Make sure students choose a small ecosystem near their home to observe.

▶ Have students visit the ecosystem once a day for a week to write down their observations.

Answer to Question

▶ Niches may include overall terms such as *producer, consumer,* and *decomposer.*

MINI-Lab ASSESSMENT

Performance: Have students describe the niche of a cat who lives in a barn.

3 CLOSE

▶Ask questions 1-3 and the **Apply** Question in the Section Review.

SECTION REVIEW ANSWERS

1. biotic—monerans, protists, fungi, plants, and animals; abiotic—soil, water, temperature, air, water, light, wind, minerals

2. A population is all the organisms of one species living together in the same place at the same time. A community is all the populations of different species that live in the same place at the same time.

3. The place where an organism lives in a community is its habitat, and its role within the community is its niche.

4. Apply: Some populations would decrease and some might increase.

5. Connect to Earth Science: The erupting volcano, with its lava flows and ash, reduced populations of wildlife and seemed to wipe out most plantlife. Within 6 months, some conifer seedlings returned and some elk had returned to the area.

Skill Builder

Students may describe themselves as humans, living in towns, inside houses, and as consumers, students, children, brothers, sisters, etc.

INQUIRY QUESTION

▶ **How can ecosystems range in size from an acorn to the Pacific Ocean?** *An ecosystem is a community interacting with the abiotic parts of its environment. Both the acorn and the Pacific Ocean have communities that interact with the abiotic parts of the environment.*

PROGRAM RESOURCES

From the **Teacher Resource Package** use:

Activity Worksheets, page 236, Mini-Lab: What organisms are found in an ecosystem?

Skillbuilder ASSESSMENT

Performance: Use this Skillbuilder to assess students' abilities to Observe their own community. Have them describe their own niche.

SECTION 26-2

PREPARATION

SECTION BACKGROUND

▶ *Symbiosis* comes from a Greek word meaning "living together."

▶ There are two main types of competition. Intraspecific competition occurs between organisms of the same species. Interspecific competition occurs between organisms of different species.

PREPLANNING

▶ Obtain a very large ball of yarn for the MINI-Lab performance assessment on page 615.

1 MOTIVATE

▶ **Bulletin Board:** Make a bulletin board entitled "Web of Life." Connect pictures of organisms with yarn to illustrate food chains and food webs. Discuss the roles of producers, consumers, and decomposers and how they are interdependent.

Cooperative Learning: Divide the class into groups of four. Use the Expert Team strategy to have students study and master competition, predation, parasitism, and mutualism.

Connect to... Earth Science

Answer: Changes in climate and human activities contribute to rapid desertification. Drought in Africa has increased markedly since 1960 and the effect on the land spreads at a rate of 5 to 6 miles per year. For a description of world deserts and related weather causes, see:

Attenborough, David. *The Atlas of the Living World.* Boston: Houghton Mifflin Co., 1989.

VideoDisc

STVS: Hydrilla Killer, Disc 6, Side 1

26-2 Biotic Relationships

New Science Words

competition
camouflage
mimicry
food chain
food web
energy pyramid

Objectives

- Identify kinds of relationships organisms have with each other.
- Describe food chains and food webs in a community.
- Explain how energy is transferred through a community.

Connect to... Earth Science

When soils are damaged by overgrazing in areas that receive little rain, a desert can form. Called desertification, desert formation is currently happening in some areas of Africa, China, and the United States. Find out more about desertification, how quickly it occurs, and the problems it causes.

Feeding Relationships

What are the biotic factors in a desert such as the one shown in Figure 26-8? Many desert animals burrow underground for protection from the sun, only venturing out after dark. The plants are cacti and other plants with leaves modified into needles to conserve water. In the desert ecosystem, the amounts of food, water, and nutrients are limited, yet the amounts of sunlight and living space are not. Many desert organisms have adapted to life under these conditions. Birds nest in cacti, whereas rodents burrow in the sand. Snakes and lizards, turtles, and many kinds of insects can be found in deserts. All of these organisms have relationships with one another. The relationships among living things are biotic relationships. Many biotic relationships are feeding relationships. These relationships include competition for resources and predator-prey relationships.

Figure 26-8. Desert organisms compete with one another for food, water, and living space.

In the desert and in every other ecosystem, organisms compete for the same food, water, living space, and other resources. **Competition** is the contest among organisms to obtain all they need to survive. Competition determines the size and location of populations, and can occur between members of the same species. Members of one species often compete for nesting sites, mates, or feeding spots. Two robins will fight over one worm in the spring! If the competition is too strong, individuals may move into less populated areas.

OPTIONS

Meeting Different Ability Levels

For Section 26-2, use the following **Teacher Resource Masters** depending upon individual students' needs.

◆ **Study Guide Master** for all students.

● **Reinforcement Master** for students of average and above average ability levels.

▲ **Enrichment Master** for above average students.

Additional Teacher Resource Package masters are listed in the OPTIONS box throughout the section. The additional masters are appropriate for all students.

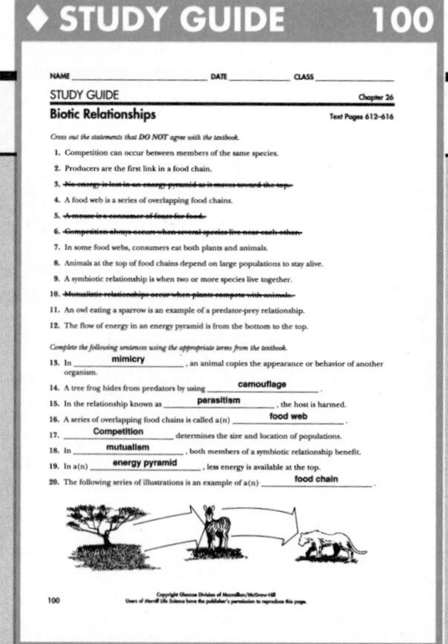

◆ STUDY GUIDE 100

NAME ______ DATE ______ CLASS ______

STUDY GUIDE Chapter 26

Biotic Relationships Text Pages 612-616

Cross out the statements that DO NOT agree with the textbook.

1. Competition can occur between members of the same species.
2. Producers are the first link in a food chain.
3. ~~No energy is lost in an energy pyramid as it moves toward the top.~~
4. A food web is a series of overlapping food chains.
5. ~~A mouse is a consumer of foxes for food.~~
6. ~~Competition always occurs when several species live near each other.~~
7. In some food webs, consumers eat both plants and animals.
8. Animals at the top of food chains depend on large populations to stay alive.
9. A symbiotic relationship is when two or more species live together.
10. ~~Mutualistic relationships occur when plants compete with animals.~~
11. An owl eating a sparrow is an example of a predator-prey relationship.
12. The flow of energy in an energy pyramid is from the bottom to the top.

Complete the following sentences using the appropriate terms from the textbook.

13. In **mimicry**, an animal copies the appearance or behavior of another organism.
14. A tree frog hides from predators by using **camouflage**.
15. In the relationship known as **parasitism**, the host is harmed.
16. A series of overlapping food chains is called a(n) **food web**.
17. **Competition** determines the size and location of populations.
18. In **mutualism**, both members of a symbiotic relationship benefit.
19. In a(n) **energy pyramid**, less energy is available at the top.
20. The following series of illustrations is an example of a(n) **food chain**.

100

Another kind of competition takes place between two different species. Both the owl and the fox catch and eat mice. Suppose half of the mouse population in a field suddenly died from a disease. What would happen to the owl and the fox? As long as there is a large mouse population, the owl and the fox can live in the same community. But they become competitors for food when the mouse population is reduced. In this case, the owl and the fox become competitors because they occupy the same niche—consumer of mice as food. Two organisms cannot occupy the same niche for long because one will eventually push the other out of that niche. In our example, the owl and the fox both feed on a wide variety of food organisms, so a smaller mouse population probably won't result in death for either. However, the owl and the fox will continue to compete for the remaining food sources in the community.

One important feeding relationship found in communities is the predator-prey relationship. In the oak-hickory forest community, owls and foxes are the predators, and mice are the prey. Animals that are prey can avoid being eaten by hiding, escaping, or defending themselves against predators. **Camouflage** is an adaptation that allows an animal to hide by blending into its surroundings. Insects such as the thorn bug in Figure 26-9 often can't be seen until they move.

Another defense against predators is mimicry. In **mimicry,** an animal copies the appearance or behavior of another organism.

Figure 26-9. Insects such as thorn bugs, leaf bugs, and stick insects resemble real thorns, leaves, and sticks.

Figure 26-10. Birds learn not to eat monarch butterflies because they are bad-tasting. Viceroys (left) are protected by their resemblence to monarchs (right). Birds avoid viceroys, too.

TYING TO PREVIOUS KNOWLEDGE: Review the definitions of *producers, consumers,* and *decomposers* from Chapter 3. Review the definition of *energy* as the ability to do work. Solar energy is transformed into chemical energy that is stored in food. Through chemical changes in cells, this energy is released for use by an organism. Review metabolism from Chapter 3.

2 TEACH

Key Concepts are highlighted.

CONCEPT DEVELOPMENT

▶ Discuss the concept of competition as it applies to success in sports and business. Have students consider the usefulness of cooperation in these areas. Relate the ideas discussed to competition and cooperation in nature.

▶ Point out that an organism's method of feeding is part of its niche in a community.

CROSS CURRICULUM

▶ **Art:** Have students research predator-prey relationships and make drawings to illustrate them.

● REINFORCEMENT 100

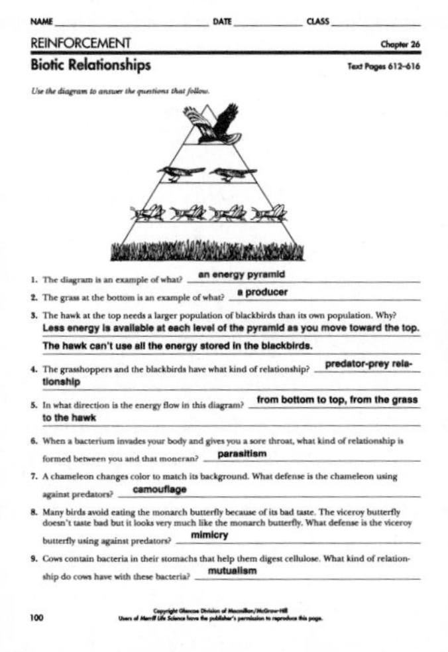

NAME ______ DATE ______ CLASS ______

REINFORCEMENT Chapter 26

Biotic Relationships Text Pages 612–616

Use the diagram to answer the questions that follow.

1. The diagram is an example of what? **an energy pyramid**
2. The grass at the bottom is an example of what? **a producer**
3. The hawk at the top needs a larger population of blackbirds than its own population. Why? **Less energy is available at each level of the pyramid as you move toward the top. The hawk can't use all the energy stored in the blackbirds.**
4. The grasshoppers and the blackbirds have what kind of relationship? **predator-prey relationship**
5. In what direction is the energy flow in this diagram? **from bottom to top, from the grass to the hawk**
6. When a bacterium invades your body and gives you a sore throat, what kind of relationship is formed between you and that moneran? **parasitism**
7. A chameleon changes color to match its background. What defense is the chameleon using against predators? **camouflage**
8. Many birds avoid eating the monarch butterfly because of its bad taste. The viceroy butterfly doesn't taste bad but it looks very much like the monarch butterfly. What defense is the viceroy butterfly using against predators? **mimicry**
9. Cows contain bacteria in their stomachs that help them digest cellulose. What kind of relationship do cows have with these bacteria? **mutualism**

100

▲ ENRICHMENT 100

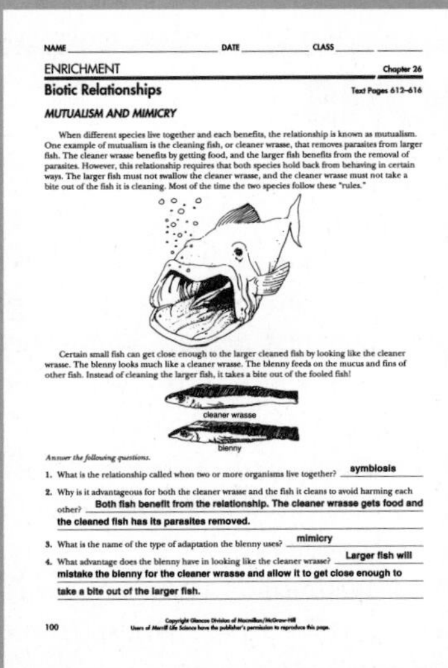

NAME ______ DATE ______ CLASS ______

ENRICHMENT Chapter 26

Biotic Relationships Text Pages 612–616

MUTUALISM AND MIMICRY

When different species live together and each benefits, the relationship is known as mutualism. One example of mutualism is the cleaning fish, or cleaner wrasse, that removes parasites from larger fish. The cleaner wrasse benefits by getting food, and the larger fish benefits from the removal of parasites. However, this relationship requires that both species hold back from behaving in certain ways. The larger fish must not swallow the cleaner wrasse, and the cleaner wrasse must not take a bite out of the fish it is cleaning. Most of the time the two species follow these "rules."

Certain small fish can get close enough to the larger cleaned fish by looking like the cleaner wrasse. The blenny looks much like a cleaner wrasse. The blenny feeds on the mucus and fins of other fish. Instead of cleaning the larger fish, it takes a bite out of the fooled fish!

Answer the following questions.

1. What is the relationship called when two or more organisms live together? **symbiosis**
2. Why is it advantageous for both the cleaner wrasse and the fish it cleans to avoid harming each other? **Both fish benefit from the relationship. The cleaner wrasse gets food and the cleaned fish has its parasites removed.**
3. What is the name of the type of adaptation the blenny uses? **mimicry**
4. What advantage does the blenny have in looking like the cleaner wrasse? **Larger fish will mistake the blenny for the cleaner wrasse and allow it to get close enough to take a bite out of the larger fish.**

100

CONCEPT DEVELOPMENT

▶Some parasites, such as tapeworms, are in continuous contact with their host. Others, such as fleas and mosquitoes, leave the host.

▶Ask students for other examples of mutualism and parasitism. Have them use reference books to find examples.

▶Ask students to name foods they eat that do not originate from green plants. Help them trace the foods they list back to the producer that began the food chain.

▶You may want to explain that herbivores are primary consumers and carnivores are secondary consumers.

▶Remind students that all living things use energy for maintenance, growth, repair, and reproduction.

CHECK FOR UNDERSTANDING

Have students diagram a food web for organisms in your area. Discuss the importance of food webs and how human activities or natural disasters disturb food webs.

RETEACH

Play the food web game. List producers, consumers, and decomposers in a food web on separate index cards. Give a card to each student. Give each producer a ball of colored yarn. Have producers roll the ball to their primary consumers, then each subsequent consumer until each chain is complete. This creates a food web. This will help students understand the interrelationships among organisms in food chains.

EXTENSION

For students who have mastered this section, use the **Reinforcement** and **Enrichment** masters or other OPTIONS provided.

REVEALING MISCONCEPTIONS

▶Students may think that parasites kill their hosts. Parasites generally do not kill their hosts. Organisms that eventually kill their hosts are called parasitoids.

▶Explain that mistletoe lives on hosts, such as oak and cedar trees. Mistletoe has chlorophyll and can make its own food, but it obtains water and nutrients from the tissues of the host.

Symbiotic Relationships in a Community

What is symbiosis?

In Figure 26-11, a bird called the oxpecker is eating insects that feed on the water buffalo. When two or more species live together, their relationship is called symbiosis. Symbiotic relationships provide food, shelter, support, or transportation for one or both organisms. Mutualism and parasitism are two types of symbiotic relationships.

Figure 26-11. The oxpecker and the water buffalo have a symbiotic relationship.

Mutualism is a symbiotic relationship in which two organisms live together and both benefit. The water buffalo and the oxpecker both benefit from their mutualistic relationship.

Parasitism is a relationship in which one organism, the parasite, is helped while the other organism, the host, is harmed. The parasite usually lives on or in the host and absorbs nutrients from the body fluids of the host. Tapeworms live as parasites in the intestines of mammals. Mistletoe is a plant that lives as a parasite on trees.

The Transfer of Energy in a Community

Many of the interactions that you have just studied are feeding relationships. Energy in food is transferred through a community by a **food chain.** In the oak-hickory forest, the trees and other producers change the energy of sunlight into chemical energy that is stored in plant leaves, stems, roots, and seeds. Consumers living in the forest community feed on these plant parts. Other consumers, such as predators, feed on these plant consumers. Decomposers break down waste materials and dead plants and animals to get energy. Food chains are a way to describe how the energy in food moves through the biotic community.

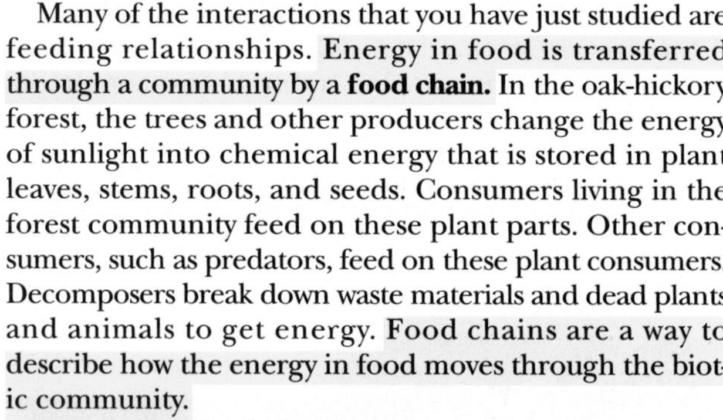

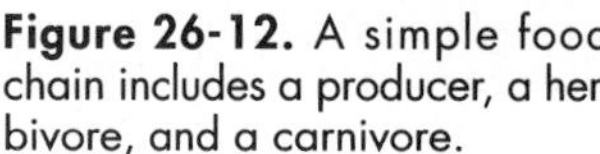

Figure 26-12. A simple food chain includes a producer, a herbivore, and a carnivore.

The simple food chain in Figure 26-12 includes the plants, the rabbit, and the hawk.

There are many food chains within a community. A consumer often feeds in a variety of food chains. Hawks, foxes, and owls all are predators that eat mice and rabbits. Mice and birds eat grains and seeds, whereas rabbits and deer both eat grass. All of these organisms feed in several food chains that overlap. A series of overlapping food chains is called a **food web.** The organisms in a food web eat or are eaten by many organisms.

OPTIONS

INQUIRY QUESTIONS

▶**A sea anemone lives on the back of a hermit crab. It hides the hermit crab, which helps protect it from predators. The hermit crab carries the sea anemone around with it and increases the area in which it can feed. What kind of symbiosis is this?** *mutualism*

▶**Why is it important that parasites not kill their hosts?** *It would result in the death of the parasites as well as the host.*

▶**What are some abiotic factors for which organisms compete?** *food, water, space, sunlight*

▶**In Chapter 9, you studied lichens. What kind of symbiosis is shown in lichens?** *mutualism*

ENRICHMENT

▶Have students construct a food web by choosing an animal and then using field guides and reference books to find out what the animal eats and what eats the animal.

Producers are the first link in any food chain. The producers provide food for themselves and the whole community. Grass, hay, and oats are examples of producers. Producers are the organisms that convert sunlight, carbon dioxide, and water into sugar and oxygen during photosynthesis. Plant-eating consumers, the herbivores, eat green plants and change some of the food stored in plants to the energy they need for life activities. Grasshoppers are herbivores that eat grass seeds. Herbivores, in turn, are eaten by meat-eating consumers, the carnivores. Woodpeckers eat grasshoppers. Some carnivores are then eaten by other carnivores. Woodpeckers, for example, are eaten by hawks. In some food webs, consumers eat both plants and animals. Consumers that eat both plants and animals are omnivores. Humans, bears, and raccoons are omnivores.

The flow of energy from grass seeds to hawks can be shown in a diagram called an **energy pyramid,** as shown

What is a food web?

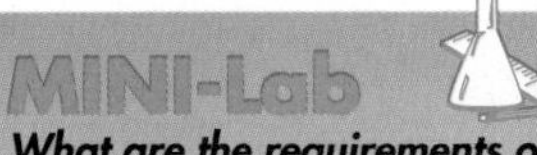

MINI-Lab

What are the requirements of ecosystems?

In your Journal, *classify* the organisms in Figure 26-13, as producers, consumers, decomposers, herbivores, carnivores, or omnivores. Next, list all the abiotic factors you can think of that are needed for these organisms to survive.

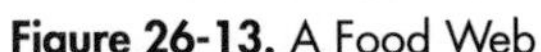

Figure 26-13. A Food Web

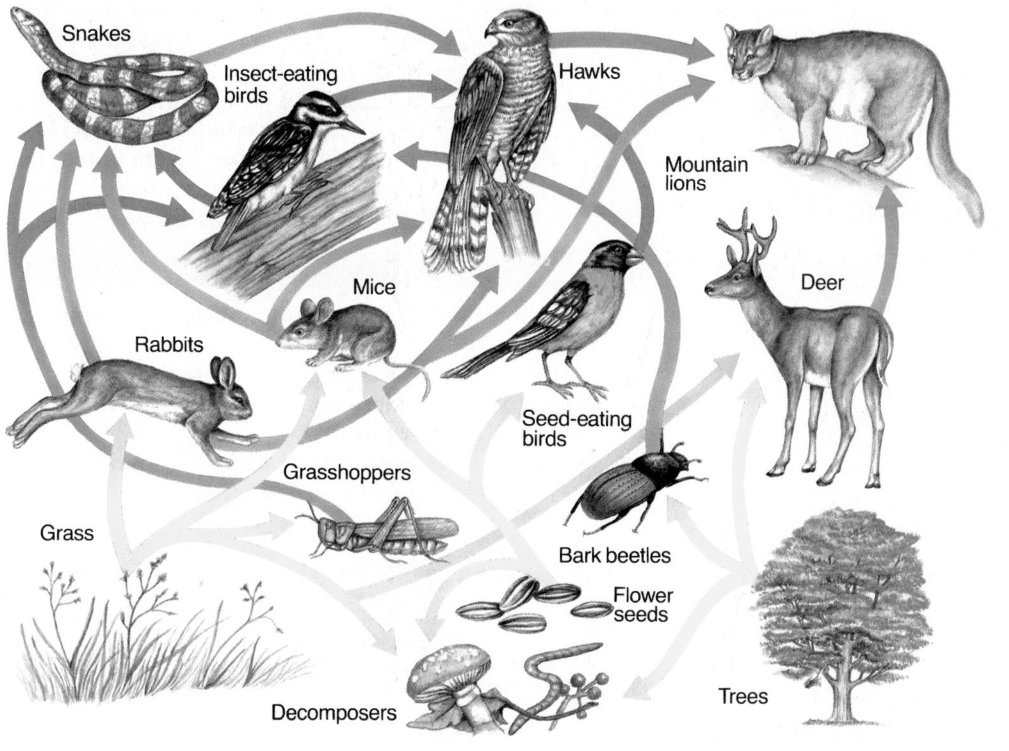

MINI-Lab

Materials: paper and pencil

Teaching Tips

► Review the meanings of *producer, consumer, herbivores, carnivores,* and *omnivores.*

► Herbivores, carnivores, and omnivores are all types of consumers.

Answers to Questions

► All green plants are producers. Everything that depends on another organism for food is a consumer.

► Organisms that eat producers directly are herbivores. Organisms that eat only meat are carnivores. Organisms that eat both are omnivores.

► Abiotic factors: sunlight, water, minerals, temperature, soil, wind

MINI-Lab

ASSESSMENT

Performance: Give each student a role to play as a deer, grass, mouse, decomposer, and so on. Have students stand in a circle and toss a ball of yarn beginning with a producer until a web is produced. Have them relate the size of the yarn ball at the beginning and end to energy loss across the system.

CONCEPT DEVELOPMENT

► Have students consider food chains that affect their lives. Then ask them to draw an energy pyramid that shows the energy relationships in that food chain. You may want to make a class bulletin board to reinforce the pyramid concept.

ASSESSMENT—ORAL

► **What is the advantage of being an omnivore?** *It makes more kinds of foods available for eating.*

► **What organisms in your home are producers?** *green plants*

► **What organisms in your home are consumers?** *people, pets—dogs, cats, fish, gerbils, etc.*

► **Do all organisms remain on the same place in the food chain throughout their lives? Explain.** *No. Tadpoles are herbivores and frogs and toads are carnivores. Many insect larvae are herbivores and then become carnivores as adults.*

PROGRAM RESOURCES

From the **Teacher Resource Package** use:

Activity Worksheets, page 237, Mini-Lab: What are the requirements of ecosystems?

Use **Laboratory Manual,** page 171, Communities, and page 175, Living Things in Soil.

CONCEPT DEVELOPMENT

▶Explain that less and less usable energy is available to larger carnivores because at each level in the energy pyramid, much energy is used for life activities and much is lost as heat to the environment.

3 CLOSE

▶Ask questions 1-2 and the **Apply** Question in the Section Review.

▶**Demonstration:** Use the classroom terrarium as an ecosystem to show producers (mosses and liverworts) and consumers (salamanders, turtles, lizards). Point out that the decomposers are bacteria and fungi.

SECTION REVIEW ANSWERS

1. competition—when two or more species compete for resources; predation—when one species is eaten by another; symbiosis—when two or more species live together

2. Mutualism is a relationship in which two organisms live together and both benefit. Parasitism is a relationship in which one organism is helped and the other is harmed.

3. Apply: Producers are the basis for all life in a community.

4. Connect to Chemistry: Biological magnification is the process in which hazardous wastes become increasingly concentrated in successive levels of the food chain. It can be seen in the environmental levels of lead.

Skill Builder

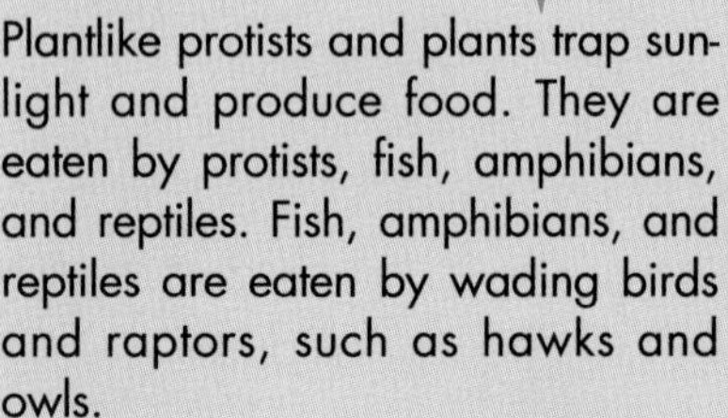

Plantlike protists and plants trap sunlight and produce food. They are eaten by protists, fish, amphibians, and reptiles. Fish, amphibians, and reptiles are eaten by wading birds and raptors, such as hawks and owls.

Skillbuilder
ASSESSMENT

Portfolio: Use this Skillbuilder to assess students' ablilities to Interpret the Scientific Illustration of an energy pyramid.

What is an energy pyramid?

in Figure 26-14. Notice that there are many more producers than herbivores, and many more herbivores than carnivores. Why do you think this is so? Less energy is available at each level of the pyramid as you move toward the top. Even though grasshoppers eat a lot of grass seed, they can't use all of the energy stored in grass seeds. In fact, only about ten percent of the energy in grass seed is available to the grasshopper. Only ten percent of the energy stored in the grasshopper is available to the woodpecker, and so on. The animals at the top of the energy pyramid depend on very large populations of producers and herbivores to stay alive.

Figure 26-14. An energy pyramid usually has three or four levels. Why aren't there more levels in an energy pyramid?

Energy pyramids are one way to show the relationships of organisms in communities. A pyramid of numbers shows how many individuals an ecosystem can support. In the example shown in Figure 26-14, a pyramid of numbers would include 175 grass seeds, 40 grasshoppers, 5 woodpeckers, and 1 hawk. Pyramids of numbers may look different from energy pyramids. One tree may support thousands of insects. What would a pyramid of numbers look like if it included one tree and one thousand insects?

SECTION REVIEW

1. Describe three relationships among organisms in a community.
2. Name and describe two examples of symbiosis.
3. **Apply:** Why are there more producers in an ecosystem than consumers?
4. **Connect to Chemistry:** Find out the meaning of biological magnification and give an example.

Skill Builder

Interpreting Scientific Illustrations

Use the energy pyramid diagram above to explain the flow of energy through a pond community. If you need help, refer to Interpreting Scientific Illustrations in the **Skill Handbook** on page 693.

OPTIONS

PROGRAM RESOURCES

From the **Teacher Resource Package** use:

Critical Thinking/Problem Solving, page 30, Problems of Species Introduction to a Community.

Transparency Masters, pages 93-94, Food Web, Energy Pyramid.

Use **Color Transparency** 47, Food Web, Energy Pyramid.

ENRICHMENT

▶Ask students to use the library and research the effects of the introduction of Russian thistle, starlings, English sparrows, gypsy moths, and kudzu into the United States.

Abiotic Factors in the Biosphere

26-3

Objectives

- Describe how materials in the biosphere are reused in a continuous cycle.
- Diagram the cycles of water, carbon, oxygen, and nitrogen in the biosphere.
- Discuss the importance of recycling these materials in the biosphere.

New Science Words

water cycle
carbon-oxygen cycle
nitrogen cycle

Cycles of Matter

Figure 26-15 shows fungi growing on a fallen tree. Fungi are decomposers. When decomposers break down waste materials and dead organisms for energy, they return carbon, oxygen, nitrogen, and other materials to the biosphere. Producers reuse these materials to make more food that is eaten by consumers. You know that energy is used as it moves through a food chain. Unlike energy, the chemical elements that make up all organisms are used over and over again. Organisms today are using the same materials that have been used since life began. Each time a substance is used, decomposed, and returned to the environment, a cycle is completed. In a cycle, the last step brings the process back to its starting point, and the cycle occurs again and again. In the biosphere, the most important materials to life on Earth are recycled. These materials are water, oxygen, carbon, and nitrogen.

Figure 26-15. Decomposers like these fungi break down dead organisms and help cycle oxygen, nitrogen, and carbon in the biosphere.

Figure 26-16. Water cycles through evaporation, condensation, and precipitation.

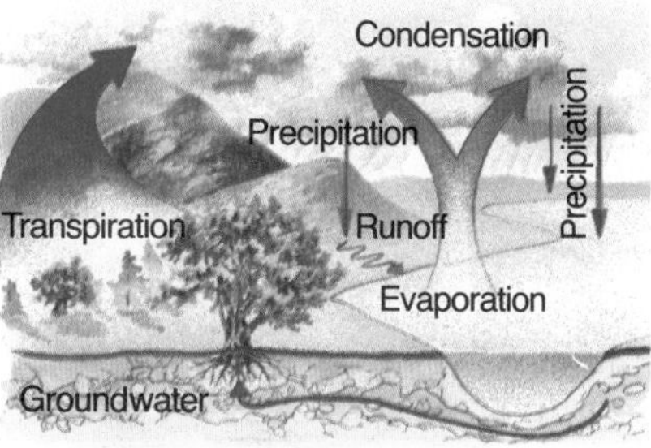

The Water Cycle

The continuous movement of water in the biosphere is called the **water cycle.** Trace the steps of the water cycle in Figure 26-16 as you read. The sun's energy causes some of the water in soil, oceans, lakes, rivers, and living organisms to evaporate and become water vapor in the air. As the water vapor in the air cools, it condenses and changes to water droplets and forms clouds. The water from the

SECTION 26-3

PREPARATION

SECTION BACKGROUND

- Less than 3 percent of the water on Earth is fresh. Most of this water is frozen in the polar ice caps. Only a very small amount of all water is available to land organisms for metabolism.
- Oxygen and carbon make up much of the body's carbohydrates, proteins, and fats. They are also involved in many chemical reactions. The oxygen-carbon cycle is driven by photosynthesis and respiration.

PREPLANNING

- Obtain binoculars, field guides, and hand lenses for Activity 26-2.

1 MOTIVATE

Cooperative Learning: Divide the class into groups of three. Use the Expert Team strategy and have students study and master the water cycle, the carbon-oxygen cycle, and the nitrogen cycle.

- Demonstrate evaporation and condensation in the water cycle by boiling water in a teakettle. Hold a cold glass plate over the steam and the steam will condense.
- Refer back to the Connect to Earth Science feature on page 612 in your discussions of the impact of abiotic factors.

VideoDisc

STVS: Cloud Chemistry, Disc 6, Side 2

OPTIONS

PROGRAM RESOURCES

From the **Teacher Resource Package** use:

Transparency Masters, pages 95-96, Water Cycle, Oxygen Cycle, Nitrogen Cycle.

Use **Color Transparency** 48, Cycles.

Science Integration Activities, 26, Carbon Dioxide and Limestone

2 TEACH

Key Concepts are highlighted.

CONCEPT DEVELOPMENT

▶ Ask the question, **Why are pollutants in water and air a long-term problem?** *They are recycled in the carbon-oxygen and water cycles.*

TECHNOLOGY

Extension: Changes in population levels will be reflected quickly in population changes. Find out about participating in a biomonitoring program.
Reference: Essman, J., and S. Zarpas. "Canaries on the Stream: Macroinvertebrates as tools for water quality studies." *The Conservationist,* May-June 1990, pp. 8-15.
Think Critically: A short lifespan allows major changes in populations in less time.

CHECK FOR UNDERSTANDING

Use overhead transparencies and ask students to describe what is occurring in the water cycle, the carbon-oxygen cycle, and the nitrogen cycle.

RETEACH

Give students unlabeled diagrams of the water, carbon-oxygen, and nitrogen cycles to label.

Figure 26-17. In the carbon-oxygen cycle, carbon and oxyen are recycled continuously during photosynthesis and respiration.

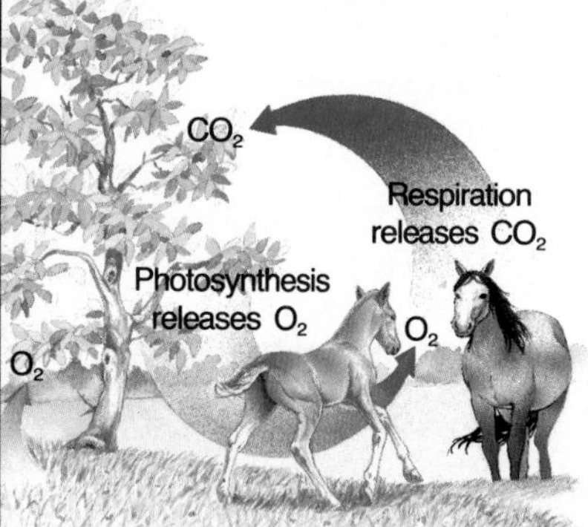

clouds returns to Earth as precipitation in the form of rain, snow, sleet, or hail. Then the cycle starts over again. You could describe the water cycle as three steps: evaporation, condensation, and precipitation.

The Carbon-Oxygen Cycle

Many other materials in the biosphere are part of cycles. The continuous movement of carbon dioxide and oxygen between the surface of Earth and the air is called the **carbon-oxygen cycle.** You can follow the carbon-oxygen cycle in Figure 26-17. During the process of photosynthesis, plants absorb carbon dioxide from the air and release oxygen. Plants, animals, and most other organisms use oxygen for respiration. Carbon dioxide is given off and returned to the cycle. Decomposers break down the carbon compounds in dead organisms and wastes and release carbon dioxide to the air. Organisms that do not decay are compressed underground. Over millions of years, they form fossil fuels such as coal, oil, and gas. When these fuels are burned, carbon dioxide again enters the air.

TECHNOLOGY

Monitoring Mayflies

The source of water pollution in streams is hard to detect because the pollutants are diluted by water. Pollutants concentrate in fish tissue, which makes detection easier, but fish move too much to be helpful in detecting the source of water pollution.

In a process called biomonitoring, scientists now evaluate the health of streams by looking at the organisms that live there. Macroinvertebrates, such as mayflies, caddisflies, and stoneflies, remain in one location and have differing tolerances to pollution. Scientists collect samples of organisms from the stream bed and count the kinds and numbers of macroinvertebrates found. Mayflies are very sensitive to pollution. Large numbers of these show a healthy stream. Aquatic earthworms and midges tolerate pollution, so streams with more of these organisms may be in trouble!

Think Critically: Most macroinvertebrates live less than a year. Why is this trait useful in biomonitoring?

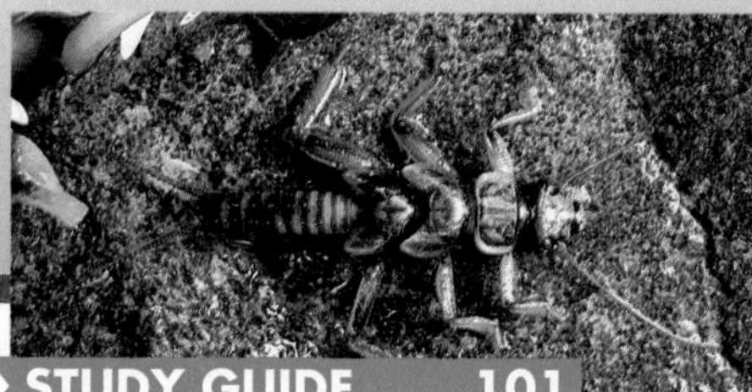

OPTIONS

Meeting Different Ability Levels

For Section 26-3, use the following **Teacher Resource Masters** depending upon individual students' needs.

◆ **Study Guide Master** for all students.

● **Reinforcement Master** for students of average and above average ability levels.

▲ **Enrichment Master** for above average students.

Additional Teacher Resource Package masters are listed in the OPTIONS box throughout the section. The additional masters are appropriate for all students.

◆ STUDY GUIDE 101

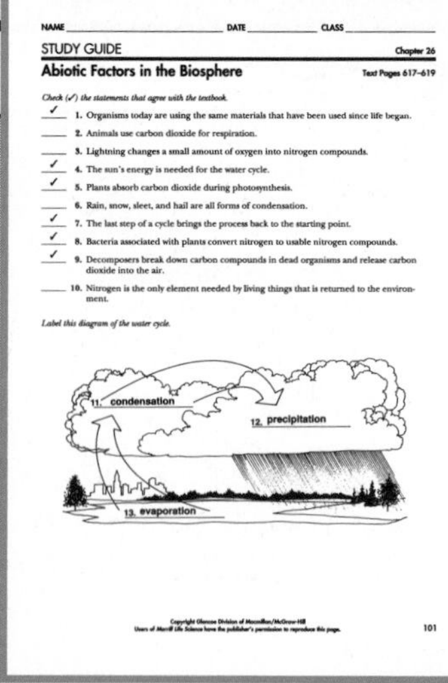

NAME ____________ DATE ________ CLASS ________

STUDY GUIDE — Chapter 26

Abiotic Factors in the Biosphere — Text Pages 617–619

Check (✓) the statements that agree with the textbook.

✓ 1. Organisms today are using the same materials that have been used since life began.
___ 2. Animals use carbon dioxide for respiration.
___ 3. Lightning changes a small amount of oxygen into nitrogen compounds.
✓ 4. The sun's energy is needed for the water cycle.
✓ 5. Plants absorb carbon dioxide during photosynthesis.
___ 6. Rain, snow, sleet, and hail are all forms of condensation.
✓ 7. The last step of a cycle brings the process back to the starting point.
✓ 8. Bacteria associated with plants convert nitrogen to usable nitrogen compounds.
✓ 9. Decomposers break down carbon compounds in dead organisms and release carbon dioxide into the air.
___ 10. Nitrogen is the only element needed by living things that is returned to the environment.

Label this diagram of the water cycle.

11. condensation
12. precipitation
13. evaporation

101

The Nitrogen Cycle

Nitrogen is an important element needed by organisms to make proteins. Even though nitrogen gas makes up 78 percent of the atmosphere, most living organisms cannot use nitrogen in this form. It has to be combined with other elements in a process called nitrogen fixation. You can see in Figure 26-18 how nitrogen is converted into usable compounds by bacteria associated with plants. A small amount is changed into nitrogen compounds by lightning. The transfer of nitrogen from the atmosphere to plants, and back to the atmosphere or directly into plants again, is the **nitrogen cycle.**

Phosphorus, sulfur, and other elements needed by living organisms are also used and returned to the environment. Just as we save aluminum, glass, and paper products to be reused, the materials that organisms need to live are recycled continuously in the biosphere.

Figure 26-18. Nitrogen can be cycled from bacteria on plant roots to plants, then to animals, and directly back to plants again as a result of decomposition.

SECTION REVIEW

1. Describe how materials move through the biosphere.
2. Describe the nitrogen cycle.
3. Explain the relationship between the cycling of oxygen and the cycling of carbon in the biosphere.
4. **Apply:** Do you think that fossil fuels are still being formed today? Why?
5. **Connect to Chemistry:** Find out about the role of phosphorus in a pond community. What happens to the pond if excess phosphorus is present?

Sequencing

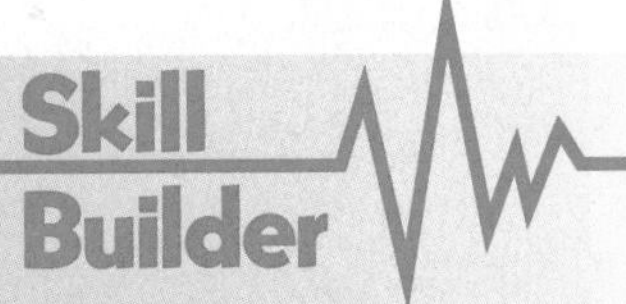

Sequence the steps in the carbon-oxygen cycle. If you need help, refer to Sequencing in the **Skill Handbook** on page 680.

● REINFORCEMENT 101

NAME ______ DATE ______ CLASS ______

REINFORCEMENT Chapter 26

Abiotic Factors in the Biosphere Text Pages 617–619

1. What is a cycle? **A cycle is a process in which the last step brings the cycle back to its starting point. A cycle occurs again and again.**
2. What are the three steps of the water cycle? **evaporation, condensation, and precipitation**
3. What causes water to evaporate? **the sun's energy**
4. Where does condensation occur? **In the atmosphere, as clouds form**

carbon dioxide
oxygen

5. What cycle does the diagram above illustrate? **the carbon-oxygen cycle**
6. What in the diagram gives off carbon dioxide? **the dog**
7. What in the diagram uses oxygen? **the dog**
8. Where and how is oxygen released in the diagram? **The bush releases oxygen through photosynthesis.**
9. What in the diagram uses carbon dioxide? **the bush**
10. What is another way in which carbon dioxide can be released from organisms? **Decomposers break down dead organisms and release carbon dioxide into the air.**
11. On the roots of clover, nitrogen is combined with other elements. What is this process called? **nitrogen fixation**
12. What other living things, besides clover, are involved in this process? **bacteria**
13. Why is this process important to other living things? **Nitrogen is needed by all living things, yet it can't be used in its pure form. Nitrogen fixation combines nitrogen with other elements into compounds that can be used.**

 101

▲ ENRICHMENT 101

NAME ______ DATE ______ CLASS ______

ENRICHMENT Chapter 26

Abiotic Factors in the Biosphere Text Pages 617–619

CARBON-OXYGEN CYCLE IN A FISH BOWL

Materials
- clear fish bowl
- distilled water at room temperature
- 2 goldfish
- goldfish food
- small *Elodea* plant

FISH FOOD

Procedure
1. Fill the bowl about three-quarters full with distilled water (tap water may be harmful to goldfish). Add the two goldfish to the bowl.
2. Over the next two days, feed the goldfish according to the instructions on the food box. Observe where the fish are spending most of their time in the bowl. What behavior do they show?
3. After two days, add the *Elodea* plant, or any other small water plant you can get at a pet store. Ask for a small plant for a goldfish bowl.
4. Over the next two days, observe the fish to see if their behavior has changed. Do they now stay in another part of the bowl?

Conclude and Apply
1. Where in the bowl did the fish tend to stay after a day or so? **The fish will tend to stay near the surface.**
2. After you added the plant, did the behavior of the fish change at all? **Most students will report that the fish began to stay longer in the deeper parts of the bowl.**
3. What does a plant give off that animals such as fish need to survive? **oxygen**
4. Why would the behavior of the fish change after the plant was added? **The plant added oxygen to the water, allowing the fish to swim deeper.**
5. What caused the fish to stay where they did at first? What did they need? **They needed the oxygen at the surface, where the water met the air.**
6. What cycle of abiotic factors does this experiment demonstrate? **the carbon-oxygen cycle**
7. What will the fish give off that will help the plant survive? **carbon dioxide**

 101

EXTENSION

For students who have mastered this section, use the **Reinforcement** and **Enrichment** masters or other OPTIONS.

3 CLOSE

▶Ask questions 1-3 and the **Apply** Question in the Section Review.

SECTION REVIEW ANSWERS

1. Decomposers break down waste materials and dead organisms for energy and return the inorganic materials to the biosphere. Producers reuse these materials to make more food that is eaten by consumers.

2. Nitrogen gas is combined with other elements in nitrogen fixation. Nitrogen is converted into usable nitrogen by bacteria associated with plants.

3. Plants absorb carbon dioxide and release oxygen in photosynthesis. Most organisms use oxygen and give off carbon dioxide in respiration.

4. Apply: Answers will vary. Some students will say no, because fossil fuels were formed during the carboniferous period. Others will say yes, in present-day bogs and marshes.

5. Connect to Chemistry: Phosphorous is needed by all organisms as a component of DNA, RNA and ATP. Excess phosphorous causes an increase in the growth of algae. When the algae dies it removes much of the oxygen from the water—affecting all other organisms in the pond.

Skill Builder

During photosynthesis, plants absorb carbon dioxide and release oxygen. Organisms use oxygen in respiration and release carbon dioxide. Decomposers break down carbon compounds in dead animals and release carbon to air.

Skillbuilder
ASSESSMENT

Performance: Use this Skillbuilder to assess students' abilities to Sequence the steps in water or nitrogen cycles.

SCIENCE & SOCIETY SECTION 26-4

PREPARATION

SECTION BACKGROUND

▶Although forest fires are usually thought of negatively, they are critical in maintaining certain types of habitats.

▶Controlled burns are used to manage certain areas and to prevent large, unpredicted, out of control fires, although local homeowners sometimes oppose them.

1 MOTIVATE

▶**Bulletin Board:** Using articles about the Yellowstone National Park fires in 1988, prepare a bulletin board showing both burned and recovering areas.

2 TEACH

Key Concepts are highlighted.

CONCEPT DEVELOPMENT

▶Forest fires aren't necessarily beneficial to wildlife, even during controlled burns. Wildlife are lost during all fires.

In Your JOURNAL

Controlled burns in national parks are controversial. Students may contact state fire marshalls or natural resource departments for information.

26-4 Friendly Fires

New Science Words

controlled burns

Objectives

- ▶ Describe the role of controlled fires in forests and why the fires are controversial.
- ▶ Identify reasons both for and against the use of controlled burns in areas where people live.

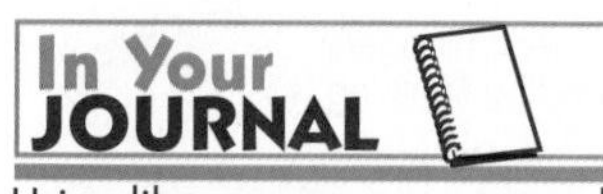

In Your JOURNAL

Using library resources, research several controlled burns. **In your Journal,** write a paragraph on your position, for or against controlled burns.

Are Forest Fires Always Harmful?

What comes to your mind first when you think of forest fires? You may think about animals such as deer running terrified through the forest, or blackened stumps of trees and scorched ground where nothing can grow. A burned forest looks so stark and damaged, it seems it will never be the same again. What most of us don't realize is that some forests actually require such fires to maintain their character. Some pine species have evolved ways to cope with periods of fire. Jack pine and lodgepole pine trees have cones that don't open to release seeds until they have been heated in a fire!

Figure 26-19. In most instances, new growth begins quickly after a forest fire.

In some types of forests, biologists conduct **controlled burns,** managed forest fires that are set periodically to control the amount of vegetation underneath the dominant forest tree species. In areas where natural fires don't occur, deep piles of dead leaves, dropped needles, and dead brush build up on the forest floor. All of this dead organic matter becomes potential fuel if a fire should start. To make sure that this organic matter doesn't build up too much, biologists sometimes start small fires to burn up the material on the forest floor. This is one way biologists hope to prevent major forest fires.

OPTIONS

Meeting Different Ability Levels

For Section 26-4, use the following **Teacher Resource Masters** depending upon individual students' needs.

◆ **Study Guide Master** for all students.

● **Reinforcement Master** for students of average and above average ability levels.

▲ **Enrichment Master** for above average students.

Additional Teacher Resource Package masters are listed in the OPTIONS box throughout the section. The additional masters are appropriate for all students.

◆ STUDY GUIDE 102

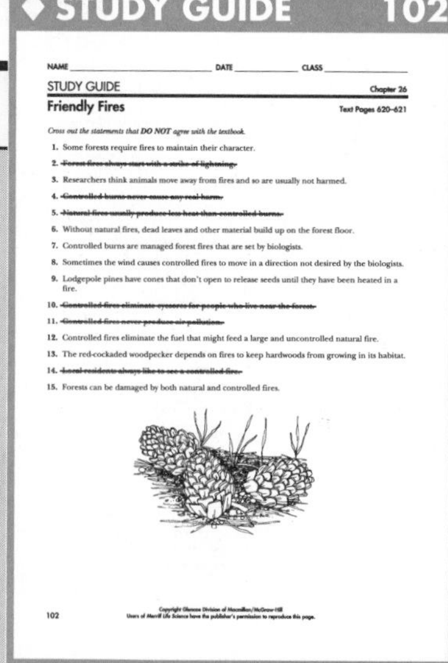

NAME ______ DATE ______ CLASS ______

STUDY GUIDE Chapter 26

Friendly Fires Text Pages 620–621

Cross out the statements that DO NOT agree with the textbook.

1. Some forests require fires to maintain their character.
2. ~~Forest fires always start with a strike of lightning.~~
3. Researchers think animals move away from fires and so are usually not harmed.
4. ~~Controlled burns never cause any real harm.~~
5. ~~Natural fires usually produce less heat than controlled burns.~~
6. Without natural fires, dead leaves and other material build up on the forest floor.
7. Controlled burns are managed forest fires that are set by biologists.
8. Sometimes the wind causes controlled fires to move in a direction not desired by the biologists.
9. Lodgepole pines have cones that don't open to release seeds until they have been heated in a fire.
10. ~~Controlled fires eliminate eyesores for people who live near the forest.~~
11. ~~Controlled fires never produce air pollution.~~
12. Controlled fires eliminate the fuel that might feed a large and uncontrolled natural fire.
13. The red-cockaded woodpecker depends on fires to keep hardwoods from growing in its habitat.
14. ~~Local residents always like to see a controlled fire.~~
15. Forests can be damaged by both natural and controlled fires.

102

What happens to the forest animals during a controlled burn? Researchers contend that wild animals are unharmed because they move away from areas on fire. Some animals, such as the rare red-cockaded woodpecker, depend upon fires to maintain their habitat. They live in open pine stands that are often invaded by hardwood trees. Fires help to keep the hardwoods from growing in the open spaces between the pines.

Controlled burns are usually less intense and produce less heat than uncontrolled forest fires. This is another reason researchers feel that it is better to have a few controlled fires periodically than a large, out-of-control fire. However, even controlled burns sometimes cause damage to areas outside of the intended area of the burn. If the wind shifts suddenly, a fire may race toward residential areas instead of staying in the forest. Homes and businesses have burned to the ground from controlled fires as well as natural forest fires. Controlled burns also result in air pollution from smoke, and newly burned areas are eyesores to those who live nearby. It may take a few years for the burned areas to begin growing again. In the meantime, local residents see damaged forest views every day.

Biologists are very careful to set fires only at times when wind conditions are right. But people who live near national forests and parks still are concerned over the need for controlled burns. They believe controlled burns are not "controlled" at all. They also aren't convinced that forest animals are unharmed as a result of these fires.

Figure 26-20. The red-cockaded woodpecker is one organism that depends on fires to maintain its habitat.

When burning takes place in the presence of enough oxygen, carbon dioxide and water are produced. Soot and smoke form when there is not enough oxygen present to burn all the carbon of the fuel. Soot is a form of carbon called graphite. Find out what other form of pure carbon is found in nature.

SECTION REVIEW

1. Name two reasons why biologists believe controlled forest fires are useful.
2. Why are local citizens often opposed to such fires?
3. **Connect to Chemistry:** Find out why there are ashes left after a piece of wood burns.

You Decide!

In popular vacation spots, people often build homes in scenic areas, such as high on a ridge top. Such spots are often in forested areas that are subject to uncontrollable forest fires. Many of these vacation homes are in locations that are difficult for forest firefighters to reach. Should people be permitted to build homes in these areas?

● REINFORCEMENT 102

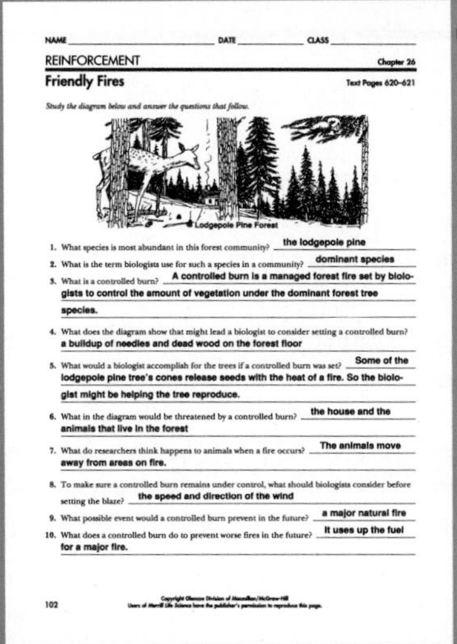

NAME ______ DATE ______ CLASS ______

REINFORCEMENT — Chapter 26

Friendly Fires — Text Pages 620–621

Study the diagram below and answer the questions that follow.

1. What species is most abundant in this forest community? **the lodgepole pine**
2. What is the term biologists use for such a species in a community? **dominant species**
3. What is a controlled burn? **A controlled burn is a managed forest fire set by biologists to control the amount of vegetation under the dominant forest tree species.**
4. What does the diagram show that might lead a biologist to consider setting a controlled burn? **a buildup of needles and dead wood on the forest floor**
5. What would a biologist accomplish for the trees if a controlled burn was set? **Some of the lodgepole pine tree's cones release seeds with the heat of a fire. So the biologist might be helping the tree reproduce.**
6. What in the diagram would be threatened by a controlled burn? **the house and the animals that live in the forest**
7. What do researchers think happens to animals when a fire occurs? **The animals move away from areas on fire.**
8. To make sure a controlled burn remains under control, what should biologists consider before setting the blaze? **the speed and direction of the wind**
9. What possible event would a controlled burn prevent in the future? **a major natural fire**
10. What does a controlled burn do to prevent worse fires in the future? **It uses up the fuel for a major fire.**

102

▲ ENRICHMENT 102

NAME ______ DATE ______ CLASS ______

ENRICHMENT — Chapter 26

Friendly Fires — Text Pages 620–621

A FIRE AT YELLOWSTONE

In the summer of 1988, a terrible fire occurred at Yellowstone National Park on the Wyoming-Montana border. This park is visited by thousands of Americans every year who think of the area as one of the most beautiful in the world. Yellowstone has many great sights, including the geyser Old Faithful. When the fire began, the National Park Service decided to let it burn. As the fire grew out of control, many places were threatened with destruction, including Old Faithful. The whole country took notice of the fire and of the debate about fire policy. Some people argued that the fire should be put out immediately. Others argued that the fire could be helpful.

Use the resources in your library to research the events and issues of this fire. Find magazine articles that discuss the events of the fire, the debate that raged along with it, and the outcome a year or two later. You can find magazine articles by using your library's computer search service or by using the recent volumes of the *Readers' Guide to Periodical Literature.* If you do not know how to use these resources, ask a librarian.

Write a report from articles you find. In your report, try to answer the following questions.

- When did the Yellowstone fire begin?
- What caused the fire?
- Was there more than one fire at Yellowstone?
- What was the reaction of the local residents?
- What was the policy of the National Park Service concerning this fire?
- Did the Park Service's policy change as the fire burned?
- How was the fire put out?
- How much destruction did the fire do to the park, its buildings, and its animals?
- What were the conclusions of the scientists and others a year later about the harmfulness of this fire?

102

CHECK FOR UNDERSTANDING

Have students write a paragraph on how they would react to a controlled burn near their homes. Make sure they can support their views.

RETEACH

Present a filmstrip or video on forest fires or fire ecology. Write to the U.S. National Park Service for information and posters to present to the class.

EXTENSION

For students who have mastered this section, use the **Reinforcement** and **Enrichment** masters or other OPTIONS.

3 CLOSE

► Ask questions 1-2 in the Section Review.

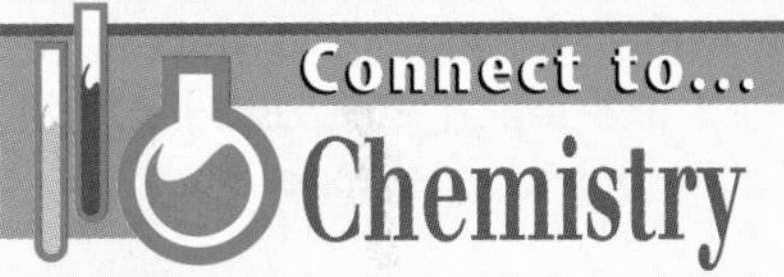

Answer: Graphite and diamond are two forms of pure carbon found in nature. Each has a distinctly different molecular structure.

SECTION REVIEW ANSWERS

1. Controlled fires help prevent major forest fires and help to maintain habitat for particular wildlife species.

2. They fear the burns will go out of control and that wildlife will be harmed.

3. Connect to Chemistry: The ash, generally a mixture of minerals, is present in the wood but will not combine with the oxygen.

YOU DECIDE!

SCIENCE & SOCIETY

Answers will vary. Ask students to back up their positions with logical support.

PROGRAM RESOURCES

From the **Teacher Resource Package** use:

Science and Society, page 30, Fire Prevention in Redwood National Park.

Activity Worksheets, page 5, Flex Your Brain.

ACTIVITY 26-2

40 minutes

OBJECTIVE: Conduct a study of an ecosystem and **determine** how the living and nonliving factors interact.

PROCESS SKILLS applied in this activity:
▶**Measuring** in Procedure Step 2.
▶**Mapping** in Procedure Step 3.
▶**Observing** in Procedure Steps 4 and 6.
▶**Recording** in Procedure Steps 5 and 6.
▶**Identifying** in Procedure Step 5.
▶**Inferring** in Analyze Question 4, Conclude and Apply Questions 7-12.

COOPERATIVE LEARNING Use Science Investigation groups of five. Assign one student to gather the materials, one to read, one to record and report, one as safety officer, and one as timekeeper.

TEACHING THE ACTIVITY

Troubleshooting: Approve the ecosystems chosen by the students before they begin their observations. Make sure they have chosen an appropriate area.

Activity ASSESSMENT

Performance: To further assess students' understanding of ecosystems, see USING LAB SKILLS, Question 12, p. 624.

ACTIVITY 26-2 Studying an Ecosystem

***Problem:** How do you study an ecosystem?*

Materials

- graph paper
- notebook
- pencil
- binoculars
- field guides
- hand lens

Procedure

1. Choose a natural community near your school or home to be your ecosystem for study. You may choose to study a pond, a forest area in a park, a rotten log, or an area around your school.
2. Decide the boundaries of the ecosystem you are studying. Make the ecosystem a size you think you can study well.
3. Make a map or drawing of the ecosystem on graph paper.
4. *Observe* the organisms that live in the ecosystem. Use a hand lens to study small creatures. Use binoculars to study organisms you cannot get near. Look for evidence, such as tracks or feathers, of organisms you cannot see.
5. Record your observations in a table like the one shown. Make drawings. Use field guides to identify the organisms.
6. Visit the ecosystem as many times as you can and at different times of day for four weeks. Make observations and record them. Pay close attention to the relationships among organisms. Also note how the organisms interact with the nonliving environment.

Data and Observations

Date	Organisms Observed	Comments

Analyze

1. What is the ecosystem you chose?
2. Describe the nonliving environment of the ecosystem.
3. How many populations did you count?
4. Which organisms in the ecosystem are producers, consumers, or decomposers?
5. What evidence of competition did you find?

Conclude and Apply

6. *Diagram* a food web in the ecosystem.
7. *Predict* what might happen if a population of producers was removed from the community.
8. *Predict* what might happen if either the predators or the decomposers were removed from the community.
9. *Identify* relationships among the organisms in the ecosystem, such as predator-prey, mutualism, or parasitism.
10. *Infer* how the nonliving environment is important in the ecosystem.
11. *Predict* what might happen if the nonliving environment changed suddenly.
12. Based on your observations, what conclusions can you make about studying an ecosystem?

ANSWERS TO QUESTIONS

1-6. Answers will vary depending on the ecosystem chosen by each group.
7. The consumers that use the producers for food might die.
8. Without predators, the community would soon become overpopulated with animals such as insects and rodents. Without decomposers, the cycling of matter would stop and the community would become diseased.
9. Answers will vary.
10. The nonliving environment provides energy, air, water, soil, and other needed elements for living organisms.
11. If the nonliving environment changed suddenly, the living organisms would be affected and might die.
12. An ecosystem is studied by observing living organisms and the relationships between them and their nonliving environment.

PROGRAM RESOURCES

From the **Teacher Resource Package** use:
Activity Worksheets, pages 232-233, Activity 26-2, Studying an Ecosystem.

CHAPTER
REVIEW

SUMMARY

26-1: Organisms and Their Environments

1. The biosphere is the part of Earth where all life is found, and it consists of biotic (living) factors as well as abiotic (nonliving) factors such as air, soil, water, and sunlight.
2. Populations are made up of all the members of a species living in the same place at the same time. A community includes all the populations of the area. The community and the abiotic factors make up the ecosystem.
3. An organism lives in its habitat within a community. The role or job of an organism within a community is its niche.

26-2: Biotic Relationships

1. Organisms have relationships with each other. These relationships include feeding relationships and symbiotic relationships.
2. Energy in food is transferred through a community in food chains. Overlapping food chains are food webs. Organisms are part of many food chains or webs in a community.
3. Energy stored in food moves through the community from producers to consumers to decomposers. At each level only a small amount of the energy consumed is available to the next level of organism.

26-3: Abiotic Factors in the Biosphere

1. Materials in the biosphere are cycled in order for organisms to survive.
2. Water, carbon, oxygen, and nitrogen are recycled in the environment.
3. Many of the materials needed by organisms are recycled.

26-4: Science and Society: Friendly Fires

1. Controlled forest fires are set to reduce the risk of uncontrolled fires.
2. Even controlled fires may become uncontrolled and damage areas where people live.

KEY SCIENCE WORDS

a. **abiotic factors**
b. **biosphere**
c. **biotic factors**
d. **camouflage**
e. **carbon-oxygen cycle**
f. **community**
g. **competition**
h. **controlled burns**
i. **ecology**
j. **ecosystem**
k. **energy pyramid**
l. **food chain**
m. **food web**
n. **habitat**
o. **mimicry**
p. **niche**
q. **nitrogen cycle**
r. **population**
s. **population density**
t. **water cycle**

UNDERSTANDING VOCABULARY

Match each phrase with the correct term from the list of Key Science Words.

1. living things in the environment
2. number of organisms in an area
3. all parts of Earth with life
4. job of a species in an environment
5. the cycle that includes evaporation, condensation, and precipitation
6. copying another organism's behavior or appearance
7. all the populations in an ecosystem
8. interaction of abiotic and biotic factors
9. series of overlapping food chains
10. where an organism lives in an ecosystem

OPTIONS

ASSESSMENT

To assess student understanding of material in this chapter, use the resources listed.

COOPERATIVE LEARNING

Consider using cooperative learning in the THINK AND WRITE CRITICALLY, APPLY, and MORE SKILL BUILDERS sections of the Chapter Review.

PROGRAM RESOURCES

From the **Teacher Resource Package** use:
Chapter Review, pages 55-56.
Chapter and Unit Tests, pages 176-179, Chapter Test.

CHAPTER
REVIEW

SUMMARY

Have students read the summary statements to review the major concepts of the chapter.

UNDERSTANDING VOCABULARY

1. c
2. s
3. b
4. p
5. t
6. o
7. f
8. j
9. m
10. n

ASSESSMENT

Portfolio

Encourage students to place in their portfolios one or two items of what they consider to be their best work. For each item, ask students to explain why that item was chosen and what they learned from it. Items might be selected from the following.

- For Your Mainstreamed Students acting, p. 604
- Check For Understanding food web diagram, p. 614
- In Your Journal research on controlled burns, p. 620

Performance

Additional performance assessments may be found in *Performance Assessment* and *Science Integration Activities* that accompany **Merrill Life Science.** Performance Task Assessment Lists and rubrics for evaluating these activities and other products generated throughout the chapter can be found in Glencoe's *Performance Assessment in Middle School Science.*

CHAPTER REVIEW

CHECKING CONCEPTS

1. a
2. d
3. c
4. d
5. d
6. d
7. b
8. d
9. d
10. d

USING LAB SKILLS

ASSESSMENT

Use these alternate lab exercises to assess students' understanding of the skills used in this chapter.

11. Students should maintain ongoing population counts and descriptions to compare with the results of Activity 26-1. Be sure the students identify the source of their new populations.

12. Abiotic factors: water, air, gravel; Biotic factors: Elodea plant and guppy

THINK AND WRITE CRITICALLY

13. If we ate producers, we would be able to get rid of farm animals and decrease use of chemicals on food crops.

14. Any reasonable answer that shows a food chain will do.

15. Students may choose any relationship to describe as long as they can explain how each organism involved benefits.

APPLY

16. More organisms would put pressure on food and space requirements, forcing some organisms to fight, migrate, or starve.

17. parasitism

CHAPTER REVIEW

CHECKING CONCEPTS

Choose the word or phrase that completes the sentence.

1. All are abiotic factors except _______.
a. animals **c.** sunlight
b. air **d.** soil

2. A coral reef and an oak-hickory forest are examples of a(n) _______.
a. niche **c.** population
b. habitat **d.** ecosystem

3. The _______ is made up of all populations in an area.
a. niche **c.** community
b. habitat **d.** dominant species

4. The number of individuals in a given area is called population _______.
a. clumping **c.** spacing
b. size **d.** density

5. An example of a consumer is _______.
a. tree **c.** grass
b. moss **d.** rabbit

6. In an ecosystem, _______ get the most energy.
a. omnivores **c.** decomposers
b. herbivores **d.** producers

7. _______ is a relationship in which one organism is helped and the other harmed.
a. Mutualism **c.** Mimicry
b. Parasitism **d.** Camouflage

8. Photosynthesis, respiration, and the burning of fuels are all directly involved in the _______.
a. water cycle **c.** phosphorus cycle
b. nitrogen cycle **d.** carbon-oxygen cycle

9. Materials that are cycled in the biosphere include all of the following except _______.
a. nitrogen **c.** phosphorus
b. sulfur **d.** energy

10. Controlled burns of forests are needed _______.
a. for normal growth **c.** to thin out brush
b. to improve habitats **d.** to decrease dead organic matter

USING LAB SKILLS

11. In Activity 26-1, you determined how a population changes over time. Repeat this experiment using a different type of pond vegetation. Compare the results of your two activities.

12. In Activity 26-2 on page 622 you studied an ecosystem. Using a covered 2-L container, gravel, aged tap water, one Elodea plant, and one guppy, make an ecosystem. What are the abiotic factors in your ecosystem? What are the biotic factors?

THINK AND WRITE CRITICALLY

Answer the following questions in your Journal using complete sentences.

13. What would be the advantages to humans if they ate food lower on the food chain?

14. Use Figure 26-7 on page 611 and describe two food chains of the coral reef ecosystem.

15. Explain the advantages of a mutualistic relationship by using an example.

APPLY

16. Explain the changes that would occur in biotic relationships if the density of a population increased dramatically.

17. A maggot grows within a gall formed on the goldenrod plant, eventually maturing and emerging as a fly while the goldenrod dies. What kind of relationship is this?

18. What would happen to an ecosystem if decomposers were removed?
19. Explain why energy flow is described as a pyramid.
20. Explain how photosynthesis and respiration are opposite reactions in the carbon-oxygen cycle.

MORE SKILL BUILDERS

If you need help, refer to the Skill Handbook.

1. **Classifying:** Classify each as producer, herbivore, carnivore, or omnivore: cow, deer, grass, shark, rabbit, bear, green algae, lion, sunflower.
2. **Making and Using Graphs:** Use the following data to graph the density of a deer population over the years. Plot the number of deer on the y-axis and years on the x-axis. Explain what may have happened to change the population density.

Year	Deer (in thousands) per 400 hectares
1905	5.7
1915	35.7
1920	142.9
1925	85.7
1935	25.7

3. **Concept Mapping:** Use this information to draw a food web of organisms living in a goldenrod field.
   Goldenrod - sap is eaten by aphids
   - nectar eaten by bees
   - pollen eaten by beetles
   - leaves eaten by beetles
   Stinkbugs eat beetles
   Spiders eat aphids
   Assassin bugs eat bees
4. **Classifying:** Classify each event in the water cycle as the result of either precipitation or evaporation:
   a. A puddle is gone after the rain.
   b. Rain falls.
   c. Snow covers the mountainside.
   d. A lake becomes shallower.
5. **Concept Mapping:** Fill in the correct terms to show the nitrogen cycle.

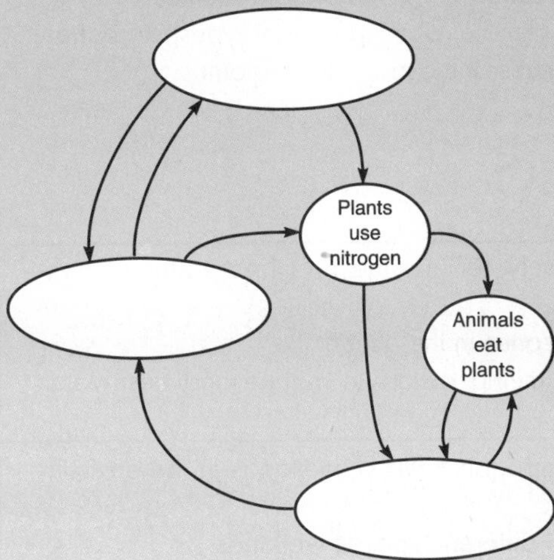

PROJECTS

1. Continue observing the ecosystem from Activity 26-2. Make observations several times a month for a year. Record your observations in your Journal.
2. Research the phosphorus cycle. Find out what role phosphorus plays in an algal bloom.

18. Dead organic matter would build up on the ground and tie up all the elements that are normally recycled in nature.
19. Energy is lost at a constant rate from one level to another in a step-by-step fashion.
20. Photosynthesis uses carbon dioxide and releases oxygen; respiration uses oxygen and releases carbon dioxide.

MORE SKILL BUILDERS

1. Classifying: producer—grass, green algae, sunflower; herbivore—cow, deer, rabbit; carnivore—shark, lion; omnivore—bear
2. Graphing: The deer population built up so high that the deer consumed all the available food. The next generations lose animals to starvation.
3. Concept Mapping:

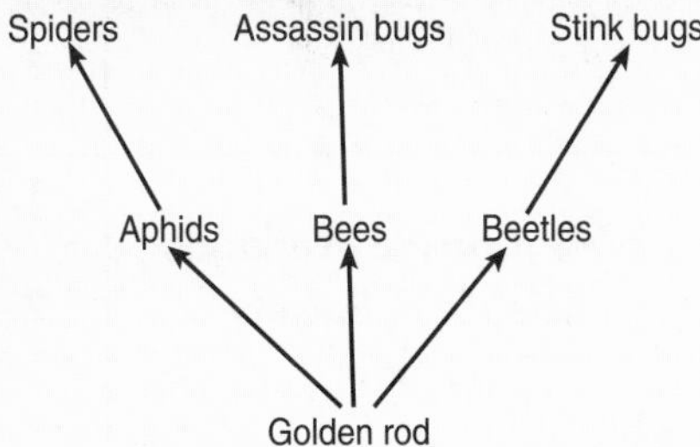

4. Classifying: evaporation, precipitation, precipitation, evaporation
5. Concept Mapping:

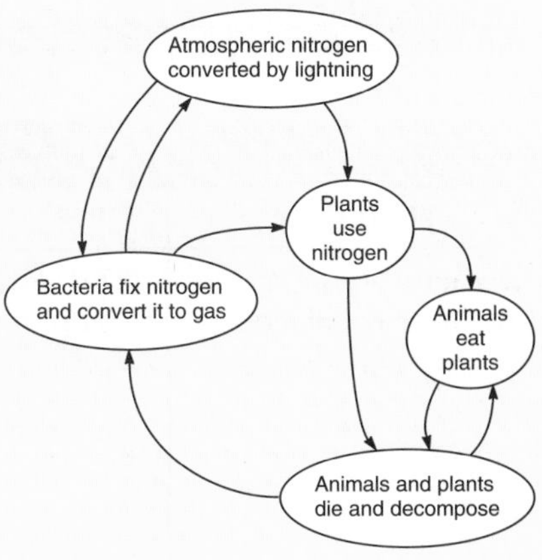

CHAPTER 27 Biomes

CHAPTER SECTION	OBJECTIVES	ACTIVITIES
27-1 Factors That Affect Biomes (3 days)	1. **Identify** and **describe** limiting factors. 2. **Define** succession and differentiate between primary and secondary succession. 3. **Describe** climax communities.	**Activity 27-1:** *Interpreting a Map,* p. 630 **MINI-Lab:** *How do communities change?* p. 631
27-2 Land Biomes (2 days)	1. **Define** biomes. 2. **Identify** the location of the major land biomes. 3. **Describe** the climate, dominant plant types, and characteristic animals of the major land biomes.	**MINI-Lab:** *Is there tundra in Africa?* p. 633
27-3 Water Ecosystems (1 day)	1. **Distinguish** between marine and freshwater ecosystems. 2. **Describe** the zones in the ocean. 3. **Identify** the limiting factors in marine and freshwater ecosystems.	
27-4 Coastal Wetlands Science & Society (1 day)	1. **List** several important roles of wetlands in the environment. 2. **Name** several different types of wetlands.	**Activity 27-2:** *Designing an Experiment (Investigating a Limiting Factor),* p. 642
Chapter Review		

ACTIVITY MATERIALS

FIND OUT	ACTIVITIES		MINI-LABS	
Page 627 globe or world map	**27-1 Interpreting a Map, p. 630** pencil and paper	**27-2 Designing an Experiment, p. 642** one package of 50 bean seeds 2 planters water soil 2 labels small spade or spoon	**How do communities change? p. 631** data table	**Is there tundra in Africa? p. 633** world biome map or world globe

CHAPTER FEATURES	TEACHER RESOURCE PACKAGE	OTHER RESOURCES
Skill Builder, *Hypothesizing,* p. 631	**Ability Level Worksheets** ◆ ***Study Guide,*** p. 103 ● ***Reinforcement,*** p. 103 ▲ ***Enrichment,*** p. 103 **Activity Worksheets,** pp. 239, 240, 245	**Lab Manual:** Limiting Factors, pp. 177-178 **STVS:** Disc 6, Side 1
Technology: *Life in a Glass World,* p. 636 **Skill Builder,** *Concept Mapping,* p. 636	**Ability Level Worksheets** ◆ ***Study Guide,*** p. 104 ● ***Reinforcement,*** p. 104 ▲ ***Enrichment,*** p. 104 **Activity Worksheets,** p. 246 **Critical Thinking/Problem Solving,** p. 31 **Concept Mapping,** p. 59 **Cross-Curricular Connections,** p. 31 **Transparency Masters,** pp. 97-98	**Color Transparency 49,** Biome Map **STVS:** Disc 6, Side 1
Problem Solving: *What Caused the Fish to Die?* p. 639 **Skill Builder,** *Hypothesizing,* p. 639	**Ability Level Worksheets** ◆ ***Study Guide,*** p. 105 ● ***Reinforcement,*** p. 105 ▲ ***Enrichment,*** p. 105	**Science Integration Activity 27**
You Decide! p. 641	**Ability Level Worksheets** ◆ ***Study Guide,*** p. 106 ● ***Reinforcement,*** p. 106 ▲ ***Enrichment,*** p. 106 **Activity Worksheets,** pp. 241-242 **Science and Society,** p. 31	**STVS:** Disc 6, Side 1
Summary Key Science Words Understanding Vocabulary Checking Concepts Using Lab Skills Think & Write Critically Apply More Skill Builders Projects	ASSESSMENT RESOURCES **Chapter Review,** pp. 57, 58 **Chapter Test,** pp. 180-183 **Performance Assessment in Middle School Science (PAMSS)**	**Chapter Review Software** **Test Bank** **Alternate Assessment** **Performance Assessment**

◆ **Basic** ● **Average** ▲ **Advanced**

ADDITIONAL MATERIALS

SOFTWARE	AUDIOVISUAL	BOOKS/MAGAZINES
Niche-Ecological Game/Simulation, Diversified Educational Enterprises. *Biomes,* Collamore Educational Publishing. *Oh, Deer!,* Queue.	*Arctic Oasis,* film, Beacon. *Life in the Desert,* film, EBEC. *Oceans,* film, Time-Life. *Seas of Grass,* film, Time-Life. *The Margins of the Land,* film, Time-Life. *The Northern Forests,* film, Time-Life. *Life in a Tropical Rain Forest,* film, BFA. *Prairie Coulee,* video, LCA. *Prairie Slough,* video, LCA. *Africa's Stolen River,* laserdisc, Image Entertainment. *Creatures of the Namib Desert,* laserdisc, Image Entertainment. *Creatures of the Mangrove,* laserdisc, Image Entertainment.	Colinvaux, Paul. *Ecology.* NY: John Wiley and Sons, Inc., 1986. Prakash, Ishwar. ed. *Desert Ecology.* NY: State Mutual Book and Periodical Service, Ltd., 1988. Simon, Noel. *Vanishing Habitats.* NY: Franklin Watts, Inc., 1987. Thurman, Harold V. *Essentials of Oceanography.* 3rd ed. Columbus, OH: Merrill Publishing, 1990.

CHAPTER

27

BIOMES

THEME DEVELOPMENT: Evolution, ecology, and homeostasis are the themes of this chapter. The chapter focuses on adaptations of organisms for living in different biomes. Interrelationships between biotic and abiotic factors are shown throughout the chapter. A climax community shows homeostasis.

CHAPTER OVERVIEW

▶ **Section 27-1:** This section defines limiting factors and explains how they affect populations. Primary succession and secondary succession are differentiated. The final stage of succession, the climax community, is discussed.

▶ **Section 27-2:** Section 27-2 identifies the location of the six major land biomes and describes the abiotic and biotic factors of each one.

▶ **Section 27-3:** Characteristics of marine and freshwater ecosystems are described. Limiting factors of marine and freshwater ecosystems are considered. Zones in the ocean and the two basic types of freshwater ecosystems are discussed.

▶ **Section 27-4: Science and Society:** Wetlands are defined and their functions are explored. The You Decide feature asks students if developers should be allowed to drain wetlands.

CHAPTER VOCABULARY

limiting factor	tundra
succession	permafrost
primary succession	littoral zone
secondary succession	sublittoral zone
climax community	pelagic zone
climate	estuary
biomes	wetland
	water table

CHAPTER

27 Biomes

OPTIONS

For Your Gifted Students

▶ Students can write a story with a setting in a particular biome discussed in the chapter. The setting should influence the characters and the plot of the story.

▶ Have students prepare a documentary on one of the biomes. They should write a script that gives great detail about the area in an interesting way. They can videotape the presentation using pictures or drawings to illustrate the important features of the biome.

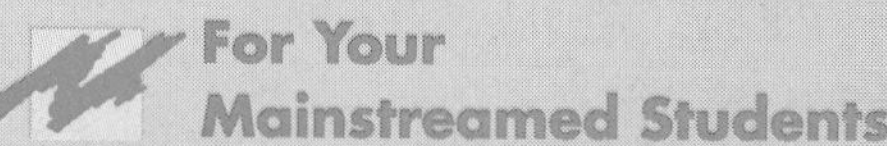

For Your Mainstreamed Students

▶ Students can plant a cactus dish garden, paying particular attention to how requirements for soil, water, and light differ from those of other plants.

▶ Have students choose one of the biomes discussed and make a mural or a mobile that shows the features of the area. They should label these features.

Do you live in an area that has pine forests, mountains, and lakes? Do you live near the ocean? You know that all life exists in the biosphere. But do you know where in the biosphere you live?

FIND OUT!

Do this simple activity to find out what part of the biosphere you live in.

Study a globe or a world map. Locate the country, state, and city where you live. What is the latitude of your state? Locate the equator. *Determine* where you live in relation to the equator. Do you have four seasons each year? What is the climate where you live?

Gearing Up

Previewing the Chapter

Use this outline to help you focus on important ideas in this chapter.

Section 27-1 Factors That Affect Biomes
- Limiting Factors
- Succession

Section 27-2 Land Biomes
- Land Biomes
- Tundra
- Northern Coniferous Forests
- Deciduous Forests
- Grasslands
- Deserts
- Tropical Rain Forests

Section 27-3 Water Ecosystems
- Marine Ecosystems
- Freshwater Ecosystems

Section 27-4 Science and Society
Coastal Wetlands
- Wetlands or Wastelands?

Previewing Science Skills

- In the **Skill Builders**, you will hypothesize and make a concept map.
- In the **Activities**, you will study maps, observe, use a table, experiment and graph.
- In the **MINI-Labs**, you will hypothesize and interpret data.

What's next?

Now that you have located where you live, learn how the climate, soil, and other abiotic factors of an area help determine what living things are found there and where they are located.

INTRODUCING THE CHAPTER

Use the Find Out activity to introduce students to biomes. Explain that they will be learning more about the biosphere in which they live. They will learn how climate, soil, and other abiotic factors determine where living organisms are found.

FIND OUT!

Materials: one globe or world map for every four students

Cooperative Learning: Use the Science Investigation strategy. Place students into groups of four. Have them work together to answer the questions.

Teaching Tips

- You may need to explain that latitude is the distance north or south of the equator and is measured in degrees.
- World maps can be used, but students can see where they live in relation to other parts of the world better with a globe.

Gearing Up

Have students study the Gearing Up feature to familiarize themselves with the chapter. Discuss the relationships of the topics in the outline.

What's Next?

Before beginning the first section, make sure students understand the connection between the Find Out activity and the topics to follow.

ASSESSMENT OPTIONS

PORTFOLIO

Refer to page 643 for suggested items that students might select for their portfolios.

PERFORMANCE ASSESSMENT

See page 643 for additional Performance Assessment options.

Process

Skillbuilders, pp. 631, 639
MINI-Lab, pp. 631, 633
Activities 27-1, p. 630; 27-2, p. 642
Using Lab Skills, p. 644

CONTENT ASSESSMENT

Assessment—Oral, pp. 631, 635, 637
Skillbuilder, pp. 631, 636
Section Reviews, pp. 631, 636, 639, 641
Chapter Review, pp. 643-645
Mini Quizzes, pp. 634, 638

GROUP ASSESSMENT

Opportunities for group assessment occur with Cooperative Learning Strategies and Flex Your Brain Activities.

SECTION 27-1

PREPARATION

SECTION BACKGROUND

▶The most important cause of succession is the altering of the physical environment by the community itself. Communities tend to alter the area in which they live in such a way as to make it less favorable for themselves and more favorable for other communities.

PREPLANNING

▶Make copies of the succession sequence for the Reteach activity.

1 MOTIVATE

▶Find a decaying log with fungi, slime molds, lichens, mosses, ferns, and plant seedlings on it. Display the log and ask students how it illustrates succession.

FLEX Your Brain

Use the Flex Your Brain activity to have students explore LIMITING FACTORS.

Flex Your Brain

ASSESSMENT

Portfolio: Use the Flex Your Brain to reinforce critical thinking and problem solving skills.

OBJECTIVES AND SCIENCE WORDS: Have students review the objectives and science words throughout the chapter to become familiar with each section.

VideoDisc

STVS: Saving the Spotted Owl, Disc 6, Side 1

27-1 Factors That Affect Biomes

New Science Words

limiting factor
succession
primary succession
secondary succession
climax community

Objectives

- ▶ **Identify and describe limiting factors.**
- ▶ **Define succession and differentiate between primary and secondary succession.**
- ▶ **Describe climax communities.**

Limiting Factors

Have you ever seen a lobster like the one in Figure 27-1a? Lobsters are found only in marine environments. The animal pictured below looks like a lobster, but it is a freshwater crayfish. What would happen to the lobster if you placed it in a freshwater stream? What would happen to the crayfish if you placed it in the ocean?

Most organisms live in a place that is, for them, the best place to live. Each species has adapted to a set of biotic and abiotic factors in its environment. Lobsters have adapted to a marine environment and cannot survive in fresh water. For lobsters, fresh water is a limiting factor. A **limiting factor** is a condition that determines the survival of an organism, population, or species in its environment. Any factor that limits the number, distribution, reproduction, or existence of organisms is a limiting factor. What are other limiting factors for the lobster?

In an oak-hickory forest, the mouse population is limited by the amount of thistle seed available for food, and the availability of mates and nesting sites. Competition among the mice for seeds, mates, or nesting sites, as well as predation by owls and hawks, are biotic factors that may limit the size of the mouse population. Abiotic factors that may become limiting include temperature, amount of sunlight, water, wind, elevation, or currents.

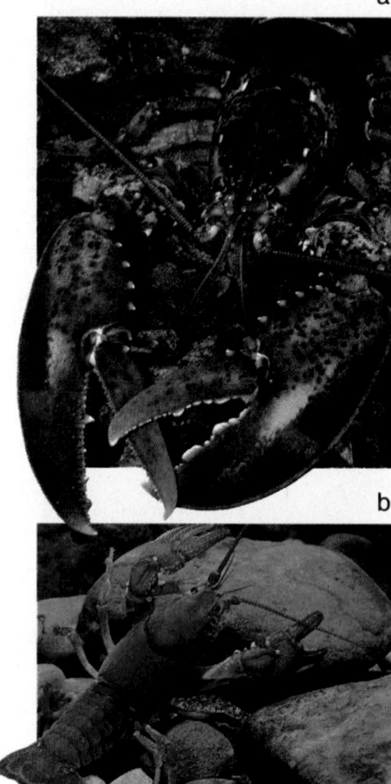

Figure 27-1. A lobster (a) is adapted to a marine environment, whereas crayfish (b) live in freshwater ecosystems.

Succession

Look at the field with scattered pine tree seedlings in Figure 27-2. In a few years, these seedlings will become

OPTIONS

Meeting Different Ability Levels

For Section 27-1, use the following **Teacher Resource Masters** depending upon individual students' needs.

◆ **Study Guide Master** for all students.
● **Reinforcement Master** for students of average and above average ability levels.
▲ **Enrichment Master** for above average students.

Additional Teacher Resource Package masters are listed in the OPTIONS box throughout the section. The additional masters are appropriate for all students.

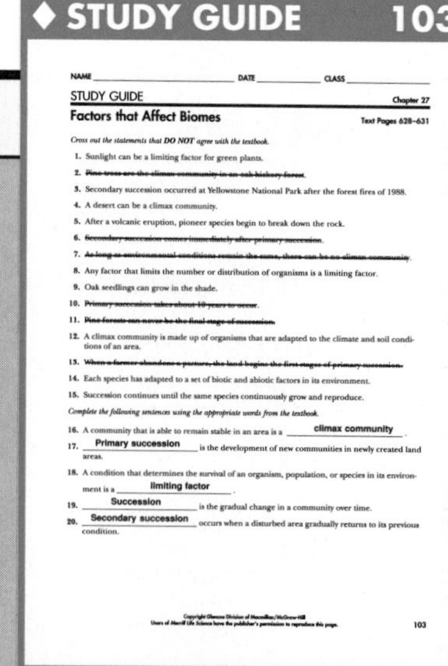

◆ STUDY GUIDE 103

NAME ____________ DATE ________ CLASS ________

STUDY GUIDE — Chapter 27

Factors that Affect Biomes — Text Pages 628–631

Cross out the statements that ***DO NOT*** *agree with the textbook.*

1. Sunlight can be a limiting factor for green plants.
2. ~~Pine trees are the climax community in an oak-hickory forest.~~
3. Secondary succession occurred at Yellowstone National Park after the forest fires of 1988.
4. A desert can be a climax community.
5. After a volcanic eruption, pioneer species begin to break down the rock.
6. ~~Secondary succession comes immediately after primary succession.~~
7. ~~As long as environmental conditions remain the same, there can be no climax community.~~
8. Any factor that limits the number or distribution of organisms is a limiting factor.
9. Oak seedlings can grow in the shade.
10. ~~Primary succession takes about 10 years to occur.~~
11. ~~Pine forests can never be the final stage of succession.~~
12. A climax community is made up of organisms that are adapted to the climate and soil conditions of an area.
13. ~~When a farmer abandons a pasture, the land begins the first stages of primary succession.~~
14. Each species has adapted to a set of biotic and abiotic factors in its environment.
15. Succession continues until the same species continuously grow and reproduce.

Complete the following sentences using the appropriate words from the textbook.

16. A community that is able to remain stable in an area is a __climax community__.
17. __Primary succession__ is the development of new communities in newly created land areas.
18. A condition that determines the survival of an organism, population, or species in its environment is a __limiting factor__.
19. __Succession__ is the gradual change in a community over time.
20. __Secondary succession__ occurs when a disturbed area gradually returns to its previous condition.

103

trees, and eventually a pine forest will stand where the field is now. Birds, mice, squirrels, chipmunks, and many other animals and plants will grow in the forest. Seedlings of tree species such as oaks, hickories, and elms may begin to grow in the shade underneath the pines. As the pine trees mature and die, the forest may become an oak-hickory forest. This gradual change in a community over time is called **succession.** Succession occurs in a community in an orderly process. Ecologists have identified two types of succession: primary succession and secondary succession.

Primary Succession

On a volcanic island, lava flows out of the volcano, cools, and forms new land. Many islands were formed this way. Over time, dust and ash settle on the lava rocks. Soil particles, bacteria, pollen, spores, and plant seeds may be carried to the island by wind, birds, or water. Pioneer species such as mosses begin to grow and break down the rock. Dead plants add organic materials, and soil begins to form. Eventually plant communities develop, and animals that eat these plants move in. The volcanic island is undergoing primary succession. **Primary succession** is the development of new communities in newly created land areas. Volcanic islands, exposed coral reefs, and human-made ponds or reservoirs undergo primary succession. Primary succession may occur over hundreds or thousands of years.

Secondary Succession

The succession of a community is also disturbed by fire, floods, earthquakes, and human activities such as logging or mining. If the disturbed area is then abandoned or left alone, succession may begin again. A pasture abandoned by a farmer may be invaded by grasses, shrubs, and trees in a process called secondary succession. In **secondary succession,** a disturbed area gradually returns to its previous condition. After fires destroyed parts of Yellowstone National Park in 1988, new plants began to grow through the ashes on the forest floor. Ecologists are studying this case of secondary succession very carefully to see what organisms return to the disturbed areas, and in what order.

Figure 27-2. Succession in an open field begins with scattered pine seedlings. Years later the field is a pine forest, and eventually it becomes an oak-hickory forest.

The eruption of Mount St. Helens killed practically all the local plant and animal life. Find out what plants and animals have returned to the area.

● REINFORCEMENT 103

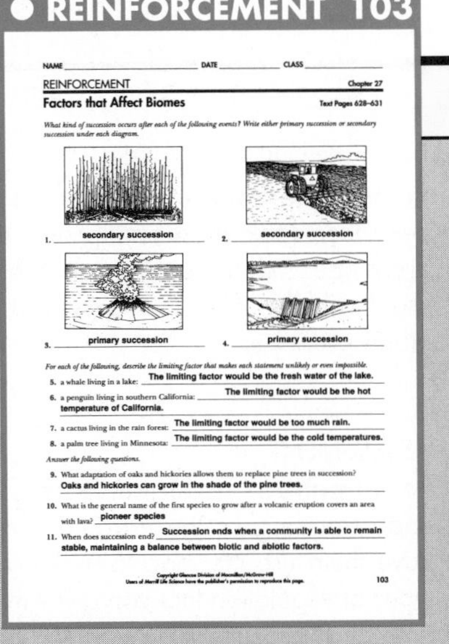

NAME ______ DATE ______ CLASS ______

REINFORCEMENT — Chapter 27

Factors that Affect Biomes — Text Pages 628–631

What kind of succession occurs after each of the following events? Write either primary succession or secondary succession under each diagram.

1. secondary succession
2. secondary succession
3. primary succession
4. primary succession

For each of the following, describe the limiting factor that makes each statement unlikely or even impossible.

5. a whale living in a lake: **The limiting factor would be the fresh water of the lake.**
6. a penguin living in southern California: **The limiting factor would be the hot temperature of California.**
7. a cactus living in the rain forest: **The limiting factor would be too much rain.**
8. a palm tree living in Minnesota: **The limiting factor would be the cold temperatures.**

Answer the following questions.

9. What adaptation of oaks and hickories allows them to replace pine trees in succession? **Oaks and hickories can grow in the shade of the pine trees.**
10. What is the general name of the first species to grow after a volcanic eruption covers an area with lava? **pioneer species**
11. When does succession end? **Succession ends when a community is able to remain stable, maintaining a balance between biotic and abiotic factors.**

103

▲ ENRICHMENT 103

NAME ______ DATE ______ CLASS ______

ENRICHMENT — Chapter 27

Factors that Affect Biomes — Text Pages 628–631

SUCCESSION

The four pictures below show succession in a small area in New York. You can think of these pictures as four photographs, with many years in between. Notice that in each picture, the plants have changed and so have the animals. Use the pictures to answer the questions below.

Stage 1 — field sparrow, sparrow, meadowlark, grasshopper sparrow, rabbit, meadow mouse, short-tailed shrew — Grasses and low shrubs

Stage 2 — robin, red fox, white-tailed deer, red squirrel, rabbit, ruffled grouse, white footed mouse, short-tailed shrew — High shrubs and shrub trees

Stage 3 — junco, robin, red fox, white-tailed deer, ruffled grouse, red squirrel, short-tailed shrew, white-footed mouse — Low trees

Stage 4 — veery, red fox, white-footed mouse, short-tailed shrew — High trees

1. In which stage is grass mostly found? **Stage 1**
2. The grasshopper sparrow is so named because of what it eats. Why do you think it is only found in Stage 1? **Grasshoppers are only found in Stage 1 because they live in the grass. The grasshopper sparrow would have food to survive only in Stage 1.**
3. By Stage 4, what type of vegetation has taken over? **high trees**
4. How did the trees change the environment, making it less favorable for the grasses to grow? **The trees took up space on the ground and also began to block out the sunlight.**
5. Which animal survived in all stages of this succession? Why? **the short-tailed shrew; Answers will vary, but should include because in all four stages there is food, shelter, and an appropriate environment for the short-tailed shrew.**

103

2 TEACH

Key Concepts are highlighted.

CONCEPT DEVELOPMENT

▶Take students on a field trip to see the stages of succession. Stages of succession can be seen in ponds, woods, road cuts, power line clearings, and outcrops. Allow students to identify the different stages of succession.

REVEALING MISCONCEPTIONS

▶Students may think that communities change because of time. Make sure they understand that communities change over time because organisms change the physical environment, making it suitable for other organisms.

CROSS CURRICULUM

▶**Art:** Have students make sequence drawings that show succession.

Answer: Pioneer plants and some pine have taken hold. Some deer began grazing within six months of the eruption.

CHECK FOR UNDERSTANDING

Have students identify how each of the following is undergoing succession: a corn field after a flood, a volcanic eruption, a newly formed sand bar in a river.

RETEACH

Arrange the stages of succession in a pond community or a forest community out of sequence in a diagram. Photocopy the diagram and have students use numbers to reorder the pictures in the proper sequence.

EXTENSION

For students who have mastered this section, use the **Reinforcement** and **Enrichment** masters or other OPTIONS provided.

ACTIVITY 27-1
25 minutes

OBJECTIVE: Use a table to determine what plants can grow in the hardiness zones shown on the map.

PROCESS SKILLS applied in this activity:
▶**Observing** in Procedure Step 1.
▶**Using a Map** in Procedure Step 3.
▶**Using a Table** in Procedure Steps 2 and 3.
▶**Inferring** in Conclude and Apply Questions 6-9.

COOPERATIVE LEARNING
Use the Paired Partners strategy.

TEACHING THE ACTIVITY

Troubleshooting: Students may have trouble locating zones in the beginning. You can enlarge the map, make a transparency, and use it on a microprojector to help get them started.

▶Display samples of seed packets. Many seed companies provide hardiness zone maps on the back of their seed packets.

PROGRAM RESOURCES

From the **Teacher Resource Package** use:

Activity Worksheets, pages 239-240, Activity 27-1, Interpreting a Map.

ACTIVITY 27-1 Interpreting a Map

Problem: *How do you use a hardiness zone map?*

Materials
- pencil and paper

Procedure
1. Study the hardiness zone map below. Hardiness zone maps show places of similar climates and vegetation type.
2. *Examine* the table under Data and Observations. This gives the northern limit to which some plants can live.
3. *Use the map and the table* to answer the questions that follow.

Data and Observations

Plant	Northern Zone Limit
White pine	2
Forsythia	2
Azalea	3
Rhododendron	2
Pyracantha	4
Chinese holly	3
Passion flower	5 and 6

Note: Zone 1 is coldest; zone 6 is warmest.

Analyze
1. In which zone do you live?
2. Which zone crosses the most states?
3. Which plants grow the farthest north?
4. Which plants cannot grow in Canada?
5. What are the states in zone 6?

Conclude and Apply
6. Could people living in Ontario use azaleas in their landscapes?
7. Could Chinese holly grow in Oklahoma?
8. How would a gardener use the map?
9. How do you use a hardiness zone map?

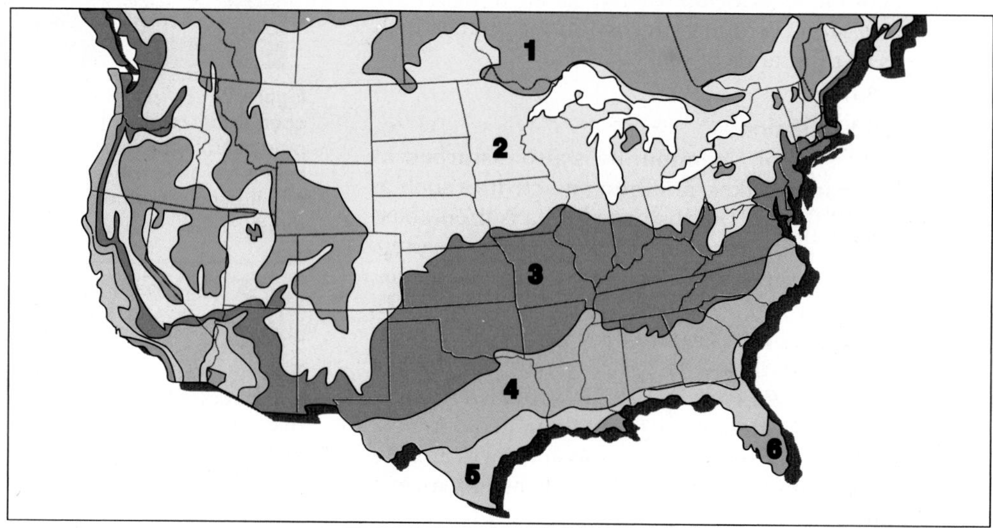

ANSWERS TO QUESTIONS

1. Answers will vary.
2. zone 2
3. white pine and *Forsythia*
4. *Pyracantha,* passion flower
5. parts of Florida, California, Louisiana, and Arizona
6. no
7. yes
8. to help order plants that will grow well in his or her area
9. Find the zone in which a plant grows best, then look on the map and find which areas are within that zone. A table may be used to show the zones in which particular plants grow best.

Activity ASSESSMENT

Performance: To further assess students' understanding of the hardiness zone map, have them use the map to make a list of the types of vegetation that can grow in their local area and name the zone in which they live.

Climax Community

What do a pine forest, oak-hickory forest, and desert have in common? Any of these ecosystems may be the final stage of succession in a particular area on Earth. Succession in any one place continues until it reaches a stage where the same species continue to grow and reproduce over and over. A community that is able to maintain a balance between abiotic and biotic factors and remain stable is a **climax community.** For any area, the type of climax community that develops is determined by the limiting factors of that environment. Where rainfall is limited and daytime temperatures are very hot, a desert may be the climax community. A climax community is made up of organisms that are adapted to the climate and soil conditions. You can see why climax communities are not the same everywhere.

During each stage of succession, the members of the community change the environment, making it more or less favorable for the present populations. A sunny field favors the growth of pine seedlings, but a pine forest does not. A pine forest favors oak and hickory seedlings. Oak and hickory forests, however, favor oak and hickory seedlings. Oak and hickory forests often are climax communities. A climax community will survive as long as environmental conditions remain the same.

MINI-Lab

How do communities change?
Use the data below to explain how tree species changed over 50 years in an area that started as an abandoned field and became a forest.

	Age of Area (years)		
Tree Species	5	20	50
Black Cherry		✓	✓
Butternut Hickory			✓
Persimmon	✓	✓	
Sassafras	✓	✓	
Shagbark Hickory			✓
Red Oak			✓
White Oak			✓
Yellow Chestnut			✓
White Ash		✓	✓
Winged Elm		✓	✓

Explain why some trees are no longer present after 50 years and why some trees did not appear until 20 or 50 years had gone by. Be sure to give the names of the trees in your answer.

SECTION REVIEW

1. Describe limiting factors and give three examples.
2. What is meant by the term *climax community*?
3. **Apply:** A beaver builds a dam across a stream and creates a pond. Would succession in this pond be primary or secondary succession? Why?
4. **Connect to Chemistry:** Oxygen is an abiotic factor that may become limiting. In what forms does oxygen occur in aquatic environments and on land?

Hypothesizing

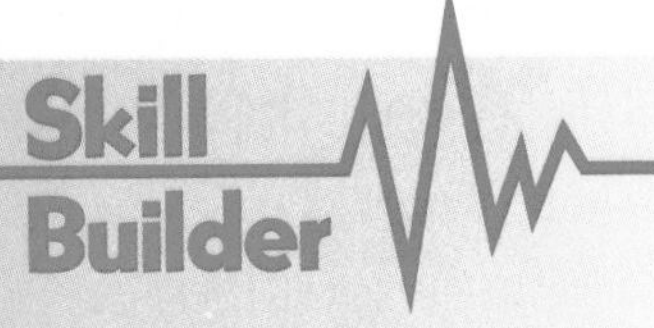

Make a hypothesis as to what would happen to succession in a pond if the pond owner removed all the cattails and reeds from around the pond edges every summer. If you need help, refer to Hypothesizing in the **Skill Handbook** on page 686.

MINI-Lab
Materials: data table
Answer: Persimmon and sassafras are able to survive in an open area and die out when shaded by other trees. Black cherry, white ash, and winged elm need enriched soil and can tolerate shade. Butternut hickory; shagbark hickory; and red, white, and chestnut oak cannot survive until other trees have become established.

MINI-Lab
ASSESSMENT
Performance: To further assess students' understanding of succession, see USING LAB SKILLS, question 12, p. 644.

3 CLOSE

▶Ask questions 1-2 and the **Apply** Question in the Section Review.

SECTION REVIEW ANSWERS

1. Student answers should reflect information on page 628.
2. a community that remains stable
3. Apply: Succession in a newly created pond would be primary succession because the pond did not exist prior to the building of a dam.
4. Connect to Chemistry: It occurs as oxygen molecules in the atmosphere. In aquatic environments, oxygen is dissolved in water.

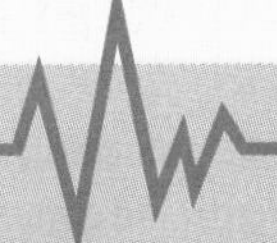

Students should hypothesize that removing cattails and weeds will prevent normal succession from taking place.

Skillbuilder
ASSESSMENT
Performance: Use this Skillbuilder to assess students' abilities to Hypothesize the effects of adding a new organism to a pond community.

OPTIONS

ASSESSMENT—ORAL

▶**Explain what would happen if gardeners did not remove weeds from their gardens.** *The garden would soon have a community of new plants leading to succession. Weeding a garden disrupts succession.*

▶**What happens to a community after a natural disaster such as a fire?** *The land goes through the process of succession.*

▶**What happens to the animals in an area when the plants change through succession?** As new plants arrive, animals adapted to the new conditions are able to move in.

PROGRAM RESOURCES

From the **Teacher Resource Package** use:
Activity Worksheets, page 245, Mini-Lab: How do communities change?
Use **Laboratory Manual,** pages 177-178, Limiting Factors.

SECTION 27-2

PREPARATION

SECTION BACKGROUND

▶Ecologists differ as to the number of biomes that should be recognized because biomes gradually merge into one another.

PREPLANNING

▶Obtain slides or pictures of different biomes for the Motivate activity.

1 MOTIVATE

▶**Discussion:** Have students look at the biome map in Figure 27-3 and locate the biome in which they live. Have them make a list of smaller ecosystems within this biome.

▶Obtain pictures or slides of the different biomes and the plants and animals that are found in each. Pictures can be used to make a bulletin board.

TYING TO PREVIOUS KNOWLEDGE: Review the concepts of adaptation and the conditions necessary for life from Chapter 1.

2 TEACH

Key Concepts are highlighted.

CONCEPT DEVELOPMENT

Cooperative Learning: Group students into teams of six. Assign each team one of the six major biomes. Have them use the Expert Teams strategy to study and master the biomes.

▶Explain the difference between *weather* and *climate*. *Climate* refers to averages of weather conditions.

27-2 Land Biomes

New Science Words

climate
biomes
tundra
permafrost

Objectives

- ▶ Define biomes.
- ▶ Identify the location of the major land biomes.
- ▶ Describe the climate, dominant plant types, and characteristic animals of the major land biomes.

Land Biomes

What are biomes?

Imagine that you could get on a train at the equator and travel north or south to the poles. What changes in the environment could you see from the train's windows? Palm trees and banana trees are found at the equator, whereas magnolias, laurels, and pecan trees are found in temperate climates. Forests of pine, spruce, and firs would be seen in colder regions near the poles. At the poles, trees would disappear altogether! These changes in forest communities as you travel away from the equator are the result of gradual changes in climate. **Climate** is the average condition of the weather in an area, as determined by rainfall, temperature, and amount of sunlight—all abiotic factors. These abiotic factors as well as other limiting factors help determine what species can survive in a particular environment. These factors also help determine the structure of climax communities all over the world. Large geographic areas that have similar climates and climax communities are called **biomes** (BI ohmz). The 1
major biomes on Earth are shown in Figure 27-3. Each major biome shares similar climate conditions, but the specific organisms found in each community will be different. Deserts in the United States and Africa look very similar, but the organisms are not the same.

Figure 27-3. The locations of the six major land biomes on Earth can be identified by using the key below.

Tundra
Northern coniferous forest
Deciduous forest
Grassland
Tropical rain forest
Desert

OPTIONS

Meeting Different Ability Levels

For Section 27-2, use the following **Teacher Resource Masters** depending upon individual students' needs.

◆ **Study Guide Master** for all students.

● **Reinforcement Master** for students of average and above average ability levels.

▲ **Enrichment Master** for above average students.

Additional Teacher Resource Package masters are listed in the OPTIONS box throughout the section. The additional masters are appropriate for all students.

◆ STUDY GUIDE 104

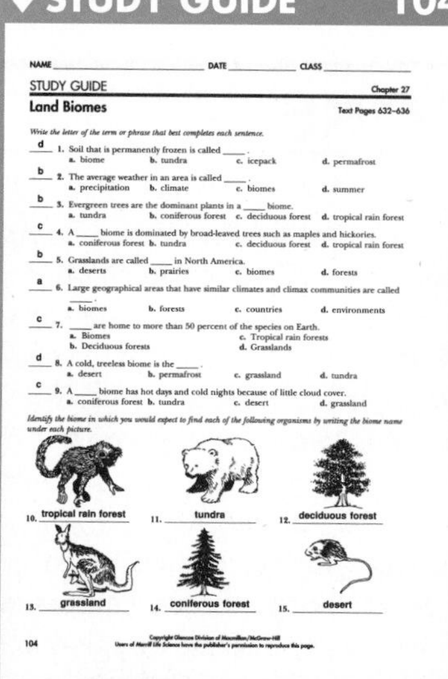
NAME ______ DATE ______ CLASS ______

STUDY GUIDE Chapter 27

Land Biomes Text Pages 632-636

Write the letter of the term or phrase that best completes each sentence.

d 1. Soil that is permanently frozen is called ____.
a. biome b. tundra c. icepack d. permafrost

b 2. The average weather in an area is called ____.
a. precipitation b. climate c. biomes d. summer

b 3. Evergreen trees are the dominant plants in a ____ biome.
a. tundra b. coniferous forest c. deciduous forest d. tropical rain forest

c 4. A ____ biome is dominated by broad-leaved trees such as maples and hickories.
a. coniferous forest b. tundra c. deciduous forest d. tropical rain forest

b 5. Grasslands are called ____ in North America.
a. deserts b. prairies c. biomes d. forests

a 6. Large geographical areas that have similar climates and climax communities are called ____.
a. biomes b. forests c. countries d. environments

c 7. ____ are home to more than 50 percent of the species on Earth.
a. Biomes b. Deciduous forests c. Tropical rain forests d. Grasslands

d 8. A cold, treeless biome is the ____.
a. desert b. permafrost c. grassland d. tundra

c 9. A ____ biome has hot days and cold nights because of little cloud cover.
a. coniferous forest b. tundra c. desert d. grassland

Identify the biome in which you would expect to find each of the following organisms by writing the biome name under each picture.

10. tropical rain forest 11. tundra 12. deciduous forest
13. grassland 14. coniferous forest 15. desert

104

Tundra

Can you picture a region that receives little precipitation yet is covered with snow and ice for nine months each year? The **tundra** (TUN drah) is a cold, dry, treeless biome where the sun barely rises during the six to nine winter months. It is located in the far northern parts of North America, Europe, and Asia and is the most continuous of Earth's biomes. Nearly 20 percent of Earth's land area is covered by tundra. ②

Low temperatures and a short growing season are the major limiting factors in the tundra. Precipitation in the tundra averages less than 25 cm yearly. Temperatures drop to minus 40°C and keep the ground frozen for most of the year. In the short summer months, the snow melts and the upper layer of soil thaws. The soil underneath that remains permanently frozen is called **permafrost.** Permafrost prevents soil from absorbing water.

What kind of organisms live in such a harsh environment? Tundra plants include lichens, mosses, and flowering plants such as cranberries. Insects such as mosquitoes and blackflies grow in vast numbers. Ducks, geese, and songbirds feed on the abundant insects. Tundra mammals include caribou, reindeer, musk oxen, and arctic foxes. Many animals come for the summer, then migrate to warmer regions in the fall. Other animals remain throughout the year. Arctic hares, polar bears, and snowy owls have white fur and feathers that provide camouflage in the snow. Why is white fur an advantage for a predator like the polar bear in the tundra?

Northern Coniferous Forests

The northern *coniferous forest biome* has as its dominant plants cone-bearing evergreen trees such as pines, firs, spruces, and cedars. These forests cover large parts of northern Europe, Asia, and North America. Winters are long and cold with heavy snows. Even though summers are short, they are longer and warmer than summers in the tundra. In these forests, the ground thaws in the summer, making it possible for trees to grow. Rainfall in this biome ranges from 50 to 125 cm yearly. ②

Because coniferous forests are dense, little light penetrates to the soil, so there are few shrubs or grasses in this biome. These forests are very productive and play an

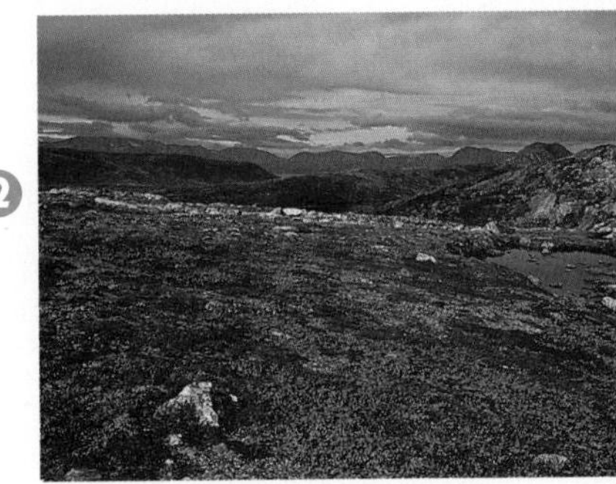

Figure 27-4. Land is so flat in the tundra that water does not drain away, and the permafrost prevents water absorption. This results in marshy conditions in the summer.

MINI-Lab

Is there tundra in Africa?

Using the land biome map on page 632, make a **hypothesis** as to whether tundra exists in Africa. Look at a globe and *compare* the latitude where tundra is found in the northern hemisphere with the same latitude in the southern hemisphere. How can you tell if your **hypothesis** is correct?

Figure 27-5. It is very dark in a coniferous forest, with deep layers of needles on the forest floor.

CONCEPT DEVELOPMENT

▶ Use this chapter as an opportunity to teach geography.

MINI-Lab

Materials: a world biome map and a globe

Teaching Tips

▶ Make sure students trace latitudes 60° and 75° on both northern and southern halves of the globe to see what land areas fall between those latitudes.

▶ Point out what land areas correspond to 30°S in the northern hemisphere.

Answer to Question: Students' hypotheses should state that there is no tundra in Africa. They should see from the globe that tundra exists only near the Arctic Circle, between latitudes 60°N and 75°N. Africa ends at approximately 30°S latitude. Only Antarctica exists at a latitude of 60° to 75°S. Antarctica is frozen year-round and thus is not tundra.

MINI-Lab ASSESSMENT

Performance: To further assess students' understanding of biomes, have them use the biome map to infer why there is no coniferous forest or tundra in the southern hemisphere.

VideoDisc

STVS: Research in the Pinelands, Disc 6, Side 1

● REINFORCEMENT 104

NAME ______ DATE ______ CLASS ______

REINFORCEMENT — Chapter 27

Land Biomes — Text Pages 632–636

Complete the table below using information in your textbook.

Biome	Climate	Dominant plants	Characteristic animals
tundra	cold; dry	lichens, moss	arctic hares, polar bears, reindeer, caribou, snowy owls
coniferous forest	cold winters; short, warm summers	pines, firs, spruces, cedars	moose, bears, wolves, lynx
deciduous forest	cold winters; hot summers; mild springs and falls	maples, oaks, hickories	deer, foxes, raccoons, squirrels, black bears
grassland	varies from hot year-round to cold winters and hot summers; little rainfall	grasses and grains	bison, antelope, zebras, giraffes, kangaroos
desert	hot days and cold nights; very dry	cacti	kangaroo rats, mice, gerbils
tropical rain forest	hot; wet	very diverse	insects, monkeys, birds, large cats

104

▲ ENRICHMENT 104

NAME ______ DATE ______ CLASS ______

ENRICHMENT — Chapter 27

Land Biomes — Text Pages 632–636

IDENTIFYING BIOMES

Use an atlas to place each of these 12 cities on the world map below. Then, compare the map you have made with the map in your textbook of the major biomes. In which biome does each city belong? Look at the information in the textbook about precipitation in each of the land biomes. Complete the table by identifying the biome each city is in.

City	Yearly Precipitation (cm)	Biome
Addis Ababa, Ethiopia	123.7	deciduous forest
Anchorage, Alaska	38.6	coniferous forest
Beijing, China	63.5	deciduous forest
Buenos Aires, Argentina	95	grassland
Capetown, South Africa	50.8	deciduous forest
Damascus, Syria	21.9	desert
Lagos, Nigeria	183.6	tropical rain forest
Madrid, Spain	41.9	grassland
Manila, Philippines	208.3	tropical rain forest
Paris, France	56.6	deciduous forest
Phoenix, Arizona	18.0	desert
Sydney, Australia	117.6	deciduous forest

Anchorage
Phoenix
Madrid
Paris
Beijing
Damascus
Manila
Lagos
Addis Ababa
Sydney
Buenos Aires
Capetown

104

CONCEPT DEVELOPMENT

▶ Point out that loss of leaves during the winter or dry season helps deciduous trees conserve water because the leaves on the ground cut down on loss of water from soil surface.

▶ Take students on a field trip to a deciduous forest. Have them observe the layered vegetation. Herbaceous plants grow on the forest floor. Shrubs and small trees grow beneath the canopy and form the understory layer. Some trees grow taller than the level of the canopy and form the emergent layer.

▶ Explain that grasslands characterized by scattered trees are called savannas.

▶ Different grasses grow well in different regions of the United States. Have students visit a local nursery or lawn care center to find out the variety of grass most suited for your area.

TEACHER F.Y.I.

▶ At one time, the grassland biome covered nearly half of the land on Earth. Today, most of the grassland is used by humans for growing crops, raising animals, and housing.

MINI QUIZ

Use the Mini Quiz to check students' recall of chapter content.

1. **_____ are large geographic areas with similar climates and climax communities.** *Biomes*
2. **What are the six major land biomes on Earth?** *tundra, northern coniferous forests, deciduous forests, grasslands, deserts, tropical rain forests*

Answer: The ice crystals take up extra room, piercing cell membranes, and destroying the cells.

Freezing temperatures can be harmful to living things because cells are made up largely of water. As water freezes, it expands. Explain what happens to cells when freezing temperatures cause ice crystals to form in them.

Figure 27-6. Deciduous forests like this once covered all of Europe and the eastern United States.

Figure 27-7. Grasslands may be very hot in the summer and very cold in the winter. Grassland species are adapted to these variations in temperatures.

important role in the timber industry. Much of the lumber used for construction comes from coniferous forests.

Coniferous forest mammals, such as moose, bears, wolves, and lynx, depend on stored body fat during the cold months. Some animals migrate to warmer climates in the fall and return in the spring. Other animals hibernate six to eight months of the year.

Deciduous Forests

Do you live in an area where the autumn leaves change color? A *deciduous forest biome* is dominated by trees that 2
have broad thin leaves, such as beeches, maples, oaks, hickories, sycamores, and elms. This biome has four seasons with about 75 to 150 cm of rainfall distributed evenly throughout the year. Winters are cold, summers are long and warm, springs and falls are mild. The climate and rich soil support large numbers of plant and animal species. Deciduous forests are found in the eastern United States, Europe, and parts of Asia, South America, Africa, and Australia.

White-tailed deer, foxes, raccoons, opossums, and squirrels are typical mammals that live in the North American deciduous forest. Black bears are found in some regions. Mice, snakes, and other animals are found on the forest floor. Some animals hibernate during cold winters. Others, including many bird species, migrate to warmer climates for the winter.

Grasslands

About one-fourth of the land surface on Earth is covered by grassland biomes. The *grassland biome* is 2
dominated by grasses. These biomes occur at about the same latitude as deciduous forests, but they do not receive enough precipitation to support trees. Rainfall in grasslands may be only 25 to 75 cm per year. Temperatures range from very hot in the summer to below freezing in the winter. Grasslands are called prairies in North America, pampas in South America, savannas or veldts in Africa, and steppes in Asia.

Many grasslands are subject to periodic fires. Grasses in these biomes have roots that can send out shoots to form new plants. Extensive root systems absorb water over large areas when it rains.

OPTIONS

ASSESSMENT—ORAL

▶ **Why do you think coniferous forests are sometimes called the "spruce-moose" biome?** *These are the dominant species in the coniferous forest.*

▶ **Why can people live in a number of biomes?** *They can create the conditions necessary to live in any biome.*

▶ **Why does the tundra become soggy in the summer?** *The permafrost does not allow the melted snow to soak into the ground.*

▶ **Why do birds migrate to the tundra in summer?** *They feed on the vast number of insects.*

▶ **How are conifers better adapted for a shorter growing season than broad-leaved deciduous trees?** *Deciduous trees cannot begin photosynthesis until new leaves are opened.*

ENRICHMENT

▶ Have students research and report on the Dust Bowl.

Some grasslands support large numbers of grazing animals. The grazing mammals in North America include bison and antelope. Zebras, wildebeasts, and giraffes are found in Africa, and kangaroos graze in Australia.

Grasslands are very important to humans. Most of the grains used for human food, such as wheat and rye, are grown in grassland areas.

Deserts

What is a desert biome?

A *desert biome* receives less than 25 cm of rainfall a year and has a high rate of evaporation. Most deserts have little cloud cover to hold the heat, so it is quickly lost once the sun goes down. The temperature may drop as much as 40°C in 12 hours. This results in hot days and cold nights. In some deserts rain may not fall for more than a year, and then a huge thunderstorm may dump more than 12 cm all at once. 2

Populations of organisms that live in deserts are adapted to the climate. Whenever sufficient rain falls, annual plants grow, flower, and produce seeds in a short time. Then they remain inactive during the long dry periods. Some plants store water in their thick stems and leaves. The saguaro cacti of Arizona and Mexico store water and carry out photosynthesis in thickened stems. The leaves or stems of many plants have sharp protective spines to protect them from thirsty animals.

Desert animal populations are also adapted to their environment. Kangaroo rats, pocket mice, and gerbils get water from the seeds they eat. Some kangaroo rats store seeds in their burrows to soak up the moisture found there. Blood vessels in the large ears of jackrabbits and foxes circulate blood to cool the animals.

Figure 27-8. Many desert plants have extensive root systems to collect as much water as possible when it rains. A single saguaro cactus can store as much as 900 liters of water.

Tropical Rain Forests

When you think of the tropics, do you imagine a jungle? A *tropical rain forest biome* has hot, wet weather throughout the year and the greatest variety of organisms on Earth. Rainfall averages 200 to 225 cm per year. High temperatures and abundant rainfall contribute to high humidity. Tropical rain forests are found in South America, Africa, and Southeast Asia. 2

Tropical rain forests contain more than 50 percent of all species of organisms on Earth. The soil is often not

In Your JOURNAL

The International Children's rainforest in Monteverde, Costa Rica, protects more than 5000 hectares of tropical rain forest. Find out how this rain forest was protected by children from the United States, Sweden, Canada, England, and Japan. Write a report **in your Journal** to share with your class.

CONCEPT DEVELOPMENT

▶**Demonstration:** Obtain a cactus plant and carefully cut off the tip of one part. Ask students what adaptations it has for living in the desert.

▶Point out that Earth's deserts are increasing due to cutting of trees and overgrazing by domestic animals. As a result of land misuse, the desert biome has spread.

▶Explain that rain forests have much more rainfall, many more species, but much poorer soil than deciduous forests.

▶There are two types of rain forest biomes. The tropical rain forest is found near the equator. The temperate rain forest extends along the west coast of North America from central California to southern Alaska. It is characterized by moderate temperatures and high humidity. Plant life includes redwoods and Sitka spruce.

In Your JOURNAL

The land has been purchased and taken out of the cycle where it would be clear-cut.

CHECK FOR UNDERSTANDING

Make a table showing precipitation, temperature, common plants, and common animals for each biome. Have students write in the name of each biome that is described.

RETEACH

Show pictures of plants and animals from each biome. Ask students to give the name of the biome in which the animal or plant lives.

EXTENSION

For students who have mastered this section, use the **Reinforcement** and **Enrichment** masters or other OPTIONS provided.

ASSESSMENT—ORAL

▶**How do animals endure the heat in some deserts?** *They stay in their burrows during the day and come out at night.*

▶**Infer why most rain forest animals live in trees.** *More food is available in the trees than on the ground.*

▶**Why do few plants grow on the forest floor in the tropical rain forest?** *The broad leaves of the tall trees shade the forest floor, making it difficult for low-growing plants to receive enough sunlight to grow.*

▶**Would any place on the rain forest floor ever have thick vegetation?** *Where sunlight reaches the forest floor, such as a clearing, an area where a tree has fallen, or along a river bank would have thick vegetation.*

ENRICHMENT

▶Research and report on the adaptations of a desert animal or plant that enable it to live and reproduce in the desert.

TECHNOLOGY

Think Critically: Food production, waste removal, air and water quality, and competition among species.
Reference: Stover, Dawn. "Inside Biosphere II." *Popular Science,* Nov. 1990, pp. 54-59.

3 CLOSE

▶Ask questions 1-3 and the **Apply** Question in the Section Review.
▶Have students make a food web for a tropical rain forest.

SECTION REVIEW ANSWERS

1. Biomes are large geographic areas that have similar climates and climax communities. The six major biomes are tundra, coniferous forest, deciduous forest, grassland, desert, and tropical rain forest.
2. The dominant plants in grasslands are grasses. In coniferous forests, the dominant plants are cone-bearing evergreen trees. Grasslands do not have enough precipitation to support trees.
3. Small mammals get water from the seeds they eat. Blood vessels in the large ears circulate blood to cool the animals.
4. Apply: Permafrost does not allow water to reach the roots of trees.
5. Connect to Earth Science: the angle at which the sunlight strikes Earth

Skill Builder

The concept map should include land biomes at the top, divided into the six major land biomes. Each biome should then divide into plant and animal sections. Lists under plant and animal headings should include two or three for each biome.

Skillbuilder
ASSESSMENT
Portfolio: Use this Skillbuilder to assess students' abilities to form a Concept Map of plant and animal species found in each biome.

TECHNOLOGY

Life in a Glass World

Biosphere II is the name given to a unique greenhouse in Arizona that is the world's largest closed ecosystem. Eight volunteers live, work, study, and play inside Biosphere II and get all the food they need from the six biomes represented there. The biomes include an agricultural biome, a desert, a marsh, an ocean, a savanna, and a rain forest. The intensive agricultural biome has food crops, fish ponds, and domestic animals. Water in Biosphere II recycles in the rain forest, although some of the total 6.6 million liters flows into the savanna.

What can scientists learn from Biosphere II? It may help scientists better understand life on Earth, provide a model for future space colonies, and allow scientists to test some new technologies.

Think Critically: What problems do you need to solve in a closed ecosystem?

very fertile because most nutrients are taken up by plants as soon as they become available through decay.

Animal life is very diverse in the rain forest, with insects, birds, monkeys, and large cats. More than 50 percent of the mammals in rain forests live in the trees. Rain forest reptiles include chameleons, snakes, iguanas, and geckos. Many scientists are studying the remaining rain forests to learn what new organisms may be found there.

SECTION REVIEW

1. Define *biome* and list the six major biomes on Earth.
2. How do grasslands differ from coniferous forests?
3. What kinds of adaptations have desert animals made to survive without much water?
4. **Apply:** Why don't large trees grow in the tundra?
5. **Connect to Earth Science:** What determines the amount of solar energy received at a particular location on Earth?

Skill Builder — Concept Mapping

Make a concept map of plant and animal species found in each land biome. If you need help, refer to Concept Mapping in the **Skill Handbook** on pages 688 and 689.

OPTIONS

INQUIRY QUESTIONS

▶**What is the average weather in an area called?** *climate*
▶**What is the permanently frozen layer of soil in the tundra called?** *permafrost*
▶**Where does much of the lumber for construction come from?** *coniferous forest*
▶**What biome is subject to periodic fires?** *grassland*
▶**Where are more than 50 percent of all species of organisms on Earth found?** *tropical rain forest*

PROGRAM RESOURCES

From the **Teacher Resource Package** use:
Activity Worksheets, page 246, MiniLab: Is there tundra in Africa?
Concept Mapping, pages 59-60.
Critical Thinking/Problem Solving, page 31, Development in the Rockies.
Transparency Masters, pages 97-98, Biome Map.
Cross-Curricular Connections, page 31, Using a Globe.
Use **Color Transparency** 49, Biome Map.

Water Ecosystems

27-3

Objectives

- **Distinguish between marine and freshwater ecosystems.**
- **Describe the zones in the ocean.**
- **Identify the limiting factors present in marine and freshwater ecosystems.**

New Science Words

littoral zone
sublittoral zone
pelagic zone
estuary

Marine Ecosystems

If you had to name the largest biome on Earth, would you answer the ocean? The major land biomes are surrounded by the oceans. The *marine ecosystem* is a continuous body of water that covers more than 70 percent of Earth's surface.

The amount of salt present is the major limiting factor in water ecosystems. Ocean water may contain as much as 3.5 percent salt, whereas fresh water contains less than 0.005 percent salt. Other limiting factors in marine ecosystems are temperature, light, the amount of dissolved oxygen, and water pressure. Light and temperature decrease as the depth of the water increases, but water pressure increases with depth.

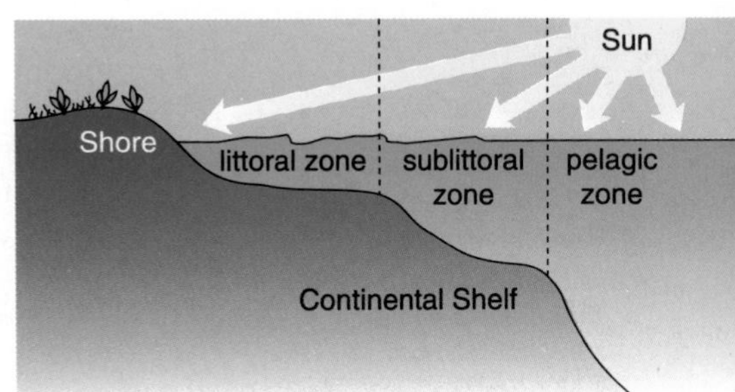

Figure 27-9. Most marine life is found in the littoral and sublittoral zones of the ocean where sunlight penetrates.

Figure 27-9 shows the major zones of the ocean. The zone along the shore is the **littoral** (LIHT uh rul) **zone.**
1 During high tide it is covered with water, and during low tide it is exposed to the air. Populations living in the littoral zone are adapted to the wet and dry conditions and to the force of the waves hitting the shore. Sea stars, sea anemones, and mussels cling to rocks. At low tide, mussels close their shells and crabs burrow into the moist sand to keep from drying out. The organisms that live there have to be able to withstand varying temperatures, water levels, and salt levels.

Figure 27-10. Organisms that live in the littoral zone are adapted to both wet and dry conditions.

The part of the ocean floor that slopes downward is the continental shelf. The shallow water above the continental shelf is called the **sublittoral zone.** This zone has enough light and nutrients to support a wide variety of

SECTION 27-3

PREPARATION

SECTION BACKGROUND

▶More than 95 percent of Earth's water is found in oceans. The remaining 5 percent is fresh water.

▶Review cyanobacteria, protists, and green algae from Chapters 8 and 9 and their importance in the food chains of marine and freshwater ecosystems.

1 MOTIVATE

▶**Bulletin Board:** Have a bulletin board with pictures of marine and freshwater ecosystems and of the organisms in each one ready to refer to as you begin this section.

2 TEACH

Key Concepts are highlighted.

CONCEPT DEVELOPMENT

▶Discuss environmental factors that affect life in the zone along the ocean's shore. Include such factors as drying out, differences between air and water temperatures, and wave action.

▶Point out that relatively few plant species exist in the ocean. Mangrove trees, eelgrass, and sea grapes grow in water along the coast.

ASSESSMENT—ORAL

▶**Why are there no green plants at the bottom of the ocean?** *Sunlight for photosynthesis does not reach the bottom of the ocean.*

▶**Why are the same kinds of one-celled organisms found in lakes and ponds around the world?** *Freshwater ecosystems have similar abiotic factors.*

▶**What would happen if all the estuaries along the coast were destroyed?** *Accept all reasonable responses, including significant changes in shellfish populations and the food webs in which they participate.*

ENRICHMENT

▶Have students prepare a report on occupations related to oceanography and present their report to the class.

▶Have students report on the commercial aspects of food production in estuaries.

FLEX Your Brain

Use the Flex Your Brain activity to have students explore WATER ECOSYSTEMS.

FLEX YOUR BRAIN
ASSESSMENT
Portfolio: Use the Flex Your Brain activity to reinforce critical thinking and problem solving skills.

CHECK FOR UNDERSTANDING
Use the Mini Quiz to check for understanding.

RETEACH
Write on the chalkboard a description of the three major ocean zones. Have students identify each one.

EXTENSION
For students who have mastered this section, use the **Reinforcement** and **Enrichment** masters or other OPTIONS.

Answer: Water provides buoyancy for organisms, which allows easy movement and little energy use.

MINI QUIZ

Use the Mini Quiz to check students' recall of chapter content.

1 **During low tide, the _____ zone is exposed to the air.** *littoral*
2 **A(n) _____ is formed where a river meets the ocean.** *estuary*
3 **Organisms in _____ are adapted to a narrow range of water temperature.** *fresh water ecosystems*

Figure 27-11. Most marine fish caught for food or sport were hatched in estuaries like this.

organisms. Most life in the ocean is found in the littoral and sublittoral zones. Microscopic organisms that float or swim near the surface of the water include protozoans, algae, and invertebrates. Shrimp, oysters, and fish all feed on these organisms and are eaten in turn by other fish, squid, some whales, walruses, and seals.

The open ocean just beyond the sublittoral zone is the **pelagic zone.** Sunlight filters down to a depth of about 200 meters. Below 200 meters this zone is dark and cold. Water pressure is extremely high. Some animals living here move up to feed, and others eat dead organisms that filter down from the water above.

2 The area where a freshwater river or stream meets the ocean is called an **estuary.** Salt marshes, deltas, and mud flats may all be estuaries. The water is shallow in estuaries, allowing light to penetrate to the bottom. Rivers bring nutrients washed from the land to algae, marsh grasses, and other plants growing along the shore. The continuous mixing of fresh water with ocean water keeps nutrients in the estuary and removes wastes from it. Oysters, clams, mussels, and snails are abundant. Estuaries serve as breeding grounds for many marine organisms such as fish, arthropods, and mollusks. Shore birds nest and raise their young in estuaries because of abundant food resources.

Find out why organisms in the oceans use less energy in moving around than you do.

Freshwater Ecosystems

What is a freshwater ecosystem?

Most of you probably are familiar with a freshwater ecosystem like the one in Figure 27-12. *Freshwater ecosystems* include lakes, ponds, streams, rivers, swamps, and bogs.

The types of organisms found in freshwater ecosystems depend largely on temperature, current, and the amounts of oxygen and minerals dissolved in the water.

Water temperature in freshwater ecosystems does not vary as much as the temperature of air. As a result, pop-
3 ulations that live in water are adapted to a narrow temperature range, and many cannot survive freezing or extreme heat. The primary limiting factor in freshwater ecosystems is the amount of oxygen dissolved in the water. A fast-moving stream has more oxygen in it than a stagnant pond. Cold water can hold more oxygen, yet there may be fewer organisms in cold water because temperature is also a limiting factor.

Figure 27-12. Fast-moving water like this contains far fewer organisms than slow-moving streams.

Lakes and ponds are classified according to the amount of organic matter. Those that are rich in organic matter

OPTIONS

Meeting Different Ability Levels

For Section 27-3, use the following **Teacher Resource Masters** depending upon individual students' needs.

◆ **Study Guide Master** for all students.

● **Reinforcement Master** for students of average and above average ability levels.

▲ **Enrichment Master** for above average students.

Additional Teacher Resource Package masters are listed in the OPTIONS box throughout the section. The additional masters are appropriate for all students.

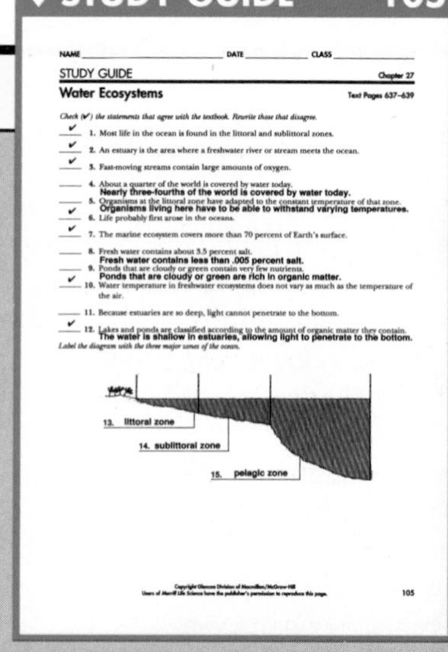

◆ STUDY GUIDE 105

NAME _____ DATE _____ CLASS _____

STUDY GUIDE — Chapter 27

Water Ecosystems — Text Pages 637–639

Check (✔) the statements that agree with the textbook. Rewrite those that disagree.

✔ 1. Most life in the ocean is found in the littoral and sublittoral zones.
✔ 2. An estuary is the area where a freshwater river or stream meets the ocean.
✔ 3. Fast-moving streams contain large amounts of oxygen.
___ 4. About a quarter of the world is covered by water today. **Nearly three-fourths of the world is covered by water today.**
___ 5. Organisms at the littoral zone have adapted to the constant temperature of that zone. **Organisms living here have to be able to withstand varying temperatures.**
✔ 6. Life probably first arose in the oceans.
✔ 7. The marine ecosystem covers more than 70 percent of Earth's surface.
___ 8. Fresh water contains about 3.5 percent salt. **Fresh water contains less than .005 percent salt.**
___ 9. Ponds that are cloudy or green contain very few nutrients. **Ponds that are cloudy or green are rich in organic matter.**
✔ 10. Water temperature in freshwater ecosystems does not vary as much as the temperature of the air.
___ 11. Because estuaries are so deep, light cannot penetrate to the bottom. **The water is shallow in estuaries, allowing light to penetrate to the bottom.**
✔ 12. Lakes and ponds are classified according to the amount of organic matter they contain.

Label the diagram with the three major zones of the ocean.

13. littoral zone
14. sublittoral zone
15. pelagic zone

Copyright Glencoe Division of Macmillan/McGraw-Hill
Users of Merrill Life Science have the publisher's permission to reproduce this page.

105

PROBLEM SOLVING

What Caused the Fish to Die?

Fish breathe by allowing water to flow over the many blood vessels in their gills. As the water passes over the blood vessels, oxygen from the water is absorbed and carbon dioxide is released.

Recently, several hundred thousand fish and other marine creatures were killed along the Gulf Coast because there was not enough oxygen in the water. Unfavorable weather conditions caused the problem. After several days of high heat and calm water, the oxygen content of the water decreased to a level that could not sustain wildlife. As the water warmed up, it lost its ability to hold oxygen. The calm water also contributed to the low oxygen levels. Turbulent water traps air bubbles, and gases from the air dissolve in the water. What caused the fish to die?

Think Critically: Why do aquariums use air pumps to bubble air through the water?

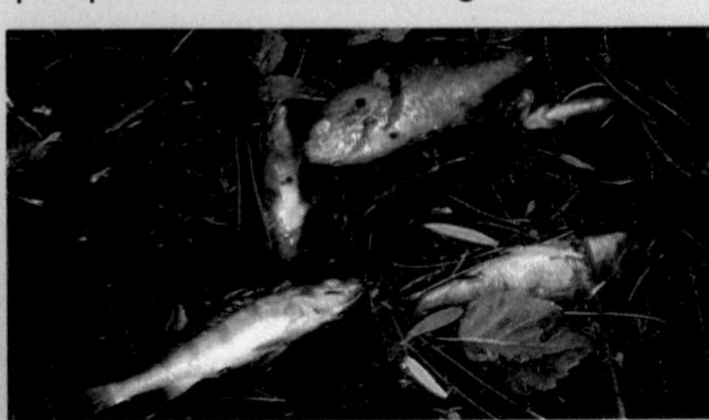

usually have cloudy or green waters. Those that contain little organic matter have water that is clearer, is blue in color, and has a sandy or rocky bottom. Which type of pond has more nutrients? Ponds with lots of organic matter have more nutrients and can support more species.

EcoTip

Boat engines may leak oil into estuary waters. Make sure boat engines are in good working order before you anchor near an estuary.

SECTION REVIEW

1. What is the main difference between marine and freshwater biomes?
2. What are the most important limiting factors in marine ecosystems? In freshwater ecosystems?
3. Name and describe the zones in the ocean.
4. **Apply:** Developers often fill in estuaries to build homes on the shoreline. Why should this concern you?
5. **Connect to Physics:** Find out why you feel lighter in the ocean than in fresh water.

Hypothesizing

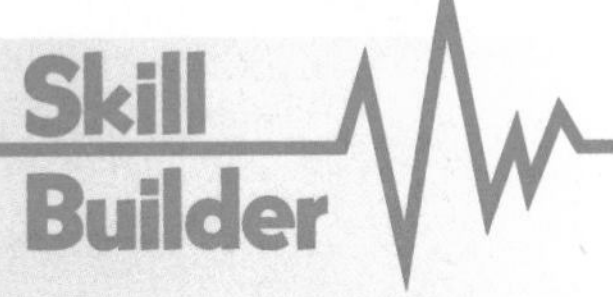

Hypothesize why some fish that live in the pelagic zone of the ocean have no eyes. If you need help, refer to Hypothesizing in the **Skill Handbook** on page 686.

● REINFORCEMENT 105

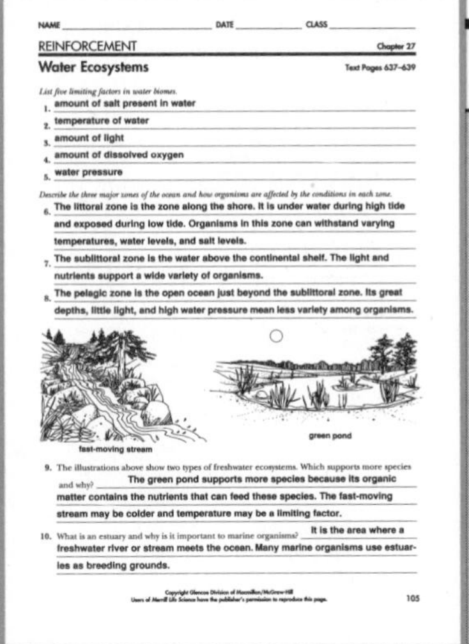

NAME ________ DATE ________ CLASS ________

REINFORCEMENT — Chapter 27

Water Ecosystems — Text Pages 637–639

List five limiting factors in water biomes.

1. amount of salt present in water
2. temperature of water
3. amount of light
4. amount of dissolved oxygen
5. water pressure

Describe the three major zones of the ocean and how organisms are affected by the conditions in each zone.

6. The littoral zone is the zone along the shore. It is under water during high tide and exposed during low tide. Organisms in this zone can withstand varying temperatures, water levels, and salt levels.
7. The sublittoral zone is the water above the continental shelf. The light and nutrients support a wide variety of organisms.
8. The pelagic zone is the open ocean just beyond the sublittoral zone. Its great depths, little light, and high water pressure mean less variety among organisms.

fast-moving stream — green pond

9. The illustrations above show two types of freshwater ecosystems. Which supports more species and why? The green pond supports more species because its organic matter contains the nutrients that can feed these species. The fast-moving stream may be colder and temperature may be a limiting factor.
10. What is an estuary and why is it important to marine organisms? It is the area where a freshwater river or stream meets the ocean. Many marine organisms use estuaries as breeding grounds.

 105

▲ ENRICHMENT 105

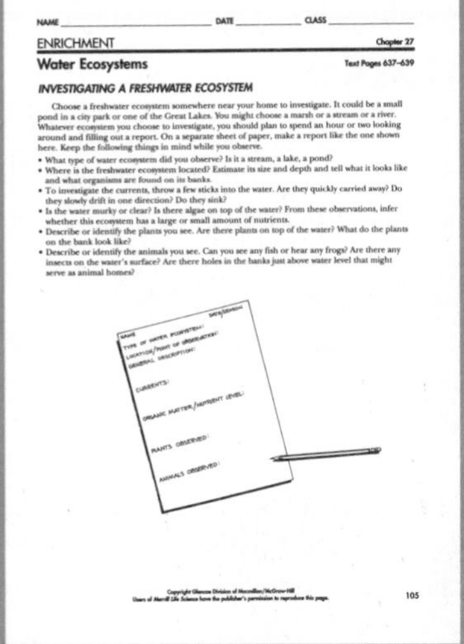

NAME ________ DATE ________ CLASS ________

ENRICHMENT — Chapter 27

Water Ecosystems — Text Pages 637–639

INVESTIGATING A FRESHWATER ECOSYSTEM

Choose a freshwater ecosystem somewhere near your home to investigate. It could be a small pond in a city park or one of the Great Lakes. You might choose a marsh or a stream or a river. Whatever ecosystem you choose to investigate, you should plan to spend an hour or two looking around and filling out a report. On a separate sheet of paper, make a report like the one shown here. Keep the following things in mind while you observe.

- What type of water ecosystem did you observe? Is it a stream, a lake, a pond?
- Where is the freshwater ecosystem located? Estimate its size and depth and tell what it looks like and what organisms are found on its banks.
- To investigate the currents, throw a few sticks into the water. Are they quickly carried away? Do they slowly drift in one direction? Do they sink?
- Is the water murky or clear? Is there algae on top of the water? From these observations, infer whether this ecosystem has a large or small amount of nutrients.
- Describe or identify the plants you see. Are there plants on top of the water? What do the plants on the bank look like?
- Describe or identify the animals you see. Can you see any fish or hear any frogs? Are there any insects on the water's surface? Are there holes in the banks just above water level that might serve as animal homes?

 105

PROBLEM SOLVING

Unfavorable weather conditions resulted in low oxygen levels.

Think Critically: Bubbling air through water allows oxygen from the air to enter water.

3 CLOSE

►Ask questions 1-3 and the **Apply** Question in the Section Review.

SECTION REVIEW ANSWERS

1. the percent of salt

2. marine—salt, temperature, light, amount of dissolved oxygen, water pressure; freshwater—temperature, current, amount of oxygen and dissolved minerals

3. The littoral zone is the zone along the shore. The shallow water above the continental shelf is the sublittoral zone. The open ocean is the pelagic zone.

4. Apply: Marine fish and other organisms breed and feed in estuaries.

5. Connect to Physics: Salt water is more dense than fresh water and buoys you up more.

Skill Builder

Organisms that live below 200 meters in the pelagic zone are in darkness and have no need for eyes.

Skillbuilder
ASSESSMENT

Performance: Assess students' abilities to Hypothesize by having them consider why flounder have two eyes on one side of their heads and none on the other.

PROGRAM RESOURCES

From the **Teacher Resource Package** use:

Activity Worksheets, page 5, Flex Your Brain.

Science Integration Activities, 27, Waves, Currents and Coastal Features.

Science Integration Activities is found in **Merrill Life Science Teacher Resource Manual.**

SECTION 27-4

PREPARATION

SECTION BACKGROUND

► Certain plants in wetlands remove toxic metal substances from water.
► Find wetlands in your state.

1 MOTIVATE

► Investigate a local wetland or create your own wetland in a classroom terrarium.
► Many animals need wetlands to reproduce. Have students name some animals that need water for their offspring to reproduce.

2 TEACH

Key Concepts are highlighted.

CONCEPT DEVELOPMENT

► Have students discuss the impact of destroyed wetlands on food webs there.

In Your JOURNAL

Student descriptions can be generic for any wetland, or can describe a wetland in your state.

SCIENCE & SOCIETY 27-4 Coastal Wetlands

New Science Words

wetland
water table

Objectives

► List several important roles of wetlands in the environment.
► Name several different types of wetlands.

In Your JOURNAL

Wetlands are interesting and exciting places to visit. **In your Journal,** write a travel brochure advertising a tour of a wetland. Describe what persons will see on the tour. You may wish to illustrate the brochure with drawings.

Wetlands or Wastelands?

Have you ever traveled along a coastline and noticed large wet areas? Many of us think of these areas as being useless, unattractive breeding grounds for mosquitoes. This type of area is known as a wetland. A **wetland** is an area of land that, for at least part of the year, is covered by shallow water. There are many different types of wetlands, including coastal marshes, swamps, and bogs.

Wetlands are unique systems that are home to many species of animals and plants. Snakes, fish, turtles, snails, insects, beavers, raccoons, and a wide variety of insects are often found in wetlands. They are all dependent on the many wetland plants for both food and shelter. Wetlands also provide a resting and feeding place for birds migrating during the fall and spring, and for shore birds all year.

Figure 27-13. Wetlands may include estuaries, where fresh water mixes with salt water, or freshwater ecosystems.

Wetlands that are located along the coast are very important in providing food and jobs. Ninety percent of the fish harvested each year in the United States spent at least part of their lives in coastal marshes. People who live near these wetlands rely on the fishing industry for jobs. In addition, the vast coastal marshes serve as protection against storms and floods. If not protected

OPTIONS

Meeting Different Ability Levels

For Section 27-4, use the following **Teacher Resource Masters** depending upon individual students' needs.

◆ **Study Guide Master** for all students.
● **Reinforcement Master** for students of average and above average ability levels.
▲ **Enrichment Master** for above average students.

Additional Teacher Resource Package masters are listed in the OPTIONS box throughout the section. The additional masters are appropriate for all students.

◆ STUDY GUIDE 106

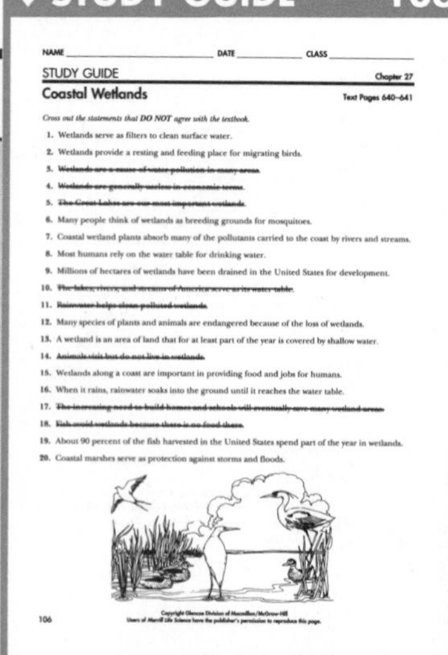

NAME ______ DATE ______ CLASS ______

STUDY GUIDE — Chapter 27

Coastal Wetlands — Text Pages 640-641

Cross out the statements that DO NOT agree with the textbook.

1. Wetlands serve as filters to clean surface water.
2. Wetlands provide a resting and feeding place for migrating birds.
3. ~~Wetlands are a cause of water pollution in many areas.~~
4. ~~Wetlands are generally useless in economic terms.~~
5. ~~The Great Lakes are our most important wetlands.~~
6. Many people think of wetlands as breeding grounds for mosquitoes.
7. Coastal wetland plants absorb many of the pollutants carried to the coast by rivers and streams.
8. Most humans rely on the water table for drinking water.
9. Millions of hectares of wetlands have been drained in the United States for development.
10. ~~The lakes, rivers, and streams of America serve as its water table.~~
11. ~~Rainwater helps clean polluted wetlands.~~
12. Many species of plants and animals are endangered because of the loss of wetlands.
13. A wetland is an area of land that for at least part of the year is covered by shallow water.
14. ~~Animals visit but do not live in wetlands.~~
15. Wetlands along a coast are important in providing food and jobs for humans.
16. When it rains, rainwater soaks into the ground until it reaches the water table.
17. ~~The increasing need to build homes and schools will eventually save many wetland areas.~~
18. ~~Fish avoid wetlands because there is no food there.~~
19. About 90 percent of the fish harvested in the United States spend part of the year in wetlands.
20. Coastal marshes serve as protection against storms and floods.

106

from the rough open waters of the ocean, both coastal cities and rural areas might suffer great damage from high water and the powerful force of the waves.

Wetlands also serve as filters to clean surface water before it enters the water table. The **water table** is the underground water supply most humans rely upon for drinking water. When it rains, the rainwater soaks into the ground until it reaches the water table. But rain also washes pollutants into surface water such as streams. Where streams empty into the ocean, coastal wetland plants and soil absorb these pollutants, preventing them from moving into the ocean itself.

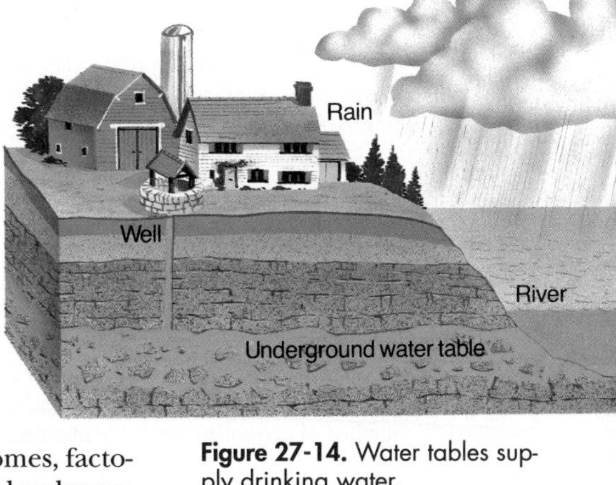

Figure 27-14. Water tables supply drinking water.

With the increasing need to build more homes, factories, schools, stores, and parking lots, land developers often choose wetland areas to develop because they appear to serve no practical use. To most people, wetlands seem useless. In the United States alone, millions of hectares of wetlands have already been drained for development. Many species of plants and animals are now endangered due to the loss of this wetland habitat.

Environmentalists are working with local governments to develop solutions. All over the country, wetland protection plans are being developed to preserve the remaining wetlands. However, as the population continues to grow, there is continuing pressure to develop wetlands. A balanced development of coastal areas may require compromise on both sides of the wetlands debate.

Did You Know?

In southern Florida, increased human population has resulted in loss of 60 percent of the original wetlands. Ninety-five percent of the wading birds in the Everglades have disappeared.

SECTION REVIEW

1. What are two important roles of wetlands?
2. **Connect to Chemistry:** Find out the salt content of the water in a coastal wetland. Does it vary? If so, when and why?

You Decide!

In some states, developers are required to build a new wetland for any that they might destroy during development. However, recent research shows that such recreated wetlands are no replacement for the natural wetlands that required years and years for nature to build. Should developers be allowed to drain wetlands for any purpose?

● REINFORCEMENT 106

NAME ______ DATE ______ CLASS ______

REINFORCEMENT — Chapter 27

Coastal Wetlands — Text Pages 640–641

Answer the following questions.

1. What is a wetland? **an area which, for at least part of the year, is covered by shallow water**
2. What are three different types of wetlands? **coastal marshes, swamps, and bogs**

Farmland
Farmland
docks
SAFE HARBOR
grocery
hardware
docks
high school
city hall
ALLEN RIVER
Oceanview development
Bigsby Swamp
Land's end
OCEAN
N W E S

Use the map of the fictional town of Oceanside to answer the following questions.

3. What type of wetland is Bigsby Swamp? **coastal marsh**
4. What probably breeds in Bigsby Swamp that bothers the townspeople? **mosquitoes**
5. What part of town looks like it is moving into the swamp? **Oceanview Development**
6. How does the swamp protect the town of Oceanside? **It protects it from storms, floods, and high ocean water.**
7. How does Bigsby Swamp remove pollutants that come down the Allen River? **The plants and the soil of the swamp absorb the pollutants carried by the river.**
8. Every morning fishing boats leave the town's docks to fish in the open ocean. How is Bigsby Swamp a benefit to the owners of those boats? **Many of the fish they will catch breed in coastal wetlands. Without the wetlands, fishing would not be as profitable.**

106

▲ ENRICHMENT 106

NAME ______ DATE ______ CLASS ______

ENRICHMENT — Chapter 27

Coastal Wetlands — Text Pages 640–641

THE EVERGLADES

The Florida Everglades are unique—there is nothing else quite like them in the world. They are a habitat for a tremendous number of organisms, and they provide a breeding ground for many more species. Despite their environmental importance, these wetlands have been under attack by human development for decades. Some scientists are concerned that the Everglades may have been irreversibly harmed. We might lose these wetlands forever.

Use library resources to write a report on the Everglades. Look in the library's computer or card catalog for books on this topic. Check the *Readers' Guide to Periodical Literature* or the library's computer search system for recent magazine articles on the Everglades. Use at least three books or articles to write your report. List these sources at the end of your report.
Try to answer these questions in your report:

- Where are the Everglades?
- How much area do the Everglades cover?
- What lake and river system feeds the Everglades?
- What animals live in the Everglades?
- What are some animals that use the Everglades as breeding grounds?
- How much area has the Everglades lost in recent years?
- What major city has been spreading into the area that once was part of the Everglades?
- What was the role of the Army Corps of Engineers in changing the environment in the Everglades?
- How is water that goes into the Everglades controlled?
- What have environmentalists been saying about the danger of losing the Everglades?
- At the end of your report, write a paragraph expressing your own opinions about the threat to the Everglades and what should be done.

Florida
Everglades National Park

106

CHECK FOR UNDERSTANDING

▶Ask question 1 and discuss question 2 in the Section Review.

▶Invite a wetland ecologist or naturalist to speak to the class on wetlands and their role in the natural world.

RETEACH

Have students draw what they think a wetland looks like and include at least five wetland animals.

EXTENSION

For students who have mastered this section, use the **Reinforcement** and **Enrichment** masters or other OPTIONS provided.

3 CLOSE

▶Ask students to investigate proper care for amphibians or reptiles found in wetlands. Put them in charge of maintaining these animals in the classroom.

SECTION REVIEW ANSWERS

1. habitat for many organisms; filter pollutants from drinking water

2. Connect to Chemistry: The salt content is less than ocean water. It varies with the tide, amount of rainfall, temperature, and other conditions.

YOU DECIDE!

Answers will vary. Students may feel that wetlands are breeding grounds for mosquitos and other insects harmful to humans. Remind students that most insects are eaten by birds and bats.

PROGRAM RESOURCES

From the **Teacher Resource Package** use:

Science and Society, page 31, Conserving Our Wetlands.

VideoDisc

STVS: Unusual Estuary, Disc 6, Side 1

ACTIVITY 27-2

OBJECTIVE: Design and carry out an experiment to determine how water affects the germination of bean seeds.
Time: One class period for designing and setting up the experiment, 10 minutes each day for 10 days for observing.

PROCESS SKILLS applied in this activity are **observing, inferring,** and **interpreting data.**

PREPARATION

- Square milk cartons with one side cut out and the ends stapled together may be used for planters.
- Other kinds of seeds may be used.
- **Caution** the students not to plant the seeds too deep in the soil.

Cooperative Learning: Divide the class into Science Investigation Teams.

THINKING CRITICALLY

Populations would overproduce if there were no limiting factors. Temperature, predators, lack of food or space are other limiting factors.

TEACHING THE ACTIVITY

*Refer to the **Activity Worksheets** for additional information and teaching strategies.*

- Use bean seeds to investigate other limiting factors such as light and soil conditions.
- **Caution** students not to overwater the planter.

SUMMING UP/SHARING RESULTS

- Usually from 20 to 23 seeds will germinate in planter A. Probably no seeds will germinate in planter B.
- The variable in the experiment was water. Planter B was the control.
- The factors that were kept the same were the number of seeds, the kind of soil, spacing, depth of seeds in soil, light, air, and temperature.
- The limiting factor in this activity was water.

ACTIVITY 27-2 DESIGNING AN EXPERIMENT
Investigating a Limiting Factor

What happens if you forget to water and feed a plant or animal? You know from this chapter that a limiting factor is a condition that determines the survival of an organism, a population, or species in its environment.

Getting Started

You need to determine a way to find out if water affects the germination of bean seeds.

Thinking Critically

What would happen in ecosystems if there were no limiting factors? List some limiting factors that reduce the number of organisms in a population.

Materials

Your cooperative group will use:

- 1 package of 50 bean seeds
- 2 planters
- water
- soil
- 2 labels
- small spade or spoon

Try It!

1. Place 5 cm of soil in each of the 2 planters.
2. Select 25 bean seeds from the package.
3. Plant 25 of the seeds about 0.5 cm deep, evenly spaced in one planter. Plant the remaining bean seeds in the other planter. Label the planters A and B.
4. Place the planters in an area where they will receive the same amount of light and air.
5. Water planter A with a measured amount of water. Do not water planter B.
6. *Observe* the seeds daily for 10 days. Record the total number of seeds that germinate and any other observations.

Summing Up/Sharing Results

- How many seeds germinated in planter A? Planter B?
- What was the variable in the experiment?
- Which planter was the control?
- What factors did you try to keep the same?
- What do bean seeds need to germinate?
- What was the limiting factor in the experiment?

Going Further!

Based on this activity, **design an experiment** that helps answer a question that occurred from your work.

GOING FURTHER!

If the watered seeds did not germinate well, students could design a new experiment using another kind of seed or different amounts of water.

Activity ASSESSMENT

Performance: To further assess students' understanding of limiting factors, have them design an experiment to test the effect of temperature on the germination of bean seeds.

PROGRAM RESOURCES

From the **Teacher Resource Package** use:
Activity Worksheets, pages 241-242, Activity 27-2, Investigating a Limiting Factor.

CHAPTER
REVIEW

SUMMARY

27-1: Factors That Affect Biomes
1. Limiting factors such as water, sunlight, and temperature determine the survival of organisms, populations, and species in the environment.
2. Over time, communities change in an orderly process called succession. Succession results in a balanced ecosystem called a climax community.
3. Primary succession occurs on new land, whereas secondary succession returns disturbed areas to a more stable climax community.

27-2: Land Biomes
1. Biomes are large geographic areas with similar climates and climax communities.
2. Land biomes include tundra, coniferous and deciduous forests, grasslands, deserts, and tropical rain forests.
3. Biomes share climate conditions, but the specific organisms found in each may differ.

27-3: Water Ecosystems
1. Marine ecosystems and freshwater ecosystems differ in the organisms that live there. This is due to the amount of salt present in the water.
2. Marine organisms have adapted to a salty environment and live in three different ocean zones—the littoral, sublittoral, and pelagic zones.
3. Freshwater organisms have adapted to life in still or moving water.

27-4: Science and Society: Coastal Wetlands
1. Wetlands are home to unique organisms, serve as nesting and feeding sites for many different animal species, and filter pollutants from surface waters.
2. Wetlands include coastal marshes, swamps, and bogs.

KEY SCIENCE WORDS

a. **biomes**
b. **climate**
c. **climax community**
d. **estuary**
e. **limiting factor**
f. **littoral zone**
g. **pelagic zone**
h. **permafrost**
i. **primary succession**
j. **secondary succession**
k. **sublittoral zone**
l. **succession**
m. **tundra**
n. **water table**
o. **wetland**

UNDERSTANDING VOCABULARY

Match each phrase with the correct term from the list of Key Science Words.

1. a condition that determines the survival of an organism
2. area where fresh water and salt water mix
3. underground water supply
4. frozen soil of the tundra
5. shallow water above the continental shelf
6. gradual change in a community over time
7. a stable, balanced community
8. ocean zone along the shore
9. areas of similar climate and communities
10. an area of land periodically covered by water

OPTIONS

ASSESSMENT
To assess student understanding of material in this chapter, use the resources listed.

COOPERATIVE LEARNING
Consider using cooperative learning in the THINK AND WRITE CRITICALLY, APPLY, and MORE SKILL BUILDERS sections of the Chapter Review.

PROGRAM RESOURCES
From the **Teacher Resource Package** use:
Chapter Review, pages 57-58.
Chapter and Unit Tests, pages 180-183, Chapter Test.

CHAPTER
REVIEW

SUMMARY
Have students read the summary statements to review the major concepts of the chapter.

UNDERSTANDING VOCABULARY

1. e
2. d
3. n
4. h
5. k
6. l
7. c
8. f
9. a
10. o

ASSESSMENT
Portfolio
Encourage students to place in their portfolios one or two items of what they consider to be their best work. For each item, ask students to explain why that item was chosen and what they learned from it. Items might be selected from the following.
- For Your Gifted Students story set in a particular biome, p. 626
- Cross Curriculum—Art drawings of succession, p. 629
- You Decide! opinion on wetlands, p. 641

Performance
Additional performance assessments may be found in *Performance Assessment* and *Science Integration Activities* that accompany **Merrill Life Science.** Performance Task Assessment Lists and rubrics for evaluating these activities and other products generated throughout the chapter can be found in Glencoe's *Performance Assessment in Middle School Science.*

CHAPTER

REVIEW

CHECKING CONCEPTS

1. d
2. a
3. c
4. d
5. d
6. c
7. d
8. c
9. c
10. b

USING LAB SKILLS

ASSESSMENT

Use these alternate lab exercises to assess students' understanding of skills used in this chapter.

11. The current biomes would change into different biomes.

12. Populations will increase and change.

THINK AND WRITE CRITICALLY

13. In primary succession, new organisms occur where none lived before. In secondary succession, one community is replaced by another after a disturbance.

14. Coastal wetlands can sustain fluctuations in water levels and remain stable. The littoral zone is a place of continual change. Organisms that live in the littoral zone have adapted to these continual changes.

15. When organisms in the rain forest die, they decompose rapidly. The nutrients are taken up by the plants quickly, so few nutrients and little organic matter remain to build soil.

APPLY

16. During cold and snowy winter months it is difficult to find food, so animals hibernate to conserve energy.

17. Climate is the result of rainfall, temperature, and sunlight—all abiotic factors that limit which species can survive in a particular area.

CHAPTER

REVIEW

CHECKING CONCEPTS

Choose the word or phrase that completes the sentence.

1. The climate of an area is determined by ______.
   a. limiting factors **c.** biotic factors
   b. succession **d.** abiotic factors

2. ______ return(s) a burned forest to its previous condition.
   a. Secondary succession
   b. Limiting factors
   c. Primary succession
   d. A climax community

3. The ______ is a treeless, cold biome.
   a. grassland **c.** tundra
   b. desert **d.** permafrost

4. All are examples of grasslands except ______.
   a. pampas **c.** savannas
   b. steppes **d.** estuaries

5. Swamps, bogs, and marshes are examples of ______.
   a. grasslands **c.** coniferous forests
   b. deserts **d.** wetlands

6. Oysters and clams have adapted to the mild waters of the ______.
   a. pelagic zone **c.** estuary
   b. sublittoral zone **d.** littoral zone

7. The major limiting factor in freshwater ecosystems is ______.
   a. current **c.** dissolved oxygen
   b. temperature **d.** rainfall

8. The ______ has the most variety of organisms of any biome.
   a. desert **c.** tropical rain forest
   b. tundra **d.** deciduous forest

9. A(n) ______ is the end result of succession.
   a. wetland **c.** climax community
   b. limiting factor **d.** estuary

10. Trees are parts of the communities of each biome except ______.
   a. deciduous forest **c.** tropical rain forest
   b. tundra **d.** coniferous forest

USING LAB SKILLS

11. In Activity 27-1 on page 630 you interpreted a hardiness zone map and learned that organisms live only in certain areas. What might be the effect on today's growing areas if all climates became warmer?

12. In the MINI-Lab on page 631 you studied how communities change. Obtain a jar of pond water and observe it over a period of several weeks. Cover the jar and place it in an area where it will receive light. Use a microscope to observe water from the jar twice each week. Record the changes you observe.

THINK AND WRITE CRITICALLY

Answer the following questions in your Journal using complete sentences.

13. Compare primary and secondary succession.

14. What is the difference between coastal wetlands and the littoral zones of marine ecosystems?

15. Why are soils in rain forests not very fertile?

APPLY

16. Why do mammals of the coniferous forest hibernate?

17. How does climate help determine what species live in communities?

18. What special adaptations might plants and animals that live in fast-moving water have?
19. Northern coniferous forests are climax communities. Why don't they change through succession to oak-hickory forests?
20. Make a food chain that could be found in a deciduous forest.

MORE SKILL BUILDERS

If you need help, refer to the Skill Handbook.

1. **Comparing and Contrasting:** Make a chart to compare and contrast these biomes.

Biome	Weather	Plants	Animals
Tundra	cold, dry	mosses	insects
Grasslands	cold, hot	grasses	bison
Deserts	hot, dry	cactus	rodents
Tropical Rain Forests	hot, wet	diverse plants	monkeys

2. **Concept Mapping:** Make a concept map for water ecosystems. Include these terms: *marine ecosystems, freshwater ecosystems, littoral zone, pelagic zone, sublittoral zone, lake, pond, river, stream.*
3. **Making and Using Graphs:** Make a bar graph of the amount of rainfall per year in each biome.

Biome	Rainfall/Year
Deciduous forests	100 cm
Tropical rain forests	225 cm
Grasslands	50 cm
Deserts	20 cm

4. **Comparing and Contrasting:** Make a chart to compare and contrast the plants and animals of the northern coniferous forests and deciduous forests.
5. **Interpreting Data:** Interpret the data given and decide which area of a stream has more varieties of living things and explain why.

Organism	Number in Pools	Number in Riffles
Midgefly larva	0	1000
Caddisfly larva	2	70
Snail	2	12
Water penny	1	7
Mayfly larva	1	250
Stonefly larva	1	61
Horsefly larva	1	8
Water strider	8	1
Dragonfly larva	2	7
Water diving beetle	3	1
Whirligig beetle	5	2

A riffle is a small wave. Find the total number of organisms in each area.

PROJECTS

1. Do you have relatives, friends, or a pen pal who lives in a different biome than you do? Each of you can make a list of common plants and animals in your biome and exchange the lists.
2. Research the Atacama Desert of Chile. Find out how much rainfall the desert receives in one year. Report to the class on your findings.
3. Find out where the grains that humans depend on for food originated, and in what type of biome. Include grains such as corn, wheat, rice, barley, rye, and sorghum in your answer.

18. They may need ways to hold on to rocks; or they may be flat, or have streamlined bodies.
19. The abiotic factors in coniferous forests are limiting to oaks and hickories.
20. acorns → mice → foxes;
acorns → mice → snakes

MORE SKILL BUILDERS

1. Comparing and Contrasting: Fill in chart as shown.
2. Concept Mapping: See below.
3. Graphing:

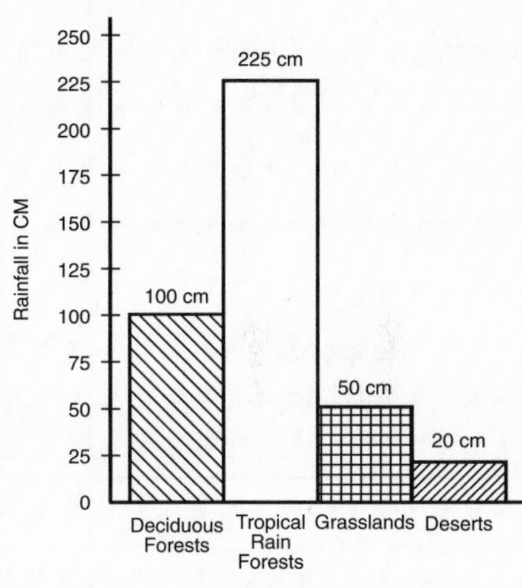

4. Comparing and Contrasting:

Northern Coniferous Forest

Plants	Animals
hemlock, spruce, pine, cedar, fir	moose, elk, bear, lynx

Deciduous Forest

Plants	Animals
beech, maple, oak, hickory, elm, sycamore	deer, fox, racoon, squirrel, bear, opossum

5. Interpreting Data: Riffles have more organisms than pools because there are more places to live, hide, and attach to. Oxygen is more available in places where water flows than in pools. Total number of organisms in pools: 26. Total number of organisms in riffles: 1419.

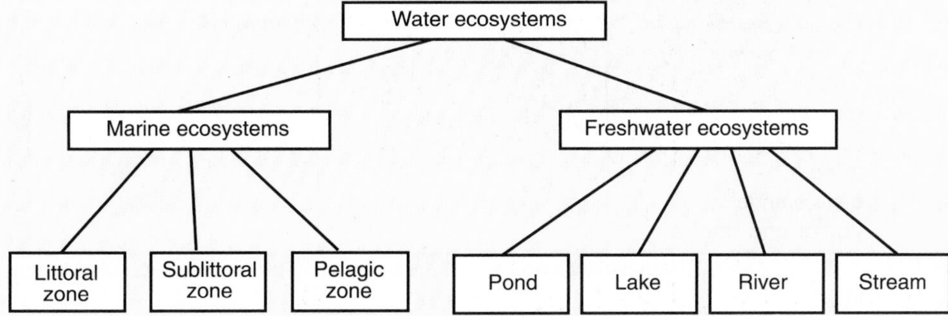

CHAPTER

28 Resources and the Environment

CHAPTER SECTION	OBJECTIVES	ACTIVITIES
28-1 Natural Resources (3 days)	1. **Identify** natural resources and describe how they are important to living organisms. 2. **Distinguish** between renewable and nonrenewable resources and give three examples of each. 3. **Describe** how resources are affected by human use.	**MINI-Lab:** *How much garbage do you produce?* p. 650 **Activity 28-1:** *Designing an Experiment (Investigating Biodegradable Substances)*, p. 652
28-2 Conservation and Protection (2 days)	1. **Define** *conservation* and how it relates to natural resources. 2. **Describe** how renewable resources can be conserved and protected. 3. **Identify** ways to reuse nonrenewable resources.	
28-3 Future Responsibility (2 days)	1. **Explain** the importance of environmental management. 2. **Identify** ways in which individuals can contribute to environmental management. 3. **Describe** ways in which individuals can conserve and protect natural resources.	**MINI-Lab:** *How many families in your school recycle?* p. 659
28-4 Earth in 2030 Science & Society (1 day)	1. **Identify** three issues that concern environmental activists. 2. **Describe** ways to improve environmental conditions on Earth.	**Activity 28-2:** *Identifying Air Pollution Sites*, p. 662
Chapter Review		

ACTIVITY MATERIALS

FIND OUT	ACTIVITIES		MINI-LABS	
Page 647 beaker coffee filter small bowl garden soil (3-4 tablespoons) stirring rod	**28-1 Designing an Experiment, p. 652** 2 clay pots 2 petri dishes 2 small aluminum pans potting soil sand or gravel biodegradable and nonbiodegradable materials 2 labels	**28-2 Identifying Air Pollution Sites, p. 662** 8 strips of cardboard 5 cm × 20 cm 4 strips of waxed paper 5 cm × 20 cm scissors hole punch string petroleum jelly stapler and staples	**How much garbage do you produce? p. 650** paper or plastic bag scale	**How many families in your school recycle? p. 659** paper pencil survey form

CHAPTER FEATURES	TEACHER RESOURCE PACKAGE	OTHER RESOURCES
Problem Solving, *Sharing a Technology,* p. 650 **Skill Builder,** *Hypothesizing,* p. 653	**Ability Level Worksheets** ◆ ***Study Guide,*** p. 107 ● ***Reinforcement,*** p. 107 ▲ ***Enrichment,*** p. 107 **Activity Worksheets,** pp. 248, 249, 254 **Transparency Masters,** pp. 99-100	**Lab Manual:** Acid Rain, p. 181; Water Pollution, p. 185 **Color Transparency 50,** Greenhouse Effect **STVS:** Disc 6, Side 1; Side 2
Technology: *Test Tube Tigers,* p. 657 **Skill Builder,** *Concept Mapping,* p. 657	**Ability Level Worksheets** ◆ ***Study Guide,*** p. 108 ● ***Reinforcement,*** p. 108 ▲ ***Enrichment,*** p. 108 **Critical Thinking/Problem Solving,** p. 32 **Concept Mapping,** p. 61 **Cross-Curricular Connections,** p. 32 **Technology,** pp. 19-20	**STVS:** Disc 2, Side 2; Disc 6, Side 1 **Science Integration Activity 28**
Skill Builder, *Recognizing Cause and Effect,* p. 659	**Ability Level Worksheets** ◆ ***Study Guide,*** p. 109 ● ***Reinforcement,*** p. 109 ▲ ***Enrichment,*** p. 109 **Activity Worksheets,** p. 255	**STVS:** Disc 6, Side 1
You Decide! p. 661	**Ability Level Worksheets** ◆ ***Study Guide,*** p. 110 ● ***Reinforcement,*** p. 110 ▲ ***Enrichment,*** p. 110 **Activity Worksheets,** pp. 250-251 **Science and Society,** p. 32	**Lab Manual:** Home Energy Needs, p. 179 **Lab Manual:** Living Space, p. 189 **STVS:** Disc 6, Side 1; Side 2
Summary Key Science Words Understanding Vocabulary Checking Concepts Using Lab Skills Think & Write Critically Apply More Skill Builders Projects	ASSESSMENT RESOURCES **Chapter Review,** pp. 59-60 **Chapter Test,** pp. 184-187 **Unit Test,** pp. 188-189 **Performance Assessment in Middle School Science (PAMSS)**	**Chapter Review Software** **Test Bank** **Alternate Assessment** **Performance Assessment**

◆ **Basic** ● **Average** ▲ **Advanced**

ADDITIONAL MATERIALS

SOFTWARE	AUDIOVISUAL	BOOKS/MAGAZINES
Energy Search, McGraw-Hill. *The Lio Project,* Queue. *The Galactic Zoo,* Focus.	*Problems of Conservation:* Forest and Range, film, EBEC. *Problems of Conservation:* Wildlife, film, EBEC. *The Garbage Explosion,* film, EBEC. *The Water Crisis,* film, Time-Life. *Loons of Amisk,* video, LCA. *The Animals Are Crying,* video, LCA. *Last Stronghold of the Eagles,* video, LCA. *Project Puffin,* video, LCA. *Plight of the Condor,* video, Coronet/MTI.	Bailey, James A. *Principles of Wildlife Management.* NY: John Wiley and Sons, Inc., 1984. Bybee, Roger W. *Human Ecology: A Perspective for Biology Education.* Reston, VA: National Association of Biology Teachers, Inc., 1984. McKee, Russell. "Tombstones of a Lost Forest." *Audubon,* March 1988, pp. 62-73.

CHAPTER

28

RESOURCES AND THE ENVIRONMENT

THEME DEVELOPMENT: How humans affect homeostasis of the environment through the use of environmental resources is the theme of this chapter.

CHAPTER OVERVIEW

- **Section 28-1:** This section defines natural resources, explains how renewable and nonrenewable resources are different, and describes how resources are affected by human use.
- **Section 28-2:** Section 28-2 defines conservation, describes how renewable resources can be conserved, and identifies ways to recycle nonrenewable resources.
- **Section 28-3:** This section describes the responsibility of individuals to the environment.
- **Section 28-4: Science and Society:** This section describes the future of Earth and the role of environmental activists in shaping that future.

CHAPTER VOCABULARY

natural resources
renewable resources
nonrenewable resources
greenhouse effect
solid wastes
erosion
conservation
reforestation
wildlife preserve
environmental management
environmental activists

CHAPTER

28 Resources and the Environment

646

OPTIONS

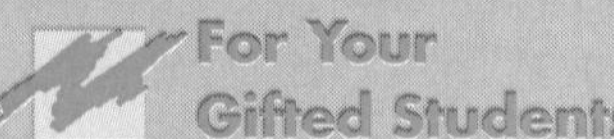

For Your Gifted Students

- Students can create a sculpture using items that might otherwise be thrown away. Their design might include cans, scrap paper, milk cartons, plastic, polystyrene cups, and so on.
- Have students make a game that promotes wise use of resources or that raises other environmental issues. Game cards can give bonus moves or points for environmentally wise activities.

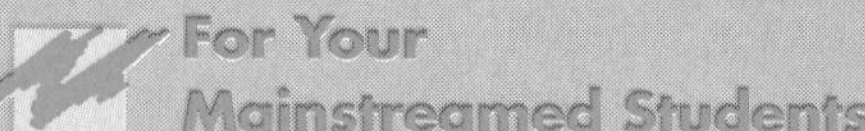

For Your Mainstreamed Students

- Ask students to brainstorm a list of things that are being done today to protect the environment. They can name people or organizations that are involved in each example.
- Students can visit an appliance store or look through a catalog to see what types of energy-saving appliances are available. Have them compare the amount of energy saved in different models.

Water is essential for all life on Earth. You know that the water used for washing dishes will go down the drain. But then where does it go? Water may go into a sewer system and enter a water treatment plant. How does today's wastewater become tomorrow's clean water? How is water cleaned so it can be used again?

FIND OUT!

Do this simple activity to find out how water can be cleaned.

Stir some garden soil and plant material into a beaker filled with water. When the water looks muddy, pour most of it through a coffee filter into a bowl. *Compare* the water in the beaker with the water in the bowl. Filters can remove some kinds of wastes from water, but do they remove all wastes? How would you remove other wastes from the filtered water?

Gearing Up

Previewing the Chapter

Use this outline to help you focus on important ideas in this chapter.

Previewing Science Skills

- In the **Skill Builders**, you will hypothesize, make concept maps, and recognize cause and effect.
- **In the Activities, you will hypothesize, experiment, and analyze.**
- In the **MINI-Labs**, you will survey, organize information, and collect and interpret data.

What's next?

Now you know that wastewater can be partially cleaned by filtering. In this chapter, you'll learn about the impact humans have on natural resources and how you can help conserve these resources.

INTRODUCING THE CHAPTER

Use the Find Out activity to show students that water is a natural resource from which waste materials can be removed. Inform students that they will learn the importance of natural resources and how to conserve them.

FIND OUT!

Preparation: The day before the activity, gather beakers, coffee filters, small bowls, and stirring rods. Mix garden soil with decaying plant leaves and other plant material.

Materials: Each group should have one beaker, one coffee filter, one small bowl, three or four tablespoons of garden soil mixture, and a stirring rod.

Cooperative Learning: Use the Science Investigation strategy. Divide the class into groups of four. Have one student gather materials, one read the activity, one record, and one check on safety and time.

Teaching Tips

- Remind students to pour the water into the coffee filter slowly so that dirty water does not overflow into the filtered water.

Gearing Up

Have students study the Gearing Up feature to familiarize themselves with the chapter. Discuss the relationships of the topics in the outline.

What's Next?

Before beginning the first section, make sure students understand the connection between the Find Out activity and the topics to follow.

ASSESSMENT OPTIONS

PORTFOLIO

Refer to page 663 for suggested items that students might select for their portfolios.

PERFORMANCE ASSESSMENT

See page 663 for additional Performance Assessment options.

Process

Skillbuilders, pp. 653, 659
MINI-Lab, p. 650
Activities 28-1, p. 652; 28-2, p. 662
Using Lab Skills, p. 664

CONTENT ASSESSMENT

Assessment—Oral, pp. 651, 656
Skillbuilder, p. 657
Section Reviews, pp. 653, 657, 659, 661
Chapter Review, pp. 663-665

GROUP ASSESSMENT

Opportunities for group assessment occur with Cooperative Learning Strategies and Flex Your Brain Activities.

SECTION 28-1

PREPARATION

SECTION BACKGROUND

▶ The world's coal supply is much larger than supplies of petroleum and natural gas.

▶ Tidal energy is one type of renewable resource that is being studied by scientists.

PREPLANNING

▶ Obtain petri dish lids, potting soil, clay pots, aluminum pie pans, scissors, metric rulers, and sand or gravel for Activity 28-1.

1 MOTIVATE

▶ **Bulletin Board:** Collect pictures of renewable resources (trees, wildlife, soil) and nonrenewable resources (coal, oil, metals, minerals). Title the bulletin board Natural Resources with the subtitles Renewable Resources and Nonrenewable Resources. Place the pictures under the correct titles.

TYING TO PREVIOUS KNOWLEDGE: Review with the students the definitions of *habitat,* the *water cycle,* and the *carbon-oxygen cycle* from Chapter 26.

OBJECTIVES AND SCIENCE WORDS: Have students review the objectives and science words in this chapter as they study each section.

28-1 Natural Resources

New Science Words

natural resources
renewable resources
nonrenewable resources
greenhouse effect
solid wastes
erosion

Objectives

- ▶ **Identify natural resources and describe how they are important to living organisms.**
- ▶ **Distinguish between renewable and nonrenewable resources and give three examples of each.**
- ▶ **Describe how resources are affected by human use.**

What Are Natural Resources?

Mountain gorillas live in family groups high in the mountains of Rwanda and Uganda in Africa. They eat fruits and leaves and drink water from clear mountain streams. At night, these gorillas bend branches of trees down to make nests to sleep on. How do you get ready for bed? You brush your teeth with a plastic toothbrush, take a drink of water, then slip between cotton sheets on your bed. Do you have anything in common with mountain gorillas? Both humans and gorillas rely on natural resources in their environment. **Natural resources** are the raw materials in the environment that organisms use for survival. Natural resources include biotic and abiotic materials such as water, soil, air, forests, wildlife, and minerals that organisms use in their daily lives. The food you eat, the clothes you wear, and the roof over your head all are necessary for your survival, and all came from natural resources.

What are natural resources?

Figure 28-1. Nonrenewable resources such as petroleum are available only in limited amounts.

The cotton used to make your sheets is a renewable resource. **Renewable resources** are those natural resources that can be replaced by nature over a period of time. Trees, wildlife, and soil are renewable resources. Cotton for sheets is a renewable resource because cotton plants can be planted every year. Air, water, and sunlight are renewable resources that all life depends upon.

Nonrenewable resources are those natural resources that are available only in limited amounts and cannot be replaced by nature. Many of the products you depend on every day are made from nonrenewable resources. Plastic toothbrushes, polystyrene cups, and gasoline all are made from petroleum, a fossil fuel. Fossil fuels are fuels made up of organisms that lived and died millions of years ago.

OPTIONS

Meeting Different Ability Levels

For Section 28-1, use the following **Teacher Resource Masters** depending upon individual students' needs.

◆ **Study Guide Master** for all students.

● **Reinforcement Master** for students of average and above average ability levels.

▲ **Enrichment Master** for above average students.

Additional Teacher Resource Package masters are listed in the OPTIONS box throughout the section. The additional masters are appropriate for all students.

◆ STUDY GUIDE 107

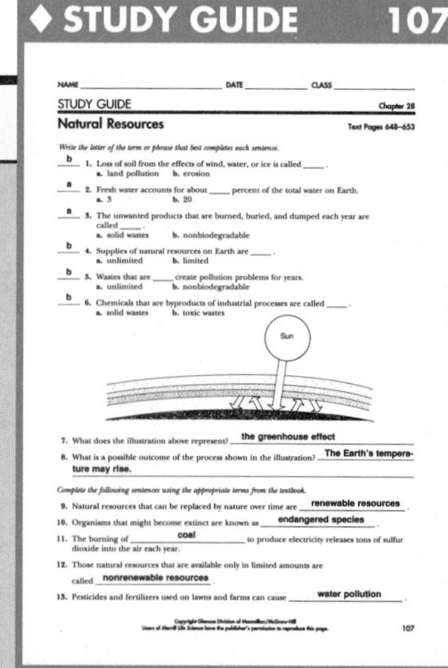

NAME ______ DATE ______ CLASS ______

STUDY GUIDE Chapter 28

Natural Resources Text Pages 648–653

Write the letter of the term or phrase that best completes each sentence.

b 1. Loss of soil from the effects of wind, water, or ice is called ____.
a. land pollution b. erosion

a 2. Fresh water accounts for about ____ percent of the total water on Earth.
a. 3 b. 20

a 3. The unwanted products that are burned, buried, and dumped each year are called ____.
a. solid wastes b. nonbiodegradable

b 4. Supplies of natural resources on Earth are ____.
a. unlimited b. limited

b 5. Wastes that are ____ create pollution problems for years.
a. unlimited b. nonbiodegradable

b 6. Chemicals that are byproducts of industrial processes are called ____.
a. solid wastes b. toxic wastes

7. What does the illustration above represent? the greenhouse effect

8. What is a possible outcome of the process shown in the illustration? The Earth's temperature may rise.

Complete the following sentences using the appropriate terms from the textbook.

9. Natural resources that can be replaced by nature over time are renewable resources.

10. Organisms that might become extinct are known as endangered species.

11. The burning of coal to produce electricity releases tons of sulfur dioxide into the air each year.

12. Those natural resources that are available only in limited amounts are called nonrenewable resources.

13. Pesticides and fertilizers used on lawns and farms can cause water pollution.

107

Petroleum, coal, peat, and natural gas are fossil fuels. Some minerals that are recycled very slowly, such as phosphorus, are also called nonrenewable resources.

Resources Are Limited

Have you ever rushed to the music store to buy a tape advertised in the paper, only to find the store is sold out when you arrive? If you looked at the advertisement carefully, you might see the words, "Supplies are limited." Supplies of natural resources on Earth are limited, too. Natural resources on Earth are not distributed evenly. You already know from Chapter 27 that deserts are biomes with very few natural resources. Today, many of Earth's natural resources are being damaged as a result of human activities. Air pollution, water pollution, and the problem of solid waste disposal are daily headlines. Soil, forests, and wildlife are natural resources in danger because of human actions. These problems are discussed in the following sections.

Air Pollution

Air pollution occurs when the air is polluted by gases from vehicles or power plants that burn fossil fuels. Air can be polluted by carbon monoxide, nitrogen oxide, sulfur dioxide, hydrocarbons, and by tiny particles of ash, lead, dust, or soot. Some air pollution occurs naturally, for example, when gases are released from a volcano. Most air pollution results from the burning of fossil fuels in power plants and in automobiles. The burning of coal to produce electricity releases tons of sulfur dioxide into the air each year in the United States. Sulfur dioxide and nitrogen oxide combine with water vapor in the air to produce acid rain.

Air pollution can also be caused by the burning of forests and grasslands. In Brazil, Indonesia, and India, farmers often clear land in this way. This burning also releases carbon dioxide into the atmosphere. Carbon dioxide and other gases form a layer around Earth that acts like a wall of windows in a greenhouse. The gases allow sunlight to pass through and warm Earth's surface. Heat and reflected sunlight radiate back into this gas layer and are trapped there, as shown in Figure 28-2. Scientists have found evidence that the **greenhouse effect** may be causing a rise in Earth's temperature.

In addition to carbon dioxide, water vapor, methane, nitrous oxide, and chlorofluorocarbons (CFCs) increase the greenhouse effect. CFCs are compounds that are used as coolants, solvents, and propellants in aerosol cans. How do CFCs affect the environment?

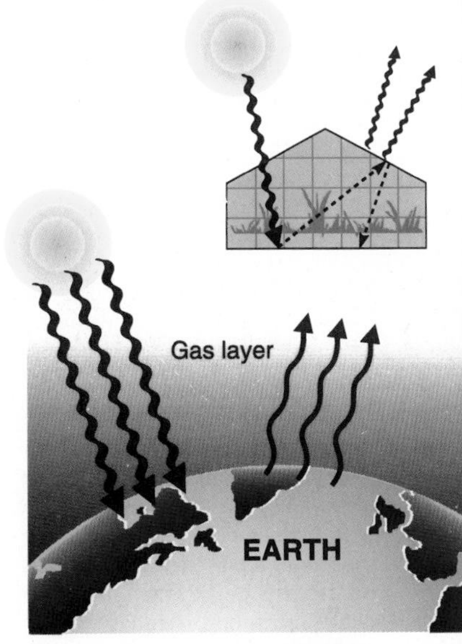

Figure 28-2. The greenhouse effect occurs when large amounts of carbon dioxide and other gases trap heat near the surface of Earth.

● REINFORCEMENT 107

NAME _____ DATE _____ CLASS _____

REINFORCEMENT — Chapter 28

Natural Resources — Text Pages 648–653

Answer the following questions using information from the textbook.

1. What are renewable resources? **natural resources that can be replaced by nature over time**
2. What are nonrenewable resources? **natural resources that are available only in limited amounts and cannot be replaced by nature**
3. Circle the renewable resources in the following list.
   sunlight (circled) gasoline cotton (circled)
   oil water (circled) coal
4. Name three fossil fuels. **coal, petroleum, peat, natural gas**
5. Why are fossil fuels called by that name? **They are made from organisms that lived and died millions of years ago.**
6. Draw and label an illustration of the greenhouse effect.
   CARBON DIOXIDE AND OTHER GASES · TRAPPED HEAT · SUN · EARTH
7. Why are biodegradable wastes preferable to nonbiodegradable wastes? **Biodegradable wastes can be broken down into their chemical components, while nonbiodegradable wastes cannot. Therefore, nonbiodegradable wastes can cause pollution problems for years.**
8. What is erosion? **loss of soil from the effects of wind, water, or ice**
9. How can cutting down trees cause erosion? **Without the trees, there are no roots to hold the soil. Rains can then wash the soil away.**
10. Why do human activities decrease populations of wild species? **Human activities change the habitats of species so fast that the species are unable to adapt and their populations decrease.**

 107

▲ ENRICHMENT 107

NAME _____ DATE _____ CLASS _____

ENRICHMENT — Chapter 28

Natural Resources — Text Pages 648–653

CARBON DIOXIDE EMISSIONS

Environmental organizations and scientists around the world collect information about what is being released into Earth's atmosphere. The substances released into the atmosphere are called emissions. As you read in your textbook, carbon dioxide emissions are especially important because they contribute to the greenhouse effect.

Carbon dioxide emissions can be divided into two types. Biological emissions are caused by cutting down and burning forests and grasslands. Industrial emissions are caused by burning fossil fuels in cars and factories.

Table A shows how many million metric tons of carbon dioxide have been released into the atmosphere in certain years. Table B shows what country or region these emissions came from in 1980 and 1985.

Using information from the tables, answer the questions below.

Table A
Carbon Dioxide Emissions per Year (million metric tons of carbon)

Year	Biological	Industrial
1950	796	1639
1955	1018	2050
1960	1349	2586
1965	1576	3154
1970	1700	4116
1975	1695	4660
1980	1691	5255
1985	(unknown)	5336

Table B
Percentage of World Total of Carbon Dioxide Emissions by Selected Countries and Regions

Country or Region	Biological (in 1980)	Industrial (in 1985)
United States	less than 1%	23%
United Kingdom	less than 1%	3%
U.S.S.R.	5%	19%
China	5%	10%
Japan	less than 3%	5%
Latin America	46%	6%
Africa	16%	3%
Middle East	0%	4%
South & Southeast Asia	25%	3%

1. How many million metric tons of carbon dioxide were released by industry in 1950? **1639** How many in 1985? **5336**
2. What happened to industrial emissions between 1950 and 1985? **Industrial emissions have steadily increased through the years.**
3. How many million metric tons of carbon dioxide were released by burning forests and grasslands in 1950? **796** How many were released in 1980? **1691**
4. Which type of carbon dioxide emissions is a greater problem, industrial or biological emissions? **Industrial emissions**
5. Which country or region accounted for the greatest percentage of industrial emissions in 1985? **the United States**

 107

2 TEACH

Key Concepts are highlighted.

CONCEPT DEVELOPMENT

▶ Make sure that students understand that petroleum is not only a source of fuel and used in the manufacture of plastics, but it is also used in the manufacture of fertilizers, pesticides, and other products.

▶ The ozone layer is a layer of gases around Earth. Ozone, a molecule made of three oxygen atoms, traps excess ultraviolet radiation and prevents it from reaching Earth.

▶ The chlorofluorocarbons (CFCs) used in insulation, coolants, and spray cans absorb ultraviolet light and release chlorine atoms which react with and destroy the ozone layer.

▶ Have students make a list of all the petroleum-based products they can think of so they will understand why the use of fossil fuels is so important.

Connect to... Chemistry

Answer: CFCs rise high into the atmosphere where ultraviolet light breaks them down and chlorine gas is given off. The chlorine breaks up the ozone layer that shields Earth's surface from ultraviolet light from the sun.

VideoDisc

STVS: Modeling Pollutants, Disc 6, Side 2
STVS: Arctic Haze, Disc 6, Side 2
STVS: Bald Eagles, Disc 6, Side 1

PROBLEM SOLVING

Think Critically: A drip method insures a steady supply of water to plants.

CONCEPT DEVELOPMENT

▶ **Activity:** Place clean baby food jars in different areas and collect rainwater. Test the water with pH paper. Compare the test results with the pH of distilled water.

▶ Have students list the uses of water and determine which uses are due to technological innovations, such as irrigation.

▶ Show pictures and discuss the pros and cons of building dams and reservoirs. Students may not realize that a valley must be flooded in order to form a reservoir.

MINI-Lab

Troubleshooting: Make sure students have divided the total amount of garbage produced by the number of people in their households.
Answer: Student answers will vary.

MINI-Lab
ASSESSMENT

Performance: To further assess students' understanding of how much waste they produce, have them combine figures to get a class average and then compare their own output to the average.

PROBLEM SOLVING

Sharing a Technology

Depending on where you live in the world, you may not think twice about planting a garden. Surely, there will be enough water.

But in many areas of the world, very little rain falls to make plants grow.

Israel is located in an arid part of the world. To supply plants with needed water, the Israelis have developed drip irrigation. Water from a well travels in pipes to thin drip lines that lie along plant rows, then drips out through small holes. Plants receive water throughout their growth period.

Similar dry conditions existed on the Navajo Reservation in Arizona. In 1984, the Navajos asked the Israelis to share their irrigation technology. As a result, Navajo farms now produce a bountiful variety of fruits and vegetables.

Think Critically: Why would continuous drip irrigation be better for roots than flooding the fields when they get dry?

MINI-Lab

How much garbage do you produce?

Ask your family to help you collect everything that is thrown away in your household in one week. Collect garbage in paper or plastic bags. At the end of the week, carefully weigh each bag. Add the weights to find the total amount of garbage, then divide by the number of people in your household. *Interpret* your data to find out how much garbage you produce in one week.

Water Pollution

In many places on Earth, water is the most scarce resource. Most of the fresh water on Earth is in the oceans or locked in polar ice caps. Fresh water accounts for only three percent of the total amount of water on Earth. Of that three percent, only 0.003 percent is clean and safe, and available for human consumption.

Water can be polluted by oil, industrial wastes, sewage, bacteria, sediments, solid wastes, and even heat. When power plants use water from rivers for cooling purposes, the water may be returned to the river several degrees hotter than it was originally. Organisms in the river cannot adjust to such quick changes in water temperature, and they may die. Water can also be polluted by pesticides and fertilizers that are used on farms and homeowners' lawns. Rain washes these chemicals out of the soil and into nearby water sources. Even though water is a renewable resource, it is expensive to clean polluted water.

Land Pollution

Did you drink a soft drink yesterday? If you did, what did you do with the can or bottle when you were finished? If you threw it in the trash, you were contributing to land pollution. Magazines, newspapers, plastic bags, glass bot-

OPTIONS

INQUIRY QUESTIONS

▶ **What are some natural air pollutants?** *pollen, soil dust, volcanic ash*

▶ **What could limit natural resources of air and water?** *wastes and pollution*

▶ **With the world's population increasing at its current rate, what are some possible problems you will have with renewable and nonrenewable resources during your lifetime?** *Answers will vary but may include shortage of resources, mandatory recycling, rationing of resources, and so on.*

PROGRAM RESOURCES

From the **Teacher Resource Package** use:

Activity Worksheets, page 254, Mini-Lab: How much garbage do you produce?

tles, aluminum cans, grass clippings, and leftover foods that are thrown away end up as solid wastes. **Solid wastes** are the unwanted products that are burned, buried, and dumped each year all over the world. Every day, each person in the United States throws away 1.8 kg of solid wastes. Most of this solid waste ends up in sanitary landfills.

What happens to the garbage in sanitary landfills? Grass cuttings, animal wastes, newspapers, and dead leaves are broken down by decomposers in the soil. Wastes that can be broken down into their chemical components are called biodegradable. Waste products that cannot be broken down by natural processes, such as aluminum cans and old tires, are called nonbiodegradable. Wastes that are nonbiodegradable create pollution problems for years.

Toxic wastes are chemicals that are byproducts of industrial processes. Many are known to cause cancer, birth defects, and other health problems. Some wastes are packed in steel drums. If the drums are not sealed properly, chemicals leak out and pollute both soil and water.

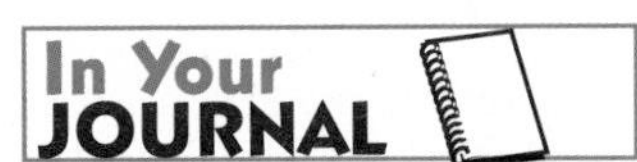

In your Journal, write an article for the local newspaper about the wise use of water resources in your area. Discuss ways to conserve water and prevent pollution.

Soil and Forest Resources

Much of the land in Nepal lies on steep mountain slopes, where fuel for cooking is scarce. Families go higher and higher up into the mountains searching for fuel. They cut trees and bring them home, leaving the slopes bare. When the rainy season comes, there are no tree roots to absorb water and hold soil in place. Water washes soil away. **Erosion** is the loss of soil from the effects of wind, water, or ice. Rain washes away the fertile topsoil, and trees are unable to grow on the barren mountain slopes.

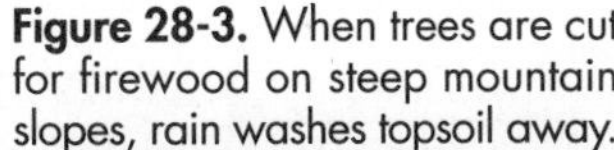

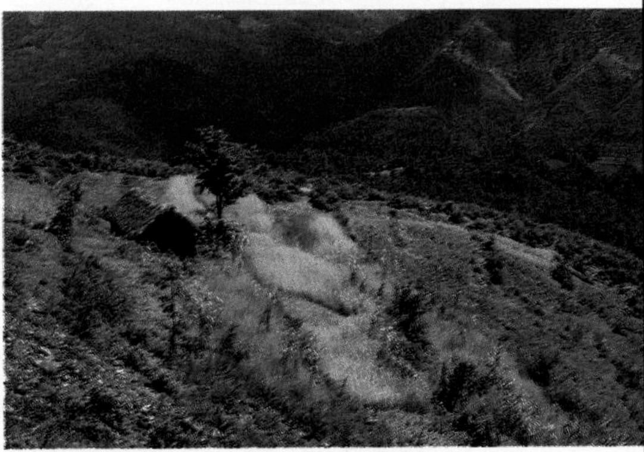

Figure 28-3. When trees are cut for firewood on steep mountain slopes, rain washes topsoil away.

Poor farming and forestry practices such as those described above often result in erosion of topsoil. It takes between 500 and 1000 years for 2.5 centimeters of topsoil to form. If rain falls after a farmer plows but before crops begin to grow, erosion may occur. Erosion also occurs when the timber in a forest is harvested by a method called clear-cutting. In this method, all the trees in a section of forest are cut down and dragged out. Not only is the soil left unprotected, but the habitats of all the organisms that live in or on the trees have been disturbed.

CONCEPT DEVELOPMENT

▶**Demonstration:** Bring a container of trash to class. Include a glass bottle, an aluminum can, a twist tie, plastic wrappers, paper, coffee filter with grounds, vegetable peelings, egg shells, and other household items. Wear plastic gloves and remove each piece from the container. Discuss which come from renewable resources and which come from nonrenewable resources.

▶**Activity:** Have each student choose a product already on the market and design a new package that is attractive and will conserve resources.

▶Explain that not all hazardous wastes come from large industries. Many harmful chemicals, such as paint removers and furniture cleaners, come from households.

▶Each year erosion carries away almost 3 billion tons of soil from the farmlands in the United States.

▶Terracing is a method of changing a hillside into broad, flat steps that help hold water. Terrace farming has been used in some parts of the world for thousands of years.

▶No-till farming is a method in which the remains of previous crops are not plowed under. New crops are planted directly in the unplowed field.

▶Invite a forester to the class to explain methods of forest conservation.

▶Stress that by careful forest management, a mature forest can be harvested without being harmed.

▶Changing temperatures, volcanic eruptions, competition from other species, and other natural happenings have caused many species to become extinct in the last 200 million years.

In Your JOURNAL

Articles will vary but should include not just things individuals can do to conserve water, but realistic things communities can do.

ASSESSMENT—ORAL

▶**What can people do to reduce the amount of waste they produce?** *Answers will vary but should include smart shopping and reusing.*

▶**How can we conserve our forest resources?** *plant trees, recycle paper products, and use fewer paper products*

▶**Why is flat land in less danger of eroding than hilly land?** *The water has no place to go.*

ENRICHMENT

▶Have students research John Muir and his work with conservation.

▶Have students research and report on strip mining and land reclamation in parts of the United States.

▶Have students research and report on aquifers.

▶Have students research and report on PCBs (polychlorinated biphenyls), toxic wastes produced in the making of paints, inks, and electrical insulators.

ACTIVITY 28-1

OBJECTIVE: Design and carry out an experiment using substances found in garbage to determine which materials are biodegradable and which are nonbiodegradable.

Time: One class period to set up the activity; 10 minutes two times a week for 4-6 weeks to collect data; one-half period to summarize the results.

PROCESS SKILLS applied in this activity are **observing, interpreting data,** and **inferring.**

PREPARATION

- Divide the materials into 2 cm squares for class use. You may have students collect materials found in garbage.
- Discuss differences between organic and inorganic materials and the role of bacterial and fungi in decomposition.

Cooperative Learning: Divide the class into Science Investigation Teams.

THINKING CRITICALLY

Student answers may vary. Biodegradable substances may include fruit and vegetables, eggshells, bread, packaging materials, coffee grounds, leaves, nuts, water-soluble plastics, and newspapers. Nonbiodegradable substances may include glass, aluminum, tin, and plastic. Bacteria and fungi cause substances to decompose naturally.

TEACHING THE ACTIVITY

*Refer to the **Activity Worksheets** for additional information and teaching strategies.*

- Alternate materials: Square milk cartons with the side cut out and the ends stapled together may be used instead of clay pots. Cartons or pots may be enclosed in a plastic bag and tied. Soil in which the materials have been placed must be well covered.
- Stir the soil to look at the items once every 2 or 3 days for 4-6 weeks.

SUMMING UP/SHARING RESULTS

- Leaves, newspapers, vegetable peelings
- Mold appears as decay occurs; an odor may be observed as the result of decomposition by bacteria.
- Most plastic, aluminum, tin, and glass; most other substances
- They pollute the environment.

GOING FURTHER!

When possible, avoid using materials that are not recyclable and dispose of nonbiodegradable materials in ways provided by your community. People can recycle materials and use biodegradable materials.

Activity
ASSESSMENT

Performance: To further assess students' understanding of biodegradable materials, see USING LAB SKILLS, Question 11, on page 664.

PROGRAM RESOURCES

From the **Teacher Resource Package** use:
Activity Worksheets, pages 248-249, Activity 28-1, Investigating Biodegradable Substances.

ACTIVITY 28-1

DESIGNING AN EXPERIMENT
Investigating Biodegradable Substances

The disposal of solid waste has become a serious problem for many communities. A common means of disposing of garbage is to add decomposed materials to landfills. What is happening to landfills?

Getting Started

In this activity, you are to devise a method to determine which substances are biodegradable.

Thinking Critically

Make a list of substances found in garbage that are biodegradable and nonbiodegradable. What causes substances to decompose naturally in the environment?

Materials  

Your cooperative group will use:

- 2 clay pots
- 2 petri dishes
- 2 aluminum pie tins
- potting soil
- sand or gravel
- biodegradable and nonbiodegradable materials
- 2 labels

Try It!

1. Obtain 8 kinds of waste materials from your teacher. *Predict* which materials are biodegradable and which materials are nonbiodegradable. Separate the two types of materials.
2. Place layers of sand or gravel in the bottom of two clay pots. Fill each pot with moistened potting soil to within 1.5 cm of the top.
3. Into one pot, place the materials that your group predicted were biodegradable on top of the soil. Cover the materials with a thin layer of soil. Label the pot *biodegradable.*
4. Into the second pot, place the materials that your group predicted were nonbiodegradable on top of the soil. Cover with a thick layer of soil. Label the pot *nonbiodegradable.*
5. Cover each pot with the lid of a petri dish. Make sure it fits tightly over the top of each pot. Place each pot in an aluminum pie pan. Add water to the pan. The water will rise through the holes in each pot and keep the contents moist.
6. *Observe* the two pots two times each week for four weeks. Record your observations.

Summing Up/Sharing Results

- Which substances decomposed first?
- What kinds of organisms were you able to observe?
- Which substances were biodegradable? Nonbiodegradable?
- How do substances that are not biodegradable affect the environment?

Going Further!

How can you use what you have learned in this activity? How can people reduce the amount of waste they produce?

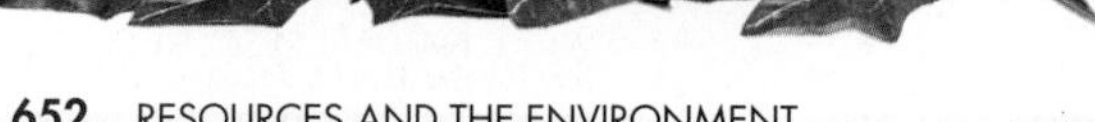

Wildlife Resources

Have you heard of the California condor? Only about 40 of these huge birds still exist, and most of them are in captivity. Condors are in danger of becoming extinct because there are so few of them left. Many species of plants and animals today also are in danger of becoming extinct because they are hunted, or because their habitats have been destroyed. Species that are in danger of becoming extinct are called endangered species.

Populations of many wild species have been decreasing, often as a result of human activities. Developers have filled in estuaries and wetlands to build housing developments, farmers have plowed grasslands to plant crops, and lumber companies have cleared forests for timber. Some wildlife species are able to survive changes like these. Raccoons often do well in housing developments, as long as they can get into garbage cans! But many times these changes in wildlife habitat are so fast that species are unable to adapt. As a result, populations decline as individual organisms die or move on to better areas.

Plant species can be important sources of food and medicines, yet many plant species are endangered also. There may be thousands of plant species in tropical rain forests that have not been studied yet, and their uses are unknown. This is one reason that scientists would like rain forest destruction to stop.

Figure 28-4. California condors are endangered as a result of habitat loss.

Why can't species adapt to changes in their habitat?

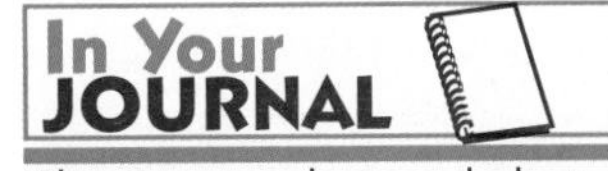

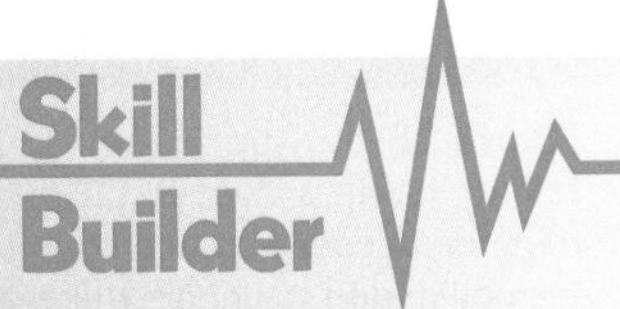

Choose an endangered plant or animal and research it. **In your Journal,** describe how this species became endangered.

SECTION REVIEW

1. What are natural resources?
2. Differentiate between renewable and nonrenewable resources.
3. Explain the greenhouse effect.
4. **Apply:** How have human activities increased the number of extinct plants and animals?
5. **Connect to Earth Science:** Find out how the cutting of trees affects global warming.

Skill Builder: Hypothesizing

Hypothesize why a population of raccoons is able to survive changes in its habitat. Explain. If you need help, refer to Hypothesizing in the **Skill Handbook** on page 686.

OPTIONS

PROGRAM RESOURCES

From the **Teacher Resource Package** use:

Activity Worksheets, pages 248-249, 28-1 Designing an Experiment.

Transparency Masters, pages 99-100, Greenhouse Effect.

Use **Color Transparency** 50, Greenhouse Effect.

Use **Laboratory Manual,** page 181, Acid Rain.

Use **Laboratory Manual,** page 185, Water Pollution.

3 CLOSE

CHECK FOR UNDERSTANDING

▶ Ask questions 1-3 and the **Apply** Question in the Section Review.

RETEACH

Have students list reasons renewable resources must be used wisely and nonrenewable resources must be conserved.

EXTENSION

For students who have mastered this section, use the **Reinforcement** and **Enrichment** masters or other OPTIONS provided.

SECTION REVIEW ANSWERS

1. Natural resources are the raw materials that organisms use for survival.
2. Renewable resources can be replaced by nature. Nonrenewable resources can't.
3. Answers should reflect information on page 649.
4. Apply: Humans have hunted and killed organisms directly or have changed their habitat so they are unable to live there.
5. Connect to Earth Science: Trees take in carbon dioxide naturally as they grow. When trees are removed, the carbon dioxide they would have removed is left in the atmosphere. Many of the trees that are cut are burned and this releases more carbon dioxide into the atmosphere.

Skill Builder

Raccoons are omnivores able to utilize many kinds of food. Omnivores are better able to adapt to changes in their habitat because they can alter what they eat more easily than other organisms.

Skillbuilder ASSESSMENT

Performance: Use this Skillbuilder to assess students' abilities to Hypothesize about the effect of change in habitat on survival. Have students suggest a change that would affect survival of an organism.

SECTION 28-2

PREPARATION

SECTION BACKGROUND

▶ In some areas, garbage is becoming a major source of fuel. Glass and metals are removed from organic material in the garbage, which is then burned to produce methane gas that can be used for heating and cooking.

1 MOTIVATE

▶ **Bulletin Board:** Make a bulletin board entitled How Humans Affect the Environment. Collect pictures of dams, large farms, shopping centers, highways, strip mining, and other subjects that would be appropriate to display.

FLEX Your Brain

Use the Flex Your Brain activity to have students explore what they know about CONSERVATION.

Flex Your Brain
ASSESSMENT
Portfolio: Use the Flex Your Brain activity to reinforce critical thinking and problem solving skills.

TYING TO PREVIOUS KNOWLEDGE: Ask students to list how they affect the environment in their everyday lives. They should think of the transportation they use, waste they create, and water they use.

28-2 Conservation and Protection

New Science Words

conservation
reforestation
wildlife preserve

Objectives

▶ Define conservation and how it relates to natural resources.
▶ Describe how renewable resources can be conserved and protected.
▶ Identify ways to reuse nonrenewable resources.

What Is Conservation?

For thousands of years, farmers in steep mountain regions of Nepal, Myanmar, Indonesia, and the Philippines have terraced the sides of mountains to grow rice. They learned that terraces prevent soil erosion and hold water needed for rice production. These farmers
1 practice soil and water conservation. **Conservation** is the wise and careful use of Earth's resources.

When humans use natural resources wisely, they can live in harmony with the community and environment around them. Because the human population continues to grow, people are becoming more aware of how important it is to conserve and protect our natural resources. There are many ways to balance the human need for natural resources with the availability of those resources. How can people clean up the air and water we rely on for life? What can be done about waste disposal? In this section, you'll see some ways people can make better use of natural resources.

Figure 28-5. Kinetic energy in falling water is used to produce other forms of energy, such as electric power and mechanical energy.

Cleaning Up the Air

What kind of gasoline does your family's car use? If it is a late model car, it probably uses unleaded gas. Lead was a common component of most gasoline until 1970 when the Clean Air Act was passed in the United States. Since that year, all new cars in the U.S. have been designed to use only unleaded gas. The Clean Air Act of 1990 requires industries to reduce their sulfur dioxide emissions by 40 percent by the year 2000.

OPTIONS

Meeting Different Ability Levels

For Section 28-2, use the following **Teacher Resource Masters** depending upon individual students' needs.

◆ **Study Guide Master** for all students.
● **Reinforcement Master** for students of average and above average ability levels.
▲ **Enrichment Master** for above average students.

Additional Teacher Resource Package masters are listed in the OPTIONS box throughout the section. The additional masters are appropriate for all students.

◆ STUDY GUIDE 108

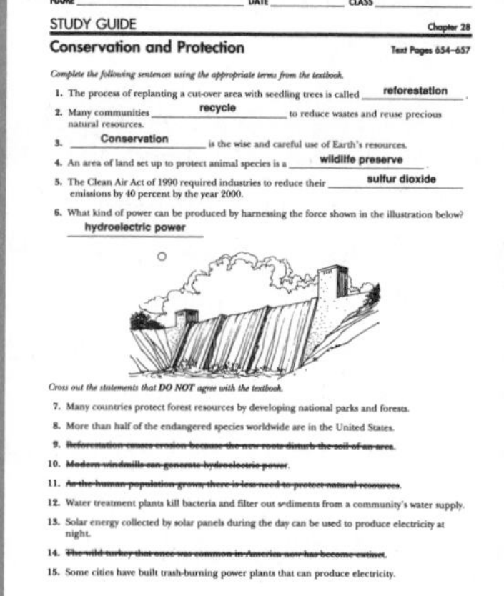

NAME ____ DATE ____ CLASS ____

STUDY GUIDE — Chapter 28

Conservation and Protection — Text Pages 654–657

Complete the following sentences using the appropriate terms from the textbook.

1. The process of replanting a cut-over area with seedling trees is called **reforestation**.
2. Many communities **recycle** to reduce wastes and reuse precious natural resources.
3. **Conservation** is the wise and careful use of Earth's resources.
4. An area of land set up to protect animal species is a **wildlife preserve**.
5. The Clean Air Act of 1990 required industries to reduce their **sulfur dioxide** emissions by 40 percent by the year 2000.
6. What kind of power can be produced by harnessing the force shown in the illustration below? **hydroelectric power**

*Cross out the statements that **DO NOT** agree with the textbook.*

7. Many countries protect forest resources by developing national parks and forests.
8. More than half of the endangered species worldwide are in the United States.
9. ~~Reforestation causes erosion because the new roots disturb the soil of an area.~~
10. ~~Modern windmills can generate hydroelectric power.~~
11. ~~As the human population grows, there is less need to protect natural resources.~~
12. Water treatment plants kill bacteria and filter out sediments from a community's water supply.
13. Solar energy collected by solar panels during the day can be used to produce electricity at night.
14. ~~The wild turkey that once was common in America now has become extinct.~~
15. Some cities have built trash-burning power plants that can produce electricity.

108

Reducing acid rain is a major focus of international groups today. At least 34 countries have agreed to work on solutions to the worldwide problem of air pollution. One way to reduce sulfur dioxide emissions is to change to another, renewable source of energy. In the desert, solar panels are now used to collect solar energy during the day. At night, this stored energy is used to produce electricity. Modern windmills like those shown in Figure 28-6 generate electricity that can be stored in batteries. Hydroelectric power has been used for many years all over the world. For example, the force of falling water turns turbines to generate electricity used by the city of Niagara Falls, New York. In the future, each country or region in a country may rely on local sources of energy rather than power plants operating far away. What renewable energy sources may be found where you live?

Cleaning Up the Water

When you brush your teeth, do you leave the water running? If you do, you are wasting water! Even though water is Earth's most abundant resource, safe, clean drinking water is scarce. Many communities must spend enormous amounts of money cleaning up water or finding clean water sources. Nearly every large community has water treatment plants that kill bacteria and filter out sediments from the water supply. Some communities reuse water. For example, water used to wash clothes could be recycled and used for washing streets or cars. Some communities have programs that help people learn to conserve water at home.

Farmers conserve soil and water by changing the way they farm. Contour plowing allows farmers to plow across fields to slow down erosion. Terracing is a way to prevent erosion in hilly areas. Today, many farmers no longer plow their fields each spring. Instead, they plant new seeds right through the leftover stubble of the last crop. Farmers can also reduce water pollution by conserving soil and reducing the need for fertilizers and pesticides.

Waste Disposal

Have you ever helped clean up a park or roadside? Wastes are natural resources in different forms. Newspapers, glass bottles, aluminum cans, and even plastic milk cartons are made from natural resources. Paper

Figure 28-6. Modern windmills harness wind energy in places where the wind blows constantly and steadily.

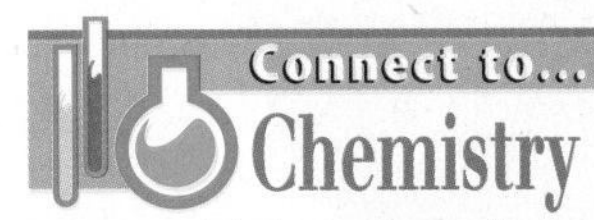

You've read about pH on shampoo bottles. The acidity of a solution is determined by using the pH scale. Substances with a pH lower than seven are acids. The lower the pH number, the stronger the acid. Substances with a pH above seven are bases. The natural pH of rainwater is about 5.0 in the eastern United States and between 5.3 and 6.5 in the western states. Find out what the pH of acid rain is.

● REINFORCEMENT 108

NAME ______ DATE ______ CLASS ______

REINFORCEMENT — Chapter 28

Conservation and Protection — Text Pages 654–657

Answer the following questions using information from the textbook.

1. What is conservation? **the wise and careful use of Earth's resources**
2. What does human population growth have to do with the need for conservation? **As the population grows, limited natural resources are used at a greater rate. Thus, the need to preserve and protect these resources becomes greater.**
3. How might the Clean Air Act of 1990 help reduce acid rain? **The Clean Air Act of 1990 requires that industries reduce sulfur dioxide emissions by 40 percent by the year 2000. Sulfur dioxide emissions contribute to acid rain.**
4. What are three renewable energy sources used by some communities to produce electricity? **solar panels, modern windmills, and hydroelectric power**
5. What does a water treatment plant do? **It kills bacteria and filters out sediments from a community's water supply.**
6. Why does it make sense to recycle glass, aluminum, and plastic products? **These products are made from limited nonrenewable resources.**
7. What are two farming methods used to prevent erosion? **contour plowing and terracing**
8. What is a wildlife preserve? **an area of land set up to protect wildlife species**
9. What do the three animals below have in common? **They were all near extinction but today are found in large numbers.**

alligator — eagle — wild turkey

108

▲ ENRICHMENT 108

NAME ______ DATE ______ CLASS ______

ENRICHMENT — Chapter 28

Conservation and Protection — Text Pages 654–657

DO YOU WASTE WATER?

Most people think water is in abundant supply. It rains from the sky, fills the oceans, and runs down rivers. But what many of us don't know is that water must be in a certain form to be used for cooking, drinking, and bathing. The water we use must be free of most salt, bacteria, and sediments. Safe, clean drinking water is actually scarce. Water purification plants clean the water we drink, but they sometimes have trouble keeping up with the demand for clean water. Do you waste water?

Procedure

1. Over a weekend, keep track of the water used in your home. Ask other family members to help you. They should report to you how they have used water each day.
2. Use the table below to help you make an estimate of how much water is used in your home over one weekend. For example, the table shows that we use 5 gallons per minute when taking a shower. If you take a 10-minute shower, you have used 50 gallons of water.
3. On a separate sheet of paper, keep track of your water use as well as everyone else's use in your home. On Monday morning, add all the figures together to get the weekend total. Then divide that number by the number of people in your household to get the average amount of water used per person over a weekend.

Activity	Water Used
taking a shower	5 gallons per minute
brushing teeth with water running	5 gallons per brushing
taking a bath	40 gallons per bath
cooking	3 gallons per meal
washing clothes	30 gallons per load
washing dishes with running water (by hand)	30 gallons per load
washing dishes (in dishwasher)	20 gallons per load
flushing a toilet	5 gallons per flush

Conclude and Apply

1. How much water did you use? **Answers will vary.**
2. How much did your entire family use? **Answers will vary.**
3. What was the average amount of water used per person? **Answers will vary.**
4. Which activity used the most water over the weekend? **Answers will vary, but may include showering or washing laundry.**
5. What could you and your family do to save water? **Answers will vary, but may include taking shorter showers and turning off the water while brushing teeth.**

108

2 TEACH

Key Concepts are highlighted.

CONCEPT DEVELOPMENT

▶ Explain that there are special devices that can be installed in faucets to help reduce water usage or to help reduce pollutants by charcoal filtering.

▶ In the 1960s, Lake Erie was too polluted for swimming or fishing, but today the lake is cleaner due to governmental cleanup efforts and new laws that require industry to curtail pollution of lake water.

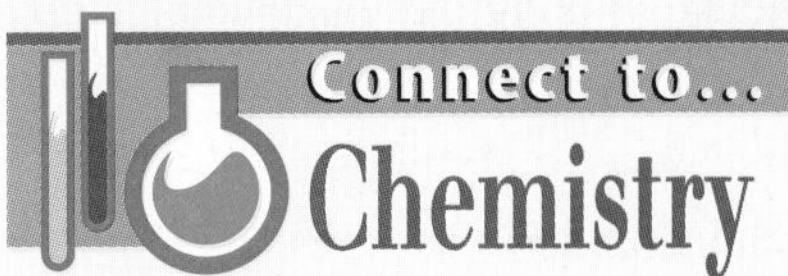

Answer: Acid rain can have a pH of less than 5.0.

CROSS CURRICULUM

▶ **Language Arts:** Have students write to their state legislators to find out what local and state laws relate to pollution and what agencies are responsible for enforcement.

▶ Humans need about 15 to 20 L of water daily to fulfill basic needs. In the United States, each person uses about 800 L each day.

CHECK FOR UNDERSTANDING

Have students write a paragraph explaining how to care for Earth's resources.

RETEACH

Review the causes and effects of acid rain and air and water pollution.

EXTENSION

For students who have mastered this section, use the **Reinforcement** and **Enrichment** masters or other OPTIONS provided.

CONCEPT DEVELOPMENT

▶ Certain methods of cutting down trees help conserve forests. Improvement cutting is the practice of removing unhealthy trees and those with little commercial value. Selective cutting removes only mature trees, leaving the young ones to grow.

▶ Display a state map on the bulletin board and highlight state park and wildlife preserves. Ask how many students have visited these areas.

REVEALING MISCONCEPTIONS

▶ When people use herbicides and pesticides to get rid of pests, the poisons often accidently kill many nonpest species of plants and animals.

MINI QUIZ

Use the Mini Quiz to check students' recall of chapter content.

1. **The wise and careful use of Earth's resources is _____ .** *conservation*
2. **An area of land set up to protect wildlife species is a _____ .** *wildlife preserve*
3. **The process of replanting an area with trees is _____ .** *reforestation*

VideoDisc

STVS: Wind Power, Disc 2, Side 2

STVS: Evaluating Artificial Reefs, Disc 6, Side 1

STVS: Shorebird Preserves, Disc 6, Side 1

products come from trees and can be recycled. Glass, aluminum, and plastic are made from nonrenewable resources.

Today, communities all over the world recycle to reduce wastes and reuse precious natural resources. Some cities collect grass clippings and dead leaves and take them to a community composting facility. Other cities have built trash-burning power plants that use trash to produce electricity. Recycling plastics is a potential industry. In Chicago, one company recycles 2 million plastic milk jugs a year into "plastic lumber."

Did You Know?

Today, the United States has 300 million hectares of forested lands, 20 percent more trees than it had 20 years ago.

Protecting Soil and Forest Resources

Lumber companies in the United States practice selective cutting in which the largest, shade-producing trees are removed from a forest. This provides smaller trees with more sunshine so they grow at an even faster rate. **Reforestation** is the process of replanting a cut-over area 3
with seedling trees. Reforestation helps to reduce soil erosion because the new trees absorb rainwater, and their roots hold soil in place.

Many countries have made efforts to protect forest resources by developing national parks and forests. In Nepal and India, forestry officials are educating farmers about reforestation and helping people create their own forests to harvest for firewood. Fast-growing tree seedlings are planted in community forests so that local people learn how to care for and manage forest resources.

Figure 28-7. Some endangered species may recover and increase in population size once laws are made to protect them and their habitats.

Protecting Wildlife

Even though extinction is a natural process, the rate of extinction of wildlife today is very high. Countries such as Kenya and Tanzania protect endangered species by creating national parks and wildlife preserves. A **wildlife** 2
preserve is an area of land set up to protect wildlife species. Hunting and fishing are prohibited in wildlife preserves to ensure that endangered species come to no harm. Out of more than 1000 species of endangered wildlife worldwide, more than half are in the United States.

Not every endangered species becomes extinct. In fact, since 1900 several endangered species in the United States have recovered as a result of laws that protect these species and their habitats. The wild turkey, pronghorn antelope,

OPTIONS

ASSESSMENT—ORAL

▶ **What does it mean to recycle wastes?** *reusing waste materials to manufacture new products*

▶ **What could you do if you discovered that someone was dumping garbage illegally in your area?** *report it to the local law enforcement agents*

▶ **Why is it safer to destroy aphids with ladybird beetles than with a pesticide?** *They do not harm other species in the area.*

▶ **What are ways that people affect wildlife?** *destroying their natural habitats, using pesticides, and overhunting*

▶ **What might require people to find new food crops?** *widespread plant disease, nuclear war, widespread drought, or climate change*

TECHNOLOGY

Test Tube Tigers

Wild Bengal tigers have been pushed into isolated areas by human encroachment into their habitat. This separates potential mates and increases the chance of inbreeding. Captive populations also have breeding problems. Nicole, a captive Siberian tigress, became the first surrogate tiger mother when she gave birth to cubs conceived through in vitro fertilization.

Nicole and two Bengal tigresses were given an injection to induce ovulation. Ripe eggs were extracted from the Bengal tigresses and mixed in a petri dish with sperm from a white tiger. Fifteen embryos were implanted in Nicole. She gave birth to three Bengal tiger cubs. The male cub and one female did not live long due to medical problems, but the second female is fine.

Think Critically: How could in vitro fertilization be used to help wild populations?

bison, bald eagle, and American alligator are five species that have gone from near extinction to large numbers of individuals today. Wildlife preserves, new conservation laws, and public education programs help to conserve wildlife and protect wildlife habitats.

SECTION REVIEW

1. Define two conservation methods used by farmers.
2. List three sources of energy that are renewable.
3. Why can recycling aluminum conserve resources?
4. How does reforestation conserve soil and forest resources?
5. **Apply:** Describe three things that you can do to conserve and protect natural resources.
6. **Connect to Chemistry:** Find out how acid rain is formed.

Concept Mapping

Skill Builder

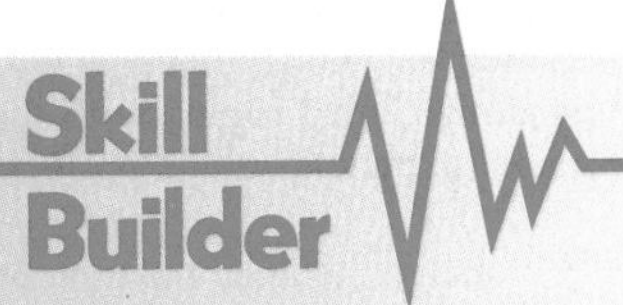

Draw a concept map that shows events that occur when a forest is clear-cut. Include the terms: *clear-cut, soil erodes, animal species leave, rainfall, no topsoil left, land unable to support plant life.* If you need help, refer to Concept Mapping in the **Skill Handbook** on pages 688 and 689.

ENRICHMENT

▶ Have students list the names of organizations that are trying to preserve wildlife in their area.

▶ Have students research and report on the passenger pigeon and how it became extinct.

PROGRAM RESOURCES

From the **Teacher Resource Package** use:

Activity Worksheets, page 5, Flex Your Brain.

Critical Thinking/Problem Solving, page 32, Protecting Our Natural Resources.

Concept Mapping, pages 61-62.

Cross-Curricular Connections, page 32, Solar Energy.

Technology, pages 19-20, Solar Power Using Photosynthesis as a Model.

TECHNOLOGY

Extension: Research and list animals in your area that are on the endangered species list. Do a report on one animal.

Reference: Stolzenburg, W. "Battling Extinction with Test-tube Tigers." *Science News*, May 26, 1990, p. 327.

Think Critically: It could be used to increase the genetic pool for reproduction by moving genetic material from one population to another.

3 CLOSE

▶ Ask questions 1-4 and the **Apply** Question in the Section Review.

SECTION REVIEW ANSWERS

1. contour plowing, terracing, planting through stubble of last crop
2. solar energy, wind energy, hydroelectric power
3. Aluminum is a nonrenewable resource. Recycling keeps it available for continued use.
4. The new trees absorb rainwater and their roots hold soil in place.
5. **Apply:** recycle nonrenewable resources, turn off water, use unleaded gas, and so on
6. **Connect to Chemistry:** Acid rain is formed when sulfur dioxide from coal-burning power plants combines with moisture in the air to form sulfuric acid and when nitrogen oxide from car exhausts combines with moisture in the air to form nitric acid.

Skill Builder

Map should show the items listed in the following order: clear-cut, animal species leave, rainfall, soil erodes, no topsoil left, land unable to support plant life.

Skillbuilder

ASSESSMENT

Portfolio: Use this Skillbuilder to assess students' abilities to form a Concept Map of events following clear-cutting of a forest.

SECTION 28-3

PREPARATION

SECTION BACKGROUND

▶ The future of all the people on Earth depends on what we do today.

1 MOTIVATE

▶ **Bulletin Board:** Display pictures on the bulletin board that show how people are changing the environment.

2 TEACH

Key Concepts are highlighted.

MINI-Lab

Materials: paper, pencils, survey form

Troubleshooting: Make sure students develop a survey form as a group and that they ask about oil from cars and old tires.

Answer: Students should be able to come up with a format for a survey and a way to tabulate results.

CHECK FOR UNDERSTANDING

To check for understanding, ask students to make a plan for environmental management around their home.

28-3 Future Responsibility

New Science Words

environmental management

Objectives

- ▶ Explain the importance of environmental management.
- ▶ Identify ways in which individuals can contribute to environmental management.
- ▶ Describe ways in which individuals can conserve and protect natural resources.

Environmental Management

If you are an architect thinking about building on a hilly site, what do you need to know? You need to know the soil type and whether there is rock underneath to support the building's weight. If there will be landscaping around the finished building, you may want to preserve the large trees that are growing there. If there is a stream at the bottom of the hill, you may have to control erosion during construction. Looking carefully at natural resources and thinking about how your actions will affect the site is one example of environmental management. **Environmental management** is the use of methods that conserve resources and protect ecosystems.

With so many people on Earth, natural resources and ecosystems are stressed. Whenever a development is planned, its effects on the environment must be considered and weighed against the benefits. When the Alaska oil pipeline was built, many people were concerned about damage to wildlife populations and the permafrost. These environmental hazards had to be weighed against the benefits of the pipeline. In many cases, changes made at the planning stages can reduce or eliminate damage to the environment.

Figure 28-8. In planning for development, the needs of wildlife must be considered along with the potential benefits of the project.

Individual Responsibility

Have you ever bought an apple at a grocery store and bitten into it, only to find a worm? Chances are this has never happened to you because commercial apple growers spray pesticides on their trees to prevent damage from pests. Would you purchase apples with wormholes if they

OPTIONS

Meeting Different Ability Levels

For Section 28-3, use the following **Teacher Resource Masters** depending upon individual students' needs.

◆ **Study Guide Master** for all students.

● **Reinforcement Master** for students of average and above average ability levels.

▲ **Enrichment Master** for above average students.

Additional Teacher Resource Package masters are listed in the OPTIONS box throughout the section. The additional masters are appropriate for all students.

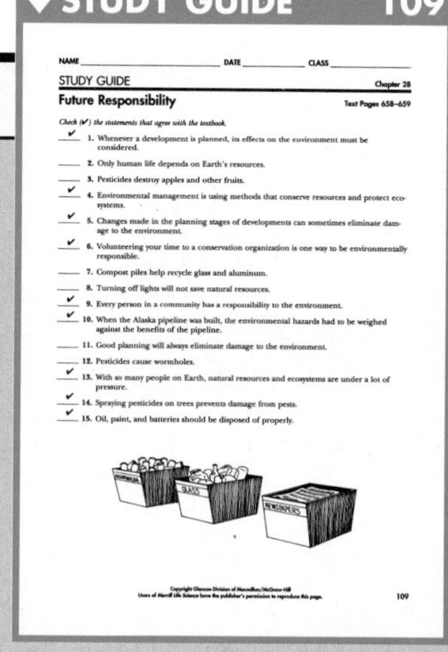

◆ STUDY GUIDE 109

NAME ______ DATE ______ CLASS ______

STUDY GUIDE Chapter 28

Future Responsibility Text Pages 658–659

Check (✔) the statements that agree with the textbook.

✔ 1. Whenever a development is planned, its effects on the environment must be considered.
___ 2. Only human life depends on Earth's resources.
___ 3. Pesticides destroy apples and other fruits.
✔ 4. Environmental management is using methods that conserve resources and protect ecosystems.
✔ 5. Changes made in the planning stages of developments can sometimes eliminate damage to the environment.
✔ 6. Volunteering your time to a conservation organization is one way to be environmentally responsible.
___ 7. Compost piles help recycle glass and aluminum.
___ 8. Turning off lights will not save natural resources.
✔ 9. Every person in a community has a responsibility to the environment.
✔ 10. When the Alaska pipeline was built, the environmental hazards had to be weighed against the benefits of the pipeline.
___ 11. Good planning will always eliminate damage to the environment.
___ 12. Pesticides cause wormholes.
✔ 13. With so many people on Earth, natural resources and ecosystems are under a lot of pressure.
✔ 14. Spraying pesticides on trees prevents damage from pests.
✔ 15. Oil, paint, and batteries should be disposed of properly.

109

were grown without chemicals? What are you willing to do to help protect and conserve natural resources?

In your community the police, sanitation workers, school teachers, firefighters, and homeowners all have responsibilities. You have the responsibilities of going to school, doing homework, and helping at home. Everyone must obey laws. But in addition to these, every person in your community also has responsibilities to the environment in which he or she lives.

What can you do to help? Help your family separate household garbage and recycle glass, plastic, aluminum, and paper. Use leftover foods to make a compost pile, and use the compost to improve your garden's soil. Take your shopping bag to the grocery store, and don't buy prepackaged foods. Always turn off the lights when you leave a room. These are things you can do to save energy, use resources wisely, and improve the environment.

In your community, you can volunteer your time to a conservation or environmental organization. If your school or community doesn't have a recycling program, start one! Begin a tree planting program, or ask your class to adopt a stretch of highway. Make sure that used oil, paint, and batteries are disposed of properly. Remember—there is only one Earth, and all life depends on Earth's limited resources. If each person conserves and protects these resources, life on Earth will be able to continue.

MINI-Lab

How many families in your school recycle?

With the help of your teacher and approval of your principal, develop a survey that can be sent home asking families if, what, and where they recycle. Set a deadline for the return of the papers. *Organize* a way to tally the *data* and be sure to report to the families on the results. Try to use recycled paper for this activity!

SECTION REVIEW

1. Why is environmental management important?
2. Name four things you can do to help conserve and protect natural resources.
3. **Apply:** If building a house on a site will result in erosion that will affect a stream, what could the builder do to eliminate the erosion? Explain.
4. **Connect to Chemistry:** How might a pesticide such as DDT be present in species other than the crop-eating insects it was intended to kill?

Recognizing Cause and Effect

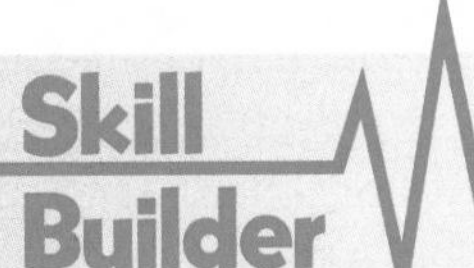

List as many causes and effects of pollution in your community as you can. Suggest ways each source could be reduced or eliminated. If you need help, refer to Recognizing Cause and Effect in the **Skill Handbook** on page 683.

● REINFORCEMENT 109

NAME ______ DATE ______ CLASS ______

REINFORCEMENT — Chapter 28

Future Responsibility — Text Pages 658–659

Answer the following questions using information from the textbook.

1. What is environmental management? **the use of methods that conserve resources and protect ecosystems**

SITE OF PROPOSED HIGHWAY

2. As the proposed highway through the Sierra Forest is planned, what should be considered? **The environmental hazards or damage should be weighed against the benefits of the highway.**
3. What might be damaged if this road is built? **the forest ecosystem and the organisms in it**
4. Why were people concerned about the construction of the Alaska oil pipeline? **They were concerned about damage to wildlife and the permafrost.**
5. Why do fruit growers use pesticides? **Pesticides prevent damage to the fruit from pests.**
6. If pesticides were not used, how would consumers have to change their attitudes? **They would have to become more willing to buy damaged fruit.**
7. In deciding to use pesticides, what two things are considered by the farmers and sellers of fruit? **They weigh the damage done to the environment by pesticides against the benefits of undamaged fruit to the consumer.**
8. Name three ways that you can be environmentally responsible. **recycle glass, plastic, aluminum, and paper; use leftovers to make a compost pile; take your own bag to the grocery store; don't buy pre-packaged foods; turn off lights when leaving a room; volunteer for a conservation organization**

 109

▲ ENRICHMENT 109

NAME ______ DATE ______ CLASS ______

ENRICHMENT — Chapter 28

Future Responsibility — Text Pages 658–659

PLANT A TREE TO SAVE THE EARTH

If you could only do one thing to help improve conditions on Earth, planting a tree would be an excellent choice. For many years, people have been clearing forests for farmland and using the wood for fuel and to build homes. Between 1950 and 1980, the forested parts of Earth were reduced by about 25 percent.

Cutting down these trees adds carbon dioxide to the atmosphere. As the trees rot or burn, carbon dioxide is released into the atmosphere. Live trees and plants use carbon dioxide during photosynthesis, but with many trees gone, the carbon dioxide remains in the atmosphere. If you planted a tree, that tree would use carbon dioxide and release oxygen during photosynthesis. Trees not only look beautiful, but they help clean the air.

Materials
- small tree
- shovel
- topsoil or peat moss
- water
- mulch

Procedure
1. Buy a small tree at a local nursery, arboretum, botanical garden, or garden store. Some places, such as your local arboretum, may give you a tree to plant. Be sure to choose a variety that grows well in your climate.
2. Ask how to plant the tree when you get it. The best times to plant trees are spring and fall.
3. Choose a sunny spot for your tree. Some trees prefer shade, but most need a sunny, uncrowded area to grow well.
4. With the shovel, dig a hole deep enough for the roots of the tree. The hole should be large enough for the roots to grow easily, but be sure not to plant the tree too deep. Loosen the soil on the bottom and the sides of the hole. This will help the roots grow more easily.
5. Place the tree in the hole and fill in around the sides with a mix of the soil you took out of the hole and topsoil or peat moss. Pat the soil down to make it firm and cover the top with about an inch of the soil mixture. Pat this soil down too.
6. Put a mulch of shredded bark, leaves, or straw over the soil to help keep in moisture.
7. Water the tree thoroughly. Water it once a week during the spring and summer months for the first two years.

Caring for your tree as it grows can be fun. You should stake it to protect it from heavy winds. You may want to put a small fence around it to protect it from animals. Find out how and when to fertilize your tree. Sometimes no matter what you do, a small tree will not live. If that happens, try to figure out why and plant another one. Remember, this is your environment and you have a responsibility for the future.

If you want to find out more about how you can help plant trees in your community, you can write to the American Forestry Association, Global Relief Program, P.O. Box 2000, Washington, D.C., 20013.

 109

RETEACH

Have students write a paper about what their individual responsibilities are toward their environment.

EXTENSION

For students who have mastered this section, use the **Reinforcement** and **Enrichment** masters or other OPTIONS.

3 CLOSE

► Ask questions 1-2 and the **Apply** Question in the Section Review.

SECTION REVIEW ANSWERS

1. Natural resources are under pressure from overpopulation. Humans must conserve and recycle resources to survive on Earth.

2. separate garbage for recycling, make a compost pile, use reusable shopping bags, turn off lights, etc.

3. Apply: The builder could terrace the site to hold soil and water, prevent erosion by not disturbing the soil and leaving trees on the site, or build during a season with little rainfall.

4. Connect to Chemistry: The pesticide accumulates in lower organisms and is passed on to predators higher in the food chain. It could be passed to humans that drank the water or ate fish that consumed plankton.

Skill Builder

Answers will vary. Students may list automobiles, factories, power plants, and so on.

Skillbuilder

ASSESSMENT

Performance: Use this Skillbuilder to assess students' abilities to Recognize the Causes and Effects of pollution in their community and make specific lists.

PROGRAM RESOURCES

From the **Teacher Resource Package** use:

Activity Worksheets, page 255, Mini-Lab: How many families in your school recycle?

SECTION 28-4

PREPARATION

SECTION BACKGROUND

▶ Environmentalists are interested in preserving the environment for future generations by recycling and using energy resources that will not harm the environment.

1 MOTIVATE

▶ Have a local recycling company send someone to talk to the class.

2 TEACH

Key Concepts are highlighted.

FLEX Your Brain

Use the Flex Your Brain activity to explore PRESERVE EARTH.

Flex Your Brain
ASSESSMENT
Performance: Use the Flex Your Brain activity to reinforce critical-thinking and problem-solving skills.

VideoDisc
STVS: Bird Sanctuary, Disc 6, Side 1
STVS: Treating Acid Lakes, Disc 6, Side 2

28-4 Earth in 2030

New Science Words

environmental activists

Objectives

- ▶ Identify three issues that concern environmental activists.
- ▶ Describe ways to improve environmental conditions on Earth.

The Future: Grand or Grim?

Pretend you're in a time machine heading for the year 2030. You are hoping for the best, but you know that when you left the 1990s things were looking grim. Earth's population had nearly doubled between 1960 and 1990. More than 400 hectares of forest were being destroyed each day. Pollutants in the air were causing global warming and acid rain. In the United States, people were throwing away 160 million tons of solid waste every year. They were not thinking about the environment of future generations.

As you leave the time machine in the year 2030, you are very afraid of what you might find. Will the world still exist? What if you can't return? To your surprise, Earth is beautiful. It soon becomes apparent why. Huge numbers of people became **environmental activists,** people who worked to conserve Earth's resources and protect the environment.

The plastic containers that you sent to the recycling center are still here, although you don't recognize them. They've been molded into building material that is used to make park benches. No coal, oil, or natural gas is sold. Nuclear energy is not avail-

OPTIONS

Meeting Different Ability Levels

For Section 28-4, use the following **Teacher Resource Masters** depending upon individual students' needs.

◆ **Study Guide Master** for all students.

● **Reinforcement Master** for students of average and above average ability levels.

▲ **Enrichment Master** for above average students.

Additional Teacher Resource Package masters are listed in the OPTIONS box throughout the section. The additional masters are appropriate for all students.

◆ STUDY GUIDE 110

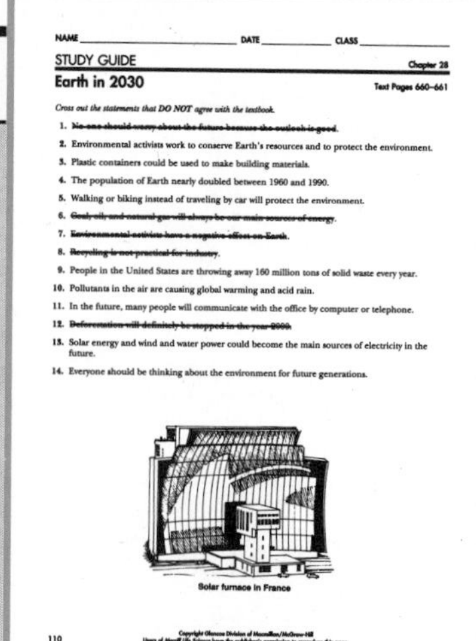

NAME ____________ DATE ________ CLASS ________

STUDY GUIDE — Chapter 28

Earth in 2030 — Text Pages 660–661

Cross out the statements that DO NOT agree with the textbook.

1. ~~No one should worry about the future because the outlook is good.~~
2. Environmental activists work to conserve Earth's resources and to protect the environment.
3. Plastic containers could be used to make building materials.
4. The population of Earth nearly doubled between 1960 and 1990.
5. Walking or biking instead of traveling by car will protect the environment.
6. ~~Coal, oil, and natural gas will always be our main sources of energy.~~
7. ~~Environmental activists have a negative effect on Earth.~~
8. ~~Recycling is not practical for industry.~~
9. People in the United States are throwing away 160 million tons of solid waste every year.
10. Pollutants in the air are causing global warming and acid rain.
11. In the future, many people will communicate with the office by computer or telephone.
12. ~~Deforestation will definitely be stopped in the year 2000.~~
13. Solar energy and wind and water power could become the main sources of electricity in the future.
14. Everyone should be thinking about the environment for future generations.

Solar furnace in France

110

able either. Solar energy and wind and water power are the main sources of electricity now. Homes are super-insulated to prevent cooling or heating. People don't drive to work, but walk or bike instead. Many people work in their homes and communicate with the office by computer or telephone. Even shopping is done from the home through computers. Air quality is good because deforestation was stopped in the year 2000. Landfills are no longer used, and everything is recycled over and over again. Nearly all material used in industry is made from recycled products, including the clock that tells you it is time to head back to the 1990s!

Plutonium is a radioactive element used as a fuel in nuclear reactors. Find out some of the characteristics of this fuel as they affect the human body.

SECTION REVIEW

1. Name three environmental crises that need our immediate attention.
2. What are three things that you can begin doing today that will affect Earth tomorrow?
3. **Connect to Physics:** Give examples of renewable energy sources.

You Decide!

In the United States, most energy for heat and electricity is produced by burning fossil fuels. Nuclear power generation is a much cleaner process, yet it results in radioactive wastes such as plutonium. Most used plutonium is stored in vats of water at power plants because there is no material that can be used to make containers to hold plutonium for the 240 000 years it is radioactive. Should the United States switch from fossil fuels to nuclear power for its energy needs?

3 CLOSE

CHECK FOR UNDERSTANDING

▶ Ask questions 1-2 in the Section Review.

RETEACH

Ask a representative of the National Audubon Society or Sierra Club to talk to the class about the environmental problems that our world is facing.

EXTENSION

For students who have mastered this section, use the **Reinforcement** and **Enrichment** masters or other OPTIONS.

Answer: Radioactive plutonium can collect in bones and change production of white blood cells.

SECTION REVIEW ANSWERS

1. global warming, acid rain, solid waste disposal
2. recycling; using solar, wind, or water power for energy needs; walking or riding a bike rather than driving
3. Connect to Physics: sun, wind, water, and geothermal energy from magma beneath volcanoes

Answers will vary. Accept all reasonable answers.

● REINFORCEMENT 110

NAME ________ DATE ________ CLASS ________

REINFORCEMENT — Chapter 28

Earth in 2030 — Text Pages 660-661

Circle the number of the actions that would make the environment a better place for future generations to live.

(1.) recycling products made from nonrenewable resources
2. using coal for power in all factories
3. building larger sanitary landfills
(4.) using solar energy to produce electricity
(5.) walking or biking to work or school
6. increasing global warming
(7.) insulating homes and work places
(8.) stopping deforestation
9. increasing Earth's human population
(10.) becoming an environmental activist

Answer the following questions using the information from the textbook.

11. Who are environmental activists? **people who work to conserve Earth's resources and protect the environment**
12. What might Earth look like in 2030 if no one is environmentally responsible? **Answers will vary. Students should mention the destruction caused by acid rain, the effects of global warming, the accumulation of solid wastes, the decline of many species, and the pollution of air and water.**

110

▲ ENRICHMENT 110

NAME ________ DATE ________ CLASS ________

ENRICHMENT — Chapter 28

Earth 2030 — Text Pages 660-661

BECOME AN ENVIRONMENTAL ACTIVIST

Many Americans, as well as people from around the world, have formed organizations that work to protect the environment. Each organization has a slightly different focus. Some, such as the National Wildlife Federation, have programs that work to protect the environment in many ways. Others, such as the Rainforest Action Network, focus on one environmental issue. The type of organization that interests you depends on your beliefs.

If you are interested in becoming an environmental activist, write to two or three of the organizations listed below. In your letter, tell them that you are a student who wants to help the environment. Ask them to send you literature about what the organization does and how you might join.

Environmental Defense Fund Inc.
257 Park South
New York, NY 10010
Membership: 100 000

Friends of the Earth
218 D Street SE
Washington, D.C. 20003
Membership: 12 000

Greenpeace, USA, Inc.
1436 U Street NW
Washington, D.C. 20009
Membership: 3 million

Izaak Walton League of America
1401 Wilson Blvd., Level B
Arlington, VA 22209
Membership: 50 000

National Audubon Society
950 3rd Avenue
New York, NY 10022
Membership: 550 000

National Wildlife Federation
1400 16th Street NW
Washington, D.C. 20036-2266
Membership: 5 million

The Nature Conservancy
1815 North Lynn Street
Arlington, VA 22209-2016
Membership: 550 000

Rainforest Action Network
301 Broadway, Suite A
San Francisco, CA 94133
Membership: 20 000

Sierra Club
730 Polk Street
San Francisco, CA 94109
Membership: 500 000

The Wilderness Society
1400 I Street NW, 10th Floor
Washington, D.C. 20005
Membership: 220 000

World Wildlife Fund
1250 24th Street NW
Washington, D.C. 20037
Membership: 600 000

110

PROGRAM RESOURCES

From the **Teacher Resource Package** use:

Activity Worksheets, page 5, Flex Your Brain.

Science and Society, page 32, How can you help save the Earth?

Use **Laboratory Manual,** page 179, Home Energy Needs.

Use **Laboratory Manual,** page 189, Living Space.

ACTIVITY 28-2
40 minutes

OBJECTIVE: Construct a device to measure air pollution and **observe** the amounts of particulates in different areas.

PROCESS SKILLS applied in this activity:
▶ **Recording** in Procedure Steps 7 and 10.
▶ **Hypothesizing** in Procedure Step 8.
▶ **Examining** in Procedure Step 10.
▶ **Inferring** in Conclude and Apply Questions 6 and 7.

COOPERATIVE LEARNING Use the Science Investigation strategy. Have one student gather the materials, one student read the activity, one student record and report, and one student check for safety and be the timekeeper.

TEACHING THE ACTIVITY
Alternate Materials: Glass slides can be covered with petroleum jelly and placed in different locations if there is not enough time to prepare the strips.
Troubleshooting: To prevent contamination, caution students not to let the petroleum jelly on the waxed paper touch anything and not to touch it.
▶ Have the students hang the strips downwind and upwind from chimneys.
▶ Discuss particulate air pollution with the students.

ANSWERS TO QUESTIONS
1-5. Answers will vary.
6. More particulates will be found on a windy day because particles will be blown from surrounding areas to the location of the strip.
7. Answers will vary. Students may suggest harmful gases, acid rain, and so on.

Activity ASSESSMENT
Performance: To further assess students' understanding of air pollution, see USING LAB SKILLS, Question 12, page 664.

ACTIVITY 28-2 Identifying Air Pollution Sites

Problem: *Where can you find air polluted with particulates?*

Material 
- 8 strips of cardboard 5 cm × 20 cm
- 4 strips of waxed paper 5 cm × 20 cm
- scissors
- hole punch
- string
- petroleum jelly
- stapler and staples

Procedure
1. Draw circles 3 cm in diameter in each of 4 strips of cardboard. Cut out the circles. Label each 1, 2, 3, or 4.
2. Cover one side of a waxed paper strip with petroleum jelly.
3. Place the waxed paper strip behind the circles on the cardboard strip with the petroleum jelly facing out. Tape in place.
4. Cover the back of the waxed paper strip with a second cardboard strip. Staple the cardboard strips together.
5. Punch a hole in one end of the double cardboard strip. Place a piece of string through the hole and make a loop in it.
6. Hang each of the strips in a different location. Choose one location that is inside a building. Handle the strips carefully so that you do not touch the petroleum jelly in the circles. Tie each strip to an object with the string so that it will not blow away.
7. Record the locations.
8. **Hypothesize** which location you expect to have the most particulates.
9. Wait two days and collect the strips. Use care not to smudge the petroleum-coated sides.
10. Examine the circles on each strip with a hand lens. Record your observations.

Data and Observations Sample Data

Strip	Location	Observations
1		Answers will depend
2		on the sites chosen by
3		students for observation.
4		

Analyze
1. What particulates did you *observe* on the strips hung outside?
2. What particulates did you observe on the strip hung indoors?
3. Which site had the most air particulates?
4. Which site had the least air particulates?
5. How did your results *compare* with your **hypothesis?**

Conclude and Apply
6. Would you expect to find more or fewer particulates on a windy day? Why?
7. What types of air pollution cannot be detected on the strips?

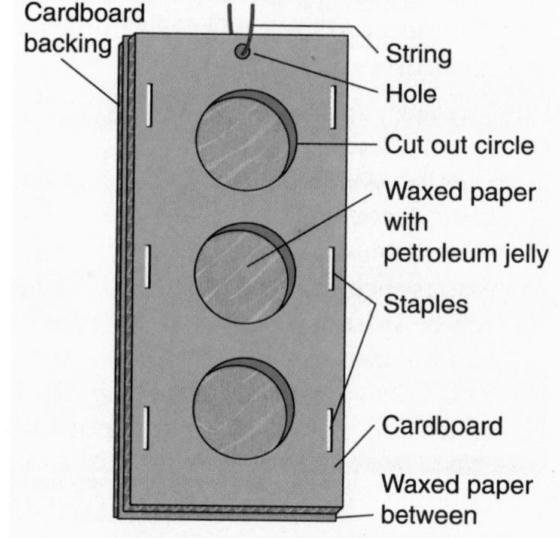

PROGRAM RESOURCES
From the **Teacher Resource Package** use:
Activity Worksheets, pages 250-251, Activity 28-2, Identifying Air Pollution Sites.

REVIEW

SUMMARY

28-1: Natural Resources

1. All living things need natural resources such as air and water to stay alive.
2. Renewable resources can be replaced by nature. Nonrenewable resources are only available in limited amounts.
3. Human use of resources causes pollution of air, water, and land.

28-2: Conservation and Protection

1. Conservation is the wise and careful use of natural resources.
2. Renewable resources are conserved through conservation and protection.
3. Nonrenewable resources are limited but can be recycled to be used again.

28-3: Future Responsibility

1. Environmental management uses methods to conserve resources and protect ecosystems.
2. Individuals can contribute to environmental management by giving consideration to the environment when development is planned.
3. Individuals can conserve and protect natural resources by recycling, using resources wisely, and becoming active members of environmental organizations.

28-4: Science and Society: Earth in 2030

1. Environmental activists work to conserve Earth's resources and protect the environment.
2. Finding renewable sources of energy and recycling nonrenewable resources lead to a better environment.

KEY SCIENCE WORDS

a. **conservation**
b. **environmental activists**
c. **environmental management**
d. **erosion**
e. **greenhouse effect**
f. **natural resources**
g. **nonrenewable resources**
h. **reforestation**
i. **renewable resources**
j. **solid wastes**
k. **wildlife preserve**

UNDERSTANDING VOCABULARY

Match each phrase with the correct term from the list of Key Science Words.

1. loss of soil
2. land area for protected wildlife
3. resources that can be replaced by nature
4. raw materials in environment needed by organisms
5. unwanted waste products
6. resources that are limited
7. replanting after trees have been cut
8. retaining heat by Earth's gas layer
9. wise and careful use of Earth's resources
10. the wise and careful use of Earth's resources

OPTIONS

ASSESSMENT

To assess student understanding of material in this chapter, use the resources listed.

COOPERATIVE LEARNING

Consider using cooperative learning in the THINK AND WRITE CRITICALLY, APPLY, and MORE SKILL BUILDERS sections of the Chapter Review.

PROGRAM RESOURCES

From the **Teacher Resource Package** use:

Chapter Review, pages 59-60.

Chapter and Unit Tests, pages 184-187, Chapter Test.

Chapter and Unit Tests, pages 188-189, Unit Test.

CHAPTER REVIEW

SUMMARY

Have students read the summary statements to review the major concepts of the chapter.

UNDERSTANDING VOCABULARY

1. d	**6.** g
2. k	**7.** h
3. i	**8.** e
4. f	**9.** c
5. j	**10.** a

ASSESSMENT

Portfolio

Encourage students to place in their portfolios one or two items of what they consider to be their best work. For each item, ask students to explain why that item was chosen and what they learned from it. Items might be selected from the following.

- For Your Gifted Students game on environmental issues, p. 646
- Enrichment report on strip mining, p. 651
- Skillbuilder hypothesis on raccoon survival, p. 653

Performance

Additional performance assessments may be found in *Performance Assessment* and *Science Integration Activities* that accompany **Merrill Life Science.** Performance Task Assessment Lists and rubrics for evaluating these activities and other products generated throughout the chapter can be found in Glencoe's *Performance Assessment in Middle School Science.*

CHAPTER

REVIEW

CHECKING CONCEPTS

1. d	**6.** b
2. a	**7.** d
3. d	**8.** d
4. a	**9.** d
5. d	**10.** a

USING LAB SKILLS

ASSESSMENT

Use these alternate lab exercises to assess students' understanding of skills used in this chapter.

11. Students may design experiments to show what conditions promote rapid breakdown of some substances, such as carrot peelings and tea leaves. The other items will not show breakdown. An alternate approach would be to have students compare different so-called biodegradable plastic bags. Some do break up when exposed to sunlight.

12. a film of tiny particles; particles in the air dissolved in the precipitation and reappeared when the precipitation evaporated

THINK AND WRITE CRITICALLY

13. Renewable resources can be replaced by nature over time. Nonrenewable resources cannot be replaced by nature.

14. Decomposers break down biodegradable resources into their chemical components and recycle them.

15. Environmental activists can lobby to pass laws to enforce recycling of natural resources, create wildlife preserves, and help clean up air and water pollution.

16. Topsoil is necessary for plants and microorganisms to grow. Without it, no nutrients are contained in the soil.

CHAPTER

REVIEW

CHECKING CONCEPTS

Choose the word or phrase that completes the sentence.

1. Fossil fuels are ____.
- **a.** renewable resources
- **b.** responsible for most land pollution
- **c.** responsible for most water pollution
- **d.** formed from dead organisms

2. Nonrenewable resources include ______.
- **a.** gold **c.** wind energy
- **b.** solar energy **d.** trees

3. Air pollution is caused by all of the following except ______.
- **a.** carbon monoxide **c.** sulfur dioxide
- **b.** nitrogen oxide **d.** oxygen

4. Trapping of ______ causes the greenhouse effect.
- **a.** carbon dioxide **c.** sulfur dioxide
- **b.** carbon monoxide **d.** nitrogen oxide

5. Only ______ percent of water on Earth is clean and safe, and available for human consumption.
- **a.** 3 **c.** 0.3
- **b.** 0.03 **d.** 0.003

6. ______ is (are) nonbiodegradable waste.
- **a.** Grass **c.** Leaves
- **b.** Aluminum cans **d.** Vegetable peelings

7. ______ are in danger of becoming extinct.
- **a.** Natural resources
- **b.** Solid wastes
- **c.** Wildlife preserves
- **d.** Endangered species

8. Problems with sanitary landfills include all of the following except _____.
- **a.** leaking drums
- **b.** containing toxic wastes
- **c.** little room left
- **d.** containing biodegradable wastes

9. ______ can cause erosion of soil.
- **a.** Fertilizers **c.** Reforestation
- **b.** Selective cutting **d.** Clear-cutting

10. An example of an extinct animal is ______.
- **a.** dodo bird **c.** American alligator
- **b.** wild turkey **d.** bald eagle

USING LAB SKILLS

11. In Activity 28-1, you studied biodegradable substances. Design an experiment to determine whether the following items are biodegradable: carrot scrapings, coffee grounds, tea leaves, paper bag, plastic foam cup, and plastic wrap from a package.

12. In Activity 28-2 on page 662, you identified air pollution sites. Wash and rinse a glass container in distilled water available at drug stores. Collect rainwater in the container and then allow the water to evaporate. What do you see? How did it get there?

THINK AND WRITE CRITICALLY

Answer the following questions in your Journal using complete sentences.

13. Distinguish between renewable and nonrenewable resources.

14. What is the importance of decomposers in a sanitary landfill?
15. How can environmental activists affect how resources are used?
16. What happens to land that has lost topsoil?

APPLY

17. Why is it beneficial to grow another crop on soil after the major crop has been harvested?
18. Explain the possible problems pollutants cause to water- and land-dwelling organisms.
19. Explain how car pooling saves fossil fuel.
20. What are the advantages gained by selective cutting rather than clear-cutting a forest?
21. Why is it important to do research now on solar energy, wind power, and water power?

MORE SKILL BUILDERS

If you need help, refer to the Skill Handbook.

1. **Classifying:** Classify each of the following as renewable or nonrenewable: copper, gold, trees, iron, wildlife, fossil fuels, cotton.
2. **Sequencing:** Sequence the making and use of an aluminum can. Then recycle it!
3. **Using Variables, Constants, and Controls:** In an experiment, show differences in air pollutants produced by old and new cars.
4. **Concept Mapping:** Complete an events chain concept map using these occurrences:
   - organisms died millions of years ago forming. . .
   - fossil fuels such as petroleum, which is used to make plastic bags, cups, and . . .
   - gasoline which is burned by cars releasing the gases carbon monoxide and . . .
   - nitrogen oxide, which pollutes the air and returns to Earth in . . .
   - acid rain, which pollutes rivers, lakes, and forests, causing death of many organisms.

PROJECTS

1. Write a report on the recovery of animals that were endangered. How were they saved?
2. Find out what recycled items are available in your community and how you would be able to tell they are made of recycled materials.

APPLY

17. A cover crop prevents erosion and retains soil which allows future crops to grow.
18. Pollutants cause destruction of habitat, diseases, and death to organisms.
19. Car pooling saves gasoline by using gas for only one car to transport up to six people.
20. Selective cutting removes the largest trees that shade out smaller trees. There is more sunlight for smaller trees to grow. Clear-cutting destroys large areas of a forest, resulting in soil erosion. In selective cutting, soil remains in the forest.
21. Eventually, fossil fuels will be used up. In the future, all energy needs will have to be met through renewable resources.

MORE SKILL BUILDERS

1. Classifying: renewable: trees, cotton, wildlife
nonrenewable: fossil fuels, copper, gold, iron
2. Sequencing: Aluminum is dug out of the ground → processed in a plant → made into a can → filled with soda pop → emptied by a consumer → thrown into a recycling bin → taken to plant and reprocessed into another can.
3. Using Variables, Constants and Controls: One design might include identifying emitted particulates by briefly holding up a white sheet behind a parked, running car. (This experiment should not be tried by students.)

4. Concept Mapping:

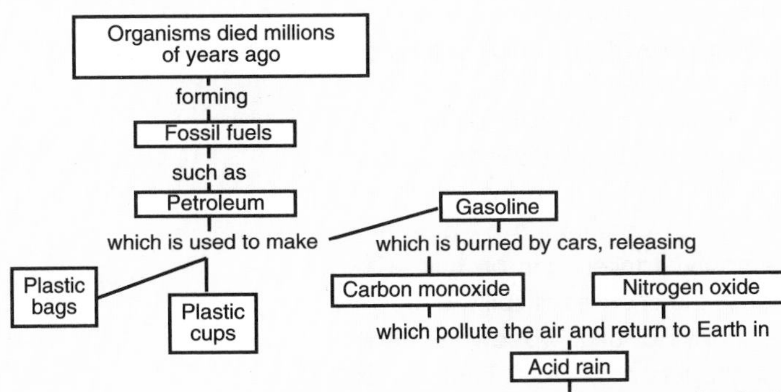

Objective

In this unit ending feature, the unit topic, "Ecology," is extended into other disciplines. Students will see how living things interact with their environment.

Motivate

Cooperative Learning: Have students work in Expert Teams. Assign one Connection to each group of students. Have each group research the geographic location of the Connection and report on its ecology.

Teaching Tips

- Tell students to think about the interrelationships of living things as they read this feature.
- Have students list some dangerous environmental conditions and suggest how these might be prevented.

Wrap-Up

Conclude this lesson by having students suggest ways to help local people prevent erosion, pollution, and the destruction of rain forests.

SOCIAL STUDIES

Background: The rain forests of Brazil contain iron ore, manganese, bauxite, and gold. Many of Brazil's poor homestead in the rain forests. Brazil sees rain forests as natural resources to be used.
Discussion: Ecologists worry that cutting the rain forests will change the world's climate. How can countries balance economic needs and population growth with environmental protection?
Answer to Question: With international help, some of the rain forests could be converted into national parks.
Extension: Have students visit a local city, county, state, or national park and talk to park managers about how the parks were created.

UNIT 8

GLOBAL CONNECTIONS

Ecology

In this unit, you have studied how living things interact with each other. Now find out how that interaction is connected to other subjects and specific places in the world.

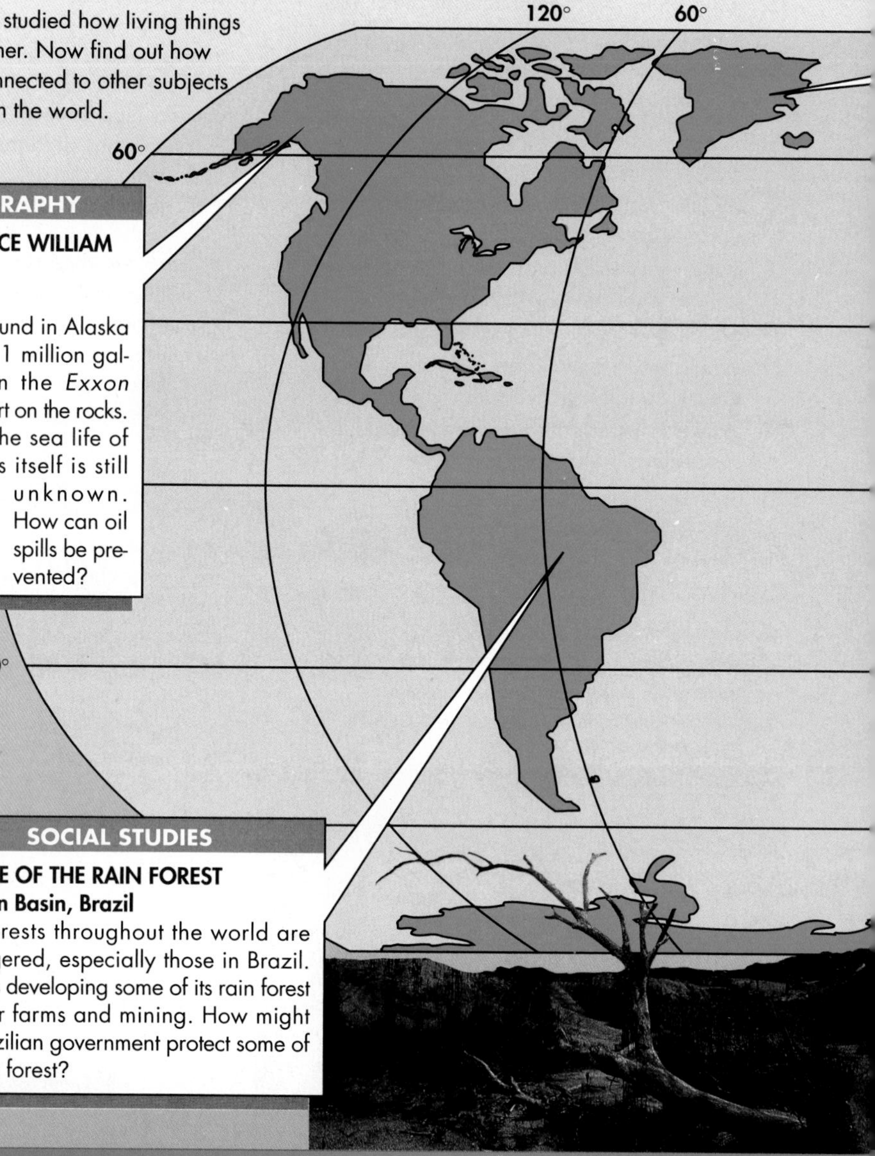

OCEANOGRAPHY

OIL SPILL AT PRINCE WILLIAM SOUND
Valdez, Alaska
Prince William Sound in Alaska received nearly 11 million gallons of oil when the *Exxon Valdez* ripped apart on the rocks. When and how the sea life of this area restores itself is still unknown. How can oil spills be prevented?

SOCIAL STUDIES

DECLINE OF THE RAIN FOREST
Amazon Basin, Brazil
Rain forests throughout the world are endangered, especially those in Brazil. Brazil is developing some of its rain forest land for farms and mining. How might the Brazilian government protect some of the rain forest?

666

OCEANOGRAPHY

Background: Crude oil eliminates the protection that natural oils give to seabirds and other animals. Animals coated with crude oil either drown or die of exposure. Other animals ingest the petroleum.
Discussion: Discuss the problems of oil spill cleanup and the fact that some microorganisms break down oil biologically.
Answer to Question: Oil can be shipped by pipeline, though it is much more costly. Tankers can be built with "double bottoms" to reduce the chances of spills.
Extension: Have students visit a local wastewater treatment plant. Ask plant managers how water treatment is related to bioremediation.

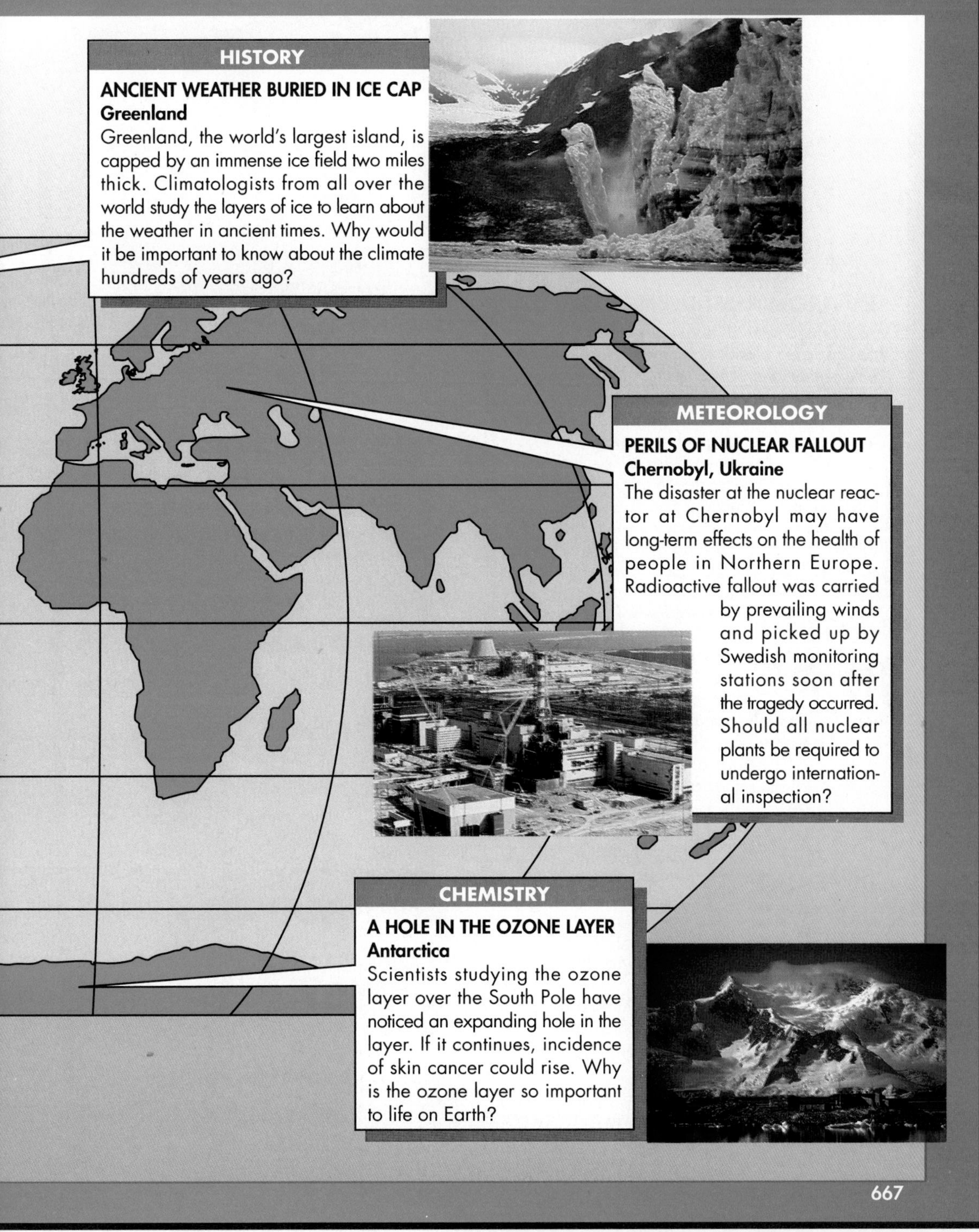

HISTORY

ANCIENT WEATHER BURIED IN ICE CAP
Greenland
Greenland, the world's largest island, is capped by an immense ice field two miles thick. Climatologists from all over the world study the layers of ice to learn about the weather in ancient times. Why would it be important to know about the climate hundreds of years ago?

METEOROLOGY

PERILS OF NUCLEAR FALLOUT
Chernobyl, Ukraine
The disaster at the nuclear reactor at Chernobyl may have long-term effects on the health of people in Northern Europe. Radioactive fallout was carried by prevailing winds and picked up by Swedish monitoring stations soon after the tragedy occurred. Should all nuclear plants be required to undergo international inspection?

CHEMISTRY

A HOLE IN THE OZONE LAYER
Antarctica
Scientists studying the ozone layer over the South Pole have noticed an expanding hole in the layer. If it continues, incidence of skin cancer could rise. Why is the ozone layer so important to life on Earth?

667

METEOROLOGY

Background: Reindeer herds in Sweden and Norway were showered with radioactive fallout after the Chernobyl disaster. Lichens, the deer population's major food source, were also contaminated.
Discussion: Nuclear power offers a substitute for oil. Both cause environmental problems. Should we continue to use nuclear power or search for something else?
Answer to Question: International inspection of all nuclear plants might raise safety standards and ensure proper operation.
Extension: Have students make an energy pyramid that shows accumulation of pollutants in the food chain. Ask, "Which organisms have the greatest amount of stored pollutants?"

HISTORY

Background: Greenland is covered by an ice sheet that has been preserved for thousands of years. Dust falling on the shield and the thickness of the layers provide information about the climatic history of Earth.
Discussion: Ask students why it is important that we learn about climatic history.
Answer to Question: If scientists can see how weather patterns and climate changed over time, they will be better able to predict future changes. Also, such study enables scientists to see if warming and cooling trends are cyclical or caused by human activities.
Extension: Have students interview senior citizens who have always lived in the community. They should ask how weather has changed over the years.

CHEMISTRY

Background: In the upper atmosphere, ozone prevents the sun's damaging ultraviolet radiation from reaching Earth. The ozone layer over Antarctica is growing thinner each year.
Discussion: As the protective ozone layer is depleted, incidence of skin cancer is rising. How can people protect themselves against UV rays?
Answer to Question: The ozone layer protects Earth from damaging ultraviolet rays.
Extension: Ask students if they have ever had a painful sunburn. Do they sunbathe now? Do they use a sunblock? Lead a discussion about damage from ultraviolet rays and how to protect skin.

ENVIRONMENTAL ENGINEER

Background: Environmental engineering is a relatively new technology. Students will need a background in math and science, particularly in environmental problems. Persons interested in such a career will need to have at least a bachelor's degree in engineering with a specialty in environmental science.

Related Career	Education
Environmental Engineer	BS degree
Chemical Engineer	BS degree
Soil Scientist	BS degree

Career Issue: Manufacturers are pressured to use environmentally-safe production methods. Many object to the extra cost. These costs are usually passed on to consumers.

What do you think? Discuss whether consumers should pay for improved methods of manufacturing.

MARINE ANIMAL TRAINER

Background: A marine animal trainer works with dolphins, seals, sea lions, and killer whales. A marine animal trainer needs an understanding of animal behavior, as well as patience and kindness. Interested students need a high school diploma and background in biology. Actual training can be learned on the job.

Related Career	Education
Marine Animal Trainer	High school
Fish Farmer	2 yr. col.
Aquarist	2 yr. col.

Career Issue: Environmentalists feel that incarceration of marine animals is inhumane and does not allow enough freedom for the animals.

What do you think? Discuss whether large marine animals should be captured and placed in marine parks and aquariums.

CAREERS

ENVIRONMENTAL ENGINEER

Environmental engineers apply the theories of science and mathematics to solve real environmental problems. They may design special equipment to study ocean depths or probe outer space. A bachelor's degree in engineering is needed for beginning engineering jobs. Environmental engineering is a new technology and requires a degree in engineering, with environmental concerns as a specialty.

Students interested in environmental engineering should take courses in math and science. They will also need to have a knowledge of environmental problems. They may work for a state or federal environmental agency, for a water purification plant, or for a company that cleans up environmental accidents.

For Additional Information

Contact JETS-Guidance, 1420 King St., Suite 405, Alexandria, VA 22314.

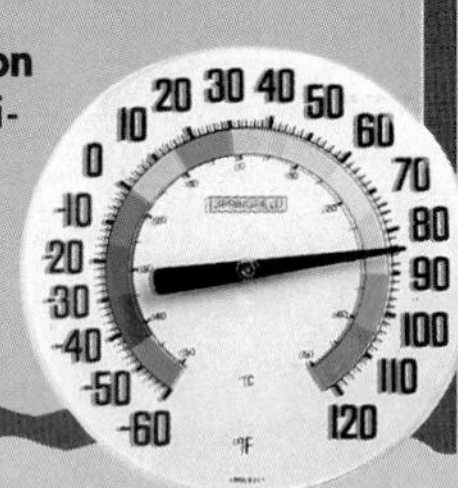

MARINE ANIMAL TRAINER

As marine parks have become more popular, the need for people to train, exhibit, and care for the animal performers has increased. Dolphins, killer whales, and seals don't naturally perform tricks. These wild animals must be taught how to perform, a task that can take months.

A person who wants to become a *marine animal trainer* should have an interest in animals and their behavior. Most animal trainers learn on-the-job as apprentices. They should have at least a high school diploma with credits in biology.

For Additional Information

Contact the International Marine Animals Trainer Association, Brookfield Zoo, Brookfield, IL 60513.

UNIT READINGS

▶Carson, Rachel. *Silent Spring.* New York: Macmillan, 1969.
▶Grove, Noel. "Air: An Atmosphere of Uncertainty." *National Geographic,* April 1987.
▶Stover, Dawn. "Inside Biosphere." *Popular Science,* November 1990.

668

UNIT READINGS

Background

▶ *Silent Spring* is a key work in the history of the environmental movement. It was the first book to alert people to the dangers of pesticides.

▶ "Inside Biosphere" is a highly readable illustrated article on Biosphere II.

▶ "Air: An Atmosphere of Uncertainty" is a beautifully illustrated and comprehensive treatment of the global problem of air pollution. Smog, acid rain, fallout, and global warming are all covered.

More Readings

1. Findley, Rowe. "Will We Save Our Own?" *National Geographic,* September, 1990.
2. Fisher, Arthur. "Soviet Space Odyssey." *Popular Science,* January 1991.
3. Cobb, Charles E. "Living With Radiation." *National Geographic,* April 1989.

SCIENCE & ART

Abuelitos Piscando Napolitos

(Grandparents Picking Prickly Pears)

by Carmen Lomas Garza

The painting, *Abuelitos Piscando Napolitos,* shows the harvest of prickly pear cactus fruits, which are used in Southwestern cooking. The artist, Carmen Lomas Garza, grew up in Kingsville, Texas. The painting is one of several works that represent Garza's response to the Latino movement. The artist wanted to show the culture of Mexican-Americans "in fine art form," as well as portraying the things that are important, beautiful, and moving to the culture. In painting these pictures, she has said that she wished to portray events that are meaningful to all Mexican-American culture.

From the picture, what can you tell about the climate of this area? The prickly pear cactus is common in the American Southwest. A number of desert animals, including the pack rat, use prickly pear cacti for food and a source of water.

Spines on cacti take the place of leaves and help to discourage predators. The fleshy pads that make up the prickly pear and other cacti are actually stems whose spongy tissues hold moisture. In the brief rainy season, these stems expand quickly as the cactus' shallow, but extensive, roots absorb the rainfall.

A number of desert animals use the spiny cacti to protect themselves. The jumping cholla (choy-yuh) cactus has stems that break off at the slightest touch. Large animals find it very unpleasant. But cactus wrens prefer to nest in chollas because they use the cholla stems to protect the entrance to their nests.

Carmen Garza inherited her talent from her grandmother who was a skilled needleworker and from her mother, who was an artist. As a teenager, she decided to become an artist. She pursued her interest in art in college and received master's degrees in both art and education.

In Your Own Words

▶ Think of an area near where you live that you and your family enjoy. It might be picnic area, a public garden, or a park. Draw a picture of the place and write a paragraph on how it relates to the ecology of the area in which you live.

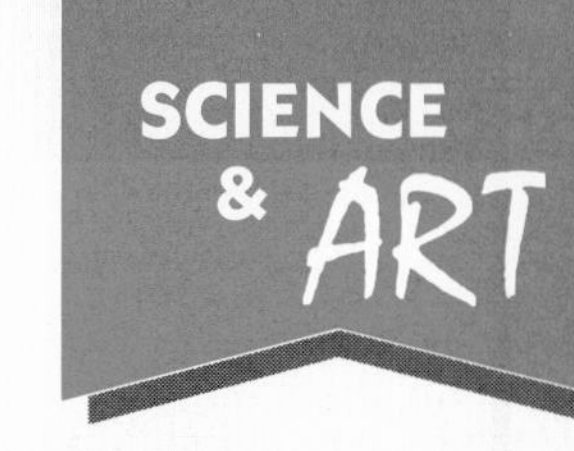

Source: *Hispanic Art in the United States: Thirty Contemporary Painters and Sculptors.* Cross River, NY: Cross River Press, Ltd., 1987.

Biography: Carmen Lomas Garza was born in Kingsville, Texas, in 1948. Her mother's family has lived in Texas for many generations, working as ranch hands and on the railroads. Carmen Garza graduated from Texas Arts and Industries University in Kingsville in 1972 with a B.S. in art education. She continued her education at the Juarez-Lincoln Campus of Antioch Graduate School in Austin, where she earned a master's degree in education in 1973. She then went on to Washington State University at Pullman, and San Francisco State University, where she received a master's degree in art in 1979. Her work has been exhibited in galleries in Texas, Pennsylvania, and New York.

TEACHING STRATEGY

Have each student choose a particular ecological area such as the desert, rain forest, tundra, coast, etc. Have them research the ways in which the ecology of an area may influence the culture of its people. Have selected students present their findings to the rest of the class for discussion.

Cooperative Learning: Have groups of students research Southwestern cooking. Have them locate recipes that use the fruit of the prickly pear cactus. If possible, obtain some prickly pears from your local market.

Other Works:

▶ Other works and articles on ecology include: Leopold, Aldo. *A Sand County Almanac.* Oxford University Press, 1949; and Gibbons, Boyd. "A Durable Scale of Values." *National Geographic,* November, 1981.

Classics

▶ Odum, Eugene. *Fundamentals of Ecology,* Philadelphia: Saunders, 1971, is the classic college-level introductory text on this topic.

▶ Carson, Rachel. *The Sea Around Us.* New York: New American Library, 1954. This book describes physical oceanography and the effect of the ocean on life and the land.

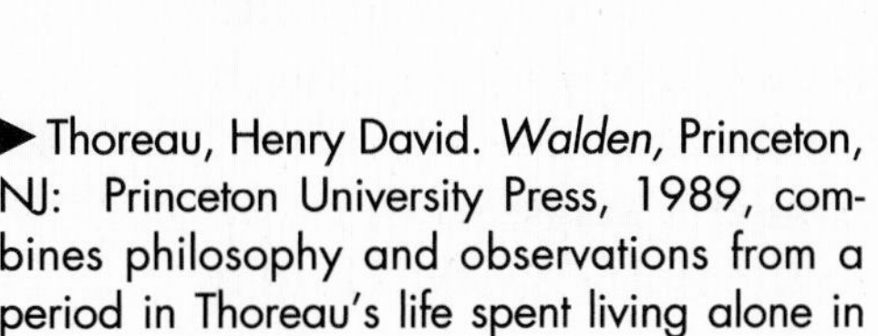

▶ Thoreau, Henry David. *Walden,* Princeton, NJ: Princeton University Press, 1989, combines philosophy and observations from a period in Thoreau's life spent living alone in a cabin on Walden Pond.

APPENDIX A

Care and Use of a Microscope

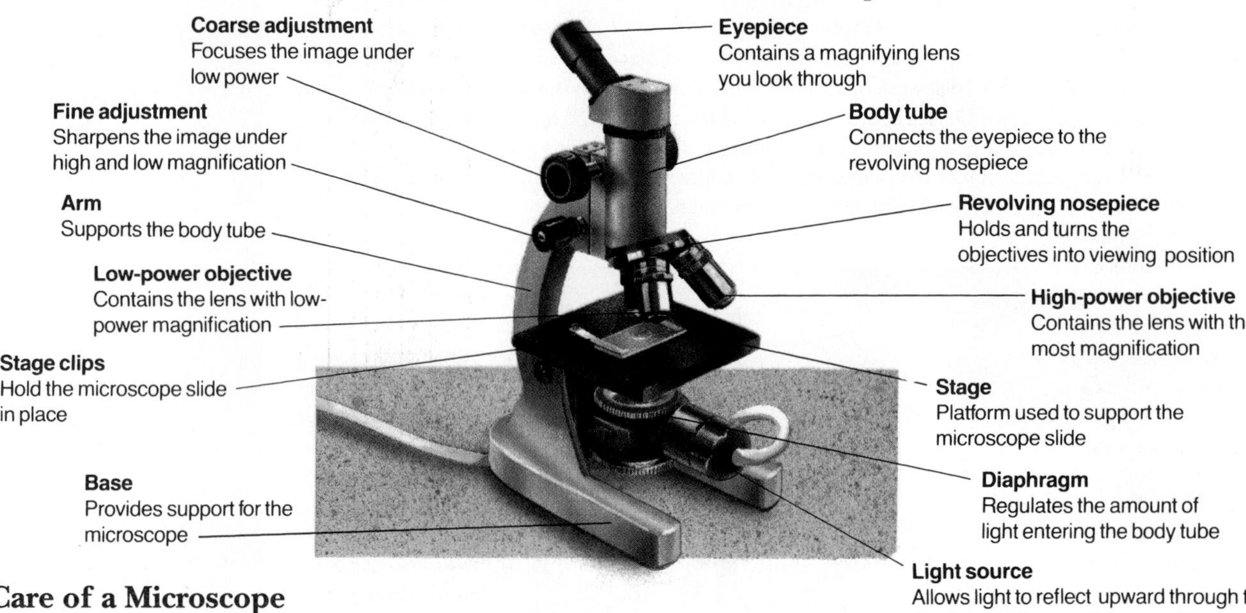

Care of a Microscope

1. Always carry the microscope holding the arm with one hand and supporting the base with the other hand.
2. Don't touch the lenses with your finger.
3. Never lower the coarse adjustment knob when looking through the eyepiece lens.
4. Always focus first with the low-power objective.
5. Don't use the coarse adjustment knob when the high-power objective is in place.
6. Store the microscope covered.

Using a Microscope

1. Place the microscope on a flat surface that is clear of objects. The arm should be toward you.
2. Look through the eyepiece. Adjust the diaphragm so that light comes through the opening in the stage.
3. Place a slide on the stage so that the specimen is in the field of view. Hold it firmly in place by using the stage clips.
4. Always focus first with the coarse adjustment and the low-power objective lens. Once the object is in focus on low power, turn the nosepiece until the high-power objective is in place. Use ONLY the fine adjustment to focus with this lens.

Making a Wet Mount Slide

1. Carefully place the item you want to look at in the center of a clean glass slide. Make sure the sample is thin enough for light to pass through.
2. Use a dropper to place one or two drops of water on the sample.
3. Hold a clean coverslip by the edges and place it at one edge of the drop of water. Slowly lower the coverslip onto the drop of water until it lies flat.
4. If you have too much water or a lot of air bubbles, touch the edge of a paper towel to the edge of the coverslip to draw off extra water and force air out.

APPENDIX B

Table B-1

SI/METRIC TO ENGLISH CONVERSIONS			
	When you want to convert:	**Multiply by:**	**To find:**
Length	inches	2.54	centimeters
	centimeters	0.39	inches
	feet	0.30	meters
	meters	3.28	feet
	yards	0.91	meters
	meters	1.09	yards
	miles	1.61	kilometers
	kilometers	0.62	miles
Mass and Weight*	ounces	28.35	grams
	grams	0.035	ounces
	pounds	0.45	kilograms
	kilograms	2.2	pounds
	tons (short)	0.91	tonnes (metric tons)
	tonnes (metric tons)	1.10	tons (short)
	pounds	4.45	newtons
	newtons	0.23	pounds
Volume	cubic inches	16.38	cubic centimeters
	cubic centimeters	0.06	cubic inches
	cubic feet	0.028	cubic meters
	cubic meters	35.3	cubic feet
	liters	1.06	quarts
	liters	0.26	gallons
	gallons	3.785	liters
Area	square inches	6.45	square centimeters
	square centimeters	0.155	square inches
	square feet	0.09	square meters
	square meters	10.76	square feet
	square miles	2.59	square kilometers
	square kilometers	0.39	square miles
	hectares	2.47	acres
	acres	0.40	hectares
Temperature	Fahrenheit	5/9 (°F – 32)	Celsius
	Celsius	9/5 °C + 32	Fahrenheit

*Weight as measured in standard Earth gravity

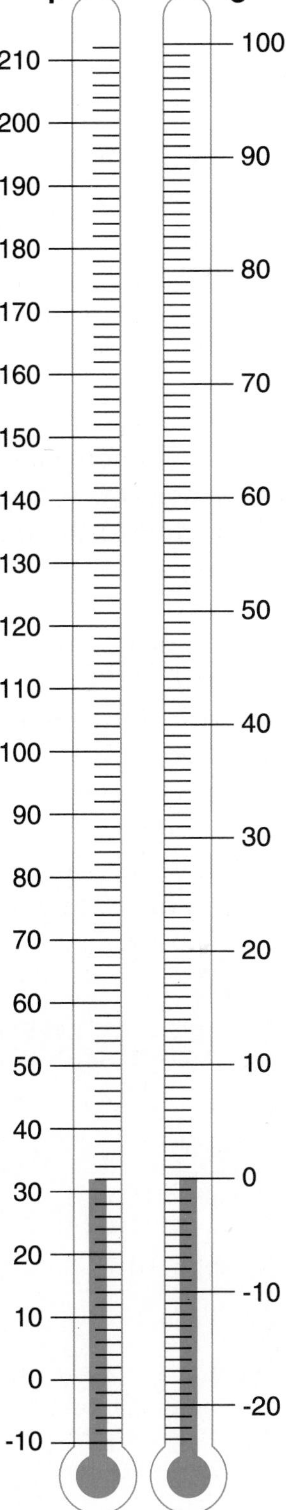

APPENDIX C

Safety in the Classroom

1. Always obtain your teacher's permission to begin an activity.
2. Study the procedure. If you have questions, ask your teacher. Understand any safety symbols shown.
3. Use the safety equipment provided for you. Goggles and a safety apron should be worn when any investigation calls for using chemicals.
4. When you are heating a test tube, always slant it so the mouth points away from you and others.
5. Never eat or drink in the lab. Never inhale chemicals. Do not taste any substance or draw any material into a tube with your mouth.
6. If you spill any chemical, wash it off immediately with water. Report the spill immediately to your teacher.
7. Know the location and proper use of the fire extinguisher, safety shower, fire blanket, first aid kit, and fire alarm.
8. Keep all materials away from open flames. Tie back long hair.
9. If a fire should break out in the classroom, or if your clothing should catch fire, smother it with the fire blanket or a coat, or get under a safety shower. **NEVER RUN.**
10. Report any accident or injury, no matter how small, to your teacher.

Follow these procedures as you clean up your work area.

1. Turn off the water and gas. Disconnect electrical devices.
2. Return materials to their places.
3. Dispose of chemicals and other materials as directed by your teacher. Place broken glass and solid substances in the proper containers. Never discard materials in the sink.
4. Clean your work area.
5. Wash your hands thoroughly after working in the laboratory.

Table C-1

FIRST AID	
Injury	**Safe response**
Burns	Apply cold water. Call your teacher immediately.
Cuts and bruises	Stop any bleeding by applying direct pressure. Cover cuts with a clean dressing. Apply cold compresses to bruises. Call your teacher immediately.
Fainting	Leave the person lying down. Loosen any tight clothing and keep crowds away. Call your teacher immediately.
Foreign matter in eye	Flush with plenty of water. Use eyewash bottle or fountain.
Poisoning	Note the suspected poisoning agent and call your teacher immediately.
Any spills on skin	Flush with large amounts of water or use safety shower. Call your teacher immediately.

SAFETY SYMBOLS

Table C-2

SAFETY SYMBOLS	
DISPOSAL ALERT This symbol appears when care must be taken to dispose of materials properly.	**ANIMAL SAFETY** This symbol appears whenever live animals are studied and the safety of the animals and the students must be ensured.
BIOLOGICAL HAZARD This symbol appears when there is danger involving bacteria, fungi, or protists.	**RADIOACTIVE SAFETY** This symbol appears when radioactive materials are used.
OPEN FLAME ALERT This symbol appears when use of an open flame could cause a fire or an explosion.	**CLOTHING PROTECTION SAFETY** This symbol appears when substances used could stain or burn clothing.
THERMAL SAFETY This symbol appears as a reminder to use caution when handling hot objects.	**FIRE SAFETY** This symbol appears when care should be taken around open flames.
SHARP OBJECT SAFETY This symbol appears when a danger of cuts or punctures caused by the use of sharp objects exists.	**EXPLOSION SAFETY** This symbol appears when the misuse of chemicals could cause an explosion.
FUME SAFETY This symbol appears when chemicals or chemical reactions could cause dangerous fumes.	**EYE SAFETY** This symbol appears when a danger to the eyes exists. Safety goggles should be worn when this symbol appears.
ELECTRICAL SAFETY This symbol appears when care should be taken when using electrical equipment.	**POISON SAFETY** This symbol appears when poisonous substances are used.
PLANT SAFETY This symbol appears when poisonous plants or plants with thorns are handled.	**CHEMICAL SAFETY** This symbol appears when chemicals used can cause burns or are poisonous if absorbed through the skin.

APPENDIX D

Classification

Merrill Life Science uses a five kingdom system of classification. In this system there is one kingdom of organisms that are prokaryotes, organisms that lack a true nucleus and organelles—Kingdom Monera. There are four kingdoms of eukaryotes: Kingdom Protista, Kingdom Fungi, the Plant Kingdom, and the Animal Kingdom.

All the phyla and divisions of organisms discussed in this textbook are outlined in this Appendix.

Kingdom Monera

Phylum Cyanobacteria: one-celled prokaryotes; make their own food, contain chlorophyll, some species form colonies, most are blue-green

Bacteria: one-celled prokaryotes; most absorb food from their surroundings; many are parasites; round, spiral, or rod shaped

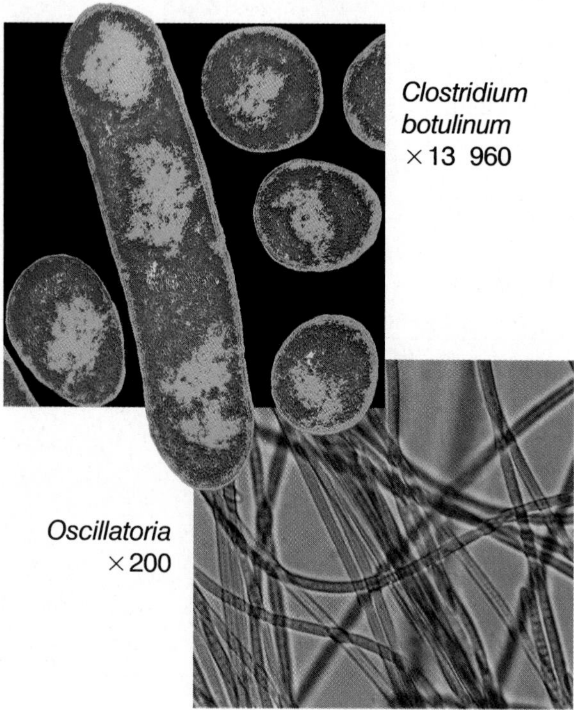

Clostridium botulinum ×13 960

Oscillatoria ×200

Kingdom Protista

Phylum Euglenophyta: one-celled; can photosynthesize or take in food; most have one flagellum

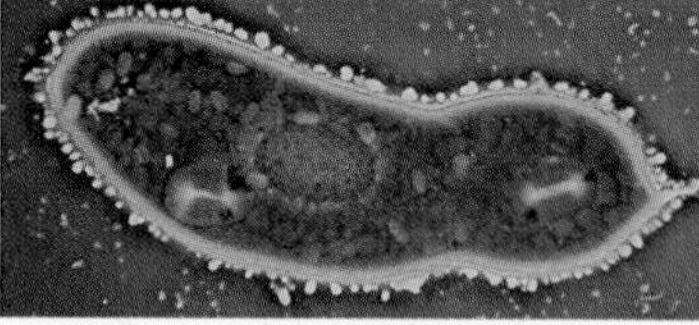

Euglena oxyuris

Phylum Crysophyta: most are one-celled; make their own food through photosynthesis; golden-brown pigments mask chlorophyll; diatoms

Phylum Pyrrophyta: one-celled; make their own food through photosynthesis; contain red pigments and have two flagella; dinoflagellates

Phylum Chlorophyta: one-celled, many-celled, or colonies; contain chlorophyll and make their own food; live on land, in fresh water or salt water; green algae

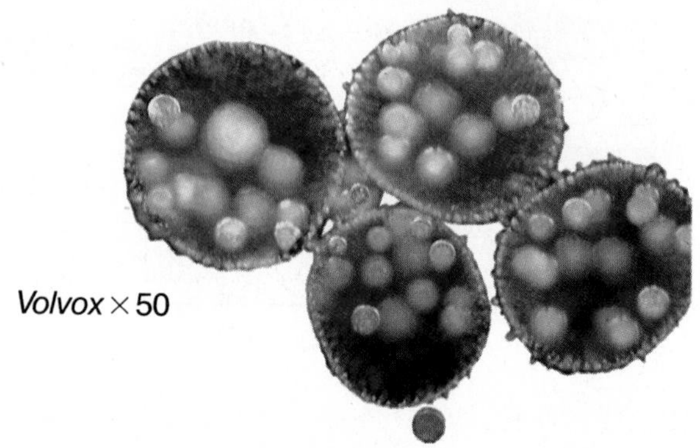

Volvox ×50

Phylum Rhodophyta: most are many-celled and photosynthetic; contain red pigments; most live in deep saltwater environments; red algae

Phylum Phaeophyta: most are many-celled and photosynthetic; contain brown pigments; most live in saltwater environments; brown algae

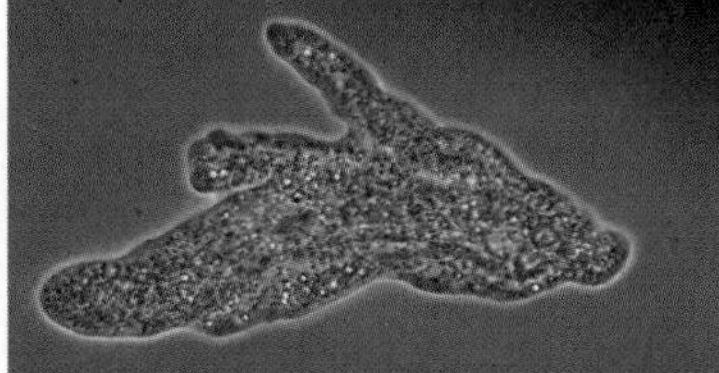

Amoeba discoides

Phylum Sarcodina: one-celled; take in food; move by means of pseudopods; free-living or parasitic; sarcodines, amoebas

Phylum Mastigophora: one-celled; take in food; have two or more flagella; free living or parasitic; flagellates

Phylum Ciliophora: one-celled; take in food; have large numbers of cilia; ciliates

Phylum Sporozoa: one-celled; take in food; no means of movement; parasites in animals; sporozoans

Phylum Myxomycetes, Phylum Acrasiomycota: one- or many-celled; absorb food; change form during life cycle; cellular and plasmodial slime molds

Phylum Oomycota: live in water or on land; one- or many-celled parasites; absorb dead organic matter; cause diseases in plants and animals; water molds and mildews

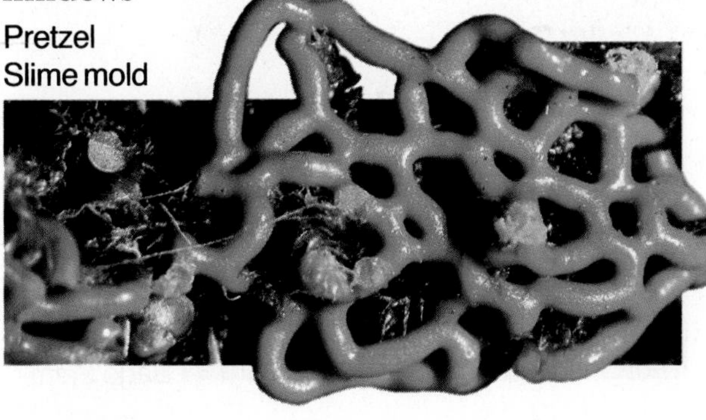

Pretzel Slime mold

Kingdom Fungi

Division Zygomycota: many-celled; absorb food; spores are produced in sporangia; zygote fungi

Division Ascomycota: one- and many-celled; absorb food; spores produced in asci; sac fungi

Division Basidiomycota: many-celled; absorb food; spores produced in basidia; club fungi

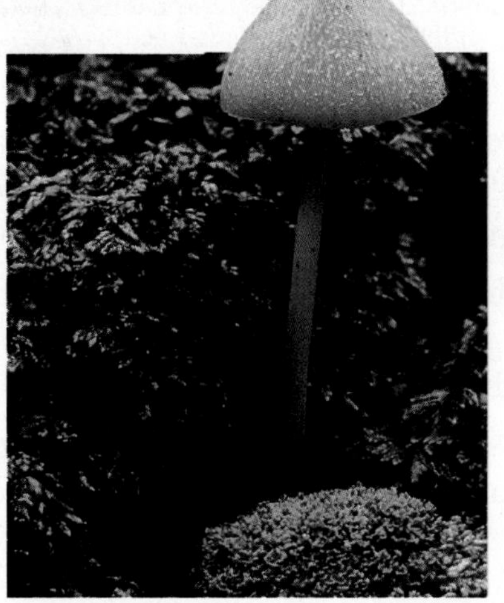

Mushroom

Division Deuteromycota: members with unknown reproductive structures; imperfect fungi

Lichens: organism formed by symbiotic relationship between an ascomycote or a basidiomycote and a green alga or a cyanobacterium; fungus provides protection and the algae or cyanobacterium provides food

Old Man's Beard lichen

Plant Kingdom

Division Bryophyta: non vascular plants that reproduce by spores produced in capsules; many-celled; green; grow in moist land environments; mosses and liverworts

Liverwort

Spore Plants

Division Lycophyta: many-celled vascular plants; spores produced in cones; live on land; are photosynthetic; club mosses

Division Sphenophyta: vascular plants with ribbed and jointed stems; scalelike leaves; spores produced in cones; horsetails

Division Pterophyta: vascular plants with feathery leaves called fronds; spores produced in clusters of sporangia called sori; live on land or in water; ferns

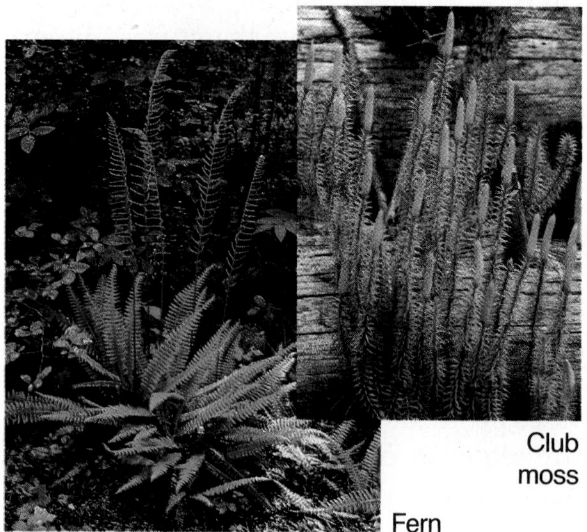

Club moss

Fern

Seed Plants

Division Ginkgophyta: deciduous gymnosperms; only one living species called the maiden hair tree; fan-shaped leaves with branching veins; reproduces with seeds; ginkgos

Division Cycadophyta: palmlike gymnosperms; large compound leaves; produce seeds in cones; cycads

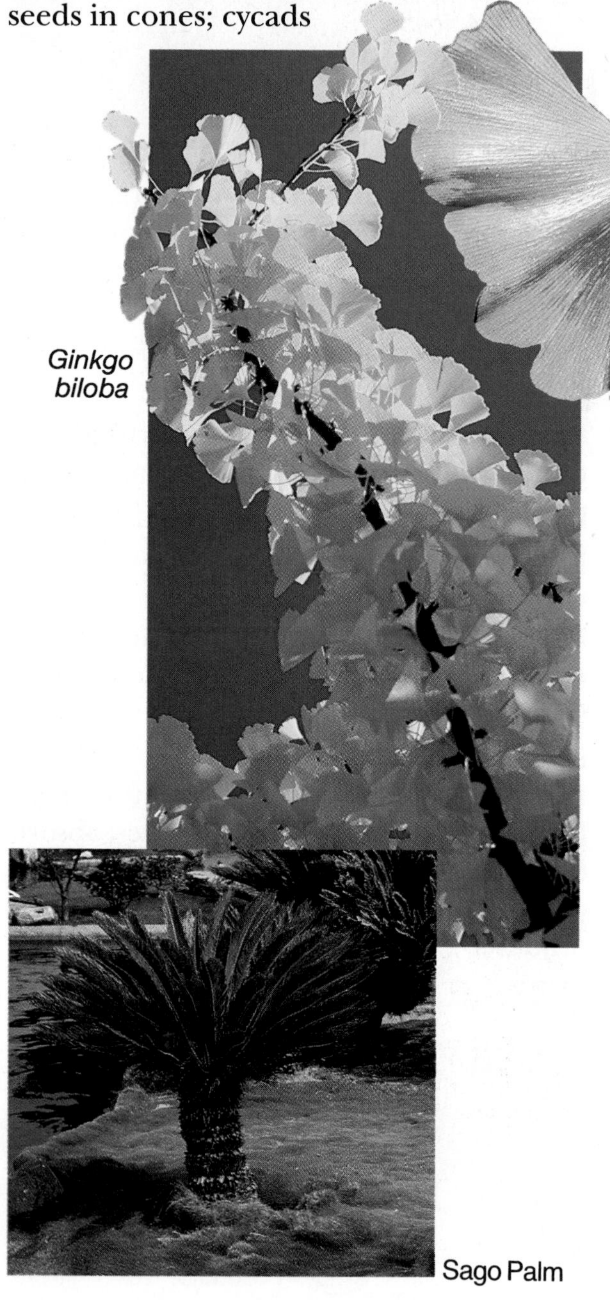

Ginkgo biloba

Sago Palm

Division Coniferophyta: deciduous or evergreen gymnosperms; trees or shrubs; needlelike or scalelike leaves; seeds produced in cones; conifers

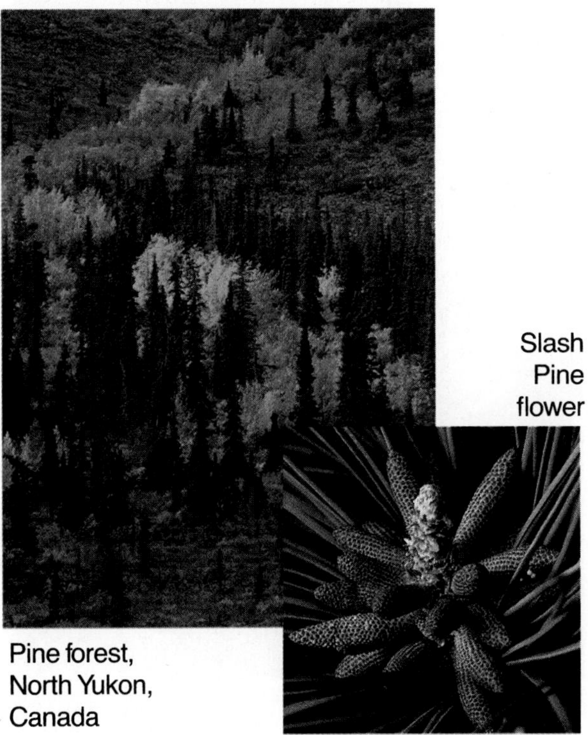
Slash Pine flower

Pine forest, North Yukon, Canada

Division Gnetophyta: shrubs or woody vines; seeds produced in cones; division contains only three genera; gnetum

Welwitschia mirabilis

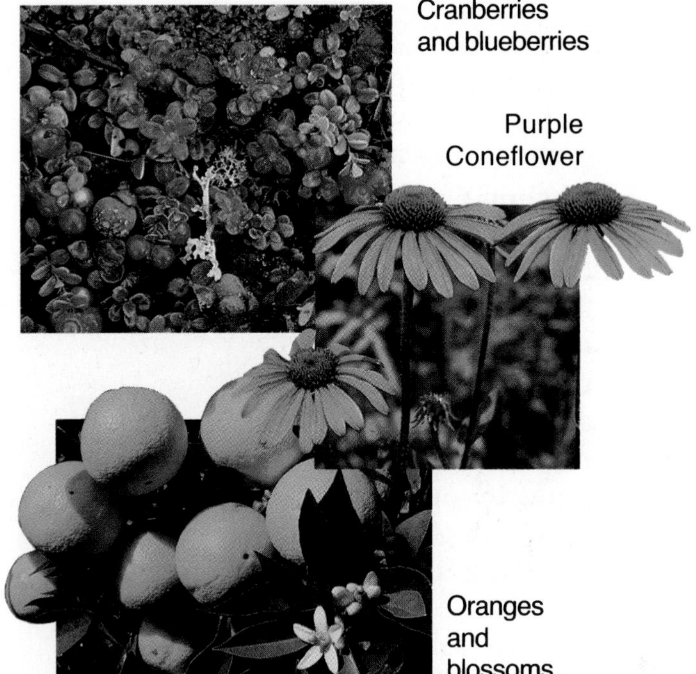
Cranberries and blueberries

Purple Coneflower

Oranges and blossoms

Division Anthophyta: dominant group of plants; ovules protected at fertilization by an ovary; sperm carried to ovules by pollen tube; produce flowers and seeds in fruits; flowering plants

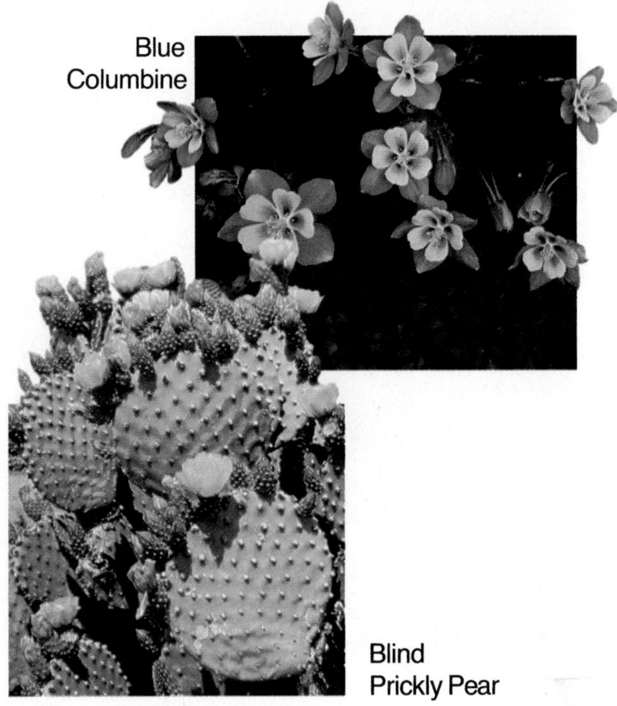
Blue Columbine

Blind Prickly Pear

Animal Kingdom

Phylum Porifera: aquatic organisms that lack true tissues and organs; they are asymmetrical and sessile; sponges

Phylum Cnidaria: radially symmetrical organisms with a digestive cavity with one opening; most have tentacles armed with stinging cells; live in aquatic environments singly or in colonies; includes jellyfish, corals, hydra, and sea anemones

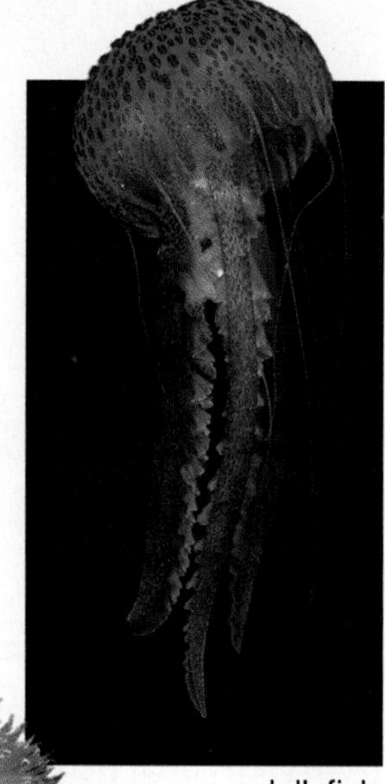

Jellyfish

Frilled Anemone

Phylum Platyhelminthes: bilaterally symmetrical worms with flattened bodies; digestive system has one opening; parasitic and free-living species; flatworms

Flatworm

Phylum Nematoda: round bilaterally symmetrical body; digestive system with two openings; some free-living forms but mostly parasitic; roundworms

Phylum Mollusca: soft-bodied animals, many with a hard shell; a mantle covers the soft body; aquatic and terrestrial species; includes clams, snails, squid, and octopuses

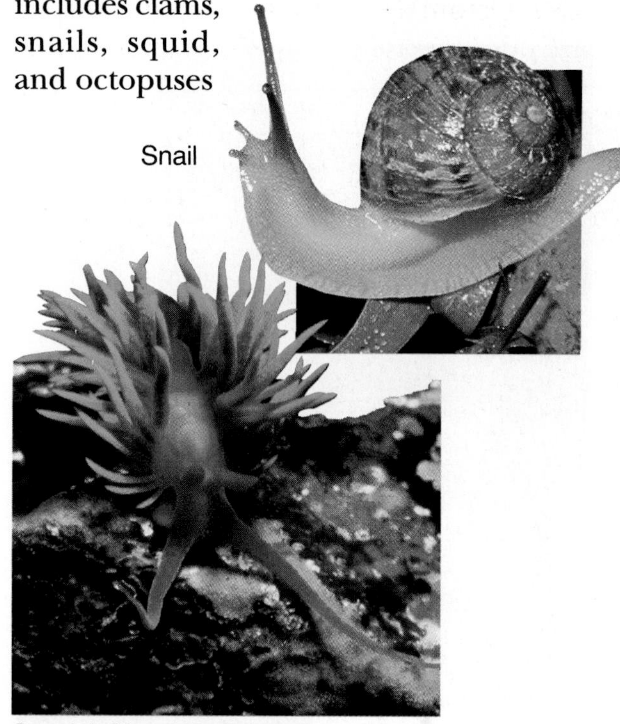

Snail

Spanish Shawl Nudibranch

Phylum Annelida: bilaterally symmetrical worms with round segmented bodies; terrestrial and aquatic species; well-developed body systems; includes earthworms, leeches, and marine polychaetes

Christmas Tree worm

Phylum Arthropoda: largest phylum of organisms that have segmented bodies with pairs of jointed appendages, and a hard exoskeleton; terrestrial and aquatic species; includes insects crustaceans, spiders, and horseshoe crabs

Swallowtail butterfly

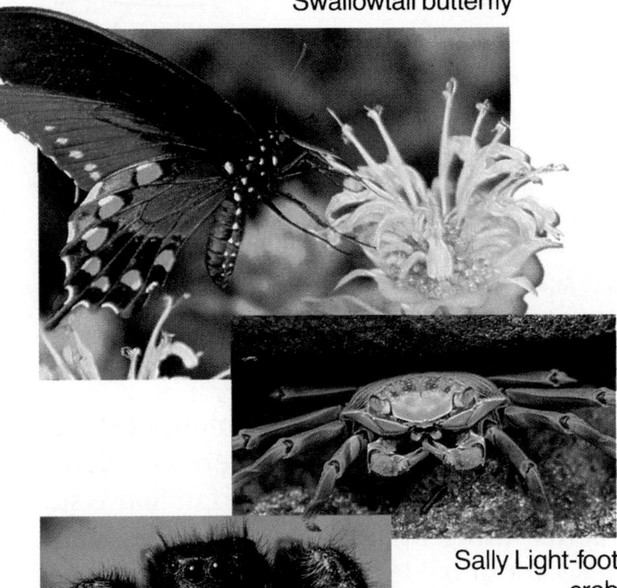

Sally Light-foot crab

Jumping spider

Phylum Echinodermata: saltwater organisms with spiny or leathery skin; water-vascular system with tube feet; radial symmetry; includes starfish, sand dollars, and sea urchins

Brittle stars

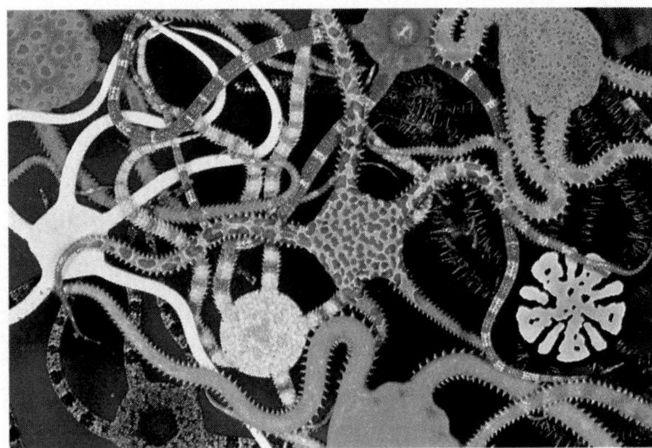

Seahorse

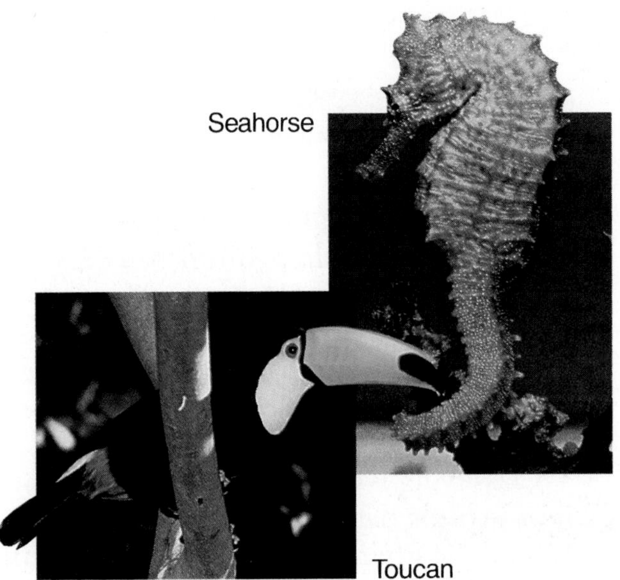

Toucan

Phylum Chordata: organisms with internal skeletons, specialized body systems, and paired appendages; all at some time have a notochord, dorsal nerve cord, gill slits, and a tail; include fish, amphibians, reptiles, birds, and mammals

Peninsula turtles

Mare and foal

Organizing Information

Classifying

You may not realize it, but you impose order on the world around you. If your shirts hang in the closet together, your socks take up a corner of a dresser drawer, or your favorite audio cassettes are stacked together, you have used the skill of classifying.

Classifying is grouping objects or events into groups based on common features. When classifying, you first make careful observations of the group of items to be classified. Select one feature that is shared by some items in the group but not others. Place the items that share the same feature in a subgroup. Place other items in subgroups based on other shared features. After you decide on the first feature that divides the items into subgroups, examine the items for other features and further divide each subgroup into smaller and smaller groups until the items have no features in common.

How would you classify these socks?

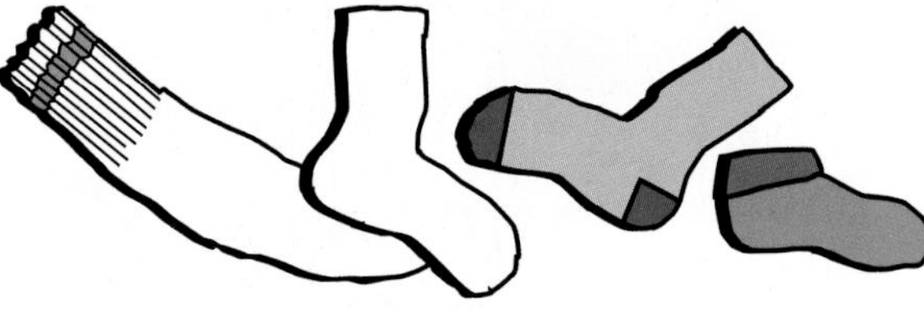

Classify the socks based on observable features. You might classify sport socks in one subgroup and dress socks in another. The sport socks could be subdivided into a white subgroup and a striped subgroup. Note that for each feature selected, each sock only fits into one subgroup. Keep selecting features until all the socks are classified. The chart shows one classification.

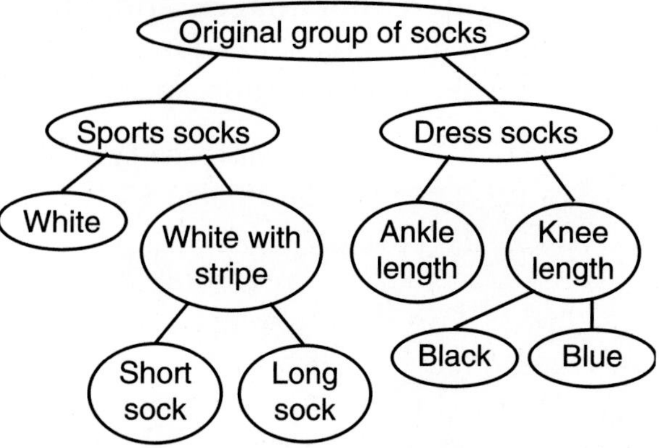

Remember, when you classify, you are grouping objects or events for a purpose.

Sequencing

A sequence is an arrangement of things or events in a particular order. A common sequence with which you are familiar is students sitting in alphabetical order. Think also about baking chocolate chip cookies. Certain steps have to be followed in order for the cookies to taste good.

When you are asked to sequence things or events, you must first identify what comes first. You then decide what should come second. Continue to choose things or events until they are all in order. Then, go back over the sequence to make sure each thing or event logically leads to the next.

Suppose you wanted to watch a rented movie that just came out on videotape. What sequence of events would you have to follow to watch the movie? You would first turn the television set to Channel 3 or 4. You would then turn the videotape player on and insert the tape. Once the tape has started playing, you would adjust the sound and picture. Then, when the movie is over, you would rewind the tape and return it to the store.

Outlining

Have you ever wondered why teachers ask students to outline what they read? The purpose of outlining is to show the relationships between main ideas and information about the main ideas. Outlining can help you organize, remember, and review written material.

When you are asked to outline, you must first find a group of words that summarizes the main idea. This group of words corresponds to the Roman numerals in an outline. Next, determine what is said about the main idea. Ideas of equal importance are grouped together and are given capital letters. Ideas of equal importance are further broken down and given numbers and letters.

To get an idea how to outline, compare the following outline with your textbook.

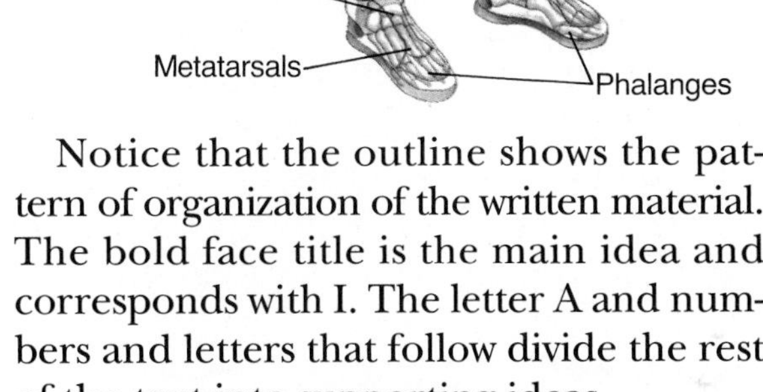

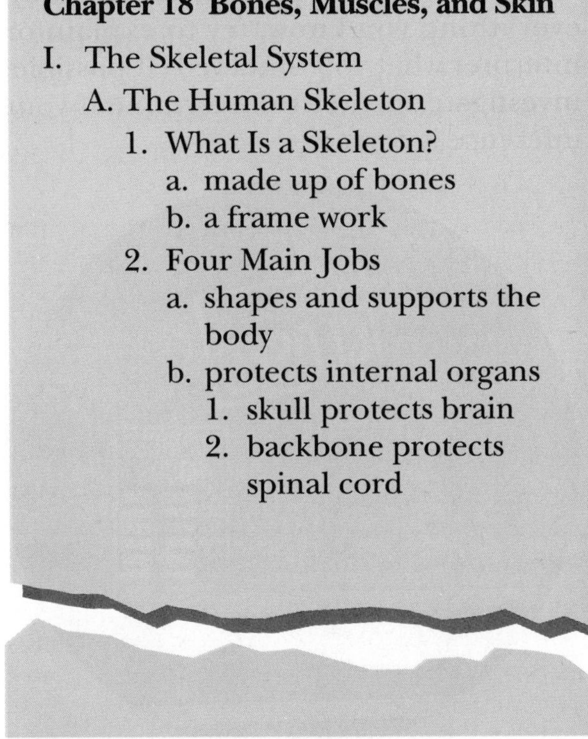

Chapter 18 Bones, Muscles, and Skin

I. The Skeletal System
  A. The Human Skeleton
    1. What Is a Skeleton?
      a. made up of bones
      b. a frame work
    2. Four Main Jobs
      a. shapes and supports the body
      b. protects internal organs
        1. skull protects brain
        2. backbone protects spinal cord

Notice that the outline shows the pattern of organization of the written material. The bold face title is the main idea and corresponds with I. The letter A and numbers and letters that follow divide the rest of the text into supporting ideas.

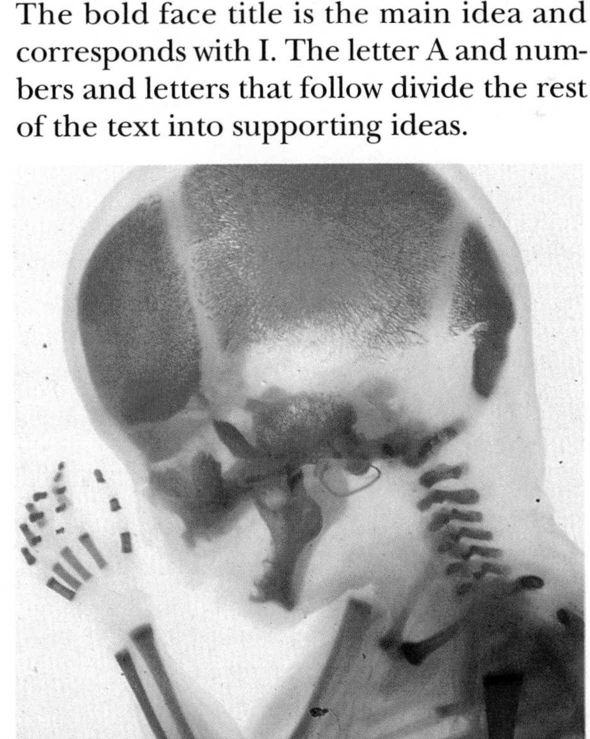

Thinking Critically

Observing and Inferring

Imagine that you have just finished a volleyball game with your friends. You hurry home to get a cold drink. Opening the refrigerator, you see a jug of orange juice at the back of the top shelf. The jug feels cold as you grasp it. "Ah, just what I need," you think. You hear the tone rise as you pour the juice into a tall glass. When you quickly down the drink, you smell the oranges and enjoy the tart taste in your mouth.

As you imagined yourself in the story, you used your senses to make observations. The basis of all scientific investigation is observation. Scientists are careful to make their observations accurate. Instruments, such as microscopes or telescopes, are used to extend their senses.

Instruments are used to make measurements. When observations involve measurements, they are called quantitative observations. Because measurements are easy to communicate and provide a concrete means of comparing collected data, scientists use them whenever possible.

When you make observations in science, you may find it helpful first to examine the entire object or situation. Then, look carefully for details using your sense of sight. Write down everything you see before using another sense to make additional observations. Continue until you have used all five senses.

Scientists often use their observations to make inferences. An inference is an attempt to explain or interpret observations or to determine what caused the events you observed. For example, if you observed a CLOSED sign in a store window around noon, you might infer the owner is taking a lunch break. But, perhaps the owner has a doctor's appointment or has taken the day off to go fishing. The only way to be sure your inference is correct is to investigate further.

When making an inference, be certain to make accurate observations and to record them carefully. Then, based on everything you know, try to explain or interpret what you observed. If possible, investigate further to determine if your inference is correct.

Comparing and Contrasting

Observations can be analyzed and then organized by noting the similarities and differences between two or more objects or situations. When you examine objects or situations to determine similarities, you are comparing. Contrasting is looking at similar objects or situations for differences.

Suppose you were asked to compare and contrast a grasshopper and a dragonfly. You would start by examining your observations. You then divide a piece of paper into two columns. List ways the insects are similar in one column and ways they are different in the other. After completing your lists, you report your findings in a table or in a graph.

Similarities you might point out are that both have three body parts, two pairs of wings, and chewing mouthparts. Differences include grasshoppers cause damage to crops but dragonflies destroy harmful insects or the grasshopper flies short distances but the dragonfly can fly long distances.

Recognizing Cause and Effect

Have you ever observed something happen and then tried to figure out why or how it came about? If so, you have observed an event and inferred a reason for the event. The event or result of action is an effect, and the reason for the event is the cause.

Suppose that every time your teacher fed fish in a classroom aquarium, she tapped the food container on the edge. Then, one day she tapped the edge of the aquarium to make a point about an ecology lesson. You observe the fish swim to the surface of the aquarium to feed.

What is the effect and what would you infer would be the cause? The effect is the fish swimming to the surface of the aquarium. You might infer the cause to be the teacher tapping on the edge of the aquarium. In determining cause and effect, you have made a logical inference based on careful observations.

Perhaps, the fish swam to the surface because they reacted to the teacher's waving hand or for some other reason. When scientists are unsure of the cause for a certain event, they often design controlled experiments to determine what caused their observations. Although you have made a sound judgment, you would have to perform an experiment to be certain that it was the tapping that caused the effect you observed.

Designing an Experiment

Designing an experiment allows you to develop your scientific skills through investigations. Such skills include measuring in SI, hypothesizing, using variables, constants and controls, and interpreting data.

Measuring in SI

You are probably familiar with the metric system of measurement. The metric system is a uniform system of measurement developed by scientists in 1795. The development of the metric system helped scientists avoid problems with different units of measurement by providing an international standard. A modern form of the metric system called the International System, or SI, was adopted for worldwide use in 1960.

The metric system is easy to use because it names units systematically and has a decimal base. For example, meter is the base unit for measuring length, gram for measuring mass, and liter for measuring volume. Unit sizes vary by multiples of ten. When changing from smaller units to larger, you divide by ten. When changing from larger units to smaller, multiply by ten. Prefixes are used to name larger and smaller units. The following table lists common metric prefixes and their meanings.

METRIC PREFIXES

Prefix	Symbol		Meaning
kilo-	k	1 000	thousand
hecto-	h	100	hundred
deka	da	10	ten
deci-	d	0.1	tenth
centi	c	0.01	hundredth
milli-	m	0.001	thousandth

Do you see how the prefix *kilo-* attached to the unit *gram* is *kilogram* or 1000 grams, or the prefix *deci-* attached to the unit *meter* is *decimeter* or one tenth (0.1) of a meter?

The meter is the SI unit used to measure distance. To visualize the length of a meter, think of a baseball bat. A baseball bat is about one meter long. When measuring smaller distances, the meter is divided into smaller units called centimeters and millimeters. A centimeter is one hundredth (0.01) of a meter, which is about the size of the width of the fingernail on your little finger. A millimeter is one thousandth of a meter (0.001), about the thickness of a dime.

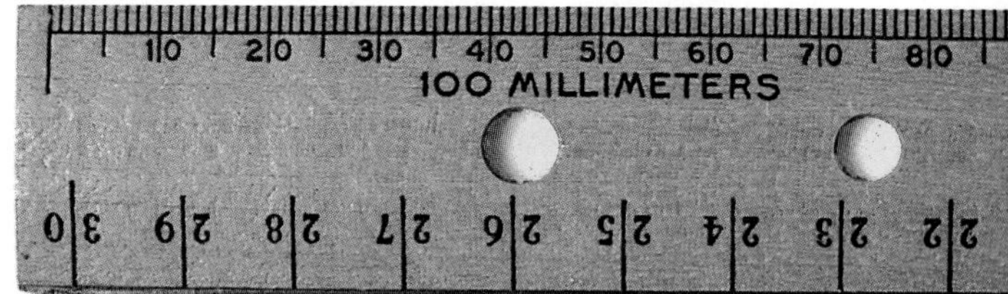

Most metersticks and metric rulers have lines indicating centimeters and millimeters. Look at the illustration. The centimeter lines are the longer numbered lines and the shorter lines between the centimeter lines are millimeter lines.

When using a metric ruler, you must first decide on a unit of measurement. You then line up the 0 centimeter mark with the end of the object being measured, and read the number of the unit where the object ends.

Units of length are also used to measure the surface area. The standard unit of area is the square meter (m^2), or a square one meter long on each side. Similarly, a square centimeter (cm^2) is a square one centimeter long on each side. Surface area is determined by multiplying the number of units in length times the number of units in width.

The volume of rectangular solids is also calculated using units of length. The cubic meter (m^3) is the standard SI unit of volume. A cubic meter is a cube one meter on a side. You can determine the volume of rectangular solids by multiplying length times width times height.

Liquid volume is measured using a unit called a liter. You are probably familiar with a two-liter soft drink bottle. One liter is about one-half of the two-liter bottle. A liter has the volume of 1000 cubic centimeters. Since the prefix *milli-* means thousandth (0.001), a milliliter equals one cubic centimeter. One milliliter of liquid would fill a cube measuring one centimeter on each side.

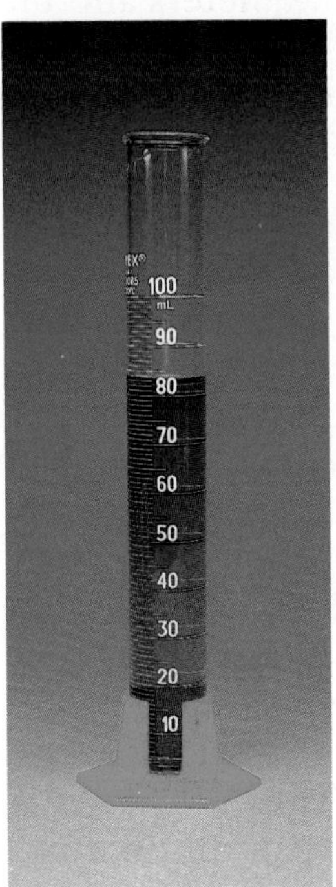

During science activities you will measure liquids using beakers and graduated cylinders marked in milliliters. A graduated cylinder is a tall cylindrical container marked with lines from bottom to top. Each graduation usually represents one milliliter.

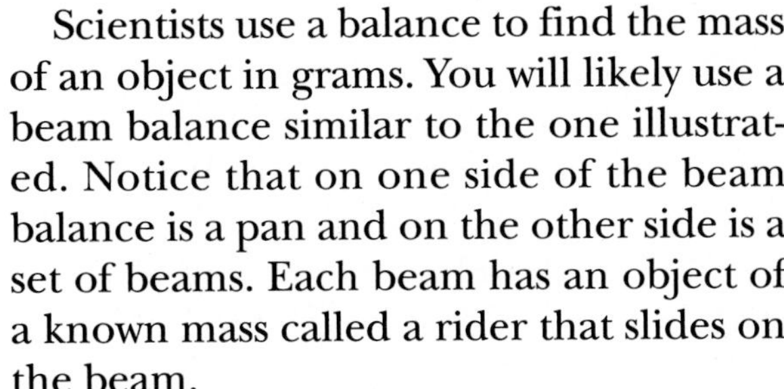

Scientists use a balance to find the mass of an object in grams. You will likely use a beam balance similar to the one illustrated. Notice that on one side of the beam balance is a pan and on the other side is a set of beams. Each beam has an object of a known mass called a rider that slides on the beam.

You must be careful when using a balance. When carrying the balance, hold the beam support with one hand and place the other hand under the balance. Also, be careful what you place on the pan. Never place a hot object or pour chemicals directly on the pan. Determine the mass of a suitable container and place dry or liquid chemicals into the container. Then determine the mass of the container and the chemicals. Finally, calculate the mass of the chemicals by subtracting the mass of the empty container.

Before you find the mass of an object, you must set the balance to zero by sliding all the riders back to the zero point. Check the pointer to make sure it swings an equal distance above and below the zero point on the scale. If the swing is unequal, find and turn the adjusting screw until you have an equal swing.

You are now ready to use the balance to find the mass of the object. Place the object on the pan. Slide the rider with the largest mass along the beams until the pointer drops below the zero point. Then move it back one notch. Repeat the process on each beam until the pointer swings an equal distance above and below the zero point. Read the masses indicated on the beams. The sum of the masses is the mass of the object.

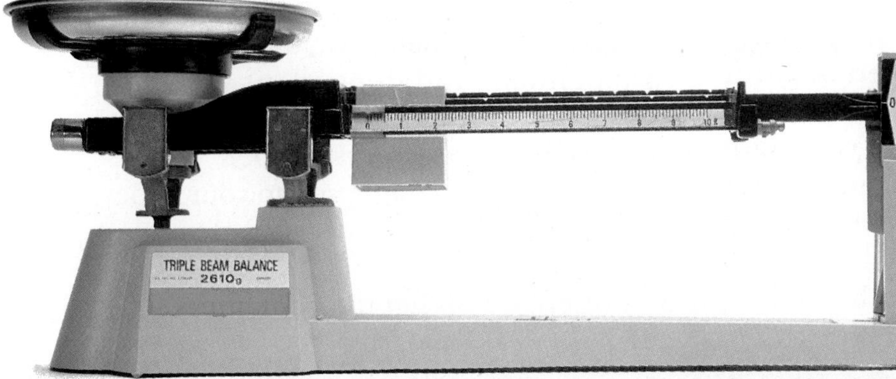

Hypothesizing

What would you do if the combination lock on your locker didn't work? Would you try the combination again? Would you check to make sure you had the right locker? You would probably try several possible solutions until you managed to open the locker.

Scientists generally use experiments to solve problems and answer questions. An experiment is a method of solving a problem in which scientists use an organized process to attempt to answer a question.

Experimentation involves defining a problem and formulating and testing hypotheses, or proposed solutions, to the problem. Each proposed solution is tested during an experiment which includes making careful observations and collecting data. After analysis of the collected data, a conclusion is formed and compared to the hypothesis.

Imagine it's after school, and you are changing clothes. You notice a brownish-black spot on a favorite shirt. Your problem is how to remove the stain from the shirt without damaging the shirt. You think that soap and water will remove the stain. You have made a hypothesis, or proposed a solution to the problem. But, making a hypothesis is not enough. A hypothesis has to be something you can test. You try soap and water, but the stain doesn't budge.

Then you decide that you need to use a stronger solvent than water. You have revised your hypothesis based on your observations. The new hypothesis is still only a proposed solution until you test it and examine the results. If the test removes the stain, the hypothesis is accepted. But, if the test doesn't remove the stain, you will have to revise and refine your hypothesis again.

Using Variables, Constants, and Controls

When scientists perform experiments, they are careful to manipulate or change only one condition and keep all other conditions in the experiment the same. The condition that is manipulated is called the independent variable. The conditions that are kept the same during an experiment are called constants. The dependent variable is any change that results from manipulating the independent variable.

Scientists can only know that the independent variable caused the change in the dependent variable if they keep all other factors the same in an experiment. Scientists also use controls to be certain that the observed changes were a result of manipulation of the independent variable. A control is a sample that is treated exactly like the experimental group except that the independent variable is not applied to the control. After the experiment, the change in the dependent variable of the control sample is compared with any change in the experimental group. This allows you to see the effect of the independent variable.

Suppose you watch your guppies one morning and they don't seem as active as usual. You check the aquarium and notice that the aquarium heater is not working. You wonder how water temperature affects the activity level of guppies and decide to design an experiment. What would be your independent and dependent variables, constants, and control in your experiment? What would your hypothesis be?

This is how you might set up your experiment. Obtain several identical clear glass containers, and fill them with the same amount of water. Let the containers set.

On the day of your experiment, fill a container with an amount of aquarium water equal to that in the test containers. After measuring and recording the water temperature, heat and cool the other containers adjusting the water temperatures in the test containers so that two have higher temperatures and two have lower temperatures than the aquarium water temperature.

Place a guppy in a container. Count the number of horizontal and vertical movements the guppy makes during five minutes. Repeat with the same guppy at each temperature. Record the data in a table.

Number of Guppy Movements		
Container	**Temperature (°C)**	**Number of movements**
Aquarium Water	38	56
A	40	61
B	42	70
C	36	46
D	34	42

What are the independent and dependent variables in the experiment? Because you are changing the temperatures of the water, the independent variable is the water temperature. Since the dependent variable is any change that results from the independent variable, the dependent variable is the number of movements the guppy makes during five minutes.

What factors are constants in the experiment? The constants are using the same size and shape containers, filling them with equal amounts of water, and counting the number of movements during the same amount of time. What was the purpose of counting the number of movements of a guppy in an identical container filled with aquarium water? The container of aquarium water is the control. The number of movements of the guppy in the aquarium water will be used to compare the movements of the guppy in water of different temperatures.

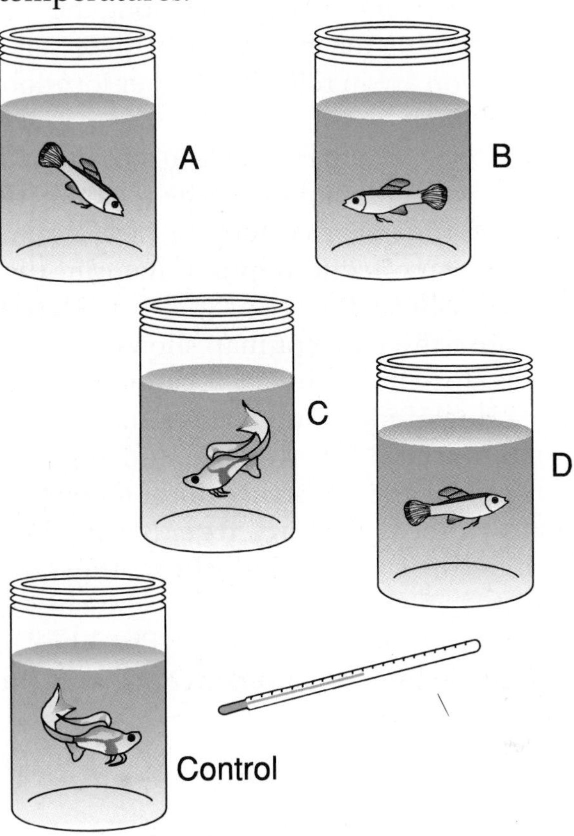

Interpreting Data

After doing a controlled experiment, you must analyze and interpret the collected data, form a conclusion, and compare the conclusion to your hypothesis. Analyze and interpret the data in the table. What conclusion did you form? The data indicate that the higher the temperature the greater the number of movements of the guppy. How does the conclusion compare with your hypothesis? Was it supported by the experiment or not?

Graphic Organizers

Concept Mapping

If you were taking an automobile trip, you would probably take along a road map. The road map shows your location, your destination, and other places along the way. By examining the map, you can understand where you are in relation to other locations on the map.

A concept map is similar to a road map. But, a concept map shows the relationship among ideas (or concepts) rather than places. A concept map is a diagram that visually shows how concepts are related. Because the concept map shows the relationships among ideas, it can clarify the meaning of ideas and terms and help you to understand what you are studying.

Look at the construction of a concept map called a **network tree.** Notice how some words are circled while others are written on connecting lines. The circled words are science concepts. The lines in the map show related concepts, and the words written on them describe relationships between the concepts. A network tree can also show more complex relationships between the concepts. For example, because chemical processes occur in plants and animals, a line labeled "affected by" could be drawn from plants and animals to chemistry. Another example of a relationship that crosses branches would be a line connecting Earth changes and matter and energy labeled "caused by interactions of." Earth changes are caused by interactions of matter and energy.

When you are asked to construct a network tree, state the topic and select the major concepts. Find related concepts and put them in order from general to specific. Branch the related concepts from the major concept and describe the relationships on the lines. Continue to write the more specific concepts. Write the relationships between the concepts on the lines until all concepts are mapped. Examine the concept map for relationships that cross branches, and add them to the concept map.

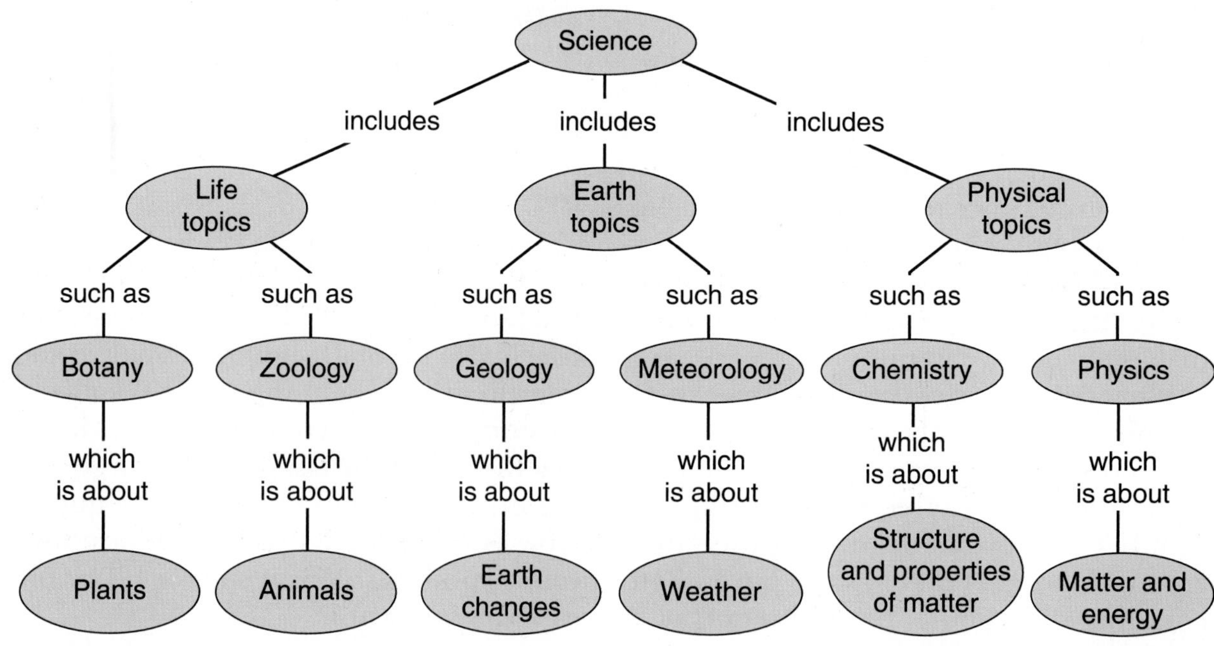

An **events chain** is another type of concept map. An events chain map is used to describe ideas in order. In science, an events chain can be used to describe a sequence of events, the steps in a procedure, or the stages of a process.

lemon

When making an events chain, you first must find the one event that starts the chain. This event is called the initiating event. You then find the next event in the chain and continue until you reach an outcome. Suppose your mother asked you to wash the dinner dishes. An events chain map might look like the one below. Notice that connecting words may not be necessary.

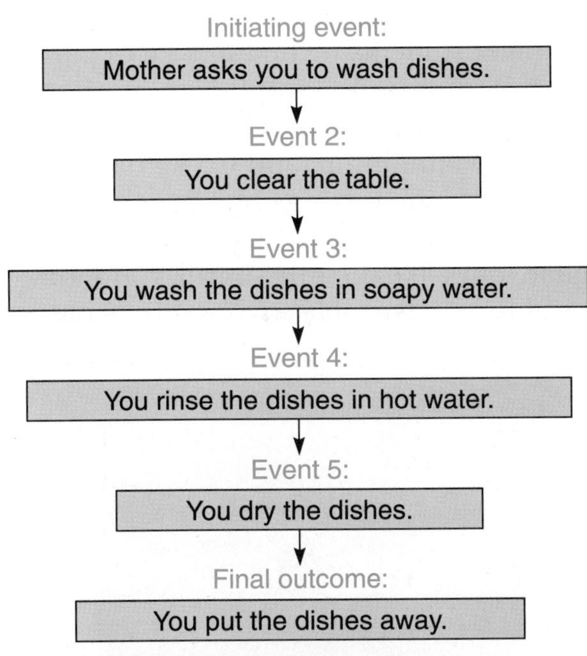
Initiating event:
Mother asks you to wash dishes.
Event 2:
You clear the table.
Event 3:
You wash the dishes in soapy water.
Event 4:
You rinse the dishes in hot water.
Event 5:
You dry the dishes.
Final outcome:
You put the dishes away.

A **cycle concept map** is a special type of events chain map. In a cycle concept map, the series of events do not produce a final outcome. The last event in the chain relates back to the initiating event.

As in the events chain map, you first decide on an initiating event and then list each important event in order. Since there is no outcome and the last event relates back to the initiating event, the cycle repeats itself. Look at the cycle map of the stages of complete insect metamorphosis.

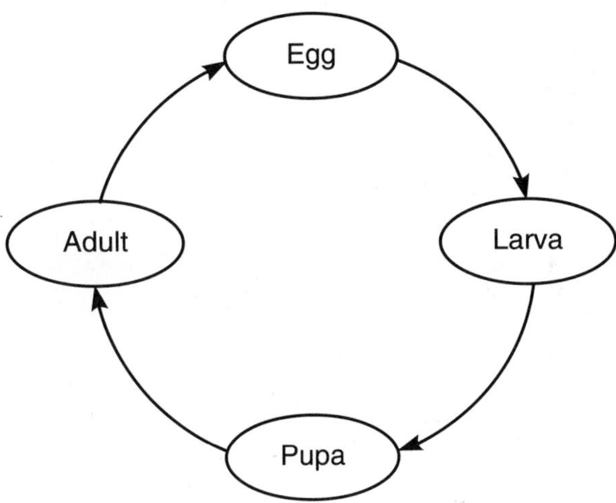
Complete Insect Metamorphosis
Egg
Larva
Pupa
Adult

There is usually not one correct way to create a concept map. As you are constructing a map, you may discover other ways to construct the map that show the relationships between concepts better. If you do discover what you think is a better way to create a concept map, do not hesitate to change it.

Concept maps are useful in understanding the ideas you have read about. As you construct a map, you are constructing knowledge and learning. Once concept maps are constructed, you can use them to review and study and to test your knowledge.

SKILL HANDBOOK 689

Making and Using Tables

Browse through your textbook and you will notice many tables both in the text and in the activities. The tables in the text arrange information in such a way that it is easier for you to understand. Also, many activities in your text have tables to complete as you do the activity. Activity tables will help you organize the data you collect during the activity so that it can be interpreted easily.

Most tables have a title telling you what is being presented. The table itself is divided into columns and rows. The column titles list items to be compared. The rows headings list the specific characteristics being compared. Within the grid of the table the collected data is recorded. Look at the following table:

EFFECT OF EXERCISE ON HEARTBEAT RATE

Pulse taken	Heartbeat rate individual	Heartbeat rate class average
at rest	73	72
after exercise	110	112
1 minute rest after exercise	94	90
5 minute rest after exercise	76	75

What is the title of this table? The title is "Effect of Exercise on Heart Rate." What items are being compared? The heart rates for an individual and the class average are being compared at rest and for several durations after exercise.

What is the average heart rate of the class 1 minute after exercise? To find the answer you must locate the column labeled "class average" and the row "1-minute rest after exercise." The data contained in the box where the column and row intersect is the answer. Did you answer 90? Whose heart rate was 110 after exercise? If you answered the individual, you have an understanding of how to use a table.

RECYCLED MATERIALS

Day of Week	Paper (kg)	Aluminum (kg)	Plastic (kg)
Mon.	4	2	0.5
Wed.	3.5	1.5	0.5
Fri.	3	1	1.5

To make a table, you simply list the items compared in columns and the characteristics compared in rows. Make a table and record the data comparing the mass of recycled materials collected by a class. On Monday, students turned in 4 kg of paper, 2 kg of aluminum, and 0.5 kg of plastic. Wednesday, they turned in 3.5 kg of paper, 1.5 kg of aluminum, and 0.5 kg of plastic. On Friday, the totals were 3 kg of paper, 1 kg of aluminum, and 1.5 kg of plastic. If your table looks like the one shown, you should be able to make tables to organize data.

Making and Using Graphs

After scientists organize data in tables, they often display the data in graphs. A graph is a diagram that shows a comparison between variables. Since graphs show a picture of collected data, they make interpretation and analysis of the data easier. The three basic types of graphs used in science are the line graph, bar graph, and pie graph.

A line graph is used to show the relationship between two variables. The variables being compared go on two axes of the graph. The independent variable always goes on the horizontal axis, called the *x*-axis. The dependent variable always goes on the vertical axis or *y*-axis.

Suppose a school started a peer study program with a class of students to see how it affected their science grades.

AVERAGE GRADES OF STUDENTS IN STUDY PROGRAM

Grading Period	Average Science Grade
First	81
Second	85
Third	86
Fourth	89

You could make a graph of the grades of students in the program over a period of time. The grading period is the independent variable and should be placed on the *x*-axis of your graph. The average grade of the students in the program is the dependent variable and would go on the *y*-axis.

After drawing your axes, you would label each axis with a scale. The *x*-axis simply lists the grading periods. To make a scale of grades on the *y*-axis, you must look at the data values. Since the lowest grade was 81 and the highest was 89, you know that you will have to start numbering at least at 81 and go through 89. You decide to start numbering at 80 and number by twos through 90.

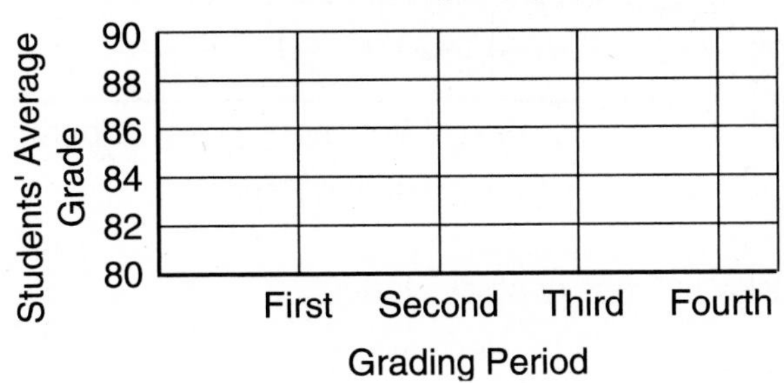

Next, plot the data points. The first pair of data you want to plot is the first grading period and 81. Locate "First" on the *x*-axis and 81 on the *y*-axis. Where an imaginary vertical line from the *x*-axis and an imaginary horizontal line from the *y*-axis would meet, place the first data point. Place the other data points the same way. After all the points are plotted, connect them with a smooth line.

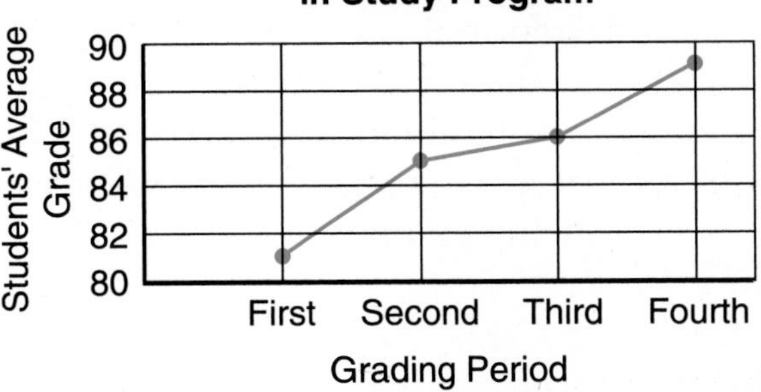

What if you wanted to compare the average grades of the class in the study group with the grades of another class? The data of the other class can be plotted on the same graph to make the comparison. You must include a key with two different lines each indicating a different set of data.

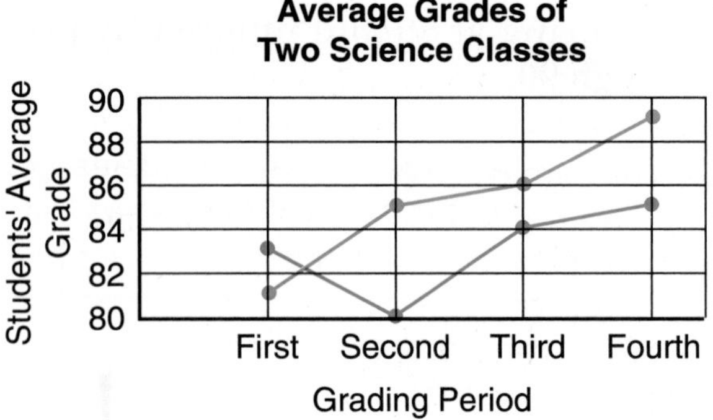

KEY Class or study students
Regular class

Bar graphs are similar to line graphs, except they are used to show comparisons between data or to display data that does not continuously change. In a bar graph, thick bars show the relationships between data rather than data points.

To make a bar graph, set up the *x*-axis and *y*-axis as you did for the line graph. The data is plotted by drawing thick bars from the *x*-axis up to an imaginary point where the *y*-axis would intersect the bar if it was extended.

Look at the bar graph comparing the wing vibration rates for different insects. The independent variable is the type of insect, and the dependent variable is the number of wing vibrations per second. The number of wing vibrations for different insects is being compared.

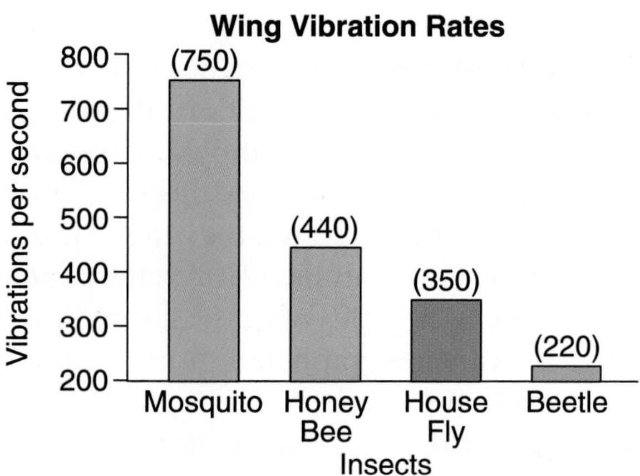

A pie graph uses a circle divided into sections to display data. Each section represents part of the whole. When all the sections are placed together, they equal 100 percent of the whole.

Suppose you wanted to make a pie graph to show the number of seeds that germinated in a package. You would have to determine the total number of seeds and the number of seeds that germinated out of the total. You count the seeds and find that there are 143 seeds in the package. Therefore, the whole pie will represent this amount.

You plant the seeds and determine that 129 seeds germinate. The group of seeds that germinated will make up one section of the pie graph, and the group of seeds that did not germinate will make up another section.

To find out how much of the pie each section should take, you must divide the number of seeds in each section by the total number of seeds. You then multiply your answer by 360, the number of degrees in a circle. Round your answer to the nearest whole number. The number of seeds that germinated would be determined as follows:

$$\frac{129}{143} \times 360 = 324.75 \text{ or } 325 \text{ degrees}$$

To plot this group on the pie graph, you need a compass and a protractor. Use the compass to draw a circle. Then, draw a straight line from the center to the edge of the circle. Place your protractor on this line and use it to mark a point on the edge of the circle at 325 degrees. Connect this point with a straight line to the center of the circle. This is the section for the group of seeds that germinated. The other section represents the group of seeds that did not germinate. Complete the graph by labeling the sections of your graph and giving the graph a title.

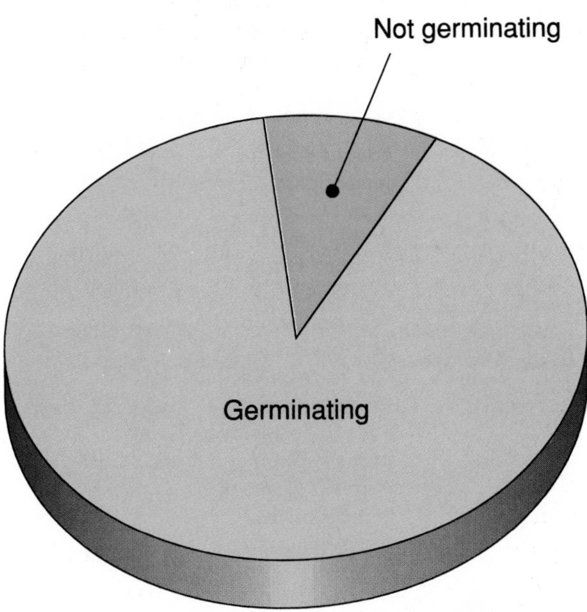

Seed Germination Rate

Interpreting Scientific Illustrations

Illustrations are included in your science textbook to help you understand, interpret, and remember what you read. Whenever you encounter an illustration, examine it carefully and read the caption. The caption is a brief comment that explains or identifies the illustrations.

Some illustrations are designed to show you how the internal parts of a structure are arranged. Look at the illustrations of the stem in Figure 11-3 on page 256. The stems have been cut so that they show sections that are at right angles to the length of each stem. This type of illustration is called a cross section. If the stems were cut lengthwise in half, the sections would be called longitudinal sections.

Symmetry refers to a similarity or likeness of parts. Many organisms and objects have symmetry. When something can be divided into two similar parts lengthwise, it has bilateral symmetry. Look at the illustrations in Figure 13-3 on page 304. The right side of the butterfly looks very similar to the left side. It has bilateral symmetry.

Other organisms and objects have radial symmetry. Radial symmetry is the arrangement of similar parts around a central point. The sand dollar in Figure 13-3 has radial symmetry. See how it can be divided anywhere through the center into similar parts.

Some organisms and objects can not be divided into two similar parts. If an organism or object can not be divided, it is asymmetrical. Think about an amoeba. Regardless of how you try to divide an amoeba, you can not divide it into two parts that look alike.

Now look at the section of the dicot stem on page 256 again. Can you see that it has radial symmetry in cross section? If you were to make a longitudinal section through the the dicot stem, you would see that it would have bilateral symmetry as well.

As you study illustrations, you will sometimes see terms that refer to the orientation of an organism. The word dorsal refers to the upper side or back of an animal. Ventral refers to the lower side or belly of the animal. The illustration of the planarian in Figure 13-12 shows a dorsal side. Figure 14-12 shows a ventral side.

PROJECT 1

OVERVIEW

OBJECTIVE: Students will *research* the cumulative nature of theory building and the feedback between knowledge and technology through a study of the development of the cell theory and the instruments that assisted in its evolution.

SUMMARY

In this project, students will research scientists who have contributed to our knowledge of the cell. Students will put together social, technological, and scientific factors that influence contributions to scientific knowledge. They will share this information in a creative manner and attempt to look at the influence that the changing state of technology has on the evolution or development of scientific knowledge about a particular subject.

TIME REQUIRED

This project is designed to be worked on throughout Unit 1. One period is required to set up research groups and choose topics. At the end, one period is required to present scenarios. The rest of the work may be interspersed with other lessons over a period of two to three weeks.

PREPARATION

- Establish cooperative groups of 4 or 5 students.
- Be aware of sources on various scientists involved in cell research.
- Oversee materials for costumes and props.

PROJECT 1

IN SEARCH OF CELLS

Even though cells make up ourselves and every living thing around us, we have not always known that cells are there. Some organisms, like protozoa, consist of only one cell, while other organisms, like people, consist of trillions of cells. While some cells, such as egg yolks, are large enough to be seen directly, most cells can not be seen without using special equipment. Centuries ago, scientists speculated about the nature of life. They weren't sure what organisms were made up of. Scientists were like detectives, trying to reconstruct a crime based only on clues. As technologies such as the microscope developed, scientists finally were able to observe cells. Cell research has sometimes involved reconstructing the nature of cells from looking at the activities and materials cells produced, rather than looking at cells themselves. As technology has improved, researchers have been able to observe more and more details about cells. Technology has taken us from the limitations of the hand lens to being able to see atoms.

Cell Detectives

Many different scientists have contributed to our present knowledge of the cell. Explore these scientists' lives and research to see how they contributed to our current understanding of cells. A list of some cell researchers is given below. Working in groups, choose one individual or a pair of individuals to research.

Eduard Buchner
Jean-Baptiste Carnoy
Albert Claude
Jewel Plummer Cobb
Rosalind Franklin
Camillo Golgi
Ross Harrison and Alexis Carrel
Sir Hans Krebs
Rita Levi-Montalcini
Jacob Monod
Mikhail Sememovich Tswett
James Watson and Francis Crick
August Weismann
Rudolf Virchow

694

REFERENCES

Brown, Walter V. and Eldridge M. Berthe. *Textbook of Cytology.* St. Louis: C.V. Mosby Co., 1974.

Curtis, Helena and N. Sue Barnes. *Biology.* New York: Worth Publishing, Inc., 1989.

de Duve, Christian. *A Guided Tour of the Living Cell Vols.* 1 and 2. New York: Scientific American Library, W.H. Freeman and Co., 1984.

Watson, James, Bruce Alberts, Dennis Bray, Julian Lewis, Martin Ruff, and Ruth Roberts. *Molecular Biology of the Cell,* 2nd Ed. New York: Garland Publishing Co., 1989.

For each subject, research the following information:

1. What was already known about the cell by the time your researcher was alive?
2. What kind of equipment did he or she have to work with?
3. What were the limitations of the equipment available?
4. What was taking place socially or politically at the time your researcher was living? How did he or she dress? How did scientists communicate with each other at that time?
5. What contribution did your researcher make to the knowledge of cells?
6. How did this person arrive at his or her contribution?

Using Your Research

After you have completed your research, write a scenario using the people in your group. This scenario should answer the research questions listed above. Design costumes that reflect the lifestyle of the scientists researched. The period of time, social class, and geographic location should all influence the costumes you design. Props should include models of the equipment available only at that particular time. Each group can present its scenario in chronological order. As a class, discuss how the scenarios show a history of cell research, and how they show a history of technological development as well.

Extension

Knowledge of the cell is far from complete. More and more as technology improves, new questions are asked. Today, a great deal of cell research centers on looking for solutions to various diseases that affect human life. What makes cells become cancerous? How can HIV be stopped? Will gene therapy be effective? Investigate the current research on muscular dystrophy, AIDS, cancer, or other diseases. How does this research make use of cell theory? What contributions might this research make to a better understanding of cells and how they work? Will you be the person in the next scenario?

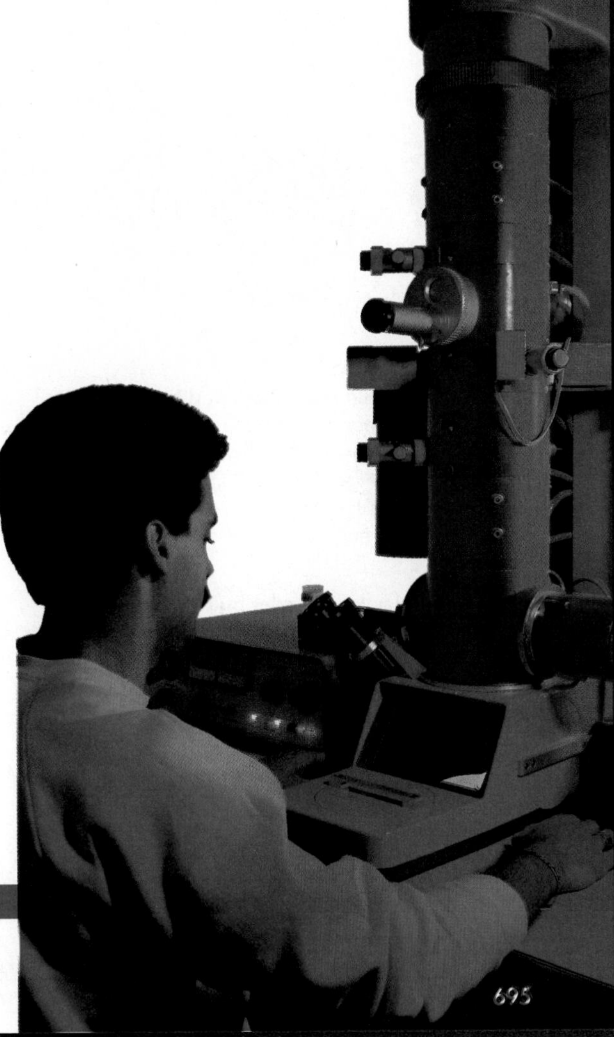

TEACHING NOTES

• Before beginning the cell research, show the class a folder. Tell them that you have the picture of an animal in the folder, and have them guess what animal it is. Use an unusual animal and give them some time to guess. They may not guess it at all. Now try the same thing, but give each group a different clue to the animal's identity. Have each group present its clue and then allow time to guess. If you have four or five groups, each with a different clue, students should be able to identify the animal this time. Discuss how particular clues influenced students' guesses. Compare this to the way scientific knowledge builds on previous ideas, modifying them as more information becomes available.

• Help students to recognize characteristics that are helpful in this approach to knowledge — openmindedness, creativity, diverse thinking, methodical approach, and so on.

• Help students determine which scientist(s) each group will select to investigate.

• Remind students that they will end their study of the cell with a chronological presentation, in period costume, with period instruments, and an explanation of how much they could know based on instruments of that period.

• Discuss what other instruments, besides microscopes, make modern knowledge of the cell possible (computers, cameras).

EXTENSION

Technology is now able to provide artificial tissues, organs, and limbs. Students may research these synthetic replacements to determine their limitations and uses. Students may wish to write a science fiction story about a time in the future when any human body part can be replaced with artificial parts. Would this be a good idea or not? What effect would this have on human lifestyle and life span? Would we have to rethink our definitions of robots and humans?

PROJECT 2

OVERVIEW

OBJECTIVES: Students will *measure, record data, analyze* statistics, and *design* a way to display data and conclusions about a human genetic characteristic.

SUMMARY

In this project, students will collect data on foot length, and then find the mean, median, and mode for foot length in different age groupings for males and for females. The class will then display and discuss these statistics.

Because this is a characteristic controlled by more than one gene, students should arrive at a bell-shaped curve on a graph. Through collection of large amounts of data, students will experience more accurate results than with a small sample.

TIME REQUIRED

This project is designed to be worked on throughout Chapter 5. The students will need a period to plan and establish data charts. They will need time outside the class to collect their data. They will need another period to design and construct their statistical display. The transfer of data and statistical analysis may be done in an additional period, or may be interspersed with other lessons.

PREPARATION

- Establish cooperative groups of 4 or 5 students.
- Assemble materials: paper strips and pencils to make rulers, paper for charts, materials for display.

PROJECT 2

KEEP THOSE TOES A-TAPPING!

Genetics influences the way you look, the way you act, and your health. Some genetically influenced characteristics are very noticeable. Skin color, hair color, eye color, and height are frequently used to describe people. Other genetically influenced characteristics are not thought about as much. Do you have attached or free earlobes? Can you taste PTC? Do you have a "hitchhiker's thumb?" Since it is hard to study a number of genetic traits all at once, this project focuses on gathering data on only one characteristic—the length of feet.

The length of the human foot is influenced by multiple genetic factors. To find out about this unique trait, first you will gather data. Then analyze it, and finally design a method to display results.

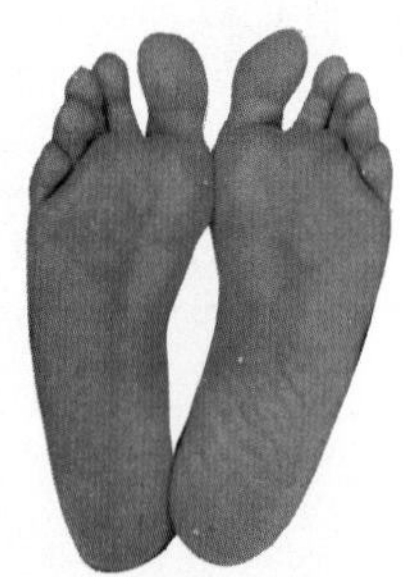

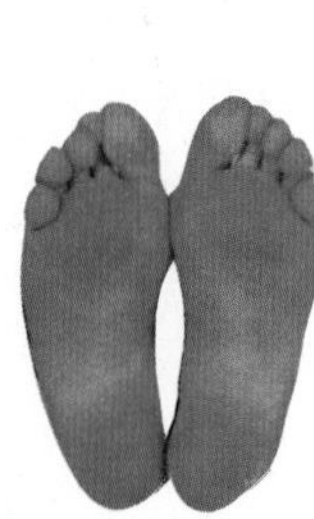

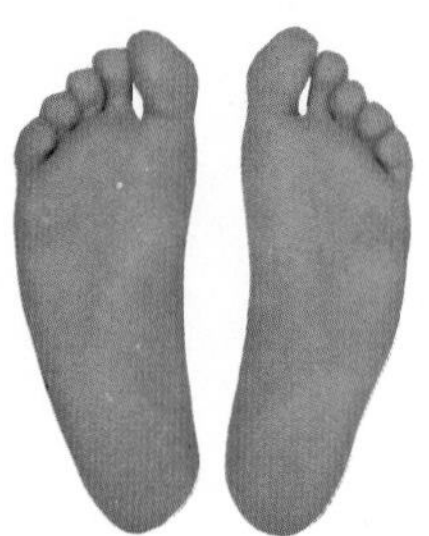

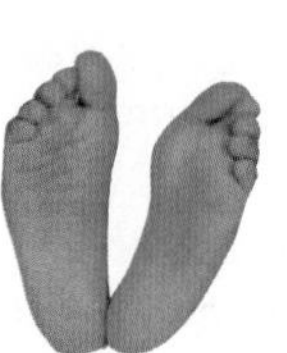

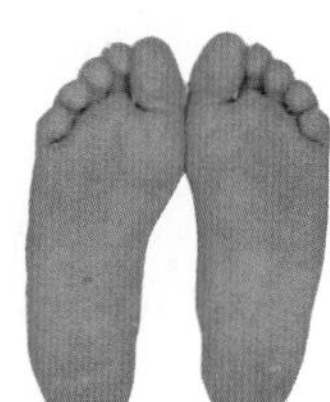

Collecting Data

Now comes the fun part. Work in a group. Each group will need to make a chart to record data. The chart will need to have horizontal heads that read Age, Sex, and Foot length (left/right) across the top. Down the side will be a number for each pair of feet you measure.

You can make a flexible ruler by placing a strip of paper over a plastic ruler with raised markings in centimeters. Rub over the ruler with your pencil and the markings will be transferred to your paper ruler.

Each group should measure the feet of *at least* 10 people. These people should be a variety of ages. Include both males and females.

Compiling Data

As a class, decide how you are going to divide the age groups. Are you going to make three-year groupings (ages 0-2, 3-5, etc.) or five year groupings (0-4, 5-9, etc.)? Are you going to make smaller groupings for young children, but increase the age increments to 10 years for adults? Each group will now be assigned a particular age grouping and will make a chart (like the one used to record the data) for just that age group. Once these age group charts are ready, each group will fill in the data they collected on the appropriate age charts.

Data and Statistics

Once everyone has recorded the data they measured, each group is ready to statistically

REFERENCES

Burns, George W. *The Science of Genetics.* New York: Macmillan Co., 1969.

Gould, Stephen Jay. *Hen's Teeth and Horses' Toes.* New York: W.W. Norton and Co., 1983.

Leigh, Julia and David Savold. *The Day that Lightening Chased the Housewife and Other Mysteries of Science.* New York: Harper and Row—Perennial Library, 1989.

Sutton, Caroline. *How Do They Do That?* New York: Quill Press, 1982.

analyze their particular age group chart. Calculate the mean, median, and mode of foot length for males and for females in a particular age group. The mean of male foot length is found by adding all the lengths of the male foot measurements in the age group and dividing by the number of male subjects in that age group. The same process is followed to find the mean of female foot length, but only the data on females is used. The foot lengths for all

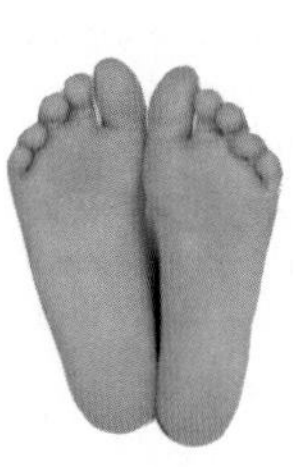

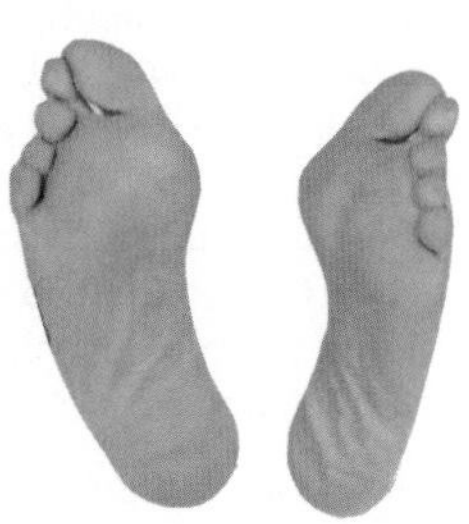

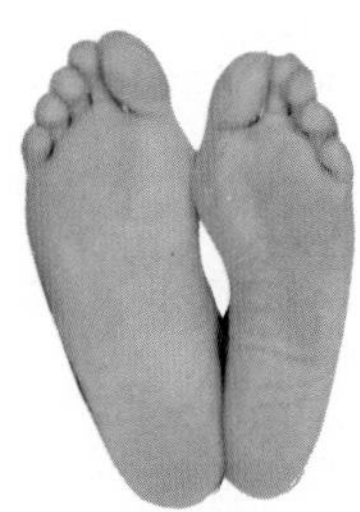

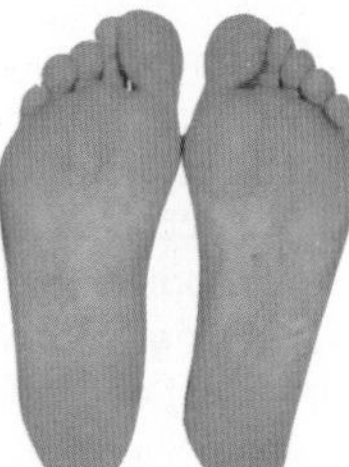

males within the age group should be listed in order from smallest to longest. The median of male foot length can then be found by finding the length in the exact middle of the list. (If there is an even number, the median is calculated by averaging the two middle lengths). The mode is the foot length that appears most frequently in the data. The same process is applied to the female foot length data to find female median foot length and the mode for female foot length.

Data Display

As a class, decide how to display the data you have analyzed so that the differences between males and females and among different age groupings can be easily seen. Make construction paper feet to display your statistics. Try to come up with a creative way of showing how foot length changes with age. You might also want to show the variation in foot size within age groupings.

Extension

Ask a local podiatrist for charts that list the average foot size for various age groups. See how close your results come to these published averages. Check with several local shoe stores to find out what sizes sell the most.

TEACHING NOTES

- This project examines one trait governed by more than one gene. While the project can begin as you start Chapter 5, the principle of polygenic inheritance is covered in Section 5-2.
- Before dividing into groups, show the class a picture of an individual from a magazine. Have the class describe this person using traits. Write the descriptive words on the board. Discuss how genetics affects some of the traits listed (pierced ears and salon-colored hair are not inherited). What other characteristics do genetics affect? A hitchhiker's thumb is one that bends outward; attached ear lobes have no free-hanging lobe.
- Students should make their measurements from the top of the big toe to the heel.
- Discuss how large amounts of data make the outcome of an investigation more reliable.

EXTENSION

What are some genetic traits that we can't observe just by looking at a person (e.g., blood type, PTC, etc.)? The ability to taste phenylthiocarbamide (PTC) is determined by placing a test paper in your mouth.

Assess students' ability to analyze a genetic trait by having them use PTC test papers to see if they are "tasters" or "nontasters." Volunteers could also test their parents and other relatives and try to do a pedigree analysis of the trait in their family. See page 119 for symbols used in drawing pedigrees. **CAUTION:** When doing pedigrees, use volunteers after parental permission to be sensitive to adoptive situations.

Being a taster is determined by a dominant allele (T), so a taster may have the genotype TT or Tt. A nontaster can only have the genotype tt. By tracing how the alleles must be passed from generation to generation, students should be able to determine the genotypes for some of their relatives.

PROJECT 3

OVERVIEW

OBJECTIVE: Students will *observe* a current concern, and apply and *evaluate* a technological solution to that concern.

SUMMARY

In this project, students will learn about composting as a means of organic waster disposal and will experiment with variables involved in the composting process to build an effective compost pile.

TIME REQUIRED

A minimum of three weeks is required to produce compost. Two full periods are required to research and plan how and where to construct compost piles. One full period is needed at the end to evaluate the data collected from the compost piles. A few minutes will be needed daily to record observations, collect temperature data, and, when necessary, turn the compost piles.

PREPARATION

- Establish cooperative groups of 4 or 5 students.
- disposable plastic gloves
- Help assemble materials that students decide they need, such as a glass jar, plastic aquarium, or window screens for compost pile soil, fertilizer, water, and organic wastes for composting.

PROJECT 3

DON'T TOSS IT, COMPOST IT!

Have you ever thought about what happens to all the trash—paper cups, napkins, paper plates and plastic, spoons—after a football game or a dance? Of course, the first stop is a trash can or dumpster. This trash usually is picked up and trucked to a landfill. But our landfills are filling up, and we are running out of room for more. Some people are even suggesting that we send our garbage into space—littering on a cosmic scale!

The Compost Solution

Many communities try to cut down on landfill problems through recycling and composting. Your community may already recycle aluminum cans, glass, plastic, and paper products. Your family can also compost kitchen scraps and garden wastes. A compost heap allows organic garbage to break down and decay—a natural process that occurs when living things die. As these organic materials are broken down by bacteria and fungi, nutrients are released into the environment and can be used by other living organisms. When you compost, you are taking part in a process that has been going on for millions of years.

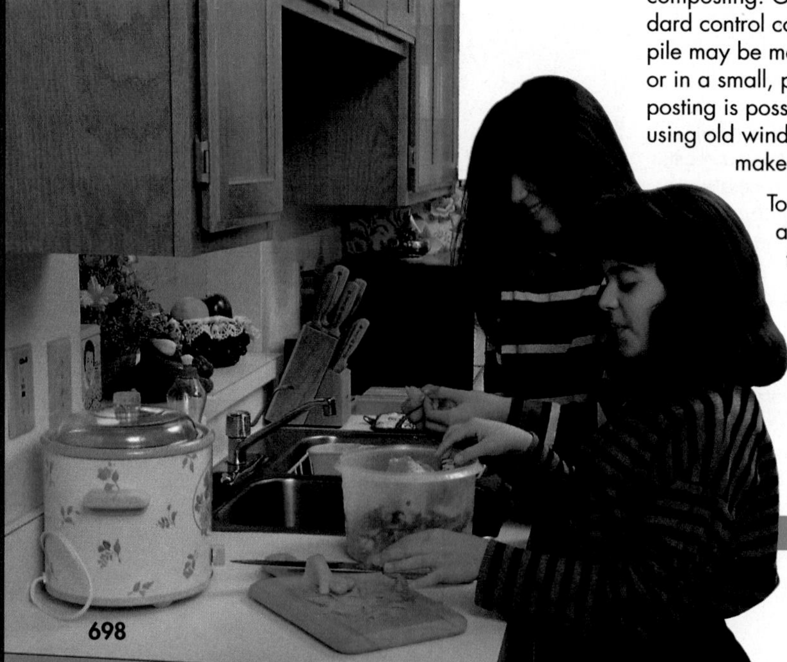

How It Works

Bacteria and fungi start the composting process by breaking down large organic materials. This releases heat as well as nutrients. The heat "cooks" and decomposes more material. A compost pile about 1 cubic meter in size works best to generate the heat required to make the process work. The pile needs to be turned or sifted regularly to keep material in contact with the warm center. The pile also needs to be kept moist and given plenty of air to keep the bacteria and fungi active.

Making a Compost Pile

Work in cooperative groups to learn about composting. One group could set up a standard control compost pile. A small, compost pile may be made in a large-mouthed glass jar or in a small, plastic aquarium. If outdoor composting is possible, you can make an enclosure using old window screens. Use four screens to make an open cube.

To build a control compost pile, alternate layers of soil (1 inch thick), a layer of kitchen or garden waste (2 inches thick), a sprinkle of fertilizer containing nitrogen, and a sprinkle of water. End with a top layer of soil. Although many types of wastes will decompose, do not add meat or fat—they can

698

REFERENCES

Campbell, Stu. *Let It Rot! The Gardener's Guide to Composting*. Pownall, VT: Storey Communications, Inc., 1990.

"How to Build a Compost Heap." Rutgers Cooperative Extension of Morris County, CN 900, Morristown, NJ 07960.

Jones, Linda L. Cronin. "Strike It Rich with Classroom Compost." *The American Biology Teacher*, Vol. 54, No. 7,. October, 1992. pp. 420-424.

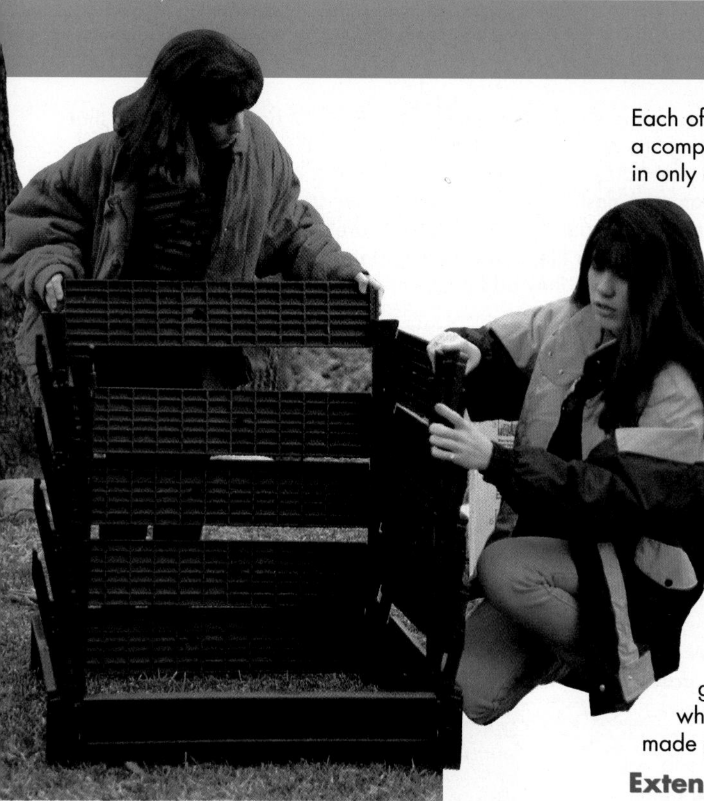

Each of the other student groups should design a compost pile that varies from the control pile in only one way. The amount of light, heat, or water the pile receives could be varied. The pile could be turned more or less frequently than the control pile. Each group should **hypothesize** how one change will affect the compost. "If we wet the pile well, then compost will be made more quickly." "If we don't turn the pile each week, it will make compost more quickly."

Analyzing the Differences

Each group will record the changes in their compost pile by finding its temperature and observing its smell, texture, and appearance. Make all observations of the compost pile before turning it with the fork or shovel. At the end of the experiment, groups should compare notes to decide which compost technique and recipe made good compost.

Extension

Find out how your community's trash and garbage is processed. Does the community compost grass clippings, or are clippings bagged to be picked up with the trash? Almost one quarter of the space in landfills in the United States is currently taken up by yard clippings! What advantages does composting have over landfills for organic materials? Perhaps some students would like to set up an on-going school composting program.

attract animals. Any type of grass clippings, leaves, or fruit or vegetable peels work well. Ash from a fireplace and sawdust are good additions also. Make sure that the soil you use is not sterilized potting soil purchased from a store. After the pile is completed, use a shovel or garden fork to turn the compost once a week. This brings in oxygen and brings new material into the warm center. Your compost is ready when the materials have become uniformly brown and crumbly.

TEACHING NOTES

- Before students begin to research compost piles, ask them to think about the trash generated by the school. Discuss the kinds of trash, and the amount of trash. Take the class to the school dumpster. With students wearing disposable gloves, get a general idea of the different kinds of trash in the dumpster.
- Broaden the discussion to think about the trash from homes. Is it different from "school trash"?
- Discuss places that have special trash problems—hospitals would be a good example of this, as would groceries and businesses using substances with radiation. Discuss ways that trash can be handled.
- Obtain permission to build one or more compost heaps on school grounds. Have students offer the school a proposal on what the compost can be used for—such as plantings on the school grounds.

EXTENSION

Some students may wish to do some research on trash solutions. Although landfills have improved and are constructed to prevent leakage of harmful chemicals into surrounding land and water, they are still non-aerobic and slow. In some instances, they even stop the decay process. They must also be vented or methane gas may build up and cause problems.

Other students may wish to try experiments growing plants using different mixtures of soil and compost to find how the soil to compost ratio affects plant growth.

PROJECT 4

OVERVIEW

OBJECTIVE: Students will *make a model* using information gained through observation and research.

SUMMARY

In this project, students will research various types of animal shelters and make a model of one. At least one group should be encouraged to build a life-size eagle's nest. They will also look at how the shelter reflects the needs of the animal and the materials available and how it is an adaptation to that environment for the animal involved. After building the model, students will have an opportunity to observe other animal home models.

TIME REQUIRED

This project is designed to be worked on throughout Unit 4. One period is needed to allow groups to select an animal home to research. One period is needed for the model presentations. Actual building of models will have to be interspersed with lessons during the unit.

PREPARATION

- Establish cooperative groups of 4 or 5 students.
- Collect sources of information on animal homes.
- Plan on where students can collect materials for use in their models. Students will be using mostly plant materials.
- For larger models, such as a life-size eagle's nest, obtain permission from the school ar a local park where the nest can be constructed and be undisturbed.

PROJECT 4 ANIMAL ENGINEERS

Have you ever found an abandoned animal home? You may not have thought of it that way, but a bird's nest, a hornet's nest, an anthill, or a rabbit burrow is a carefully engineered home. Just like your home, these homes provide shelter and protection from the environment and predators. A nest is usually made up of twigs. Nest sizes differ. A hummingbird nest is the size of a quarter. Bald eagle's nests are very large and may be 2 meters in diameter. Eagles mate for life, and a nesting pair may return to the same nest for as long as 30 years. Each year, they add a little more to their nest, and it grows.

Building an Animal Home

Animals use all types of materials in their homes. Animals build with twigs, mud, grass, hair, fur, stone, wood, or anything else they can find. Ospreys, large birds that live near the water, may use discarded fishing line to build nests, and frequently use seaweed. Some osprey nests have included bicycle tires and rubber gloves! Eagles in contrast, are very particular about what goes into their nests. Their nests are huge and made of sticks no less than one inch in diameter. Then they are lined with evergreens. These nests weight hundreds of pounds. They are built to last many years. Built in trees, an eagles nest is exposed to severe weather conditions.

Some animal homes are designed to blend into the environment. Prairie dog towns are basically underground. Other animals, like the magpie, a bird, like to collect shiny objects to "decorate" their homes.

Work in cooperative groups to pick an animal home to research. What materials does the animal use? How big is the home? What kind of environment is involved? What is the actual building process the animal uses? How long is the home expected to last? After researching a particular animal home, build a nest or burrow using the same materials and dimensions that the animal uses.

700

REFERENCES

Boardman, Richard. *Animal Colonies: Development and Function Through Time.* New York: Van Nostrand Reinhold, 1973.

Cassidy, James ed. *Book of North American Birds.* Pleasantville, NY: The Reader's Digest Association, Inc., 1990.

McDonnell, Janet. *Animal Builders.* Mankato, MN: Child's World, 1989.

The Model Home Show

Plan a display of your model homes that shows what is special about this animal home. Show how it fits into the particular environment in which it is found. How does it offer protection and shelter for the animal involved? Does the environment influence the materials, and therefore, the structure of the animal house? Think about the wind and rain that some structures have to withstand.

Extension

Find out about animals that live in your area. What animals have you observed? If possible, go on a walk with a naturalist to learn more about the wildlife in your area. You may wish to focus particularly on bird nests. What types of materials do birds in your area use? Do they use twigs? Are the nests lined with feathers? Take pictures or make sketches of any nests you observe in your area. See if you can decide what kind of bird is using the nest just by its size and makeup.

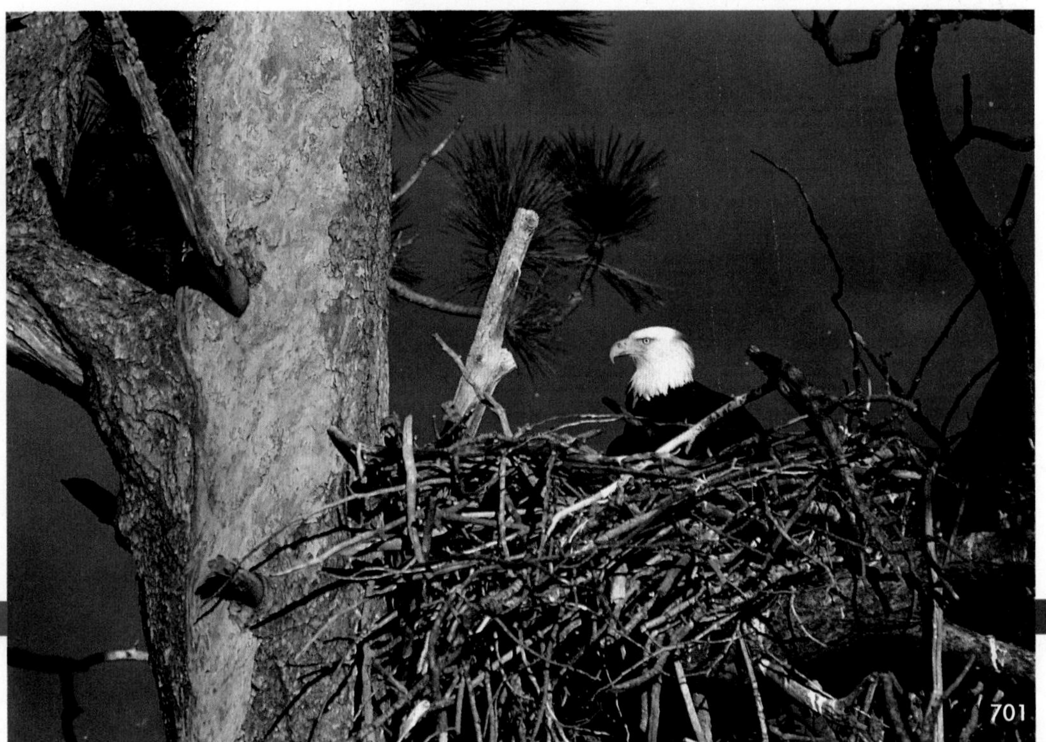

TEACHING NOTES

• Before having students choose animal homes to research, have them brainstorm some they are familiar with. List these on the chalkboard and have students analyze features common to all homes on the list. Help students to see that animal homes are affected by the materials available and the stresses of the environment. Draw from students the fact that plant materials and Earth-related structures and materials (caves, soil burrows, and so on) are what animals draw on for building shelters. Discuss the needs that all living organisms have, bringing the focus onto the needs of shelter and protection, and how they help species survive. Discuss the effects of destruction of nesting sites on populations.

• Ask students about the specific shelter requirements of the animals they have named. For example, a reptile will seek out a shelter that helps it maintain a relatively stable body temperature. A desert dweller needs to have a shelter that provides a lower temperature and shade. An animal in a very cold climate might need a shelter that provides warmth.

EXTENSION

To assess students' understanding of the materials and engineering that goes into animal-built shelters, have them construct nesting sites for various birds, such as a nest cone for mourning doves, or a bluebird box. Nesting materials can also be left out for the use of nearby birds. Hair from a hairbrush, string or yarn, and other supplies are useful to many birds. Interested students may also want to try constructing an animal burrow. A piece of clear plastic can be put on one side of the burrow model.

PROJECT 5

OVERVIEW

OBJECTIVE: Students will *apply* a particular theory to research and construct a *model.*

SUMMARY

In this project, students will attempt to research a prehistoric life form based on a particular theoretical perspective. They will apply this research to build a model to represent a particular dinosaur. After doing this research and building the model, students will discuss the relationship between the apparent abilities of the animal and its body structure.

TIME REQUIRED

This project is designed to be worked on throughout Unit 5. One to two periods are required to set up the project. Students will need time throughout the unit to work on research, construction, and presentation of dinosaur information.

PREPARATION

- Establish cooperative groups of 4 or 5 students.
- Make available valid information sources on dinosaurs and materials for dinosaur and diorama construction.

PROJECT 5

WAS IT A REPTILE? WAS IT A BIRD?

Dinosaurs are popular attractions in stories, as action figures, on TV shows, and in museum displays. We have lots of pieces of evidence from dinosaur times, but this evidence can be woven into a variety of theories about the nature of dinosaurs, dinosaur lifestyles, and the cause of dinosaurs' demise. Originally, most scientists accepted the idea that dinosaurs were cold-blooded reptiles, slow-moving, and not terribly bright. Today, however, new evidence is causing scientists to question some of these ideas. What are some of the current controversies?

Theories in Conflict

Scientists have many questions about dinosaurs.

- Were dinosaurs cold-blooded or warm-blooded?
- Were the dinosaurs more closely related to reptiles or to birds?
- Were their bones solid like reptile bones or hollow inside like bird bones?
- Were dinosaurs slow-moving, or could some of them move quickly?
- Impressions of dinosaur skin show gland cells. Were these sweat glands or poison glands?
- Did dinosaurs live in hot, humid areas or hot, dry areas?
- Are bones that are found together from one dinosaur or from many? How can scientists tell?
- Are the eggs that are found with dinosaur bones their dinner or their off-spring?
- What caused dinosaurs to disappear? Geological processes such as volcanoes, or comets from space?

Using a Theory to Design a Dinosaur

Jack Horner from the Museum of the Rockies in Bozman, Montana thinks that dinosaurs were more closely related to birds than to reptiles. He has found that many dinosaur bones appear to be hollow. By using the CAT scan X ray, he has determined that the skull and other bones contain many holes for blood vessels similar to those found in warm-blooded animals like birds. These ideas are still controversial, but you can use Dr. Horner's theory to construct a dinosaur yourself.

702

REFERENCES

Herner, John R. and James Gorman. *Digging Dinosaurs.* New York: Workman Publishers, 1988.

Lessem, Don. "Skinning the Dinosaur." *Discover,* March 1989, pp. 38-44.

Robbins, Jim. "The Real Jurrasic Park." *Discover,* March 1991, pp. 52-59.

Trefil, James. "Whale Feet." *Discover,* May 1991, pp. 44-48.

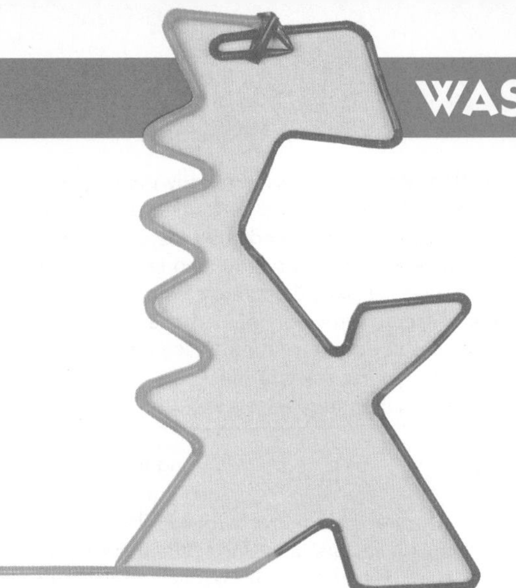

WAS IT REALLY WARM?

In a small group, look at books on dinosaurs and find one particular dinosaur that you can construct a model of. Make sure there is a diagram of the bone structure to give you guidance. Decide what materials you will use. Straws are hollow and lightweight like bird bones would be. Use bendable straws. The straws can be joined with tape.

Study the environments in which your dinosaur seems to have lived. Build your dinosaur and a realistic diorama showing its environment and something about its lifestyle.

Presenting Your Information

Prepare a brief description to explain the model and diorama. Arrange a day for groups to share these projects with each other. Look for connections between the various forms researched. Can a food web be constructed to connect the various dinosaurs?

Extension

Discuss whether you agree with Dr. Horner's theory or not. What differences in dinosaur lifestyle are involved in accepting or rejecting Dr. Horner's theory? Do you have any different theory about the nature of dinosaurs? What evidence currently supports this theory? What further evidence would be needed to support your theory?

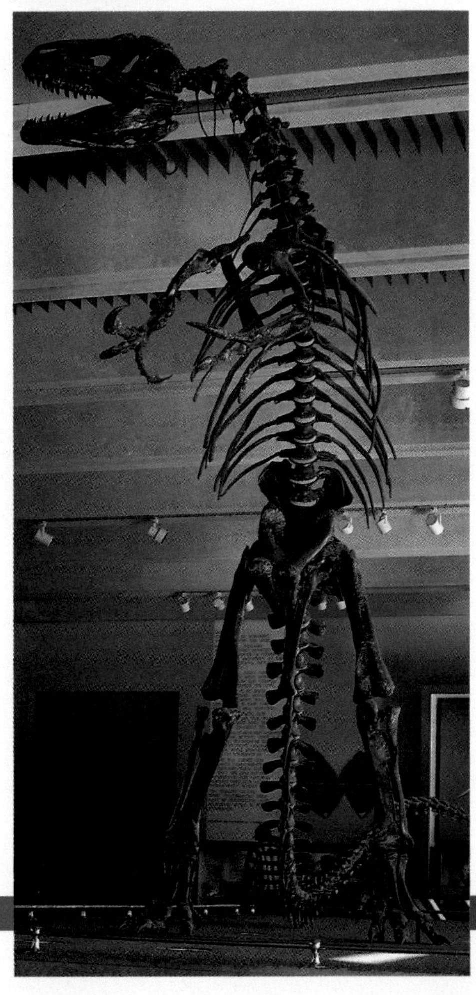

TEACHING NOTES

- Before students begin their research, ask them to talk about how dinosaurs are represented in stories, movies, TV shows, stuffed toys, etc. Discuss why people seem to have such an interest in dinosaurs, and how representations are not always scientifically based or accurate.
- Focus on bone structure and why animals stand on hind legs or need to be on all fours to support themselves.
- Discuss how bones are constructed to withstand stress. Consider getting input from a structural engineer as to how some structures are supported. Discuss how this relates to dinosaurs.
- Focus on differences between warm-blooded and cold-blooded animals and their needs. Discuss current research on why it is now thought that dinosaurs were warm-blooded instead of the long held idea that they were more reptilian, or cold-blooded in nature.

EXTENSION

One major area of current research focuses on what caused the dinosaurs to become extinct. Students could look at the various theories suggested to explain the major extinctions of dinosaurs that appear in the fossil record. What kinds of evidence support each particular theory? What further evidence would be helpful in deciding among the various theories?

Another area of controversy is the idea of remnant dinosaurs. The most famous example of this would be the Loch Ness monster.

PROJECT 6

OVERVIEW

OBJECTIVE: Students will *practice recording observations,* and *analyzing* the *data* produced through the use of graphs.

SUMMARY

In this activity, students will measure and record pulse and breathing rate in a resting state, a state of normal activity, and after exercise. They will also chart changes in performance produced by training for an athletic event. This information will then be analyzed and students will explain the results using their knowledge of various organ systems.

TIME REQUIRED

This project is designed to be worked on throughout Unit 6. One period is required to set up the Olympic competition. One period a week is needed for training and for measuring performance. Students will need one period to analyze and explain their findings.

PREPARATION

- Establish cooperative groups of 4 or 5 students.
- Obtain ropes, balls, space for Olympic events, paper to record data, and graph paper.

PROJECT 6

ON YOUR MARK! GET SET! GO!

The Olympics is exciting sports competition, showing the human body performing at its peak. Many organ systems must work together in an athletic performance. The brain receives input from the sensory organs and directs and coordinates the action. Nerves carry the brain's directions, and stimulate the muscles to contract and to move the muscular and skeletal systems. In addition, nerves stimulate the circulatory and respiratory systems to speed the flow of blood with its increased supply of oxygen to the active muscles. The endocrine system also regulates performance through the hormones it releases. Adrenalin, for example, is released during times of stress, and may give people greater strength, speed, and quickness than usual—certainly useful to athletes in a tough competition!

Classroom Olympics

How fast can you run the 100-meter dash? Does training affect an athlete's ability to perform? For this project, you will plan, train, and participate in your own Olympic events. Working in groups, students will decide on physical activities that they would like to include in the classroom games. Each group is responsible for planning one event and demonstrating the event for the class. After the events are decided, each group will decide who will run the group's event and who will participate in the events planned by other groups.

Training

Like Olympic athletes, the classroom athletes will need to train. Unlike Olympic athletes, you are competing against your own record. First you want to look at your current performance level in your event. Have a partner record your level of achievement in the event so that you can compare your later performances to this baseline. Now you are ready to train. Have your partner measure your body functions, such as pulse rate and breaths per minute, before you begin. Record these measurements—they will show the normal speed of your circulatory and respiratory systems. Sit quietly for 5 minutes and record the same measurements. These will show your resting rates. Now train in your event for 5 minutes and record these measurements again. Have your circulatory and respiratory systems slowed down or speeded up?

When you train, be consistent in the way you measure performance. For example, if you are

704

REFERENCES

Mallon, Bill. *The Olympic Record Book.* Hamden, CT: Garland Pub., 1988.

Wallechinsky, David. *The Complete Book of the Olympics,* Rev. ed. New York: Penguin, 1988.

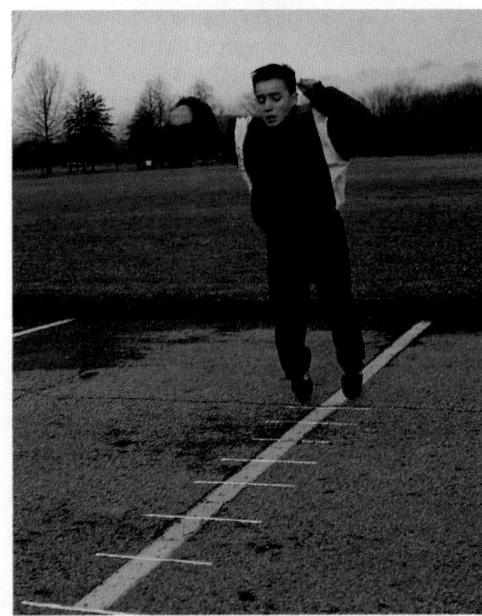

doing a running event, time your runs on the same course each time. If you are jumping rope, you might measure how long you can jump rope without tripping up. When your training time is finished, switch jobs and let your partner train while you record. Although you will have some time to train in school, you may want to keep practicing at home—or at least do some conditioning. Athletes do conditioning exercises such as running to increase their strength and endurance in general, rather than practice a particular skill.

Graphing Your Results

Within your group, analyze the data you have collected on performance. On the *x*-axis, record the number of each trial. On the *y*-axis, record your performance in seconds, minutes, meters, number of jumps, etc. Use another graph to chart the pattern of your pulse and breathing rates. Has your level of performance changed at all? Have your pulse and breathing rates changed at all as your training progressed? Think about the organ systems involved in your event and explain the changes or lack of change.

Extension

You may wish to investigate the effect of learning on performance. Groups can design games or tasks that require learning a strategy and see how practice affects performance. One example of this kind of task would involve numbering small pieces of paper and having students pick them up in order.

TEACHING NOTES

• Begin by asking students what sports events they like to play in or watch. Ask if anyone watches the Olympic competitions. What events do they particularly like? Are there any events that do not get much TV coverage on American stations? Discuss why some events are more heavily televised in the United States than others. Discuss what countries appear to have more success in certain events than other countries. For example, Russian gymnasts have been very successful, and winning Olympic cross-country skiers are frequently from just a few European countries. Why do athletes from these countries have such success? Factors can include things such as availability of snow for practice, cost of the sport, emphasis on training and support of athletes, and so on.

EXTENSION

Students can investigate Olympic records for various events. Why are records continually being broken? Are there any records that have not changed much over time? What factors influence how quickly an athletic record is likely to be broken?

PROJECT 7

OVERVIEW

OBJECTIVE: Students will practice sound research techniques, *collect data* on a real life concern, and *analyze,* present, and act upon it appropriately.

SUMMARY

In this activity, students will use a variety of research techniques to learn about people's perspectives on foods and health. Students may research:

- people's attitudes using a questionnaire
- buying habits by observing grocery store behavior
- taste perceptions using a blind taste test
- food options at school by surveying available resources
- advertising strategies by analyzing media influence.

After collecting the data, students will analyze and summarize the data, and discuss ways to use this data to promote healthy eating.

TIME REQUIRED

This project is designed to be worked on throughout Unit 7. One to two periods are required to set up the project, then group time should be interspersed with other lessons.

PREPARATION

- Establish cooperative groups of 4 or 5 students.
- Assess media sources needed.
- Obtain permission of grocer and cooperation of cafeteria staff.
- List ingredients for taste test recipes.

PROJECT 7 TO YOUR HEALTH!

What is a healthy lifestyle? Rest, exercise, and even your outlook on life—whether you are optimistic or pessimistic—all play a part. Having a good diet is important too, but not everyone agrees on what a healthy diet is. Different people also require different amounts of foods. Athletes frequently "eat to win," selecting meals that are high in carbohydrates. Males and females need different amounts of calcium. Body image affects diet greatly. Forty percent of 4th grade girls in the US are on some diet, frequently self-imposed, to either gain or lose weight. Some of these diets are based on healthy principles and make use of nutritious, low-fat foods. Others may be unhealthy, like those that rely almost exclusively on only one particular type of food.

Gathering Data

Work in groups of four or five. Each group should decide on a question to investigate people's attitudes about healthy eating habits, and devise a way to research these attitudes. Some examples of possible research approaches include:

1. Set up and administer a blind taste test to 50 people.
   Prepare the same recipe for banana bread or corn bread two ways. First, prepare the bread using a standard recipe. Then, make a "healthy" substitution of fruit for sugar, or canola oil for butter. Don't tell them the difference. See which recipe people prefer. Do different age groups have different preferences? Do they even notice a difference? Do males and females?

2. Research the focus of food advertising. Collect samples of advertising from various media sources and analyze the messages advertisers send to consumers about health and food. Do the messages appear to target any groups in particular? How can you tell?

706

REFERENCES

The Wellness Encyclopedia. Boston, MA: Houghton Mifflin Co., 1991.

3. Observe grocery store behavior.
After obtaining permission from the store owner or manager, students can map the layout of food in a grocery store. The manager may be interviewed to discover why the store places items in particular places, and how "specials" are chosen and displayed. Students can also observe how people circulate through the store. Again, look for patterns by age group, sex, time of day, etc.

4. Design and give a questionnaire to 50 people.
Decide on background information that might help you to categorize responses.
For example:
Male or female?
Age?
Occupation?
Hobbies?
Choose "yes" or "no" questions about what people actually eat and what they think they should eat.
For example:
Are you on a diet?
Do you think you can get healthy food at a fast food restaurant?
Do you exercise on a regular basis?

Make sure the questions you choose are clear and will result in enough information for analysis. Ask for the information you want. Decide how many people you want to survey and make copies of the questionnaire. Decide when and where the survey will be done and who will conduct it. You may want to survey different groups (e.g., athletes, different age groups, etc.) to see if their answers vary.

5. Survey the food choices at your school. Analyze the daily menu for the cafeteria and the contents of any available snack machines. Do these foods encourage healthy eating? Students might also survey actual consumption patterns by randomly observing one table's occupants each day.

Presenting the Data

Each group must decide:

How will the data be summarized—a chart? A graph? A video-tape? A report? How will you present your results? A poster? A verbal report? A written summary?

What Can You Do?

Discuss the results of the research as a class. Use what you have learned to make an ad campaign for healthy eating. This campaign could include posters, "radio ads" for school announcements, and print ads for student newsletters. Be creative!

TEACHING NOTES

- Before students begin their research, ask them to talk about diets they have used or heard about. Discuss why dieting is a major concern for many young people. One out of four college freshman women have an eating disorder of some kind. Point out the connections between dieting, body image, and social acceptance. Encourage students to see the difference between a healthy diet and a diet to achieve a particular body image (where the diet may or may not be healthy).
- Explain the importance of obtaining answers from a questionnaire without prejudicing the person being questioned. Make questions simple and clear.
- Explain that a blind taste test does not call out the difference between two items, but seeks to have the customer respond to the differences without being informed of specific differences. The two items compared should be nearly identical, with but one variable changed. One suggestion is to use a recipe for banana bread or a basic cookie where dry milk is substituted for half the sugar required.

EXTENSION

Students could plan a health party and prepare and share healthy snack foods.

Students can also encourage healthy eating by recommending changes in food options offered at school, or by organizing health awareness activities to share with other classes.

PROJECT 8

OVERVIEW

OBJECTIVE: Students will *design and construct a map* that shows facts about different biomes.

SUMMARY

In this project, students will work in groups to plan and construct a map of the world that will display physical features, political boundaries, plants and animals found in a particular part of the world, and other interesting information as they study Unit 8. Groups will relate the geographic features to characteristics uncovered in their research. Finally the groups will display their maps and challenge classmates with questions that can be answered using the maps and research notes.

TIME REQUIRED

Students will need parts of class periods for one week to design and construct their basic map. Research time in the library can be given throughout the unit to gather information on the various parts of the world. Students will need a full period to choose their geographic "clues" and to play the detective game.

PREPARATION

- Establish cooperative groups of 4 or 5 students.
- Assign students to assemble materials such as posterboard or paper for map, rulers to make the map to scale, colored pens, pencils, or crayons, 3″ × 5″ notecards, resource materials with information on various countries.

PROJECT 8

WHERE IN THE WORLD?

The National Geographic Society defines geography as "a field of knowledge that deals with the Earth and all the life on it." Geography, then, is much more than maps and places. To know the geography of a place means to understand the physical, cultural, political, economic, historical, and environmental facts about it.

Making a Map

Divide your class into cooperative groups, and have each group design and make a world map. Each group may make a map or all the groups in the class may make one huge map. Look at a world atlas before you begin to see how cartographers (map makers) design maps. What kinds of things do they include? Decide how large you want your map to be. Size your map to scale so that the relative sizes of the continents and countries are in proportion to one another. Use different colors to distinguish countries from each other. Include a legend to identify mountains, deserts, bodies of water, and other physical features. Plan to use magazine photographs and drawings to illustrate your map.

Researching Your World

Each group should research just one part of the world. Use natural history magazines, library references, and atlases to find interesting facts about your region that your classmates won't already know. Look for features that are unique to your part of the world. For example, your region may include the largest desert or the largest body of water. It may have the least populated country or the country with the greatest rainfall. Does it include countries that

708

REFERENCES

Encyclopedia of World Geography. New York: Dorset Press, 1989.

Espenshade, Edward B., ed. *Goode's World Atlas,* 18th Edition. Chicago, IL: Rand McNally, 1990.

have changed names? Place names frequently change as political situations change. The Soviet Union has separated into many smaller republics that have reclaimed their old names. The country that was once Upper Volta changed its name to Burkino Faso after it gained independence.

Put the World Together

After your group has finished its research, you will have more information than you can include on one map. Decide what will go on the map and how the class groups will share what they've learned.

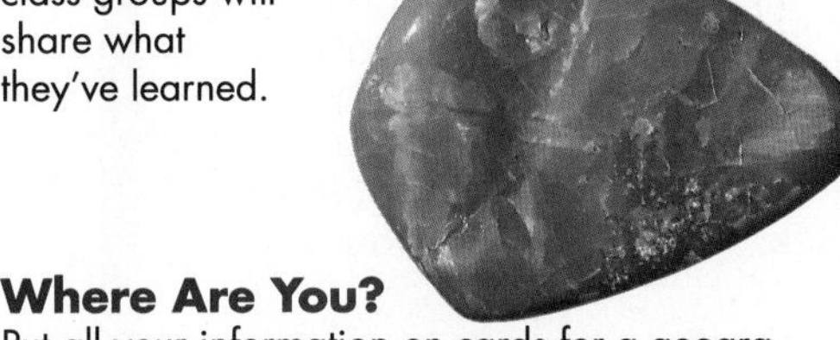

Where Are You?

Put all your information on cards for a geography game. Have each group write ten "clues" using their cards. After all groups have assembled their clues, begin the game. One group should share a clue and another class group, working as a team, should figure out where in the world or what in the world is being described. Work out a scoring system. Several example clues are given in the next column.

- You are looking for a country that has the largest movie industry. Where should you look?
- You are looking for a place where there are brown bears, white bears, and where the Inuit people live. Where should you look?
- You are in the country that contains the world's largest opal mines. Where are you?

An Extension

Is It Imported?

How interdependent are the countries of the world? Investigate where manufactured items come from. Survey 20 to 30 items at home to find out where they were made. Chart your findings to show what countries are represented. Look at labels on food, and tags on clothing, appliances, games, and electronics. Locate these places on your map.

TEACHING NOTES

- Ask students why they think they like a particular place. Is it the physical features it has? The weather? The plants and animals? The people? The food? Point out that all these things are part of that area's geography. How does geography affect relationships between countries and cultures? What does geography affect? (trade, politics, materials availability) Does geography make interaction with other countries easy, or hard?
- Have students decide if they want separate group world maps, or one large map.
- Have students discuss their section of the world in terms of its biomes.

EXTENSION

Multinational corporations are becoming more and more common. Students may wish to choose a particular multinational company and look at the goods it produces. They could map where the different stages of manufacture and distribution take place. For example, raw materials may be shipped from one country to be worked on in factories in another country. The product may then be shipped to yet another country to be sold. Why do these companies operate in this fashion? What geographic factors influence this process?

GLOSSARY

This glossary defines each key term that appears in **bold type** in the text. It also shows the page number where you can find the word used. Some other terms that you may need to look up are included, too. We also show how to pronounce some of the words. A key to pronunciation is in the table below.

PRONUNCIATION KEY		
a . . . **b**a**ck** (bak)	oh . . . g**o** (goh)	sh . . . **sh**elf (shelf)
ay . . . d**a**y (day)	aw . . . s**o**ft (sawft)	ch . . . na**t**ure (nay chur)
ah . . . f**a**ther (fahth ur)	or . . . **or**bit (or but)	g . . . **g**ift (gihft)
ow . . . fl**ow**er (flow ur)	oy . . . c**oi**n (coyn)	j . . . **g**em (jem)
ar . . . c**ar** (car)	oo . . . f**oo**t (foot)	ing . . . s**ing** (sing)
e . . . l**e**ss (les)	ew . . . f**oo**d (fewd)	zh . . . vi**s**ion (vihzh un)
ee . . . l**ea**f (leef)	yoo . . . p**u**re (pyoor)	k . . . **c**a**k**e (kayk)
ih . . . tr**i**p (trihp)	yew . . . f**ew** (fyew)	s . . . **s**eed, **c**ent (seed, sent)
i (i + con + e) . . . **i**dea (i dee uh), l**i**fe (life)	uh . . . comm**a** (cahm uh)	z . . . **z**one, rai**s**e (zohn, rayz)
	u (+ con) . . . flow**er** (flow ur)	

abiotic (ay bi AHT ihk) **factors:** nonliving things in the environment, including soil, water, temperature, air, light, wind, and minerals. (607)
acid rain: rain that combines with sulfur dioxide and nitrogen oxide in the air to form sulfuric acid and nitric acid and then falls to Earth. (268)
active immunity: immunity caused by the body making its own antibodies. (564)
active transport: movement of material through a cell membrane with the use of energy. (58)
adaptation: any characteristic of an organism that helps it to survive in its environment. (8)
adolescence: the period in human growth that begins around ages 12 to 14 and lasts until age 20. (542)
adulthood: the period in human growth that begins with the end of adolescence (age 20) and extends to old age. (542)
aerobic: requiring oxygen for respiration. (190)
aggression: a forceful act used to dominate or control another animal. (401)
AIDS: acquired immune deficiency syndrome, a fatal communicable disease caused by an RNA virus; spread by sexual contact, by use of contaminated needles, through the placenta or by nursing by an AIDS - infected mother. (186)
algae: plantlike protists that contain chlorophyll and make their own food; they have no roots, stems, or leaves and live in or near water. (203)
alleles (uh LEE LZ): the different forms a gene may have for a trait. (106)
allergens: substances that cause an allergic response in the body. (572)
allergy: a strong reaction of the immune system to a foreign substance. (572)
alternation of generations: in some protists and in plants, a cycle that alternates between sporophyte and gametophyte generations. (238)
alveoli (al VE uh li): in the lungs, clusters of tiny thin-walled sacs where oxygen and carbon dioxide are exchanged. (490)
Alzheimer's disease: a disease of unknown cause in which nerve cells in the brain fail to communicate with each other, causing memory loss. (520)
amino (uh MEE noh) **acids:** the building blocks of proteins. (443)
amniote egg: a type of egg that provides a complete environment for the developing embryo. (364)
amniotic sac: a thin membrane surrounding an embryo, filled with amnion (a clear liquid). (537)
amphibian: a cold-blooded vertebrate whose permeable skin, strong skeleton, and hibernation or estivation abilities adapt it to life in water and on land; examples are frogs, toads, salamanders. (356)
anabolic steriods: drugs similar to testosterone; increase muscle mass; thought to improve athletic performance. (430)
anaerobic: describes organisms that can live in environments without oxygen. (190)
angiosperm: a vascular plant in which the seed is enclosed in a fruit, such as an apple. (256)

anorexia nervosa: an eating disorder in which a person eats very little to avoid gaining weight; it causes extreme weight loss that can damage organs and cause death. (456)
antibiotic: a drug used to kill bacteria. (194)
antibody: a substance made by the body in response to an antigen. (564)
antigens: proteins and chemicals that are foreign to the body. (564)
antiseptic: a substance that kills bacteria on living tissue. (559)
anus: an opening at the end of the digestive tract through which solid wastes leave the body. (317)
appendages: structures growing from the body, such as arms and legs. (332)
arteries: vessels that move blood from the heart. (466)
Arthropoda (AR thruh pahd uh): largest animal phylum; includes insects, shrimp, spiders, and centipedes. (332)
asci: the spore-producing sacs of sac fungi. (213)
asexual reproduction: reproduction in which a new organism is produced from a single parent. (80)
asthma: a disorder in which the bronchial tubes contract rapidly, causing shortness of breath or coughing; caused by smoking, allergy, or stress. (493)
atherosclerosis (ath uh roh skluh ROH sus): a circulatory system disorder in which fatty deposits form on artery walls and restrict blood flow. (468)
atria (AY tree uh): the two upper chambers of the human heart *(singular:* atrium). (464)
autologous blood transfer: the use of one's own blood for transfusion. (478)
auxin (AHKS ihn): a plant hormone that causes plants to grow or bend toward light. (283)
axon: the part of a neuron that carries messages away from the cell body. (509)

B

basidium: the club-shaped spore-producing structure of club fungi. (214)
behavior: the way an organism acts toward its environment. (394)
bilateral symmetry: describes animals that have body parts arranged in the same way on both sides of their body. (304)
binomial nomenclature: the system of naming all organisms with two words; *Homo sapiens.* (157)
biodegradable: a substance that easily decomposes in the environment, usually by bacteria or fungi breaking it down into its basic elements. (66)
biogenesis: the theory that living things can be produced only from other living things. (11)
biological indicators: organisms that reflect the condition of the environment. (361)
biome (BI ohm): a large geographic area that has similar climates and climax communities. (632)
biosphere: all areas of Earth where life exists, including air, land, and water. (606)
biotic (bi AHT ihk) **factors:** living things in the environment, such as plants and animals. (606)
bladder: a bag-shaped muscular organ that stores urine until it leaves the body. (499)
blood pressure: pressure of blood in arteries. (467)
bog: low-lying, spongy, wet ground composed mainly of dead and decaying plants. (246)
brainstem: the part of the brain extending from the cerebrum to the spinal cord; controls involuntary muscle movements. (511)
bronchi: two tubes off the trachea that carry air into the lungs *(singular:* bronchus). (490)
budding: a type of asexual reproduction in which a new organism grows from the body of its parent; for example, yeast reproduces this way. (213)
bulimia (buh LEE mee yuh): an eating disorder in which a person eats huge amounts of food and then vomits to prevent weight gain. (456)

C

cambium (KAM bee um): tissue in vascular plants that grows to produce new xylem and phloem cells. (260)
camouflage: an adaptation that allows an animal to hide by blending into its surroundings. (613)
cancer: a major chronic disease resulting from uncontrolled cell growth. (571)
capillaries: microscopic blood vessels that connect arteries and veins; nutrients, oxygen, and wastes are exchanged through their one-cell-thick walls. (466)
carbohydrates: organic compounds that provide most of the body's energy; made up of carbon, hydrogen, and oxygen. (56, 443)
carbon-oxygen cycle: the continuous movement of carbon dioxide and oxygen between Earth's surface and the air. (618)
cardiac muscle: involuntary muscle tissue found only in the heart. (427)
carnivore: an animal that eats only other animals, such as a cat. (380)
carrying capacity: the number of animals that can obtain food and shelter from the environment in a given area; the ability of an environment to support its animal population. (407)
cartilage: a thick, tough, smooth, flexible tissue that is harder than flesh but softer than bone (like your ears); it covers the ends of bones to allow movement and cushion shock. (352, 422)
catalase: enzyme that breaks down hydrogen peroxide. (545)
cell: the smallest unit of an organism that can perform life functions. (6)

cell membrane: the selectively permeable outer boundary of a cell that allows food and oxygen to move into the cell and wastes to leave it. (37)

cell theory: a major theory of life science: all organisms consist of cells; cells are basic units of structure and function in organisms; and all cells come from other cells. (34)

cell wall: a rigid structure made of cellulose that surrounds a plant cell membrane to support and protect it. (41)

cellular respiration: the process by which cells in some organisms release energy from sugar in the presence of oxygen. (64)

cellulose: the organic compound that forms plant cell walls; made up of long chains of sugar molecules. (232)

central nervous system: part of the body's control system, including the brain and spinal cord. (510)

centromere: a knotlike region that holds double-stranded chromosomes together. (76)

cerebellum: the part of the brain that coordinates muscle movements. (511)

cerebrum: the largest of the three parts of the brain; interprets impulses from the senses, stores memory, and controls voluntary muscles. (510)

chemical digestion: processing of food by fluids in the mouth, stomach, and small intestine, to break down large molecules into smaller ones that can be absorbed by cells. (451)

chemotherapy: the use of chemicals to destroy cancer cells. (571)

childhood: the period in human growth from age one to age 12. (541)

chitin: a strong, flexible carbohydrate forming the cell walls of hyphae and found in the body-covering and wings of insects. (211)

chloroplast: a plant cell organelle in which light energy is changed into chemical energy in the form of sugar during the process of photosynthesis. (41)

chordate: a member of the animal phylum Chordata that has a notochord, a hollow dorsal nerve cord, and gill slits at some time in its life cycle. (350)

chromatin: long strands of DNA in a cell's nucleus; chromatin coils into the form of chromosomes when a cell divides. (38)

chromosomes: threadlike strands of DNA and protein in a cell nucleus that carry the code for the inherited characteristics of an organism. (76)

chronic bronchitis: a disease in which too much mucus is produced in the bronchial tubes. (493)

chronic disease: a noncommunicable disease that lasts a long time. (570)

chyme (KIME): the thin, watery liquid in the digestive tract; partially digested food. (453)

cilia: short, hairlike structures that extend from the cell membrane and help tiny organisms move; found in respiratory passages. (207)

circadian rhythm: behavior that recurs on a 24-hour cycle. (404)

class: the third-highest taxonomical category, below a phylum (animals) or division (plants). (163)

classify: to group ideas, information, or objects, based on their similarities. (156)

climate: the average weather in an area, determined by rainfall, temperature, and sunlight. (632)

climax community: a stable community with a balance of abiotic and biotic factors. (631)

closed circulatory system: a blood-circulation system using vessels to transport blood to the internal organs, as in humans. (326)

cnidarians (ni DAIR ee uhnz): a phylum of animals having stinging cells. (311)

cocaine: a highly addictive stimulant drug made from the leaves of the South American coca plant. (589)

cochlea (KOH klee uh): a sound-sensitive structure in the inner ear; shaped like a snail's shell. (516)

coevolution: occurs when two species evolve structures and behaviors in response to changes in each other over a long period of time. (286)

cold-blooded: describes an animal whose body temperature varies with the temperature of its surroundings. (351)

collar cells: cells that line the inside of a sponge and help water move through the sponge. (309)

communicable disease: a disease that is transmitted from one organism to another. (560)

communication: an exchange of information among animals, by means of cries, movements, touch, speech, pheromones, and so on. (403)

community: all the populations of different species that live in the same place at the same time. (610)

competition: the striving among organisms for the existing food, water, and living space available in an ecosystem; competition determines the size and location of populations. (612)

compound light microscope: an instrument that uses lenses to magnify objects. (31)

conditioning: modifying behavior so that a response previously associated with one stimulus becomes associated with another stimulus. (398)

conservation: the wise and careful use of Earth's resources. (654)

consumers: organisms that can't make their own food. (63)

contour feathers: on birds, the strong feathers that give birds their coloring and sleek shape. (372)

control: in an experiment, the standard for comparison. (15)

controlled burns: managed forest fires set peri-odically to control vegetation and plant debris underneath the dominant forest tree species. (620)

coronary circulation: the flow of blood to heart tissues. (467)

courtship behavior: behaviors that help males and females of a species prepare for mating. (402)

crack: a strong form of cocaine. (590)

crop: in an earthworm, a sac in the digestive system that stores soil eaten by the worm. (328)

cuticle: a waxy protective layer on stems and leaves of a plant that helps the plant conserve water. (233)

cyanobacteria (si AN oh bak TEER ee uh): bacteria that make their own food from sunlight; often a blue-green color. (191)

cyclic behaviors: innate behaviors that occur in a repeating pattern. (404)

cyst: a young, parasitic worm with a protective covering. (318)

cytoplasm (SI toh plaz uhm): the fluid material inside the cell membrane that contains structures that carry out life processes. (38)

day-neutral plant: a plant that isn't very sensitive to the length of darkness and thus can flower over a wide range of days. (284)

dendrite: the portion of a neuron that receives messages and sends them to the cell body. (509)

depressant: a substance that slows the central nervous system, affecting judgment, reasoning, memory, and motor control. (584)

dermis: the inner layer of skin; it contains blood vessels, nerves, and sweat glands. (433)

development: the changes living things undergo as they grow. (8)

diaphragm (DI uh fram): a muscle beneath the lungs that helps move air in and out of the body. (490)

dichotomous key: a step-by-step guide to identifying an organism, requiring the choice of one of two descriptions at each step. (168)

dicots (DI kahts): angiosperms having two seed leaves inside their seeds; for example, maple trees. (256)

diffusion: the movement of molecules from where they are concentrated to where they are less concentrated; a type of passive transport in cells. (58)

digestion: the process that breaks down food into small molecules that can be absorbed. (451)

diploid chromosome number: in organisms that reproduce sexually, the total number of paired chromosomes in a body cell of the organism. (83)

disinfectant: a substance that kills bacteria on nonliving objects. (559)

division: the second-highest of the taxonomical categories in the plant kingdom (in the animal kingdom, the word *phylum* replaces division.) (163)

DNA: *d*eoxyribo*n*ucleic *a*cid; an acid in the nuclei of cells that codes and stores genetic information; consists of strands of molecules that control cell activities using coded instructions. (86)

dominant: the form of a trait that appears to dominate or mask another form of the same trait. (108)

dorsal nerve cord: a bundle of nerves that lies above the notochord in a chordate animal; the spinal cord in most vertebrates. (350)

down feathers: on birds, the soft, fluffy insulating feathers that cover their skin. (373)

drug: any chemical substance that changes the way a person thinks, feels, or acts. (580)

drug abuse: deliberate use of a drug for other than its intended purpose. (580)

drug misuse: use of a drug for the purpose it was made, but taking it improperly. (580)

E

echinoderms (ih KI nuh durmz): spiny-skinned invertebrate animals that live on the seafloor, including sea stars and sand dollars. (342)

ecology: the study of relationships between organisms and their environments. (23, 606)

ecosystem: a community interacting with the nonliving (abiotic) parts of its environment. (611)

egg: in organisms that reproduce sexually, the gamete from the female parent. (82)

electron microscope: an instrument using beams of electrons that magnifies objects too small to be seen with a light microscope. (32)

embryo: a fertilized egg during early growth; in humans, a fertilized egg during the first two months of pregnancy. (537)

embryology: the study of embryos, which are the earliest stages of an organism's development. (142)

emphysema (em fuh SEE muh): a disease in which the alveoli in the lungs lose their ability to expand and contract, mostly caused by smoking. (492)

endangered species: one whose population is so small that it is in danger of extinction. (144)

endocytosis: a process by which cells transport a large body, such as a large protein molecule, through the cell membrane into the cytoplasm. (60)

endoplasmic reticulum (end oh PLAZ mihk re TIK yew luhm): a cell organelle consisting of folded membranes that move materials around within the cell. (39)

endoskeleton: the internal skeleton of an organism that supports and protects the internal organs and provides a frame for muscles. (351)

endospore (EN doh spor): a thick-walled cell that some bacteria produce around themselves, especially for protection from heat and drought. (194)

energy pyramid: a model of the flow of energy in a food chain; less energy is available at each level of the pyramid as you move toward the top. (615)

environmental activists: people who work to conserve Earth's resources and protect the environment. (660)

environmental management: conserving resources and protecting ecosystems by carefully managing their use. (658)

enzyme (EN zim): a protein that speeds chemical reactions in cells without being changed itself. (57)
epidermis: the surface or outer layer of the two layers of skin. (432)
epiphyte: a plant that grows on other plants for physical support. (242)
equilibrium: condition in which the number of molecules of a substance is maintained at the same amount throughout a space. (59)
erosion: the wearing away of soil by wind, water, and ice. (651)
estivation: an adaptation for survival in hot, dry weather during which an animal becomes inactive and all body processes slow down. (356)
estuary: the area where a freshwater river or stream meets the ocean; examples are salt marshes, deltas, and mud flats. (638)
ethnobotany (eth noh BAH tuhn ee): the study of tribal people and their use of plants. (288)
eukaryote: an organism or cells with a formed nucleus and organelles surrounded by membranes. (161)
evolution: changes that occur over time in the hereditary features of organisms. (130)
exocytosis: a process by which a cell moves large molecules out through the cell membrane. (60)
exoskeleton: on all arthropods, the hard, lightweight external covering that shields, supports, and protects their bodies. (333)
extinction: the dying out of an entire species. (144)

family: the fifth-highest taxonomical category, below an order. (163)
fats: nutrients that release energy; also help the body store some vitamins. (444)
fermentation: a process by which yeast and some bacteria release energy needed for life processes by breaking down glucose into alcohol and carbon dioxide, without the use of oxygen. (65, 584)
fertilization: in organisms that reproduce sexually, the fusion of two gametes. (83)
fetus (FEE tus): a developing embryo; in humans, after the first two months of pregnancy. (538)
filter feeder: an organism that obtains food by filtering it from the water in which it lives. (309)
fins: fanlike structures on fish; adapted for steering, balancing, and moving through the water. (351)
fish: a cold-blooded vertebrate whose gills, fins, and scales adapt it to living in water. (351)
fission: a reproduction method used by bacteria in which one divides to form two bacteria having identical genetic material. (189)
flagellum (fluh JELL uhm): a whiplike tail on bacteria and some protists that helps them move through a moist environment. (189)
food chain: the feeding relationships that transfer energy through a community; starts with producers and moves through herbivores to carnivores. (614)
food group: a group of foods containing the same nutrients; for example, the meat group. (448)
food web: a series of overlapping food chains. (614)
fossils: remains of life from an earlier time. (137)
fracture: a break in a bone. (422)
free-living: organisms that find their own food and place to live without depending on other organisms; the opposite of a parasite. (315)
frond: the leaf of a fern. (244)

gametophyte: the form of a plant that produces male and/or female gametes. (237)
gene: the region of DNA that directs the making of a specific protein, thus controlling traits that are passed to offspring. (89)
genetic engineering: experimental methods for altering genes in offspring, to produce desirable traits or eliminate undesirable traits. (121)
genetics: the study of how a gene affects the traits of an offspring. (106)
genome: a map of the location of individual genes on every chromosome of an individual. (122)
genotype: the combination of dominant and/or recessive genes (alleles) present in the cells of an organism. (110)
genus: the second-largest taxonomic category. (157)
gestation period: the time between fertilization and the birth of an offspring. (381)
gill: in water-dwelling creatures, a breathing organ that extracts oxygen from the water. (324)
gill slits: paired openings in the throat; in fish, they develop into gills for breathing under water. (351)
gizzard: in an earthworm, a muscular structure in the digestive system that grinds up soil. (328)
Golgi (GAWL jee) **bodies:** cell organelles consisting of stacks of membrane-covered sacs that pinch off and move proteins to the outside of the cell. (39)
gradualism: a model that describes evolution as working steadily, gradually, and continuously to slowly change species into new species. (135)
greenhouse effect: the process by which heat radiated from Earth's surface is trapped and reflected back to Earth by gases in the atmosphere, causing worldwide temperatures to increase. (649)
guard cells: in a plant leaf, cells that surround the stomata to open and close them. (261)
gymnosperm: a vascular plant that produces seeds on the scales of cones, such as pine cones. (255)

habitat: the place where an organism lives within a community. (610)
hallucinogens (huh LEWS un uh junz): drugs that affect the user's perception of the world, causing reality to be confused with fantasy. (591)
haploid chromosome number: in organisms that reproduce sexually, the number of chromosomes in a gamete. (83)
hashish: a drug that is more concentrated and powerful than marijuana. (588)
hemoglobin: a chemical in red blood cells that carries oxygen to tissues and gives blood its red color. (472)
herbivore: an animal that eats only plants. (380)
heredity: passing traits from parent to offspring. (106)
hermaphrodite: an animal that produces both sperm and eggs. (310)
heroin: a powerful pain-relieving drug made from morphine; it is highly addictive. (592)
heterozygous (het uh roh ZI gus): an organism that has two different alleles for a trait. (110)
hibernation: an adaptation for winter survival during which an animal becomes inactive and all body processes slow down. (356)
homeostasis: an organism's or cell's constant adjustment to maintain stable conditions in itself, despite changes in its environment. (7)
hominids: earliest humanlike primates; eat both meat and vegetables and walk upright on two feet. (147)
Homo sapiens: our own species, a hominid primate mammal. (148)
homologous: body parts in different species but which are similar in origin and structure, such as limbs on different animals. (141)
homologous blood transfer: blood taken from one person and transfused into another. (478)
homozygous (ho muh ZI gus): an organism that has two identical alleles for a trait. (110)
hormones: secretions from endocrine glands that control certain body activities. (522)
hypertension (hi pur TEN chun): a circulatory disorder in which blood pressure is too high. (468)
hyphae (HI fee): masses of threadlike structures that form the body of a fungus. (211)
hypothesis: a prediction that can be tested. (14)

immovable joint: a juncture of two or more bones, made so that no movement occurs. (423)
immune system: a complex group of defenses that the body uses to fight disease. (563)
imprinting: a type of learning in which an animal forms a social attachment to another organism during a specific period after birth or hatching. (397)
incomplete dominance: a condition that exists when both alleles for a trait are expressed. (114)
incubate: to keep an egg or newborn animal warm; the heat helps eggs develop until they hatch and helps young to survive. (372)
infancy: the period in human growth from four weeks after birth to one year. (541)
innate behavior: behavior that an organism has when it is born and does not have to learn. (394)
insight: a form of reasoning that enables animals to use past experiences to solve new problems. (399)
instinct: a complex pattern of innate behavior involving multiple actions. (395)
interneurons: nerve cells throughout the brain and spinal cord that transmit impulses from sensory neurons to motor neurons. (509)
invertebrate: an animal without a backbone. (303)
involuntary muscle: a muscle that can't be consciously controlled, such as heart and digestive muscles. (426)

joint: any place where two or more bones meet. (423)

kidneys: two organs that filter blood to produce the waste liquid called urine. (498)
kingdom: the highest of the taxonomic categories, having the largest number of species; kingdoms include all other categories. (156)

larva: a stage in development between egg and adult. (310)
larynx: the structure between the pharynx and trachea to which the vocal cords are attached. (489)
latent virus: a virus that becomes part of a cell's DNA and lies dormant until something causes it to become active, whereupon it destroys the cell and makes new viruses. (183)
law: a rule that describes a pattern in nature and what will happen under specific conditions. (16)
learning: the process of developing a behavior through experience; an organism repeats behaviors that fulfill a need and avoids those that achieve nothing or cause pain. (396)
lichen: a fungus living in a mutualistic relationship with green algae. (215)
life span: the length of time an organism lives. (8)
ligament: a tough band of tissue that holds bones together. (423)

limiting factor: a condition, such as the quality or amount of water or food, that determines whether an organism, population, or species survives in its environment. (628)

lipid: an organic compound that stores and releases large amounts of energy; fats or oils. (57)

littoral (LIHT uh rul) **zone:** the ocean zone along the shore that is covered with water during high tide and exposed to air during low tide. (637)

long-day plant: a plant that requires short nights (long days) to flower. (284)

lymph (LIHMF): fluid in body tissues made up of water, dissolved substances, and lymphocytes. (480)

lymphatic (lihm FAT ihk) **system:** collects fluid from body tissues and returns it to the blood through lymphatic vessels. (480)

lymph node: structures throughout the body that filter microorganisms and foreign material from lymph before it returns to blood. (481)

lymphocyte (LIHM fuh site): a white blood cell that fights disease-causing antigens by engulfing and digesting them. (481, 566)

lysosome (LI suh sohm): an organelle with chemicals that digest waste and worn-out cell parts. (40)

mammal: a warm-blooded vertebrate with insulating hair and mammary glands. (379)

mammary glands: glands in female mammals that produce milk for feeding young. (379)

manatee: a large, saltwater mammal living in warm coastal waters; commonly called a sea cow. (386)

mantle: in a mollusk, the outside covering of the soft body; it secretes chemicals that become the shell or protects the body if no shell exists. (324)

marijuana: a drug made of stems, leaves, flowers, and seeds of Indian hemp; also called pot. (588)

marrow: a red or yellow fatty tissue in certain bones that produces red and white blood cells. (422)

marsupial: a mammal with a pouch on its abdomen for carrying and nursing its young. (381)

mechanical digestion: processing of food by chewing, by churning, and by the action of bile, to break it into smaller particles. (451)

medusa: a cnidarian (stinging-celled animal) that is bell-shaped and free-swimming. (312)

meiosis: the division of the cell nucleus to produce sex cells (gametes). (82)

melanin (MEL uh nuhn): a chemical produced by the epidermis that gives skin its color. (433)

menopause: in older females, the period of years when the menstrual cycle becomes irregular and eventually stops. (533)

menstrual (MEN strul) **cycle:** in females, the cycle of egg production and menstruation. (532)

menstruation (men STRAY shun): in females, the monthly discharge through the vagina of the lining of the uterus and some blood if fertilization of an egg has not occurred. (533)

metabolism: all of the changes in an organism that enable it to live, grow, and reproduce. (63)

metamorphosis: in insects, the changes of form during the life cycle: egg, larva, pupa, adult. (337)

methamphetamine (meth am FET uh men): a very powerful stimulant. (587)

migration: the instinctive seasonal movement of animals, such as birds flying south for the winter. (405)

mimicry: a defense against predators in which an animal copies the appearance or behavior of another organism. (613)

minerals: inorganic nutrients that regulate chemical reactions in the body, such as building cells and sending nerve impulses. (446)

mitochondria (mi toh KAWN dree uh): a cell organelle that breaks down food molecules and releases energy. (40)

mitosis: the process by which a nucleus divides into two nuclei, each containing the same number of chromosomes that the parent cell had. (75)

mollusk: a soft-bodied invertebrate that usually has a shell, such as a snail. (324)

molting: in animals, the periodic shedding and replacing of the old body covering, such as skin or an exoskeleton. (333)

monocots (MAHN uh kahts): angiosperms having a single seed leaf inside their seeds. (256)

monotreme: a mammal that lays eggs having a tough, leathery shell; for example, the platypus. (382)

motivation: some stimulus within an animal that causes it to act. (398)

motor neurons: nerve cells that conduct impulses from the brain or spinal cord to muscles or glands throughout the body. (509)

movable joint: a place where two or more bones meet or are attached and allow movement. (423)

multiple alleles: having more than two alleles that control a trait. (115)

muscle: tissue that can relax and contract to allow movement. (426)

mutation: any permanent change in the genetic material (DNA) of a cell. (91)

mutualistic: a relationship in which two organisms live together; both benefit in some way. (215)

natural resources: raw materials in the environment that organisms use for survival, such as air, water, soil, and forests. (648)

natural selection: Darwin's theory that organisms best adapted with traits for an environment will survive to pass on those traits to their offspring. (132)

nephron (NEF rahn): the filtering unit of the kidney. (499)

neurons: nerve cells that carry impulses throughout the body. (509)

niche: the role of a species within a community. (610)

nitrogen cycle: the continuous movement of nitrogen from the atmosphere, to plants, and back to the atmosphere (or directly into plants) again. (619)

nitrogen dioxide: a major air-polluting chemical in cigarette smoke and automobile exhaust that is harmful when inhaled. (496)
nitrogen-fixing bacteria: bacteria that change nitrogen in the air into nitrogen compounds that are useful to plants and animals. (194)
noncommunicable disease: a disease that cannot be spread from one organism to another. (570)
nonrenewable resources: natural resources that are available only in limited amounts and are not being replaced by nature. (648)
nonvascular plant: a plant lacking tube-like vessels to transport food and water; it absorbs water directly through its cell membranes. (234)
notochord: a flexible, rodlike structure along the dorsal side (back) of a chordate animal; the backbone in vertebrates. (350)
nucleic acids: large organic molecules that store important information in cells and direct their activities; two types are DNA and RNA. (57)
nucleus: a structure inside a cell that directs the cell's activities; contains chromatin; chromosomes in a dividing cell. (38)
nutrients: the chemicals in foods that provide energy and materials for life activities. (442)

olfactory cells: nerve cells in the nasal passages that respond to chemical stimuli in air. (517)
omnivore: an animal that derives its energy from both plants and other animals. (380)
open circulatory system: a blood-circulation system that lacks vessels and instead bathes internal organs in blood, as in mollusks. (324)
order: the fourth-highest taxonomical category, below a class. (163)
organ: a structure made up of different types of tissues that work together to do a specific job; for example, the heart. (45)
organ system: a group of organs working together for a specific job; for example, the digestive system. (45)
organelles: membrane-bound structures within the cytoplasm of eukaryotic cells; organelles break down food, move wastes, and store materials. (38)
organism: a living thing that is made of one or more cells, uses energy, moves, responds to its environment, reproduces, adapts, and has a life span. (6)
osmosis: the passive transport of water through a cell membrane by diffusion. (59)
ovary: in angiosperms, the swollen base of the pistil, where ovules form (264); in female animals, the organ that produces ova, or egg cells. (531)
ovulation (ahv yuh LAY shun): in human females, the monthly release of an egg from an ovary. (531)
ovule (OHV yewl): the reproductive part of a female plant that contains the eggs. (262)

palisade (pal uh SAYD) **layer:** in a plant leaf, rows of closely packed cells that are near the surface and contain chlorophyll. (261)
parasite: usually an organism, but anything that obtains food from a host organism and at the same time harms the host organism. (183)
passive immunity: immunity caused by antibodies introduced from an outside source. (564)
passive transport: movement of material through a cell membrane without the use of energy. (58)
pasteurization: heating food to a temperature below boiling long enough to kill most bacteria. (559)
pathogen: any organism that causes disease. (194)
pedigree: a diagram that traces the history of a trait in a family. (119)
pelagic zone: the open ocean just beyond the sublittoral zone. (638)
periosteum (per ee AHS tee uhm): a tight-fitting membrane covering bone. (421)
peripheral nervous system: part of the nervous system, including cranial nerves and spinal nerves that connect the brain and spinal cord to other body parts. (510)
peristalsis (pe ruh STAHL sihs): contractions that move food through the digestive system. (452)
permafrost: in tundra areas of Earth, the deeper soil that remains permanently frozen underneath the topsoil layer. (633)
pesticide: a chemical that kills undesirable plant or animal pests. (340)
pharynx (FER ingks): a tubelike passageway for both food and air, located between the nasal cavity and the esophagus. (489)
phenotype (FEE nuh tipe): a physical trait in an organism; resulting from its genetic makeup. (110)
phloem (FLOW em): tubular cells in vascular plants that move food from leaves and stems to other parts of the plant for use or storage. (260)
photoperiodism: the flowering response of a plant to change in the length of day and night. (284)
photosynthesis (foht oh SIHN thuh sus): a chemical reaction used by producers, such as green plants, to produce food; light energy is used to produce chemical energy, converting carbon dioxide and water into sugar and oxygen. (63, 278)
phylogeny: the evolutionary history of an organism. (160)
phylum: the second-highest of the taxonomical categories in the animal kingdom. (163)
physical dependence: taking a drug because the body has a chemical need for it and can't function without it; also called addiction. (581)
pioneer species: the first plants to grow in a new or disturbed area; their decay creates material on which other plant species grow. (239)
pistil (PIHS tul): the female reproductive organ of a flower. (264)

placenta: part of the sac that surrounds an embryo in mammals; attached to the uterus; many blood vessels provide life support for an embryo. (381)

placental mammal: a mammal whose young develop inside the female in a uterus. (381)

plasma: the liquid part of blood, made mostly of water but also containing dissolved nutrients, minerals, and gases such as oxygen and carbon dioxide. (472)

platelet: a cell fragment that helps blood clot. (472)

poaching: illegal hunting of animals. (387)

pollen grain (PAHL un GRAYN): the reproductive part of a male plant that contains the sperm. (262)

pollen tube: a hollow structure that grows from a pollen grain into an ovule, through which a sperm swims into the ovule for fertilization. (262)

pollination: the transfer of pollen grains from the stamen to the ovule. (264)

polygenic (pahl ih JEHN ihk) **inheritance:** groups of gene pairs act together to produce a trait. (116)

polyp: a cnidarian (stinging-celled animal) that is tube-shaped and sessile. (312)

population: organisms of one species living in the same place at the same time. (134, 608)

population density: the number of individuals per unit of living space. (608)

Porifera: the phylum to which sponges belong; the name means "pore-bearing." (308)

predator: an organism that kills and eats prey to obtain energy. (352)

preening: a behavior of birds in which a bird uses its beak to rub oil over its feathers to condition them and make them water repellent. (373)

pregnancy: period between fertilization and birth. (536)

prey: an organism that is eaten by a predator. (352)

primary succession: the development of new communities in newly created land areas, such as new volcanic islands. (629)

primates: the group of mammals to which monkeys, apes, and humans belong. (146)

probability: a branch of science that determines the likelihood of something happening. (109)

producers: green plants that make their own food by photosynthesis. (63)

prokaryote: single-celled organism that doesn't have an organized nucleus or organelles. (161)

protein: an organic compound made up of amino acids; used throughout the body for growth and to replace and repair cells. (57, 443)

prothallus: the structure in ferns that produces gametes. (244)

protists: members of the Kingdom Protista; simple organisms having cells with a nucleus; probable ancestors of the fungi, plant, and animal. (202)

protozoa (proht uh ZOH uh): one-celled animal-like protists; many are parasites. (206)

pseudopod (SEWD uh pahd): a footlike extension of cytoplasm used by some organisms to move and to trap food. (206)

psychological dependence: taking a drug only because you think that you need it. (581)

pulmonary (PUL mo ner e) **circulation:** the flow of blood from the heart, to the lungs, and back to the heart. (465)

punctuated equilibrium: a model that describes evolution as working rapidly, changing species to new species by the mutation of a few genes. (135)

Punnett square: a tool that shows how genes can combine; used to predict results in genetics. (110)

purebred: an organism that always produces the same traits in its offspring. (108)

radial symmetry: describes animals that have body parts arranged in a circle around a central point, similar to a bicycle wheel. (303)

radioactive elements: elements whose atoms give off radiation, a form of atomic energy. (138)

radula: in mollusks, a tongue-like organ with rows of teeth that scrape and tear food. (325)

recessive: the form of a trait that appears least often in offspring; for example, left-handedness is recessive compared to right-handedness. (108)

reflex: a type of innate behavior that is an automatic response to a stimulus; not involving the brain. (395, 512)

reforestation: replanting an area with seedling trees. (656)

regeneration: a type of asexual reproduction in which a whole new organism grows from just a part of the parent organism; a sea star. (309)

rejection: a process whereby the immune system attacks a transplanted organ. (46)

relative dating: estimating the age of a fossil by comparing it to younger fossils in the rock layer above and to older fossils in the rock layer below. (138)

renewable resources: natural resources that constantly are being replaced by nature. (648)

reptile: a cold-blooded vertebrate that has dry, scaly skin and that lays eggs covered with a leathery shell; examples are lizards, snakes, turtles. (363)

response: the reaction to a stimulus. (7)

retina: light-sensitive tissue at back of the eye. (515)

rhizoids: rootlike filaments containing only a few long cells; they hold moss plants in place. (236)

rhizome: the underground stem of a fern. (244)

ribosome: a cell organelle on which protein is made. (39)

RNA: ribonucleic acid; carries codes for making proteins. (90)

saliva: the watery substance produced in the mouth that begins the chemical digestion of food. (452)

saprophyte (SA proh fite): any organism that uses dead material as a food and energy source. (193)
scales: hard, thin, overlapping plates that cover and protect a fish's body. (351)
scientific methods: problem-solving procedures used by scientists: define the problem, make a hypothesis, test the hypothesis, analyze the results, and draw conclusions. (14)
secondary succession: the gradual return of a disturbed area to its former communities. (629)
sedimentary rock: a type of rock formed when fine particles like mud and sand settle out of water and become cemented together. (137)
semen: the mixture of sperm and a fluid that nourishes the sperm and helps them to move. (531)
sensory neurons: nerve cells that transmit stim-uli from receptors to the brain or spinal cord. (509)
sessile: describes organisms, such as trees, that remain attached to one place during their lifetime. (308)
setae (SEE tee): in a segmented worm, bristle-like structures on the outside of the body that help it grip soil and move. (327)
sex-linked gene: an allele inherited on a sex chromosome. (119)
sexually transmitted diseases (STDs): diseases transmitted during sexual contact. (561)
sexual reproduction: a type of reproduction in which a new organism is produced by combining sex cells from two parents. (81)
short-day plant: a plant that requires long nights (short days) to flower. (284)
skeletal muscles: voluntary muscle tissue attached to bones; the muscles that makes the bones move. (427)
skeletal system: the body's network of bones, which form a rigid frame to support the body, protect internal organs, generate red blood cells, and store calcium and phosphorus. (420)
smooth muscle: involuntary muscle tissue that forms the walls of the stomach, intestines, uterus, and blood vessels. (427)
social behavior: interactions among organisms of the same species, including mating, caring for young, protection, getting food, and claiming territory. (403)
society: a group of animals of the same species living and working together in an organized way. (403)
solid wastes: solid products that can be recycled, burned, buried, or dumped. (651)
sori: structures that produce spores on the underside of fern fronds (*singular:* sorus). (244)
species: organisms whose members are alike and successfully reproduce among themselves. (130, 157)
species diversity: the great variety of plant and animal species on Earth. (164)
sperm: the gamete or reproductive cell from the male parent, produced by the testes. (82, 530)
spiracles: in an arthropod, openings in the abdomen that allow oxygen and carbon dioxide to move into and out of the lungs. (333)
spontaneous generation: a belief that living things come from nonliving matter; for example, that frogs arise spontaneously from mud. (10)
sporangia: the round, spore-producing cases of zygote fungi. (212)
spore: a resistant reproductive cell that forms new organisms without fertilization; in fungi, ferns and some protists. (212)
sporophyte: a capsule in which spores are produced by meiosis in plants such as mosses. (237)
stamen (STAY mun): the male reproductive organ of a flower. (264)
stimulant: a substance that stimulates the central nervous system and speeds up other body systems; for example, the caffeine in coffee, tea, chocolate, and cola drinks. (583)
stimulus: anything an organism responds to, such as sound, light, heat, vibration, odor, movement, hunger, thirst, and so on. (7, 282)
stomata: in a plant leaf, small pores in the surface that allow carbon dioxide, water, and oxygen to enter and leave (*singular:* stoma). (261)
sublittoral zone: the shallow water above the continental shelf. (637)
succession: a gradual, orderly change of species in a community over time. (629)
symbiosis: a condition where two organisms live together for mutual benefit. (215)
synapse: the small space between two neurons across which impulses can travel. (510)
systemic (sihs TEM ihk) **circulation:** the flow of blood from the heart, to all body tissues (except lungs and heart), and back to the heart. (467)

T

target tissue: tissue affected by hormones. (522)
taste buds: tissue located on the tongue that responds to chemical stimuli. (517)
taxonomy (tak SAHN uh mee): the science of grouping and naming organisms. (156)
technology: the application of scientific knowledge to improve the quality of human life. (22)
tendon: a thick band of tissue that attaches a muscle to a bone. (427)
tentacles: the armlike structures that surround the mouths of some organisms and help them to capture food. (311)
territory: an area that an animal defends from other members of the same species. (401)
testes: organs in males that produce sperm. (530)
theory: description of nature; subject to change when evidence changes. (16)
tissue: similar cells that do the same sort of work; for example, all muscle tissue contracts. (45)
tolerance: occurs when the body adjusts to a drug and needs increasing doses to produce the same effect. (581)

toxin: a poison produced by disease-causing organisms (pathogens). (194)
trachea: a tube that carries air to the bronchi. (489)
transgenic organisms: organisms that contain genetic information from another species. (92)
transpiration: in plants, loss of water vapor through the stomata of a leaf. (277)
trial and error: behavior that is modified by experience. (397)
tropism: the response of a plant to a stimulus. (282)
tube feet: in an echinoderm, structures attached to the water vascular system that act like suction cups and help the echinoderm to move, feed, get oxygen, and get rid of waste. (342)
tumor: an abnormal growth of tissue. (571)
tundra (TUN drah): a cold, dry, treeless biome where the sun is barely visible during the six to nine months of winter. (633)

umbilical cord: a bundle of blood vessels connecting the placenta to the embryo's navel in mammals; carries nutrients and oxygen to the embryo. (381)
ureters (YOOR ut urz): tubes that lead from each kidney to the bladder. (499)
urethra (yoo REE thruh): a tube leading from the bladder to the outside of the body. (499)
urinary system: a system of excretory organs that rids blood of wastes, excess water, and excess salts. (498)
urine: waste liquid collected by the kidneys and containing water, salts, and other wastes. (499)
uterus: in females, the pear-shaped, muscular organ in which a fertilized egg develops into a baby; also called the womb. (531)

vaccination: administration of a weakened virus to develop immunity against a disease. (568)
vaccine: a solution made from a killed or weakened virus; causes artificial immunity. (184)
vacuole: a storage area in a cell for water, food, or waste products. (41)
vagina: in females, the passageway that leads from the uterus to outside of the female's body; also called the birth canal because offspring pass through it when being born. (531)
variable: in an experiment, the factor tested. (15)
variation: the occurrence of an inherited trait that makes an individual different from other members within the same species. (133)
vascular plant: a plant having tube-like vessels that transport food and water to its cells. (234)
vascular tissue: in vascular plants, tissue made up of long cells that form tubes. (241)
veins: vessels that move blood toward the heart, carrying wastes. (466)
ventricles (VEN trih kulz): the two lower chambers of the human heart. (464)
vertebrate: animal with a backbone. (303)
vestigial structure: a body part that is reduced in size and is no longer used by the organism. (141)
villi (VIHL i): tiny, fingerlike projections on the inner surface of the small intestine. (454)
virus: a microscopic particle made of either a DNA or an RNA core and covered with a protein coat; it infects host cells in order to reproduce. (182)
vitamins: organic nutrients that promote growth, regulate body functions, and help the body use other nutrients. (445)
voluntary muscle: a muscle you can control, such as arm and leg muscles. (426)

warm-blooded: describes an animal whose body temperature stays the same, regardless of temperature changes in its environment. (351)
water cycle: the continuous movement of water in the biosphere through evaporation, condensation, and precipitation. (617)
water table: the upper surface of underground water; this underground water supplies drinking water for most humans. (641)
water-vascular system: a network of water-filled canals, unique in echinoderms. (342)
wetland: an area of land covered by shallow water at least part of the year. (640)
wildlife preserve: an area of land set aside to protect wildlife species. (656)
withdrawal: an illness that occurs when a person stops taking a drug on which the person is physically dependent. (581)

xylem (ZI lum): tubular vessels in vascular plants that transport water and minerals from the roots up through the plant. (260)

zygote: in organisms that reproduce sexually, the cell that forms in fertilization. (83)

GLOSSARY/GLOSARIO

This glossary defines each key term that appears in **bold type** in the text. It also shows the page number where you can find the word used. Some other terms that you may need to look up are included, too.

abiotic factors/factores abióticos: Seres no vivientes en el medio ambiente, incluyendo la tierra, el agua, la temperatura, el aire, la luz, el viento y los minerales. (Pág. 607)
acid rain/lluvia ácida: Lluvia que se combina con dióxido de sulfuro y óxido de nitrógeno en el aire para formar ácido sulfúrico y ácido nítrico que caen sobre la Tierra. (Pág. 268)
active immunity/inmunidad activa: Ocurre cuando el cuerpo crea anticuerpos en respuesta a los antígenos. (Pág. 564)
active transport/transporte activo: Cuando se utiliza energía para mover materiales a través de una membrana celular. (Pág. 58)
adaptation/adaptación: Las características que posee un organismo que le ayudan a sobrevivir en su medio ambiente. (Pág. 8)
adolescence/adolescencia: Período de crecimiento humano que comienza alrededor de los 12 a los 14 años y dura hasta los 20 años. (Pág. 542)
adulthood/edad adulta: Período de crecimiento humano que comienza al final de la adolescencia (20 años de edad) y se extiende hasta la vejez. (Pág. 542)
aerobic/aeróbico: Organismo que requiere oxígeno para la respiración. (Pág. 190)
aggression/agresión: Acción de usar fuerza para dominar o controlar a otro animal. (Pág. 401)
AIDS/SIDA: Síndrome de inmunodeficiencia adquirida, una enfermedad transmisible causada por un virus RNA; se contagia por medio del contacto sexual, del uso de agujas contaminadas, a través de la placenta o por una madre infectada que amamanta a su nene. (Pág. 186)
algae/algas: Protistas que parecen plantas, contienen clorofila y fabrican su propio alimento; no tienen ni raíces, ni tallos, ni hojas y viven en o cerca del agua. (Pág. 203)
alleles/alelos: Las diferentes formas que un gene puede tener para un rasgo. (Pág. 106)
allergens/alergenos: Sustancias que ocasionan una reacción alérgica en el cuerpo. (Pág. 572)
allergy/alergia: Reacción fuerte del sistema inmunológico en respuesta a una sustancia extraña. (Pág. 572)
alternation of generations/alternación de generaciones: Ciclo que alterna entre las generaciones de esporofitos y gametofitos, en algunos protistas y plantas. (Pág. 238)
alveoli/alvéolos: Conjunto de saquitos con paredes delgadas que se encuentran en los pulmones y en donde se intercambian el oxígeno y el dióxido de carbono. (Pág. 490)
Alzheimer's disease/enfermedad de Alzheimer: Enfermedad de origen desconocido en la cual las células nerviosas del encéfalo dejan de comunicarse unas con otras, ocasionando pérdida de la memoria. (Pág. 520)
amino acids/aminoácidos: Sustancias básicas de las proteínas. (Pág. 443)
amniote egg/huevo amniótico: Tipo de huevo que provee un ambiente completo para el desarrollo del embrión. (Pág. 364)
amniotic sac/bolsa amniótica: Membrana fina que rodea al embrión, llena de líquido amniótico (un líquido claro). (Pág. 536)
amphibian/anfibio: Animal vertebrado de sangre fría cuya piel permeable, esqueleto fuerte y habilidad para la hibernación o estivación lo adaptan para vivir en el agua o en la tierra; algunos ejemplos son las ranas, los sapos y las salamandras. (Pág. 356)
anabolic steroids/esteroides anabólicos: Drogas que contienen una variedad de la hormona testosterona. (Pág. 430)
anaerobic/anaeróbico: Describe a los organismos que pueden vivir en ambientes sin oxígeno. (Pág. 190)
angiosperm/angiosperma: Planta vascular en la cual la semilla se encuentra rodeada por el fruto, como por ejemplo, la manzana. (Pág. 256)

anorexia nervosa/anorexia nervosa: Trastorno alimenticio en que la persona come muy poco para evitar el aumento de peso; causa disminución extrema de peso, la cual puede causar daño a los órganos y hasta causar la muerte. (Pág. 456)

antibiotic/antibiótico: Una droga que se usa para matar las bacterias. (Pág. 194)

antibody/anticuerpo: Sustancia que forma el cuerpo en respuesta a los antígenos. (Pág. 564)

antigens/antígenos: Proteínas y sustancias químicas que son extrañas para el cuerpo. (Pág. 564)

antiseptic/antiséptico: Sustancia que mata las bacterias en el tejido vivo. (Pág. 559)

anus/ano: Abertura al final del sistema digestivo a través del cual salen los excrementos del cuerpo. (Pág. 317)

appendages/apéndices: Estructuras que crecen del tronco del cuerpo, como por ejemplo, los brazos y las piernas. (Pág. 332)

arteries/arterias: Vasos sanguíneos que mueven sangre desde el corazón. (Pág. 465)

Arthropoda/Artropoda: El fílum más grande de los animales; incluye insectos, camarones, arañas y ciempiés. (Pág. 332)

asci/asca: Sacos productores de esporas en los ascomicetos. (Pág. 213)

asexual reproduction/reproducción asexual: Reproducción en la que se produce un nuevo organismo de un solo padre. (Pág. 80)

asthma/asma: Trastorno en que los bronquios se contraen rápidamente, causando falta de aire o tos; lo ocasiona el fumar, las alergias o el estrés. (Pág. 493)

atherosclerosis/aterosclerosis: Trastorno del sistema circulatorio en que se acumulan depósitos grasos en las paredes arteriales y limitan el flujo de la sangre. (Pág. 468)

atria/aurículas: Las dos cámaras superiores del corazón humano. (Pág. 464)

autologous blood transfer/transferencia de sangre propia: Uso de la sangre propia para una transfusión. (Pág. 478)

auxin/auxina: Hormona vegetal que hace que las plantas crezcan o se doblen hacia la luz. (Pág. 283)

axon/axón: Parte de la neurona que lleva los mensajes de las células del cuerpo. (Pág. 509)

B

basidium/basidio: Estructuras en forma de bastos que producen esporas en los hongos basidiomicetos. (Pág. 214)

behavior/comportamiento: La manera en que un organismo actúa hacia su ambiente. (Pág. 394)

bilateral symmetry/simetría bilateral: Se dice de los animales que tienen las partes del cuerpo distribuidas en la misma forma a ambos lados del cuerpo. (Pág. 304)

binomial nomenclature/nomenclatura binaria: El sistema de nombrar todos los organismos con dos palabras; *Homo sapiens.* (Pág. 157)

biodegradable/biodegradable: La sustancia que se descompone fácilmente en sus elementos básicos, en el medio ambiente, debido a la intervención de las bacterias o los hongos. (Pág. 66)

biogenesis/biogénesis: La teoría de que los seres vivientes solo pueden provenir de otros seres vivientes. (Pág. 11)

biological indicators/índices biológicos: Organismos que reflejan la condición del medio ambiente. (Pág. 361)

biome/bioma: Área geográfica grande que tiene un clima y comunidades clímax similares. (Pág. 632)

biosphere/biosfera: Todos los lugares de la Tierra en donde existe la vida, incluye el aire, el agua y la tierra. (Pág. 606)

biotic factors/factores bióticos: Seres vivientes en el ambiente, tales como las plantas y los animales. (Pág. 606)

bladder/vejiga: Órgano muscular en forma de saco que almacena la orina hasta que sale del cuerpo. (Pág. 499)

blood pressure/presión sanguínea: Presión sanguínea en las arterias. (Pág. 467)

bog/pantano: Tierra baja y pantanosa compuesta principalmente de plantas muertas y en proceso de descomposición. (Pág. 246)

brainstem/bulbo raquídeo: Parte del encéfalo que se extiende desde el cerebro hasta la médula espinal; controla el movimiento involuntario de los músculos. (Pág. 511)

budding/gemación: Un tipo de reproducción asexual en el que un nuevo organismo crece del cuerpo de la célula madre; por ejemplo, la levadura se reproduce de esta forma. (Pág. 213)

bulimia/bulimia: Trastorno alimenticio en que la persona consume grandes cantidades de alimentos y luego los vomita para no aumentar de peso. (Pág. 456)

C

cambium/cambium: Tejido en plantas vasculares que crece para producir células de xilema y de floema nuevas. (Pág. 260)

camouflage/camuflaje: Adaptación que permite que un animal armonice con su medio ambiente para poderse esconder. (Pág. 613)

cancer/cáncer: Enfermedad seria y crónica que resulta del crecimiento descontrolado de las células. (Pág. 571)
capillaries/capilares: Vasos sanguíneos microscópicos que conectan las arterias y las venas; alimentos, oxígeno y materiales de desecho se intercambian a través de sus paredes delgadas (de una célula de grosor). (Pág. 466)
carbohydrates/carbohidratos: Compuestos orgánicos que proveen la mayor parte de la energía corporal; están hechos de carbono, hidrógeno y oxígeno. (Págs. 56, 443)
carbon-oxygen cycle/ciclo del carbono-oxígeno: Movimiento continuo del dióxido de carbono y el oxígeno entre la atmósfera de la Tierra y el aire. (Pág. 618)
cardiac muscle/músculo cardíaco: Tejido muscular involuntario que solo se encuentra en el corazón. (Pág. 427)
carnivore/carnívoro: Animal que solo come otros animales, como por ejemplo, el gato. (Pág. 380)
carrying capacity/capacidad de sustento: El número de animales que pueden obtener alimentos y refugio del medio ambiente en un área determinada; la capacidad del medio ambiente para sostener su población animal. (Pág. 407)
cartilage/cartílago: Tejido liso, grueso y flexible que es más duro que la carne, pero más blando que el hueso (por ejemplo, las orejas); cubre los extremos de los huesos, permite el movimiento y amortigua los golpes. (Págs. 352, 422)
catalase/catalasa: Enzima que descompone el peróxido de hidrógeno. (Pág. 545)
cell/célula: La unidad más pequeña de que está compuesto un organismo y la cual puede llevar a cabo funciones vitales. (Pág. 6)
cell membrane/membrana celular: La parte externa, y selectivamente permeable, que permite la entrada de alimento y agua y la salida de los desechos. (Pág. 37)
cell theory/teoría celular: La importante teoría de la ciencia biológica: Todos los organismos están compuestos de células; las células son las unidades básicas estructurales y funcionales de los organismos; y cada célula proviene de otra célula. (Pág. 34)
cellulose/celulosa: Compuesto orgánico que forma las paredes celulares de las plantas; está compuesta de cadenas largas de moléculas de azúcar. (Pág. 232)
cell wall/pared celular: La estructura rígida compuesta de celulosa que rodea la membrana celular de las plantas y que proporciona protección y soporte a las mismas. (Pág. 41)
central nervous system/sistema nervioso central: Parte del sistema de control del cuerpo, incluye el encéfalo y la médula espinal. (Pág. 510)
centromere/centrosoma: Una región en forma de lazo que sujeta las dobles fibras de los cromosomas. (Pág. 76)
cerebellum/cerebelo: Parte del encéfalo que coordina el movimiento de los músculos. (Pág. 511)
cerebrum/cerebro: La parte más grande de las tres que comprenden el encéfalo; interpreta los impulsos de los sentidos, almacena la memoria y controla los músculos voluntarios. (Pág. 511)
chemical digestion/digestión química: Procesamiento de alimentos por medio de fluidos en la boca, el estómago y el intestino delgado para romper moléculas grandes en moléculas más pequeñas que pueden se absorbidas por las células. (Pág. 451)
chemotherapy/quimoterapia: Uso de sustancias químicas para destruir las células cancerosas. (Pág. 571)
childhood/niñez: Período de crecimiento humano entre uno y doce años. (Pág. 541)
chitin/quitina: Carbohidrato duro y flexible del cual están formadas las paredes celulares de las hifas; se encuentra en la cubierta del cuerpo y en las alas de los insectos. (Pág. 211)
chloroplasts/cloroplastos: Organelos celulares que durante el proceso de fotosíntesis transforman la energía solar en energía química en forma de azúcar. (Pág. 41)
chordate/cordado: Un miembro del fílum Chordata de animales que tiene un notocordio, un cordón nervioso dorsal hueco y una hendidura branquial en algún momento de su ciclo vital. (Pág. 350)
chromatin/cromatina: Fibras largas de DNA en el núcleo de una célula; al dividirse la célula, la cromatina se enrosca en la forma de los cromosomas. (Pág. 38)
chromosomes/cromosomas: Fibras largas y delgadas de DNA y proteína en el núcleo de una célula, las cuales llevan el código para las características heredadas del organismo. (Pág. 76)
chronic bronchitis/bronquitis crónica: Enfermedad en que se produce demasiada mucosidad en los bronquios. (Pág. 493)
chronic disease/enfermedad crónica: Enfermedad no contagiosa que dura mucho tiempo. (Pág. 570)
chyme/quimo: Líquido claro y acuoso en el sistema digestivo; alimentos parcialmente digeridos. (Pág. 453)
cilia/cilios: Estructuras cortas parecidas a unos vellos que se extienden de la membrana celular y ayudan en la locomoción de organismos pequeños. (Pág. 207)
circadian rhythm/ritmo circadiano: Comportamiento que ocurre regularmente en un período de 24 horas. (Pág. 404)
class/clase: La tercera categoría más alta de la taxonomía, debajo del fílum (en los animales) o división (en las plantas). (Pág. 163)
classify/clasificar: Agrupar ideas, información u objetos basándose en lo que tienen semejante. (Pág. 156)
climate/clima: Tiempo promedio de un área, lo determinan la cantidad de lluvia, la temperatura y la cantidad de luz solar. (Pág. 632)

climax community/comunidad clímax: Comunidad estable con un balance de factores bióticos y abióticos. (Pág. 631)
closed circulatory system/sistema circulatorio cerrado: Sistema circulatorio de la sangre que posee vasos para transportar la sangre a los órganos internos, como por ejemplo, el de los seres humanos. (Pág. 326)
cnidarian/celentéreo: Fílum de animales que poseen células que pican. (Pág. 311)
cocaine/cocaína: Droga estimulante muy adictiva que se extrae de las hojas de la planta de la coca suramericana. (Pág. 589)
cochlea/cóclea: Estructura en el oído interno sensible al sonido; tiene forma de caracol. (Pág. 516)
coevolution/coevolución: Ocurre cuando dos especies evolucionan estructuras y comportamientos en respuesta a los cambios en cada una de ellas por un largo período de tiempo. (Pág. 286)
cold-blooded/de sangre fría: Describe animales cuya temperatura corporal cambia de acuerdo a la temperatura de su medio ambiente. (Pág. 351)
collar cells/células del cuello: Células que cubren la parte interior de las esponjas y facilitan el movimiento del agua a través de las mismas. (Pág. 309)
communicable disease/enfermedad contagiosa: Enfermedad que es transmitida de un organismo a otros. (Pág. 560)
communication/comunicación: Intercambio de información entre los animales, por medio de aullidos, movimientos, tacto, lenguaje o feramonas. (Pág. 403)
community/comunidad: Todas las poblaciones de diferentes especies que viven en el mismo lugar al mismo tiempo. (Pág. 610)
competition/competencia: Esfuerzo entre los organismos por obtener el alimento, el agua y el espacio que existe en un ecosistema; la competencia determina el tamaño y la localización de una población. (Pág. 612)
compound light microscope/microscopio de luz: Un instrumento que utiliza lentes para magnificar los objetos. (Pág. 31)
conditioning/acondicionamiento: Modificación del comportamiento de manera que la reacción que previamente se asociaba con un estímulo, ahora se asocia con otro. (Pág. 398)
conservation/conservación: Uso prudente y cuidadoso de los recursos de la Tierra. (Pág. 654)
consumers/consumidores: Organismos que no pueden producir su propio alimento. (Pág. 63)
contour feathers/plumas de contorno: Plumas fuertes que les dan a las aves sus bellos coloridos y formas. (Pág. 372)
control/control: En un experimento, el estándar que se emplea para hacer comparaciones. (Pág. 15)
controlled burns/incendios controlados: Incendios forestales que se prenden periódicamente con el propósito de controlar la vegetación y los desechos vegetales debajo de las especies dominantes de árboles. (Pág. 620)
coronary circulation/circulación coronaria: Flujo de sangre a los tejidos del corazón. (Pág. 467)
courtship behavior/comportamiento de cortejo: Comportamiento que ayuda a las hembras y a los machos de una especie a prepararse para el apareamiento. (Pág. 402)
crack/crack: Forma poderosa de la cocaína. (Pág. 590)
crop/buche: Saco en el sistema digestivo de la lombriz de tierra, en el cual almacena la tierra que se come. (Pág. 328)
cuticle/cutícula: Capa cerosa protectora en los tallos y en las hojas de las plantas que las ayuda a conservar el agua. (Pág. 233)
cyclic behavior/comportamiento cíclico: Comportamiento innato que ocurre en una forma repetitiva. (Pág. 404)
cyst/quiste: Gusano parasítico joven con una cubierta protectora. (Pág. 318)
cytoplasm/citoplasma: El material líquido dentro de la membrana celular que contiene estructuras que llevan a cabo procesos vitales. (Pág. 38)

D

day-neutral plant/planta de día neutral: Planta que no es afectada por la duración de la noche y por lo tanto florece bajo una amplia gama de horas de luz solar. (Pág. 284)
dendrite/dendrita: Parte de la neurona que recibe mensajes y los transmite a las células del cuerpo. (Pág. 509)
depressant/depresivo: Sustancia que retarda el sistema nervioso central, afecta el razonamiento, la memoria y las funciones motoras. (Pág. 584)
dermis/dermis: La capa inferior de la piel; contiene vasos sanguíneos, nervios y glándulas sudoríparas. (Pág. 433)
development/desarrollo: Los cambios por los que pasa un ser viviente. (Pág. 8)
diaphragm/diafragma: Músculo debajo de los pulmones que ayuda a mover el aire hacia adentro y fuera del cuerpo. (Pág. 490)
dichotomous key/clave dicotómica: Una guía, paso por paso, para la identificación de organismos, la cual requiere escoger una descripción de las dos que se dan en cada paso. (Pág. 168)
dicots/dicotiledóneas: Plantas angiospermas con dos cotiledones dentro de la semilla; como por ejemplo, los árboles de arce. (Pág. 256)
diffusion/difusión: Movimiento molecular de tipo pasivo desde un lugar en donde hay una mayor concentración hacia otro lugar donde hay menor concentración; un tipo de transporte pasivo en las células. (Pág. 58)

digestion/digestión: Proceso que rompe los alimentos en pequeñas moléculas que pueden ser absorbidas por el cuerpo. (Pág. 451)

disinfectant/desinfectante: Sustancia que mata las bacterias en los objetos inertes. (Pág. 559)

division/división: La segunda categoría más alta de la taxonomía en el reino de las plantas (en el reino animal, la palabra *fílum* reemplaza la división). (Pág. 163)

DNA/DNA: Ácido desoxirribonucleico; un ácido en los núcleos de las células que codifica y almacena la información genética; consiste en filamentos de moléculas que controlan las actividades celulares por medio del uso de instrucciones codificadas. (Pág. 86)

dominant/dominante: La forma de un rasgo que parece dominar u ocultar otra forma del mismo rasgo. (Pág. 108)

dorsal nerve cord/cordón nervioso dorsal: En los animales cordados, manojo de nervios localizado sobre el notocordio; la espina dorsal en la mayoría de los vertebrados. (Pág. 350)

down feathers/plumón: En las aves, tipo de pluma suave y esponjosa que les cubren su piel y les sirve de aislante. (Pág. 373)

drug/droga: Cualquier sustancia química que cambia la forma en que una persona se siente, piensa o actúa. (Pág. 580)

drug abuse/abuso de las drogas: Uso intencional de una droga para otro propósito y no para el que fue designada la droga. (Pág. 580)

drug misuse/uso inapropiado de las drogas: Uso de una droga para el propósito que se recetó, pero en forma inapropiada. (Pág. 580)

E

echinoderms/equinodermos: Animales invertebrados con espinas que viven en el fondo del mar, incluyen las estrellas de mar y las holoturias. (Pág. 342)

ecology/ecología: El estudio de las relaciones entre organismos y su ambiente. (Págs. 23, 606)

ecosystem/ecosistema: La interacción entre una comunidad y las partes no vivas (abióticas) de su medio ambiente. (Pág. 611)

egg/óvulo: En los organismos que se reproducen sexualmente, el gameto o célula reproductora de la hembra. (Pág. 82)

electron microscope/microscopio electrónico: Un instrumento que utiliza rayos de electrones para magnificar los objetos que son demasiado pequeños para verse con el microscopio de luz. (Pág. 32)

embryo/embrión: En los seres humanos, primera etapa de crecimiento de un óvulo fecundado; los dos primeros meses de embarazo. (Pág. 537)

embryology/embriología: El estudio de los embriones, los cuales son una de las primeras etapas del desarrollo de un organismo. (Pág. 142)

emphysema/enfisema: Enfermedad en que los alvéolos pierden la habilidad de expandirse y contraerse, causada principalmente por el fumar. (Pág. 492)

endangered species/especies en peligro de extinción: Una especie cuya población es muy pequeña y está en peligro de desaparecer. (Pág. 144)

endocytosis/endocitosis: Proceso por medio del cual las células transportan una masa de un material, como por ejemplo, una molécula grande de proteína a través de la membrana celular hasta el citoplasma. (Pág. 60)

endoplasmic reticulum/retículo endoplásmico: Un organelo celular que consiste de membranas dobladas las cuales transportan materiales dentro de la célula. (Pág. 39)

endoskeleton/endoesqueleto: El esqueleto interno de un organismo, que le da soporte y protege los órganos internos y también provee una estructura para los músculos. (Pág. 351)

endospores/endoesporas: Pared celular gruesa con la cual se rodean algunas bacterias, especialmente para protegerse del calor y de la sequía. (Pág. 194)

energy pyramid/pirámide de energía: Modelo del flujo de energía en una cadena alimenticia; hay menos energía accesible a medida que se avanza hacia el pico de la pirámide. (Pág. 615)

environmental activist/activista ambiental: Persona que trabaja para conservar los recursos terrestres y proteger el medio ambiente. (Pág. 660)

environmental management/administración del ambiente: Conservación y protección de los ecosistemas terrestres por medio de la administración cuidadosa de sus usos. (Pág. 658)

enzyme/enzima: La proteína que, sin ser alterada, acelera las reacciones químicas en las células. (Pág. 57)

epidermis/epidermis: La capa superficial o externa de las dos capas de la piel. (Pág. 432)

epiphyte/epífita: Planta que crece sobre otras plantas que le proveen soporte. (Pág. 242)

equilibrium/equilibrio: Condición en la cual el número de moléculas en una sustancia se mantiene constante a través de cierto espacio. (Pág. 59)

erosion/erosión: Desgaste del terreno causado por el viento, el agua y el hielo. (Pág. 651)

estivation/estivación: Adaptación para sobrevivir en tiempo seco y caluroso en la cual el animal se vuelve inactivo y ocurre una disminución de todos los procesos corporales. (Pág. 356)

estuary/estuario: Área en donde una corriente de agua dulce o un río desemboca en el océano; como por ejemplo, las ciénagas marinas, los deltas y las tierras bajas que quedan inundadas por la marea alta. (Pág. 638)

ethnobotany/etnobotánica: El estudio de gentes de tribus y de cómo estas gentes utilizan las plantas. (Pág. 288)
eukaryote/eucariota: Organismo o células con núcleo y con organelos que están rodeados por una membrana. (Pág. 161)
evolution/evolución: Cambios que ocurren con el paso del tiempo en los rasgos hereditarios de los organismos. (Pág. 130)
exocytosis/exocitosis: El proceso por medio del cual la célula expulsa moléculas grandes a través de la membrana celular. (Pág. 60)
exoskeleton/exoesqueleto: Cubierta dura y liviana en el exterior de los artrópodos, la cual los protege y les da soporte a sus cuerpos. (Pág. 333)
extinction/extinción: La muerte de una especie completa. (Pág. 144)

F

family/familia: La quinta categoría más alta de la taxonomía, debajo del orden. (Pág. 163)
fats/grasas: Nutrimientos que liberan energía; también ayudan al cuerpo a almacenar algunas vitaminas. (Pág. 444)
fermentation/fermentación: Proceso por medio del cual la levadura y algunas bacterias liberan la energía necesaria para sus procesos vitales, al descomponer la glucosa en alcohol y dióxido de carbono, en la ausencia del oxígeno. (Págs. 65, 584)
fertilization/fecundación: La fusión de dos gametos en los organismos que se reproducen sexualmente. (Pág. 83)
fetus/feto: Embrión en etapa de desarrollo; en los seres humanos, embrión después de los primeros dos meses de embarazo. (Pág. 538)
filter feeder/alimentador filtrador: Organismo que obtiene su alimento al filtrarlo del agua en la cual vive. (Pág. 309)
fins/aletas: En los peces, estructuras en forma de abanico; adaptadas para el movimiento, equilibrio y dirección del movimiento en el agua. (Pág. 351)
fish/pez: Animal vertebrado de sangre fría cuyas branquias, aletas y escamas lo ayudan a adaptarse para vivir en el agua. (Pág. 351)
fission/fisión: Un método de reproducción que usan las bacterias, en el que la bacteria se divide para formar dos bacterias con material genético idéntico. (Pág. 189)
flagellum/flagelo: Una cola parecida a un látigo en las bacterias y algunos protistas que las ayudan a moverse en ambientes húmedos. (Pág. 189)
food chain/cadena alimenticia: Relación en la forma de alimentación que transfiere energía a través de una comunidad, empezando con los productores y avanzando hacia los animales herbívoros hasta los animales carnívoros. (Pág. 614)
food group/grupo de alimentos: Un grupo de alimentos que contienen los mismos nutrimientos, por ejemplo, el grupo de las carnes. (Pág. 448)
food web/red alimenticia: Serie de cadenas alimenticias superpuestas. (Pág. 614)
fossils/fósiles: Restos de vidas que existieron en tiempos pasados. (Pág. 137)
fracture/fractura: El rompimiento de un hueso. (Pág. 422)
free-living/de vida libre: Organismos que encuentran su propio alimento y lugar para vivir sin depender de otros organismos; lo opuesto de los parásitos. (Pág. 315)
frond/fronda: La hoja de un helecho. (Pág. 244)

G

gametophyte/gametofito: La forma de una planta que produce gametos masculinos y/o femeninos. (Pág. 237)
gene/gene: Región del DNA que dirige la producción de una cierta proteína, así controla las características que se pasan a la progenie. (Pág. 89)
genetic engineering/ingeniería genética: Métodos experimentales que se usan para alterar los genes de la progenie y así controlar los rasgos que se pasan de los padres a la progenie. (Pág. 121)
genetics/genética: El estudio de cómo un gene afecta los rasgos de la progenie. (Pág. 106)
genome/genoma: Un mapa de la localización de los genes individuales en cada cromosoma de un individuo. (Pág. 122)
genotype/genetipo: La combinación de genes dominantes y/o de genes recesivos (alelos) que están presentes en las células de un organismo. (Pág. 110)
genus/genio: La segunda categoría más grande de la taxonomía. (Pág. 157)
gestation period/período de gestación: Tiempo que transcurre entre la fecundación del huevo y el nacimiento de la progenie. (Pág. 381)
gills/branquias: En animales acuáticos, el órgano que emplean para extraer el oxígeno del agua. (Pág. 324)
gill slits/hendiduras branquiales: Un par de aberturas en la garganta; se desarrollan en branquias en los peces, las cuales usan para respirar en el agua. (Pág. 351)
gizzard/molleja: Estructura muscular en el sistema digestivo de la lombriz de tierra, la cual muele la tierra. (Pág. 328)
Golgi bodies/cuerpos de Golgi: Organelos celulares que consisten en capas de saquitos membranosos

los cuales comprimen las proteínas y las mueven fuera de la célula. (Pág. 39)
gradualism/proceder gradualmente: Un modelo que describe la evolución como algo que ha ocurrido fija, gradual y continuamente para cambiar las especies en nuevas especies. (Pág. 135)
greenhouse effect/efecto de invernadero: Proceso por el cual el calor que irradia la superficie terrestre queda atrapado por los gases atmosféricos y se vuelve a reflejar a la Tierra, provocando un aumento en la temperatura global. (Pág. 649)
guard cells/células guardianas: Células que rodean el estoma para abrirlo y cerrarlo en las hojas de una planta. (Pág. 261)
gymnosperm/gimnosperma: Planta vascular que produce semillas en las escamas de los conos, como por ejemplo, los conos de los pinos. (Pág. 255)

habitat/hábitat: Lugar donde vive un organismo dentro de la comunidad. (Pág. 610)
hallucinogens/alucinógenos: Drogas que distorcionan la percepción, haciendo que el usuario se confunda entre la realidad y la fantasía. (Pág. 591)
haploid chromosome number/número haploide de cromosomas: Número de cromosomas en un gameto de un organismo que se reproduce sexualmente. (Pág. 83)
hashish/hachís: Droga más concentrada y poderosa que la mariguana. (Pág. 588)
hemoglobin/hemoglobina: Sustancia química en los glóbulos rojos de la sangre que lleva oxígeno a los tejidos y le da el color rojo a la sangre. (Pág. 472)
herbivore/herbívoro: Animal que solo come plantas. (Pág. 380)
heredity/herencia: Pasar los rasgos de los padres a la progenie. (Pág. 106)
hermaphrodite/hermafrodita: Animal que produce tanto espermatozoides como huevos. (Pág. 310)
heroin/heroína: Droga poderosa para aliviar el dolor, altamente adictiva, que se extrae de la morfina. (Pág. 592)
heterozygous/heterocigoso: Un organismo que tiene dos alelos diferentes para un rasgo. (Pág. 110)
hibernation/hibernación: Adaptación para sobrevivir el invierno, durante la cual el animal se vuelve inactivo y ocurre una disminución de todos los procesos corporales. (Pág. 356)
homeostasis/homeostasis: La adaptación constante de un organismo o de una célula para mantenerse en una condición estable, a pesar de los cambios en el medio ambiente. (Pág. 7)
hominids/homínidos: Los primeros primates que se parecían a los seres humanos actuales; comían tanto vegetales como animales y caminaban verticalmente en dos pies. (Pág. 147)
Homo sapiens/Homo sapiens: Nuestra propia especie, un mamífero primate homínido. (Pág. 148)
homologous/homólogo: Partes del cuerpo en especies diferentes, pero que tienen origen y estructuras similares, tales como las extremidades en los diferentes animales. (Pág. 141)
homologous blood transfer/transferencia de sangre homóloga: Involucra la extracción de sangre de una persona para transferírsela a otra. (Pág. 478)
homozygous/homocigoso: Un organismo que tiene dos alelos idénticos para un rasgo. (Pág. 110)
hormones/hormonas: Secreciones de las glándulas endocrinas que controlan ciertas actividades del cuerpo. (Pág. 522)
hypertension/hipertensión: Trastorno circulatorio en que la presión sanguínea es muy alta. (Pág. 468)
hyphae/hifa: Multitud de estructuras en forma de hilo que forman el cuerpo de los hongos. (Pág. 211)
hypothesis/hipótesis: Una predicción que se puede comprobar. (Pág. 14)

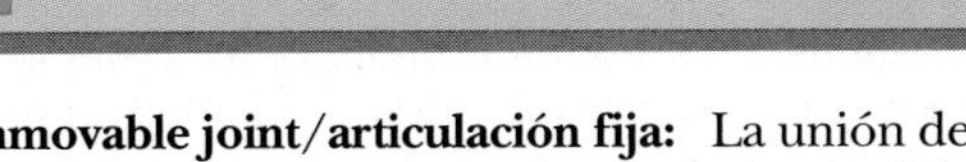

immovable joint/articulación fija: La unión de dos o más huesos que no permite movimiento. (Pág. 423)
immune system/sistema inmunológico: Conjunto complejo de defensas que usa el cuerpo para combatir las enfermedades. (Pág. 563)
imprinting/impronta: Tipo de aprendizaje en que un animal forma un lazo social con otro animal durante un período crítico: acabando de nacer o al salir del cascarón. (Pág. 397)
incomplete dominance/dominancia incompleta: La condición que existe cuando se expresan ambos alelos para un rasgo. (Pág. 114)
incubate/incubar: Mantener abrigado un huevo o un animal recién nacido; el calor ayuda al desarrollo de los huevos hasta que sale el animal del cascarón y ayuda a la sobrevivencia del animal joven. (Pág. 372)
infancy/infancia: Período de crecimiento humano que va desde las cuatro semanas después del nacimiento hasta el año. (Pág. 541)
innate behavior/comportamiento innato: Comportamiento con que nace un organismo, el cual no tiene que ser aprendido. (Pág. 394)
insight/discernimiento: Forma de razonamiento que les permite a los animales usar las experiencias anteriores para resolver problemas nuevos. (Pág. 399)

instinct/instinto: Patrón complejo de comportamiento innato que involucra acciones múltiples. (Pág. 395)

interneurons/neuronas internunciales: Células nerviosas a través del encéfalo y la médula espinal, las cuales transmiten impulsos de las neuronas sensoriales a las neuronas motoras. (Pág. 509)

invertebrates/invertebrados: Animales sin espina dorsal. (Pág. 303)

involuntary muscles/músculos involuntarios: Músculos que no pueden controlarse conscientemente, como por ejemplo, el músculo del corazón y el sistema digestivo. (Pág. 426)

joint/articulación: Cualquier lugar en donde se encuentran dos huesos. (Pág. 423)

kidneys/riñones: Dos órganos que filtran la sangre y producen el desperdicio líquido conocido como la orina. (Pág. 498)

kingdom/reino: La más alta categoría de la taxonomía, la cual tiene el número más grande de especies; los reinos incluyen todas las otras categorías. (Pág. 156)

larva/larva: Etapa del desarrollo entre el huevo y el adulto. (Pág. 310)

larynx/laringe: Estructura entre la faringe y la tráquea a la cual están conectadas las cuerdas vocales. (Pág. 489)

latent virus/virus latente: Un virus que llega a formar parte del DNA de la célula y que se queda inactivo hasta que algo ocasiona su actividad, entonces destruye la célula y reproduce nuevos virus. (Pág. 183)

law/ley: Una regla que describe un patrón en la naturaleza y lo que suele ocurrir dadas ciertas condiciones. (Pág. 16)

learning/aprendizaje: Proceso por medio del cual se desarrolla un comportamiento por medio de la experiencia; un organismo repite aquellos comportamientos que satisfacen necesidades y evita aquellos que no le proporcionan nada o que le causan dolor. (Pág. 396)

lichen/liquen: Hongo que vive en una relación mutualista con un alga verde. (Pág. 215)

life span/duración de la vida: El período de tiempo que vive un organismo. (Pág. 8)

ligament/ligamento: Banda dura de tejido que mantiene unidos los huesos. (Pág. 423)

limiting factor/factor limitante: Una condición, como por ejemplo la cantidad de agua o alimentos, que determina si un organismo, una población o una especie sobrevive en su ambiente. (Pág. 628)

lipid/lípido: Compuesto orgánico que almacena y libera grandes cantidades de grasas o aceites que proporcionan energía. (Pág. 57)

littoral zone/zona del litoral: Zona del océano a lo largo de la costa, que está cubierta de agua durante la marea alta y expuesta al aire durante la marea baja. (Pág. 637)

long-day plant/planta de día largo: Planta que requiere noches cortas y días largos para florecer. (Pág. 284)

lymph/linfa: Fluido en los tejidos del cuerpo compuesto de agua, sustancias disueltas y linfocitos. (Pág. 480)

lymph nodes/ganglios linfáticos: Estructuras a través del cuerpo que filtran los microorganismos y los materiales extraños de la linfa antes de devolverla a la sangre. (Pág. 481)

lymphatic system/sistema linfático: Recoge el fluido de los tejidos del cuerpo y lo devuelve a la sangre a través de los vasos linfáticos. (Pág. 480)

lymphocyte/linfocito: Glóbulo blanco de la sangre que combate los antígenos que causan enfermedades al rodearlos e ingerirlos. (Págs. 481, 566)

lysosome/lisosoma: Un organelo que contiene sustancias químicas que digieren los desperdicios y las partes celulares desgastadas. (Pág. 40)

mammal/mamífero: Animal vertebrado de sangre caliente que tiene pelo que lo aisla y glándulas mamarias. (Pág. 379)

mammary glands/glándulas mamarias: Glándulas en las hembras mamíferas que producen leche para alimentar a las crías. (Pág. 379)

manatee/manatí: Mamífero marino de gran tamaño que vive en las aguas cálidas costeras; se llama comúnmente vaca marina. (Pág. 386)

mantle/manto: La cubierta exterior que cubre el cuerpo de los moluscos; secreta una sustancia química la cual se convierte en el caparazón o les protege el cuerpo si carecen de él. (Pág. 324)
marijuana/mariguana: Droga que se fabrica de los tallos, las hojas, las flores y las semillas de la planta de cáñamo, también se le conoce con el nombre de porro. (Pág. 588)
marrow/médula: Tejido graso rojo o amarillo en ciertos huesos, el cual produce los glóbulos rojos y blancos de la sangre. (Pág. 422)
marsupial/marsupial: Mamífero con una bolsa abdominal donde guarda y amamanta las crías. (Pág. 381)
mechanical digestion/digestión mecánica: Procesamiento de alimentos masticándolos, revolviéndolos y por acción de la bilis para romperlos en pedacitos más pequeños. (Pág. 451)
medusa/medusa: Animal celenterado que tiene forma de campana y es un nadador libre. (Pág. 312)
meiosis/meiosis: La división del núcleo de la célula para producir células sexuales (gametos). (Pág. 82)
melanin/melanina: Sustancia química producida por la epidermis, la cual le da el color a la piel. (Pág. 433)
menopause/menopausia: En las hembras de edad avanzada, período de años en que el ciclo menstrual es irregular y eventualmente cesa. (Pág. 533)
menstrual cycle/ciclo menstrual: En las hembras, ciclo de producción del óvulo y la menstruación. (Pág. 532)
menstruation/menstruación: En las hembras, flujo mensual de expulsión del forro uterino y sangre a través de la vagina si no ha ocurrido la fecundación del huevo. (Pág. 533)
metabolism/metabolismo: Todos los cambios en un organismo que le permiten vivir, crecer y reproducirse. (Pág. 63)
metamorphosis/metamorfosis: Cambios de forma que ocurren durante el ciclo de vida de los insectos: huevo, larva, ninfa, adulto. (Pág. 337)
methamphetamine/metanfetamina: Un estimulante muy poderoso. (Pág. 587)
migration/migración: Movimiento instintivo de ciertos animales asociado con las estaciones, como por ejemplo, el que las aves vuelen al sur en el invierno. (Pág. 405)
mimicry/mimetismo: Defensa contra predadores en que un animal copia o imita la apariencia o el comportamiento de otro organismo. (Pág. 613)
minerals/minerales: Nutrimientos inorgánicos que regulan las reacciones químicas en el cuerpo, tales como la formación de células y la transmisión de impulsos nerviosos. (Pág. 446)
mitochondria/mitocondria: Un organelo celular que digiere las moléculas de alimento y libera energía. (Pág. 40)
mitosis/mitosis: El proceso mediante el cual el núcleo se divide en dos núcleos, cada uno de estos contiene el mismo número de cromosomas que tenía la célula madre. (Pág. 75)
mollusk/molusco: Un animal invertebrado de cuerpo blando que normalmente tiene un caparazón, como por ejemplo, un caracol. (Pág. 324)
molting/muda: Cambio periódico y reemplazo de la cubierta vieja de los cuerpos de los animales, como por ejemplo, la piel o el exoesqueleto. (Pág. 333)
monocots/monocotiledóneas: Plantas angiospermas con un solo cotiledón dentro de la semilla. (Pág. 256)
monotreme/monotrema: Mamífero que pone huevos con cáscara dura y correosa; como por ejemplo el ornitorrinco. (Pág. 382)
motivation/motivación: Algún estímulo interior que causa que un animal actúe. (Pág. 398)
motor neurons/neuronas motoras: Células nerviosas que conducen los impulsos desde el encéfalo o la médula espinal hasta los músculos o glándulas a través del cuerpo. (Pág. 509)
movable joint/articulación móvil: Lugar en donde se encuentran o están pegados dos o más huesos y permiten el movimiento. (Pág. 423)
multiple alleles/alelos múltiples: Tener más de dos alelos que controlan un rasgo. (Pág. 115)
muscle/músculo: Tejido que se relaja y se contrae para permitir el movimiento. (Pág. 426)
mutation/mutación: Cualquier cambio permanente en el material genético (DNA) de una célula. (Pág. 91)
mutualistic/mutualística: Relación en la cual dos organismos viven juntos y ambos se benefician de alguna forma. (Pág. 215)

natural resources/recursos naturales: Materias primas en el medio ambiente que los organismos utilizan para la sobrevivencia, incluyen el aire, el agua, la tierra y los bosques. (Pág. 648)
natural selection/selección natural: Teoría de Darwin que dice que los organismos mejor adaptados, con rasgos para sobrevivir en cierto medio ambiente son los que sobreviven para pasar esos rasgos a la progenie. (Pág. 132)
nephron/nefrón: Especie de filtro en los riñones. (Pág. 499)
neuron/neurona: Célula nerviosa que lleva impulsos a través de todo el cuerpo. (Pág. 509)
niche/nicho: El papel que desempeña una especie dentro de una comunidad. (Pág. 610)
nitrogen cycle/ciclo del nitrógeno: Movimiento continuo del nitrógeno desde la atmósfera a las plantas y de nuevo a la atmósfera (o directamente a las plantas). (Pág. 619)
nitrogen dioxide/dióxido de nitrógeno: Contaminante principal del aire, el cual es peligroso

inhalar, expelido por el humo de cigarrillo y los gases de los automóviles. (Pág. 496)

nitrogen-fixing bacteria/bacteria nitrificante: Bacteria que cambia el oxígeno del aire en compuestos nítricos que son útiles para las plantas y los animales. (Pág. 194)

noncommunicable disease/enfermedad no contagiosa: Enfermedad que no se puede contagiar de un organismo a otro. (Pág. 570)

nonrenewable resources/recursos no renovables: Recursos naturales que están disponibles solo en cantidades limitadas y que la naturaleza no reemplaza rápidamente. (Pág. 648)

nonvascular plants/plantas no vasculares: Plantas que carecen de tubos para transportar el alimento y el agua; estas plantas absorben el agua directamente a través de sus membranas celulares. (Pág. 234)

notochord/notocordio: Estructura flexible en forma de bastón a lo largo del lado dorsal (espalda) de un animal cordado; la espina dorsal en algunos vertebrados. (Pág. 350)

nucleic acids/ácidos nucleicos: Moléculas orgánicas grandes que además de almacenar información importante en las células, gobiernan sus actividades; dos tipos son el DNA y el RNA. (Pág. 57)

nucleus/núcleo: La estructura dentro de la célula que controla las actividades de la célula; contiene cromatina; cromosomas en una célula en el proceso de reproducción. (Pág. 38)

nutrients/nutrimientos: Sustancias químicas en los alimentos que proveen energía y materiales para las actividades vitales. (Pág. 442)

olfactory cells/células olfativas: Células nerviosas en las cavidades nasales que responden a los estímulos químicos en el aire. (Pág. 517)

omnivore/omnívoro: Animal que obtiene su energía tanto de plantas como de otros animales. (Pág. 380)

open circulary system/sistema circulatorio abierto: Sistema circulatorio que no posee vasos y, en su lugar, baña los órganos internos con la sangre, como por ejemplo, en los moluscos. (Pág. 324)

order/orden: La cuarta categoría más alta de la taxonomía, debajo de la clase. (Pág. 163)

organ/órgano: La estructura compuesta de diferentes tipos de tejidos, que funcionan juntos para realizar una tarea específica; por ejemplo, el corazón. (Pág. 45)

organ system/sistema de órganos: Un grupo de órganos que funcionan juntos para realizar una tarea específica; por ejemplo, el sistema digestivo. (Pág. 45)

organelles/organelos: Estructuras dentro del citoplasma de células eucariotas, los cuales digieren el alimento, eliminan los desechos y almacenan los materiales. (Pág. 38)

organism/organismo: Un ser viviente compuesto de una o más células, usa energía, se mueve, reacciona y se adapta a su medio ambiente, sereproduce y vive un determinado período de tiempo. (Pág. 6)

osmosis/ósmosis: El transporte pasivo de agua a través de una membrana celular por medio de la difusión. (Pág. 59)

ovary/ovario: Parte hinchada de la base del pistilo donde se forman los óvulos de las angiospermas (Pág. 264); órgano que produce los huevos en los animales. (Pág. 531)

ovulation/ovulación: Expulsión mensual de un óvulo de los ovarios de las hembras. (Pág. 531)

ovule/óvulo: La parte reproductora de la planta hembra que contiene los huevos. (Pág. 262)

palisade layer/capa de empalizada: Filas de células empacadas densamente cerca de la superficie, las cuales contienen clorofila en las hojas de una planta. (Pág. 261)

parasite/parásito: Generalmente un organismo, pero también cualquier cosa que se alimenta del organismo huésped y al mismo tiempo causa daño al mismo. (Pág. 183)

passive immunity/inmunidad pasiva: Inmunidad que causan los anticuerpos que se introducen al cuerpo de fuentes externas. (Pág. 564)

passive transport/transporte pasivo: Movimiento de materiales a través de una membrana, pero sin usar energía. (Pág. 58)

pasteurization/pasteurización: Calentamiento de los alimentos a una temperatura un poco menor que el punto de ebullición, pero por suficiente tiempo para matar la mayor parte de las bacterias. (Pág. 559)

pathogen/patógeno: Cualquier organismo que causa enfermedad. (Pág. 194)

pedigree/pedigree o árbol genealógico: Un diagrama que traza la historia de un rasgo en una familia. (Pág. 119)

pelagic zone/zona pelágica: Área de mar abierto más allá de la zona sublitoral. (Pág. 638)

periosteum/periósteo: Membrana apretada que forra los huesos. (Pág. 421)

peripheral nervous system/sistema nervioso periférico: Parte del sistema nervioso que comprende los nervios craneales y los espinales que conectan el encéfalo y la médula espinal a las otras partes del cuerpo. (Pág. 510)

peristalsis/peristalsis: Contracciones que mueven los alimentos a través del sistema digestivo. (Pág. 452)

permafrost/permagel: En las regiones de la tundra, tierra que permanece profundamente congelada debajo de la superficie. (Pág. 633)

pesticide/insecticida: Sustancia química que mata insectos perjudiciales indeseables. (Pág. 340)

pharynx/faringe: Pasaje tubular, para el aire y los alimentos, que se encuentra situado entre la cavidad nasal y el esófago. (Pág. 489)

phenotype/fenotipo: Un rasgo físico en un organismo; que resulta de su composición genética. (Pág. 110)

phloem/floema: Células tubulares en las plantas vasculares que transportan el alimento desde las hojas y tallos a las otras partes de la planta para ser almacenado o usado. (Pág. 260)

photoperiodism/fotoperiodicidad: La respuesta de una planta hacia la floración, la planta responde a la duración de las horas de luz y de oscuridad en un día. (Pág. 284)

photosynthesis/fotosíntesis: La reacción química que utilizan los productores, tales como las plantas verdes, para producir el alimento; dicha reacción transforma la energía solar a energía química y convierte el dióxido de carbono y el agua en azúcar y oxígeno. (Págs. 63, 278)

phylogeny/filogenia: La historia de la evolución de un organismo. (Pág. 160)

phylum/fílum: La segunda categoría más alta de la taxonomía en el reino animal. (Pág. 163)

physical dependence/dependencia física: Uso de una droga porque el cuerpo tiene una dependencia química y no puede funcionar sin ella; también se llama narcomanía o adicción a las drogas. (Pág. 581)

pioneer species/especie precursora: Las primeras plantas que crecen en un área nueva o perturbada; su descomposición produce el material que necesitan otras plantas para crecer. (Pág. 239)

pistil/pistilo: Órgano reproductor femenino de la flor. (Pág. 264)

placenta/placenta: Parte de la bolsa que rodea el embrión de los mamíferos; pegada al útero; muchos vasos sanguíneos proveen sustento al embrión. (Pág. 381)

placental mammal/mamífero placentario: Animal cuyas crías se desarrollan dentro del útero materno. (Pág. 381)

plasma/plasma: Parte líquida de la sangre, formada principalmente de agua, pero también contiene alimentos disueltos, minerales y gases tales como el oxígeno y el dióxido de carbono. (Pág. 472)

platelet/plaqueta: Fragmento celular que ayuda en la coagulación de la sangre. (Pág. 472)

poaching/caza o pesca ilegal: Caza o pesca ilegal de animales. (Pág. 387)

pollen grains/granos de polen: La parte reproductora de la planta macho que contiene el espermatozoide. (Pág. 262)

pollen tube/tubo de polen: Estructura hueca que crece del grano de polen en el óvulo, a través de la cual el espermatozoide nada dentro del óvulo para la fecundación. (Pág. 262)

pollination/polinización: Transferencia de los granos de polen desde el estambre hasta el óvulo. (Pág. 264)

polygenic inheritance/herencia poligénica: Grupos de pares de genes que actúan juntos para producir un rasgo. (Pág. 116)

polyp/pólipo: Animal celenterado que tiene forma de tubo y es sésil. (Pág. 312)

population/población: Organismos de una misma especie que viven en el mismo lugar al mismo tiempo. (Págs. 134, 608)

population density/densidad de población: Número de organismos por unidad de espacio habitable. (Pág. 608)

Porifera/Porifera: Fílum al cual pertenecen las esponjas; el nombre significa: "con poros". (Pág. 308)

predator/predador: Organismo que mata y se come a la presa para obtener energía. (Pág. 352)

preening/arreglarse las plumas con el pico: Comportamiento que tienen las aves de usar el pico para frotarse aceite sobre las plumas para hacerlas impermeables al agua. (Pág. 373)

pregnancy/embarazo: Período de tiempo entre la fecundación del huevo y el nacimiento del nene. (Pág. 536)

prey/presa: Organismo que es consumido por un predador. (Pág. 352)

primary succession/sucesión primaria: Desarrollo de nuevas comunidades en las nuevas áreas de terreno, tales como las islas volcánicas nuevas. (Pág. 629)

primates/primates: El grupo de mamíferos al cual pertenecen los monos, simios y los seres humanos. (Pág. 146)

probability/probabilidad: Una rama de la ciencia que determina si es posible que algo suceda. (Pág. 109)

producers/productoras: Plantas verdes que producen su propio alimento por medio de la fotosíntesis. (Pág. 63)

prokaryote/procariota: Organismo de una sola célula que no tiene un núcleo organizado ni tiene organelos. (Pág. 161)

protein/proteína: Compuesto orgánico compuesto de aminoácidos; lo usa el cuerpo para crecer y para reparar y reemplazar las células. (Págs. 57, 443)

prothallus/prótalo: La estructura en los helechos que produce los gametos. (Pág. 244)

protists/protistas: Miembros del reino protista; organismos simples cuyas células poseen núcleo; posiblemente los antecesores de los hongos, las plantas y los animales. (Pág. 202)

protozoa/protozoario: Protistas unicelulares que se parecen a los animales; la mayoría son parásitos. (Pág. 206)

pseudopod/seudópodo: Una extensión del citoplasma parecida a una pata, la cual usan algunos organismos para movilizarse y para atrapar sus alimentos. (Pág. 206)

psychological dependence/dependencia psicológica: Uso de una droga solo porque la persona piensa que la necesita. (Pág. 581)

pulmonary circulation/circulación pulmonar: Flujo de sangre desde el corazón hasta los pulmones y de vuelta al corazón. (Pág. 466)

punctuated equilibrium/equilibrio interrumpido: Un modelo que describe la evolución como algo que ha ocurrido rápidamente para cambiar viejas especies en nuevas especies por medio de la mutación de unos cuantos genes. (Pág. 135)

Punnett square/cuadrado de Punnett: Un instrumento que muestra cómo se pueden combinar los genes; se usa para predecir los resultados en la genética. (Pág. 110)

purebred/de raza pura: Un organismo que siempre produce los mismos rasgos en su progenie. (Pág. 108)

R

radial symmetry/simetría radial: Se dice de los animales que tienen las partes del cuerpo distribuidas alrededor del eje de un círculo, a semejanza de una rueda de bicicleta. (Pág. 303)

radioactive elements/elementos radiactivos: Elementos cuyos átomos emiten radiación, una forma de energía atómica. (Pág. 138)

radula/rádula: En los moluscos, órgano, parecido a una lengua, con hileras de dientes que raspan y desgarran los alimentos. (Pág. 325)

recessive/recesivo: La forma de un rasgo que menos aparece en la progenie; por ejemplo, el ser zurdo es un rasgo recesivo, comparado con el ser diestro. (Pág. 108)

reflex/reflejo: Tipo de comportamiento innato que es una respuesta automática a un estímulo; no involucra el encéfalo. (Págs. 512, 395)

reforestation/reforestación: Plantar de nuevo un área con árboles jóvenes. (Pág. 656)

regeneration/regeneración: Tipo de reproducción asexual en el cual un organismo entero nuevo crece de solo una parte del organismo padre. (Pág. 309)

rejection/rechazo: Proceso en el cual el sistema inmunológico ataca el órgano trasplantado porque este es algo extraño para el cuerpo. (Pág. 46)

relative dating/datación relativa: La estimación de la edad de un fósil por medio de la comparación con fósiles más jóvenes en la capa de roca superior y con fósiles más viejos en la capa de roca inferior. (Pág. 138)

renewable resources/recursos renovables: Recursos naturales que la naturaleza reemplaza constantemente. (Pág. 648)

reptile/reptil: Animal vertebrado de sangre fría y piel seca y escamosa que pone huevos cubiertos con una cáscara correosa; algunos ejemplos son los lagartos, las serpientes y las tortugas. (Pág. 363)

response/respuesta: La reacción a un estímulo. (Pág. 7)

retina/retina: Tejido sensible a la luz en la parte trasera del ojo. (Pág. 515)

rhizoids/rizoides: Estructuras filamentosas que parecen raíces y solo contienen unas cuantas células largas; sostienen los musgos en su sitio. (Pág. 236)

rhizome/rizoma: El tallo subterráneo de un helecho. (Pág. 244)

ribosomes/ribosomas: Un organelo celular que produce proteína. (Pág. 39)

S

saliva/saliva: Sustancia acuosa producida en la boca que comienza la digestión química de los alimentos. (Pág. 452)

saprophyte/saprófago: Cualquier organismo que usa material muerto como alimento y fuente de energía. (Pág. 193)

scales/escamas: Placas delgadas y duras que superpuestas cubren y protegen el cuerpo de los peces. (Pág. 351)

scientific method/método científico: Procedimiento usado por los científicos para resolver problemas: definir el problema, formar una hipótesis, comprobar la hipótesis, analizar los resultados y sacar conclusiones. (Pág. 14)

secondary succession/sucesión secundaria: Retorno gradual de un área perturbada a sus comunidades pasadas. (Pág. 629)

sedimentary rock/roca sedimentaria: Un tipo de roca que se forma por el asentamiento de partículas finas de barro y arena que luego se cementan en una roca. (Pág. 137)

semen/semen: Mezcla de espermatozoides y fluido que los nutre y les facilita la locomoción. (Pág. 531)

sensory neurons/neuronas sensoriales: Células nerviosas que transmiten estímulos de los receptores al encéfalo o la médula espinal. (Pág. 509)

sessile/sésil: Describe los organismos, como los árboles que permanecen anclados a un sitio durante todas sus vidas. (Pág. 308)

setae/seta: Estructuras que parecen cerdas en el exterior de los gusanos segmentados, las cuales les ayudan a agarrarse a la tierra para poder moverse. (Pág. 327)

sex-linked gene/gene ligado al sexo: Un alelo que se hereda en los cromosomas que determinan el sexo. (Pág. 119)
sexual reproduction/reproducción sexual: El tipo de reproducción en que se produce un nuevo organismo de la combinación de células sexuales de ambos padres. (Pág. 81)
sexually transmitted disease/enfermedad transmitida sexualmente: Enfermedad que se transmite durante el contacto sexual. (Pág. 561)
short-day plant/planta de día corto: Planta que requiere noches largas y días cortos para florecer. (Pág. 284)
skeletal muscles/músculos del esqueleto: Tejido muscular voluntario conectado a los huesos; los músculos que permiten que los huesos se muevan. (Pág. 427)
skeletal system/sistema del esqueleto: Conjunto de huesos del cuerpo, el cual forma un marco rígido que da apoyo al cuerpo, protege los órganos internos, produce los glóbulos rojos y almacena el calcio y el fósforo del cuerpo. (Pág. 420)
smooth muscle/músculo liso: Tejido muscular involuntario que forma las paredes del estómago, los intestinos, el útero y los vasos sanguíneos. (Pág. 427)
social behavior/comportamiento social: Interacciones entre los organismos de una misma especie, que incluyen el apareamiento, el cuidado de las crías, protección, obtención de alimentos y defensa del territorio. (Pág. 403)
society/sociedad: Grupo de animales de la misma especie que viven y trabajan juntos en una forma organizada. (Pág. 403)
solid wastes/desperdicios sólidos: Productos sólidos que se pueden reciclar, quemar o botar. (Pág. 651)
sori/soros: Estructuras que producen esporas en el envés de las frondas. (Pág. 244)
species/especies: Organismos cuyos miembros son iguales y se reproducen exitosamente entre sí. (Págs. 130, 157)
species diversity/diversidad de la especie: La gran variedad de especies de plantas y animales sobre la Tierra. (Pág. 164)
sperm/espermatozoide: El gameto o célula reproductora del macho, producida en los testículos. (Págs. 82, 530)
spiracle/espiráculo: Abertura en el abdomen de los artrópodos que permite la entrada y la salida de oxígeno y de dióxido de carbono en los pulmones. (Pág. 333)
spontaneous generation/generación espontánea: La creencia de que los seres vivientes provenían de materia no viva; por ejemplo, de que los sapos surgían espontáneamente del barro. (Pág. 10)
sporangia/esporangios: Envolturas redondas que producen esporas en los hongos cigotos. (Pág. 212)
spore/espora: Célula reproductora resistente que forma nuevos organismos, sin intervenir la fecundación, en los hongos, los helechos y algunos protistas. (Pág. 212)
sporophyte/esporofito: Cápsula en la cual se producen esporas durante la meiosis, en plantas como los musgos. (Pág. 237)
stamen/estambre: Órgano reproductor masculino de la flor. (Pág. 264)
stimulant/estimulante: Sustancia, como la cafeína (en el café, el té, el chocolate y los refrescos), que estimula el sistema nervioso central y acelera los otros sistemas del cuerpo. (Pág. 583)
stimulus/estímulo: Cualquier cosa a la cual responde un organismo, como por ejemplo, el sonido, la luz, el calor, la vibración, el olor, el movimiento, el hambre, la sed, etc. (Págs. 7, 282)
stomata/estoma: Especie de poro en la superficie de la hoja que permite la entrada y la salida del dióxido de carbono, del oxígeno y del agua. (Pág. 261)
sublittoral zone/zona sublitoral: Zona de agua poco profunda sobre la plataforma continental. (Pág. 637)
succession/sucesión: Cambio gradual y ordenado de las especies en una comunidad a través del tiempo. (Pág. 629)
symbiosis/simbiosis: Condición en la cual dos organismos viven juntos para beneficio mutuo. (Pág. 215)
synapse/sinapsis: Pequeño espacio entre las neuronas, a través del cual se pueden transmitir los impulsos. (Pág. 510)
systemic circulation/circulación sistémica: Flujo de sangre desde el corazón hasta todos los tejidos del cuerpo (excepto los del corazón y de los pulmones), y de vuelta al corazón. (Pág. 467)

target tissue/tejido asignado: Tejido afectado por hormonas. (Pág. 522)
taste buds/papilas gustativas: Tejido localizado en la lengua que responde a los estímulos químicos. (Pág. 517)
taxonomy/taxonomía: La ciencia que trata con la agrupación y dar nombres a los organismos. (Pág. 156)
technology/tecnología: La aplicación del conocimiento científico para mejorar la calidad de la vida humana. (Pág. 22)
tendon/tendón: Banda gruesa de tejido que conecta los músculos a los huesos. (Pág. 427)
tentacles/tentáculos: Estructuras que parecen brazos y que se encuentran alrededor de la boca de algunos organismos y les ayudan a atrapar los alimentos. (Pág. 311)

territory/territorio: Área que un animal defiende de los otros miembros de la misma especie. (Pág. 401)
testes/testículos: Órganos en el sistema reproductor del hombre que producen los espermatozoides. (Pág. 530)
theory/teoría: Descripción de la naturaleza, la cual cambia cuando la evidencia cambia. (Pág. 16)
tissues/tejidos: Células parecidas que realizan la misma función; por ejemplo, todos los tejidos musculares se contraen. (Pág. 45)
tolerance/tolerancia: Ocurre cuando el cuerpo se acostumbra a una droga y necesita un aumento paulatino de la dosis para producir el mismo efecto. (Pág. 581)
toxin/toxina: Un veneno producido por organismos que causan enfermedades (patógenos). (Pág. 194)
trachea/tráquea: Tubo que lleva el aire a los bronquios. (Pág. 489)
transgenic organisms/organismos transgénicos: Organismos que contienen información genética de otras especies. (Pág. 92)
transpiration/transpiración: Pérdida del vapor de agua a través del estoma de la hoja en las plantas. (Pág. 277)
trial and error/comportamiento de tanteos: Comportamiento que es modificado por la experiencia. (Pág. 397)
tropism/tropismo: La respuesta de una planta a un estímulo. (Pág. 282)
tube feet/patas tubulares: Estructuras pegadas al sistema vascular acuífero de los equinodermos, las cuales actúan como copas de succión y ayudan a los equinodermos a moverse, alimentarse, obtener oxígeno y deshacerse de los desperdicios. (Pág. 342)
tumor/tumor: Crecimiento anormal del tejido. (Pág. 571)
tundra/tundra: Bioma frío, seco y sin árboles en donde el Sol casi no brilla durante los seis o nueve meses de invierno. (Pág. 633)

U

umbilical cord/cordón umbilical: En los mamíferos, manojo de vasos sanguíneos que conecta la placenta al ombligo y que transporta oxígeno y nutrimiento al embrión. (Pág. 381)
ureters/uréteres: Tubos que van desde los riñones a la vejiga. (Pág. 499)
urethra/uretra: Tubo que conecta la vejiga con el exterior del cuerpo. (Pág. 499)
urinary system/sistema urinario: Sistema de excreción que se deshace de los desperdicios de la sangre, del exceso de agua y de sales. (Pág. 498)
urine/orina: Desperdicio líquido colectado por los riñones, el cual contiene agua, sales y otros desperdicios. (Pág. 499)
uterus/útero: En las hembras, órgano muscular en forma de pera en donde el huevo fecundado se desarrolla en un nene; también se le llama la matriz. (Pág. 531)

V

vaccination/vacunación: Administración de un virus debilitado para desarrollar inmunidad contra una enfermedad. (Pág. 568)
vaccine/vacuna: Una solución hecha de un virus muerto o debilitado; causa inmunidad artificial. (Pág. 184)
vacuole/vacuola: Un lugar en la célula donde se almacenan el agua, los alimentos o los desechos. (Pág.41)
vagina/vagina: En las hembras, canal que va del útero al exterior del cuerpo; también denominado el canal de nacimiento porque la progenie pasa por él al nacer. (Pág. 531)
variable/variable: El factor que se comprueba en un experimento. (Pág. 15)
variation/variación: La ocurrencia de un rasgo heredado que hace que un organismo sea diferente a los otros miembros de la misma especie. (Pág. 133)
vascular plants/plantas vasculares: Plantas que poseen unas estructuras en forma de tubos, las cuales transportan el alimento y el agua a las células de la planta. (Pág. 234)
vascular tissue/tejido vascular: Tejido compuesto de células alargadas que forman los tubos en las plantas vasculares. (Pág. 241)
veins/venas: Vasos sanguíneos que mueven sangre hacia el corazón, acarreando los materiales de desecho. (Pág. 466)
ventricles/ventrículos: Las dos cámaras inferiores del corazón humano. (Pág. 464)
vertebrates/vertebrados: Animales con espina dorsal. (Pág. 303)
vestigial structure/estructura atrofiada: Una parte del cuerpo que ha sido reducida en tamaño y que el organismo no usa más. (Pág. 141)
villi/vellosidad intestinal: Proyecciones finas y diminutas en la parte interior del intestino delgado. (Pág. 454)
virus/virus: Una partícula microscópica cuyo centro está compuesto ya sea de DNA o de RNA y cubierta de una capa de proteína; el virus infecta a las células huéspedes para poder reproducirse. (Pág. 182)
vitamins/vitaminas: Nutrimientos orgánicos que promueven el crecimiento, regulan las funciones del

cuerpo y ayudan al cuerpo a usar otros nutrimientos. (Pág. 445)

voluntary muscles/músculos voluntarios: Músculos que se pueden controlar, por ejemplo, los músculos de los brazos o de las piernas. (Pág. 426)

warm-blooded/de sangre caliente: Describe animales cuya temperatura corporal se mantiene constante sin importar cuál sea la temperatura del ambiente. (Pág. 351)

water cycle/ciclo del agua: Movimiento continuo del agua en la biosfera por medio de la evaporación, la condensación y la precipitación. (Pág. 617)

water table/nivel hidrostático: Parte superior del agua subterránea; esta fuente de agua provee agua potable para la mayor parte de los seres humanos. (Pág. 641)

water-vascular system/sistema vascular acuífero: Sistema de canales llenos de agua distintivo de los equinodermos. (Pág. 342)

wetland/marisma: Terreno que por parte del año está cubierto de aguas poco profundas; por ejemplo, los pantanos y las ciénagas. (Pág. 640)

wildlife preserve/reserva de fauna: Área de terreno reservada para la protección de distintas especies de fauna. (Pág. 656)

withdrawal/síndrome de abstinencia: Enfermedad que ocurre cuando una persona deja de tomar una droga a la cual el cuerpo se había habituado. (Pág. 581)

xylem/xilema: Tubos que transportan el agua y los minerales desde la raíz a través de toda la planta. (Pág. 260)

zygote/cigoto: En los organismos que se reproducen sexualmente, célula que resulta durante la fecundación. (Pág. 83)

INDEX

The Index for *Merrill Life Science* will help you locate major topics in the book quickly and easily. Each entry in the Index is followed by the numbers of the pages on which the entry is discussed. A page number given in **boldface type** indicates the page on which that entry is defined. A page number given in *italic type* indicates a page on which the entry is used in an illustration or photograph. The abbreviation *act.* indicates a page on which the entry is used in an Activity.

R

PHOTO CREDITS

Cover, ©Francis & Donna Caldwell/Global Pictures; **v,** Runk/Schoenberger from Grant Heilman; **vi,** Michael & Patricia Fogden; **vii,** (t) Dwight R. Kuhn, (b) Merlin D. Tuttle/Bat Conservation Internat'l./Photo Researchers; **viii,** (t) Runk/Schoenberger from Grant Heilman, (b) Animals Animals/Bates Littlehales; **ix,** William J. Weber; **x,** Hickson & Associates; **xi,** (t) Horticultural Photography/Corvallis, Oregon, (c) SuperStock, (b) Brian Brake/Photo Researchers; **xii,** (t) G.R. Roberts, (b) Lucian Niemeyer/LNS Arts; **xiii,** Dwight R. Kuhn; **xiv,** Elaine Shay; **xv,** Animals Animals/Zig Leszczynski; **xvi,** Jeff Foott; **xvii,** Runk/Schoenberger from Grant Heilman; **xviii,** StudiOhio; **xix, xx,** Doug Martin; **2-3,** R.S. Virdee from Grant Heilman, (inset) Eric V. Grave/Photo Researchers; **4,** Doug Martin; **5,** Aaron Haupt; **6,** Larry Mulvehill/FPG; **7,** (t) Alan Carey, (b) Robert & Linda Mitchell; **11,** Aaron Haupt/Glencoe; **12,** E.R. Degginger; **13,** (t) Kjell Sandved, (c) Oxford Molecular Biophysics Laboratory/Science Photo Library/Photo Researchers, (b) Larry Lefever from Grant Heilman; **14,** ©Astrid & Frieder Michler/Peter Arnold, Inc., (inset) Doug Martin; **15,** Biophoto Associates/Photo Researchers; **16,** Jim Runningen; **17,** (t) Ken Frick, (b) SuperStock; **19,** ©Manfred Kage/Peter Arnold, Inc., (inset) Ken Frick; **20,** (tl) Ted Rice, (r) Gerard Photography, (bl) Aaron Haupt; **22,** file photo; **23,** Janet L. Adams; **24,** Matt Meadows; **26,** Lee Kuhn/FPG; **27,** Lloyd Lemmerman; **28,** Dr. Jeremy Burgess/Science Photo Library/Photo Researchers; **29,** StudiOhio; **30,** (l) E.R. Degginger, (r) Pierre Berger/Photo Researchers; **31,** Dwight R. Kuhn; **32,** Dr. Jeremy Burgess/Science Photo Library/Photo Researchers; **33,** (t) courtesy Digital Instruments, (b) The Bettmann Archive; **34,** Dr. Lloyd M. Beidler/Science Photo Library/Photo Researchers; **35,** Michael Abbey/Science Source/Photo Researchers; **36,** (l) E.R. Degginger, (r) Dwight R. Kuhn; **38,** Don Fawcett/Photo Researchers; **39,** (t) Don Fawcett/Photo Researchers, (b) Biophoto Associates/Photo Researchers; **40,** (t) CNRI/Science Photo Library/Photo Researchers, (b) Robert & Linda Mitchell; **42,** CNRI/Science Photo Library/Photo Researchers; **46,** StudiOhio; **47,** Tim Courlas; **48,** Aaron Haupt; **51,** StudiOhio; **52,** Howard DeCruyenaere; **53,** Ken Frick; **55,** Aaron Haupt; **56,** (t) Larry Pierce/StockPhotos, Inc., (b) Ken Frick; **58, 61,** Ken Frick; **62, 63, 64,** Aaron Haupt; **65,** David Madison; **66,** Brent Turner/BLT Productions; **67,** Aaron Haupt; **68,** Doug Martin; **70,** Aaron Haupt/Glencoe; **71,** Elaine Shay; **72,** ©Lennart Nilsson/THE BODY VICTORIOUS/Dell Publishing; **73,** Ken Frick; **74,** Bruce Iverson; **75,** A. Cruickshank/VIREO; **76,** (t) Biophoto Associates/Photo Researchers, (b) Michael Abbey/Science Source/Photo Researchers; **77,** (l,c) Michael Abbey/Science Source/Photo Researchers, (r) Runk/Schoenberger from Grant Heilman; **78,** Focus on Sports; **79,** Bruce Iverson; **80,** ©M.I. Walker/Science Source; **81,** Dwight R. Kuhn; **82,** ©SIU/Peter Arnold, Inc.; **86,** Pictorial Parade; **87,** ©Nelson Max/LLNL/Peter Arnold, Inc.; **91,** E.R. Degginger; **92,** First Image; **93,** Runk/Schoenberger from Grant Heilman; **94,** KS Studios; **97,** Doug Martin; **98,** (t) ©Biophoto Associates/Science Source/Photo Researchers, (b) Dann Blackwood/Woods Hole Oceanographic Institute; **99,** (t) SuperStock, (c) Marty Snyderman, (b) Tom McHugh/Photo Researchers; **100,** (tl,tr) First Image, (bl) Roger K. Burnard, (br) Sinclair Stammers/Science Photo Library/Photo Researchers; **101,** Teri McNew; **102-103,** E.R. Degginger, (inset) Richard Small/Photo Op; **104,** Animals Animals/Zig Leszczynski; **107,** The Bettmann Archive; **108,** Animals Animals/Robert Pearcy; **111,** Glencoe photo; **112,** Runk/Schoenberger from Grant Heilman; **113,** Bud Fowle; **116,** Daniel Erickson; **117,** Omikron/Photo Researchers; **118,** Don Kelly from Grant Heilman; **119,** E.R. Degginger; **120,** Don Kelly from Grant Heilman; **121,** ©Jim Olive/Peter Arnold, Inc.; **123,** Dan McCoy from Rainbow; **128,** Ron Mellot; **129,** Aaron Haupt; **131,** (t) Doug Martin, (b) Stephen J. Krasemann/Photo Researchers; **132,** (t,bl) Wolfgang Kaehler, (c) SuperStock, (br) Alton Biggs; **133,** William R. King; **134,** (l) David M. Dennis, (r) E.R. Degginger; **136,** E.R. Degginger; **137,** (l) Ted Clutter/Photo Researchers, (c) Robert & Linda Mitchell, (r) J. Koivula/Science Source/Photo Researchers; **140,** Tim Davis/Photo Researchers; **144,** E.R. Degginger; **145,** Lynn M. Stone; **146,** Frans Lanting/Minden Pictures; **147,** (l) Lynn M. Stone, (r) Animals Animals/Doug Wechsler, (b) Cleveland Museum of Natural History; **148,** (l) John Cunningham/Visuals Unlimited, (r) E.R. Degginger; **149,** (l) The Bettmann Archive, (r) E.R. Degginger; **150,** Nigel Cattlin/Earth Scenes; **154,** Doug Martin; **155,** Lynn M. Stone; **157,** (t) Jean Wentworth, (b,c) E.R. Degginger; **158,** Ken Frick; **160,** E.R. Degginger; **161,** (t) Animals Animals/Zig Leszczynski, (c) Biological Photo Service, (b) Don Fawcett/Science Source/Photo Researchers; **163,** Lynn M. Stone; **164,** Frans Lanting/Minden Pictures; **165,** John Elk, III; **166,** (l) Steven Holt/VIREO, (c) B. Gadsby/Academy of Natural Sciences/VIREO, (r) C.H. Greenewalt/Academy of Natural Sciences/VIREO; **167,** (l) Marty Snyderman, (r) Lynn M. Stone; **168,** Doug Martin; **170,** (l) Animals Animals/E.R. Degginger, (r) Leonard Lee Rue III/Photo Researchers, Inc.; **172,** E.R. Degginger; **173,** (l) Leonard Lee Rue III/Photo Researchers, (c) Priscilla Connell: PHOTO/NATS, Inc., (r) Charlie Ott/Photo Researchers; **174,** (l) M. Serraillier/Photo Researchers, (r) UPI/Bettmann; **175,** (t) The Bettmann Archive, (c) Tom McHugh/Steinhart Aquarium/Photo Researchers, (b) UPI/Bettmann; **176,** (tl) Tim Courlas, (tr) Bob Daemmrich/Stock Boston, (bl) Doug Martin/Glencoe, (br) Lynn M. Stone; **177,** Alton Biggs; **178-179,** Karen Jettmar/Gamma-Liaison Network, (inset) ©Manfred Kage/Peter Arnold, Inc.; **180,** USDA/Science Source/Photo Researchers; **181,** ©Biophoto Associates/Science Source; **182,** B. Heggeler/Biozentrum, University of Basel/Science Photo Library/Photo Researchers; **184,** Culver Pictures; **185,** SuperStock; **186,** ©Lennart Nilsson/Boehringer Ingelheim International GmbH; **187,** First Image; **188,** Aaron Haupt; **189,** (t) Science Photo Library/Photo Researchers, (b) CNRI/Science Photo Library/Photo Researchers; **190,** (t) Michael Collier, (b) file photo; **191,** ©Tom E. Adams/Peter Arnold, Inc.; **193,** (l) Richard Gross, (r) John Colwell from Grant Heilman; **194,** Alfred Pasieka/Science Photo Library/Photo Researchers; **195,** Gary Milburn/Tom Stack & Associates; **196,** ©Bill Tijerina; **198,** Dr. Tony Brain/Science Photo Library/Photo Researchers; **199,** CNRI/Science Photo Library/Custom Medical Stock Photo; **200,** E.R. Degginger; **201,** Elaine Shay; **203,** (t) Biophoto Associates/Science Source/Photo Researchers, (b) Jan Hinsch/Science Photo Library/Photo Researchers; **204,** (t) Biophoto Associates/Photo Researchers, (bc) Runk/Schoenberger from Grant Heilman; **206,** (l) G.R. Roberts, (r) Bruce Iverson; **207,** (t) John Durham/Science Photo Library/Photo Researchers, (b) Eric Gray/Photo Researchers; **208,** (t) Custom Medical Stock Photo, (b) DRK Photos; **209,** Alan Pitcairn from Grant Heilman; **210,** Runk/Schoenberger from Grant Heilman; **211,** Gwen Fidler; **212,** (t) Michael & Patricia Fogden, (b) Barry L. Runk from Grant Heilman; **213,** (t) ©David Scharf/Peter Arnold, Inc., (b) Custom Medical Stock Photo; **214,** (l) William D. Popejoy, (r) David M. Dennis; **215,** (l) Alvin E. Staffan, (r) Gwen Fidler; **216,** Gene Frazier; **217,** Doug Martin; **218,** Runk/Schoenberger from Grant Heilman; **220,** ©Kevin Schafer/Peter Arnold, Inc.; **222,** (t) Gary Retherford/Photo Researchers, (b) The Bettmann Archive; **223,** (t) The Bettmann Archive, (c) F. Jalain/Photo Researchers, (b) Herman Emmet/Photo Researchers; **224,** (tl) Doug Martin, (tr) Jean Chemelli, (bl,br) file photo; **225,** file photo; **226-227,** Raphael Gaillarde/Gamma-Liaison, (inset) Earth Scenes/Michael Fodgen; **228,** ©David Cavagnaro/Peter Arnold, Inc.; **229,** Elaine Shay; **230,** E.R. Degginger; **231,** G.R. Roberts; **232,** (l) ©Biophoto Association/Science Source, (r) E.R. Degginger; **233,** ©1978 Michael Lustbader/Photo Researchers; **235,** E.R. Degginger; **236,** (l) E.R. Degginger, (r) Robert & Linda Mitchell; **237,** E.R. Degginger; **239,** Earth Scenes/Breck P. Kent; **240,** (t) Robert & Linda Mitchell, (b) E.R. Degginger; **242,** (tl,b) E.R. Degginger, (tr) Earth Scenes/Oxford Scientific Films; **243,** E.R. Degginger; **244,** Aaron Haupt; **245,** ©A. Peter Margosian: PHOTO/NATS, Inc.; **246,** G.R. Roberts; **248,** Doug Martin; **249,** Alvin E. Staffan; **250,** Steve Lissau; **252,** Serraillier/Rapho/Photo Researchers; **253,** Mary Lou Uttermohlen; **254,** Elaine Shay; **255,** (t,bl,bc) E.R. Degginger, (br) Robert & Linda Mitchell; **257,** John Elk, III; **258,** G.R. Roberts; **259,** Dwight R. Kuhn; **260,** Earth Scenes/George Bernard; **262,** Wolfgang Kaehler; **263,** Larry Minden/Minden Pictures; **264,** Dwight R. Kuhn; **265,** (tl) Earth Scenes/C.C. Lockwood, (tc) Robert & Linda Mitchell, (tr) Dwight R. Kuhn, (b) Russ Lappa; **267,** E.R. Degginger; **268,** John Elk, III; **269,** E.R. Degginger; **270,** Runk/Schoenberger from Grant Heilman, (inset) Bruce Iverson; **273,** Dwight R. Kuhn; **274,** E.R. Degginger; **275,** Matt Meadows; **277,** (l,r) Dwight R. Kuhn, (b) Lynn M. Stone; **278,** Dr. James Utzinger/OSU; **279,** (l) Marty Snyderman, (r) Michael Fogden; **281,** Doug Martin; **282,** (t,b) Earth Scenes/John L. Pontier; **283,** (t) Gregory Scott, (b) file photo; **284,** (t) Richard Gross, (b) J. Howard/Photo Researchers,Inc.; **285,** (l) G.R. Roberts, (c) Animals Animals/Donald Specker, (r) Lynn M. Stone; **286,** (t) Merlin D. Tuttle/Bat Conservation Internat'l./Photo Researchers, (b) Robert Lee/Photo Researchers; **287,** (t) Wolfgang Kaehler, (c,b) Alton Biggs; **288,** Runk/Schoenberger from Grant Heilman; **289,** ©Michael J. Balick/Peter Arnold, Inc.; **292,** G.R. Roberts; **293,** Wolfgang Kaehler; **294,** (t) Martin Bond/Science Photo Library/Photo Researchers, (b) Michael Collier; **295,** (t) Tom McHugh/Photo Researchers, (c) Dan Guravich/Photo Researchers, (b) ©Kevin Schafer/Peter Arnold, Inc.; **296,** (tl) Lowell Georgia/Photo Researchers, (tr) First Image, (bl) Steve Lissau, (br) Aaron Haupt; **297,** Bob Winsett/Tom Stack & Associates; **298-299,** Robert & Linda Mitchell, (inset) Porterfield-Chickering/Photo Researchers; **300,** Nancy Sefton; **301,** Marty Snyderman; **302,** Paul W. Nesbit; **306,** Jim Cox/The Salk Institute; **307,** Dan McCoy from Rainbow; **308,** Nancy Sefton; **310,** E.R. Degginger; **311,** (t) Runk/Schoenberger from Grant Heilman, (b) Marty Snyderman; **313,** Nancy Sefton; **314,** Grant Heilman Photography; **315,** Robert & Linda Mitchell; **316,** Harris Biological Supplies, Ltd.; **317,** (t) V. Grave/Photo Researchers, (b) Alan Carey; **318,** Robert & Linda Mitchell; **320,** Biophoto Associates/Science Source/Photo Researchers; **321,** Ruth Dixon; **322,** Animals Animals/Fred Whitehead; **323,** E.R. Degginger; **325,** (t) Geri Murphy, (c) Al Grotell, (b) Marty Snyderman; **326,** Frank Lerner; **327,** (t) Geri Murphy, (b) Marty Snyderman; **328,** E.R. Degginger; **329,** Runk/Schoenberger from Grant Heilman; **330,** (l) Glencoe photo, (r) L. Boilly/Photo Researchers; **331,** Doug Martin; **333,** E.R. Degginger; **334,** (t) E.R. Degginger, (b) Grant Heilman Photography, (inset) Science Photo Library/Photo Researchers; **339,** E.R. Degginger; **340,** Animals Animals/Keith Gillett; **341,** John Elk, III; **342,** Nancy Sefton; **343,** (t) Stan Elems/Visuals Unlimited, (cl) Al Grotell, (cr) Nancy Sefton, (bl) Joey Jacques, (br) Marty Snyderman; **344,** Nancy Sefton; **346,** Marty Snyderman; **348,** E.R. Degginger; **349,** Matt Meadows; **350,** (t) Al Grotell, (b) Animals Animals/Oxford Scientific Films; **352,** (t) Animals Animals/Zig Leszczynski; (bl) Al Grotell, (br) Marty Snyderman; **353,** Al Grotell; **354,** Animals Animals/Bates Littlehales; **355;** Matt Meadows; **356,** (t) Animals Animals/Zig Leszczynski, (b) David M. Dennis; **357,** Animals Animals/Michael P. Gadomski; **359,** Animals Animals/Michael Fogden; **360,** Wolfgang Kaehler; **361,** (t) Animals

744

Animals/C.W. Schwartz, (b) Suzanne & Joseph Collins/Photo Researchers; **362,** E.R. Degginger; **363,** (t) Animals Animals/Patti Murray, (b) Animals Animals/Joe McDonald; **364,** Animals Animals/Zig Leszczynski; **365,** (tl) Animals Animals/Patricia Caufield, (tc) Animals Animals/Joe McDonald, (tr,cc,bc) Lynn M. Stone, (ct) Animals Animals/E.R. Degginger, (cb,br) Animals Animals/Zig Leszczynski (bl) Wolfgang Kaehler; **366,** Animals Animals/Zig Leszczynski; **368,** Tom Stack/Tom Stack & Associates; **370,** SuperStock; **371,** Aaron Haupt; **372,** B. Chudleish/VIREO; **373,** (t) M.P. Kahl/VIREO, (b) Alan Carey; **374,** (l) Roger K. Burnard, (r) Wolfgang Kaehler; **376,** (tl) Superstock, (tr) Lynn M. Stone, (cl) Animals Animals/Ken Cole, (cr) Animals Animals/Francis Lepine, (b) SuperStock; **378,** Wolfgang Kaehler; **379,** Robert & Linda Mitchell; **380,** (l) Johnny Johnson, (c) Grant Heilman Photography, (r) SuperStock; **381,** (t) Runk/Schoenberger/Grant Heilman, (bl) E.R. Degginger, (br) Animals Animals/Mickey Gibson; **382,** Breck P. Kent; **383,** (left, t to b) Merlin Tuttle/Photo Researchers, Animals Animals/Brian Milne, Lynn M. Stone, Alvin E. Staffan, Roger K. Burnard, (right, t to b) William J. Weber, Marty Snyderman, Animals Animals/Mickey Gibson; **384,** (t) Rod Furgason/Unicorn Stock Photos, (b) Animals Animals/Anup & Manoj Shah; **385,** (l) SuperStock, (r) Jeff Foott; **386,387,** Marty Snyderman; **389,** Visuals Unlimited; **390,** Animals Animals/Zig Leszczynski; **392,** Barry Runk from Grant Heilman; **393,** Doug Martin; **394,** Lee F. Snyder/Photo Researchers; **395,** (t) Stephen Dalton/Photo Researchers, (bl) Mark Phillips/Photo Researchers, (br) Mitch Reardon/Photo Researchers; **396,** (t) Animals Animals/David Thompson,Oxford Scientific Films, (b) Johnny Johnson; **397,** Nina Leen, LIFE MAGAZINE ©Time Warner Inc.; **398,** (t) Brent Turner/BLT Productions, (b) E.R. Degginger; **401,** Jeff Foott; **402,** (l) Paul Brown, (r) Animals Animals/Leonard Lee Rue III; **403,** Jim Brandenburg/Minden Pictures; **404,** (t) Jess R. Lee/Photo Researchers, (b) Research Center of Georgia State University and the Yerkes Regional Primate Research Center of Emory University, Atlanta; **406,** Frans Lanting/Minden Pictures; **407,** Roger K. Burnard; **410,** E.R. Degginger; **411,** Lang Elliott/Cornell Laboratory of Ornithology; **412,** (t) Scott Camazine/Photo Researchers, (b) Lynn M. Stone; **413,** (t) Art Resource/New York, (c) Bildarchiv Okapia/Photo Researchers, (b) Jeff Foott/Tom Stack & Associates; **414,** (tl) ©1991 Norbert Wu, (tr) First Image, (bl) Cliff Beaver, (br) Kevin Barry; **415,** Michael Tropea; **416-417,** ©Lennart Nilsson/BEHOLD MAN/Little, Brown and Company, (inset) Aaron Haupt; **418,** David Madison; **419,** StudiOhio; **421,** Runk/Schoenberger from Grant Heilman; **422,** (t) ©Lennart Nilsson/BEHOLD MAN/Little, Brown and Company, (b) Princess Margaret Rose Orthopaedic Hospital/Science Photo Library/Custom Medical Stock Photo; **423,** Hickson & Associates; **424,** Science Photo Library/Custom Medical Stock Photo; **427,** (l,r) Runk/Schoenberger from Grant Heilman, (c) Michael Abbey/Science Source/Photo Researchers; **429,** Urike Welsch/Photo Researchers; **430, 431,** David Madison; **432,** CNRI/Science Photo Library/Photo Researchers; **435,** Johnny Johnson; **436,** Doug Martin; **439,** Custom Medical Stock Photo; **440,** ©Lennart Nilsson; **441,** Ken Frick; **442,** Wolfgang Kaehler; **443,** (l) Grant Heilman Photography, (r) Kurt Thorson; **444,** Bruce Iverson; **445,** Hickson & Associates; **446,** David Madison; **448,** (l,cr) Elaine Shay/Glencoe, (cl,r) Jack Sekowski; **449,** Randy Schieber; **450,** Doug Martin; **452,** Bruce Iverson; **453,** ©Lennart Nilsson/BEHOLD MAN/Little, Brown and Company; **454,** Fawcett/Hirokawa/Heuser/Science Source/Photo Researchers; **455,** Grant Heilman Photography; **457,** Tim Cairns; **458,** Doug Martin; **462,** David Madison; **463,** StudiOhio; **464,** Doug Martin; **468,** Ray Simons/Photo Researchers; **470,** StudiOhio; **471,** Doug Martin; **472,** (l) ©Lennart Nilsson/THE INCREDIBLE MACHINE/Boheringer Ingelheim International GmbH/National Geographic Society, (r) ©Lennart Nilsson/THE BODY VICTORIOUS/Boehringer Ingelheim International GmbH/Dell Publishing; **473,** (l) NIBSC/Science Photo Library/Photo Researchers, (r) Richard B. Levine; **474,** First Image; **475,** ©Farrell Greham/Photo Researchers; **476,** Prof. Aaron Polliack/Science Photo Library/Photo Researchers; **478,** Doug Martin; **482,** Donna Bise/Gamma Liaison; **483,** StudiOhio; **484,** Aaron Haupt/Glencoe; **485,** StudiOhio; **486,** David Madison; **487,** StudiOhio; **492,** Custom Medical Stock Photo; **493,** (l) StudiOhio, (r) ©Lennart Nilsson/THE INCREDIBLE MACHINE/Boehringer Ingelheim International GmbH/National Geographic Society; **494,** American Cancer Society; **495,** KS Studios, (inset) Doug Martin; **497,** (t) Randy Schieber, (b) StudiOhio; **499,** Spinoff Magazine, 1989/NASA; **501,** Custom Medical Stock Photo; **502, 503,** Doug Martin; **504,** Aaron Haupt/Glencoe; **506,** ©Lennart Nilsson/BEHOLD MAN/Little, Brown and Company; **509,** Custom Medical Stock Photo; **510,** Don Fawcett/Science Source/Photo Researchers; **511,** John Meyer/Custom Medical Stock Photo; **517,** ©Lennart Nilsson/BEHOLD MAN/Little, Brown and Company; **519,** Ken Frick; **520,** Randy Schieber; **522,** Focus on Sports; **524,** Brent Turner/BLT Productions; **526,** David Madison; **527,** Custom Medical Stock Photo; **528,** Howard DeCruyenaere; **529,** Ken Frick; **531,** C. Edelmann/La Villette/Photo Researchers; **535,** ©Lennart Nilsson/THE INCREDIBLE MACHINE/National Geographic Society; **536,** First Image; **538,** (t) Saturn Scientific, Inc., (b) CNRI/Science Photo Library/Photo Researchers; **539,** Charles Zirkle; **541,** (l) SuperStock, (r) Bud Fowle; **542,** Doug Martin; **543,** Bud Fowle; **544,** Brent Turner/BLT Productions; **545,** David Madison; **549,** Aaron Haupt; **550,** (l) John Running/Stock Boston, (b) SuperStock; **551,** (t) Historical Picture Services, (c) Roger K. Burnard, (b) Science Photo Library/Photo Researchers; **552,** (tl) First Image, (tr) Hickson & Associates, (bl) William McCoy from Rainbow, (inset) StudiOhio; **553,** ©Maria Mulas; **554-555,** ©Manfred Kage/Peter Arnold, Inc., (inset) Custom Medical Stock Photo; **556,** ©David Scharf/Peter Arnold, Inc.; **557,** StudiOhio; **558,** (l) Bruce Iverson, (r) Mary Evans Picture Library/Photo Researchers; **559,** (t) Van Bucher/Photo Researchers, (b) Doug Martin; **560,** Lindsay Gerard/Glencoe; **561,** (l) CNRI/Science Photo Library/Photo Researchers, (r) CNRI/Science Photo Library/Photo Researchers; **562,** Doug Martin; **564,** Tim Courlas; **565,** Biomedical Communications/Photo Researchers; **566,** CDC/Science Source/Photo Researchers; **567,** Doug Martin; **568,** Culver Pictures, Inc.; **569,** Biophoto Associates/Science Source/Photo Researchers; **570,** (t) Thompson/Stammers/Photo Researchers, (b) CNRI/Science Photo Library/Photo Researchers; **571,** Aaron Haupt/Glencoe; **572,** (l) Ken Frick, (c) Horticultural Photography, Corvallis, Oregon, (r) Randy Trine; **573,** courtesy The Navajo Nation; **574,** Matt Meadows; **577,** Cecil H. Fox/Science Source/Photo Researchers; **578,** Doug Martin; **579,** J. Swarwtz; **580,** Aaron Haupt; **581,** Medfusion Inc./A Medex Inc. Company; **582,** Aaron Haupt; **583,** Eric Hoffhines; **586,** Aaron Haupt/Glencoe; **587,** SuperStock; **588,** J.R. Schnelzer; **589,** (t) David L. Perry, (c) J.R. Schnelzer, (b) Anne Heimann/The Stock Market; **590,** Larry Hamill; **591,** (l) Custom Medical Stock Photo, (r) Runk/Schoenberger from Grant Heilman; **592,** Doug Martin; **593,** Carolina Biological Supply Company; **594,** Bruce Iverson; **595,** Doug Martin; **596,** David M. Dennis; **597,** Aaron Haupt; **598,** (t) Holly Kuper, (b) SuperStock; **599,** (t) Commercial Image, (c) Owen Franken/Stock Boston, (b) Robert Caputo/Stock Boston; **600,** (tl) First Image, (tr) First Image, (bl) Tim Courlas, (br) StudiOhio; **601,** The Bettmann Archive; **602-603,** Robert & Linda Mitchell, (inset) Terry Donnelly/Tom Stack & Associates; **604,** E.R. Degginger; **605,** Brian Brake/Photo Researchers; **606,** Frederick J. Grassle/Woods Hole Oceanographic Institution; **607,** (t) Galen Rowell/Mountain Light, (b) Ken Frick; **608,** (l) Johanna Schumacher/VIREO, (c) E.R. Degginger, (r) J. Ramey/Stockphotos, Inc.; **609,** ©Matt Meadows; **610,** Laurence Pringle/Photo Researchers; **611,612,613,614,617,** E.R. Degginger; **618,** Runk/Schoenberger from Grant Heilman; **620,** Jeff Foott; **621,** ©Julia Sims/Peter Arnold, Inc.; **622,** E.R. Degginger; **626,** Jeff Foott; **627,** Aaron Haupt; **628,** (t) E.R. Degginger, (b) Alvin E. Staffan; **633,** (t) Wolfgang Kaehler, (b) Aaron Haupt/Glencoe; **634,** (t) Peter & Stephen Maslowski/Photo Researchers, (c) E.R. Degginger, (b) Jeff Foott; **635,** (t) E.R. Degginger, (b) Earth Scenes/Michael Fogden; **636,** Jeff Topping/©Space Biospheres Venture; **637,** William D. Popejoy; **638,** (t) Jeff Foott, (b) E.R. Degginger; **639,** E.R. Degginger; **640,** Lucian Niemeyer/LNS Arts; **646,** J. Ramey/Stockphotos Inc.; **647,** Hickson & Associates; **648,** Ronald Johnson/Stockphotos, Inc.; **650,** Bobbe Wolf/International Stock Photo; **651,** Earth Scenes/Mike Andrews; **652,** Ken Frick; **653,** Photo Researchers; **654,** Doug Martin; **655,** E.R. Degginger; **656,** Jeff Foott; **657,** Kjell Sandved; **658,** Johnny Johnson; **660,** ©Space Biospheres Venture; **661,** SuperStock; **662,** E.R. Degginger; **664,** Aaron Haupt/Glencoe; **665,** Weslet Bocxe/Photo Researchers; **666,** (t) T. Kitchin/Tom Stack & Associates, (b) Lionel Delevingne/Stock Boston; **667,** (t) Mark Newman/Tom Stack & Associates, (c) TASS from Sovfoto, (b) Robert Hernandez/Photo Researchers; **668,** (tl) First Image, (tr) Lawrence Migdale/Stock Boston, (bl) Tim Courlas, (br) Roger K. Burnard; **669,** Carmen Lomas Garza; **674,** (tl) A.B. Dowsett/Science Photo Library/Photo Researchers, (tr) Biophoto Associates/Photo Researchers, (b) Runk/Schoenberger from Grant Heilman; **675,** (tl) Harris Biological Supplies, (tr) David Scharf/Peter Arnold, Inc., (cr) Gustav Verderber, (bl) David M. Dennis/Tom Stack & Associates, (br) Lynn M. Stone; **676,** (tl) Alton Biggs, (cl) E.R. Degginger, (cr,and inset) Unicorn Stock Photos, (bl) David M. Dennis, (br) Kenneth Fink/Photo Researchers; **677,** (tl) SuperStock, (cl) William Westwater, (bl) Roger Caras/Photo Researchers, (right top to bottom) Johnny Johnson, Alvin E. Staffan, Glencoe photo, Sean Warren, Gwen Fidler; **678,** (tl) Bill Curtsinger/Photo Researchers, (tr) Gene Frazier, (cl) Breck P. Kent, (cr,bl) ©1991 Norbert Wu, (br) Joey Jacques; **679,** (left top to bottom) Hutchins, Wolfgang Kaehler, Lynn M. Stone, Kjell Sandved, (right top to bottom) Al Grotell, Jim Rorabaugh, Lynn M. Stone, Jean Wentworth; **681,** Carolina Biological Supply Company; **682,** (t) Ken Frick, (b) Lloyd Lemmerman; **683,** (tl) Lynn M. Stone, (tr) Alvin E. Staffan, (b) Janet L. Adams; **684,** Ken Frick; **685,** Doug Martin; **689,690,** Ken Frick; **694,695,** Matt Meadows; **696-697,** (c) Hickson & Associates; **696,** (b) Hickson & Associates; **697,** (b) Ken Frick; **698,699,** Matt Meadows; **700,** (l) Wendy Shattil & Bob Rozinski/Tom Stack and Associates, (r) Matt Meadows; **701,** Tom and Pat Leeson; **702,** Tim Courlas; **703,** (l) Hickson & Associates, (r) Tom McHugh/Photo Researchers; **704,** David Madison; **705,706,707,** Matt Meadows; **708,** Gerard Photography; **709,** (t) Doug Martin, (c) D. C. H. Plowes, (b) David R. Frazier.

745